Brian Street

Maximal Subellipticity

De Gruyter Studies in Mathematics

Volume 93

Brian Street

Maximal Subellipticity

DE GRUYTER

Mathematics Subject Classification 2020
Primary: 35-02, 35H20, 35B65; Secondary: 35G05, 35G20, 35J70, 35R03, 42B20, 42B25, 42B35, 42B37, 53C17

Author
Prof. Brian Street
University of Wisconsin – Madison
Department of Mathematics
480 Lincoln Drive
Madison, WI 53706-1388
USA
street@math.wisc.edu

ISBN 978-3-11-108517-3
e-ISBN (PDF) 978-3-11-108564-7
e-ISBN (EPUB) 978-3-11-108594-4
ISSN 0179-0986

Library of Congress Control Number: 2023932713

Bibliographic information published by the Deutsche Nationalbibliothek
The Deutsche Nationalbibliothek lists this publication in the Deutsche Nationalbibliografie; detailed bibliographic data are available on the Internet at http://dnb.dnb.de.

Typesetting: VTeX UAB, Lithuania
Printing and binding: CPI books GmbH, Leck

www.degruyter.com

Contents

1 Introduction

Elliptic partial differential equations (PDEs) hold a special place among PDEs: sharp results are known for very general linear and even fully nonlinear elliptic PDEs. Many of the classical techniques from harmonic analysis were first developed to study these sharp results; the study of elliptic PDEs leans heavily on the Fourier transform and Riemannian geometry. Starting with the work of Hörmander, Kohn, Folland, Stein, and Rothschild in the 1960s and 1970s, a far-reaching generalization of ellipticity was introduced, now known as *maximal subellipticity* or *maximal hypoellipticity*.[1] Maximally subelliptic operators generalize elliptic operators and can be significantly more degenerate. In the intervening years, many authors have adapted results from elliptic PDEs to various special cases of maximally subelliptic PDEs.

One of the central problems in the study of PDEs is to establish regularity theorems, for both linear and nonlinear PDEs. Let $\mathscr{X}^{\mathbf{s}}$ denote a family of spaces indexed by a regularity parameter $\mathbf{s} \in \mathbb{R}$ (for example, $\mathscr{X}^{\mathbf{s}}$ could denote the L^p Sobolev space of order $\mathbf{s}$). A linear regularity theorem near a point x_0 for a linear partial differential operator $\mathscr{P}$ is a result of the form

$$\mathscr{P}u \in \mathscr{X}^{\mathbf{s}} \text{ near } x_0 \implies u \in \mathscr{X}^{\mathbf{t}} \text{ near } x_0, \tag{\textsc{Linear}}$$

for some $\mathbf{s}, \mathbf{t} \in \mathbb{R}$.

Nonlinear regularity theorems are also of interest. Fix $\kappa \in \{1, 2, 3, \ldots\}$. For a function u, we write $\{\partial_x^\alpha u\}_{|\alpha| \leq \kappa}$ for the vector whose components are given by $\partial_x^\alpha u$, where $|\alpha| \leq \kappa$. For a given function $F(x, \zeta)$, a *nonlinear* regularity theorem near a point x_0 is a result of the form

$$F(x, \{\partial_x^\alpha u(x)\}_{|\alpha| \leq \kappa}) \in \mathscr{X}^{\mathbf{s}} \text{ near } x_0 \implies u \in \mathscr{X}^{\mathbf{t}} \text{ near } x_0. \tag{\textsc{Nonlinear}}$$

Results like (Linear) and (Nonlinear) can be very difficult to establish. One often has to work with specific properties of the partial differential operators, and abstract general results are rare. In contrast to these difficulties, when considering *elliptic* PDEs, both (Linear) and (Nonlinear) are well understood and sharp results are known. One of the main goals of this text is to establish sharp results of the form (Linear) and (Nonlinear) for general maximally subelliptic PDEs, thereby generalizing the sharp results for general elliptic PDEs to a much larger class of PDEs.

The development of this regularity theory will take us through many generalizations of classical notions from the area of elliptic PDEs. The Fourier transform is not a

---

1 The PDEs studied in this text have been called both *maximally subelliptic* and *maximally hypoelliptic*. We use the more descriptive, but less common, terminology maximally subelliptic; see Remark 1.6.2 for some comments on this choice. These operators are subelliptic (see (1.18) and Corollary 8.2.5) and therefore hypoelliptic (see Definition 2.2.37 and Proposition 2.2.38).

https://doi.org/10.1515/9783111085647-001

decisive tool, but many ideas from elliptic theory have generalizations to the maximally subelliptic setting:
- Riemannian geometry is replaced by a Carnot–Carathéodory (or sub-Riemannian) geometry,
- Besov and Triebel–Lizorkin spaces have generalizations which are adapted to the Carnot–Carathéodory geometry,
- pseudo-differential operators are replaced by singular integral operators adapted to the Carnot–Carathéodory geometry.

While most of the main results of this text are new, they are based on previous work of many authors. At the end of most chapters, we include a section "Further reading and references," which attempts to give a history of the ideas we use and also points the reader to references on related subjects which we do not directly cover.

This text focuses on generalizing several aspects of elliptic theory to the maximally subelliptic setting. However, there are many parts of elliptic theory which we do not address; given almost any result from the vast elliptic theory, one can ask whether a generalization holds for maximally subelliptic equations. Usually, the right generalization is not immediately clear: one needs to find the right function spaces, operators, and geometry to use. One also usually needs a completely new proof. Nevertheless, it is often possible to find the right generalization (see Sections 8.10 and 9.5 for many examples from the literature). It seems likely that this will be a fruitful area of study for years to come.

The author was partially supported by National Science Foundation grants DMS-1764265 and DMS-2153069. The opinions, findings, and conclusions expressed are those of the author and do not necessarily reflect the views of the National Science Foundation.

1.1 Basic definitions

Let $\mathfrak{M}$ be a smooth manifold[2] and let Vol be any smooth, strictly positive density on $\mathfrak{M}$ (see Definition 3.1.5); the definitions that follow do not depend on the choice of Vol, which plays the role of "Lebesgue measure" on $\mathfrak{M}$. Let $\mathbb{N} = \{0, 1, 2, \ldots\}$ and $\mathbb{N}_+ = \{1, 2, 3, \ldots\}$.

Definition 1.1.1. Let $W = \{W_1, \ldots, W_r\}$ be a finite collection of smooth vector fields on $\mathfrak{M}$. For $m \in \mathbb{N}_+$ and $x \in \mathfrak{M}$, we say W satisfies Hörmander's condition of order m at x if

2 All manifolds in this text are manifolds without boundary. Our main results are local, so $\mathfrak{M}$ can usually be taken to be an open subset of $\mathbb{R}^n$. Nevertheless, it is often useful to work in the abstract setting of manifolds, since many proofs exploit changing coordinates in a quantitative way. See Section 3.3 for a first example of this and Appendix A for more details.

the commutators of $W_1, \dots, W_r$ up to order m span $T_x\mathfrak{M}$, that is, if

$$W_1(x), W_2(x), \dots, W_r(x), \dots, [W_i, W_j](x), \dots, [W_i, [W_j, W_k]](x),$$
$$\dots, \dots, \text{commutators of order } m \text{ evaluated at } x$$

span $T_x\mathfrak{M}$. We say W satisfies Hörmander's condition at x if $\exists m \in \mathbb{N}_+$ such that W satisfies Hörmander's condition of order m at x. We say W satisfies Hörmander's condition of order m on $\mathfrak{M}$ if $\forall x \in \mathfrak{M}$, W satisfies Hörmander's condition of order m at x. We say W satisfies Hörmander's condition on $\mathfrak{M}$ if $\forall x \in \mathfrak{M}$, W satisfies Hörmander's condition at x.

Example 1.1.2. (i) With $\mathfrak{M} = \mathbb{R}^2$, let $W_1 = \partial_x$ and $W_2 = x\partial_y$. Note that $\operatorname{span}\{W_1(0), W_2(0)\} \neq T_0\mathbb{R}^2$. However, $[W_1, W_2] = \partial_y$, so $\operatorname{span}\{W_1(0), W_2(0), [W_1, W_2](0)\} = T_0\mathbb{R}^2$. Thus, $\{W_1, W_2\}$ satisfies Hörmander's condition of order 2 at 0. In fact, $\{W_1, W_2\}$ satisfies Hörmander's condition of order 2 on $\mathbb{R}^2$.

(ii) With $\mathfrak{M} = \mathbb{R}^2$, let $W_1 = \partial_x$ and $W_2 = x^2\partial_y$. In this case $[W_1, [W_1, W_2]] = 2\partial_y$, and $W = \{W_1, W_2\}$ satisfies Hörmander's condition of order 3 on $\mathbb{R}^2$. More generally, $\{\partial_x, x^m\partial_y\}$ satisfies Hörmander's condition of order $m+1$ on $\mathbb{R}^2$.

(iii) On $\mathbb{R}^3$, with coordinates (x, y, t), let $W_1 = \partial_x + 2y\partial_t$ and $W_2 = \partial_y - 2x\partial_t$. In this case $W_1(\zeta)$ and $W_2(\zeta)$ do not span $T_\zeta\mathbb{R}^3$ for any ζ. However, $[W_1, W_2] = -4\partial_t$, and therefore $W_1(\zeta), W_2(\zeta), [W_1, W_2](\zeta)$ span $T_\zeta\mathbb{R}^3$ for every $\zeta \in \mathbb{R}^3$. We conclude $W = \{W_1, W_2\}$ satisfies Hörmander's condition of order 2 on $\mathbb{R}^3$. In this case, W_1 and W_2 are left invariant vector fields on the Heisenberg group; see [216, Chapters XII and XIII] for more information on the Heisenberg group. The Heisenberg group is a simple example of a graded nilpotent Lie group, and such groups play an important role in the theory of maximal subellipticity (see Sections 4.5.3 and 8.8).

(iv) On $\mathbb{R}^n$, let $W_1 = \partial_{x_1}, \dots, W_n = \partial_{x_n}$. Then $W = \{W_1, \dots, W_n\}$ span the tangent space at every point and therefore satisfy Hörmander's condition of order 1. This example plays an important role when we see elliptic operators as a special case of maximally subelliptic operators in Example 1.1.10 (i).

(v) The above examples are all homogeneous under appropriately chosen dilations; for example the vector fields in (i) are homogeneous under the dilations $(x, y) \mapsto (\delta x, \delta^2 y)$. Usually, there will be no such choice of dilations. For example, with $\mathfrak{M} = \mathbb{R}^2$, let $W_1 = \partial_x$ and $W_2 = (x + x^2)\partial_y$. Then $W = \{W_1, W_2\}$ satisfies Hörmander's condition of order 2 on $\mathbb{R}^2$, but the vector fields are not homogeneous with respect to dilations. To deal with general Hörmander vector fields, we work with more abstract scaling maps instead of simple dilations; see Section 1.9.

(vi) On $\mathbb{R}^3$, let $W_1 = \partial_x$ and $W_2 = x\partial_y$. Then W_1 and W_2 do not satisfy Hörmander's condition at any point. However, for each $t_0 \in \mathbb{R}$, if we restrict our attention to the submanifold $L_{t_0} := \{(x, y, t_0) : x, y \in \mathbb{R}\}$, then W_1 and W_2 are tangent to L_{t_0} and satisfy Hörmander's condition of order 2 on L_{t_0}. This idea can be generalized using the classical Frobenius theorem (see Section 3.1.3 and Remark 3.4.11).

(vii) On $\mathbb{R}^2$, let $W_1 = \partial_x$ and $W_2 = e^{-1/x^2}\partial_y$. Then $W = \{W_1, W_2\}$ does not satisfy Hörmander's condition at any point of the form $(0,y) \in \mathbb{R}^2$. Unlike (vi), there is not a submanifold passing through $(0,y)$ such that W_1 and W_2 are tangent to the submanifold and satisfy Hörmander's condition.

Given a set of vector fields satisfying Hörmander's condition, we will often pair each vector field with a "formal degree," which is an element of $\mathbb{N}_+$.

Definition 1.1.3. We call a finite set $(W, \vec{d}) = \{(W_1, \vec{d}_1), \ldots, (W_r, \vec{d}_r)\}$, where each W_j is a smooth vector field on $\mathfrak{M}$ and each $\vec{d}_j \in \mathbb{N}_+$, a finite set of vector fields with formal degrees. If $W = \{W_1, \ldots, W_r\}$ satisfies Hörmander's condition, we call $(W, \vec{d})$ a set of Hörmander vector fields with formal degrees.

Let $(W, \vec{d}) = \{(W_1, \vec{d}_1), \ldots, (W_r, \vec{d}_r)\}$ be a set of Hörmander vector fields with formal degrees on $\mathfrak{M}$. Informally, we will treat W_j as a differential operator of "degree" $\vec{d}_j$, even though W_j is a differential operator of order 1 in the usual sense. The case $\vec{d}_1 = \vec{d}_2 = \cdots = \vec{d}_r = 1$ is particularly interesting, and the reader may find it enlightening to interpret the definitions in this special case. When we consider this case, we will write $(W, 1)$ instead of $(W, \vec{d})$.

Because the vector fields $W_1, \ldots, W_r$ do not necessarily commute, when we consider polynomials in these vector fields we use noncommutative polynomials. For this, we use the following notation.

Definition 1.1.4. Let $\alpha = (\alpha_1, \ldots, \alpha_L) \in \{1, \ldots, r\}^L$ be a list of elements of $\{1, \ldots, r\}$. We set $W^\alpha := W_{\alpha_1} W_{\alpha_2} \cdots W_{\alpha_L}$, $|\alpha| := L$, and $\deg_{\vec{d}}(\alpha) := \vec{d}_{\alpha_1} + \vec{d}_{\alpha_2} + \cdots + \vec{d}_{\alpha_L}$. We call such a list α an *ordered multi-index*.

Remark 1.1.5. We treat $[W_j, W_k] = W_j W_k - W_k W_j$ as a differential operator of degree $\vec{d}_j + \vec{d}_k$. For example, when $(W_1, \vec{d}_1) = (\partial_x, 1)$ and $(W_2, \vec{d}_2) = (x\partial_y, 1)$, we will treat $\partial_y = [W_1, W_2]$ as a differential operator of degree 2 (see Example 1.6.1). Thus, even though ∂_y is a differential operator of *order* 1 in the classical sense, in the theory which follows we treat it as an operator of degree 2. In particular, the vector fields W_1 and W_2 do not even approximately commute: $[W_1, W_2] = W_1 W_2 - W_2 W_1$ is a partial differential operator of the same degree as $W_1 W_2$. This is in contrast to elliptic theory, where $[\partial_x, \partial_y] = 0$. This difference underlies nearly all the difficulties when studying maximally subelliptic operators. Nearly every proof from classical elliptic theory at some point, either implicitly or explicitly, uses the commutative nature of the elliptic setting. Because of this, these proofs do not translate to the maximally subelliptic setting and new ideas are needed.

In what follows, $\kappa \in \mathbb{N}_+$ will be such that $\vec{d}_j$ divides κ, $1 \le j \le r$, and $n_j := \kappa / \vec{d}_j \in \mathbb{N}_+$. We write $A \Subset B$ if A is a relatively compact subset of B.

1.1.1 Linear maximally subelliptic operators

Fix $\mathbf{D}_1, \mathbf{D}_2 \in \mathbb{N}_+$. Given a partial differential operator, $\mathscr{P}$, with coefficients in $C^\infty_{\mathrm{loc}}(\mathfrak{M}; \mathbb{M}^{\mathbf{D}_1\times\mathbf{D}_2}(\mathbb{C}))$,[3] we write $\mathscr{P}$ in the form[4]

$$\mathscr{P} = \sum_{\deg_{\vec{d}}(\alpha)\leq\kappa} a_\alpha(x)W^\alpha, \quad a_\alpha \in C^\infty_{\mathrm{loc}}(\mathfrak{M}; \mathbb{M}^{\mathbf{D}_1\times\mathbf{D}_2}(\mathbb{C})). \tag{1.1}$$

Recall $\vec{d}_j$ is assumed to divide κ, $\forall j$. By Hörmander's condition, *every* partial differential operator with smooth coefficients is of the form (1.1).

Remark 1.1.6. The form (1.1) is not unique. For example, if $W_1 = \partial_x$ and $W_2 = \partial_y$, then we can write $0 = [W_1, W_2] = W_1W_2 - W_2W_1$. We always pick a representation of the form (1.1) with the least possible κ (for a fixed choice of $(W, \vec{d})$). The value of κ depends on $(W, \vec{d})$ and can be considered the degree of $\mathscr{P}$ with respect to $(W, \vec{d})$.

Definition 1.1.7. We say that $\mathscr{P}$ given by (1.1) is maximally subelliptic of degree κ with respect to $(W, \vec{d})$ on $\mathfrak{M}$, if for every relatively compact, open set $\Omega \Subset \mathfrak{M}$, there exists $C_\Omega \geq 0$, such that $\forall f \in C^\infty_0(\Omega; \mathbb{C}^{\mathbf{D}_2})$,

$$\sum_{j=1}^r \|W_j^{n_j} f\|_{L^2(\mathfrak{M},\mathrm{Vol};\mathbb{C}^{\mathbf{D}_2})} \leq C_\Omega\big(\|\mathscr{P}f\|_{L^2(\mathfrak{M},\mathrm{Vol};\mathbb{C}^{\mathbf{D}_1})} + \|f\|_{L^2(\mathfrak{M},\mathrm{Vol};\mathbb{C}^{\mathbf{D}_2})}\big).$$

When $\mathfrak{M}$ is clear from the context, we say $\mathscr{P}$ is maximally subelliptic of degree κ with respect to $(W, \vec{d})$. Recall, $n_j = \kappa/\vec{d}_j \in \mathbb{N}_+$.

Remark 1.1.8. Given $(W, \vec{d})$, Definition 1.1.7 defines a class of operators which are maximally subelliptic with respect to $(W, \vec{d})$. If one changes $(W, \vec{d})$, the class of operators may change as well.

Remark 1.1.9. The density Vol does not play a role in Definition 1.1.7: one obtains an equivalent definition with any choice of smooth, strictly positive density.

Given a partial differential operator with smooth coefficients, $\mathscr{P}$, to show $\mathscr{P}$ is maximally subelliptic, one needs to:
- find the right choice $(W, \vec{d})$ of Hörmander vector fields with formal degrees,
- find the minimal κ such that $\mathscr{P}$ can be written in the form (1.1), with this choice of $(W, \vec{d})$,
- prove that $\mathscr{P}$ is maximally subelliptic of degree κ with respect to $(W, \vec{d})$.

---

3 See Sections 2.1 and 3.1 for a discussion of C^∞_0 (the space of smooth functions with compact support), C^∞_{loc} (the space of locally smooth functions), and C^∞ (the space of bounded smooth functions, all of whose derivatives are bounded).

4 In this text we work with trivial vector bundles, though most definitions and results can be easily extended to general vector bundles – see Section 8.6. Our main results are local in nature, so general vector bundles do not give much additional generality.

This may not even be possible: not every partial differential operator can be seen as maximally subelliptic with respect to some (W, d).[5] Even when it is possible, the above steps are not necessarily easy (however, there have been some recent advances in recognizing maximally subelliptic operators – see Section 8.8). Once a partial differential operator is shown to be maximally subelliptic, many sharp results follow from the theory in this text.

Example 1.1.10. In the examples that follow, $(W, d) = \{(W_1, d_1), \ldots, (W_r, d_r)\}$ will denote any set of Hörmander vector fields with formal degrees on $\mathfrak{M}$. We write $(W, 1)$ for the special case where $d_1 = d_2 = \cdots = d_r = 1$. Recall that $n_j = \kappa / d_j \in \mathbb{N}_+$.

(i) With $\mathfrak{M} = \mathbb{R}^n$, let $(\partial, 1) := \{(\partial_{x_1}, 1), \ldots, (\partial_{x_n}, 1)\}$. Then $\mathscr{P}$ is maximally subelliptic of degree κ with respect to $(\partial, 1)$ if and only if $\mathscr{P}$ is locally elliptic of degree κ in the classical sense (see Definition 2.7.1). In particular, the Laplacian

$$\triangle = \partial_{x_1}^* \partial_{x_1} + \cdots + \partial_{x_n}^* \partial_{x_n} = -\partial_{x_1}^2 - \partial_{x_2}^2 - \cdots - \partial_{x_n}^2$$

is maximally subelliptic of degree 2 with respect to $(\partial, 1)$.

(ii) The “gradient”

$$\nabla_{\kappa,W} f := (W_1^{n_1} f, W_2^{n_2} f, \ldots, W_r^{n_r} f)$$

is maximally subelliptic of degree κ with respect to (W, d) on $\mathfrak{M}$; this follows immediately from Definition 1.1.7.

(iii) The “sub-Laplacian”

$$\mathscr{L} := \nabla_{\kappa,W}^* \nabla_{\kappa,W} = \sum_{j=1}^{r} (W_j^{n_j})^* W_j^{n_j}$$

is maximally subelliptic of degree κ with respect to (W, d) on $\mathfrak{M}$; here $*$ denotes the formal $L^2(\mathfrak{M}, \mathrm{Vol})$ adjoint. The maximal subellipticity of $\mathscr{L}$ follows from (ii) and Theorem 8.1.1 (i) ⇒ (vi). In the special case where (W, d) is replaced by $(W, 1)$ and $\kappa = 1$, the maximal subellipticity of $\mathscr{L}$ was shown by Rothschild and Stein [200]; in that case $\mathscr{L}$ is called Hörmander's sub-Laplacian. Just as the Laplacian plays an important role in the study of elliptic PDEs, the sub-Laplacian plays an important role in the study of maximally subelliptic PDEs (see Section 4.5). The sub-Laplacian can be used to give several interesting examples of maximally subelliptic operators:

(a) The operator

$$\partial_x^2 + (x\partial_y)^2 = \partial_x^2 + x^2 \partial_y^2 \tag{1.2}$$

5 For example, all maximally subelliptic operators are subelliptic (see (1.18) and Corollary 8.2.5), while not every partial differential operator is subelliptic.

is maximally subelliptic of degree 2 with respect to $\{(\partial_x, 1), (x\partial_y, 1)\}$ on $\mathbb{R}^2$. Note that this operator is not elliptic at $x = 0$. See also Example 1.1.2 (i). The operator in (1.2) is known as the Grushin operator.

(b) More generally, for $n_1, n_2 \in \mathbb{N}_+$, the operator

$$(-1)^{n_1}\partial_x^{2n_1} + (-1)^{n_2}(x\partial_y)^{2n_2} = (-1)^{n_1}\partial_x^{2n_1} + (-1)^{n_2}x^{2n_2}\partial_y^{2n_2}$$

is maximally subelliptic of degree $2n_1n_2$ with respect to $\{(\partial_x, n_2), (x\partial_y, n_1)\}$ on $\mathbb{R}^2$.

(c) Item (a) can be generalized in another way:

$$\partial_x^2 + (x^m\partial_y)^2 = \partial_x^2 + x^{2m}\partial_y^2 \tag{1.3}$$

is maximally subelliptic of degree 2 with respect to $\{(\partial_x, 1), (x^m\partial_y, 1)\}$ on $\mathbb{R}^2$. See also Example 1.1.2 (ii). The operator in (1.3) is known as a higher-order Grushin operator.

(d) The operator

$$(\partial_x + 2y\partial_t)^2 + (\partial_y - 2x\partial_t)^2 \tag{1.4}$$

is maximally subelliptic of degree 2 with respect to $\{(\partial_x + 2y\partial_t, 1), (\partial_y - 2x\partial_t, 1)\}$ on $\mathbb{R}^3$; see also Example 1.1.2 (iii). Note that the operator in (1.4) is **nowhere elliptic**. In fact, we will see that maximally subelliptic operators are usually nowhere elliptic. For example, in the special case where $d_1 = \cdots = d_r = 1$, the following are equivalent for a fixed $x_0 \in \mathfrak{M}$ and an operator $\mathcal{P}$ written in the form (1.1):

- $\mathcal{P}$ is elliptic of degree κ near x_0,
- $\mathcal{P}$ is maximally subelliptic of degree κ with respect to $(W, 1)$ near x_0 and $W = \{W_1, \dots, W_r\}$ satisfies Hörmander's condition of order 1 at x_0.

Moreover, the higher the order of Hörmander's condition at x_0, the more degenerate maximally subelliptic operators are at x_0. See Remark 8.2.7. Because of this, classical elliptic theory is not useful to study general maximally subelliptic operators.

(iv) On $\mathbb{R}^{n+1}$ with coordinates $(t, x_1, x_2, \dots, x_n)$, the heat operator

$$\partial_t + \partial_{x_1}^2 + \partial_{x_2}^2 + \cdots + \partial_{x_n}^2 \tag{1.5}$$

is maximally subelliptic of degree 2 with respect to $\{(\partial_t, 2), (\partial_{x_1}, 1), (\partial_{x_2}, 1), \dots, (\partial_{x_n}, 1)\}$ on $\mathbb{R}^{n+1}$. More generally, locally parabolic partial differential operators are the special case of maximally subelliptic operators when we choose $(W, d) = \{(\partial_t, 2), (\partial_{x_1}, 1), (\partial_{x_2}, 1), \dots, (\partial_{x_n}, 1)\}$.

(v) Let $W = \{W_1, \dots, W_r\}$ be vector fields satisfying Hörmander's condition on $\mathfrak{M}$. Generalizing (1.5), Rothschild and Stein [200, § 18] showed that

$$W_1 - W_2^* W_2 - W_3^* W_3 - \cdots - W_r^* W_r \tag{1.6}$$

is maximally subelliptic of degree 2 with respect to $\{(W_1, 2), (W_2, 1), (W_3, 1), \dots, (W_r, 1)\}$ on $\mathfrak{M}$. Operators of the type (1.6) arise in the study of Kolmogorov–Fokker–Planck equations and account for one of the main original motivating examples for studying maximally subelliptic operators; see [16, Chapter 2].

(vi) Let $W = \{W_1, \dots, W_r\}$ be vector fields satisfying Hörmander's condition. Let $a_1 \in \mathbb{N}_+$ be odd, let $a_2, \dots, a_r \in \mathbb{N}_+$ be even, and let $\kappa \in \mathbb{N}_+$ be the least common multiple of $a_1, \dots, a_r$. Generalizing (v), Androulidakis, Moshen, and Yuncken [2, Proposition 4.1] showed that the operator

$$W_1^{a_1} + \sum_{j=2}^{r} (-1)^{a_j/2} W_j^{a_j}$$

is maximally subelliptic of degree κ with respect to the Hörmander vector fields with formal degrees $(W, \vec{d}) = \{(W_1, \kappa/a_1), \dots, (W_r, \kappa/a_r)\}$.

(vii) Let $a_{i,j}(x) : \mathfrak{M} \to \mathbb{M}^{r\times r}(\mathbb{C})$ be a self-adjoint matrix depending smoothly on $x \in \mathfrak{M}$ which is strictly positive definite at every point. Then the operator

$$u \mapsto \sum_{i,j} a_{i,j}(x) W_i W_j u(x)$$

is maximally subelliptic of degree 2 with respect to $(W, 1)$ on $\mathfrak{M}$. See Example 8.9.2.

(viii) The operator in (vii) can be generalized to higher-order operators. Indeed, let $a_{i,j}(x) : \mathfrak{M} \to \mathbb{M}^{r\times r}(\mathbb{C})$ be a self-adjoint matrix depending smoothly on $x \in \mathfrak{M}$ which is strictly positive definite at every point. Then the operator

$$u \mapsto \sum_{i,j} a_{i,j}(x) W_i^{n_i} W_j^{n_j} u(x)$$

is maximally subelliptic of degree 2κ with respect to $(W, \vec{d})$ on $\mathfrak{M}$. One can also add lower-order terms. See Proposition 8.9.1 for details. For a different generalization of (vii) to higher-order operators, see Proposition 8.9.4.

(ix) Consider the Laplacian on $\mathbb{R}^2$: $\triangle = -\partial_x^2 - \partial_y^2$. As described in (i), $\triangle$ is maximally subelliptic of degree 2 with respect to $\{(\partial_x, 1), (\partial_y, 1)\}$ on $\mathbb{R}^2$. Consider, instead, the Hörmander vector fields with formal degrees $(W, \vec{d}) = \{(\partial_x, 1), (x\partial_y, 1)\}$. Then, on any neighborhood of $(0, 0)$, $\triangle$ is a polynomial of degree 4 in ∂_x and $x\partial_y$, and is not maximally subelliptic of degree 4 (or of any degree) with respect to $(W, \vec{d})$ on any neighborhood of $(0, 0)$. This highlights the fact that maximal subellipticity depends on the choice of Hörmander vector fields with formal degrees. Depending on the choice of $(W, \vec{d})$, elliptic operators may not be maximally subelliptic with respect to $(W, \vec{d})$ (see also the discussion in (iii), part (d)). Thus, ellipticity is not a stronger condition than maximal subellipticity; it is a special case of maximal subellipticity.

Maximally subelliptic operators in general, and the sub-Laplacian in particular, arise in many settings, including probability theory, several complex variables, and stochastic PDEs. The text [16] gives an excellent history of the subject; see also Sections 8.10 and 9.5 for a further discussion and many references.

1.1.2 Nonlinear maximally subelliptic operators

Fix $\mathbf{D}_1, \mathbf{D}_2 \in \mathbb{N}_+$ and let

$$N := \mathbf{D}_2 \times \#\{\alpha \text{ an ordered multi-index} : \deg_{\vec{d}}(\alpha) \leq \kappa\}. \tag{1.7}$$

Given a sufficiently smooth function $u = (u_1, \dots, u_{\mathbf{D}_2}) : \mathfrak{M} \to \mathbb{R}^{\mathbf{D}_2}$, we define

$$\{W^\alpha u\}_{\deg_{\vec{d}}(\alpha)\leq\kappa} : \mathfrak{M} \to \mathbb{R}^N$$

to be the vector-valued function whose components are given by $W^\alpha u_j$, $\deg_{\vec{d}}(\alpha) \leq \kappa$ and $1 \leq j \leq \mathbf{D}_2$.

Let $F(x, \zeta) : \mathfrak{M} \times \mathbb{R}^N \to \mathbb{R}^{\mathbf{D}_1}$ be a function.[6] Given a nonlinear PDE, we write it in the form

$$F(x, \{W^\alpha u(x)\}_{\deg_{\vec{d}}(\alpha)\leq\kappa}) = g(x). \tag{1.8}$$

By Hörmander's condition, *every* nonlinear PDE can be written in the form (1.8).

For a fixed $x_0 \in \mathfrak{M}$ and $u : \mathfrak{M} \to \mathbb{R}^{\mathbf{D}_2}$ the operator

$$\mathscr{P}_{u,x_0} v := d_\zeta F(x_0, \{W^\alpha u(x_0)\}_{\deg_{\vec{d}}(\alpha)\leq\kappa})\{W^\alpha v\}_{\deg_{\vec{d}}(\alpha)\leq\kappa} \tag{1.9}$$

is a partial differential operator of the form

$$\mathscr{P}_{u,x_0} v = \sum_{\deg_{\vec{d}}(\alpha)\leq\kappa} a_{\alpha,u,x_0} W^\alpha v,$$

where $a_{\alpha,u,x_0} \in \mathbb{M}^{\mathbf{D}_1\times\mathbf{D}_2}(\mathbb{R})$ are *constant* matrices which depend on x_0 and a finite number of Taylor coefficients of u at x_0.

6 For the purposes of this introduction, we assume that all functions are smooth enough that every expression makes sense.

Definition 1.1.11. Fix $x_0 \in \mathfrak{M}$, $u : \mathfrak{M} \to \mathbb{R}^{\mathbf{D}_2}$, and $F : \mathfrak{M} \times \mathbb{R}^N \to \mathbb{R}^{\mathbf{D}_1}$; we assume both u and F are sufficiently smooth so that all objects which follow make sense. We say the nonlinear partial differential operator

$$u \mapsto F(x, \{W^\alpha u(x)\}_{\deg_{\vec{d}}(\alpha) \le \kappa}) \tag{1.10}$$

is maximally subelliptic at (x_0, u) of degree κ with respect to $(W, \vec{d})$ if there exists an open neighborhood $U \subseteq \mathfrak{M}$ of x_0 such that the linear partial differential operator $\mathscr{P}_{u,x_0}$ given by (1.9) is maximally subelliptic of degree κ with respect to $(W, \vec{d})$ on U.

We do not assume the function $F(x, \zeta)$ in Definition 1.1.11 is smooth. In particular, by considering the case where $F(x, \zeta)$ is linear in ζ but not smooth in x, we will be able to study linear maximally subelliptic operators with non-smooth coefficients as a special case of Definition 1.1.11. This will allow us to conclude results like interior Schauder estimates by using a more general fully nonlinear theory. See, for example, Corollary 9.1.10.

Because the definition of a fully nonlinear equation being maximally subelliptic is just that a "frozen coefficient" version of the linearization is maximally subelliptic, once one has examples of linear maximally subelliptic operators (as in Example 1.1.10) it is simple to write down examples of nonlinear maximally subelliptic operators. Moreover, important examples of fully nonlinear elliptic operators have generalizations to the maximally subelliptic setting, as the next examples show.

Example 1.1.12 (The Monge–Ampère equation). An important nonlinear PDE is the Monge–Ampère equation. For a function $u : \mathbb{R}^n \to \mathbb{R}$, we consider the Hessian $Hu = (\partial_{x_i}\partial_{x_j} u(x))_{1\le i,j\le n}$. If $Hu(x_0) > 0$, i. e., $Hu(x_0)$ is a strictly positive definite matrix, then the fully nonlinear operator $u \mapsto \det Hu$ is elliptic at (x_0, u); this is known as the Monge–Ampère equation. Note that the condition $Hu(x_0) > 0$ implies u is strictly convex near x_0 (provided u is at least C^2).

(i) As first noted in [157] and [57], a natural generalization of the Monge–Ampère equation to the maximally subelliptic setting is as follows. Let $(W, 1) = \{(W_1, 1), \ldots, (W_r, 1)\}$ be Hörmander vector fields each paired with formal degree 1. For a function $u : \mathfrak{M} \to \mathbb{R}$, define a symmetric $r \times r$ matrix $\mathscr{W} u(x)$ whose (i,j) component is given by $\frac{1}{2}(W_i W_j + W_j W_i)u(x)$. If $\mathscr{W} u(x_0) > 0$ (i. e., if the matrix is strictly positive definite at x_0), then $u \mapsto \det \mathscr{W} u$ is maximally subelliptic at (x_0, u) of degree 2 with respect to $(W, 1)$ – this is proved, and more details are given, in Section 9.4.2. By choosing $W_1, \ldots, W_r$ to satisfy Hörmander's condition of a very high order, this fully nonlinear equation is very "degenerate;" see the discussion in Example 1.1.10 (d).

(ii) The maximally subelliptic Monge–Ampère equation from (i) can be further generalized to higher-order fully nonlinear equations. Let $(W, \vec{d}) = \{(W_1, \vec{d}_1), \ldots, (W_r, \vec{d}_r)\}$ be Hörmander vector fields with formal degrees. As above, fix $\kappa \in \mathbb{N}_+$ such that $\vec{d}_j$ divides κ, $\forall j$, and set $n_j := \kappa / \vec{d}_j \in \mathbb{N}_+$. For a function $u : \mathfrak{M} \to \mathbb{R}$, define a symmetric $r \times r$ matrix $\mathscr{W}[\kappa]u(x)$ whose (i,j) component is given by $\frac{1}{2}(W_i^{n_i} W_j^{n_j} + W_j^{n_j} W_i^{n_i})u(x)$. If

$\mathscr{W}[\kappa]u(x_0) > 0$, then $u \mapsto \det \mathscr{W}[\kappa]u$ is maximally subelliptic at (x_0, u) of degree 2κ with respect to $(W, \vec{d})$ – this is proved, and more details are given, in Section 9.4.3.

(iii) There is a different generalization (i) to higher-order operators. With the same setting as in (ii), for a function $u : \mathfrak{M} \to \mathbb{R}$ define a symmetric, square matrix, $\widehat{\mathscr{W}}[\kappa]u(x)$, whose components are parameterized by (α, β), where $\deg_{\vec{d}}(\alpha) = \deg_{\vec{d}}(\beta) = \kappa$, by

$$(\widehat{\mathscr{W}}[\kappa]u(x))_{\alpha,\beta} = \frac{1}{2}(W^\alpha W^\beta u(x) + W^\beta W^\alpha u(x)), \quad \deg_{\vec{d}}(\alpha) = \deg_{\vec{d}}(\beta) = \kappa.$$

If $\widehat{\mathscr{W}}[\kappa]u(x_0) > 0$, then $u \mapsto \det \widehat{\mathscr{W}}[\kappa]u$ is maximally subelliptic at (x_0, u) of degree 2κ with respect to $(W, \vec{d})$ – this is proved, and more details are given, in Section 9.4.3.

Example 1.1.13. The ellipticity of the classical Monge–Ampère equation is just one example of a second-order fully nonlinear equation whose linearization is elliptic because its principal part is given by a strictly positive definite quadratic form – a phenomenon which often appears in elliptic theory. This phenomenon can be generalized to the maximally subelliptic setting as follows. Let $(W, 1) = \{(W_1, 1), \dots, (W_r, 1)\}$ be Hörmander vector fields, each paired with formal degree 1. Consider a nonlinear partial differential operator of the form

$$u \mapsto F(x, \{W^\alpha u(x)\}_{|\alpha| \le 2}), \tag{1.11}$$

where F is a real-valued function. We consider $F = F(x, \zeta)$ as a function of $x \in \mathfrak{M}$ and $\zeta \in \mathbb{R}^{r^2+r+1}$: $\zeta = (\zeta_0, \zeta_1, \dots, \zeta_r, \dots, \zeta_{i,j}, \dots, \zeta_{r,r})$, where, for example, the coordinate $\zeta_{i,j}$ corresponds to $W_i W_j u$. For a fixed $u : \mathfrak{M} \to \mathbb{R}$ and $x_0 \in \mathfrak{M}$, if the $r \times r$ matrix

$$\begin{pmatrix} \partial_{\zeta_{1,1}} F(x_0, \{W^\alpha u(x_0)\}_{|\alpha| \le 2}) & \cdots & \partial_{\zeta_{1,r}} F(x_0, \{W^\alpha u(x_0)\}_{|\alpha| \le 2}) \\ \vdots & \ddots & \vdots \\ \partial_{\zeta_{r,1}} F(x_0, \{W^\alpha u(x_0)\}_{|\alpha| \le 2}) & \cdots & \partial_{\zeta_{r,r}} F(x_0, \{W^\alpha u(x_0)\}_{|\alpha| \le 2}) \end{pmatrix}$$

is symmetric and strictly positive definite, then (1.11) is maximally subelliptic at (x_0, u) of degree 2 with respect to $(W, 1)$. See Section 9.4.1 for details. Moreover, similar to Example 1.1.12 (ii) and (iii), this idea can be generalized to higher-order operators. See Section 9.4.4 for details.

See Section 9.5 for several other settings where nonlinear maximally subelliptic and other related nonlinear PDEs arise.

Remark 1.1.14. One main goal of this text is to study the regularity theory of classical solutions to general fully nonlinear maximally subelliptic equations of arbitrary order. Because we work with general equations, we do not introduce any notion of a weak solution.

1.2 Background: elliptic theory

All the results of this text have classical versions for elliptic equations. We describe the main results we generalize here; these topics are discussed in more detail in Chapter 2. We work in $\mathbb{R}^n$.

The sharp regularity theory of elliptic operators is often described in terms of the classical Besov ($\mathscr{B}^{\mathbf{s}}_{p,q}(\mathbb{R}^n)$) and Triebel–Lizorkin ($\mathscr{F}^{\mathbf{s}}_{p,q}(\mathbb{R}^n)$) spaces; here $\mathbf{s} \in \mathbb{R}$ and $p, q \in (0, \infty]$. However, for the purposes of this text, we consider only the simplest setting when $p, q \in [1, \infty]$ in general and restrict our attention to $p \in (1, \infty)$ for $\mathscr{F}^{\mathbf{s}}_{p,q}$. A unifying aspect of these spaces is that ∂_{x_j} acts as a differential operator of order 1: $\partial_{x_j} : \mathscr{X}^{\mathbf{s}} \to \mathscr{X}^{\mathbf{s}-1}$, where $\mathscr{X}^{\mathbf{s}}$ is either $\mathscr{B}^{\mathbf{s}}_{p,q}(\mathbb{R}^n)$ or $\mathscr{F}^{\mathbf{s}}_{p,q}(\mathbb{R}^n)$. See Section 2.4 for a more complete discussion.

While these spaces have a technical definition, many of the classical spaces from harmonic analysis can be seen as special cases. For example, $\mathscr{F}^{\mathbf{s}}_{p,2}(\mathbb{R}^n)$ is the L^p Sobolev space of order $\mathbf{s} \in \mathbb{R}$. Also, for $\mathbf{s} > 0$, $\mathscr{C}^{\mathbf{s}}(\mathbb{R}^n) := \mathscr{B}^{\mathbf{s}}_{\infty,\infty}(\mathbb{R}^n)$ is the so-called Zygmund–Hölder space of order $\mathbf{s}$. When $\mathbf{s} \in (0,\infty)\setminus\mathbb{N}_+$, $\mathscr{C}^{\mathbf{s}}(\mathbb{R}^n) = C^{\lfloor \mathbf{s} \rfloor, \mathbf{s}-\lfloor \mathbf{s} \rfloor}(\mathbb{R}^n)$ is the usual Hölder space of order $\mathbf{s}$. However, when $\mathbf{s} \in \mathbb{N}_+$, $\mathscr{C}^{\mathbf{s}}(\mathbb{R}^n)$ is not a Hölder space: $C^{m+1}(\mathbb{R}^n) \subsetneq C^{m,1}(\mathbb{R}^n) \subsetneq \mathscr{C}^{m+1}(\mathbb{R}^n)$; see Section 2.5 for a detailed description of $\mathscr{C}^{\mathbf{s}}(\mathbb{R}^n)$.

Some key results we generalize from the elliptic setting are:

- **Parametrices:** Let $\mathscr{P}$ be an elliptic partial differential operator of order κ with smooth coefficients. Then there is an operator T which is locally a pseudo-differential operator of order $-\kappa$ such that

$$T\mathscr{P} \equiv I \mod C^{\infty}_{\mathrm{loc}}.$$

  See Theorem 2.7.2 (g) for a more detailed statement and Proposition 2.2.27 for a proof of a similar result.
- **Linear regularity:** Let $\mathscr{P}$ be an elliptic partial differential operator of order κ on $\mathbb{R}^n$, with smooth coefficients. Then if $\mathscr{X}^{\mathbf{s}}$ denotes any one of the scales of spaces $\mathscr{B}^{\mathbf{s}}_{p,q}(\mathbb{R}^n)$ or $\mathscr{F}^{\mathbf{s}}_{p,q}(\mathbb{R}^n)$, we have, for any distribution u and any point $x_0 \in \mathbb{R}^n$,

$$\mathscr{P}u \in \mathscr{X}^{\mathbf{s}} \text{ near } x_0 \iff u \in \mathscr{X}^{\mathbf{s}+\kappa} \text{ near } x_0.$$

  See Theorem 2.7.2 (e) for a more precise statement.
- **Fully nonlinear regularity:** Fix $\mathbf{s} > \mathbf{r} > 0$ and let $F(x, \zeta)$ be a function which is $\mathscr{C}^{\mathbf{s}}$ in the x variable and $C^{\infty}_{\mathrm{loc}}$ in the ζ variable.[7] Fix $x_0 \in \mathbb{R}^n$ and suppose $u \in \mathscr{C}^{\mathbf{r}+\kappa}(\mathbb{R}^n)$ is such that $F(x, \{\partial^{\alpha}_x u(x)\}_{|\alpha|\leq\kappa})$ is elliptic near x_0 and u. That is, suppose that the constant coefficient, linear partial differential operator

$$\mathscr{P}_{u,x_0} v := d_{\zeta} F(x_0, \{\partial_x u(x_0)\}_{|\alpha|\leq\kappa})\{\partial^{\alpha}_x v\}_{|\alpha|\leq\kappa}$$

7 We only require finite regularity for $F(x, \zeta)$ in the ζ variable, though we do not make that precise here. See Theorem 2.8.3 for more details.

is elliptic of order κ. Then

$$F(x, \{\partial_x^\alpha u(x)\}_{|\alpha|\leq\kappa}) \in \mathscr{C}^{\mathbf{s}}(\mathbb{R}^n) \text{ near } x_0 \iff u \in \mathscr{C}^{\mathbf{s}+\kappa}(\mathbb{R}^n) \text{ near } x_0.$$

See Theorem 2.8.3 for a more general and precise statement.

- **Interior Schauder estimates:** Fix $\mathbf{s} > \mathbf{r} > 0$ and let $\mathscr{P}$ be an elliptic partial differential operator of order κ with $\mathscr{C}^{\mathbf{s}}$ coefficients. Then, for $x_0 \in \mathbb{R}^n$ and $u \in \mathscr{C}^{\mathbf{r}+\kappa}(\mathbb{R}^n)$,

$$\mathscr{P}u \in \mathscr{C}^{\mathbf{s}}(\mathbb{R}^n) \text{ near } x_0 \iff u \in \mathscr{C}^{\mathbf{s}+\kappa}(\mathbb{R}^n) \text{ near } x_0.$$

  This is the special case of the fully nonlinear result described above where we take $F(x,\zeta)$ to be linear in ζ. See Corollary 2.8.5 for a more precise statement.

- **Gaussian bounds:** Let $\mathscr{P}$ be an elliptic partial differential operator of order κ with smooth coefficients. Let $\mathscr{L}$ be any non-negative self-adjoint extension of the non-negative symmetric operator $\mathscr{P}^*\mathscr{P}$. Then, for $t > 0$, the heat operator $e^{-t\mathscr{L}}$ is given by integration against a smooth function, which we denote by $e^{-t\mathscr{L}}(x,y)$. Moreover, we have local Gaussian bounds for this heat kernel: for every compact set $\mathcal{K} \Subset \mathbb{R}^n$, $\exists c = c(\mathcal{K}) > 0$, $\forall \alpha, \beta \in \mathbb{N}^n$, $\forall s \in \mathbb{N}$, $\exists C = C(\alpha,\beta,s,\mathcal{K}) \geq 0$, so that $\forall t \in (0,1]$, $\forall x,y \in \mathcal{K}$,

$$|\partial_t^s \partial_x^\alpha \partial_y^\beta e^{-t\mathscr{L}}(x,y)| \leq C(|x-y| + t^{\frac{1}{2\kappa}})^{-n-|\alpha|-|\beta|-2\kappa s} \exp\left(-c\left(\frac{|x-y|^{2\kappa}}{t}\right)^{1/(2\kappa-1)}\right).$$

  See Theorem 2.7.2 (i) for more details.

In fact, Theorem 2.7.2 shows that more precise restatements of the parametrices, linear regularity, and Gaussian bounds described above are actually equivalent to ellipticity.

As described in the above bullet points on parametrices and local regularity, linear elliptic operators with smooth coefficients are locally *left invertible* modulo C^∞_{loc} as maps $\mathscr{X}^{\mathbf{s}+\kappa} \to \mathscr{X}^{\mathbf{s}}$, where $\mathscr{X}^{\mathbf{s}}$ denotes either a scale of Besov or Triebel–Lizorkin spaces. In particular, they are locally left invertible modulo C^∞_{loc} as maps $\mathscr{C}^{\mathbf{s}+\kappa}(\mathbb{R}^n) \to \mathscr{C}^{\mathbf{s}}(\mathbb{R}^n)$. As we will see in the proof of Theorem 2.8.7, one can use this invertibility to apply ideas similar to the contraction mapping principle (along with some careful covering lemmas) to deduce the fully nonlinear regularity described above. Elliptic operators are the only partial differential operators which have this left invertibility on the classical spaces, which prevents such a simple argument from directly working for any operators which are not elliptic. To avoid this problem, when we move to studying more general maximally subelliptic operators, we introduce generalizations of the Besov and Triebel–Lizorkin spaces, adapted to the operator in question, on which it is left invertible modulo smooth functions.

1.3 Carnot–Carathéodory geometry

Intimately tied to a maximally subelliptic operator is a metric geometry on $\mathfrak{M}$ induced by the Hörmander vector fields with formal degrees $(W, \vec{d})$, known as a Carnot–Carathéodory (or sub-Riemannian) geometry. This geometry plays the role for maximal subellipticity with respect to $(W, \vec{d})$ that Riemannian geometry plays for ellipticity.

Let $\mathfrak{M}$ be a connected smooth manifold of dimension ≥ 1 and let $W = \{W_1, \ldots, W_r\}$ be a finite set of smooth vector fields on $\mathfrak{M}$. We define the Carnot–Carathéodory ball of unit radius, centered at $x \in \mathfrak{M}$, by

$$B_W(x) := \left\{ y \in \mathfrak{M} \;\middle|\; \begin{array}{l} \exists \gamma : [0,1] \to \mathfrak{M}, \gamma(0) = x, \gamma(1) = y, \\ \gamma \quad \text{is absolutely continuous}, \\ \gamma'(t) = \sum_{j=1}^{r} a_j(t) W_j(\gamma(t)) \quad \text{almost everywhere}, \\ a_j \in L^\infty([0,1]), \left\| \sum_{j=1}^{r} |a_j|^2 \right\|_{L^\infty([0,1])} < 1 \end{array} \right\}. \tag{1.12}$$

In particular, since γ is absolutely continuous in (1.12), we have in local coordinates $\gamma(t_1) = \gamma(t_0) + \int_{t_0}^{t_1} \sum_{j=1}^{r} a_j(t) W_j(\gamma(t))\, dt$.

To define balls of other radii, we scale the vector fields. We use a list of vector fields with formal degrees $(W, \vec{d}) := \{(W_1, \vec{d}_1), \ldots, (W_r, \vec{d}_r)\}$, with $\vec{d}_j \in \mathbb{N}_+$. For $\delta > 0$ set $\delta^{\vec{d}} W := \{\delta^{\vec{d}_1} W_1, \ldots, \delta^{\vec{d}_r} W_r\}$. We define the Carnot–Carathéodory ball of radius $\delta > 0$, centered at $x \in \mathfrak{M}$, by

$$B_{(W,\vec{d})}(x, \delta) := B_{\delta^{\vec{d}} W}(x). \tag{1.13}$$

Corresponding to the balls $B_{(W,\vec{d})}(x, \delta)$ is the Carnot–Carathéodory distance

$$\rho_{(W,\vec{d})}(x,y) := \inf\{\delta > 0 : y \in B_{(W,\vec{d})}(x, \delta)\}. \tag{1.14}$$

It is not hard to see that $\rho_{(W,\vec{d})}$ is an extended metric on $\mathfrak{M}$ (it may take the value $+\infty$, but otherwise it satisfies the usual axioms for a metric).

A main setting in which we study these Carnot–Carathéodory balls is when $(W, \vec{d})$ are Hörmander vector fields with formal degrees (see Definition 1.1.3). In this case, a classical theorem of Chow [45] (see Lemma 3.1.7) implies that $\rho_{(W,\vec{d})}$ is a metric on $\mathfrak{M}$ (i. e., $\rho_{(W,\vec{d})}(x,y) < \infty$, $\forall x, y \in \mathfrak{M}$), and the metric topology on $\mathfrak{M}$ induced by $\rho_{(W,\vec{d})}$ agrees with the topology on $\mathfrak{M}$ (as a manifold) – see Lemma 3.1.7. In this setting, Nagel, Stein, and Wainger [189] proved the important inequality for small $\delta > 0$

$$\mathrm{Vol}(B_{(W,\vec{d})}(x, 2\delta)) \lesssim \mathrm{Vol}(B_{(W,\vec{d})}(x, \delta)), \tag{1.15}$$

Table 1.1: Analogs between ellipticity and maximal subellipticity.

Elliptic	Maximally subelliptic
Riemannian geometry	Carnot–Carathéodory (CC) geometry
Pseudo-differential operators	Singular integral operators adapted to the CC geometry
Classical (Riemannian) function spaces	Function spaces adapted to the CC geometry
Gaussian bounds for heat equations using a Riemannian metric	Gaussian bounds for heat equations using the CC metric
Standard dilation maps on $\mathbb{R}^n$	Abstract scaling maps adapted to the CC geometry
Fourier transform	No single analog

uniformly as x ranges over a compact set. Formula (1.15) implies that the balls $B_{(W,d)}(x,\delta)$ when paired with the smooth, strictly positive density Vol locally give $\mathfrak{M}$ the structure of a space of homogeneous type in the sense of Coifman and Weiss [52]. This gives access to the robust Calderón–Zygmund theory, which is an essential tool of this text (see, for example, Section 1.4).

See Table 1.1 for some of the many ways Carnot–Carathéodory geometry arises when studying maximally subelliptic operators.

1.4 Singular integrals

Unlike the case of elliptic operators, standard pseudo-differential operators and the Fourier transform are not decisive tools when studying maximally subelliptic operators. Since the work of Folland and Stein [91], Folland [90], Rothschild and Stein [200], and Nagel, Stein, and Wainger [189], it has been clear that singular integral operators adapted to the Carnot–Carathéodory geometry can often take the place of pseudo-differential operators. The main results of this text are local, so it suffices to consider operators whose Schwartz kernels have compact support.

For $t \in \mathbb{R}$, in Chapter 5 we define a class of singular integral operators $\mathscr{A}^t(W,d)$ whose Schwartz kernels have compact support. These satisfy the following properties:

- $\mathscr{A}^t(W,d)$ is a filtered algebra: For $T \in \mathscr{A}^t(W,d)$ and $S \in \mathscr{A}^s(W,d)$, we have $TS \in \mathscr{A}^{t+s}(W,d)$.
- Operators in $\mathscr{A}^t(W,d)$ have good boundedness properties on function spaces adapted to (W,d). See Section 1.5.
- Left parametrices for maximally subelliptic operators of degree κ are locally in $\mathscr{A}^{-\kappa}(W,d)$. See Section 1.6.
- If $\psi \in C_0^\infty(\mathfrak{M})$, then $\mathrm{Mult}[\psi]W_j \in \mathscr{A}^{d_j}(W,d)$, where $\mathrm{Mult}[\psi] : f \mapsto \psi f$. More generally, if $\mathscr{P}$ is a partial differential operator of the form (1.1), then $\mathrm{Mult}[\psi]\mathscr{P} \in \mathscr{A}^\kappa(W,d)$.

- Operators in $\mathscr{A}^t(W, \vec{d})$ are pseudo-local: their Schwartz kernels are smooth off the diagonal.

Remark 1.4.1. If $T \in \mathscr{A}^t(W, \vec{d})$ and $S \in \mathscr{A}^s(W, \vec{d})$, then by the filtered algebra property we have $[T, S] = TS - ST \in \mathscr{A}^{t+s}(W, \vec{d})$; however, unlike pseudo-differential operators we do not have $[T, S] \in \mathscr{A}^{t+s-1}(W, \vec{d})$. This is unavoidable: we wish to treat $[W_j, W_k]$ as a differential operator of degree $\vec{d}_j + \vec{d}_k$ (see Remark 1.1.5). Because of this, most proofs from elliptic theory which rely on pseudo-differential operators do not directly translate to the maximally subelliptic setting.

We defer the technical definition of $\mathscr{A}^t(W, \vec{d})$ to Section 5.2.1 (where, in fact, four equivalent definitions are given). However, $T \in \mathscr{A}^0(W, \vec{d})$ implies the following:
- The Schwartz kernel $T(x, z)$ of T agrees with a smooth function (which we again denote by $T(x, z)$) away from the diagonal $x = z$. Moreover, for all ordered multi-indices α, β, we have

$$|W_x^\alpha W_z^\beta T(x, z)| \leq C_{\alpha,\beta} \frac{\rho_{(W,\vec{d})}(x, z)^{-\deg_{\vec{d}}(\alpha) - \deg_{\vec{d}}(\beta)}}{\mathrm{Vol}(B_{(W,\vec{d})}(x, \rho_{(W,\vec{d})}(x, z))) \wedge 1},$$

  where W_x denotes the list of vector fields $W_1, \ldots, W_r$ acting as partial differential operators in the x variable, and similarly for W_z.
- T has good bounds when acting on certain bump functions which are adapted to the Carnot–Carathéodory geometry.

In particular, T is a Calderón–Zygmund singular integral operator with respect to the local space of homogeneous type on $\mathfrak{M}$ induced by the Carnot–Carathéodory metric $\rho_{(W,\vec{d})}$. The classical Calderón–Zygmund theory can then be used to show that T extends to a bounded operator on $L^p(\mathfrak{M}, \mathrm{Vol})$, $1 < p < \infty$. This boundedness is an important tool in our study of the regularity theory of maximally subelliptic operators.

While the singular integrals above are our main replacement for pseudo-differential operators, there is a subclass of these singular integrals which is somewhat more analogous to the classical approach to pseudo-differential operators; we call this subclass (generalized) pseudo-differential operators.[8] We present and study these operators in Chapter 4. Unfortunately, though they do have some important applications, these generalized pseudo-differential operators do not seem robust enough to establish all the results we need, so we instead turn to the more general singular integral operators in Chapter 5.

---

8 This class of operators appears in the Rothschild and Stein lifting procedure, which is very useful, but has some limitations. See Section 4.5.7.

1.5 Function spaces

The sharp regularity theory of maximally subelliptic operators is in terms of Besov ($\mathscr{B}^{\mathbf{s}}_{p,q}(W, \vec{d})$) and Triebel–Lizorkin ($\mathscr{F}^{\mathbf{s}}_{p,q}(W, \vec{d})$) spaces which are adapted to $(W, \vec{d})$;[9] here $\mathbf{s} \in \mathbb{R}$. We define and study these spaces in Chapter 6. The two most important special cases are $\mathscr{F}^{\mathbf{s}}_{p,2}(W, \vec{d})$, which can be viewed as L^p Sobolev spaces adapted to $(W, \vec{d})$, and $\mathscr{C}^{\mathbf{s}}(W, \vec{d}) := \mathscr{B}^{\mathbf{s}}_{\infty,\infty}(W, \vec{d})$, $\mathbf{s} > 0$, which can be viewed as a Zygmund–Hölder space adapted to $(W, \vec{d})$. In Chapters 6 and 7 we establish the following properties for $\mathscr{X}^{\mathbf{s}}(W, \vec{d}) \in \{\mathscr{B}^{\mathbf{s}}_{p,q}(W, \vec{d}), \mathscr{F}^{\mathbf{s}}_{p,q}(W, \vec{d})\}$:

- W_j acts as a differential operator of degree $\vec{d}_j$. More precisely, $W_j : \mathscr{X}^{\mathbf{s}}(W, \vec{d}) \to \mathscr{X}^{\mathbf{s}-\vec{d}_j}(W, \vec{d})$.
- If $T \in \mathscr{A}^{\mathbf{t}}(W, \vec{d})$, then $T : \mathscr{X}^{\mathbf{s}}(W, \vec{d}) \to \mathscr{X}^{\mathbf{s}-\mathbf{t}}(W, \vec{d})$.
- $\mathscr{F}^{\mathbf{s}}_{p,2}(W, \vec{d})$ can be viewed as the L^p Sobolev space of order $\mathbf{s} \in \mathbb{R}$ with respect to $(W, \vec{d})$, for $1 < p < \infty$. More precisely, let $\kappa \in \mathbb{N}$ be such that $\vec{d}_j$ divides κ, $1 \leq j \leq r$. Then, for a distribution u with compact support,
$$u \in \mathscr{F}^{\kappa}_{p,2}(W, \vec{d}) \iff W^{\alpha}u \in L^p(\mathfrak{M}, \mathrm{Vol}), \forall \deg_{\vec{d}}(\alpha) \leq \kappa.$$
In particular, $\mathscr{F}^{0}_{p,2}(W, \vec{d})$ is locally equal to $L^p(\mathfrak{M}, \mathrm{Vol})$.
- For $\mathbf{s} \in (0, 1)$, the space $\mathscr{C}^{\mathbf{s}}(W, \vec{d})$ is locally equal to the Hölder space ($C^{0,\mathbf{s}}(W, \vec{d})$) with respect to the Carnot–Carathéodory metric $\rho_{(W,\vec{d})}$:
$$\|f\|_{C^{0,\mathbf{s}}(W,\vec{d})} := \|f\|_{C(\mathfrak{M})} + \sup_{\substack{x,y \in \mathfrak{M} \\ x \neq y}} \rho_{(W,\vec{d})}(x, y)^{-\mathbf{s}} |f(x) - f(y)|.$$
Moreover, in the special case where $\vec{d}_1 = \vec{d}_2 = \cdots = \vec{d}_r = 1$ and $\mathbf{s} \in (0, \infty) \setminus \mathbb{N}_+$, $\mathscr{C}^{\mathbf{s}}(W, 1)$ is locally equal to the naturally defined Hölder space $C^{\lfloor \mathbf{s} \rfloor, \mathbf{s} - \lfloor \mathbf{s} \rfloor}(W, 1)$ with norm
$$\|f\|_{C^{m,t}(W,1)} := \sum_{|\alpha| \leq m} \|W^{\alpha} f\|_{C^{0,t}(W,1)}, \quad m \in \mathbb{N},\ t \in [0, 1].$$
See Theorem 7.7.23 for details.
- When $\mathfrak{M} = \mathbb{R}^n$, with $(\partial, 1) = \{(\partial_{x_1}, 1), \ldots, (\partial_{x_n}, 1)\}$, $\mathscr{B}^{\mathbf{s}}_{p,q}(\partial, 1)$ and $\mathscr{F}^{\mathbf{s}}_{p,q}(\partial, 1)$ are locally equal to their classical counterparts $\mathscr{B}^{\mathbf{s}}_{p,q}(\mathbb{R}^n)$ and $\mathscr{F}^{\mathbf{s}}_{p,q}(\mathbb{R}^n)$; see Section 6.6.1 for details.

One interesting aspect of our theory is that we define the Besov and Triebel–Lizorkin spaces for all regularity parameters $\mathbf{s} \in \mathbb{R}$ in such a way that we can concretely under-

9 In defining these spaces, we focus only on distributions with compact support. This is sufficient since the main results of this text are local. Though it is likely not necessary, we also restrict our attention to $p, q \in [1, \infty]$ in general and to $p \in (1, \infty)$ when considering $\mathscr{F}^{\mathbf{s}}_{p,q}$.

stand these spaces – for example we can characterize $\mathscr{C}^{\mathbf{s}}(W, \vec{d})$ in terms of differences. In the literature there are two typical situations similar to ours:

- Theories which define $\mathscr{B}^{\mathbf{s}}_{p,q}$ and/or $\mathscr{F}^{\mathbf{s}}_{p,q}$ for all $\mathbf{s} \in \mathbb{R}$. Such theories usually use some sort of Lie group structure to help define the spaces (e. g., for the classical spaces, $\mathbb{R}^n$ is used).
- Theories which define $\mathscr{B}^{\mathbf{s}}_{p,q}$ and/or $\mathscr{F}^{\mathbf{s}}_{p,q}$ for $|\mathbf{s}|$ small (e. g., $|\mathbf{s}| < 1$). With the restriction that $\mathbf{s}$ must be small, one can define these function spaces for much more general underlying spaces.

See Section 6.14 for many references. In our setting, we define and study the Besov and Triebel–Lizorkin spaces *for all* regularity parameters $\mathbf{s} \in \mathbb{R}$, even though there is no group structure on which to rely.

1.6 Linear theory

Let $\kappa \in \mathbb{N}_+$ be such that $\vec{d}_j$ divides κ, $1 \le j \le r$, and let $\mathscr{P} = \sum_{\deg_{\vec{d}}(\alpha) \le \kappa} a_\alpha(x) W^\alpha$ be a partial differential operator with smooth coefficients written in the form (1.1). If $\mathscr{P}$ is maximally subelliptic of degree κ with respect to $(W, \vec{d})$ on $\mathfrak{M}$, then in Theorem 8.1.1 we establish the following.

- **Parametrices:** There is an operator T, which is locally in $\mathscr{A}^{-\kappa}(W, \vec{d})$, such that
$$T\mathscr{P} \equiv I \mod C^\infty_{\mathrm{loc}}(\mathfrak{M}).$$
- **Regularity:** If $\mathscr{X}^{\mathbf{s}}(W, \vec{d})$ denotes any one of the scales of spaces $\mathscr{B}^{\mathbf{s}}_{p,q}(W, \vec{d})$ or $\mathscr{F}^{\mathbf{s}}_{p,q}(W, \vec{d})$, then for any distribution u and any point $x_0 \in \mathfrak{M}$,
$$\mathscr{P}u \in \mathscr{X}^{\mathbf{s}}(W, \vec{d}) \text{ near } x_0 \iff u \in \mathscr{X}^{\mathbf{s}+\kappa}(W, \vec{d}) \text{ near } x_0. \tag{1.16}$$
- **Gaussian bounds:** Let $\mathscr{L}$ be any non-negative self-adjoint extension of the non-negative symmetric operator $\mathscr{P}^* \mathscr{P}$. Then for $t > 0$, the heat operator $e^{-t\mathscr{L}}$ is given by integration against a smooth function, which we denote by $e^{-t\mathscr{L}}(x, y)$. Moreover, we have local Gaussian bounds for this heat kernel in terms of the metric $\rho_{(W,\vec{d})}$: for every $\mathcal{K} \Subset \mathfrak{M}$ compact, there exists $c > 0$, such that for all ordered multi-indices α, β and all $s \in \mathbb{N}$, there exists $C \ge 0$, $\forall t \in (0, 1]$, $\forall x, y \in \mathcal{K}$, so that
$$\begin{aligned} |\partial_t^s W_x^\alpha W_y^\beta e^{-t\mathscr{L}}(x,y)| &\le C(\rho_{(W,\vec{d})}(x,y) + t^{\frac{1}{2\kappa}})^{-\deg_{\vec{d}}(\alpha) - \deg_{\vec{d}}(\beta) - 2\kappa s} \\ &\quad \times \exp\left(-c\left(\frac{\rho_{(W,\vec{d})}(x,y)^{2\kappa}}{t}\right)^{\frac{1}{2\kappa-1}}\right)\left(\mathrm{Vol}(B_{(W,\vec{d})}(x, \rho_{(W,\vec{d})}(x,y) + t^{\frac{1}{2\kappa}})) \wedge 1\right)^{-1}. \end{aligned}$$

See Theorem 8.1.1 for more precise versions of the above. Moreover, once made precise, Theorem 8.1.1 shows that each of the above conditions are *equivalent* to maximal subellipticity, and moreover gives several other equivalent conditions, including:

- $\mathscr{P}_0 := \sum_{\deg(\alpha)=\kappa} a_\alpha W^\alpha$ is maximally subelliptic of degree κ with respect to $(W, \vec{d})$ on $\mathfrak{M}$.
- For all $x_0 \in \mathfrak{M}$, there exists an open neighborhood $U \subseteq \mathfrak{M}$ of x_0 such that the frozen coefficient operator $\sum_{\deg_{\vec{d}}(\alpha)\leq\kappa} a_\alpha(x_0) W^\alpha$ is maximally subelliptic of degree κ with respect to $(W, \vec{d})$ on U.
- $\mathscr{P}^* \mathscr{P}$ is maximally subelliptic of degree 2κ with respect to $(W, \vec{d})$ on $\mathfrak{M}$.

Example 1.6.1. Let $\mathscr{P} = \partial_x^2 + x^2\partial_y^2$. Then $\mathscr{P}$ is maximally subelliptic of degree 2 with respect to $(W, \vec{d}) := \{(\partial_x, 1), (x\partial_y, 1)\}$ on $\mathbb{R}^2$; see Example 1.1.10 (iii). Applying (1.16) in the case $\mathscr{X}^0(W, \vec{d}) = \mathscr{F}^0_{p,2}(W, \vec{d}) = L^p(\mathbb{R}^2)$, $1 < p < \infty$, we see that for $\zeta \in \mathbb{R}^2$,

$$\mathscr{P}u \in L^p \text{ near } \zeta \iff u, \partial_x u, x\partial_y u, \partial_x^2 u, x^2\partial_y^2 u, \partial_x x\partial_y u, x\partial_y\partial_x u \in L^p \text{ near } \zeta.$$

In particular, $\mathscr{P}u \in L^p$ near ζ implies $\partial_y u = [\partial_x, x\partial_y]u \in L^p$ near ζ, but **does not imply** $\partial_y^2 u \in L^p$ near ζ. This is why we treat ∂_y as a differential operator of degree 2 in this setting; see Remark 1.1.5.

The regularity theory described above is clearly sharp (the conclusion (1.16) is an "if and only if"). However, it is also of interest to study the regularity theory of maximally subelliptic operators with respect to other geometries; for example, in Section 8.2.2 we present the sharp regularity theory of maximally subelliptic operators with respect to the standard Besov and Triebel–Lizorkin spaces on $\mathfrak{M}$, here denoted by $\mathscr{B}^{\mathbf{s}}_{p,q,\mathrm{std}}$ and $\mathscr{F}^{\mathbf{s}}_{p,q,\mathrm{std}}$ (in any local coordinate system, $\mathscr{B}^{\mathbf{s}}_{p,q,\mathrm{std}}$ and $\mathscr{F}^{\mathbf{s}}_{p,q,\mathrm{std}}$ are locally equal to their classical counterparts $\mathscr{B}^{\mathbf{s}}_{p,q}(\mathbb{R}^n)$ and $\mathscr{F}^{\mathbf{s}}_{p,q}(\mathbb{R}^n)$). In Corollary 8.2.3, the following is shown, with $\mathscr{X}^{\mathbf{s}}_{\mathrm{std}}$ equal to either $\mathscr{B}^{\mathbf{s}}_{p,q,\mathrm{std}}$ or $\mathscr{F}^{\mathbf{s}}_{p,q,\mathrm{std}}$:

> Suppose $\mathscr{P}$, given by (1.1), is maximally subelliptic of degree κ with respect to $(W, \vec{d})$ on $\mathfrak{M}$. Then, for any distribution u and $x_0 \in \mathfrak{M}$,
>
> $$\mathscr{P}u \in \mathscr{X}^{\mathbf{s}}_{\mathrm{std}} \text{ near } x_0 \implies u \in \mathscr{X}^{\mathbf{s}+\kappa/\lambda_{\mathrm{std}}}_{\mathrm{std}} \text{ near } x_0, \tag{1.17}$$
>
> where we have an explicit formula for $\lambda_{\mathrm{std}} = \lambda_{\mathrm{std}}(x_0, (W, \vec{d})) > 0$. For example, in the special case $\vec{d}_1 = \vec{d}_2 = \cdots = \vec{d}_r = 1$, $\lambda_{\mathrm{std}} = m$, where $W_1, \ldots, W_r$ satisfy Hörmander's condition of order exactly m at x_0; see Definition 6.6.15 for the formula for $\lambda_{\mathrm{std}}(x_0, (W, \vec{d}))$ in the general setting. Furthermore, Corollary 8.2.3 shows that this result is sharp: there is no better regularity result in terms of the classical Besov and Triebel–Lizorkin spaces (see Corollary 8.2.3 for a precise statement of this sharpness). See Corollary 8.2.9 for some more regularity properties in terms of the standard Besov and Triebel–Lizorkin spaces.

In particular, maximally subelliptic operators are subelliptic. Indeed, for $1 < p < \infty$, we may take $\mathscr{X}^{\mathbf{s}}_{\mathrm{std}} = \mathscr{F}^{\mathbf{s}}_{p,2,\mathrm{std}} = L^p_{\mathbf{s}}$, the standard L^p Sobolev space of order $\mathbf{s} \in \mathbb{R}$. In this special case, (1.17) becomes

$$\mathscr{P}u \in L^p_{\mathbf{s}} \text{ near } x_0 \implies u \in L^p_{\mathbf{s}+\kappa/\lambda_{\text{std}}} \text{ near } x_0, \tag{1.18}$$

and this subelliptic gain is sharp. See Corollary 8.2.5.

See Section 8.2 for additional regularity properties of maximally subelliptic operators. In particular, we study the regularity theory with respect to function spaces $\mathscr{X}^{\mathbf{s}}(Z, \vec{d})$, where $(Z, \vec{d})$ is another set of Hörmander vector fields with formal degrees.

Remark 1.6.2. The above discussion justifies the name maximally subelliptic. Indeed, (1.18) shows maximally subelliptic operators are subelliptic (Definition 2.2.32); informally, u is strictly more regular than $\mathscr{P}u$. Moreover, (1.16) shows that u is more regular than $\mathscr{P}u$ by the clearly maximal possible amount, and hence *maximally subelliptic*. See Corollary 8.2.9 (i) for another perspective on this maximal gain using the standard L^p Sobolev spaces. Because maximally subelliptic operators are subelliptic, they are also hypoelliptic (see Definition 2.2.37 and Proposition 2.2.38); though hypoellipticity is a more qualitative concept than subellipticity, and it is perhaps less clear what "maximally hypoelliptic" should mean. This is why we have opted to call the operators in this text maximally subelliptic rather than maximally hypoelliptic.

1.7 Fully nonlinear theory

The sharp regularity theory for fully nonlinear, maximally subelliptic PDEs is in terms of the Zygmund–Hölder spaces adapted to $(W, \vec{d})$: $\mathscr{C}^{\mathbf{s}}(W, \vec{d}) = \mathscr{B}^{\mathbf{s}}_{\infty,\infty}(W, \vec{d})$. Important for the nonlinear theory is that $\mathscr{C}^{\mathbf{s}}(W, \vec{d})$ forms an algebra, and in fact we can understand compositions of classical Zygmund–Hölder functions with functions in $\mathscr{C}^{\mathbf{s}}(W, \vec{d})$ – see Sections 7.4 and 7.5.

Fix $\kappa \in \mathbb{N}_+$, and let

$$u \mapsto F(x, \{W^\alpha u(x)\}_{\deg_{\vec{d}}(\alpha) \le \kappa})$$

be a fully nonlinear equation of the type described in Section 1.1.2.

Fully nonlinear regularity Fix $\mathbf{s} > \mathbf{r} > 0$, $x_0 \in \mathfrak{M}$, and a function $u : \mathfrak{M} \to \mathbb{R}^{\mathrm{D}_2}$. Suppose:

(a) $u \in \mathscr{C}^{\mathbf{r}+\kappa}(W, \vec{d})$ near x_0,
(b) $F(x, \zeta)$ is $\mathscr{C}^{\mathbf{s}}(W, \vec{d})$ in the x variable and C^∞ in the ζ variable, for x near x_0,
(c) $u \mapsto F(x, \{W^\alpha u(x)\}_{\deg_{\vec{d}}(\alpha) \le \kappa})$ is maximally subelliptic at (x_0, u) of degree κ with respect to $(W, \vec{d})$ in the sense of Definition 1.1.11.

Then

$$F(x, \{W^\alpha u(x)\}_{\deg_{\vec{d}}(\alpha) \le \kappa}) \in \mathscr{C}^{\mathbf{s}}(W, \vec{d}) \text{ near } x_0 \iff u \in \mathscr{C}^{\mathbf{s}+\kappa}(W, \vec{d}) \text{ near } x_0. \tag{1.19}$$

See Theorem 9.1.2 for a more precise and more general version of this result, where $F(x,\zeta)$ is only assumed to have finite regularity in the ζ variable. In the above result, $\mathbf{r}$ should be thought of as small. Thus, (a) implies that $W^\alpha u$ is continuous near x_0, $\forall \deg_{d}(\alpha) \leq \kappa$, but it does not imply much more than that. The above result shows that even though u is a priori assumed to have low regularity, since it satisfies a maximally subelliptic equation it must have a higher regularity. Since the conclusion (1.19) is an "if and only if," it is clearly sharp.

Regularity in classical Zygmund–Hölder spaces Even though (1.19) gives a sharp regularity result, it is also of interest to understand the regularity theory with respect to the classical Zygmund–Hölder spaces. Fix $\mathbf{s} > \mathbf{r} > 0$, $x_0 \in \mathfrak{M}$, and a function $u : \mathfrak{M} \to \mathbb{R}^{\mathbf{D}_2}$. Let $\mathscr{C}^{\mathbf{s}}_{\mathrm{std}}$ denote the standard (local) Zygmund–Hölder space of order $\mathbf{s}$ on $\mathfrak{M}$.[10] Suppose:
- $W^\alpha u \in \mathscr{C}^{\mathbf{r}}_{\mathrm{std}}$, $\forall \deg_{d}(\alpha) \leq \kappa$,
- $F(x,\zeta)$ is $\mathscr{C}^{\mathbf{s}}_{\mathrm{std}}$ in the x variable and C^∞ in the ζ variable, for x near x_0,
- $u \mapsto F(x, \{W^\alpha u(x)\}_{\deg_{d}(\alpha)\leq\kappa})$ is maximally subelliptic at (x_0, u) of degree κ with respect to (W, d) in the sense of Definition 1.1.11.

Then, $\forall \epsilon > 0$,

$$F(x, \{W^\alpha u(x)\}_{\deg_{d}(\alpha)\leq\kappa}) \in \mathscr{C}^{\mathbf{s}}_{\mathrm{std}} \text{ near } x_0 \implies u \in \mathscr{C}^{\mathbf{s}+\kappa/\lambda_{\mathrm{std}}-\epsilon}_{\mathrm{std}} \text{ near } x_0, \tag{1.20}$$

where $\lambda_{\mathrm{std}} = \lambda_{\mathrm{std}}(x_0, (W, d))$ is the same constant as in (1.17).[11] See Corollary 9.1.8 for a more precise and more general version of this result and for other related regularity properties. Except for the loss of an arbitrarily small $\epsilon > 0$, (1.20) is sharp – indeed, as described in Corollary 8.2.3, it is sharp for linear equations. We do not know if this loss of ϵ is necessary.

See Section 9.1.1 for a further discussion of these results, along with a statement of regularity theory in terms of the Zygmund–Hölder spaces $\mathscr{C}^{\mathbf{s}}(Z, d)$, with respect to a different set of Hörmander vector fields with formal degrees (Z, d). See Section 9.3 for a description of blowup near the boundary for solutions in terms of weighted Zygmund–Hölder norms.

1.7.1 The inverse function theorem

Much like classical elliptic theory, a main source of intuition for our proof of the fully nonlinear regularity theory comes from a standard proof of the inverse function theo-

10 In local coordinates, $\mathscr{C}^{\mathbf{s}}_{\mathrm{std}} = \mathscr{C}^{\mathbf{s}}(\mathbb{R}^n)$.

11 In particular, in the special case where $d_1 = \cdots = d_r = 1$, we have $\lambda_{\mathrm{std}} = m$, where $W_1, \ldots, W_r$ satisfy Hörmander's condition of order exactly m at x_0. See Definition 6.6.15 for the formula for $\lambda_{\mathrm{std}}(x_0, (W, d))$ in the general setting.

rem in Banach spaces: we use a fixed point argument to prove regularity of fully nonlinear equations. We present this proof of the inverse function theorem, and then describe how it informs our proof of (1.19). Here $\mathscr{X}$ and $\mathscr{Y}$ will both denote Banach spaces. Given a function $f : U \to \mathscr{Y}$, where $U \subseteq \mathscr{X}$ is open, we say f is C^k if it is k times continuously Fréchet differentiable.

Theorem 1.7.1 (Banach space inverse function theorem). *Suppose $f : U \to \mathscr{Y}$ is C^k, where $k \geq 1$, and $U \subseteq \mathscr{X}$ is open. Fix $x_0 \in U$. If $df(x_0) : \mathscr{X} \to \mathscr{Y}$ has a bounded left inverse, $\mathscr{Y} \to \mathscr{X}$, then there exists a ball $B \subseteq U$, centered at x_0, and an open set $V \subseteq \mathscr{Y}$ such that $f|_B : B \to V$ has a left inverse $g : V \to B$ which is C^k.*

Remark 1.7.2. In short, Theorem 1.7.1 says that if $df(x_0)$ is left invertible, then f is left invertible near x_0. A similar proof shows that if $df(x_0)$ is invertible, then f is invertible near x_0. A similar result can also be stated when $df(x_0)$ is right invertible. However, it is the proof of the case where $df(x_0)$ is left invertible which is most informative for our applications.

Lemma 1.7.3. *Suppose $U \subseteq \mathscr{X}$ is an open set, $\alpha \in (0,1)$, $r : U \to \mathscr{X}$, $F(x) = x + r(x)$ for $x \in U$, and r satisfies*

$$\|r(x) - r(y)\|_{\mathscr{X}} \leq \alpha \|x - y\|_{\mathscr{X}}, \quad \forall x, y \in U. \tag{1.21}$$

Then $F(U)$ is open in $\mathscr{X}$ and $F : U \to F(U)$ is a homeomorphism.

Proof. First note that from (1.21), we have

$$\begin{aligned} \|x - y\|_{\mathscr{X}} &= \|F(x) - F(y) - (r(x) - r(y))\|_{\mathscr{X}} \\ &\leq \|F(x) - F(y)\|_{\mathscr{X}} + \|r(x) - r(y)\|_{\mathscr{X}} \\ &\leq \|F(x) - F(y)\|_{\mathscr{X}} + \alpha\|x - y\|_{\mathscr{X}}, \end{aligned}$$

from which it follows that $\|x-y\|_{\mathscr{X}} \leq (1-\alpha)^{-1}\|F(x)-F(y)\|_{\mathscr{X}}$, and therefore F is injective on U. Let $V := F(U)$ and set $G := F^{-1} : V \to U$, which exists since F is injective.

We next show V is open. Take $z_0 = F(x_0) = x_0 + r(x_0) \in V$; we wish to show that for z close to z_0, there exists $x \in U$ with $F(x) = z$, or equivalently $x = z - r(x)$.

Fix $\delta > 0$ such that $\overline{B_{\mathscr{X}}(x_0, \delta)} \subseteq U$ and $z \in B_{\mathscr{X}}(z_0, (1-\alpha)\delta)$, where $B_{\mathscr{X}}(x, \delta)$ denotes the ball of radius δ, centered at x, in $\mathscr{X}$. Define

$$\mathscr{T}_z(x) := z - r(x).$$

We wish to find $x \in U$ such that $\mathscr{T}_z(x) = x$. We will do this by showing that $\mathscr{T}_z : \overline{B_{\mathscr{X}}(x_0, \delta)} \to \overline{B_{\mathscr{X}}(x_0, \delta)}$ is a strict contraction. Using (1.21), for $x \in \overline{B_{\mathscr{X}}(x_0, \delta)}$,

$$\begin{aligned} \|\mathscr{T}_z(x) - x_0\|_{\mathscr{X}} &= \|\mathscr{T}_z(x) - \mathscr{T}_{z_0}(x_0)\|_{\mathscr{X}} = \|z - z_0 - (r(x) - r(x_0))\|_{\mathscr{X}} \\ &\leq \|z - z_0\|_{\mathscr{X}} + \alpha\|x - x_0\|_{\mathscr{X}} < (1-\alpha)\delta + \alpha\delta = \delta, \end{aligned}$$

showing that $\mathscr{T}_z : \overline{B_{\mathscr{X}}(x_0,\delta)} \to \overline{B_{\mathscr{X}}(x_0,\delta)}$. Also, using (1.21),

$$\|\mathscr{T}_z(x) - \mathscr{T}_z(y)\|_{\mathscr{X}} = \|r(x) - r(y)\|_{\mathscr{X}} \le \alpha\|x - y\|_{\mathscr{X}}, \tag{1.22}$$

so $\mathscr{T}_z$ is a strict contraction. The contraction mapping principle shows that there is a unique fixed point $x \in \overline{B_{\mathscr{X}}(x_0,\delta)} \subseteq U$ such that $\mathscr{T}_z(x) = x$. Since

$$x = \mathscr{T}_z(x) = z - F(x) + x,$$

we see that $F(x) = z$, i. e., $G(z) = x$. This also shows $B(z_0,(1-\alpha)\delta) \subseteq F(U) = V$, and therefore V is open.

Finally, we wish to show that G is continuous. Using that $G(z)$ is a fixed point of the map $\mathscr{T}_z$ and (1.22), for $z, w \in V$ we have

$$\begin{aligned}\|G(z) - G(w)\|_{\mathscr{X}} &= \|\mathscr{T}_z(G(z)) - \mathscr{T}_w(G(w))\|_{\mathscr{X}}\\ &\le \|\mathscr{T}_z(G(z)) - \mathscr{T}_w(G(z))\|_{\mathscr{X}} + \|\mathscr{T}_w(G(z)) - \mathscr{T}_w(G(w))\|_{\mathscr{X}}\\ &\le \|\mathscr{T}_z(G(z)) - \mathscr{T}_w(G(z))\|_{\mathscr{X}} + \alpha\|G(z) - G(w)\|_{\mathscr{X}}.\end{aligned}$$

We conclude that

$$\|G(z) - G(w)\|_{\mathscr{X}} \le (1-\alpha)^{-1}\|\mathscr{T}_z(G(z)) - \mathscr{T}_w(G(z))\|_{\mathscr{X}} = (1-\alpha)^{-1}\|z - w\|_{\mathscr{X}}.$$

It follows that $G = F^{-1} : V \to B$ is continuous, completing the proof. □

Proof of Theorem 1.7.1. Let L_{x_0} be a left inverse to $df(x_0)$. Define $F(x) := L_{x_0} f(x) : \mathscr{X} \to \mathscr{X}$ and set $r(x) := F(x) - x$. Note that $dF(x_0) = I$, where I denotes the identity map $\mathscr{X} \to \mathscr{X}$. Thus, $dr(x_0) = 0$.

By choosing a small enough open ball, $B \subseteq \mathscr{X}$, centered at x_0, we have $\sup_{x\in B}\|dr(x)\|_{\mathscr{X}\to\mathscr{X}} \le 1/2$; therefore, for $x, y \in B$, $\|r(x) - r(y)\|_{\mathscr{X}} \le \frac{1}{2}\|x - y\|_{\mathscr{X}}$.

Lemma 1.7.3 applies to $F|_B : B \to \mathscr{X}$ to show that $F(B) \subseteq \mathscr{X}$ is open and F has a continuous inverse $G : F(B) \to B$. Set $V := L_{x_0}^{-1}(B)$ (the set theoretic pre-image), which is open in $\mathscr{Y}$, and set $g(y) := G(L_{x_0}(y)) : V \to B$. Note that for $x \in B$, we have $g(f(x)) = G(F(x)) = x$.

Finally, it follows from the chain rule that G is C^k; therefore, g is C^k as well. □

A recap of the proof of Theorem 1.7.1 is given now. To find a local left inverse to f, we apply the contraction mapping principle to $L_{x_0} f(x)$, where L_{x_0} is a left inverse for $df(x_0)$.

When we consider a nonlinear equation written in the form

$$F(x, \{W^\alpha u(x)\}_{\deg_{\vec{d}}(\alpha)\le\kappa}), \tag{1.23}$$

understanding regularity theory for u near x_0 is akin to finding a left inverse, near u and x_0, modulo smooth functions, for the map

$$v \mapsto F(x, \{W^\alpha v(x)\}_{\deg_{\vec{d}}(\alpha)\leq\kappa}), \quad \mathscr{C}^{\mathbf{s}+\kappa}(W, \vec{d}) \to \mathscr{C}^{\mathbf{s}}(W, \vec{d}). \tag{1.24}$$

The assumption that (1.23) is maximally subelliptic at (x_0, u) of degree κ with respect to $(W, \vec{d})$ is the same as saying $\mathscr{P}_{x_0,u}$, given by (1.9), is maximally subelliptic of degree κ with respect to $(W, \vec{d})$ on a neighborhood of x_0. $\mathscr{P}_{x_0,u}$ is not exactly the derivative of the map (1.24), but near x_0 is a perturbation of this derivative. As described in Section 1.6, since $\mathscr{P}_{x_0,u}$ is maximally subelliptic, it has a local left parametrix $T_{x_0,u}$ which is locally in $\mathscr{A}^{-\kappa}(W, \vec{d})$. This parametrix takes $\mathscr{C}^{\mathbf{s}}(W, \vec{d}) \to \mathscr{C}^{\mathbf{s}+\kappa}(W, \vec{d})$ (locally). Following the proof of Theorem 1.7.1, we will prove the regularity of u by applying the contraction mapping principle to a localization of

$$v \mapsto T_{x_0,u}F(x, \{W^\alpha v(x)\}_{\deg_{\vec{d}}(\alpha)\leq\kappa}),$$

which takes $\mathscr{C}^{\mathbf{s}+\kappa}(W, \vec{d}) \to \mathscr{C}^{\mathbf{s}+\kappa}(W, \vec{d})$.

Unfortunately, the proof is not quite as simple as the proof of Theorem 1.7.1. One main issue is that $T_{x_0,u}$ is not actually an inverse for $\mathscr{P}_{x_0,u}$, and we need to introduce cut-off functions to localize all our computations. Due to these cut-off functions, the contraction mapping principle does not directly apply. However, by employing a covering argument inspired by a work of Simon [212], we are able to get around these issues. See Section 9.2 for details.

Remark 1.7.4. There are other inverse function theorems which often appear in the study of PDEs. Notably, the Nash–Moser inverse function theorem [105] is a powerful tool which applies to a wide class of fully nonlinear PDEs, a much larger class of PDEs than the Banach space inverse function theorem (Theorem 1.7.1) applies to. However, when it applies, the Banach space inverse function theorem often yields sharper and stronger results than the Nash–Moser inverse function theorem does in general. A main theme of this text, especially in Chapter 9, is that ideas from the Banach space inverse function theorem apply to maximally subelliptic equations, once the right spaces are chosen, and sharp results can be deduced.

1.8 Multi-parameter theory

When proving results for maximally subelliptic operators with respect to the adapted Besov ($\mathscr{B}^{\mathbf{s}}_{p,q}(W, \vec{d})$) and Triebel–Lizorkin ($\mathscr{F}^{\mathbf{s}}_{p,q}(W, \vec{d})$) spaces we employ the classical (single-parameter) Calderón–Zygmund theory to understand the parametrix and function spaces. However, when we prove results with respect to function spaces which are not adapted to the vector fields $(W, \vec{d})$ (for example when we prove results concerning the standard Besov [$\mathscr{B}^{s}_{p,q,\mathrm{std}}$] and Triebel–Lizorkin [$\mathscr{F}^{s}_{p,q,\mathrm{std}}$] spaces), classical Calderón–Zygmund theory is no longer sufficient because more than one geometry is relevant.

To deal with this issue, we turn to multi-parameter singular integrals and function spaces of the type originally studied in [220]. For example, when we study the regularity theory of fully nonlinear maximally subelliptic equations with respect to the standard Zygmund–Hölder spaces, $\mathscr{C}^{\bullet}_{\mathrm{std}}$, we employ an algebra of singular integral operators which (locally) contains both the parametrices for maximally subelliptic operators and standard pseudo-differential operators. This algebra is filtered by $\mathbb{R}^2$ denoting the two regularity parameters: one for the Carnot–Carathéorody geometry adapted to $(W, d\!\!\!/)$ and one for the standard Riemannian geometry.

Fix $\nu \in \mathbb{N}_+$. We will now discuss the ν-parameter theory. We are given ν finite sets of vector fields with formal degrees:

$$(W^\mu, d\!\!\!/^\mu) := \{(W_1^\mu, d\!\!\!/_1^\mu), \dots, (W_{r_\mu}^\mu, d\!\!\!/_{r_\mu}^\mu)\}, \quad 1 \le \mu \le \nu,$$

where each $d\!\!\!/_j^\mu \in \mathbb{N}_+$ and $W_{j_\mu}^\mu$ is a smooth vector field; we will assume various conditions on these sets, depending on the context. For each μ, the balls $B_{(W^\mu, d\!\!\!/^\mu)}(x, \delta_\mu)$, given by (1.13), induce a geometry on $\mathfrak{M}$. These ν geometries might interact in complicated ways.

Let $\mathfrak{e}_1, \dots, \mathfrak{e}_\nu$ denote the standard basis in $\mathbb{R}^\nu$. Define

$$\begin{aligned}(W, \vec{d\!\!\!/}) &= \{(W_1, \vec{d\!\!\!/}_1), \dots, (W_r, \vec{d\!\!\!/}_r)\} \\ &:= \{(W_j^\mu, d\!\!\!/_j^\mu \mathfrak{e}_\mu) : 1 \le \mu \le \nu, 1 \le j \le r_\mu\},\end{aligned} \tag{1.25}$$

so that $(W, \vec{d\!\!\!/})$ is a list of smooth vector fields, W_j, each paired with a degree $\vec{d\!\!\!/}_j \in \mathbb{N}^\nu \setminus \{0\}$.

Usually, we will consider the case where $(W^1, d\!\!\!/^1)$ are Hörmander vector fields with formal degrees, and we will study maximally subelliptic operators with respect to $(W^1, d\!\!\!/^1)$.

Example 1.8.1. Let $\mathfrak{M} = \mathbb{R}^n$ and suppose $(W^1, d\!\!\!/^1)$ are Hörmander vector fields with formal degrees on $\mathbb{R}^n$. Given an either linear or fully nonlinear, maximally subelliptic operator with respect to $(W^1, d\!\!\!/^1)$, if we wish to study regularity theory with respect to the classical Besov and Triebel–Lizorkin spaces, we take $\nu = 2$ and $(W^2, d\!\!\!/^2) = \{(\partial_{x_1}, 1), \dots, (\partial_{x_n}, 1)\}$.

We always assume that $(W^1, d\!\!\!/^1), \dots, (W^\nu, d\!\!\!/^\nu)$ "locally weakly approximately commute" as in Definition 3.8.5. Under an additional finite type condition on each of the $(W^\mu, d\!\!\!/^\mu)$ we introduce and study the following:

- In Chapter 6, we introduce Besov ($\mathscr{B}^{\vec{s}}_{p,q}(W, \vec{d\!\!\!/})$) and Triebel–Lizorkin ($\mathscr{F}^{\vec{s}}_{p,q}(W, \vec{d\!\!\!/})$) spaces indexed by a regularity parameter $\vec{s} \in \mathbb{R}^\nu$. These are defined in such a way that $W_j : \mathscr{X}^{\vec{s}}(W, \vec{d\!\!\!/}) \to \mathscr{X}^{\vec{s}-\vec{d\!\!\!/}_j}(W, \vec{d\!\!\!/})$; i. e., W_j acts as a differential operator of degree $\vec{d\!\!\!/}_j \in \mathbb{N}^\nu$.

- In Chapter 5, we introduce two filtered algebras of singular integral operators indexed by $\vec{s} \in \mathbb{R}^\nu$: $\mathscr{A}^{\vec{s}}(W, \vec{d})$ and $\widetilde{\mathscr{A}}^{\vec{s}}(W, \vec{d})$.
  - If $T \in \mathscr{A}^{\vec{t}}(W, \vec{d})$ and $S \in \mathscr{A}^{\vec{s}}(W, \vec{d})$, then $TS \in \mathscr{A}^{\vec{s}+\vec{t}}(W, \vec{d})$; a similar result can be obtained with $\mathscr{A}^{\bullet}(W, \vec{d})$ replaced by $\widetilde{\mathscr{A}}^{\bullet}(W, \vec{d})$.
  - $\mathscr{A}^{\vec{s}}(W, \vec{d})$ is a generalization of the Calderón–Zygmund theory we use in the single-parameter case, and is only defined in the special case where all $(W^\mu, \vec{d}^\mu)$ are Hörmander vector fields with formal degrees, which pairwise "locally weakly approximately commute."
  - $\mathscr{A}^{\vec{s}}(W, \vec{d}) \subseteq \widetilde{\mathscr{A}}^{\vec{s}}(W, \vec{d})$ and moreover $\widetilde{\mathscr{A}}^{\vec{s}}(W, \vec{d})$ is defined even when $(W^\mu, \vec{d}^\mu)$ do not satisfy Hörmander's condition. This is important because the assumption of Hörmander's condition for all $(W^\mu, \vec{d}^\mu)$ is not quantitatively preserved under our scaling arguments. $\widetilde{\mathscr{A}}^{\vec{s}}(W, \vec{d})$ provides an algebra with good scaling properties.
  - If $\mathscr{P}$ is a maximally subelliptic operator of degree κ with respect to $(W^1, \vec{d}^1)$ on $\mathfrak{M}$, then $\mathscr{P}$ has a left parametrix which is locally in $\widetilde{\mathscr{A}}^{-\kappa \mathfrak{e}_1}(W, \vec{d})$ and locally in $\mathscr{A}^{-\kappa \mathfrak{e}_1}(W, \vec{d})$ when it is defined.
  - When $\mathfrak{M} = \mathbb{R}^n$, if $(W^\mu, \vec{d}^\mu) = \{(\partial_{x_1}, 1), \ldots, (\partial_{x_n}, 1)\}$, then standard pseudodifferential operators of order $k \in \mathbb{R}$ with compact support are in $\widetilde{\mathscr{A}}^{k\mathfrak{e}_\mu}(W, \vec{d})$ and in $\mathscr{A}^{k\mathfrak{e}_\mu}(W, \vec{d})$ when it is defined. See Section 5.8.2.
  - If $T \in \widetilde{\mathscr{A}}^{\vec{t}}(W, \vec{d})$, then $T : \mathscr{X}^{\vec{s}}(W, \vec{d}) \to \mathscr{X}^{\vec{s}-\vec{t}}(W, \vec{d})$, where $\mathscr{X}^{\vec{s}}$ denotes either $\mathscr{B}^{\vec{s}}_{p,q}$ or $\mathscr{F}^{\vec{s}}_{p,q}$.

The multi-parameter theory is useful when studying both linear and nonlinear equations, but its use is especially transparent in nonlinear theory. We describe here how it can be used to prove (1.20); the same ideas are used to prove more general results in Chapter 9.

Let $(W^1, \vec{d}^1) = \{(W^1_1, \vec{d}^1_1), \ldots, (W^1_{r_1}, \vec{d}^1_{r_1})\}$ be Hörmander vector fields with (single-parameter) formal degrees on $\mathbb{R}^n$. Consider a nonlinear operator of the form

$$v \mapsto F\big(x, \{(W^1)^\alpha v(x)\}_{\deg_{\vec{d}^1}(\alpha) \leq \kappa}\big) \tag{1.26}$$

which is maximally subelliptic at (x_0, u) of degree κ with respect to $(W^1, \vec{d}^1)$. For simplicity in this section we take $F(x, \zeta) \in C^\infty$, though we deal with finite smoothness when we present this result in Corollary 9.1.8.

Suppose we wish to prove (1.20) using a contraction mapping argument as described in Section 1.7.1. We immediately run into a problem. Thought of as a map on the standard Zygmund–Hölder spaces, the best we can say in general is that (1.26) maps $\mathscr{C}^{\mathbf{s}+\kappa}_{\mathrm{std}} \to \mathscr{C}^{\mathbf{s}}_{\mathrm{std}}$, near x_0. However, if we define $\mathscr{P}_{u,x_0}$ by (1.9) and let T_{u,x_0} be a left parametrix for $\mathscr{P}_{u,x_0}$ near x_0, then (by (1.17)) the best we can say is that $T_{u,x_0} : \mathscr{C}^{\mathbf{s}}_{\mathrm{std}} \to \mathscr{C}^{\mathbf{s}+\kappa/\lambda_{\mathrm{std}}}_{\mathrm{std}}$. We conclude that

$$v \mapsto T_{u,x_0} F\big(x, \{(W^1)^\alpha v(x)\}_{\deg_{\vec{d}}(\alpha) \leq \kappa}\big) : \mathscr{C}^{\mathbf{s}+\kappa}_{\mathrm{std}} \to \mathscr{C}^{\mathbf{s}+\kappa/\lambda_{\mathrm{std}}}_{\mathrm{std}}, \quad \text{near } x_0. \tag{1.27}$$

Unless we are in the elliptic setting we have $\lambda_{\mathrm{std}} > 1$, so (1.27) maps from one space to a different space, and we cannot use the contraction mapping principle.

Instead, we work with a multi-parameter theory. We set $(W^2, \vec{d}^2) := \{(\partial_{x_1}, 1), \ldots, (\partial_{x_n}, 1)\}$, $\nu = 2$, and define $(W, \vec{d})$ as in (1.25). For $\vec{\mathbf{s}} \in (0, \infty)^2$, set $\mathscr{C}^{\vec{\mathbf{s}}}(W, \vec{d}) := \mathscr{B}^{\vec{\mathbf{s}}}_{\infty,\infty}(W, \vec{d})$. Then, we will show (1.26) maps $\mathscr{C}^{\vec{\mathbf{s}}+\kappa e_1}(W, \vec{d}) \to \mathscr{C}^{\vec{\mathbf{s}}}(W, \vec{d})$, near x_0, and $T_{u,x_0} : \mathscr{C}^{\vec{\mathbf{s}}}(W, \vec{d}) \to \mathscr{C}^{\vec{\mathbf{s}}+\kappa e_1}(W, \vec{d})$. In particular, we have

$$v \mapsto T_{u,x_0} F(x, \{(W^1)^\alpha v(x)\}_{\deg_d(\alpha) \le \kappa}) : \mathscr{C}^{\vec{\mathbf{s}}+\kappa e_1}(W, \vec{d}) \to \mathscr{C}^{\vec{\mathbf{s}}+\kappa e_1}(W, \vec{d}), \quad \text{near } x_0, \tag{1.28}$$

and we will be able to apply a contraction mapping type argument to (1.28) in the Banach space $\mathscr{C}^{\vec{\mathbf{s}}+\kappa e_1}(W, \vec{d})$.

To conclude (1.20) we argue roughly as follows (see the proof of Proposition 9.1.6 for full details in a more general case):

- We start with $u \in \mathscr{C}^{\mathbf{r}}_{\mathrm{std}}$ such that $F(x, \{W^\alpha u(x)\}_{\deg_d(\alpha) \le \kappa}) \in \mathscr{C}^{\mathbf{s}}_{\mathrm{std}}$.
- Fix $\epsilon \in (0, \mathbf{r})$ small. We show that there exist $\epsilon' > 0$ with:
  - $u \in \mathscr{C}^{\mathbf{r}}_{\mathrm{std}} \subseteq \mathscr{C}^{(\epsilon', \mathbf{r}-\epsilon)}(W, \vec{d})$ near x_0,
  - $F(x, \{W^\alpha u(x)\}_{\deg_d(\alpha) \le \kappa}) \in \mathscr{C}^{\mathbf{s}}_{\mathrm{std}} \subseteq \mathscr{C}^{(\epsilon', \mathbf{s}-\epsilon)}(W, \vec{d})$ near x_0.
- Using (1.28) we apply a contraction mapping type argument in $\mathscr{C}^{(\epsilon'+\kappa, \mathbf{s}-\epsilon)}(W, \vec{d})$ to show $u \in \mathscr{C}^{(\epsilon'+\kappa, \mathbf{s}-\epsilon)}(W, \vec{d})$ near x_0. This is the main (and most difficult) step of the proof.
- We then show that

  $$u \in \mathscr{C}^{(\epsilon'+\kappa, \mathbf{s}-\epsilon)}(W, \vec{d}) \subseteq \mathscr{C}^{(\epsilon', \mathbf{s}+\kappa/\lambda_{\mathrm{std}}-\epsilon)}(W, \vec{d}) \subseteq \mathscr{C}^{\mathbf{s}+\kappa/\lambda_{\mathrm{std}}-\epsilon}_{\mathrm{std}},$$

  near x_0.

In particular, the loss of ϵ derivatives comes from the conversion between the single-parameter Zygmund–Hölder spaces and the multi-parameter Zygmund–Hölder spaces. When working directly with multi-parameter Zygmund–Hölder spaces, our regularity results are sharp and generalize (1.19); see Theorem 9.1.14.

1.9 The main tool: scaling

While most tools from classical elliptic theory (e. g., the Fourier transform) do not translate to the maximally subelliptic setting, there is one tool which does: scaling. Scaling for elliptic equations is very simple: we use the standard dilation maps on $\mathbb{R}^n$. When we turn to the maximally subelliptic setting, we cannot write down such a simple formula in general, and instead satisfy ourselves by establishing that scaling maps with appropriate properties exist.

1.9.1 Scaling for elliptic operators

We begin by explaining how scaling works for elliptic operators, which is very simple, and a standard tool in elliptic theory. For each $x \in \mathbb{R}^n$ and $\delta > 0$, we define the scaling map

$$\Phi_{x,\delta}(t) = x + \delta t : B^n(1) \xrightarrow{\sim} B^n(x,\delta),$$

where $B^n(x,\delta) := \{y \in \mathbb{R}^n : |x-y| < \delta\}$ and $B^n(1) := B^n(0,1)$.

Let $\mathscr{E} := \sum_{|\alpha|\leq\kappa} a_\alpha(x)\partial_x^\alpha$ be an elliptic operator on $\mathbb{R}^n$ with smooth coefficients. Set

$$\mathscr{E}_{x,\delta} := \Phi_{x,\delta}^* \delta^\kappa \mathscr{E}(\Phi_{x,\delta})_* = \sum_{|\alpha|\leq\kappa} a_\alpha(x+\delta t)\delta^{\kappa-|\alpha|}\partial_t.$$

It is not hard to see that $\mathscr{E}_{x,\delta}$ is an elliptic operator on $B^n(1)$, *uniformly* as x ranges over compact sets and $\delta \in (0,1]$. Because of this, if we want to study the elliptic operator $\mathscr{E}$ on a small ball $B^n(x,\delta)$, it suffices to instead study a different elliptic operator, $\mathscr{E}_{x,\delta}$, on the unit ball $B^n(1)$. On the Fourier transform side, this means that if we want to study high frequencies for an elliptic operator, it suffices to instead study low frequencies for a different elliptic operator. It is this invariance under scaling that makes the classical Calderón–Zygmund theory on $\mathbb{R}^n$ an important tool when studying elliptic operators.

1.9.2 Scaling for maximally subelliptic operators

Let $(W, \vec{d}) = \{(W_1, \vec{d}_1), \dots, (W_r, \vec{d}_r)\}$ be Hörmander vector fields with formal degrees on $\mathfrak{M}$. Fix a compact set $\mathcal{K} \Subset \mathfrak{M}$. For each $x \in \mathcal{K}$ and $\delta \in (0,1]$, Nagel, Stein, and Wainger [189] introduced a scaling map,

$$\Phi_{x,\delta} : B^n(1) \xrightarrow{\sim} \Phi_{x,\delta}(B^n(1)) \approx B_{(W,\vec{d})}(x,\delta),$$

where $\Phi_{x,\delta}(B^n(1))$ is not exactly equal to $B_{(W,\vec{d})}(x,\delta)$, but is comparable to it.

Let $\mathscr{P} := \sum_{\deg_{\vec{d}}(\alpha)\leq\kappa} a_\alpha(x)W^\alpha$ be maximally subelliptic of degree κ with respect to $(W, \vec{d})$ on $\mathfrak{M}$. We have the following:

- Define $W_j^{x,\delta} := \Phi_{x,\delta}^* \delta^{\vec{d}_j} W_j$ and

$$(W^{x,\delta}, \vec{d}) := \{(W_1^{x,\delta}, \vec{d}_1), \dots, (W_r^{x,\delta}, \vec{d}_r)\}.$$

  Then $(W^{x,\delta}, \vec{d})$ are Hörmander vector fields with formal degrees *uniformly* for $x \in \mathcal{K}$ and $\delta \in (0,1]$.
- Set $\mathscr{P}_{x,\delta} := \Phi_{x,\delta}^* \delta^\kappa \mathscr{P}(\Phi_{x,\delta})_*$. Then $\mathscr{P}_{x,\delta}$ is maximally subelliptic of degree κ with respect to $(W^{x,\delta}, \vec{d})$ on $B^n(1)$, *uniformly* for $x \in \mathcal{K}$ and $\delta \in (0,1]$.

See Theorem 3.3.7 for a much more detailed discussion of these maps and Section 3.3.1 for a more detailed discussion of scaling maximally subelliptic operators; the main point is that every estimate we need for our proofs holds uniformly for $x \in \mathcal{K}$ and $\delta \in (0,1]$.

Because of the above, and just as in the elliptic setting, to study a maximally subelliptic operator at a small (Carnot–Carathéodory) scale $B_{(W,d)}(x,\delta)$, it suffices to instead study a different maximally subelliptic operator at the unit scale $B^n(1)$. This is the key fact needed to begin applying the Calderón–Zygmund theory, with respect to the Carnot–Carathéodory metric, to these operators.

In our proofs, we iterate these scaling ideas several times, leading to several "layers" of scaling. In elliptic theory, adding such layers adds no complications because the scaling maps are so simple. However, in our setting, this means that we need results which are more quantitative – uniformity on compact sets is not sufficient. To achieve these more quantitative results and to allow us to study the multi-parameter theory described in Section 1.8, we use recent work on these scaling maps [228], which is based on an idea of Tao and Wright [232]. This technical, quantitative result, described in Section 3.6 and Appendix A, is the central tool in this text and forms the foundation for the proof of every main result.

1.10 Outline

- **Chapter 2:** We describe classical results from elliptic theory which we generalize in later chapters. Many of our proofs are much simpler (and well known) in the classical setting. The proofs given in Chapter 2 may help the reader better understand the logical structure of the more abstract proofs given in the maximally subelliptic setting.
- **Chapter 3:** We study vector fields and Carnot–Carathéodory geometry. This chapter is the toolbox we need when studying maximally subelliptic operators. In particular, we define and study the scaling maps described in Section 1.9. The reader interested in understanding the proofs for maximally subelliptic operators might only skim this chapter on a first pass, referring back whenever a particular tool is needed.
- **Chapter 4:** We introduce generalized pseudo-differential operators corresponding to a finite list of vector fields with multi-parameter formal degrees as discussed in Section 1.8. These pseudo-differential operators provide several useful examples of the singular integral operators studied in Chapter 5.
- **Chapter 5:** We study singular integral operators, both in the single-parameter case described in Section 1.4 and in the multi-parameter case described in Section 1.8.
- **Chapter 6:** We study the Besov and Triebel–Lizorkin spaces described in Section 1.5, both in the single-parameter case and in the multi-parameter case described in Section 1.8.

- **Chapter 7:** We focus on the Zygmund–Hölder spaces $\mathscr{C}^{\vec{s}} = \mathscr{B}^{\vec{s}}_{\infty,\infty}$. We study compositions of functions as well as characterizations of these spaces in terms of differences.
- **Chapter 8:** We develop the linear theory of maximally subelliptic operators.
- **Chapter 9:** We develop the fully nonlinear theory of maximally subelliptic operators.
- **Appendix A:** We present the proof from [228] of the main technical theorem underlying all the scaling results in Chapter 3. The results in Appendix A form the heart of every main proof in this text.

2 Ellipticity

Maximally subelliptic partial differential operators are a generalization of elliptic partial differential operators, and all the main results in this text are generalizations of classical results concerning elliptic operators. Elliptic operators may be the most well-understood class of partial differential operators: not only is the linear theory of elliptic PDEs well understood, but many sharp results are known for general fully nonlinear elliptic PDEs.

In the classical study of elliptic PDEs, the Fourier transform and the group structure on $\mathbb{R}^n$ play a central role. In the more general setting of maximally subelliptic operators, these tools are not as applicable. Thus, while we will see that many of the classical results from linear and nonlinear elliptic PDEs generalize to the case of maximally subelliptic PDEs, the classical proofs do not generalize. In this chapter, we will reframe these classical results in such a way that the group structure and the Fourier transform play a secondary role. These reframed results and methods will generalize to maximally subelliptic operators more directly.

Set $D_x = (D_{x_1}, \ldots, D_{x_n}) := \frac{1}{2\pi i}\partial_x = \frac{1}{2\pi i}(\partial_{x_1}, \ldots, \partial_{x_n})$, where $\partial_{x_j} = \frac{\partial}{\partial x_j}$. Consider a linear partial differential operator of order m on $\mathbb{R}^n$ with smooth coefficients:

$$\mathscr{P} := \sum_{|\alpha| \le m} a_\alpha(x) D_x^\alpha, \quad a_\alpha : \mathbb{R}^n \to \mathbb{C} \text{ smooth.}$$

Corresponding to $\mathscr{P}$ is its characteristic polynomial or symbol

$$P(x, \xi) := \sum_{|\alpha| \le m} a_\alpha(x) \xi^\alpha,$$

with principal symbol $P_m(x, \xi) := \sum_{|\alpha|=m} a_\alpha(x)\xi^\alpha$.

Definition 2.0.1. We say $\mathscr{P}$ is elliptic near $x_0 \in \mathbb{R}^n$ if

$$P_m(x_0, \xi) = 0 \iff \xi = 0.$$

For $\Omega \subseteq \mathbb{R}^n$ open, we say $\mathscr{P}$ is elliptic on Ω if $\mathscr{P}$ is elliptic near x_0, $\forall x_0 \in \Omega$.

The key result that we will focus on, and the reason fully nonlinear elliptic PDEs are so well understood, is that elliptic operators are locally left invertible modulo smooth functions on many spaces. For example, suppose $\mathscr{P}$ is a partial differential operator of order m, which is elliptic near x_0. Let u be a distribution on $\mathbb{R}^n$. Then letting L^p_s denote the L^p Sobolev space of order $s \in \mathbb{R}$ (see Definition 2.2.11), a standard result is, for $1 < p < \infty$,

$$\mathscr{P}u \in L^p_s \text{ near } x_0 \iff u \in L^p_{s+m} \text{ near } x_0;$$

https://doi.org/10.1515/9783111085647-002

see Proposition 2.2.35. It is not just Sobolev spaces which enjoy this property. Indeed, for $l \in \mathbb{N}$ and $r \in [0,1]$, let $C^{l,r}$ denote the standard Hölder space on $\mathbb{R}^n$ (see (2.44)). Then, for $l \in \mathbb{N}$ and $r \in (0,1)$,

$$\mathscr{P}u \in C^{l,r} \text{ near } x_0 \iff u \in C^{l+m,r} \text{ near } x_0. \tag{2.1}$$

This does not hold for $r \in \{0,1\}$; however, one can generalize (2.1) to integer exponents by using the classical Zygmund–Hölder spaces (see Section 2.5). We denote these by $\mathscr{C}^s$, $s > 0$. We have $C^{l,r} = \mathscr{C}^{l+r}$ for $r \notin \{0,1\}$ and

$$\mathscr{P}u \in \mathscr{C}^s \text{ near } x_0 \iff u \in \mathscr{C}^{s+m} \text{ near } x_0;$$

see Theorem 2.7.2 (e).

There is a more general framework to understand the above results: the Besov and Triebel–Lizorkin spaces (see Section 2.4). We denote the scale of Besov spaces by $\mathscr{B}^s_{p,q}$, where $1 \le p, q \le \infty$, and $s \in \mathbb{R}$, and the scale of Triebel–Lizorkin spaces by $\mathscr{F}^s_{p,q}$, where $1 < p < \infty$, $1 < q \le \infty$, and $s \in \mathbb{R}$.[1] We have $L^p_s = \mathscr{F}^s_{p,2}$ for $1 < p < \infty$ and $s \in \mathbb{R}$, and $\mathscr{C}^s = \mathscr{B}^s_{\infty,\infty}$ and $s > 0$. Thus, one many generalize the above results to

$$\begin{aligned} \mathscr{P}u \in \mathscr{B}^s_{p,q} \text{ near } x_0 &\iff u \in \mathscr{B}^{s+m}_{p,q} \text{ near } x_0, \\ \mathscr{P}u \in \mathscr{F}^s_{p,q} \text{ near } x_0 &\iff u \in \mathscr{F}^{s+m}_{p,q} \text{ near } x_0; \end{aligned}$$

see Theorem 2.7.2 (e). It is these results which will naturally generalize to the case of maximally subelliptic operators.

Example 2.0.2. The Laplacian on $\mathbb{R}^n$, $\triangle = -\partial_{x_1}^2 - \partial_{x_2}^2 - \cdots - \partial_{x_n}^2$, is an elliptic operator. Its principal symbol is $4\pi^2|\xi|^2$.

2.1 Some basic function spaces

For any topological space Ω, we let $C(\Omega)$ denote the space of bounded, continuous functions $f : \Omega \to \mathbb{C}$ endowed with the norm

$$\|f\|_{C(\Omega)} := \sup_{x\in\Omega} |f(x)|.$$

With this norm, $C(\Omega)$ is a Banach space.

Let $\Omega \subseteq \mathbb{R}^n$ be an open set. For $m \in \mathbb{N}$ let

$$\begin{aligned} C^m(\Omega) &:= \{f \in C(\Omega) : \partial_x^\alpha f \in C(\Omega), \forall |\alpha| \le m\}, \\ \|f\|_{C^m(\Omega)} &:= \sum_{|\alpha|\le m} \|\partial_x^\alpha f\|_{C(\Omega)}, \end{aligned} \tag{2.2}$$

---

1 It is possible to define these spaces for all positive p, q; see Remark 2.4.6.

so that $C^m(\Omega)$ is a Banach space. Let $C^m_{\mathrm{loc}}(\Omega)$ be the space of those $f : \Omega \to \mathbb{C}$ such that $\forall x \in \Omega$, there exists an open neighborhood V of x such that $f|_V \in C^m(V)$; $C^m_{\mathrm{loc}}(\Omega)$ is a Fréchet space (see [238, Chapter 10]). Similarly, for any function space we define the local version by adding a subscript loc. We let

$$C^\infty(\Omega) := \bigcap_{m\in\mathbb{N}} C^m(\Omega), \quad C^\infty_{\mathrm{loc}}(\Omega) := \bigcap_{m\in\mathbb{N}} C^m_{\mathrm{loc}}(\Omega), \tag{2.3}$$

and give these spaces the usual Fréchet topologies (see [238, Chapter 10]).

We let $C_0^\infty(\Omega)$ be the space of all $f \in C^\infty(\Omega)$ such that $\mathrm{supp}(f) \Subset \Omega$. Furthermore, we give $C_0^\infty(\Omega)$ the structure of a locally convex topological vector space in the usual way (see [238, Chapter 13, Example 2]).

For all of the above spaces, we define the analogous spaces of functions taking values in a Banach space $\mathscr{X}$ in the obvious way, and denote these spaces by $C(\Omega;\mathscr{X})$, $C^m(\Omega;\mathscr{X})$, $C^m_{\mathrm{loc}}(\Omega;\mathscr{X})$, $C_0^\infty(\Omega;\mathscr{X})$, etc.

We let $\mathscr{S}(\mathbb{R}^n)$ denote the space of Schwartz functions on $\mathbb{R}^n$, that is, the space of all $f \in C^\infty(\mathbb{R}^n)$ such that for each $\alpha, \beta \in \mathbb{N}^n$ the following semi-norm is finite:

$$\|f\|_{\alpha,\beta} := \sup_{x\in\mathbb{R}^n} |x^\alpha \partial_x^\beta f(x)|.$$

We give $\mathscr{S}(\mathbb{R}^n)$ the coarsest topology such that all such semi-norms are continuous.

Often, given a locally convex topological vector space, $\mathscr{X}$, we are interested in the bounded subsets of $\mathscr{X}$, defined as follows.

Definition 2.1.1. A subset $\mathcal{B} \subseteq \mathscr{X}$ of a locally convex topological vector space $\mathscr{X}$ is said to be a bounded subset of $\mathscr{X}$ if for every continuous semi-norm $\|\cdot\|$ on $\mathscr{X}$ we have

$$\sup_{f\in\mathcal{B}} \|f\| < \infty.$$

Remark 2.1.2. For most of the above spaces, the bounded subsets are easy to understand directly from the definition. However, the semi-norms which induce the topology on $C_0^\infty(\Omega)$ are a bit more complicated. Fortunately, the bounded subsets have a simple characterization: $\mathcal{B} \subset C_0^\infty(\Omega)$ is bounded if and only if the following two conditions hold:
- $\bigcup_{f\in\mathcal{B}} \mathrm{supp}(f) \Subset \Omega$,
- for every $m \in \mathbb{N}$, $\sup_{f\in\mathcal{B}} \|f\|_{C^m} < \infty$.

For a proof of this fact, see [238, Proposition 14.6 and Chapter 13, Example 2]. In $C_0^\infty(\Omega)$, the bounded sets are precisely the pre-compact sets.

Given a locally convex topological vector space $\mathscr{X}$, we denote by $\mathscr{X}'$ its dual. Thus, $C_0^\infty(\Omega)'$ is the usual space of distributions on Ω, $C^\infty_{\mathrm{loc}}(\Omega)'$ is the space of distributions with compact support in Ω (see [238, Theorem 24.2]), and $\mathscr{S}(\mathbb{R}^n)'$ is the space of tempered distributions. We give these spaces the usual weak dual topology.

Given a locally integrable function $f : \Omega \to \mathbb{C}$, we obtain a distribution by

$$\phi \mapsto \int f(x)\phi(x)\, dx. \tag{2.4}$$

Thus, given a function, we may treat it as a distribution. Given an arbitrary distribution $\lambda \in C_0^\infty(\Omega)'$ we abuse notation and for $\phi \in C_0^\infty(\Omega)$, we write

$$\lambda(\phi) = \int \lambda(x)\phi(x)\, dx,$$

whether or not λ is given by integration against a locally integrable function. If we say that λ agrees with a locally integrable function, this means that there exists f such that λ is given by (2.4). In this case, we identify λ with f, and may then treat it as a function.

If $\mathscr{X}_1$ and $\mathscr{X}_2$ are two locally convex topological vector spaces and we write $T : \mathscr{X}_1 \to \mathscr{X}_2$, this means that T is a continuous linear map from $\mathscr{X}_1$ to $\mathscr{X}_2$. We write $\mathrm{Hom}(\mathscr{X}_1, \mathscr{X}_2)$ for the space of all such continuous linear maps. The Schwartz kernel theorem gives a convenient characterization of certain spaces of linear maps.

Theorem 2.1.3 (The Schwartz kernel theorem). *Given $K(x,y) \in C_0^\infty(\Omega\times\Omega)'$, we obtain $T \in \mathrm{Hom}(C_0^\infty(\Omega), C_0^\infty(\Omega)')$ by, for $\phi_1, \phi_2 \in C_0^\infty(\Omega)$,*

$$\int \phi_2(x)(T\phi_1)(x)\, dx = \iint \phi_2(x)\phi_1(y)K(x,y)\, dxdy.$$

We similarly define maps $C_{\mathrm{loc}}^\infty(\Omega \times \Omega)' \to \mathrm{Hom}(C_{\mathrm{loc}}^\infty(\Omega), C_{\mathrm{loc}}^\infty(\Omega)')$ and $\mathscr{S}(\mathbb{R}^n \times \mathbb{R}^n)' \to \mathrm{Hom}(\mathscr{S}(\mathbb{R}^n), \mathscr{S}(\mathbb{R}^n)')$. The above maps are all bijections. The distribution K is called the Schwartz kernel of the operator T.

Henceforth, we identify operators with their Schwartz kernel. Thus, given $T : C_0^\infty(\Omega) \to C_0^\infty(\Omega)'$, we sometimes treat T as a distribution $T(x,y) \in C_0^\infty(\Omega\times\Omega)'$, and given a distribution $T(x,y) \in C_0^\infty(\Omega \times \Omega)'$, we sometimes treat T as an operator $T : C_0^\infty(\Omega) \to C_0^\infty(\Omega)'$. Moreover, given a locally integrable function $F(x,y) : \Omega \times \Omega \to \mathbb{C}$, we identify F with a distribution on $\Omega \times \Omega$ via (2.4), and therefore with a map $F : C_0^\infty(\Omega) \to C_0^\infty(\Omega)'$. For a proof of the Schwartz kernel theorem, as well as an elegant description of a more general result, see [238, Chapter 51].

Definition 2.1.4. Given two locally convex topological vector spaces $\mathscr{X}_1$ and $\mathscr{X}_2$, the topology of bounded convergence on $\mathrm{Hom}(\mathscr{X}_1, \mathscr{X}_2)$ can be defined as follows. Given a bounded set $\mathcal{B} \subset \mathscr{X}_1$ and a continuous semi-norm $|\cdot|$ on $\mathscr{X}_2$, define a semi-norm on $\mathrm{Hom}(\mathscr{X}_1, \mathscr{X}_2)$ by

$$\|T\| := \sup_{f\in\mathcal{B}} |Tf|.$$

The topology of bounded convergence is the coarsest topology such that all of the above semi-norms are continuous.

If $(\mathfrak{M}, \mu)$ is a measure space, we define $L^p(\mathfrak{M}, \mu)$ for $p \in [1, \infty]$ in the usual way. When $\mathfrak{M}$ is an open subset of $\mathbb{R}^n$, we give $\mathfrak{M}$ Lebesgue measure unless otherwise mentioned. So, for example, $L^p(\mathbb{R}^n)$ denotes the usual L^p space on $\mathbb{R}^n$.

Definition 2.1.5. For a distribution $\lambda \in C_0^\infty(\Omega)'$, we define $\overline{\lambda}$ by

$$\overline{\lambda}(\phi) = \overline{\lambda(\overline{\phi})}, \quad \phi \in C_0^\infty(\Omega).$$

Definition 2.1.6. Given any operator $T \in \operatorname{Hom}(C_0^\infty(\Omega), C_0^\infty(\Omega)')$, we define the formal L^2 adjoint of T, denoted by T^*, where $T^* \in \operatorname{Hom}(C_0^\infty(\Omega), C_0^\infty(\Omega)')$, by

$$\int \psi(x)\overline{(T\phi)(x)}\, dx = \int (T^*\psi)(x)\overline{\phi(x)}\, dx, \quad \phi, \psi \in C_0^\infty(\Omega).$$

We use a similar definition with $C_0^\infty(\Omega)$ replaced with $C_{\mathrm{loc}}^\infty(\Omega)$ or $\mathscr{S}(\Omega)$ throughout.

2.2 Pseudo-differential operators and the Fourier transform

A central tool in the study of elliptic PDEs is the Fourier transform.

Definition 2.2.1. Given $f \in \mathscr{S}(\mathbb{R}^n)$, we define the Fourier transform of f, $\hat{f}(\xi) : \mathbb{R}^n \to \mathbb{C}$, by

$$\hat{f}(\xi) := \int f(x) e^{-2\pi i x \cdot \xi}\, dx.$$

We have the following well-known result.

Proposition 2.2.2. *The Fourier transform $f \mapsto \hat{f}$ is an automorphism of the Fréchet space $\mathscr{S}(\mathbb{R}^n)$ with inverse $g \mapsto \check{g}$ given by*

$$\check{g}(x) := \int g(\xi) e^{2\pi i x \cdot \xi}\, d\xi.$$

The Fourier transform allows us to introduce another key tool in the study of elliptic operators: pseudo-differential operators. To a tempered distribution $a \in \mathscr{S}(\mathbb{R}^n \times \mathbb{R}^n)'$, we may associate a continuous linear map $a(x, D) : \mathscr{S}(\mathbb{R}^n) \to \mathscr{S}(\mathbb{R}^n)'$ by

$$\int g(x)(a(x, D)f)(x)\, dx := \iint g(x) e^{2\pi i x \cdot \xi} \hat{f}(\xi) a(x, \xi)\, dx d\xi.$$

An analog of the Schwartz kernel theorem (Theorem 2.1.3) states that the map $a \mapsto a(x, D)$ is a bijection $\mathscr{S}(\mathbb{R}^n \times \mathbb{R}^n)' \to \operatorname{Hom}(\mathscr{S}(\mathbb{R}^n), \mathscr{S}(\mathbb{R}^n)')$. The tempered distribution $a(x, \xi)$ is called the symbol of the operator $a(x, D)$ and for appropriate choices of a, operators written in this way are called pseudo-differential operators.

We now introduce the class of symbols which we will use. The class that follows is the mostly commonly used class of symbols, and we refer to them as “standard symbols.”

Definition 2.2.3. Fix $m \in \mathbb{R}$. The space of standard symbols of order m, S^m, is the Fréchet space consisting of functions $a \in C^\infty_{\mathrm{loc}}(\mathbb{R}^n \times \mathbb{R}^n)$, which satisfy, for all multi-indices $\alpha, \beta \in \mathbb{N}^n$,

$$|\partial_x^\alpha \partial_\xi^\beta a(x,\xi)| \leq C_{\alpha,\beta}(1+|\xi|)^{m-|\beta|}.$$

We give S^m the coarsest topology such that the least possible $C_{\alpha,\beta}$ defines a continuous semi-norm on S^m (for each α, β). Operators of the form $a(x, D)$ for $a \in S^m$ are called pseudo-differential operators of order m.

Remark 2.2.4. It can be shown that for $a \in S^m$, $a(x,D) : \mathscr{S}(\mathbb{R}^n) \to \mathscr{S}(\mathbb{R}^n)$.

The first example to consider, and the one that justifies the name, is that of linear partial differential operators with smooth coefficients.

Example 2.2.5. Consider a linear partial differential operator of order $m \in \mathbb{N}_+$, $P(x,D)$, of the form

$$P(x,D)f(x) = \sum_{|\alpha|\leq m} a_\alpha(x) D^\alpha f(x),$$

where $a_\alpha \in C^\infty(\mathbb{R}^n)$ (see (2.3)). Then $P(x, D)$ is a pseudo-differential operator of order m, with symbol $P(x,\xi) = \sum_{|\alpha|\leq m} a_\alpha(x)\xi^\alpha$.

We state the next lemma without proof.

Lemma 2.2.6. *Let $a(x,\xi) \in S^m$ and let $a(x,D)^*$ be the formal L^2 adjoint of $a(x,D)$ (see Definition 2.1.6). Then $a(x,D)^*$ is a pseudo-differential operator of order m. Moreover, one can obtain an asymptotic expansion for the symbol of $a(x,D)^*$ in terms of a; see* [216, *Chapter VI, Section* 6.2].

Remark 2.2.7. Combining Lemma 2.2.6 and Remark 2.2.4, for $a \in S^m$ we may extend $a(x,D)$ to be a map $a(x,D) : \mathscr{S}(\mathbb{R}^n)' \to \mathscr{S}(\mathbb{R}^n)'$ by duality:

$$\int (a(x,D)f)(x)\overline{g(x)}\,dx = \int f(x)\overline{(a(x,D)^*g)(x)}\,dx, \quad f \in \mathscr{S}(\mathbb{R}^n)', g \in \mathscr{S}(\mathbb{R}^n).$$

The two central results concerning pseudo-differential operators which we use are that they form a filtered algebra and that they are bounded on appropriate L^p Sobolev spaces. We turn to stating these results in detail, though we do not include all the proofs. For further discussion and proofs, see [216, Chapter VI].

Theorem 2.2.8 (The calculus of pseudo-differential operators). *Suppose $a \in S^{m_1}$ and $b \in S^{m_2}$. We consider the operator $a(x,D)b(x,D) : \mathscr{S}(\mathbb{R}^n) \to \mathscr{S}(\mathbb{R}^n)$. There exists $c \in S^{m_1+m_2}$ such that $c(x,D) = a(x,D)b(x,D)$. Furthermore, for every $N \in \mathbb{N}$,*

$$c - \sum_{|\alpha|<N} \frac{(2\pi i)^{-|\alpha|}}{\alpha!}(\partial_\xi^\alpha a)(\partial_x^\alpha b) \in S^{m_1+m_2-N}.$$

Remark 2.2.9. Taking $N = 1$ in Theorem 2.2.8, we have $c(x,\xi) \equiv a(x,\xi)b(x,\xi) \mod S^{m_1+m_2-1}$, and therefore $[a(x,D), b(x,D)]$ is a pseudo-differential operator of order m_1+m_2-1. Informally, pseudo-differential operators commute modulo lower-order terms. This is a key property in the classical study of elliptic PDEs, and we will lose this property when moving to the more general setting of maximally subelliptic PDEs. See Remark 1.4.1.

Definition 2.2.10. For $s \in \mathbb{R}$, we let Λ^s denote the pseudo-differential operator of order s with symbol $(1+|\xi|^2)^{s/2}$. It is immediate to verify that $\Lambda^s\Lambda^t = \Lambda^{s+t}$. As such, $\Lambda^s : \mathscr{S}(\mathbb{R}^n) \to \mathscr{S}(\mathbb{R}^n)$ is an automorphism with inverse Λ^{-s}.

Definition 2.2.11. For $p \in (1,\infty)$ and $s \in \mathbb{R}$, we define $L^p_s(\mathbb{R}^n)$, the standard L^p Sobolev space of order s on $\mathbb{R}^n$, to be the space of all $f \in \mathscr{S}(\mathbb{R}^n)'$ such that $\Lambda^s f \in L^p$. We endow this space with the norm

$$\|f\|_{L^p_s} := \|\Lambda^s f\|_{L^p(\mathbb{R}^n)}.$$

By convention, we take $\|f\|_{L^p_s} = \infty$ if $f \notin L^p_s(\mathbb{R}^n)$.

Theorem 2.2.12. *Let $a \in S^m$. Then $a(x,D)$ extends to a continuous map $a(x,D) : L^p_s \to L^p_{s-m}$, $p \in (1,\infty)$, $s \in \mathbb{R}$. In particular, pseudo-differential operators of order* 0 *are bounded on $L^p(\mathbb{R}^n)$, $1 < p < \infty$.*

We will see Theorem 2.2.12 as part of a more general result (see Corollary 2.4.8).

2.2.1 Pseudo-differential operators without the Fourier transform

When we turn to maximally subelliptic operators, the Fourier transform is no longer a decisive tool, and pseudo-differential operators do not play a central role. Instead, we will have several types of singular integral operators which will play the role of pseudo-differential operators in various settings. As a first step, we need to characterize pseudo-differential operators in a way which does not use the Fourier transform. We do this in two ways, both of which motivate later definitions.

For the first way, we follow the presentation in [216, Chapter VI, Section 7.4]. The idea is as follows. For $a(x,\xi) \in S^m$ let $\check{a}(x,z)$ denote the inverse Fourier transform of a in the ξ variable. This tempered distribution is from the following class of kernels.

Definition 2.2.13. We say $K(x,z) \in \mathscr{S}(\mathbb{R}^n \times \mathbb{R}^n)'$ is a kernel of order $m \in \mathbb{R}$ if the following conditions hold:

– (Growth condition) $K(x,z)$ agrees with a C^∞ function on the set $\{(x,z) : z \neq 0\}$, and for all multi-indices $\alpha, \beta \in \mathbb{N}^n$ and every $N \geq 0$ so that $n+m+|\alpha|+N > 0$,

$$|\partial_x^\beta \partial_z^\alpha K(x,z)| \leq C_{\alpha,\beta,N}|z|^{-n-m-|\alpha|-N}.$$

- (Cancelation condition) For $m \geq 0$, we assume for every bounded set $\mathcal{B} \subset C_0^\infty(\mathbb{R}^n)$ and every multi-index $\alpha \in \mathbb{N}^n$

$$\sup_{x \in \mathbb{R}^n} \sup_{R \geq 1} \sup_{\phi \in \mathcal{B}} R^{-m} \left| \int \partial_x^\alpha K(x,z)\phi(Rz)\, dz \right| < \infty.$$

- If $m < 0$, the growth condition implies that for $\phi \in C_0^\infty(\mathbb{R}^n)$, the limit

$$\lim_{\epsilon \downarrow 0} \int_{|z|>\epsilon} K(x,z)\phi(z)\, dz$$

converges for each $x \in \mathbb{R}^n$. We assume that $\int K(x,z)\phi(z)\, dz$ agrees with this limit.

Proposition 2.2.14 (See Chapter VI, Section 7.4 of [216]). *$a \mapsto \check{a}$ is a bijection between S^m and kernels of order m.*

Remark 2.2.15. In light of Proposition 2.2.14, we may write pseudo-differential operators in another way. Namely, for $a \in S^m$,

$$a(x,D)f(x) = \int f(x-z)\check{a}(x,z)\, dz, \quad f \in \mathcal{S}(\mathbb{R}^n),$$

where $\check{a}(x,z)$ is a kernel of order m. We will use this idea to generalize some aspects of pseudo-differential operators to settings which have no immediate analog of the Fourier transform.

Example 2.2.16. A simple example to keep in mind is the Laplacian on $\mathbb{R}^2$, which has Green's function $K(x-y) = -\frac{1}{2\pi}\log(|x-y|)$. If $\phi \in C_0^\infty(\mathbb{R}^n)$ and we consider the distribution $\phi(x)K(x-y)\phi(y)$, then this is the Schwartz kernel of a pseudo-differential operator of order -2. Here, $n = 2$ and $m = -2$. This demonstrates the need for N in the growth condition in Definition 2.2.13. We will see this phenomenon several times in this text.

The second way to view pseudo-differential operators without using the Fourier transform utilizes a Littlewood–Paley decomposition. To describe this, we need to introduce some more function spaces on $\mathbb{R}^n$.

Definition 2.2.17. We let $\mathcal{S}_0(\mathbb{R}^n)$ denote the space of all $f \in \mathcal{S}(\mathbb{R}^n)$ such that

$$\int t^\alpha f(t)\, dt = 0, \quad \forall \alpha \in \mathbb{N}^n.$$

$\mathcal{S}_0(\mathbb{R}^n)$ is called the space of Schwartz functions, all of whose moments vanish. It is a closed subspace of $\mathcal{S}(\mathbb{R}^n)$, and as such inherits a Fréchet topology from $\mathcal{S}(\mathbb{R}^n)$.

Lemma 2.2.18. *Let $f \in \mathcal{S}(\mathbb{R}^n)$. Then $f \in \mathcal{S}_0(\mathbb{R}^n)$ if and only if $\forall \alpha \in \mathbb{N}^n$, $\partial_\xi^\alpha \hat{f}(0) = 0$.*

Proof. This is immediate from the definitions. □

Corollary 2.2.19. *Fix $s \in \mathbb{R}$. The map $\triangle^s : \mathscr{S}_0(\mathbb{R}^n) \to \mathscr{S}_0(\mathbb{R}^n)$ given by $\triangle^s : f \mapsto (|2\pi\xi|^{2s}\hat{f}(\xi))^\vee$ is an automorphism.*

Proof. It follows easily from Lemma 2.2.18 that for $f \in \mathscr{S}_0(\mathbb{R}^n)$, $\triangle^s f \in \mathscr{S}_0(\mathbb{R}^n)$. The closed graph theorem shows that $\triangle^s$ is continuous. The continuous inverse of $\triangle^s$ is $\triangle^{-s}$. □

Corollary 2.2.19 gives a characterization of $\mathscr{S}_0(\mathbb{R}^n)$ which will be useful later.

Corollary 2.2.20. *Let $\mathcal{T} \subseteq \mathscr{S}(\mathbb{R}^n)$ be the largest subset such that $\forall f \in \mathcal{T}$*

$$f = \sum_{j=1}^{n} \partial_{t_j} f_a, \quad f_a \in \mathcal{T}. \tag{2.5}$$

Then $\mathcal{T} = \mathscr{S}_0(\mathbb{R}^n)$.

Proof. For $f \in \mathscr{S}_0(\mathbb{R}^n)$, we have

$$f = \triangle\triangle^{-1} f = \sum_{j=1}^{n} \partial_{t_j}(-\partial_{t_j} \triangle^{-1} f).$$

It follows from Corollary 2.2.19 that $(-\partial_{t_j} \triangle^{-1} f) \in \mathscr{S}_0(\mathbb{R}^n)$. We conclude $\mathscr{S}_0(\mathbb{R}^n) \subseteq \mathcal{T}$.

Suppose $f \in \mathcal{T}$ and fix a multi-index $\alpha \in \mathbb{N}^n$. Applying (2.5) $|\alpha| + 1$ times, we have

$$f = \sum_{|\beta|=|\alpha|+1} \partial_t^\beta f_\beta, \quad f_\beta \in \mathcal{T} \subseteq \mathscr{S}(\mathbb{R}^n).$$

Thus,

$$\int t^\alpha f(t)\, dt = \sum_{|\beta|=|\alpha|+1} \int t^\alpha \partial_t^\beta f_\beta(t)\, dt = 0.$$

Since $\alpha \in \mathbb{N}^n$ was arbitrary, we conclude $f \in \mathscr{S}_0(\mathbb{R}^n)$. This shows $\mathcal{T} \subseteq \mathscr{S}_0(\mathbb{R}^n)$ and completes the proof. □

Definition 2.2.21. We let $C^\infty(\mathbb{R}^n)\widehat{\otimes}\mathscr{S}(\mathbb{R}^n)$ be the space of all $\varsigma(x,t) \in C^\infty(\mathbb{R}^n \times \mathbb{R}^n)$ such that for each $\alpha, \beta, \gamma \in \mathbb{N}^n$, the following semi-norm is finite:

$$\|\varsigma\|_{\alpha,\beta,\gamma} := \sup_{x,t\in\mathbb{R}^n} |t^\alpha \partial_t^\beta \partial_x^\gamma \varsigma(x,t)|.$$

We give $C^\infty(\mathbb{R}^n)\widehat{\otimes}\mathscr{S}(\mathbb{R}^n)$ the coarsest topology such that all such semi-norms are continuous; with this topology $C^\infty(\mathbb{R}^n)\widehat{\otimes}\mathscr{S}(\mathbb{R}^n)$ is a Fréchet space.

Remark 2.2.22. The notation $C^\infty(\mathbb{R}^n)\widehat{\otimes}\mathscr{S}(\mathbb{R}^n)$ is justified because the space agrees with the completed tensor product of the nuclear space $\mathscr{S}(\mathbb{R}^n)$ and the Fréchet space $C^\infty(\mathbb{R}^n)$. Whenever we use the symbol $\widehat{\otimes}$ it denotes the completed tensor product where one of the two factors is nuclear. We will not use any details of the theory of nuclear spaces, though.

For the interested reader, [238, Part III] is an excellent introduction to the theory of such tensor products.

Definition 2.2.23. We let $C^\infty(\mathbb{R}^n)\widehat{\otimes}\mathscr{S}_0(\mathbb{R}^n)$ be the space of those $\varsigma(x,t) \in C^\infty(\mathbb{R}^n)\widehat{\otimes}\mathscr{S}(\mathbb{R}^n)$ such that for all $\alpha \in \mathbb{N}^n$,

$$\int t^\alpha \varsigma(x,t)\,dt = 0, \quad \forall x \in \mathbb{R}^n.$$

$C^\infty(\mathbb{R}^n)\widehat{\otimes}\mathscr{S}_0(\mathbb{R}^n)$ is a closed subspace of $C^\infty(\mathbb{R}^n)\widehat{\otimes}\mathscr{S}(\mathbb{R}^n)$ and inherits its Fréchet topology.

For $\varsigma(x,t) \in C^\infty(\mathbb{R}^n)\widehat{\otimes}\mathscr{S}(\mathbb{R}^n)$, let $\hat{\varsigma}(x,\xi)$ denote the Fourier transform of ς in the t variable. Note that $\varsigma \mapsto \hat{\varsigma}$ is an automorphism of $C^\infty(\mathbb{R}^n)\widehat{\otimes}\mathscr{S}(\mathbb{R}^n)$ with inverse $\varsigma \mapsto \check{\varsigma}$, where $\check{\varsigma}$ denotes the inverse Fourier transform in the second variable.

We have analogs of Lemma 2.2.18 and Corollary 2.2.19 for $C^\infty(\mathbb{R}^n)\widehat{\otimes}\mathscr{S}_0(\mathbb{R}^n)$, which we state without proof, since the proofs are nearly identical.

Lemma 2.2.24. *Let $\varsigma \in C^\infty(\mathbb{R}^n)\widehat{\otimes}\mathscr{S}(\mathbb{R}^n)$. Then $\varsigma \in C^\infty(\mathbb{R}^n)\widehat{\otimes}\mathscr{S}_0(\mathbb{R}^n)$ if and only if $\forall \alpha \in \mathbb{N}^n$, $\partial_\xi^\alpha \hat{\varsigma}(x,\xi)\big|_{\xi=0} = 0$.*

Corollary 2.2.25. *Fix $s \in \mathbb{R}$. The map*

$$\triangle_t^s : C^\infty(\mathbb{R}^n)\widehat{\otimes}\mathscr{S}_0(\mathbb{R}^n) \to C^\infty(\mathbb{R}^n)\widehat{\otimes}\mathscr{S}_0(\mathbb{R}^n)$$

given by $\triangle_t^s : \varsigma(x,t) \mapsto (|2\pi\xi|^{2s}\hat{\varsigma}(x,\xi))^\vee$ is an automorphism.

For $\varsigma(x,t) \in C^\infty(\mathbb{R}^n)\widehat{\otimes}\mathscr{S}(\mathbb{R}^n)$ and $j \in \mathbb{N}$, let $\mathrm{Dil}_{2^j}(\varsigma)(x,t) := 2^{jn}\varsigma(x,2^j t) \in C^\infty(\mathbb{R}^n)\widehat{\otimes}\mathscr{S}(\mathbb{R}^n)$.

Theorem 2.2.26. *Let $a(x,\xi) \in \mathscr{S}(\mathbb{R}^n \times \mathbb{R}^n)'$ and fix $m \in \mathbb{R}$. The following are equivalent:*
(i) *$\check{a}$ is a kernel of order m,*
(ii) *$a \in S^m$,*
(iii) *there is a bounded set $\{\varsigma_j : j \in \mathbb{N}\} \subset C^\infty(\mathbb{R}^n)\widehat{\otimes}\mathscr{S}(\mathbb{R}^n)$ with $\varsigma_j \in C^\infty(\mathbb{R}^n)\widehat{\otimes}\mathscr{S}_0(\mathbb{R}^n)$ for $j > 0$ such that*

$$\check{a}(x,t) = \sum_{j\in\mathbb{N}} 2^{jm}\,\mathrm{Dil}_{2^j}(\varsigma_j)(x,t),$$

where the sum is taken in the sense of $\mathscr{S}(\mathbb{R}^n \times \mathbb{R}^n)'$ (and every such sum converges in this sense).

Before we turn to the proof of Theorem 2.2.26 we need some new notation. For two real numbers $a, b \in \mathbb{R}$, we write $a \vee b = \max\{a,b\}$ and $a \wedge b = \min\{a,b\}$.

Proof of Theorem 2.2.26. (i) ⇔ (ii): This is Proposition 2.2.14.

(iii) ⇒ (ii): Suppose $\{\varsigma_j : j \in \mathbb{N}\} \subset C^\infty(\mathbb{R}^n)\widehat{\otimes}\mathscr{S}(\mathbb{R}^n)$ is a bounded set with $\varsigma_j \in C^\infty(\mathbb{R}^n)\widehat{\otimes}\mathscr{S}_0(\mathbb{R}^n)$ for $j > 0$. Because the Fourier transform of $\mathrm{Dil}_{2^j}(\varsigma_j)$ is $\hat{\varsigma}_j(x, 2^{-j}\xi)$, we consider the sum

$$a(x,\xi) = \sum_{j\in\mathbb{N}} 2^{jm}\hat{\varsigma}_j(x, 2^{-j}\xi).$$

We will show that this sum converges in the sense of tempered distributions (and in $C^\infty_{\mathrm{loc}}(\mathbb{R}^n \times \mathbb{R}^n)$) to a symbol $a(x,\xi) \in S^m$. In particular, we wish to show

$$\left|\partial_x^\alpha \partial_\xi^\beta \sum_{j\in\mathbb{N}} 2^{jm}\hat{\varsigma}_j(x, 2^{-j}\xi)\right| \lesssim (1+|\xi|)^{m-|\beta|}. \tag{2.6}$$

Note that since $\partial_x^\alpha\partial_\xi^\beta \hat{\varsigma}_j(x,2^{-j}\xi) = 2^{-j|\beta|}(\partial_x^\alpha\partial_\xi^\beta\hat{\varsigma}_j)(x,2^{-j}\xi)$ and $\partial_x^\alpha\partial_\xi^\beta\hat{\varsigma}_j$ is of the same form as $\hat{\varsigma}_j$, it suffices to prove (2.6) with $\alpha = \beta = 0$ (by replacing m with $m - |\beta|$). We will show that

$$\sum_{j\in\mathbb{N}_+} (1+|\xi|)^{-m} 2^{jm}\hat{\varsigma}_j(x, 2^{-j}\xi) \tag{2.7}$$

converges uniformly in x and ξ. It then immediately follows that the same is true with $\mathbb{N}_+$ replaced by $\mathbb{N}$, and this will complete the proof of (ii).

Set $L = |m| + 1$. We claim, for $j \in \mathbb{N}_+$,

$$|\hat{\varsigma}_j(x,\xi)| \lesssim (1+|\xi|)^{-L} \wedge |\xi|^L. \tag{2.8}$$

Indeed,

$$|\hat{\varsigma}_j(x,\xi)| \lesssim (1+|\xi|)^{-L},$$

which follows immediately from the fact that $\{\hat{\varsigma}_j : j \in \mathbb{N}\} \subset C^\infty(\mathbb{R}^n)\widehat{\otimes}\mathscr{S}(\mathbb{R}^n)$ is a bounded set. To see that

$$|\hat{\varsigma}_j(x,\xi)| \lesssim |\xi|^L, \tag{2.9}$$

note that $\varsigma_j = \triangle_t^{L/2}(\triangle_t^{-L/2}\varsigma_j)$, where $\{\triangle_t^{-L/2}\varsigma_j : j \in \mathbb{N}_+\} \subset C^\infty(\mathbb{R}^n)\widehat{\otimes}\mathscr{S}_0(\mathbb{R}^n)$ is a bounded set, by Corollary 2.2.25. Since

$$\hat{\varsigma}_j = |2\pi\xi|^L(\triangle_t^{L/2}\varsigma_j)^\wedge,$$

(2.9) follows.

We separate (2.7) into two parts: $\sum_{2\le 2^j<|\xi|}$ and $\sum_{2^j\ge|\xi|\vee 2}$. Using (2.8), we have

$$\begin{aligned}\sum_{|\xi|\vee 2\le 2^j}(1+|\xi|)^{-m}2^{jm}|\hat{\varsigma}_j(x,2^{-j}\xi)| &\lesssim \sum_{|\xi|\vee 2\le 2^j}(1+|\xi|)^{-m}2^{jm}2^{-Lj}|\xi|^L\\ &= \sum_{|\xi|\vee 2\le 2^j}(1+|\xi|)^{-m}2^{j(m-|m|-1)}|\xi|^{|m|+1}\lesssim 1.\end{aligned}$$

Using (2.8) again, we have

$$\begin{aligned}\sum_{2\le 2^j<|\xi|}(1+|\xi|)^{-m}2^{jm}|\hat{\varsigma}_j(x,2^{-j}\xi)| &\lesssim \sum_{2\le 2^j<|\xi|}(1+|\xi|)^{-m}2^{jm}(2^{-j}|\xi|)^{-L}\\ &= \sum_{2\le 2^j<|\xi|}(1+|\xi|)^{-m}2^{j(m+|m|+1)}|\xi|^{-|m|-1}\lesssim 1.\end{aligned}$$

This completes the proof that (2.7) converges uniformly in ξ and x, and therefore completes the proof of (ii).

(ii) ⇒ (iii): Let $a(x,\xi)\in S^m$. Fix $\phi(\xi)\in C_0^\infty(\mathbb{R}^n)$ with ϕ equal to 1 on a neighborhood of 0. Set $\psi(\xi)=\phi(\xi)-\phi(2\xi)$ so that ψ equals 0 on a neighborhood of 0 and $1=\phi(\xi)+\sum_{j\ge 1}\psi(2^{-j}\xi)$. Define $\hat{\varsigma}_j$ by

$$2^{jm}\hat{\varsigma}_j(x,2^{-j}\xi)=\begin{cases}\psi(2^{-j}\xi)a(x,\xi), & j\ge 1,\\ \phi(\xi)a(x,\xi), & j=0.\end{cases}$$

Note that $a(x,\xi)=\sum_{j\in\mathbb{N}}2^{jm}\hat{\varsigma}_j(x,2^{-j}\xi)$ (with convergence pointwise and in the sense of tempered distributions). We let ς_j be the inverse Fourier transform in the second variable of $\hat{\varsigma}_j$. Thus, $\check{a}(x,t)=\sum_{j\in\mathbb{N}}2^{jm}\operatorname{Dil}_{2^j}(\varsigma_j)(x,t)$ (where this sum converges in the sense of tempered distributions). It follows easily from the definitions that $\{\hat{\varsigma}_j : j\in\mathbb{N}\}\subset C^\infty(\mathbb{R}^n)\widehat{\otimes}\mathscr{S}(\mathbb{R}^n)$ is a bounded set, and therefore $\{\varsigma_j : j\in\mathbb{N}\}\subset C^\infty(\mathbb{R}^n)\widehat{\otimes}\mathscr{S}(\mathbb{R}^n)$ is a bounded set. Since ψ equals 0 on a neighborhood of $\xi=0$, the same is true of $\hat{\varsigma}_j(x,\xi)$ for $j\ge 1$ and therefore $\varsigma_j\in C^\infty(\mathbb{R}^n)\widehat{\otimes}\mathscr{S}_0(\mathbb{R}^n)$ for $j\ge 1$ by Lemma 2.2.24. This completes the proof of (iii). □

2.2.2 Elliptic operators and pseudo-differential operators

The calculus of pseudo-differential operators (Theorem 2.2.8) is a powerful tool for studying elliptic partial differential operators. In this section, we describe some simple consequences of this calculus.

Fix an open set $\Omega\subseteq\mathbb{R}^n$ and let

$$P(x,D)=\sum_{|\alpha|\le m}a_\alpha(x)\partial_x^\alpha$$

be a partial differential operator of order $m \in \mathbb{N}_+$, with smooth coefficients $a_\alpha \in C^\infty_{\mathrm{loc}}(\Omega)$.

For a function $\phi \in C_0^\infty(\mathbb{R}^n)$ we define the operator $\mathrm{Mult}[\phi] : f \mapsto \phi f$, the operator given by multiplication by ϕ. For two functions $\phi_1, \phi_2 \in C_0^\infty(\Omega)$, we write $\phi_1 \prec \phi_2$ if $\phi_2 \equiv 1$ on a neighborhood of $\mathrm{supp}(\phi_1)$.

Proposition 2.2.27. *Suppose $P(x, D)$ is elliptic near $x_0 \in \Omega$. Then there exists an open neighborhood $V \subseteq \Omega$ of x_0 such that the following holds. For all $\phi_1, \phi_2, \phi_3 \in C_0^\infty(V)$ with $\phi_1 \prec \phi_2 \prec \phi_3$ and for all $N \in \mathbb{N}_+$, there exists $b_{N,\phi_1,\phi_2}(x, \xi) \in S^{-m}$ such that*

$$\mathrm{Mult}[\phi_1] b_{N,\phi_1,\phi_2}(x, D) \,\mathrm{Mult}[\phi_2] P(x, D) = \mathrm{Mult}[\phi_1] + \mathrm{Mult}[\phi_1] R_N \,\mathrm{Mult}[\phi_3], \tag{2.10}$$

where R_N is a pseudo-differential operator of order $-N$.

Proof. Note that it suffices to prove the result with ϕ_3 replaced by the constant function 1. Indeed, if we have the result with ϕ_3 replaced by 1, then multiplying both sides on the right by $\mathrm{Mult}[\phi_3]$ and using $\mathrm{Mult}[\phi_2]P(x, D)\,\mathrm{Mult}[\phi_3] = \mathrm{Mult}[\phi_2]P(x, D)$ and $\mathrm{Mult}[\phi_1]\,\mathrm{Mult}[\phi_3] = \mathrm{Mult}[\phi_1]$, the result follows.

Since $P(x, \xi)$ is elliptic near x_0, it is easy to see that there exists an open neighborhood $V \subseteq \Omega$ of x_0 such that there exist $R > 0$, $C > 0$ with

$$|P(x, \xi)| \geq C|\xi|^m, \quad \forall |\xi| \geq R, x \in V.$$

Fix $\phi_1, \phi_2 \in C_0^\infty(V)$ with $\phi_1, \prec \phi_2$. We claim that (2.10) holds with ϕ_3 replaced by the constant function 1, and this will complete the proof. We prove this by induction on N.

For the base case, $N = 1$, let $\eta \in C_0^\infty(\mathbb{R}^n)$ equal 1 on $\overline{B^n(R)} = \{\xi \in \mathbb{R}^n : |\xi| \leq R\}$ and set

$$b_{1,\phi_2}(x, \xi) = (1 - \eta(\xi))\phi_2(x)P(x, \xi)^{-1}.$$

It is immediate to verify that $b_{1,\phi_2} \in S^{-m}$, and Theorem 2.2.8 shows

$$\mathrm{Mult}[\phi_1] b_{1,\phi_2}(x, D)\,\mathrm{Mult}[\phi_2] P(x, D) = \mathrm{Mult}[\phi_1] + \mathrm{Mult}[\phi_1] R_1,$$

where R_1 is a pseudo-differential operator of order -1. This completes the proof of the base case.

Suppose now we have (2.10) (with ϕ_3 replaced by 1) for N and we wish to prove it for $N + 1$. Set

$$b_{N+1,\phi_1,\phi_2}(x, D) = b_N(x, D) - R_N b_1(x, D),$$

so that $b_{N+1,\phi_1,\phi_2} \in S^{-m}$. We have

$$\begin{aligned} &\mathrm{Mult}[\phi_1] b_{N+1,\phi_1,\phi_2}(x, D)\,\mathrm{Mult}[\phi_2] P(x, D) \\ &= \mathrm{Mult}[\phi_1] + \mathrm{Mult}[\phi_1] R_N - \mathrm{Mult}[\phi_1] R_N b_{1,\phi_2}(x, D)\,\mathrm{Mult}[\phi_2] P(x, D). \end{aligned} \tag{2.11}$$

It follows easily from Theorem 2.2.8 and the definition of b_{1,ϕ_2} that

$$\mathrm{Mult}[\phi_1]R_N - \mathrm{Mult}[\phi_1]R_N b_{1,\phi_2}(x,D)\,\mathrm{Mult}[\phi_2]P(x,D) = \mathrm{Mult}[\phi_1]R_{N+1},$$

where R_{N+1} is a pseudo-differential operator of order $-N-1$. Combining this with (2.11) completes the proof. □

Remark 2.2.28. One can improve Proposition 2.2.27 to find $b_{\infty,\phi_1,\phi_2} \in S^{-m}$ so that

$$\mathrm{Mult}[\phi_1]b_{\infty,\phi_1,\phi_2}(x,D)\,\mathrm{Mult}[\phi_2]P(x,D) = \mathrm{Mult}[\phi_1] + \mathrm{Mult}[\phi_1]R_\infty\,\mathrm{Mult}[\phi_3]$$

and R_∞ is a pseudo-differential operator whose symbol is in $\bigcap_{N\in\mathbb{R}} S^N$. See [216, Chapter VI, Section 7.12].

Using Proposition 2.2.27, we can establish the sharp regularity theory for elliptic operators on appropriate L^p Sobolev spaces (see Remark 2.2.31), as made precise in the next definition and corollary.

Definition 2.2.29. For $u \in C_0^\infty(\Omega)'$ and $x_0 \in \Omega$, we say $u \in L_s^p$ near x_0 if there exists $\phi \in C_0^\infty(\Omega)$ with $\phi \equiv 1$ on a neighborhood of x_0 such that $\phi u \in L_s^p(\mathbb{R}^n)$.

Corollary 2.2.30. *Suppose $P(x,D)$ is elliptic near x_0 and $u \in C_0^\infty(\Omega)'$ satisfies $P(x,D)u \in L_s^p$ near x_0, for some $p \in (1,\infty)$, $s \in \mathbb{R}$. Then $u \in L_{s+m}^p$ near x_0.*

Proof. Let $\psi \in C_0^\infty(\Omega)$ equal 1 on a neighborhood of x_0. We may, without loss of generality, replace u with ψu and therefore assume $u \in C_{\mathrm{loc}}^\infty(\Omega)' \subseteq C_{\mathrm{loc}}^\infty(\mathbb{R}^n)'$. It is well known that $C_{\mathrm{loc}}^\infty(\mathbb{R}^n)' \subset \bigcup_{t\in\mathbb{R}} L_t^p$, and we may therefore assume $u \in L_{-M}^p$ for some M.

Take $\phi_1, \phi_2, \phi_3 \in C_0^\infty(\Omega)$ such that ϕ_1 equals 1 on a neighborhood of x_0, $\phi_1 \prec \phi_2 \prec \phi_3$ and $\phi_2 P(x,D)u \in L_s^p$. For $N \in \mathbb{N}_+$, let $b_N = b_{N,\phi_1,\phi_2} \in S^{-m}$ be as in Proposition 2.2.27. Then

$$\mathrm{Mult}[\phi_1]u = \mathrm{Mult}[\phi_1]b_N(x,D)\,\mathrm{Mult}[\phi_2]P(x,D)u - \mathrm{Mult}[\phi_1]R_N\,\mathrm{Mult}[\phi_3]u.$$

By Theorem 2.2.12, $\mathrm{Mult}[\phi_1]b_N(x,D)\,\mathrm{Mult}[\phi_2]P(x,D)u \in L_{s+m}^p$ and $\mathrm{Mult}[\phi_1]R_N\,\mathrm{Mult}[\phi_3]u \in L_{N-M}^p$. Taking $N \geq s+m-M$, it follows that $\mathrm{Mult}[\phi_1]u \in L_{s+m}^p$, completing the proof. □

Remark 2.2.31. Corollary 2.2.30 shows that if $P(x,D)$ is elliptic near x_0 and $P(x,D)u \in L_s^p$ near x_0, then $u \in L_{s+m}^p$ near x_0. The converse is true as well: if $u \in L_{s+m}^p$ near x_0, then $P(x,D)u \in L_s^p$ near x_0. This follows because $P(x,D)$ is a differential operator of order m. Thus, as far as regularity is concerned, we can consider elliptic operators of order m as being left invertible as maps from L_{s+m}^p to L_s^p. They may not be actually left invertible – elliptic operators may have non-trivial kernels. However, all functions in the kernel must be smooth (this follows from Corollary 2.2.30). In many proofs we will work modulo smooth functions and can often treat such operators as invertible.

A very similar proof to Corollary 2.2.30 shows that elliptic operators are "subelliptic." To describe this, we introduce a definition.

Definition 2.2.32. Let $T \in \mathrm{Hom}(C_0^\infty(\Omega)', C_0^\infty(\Omega)')$. For $p \in (1,\infty)$ we say T is L^p subelliptic of order $\epsilon > 0$ on an open set $V \subseteq \Omega$ if $\forall \phi_1, \phi_2 \in C_0^\infty(V)$, with $\phi_1 \prec \phi_2$, we have the following estimate $\forall s \in \mathbb{R}$:

$$\|\phi_1 u\|_{L^p_{s+\epsilon}} \le C_{\phi_1,\phi_2,s}(\|\phi_2 Tu\|_{L^p_s} + \|\phi_2 u\|_{L^p_s}), \quad \forall u \in C_0^\infty(\Omega)', \tag{2.12}$$

where if the right-hand side is finite, then the left-hand side is also finite. We say T is L^p subelliptic on V if T is L^p subelliptic of order ϵ on V for some $\epsilon > 0$. We say T is subelliptic on V if T is L^2 subelliptic on V.

Subellipticity automatically implies a stronger estimate by iterating (2.12).

Lemma 2.2.33. *Suppose $T \in \mathrm{Hom}(C_0^\infty(\Omega)', C_0^\infty(\Omega)')$, $p \in (1,\infty)$, and T is L^p subelliptic of order $\epsilon > 0$ on an open set $V \subseteq \Omega$. Then $\forall \phi_1, \phi_2 \in C_0^\infty(V)$, with $\phi_1 \prec \phi_2$, we have the following estimate $\forall s \in \mathbb{R}$, $N \ge 0$:*

$$\|\phi_1 u\|_{L^p_{s+\epsilon}} \le C_{\phi_1,\phi_2,s,N}(\|\phi_2 Tu\|_{L^p_s} + \|\phi_2 u\|_{L^p_{s-N}}), \quad \forall u \in C_0^\infty(\Omega)', \tag{2.13}$$

where if the right-hand side is finite, then the left-hand side is also finite.

Before we begin the proof of Lemma 2.2.33 we introduce some new notation. We write $A \lesssim B$ for $A \le CB$, where C is a constant which only depends on certain parameters. It will be clear from the context what parameters C is allowed to depend on; for example, in the proof of Lemma 2.2.33, we will allow C to depend on the same things as the constant in (2.13).

Proof. We prove by induction on $N \in \mathbb{N}$ that $\forall \phi_1, \phi_2 \in C_0^\infty(V)$, with $\phi_1 \prec \phi_2$, and all $s \in \mathbb{R}$, $N \ge 0$,

$$\|\phi_1 u\|_{L^p_{s+\epsilon}} \lesssim \|\phi_2 Tu\|_{L^p_s} + \|\phi_2 u\|_{L^p_{s-N\epsilon}}, \quad \forall u \in C_0^\infty(\Omega)', \tag{2.14}$$

which will complete the proof by taking N large.

The base case, $N = 0$, of (2.14) is the assumption of the lemma. We assume (2.14) for $N \in \mathbb{N}$ and prove it for $N+1$. Fix $\phi_1, \phi_2 \in C_0^\infty(V)$ with $\phi_1 \prec \phi_2$ and let $\psi \in C_0^\infty(V)$ with $\phi_1 \prec \psi \prec \phi_2$. Applying the inductive hypothesis with ϕ_2 replaced by ψ, we have

$$\|\phi_1 u\|_{L^p_{s+\epsilon}} \lesssim \|\psi Tu\|_{L^p_s} + \|\psi u\|_{L^p_{s-N\epsilon}}. \tag{2.15}$$

Applying the assumption (2.12) we see

$$\|\psi u\|_{L^p_{s-N\epsilon}} \lesssim \|\phi_2 Tu\|_{L^p_{s-(N+1)\epsilon}} + \|\phi_2 u\|_{L^p_{s-(N+1)\epsilon}}. \tag{2.16}$$

Using $\|\psi Tu\|_{L^p_s} \lesssim \|\phi_2 Tu\|_{L^p_s}$ and $\|\phi_2 Tu\|_{L^p_{s-(N+1)\epsilon}} \lesssim \|\phi_2 Tu\|_{L^p_s}$, combining (2.15) and (2.16) gives (2.14) with N replaced by $N+1$, completing the proof. □

Proposition 2.2.34. *Suppose $P(x,D)$ is elliptic near x_0. Then there is an open neighborhood $V \subseteq \Omega$ of x_0 such that $P(x,D)$ is L^p subelliptic of order m on V, $\forall p \in (1,\infty)$.*

Proof. Let $V \subseteq \Omega$ be the open neighborhood of x_0 from Proposition 2.2.27. Fix $\phi_1, \phi_2, \phi_3 \in C_0^\infty(V)$ with $\phi_1 \prec \phi_2 \prec \phi_3$. For $N \in \mathbb{N}$, let $b_N = b_{N,\phi_1,\phi_2} \in S^{-m}$ be as in Proposition 2.2.27. We have, using Theorem 2.2.12, $\forall u \in C_0^\infty(\Omega)'$,

$$\|\phi_1 b_N(x,D)\phi_2 u\|_{L^p_{s+m}} \lesssim \|\phi_2 u\|_{L^p_s}. \tag{2.17}$$

Replacing u with $P(x,D)u$ in (2.17), we have

$$\|\phi_1 b_N(x,D)\phi_2 P(x,D)u\|_{L^p_{s+m}} \lesssim \|\phi_2 P(x,D)u\|_{L^p_s}. \tag{2.18}$$

Proposition 2.2.27 gives $\phi_1 b_N(x,D)\phi_2 P(x,D)u = \phi_1 u + \phi_1 R_N \phi_3 u$, where R_N is a pseudodifferential operator of order $-N$. Plugging this into (2.18) and using Theorem 2.2.12, we have

$$\|\phi_1 u\|_{L^p_{s+m}} \lesssim \|\phi_2 P(x,D)u\|_{L^p_s} + \|\phi_3 u\|_{L^p_{s+m-N}}.$$

Since N was arbitrary and $\|\phi_2 P(x,D)u\|_{L^p_s} \lesssim \|\phi_3 P(x,D)u\|_{L^p_s}$, this completes the proof of L^p subellipticity (with ϕ_2 replaced by ϕ_3 in Definition 2.2.32). □

A similar proof yields the following.

Proposition 2.2.35. *Suppose $P(x,D)$ is elliptic on Ω. Then $P(x,D)$ is L^p subelliptic of order m on Ω, $\forall p \in (1,\infty)$.*

Remark 2.2.36. Proposition 2.2.35 is optimal in the sense that a differential operator of order m can never be L^p subelliptic of order ϵ for $\epsilon > m$ (as can be immediately seen from the definition). In fact, being subelliptic of order m characterizes elliptic operators (see Theorem 2.7.2 (b)).

There is a third, even weaker condition: hypoellipticity.

Definition 2.2.37. Let $T \in \mathrm{Hom}(C_0^\infty(\Omega)', C_0^\infty(\Omega)')$. We say T is hypoelliptic on Ω if the following holds. Suppose $V \subseteq \Omega$ is an open set and $u \in C_0^\infty(\Omega)'$ satisfies $Tu\big|_V \in C^\infty_{\mathrm{loc}}(V)$. Then $u\big|_V \in C^\infty_{\mathrm{loc}}(V)$.

Proposition 2.2.38. *Suppose T is L^p subelliptic on Ω for some $p \in (1,\infty)$. Then T is hypoelliptic on Ω.*

Proof. Let $V \subseteq \Omega$ be open, let $u \in C_0^\infty(\Omega)'$, and suppose $Tu\big|_V \in C^\infty_{\mathrm{loc}}(V)$. Fix $x_0 \in V$. We wish to show that there is a neighborhood $U \subseteq V$ of x_0 such that $u\big|_U \in C^\infty_{\mathrm{loc}}(U)$, and this will complete the proof.

Take $\phi_1, \phi_2 \in C_0^\infty(V)$, with ϕ_1 equal to 1 on a neighborhood of x_0, and $\phi_1 \prec \phi_2$. By subellipticity and Lemma 2.2.33, we have, for some $\epsilon > 0$ and $\forall s \in \mathbb{R}$, $N \in \mathbb{N}$,

$$\|\phi_1 u\|_{L^p_{s+\epsilon}} \leq C_{\phi_1,\phi_2,s,N}\big(\|\phi_2 Tu\|_{L^p_s} + \|\phi_2 u\|_{L^p_{s-N}}\big).$$

Since $\phi_2 Tu \in C_0^\infty(\mathbb{R}^n)$, by hypothesis, $\|\phi_2 Tu\|_{L^p_s} < \infty$, $\forall s \in \mathbb{R}$. By taking N sufficiently large, $\|\phi_2 u\|_{L^p_{s-N}} < \infty$ (as this it true for any distribution with compact support, for some N). Thus, $\phi_1 u \in \bigcap_{s\in\mathbb{R}} L^p_s$ and the Sobolev embedding theorem implies $\phi_1 u \in C_0^\infty(\mathbb{R}^n)$. Since ϕ_1 equals 1 on a neighborhood of x_0, the result follows. □

Propositions 2.2.35 and 2.2.38 can be succinctly restated as the following implications:

$$\text{ellipticity} \Rightarrow \text{subellipticity} \Rightarrow \text{hypoellipticity}.$$

None of the reverse implications hold, in general, even for partial differential operators.[2] The goal of this text is to study a class of operators which are subelliptic and which generalize elliptic operators: the maximally subelliptic operators. Elliptic operators are a special case of maximally subelliptic operators (see Example 1.1.10 (i)) and

$$\text{maximal subellipticity} \Rightarrow \text{subellipticity} \Rightarrow \text{hypoellipticity}.$$

2.3 Singular integral operators

The theory of pseudo-differential operators is closely tied to the group structure on $\mathbb{R}^n$. In the study of maximally subelliptic operators, we need to move beyond this group structure, and to do this we turn to the Calderón–Zygmund theory of singular integrals. We introduce a filtered algebra of singular integral operators which contains pseudo-differential operators, but is less tied to the group structure of $\mathbb{R}^n$. The ideas here are amenable to the generalizations we require in later chapters.

Following Theorem 2.2.26 we define this algebra in three equivalent ways. The first way will be the most familiar to readers who have studied singular integrals. However, it is the later two ways which are more important for our generalizations and applications.

Definition 2.3.1. We say $T \in \mathrm{Hom}(\mathscr{S}(\mathbb{R}^n), C^\infty_{\mathrm{loc}}(\mathbb{R}^n))$ is a singular integral operator of order $s \in \mathbb{R}$ if:

- (Growth condition) $T(x,y)$ agrees with a C^∞_{loc} function on $\{(x,y) : x \neq y\}$ and for all multi-indices $\alpha, \beta \in \mathbb{N}^n$ and all $N \geq 0$ such that $n + s + |\alpha| + |\beta| + N > 0$,

$$\big|\partial_x^\alpha \partial_y^\beta T(x,y)\big| \leq C_{\alpha,\beta,M} |x-y|^{-n-s-|\alpha|-|\beta|-N}.$$

2 See Remark 8.2.7, where it is shown that maximally subelliptic operators are subelliptic but very rarely elliptic. See [143] for examples of partial differential operators which are hypoelliptic but far from subelliptic.

- (Cancelation condition) For all bounded sets $\mathcal{B} \subset C_0^\infty(\mathbb{R}^n)$, all multi-indices $\alpha \in \mathbb{N}^n$, and all $N \geq 0$ such that $n + s + |\alpha| + N > 0$, there exists $C_{\mathcal{B},\alpha,N} \geq 0$ such that the following holds. For $\phi \in \mathcal{B}$ and $z \in \mathbb{R}^n$, set $\phi_{R,z}(x) := \Phi(R(x - z))$. We assume

$$\sup_{\phi \in \mathcal{B}} \sup_{R \geq 1} \sup_{x,z \in \mathbb{R}^n} R^{-s-|\alpha|-N} |\partial_x^\alpha T \phi_{R,z}(x)| \leq C_{\mathcal{B},\alpha,N}.$$

We also assume that the same holds with T replaced by its formal L^2 adjoint T^* (see Definition 2.1.6).

Remark 2.3.2. A singular integral operator of order $s \in \mathbb{R}$ is a priori an operator in $\mathrm{Hom}(\mathscr{S}(\mathbb{R}^n), C_{\mathrm{loc}}^\infty(\mathbb{R}^n))$. However, we will see that, in fact, $T \in \mathrm{Hom}(\mathscr{S}(\mathbb{R}^n), \mathscr{S}(\mathbb{R}^n))$ and that T extends to an operator $T \in \mathrm{Hom}(\mathscr{S}(\mathbb{R}^n)', \mathscr{S}(\mathbb{R}^n)')$. See Remark 2.3.29.

The equivalent characterizations of Definition 2.3.1 which we use are closely tied to a Littlewood–Paley decomposition of the operators. To discuss this, we need to introduce a new kind of operator.

Definition 2.3.3. We say $\mathcal{F} \subset C^\infty(\mathbb{R}^n \times \mathbb{R}^n) \times (0, 1]$ is a bounded set of pre-elementary operators if $\forall m \geq 0$, $\forall \alpha, \beta \in \mathbb{N}^n$, $\exists C_{m,\alpha,\beta} \geq 0$ such that

$$\sup_{(F,2^{-j}) \in \mathcal{F}} \left|(2^{-j}\partial_x)^\alpha (2^{-j}\partial_z)^\beta F(x, z)\right| \leq C_{m,\alpha,\beta} 2^{jn} (1 + 2^j|x - z|)^{-m}.$$

Example 2.3.4. If $\mathcal{B} \subset \mathscr{S}(\mathbb{R}^n)$ is a bounded set, then

$$\{(2^{jn}\phi(2^j(x - z)), 2^{-j}) \mid \phi \in \mathcal{B}, j \in [0, \infty)\}$$

is a bounded set of pre-elementary operators.

Definition 2.3.5. We define the set of bounded sets of elementary operators, $\mathcal{G}$, to be the largest set of subsets of $C^\infty(\mathbb{R}^n \times \mathbb{R}^n) \times (0, 1]$ such that $\forall \mathcal{E} \in \mathcal{G}$:
- $\mathcal{E}$ is a bounded set of pre-elementary operators,
- $\forall (E, 2^{-j}) \in \mathcal{E}$,

$$E = \sum_{|\alpha|,|\beta| \leq 1} 2^{-(2-|\alpha|-|\beta|)j} (2^{-j}\partial_x)^\alpha E_{\alpha,\beta} (2^{-j}\partial_x)^\beta, \tag{2.19}$$

where we are treating E and $E_{\alpha,\beta}$ as operators (by identifying an operator with its Schwartz kernel), and

$$\{(E_{\alpha,\beta}, 2^{-j}) : (E, 2^{-j}) \in \mathcal{E}, |\alpha|, |\beta| \leq 1\} \in \mathcal{G}.$$

Remark 2.3.6. In Definition 2.3.5, $\mathcal{G}$ is defined to be the *largest* set satisfying certain axioms. There is a unique largest such set, since if $\{\mathcal{G}_a\}$ is the collection of all sets satisfying

the axioms, then $\bigcup_a \mathscr{G}_a$ also satisfies the axioms. In practice, we will show that a set $\mathcal{E}$ is a bounded set of elementary operators by showing that:
- $\mathcal{E}$ is a bounded set of pre-elementary operators, and
- for each $(E, 2^{-j}) \in \mathcal{E}$, E can be written in the form (2.19), where each $E_{\alpha,\beta}$ is "of the same form" as E.

See Example 2.3.8 for a simple version of this argument.

Remark 2.3.7. Note that (2.19) is trivial when $j = 0$ by taking $E_{0,0} = E$ and $E_{\alpha,\beta} = 0$ for $|\alpha| + |\beta| > 0$.

Example 2.3.8. – If $\mathcal{B}_1 \subset \mathscr{S}_0(\mathbb{R}^n)$ is a bounded set, then

$$\{(2^{jn}\phi(2^j(x - z)), 2^{-j}) : \phi \in \mathcal{B}_1, j \in [0, \infty)\} \tag{2.20}$$

is a bounded set of elementary operators. Indeed, the proof of Corollary 2.2.20 shows that for each $\phi \in \mathcal{B}_1$, we have

$$\phi(x - z) = \sum_{|\alpha|,|\beta|=1} \partial_x^\alpha \partial_z^\beta \phi_{\alpha,\beta}(x - z),$$

where

$$\{\phi_{\alpha,\beta} : \phi \in \mathcal{B}_1, |\alpha|, |\beta| = 1\} \subseteq \mathscr{S}_0(\mathbb{R}^n)$$

is a bounded set. From here, it follows easily that (2.20) is a bounded set of elementary operators. See Lemma 2.3.17 for a more general version of this fact.
- If $\mathcal{B}_2 \subset \mathscr{S}(\mathbb{R}^n)$ is a bounded set, then

$$\{(\phi(x - z), 2^{-0}) : \phi \in \mathcal{B}_2\}$$

is a bounded set of elementary operators (see Remark 2.3.7).

Remark 2.3.9. See [220, Section 1.2] for a closely related perspective on singular integrals, which makes the relationship between $\mathscr{S}_0$ and bounded sets of elementary operators even more clear.

Theorem 2.3.10. *Let $T \in \mathrm{Hom}(\mathscr{S}(\mathbb{R}^n), C^\infty_{\mathrm{loc}}(\mathbb{R}^n))$ and fix $s \in \mathbb{R}$. The following are equivalent:*
(i) *T is a singular integral operator of order s.*
(ii) *For every bounded set of elementary operators $\mathcal{E}$,*

$$\{(2^{-js}TE, 2^{-j}) : (E, 2^{-j}) \in \mathcal{E}\}$$

is a bounded set of elementary operators.

(iii) *There exists a bounded set of elementary operators* $\{(E_j, 2^{-j}) : j \in \mathbb{N}\}$ *such that*

$$T = \sum_{j\in\mathbb{N}} 2^{js} E_j,$$

where this sum converges in the sense of tempered distributions (we will see in Lemma 2.3.20 *that every such sum converges in the sense of tempered distributions).*

From Theorem 2.3.10 we immediately see that singular integral operators form a filtered algebra.

Corollary 2.3.11. *Let T be an operator of order $s_1 \in \mathbb{R}$ and let S be an operator of order $s_2 \in \mathbb{R}$. Then TS is an operator of order $s_1 + s_2$.*

Proof. This follows immediately from Theorem 2.3.10 (ii). □

Remark 2.3.12. It follows immediately from either Theorem 2.3.10 (i) or (ii) that ∂_x^α is a singular integral operator of order $|\alpha|$.

Corollary 2.3.13. *Singular integral operators of order* 0 *extend to bounded operators on* $L^p(\mathbb{R}^n)$, $1 < p < \infty$.

Corollary 2.3.13 can be deduced from Definition 2.3.1 and the $T(1)$ theorem of David and Journé [60]; see the presentation in [216, Chapter VII, § 3.2]. However, we defer the proof to later in the section, where we will see it as a relatively elementary consequence of Theorem 2.3.10.

The rest of this section is devoted to Theorem 2.3.10 and Corollary 2.3.13. For Theorem 2.3.10, we prove only (ii) ⇔ (iii) as that is the equivalence that is of central importance to our later generalizations. The remaining equivalence, (i) ⇔ (iii), is proved in an essentially more general setting in Theorem 5.2.12, and we refer the reader there for details.

Our first result says that pseudo-differential operators of order t are singular integral operators of order t.

Proposition 2.3.14. *Let $a(x, D)$ be a pseudo-differential operator of order $s \in \mathbb{R}$. Then $a(x, D)$ satisfies the conclusions of Theorem* 2.3.10 (iii).

Remark 2.3.15. Proposition 2.3.14 would follow immediately from Theorem 2.2.26 (ii) ⇒ (i) combined with Theorem 2.3.10 (i) ⇒ (iii). However, this is circular: Proposition 2.3.14 is central to the proof of Theorem 2.3.10.

Remark 2.3.16. Using Proposition 2.3.14 and Corollary 2.3.13, one can “move derivatives past” singular integral operators. Indeed, suppose T is a singular integral operator of order s. Then

$$\partial_{x_j} T = \partial_{x_j} T(1+\triangle)^{-1} - \sum_{k=1}^{n} (\partial_{x_j} T(I+\triangle)^{-1}\partial_{x_k})\partial_{x_k} = S_0^{(s-1)} + \sum_{k=1}^{n} S_k^{(s)} \partial_{x_k},$$

where each $S_k^{(s)}$ is a singular integral operator of order s and $S_0^{(s-1)}$ is a singular integral operator of order $s-1$. However, the best one can say in general is that $[\partial_{x_j}, T]$ is a singular integral operator of order $s+1$; this is in contrast to pseudo-differential operators where the commutator is an operator of order s (see Theorem 2.2.8). This lack of commutativity is a major inconvenience when studying elliptic PDEs; however it is an essential feature in our study of maximally subelliptic PDEs (see Remark 1.4.1).

Lemma 2.3.17. *Let* $\{\varsigma_j(x,t) : j \in \mathbb{N}\} \subset C^\infty(\mathbb{R}^n)\widehat{\otimes}\mathscr{S}(\mathbb{R}^n)$ *be a bounded set with* $\varsigma_j(x,t) \in C^\infty(\mathbb{R}^n)\widehat{\otimes}\mathscr{S}_0(\mathbb{R}^n)$ *for* $j > 0$. *Then*

$$\mathcal{E} := \{(2^{jn}\varsigma_j(x, 2^j(x-y)), 2^{-j}) : j \in \mathbb{N}\}$$

is a bounded set of elementary operators.

Proof. It follows immediately from the definitions that $\mathcal{E}$ is a bounded set of pre-elementary operators. Thus, it suffices to show that each $2^{jn}\varsigma_j(x, 2^j(x-y))$ is an appropriate sum of derivatives of operators of the same form. Set

$$\tilde{\varsigma}_j(x,t) := \begin{cases} \triangle_t^{-1}\varsigma_j(x,t), & j \geq 1, \\ \varsigma_0(x,t), & j = 0. \end{cases}$$

By Corollary 2.2.25, $\{\tilde{\varsigma}_j : j \in \mathbb{N}\} \subset C^\infty(\mathbb{R}^n)\widehat{\otimes}\mathscr{S}(\mathbb{R}^n)$ is a bounded set and $\tilde{\varsigma}_j \in C^\infty(\mathbb{R}^n)\widehat{\otimes}\mathscr{S}_0(\mathbb{R}^n)$ for $j \geq 1$. The same is therefore also true of $\{\partial_{x_k}\tilde{\varsigma}_j(x,t) : j \in \mathbb{N}\}$. Furthermore, we have, for $j \geq 1$,

$$2^{nj}\varsigma_j(x, 2^j(x-y)) = \sum_{k=1}^n 2^{nj}(2^{-j}\partial_{x_k})(2^{-j}\partial_{y_k})\tilde{\varsigma}_j(x, 2^j(x-y)) - \sum_{k=1}^n 2^{nj-j}(2^{-j}\partial_{y_k})(\partial_{x_k}\tilde{\varsigma}_j)(x, x-y).$$

This sees $2^{nj}\varsigma_j(x, 2^j(x-y))$ as an appropriate sum of derivatives of operators of the same form as in (2.19). Since (2.19) is trivial when $j = 0$ (Remark 2.3.7), this completes the proof. □

Proof of Proposition 2.3.14. Let $a \in S^s$. Then by Theorem 2.2.26 (iii) we may write

$$\check{a}(x,t) = \sum_{j\in\mathbb{N}} \mathrm{Dil}_{2^j}(\varsigma_j)(x,t),$$

with convergence in the sense of tempered distributions, where $\{\varsigma_j : j \in \mathbb{N}\} \subset C^\infty(\mathbb{R}^n)\widehat{\otimes}\mathscr{S}(\mathbb{R}^n)$ with $\varsigma_j \in C^\infty(\mathbb{R}^n)\widehat{\otimes}\mathscr{S}_0(\mathbb{R}^n)$ for $j > 0$. The Schwartz kernel of $a(x,D)$ is given by $\check{a}(x, x-y)$ and therefore we have

$$a(x,D) = \sum_{j\in\mathbb{N}} \mathrm{Dil}_{2^j}(\varsigma_j)(x, x-y),$$

with convergence in the sense of tempered distributions. Lemma 2.3.17 shows that $\{(\mathrm{Dil}_{2^j}(\varsigma_j)(x, x-y), 2^{-j}) : j \in \mathbb{N}\}$ is a bounded set of elementary operators, completing the proof. □

Proposition 2.3.18. *Let $\mathcal{E}$ be a bounded set of elementary operators. Then:*

(a) *If $\psi \in C^\infty(\mathbb{R}^n)$, then*

$$\{(\mathrm{Mult}[\psi]E, 2^{-j}), (E\,\mathrm{Mult}[\psi], 2^{-j}) : (E, 2^{-j}) \in \mathcal{E}\}$$

is a bounded set of elementary operators.

(b) *$\{(E^*, 2^{-j}) : (E, 2^{-j}) \in \mathcal{E}\}$ is a bounded set of elementary operators.*

(c) *Fix a multi-index $\alpha \in \mathbb{N}^n$. Then*

$$\{((2^{-j}\partial_x)^\alpha E, 2^{-j}), (E(2^{-j}\partial_x)^\alpha, 2^{-j}) : (E, 2^{-j}) \in \mathcal{E}\}$$

is a bounded set of elementary operators.

(d) *For all $N \in \mathbb{N}$, $(E, 2^{-j}) \in \mathcal{E}$ can be written as*

$$E = \sum_{|\alpha|\le N} 2^{-j(N-|\alpha|)}(2^{-j}\partial_x)^\alpha E_\alpha,$$

where $\{(E_\alpha, 2^{-j}) : (E, 2^{-j}) \in \mathcal{E}, |\alpha| \le N\}$ is a bounded set of elementary operators. Similarly, each $(E, 2^{-j}) \in sE$ can be written as

$$E = \sum_{|\alpha|\le N} 2^{-j(N-|\alpha|)}\widetilde{E}_\alpha(2^{-j}\partial_x)^\alpha,$$

where $\{(\widetilde{E}_\alpha, 2^{-j}) : (E, 2^{-j}) \in \mathcal{E}, |\alpha| \le N\}$ is a bounded set of elementary operators.

Proof. This follows easily from the definitions. See Proposition 5.5.5 for a similar result in a more general context. □

Lemma 2.3.19. *For every $m > n$, $\forall j_1, j_2 \in [0, \infty)$,*

$$\int 2^{nj_1}(1 + 2^{j_1}|x-y|)^{-m} 2^{nj_2}(1 + 2^{j_2}|y-z|)^{-m}\, dy \le C_{m,n} 2^{n(j_1\wedge j_2)}(1 + 2^{j_1\wedge j_2}|x-z|)^{-m}.$$

Proof. By symmetry in x and z, it suffices to prove the case where $j_2 \le j_1$. By a simple dilation argument, it suffices to prove the case where $j_2 = 0$. By a translation, it suffices to prove the case $x = 0$. Thus, it suffices to show that, $\forall j \in [0, \infty)$,

$$\int 2^{nj}(1 + 2^j|y|)^{-m}(1 + |y-z|)^{-m}\, dy \le C_{m,n}(1 + |z|)^{-m}.$$

We leave this elementary estimate to the reader. □

Lemma 2.3.20. *Fix $s \in \mathbb{R}$ and let $\{(E_j, 2^{-j}) : j \in \mathbb{N}\}$ be a bounded set of elementary operators. Then the sum*

$$\sum_{j \in \mathbb{N}} 2^{js} E_j$$

converges in the topology of bounded convergence on $\mathrm{Hom}(\mathscr{S}(\mathbb{R}^n), \mathscr{S}(\mathbb{R}^n))$ (and therefore converges in $\mathscr{S}(\mathbb{R}^n \times \mathbb{R}^n)'$).

Proof. Let $\mathcal{B} \subset \mathscr{S}(\mathbb{R}^n)$ be a bounded set. We will show that for all multi-indices $\alpha \in \mathbb{N}^n$, all $m \in \mathbb{N}$, and all $f \in \mathcal{B}$,

$$\sum_{j \in \mathbb{N}} 2^{js}(1 + |x|)^m \partial_x^\alpha E_j f(x) \tag{2.21}$$

converges absolutely and uniformly for $x \in \mathbb{R}^n$ and $f \in \mathcal{B}$, and the result will follow. Note that

$$\sum_{j \in \mathbb{N}} 2^{js}(1 + |x|)^m \partial_x^\alpha E_j f(x) = \sum_{j \in \mathbb{N}} 2^{j(s+|\alpha|)}(1 + |x|)^m (2^{-j}\partial_x)^\alpha E_j f(x).$$

Since $\{((2^{-j}\partial_x)^\alpha E_j, 2^{-j}) : j \in \mathbb{N}\}$ is a bounded set of elementary operators by Proposition 2.3.18 (c), we see that it suffices to prove the convergence of (2.21) in the case $\alpha = 0$ (by replacing s with $s + \alpha$).

Fix $N \in \mathbb{N}$. By Proposition 2.3.18 (d) we have

$$\sum_{j \in \mathbb{N}} (1 + |x|)^m 2^{js} E_j f = \sum_{j \in \mathbb{N}} \sum_{|\beta| \le N} 2^{j(s-N)} E_{j,\beta} \partial_x^\beta f.$$

By taking $N \ge s + 1$ and replacing f with $\partial_x^\beta f$, we see that it suffices to prove the convergence of (2.21) in the case $\alpha = 0$ and $s = -1$.

By taking $z = 0$ in Lemma 2.3.19, we see that for $f \in \mathcal{B}$ and $(E, 2^{-j}) \in \mathcal{E}$,

$$|E_j f(x)| \lesssim (1 + |x|)^{-m}.$$

Thus,

$$\sum_{j \in \mathbb{N}} |2^{-j}(1 + |x|)^m E_j f(x)| \lesssim \sum_{j \in \mathbb{N}} 2^{-j}.$$

The desired convergence of (2.21) follows (in the case $\alpha = 0$, $s = -1$), completing the proof. □

Corollary 2.3.21. *Let $I \in \mathrm{Hom}(\mathscr{S}(\mathbb{R}^n), \mathscr{S}(\mathbb{R}^n))$ be the identity operator. Then there exists a bounded set of elementary operators $\{(E_j, 2^{-j}) : j \in \mathbb{N}\}$ such that*

$$I = \sum_{j\in\mathbb{N}} E_j,$$

where the sum is taken in the sense of the topology of bounded convergence on $\mathrm{Hom}(\mathscr{S}(\mathbb{R}^n), \mathscr{S}(\mathbb{R}^n))$.

Proof. Since I is a pseudo-differential operator of order 0, this follows from Proposition 2.3.14 and Lemma 2.3.20. □

Proof of Theorem 2.3.10 (ii) ⇒ (iii). By Corollary 2.3.21, we may write the identity operator, I, as

$$I = \sum_{j\in\mathbb{N}} E_j,$$

where $\{(E_j, 2^{-j}) : j \in \mathbb{N}\}$ is a bounded set of elementary operators, and this sum converges in the topology of bounded convergence on $\mathrm{Hom}(\mathscr{S}(\mathbb{R}^n), \mathscr{S}(\mathbb{R}^n))$. Thus,

$$T = TI = \sum_{j\in\mathbb{N}} 2^{js}(2^{-js}TE_j).$$

Since $\{(2^{-js}TE_j, 2^{-j}) : j \in \mathbb{N}\}$ is a bounded set of elementary operators, by hypothesis, this completes the proof. □

To prove Theorem 2.3.10 (iii) ⇒ (ii), we use the next proposition, which is central to our analysis.

Proposition 2.3.22. *Let $\mathcal{E}$ be a bounded set of elementary operators. Then, $\forall N \in \mathbb{N}$, the set*

$$\{(2^{N|j_1-j_2|}E_1E_1, 2^{-j_1}), (2^{N|j_1-j_2|}E_1E_2, 2^{-j_2}) : (E_1, 2^{-j_1}), (E_2, 2^{-j_2}) \in \mathcal{E}\}$$

is a bounded set of elementary operators.

To prove Proposition 2.3.22 we need two lemmas.

Lemma 2.3.23. *Let $\mathcal{E}$ be a bounded set of pre-elementary operators. Then, $\forall \alpha, \beta \in \mathbb{N}^n$, $m \in \mathbb{N}$, $\exists C = C(m, \alpha, \beta) \geq 0$, such that $\forall (F_1, 2^{-j_1}), (F_2, 2^{-j_2}) \in \mathcal{F}$, and letting $k \in \{j_1, j_2\}$,*

$$2^{-(m+n+|\alpha|\vee|\beta|)|j_1-j_2|}\left|(2^{-k}\partial_x)^\alpha(2^{-k}\partial_z)^\beta[F_1F_2](x,z)\right| \leq C2^{kn}(1 + 2^k|x-z|)^{-m},$$

where $[F_1F_2](x, z)$ denotes the Schwartz kernel of the operator F_1F_2.

Proof. We prove the result for $k = j_2$; the proof for $k = j_1$ is similar and we leave it to the reader. Thus, we wish to show

$$2^{-(m+n+|\alpha|)|j_1-j_2|}\left|(2^{-j_2}\partial_x)^\alpha(2^{-j_2}\partial_z)^\beta[F_1F_2](x,z)\right| \lesssim 2^{j_2n}(1 + 2^{j_2}|x-z|)^{-m};$$

it therefore suffices to show

$$2^{-(m+n)|j_1-j_2|}\left|(2^{-j_1}\partial_x)^\alpha(2^{-j_2}\partial_z)^\beta[F_1F_2](x,z)\right| \lesssim 2^{j_2n}(1+2^{j_2}|x-z|)^{-m}. \tag{2.22}$$

But Lemma 2.3.19 shows

$$\begin{aligned}&2^{-(m+n)|j_1-j_2|}\left|(2^{-j_1}\partial_x)^\alpha(2^{-j_2}\partial_z)^\beta[F_1F_2](x,z)\right|\\ &\lesssim 2^{-(m+n)|j_1-j_2|}2^{n(j_1\wedge j_2)}(1+2^{j_1\wedge j_2}|x-z|)^{-m}.\end{aligned}$$

Formula (2.22) follows, completing the proof. □

Lemma 2.3.24. *Let $\mathcal{E}$ be a bounded set of elementary operators. Then, $\forall N\in\mathbb{N}$, the set*

$$\{(2^{N|j_1-j_2|}E_1E_1,2^{-j_1}),(2^{N|j_1-j_2|}E_1E_2,2^{-j_2}):(E_1,2^{-j_1}),(E_2,2^{-j_2})\in\mathcal{E}\}$$

is a bounded set of pre-elementary operators.

Proof. Fix $N_0\in\mathbb{N}$. We will show that

$$\{(2^{N_0|j_1-j_2|}E_1E_1,2^{-j_1}):(E_1,2^{-j_1}),(E_2,2^{-j_2})\in\mathcal{E}\}$$

is a bounded set of pre-elementary operators. The proof of the result with j_1 replaced by j_2 is similar and we leave it to the reader.

Fix $m\in\mathbb{N}$ and let $\alpha,\beta\in\mathbb{N}^n$. We wish to show

$$2^{N_0|j_1-j_2|}\left|(2^{-j_1}\partial_x)^\alpha(2^{-j_1}\partial_z)^\beta[E_1E_2](x,z)\right| \lesssim 2^{nj_1}(1+2^{j_1}|x-z|)^{-m}. \tag{2.23}$$

Set $N:=m+n+|\alpha|\vee|\beta|$. We separate the proof into two cases. When $j_1\le j_2$, we apply Proposition 2.3.18 (d) to write

$$E_2=\sum_{|\gamma|\le N+N_0}2^{j_2(|\gamma|-N-N_0)}(2^{-j_2}\partial_x)^\gamma E_{2,\gamma},$$

where $\{(E_{2,\gamma},2^{-j_2}):(E_2,2^{-j_2})\in\mathcal{E},|\gamma|\le N+N_0\}$ is a bounded set of elementary operators. Thus,

$$\begin{aligned}E_1E_2&=\sum_{|\gamma|\le N+N_0}2^{-j_2(N+N_0-|\gamma|)}2^{-|\gamma||j_1-j_2|}E_1(2^{-j_1}\partial_x)^\gamma E_{2,\gamma}\\ &=\sum_{|\gamma|\le N+N_0}2^{-|j_1-j_2|(N+N_0)}E_{1,\gamma}E_{2,\gamma},\end{aligned} \tag{2.24}$$

where $E_{1,\gamma}=2^{-j_1(N+N_1-|\gamma|)}E_1(2^{-j_1}\partial_x)^\alpha$, and therefore by Proposition 2.3.18 (c), $\{(E_{1,\gamma},2^{-j_1}):(E,2^{-j})\in\mathcal{E},|\gamma|\le N+N_0\}$ is a bounded set of elementary operators. Plugging (2.24) into (2.23), we see that it suffices to prove (2.23) with N_0 replaced by $-m-n-|\alpha|\wedge|\beta|$.

With this replacement, (2.23) follows directly from Lemma 2.3.23, which completes the proof in the case $j_1 \le j_2$.

When $j_1 \ge j_2$ the proof is similar, except that we instead apply Proposition 2.3.18 (d) to E_1 to write

$$E_1 = \sum_{|\gamma| \le N+N_0} 2^{j_1(|\gamma|-N-N_0)} E_{1,\gamma}(2^{-j_1}\partial_x)^{\gamma}.$$

From here the proof is similar, and we leave the details to the reader. □

Proof of Proposition 2.3.22. We prove that

$$\mathcal{E}_N := \{(2^{N|j_1-j_2|}E_1E_2, 2^{-j_1}) : (E, 2^{-j_1}), (E, 2^{-j_2})\}$$

is a bounded set of elementary operators. The proof for the result with j_1 replaced by j_2 is similar, and we leave it to the reader.

Lemma 2.3.24 shows that $\mathcal{E}_N$ is a bounded set of pre-elementary operators. The result will follow once we show that $2^{N|j_1-j_2|}E_1E_2$ is a sum of derivatives of operators of the same form as in (2.19).

Using Proposition 2.3.18 (d) we have

$$\begin{aligned} 2^{N|j_1-j_2|}E_1E_2 &= \sum_{|\alpha|,|\beta|\le 1} 2^{N|j_1-j_2|-(1-|\alpha|)j_1-(1-|\beta|)j_2}(2^{-j_1}\partial_x)^{\alpha}E_{1,\alpha}E_{2,\beta}(2^{-j_2}\partial_x)^{\beta} \\ &= \sum_{|\alpha|,|\beta|\le 1} 2^{-(2-|\alpha|-|\beta|)j_1}(2^{-j_1}\partial_x)^{\alpha}[2^{N|j_1-j_2|+(j_1-j_2)}E_{1,\alpha}E_{2,\alpha}](2^{-j_1}\partial_x)^{\beta}, \end{aligned}$$

where $\{(E_{1,\alpha}, 2^{-j_1}), (E_{2,\beta}, 2^{-j_2}) : (E_1, 2^{-j_1}), (E_2, 2^{-j_2}) \in \mathcal{E}, |\alpha|, |\beta| \le 1\}$ is a bounded set of elementary operators. This completes the proof since $2^{N|j_1-j_2|+(j_1-j_2)}E_{1,\alpha}E_{2,\alpha}$ is of the same form as $2^{N|j_1-j_2|}E_1E_2$ (with a different choice of N). □

Proof of Theorem 2.3.10 (iii) ⇒ (ii). Fix $s \in \mathbb{R}$ and let $T = \sum_{j\in\mathbb{N}} 2^{js}E_j$, where $\{(E_j, 2^{-j}) : j \in \mathbb{N}\}$ is a bounded set of elementary operators. Let $\mathcal{E}$ be a bounded set of elementary operators. For $(E, 2^{-k}) \in \mathcal{E}$ and $j \in \mathbb{N}$, set $\widetilde{E}_j := 2^{(j-k)s+|j-k|}E_jE$, so that

$$\{(\widetilde{E}_j, 2^{-k}) : j \in \mathbb{N}, (E, 2^{-k}) \in \mathcal{E}\}$$

is a bounded set of elementary operators by Proposition 2.3.22. Set $\widetilde{E} := \sum_{j\in\mathbb{N}} 2^{-|j-k|}\widetilde{E}_j$, so that $\{(\widetilde{E}, 2^{-k}) : (E, 2^{-k}) \in \mathcal{E}\}$ is a bounded set of elementary operators. We have

$$2^{-ks}TE = \sum_{j\in\mathbb{N}} 2^{(j-k)s}E_jE = \sum_{j\in\mathbb{N}} 2^{-|j-k|}\widetilde{E}_j = \widetilde{E}.$$

Thus, $\{(2^{-ks}TE, 2^{-k}) : (E, 2^{-k}) \in \mathcal{E}\}$ is a bounded set of elementary operators, completing the proof. □

Finally, we turn to the proof of Corollary 2.3.13 without using the $T(1)$ theorem. For this, we employ the Cotlar–Stein lemma.

Lemma 2.3.25 (The Cotlar–Stein lemma). *Let $\mathcal{H}_1$ and $\mathcal{H}_2$ be Hilbert spaces. For $j \in \mathbb{N}$, let $T_j : \mathcal{H}_1 \to \mathcal{H}_2$ be bounded operators such that*

$$\sup_{j\in\mathbb{N}} \sum_{k\in\mathbb{N}} \|T_j^* T_k\|_{\mathcal{H}_1\to\mathcal{H}_1}^{\frac{1}{2}} \le A < \infty,$$

$$\sup_{j\in\mathbb{N}} \sum_{k\in\mathbb{N}} \|T_j T_k^*\|_{\mathcal{H}_2\to\mathcal{H}_2}^{\frac{1}{2}} \le B < \infty.$$

Then the sum $\sum_{j\in\mathbb{N}} T_j$ converges in the strong operator topology on $\mathrm{Hom}(\mathcal{H}_1, \mathcal{H}_2)$ to a bounded operator $\mathcal{H}_1 \to \mathcal{H}_2$ and

$$\Big\|\sum_{j\in\mathbb{N}} T_j\Big\|_{\mathcal{H}_1\to\mathcal{H}_2} \le \sqrt{AB}.$$

For a proof of Lemma 2.3.25, see [216, Chapter VII, § 2]. The statement in that reference is slightly different from the statement here, though the proof can be adapted to prove the result as we have stated it.

Lemma 2.3.26. *Let $\mathcal{E}$ be a bounded set of pre-elementary operators. Then*

$$\sup_{p\in[1,\infty]} \sup_{(E,2^{-j})\in\mathcal{E}} \|E\|_{L^p\to L^p} < \infty.$$

Proof. First we claim

$$\sup_{(E,2^{-j})\in\mathcal{E}} \sup_{x\in\mathbb{R}^n} \int |E(x,y)|\, dy < \infty. \tag{2.25}$$

Indeed, if we take $m = n + 1$, we have

$$\int |E(x,y)|\, dy \lesssim \int 2^{jn}(1 + 2^j|x-y|)^{-m}\, dy = \int (1 + |y|)^{-m}\, dy \lesssim 1,$$

as desired. Formula (2.25) implies

$$\sup_{(E,2^{-j})\in\mathcal{E}} \|E\|_{L^\infty\to L^\infty} < \infty. \tag{2.26}$$

A similar proof shows

$$\sup_{(E,2^{-j})\in\mathcal{E}} \sup_{y\in\mathbb{R}^n} \int |E(x,y)|\, dx < \infty,$$

which implies

$$\sup_{(E,2^{-j})\in\mathcal{E}} \|E\|_{L^1\to L^1} < \infty. \tag{2.27}$$

Interpolating (2.26) and (2.27) completes the proof. □

Lemma 2.3.27. *Let T be a singular integral operator of order $s \in \mathbb{R}$. Then the formal L^2 adjoint of T, T^* (see Definition 2.1.6) is also a singular integral operator of order s.*

Proof. This can be seen directly from Definition 2.3.1. □

Lemma 2.3.28. *Let T be a singular integral operator of order 0. Then T extends to a bounded operator on $L^2(\mathbb{R}^n)$.*

Proof. By Theorem 2.3.10 (iii), we may write $T = \sum_{j\in\mathbb{N}} E_j$, where $\{(E_j, 2^{-j}) : j \in \mathbb{N}\}$ is a bounded set of elementary operators. Combining Proposition 2.3.18 (b) with Lemma 2.3.24, we see that for every $N \in \mathbb{N}$,

$$\{(2^{N|j-k|}E_j^* E_k, 2^{-j}), (2^{N|j-k|}E_j E_k^*, 2^{-j}) : j, k \in \mathbb{N}\}$$

is a bounded set of elementary operators. Lemma 2.3.26 implies

$$\|E_j^* E_k\|_{L^2\to L^2}, \|E_j E_k^*\|_{L^2\to L^2} \lesssim 2^{-|j-k|}.$$

The result now follows from the Cotlar–Stein lemma (Lemma 2.3.25). □

Proof of Corollary 2.3.13. Let T be a singular integral operator of order 0. Lemma 2.3.28 shows that T extends to a bounded operator on $L^2(\mathbb{R}^n)$. Using the growth condition from Definition 2.3.1 it is a standard result that this implies that T is weak type $(1,1)$; see [216, Chapter I, § 5].

The Marcinkiewicz interpolation theorem then shows that T extends a bounded operator on L^p, $1 < p \le 2$. Lemma 2.3.27 shows that T^* is also a singular integral operator of order 0. Thus, the same argument shows that T^* extends to a bounded operator on L^p, $1 < p \le 2$, and therefore T extends to a bounded operator on L^p, $2 \le p < \infty$, completing the proof. □

Remark 2.3.29. Let T be a singular integral operator of order $s \in \mathbb{R}$. A priori, $T \in \mathrm{Hom}(\mathscr{S}(\mathbb{R}^n), C^\infty_{\mathrm{loc}}(\mathbb{R}^n))$. However, it follows from Theorem 2.3.10 (iii) and Lemma 2.3.20 that, in fact, $T \in \mathrm{Hom}(\mathscr{S}(\mathbb{R}^n), \mathscr{S}(\mathbb{R}^n))$. Since T is a singular integral operator of order s if and only if T^* is a singular integral operator of order s (by Lemma 2.3.27), we see that T extends by duality to an operator in $\mathrm{Hom}(\mathscr{S}(\mathbb{R}^n)', \mathscr{S}(\mathbb{R}^n)')$.

2.3.1 Local, matrix-valued operators

We will often be considering results where the estimates are local. We also want to consider the possibility of operators acting on functions taking values in a finite-dimensional complex vector space. To describe this, we introduce the following definitions.

Definition 2.3.30. Fix $\mathbf{D}_1, \mathbf{D}_2 \in \mathbb{N}_+$ and let $\mathbb{M}^{\mathbf{D}_1 \times \mathbf{D}_2}(\mathbb{C})$ denote the Banach space of complex $\mathbf{D}_1 \times \mathbf{D}_2$ matrices. Fix a connected open set $\Omega \subseteq \mathbb{R}^n$.

- We say $T \in \mathrm{Hom}(C_0^\infty(\Omega; \mathbb{C}^{\mathbf{D}_1}), C_{\mathrm{loc}}^\infty(\Omega; \mathbb{C}^{\mathbf{D}_2})')$ is locally a singular integral operator of order $s \in \mathbb{R}$ if $\forall \phi_1, \phi_2 \in C_0^\infty(\Omega)$, $\mathrm{Mult}[\phi_1] T \,\mathrm{Mult}[\phi_2]$ is a $\mathbf{D}_2 \times \mathbf{D}_1$ matrix of singular integral operators of order s.
- We say $T \in \mathrm{Hom}(C_0^\infty(\Omega; \mathbb{C}^{\mathbf{D}_1}), C_{\mathrm{loc}}^\infty(\Omega; \mathbb{C}^{\mathbf{D}_2})')$ is locally a pseudo-differential operator of order $s \in \mathbb{R}$ if $\forall \phi_1, \phi_2 \in C_0^\infty(\Omega)$, $\mathrm{Mult}[\phi_1] T \,\mathrm{Mult}[\phi_2]$ is a $\mathbf{D}_2 \times \mathbf{D}_1$ matrix of pseudo-differential operators of order s.
- We say $\mathcal{E} \subset C_{\mathrm{loc}}^\infty(\Omega \times \Omega; \mathbb{M}^{\mathbf{D}_1 \times \mathbf{D}_2}(\mathbb{C}))$ is locally a bounded set of pre-elementary operators if $\forall \phi_1, \phi_2 \in C_0^\infty(\Omega)$, if we write $(F, 2^{-j}) \in \mathcal{E}$ as $(F_{l_1, l_2})_{1 \le l_1 \le \mathbf{D}_1, 1 \le l_2 \le \mathbf{D}_2}$, then

$$\{(\mathrm{Mult}[\phi_1] F_{l_1,l_2} \,\mathrm{Mult}[\phi_2], 2^{-j}) : (F, 2^{-j}) \in \mathcal{E}, 1 \le l_1 \le \mathbf{D}_1, 1 \le l_2 \le \mathbf{D}_2\}$$

  is a bounded set of pre-elementary operators.

Remark 2.3.31. The above definitions can be generalized to operators acting on sections of a vector bundle. See Section 8.6 for details in the more general maximally subelliptic setting.

2.4 Besov and Triebel–Lizorkin spaces

Besov and Triebel–Lizorkin spaces are a central tool in the classical study of elliptic PDEs: they are the function spaces on which elliptic PDEs are most well behaved. Both the classical L^p Sobolev spaces for $1 < p < \infty$ (see Proposition 2.4.9) and Hölder spaces $C^{m,r}$ for $0 < r < 1$ (see Corollary 2.5.6) are special cases, as we will see. The usual study of these spaces is intimately tied to the Fourier transform; however, when we turn to generalizations in Chapter 6 we will have to leave behind the Fourier transform altogether.

We start by describing the main definitions and results of this section, and then turn to the proofs. To begin, we need to introduce vector-valued Lebesgue spaces. Let $(\mathfrak{M}, \mu)$ be a measure space and let $\mathcal{A}$ be a countable or finite set. Let $f_j : \mathfrak{M} \to \mathbb{C}$ be a sequence of a. e. equivalence classes of measurable functions indexed by $j \in \mathcal{A}$. For $p, q \in [1, \infty]$, we set

$$\|\{f_j\}_{j \in \mathcal{A}}\|_{L^p(\mathfrak{M},\mu;\ell^q(\mathcal{A}))} := \begin{cases} \left\| \left(\sum_{j \in \mathcal{A}} |f_j(x)|^q\right)^{\frac{1}{q}} \right\|_{L^p(\mathfrak{M},\mu)}, & q \in [1, \infty), \\ \left\| \sup_{j \in \mathcal{A}} |f_j(x)| \right\|_{L^p(\mathfrak{M},\mu)}, & q = \infty, \end{cases} \tag{2.28}$$

$$\|\{f_j\}_{j \in \mathcal{A}}\|_{\ell^q(\mathcal{A};L^p(\mathfrak{M},\mu))} := \begin{cases} \left(\sum_{j \in \mathcal{A}} \|f_j\|_{L^p(\mathfrak{M},\mu)}^q\right)^{\frac{1}{q}}, & q \in [1, \infty), \\ \sup_{j \in \mathcal{A}} \|f_j\|_{L^p(\mathfrak{M},\mu)}, & q = \infty. \end{cases} \tag{2.29}$$

For $p, q \in [1,\infty]$, if $\mathcal{V}$ is either $L^p(\mathfrak{M},\mu;\ell^q(\mathcal{A}))$ or $\ell^q(\mathcal{A};L^p(\mathfrak{M},\mu))$, we let the space $\mathcal{V}$ consist of those sequences $\{f_j(x)\}_{j\in\mathcal{A}}$ such that $\|\{f_j\}_{j\in\mathcal{A}}\|_{\mathcal{V}} < \infty$. With the norm $\|\cdot\|_{\mathcal{V}}$, $\mathcal{V}$ is a Banach space.

Let $\hat{\phi} \in C_0^\infty(\mathbb{R}^n)$ be such that $\hat{\phi}(\xi) \equiv 1$ on a neighborhood of $0 \in \mathbb{R}^n$. For $j \geq 0$ set

$$\hat{\psi}_j(\xi) := \begin{cases} \hat{\phi}(\xi), & j = 0, \\ \hat{\phi}(\xi) - \hat{\phi}(2\xi), & j \geq 1, \end{cases}$$

so that $\hat{\psi}_j = \hat{\psi}_1$ for $j \geq 1$ and

$$\sum_{j=0}^{\infty} \hat{\psi}_j(2^{-j}\xi) \equiv 1.$$

For $j \geq 0$, define $D_j : \mathcal{S}(\mathbb{R}^n)' \to \mathcal{S}(\mathbb{R}^n)'$ by

$$(D_j f)^\wedge(\xi) := \hat{\psi}_j(2^{-j}\xi)\hat{f}(\xi), \tag{2.30}$$

where $\wedge$ denotes the Fourier transform. In other words, if $\psi_j \in \mathcal{S}(\mathbb{R}^n)$ is the inverse Fourier transform of $\hat{\psi}_j$, then

$$D_j f = f * \mathrm{Dil}_{2^j}(\psi_j).$$

In particular, for $f \in \mathcal{S}(\mathbb{R}^n)'$, $D_j f \in C^\infty_{\mathrm{loc}}(\mathbb{R}^n) \cap \mathcal{S}(\mathbb{R}^n)'$, and we may therefore treat $D_j f$ as a function.

Remark 2.4.1. By choosing $\hat{\phi}$ to be real-valued and even, we can ensure ψ_j is real-valued, $\forall j$. Thus, when necessary, we may assume D_j takes real-valued functions to real-valued functions.

Definition 2.4.2. For $s \in \mathbb{R}$, $p, q \in [1,\infty]$, we define the extended Besov norm,

$$\|f\|_{\mathcal{B}^s_{p,q}(\mathbb{R}^n)} := \left\|\{2^{js} D_j f\}_{j\in\mathbb{N}}\right\|_{\ell^q(\mathbb{N};L^p(\mathbb{R}^n))}, \quad f \in \mathcal{S}(\mathbb{R}^n)',$$

and the Besov space,

$$\mathcal{B}^s_{p,q}(\mathbb{R}^n) := \{f \in \mathcal{S}(\mathbb{R}^n)' : \|f\|_{\mathcal{B}^s_{p,q}(\mathbb{R}^n)} < \infty\}.$$

For $s \in \mathbb{R}$, $p \in (1,\infty)$, $q \in (1,\infty]$, we define the extended Triebel–Lizorkin norm,

$$\|f\|_{\mathcal{F}^s_{p,q}(\mathbb{R}^n)} := \left\|\{2^{js} D_j f\}_{j\in\mathbb{N}}\right\|_{L^p(\mathbb{R}^n;\ell^q(\mathbb{N}))}, \quad f \in \mathcal{S}(\mathbb{R}^n)',$$

and the Triebel–Lizorkin space,

$$\mathcal{F}^s_{p,q}(\mathbb{R}^n) := \{f \in \mathcal{S}(\mathbb{R}^n)' : \|f\|_{\mathcal{F}^s_{p,q}(\mathbb{R}^n)} < \infty\}.$$

Remark 2.4.3. It is well known that $\mathscr{B}^s_{p,q}(\mathbb{R}^n)$ and $\mathscr{F}^s_{p,q}(\mathbb{R}^n)$ are Banach spaces with the above norms; see [244, § 2.3.3].

Remark 2.4.4. If $\mathscr{X}$ is a finite-dimensional Banach space, we can similarly define the spaces $\mathscr{F}^s_{p,q}(\mathbb{R}^n;\mathscr{X})$ and $\mathscr{B}^s_{p,q}(\mathbb{R}^n;\mathscr{X})$ of distributions taking values in $\mathscr{X}$, by making the obvious modifications.

Proposition 2.4.5. *The equivalence class of the extended norms $\|\cdot\|_{\mathscr{B}^s_{p,q}(\mathbb{R}^n)}$ and $\|\cdot\|_{\mathscr{F}^s_{p,q}(\mathbb{R}^n)}$ does not depend on the choice of $\hat{\phi}$. In particular, the Banach spaces $\mathscr{B}^s_{p,q}(\mathbb{R}^n)$ and $\mathscr{F}^s_{p,q}(\mathbb{R}^n)$ do not depend on the choice of $\hat{\phi}$.*

Remark 2.4.6. It is possible to extend the definition of $\|\cdot\|_{\mathscr{B}^s_{p,q}(\mathbb{R}^n)}$ to a well-defined equivalence class of quasi-norms for $p,q\in(0,\infty]$ and to similarly extend $\|\cdot\|_{\mathscr{F}^s_{p,q}(\mathbb{R}^n)}$ to $p\in(0,\infty)$, $q\in(0,\infty]$, though the proofs in these settings require an extra twist. See [244, § 2.3.1 and § 2.3.2] for details. We will not go down this path in this text, though generalizing some of our later results in this way might be interesting for other applications.

Theorem 2.4.7. *If T is a singular integral operator of order $s\in\mathbb{R}$, then T is a bounded operator:*

$$T:\mathscr{B}^{s_0}_{p,q}(\mathbb{R}^n)\to\mathscr{B}^{s_0-s}_{p,q}(\mathbb{R}^n),\quad s_0\in\mathbb{R},\ p,q\in[1,\infty],$$
$$T:\mathscr{F}^{s_0}_{p,q}(\mathbb{R}^n)\to\mathscr{F}^{s_0-s}_{p,q}(\mathbb{R}^n),\quad s_0\in\mathbb{R},\ p\in(1,\infty),\ q\in(1,\infty].$$

Recall that Tf is defined for any $f\in\mathscr{S}(\mathbb{R}^n)'$ – see Remark 2.3.29.

Corollary 2.4.8. *If $a(x,D)$ is a pseudo-differential operator of order $s\in\mathbb{R}$, then $a(x,D)$ is a bounded operator*

$$a(x,D):\mathscr{B}^{s_0}_{p,q}(\mathbb{R}^n)\to\mathscr{B}^{s_0-s}_{p,q}(\mathbb{R}^n),\quad s_0\in\mathbb{R},\ p,q\in[1,\infty],$$
$$a(x,D):\mathscr{F}^{s_0}_{p,q}(\mathbb{R}^n)\to\mathscr{F}^{s_0-s}_{p,q}(\mathbb{R}^n),\quad s_0\in\mathbb{R},\ p\in(1,\infty),\ q\in(1,\infty].$$

Proof. Since $a(x,D)$ is a singular integral operator of order s by Proposition 2.3.14, this follows immediately from Theorem 2.4.7. □

Proposition 2.4.9. *For $p\in(1,\infty)$ and $s\in\mathbb{R}$, $\mathscr{F}^s_{p,2}(\mathbb{R}^n)=L^p_s$ and*

$$\|f\|_{\mathscr{F}^s_{p,2}(\mathbb{R}^n)}\approx\|f\|_{L^p_s},\quad \forall f\in L^p_s,$$

where the implicit constants depend on p and s, but not on f. Here, L^p_s is the L^p Sobolev space of order s (see Definition 2.2.11).

We turn to the proofs of the above results. A convenient tool is the Calderón reproducing formula. For $C_0 \in \mathbb{N}$, define

$$\widetilde{D}_j^{C_0} := \sum_{\substack{|k-j|\le C_0\\ k\in\mathbb{N}}} D_k.$$

Lemma 2.4.10 (Calderón reproducing formula). *If $C_0 = C_0(\hat{\phi}) \in \mathbb{N}$ is sufficiently large, then*

$$D_j\widetilde{D}_j^{C_0} = D_j, \quad \forall j \in \mathbb{N}.$$

Proof. Note that

$$(\widetilde{D}_j^{C_0} f)^{\wedge}(\xi) = \sum_{\substack{|k-j|\le C_0\\ k\in\mathbb{N}}} \hat{\psi}_k(2^{-k}\xi)\hat{f}(\xi).$$

Thus, $D_j\widetilde{D}_j^{C_0} = D_j$ if $\sum_{\substack{|k-j|\le C_0\\ k\in\mathbb{N}}} \hat{\psi}_k(2^{-k+j}\xi) \equiv 1$ on $\operatorname{supp}(\hat{\psi}_j)$. But,

$$\sum_{\substack{|k-j|\le C_0\\ k\in\mathbb{N}}} \hat{\psi}_k(2^{-k+j}\xi) = \begin{cases} \hat{\phi}(2^{-C_0}\xi) - \hat{\phi}(2^{C_0}\xi), & j > C_0,\\ \hat{\phi}(2^{-C_0}\xi), & j \le C_0,\end{cases}$$

and

$$\hat{\psi}_j(\xi) = \begin{cases} \hat{\phi}(\xi) - \hat{\phi}(2\xi), & j \ge 1,\\ \hat{\phi}(\xi), & j = 0.\end{cases}$$

From here the result follows from the fact that $\hat{\phi} \equiv 1$ on a neighborhood of 0. □

Henceforth, we fix $C_0 = C_0(\hat{\phi}) \in \mathbb{N}$ so large that Lemma 2.4.10 holds.

A central part of the proofs of the above results is the following technical result. For a bounded set of elementary operators $\mathcal{E}$, $s \in \mathbb{R}$, and

$$\mathscr{V} \in \{L^p(\mathbb{R}^n; \ell^q(\mathbb{N})) : p \in (1,\infty), q \in (1,\infty]\} \cup \{\ell^q(\mathbb{N}; L^p(\mathbb{R}^n)) : p, q \in [1,\infty]\},$$

define a semi-norm by

$$\|f\|_{\mathscr{V},s,\mathcal{E}} := \sup_{\{(E_j,2^{-j}):j\in\mathbb{N}\}\subseteq\mathcal{E}} \|\{2^{js}E_j f\}_{j\in\mathbb{N}}\|_{\mathscr{V}}.$$

Proposition 2.4.11. *For $p, q \in [1,\infty]$ and $s \in \mathbb{R}$, $\|\cdot\|_{\ell^q(L^p),s,\mathcal{E}}$ is a continuous semi-norm on $\mathscr{B}^s_{p,q}(\mathbb{R}^n)$ and*

$$\|f\|_{\ell^q(L^p),s,\mathcal{E}} \le C_{p,q,s,\mathcal{E}}\|f\|_{\mathscr{B}^s_{p,q}(\mathbb{R}^n)}, \quad \forall f \in \mathscr{S}(\mathbb{R}^n)'. \tag{2.31}$$

For $p \in (1,\infty)$, $q \in (1,\infty]$, and $s \in \mathbb{R}$, $\|\cdot\|_{L^p(\ell^q),s,\mathcal{E}}$ is a continuous semi-norm on $\mathscr{F}^s_{p,q}(\mathbb{R}^n)$ and

$$\|f\|_{L^p(\ell^q),s,\mathcal{E}} \le C_{p,q,s,\mathcal{E}}\|f\|_{\mathscr{F}^s_{p,q}(\mathbb{R}^n)}, \quad \forall f \in \mathscr{S}(\mathbb{R}^n)'. \tag{2.32}$$

The constants in the above depend on $\mathcal{E}$, p, q, s, and n, though they may be chosen uniformly on compact subsets of $s \in \mathbb{R}$.

Notation 2.4.12. For the rest of the section, for $s \in \mathbb{R}$, $\mathscr{X}^s$ will denote any one of the spaces

$$\mathscr{X}^s \in \{\mathscr{B}^s_{p,q}(\mathbb{R}^n) : p,q \in [1,\infty]\} \bigcup \{\mathscr{F}^s_{p,q}(\mathbb{R}^n) : p \in (1,\infty), q \in (1,\infty]\}.$$

We set

$$\mathscr{V} := \begin{cases} \ell^q(\mathbb{N}; L^p(\mathbb{R}^n)), & \text{if } \mathscr{X}^s = \mathscr{B}^s_{p,q}(\mathbb{R}^n), \\ L^p(\mathbb{R}^n; \ell^q(\mathbb{N})), & \text{if } \mathscr{X}^s = \mathscr{F}^s_{p,q}(\mathbb{R}^n). \end{cases}$$

With this notation we have

$$\|f\|_{\mathscr{X}^s} = \|f\|_{\mathscr{V},s,\{(D_j,2^{-j}):j\in\mathbb{N}\}} = \left\|\{2^{js}D_j f\}_{j\in\mathbb{N}}\right\|_{\mathscr{V}}.$$

To prove Proposition 2.4.11 we require the next lemma.

Lemma 2.4.13. *Let $\mathcal{E}$ be a bounded set of pre-elementary operators. Let $\{(E_j, 2^{-j}) : j \in \mathbb{N}\} \subseteq \mathcal{E}$ and define an operator on sequences of $L^1_{\mathrm{loc}}(\mathbb{R}^n)$ functions $\{f_j(x)\}_{j\in\mathbb{N}}$ by*

$$\mathcal{T}\{f_j\}_{j\in\mathbb{N}} = \{E_j f_j\}_{j\in\mathbb{N}}.$$

For

$$\mathcal{V} \in \{L^p(\mathbb{R}^n; \ell^q(\mathbb{N})) : p \in (1,\infty), q \in (1,\infty]\} \cup \{\ell^q(\mathbb{N}; L^p(\mathbb{R}^n)) : p,q \in [1,\infty]\},$$

we have $\mathcal{T} \in \mathrm{Hom}(\mathcal{V},\mathcal{V})$ and

$$\|\mathcal{T}\|_{\mathcal{V}\to\mathcal{V}} \le C_{\mathcal{E},\mathcal{V}}.$$

Before we prove Lemma 2.4.13 we introduce some preliminary results. For a function $f \in L^1_{\mathrm{loc}}(\mathbb{R}^n)$, we use the Hardy–Littlewood maximal function defined by

$$\mathcal{M}f(x) = \sup_{r>0} \frac{1}{\mathrm{Vol}(B^n(x,r))} \int_{B^n(x,r)} |f(y)|\, dy, \tag{2.33}$$

where $B^n(x,r) = \{y \in \mathbb{R}^n : |x-y| < r\}$ and $\mathrm{Vol}(A)$ denotes the Lebesgue measure of A.

Lemma 2.4.14. *For $p \in (1,\infty)$ and $q \in (1,\infty]$,*

$$\|\{\mathcal{M}f_j\}_{j\in\mathbb{N}}\|_{L^p(\mathbb{R}^n;\ell^q(\mathbb{N}))} \leq C_{p,q,n}\|\{f_j\}_{j\in\mathbb{N}}\|_{L^p(\mathbb{R}^n;\ell^q(\mathbb{N}))},$$

where if the right-hand side is finite, then the left-hand side is finite.

Proof. This is due to C. Fefferman and E. M. Stein [83]. See also [216, Chapter II, § 1.3.1]. □

Lemma 2.4.15. *Let $\mathcal{E}$ be a bounded set of pre-elementary operators. Then there exists $C = C(\mathcal{E}) \geq 0$ such that $\forall (E, 2^{-j}) \in \mathcal{E}$ and all $f \in L^1_{\mathrm{loc}}$,*

$$|Ef(x)| \leq C\mathcal{M}f(x).$$

Proof. This follows easily from the definitions. □

Proof of Lemma 2.4.13. Using Lemmas 2.4.14 and 2.4.15, for $p \in (1,\infty)$, $q \in (1,\infty]$ we have

$$\|\mathcal{T}\{f_j\}_{j\in\mathbb{N}}\|_{L^p(\ell^q)} = \|\{E_jf_j\}_{j\in\mathbb{N}}\|_{L^p(\ell^q)} \lesssim \|\{\mathcal{M}f_j\}_{j\in\mathbb{N}}\|_{L^p(\ell^q)} \lesssim \|\{f_j\}_{j\in\mathbb{N}}\|_{L^p(\ell^q)}.$$

Using Lemma 2.3.26, for $p, q \in [1,\infty]$ we have

$$\|\mathcal{T}\{f_j\}_{j\in\mathbb{N}}\|_{\ell^q(L^p)} = \|\{E_jf_j\}_{j\in\mathbb{N}}\|_{\ell^q(L^p)} \lesssim \|\{f_j\}_{j\in\mathbb{N}}\|_{\ell^q(L^p)},$$

completing the proof. □

In the remainder of this section, we will have several operators indexed by $j \in \mathbb{N}$; for example, we may have E_j for $j \in \mathbb{N}$. For such a sequence of operators, we define $E_j = 0$ for $j \in \mathbb{Z} \setminus \mathbb{N}$.

Lemma 2.4.16. *Let $\mathcal{E}$ be a bounded set of elementary operators and let*

$$\{(E^1_j, 2^{-j}) : j \in \mathbb{N}\}, \{(E^2_j, 2^{-j}) : j \in \mathbb{N}\} \subseteq \mathcal{E}.$$

For $k \in \mathbb{Z}$, set

$$\mathcal{T}_k\{f_j\}_{j\in\mathbb{N}} := \{E^1_j E^2_{j+k} f_j\}_{j\in\mathbb{N}}.$$

Then, $\forall N \in \mathbb{N}$,

$$\forall \mathcal{V} \in \{L^p(\mathbb{R}^n;\ell^q(\mathbb{N})) : p \in (1,\infty), q \in (1,\infty]\} \cup \{\ell^q(\mathbb{N}; L^p(\mathbb{R}^n)) : p, q \in [1,\infty]\},$$

we have $\mathcal{T}_k \in \mathrm{Hom}(\mathcal{V},\mathcal{V})$, $\forall k \in \mathbb{Z}$, and there exists $C_{\mathcal{E},N,\mathcal{V}} \geq 0$ such that $\forall k \in \mathbb{Z}$

$$\|\mathcal{T}_k\|_{\mathcal{V}\to\mathcal{V}} \leq C_{\mathcal{E},N,\mathcal{V}} 2^{-N|k|}.$$

Proof. Lemma 2.3.24 shows that, $\forall N \in \mathbb{N}$,

$$\{(2^{N|k|}F_1F_2, 2^{-j}) : (F_1, 2^{-j}), (F_2, 2^{-(j+k)}) \in \mathcal{E}\}$$

is a bounded set of elementary operators. From here, the result follows from Lemma 2.4.13. □

Lemma 2.4.17. *Let $\mathcal{E}$ be a bounded set of pre-elementary operators and let $f \in \mathscr{S}(\mathbb{R}^n)'$. Then, $\forall (E, 2^{-k}) \in \mathcal{E}$,*

$$E \sum_{j\in\mathbb{N}} D_j f = \sum_{j\in\mathbb{N}} ED_j f(x) = Ef(x),$$

where this sum converges pointwise. Here, D_j are the operators from (2.30).

Proof. It follows easily from the definitions that for $g \in \mathscr{S}(\mathbb{R}^n)$, $\sum_{j\in\mathbb{N}} D_j^* g = g$, with convergence in $\mathscr{S}(\mathbb{R}^n)$. Thus, by duality, $\sum_{j\in\mathbb{N}} D_j f = f$, with convergence in $\mathscr{S}(\mathbb{R}^n)'$. Since, for x fixed, $E(x, \cdot) \in \mathscr{S}(\mathbb{R}^n)$ and $ED_j f(x) = \int E(x,y)(D_j f)(y)\, dy$, the result follows. □

Proof of Proposition 2.4.11. We prove (2.31) and (2.32) simultaneously, by using Notation 2.4.12.

Let $\{(E_j, 2^{-j}) : j \in \mathbb{N}\} \subseteq \mathcal{E}$. Using Lemmas 2.4.10 and 2.4.17, we have

$$\begin{aligned} \|\{2^{js}E_j f\}_{j\in\mathbb{N}}\|_{\mathcal{V}} &= \left\| \left\{ \sum_{k\in\mathbb{Z}} 2^{js} E_j D_{j+k} f \right\}_{j\in\mathbb{N}} \right\|_{\mathcal{V}} \\ &\leq \sum_{k\in\mathbb{Z}} \|\{2^{js}E_j D_{j+k} f\}_{j\in\mathbb{N}}\|_{\mathcal{V}} = \sum_{k\in\mathbb{Z}} \|\{2^{js}E_j D_{j+k} \widetilde{D}_{j+k}^{C_0} f\}_{j\in\mathbb{N}}\|_{\mathcal{V}}, \end{aligned} \tag{2.34}$$

where the $\leq$ inequality uses the triangle inequality, Fatou's lemma, and the pointwise convergence of the sum (Lemma 2.4.17).

Define, for $k \in \mathbb{Z}$, the vector-valued operator $\mathcal{T}_k$ by

$$\mathcal{T}_k\{f_j\}_{j\in\mathbb{N}} := \{2^{|k|-ks} E_j D_{j+k} f_j\}_{j\in\mathbb{N}}.$$

Since $\{(D_j, 2^{-j}) : j \in \mathbb{N}\}$ is a bounded set of elementary operators (see Example 2.3.8), Lemma 2.4.16 shows

$$\|\mathcal{T}_k\|_{\mathcal{V}\to\mathcal{V}} \lesssim 1.$$

Thus,

$$
\begin{aligned}
\sum_{k\in\mathbb{Z}} \|\{2^{js}E_jD_{j+k}\widetilde{D}^{C_0}_{j+k}f\}_{j\in\mathbb{N}}\|_{\mathscr{V}} &= \sum_{k\in\mathbb{Z}} 2^{-|k|}\|\mathcal{T}_k\{2^{(j+k)s}\widetilde{D}^{C_0}_{j+k}f\}_{j\in\mathbb{N}}\|_{\mathscr{V}} \\
&\lesssim \sum_{k\in\mathbb{Z}} 2^{-|k|}\|\{2^{(j+k)s}\widetilde{D}^{C_0}_{j+k}f\}_{j\in\mathbb{N}}\|_{\mathscr{V}} \le \sum_{k\in\mathbb{Z}} 2^{-|k|}\|\{2^{js}\widetilde{D}^{C_0}_{j}f\}_{j\in\mathbb{N}}\|_{\mathscr{V}} \\
&\lesssim \sum_{k\in\mathbb{Z}} 2^{-|k|}\|\{2^{js}D_jf\}_{j\in\mathbb{N}}\|_{\mathscr{V}} \lesssim \|f\|_{\mathscr{X}^s},
\end{aligned}
\tag{2.35}
$$

where the final $\lesssim$ estimate used the definition of $\widetilde{D}^{C_0}_j$. Combining (2.34) and (2.35) and taking the supremum over all $\{(E_j, 2^{-j}) : j \in \mathbb{N}\} \subseteq \mathcal{E}$ shows

$$
\|f\|_{\mathscr{V},s,\mathcal{E}} = \sup_{\{(E_j,2^{-j}):j\in\mathbb{N}\}\subseteq\mathcal{E}} \|\{2^{js}E_jf\}_{j\in\mathbb{N}}\|_{\mathscr{V}} \lesssim \|f\|_{\mathscr{X}^s},
$$

completing the proof. □

Proposition 2.4.11 leads to the following characterization of $\mathscr{X}^s$.

Proposition 2.4.18. *For $s \in \mathbb{R}$ and $f \in \mathscr{S}(\mathbb{R}^n)'$, the following are equivalent:*
(i) $f \in \mathscr{X}^s$,
(ii) *for every bounded set of elementary operators $\mathcal{E}$, $\|f\|_{\mathscr{V},s,\mathcal{E}} < \infty$.*

Proof. (i) ⇒ (ii) follows from Proposition 2.4.11. For (ii) ⇒ (i), note that $\{(D_j, 2^{-j}) : j \in \mathbb{N}\}$ is a bounded set of elementary operators (see Example 2.3.8) and

$$
\|f\|_{\mathscr{X}^s} = \|f\|_{\mathscr{V},s,\{(D_j,2^{-j}):j\in\mathbb{N}\}}.
$$
□

Proof of Proposition 2.4.5. Let $\hat{\gamma} \in C_0^\infty(\mathbb{R}^n)$ satisfy $\hat{\gamma}(\xi) \equiv 1$ on a neighborhood of 0. Define $\overline{D}_j$ in terms of $\hat{\gamma}$, just as D_j was defined in terms of $\hat{\phi}$. The statement of the proposition is equivalent to

$$
\|f\|_{\mathscr{V},s,\{(\overline{D}_j,2^{-j}):j\in\mathbb{N}\}} \approx \|f\|_{\mathscr{V},s,\{(D_j,2^{-j}):j\in\mathbb{N}\}}.
\tag{2.36}
$$

Since the assumptions of (2.36) are symmetric in $\overline{D}_j$ and D_j, it suffices to show just the $\lesssim$ part of (2.36). Since $\{(\overline{D}_j, 2^{-j}) : j \in \mathbb{N}\}$ is a bounded set of elementary operators (see Example 2.3.8), this follows from Proposition 2.4.11, completing the proof. □

Next we turn to the proof of Theorem 2.4.7. We will use the next lemma.

Lemma 2.4.19. *Let T be a singular integral operator of order $s \in \mathbb{R}$ and let $\mathcal{E}$ be a bounded set of elementary operators. Then*

$$
\{(2^{-js}ET, 2^{-j}) : (E, 2^{-j}) \in \mathcal{E}\}
$$

is a bounded set of elementary operators.

Proof. By Lemma 2.3.27, T^* is a singular integral operator of order s. By Proposition 2.3.18 (b), $\{(E^*, 2^{-j}) : (E, 2^{-j}) \in \mathcal{E}\}$ is a bounded set of elementary operators. From here, Theorem 2.3.10 (ii) implies that

$$\{(2^{-js}T^*E^*, 2^{-j}) : (E, 2^{-j}) \in \mathcal{E}\}$$

is a bounded set of elementary operators. Proposition 2.3.18 (b) applied to this set shows

$$\{(2^{-js}ET, 2^{-j}) : (E, 2^{-j}) \in \mathcal{E}\}$$

is a bounded set of elementary operators, completing the proof. □

Proof of Theorem 2.4.7. Using Notation 2.4.12, we wish to show that $T : \mathscr{X}^{s_0} \to \mathscr{X}^{s_0 - s}$ is bounded. Set $E_j := 2^{-js}D_jT$. Since $\{(D_j, 2^{-j}) : j \in \mathbb{N}\}$ is a bounded set of elementary operators (see Example 2.3.8), Lemma 2.4.19 shows that $\{(E_j, 2^{-j}) : j \in \mathbb{N}\}$ is a bounded set of elementary operators. We have, for $f \in \mathscr{S}(\mathbb{R}^n)'$,

$$\begin{aligned}\|Tf\|_{\mathscr{X}^{s-s_0}} &= \left\|\{2^{j(s_0-s)}D_jTf\}_{j\in\mathbb{N}}\right\|_{\mathscr{V}} = \left\|\{2^{js_0}E_jf\}_{j\in\mathbb{N}}\right\|_{\mathscr{V}} \\ &= \|f\|_{\mathscr{V},s,\{(E_j,2^{-j}):j\in\mathbb{N}\}} \lesssim \|f\|_{\mathscr{X}^s},\end{aligned}$$

where the last inequality follows from Proposition 2.4.11 and completes the proof. □

We turn to the proof of Proposition 2.4.9. We begin with the case $s = 0$, where $L^p_0 = L^p$. First, we introduce the Khintchine inequality, which we state without proof.

Theorem 2.4.20 (The Khintchine inequality – see [103]). *Let $\{\epsilon_j\}_{j=0}^{\infty}$ be a sequence of i. i. d. random variables with mean zero taking values in ± 1. Let $x_0, x_1, x_2, \dots$ be a sequence of complex numbers with $\sum |x_j|^2 < \infty$ and fix $0 < p < \infty$. There are constants A_p and B_p, depending only on p, such that*

$$A_p\left(\sum_{j=0}^{\infty}|x_j|^2\right)^{\frac{1}{2}} \le \left(\mathbb{E}\left|\sum_{j=0}^{\infty}\epsilon_j x_j\right|^p\right)^{\frac{1}{p}} \le B_p\left(\sum_{j=0}^{\infty}|x_j|^2\right)^{\frac{1}{2}}.$$

Remark 2.4.21. The use of the Khintchine inequality in this and similar settings is very common. See, for instance, [216, page 267] and [213, Chapter 4, Section 5].

Lemma 2.4.22. *We have, for $1 < p < \infty$,*

$$\left\|\{D_jf\}_{j\in\mathbb{N}}\right\|_{L^p(\ell^2)} \lesssim \|f\|_{L^p(\mathbb{R}^n)}, \quad \forall f \in L^p(\mathbb{R}^n), \tag{2.37}$$

where the implicit constant depends on $p \in (1, \infty)$ and n.

Proof. Fix $p \in (1, \infty)$. Given any sequence $\epsilon_j \in \{-1, 1\}$, set $T_\epsilon := \sum_{j\in\mathbb{N}} \epsilon_j D_j$. Using the fact that $\{(D_j, 2^{-j}) : j \in \mathbb{N}\}$ is a bounded set of elementary operators (see Example 2.3.8), Corollary 2.3.13 shows that for $f \in L^p$, $\|T_\epsilon f\|_{L^p} \lesssim \|f\|_{L^p}$. Though it is not stated there, the same proof shows that this is uniform in ϵ:

$$\sup_{\epsilon_j\in\{-1,1\}} \|T_\epsilon f\|_{L^p} \lesssim \|f\|_{L^p}. \tag{2.38}$$

Using the Khintchine inequality (Theorem 2.4.20), we have with ϵ_j a sequence of i. i. d. random variables of mean zero taking values in ± 1,

$$\begin{aligned}\|\{D_j f\}_{j\in\mathbb{N}}\|_{L^p(\ell^2)} &\approx \left\| \left(\mathbb{E} \left| \sum_{j=0}^{\infty} \epsilon_j D_j f \right|^p \right)^{\frac{1}{p}} \right\|_{L^p} \\ &= \left(\mathbb{E} \left\| \sum_{j=0}^{\infty} \epsilon_j D_j f \right\|_{L^p}^p \right)^{\frac{1}{p}} \leq \sup_{\epsilon_j\in\{-1,1\}} \|T_\epsilon f\|_{L^p} \lesssim \|f\|_{L^p},\end{aligned}$$

where the last inequality follows from (2.38), completing the proof. □

Lemma 2.4.23. *For $1 < p < \infty$, $L^p \subseteq \mathscr{F}^0_{p,2}(\mathbb{R}^n)$ and*

$$\|f\|_{\mathscr{F}^0_{p,2}(\mathbb{R}^n)} \lesssim \|f\|_{L^p}, \quad \forall f \in L^p, \tag{2.39}$$

where the implicit constant depends on $p \in (1,\infty)$ and n.

Proof. Formula (2.39) is just a restatement of (2.37). The result follows. □

Lemma 2.4.24. *For $p \in (1,\infty)$, $\mathscr{F}^0_{p,2}(\mathbb{R}^n) \subseteq L^p$ and $\|f\|_{L^p} \lesssim \|f\|_{\mathscr{F}^0_{p,2}(\mathbb{R}^n)}$, $\forall f \in \mathscr{F}^0_{p,s}(\mathbb{R}^n)$, where the implicit constant depends on $p \in (1,\infty)$ and n.*

Proof. We take $\langle g, f\rangle = \int g(x)\overline{f(x)}\,dx$; here if $f \in \mathscr{S}(\mathbb{R}^n)'$, $\overline{f}$ is defined as in Definition 2.1.5.

Suppose $f \in \mathscr{F}^0_{p,2}(\mathbb{R}^n)$. Let $q \in (1,\infty)$ be dual to p (i. e., $p^{-1} + q^{-1} = 1$). We will show that $\forall g \in \mathscr{S}(\mathbb{R}^n)$,

$$|\langle g, f\rangle| \lesssim \|g\|_{L^q} \|f\|_{\mathscr{F}^0_{p,2}(\mathbb{R}^n)},$$

and the result will follow. Since $\sum_{j\in\mathbb{N}} D_j g = g$, with convergence in $\mathscr{S}(\mathbb{R}^n)$, we have, using Lemma 2.4.10,

$$\begin{aligned}|\langle g, f\rangle| &= \left| \sum_{j\in\mathbb{N}} \langle D_j g, f\rangle \right| = \left| \sum_{j\in\mathbb{N}} \langle D_j \widetilde{D}_j^{C_0} g, f\rangle \right| = \left| \sum_{j\in\mathbb{N}} \langle \widetilde{D}_j^{C_0} g, D_j^* f\rangle \right| \\ &\leq \int \sum_{j\in\mathbb{N}} |\widetilde{D}_j^{C_0} g(x)| |D_j^* f(x)|\,dx \leq \int \left(\sum_{j\in\mathbb{N}} |\widetilde{D}_j^{C_0} g(x)|^2 \right)^{\frac{1}{2}} \left(\sum_{j\in\mathbb{N}} |D_j^* f(x)|^2 \right)^{\frac{1}{2}} dx \\ &\leq \|\{\widetilde{D}_j^{C_0} g\}_{j\in\mathbb{N}}\|_{L^q(\ell^2)} \|\{D_j^* f\}_{j\in\mathbb{N}}\|_{L^p(\ell^2)} \lesssim \|\{D_j g\}_{j\in\mathbb{N}}\|_{L^q(\ell^2)} \|\{D_j^* f\}_{j\in\mathbb{N}}\|_{L^p(\ell^2)} \\ &\lesssim \|g\|_{L^q} \|f\|_{\mathscr{F}^0_{p,2}(\mathbb{R}^n)},\end{aligned}$$

where the first $\lesssim$ estimate used the definition of $\widetilde{D}_j^{C_0}$ and the second used Lemma 2.4.22 and (2.32). □

Proof of Proposition 2.4.9. The case $s = 0$ follows by combining Lemmas 2.4.24 and 2.4.23. Fix $s \in \mathbb{R}$. Let Λ^s be the pseudo-differential operator of order s from Definition 2.2.10. By Definition 2.2.11, Λ^s is an isomorphism $L_s^p \to L^p$, $1 < p < \infty$. By Corollary 2.4.8, $\Lambda^s : \mathscr{F}_{p,2}^s(\mathbb{R}^n) \to \mathscr{F}_{p,2}^0(\mathbb{R}^n) = L^p$ is an isomorphism with continuous inverse Λ^{-s}. We conclude that the identity map $I = \Lambda^{-s}\Lambda^s$ is an isomorphism $L_s^p \to \mathscr{F}_{p,2}^s(\mathbb{R}^n)$, completing the proof. □

We close this section with the following useful proposition.

Proposition 2.4.25. *For $f \in \mathscr{S}(\mathbb{R}^n)'$, the following are equivalent:*
(i) $f \in \mathscr{X}^{s+1}$,
(ii) $f \in \mathscr{X}^s$ *and* $\partial_{x_j} f \in \mathscr{X}^s$, $1 \le j \le n$,
(iii) $f \in \mathscr{X}^{s-1}$ *and* $\partial_{x_j}^2 f \in \mathscr{X}^{s-1}$, $1 \le j \le n$.

In this case, we have

$$\|f\|_{\mathscr{X}^{s+1}} \approx \|f\|_{\mathscr{X}^s} + \sum_{j=1}^n \|\partial_{x_j} f\|_{\mathscr{X}^s} \approx \|f\|_{\mathscr{X}^{s-1}} + \sum_{j=1}^n \|\partial_{x_j}^2 f\|_{\mathscr{X}^{s-1}}. \tag{2.40}$$

Proof. (i) ⇒ (ii): Suppose $f \in \mathscr{X}^{s+1}$. Clearly $f \in \mathscr{X}^s$. Since ∂_{x_j} is a pseudo-differential operator of order 1, Corollary 2.4.8 shows that $\partial_{x_j} f \in \mathscr{X}^s$ and

$$\|f\|_{\mathscr{X}^s} + \sum_{j=1}^n \|\partial_{x_j} f\|_{\mathscr{X}^s} \lesssim \|f\|_{\mathscr{X}^{s+1}}. \tag{2.41}$$

(ii) ⇒ (iii): Suppose $f \in \mathscr{X}^s$ and $\partial_{x_j} f \in \mathscr{X}^s$, $1 \le j \le n$. Since ∂_{x_j} is a pseudo-differential operator of order 1, Corollary 2.4.8 shows $\partial_{x_j}^2 f \in \mathscr{X}^{s-1}$ and

$$\|f\|_{\mathscr{X}^{s-1}} + \sum_{j=1}^n \|\partial_{x_j}^2 f\|_{\mathscr{X}^{s-1}} \lesssim \|f\|_{\mathscr{X}^s} + \sum_{j=1}^n \|\partial_{x_j} f\|_{\mathscr{X}^s}. \tag{2.42}$$

(iii) ⇒ (i): Suppose $f \in \mathscr{X}^{s-1}$ and $\partial_{x_j}^2 f \in \mathscr{X}^{s-1}$, $1 \le j \le n$. Since $(1 + \triangle)^{-1} = \Lambda^{-2}$ is a pseudo-differential operator of order -2, Corollary 2.4.8 shows $f = \Lambda^{-2}(1 + \triangle)f \in \mathscr{X}^{s+1}$, and

$$\|f\|_{\mathscr{X}^{s+1}} \lesssim \|(1 + \triangle)f\|_{\mathscr{X}^{s-1}} \lesssim \|f\|_{\mathscr{X}^{s-1}} + \sum_{j=1}^n \|\partial_{x_j}^2 f\|_{\mathscr{X}^{s-1}}. \tag{2.43}$$

Formula (2.40) follows from (2.41), (2.42), and (2.43). □

2.5 Zygmund–Hölder spaces

We now turn to one of the most important classes of spaces we consider: the Zygmund–Hölder spaces. Recall the classical Hölder spaces on $\mathbb{R}^n$. For $r \in [0,1]$ and $f : \mathbb{R}^n \to \mathbb{C}$, we set

$$\|f\|_{C^{0,r}} := \|f\|_{C(\mathbb{R}^n)} + \sup_{x \neq y \in \mathbb{R}^n} |x-y|^{-r} |f(x) - f(y)|,$$

and for $m \in \mathbb{N}$,

$$\|f\|_{C^{m,r}} := \sum_{|\alpha| \leq m} \|\partial_x^\alpha f\|_{C^{0,r}}.$$

We set

$$C^{m,r}(\mathbb{R}^n) := \{f \in C^m(\mathbb{R}^n) : \|f\|_{C^{m,r}} < \infty\}. \tag{2.44}$$

As is well known, and as we will see, Hölder spaces with non-integer exponents (i. e., when $r \in (0,1)$) play a central role in the study of nonlinear elliptic PDEs. However, when $r \in \{0,1\}$, the Hölder space is often not the right space to use. Instead, we follow Zygmund's lead [255], and use the Zygmund–Hölder spaces. Thus, for $s \in (0,\infty)$, we will define the Zygmund–Hölder space $\mathscr{C}^s(\mathbb{R}^n)$. We will see (Corollary 2.5.6) that for $m \in \mathbb{N}$ and $r \in (0,1)$, $C^{m,r}(\mathbb{R}^n) = \mathscr{C}^{m+r}(\mathbb{R}^n)$. However, for $r \in \{0,1\}$, these spaces differ. We have, for $m \in \mathbb{N}$,

$$C^{m+1,0}(\mathbb{R}^n) \subsetneq C^{m,1}(\mathbb{R}^n) \subsetneq \mathscr{C}^{m+1}(\mathbb{R}^n).$$

We will give a few equivalent characterizations of the space $\mathscr{C}^s(\mathbb{R}^n)$. We begin by defining it as follows.

Definition 2.5.1. For $s > 0$, we define the Zygmund–Hölder space $\mathscr{C}^s(\mathbb{R}^n) := \mathscr{B}^s_{\infty,\infty}(\mathbb{R}^n)$ and $\|\cdot\|_{\mathscr{C}^s(\mathbb{R}^n)} := \|\cdot\|_{\mathscr{B}^s_{\infty,\infty}(\mathbb{R}^n)}$. If $\mathscr{X}$ is a finite-dimensional Banach space, we similarly define $\mathscr{C}^s(\mathbb{R}^n;\mathscr{X}) := \mathscr{B}^s_{\infty,\infty}(\mathbb{R}^n;\mathscr{X})$ and $\|\cdot\|_{\mathscr{C}^s(\mathbb{R}^n;\mathscr{X})} := \|\cdot\|_{\mathscr{B}^s_{\infty,\infty}(\mathbb{R}^n;\mathscr{X})}$.

$\mathscr{C}^s(\mathbb{R}^n)$ can be given a more elementary characterization in terms of differences. For $f : \mathbb{R}^n \to \mathbb{C}$ and $v \in \mathbb{R}^n$, set

$$\mathrm{Diff}_v f(x) := f(x+v) - f(x).$$

For the next definition, recall that $\mathfrak{e}_1, \dots, \mathfrak{e}_n$ denotes the standard basis in $\mathbb{R}^n$.

Definition 2.5.2. For $s \in (0,\infty)$ and $l \in \mathbb{N}$ with $l > s$, define

$$\|f\|_{\mathscr{C}^s_l} := \sum_{l'=0}^{l} \max_{1 \leq k \leq n} \sup_{\delta \in (0,1]} \sup_{x \in \mathbb{R}^n} \delta^{-sl'/l} \left|\mathrm{Diff}^{l'}_{\delta \mathfrak{e}_k} f(x)\right|,$$

where $\mathrm{Diff}_v^{l'}$ denotes the operator Diff_v applied l' times. We set

$$\mathscr{C}_l^s(\mathbb{R}^n) := \{f \in C(\mathbb{R}^n) : \|f\|_{\mathscr{C}_l^s} < \infty\}.$$

The main result of this section is the next proposition.

Proposition 2.5.3. *For $s \in (0,\infty)$ and $l \in \mathbb{N}$ with $l > s$, we have $\mathscr{C}^s(\mathbb{R}^n) = \mathscr{C}_l^s(\mathbb{R}^n)$ and*

$$\|f\|_{\mathscr{C}^s(\mathbb{R}^n)} \approx \|f\|_{\mathscr{C}_l^s(\mathbb{R}^n)}, \quad \forall f \in \mathscr{C}^s(\mathbb{R}^n),$$

where the implicit constants depend on s, l, and n. In particular, $\mathscr{C}_l^s(\mathbb{R}^n)$ does not depend on the choice of l, so long as $l > s$.

Remark 2.5.4. For another, closely related, characterization of $\mathscr{C}^s(\mathbb{R}^n)$, see Remark 2.5.13.

Remark 2.5.5. Another equivalent norm for $\mathscr{C}^s(\mathbb{R}^n)$ which is commonly seen is, for $s \in (0,2)$,

$$\sup_{x\in\mathbb{R}^n} |f(x)| + \sup_{x\in\mathbb{R}^n} \sup_{0\neq h\in\mathbb{R}^n} |h|^{-s} |\mathrm{Diff}_h^2 f(x)|.$$

This definition can be extended to all $s > 0$ by using Corollary 2.5.7. That this norm is equivalent to $\|\cdot\|_{\mathscr{C}^s}$ follows easily from Remark 2.5.13.

Before we prove Proposition 2.5.3, we use it to see some simple corollaries.

Corollary 2.5.6. *For $m \in \mathbb{N}$, $r \in (0,1)$, we have $\mathscr{C}^{m+r}(\mathbb{R}^n) = C^{m,r}(\mathbb{R}^n)$ and*

$$\|f\|_{\mathscr{C}^{m+r}} \approx \|f\|_{C^{m,r}}, \quad \forall f \in C^{m,r}(\mathbb{R}^n),$$

where the implicit constants depend on m, r, and n.

Proof. We begin with the case $m = 0$. Indeed, in this case, $\mathscr{C}^r = \mathscr{C}_1^r = C^{0,r}$, by Proposition 2.5.3 and the definition of $\mathscr{C}_1^r$. Furthermore, also directly from the definition, $\|\cdot\|_{\mathscr{C}_1^r} \approx \|\cdot\|_{C^{0,r}}$. This completes the case $m = 0$.

When $m \in \mathbb{N}_+$, we use the fact that $\mathscr{C}^{m+r} = \mathscr{B}_{\infty,\infty}^{m+r}(\mathbb{R}^n)$ (with equality of norms) and apply Proposition 2.4.25 m times to see

$$\|f\|_{\mathscr{C}^{m+r}} \approx \sum_{|\alpha|\leq m} \|\partial_x^\alpha f\|_{\mathscr{C}^r} \approx \sum_{|\alpha|\leq m} \|\partial_x^\alpha f\|_{\mathscr{C}_1^r} \approx \sum_{|\alpha|\leq m} \|\partial_x^\alpha f\|_{C^{0,r}} = \|f\|_{C^{m,r}}.$$

A similar proof shows that $\mathscr{C}^{m+r} = C^{m,r}$. □

Corollary 2.5.7. *For $m \in \mathbb{N}$ and $r \in (0,2)$,*

$$\mathscr{C}^{m+r}(\mathbb{R}^n) = \{f \in C^m(\mathbb{R}^n) : \partial_x^\alpha f \in \mathscr{C}_2^r, \forall |\alpha| \leq m\}$$

and

$$\|f\|_{\mathscr{C}^{m+r}} \approx \sum_{|\alpha|\le m} \|\partial_x^\alpha f\|_{\mathscr{C}_2^r},$$

where the implicit constants depend on m, r, and n.

Proof. Using the fact that $\mathscr{C}^{m+r} = \mathscr{B}^{m+r}_{\infty,\infty}(\mathbb{R}^n)$ (with equality of norms), the result with $\mathscr{C}_2^r$ replaced by $\mathscr{C}^r$ follows from m applications of Proposition 2.4.25. Since $r \in (0,2)$, the result now follows from Proposition 2.5.3. □

Remark 2.5.8. The characterization of $\mathscr{C}^{m+r}$ given in Corollary 2.5.7 is the one most commonly used.

We now turn to the proof of Proposition 2.5.3.

Lemma 2.5.9. *If $g \in C^l(\mathbb{R}^n)$, $1 \le k \le n$, and $\delta \in (0,1]$,*

$$\sup_{x\in\mathbb{R}^n} |\mathrm{Diff}^l_{\delta e_k} g(x)| \lesssim \Big(\sup_{x\in\mathbb{R}^n} |(\delta\partial_{x_k})^l g(x)|\Big) \wedge \|g\|_{L^\infty}.$$

Proof. It is clear that

$$\sup_{x\in\mathbb{R}^n} |\mathrm{Diff}^l_{\delta e_k} g(x)| \lesssim \|g\|_{L^\infty}.$$

To see that

$$\sup_{x\in\mathbb{R}^n} |\mathrm{Diff}^l_{\delta e_k} g(x)| \lesssim \sup_{x\in\mathbb{R}^n} |(\delta\partial_{x_k})^l g(x)|, \tag{2.45}$$

note that the fundamental theorem of calculus gives

$$\mathrm{Diff}_{\delta e_k} g(x) = \int_0^1 (\delta\partial_{x_k} g)(x + \delta t e_k)\, dt,$$

and therefore,

$$\mathrm{Diff}^l_{\delta e_k} g(x) = \int_{[0,1]^l} ((\delta\partial_{x_k})^l g)\left(x + \delta \sum_{r=1}^l t_r\right) dt.$$

The estimate (2.45) follows, completing the proof. □

Lemma 2.5.10. *Let $l \in \mathbb{N}_+$. Then, for all $s \in (0,l)$, we have $\mathscr{C}^s \subseteq \mathscr{C}^s_l$ and $\|f\|_{\mathscr{C}^s_l} \lesssim \|f\|_{\mathscr{C}^s}$, $\forall f \in \mathscr{C}^s$.*

Proof. We proceed by induction on $l \in \mathbb{N}_+$. That is, suppose we know that the result holds for all $l' \in \mathbb{N}_+$ with $l' < l$. We will then prove the result for l. Because of how we have set up the induction, we do not require a base case (the case $l = 1$ is also included in the inductive step). Fix $s \in (0, l)$.

Suppose $f \in \mathscr{C}^s$. Then $\|D_j f\|_{L^\infty} \leq 2^{-js}\|f\|_{\mathscr{C}^s}$, $\forall j \in \mathbb{N}$, where D_j is as in (2.30). In particular, since $f = \sum_{j\in\mathbb{N}} D_j f$ in the sense of $\mathscr{S}(\mathbb{R}^n)'$, this sum in fact converges uniformly and therefore $f \in C(\mathbb{R}^n)$ and $\|f\|_{C(\mathbb{R}^n)} \lesssim \|f\|_{\mathscr{C}^s}$.

For $0 < l' < l$ we have, by the inductive hypothesis,

$$\max_{1\leq k\leq n} \sup_{\delta\in(0,1]} \sup_{x\in\mathbb{R}^n} \delta^{-sl'/l}\left|\mathrm{Diff}^{l'}_{\delta e_k} f(x)\right| \lesssim \|f\|_{\mathscr{C}^{sl'/l}} \leq \|f\|_{\mathscr{C}^s},$$

where we have used $l > s$ and therefore $l' > sl'/l$. Thus, it suffices to show that for $\delta \in (0, 1]$, $1 \leq k \leq n$,

$$\left|\mathrm{Diff}^{l}_{\delta e_k} f(x)\right| \lesssim \delta^s \|f\|_{\mathscr{C}^s}. \tag{2.46}$$

Without loss of generality, we assume $\|f\|_{\mathscr{C}^s} = 1$.

Set $E_j := (2^{-j}\partial_{x_k})^l D_j$, so that $\{(E_j, 2^{-j}) : j \in \mathbb{N}\}$ is a bounded set of elementary operators by Proposition 2.3.18 (c). By Proposition 2.4.11, $\forall j \in \mathbb{N}$, $\|E_j f\|_{L^\infty} \lesssim 2^{-js}\|f\|_{\mathscr{B}^s_{\infty,\infty}(\mathbb{R}^n)} \lesssim 2^{-js}$. By Lemma 2.5.9 we have

$$\begin{aligned}\left|\mathrm{Diff}^{l}_{\delta e_k} D_j f(x)\right| &\lesssim \|(\delta\partial_{x_k})^l D_j f\|_{L^\infty} \wedge \|D_j f\|_{L^\infty}\\ &= ((\delta 2^j)^l \|E_j f\|_{L^\infty}) \wedge \|D_j f\|_{L^\infty} \lesssim 2^{-js}((\delta 2^j)^l \wedge 1).\end{aligned}$$

Thus,

$$\begin{aligned}\left|\mathrm{Diff}^{l}_{\delta e_k} f(x)\right| &\leq \sum_{j\in\mathbb{N}} \left|\mathrm{Diff}^{l}_{\delta e_k} D_j f(x)\right| \lesssim \sum_{j\in\mathbb{N}} 2^{-js}((\delta 2^j)^l \wedge 1)\\ &= \sum_{2^{-j}\geq\delta} 2^{j(l-s)}\delta^l + \sum_{2^{-j}<\delta} 2^{-js} \approx \delta^s,\end{aligned}$$

where in the estimate, we have used $l > s$. This proves (2.46) and completes the proof. □

We now turn to the containment $\mathscr{C}^s_l \subseteq \mathscr{C}^s$. To prove this, we are more precise about the choice of $\hat{\phi}$ in the definition of D_j. We use the next lemma.

Lemma 2.5.11. *Suppose $f \in \mathscr{S}_0(\mathbb{R}^n)$ with $\operatorname{supp}\hat{f} \subset B^n(1)$. Then, $\forall l \in \mathbb{N}$ there exist $g_{j,l} \in \mathscr{S}_0(\mathbb{R}^n)$, $j = 1, \ldots n$, such that*

$$f = \sum_{j=1}^{n} \mathrm{Diff}^{l}_{e_j} g_{j,l}.$$

Proof. Fix $l \in \mathbb{N}$. By Lemma 2.2.18, we may write

$$\hat{f}(\xi) = \sum_{j=1}^{n} (2\pi i \xi_j)^l \hat{h}_{j,l}(\xi), \tag{2.47}$$

where $\hat{h}_{j,l}(\xi) \in \mathscr{S}(\mathbb{R}^n)$, $\operatorname{supp}(\hat{h}_{j,l}) \subset B^n(1)$, and $\hat{h}_{j,l}(\xi)$ vanishes to infinite order at 0.

We set

$$\hat{g}_{j,l}(\xi) := \left(\frac{2\pi i \xi_j}{e^{-2\pi i \xi_j} - 1} \right)^l \hat{h}_{j,l}(\xi). \tag{2.48}$$

Since $\hat{h}_{j,k}$ has compact support in $B^n(1)$ and vanishes to infinite order at 0, it follows that $\hat{g}_{j,l} \in \mathscr{S}(\mathbb{R}^n)$ and vanishes to infinite order at 0. Therefore, its inverse Fourier transform, $g_{j,l}$, is in $\mathscr{S}_0(\mathbb{R}^n)$ (Lemma 2.2.18). Combining (2.47) and (2.48), we have

$$\hat{f}(\xi) = \sum_{j=1}^{n} (e^{-2\pi i \xi_j} - 1)^l \hat{g}_{j,l}(\xi).$$

Taking the inverse Fourier transform yields the result. □

Fix $\hat{\phi} \in C_0^\infty(B^n(1))$ with $\hat{\phi} \equiv 1$ on a neighborhood of 0. Define D_j as in (2.30). Recall that $\sum_{j \in \mathbb{N}} D_j = I$ and $\{(D_j, 2^{-j}) : j \in \mathbb{N}\}$ is a bounded set of elementary operators (see Example 2.3.8).

Lemma 2.5.12. *Let $l \in \mathbb{N}_+$. Then, for all $s > 0$, we have $\mathscr{C}_l^s \subseteq \mathscr{C}^s$ and $\|f\|_{\mathscr{C}^s} \lesssim \|f\|_{\mathscr{C}_l^s}$, $\forall f \in \mathscr{C}_l^s$.*

Proof. We wish to show $\|f\|_{\mathscr{C}^s} \lesssim \|f\|_{\mathscr{C}_l^s}$, $\forall f \in \mathscr{C}_l^s$. Let $\psi \in \mathscr{S}(\mathbb{R}^n)$ be defined by $\hat{\psi}(\xi) := \hat{\phi}(\xi) - \hat{\phi}(2\xi)$, so that $\psi \in \mathscr{S}_0(\mathbb{R}^n)$ and $D_j f = f * \operatorname{Dil}_\psi(2^j)$, for $j \geq 1$. By Lemma 2.5.11,

$$\psi = \sum_{k=1}^{n} \operatorname{Diff}_{e_k}^l \gamma_k,$$

where $\gamma_k \in \mathscr{S}_0(\mathbb{R}^n)$.

We will show, for $f \in \mathscr{C}_l^s$, that

$$\|D_j f\|_{L^\infty} \lesssim 2^{-js} \|f\|_{\mathscr{C}_l^s}. \tag{2.49}$$

This implies $\|f\|_{\mathscr{C}^s} \lesssim \|f\|_{\mathscr{C}_l^s}$, and will complete the proof. The estimate (2.49) is easy when $j = 0$; indeed, $\|D_0 f\|_{L^\infty} \leq \|f\|_{L^\infty} \|\phi\|_{L^1} \lesssim \|f\|_{L^\infty} \leq \|f\|_{\mathscr{C}_l^s}$. For $j \geq 1$, we have

$$\begin{aligned}
|D_j f(x)| &= \left|\int f(x-y)\,\mathrm{Dil}_{2^j}(\psi)(y)\,dy\right| \leq \sum_{k=1}^{n} \left|\int f(x-y)\,\mathrm{Dil}_{2^j}(\mathrm{Diff}^l_{\mathfrak{e}_k}\gamma_k)(y)\,dy\right| \\
&= \sum_{k=1}^{n} \left|\int f(x-y)\,\mathrm{Diff}^l_{2^{-j}\mathfrak{e}_k}\mathrm{Dil}_{2^j}(\gamma_k)(y)\,dy\right| \\
&= \sum_{k=1}^{n} \left|\int (\mathrm{Diff}^l_{2^{-j}\mathfrak{e}_k} f)(x-y)\,\mathrm{Dil}_{2^j}(\gamma_k)(y)\,dy\right| \\
&\leq \sum_{k=1}^{n} \|\mathrm{Diff}^l_{2^{-j}\mathfrak{e}_k} f\|_{L^\infty}\|\gamma_k\|_{L^1} \\
&\lesssim 2^{-ls}\|f\|_{\mathscr{C}^s_l}.
\end{aligned}$$

This proves (2.49) and completes the proof. □

Proof of Proposition 2.5.3. This follows by combining Lemmas 2.5.10 and 2.5.12. □

Remark 2.5.13. The proof of Proposition 2.5.3 gives another characterization of $\mathscr{C}^s(\mathbb{R}^n)$. Indeed, let $l > s$. Then $\mathscr{C}^s(\mathbb{R}^n)$ consists of those $f \in C(\mathbb{R}^n)$ such that the following norm is finite:

$$\|f\|_{C(\mathbb{R}^n)} + \max_{1\leq k\leq n}\sup_{\delta\in(0,1]}\sup_{x\in\mathbb{R}}\delta^{-s}|\mathrm{Diff}^l_{\delta\mathfrak{e}_k} f(x)|.$$

Moreover, this defines an equivalent norm on $\mathscr{C}^s(\mathbb{R}^n)$.

Our first clue that $\mathscr{C}^s(\mathbb{R}^n)$ will be useful in the nonlinear setting is that it forms an algebra.

Lemma 2.5.14. *$\mathscr{C}^s(\mathbb{R}^n)$ is an algebra. Moreover,*

$$\|fg\|_{\mathscr{C}^s} \leq C_{s,n}\|f\|_{\mathscr{C}^s}\|g\|_{\mathscr{C}^s}, \quad \forall f,g \in \mathscr{C}^s(\mathbb{R}^n). \tag{2.50}$$

Remark 2.5.15. For applications to PDEs, one often requires a stronger estimate than (2.50), namely, the *tame estimate*:

$$\|fg\|_{\mathscr{C}^s} \lesssim \|f\|_{\mathscr{C}^s}\|g\|_{L^\infty} + \|f\|_{L^\infty}\|g\|_{\mathscr{C}^s}.$$

We prove this tame estimate in Theorem 2.6.2 (iii) using paraproducts. For now, we content ourselves with the much easier non-tame estimate (2.50).

Proof of Lemma 2.5.14. For $v \in \mathbb{R}^n$, let $\tau_v f(x) = f(x+v)$. Note that

$$\mathrm{Diff}^l_v(fg) = \sum_{j=0}^{l}\binom{l}{j}(\tau^j_v\,\mathrm{Diff}^{l-j}_v f)(\mathrm{Diff}^j_v g).$$

Letting $f,g \in \mathscr{C}^s(\mathbb{R}^n)$, $l > s$, $\delta \in (0,1]$, $1 \leq k \leq n$, and $0 \leq l' \leq l$, we have

$$\begin{aligned}|\mathrm{Diff}^{l'}_{\delta e_k}(fg)(x)| &\lesssim \sum_{r=0}^{l'} |\tau^r_{\delta e_k}\,\mathrm{Diff}^{l'-r}_{\delta e_k} f(x)||\mathrm{Diff}^r_{\delta e_k} g(x)| \\ &\lesssim \sum_{r=0}^{l'} \|\,\mathrm{Diff}^{l'-r}_{\delta e_k} f\|_{L^\infty}\|\,\mathrm{Diff}^r_{\delta e_k} g\|_{L^\infty} \\ &\lesssim \sum_{r=0}^{l'} \delta^{s(l'-r)/l}\delta^{sr/l}\|f\|_{\mathscr{C}^s_l}\|g\|_{\mathscr{C}^s_l} \lesssim \delta^{sl'/l}\|f\|_{\mathscr{C}^s_l}\|g\|_{\mathscr{C}^s_l}.\end{aligned}$$

It follows that $\|fg\|_{\mathscr{C}^s_l} \lesssim \|f\|_{\mathscr{C}^s_l}\|g\|_{\mathscr{C}^s_l}$. Applying Proposition 2.5.3 completes the proof. □

2.5.1 Decomposition into smooth functions

One of the most convenient ways to study functions in $\mathscr{C}^s(\mathbb{R}^n)$ is to decompose them into a sum of smooth functions. The next proposition gives such a decomposition.

Proposition 2.5.16. *Let $s > 0$. For $f \in \mathscr{S}(\mathbb{R}^n)'$, the following are equivalent:*
(i) $f \in \mathscr{C}^s(\mathbb{R}^n)$.
(ii) *There exists a sequence $\{f_j\}_{j\in\mathbb{N}} \subset C^\infty(\mathbb{R}^n)$ such that for every multi-index α,*

$$\sup_{j\in\mathbb{N}} 2^{js}\|(2^{-j}\partial_x)^\alpha f_j\|_{L^\infty(\mathbb{R}^n)} < \infty, \tag{2.51}$$

such that $f = \sum_{j\in\mathbb{N}} f_j$, with convergence in $C(\mathbb{R}^n)$.

In this case, for $M > s$ there exists $C = C(s) \geq 0$ such that

$$\|f\|_{\mathscr{C}^s(\mathbb{R}^n)} \leq C \sum_{|\alpha|\leq M} \sup_{j\in\mathbb{N}} 2^{js}\|(2^{-j}\partial_x)^\alpha f_j\|_{L^\infty(\mathbb{R}^n)}. \tag{2.52}$$

Furthermore, f_j may be chosen such that for every multi-index α, there exists $C_{\alpha,s} \geq 0$ such that

$$\sup_{j\in\mathbb{N}} 2^{js}\|(2^{-j}\partial_x)^\alpha f_j\|_{L^\infty(\mathbb{R}^n)} \leq C_{\alpha,s}\|f\|_{\mathscr{C}^s(\mathbb{R}^n)}. \tag{2.53}$$

Proof. Suppose (i) holds, i. e., $f \in \mathscr{C}^s(\mathbb{R}^n)$. Set $f_j := D_j f$ and define $E_{j,\alpha} := (2^{-j}\partial_x)^\alpha D_j$. Since $\{(D_j, 2^{-j}) : j \in \mathbb{N}\}$ is a bounded set of elementary operators (see Example 2.3.8), Proposition 2.3.18 (c) implies that $\{(E_{j,\alpha}, 2^{-j}) : j \in \mathbb{N}\}$ is also a bounded set of elementary operators. Therefore, by Proposition 2.4.11, using $\mathscr{C}^s(\mathbb{R}^n) = \mathscr{B}^s_{\infty,\infty}(\mathbb{R}^n)$,

$$\|(2^{-j}\partial_x)^\alpha f_j\|_{L^\infty(\mathbb{R}^n)} = \|E_{j,\alpha} f_j\|_{L^\infty(\mathbb{R}^n)} \lesssim 2^{-js}\|f\|_{\mathscr{C}^s(\mathbb{R}^n)}.$$

This shows that $f_j \in C^\infty(\mathbb{R}^n)$ and (2.53) holds (and therefore (2.51) holds). We already know that $\sum_{j\in\mathbb{N}} D_j f = f$, with convergence in $\mathscr{S}(\mathbb{R}^n)'$; (2.51) with $|\alpha| = 0$ implies that $\sum_{j\in\mathbb{N}} f_j$ converges in $C(\mathbb{R}^n)$. This establishes (ii).

Conversely, suppose (ii) holds so that $f = \sum_{j\in\mathbb{N}} f_j$, where f_j satisfies (2.51). Fix $M > s$. We wish to show

$$\sup_{j\in\mathbb{N}} 2^{js}\|D_j f\|_{L^\infty(\mathbb{R}^n)} \lesssim \sum_{|\alpha|\le M} \sup_{j\in\mathbb{N}} 2^{js}\|(2^{-j}\partial_x)^\alpha f_j\|_{L^\infty(\mathbb{R}^n)}. \tag{2.54}$$

Indeed, (2.54) is equivalent to (2.52), which implies (i).

Consider,

$$2^{js}\|D_j f\|_{L^\infty(\mathbb{R}^n)} \le 2^{js}\sum_{j\ge k}\|D_j f_k\|_{L^\infty(\mathbb{R}^n)} + 2^{js}\sum_{j<k}\|D_j f_k\|_{L^\infty(\mathbb{R}^n)}. \tag{2.55}$$

We bound the two terms on the right-hand side of (2.55) separately. For the second term, using Lemma 2.3.26 we obtain

$$\begin{aligned} 2^{js}\sum_{j<k}\|D_j f_k\|_{L^\infty(\mathbb{R}^n)} &\lesssim 2^{js}\sum_{j<k}\|f_k\|_{L^\infty(\mathbb{R}^n)} \\ &\le \Big(\sup_k 2^{ks}\|f_k\|_{L^\infty(\mathbb{R}^n)}\Big)\sum_{j<k}2^{s(j-k)} \lesssim \sup_k 2^{ks}\|f_k\|_{L^\infty(\mathbb{R}^n)}. \end{aligned} \tag{2.56}$$

For the first term on the right-hand side of (2.55), by Proposition 2.3.18 (d), we may write

$$D_j = \sum_{|\alpha|\le M} 2^{-j(M-|\alpha|)}E_{j,\alpha}(2^{-j}\partial_x)^\alpha,$$

where $\{(E_{j,\alpha}, 2^{-j}) : j \in \mathbb{N}\}$ is a bounded set of elementary operators. Thus, using Lemma 2.3.26, we have

$$\begin{aligned} 2^{js}\sum_{j\ge k}\|D_j f_k\|_{L^\infty(\mathbb{R}^n)} &\le 2^{js}\sum_{j\ge k}\sum_{|\alpha|\le M} 2^{-Mj}\|E_{j,\alpha}\partial_x^\alpha f_k\|_{L^\infty(\mathbb{R}^n)} \\ &\lesssim 2^{js}\sum_{j\ge k}\sum_{|\alpha|\le M} 2^{-Mj+|\alpha|k}\|(2^{-k}\partial_x)^\alpha f_k\|_{L^\infty(\mathbb{R}^n)} \\ &\le \Big(\sum_{|\alpha|\le M}\sup_k 2^{ks}\|(2^{-k}\partial_x)^\alpha f_k\|_{L^\infty(\mathbb{R}^n)}\Big)\sum_{j\ge k}2^{s(j-k)}2^{M(k-j)} \\ &\lesssim \sum_{|\alpha|\le M}\sup_k 2^{ks}\|(2^{-k}\partial_x)^\alpha f_k\|_{L^\infty(\mathbb{R}^n)}, \end{aligned} \tag{2.57}$$

where the last line used $M > s$. Plugging (2.56) and (2.57) into (2.55) establishes (2.54) and completes the proof. □

2.6 Compositions and product Zygmund–Hölder spaces

Fix $N \in \mathbb{N}_+$ and $s > 0$. For a function $u \in \mathscr{C}^s(\mathbb{R}^n;\mathbb{R}^N)$,[3] we will consider compositions of the form $F(x,u(x))$, where $F(x,\zeta) : \mathbb{R}^n \times \mathbb{R}^N \to \mathbb{C}$ is a sufficiently smooth function.

Informally, the class of functions $F(x,\zeta)$ are $\mathscr{C}^s(\mathbb{R}^n)$ in the x variable and $\mathscr{C}^t(\mathbb{R}^N)$ in the ζ variable, where $s,t > 0$. To rigorously define this product Zygmund–Hölder space, let $\tilde{D}_l$ be defined just like D_j (see (2.30)) but with n replaced by N. In particular, $\sum_{l\in\mathbb{N}} \tilde{D}_j = I$ on $\mathscr{S}(\mathbb{R}^N)'$. For $j,l \in \mathbb{N}$, set $\widehat{D}_{j,l} := D_j \otimes \tilde{D}_l$.

Definition 2.6.1. For $s,t > 0$ we let $\mathscr{C}^{s,t}(\mathbb{R}^n \times \mathbb{R}^N)$ consist of those tempered distributions $F \in \mathscr{S}(\mathbb{R}^n \times \mathbb{R}^N)'$ such that the following norm is finite:

$$\|F\|_{\mathscr{C}^{s,t}(\mathbb{R}^n\times\mathbb{R}^N)} := \sup_{j,l} 2^{js+lt} \|\widehat{D}_{j,l}F\|_{L^\infty(\mathbb{R}^n\times\mathbb{R}^N)}.$$

For $M \in \mathbb{N}_+$, we define the vector-valued space $\mathscr{C}^{s,t}(\mathbb{R}^n \times \mathbb{R}^N;\mathbb{C}^M)$ to consist of those tempered distributions $F = (F_1,\ldots,F_M)$ with each $F_k \in \mathscr{C}^{s,t}(\mathbb{R}^n \times \mathbb{R}^N)$. We set

$$\|F\|_{\mathscr{C}^{s,t}(\mathbb{R}^n\times\mathbb{R}^N;\mathbb{C}^M)} := \sum_{k=1}^{M} \|F_k\|_{\mathscr{C}^{s,t}(\mathbb{R}^n\times\mathbb{R}^N)}.$$

To study fully nonlinear elliptic PDEs, we use the next composition theorem.

Theorem 2.6.2. *Fix $s,t > 0$ and let $F \in \mathscr{C}^{s,t}(\mathbb{R}^n \times \mathbb{R}^N)$, $u,v \in \mathscr{C}^s(\mathbb{R}^n;\mathbb{R}^N)$, and $w \in \mathscr{C}^s(\mathbb{R}^n)$. Then:*

(i) *If $t > \lfloor s \rfloor + 2$, then*

$$F(x,u(x)+\zeta) \in \mathscr{C}^{s,t-\lfloor s\rfloor-2}(\mathbb{R}^n \times \mathbb{R}^N)$$

and

$$\|F(x,u(x)+\zeta)\|_{\mathscr{C}^{s,t-\lfloor s\rfloor-2}} \lesssim \|F\|_{\mathscr{C}^{s,t}}(1+\|u\|_{\mathscr{C}^s})(1+\|u\|_{L^\infty})^{\lfloor s\rfloor+1}.$$

(ii) *If $t > \lfloor s \rfloor + 2$, then*

$$F(x,u(x)) \in \mathscr{C}^s(\mathbb{R}^n)$$

and

$$\|F(x,u(x))\|_{\mathscr{C}^s} \lesssim \|F\|_{\mathscr{C}^{s,t}}(1+\|u\|_{\mathscr{C}^s})(1+\|u\|_{L^\infty})^{\lfloor s\rfloor+1}.$$

---

3 Here we are considering functions taking values in the real Banach space $\mathbb{R}^N$. In most of the text, we use complex Banach spaces, but for the purposes of this section the distinction is not relevant: every element of $\mathbb{R}^N$ can be viewed as an element of $\mathbb{C}^N$, and every element of $\mathbb{C}^N$ can be viewed as an element of $\mathbb{R}^{2N}$. See also Remark 2.4.1.

(iii) *We have* $u \cdot v \in \mathscr{C}^s(\mathbb{R}^n)$ *and*

$$\|u \cdot v\|_{\mathscr{C}^s} \lesssim \|u\|_{\mathscr{C}^s}\|v\|_{L^\infty} + \|u\|_{L^\infty}\|v\|_{\mathscr{C}^s}.$$

(iv) *If* $t > \lfloor s \rfloor + 2$, *we have* $F(x, u(x) + \zeta)w(x) \in \mathscr{C}^{s,t-\lfloor s \rfloor - 2}$ *and*

$$\|F(x, u(x) + \zeta)w(x)\|_{\mathscr{C}^{s,t-\lfloor s\rfloor-2}} \lesssim \|F\|_{\mathscr{C}^{s,t}}(\|w\|_{\mathscr{C}^s} + \|u\|_{\mathscr{C}^s}\|w\|_{L^\infty})(1 + \|u\|_{L^\infty})^{\lfloor s \rfloor + 1}.$$

(v) *If* $t > \lfloor s \rfloor + 3$, *then*

$$\begin{aligned}&\|F(x, u(x) + \zeta) - F(x, v(x) + \zeta)\|_{\mathscr{C}^{s,t-\lfloor s\rfloor-3}} \\ &\quad \lesssim \|F\|_{\mathscr{C}^{s,t}}(\|u - v\|_{\mathscr{C}^s} + (\|u\|_{\mathscr{C}^s} + \|v\|_{\mathscr{C}^s})\|u - v\|_{L^\infty})(1 + \|u\|_{L^\infty} + \|v\|_{L^\infty})^{\lfloor s \rfloor + 1}.\end{aligned}$$

(vi) *If* $t > \lfloor s \rfloor + 3$, *then*

$$\begin{aligned}&\|F(x, u(x)) - F(x, v(x))\|_{\mathscr{C}^s} \\ &\quad \lesssim \|F\|_{\mathscr{C}^{s,t}}(\|u - v\|_{\mathscr{C}^s} + (\|u\|_{\mathscr{C}^s} + \|v\|_{\mathscr{C}^s})\|u - v\|_{L^\infty})(1 + \|u\|_{L^\infty} + \|v\|_{L^\infty})^{\lfloor s \rfloor + 1}.\end{aligned}$$

(vii) *Fix* $L \in \mathbb{N}_+$ *and suppose* $\partial_\zeta^\beta F(x, 0) \equiv 0$, $\forall |\beta| < L$. *If* $t > \lfloor s \rfloor + 2 + L$, *then*

$$\|F(x, u(x))\|_{\mathscr{C}^s} \lesssim \|F\|_{\mathscr{C}^{s,t}}(1 + \|u\|_{L^\infty})^{\lfloor s \rfloor + 2}\|u\|_{L^\infty}^{L-1}\|u\|_{\mathscr{C}^s}.$$

(viii) *Fix* $L \in \mathbb{N}$, $L \geq 2$ *and suppose* $\partial_\zeta^\beta F(x, 0) \equiv 0$, $\forall 0 < |\beta| \leq L$ *and* $t > \lfloor s \rfloor + 2 + L$. *Then*

$$\begin{aligned}\|F(x, u(x)) - F(x, v(x))\|_{\mathscr{C}^s} &\lesssim \|F\|_{\mathscr{C}^{s,t}}(1 + \|u\|_{L^\infty} + \|v\|_{L^\infty})^{\lfloor s \rfloor + 2} \\ &\qquad \times (\|u\|_{L^\infty} + \|v\|_{L^\infty})^{L-2}(\|u\|_{\mathscr{C}^s} + \|v\|_{\mathscr{C}^s})\|u - v\|_{L^\infty} \\ &\qquad + \|F\|_{\mathscr{C}^{s,t}}(\|u\|_{L^\infty} + \|v\|_{L^\infty})^{L-1}\|u - v\|_{\mathscr{C}^s}.\end{aligned}$$

The implicit constants do not depend on F, u, v, *or* w, *but may depend on* s, t, n, *and* N.

Remark 2.6.3. Of particular interest is Theorem 2.6.2 (vi), which, roughly, says the map $u \mapsto F(x, u(x))$ is Lipschitz in u. This will allow us to use arguments similar to the contraction mapping principle when studying fully nonlinear elliptic PDEs.

Remark 2.6.4. Theorem 2.6.2 assumes more regularity than is necessary in the ζ variable (i. e., the results are not sharp in t). We proceed with these non-sharp results because the proof of Theorem 2.6.2 easily generalizes to the more general situation we consider in Section 7.5. Understanding the sharp regularity needed in the ζ variable can be somewhat complicated; see, e. g., [14].

Remark 2.6.5. All the estimates in Theorem 2.6.2 are *tame* in the sense that each term on the right-hand side of the inequalities only has one factor of $\|\cdot\|_{\mathscr{C}^s(W,\vec{d})}$; all the other factors are $\|\cdot\|_{L^\infty}$. Tame estimates are an important tool in the field of nonlinear PDEs (see, for example, [105]) and are an essential tool in our study of fully nonlinear PDEs.

Before we prove Theorem 2.6.2, we state (without proof) several basic properties of $\mathscr{C}^{s,t}(\mathbb{R}^n \times \mathbb{R}^N)$; the proofs are very similar to the corresponding proofs for $\mathscr{C}^s(\mathbb{R}^n)$.

Proposition 2.6.6. *For $s, t > 0$, $\mathscr{C}^{s,t}(\mathbb{R}^n \times \mathbb{R}^N)$ is a Banach space when endowed with the norm $\|\cdot\|_{\mathscr{C}^{s,t}(\mathbb{R}^n\times\mathbb{R}^N)}$. Neither the space $\mathscr{C}^{s,t}(\mathbb{R}^n \times \mathbb{R}^N)$ nor the equivalence class of $\|\cdot\|_{\mathscr{C}^{s,t}(\mathbb{R}^n\times\mathbb{R}^N)}$ depends on the choices we made when defining D_j and $\widetilde{D}_l$.*

Proposition 2.6.7. *Let $s, t > 0$. For $F \in \mathscr{S}(\mathbb{R}^n \times \mathbb{R}^N)'$, the following are equivalent:*
(i) *We have $F \in \mathscr{C}^{s,t}(\mathbb{R}^n \times \mathbb{R}^N)$.*
(ii) *There exists a sequence $\{F_{j,l}\}_{j,l\in\mathbb{N}} \subset C^\infty(\mathbb{R}^n \times \mathbb{R}^N)$ such that for all multi-indices α, β,*

$$\sup_{j,l\in\mathbb{N}} 2^{js}2^{lt}\|(2^{-j}\partial_x)^\alpha(2^{-l}\partial_x)^\beta F_{j,k}\|_{L^\infty(\mathbb{R}^n\times\mathbb{R}^N)} < \infty,$$

and $F = \sum_{j,l\in\mathbb{N}} F_{j,l}$ with convergence in $C(\mathbb{R}^n \times \mathbb{R}^N)$.

In this case, there exist $M = M(s,t) \in \mathbb{N}$ and $C_{s,t} \geq 0$ such that

$$\|F\|_{\mathscr{C}^{s,t}(\mathbb{R}^n\times\mathbb{R}^N)} \leq C_{s,t} \sum_{|\alpha|,|\beta|\leq M} \sup_{j,l\in\mathbb{N}} 2^{js}2^{lt}\|(2^{-j}\partial_x)^\alpha(2^{-l}\partial_x)^\beta F_{j,k}\|_{L^\infty(\mathbb{R}^n\times\mathbb{R}^N)}.$$

Furthermore, $F_{j,l}$ can be chosen such that $\forall \alpha, \beta$, there exists $C_{s,t,\alpha,\beta} \geq 0$,

$$\begin{aligned}\sup_{j,l\in\mathbb{N}} 2^{js}2^{lt}&\|(2^{-j}\partial_x)^\alpha(2^{-l}\partial_x)^\beta F_{j,k}\|_{L^\infty(\mathbb{R}^n\times\mathbb{R}^N)} \\ &\leq C_{s,t,\alpha,\beta}\|F\|_{\mathscr{C}^{s,t}(\mathbb{R}^n\times\mathbb{R}^N)}.\end{aligned} \tag{2.58}$$

Proposition 2.6.8. *Let $s > 0$, let $\beta \in \mathbb{N}^N$ be a multi-index, and let $t > |\beta|$. Then $F(x,\zeta) \mapsto \partial_\zeta^\beta F(x,\zeta)$ is continuous, $\mathscr{C}^{s,t}(\mathbb{R}^n \times \mathbb{R}^N) \to \mathscr{C}^{s,t-|\beta|}(\mathbb{R}^n \times \mathbb{R}^N)$.*

Lemma 2.6.9. *For $s, t > 0$, we have:*
(i) *The map $F(x,\zeta) \mapsto F(x,0)$ is continuous, $\mathscr{C}^{s,t}(\mathbb{R}^n \times \mathbb{R}^N) \to \mathscr{C}^s(\mathbb{R}^n)$.*
(ii) *The map $F(x,\zeta_1,\ldots,\zeta_N) \mapsto F(x,\zeta_1,\ldots,\zeta_{N-1},0)$ is continuous, $\mathscr{C}^{s,t}(\mathbb{R}^n \times \mathbb{R}^N) \to \mathscr{C}^{s,t}(\mathbb{R}^n \times \mathbb{R}^{N-1})$.*
(iii) *For $x_0 \in \mathbb{R}^n$, the map $F(x,\zeta) \mapsto F(x_0,\zeta)$ is continuous, $\mathscr{C}^{s,t}(\mathbb{R}^n \times \mathbb{R}^N) \to \mathscr{C}^t(\mathbb{R}^N)$.*

Proof. This follows easily from Propositions 2.5.16 and 2.6.7. □

With the above facts in hand, we turn to sketching the proof of Theorem 2.6.2. In Section 7.5 we repeat this proof in a more difficult setting and include all the details.

Lemma 2.6.10. *Let $s, t > 0$ and set $M_1 := \lfloor s \rfloor + 1$ and $M_2 := \lfloor t \rfloor + 1$. Suppose $\forall j, k, l \in \mathbb{N}$, we are given $H_{j,k,l}(x,\zeta) \in C^\infty(\mathbb{R}^n \times \mathbb{R}^N)$ such that $\forall |\alpha| \leq M_1$ and $|\beta| \leq M_2$,*

$$\|(2^{-(k\vee j)}\partial_x)^\alpha(2^{-l}\partial_\zeta)^\beta H_{j,k,l}\|_{L^\infty(\mathbb{R}^n\times\mathbb{R}^N)} \leq A2^{-ks}2^{-js}2^{-lt},$$

for some $A \geq 0$. Set $H(x,\zeta) := \sum_{j,k,l\in\mathbb{N}} H_{j,k,l}(x,\zeta)$. Then $H \in \mathscr{C}^{s,t}(\mathbb{R}^n \times \mathbb{R}^N)$ and

$$\|H\|_{\mathscr{C}^{s,t}(\mathbb{R}^n\times\mathbb{R}^N)} \leq C_{s,t,n,N}A, \tag{2.59}$$

for some $C_{s,t,n,N} \geq 0$.

Proof. We will show for $j, k, k', l, l' \in \mathbb{N}$ that

$$\|\widehat{D}_{k,l}H_{j,k',l'}\|_{L^\infty(\mathbb{R}^n\times\mathbb{R}^N)} \lesssim A2^{-M_1((j\vee k\vee k')-(j\vee k'))}2^{-M_2((l\vee l')-l')}2^{-js}2^{-k's}2^{-l't}. \tag{2.60}$$

First we see why (2.60) will complete the proof. Indeed, using (2.60), we have

$$\begin{aligned}\|\widehat{D}_{k,l}H\|_{L^\infty(\mathbb{R}^n\times\mathbb{R}^N)} &\leq \sum_{j,k',l'\in\mathbb{N}} \|\widehat{D}_{k,l}H_{j,k',l'}\|_{L^\infty}\\ &\lesssim A \sum_{j,k',l'\in\mathbb{N}} 2^{-M_1((j\vee k\vee k')-(j\vee k'))}2^{-M_2((l\vee l')-l')}2^{-js}2^{-k's}2^{-l't} \lesssim A2^{-ks}2^{-lt},\end{aligned}$$

where the final estimate is elementary and uses $M_1 > s$ and $M_2 > t$ – see Lemma 7.3.2. This estimate is equivalent to (2.59), which then completes the proof.

We turn to the proof of (2.60), which is separated into several cases. Using Proposition 2.3.18 (d) we may write

$$D_k = \sum_{|\alpha|\leq M_1} 2^{-k(M_1-|\alpha|)}E_{k,\alpha}(2^{-k}\partial_x)^\alpha, \tag{2.61}$$

$$\widetilde{D}_l = \sum_{|\beta|\leq M_1} 2^{-l(M_2-|\beta|)}E_{l,\beta}(2^{-l}\partial_\zeta)^\beta, \tag{2.62}$$

where $\{(E_{k,\alpha}, 2^{-k}) : k \in \mathbb{N}, |\alpha| \leq M_1\}$ and $\{(\widetilde{E}_{l,\beta}, 2^{-l}) : l \in \mathbb{N}, |\beta| \leq M_2\}$ are bounded sets of elementary operators on $\mathbb{R}^n$ and $\mathbb{R}^N$, respectively. In particular, using Lemma 2.3.26, we have

$$\|D_k\|_{L^\infty\to L^\infty}, \|\widetilde{D}_l\|_{L^\infty\to L^\infty}, \|E_{k,\alpha}\|_{L^\infty\to L^\infty}, \|\widetilde{E}_{l,\beta}\|_{L^\infty\to L^\infty} \lesssim 1, \tag{2.63}$$

$\forall k, l \in \mathbb{N}$, $|\alpha| \leq M_1$, $|\beta| \leq M_2$.

First, we consider the case where $k' \vee j = k \vee k' \vee j$ and $l' = l \vee l'$. Thus, using (2.63) and the fact that $\widehat{D}_{k,l} = D_k \otimes \widetilde{D}_l$, we have

$$\|\widehat{D}_{k,l}H_{j,k',l'}\|_{L^\infty} \lesssim \|H_{j,k',l'}\|_{L^\infty} \leq A2^{-js}2^{-k's}2^{-l't},$$

establishing (2.60) in this case.

We next consider the case where $k > k' \vee j$ but $l \leq l'$. Using (2.61) and (2.63), we have

$$\begin{aligned}\|\widehat{D}_{k,l}H_{j,k',l'}\|_{L^\infty} &\leq \sum_{|\alpha|\leq M_1} 2^{-(M_1-|\alpha|)k}\|(E_{k,\alpha}\otimes\widetilde{D}_l)(2^{-k}\partial_x)^\alpha H_{j,k',l'}\|_{L^\infty}\\ &\lesssim \sum_{|\alpha|\leq M_1} 2^{-(M_1-|\alpha|)k}\|(2^{-k}\partial_x)^\alpha H_{j,k',l'}\|_{L^\infty}\\ &\lesssim \sum_{|\alpha|\leq M_1} 2^{-(M_1-|\alpha|)k}2^{-|\alpha|(k-(k'\vee j))}\|(2^{-(k'\vee j)}\partial_x)^\alpha H_{j,k',l'}\|_{L^\infty}\\ &\lesssim A2^{-M_1(k-(k'\vee j))}2^{-k's}2^{-js}2^{-l't},\end{aligned}$$

which establishes (2.60) in this case.

We next consider the case where $k > k' \vee j$ and $l > l'$. Using (2.61), (2.62), and (2.63), we have

$$\begin{aligned}&\|\widehat{D}_{k,l}H_{j,k',l'}\|_{L^\infty}\\ &\leq \sum_{\substack{|\alpha|\leq M_1\\|\beta|\leq M_2}} 2^{-(M_1-|\alpha|)k}2^{-l(M_2-|\beta|)}\|(E_{k,\alpha}\otimes\widetilde{E}_{l,\beta})(2^{-k}\partial_x)^\alpha(2^{-l}\partial_\zeta)^\beta H_{j,k',l'}\|_{L^\infty}\\ &\lesssim \sum_{\substack{|\alpha|\leq M_1\\|\beta|\leq M_2}} 2^{-(M_1-|\alpha|)k}2^{-l(M_2-|\beta|)}\|(2^{-k}\partial_x)^\alpha(2^{-l}\partial_\zeta)^\beta H_{j,k',l'}\|_{L^\infty}\\ &= \sum_{\substack{|\alpha|\leq M_1\\|\beta|\leq M_2}} 2^{-(M_1-|\alpha|)k}2^{-l(M_2-|\beta|)}2^{-|\alpha|(k-(k'\vee j))}2^{-|\beta|(l-l')}\|(2^{-(k'\vee j)}\partial_x)^\alpha(2^{-l'}\partial_\zeta)^\beta H_{j,k',l'}\|_{L^\infty}\\ &\lesssim A2^{-M_1(k-(k'\vee j))}2^{-M_2(l-l')}2^{-k's}2^{-js}2^{-l't},\end{aligned}$$

which establishes (2.60) in this case.

Finally, we consider the case where $k \leq k' \vee j$ but $l > l'$. Using (2.62) and (2.63), we have

$$\begin{aligned}\|\widetilde{D}_{k,l}H_{j,k',l'}\|_{L^\infty} &\leq \sum_{|\beta|\leq M_2} 2^{-(M_2-|\beta|)l}\|(D_k\otimes\widetilde{E}_{l,\beta})(2^{-l}\partial_\zeta)^\beta H_{j,k',l'}\|_{L^\infty}\\ &\lesssim \sum_{|\beta|\leq M_2} 2^{-(M_2-|\beta|)l}\|(2^{-l}\partial_\zeta)^\beta H_{j,k',l'}\|_{L^\infty}\\ &\lesssim \sum_{|\beta|\leq M_2} 2^{-(M_2-|\beta|)l}2^{-|\beta|(l-l')}\|(2^{-l'}\partial_\zeta)^\beta H_{j,k',l'}\|_{L^\infty}\\ &\lesssim A2^{-M_2(l-l')}2^{-js}2^{-k's}2^{-l't},\end{aligned}$$

establishing (2.60) in this case, completing the proof. □

Proof of Theorem 2.6.2. (i): We begin with the proof of (i), which contains most of the difficulties. To prove this, we wish to decompose $F(x, u(x) + \zeta)$ as a sum as in Lemma 2.6.10. First, we decompose $F = \sum_{k,l\in\mathbb{N}} F_{k,l}$ as in Proposition 2.6.7, where $F_{k,l}$ satisfy (2.58). Next, take $P_J := \sum_{j\leq J} D_j$ so that $P_J f = f * \mathrm{Dil}_{2^J}(\phi)$ (see the definition of D_j in (2.30)). In particular, for $u \in \mathscr{C}^s(\mathbb{R}^n)$, we have $P_J u \to u$ in $C(\mathbb{R}^n)$. The decomposition of $F(x, u(x) + \zeta)$ we use is given by

$$\begin{aligned}
F(x, u(x) + \zeta) &= \lim_{J\to\infty} F(x, P_J u(x) + \zeta) \\
&= F(x, D_0 u(x) + \zeta) + \sum_{j=1}^{\infty} (F(x, P_j u(x) + \zeta) - F(x, P_{j-1} u(x) + \zeta)) \\
&= \sum_{k,l\in\mathbb{N}} F_{k,l}(x, D_0 u(x) + \zeta) \\
&\quad + \sum_{k,l\in\mathbb{N}} \sum_{j=1}^{\infty} (F_{k,l}(x, P_j u(x) + \zeta) - F_{k,l}(x, P_{j-1} u(x) + \zeta)) \\
&=: \sum_{j,k,l\in\mathbb{N}} H_{j,k,l}(x, \zeta),
\end{aligned}$$

where

$$H_{j,k,l} = \begin{cases} F_{k,l}(x, D_0 u(x) + \zeta), & j = 0, \\ F_{k,l}(x, P_j u(x) + \zeta) - F_{k,l}(x, P_{j-1} u(x) + \zeta), & j \geq 1. \end{cases}$$

To prove (i), we will show

$$\begin{aligned}
&\|(2^{-(k\vee j)}\partial_x)^\alpha (2^{-l}\partial_\zeta)^\beta H_{j,k,l}\|_{L^\infty} \\
&\quad \leq C_{\alpha,\beta} \|F\|_{\mathscr{C}^{s,t}(\mathbb{R}^n\times\mathbb{R}^N)} 2^{-ks} 2^{-js} 2^{-l(t-|\alpha|-1)} (1 + \|u\|_{\mathscr{C}^s})(1 + \|u\|_{L^\infty})^{|\alpha|}.
\end{aligned} \tag{2.64}$$

To see why (2.64) completes the proof of (i), set $M_1 := \lfloor s \rfloor + 1$ and $M_2 := \lfloor t \rfloor - \lfloor s \rfloor - 1$. We have by (2.64), $\forall |\alpha| \leq M_1$, $\forall |\beta| \leq M_2$,

$$\begin{aligned}
&\|(2^{-(k\vee j)}\partial_x)^\alpha (2^{-l}\partial_\zeta)^\beta H_{j,k,l}\|_{L^\infty} \\
&\quad \lesssim \|F\|_{\mathscr{C}^{s,t}(\mathbb{R}^n\times\mathbb{R}^N)} 2^{-ks} 2^{-js} 2^{-l(t-\lfloor s \rfloor-2)} (1 + \|u\|_{\mathscr{C}^s})(1 + \|u\|_{L^\infty})^{\lfloor s \rfloor+1}.
\end{aligned}$$

This shows that $H_{j,k,l}$ satisfies the assumptions of Lemma 2.6.10 with t replaced by $t - \lfloor s \rfloor - 2$. The conclusion of Lemma 2.6.10 gives (i).

We turn to proving (2.64). When $j = 0$, proving (2.64) is straightforward, and we leave it to the reader. When $j \geq 1$, we use

$$\begin{aligned}
H_{j,k,l}(x, \zeta) &= F_{k,l}(x, P_j u(x) + \zeta) - F_{k,l}(x, P_{j-1} u(x) + \zeta) \\
&= \int_0^1 \partial_\sigma (F_{k,l}(x, (1-\sigma) P_j u(x) + \sigma P_{j-1} u(x) + \zeta))\, d\sigma \\
&= \int_0^1 d_\zeta F_{k,l}(x, (1-\sigma) P_j u(x) + \sigma P_{j-1} u(x) + \zeta)\, d\sigma D_{j-1} u(x).
\end{aligned}$$

Thus, to prove (2.64) it suffices to show that

$$\begin{aligned}&\sup_{\sigma\in[0,1]} \|(2^{-(k\vee j)}\partial_x)^\alpha(2^{-l}\partial_\zeta)^\beta(d_\zeta F_{k,l})(x,(1-\sigma)P_ju(x)+\sigma P_{j-1}u(x)+\zeta)D_{j-1}u(x)\|_{L^\infty}\\&\quad\lesssim \|F\|_{\mathscr{C}^{s,t}(\mathbb{R}^n\times\mathbb{R}^N)}2^{-ks}2^{-js}2^{-l(t-|\alpha|-1)}(1+\|u\|_{\mathscr{C}^s})(1+\|u\|_{L^\infty})^{|\alpha|}.\end{aligned} \tag{2.65}$$

Proving the estimate (2.65) is straightforward, and we leave it to the reader (see the proof of the analogous result Theorem 7.5.2).

(ii): Since $F(x,u(x)) = F(x,u(x)+\zeta)|_{\zeta=0}$, (ii) follows by combining (i) and Lemma 2.6.9 (i).

(iii): We first consider the case where $\|u\|_{L^\infty} = \|v\|_{L^\infty} = 1$. Fix $\phi(\zeta_1,\zeta_2)\in C_0^\infty(\mathbb{R}^N\times\mathbb{R}^N)$ with $\phi(\zeta_1,\zeta_2)\equiv 1$ if $|\zeta_1|,|\zeta_2|\le 1$. Set $F(x,\zeta) := \phi(\zeta_1,\zeta_2)\zeta_1\cdot\zeta_2 \in C_0^\infty(\mathbb{R}^{2N})$. Since $\phi\in C_0^\infty(\mathbb{R}^{2N})$, it follows that $F(x,\zeta)\in\mathscr{C}^{s,t}(\mathbb{R}^n\times\mathbb{R}^{2N})$, $\forall t>0$. Using the fact that $\|u\|_{L^\infty} = \|v\|_{L^\infty} = 1$, we have $u(x)\cdot v(x) = F(u(x),v(x))$ and (ii) implies

$$\begin{aligned}\|u\cdot v\|_{\mathscr{C}^s} &= \|F(x,u(x),v(x))\|_{\mathscr{C}^s}\\&\lesssim \|F\|_{\mathscr{C}^{s,\lfloor s\rfloor+3}}(1+\|u\|_{\mathscr{C}^s}+\|v\|_{\mathscr{C}^s})(1+\|u\|_{L^\infty}+\|v\|_{L^\infty})\\&\lesssim 1+\|u\|_{\mathscr{C}^s}+\|v\|_{\mathscr{C}^s}.\end{aligned} \tag{2.66}$$

For general (not identically zero) $u,v\in\mathscr{C}^s(\mathbb{R}^n;\mathbb{R}^N)$, we apply (2.66) to $u/\|u\|_{L^\infty}$ and $v/\|v\|_{L^\infty}$ to see

$$\begin{aligned}\|u\cdot v\|_{\mathscr{C}^s} &\lesssim \|u\|_{L^\infty}\|v\|_{L^\infty}+\|u\|_{\mathscr{C}^s}\|v\|_{L^\infty}+\|u\|_{L^\infty}\|v\|_{\mathscr{C}^s}\\&\lesssim \|u\|_{\mathscr{C}^s}\|v\|_{L^\infty}+\|u\|_{L^\infty}\|v\|_{\mathscr{C}^s},\end{aligned}$$

where the final inequality uses $\|u\|_{L^\infty}\lesssim\|u\|_{\mathscr{C}^s}$.

(iv): First we prove the result when $\|w\|_{L^\infty}=1$. Let $\phi(\zeta')\in C_0^\infty(\mathbb{R})$ satisfy $\phi\equiv 1$ for $|\zeta'|\le 1$. Set

$$\widetilde{F}(x,\zeta,\zeta') := F(x,\zeta)\phi(\zeta')\zeta'.$$

It follows easily from Proposition 2.6.7 that $\widetilde{F}\in\mathscr{C}^{s,t}(\mathbb{R}^n\times\mathbb{R}^{N+1})$ and $\|\widetilde{F}\|_{\mathscr{C}^{s,t}(\mathbb{R}^n\times\mathbb{R}^{N+1})}\lesssim \|F\|_{\mathscr{C}^{s,t}(\mathbb{R}^n\times\mathbb{R}^N)}$.

When $\|w\|_{L^\infty}=1$, we have $F(x,u(x)+\zeta)w(x) = \widetilde{F}(x,u(x)+\zeta,w(x)+\zeta')|_{\zeta'=0}$. Thus, applying (i) and Lemma 2.6.9 (ii), we have

$$\begin{aligned}&\|F(x,u(x)+\zeta)w(x)\|_{\mathscr{C}^{s,t-\lfloor s\rfloor-2}(\mathbb{R}^n\times\mathbb{R}^N)}\\&\quad= \|\widetilde{F}(x,u(x)+\zeta,w(x)+\zeta')|_{\zeta'=0}\|_{\mathscr{C}^{s,t-\lfloor s\rfloor-2}(\mathbb{R}^n\times\mathbb{R}^N)}\\&\quad\lesssim \|\widetilde{F}(x,u(x)+\zeta,w(x)+\zeta')\|_{\mathscr{C}^{s,t-\lfloor s\rfloor-2}(\mathbb{R}^n\times\mathbb{R}^{N+1})}\\&\quad\lesssim \|F\|_{\mathscr{C}^{s,t}}(1+\|u\|_{\mathscr{C}^s}+\|w\|_{\mathscr{C}^s})(1+\|u\|_{L^\infty}+\|w\|_{L^\infty})^{\lfloor s\rfloor+1}\\&\quad\lesssim \|F\|_{\mathscr{C}^{s,t}}(1+\|u\|_{\mathscr{C}^s}+\|w\|_{\mathscr{C}^s})(1+\|u\|_{L^\infty})^{\lfloor s\rfloor+1}.\end{aligned} \tag{2.67}$$

For general, not identically zero, $w \in \mathscr{C}^s(\mathbb{R}^n)$, we apply (2.67) with w replaced by $w/\|w\|_{L^\infty}$ to see

$$\begin{aligned}
&\|F(x, u(x) + \zeta)w(x)\|_{\mathscr{C}^{s,t-\lfloor s\rfloor-2}(\mathbb{R}^n\times\mathbb{R}^N)} \\
&\quad\lesssim \|F\|_{\mathscr{C}^{s,t}}\big(\|w\|_{L^\infty} + \|w\|_{L^\infty}\|u\|_{\mathscr{C}^s} + \|w\|_{\mathscr{C}^s}\big)\big(1 + \|u\|_{L^\infty}\big)^{\lfloor s\rfloor+1} \\
&\quad\lesssim \|F\|_{\mathscr{C}^{s,t}}\big(\|w\|_{L^\infty}\|u\|_{\mathscr{C}^s} + \|w\|_{\mathscr{C}^s}\big)\big(1 + \|u\|_{L^\infty}\big)^{\lfloor s\rfloor+1},
\end{aligned}$$

where the final inequality used $\|w\|_{L^\infty} \lesssim \|w\|_{\mathscr{C}^s}$.

(v): By Proposition 2.6.8, the components of the vector $d_\zeta F$ satisfy the same assumptions as F, but with t replaced by $t-1$. Using (iv), we have

$$\begin{aligned}
&\|F(x, u(x) + \zeta) - F(x, v(x) + \zeta)\|_{\mathscr{C}^{s,t-\lfloor s\rfloor-3}} \\
&\quad= \left\| \int_0^1 d_\zeta F(x, \sigma u(x) + (1-\sigma)v(x) + \zeta)(u(x) - v(x))\, d\sigma \right\|_{\mathscr{C}^{s,t-\lfloor s\rfloor-3}} \\
&\quad\le \int_0^1 \|d_\zeta F(x, \sigma u(x) + (1-\sigma)v(x) + \zeta)(u(x) - v(x))\|_{\mathscr{C}^{s,t-\lfloor s\rfloor-3}}\, d\sigma \\
&\quad\le \sup_{\sigma\in[0,1]} \|d_\zeta F(x, \sigma u(x) + (1-\sigma)v(x) + \zeta)(u(x) - v(x))\|_{\mathscr{C}^{s,t-\lfloor s\rfloor-3}} \\
&\quad\lesssim \|d_\zeta F\|_{\mathscr{C}^{s,t-1}}\big(\|u-v\|_{\mathscr{C}^s} + \|\sigma u + (1-\sigma)v\|_{\mathscr{C}^s}\|u-v\|_{L^\infty}\big) \\
&\qquad\times \big(1 + \|\sigma u + (1-\sigma)v\|_{L^\infty}\big)^{\lfloor s\rfloor+1} \\
&\quad\lesssim \|F\|_{\mathscr{C}^{s,t}}\big(\|u-v\|_{\mathscr{C}^s} + (\|u\|_{\mathscr{C}^s} + \|v\|_{\mathscr{C}^s})\|u-v\|_{L^\infty}\big) \\
&\qquad\times \big(1 + \|u\|_{L^\infty} + \|v\|_{L^\infty}\big)^{\lfloor s\rfloor+1},
\end{aligned}$$

where the last inequality uses Proposition 2.6.8.

(vi): Item (vi) follows by combining (v) and Lemma 2.6.9 (i).

(vii): We prove (vii) by induction on L and begin with the base case, $L = 1$. When $L = 1$, using the fact that $F(x, 0) \equiv 0$, we have $F(x, u(x)) = F(x, u(x)) - F(x, 0)$. From here, the result when $L = 1$ follows from the case $v = 0$ of (vi).

We assume (vii) for some $L \in \mathbb{N}_+$ and prove it for $L+1$; thus we assume the hypotheses with L replaced by $L+1$. Using the fact that $F(x, 0) \equiv 0$ and using (iii), we have

$$\begin{aligned}
\|F(x, u(x))\|_{\mathscr{C}^s} &= \|F(x, u(x)) - F(x, v(x))\|_{\mathscr{C}^s} \\
&= \left\| \int_0^1 d_\zeta F(x, \sigma u(x))u(x)\, d\sigma \right\|_{\mathscr{C}^s} \\
&\le \int_0^1 \|d_\zeta F(x, \sigma u(x))u(x)\|_{\mathscr{C}^s}\, d\sigma \qquad (2.68)
\end{aligned}$$

$$\begin{aligned}
&\lesssim \sup_{\sigma\in[0,1]} \left\| d_\zeta F(x,\sigma u(x))u(x) \right\|_{\mathscr{C}^s} \\
&\lesssim \sup_{\sigma\in[0,1]} \left\| d_\zeta F(x,\sigma u(x)) \right\|_{\mathscr{C}^s} \|u\|_{L^\infty} + \sup_{\sigma\in[0,1]} \|d_\zeta F(x,\sigma u(x))\|_{L^\infty} \|u\|_{\mathscr{C}^s} \\
&=: (\mathrm{I}) + (\mathrm{II}),
\end{aligned}$$

where we applied (iii) to the various summands of the matrix multiplication $d_\zeta F(x,\sigma u(x))u(x)$.

To bound (I) we use the inductive hypothesis applied to $d_\zeta F$ with t replaced by $t-1 > \lfloor s \rfloor + 2 + L$ and Proposition 2.6.8 to see

$$\begin{aligned}
(\mathrm{I}) &= \sup_{\sigma\in[0,1]} \left\| d_\zeta F(x,\sigma u(x)) \right\|_{\mathscr{C}^s} \|u\|_{L^\infty} \\
&\lesssim \sup_{\sigma\in[0,1]} \|d_\zeta F\|_{\mathscr{C}^{s,t-1}} (1 + \|\sigma u\|_{L^\infty})^{\lfloor s \rfloor + 2} \|\sigma u\|_{L^\infty}^{L-1} \|\sigma u\|_{\mathscr{C}^s} \|u\|_{L^\infty} \qquad (2.69) \\
&\lesssim \|F\|_{\mathscr{C}^{s,t}} (1 + \|u\|_{L^\infty})^{\lfloor s \rfloor + 2} \|u\|_{L^\infty}^{L} \|u\|_{\mathscr{C}^s}.
\end{aligned}$$

For (II) we use the fact that $\partial_\zeta^\beta d_\zeta F(x,0) \equiv 0$, $\forall |\beta| < L$, the fact that $t > L+1$, and Proposition 2.6.8 to see that

$$\begin{aligned}
\sup_{\sigma\in[0,1]} \left\| d_\zeta F(x,\sigma u(x)) \right\|_{L^\infty} &\lesssim \sup_{\sigma\in[0,1]} \sum_{|\beta|=L} \left\| \partial_\zeta^L d_\zeta F(x,\zeta) \right\|_{L^\infty} \|\sigma u(x)\|_{L^\infty}^L \\
&\lesssim \|F\|_{\mathscr{C}^{s,t}} \|u\|_{L^\infty}^L.
\end{aligned}$$

We conclude that

$$(\mathrm{II}) \lesssim \|F\|_{\mathscr{C}^{s,t}} \|u\|_{L^\infty}^L \|u\|_{\mathscr{C}^s}. \qquad (2.70)$$

Using (2.69) and (2.70) to bound the right-hand side of (2.68) completes the inductive step.

(viii): We have, using (iii),

$$\begin{aligned}
\|F(x,u(x)) - F(x,v(x))\|_{\mathscr{C}^s} &= \left\| \int_0^1 d_\zeta F(x,\sigma u(x) + (1-\sigma)v(x))(u(x)-v(x))\, d\sigma \right\|_{\mathscr{C}^s} \\
&\le \int_0^1 \left\| d_\zeta F(x,\sigma u(x) + (1-\sigma)v(x))(u(x)-v(x)) \right\|_{\mathscr{C}^s} d\sigma \\
&\le \sup_{\sigma\in[0,1]} \left\| d_\zeta F(x,\sigma u(x) + (1-\sigma)v(x))(u(x)-v(x)) \right\|_{\mathscr{C}^s} \qquad (2.71) \\
&\lesssim \sup_{\sigma\in[0,1]} \|d_\zeta F(x,\sigma u(x) + (1-\sigma)v(x))\|_{\mathscr{C}^s} \|u-v\|_{L^\infty} \\
&\quad + \sup_{\sigma\in[0,1]} \|d_\zeta F(x,\sigma u(x) + (1-\sigma)v(x))\|_{L^\infty} \|u-v\|_{\mathscr{C}^s} \\
&=: (\mathrm{III}) + (\mathrm{IV}),
\end{aligned}$$

where we applied (iii) to the various summands of the matrix multiplication $d_\zeta F(x, \sigma u(x) + (1-\sigma)v(x))(u(x) - v(x))$.

For (III), we have $\partial_\zeta^\beta d_\zeta F(x,0) \equiv 0, \forall |\beta| \leq L-1$, and we may therefore apply (vii) with L replaced by $L-1$ and t replaced by $t-1$ to see

$$\begin{aligned}
&\sup_{\sigma\in[0,1]} \|d_\zeta F(x, \sigma u(x) + (1-\sigma)v(x))\|_{\mathscr{C}^s} \\
&\quad\lesssim \sup_{\sigma\in[0,1]} \|d_\zeta F\|_{\mathscr{C}^{s,t-1}} (1 + \|\sigma u + (1-\sigma)v\|_{L^\infty})^{\lfloor s\rfloor+2} \|\sigma u + (1-\sigma)v\|_{L^\infty}^{L-2} \|\sigma u + (1-\sigma)v\|_{\mathscr{C}^s} \\
&\quad\lesssim \|F\|_{\mathscr{C}^{s,t}} (1 + \|u\|_{L^\infty} + \|v\|_{L^\infty})^{\lfloor s\rfloor+2} (\|u\|_{L^\infty} + \|v\|_{L^\infty})^{L-2} (\|u\|_{\mathscr{C}^s} + \|v\|_{\mathscr{C}^s}),
\end{aligned}$$

where the final estimate used Proposition 2.6.8. It follows that

$$(\mathrm{III}) \lesssim \|F\|_{\mathscr{C}^{s,t}} (1 + \|u\|_{L^\infty} + \|v\|_{L^\infty})^{\lfloor s\rfloor+\nu+1} (\|u\|_{L^\infty} + \|v\|_{L^\infty})^{L-2} (\|u\|_{\mathscr{C}^s} + \|v\|_{\mathscr{C}^s}) \|u - v\|_{L^\infty}. \tag{2.72}$$

For (IV), we use the fact that $\partial_\zeta^\beta d_\zeta f(x,0) \equiv 0, \forall |\beta| < L-1$ to see

$$\begin{aligned}
&\sup_{\sigma\in[0,1]} \|d_\zeta F(x, \sigma u(x) + (1-\sigma)v(x))\|_{L^\infty} \\
&\quad\lesssim \sup_{\sigma\in[0,1]} \sum_{|\beta|=L-1} \|\partial_\zeta^\beta d_\zeta F(x, \sigma u(x) + (1-\sigma)v(x))\|_{L^\infty} \|\sigma u + (1-\sigma)v\|_{L^\infty}^{L-1} \\
&\quad\lesssim \|F\|_{\mathscr{C}^{s,t}} (\|u\|_{L^\infty} + \|v\|_{L^\infty})^{L-1},
\end{aligned}$$

where in the last inequality we used $t > L$ and Proposition 2.6.8. It follows that

$$(\mathrm{IV}) \lesssim \|F\|_{\mathscr{C}^{s,t}} (\|u\|_{L^\infty} + \|v\|_{L^\infty})^{L-1} \|u - v\|_{\mathscr{C}^s}. \tag{2.73}$$

Using (2.72) and (2.73) to bound the right-hand side of (2.71) completes the proof. □

2.7 Linear elliptic operators

Our main result about linear maximally subelliptic operators with smooth coefficients (Theorem 8.1.1) is a theorem giving several equivalent characterizations of such operators. Here, we state the special case of this theorem in the case of elliptic operators, a setting where the theorem is well known.

Fix a connected open set $\Omega \subseteq \mathbb{R}^n$, $\mathbf{D}_1, \mathbf{D}_2, \kappa \in \mathbb{N}_+$, and let

$$\mathscr{P} := \sum_{|\alpha|\leq\kappa} a_\alpha(x)\partial_x^\alpha, \quad a_\alpha \in C_{\mathrm{loc}}^\infty(\Omega; \mathbb{M}^{\mathbf{D}_1\times\mathbf{D}_2}(\mathbb{C})).$$

Definition 2.7.1. Let

$$P_\kappa(x,\xi) := \sum_{|\alpha|=\kappa} a_\alpha(x)(2\pi i)^{|\alpha|}\xi^\alpha. \tag{2.74}$$

We say $\mathscr{P}$ is elliptic near $x_0 \in \Omega$ if the $\mathbf{D}_1 \times \mathbf{D}_2$ matrix $P_\kappa(x_0, \xi)$ is left invertible $\forall \xi \neq 0$. Note that this requires $\mathbf{D}_1 \geq \mathbf{D}_2$. We say $\mathscr{P}$ is elliptic on Ω if $\mathscr{P}$ is elliptic near x_0, $\forall x_0 \in \Omega$.

Letting $\mathscr{P}^* \mathscr{P}$, where $\mathscr{P}^*$ is the formal adjoint of $\mathscr{P}$, we have

$$\mathscr{P}^* \mathscr{P} = \sum_{|\alpha| \leq 2\kappa} b_\alpha(x) \partial_x^\alpha, \quad b_\alpha \in C^\infty_{\mathrm{loc}}(\Omega; \mathbb{M}^{\mathbf{D}_2 \times \mathbf{D}_2}(\mathbb{C})).$$

It makes sense to ask whether $\mathscr{P}^* \mathscr{P}$ is elliptic, since it is a partial differential operator of order 2κ. We sometimes consider $\mathscr{P}^* \mathscr{P}$ as an unbounded operator on $L^2(\Omega)$, with dense domain $C_0^\infty(\Omega)$; then $\mathscr{P}^* \mathscr{P}$ is a non-negative symmetric operator.

For two operators $T, S \in \mathrm{Hom}(C_0^\infty(\Omega; \mathbb{C}^{\mathbf{D}_1}), C_0^\infty(\Omega; \mathbb{C}^{\mathbf{D}_2})')$, we write

$$T \equiv S \mod C^\infty_{\mathrm{loc}}(\Omega \times \Omega; \mathbb{M}^{\mathbf{D}_2 \times \mathbf{D}_1}(\mathbb{C})),$$

if the Schwartz kernel of $T - S$ is given by integration against a function in $C^\infty_{\mathrm{loc}}(\Omega \times \Omega; \mathbb{M}^{\mathbf{D}_2 \times \mathbf{D}_1}(\mathbb{C}))$, that is,

$$\int g(x) \cdot (T - S) f(x)\, dx = \iint g(x) \cdot (T - S)(x, y) f(y)\, dy,$$

$\forall f \in C_0^\infty(\Omega; \mathbb{C}^{\mathbf{D}_1})$, $\forall g \in C_0^\infty(\Omega; \mathbb{C}^{\mathbf{D}_2})$, where $T - S \in C^\infty_{\mathrm{loc}}(\Omega \times \Omega; \mathbb{M}^{\mathbf{D}_2 \times \mathbf{D}_1}(\mathbb{C}))$.

We now state a basic theorem regarding linear elliptic partial differential operators. All aspects of this theorem are well known, and some are trivial; we will later prove a generalization for maximally subelliptic operators (see Theorem 8.1.1). In this theorem, and in the rest of the text, for a matrix M, we write $|M|$ for the usual operator norm.

Theorem 2.7.2. *Let $\mathscr{P}$ be given by* (2.74). *The following are equivalent:*

(a) *$\mathscr{P}$ is elliptic on Ω.*

(b) *For every pre-compact, open set $\Omega_1 \Subset \Omega$, there exists $C_{\Omega_1} \geq 0$ such that $\forall f \in C_0^\infty(\Omega_1; \mathbb{C}^{\mathbf{D}_2})$,*

$$\sum_{j=1}^n \|\partial_{x_j}^\kappa f\|_{L^2(\Omega_1; \mathbb{C}^{\mathbf{D}_2})} \leq C_{\Omega_1} \big(\|\mathscr{P} f\|_{L^2(\Omega_1; \mathbb{C}^{\mathbf{D}_1})} + \|f\|_{L^2(\Omega_1; \mathbb{C}^{\mathbf{D}_2})} \big).$$

(c) *$\mathscr{P}_0 := \sum_{|\alpha| = \kappa} a_\alpha(x) \partial_x^\alpha$ is elliptic on Ω.*

(d) *The frozen coefficient operator $\sum_{|\alpha| \leq \kappa} a_\alpha(x_0) \partial_x^\alpha$ is elliptic on Ω, $\forall x_0 \in \Omega$.*

(e) *For all $\phi_1, \phi_2 \in C_0^\infty(\Omega)$ with $\phi_1 \prec \phi_2$, the following holds $\forall u \in C_0^\infty(\Omega; \mathbb{C}^{\mathbf{D}_2})'$:*

- *for $p, q \in [1, \infty]$ and $s \in \mathbb{R}$, if $\phi_2 \mathscr{P} u \in \mathscr{B}^s_{p,q}(\mathbb{R}^n)$, then $\phi_1 u \in \mathscr{B}^{s+\kappa}_{p,q}(\mathbb{R}^n)$, and for such u*

$$\|\phi_1 u\|_{\mathscr{B}^{s+\kappa}_{p,q}(\mathbb{R}^n)} \leq C_{s,p,q,\phi_1,\phi_2,\mathscr{P}} \big(\|\phi_2 \mathscr{P} u\|_{\mathscr{B}^s_{p,q}(\mathbb{R}^n)} + \|\phi_2 u\|_{\mathscr{B}^s_{p,q}(\mathbb{R}^n)} \big),$$

- *for $p \in (1, \infty)$, $q \in (1, \infty]$, and $s \in \mathbb{R}$, if $\phi_2 \mathscr{P} u \in \mathscr{F}^s_{p,q}(\mathbb{R}^n)$, then $\phi_1 u \in \mathscr{F}^{s+\kappa}_{p,q}(\mathbb{R}^n)$, and for such u*

$$\|\phi_1 u\|_{\mathscr{F}^{s+\kappa}_{p,q}(\mathbb{R}^n)} \le C_{s,p,q,\phi_1,\phi_2,\mathscr{P}}\big(\|\phi_2 \mathscr{P} u\|_{\mathscr{F}^{s}_{p,q}(\mathbb{R}^n)} + \|\phi_2 u\|_{\mathscr{F}^{s}_{p,q}(\mathbb{R}^n)}\big).$$

(f) *$\mathscr{P}^* \mathscr{P}$ is elliptic on Ω.*

(g) *There exists an operator $T \in \mathrm{Hom}(C_0^\infty(\Omega; \mathbb{C}^{\mathbf{D}_1}), C^\infty_{\mathrm{loc}}(\Omega; \mathbb{C}^{\mathbf{D}_2}))$ which is locally a singular integral operator of order $-\kappa$ (see Definitions* 2.3.30 *and* 2.3.1 *for this concept), such that*

$$T\mathscr{P} \equiv I \mod C^\infty_{\mathrm{loc}}(\Omega \times \Omega; \mathbb{M}^{\mathbf{D}_2 \times \mathbf{D}_2}(\mathbb{C})).$$

In fact, T is locally a pseudo-differential operator of order $-\kappa$ (see Definitions 2.3.30 *and* 2.2.3*).*

(h) *There exists an operator $S \in \mathrm{Hom}(C_0^\infty(\Omega; \mathbb{C}^{\mathbf{D}_2}), C^\infty_{\mathrm{loc}}(\Omega; \mathbb{C}^{\mathbf{D}_2}))$ which is locally a singular integral operator of order -2κ such that*

$$S\mathscr{P}^* \mathscr{P}, \mathscr{P}^* \mathscr{P} S \equiv I \mod C^\infty_{\mathrm{loc}}(\Omega \times \Omega; \mathbb{M}^{\mathbf{D}_2 \times \mathbf{D}_2}(\mathbb{C})).$$

In fact, S is locally a pseudo-differential operator of order -2κ.

(i) *For every non-negative self-adjoint extension $\mathscr{L}$ of $\mathscr{P}^* \mathscr{P}$ and $\forall t > 0$, $e^{-t\mathscr{L}}$ is given by integration against a $C^\infty_{\mathrm{loc}}(\Omega \times \Omega; \mathbb{M}^{\mathbf{D}_2 \times \mathbf{D}_2}(\mathbb{C}))$ function (denoted by $e^{-t\mathscr{L}}(x,y)$), which satisfies the following. For every $\Omega_1 \Subset \Omega$, $\exists c = c(\Omega_1) > 0$, $\forall \alpha, \beta \in \mathbb{N}^n$, $\forall s \in \mathbb{N}$, $\exists C = C(\alpha, \beta, s, \Omega_1) \ge 0$, $\forall t \in (0,1]$, $\forall x, y \in \Omega_1$, such that*

$$\big|\partial_t^s \partial_x^\alpha \partial_y^\beta e^{-t\mathscr{L}}(x,y)\big| \le C\big(|x-y| + t^{\frac{1}{2\kappa}}\big)^{-n-|\alpha|-|\beta|-2\kappa s} \exp\left(-c\left(\frac{|x-y|^{2\kappa}}{t}\right)^{1/(2\kappa-1)}\right).$$

(j) *There exists a non-negative self-adjoint extension $\mathscr{L}$ of $\mathscr{P}^* \mathscr{P}$ satisfying the conclusions of* (i).

(k) *For every non-negative self-adjoint extension $\mathscr{L}$ of $\mathscr{P}^* \mathscr{P}$ and $\forall t > 0$, $e^{-t\mathscr{L}}$ is given by integration against a $C^\infty_{\mathrm{loc}}(\Omega \times \Omega; \mathbb{M}^{\mathbf{D}_2 \times \mathbf{D}_2}(\mathbb{C}))$ function (denoted by $e^{-t\mathscr{L}}(x,y)$), which satisfies the following. For every $\Omega_1 \Subset \Omega$, $\exists c = c(\Omega_1) > 0$, $\forall \alpha, \beta \in \mathbb{N}^n$, $\forall l, s \in \mathbb{N}$, $\exists C = C(\alpha, \beta, s, l) \ge 0$, $\forall t \in (0,1]$, $\forall x, y \in \Omega_1$, such that*

$$\big|\partial_t^s \partial_x^\alpha \partial_y^\beta e^{-t\mathscr{L}}(x,y)\big| \le C t^{-|\alpha|/2\kappa - |\beta|/2\kappa - s}\big(|x-y| + t^{\frac{1}{2\kappa}}\big)^{-n}\big(1 + t^{-1/2\kappa}|x-y|\big)^{-l}.$$

(l) *There exists a non-negative self-adjoint extension $\mathscr{L}$ of $\mathscr{P}^* \mathscr{P}$ satisfying the conclusions of* (k).

The main aspects of Theorem 2.7.2 are a special case of Theorem 8.1.1 and we refer the reader there for proofs.

Remark 2.7.3. Theorem 2.7.2 concerns elliptic operators acting on vector-valued functions. One can easily generalize this theorem to operators acting on sections of a finite-dimensional vector bundle. See Section 8.6 for a more general version of this for maximally subelliptic operators.

2.8 Nonlinear elliptic equations

It follows from Theorem 2.7.2 that elliptic partial differential operators of order κ are locally left invertible, modulo smooth functions as maps from $\mathscr{C}^{\mathbf{s}+\kappa}(\mathbb{R}^n) \to \mathscr{C}^{\mathbf{s}}(\mathbb{R}^n)$. In this section, we use this fact to deduce the sharp regularity theory for fully nonlinear elliptic equations.

Fix $\mathbf{D}_1, \mathbf{D}_2, \kappa \in \mathbb{N}_+$ and set

$$N := \mathbf{D}_2 \times \#\{\alpha \in \mathbb{N}^n : |\alpha| \le \kappa\}.$$

Given $u = (u_1, \dots, u_{\mathbf{D}_2}) : \mathbb{R}^n \to \mathbb{R}^{\mathbf{D}_2}$, we define

$$\{\partial_x^\alpha u\}_{|\alpha|\le\kappa} : \mathbb{R}^n \to \mathbb{R}^N$$

to be the vector whose components are given by $\partial_x^\alpha u_j$, $1 \le j \le \mathbf{D}_2$, $|\alpha| \le \kappa$.

Let $F(x,\zeta) : \mathbb{R}^n \times \mathbb{R}^N \to \mathbb{R}^{\mathbf{D}_1}$ be a function. For a fixed $x_0 \in \mathbb{R}^n$ and $u : \mathbb{R}^n \to \mathbb{R}^{\mathbf{D}_2}$,

$$v \mapsto d_\zeta F(x_0, \{\partial_x^\alpha u(x_0)\}_{|\alpha|\le\kappa})\{\partial_x^\alpha v(x)\}_{|\alpha|\le\kappa}$$

is a constant coefficient partial differential operator of degree κ of the form

$$v \mapsto \sum_{|\alpha|\le\kappa} a_{\alpha,x_0,u}\partial_x^\alpha v,$$

where $a_{\alpha,x_0,u} \in \mathbb{M}^{\mathbf{D}_1\times\mathbf{D}_2}(\mathbb{R})$ are constant matrices which only depend on the Taylor coefficients of u at x_0 up to order κ.

Definition 2.8.1. Let $x_0 \in \mathbb{R}^n$, $u : \mathbb{R}^n \to \mathbb{R}^{\mathbf{D}_2}$, and $F(x,\zeta) : \mathbb{R}^n \times \mathbb{R}^N \to \mathbb{R}^{\mathbf{D}_1}$; we assume that both u and F are sufficiently smooth for all objects that follow to make sense. We say the nonlinear partial differential operator

$$F(x, \{\partial_x^\alpha u(x)\}_{|\alpha|\le\kappa})$$

is elliptic at (x_0, u) of degree κ if the constant coefficient linear partial differential operator of degree κ

$$v \mapsto d_\zeta F(x_0, \{\partial_x^\alpha u(x_0)\}_{|\alpha|\le\kappa})\{\partial_x^\alpha u(x)\}$$

is elliptic.

Definition 2.8.2.
- For $\mathbf{r} > 0$, we say $u \in \mathscr{C}^{\mathbf{r}}$ near x_0 if there exists $\psi \in C_0^\infty(\mathbb{R}^n)$ with $\psi \equiv 1$ on a neighborhood of x_0 such that $\psi u \in \mathscr{C}^{\mathbf{r}}(\mathbb{R}^n)$.
- For $\mathbf{r}, \mathbf{t} > 0$, we say $F(x,\zeta) \in \mathscr{C}^{\mathbf{r},\mathbf{t}}$ near x_0 if there exists $\psi \in C_0^\infty(\mathbb{R}^n)$ with $\psi \equiv 1$ on a neighborhood of x_0 such that $\psi(x)F(x,\zeta) \in \mathscr{C}^{\mathbf{r},\mathbf{t}}(\mathbb{R}^n \times \mathbb{R}^N)$.

The main result of this section (Theorem 2.8.7) is a quantitative theorem. It has, as an immediate consequence, the next qualitative theorem.

Theorem 2.8.3 (Main qualitative theorem). *Fix* $\mathbf{s} > \mathbf{r} > 0$, $\mathbf{t} > 2\lfloor\mathbf{s}\rfloor + 6$, *and* $x_0 \in \mathbb{R}^n$. *Suppose that:*
- $F \in \mathscr{C}^{\mathbf{s},\mathbf{t}}$ *near* x_0.
- $u \in \mathscr{C}^{\mathbf{r}+\kappa}$ *near* x_0.
- $F(x, \{\partial_x^\alpha u(x)\}_{|\alpha|\leq\kappa}) \in \mathscr{C}^{\mathbf{s}}$ *near* x_0.
- $F(x, \{\partial_x^\alpha u(x)\}_{|\alpha|\leq\kappa})$ *is elliptic of order* κ *at* (x_0, u).

Then $u \in \mathscr{C}^{\mathbf{s}+\kappa}$ *near* x_0.

Remark 2.8.4. Theorem 2.8.3 is not sharp with respect to the regularity of $F(x,\zeta)$ in the ζ variable, i. e., it is not sharp in $\mathbf{t}$. However, it is sharp in $\mathbf{s}$, and $\mathbf{r}$ (the a priori regularity of u) does not play a role in the conclusion.

Theorem 2.8.3 has, as an easy consequence, the classical interior Schauder estimates.

Corollary 2.8.5 (Qualitative Schauder estimates). *Fix* $\mathbf{s} > \mathbf{r} > 0$ *and let* $\mathscr{P}$ *be a linear partial differential operator of the form*

$$\mathscr{P}u(x) := \sum_{|\alpha|\leq\kappa} a_\alpha(x)\partial_x^\alpha u(x),$$

where $a_\alpha : \mathbb{R}^n \to \mathbb{M}^{\mathbf{D}_1\times\mathbf{D}_2}(\mathbb{R})$. *Fix* $x_0 \in \mathbb{R}^n$ *and suppose that:*
- $a_\alpha \in \mathscr{C}^{\mathbf{s}}$ *near* x_0, *for all* $|\alpha| \leq \kappa$.
- $u \in \mathscr{C}^{\mathbf{r}+\kappa}$ *near* x_0.
- $\mathscr{P}u \in \mathscr{C}^{\mathbf{s}}$ *near* x_0.
- *Set* $\mathscr{P}_{x_0} := \sum_{|\alpha|\leq\kappa} a_\alpha(x_0)\partial_x^\alpha$; *we suppose* $\mathscr{P}_{x_0}$ *is elliptic of order* κ.

Then $u \in \mathscr{C}^{\mathbf{s}+\kappa}$ *near* x_0.

Proof. Take $\psi \in C_0^\infty(\mathbb{R}^n)$, with $\psi \equiv 1$ on a neighborhood of x_0 and such that $\psi u \in \mathscr{C}^{\mathbf{r}+\kappa}(\mathbb{R}^n)$. Dividing u by $\sum_{|\alpha|\leq\kappa} \|\partial_x^\alpha \psi u\|_{L^\infty}$, we see that we may (without loss of generality) assume that $|\partial_x^\alpha u(x)| \leq 1$ for x near x_0, $\forall|\alpha| \leq \kappa$.

Fix $\phi(\zeta) \in C_0^\infty(\mathbb{R}^N)$ with $\phi(\zeta) \equiv 1$ on a neighborhood of $\{\zeta : |\zeta_\alpha| \leq 2\}$. Define

$$F(x,\zeta) := \sum_{|\alpha|\leq\kappa} a_\alpha(x)\zeta_\alpha\phi(\zeta).$$

Clearly, $F(x,\zeta) \in \mathscr{C}^{\mathbf{s},2\lfloor\mathbf{s}\rfloor+7}$ near x_0. Moreover, since $|\partial_x^\alpha u(x)| \leq 1$ near x_0, $\forall|\alpha| \leq \kappa$, we have

$$F(x, \{\partial_x^\alpha u(x)\}_{|\alpha|\leq\kappa}) = \mathscr{P}u(x) \in \mathscr{C}^{\mathbf{s}}, \quad x \text{ near } x_0,$$

and

$$d_\zeta F(x_0, \{\partial_x^\alpha u(x_0)\}_{|\alpha|\le\kappa})\{\partial_x^\alpha\}_{|\alpha|\le\kappa} = \mathscr{P}_{x_0}.$$

Thus, Theorem 2.8.3 applies to show $u \in \mathscr{C}^{\mathbf{s}+\kappa}$ near x_0, completing the proof. □

Remark 2.8.6. The a priori assumption that $u \in \mathscr{C}^{\mathbf{r}+\kappa}$ near x_0 in Corollary 2.8.5 can be weakened considerably by an easier reprise of the proof of Theorem 2.8.3, using the fact that the PDE in Corollary 2.8.5 is linear; see [212]. Thus, seeing Schauder estimates as a consequence of estimates for fully nonlinear equations (as in Theorem 2.8.3) introduces unnecessary difficulties and assumptions. A main focus of this text is to study fully nonlinear equations, so we do not discuss such improvements here.

Theorem 2.8.3 follows easily from the next quantitative theorem.

Theorem 2.8.7. *Let* $\mathbf{s} > \mathbf{r} > 0$ *and* $\sigma > 0$. *Suppose that:*
- $F \in \mathscr{C}^{\mathbf{s},2\lfloor\mathbf{s}\rfloor+6+\sigma}(\mathbb{R}^n \times \mathbb{R}^N; \mathbb{R}^{\mathbf{D}_1})$.
- $u \in \mathscr{C}^{\mathbf{r}+\kappa}(\mathbb{R}^n; \mathbb{R}^{\mathbf{D}_2})$.
- $g \in \mathscr{C}^{\mathbf{s}}(\mathbb{R}^n; \mathbb{R}^{\mathbf{D}_1})$.
- $F(x, \{\partial_x^\alpha u(x)\}_{|\alpha|\le\kappa}) = g(x)$, $\forall x \in B^n(3/4)$.
- *Let* $\mathscr{P}v := d_\zeta F(0, \{\partial_x^\alpha u(0)\}_{|\alpha|\le\kappa})\{\partial_x^\alpha v\}_{|\alpha|\le\kappa}$, *so that* $\mathscr{P}$ *is a constant coefficient partial differential operator of order* κ *with coefficients in* $\mathbb{M}^{\mathbf{D}_1\times\mathbf{D}_2}(\mathbb{R})$. *We assume that* $\mathscr{P}$ *is elliptic, i. e.,* $\exists A \ge 0$ *such that*

$$\sum_{j=1}^{n} \|\partial_{x_j}^\kappa f\|_{L^2} \le A(\|\mathscr{P}f\|_{L^2} + \|f\|_{L^2}), \quad \forall f \in C_0^\infty(\mathbb{R}^n; \mathbb{C}^{\mathbf{D}_2}). \tag{2.75}$$

Let $\psi_0 \in C_0^\infty(B^n(2/3))$ *and for* $\delta > 0$ *set* $\psi_0^\delta(x) := \psi_0(\delta^{-1}x)$. *Then* $\exists \delta \in (0,1]$, *depending only on* $n, A, \mathbf{D}_1, \mathbf{D}_2, \mathbf{s}, \mathbf{r}, \sigma, \kappa$ *and upper bounds for* $\|F\|_{\mathscr{C}^{\mathbf{s},2\lfloor\mathbf{s}\rfloor+6+\sigma}}$, $\|g\|_{\mathscr{C}^{\mathbf{s}}}$, $\|u\|_{\mathscr{C}^{\mathbf{r}+\kappa}}$ *such that* $\psi_0^\delta u \in \mathscr{C}^{\mathbf{s}+\kappa}$ *and*

$$\|\psi_0^\delta u\|_{\mathscr{C}^{\mathbf{s}+\kappa}} \le C_5,$$

where $C_5 \ge 0$ *can be chosen to depend only on* $n, A, \mathbf{D}_1, \mathbf{D}_2, \mathbf{s}, \mathbf{r}, \kappa, \sigma, \psi_0$, *and upper bounds for* $\|F\|_{\mathscr{C}^{\mathbf{s},2\lfloor\mathbf{s}\rfloor+6+\sigma}}$, $\|g\|_{\mathscr{C}^{\mathbf{s}}}$, $\|u\|_{\mathscr{C}^{\mathbf{r}+\kappa}}$.

We sketch the proof of Theorem 2.8.7; a full proof with all details of a more general result is given in the maximally subelliptic setting (see Theorem 9.2.1). There are many proofs of results like Theorem 2.8.7 in the literature (see [233, Chapter 14] for several approaches). However, these proofs do not directly generalize to the setting of maximally subelliptic PDEs (see Remark 1.1.5). Instead, we adapt a proof due to Simon [212] to this setting; while the proof we present is somewhat more complicated than other proofs of Theorem 2.8.7, it has the advantage of also working in the more general setting of maximally subelliptic operators.

We write $\mathfrak{D}^\kappa := \{\partial_x^\alpha\}_{|\alpha|\le\kappa}$ so that $\mathfrak{D}^\kappa u = \{\partial_x^\alpha u\}_{|\alpha|\le\kappa}$. Throughout this section, we take $\mathbf{s} > \mathbf{r} > 0$ and $\sigma > 0$.

We begin by reducing Theorem 2.8.7 to the case of a perturbation of a constant coefficient linear elliptic operator. We do this in two steps, which can be found in Section 2.8.1 and Section 2.8.2; this reduction is straightforward. Then, once we have reduced to the case of a perturbation of a constant coefficient linear elliptic operator, in Section 2.8.3 we use ideas from [212] to deduce the result.

2.8.1 Reduction I

To prove Theorem 2.8.7, we reduce it to the following, simpler proposition.

Proposition 2.8.8. *Let $C_3 \ge 0$ be given. There exists $\epsilon_2 = \epsilon_2(C_3, n, \mathbf{D}_1, \mathbf{D}_2, \kappa, \mathbf{s}, \mathbf{r}, \sigma) > 0$ sufficiently small such that the following holds. Suppose that:*

- $G \in \mathscr{C}^{\lfloor \mathbf{s}\rfloor+4+\sigma}(\mathbb{R}^N; \mathbb{R}^{\mathbf{D}_1})$ *with* $\|G\|_{\mathscr{C}^{\lfloor \mathbf{s}\rfloor+4+\sigma}} \le C_3$.
- $u \in \mathscr{C}^{\mathbf{r}+\kappa}(\mathbb{R}^n; \mathbb{R}^{\mathbf{D}_2})$ *with* $\|u\|_{\mathscr{C}^{\mathbf{r}+\kappa}} \le \epsilon_2$.
- $R_3 \in \mathscr{C}^{\mathbf{s},\lfloor \mathbf{s}\rfloor+3+\sigma}(\mathbb{R}^n \times \mathbb{R}^N; \mathbb{R}^{\mathbf{D}_1})$ *with* $\|R_3\|_{\mathscr{C}^{\mathbf{s},\lfloor \mathbf{s}\rfloor+3+\sigma}} \le \epsilon_2$.
- *Let* $\mathscr{P} := dG(\mathfrak{D}^\kappa u(0))\mathfrak{D}^\kappa$. *We assume $\mathscr{P}$ is elliptic of degree κ, i. e., $\exists A_2 \ge 0$ with*

$$\sum_{j=1}^n \|\partial_{x_j}^\kappa f\|_{L^2} \le A_2(\|\mathscr{P} f\|_{L^2} + \|f\|_{L^2}), \quad \forall f \in C_0^\infty(\mathbb{R}^n; \mathbb{C}^{\mathbf{D}_2}).$$

- $G(\mathfrak{D}^\kappa u(x)) + R_3(x, \mathfrak{D}^\kappa u(x)) = 0, \forall x \in B^n(13/16)$.

Then, $\forall \psi_0 \in C_0^\infty(B^n(2/3))$, $\psi_0 u \in \mathscr{C}^{\mathbf{s}+\kappa}(\mathbb{R}^n; \mathbb{R}^{\mathbf{D}_2})$ and $\|\psi_0 u\|_{\mathscr{C}^{\mathbf{s}+\kappa}} \le C_4$, where $C_4 \ge 0$ can be chosen to depend only on C_3, n, A_2, $\mathbf{D}_1$, $\mathbf{D}_2$, $\mathbf{s}$, κ, $\mathbf{r}$, σ, and ψ_0.

The rest of this section is devoted to sketching the reduction of Theorem 2.8.7 to Proposition 2.8.8; we henceforth assume the setting of Theorem 2.8.7. By replacing $F(x,\zeta)$ with $F(x,\zeta) - g(x)$, it suffices to consider only the case $g = 0$; we henceforth assume $g = 0$.

Take $J \in \mathbb{N}_+$ large, to be chosen later. Set

$$u_J := \sum_{0\le j<J} D_j u, \quad v_J := 2^{\kappa J} \sum_{j\ge J} D_j u. \tag{2.76}$$

Here D_j are defined by (2.30); they are the usual Littlewood–Paley projections satisfying $\sum_{j\in\mathbb{N}} D_j = I$. In particular, we have

$$u_J + 2^{-\kappa J} v_J = \sum_{j\in\mathbb{N}} D_j u = u.$$

Set

$$G_J(\zeta) := F(0, \mathfrak{D}^\kappa u_J(0) + \{2^{-(\kappa-|\alpha|)J}\zeta_\alpha\}_{|\alpha|\le\kappa}),$$
$$H_J(x,\zeta) := F(x, \mathfrak{D}^\kappa u_J(x) + \{2^{-(\kappa-|\alpha|)J}\zeta_\alpha\}_{|\alpha|\le\kappa}) - F(0, \mathfrak{D}^\kappa u_J(0) + \{2^{-(\kappa-|\alpha|)J}\zeta_\alpha\}_{|\alpha|\le\kappa}).$$

Using (2.76), we have

$$\begin{aligned} &G_J(\{(2^{-J}\partial_x)^\alpha v_J(x)\}_{|\alpha|\le\kappa}) + H_J(x, \{(2^{-J}\partial_x)^\alpha v_J(x)\}_{|\alpha|\le\kappa}) \\ &\quad = F(x, \{\partial_x^\alpha u(x)\}_{|\alpha|\le\kappa}) = 0, \quad \forall x \in B^n(3/4), \end{aligned} \tag{2.77}$$

where in the last equality we used the assumptions of Theorem 2.8.7 and the fact that we have already reduced to the case $g = 0$.

Our goal is to rescale (2.77) and apply Proposition 2.8.8 to the rescaled equation. Fix $\psi \in C_0^\infty(B^n(1))$ with $\psi \equiv 1$ on $B^n(15/16)$. Set

$$\hat{v}_J(y) := \psi(y) v_J(2^{-J}y), \quad \hat{H}_J(y,\zeta) := \psi(y) H_J(2^{-J}y, \zeta).$$

Then (2.77) implies

$$G_J(\mathfrak{D}^\kappa \hat{v}_J(y)) + \hat{H}_J(y, \mathfrak{D}^\kappa \hat{v}_J(y)) = 0, \quad \forall y \in B^n(3/4). \tag{2.78}$$

We will apply Proposition 2.8.8 with u replaced by $\hat{v}_J$ (for J large), G replaced by G_J, and R_3 replaced by $\hat{H}_J$.

An elementary proof, which we leave to the reader, shows that

$$\|G_J\|_{\mathscr{C}^{\lfloor \mathbf{s}\rfloor+4+\sigma}} \lesssim 1, \tag{2.79}$$
$$\|\hat{H}_J\|_{\mathscr{C}^{\mathbf{s},\lfloor \mathbf{s}\rfloor+3+\sigma}} \lesssim 2^{-J(\mathbf{r}\wedge 1)}, \tag{2.80}$$
$$\|\hat{v}_J\|_{\mathscr{C}^{\mathbf{r}+\kappa}} \lesssim 2^{-J\mathbf{r}}. \tag{2.81}$$

Let $\mathscr{P}$ be as in the statement of Theorem 2.8.7 and write $\mathscr{P} := \sum_{|\alpha|\le\kappa} a_\alpha \partial_x^\alpha$, where $a_\alpha \in \mathbb{M}^{\mathbf{D}_1\times\mathbf{D}_2}(\mathbb{R})$. Define

$$\mathscr{P}_J := \sum_{|\alpha|\le\kappa} a_\alpha 2^{-J(\kappa-|\alpha|)}\partial_x^\alpha.$$

Multiplying (2.75) by $2^{-\kappa J}$, we have, $\forall f \in C_0^\infty(\mathbb{R}^n; \mathbb{C}^{\mathbf{D}_2})$,

$$\begin{aligned} \sum_{j=1}^n \|(2^{-J}\partial_{x_j})^\kappa f\|_{L^2} &\le A(\|2^{-\kappa J}\mathscr{P} f\|_{L^2} + 2^{-\kappa J}\|f\|_{L^2}) \\ &\le A(\|2^{-\kappa J}\mathscr{P} f\|_{L^2} + \|f\|_{L^2}). \end{aligned} \tag{2.82}$$

Putting $f(x) = g(2^J x)$ in (2.82) and changing variables shows, $\forall g \in C_0^\infty(\mathbb{R}^n; \mathbb{C}^{\mathbf{D}_2})$,

$$\sum_{j=1}^{n} \|\partial_{x_j}^{\kappa} g\|_{L^2} \leq A(\|\mathscr{P}_J g\|_{L^2} + \|g\|_{L^2}), \tag{2.83}$$

i. e., $\mathscr{P}_J$ is elliptic with the same constant A as $\mathscr{P}$.

Finally, note, using (2.76),

$$\begin{aligned} dG_J(\mathfrak{D}^\kappa \hat{v}_J(0))\{\zeta_\alpha\}_{|\alpha|\leq\kappa} &= dG_J(\{2^{J|\alpha|}\partial_x^\alpha v_J(0)\}_{|\alpha|\leq\kappa})\{\zeta_\alpha\}_{|\alpha|\leq\kappa} \\ &= dF(0, \mathfrak{D}^\kappa u(0))\{2^{-(\kappa-|\alpha|)J}\zeta_\alpha\}_{|\alpha|\leq\kappa} = \sum_{|\alpha|\leq\kappa} a_\alpha 2^{-(\kappa-|\alpha|)J}\zeta_\alpha. \end{aligned}$$

Thus,

$$dG_J(\mathfrak{D}^\kappa \hat{v}_J(0))\mathfrak{D}^\kappa = \mathscr{P}_J.$$

By (2.79), there exists $C_3 \approx 1$ with $\|G_J\|_{\mathscr{C}^{\lfloor s\rfloor+4+\sigma}} \leq C_3$. Let $\epsilon_2 > 0$ be as in Proposition 2.8.8 with this choice of C_3 and with A_2 replaced by A. By (2.80) and (2.81), if we take $J \approx 1$ sufficiently large, we have $\|\hat{H}_J\|_{\mathscr{C}^{s,\lfloor s\rfloor+3+\sigma}} \leq \epsilon_2$ and $\|\hat{v}_J\|_{\mathscr{C}^{r+\kappa}} \leq \epsilon_2$. Using this and the above remarks, we have shown that all the hypotheses of Proposition 2.8.8 hold for equation (2.78). The conclusion of Proposition 2.8.8 in this case is

$$\|\psi_0 \hat{v}_J\|_{\mathscr{C}^{s+\kappa}} \lesssim 1. \tag{2.84}$$

Since $\psi_0 \prec \psi$, (2.84) is equivalent to

$$\|\psi_0(\cdot)v_J(2^{-J}\cdot)\|_{\mathscr{C}^{s+\kappa}} \lesssim 1. \tag{2.85}$$

Since $J \approx 1$, (2.85) is equivalent to

$$\|\psi_0(2^J\cdot)v_J(\cdot)\|_{\mathscr{C}^{s+\kappa}} \lesssim 1. \tag{2.86}$$

Also, since $J \approx 1$, we have $u_J \in C^\infty(\mathbb{R}^n)$ and $\|u_J\|_{C^L} \lesssim 1, \forall L \in \mathbb{N}$. In particular, $\|u_J\|_{\mathscr{C}^{s+\kappa}} \lesssim 1$, and therefore

$$\|\psi_0(2^J\cdot)u_J(\cdot)\|_{\mathscr{C}^{s+\kappa}} \lesssim 1. \tag{2.87}$$

Since $u = u_J + 2^{-\kappa J}v_J$, combining (2.86) and (2.87) shows

$$\|\psi_0(2^J\cdot)u(\cdot)\|_{\mathscr{C}^{s+\kappa}} \lesssim 1.$$

Theorem 2.8.7 follows with $\delta = 2^{-J}$.

2.8.2 Reduction II

To prove Proposition 2.8.8, we reduce it to the following, simpler proposition.

Proposition 2.8.9. *Fix $A, B, C_1 \geq 0$. There exists $\epsilon_1 = \epsilon_1(n, A, B, C_1, \mathbf{s}, \mathbf{r}, \sigma, \kappa, n, \mathbf{D}_1, \mathbf{D}_2) > 0$ such that the following holds. Let $\mathscr{P} = \sum_{|\alpha|\leq\kappa} a_\alpha \partial_x^\alpha$, where $a_\alpha \in \mathbb{M}^{\mathbf{D}_1\times\mathbf{D}_2}(\mathbb{R})$ and $|a_\alpha| \leq B$, $\forall \alpha$. Suppose $\mathscr{P}$ is elliptic in the following quantitative sense:*

$$\sum_{j=1}^{n} \|\partial_{x_j}^\kappa f\|_{L^2} \leq A(\|\mathscr{P}f\|_{L^2} + \|f\|_{L^2}), \quad \forall f \in C_0^\infty(\mathbb{R}^n; \mathbb{C}^{\mathbf{D}_2}).$$

Suppose $R_1(x,\zeta) \in \mathscr{C}^{\mathbf{s},\lfloor\mathbf{s}\rfloor+3+\sigma}(\mathbb{R}^n \times \mathbb{R}^N; \mathbb{R}^{\mathbf{D}_1})$, $R_2 \in \mathscr{C}^{\mathbf{s},\lfloor\mathbf{s}\rfloor+4+\sigma}(\mathbb{R}^n \times \mathbb{R}^N; \mathbb{R}^{\mathbf{D}_1})$, and $u \in \mathscr{C}^{\mathbf{r}+\kappa}(\mathbb{R}^n; \mathbb{R}^{\mathbf{D}_2})$ with

$$\|R_1\|_{\mathscr{C}^{\mathbf{s},\lfloor\mathbf{s}\rfloor+3+\sigma}} \leq \epsilon_1, \quad \|R_2\|_{\mathscr{C}^{\mathbf{s},\lfloor\mathbf{s}\rfloor+4+\sigma}} \leq C_1, \quad \|u\|_{\mathscr{C}^{\mathbf{r}+\kappa}} \leq \epsilon_1,$$

and $\partial_\zeta^\beta R_2(x,0) \equiv 0$, $\forall |\beta| \leq 1$. Suppose

$$\mathscr{P}u(x) = R_1(x, \mathfrak{D}^\kappa u(x)) + R_2(x, \mathfrak{D}^\kappa u(x)), \quad \forall x \in B^n(13/16). \tag{2.88}$$

Then, $\forall \psi_0 \in C_0^\infty(B^n(2/3))$, we have $\psi_0 u \in \mathscr{C}^{\mathbf{s}+\kappa}(\mathbb{R}^n; \mathbb{R}^{\mathbf{D}_2})$ and

$$\|\psi_0 u\|_{\mathscr{C}^{\mathbf{s}+\kappa}} \leq C_2,$$

where $C_2 = C_2(n, A, B, C_1, \mathbf{s}, \mathbf{r}, \sigma, \kappa, n, \mathbf{D}_1, \mathbf{D}_2, \psi_0) \geq 0$.

The rest of this section is devoted to the reduction of Proposition 2.8.8 to Proposition 2.8.9; thus, we assume the setting of Proposition 2.8.8. Our goal is to choose $\epsilon_2 > 0$ small (as in Proposition 2.8.8) so that the assumptions of Proposition 2.8.9 hold.

The hypotheses of Proposition 2.8.8 give, for $f \in C_0^\infty(\mathbb{R}^n; \mathbb{C}^{\mathbf{D}_2})$,

$$\begin{aligned}\sum_{|\alpha|\leq\kappa} \|\partial_x^\alpha f\|_{L^2} &\lesssim \sum_{j=1}^{n} \|\partial_{x_j}^\kappa f\|_{L^2} + \|f\|_{L^2} \lesssim \|dG(\mathfrak{D}^\kappa u(0))\mathfrak{D}^\kappa f\|_{L^2} + \|f\|_{L^2} \\ &\lesssim \|dG(0)\mathfrak{D}^\kappa f\|_{L^2} + \|f\|_{L^2} + |dG(\mathfrak{D}^\kappa u(0)) - dG(0)| \sum_{|\alpha|\leq\kappa} \|\partial_x^\alpha f\|_{L^2}.\end{aligned} \tag{2.89}$$

Since $\|\mathfrak{D}^\kappa u\|_{L^\infty} \lesssim \|u\|_{\mathscr{C}^{\mathbf{r}+\kappa}} \leq \epsilon_2$, (2.89) implies that there exists $A_3 \approx 1$ with

$$\sum_{|\alpha|\leq\kappa} \|\partial_x^\alpha f\|_{L^2} \leq A_3(\|dG(0)\mathfrak{D}^\kappa f\|_{L^2} + \|f\|_{L^2}) + A_3\epsilon_2 \sum_{|\alpha|\leq\kappa} \|\partial_x^\alpha f\|_{L^2}. \tag{2.90}$$

If we choose $\epsilon_2 \leq 1/2A_3$, (2.90) implies, for $f \in C_0^\infty(\mathbb{R}^n; \mathbb{C}^{\mathbf{D}_2})$,

$$\sum_{|\alpha|\leq\kappa} \|\partial_x^\alpha f\|_{L^2} \leq 2A_3(\|dG(0)\mathfrak{D}^\kappa f\|_{L^2} + \|f\|_{L^2}). \tag{2.91}$$

Consider we have

$$\left|G(\mathfrak{D}^{\kappa}u(0)) - G(0)\right| \lesssim \|\mathfrak{D}^{\kappa}u\|_{L^{\infty}} \lesssim \|u\|_{\mathscr{C}^{r+\kappa}} \lesssim \epsilon_2. \tag{2.92}$$

But we also have, by the hypotheses of Proposition 2.8.8,

$$\begin{aligned}\left|G(\mathfrak{D}^{\kappa}u(0))\right| &\leq \left|G(\mathfrak{D}^{\kappa}u(0)) + R_3(0,\mathfrak{D}^{\kappa}u(0))\right| + \|R_3\|_{L^{\infty}}\\ &= 0 + \|R_3\|_{L^{\infty}} \lesssim \epsilon_2.\end{aligned} \tag{2.93}$$

Combining (2.92) and (2.93) gives

$$|G(0)| \lesssim \epsilon_2. \tag{2.94}$$

Define $E(\zeta)$ by $G(\zeta) := G(0) + d_{\zeta}G(0) + E(\zeta)$, so that $E(0) = 0$, $d_{\zeta}E(0) = 0$, and

$$\|E\|_{\mathscr{C}^{\lfloor s\rfloor+4+\sigma}} \lesssim 1. \tag{2.95}$$

Set $\mathscr{P} := d_{\zeta}G(0)\mathfrak{D}^{\kappa}$. Rewriting the assumption that $G(\mathfrak{D}^{\kappa}u(x)) + R_3(x,\mathfrak{D}^{\kappa}u(x)) = 0$, $\forall x \in B^n(13/16)$, we have

$$\mathscr{P}u = -G(0) - R_3(x,\mathfrak{D}^{\kappa}u(x)) - E(\mathfrak{D}^{\kappa}u(x)), \quad \forall x \in B^n(13/16). \tag{2.96}$$

Fix $\psi_1 \in C_0^{\infty}(B^n(1))$ with $\psi_1 \equiv 1$ on $B^n(13/16)$. Set

$$R_1(x,\zeta) := \psi_1(x)\big(G(0) + R_3(x,\zeta)\big), \quad R_2(x,\zeta) := -\psi_1(x)E(\zeta).$$

Since $E(0) = 0$ and $dE(0) = 0$, we have $\partial_{\zeta}^{\beta}R_2(x,0) = 0$, $\forall|\beta| \leq 1$. Formula (2.96) implies

$$\mathscr{P}u = R_1(x,\mathfrak{D}^{\kappa}u(x)) + R_2(x,\mathfrak{D}^{\kappa}u(x)), \quad \forall x \in B^n(13/16). \tag{2.97}$$

Furthermore, using (2.94), we have

$$\|R_1\|_{\mathscr{C}^{s,\lfloor s\rfloor+3+\sigma}} \lesssim |G(0)| + \|R_3\|_{\mathscr{C}^{s,\lfloor s\rfloor+3+\sigma}} \lesssim \epsilon_2,$$

and using (2.95), we have

$$\|R_2\|_{\mathscr{C}^{s,\lfloor s\rfloor+4+\sigma}} \lesssim \|E\|_{\mathscr{C}^{\lfloor s\rfloor+4+\sigma}} \lesssim 1.$$

Finally, we have $|d_{\zeta}G(0)| \leq \|G\|_{\mathscr{C}^{\lfloor s\rfloor+4+\sigma}} \lesssim 1$. Thus, if $\mathscr{P} = \sum_{|\alpha|\leq\kappa} a_{\alpha}\partial_x^{\alpha}$, we have $|a_{\alpha}| \lesssim 1$, $\forall\alpha$.

Thus, by taking $\epsilon_2 > 0$ small, all the hypotheses of Proposition 2.8.9 hold (the ellipticity of $\mathscr{P}$ is (2.91)). The conclusion of Proposition 2.8.9, in this case, is exactly the conclusion of Proposition 2.8.8, completing the proof of Proposition 2.8.8.

2.8.3 Completion of the proof

In this section we complete the proof of Theorem 2.8.7 by proving Proposition 2.8.9.

For each $\delta > 0$, we define a norm on $\mathscr{C}^{\mathbf{s}}(\mathbb{R}^n)$ given by

$$\|f\|_{\mathscr{C}^{\mathbf{s}}(\delta)} := \|f(\delta\cdot)\|_{\mathscr{C}^{\mathbf{s}}(\mathbb{R}^n)},$$

and more generally, for $\mathbf{s} \in \mathbb{R}$, we set

$$\|f\|_{\mathscr{B}^{\mathbf{s}}_{\infty,\infty}(\delta)} := \|f(\delta\cdot)\|_{\mathscr{B}^{\mathbf{s}}_{\infty,\infty}(\mathbb{R}^n)}.$$

In particular, $\|f\|_{\mathscr{C}^{\mathbf{s}}} = \|f\|_{\mathscr{C}^{\mathbf{s}}(1)}$ and if $\delta_1 \approx \delta_2$, then $\|f\|_{\mathscr{C}^{\mathbf{s}}(\delta_1)} \approx \|f\|_{\mathscr{C}^{\mathbf{s}}(\delta_2)}$.

We write $B^n(x,\delta) := \{y \in \mathbb{R}^n : |x-y| < \delta\}$. For the proof we require scaled bump functions near each point. Fix $\phi_{0,1} \in C_0^\infty(B^n(3/4))$ with $\phi_{0,1} \equiv 1$ on $B^n(2/3)$. For $x \in \mathbb{R}^n$ and $\delta > 0$, set $\phi_{x,\delta}(y) := \phi(\delta^{-1}(y-x))$.

Proposition 2.8.10. *There exists $N_1 \in \mathbb{N}$, depending only on n, such that the following hold:*

(i) *For all $\mathbf{s} > 0$, $\exists C_{n,\mathbf{s}} \geq 0$, $\forall x \in \mathbb{R}^n$, $\delta > 0$, there exist $x_1, \dots, x_{N_1} \in B^n(x,\delta)$ with $B^n(x_j, \delta/4) \subseteq B^n(x,\delta)$ and*

$$\|\phi_{x,\delta} u\|_{\mathscr{C}^{\mathbf{s}}(\delta)} \leq C_{n,\mathbf{s}} \sum_{j=1}^{N_1} \|\phi_{x_j,\delta/32} u\|_{\mathscr{C}^{\mathbf{s}}(\delta)}, \quad \forall u \in C_0^\infty(\mathbb{R}^n)',$$

where if the right-hand side is finite, so is the left-hand side.

(ii) *For all $\mathbf{s} > 0$, $\forall \psi \in C_0^\infty(B^n(2/3))$, $\exists C_{n,\mathbf{s},\psi} \geq 0$, $\exists x_1, \dots, x_{N_1} \in B^n(2/3)$ with $B^n(x_j, 1/4) \subseteq B^n(3/4)$ such that*

$$\|\psi u\|_{\mathscr{C}^{\mathbf{s}}} \leq C_{n,\mathbf{s},\psi} \sum_{j=1}^{N_1} \|\phi_{x_j,1/32} u\|_{\mathscr{C}^{\mathbf{s}}}, \quad \forall u \in C_0^\infty(\mathbb{R}^n)',$$

where if the right-hand side is finite, so is the left-hand side.

Proof. (i): By translating and scaling, it suffices to prove (i) for $x = 0$ and $\delta = 1$. Cover $\overline{B^n(0,3/4)}$ by a finite collection of balls of the form $B^n(x_1, 1/64), \dots, B^n(x_{N_1}, 1/64)$, with $x_j \in B^n(3/4)$; note that $B^n(x_j, 1/4) \subseteq B^n(0,1)$. Let $\psi_{x_j} \in C_0^\infty(B^n(x_j, 1/64))$ be a partition of unity on $B^n(3/4)$ subordinate to this cover, i. e., $\sum_{j=1}^{N_1} \psi_{x_j} \equiv 1$ on $B^n(0,3/4)$. Then we have

$$\begin{aligned}
\|\phi_{0,1} u\|_{\mathscr{C}^{\mathbf{s}}} &\leq \sum_{j=1}^{N_1} \|\phi_{0,1}\psi_{x_j} u\|_{\mathscr{C}^{\mathbf{s}}} \lesssim \sum_{j=1}^{N_1} \|\psi_{x_j} u\|_{\mathscr{C}^{\mathbf{s}}} \\
&= \sum_{j=1}^{N_1} \|\psi_{x_j}\phi_{x_j,1/32} u\|_{\mathscr{C}^{\mathbf{s}}} \lesssim \sum_{j=1}^{N_1} \|\phi_{x_j,1/32} u\|_{\mathscr{C}^{\mathbf{s}}},
\end{aligned}$$

completing the proof of (i).

(ii): Since $\psi \prec \phi_{0,1}$, we have, by (i) with $x = 0$ and $\delta = 1$,

$$\|\psi u\|_{\mathscr{C}^s} = \|\psi\phi_{0,1}u\|_{\mathscr{C}^s} \lesssim \|\phi_{0,1}u\|_{\mathscr{C}^s} \lesssim \sum_{j=1}^{N_1} \|\phi_{x_j,1/32}u\|_{\mathscr{C}^s}.$$ □

We henceforth take the same setting and notation as in Proposition 2.8.9. Our goal is to choose $\epsilon_1 > 0$ sufficiently small so that the conclusions of Proposition 2.8.9 hold. By replacing u with $\tilde{\psi}u$, where $\tilde{\psi} \in C_0^\infty(B^n(1))$ and $\tilde{\psi} \equiv 1$ on $B^n(31/32)$, we may henceforth assume $\operatorname{supp}(u) \subset B^n(1)$.

Lemma 2.8.11. *For $x \in \mathbb{R}^n$ and $\delta \in (0,1]$,*

$$\|\phi_{x,\delta/32}v\|_{\mathscr{C}^{s+\kappa}(\delta)} \lesssim \|\phi_{x,\delta/16}\delta^{2\kappa}\mathscr{P}^*\mathscr{P}v\|_{\mathscr{B}^{s-\kappa}_{\infty,\infty}(\delta)} + \|\phi_{x,\delta/16}v\|_{L^\infty}, \quad \forall v \in C_0^\infty(\mathbb{R}^n;\mathbb{R}^{\mathbf{D}_2})', \tag{2.98}$$

where if the right-hand side is finite, so is the left-hand side.

Proof. Write $\mathscr{P}^*\mathscr{P} = \sum_{|\alpha|\le 2\kappa} b_\alpha \partial_x^\alpha$, so that $b_\alpha \in \mathbb{M}^{\mathbf{D}_2\times\mathbf{D}_2}(\mathbb{R})$ and $|b_\alpha| \lesssim 1$. For $\delta \in (0,1]$, set $\mathscr{L}_\delta := \sum_{|\alpha|\le 2\kappa} b_\alpha \delta^{2\kappa-|\alpha|}\partial_x^\alpha$.

Since $\mathscr{P}$ is elliptic, the same is true of $\mathscr{P}^*\mathscr{P}$, and in fact,

$$\sum_{j=1}^{n} \|\partial_{x_j}^{2\kappa} f\|_{L^2} \lesssim \|\mathscr{P}^*\mathscr{P}f\|_{L^2} + \|f\|_{L^2}, \quad \forall f \in C_0^\infty(\mathbb{R}^n;\mathbb{C}^{\mathbf{D}_2}). \tag{2.99}$$

Multiplying both sides of (2.99) by $\delta^{2\kappa}$, we get

$$\sum_{j=1}^{n} \|(\delta\partial_{x_j})^{2\kappa} f\|_{L^2} \lesssim \|\delta^{2\kappa}\mathscr{P}^*\mathscr{P}f\|_{L^2} + \|f\|_{L^2}, \quad \forall f \in C_0^\infty(\mathbb{R}^n;\mathbb{C}^{\mathbf{D}_2}). \tag{2.100}$$

Replacing f with $g(\delta^{-1}\cdot)$ in (2.100) and changing variables, we get

$$\sum_{j=1}^{n} \|\partial_{x_j}^{2\kappa} g\|_{L^2} \lesssim \|\mathscr{L}_\delta g\|_{L^2} + \|g\|_{L^2}, \quad \forall g \in C_0^\infty(\mathbb{R}^n;\mathbb{C}^{\mathbf{D}_2}),$$

that is, $\mathscr{L}_\delta$ is elliptic (uniformly in $\delta \in (0,1]$ and any other relevant parameters).

Since $\phi_{0,1/32} \prec \phi_{0,1/16}$, standard elliptic estimates (see Theorem 2.7.2 (e)) show that

$$\|\phi_{0,1/32}v\|_{\mathscr{C}^{s+\kappa}(1)} \lesssim \|\phi_{0,1/16}\mathscr{L}_\delta v\|_{\mathscr{B}^{s-\kappa}_{\infty,\infty}(1)} + \|\phi_{0,1/16}v\|_{L^\infty}, \quad \forall v \in C_0^\infty(\mathbb{R}^n;\mathbb{R}^{\mathbf{D}_2})'. \tag{2.101}$$

Translating and dilating (2.101) gives (2.98) and completes the proof. □

Fix $\psi_1, \psi_2, \psi_3 \in C_0^\infty(B^n(13/16))$ with $\psi_3 \prec \psi_2 \prec \psi_1$ and $\psi_3 \equiv 1$ on a neighborhood of $\overline{B^n(3/4)}$. Since $\mathscr{P}$ is elliptic, Theorem 2.7.2 (a) ⇒ (h) implies that there exists an operator T which is locally a singular integral operator of order -2κ such that

$$T\mathscr{P}^*\mathscr{P}, \mathscr{P}^*\mathscr{P}T \equiv I \mod C^\infty_{\mathrm{loc}}(\mathbb{R}^n \times \mathbb{R}^n; \mathbb{M}^{\mathbf{D}_2\times\mathbf{D}_2}(\mathbb{C})). \tag{2.102}$$

In fact, since $\mathscr{P}$ is a constant coefficient operator, we can choose T to be a multiplier, but the details of T are not important for our purposes; all we need are some basic properties of T like the boundedness on Besov spaces given by Theorem 2.4.7.

Using (2.88), we define

$$\begin{aligned} V_\infty &:= \psi_2 T\mathscr{P}^*\psi_1(x)[R_1(x, \mathfrak{D}^\kappa u(x)) + R_2(x, \mathfrak{D}^\kappa u(x))] \\ &= \psi_2 T\mathscr{P}^*\psi_1\mathscr{P}u = \psi_2 T\mathscr{P}^*\mathscr{P}u + \psi_2 T\mathscr{P}^*(1-\psi_1)\mathscr{P}u \\ &= \psi_2 u + \psi_2 e_\infty, \end{aligned} \tag{2.103}$$

where, by the pseudo-locality of T (see the growth condition in Definition 2.3.1) and (2.102), we have $e_\infty \in C_0^\infty(B^n(1))$. Moreover, we have

$$\|e_\infty\|_{\mathscr{C}^{\mathbf{r}+\kappa}}, \|e_\infty\|_{\mathscr{C}^{\mathbf{s}+\kappa}} \lesssim \|u\|_{\mathscr{C}^{\mathbf{r}+\kappa}} \lesssim \epsilon_1. \tag{2.104}$$

Since T is locally a singular integral operator of -2κ, $T\mathscr{P}^*\psi_1\mathscr{P}$ is locally a singular integral operator of order 0. Thus, Theorem 2.4.7 combined with the fact that we are assuming $\mathrm{supp}(u) \subset B^n(1)$ implies

$$\|V_\infty\|_{\mathscr{C}^{\mathbf{r}+\kappa}} = \|\psi_2 T\mathscr{P}^*\psi_1\mathscr{P}u\|_{\mathscr{C}^{\mathbf{r}+\kappa}} \lesssim \|u\|_{\mathscr{C}^{\mathbf{r}+\kappa}} \lesssim \epsilon_1,$$

that is, $\exists C_3 \approx 1$ with

$$\|V_\infty\|_{\mathscr{C}^{\mathbf{r}+\kappa}} \le C_3\epsilon_1. \tag{2.105}$$

Set $H := u - V_\infty$, so that

$$\|H\|_{\mathscr{C}^{\mathbf{r}+\kappa}} \le \|u\|_{\mathscr{C}^{\mathbf{r}+\kappa}} + \|V_\infty\|_{\mathscr{C}^{\mathbf{r}+\kappa}} \lesssim \epsilon_1.$$

Also, by (2.103) we have $\psi_3 H = \psi_3 e_\infty$, and therefore using (2.104) we have

$$\|\psi_3 H\|_{\mathscr{C}^{\mathbf{s}+\kappa}} \lesssim \|e_\infty\|_{\mathscr{C}^{\mathbf{s}+\kappa}} \lesssim \epsilon_1. \tag{2.106}$$

For $V \in \mathscr{C}^{\mathbf{r}+\kappa}$, define

$$\mathscr{T}(V) := \psi_2 T\mathscr{P}^*\psi_1(x)[R_1(x, \mathfrak{D}^\kappa(V+H)(x)) + R_2(x, \mathfrak{D}^\kappa(V+H)(x))].$$

It follows from the fact that $H \in \mathscr{C}^{\mathbf{r}+\kappa}$, Theorem 2.4.7, and Theorem 2.6.2 that $\mathscr{T} : \mathscr{C}^{\mathbf{r}+\kappa} \to \mathscr{C}^{\mathbf{r}+\kappa}$. By the definition of V_∞ (see (2.103)), we have $\mathscr{T}(V_\infty) = V_\infty$.

Let $D \in [C_3 \vee 1, 1/\epsilon_1]$ be a fixed number, to be chosen later (we will pick $D \approx 1$). Define

$$\mathscr{M}_{D,\epsilon_1,\mathbf{r}} := \{V \in \mathscr{C}^{\mathbf{r}+\kappa} : \|V\|_{\mathscr{C}^{\mathbf{r}+\kappa}} \le D\epsilon_1\}.$$

Note that $\mathscr{M}_{D,\epsilon_1,\mathbf{r}}$ is a closed subspace of the Banach space $\mathscr{C}^{\mathbf{r}+\kappa}$, and therefore it is a complete metric space with distance induced by $\|\cdot\|_{\mathscr{C}^{\mathbf{r}+\kappa}}$. Furthermore, since $D \geq C_3$, (2.105) shows $V_\infty \in \mathscr{M}_{D,\epsilon_1,\mathbf{r}}$. We will show that $\mathscr{T} : \mathscr{M}_{D,\epsilon_1,\mathbf{r}} \to \mathscr{M}_{D,\epsilon_1,\mathbf{r}}$ and that it is a strict contraction (with appropriate choices of D and ϵ_1); it will then follow that V_∞ is the unique fixed point of $\mathscr{T}$ in $\mathscr{M}_{D,\epsilon_1,\mathbf{r}}$.

Using Theorem 2.4.7, Theorem 2.6.2 (ii) applied to R_1, and Theorem 2.6.2 (vii) with $L = 2$ applied to R_2, we have, for $V \in \mathscr{M}_{D,\epsilon_1,\mathbf{r}}$,

$$\begin{aligned}
\|\mathscr{T}(V)\|_{\mathscr{C}^{\mathbf{r}+\kappa}} &= \left\|\psi_2 T \mathscr{P}^* \psi_1(R_1(x,\mathfrak{D}^\kappa(V+H)(x)) + R_2(x,\mathfrak{D}^\kappa(V+H)(x)))\right\|_{\mathscr{C}^{\mathbf{r}+\kappa}} \\
&\lesssim \left\|R_1(x,\mathfrak{D}^\kappa(V+H)(x))\right\|_{\mathscr{C}^{\mathbf{r}}} + \left\|R_2(x,\mathfrak{D}^\kappa(V+H)(x))\right\|_{\mathscr{C}^{\mathbf{r}}} \\
&\lesssim \|R_1\|_{\mathscr{C}^{\mathbf{r},\lfloor\mathbf{r}\rfloor+2+\sigma}}(1+\|\mathfrak{D}^\kappa(V+H)\|_{\mathscr{C}^{\mathbf{r}}})(1+\|\mathfrak{D}^\kappa(V+H)\|_{L^\infty})^{\lfloor\mathbf{r}\rfloor+1} \\
&\quad + \|R_2\|_{\mathscr{C}^{\mathbf{r},\lfloor\mathbf{r}\rfloor+4+\sigma}}(1+\|\mathfrak{D}^\kappa(V+H)\|_{L^\infty})^{\lfloor\mathbf{r}\rfloor+2}\|\mathfrak{D}^\kappa(V+H)\|_{L^\infty}\|\mathfrak{D}^\kappa(V+H)\|_{\mathscr{C}^{\mathbf{r}}} \\
&\lesssim \epsilon_1(1+D\epsilon_1)(1+D\epsilon_1)^{\lfloor\mathbf{r}\rfloor+1} + (1+D\epsilon_1)^{\lfloor\mathbf{r}\rfloor+2}(D\epsilon_1)^2 \lesssim \epsilon_1 + D^2\epsilon_1^2,
\end{aligned}$$

that is, $\exists C_4 \approx 1$ with

$$\|\mathscr{T}(V)\|_{\mathscr{C}^{\mathbf{r}+\kappa}} \leq C_4(\epsilon_1 + D^2\epsilon_1^2), \quad \forall V \in \mathscr{M}_{D,\epsilon_1,\mathbf{r}}. \tag{2.107}$$

Take $D := \max\{C_3, 2C_4, 1\}$, so that $D \approx 1$, and take $\epsilon_1 \in (0, 1/D^2]$ to be chosen later. Then $C_4(\epsilon_1 + \epsilon_1^2 D^2) \leq D\epsilon_1$ and (2.107) shows that

$$\|\mathscr{T}(V)\|_{\mathscr{C}^{\mathbf{r}+\kappa}} \leq D\epsilon_1, \quad \forall V \in \mathscr{M}_{D,\epsilon_1,\mathbf{r}}.$$

We conclude that $\mathscr{T} : \mathscr{M}_{D,\epsilon_1,\mathbf{r}} \to \mathscr{M}_{D,\epsilon_1,\mathbf{r}}$.

Now that we have chosen $D \approx 1$, we have, for $V \in \mathscr{M}_{D,\epsilon_1,\mathbf{r}}$, $\|V\|_{\mathscr{C}^{\mathbf{r}+\kappa}} \leq D\epsilon_1 \lesssim \epsilon_1$. Using Theorem 2.6.2 (vi) applied to R_1 and Theorem 2.6.2 (viii) with $L = 2$ applied to R_2, we have, for $V_1, V_2 \in \mathscr{M}_{D,\epsilon_1,\mathbf{r}}$,

$$\begin{aligned}
&\|\mathscr{T}(V_1) - \mathscr{T}(V_2)\|_{\mathscr{C}^{\mathbf{r}+\kappa}} \\
&\quad = \|\psi_2 T \mathscr{P}^* \psi_1(R_1(x,\mathfrak{D}^\kappa(V_1+H)(x)) - R_1(x,\mathfrak{D}^\kappa(V_2+H)(x)) \\
&\qquad + R_2(x,\mathfrak{D}^\kappa(V_1+H)(x)) - R_2(x,\mathfrak{D}^\kappa(V_2+H)(x)))\|_{\mathscr{C}^{\mathbf{r}+\kappa}} \\
&\quad \lesssim \|R_1(x,\mathfrak{D}^\kappa(V_1+H)(x)) - R_1(x,\mathfrak{D}^\kappa(V_2+H)(x))\|_{\mathscr{C}^{\mathbf{r}}} \\
&\qquad + \|R_2(x,\mathfrak{D}^\kappa(V_1+H)(x)) - R_2(x,\mathfrak{D}^\kappa(V_2+H)(x))\|_{\mathscr{C}^{\mathbf{r}}} \\
&\quad \lesssim \|R_1\|_{\mathscr{C}^{\mathbf{r},\lfloor\mathbf{r}\rfloor+3+\sigma}}(\|\mathfrak{D}^\kappa(V_1-V_2)\|_{\mathscr{C}^{\mathbf{r}}} \\
&\qquad + (\|\mathfrak{D}^\kappa(V_1+H)\|_{\mathscr{C}^{\mathbf{r}}} + \|\mathfrak{D}^\kappa(V_2+H)\|_{\mathscr{C}^{\mathbf{r}}})\|\mathfrak{D}^\kappa(V_1-V_2)\|_{L^\infty}) \\
&\qquad\quad \times (1+\|\mathfrak{D}^\kappa(V_1+H)\|_{L^\infty} + \|\mathfrak{D}^\kappa(V_2+H)\|_{L^\infty})^{\lfloor\mathbf{r}\rfloor+1} \\
&\qquad + \|R_2\|_{\mathscr{C}^{\mathbf{r},\lfloor\mathbf{r}\rfloor+4+\sigma}}(1+\|\mathfrak{D}^\kappa(V_1+H)\|_{L^\infty} + \|\mathfrak{D}^\kappa(V_2+H)\|_{L^\infty})^{\lfloor\mathbf{r}\rfloor+2} \\
&\qquad\quad \times (\|\mathfrak{D}^\kappa(V_1+H)\|_{\mathscr{C}^{\mathbf{r}}} + \|\mathfrak{D}^\kappa(V_2+H)\|_{\mathscr{C}^{\mathbf{r}}})\|\mathfrak{D}^\kappa(V_1-V_2)\|_{L^\infty} \\
&\qquad + \|R_2\|_{\mathscr{C}^{\mathbf{r},\lfloor\mathbf{r}\rfloor+4+\sigma}}(\|\mathfrak{D}^\kappa(V_1+H)\|_{L^\infty} + \|\mathfrak{D}^\kappa(V_2+H)\|_{L^\infty})\|\mathfrak{D}^\kappa(V_1-V_2)\|_{\mathscr{C}^{\mathbf{r}}}
\end{aligned}$$

$$\begin{aligned}&\lesssim \epsilon_1(\|V_1 - V_2\|_{\mathscr{C}^{\mathbf{r}+\kappa}} + \epsilon_1\|\mathfrak{D}^\kappa(V_1 - V_2)\|_{L^\infty})(1+\epsilon_1)^{\lfloor\mathbf{r}\rfloor+1}\\&\quad + (1+\epsilon_1)^{\lfloor\mathbf{r}\rfloor+2}\epsilon_1\|\mathfrak{D}^\kappa(V_1 - V_2)\|_{L^\infty}\\&\quad + \epsilon_1\|\mathfrak{D}^\kappa(V_1 - V_2)\|_{L^\infty}\\&\lesssim \epsilon_1\|V_1 - V_2\|_{\mathscr{C}^{\mathbf{r}+\kappa}} + \epsilon_1\|\mathfrak{D}^\kappa(V_1 - V_2)\|_{L^\infty}\\&\lesssim \epsilon_1\|V_1 - V_2\|_{\mathscr{C}^{\mathbf{r}+\kappa}}.\end{aligned}$$

We conclude that $\exists C_5 \approx 1$ with

$$\|\mathscr{T}(V_1) - \mathscr{T}(V_2)\|_{\mathscr{C}^{\mathbf{r}+\kappa}} \leq C_5\epsilon_1\|V_1 - V_2\|_{\mathscr{C}^{\mathbf{r}+\kappa}}.$$

In particular, we will take $\epsilon_1 \in (0, \min\{1/2C_5, 1/2D\}]$, so that

$$\|\mathscr{T}(V_1) - \mathscr{T}(V_2)\|_{\mathscr{C}^{\mathbf{r}+\kappa}} \leq \frac{1}{2}\|V_1 - V_2\|_{\mathscr{C}^{\mathbf{r}+\kappa}}, \quad \forall V_1, V_2 \in \mathscr{M}_{D,\epsilon_1,\mathbf{r}}.$$

Thus, $\mathscr{T} : \mathscr{M}_{D,\epsilon_1,\mathbf{r}} \to \mathscr{M}_{D,\epsilon_1,\mathbf{r}}$ is a strict contraction with unique fixed point V_∞.

Remark 2.8.12. At this point it is tempting to try to show that $\mathscr{T}$ is also a strict contraction on $\mathscr{M}_{D,\epsilon',\mathbf{s}}$, for some small $\epsilon' > 0$, and then conclude that V_∞ must be the fixed point of this contraction and therefore $V_\infty \in \mathscr{C}^{\mathbf{s}+\kappa}$. Unfortunately, we only know an estimate like $\|\psi_3 H\|_{\mathscr{C}^{\mathbf{s}}} \lesssim \epsilon_1$, where we would need a stronger estimate like $\|\psi_1 H\|_{\mathscr{C}^{\mathbf{s}}} \lesssim \epsilon_1$ to make such an argument work. In short, the localizations get in the way of directly applying the contraction mapping principle to $\mathscr{T}$ on functions in $\mathscr{C}^{\mathbf{s}+\kappa}$. To get around this issue, we apply a clever covering type argument adapted from a work of Simon [212].

Set $V_0 := 0$ and for $j \geq 1$ let $V_j := \mathscr{T}(V_{j-1})$. We have established $V_j \in \mathscr{M}_{D,\epsilon_1,\mathbf{r}}$, $\forall j$, and the contraction mapping principle shows that $V_j \xrightarrow{j\to\infty} V_\infty$ in $\mathscr{C}^{\mathbf{r}+\kappa}$.

Fix $x_0 \in \mathbb{R}^n$ and $\delta \in (0,1]$ such that $B^n(x_0, \delta/4) \subseteq B^n(3/4)$. Applying Lemma 2.8.11, we see that

$$\|\phi_{x_0,\delta/32}V_{j+1}\|_{\mathscr{C}^{\mathbf{s}+\kappa}(\delta)} \lesssim \|\phi_{x_0,\delta/16}\delta^{2\kappa}\mathscr{P}^*\mathscr{P}V_{j+1}\|_{\mathscr{B}^{\mathbf{s}-\kappa}_{\infty,\infty}(\delta)} + \|\phi_{x_0,\delta/16}V_{j+1}\|_{L^\infty}. \tag{2.108}$$

We will estimate the right-hand side of (2.108).

For the second term on the right-hand side of (2.108), we have

$$\|\phi_{x_0,\delta/16}V_{j+1}\|_{L^\infty} \leq \|V_{j+1}\|_{L^\infty} \lesssim \|V_{j+1}\|_{\mathscr{C}^{\mathbf{r}+\kappa}} \lesssim \epsilon_1. \tag{2.109}$$

For the first term on the right-hand side of (2.108), we use $\operatorname{supp}(\phi_{x_0,\delta/16}) \subseteq B^n(x_0, \delta/4) \subseteq B^n(3/4)$; thus by the choice of ψ_2 we have $\phi_{x_0,\delta/16} \prec \psi_2$. So, we have

$$\begin{aligned}\phi_{x_0,\delta/16}\mathscr{P}^*\mathscr{P}\operatorname{Mult}[\psi_2]T\mathscr{P}^*\operatorname{Mult}[\psi_1] &= \phi_{x_0,\delta/16}\mathscr{P}^*\mathscr{P}T\mathscr{P}^*\operatorname{Mult}[\psi_1]\\&= \phi_{x_0,\delta/16}\mathscr{P}^*\operatorname{Mult}[\psi_1] + \phi_{x_0,\delta/16}E_\infty\operatorname{Mult}[\psi_1],\end{aligned}$$

where $E_\infty \in C_0^\infty(B^n(1) \times B^n(1))$ with $\|E_\infty\|_{C^L} \lesssim_L 1$, $\forall L \in \mathbb{N}$. Thus, we have

$$
\begin{aligned}
&\|\phi_{x_0,\delta/16}\delta^{2\kappa}\mathscr{P}^*\mathscr{P}V_{j+1}\|_{\mathscr{B}^{s-\kappa}_{\infty,\infty}(\delta)} \\
&\quad = \|\phi_{x_0,\delta/16}\delta^{2\kappa}\mathscr{P}^*\mathscr{P}\mathscr{T}(V_j)\|_{\mathscr{B}^{s-\kappa}_{\infty,\infty}(\delta)} \\
&\quad = \big\|\phi_{x_0,\delta/16}\delta^{2\kappa}\mathscr{P}^*\mathscr{P}\psi_2 T\mathscr{P}^*\psi_1(R_1(x,\mathfrak{D}^\kappa(V_j+H)(x)) + R_2(x,\mathfrak{D}^\kappa(V_j+H)(x)))\big\|_{\mathscr{B}^{s-\kappa}_{\infty,\infty}(\delta)} \\
&\quad \le \big\|\phi_{x_0,\delta/16}\delta^{2\kappa}\mathscr{P}^*\psi_1(R_1(x,\mathfrak{D}^\kappa(V_j+H)(x)) + R_2(x,\mathfrak{D}^\kappa(V_j+H)(x)))\big\|_{\mathscr{B}^{s-\kappa}_{\infty,\infty}(\delta)} \\
&\qquad + \big\|\phi_{x_0,\delta/16}\delta^{2\kappa}E_\infty\psi_1(R_1(x,\mathfrak{D}^\kappa(V_j+H)(x)) + R_2(x,\mathfrak{D}^\kappa(V_j+H)(x)))\big\|_{\mathscr{B}^{s-\kappa}_{\infty,\infty}(\delta)} \\
&\quad =: \text{(I)} + \text{(II)}.
\end{aligned}
\tag{2.110}
$$

To bound (II), we use the fact that for some $L \approx 1$,

$$
\begin{aligned}
\text{(II)} &\lesssim \delta^{2\kappa}\|E_\infty\|_{C^L}\|R_1(x,\mathfrak{D}^\kappa(V_j+H)(x)) + R_2(x,\mathfrak{D}^\kappa(V_j+H)(x))\|_{L^\infty} \\
&\lesssim \delta^{2\kappa}(\|R_1(x,\mathfrak{D}^\kappa(V_j+H)(x))\|_{L^\infty} + \|R_2(x,\mathfrak{D}^\kappa(V_j+H)(x))\|_{L^\infty}).
\end{aligned}
\tag{2.111}
$$

We have

$$
\|R_1(x,\mathfrak{D}^\kappa(V_j+H)(x))\|_{L^\infty} \le \|R_1\|_{L^\infty} \lesssim \|R_1\|_{\mathscr{C}^{s,\lfloor s\rfloor+3+\sigma}} \le \epsilon_1. \tag{2.112}
$$

Since $R_2(x,0) \equiv 0$, we have

$$
\begin{aligned}
\|R_2(x,\mathfrak{D}^\kappa(V_j+H)(x))\|_{L^\infty} &\lesssim \sum_{|\beta|\le 1}\|\partial_\zeta^\beta R_2(x,\zeta)\|_{L^\infty}\|\mathfrak{D}^\kappa(V_j+H)\|_{L^\infty} \\
&\lesssim \sum_{|\beta|\le 1}\|\partial_\zeta^\beta R_2\|_{\mathscr{C}^{s,\lfloor s\rfloor+3+\sigma}}\|\mathfrak{D}^\kappa(V_j+H)\|_{L^\infty} \\
&\lesssim \|R_2\|_{\mathscr{C}^{s,\lfloor s\rfloor+4+\sigma}}\|V_j+H\|_{\mathscr{C}^{r+\kappa}} \lesssim \epsilon_1.
\end{aligned}
\tag{2.113}
$$

Using (2.112) and (2.113) to estimate the right-hand side of (2.111) shows that

$$
\text{(II)} \lesssim \delta^{2\kappa}\epsilon_1 \lesssim \epsilon_1. \tag{2.114}
$$

For (I), we use the fact that $\phi_{x_0,\delta/16} \prec \phi_{x_0,\delta/8}$ to see that

$$
\begin{aligned}
\text{(I)} &= \Big\|\phi_{x_0,\delta/16}\delta^{2\kappa}\mathscr{P}^*\psi_1(R_1(x,\phi_{x_0,\delta/8}(x)\mathfrak{D}^\kappa(V_j+H)(x)) \\
&\qquad + R_2(x,\phi_{x_0,\delta/8}(x)\mathfrak{D}^\kappa(V_j+H)(x)))\Big\|_{\mathscr{B}^{s-\kappa}_{\infty,\infty}(\delta)} \\
&\lesssim \delta^\kappa\Big\|\phi_{x_0,\delta/8}(R_1(x,\phi_{x_0,\delta/8}(x)\mathfrak{D}^\kappa(V_j+H)(x)) \\
&\qquad + R_2(x,\phi_{x_0,\delta/8}(x)\mathfrak{D}^\kappa(V_j+H)(x)))\Big\|_{\mathscr{C}^s(\delta)} \\
&\le \delta^\kappa\|\phi_{x_0,\delta/8}R_1(x,\phi_{x_0,\delta/8}(x)\mathfrak{D}^\kappa(V_j+H)(x))\|_{\mathscr{C}^s(\delta)} \\
&\qquad + \delta^\kappa\|\phi_{x_0,\delta/8}R_2(x,\phi_{x_0,\delta/8}(x)\mathfrak{D}^\kappa(V_j+H)(x))\|_{\mathscr{C}^s(\delta)} \\
&=: \text{(III)} + \text{(IV)},
\end{aligned}
\tag{2.115}
$$

where we used $\phi_{x_0,\delta/8} \prec \psi_1$ (since $\operatorname{supp}(\phi_{x_0,\delta/8}) \subseteq B^n(x_0,\delta/4) \subseteq B^n(3/4)$).

We turn to estimating (III) and (IV); this is the key place where the tame estimates from Theorem 2.6.2 arise (see Remark 2.6.5). Indeed, important for bounding (III) and (IV) is the following estimate:

$$\begin{aligned}\delta^\kappa \|\phi_{x_0,\delta/8}\mathfrak{D}^\kappa(V_j+H)\|_{\mathscr{C}^s(\delta)} &\lesssim \sum_{|\alpha|\le\kappa} \|\phi_{x_0,\delta/8}(\delta\partial_x)^\alpha(V_j+H)\|_{\mathscr{C}^s(\delta)} \\ &= \sum_{|\alpha|\le\kappa} \|\phi_{x_0/\delta,1/8}\partial_x^\alpha(V_j(\delta\cdot)+H(\delta\cdot))\|_{\mathscr{C}^s(1)} \\ &\lesssim \|\phi_{x_0/\delta,1/4}(V_j(\delta\cdot)+H(\delta\cdot))\|_{\mathscr{C}^{s+\kappa}(1)} \\ &= \|\phi_{x_0,\delta/4}(V_j+H)\|_{\mathscr{C}^{s+\kappa}(\delta)}.\end{aligned} \tag{2.116}$$

The idea in (2.116) is that we require the factor δ^κ on the left-hand side to achieve the needed estimate. In contrast, to estimate a similar L^∞ norm we do not require the factor of δ^κ. Indeed, we have

$$\|\phi_{x_0,\delta/8}\mathfrak{D}^\kappa(V_j+H)\|_{L^\infty} \lesssim \|\mathfrak{D}^\kappa(V_j+H)\|_{L^\infty} \lesssim \|V_j+H\|_{\mathscr{C}^{r+\kappa}} \lesssim \epsilon_1.$$

We also require another estimate. Since $\operatorname{supp}(\phi_{x_0,\delta/4}) \subseteq B^n(x_0,\delta/4) \subseteq B^n(3/4)$, we have $\phi_{x_0,\delta/4}\psi_3 = \phi_{x_0,\delta/4}$. Therefore,

$$\begin{aligned}\|\phi_{x_0,\delta/4}H\|_{\mathscr{C}^{s+\kappa}(\delta)} &= \|\phi_{x_0,\delta/4}\psi_3 H\|_{\mathscr{C}^{s+\kappa}(\delta)} = \|\phi_{x_0/\delta,1/4}\psi_3(\delta\cdot)H(\delta\cdot)\|_{\mathscr{C}^{s+\kappa}(1)} \\ &\lesssim \|\psi_3(\delta\cdot)H(\delta\cdot)\|_{\mathscr{C}^{s+\kappa}(1)} = \|\psi_3 H\|_{\mathscr{C}^{s+\kappa}(\delta)} \lesssim \|\psi_3 H\|_{\mathscr{C}^{s+\kappa}(1)} \lesssim \epsilon_1.\end{aligned} \tag{2.117}$$

Combining (2.116) and (2.117), we get

$$\delta^\kappa \|\phi_{x_0,\delta/8}\mathfrak{D}^\kappa(V_j+H)\|_{\mathscr{C}^s(\delta)} \lesssim \|\phi_{x_0,\delta/4}V_j\|_{\mathscr{C}^{s+\kappa}(\delta)} + \epsilon_1.$$

In the next estimates, we use the above inequalities freely.

We have, using Theorem 2.6.2 (ii),

$$\begin{aligned}\text{(III)} &= \delta^\kappa \left\|\phi_{x_0,\delta/8}R_1(x,\phi_{x_0,\delta/8}(x)\mathfrak{D}^\kappa(V_j+H)(x))\right\|_{\mathscr{C}^s(\delta)} \\ &= \delta^\kappa \left\|\phi_{x_0/\delta,1/8}R_1(\delta x,\phi_{x_0/\delta,1/8}(x)\mathfrak{D}^\kappa(V_j+H)(\delta x))\right\|_{\mathscr{C}^s(1)} \\ &\lesssim \|R_1(\delta x,\zeta)\|_{\mathscr{C}^{s,\lfloor s\rfloor+2+\sigma}}\big(\delta^\kappa + \delta^\kappa\|\phi_{x_0/\delta,1/8}\mathfrak{D}^\kappa(V_j+H)(\delta x)\|_{\mathscr{C}^s(1)}\big) \\ &\quad\times \big(1+\|\phi_{x_0/\delta,1/8}\mathfrak{D}^\kappa(V_j+H)(\delta x)\|_{L^\infty}\big)^{\lfloor s\rfloor+1} \\ &\lesssim \|R_1\|_{\mathscr{C}^{s,\lfloor s\rfloor+2+\sigma}}\big(\delta^\kappa + \delta^\kappa\|\phi_{x_0,\delta/8}\mathfrak{D}^\kappa(V_j+H)\|_{\mathscr{C}^s(\delta)}\big) \\ &\quad\times \big(1+\|\phi_{x_0,\delta/8}\mathfrak{D}^\kappa(V_j+H)\|_{L^\infty}\big)^{\lfloor s\rfloor+1} \\ &\lesssim \epsilon_1(1+\|\phi_{x_0,\delta/4}V_j\|_{\mathscr{C}^{s+\kappa}(\delta)}+\epsilon_1)(1+\epsilon_1)^{\lfloor s\rfloor+1} \\ &\lesssim \epsilon_1 + \epsilon_1\|\phi_{x_0,\delta/4}V_j\|_{\mathscr{C}^{s+\kappa}(\delta)}.\end{aligned}$$

Using a similar argument for (IV), but with Theorem 2.6.2 (vii) in place of Theorem 2.6.2 (ii), we have

$$\begin{aligned}
(\mathrm{IV}) &= \delta^{\kappa}\|\phi_{x_0,\delta/8}R_2(x,\phi_{x_0,\delta/8}(x)\mathfrak{D}^{\kappa}(V_j+H)(x))\|_{\mathscr{C}^s(\delta)}\\
&\lesssim \|R_2\|_{\mathscr{C}^{s,\lfloor s\rfloor+4+\sigma}}(1+\|\phi_{x_0,\delta/8}\mathfrak{D}^{\kappa}(V_j+H)\|_{L^\infty})^{\lfloor s\rfloor+2}\\
&\quad\times\|\phi_{x_0,\delta/8}\mathfrak{D}^{\kappa}(V_j+H)\|_{L^\infty}\delta^{\kappa}\|\phi_{x_0,\delta/8}\mathfrak{D}^{\kappa}(V_j+H)\|_{\mathscr{C}^s(\delta)}\\
&\lesssim (1+\epsilon_1)^{\lfloor s\rfloor+2}\epsilon_1(\|\phi_{x_0,\delta/4}V_j\|_{\mathscr{C}^{s+\kappa}(\delta)}+\epsilon_1)\\
&\lesssim \epsilon_1^2+\epsilon_1\|\phi_{x_0,\delta/4}V_j\|_{\mathscr{C}^{s+\kappa}(\delta)}.
\end{aligned}$$

Plugging the above estimates into (2.115) gives

$$(\mathrm{I})\lesssim \epsilon_1+\epsilon_1\|\phi_{x_0,\delta/4}V_j\|_{\mathscr{C}^{s+\kappa}(\delta)}. \tag{2.118}$$

Using (2.118) and (2.111) to bound the right-hand side of (2.110) gives

$$\|\phi_{x_0,\delta/16}\delta^{2\kappa}\mathscr{P}^*\mathscr{P}V_{j+1}\|_{\mathscr{B}^{s-\kappa}_{\infty,\infty}(\delta)}\lesssim \epsilon_1+\epsilon_1\|\phi_{x_0,\delta/4}V_j\|_{\mathscr{C}^{s+\kappa}(\delta)}. \tag{2.119}$$

Using (2.119) and (2.109) to bound the right-hand side of (2.108) shows that for $x_0\in\mathbb{R}^n$, $\delta\in(0,1]$ such that $B^n(x_0,\delta/4)\subseteq B^n(3/4)$, we have

$$\begin{aligned}
\|\phi_{x_0,\delta/32}V_{j+1}\|_{\mathscr{C}^{s+\kappa}(\delta)} &\lesssim \epsilon_1+\epsilon_1\|\phi_{x_0,\delta/4}V_j\|_{\mathscr{C}^{s+\kappa}(\delta)}\\
&\approx \epsilon_1+\epsilon_1\|\phi_{x_0,\delta/4}V_j\|_{\mathscr{C}^{s+\kappa}(\delta/4)}.
\end{aligned} \tag{2.120}$$

Set, for $j\in\mathbb{N}$,

$$Q_j := \sup_{\substack{\delta\in(0,1],x\in\mathbb{R}^n\\ B^n(x,\delta)\subseteq B^n(3/4)}}\|\phi_{x,\delta}V_j\|_{\mathscr{C}^{s+\kappa}(\delta)}\in[0,\infty].$$

Since $V_0=0$, we have $Q_0=0$.

For $x_0\in B^n(3/4)$ and $\delta\in(0,1]$ such that $B^n(x_0,\delta/4)\subseteq B^n(3/4)$, (2.120) implies

$$\|\phi_{x_0,\delta/32}V_{j+1}\|_{\mathscr{C}^{s+\kappa}(\delta)}\lesssim \epsilon_1+\epsilon_1 Q_j. \tag{2.121}$$

Take $x\in B^n(3/4)$ and $\delta\in(0,1]$ such that $B^n(x,\delta)\subseteq B^n(3/4)$. Take $N_1\approx 1$ and $x_1,\dots,x_{N_1}$ as in Proposition 2.8.10 (i). In particular, we have $B^n(x_j,\delta/4)\subseteq B^n(x,\delta)\subseteq B^n(3/4)$, and therefore by (2.121), we have

$$\|\phi_{x_l,\delta/32}V_{j+1}\|_{\mathscr{C}^{s+\kappa}(\delta)}\lesssim \epsilon_1+\epsilon_1 Q_j,\quad 1\le l\le N_1. \tag{2.122}$$

Proposition 2.8.10 (i) gives

$$\|\phi_{x,\delta} V_{j+1}\|_{\mathscr{C}^{s+\kappa}(\delta)} \lesssim \sum_{l=1}^{N_1} \|\phi_{x_l,\delta/32} V_{j+1}\|_{\mathscr{C}^{s+\kappa}(\delta)} \lesssim \epsilon_1 + \epsilon_1 Q_j, \tag{2.123}$$

where the final estimate used (2.122) and the fact that $N_1 \approx 1$. Taking the supremum over all such x and δ in (2.123) gives

$$Q_{j+1} \lesssim \epsilon_1 + \epsilon_1 Q_j,$$

that is, $\exists C_6 \approx 1$ with $Q_{j+1} \le C_6(\epsilon_1 + \epsilon_1 Q_j)$. We take $\epsilon_1 := \min\{1/2C_6, 1/2C_5, 1/D^2\}$. Then we have

$$Q_{j+1} \le \frac{1}{2} + \frac{Q_j}{2}.$$

A simple induction, using the fact that $Q_0 = 0$, shows that $Q_j \le 1$, $\forall j \in \mathbb{N}$.

Take $\psi_0 \in C_0^\infty(B^n(2/3))$ and take $x_1, \ldots, x_{N_1}$ as in Proposition 2.8.10 (ii). We have

$$\|\psi_0 V_j\|_{\mathscr{C}^{s+\kappa}} \lesssim \sum_{l=1}^{N_1} \|\phi_{x_l,1/32} V_j\|_{\mathscr{C}^{s+\kappa}} \approx \sum_{l=1}^{N_1} \|\phi_{x_l,1/32} V_j\|_{\mathscr{C}^{s+\kappa}(1/32)} \le N_1 Q_j \lesssim 1,$$

that is,

$$\|D_l \psi_0 V_j\|_{L^\infty} \lesssim 2^{-l(\mathbf{s}+\kappa)}, \quad \forall l \in \mathbb{N}. \tag{2.124}$$

Since $V_j \xrightarrow{j\to\infty} V_\infty$ in $\mathscr{C}^{\mathbf{r}+\kappa}$, we have $V_j \to V_\infty$ in L^∞. Since $D_l : L^\infty \to L^\infty$ is bounded, taking the limit as $j \to \infty$ in (2.124) gives

$$\|D_l \psi_0 V_\infty\|_{L^\infty} \lesssim 2^{-l(\mathbf{s}+\kappa)}, \quad \forall l \in \mathbb{N}.$$

This is equivalent to $\|\psi_0 V_\infty\|_{\mathscr{C}^{s+\kappa}} \lesssim 1$. We have

$$\begin{aligned}\|\psi_0 u\|_{\mathscr{C}^{s+\kappa}} &\le \|\psi_0 V_\infty\|_{\mathscr{C}^{s+\kappa}} + \|\psi_0 H\|_{\mathscr{C}^{s+\kappa}} \\ &\lesssim \|\psi_0 V_\infty\|_{\mathscr{C}^{s+\kappa}} + \|\psi_3 H\|_{\mathscr{C}^{s+\kappa}} \lesssim 1,\end{aligned}$$

where the last inequality used (2.106), completing the proof.

2.8.4 Weighted estimates near the boundary

Let $\Omega \subseteq \mathbb{R}^n$ be an open set. If a nonlinear elliptic equation holds on Ω,

$$F(x, \{\partial_x^\alpha u(x)\}_{|\alpha|\le\kappa}) = g(x), \quad x \in \Omega,$$

then the estimates on $\|u\|_{\mathscr{C}^{s+\kappa}}$ implied by Theorem 2.8.3 blow up near $\partial\Omega$. To make this blowup precise we need to introduce weighted analogs of the Zygmund–Hölder norms $\|\cdot\|_{\mathscr{C}^s_l}$ defined in Definition 2.5.2.

For $v \in \mathbb{R}^n$ and $l \in \mathbb{N}$, set

$$\Omega_{v,l} := \{x \in \Omega : x, x+v, x+2v, \ldots, x+lv \in \Omega\}.$$

Note that for $f \in C_{\mathrm{loc}}(\Omega)$, $\mathrm{Diff}_v^l f(x)$ is defined for $x \in \Omega_{v,l}$.

Definition 2.8.13. For $l \in \mathbb{N}_+$, $\mathbf{s} \in (0,l)$, and $f \in C_{\mathrm{loc}}(\Omega)$, we define the weighted interior norm

$$\|f\|_{\widehat{\mathscr{C}}^{\mathbf{s}}_l(\Omega)} := \sum_{l'=0}^{l} \max_{1\leq k\leq n} \sup_{\delta\in(0,1]} \sup_{x\in\Omega_{\delta e_k, l'}} (\mathrm{dist}(x,\partial\Omega)\wedge 1)^{\mathbf{s}l'/l} \delta^{-\mathbf{s}l'/l} \left|\mathrm{Diff}_{\delta e_k}^{l'} f(x)\right|,$$

where $\mathrm{dist}(x,\partial\Omega) = \inf\{|x-y| : y \in \partial\Omega\}$.

Theorem 2.8.14. *Let $\mathbf{s} > \mathbf{r} > 0$, $\sigma > 0$. Suppose that:*

- $F(x,\zeta) \in \mathscr{C}^{\mathbf{s},3\lfloor\mathbf{s}\rfloor+8+\sigma}(\mathbb{R}^n \times \mathbb{R}^N; \mathbb{R}^{\mathbf{D}_1})$.
- $u \in \mathscr{C}^{\mathbf{r}+\kappa}(\mathbb{R}^n; \mathbb{R}^{\mathbf{D}_2})$.
- $g \in \mathscr{C}^{\mathbf{s}}(\mathbb{R}^n; \mathbb{R}^{\mathbf{D}_1})$.
- $F(x, \{\partial_x^\alpha u(x)\}_{|\alpha|\leq\kappa}) = g(x)$, $\forall x \in \Omega$; *since the functions involved are continuous, this equality in fact holds on $\overline{\Omega}$.*
- *For each $x_0 \in \Omega$, define the constant coefficient, linear differential operator*

$$\mathscr{P}_{x_0} v := d_\zeta F(x_0, \{\partial_x^\alpha u(x_0)\}_{|\alpha|\leq\kappa})\{\partial_x^\alpha v\}_{|\alpha|\leq\kappa}. \tag{2.125}$$

We assume $\mathscr{P}_{x_0}$ is elliptic, uniformly for $x_0 \in \Omega$, that is, $\exists A \geq 0$, $\forall x_0 \in \Omega$,

$$\sum_{j=1}^{n} \|\partial_{x_j}^\kappa f\|_{L^2} \leq A(\|\mathscr{P}_{x_0} f\|_{L^2} + \|f\|_{L^2}), \quad \forall f \in C_0^\infty(\mathbb{R}^n; \mathbb{R}^{\mathbf{D}_2}). \tag{2.126}$$

Then, $\forall l \in \mathbb{N}_+$ with $l > \mathbf{s}+\kappa$, $\exists C \geq 0$ such that

$$\|u\|_{\widehat{\mathscr{C}}^{\mathbf{s}}_l(\Omega)} \leq C.$$

Here, $C \geq 0$ can be chosen to depend only on n, $\mathbf{s}$, l, $\mathbf{r}$, κ, σ, A, $\mathbf{D}_1$, $\mathbf{D}_2$, and upper bounds for $\|u\|_{\mathscr{C}^{\mathbf{r}+\kappa}}$, $\|g\|_{\mathscr{C}^{\mathbf{s}}}$, and $\|F\|_{\mathscr{C}^{\mathbf{s},3\lfloor\mathbf{s}\rfloor+8+\sigma}}$.

Remark 2.8.15. Results like Theorem 2.8.14 are often stated in terms of weighted Hölder norms, $C^{m,\alpha}$, where $\alpha \notin \{0,1\}$. By using weighted Zygmund–Hölder norms one can state the results for all $\mathbf{s} > 0$, not just $\mathbf{s} \in (0,\infty)\setminus\mathbb{N}$. Using results like Corollary 2.5.6, one can easily show that Theorem 2.8.14 implies a corresponding result about weighted Hölder norms with non-integer exponents.

The rest of this section is devoted to a sketch of the proof of Theorem 2.8.14. A proof with all details of a more general result in the maximally subelliptic setting is given in Section 9.3.

By replacing $F(x,\zeta)$ by $F(x,\zeta) - g(x)$, we may henceforth assume $g = 0$. For $J \in [0,\infty)$, set

$$u_J := \sum_{0\le j<J} D_j u, \quad v_J := 2^{\kappa J} \sum_{j\ge J} D_j u,$$

so that

$$u_J + 2^{-\kappa J} v_J = \sum_{j\in\mathbb{N}} D_j u = u. \tag{2.127}$$

Set

$$F_J(x,\zeta) := F(x, \{\partial_x^\alpha u_J(x)\}_{|\alpha|\le\kappa} + \{2^{-(\kappa-|\alpha|)J}\zeta_\alpha\}_{|\alpha|\le\kappa}). \tag{2.128}$$

Then, using (2.127), we have

$$F_J(x, \{(2^{-J}\partial_x)^\alpha v_J(x)\}_{|\alpha|\le\kappa}) = F(x, \{\partial_x^\alpha u(x)\}_{|\alpha|\le\kappa}) = 0, \quad \forall x\in\Omega. \tag{2.129}$$

Fix $\xi_0 \in (0,1]$ a small number to be chosen later; we will choose $\xi_0 \approx 1$. Fix $\delta > 0$, $0 \le l' \le l$, $1 \le k \le n$, and $x_0 \in \Omega_{\xi_0\delta\mathfrak{e}_k,l'}$ and set $2^{-J_0} := \operatorname{dist}(x_0,\partial\Omega)\wedge 1$. To prove the statement of the theorem, it suffices to show that (for some $\xi_0 \approx 1$)

$$2^{-J_0(\mathrm{s}+\kappa)l'/l}\delta^{-(\mathrm{s}+\kappa)l'/l}\left|\mathrm{Diff}^{l'}_{\xi_0\delta\mathfrak{e}_k} u(x_0)\right| \lesssim 1. \tag{2.130}$$

If $\delta \ge 2^{-J_0}$, then we have

$$2^{-J_0(\mathrm{s}+\kappa)l'/l}\delta^{-(\mathrm{s}+\kappa)l'/l}\left|\mathrm{Diff}^{l'}_{\xi_0\delta\mathfrak{e}_k} u(x_0)\right| \lesssim \|u\|_{L^\infty} \lesssim \|u\|_{\mathscr{C}^{\mathrm{r}+\kappa}} \lesssim 1,$$

establishing (2.130). If $l' = 0$, then δ and 2^{-J_0} do not play a role in (2.130) and the same proof establishes (2.130).

We turn to proving (2.130) in the remaining case where $2^{-J_0} > \delta$ and $1 \le l' \le l$. Take $\psi \in C_0^\infty(B^n(1))$ with $\psi \equiv 1$ on $B^n(3/4)$. Set

$$\hat{v}_{x_0,J_0}(y) := v_{J_0}(2^{-J_0}y + x_0), \quad \hat{u}_{x_0,J_0}(y) := \psi(y)u_{J_0}(2^{-J_0}y + x_0),$$
$$\widehat{F}_{x_0,J_0}(y,\zeta) := F_{J_0}(2^{-J_0}y + \zeta).$$

It is straightforward to show

$$\|\hat{u}_{x_0,J_0}\|_{\mathscr{C}^{s+\kappa}} \lesssim \|u\|_{\mathscr{C}^{r+\kappa}} \lesssim 1, \quad \|\hat{v}_{x_0,J_0}\|_{\mathscr{C}^{r+\kappa}} \lesssim \|u\|_{\mathscr{C}^{r+\kappa}} \lesssim 1, \tag{2.131}$$

and using Theorem 2.6.2 (i),

$$\|\widehat{F}_{x_0,J_0}\|_{\mathscr{C}^{s,2\lfloor s\rfloor+6+\sigma}} \lesssim \|F\|_{\mathscr{C}^{s,3\lfloor s\rfloor+8+\sigma}}. \tag{2.132}$$

Rescaling (2.129) shows

$$\widehat{F}_{x_0,J_0}(y, \{\partial_y^\alpha \hat{v}_{x_0,J_0}(y)\}_{|\alpha|\le\kappa}) = 0, \quad \forall y \in B^n(3/4). \tag{2.133}$$

Our goal is to apply Theorem 2.8.7 to (2.133). To do so, we need to show (2.133) is elliptic, uniformly in $x_0 \in \Omega$. That is, define

$$\widehat{\mathscr{P}}_{x_0,J_0} := d_\zeta \widehat{F}_{x_0,J_0}(0, \{\partial_y^\alpha \hat{v}_{x_0,J_0}(0)\}_{|\alpha|\le\kappa})\{\partial_y^\alpha\}_{|\alpha|\le\kappa}.$$

We with to show $\widehat{\mathscr{P}}_{x_0,J_0}$ is elliptic of degree κ, uniformly for $x_0 \in \Omega$.

Recall $\mathscr{P}_{x_0}$ from (2.125). From its definition, $\mathscr{P}_{x_0}$ is of the form

$$\mathscr{P}_{x_0} = \sum_{|\alpha|\le\kappa} a_{\alpha,x_0}\partial_x^\alpha, \quad a_{\alpha,x_0} \in \mathbb{M}^{\mathbf{D}_1\times\mathbf{D}_2}(\mathbb{R}),$$

and we have

$$d_\zeta F(x_0, \{\partial_x^\alpha u(x_0)\}_{|\alpha|\le\kappa})\{\zeta_\alpha\}_{|\alpha|\le\kappa} = \sum_{|\alpha|\le\kappa} a_{\alpha,x_0}\zeta_\alpha.$$

Using (2.127) and (2.128), we have

$$\begin{aligned}
\widehat{\mathscr{P}}_{x_0,J_0} w &= d_\zeta \widehat{F}_{x_0,J_0}(0, \{\partial_y^\alpha \hat{v}_{x_0,J_0}(y)\}_{|\alpha|\le\kappa})\{\partial_y^\alpha w\}_{|\alpha|\le\kappa} \\
&= d_\zeta F_{J_0}(x_0, \{(2^{-J_0}\partial_x)^\alpha v_{J_0}(x_0)\})\{\partial_y^\alpha w\}_{|\alpha|\le\kappa} \\
&= d_\zeta F(x_0, \{\partial_x^\alpha u(x_0)\}_{|\alpha|\le\kappa})\{2^{-(\kappa-|\alpha|)J_0}\partial_y^\alpha w\}_{|\alpha|\le\kappa} \\
&= \sum_{|\alpha|\le\kappa} 2^{-(\kappa-|\alpha|)J_0} a_{\alpha,x_0}\partial_y^\alpha w.
\end{aligned}$$

Multiplying (2.126) by $2^{-\kappa J_0}$ gives, $\forall f \in C_0^\infty(\mathbb{R}^n; \mathbb{R}^{\mathbf{D}_2})$,

$$\begin{aligned}
\sum_{j=1}^n \|(2^{-J_0}\partial_{x_j})^\kappa f\|_{L^2} &\le A(\|2^{-\kappa J_0}\mathscr{P}_{x_0} f\|_{L^2} + 2^{-\kappa J_0}\|f\|_{L^2}) \\
&\le A(\|2^{-\kappa J_0}\mathscr{P}_{x_0} f\|_{L^2} + \|f\|_{L^2}).
\end{aligned} \tag{2.134}$$

Applying (2.134) to $f(x) = g(2^{J_0}x)$ and changing variables shows

$$\sum_{j=1}^{n} \|\partial_{y_j}^{\kappa} g\|_{L^2} \leq A(\|\widehat{\mathscr{P}}_{x_0,J_0} g\|_{L^2} + \|g\|_{L^2}), \quad \forall g \in C_0^\infty(\mathbb{R}^n; \mathbb{R}^{\mathbf{D}_2}).$$

In particular, $\widehat{\mathscr{P}}_{x_0,J_0}$ is elliptic, uniformly for $x_0 \in \Omega$.

In light of the above, Theorem 2.8.7 applies to (2.133). Let $\psi_0 \in C_0^\infty(B^n(2/3))$ with $\psi_0 \equiv 1$ on $B^n(1/2)$ and for $\delta_1 > 0$ set $\psi_0^{\delta_1}(y) := \psi_0(\delta_1^{-1}y)$, so that $\psi_0^{\delta_1} \equiv 1$ on $B^n(\delta_1/2)$. The conclusion of Theorem 2.8.7, when applied to (2.133), shows that there exists $\delta_1 \in (0,1]$, $\delta_1 \approx 1$, with

$$\|\psi_0^{\delta_1} \hat{v}_{x_0,J_0}\|_{\mathscr{C}^{s+\kappa}} \lesssim 1. \tag{2.135}$$

Recall that, by (2.131), we also have

$$\|\hat{u}_{x_0,J_0}\|_{\mathscr{C}^{s+\kappa}} \lesssim 1. \tag{2.136}$$

By Proposition 2.5.3, (2.135) and (2.136) imply

$$\|\psi_0^{\delta_1} \hat{v}_{x_0,J_0}\|_{\mathscr{C}_l^{s+\kappa}}, \|\hat{u}_{x_0,J_0}\|_{\mathscr{C}_l^{s+\kappa}} \lesssim 1.$$

In particular,

$$(2^{J_0}\delta)^{-(s+\kappa)l'/l} |\mathrm{Diff}_{2^{J_0}\delta\xi_0\mathfrak{e}_k}^{l'} (\psi_0^{\delta_1} \hat{v}_{x_0,J_0})(0)| \lesssim 1, \tag{2.137}$$

$$(2^{J_0}\delta)^{-(s+\kappa)l'/l} |\mathrm{Diff}_{2^{J_0}\delta\xi_0\mathfrak{e}_k}^{l'} \hat{u}_{x_0,J_0}(0)| \lesssim 1. \tag{2.138}$$

By taking $\xi_0 := \delta_1/2l \approx 1$ and using $2^{J_0}\delta \leq 1$, we see that

$$\psi_0^{\delta_1}(2^{J_0}\delta\xi_0 l'' \mathfrak{e}_k) = 1, \quad \forall 0 \leq l'' \leq l, \tag{2.139}$$

and since $\psi \equiv 1$ on $B^n(3/4)$,

$$\psi(2^{J_0}\delta\xi_0 l'' \mathfrak{e}_k) = 1, \quad \forall 0 \leq l'' \leq l. \tag{2.140}$$

Using (2.137) and (2.139) shows

$$(2^{J_0}\delta)^{-(s+\kappa)l'/l} |\mathrm{Diff}_{\delta\xi_0\mathfrak{e}_k}^{l'} v_{J_0}(x_0)| = (2^{J_0}\delta)^{-(s+\kappa)l'/l} |\mathrm{Diff}_{2^{J_0}\delta\xi_0\mathfrak{e}_k}^{l'} \hat{v}_{x_0,J_0}(0)| \lesssim 1. \tag{2.141}$$

Using (2.138) and (2.140) shows

$$(2^{J_0}\delta)^{-(s+\kappa)l'/l} |\mathrm{Diff}_{\delta\xi_0\mathfrak{e}_k}^{l'} u_{J_0}(x_0)| = (2^{J_0}\delta)^{-(s+\kappa)l'/l} |\mathrm{Diff}_{2^{J_0}\delta\xi_0\mathfrak{e}_k}^{l'} \hat{u}_{x_0,J_0}(0)| \lesssim 1. \tag{2.142}$$

Since $u = u_{J_0} + 2^{-\kappa J_0} v_{J_0}$ (see (2.127)), (2.130) follows by combining (2.141) and (2.142), completing the proof.

2.9 Further reading and references

The prototypical example of a singular integral operator, the Hilbert transform, first arose in Hilbert's work on what is now known as the Riemann–Hilbert problem. Hilbert's proof was published by Weyl [250]. These results were later improved by Schur [208]. The results of Hilbert, Weyl, and Schur were restricted to L^2; Marcel Riesz extended these results to L^p $(1 < p < \infty)$ [204]. These references were focused on "complex analysis methods," and did not generalize to higher dimensions. Besikovitch [10], Titchmarsh [237], and Marcinkiewicz [166] offered a "real-variable" analysis of the Hilbert transform.

The real-variable analysis of the Hilbert transform was a main motivating example for Calderón and Zygmund when they introduced the first iteration of what are now known as Calderón–Zygmund singular integrals [27]. The proof of Corollary 2.3.13 uses their methods; the interpolation theorem used in that proof (the "Marcinkiewicz interpolation theorem") was proved by Marcinkiewicz [167]. Also used in Corollary 2.3.13 was the Cotlar–Stein lemma (Lemma 2.3.25). The special case of the Cotlar–Stein lemma when T_j are all assumed to be self-adjoint and to pairwise commute was originally proved by Cotlar [54]. Lemma 2.3.25 in its full generality is due to E. M. Stein; see [216, Chapter VII, § 2]. Though it may not be immediately apparent, these methods form the foundation for many of the L^p estimates later in this text.

The concept of a pseudo-differential operator is rooted in the work of Marcinkiewicz [167], later work by Calderón and Zygmund, and the work of Seeley in his thesis [209]. This was follows by further work of Seeley [210] and culminated in the work of Unterberger and Bokobza [247, 246], Kohn and Nirenberg [144], and Hörmander [124]. It was these last two references that first exhibited the theory of pseudo-differential operators as covered in Section 2.2, though our presentation more closely follows the one from [216, Chapter VI]. As mentioned in Section 2.2.1, the generalizations of pseudo-differential operators in Chapter 4 cannot use the Fourier transform, and are more closely related to the earlier works mentioned above.

The theory of non-translation invariant, non-homogeneous Calderón–Zygmund singular integral operators (roughly speaking, singular integral operators of order 0 from Definition 2.3.1) was the work of many authors. A systematic approach, working in more general "spaces of homogeneous type" and with much less regular kernels, was developed by Coifman and Weiss [52]. Definition 2.3.1 (at least for singular integrals of order $> -n$) was taken from much more recent work of Nagel, Rosay, Stein, and Wainger [182] and Koenig [140], who worked in the more general setting described in Sections 5.1.1 and 5.2.1 (see Section 5.13 for further comments on their work). Results like the characterization of singular integral operators given in Theorem 2.3.10 can be found in [220, Chapter I], though they are closely related to much earlier and well-known methods – for example Theorem 2.2.26. A key idea of Theorem 2.3.10 – to characterize singular integrals in terms of their action on certain special functions or operators (e. g., Theorem 2.3.10 (ii)) – was first pointed out to the author in a graduate course given by E. M. Stein at Princeton University in 2007. The concept of defining the singular integrals partially in terms of a "cancelation condition" (as in Definition 2.3.1) is closely related to the hypotheses of the $T(1)$ theorem of David and Journé [60]; see also the presentation in [216, pages 293–294]. This idea was further championed by E. M. Stein; see, e. g., [216, page 248].

The decomposition of singular integral operators as in Theorem 2.3.10 (iii) is called a Littlewood–Paley decomposition of the operator, named so because the first place where similar decompositions appeared was in the work of Littlewood and Paley on Fourier series [149–151, 194]. These ideas were later worked on by Zygmund and Marcinkiewicz, but were moved to higher dimensions and used in greater generality by E. M. Stein (see, e. g., [214]). Such decompositions are now common in the theory of pseudo-differential and singular integral operators; see for example Theorem 2.2.26 (iii).

The Hardy–Littlewood maximal function (2.33) was introduced by Hardy and Littlewood, and the vector-valued result we used (Lemma 2.4.14) is due to C. Fefferman and E. M. Stein [83].

For more history on some of the above topics, see the expository articles of E. M. Stein [215, 217] along with Stein's book [216]. In particular, the case $s = 0$ of Proposition 2.4.9 is often known as a square function estimate; [215] contains a detailed history of such square functions.

The Besov spaces $\mathscr{B}^s_{p,q}(\mathbb{R}^n)$ were introduced by Peetre [195] (which was preceded by some earlier ideas of Hörmander in the case $p = q = 2$ [123]). The Triebel–Lizorkin spaces $\mathscr{F}^s_{p,q}(\mathbb{R}^n)$ were introduced by Triebel [240] with closely related definitions given by Lizorkin [152, 153]. Besov and Triebel–Lizorkin spaces unify a number of classical function spaces. Two of these are of particular interest to us: the Sobolev spaces L^p_s (Definition 2.2.11) and the Zygmund–Hölder spaces $\mathscr{C}^s$ (Section 2.5). The Sobolev spaces were introduced by Aronszajn and Smith [4] and Calderón [26]. The Zygmund–Hölder spaces were introduced by Zygmund [255]. For a detailed discussion of the spaces generalized by the Besov and Triebel–Lizorkin spaces and the history of these spaces, we refer the reader to [244, § 2.2.1 and § 2.3.5]. The theory of these spaces is extensive (see, for example, the books by Triebel [244, 242, 243, 241]). We have only described the very basic definitions and results in this chapter, as that is all we generalize to the maximally subelliptic setting. However, it would be interesting to generalize other aspects of these theories to the maximally subelliptic setting.

The theory of linear elliptic operators described in Section 2.7 is classical. Theorem 2.7.2 is well known, in some form or another, to any expert in the field.

The history of results like Theorem 2.8.3 is long, and many improvements are known. The expository article [20] contains a detailed history. There are several known proofs of Theorem 2.8.3; however, most of them do not easily generalize to the maximally subelliptic setting. This is usually because they implicitly rely on the basic fact that ∂_{x_j} and ∂_{x_k} commute. For example, one standard approach can be found on pages 139 and 140 of [233] in the proof of Theorem 4.6. There is it used that derivatives and differences commute, a property which we do not have when we move to the maximally subelliptic setting in Chapter 9.

3 Vector fields and Carnot–Carathéodory geometry

Central to the study of maximally subelliptic operators is the Carnot–Carathéodory geometry induced by the vector fields with formal degrees $(W, \vec{d}) = \{(W_1, \vec{d}_1), \ldots, (W_r, \vec{d}_r)\}$, with balls $B_{(W,\vec{d})}(x, \delta)$ defined in (1.13) and associated metric $\rho_{(W,\vec{d})}(x,y)$ defined in (1.14).

In this chapter we present the main tools we need to study these geometries and vector fields. A key goal is to define and study the associated scaling maps as discussed in Section 1.9. We study these scaling maps both in the single-parameter case (Section 3.3) and in a more general multi-parameter setting (Section 3.5). These theorems about scaling (along with other, more quantitative versions we need) all follow from a single, abstract, technical theorem of the author and Stovall [228], which is presented in Section 3.6 and proved in Appendix A.

Since this chapter is the toolbox for the rest of the text, the reader wishing to understand the main ideas might start by reading up through Section 3.3, and then referring back to this chapter whenever a tool or topic is needed.

3.1 Manifolds

We assume the reader is familiar with the basic definitions and results concerning manifolds and vector fields; for background see [147]. In this section, we fix notation and review some concepts which, though they are standard, may not be immediately recalled by readers who only know the basic theory of manifolds.

For $L \in \mathbb{N}$ and an open subset $\Omega \subseteq \mathbb{R}^n$, we defined the Banach space $C^L(\Omega)$ of L-times differential functions whose derivatives up to order L were bounded (see (2.2)); we similarly defined the Fréchet space $C^\infty(\Omega)$. We also defined the Fréchet space $C^L_{\mathrm{loc}}(\Omega)$ to be the space of functions which were locally in C^L.

Let $\mathfrak{M}$ be a connected C^∞ manifold (all manifolds in this text are assumed to be second countable). There is no intrinsic definition of a Banach space $C^L(\mathfrak{M})$, unless $\mathfrak{M}$ is compact. However, there is an intrinsic definition of $C^L_{\mathrm{loc}}(\mathfrak{M})$. Thus, whenever we speak of functions on a manifold, we will always use the local spaces $C^L_{\mathrm{loc}}(\mathfrak{M})$, and we will not use any norm of the form $\|f\|_{C^L(\mathfrak{M})}$, as there is no intrinsic definition of this norm. We let $C^\infty_0(\mathfrak{M})$ denote the locally convex topological vector space consisting of those functions in $C^\infty_{\mathrm{loc}}(\mathfrak{M})$ which have compact support. As before we denote by $C^\infty_0(\mathfrak{M})'$ the space of distributions on $\mathfrak{M}$ and by $C^\infty_{\mathrm{loc}}(\mathfrak{M})'$ the space of distributions with compact support.

Definition 3.1.1. We say $\Phi : U \to \mathfrak{M}$ is a smooth coordinate system if $U \subseteq \mathbb{R}^n$ and $\Phi(U) \subseteq \mathfrak{M}$ are open and $\Phi : U \to \Phi(U)$ is a C^∞_{loc} diffeomorphism.

Let $\mathscr{V}$ denote a C^∞ vector bundle over $\mathfrak{M}$. We let $C^\infty_{\mathrm{loc}}(\mathfrak{M}; \mathscr{V})$ denote the space of C^∞_{loc} sections of $\mathscr{V}$. If $\mathscr{V}$ is a finite-dimensional vector space, we let $C^\infty_{\mathrm{loc}}(\mathfrak{M}; \mathscr{V})$ denote the space of C^∞_{loc} sections of the trivial vector bundle $\mathfrak{M} \times \mathscr{V}$. In particular, $C^\infty_{\mathrm{loc}}(\mathfrak{M}; T\mathfrak{M})$ denotes the space of smooth vector fields on $\mathfrak{M}$. Let $\Omega \subseteq \mathbb{R}^n$ be an open set. Then $T\Omega$ can be

https://doi.org/10.1515/9783111085647-003

identified with the trivial vector bundle $\Omega \times \mathbb{R}^n$ and therefore $C^\infty_{\mathrm{loc}}(\Omega; T\Omega) \cong C^\infty_{\mathrm{loc}}(\Omega; \mathbb{R}^n)$. Thus, for a vector field Y on Ω, it makes sense to ask if $Y \in C^L(\Omega; \mathbb{R}^n)$. We do this by identifying $Y = \sum_{j=1}^n a_j \partial_{x_j}$ with the function $a(x) = (a_1(x), \ldots, a_n(x))$. We then define $\|Y\|_{C^L(\Omega;\mathbb{R}^n)} := \|a\|_{C^L(\Omega;\mathbb{R}^n)}$, and we say $Y \in C^L(\Omega; \mathbb{R}^n)$ if and only if $a \in C^L(\Omega; \mathbb{R}^n)$. Such a norm and space are only defined for vector fields on open subsets of $\mathbb{R}^n$ and are not defined on an abstract manifold. This is an important distinction in many of our proofs, where we will be given vector fields on an abstract manifold, and we will find a coordinate system in which they are "normalized"; see Section 3.3.

Suppose $\mathfrak{M}$ and $\mathfrak{N}$ are C^∞ manifolds and $\Phi : \mathfrak{M} \to \mathfrak{N}$ is a C^∞_{loc} map. We write Φ^* for the pullback via Φ and Φ_* for the pushforward via Φ. If $\Phi : \mathfrak{M} \to \mathfrak{N}$ is a C^∞_{loc} diffeomorphism, then Φ^* and Φ_* are inverses to each other, whenever they are defined.

Example 3.1.2. Suppose $\Phi : \mathfrak{M} \to \mathfrak{N}$ is a C^∞_{loc} diffeomorphism. Then:
- If $f \in C^\infty_{\mathrm{loc}}(\mathfrak{N})$, then $\Phi^* f = f \circ \Phi \in C^\infty_{\mathrm{loc}}(\mathfrak{M})$.
- If $g \in C^\infty_{\mathrm{loc}}(\mathfrak{M})$, then $\Phi_* g = g \circ \Phi^{-1} \in C^\infty_{\mathrm{loc}}(\mathfrak{N})$.
- If $X \in C^\infty_{\mathrm{loc}}(\mathfrak{N}; T\mathfrak{N})$, then $\Phi^* X \in C^\infty_{\mathrm{loc}}(\mathfrak{M}; T\mathfrak{M})$ is defined by

$$\Phi^* X(m) = (d\Phi(m))^{-1}(X(\Phi(m))) \in T_m\mathfrak{M}.$$

- If $Y \in C^\infty_{\mathrm{loc}}(\mathfrak{M}; T\mathfrak{M})$, then $\Phi_* Y \in C^\infty_{\mathrm{loc}}(\mathfrak{N}; T\mathfrak{N})$ is defined by

$$\Phi_* Y(n) = d\Phi(\Phi^{-1}(n))(Y(\Phi^{-1}(n))) \in T_n\mathfrak{N}.$$

Remark 3.1.3. If we consider vector fields as first-order partial differential operators and $\Phi : \mathfrak{M} \to \mathfrak{N}$ is a C^∞_{loc} diffeomorphism, then for $X \in C^\infty_{\mathrm{loc}}(\mathfrak{N}; T\mathfrak{N})$ we have

$$\Phi^*(X(\Phi_* g)) = (\Phi^* X)g, \quad \forall g \in C^\infty_{\mathrm{loc}}(\mathfrak{M}).$$

We will have occasion to use densities on a manifold. Given a real vector space $\mathscr{V}$ of dimension n, we let $|\mathscr{V}|$ denote the space of all maps

$$\sigma : \underbrace{\mathscr{V} \times \mathscr{V} \times \cdots \times \mathscr{V}}_{n \text{ factors}} \to \mathbb{R}$$

such that for any linear map $A : \mathscr{V} \to \mathscr{V}$ we have $\sigma(Av_1, \ldots, Av_n) = |\det A|\sigma(v_1, \ldots, v_n)$. There is a canonical element of $|\mathbb{R}^n|$, called the Lebesgue density, given by

$$\sigma_{\mathrm{Leb}}(v_1, \ldots, v_n) = |\det(v_1|\cdots|v_n)|, \quad v_1, \ldots, v_n \in \mathbb{R}^n,$$

where $(v_1|\cdots|v_n)$ denotes the $n \times n$ matrix with columns $v_1, \ldots, v_n$.

Definition 3.1.4. A density on $\mathfrak{M}$ is a map σ, which assigns to each $x \in \mathfrak{M}$ an element $\sigma(x) \in |T_x\mathfrak{M}|$.

Let $\mathfrak{M}$ and $\mathfrak{N}$ be two manifolds. Given a C^∞_{loc} map $\Phi : \mathfrak{M} \to \mathfrak{N}$ and a density σ on $\mathfrak{N}$, we define a density $\Phi^*\sigma$ on $\mathfrak{M}$ by

$$(\Phi^*\sigma)(m)(v_1,\dots,v_n) = \sigma(\Phi(m))(d\Phi(m)v_1,\dots,d\Phi(m)v_n), \quad v_1,\dots,v_n \in T_m\mathfrak{M}. \tag{3.1}$$

Definition 3.1.5. Let $\mathfrak{M}$ be a C^∞ manifold of dimension n and let σ a density on $\mathfrak{M}$. We say σ is a smooth density if the following holds. For every smooth coordinate system $\Phi : U \to \Phi(U)$, where $U \subseteq \mathbb{R}^n$ is open, we have

$$\Phi^*\sigma = f\sigma_{\mathrm{Leb}},$$

where $f \in C^\infty_{\mathrm{loc}}(U)$. We say σ is a strictly positive density if $f > 0$ for all such Φ.

Given a strictly positive, smooth density σ on $\mathfrak{M}$, σ induces a non-negative Borel measure on $\mathfrak{M}$ which is strictly positive on every open set. Indeed, this may be defined locally as follows. If $S \subseteq \mathfrak{M}$ is a Borel set such that there exists a smooth coordinate system $\Phi : U \to \Phi(U)$, we define

$$\sigma(S) := \int_{\Phi^{-1}(S)} f(x)\,dx,$$

where $\Phi^*\sigma = f\sigma_{\mathrm{Leb}}$ and dx denotes the Lebesgue measure on $\mathbb{R}^n$. Henceforth, we identify σ with this measure. For more details on densities, we refer the reader to the particularly lucid presentation in [100] (see also [191], where densities are called 1-densities).

3.1.1 The exponential map

Let $\mathfrak{M}$ be a C^∞ manifold and let $X \in C^\infty_{\mathrm{loc}}(\mathfrak{M}; T\mathfrak{M})$ be a smooth vector field on $\mathfrak{M}$. For $x \in \mathfrak{M}$ and $t \in \mathbb{R}$ small, we define

$$e^{tX}x = E(t),$$

where $E(t)$ is the unique solution to the ordinary differential equation (ODE)

$$\frac{d}{dt}E(t) = X(E(t)), \quad E(0) = x.$$

Classical results from the field of ODEs show that $(t,x) \mapsto e^{tX}x$ defines a smooth map $U \to M$, where U is an open neighborhood of $\{0\} \times \mathfrak{M} \subset \mathbb{R} \times \mathfrak{M}$. Furthermore, we have the identity $e^{-tX}e^{tX}x = e^{0X}x = x$ for (t,x) in an open neighborhood of $\{0\} \times \mathfrak{M} \subset \mathbb{R} \times \mathfrak{M}$. It follows from the definition that

$$\frac{d}{dt}f(e^{tX}x) = (Xf)(e^{tX}x). \tag{3.2}$$

Let $\Phi_t(x) := e^{tX}x$. By the above remarks, given a relatively compact, open set $\Omega \Subset \mathfrak{M}$, for $t \in \mathbb{R}$ sufficiently small (depending on Ω), $\Phi_t : \Omega \to \Phi_t(\Omega)$ is a C^∞_{loc} diffeomorphism with inverse Φ_{-t}. We define the Lie derivative with respect to X by $\mathrm{Lie}_X = \frac{d}{dt}\big|_{t=0}\Phi_t^*$.

Example 3.1.6. For $f \in C^\infty_{\mathrm{loc}}(\mathfrak{M})$, $\mathrm{Lie}_X f = Xf$. For $Y \in C^\infty_{\mathrm{loc}}(\mathfrak{M}; T\mathfrak{M})$, $\mathrm{Lie}_X Y = [X, Y]$. For a smooth density σ on $\mathfrak{M}$, $\mathrm{Lie}_X \sigma = \frac{d}{dt}\big|_{t=0}\Phi_t^*\sigma$, where $\Phi_t^*\sigma$ is defined by (3.1); in particular, if $\mathfrak{M} = \mathbb{R}^n$, then $\mathrm{Lie}_X(f\sigma_{\mathrm{Leb}}) = (Xf + f\nabla\cdot X)\sigma_{\mathrm{Leb}}$, where $\nabla\cdot X$ denotes the divergence of X.

We will have many uses for the exponential map in this text. We begin with a simple example, following the work of Nagel, Stein, and Wainger [189].

Lemma 3.1.7. *Let $(W, \vec{d}) = \{(W_1, \vec{d}_1), \dots, (W_r, \vec{d}_r)\} \subset C^\infty_{\mathrm{loc}}(\mathfrak{M}; T\mathfrak{M}) \times \mathbb{N}_+$ be a set of Hörmander vector fields with formal degrees on the connected C^∞ manifold $\mathfrak{M}$. Then $\rho_{(W,d)}$ is a metric, and the metric topology induced by $\rho_{(W,d)}$ is the same as the underlying topology on $\mathfrak{M}$.*

Proof. $\rho_{(W,d)}$ is easily seen to be symmetric and satisfy the triangle inequality. If U is an open neighborhood of $x \in \mathfrak{M}$, then the Picard–Lindelöf theorem shows that there is a $\delta > 0$ such that $B_{(W,d)}(x, \delta) \subseteq U$. Thus, the topology induced by $\rho_{(W,d)}$ is finer than the topology on $\mathfrak{M}$; this also implies $\rho_{(W,d)}(x, y) > 0$ for $x \neq y$.

Next, we will show that the topology on $\mathfrak{M}$ is finer than the metric topology induced by $\rho_{(W,d)}$. Fix $x \in \mathfrak{M}$ and $\delta > 0$. We wish to show that there is an open set U with $x \in U \subseteq B_{(W,d)}(x, \delta)$. For smooth vector fields $S_1, S_2, \dots \in C^\infty_{\mathrm{loc}}(\mathfrak{M}; T\mathfrak{M})$ and $a \in \mathbb{R}$ small, set

$$C_1(a, S_1) := e^{aS_1}, \quad C_2(a, S_1, S_2) := e^{aS_1}e^{aS_2}e^{-aS_1}e^{-aS_2},$$

and more generally,

$$C_l(a, S_1, \dots, S_l) := e^{aS_1}C_{l-1}(a, S_2, \dots, S_{l-1})e^{-aS_1}(C_{l-1}(a, S_2, \dots, S_l))^{-1}.$$

Standard results (see Section 3.1.2) show that

$$\begin{aligned} \partial_a^l\big|_{a=0}C_l(a, S_1, \dots, S_l)x &= [S_1, [S_2, \dots, [S_{l-1}, S_l]\cdots]](x), \\ \partial_a^k\big|_{a=0}C_l(a, S_1, \dots, S_l)x &= 0, \quad \forall k < l. \end{aligned}$$

Pick $Y_1, \dots, Y_n$ of the form

$$Y_j = [W_{l_1^j}, [W_{l_2^j}, \dots, [W_{l^j_{k_j-1}}, W_{l_{k_j}}]\cdots]]$$

so that $Y_1(x), \dots, Y_n(x)$ form a basis for $T_x\mathfrak{M}$ (this is always possible since W satisfies Hörmander's condition). Set, for $t \in \mathbb{R}$ with t small,

$$C_j(t) := \begin{cases} C_{k_j}(t^{\frac{1}{k_j}}, W_{l_1^j}, W_{l_2^j}, \dots, W_{l_{k_j}^j}), & k_j \text{ odd or } k_j \text{ even and } t \geq 0, \\ C_{k_j}(|t|^{\frac{1}{k_j}}, W_{l_1^j}, W_{l_2^j}, \dots, W_{l_{k_j}^j})^{-1}, & k_j \text{ even and } t < 0. \end{cases}$$

Set

$$D(t_1, \dots, t_n) := C_1(t_1)C_2(t_2)\cdots C_n(t_n)x.$$

Note that $D(0) = x$. It is easy to see that D is C^1_{loc} on a neighborhood of $0 \in \mathbb{R}^n$. Furthermore, since $\partial_{t_j}|_{t=0}D(t) = Y_j(x)$, we see that $dD(0)$ is an invertible map. The inverse function theorem applies to show that if $\epsilon > 0$ is sufficiently small, $D(B^n(\epsilon))$ is open. Furthermore, if $\epsilon > 0$ is small enough, it is easy to see that $D(B^n(\epsilon)) \subseteq B_{(W,d)}(x, \delta)$. Since $x = D(0) \in D(B^n(\epsilon))$, it follows that the topology on $\mathfrak{M}$ is finer than the topology induced by $\rho_{(W,d)}$. Thus, the topology induced by $\rho_{(W,d)}$ agrees with the topology on $\mathfrak{M}$.

To see that $\rho_{(W,d)}$ is a metric, it remains to show that $\rho_{(W,d)}(x,y) < \infty$, $\forall x, y$. We have already shown that for each x, $\{y \in \mathfrak{M} : \rho_{(W,d)}(x,y) < \infty\}$ is an open set, and by the same proof its complement is open. Since $\mathfrak{M}$ is connected, $\{y \in \mathfrak{M} : \rho_{(W,d)}(x,y) < \infty\} = \mathfrak{M}$, completing the proof. □

3.1.2 The Baker–Campbell–Hausdorff formula

The Baker–Campbell–Hausdorff formula is a formal series, where if one considers the algebra of formal series in two non-commuting indeterminants X, Y, then $e^X e^Y = e^{\mathrm{BCH}(X,Y)}$, where $\mathrm{BCH}(X,Y)$ is an infinite formal (Lie) series

$$\begin{aligned} \mathrm{BCH}(X,Y) = X + Y + \frac{1}{2}[X,Y] + \frac{1}{12}[X,[X,Y]] + \frac{1}{12}[Y,[Y,X]] \\ + \text{commutators of order } 3 + \text{commutators of order } 4 + \cdots. \end{aligned}$$

This formal series is universal (the coefficients are universal constants). It is possible to write down an explicit formula for the coefficients (see, for example, [12]), though what is important for our applications is that it is a universal formula.

Let $X \in C^\infty_{\mathrm{loc}}(\mathfrak{M}; T\mathfrak{M})$ be a smooth vector field and let $f \in C^\infty_{\mathrm{loc}}(\mathfrak{M})$. Define

$$e^{tX} f(x) := f(e^{tX}x).$$

Then $(t,x) \mapsto e^{tX}f(x) \in C^\infty_{\mathrm{loc}}(U)$, where U is an open neighborhood of $\{0\} \times \mathfrak{M}$ in $\mathbb{R} \times \mathfrak{M}$. Repeatedly using (3.2), we see that the Taylor series in the t variable of $e^{tX}f(x)$ can be computed as

$$e^{tX}f(x) \sim \sum_{j=0}^{\infty} \frac{(tX)^j}{j!} f(x).$$

As a consequence, if $Y \in C^\infty_{\mathrm{loc}}(\mathfrak{M}; T\mathfrak{M})$, then as a formal power series in t and s, we have

$$f(e^{sY} e^{tX} x) = e^{tX} e^{sY} f(x) \sim e^{\mathrm{BCH}(tX, sY)} f(x), \tag{3.3}$$

where the right-hand side is only interpreted as a formal power series. If sufficiently high-order commutators of X and Y are zero, then $\mathrm{BCH}(tX, sY)$ is a finite sum and (3.3) is an equality of functions, not just of formal sums.

One application of (3.3) is that it allows us to compute the Taylor series in the s variable of $X(f(e^{sY}x))$. Indeed,

$$Xf(e^{sY}x) = \frac{d}{dt}\bigg|_{t=0} f(e^{sY} e^{tX} x) \sim \frac{d}{dt}\bigg|_{t=0} e^{\mathrm{BCH}(tX,sY)} f(x).$$

3.1.3 The Frobenius theorem

Closely tied to Carnot–Carathéodory geometry is a quantitative version of the classical Frobenius theorem. In this section, we describe this classical theorem in the form which is of most use to us. The presentation here is a special case of the theory developed by Sussman [231], and we refer the reader to that paper for more details.

Definition 3.1.8. Let $\mathfrak{N}$ and $\mathfrak{M}$ be C^∞ manifolds. An immersion of $\mathfrak{N}$ in $\mathfrak{M}$ is a smooth map $\mathfrak{N} \to \mathfrak{M}$ whose differential is everywhere injective.

Definition 3.1.9. Let $\mathfrak{M}$ be a C^∞ manifold. An injectively immersed submanifold $\mathfrak{N}$ of $\mathfrak{M}$ is a smooth manifold $\mathfrak{N}$ along with an injective immersion $i : \mathfrak{N} \hookrightarrow \mathfrak{M}$.

Remark 3.1.10. If $\mathfrak{N} \hookrightarrow \mathfrak{M}$ is an injectively immersed submanifold of $\mathfrak{M}$, then we identify $\mathfrak{N}$ with its image under the immersion, and therefore think of $\mathfrak{N}$ as a *subset* of $\mathfrak{M}$. However, $\mathfrak{N}$ need not be a topological subspace of $\mathfrak{M}$. See Example 3.1.19.

Remark 3.1.11. If $i : \mathfrak{N} \hookrightarrow \mathfrak{M}$ is an injectively immersed submanifold, for $x \in \mathfrak{N}$, we may identify $T_x\mathfrak{N}$ with its image under $di(x) : T_x\mathfrak{N} \hookrightarrow T_{i(x)}\mathfrak{M}$. Henceforth, we write $T_x\mathfrak{N}$ to denote this subspace of $T_{i(x)}\mathfrak{M}$.

Note that $C^\infty_{\mathrm{loc}}(\mathfrak{M}; T\mathfrak{M})$ is a $C^\infty_{\mathrm{loc}}(\mathfrak{M})$ module: given $f, g \in C^\infty_{\mathrm{loc}}(\mathfrak{M})$ and $X, Y \in C^\infty_{\mathrm{loc}}(\mathfrak{M}; T\mathfrak{M})$, $fX + gY \in C^\infty_{\mathrm{loc}}(\mathfrak{M}; T\mathfrak{M})$. For the next three definitions, let $\mathcal{D} \subseteq C^\infty_{\mathrm{loc}}(\mathfrak{M}; T\mathfrak{M})$ be a $C^\infty_{\mathrm{loc}}(\mathfrak{M})$-submodule.

Definition 3.1.12. For $x \in \mathfrak{M}$, set $\mathcal{D}(x) := \mathrm{span}\{X(x) : X \in \mathcal{D}\}$.

Definition 3.1.13. We say $\mathcal{D}$ is locally finitely generated as a $C^\infty_{\mathrm{loc}}(\mathfrak{M})$-module if for each $x \in \mathfrak{M}$, there is a neighborhood $V \subseteq \mathfrak{M}$ of x and a finite collection $Z_1, \ldots, Z_q \in \mathcal{D}$ such that for every $Y \in \mathcal{D}$, there exist $c_1, \ldots, c_q \in C^\infty_{\mathrm{loc}}(V)$ with $Y|_V = \sum_{j=1}^q c_j Z_j|_V$.

Definition 3.1.14. We say $\mathcal{D}$ is involutive if $\forall X, Y \in \mathcal{D}$, $[X, Y] \in \mathcal{D}$, that is, if $\mathcal{D}$ is a Lie subalgebra of $C^\infty_{\mathrm{loc}}(\mathfrak{M}; T\mathfrak{M})$.

Theorem 3.1.15 (The Frobenius theorem; see Theorem 4.2 in [231]). *Let $\mathcal{D} \subseteq C^\infty_{\mathrm{loc}}(\mathfrak{M}; T\mathfrak{M})$ be a $C^\infty_{\mathrm{loc}}(\mathfrak{M})$-submodule which is involutive and locally finitely generated as a $C^\infty_{\mathrm{loc}}(\mathfrak{M})$-module. Then, for each point $x \in \mathfrak{M}$, there is a unique, maximal, connected, injectively immersed submanifold $\mathfrak{L} \hookrightarrow \mathfrak{M}$, with $x \in \mathfrak{L}$ and $\forall y \in \mathfrak{L}$, $T_y\mathfrak{L} = \mathcal{D}(y)$. The set of all such $\mathfrak{L}$, as $x \in \mathfrak{M}$ varies, forms a partition of $\mathfrak{M}$ into disjoint injectively immersed submanifolds.*

Definition 3.1.16. In the setting of Theorem 3.1.15, we say $\mathfrak{L}$ is the leaf passing through x, and we refer to the conclusion by saying that $\mathcal{D}$ foliates $\mathfrak{M}$ into leaves.

Remark 3.1.17. Under the conditions of Theorem 3.1.15, the vector fields in $\mathcal{D}$ are tangent to each leaf. Thus, we may treat them as vector fields on the leaves. By the conclusion of Theorem 3.1.15, these vector fields span the tangent space to the leaf at every point.

Example 3.1.18. Consider $\mathbb{R}^2$ with the usual coordinates (x, y). Let $\mathcal{D} = C^\infty_{\mathrm{loc}}(\mathbb{R}^2)\{\partial_y\}$. Then the leaf passing through (x_0, y_0) is

$$\{(x_0, y) : y \in \mathbb{R}\}.$$

Thus, in this case, the Frobenius theorem decomposes $\mathbb{R}^2$ into its usual product structure $\mathbb{R} \times \mathbb{R}$.

Example 3.1.19. Let $\mathfrak{M} = \mathbb{R}/\mathbb{Z} \times \mathbb{R}/\mathbb{Z}$ be the torus with inherited coordinates (x, y) from $\mathbb{R} \times \mathbb{R}$. Fix $\theta \in \mathbb{R} \setminus \mathbb{Q}$ an irrational number. Let $\mathcal{D} = C^\infty_{\mathrm{loc}}(\mathfrak{M})\{\partial_x + \theta\partial_y\}$. In this case the leaves are one-dimensional, dense subsets of $\mathfrak{M}$ and carry a topology strictly finer than the subspace topology.

Example 3.1.20. Consider the case where $\mathcal{D} = C^\infty_{\mathrm{loc}}(\mathfrak{M}; T\mathfrak{M})$. In this case, there is only one leaf, and that leaf equals $\mathfrak{M}$. While this example is trivial from the perspective of Theorem 3.1.15, in Section 3.6 we introduce a quantitative version of the Frobenius theorem where this example is not trivial. In fact, from a quantitative perspective, this example is the most important one in this text.

Suppose $\mathcal{D}$ satisfies the conditions of Theorem 3.1.15 and let $\mathfrak{L}_x$ denote the leaf passing through $x \in \mathfrak{M}$. Note

$$\dim \mathfrak{L}_x = \dim T_x\mathfrak{L}_x = \dim \mathcal{D}(x).$$

Nowhere did we assume $\dim \mathcal{D}(x)$ is constant in x, and therefore the dimensions of the leaves may vary from point to point. To discuss this, we introduce a definition.

Definition 3.1.21. We say that $x \in \mathfrak{M}$ is a singular point of $\mathcal{D}$ if $\dim \mathcal{D}(x)$ is not constant on any neighborhood of x.

Example 3.1.22. Let $\mathfrak{M} = \mathbb{R}^2$ with the usual coordinates (x, y). We set $\mathcal{D} = C^\infty_{\mathrm{loc}}(\mathbb{R}^2)\{x\partial_y\}$. Note that

$$\dim \mathcal{D}(x,y) = \begin{cases} 1 & \text{if } x \neq 0, \\ 0 & \text{if } x = 0. \end{cases}$$

If $(x_0, y_0) \in \mathbb{R}^2$ has $x_0 \neq 0$, then the leaf passing through (x_0, y_0) is $\{(x_0, y) : y \in \mathbb{R}\}$. On the other hand, the leaf passing through $(0, y_0)$ is the point $\{(0, y_0)\}$. In this case, each point of the form $(0, y_0)$ is a singular point of $\mathcal{D}$. The map $(x,y) \mapsto \dim \mathcal{D}(x,y)$ is not continuous at the singular points.

Suppose $\mathcal{D}$ satisfies the conditions of Theorem 3.1.15 and let $\mathfrak{L}_x$ denote the leaf passing through $x \in \mathfrak{M}$. As we saw in Example 3.1.22, the map $x \mapsto \dim \mathfrak{L}_x = \dim \mathcal{D}(x)$ may not be continuous $\mathfrak{M} \to \mathbb{N}$. However, it is lower semi-continuous, as the next result shows.

Lemma 3.1.23. *Let $\mathcal{D} \subseteq C^\infty_{\mathrm{loc}}(\mathfrak{M}; T\mathfrak{M})$ be a $C^\infty_{\mathrm{loc}}(\mathfrak{M})$-submodule which is locally finitely generated as a $C^\infty_{\mathrm{loc}}(\mathfrak{M})$-module. Then $x \mapsto \dim \mathcal{D}(x)$, $\mathfrak{M} \to \mathbb{N}$ is lower semi-continuous.*

Proof. We will show $\{x : \dim \mathcal{D}(x) > r\} \subseteq \mathfrak{M}$ is open, $\forall r \in \mathbb{R}$. Letting $m - 1 = \lfloor r \rfloor \vee -1$, it suffices to show that $\{x : \dim \mathcal{D}(x) > m - 1\}$ is open. The case where $m - 1 = -1$ is trivial, so we assume $m - 1 \geq 0$. Fix a point $x_0 \in \{x : \dim \mathcal{D}(x) > m - 1\}$. Let $V \subseteq \mathfrak{M}$ be an open neighborhood of x_0 so small that there is a smooth coordinate system $\Phi : U \xrightarrow{\sim} V$, with $U \subseteq \mathbb{R}^n$ open, and such that $\mathcal{D}$ is finitely generated on V as a $C^\infty_{\mathrm{loc}}(V)$-module. Thus, we have $Z_1, \ldots, Z_q \in \mathcal{D}$ such that $Z_1, \ldots, Z_q$ generate $\mathcal{D}$ on V as a $C^\infty_{\mathrm{loc}}(V)$-module. Note that for $y \in V$, $\mathcal{D}(y) = \mathrm{span}\{Z_1(y), \ldots, Z_q(y)\}$, and therefore,

$$\begin{aligned} &\{u \in U : \max\{|\det A| : A \text{ is an } m \times m \text{ submatrix of } (\Phi^* Z_1(u)| \cdots |\Phi^* Z_q(u))\} > 0\} \\ &\quad = \Phi^{-1}(\{x : \dim \mathcal{D}(x) > m - 1\}), \end{aligned}$$

where each $\Phi^* Z_j(u)$ is treated as an n-dimensional column vector. The left-hand side of the above equation is clearly an open neighborhood of $\Phi^{-1}(x_0) \in U$, and the result follows. □

Remark 3.1.24. The leaves of a foliation are manifolds and are therefore defined via an atlas of coordinate charts. In the usual proofs of the Frobenius theorem, one does not have good quantitative control of these coordinate charts near a singular point. In this text, we require better quantitative control than the standard proofs yield; in particular, we require uniform control near singular points. See Remark 3.5.2 for a further discussion of the sort of control we use.

3.2 The unit scale

In this text, scaling is the process of changing coordinates to turn a small scale into the unit scale. In this section, we describe the unit scale, which will always take place on $B^n(1)$, the unit ball in $\mathbb{R}^n$.

Let $(W, d) = \{(W_1, d_1), \dots, (W_r, d_r)\} \subset C^\infty(B^n(1); \mathbb{R}^n) \times \mathbb{N}_+$ be C^∞ vector fields on $B^n(1)$ paired with formal degrees. We assume $W = \{W_1, \dots, W_r\}$ satisfies Hörmander's condition of order $m \in \mathbb{N}_+$. Let $X_1, \dots, X_q$ be an enumeration of all commutators of $W_1, \dots, W_r$ up to order m, so that, by hypothesis, $X_1(x), \dots, X_q(x)$ span $T_x B^n(1)$, $\forall x \in B^n(1)$. We assume, moreover, that

$$\tau := \inf_{x \in B^n(1)} \max_{j_1, \dots, j_n \in \{1, \dots, q\}} \left|\det(X_{j_1}(x)| \cdots |X_{j_n}(x))\right| > 0. \tag{3.4}$$

For each $L \in \mathbb{N}$, fix $C_L \geq 0$ so that

$$\|W_j\|_{C^L(B^n(1);\mathbb{R}^n)} \leq C_L, \quad 1 \leq j \leq r.$$

Definition 3.2.1. For a parameter ι, we say C is an ι-unit-admissible constant if there exists $L \in \mathbb{N}$, depending only on ι, and upper bounds for $m, r, \max_{1 \leq j \leq r} d_j$, such that C can be chosen to depend only on ι and upper bounds for $n, m, r, \max_{1 \leq j \leq r} d_j, C_L$, and τ^{-1}.

Definition 3.2.2. Let $\mathcal{I}$ be a set, and for each $\kappa \in \mathcal{I}$, suppose $(W^\kappa, d^\kappa) = \{(W_1^\kappa, d_1^\kappa), \dots, (W_{r_\kappa}^\kappa, d_{r_\kappa}^\kappa)\}$ are C^∞ vector fields with formal degrees on $B^{n_\kappa}(1)$. We say (W^κ, d^κ) are Hörmander vector fields with formal degrees at the unit scale, uniformly in κ, if ι-unit-admissible constants (as in Definition 3.2.1) can be chosen independent of κ, that is, if the parameters $n, m, r, \max_{1 \leq j \leq r} d_j, C_L$, and τ^{-1} described there are bounded uniformly in κ.

Example 3.2.3. For $\delta > 0$, consider the vector fields with formal degrees, on $\mathbb{R}^n$, $\{(\delta\partial_{x_1}, 1), \dots, (\delta\partial_{x_n}, 1)\}$. The parameters C_L and τ^{-1} are not bounded uniformly in δ; indeed, $C_L \approx \delta$ and $\tau \approx \delta^n$. Thus, these vector fields with formal degrees are not Hörmander vector fields with formal degrees at the unit scale, uniformly in δ. However, for $x \in \mathbb{R}^n$ and $\delta > 0$, define the map $\Phi_{x,\delta} : B^n(1) \to (x + B^n(\delta))$ by $\Phi_{x,\delta}(t) = x + \delta t$. Then $\Phi_{x,\delta}^* \delta\partial_{x_j} = \partial_{t_j}$. Therefore, $\{(\Phi_{x,\delta}^* \delta\partial_{x_1}, 1), \dots, (\Phi_{x,\delta}^* \delta\partial_{x_n}, 1)\}$ are Hörmander vector fields with formal degrees at the unit scale, uniformly in δ. Thus, we have used $\Phi_{x,\delta}$ to change coordinates that turned vector fields at "scale δ" into vector fields at the "unit scale". Theorems 3.3.1, 3.3.7, 3.5.1, and 3.5.4 are more complicated versions of this simple idea.

Whenever we prove a result about Hörmander vector fields with formal degrees (W, d), the result will be uniform in the following sense. When we apply the result to vector fields (W^κ, d^κ) which are Hörmander vector fields with formal degrees at the unit scale, uniformly in κ, the estimates proved will all hold uniformly in κ. This is an important point, because many of our later results will apply earlier results infinitely many times in their proofs, and it is important that the estimates are uniform over these applications. In particular, by proving a uniform result at the unit scale, rescaling techniques will allow us to prove a result at every scale, uniformly in the scale. For example, we have the next result at the unit scale, which we will later use to prove results at every scale (see the proof of Theorem 3.5.1 (b)).

Lemma 3.2.4. *Let (W, d) be Hörmander vector fields with formal degrees at the unit scale. Then there exists a 0-unit-admissible constant $\eta > 0$ such that*

$$B^n(\eta) \subseteq B_W(0) = B_{(W,d)}(0, 1).$$

Proof. Lemma 3.1.7 shows that such an $\eta > 0$ exists. By keeping track of the estimates in the proof of that lemma, it is easy to see that η can be taken to be a 0-unit-admissible constant. □

3.3 Scaling and Hörmander vector fields

Let $\mathfrak{M}$ be a connected C^∞ manifold and let Vol be a strictly positive C^∞_{loc} density on $\mathfrak{M}$. Let $(W, d) \subset C^\infty_{\mathrm{loc}}(\mathfrak{M}; T\mathfrak{M}) \times \mathbb{N}_+$ be Hörmander vector fields with formal degrees.

Fix $\mathcal{K} \Subset \mathfrak{M}$ a compact set and $\Omega_1 \Subset \mathfrak{M}$ a relatively compact, open set, with $\mathcal{K} \Subset \Omega_1 \Subset \mathfrak{M}$. In what follows we write $A \lesssim B$ for $A \leq CB$, where $C \geq 0$ does not depend on $x, y \in \mathcal{K}$ or $\delta > 0$. C may depend on any of the other ingredients in the problem. We write $A \approx B$ for $A \lesssim B$ and $B \lesssim A$. All of the estimates and constants below depend on the particular choice of (W, d) and on the choice of $\mathcal{K}$ and Ω_1.

We state the main theorem of this section twice, beginning with a simpler version. We defer proofs of these results to Section 3.7.

Theorem 3.3.1. *There exist $\delta_0 = \delta_0(\mathcal{K}, \Omega_1) \in (0, 1]$ and $\xi_3 = \xi_3(\mathcal{K}, \Omega_1) \in (0, 1]$ such that the following hold:*

(a) $\forall x \in \mathcal{K}$, $\delta \in (0, \delta_0]$, $B_{(W,d)}(x, \delta) \subseteq \Omega_1$ *and* $B_{(W,d)}(x, \delta)$ *is open in* $\mathfrak{M}$.

(b) $\forall x \in \mathcal{K}$, $\delta \in (0, \delta_0]$, $\mathrm{Vol}(B_{(W,d)}(x, 2\delta)) \lesssim \mathrm{Vol}(B_{(W,d)}(x, \delta))$. $\forall x \in \mathcal{K}$, $\delta > 0$, $\mathrm{Vol}(B_{(W,d)}(x, 2\delta) \wedge 1 \lesssim \mathrm{Vol}(B_{(W,d)}(x, \delta)) \wedge 1$.

For $x \in \mathcal{K}$, $\delta \in (0, 1]$, there exists a map $\Phi_{x,\delta} : B^n(1) \to \mathfrak{M}$ such that:

(c) $\Phi_{x,\delta}(0) = x$.

(d) $\Phi_{x,\delta}$ *is a smooth coordinate system, that is,* $\Phi_{x,\delta}(B^n(1)) \subseteq \mathfrak{M}$ *is open and* $\Phi_{x,\delta} : B^n(1) \to \Phi_{x,\delta}(B^n(1))$ *is a* C^∞_{loc} *diffeomorphism.*

(e) $B_{(W,d)}(x, \xi_3\delta) \subseteq \Phi_{x,\delta}(B^n(1)) \subseteq B_{(W,d)}(x, \delta)$.

For $x \in \mathcal{K}$, $\delta \in (0, 1]$, set $W_j^{x,\delta} := \Phi_{x,\delta}^* \delta^{d_j} W_j$. Let $(W^{x,\delta}, d) = \{(W_1^{x,\delta}, d_1), \ldots, (W_r^{x,\delta}, d_r)\} \subset C^\infty(B^n(1); TB^n(1)) \times \mathbb{N}_+$.

(f) $(W^{x,\delta}, d)$ *are Hörmander vector fields with formal degrees at the unit scale, uniformly for* $x \in \mathcal{K}$ *and* $\delta \in (0, 1]$*; see Definition* 3.2.2.

Remark 3.3.2. The most important aspect of Theorem 3.3.1 is (f); this sees the maps $\Phi_{x,\delta}$ as scaling maps which take the case where δ is small and changes coordinates to turn it into the case $\delta = 1$. Thus, once we prove a result for Hörmander vector fields with formal degrees at the unit scale, we obtain a result at every small scale by pushing the result forward via $\Phi_{x,\delta}$. Another important aspect is (b), which is the main inequality needed to show that the balls $B_{(W,\vec{d})}(x,\delta)$, when paired with Vol, locally give $\mathfrak{M}$ the structure of a space of homogeneous type in the sense of Coifman and Weiss [52].

Example 3.3.3. To see the maps $\Phi_{x,\delta}$ as scaling maps, it is useful to consider some simple special cases.

- When $\mathfrak{M} = \mathbb{R}^n$ and $(W,\vec{d}) = \{(\partial_{x_1},1),\dots,(\partial_{x_n},1)\}$, we may take $\Phi_{x,\delta}(t_1,\dots,t_n) = (x_1+\delta t_1,\dots,x_n+\delta t_n)$. In this case we have the simple identity, $\Phi^*_{x,\delta}\delta\partial_{x_j} = \partial_{t_j}$. Thus, in the "elliptic setting," the maps $\Phi_{x,\delta}$ are nothing more than the usual dilation maps.
- With $\mathfrak{M} = \mathbb{R}^2$ and $(W,\vec{d}) = \{(\partial_{x_1},1),(x_1\partial_{x_2},1)\}$, the map $\Phi_{x,\delta}$ depends in a non-trivial way on the base point x. Indeed, when $x=0$, we may take
$$\Phi_{0,\delta}(t_1,t_2) = (\delta t_1,\delta^2 t_2).$$
Then we have $\Phi^*_{0,\delta}\delta\partial_{x_1} = \partial_{t_1}$ and $\Phi^*_{0,\delta}\delta x_1\partial_{x_2} = t_1\partial_{t_2}$. However, when $x_1 = 1$ and $\delta \ll 1$, we instead take $\Phi_{(1,0),\delta}(t_1,t_2) = (\delta t_1,\delta t_2)$. The map $\Phi_{x,\delta}$ from Theorem 3.3.1 correctly interpolates between these two choices in a non-trivial way which depends on both δ and x_1.

We require more detailed information than Theorem 3.3.1 gives, and for this we introduce some new notation.

Definition 3.3.4. Let $\mathcal{S} \subseteq C^\infty_{\mathrm{loc}}(\mathfrak{M};T\mathfrak{M})\times\mathbb{N}_+$. We let $\mathrm{Gen}(\mathcal{S}) \subseteq C^\infty_{\mathrm{loc}}(\mathfrak{M};T\mathfrak{M})\times\mathbb{N}_+$ be the smallest subset of $C^\infty_{\mathrm{loc}}(\mathfrak{M};T\mathfrak{M})\times\mathbb{N}_+$ such that:

- $\forall (X_1,d_1),(X_2,d_2)\in\mathrm{Gen}(\mathcal{S})$, $([X_1,X_2],d_1+d_2)\in\mathrm{Gen}(\mathcal{S})$.
- $\mathcal{S}\subseteq\mathrm{Gen}(\mathcal{S})$.

For the remainder of the section, fix $\Omega_2 \Subset \mathfrak{M}$, a relatively compact, open set, with $\mathcal{K}\Subset\Omega_1\Subset\Omega_2\Subset\mathfrak{M}$.

Lemma 3.3.5. *There exists a finite set* $(X,d) = \{(X_1,d_1),\dots,(X_q,d_q)\}\subset\mathrm{Gen}((W,\vec{d}))$ *such that* $(W,\vec{d})\subseteq(X,d)$ *and* $\forall 1\le j,k\le q$,

$$[X_j,X_k] = \sum_{d_l\le d_j+d_k} c^l_{j,k}X_l, \tag{3.5}$$

where $c^l_{j,k}\in C^\infty_{\mathrm{loc}}(\Omega_2)$.

Proof. Since $W=\{W_1,\dots,W_r\}$ satisfies Hörmander's condition on $\mathfrak{M}$, by the compactness of $\overline{\Omega_2}$, there exists $m\in\mathbb{N}_+$ so that W satisfies Hörmander's condition of order m on $\overline{\Omega_2}$. Set

$$(X,d) := \Big\{(Z,e)\in\mathrm{Gen}((W,d)) : e\le m\max_{1\le j\le r}\vec{d}_j\Big\}.$$

Note that $(X, d) = \{(X_1, d_1), \dots, (X_q, d_q)\}$ is a finite set and $X_1(x), \dots, X_q(x)$ span $T_x\Omega_2$, $\forall x \in \Omega_2$ (by Hörmander's condition of order m).

We claim (X, d) satisfies the conclusions of the lemma. It is clear that $(W, \vec{d}) \subseteq (X, d)$, so we only need to verify (3.5). For $1 \le j, k \le q$, there are two cases. If $d_j + d_k \le m \max_{1\le l\le r} \vec{d}_l$, then $([X_j, X_k], d_j + d_k) \in (X, d)$ and (3.5) follows trivially (with $c_{j,k}^l$ constants in $\{0, 1\}$). If $d_j + d_k > m \max_{1\le l\le r} \vec{d}_l \ge \max_l d_l$, then using the fact that $X_1(x), \dots, X_q(x)$ span $T_x\Omega_2$, $\forall x \in \Omega_2$, we have

$$[X_j, X_k] = \sum_{l=1}^{q} c_{j,k}^l X_l = \sum_{d_l \le d_j + d_k} c_{j,k}^l X_l, \quad c_{j,k}^l \in C^\infty_{\mathrm{loc}}(\Omega_2),$$

completing the proof. □

Example 3.3.6. When $\mathfrak{M} = \mathbb{R}^2$ and $(W, \vec{d}) = \{(\partial_x, 1), (x\partial_y, 1)\}$, we may take $(X, d) = \{(\partial_x, 1), (x\partial_y, 1), (\partial_y, 2)\}$. This encodes the fact that we consider $\partial_y = [\partial_x, x\partial_y]$ as a differential operator of degree 2 in this setting. See Remark 1.1.5.

Henceforth, we let $(X, d) = \{(X_1, d_1), \dots, (X_q, d_q)\}$ be as in the conclusion of Lemma 3.3.5 (there are many possible choices of (X, d), but for what follows it is irrelevant which choice is used). For $x \in \mathcal{K}$, $\delta > 0$, set

$$\Lambda(x, \delta) := \max_{j_1, \dots, j_n \in \{1, \dots, q\}} \mathrm{Vol}(x)(\delta^{d_{j_1}} X_{j_1}(x), \dots, \delta^{d_{j_n}} X_{j_n}(x)).$$

Note that $\Lambda(x, \delta) > 0$ since $X_1(x), \dots, X_q(x)$ span $T_x\Omega_2$ and Vol is a strictly positive density.

Theorem 3.3.7. *There exist $\delta_0 = \delta_0(\mathcal{K}, \Omega_1) \in (0, 1]$ and $\xi_3 = \xi_3(\mathcal{K}, \Omega_1) \in (0, 1]$ such that the following hold:*

(a) $\forall x \in \mathcal{K}$, $\delta \in (0, \delta_0]$, $B_{(W,\vec{d})}(x, \delta) \subseteq B_{(X,d)}(x, \delta) \subseteq \Omega_1$, *and* $B_{(W,\vec{d})}(x, \delta)$ *and* $B_{(X,d)}(x, \delta)$ *are open in* $\mathfrak{M}$.

(b) $\forall x \in \mathcal{K}$, $\delta \in (0, 1]$, $B_{(X,d)}(x, \xi_3\delta) \subseteq B_{(W,\vec{d})}(x, \delta) \subseteq B_{(X,d)}(x, \delta)$.

(c) $\forall x \in \mathcal{K}$, $\delta \in (0, \delta_0]$, $\mathrm{Vol}(B_{(W,\vec{d})}(x, \delta)) \approx \mathrm{Vol}(B_{(X,d)}(x, \delta)) \approx \Lambda(x, \delta)$.

(d) $\forall x \in \mathcal{K}$, $\delta > 0$, $\mathrm{Vol}(B_{(W,\vec{d})}(x, \delta)) \wedge 1 \approx \mathrm{Vol}(B_{(X,d)}(x, \delta)) \wedge 1 \approx \Lambda(x, \delta) \wedge 1$.

(e) $\forall x \in \mathcal{K}$, $\delta \in (0, \delta_0]$, $\mathrm{Vol}(B_{(W,\vec{d})}(x, 2\delta)) \lesssim \mathrm{Vol}(B_{(W,\vec{d})}(x, \delta))$ *and* $\mathrm{Vol}(B_{(X,d)}(x, 2\delta)) \lesssim \mathrm{Vol}(B_{(X,d)}(x, \delta))$.

(f) $\forall x \in \mathcal{K}$, $\delta > 0$, $\mathrm{Vol}(B_{(W,\vec{d})}(x, 2\delta)) \wedge 1 \lesssim \mathrm{Vol}(B_{(W,\vec{d})}(x, \delta)) \wedge 1$ *and* $\mathrm{Vol}(B_{(X,d)}(x, 2\delta)) \wedge 1 \lesssim \mathrm{Vol}(B_{(X,d)}(x, \delta)) \wedge 1$.

For all $x \in \mathcal{K}$, $\delta \in (0, 1]$, there exists $\Phi_{x,\delta} : B^n(1) \to B_{(X,d)}(x, \delta) \cap \Omega_1$ such that:

(g) $\Phi_{x,\delta}(0) = x$.

(h) $\Phi_{x,\delta}$ *is a smooth coordinate system, that is,* $\Phi_{x,\delta}(B^n(1)) \subseteq \mathfrak{M}$ *is open and* $\Phi_{x,\delta} : B^n(1) \to \Phi_{x,\delta}(B^n(1))$ *is a* C^∞_{loc} *diffeomorphism.*

(i) $\forall x \in \mathcal{K}$, $\delta \in (0, 1]$, $B_{(W,\vec{d})}(x, \xi_3\delta) \subseteq B_{(X,d)}(x, \xi_3\delta) \subseteq \Phi_{x,\delta}(B^n(1/2)) \subseteq \Phi_{x,\delta}(B^n(1)) \subseteq B_{(X,d)}(x, \delta)$.

Set $W_j^{x,\delta} := \Phi_{x,\delta}^* \delta^{d_j} W_j$ *and* $X_j^{x,\delta} := \Phi_{x,\delta}^* \delta^{d_j} X_j$. *Let* $(W^{x,\delta}, d) = \{(W_1^{x,\delta}, d_1), \ldots, (W_r^{x,\delta}, d_r)\}$, $W^{x,\delta} = \{W_1^{x,\delta}, \ldots, W_r^{x,\delta}\}$, *and* $X^{x,\delta} := \{X_1^{x,\delta}, \ldots, X_q^{x,\delta}\}$.

(j) $W_j^{x,\delta}$ *and* $X_k^{x,\delta}$ *are* C^∞, *uniformly for* $x \in \mathcal{K}$ *and* $\delta \in (0,1]$ *in the sense that* $\forall L \in \mathbb{N}$, $1 \le j \le r, 1 \le k \le q$,

$$\sup_{\substack{x\in\mathcal{K}\\ \delta\in(0,1]}} \|W_j^{x,\delta}\|_{C^L(B^n(1);\mathbb{R}^n)} < \infty, \quad \sup_{\substack{x\in\mathcal{K}\\ \delta\in(0,1]}} \|X_k^{x,\delta}\|_{C^L(B^n(1);\mathbb{R}^n)} < \infty.$$

(k) $X_1^{x,\delta}(u), \ldots, X_q^{x,\delta}(u)$ *span* $T_u B^n(1)$ *uniformly in* $x \in \mathcal{K}$, $\delta \in (0,1]$, $u \in B^n(1)$ *in the sense that*

$$\inf_{\substack{x\in\mathcal{K}\\ \delta\in(0,1]}} \max_{j_1,\ldots,j_n\in\{1,\ldots,q\}} \inf_{u\in B^n(1)} \left|\det(X_{j_1}^{x,\delta}(u)|\cdots|X_{j_n}^{x,\delta}(u))\right| > 0.$$

(l) $(W^{x,\delta}, d)$ *are Hörmander vector fields with formal degrees at the unit scale, uniformly for* $x \in \mathcal{K}$ *and* $\delta \in (0,1]$.

(m) $\forall x \in \mathcal{K}$, $\delta \in (0,1]$,

$$\|f\|_{C^L(B^n(1))} \approx \sum_{|\alpha|\le L} \|(X^{x,\delta})^\alpha f\|_{C(B^n(1))}, \quad \forall f \in C(B^n(1)),$$

where the implicit constants depend on $L \in \mathbb{N}$, *but not on* x, δ, *or* f. *Here, we are using ordered multi-index notation; see Definition* 1.1.4.

For $x \in \mathcal{K}$, $\delta \in (0,1]$, *define* $h_{x,\delta} \in C^\infty(B^n(1))$ *by* $\Phi_{x,\delta}^* \mathrm{Vol} = \Lambda(x,\delta) h_{x,\delta} \sigma_{\mathrm{Leb}}$.

(n) $\forall x \in \mathcal{K}$, $\delta \in (0,1]$, $u \in B^n(1)$, $h_{x,\delta}(u) \approx 1$ *(with implicit constants independent of* u, x, δ*), and* $\forall L \in \mathbb{N}$,

$$\sup_{\substack{x\in\mathcal{K}\\ \delta\in(0,1]}} \|h_{x,\delta}\|_{C^L(B^n(1))} < \infty.$$

Corollary 3.3.8. *For each* $x \in \mathcal{K}$, $\delta \in (0,1]$, *there exists* $\phi_{x,\delta} \in C_0^\infty(\Omega_1)$ *such that*

(i) $\mathrm{supp}(\phi_{x,\delta}) \subseteq B_{(W,d)}(x,\delta) \cap \Omega_1$.

(ii) $\phi_{x,\delta} \equiv 1$ *on a neighborhood of the closure of* $B_{(W,d)}(x, \xi_3^2\delta)$, *where* $\xi_3 > 0$ *is as in Theorem* 3.3.7.

(iii) *For every* $N \in \mathbb{N}$,

$$\sup_{\substack{x,\in\mathcal{K}\\ \delta\in(0,1]}} \sum_{|\alpha|\le N} \|(\delta^d W)^\alpha \phi_{x,\delta}\|_{C(\mathfrak{M})} < \infty.$$

Corollary 3.3.9. *Let* $\Omega \Subset \mathfrak{M}$. *Then there exist* $C = C(\Omega) \ge 0$, $\delta_1 = \delta_1(\Omega) > 0$, *and* $Q_1 = Q_1(\Omega)$, $Q_2 = Q_2(\Omega)$, *with* $n \le Q_1 \le Q_2$, *such that the following hold:*

(i) $\forall x \in \Omega$,

$$\mathrm{Vol}(B_{(W,d)}(x,\delta_1)) \approx \mathrm{Vol}(B_{(X,d)}(x,\delta_1)) \approx 1.$$

(ii) $\forall x \in \Omega$, $\delta \in (0,\delta_1]$, $\xi \in (0,1]$,

$$\begin{aligned}\xi^{-Q_1}\,\mathrm{Vol}(B_{(W,d)}(x,\xi\delta)) &\lesssim \mathrm{Vol}(B_{(W,d)}(x,\delta)) \\ &\lesssim \xi^{-Q_2}\,\mathrm{Vol}(B_{(W,d)}(x,\xi\delta)),\end{aligned}$$

and the same estimates hold with (W, d) *replaced by* (X, d).

(iii) $\forall x \in \Omega$, $\delta > 0$, $\xi \in (0,1]$,

$$\begin{aligned}(\xi^{-Q_1}\,\mathrm{Vol}(B_{(W,d)}(x,\xi\delta))) \wedge 1 &\lesssim \mathrm{Vol}(B_{(W,d)}(x,\delta)) \wedge 1 \\ &\lesssim (\xi^{-Q_2}\,\mathrm{Vol}(B_{(W,d)}(x,\xi\delta))) \wedge 1,\end{aligned}$$

and the same estimates hold with (W, d) *replaced by* (X, d).

Remark 3.3.10. If $(W, d) \subset C^\infty(B^n(1); \mathbb{R}^n)\times\mathbb{N}_+$ are Hörmander vector fields with formal degrees at the unit scale, as in Section 3.2, and if one fixes sets $\mathcal{K} \Subset \Omega_1 \Subset B^n(1)$, then all of the constants in this section can be taken to be unit-admissible constants in the sense of Definition 3.2.1. So, for example, if we write $A \lesssim B$, this can be taken to mean $A \le CB$, where C is a unit-admissible constant (also depending on the choice of $\mathcal{K}$ and Ω_1).

3.3.1 Scaling and maximal subellipticity

We are prepared to describe one of the main ideas in this text: maximal subellipticity is quantitatively preserved under the scaling maps $\Phi_{x,\delta}$ from Theorem 3.3.7. Let

$$\mathscr{P} = \sum_{\deg_d(\alpha)\le\kappa} a_\alpha(x) W^\alpha, \quad a_\alpha \in C^\infty_{\mathrm{loc}}(\mathfrak{M}; \mathbb{M}^{\mathbf{D}_1\times\mathbf{D}_2}(\mathbb{C})),$$

where $\kappa \in \mathbb{N}_+$ is such that d_j divides κ, $\forall j$. Suppose that $\mathscr{P}$ is maximally subelliptic of degree κ with respect to (W, d) on $\mathfrak{M}$.

We take the same setting as in Theorem 3.3.7. In particular, since $\Omega_1 \Subset \mathfrak{M}$ is relatively compact and $\mathscr{P}$ is maximally subelliptic, there exists $A \ge 0$ such that, with $n_j := \kappa/d_j \in \mathbb{N}_+$,

$$\sum_{j=1}^{r} \|W_j^{n_j} f\|_{L^2(\mathfrak{M},\mathrm{Vol};\mathbb{C}^{\mathbf{D}_2})} \le A(\|\mathscr{P} f\|_{L^2(\mathfrak{M},\mathrm{Vol};\mathbb{C}^{\mathbf{D}_1})} + \|f\|_{L^2(\mathfrak{M},\mathrm{Vol};\mathbb{C}^{\mathbf{D}_2})}), \tag{3.6}$$

$\forall f \in C_0^\infty(\Omega_1; \mathbb{C}^{\mathbf{D}_2})$. For $x \in \mathcal{K}$, $\delta \in (0,1]$, set

$$\mathscr{P}^{x,\delta} := \Phi_{x,\delta}^{*}\delta^{\kappa}\mathscr{P}(\Phi_{x,\delta})_{*} = \sum_{\deg_{d}(\alpha)\leq\kappa} b_{\alpha}^{x,\delta}(W^{x,\delta})^{\alpha},$$

where $b_{\alpha}^{x,\delta} = \delta^{\kappa-\deg_{d}(\alpha)}a_{\alpha}\circ\Phi_{x,\delta} \in C^{\infty}(B^{n}(1); \mathbb{M}^{\mathbf{D}_1\times\mathbf{D}_2}(\mathbb{C}))$.

Proposition 3.3.11. *$\mathscr{P}^{x,\delta}$ is maximally subelliptic of degree κ with respect to $(W^{x,\delta}, \vec{d})$, uniformly for $x \in \mathcal{K}$ and $\delta \in (0,1]$ in the sense that:*

(i) *We have*

$$\sum_{j=1}^{r}\|(W_{j}^{x,\delta})^{n_j}f\|_{L^2(B^n(1),h_{x,\delta}\sigma_{\mathrm{Leb}};\mathbb{C}^{\mathbf{D}_2})} \leq A\big(\|\mathscr{P}^{x,\delta}f\|_{L^2(B^n(1),h_{x,\delta}\sigma_{\mathrm{Leb}};\mathbb{C}^{\mathbf{D}_1})} + \|f\|_{L^2(B^n(1),h_{x,\delta}\sigma_{\mathrm{Leb}};\mathbb{C}^{\mathbf{D}_2})}\big),$$

$\forall f \in C_0^{\infty}(B^n(1); \mathbb{C}^{\mathbf{D}_2})$, *where this is the same constant A as in* (3.6).

(ii) *For every $L \in \mathbb{N}$, we have*

$$\sup_{\substack{x\in\mathcal{K}\\ \delta\in(0,1]}} \max_{\deg_{d}(\alpha)\leq\kappa} \|b_{\alpha}\|_{C^{L}(B^n(1);\mathbb{M}^{\mathbf{D}_1\times\mathbf{D}_2})} < \infty.$$

Remark 3.3.12. Proposition 3.3.11 shows that $\mathscr{P}^{x,\delta}$ is maximally subelliptic of degree κ with respect to $(W^{x,\delta}, \vec{d})$ on $B^n(1)$, in a way which is uniform in $x \in \mathcal{K}$ and $\delta \in (0,1]$. Theorem 3.3.7 (l) shows that $(W^{x,\delta}, \vec{d})$ are Hörmander vector fields at the unit scale, uniformly for $x \in \mathcal{K}$ and $\delta \in (0,1]$, and Theorem 3.3.7 (n) shows that $h_{x,\delta}$ is smooth and bounded away from 0, uniformly for $x \in \mathcal{K}$ and $\delta \in (0,1]$. Thus, once we prove a result for maximally subelliptic operators, that result will hold for $\mathscr{P}^{x,\delta}$, uniformly for $x \in \mathcal{K}$ and $\delta \in (0,1]$. By then pushing the result forward via $\Phi_{x,\delta}$, we will obtain results for $\mathscr{P}$ on $\Phi_{x,\delta}(B^n(1))$, which hold uniformly for all $x \in \mathcal{K}$ and $\delta \in (0,1]$. It follows from Theorem 3.3.7 (i) that $\Phi_{x,\delta}(B^n(1))$ is comparable to $B_{(W,\vec{d})}(x,\delta)$. Thus, by proving a result about maximally subelliptic operators at the unit scale, we will be able to conclude results about maximally subelliptic operators at every small Carnot–Carathéodory scale, $\delta \in (0,1]$, uniformly in the scale.

Proof of Proposition 3.3.11. (i): Multiplying (3.6) by δ^{κ}, we see that $\forall f \in C_0^{\infty}(\Omega_1; \mathbb{C}^{\mathbf{D}_2})$,

$$\begin{aligned}\sum_{j=1}^{r}\|(\delta^{d_j}W_j)^{n_j}f\|_{L^2(\mathfrak{M},\mathrm{Vol};\mathbb{C}^{\mathbf{D}_2})} &\leq A\big(\|\delta^{\kappa}\mathscr{P}f\|_{L^2(\mathfrak{M},\mathrm{Vol};\mathbb{C}^{\mathbf{D}_1})} + \delta^{\kappa}\|f\|_{L^2(\mathfrak{M},\mathrm{Vol};\mathbb{C}^{\mathbf{D}_2})}\big)\\ &\leq A\big(\|\delta^{\kappa}\mathscr{P}f\|_{L^2(\mathfrak{M},\mathrm{Vol};\mathbb{C}^{\mathbf{D}_1})} + \|f\|_{L^2(\mathfrak{M},\mathrm{Vol};\mathbb{C}^{\mathbf{D}_2})}\big).\end{aligned} \tag{3.7}$$

Let $g \in C_0^{\infty}(B^n(1); \mathbb{C}^{\mathbf{D}_2})$. Theorem 3.3.7 (h) shows $f := g\circ\Phi_{x,\delta} \in C_0^{\infty}(\Omega_1; \mathbb{C}^{\mathbf{D}_2})$. Applying (3.7) with this choice of g, using $W_j^{x,\delta} = \Phi_{x,\delta}\delta^{d_j}W_j$, and changing variables, we see that

$$\sum_{j=1}^{r}\|(W_j^{x,\delta})^{n_j}g\|_{L^2(B^n(1),\Phi_{x,\delta}^{*}\mathrm{Vol})} \leq A\big(\|\mathscr{P}^{x,\delta}g\|_{L^2(B^n(1),\Phi_{x,\delta}^{*}\mathrm{Vol})} + \|g\|_{L^2(B^n(1),\Phi_{x,\delta}^{*}\mathrm{Vol})}\big). \tag{3.8}$$

Finally, using that $\Phi_{x,\delta}^* \operatorname{Vol} = \Lambda(x,\delta) h_{x,\delta} \sigma_{\mathrm{Leb}}$, multiplying both sides of (3.8) by $\Lambda(x,\delta)^{-\frac{1}{2}}$ gives (i).

(ii): Using Theorem 3.3.7 (m), we have, for each α with $\deg_{\vec{d}}(\alpha) \le \kappa$ and each $x \in \mathcal{K}$, $\delta \in (0,1]$,

$$\begin{aligned} \|b_\alpha^{x,\delta}\|_{C^L(B^n(1))} &\approx \sum_{|\beta|\le L} \|(X^{x,\delta})^\beta b_\alpha^{x,\delta}\|_{C(B^n(1))} \\ &= \sum_{|\beta|\le L} \delta^{\kappa-\deg_{\vec{d}}(\beta)} \|(\delta^d X)^\beta a_\alpha\|_{C(\Phi_{x,\delta}(B^n(1)))} \le \sum_{|\beta|\le L} \|X^\beta a_\alpha\|_{C(\Omega_1)}. \end{aligned} \tag{3.9}$$

Since Ω_1 is relatively compact and a_α and $X_1, \dots, X_q$ are smooth, we have

$$\sum_{|\beta|\le L} \|X^\beta a_\alpha\|_{C(\Omega_1)} < \infty.$$

We conclude the right-hand side of (3.9) is finite and does not depend on $x \in \mathcal{K}$ or $\delta \in (0,1]$. Item (ii) follows. □

3.4 Finitely generated setting

As we will see in later sections, the assumption of Hörmander's condition is not essential to obtain many of our results concerning Carnot–Carathéodory geometry and associated scaling maps. This is important when we study two different Carnot–Carathéodory geometries simultaneously, as Hörmander's condition will not be quantitatively preserved under the adapted rescaling (though it will be qualitatively preserved).

The key requirement is the conclusion of Lemma 3.3.5. In this section, we investigate this condition. For later applications, we fix $\nu \in \mathbb{N}_+$ and associate to the vector fields multi-parameter formal degrees which are elements of $\mathbb{N}^\nu \setminus \{0\}$. Throughout this section, we let $\mathfrak{M}$ be a connected C^∞ manifold.

Definition 3.4.1. Let $\mathcal{S} \subseteq C^\infty_{\mathrm{loc}}(\mathfrak{M}; T\mathfrak{M}) \times (\mathbb{N}^\nu \setminus \{0\})$. For $\Omega \subseteq \mathfrak{M}$ open, we say $\mathcal{S}$ is finitely generated on Ω if there is a finite subset $\mathcal{F} \subseteq \mathcal{S}$ such that $\forall (X_0, \vec{d}_0) \in \mathcal{S}$,

$$X_0 = \sum_{\substack{(X_1,\vec{d}_1)\in\mathcal{F} \\ \vec{d}_1 \le \vec{d}_0}} c_{(X_1,\vec{d}_1)} X_1, \quad c_{(X_1,\vec{d}_1)} \in C^\infty_{\mathrm{loc}}(\Omega), \tag{3.10}$$

where for the vectors $\vec{d}_0$ and $\vec{d}_1$ we have written $\vec{d}_0 \le \vec{d}_1$ to mean the inequality holds for each component: $\vec{d}_0^\mu \le \vec{d}_1^\mu$, $\forall \mu \in \{1, \dots, \nu\}$. If we wish to make the choice of $\mathcal{F}$ explicit, we say $\mathcal{S}$ is finitely generated by $\mathcal{F}$ on Ω.

Remark 3.4.2. In Definition 3.4.1, and in the rest of this text, we have drawn an $\vec{}$ over $\vec{d}$ to remind the reader that $\vec{d}$ is a vector.

Definition 3.4.3. Let $\mathcal{S} \subseteq C^\infty_{\mathrm{loc}}(\mathfrak{M}; T\mathfrak{M}) \times (\mathbb{N}^\nu \setminus \{0\})$. For $\Omega \subseteq \mathfrak{M}$ open, we say $\mathcal{S}$ is linearly finitely generated on Ω if $\mathcal{S}$ is finitely generated by $\mathcal{F}$ on Ω, where for each $(X, d) \in \mathcal{F}$, d is non-zero in only one component. If we wish to make the choice of $\mathcal{F}$ explicit, we say $\mathcal{S}$ is linearly finitely generated by $\mathcal{F}$ on Ω.

Remark 3.4.4. When $\nu = 1$, finitely generated and linearly finitely generated are equivalent.

Definition 3.4.5. Let $\mathcal{S} \subseteq C^\infty_{\mathrm{loc}}(\mathfrak{M}; T\mathfrak{M}) \times (\mathbb{N}^\nu \setminus \{0\})$. We say $\mathcal{S}$ is locally finitely generated (respectively, locally linearly finitely generated) if for every relatively compact open set $\Omega \Subset \mathfrak{M}$, $\mathcal{S}$ is finitely generated (respectively, linearly finitely generated) on Ω. If we wish to make $\mathfrak{M}$ explicit, we say $\mathcal{S}$ is locally finitely generated on $\mathfrak{M}$ (respectively, locally linearly finitely generated on $\mathfrak{M}$).

Lemma 3.4.6. *Suppose $\mathcal{S} \subseteq C^\infty_{\mathrm{loc}}(\mathfrak{M}; T\mathfrak{M}) \times (\mathbb{N}^\nu \setminus \{0\})$ and let $\mathcal{D} \subseteq C^\infty_{\mathrm{loc}}(\mathfrak{M}; T\mathfrak{M})$ be the $C^\infty_{\mathrm{loc}}(\mathfrak{M})$-module generated by $\{X : (X, \vec{d}) \in \mathcal{S}\}$. If $\mathcal{S}$ is locally finitely generated, then $\mathcal{D}$ is locally finitely generated as a $C^\infty_{\mathrm{loc}}(\mathfrak{M})$-module.*

Proof. This follows immediately from the definitions. □

Next, we adapt Definition 3.3.4 to the multi-parameter setting.

Definition 3.4.7. Let $\mathcal{S} \subseteq C^\infty_{\mathrm{loc}}(\mathfrak{M}; T\mathfrak{M}) \times (\mathbb{N}^\nu \setminus \{0\})$. We let $\mathrm{Gen}(\mathcal{S}) \subseteq C^\infty_{\mathrm{loc}}(\mathfrak{M}; T\mathfrak{M}) \times (\mathbb{N}^\nu \setminus \{0\})$ be the smallest subset of $C^\infty_{\mathrm{loc}}(\mathfrak{M}; T\mathfrak{M}) \times (\mathbb{N}^\nu \setminus \{0\})$ such that:
- $\forall (X_1, \vec{d}_1), (X_2, \vec{d}_2) \in \mathrm{Gen}(\mathcal{S})$, $([X_1, X_2], \vec{d}_1 + \vec{d}_2) \in \mathrm{Gen}(\mathcal{S})$.
- $\mathcal{S} \subseteq \mathrm{Gen}(\mathcal{S})$.

Example 3.4.8. In this text we are often given a set $\mathcal{S} \subseteq C^\infty_{\mathrm{loc}}(\mathfrak{M}; T\mathfrak{M}) \times (\mathbb{N}^\nu \setminus \{0\})$ and we will be interested in whether $\mathrm{Gen}(\mathcal{S})$ is locally finitely generated. It is instructive to see an example where $\mathrm{Gen}(\mathcal{S})$ is not locally finitely generated. On $\mathfrak{M} = \mathbb{R}^2$, let $X_1 = \partial_x$, $X_2 = e^{-1/x^2}\partial_y$, and $\mathcal{S} = \{(X_1, 1), (X_2, 1)\}$. Let Ω be any neighborhood of $0 \in \mathbb{R}^2$. Then $\mathrm{Gen}(\mathcal{S})$ is not finitely generated on Ω. Indeed, any commutator of the form

$$[X_1, [X_1, [X_1, \ldots, [X_1, X_2] \cdots]]]$$

is not spanned (with bounded coefficients) on any neighborhood of 0 by commutators with fewer terms. This example is in contrast to Proposition 3.4.15, which shows that such a phenomenon cannot occur with real analytic vector fields.

Example 3.4.9. An important example where $\mathrm{Gen}(\mathcal{S})$ is finitely generated but not linearly finitely generated comes from the Heisenberg group. The Heisenberg group, $\mathbb{H}^1$, has a three-dimensional Lie algebra spanned by vector fields X, Y, T, where $[X, Y] = T$ and T is in the center. As a manifold, $\mathbb{H}^1 \cong \mathbb{R}^3$ and the vector fields X, Y, T form a basis for the tangent space at every point. Set $\mathcal{S} := \{(X, (1, 0)), (Y, (0, 1))\}$. Then, for any non-empty open set $\Omega \subset \mathbb{H}^1$, it is immediate to see that $\mathrm{Gen}(\mathcal{S})$ is finitely generated on Ω, but not linearly finitely generated on Ω. A main point here is that the vector fields X and Y are far from commuting. This example is in contrast to Proposition 3.8.8, which shows that if the vector fields "weakly approximately commute," then $\mathrm{Gen}(\mathcal{S})$ is linearly finitely generated.

Lemma 3.4.10. *Let $\mathcal{S} \subseteq C^{\infty}_{\mathrm{loc}}(\mathfrak{M}; T\mathfrak{M}) \times (\mathbb{N}^{\nu} \setminus \{0\})$ and let $\mathcal{D}$ be the $C^{\infty}_{\mathrm{loc}}(\mathfrak{M})$-module generated by $\{X : (X, d) \in \mathrm{Gen}(\mathcal{S})\}$. Then $\mathcal{D}$ is involutive.*

Proof. This follows immediately from the definitions. □

Remark 3.4.11. Let $\mathcal{S} \subseteq C^{\infty}_{\mathrm{loc}}(\mathfrak{M}; T\mathfrak{M}) \times (\mathbb{N}^{\nu} \setminus \{0\})$. If $\mathrm{Gen}(\mathcal{S})$ is locally finitely generated, then Lemmas 3.4.6 and 3.4.10 show that the Frobenius theorem (Theorem 3.1.15) applies with $\mathcal{D}$ equal to the $C^{\infty}_{\mathrm{loc}}(\mathfrak{M})$-module generated by $\{X : (X, \vec{d}) \in \mathrm{Gen}(\mathcal{S})\}$. Thus, $\mathfrak{M}$ is foliated into leaves, and the tangent space to the leaf passing through $y \in \mathfrak{M}$ at y is equal to $\mathrm{span}\{X(y) : (X, \vec{d}) \in \mathrm{Gen}(\mathcal{S})\}$. In other words, $\{X : (X, \vec{d}) \in \mathcal{S}\}$ satisfies Hörmander's condition on each leaf.

Remark 3.4.12. Suppose $\mathrm{Gen}(\mathcal{S})$ is finitely generated by

$$(X, \vec{d}) = \{(X_1, \vec{d}_1), \dots, (X_q, \vec{d}_q)\} \subseteq \mathrm{Gen}(\mathcal{S})$$

on Ω. Then, since $([X_j, X_k], \vec{d}_j + \vec{d}_k) \in \mathrm{Gen}(\mathcal{S})$, we have

$$[X_j, X_k] = \sum_{\vec{d}_l \le \vec{d}_j + \vec{d}_k} c^l_{j,k} X_l, \quad c^l_{j,k} \in C^{\infty}_{\mathrm{loc}}(\Omega).$$

In light of Remark 3.4.12, Lemma 3.3.5 can be restated as follows: if $(W, \vec{d}) \subset C^{\infty}_{\mathrm{loc}}(\mathfrak{M}; T\mathfrak{M}) \times \mathbb{N}_+$ are Hörmander vector fields with formal degrees, then $\mathrm{Gen}((W, \vec{d}))$ is locally finitely generated. The next definition and result generalize this to the multi-parameter setting.

Definition 3.4.13. Let $\mathcal{S}_1 \subseteq C^{\infty}_{\mathrm{loc}}(\mathfrak{M}; T\mathfrak{M}) \times (\mathbb{N}^{\nu_1} \setminus \{0\})$ and $\mathcal{S}_2 \subseteq C^{\infty}_{\mathrm{loc}}(\mathfrak{M}; T\mathfrak{M}) \times (\mathbb{N}^{\nu_2} \setminus \{0\})$. We let

$$\begin{aligned} \mathcal{S}_1 \boxtimes \mathcal{S}_2 &:= \{(X_1, (d_1, 0_{\nu_2})) : (X_1, d_1) \in \mathcal{S}_1\} \cup \{(X_2, (0_{\nu_1}, d_2)) : (X_2, d_2) \in \mathcal{S}_2\} \\ &\subseteq C^{\infty}_{\mathrm{loc}}(\mathfrak{M}; T\mathfrak{M}) \times (\mathbb{N}^{n_1+\nu_2} \setminus \{0\}). \end{aligned}$$

Proposition 3.4.14. *For $\mu \in \{1, \dots, \nu\}$, let*

$$(W^{\mu}, \vec{d}^{\mu}) = \{(W^{\mu}_1, \vec{d}^{\mu}_1), \dots, (W^{\mu}_{r_\mu}, \vec{d}^{\mu}_{r_\mu})\} \subset C^{\infty}_{\mathrm{loc}}(\mathfrak{M}; T\mathfrak{M}) \times \mathbb{N}_+$$

be Hörmander vector fields with formal degrees. Set

$$\begin{aligned} (W, \vec{d}) &= \{(W_1, \vec{d}_1), \dots, (W_r, \vec{d}_r)\} := (W^1, \vec{d}^1) \boxtimes (W^2, \vec{d}^2) \boxtimes \cdots \boxtimes (W^{\nu}, \vec{d}^{\nu}) \\ &= \{(W^{\mu}_j, \vec{d}^{\mu}_j \mathfrak{e}_\mu) : \mu \in \{1, \dots, \nu\}, 1 \le j \le q_\mu\} \subset C^{\infty}_{\mathrm{loc}}(\mathfrak{M}; T\mathfrak{M}) \times (\mathbb{N}^{\nu} \setminus \{0\}), \end{aligned}$$

where $\mathfrak{e}_1, \dots, \mathfrak{e}_\nu$ *denotes the standard basis of* $\mathbb{R}^\nu$*. Then* $\mathrm{Gen}((W, \vec{d}))$ *is locally finitely generated.*

Proof. Fix $\Omega \Subset \mathfrak{M}$. Since $W^\mu = \{W_1^\mu, \dots, W_{r_\mu}^\mu\}$ satisfies Hörmander's condition on $\mathfrak{M}$, for each $\mu \in \{1, \dots, \nu\}$, by the compactness of $\overline{\Omega}$ there exists $m \in \mathbb{N}_+$ so that W^μ satisfies Hörmander's condition of order m on $\overline{\Omega}$, for all $\mu \in \{1, \dots, \nu\}$. Set

$$\mathcal{F} := \left\{(Z, \vec{e}) \in \mathrm{Gen}((W, \vec{d})) : |e|_\infty \le m \max_{1\le\mu\le\nu} \max_{1\le j\le r_\mu} d_j^\mu \right\}.$$

Note that $\mathcal{F}$ is a finite set. Enumerate $\mathcal{F} = \{(X_1, \vec{d}_1), \dots, (X_q, \vec{d}_q)\}$. Here, we have used $|e|_\infty$ to denote the ℓ^∞ norm of the vector e. Since (W^μ, d^μ) satisfies Hörmander's condition of order m on $\overline{\Omega}$, we have $\forall \mu \in \{1, \dots, \nu\}$, $x \in \Omega$,

$$\mathrm{span}\{X_j(x) : d_j \text{ is non-zero in only the } \mu \text{ component}\} = T_x\Omega. \tag{3.11}$$

We claim $\mathrm{Gen}((W, \vec{d}))$ is finitely generated by $\mathcal{F}$ on Ω. Indeed, take $(X_0, \vec{d}_0) \in \mathrm{Gen}((W, \vec{d}))$. If $|\vec{d}_0|_\infty \le m \max_{1\le\mu\le\nu} \max_{1\le j\le r_\mu} d_j^\mu$, then $(X_0, \vec{d}_0) \in \mathcal{F}$ and (3.10) holds trivially. Otherwise, there exists μ with $\vec{d}_0^\mu > m \max_{1\le\mu\le\nu} \max_{1\le j\le r_\mu} d_j^\mu$. Using (3.11), we may write

$$X_0 = \sum_{\substack{\vec{d}_l \text{ is non-zero in only} \\ \text{the } \mu \text{ component}}} c_l X_l, \quad c_l \in C_{\mathrm{loc}}^\infty(\Omega).$$

Since $\vec{d}_l^\mu \le |\vec{d}_l|_\infty < \vec{d}_0^\mu$, for $1 \le l \le q$, (3.10) follows, completing the proof. □

Proposition 3.4.14 used Hörmander's condition to prove that $\mathrm{Gen}((W, \vec{d}))$ is locally finitely generated. However, Hörmander's condition is not always necessary for such a conclusion. A particularly compelling case is when the vector fields are real analytic.

Proposition 3.4.15. *Let* $\mathfrak{M}$ *be a real analytic manifold, and let* $\mathcal{S} \subset C_{\mathrm{loc}}^\omega(\mathfrak{M}; T\mathfrak{M}) \times \mathbb{N}^\nu \setminus \{0\}$ *be a finite set of real analytic vector fields paired with ν-parameter formal degrees. Then* $\mathrm{Gen}(\mathcal{S})$ *is locally finitely generated.*

To prove Proposition 3.4.15 we use the next lemma.

Lemma 3.4.16. *Let* $\mathcal{S}$ *be as in Proposition* 3.4.15*. Then, for every* $x \in \mathfrak{M}$*, there is a finite set* $\mathcal{F}_x \subset \mathrm{Gen}(\mathcal{S})$ *such that for each* $(X_0, \vec{d}_0) \in \mathrm{Gen}(\mathcal{S})$*, there is an open neighborhood* $V_{(X_0,\vec{d}_0),x}$ *of* x *with*

$$X_0 = \sum_{\substack{(X_1,\vec{d}_1)\in\mathcal{F}_x \\ \vec{d}_1\le\vec{d}_0}} c_{x,(X_0,\vec{d}_0)}^{(X_1,\vec{d}_1)} X_1, \quad c_{x,(X_0,\vec{d}_0)}^{(X_1,\vec{d}_1)} \in C_{\mathrm{loc}}^\infty(V_{(X_0,\vec{d}_0),x}).$$

Proof of Proposition 3.4.15 *given Lemma* 3.4.16. Let $\mathcal{F}_x$ be as in Lemma 3.4.16, let $M_x := \max\{|\vec{d}_1|_\infty : (X_1, \vec{d}_1) \in \mathcal{F}_x \cup \mathcal{S}\}$, and let $\mathcal{F}'_x := \{(X_1, \vec{d}_1) : |\vec{d}_1|_\infty \leq M_x\}$. Note that $\mathcal{F}'_x$ is a finite set and $\mathcal{S} \cup \mathcal{F}_x \subseteq \mathcal{F}'_x$. Let $V_{(X_0,\vec{d}_0),x}$ be as in Lemma 3.4.16 and set

$$V_x := \bigcap_{(X_1,\vec{d}_1),(X_2,\vec{d}_2)\in\mathcal{F}'_x} V_{([X_1,X_2],\vec{d}_1+\vec{d}_2),x}.$$

By the definition of V_x, we have, for $(X_1, \vec{d}_1), (X_2, \vec{d}_2) \in \mathcal{F}'_x$,

$$[X_1, X_2] = \sum_{\substack{(X_3,\vec{d}_3)\in\mathcal{F}'_x \\ \vec{d}_3\leq\vec{d}_1+\vec{d}_2}} c^{(X_3,\vec{d}_3)}_{(X_1,\vec{d}_1),(X_2,\vec{d}_2)} X_3, \quad c^{(X_3,\vec{d}_3)}_{(X_1,\vec{d}_1),(X_2,\vec{d}_2)} \in C_0^\infty(V_x). \tag{3.12}$$

Let $\mathcal{T}_x$ denote the set of all $(X_0, \vec{d}_0) \in \mathrm{Gen}(\mathcal{S})$ such that

$$X_0 = \sum_{\substack{(X_1,\vec{d}_1)\in\mathcal{F}'_x \\ \vec{d}_1\leq\vec{d}_0}} c^{(X_1,\vec{d}_1)}_{(X_0,\vec{d}_0)} X_1, \quad c^{(X_1,\vec{d}_1)}_{(X_0,\vec{d}_0)} \in C^\infty_{\mathrm{loc}}(V_x). \tag{3.13}$$

We claim $\mathcal{T}_x = \mathrm{Gen}(\mathcal{S})$. Indeed, (3.12) shows that $\mathcal{S} \subseteq \mathcal{F}'_x \subseteq \mathcal{T}_x$. Also, if $(X_1, \vec{d}_1), (X_2, \vec{d}_2) \in \mathcal{T}_x$, it follows from (3.12) and (3.13) that $([X_1,X_2], \vec{d}_1 + \vec{d}_2) \in \mathcal{T}_x$. Thus, $\mathcal{T}_x = \mathrm{Gen}(\mathcal{S})$.

Fix Ω open with $\Omega \Subset \mathfrak{M}$. $\{V_x : x \in \overline{\Omega}\}$ is an open cover for the compact set $\overline{\Omega}$. We extract a finite subcover $V_{x_1}, \ldots, V_{x_N}$. Let $M := \max\{M_{x_j} : 1 \leq j \leq N\}$ and set $\mathcal{F} := \{(X_0, \vec{d}_0) :\in \mathrm{Gen}(\mathcal{S}) : |\vec{d}_0| \leq M\}$. Note that $\mathcal{S} \subseteq \mathcal{F}_{x_j} \subseteq \mathcal{F}$, for $1 \leq j \leq N$, and $\mathcal{F}$ is a finite set.

Since $\mathcal{T}_{x_j} = \mathrm{Gen}(\mathcal{S})$, a simple partition of unity argument together with the definition of $\mathcal{T}_{x_j}$ (see (3.13)) shows that $\forall (X_0, \vec{d}_0) \in \mathrm{Gen}(\mathcal{S})$,

$$X_0 = \sum_{\substack{(X_1,\vec{d}_1) \\ \vec{d}_1\leq\vec{d}_0}} c^{(X_1,\vec{d}_1)}_{(X_0,\vec{d}_0)} X_1, \quad c^{(X_1,\vec{d}_1)}_{(X_0,\vec{d}_0)} \in C^\infty_{\mathrm{loc}}(\Omega).$$

This completes the proof. □

We turn to the proof of Lemma 3.4.16. Since it is a local result, it suffices to work on a neighborhood of $0 \in \mathbb{R}^n$. We write $f : \mathbb{R}^n_0 \to \mathbb{R}^m$ to denote that f is a germ of a function defined near $0 \in \mathbb{R}^n$. Let

$$\mathcal{A}_n := \{f : \mathbb{R}^n_0 \to \mathbb{R} \mid f \text{ is real analytic}\},$$
$$\mathcal{A}^m_n := \{f : \mathbb{R}^n_0 \to \mathbb{R}^m \mid f \text{ is real analytic}\}.$$

Note that $\mathcal{A}^m_n$ can be identified with the m-fold Cartesian product of $\mathcal{A}_n$ and $\mathcal{A}^n_n$ can be identified with the set of germs of real analytic vector fields defined near $0 \in \mathbb{R}^n$.

Lemma 3.4.17. *The ring $\mathcal{A}_n$ is Noetherian.*

Comments on the proof. This is a simple consequence of the Weierstrass preparation theorem; see page 148 of [252]. The proof in [252] is for the formal power series ring; however, as mentioned on page 130 of [252], the proof also works on the ring of convergent power series, i. e., the ring of power series with some positive radius of convergence. This is exactly the ring $\mathcal{A}_n$. □

Lemma 3.4.18. *The module $\mathcal{A}_n^m$ is a Noetherian $\mathcal{A}_n$-module.*

Comments on the proof. Since $\mathcal{A}_n$ is a Noetherian $\mathcal{A}_n$-module by Lemma 3.4.17, $\mathcal{A}_n^m$ is a finite direct sum of Noetherian $\mathcal{A}_n$-modules. It is therefore a Noetherian $\mathcal{A}_n$-module (see [98, Corollary 1.3]). □

Lemma 3.4.19. *Let $\mathcal{T} \subseteq \mathcal{A}_n^n \times \mathbb{N}^\nu$. Then there exists a finite set $\mathcal{F} \subseteq \mathcal{T}$ such that for every $(X_0, \vec{d}_0) \in \mathcal{T}$,*

$$X_0 = \sum_{\substack{(X_1,\vec{d}_1)\in\mathcal{F} \\ \vec{d}_1 \leq \vec{d}_0}} c_{(X_0,\vec{d}_0)}^{(X_1,\vec{d}_1)} X_1, \quad c_{(X_0,\vec{d}_0)}^{(X_1,\vec{d}_1)} \in \mathcal{A}_n.$$

Proof. Define a map $\iota : \mathcal{A}_n^n \times \mathbb{N}^\nu \to \mathcal{A}_{n+\nu}^n$ by $\iota(X_1, \vec{d}_1) = t^{\vec{d}_1} X_1(x)$, where $t \in \mathbb{R}^\nu$, $x \in \mathbb{R}^n$. Let M be the submodule of $\mathcal{A}_{n+\nu}^n$ generated by $\iota\mathcal{T}$. M is finitely generated by Lemma 3.4.18. Let $\mathcal{F} \subseteq \mathcal{T}$ be a finite set so that $\iota\mathcal{F}$ generates M. We will show that $\mathcal{F}$ satisfies the conclusions of the lemma.

Let $(X_0, \vec{d}_0) \in \mathcal{T}$. Since $t^{\vec{d}_0} X_0 \in M$, we have

$$t^{\vec{d}_0} X_0(x) = \sum_{(X_1,\vec{d}_1)\in\mathcal{F}} \hat{c}_{(X_0,\vec{d}_0)}^{(X_1,\vec{d}_1)}(t,x) t^{\vec{d}_1} X_1(x)$$

on a neighborhood of $(0, 0) \in \mathbb{R} \times \mathbb{R}^n$. Applying $\frac{1}{\vec{d}_0!}\partial_t^{\vec{d}_0}\big|_{t=0}$ to both sides and using that $\frac{1}{\vec{d}_0!}\partial_t^{\vec{d}_0}\big|_{t=0} \hat{c}_{(X_0,\vec{d}_0)}^{(X_1,\vec{d}_1)}(t,x) t^{\vec{d}_1} = 0$ if $\vec{d}_1 \not\leq \vec{d}_0$, we have

$$X_0(x) = \sum_{\substack{(X_1,\vec{d}_1)\in\mathcal{F} \\ \vec{d}_1 \leq \vec{d}_0}} \left[\frac{1}{\vec{d}_0!} \partial_t^{\vec{d}_0} \bigg|_{t=0} \hat{c}_{(X_0,\vec{d}_0)}^{(X_1,\vec{d}_1)}(t,x) t^{\vec{d}_1} \right] X_1(x).$$

The result follows. □

Proof of Lemma 3.4.16. Since the result is local, it suffices to prove the result when $\mathfrak{M}$ is an open neighborhood of $0 \in \mathbb{R}^n$ and $x = 0$. Applying Lemma 3.4.19 to $\mathcal{T} := \mathrm{Gen}(\mathcal{S})$, the result follows. □

3.5 Multi-parameter Carnot–Carathéodory geometry

Fix $\nu \in \mathbb{N}_+$ and let $\mathfrak{M}$ be a connected C^∞ manifold of dimension n. Let

$$(W, \vec{d}) = \{(W_1, \vec{d}_1), \dots, (W_r, \vec{d}_r)\} \subset C^\infty_{\mathrm{loc}}(\mathfrak{M}; T\mathfrak{M}) \times (\mathbb{N}^\nu \setminus \{0\}) \tag{3.14}$$

be a finite set. Our main assumption in this section, which we assume throughout, is the following.

Main Assumption: $\mathrm{Gen}((W, \vec{d}))$ is locally finitely generated on $\mathfrak{M}$.

Note that we do not assume that $W_1, \dots, W_r$ satisfy Hörmander's condition. However, Remark 3.4.11 shows that the Frobenius theorem (Theorem 3.1.15) applies to foliate $\mathfrak{M}$ into leaves: through each point x, there is a unique, maximal, connected, injectively immersed submanifold $\mathrm{Leaf}_x \hookrightarrow \mathfrak{M}$ such that $W_1, \dots, W_r$ are tangent to Leaf_x and satisfy Hörmander's condition on Leaf_x. Set $N(x) := \dim \mathrm{Leaf}_x$; Lemma 3.1.23 shows $x \mapsto N(x)$ is lower semi-continuous.

Fix $\Omega \Subset \mathfrak{M}$, an open, relatively compact set. Pick a finite set $(X, \vec{d}) = \{(X_1, \vec{d}_1), \dots, (X_q, \vec{d}_q)\} \subset \mathrm{Gen}((W, \vec{d}))$ so that $\mathrm{Gen}((W, \vec{d}))$ is finitely generated by $(X, \vec{d})$ on Ω and $(W, \vec{d}) \subseteq (X, \vec{d})$. Note that, $\forall x, y \in \Omega$, $T_y \mathrm{Leaf}_x = \mathrm{span}\{X_1(y), \dots, X_q(y)\}$.

For $\delta \in [0, \infty)^\nu$, we set $\delta^{\vec{d}} W = \{\delta^{\vec{d}_1} W_1, \dots, \delta^{\vec{d}_r} W_r\}$, where $\delta^{\vec{d}_j}$ is defined via standard multi-index notation. We similarly set $\delta^{\vec{d}} X = \{\delta^{\vec{d}_1} X_1, \dots, \delta^{\vec{d}_q} X_q\}$. For $x \in \mathfrak{M}$, we define

$$B_{(W,\vec{d})}(x, \delta) := B_{\delta^{\vec{d}} W}(x), \quad B_{(X,\vec{d})}(x, \delta) := B_{\delta^{\vec{d}} X}(x),$$

where $B_{\delta^{\vec{d}} W}(x)$ and $B_{\delta^{\vec{d}} X}(x)$ are defined via (1.12).

Fix sets $\mathcal{K} \Subset \Omega_1 \Subset \Omega$ with $\mathcal{K}$ compact and Ω_1 open and relatively compact in Ω. In what follows, we write $A \lesssim B$ for $A \le CB$, where $C \ge 0$ does not depend on $x \in \mathcal{K}$ or $\delta \in (0, 1]^\nu$.

Theorem 3.5.1. *There exists $\delta_0 = \delta_0(\mathcal{K}, \Omega_1) \in (0, 1]$ and $\xi_3 = \xi_3(\mathcal{K}, \Omega_1) \in (0, 1]$ such that the following hold:*

(a) $\forall x \in \mathcal{K}$, $\delta \in (0, \delta_0]^\nu \setminus \{0\}$, *we have* $B_{(W,\vec{d})}(x, \delta) \subseteq B_{(X,\vec{d})}(x, \delta) \subseteq \Omega_1 \cap \mathrm{Leaf}_x$, *and* $B_{(W,\vec{d})}(x, \delta)$ *and* $B_{(X,\vec{d})}(x, \Omega_1)$ *are open in* Leaf_x.

(b) $\forall x \in \mathcal{K}$, $\delta \in (0, 1]^\nu$, *we have* $B_{(X,\vec{d})}(x, \xi_3 \delta) \subseteq B_{(W,\vec{d})}(x, \delta) \subseteq B_{(X,\vec{d})}(x, \delta)$.

For all $x \in \mathcal{K}$, $\delta \in (0, 1]^\nu$, there exists $\Phi_{x,\delta} : B^{N(x)}(1) \to B_{(X,\vec{d})}(x, \delta) \cap \Omega_1$ such that:

(c) $\Phi_{x,\delta}(0) = 0$.

(d) $\Phi_{x,\delta}$ *is a smooth coordinate system on* Leaf_x.

(e) $B_{(W,\vec{d})}(x, \xi_3 \delta) \subseteq B_{(X,\vec{d})}(x, \xi_3 \delta) \subseteq \Phi_{x,\delta}(B^{N(x)}(1/2)) \subseteq \Phi_{x,\delta}(B^{N(x)}(1)) \subseteq B_{(X,\vec{d})}(x, \delta) \cap \Omega_1 \subseteq \mathrm{Leaf}_x \cap \Omega_1$.

(f) $\forall \delta \in (0, 1]^\nu$ *and* $m \in \mathbb{N}$, *the map* $(x, u) \mapsto \Phi_{x,\delta}(u)$ *taking* $\{x \in \mathcal{K} : N(x) = m\} \times B^m(1) \to \mathfrak{M}$ *is Borel measurable.*

Set $W_j^{x,\delta} := \Phi_{x,\delta}^* \delta^{\vec{d}_j} W_j$ *and* $X_j^{x,\delta} := \Phi_{x,\delta}^* \delta^{\vec{d}_j} X_j$. *Let*

$$W^{x,\delta} = \{W_1^{x,\delta}, \dots, W_r^{x,\delta}\}, \quad X^{x,\delta} := \{X_1^{x,\delta}, \dots, X_q^{x,\delta}\},$$
$$(W^{x,\delta}, |\vec{d}|_1) := \{(W_1^{x,\delta}, |\vec{d}_1|_1), \dots, (W_r^{x,\delta}, |\vec{d}_r|_1)\},$$

where $|\vec{d}_j|_1$ *denotes the* ℓ^1 *norm of the vector* $\vec{d}_j$.

(g) $W_j^{x,\delta}$ *and* $X_k^{x,\delta}$ *are* C^∞, *uniformly for* $x \in \mathcal{K}$ *and* $\delta \in (0,1]^\nu$ *in the sense that* $\forall L \in \mathbb{N}$, $1 \le j \le r$, $1 \le k \le q$,

$$\sup_{\substack{x\in\mathcal{K}\\ \delta\in(0,1]^\nu}} \|W_j^{x,\delta}\|_{C^L(B^{N(x)}(1);\mathbb{R}^{N(x)})} < \infty, \qquad \sup_{\substack{x\in\mathcal{K}\\ \delta\in(0,1]^\nu}} \|X_k^{x,\delta}\|_{C^L(B^{N(x)}(1);\mathbb{R}^{N(x)})} < \infty.$$

(h) $X_1^{x,\delta}(u), \dots, X_q^{x,\delta}(u)$ *span* $T_u B^{N(x)}(1)$, *uniformly for* $x \in \mathcal{K}$, $\delta \in (0,1]^\nu$, *and* $u \in B^{N(x)}(1)$ *in the sense that*

$$\inf_{\substack{x\in\mathcal{K}\\ \delta\in(0,1]^\nu}} \max_{j_1,\dots,j_{N(x)}\in\{1,\dots,q\}} \inf_{u\in B^{N(x)}(1)} \left|\det(X_{j_1}^{x,\delta}(u)|\cdots|X_{j_{N(x)}}^{x,\delta}(u))\right| > 0,$$

where each $X_j^{x,\delta}(u)$ *is treated as a column vector in* $\mathbb{R}^{N(x)}$.

(i) $(W^{x,\delta}, |\vec{d}|_1)$ *are Hörmander vector fields with formal degrees at the unit scale, uniformly for* $x \in \mathcal{K}$ *and* $\delta \in (0,1]^\nu$.

(j) *For all* $L \in \mathbb{N}$,

$$\|f\|_{C^L(B^{N(x)}(1))} \approx \sum_{|\alpha|\le L} \|(X^{x,\delta})^\alpha f\|_{C(B^{N(x)}(1))}, \quad \forall x \in \mathcal{K}, \delta \in (0,1]^\nu,$$

where the implicit constants depend on L, *but not on* f, x, *or* δ.

We defer the proof of Theorem 3.5.1 to Section 3.7.

Remark 3.5.2. The map $\Phi_{x,\delta}$ in Theorem 3.5.1 is a coordinate chart on a neighborhood of x in Leaf_X. Theorem 3.5.1 shows that this coordinate chart has good quantitative estimates, uniformly for $x \in \mathcal{K}$ and $\delta \in (0,1]^\nu$. In particular, these quantitative estimates remain uniform if x approaches a singular point. The main aspects of this uniformity are Theorem 3.5.1 (g), (h), and (i).

Remark 3.5.3. When working with multi-parameter balls, it is often more convenient to write $\delta \in [0,\infty)^\nu$ as $\delta = 2^{-j}$, where $j \in (-\infty,\infty]^\nu$. This is defined by $(\delta_1,\dots,\delta_\nu) = (2^{-j_1},\dots,2^{-j_\nu})$, where $2^{-\infty} = 0$. For $j,k \in (-\infty,\infty]^\nu$ we write $j \le k$ to mean $j_\mu \le k_\mu$ for $\mu \in \{1,\dots,\nu\}$ and we write $j \wedge k = (j_1 \wedge k_1, \dots, j_\nu \wedge k_\nu) \in (-\infty,\infty]^\nu$.

3.5.1 An important special case: Hörmander's condition

In this section, we describe an important special case of Theorem 3.5.1. As above, we let $\mathfrak{M}$ be a connected C^∞ manifold of dimension n. We let Vol be a smooth, strictly positive density on $\mathfrak{M}$. Fix $\nu \in \mathbb{N}_+$ and for each $\mu \in \{1,\dots,\nu\}$, let

$$(W^\mu, d^\mu) = \{(W_1^\mu, d_1^\mu),\dots,(W_{r_\mu}^\mu, d_{r_\mu}^\mu)\} \subset C^\infty_{\mathrm{loc}}(\mathfrak{M}; T\mathfrak{M}) \times \mathbb{N}_+$$

be Hörmander vector fields with single-parameter formal degrees. We define

$$\begin{aligned}(W,\vec{d}) &= \{(W_1,\vec{d}_1),\dots,(W_r,\vec{d}_r)\} := (W^1,d^1)\boxtimes(W^2,d^2)\boxtimes\cdots\boxtimes(W^\nu,d^\nu)\\ &= \{(W_j^\mu, d_j^\mu \mathfrak{e}_\mu) : \mu\in\{1,\dots,\nu\}, 1\le j\le r_\mu\} \subset C^\infty_{\mathrm{loc}}(\mathfrak{M};T\mathfrak{M})\times(\mathbb{N}^\nu\setminus\{0\}),\end{aligned}$$

where $\mathfrak{e}_1,\dots,\mathfrak{e}_\nu$ denotes the standard basis of $\mathbb{R}^\nu$. By Proposition 3.4.14, $\mathrm{Gen}((W,\vec{d}))$ is locally finitely generated on $\mathfrak{M}$. Thus, Theorem 3.5.1 applies in this setting. In this case, $N(x) = n$ and $\mathrm{Leaf}_x = \mathfrak{M}$, $\forall x \in \mathfrak{M}$.

We take $\mathcal{K} \Subset \Omega_1 \Subset \Omega \Subset \mathfrak{M}$ and $(X,\vec{d})$ as in Theorem 3.5.1. For $x\in\mathcal{K}$, $\delta\in(0,1]^\nu$, set

$$\Lambda(x,\delta) := \max_{j_1,\dots,j_n\in\{1,\dots,q\}} \mathrm{Vol}(x)(\delta^{\vec{d}_{j_1}}X_{j_1}(x),\dots,\delta^{\vec{d}_{j_n}}X_{j_n}(x)). \tag{3.15}$$

Note that $\Lambda(x,\delta) > 0$, $\forall x\in\mathcal{K}$ and $\delta\in(0,1]^\nu$. We have the following theorem.

Theorem 3.5.4. *Theorem 3.5.1 holds with $N(x) = n$ and $\mathrm{Leaf}_x = \mathfrak{M}$, $\forall x\in\mathfrak{M}$. Moreover, we also have:*

(k) $\forall x\in\mathcal{K}$, $\delta\in(0,\delta_0]^\nu$, *we have* $\mathrm{Vol}(B_{(W,\vec{d})}(x,\delta)) \approx \mathrm{Vol}(B_{(X,\vec{d})}(x,\delta)) \approx \Lambda(x,\delta)$.

(l) $\forall x\in\mathcal{K}$, $\delta\in(0,\infty)^\nu$, *we have* $\mathrm{Vol}(B_{(W,\vec{d})}(x,\delta))\wedge 1 \approx \mathrm{Vol}(B_{(X,\vec{d})}(x,\delta))\wedge 1 \approx \Lambda(x,\delta)\wedge 1$.

(m) $\forall x \in \mathcal{K}$, $\delta \in (0,\delta_0]^\nu$, *we have* $\mathrm{Vol}(B_{(W,\vec{d})}(x,2\delta)) \lesssim \mathrm{Vol}(B_{(W,\vec{d})}(x,\delta))$ *and* $\mathrm{Vol}(B_{(X,\vec{d})}(x,2\delta)) \lesssim \mathrm{Vol}(B_{(X,\vec{d})}(x,\delta))$.

(n) $\forall x \in \mathcal{K}$, $\delta \in (0,\infty)^\nu$, *we have* $\mathrm{Vol}(B_{(W,\vec{d})}(x,2\delta))\wedge 1 \lesssim \mathrm{Vol}(B_{(W,\vec{d})}(x,\delta))\wedge 1$ *and* $\mathrm{Vol}(B_{(X,\vec{d})}(x,2\delta))\wedge 1 \lesssim \mathrm{Vol}(B_{(X,\vec{d})}(x,\delta))\wedge 1$.

For $x\in\mathcal{K}$ and $\delta\in(0,1]^\nu$, define $h_{x,\delta}\in C^\infty(B^n(1))$ by $\Phi_{x,\delta}^*\mathrm{Vol} = \Lambda(x,\delta)h_{x,\delta}\sigma_{\mathrm{Leb}}$.

(o) $\forall x\in\mathcal{K}$, $\delta\in(0,1]^\nu$, $u\in B^n(1)$, $h_{x,\delta}(u)\approx 1$, *with implicit constants independent of x, δ, and u.*

(p) *For all $L\in\mathbb{N}$,*

$$\sup_{\substack{x\in\mathcal{K}\\ \delta\in(0,1]^\nu}} \|h_{x,\delta}\|_{C^L(B^n(1))} < \infty.$$

We defer the proof of Theorem 3.5.4 to Section 3.7.

Example 3.5.5. Consider the following very simple special case. We let $\mathfrak{M} = \mathbb{R}$, $\nu = 2$, $(W^1,d^1) = \{(W_1^1,d_1^1)\} := \{(\partial_x,1)\}$, and $(W^2,d^2) = \{(W_1^1,d_1^2)\} := \{(\partial_x,1)\}$. Let $\delta = (\delta_1,\delta_2)\in$

$(0,1]^2$. This satisfies the conditions of Theorem 3.5.4. If $\delta_1 \geq \delta_2$, then the map $\Phi_{x,\delta}$ in Theorems 3.5.1 and 3.5.4 can be taken to be $\Phi_{x,\delta}(t) = x + \delta_1 t$. In this case, $\Phi_{x,\delta}^* \delta_1^{d_1^1} W_1^1 = \partial_t$ and $\Phi_{x,\delta}^* \delta_2^{d_1^2} W_1^2 = (\delta_2/\delta_1)\partial_t$. Thus, if $\delta_2 \ll \delta_1$, then while $\Phi_{x,\delta}^* \delta_2^{d_1^2} W_1^2$ satisfies Hörmander's condition (in fact, it spans the tangent space at every point), it does not do so uniformly in δ. Therefore, the conditions of Theorem 3.5.4 are not quantitatively invariant under rescaling. However, the conditions of Theorem 3.5.1 are quantitatively invariant under rescaling. Even though our main motivations come from the setting of Theorem 3.5.4, because we will often need to rescale and prove quantitative results, it is often best to work in the more general setting of Theorem 3.5.1, since it is quantitatively preserved under rescalings.

3.5.2 Dropping parameters

Theorems 3.5.1 and 3.5.4 consider only $\delta \in (0,\infty)^\nu$. In fact, similar results hold for $\delta \in [0,\infty)^\nu \setminus \{0\}$. This is because when some of the coordinates of δ are equal to 0, the balls $B_{(W,\vec{d})}(x,\delta)$ can be viewed as lower-parameter balls where none of the coordinates of δ are equal to 0. This section is devoted to explaining this.

Fix $\emptyset \neq E \subseteq \{1,\dots,\nu\}$. Then under the assumptions of either Theorem 3.5.1 or Theorem 3.5.4, the list of sets of vector fields with formal degrees

$$(W^\mu, \vec{d}^\mu), \quad \mu \in E,$$

satisfies the same assumptions as

$$(W^\mu, \vec{d}^\mu), \quad \mu \in \{1,\dots,\nu\},$$

but with ν replaced by $|E|$. We let

$$(W^E, \vec{d}^E) \subset C^\infty_{\mathrm{loc}}(\mathfrak{M}; T\mathfrak{M}) \times \mathbb{N}^{|E|}$$

be as in (3.14), but with $(W^1, \vec{d}^1), \dots, (W^\nu, \vec{d}^\nu)$ replaced by $(W^\mu, \vec{d}^\mu)$, $\mu \in E$.

Moreover, suppose $\delta \in [0,\infty)^\nu \setminus \{0\}$. Set $E := \{\mu \in \{1,\dots,\nu\} : \delta_\mu \neq 0\}$ and let $\delta_E := (\delta_\mu)_{\mu\in E} \in (0,\infty)^{|E|}$ be the non-zero components of δ. Then,

$$B_{(W^E,\vec{d}^E)}(x,\delta_E) = B_{(W,\vec{d})}(x,\delta).$$

Because of this, we can obtain versions of Theorems 3.5.1 and 3.5.4 for the balls $B_{(W,\vec{d})}(x,\delta)$, when some of the coordinates of δ are 0, by applying those results to $(W^E, \vec{d}^E)$.

3.6 The quantitative coordinate system

All of the main scaling theorems of this chapter follow from a single, abstract theorem concerning the existence of a coordinate system with good quantitative estimates. This theorem is due to the author and Stovall [228], and we present it here. An important aspect of the theory in this section is that all assumptions and estimates remain unchanged when setting is transformed under an arbitrary C^∞_{loc} diffeomorphism: every estimate is "coordinate-free." This comes up in Remark 3.6.1; see Remarks 3.6.6 and A.2.3 for a further discussion of this invariance.

Let $\mathfrak{M}$ be a C^∞ manifold of dimension n and let $X = \{X_1, \dots, X_q\} \subset C^\infty_{\mathrm{loc}}(\mathfrak{M}; T\mathfrak{M})$ be a finite set of smooth vector fields on $\mathfrak{M}$ such that $\forall x \in \mathfrak{M}$, $T_x\mathfrak{M} = \operatorname{span}\{X_1(x), \dots, X_q(x)\}$. Let Vol be a smooth, strictly positive density on $\mathfrak{M}$. For $\delta > 0$ and $x \in \mathfrak{M}$, set $B_X(x,\delta) := B_{\{\delta X_1,\dots,\delta X_q\}}(x)$, where the latter ball is defined by (1.12).

Remark 3.6.1. When we use the results of this section to prove results like Theorem 3.3.7, we apply the theory of this section with $X_1, \dots, X_q$ replaced by $\delta^{d_1}X_1, \dots, \delta^{d_q}X_q$, where $X_1, \dots, X_q$ satisfy certain assumptions and δ is small. We require that the results we prove are uniform in δ (and other parameters) in such applications. However, since there is no parameter like δ in this section, we instead proceed by keeping careful track of what each estimate does depend on. This will allow us to see that the estimates do not depend on the relevant parameters (like δ) in our applications. Unfortunately, this makes the statements in this section somewhat technical.

Fix $x_0 \in \mathfrak{M}$ and $\xi > 0$. We write

$$[X_j, X_k] = \sum_{l=1}^{q} c^l_{j,k} X_l,$$

and we assume $c^l_{j,k} \in C(B_X(x_0,\xi))$. The ball $B_X(x_0,\xi)$ is open in $\mathfrak{M}$ (see Lemma 3.1.7) and we may therefore treat $B_X(x_0,\xi)$ as a manifold.

Definition 3.6.2. Let Z be a one-dimensional real vector space. For $z_1, z_2 \in Z$, $z_2 \neq 0$, we set $z_1/z_2 := \lambda(z_1)/\lambda(z_2) \in \mathbb{R}$, where $\lambda : Z \to \mathbb{R}$ is any non-zero linear map. It is easy to see that z_1/z_2 does not depend on the choice of λ.

Fix $\zeta \in (0,1]$. Reorder $X_1, \dots, X_q$ so that

$$\max_{j_1,\dots,j_n \in \{1,\dots,q\}} \left| \frac{X_{j_1}(x_0) \wedge X_{j_2}(x_0) \wedge \cdots \wedge X_{j_n}(x_0)}{X_1(x_0) \wedge X_2(x_2) \wedge \cdots \wedge X_n(x_0)} \right| \leq \zeta^{-1}. \tag{3.16}$$

This quotient is defined by Definition 3.6.2, where we have used the fact that $\bigwedge^n T_{x_0}\mathfrak{M}$ is a one-dimensional vector space, since $T_{x_0}\mathfrak{M}$ is an n-dimensional vector space. Note that we may always reorder $X_1, \dots, X_q$ in this way: it is always possible to pick $X_1, \dots, X_n$ so that the left-hand side of (3.16) equals 1, though we will require the flexibility to take $\zeta < 1$ in our applications. See Remark 3.6.9 for a further comment on this quantity.

For $1 \leq j \leq q$, define $f_j \in C^\infty_{\mathrm{loc}}(\mathfrak{M})$ by $\mathrm{Lie}_{X_j} \mathrm{Vol} = f_j \mathrm{Vol}$.

Definition 3.6.3. For $x \in \mathfrak{M}$, $\eta > 0$, and $U \subseteq \mathfrak{M}$, we say X satisfies $\mathcal{C}(x, \eta, U)$ if for every $a \in B^q(\eta)$ the expression

$$e^{a_1 X_1 + \cdots + a_q X_q} x$$

exists in U. More precisely, consider the differential equation

$$\partial_r E(r) = a_1 X_1(E(r)) + \cdots + a_q X_q(E(r)), \quad E(0) = x.$$

We assume that a solution to this differential equation exists up to $r = 1$, $E : [0,1] \to U$. We have $E(r) = e^{r a_1 X_1 + \cdots + r a_q X_q} x$.

We require two more technical quantities related to X, though the reader may wish to skip these on a first reading:

- Let $\eta \in (0,1]$ be such that $\{X_1, \ldots, X_n\}$ satisfies $\mathcal{C}(x_0, \eta, \mathfrak{M})$.
- Let $\tau_0 \in (0,1]$ be such that for $\tau \in (0, \tau_0]$ the following holds: if $z \in B_{\{X_1,\ldots,X_n\}}(x_0, \xi)$ is such that $\{X_1, \ldots, X_n\}$ satisfies $\mathcal{C}(z, \tau, B_{\{X_1,\ldots,X_n\}}(x_0, \xi))$, $t \in B^n(\tau)$ is such that $e^{t_1 X_1 + \cdots + t_n X_n} z = z$, and $X_1(z), \ldots, X_n(z)$ are linearly independent, then $t = 0$.

Such an η and τ_0 always exist (by possibly shrinking $\xi > 0$) – see Proposition 3.6.7 and Remark 3.6.8. However, our quantitative estimates will depend on η and τ_0.

Definition 3.6.4. For $L \in \mathbb{N}$, we say C is an L-admissible constant if C can be chosen to depend only on upper bounds for the following quantities, which we assume to be finite:

- $\sum_{|\alpha| \leq L} \|X^\alpha c^l_{j,k}\|_{C(B_X(x_0,\xi))}$, $1 \leq j, k, l \leq q$,
- $L, q, \eta^{-1}, \tau_0^{-1}, \zeta^{-1}$, and ξ^{-1}.

We say C is an L, Vol-admissible constant if C is an L-admissible constant which is also allowed to depend on upper bounds for the following quantity, which we assume to be finite:

$$\sum_{|\alpha| \leq L} \|X^\alpha f_j\|_{C(B_X(x_0,\xi))}, \quad 1 \leq j \leq q.$$

We write $A \lesssim_L B$ for $A \leq CB$, where C is a positive L-admissible constant and write $A \approx_L B$ for $A \lesssim_L B$ and $B \lesssim_L A$. We similarly define $A \lesssim_{L,\mathrm{Vol}} B$ and $A \approx_{L,\mathrm{Vol}} B$.

Because $\{X_1, \ldots, X_n\}$ satisfies $\mathcal{C}(x_0, \eta, \mathfrak{M})$, by hypothesis, we may define the map, for $t \in B^n(\eta)$,

$$\Phi(t) := e^{t_1 X_1 + \cdots + t_n X_n} x_0. \tag{3.17}$$

Let $\eta_0 := \min\{\eta, \xi\}$ so that $\Phi : B^n(\eta_0) \to B_X(x_0, \xi)$. Note that $\Phi \in C^\infty_{\mathrm{loc}}$ since $X_1, \ldots, X_q \in C^\infty_{\mathrm{loc}}(\mathfrak{M}; T\mathfrak{M})$.

Theorem 3.6.5. *There exist 1-admissible constants $\eta_1, \xi_2 \in (0,1]$ such that:*
(a) *$\Phi(B^n(\eta_1))$ is an open subset of $B_X(x_0,\xi)$ and is therefore a submanifold of $\mathfrak{M}$.*
(b) *$\Phi : B^n(\eta_1) \to \Phi(B^n(\eta_1))$ is a C^∞_{loc} diffeomorphism.*
(c) *$B_X(x_0,\xi_2) \subseteq \Phi(B^n(\eta_1/2)) \subseteq \Phi(B^n(\eta_1)) \subseteq B_X(x_0,\xi)$.*

Let $Y_j = \Phi^* X_j$ and write

$$\begin{bmatrix} Y_1 \\ Y_2 \\ \vdots \\ Y_n \end{bmatrix} = (I+A) \begin{bmatrix} \partial_{t_1} \\ \partial_{t_2} \\ \vdots \\ \partial_{t_n} \end{bmatrix},$$

where $A \in C^\infty(B^n(\eta_1); \mathbb{M}^{n\times n})$.
(d) *$A(0) = 0$ and $\sup_{t\in B^n(\eta_1)} \|A\|_{\mathbb{M}^{n\times n}} \le \frac{1}{2}$.*
(e) *We have the following quantitative regularity for Y_j, $1 \le j \le q$:*

$$\|Y_j\|_{C^L(B^n(\eta_1);\mathbb{R}^n)} \lesssim_L 1.$$

(f) *We have the following equivalence of norms for $f \in C(B^n(1))$:*

$$\|f\|_{C^L(B^n(\eta_1))} \approx_{(L-1)\vee 0} \sum_{|\alpha|\le L} \|Y^\alpha f\|_{C(B^n(\eta_1))}.$$

Define $h \in C^\infty(B^n(\eta_1))$ by

$$\Phi^* \operatorname{Vol} = \Big(\max_{j_1,\dots,j_n \in\{1,\dots,q\}} \operatorname{Vol}(x_0)(X_{j_1}(x_0),\dots,X_{j_n}(x_0))\Big) h \sigma_{\mathrm{Leb}}.$$

(g) *$h(t) \approx_{0,\mathrm{Vol}} 1$, $\forall t \in B^n(\eta_1)$, with implicit constant independent of t.*
(h) *We have the following quantitative regularity of h: for all $L \in \mathbb{N}$,*

$$\|h\|_{C^L(B^n(\eta_1))} \lesssim_{L,\mathrm{Vol}} 1.$$

Remark 3.6.6. Note that the estimates in Theorem 3.6.5 do not depend on the norms of the coefficients of the vector fields $X_1,\dots,X_q$ in any coordinate system. In fact, the quantitative estimates are invariant under conjugating the setting by an arbitrary C^∞_{loc} diffeomorphism. More is true: in Appendix A we prove a generalization of this result which is invariant under arbitrary C^2_{loc} diffeomorphisms; thus, one does not even need to assume $X_1,\dots,X_q$ are C^∞_{loc} in order to achieve the above result. In particular, one can conclude that $Y_1,\dots,Y_q$ are C^∞_{loc} even in some cases when $X_1,\dots,X_q$ are merely C^1_{loc}. See Appendix A, and in particular Remark A.2.3, for more on this.

Proposition 3.6.7. *$\eta > 0$ and $\tau_0 > 0$ satisfying the above assumptions always exist. Indeed, let $\mathcal{K} \Subset \mathfrak{M}$ be a compact set. Then:*

- $\exists \eta \in (0,1]$ *such that* $\forall x_0 \in \mathcal{K}$, X *satisfies* $\mathcal{C}(x, \eta, \mathfrak{M})$.
- $\exists \tau_0 \in (0,1]$ *such that* $\forall \theta \in \mathbb{R}^q$ *with* $|\theta| = 1$, *if* $x \in \mathcal{K}$ *is such that* $\theta_1 X_1(x) + \cdots + \theta_q X_q(x) \neq 0$, *then* $\forall r \in (0, \tau_0]$,

$$e^{r\theta_1 X_1 + \cdots + r\theta_q X_q} x \neq x.$$

Remark 3.6.8. Proposition 3.6.7 immediately implies that $\eta \in (0,1]$, as in the above assumptions, exists. By possibly shrinking $\xi > 0$, $B_{\{X_1,\ldots,X_n\}}(x_0, \xi)$ is pre-compact in $\mathfrak{M}$; the existence of $\tau_0 \in (0,1]$ follows by applying Proposition 3.6.7 to the closure of this set.

Remark 3.6.9. Let $\mathscr{V}$ be an n-dimensional real vector space, let $v_1, \ldots, v_n \in \mathscr{V}$ be a basis for $\mathscr{V}$, and let $w_1, \ldots, w_n \in \mathscr{V}$. Then, as described above, Definition 3.6.2 defines the quantity

$$\frac{w_1 \wedge w_2 \wedge \cdots \wedge w_n}{v_1 \wedge v_2 \wedge \cdots \wedge v_n} \in \mathbb{R}. \tag{3.18}$$

This quantity can be described in a much more concrete way. Indeed, pick any basis for $\mathscr{V}$ and let $V_1, \ldots, V_n$ and $W_1, \ldots, W_n$ be the column vectors given by writing $v_1, \ldots, v_n$ and $w_1, \ldots, w_n$ in this basis. Then

$$\frac{w_1 \wedge w_2 \wedge \cdots \wedge w_n}{v_1 \wedge v_2 \wedge \cdots \wedge v_n} = \frac{\det(W_1|W_2|\cdots|W_n)}{\det(V_1|V_2|\cdots|V_n)}.$$

The reason we have chosen the more abstract notation (3.18) is that it makes it clear that this quantity does not depend on the choice of basis or coordinate system.

For proofs of the results in this section, see Appendix A.

3.7 Proofs of scaling results

In this section, we use Theorem 3.6.5 to prove Theorems 3.3.1, 3.3.7, 3.5.1, and 3.5.4 and Corollaries 3.3.8 and 3.3.9.

Theorem 3.3.1 is a special case of Theorem 3.3.7, which, in turn, is a special case of Theorem 3.5.4. We prove Theorems 3.5.1 and 3.5.4 simultaneously.

Proof of Theorems 3.5.1 *and* 3.5.4. Most of this proof is in the more general setting of Theorem 3.5.1, though we will occasionally work in the special setting of Theorem 3.5.4.

We will apply Theorem 3.6.5 with x_0 replaced by $x \in \mathcal{K}$, $\mathfrak{M}$ replaced by $\mathrm{Leaf}_x \cap \Omega$, and $X = \{X_1, \ldots, X_q\}$ replaced by $\delta^{\vec{d}} X = \{\delta^{\vec{d}_1} X_1, \ldots, \delta^{\vec{d}_q} X_q\}$ for $\delta \in (0,1]^\nu$. We need to show that the hypotheses of Theorem 3.6.5 hold uniformly for $x \in \mathcal{K}$ and $\delta \in (0,1]^\nu$, with these choices. More precisely, we will show that L-admissible constants and L, Vol-admissible constants, as in Definition 3.6.4, can be chosen independent of $x \in \mathcal{K}$ and $\delta \in (0,1]^\nu$.

By the Picard–Lindelöf theorem, we may take $\xi \in (0,1]$ so small that $\forall x \in \mathcal{K}$, $B_X(x,\xi) \subseteq \Omega_1$. We will choose $\delta_0 \in (0,\xi]$. Since $\delta_0 \leq \xi$, we have, $\forall \delta \in (0,\delta_0]^\nu$, $x \in \mathcal{K}$,

$$B_{(X,\vec{d})}(x,\delta) = B_{\delta^{\vec{d}}X}(x) \subseteq B_X(x,\xi) \subseteq \Omega_1. \tag{3.19}$$

For any x and δ, we have $B_{(W,\vec{d})}(x,\delta) \subseteq B_{(X,\vec{d})}(x,\delta)$ (since $(W,\vec{d}) \subseteq (X,\vec{d})$). This, combined with (3.19), Lemma 3.1.7, and the immediate fact that $B_X(x,\xi) \subseteq \mathrm{Leaf}_x$, proves (a).

For $x \in \mathcal{K}$ and $\delta \in (0,1]^\nu$, pick $j_1 = j_1(x,\delta), \ldots, j_{N(x)} = j_{N(x)}(x,\delta) \in \{1,\ldots,q\}$ such that

$$\max_{k_1,\ldots,k_{N(x)} \in \{1,\ldots,q\}} \left| \frac{(\delta^{\vec{d}_{k_1}} X_{k_1}(x)) \wedge \cdots \wedge (\delta^{\vec{d}_{k_{N(x)}}} X_{k_{N(x)}}(x))}{(\delta^{\vec{d}_{j_1}} X_{j_1}(x)) \wedge \cdots \wedge (\delta^{\vec{d}_{j_{N(x)}}} X_{j_{N(x)}}(x))} \right| > 1/2. \tag{3.20}$$

It is always possible to pick $j_1(x,\delta), \ldots, j_{N(x)}(x,\delta)$ as in (3.20); indeed, we may pick them so that the left-hand side equals 1 (see also Remark 3.6.9). We require the choice of $j_1,\ldots,j_n$ to be Borel measurable in x. Indeed, fix $m \in \mathbb{N}$, and note that the set $\{x \in \mathcal{K} : N(x) = m\}$ is measurable by Lemma 3.1.23 (the function $N(x)$ is lower semi-continuous). For each $x \in \{x \in \mathcal{K} : N(x) = m\}$, we pick $j_1(x,\delta), \ldots, j_m(x,\delta)$ as in (3.20) such that the map $x \mapsto j_l(x,\delta)$, $\{x \in \mathcal{K} : N(x) = m\} \to \mathbb{N}$ is Borel measurable (this is possible since the left-hand side of (3.20) is continuous in x for a fixed $k_1,\ldots,k_m,j_1,\ldots,j_m$).

We will apply Theorem 3.6.5 with $X_1,\ldots,X_n$ replaced by

$$\delta^{\vec{d}_{j_1(x,\delta)}} X_{j_1(x,\delta)}, \ldots, \delta^{\vec{d}_{j_{N(x)}(x,\delta)}} X_{j_{N(x)}(x,\delta)}; \tag{3.21}$$

thus, we are taking $\zeta = 1/2$ in that theorem.

By Proposition 3.6.7, with $\mathfrak{M}$ replaced by Ω_1, we can find $\eta = \eta(\mathcal{K},\Omega_1) \in (0,1]$ and $\tau_0 = \tau_0(\mathcal{K},\Omega_1) \in (0,1]$ satisfying the conclusions of that proposition when applied to X. It follows immediately that the same η and τ_0 satisfy the hypotheses of Theorem 3.6.5 with $x \in \mathcal{K}$ and X replaced by $\delta^{\vec{d}}X$ for $\delta \in (0,1]^\nu$ (and $X_1,\ldots,X_n$ replaced by (3.21)).

Since $\mathrm{Gen}((W,\vec{d}))$ is finitely generated by $(X,\vec{d})$ on Ω, Remark 3.4.12 shows that

$$[X_j, X_k] = \sum_{\vec{d}_l \leq \vec{d}_j + \vec{d}_k} c_{j,k}^l X_l, \quad c_{j,k}^l \in C^\infty_{\mathrm{loc}}(\Omega). \tag{3.22}$$

Set, for $\delta \in (0,1]^\nu$,

$$c_{j,k}^{l,\delta} := \begin{cases} \delta^{\vec{d}_j + \vec{d}_k - \vec{d}_l} c_{j,k}^l & \text{if } \vec{d}_j + \vec{d}_k \geq \vec{d}_l, \\ 0 & \text{otherwise.} \end{cases}$$

Note that $c_{j,k}^{l,\delta} \in C^\infty_{\mathrm{loc}}(\Omega)$, uniformly for $\delta \in (0,1]^\nu$. Multiplying both sides of (3.22) by $\delta^{\vec{d}_j + \vec{d}_k}$, we have

$$[\delta^{\vec{d}_j}X_j, \delta^{\vec{d}_k}X_k] = \sum_{l=1}^{q} c_{j,k}^{l,\delta}\delta^{\vec{d}_l}X_l.$$

In our application of Theorem 3.6.5, $c_{j,k}^{l,\delta}$ plays the role of $c_{j,k}^{l}$. When considering the special case of Theorem 3.5.4, let $f_j \in C^\infty_{\mathrm{loc}}(\mathfrak{M})$ be defined by $\mathrm{Lie}_{X_j}\,\mathrm{Vol} = f_j\,\mathrm{Vol}$ (such an f_j exists, since Vol is a smooth, strictly positive density). Let $f_j^\delta := \delta^{\vec{d}_j}f_j$, so that $\mathrm{Lie}_{\delta^{\vec{d}_j}X_j}\,\mathrm{Vol} = f_j^\delta\,\mathrm{Vol}$. Note that $f_j^\delta \in C^\infty_{\mathrm{loc}}(\mathfrak{M})$, uniformly for $\delta \in (0,1]^\nu$.

For every $L \in \mathbb{N}$, we have

$$\begin{aligned}\sup_{\substack{\delta\in(0,1]^\nu\\ x\in\mathcal{K}}}\max_{j,k,l\in\{1,\dots,q\}}\sum_{|\alpha|\le L}\|(\delta^{\vec{d}}X)^\alpha c_{j,k}^{l,\delta}\|_{C(B_{\delta^{\vec{d}}X}(x,\xi))} &\le \sup_{\delta\in(0,1]^\nu}\max_{j,k,l\in\{1,\dots,q\}}\sum_{|\alpha|\le L}\|(\delta^{\vec{d}}X)^\alpha c_{j,k}^{l,\delta}\|_{C(\Omega_1)}\\ &\le \max_{\vec{d}_l\le\vec{d}_j+\vec{d}_k}\sum_{|\alpha|\le L}\|X^\alpha c_{j,k}^{l}\|_{C(\Omega_1)} < \infty.\end{aligned}$$

Similarly in the special case of Theorem 3.5.4, we have

$$\sup_{\substack{\delta\in(0,1]^\nu\\ x\in\mathcal{K}}}\max_{j\in\{1,\dots,q\}}\sum_{|\alpha|\le L}\|(\delta^{\vec{d}}X)^\alpha f_j^\delta\|_{C(B_{\delta^{\vec{d}}X}(x,\xi))} \le \max_{j\in\{1,\dots,q\}}\sum_{|\alpha|\le L}\|X^\alpha f_j\|_{C(\Omega_1)} < \infty.$$

The above shows that L-admissible constants for any $L \in \mathbb{N}$ and L, Vol-admissible constants in the special case of Theorem 3.5.4 as in Definition 3.6.4 can be chosen independent of $x \in \mathcal{K}$ and $\delta \in (0,1]^\nu$. Thus, Theorem 3.6.5 applies with x_0 replaced by $x \in \mathcal{K}$ and X replaced by $\delta^{\vec{d}}X$ for $\delta \in (0,1]^\nu$, and any constants in that theorem can be chosen uniformly for $x \in \mathcal{K}$ and $\delta \in (0,1]^\nu$.

Let $\eta_1 > 0$ be as in Theorem 3.6.5; as described above, η_1 can be chosen independent of $x \in \mathcal{K}$ and $\delta \in (0,1]^\nu$. Let $\widetilde{\Phi}_{x,\delta} : B^{N(x)}(\eta_1) \to B_{\delta^{\vec{d}}X}(x,\xi) \subseteq B_{(X,\vec{d})}(x,\delta) \cap \Omega_1$ be the map Φ from Theorem 3.6.5, that is,

$$\widetilde{\Phi}_{x,\delta}(t) = \exp(t_1\delta^{\vec{d}_{j_1(x,\delta)}}X_{j_1(x,\delta)} + \cdots + t_{N(x)}\delta^{\vec{d}_{j_{N(x)}(x,\delta)}}X_{j_{N(x)}(x,\delta)})x.$$

Clearly, $\Phi_{x,\delta}(0) = x$. The Borel measurability of $j_l(x,\delta)$ along with standard theorems from the field of ODEs shows that for each $m \in \mathbb{N}$ and $\delta \in (0,1]^\nu$, the map

$$(x,t) \mapsto \widetilde{\Phi}_{x,\delta}(t), \quad \{x \in \mathcal{K} : N(x) = m\} \times B^m(\eta_1) \to \mathfrak{M}$$

is Borel measurable. By Theorem 3.6.5 (a) and (b), $\widetilde{\Phi}_{x,\delta}$ is a smooth coordinate system on Leaf_x.

Set $\Phi_{x,\delta}(t) := \widetilde{\Phi}_{x,\delta}(\eta_1 t) : B^{N(x)}(1) \to B_{(X,\vec{d})}(x,\delta) \cap \Omega_1$. The results (c), (d), and (f) now follow from the corresponding results for $\widetilde{\Phi}_{x,\delta}$ described above.

Taking $\xi_2 \in (0,1]$ as in Theorem 3.6.5 and using Theorem 3.6.5 (c) and the fact that $(W,\vec{\mathsf{d}}) \subseteq (X,\vec{d})$, we have, for all $x \in \mathcal{K}$, $\delta \in (0,1]^\nu$, and $\xi' \in (0,\xi_2]$,

$$\begin{aligned} B_{(W,\vec{d})}(x,\xi'\delta) &\subseteq B_{(X,\vec{d})}(x,\xi'\delta) \subseteq B_{(X,\vec{d})}(x,\xi_2\delta) \subseteq B_{\delta^{\vec{d}}X}(x,\xi_2) \\ &\subseteq \widetilde{\Phi}_{x,\delta}(B^{N(x)}(\eta_1/2)) = \Phi_{x,\delta}(B^{N(x)}(1/2)) \subseteq \Phi_{x,\delta}(B^{N(x)}(1)) \\ &= \widetilde{\Phi}_{x,\delta}(B^{N(x)}(\eta_1)) \subseteq B_{\delta^{\vec{d}}X}(x,\xi) \subseteq B_{(X,\vec{d})}(x,\delta). \end{aligned} \tag{3.23}$$

We will later pick $\xi_3 \in (0,\xi_2]$, and item (e) will follow from (3.23) with $\xi' = \xi_3$.

Note that for any vector field Z,

$$\Phi_{x,\delta}^* Z = \eta_1^{-1}\widetilde{\Phi}_{x,\delta}^* Z. \tag{3.24}$$

Since $\eta_1 \approx 1$, (3.24) allows us to easily turn estimates about $\widetilde{\Phi}_{x,\delta}^* Z$ into estimates about $\Phi_{x,\delta}^* Z$. Theorem 3.6.5 (e) combined with (3.24) shows that for $1 \le k \le q$ and $L \in \mathbb{N}$,

$$\|X_k^{x,\delta}\|_{C^L(B^{N(x)}(1))} \lesssim 1.$$

Since each $W_j^{x,\delta}$ is of the form $X_k^{x,\delta}$ for some k, item (g) follows.

Theorem 3.6.5 (d) shows that the matrix $I + A(t)$ from that theorem is invertible with $\det(I + A(t)) \gtrsim 1$, and therefore using (3.24) we have

$$\begin{aligned} &\inf_{u\in B^{N(x)}(1)} \left|\det(X_{j_1(x,\delta)}^{x,\delta}(u)|\cdots|X_{j_{N(x)}(x,\delta)}^{x,\delta}(u))\right| \\ &\approx \inf_{u\in B^{N(x)}(1)} \left|\det(\widetilde{\Phi}_{x,\delta}^*\delta^{d_{j_1}}X_{j_1}(u)|\cdots|\widetilde{\Phi}_{x,\delta}^*\delta^{d_{j_{N(x)}}}X_{j_{N(x)}}(u))\right| \\ &= \inf_{u\in B^{N(x)}(1)} \left|\det(I + A(u))\right| \approx 1. \end{aligned}$$

Item (h) follows.

Since each $(X_k, \vec{d}_k) \in \mathrm{Gen}((W,\vec{d}))$, we see that $\delta^{\vec{d}_j}X_j$ is a fixed commutator of the vector fields $\delta^{\vec{d}}W$. Thus, each $X_j^{x,\delta}$ is the same commutator of the vector fields $W^{x,\delta}$ (since the pullback of the commutator is the commutator of the pullback). From this, (i) follows from (g) and (h).

Item (j) follows immediately from (3.24) and Theorem 3.6.5 (f).

We turn to (b). We always have $B_{(W,\vec{d})}(x,\delta) \subseteq B_{(X,\vec{d})}(x,\delta)$ for any x and δ, since $(W,\vec{d}) \subseteq (X,\vec{d})$. By (3.23) with $\xi' = \xi_2$, we have, for $x \in \mathcal{K}$ and $\delta \in (0,1]^\nu$,

$$B_{(\xi_2\delta)^{\vec{d}}W}(x) = B_{(W,\vec{d})}(x,\xi_2\delta) \subseteq \Phi_{x,\delta}(B^{N(x)}(1/2)).$$

Pulling this equation back via $\Phi_{x,\delta}$, we have

$$B_{\xi_2^{|\vec{d}|_1}W^{x,\delta}}(0) = B_{\Phi_{x,\delta}^*(\xi_2\delta)^{\vec{d}}W}(x) \subseteq B^{N(x)}(1/2).$$

By (i) and the fact that $\xi_2 \approx 1$, we see that $(\xi_2^{|\vec{d}|_1}W^{x,\delta}, |\vec{d}|_1)$ are Hörmander vector fields with formal degrees at the unit scale, uniformly for $x \in \mathcal{K}$ and $\delta \in (0,1]^\nu$. Lemma 3.2.4 shows that there exists $\eta_2 \in (0,1/2]$ with η_2 independent of $x \in \mathcal{K}$ and $\delta \in (0,1]$ such that

$$B^{N(x)}(\eta_2) \subseteq B_{\xi_2^{|\vec{d}|_1} W^{x,\delta}}(0) = B_{\Phi^*_{x,\delta}(\xi_2\delta)^{\vec{d}} W}(x) \subseteq B^{N(x)}(1/2).$$

Using (g), the Picard–Lindelöf theorem shows that there exists $\xi_3 \in (0, \xi_2]$, independent of $x \in \mathcal{K}$ and $\delta \in (0,1]^\nu$, so small that

$$B_{(X^{x,\delta}, |\vec{d}|_1)}(0, \xi_3) \subseteq B^{N(x)}(\eta_2) \subseteq B_{\xi_2^{|\vec{d}|_1} W^{x,\delta}}(0).$$

Pushing this forward via $\Phi_{x,\delta}$, we have

$$\begin{aligned} B_{(X,\vec{d})}(x, \xi_3\delta) = B_{(\delta^{\vec{d}} X, |\vec{d}|_1|)}(x, \xi_3) &= \Phi_{x,\delta}(B_{(X^{x,\delta}, |\vec{d}|_1)}(0, \xi_3)) \subseteq \Phi_{x,\delta}(B^n(\eta_2)) \\ \subseteq \Phi_{x,\delta}(B_{\xi_2^{|\vec{d}|_1} W^{x,\delta}}(0)) &= B_{(W,\vec{d})}(x, \xi_2\delta) \subseteq B_{(W,\vec{d})}(x, \delta), \end{aligned} \tag{3.25}$$

where in the last containment, we have used $\xi_2 \le 1$. This completes the proof of (b).

We have completed the proof of Theorem 3.5.1. For the remainder of this proof, we assume the additional assumptions of Theorem 3.5.4 – note that $N(x) = n$, $\forall x$ in this case. Item (o) follows from Theorem 3.6.5 (g) and item (p) follows from Theorem 3.6.5 (h).

Using (3.25) and the fact that $B_{(W,\vec{d})}(x, \xi_3\delta) \subseteq B_{(X,\vec{d})}(x, \xi_3\delta)$, we have, for $x \in \mathcal{K}$ and $\delta \in (0,1]^\nu$,

$$\begin{aligned} \operatorname{Vol}(B_{(W,\vec{d})}(x, \xi_3\delta)) \le \operatorname{Vol}(B_{(X,\vec{d})}(x, \xi_3\delta)) &\le \operatorname{Vol}(\Phi_{x,\delta}(B^n(1))) \\ &= \Lambda(x,\delta) \int_{B^n(1)} h_{x,\delta}(t)\, dt \approx \Lambda(x,\delta) \approx \Lambda(x, \xi_3\delta), \end{aligned} \tag{3.26}$$

where in the second to last estimate we have used (o) and in the last estimate we have used $\xi_3 \approx 1$ and the formula for Λ. Similarly, again using (3.25), $\eta_2 \approx 1$, and (o), we have

$$\begin{aligned} \Lambda(x,\delta) \approx \Lambda(x,\delta) \int_{B^n(\eta_2)} h_{x,\delta}(t)\, dt &= \operatorname{Vol}(\Phi_{x,\delta}(B^n(\eta_2))) \\ &\le \operatorname{Vol}(B_{(W,\vec{d})}(x,\delta)) \le \operatorname{Vol}(B_{(X,\vec{d})}(x,\delta)). \end{aligned} \tag{3.27}$$

Combining (3.26) and (3.27), we have, for $\delta \in (0, \xi_3]^\nu$,

$$\operatorname{Vol}(B_{(W,\vec{d})}(x,\delta)) \approx \operatorname{Vol}(B_{(X,\vec{d})}(x,\delta)) \approx \Lambda(x,\delta). \tag{3.28}$$

Thus, by taking $\delta \in (0, \xi_3]$, (k) follows.

Next, we turn to (l). When $\delta \in (0, \delta_0]^\nu$, (l) follows from (k). Suppose $\delta = (\delta_1, \dots, \delta_\nu) \in (0,\infty)^\nu$ with $\delta_\mu > \delta_0$ for some μ. Then, using (k), we have

$$\begin{aligned} \operatorname{Vol}(B_{(W,\vec{d})}(x,\delta)) &\ge \operatorname{Vol}(B_{(W,\vec{d})}(x, \delta_0 \mathfrak{e}_\mu)) \\ &\approx \Lambda(x, \delta_0 \mathfrak{e}_\mu) \ge \inf_{x \in \mathcal{K}} \min_{1 \le \mu' \le \nu} \Lambda(x', \delta_0 \mathfrak{e}_{\mu'}) \gtrsim 1, \end{aligned}$$

where in the last estimate, we have used the compactness of $\mathcal{K}$, the formula for Λ, the fact that $\delta_0 \approx 1$, and the fact that $X_1^{\mu'}, \ldots, X_{q'_\mu}^{\mu'}$ span the tangent space at every point of $\Omega \Supset \mathcal{K}$, $\forall \mu' \in \{1, \ldots, \nu\}$. We also have

$$\mathrm{Vol}(B_{(X,\vec{d})}(x,\delta)) \geq \mathrm{Vol}(B_{(W,\vec{d})}(x,\delta)) \gtrsim 1.$$

Finally, using the formula for Λ, we have

$$\Lambda(x,\delta) \geq \inf_{x \in \mathcal{K}} \min_{1 \leq \mu' \leq \nu} \Lambda(x', \delta_0 \mathfrak{e}_{\mu'}) \gtrsim 1.$$

Combining the above three estimates shows that

$$\mathrm{Vol}(B_{(W,\vec{d})}(x,\delta)) \wedge 1 \approx 1, \quad \mathrm{Vol}(B_{(X,\vec{d})}(x,\delta)) \wedge 1 \approx 1, \quad \Lambda(x,\delta) \wedge 1 \approx 1,$$

completing the proof of (l).

Taking $\delta_0 := \min\{\xi, \xi_3/2\}$, we have by (3.28), for $x \in \mathcal{K}$ and $\delta \in (0, \delta_0]^\nu$,

$$\mathrm{Vol}(B_{(W,\vec{d})}(x,2\delta)) \approx \Lambda(x,2\delta) \approx \Lambda(x,\delta) \approx \mathrm{Vol}(B_{(W,\vec{d})}(x,\delta)),$$

where we have used the formula for Λ to see that $\Lambda(x,2\delta) \approx \Lambda(x,\delta)$. A similar proof gives the same estimate with $(W,\vec{d})$ replaced by $(X,\vec{d})$. This proves (m). A similar proof using (l) in place of (3.28) proves (n). □

Proof of Corollary 3.3.8. Fix $\phi \in C_0^\infty(B^n(1))$ with $\phi \equiv 1$ on $B^n(3/4)$. Set

$$\phi_{x,\delta}(t) := \phi(\Phi_{x,\xi_3\delta}(t)) \in C_0^\infty(\Omega_1),$$

where $\Phi_{x,\delta}$ and ξ_3 are as in Theorem 3.3.7. By Theorem 3.3.7 (b) we have

$$\mathrm{supp}(\phi_{x,\delta}) \subseteq \Phi_{x,\xi_3\delta}(B^n(1)) \subseteq B_{(X,d)}(x,\xi_3\delta) \cap \Omega_1 \subseteq B_{(W,d)}(x,\delta) \cap \Omega_1,$$

proving (i).

By definition, $\phi_{x,\delta}$ equals 1 on $\Phi_{x,\xi_3\delta}(B^n(3/4))$, and therefore it equals 1 on a neighborhood of the closure of $\Phi_{x,\xi_3\delta}(B^n(1/2)) \supseteq B_{(W,d)}(x,\xi_3^2\delta)$, where the containment follows from Theorem 3.3.7 (i). This proves (ii).

Fix $N \in \mathbb{N}$. By Theorem 3.3.7 (j),

$$\sum_{|\alpha| \leq N} \|(W^{x,\xi_3\delta})^\alpha \phi\|_{C(B^n(1))} \lesssim 1.$$

Pushing this equation forward via $\Phi_{x,\xi_3\delta}$, we have

$$\sum_{|\alpha| \leq N} \|((\xi_3\delta)^d W)^\alpha \phi_{x,\delta}\|_{C(\Phi_{x,\xi_3\delta}(B^n(1)))} \lesssim 1.$$

Since $\xi_3 \approx 1$ and $\mathrm{supp}(\phi_{x,\delta}) \subseteq \Phi_{x,\xi_3\delta}(B^n(1))$, (iii) follows, completing the proof. □

Finally, we turn to the proof of Corollary 3.3.9. This is a simple consequence of Theorem 3.3.7.

Proof of Corollary 3.3.9. Set $\mathcal{K} := \overline{\Omega}$ and fix $\Omega_1 \Subset \mathfrak{M}$ open with $\mathcal{K} \Subset \Omega_1$. We apply Theorem 3.3.7 with this choice of $\mathcal{K}$ and Ω_1. Set $\delta_1 := \delta_0$, where δ_0 is as in Theorem 3.3.7. By Theorem 3.3.7 (c), we have, for $x \in \mathcal{K}$, $\delta \in (0, \delta_1]$,

$$\operatorname{Vol}(B_{(W,d)}(x,\delta) \approx \operatorname{Vol}(B_{(X,d)}(x,\delta) \approx \Lambda(x,\delta). \tag{3.29}$$

(i): Since $x \mapsto \Lambda(x,\delta_1)$ and $\Lambda(x,\delta_1)$ is never 0 on Ω_1, we see that

$$\inf_{x\in\mathcal{K}} \Lambda(x,\delta_1) > 0. \tag{3.30}$$

Item (i) now follows from (3.29).

(ii): By the formula for $\Lambda(x,\delta)$, there exist $Q_1 = Q_1(\Omega)$, $Q_2 = Q_2(\Omega)$ with $n \le Q_1 \le Q_2$ such that for all $\delta > 0$, $\xi \in (0,1]$,

$$\xi^{-Q_1}\Lambda(x,\xi\delta) \lesssim \Lambda(x,\delta) \lesssim \xi^{-Q_2}\Lambda(x,\xi\delta).$$

Item (i) now follows from (3.29).

(iii): We begin with the first inequality. We prove only the result for $(W, \vec{d})$; the same proof works with (X, d) in place of $(W, \vec{d})$. If $\delta \in (0,\delta_1]$, then by (ii),

$$\xi^{-Q_1}\operatorname{Vol}(B_{(W,d)}(x,\xi\delta)) \lesssim \operatorname{Vol}(B_{(W,d)}(x,\delta)),$$

and it follows that $(\xi^{-Q_1}\operatorname{Vol}(B_{(W,d)}(x,\xi\delta))) \wedge 1 \lesssim \operatorname{Vol}(B_{(W,d)}(x,\delta)) \wedge 1$. If $\delta > \delta_0$, then by (i), $\operatorname{Vol}(B_{(W,d)}(x,\delta)) \wedge 1 \approx 1$, and the first inequality in (iii) follows.

We turn to the second inequality in (iii). If $\delta \in (0,\delta_1]$, then by (ii), $\operatorname{Vol}(B_{(W,d)}(x,\delta)) \lesssim \xi^{-Q_2}\operatorname{Vol}(B_{(W,d)}(B_{(W,d)}(x,\xi\delta))$, and the second inequality follows. If $\delta > \delta_1$, then the result will follow once we show $\xi^{-Q_2}\operatorname{Vol}(B_{(W,d)}(B_{(W,d)}(x,\xi\delta)) \gtrsim 1$. Indeed, we have, using (3.29) and possibly increasing Q_2,

$$\begin{aligned}\xi^{-Q_2}\operatorname{Vol}(B_{(W,d)}(B_{(W,d)}(x,\xi\delta)) &\ge \xi^{-Q_2}\operatorname{Vol}(B_{(W,d)}(B_{(W,d)}(x,\xi\delta_1))\\ &\approx \xi^{-Q_2}\Lambda(x,\xi\delta_1) \gtrsim \Lambda(x,\delta_1) \gtrsim 1,\end{aligned}$$

where $\xi^{-Q_2}\Lambda(x,\xi\delta_1) \gtrsim \Lambda(x,\delta_1)$ follows from the formula for Λ (by taking Q_2 large enough), and $\Lambda(x,\delta_1) \gtrsim 1$ follows from (3.30). □

3.8 Approximately commuting vector fields

Fix $\nu_1, \nu_2 \in \mathbb{N}_+$. Let $\mathcal{S}_1 \subseteq C^\infty_{\mathrm{loc}}(\mathfrak{M}; T\mathfrak{M}) \times (\mathbb{N}^{\nu_1} \setminus \{0\})$ and $\mathcal{S}_2 \subseteq C^\infty_{\mathrm{loc}}(\mathfrak{M}; T\mathfrak{M}) \times (\mathbb{N}^{\nu_2} \setminus \{0\})$. In this text we define various objects related to $\mathcal{S}_1$ and $\mathcal{S}_2$ (for example, the Carnot–Carathéodory balls defined in Section 3.5, the singular integrals defined in Chapter 5, or the function spaces defined in Chapter 5). A main theme is that these objects associated with $\mathcal{S}_1$ behave well with the same objects associated with $\mathcal{S}_2$ provided $\mathcal{S}_1$ and $\mathcal{S}_2$ "locally weakly approximately commute," a notion that we make precise in this section (see Definition 3.8.5). Throughout this section, $\Omega \Subset \mathfrak{M}$ is an open, relatively compact set.

Definition 3.8.1. We say $\mathcal{S}_1$ and $\mathcal{S}_2$ strongly approximately commute on Ω if $\forall (Z_1, \vec{d}_1) \in \mathcal{S}_1$ and $(Z_2, \vec{d}_2) \in \mathcal{S}_2$, there are finite sets

$$\mathcal{F}_1 \subseteq \{(Y, \vec{d}) \in \mathcal{S}_1 : \vec{d} \leq \vec{d}_1\}, \quad \mathcal{F}_2 \subseteq \{(Y, \vec{d}) \in \mathcal{S}_2 : \vec{d} \leq \vec{d}_2\},$$

such that

$$[Z_1, Z_2] = \sum_{(Y,\vec{d}) \in \mathcal{F}_1} a_{(Y,\vec{d})} Y + \sum_{(Y,\vec{d}) \in \mathcal{F}_2} b_{(Y,\vec{d})} Y, \quad a_{(Y,\vec{d})}, b_{(Y,\vec{d})} \in C^\infty_{\mathrm{loc}}(\Omega). \tag{3.31}$$

We say $\mathcal{S}_1$ and $\mathcal{S}_2$ locally strongly approximately commute if $\mathcal{S}_1$ and $\mathcal{S}_2$ strongly approximately commute on Ω, for all $\Omega \Subset \mathfrak{M}$ open.

Example 3.8.2. In the case $\nu_2 = 1$, let $\mathcal{S}_1 \subseteq C^\infty_{\mathrm{loc}}(\mathbb{R}^n; T\mathbb{R}^n) \times (\mathbb{N}^{\nu_1} \setminus \{0\})$ be any set of vector fields with formal degrees on $\mathbb{R}^n$. Let $\mathcal{S}_2 = \{(\partial_{x_1}, 1), \ldots, (\partial_{x_n}, 1)\} \subset C^\infty_{\mathrm{loc}}(\mathbb{R}^n; T\mathbb{R}^n) \times \mathbb{N}_+$. Then $\mathcal{S}_1$ and $\mathcal{S}_2$ locally strongly approximately commute. Indeed, for $(Z_1, \vec{d}_1) \in \mathcal{S}_1$, we have

$$[Z_1, \partial_{x_k}] = \sum_{l=1}^{n} a^l_{(Z_1,\vec{d}_1),k} \partial_{x_l}, \quad a^l_{(Z_1,\vec{d}_1),k} \in C^\infty_{\mathrm{loc}}(\mathbb{R}^n).$$

We generalize Example 3.8.2 via the next lemma.

Lemma 3.8.3. *Let $\mathcal{S}_1 \subseteq C^\infty_{\mathrm{loc}}(\mathfrak{M}; T\mathfrak{M}) \times (\mathbb{N}^{\nu_1} \setminus \{0\})$ be any set of vector fields with formal degrees on $\mathbb{R}^n$. Let $\mathcal{S}_2 \subseteq C^\infty_{\mathrm{loc}}(\mathfrak{M}; T\mathfrak{M}) \times \mathbb{N}_+$ be a set of vector fields paired with single-parameter formal degrees. Suppose*

$$T_x\mathfrak{M} = \mathrm{span}\{Z_2(x) : (Z_2, 1) \in \mathcal{S}_2\}, \quad \forall x \in \mathfrak{M}.$$

Then $\mathcal{S}_1$ and $\mathcal{S}_2$ locally strongly approximately commute.

Proof. Let $\Omega \Subset \mathfrak{M}$ be an open set. Since $\overline{\Omega}$ is compact, there is a finite set $\mathcal{F}_2 \subseteq \mathcal{S}_2$ such that

$$T_x\mathfrak{M} = \mathrm{span}\{Y(x) : (Y, 1) \in \mathcal{F}_2\}, \quad \forall x \in \overline{\Omega}. \tag{3.32}$$

Let $(Z_1, \vec{d}_1) \in \mathcal{S}_1$ and $(Z_2, \vec{d}_2) \in \mathcal{S}_2$. Since $[Z_1, Z_2]$ is a smooth vector field, (3.32) implies

$$[Z_1, Z_2] = \sum_{(Y,1) \in \mathcal{F}_2} a_{(Y,1)} Y, \quad a_{(Y,1)} \in C^\infty_{\mathrm{loc}}(\mathfrak{M}).$$

This establishes (3.31) with $\mathcal{F}_1 = \emptyset$, proving that $\mathcal{S}_1$ and $\mathcal{S}_2$ strongly approximately commute on Ω. Since $\Omega \Subset \mathfrak{M}$ was an arbitrary relatively compact open set, we conclude $\mathcal{S}_1$ and $\mathcal{S}_2$ locally strongly approximately commute. □

Example 3.8.4. Let $\mathfrak{M}$ be a Lie group. Suppose the vector fields in $\mathcal{S}_1$ are left invariant vector fields and the vector fields in $\mathcal{S}_2$ are right invariant vector fields. Then $\mathcal{S}_1$ and $\mathcal{S}_2$ locally strongly approximately commute. Indeed, $[Z_1, Z_2] = 0$, for all $(Z_1, \vec{d}_1) \in \mathcal{S}_1$ and $(Z_2, \vec{d}_2) \in \mathcal{S}_2$.

Definition 3.8.5. We say $\mathcal{S}_1$ and $\mathcal{S}_2$ weakly approximately commute on Ω if $\mathrm{Gen}(\mathcal{S}_1)$ and $\mathrm{Gen}(\mathcal{S}_2)$ strongly approximately commute on Ω. We say $\mathcal{S}_1$ and $\mathcal{S}_2$ locally weakly approximately commute if $\mathcal{S}_1$ and $\mathcal{S}_2$ weakly approximately commute on Ω, for all open, relatively compact sets $\Omega \Subset \mathfrak{M}$.

Proposition 3.8.6. (a) *Suppose $\mathcal{S}_1$ and $\mathcal{S}_2$ strongly approximately commute on Ω. Then $\mathcal{S}_1$ and $\mathcal{S}_2$ weakly approximately commute on Ω.*
(b) *Suppose $\mathcal{S}_1$ and $\mathcal{S}_2$ locally strongly approximately commute. Then $\mathcal{S}_1$ and $\mathcal{S}_2$ locally weakly approximately commute.*

The proof of Proposition 3.8.6 uses the next lemma.

Lemma 3.8.7. *Suppose $\mathcal{S}_1$ and $\mathcal{S}_2$ strongly approximately commute on Ω. Then* $\mathrm{Gen}(\mathcal{S}_1)$ *and $\mathcal{S}_2$ strongly approximately commute on Ω.*

Proof. Let $\widehat{\mathcal{S}_1} \subseteq \mathrm{Gen}(\mathcal{S}_1)$ be the set of all $(Z_1, \vec{d}_1) \in \mathrm{Gen}(\mathcal{S}_1)$ such that $\forall (Z_2, \vec{d}_2) \in \mathcal{S}_2$ there are finite sets

$$\mathcal{F}_1 \subseteq \{(Y, \vec{d}) \in \mathrm{Gen}(\mathcal{S}_1) : \vec{d} \leq \vec{d}_1\}, \quad \mathcal{F}_2 \subseteq \{(Y, \vec{d}) \in \mathcal{S}_2 : \vec{d} \leq \vec{d}_2\},$$

with

$$[Z_1, Z_2] = \sum_{(Y,\vec{d}) \in \mathcal{F}_1} a_{(Y,\vec{d})} Y + \sum_{(Y,\vec{d}) \in \mathcal{F}_2} b_{(Y,\vec{d})} Y, \quad a_{(Y,\vec{d})}, b_{(Y,\vec{d})} \in C^\infty_{\mathrm{loc}}(\Omega).$$

The assumption of the lemma implies $\mathcal{S}_1 \subseteq \widehat{\mathcal{S}_1}$. Our goal is to show $\widehat{\mathcal{S}_1} = \mathrm{Gen}(\mathcal{S}_1)$, which is equivalent to the conclusion of the lemma. Since $\mathcal{S}_1 \subseteq \widehat{\mathcal{S}_1} \subseteq \mathrm{Gen}(\mathcal{S}_1)$, it suffices to show that if $(Z_1, \vec{d}_1), (X_1, \vec{d}_1) \in \widehat{\mathcal{S}_1}$, then $([Z_1, X_1], \vec{d}_1 + \vec{d}_1) \in \widehat{\mathcal{S}_1}$. Thus, we wish to consider, for $(Z_2, \vec{d}_2) \in \mathcal{S}_2$,

$$[[Z_1, X_1], Z_2] = [[Z_1, Z_2], X_1] + [Z_1, [Z_2, X_1]]. \tag{3.33}$$

The two terms on the right-hand side of (3.33) are similar, so we only show that the first is of the desired form, that is, we wish to show that there are finite sets

$$\widehat{\mathcal{F}}_1 \subseteq \{(Y, \vec{d}) \in \mathrm{Gen}(\mathcal{S}_1) : \vec{d} \leq \vec{d}_1 + \vec{d}_2\}, \quad \widehat{\mathcal{F}}_2 \subseteq \{(Y, \vec{d}) \in \mathcal{S}_2 : \vec{d} \leq \vec{d}_2\}$$

such that

$$[[Z_1, Z_2], X_1] = \sum_{(Y,\vec{d})\in\widehat{\mathcal{F}}_1} a_{(Y,\vec{d})} Y + \sum_{(Y,\vec{d})\in\widehat{\mathcal{F}}_2} b_{(Y,\vec{d})} Y, \quad a_{(Y,\vec{d})}, b_{(Y,\vec{d})} \in C^\infty_{\mathrm{loc}}(\Omega). \tag{3.34}$$

By the hypothesis that $(Z_1, \vec{d}_1) \in \widehat{\mathcal{S}}_1$, there exist finite sets $\mathcal{F}_1 \subseteq \{(Y, \vec{d}) \in \mathrm{Gen}(\mathcal{S}_1) : \vec{d} \le \vec{d}_1\}$ and $\mathcal{F}_2 \subseteq \{(Y, \vec{d}) \in \mathcal{S}_2 : \vec{d} \le \vec{d}_2\}$ such that

$$[Z_1, Z_2] = \sum_{(Y,\vec{d})\in\mathcal{F}_1} a_{(Y,\vec{d})} Y + \sum_{(Y,\vec{d})\in\mathcal{F}_2} b_{(Y,\vec{d})} Y, \quad a_{(Y,\vec{d})}, b_{(Y,\vec{d})} \in C^\infty_{\mathrm{loc}}(\Omega).$$

Thus, we have

$$\begin{aligned} [[Z_1, Z_2], X_1] = &- \sum_{(Y,\vec{d})\in\mathcal{F}_1} (X_1 a_{(Y,\vec{d})}) Y - \sum_{(Y,\vec{d})\in\mathcal{F}_2} (X_1 b_{(Y,\vec{d})}) Y \\ &+ \sum_{(Y,\vec{d})\in\mathcal{F}_1} [Y, X_1] + \sum_{(Y,\vec{d})\in\mathcal{F}_2} b_{(Y,\vec{d})} [Y, X_1]. \end{aligned} \tag{3.35}$$

The first two terms on the right-hand side of (3.35) are of the desired form from (3.34). That the last term on the right-hand side of (3.35) is of the desired form follows from the fact that $(X_1, \vec{d}_1) \in \widehat{\mathcal{S}}_1$ and for $(Y, \vec{d}) \in \mathcal{F}_2$ we have $\vec{d}_1 + \vec{d} \le \vec{d}_1 + \vec{d}_2$. Finally, for the third term on the right-hand side of (3.35), note that for $(Y, \vec{d}) \in \mathcal{F}_1$, we have $([Y, X_1], \vec{d} + \vec{d}_1) \in \mathrm{Gen}(\mathcal{S}_1)$ with $\vec{d} + \vec{d}_1 \le \vec{d}_1 + \vec{d}_1$, and therefore the third term is of the desired form as well. We conclude $([Z_1, X_1], \vec{d}_1 + \vec{d}_1) \in \widehat{\mathcal{S}}_1$, completing the proof. □

Proof of Proposition 3.8.6. (a): By Lemma 3.8.7, $\mathrm{Gen}(\mathcal{S}_1)$ and $\mathcal{S}_2$ strongly approximately commute on Ω. Another application of Lemma 3.8.7 then shows that $\mathrm{Gen}(\mathcal{S}_1)$ and $\mathrm{Gen}(\mathcal{S}_2)$ strongly approximately commute on Ω, completing the proof of (a).

(b): This follows immediately from (a). □

The way we will usually obtain a set $\mathcal{S} \subseteq C^\infty_{\mathrm{loc}}(\mathfrak{M}; T\mathfrak{M}) \times (\mathbb{N}^\nu \setminus \{0\})$ is by creating it out of sets of vector fields with single-parameter formal degrees as follows. Given $\mathcal{S}_1, \ldots, \mathcal{S}_\nu \subseteq C^\infty_{\mathrm{loc}}(\mathfrak{M}; T\mathfrak{M}) \times \mathbb{N}_+$, we set

$$\begin{aligned} \mathcal{S} &:= \mathcal{S}_1 \boxtimes \mathcal{S}_2 \boxtimes \cdots \boxtimes \mathcal{S}_\nu \\ &= \{(X_\mu, d_\mu \mathfrak{e}_\mu) : \mu \in \{1, \ldots, \nu\}, (X_\mu, d_\mu) \in \mathcal{S}_\mu\} \\ &\subset C^\infty_{\mathrm{loc}}(\mathfrak{M}; T\mathfrak{M}) \times \mathbb{N}^\nu \setminus \{0\}. \end{aligned} \tag{3.36}$$

Proposition 3.8.8. *Let* $\mathcal{S}_1, \ldots, \mathcal{S}_\nu \subseteq C^\infty_{\mathrm{loc}}(\mathfrak{M}; T\mathfrak{M}) \times \mathbb{N}_+$ *and define* $\mathcal{S} = \mathcal{S}_1 \boxtimes \mathcal{S}_2 \boxtimes \cdots \boxtimes \mathcal{S}_\nu$ *as in* (3.36).

(a) *Suppose* $\mathcal{S}_1, \ldots, \mathcal{S}_\nu$ *pairwise weakly approximately commute on* Ω. *For each* $\mu \in \{1, \ldots, \nu\}$, *suppose* $\mathrm{Gen}(\mathcal{S}_\mu)$ *is finitely generated on* Ω *by* $\mathcal{F}_\mu \subset \mathrm{Gen}(\mathcal{S}_\mu)$. *Then* $\mathrm{Gen}(\mathcal{S})$ *is linearly finitely generated on* Ω *by*

$$\mathcal{F} := \mathcal{F}_1 \boxtimes \mathcal{F}_2 \boxtimes \cdots \boxtimes \mathcal{F}_\nu = \{(X_\mu, d_\mu \mathfrak{e}_\mu) : \mu \in \{1, \dots, \nu\}, (X_\mu, d_\mu) \in \mathcal{F}_\mu\} \subseteq \mathcal{S}.$$

(b) *Suppose* $\mathcal{S}_1, \dots, \mathcal{S}_\nu$ *pairwise locally weakly approximately commute and suppose for each* $\mu \in \{1, \dots, \nu\}$ *that* $\mathrm{Gen}(\mathcal{S}_\mu)$ *is locally finitely generated. Then* $\mathrm{Gen}(\mathcal{S})$ *is locally linearly finitely generated.*

Proof. Item (b) follows immediately from (a), so we prove only (a).

Clearly $\mathcal{F}$ is a finite set and for each $(X_0, \vec{d}_0) \in \mathcal{F}$, $\vec{d}_0$ is non-zero in only one component. We will show that $\mathrm{Gen}(\mathcal{S})$ is finitely generated by $\mathcal{F}$ on Ω, which will complete the proof.

Let $\mathcal{T}$ be the set of all those $(X_0, \vec{d}_0) \in \mathrm{Gen}(\mathcal{S})$ such that

$$X_0 = \sum_{\substack{(X_1, \vec{d}_1) \in \mathcal{F} \\ \vec{d}_1 \leq \vec{d}_0}} c_{(X_0, \vec{d}_0)}^{(X_1, \vec{d}_1)} X_1, \quad c_{(X_0, \vec{d}_0)}^{(X_1, \vec{d}_1)} \in C^\infty_{\mathrm{loc}}(\Omega). \tag{3.37}$$

We wish to show that $\mathcal{T} = \mathrm{Gen}(\mathcal{S})$. By the definition of $\mathcal{S}$ and $\mathcal{F}$, we have $\mathcal{S} \subseteq \mathcal{T}$.

Suppose $(X_1, \vec{d}_1), (X_2, \vec{d}_2) \in \mathcal{F}$. We claim $([X_1, X_2], \vec{d}_1 + \vec{d}_2) \in \mathcal{T}$. Indeed, $(X_1, \vec{d}_1) = (X_{\mu_1}, d_{\mu_1} \mathfrak{e}_{\mu_1})$ and $(X_2, \vec{d}_2) = (X'_{\mu_2}, d'_{\mu_2} \mathfrak{e}_{\mu_2})$ for some $(X_{\mu_1}, d_{\mu_1}) \in \mathrm{Gen}(\mathcal{S}_{\mu_1})$ and $(X'_{\mu_2}, d'_{\mu_2}) \in \mathrm{Gen}(\mathcal{S}_{\mu_2})$. There are two cases. If $\mu_1 = \mu_2$, then since $([X_{\mu_1}, X'_{\mu_2}], d_{\mu_1} + d'_{\mu_2}) \in \mathrm{Gen}(\mathcal{S}_{\mu_1})$ and $\mathrm{Gen}(\mathcal{S}_{\mu_1})$ is finitely generated by $\mathcal{F}_{\mu_1}$ on Ω, we have

$$[X_{\mu_1}, X'_{\mu_2}] = \sum_{\substack{(X''_{\mu_1}, d''_{\mu_1}) \in \mathcal{F}_{\mu_1} \\ d''_{\mu_1} \leq d_{\mu_1} + d'_{\mu_2}}} c_{(X''_{\mu_1}, d''_{\mu_1})} X''_{\mu_1}, \quad c_{(X''_{\mu_1}, d''_{\mu_1})} \in C^\infty_{\mathrm{loc}}(\Omega).$$

From here it follows immediately from the definitions that $([X_1, X_2], \vec{d}_1 + \vec{d}_2) = ([X_{\mu_1}, X'_{\mu_2}], (d_{\mu_1} + d'_{\mu_2}) \mathfrak{e}_{\mu_1}) \in \mathcal{T}$. If $\mu_1 \neq \mu_2$, then since $\mathrm{Gen}(\mathcal{S}_{\mu_1})$ and $\mathrm{Gen}(\mathcal{S}_{\mu_2})$ strongly approximately commute on Ω, using the fact that $\mathrm{Gen}(\mathcal{S}_{\mu_1})$ and $\mathrm{Gen}(\mathcal{S}_{\mu_2})$ are finitely generated by $\mathcal{F}_1$ and $\mathcal{F}_2$ (respectively) on Ω, we have

$$[X_{\mu_1}, X'_{\mu_2}] = \sum_{\substack{(X''_{\mu_1}, d''_{\mu_1}) \in \mathcal{F}_{\mu_1} \\ d''_{\mu_1} \leq d_{\mu_1}}} b_{(X''_{\mu_1}, d''_{\mu_1})} X''_{\mu_1} + \sum_{\substack{(X'''_{\mu_2}, d'''_{\mu_2}) \in \mathcal{F}_{\mu_2} \\ d'''_{\mu_2} \leq d'_{\mu_2}}} c_{(X'''_{\mu_2}, d'''_{\mu_2})} X'''_{\mu_2},$$
$$b_{(X''_{\mu_1}, d''_{\mu_1})}, c_{(X'''_{\mu_2}, d'''_{\mu_2})} \in C^\infty_{\mathrm{loc}}(\Omega).$$

From here it follows immediately from the definitions that $([X_1, X_2], \vec{d}_1 + \vec{d}_2) = ([X_{\mu_1}, X'_{\mu_2}], d_{\mu_1} \mathfrak{e}_{\mu_1} + d'_{\mu_2} \mathfrak{e}_{\mu_2}) \in \mathcal{T}$.

Now suppose $(X_1, \vec{d}_1), (X_2, \vec{d}_2) \in \mathcal{T}$. We claim $([X_1, X_2], \vec{d}_1 + \vec{d}_2) \in \mathcal{T}$. Indeed, this follows immediately from the definition of $\mathcal{T}$, (3.37), and the fact that we just proved that if $(X'_1, \vec{d}'_1), (X_2, \vec{d}'_2) \in \mathcal{F}$, then $([X'_1, X'_2], \vec{d}'_1 + \vec{d}'_2) \in \mathcal{T}$. Combining this with $\mathcal{S} \subseteq \mathcal{T} \subseteq \mathrm{Gen}(\mathcal{S})$, it follows that $\mathcal{T} = \mathrm{Gen}(\mathcal{S})$.

Saying that $\mathcal{T} = \mathrm{Gen}(\mathcal{S})$ is equivalent to saying that $\mathrm{Gen}(\mathcal{S})$ is finitely generated by $\mathcal{F}$ on Ω, which completes the proof. □

Example 3.8.9. This is an important example for our applications. Suppose $(W^1, d^1) \subset C^\infty_{\mathrm{loc}}(\Omega; T\Omega) \times \mathbb{N}_+$ are Hörmander vector fields with formal degrees on an open set $\Omega \subseteq \mathbb{R}^n$ (see Definition 1.1.3). Let $(W^2, d^2) = \{(\partial_{x_1}, 1), \ldots, (\partial_{x_n}, 1)\}$. By Example 3.8.2, (W^1, d^1) and (W^2, d^2) locally strongly approximately commute, and by Proposition 3.8.6, (W^1, d^1) and (W^2, d^2) locally weakly approximately commute. By Proposition 3.4.14 (in the case $\nu = 1$), $\mathrm{Gen}((W^1, d^1))$ and $\mathrm{Gen}((W^2, d^2))$ are both locally finitely generated. Set $(W, \vec{d}) := (W^1, d^1) \boxtimes (W^2, d^2) \subset C^\infty_{\mathrm{loc}}(\Omega; T\Omega) \times (\mathbb{N}^2 \setminus \{0\})$. Combining the above with Proposition 3.8.8 shows that $\mathrm{Gen}((W, \vec{d}))$ is locally linearly finitely generated.

Proposition 3.8.10. *Suppose that for $j = 1, 2, 3$, $\mathcal{S}_j \subseteq C^\infty_{\mathrm{loc}}(\mathfrak{M}; T\mathfrak{M}) \times (\mathbb{N}^{\nu_j} \setminus \{0\})$, where $\nu_j \in \mathbb{N}_+$. Define*

$$\mathcal{S}_0 := \mathcal{S}_1 \boxtimes \mathcal{S}_2 \subseteq C^\infty_{\mathrm{loc}}(\mathfrak{M}; T\mathfrak{M}) \times (\mathbb{N}^{\nu_1+\nu_2} \setminus \{0\}).$$

Then we have the following:

(i) *If, for $j = 1, 2$, $\mathcal{S}_j$ and $\mathcal{S}_3$ strongly approximately commute on Ω, then $\mathcal{S}_0$ and $\mathcal{S}_3$ strongly approximately commute on Ω.*

(ii) *If, for $j = 1, 2$, $\mathcal{S}_j$ and $\mathcal{S}_3$ locally strongly approximately commute, then $\mathcal{S}_0$ and $\mathcal{S}_3$ locally strongly approximately commute.*

(iii) *If, for $j = 1, 2$, $\mathcal{S}_j$ and $\mathcal{S}_3$ weakly approximately commute on Ω, then $\mathcal{S}_0$ and $\mathcal{S}_3$ weakly approximately commute on Ω.*

(iv) *If, for $j = 1, 2$, $\mathcal{S}_j$ and $\mathcal{S}_3$ locally weakly approximately commute, then $\mathcal{S}_0$ and $\mathcal{S}_3$ locally weakly approximately commute.*

Proof. Item (ii) follows immediately from (i) and item (iv) follows immediately from (iii), so we prove only (i) and (iii).

(i): Let $(Z_0, \vec{d}_0) \in \mathcal{S}_0$ and $(Z_3, \vec{d}_3) \in \mathcal{S}_3$. We wish to show that there are finite sets

$$\mathcal{F}_0 \subseteq \{(Y, \vec{d}) \in \mathcal{S}_0 : \vec{d} \le \vec{d}_0\}, \quad \mathcal{F}_3 \subseteq \{(Y, \vec{d}) \in \mathcal{S}_3 : \vec{d} \le \vec{d}_3\}$$

such that

$$[Z_0, Z_3] = \sum_{(Y,\vec{d}) \in \mathcal{F}_0} a_{(Y,\vec{d})} Y + \sum_{(Y,\vec{d}) \in \mathcal{F}_3} b_{(Y,\vec{d})} Y, \quad a_{(Y,\vec{d})}, b_{(Y,\vec{d})} \in C^\infty_{\mathrm{loc}}(\Omega). \tag{3.38}$$

There are two cases: either $(Z_0, \vec{d}_0) = (Z_1, (\vec{d}_1, 0_{\nu_2}))$ for some $(Z_1, \vec{d}_1) \in \mathcal{S}_1$ or $(Z_0, \vec{d}_0) = (Z_2, (0_{\nu_1}, \vec{d}_2))$ for some $(Z_2, \vec{d}_2) \in \mathcal{S}_2$. We consider only the first case, where $(Z_0, \vec{d}_0) = (Z_1, (\vec{d}_1, 0_{\nu_2}))$ for some $(Z_1, \vec{d}_1) \in \mathcal{S}_1$; the other case is similar. In this case, by the assumption that $\mathcal{S}_1$ and $\mathcal{S}_3$ strongly approximately commute on Ω, there exist finite sets

$$\mathcal{F}_1 \subseteq \{(Y, \vec{d}) \in \mathcal{S}_1 : \vec{d} \le \vec{d}_1\}, \quad \mathcal{F}_3 \subseteq \{(Y, \vec{d}) \in \mathcal{S}_3 : \vec{d} \le \vec{d}_3\}$$

with

$$[Z_1, Z_3] = \sum_{(Y,\vec{d})\in\mathcal{F}_0} a_{(Y,\vec{d})} Y + \sum_{(Y,\vec{d})\in\mathcal{F}_3} b_{(Y,\vec{d})} Y, \quad a_{(Y,\vec{d})}, b_{(Y,\vec{d})} \in C^\infty_{\mathrm{loc}}(\Omega).$$

Setting

$$\mathcal{F}_0 := \{(Y, (\vec{d}, 0_{\nu_2})) : (Y, \vec{d}) \in \mathcal{F}_1\} \subseteq \{(Y, \vec{d}) \in \mathcal{S}_0 : \vec{d} \leq \vec{d}_0\}$$

establishes (3.38) and completes the proof of (i).

(iii): The assumption can be restated as follows: $\mathrm{Gen}(\mathcal{S}_j)$ and $\mathrm{Gen}(\mathcal{S}_3)$ strongly approximately commute on Ω, for $j = 1, 2$. Set

$$\mathcal{T} := \mathrm{Gen}(\mathcal{S}_1) \boxtimes \mathrm{Gen}(\mathcal{S}_2) \subseteq \mathrm{Gen}(\mathcal{S}_0).$$

By (i), $\mathcal{T}$ and $\mathrm{Gen}(\mathcal{S}_3)$ strongly approximately commute on Ω. By Proposition 3.8.6 (a), $\mathcal{T}$ and $\mathrm{Gen}(\mathcal{S}_3)$ weakly approximately commute on Ω, i. e., $\mathrm{Gen}(\mathcal{T})$ and $\mathrm{Gen}(\mathcal{S}_3)$ strongly approximately commute on Ω.

Since $\mathcal{S}_0 \subseteq \mathcal{T} \subseteq \mathrm{Gen}(\mathcal{S}_0)$, we have $\mathrm{Gen}(\mathcal{T}) = \mathrm{Gen}(\mathcal{S}_0)$. We conclude that $\mathrm{Gen}(\mathcal{S}_0)$ and $\mathrm{Gen}(\mathcal{S}_3)$ strongly approximately commute on Ω, i. e., $\mathcal{S}_0$ and $\mathcal{S}_3$ weakly approximately commute on Ω. □

Corollary 3.8.11. *Fix $\nu_1, \nu_2 \in \mathbb{N}_+$. Let*

$$\mathcal{S}_1, \ldots, \mathcal{S}_{\nu_1}, \mathcal{T}_1, \ldots, \mathcal{T}_{\nu_2} \subseteq C^\infty_{\mathrm{loc}}(\mathfrak{M}; T\mathfrak{M}) \times \mathbb{N}_+.$$

Suppose that $\forall \mu_1 \in \{1, \ldots, \nu_1\}$ and $\forall \mu_2 \in \{1, \ldots, \nu_2\}$, $\mathcal{S}_{\mu_1}$ and $\mathcal{T}_{\mu_2}$ locally weakly approximately commute. Set

$$\mathcal{S}_0 := \mathcal{S}_1 \boxtimes \mathcal{S}_2 \boxtimes \cdots \boxtimes \mathcal{S}_{\nu_1} \subseteq C^\infty_{\mathrm{loc}}(\mathfrak{M}; T\mathfrak{M}) \times (\mathbb{N}^{\nu_1} \setminus \{0\}),$$
$$\mathcal{T}_0 := \mathcal{T}_1 \boxtimes \mathcal{T}_2 \boxtimes \cdots \boxtimes \mathcal{T}_{\nu_2} \subseteq C^\infty_{\mathrm{loc}}(\mathfrak{M}; T\mathfrak{M}) \times (\mathbb{N}^{\nu_2} \setminus \{0\}).$$

Then $\mathcal{S}_0$ and $\mathcal{T}_0$ locally weakly approximately commute.

Proof. For $\mu_1 \in \{1, \ldots, \nu_1\}$ and $\mu_2 \in \{1, \ldots, \nu_2\}$, set

$$\mathcal{S}^{\mu_1} := \mathcal{S}_1 \boxtimes \mathcal{S}_2 \boxtimes \cdots \boxtimes \mathcal{S}_{\mu_1} \subseteq C^\infty_{\mathrm{loc}}(\mathfrak{M}; T\mathfrak{M}) \times (\mathbb{N}^{\mu_1} \setminus \{0\}),$$
$$\mathcal{T}^{\mu_2} := \mathcal{T}_1 \boxtimes \mathcal{T}_2 \boxtimes \cdots \boxtimes \mathcal{T}_{\mu_2} \subseteq C^\infty_{\mathrm{loc}}(\mathfrak{M}; T\mathfrak{M}) \times (\mathbb{N}^{\mu_2} \setminus \{0\}).$$

First claim: For all $\mu_1 \in \{1, \ldots, \nu_1\}$ and all $\mu_2 \in \{1, \ldots, \nu_2\}$, $\mathcal{S}^{\mu_1}$ and $\mathcal{T}_{\mu_2}$ locally weakly approximately commute.

We prove the first claim by induction on μ_1. Since $\mathcal{S}^1 = \mathcal{S}_1$, the base case ($\mu_1 = 1$) follows from the hypotheses of the corollary. We assume the first claim for some $\mu_1 \in \{1, \ldots, \nu_1 - 1\}$ and prove it for $\mu_1 + 1$. The inductive hypothesis implies that $\mathcal{S}^{\mu_1}$ and

$\mathcal{T}_{\mu_2}$ locally weakly approximately commute for all $\mu_2 \in \{1,\dots,\nu_2\}$. The hypotheses of the corollary imply that $\mathcal{S}_{\mu_1+1}$ and $\mathcal{T}_{\mu_2}$ locally weakly approximately commute for all $\mu_2 \in \{1,\dots,\nu_2\}$. Proposition 3.8.10 (iv) with $\mathcal{S}_1$ in that proposition replaced with $\mathcal{S}^{\mu_1}$, $\mathcal{S}_2$ replaced with $\mathcal{S}_{\mu_1+1}$, and $\mathcal{S}_3$ replaced with $\mathcal{T}_{\mu_2}$ implies that $\mathcal{S}^{\mu_1+1}$ and $\mathcal{T}_{\mu_2}$ locally weakly approximately commute, for all $\mu_2 \in \{1,\dots,\nu_2\}$. This completes the inductive step and completes the proof of the first claim.

Second claim: For all $\mu_2 \in \{1,\dots,\nu_2\}$, $\mathcal{S}^{\nu_1}$ and $\mathcal{T}^{\mu_2}$ locally weakly approximately commute.

We prove the second claim by induction on μ_2. Since $\mathcal{T}^1 = \mathcal{T}_1$, the base case ($\mu_2 = 1$) follows from the first claim. We assume the second claim for some $\mu_2 \in \{1,\dots,\nu_2 - 1\}$ and prove it for $\mu_2 + 1$. The inductive hypothesis implies that $\mathcal{S}^{\nu_1}$ and $\mathcal{T}^{\mu_2}$ weakly locally approximately commute, and the first claim implies $\mathcal{S}^{\nu_1}$ and $\mathcal{T}_{\mu_2+1}$ weakly locally approximately commute. Proposition 3.8.10 (iv) with $\mathcal{S}_1$ in that proposition replaced with $\mathcal{T}^{\mu_2}$, $\mathcal{S}_2$ replaced with $\mathcal{T}_{\mu_2+1}$, and $\mathcal{S}_3$ replaced with $\mathcal{S}^{\nu_1}$ implies that $\mathcal{S}^{\nu_1}$ and $\mathcal{T}^{\mu_2+1}$ weakly locally approximately commute. This completes the inductive step and completes the proof of the second claim.

Since $\mathcal{S}_0 = \mathcal{S}^{\nu_1}$ and $\mathcal{T}_0 = \mathcal{T}^{\nu_2}$, the corollary follows from the second claim. □

3.9 Integrating over leaves

Let $\mathfrak{M}$ be a connected C^∞ manifold, with smooth strictly positive density Vol. Fix $\nu \in \mathbb{N}_+$ and let

$$(W, \vec{d}) = \{(W_1, \vec{d}_1), \dots, (W_r, \vec{d}_r)\} \subset C^\infty_{\mathrm{loc}}(\mathfrak{M}; T\mathfrak{M}) \times (\mathbb{N}^\nu \setminus \{0\})$$

be a finite set. Our main assumption is the following.

Main assumption: $\mathrm{Gen}((W, \vec{d}))$ is locally finitely generated on $\mathfrak{M}$.

Theorem 3.5.1 applies in this setting, though that theorem does not give any relationship between the Carnot–Carathéodory geometry on each leaf and the density Vol, which is defined on the total space $\mathfrak{M}$. The main result of this section gives a way to estimate an integral over the total space by integrals over leaves.

Fix open sets $\Omega_3 \Subset \Omega_4 \Subset \mathcal{K} \Subset \Omega_1 \Subset \Omega \Subset \mathfrak{M}$, with Ω, Ω_1, Ω_3, and Ω_4 open and relatively compact and $\mathcal{K}$ compact. Let $\mathrm{Gen}((W, \vec{d}))$ be finitely generated by $(X, \vec{d}) = \{(X_1, \vec{d}_1), \dots, (X_q, \vec{d}_q)\}$ on Ω. Theorem 3.5.1 applies with this choice of $\mathcal{K}$, Ω_1, and Ω. We let $\Phi_{x,\delta} : B^{N(x)}(1) \to \mathfrak{M}$ be as in that theorem. The main result of this section is the following proposition.

Proposition 3.9.1. *There exists $\eta_1 = \eta_1(\mathcal{K}, \Omega_3, \Omega_4, (X, \vec{d})) \in (0,1]$ such that $\forall \delta \in (0,1]^\nu$, $\eta' \in (0, \eta_1]$, and $f \in C(\mathfrak{M})$ with $\mathrm{supp}(f) \subseteq \Omega_3$ and $f \geq 0$,*

$$\int_{\Omega_3} f(x)\, d\,\mathrm{Vol}(x) \approx \int_{\Omega_4} \int_{B^{N(x)}(\eta')} f(\Phi_{x,\delta}(u))\, du\, d\,\mathrm{Vol}(x),$$

where the implicit constants depend on η'. Here, we have used Theorem 3.5.1 (f) to see that the integral on the right-hand side makes sense.

To prove Proposition 3.9.1, we require two lemmas.

Lemma 3.9.2. *There exists $\eta_2 = \eta_2(\mathcal{K}, \Omega_3, \Omega_4, (X, \vec{d})) \in (0,1]$ such that $\forall \eta' \in (0, \eta_2]$ and all $f \in C(\mathfrak{M})$ with $\operatorname{supp}(f) \subseteq \Omega_3$ and $f \geq 0$, we have $\forall \delta \in (0,1]^\nu$,*

$$\int_{\Omega_3} f(x)\, d\operatorname{Vol}(x) \approx \frac{1}{(\eta')^q} \int_{\Omega_4} \int_{B^q(\eta')} f(e^{t\cdot\delta^{\vec{d}}X}x)\, dt\, d\operatorname{Vol}(x).$$

Proof. If $\eta_2 = \eta_2(\mathcal{K}, \Omega_3, \Omega_4, (X, \vec{d})) \in (0,1]$ is sufficiently small, then for $t \in B^q(\eta_2)$, we have $e^{t\cdot X}\Omega_4 \supseteq \Omega_3$. Furthermore, since $d_x e^{t\cdot X}\big|_{t=0} = I$, if η_2 is sufficiently small and $t \in B^q(\eta_2)$, a change of variables shows that

$$\int_{\Omega_3} f(x)\, d\operatorname{Vol}(x) \approx \int_{\Omega_4} f(e^{t\cdot X}x)\, d\operatorname{Vol}(x), \quad \forall t \in B^q(\eta_2).$$

In particular, by replacing t with $(\delta^{\vec{d}_1}t_1, \ldots, \delta^{\vec{d}_q}t_q)$, we have, for $\delta \in (0,1]^\nu$,

$$\int_{\Omega_3} f(x)\, d\operatorname{Vol}(x) \approx \int_{\Omega_4} f(e^{t\cdot\delta^{\vec{d}}X}x)\, d\operatorname{Vol}(x), \quad \forall t \in B^q(\eta_2).$$

Averaging both sides of this equation over $B^q(\eta')$ yields the result. □

Lemma 3.9.3. *Let η_2 be as in Lemma 3.9.2. There exist $\eta_1 = \eta_1(\mathcal{K}, \Omega_3, \Omega_4, (X, \vec{d})) \in (0,1]$ and $\eta_3 = \eta_3(\mathcal{K}, \Omega_3, \Omega_4, (X, \vec{d})) \in (0, \eta_2]$ such that $\forall \eta' \in (0, \eta_1]$, there exists $\eta_4 = \eta_4(\eta', \mathcal{K}, \Omega_3, \Omega_4, (X, \vec{d})) \in (0, \eta_3]$ such that the following holds. For all $f \in C(\mathfrak{M})$ with $f \geq 0$ and for all $\delta \in (0,1]^\nu$, $x \in \Omega_4$,*

$$\int_{B^q(\eta_4)} f(e^{t\cdot\delta^{\vec{d}}X}x)\, dt \lesssim \int_{B^{N(x)}(\eta')} f \circ \Phi_{x,\delta}(u)\, du \lesssim \int_{B^q(\eta_3)} f(e^{t\cdot\delta^{\vec{d}}X}x)\, dt.$$

Proof. Theorem 3.5.1 (g) shows that

$$\|\Phi_{x,\delta}^* \delta^{\vec{d}_j} X_j\|_{C^2(B^{N(x)}(1);\mathbb{R}^{N(x)})} \lesssim 1. \tag{3.39}$$

Let $\Theta_{x,\delta}(t) := e^{t\cdot\Phi_{x,\delta}^*\delta^{\vec{d}}X}0$. Using (3.39), standard results from the field of ODEs imply that if $\eta_5 \gtrsim 1$ is sufficiently small,

$$\|\Theta_{x,\delta}\|_{C^2(B^{N(x)}(\eta_5);\mathbb{R}^{N(x)})} \lesssim 1. \tag{3.40}$$

Since

$$\partial_{t_j}\big|_{t=0}\Theta_{X,\delta}(t) = \Phi^*_{X,\delta}\delta^{d_j}X_j(0),$$

Theorem 3.5.1 (h) shows that we can decompose the t variable into (t^1, t^2), where $t^1 = (t_{j_1}, t_{j_2}, \ldots, t_{j_{N(x)}})$, t^2 denotes the rest of the variables in t, and

$$\left|\det d_{t^1}\big|_{t=0}\Theta(t)\right| \approx 1. \tag{3.41}$$

Combining this with (3.40) shows that there exists $\eta_3 = \eta_3(\mathcal{K}, \Omega_3, \Omega_4, (X, \vec{d})) \in (0, \eta_2]$ such that $\forall t \in B^q(\eta_3)$,

$$\left|\det d_{t^1}\big|_{t=0}\Theta_{X,\delta}(t)\right| \approx 1. \tag{3.42}$$

Using the inverse function theorem, (3.39), (3.40), and the fact that $\Theta_{X,\delta}(0) = 0$, we can show that there exists $\eta_1 = \eta_1(\mathcal{K}, \Omega_3, \Omega_4, (X, \vec{d})) \in (0, 1]$ with

$$B^{N(x)}(\eta_1) \subseteq \Theta_{X,\delta}(B^q(\eta_3)). \tag{3.43}$$

Given $\eta' \in (0, \eta_1]$, the Picard–Lindelöf theorem and (3.39) show that there exists $\eta_4 = \eta_4(\eta', \mathcal{K}, \Omega_3, \Omega_4, (X, \vec{d})) \in (0, \eta_3]$ such that

$$\Theta_{X,\delta}(t) = e^{t\cdot\Phi^*_{X,\delta}\delta^d X}0 \in B^{N(x)}(\eta_1), \quad \forall t \in B^q(\eta_4). \tag{3.44}$$

Since $x = \Phi_{x,\delta}(0)$ (see Theorem 3.5.1 (c)), we have by a change of variables, using (3.42), (3.43), and (3.44),

$$\begin{aligned}\int\limits_{B^q(\eta_4)} f(e^{t\cdot\delta^d X}x)\,dt &= \int\limits_{B^q(\eta_4)} f(e^{t\cdot\delta^d X}\Phi_{x,\delta}(0))\,dt = \int\limits_{B^q(\eta_4)} f\circ\Phi_{x,\delta}(\Theta_{x,\delta}(t))\,dt \\ &\lesssim \int\limits_{B^{N(x)}(\eta')} f\circ\Phi_{x,\delta}(u)\,du \lesssim \int\limits_{B^q(\eta_3)} f\circ\Phi_{x,\delta}(\Theta_{x,\delta}(t))\,dt = \int\limits_{B^q(\eta_3)} f(e^{t\cdot\delta^d X}x)\,dt,\end{aligned}$$

completing the proof. □

Proof of Proposition 3.9.1. Applying Lemmas 3.9.2 and 3.9.3, we have with η_1, η_3, and η_4 as in those lemmas, for $f \in C(\mathfrak{M})$ with $\operatorname{supp}(f) \subseteq \Omega_3$ and $f \geq 0$, $\forall \eta' \in (0, \eta_1]$,

$$\begin{aligned}\int\limits_{\Omega_3} f(x)\,d\mathrm{Vol}(x) &\approx \int\limits_{\Omega_4}\int\limits_{B^q(\eta_4)} f(e^{t\cdot\delta^d X}x)\,dt\,d\mathrm{Vol}(x) \\ &\lesssim \int\limits_{\Omega_4}\int\limits_{B^{N(x)}(\eta')} f\circ\Phi_{x,\delta}(u)\,du\,d\mathrm{Vol}(x) \lesssim \int\limits_{\Omega_4}\int\limits_{B^q(\eta_3)} f(e^{t\cdot\delta^d X}x)\,dt\,d\mathrm{Vol}(x) \\ &\approx \int\limits_{\Omega_3} f(x)\,d\mathrm{Vol}(x),\end{aligned}$$

where the implicit constants depend on $\eta' \in (0, \eta_1]$. □

3.10 Integrals and approximately commuting vector fields

Let $\mathfrak{M}$ be a connected C^∞ manifold with smooth, strictly positive density Vol. Fix $\nu \in \mathbb{N}_+$. For each $\mu \in \{1,\ldots,\nu\}$, let

$$(W^\mu, \vec{d}^\mu) = \{(W_1^\mu, \vec{d}_1^\mu), \ldots, (W_{r_\mu}^\mu, \vec{d}_{r_\mu}^\mu)\} \subset C^\infty_{\mathrm{loc}}(\mathfrak{M}; T\mathfrak{M}) \times \mathbb{N}_+$$

be Hörmander vector fields with formal degrees. Set

$$\begin{aligned}(W, \vec{d}) = \{(W_1, \vec{d}_1), \ldots, (W_r, \vec{d}_r)\} &:= (W^1, \vec{d}^1) \boxtimes (W^2, \vec{d}^2) \boxtimes \cdots \boxtimes (W^\nu, \vec{d}^\nu)\\ &:= \{(W_j^\mu, \vec{d}_j^\mu \mathfrak{e}_\mu) : \mu \in \{1,\ldots,\nu\}, j \in \{1,\ldots,q_\mu\}\} \subset C^\infty_{\mathrm{loc}}(\mathfrak{M}; T\mathfrak{M}) \times (\mathbb{N}^\nu \setminus \{0\}),\end{aligned}$$

where $\mathfrak{e}_1, \ldots, \mathfrak{e}_\nu$ denotes the standard basis of $\mathbb{R}^\nu$.

Main assumption: In this section, we assume the lists $(W^\mu, \vec{d}^\mu)$ pairwise locally weakly approximately commute, $\mu \in \{1,\ldots,\nu\}$.

The main result of this section is the following proposition; in it we write $\delta \in [0,\infty)^\nu$ as $\delta = 2^{-j}$, where $j \in (-\infty,\infty]^\nu$. See Remark 3.5.3.

Proposition 3.10.1. *Fix an open, relatively compact set $\Omega \Subset \mathfrak{M}$. Then there exists $D = D((W,\vec{d}),\Omega) \geq 0$ such that $\forall j,k \in (-\infty,\infty]^\nu \setminus \{(\infty,\ldots,\infty)\}$, $\forall x,z \in \Omega$,*

$$\begin{aligned}&\int_\Omega \frac{\chi_{B_{(W,\vec{d})}(x,2^{-j})}(y)}{(\mathrm{Vol}(B_{(W,\vec{d})}(x,2^{-j})) \wedge 1)} \frac{\chi_{B_{(W,\vec{d})}(y,2^{-k})}(z)}{(\mathrm{Vol}(B_{(W,\vec{d})}(y,2^{-k})) \wedge 1)}\, d\mathrm{Vol}(y)\\ &\quad \leq D \frac{\chi_{B_{(W,\vec{d})}(x,2\cdot 2^{-j\wedge k})}(z)}{\mathrm{Vol}(B_{(W,\vec{d})}(x,2^{-j\wedge k})) \wedge 1}.\end{aligned} \tag{3.45}$$

The rest of this section is devoted to the proof of Proposition 3.10.1. Fix $\Omega \Subset \Omega_1 \Subset \Omega_2 \Subset \Omega_3 \Subset \mathfrak{M}$, with Ω_1, Ω_2, and Ω_3 open and relatively compact.

By Proposition 3.4.14 (applied with $\nu = 1$), each $\mathrm{Gen}((W^\mu, \vec{d}^\mu))$ is finitely generated on Ω_3 by some $(X^\mu, d^\mu) = \{(X_1^\mu, d_1^\mu), \ldots, (X_{q_\mu}^\mu, d_{q_\mu}^\mu)\} \subset \mathrm{Gen}((W^\mu, \vec{d}^\mu))$. We define

$$\begin{aligned}(X, \vec{d}) = \{(X_1, \vec{d}_1), \ldots, (X_q, \vec{d}_q)\} &:= (X^1, d^1) \boxtimes (X^2, d^2) \boxtimes \cdots \boxtimes (X^\nu, d^\nu)\\ &\{(X_j^\mu, d_j^\mu \mathfrak{e}_\mu) : \mu \in \{1,\ldots,\nu\}, 1 \leq j \leq q_\mu\}.\end{aligned}$$

By Proposition 3.8.8, $\mathrm{Gen}((W,\vec{d}))$ is linearly finitely generated by $(X,\vec{d})$ on Ω_3.

For $a > 0$ and $j \in [0,\infty]^\nu$, set

$$A^a_{2^{-j}} f(x) = \int_{B^q(a)} f(e^{-t^1 \cdot 2^{-j_1 d^1} X^1} e^{-t^2 \cdot 2^{-j_2 d^2} X^2} \cdots e^{-t^\nu \cdot 2^{-j_\nu d^\nu} X^\nu} x)\, dt,$$

where $t = (t^1, \dots, t^\nu) \in \mathbb{R}^{q_1} \times \cdots \times \mathbb{R}^{q_\nu} = \mathbb{R}^q$ and $2^{-j_\mu d^\mu} X^\mu = (2^{-j_\mu d_1^\mu} X_1^\mu, \dots, 2^{-j_\mu d^\mu} X_{q_\mu}^\mu)$ with $2^{-\infty} = 0$. For $a > 0$ sufficiently small, we treat $A_{2^{-j}}^a$ as an operator in $\mathrm{Hom}(C_0^\infty(\Omega_2), C_{\mathrm{loc}}^\infty(\Omega_3))$, and we write the Schwartz kernel as $A_{2^{-j}}^a(x,y)$.

Lemma 3.10.2. *Fix $\xi_4 > 0$. Then there exists $a_0 = a_0(\xi_4, \Omega_1, \Omega_2, (X, \vec{d})) > 0$ such that $\forall a \in (0, a_0]$, $\exists \xi_5 = \xi_5(a, \Omega_1, \Omega_2, (X, \vec{d})) > 0$ such that the following hold:*

(i) *$\forall \delta \in [0,1]^\nu \setminus \{0\}$, $\forall x \in \Omega_1$, $\operatorname{supp} A_\delta^a(x, \cdot) \Subset \Omega_2$, and $A_\delta^a|_{\Omega_1 \times \Omega_2}$ is a bounded measurable function.*

(ii) *There exists $C_1 = C_1(a, \Omega_1, \Omega_2, (X, \vec{d})) \geq 0$ such that $\forall x \in \Omega_1, y \in \Omega_2$, $\delta \in [0,1]^\nu \setminus \{0\}$,*

$$\frac{\chi_{B_{(W,\vec{d})}(x, \xi_5 \delta)}(y)}{\mathrm{Vol}(B_{(W,\vec{d})}(x, \xi_5 \delta)) \wedge 1} \leq C_1 A_\delta^a(x,y).$$

(iii) *There exists $C_2 = C_2(a, \Omega_1, \Omega_2, (X, \vec{d})) \geq 0$ such that $\forall x \in \Omega_1, y \in \Omega_2$, $\delta \in [0,1]^\nu \setminus \{0\}$,*

$$A_\delta^a(x,y) \leq C_2 \frac{\chi_{B_{(W,\vec{d})}(x, \xi_4 \delta)}(y)}{\mathrm{Vol}(B_{(W,\vec{d})}(x, \xi_4 \delta)) \wedge 1}.$$

(iv) *There exists $C_3 = C_3(a, \Omega_1, \Omega_2, (X, \vec{d})) \geq 0$ such that $\forall 2^{-j}, 2^{-k} \in [0,1]^\nu$ with $2^{-j \wedge k} \neq 0$, $\forall x \in \Omega_1, y \in \Omega_2$,*

$$[A_{2^{-j}}^a A_{2^{-k}}^a](x,y) \leq C_3 \frac{\chi_{B_{(W,\vec{d})}(x, \xi_4 2^{-j \wedge k})}(y)}{\mathrm{Vol}(B_{(W,\vec{d})}(x, \xi_4 2^{-j \wedge k})) \wedge 1}.$$

Proof. For (i), (ii), and (iii) we claim that it suffices to consider only the case $\delta \in (0,1]^\nu$, i. e., δ is non-zero in every component. Indeed, if $\delta \in [0,1]^\nu \setminus \{0\}$, then if we set $E = \{\mu : \delta_\mu \neq 0\}$, the above results with δ are equivalent to the same results applied to $(W^E, \vec{d}^E)$ with δ replaced by δ_E, where $(W^E, \vec{d}^E)$ and $\delta_E \in (0,1]^{|E|}$ are as in Section 3.5.2. Similarly, it suffices to prove (iv) in the case where $j \wedge k$ is not infinite in any component. We henceforth assume $\delta \in (0,1]^\nu$ in (i), (ii), and (iii) and $j \wedge k$ is not infinite in any component in (iv).

For $t \in \mathbb{R}^q$ with t small, set

$$\Gamma_t(x) = e^{-t^1 \cdot X^1} e^{-t^2 \cdot X^2} \cdots e^{-t^\nu \cdot X^\nu} x$$

so that

$$A_\delta^a f(x) = \int_{B^q(a)} f(\Gamma_{\delta^{\vec{d}} t}(x))\, dt$$

and $\delta^{\vec{d}} t = (\delta^{\vec{d}_1} t_1, \dots, \delta^{\vec{d}_q} t_q)$.

Fix $\Omega_{1.5}$ open with $\Omega_1 \Subset \Omega_{1.5} \Subset \Omega_2$. We apply Theorem 3.5.1 with $\mathcal{K}$ replaced by $\overline{\Omega_1}$ and Ω replaced by $\Omega_{1.5}$. Without loss of generality (using Theorem 3.5.4 (n)) we may shrink

ξ_4 so that $\xi_4 \le \xi_3 \wedge \delta_0 \wedge 1$ (where ξ_3 and δ_0 are as in Theorem 3.5.1). In particular, by Theorem 3.5.1 (a), we have $B_{(X,\vec{d})}(x, \xi_4\delta) \subseteq \Omega_{1.5}$, $\forall x \in \Omega_1$, $\delta \in (0,1]^\nu$.

If $0 < a_0 \le \xi_4/\nu$ is small enough that $\Gamma_t(x)$ is defined $\forall t \in B^q(a)$, $x \in \Omega_1$, then we have for such t and x, $\Gamma_{\delta^{\vec{d}}t}(x) \in B_{(X,\vec{d})}(x, \xi_4\delta) \subseteq \Omega_{1.5}$. And therefore, for $x \in \Omega_1$ and $a \in (0, a_0]$, $\operatorname{supp} A_\delta^a(x,\cdot) \subseteq \overline{\Omega_{1.5}} \subseteq \Omega_2$.

Note that

$$\partial_{t_j^\mu}\Gamma_t(x) = X_j^\mu(x).$$

Since, for each $\mu \in \{1,\dots,\nu\}$, the vector fields $X_1^\mu(x),\dots,X_{q_\mu}^\mu(x)$ span the tangent space at every point of Ω_3 (because W^μ satisfies Hörmander's condition), $d_t|_{t=0}\Gamma_t(x) : \mathbb{R}^q \to T_x\mathfrak{M}$ is surjective for each $x \in \Omega_3$. Since $\Gamma_t(x)$ is smooth in both variables (see Section 3.1.1), the inverse function theorem along with a simple change of variables shows that if $a_0 > 0$ is sufficiently small and $a \in (0, a_0]$, then $A_\delta^a(x,y)|_{\Omega_1\times\Omega_2}$ is a bounded measurable function. This completes the proof of (i).

Next, we prove (iv); recall that we are assuming without loss of generality that $j \wedge k \in \mathbb{R}^\nu$. Let $l = j \wedge k$, $\hat{j} = j - l$, and $\hat{k} = k - l$; note that $\hat{j} \wedge \hat{k} = 0 \in \mathbb{R}^\nu$. Let $\Phi_{x,2^{-l}}$ be the map from Theorem 3.5.1. Let $X_p^{m,x,2^{-l}} := \Phi_{x,\delta}^* 2^{-l_\mu d_p^\mu} X_p^\mu$. By Theorem 3.5.1 (h), we may pick n of the vector fields from $\{X_p^{\mu,x,2^{-l}} : \mu \in \{1,\dots,\nu\}, 1 \le p \le q_\mu\}$ and call them $Y_1^{x,2^{-l}},\dots,Y_n^{x,2^{-l}}$, with

$$\inf_{u\in B^n(1)} \left|\det(Y_1^{x,2^{-l}}(u)|\cdots|Y_n^{x,2^{-l}}(u))\right| \gtrsim 1. \tag{3.46}$$

Using the fact that $\Phi_{x,2^{-l}}(0) = x$ (Theorem 3.5.1 (c)), we have

$$\begin{aligned}
[A_{2^{-j}}^a A_{2^{-k}}^a f](x) &= \int_{B^q(a)}\int_{B^q(a)} f(e^{-t^1\cdot 2^{-k_1 d^1}X^1} e^{-t^2\cdot 2^{-k_2 d^2}X^2}\cdots e^{-t^\nu\cdot 2^{-k_\nu d^\nu}X^\nu} \\
&\qquad e^{-s^1\cdot 2^{-j_1 d^1}X^1} e^{-s^2\cdot 2^{-j_2 d^2}X^2}\cdots e^{-s^\nu\cdot 2^{-j_\nu d^\nu}X^\nu}\Phi_{x,2^{-l}}(0))\,ds\,dt \\
&= \int_{B^q(a)}\int_{B^q(a)} f\circ\Phi_{x,2^{-l}}(e^{-t^1\cdot 2^{-\hat{k}_1 d^1}X^{1,x,2^{-l}}} e^{-t^2\cdot 2^{-\hat{k}_2 d^2}X^{2,x,2^{-l}}}\cdots e^{-t^\nu\cdot 2^{-\hat{k}_\nu d^\nu}X^{\nu,x,2^{-l}}} \\
&\qquad e^{-s^1\cdot 2^{-\hat{j}_1 d^1}X^{1,x,2^{-l}}} e^{-s^2\cdot 2^{-\hat{j}_2 d^2}X^{2,x,2^{-l}}}\cdots e^{-s^\nu\cdot 2^{-\hat{j}_\nu d^\nu}X^{\nu,x,2^{-l}}}0)\,ds\,dt \\
&=: \int_{B^q(a)}\int_{B^q(a)} f\circ\Phi_{x,2^{-l}}(\Theta_{\hat{j},\hat{k},x,l}(s,t))\,ds\,dt,
\end{aligned}$$

where

$$\begin{aligned}
\Theta_{\hat{j},\hat{k},x,l}(s,t) = {}& e^{-t^1\cdot 2^{-\hat{k}_1 d^1}X^{1,x,2^{-l}}} e^{-t^2\cdot 2^{-\hat{k}_2 d^2}X^{2,x,2^{-l}}}\cdots e^{-t^\nu\cdot 2^{-\hat{k}_\nu d^\nu}X^{\nu,x,2^{-l}}} \\
& e^{-s^1\cdot 2^{-\hat{j}_1 d^1}X^{1,x,2^{-l}}} e^{-s^2\cdot 2^{-\hat{j}_2 d^2}X^{2,x,2^{-l}}}\cdots e^{-s^\nu\cdot 2^{-\hat{j}_\nu d^\nu}X^{\nu,x,2^{-l}}}0.
\end{aligned}$$

Note that

$$\partial_{s_p^\mu}\Theta_{\hat{j},\hat{k},x,l}\big|_{s=t=0} = -2^{-\hat{j}_\mu d_p^\mu} X_p^{\mu,x,2^{-l}}(0), \quad \partial_{t_p^\mu}\Theta_{\hat{j},\hat{k},x,l}\big|_{s=t=0} = -2^{-\hat{k}_\mu d_p^\mu} X_p^{\mu,x,2^{-l}}(0).$$

Thus, since $\hat{j} \wedge \hat{k} = 0$, each of the vectors $Y_1^{x,2^{-l}}(0), \dots, Y_n^{x,2^{-l}}(0)$ from (3.46) appears as either $-\partial_{s_p^\mu}\Theta_{\hat{j},\hat{k},x,l}|_{s=t=0}$ or $-\partial_{t_p^\mu}\Theta_{\hat{j},\hat{k},x,l}|_{s=t=0}$ for some p and μ. Let $u^1 \in \mathbb{R}^n$ denote those coordinates of (s,t) corresponding to the vectors $Y_1^{x,2^{-l}}(0), \dots, Y_n^{x,2^{-l}}(0)$ and let u^2 denote the rest of the coordinates of (s,t), so that

$$-\partial_{u_p^1}\Theta_{\hat{j},\hat{k},l}(u^1,u^2)\big|_{u^1=u^2=0} = Y_p^{x,2^{-l}}(0).$$

Combining this with (3.46), we have

$$\left|\det d_{u^1}\Theta_{\hat{j},\hat{k},x,l}(0,0)\right| \gtrsim 1. \tag{3.47}$$

Theorem 3.5.1 (g) shows that $\|X_p^{\mu,x,2^{-l}}\|_{C^1(B^n(1);\mathbb{R}^n)} \lesssim 1$; thus, if $a_0 \gtrsim 1$ is sufficiently small, standard theorems from the field of ODEs show that

$$\left\|\Theta_{\hat{j},\hat{k},x,l}\right\|_{C^1(B^q(a)\times B^q(a);\mathbb{R}^n)} \lesssim 1, \quad \forall a \in (0,a_0]. \tag{3.48}$$

By Lemma 3.2.4, Theorem 3.5.1 (i), and the fact that $\xi_4 \approx 1$, we may pick $\eta_0 \gtrsim 1$ so small that

$$B^n(\eta_0) \subseteq B_{\xi_4^{|\vec{d}|_1} W^{x,2^{-l}}}(0). = B_{(W^{x,2^{-l}},|\vec{d}|_1)}(0,\xi_4/2),$$

so that

$$\Phi_{x,2^{-l}}(B^n(\eta_0)) \subseteq B_{(W,\vec{d})}(x,\xi_4 2^{-l}). \tag{3.49}$$

Applying a change of variables in the u^1 variable, with $v = \Theta_{\hat{j},\hat{k},x,l}(u^1,u^2)$, (3.47) and (3.48) show that if $a_0 = a_0(\eta_0,(X,\vec{d}),\Omega_1,\Omega_2) > 0$ is sufficiently small and $a \in (0,a_0]$, then

$$\begin{aligned}&\int\limits_{B^q(a)\times B^q(a)} f\circ\Phi_{x,2^{-l}}(\Theta_{\hat{j},\hat{k},x,l}(s,t))\,ds\,dt\\ &\quad = \int\limits_{B^n(\eta_0)}\int\limits_{B^q(2a)} f\circ\Phi_{x,2^{-l}}(v) J_{a,\hat{j},\hat{k},x,l}(v,u^2)\,du^2\,dv,\end{aligned}$$

where $|J_{a,\hat{j},\hat{k},x,l}(v,u^2)| \lesssim 1$.

Let $\Lambda(x, 2^{-l})$ and $h_{x,2^{-l}}$ be as in Theorem 3.5.4. We have

$$\begin{aligned}&\int_{B^n(\eta_0)}\int_{B^q(2a)} f \circ \Phi_{x,2^{-l}}(v) J_{a,\hat{j},\hat{k},x,l}(v, u^2)\, du^2\, dv \\ &= \Lambda(x, 2^{-l})^{-1} \int_{B^n(\eta_0)}\int_{B^q(2a)} f \circ \Phi_{x,2^{-l}}(v) h_{x,2^{-l}}(v)^{-1} J_{a,\hat{j},\hat{k},x,l}(v, u^2)\, du^2 \Lambda(x, 2^{-l}) h_{x,2^{-l}}(v)\, dv.\end{aligned}$$

Since $\Phi^*_{x,2^{-l}} \mathrm{Vol} = \Lambda(x, 2^{-l}) h_{x,2^{-l}} \sigma_{\mathrm{Leb}}$, applying a change of variables $y = \Phi_{x,2^{-l}}(v)$ gives

$$\begin{aligned}&\Lambda(x, 2^{-l})^{-1} \int_{B^n(\eta_0)}\int_{B^q(2a)} f \circ \Phi_{x,2^{-l}}(v) h_{x,2^{-l}}(v)^{-1} J_{a,\hat{j},\hat{k},x,l}(v, u^2)\, du^2 \Lambda(x, 2^{-l}) h_{x,2^{-l}}(v)\, dv \\ &= \Lambda(x, 2^{-l})^{-1} \int_{\Phi_{x,2^{-l}}(B^n(\eta_0))}\int_{B^q(2a)} f(y) h_{x,2^{-l}}(\Phi^{-1}_{x,2^{-l}}(y))^{-1} J_{a,\hat{j},\hat{k},x,l}(\Phi^{-1}_{x,2^{-l}}(y), u^2)\, du^2\, d\mathrm{Vol}(y).\end{aligned}$$

Combining the above equations yields

$$\begin{aligned}A^a_{2^{-j}} A^a_{2^{-k}}(x, y) = {}& \chi_{\Phi_{x,2^{-j}}(B^n(\eta_0))}(y) \Lambda(x, 2^{-l})^{-1} \\ & \times h_{x,2^{-l}} \circ \Phi^{-1}_{x,2^{-l}}(y)^{-1} \int_{B^q(2a)} J_{a,\hat{j},\hat{k},x,l}(\Phi^{-1}_{x,2^{-l}}(y), u^2)\, du^2.\end{aligned}$$

Since $|J_{a,\hat{j},\hat{k},x,l}| \lesssim 1$, $h_{x,2^{-l}} \approx 1$ (Theorem 3.5.4 (o)), using (3.49), we have

$$\begin{aligned}|A^a_{2^{-j}} A^a_{2^{-k}}(x, y)| &\lesssim \Lambda(x, 2^{-l})^{-1} \chi_{B_{(W,\vec{d})}(x,\xi_4 2^{-l})}(y) \lesssim \frac{\chi_{B_{(W,\vec{d})}(x,\xi_4 2^{-l})}(y)}{\Lambda(x, 2^{-l}) \wedge 1} \\ &\approx \frac{\chi_{B_{(W,\vec{d})}(x,\xi_4 2^{-l})}(y)}{\mathrm{Vol}(B_{(W,\vec{d})}(x, 2^{-l})) \wedge 1} \lesssim \frac{\chi_{B_{(W,\vec{d})}(x,\xi_4 2^{-l})}(y)}{\mathrm{Vol}(B_{(W,\vec{d})}(x, \xi_4 2^{-l})) \wedge 1},\end{aligned}$$

where in the second to last estimate we have used Theorem 3.5.4 (l) and in the last estimate we have used the fact that $\xi_4 \leq 1$. This completes the proof of (iv).

Item (iii) follows from (iv). Indeed, when $j = (\infty, \dots, \infty)$, $A^a_{2^{-j}} f(x) = |B^q(a)| f(x)$. Thus, (iv) with $2^{-j} = 0$ and $2^{-k} = \delta$ implies (iii).

Finally, we turn to (ii), which has a very similar proof to (iv). Again, we take $\Phi_{x,\delta}$ as in Theorem 3.5.4, and we let $X_j^{\mu,x,\delta} := \Phi^*_{x,\delta} \delta_\mu^{d_j^\mu} X_j^\mu$. Using the fact that $\Phi_{x,\delta}(0) = x$ (Theorem 3.5.1 (c)), we have

$$\begin{aligned}A_\delta^a f(x) &= \int\limits_{B^q(a)} f(e^{-t^1\cdot\delta_1^{d^1}X^1}e^{-t^2\cdot\delta_2^{d^2}X^2}\cdots e^{-t^\nu\cdot\delta_\nu^{d^\nu}X^\nu}\Phi_{x,\delta}(0))\,dt \\ &= \int\limits_{B^q(a)} f\circ\Phi_{x,\delta}(e^{-t^1\cdot X^{1,x,\delta}}e^{-t^2\cdot X^{2,x,\delta}}\cdots e^{-t^\nu\cdot X^{\nu,x,\delta}}0)\,dt \\ &=: \int\limits_{B^q(a)} f\circ\Phi_{x,\delta}(\Theta_{x,\delta}(t))\,dt,\end{aligned}$$

where

$$\Theta_{x,\delta}(t) = e^{-t^1\cdot X^{1,x,\delta}}e^{-t^2\cdot X^{2,x,\delta}}\cdots e^{-t^\nu\cdot X^{\nu,x,\delta}}0.$$

Note that

$$\partial_{t_j^\mu}\Theta_{x,\delta}(t)\big|_{t=0} = -X_j^{\mu,x,\delta}(0),$$

and therefore by Theorem 3.5.1 (h),

$$\max_{l_1,\dots,l_n\in\{1,\dots,q\}}\left|\det(\partial_{t_{l_1}}\Theta_{x,\delta}(0)|\cdots|\partial_{t_{l_n}}\Theta_{x,\delta}(0))\right| \gtrsim 1. \tag{3.50}$$

Theorem 3.5.1 (g) shows that $\|X_j^{\mu,x,2^{-l}}\|_{C^1(B^n(1);\mathbb{R}^n)} \lesssim 1$; thus, if $a_0 \gtrsim 1$ is sufficiently small, standard theorems from the field of ODEs show that

$$\left\|\Theta_{x,\delta}\right\|_{C^1(B^q(a);\mathbb{R}^n)} \lesssim 1, \quad \forall a\in(0,a_0]. \tag{3.51}$$

Picking $l_1,\dots,l_n$ to achieve the max in (3.50) and applying a change of variables in the $(t_{l_1},\dots,t_{l_n})$ variable, (3.50), (3.51), and the fact that $\Theta_{x,\delta}(0)=0$ show that

$$\int\limits_{B^q(a)} f\circ\Phi_{x,\delta}(\Theta_{x,\delta}(t))\,dt = \int\limits_{B^n(1)} f\circ\Phi_{x,\delta,a}(v)J_{x,\delta,a}(v)\,dv, \tag{3.52}$$

where $J_{x,\delta,a}(v)\geq 0$ and there exists $\eta_0=\eta_0(a,(X,\vec{d}),\Omega_1,\Omega_2)>0$ such that $J_{x,\delta,a}(v)\gtrsim 1$ on $B^n(\eta_0)$. Taking $\Lambda(x,\delta)$ and $h_{x,\delta}$ as in Theorem 3.5.4 and using (3.52),

$$A_\delta^a f(x) = \Lambda(x,\delta)^{-1}\int\limits_{B^n(1)} f\circ\Phi_{x,\delta}(v)h_{x,\delta}(v)^{-1}J_{x,\delta,a}(v)h_{x,\delta}(v)\Lambda(x,\delta)\,dv.$$

Since $\Phi_{x,\delta}^*\operatorname{Vol} = \Lambda(x,\delta)h_{x,\delta}\sigma_{\mathrm{Leb}}$, this implies

$$A_\delta^a f(x) = \Lambda(x,\delta)^{-1}\int\limits_{\Phi_{x,\delta}(B^n(1))} f(y)h_{x,\delta}(\Phi_{x,\delta}^{-1}(y))J_{x,\delta,a}(\Phi_{x,\delta}^{-1}(y))\,d\operatorname{Vol}(y).$$

Thus, we see

$$A_\delta^a(x,y) = \Lambda(x,\delta)^{-1}\chi_{\Phi_{x,\delta}(B^n(1))}(y)h_{x,\delta}(\Phi_{x,\delta}^{-1}(y))J_{x,\delta,a}(\Phi_{x,\delta}^{-1}(y)).$$

Using Theorem 3.5.1 (g), we may pick $\xi_5 = \xi_5(\eta_0, (X,\vec{d}), \Omega_1, \Omega_2) \in (0, \xi_4] > 0$ so small that

$$B_{(X^{x,\delta},|\vec{d}|_1)}(0,\xi_5) \subseteq B^n(\eta_0/2).$$

This implies that

$$B_{(W,\vec{d})}(x,\xi_5\delta) \subseteq B_{(X,\vec{d})}(x,\xi_5\delta) \subseteq \Phi_{x,\delta}(B^n(\eta_0)).$$

Since $J_{x,\delta}(v) \gtrsim 1$ on $B^n(\eta_0)$ and $h_{x,\delta} \approx 1$ (Theorem 3.5.4 (o)), this implies

$$A_\delta^a(x,y) \gtrsim \Lambda(x,\delta)^{-1}, \quad y \in B_{(W,\vec{d})}(x,\xi_5\delta). \tag{3.53}$$

Since Ω_1 is compact, the formula for Λ gives $\sup_{\substack{x\in\overline{\Omega_1}\\ \delta\in(0,1]^\nu}} \Lambda(x,\delta) \lesssim 1$. Thus,

$$\Lambda(x,\delta) \approx \Lambda(x,\delta)\wedge 1 \approx \mathrm{Vol}(B_{(W,\vec{d})}(x,\delta))\wedge 1, \tag{3.54}$$

where the last estimate uses Theorem 3.5.4 (l). Since $\xi_5 \approx 1$, Theorem 3.5.4 (n) shows that

$$\mathrm{Vol}(B_{(W,\vec{d})}(x,\delta))\wedge 1 \approx \mathrm{Vol}(B_{(W,\vec{d})}(x,\xi_5\delta))\wedge 1. \tag{3.55}$$

Plugging (3.54) and (3.55) into (3.53) yields

$$A_\delta^a(x,y) \gtrsim (\mathrm{Vol}(B_{(W,\vec{d})}(x,\xi_5\delta))\wedge 1)^{-1}, \quad y \in B_{(W,\vec{d})}(x,\xi_5\delta),$$

which completes the proof of (ii). □

Proof of Proposition 3.10.1. Since $B_{(W,\vec{d})}(x,2^{-j}), B_{(W,\vec{d})}(x,2^{-k}) \subseteq B_{(W,\vec{d})}(x,2^{-j\wedge k})$ and the balls $r \mapsto B_{(W,\vec{d})}(x,r2^{-j})$ are metric balls, it follows that the integral on the left-hand side of (3.45) equals zero unless $z \in B_{(W,\vec{d})}(x,2\cdot 2^{-j\wedge k})$. Thus, to prove (3.45) it suffices to show that for $x,z \in \Omega$,

$$\begin{aligned}\int_\Omega &\frac{\chi_{B_{(W,\vec{d})}(x,2^{-j})}(y)}{(\mathrm{Vol}(B_{(W,\vec{d})}(x,2^{-j}))\wedge 1)}\frac{\chi_{B_{(W,\vec{d})}(y,2^{-k})}(z)}{(\mathrm{Vol}(B_{(W,\vec{d})}(y,2^{-k}))\wedge 1)}\,d\,\mathrm{Vol}(y)\\ &\lesssim \frac{1}{\mathrm{Vol}(B_{(W,\vec{d})}(x,2^{-j\wedge k}))\wedge 1}.\end{aligned} \tag{3.56}$$

Fix Ω_1, Ω_2, and Ω_3 open with $\Omega \Subset \Omega_1 \Subset \Omega_2 \Subset \Omega_3 \Subset \mathfrak{M}$. We apply Lemma 3.10.2 with $\xi_4 = 1$ and take $a_0 > 0$ and $\xi_5 > 0$ (with $a = a_0$) as in that lemma. We separate the proof of (3.56) into two cases.

The first case is when $|2^{-j}|_\infty, |2^{-k}|_\infty \le \xi_5$. In this case, set $2^{-j} = \xi_5 2^{-\hat{j}}$ and $2^{-k} = \xi_5 2^{-\hat{k}}$, with $2^{-\hat{j}}, 2^{-\hat{k}} \in (0,1]^\nu$. We have, for $x, y, z \in \Omega$, by Lemma 3.10.2 (ii),

$$\frac{\chi_{B_{(W,\vec{d})}(x,2^{-j})}(y)}{\operatorname{Vol}(B_{(W,\vec{d})}(x,2^{-j}))\wedge 1} \lesssim A^{a_0}_{2^{-\hat{j}}}(x,y), \qquad \frac{\chi_{B_{(W,\vec{d})}(y,2^{-k})}(z)}{\operatorname{Vol}(B_{(W,\vec{d})}(y,2^{-k}))\wedge 1} \lesssim A^{a_0}_{2^{-\hat{k}}}(y,z).$$

Thus,

$$\begin{aligned}&\int_\Omega \frac{\chi_{B_{(W,\vec{d})}(x,2^{-j})}(y)}{(\operatorname{Vol}(B_{(W,\vec{d})}(x,2^{-j}))\wedge 1)} \frac{\chi_{B_{(W,\vec{d})}(y,2^{-k})}(z)}{(\operatorname{Vol}(B_{(W,\vec{d})}(y,2^{-k}))\wedge 1)}\, d\operatorname{Vol}(y) \\ &\quad \lesssim [A^{a_0}_{2^{-\hat{j}}} A^{a_0}_{2^{-\hat{k}}}](x,z) \lesssim (\operatorname{Vol}(B_{(W,\vec{d})}(x,2^{-\hat{j}\wedge\hat{k}}))\wedge 1)^{-1}.\end{aligned} \tag{3.57}$$

Since $\xi_5 \approx 1$, Theorem 3.5.4 (n) shows that

$$\operatorname{Vol}(B_{(W,\vec{d})}(x,2^{-\hat{j}\wedge\hat{k}}))\wedge 1 \approx \operatorname{Vol}(B_{(W,\vec{d})}(x,2^{-j\wedge k}))\wedge 1.$$

Combining this with (3.57) proves (3.56) and completes the case where $|2^{-j}|_\infty, |2^{-k}|_\infty \le \xi_5$.

We turn to the case where either $|2^{-j}|_\infty > \xi_5$ or $|2^{-k}|_\infty > \xi_5$. First, we claim that for any $\delta \in [0,\infty)^\nu \setminus \{0\}$, $x, z \in \Omega$,

$$\frac{\chi_{B_{(W,\vec{d})}(x,\delta)}(z)}{\operatorname{Vol}(B_{(W,\vec{d})}(x,\delta))\wedge 1} \approx \frac{\chi_{B_{(W,\vec{d})}(z,\delta)}(x)}{\operatorname{Vol}(B_{(W,\vec{d})}(z,\delta))\wedge 1}. \tag{3.58}$$

To prove (3.58), it suffices to consider the case where $\delta \in (0,\infty)^\nu$ by the discussion in Section 3.5.2. For $z \in B_{(W,\vec{d})}(x,\delta)$, $B_{(W,\vec{d})}(x,\delta) \subseteq B_{(W,\vec{d})}(z,3\delta)$, and therefore,

$$\begin{aligned}\frac{\chi_{B_{(W,\vec{d})}(x,\delta)}(z)}{\operatorname{Vol}(B_{(W,\vec{d})}(x,\delta))\wedge 1} &\ge \frac{\chi_{B_{(W,\vec{d})}(x,\delta)}(z)}{\operatorname{Vol}(B_{(W,\vec{d})}(z,3\delta))\wedge 1} \\ &= \frac{\chi_{B_{(W,\vec{d})}(z,\delta)}(x)}{\operatorname{Vol}(B_{(W,\vec{d})}(z,3\delta))\wedge 1} \approx \frac{\chi_{B_{(W,\vec{d})}(z,\delta)}(x)}{\operatorname{Vol}(B_{(W,\vec{d})}(z,\delta))\wedge 1},\end{aligned}$$

where the last estimate uses Theorem 3.5.4 (n). This proves the $\gtrsim$ part of (3.58); the $\lesssim$ part follows by reversing the roles of x and z.

Formula (3.58) implies that (3.45) is symmetric in x and z, so that it suffices to prove (3.56) in the case where $|2^{-k}|_\infty > \xi_5$. Because our assumptions are symmetric in the parameters $1, \dots, \nu$, it suffices to consider only the case $2^{-k_1} > \xi_5$. Since $2^{-k_1} > \xi_5$, we have

$$\operatorname{Vol}(B_{(W,\vec{d})}(x,2^{-k})) \ge \operatorname{Vol}(B_{(W^1,d^1)}(x,2^{-k_1})) \ge \operatorname{Vol}(B_{(W^1,d^1)}(x,\xi_5)) \gtrsim 1, \tag{3.59}$$

and therefore

$$\mathrm{Vol}(B_{(W,\vec{d})}(x, 2^{-j\wedge k})) \geq \mathrm{Vol}(B_{(W,\vec{d})}(x, 2^{-k})) \gtrsim 1. \tag{3.60}$$

Using (3.59) and (3.60), we see that

$$\begin{aligned}
&\int_{\Omega} \frac{\chi_{B_{(W,\vec{d})}(x,2^{-j})}(y)}{(\mathrm{Vol}(B_{(W,\vec{d})}(x, 2^{-j})) \wedge 1)} \frac{\chi_{B_{(W,\vec{d})}(y,2^{-k})}(z)}{(\mathrm{Vol}(B_{(W,\vec{d})}(y, 2^{-k})) \wedge 1)} \, d\,\mathrm{Vol}(y) \\
&\lesssim \int_{\Omega} \frac{\chi_{B_{(W,\vec{d})}(x,2^{-j})}(y)}{\mathrm{Vol}(B_{(W,\vec{d})}(x, 2^{-j})) \wedge 1} \, d\,\mathrm{Vol}(y) \\
&= \frac{\mathrm{Vol}(B_{(W,\vec{d})}(x, 2^{-j}) \cap \Omega)}{\mathrm{Vol}(B_{(W,\vec{d})}(x, 2^{-j})) \wedge 1} \lesssim 1 \lesssim \frac{1}{\mathrm{Vol}(B_{(W,\vec{d})}(x, 2^{-j\wedge k})) \wedge 1},
\end{aligned}$$

where we have used $\mathrm{Vol}(\Omega) < \infty$, since $\overline{\Omega}$ is compact. This completes the proof. □

3.11 Maximal functions

In this section, we present maximal functions adapted to some multi-parameter Carnot–Carathéodory geometries. Let $\mathfrak{M}$ be a connected C^∞ manifold, with smooth, strictly positive density Vol. Fix $\nu \in \mathbb{N}_+$, and for each $\mu \in \{1, \ldots, \nu\}$, let

$$(W^\mu, d^\mu) = \{(W_1^\mu, d_1^\mu), \ldots, (W_{r_\mu}^\mu, d_{r_\mu}^\mu)\} \subset C^\infty_{\mathrm{loc}}(\mathfrak{M}; T\mathfrak{M}) \times \mathbb{N}_+.$$

Set

$$\begin{aligned}
(W, \vec{d}) = \{(W_1, \vec{d}_1), \ldots, (W_r, \vec{d}_r)\} &:= (W^1, d^1) \boxtimes (W^2, d^2) \cdots \boxtimes (W^\nu, d^\nu) \\
&= \{(W_j^\mu, d_j^\mu \mathfrak{e}_\mu) : \mu \in \{1, \ldots, \nu\}, 1 \leq j \leq r_\mu\} \subset C^\infty_{\mathrm{loc}}(\mathfrak{M}; T\mathfrak{M}) \times (\mathbb{N}^\nu \setminus \{0\}).
\end{aligned} \tag{3.61}$$

We study maximal functions in two situations:

The finitely generated setting: In this setting, we assume $\mathrm{Gen}((W^\mu, d^\mu))$ is locally finitely generated on $\mathfrak{M}$ for each $\mu \in \{1, \ldots, \nu\}$.

The Hörmander setting: In this setting, we assume (W^μ, d^μ) are Hörmander vector fields with formal degrees for each $\mu \in \{1, \ldots, \nu\}$ and that $(W^1, d^1), \ldots, (W^\nu, d^\nu)$ pairwise locally weakly approximately commute.

By Proposition 3.4.14 (with $\nu = 1$), the Hörmander setting is a special case of the finitely generated setting. The maximal functions we present in the Hörmander setting will perhaps seems a little more familiar to the reader, though as we will see, they are essentially a special case of the maximal functions we present in the finitely generated setting.

3.11.1 The finitely generated setting

In this section we assume the assumptions from the **finitely generated setting** described above.

Fix $\mathcal{K} \Subset \Omega_1 \Subset \Omega \Subset \mathfrak{M}$, with Ω_1 and Ω open and relatively compact and $\mathcal{K}$ compact. For each $\mu \in \{1,\dots,\nu\}$, pick

$$(X^\mu, d^\mu) = \{(X_1^\mu, d_1^\mu), \dots, (X_{q_\mu}^\mu, d_{q_\mu}^\mu)\} \subset \mathrm{Gen}((W^\mu, \vec{d}^\mu))$$

so that $\mathrm{Gen}((W^\mu, \vec{d}^\mu))$ is finitely generated by (X^μ, d^μ) on Ω.

Set

$$\begin{aligned}(X, \vec{d}) &= \{(X_1, \vec{d}_1), \dots, (X_q, \vec{d}_q)\} := (X^1, d^1) \boxtimes (X^2, d^2) \boxtimes \cdots \boxtimes (X^\nu, d^\nu) \\ &= \{(X_j^\mu, d_j^\mu \mathfrak{e}_\mu) : \mu \in \{1,\dots,\nu\}, 1 \le j \le q_\mu\} \subset C^\infty_{\mathrm{loc}}(\mathfrak{M}; T\mathfrak{M}) \times (\mathbb{N}^\nu \setminus \{0\}).\end{aligned} \tag{3.62}$$

Let $a = a(\mathcal{K}, \Omega_1, (X, \vec{d})) > 0$ be a small number, to be chosen later. For $t \in \mathbb{R}^q$, we write $t = (t^1, t^2, \dots, t^\nu) \in \mathbb{R}^{q_1} \times \mathbb{R}^{q_2} \times \cdots \times \mathbb{R}^{q_\nu}$. For $\delta = (\delta_1, \dots, \delta_\nu) \in (0,1]^\nu$, set

$$A^a_{\delta,(X,\vec{d})} f(x) := \int_{B^q(a)} f(e^{-t^1 \cdot \delta^{d_1} X^1} e^{-t^2 \cdot \delta^{d^2} X^2} \cdots e^{-t^\nu \cdot \delta^{d^\nu} X^\nu} x)\, dt, \tag{3.63}$$

where, as before, $\delta_\mu^{d^\mu} X^\mu = (\delta_\mu^{d_1^\mu} X_1^\mu, \dots, \delta_\mu^{d_{q_\mu}^\mu} X_{q_\mu}^\mu)$.

Lemma 3.11.1. *If $a = a(\mathcal{K}, \Omega_1, (X, \vec{d})) > 0$ is sufficiently small, then for $\delta \in (0,1]^\nu$, $A^a_{\delta,(X,\vec{d})}$ extends to a continuous map $A^a_{\delta,(X,\vec{d})} : L^p(\mathcal{K}, \mathrm{Vol}) \to L^p(\mathfrak{M}, \mathrm{Vol})$, $p \in [1,\infty]$, and $\forall f \in L^p(\mathcal{K}, \mathrm{Vol})$, $\mathrm{supp}(A^a_{\delta,(X,\vec{d})} f) \subseteq \Omega_1$. Finally,*

$$\sup_{p \in [1,\infty]} \sup_{\delta \in (0,1]^\nu} \|A^a_{\delta,(X,\vec{d})}\|_{L^p \to L^p} < \infty. \tag{3.64}$$

We prove Lemma 3.11.1 below. In light of Lemma 3.11.1, $A^a_{\delta,(X,\vec{d})} f(x)$ is defined almost everywhere, for $f \in L^1(\mathcal{K}, \mathrm{Vol})$. For such f, we set

$$\widetilde{\mathcal{M}}_{(X,\vec{d}),\mathcal{K},a} := \sup_{l \in \mathbb{N}^\nu} A^a_{2^{-l},(X,\vec{d})} |f|(x). \tag{3.65}$$

The main result of this section is the following.

Theorem 3.11.2. *If $a = a(\mathcal{K}, \Omega_1, (X, \vec{d})) > 0$ is sufficiently small, then $\widetilde{\mathcal{M}}_{(X,\vec{d}),\mathcal{K},a} : L^p(\mathcal{K}, \mathrm{Vol}) \to L^p(\Omega_1, \mathrm{Vol})$, for $p \in (1,\infty]$. In particular, for $f \in L^p(\mathcal{K}, \mathrm{Vol})$, $\mathrm{supp}(\widetilde{\mathcal{M}}_{(X,\vec{d}),\mathcal{K},a} f) \subseteq \Omega_1$. Also, for $p \in (1,\infty]$,*

$$\|\widetilde{\mathcal{M}}_{(X,\vec{d}),\mathcal{K},a} f\|_{L^p(\Omega_1,\mathrm{Vol})} \lesssim \|f\|_{L^p(\mathcal{K},\mathrm{Vol})}, \quad f \in L^p(\mathcal{K}, \mathrm{Vol}), \tag{3.66}$$

where the implicit constant depends on $p \in (1,\infty]$ but not on f. Moreover, for $p \in (1,\infty)$, $q \in (1,\infty]$,

$$\widetilde{\mathcal{M}}_{(X,\vec{d}),\mathcal{K},a} : L^p(\mathcal{K}, \mathrm{Vol}; \ell^q(\mathbb{N})) \to L^p(\Omega_1, \mathrm{Vol}; \ell^q(\mathbb{N})),$$

and for $p \in (1,\infty)$, $q \in (1,\infty]$,

$$\begin{aligned} \|\{\widetilde{\mathcal{M}}_{(X,\vec{d}),\mathcal{K},a} f_j\}_{j\in\mathbb{N}}\|_{L^p(\Omega_1,\mathrm{Vol};\ell^q(\mathbb{N}))} &\lesssim \|\{f_j\}_{j\in\mathbb{N}}\|_{L^p(\mathcal{K},\mathrm{Vol};\ell^q(\mathbb{N}))}, \\ \{f_j\}_{j\in\mathbb{N}} &\in L^p(\mathcal{K}, \mathrm{Vol}; \ell^q(\mathbb{N})), \end{aligned} \tag{3.67}$$

where the implicit constant depends on p and q but not on f_j.

The rest of this section is devoted to the proof of Theorem 3.11.2. We begin with a result which implies Lemma 3.11.1.

Lemma 3.11.3. *For each $\mu \in \{1,\dots,\nu\}$, let $Z_1^\mu, \dots, Z_{q_\mu}^\mu \in C^\infty_{\mathrm{loc}}(\mathfrak{M}; T\mathfrak{M})$. Let Z^μ denote the vector of vector fields $Z^\mu = (Z_1^\mu, \dots, Z_{q_\mu}^\mu)$. Then there exists $a = a(Z^1,\dots,Z^\nu,\mathcal{K},\Omega_1) > 0$ such that the following holds $\forall \zeta \in L^1(B^{q_1+q_2+\dots+q_\nu}(a))$. For $f \in C(\mathfrak{M})$ with $\mathrm{supp}(f) \subseteq \mathcal{K}$, set*

$$S_\zeta f(x) := \int f(e^{-t^1\cdot Z^q} e^{-t^2\cdot Z^2} \cdots e^{-t^\nu\cdot Z^\nu} x)\zeta(t^1,\dots t^\nu)\, dt^1\, dt^2 \cdots dt^\nu.$$

Then S_ζ extends to a bounded operator $S_\zeta : L^p(\mathcal{K}, \mathrm{Vol}) \to L^p(\mathfrak{M}, \mathrm{Vol})$, for $p \in [1,\infty]$. Moreover, $\mathrm{supp}(S_\delta f) \subseteq \Omega_1$, $\forall f \in L^p(\mathcal{K}, \mathrm{Vol})$. Finally,

$$\|S_\zeta\|_{L^p(\mathcal{K},\mathrm{Vol})\to L^p(\Omega_1,\mathrm{Vol})} \leq C\|\zeta\|_{L^1}, \tag{3.68}$$

where $C \geq 0$ does not depend on $p \in [1,\infty]$ or ζ, though it may depend on any other ingredient.

Proof. Set $q = q_1 + q_2 + \dots + q_\nu$. Fix Ω_{-1}, Ω_0 open with $\mathcal{K} \Subset \Omega_{-1} \Subset \Omega_0 \Subset \Omega_1$. For $t = (t^1,\dots,t^\nu) \in \mathbb{R}^q \cong \mathbb{R}^{q_1} \times \mathbb{R}^{q_2} \times \dots \times \mathbb{R}^{q_\nu}$, let $\Gamma_t(x) := e^{-t^1\cdot Z^1} e^{-t^2\cdot Z^2} \cdots e^{-t^\nu\cdot Z^\nu} x$. For $a > 0$ sufficiently small and $t \in B^q(a)$, the Picard–Lindelöf theorem shows that $\Gamma_t(x) \in \Omega_0$ is defined for all $x \in \Omega_{-1}$. Standard results from the field of ODEs show that $\Gamma_t(x)$ is C^∞ in both variables.

Since $d_x\Gamma_t(x)|_{t=0} = I$, if a is sufficiently small and $t \in B^q(a)$, a simple change of variables shows that for any $g \in C(\mathfrak{M})$ and $t \in B^q(a)$,

$$\int_{\Omega_{-1}} g(\Gamma_t(x))\, d\,\mathrm{Vol}(x) = \int_{\Omega} g(u)J(t,u)\, d\,\mathrm{Vol}(u),$$

where $\sup_{t,u} |J(t,u)| < \infty$. Thus, for $f \in C(\mathfrak{M})$ with $\mathrm{supp}(f) \subseteq \Omega_{-1}$, we have $\mathrm{supp}(S_\zeta(f)) \subseteq \overline{\Omega_0} \subseteq \Omega_1$, and

$$\int_{\Omega_1} |S_\zeta f(x)|\, d\,\mathrm{Vol}(x) \le \int_{\Omega_1}\int |f(\Gamma_t(x))||\zeta(t)|\, dt\, d\,\mathrm{Vol}(x),$$

$$\int_{\Omega_1}\int |g(u)||J(t,u)||\zeta(t)|\, dt\, d\,\mathrm{Vol}(x) \lesssim \|g\|_{L^1(\Omega_{-1},\mathrm{Vol})}\|\zeta\|_{L^1}.$$

Thus, S_ζ extends to a bounded operator $L^1(\Omega_{-1},\mathrm{Vol}) \to L^1(\Omega_1,\mathrm{Vol})$ and (3.68) holds for $p = 1$; in particular, this extended operator is well-defined on almost everywhere equivalence classes of functions in $L^1(\Omega_{-1},\mathrm{Vol})$.

For $f \in L^\infty(\Omega_{-1},\mathrm{Vol}) \subseteq L^1(\Omega_{-1},\mathrm{Vol})$, $S_\zeta f$ is, therefore, defined almost everywhere. Furthermore, we may take $f_k \in C(\Omega_{-1})$ with $f_k \to f$ in L^1 and $\|f_k\|_{L^\infty} \le \|f\|_{L^\infty}$, $\forall k$. By taking a subsequence, we may ensure $f_k \to f$ almost everywhere and $S_\zeta f_k \to S_\zeta f$ almost everywhere. We have, for almost every x,

$$\begin{aligned}|S_\zeta f(x)| &= \lim_{k\to\infty} |S_\zeta f_k(x)| \le \liminf_{k\to\infty} \int |f_k(\Gamma_t(x))||\zeta(t)|\, dt\\ &\le \liminf_{k\to\infty} \|f_k\|_{L^\infty}\|\zeta\|_{L^1} \le \|f\|_{L^\infty}\|\zeta\|_{L^1}.\end{aligned}$$

This proves the lemma for $p = \infty$. The Riesz–Thorin interpolation theorem completes the proof. □

Proof of Lemma 3.11.1. We have

$$\begin{aligned}A^a_{\delta,(X,\vec{d})}f(x) &= \int_{B^q(a)} f(e^{-t^1\cdot\delta^{d_1}X^1}e^{-t^2\cdot\delta^{d^2}X^2}\cdots e^{-t^\nu\cdot\delta^{d^\nu}X^\nu}x)\, dt\\ &= \int f(e^{-t^1\cdot X^1}e^{-t^2\cdot X^2}\cdots e^{-t^\nu\cdot X^\nu}x)\delta_1^{-|d^1|_1}\delta_2^{-|d^2|_1}\cdots\delta_\nu^{-|d^\nu|_1}\\ &\quad\times \chi_{B^q(a)}(\delta_1^{-d^1}t^1,\dots,\delta_\nu^{-d^\nu}t^\nu)\, dt\\ &=: \int f(e^{-t^1\cdot X^1}e^{-t^2\cdot X^2}\cdots e^{-t^\nu\cdot X^\nu}x)\zeta_\delta(t)\, dt,\end{aligned}$$

where $\delta_\mu^{-d^\mu}t^\mu = (\delta_\mu^{-d_1^\mu}t_1^\mu,\dots,\delta_\mu^{-d_{q_\mu}^\mu}t_{q_\mu}^\mu)$. The result now follows from Lemma 3.11.3. □

Lemma 3.11.4. *It suffices to prove Theorem* 3.11.2 *in the special case* $\nu = 1$.

Proof. Suppose we know Theorem 3.11.2 in the case $\nu = 1$. Note that the hypotheses of Theorem 3.11.2 hold with $(W,\vec{d})$ and $(X,\vec{d})$ replaced by (W^μ, d^μ) and (X^μ, d^μ), for each $\mu \in \{1,\dots,\nu\}$.

Fix open sets

$$\mathcal{K} \Subset \Omega_1^\nu \Subset \Omega_2^\nu \Subset \Omega_1^{\nu-1} \Subset \Omega_2^{\nu-1} \Subset \cdots \Subset \Omega_1^1 \Subset \Omega_2^1 \Subset \Omega_1.$$

Let $a_\mu = a_\mu(\overline{\Omega_1}^\mu, \Omega_2^\mu, (X^\mu, d^\mu)) > 0$ be as in the $\nu = 1$ case of Theorem 3.11.2 applied to (X^μ, d^μ). Note that if $a = a(\mathcal{K}, \Omega_1, (X, \vec{d})) > 0$ is small enough, we have, for $\delta = (\delta_1, \dots, \delta_\nu) \in (0,1]^\nu$,

$$\begin{aligned} A^a_{\delta,(X,\vec{d})}|f|(x) &= \int_{B^q(a)} |f(e^{-t^1\cdot\delta^{d_1}X^1} e^{-t^2\cdot\delta^{d^2}X^2} \cdots e^{-t^\nu\cdot\delta^{d^\nu}X^\nu}x)|\,dt \\ &\le \int_{B^{q_1}(a_1)}\int_{B^{q_2}(a_2)} \cdots \int_{B^{q_\nu}(a_\nu)} |f(e^{-t^1\cdot\delta^{d_1}X^1} e^{-t^2\cdot\delta^{d^2}X^2} \cdots e^{-t^\nu\cdot\delta^{d^\nu}X^\nu}x)|\,dt^1\,dt^2 \cdots dt^\nu \\ &= A^{a_1}_{\delta_1,(X^1,d^1)} A^{a_2}_{\delta_2,(X^2,d^2)} \cdots A^{a_\nu}_{\delta_\nu,(X^\nu,d^\nu)}|f|(x). \end{aligned}$$

Moreover, if $\mathrm{supp}(f) \subseteq \mathcal{K}$, then

$$\mathrm{supp}(A^{a_\mu}_{\delta_\mu,(X^\mu,d^\mu)} A^{a_{\mu+1}}_{\delta_{\mu+1},(X^\mu,d^\mu)} \cdots A^{a_\nu}_{\delta_\nu,(X^\nu,d^\nu)} f) \subset \Omega_1^{\mu-1}.$$

Thus, if $\mathrm{supp}(f) \subseteq \mathcal{K}$, $j \in \mathbb{N}^\nu$,

$$A^a_{2^{-j},(X,\vec{d})}|f|(x) \le \widetilde{\mathcal{M}}_{(X^1,d^1),\overline{\Omega_1^1},a_1} \widetilde{\mathcal{M}}_{(X^2,d^2),\overline{\Omega_1^2},a_2} \cdots \widetilde{\mathcal{M}}_{(X^\nu,d^\nu),\overline{\Omega_1^\nu},a_\nu} f(x),$$

and therefore,

$$\widetilde{\mathcal{M}}_{(X,\vec{d}),\mathcal{K},a} f(x) \le \widetilde{\mathcal{M}}_{(X^1,d^1),\overline{\Omega_1^1},a_1} \widetilde{\mathcal{M}}_{(X^2,d^2),\overline{\Omega_1^2},a_2} \cdots \widetilde{\mathcal{M}}_{(X^\nu,d^\nu),\overline{\Omega_1^\nu},a_\nu} f(x).$$

From here the bounds for $\widetilde{\mathcal{M}}_{(X,\vec{d}),\mathcal{K},a}$ follow from the case $\nu = 1$ applied to each $\widetilde{\mathcal{M}}_{(X^\mu,d^\mu),\overline{\Omega_1^\mu},a_\mu}$. □

In light of Lemma 3.11.4, it suffices to prove Theorem 3.11.2 in the case $\nu = 1$. We do this in two steps. In the first, we prove a quantitative result on the unit ball in $\mathbb{R}^N$. Then we combine this quantitative result with Proposition 3.9.1 to complete the proof.

3.11.1.1 A result on the unit ball

Fix $N \in \mathbb{N}_+$ and let $(Y, \vec{d}) = \{(Y_1, \vec{d}_1), \dots, (Y_q, \vec{d}_q)\} \subset C^\infty(B^n(1)) \times \mathbb{N}_+$ be C^∞ vector fields on $B^N(1)$, paired with formal degrees. We suppose

$$\inf_{u\in B^N(1)} \max_{j_1,\dots,j_N\in\{1,\dots q\}} |\det(Y_{j_1}(u)|\cdots|Y_{j_N}(u))| \ge \gamma_0 > 0. \tag{3.69}$$

Fix $\eta_4 \in (0,1/2]$ and $\eta_3 \in (0,\eta_4)$. We will study the maximal function $\widetilde{\mathcal{M}}_{(Y,\vec{d}),\overline{B^N(\eta_3)},a}$ defined by (3.65) in the case $\nu = 1$.

What is crucial is keeping track of what our estimates depend on. For this, we introduce the following definition, for use just in this section.

Definition 3.11.5. For a parameter ι, we say $C \in \mathbb{R}$ is an ι-admissible constant if C can be chosen to depend only on ι and upper bounds for N, q, γ_0^{-1}, $(\eta_4 - \eta_3)^{-1}$, $\max_{1\le j\le q} \check{d}_j$, and

$$\max_{j\in\{1,\dots,q\}} \|Y_j\|_{C^L(B^q(1);\mathbb{R}^N)},$$

where L can be chosen to depend only on ι and upper bounds for N and q. We write $A \lesssim_\iota B$ for $A \le CB$, where C is an ι-admissible constant. We say C is an admissible constant if it is a 0-admissible constant.

The main result of this section is the next proposition.

Proposition 3.11.6. *There is an admissible constant $a \in (0,1]$ such that for $p \in (1,\infty]$, we have*

$$\|\widetilde{\mathcal{M}}_{(Y,\check{d}),\overline{B^N(\eta_3)},a} f\|_{L^p(B^N(\eta_4),\sigma_{\mathrm{Leb}})} \le C_p \|f\|_{L^p}, \quad \forall f \in L^1(B^N(\eta_3);\sigma_{\mathrm{Leb}}),$$

where $C_p \ge 0$ is a p-admissible constant. Also, for $p \in (1,\infty)$, $q \in (1,\infty]$,

$$\|\{\widetilde{\mathcal{M}}_{(Y,\check{d}),\overline{B^N(\eta_3)},a} f_j\}_{j\in\mathbb{N}}\|_{L^p(B^N(\eta_4),\sigma_{\mathrm{Leb}};\ell^q(\mathbb{N}))} \le C_{p,q} \|\{f_j\}_{j\in\mathbb{N}}\|_{L^p(B^N(\eta_3),\sigma_{\mathrm{Leb}};\ell^q(\mathbb{N}))},$$
$$\{f_j\}_{j\in\mathbb{N}} \in L^p(B^N(\eta_3),\sigma_{\mathrm{Leb}};\ell^q(\mathbb{N})),$$

where $C_{p,q} \ge 0$ is a p,q-admissible constant.

The rest of this section is devoted to the proof of Proposition 3.11.6. To do this, we introduce a more standard maximal function. For $f \in L^1(B^N(\eta_3);\sigma_{\mathrm{Leb}})$ and $\delta_1 \in (0,1]$, we set

$$\widehat{\mathcal{M}}_{(Y,\check{d}),\delta_1} f(x) := \sup_{\delta\in(0,\delta_1]} \frac{1}{\mathrm{Vol}(B_{(Y,\check{d})}(x,\delta))} \int_{B_{(Y,\check{d})}(x,\delta)} |f(u)|\,du.$$

Proposition 3.11.7. *There exists an admissible constant $\delta_1 > 0$ such that for $p \in (1,\infty]$, we have*

$$\|\widehat{\mathcal{M}}_{(Y,\check{d}),\delta_1} f\|_{L^p} \le C_p \|f\|_{L^p}, \quad \forall f \in L^1(B^N(\eta_3);\sigma_{\mathrm{Leb}}), \tag{3.70}$$

where $C_p \ge 0$ is a p-admissible constant. Also, for $p \in (1,\infty)$, $q \in (1,\infty]$,

$$\|\{\widehat{\mathcal{M}}_{(Y,\check{d}),\delta_1} f_j\}_{j\in\mathbb{N}}\|_{L^p(B^N(\eta_4),\sigma_{\mathrm{Leb}};\ell^q(\mathbb{N}))} \le C_{p,q} \|\{f_j\}_{j\in\mathbb{N}}\|_{L^p(B^N(\eta_3),\sigma_{\mathrm{Leb}};\ell^q(\mathbb{N}))}, \tag{3.71}$$
$$\{f_j\}_{j\in\mathbb{N}} \in L^p(B^N(\eta_3),\sigma_{\mathrm{Leb}};\ell^q(\mathbb{N})), \tag{3.72}$$

where $C_{p,q} \ge 0$ is a p,q-admissible constant.

Proof. Because $Y_1, \ldots, Y_q$ span the tangent space to $B^N(1)$ at every point, they satisfy Hörmander's condition of order 1 on $B^N(1)$. In particular, (3.69) together with the definition of admissible constants implies

$$[Y_j, Y_k] = \sum_{l=1}^{q} c_{j,k}^l Y_l, \quad c_{j,k}^l Y_l, \quad c_{j,k}^l \in C^\infty(B^N(1)),$$

with $\|c_{j,k}^l\|_{C^L(B^N(1))} \lesssim_L 1$, $\forall L \in \mathbb{N}$. Thus, Theorem 3.3.7 applies with $(W, \vec{d}) = (X, d) := (Y, \vec{d})$, $\mathcal{K} = \overline{B^N(\eta_4)}$, and $\Omega = B^N((1 + \eta_4)/2)$. Moreover, by tracing through the proof, it is immediate to see that all of the constants in that theorem can be taken to be admissible constants in the sense of Definition 3.11.5.

In particular, by Theorem 3.3.7 (e), we have

$$\sigma_{\mathrm{Leb}}(B_{(Y,\vec{d})}(x, 2\delta)) \lesssim \sigma_{\mathrm{Leb}}(B_{(Y,\vec{d})}(x, \delta)), \quad x \in B^N(x, 2\delta), \delta \in (0, \delta_0], \tag{3.73}$$

where $\delta_0 > 0$ is the admissible constant from Theorem 3.3.7.

Formula (3.73) is the main inequality needed to show that the balls $B_{(Y,\vec{d})}(x, \delta)$, $\delta \in (0, 1]$, when paired with σ_{Leb}, locally give $B^N(\eta_4)$ the structure of a space of homogeneous type in the sense of Coifman and Weiss [52]. By taking $\delta_1 = \delta_0$, the result now follows from standard methods: (3.70) follows from the methods in [216, Chapter I, Section 3.1] and (3.71) follows from the methods in [83]. □

Proof of Proposition 3.11.6. Let $\delta_1 \in (0, 1]$ be the admissible constant from Proposition 3.11.7. Similar to the remarks in the proof of Proposition 3.11.7, Lemma 3.10.2 applies with $\xi_4 = \delta_1$, $\nu = 1$, $(W, \vec{d}) = (X, \vec{d}) := (Y, \vec{d})$, $\Omega_1 = B^N(\eta_4)$, and $\Omega_2 = B^N((1 + \eta_4)/2)$; moreover, by tracing through the proof it is easy to see that each of the constants in Lemma 3.10.2 can be chosen to be an admissible constant in the sense of Definition 3.11.5. We let $a = a_0 \wedge 1$, where $a_0 > 0$ is as in Lemma 3.10.2.

By Lemma 3.10.2 (iii), we have, for $x \in B^N(\eta_4) = \Omega_1$ and $f \in L^1(B^N(\eta_3))$,

$$A^a_{\delta,(Y,\vec{d})}|f|(x) \lesssim \widehat{\mathcal{M}}_{(Y,\vec{d}),\delta_1} f(x),$$

and therefore, for $x \in B^N(\eta_4)$ and $f \in L^1(B^N(\eta_3))$,

$$\widetilde{\mathcal{M}}_{(Y,\vec{d}),\overline{B^N(\eta_3)},a} f(x) \lesssim \widehat{\mathcal{M}}_{(Y,\vec{d}),\delta_1} f(x).$$

From here the result follows immediately from Proposition 3.11.7. □

3.11.1.2 Proof of Theorem 3.11.2

In this section, we use Propositions 3.9.1 and 3.11.6 to complete the proof of Theorem 3.11.2. As described in Lemma 3.11.4, it suffices to consider only the case $\nu = 1$.

Thus, we assume we are given

$$(W, \vec{d}) = \{(W_1, \vec{d}_1), \ldots, (W_r, \vec{d}_r)\} \subset C^\infty_{\mathrm{loc}}(\mathfrak{M}; T\mathfrak{M}) \times \mathbb{N}_+,$$

such that $\mathrm{Gen}((W, \vec{d}))$ is locally finitely generated on $\mathfrak{M}$. We pick

$$(X, d) = \{(X_1, d_1), \ldots, (X_q, d_q)\} \subset \mathrm{Gen}((W, \vec{d}))$$

such that $(W, \vec{d})$ is finitely generated by (X, d) on Ω.

Theorem 3.5.1 applies (with $\nu = 1$) to these vector fields. We let $\Phi_{x,\delta}$, $N(x)$, and

$$X^{x,\delta} = (X_1^{x,\delta}, \ldots, X_q^{x,\delta}) \subset C^\infty(B^{N(x)}(1); TB^{N(x)}(1)) \times \mathbb{N}_+$$

be as in that theorem, for $x \in \mathcal{K}$, $\delta \in (0,1]$. For $x \in \mathcal{K}$, $\delta \in (0,1]$, define $A^a_{\delta,(X^{x,1},d)}$ by the $\nu = 1$ case of (3.63).

Taking $(Y, \vec{d}) = (X^{x,\delta}, d)$, it follows from Theorem 3.5.1 (g) and (h) that admissible constants as defined in Definition 3.11.5 can be chosen independent of $x \in \mathcal{K}$. This leads us to the next result.

Lemma 3.11.8. *Let $\eta_3, \eta_4 \in (0, 1/2]$, $\eta_3 < \eta_4$. There exists $a = a((X, d), \mathcal{K}, \Omega_1, \eta_3, \eta_4) \in (0, 1]$ such that the following holds. For $p \in (1, \infty]$ there exists $C = C(p, (X, d), \mathcal{K}, \Omega_1, \eta_3, \eta_4) \geq 0$ such that $\forall x \in \mathcal{K}$,*

$$\Big\| \sup_{l \in \mathbb{N}} A^a_{2^{-l},(X^{x,1},d)} |g| \Big\|_{L^p(B^{N(x)}(\eta_4))} \leq \|g\|_{L^p(B^{N(x)}(\eta_3))}, \quad \forall g \in L^p(B^{N(x)}(\eta_3)).$$

For $p \in (1, \infty)$, $q \in (1, \infty]$, there exists $C = C(p, q, (X, d), \mathcal{K}, \Omega_1, \eta_3, \eta_4) \geq 0$ such that $\forall x \in \mathcal{K}$,

$$\Big\| \Big\{ \sup_{l \in \mathbb{N}} A^a_{2^{-l},(X^{x,1},d)} |g_j| \Big\}_{j \in \mathbb{N}} \Big\|_{L^p(B^{N(x)}(\eta_4), \sigma_{\mathrm{Leb}}; \ell^q(\mathbb{N}))} \leq C \|\{g_j\}_{j \in \mathbb{N}}\|_{L^p(B^{N(x)}(\eta_3), \sigma_{\mathrm{Leb}}; \ell^q(\mathbb{N}))},$$
$$\{g_j\}_{j \in \mathbb{N}} \in L^p(B^{N(x)}(\eta_3), \sigma_{\mathrm{Leb}}; \ell^q(\mathbb{N})).$$

Proof. Note that for $g \in L^1(B^{N(x)}(\eta_3))$,

$$\widetilde{\mathcal{M}}_{(X^{x,1},d),\overline{B^N(\eta_3)},a} g(u) = \sup_{l \in \mathbb{N}} A^a_{2^{-l},(X^{x,1},d)} |g|(u).$$

Thus, this result follows from Proposition 3.11.6 applied with $(Y, \vec{d}) = (X^{x,1}, d)$. The main point here, as discussed above, is that admissible constants as defined in Definition 3.11.5 can be chosen independent of $x \in \mathcal{K}$, and therefore the conclusions of Proposition 3.11.6 are uniform in $x \in \mathcal{K}$. □

Proof of Theorem 3.11.2. We will pick $a = a(\mathcal{K}, \Omega_1, (X, d)) > 0$ in the proof; we will choose it small enough that Lemma 3.11.1 holds with our choice of a. In particular, if $\mathrm{supp}(f) \subseteq \mathcal{K}$, then $\mathrm{supp}(A^a_{\delta,(X,d)}) \subseteq \Omega_1$.

When $p = \infty$, (3.66) follows immediately from the $p = \infty$ case of (3.64). Thus, it suffices to consider only the case $p \in (1, \infty)$. By possibly increasing the size of $\mathcal{K}$, it suffices to consider the case where $\mathcal{K} = \overline{\Omega_0}$ and Ω_0 is an open set with $\Omega_0 \Subset \Omega_1$. In this case, by a density argument, it suffices to prove (3.66) and (3.67) in the case where f and f_j are continuous. We henceforth assume these functions are continuous.

Fix a compact set $\mathcal{K}'$ and open sets Ω_1', Ω_3, and Ω_4 with $\mathcal{K} \Subset \Omega_1 \Subset \Omega_3 \Subset \Omega_4 \Subset \mathcal{K}' \Subset \Omega_1' \Subset \mathfrak{M}$. We will apply our previous results (e. g., Lemma 3.11.8) with $\mathcal{K}$ and Ω_1 replaced by $\mathcal{K}'$ and Ω_1', respectively.

Let $\eta_1 = \eta_1(\mathcal{K}', \Omega_3, \Omega_4, (X, d)) \in (0, 1]$ be as in Proposition 3.9.1. Set $\eta_3 := \eta_1/4$ and $\eta_4 = \eta_1/2$. Take $a = a((X, d), \mathcal{K}', \Omega_1', \eta_3, \eta_4) \in (0, 1]$ small enough that Lemma 3.11.8 holds.

Note that for $f \in C(\mathfrak{M})$ and $x \in \mathcal{K}$, we have

$$\begin{aligned}(A^a_{\delta,(X,d)} f) \circ \Phi_{x,1}(u) &= \int_{B^q(a)} f(e^{-t\cdot\delta^d X} \Phi_{x,1}(u))\, dt \\ &= \int_{B^q(a)} f \circ \Phi_{x,1}(e^{-t\cdot\delta^d X^{x,1}} u)\, dt = A^a_{\delta,(X^{x,1},d)}(f \circ \Phi_{x,1}).\end{aligned}$$

Fix $M \in \mathbb{N}_+$. For $p \in (1, \infty)$ and $f \in C(\mathfrak{M})$ with $\operatorname{supp}(f) \subseteq \mathcal{K}$, using Proposition 3.9.1 and Lemma 3.11.8, we have with implicit constants independent of $M \in \mathbb{N}$

$$\begin{aligned}\Big\| \sup_{l \in \{0,\dots,M\}} A^a_{2^{-l},(X,d)} |f| \Big\|^p_{L^p(\Omega_1,\mathrm{Vol})} &\approx \int_{\Omega_4} \int_{B^{N(x)}(\eta_3)} \sup_{l\in\{0,\dots M\}} \big|(A^a_{2^{-l},(X,d)} f)(\Phi_{x,1}(u))\big|^p \, du\, d\,\mathrm{Vol}(x) \\ &= \int_{\Omega_4} \int_{B^{N(x)}(\eta_3)} \sup_{l\in\{0,\dots M\}} \big|A^a_{2^{-l},(X^{x,1},d)}(f \circ \Phi_{x,1})(u)\big|^p \, du\, d\,\mathrm{Vol}(x) \\ &\lesssim \int_{\Omega_4} \int_{B^{N(x)}(\eta_4)} |f \circ \Phi_{x,1}(u)| \, du\, d\,\mathrm{Vol}(x) \\ &\approx \int_{\Omega_3} |f(x)|^p \, d\,\mathrm{Vol}(x) = \|f\|^p_{L^p}.\end{aligned}$$

Taking the limit $M \to \infty$ and applying the monotone convergence theorem, (3.66) is proved.

A very similar proof gives (3.67) in the case where all but finitely many of the f_j are identically zero. The monotone convergence theorem then gives (3.67) in the general case. □

3.11.2 The Hörmander setting

In this section, we introduce a maximal function in the "Hörmander setting" described above. Thus, for each $\mu \in \{1, \dots, \nu\}$, we are given Hörmander vector fields with formal degrees

$$(W^\mu, \vec{d}^\mu) = \{(W_1^\mu, \vec{d}_1^\mu), \dots, (W_{r_\mu}^\mu, \vec{d}_{r_\mu}^\mu)\} \subset C_{\mathrm{loc}}^\infty(\mathfrak{M}; T\mathfrak{M}) \times \mathbb{N}_+.$$

We assume that $(W^1, \vec{d}^1), \dots, (W^\nu, \vec{d}^\nu)$ pairwise locally weakly approximately commute. We define $(W, \vec{d})$ by (3.61). For $f \in L^1(\mathfrak{M}, \mathrm{Vol})$, set

$$\mathcal{M}_{(W,\vec{d})} f(x) := \sup_{\delta \in (0,\infty)^\nu} \frac{1}{\mathrm{Vol}(B_{(W,\vec{d})}(x,\delta)) \wedge 1} \int_{B_{(W,\vec{d})}(x,\delta)} |f(y)| \, d\,\mathrm{Vol}(y).$$

Theorem 3.11.9. *Fix a relatively compact, open set $\Omega \Subset \mathfrak{M}$. Then, for $p \in (1,\infty]$, there exists $C_{p,\Omega,(W,\vec{d})} \geq 0$ such that*

$$\|\mathcal{M}_{(W,\vec{d})} f\|_{L^p(\Omega,\mathrm{Vol})} \leq C_{p,\Omega,(W,\vec{d})} \|f\|_{L^p(\Omega,\mathrm{Vol})}, \quad \forall f \in L^p(\Omega, \mathrm{Vol}).$$

Also, for $p \in (1,\infty)$, $q \in (1,\infty]$, there exists $C_{p,q,\Omega,(W,\vec{d})} \geq 0$ such that

$$\|\{\mathcal{M}_{(W,\vec{d})} f_j\}_{j\in\mathbb{N}}\|_{L^p(\Omega,\mathrm{Vol};\ell^q(\mathbb{N}))} \leq C_{p,q,\Omega,(W,\vec{d})} \|\{f_j\}_{j\in\mathbb{N}}\|_{L^p(\Omega,\mathrm{Vol};\ell^q(\mathbb{N}))},$$
$$\forall \{f_j\}_{j\in\mathbb{N}} \in L^p(\Omega, \mathrm{Vol}; \ell^q(\mathbb{N})).$$

Proof. Let $\xi_5 \in (0,1]$ be a small number, to be chosen later. Define

$$\mathcal{M}_{(W,\vec{d}),\xi_5} f(x) := \sup_{j\in\mathbb{N}^\nu} \frac{1}{\mathrm{Vol}(B_{(W,\vec{d})}(x,\xi_5 2^{-j})) \wedge 1} \int_{B_{(W,\vec{d})}(x,\xi_5 2^{-j})} |f(y)| \, d\,\mathrm{Vol}(y).$$

We claim, for $f \in L^1(\Omega, \mathrm{Vol})$ and $x \in \Omega$,

$$\mathcal{M}_{(W,\vec{d})} f(x) \lesssim \mathcal{M}_{(W,\vec{d}),\xi_5} f(x) + \chi_\Omega(x) \int |f(y)| \, d\,\mathrm{Vol}(y). \tag{3.74}$$

Indeed, we will show, for such x and f and for $\delta \in (0,\infty)^\nu$,

$$\begin{aligned} &\frac{1}{\mathrm{Vol}(B_{(W,\vec{d})}(x,\delta)) \wedge 1} \int_{B_{(W,\vec{d})}(x,\delta)} |f(y)| \, d\,\mathrm{Vol}(y) \\ &\quad \lesssim \mathcal{M}_{(W,\vec{d}),\xi_5} f(x) + \chi_\Omega(x) \int |f(y)| \, d\,\mathrm{Vol}(y), \end{aligned} \tag{3.75}$$

which immediately implies (3.74), by taking the supremum over δ. We separate the proof of (3.75) into two cases. The first is when $|\delta|_\infty \leq \xi_5$. In this case, pick $l = (l_1, \dots, l_\nu) \in \mathbb{N}^\nu$ such that $\forall \mu \in \{1, \dots, \nu\}$, we have $\xi_5 2^{-l_\mu - 1} \leq \delta_\mu \leq \xi_5 2^{-l_\mu}$. Then

$$\int_{B_{(W,\vec{d})}(x,\delta)} |f(y)| d\,\mathrm{Vol}(y) \leq \int_{B_{(W,\vec{d})}(x,\xi_5 2^{-l})} |f(y)| \, d\,\mathrm{Vol}(y), \tag{3.76}$$

and using Theorem 3.5.4 (n),

$$\mathrm{Vol}(B_{(W,\vec{d})}(x,\xi_5 2^{-l}))\wedge 1 \approx \mathrm{Vol}(B_{(W,\vec{d})}(x,\delta))\wedge 1. \tag{3.77}$$

Combining (3.76) and (3.77) shows

$$\begin{aligned}\frac{1}{\mathrm{Vol}(B_{(W,\vec{d})}(x,\delta))\wedge 1}&\int\limits_{B_{(W,\vec{d})}(x,\delta)} |f(y)|\,d\,\mathrm{Vol}(y)\\ &\lesssim \frac{1}{\mathrm{Vol}(B_{(W,\vec{d})}(x,\xi_5 2^{-l}))\wedge 1}\int\limits_{B_{(W,\vec{d})}(x,\xi_5 2^{-l})} |f(y)|\,d\,\mathrm{Vol}(y)\\ &\le \mathcal{M}_{(W,\vec{d}),\xi_5} f(x),\end{aligned}$$

completing the proof of (3.75) when $|\delta|_\infty \le \xi_5$.

Suppose $|\delta|_\infty > \xi_5$. Without loss of generality, we assume $\delta_1 > \xi_5$. We apply Theorem 3.5.4 with Ω replaced by some $\Omega' \Subset \mathfrak{M}$ with $\Omega \Subset \Omega'$. Using Theorem 3.5.4 (l) and the formula for Λ from that theorem, we have

$$\begin{aligned}\mathrm{Vol}(B_{(W,\vec{d})}(x,\delta))\wedge 1 &\approx \Lambda(x,\delta)\wedge 1 \ge \Lambda(x,(\xi_5,0,0,\dots,0))\wedge 1\\ &\ge \inf_{y\in\Omega'} \Lambda(y,(\xi_5,0,0,\dots,0))\wedge 1 \gtrsim 1.\end{aligned}$$

Therefore,

$$\frac{1}{\mathrm{Vol}(B_{(W,\vec{d})}(x,\delta))\wedge 1}\int\limits_{B_{(W,\vec{d})}(x,\delta)} |f(y)|\,d\,\mathrm{Vol}(y) \lesssim \int |f(y)|\,dy.$$

Since we are only considering $x \in \Omega$, this proves (3.75) and completes the proof of (3.74).

Thus, it suffices to estimate the two terms on the right-hand side of (3.74). We begin with the second term. Using that $\mathrm{Vol}(\Omega) < \infty$, since Ω is relatively compact, we have $\mathrm{Vol}(\Omega) \lesssim 1$, since our estimates are allowed to depend on Ω. We have, for $p \in [1,\infty]$ and $f \in L^1(\Omega,\mathrm{Vol})$, using Hölder's inequality,

$$\left\| \left(\int\limits_{\Omega} |f(y)|\,d\,\mathrm{Vol}(y) \right)\chi_\Omega \right\|_{L^p(\Omega,\mathrm{Vol})} \lesssim \|f\|_{L^1(\Omega,\mathrm{Vol})} \lesssim \|f\|_{L^p(\Omega,\mathrm{Vol})}.$$

Similarly, using the triangle inequality and Hölder's inequality, for $p,q \in [1,\infty]$ and $\{f_j\}_{j\in\mathbb{N}} \in L^1(\Omega,\mathrm{Vol};\ell^q(\mathbb{N}))$, we have

$$
\begin{aligned}
&\left\| \left\{ \left(\int_\Omega |f_j(y)| \, d\,\mathrm{Vol}(y) \right) \chi_\Omega \right\}_{j\in\mathbb{N}} \right\|_{L^p(\Omega,\mathrm{Vol};\ell^q(\mathbb{N}))} \\
&\quad \le \left\| \int_\Omega \left(\sum_{j\in\mathbb{N}} |f_j(y)|^q \right)^{\frac{1}{q}} d\,\mathrm{Vol}(y) \right\|_{L^p(\Omega,\mathrm{Vol})} \approx \left\| \left(\sum_{j\in\mathbb{N}} |f_j|^q \right)^{\frac{1}{q}} \right\|_{L^1(\Omega,\mathrm{Vol})} \\
&\quad \lesssim \|\{f_j\}_{j\in\mathbb{N}}\|_{L^p(\Omega,\mathrm{Vol};\ell^q(\mathbb{N}))}.
\end{aligned}
$$

The completes the proof of the desired estimates for the second term on the right-hand side of (3.74).

Finally, we estimate the first term on the right-hand side of (3.74). For this we need to choose $\xi_5 > 0$. Fix relatively compact open sets $\Omega_1, \Omega_2, \Omega_1'$, and Ω' and a compact set $\mathcal{K}$ with

$$\Omega \Subset \Omega_1 \Subset \Omega_2 \Subset \mathcal{K} \Subset \Omega_1' \Subset \Omega' \Subset \mathfrak{M}.$$

By Proposition 3.4.14 in the case $\nu = 1$, we may pick $(X^\mu, d^\mu) \subset \mathrm{Gen}((W^\mu, \vec{d}^\mu))$ such that $\mathrm{Gen}((W^\mu, \vec{d}^\mu))$ is finitely generated by (X^μ, d^μ) on Ω'. We define $(X, \vec{d})$ by (3.62) and $A^a_{\delta,(X,\vec{d})}$ by (3.63). Take $a = a(\mathcal{K}, \Omega_1', (X, \vec{d})) > 0$ as in Theorem 3.11.2.

Let $a_0 = a_0(1, \Omega_1, \Omega_2, (X, \vec{d})) > 0$ be as in Lemma 3.10.2 (with $\xi_4 = 1$ in that lemma); by replacing a with $a \wedge a_0$, we may assume $a \in (0, a_0]$. Let $\xi_5 = \xi_5(a, \Omega_1, \Omega_2, (X, \vec{d})) > 0$ be as in Lemma 3.10.2. Using Lemma 3.10.2 (ii), we have, for $l \in \mathbb{N}^\nu$, $x \in \Omega_1$,

$$
\begin{aligned}
&\frac{1}{\mathrm{Vol}(B_{(W,\vec{d})}(x, \xi_5 2^{-l})) \wedge 1} \int_{B_{(W,\vec{d})}(x,\xi_5 2^{-l})} |f(y)| \, d\,\mathrm{Vol}(y) \\
&\quad \lesssim A^a_{2^{-l},(X,\vec{d})} |f|(x) \le \widetilde{\mathcal{M}}_{(X,\vec{d}),\mathcal{K},a} f(x).
\end{aligned}
$$

Taking the supremum over $l \in \mathbb{N}^\nu$, we have $\mathcal{M}_{(W,\vec{d}),\xi_5} f(x) \lesssim \widetilde{\mathcal{M}}_{(X,\vec{d}),\mathcal{K},a} f(x)$, for $x \in \Omega$. The desired estimates for $\mathcal{M}_{(W,\vec{d}),\xi_5}$ follow immediately from Theorem 3.11.2, which completes the proof. □

3.12 Approximately commuting partial differential operators

Let $\nu_1, \nu_2 \in \mathbb{N}_+$ and for $j = 1, 2$, $\mathcal{S}_j \subseteq C^\infty_{\mathrm{loc}}(\mathfrak{M}; T\mathfrak{M}) \times (\mathbb{N}^{\nu_j} \setminus \{0\})$. If $\mathcal{S}_1$ and $\mathcal{S}_2$ locally weakly approximately commute, the main result of this section shows that partial differential operators which are polynomials in the vector fields from $\mathcal{S}_1$ and $\mathcal{S}_2$ also "approximately commute." We turn to making this precise. Throughout this section, $\Omega \Subset \mathfrak{M}$ will denote a relatively compact, open set. Also, ν_1, ν_2, ν will denote elements of $\mathbb{N}_+$ and $\mathcal{S}_j \subseteq C^\infty_{\mathrm{loc}}(\mathfrak{M}; T\mathfrak{M}) \times (\mathbb{N}^{\nu_j} \setminus \{0\})$, for $j = 1, 2$ and $\mathcal{S} \subseteq C^\infty_{\mathrm{loc}}(\mathfrak{M}; T\mathfrak{M}) \times (\mathbb{N}^\nu \setminus \{0\})$.

Definition 3.12.1. Given $\kappa \in \mathbb{Z}^\nu$ and $k, L \in \mathbb{Z}$, we say a partial differential operator on Ω with smooth coefficients, $\mathscr{P}$, is an $\mathcal{S}$ partial differential operator on Ω of degree $\leq \kappa$, length $\leq k$, and order $\leq L$ if:

- If $\kappa \notin \mathbb{N}^\nu$, or $k \leq -1$, or $L \leq -1$, then $\mathscr{P} = 0$.
- If $\kappa \in \mathbb{N}^\nu$ and $k, L \in \mathbb{N}$, then $\mathscr{P}$ can be written as a finite sum of terms of the form

$$\mathrm{Mult}[g] Z_1 Z_2 \cdots Z_K,$$

where $g \in C^\infty_{\mathrm{loc}}(\Omega)$, $K \leq L$, and $(Z_1, \vec{d}_1), \ldots, (Z_K, \vec{d}_k) \in \mathcal{S}$ with $\vec{d}_1 + \vec{d}_2 + \cdots + \vec{d}_K \leq \kappa$ and $|\vec{d}_1 + \vec{d}_2 + \cdots + \vec{d}_K|_1 \leq k$.

We say $\mathscr{P}$ is an $\mathcal{S}$ partial differential operator on Ω of degree $\leq \kappa$ and length $\leq k$ if there is an $L \in \mathbb{Z}$ such that $\mathscr{P}$ is an $\mathcal{S}$ partial differential operator on Ω of degree $\leq \kappa$, length $\leq k$, and order $\leq L$. We say $\mathscr{P}$ is an $\mathcal{S}$ partial differential operator on Ω of degree $\leq \kappa$ if there is a $k \in \mathbb{Z}$ such that $\mathscr{P}$ is an $\mathcal{S}$ partial differential operator of degree $\leq \kappa$ and length $\leq k$.

Remark 3.12.2. Note that if $\mathscr{P}$ is an $\mathcal{S}$ partial differential operator on Ω of degree $\leq \kappa \in \mathbb{N}^\nu$, then it is a partial differential operator on Ω of degree $\leq \kappa$ and length $\leq |\kappa|_1$.

Remark 3.12.3. We will sometimes speak of $\mathcal{S}$ partial differential operators on Ω of degree $\leq \kappa$, where $\kappa \in \mathbb{R}^\nu$. The definition is the same, and is equivalent to replacing $\kappa = (\kappa_1, \ldots, \kappa_\nu)$ with $(\lfloor \kappa_1 \rfloor, \ldots, \lfloor \kappa_\nu \rfloor) \in \mathbb{Z}^\nu$.

Lemma 3.12.4. *Suppose, for $j = 1, 2$, $\mathscr{P}_j$ is an $\mathcal{S}$ partial differential operator on Ω of degree $\leq \kappa_j$, length $\leq k_j$, and order $\leq L_j$. Then $\mathscr{P}_1 \mathscr{P}_2$ is an $\mathcal{S}$ partial differential operator on Ω of degree $\leq \kappa_1 + \kappa_2$, length $\leq k_1 + k_2$, and order $\leq L_1 + L_2$.*

Proof. If $\kappa_1 \in \mathbb{Z}^\nu \setminus \mathbb{N}^\nu$, $\kappa_2 \in \mathbb{Z}^\nu \setminus \mathbb{N}^\nu$, $k_1 \leq -1$, $k_2 \leq -1$, $L_1 \leq -1$, or $L_2 \leq -1$, then $\mathscr{P}_1 \mathscr{P}_2 = 0$ and the result is trivial. Thus, we may assume $\kappa_1, \kappa_2 \in \mathbb{N}^\nu$ and $k_1, k_2, L_1, L_2 \in \mathbb{N}$.

By the definition, $\mathscr{P}_1$ can be written as a finite sum of terms of the form

$$\mathrm{Mult}[g] Z_1 Z_2 \cdots Z_{K_1}, \tag{3.78}$$

where $g \in C^\infty_{\mathrm{loc}}(\Omega)$, $K_1 \leq L_1$, and $(Z_1, \vec{d}_1), \ldots, (Z_{K_1}, \vec{d}_{K_1}) \in \mathcal{S}$ with $\vec{d}_1 + \cdots + \vec{d}_{K_1} \leq \kappa_1$ and $|\vec{d}_1 + \cdots + \vec{d}_{K_1}|_1 \leq k_1$. And $\mathscr{P}_2$ can be written as a finite sum of terms of the form

$$\mathrm{Mult}[f] Y_1 Y_2 \cdots Y_{K_2}, \tag{3.79}$$

where $f \in C^\infty_{\mathrm{loc}}(\Omega)$, $K_2 \leq L_2$, and $(Y_1, \vec{d}_1), \ldots, (Y_{K_2}, \vec{d}_{K_2}) \in \mathcal{S}$ with $\vec{d}_1 + \cdots + \vec{d}_{K_2} \leq \kappa_2$ and $|\vec{d}_1 + \cdots + \vec{d}_{K_2}|_1 \leq k_2$.

It suffices to prove the case where $\mathscr{P}_1$ is given by a single term of the form (3.78) and $\mathscr{P}_2$ is given by a single term of the form (3.79). In this case, $\mathscr{P}_1 \mathscr{P}_2$ can be written as a finite sum of terms of the form

$$\mathrm{Mult}[h] Z_{l_1} Z_{l_2} \cdots Z_{l_R} Y_1 Y_2 \cdots Y_{K_2},$$

where $1 \leq l_1 < l_2 < \cdots < l_R \leq K_1$ and $h \in C^\infty_{\mathrm{loc}}(\Omega)$. This is clearly an $\mathcal{S}$ partial differential operator of degree $\leq \kappa_1 + \kappa_2$, length $\leq k_1 + k_2$, and order $\leq L_1 + L_2$, completing the proof. □

Lemma 3.12.5. *Suppose $\mathscr{P}$ is an $\mathcal{S}$ partial differential operator on Ω of degree $\leq \kappa$, length $\leq k$, and order $\leq L$. Then $\mathscr{P}^*$ is an $\mathcal{S}$ partial differential operator on Ω of degree $\leq \kappa$, length $\leq k$, and order $\leq L$, where $\mathscr{P}^*$ denotes the formal L^2 adjoint of $\mathscr{P}$, thought of as an operator on $C^\infty_0(\Omega)$.*

Proof. If $\kappa \in \mathbb{Z}^\nu \setminus \mathbb{N}^\nu$, $k \leq -1$, or $L \leq -1$, then $\mathscr{P} = 0$ and the result is trivial. Thus, we may assume $\kappa \in \mathbb{N}^\nu$ and $k, L \in \mathbb{N}$.

By definition, $\mathscr{P}$ can be written as a finite sum of terms of the form

$$\mathrm{Mult}[g] Z_1 Z_2 \cdots Z_K, \tag{3.80}$$

where $g \in C^\infty_{\mathrm{loc}}(\Omega)$, $K \leq L$, and $(Z_1, \vec{d}_1), \ldots, (Z_K, \vec{d}_K) \in \mathcal{S}$ with $\vec{d}_1 + \cdots + \vec{d}_{K_1} \leq \kappa$ and $|\vec{d}_1 + \cdots + \vec{d}_{K_1}|_1 \leq k$.

It suffices to consider the case where $\mathscr{P}$ is given by a single term of the form (3.80). The formal adjoint of (3.80) is given by

$$Z_K^* Z_{K-1}^* \cdots Z_1^* \,\mathrm{Mult}[\overline{g}]. \tag{3.81}$$

Since $Z_j^* = -Z_j + \mathrm{Mult}[f_j]$, for some $f_j \in C^\infty_{\mathrm{loc}}(\mathfrak{M})$, we see that each Z_j^* is an $\mathcal{S}$ partial differential operator on Ω of degree $\leq \vec{d}_j$, length $\leq |\vec{d}_j|_1$, and order ≤ 1. Combining this with the fact that $\mathrm{Mult}[\overline{g}]$ is an $\mathcal{S}$ partial differential operator on Ω of degree $\leq 0 \in \mathbb{N}^\nu$, length $\leq 0 \in \mathbb{N}$, and order $\leq 0 \in \mathbb{N}$, Lemma 3.12.4 implies that the operator in (3.81) is an $\mathcal{S}$ partial differential operator on Ω of degree $\leq \vec{d}_1 + \cdots + \vec{d}_K \leq \kappa$, length $\leq |\vec{d}_1|_1 + \cdots + |\vec{d}_K|_1 \leq k$, and order $\leq K \leq L$, completing the proof. □

The main result of this section is the next theorem.

Theorem 3.12.6. *Suppose $\mathcal{S}_1$ and $\mathcal{S}_2$ weakly approximately commute on Ω. Let $\kappa_1 \in \mathbb{Z}^{\nu_1}$, $\kappa_2 \in \mathbb{Z}^{\nu_2}$, and $k_1, k_2 \in \mathbb{Z}$. For $j = 1, 2$, let $\mathscr{P}_j^{(\kappa_j, k_j)}$ be an $\mathcal{S}_j$ partial differential operator on Ω of degree $\leq \kappa_j$ and length $\leq k_j$. Then*

$$\mathscr{P}_1^{(\kappa_1,k_1)} \mathscr{P}_2^{(\kappa_2,k_2)} = \mathscr{P}_2^{(\kappa_2,k_2)} \mathscr{P}_1^{(\kappa_1,k_1)} + \mathrm{Error},$$

where Error *is a finite sum of operators of the form $\widetilde{\mathscr{P}}_2^{(\kappa_2,k_2-1)} \widetilde{\mathscr{P}}_1^{(\kappa_1,k_1)}$ and $\widetilde{\mathscr{P}}_2^{(\kappa_2,k_2)} \widetilde{\mathscr{P}}_1^{(\kappa_1,k_1-1)}$, where each $\widetilde{\mathscr{P}}_j^{(\kappa_j,l)}$ is an $\mathcal{S}_j$ partial differential operator on Ω of degree $\leq \kappa_j$ and length $\leq l$.*

Before we prove Theorem 3.12.6, we introduce three corollaries.

Corollary 3.12.7. *Suppose $\mathcal{S}_1$ and $\mathcal{S}_2$ weakly approximately commute on Ω. Let $\kappa_1 \in \mathbb{Z}^{\nu_1}$ and $\kappa_2 \in \mathbb{Z}^{\nu_2}$. For $j = 1, 2$, let $\mathscr{P}_j^{(\kappa_j)}$ be an $\mathcal{S}_j$ partial differential operator on Ω of degree $\leq \kappa_j$. Then*

$$\mathscr{P}_1^{(\kappa_1)} \mathscr{P}_2^{(\kappa_2)} = \mathscr{P}_2^{(\kappa_2)} \mathscr{P}_1^{(\kappa_1)} + \text{Error},$$

where Error *is a finite sum of operators of the form $\widetilde{\mathscr{P}}_2^{(\kappa_2,|\kappa_2|_1-1)} \widetilde{\mathscr{P}}_1^{(\kappa_1,|\kappa_1|_1)}$ and $\widetilde{\mathscr{P}}_2^{(\kappa_2,|\kappa_2|_1)} \widetilde{\mathscr{P}}_1^{(\kappa_1,|\kappa_1|_1-1)}$, where each $\widetilde{\mathscr{P}}_j^{(\kappa_j,l)}$ is an $\mathcal{S}_j$ partial differential operator on Ω of degree $\leq \kappa_j$ and length $\leq l$.*

Proof. Since $\mathscr{P}_j$ is an $\mathcal{S}_j$ partial differential operator on Ω of degree $\leq \kappa_j$ if and only if it is an $\mathcal{S}_j$ partial differential operator on Ω of degree $\leq \kappa_j$ and length $\leq |\kappa_j|_1$ (see Remark 3.12.2) this follows immediately from Theorem 3.12.6. □

Corollary 3.12.8. *Suppose $\mathcal{S}_1$ and $\mathcal{S}_2$ weakly approximately commute on Ω. Let $\kappa_1 \in \mathbb{Z}^{\nu_1}$ and $\kappa_2 \in \mathbb{Z}^{\nu_2}$. For $j = 1, 2$, let $\mathscr{P}_j^{(\kappa_j)}$ be an $\mathcal{S}_j$ partial differential operator on Ω of degree $\leq \kappa_j$. Then $\mathscr{P}_1^{(\kappa_1)} \mathscr{P}_2^{(\kappa_2)}$ can be written as a finite sum of operators of the form $\widetilde{\mathscr{P}}_2^{(\kappa_2)} \widetilde{\mathscr{P}}_1^{(\kappa_1)}$, where each $\widetilde{\mathscr{P}}_j^{(\kappa_j)}$ is an $\mathcal{S}_j$ partial differential operator on Ω of degree $\leq \kappa_j$.*

Proof. This follows immediately from Corollary 3.12.7. □

Corollary 3.12.9. *Suppose $\nu_1 = \nu_2 = 1$ and suppose $\mathcal{S}_1$ and $\mathcal{S}_2$ weakly approximately commute on Ω. Let $\kappa_1, \kappa_2 \in \mathbb{Z}$ and for $j = 1, 2$, let $\mathscr{P}_j^{(\kappa_j)}$ be an $\mathcal{S}_j$ partial differential operator on Ω of degree $\leq \kappa_j$. Then*

$$\mathscr{P}_1^{(\kappa_1)} \mathscr{P}_2^{(\kappa_2)} = \mathscr{P}_2^{(\kappa_2)} \mathscr{P}_1^{(\kappa_1)} + \text{Error},$$

where Error *is a finite sum of operators of the form $\widetilde{\mathscr{P}}^{(\kappa_2-1)} \widetilde{\mathscr{P}}^{(\kappa_1)}$ and $\widetilde{\mathscr{P}}^{(\kappa_2)} \widetilde{\mathscr{P}}^{(\kappa_1-1)}$, where each $\widetilde{\mathscr{P}}_j^{(\gamma)}$ is an $\mathcal{S}_j$ partial differential operator on Ω of degree $\leq \gamma$.*

Proof. When $\nu = 1$, $\mathscr{P}$ is an $\mathcal{S}$ partial differential operator on Ω of degree $\leq \kappa$ and length $\leq k$ if and only if $\mathscr{P}$ is an $\mathcal{S}$ partial differential operator on Ω of degree $\leq \min\{\kappa, k\}$ (see Definition 3.12.1). Using this, the result follows immediately from Corollary 3.12.7. □

We now turn to the proof of Theorem 3.12.6. To prove it, we introduce some new notation. For $j = 1, 2$, we write $\mathscr{P}_j^{(\kappa_j,k_j,L_j)}$ to denote a particular (fixed) $\mathcal{S}_j$ partial differential operator on Ω of degree $\leq \kappa_j$, length $\leq k_j$ and order $\leq L_j$. We write $\widetilde{\mathscr{P}}_j^{(\kappa_j,k_j,L_j)}$ to denote an $\mathcal{S}_j$ partial differential operator on Ω of degree $\leq \kappa_j$, length $\leq k_j$, and order $\leq L_j$ which may change from line to line. If we have a term which involves $\widetilde{\mathscr{P}}_1^{(\kappa_1,k_1,L_1)}$ or $\widetilde{\mathscr{P}}_2^{(\kappa_2,k_2,L_2)}$ and we write $\sum$ in front of it, we mean a finite sum of terms of that form, where the operators $\widetilde{\mathscr{P}}_1^{(\kappa_1,k_1,L_1)}$ and $\widetilde{\mathscr{P}}_2^{(\kappa_2,k_2,L_2)}$ vary in the sum: only the terms with a ~ over them vary in the sum. For example,

$$\sum \mathscr{P}_1^{(\kappa_1,k_1,L_1)} \widetilde{\mathscr{P}}_2^{(\kappa_2,k_2,L_2)} \tag{3.82}$$

means there is a finite collection $\mathscr{P}_{2,1}^{(\kappa_2,k_2,L_2)},\ldots,\mathscr{P}_{2,Q}^{(\kappa_2,k_2,L_2)}$ of $\mathcal{S}_2$ partial differential operators on Ω of degree $\le \kappa_2$, length $\le k_2$, and order $\le L_2$ such that the expression in (3.82) equals

$$\sum_{j=1}^{Q} \mathscr{P}_1^{(\kappa_1,k_1,L_1)}\mathscr{P}_{2,j}^{(\kappa_2,k_2,L_2)}.$$

Using this new notation, we begin by proving Theorem 3.12.6 in the special case where $\mathcal{S}_1$ and $\mathcal{S}_2$ strongly approximately commute, as stated in the next proposition.

Proposition 3.12.10. *Suppose $\mathcal{S}_1$ and $\mathcal{S}_2$ strongly approximately commute on Ω. Then*

$$\begin{aligned}\mathscr{P}_1^{(\kappa_1,k_1,L_1)}\mathscr{P}_2^{(\kappa_2,k_2,L_2)} &= \mathscr{P}_2^{(\kappa_2,k_2,L_2)}\mathscr{P}_1^{(\kappa_1,k_1,L_1)}\\ &\quad+\sum \widetilde{\mathscr{P}}_2^{(\kappa_2,k_2-1,L_2-1)}\widetilde{\mathscr{P}}_1^{(\kappa_1,k_1,L_1)}\\ &\quad+\sum \widetilde{\mathscr{P}}_2^{(\kappa_2,k_2,L_2)}\widetilde{\mathscr{P}}_1^{(\kappa_1,k_1-1,L_1-1)}.\end{aligned}\tag{3.83}$$

Proof. If $\kappa_1 \in \mathbb{Z}^{\nu_1}\setminus\mathbb{N}^{\nu_1}$, $k_1 \le -1$, or $L_1 \le -1$, then $\mathscr{P}_1^{(\kappa_1,k_1,L_1)} = 0$ and the result is trivial. Similarly if $\kappa_2 \in \mathbb{Z}^{\nu_2}\setminus\mathbb{N}^{\nu_2}$, $k_2 \le -1$, or $L_2 \le -1$, then $\mathscr{P}_1^{(\kappa_2,k_2,L_2)} = 0$ and the result is trivial. We henceforth assume $\kappa_1 \in \mathbb{N}^{\nu_1}$, $\kappa_2 \in \mathbb{N}^{\nu_2}$, and $k_1, k_2, L_1, L_2 \in \mathbb{N}$.

We first consider the case $L_2 = 0$. In this case, $\mathscr{P}_2^{(\kappa_2,k_2,0)} = \mathrm{Mult}[g_2]$ for some $g_2 \in C^\infty_{\mathrm{loc}}(\Omega)$. By definition, $\mathscr{P}_1^{(\kappa_1,k_1,L_1)}$ can be written as a finite sum of terms of the form

$$\mathrm{Mult}[g_1]Z_1Z_2\cdots Z_K,$$

where $g_1 \in C^\infty_{\mathrm{loc}}(\Omega)$, $K \le L_1$, $(Z_1,\vec{d}_1),\ldots,(Z_K,\vec{d}_K) \in \mathcal{S}_1$, with $\vec{d}_1+\cdots+\vec{d}_K \le \kappa_1$ and $|\vec{d}_1+\cdots+\vec{d}_K|_1 \le k_1$. Note that

$$\begin{aligned}\mathrm{Mult}[g_1]Z_1Z_2\cdots Z_K\mathscr{P}_2^{(\kappa_2,k_2,0)} &= \mathrm{Mult}[g_1]Z_1Z_2\cdots Z_K\,\mathrm{Mult}[g_2]\\ &= \mathrm{Mult}[g_2]\,\mathrm{Mult}[g_1]Z_1Z_2\cdots Z_K + \mathrm{Error}\\ &= \mathscr{P}_2^{(\kappa_2,k_2,0)}\,\mathrm{Mult}[g_1]Z_1Z_2\cdots Z_K + \mathrm{Error},\end{aligned}$$

where Error is a finite sum of terms of the form

$$\mathrm{Mult}[g_3]Z_{l_1}Z_{l_2}\cdots Z_{l_R},\quad g\in C^\infty_{\mathrm{loc}}(\Omega), R<K\le L_1, 1\le l_1<l_2<\cdots<l_R\le K.$$

Since $|\vec{d}_{l_1}+\cdots+\vec{d}_{l_R}|_1 < |\vec{d}_1+\cdots+\vec{d}_K|_1 \le k_1$, we see that Error is of the form $\widetilde{\mathscr{P}}_1^{(\kappa_1,k_1-1,L_1-1)}$. We conclude that

$$\begin{aligned}\mathscr{P}_1^{(\kappa_1,k_1,L_1)}\mathscr{P}_2^{(\kappa_2,k_2,L_2)} &= \mathscr{P}_2^{(\kappa_2,k_2,L_2)}\mathscr{P}_1^{(\kappa_1,k_1,L_1)} + \sum \mathrm{Mult}[1]\widetilde{\mathscr{P}}_1^{(\kappa_1,k_1-1,L_1-1)}\\ &= \mathscr{P}_2^{(\kappa_2,k_2,L_2)}\mathscr{P}_1^{(\kappa_1,k_1,L_1)} + \sum \widetilde{\mathscr{P}}_2^{(\kappa_2,k_2,0)}\widetilde{\mathscr{P}}_1^{(\kappa_1,k_1-1,L_1-1)},\end{aligned}$$

completing the proof when $L_2 = 0$. A similar proof gives the result when $L_1 = 0$, and we leave the details to the interested reader.

If $k_1 = 0$ and $L_1 > 0$, then $\mathscr{P}_1^{(\kappa_1,0,L_1)}$ is of the form $\mathscr{P}_1^{(\kappa_1,0,0)}$ (see Definition 3.12.1) and therefore the case $k_1 = 0$ follows from the already established case $L_1 = 0$. Similarly, the case $k_2 = 0$ follows from the already established case $L_2 = 0$. Thus, we may henceforth assume $L_1, L_2, k_1, k_2 \in \mathbb{N}_+$.

Next, we consider the case $L_1 = L_2 = 1$ and $k_1, k_2 \in \mathbb{N}_+$. In this case,

$$\mathscr{P}_1^{(\kappa_1,k_1,1)} = \sum_{l=1}^{R_1} \mathrm{Mult}[g_l]Y_l + \mathrm{Mult}[g_0], \quad (Y_l, \vec{d}_l) \in \mathcal{S}_1, \vec{d}_l \leq \kappa_1, g_l \in C^\infty_{\mathrm{loc}}(\Omega), \tag{3.84}$$

$$\mathscr{P}_2^{(\kappa_2,k_2,1)} = \sum_{l=1}^{R_2} \mathrm{Mult}[f_l]Z_l + \mathrm{Mult}[f_0], \quad (Z_l, \vec{d}_l) \in \mathcal{S}_2, \vec{d}_l \leq \kappa_2, f_l \in C^\infty_{\mathrm{loc}}(\Omega). \tag{3.85}$$

It suffices to consider the cases where $\mathscr{P}_1^{(\kappa_1,k_1,1)}$ is given by one of the terms on the right-hand side of (3.84) and $\mathscr{P}_2^{(\kappa_2,k_2,1)}$ is given by one of the terms on the right-hand side of (3.85). If $\mathscr{P}_1^{(\kappa_1,k_1,1)} = \mathrm{Mult}[g_0]$, then it is of the form $\mathscr{P}_1^{(\kappa_1,k_1,0)}$ and the result follows from the already proved case $L_1 = 0$. Similarly, if $\mathscr{P}_2^{(\kappa_2,k_2,1)} = \mathrm{Mult}[f_0]$, then it is of the form $\mathscr{P}_2^{(\kappa_2,k_2,0)}$, and the result follows from the already proved case $L_2 = 0$. Thus, it suffices to assume that $\mathscr{P}_1^{(\kappa_1,k_1,1)}$ and $\mathscr{P}_2^{(\kappa_2,k_2,1)}$ have the following forms:

$$\mathscr{P}_1^{(\kappa_1,k_1,1)} = \mathrm{Mult}[g_1]Y_1, \quad (Y_1, \vec{d}_1) \in \mathcal{S}_1, \vec{d}_1 \leq \kappa_1, g_1 \in C^\infty_{\mathrm{loc}}(\Omega), \tag{3.86}$$

$$\mathscr{P}_2^{(\kappa_2,k_2,1)} = \mathrm{Mult}[f_1]Z_1, \quad (Z_1, \vec{d}_1) \in \mathcal{S}_2, \vec{d}_1 \leq \kappa_2, f_1 \in C^\infty_{\mathrm{loc}}(\Omega). \tag{3.87}$$

Using the assumption that $\mathcal{S}_1$ and $\mathcal{S}_2$ strongly approximately commute on Ω, there are finite sets

$$\mathcal{F}_1 \subseteq \{(Y_0, \vec{d}_0) \in \mathcal{S}_1 : \vec{d}_0 \leq \vec{d}_1\}, \quad \mathcal{F}_2 \subseteq \{(Z_0, \vec{d}_0) \in \mathcal{S}_2 : \vec{d}_0 \leq \vec{d}_1\}$$

such that

$$[Y_1, Z_1] = \sum_{(Y_0,\vec{d}_0)\in\mathcal{F}_1} a_{(Y_0,\vec{d}_0)} Y_0 + \sum_{(Z_0,\vec{d}_0)\in\mathcal{F}_2} b_{(Z_0,\vec{d}_0)} Z_0, \quad a_{(Y_0,\vec{d}_0)}, b_{(Z_0,\vec{d}_0)} \in C^\infty_{\mathrm{loc}}(\Omega).$$

Note that, since $\vec{d}_1 \leq \kappa_1$ and $|\vec{d}_1|_1 \leq k_1$, each $(Y_0, \vec{d}_0) \in \mathcal{F}_1$ is of the form $\overline{\mathscr{P}}_1^{(\kappa_1,k_1,1)}$. Similarly, each $(Z_0, \vec{d}_0) \in \mathcal{F}_2$ is of the form $\overline{\mathscr{P}}_2^{(\kappa_2,k_2,1)}$. We have, assuming $\mathscr{P}_1^{(\kappa_1,k_1,1)}$ and $\mathscr{P}_2^{(\kappa_2,k_2,1)}$ are given by (3.86) and (3.87),

$$\begin{aligned}
\mathscr{P}_1^{(\kappa_1,k_1,1)} \mathscr{P}_2^{(\kappa_2,k_2,1)} &= \mathrm{Mult}[g_1]Y_1 \,\mathrm{Mult}[f_1]Z_1 \\
&= \mathrm{Mult}[f_1]Z_1 \,\mathrm{Mult}[g_1]Y_1 + \mathrm{Mult}[g_1(Y_1 f_1)]Z_1 \\
&\quad - \mathrm{Mult}[f_1(Z_1 g_1)]Y_1 + \mathrm{Mult}[g_1 f_1][Y_1, Z_1]
\end{aligned}$$

$$
\begin{aligned}
&= \mathscr{P}_2^{(\kappa_2,k_2,1)} \mathscr{P}_1^{(\kappa_1,k_1,1)} + \mathrm{Mult}[g_1(Y_1 f)]Z_1 - \mathrm{Mult}[f_1(Z_1 g_1)]Y_1 \\
&\quad + \sum_{(Y_0,\vec{d}_0)\in\mathcal{F}_1} \mathrm{Mult}[f_1 g_1 a_{(Y_0,\vec{d}_0)}]Y_0 + \sum_{(Z_0,\vec{d}_0)\in\mathcal{F}_2} \mathrm{Mult}[f_1 g_1 b_{(Z_0,\vec{d}_0)}]Z_0 \\
&= \mathscr{P}_2^{(\kappa_2,k_2,1)} \mathscr{P}_1^{(\kappa_1,k_1,1)} + \sum \widetilde{\mathscr{P}}_2^{(\kappa_2,k_2,1)} \mathrm{Mult}[1] + \sum \mathrm{Mult}[1] \widetilde{\mathscr{P}}_1^{(\kappa_1,k_1,1)} \\
&= \mathscr{P}_2^{(\kappa_2,k_2,1)} \mathscr{P}_1^{(\kappa_1,k_1,1)} + \sum \widetilde{\mathscr{P}}_2^{(\kappa_2,k_2,1)} \widetilde{\mathscr{P}}_1^{(\kappa_1,k_1-1,0)} + \sum \widetilde{\mathscr{P}}_2^{(\kappa_2,k_2-1,0)} \widetilde{\mathscr{P}}_1^{(\kappa_1,k_1,1)},
\end{aligned}
$$

where in the last equality, we have used the fact that Mult[1] is of the form $\widetilde{\mathscr{P}}_1^{(\kappa_1,k_1-1,0)}$ and of the form $\widetilde{\mathscr{P}}_2^{(\kappa_2,k_2-1,0)}$ (which uses the fact that $\kappa_1 \in \mathbb{N}^{\nu_1}$, $\kappa_1 \in \mathbb{N}^{\nu_2}$, and $k_1, k_2 \geq 1$). This completes the proof of the case $(L_1, L_2) = (1,1)$ and $k_1, k_2 \in \mathbb{N}_+$.

Combining all of the above arguments, we have proved the result for all (L_1, L_2), with $L_1 \leq 1$, $L_2 \leq 1$, and $k_1, k_2 \in \mathbb{Z}$. We proceed by induction on (L_1, L_2), the base case being $(L_1, L_2) = (1,1)$, which we have already shown. Thus, fix (L_1, L_2) with $L_1 \geq 1$, $L_2 \geq 1$, and $L_1 + L_2 > 2$; we wish to prove the result for this L_1 and L_2, and as described above we may assume $\kappa_1 \in \mathbb{N}^{\nu_1}$, $\kappa_2 \in \mathbb{N}^{\nu_2}$, and $k_1, k_2 \in \mathbb{N}_+$. We suppose we know the result for all (L_1', L_2') with $L_1' \leq L_1$, $L_2' \leq L_2$, and $L_1' + L_2' < L_1 + L_2$ (and all $\kappa_1 \in \mathbb{Z}^{\nu_1}$, $\kappa_2 \in \mathbb{Z}^{\nu_2}$, $k_1, k_2 \in \mathbb{Z}$). By the inductive hypothesis, for such L_1' and L_1' (and all $\kappa_1' \in \mathbb{Z}^{\nu_1}$, $\kappa_2' \in \mathbb{Z}^{\nu_2}$, and $k_1', k_2' \in \mathbb{Z}$), we have

$$
\widetilde{\mathscr{P}}_1^{(\kappa_1',k_1',L_1')} \widetilde{\mathscr{P}}_2^{(\kappa_2',k_2',L_2')} = \sum \widetilde{\mathscr{P}}_2^{(\kappa_2',k_2',L_2')} \widetilde{\mathscr{P}}_1^{(\kappa_1',k_1',L_1')}, \tag{3.88}
$$

where we recall that the terms with a ~ over them may vary from term to term.

By the definitions (see Definition 3.12.1), $\mathscr{P}_1^{(\kappa_1,k_1,L_1)}$ is a finite sum of operators of the form

$$
\mathrm{Mult}[g] Y_1 Y_2 \cdots Y_K, \tag{3.89}
$$

where $K \leq L_1$, $(Y_1, \vec{d}_1), \ldots, (Y_K, \vec{d}_K) \in \mathcal{S}_1$, $\vec{d}_1 + \cdots + \vec{d}_K \leq \kappa_1$, and $|\vec{d}_1 + \cdots + \vec{d}_K|_1 \leq k_1$. When $K \geq 1$, we may write (3.89) as $(\mathrm{Mult}[g] Y_1 Y_2 \cdots Y_{K-1}) Y_K$. When $K = 0$, we may write (3.89) as $\mathrm{Mult}[g]\,\mathrm{Mult}[1]$. This shows that in either case, we may write $\mathscr{P}_1^{(\kappa_1,k_1,L_1)}$ as a finite sum of terms of the form

$$
\mathscr{P}_1^{(\kappa_{1,1},k_{1,1},L_1-1)} \mathscr{P}_1^{(\kappa_{1,2},k_{1,2},1)}, \quad \kappa_{1,1} + \kappa_{1,2} = \kappa_1,\ k_{1,1} + k_{1,2} = k_1. \tag{3.90}
$$

Similarly, $\mathscr{P}_2^{(\kappa_2,k_2,L_2)}$ can be written as a finite sum of terms of the form

$$
\mathscr{P}_2^{(\kappa_{2,1},k_{2,1},L_2-1)} \mathscr{P}_2^{(\kappa_{2,2},k_{2,2},1)}, \quad \kappa_{2,1} + \kappa_{2,2} = \kappa_1,\ k_{2,1} + k_{2,2} = k_1. \tag{3.91}
$$

Thus, it suffices to prove the result in the special case where $\mathscr{P}_1^{(\kappa_1,k_1,L_1)}$ is a single term of the form (3.90) and $\mathscr{P}_2^{(\kappa_2,k_2,L_2)}$ is a single term of the form (3.91).

Using the inductive hypothesis, we have

$$\begin{aligned}
\mathscr{P}_1^{(\kappa_1,k_1,L_1)}\mathscr{P}_2^{(\kappa_2,k_2,L_2)} &= \mathscr{P}_1^{(\kappa_{1,1},k_{1,1},L_1-1)}\mathscr{P}_1^{(\kappa_{1,2},k_{1,2},1)}\mathscr{P}_2^{(\kappa_{2,1},k_{2,1},L_2-1)}\mathscr{P}_2^{(\kappa_{2,2},k_{2,2},1)}\\
&= \mathscr{P}_1^{(\kappa_{1,1},k_{1,1},L_1-1)}\mathscr{P}_2^{(\kappa_{2,1},k_{2,1},L_2-1)}\mathscr{P}_1^{(\kappa_{1,2},k_{1,2},1)}\mathscr{P}_2^{(\kappa_{2,2},k_{2,2},1)}\\
&\quad+\sum \mathscr{P}_1^{(\kappa_{1,1},k_{1,1},L_1-1)}\widetilde{\mathscr{P}}_2^{(\kappa_{2,1},k_{2,1}-1,L_2-2)}\widetilde{\mathscr{P}}_1^{(\kappa_{1,2},k_{1,2},1)}\mathscr{P}_2^{(\kappa_{2,2},k_{2,2},1)}\\
&\quad+\sum \mathscr{P}_1^{(\kappa_{1,1},k_{1,1},L_1-1)}\widetilde{\mathscr{P}}_2^{(\kappa_{2,1},k_{2,1},L_2-1)}\widetilde{\mathscr{P}}_1^{(\kappa_{1,2},k_{1,2}-1,0)}\mathscr{P}_2^{(\kappa_{2,2},k_{2,2},1)}\\
&=: \text{(I)} + \text{(II)} + \text{(III)}.
\end{aligned}$$

Using repeated application of (3.88) and Lemma 3.12.4, we have

$$\begin{aligned}
\text{(II)} &= \sum \widetilde{\mathscr{P}}_2^{(\kappa_{2,1},k_{2,1}-1,L_2-2)}\widetilde{\mathscr{P}}_2^{(\kappa_{2,2},k_{2,2},1)}\widetilde{\mathscr{P}}_1^{(\kappa_{1,1},k_{1,1},L_1-1)}\widetilde{\mathscr{P}}_1^{(\kappa_{1,2},k_{1,2},1)}\\
&= \sum \widetilde{\mathscr{P}}_2^{(\kappa_2,k_2-1,L_2-1)}\widetilde{\mathscr{P}}_1^{(\kappa_1,k_1,L_1)},
\end{aligned}$$

which is of the desired form for our error term in (3.83). Similarly, we have

$$\begin{aligned}
\text{(III)} &= \sum \widetilde{\mathscr{P}}_2^{(\kappa_{2,1},k_{2,1},L_2-1)}\widetilde{\mathscr{P}}_2^{(\kappa_{2,2},k_{2,2},1)}\widetilde{\mathscr{P}}_1^{(\kappa_{1,1},k_{1,1},L_1-1)}\widetilde{\mathscr{P}}_1^{(\kappa_{1,2},k_{1,2}-1,0)}\\
&= \sum \widetilde{\mathscr{P}}_2^{(\kappa_2,k_2,L_2)}\widetilde{\mathscr{P}}_1^{(\kappa_1,k_1-1,L_1-1)},
\end{aligned}$$

which is also of the desired form for the error term in (3.83).

Turning to (I), we have by the base case $(L_1', L_2') = (1,1)$

$$\begin{aligned}
\text{(I)} &= \mathscr{P}_1^{(\kappa_{1,1},k_{1,1},L_1-1)}\mathscr{P}_2^{(\kappa_{2,1},k_{2,1},L_2-1)}\mathscr{P}_2^{(\kappa_{2,2},k_{2,2},1)}\mathscr{P}_1^{(\kappa_{1,2},k_{1,2},1)}\\
&\quad+\sum \mathscr{P}_1^{(\kappa_{1,1},k_{1,1},L_1-1)}\mathscr{P}_2^{(\kappa_{2,1},k_{2,1},L_2-1)}\widetilde{\mathscr{P}}_2^{(\kappa_{2,2},k_{2,2}-1,0)}\widetilde{\mathscr{P}}_1^{(\kappa_{1,2},k_{1,2},1)}\\
&\quad+\sum \mathscr{P}_1^{(\kappa_{1,1},k_{1,1},L_1-1)}\mathscr{P}_2^{(\kappa_{2,1},k_{2,1},L_2-1)}\widetilde{\mathscr{P}}_2^{(\kappa_{2,2},k_{2,2},1)}\widetilde{\mathscr{P}}_1^{(\kappa_{1,2},k_{1,2}-1,0)}\\
&=: \text{(IV)} + \text{(V)} + \text{(VI)}.
\end{aligned}$$

Using repeated application of (3.88) and Lemma 3.12.4, we have

$$\begin{aligned}
\text{(V)} &= \sum \widetilde{\mathscr{P}}_2^{(\kappa_{2,1},k_{2,1},L_2-1)}\widetilde{\mathscr{P}}_2^{(\kappa_{2,2},k_{2,2}-1,0)}\widetilde{\mathscr{P}}_1^{(\kappa_{1,1},k_{1,1},L_1-1)}\widetilde{\mathscr{P}}_1^{(\kappa_{1,2},k_{1,2},1)}\\
&= \sum \widetilde{\mathscr{P}}_2^{(\kappa_2,k_2-1,L_2-1)}\widetilde{\mathscr{P}}_1^{(\kappa_1,k_1,L_1)},
\end{aligned}$$

which is of the desired form for our error term in (3.83). Similarly, we have

$$\begin{aligned}
\text{(VI)} &= \sum \widetilde{\mathscr{P}}_2^{(\kappa_{2,1},k_{2,1},L_2-1)}\widetilde{\mathscr{P}}_2^{(\kappa_{2,2},k_{2,2},1)}\widetilde{\mathscr{P}}_1^{(\kappa_{1,1},k_{1,1},L_1-1)}\widetilde{\mathscr{P}}_1^{(\kappa_{1,2},k_{1,2}-1,0)}\\
&= \widetilde{\mathscr{P}}_2^{(\kappa_2,k_2,L_2)}\widetilde{\mathscr{P}}_2^{(\kappa_1,k_1-1,L_1-1)},
\end{aligned}$$

which is also of the desired form for the error term in (3.83).

Turning to (IV), using the inductive hypothesis, Lemma 3.12.4, and the assumption that $\mathscr{P}_1^{(\kappa_1,k_1,L_1)}$ and $\mathscr{P}_2^{(\kappa_2,k_2,L_2)}$ are given by (3.90) and (3.91), we have

$$
\begin{aligned}
(\mathrm{IV}) &= \mathscr{P}_1^{(\kappa_{1,1},k_{1,1},L_1-1)} \mathscr{P}_2^{(\kappa_2,k_2,L_2)} \mathscr{P}_1^{(\kappa_{1,2},k_{1,2},1)} \\
&= \mathscr{P}_2^{(\kappa_2,k_2,L_2)} \mathscr{P}_1^{(\kappa_{1,1},k_{1,1},L_1-1)} \mathscr{P}_1^{(\kappa_{1,2},k_{1,2},1)} \\
&\quad + \sum \widetilde{\mathscr{P}}_2^{(\kappa_2,k_2-1,L_2-1)} \widetilde{\mathscr{P}}_1^{(\kappa_{1,1},k_{1,1},L_1-1)} \mathscr{P}_1^{(\kappa_{1,2},k_{1,2},1)} \\
&\quad + \sum \widetilde{\mathscr{P}}_2^{(\kappa_2,k_2,L_2)} \widetilde{\mathscr{P}}_1^{(\kappa_{1,1},k_{1,1}-1,L_1-2)} \mathscr{P}_1^{(\kappa_{1,2},k_{1,2},1)} \\
&= \mathscr{P}_2^{(\kappa_2,k_2,L_2)} \mathscr{P}_1^{(\kappa_1,k_1,L_1)} + \sum \widetilde{\mathscr{P}}_2^{(\kappa_2,k_2,L_2)} \widetilde{\mathscr{P}}_1^{(\kappa_1,k_1-1,L_1-1)} \\
&\quad + \sum \widetilde{\mathscr{P}}_2^{(\kappa_2,k_2-1,L_2-1)} \widetilde{\mathscr{P}}_1^{(\kappa_1,k_1,L_1)},
\end{aligned}
$$

which is of the desired form (3.83) and completes the proof. □

Lemma 3.12.11. *$\mathscr{P}$ is an $\mathcal{S}$ partial differential operator on Ω of degree $\le \kappa$ and length $\le k$ if and only if $\mathscr{P}$ is a* $\mathrm{Gen}(\mathcal{S})$ *partial differential operator on Ω of degree $\le \kappa$ and length $\le k$.*

Proof. If $\kappa \in \mathbb{Z}^\nu \setminus \mathbb{N}^\nu$ or $k \in \mathbb{Z} \setminus \mathbb{N}$, then $\mathscr{P} = 0$ and the result is trivial. We henceforth assume $\kappa \in \mathbb{N}^\nu$ and $k \in \mathbb{N}$.

Since $\mathcal{S} \subseteq \mathrm{Gen}(\mathcal{S})$, if $\mathscr{P}$ is an $\mathcal{S}$ partial differential operator on Ω of degree $\le \kappa$ and length $\le k$, then it is a $\mathrm{Gen}(\mathcal{S})$ partial differential operator on Ω of degree $\le \kappa$ and length $\le k$.

Suppose $\mathscr{P}$ is a $\mathrm{Gen}(\mathcal{S})$ partial differential operator on Ω of degree $\le \kappa$ and length $\le k$. Then $\mathscr{P}$ can be written as a finite sum of terms of the form

$$\mathrm{Mult}[g] Z_1 Z_2 \cdots Z_K, \tag{3.92}$$

where $(Z_1, \vec{d}_1), \dots, (Z_K, \vec{d}_K) \in \mathrm{Gen}(\mathcal{S})$, $\vec{d}_1 + \cdots + \vec{d}_K \le \kappa$, and $|\vec{d}_1 + \cdots + \vec{d}_K|_1 \le k$. By the definition of $\mathrm{Gen}(\mathcal{S})$, each Z_j is an $\mathcal{S}$ partial differential operator on Ω of degree $\le \vec{d}_j$ and length $\le |\vec{d}_j|_1$. Lemma 3.12.4 then implies that terms of the form (3.92) are $\mathcal{S}$ partial differential operators on Ω of degree $\le \kappa$ and length $\le k$, and therefore the same is true of $\mathscr{P}$, completing the proof. □

Remark 3.12.12. Lemma 3.12.11 shows that the degree and length of an $\mathcal{S}$ partial differential operator on Ω do not depend on whether one uses $\mathcal{S}$ or $\mathrm{Gen}(\mathcal{S})$. However, the *order* does depend on whether one uses $\mathcal{S}$ or $\mathrm{Gen}(\mathcal{S})$.

Proof of Theorem 3.12.6. By Lemma 3.12.11, it suffices to prove the result with $\mathcal{S}_1$ and $\mathcal{S}_2$ replaced by $\mathrm{Gen}(\mathcal{S}_1)$ and $\mathrm{Gen}(\mathcal{S}_2)$. Since $\mathcal{S}_1$ and $\mathcal{S}_2$ weakly approximately commute on Ω, $\mathrm{Gen}(\mathcal{S}_1)$ and $\mathrm{Gen}(\mathcal{S}_2)$ strongly approximately commute on Ω. The result now follows from Proposition 3.12.10. □

Theorem 3.12.6 has a consequence which will be of use to us.

Proposition 3.12.13. *Suppose $\mathcal{S}_1$ and $\mathcal{S}_2$ weakly approximately commute on Ω. Set $\nu = \nu_1 + \nu_2$ and*

$$\mathcal{S} := \mathcal{S}_1 \boxtimes \mathcal{S}_2 \subseteq C^\infty_{\mathrm{loc}}(\mathfrak{M}; T\mathfrak{M}) \times (\mathbb{N}^\nu \setminus \{0\}).$$

Let $\kappa = (\kappa_1, \kappa_2) \in \mathbb{Z}^{\nu_1} \times \mathbb{Z}^{\nu_2} \cong \mathbb{Z}^{\nu}$. Suppose $\mathscr{P}^{(\kappa)}$ is an $\mathcal{S}$ partial differential operator on Ω of degree $\leq \kappa$. Then $\mathscr{P}^{(\kappa)}$ can be written as a finite sum of operators of the form $\widetilde{\mathscr{P}}_1^{(\kappa_1)} \widetilde{\mathscr{P}}_2^{(\kappa_2)}$, where each $\widetilde{\mathscr{P}}_j^{(\kappa_j)}$ is an $\mathcal{S}_j$ partial differential operator on Ω of degree $\leq \kappa_j$.

Proof. If $\kappa \in \mathbb{Z}^{\nu} \setminus \mathbb{N}^{\nu}$, then $\mathscr{P}^{(\kappa)} = 0$ and the result is trivial. We henceforth assume $\kappa \in \mathbb{N}^{\nu}$.

We claim that $\mathscr{P}^{(\kappa)}$ can be written as a finite sum of operators of the form

$$\mathscr{P}_1^{(\sigma_1)} \mathscr{P}_2^{(\tau_1)} \mathscr{P}_1^{(\sigma_2)} \mathscr{P}_2^{(\tau_2)} \cdots \mathscr{P}_1^{(\sigma_L)} \mathscr{P}_2^{(\tau_L)}, \tag{3.93}$$

for some $L \geq 1$, where $\sigma_1, \ldots, \sigma_L \in \mathbb{N}^{\nu_1}$, $\tau_1, \ldots, \tau_L \in \mathbb{N}^{\nu_2}$, $\sigma_1 + \cdots + \sigma_L \leq \kappa_1$, $\tau_1 + \cdots + \tau_L \leq \kappa_2$, each $\mathscr{P}_1^{(\sigma_j)}$ is an $\mathcal{S}_1$ partial differential operator on Ω of degree $\leq \sigma_j$, and each $\mathscr{P}_2^{(\tau_j)}$ is an $\mathcal{S}_2$ partial differential operator on Ω of degree $\leq \tau_j$. Indeed, by Definition 3.12.1, $\mathscr{P}^{(\kappa)}$ can be written as a finite sum of operators of the form

$$\mathrm{Mult}[g] \mathscr{P}_2^{(\tau_1)} \mathscr{P}_1^{(\sigma_2)} \mathscr{P}_2^{(\tau_2)} \cdots \mathscr{P}_1^{(\sigma_L)} \mathscr{P}_2^{(\tau_L)},$$

where $g \in C^{\infty}_{\mathrm{loc}}(\Omega)$, $(\sigma_2 + \cdots + \sigma_L, \tau_1 + \cdots + \tau_L) = (\kappa_1, \kappa_2) = \kappa$, and each $\mathscr{P}_1^{(\sigma_j)}$ is of the form $Y_1 Y_2 \cdots Y_R$ for some $R \geq 0$ (where $\mathscr{P}_1^{(\sigma_j)} = \mathrm{Mult}[1]$ if $R = 0$), where $(Y_1, \vec{d}_1), \ldots, (Y_R, \vec{d}_R) \in \mathcal{S}_1$ with $\vec{d}_1 + \cdots + \vec{d}_R \leq \sigma_j$. Similarly, each $\mathscr{P}_2^{(\tau_j)}$ is of the form $Z_1 Z_2 \cdots Z_R$ (for some $R \geq 0$), where $(Z_1, \vec{d}_1), \ldots, (Z_R, \vec{d}_R) \in \mathcal{S}_2$ with $\vec{d}_1 + \cdots + \vec{d}_R \leq \tau_j$.

Setting $\sigma_1 = 0$ and $\mathscr{P}_1^{(\sigma_1)} = \mathrm{Mult}[g]$, we see that each $\mathscr{P}_1^{(\sigma_j)}$ is an $\mathcal{S}_1$ partial differential operator on Ω of degree $\leq \sigma_j$ and each $\mathscr{P}_2^{(\tau_j)}$ is an $\mathcal{S}_2$ partial differential operator on Ω of degree $\leq \tau_j$. This establishes that $\mathscr{P}^{(\kappa)}$ can be written as a sum of terms of the form (3.93).

Thus, it suffices to prove the result when $\mathscr{P}^{(\kappa)}$ is given by a single term of the form (3.93). In that case, by repeated application of Corollary 3.12.8, $\mathscr{P}^{(\kappa)}$ can be written as a finite sum of terms of the form

$$\widetilde{\mathscr{P}}_1^{(\sigma_1)} \widetilde{\mathscr{P}}_1^{(\sigma_2)} \cdots \widetilde{\mathscr{P}}_1^{(\sigma_L)} \widetilde{\mathscr{P}}_2^{(\tau_1)} \widetilde{\mathscr{P}}_2^{(\tau_2)} \cdots \widetilde{\mathscr{P}}_2^{(\tau_L)}. \tag{3.94}$$

Lemma 3.12.4 shows that (3.94) is of the form $\widetilde{\mathscr{P}}_1^{(\kappa_1)} \widetilde{\mathscr{P}}_2^{(\kappa_2)}$, completing the proof. □

3.13 Filtered modules of vector fields

Many of the definitions in this chapter can be restated in the language of filtered modules. This can be useful when understanding certain qualitative properties of the definitions, though is not as useful in the quantitative analysis needed for our applications. We describe this perspective in this section.

Our modules will be filtered by the semi-group $\mathbb{N}^{\nu}$, with the partial ordering for $a, b \in \mathbb{N}^{\nu}$, $a \leq b$ if $a_{\mu} \leq b_{\mu}$ for all $\mu \in \{1, \ldots, \nu\}$. We will sometimes use $[0, \infty)^{\nu}$ in place of $\mathbb{N}^{\nu}$, with the same definitions.

Definition 3.13.1. Let R be a commutative ring. An $\mathbb{N}^\nu$-filtration $F_\cdot$ of an R-module M assigns to each $D \in \mathbb{N}^\nu$ an R-submodule F_D of M, satisfying $D_1 \leq D_2 \Rightarrow F_{D_1} \subseteq F_{D_2}$.

Remark 3.13.2. In Definition 3.13.1, we do not assume that a filtration satisfies $\bigcup_{D\in\mathbb{N}^\nu} F_D = M$.

Definition 3.13.3. Let $F_\cdot$ be an $\mathbb{N}^\nu$-filtration of an R-module M. We say $F_\cdot$ is finitely generated if there is a finite set $\mathcal{F} \subseteq \mathfrak{M}$ such that $F_D \cap \mathcal{F}$ generates F_D as an R-module.

Definition 3.13.4. A $\mathbb{N}^\nu$ filtration $F_\cdot$ of a Lie algebra A assigns to each $D \in \mathbb{N}^\nu$ a vector subspace F_D of A, such that $D_1 \leq D_2 \Rightarrow F_{D_1} \subseteq F_{D_2}$ and $[F_{D_1}, F_{D_2}] \subseteq F_{D_1+D_2}$, $\forall D_1, D_2 \in \mathbb{N}^\nu$.

Definition 3.13.5. Given two $\mathbb{N}^\nu$-filtrations $F_\cdot$ and $G_\cdot$ of a module M or a Lie algebra A, we say $F_\cdot \subseteq G_\cdot$ if $F_D \subseteq G_D$, $\forall D \in \mathbb{N}^\nu$. We say $F_\cdot = G_\cdot$ if $F_D = G_D$ for all $D \in \mathbb{N}^\nu$.

Let $\mathfrak{M}$ be a smooth manifold. For open sets $\Omega \subseteq \mathfrak{M}$, the map $\Omega \mapsto C^\infty_{\mathrm{loc}}(\Omega)$ gives a sheaf of rings on $\mathfrak{M}$. The map $\Omega \mapsto C^\infty_{\mathrm{loc}}(\Omega; T\Omega)$ gives a sheaf of modules over this sheaf of rings. This is also a sheaf of Lie algebras via the usual Lie bracket of vector fields.

Definition 3.13.6. An $\mathbb{N}^\nu$-filtration of vector fields on $\mathfrak{M}$, $\Gamma_\cdot(\cdot)$, is for each open set $\Omega \subseteq \mathfrak{M}$ a filtration, $\Gamma_\cdot(\Omega)$, of the module $C^\infty_{\mathrm{loc}}(\Omega; T\Omega)$ such that for each $D \in \mathbb{N}^\nu$, $\Omega \mapsto \Gamma_D(\Omega)$ is a subpresheaf of the sheaf $\Omega \mapsto C^\infty_{\mathrm{loc}}(\Omega; T\Omega)$. We say Γ is a Lie algebra $\mathbb{N}^\nu$-filtration of vector fields on $\mathfrak{M}$ if in addition each $\Gamma_\cdot(\Omega)$ is a filtration of the Lie algebra $C^\infty_{\mathrm{loc}}(\Omega; T\Omega)$.

Definition 3.13.7. Let Γ be an $\mathbb{N}^\nu$-filtration of vector fields on $\mathfrak{M}$. We say Γ is locally finitely generated if for each relatively compact, open set $\Omega \Subset \mathfrak{M}$, $\Gamma_\cdot(\Omega)$ is finitely generated.

Definition 3.13.8. Given a set $\mathcal{S} \subseteq C^\infty_{\mathrm{loc}}(\mathfrak{M}; T\mathfrak{M}) \times \mathbb{N}^\nu$, we obtain a filtration of vector fields on $\mathfrak{M}$ defined by, for $D \in \mathbb{N}^\nu$,

$$\Gamma_D(\Omega, \mathcal{S}) := C^\infty_{\mathrm{loc}}(\Omega) \text{ module generated by } \{X\big|_\Omega : (X, d) \in \mathcal{S}, d \leq D\}.$$

As we will see, many of the properties we study about sets $\mathcal{S} \subseteq C^\infty_{\mathrm{loc}}(\mathfrak{M}; T\mathfrak{M}) \times \mathbb{N}^\nu$ only depend on the associated filtration of vector fields $\Gamma_\cdot(\cdot, \mathcal{S})$. For example, we have the follow lemma, whose simple proof we leave to the interested reader.

Lemma 3.13.9. *Let $\mathcal{S} \subseteq C^\infty_{\mathrm{loc}}(\mathfrak{M}; T\mathfrak{M}) \times \mathbb{N}^\nu$. Then:*
(i) *$\mathcal{S}$ is strongly locally finitely generated if and only if $\Gamma_\cdot(\cdot, \mathcal{S})$ is locally finitely generated. $\mathcal{S}$ is strongly finitely generated on Ω if and only if $\Gamma_\cdot(\Omega, \mathcal{S})$ is finitely generated. Moreover, for $\Omega \Subset \mathfrak{M}$ open, if $\mathcal{S}$ is finitely generated on Ω by $\mathcal{F} \subseteq \mathcal{S}$, then $\Gamma_D(\Omega, \mathcal{S})$ is finitely generated by $\{X|_\Omega : (X, d) \in \mathcal{F}, d \leq D\}$.*
(ii) *$\Gamma_\cdot(\cdot, \mathrm{Gen}(\mathcal{S}))$ is the Lie algebra $\mathbb{N}^\nu$-filtration of vector fields on $\mathfrak{M}$ generated by $\Gamma_\cdot(\cdot, \mathcal{S})$. More precisely, $\Gamma_\cdot(\cdot, \mathrm{Gen}(\mathcal{S}))$ is the smallest Lie algebra $\mathbb{N}^\nu$-filtration of vector fields on $\mathfrak{M}$ which contains $\Gamma_\cdot(\cdot, \mathcal{S})$.*

(iii) *Let* $\mathcal{S}_1 \subseteq C^\infty_{\mathrm{loc}}(\mathfrak{M}; T\mathfrak{M}) \times \mathbb{N}^{\nu_1}$ *and* $\mathcal{S}_2 \subseteq C^\infty_{\mathrm{loc}}(\mathfrak{M}; T\mathfrak{M}) \times \mathbb{N}^{\nu_2}$. *Then* $\mathcal{S}_1$ *and* $\mathcal{S}_2$ *weakly approximately commute on* $\Omega \Subset \mathfrak{M}$ *if and only if the following holds. If*

$$\mathcal{S} := \{(X_1, (d_1, 0_{\nu_2})), (X_2, (0_{\nu_1}, d_2)) : (X_1, d_1) \in \mathrm{Gen}(\mathcal{S}_1), (X_2, d_2) \in \mathrm{Gen}(\mathcal{S}_2)\},$$

then $\Gamma_\cdot(\Omega, \mathcal{S})$ *is a Lie algebra filtration.*

3.14 Control of vector fields

Fix $\nu_1, \nu_2 \in \mathbb{N}_+$ and for $j = 1, 2$ let $\mathcal{S}_j \subseteq C^\infty_{\mathrm{loc}}(\mathfrak{M}; T\mathfrak{M}) \times (\mathbb{N}^{\nu_j} \setminus \{0\})$. If $(X_2, \vec{d}_2) \in \mathcal{S}_2$, we think of X_2 as a differential operator of "degree" $\vec{d}_j \in \mathbb{N}^{\nu_2}$. The goal of this section is to describe a situation in which we can "trade" such vector fields in $\mathcal{S}_2$ for vector fields in $\mathcal{S}_1$.

Fix a $\nu_2 \times \nu_1$ matrix λ with components in $[0, \infty]$. We use the convention that $0 \cdot \infty = 0$.

Definition 3.14.1. We define $\lambda^\top(\mathcal{S}_2) \subseteq C^\infty_{\mathrm{loc}}(\mathfrak{M}; T\mathfrak{M}) \times [0, \infty)^\nu$ to be the set of all $(X_2, \lambda^\top(\vec{d}_2))$, where $(X_2, \vec{d}_2) \in \mathcal{S}_2$ and $\lambda^\top(\vec{d}_2)$ is not infinite in any component.

Definition 3.14.2. We say $\mathcal{S}_1$ strongly λ-controls $\mathcal{S}_2$ on Ω if $\forall (Y_2, \vec{d}_2) \in \lambda^\top(\mathcal{S}_2)$, we have

$$Y_2 \in C^\infty_{\mathrm{loc}}(\Omega) \text{ module generated by } \{X_1 : (X_1, \vec{d}_1) \in \mathcal{S}_1, \vec{d}_1 \leq \vec{d}_2\}.$$

We say $\mathcal{S}_1$ locally strongly λ-controls $\mathcal{S}_2$ on $\mathfrak{M}$ if $\mathcal{S}_1$ strongly λ-controls $\mathcal{S}_2$ on Ω, for all $\Omega \Subset \mathfrak{M}$ open. When $\nu_1 = \nu_2$, we say $\mathcal{S}_1$ strongly controls $\mathcal{S}_2$ on Ω if $\mathcal{S}_1$ strongly I-controls $\mathcal{S}_2$ on Ω, where I is the identity matrix. We say $\mathcal{S}_1$ locally strongly controls $\mathcal{S}_2$ on $\mathfrak{M}$ if $\mathcal{S}_1$ locally strongly I-controls $\mathcal{S}_2$ on $\mathfrak{M}$.

Definition 3.14.3. We say $\mathcal{S}_1$ weakly λ-controls $\mathcal{S}_2$ on Ω if $\mathrm{Gen}(\mathcal{S}_1)$ strongly λ-controls $\mathcal{S}_2$ on Ω. We say $\mathcal{S}_1$ locally weakly λ-controls $\mathcal{S}_2$ on $\mathfrak{M}$ if $\mathcal{S}_1$ weakly λ-controls $\mathcal{S}_2$ on Ω, for all $\Omega \Subset \mathfrak{M}$ open. When $\nu_1 = \nu_2$, we say $\mathcal{S}_1$ weakly controls $\mathcal{S}_2$ on Ω if $\mathcal{S}_1$ weakly I-controls $\mathcal{S}_2$ on Ω, where I is the identity matrix. We say $\mathcal{S}_1$ locally weakly controls $\mathcal{S}_2$ on $\mathfrak{M}$ if $\mathcal{S}_1$ locally weakly I-controls $\mathcal{S}_2$ on $\mathfrak{M}$.

Example 3.14.4. Let $(W, 1) = \{(W_1, 1), \ldots, (W_r, 1)\}$ be such that $W_1, \ldots, W_{r_1}$ satisfy Hörmander's condition of order m on Ω. Let $(Z, 1) = \{(Z_1, 1), \ldots, (Z_s, 1)\}$ be such that $Z_1(x), \ldots, Z_s(x)$ span $T_x\Omega$, $\forall x \in \Omega$ (i. e., $Z_1, \ldots, Z_s$ satisfy Hörmander's condition of order 1 on Ω). Then $(Z, 1)$ strongly controls $(W, 1)$ on Ω, and $(W, 1)$ weakly m-controls $(Z, 1)$ on Ω. For a generalization of this example, see the discussion of λ_{std} and Λ_{std} (Definition 6.6.15) in Section 6.6.2.

Definition 3.14.5. When $\nu_1 = \nu_2$, we say $\mathcal{S}_1$ and $\mathcal{S}_2$ are weakly (respectively, strongly) equivalent on Ω if $\mathcal{S}_1$ weakly (respectively, strongly) controls $\mathcal{S}_2$ on Ω and $\mathcal{S}_2$ weakly (respectively, strongly) controls $\mathcal{S}_1$ on Ω. We say $\mathcal{S}_1$ and $\mathcal{S}_2$ are locally weakly (respectively, strongly) equivalent on $\mathfrak{M}$ if $\mathcal{S}_1$ and $\mathcal{S}_2$ are weakly (respectively, strongly) equivalent on Ω, $\forall \Omega \Subset \mathfrak{M}$ open.

Remark 3.14.6. Using the notation from Section 3.13 we have the following restatements of the above definitions:

- $\mathcal{S}_1$ strongly λ-controls $\mathcal{S}_2$ on Ω if and only if $\Gamma_\cdot(\Omega, \lambda^\top(\mathcal{S}_2)) \subseteq \Gamma_\cdot(\Omega, \mathcal{S}_1)$. Here, we are using filtration indexed by $[0,\infty)^\nu$ instead of $\mathbb{N}^\nu$, though the definitions are the same.
- $\mathcal{S}_1$ weakly λ-controls $\mathcal{S}_2$ on Ω if and only if $\Gamma_\cdot(\Omega, \lambda^\top(\mathcal{S}_2)) \subseteq \Gamma_\cdot(\Omega, \mathrm{Gen}(\mathcal{S}_1))$.
- $\mathcal{S}_1$ and $\mathcal{S}_2$ are strongly equivalent on Ω if and only if $\Gamma_\cdot(\Omega, \mathcal{S}_2) = \Gamma_\cdot(\Omega, \mathcal{S}_1)$.
- $\mathcal{S}_1$ and $\mathcal{S}_2$ are weakly equivalent on Ω if and only if $\Gamma_\cdot(\Omega, \mathrm{Gen}(\mathcal{S}_2)) = \Gamma_\cdot(\Omega, \mathrm{Gen}(\mathcal{S}_1))$, that is, if and only if the filtered Lie algebra generated by $\Gamma_\cdot(\Omega, \mathcal{S}_2)$ equals the filtered Lie algebra generated by $\Gamma_\cdot(\Omega, \mathcal{S}_1)$ – see Lemma 3.13.9 (ii).

Remark 3.14.7. Weak and strong control on Ω each give a partial ordering on subsets of $C^\infty_{\mathrm{loc}}(\mathfrak{M}; T\mathfrak{M}) \times (\mathbb{N}^\nu \setminus \{0\})$. Weak and strong equivalence on Ω each give an equivalence relation on such subsets. This follows immediately from Remark 3.14.6.

Lemma 3.14.8. *$\mathcal{S}_1$ weakly λ-controls $\mathcal{S}_2$ on Ω if and only if $\mathcal{S}_1$ weakly λ-controls* $\mathrm{Gen}(\mathcal{S}_2)$ *on Ω.*

Proof. The "if" part of the lemma is clear, so we focus on the "only if" part. Suppose $\mathcal{S}_1$ weakly λ-controls $\mathcal{S}_2$ on Ω. Then, by Remark 3.14.6,

$$\Gamma_\cdot(\Omega, \lambda^\top(\mathcal{S}_2)) \subseteq \Gamma_\cdot(\Omega, \mathrm{Gen}(\mathcal{S}_1)).$$

Since $\Gamma_\cdot(\Omega, \mathrm{Gen}(\mathcal{S}_1))$ is a Lie algebra filtration, this shows that $\Gamma_\cdot(\Omega, \mathrm{Gen}(\mathcal{S}_1))$ contains the Lie algebra filtration generated by $\Gamma_\cdot(\Omega, \lambda^\top(\mathcal{S}_2))$. It is easy to see that this is exactly $\Gamma_\cdot(\Omega, \lambda^\top(\mathrm{Gen}(\mathcal{S}_2)))$, completing the proof. □

Lemma 3.14.9. *For $j = 1, 2$, let $\mathcal{S}_j', \mathcal{F}_j \subseteq C^\infty_{\mathrm{loc}}(\mathfrak{M}; T\mathfrak{M}) \times (\mathbb{N}^{\nu_j} \setminus \{0\})$ with $\mathcal{F}_j$ finite.*

(i) *If $\mathcal{S}_1$ and $\mathcal{S}_1'$ are strongly equivalent on Ω and $\mathcal{S}_2$ and $\mathcal{S}_2'$ are strongly equivalent on Ω, then $\mathcal{S}_1$ strongly λ-controls $\mathcal{S}_2$ on Ω if and only if $\mathcal{S}_1'$ strongly λ-controls $\mathcal{S}_2'$ on Ω.*
(ii) *If $\mathcal{S}_1$ and $\mathcal{S}_1'$ are weakly equivalent on Ω and $\mathcal{S}_2$ and $\mathcal{S}_2'$ are weakly equivalent on Ω, then $\mathcal{S}_1$ weakly λ-controls $\mathcal{S}_2$ on Ω if and only if $\mathcal{S}_1'$ weakly λ-controls $\mathcal{S}_2'$ on Ω.*
(iii) *Suppose $\mathcal{S}_1$ is finitely generated by $\mathcal{F}_1 \subseteq \mathcal{S}_1$ on Ω. Then $\mathcal{S}_1$ strongly λ-controls $\mathcal{S}_2$ on Ω if and only if $\mathcal{F}_1$ strongly λ-controls $\mathcal{S}_2$ on Ω.*
(iv) *Suppose $\mathcal{S}_2$ is finitely generated by $\mathcal{F}_2 \subseteq \mathcal{S}_2$ on Ω. Then $\mathcal{S}_1$ strongly λ-controls $\mathcal{S}_2$ on Ω if and only if $\mathcal{S}_1$ strongly λ-controls $\mathcal{F}_2$ on Ω.*
(v) *Suppose* $\mathrm{Gen}(\mathcal{S}_1)$ *is finitely generated by $\mathcal{F}_1 \subset \mathrm{Gen}(\mathcal{S}_1)$ on Ω. Then $\mathcal{S}_1$ weakly λ-controls $\mathcal{S}_2$ on Ω if and only if $\mathcal{F}_1$ strongly λ-controls $\mathcal{S}_2$ on Ω.*
(vi) *Suppose* $\mathrm{Gen}(\mathcal{S}_2)$ *is finitely generated by $\mathcal{F}_2 \subset \mathrm{Gen}(\mathcal{S}_2)$ on Ω. Then $\mathcal{S}_1$ weakly λ-controls $\mathcal{S}_2$ on Ω if and only if $\mathcal{S}_1$ weakly λ-controls $\mathcal{F}_2$ on Ω.*

Proof. We use Remark 3.14.6 freely in this proof.

(i): That $\mathcal{S}_j$ and $\mathcal{S}_j'$ are strongly equivalent on Ω is equivalent to $\Gamma_\cdot(\Omega, \mathcal{S}_j) = \Gamma_\cdot(\Omega, \mathcal{S}_j')$. It follows that $\Gamma_\cdot(\Omega, \lambda^t(\mathcal{S}_2)) = \Gamma_\cdot(\Omega, \lambda^t(\mathcal{S}_2'))$. That $\mathcal{S}_1$ strongly λ-controls $\mathcal{S}_2$ is equivalent to $\Gamma_\cdot(\Omega, \lambda^t(\mathcal{S}_2)) \subseteq \Gamma_\cdot(\Omega, \mathcal{S}_1)$, and similarly for $\mathcal{S}_j$ replaced with $\mathcal{S}_j'$. The result follows.

(ii): By Lemma 3.14.8, $\mathcal{S}_1$ weakly λ-controls $\mathcal{S}_2$ if and only if $\mathcal{S}_1$ weakly λ-controls $\mathrm{Gen}(\mathcal{S}_2)$. The result follows by applying (i) with $\mathcal{S}_1, \mathcal{S}_2, \mathcal{S}_1', \mathcal{S}_2'$ replaced with $\mathrm{Gen}(\mathcal{S}_1)$, $\mathrm{Gen}(\mathcal{S}_2), \mathrm{Gen}(\mathcal{S}_1'), \mathrm{Gen}(\mathcal{S}_2')$.

(iii) and (iv): Since $\mathcal{S}$ is finitely generated by $\mathcal{F} \subseteq \mathcal{S}$ on Ω if and only if $\mathcal{F} \subseteq \mathcal{S}$ is finite and $\mathcal{S}$ and $\mathcal{F}$ are strongly equivalent on Ω, these results follow immediately from (i).

(v): This follows from (iii) by replacing $\mathcal{S}_1$ with $\mathrm{Gen}(\mathcal{S}_1)$.

(vi): By Lemma 3.14.8, $\mathcal{S}_1$ weakly λ-controls $\mathcal{S}_2$ if and only if $\mathcal{S}_1$ weakly λ-controls $\mathrm{Gen}(\mathcal{S}_2)$. Using this the result follows from (iv) with $\mathcal{S}_1$ and $\mathcal{S}_2$ replaced by $\mathrm{Gen}(\mathcal{S}_1)$ and $\mathrm{Gen}(\mathcal{S}_2)$. □

The most important definitions of this text only depend on the weak equivalence class of $\mathcal{S}$. For example, we have the next result.

Proposition 3.14.10. *Let $\mathcal{S}_1, \mathcal{S}_2, \mathcal{S}_1', \mathcal{S}_2' \subseteq C^\infty_{\mathrm{loc}}(\mathfrak{M}; T\mathfrak{M}) \times (\mathbb{N}^\nu \setminus \{0\})$. Suppose $\mathcal{S}_1$ and $\mathcal{S}_1'$ are weakly equivalent on Ω and $\mathcal{S}_2$ and $\mathcal{S}_2'$ are weakly equivalent on Ω. Then $\mathcal{S}_1$ and $\mathcal{S}_2$ weakly approximately commute on Ω if and only if $\mathcal{S}_1'$ and $\mathcal{S}_2'$ weakly approximately commute on Ω.*

Proof. Set

$$\mathcal{S} := \mathrm{Gen}(\mathcal{S}_1) \boxtimes \mathrm{Gen}(\mathcal{S}_2), \quad \mathcal{S}' := \mathrm{Gen}(\mathcal{S}_1') \boxtimes \mathrm{Gen}(\mathcal{S}_2').$$

Suppose $\mathcal{S}_1$ and $\mathcal{S}_2$ weakly approximately commute on Ω. We will show that $\mathcal{S}_1'$ and $\mathcal{S}_2'$ weakly approximately commute on Ω; the reverse implication will then follow by reversing the roles of $\mathcal{S}_1$ and $\mathcal{S}_2$ with $\mathcal{S}_1'$ and $\mathcal{S}_2'$. That $\mathcal{S}_1$ and $\mathcal{S}_2$ weakly approximately commute on Ω is easily seen to be equivalent to $\Gamma_\cdot(\Omega, \mathcal{S}) = \Gamma_\cdot(\Omega, \mathrm{Gen}(\mathcal{S}))$.

Since $\mathcal{S}_1$ and $\mathcal{S}_1'$ are weakly equivalent on Ω, Remark 3.14.6 shows that $\Gamma_\cdot(\Omega, \mathrm{Gen}(\mathcal{S}_1)) = \Gamma_\cdot(\Omega, \mathrm{Gen}(\mathcal{S}_1'))$ and similarly $\Gamma_\cdot(\Omega, \mathrm{Gen}(\mathcal{S}_2)) = \Gamma_\cdot(\Omega, \mathrm{Gen}(\mathcal{S}_2'))$. It follows that $\Gamma_\cdot(\Omega, \mathcal{S}) = \Gamma_\cdot(\Omega, \mathcal{S}')$. Since $\Gamma_\cdot(\Omega, \mathcal{S}) = \Gamma_\cdot(\Omega, \mathrm{Gen}(\mathcal{S}))$, we know that $\Gamma_\cdot(\Omega, \mathcal{S}) = \Gamma_\cdot(\Omega, \mathcal{S}')$ is a Lie algebra filtration (see Lemma 3.13.9 (ii)), and therefore $\Gamma_\cdot(\Omega, \mathcal{S}') = \Gamma_\cdot(\Omega, \mathrm{Gen}(\mathcal{S}'))$. This is equivalent to $\mathcal{S}_1'$ and $\mathcal{S}_2'$ weakly approximately commuting on Ω, completing the proof. □

3.14.1 Further comments on equivalence

The main setting we will be considering in this text is as follows. Fix $\nu \in \mathbb{N}_+$. For each $\mu \in \{1, \ldots, \nu\}$, we are given a finite set $(W^\mu, \vec{d}^\mu) = \{(W_1^\mu, \vec{d}_1^\mu), \ldots, (W_{r_\mu}^\mu, \vec{d}_{r_\mu}^\mu)\} \subset C^\infty_{\mathrm{loc}}(\mathfrak{M}; T\mathfrak{M}) \times \mathbb{N}_+$. We assume that for each μ, $\mathrm{Gen}((W^\mu, \vec{d}^\mu))$ is locally finitely generated on $\mathfrak{M}$, and we assume that $(W^1, \vec{d}^1), \ldots, (W^\nu, \vec{d}^\nu)$ pairwise weakly approximately commute on $\mathfrak{M}$. Set

$$\begin{aligned}(W, \vec{d}) = \{(W_1, \vec{d}_1), \ldots, (W_r, \vec{d}_r)\} &:= (W^1, \vec{d}^1) \boxtimes (W^2, \vec{d}^2) \boxtimes \cdots \boxtimes (W^\nu, \vec{d}^\nu) \\ &= \{(W_j^\mu, \vec{d}_j^\mu \mathfrak{e}_\mu) : \mu \in \{1, \ldots, \nu\}, 1 \le j \le r_\mu\} \subset C^\infty_{\mathrm{loc}}(\mathfrak{M}; T\mathfrak{M}) \times (\mathbb{N}^\nu \setminus \{0\}).\end{aligned} \tag{3.95}$$

Fix an open, relatively compact set $\Omega \Subset \mathfrak{M}$. For each $\mu \in \{1,\dots,\nu\}$, let $(Z^\mu, \vec{d}^\mu) = \{(Z_1^\mu, \vec{d}_1^\mu), \dots, (Z_{s_\mu}^\mu, \vec{d}_{s_\mu}^\mu)\} \subset C_{\mathrm{loc}}^\infty(\mathfrak{M}; T\mathfrak{M}) \times \mathbb{N}_+$. We assume $(W^\mu, \vec{d}^\mu)$ is weakly equivalent to $(Z^\mu, \vec{d}^\mu)$ on Ω. Define $(Z, \vec{d})$ in the same way as $(W, \vec{d})$ (see (3.95)), with each $(W^\mu, \vec{d}^\mu)$ replaced by $(Z^\mu, \vec{d}^\mu)$. The goal of this section is to show that $(W, \vec{d})$ can be replaced by $(Z, \vec{d})$ in many settings. We will often be given a set like $(W, \vec{d})$ and we will define various objects associated to this set; we will see that many of these objects can be equivalently defined using $(Z, \vec{d})$.

Let $(X^\mu, d^\mu) = \{(X_1^\mu, d_1^\mu), \dots, (X_{q_\mu}^\mu, d_{q_\mu}^\mu)\} \subset \mathrm{Gen}((W^\mu, \vec{d}^\mu))$ be such that $(W^\mu, \vec{d}^\mu) \subseteq (X^\mu, d^\mu)$ and $\mathrm{Gen}((W^\mu, \vec{d}^\mu))$ is finitely generated by (X^μ, d^μ) on Ω. Set

$$\begin{aligned}(X, \vec{d}) &= \{(X_1, \vec{d}_1), \dots, (X_q, \vec{d}_q)\} := (X^1, d^1) \boxtimes (X^2, d^2) \boxtimes \cdots \boxtimes (X^\nu, d^\nu)\\ &= \{(X_j^\mu, d_j^\mu \mathfrak{e}_\mu) : \mu \in \{1,\dots,\nu\}, 1 \le j \le q_\mu\} \subset C_{\mathrm{loc}}^\infty(\mathfrak{M}; T\mathfrak{M}) \times (\mathbb{N}^\nu \setminus \{0\}).\end{aligned}$$

By Proposition 3.8.8, $\mathrm{Gen}((W, \vec{d}))$ is finitely generated by $(X, \vec{d})$ on Ω.

Lemma 3.14.11. *For each $\mu \in \{1,\dots,\nu\}$, $\mathrm{Gen}((Z^\mu, \vec{d}^\mu))$ is finitely generated on Ω and $(Z^1, \vec{d}^1), \dots, (Z^\nu, \vec{d}^\nu)$ pairwise weakly approximately commute on Ω. Finally, $(W, \vec{d})$ and $(Z, \vec{d})$ are weakly equivalent on Ω.*

Proof. Using Lemma 3.13.9 (i), that $(W^\mu, \vec{d}^\mu)$ is weakly finitely generated on Ω is equivalent to $\Gamma_\cdot(\Omega, \mathrm{Gen}((W^\mu, \vec{d}^\mu)))$ being finitely generated. Since $\Gamma_\cdot(\Omega, \mathrm{Gen}((W^\mu, \vec{d}^\mu))) = \Gamma_\cdot(\Omega, \mathrm{Gen}((Z^\mu, \vec{d}^\mu)))$, by hypothesis (see Remark 3.14.6), it follows that $\mathrm{Gen}((Z^\mu, \vec{d}^\mu))$ is finitely generated on Ω.

Proposition 3.14.10 implies that $(Z^1, \vec{d}^1), \dots, (Z^\nu, \vec{d}^\nu)$ pairwise weakly approximately commute on Ω.

Since $\Gamma_\cdot(\Omega, \mathrm{Gen}((W^\mu, \vec{d}^\mu))) = \Gamma_\cdot(\Omega, \mathrm{Gen}((Z^\mu, \vec{d}^\mu)))$ for each μ, it follows easily that $\Gamma_\cdot(\Omega, \mathrm{Gen}((W, \vec{d}))) = \Gamma_\cdot(\Omega, \mathrm{Gen}((Z, \vec{d})))$. Indeed, $\Gamma_\cdot(\Omega, \mathrm{Gen}((W, \vec{d})))$ is nothing more than the filtered Lie algebra generated by the filtered $C_{\mathrm{loc}}^\infty(\Omega)$ module generated by

$$\{(V_j^\mu, f_j^\mu \mathfrak{e}_\mu) : \mu \in \{1,\dots,\nu\}, (V_j^\mu, f_j^\mu) \in \Gamma_\cdot(\Omega, \mathrm{Gen}((W^\mu, \vec{d}^\mu)))\},$$

and similarly for $(W, \vec{d})$ replaced by $(Z, \vec{d})$. This shows that $(W, \vec{d})$ and $(Z, \vec{d})$ are weakly equivalent on Ω. □

Lemma 3.14.11 shows that $(Z^1, \vec{d}^1), \dots, (Z^\nu, \vec{d}^\nu)$ satisfy the same hypotheses as $(W^1, \vec{d}^1), \dots, (W^\nu, \vec{d}^\nu)$ on Ω. Let $(Y^\mu, \vec{d}^\mu) = \{(Y_1^\mu, \vec{d}_\mu^1), \dots, (Y_{p_\mu}^\mu, \vec{d}_{p_\mu}^\mu)\} \subset \mathrm{Gen}((Z^\mu, \vec{d}^\mu))$ be such that $(Z^\mu, \vec{d}^\mu) \subseteq (Y^\mu, \vec{d}^\mu)$ and $\mathrm{Gen}((Z^\mu, \vec{d}^\mu))$ is finitely generated by $(Y^\mu, \vec{d}^\mu)$ on Ω, that is, $(Y^\mu, \vec{d}^\mu)$ is to $(Z^\mu, \vec{d}^\mu)$ as (X^μ, d^μ) is to $(W^\mu, \vec{d}^\mu)$. Define $(Y, \vec{d})$ in terms of $(Y^1, \vec{d}^1), \dots, (Y^\nu, \vec{d}^\nu)$ in the same way $(X, \vec{d})$ was defined in terms of $(X^1, d^1), \dots, (X^\nu, d^\nu)$. By Proposition 3.8.8 and Lemma 3.14.11, $\mathrm{Gen}((Z, \vec{d}))$ is finitely generated on Ω by $(Y, \vec{d})$.

Lemma 3.14.12. *$(X, \vec{d})$ and $(Y, \vec{d})$ are strongly equivalent on Ω.*

Proof. Since $\mathrm{Gen}((W,\vec{d}))$ is finitely generated by $(X,\vec{d})$ on Ω, $\mathrm{Gen}((Z,\vec{d}))$ is finitely generated by $(Y,\vec{d})$ on Ω, and $(W,\vec{d})$ and $(Z,\vec{d})$ are weakly equivalent on Ω (see Lemma 3.14.11), we have

$$\Gamma_{\cdot}(\Omega,(X,\vec{d})) = \Gamma_{\cdot}(\Omega,\mathrm{Gen}((W,\vec{d}))) = \Gamma_{\cdot}(\Omega,\mathrm{Gen}((Z,\vec{d}))) = \Gamma_{\cdot}(\Omega,(Y,\vec{d})),$$

completing the proof. □

Proposition 3.14.13. *Fix $\mathcal{K} \Subset \Omega$, a compact set. Then there exists $\xi_4 = \xi_4(\mathcal{K},\Omega,(W,\vec{d}),(Z,\vec{d})) \in (0,1]$ such that for all $x \in \mathcal{K}$ and $\delta \in (0,1]^\nu$,*

$$B_{(W,\vec{d})}(x,\xi_4\delta) \subseteq B_{(Z,\vec{d})}(x,\delta) \quad \textit{and} \quad B_{(Z,\vec{d})}(x,\xi_4\delta) \subseteq B_{(W,\vec{d})}(x,\delta).$$

Proof. We will prove the containment $B_{(Z,\vec{d})}(x,\xi_4\delta) \subseteq B_{(W,\vec{d})}(x,\delta)$; because our assumptions are symmetric in $(Z,\vec{d})$ and $(W,\vec{d})$ (by Lemma 3.14.11), the other containment follows from the same proof.

Fix Ω_1 open with $\mathcal{K} \Subset \Omega_1 \Subset \Omega$ and take $\xi_5 \in (0,1]$ so small that $B_{(Y,\vec{d})}(x,\xi_5\delta) \subseteq \Omega_1$ for all $\delta \in (0,1]^\nu$ and $x \in \mathcal{K}$. We claim that there exists $C \geq 1$ with

$$B_{(Y,\vec{d})}(x,\xi_5\delta) \subseteq B_{(X,\vec{d})}(x,C\delta), \tag{3.96}$$

$\forall x \in \mathcal{K}$ and $\delta \in (0,1]^\nu$. Indeed, let $C \geq 1$ be a large constant to be chosen later and fix $y \in B_{(Y,\vec{d})}(x,\xi_5\delta)$. Then there exists an absolutely continuous $\gamma : [0,1] \to B_{(Y,\vec{d})}(x,\xi_5\delta) \subseteq \Omega_1$, with $\gamma(0) = x$, $\gamma(1) = y$, and $\gamma'(t) = \sum_j a_j(t)(\xi_5\delta)^{\vec{d}_j} Y_j(\gamma(t))$ almost everywhere, where $\sum |a_j(t)|^2 < 1$ for almost every t.

Lemma 3.14.12 shows that $(X,\vec{d})$ and $(Y,\vec{d})$ are strongly equivalent on Ω. Thus,

$$Y_j = \sum_{\vec{d}_k \leq \vec{d}_j} b_j^k X_k, \quad b_j^k \in C^\infty_{\mathrm{loc}}(\Omega).$$

Therefore,

$$\begin{aligned}\gamma'(t) &= \sum_k \Bigg(\sum_{\vec{d}_j \geq \vec{d}_k} a_j(t) b_j^k(\gamma(t)) \xi_5^{\vec{d}_j} \delta^{\vec{d}_j - \vec{d}_k} \Bigg) \delta^{\vec{d}_k} X_k(\gamma(t)) \\ &= \sum_k \Bigg(\sum_{\vec{d}_j \geq \vec{d}_k} C^{-|\vec{d}_k|_1} a_j(t) b_j^k(\gamma(t)) \xi_5^{\vec{d}_j} \delta^{\vec{d}_j - \vec{d}_k} \Bigg) (C\delta)^{\vec{d}_k} X_k(\gamma(t)) \\ &=: \sum_k \tilde{a}_k(t)(C\delta)^{\vec{d}_k} X_k(\gamma(t)).\end{aligned}$$

Since $\gamma(t) \in \Omega_1$, $\forall t$, and $\Omega_1 \Subset \Omega$, we have $|b_j^k(\gamma(t))| \lesssim 1$, for almost every t. Thus, by taking C sufficiently large (independent of δ, x, and y) we have $\sum |\tilde{a}_k(t)|^2 < 1$, that is, $y = \gamma(1) \in B_{(X,\vec{d})}(x,C\delta)$, establishing (3.96).

Letting $\xi_3 \in (0,1]$ be as in Theorem 3.5.1, Theorem 3.5.1 (b) shows that for all $x \in \mathcal{K}$ and $\delta \in (0,1]^\nu$, $B_{(X,d)}(x, \xi_3\delta) \subseteq B_{(W,\vec{d})}(x,\delta)$. Using that $(Z,\vec{d}) \subseteq (Y,\vec{d})$ and (3.96), for $x \in \mathcal{K}$ and $\delta \in (0,1]^\nu$, we have

$$B_{(Z,\vec{d})}(x, \xi_3\xi_5 C^{-1}\delta) \subseteq B_{(Y,\vec{d})}(x, \xi_3\xi_5 C^{-1}\delta) \subseteq B_{(X,\vec{d})}(x, \xi_3\delta) \subseteq B_{(W,\vec{d})}(x,\delta),$$

which completes the proof with $\xi_4 := \xi_3\xi_5 C^{-1}$. □

Proposition 3.14.14. *Suppose, in addition to the above assumptions, that for each $\mu \in \{1,\ldots,\nu\}$, $(W^\mu, \vec{d}^\mu)$ are Hörmander vector fields with formal degrees on Ω. Then, for each $\mu \in \{1,\ldots,\nu\}$, $(Z^\mu, \vec{d}^\mu)$ are Hörmander vector fields with formal degrees on Ω. Moreover, if $\mathcal{K} \Subset \mathfrak{M}$ is compact, then $\forall x \in \mathcal{K}$, $\delta \in (0,\infty)^\nu$,*

$$\mathrm{Vol}(B_{(W,\vec{d})}(x,\delta)) \wedge 1 \approx \mathrm{Vol}(B_{(Z,\vec{d})}(x,\delta)) \wedge 1,$$

where the implicit constants depend on $\mathcal{K}$, Ω, $(W,\vec{d})$, and $(Z,\vec{d})$, but not on x or δ.

Proof. That $(W^\mu, \vec{d}^\mu)$ are Hörmander vector fields with formal degrees on Ω is equivalent to the vector fields in $\Gamma_\cdot(\Omega, \mathrm{Gen}((W^\mu, \vec{d}^\mu)))$ spanning the tangent space at every point of Ω. Similarly, that $(Z^\mu, \vec{d}^\mu)$ are Hörmander vector fields with formal degrees on Ω is equivalent to the vector fields in $\Gamma_\cdot(\Omega, \mathrm{Gen}((Z^\mu, \vec{d}^\mu)))$ spanning the tangent space at every point of Ω. By hypothesis, $\Gamma_\cdot(\Omega, \mathrm{Gen}((W^\mu, \vec{d}^\mu))) = \Gamma_\cdot(\Omega, \mathrm{Gen}((Z^\mu, \vec{d}^\mu)))$, so we see that $(Z^\mu, \vec{d}^\mu)$ are Hörmander vector fields with formal degrees on Ω.

With ξ_4 as in Proposition 3.14.13, we have for $\delta \in (0,\xi_4]^\nu$, using Theorem 3.5.4 (n),

$$\mathrm{Vol}(B_{(Z,\vec{d})}(x,\delta)) \wedge 1 \le \mathrm{Vol}(B_{(W,\vec{d})}(x,\xi_4^{-1}\delta)) \wedge 1 \lesssim \mathrm{Vol}(B_{(W,\vec{d})}(x,\delta)) \wedge 1.$$

Suppose $\delta \in (0,\infty)^\nu \setminus (0,\xi_4]^\nu$. Without loss of generality, assume $\delta_1 > \xi_4$. Then, using Theorem 3.5.4 (l), we have for $x \in \mathcal{K}$

$$\mathrm{Vol}(B_{(W,\vec{d})}(x,\delta)) \wedge 1 \ge \mathrm{Vol}(B_{(W,\vec{d})}(x,\xi_4 \mathfrak{e}_1)) \wedge 1 \approx \Lambda(x,\xi_4\mathfrak{e}_1) \wedge 1 \ge \inf_{y\in\mathcal{K}} \Lambda(y,\xi_4\mathfrak{e}_1) \wedge 1 \gtrsim 1,$$

where the last line uses the formula for Λ, (3.15), and the compactness of $\mathcal{K}$. Similarly, we have $\mathrm{Vol}(B_{(Z,\vec{d})}(x,\delta)) \wedge 1 \gtrsim 1$. Thus,

$$\mathrm{Vol}(B_{(Z,\vec{d})}(x,\delta)) \wedge 1 \approx 1 \approx \mathrm{Vol}(B_{(W,\vec{d})}(x,\delta)) \wedge 1.$$

Combining the above estimates, we conclude that for $\delta \in (0,\infty)^\nu$ and $x \in \mathcal{K}$,

$$\mathrm{Vol}(B_{(Z,\vec{d})}(x,\delta)) \wedge 1 \lesssim \mathrm{Vol}(B_{(W,\vec{d})}(x,\delta)) \wedge 1.$$

The reverse inequality follows by reversing the roles of $(W,\vec{d})$ and $(Z,\vec{d})$ throughout.

□

3.15 The main multi-parameter setting

In our applications to maximally subelliptic operators, a special case of the multi-parameter setting in Section 3.5 is of central interest. In this section, we describe this setting and show that it is invariant under an adapted *single-parameter* scaling.

Let $\mathfrak{M}$ be a connected C^∞ manifold of dimension $n \in \mathbb{N}_+$ with smooth, strictly positive density Vol. Fix $\nu \geq 1$. For each $\mu \in \{1, \dots, \nu\}$, we suppose we are given a list of vector fields with single-parameter formal degrees:

$$(W^\mu, \vec{d}^\mu) := \{(W_1^\mu, \vec{d}_1^\mu), \dots, (W_{r_\mu}^\mu, \vec{d}_{r_\mu}^\mu)\} \subset C^\infty_{\mathrm{loc}}(\mathfrak{M}; T\mathfrak{M}) \times \mathbb{N}_+.$$

We suppose:

(I) $(W^1, \vec{d}^1)$ are Hörmander vector fields with formal degrees.

(II) $\mathrm{Gen}((W^\mu, \vec{d}^\mu))$ is locally finitely generated for $\mu \in \{2, \dots, \nu\}$.

(III) $(W^1, \vec{d}^1), \dots, (W^\nu, \vec{d}^\nu)$ pairwise weakly locally approximately commute on $\mathfrak{M}$.

3.15.1 The multi-parameter unit scale

To present the scaling results adapted to the above setting, we need to introduce a version of the above setting at the "unit scale," and we will rescale to this unit scale. We work on $B^n(1)$ endowed with the density $\mathrm{Vol} = h\sigma_{\mathrm{Leb}}$, where $h \in C^\infty(B^n(1); \mathbb{R})$, with $\inf_{x \in B^n(1)} h(x) > 0$. Fix $\nu \geq 1$ and assume that for each $\mu \in \{1, \dots, \nu\}$ we are given

$$(W^\mu, \vec{d}^\mu) := \{(W_1^\mu, \vec{d}_1^\mu), \dots, (W_{r_\mu}^\mu, \vec{d}_{r_\mu}^\mu)\} \subset C^\infty(B^n(1); TB^n(1)) \times \mathbb{N}_+.$$

We wish to make assumptions (I)–(III) described above quantitative.

To do this, we assume that for each $\mu \in \{1, \dots, \nu\}$ we are given a finite set

$$(X^\mu, d^\mu) = \{(X_1^\mu, d_1^\mu), \dots, (X_{q_\mu}^\mu, d_{q_\mu}^\mu)\} \subset \mathrm{Gen}((W^\mu, \vec{d}^\mu))$$

with $(W^\mu, \vec{d}^\mu) \subseteq (X^\mu, d^\mu)$ such that the following holds. Set

$$\begin{aligned}(X, \vec{d}) &= \{(X_1, \vec{d}_1), \dots, (X_q, \vec{d}_q)\} \\ &:= (X^1, d^1) \boxtimes (X^2, d^2) \boxtimes \cdots \boxtimes (X^\nu, d^\nu) \subset C^\infty(B^n(1); TB^n(1)) \times (\mathbb{N}^\nu \setminus \{0\}).\end{aligned}$$

We assume that:

(i) $[X_j, X_k] = \sum_{\vec{d}_l \leq \vec{d}_j + \vec{d}_k} c_{j,k}^l X_l$, $c_{j,k}^l \in C^\infty(B^n(1))$.

(ii) $\mathrm{span}\{X_1^1(x), \dots, X_{q_1}^1(x)\} = T_x B^n(1)$, $\forall x \in B^n(1)$, and moreover,

$$\max_{j_1, \dots, j_n \in \{1, \dots, q\}} \inf_{x \in B^n(1)} \left|\det(X_{j_1}^1(x) | \cdots | X_{j_n}^1(x))\right| > 0. \tag{3.97}$$

Item (ii) implies (I) holds and (i) implies (II) and (III) hold.

Definition 3.15.1. For a parameter ι, we say C is an ι-multi-parameter unit-admissible constant if there exists $L \in \mathbb{N}$, depending only on ι and upper bounds for q, ν, and $\max_{1\le j\le q}|d_j|_1$, such that C can be chosen to depend only on ι, upper bounds for q, ν, $\max_{1\le j\le q}|d_j|_1$, $\max_{1\le j\le q}\|X_j\|_{C^L(B^n(1);\mathbb{R}^n)}$, $\max_{1\le j,k,l\le q}\|c_{j,k}^l\|_{C^L(B^n(1))}$, and $\|h\|_{C^L(B^n(1))}$, and lower bounds > 0 for $\inf_{x\in B^n(1)} h(x)$ and the left-hand side of (3.97). If ι_0 is another parameter, we say $C = C(\iota_0)$ is an ι-multi-parameter unit-admissible constant if C is an ι-multi-parameter unit-admissible constant, which can also depend on ι_0.

3.15.2 Scaling

We work in the setting at the start of this section. Thus, we are given vector fields with formal degrees $(W^1, \vec{d}^1), \dots, (W^\nu, \vec{d}^\nu) \subset C^\infty_{\mathrm{loc}}(\mathfrak{M}; T\mathfrak{M}) \times \mathbb{N}_+$ satisfying hypotheses (I)–(III). In this section, we describe a single-parameter scaling adapted to these vector fields.

Fix $\mathcal{K} \Subset \Omega_1 \Subset \Omega_2 \Subset \mathfrak{M}$ with $\mathcal{K}$ compact and Ω_1 and Ω_2 open and relatively compact. For each $\mu \in \{1, \dots, \nu\}$, let $(X^\mu, d^\mu) \subset \mathrm{Gen}((W^\mu, \vec{d}^\mu))$ be such that $\mathrm{Gen}((W^\mu, \vec{d}^\mu))$ is finitely generated by (X^μ, d^μ) on Ω_2 and $(W^\mu, \vec{d}^\mu) \subseteq (X^\mu, d^\mu)$. This is always possible: for $2 \le \mu \le \nu$ this follows from hypothesis (II) and for $\mu = 1$ this uses hypothesis (I) and the case $\nu = 1$ of Proposition 3.4.14. Because $W_1^1, \dots, W_{r_1}^1$ satisfy Hörmander's condition, $X_1^1, \dots, X_{q_1}^1$ span the tangent space at every point of Ω_2.

Set $\lambda_1 := 1$ and for $\mu \in \{2, \dots, \nu\}$ pick $\lambda_\mu > 0$ such that for $1 \le j \le r_m u$,

$$W_j^\mu = \sum_{d_l^1 \le \lambda_\mu d_j^\mu} c_{j,\mu}^l X_l^1, \quad c_{j,\mu}^l \in C^\infty_{\mathrm{loc}}(\Omega_2); \tag{3.98}$$

this is always possible since $X_1^1, \dots, X_{q_1}^1$ span the tangent space at every point of Ω_2. Let $\lambda := (\lambda_1, \dots, \lambda_\nu) \in (0,\infty)^\nu$.

Remark 3.15.2. Formula (3.98) is a more explicit version of the fact that $(W^1, \vec{d}^1)$ weakly λ_μ controls $(W^\mu, \vec{d}^\mu)$ on Ω_2; see Definition 3.14.3.

Theorem 3.3.7 applies with $(W^1, \vec{d}^1)$ and (X^1, d^1) in place of $(W, \vec{d})$ and (X, d). Let $\Phi_{x,\delta} : B^n(1) \to B_{(X^1,d^1)}(x,\delta) \cap \Omega_1$ be as in that theorem. Then we have the following.

Theorem 3.15.3. *For $\mu \in \{1, \dots, \nu\}$, $x \in \mathcal{K}$, and $\delta \in (0,1]$ we set*

$$W_j^{\mu,x,\delta} := \Phi_{x,\delta}^* \delta^{\lambda_\mu \vec{d}_j^\mu} W_j^\mu, \quad X_k^{\mu,x,\delta} := \Phi_{x,\delta}^* \delta^{\lambda_\mu d_k^\mu} X_k^\mu,$$
$$(W^{\mu,x,\delta}, \vec{d}^\mu) := \{(W_1^{\mu,x,\delta}, \vec{d}_1^\mu), \dots, (W_{r_\mu}^{\mu,x,\delta}, \vec{d}_{r_\mu}^\mu)\} \subset C^\infty(B^n(1); TB^n(1)) \times \mathbb{N}_+,$$
$$(X^{\mu,x,\delta}, d^\mu) := \{(X_1^{\mu,x,\delta}, d_1^\mu), \dots, (X_{q_\mu}^{\mu,x,\delta}, d_{q_\mu}^\mu)\} \subset C^\infty(B^n(1); TB^n(1)) \times \mathbb{N}_+.$$

Let $h_{x,\delta}$ be the function h from this application of Theorem 3.3.7 (n). Then $(W^{1,x,\delta}, \vec{d}^1), \dots, (W^{\nu,x,\delta}, \vec{d}^\nu)$, $(X^{1,x,\delta}, d^1), \dots, (X^{\nu,x,\delta}, d^\nu)$, and $h_{x,\delta}$ satisfy the hypotheses of Section 3.15.1 uniformly for $x \in \mathcal{K}$, $\delta \in (0,1]$, in the sense that multi-parameter unit-admissible constants *(as in Definition 3.15.1) can be chosen independent of $x \in \mathcal{K}$ and $\delta \in (0,1]$. More precisely, we have:*

(a) $(W^{\mu,x,\delta}, \vec{d}^{\mu}) \subseteq (X^{\mu,x,\delta}, d^{\mu}) \subset \mathrm{Gen}((W^{\mu,x,\delta}, \vec{d}^{\mu}))$, *for* $\mu \in \{1,\dots,\nu\}$, $x \in \mathcal{K}$, $\delta \in (0,1]$.

(b) $W_j^{\mu,x,\delta}$ *and* $X_k^{\mu,x,\delta}$ *are* C^∞ *uniformly for* $x \in \mathcal{K}$ *and* $\delta \in (0,1]$. *More precisely,* $\forall L \in \mathbb{N}$, $\mu \in \{1,\dots,\nu\}$, $1 \le j \le r_\mu$, $1 \le k \le q_\mu$,

$$\sup_{\substack{x\in\mathcal{K}\\ \delta\in(0,1]}} \|W_j^{\mu,x,\delta}\|_{C^L(B^n(1);\mathbb{R}^n)} < \infty, \quad \sup_{\substack{x\in\mathcal{K}\\ \delta\in(0,1]}} \|X_k^{\mu,x,\delta}\|_{C^L(B^n(1);\mathbb{R}^n)} < \infty.$$

(c) *We have* $X_1^{1,x,\delta},\dots,X_{q_1}^{1,x,\delta}$ *span the tangent space to every point of* $B^n(1)$ *uniformly in* $x \in \mathcal{K}$ *and* $\delta \in (0,1]$ *in the sense that*

$$\inf_{x\in\mathcal{K},\delta\in(0,1]} \max_{j_1,\dots,j_n\in\{1,\dots q_1\}} \inf_{u\in B^n(1)} \left|\det(X_{j_1}^{1,x,\delta}(u),\dots,X_{j_n}^{1,x,\delta}(u))\right| > 0.$$

(d) *Define*

$$\begin{aligned}(X^{x,\delta}, \vec{d}) &= \{(X_1^{x,\delta}, \vec{d}_1),\dots,(X_q^{x,\delta}, \vec{d}_q)\}\\ &:= (X^{1,x,\delta}, d^1) \boxtimes (X^{2,x,\delta}, d^2) \boxtimes \cdots \boxtimes (X^{\nu,x,\delta}, d^\nu).\end{aligned}$$

Then, for $1 \le j,k \le q$, $x \in \mathcal{K}$, $\delta \in (0,1]$,

$$[X_j^{x,\delta}, X_k^{x,\delta}] = \sum_{\vec{d}_l \le \vec{d}_j + \vec{d}_k} c_{j,k}^{l,x,\delta} X_l^{x,\delta}, \quad c_{j,k}^{l,x,\delta} \in C^\infty(B^n(1)),$$

and for every $L \in \mathbb{N}$,

$$\sup_{\substack{x\in\mathcal{K}\\ \delta\in(0,1]}} \|c_{j,k}^{l,x,\delta}\|_{C^L(B^n(1))} < \infty. \tag{3.99}$$

(e) $\inf_{u\in B^n(1)} \inf_{x\in\mathcal{K},\delta\in(0,1]} h_{x,\delta}(u) > 0$.

(f) $\forall L \in \mathbb{N}$, $\sup_{x\in\mathcal{K},\delta\in(0,1]} \|h_{x,\delta}\|_{C^L(B^n(1))} < \infty$.

Proof. (a): Since $(W^\mu, \vec{d}^\mu) \subseteq (X^\mu, d^\mu) \subset \mathrm{Gen}((W^\mu, \vec{d}^\mu))$, we have

$$(\delta^{\lambda_\mu \vec{d}^\mu} W^\mu, \vec{d}^\mu) \subseteq (\delta^{\lambda_\mu d^\mu} X^\mu, d^\mu) \subset \mathrm{Gen}((\delta^{\lambda_\mu \vec{d}^\mu} W^\mu, \vec{d}^\mu)). \tag{3.100}$$

Pulling (3.100) back via $\Phi_{x,\delta}$ gives (a).

(b): When $\mu = 1$, (b) follows from Theorem 3.3.7 (j). Let $\mu \in \{2,\dots,\nu\}$. We first prove the result for $W_j^{\mu,x,\delta}$. Multiplying (3.98) by $\delta^{\lambda_\mu \vec{d}_j^\mu}$, we have

$$\delta^{\lambda_\mu \vec{d}_j^\mu} W_j^\mu = \sum_{d_l^1 \le \lambda_\mu \vec{d}_j^\mu} (\delta^{\lambda_\mu \vec{d}_j^\mu - d_l^1} c_{j,\mu}^l) \delta^{d_l^1} X_l^1. \tag{3.101}$$

Pulling (3.101) back via $\Phi_{x,\delta}$ we have

$$W_j^{\mu,x,\delta} = \sum_{l=1}^{q_1} c_{j,\mu}^{l,x,\delta} X_l^{1,x,\delta},$$

where

$$c_{j,\mu}^{l,x,\delta} := \begin{cases} \delta^{\lambda_\mu \vec{d}_j^\mu - d_l^1} c_{j,\mu}^l \circ \Phi_{x,\delta}, & d_l^1 \le \lambda_\mu \vec{d}_j^\mu, \\ 0, & \text{otherwise.} \end{cases}$$

Since we have already established the case $\mu = 1$, to complete the proof it suffices to show that $\forall L \in \mathbb{N}$,

$$\sup_{\substack{x\in\mathcal{K} \\ \delta\in(0,1]}} \|c_{j,\mu}^{l,x,\delta}\|_{C^L(B^n(1))} < \infty. \tag{3.102}$$

The estimate (3.102) is trivial when $d_l^1 > \lambda_\mu \vec{d}_j^\mu$. When $d_l^1 \le \lambda_\mu \vec{d}_j^\mu$, we have, using Theorem 3.3.7 (m), for $x \in \mathcal{K}$, $\delta \in (0,1]$,

$$\begin{aligned} \|c_{j,\mu}^{l,x,\delta}\|_{C^L(B^n(1))} &\approx \sum_{|\alpha|\le L} \|(X^{1,x,\delta})^\alpha c_{j,\mu}^{l,x,\delta}\|_{C(B^n(1))} \\ &= \sum_{|\alpha|\le L} \delta^{\lambda_\mu \vec{d}_j^\mu - d_l^1 + \deg_{d^1}(\alpha)} \|(X^1)^\alpha c_{j,\mu}^l\|_{C(\Phi_{x,\delta}(B^n(1)))} \\ &\lesssim \sum_{|\alpha|\le L} \|(X^1)^\alpha c_{j,\mu}^l\|_{C(\Omega_1)} \lesssim 1, \end{aligned}$$

where the final $\lesssim 1$ uses that $\Omega_1 \Subset \Omega_2$. This proves (3.102) and completes the proof of (b) for $W_j^{\mu,x,\delta}$. Since each $X_k^{\mu,x,\delta}$ is a fixed commutator (independent of x and δ) of the $W_k^{\mu,x,\delta}$, the result for $X_k^{\mu,x,\delta}$ follows as well, completing the proof of (b).

(c): This is a restatement of Theorem 3.3.7 (k).

(d): Set

$$\begin{aligned} (W, \vec{d}) &= \{(W_1, \vec{d}_1), \dots, (W_r, \vec{d}_1)\} \\ &:= (W^1, d^1) \boxtimes (W^2, d^2) \boxtimes \cdots \boxtimes (W^\nu, d^\nu) \subset C^\infty_{\mathrm{loc}}(\mathfrak{M}; T\mathfrak{M}) \times (\mathbb{N}^\nu \setminus \{0\}), \\ (X, \vec{d}) &= \{(X_1, \vec{d}_1), \dots, (X_q, \vec{d}_1)\} \\ &:= (X^1, d^1) \boxtimes (X^2, d^2) \boxtimes \cdots \boxtimes (X^\nu, d^\nu) \subset C^\infty_{\mathrm{loc}}(\mathfrak{M}; T\mathfrak{M}) \times (\mathbb{N}^\nu \setminus \{0\}). \end{aligned}$$

Clearly, $(W, \vec{d}) \subseteq (X, \vec{d})$. By Proposition 3.8.8 (a), $\mathrm{Gen}((W, \vec{d}))$ is linearly finitely generated by $(X, \vec{d})$ on Ω_2. By Remark 3.4.12, we have

$$[X_j, X_k] = \sum_{\vec{d}_l \le \vec{d}_j + \vec{d}_k} c_{j,k}^l X_l, \quad c_{j,k}^l \in C^\infty_{\mathrm{loc}}(\Omega_2). \tag{3.103}$$

Multiplying (3.103) by $\delta^{\lambda\cdot(\vec{d}_j+\vec{d}_k)}$ and pulling back via $\Phi_{x,\delta}$, we obtain

$$[X_j^{x,\delta}, X_k^{x,\delta}] = \sum_{d_l \leq d_j+d_k} c_{j,k}^{l,x,\delta} X_l^{x,\delta},$$

where $c_{j,k}^{l,x,\delta} := \delta^{\lambda\cdot(d_j+d_k-d_l)} c_{j,k}^l$. It remains to show that (3.99) holds. This can be done by an almost identical argument to the proof of (3.102); we leave the details to the interested reader.

(e) and (f): These follow from Theorem 3.3.7 (n). □

3.15.3 Quantitative scaling

After applying Theorem 3.15.3 we obtain the following objects at the unit scale: $(W^{1,x,\delta}, \vec{d}^1), \ldots, (W^{\nu,x,\delta}, \vec{d}^\nu)$, $(X^{1,x,\delta}, d^1), \ldots, (X^{\nu,x,\delta}, d^\nu)$, and $h_{x,\delta}$, which satisfy the hypotheses of Section 3.15.1 uniformly in x and δ. We can then apply Theorem 3.15.3 to these new vector fields to obtain another scaling map. Sometimes, we will need to do this process repeatedly, working with several levels of iterated scaling. To do this, we need a more quantitative version of Theorems 3.3.7 and 3.15.3. The idea is that we will start with vector fields at the multi-parameter unit scale (as in Section 3.15.1) and have a scaling result in which all estimates are multi-parameter unit-admissible constants (as in Definition 3.15.1). As with all the scaling results in this text, we will see this as a simple consequence of Theorem 3.6.5.

Thus, in this section, we take the same setting as in Section 3.15.1, so we are given $(W^1, \vec{d}^1), \ldots, (W^\nu, \vec{d}^\nu)$, $(X^1, d^1), \ldots, (X^\nu, d^\nu)$, and h satisfying the assumptions of that section (in particular, satisfying (i) and (ii)).

Notation 3.15.4. In this section, we write $A \lesssim_{\iota;\iota_0} B$ to denote $A \leq CB$, where $C = C(\iota_0) \geq 0$ is an ι-multi-parameter unit-admissible constant as in Definition 3.15.1. We write $A \approx_{\iota;\iota_0} B$ for $A \lesssim_{\iota;\iota_0} B$ and $B \lesssim_{\iota;\iota_0} A$. We write $A \lesssim_{\iota_0} B$ and $A \approx_{\iota_0} B$ for $A \lesssim_{0;\iota_0} B$ and $A \approx_{0;\iota_0} B$, respectively.

Set $\lambda_1 := 1$ and for $\mu \in \{2, \ldots, \nu\}$ take $\lambda_\mu > 0$ such that $\forall 1 \leq j \leq q_\mu$ we may write

$$W_j^\mu = \sum_{d_l^1 \leq \lambda_\mu \vec{d}_j^\mu} c_{j,\mu}^l X_l^1, \quad c_{j,\mu}^l \in C^\infty(B^n(1)), \tag{3.104}$$

and such that $\|c_{j,\mu}^l\|_{C^L(B^n(1))} \lesssim_L 1$, $\forall L \in \mathbb{N}$; this is always possible by (ii). Define

$$\begin{aligned}\Lambda(x,\delta) &:= h(x) \max_{1\leq j_1,\ldots,j_n\leq q_1} \left|\det(\delta^{d_{j_1}^1} X_{j_1}^1(x)| \cdots |\delta^{d_{j_n}^1} X_{j_n}^1(x))\right| \\ &= h(x) \max_{1\leq j_1,\ldots,j_n\leq q_1} \sigma_{\mathrm{Leb}}(\delta^{d_{j_1}^1} X_{j_1}^1(x), \ldots, \delta^{d_{j_n}^1} X_{j_n}^1(x)).\end{aligned}$$

Theorem 3.15.5. *Fix $\sigma \in (0,1)$.*[1] *There exist 0-multi-parameter unit-admissible constants $\delta_0 = \delta_0(\sigma) \in (0,1]$ and $\xi_3 = \xi_3(\sigma) \in (0,1]$ such that the following hold:*

(a) $\forall x \in \overline{B^n(\sigma)}$, $\delta \in (0,1]$,

$$B_{(X^1,d^1)}(x, \xi_3\delta) \subseteq B_{(W^1,\vec{d}^1)}(x,\delta) \subseteq B_{(X^1,d^1)}(x,\delta).$$

(b) $\forall x \in \overline{B^n(\sigma)}$, $\delta \in (0,\delta_0]$,

$$\mathrm{Vol}(B_{(W^1,\vec{d}^1)}(x,\delta)) \approx_\sigma \mathrm{Vol}(B_{(X^1,d^1)}(x,\delta)) \approx_\sigma \Lambda(x,\delta).$$

(c) $\forall x \in \overline{B^n(\sigma)}$, $\delta \in (0,\delta_0]$,

$$\mathrm{Vol}(B_{(W^1,\vec{d}^1)}(x,2\delta)) \lesssim \mathrm{Vol}(B_{(W^1,\vec{d}^1)}(x,\delta)) \quad \textit{and}$$
$$\mathrm{Vol}(B_{(X^1,d^1)}(x,2\delta)) \lesssim \mathrm{Vol}(B_{(X^1,d^1)}(x,\delta)).$$

For all $x \in \overline{B^n(\sigma)}$, $\delta \in (0,1]$, there exists a map $\Phi_{x,\delta} : B^n(1) \to B_{(X^1,d^1)}(x,\delta) \cap B^n((1+\sigma)/2)$ such that:

(d) $\Phi_{x,\delta}(0) = x$.

(e) *$\Phi_{x,\delta}$ is a smooth coordinate system, that is, $\Phi_{x,\delta}(B^n(1)) \subseteq B^n(1)$ is open and $\Phi_{x,\delta} : B^n(1) \to \Phi_{x,\delta}(B^n(1))$ is a C^∞_{loc} diffeomorphism.*

(f) $\forall x \in \overline{B^n(\sigma)}$, $\delta \in (0,1]$, $B_{(W^1,\vec{d}^1)}(x,\xi_3\delta) \subseteq B_{(X^1,d^1)}(x,\xi_3\delta) \subseteq \Phi_{x,\delta}(B^n(1/2)) \subseteq \Phi_{x,\delta}(B^n(1)) \subseteq B_{(X^1,d^1)}(x,\delta)$.

Set $W_j^{\mu,x,\delta} := \Phi_{x,\delta}^* \delta^{\lambda_\mu \vec{d}_j^\mu} W_j^\mu$ and $X_k^{\mu,x,\delta} := \Phi_{x,\delta}^* \delta^{\lambda_\mu d_k^\mu} X_k^\mu$. Let

$$(W^{\mu,x,\delta}, \vec{d}^\mu) := \{(W_1^{\mu,x,\delta}, \vec{d}_1^\mu), \ldots, (W_{r_\mu}^{\mu,x,\delta}, \vec{d}_{r_\mu}^\mu)\},$$
$$(X^{\mu,x,\delta}, d^\mu) := \{(X_1^{\mu,x,\delta}, d_1^\mu), \ldots, (X_{q_\mu}^{\mu,x,\delta}, d_{q_\mu}^\mu)\}.$$

Define $h_{x,\delta}$ by $\Phi_{x,\delta}^* h\sigma_{\mathrm{Leb}} = \Lambda(x,\delta) h_{x,\delta}\sigma_{\mathrm{Leb}}$. Then $(W^{1,x,\delta}, \vec{d}^1), \ldots, (W^{\nu,x,\delta}, \vec{d}^\nu)$, $(X^{1,x,\delta}, d^1), \ldots, (X^{\nu,x,\delta}, d^\nu)$, and $h_{x,\delta}$ satisfy the hypotheses of Section 3.15.1 quantitatively in the sense that ι-multi-parameter unit-admissible constants C with respect to these choices can be chosen to be $C = C(\sigma)$, ι-multi-parameter unit-admissible constants with respect to $(W^1, \vec{d}^1), \ldots, (W^\nu, \vec{d}^\nu)$, $(X^1, d^1), \ldots, (X^\nu, d^\nu)$, and h. More precisely, we have:

(g) $(W^{\mu,x,\delta}, \vec{d}^\mu) \subseteq (X^{\mu,x,\delta}, d^\mu) \subset \mathrm{Gen}((W^{\mu,x,\delta}, \vec{d}^\mu))$.

(h) $\forall L \in \mathbb{N}$, $x \in \overline{B^n(\sigma)}$, $\delta \in (0,1]$,

$$\|W_j^{\mu,x,\delta}\|_{C^L(B^n(1);\mathbb{R}^n)} \lesssim_{L;\sigma} 1,$$
$$\|X_k^{\mu,x,\delta}\|_{C^L(B^n(1);\mathbb{R}^n)} \lesssim_{L;\sigma} 1.$$

1 In many of our applications, we will take $\sigma = 7/8$.

(i) *For all* $x \in \overline{B^n(\sigma)}$, $\delta \in (0,1]$,

$$\max_{j_1,\dots,j_n \in \{1,\dots,q_1\}} \inf_{u \in B^n(1)} \left|\det(X^{1,x,\delta}_{j_1}(u)|\cdots|X^{1,x,\delta}_{j_n}(u))\right| \gtrsim_\sigma 1.$$

(j) $\forall x \in \overline{B^n(\sigma)}$, $\delta \in (0,1]$,

$$\|f\|_{C^L(B^n(1))} \approx_{L;\sigma} \sum_{|\alpha|\le L} \|(X^{1,x,\delta})^\alpha f\|_{C^{B^n(1)}}.$$

(k) *For* $x \in \overline{B^n(\sigma)}$ *and* $\delta \in (0,1]$, *set*

$$\begin{aligned}(X^{x,\delta}, \vec{d}) &= \{(X^{x,\delta}_1, \vec{d}_1), \dots, (X^{x,\delta}_q, \vec{d}_q)\} \\ &:= (X^{1,x,\delta}, d^1) \boxtimes (X^{2,x,\delta}, d^2) \boxtimes \cdots \boxtimes (X^{\nu,x,\delta}, d^\nu).\end{aligned}$$

Then

$$[X^{x,\delta}_j, X^{x,\delta}_k] = \sum_{\vec{d}_l \le \vec{d}_j + \vec{d}_k} c^{l,x,\delta}_{j,k} X^{x,\delta}_l, \quad c^{l,x,\delta}_{j,k} \in C^\infty(B^n(1)),$$

where $\forall L \in \mathbb{N}$,

$$\|c^{l,x,\delta}_{j,k}\|_{C^L(B^n(1))} \lesssim_{L;\sigma} 1.$$

(l) $\forall L \in \mathbb{N}$, $\|h_{x,\delta}\|_{C^L(B^n(1))} \lesssim_{L;\sigma} 1$.
(m) $h_{x,\delta}(u) \approx_\sigma 1$, $\forall u \in B^n(1)$.

To prove Theorem 3.15.5, we require the next lemma.

Lemma 3.15.6. *For each* $\mu \in \{2,\dots,\nu\}$ *and* $1 \le k \le q_\mu$, *we have*

$$X^\mu_k = \sum_{d^1_l \le \lambda_\mu d^\mu_k} c^l_{k,\mu} X_l, \quad c^l_{k,\mu} \in C^\infty(B^n(1)),$$

such that $\forall L \in \mathbb{N}$,

$$\|c^l_{k,\mu}\|_{C^L(B^n(1))} \lesssim_L 1.$$

Proof. Recursively set $\mathrm{Gen}_0((W^\mu, \vec{d}^\mu)) := (W^\mu, \vec{d}^\mu)$ and for $K \ge 1$,

$$\begin{aligned}\mathrm{Gen}_K((W^\mu, \vec{d}^\mu)) &:= \mathrm{Gen}_{K-1}((W^\mu, \vec{d}^\mu)) \\ &\quad \cup \{([Z_1, Z_2], \vec{d}_1 + \vec{d}_2) : (Z_1, \vec{d}_1), (Z_2, \vec{d}_2) \in \mathrm{Gen}_{K-1}((W^\mu, \vec{d}^\mu))\}.\end{aligned}$$

Note that

$$(X^\mu, d^\mu) \subseteq \mathrm{Gen}_{\max_{1\le k\le q_\mu} d_k^\mu}((W^\mu, \vec{d}^\mu)). \tag{3.105}$$

We will show that $\forall (Z, \vec{d}) \in \mathrm{Gen}_K((W^\mu, \vec{d}^\mu))$ we have

$$Z = \sum_{d_l^1 \le \lambda_\mu \vec{d}} b_l X_l^1, \quad b_l \in C^\infty(B^n(1)), \tag{3.106}$$

with $\|b_l\|_{C^L(B^n(1))} \lesssim_{(L,K)} 1$, $\forall L \in \mathbb{N}$, and the result will follow using (3.105).

We prove (3.106) by induction on K. The base case, $K = 0$, is (3.104). We assume (3.106) for some $K - 1 \ge 0$ and prove it for K.

Let $(Z_1, \vec{d}_1), (Z_2, \vec{d}_2) \in \mathrm{Gen}_[((W^\mu, \vec{d}^\mu))K-1]$. By the inductive hypothesis we have for $j = 1, 2$,

$$Z_j = \sum_{d_l^1 \le \lambda_\mu \vec{d}_j} b_j^l X_l^1, \quad b_j^l \in C^\infty(B^n(1)),$$

with $\|b_j^l\|_{C^L(B^n(1))} \lesssim_{L,K} 1$, $\forall L \in \mathbb{N}$. Thus, we have

$$\begin{aligned}
[Z_1, Z_2] &= \sum_{\substack{d_{l_1}^1 \le \lambda_\mu \vec{d}_1 \\ d_{l_2}^1 \le \lambda_\mu \vec{d}_2}} b_1^{l_1} b_2^{l_2} [X_{l_1}^1, X_{l_2}^2] + \sum_{\substack{d_{l_1}^1 \le \lambda_\mu \vec{d}_1 \\ d_{l_2}^1 \le \lambda_\mu \vec{d}_2}} \big(b_l^{l_1}(X_{l_1}^1 b_2^{l_2}) X_{l_2}^1 - b_2^{l_2}(X_{l_2}^1 b_1^{l_1}) X_{l_1}^1\big) \\
&= \sum_{\substack{d_{l_1}^1 \le \lambda_\mu \vec{d}_1 \\ d_{l_2}^1 \le \lambda_\mu \vec{d}_2}} b_1^{l_1} b_2^{l_2} \sum_{d_{l_3}^1 \le d_{l_1}^1 + d_{l_2}^1} c_{l_1,l_2}^{l_3} c_{l_1,l_2}^{l_3,1} X_{l_3}^1 + \sum_{\substack{d_{l_1}^1 \le \lambda_\mu \vec{d}_1 \\ d_{l_2}^1 \le \lambda_\mu \vec{d}_2}} \big(b_l^{l_1}(X_{l_1}^1 b_2^{l_2}) X_{l_2}^1 - b_2^{l_2}(X_{l_2}^1 b_1^{l_1}) X_{l_1}^1\big),
\end{aligned}$$

where $\|c_{l_1,l_2}^{l_3,1}\|_{C^L(B^n(1))} \lesssim_L 1$; here, the final equality uses hypothesis (i) from Section 3.15.1 applied when the vectors $\vec{d}_j$ and $\vec{d}_k$ are non-zero in only the first component. The above sum includes only terms of the form X_l^1 where $d_l^1 \le \lambda_\mu(\vec{d}_1 + \vec{d}_2)$. This establishes (3.106) for $([Z_1, Z_2], \vec{d}_1 + \vec{d}_2)$, completing the proof. □

Proof of Theorem 3.15.5. The heart of the proof is to apply Theorem 3.6.5 with x_0 replaced by $x \in \overline{B^n(\sigma)}$, $X_1, \dots, X_q$ replaced by $\delta^{d_1^1} X_1^1, \dots, \delta^{d_{q_1}^1} X_{q_1}^1$, Vol $:= h\sigma_{\mathrm{Leb}}$, and $\mathfrak{M} = B^n((1+\sigma)/2)$. The main point is that L-admissible constants and L, Vol-admissible constants $C \ge 0$ as in Definition 3.6.4 are $C = C(\sigma)$, L-multi-parameter unit-admissible constants as in Definition 3.15.1. To see this, we must verify that all of the quantities in Definition 3.6.4 are $\lesssim_{L;\sigma} 1$. Throughout this proof, x will denote an arbitrary element of $\overline{B^n(\sigma)}$ and δ will denote an arbitrary element of $(0, 1]$

By the Picard–Lindelöf theorem, if $\eta \gtrsim_\sigma 1$ is sufficiently small (η depending only on σ and $\sum_{j=1}^{q_1} \|X_j^1\|_{C^1(B^n(1);\mathbb{R}^n)}$), then $X_1^1, \dots, X_{q_1}^1$ satisfy $\mathcal{C}(x, \eta, B^n((1+\sigma)/2))$ (see Definition 3.6.3), and therefore $\delta^{d_1^1} X_1^1, \dots, \delta^{d_{q_1}^1} X_{q_1}^1$ also satisfy $\mathcal{C}(x, \eta, B^n((1+\sigma)/2))$. Lemma A.2.11 shows that $\tau_0 = \tau_0(\sigma) > 0$ as in the hypotheses of Theorem 3.6.5 can be chosen to be a 0-multi-parameter unit-admissible constant.

In this application of Theorem 3.6.5, we take $\xi = 1$. Applying assumption (i) from Section 3.15.1 applied when the vectors $\vec{d}_j$ and $\vec{d}_k$ are non-zero in only the first component, we have

$$[X_j^1, X_k^1] = \sum_{d_l^1 \leq d_j^1 + d_k^1} c_{j,k}^{1,l} X_l^1, \quad c_{j,k}^{l,1} \in C^\infty(B^n(1)), \tag{3.107}$$

and $\forall L \in \mathbb{N}$, $\|c_{j,k}^{1,l}\|_{C^L(B^n(1))} \lesssim_{L;1} 1$. Multiplying (3.107) by $\delta^{d_j^1 + d_k^1}$, we have

$$[\delta^{d_j^1} X_j^1, d_k^1 X_k^1] = \sum_{l=1}^{q_1} c_{j,k}^{1,l,\delta} \delta^{d_l^1} X_l^1,$$

where

$$c_{j,k}^{l,1,\delta} := \begin{cases} \delta^{d_j^1 + d_k^1 - d_l^1} c_{j,k}^{1,l}, & d_l^1 \leq d_j^1 + d_k^1, \\ 0 & \text{otherwise.} \end{cases}$$

We have, for every $L \in \mathbb{N}$,

$$\begin{aligned} \max_{j,k,l} \sum_{|\alpha| \leq L} \|(\delta^{d^1} X^1)^\alpha c_{j,k}^{1,l,\delta}\|_{C(B_{\delta^{d^1} X^1}(x))} &\leq \max_{d_l^1 \leq d_j^1 + d_k^1} \sum_{|\alpha| \leq L} \|(X^1)^\alpha c_{j,k}^l\|_{C(B^n(1))} \\ &\lesssim_{L;1} \max_{d_l^1 \leq d_j^1 + d_k^1} \sum_{|\alpha| \leq L} \|c_{j,k}^l\|_{C^{|\alpha|}(B^n(1))} \lesssim_{L;1} 1. \end{aligned}$$

Consider

$$\mathrm{Lie}_{\delta^{d_k^1} X_k} \mathrm{Vol} = (\delta^{d_k^1} X_k^1 h)\sigma_{\mathrm{Leb}} + \delta^{d_k^1} h \,\mathrm{div}(X_k)\sigma_{\mathrm{Leb}} =: f_{k,\delta}\sigma_{\mathrm{Leb}}.$$

We have, for $L \in \mathbb{N}$,

$$\sum_{|\alpha| \leq L} \|(\delta^{d^1} X^1)^\alpha f_{j,\delta}\|_{C(B_{\delta^{d^1} X}(x))} \leq \sum_{|\alpha| \leq L} \|(X^1)^\alpha X_k h\|_{C(B^n(1))} + \sum_{|\alpha| \leq L} \|(X^1)^\alpha h\|_{C(B^n(1))} \lesssim_{L;1} 1.$$

We use $\zeta = 1$ in this application of Theorem 3.6.5, so that (after reordering the vector fields) the left-hand side of (3.16) equals 1; this is always possible – see the remarks following (3.16).

Combining all of the above, we have shown L-admissible constants and L, Vol-admissible constants $C \geq 0$ as in Definition 3.6.4 are $C = C(\sigma)$, L-multi-parameter unit-admissible constants as in Definition 3.15.1. Thus, with the above choices, Theorem 3.6.5 applies. Let $\eta_1 > 0$ and let $\Psi_{x,\delta} : B^n(\eta_1) \xrightarrow{\sim} \Psi_{x,\delta}(B^n(\eta_1))$ be the map Φ from that theorem. Set $\Phi_{x,\delta}(t) := \Psi_{x,\delta}(\eta_1 t)$. From here (and using Lemma 3.15.6), the result follows just as in the proofs of Theorems 3.3.7 and 3.15.3; all the estimates are multi-parameter unit-admissible constants due to the above remarks. We leave the final details to the interested reader. □

3.16 Further reading and references

The study of Carnot–Carathéodory and sub-Riemannian geometry has a long history; we only cover the parts of that history directly relevant to the results in this chapter. The general theory of Carnot–Carahtéodory geometry began with the work of Carathéodory [33] and Chow [45]. The quantitative study of Carnot–Carathéodory geometry, more suitable to applications to PDEs, began with the work of Nagel, Stein, and Wainger [189]. They proved a result very similar to Theorem 3.3.7. In particular, they proved that Carnot–Carathéodory balls in Euclidean space give rise to a space of homogeneous type in the sense of Coifman and Weiss [52], paving the way for using techniques from harmonic analysis adapted to these geometries. Around the same time, C. Fefferman and Sanchez-Calle proved some related results [84].

Nagel, Stein, and Wainger's methods were based on Taylor series and the Baker–Campbell–Hausdorff formula, and while they can be adapted to some settings beyond the one described in Section 3.3, they have limitations when working in the multi-parameter setting or when working with vector fields with low regularity.[2] Tao and Wright [232] improved on Nagel, Stein, and Wainger's methods, when they studied two-parameter balls $B(x, \delta_1, \delta_2)$ under a weakly comparable condition $\delta_1^N \lesssim \delta_2 \lesssim \delta_1^{1/N}$, for some large N. Importantly, they replaced some of Nagel, Stein, and Wainger's Taylor series methods with more robust methods from the field of ODEs. The author later expanded on these ideas in [219], where general multi-parameter balls were studied without the weakly comparable hypothesis.

Theorem 3.6.5 was proved by the author and Stovall in [228]; it implies all of the above previous results and is stronger than they are. A main difference is the following. In previous works, all of the estimates depended on the C^m norms of the coefficients of the vector fields in some given coordinate system. Theorem 3.6.5 provides a coordinate system in which the C^m norms of the vector fields have good estimates, but the hypotheses of the theorem do not assume the existence of such a coordinate system. In fact, all of the estimates in Theorem 3.6.5 remain unchanged if the whole setting is pushed forward under an arbitrary C^∞_{loc} diffeomorphism. See Section A.5 for a further discussion of the history of these ideas and more results.

Because Nagel, Stein, and Wainger showed that single-parameter Carnot–Carathéodory balls induce a space of homogeneous type, the theory of the maximal functions in Theorem 3.11.9 in the special case $\nu = 1$ follows from their work. When the vector fields do not necessarily satisfy Hörmander's condition as in the $\nu = 1$ case of Theorem 3.11.2, the corresponding maximal function was introduced in [219]. Bounding a multi-parameter maximal function by the composition of several single-parameter

2 We present a more quantitative version of Theorem 3.3.7 in Proposition 8.3.21, which is a key step in the proof of Theorem 8.1.1. It is possible that one could obtain a similar result using the methods of [189]; however, it is much more straightforward to use Theorem 3.6.5.

maximal functions, as we did in the proof of Lemma 3.11.4, dates back to the work of Jessen, Marcinkiewicz, and Zygmund [135]. The particular multi-parameter maximal functions studied in Theorem 3.11.2 were first studied by the author in [219]. The study of vector-valued maximal operators began with the work of C. Fefferman and Stein [83]. Stronger multi-parameter maximal functions than the ones studied here were studied by the author and Stein in [226, 225]; these stronger results require a more intricate proof and are not useful for the applications in this text.

The concepts of finitely generated and linearly finitely generated, as in Definitions 3.4.1 and 3.4.3, were first introduced in this context in [221], where their relationship to developing function spaces adapted to vector fields was described. We use similar ideas in Chapter 6. The relationship between linearly finitely generated, multi-parameter singular integrals and parametrices for differential operators was first described in [220]; we will build on these ideas in Chapter 5.

For more details on, and a detailed history of, the Baker–Campbell–Hausdorff formula (sometimes called the Campbell–Hausdorff formula or the Baker–Campbell–Hausdorff–Dynkin formula) we refer the reader to [12].

The concepts of control, equivalence, and sharp control described in Section 3.14 were defined in a more general setting and in a more general way in [221]. The more general approach used in [221] might be useful for some questions like the ones discussed in this text, but this comes at the cost of more technicalities and does not change the main thrust of our results. For this reason, we chose the simpler approach in Section 3.14.

4 Pseudo-differential operators

Let $\mathfrak{M}$ be a C^∞ manifold of dimension $n \in \mathbb{N}_+$, fix $\nu \in \mathbb{N}_+$, and for each $\mu \in \{1, \ldots, \nu\}$, let

$$(X^\mu, d^\mu) = \{(X_1^\mu, d_1^\mu), \ldots, (X_{q_\mu}^\mu, d_{q_\mu}^\mu)\} \subset C^\infty_{\mathrm{loc}}(\mathfrak{M}; T\mathfrak{M}) \times \mathbb{N}_+$$

be a finite set of smooth vector fields with formal degrees.

Set

$$\begin{aligned}(X, \vec{d}) = \{(X_1, \vec{d}_1), \ldots, (X_q, \vec{d}_q)\} &:= (X^1, d^1) \boxtimes (X^2, d^2) \boxtimes \cdots \boxtimes (X^\nu, d^\nu)\\ &= \{(X_j^\mu, d_j^\mu \mathfrak{e}_\mu) : \mu \in \{1, \ldots, \nu\}, 1 \le j \le q_\mu\} \subset C^\infty_{\mathrm{loc}}(\mathfrak{M}; T\mathfrak{M}) \times (\mathbb{N}^\nu \setminus \{0\}),\end{aligned} \tag{4.1}$$

where we recall that $\mathfrak{e}_\mu \in \mathbb{R}^\nu$ is the vector which equals 1 in the μ-th component and 0 in all other components. Note that $q = q_1 + \cdots + q_\nu$ and each $\vec{d}_j$ is non-zero in precisely one component. The goal in this chapter is to define the concept of a pseudo-differential operator with respect to $(X, \vec{d})$.

Main assumption: The main assumption of this chapter, which we assume throughout, is

$$[X_j, X_k] = \sum_{\vec{d}_l \le \vec{d}_j + \vec{d}_k} c_{j,k}^l X_l, \quad c_{j,k}^l \in C^\infty_{\mathrm{loc}}(\mathfrak{M}). \tag{4.2}$$

We do not assume $X_1, \ldots, X_q$ span the tangent space at any point. Recall that $\vec{d}_l \le \vec{d}_j + \vec{d}_k$ means that the inequality holds coordinatewise.

Remark 4.0.1. The main way $(X, \vec{d})$ will arise in our applications is the following. We will be given ν sets $\mathcal{S}_1, \ldots, \mathcal{S}_\nu \subseteq C^\infty_{\mathrm{loc}}(\mathfrak{M}; T\mathfrak{M}) \times \mathbb{N}_+$ such that each $\mathrm{Gen}(\mathcal{S}_\mu)$ is locally finitely generated on $\mathfrak{M}$ and such that $\mathcal{S}_1, \ldots, \mathcal{S}_\nu$ pairwise locally weakly approximately commute. Fix $\Omega \Subset \mathfrak{M}$ open and relatively compact and let $\mathrm{Gen}(\mathcal{S}_\mu)$ be finitely generated by (X^μ, d^μ) on Ω. Then $(X, \vec{d})$ satisfies (4.2) with $\mathfrak{M}$ replaced by Ω; this follows from Proposition 3.8.8 combined with Remark 3.4.12. A particularly important example is Example 3.8.9.

Fix pre-compact open sets $\Omega_1, \Omega_2 \Subset \mathfrak{M}$ with $\Omega_1 \Subset \Omega_2$. For $a = a(\Omega_1, \Omega_2, (X, \vec{d})) > 0$ sufficiently small and $t = (t^1, \ldots, t^\nu) \in B^q(a)$ (where $t^\mu \in \mathbb{R}^{q_\mu}$), define, for $x \in \Omega_1$,

$$\Gamma(t, x) = \Gamma((t^1, \ldots, t^\nu), x) = e^{-t^1 \cdot X^1} e^{-t^2 \cdot X^2} \cdots e^{-t^\nu \cdot X^\nu} x. \tag{4.3}$$

Note that if we take $a > 0$ sufficiently small, then $\Gamma(t, x) \in \Omega_2$, $\forall x \in \Omega_1$, $t \in B^q(a)$. Following Remark 2.2.15, $(X, \vec{d})$ pseudo-differential operators will take the form

$$f \mapsto \int f(\Gamma(t, x)) K(x, t)\, dt,$$

https://doi.org/10.1515/9783111085647-004

for a class of kernels $K(x,t)$ made precise in Section 4.1.

One of our main uses for $(X,\vec{d})$ pseudo-differential operators is to create a parametrix for the so-called sub-Laplacian, which is a key tool in the study of maximally subelliptic operators. See Section 4.5.

Example 4.0.2. Consider the case where $\mathfrak{M} = \mathbb{R}^2$ and we are working with the Hörmander vector fields with formal degrees $(W, \vec{d}) = \{(\partial_x, 1), (x\partial_y, 1)\}$. In this case we take $\nu = 1$ and we may take $(X,d) = \{(X_1,d_1),(X_2,d_2),(X_3,d_3)\} = \{(\partial_1, x),(x\partial_y,1),(\partial_y,2)\}$; see Example 3.3.6. Then we are considering operators of the form

$$f \mapsto \int f(e^{-t_1X_1-t_2X_2-t_3X_3}x)K(x,(t_1,t_2,t_3))\,dt.$$

Here, the kernel $K(x,t)$ will be defined in a way which respects the fact that $X_1 = \partial_x$ and $X_2 = x\partial_y$ have formal degree 1, but $X_3 = \partial_y$ has formal degree 2. See Definitions 4.1.1 and 4.1.2.

4.1 Symbols of pseudo-differential operators

For $m \in \mathbb{R}^\nu$, we define a class of symbols $S^m(a,\Omega,\vec{d})$, which depend on a pre-compact, open set $\Omega \Subset \mathfrak{M}$, a small number $a > 0$ (to be chosen later), and the degrees d_j^μ associated with the vector fields X_j^μ (though they do not depend on the particular vector fields).

For $t^\mu = (t_1^\mu,\dots,t_{q_\mu}^\mu) \in \mathbb{R}^{q_\mu}$ we define dilations by, for $\delta_\mu > 0$,

$$\delta_\mu^{d^\mu} t^\mu = (\delta_\mu^{d_1^\mu} t_1^\mu,\dots,\delta_\mu^{d_{q_\mu}^\mu} t_{q_\mu}^\mu). \tag{4.4}$$

We define a homogeneous norm as

$$\|t^\mu\|_\mu := \left(\sum_{l=1}^{q_\mu} |t_l^\mu|^{(2\prod_{j=1}^{q_\mu} d_j^\mu)/d_l^\mu}\right)^{1/2\prod_{j=1}^{q_\mu} d_j^\mu} \approx \sum_{l=1}^{q_\mu} |t_l^\mu|^{1/d_l^\mu}. \tag{4.5}$$

Note that

$$\|\delta_\mu^{d^\mu} t^\mu\|_\mu = \delta_\mu \|t^\mu\|_\mu. \tag{4.6}$$

For a multi-index $\gamma = (\gamma_1,\dots,\gamma_{q_\mu}) \in \mathbb{N}^{q_\mu}$, set $\deg_{d^\mu}(\gamma) = \sum_{l=1}^{q_\mu} \gamma_j d_j^\mu$. Note that $(\delta_\mu^{d^\mu} t^\mu)^\gamma = \delta_\mu^{\deg_{d^\mu}(\gamma)} t^\mu$.

We also define multi-parameter dilations; for $\delta \in [0,\infty)^\nu$ and $t = (t^1,\dots,t^\nu) \in \mathbb{R}^{q_1} \times \cdots \times \mathbb{R}^{q_\nu} = \mathbb{R}^q$, we set

$$\delta^{\vec{d}} t = (\delta_1^{d^1} t^1,\dots,\delta_\nu^{d^\nu} t^\nu). \tag{4.7}$$

For $\gamma = (\gamma^1, \dots, \gamma^\nu) \in \mathbb{N}^{q_1} \times \cdots \times \mathbb{N}^{q_\nu} = \mathbb{N}^q$, we set

$$\deg_{\vec{d}}(\gamma) = \sum_{\mu} \deg_{d^\mu}(\gamma^\mu) \mathfrak{e}_\mu \in \mathbb{N}^\nu. \tag{4.8}$$

Note that $(\delta^{\vec{d}} t)^\gamma = \delta^{\deg_{\vec{d}}(\gamma)} t^\gamma$.

Definition 4.1.1. For $m = (m_1, \dots, m_\nu) \in \mathbb{R}^\nu$ and $a > 0$, we set $S^m(a, \Omega, \vec{d})$ to be the set of all $b(x, s, \xi) \in C^\infty_{\text{loc}}(\mathfrak{M} \times B^q(a) \times \mathbb{R}^q)$ such that:
(i) $\{(x, s) \in \mathfrak{M} \times B^q(a) : \exists \xi \in \mathbb{R}^q \text{ with } b(x, s, \xi) \neq 0\} \Subset \Omega \times B^q(a)$.
(ii) For every smooth coordinate system $\Phi : B^n(1) \xrightarrow{\sim} \Phi(B^n(1)) \subseteq \mathfrak{M}$ and all multi-indices $\alpha \in \mathbb{N}^n, \beta \in \mathbb{N}^q, \gamma = (\gamma_1, \dots, \gamma_\nu) \in \mathbb{N}^{q_1} \times \cdots \times \mathbb{N}^{q_\nu} \cong \mathbb{N}^q$,

$$\left|\partial_u^\alpha \partial_s^\beta \partial_\xi^\gamma b(\Phi(u), s, \xi)\right| \leq C_{\Phi,\alpha,\beta,\gamma} \prod_{\mu=1}^{\nu} (1 + \|\xi^\mu\|_\mu)^{m_\mu - \deg_{d^\mu}(\gamma_\mu)}.$$

Definition 4.1.2. For pre-compact, open sets $\Omega_1, \Omega_2 \Subset \mathfrak{M}$ with $\Omega_1 \Subset \Omega_2$, if $a = a(\Omega_1, \Omega_2, (X, \vec{d})) > 0$ is sufficiently small (how small to be chosen later, but in particular so small that $\Gamma(B^q(a) \times \Omega_1) \Subset \Omega_2$) and $b(x, s, \xi) \in S^m(a, \Omega_1, \vec{d})$, for some $m \in \mathbb{R}^\nu$, we call operators of the form

$$f \mapsto \int f(\Gamma(t, x)) \check{b}(x, t, t)\, dt, \quad C_0^\infty(\mathfrak{M}) \to C_0^\infty(\mathfrak{M}) \tag{4.9}$$

$(X, \vec{d}), a$ pseudo-differential operators of order m supported in $\Omega_1 \times \Omega_2$. Here, $\check{b}(x, s, t)$ denotes the inverse Fourier transform in the ξ variable of $b(x, s, \xi)$ (see Remark 4.1.4) and Γ is as in (4.3).

Remark 4.1.3. In Definition 4.1.2 a small number $a = a(\Omega_1, \Omega_2, (X, \vec{d})) > 0$ is used in the definition of $(X, \vec{d}), a$ pseudo-differential operators. Various results which follow require that $a = a(\Omega_1, \Omega_2, (X, \vec{d})) > 0$ be "sufficiently small." In these results, how small a has to be can depend on Ω_1, Ω_2, and the particular vector fields with formal degrees $(X, \vec{d})$. Since there are only finitely many of these results, if $a = a(\Omega_1, \Omega_2, (X, \vec{d})) > 0$ is chosen sufficiently small, all of these results hold. In our applications, we will take a that small.

Remark 4.1.4. $\check{b}(x, s, t)$ is a priori only defined as a tempered distribution, and this does not immediately give $\check{b}(x, t, t)$ a meaning. However, we define the integral in (4.9) as

$$\int f(\Gamma(t, x)) \check{b}(x, t, t)\, dt := \iint f(\Gamma(t, x)) b(x, t, \xi) e^{2\pi i t \cdot \xi}\, dt\, d\xi.$$

It is easy to see that this converges for $f \in C_0^\infty(\mathfrak{M})$ and $a > 0$ sufficiently small. Theorem 4.3.3 extends this definition to $f \in C_0^\infty(\mathfrak{M})'$.

Remark 4.1.5. $(X, \vec{d}), a$ pseudo-differential operators have many similarities with the standard pseudo-differential operators described in Definition 2.2.3. However, unlike the standard pseudo-differential operators, we do not have a robust calculus like the one in Theorem 2.2.8. In fact, a very useful fact about standard pseudo-differential operators is that if $b_1(x,D)$ and $b_2(x,D)$ are standard pseudo-differential operators of orders m_1 and m_2, then $[b_1(x,D), b_2(x,D)]$ is a standard pseudo-differential operator of order m_1+m_2-1. In particular, the commutator is of lower order than the composition $b_1(x,D)b_2(x,D)$, which is of order m_1+m_2. $(X,\vec{d}), a$ pseudo-differential operators have no such gain when considering commutators; in fact, this is a necessary part of our theory – see Proposition 4.5.31. There are some remnants of the calculus of pseudo-differential operators, however; see Proposition 4.5.31. See also Section 4.6 for some comments on recent results for related pseudo-differential operators where a calculus can be obtained.

Example 4.1.6. Let $\psi \in C_0^\infty(\Omega_1)$. Then for any $a > 0$, $\mathrm{Mult}[\psi]$ is an $(X,\vec{d}), a$ pseudo-differential operator of order 0 supported in $\Omega_1 \times \Omega_2$. Indeed, take $\eta \in C_0^\infty(B^q(a))$, with $\eta \equiv 1$ on a neighborhood of 0. Set $b(x,s,\xi) := \psi(x)\eta(s)$; it is immediate to verify that $b \in S^0(a,\Omega_1,\vec{d})$. Note that $\check{b}(x,t,t) = \psi(x)\eta(t)\delta_0(t) = \psi(x)\delta_0(t)$, where $\delta_0(t)$ denotes the Dirac δ function at 0. Thus,

$$\int f(\Gamma(t,x))\check{b}(x,t,t)\,dt = \psi(x)f(\Gamma(0,x)) = \psi(x)f(x).$$

See Proposition 6.11.3 for a more detailed discussion of this perspective for multiplication operators.

4.1.1 Connection with standard pseudo-differential operators

Formally, Definition 4.1.2 is very similar to the formula for standard pseudo-differential operators on $\mathbb{R}^n$ given in Remark 2.2.15. In this section, we make precise the connection between the two.

Let $\mathfrak{M} = \mathbb{R}^n$ and consider the vector fields on $\mathbb{R}^n$ with formal degrees $(\partial,1) = \{(\partial_{x_1},1),\ldots,(\partial_{x_n},1)\}$. $(\partial,1)$ satisfies the main assumption (4.2) (with $c_{j,k}^l = 0$ for all j,k,l). Thus, it makes sense to talk about $(\partial,1)$ pseudo-differential operators. These are not exactly the same as standard pseudo-differential operators as described in Definition 2.2.3, though they are almost the same, as the next proposition shows.

Proposition 4.1.7. *Fix relatively compact, open sets $\Omega_1 \Subset \Omega_2 \Subset \mathbb{R}^n$.*

(i) *Let $b(x,D) : \mathscr{S}(\mathbb{R}^n) \to \mathscr{S}(\mathbb{R}^n)$ be a (standard) pseudo-differential operator of order $m \in \mathbb{R}$ and let $\psi \in C_0^\infty(\Omega_1)$. Then if $a = a(\Omega_1,\Omega_2) > 0$ is sufficiently small,*

$$\mathrm{Mult}[\psi]b(x,D) = T + R,$$

where T is a $(\partial,1), a$ pseudo-differential operator of order m supported in $\Omega_1 \times \Omega_2$ and $R \in C_{\mathrm{loc}}^\infty(\mathbb{R}^n \times \mathbb{R}^n)$ (i. e., the Schwartz kernel of R is infinitely smooth).

(ii) *Let $a > 0$ and suppose T is a $(\partial, 1), a$ pseudo-differential operator of order $m \in \mathbb{R}$ supported in $\Omega_1 \times \Omega_2$. Then T is a (standard) pseudo-differential operator of order m.*

Proof. (i): Take $a = a(\Omega_1, \Omega_2) > 0$ so small that $\{x + t : x \in \Omega_1, |t| < a\} \Subset \Omega_2$. Let $\eta \in C_0^\infty(B^n(a))$ equal 1 on a neighborhood of 0. For $b \in S^m$ (where S^m is defined in Definition 2.2.3), we have

$$\begin{aligned} \mathrm{Mult}[\psi]b(x,D)f(x) &= \int f(x-t)\psi(x)\check{b}(x,t)\,dt \\ &= \int f(e^{-t\cdot\partial}x)\psi(x)\eta(t)\check{b}(x,t)\,dt + \int f(x-t)\psi(x)(1-\eta(t))\check{b}(x,t)\,dt \\ &=: Tf(x) + Rf(x), \end{aligned}$$

where we have used $e^{-t\cdot\partial}x = x - t$. Note that $Tf(x)$ is a $(\partial, 1), a$ pseudo-differential operator of order m with symbol $\eta(s)b(x,\xi) \in S^m(a, \Omega, 1)$. By Proposition 2.2.14, using the fact that $1 - \eta(t)$ is zero on a neighborhood of 0, we have $\psi(x)(1-\eta(t))\check{b}(x,t) \in C^\infty_{\mathrm{loc}}$, and therefore the Schwartz kernel of R is in C^∞_{loc}.

(ii): Since T is a $(\partial, 1), a$ pseudo-differential operator of order m supported in $\Omega_1 \times \Omega_2$, there is $b \in S^m(a, \Omega_1, 1)$ so that

$$Tf(x) = \int f(x-t)b(x,t,\xi)e^{2\pi i t\cdot\xi}\,dt\,d\xi = \int f(y)c(x,y,\xi)e^{2\pi i(y-x)\cdot\xi}\,dy\,d\xi,$$

where $c(x,y,\xi) = b(x, y-x, \xi)$. For all $\alpha, \beta, \gamma \in \mathbb{N}^n$, we have

$$|\partial_x^\alpha \partial_y^\beta \partial_\xi^\gamma c(x,y,\xi)| \le C_{\alpha,\beta,\gamma}(1+|\xi|)^{m-|\gamma|}.$$

It follows that c is a "compound symbol of order m" in the sense of [216, Chapter VI, Section 6.1], and the results there show that T is a pseudo-differential operator of order m. □

4.1.2 Littlewood–Paley decomposition of symbols

Theorem 2.2.26 (ii) ⇔ (iii) provides a decomposition of standard symbols. We require a similar decomposition for symbols in $S^m(a, \Omega, \vec{d})$. For this, we require some definitions; throughout, $a > 0$ and $\Omega \Subset \mathfrak{M}$ is open and relatively compact.

Definition 4.1.8. The space $C_0^\infty(\Omega)\widehat{\otimes}C_0^\infty(B^q(a))\widehat{\otimes}\mathscr{S}(\mathbb{R}^q)$ is the space of all $f(x,s,t) \in C^\infty_{\mathrm{loc}}(\mathfrak{M} \times B^q(a) \times \mathbb{R}^q)$ such that:

(i) $\{(x,s) \in \Omega \times B^q(a) : \exists t \in \mathbb{R}^q, f(x,s,t) \neq 0\} \Subset \Omega \times B^q(a)$.

(ii) For every smooth coordinate system $\Phi : B^n(1) \xrightarrow{\sim} \Phi(B^n(1)) \subseteq \mathfrak{M}$, $\forall m \in \mathbb{N}$, for all multi-indices $\alpha \in \mathbb{N}^n$, $\beta, \gamma \in \mathbb{N}^q$,

$$|\partial_u^\alpha \partial_s^\beta \partial_t^\gamma f(\Phi(u), s, t)| \le C_{\Phi,m,\alpha,\beta,\gamma}(1+|t|)^{-m}.$$

Remark 4.1.9. $C_0^\infty(\Omega)\widehat{\otimes}C_0^\infty(B^q(a))\widehat{\otimes}\mathscr{S}(\mathbb{R}^q)$ is a locally convex topological vector space. The topology can be understood as follows. For compact sets $\mathcal{K}_1 \Subset \Omega$ and $\mathcal{K}_2 \Subset B^q(a)$, the space $C_0^\infty(\mathcal{K}_1)\widehat{\otimes}C_0^\infty(\mathcal{K}_2)\widehat{\otimes}\mathscr{S}(\mathbb{R}^q)$ consisting of functions in $C_0^\infty(\Omega)\widehat{\otimes}C_0^\infty(B^q(a))\widehat{\otimes}\mathscr{S}(\mathbb{R}^q)$ whose support in the (x,s) variable is contained in $\mathcal{K}_1 \times \mathcal{K}_2$ can be given a Fréchet topology in the usual way. Letting $\Omega = \bigcup \mathcal{K}_1^m$ and $B^q(a) = \bigcup \mathcal{K}_2^m$, where $\mathcal{K}_1^m$ and $\mathcal{K}_2^m$ are increasing sequences of compact sets, we give $C_0^\infty(\Omega)\widehat{\otimes}C_0^\infty(B^q(a))\widehat{\otimes}\mathscr{S}(\mathbb{R}^q)$ the topology of the inductive limit of the Fréchet spaces $C_0^\infty(\mathcal{K}_1^m)\widehat{\otimes}C_0^\infty(\mathcal{K}_2^m)\widehat{\otimes}\mathscr{S}(\mathbb{R}^q)$; it is easy to see that this limit does not depend on the choice of sequences of compact sets. Thus, not only is $C_0^\infty(\Omega)\widehat{\otimes}C_0^\infty(B^q(a))\widehat{\otimes}\mathscr{S}(\mathbb{R}^q)$ a locally convex topological vector space, it is an LF space in the sense of [238, Chapter 13].

Lemma 4.1.10. *$\mathcal{B} \subset C_0^\infty(\Omega)\widehat{\otimes}C_0^\infty(B^q(a))\widehat{\otimes}\mathscr{S}(\mathbb{R}^q)$ is a bounded set if and only if:*
- $\{(x,s) \in \Omega \times B^q(a) : \exists f \in \mathcal{B}, t \in \mathbb{R}^q, f(x,s,t) \neq 0\} \Subset \Omega \times B^q(a)$.
- *The constant $C_{\Phi,m,\alpha,\beta,\gamma}$ from Definition* 4.1.8 (ii) *can be chosen independent of $f \in \mathcal{B}$.*

Proof. This follows from [238, Proposition 14.6], combined with Remark 4.1.9. □

Definition 4.1.11. For a set $E \subseteq \{1,\dots,\nu\}$, we let $C_0^\infty(\Omega)\widehat{\otimes}C_0^\infty(B^q(a))\widehat{\otimes}\mathscr{S}_E(\mathbb{R}^q)$ be the subspace of all $f \in C_0^\infty(\Omega)\widehat{\otimes}C_0^\infty(B^q(a))\widehat{\otimes}\mathscr{S}(\mathbb{R}^q)$ such that

$$\int_{\mathbb{R}^{q_\mu}} (t^\mu)^\alpha f(x,s,t)\,dt^\mu = 0, \quad \forall \mu \in E, \alpha \in \mathbb{N}^{q_\mu}.$$

$C_0^\infty(\Omega)\widehat{\otimes}C_0^\infty(B^q(a))\widehat{\otimes}\mathscr{S}_E(\mathbb{R}^q)$ is a closed subspace of $C_0^\infty(\Omega)\widehat{\otimes}C_0^\infty(B^q(a))\widehat{\otimes}\mathscr{S}(\mathbb{R}^q)$ and we give it the subspace topology.

Definition 4.1.12. Given $f(x,s,t) \in C_0^\infty(\Omega)\widehat{\otimes}C_0^\infty(B^q(a))\widehat{\otimes}\mathscr{S}(\mathbb{R}^q)$ and $2^j = (2^{j_1},\dots,2^{j_\nu}) \in (0,\infty)^\nu$, we define $\mathrm{Dil}_{2^j}^{\vec{d}}(f)(x,s,t) \in C_0^\infty(\Omega)\widehat{\otimes}C_0^\infty(B^q(a))\widehat{\otimes}\mathscr{S}(\mathbb{R}^q)$ by

$$\mathrm{Dil}_{2^j}^{\vec{d}}(f)(x,s,t) = 2^{j_1(\sum_{l=1}^{q_1} d_l^1)+j_2(\sum_{l=1}^{q_2} d_l^2)+\cdots+j_\nu(\sum_{l=1}^{q_\nu} d_l^\nu)} f(x,s,2^{j\vec{d}}t),$$

where $2^{j\vec{d}}t = (2^j)^{\vec{d}}t$ is defined by (4.7).

Remark 4.1.13. Note that $\mathrm{Dil}_{2^j}^{\vec{d}}(f)$ is defined so that $\int \mathrm{Dil}_{2^j}^{\vec{d}}(f)(x,s,t)\,dt = \int f(x,s,t)\,dt$.

The main proposition of this section is the following.

Proposition 4.1.14. *Fix $a > 0$ and $m \in \mathbb{R}^\nu$. The following are equivalent:*
(i) $b \in S^m(a,\Omega,\vec{d})$.
(ii) *There exists a bounded set*

$$\{\varsigma_j : j \in \mathbb{N}^\nu\} \subset C_0^\infty(\Omega)\widehat{\otimes}C_0^\infty(B^q(a))\widehat{\otimes}\mathscr{S}(\mathbb{R}^q)$$

such that

$$\varsigma_j \in C_0^\infty(\Omega)\widehat{\otimes}C_0^\infty(B^q(a))\widehat{\otimes}\mathscr{S}_{\{\mu:j_\mu\neq 0\}}(\mathbb{R}^q)$$

and

$$\check{b}(x,s,t) = \sum_{j\in\mathbb{N}^\nu} 2^{j\cdot m}\,\mathrm{Dil}_{2^j}^{\vec{d}}(\varsigma_j)(x,s,t),$$

where $\check{b}(x,s,t)$ denotes the inverse Fourier transform of $b(x,s,\xi)$ in the ξ variable, and the above sum converges in the weak topology on

$$(C_0^\infty(\Omega)\widehat{\otimes}C_0^\infty(B^q(a))\widehat{\otimes}\mathscr{S}(\mathbb{R}^q))'.$$

Moreover, every such sum converges in this sense.

The rest of this section is devoted to the proof of Proposition 4.1.14, which uses standard techniques. For $\varsigma(x,s,t) \in C_0^\infty(\Omega)\widehat{\otimes}C_0^\infty(B^q(a))\widehat{\otimes}\mathscr{S}(\mathbb{R}^q)$, we write $\hat{\varsigma}(x,s,\xi) \in C_0^\infty(\Omega)\widehat{\otimes}C_0^\infty(B^q(a))\widehat{\otimes}\mathscr{S}(\mathbb{R}^q)$ for the Fourier transform of ς in the t variable. It is easy to see that this gives an automorphism of $C_0^\infty(\Omega)\widehat{\otimes}C_0^\infty(B^q(a))\widehat{\otimes}\mathscr{S}(\mathbb{R}^q)$ (see also [238, Proposition 13.1]). In particular, it takes bounded subsets of $C_0^\infty(\Omega)\widehat{\otimes}C_0^\infty(B^q(a))\widehat{\otimes}\mathscr{S}(\mathbb{R}^q)$ to bounded subsets of $C_0^\infty(\Omega)\widehat{\otimes}C_0^\infty(B^q(a))\widehat{\otimes}\mathscr{S}(\mathbb{R}^q)$.

Definition 4.1.15. For a set $E \subseteq \{1,\dots,\nu\}$, we let $C_0^\infty(\Omega)\widehat{\otimes}C_0^\infty(B^q(a))\widehat{\otimes}\widehat{\mathscr{S}_E}(\mathbb{R}^q)$ be the subspace of all $f(x,s,\xi) \in C_0^\infty(\Omega)\widehat{\otimes}C_0^\infty(B^q(a))\widehat{\otimes}\mathscr{S}(\mathbb{R}^q)$ such that

$$\partial_{\xi^\mu}^\alpha f(x,s,\xi)\big|_{\xi^\mu=0} = 0, \quad \forall \mu \in E, \alpha \in \mathbb{N}^{q_\mu}.$$

$C_0^\infty(\Omega)\widehat{\otimes}C_0^\infty(B^q(a))\widehat{\otimes}\widehat{\mathscr{S}_E}(\mathbb{R}^q)$ is a closed subspace of $C_0^\infty(\Omega)\widehat{\otimes}C_0^\infty(B^q(a))\widehat{\otimes}\mathscr{S}(\mathbb{R}^q)$ and we give it the subspace topology.

Lemma 4.1.16. *The map $\varsigma \mapsto \hat{\varsigma}$ is an isomorphism $C_0^\infty(\Omega)\widehat{\otimes}C_0^\infty(B^q(a))\widehat{\otimes}\mathscr{S}_E(\mathbb{R}^q) \to C_0^\infty(\Omega)\widehat{\otimes}C_0^\infty(B^q(a))\widehat{\otimes}\widehat{\mathscr{S}_E}(\mathbb{R}^q)$.*

Proof. This follows easily from the definitions. □

Lemma 4.1.17. *Fix $s \in \mathbb{R}$ and $E \subseteq \{1,\dots,\nu\}$. Then, $\forall \mu \in E$, the maps $f(x,s,\xi) \mapsto |\xi^\mu|^s f(x,s,\xi)$ and $f(x,s,\xi) \mapsto \|\xi^\mu\|_\mu^s f(x,s,\xi)$ are automorphisms of $C_0^\infty(\Omega)\widehat{\otimes}C_0^\infty(B^q(a))\widehat{\otimes}\widehat{\mathscr{S}_E}(\mathbb{R}^q)$.*

Proof. Note $\|\xi^\mu\|_\mu$ is smooth away from $\xi^\mu = 0$ (see (4.5)). Thus, it follows that the map $f(x,s,\xi) \mapsto \|\xi^\mu\|_\mu^s f(x,s,\xi)$ is a bijection

$$C_0^\infty(\Omega)\widehat{\otimes}C_0^\infty(B^q(a))\widehat{\otimes}\widehat{\mathscr{S}_E}(\mathbb{R}^q) \to C_0^\infty(\Omega)\widehat{\otimes}C_0^\infty(B^q(a))\widehat{\otimes}\widehat{\mathscr{S}_E}(\mathbb{R}^q),$$

with inverse $f(x,s,\xi) \mapsto \|\xi^\mu\|_\mu^{-s} f(x,s,\xi)$. It is elementary to show that this map is continuous (see [238, Proposition 13.1]), and we leave the details to the interested reader. A similar proof gives the result for $f(x,s,\xi) \mapsto |\xi^\mu|^s f(x,s,\xi)$. □

Definition 4.1.18. For $s \in \mathbb{R}$, $\mu \in \{1, \ldots, \nu\}$, set

$$\triangle_\mu^s \varsigma(x, s, t) = (|2\pi\xi^\mu|^{2s} \hat{\varsigma}(x, s, \xi))^\vee,$$

where $\vee$ denotes the inverse Fourier transform in the ξ variable.

Corollary 4.1.19. *Fix $s \in \mathbb{R}$ and $E \subseteq \{1, \ldots, \nu\}$. Then $\forall E$ such that $\mu \in E$,*

$$\triangle_\mu^s : C_0^\infty(\Omega) \widehat{\otimes} C_0^\infty(B^q(a)) \widehat{\otimes} \mathscr{S}_E(\mathbb{R}^q) \to C_0^\infty(\Omega) \widehat{\otimes} C_0^\infty(B^q(a)) \widehat{\otimes} \mathscr{S}_E(\mathbb{R}^q)$$

is an automorphism.

Proof. This follows immediately from Lemmas 4.1.16 and 4.1.17. □

Proof of Proposition 4.1.14. Applying the Fourier transform in the t variable, Lemma 4.1.16 shows that (ii) is equivalent to:
(ii)' There is a bounded set

$$\{b_j(x, s, \xi) : j \in \mathbb{N}^\nu\} \subset C_0^\infty(\Omega) \widehat{\otimes} C_0^\infty(B^q(a)) \widehat{\otimes} \mathscr{S}(\mathbb{R}^q)$$

such that

$$b_j(x, s, \xi) \in C_0^\infty(\Omega) \widehat{\otimes} C_0^\infty(B^q(a)) \widehat{\otimes} \widehat{\mathscr{S}}_{\{\mu : j_\mu \neq 0\}}(\mathbb{R}^q)$$

and such that

$$b(x, s, \xi) = \sum_{j \in \mathbb{N}^\nu} 2^{j \cdot m} b_j(x, s, 2^{-j\vec{d}} \xi),$$

with convergence in the weak topology on

$$(C_0^\infty(\Omega) \widehat{\otimes} C_0^\infty(B^q(a)) \widehat{\otimes} \mathscr{S}(\mathbb{R}^q))',$$

and every such sum converges in this sense.

(ii)' ⇒ (i): Let $b_j(x, s, \xi)$ be as in (ii)'. We consider the sum

$$b(x, s, \xi) = \sum_{j \in \mathbb{N}^\nu} \sum_{j \in \mathbb{N}^\nu} 2^{j \cdot m} b_j(x, s, 2^{-j\vec{d}} \xi). \tag{4.10}$$

We will show that this sum converges in the weak topology on

$$(C_0^\infty(\Omega) \widehat{\otimes} C_0^\infty(B^q(a)) \widehat{\otimes} \mathscr{S}(\mathbb{R}^q))',$$

and converges in $C_{\text{loc}}^\infty(\mathfrak{M} \times B^q(a) \times \mathbb{R}^q)$ to a symbol $b(s, x, \xi) \in S^m(a, \Omega, \vec{d})$. That such b satisfies Definition 4.1.8 (i) is obvious. Thus, to show $b \in S^m(a, \Omega, \vec{d})$, we wish to show that for any smooth coordinate system $\Phi : B^n(1) \xrightarrow{\sim} \Phi(B^n(1)) \subseteq \mathfrak{M}$,

$$\left|\partial_u^\alpha \partial_s^\beta \partial_\xi^\gamma \sum_{j\in\mathbb{N}} 2^{j\cdot m} b_j(\Phi(u), s, 2^{-j\vec{d}}\xi)\right| \lesssim \prod_{\mu=1}^{\nu} (1+\|\xi^\mu\|_\mu)^{m_\mu - \deg_\mu(\gamma_\mu)}. \tag{4.11}$$

Note that with $\Phi^* b_j(u,s,\xi) := b_j(\Phi(u), s, \xi)$,

$$\partial_u^\alpha \partial_s^\beta \partial_\xi^\gamma b_j(\Phi(u), s, 2^{-jd}\xi) = 2^{-j\cdot\deg_d(\gamma)}(\partial_u^\alpha \partial_s^\beta \partial_\xi^\gamma \Phi^* b_j)(u, s, 2^{-jd}\xi),$$

it suffices to prove (4.11) in the case $\alpha = 0$, $\beta = 0$, and $\gamma = 0$, by replacing m with $m - \deg_{\vec{d}}(\gamma)$.

We will show that for each $E \subseteq \{1, \dots, \nu\}$,

$$\sum_{\substack{j\in\mathbb{N}^\nu \\ j_\mu>0, \mu\in E \\ j_\mu=0, \mu\notin E}} \left[\prod_{\mu=1}^{\nu} (1+\|\xi^\mu\|_\mu)^{-m_\mu} 2^{j_\mu m_\mu}\right] |b_j(x, s, 2^{-j\cdot\vec{d}}\xi)| \tag{4.12}$$

converges absolutely and is bounded independent of x, s, and ξ, which, when combined with the above remarks, implies both the desired convergence of (4.10) and the desired bound (4.11).

Set $L \geq \max_{1\leq\mu\leq\nu} |m_\mu| + 1$. We claim, with $E = \{\mu : j_\mu \neq 0\}$,

$$|\varsigma_j(x, s, \xi)| \lesssim \left[\prod_{\mu\notin E} (1+\|\xi^\mu\|_\mu)^{-L}\right] \times \left[\prod_{\mu\in E} (1+\|\xi^\mu\|_\mu)^{-L} \wedge \|\xi^\mu\|_\mu^L\right]. \tag{4.13}$$

Indeed, take $F \subseteq E$ and note that by Lemma 4.1.17, if $b_{j,F}(x, s, \xi) := \|\xi^\mu\|_\mu^{-L} b_j(x, s, \xi)$, then

$$\{b_{j,F} : j \in \mathbb{N}^\nu, j_\mu \neq 0, \forall \mu \in F\} \subset C_0^\infty(\Omega)\widehat{\otimes} C_0^\infty(B^q(a))\widehat{\otimes}\mathscr{S}(\mathbb{R}^q)$$

is a bounded set. Thus, it follows immediately from the definitions that

$$\begin{aligned} |b_j(x, s, \xi)| &= \|\xi^\mu\|_\mu^L |b_{j,F}(x, s, \xi)| \\ &\lesssim \left[\prod_{\mu\notin E} (1+\|\xi^\mu\|_\mu)^{-L}\right] \times \left[\prod_{\mu\notin F} (1+\|\xi^\mu\|_\mu)^{-L}\right] \times \left[\prod_{\mu\in F} \|\xi^\mu\|_\mu^L\right]. \end{aligned}$$

Taking the minimum of the right-hand side over $F \subseteq E$ yields (4.13).

We separate the sum (4.12) into several parts, one for each subset $F \subseteq E$. For a fixed $\xi \in \mathbb{R}^q$, we set

$$\begin{aligned} A_{F,E}(\xi) := \{j \in \mathbb{N}^\nu :& \forall \mu \notin E, j_\mu = 0, \\ & \forall \mu \in E, j_\mu > 0, \\ & \forall \mu \in F, \|\xi^\mu\|_\mu \vee 2 < 2^{j_\mu}, \\ & \forall \mu \in E \setminus F, 2 \leq 2^{j_\mu} \leq 2 \vee \|\xi^\mu\|_\mu\}. \end{aligned}$$

Note that

$$\bigcup_{F\subseteq E} A_{F,E}(\xi) = \{j \in \mathbb{N}^\nu : j_\mu = 0 \Leftrightarrow \mu \in E\}. \tag{4.14}$$

Using (4.13), we have

$$\begin{aligned}
&\sum_{j\in A_{F,E}(\xi)} \left[\prod_{\mu\in\{1,\dots,\nu\}} (1+\|\xi^\mu\|_\mu)^{-m_\mu} 2^{j_\mu}\right] |b_j(x,s,2^{-j\vec{d}}\xi)| \\
&\lesssim \sum_{j\in A_{F,E}(\xi)} \left[\prod_{\mu\in\{1,\dots,\nu\}} (1+\|\xi^\mu\|_\mu)^{-m_\mu} 2^{j m_\mu}\right] \times \left[\prod_{\mu\notin F} (1+2^{-j_\mu}\|\xi^\mu\|_\mu)^{-L}\right] \\
&\quad \times \left[\prod_{\mu\in F} (2^{-j_\mu}\|\xi^\mu\|_\mu)^{L}\right] \\
&\lesssim \sum_{j\in A_{F,E}(\xi)} \left[\prod_{\mu\in E^c} 1\right] \times \left[\prod_{\mu\in F^c\setminus E^c} (1+\|\xi^\mu\|_\mu)^{-m_\mu} 2^{j_\mu(m_\mu+L)} \|\xi^\mu\|_\mu^{-L}\right] \\
&\quad \times \left[\prod_{\mu\in F} (1+\|\xi^\mu\|_\mu)^{-m_\mu} 2^{j_\mu(m_\mu-L)} \|\xi^\mu\|_\mu^{L}\right] \\
&\lesssim \left[\prod_{\mu\in F^c\setminus E^c} (1+\|\xi^\mu\|_\mu)^{-m_\mu} \|\xi^\mu\|^{-L} \|\xi^\mu\|_\mu^{m_\mu+L}\right] \\
&\quad \times \left[\prod_{\mu\in F} (1+\|\xi^\mu\|_\mu)^{-m_\mu} \|\xi^\mu\|_\mu^{m_\mu-L} \|\xi^\mu\|_\mu^{L}\right] \\
&\lesssim 1,
\end{aligned}$$

where in the second to last estimate we have used the definition of $A_{F,E}(\xi)$ and the fact that these are geometric sums. Combining this estimate with (4.14) proves (4.12) and completes the proof of (i).

(i) $\Rightarrow$ (ii)': For each $\mu \in \{1,\dots,\nu\}$, let $\phi_\mu \in C_0^\infty(\mathbb{R}^{q_\mu})$ satisfy $\phi_\mu(\xi^\mu) = 1$ on a neighborhood of $0 \in \mathbb{R}^{q_\mu}$. Set $\psi_\mu(\xi^\mu) := \phi_\mu(\xi^\mu) - \phi_\mu(2^{d^\mu}\xi^\mu)$. Note that $1 = \phi_\mu(\xi^\mu) + \sum_{j_\mu>0} \psi_\mu(2^{-j_\mu d^\mu}\xi^\mu)$. Set

$$b_j(x,s,2^{-j\vec{d}}\xi) := 2^{-j\cdot m}\left[\prod_{\mu \text{ with } j_\mu=0} \phi_\mu(\xi^\mu)\right] \times \left[\prod_{\mu \text{ with } j_\mu>0} \psi_\mu(2^{-j_\mu d^\mu}\xi^\mu)\right] b(x,s,\xi).$$

Note that $b(x,s,\xi) = \sum_{j\in\mathbb{N}^\nu} 2^{j\cdot m} b_j(x,s,2^{-j\vec{d}}\xi)$, $\forall \xi \in \mathbb{R}^q$.

Since $\phi_\mu(\xi^\mu)$ is supported on $\|\xi^\mu\|_\mu \lesssim 1$ and $\psi_\mu(2^{-j_\mu d^\mu}\xi^\mu)$ is supported on $\|\xi^\mu\|_\mu \approx 2^{j_\mu}$, it follows easily from the definitions that

$$\{b_j(x,s,\xi) : j \in \mathbb{N}^\nu\} \subset C_0^\infty(\Omega)\widehat{\otimes} C_0^\infty(B^q(a))\widehat{\otimes}\mathscr{S}(\mathbb{R}^q)$$

is a bounded set. Furthermore, since $\psi_\mu(\xi^\mu)$ vanishes on a neighborhood of $0 \in \xi^\mu$, we see $b_j(x,s,\xi) \in C_0^\infty(\Omega)\widehat{\otimes}C_0^\infty(B^q(a))\widehat{\otimes}\widehat{\mathscr{S}}_{\{\mu:j_\mu\neq 0\}}(\mathbb{R}^q)$.

Thus, by the proof of (ii)' ⇒ (i), the sum $\sum_{j\in\mathbb{N}^\nu} 2^{j\cdot m} b_j(x,s,2^{-j\vec{d}}\xi)$ converges in $C^\infty_{\mathrm{loc}}(\mathfrak{M}\times B^q(a)\times\mathbb{R}^q)$ and in the sense of the weak topology on

$$(C_0^\infty(\Omega)\widehat{\otimes}C_0^\infty(B^q(a))\widehat{\otimes}\mathscr{S}(\mathbb{R}^q))'.$$

Since we already know that the sum converges pointwise to $b(x,s,\xi)$, the sum converges in the above senses to $b(x,s,\xi)$. This completes the proof. □

4.2 The exponential map

In this section, we revisit the exponential map defined in Section 3.1.1 and describe the relationship between our main assumption on $(X,\vec{d})$, namely (4.2), and the map Γ defined in (4.3).

Recall the degree of a multi-index $\alpha\in\mathbb{N}^q$, $\deg_{\vec{d}}(\alpha)\in\mathbb{N}^\nu$, defined in (4.8).

Definition 4.2.1. For a function $f\in C^\infty_{\mathrm{loc}}(U)$, where $U\subseteq\mathbb{R}^q\times\mathfrak{M}$ is an open neighborhood of $\{0\}\times\mathfrak{M}$, and for $\vec{d}_0\in\mathbb{Z}^\nu$, we say

$$\deg_{\vec{d}}(f)\geq\vec{d}_0$$

if $f(t,x)=\sum_{k=1}^K t^{\alpha_k} f_k(t,x)$, where $\deg_{\vec{d}}(\alpha_k)\geq\vec{d}_0$ for $1\leq k\leq K$ and $f_k\in C^\infty_{\mathrm{loc}}(U')$ for some open neighborhood $U'\subseteq\mathbb{R}^q\times\mathfrak{M}$ of $\{0\}\times\mathfrak{M}$. The inequality $\deg_{\vec{d}}(\alpha_k)\geq\vec{d}_0$ means the vector $\deg_{\vec{d}}(\alpha_k)-\vec{d}_0$ is non-negative in every component.

The main proposition of this section is the following.

Proposition 4.2.2. *For $f\in C^\infty_{\mathrm{loc}}(\mathfrak{M})$, we have the following identities for $1\leq j\leq q$:*
(i) $X_j(f(\Gamma(t,x)))=((X_j+\sum_{k=1}^q g_{1,j}^k X_k)f)(\Gamma(t,x))$.
(ii) $(X_jf)(\Gamma(t,x))=(X_j+\sum_{k=1}^q g_{2,j}^k X_k)(f(\Gamma(t,x)))$.
(iii) $\partial_{t_j} f(\Gamma(t,x))=((-X_j+\sum_{k=1}^q g_{3,j}^k X_k)f)(\Gamma(t,x))$.
(iv) $\partial_{t_j} f(\Gamma(t,x))=(-X_j+\sum_{k=1}^q g_{4,j}^k X_k)(f(\Gamma(t,x)))$.
(v) $X_j(f(\Gamma(t,x)))=(-\partial_{t_j}+\sum_{k=1}^q g_{5,j}^k \partial_{t_k})(f(\Gamma(t,x)))$.
(vi) $(X_jf)(\Gamma(t,x))=(-\partial_{t_j}+\sum_{k=1}^q g_{6,j}^k \partial_{t_k})(f(\Gamma(t,x)))$.

Here $g_{l,j}^l(t,x)\in C^\infty_{\mathrm{loc}}(U)$, $U\subseteq\mathbb{R}^q\times\mathfrak{M}$ is an open neighborhood of $\{0\}\times\mathfrak{M}$, $\deg_{\vec{d}}(g_{l,j}^k)\geq\vec{d}_k-\vec{d}_j$, and $g_{l,j}^k(0,x)=0$.

To prove Proposition 4.2.2 we introduce some matrix notation, which we will use only in this section. X will denote the column vector of vector fields $[X_1,\dots,X_q]^\top$. If $g_{l,j}^k$ is as in the conclusion of Proposition 4.2.2, we write $G_l(t,x)$ for the $q\times q$ matrix with the (j,k) component equal to $g^k_{l,j}(t,x)$, so that $G_l \in C^\infty_{\mathrm{loc}}(\mathfrak{M};\mathbb{M}^{q\times q})$. For example, Proposition 4.2.2 (i) can be rewritten as $X(f(\Gamma(t,x))) = ((I+G_1)Xf)(\Gamma(t,x))$ and (ii) can be rewritten as $(Xf)(\Gamma(t,x)) = (I+G_2)X(f(\Gamma(t,x)))$.

Lemma 4.2.3. *Let $A, B \in C^\infty_{\mathrm{loc}}(U;\mathbb{M}^{q\times q})$ be $q\times q$ matrices with components $A^k_j(t,x)$ and $B^k_j(t,x)$, respectively, where U is an open neighborhood of $\{0\}\times\mathfrak{M}$ in $\mathbb{R}^q\times\mathfrak{M}$. Suppose that*

$$\deg_{\vec{d}}(A^k_j), \deg_{\vec{d}}(B^k_j) \geq \vec{d}_k - \vec{d}_j.$$

Then if $(AB)^k_j(t,x)$ is the (j,k) component of AB, we have

$$\deg_{\vec{d}}((AB)^k_j) \geq \vec{d}_k - \vec{d}_j. \tag{4.15}$$

Also,

$$\deg_{\vec{d}}(A^k_j(t,\Gamma(t,x))) \geq \vec{d}_k - \vec{d}_j, \tag{4.16}$$

and similarly with $\Gamma(t,x)$ replaced by $\Gamma(-t,x)$. Finally, if $A(0,x)\equiv 0$, then there is an open neighborhood U_1 of $\{0\}\times\mathfrak{M}$ in $\mathbb{R}^q\times\mathfrak{M}$ such that $(I+A)^{-1}\in C^\infty_{\mathrm{loc}}(U_1;\mathbb{M}^{q\times q})$, and

$$\deg_{\vec{d}}(((I+A)^{-1}-I)^k_j) \geq \vec{d}_k - \vec{d}_j, \tag{4.17}$$

where $((I+A)^{-1}-I)^k_j$ denotes the (j,k) component of the matrix $(I+A)^{-1}-I$.

Proof. It follows immediately from the definitions that $\deg_{\vec{d}}(A^l_j B^k_l) \geq \vec{d}_l - \vec{d}_j + \vec{d}_k - \vec{d}_l = \vec{d}_k - \vec{d}_j$, and (4.15) follows. Formula (4.16) follows immediately from the definitions. For (4.17), since $A(0,x)\equiv 0$, for t sufficiently small we have $(I+A)^{-1}-I = \sum_{l=1}^\infty (-1)^l A^l$. Using (4.15), a straightforward argument completes the proof; we leave the details to the interested reader. □

The next lemma is a key step in the proof of Proposition 4.2.2.

Lemma 4.2.4. *Let $\widehat{\Gamma}(t,x) := (t,\Gamma(t,x)) \in \mathbb{R}^q\times\mathfrak{M}$. Then, for $1\leq j\leq q$,*

$$\widehat{\Gamma}_* X_j = X_j + \sum_{k=1}^q g^k_{1,j} X_k,$$

$$\widehat{\Gamma}_* \partial_{t_j} = \partial_{t_j} - X_j + \sum_{k=1}^q g^k_{3,j} X_k,$$

where $g^k_{1,j}$ and $g^k_{3,j}$ are as in Proposition 4.2.2.

To prove Lemma 4.2.4 we require a few lemmas. For the next lemma, let $Z \in C^\infty_{\mathrm{loc}}(\mathfrak{M};T\mathfrak{M})$, and for $s\in\mathbb{R}$ small, set $\Phi_s(x) = e^{sZ}x$. $\Phi_s(x)$ is defined on a neighborhood of $\{0\}\times\mathfrak{M}$ in $\mathbb{R}\times\mathfrak{M}$. Since $\Phi_0(x) = x$, given $x \in \mathfrak{M}$, for s small, there is a neighborhood V of x such that $\Phi_s|_V$ is a diffeomorphism onto its image (by the inverse function theorem). Thus, if Y is another vector field, it makes sense to consider $(\Phi_s)_* Y$, which is an element of $C^\infty_{\mathrm{loc}}(U; T\mathfrak{M})$, where U is a neighborhood of $\{0\} \times \mathfrak{M}$ in $\mathbb{R} \times \mathfrak{M}$.

Lemma 4.2.5. (i) $(\Phi_{-s})_* Z = Z$.
(ii) $\partial_s(\Phi_{-s})_* Y = (\Phi_{-s})_*([Z, Y])$.

Proof. For (i), let $f \in C^\infty_{\mathrm{loc}}(\mathfrak{M}; T\mathfrak{M})$. Then

$$\begin{aligned}((\Phi_{-s})_* Z)f &= (\Phi_{-s})_* Z\Phi^*_{-s} f(x) = (\Phi_{-s})_* Zf(e^{-sZ}x) = (\Phi_{-s})_* \partial_u|_{u=0} f(e^{-sZ}e^{uZ}x)\\ &= (\Phi_{-s})_* \partial_u|_{u=0} f(e^{uZ}e^{-sZ}x) = (\Phi_{-s})_*(Zf)(e^{-sZ}x) = Zf(x),\end{aligned}$$

proving (i).

We claim

$$\partial_s|_{s=0}(\Phi_{-s})_* Y = [Z, Y]. \tag{4.18}$$

Indeed, for $f \in C^\infty_{\mathrm{loc}}(\mathfrak{M}; T\mathfrak{M})$,

$$\begin{aligned}(\partial_s|_{s=0}(\Phi_{-s})_* Y)f(x) &= \partial_s|_{s=0}(\Phi_{-s})_* Y\Phi^*_{-s} f(x)\\ &= \partial_s|_{s=0}(\Phi_{-s})_* Yf(x) + \partial_s|_{s=0} Y\Phi^*_{-s} f(x)\\ &= \partial_s|_{s=0}(Yf)(e^{sZ}x) + \partial_s|_{s=0} Y(f(e^{-sZ}x)) = ZYf - YZf,\end{aligned}$$

proving (4.18).

Consider, using (4.18),

$$\partial_s(\Phi_{-s})_* Y = \partial_\epsilon|_{\epsilon=0}(\Phi_{-s-\epsilon})_* Y = \partial_\epsilon|_{\epsilon=0}(\Phi_{-\epsilon})_*(\Phi_{-s})_* Y = [Z, (\Phi_{-s})_* Y]. \tag{4.19}$$

For $f \in C^\infty_{\mathrm{loc}}(\mathfrak{M}; T\mathfrak{M})$, by (i) we have $Zf = (\Phi_{-s})_* Z\Phi^*_{-s} f$, and therefore,

$$\begin{aligned}f &= (Z(\Phi_{-s})_* Y\Phi^*_{-s} - (\Phi_{-s})_* Y\Phi^*_{-s} Z)f\\ &= ((\Phi_{-s})_* ZY\Phi^*_{-s} - (\Phi_{-s})_* YZ\Phi^*_{-s})f = ((\Phi_{-s})_*[Z, Y])f.\end{aligned} \tag{4.20}$$

Combining (4.19) and (4.20) proves (ii). □

For $1 \le \kappa \le \mu$, set

$$\Gamma_\kappa(t^\kappa, x) = e^{-t^\kappa \cdot X^\kappa} x,$$

and for $\epsilon \in [0, 1]$, set

$$\widehat{\Gamma}_{\epsilon,\kappa}(t, x) = \widehat{\Gamma}_{\epsilon,\kappa}((t^1, \ldots, t^\nu), x) := ((t^1, \ldots, t^\nu), \Gamma_\kappa(\epsilon t^\kappa, x)).$$

If U is a sufficiently small neighborhood of $\{0\} \times \mathfrak{M}$ in $\mathbb{R}^q \times \mathfrak{M}$, $\widehat{\Gamma}_{\epsilon,\kappa}|_U$ is a smooth diffeomorphism onto its image, for all $\epsilon \in [0,1]$, and depends smoothly on ϵ. For the rest of this section, all functions of (t,x) will be smooth and defined on some open neighborhood of $\{0\}\times\mathfrak{M}$ in $\mathbb{R}^q\times\mathfrak{M}$. All functions of (ϵ,t,x) will be smooth and defined on $[0,1]\times U$, where U is some open neighborhood of $\{0\}\times\mathfrak{M}$ in $\mathbb{R}^q\times\mathfrak{M}$. Given a function $f(\epsilon,t,x)$, we define $\deg_{\vec{d}}(f) \geq \vec{d}_0$ just as in Definition 4.2.1, but the functions f_k from that definition depend on (ϵ,t,x). In the next lemmas, we write t^κ for an element of $\mathbb{R}^{q_\kappa}$, where $\kappa \in \{1,\dots,\nu\}$. However, for a multi-index $\alpha \in \mathbb{N}^q$, we write t^α for the usual multi-index notation. α,β,γ will all denote multi-indices in $\mathbb{N}^q$, while κ will denote an element of $\{1,\dots,\nu\}$.

Lemma 4.2.6. *For* $1 \leq j \leq q$,

$$(\widehat{\Gamma}_{\epsilon,\kappa})_* X_j = \sum_{j=1}^{q} h_{j,\kappa}^{k,1}(\epsilon,t,x) X_k,$$

where $\deg_{\vec{d}}(h_{j,\kappa}^{k,1}) \geq \vec{d}_k - \vec{d}_j$.

Proof. Set $M_0 := 2\nu \max_{1\leq l\leq q} |\vec{d}_l|_1$. Fix $1 \leq k \leq q$ and consider, for $\alpha \in \mathbb{N}^q$ with $|\alpha| \leq M_0$ and $\deg_{\vec{d}}(\alpha) \geq \vec{d}_k - \vec{d}_j$, using Lemma 4.2.5 (ii), we have

$$\begin{aligned}
\partial_\epsilon(\widehat{\Gamma}_{\epsilon,\kappa})_*(t^\alpha X_k) &= (\widehat{\Gamma}_{\epsilon,\kappa})_*[t^\kappa\cdot X^\kappa, t^\alpha X_k] = \sum_{j'=1}^{q_\kappa} t_{j'}^\kappa t^\alpha (\Gamma_{\epsilon,\kappa})_*[X_{j'}^\kappa, X_k] \\
&= \sum_{j'=1}^{\kappa} t_{j'}^\kappa t^\alpha (\widehat{\Gamma}_{\epsilon,\kappa})_* \sum_{\vec{d}_l \leq d_{j'}^\kappa \mathfrak{e}_\kappa + \vec{d}_k} c_{j',k}^{l,\kappa} X_l \\
&= \sum_{j'=1}^{q_\kappa} \sum_{\vec{d}_l \leq d_{j'}^\kappa \mathfrak{e}_\kappa + \vec{d}_k} c_{j',k}^{l,\kappa}(\Gamma_{\epsilon,\kappa}(-t^\kappa,x))(\widehat{\Gamma}_{\epsilon,\kappa})_* t_{j'}^\kappa t^\alpha X_l,
\end{aligned} \tag{4.21}$$

where $c_{j',k}^{l,\kappa} \in C^\infty_{\mathrm{loc}}(\mathfrak{M})$ come from our main hypothesis (4.2). We consider the right-hand side of (4.21). Note that in the sum, the term $t^\beta := t_{j'}^\kappa t^\alpha$, where $\deg_{\vec{d}}(\beta) = \deg_{\vec{d}}(\alpha) + d_{j'}^\kappa \mathfrak{e}_\kappa \geq \vec{d}_k - \vec{d}_j + d_{j'}^\kappa \mathfrak{e}_\kappa \geq \vec{d}_l - \vec{d}_j$. Thus, X_l is multiplied by t^β, where $\deg_{\vec{d}}(\beta) \geq \vec{d}_l - \vec{d}_j$.

There are two cases. If $|\alpha| < M_0$, then $|\beta| \leq M_0$, in which case (4.21) shows $\partial_\epsilon(\widehat{\Gamma}_{\epsilon,\kappa})_*(t^\alpha X_k)$ is a C^∞_{loc} linear combination of terms of the same form as $(\widehat{\Gamma}_{\epsilon,\kappa})_*(t^\alpha X_k)$. If $|\alpha| = M_0$, then $|\beta| = M_0 + 1$ and at least one component of $\deg_{\vec{d}}(\beta)$ must be greater than M_0/ν. Thus, there exist $\gamma_1,\gamma_2 \in \mathbb{N}^q$, with $\beta = \gamma_1 + \gamma_2$, $|\gamma_1| \leq M_0$, and $\deg_{\vec{d}}(\gamma_1) \geq \vec{d}_l - \vec{d}_j$. In fact, one may take $\gamma_1 = \beta - \gamma_2 \in \mathbb{N}^q$, where $\gamma_2 \in \mathbb{N}^q$ is equal to 1 in one component, zero in all other components, and such that $\deg_{\vec{d}}(\gamma_2)$ is non-zero in the component where $\deg_{\vec{d}}(\beta)$ is greater than M_0/ν; then, in that component $\deg_{\vec{d}}(\gamma_1)$ will be $\geq \max_{1\leq l\leq q} |\vec{d}_l|_1$. Since $\deg_{\vec{d}}(\gamma_1)$ equals $\deg_{\vec{d}}(\beta)$ in all other components, it follows that $\deg_{\vec{d}}(\gamma_1) \geq \vec{d}_l - \vec{d}_j$. Thus, in this case (4.21) again shows that $\partial_\epsilon(\widehat{\Gamma}_{\epsilon,\kappa})_*(t^\alpha X_k)$ is a C^∞_{loc} linear combination of terms of the same form as $(\widehat{\Gamma}_{\epsilon,\kappa})_*(t^\alpha X_k)$.

Hence, if we consider the vector of vector fields

$$V_j(\epsilon,t,x) = ((\widehat{\Gamma}_{\epsilon,\kappa})_* t^\alpha X_k),$$

where k ranges over $1 \le k \le q$ and $\alpha \in \mathbb{N}^q$ ranges over all $|\alpha| \le M_0$ with $\deg_{\vec{d}}(\alpha) \ge \vec{d}_k - \vec{d}_j$, then V_j satisfies an ODE of the form

$$\partial_\epsilon V_j(\epsilon,t,x) = M(\epsilon,t,x)V_j(\epsilon,t,x),$$

for some smooth matrix $M(\epsilon,t,x)$. Solving this ODE, we see that

$$V_j(\epsilon,t,x) = A(\epsilon,t,x)V_j(0,t,x),$$

where $A(\epsilon,t,x)$ is some smooth matrix. Since $(\widehat{\Gamma}_{\epsilon,\kappa})_* X_j$ is one of the components of $V_j(\epsilon,t,x)$ and $(\widehat{\Gamma}_{0,\kappa})_* t^\alpha X_k = t^\alpha X_k$, the result follows. □

Lemma 4.2.7. *$(\widehat{\Gamma}_{\epsilon,\kappa})_* X_j = X_j + \sum_{k=1}^q h_{j,\kappa}^{k,2}(\epsilon,t,x)X_k$, where $\deg_{\vec{d}}(h_{j,\kappa}^{k,2}) \ge \vec{d}_k - \vec{d}_j$ and $h_{j,\kappa}^{k,2}(\epsilon,0,x) = 0$.*

Proof. Using Lemma 4.2.5 (ii) and (4.2), we have

$$\partial_\epsilon(\widehat{\Gamma}_{\epsilon,\kappa})_* X_j = (\widehat{\Gamma}_{\epsilon,\kappa})_*[t^\kappa \cdot X^\kappa, X_j] = \sum_{j'=1}^{q_\kappa} \sum_{\vec{d}_l \le d_{j'}^\kappa \mathfrak{e}_\kappa + \vec{d}_j} c_{j,k}^{l,\kappa}(\Gamma_{\epsilon,\kappa}(t,x))(\widehat{\Gamma}_{\epsilon,\kappa})_* t_{j'}^\kappa X_l, \tag{4.22}$$

where $c_{j,k}^{l,\kappa} \in C_{\mathrm{loc}}^\infty(\mathfrak{M})$. Using Lemma 4.2.6, for $\vec{d}_l \le d_{j'}^\kappa \mathfrak{e}_\kappa + \vec{d}_j$, we have

$$(\widehat{\Gamma}_{\epsilon,\kappa})_* t_{j'}^\kappa X_l = \sum_{r=1}^q t_{j'}^\kappa h_{k,\kappa}^{r,1}(\epsilon,t,x)X_r,$$

where $\deg_{\vec{d}}(h_{k,\kappa}^{r,1}) \ge \vec{d}_r - \vec{d}_l$, and therefore $\deg_{\vec{d}}(t_{j'}^\kappa h_{k,\kappa}^{r,1}) \ge \vec{d}_r - \vec{d}_l + d_{j'}^\kappa \mathfrak{e}_\kappa \ge \vec{d}_r - \vec{d}_j$. Plugging this into (4.22), we see

$$\partial_\epsilon(\widehat{\Gamma}_{\epsilon,\kappa})_* X_j = \sum_{k=1}^q h_{j,\kappa}^{k,3}(\epsilon,t,x)X_k, \tag{4.23}$$

where $h_{j,\kappa}^{k,3}(\epsilon,0,x) = 0$ and $\deg_{\vec{d}}(h_{j,\kappa}^{k,3}) \ge \vec{d}_k - \vec{d}_j$.

Integrating (4.23) and using $(\widehat{\Gamma}_{0,\kappa})_* X_j = X_j$, we see that

$$(\widehat{\Gamma}_{\epsilon,\kappa})_* X_j = X_j + \sum_{k=1}^q \left(\int_0^\epsilon h_{j,\kappa}^{k,3}(\epsilon',t,x)\,d\epsilon' \right) X_k,$$

and the result follows. □

Recall that we have separated $t \in \mathbb{R}^q$ as $t = (t^1, \ldots, t^\nu)$, where $t^\kappa \in \mathbb{R}^{q_\kappa}$. Thus, for a given $j \in \{1, \ldots, q\}$, t_j is a coordinate of precisely one t^κ.

Lemma 4.2.8. *We have*

$$(\widehat{\Gamma}_{\epsilon,\kappa})_* \partial_{t_j} = \begin{cases} \partial_{t_j}, & t_j \text{ is not a coordinate of } t^\kappa, \\ \partial_{t_j} - \epsilon X_j + \sum_{k=1}^q h_{j,\kappa}^{k,4}(\epsilon, t, x) X_k, & t_j \text{ is a coordinate of } t^\kappa, \end{cases}$$

where $\deg_{\vec{d}}(h_{j,\kappa}^{k,4}) \geq \vec{d}_k - \vec{d}_j$ *and* $h_{j,\kappa}^{k,4}(\epsilon, 0, x) = 0$.

Proof. We lift the vector fields X_k^κ to $\mathbb{R}^q \times \mathfrak{M}$, by viewing them as tangent to $\{t\} \times \mathfrak{M}$, $\forall t \in \mathbb{R}^q$, and we use the same notation X_k^κ for the lifted vector fields. Let $Y^\kappa(t, x) = t^\kappa \cdot X^\kappa(x) \in C^\infty_{\mathrm{loc}}(\mathbb{R}^q \times \mathfrak{M}; T(\mathbb{R}^q \times \mathfrak{M}))$. Using this notation, $\widehat{\Gamma}_{(\epsilon,\kappa)}(t, x) = e^{\epsilon Y^\kappa}(t, x)$. Thus, using Lemma 4.2.5 (ii),

$$\begin{aligned} \partial_\epsilon (\widehat{\Gamma}_{\epsilon,\kappa})_* \partial_{t_j} = (\widehat{\Gamma}_{\epsilon,\kappa})_* [Y^\kappa, \partial_{t_j}] &= \begin{cases} 0, & t_j \text{ is not a coordinate of } t^\kappa, \\ (\widehat{\Gamma}_{\epsilon,\kappa})_*(-X_j), & t_j \text{ is a coordinate of } t^\kappa, \end{cases} \\ &= \begin{cases} 0, & t_j \text{ is not a coordinate of } t^\kappa, \\ -X_j - \sum_{k=1}^q h_{j,\kappa}^{k,2} X_k, & t_j \text{ is a coordinate of } t^\kappa, \end{cases} \end{aligned}$$

where $h_{j,\kappa}^{k,2}(\epsilon, t, x)$ is as in Lemma 4.2.7. Using the fact that $(\widehat{\Gamma}_{0,\kappa})_* \partial_{t_j} = \partial_{t_j}$ and integrating we see

$$(\widehat{\Gamma}_{\epsilon,\kappa})_* \partial_{t_j} = \begin{cases} \partial_{t_j}, & t_j \text{ is not a coordinate of } t^\kappa, \\ \partial_{t_j} - \epsilon X_j - \sum_{k=1}^q \left(\int_0^\epsilon h_{j,\kappa}^{k,2}(\epsilon', t, x)\, d\epsilon' \right) X_k, & t_j \text{ is a coordinate of } t^\kappa. \end{cases}$$

The result follows. □

Corollary 4.2.9. *We have*

$$(\widehat{\Gamma}_{1,\kappa})_* X_j = X_j + \sum_{k=1}^q h_{j,\kappa}^{k,5}(t, x) X_k,$$

$$(\widehat{\Gamma}_{1,\kappa})_* \partial_{t_j} = \begin{cases} \partial_{t_j}, & t_j \text{ is not a coordinate of } t^\kappa, \\ \partial_{t_j} - X_j + \sum_{k=1}^q h_{j,\kappa}^{k,6}(t, x) X_k, & t_j \text{ is a coordinate of } t^\kappa, \end{cases}$$

where $h_{j,\kappa}^{k,l}(0, x) = 0$ *and* $\deg_{\vec{d}}(h_{j,\kappa}^{k,l}) \geq \vec{d}_k - \vec{d}_j$, *for* $l = 5, 6$.

Proof. This follows by taking $\epsilon = 1$ in Lemmas 4.2.7 and 4.2.8. □

Lemma 4.2.10. *For* $\kappa \in \{1, \ldots, \nu\}$, $j \in \{1, \ldots, q\}$,

$$(\widehat{\Gamma}_{1,\kappa} \circ \widehat{\Gamma}_{1,\kappa+1} \circ \cdots \circ \widehat{\Gamma}_{1,\nu})_* X_j = X_j + \sum_{k=1}^q h_{j,\kappa}^{k,7}(t, x) X_k, \tag{4.24}$$

$$(\widehat{\Gamma}_{1,\kappa} \circ \widehat{\Gamma}_{1,\kappa+1} \circ \cdots \circ \widehat{\Gamma}_{1,\nu})_* \partial_{t_j} = \begin{cases} \partial_{t_j}, & t_j \text{ is not a coordinate of } (t^\kappa, t^{\kappa+1}, \ldots, t^\nu), \\ \partial_{t_j} - X_j + \sum_{k=1}^q h_{j,\kappa}^{k,8}(t,x) X_k, & t_j \text{ is a coordinate of } (t^\kappa, t^{\kappa+1}, \ldots, t^\nu), \end{cases} \tag{4.25}$$

where $h_{j,\kappa}^{k,l}(0,x) = 0$ *and* $\deg_{\vec{d}}(h_{j,\kappa}^{k,l}) \geq \vec{d}_k - \vec{d}_j$, *for* $l = 7, 8$.

Proof. We write $H_{\kappa,l}(t,x)$ for the $q \times q$ matrix whose (j,k) component is $h_{j,\kappa}^{k,l}$, where $h_{j,\kappa}^{k,l}$ is either as in the statement of the lemma (for $l = 7, 8$) or as in Corollary 4.2.9 (for $l = 5, 6$). We freely use (4.15) from Lemma 4.2.3, so that the product of two matrices of the type $H_{\kappa,l}$ is again of the same form.

We begin with (4.24). We proceed by induction on κ; the base case, $\kappa = \nu$, follows from Corollary 4.2.9. Thus, we assume the result for κ replaced by $\kappa + 1$ and prove it for κ. By the inductive hypothesis and Corollary 4.2.9, we have

$$\begin{aligned} (\widehat{\Gamma}_{1,\kappa} \circ \cdots \circ \widehat{\Gamma}_{1,\nu})_* X &= (\widehat{\Gamma}_{1,\kappa})_* (\widehat{\Gamma}_{1,\kappa+1} \circ \cdots \circ \widehat{\Gamma}_{1,\nu})_* \\ &= (\widehat{\Gamma}_{1,\kappa})_* (I + H_{\kappa+1,7}) X = (I + H_{\kappa+1,7}(t, \Gamma_\kappa(t,x)))(I + H_{\kappa,5}(t,x)) X \\ &=: (I + H_{\kappa,7}(t,x)). \end{aligned}$$

That $H_{\kappa,7}(t,x)$ is of the desired form follows from Lemma 4.2.3; this completes the proof of (4.24).

We turn to (4.25). Let B_κ denote the $q \times q$ matrix which equals -1 in the (j,j) component if t_j is a coordinate of t^κ and is equal to 0 in all other components. Let ∂_t denote the column vector $\partial_t = [\partial_{t_1}, \ldots, \partial_{t_q}]^\top$. We claim that

$$(\widehat{\Gamma}_{1,\kappa} \circ \widehat{\Gamma}_{1,\kappa+1} \circ \cdots \circ \widehat{\Gamma}_{1,\nu})_* \partial_t = \partial_t - \left(\sum_{\kappa'=\kappa}^{\nu} B_{\kappa'} \right) X + \left(\sum_{\kappa'=\kappa}^{\nu} B_{\kappa'} H_{9,\kappa'}(t,x) \right) X, \tag{4.26}$$

where $H_{9,\kappa}(0,x) = 0$ and if $h_{j,\kappa}^{k,9}$ is the (j,k) component of $H_{9,\kappa}$, then $\deg_{\vec{d}}(h_{j,\kappa}^{k,0}) \geq \vec{d}_k - \vec{d}_j$. Formula (4.25) follows immediately from (4.26) and we prove (4.26) by induction on κ. The base case, $\kappa = \nu$, follows from Corollary 4.2.9. For the inductive step, we assume (4.26) with κ replaced by $\kappa + 1$ and prove it for κ. We have, by the inductive hypothesis and Corollary 4.2.9,

$$\begin{aligned} (\widehat{\Gamma}_{1,\kappa} \circ \cdots \circ \widehat{\Gamma}_{1,\nu})_* \partial_t &= (\widehat{\Gamma}_{1,\kappa})_* (\widehat{\Gamma}_{1,\kappa+1} \circ \cdots \circ \widehat{\Gamma}_{1,\nu})_* \partial_t \\ &= (\widehat{\Gamma}_{1,\kappa})_* \left(\partial_t - \left(\sum_{\kappa'=\kappa+1}^{\nu} B_{\kappa'} \right) X + \left(\sum_{\kappa'=\kappa+1}^{\nu} B_{\kappa'} H_{9,\kappa'} \right) X \right) \\ &= \partial_t - B_\kappa X + B_\kappa H_{\kappa,6} X - \left(\sum_{\kappa'=\kappa+1}^{\nu} B_{\kappa'} \right) (X + H_{\kappa,5} X) \\ &\quad + \left(\sum_{\kappa'=\kappa+1}^{\nu} B_{\kappa'} H_{9,\kappa'} \right) (I + H_{\kappa,5}) X. \end{aligned}$$

This is of the desired form by Lemma 4.2.3, completing the proof. □

Proof of Lemma 4.2.4. Since $\widehat{\Gamma}(t,x) = \widehat{\Gamma}_{1,1} \circ \widehat{\Gamma}_{1,2} \circ \cdots \circ \widehat{\Gamma}_{1,\nu}(t,x)$, this follows immediately from Lemma 4.2.10. □

Proof of Proposition 4.2.2. We prove parts of the proposition with $f(x)$ replaced by a more general function $F(t,x)$ which is allowed to depend on $t \in \mathbb{R}^q$.

Using Lemma 4.2.4, we have

$$\begin{aligned} X_j F(t,\Gamma(t,x)) &= \widehat{\Gamma}^* \widehat{\Gamma}_* X_j \widehat{\Gamma}^* F(t,x) = \widehat{\Gamma}^*((\widehat{\Gamma}_* X_j) F(t,x)) \\ &= \widehat{\Gamma}^*\left(\left(X_j + \sum_{k=1}^{q} g_{1,j}^k X_k\right) F\right)(t,x) = \left(\left(X_j + \sum_{k=1}^{q} g_{1,j}^k X_k\right) F\right)(t,\Gamma(t,x)), \end{aligned}$$

proving (i).

For (ii), we define G_2 by $G_2(t,\Gamma_t(x)) = (I + G_1(t,x))^{-1} - I$, that is, $G_2(t,x) = (I + G_1)^{-1}(t,\Gamma_{-t}(x)) - I$. By Lemma 4.2.3, G_2 is of the desired form, and we have, using (i),

$$((I + G_2)X)(F(t,\Gamma_t(x))) = ((I + G_1)^{-1}(I + G_1)XF)(t,\Gamma_t(x)) = (XF)(t,\Gamma_t(x)),$$

proving (ii).

For (iii), similar to the proof of (i), we have, by Lemma 4.2.4,

$$\begin{aligned} \partial_{t_j} F(t,\Gamma_t(x)) &= \widehat{\Gamma}^* \widehat{\Gamma}_* \partial_{t_j} \widehat{\Gamma}^* F(t,x) = \widehat{\Gamma}^*\left(\left(\partial_{t_j} - X_j + \sum_{k=1}^{q} g_{3,j}^k X_k\right) F\right)(t,x) \\ &= \left(\left(\partial_{t_j} - X_j + \sum_{k=1}^{q} g_{3,j}^k X_k\right) F\right)(t,\Gamma_t(x)), \end{aligned}$$

proving (iii).

For (iv), we have, by using (ii) and (iii),

$$\begin{aligned} \partial_{t_j} f(\Gamma_t(x)) &= ((-I + G_3)Xf)(\Gamma_t(x)) \\ &= (-I + G_3(t,\Gamma_t(x)))(I + G_2(t,x))X(f(\Gamma_t(x))), \end{aligned}$$

which is of the desired form by Lemma 4.2.3.

For (v), we have, by using (iv),

$$\begin{aligned} X(f(\Gamma(t,x))) &= (-I + G_4(t,x))^{-1}(I + G_4(t,x))X(f(\Gamma(t,x))) \\ &= (-I + G_4(t,x))^{-1}\partial_t(f(\Gamma(t,x))), \end{aligned}$$

which is of the desired form by Lemma 4.2.3.

For (vi), we have, by using (iii),

$$\begin{aligned} (Xf)(\Gamma_t(x)) &= (-I + G_3(t,\Gamma(t,x)))^{-1}((-I + G_3)Xf)(\Gamma(t,x)) \\ &= (-I + G_3(t,\Gamma(t,x)))^{-1}\partial_t(f(\Gamma(t,x))), \end{aligned}$$

which is of the desired form by Lemma 4.2.3. □

4.3 Littlewood–Paley decompositions

Proposition 4.1.14 gave a decomposition of symbols in $S^m(a, \Omega, \vec{d})$ into Littlewood–Paley scales. This immediately gives a corresponding decomposition of $(X, \vec{d})$ pseudo-differential operators as defined in Definition 4.1.2. In this section, we give a description of the properties of this decomposition.

In this section, fix relatively compact open sets $\Omega_1, \Omega_2 \Subset \mathfrak{M}$, with $\Omega_1 \Subset \Omega_2$, and take $a = a(\Omega_1, \Omega_2, (X, \vec{d})) > 0$ small, in particular so small that

$$\overline{\Gamma(B^q(a) \times \Omega_1)} \Subset \Omega_2.$$

Throughout this section, we may shrink $a = a(\Omega_1, \Omega_2, (X, \vec{d})) > 0$ as needed to make various results hold; we will only need to shrink a a finite number of times, so we will be left with a strictly positive number.

As in Section 4.1.2, for $j \in \mathbb{Z}^\nu$ we write $2^j = (2^{j_1}, 2^{j_2}, \dots, 2^{j_\nu}) \in (0, \infty)^\nu$. We write $2^{j \cdot \vec{d}} X = (2^j)^{\vec{d}} X = \{2^{j \cdot d_1} X_1, \dots, 2^{j \cdot d_\nu} X_\nu\}$.

Definition 4.3.1. (i) A bounded set of $(X, \vec{d})$, a pre-pseudo-differential operator scales supported in $\Omega_1 \times \Omega_2$ is a subset $\mathcal{E} \subset \mathrm{Hom}(C^\infty_{\mathrm{loc}}(\mathfrak{M}), C^\infty_0(\Omega_1)) \times (0,1]^\nu$ such that each $(E, 2^{-j}) \in \mathcal{E}$ is of the form

$$Ef(x) = \int f(\Gamma(t,x)) \, \mathrm{Dil}^{\vec{d}}_{2^j}(\varsigma)(x,t,t) \, dt,$$

where

$$\{\varsigma : (E, 2^{-j}) \in \mathcal{E}\} \subset C^\infty_0(\Omega_1) \widehat{\otimes} C^\infty_0(B^q(a)) \widehat{\otimes} \mathscr{S}(\mathbb{R}^q)$$

is a bounded set.

(ii) If, in addition, for each $(E_j, 2^{-j}) \in \mathcal{E}$, we have $\varsigma \in C^\infty_0(\Omega_1) \widehat{\otimes} C^\infty_0(B^q(a)) \widehat{\otimes} \mathscr{S}_{\{\mu : j_\mu \neq 0\}}(\mathbb{R}^q)$, we say $\mathcal{E}$ is a bounded set of $(X, \vec{d})$, a pseudo-differential operator scales supported in $\Omega_1 \times \Omega_2$.

Remark 4.3.2. Note that if $\mathcal{E}$ is a bounded set of $(X, \vec{d})$, a pre-pseudo-differential operator scales supported in $\Omega_1 \times \Omega_2$, then for each $(E, 2^{-j}) \in \mathcal{E}$, the Schwartz kernel of E is supported in $\Omega_1 \times \Omega_2$.

Theorem 4.3.3. *If $a = a(\Omega_1, \Omega_2, (X, \vec{d})) > 0$ is sufficiently small, then for $T \in \mathrm{Hom}(C^\infty_{\mathrm{loc}}(\mathfrak{M}), C^\infty_0(\Omega_1))$ the following are equivalent for $m \in \mathbb{R}^\nu$:*

(i) *T is an $(X, \vec{d})$, a pseudo-differential operator of order m supported in $\Omega_1 \times \Omega_2$.*

(ii) *There is a bounded set of $(X, \vec{d})$, a pseudo-differential operators scales supported in $\Omega_1 \times \Omega_2$, $\{(E_j, 2^{-j}) : j \in \mathbb{N}^\nu\}$, such that $T = \sum_{j \in \mathbb{N}^\nu} 2^{j \cdot m} E_j$, where this sum converges in the topology of bounded convergence on $\mathrm{Hom}(C^\infty_{\mathrm{loc}}(\mathfrak{M}), C^\infty_0(\Omega_1))$.*

Furthermore, every sum of the type described in (ii) converges in the topology of bounded convergence on $\mathrm{Hom}(C^\infty_{\mathrm{loc}}(\mathfrak{M}), C^\infty_0(\Omega_1))$ *and in the weak operator topology on* $\mathrm{Hom}(C^\infty_0(\mathfrak{M})', C^\infty_{\mathrm{loc}}(\Omega_1)')$, *where* $C^\infty_{\mathrm{loc}}(\Omega_1)'$ *is given the weak topology. In particular, every* $(X, \vec{d})$, *a pseudo-differential operator extends to an operator in* $\mathrm{Hom}(C^\infty_0(\mathfrak{M})', C^\infty_{\mathrm{loc}}(\Omega_1)')$.

We complete the proof of Theorem 4.3.3 in Section 4.3.3; we begin with some basic properties of $(X, \vec{d})$ pseudo-differential operator scales.

Lemma 4.3.4. *If* $a = a(\Omega_1, \Omega_2, (X, \vec{d})) > 0$ *is sufficiently small, the following holds.*

(i) *Let* $\mathcal{E}$ *be a bounded set of* $(X, \vec{d})$, *a pre-pseudo-differential operator scales supported in* $\Omega_1 \times \Omega_2$. *Let* α *be an ordered multi-index. Then*

$$\mathcal{E}_\alpha := \{((2^{-j\cdot\vec{d}}X)^\alpha E, 2^{-j}), (E(2^{-j\cdot\vec{d}}X)^\alpha, 2^{-j}) : (E, 2^{-j}) \in \mathcal{E}\}$$

is a bounded set $(X, \vec{d})$, *a pre-pseudo-differential operator scales supported in* $\Omega_1 \times \Omega_2$.

(ii) *If, in addition,* $\mathcal{E}$ *is a bounded set of* $(X, \vec{d})$, *a pseudo-differential operator scales supported in* $\Omega_1 \times \Omega_2$, *then* $\mathcal{E}_\alpha$ *is also a bounded set of* $(X, \vec{d})$, *a pseudo-differential operator scales supported in* $\Omega_1 \times \Omega_2$.

Proof. We prove only (ii); the proof of (i) follows from a simpler version of the same argument.

We will show that for $l \in \{1, \dots, q\}$,

$$\{(2^{-j\cdot d_l}X_l E, 2^{-j}), (E 2^{-j\cdot\vec{d}_l}X_l, 2^{-j}) : (E, 2^{-j}) \in \mathcal{E}\}$$

is a bounded set of $(X, \vec{d})$, a pseudo-differential operator scales supported in $\Omega_1 \times \Omega_2$. The result for general α then follows by repeated application of this result.

Consider, using Proposition 4.2.2 (v),

$$X_l f(\Gamma(t, x)) = \left(-\partial_{t_l} + \sum_{k=1}^{q} g^k_{5,l}\partial_{t_k}\right) f(\Gamma(t, x)),$$

where $\deg_{\vec{d}} g^k_{5,l} \geq \vec{d}_k - \vec{d}_l$ and $g^k_{5,l}(t, x)$ is defined on some open neighborhood of $\{0\} \times \mathfrak{M}$ in $\mathbb{R}^q \times \mathfrak{M}$. Treating l as fixed, we write $g^k_{5,l}(t, x) = \sum_{r=1}^{K} t^{\beta_{r,k}} g_{r,k}(t, x)$, where $g_{r,k}(t, x) \in C^\infty_{\mathrm{loc}}(B^q(a) \times \Omega_1)$ and $\deg_{\vec{d}}(\beta_{r,k}) \geq \vec{d}_k - \vec{d}_l$ (where we have taken $a > 0$ sufficiently small).

For $(E, 2^{-j}) \in \mathcal{E}$, we have

$$Ef(x) = \int f(\Gamma(t, x))\, \mathrm{Dil}^{\vec{d}}_{2^j}(\varsigma)(x, t, t)\, dt,$$

where ς is as in Definition 4.3.1. Thus, if $a > 0$ is chosen so small that $g^k_{5,l}(t, x)$ is defined and smooth for $x \in \Omega_1$ and $t \in B^q(a)$, we have by integration by parts

$$2^{-j\cdot \vec{d}_l} X_l E f(x) = \int 2^{-j\cdot \vec{d}_l} \left(\left(-\partial_{t_l} + \sum_{k=1}^{q} g_{5,l}^{k} \partial_{t_k} \right) f(\Gamma(t,x)) \right) \mathrm{Dil}_{2^j}^{\vec{d}}(\varsigma)(x,t,t)\, dt$$
$$= \int f(\Gamma(t,x))\, \mathrm{Dil}_{2^j}^{\vec{d}}(\tilde{\varsigma})(x,t,t)\, dt,$$

where

$$\tilde{\varsigma}(x,s,t) = \partial_{t_l} \varsigma(x,s,t) - \left[\sum_{r=1}^{K} \partial_{s_k} (s^{\beta_{r,k}} g_{r,k}(s,x)) \right] \varsigma(x,s,t)$$
$$- \sum_{r=1}^{K} g_{r,k}(s,x) 2^{-j\cdot(\vec{d}_l - \vec{d}_k + \deg_{\vec{d}}(\beta_{r,k}))} \partial_{t_k} t^{\beta_{r,k}} \varsigma(x,s,t).$$

Since $\vec{d}_l - \vec{d}_k + \deg_{\vec{d}}(\beta_{r,k}) \geq 0$, we have $2^{-j\cdot(\vec{d}_l - \vec{d}_k + \deg_{\vec{d}}(\beta_{r,k}))} \in (0,1]$. Note that ∂_{t_k}, multiplication by $t^{\beta_{r,k}}$, and multiplication by any smooth function $g(s,x)$ are all continuous maps $C_0^\infty(\Omega_1) \widehat{\otimes} C_0^\infty(B^q(a)) \widehat{\otimes} \mathscr{S}(\mathbb{R}^q) \to C_0^\infty(\Omega_1) \widehat{\otimes} C_0^\infty(B^q(a)) \widehat{\otimes} \mathscr{S}(\mathbb{R}^q)$, which preserve the subspaces $C_0^\infty(\Omega_1) \widehat{\otimes} C_0^\infty(B^q(a)) \widehat{\otimes} \mathscr{S}_E(\mathbb{R}^q)$, $\forall E \subseteq \{1,\dots,\nu\}$. Thus, we have $\tilde{\varsigma} \in C_0^\infty(\Omega_1) \widehat{\otimes} C_0^\infty(B^q(a)) \widehat{\otimes} \mathscr{S}_{\{\mu : j_\mu \neq 0\}}(\mathbb{R}^q)$ and

$$\{\tilde{\varsigma} : (E, 2^{-j}) \in \mathcal{E}\} \subset C_0^\infty(\Omega_1) \widehat{\otimes} C_0^\infty(B^q(a)) \widehat{\otimes} \mathscr{S}(\mathbb{R}^q)$$

is a bounded set, completing the proof that

$$\{(2^{-j\cdot \vec{d}_l} X_l E, 2^{-j}) : (E, 2^{-j}) \in \mathcal{E}\}$$

is a bounded set of $(X, \vec{d})$, a pseudo-differential operator scales supported in $\Omega_1 \times \Omega_2$.

A similar proof shows that

$$\{(E 2^{-j\cdot \vec{d}_l} X_l, 2^{-j}) : (E, 2^{-j}) \in \mathcal{E}\}$$

is a bounded set of $(X, \vec{d})$ pseudo-differential operator scales supported in $\Omega_1 \times \Omega_2$, where we use Proposition 4.2.2 (vi) in place of Proposition 4.2.2 (v). □

Along with the conclusions of Lemma 4.3.4 (ii), one of the main properties of pseudo-differential operator scales is the next proposition.

Proposition 4.3.5. *If $a = a(\Omega_1, \Omega_2, (X, \vec{d})) > 0$ is sufficiently small, the following holds. Let $\mathcal{E}$ be a bounded set of $(X, \vec{d})$, a pseudo-differential operator scales supported in $\Omega_1 \times \Omega_2$. Then, $\forall \mu \in \{1,\dots,\nu\}$, $(E, 2^{-j}) \in \mathcal{E}$,*

$$E = \sum_{\substack{|\alpha_\mu| \leq 1 \\ |\beta_\mu| \leq 1}} 2^{-j_\mu(2 - |\alpha_\mu| - |\beta_\mu|)} (2^{-j_\mu d^\mu} X^\mu)^{\alpha_\mu} E_{\mu,\alpha_\mu,\beta_\mu} (2^{-j_\mu d^\mu} X^\mu)^{\beta_\mu},$$

where

$$\{(E_{\mu,\alpha_\mu,\beta_\mu}, 2^{-j}) : (E, 2^{-j}) \in \mathcal{E}, \mu \in \{1,\dots,\nu\}, |\alpha_\mu|, |\beta_\mu| \le 1\}$$

is a bounded set of $(X, \vec{d})$, *a pseudo-differential scales supported in* $\Omega_1 \times \Omega_2$.

To prove Proposition 4.3.5, we require two lemmas.

Lemma 4.3.6. *Fix* $\mu_0 \in \{1,\dots,\nu\}$, $1 \le l \le q_{\mu_0}$, $1 \le k \le q$. *Let* $g(t,x) \in C^\infty_{\mathrm{loc}}(U)$, *where* $U \subseteq \mathbb{R}^q \times \mathfrak{M}$ *is an open neighborhood of* $\{0\} \times \mathfrak{M}$, *and suppose* g *satisfies* $\deg_{\vec{d}}(g) \ge \vec{d}_k - d_l^{\mu_0} \mathfrak{e}_{\mu_0}$. *Then if* $b = b(g, \Omega_1, \Omega_2, (X, \vec{d})) > 0$ *is small enough, the following holds.*

Let $\mathcal{B} \subset C_0^\infty(\Omega_1)\widehat{\otimes} C_0^\infty(B^q(b))\widehat{\otimes}\mathscr{S}(\mathbb{R}^q)$ *be a bounded set. For* $\varsigma \in \mathcal{B}$, $j \in \mathbb{N}^\nu$, *define operators* $F_1 = F_1[j,\varsigma]$ *and* $F_2 = F_2[j,\varsigma]$ *by*

$$F_1 f(x) = 2^{-j_{\mu_0} d_l^{\mu_0}} \int (gX_k f)(\Gamma(t,x)) \operatorname{Dil}^{\vec{d}}_{2^j}(\varsigma)(x,t,t)\, dt,$$

$$F_2 f(x) = 2^{-j_{\mu_0} d_l^{\mu_0}} \int [gX_k(f(\Gamma(t,x)))] \operatorname{Dil}^{\vec{d}}_{2^j}(\varsigma)(x,t,t)\, dt.$$

Then if $\varsigma \in C_0^\infty(\Omega_1)\widehat{\otimes} C_0^\infty(B^q(b))\widehat{\otimes}\mathscr{S}_{\{\mu: j_\mu \ne 0\}}(\mathbb{R}^q) \cap \mathcal{B}$, *we have*

$$F_1 = \sum_{|\alpha_{\mu_0}| \le 1} 2^{-j_{\mu_0}(1-|\alpha_{\mu_0}|)} E_{\alpha_{\mu_0},1} (2^{-j_{\mu_0} d^{\mu_0}} X^{\mu_0})^{\alpha_{\mu_0}}, \tag{4.27}$$

$$F_2 = \sum_{|\alpha_{\mu_0}| \le 1} 2^{-j_{\mu_0}(1-|\alpha_{\mu_0}|)} (2^{-j_{\mu_0} d^{\mu_0}} X^{\mu_0})^{\alpha_{\mu_0}} E_{\alpha_{\mu_0},2}, \tag{4.28}$$

where

$$\{(E_{\alpha_{\mu_0},1}, 2^{-j}), (E_{\alpha_{\mu_0},2}, 2^{-j}) : j \in [0,\infty)^\nu,$$
$$\varsigma \in C_0^\infty(\Omega_1)\widehat{\otimes} C_0^\infty(B^q(b))\widehat{\otimes}\mathscr{S}_{\{\mu: j_\mu \ne 0\}}(\mathbb{R}^q) \cap \mathcal{B}, |\alpha_{\mu_0}| \le 1\}$$

is a bounded set of $(X, \vec{d})$, b *pseudo-differential operator scales supported in* $\Omega_1 \times \Omega_2$.

Proof. By taking b sufficiently small, we may write

$$g(t,x) = \sum_{r=1}^{R} t^{\gamma_r} g_r(t,x),$$

where $g_r(t,x) \in C^\infty_{\mathrm{loc}}(B^q(b) \times \Omega_2)$ and $\deg_{\vec{d}}(\gamma_r) \ge \vec{d}_k - d_l^{\mu_0} \mathfrak{e}_{\mu_0}$. This decomposes the right-hand sides of (4.27) and (4.28) into R terms, and it suffices to study each of these terms separately. Thus, we henceforth take $g(t,x) = t^\gamma h(t,x)$, where $h(t,x) \in C^\infty_{\mathrm{loc}}(B^q(b) \times \Omega_2)$ and $\deg_{\vec{d}}(\gamma) \ge \vec{d}_k - d_l^{\mu_0} \mathfrak{e}_{\mu_0}$.

The formal degree $\vec{d}_k \in \mathbb{N}^\nu$ is non-zero in precisely one component, and we separate the proof into two cases. The first case is when $\vec{d}_k$ is non-zero in the μ_0 component. In this case, $(X_k, \vec{d}_k) = (X_p^{\mu_0}, d_p^{\mu_0} \mathfrak{e}_{\mu_0})$ for some $1 \le p \le q_{\mu_0}$ and $\deg_{\vec{d}}(\gamma)_{\mu_0} \ge d_p^{\mu_0} - d_l^{\mu_0}$.

For (4.27), we have

$$F_1 f(x) = 2^{-j_{\mu_0} d_l^{\mu_0}} \int (t^\gamma h X_p^{\mu_0} f)(\Gamma(t,x)) \operatorname{Dil}_{2^j}^{\vec{d}}(\varsigma)(x,t,t)\, dt$$
$$= \int (2^{-j_{\mu_0} d_p^{\mu_0}} X_p^{\mu_0} f)(\Gamma(t,x)) \operatorname{Dil}_{2^j}^{\vec{d}}(\tilde{\varsigma}_1)(x,t,t)\, dt,$$

where

$$\tilde{\varsigma}_1(x,s,t) = 2^{j_{\mu_0}(d_p^{\mu_0} - d_l^{\mu_0}) - j\cdot \deg_{\vec{d}}(\gamma)} h(s,\Gamma(s,x)) t^\gamma \varsigma(x,s,t).$$

Since $2^{j_{\mu_0}(d_p^{\mu_0} - d_l^{\mu_0}) - j\cdot \deg_{\vec{d}}(\gamma)} \in (0,1]$, we see $\tilde{\varsigma}_1 \in C_0^\infty(\Omega_1) \widehat{\otimes} C_0^\infty(B^q(b)) \widehat{\otimes} \mathscr{S}_{\{\mu: j_\mu \neq 0\}}(\mathbb{R}^q)$ and

$$\{\tilde{\varsigma}_1 : j \in [0,\infty)^\nu, \varsigma \in C_0^\infty(\Omega_1) \widehat{\otimes} C_0^\infty(B^q(b)) \widehat{\otimes} \mathscr{S}_{\{\mu: j_\mu \neq 0\}}(\mathbb{R}^q) \cap \mathcal{B}\}$$
$$\subset C_0^\infty(\Omega_1) \widehat{\otimes} C_0^\infty(B^q(b)) \widehat{\otimes} \mathscr{S}(\mathbb{R}^q)$$

is a bounded set. Formula (4.27) follows.

For (4.28), we have

$$F_2 f(x) = 2^{-j_{\mu_0} d_l^{\mu_0}} \int [t^\gamma h X_p^{\mu_0}(f(\Gamma(t,x)))] \operatorname{Dil}_{2^j}^{\vec{d}}(\varsigma)(x,t,t)\, dt$$
$$= 2^{-j_{\mu_0}} \int f(\Gamma(t,x)) \operatorname{Dil}_{2^j}^{\vec{d}}(\tilde{\varsigma}_2)(x,t,t)\, dt$$
$$+ 2^{-j_{\mu_0} d_p^{\mu_0}} X_p^{\mu_0} \int f(\Gamma(t,x)) \operatorname{Dil}_{2^j}^{\vec{d}}(\tilde{\varsigma}_3)(x,t,t)\, dt,$$

where

$$\varsigma_2(x,s,t) = -2^{-j_{\mu_0}(d_l^{\mu_0} - 1)} X_p^{\mu_0}[s^\gamma h(s,x) \varsigma(x,s,t)],$$
$$\varsigma_3(x,s,t) = 2^{j_{\mu_0}(d_p^{\mu_0} - d_l^{\mu_0}) - j\cdot \deg_{\vec{d}}(\gamma)} h(s,x) t^\gamma \varsigma(x,s,t).$$

Since $2^{-j_{\mu_0}(d_l^{\mu_0} - 1)}, 2^{j_{\mu_0}(d_p^{\mu_0} - d_l^{\mu_0}) - j\cdot \deg_{\vec{d}}(\gamma)} \in (0,1]$, it is immediate to verify that $\tilde{\varsigma}_2, \tilde{\varsigma}_3 \in C_0^\infty(\Omega_1) \widehat{\otimes} C_0^\infty(B^q(b)) \widehat{\otimes} \mathscr{S}_{\{\mu: j_\mu \neq 0\}}(\mathbb{R}^q)$ and

$$\{\tilde{\varsigma}_2, \tilde{\varsigma}_3 : j \in [0,\infty)^\nu, \varsigma \in C_0^\infty(\Omega_1) \widehat{\otimes} C_0^\infty(B^q(b)) \widehat{\otimes} \mathscr{S}_{\{\mu: j_\mu \neq 0\}}(\mathbb{R}^q) \cap \mathcal{B}\}$$
$$\subset C_0^\infty(\Omega_1) \widehat{\otimes} C_0^\infty(B^q(b)) \widehat{\otimes} \mathscr{S}(\mathbb{R}^q)$$

is a bounded set. Formula (4.28) follows.

We now consider the case where $\vec{d}_k$ is non-zero is some component other than μ_0. Thus, $\deg_{\vec{d}}(\gamma) \geq \vec{d}_k$ (since $\vec{d}_k$ is non-zero in precisely one component). For (4.27), we have

$$F_1 f(x) = 2^{-j_{\mu_0} d_l^{\mu_0}} \int (t^\gamma h X_k f)(\Gamma(t,x)) \operatorname{Dil}_{2^j}^{\vec{d}}(\varsigma)(x,t,t)\, dt$$
$$= 2^{-j_{\mu_0}} \int (2^{-j\cdot \vec{d}_k} X_k f)(\Gamma(t,x)) \operatorname{Dil}_{2^j}^{\vec{d}}(\tilde{\varsigma}_4)(x,t,t)\, dt,$$

where

$$\tilde{\varsigma}_4(x,s,t) = 2^{j\cdot\vec{d}_\kappa - j\cdot\deg_{\vec{d}}(\gamma)} 2^{-j_{\mu_0}(d_l^{\mu_0}-1)} t^\gamma h(s,x)\varsigma(x,s,t).$$

Since $2^{j\cdot\vec{d}_\kappa - j\cdot\deg_{\vec{d}}(\gamma)}2^{-j_{\mu_0}} \in (0,1]$, it follows that $\tilde{\varsigma}_4 \in C_0^\infty(\Omega_1)\widehat{\otimes}C_0^\infty(B^q(b))\widehat{\otimes}\mathscr{S}_{\{\mu: j_\mu\neq 0\}}(\mathbb{R}^q)$ and

$$\begin{aligned}&\{\tilde{\varsigma}_4 : j \in [0,\infty)^\nu, \varsigma \in C_0^\infty(\Omega_1)\widehat{\otimes}C_0^\infty(B^q(b))\widehat{\otimes}\mathscr{S}_{\{\mu: j_\mu\neq 0\}}(\mathbb{R}^q)\cap \mathcal{B}\}\\ &\quad\subset C_0^\infty(\Omega_1)\widehat{\otimes}C_0^\infty(B^q(b))\widehat{\otimes}\mathscr{S}(\mathbb{R}^q)\end{aligned}$$

is a bounded set. Thus, we have shown that

$$F_1 = 2^{-j_{\mu_0}} E(2^{-j\cdot\vec{d}_\kappa} X_\kappa),$$

where

$$\{(E, 2^{-j}) : j \in [0,\infty)^\nu, \varsigma \in C_0^\infty(\Omega_1)\widehat{\otimes}C_0^\infty(B^q(b))\widehat{\otimes}\mathscr{S}_{\{\mu: j_\mu\neq 0\}}(\mathbb{R}^q)\cap \mathcal{B}\}$$

is a bounded set of $(X,\vec{d}), b$ pseudo-differential operator scales supported in $\Omega_1 \times \Omega_2$. Applying Lemma 4.3.4 (ii) to $E(2^{-j\cdot\vec{d}_\kappa}X_\kappa)$ proves

$$F_1 = 2^{-j_{\mu_0}}\widetilde{E},$$

where

$$\{(\widetilde{E}, 2^{-j}) : j \in [0,\infty)^\nu, \varsigma \in C_0^\infty(\Omega_1)\widehat{\otimes}C_0^\infty(B^q(b))\widehat{\otimes}\mathscr{S}_{\{\mu: j_\mu\neq 0\}}(\mathbb{R}^q)\cap \mathcal{B}\}$$

is a bounded set of $(X,\vec{d}), b$ pseudo-differential operator scales supported in $\Omega_1\times\Omega_2$; (4.27) follows (with only one term in that formula: the one when $|\alpha_\mu| = 0$).

For (4.28), we have

$$\begin{aligned}F_2 f(x) &= 2^{-j_{\mu_0} d_l^{\mu_0}} \int [t^\gamma h X_\kappa(f(\Gamma(t,x)))]\,\mathrm{Dil}_{2^j}^{\vec{d}}(\varsigma)(x,t,t)\,dt\\ &= 2^{-j_{\mu_0}} \int f(\Gamma(t,x))\,\mathrm{Dil}_{2^j}^{\vec{d}}(\tilde{\varsigma}_5)(x,t,t)\,dt\\ &\quad + 2^{-j_{\mu_0}} 2^{-j\cdot\vec{d}_\kappa} X_\kappa \int f(\Gamma(t,x))\,\mathrm{Dil}_{2^j}^{\vec{d}}(\tilde{\varsigma}_6)(x,t,t)\,dt,\end{aligned}$$

where

$$\begin{aligned}\varsigma_5(x,s,t) &= -2^{-j_{\mu_0}(d_l^{\mu_0}-1)} X_\kappa[s^\gamma h(s,x)\varsigma(x,s,t)],\\ \varsigma_6(x,s,t) &= 2^{-j_{\mu_0}(d_l^{\mu_0}-1)} 2^{j\cdot\vec{d}_\kappa - \deg_{\vec{d}}(\gamma)\cdot j} h(s,x) t^\gamma \varsigma(x,s,t).\end{aligned}$$

Since $2^{-j_{\mu_0}(d_l^{\mu_0}-1)}, 2^{j\cdot\vec{d}_\kappa - \deg_{\vec{d}}(\gamma)\cdot j} \in (0,1]$, it follows that

$$\tilde{\varsigma}_5, \tilde{\varsigma}_6 \in C_0^\infty(\Omega_1)\widehat{\otimes} C_0^\infty(B^q(b))\widehat{\otimes}\mathscr{S}_{\{\mu:j_\mu\neq 0\}}(\mathbb{R}^q)$$

and

$$\{\tilde{\varsigma}_5, \tilde{\varsigma}_6 : j \in [0,\infty)^\nu, \varsigma \in C_0^\infty(\Omega_1)\widehat{\otimes} C_0^\infty(B^q(b))\widehat{\otimes}\mathscr{S}_{\{\mu:j_\mu\neq 0\}}(\mathbb{R}^q) \cap \mathcal{B}\}$$
$$\subset C_0^\infty(\Omega_1)\widehat{\otimes} C_0^\infty(B^q(b))\widehat{\otimes}\mathscr{S}(\mathbb{R}^q)$$

is a bounded set. Thus, we have shown that

$$F_2 = 2^{-j_{\mu_0}} E_1 + 2^{-j_{\mu_0}} 2^{-j\cdot \vec{d}_k} X_k E_2,$$

where

$$\{(E_1, 2^{-j}), (E_2, 2^{-j}) : j \in [0,\infty)^\nu, \varsigma \in C_0^\infty(\Omega_1)\widehat{\otimes} C_0^\infty(B^q(b))\widehat{\otimes}\mathscr{S}_{\{\mu:j_\mu\neq 0\}}(\mathbb{R}^q) \cap \mathcal{B}\}$$

is a bounded set of $(X, \vec{d}), b$ pseudo-differential operator scales supported in $\Omega_1 \times \Omega_2$. As in the proof of (4.27), applying Lemma 4.3.4 (ii) to $2^{-j\cdot\vec{d}_k} X_k E_2$ completes the proof of (4.28). □

Lemma 4.3.7. *If $a = a(\Omega_1, \Omega_2, (X, \vec{d})) > 0$ is sufficiently small, the following holds. Let $\mathcal{B} \subset C_0^\infty(\Omega_1)\widehat{\otimes} C_0^\infty(B^q(a))\widehat{\otimes}\mathscr{S}(\mathbb{R}^q)$ be a bounded set. For $\varsigma \in \mathcal{B}$, $j \in \mathbb{N}^\nu$, define an operator $F = F[j,\varsigma]$ by*

$$Ff(x) = \int f(\Gamma(t,x))\,\mathrm{Dil}_{2^j}^{\vec{d}}(\varsigma)(x,t,t)\,dt.$$

Then if $\varsigma \in C_0^\infty(\Omega_1)\widehat{\otimes} C_0^\infty(B^q(a))\widehat{\otimes}\mathscr{S}_{\{\mu:j_\mu\neq 0\}}(\mathbb{R}^q) \cap \mathcal{B}$, we have

$$F = \begin{cases} \sum_{|\alpha_{\mu_0}|\le 1} 2^{-j_{\mu_0}(1-|\alpha_{\mu_0}|)} E_{\alpha_{\mu_0},1} (2^{-j_{\mu_0} d^{\mu_0}} X^{\mu_0})^{\alpha_{\mu_0}}, \\ \sum_{|\alpha_{\mu_0}|\le 1} 2^{-j_{\mu_0}(1-|\alpha_{\mu_0}|)} (2^{-j_{\mu_0} d^{\mu_0}} X^{\mu_0})^{\alpha_{\mu_0}} E_{\alpha_{\mu_0},2}, \end{cases}$$

where

$$\{(E_{\alpha_{\mu_0},1}, 2^{-j}), (E_{\alpha_{\mu_0},2}, 2^{-j}) : |\alpha_{\mu_0}| \le 1, |\beta_{\mu_0}| \le 1, j \in [0,\infty)^\nu,$$
$$\varsigma \in C_0^\infty(\Omega_1)\widehat{\otimes} C_0^\infty(B^q(a))\widehat{\otimes}\mathscr{S}_{\{\mu:j_\mu\neq 0\}}(\mathbb{R}^q) \cap \mathcal{B}\}$$

is a bounded set of $(X, \vec{d})$, a pseudo-differential operator scales supported in $\Omega_1 \times \Omega_2$.

Proof. We take a so small that a is less than $b(g, \Omega_1, \Omega_2)$ from Lemma 4.3.6, where g can be any of the functions $g_{3,j}^k$ or $g_{4,j}^k$ from Proposition 4.2.2 (note there are only $2q^2$ such functions, and in particular, only finitely many).

Fix $j \in [0,\infty)^\nu$ and $\varsigma \in (C_0^\infty(\Omega_1)\widehat{\otimes} C_0^\infty(B^q(a))\widehat{\otimes}\mathscr{S}_{\{\mu:j_\mu\neq 0\}}(\mathbb{R}^q)) \cap \mathcal{B}$. If $j_{\mu_0} = 0$, the result is trivial, as we can take

$$E_{a_{\mu_0},1} = E_{a_{\mu_0},2} = \begin{cases} F, & |a_{\mu_0}| = 0, \\ 0, & \text{otherwise.} \end{cases}$$

Thus, we may assume $j_{\mu_0} > 0$, so we have by Corollary 4.1.19, with $\triangle_{\mu_0}$ the Laplacian in the t^{μ_0} variable,

$$\varsigma(x,s,t) = \triangle_{\mu_0}\triangle_{\mu_0}^{-1} = \sum_{l=1}^{q_{\mu_0}} -\partial_{t_l^{\mu_0}}(\partial_{t_l^{\mu_0}}\triangle_{\mu_0}^{-1}\varsigma(x,s,t)),$$

where $\partial_{t_l^{\mu_0}}\triangle_{\mu_0}^{-1}\varsigma \in C_0^\infty(\Omega_1)\widehat{\otimes}C_0^\infty(B^q(a))\widehat{\otimes}\mathscr{S}_{\{\mu:j_\mu\neq 0\}}(\mathbb{R}^q)$ and

$$\{(\partial_{t_l^{\mu_0}}\triangle_{\mu_0}^{-1}\varsigma) : \varsigma \in C_0^\infty(\Omega_1)\widehat{\otimes}C_0^\infty(B^q(a))\widehat{\otimes}\mathscr{S}_{\{\mu_0\}}(\mathbb{R}^q) \cap \mathcal{B}\} \\ \subset C_0^\infty(\Omega_1)\widehat{\otimes}C_0^\infty(B^q(a))\widehat{\otimes}\mathscr{S}(\mathbb{R}^q)$$

is a bounded set. Thus, it suffices to prove the result with $\varsigma(x,s,t)$ replaced by $-\partial_{t_l^{\mu_0}}\varsigma(x,s,t)$ for some $l \in \{1,\ldots,q_{\mu_0}\}$. We henceforth do this.

Using integration by parts we have

$$\begin{aligned} Ff(x) &= \int f(\Gamma(t,x))\,\mathrm{Dil}_{2^j}^{\vec{d}}(-\partial_{t_l^{\mu_0}}\varsigma)(x,t,t)\,dt \\ &= 2^{-j_{\mu_0}}\int f(\Gamma(t,x))\,\mathrm{Dil}_{2^j}^{\vec{d}}(\tilde{\varsigma})(x,t,t)\,dt \\ &\quad + 2^{-j_{\mu_0}d_l^{\mu_0}}\int [\partial_{t_l^{\mu_0}}(f(\Gamma(t,x)))]\,\mathrm{Dil}_{2^j}^{\vec{d}}(\varsigma)(x,t,t)\,dt \\ &=: 2^{-j_{\mu_0}}G_1f(x) + G_2f(x), \end{aligned} \tag{4.29}$$

where

$$\tilde{\varsigma}(x,s,t) = 2^{-j_{\mu_0}(d_l^{\mu_0}-1)}\partial_{s_l^{\mu_0}}\varsigma(x,s,t).$$

Since $2^{-j_{\mu_0}(d_l^{\mu_0}-1)} \in (0,1]$, it follows easily that $\tilde{\varsigma} \in C_0^\infty(\Omega_1)\widehat{\otimes}C_0^\infty(B^q(a))\widehat{\otimes}\mathscr{S}_{\{\mu:j_\mu\neq 0\}}(\mathbb{R}^q)$ and

$$\{\tilde{\varsigma} : j \in [0,\infty)^\nu, \varsigma \in C_0^\infty(\Omega_1)\widehat{\otimes}C_0^\infty(B^q(b))\widehat{\otimes}\mathscr{S}_{\{\mu:j_\mu\neq 0\}}(\mathbb{R}^q) \cap \mathcal{B}\} \\ \subset C_0^\infty(\Omega_1)\widehat{\otimes}C_0^\infty(B^q(b))\widehat{\otimes}\mathscr{S}(\mathbb{R}^q)$$

is a bounded set. Thus,

$$\{(G_1, 2^{-j}) : j \in [0,\infty)^\nu, \varsigma \in C_0^\infty(\Omega_1)\widehat{\otimes}C_0^\infty(B^q(a))\widehat{\otimes}\mathscr{S}_{\{\mu:j_\mu\neq 0\}}(\mathbb{R}^q) \cap \mathcal{B}\}$$

is a bounded set of $(X, \vec{d})$, a pseudo-differential scales supported in $\Omega_1\times\Omega_2$, and therefore the term $2^{-j_{\mu_0}}G_1$ on the right-hand side of (4.29) is of the desired form.

It remains to show that operator G_2 is of the desired form. Applying Proposition 4.2.2 (iii) and (iv), by taking l_0 so that $(X_l^{\mu_0}, d_l^{\mu_0}\mathfrak{e}_{\mu_0}) = (X_{l_0}, \vec{d}_{l_0})$,

$$G_2 f(x) = \begin{cases} 2^{-j_{\mu_0} d_l^{\mu_0}} \int ((-X_l + \sum_{k=1}^q g_{3,l_0}^k X_k) f)(\Gamma(t,x))\, \mathrm{Dil}_{2^j}^{\vec{d}}(\varsigma)(x,t,t)\, dt, \\ 2^{-j_{\mu_0} d_l^{\mu_0}} \int (-X_l + \sum_{k=1}^q g_{4,l_0}^k X_k)(f(\Gamma(t,x)))\, \mathrm{Dil}_{2^j}^{\vec{d}}(\varsigma)(x,t,t)\, dt, \end{cases}$$

$$=: \begin{cases} F_1 f(x), \\ F_2 f(x), \end{cases}$$

where $\deg_{\vec{d}}(g_{r,l_0}^k) \geq \vec{d}_k - d_l^{\mu_0}\mathfrak{e}_{\mu_0}$. Note also that for the constant function $g(t,x) = -1$ we have $\deg_{\vec{d}}(g) \geq -d_l^{\mu_0}\mathfrak{e}_{\mu_0}$. Thus, F_1 and F_2 are sums of operators of the type of the same name in Lemma 4.3.6. Lemma 4.3.6 shows these operators are of the desired form, completing the proof. □

Proof of Proposition 4.3.5. Lemma 4.3.7 is equivalent to saying

$$E = \begin{cases} \sum_{|\alpha_{\mu_0}|\leq 1} 2^{-j_{\mu_0}(1-|\alpha_{\mu_0}|)} E_{\alpha_{\mu_0},1} (2^{-j_{\mu_0} d^{\mu_0}} X^{\mu_0})^{\alpha_{\mu_0}}, \\ \sum_{|\alpha_{\mu_0}|\leq 1} 2^{-j_{\mu_0}(1-|\alpha_{\mu_0}|)} (2^{-j_{\mu_0} d^{\mu_0}} X^{\mu_0})^{\alpha_{\mu_0}} E_{\alpha_{\mu_0},2}, \end{cases} \tag{4.30}$$

where

$$\{(E_{\alpha_{\mu_0},1}, 2^{-j}), (E_{\alpha_{\mu_0},2}, 2^{-j}) : |\alpha_{\mu_0}| \leq 1, |\beta_{\mu_0}| \leq 1, j \in [0,\infty)^\nu,$$
$$\varsigma \in C_0^\infty(\Omega_1)\widehat{\otimes} C_0^\infty(B^q(a))\widehat{\otimes}\mathscr{S}_{\{\mu : j_\mu \neq 0\}}(\mathbb{R}^q) \cap \mathcal{B}\}$$

is a bounded set of $(X, \vec{d}), a$ pseudo-differential operators supported in $\Omega_1 \times \Omega_2$. The result follows by two applications of (4.30). □

4.3.1 Lebesgue space bounds

We will discuss the boundedness properties of $(X, \vec{d}), a$ pseudo-differential operators on various function spaces as part of a more general theory in Chapter 6. However, in this section, we describe the L^p boundedness properties of $(X, \vec{d}), a$ pseudo-differential operator scales, which will be an essential tool in later chapters. Fix a smooth, strictly positive density Vol on $\mathfrak{M}$.

Proposition 4.3.8. *If $a = a(\Omega_1, \Omega_2, (X, \vec{d})) > 0$ is sufficiently small, the following holds. Let $\mathcal{E}$ be a bounded set of $(X, \vec{d}), a$ pre-pseudo-differential operator scales supported in $\Omega_1 \times \Omega_2$. Then*

$$\sup_{(E,2^{-j})\in\mathcal{E}} \sup_{p\in[1,\infty]} \|E\|_{L^p(\mathfrak{M},\mathrm{Vol})\to L^p} < \infty. \tag{4.31}$$

Also, for $p \in (1,\infty)$, $q \in (1,\infty]$, there exists $C = C(p,q,\mathcal{E}) \geq 0$ such that the following holds. Let $\mathcal{E}' = \{(E_l, 2^{-j_l}) : l \in \mathbb{N}\} \subseteq \mathcal{E}$. Then

$$\sup_{\mathcal{E}'} \sup_{(E_l,2^{-j_l})\in\mathcal{E}'} \|\{E_l f_l\}_{l\in\mathbb{N}}\|_{L^p(\mathfrak{M},\mathrm{Vol};\ell^q(\mathbb{N}))} \leq C\|\{f_l\}_{l\in\mathbb{N}}\|_{L^p(\mathfrak{M},\mathrm{Vol};\ell^q(\mathbb{N}))}, \tag{4.32}$$

for all $\{f_l\}_{l\in\mathbb{N}} \in L^p(\mathfrak{M},\mathrm{Vol};\ell^q(\mathbb{N}))$, where the supremum $\sup_{\mathcal{E}'}$ is taken over all such countable subsets of $\mathcal{E}$.

This section is devoted to the proof of Proposition 4.3.8.

Lemma 4.3.9. *Let $\mathcal{B} \subset C_0^\infty(\Omega_1)\widehat{\otimes}C_0^\infty(B^q(a))\widehat{\otimes}\mathscr{S}(\mathbb{R}^q)$ be a bounded set. Then, for $j \in [0,\infty)^\nu$ and $\varsigma \in \mathcal{B}$, there exists*

$$\{\gamma_k : k \in \mathbb{N}^\nu, k \leq j\} \subset C_0^\infty(\Omega_1 \times B^q(a) \times B^q(a))$$

such that

$$\mathrm{Dil}_{2^j}^{\vec{d}}(\varsigma)(x,t,t) = \sum_{\substack{k\leq j\\ k\in\mathbb{N}^\nu}} \mathrm{Dil}_{2^k}^{\vec{d}}(\gamma_k)(x,t,t). \tag{4.33}$$

Furthermore, for every $M \in \mathbb{N}$,

$$\{2^{M|j-k|}\gamma_k : j \in [0,\infty)^\nu, k \leq j, k \in \mathbb{N}^\nu, \varsigma \in \mathcal{B}\} \subset C_0^\infty(\Omega_1 \times B^q(a) \times B^q(a)) \tag{4.34}$$

is a bounded set.

Proof. Fix $\eta \in C_0^\infty(B^q(a))$ so that $\eta \equiv 1$ on a neighborhood of the closure of

$$\{s \in B^q(a) : \exists \varsigma \in \mathcal{B}, x \in \Omega, t \in \mathbb{R}^q, \varsigma(x,s,t) \neq 0\}.$$

See Lemma 4.1.10 for why this is possible.

Take $\varsigma \in \mathcal{B}$ and $j \in [0,\infty)^\nu$. For $k \in \mathbb{N}^\nu$ with $k \leq j$, let

$$\sigma_k(t) := \sum_{\substack{p\in\{0,1\}^\nu\\ k+p\leq j}} (-1)^{p_1+p_2+\cdots+p_\nu}\eta(2^{p\cdot\vec{d}}t),$$

so that $\eta(t) = \sum_{0\leq k\leq j} \sigma_k(2^{k\cdot\vec{d}}t)$. Note that $\sigma_k(t) = 0$ if $k_\mu \leq j_\mu - 1$ and $|t^\mu|$ is sufficiently small (independent of j,k). Define, for $k \in \mathbb{N}^\nu$, $k \leq j$,

$$\gamma_k(x,s,t) := \sigma_k(t)\,\mathrm{Dil}_{2^{j-k}}^{\vec{d}}(\varsigma)(x,s,t).$$

It follows immediately from the definitions that

$$\{\gamma_k : j \in [0,\infty)^\nu, k \leq j, k \in \mathbb{N}^\nu, |j-k|_\infty < 1, \varsigma \in \mathcal{B}\} \subset C_0^\infty(\Omega_1 \times B^q(a) \times B^q(a))$$

is a bounded set, so to prove (4.34) is a bounded set it suffices to consider the case where $|j-k|_\infty \geq 1$. Take μ such that $|j-k|_\infty = j_\mu - k_\mu \geq 1$. Because $\sigma_k(t) = 0$ for $|t_\mu|$ sufficiently small, we see by Lemma 4.1.10 that for every smooth coordinate system $\Phi : B^n(1) \xrightarrow{\sim} \Phi(B^n(1)) \subseteq \mathfrak{M}$, $\forall m \in \mathbb{N}$, for all multi-indices $\alpha \in \mathbb{N}^n, \beta_1, \beta_2 \in \mathbb{N}^q$,

$$|\partial_u^\alpha \partial_s^{\beta_1} \partial_t^{\beta_2} \gamma_k(\Phi(u), s, t)| \lesssim \chi_{\{|t^\mu| \approx 1\}}(1 + |2^{(j_\mu - k_\mu) d^\mu} t^\mu|)^{-m} \lesssim 2^{-m|j-k|_\infty}.$$

Since γ_k is clearly supported in a compact set of $\Omega_1 \times B^q(a) \times B^q(a)$ independent of j, k, it follows that for every $M \in \mathbb{N}$,

$$\begin{aligned}&\{2^{M|j-k|}\gamma_k : j \in [0,\infty)^\nu, k \leq j, k \in \mathbb{N}^\nu, |j-k|_\infty \geq 1, \varsigma \in \mathcal{B}\} \\ &\quad \subset C_0^\infty(\Omega_1 \times B^q(a) \times B^q(a))\end{aligned}$$

is a bounded set, completing the proof that (4.34) is bounded.

Since

$$\operatorname{Dil}_{2^j}^{\vec{d}}(\varsigma)(x,t,t) = \eta(t) \operatorname{Dil}_{2^j}^{\vec{d}}(\varsigma)(x,t,t) = \sum_{k \leq j} \operatorname{Dil}_{2^k}^{\vec{d}}(\gamma_k)(x,t,t),$$

formula (4.33) follows, completing the proof. □

Define $A^a_{\delta,(X,\vec{d})}$ as in (3.63). Set $\mathcal{K} := \overline{\Omega_2}$ and let Ω_3 be an open, relatively compact set such that $\mathcal{K} \Subset \Omega_3 \Subset \mathfrak{M}$. In what follows, we think of Ω_3 as being chosen depending only on Ω_2.

Lemma 4.3.10. *If $a = a(\Omega_1, \Omega_2, (X, \vec{d})) > 0$, the following holds. Let $\mathcal{E}$ be a bounded set of $(X, \vec{d})$, a pre-pseudo-differential scales supported in $\Omega_1 \times \Omega_2$. Then there exists $C = C(\mathcal{E}) \geq 0$ such that for $(E, 2^{-j}) \in \mathcal{E}$,*

$$|Ef(x)| \leq C\chi_{\Omega_3} \sum_{\substack{k \leq j \\ k \in \mathbb{N}^\nu}} 2^{-|j-k|} A^a_{2^{-k},(X,\vec{d})} |\chi_{\Omega_2} f|(x), \tag{4.35}$$

$$|Ef(x)| \leq C\chi_{\Omega_3} \widetilde{\mathcal{M}}_{(X,\vec{d}),\mathcal{K},a}(\chi_{\Omega_2} f)(x). \tag{4.36}$$

Proof. For $(E, 2^{-j}) \in \mathcal{E}$, $Ef = E\chi_{\Omega_2} f$, and we have

$$Ef(x) = \int (\chi_{\Omega_2} f)(\Gamma_t(x)) \operatorname{Dil}_{2^j}^{\vec{d}}(\varsigma)(x,t,t)\, dt,$$

where

$$\{\varsigma : (E, 2^{-j}) \in \mathcal{E}\} \subset C_0^\infty(\Omega_1) \widehat{\otimes} C_0^\infty(B^q(a)) \widehat{\otimes} \mathscr{S}(\mathbb{R}^q)$$

is bounded. We decompose

$$\mathrm{Dil}^{\vec{d}}_{2^j}(\varsigma)(x,t,t) = \sum_{\substack{k\leq j\\ k\in\mathbb{N}^\nu}} \mathrm{Dil}^{\vec{d}}_{2^k}(\gamma_k)(x,t,t),$$

where $\gamma_k \in C_0^\infty(\Omega_1 \times B^q(a) \times B^q(a))$ is as in Lemma 4.3.9. We have

$$\begin{aligned}
|Ef(x)| &\leq \sum_{\substack{k\leq j\\ k\in\mathbb{N}^\nu}} \int |(\chi_{\Omega_2} f)(\Gamma(t,x))| \big|\mathrm{Dil}^{\vec{d}}_{2^k}(\gamma_k)(x,t,t)\big|\, dt \\
&= \sum_{\substack{k\leq j\\ k\in\mathbb{N}^\nu}} \int |(\chi_{\Omega_2} f)(\Gamma(2^{-j\cdot\vec{d}}t,x))| \big|\gamma_k(x,2^{-j\cdot\vec{d}}t,t)\big|\, dt \\
&\lesssim \sum_{\substack{k\leq j\\ k\in\mathbb{N}^\nu}} 2^{-|j-k|} \int\limits_{B^q(a)} |(\chi_{\Omega_2} f)(\Gamma(2^{-j\cdot\vec{d}}t,x))|\, dt \\
&= \sum_{\substack{k\leq j\\ k\in\mathbb{N}^\nu}} 2^{-|j-k|} A^a_{2^{-k},(X,\vec{d})}(\chi_{\Omega_2} f)(x),
\end{aligned}$$

where in the second to last line, we used that (4.34) is a bounded set. This implies (4.35), without the factor of χ_{Ω_3} on the right-hand side. By Lemma 3.11.1, if $a > 0$ is sufficiently small, then $\mathrm{supp}(A^a_{\delta,(X,\vec{d})}\chi_{\Omega_2} f) \subseteq \Omega_3$, and we may therefore include the term χ_{Ω_3} on the right-hand side of (4.35).

Since $\chi_{\Omega_3} A^a_{\delta,(X,\vec{d})}|\chi_{\Omega_2} f|(x) \leq \chi_{\Omega_3}\widetilde{\mathcal{M}}_{(X,\vec{d}),\mathcal{K},a}(\chi_{\Omega_2} f)(x)$, by the definition of $\widetilde{\mathcal{M}}_{(X,\vec{d}),\mathcal{K},a}$, (4.36) follows from (4.35). □

Proof of Proposition 4.3.8. By taking $a > 0$ sufficiently small, (4.31) follows from (4.35) and Lemma 3.11.1 (with Ω_1 replaced by Ω_3 in that lemma). Also, (4.32) follows from (4.36) and Theorem 3.11.2 (again, with Ω_1 replaced by Ω_3 in that theorem). □

4.3.2 Adjoints

Fix a strictly positive, smooth density Vol on $\mathfrak{M}$. The $L^2(\mathfrak{M}, \mathrm{Vol})$ adjoint of an $(X, \vec{d}), a$ pseudo-differential operator is not precisely an $(X, \vec{d}), a$ pseudo-differential operator: taking adjoints reverses the order of $(X^1, d^1), \ldots, (X^\nu, d^\nu)$. However, since our assumptions are symmetric in $(X^1, d^1), \ldots, (X^\nu, d^\nu)$, the adjoint of an $(X, \vec{d}), a$ pseudo-differential operator is "of the same form" as the original operator.

To make this precise, for a permutation σ of $\{1, \ldots, \nu\}$, let

$$(Z^1, \vec{d}^1) = (X^{\sigma(1)}, d^{\sigma(1)}), \ldots, (Z^\nu, \vec{d}^\nu) = (X^{\sigma(\nu)}, d^{\sigma(\nu)}),$$

and let $(Z, \vec{d}) \subset C^\infty_{\mathrm{loc}}(\mathfrak{M}) \times \mathbb{N}^\nu$ be given by

$$(Z, \vec{d}) = (Z^1, \vec{d}^1) \boxtimes (Z^2, \vec{d}^2) \boxtimes \cdots \boxtimes (Z^\nu, \vec{d}^\nu).$$

Then $(Z, \vec{d})$ satisfies all the same hypotheses as $(X, \vec{d})$, so it makes sense to talk about $(Z, \vec{d})$, *a* pseudo-differential operators supported on $\Omega_1 \times \Omega_2$. In the rest of this section, we take the particular permutation $\sigma(1) = \nu, \sigma(2) = \nu - 1, \ldots, \sigma(\nu) = 1$, and we define $(Z, \vec{d})$ as above. Fix a relatively compact open set $\Omega_3 \Subset \mathfrak{M}$, with $\Omega_2 \Subset \Omega_3$.

Lemma 4.3.11. *If $a = a(\Omega_1, \Omega_2, \Omega_3, (X, \vec{d})) > 0$ is sufficiently small, the following hold:*

(i) *Let $\mathcal{E}$ be a bounded set of $(X, \vec{d})$, a pre-pseudo-differential operator scales supported in $\Omega_1 \times \Omega_2$. Then*

$$\mathcal{E}^* := \{(E^*, 2^{-j}) : (E, 2^{-j}) \in \mathcal{E}\} \tag{4.37}$$

is a bounded set of $(Z, \vec{d})$, a pre-pseudo-differential scales supported in $\Omega_2 \times \Omega_3$. Here, E^* denotes the $L^2(\mathfrak{M}, \mathrm{Vol})$ adjoint of E.

(ii) *If, in addition, $\mathcal{E}$ is a bounded set of $(X, \vec{d})$, a pseudo-differential operator scales supported in $\Omega_1 \times \Omega_2$, then $\mathcal{E}^*$ is also a bounded set of $(X, \vec{d})$, a pseudo-differential operator scales supported in $\Omega_1 \times \Omega_2$.*

Proof. We prove only (ii); (i) follows by a simpler version of the same proof.

We write $\Gamma_t(x) = \Gamma(t, x)$, and for each t we think of $\Gamma_t(\cdot)$ as a map. Note that

$$\Gamma_t^{-1}(x) = e^{t^\nu X^\nu} e^{t^{\nu-1} \cdot X^{\nu-1}} \cdots e^{t^1 \cdot X^1} x,$$

that is, Γ_{-t}^{-1} is to $(Z, \vec{d})$ as Γ_t is to $(X, \vec{d})$.

Since $\Gamma(0, x) = x$, for t sufficiently small, $\Gamma(t, \cdot)|_{\Omega_2}$ is a diffeomorphism onto its image and $\Gamma(t, \Omega_1) \Subset \Omega_2$. Let $h(t, x)\,\mathrm{Vol} = \Gamma(t, \cdot)_* \mathrm{Vol}$, so that $h \in C^\infty_{\mathrm{loc}}(B^q(a) \times \Omega_2)$, provided a is chosen sufficiently small.

For each $(E, 2^{-j}) \in \mathcal{E}$, we have

$$Ef(x) = \int f(\Gamma(t, x))\, \mathrm{Dil}_{2^j}^{\vec{d}}(\varsigma)(x, t, t)\, dt,$$

where $\varsigma \in C_0^\infty(\Omega_1) \widehat{\otimes} C_0^\infty(B^q(a)) \widehat{\otimes} \mathscr{S}_{\{\mu : j_\mu \neq 0\}}(\mathbb{R}^q)$, and

$$\{\varsigma : (E, 2^{-j}) \in \mathcal{E}\} \subset C_0^\infty(\Omega_1) \widehat{\otimes} C_0^\infty(B^q(a)) \widehat{\otimes} \mathscr{S}(\mathbb{R}^q)$$

is a bounded set.

We have

$$\begin{aligned}\int (Ef)(x)\overline{g(x)}\, d\,\mathrm{Vol}(x) &= \iint f(\Gamma(t, x))\overline{g(x)}\, \mathrm{Dil}_{2^j}^{\vec{d}}(\varsigma)(x, t, t)\, d\,\mathrm{Vol}(x)\, dt \\ &= (-1)^q \iint f(v)\overline{g(\Gamma_{-t}^{-1}(v))}\, \mathrm{Dil}_{2^j}^{\vec{d}}(\varsigma)(\Gamma_{-t}^{-1}(v), -t, -t) h(-t, v)\, d\,\mathrm{Vol}(v)\, dt.\end{aligned}$$

We conclude that

$$E^* g(x) = \int g(\Gamma_{-t}^{-1}(x)) \operatorname{Dil}_{2^j}(\tilde{\varsigma})(x, t, t)\, dt,$$

where

$$\tilde{\varsigma}(x, s, t) = (-1)^q h(-s, x) \overline{\varsigma(\Gamma_{-s}^{-1}(x), -s, -t)}.$$

It is straightforward to verify that $\tilde{\varsigma} \in C_0^\infty(\Omega_2) \widehat{\otimes} C_0^\infty(B^q(a)) \widehat{\otimes} \mathscr{S}_{\{\mu: j_\mu \neq 0\}}(\mathbb{R}^q)$, and

$$\{\tilde{\varsigma} : (E, 2^{-j}) \in \mathcal{E}\} \subset C_0^\infty(\Omega_2) \widehat{\otimes} C_0^\infty(B^q(a)) \widehat{\otimes} \mathscr{S}(\mathbb{R}^q)$$

is a bounded set. Thus, if $a > 0$ is sufficiently small, (4.37) is a bounded set of $(Z, \vec{d})$, a pseudo-differential operator scales supported in $\Omega_2 \times \Omega_3$, as desired. □

4.3.3 Proof of the Littlewood–Paley decomposition

In this section, we complete the proof of Theorem 4.3.3. Fix smooth vector fields $Y_1, \ldots, Y_L \in C_{\mathrm{loc}}^\infty(\mathfrak{M}; T\mathfrak{M})$ so that $\forall x \in \Omega_2$, $\operatorname{span}\{Y_1(x), \ldots, Y_L(x)\} = T_x \Omega_2$.

Lemma 4.3.12. *If $a = a(\Omega_1, \Omega_2, (X, \vec{d})) > 0$ is sufficiently small, the following holds. Let $\mathcal{E}$ be a bounded set of $(X, \vec{d})$, a pseudo-differential operator scales supported in $\Omega_1 \times \Omega_2$. Then, for $(E, 2^{-j}) \in \mathcal{E}$, $1 \leq k \leq L$,*

$$Y_k E = E_0 + \sum_{l=1}^{L} E_l Y_l,$$

where $\{(E_0, 2^{-j}), (E_1, 2^{-j}), \ldots, (E_L, 2^{-j}) : (E, 2^{-j}), 1 \leq k \leq L \in \mathcal{E}\}$ is a bounded set of $(X, \vec{d})$, a pseudo-differential operator scales supported in $\Omega_1 \times \Omega_2$.

Proof. Let $(E, 2^{-j}) \in \mathcal{E}$. Then

$$Ef(x) = \int f(\Gamma(t, x)) \operatorname{Dil}_{2^j}^{\vec{d}}(\varsigma)(x, t, t)\, dt,$$

where $\{\varsigma : (E, 2^{-j}) \in \mathcal{E}\} \subset C_0^\infty(\Omega_1) \widehat{\otimes} C_0^\infty(B^q(a)) \widehat{\otimes} \mathscr{S}(\mathbb{R}^q)$ is a bounded set and $\varsigma \in C_0^\infty(\Omega_1) \widehat{\otimes} C_0^\infty(B^q(a)) \widehat{\otimes} \mathscr{S}_{\{\mu: j_\mu \neq 0\}}(\mathbb{R}^q)$. By the chain rule, we have

$$\begin{aligned} Y_k Ef(x) &= \int f(\Gamma(t, x)) \operatorname{Dil}_{2^j}^{\vec{d}}(Y_k \varsigma)(x, t, t)\, dt + \sum_{l=1}^{L} (Y_l f)(\Gamma(t, x)) h_l(t, x) \operatorname{Dil}_{2^j}^{\vec{d}}(\varsigma)(x, t, t)\, dt \\ &=: E_0 f(x) + \sum_{l=1}^{L} E_l Y_l f(x), \end{aligned}$$

where $h_l(t, x) \in C_{\mathrm{loc}}^\infty(B^q(a) \times \Omega_2)$. Set $\varsigma_0(x, s, t) := Y_k \varsigma(x, s, t)$ and $\varsigma_l(x, s, t) := h_l(s, x) \varsigma(x, s, t)$ for $l = 1, \ldots, L$. We have $\varsigma_l \in C_0^\infty(\Omega_1) \widehat{\otimes} C_0^\infty(B^q(a)) \widehat{\otimes} \mathscr{S}_{\{\mu: j_\mu \neq 0\}}(\mathbb{R}^q)$ and

$$\{\varsigma_l : (E, 2^{-j}) \in \mathcal{E}, 0 \le l \le L, 1 \le k \le L\} \subset C_0^\infty(\Omega_1)\widehat{\otimes} C_0^\infty(B^q(a))\widehat{\otimes}\mathscr{S}(\mathbb{R}^q)$$

is a bounded set. Since $E_l f(x) = \int f(\Gamma(t,x))\,\mathrm{Dil}_{2^j}(\varsigma_l)(x,t,t)\,dt$, the result follows. □

Lemma 4.3.13. *If $a = a(\Omega_1, \Omega_2, (X, \vec{d})) > 0$ is sufficiently small, the following holds. Let $\mathcal{E}$ be a bounded set of $(X, \vec{d})$, a pseudo-differential operator scales supported in $\Omega_1 \times \Omega_2$, and let $\mathcal{B} \subset C^\infty_{\mathrm{loc}}(\mathfrak{M})$ be a bounded set. Then, for every $M \in \mathbb{N}$ and every ordered multi-index α,*

$$\sup_{\substack{(E,2^{-j})\in\mathcal{E}\\ f\in\mathcal{B}}} 2^{M|j|_\infty} \|Y^\alpha E f\|_{L^\infty(\mathfrak{M},\mathrm{Vol})} < \infty.$$

Proof. Let $(E, 2^{-j}) \in \mathcal{E}$. Repeated application of Lemma 4.3.12 shows that

$$Y^\alpha E f = \sum_{|\beta|\le|\alpha|} E_\beta Y^\beta f,$$

where $\{(E_\beta, 2^{-j}) : (E, 2^{-j}) \in \mathcal{E}, |\beta| \le |\alpha|\}$ is a bounded set of $(X, \vec{d})$, a pseudo-differential scales. Since $\{Y^\beta f : f \in \mathcal{B}\} \subset C^\infty_{\mathrm{loc}}(\mathfrak{M})$ is a bounded set, it suffices to prove the case $|\alpha| = 0$.

Fix $\mu \in \{1, \ldots, \nu\}$ so that $j_\mu = |j|_\infty$. Applying Proposition 4.3.5 M times we see that

$$\begin{aligned} Ef &= \sum_{\substack{|\alpha_\mu|\le M\\ |\beta_\mu|\le M}} 2^{-j_\mu(2M-|\alpha_\mu|-|\beta_\mu|)} (2^{-j_\mu d^\mu} X^\mu)^{\alpha_\mu} E_{\mu,\alpha_\mu,\beta_\mu} (2^{-j_\mu d^\mu} X^\mu)^{\beta_\mu} f \\ &=: \sum_{\substack{|\alpha_\mu|\le M\\ |\beta_\mu|\le M}} 2^{-j_\mu(M-|\beta_\mu|+\deg_{d^\mu}(\beta_\mu))} 2^{-j_\mu(M-|\alpha_\mu|)} \widetilde{E}_{\mu,\alpha_\mu,\beta_\mu} f_{\beta_\mu}, \end{aligned}$$

where $\widetilde{E}_{\mu,\alpha_\mu,\beta_\mu} = (2^{-j_\mu d^\mu} X^\mu)^{\alpha_\mu} E_{\mu,\alpha_\mu,\beta_\mu}$ and $f_{\beta_\mu} = (X^\mu)^{\beta_\mu} f$. By Proposition 4.3.5, $\{(E_{\mu,\alpha_\mu,\beta_\mu}, 2^{-j}) : (E, 2^{-j}) \in \mathcal{E}\}$ is a bounded set of $(X, \vec{d})$, a pseudo-differential scales supported in $\Omega_1 \times \Omega_2$ and therefore by Lemma 4.3.4 (ii), $\{(\widetilde{E}_{\mu,\alpha_\mu,\beta_\mu}, 2^{-j}) : (E, 2^{-j}) \in \mathcal{E}\}$ is a bounded set of $(X, \vec{d})$, a pseudo-differential scales supported in $\Omega_1 \times \Omega_2$. Clearly, $\{f_{\beta_\mu} : f \in \mathcal{B}\} \subset C^\infty_{\mathrm{loc}}(\mathfrak{M})$ is a bounded set. Thus, since $2^{-j_\mu(M-|\beta_\mu|+\deg_{d^\mu}(\beta_\mu))} 2^{-j_\mu(M-|\alpha_\mu|)} \le 2^{-Mj_\mu} = 2^{-M|j|_\infty}$, we see that it suffices to prove the result in the case $M = 0$.

For the case $M = 0$, take $\phi \in C_0^\infty(\mathfrak{M})$ so that $\phi = 1$ on Ω_2. We then have, by Proposition 4.3.8,

$$\sup_{\substack{(E,2^{-j})\in\mathcal{E}\\ f\in\mathcal{B}}} \|Ef\|_{L^\infty} = \sup_{\substack{(E,2^{-j})\in\mathcal{E}\\ f\in\mathcal{B}}} \|E\phi f\|_{L^\infty} \lesssim \sup_{f\in\mathcal{B}} \|\phi f\|_{L^\infty} < \infty,$$

completing the proof. □

Proof of Theorem 4.3.3. Let $\{(E_j, 2^{-j}) : j \in \mathbb{N}^\nu\}$ be a bounded set of $(X, \vec{d})$, a pseudo-differential operator scales supported in $\Omega_1 \times \Omega_2$.

We begin by showing that $\sum_{j\in\mathbb{N}^\nu} 2^{j\cdot m} E_j$ converges in the typology of bounded convergence on $\mathrm{Hom}(C^\infty_{\mathrm{loc}}(\mathfrak{M}), C^\infty_0(\Omega_1))$. Let $\mathcal{B} \subset C^\infty_{\mathrm{loc}}(\mathfrak{M})$ be a bounded set. By the definition of bounded set of $(X, \vec{d})$, *a* pseudo-differential operator data, we have

$$\bigcup_{j\in\mathbb{N}^\nu} \bigcup_{f\in\mathcal{B}} \operatorname{supp}(E_j f) \Subset \Omega_1.$$

Thus, to prove the desired convergence, it suffices to show that for every ordered multi-index α, $\sum_{j\in\mathbb{N}^\nu} Y^\alpha E_j f(x)$ converges uniformly in x and $f \in \mathcal{B}$. This follows immediately from Lemma 4.3.13, completing the proof of the convergence in the typology of bounded convergence on $\mathrm{Hom}(C^\infty_{\mathrm{loc}}(\mathfrak{M}), C^\infty_0(\Omega_1))$.

By the definition of bounded sets of $(X, \vec{d})$, *a* pseudo-differential scales supported in $\Omega_1 \times \Omega_2$, we may pick $\phi \in C^\infty_0(\Omega_1)$ so that $\mathrm{Mult}[\phi]E_j = E_j$, $\forall j \in \mathbb{N}$. Let $f \in C^\infty_0(\mathfrak{M})'$ and consider that for $g \in C^\infty_{\mathrm{loc}}(\Omega_1)$, we have

$$\left\langle \sum_{|j|\le M} 2^{m\cdot j} E_j f, g \right\rangle_{L^2(\mathfrak{M},\mathrm{Vol})} = \left\langle f, \sum_{|j|\le M} 2^{m\cdot j} E_j^* \phi g \right\rangle_{L^2(\mathfrak{M},\mathrm{Vol})}, \tag{4.38}$$

where $\langle\cdot,\cdot\rangle_{L^2}$ denotes the (formal) L^2 inner product. By Lemma 4.3.11 (ii), $\{(E_j^*, 2^{-j}) : j \in \mathbb{N}^\nu\}$ is a bounded set of $(X, \vec{d})$, *a* pseudo-differential operator scales supported in $\Omega_2 \times \Omega_3$, for an open set Ω_3 with $\Omega_2 \Subset \Omega_3 \Subset \mathfrak{M}$. Thus, $\sum_{j\in\mathbb{N}^\nu} 2^{m\cdot j} E_j^* \phi g$ converges in $C^\infty_0(\Omega_2)$. By (4.38), we conclude that $\sum_{j\in\mathbb{N}^\nu} 2^{m\cdot j} E_j f$ converges in the weak topology on $C^\infty_{\mathrm{loc}}(\Omega_1)'$, that is, $\sum_{j\in\mathbb{N}^\nu} 2^{m\cdot j} E_j$ converges in the weak operator topology on $\mathrm{Hom}(C^\infty_0(\mathfrak{M})', C^\infty_{\mathrm{loc}}(\Omega_1)')$, where $C^\infty_{\mathrm{loc}}(\Omega_1)'$ is given the weak topology.

Now that the desired convergence of the sums has been shown, (i) ⇔ (ii) follows immediately from Proposition 4.1.14. □

4.4 Adding parameters

Let $E = \{\mu_1 < \mu_2 < \cdots < \mu_{|E|}\} \subseteq \{1, \ldots, \nu\}$. For a vector in $v \in \mathbb{R}^{|E|}$, we index the coordinates of v by $\mu \in E$, so it makes sense to write v_μ for $\mu \in E$. We consider the set of vector fields with $|E|$-parameter formal degrees by

$$\begin{aligned}(X^E, \vec{d}^E) &= \{(X_1^E \vec{d}_1^E), \ldots, (X_{q_E}^E, \vec{d}_{q_E}^E)\} \\ &:= (X^{\mu_1}, d^{\mu_1}) \boxtimes (X^{\mu_2}, d^{\mu_2}) \boxtimes \cdots \boxtimes (X^{\mu_{|E|}}, d^{\mu_{|E|}}) \subset C^\infty_{\mathrm{loc}}(\mathfrak{M}; T\mathfrak{M}) \times (\mathbb{N}^{|E|} \setminus \{0\}).\end{aligned}$$

Note that $(X^E, \vec{d}^E)$ satisfies (4.2). It therefore makes sense to talk of $(X^E, \vec{d}^E)$, *a* pseudo-differential operators.

Proposition 4.4.1. *If $a = a(\Omega_1, \Omega_2, (X, \vec{d})) > 0$ is chosen sufficiently small, then the following holds. Suppose T is an $(X^E, \vec{d}^E)$, a pseudo-differential operator of order $m^E \in \mathbb{R}^{|E|}$*

supported in $\Omega_1 \times \Omega_2$. *Let* $m \in \mathbb{R}^\nu$ *be such that* $m_\mu = m_\mu^E$, *for* $\mu \in E$ *and* $m_\mu = 0$ *for* $\mu \notin E$. *Then* T *is an* $(X, \vec{d})$, *a pseudo-differential operator of order* m *supported in* $\Omega_1 \times \Omega_2$.

Proof. Let $E = \{\mu_1 < \mu_2 < \cdots < \mu_{|E|}\}$. Set $t^E = (t^{\mu_1}, t^{\mu_2}, \ldots, t^{\mu_{|E|}})$ and define t^{E^c} similarly with E replaced by E^c. By hypothesis,

$$Tf(x) = \int f(e^{-t^{\mu_1}\cdot X^{\mu_1}} e^{-t^{\mu_2}\cdot X^{\mu_2}} \cdots e^{-t^{\mu_{|E|}}\cdot X^{\mu_{|E|}}} x)\check{b}_E(x, t^E, t^E)\, dt^E,$$

where $b_E(x, s^E, \xi^E) \in S^{m_E}(a, \Omega_1, \vec{d}^E)$. Fix $\eta(t^{E^c}) \in C_0^\infty(\mathbb{R}^{q_{E^c}})$ with $\eta(t^{E^c}) = 1$ on a neighborhood of 0, and with small support. Set

$$b(x, s, \xi) := b_E(x, s^E, \xi^E)\eta(s^{E^c}).$$

Note that if the support of η is sufficiently small, we have $b(x, s, \xi) \in S^m(a, \Omega_1, \vec{d})$, and

$$\check{b}(x, t, t) = \check{b}_E(x, t^E, t^E) \otimes \delta_0(t^{E^c}),$$

where δ_0 denotes the Dirac δ function at 0. We have

$$\begin{aligned} Tf(x) &= \int f(e^{-t^{\mu_1}\cdot X^{\mu_1}} e^{-t^{\mu_2}\cdot X^{\mu_2}} \cdots e^{-t^{\mu_{|E|}}\cdot X^{\mu_{|E|}}} x)\check{b}_E(x, t^E, t^E)\, dt^E \\ &= \int f(\Gamma(t, x))\check{b}_E(x, t^E, t^E) \otimes \delta_0(t^{E^c})\, dt = \int f(\Gamma(t, x))\check{b}(x, t, t)\, dt, \end{aligned}$$

completing the proof. □

4.5 The sub-Laplacian

Let $(W, \vec{d}) = \{(W_1, \vec{d}_1), \ldots, (W_r, \vec{d}_r)\} \subset C_{\mathrm{loc}}^\infty(\mathfrak{M}; T\mathfrak{M}) \times \mathbb{N}_+$ be such that $\mathrm{Gen}((W, \vec{d}))$ is locally finitely generated. Fix $\kappa \in \mathbb{N}_+$ such that $\vec{d}_j$ divides κ, $1 \le j \le r$. Fix a smooth, strictly positive density Vol on $\mathfrak{M}$. We define the sub-Laplacian, associated with $(W, \vec{d})$, by

$$\mathcal{L} := \sum_{j=1}^{r} (W_j^{\kappa/\vec{d}_j})^* W_j^{\kappa/\vec{d}_j}, \tag{4.39}$$

where $*$ denotes the formal $L^2(\mathfrak{M}, \mathrm{Vol})$ adjoint.

Fix relatively compact open sets $\Omega_1 \Subset \Omega_2 \Subset \mathfrak{M}$. By hypothesis, $\mathrm{Gen}((W, \vec{d}))$ is finitely generated by some $(X, d) = \{(X_1, d_1), \ldots, (X_q, d_q)\} \subset \mathrm{Gen}((W, \vec{d}))$ on Ω_2. In Section 4.5.5 we will pick a particular choice of (X, d); with this choice, the main result of this section is the following.

Theorem 4.5.1. *For every* $a > 0$ *(sufficiently small) and every* $\psi \in C_0^\infty(\Omega_1)$, *there is an* (X, d), *a pseudo-differential operator,* T_ψ, *of order* -2κ *supported in* $\Omega_1 \times \Omega_2$ *such that*

$$T_\psi \mathscr{L} = \mathrm{Mult}[\psi] + E,$$

where E is an (X, d), a pseudo-differential operator of order -1 supported in $\Omega_1 \times \Omega_2$.

This section is devoted to the proof of Theorem 4.5.1.

Remark 4.5.2. A primary example to consider is the case where $(W, \vec{d})$ are Hörmander vector fields with formal degrees (see Proposition 3.4.14). In this case, $\mathscr{L}$ is maximally subelliptic (see Example 1.1.10 (iii)), and we can obtain much more detailed information about $\mathscr{L}$ (see Theorem 8.1.1).

Remark 4.5.3. The proof of Theorem 4.5.1 uses a method of Rothschild and Stein [200], where we use a fundamental solution for a sub-Laplacian on a high-dimensional nilpotent Lie group to deduce the result. After Rothschild and Stein's work, this idea was furthered by Goodman [99], and our approach more closely follows his. One thing that sets our analysis apart from these references is that we do not assume the vector fields $W_1, \ldots, W_r$ satisfy Hörmander's condition. We only assume the weaker condition that $\mathrm{Gen}((W, \vec{d}))$ is locally finitely generated. For example, when the vector fields are real analytic, this condition always holds (Proposition 3.4.15). The key to being able to work under this weaker condition is Proposition 4.2.2.

Remark 4.5.4. Theorem 4.5.1 gives a left parametrix for $\mathscr{L}$. A similar proof gives a right parametrix, that is, there is an (X, d), a pseudo-differential operator, S_ψ, of order -2κ supported in $\Omega_1 \times \Omega_2$ such that $\mathscr{L} S_\psi = \mathrm{Mult}[\psi] + E_1$, where E_1 is an (X, d), a pseudo-differential operator of order -1 supported in $\Omega_1 \times \Omega_2$. The proof of this is a straightforward modification of the proof of Theorem 4.5.1 and we leave the details to the interested reader.

4.5.1 The sub-Laplacian is subelliptic

Let $\mathfrak{M}$ be an open subset of $\mathbb{R}^n$ and let $(W, \vec{d}) = \{(W_1, \vec{d}_1), \ldots, (W_r, \vec{d}_r)\} \subset C^\infty_{\mathrm{loc}}(\mathfrak{M}; T\mathfrak{M}) \times \mathbb{N}_+$ be Hörmander vector fields with formal degrees. Define $\mathscr{L}$ as in (4.39), where Vol is taken to be the Lebesgue measure. The main result of this section is the following.

Theorem 4.5.5. *For every relatively compact, open set $\Omega_1 \Subset \mathfrak{M}$, $\mathscr{L}$ is subelliptic on Ω_1. More precisely, there exists $\epsilon_0 = \epsilon_0(\Omega_1) > 0$ such that $\forall s \in \mathbb{R}$, $N \geq 0$, $\forall \phi_1, \phi_2 \in C^\infty_0(\Omega_1)$ with $\phi_1 \prec \phi_2$,*

$$\|\phi_1 u\|_{L^2_{s+\epsilon_0}} \leq C_{\phi_1,\phi_2,s,N}\left(\|\phi_2 \mathscr{L} u\|_{L^2_s} + \|\phi_2 u\|_{L^2_{s-N}}\right), \quad \forall u \in C^\infty_0(\mathfrak{M})',$$

where if the right-hand side is finite, so is the left-hand side.

This section is devoted to the proof of Theorem 4.5.5. The results in this section are not sharp: we do not obtain the optimal value for ϵ_0 here. However, we do obtain the optimal value for ϵ_0 in Section 8.2.2, where we also obtain results for more general spaces. The results here are an important step in obtaining these more general and sharper results.

We require some new notation. Throughout this section we write $\langle\cdot,\cdot\rangle$ for the usual $L^2(\mathbb{R}^n)$ inner product. We write Λ^s for the standard pseudo-differential operator of order s given in Definition 2.2.10. We write

$$\|f_1\|_{L^2_{s_1}} \lesssim \|f_2\|_{L^2_{s_2}} + \|f_3\|_{L^2_{-\infty}}$$

to mean that $\forall N \in \mathbb{N}$, there exists C_N such that

$$\|f_1\|_{L^2_{s_1}} \leq C_N(\|f_2\|_{L^2_{s_2}} + \|f_3\|_{L^2_{-N}}).$$

We write (l.c.) to denote "large constant" and (s.c.) to denote "small constant." An expression of the form

$$A \leq (\text{l.c.})B_1 + (\text{s.c.})B_2 \tag{4.40}$$

means that $\forall \epsilon > 0$, $\exists C_\epsilon \geq 0$ with

$$A \leq C_\epsilon B_1 + \epsilon B_2.$$

Note that $AB \leq (\text{l.c.})A + (\text{s.c.})B$. Also, if $A \leq (\text{l.c.})B + (\text{s.c.})A$, then $A \lesssim B$. We can combine the two above notations as follows. If we write

$$\|f_1\|_{L^2_{s_1}} \leq (\text{s.c.})\|f_2\|_{L^2_{s_2}} + (\text{l.c.})(\|f_3\|_{L^2_{s_3}} + \|f_4\|_{L^2_{-\infty}}),$$

this means that $\forall \epsilon > 0$, $N \in \mathbb{N}$, $\exists C_{\epsilon,N} \geq 0$, such that

$$\|f_1\|_{L^2_{s_1}} \leq \epsilon\|f_2\|_{L^2_{s_2}} + C_{\epsilon,N}(\|f_3\|_{L^2_{s_3}} + \|f_4\|_{L^2_{-N}}).$$

We write $S^{(s)}$ to denote some standard pseudo-differential operator of order s, on $\mathbb{R}^n$, which may change from line to line. In each case, the operator $S^{(s)}$ can be written down explicitly, but what is important is only that it is a pseudo-differential operator of order s. Thus, using Theorem 2.2.12, we have estimates like

$$\|S^{(s)}u\|_{L^2_{s_1}} \lesssim \|u\|_{L^2_{s_1+s}}.$$

We use the calculus of pseudo-differential operators (Theorem 2.2.8) freely in this section.

Fix relatively compact open sets $\Omega_1 \Subset \Omega_2 \Subset \Omega_3 \Subset \mathfrak{M}$. Because $W_1, \ldots, W_r$ satisfy Hörmander's condition on $\mathfrak{M}$, there exists $m \in \mathbb{N}_+$ such that $W_1, \ldots, W_r$ satisfy Hörmander's condition of order m on $\overline{\Omega_2}$; we henceforth fix the smallest such m. Fix $\psi \in C_0^\infty(\Omega_3)$ with

$\psi \equiv 1$ on a neighborhood of $\overline{\Omega_2}$. By replacing $W_1, \ldots, W_r$ with $\psi W_1, \ldots \psi W_r$, the vector fields still satisfy Hörmander's condition of order m on $\overline{\Omega_2}$, and they can be thought of as globally defined standard pseudo-differential operators on $\mathbb{R}^n$. We henceforth make this replacement.

Proposition 4.5.6. *For any $s < s_2 \in \mathbb{R}$, we have*

$$\|u\|_{L^2_s} \leq (\text{s.c.})\|u\|_{L^2_{s_2}} + (\text{l.c.})\|u\|_{L^2_{-\infty}}, \quad u \in C_0^\infty(\mathbb{R}^n).$$

To prove Proposition 4.5.6, we prove the following intermediate lemma.

Lemma 4.5.7. *Let $s < s_2$. Then*

$$\|u\|_{L^2_s} \leq (\text{s.c.})\|u\|_{L^2_{s_2}} + (\text{l.c.})\|u\|_{L^2_{2s-s_2}}, \quad u \in C_0^\infty(\mathbb{R}^n).$$

Proof. We have

$$\|u\|^2_{L^2_s} = \langle \Lambda^s u, \Lambda^s u\rangle = \langle \Lambda^{s_2} u, \Lambda^{2s-s_2} u\rangle \leq (\text{s.c.})\|u\|^2_{L^2_{s_2}} + (\text{l.c.})\|u\|_{L^2_{2s-s_2}},$$

where the last estimate follows from the Cauchy–Schwartz inequality. □

Proof of Proposition 4.5.6. We claim that for all $j \in \mathbb{N}$,

$$\|u\|_{L^2_s} \leq (\text{s.c.})\|u\|_{L^2_{s_2}} + (\text{l.c.})\|u\|_{L^2_{2^j s-(2^j-1)s_2}}, \quad u \in C_0^\infty(\mathbb{R}^n). \tag{4.41}$$

The result then follows from (4.41) by taking j large.

We prove (4.41) by induction on j. The base case ($j = 0$) is obvious (in this case, the small constant can be taken to be 0 and the large constant can be taken to be 1).

We assume we have (4.41) for j and prove it for $j + 1$. Lemma 4.5.7 shows

$$\|u\|_{L^2_{2^j s-(2^j-1)s_2}} \leq (\text{s.c.})\|u\|_{L^2_{s_2}} + (\text{l.c.})\|u\|_{L^2_{2^{j+1}s-(2^{j+1}-1)s_2}}.$$

Combining this with the inductive hypothesis proves (4.41) with j replaced by $j + 1$, completing the proof. □

The next proposition is the key step where Hörmander's condition comes into play.

Proposition 4.5.8. *For all $s \in \mathbb{R}$,*

$$\|u\|_{L^2_{s+2^{(1-m)}}} \lesssim \sum_{j=1}^{r} \|W_j u\|_{L^2_s} + \|u\|_{L^2_s}, \quad u \in C_0^\infty(\Omega_1). \tag{4.42}$$

Remark 4.5.9. $s + 2^{(1-m)}$ is not the optimal value on the left-hand side of (4.42). In fact, the optimal value is essentially $s + \frac{1}{m}$ (see Corollary 8.2.5 for precise details on what the optimal value is). The optimal estimate is a result of Rothschild and Stein [200, Theorem 12]. This follows, along with many other sharp results, from the theory of Chapter 8. However, we use the results from this chapter to prove the results in Chapter 8, and therefore the non-optimal result Proposition 4.5.8 plays an important role in establishing the optimal results.

Let $\mathcal{Y}_1 := \{W_1, \dots, W_r\}$, and for $j > 1$ recursively define

$$\mathcal{Y}_j := \{[W_l, V] : 1 \le l \le r, V \in \mathcal{Y}_{j-1}\}.$$

Each $\mathcal{Y}_j$ is a finite set and our hypothesis of Hörmander's condition implies that the vector fields in $\bigcup_{j=1}^m \mathcal{Y}_j$ span the tangent space at every point of $\overline{\Omega_2}$. To prove Proposition 4.5.8, we use the following lemma.

Lemma 4.5.10. *Let $Z_j \in \mathcal{Y}_j$. Then*

$$\|Z_j u\|_{L^2_{s+2^{1-j}-1}} \lesssim \sum_{k=1}^r \|W_k u\|_{L^2_s} + \|u\|_{L^2_s}, \quad u \in C_0^\infty(\Omega_1).$$

Proof. We prove the result by induction on j. The base case, $j = 1$, is obvious. Fix $j \ge 2$ for which we wish to prove the result. We assume the result for all lesser values of j. Take $Z_j \in \mathcal{Y}_j$ and write $Z_j = [W_l, Z_{j-1}]$ for some $1 \le l \le r$ and some $Z_{j-1} \in \mathcal{Y}_{j-1}$. We have

$$\begin{aligned} \|Z_j u\|^2_{L^2_{s+2^{1-j}-1}} &= \langle Z_j u, \Lambda^{2s+2^{2-j}-2} Z_j u\rangle \\ &= \langle W_l Z_{j-1} u, \Lambda^{2s+2^{2-j}-2} Z_j u\rangle - \langle Z_{j-1} W_l u, \Lambda^{2s+2^{2-j}-2} Z_j u\rangle. \end{aligned} \tag{4.43}$$

We bound the two terms on the right-hand side of (4.43) separately. We write $W_l^* = -W_l + \mathrm{Mult}[f]$ for some $f \in C_0^\infty(\Omega_3)$. We have, for $u \in C_0^\infty(\Omega_1)$,

$$\begin{aligned} &|\langle W_l Z_{j-1} u, \Lambda^{2s+2^{2-j}-2} Z_j u\rangle| \\ &\quad = |\langle Z_{j-1} u, W_l \Lambda^{2s+2^{2-j}-2} Z_j u\rangle - \langle Z_{j-1} u, f \Lambda^{2s+2^{2-j}-2} Z_j u\rangle| \\ &\quad \le |\langle Z_{j-1} u, \Lambda^{2s+2^{2-j}-2} Z_j W_l u\rangle| + |\langle Z_{j-1} u, S^{(2s+2^{2-j}-1)} u\rangle|, \end{aligned} \tag{4.44}$$

where in the last inequality we have used the calculus of pseudo-differential operators.

Consider

$$\begin{aligned} |\langle Z_{j-1} u, S^{(2s+2^{2-j}-1)} u\rangle| &= |\langle S^{(s+2^{2-j}-1)} Z_{j-1} u, \Lambda^s u\rangle| \\ &\lesssim \|Z_{j-1} u\|^2_{L^2_{s+2^{2-j}-1}} + \|u\|^2_{L^2_s} \lesssim \sum_{k=1}^r \|W_k u\|^2_{L^2_s} + \|u\|^2_{L^2_s}, \end{aligned} \tag{4.45}$$

where the last inequality follows from the inductive hypothesis. Also,

$$|\langle Z_{j-1}u, \Lambda^{2s+2^{2-j}-2} Z_j W_l u\rangle| \lesssim \|Z_{j-1}u\|^2_{L^2_{s+2^{2-j}-1}} + \|W_l u\|^2_{L^2_s} \lesssim \sum_{k=1}^{r} \|W_k u\|^2_{L^2_s} + \|u\|^2_{L^2_s}, \tag{4.46}$$

where in the first inequality we have used $\Lambda^{s-1} Z_j W_l u = S^{(s)} W_l u$ (since Z_j is a pseudo-differential operator of order 1) and in the second inequality we have used the inductive hypothesis. Combining (4.44), (4.45), and (4.46), we see that

$$|\langle W_l Z_{j-1}u, \Lambda^{2s+2^{2-j}-2} Z_j u\rangle| \lesssim \sum_{k=1}^{r} \|W_k u\|^2_{L^2_s} + \|u\|^2_{L^2_s}, \quad \forall u \in C_0^\infty(\Omega_1). \tag{4.47}$$

We now turn to the second term on the right-hand side of (4.43). Recall that for $j \geq 2$,

$$\begin{aligned}
|\langle Z_{j-1}W_l u, \Lambda^{2s+2^{2-j}-2} Z_j u\rangle| &\leq |\langle W_l u, Z_{j-1}\Lambda^{2s+2^{2-j}-2} Z_j u\rangle| + |\langle W_l u, S^{(2s+2^{2-j}-1)} u\rangle| \\
&\lesssim |\langle W_l u, Z_{j-1}\Lambda^{2s+2^{2-j}-2} Z_j u\rangle| + \|W_l u\|^2_{L^2_{s+2^{2-j}-1}} + \|u\|^2_{L^2_s} \\
&\lesssim |\langle W_l u, \Lambda^{2s+2^{2-j}-2} Z_j Z_{j-1} u\rangle| + |\langle W_l u, S^{(2s+2^{2-j}-1)} u\rangle| \\
&\quad + \|W_l u\|^2_{L^2_{s+2^{2-j}-1}} + \|u\|^2_{L^2_s} \\
&\lesssim |\langle W_l u, \Lambda^{2s+2^{2-j}-2} Z_j Z_{j-1} u\rangle| + \|W_l u\|^2_{L^2_{s+2^{2-j}-1}} + \|u\|^2_{L^2_s} \\
&\lesssim \|W_l u\|^2_{L^2_s} + \|Z_{j-1}u\|^2_{L^2_{s+2^{2-j}-1}} + \|u\|^2_{L^2_s} \\
&\lesssim \sum_{k=1}^{r} \|W_k u\|^2_{L^2_s} + \|u\|^2_{L^2_s},
\end{aligned} \tag{4.48}$$

where the last inequality follows from the inductive hypothesis. Combining (4.43), (4.47), and (4.48) completes the proof. □

Proof of Proposition 4.5.8. Note that for $u \in C_0^\infty(\Omega_1)$,

$$\begin{aligned}
\|u\|_{L^2_{s+2^{1-m}}} &\lesssim \sum_{j=1}^{n} \|\partial_{x_j} u\|_{L^2_{s+2^{1-m}-1}} + \|u\|_{L^2_s} \\
&\lesssim \sum_{Z \in \bigcup_{j=1}^m \mathcal{Y}_j} \|Zu\|_{L^2_{s+2^{1-m}-1}} + \|u\|_{L^s_2} \lesssim \sum_{k=1}^{r} \|W_k u\|_{L^2_s} + \|u\|_{L^2_s},
\end{aligned}$$

where the last inequality follows from Lemma 4.5.10. □

Set $n_j := \kappa / d_j$ and pick $\epsilon_0 > 0$ such that $\forall j, j' \in \{1, \ldots r\}$, $(n_j - (n_{j'} - 1))\epsilon_0 \leq 2^{-m}$.

Lemma 4.5.11. *Let $Y \in C_0^\infty(\Omega_3; T\Omega_3)$. Then, $\forall s \in \mathbb{R}$, $k \in \mathbb{N}_+$,*

$$\|Y^k u\|_{L^2_s} \leq (\text{s.c.})\|Y^{k-1}u\|_{L^2_{s+\epsilon_0}} + (\text{l.c.})(\|Y^{k+1}u\|_{L^2_{s-\epsilon_0}} + \|u\|_{L^2_{-\infty}}), \quad u \in C_0^\infty(\Omega_1).$$

Proof. Using that $Y^* = -Y + \mathrm{Mult}[f]$, where $f \in C_0^\infty(\Omega_3)$, we have, using Proposition 4.5.6,

$$\begin{aligned}
\|Y^k u\|_{L_s^2} &= \left|\langle \Lambda^{2s} Y^k u, Y^k u\rangle\right| \\
&\le \left|\langle \Lambda^{s+\epsilon_0} Y^{k-1} u, \Lambda^{s-\epsilon_0} Y^{k+1} u\rangle\right| + \left|\langle S^{(s+\epsilon_0)} Y^{k-1} u, S^{(s-\epsilon_0)} Y^k u\rangle\right| \\
&\le (\text{s.c.})\|Y^{k-1}u\|_{L^2_{s+\epsilon_0}}^2 + (\text{l.c.})(\|Y^{k+1}u\|_{L^2_{s-\epsilon_0}}^2 + \|Y^k u\|_{L^2_{s-\epsilon_0}}^2) \\
&\le (\text{s.c.})(\|Y^{k-1}u\|_{L^2_{s+\epsilon_0}}^2 + \|Y^k u\|_{L^2_s}^2) + (\text{l.c.})(\|Y^{k+1}u\|_{L^2_{s-\epsilon_0}}^2 + \|Y^k u\|_{L^2_{-\infty}}) \\
&\le (\text{s.c.})(\|Y^{k-1}u\|_{L^2_{s+\epsilon_0}}^2 + \|Y^k u\|_{L^2_s}^2) + (\text{l.c.})(\|Y^{k+1}u\|_{L^2_{s-\epsilon_0}}^2 + \|u\|_{L^2_{-\infty}}),
\end{aligned}$$

where the second to last inequality uses Proposition 4.5.6. Subtracting a small constant times $\|Y^k u\|_{L_s^2}^2$ from both sides completes the proof. □

Lemma 4.5.12. *Let $Y \in C_0^\infty(\Omega_3; T\Omega_3)$. Then, $\forall s \in \mathbb{R}$, $l \in \mathbb{N}_+$,*

$$\sum_{k=1}^{l-1} \|Y^k u\|_{L^2_{s+(l-k)\epsilon_0}} \le (\text{l.c.})(\|Y^l u\|_{L_s^2} + \|u\|_{L^2_{-\infty}}) + (\text{s.c.})\|u\|_{L^2_{s+l\epsilon_0}}, \quad u \in C_0^\infty(\Omega_1).$$

Proof. When $l = 1$, the result is trivial since the left-hand side equals 0, so we assume $l \ge 2$. We prove, for $1 \le j \le l-1$,

$$\sum_{k=j}^{l-1} \|Y^k u\|_{L^2_{s+(l-k)\epsilon_0}} \le (\text{l.c.})(\|Y^l u\|_{L_s^2} + \|u\|_{L^2_{-\infty}}) + (\text{s.c.})\|Y^{j-1} u\|_{L^2_{s+(l-(j-1))\epsilon_0}}, \tag{4.49}$$

and the result will follow by taking $j = 1$. We prove (4.49) by induction on j. The base case, $j = l-1$, follows directly from Lemma 4.5.11. We assume (4.49) with j replaced by $j+1$ and prove it for j. Using Lemma 4.5.11 and then the inductive hypothesis, we have

$$\begin{aligned}
\sum_{k=j}^{l-1} \|Y^k u\|_{L^2_{s+(l-k)\epsilon_0}} &\le \sum_{k=j+1}^{l-1} \|Y^k u\|_{L^2_{s+(l-k)\epsilon_0}} + (\text{l.c.})(\|Y^{j+1}u\|_{L^2_{s+(l-(j+1))\epsilon_0}} + \|u\|_{L^2_{-\infty}}) \\
&\quad + (\text{s.c.})\|Y^{j-1}u\|_{L^2_{s+(l-(j-1))\epsilon_0}} \\
&\le (\text{l.c.})(\|Y^l u\|_{L_s^2} + \|u\|_{L^2_{-\infty}}) \\
&\quad + (\text{s.c.})(\|Y^j u\|_{L^2_{s+(l-j)\epsilon_0}} + \|Y^{j-1}u\|_{L^2_{s+(l-(j-1))\epsilon_0}}).
\end{aligned}$$

Subtracting a small constant times $\|Y^j u\|_{L^2_{s+(l-j)\epsilon_0}}$ from both sides yields (4.49) and completes the proof. □

Lemma 4.5.13. *For $s \in \mathbb{R}$,*

$$\sum_{j=1}^{r} \sum_{l=0}^{n_j-1} \|W_j^l u\|_{L^2_{s+(n_j-l)\epsilon_0}} \le \sum_{j=1}^{r} \|W_j^{n_j} u\|_{L_s^2} + \|u\|_{L^2_{-\infty}}, \quad u \in C_0^\infty(\Omega_1).$$

Proof. By Lemma 4.5.12, we have

$$\sum_{j=1}^{r}\sum_{l=0}^{n_j-1}\|W_j^l u\|_{L^2_{s+(n_j-l)\epsilon_0}} \lesssim \sum_{j=1}^{r}\|W_j^{n_j}u\|_{L^2_s} + \sum_{j=1}^{r}\|u\|_{L^2_{s+n_j\epsilon_0}}. \tag{4.50}$$

Using Proposition 4.5.8, the fact that ϵ_0 satisfies $n_j\epsilon_0 - 2^{1-m} \leq (n_{j'}-1)\epsilon_0 - 2^{-m}$, $\forall j,j'$, and Proposition 4.5.6, we have

$$\begin{aligned}\sum_{j=1}^{r}\|u\|_{L^2_{s+n_j\epsilon_0}} &\lesssim \sum_{j=1}^{r}\sum_{j'=1}^{r}\|W_{j'}u\|_{L^2_{s+n_j\epsilon_0-2^{1-m}}} + \sum_{j=1}^{r}\|u\|_{L^2_{s+n_j\epsilon_0-2^{1-m}}}\\ &\lesssim \sum_{j'=1}^{r}\|W_{j'}u\|_{L^2_{s+(n_{j'}-1)\epsilon_0-2^{-m}}} + \sum_{j=1}^{r}\|u\|_{L^2_{s+n_j\epsilon_0-2^{1-m}}}\\ &\leq (\text{s.c.})\left(\sum_{j'=1}^{r}\|W_{j'}u\|_{L^2_{s+(n_{j'}-1)\epsilon_0}} + \sum_{j=1}^{r}\|u\|_{L^2_{s+n_j\epsilon_0}}\right) + (\text{l.c.})\|u\|_{L^2_{-\infty}}.\end{aligned} \tag{4.51}$$

Plugging (4.51) into (4.50), we have

$$\begin{aligned}\sum_{j=1}^{r}\sum_{l=0}^{n_j-1}\|W_j^l u\|_{L^2_{s+(n_j-l)\epsilon_0}} &\leq (\text{l.c.})\left(\sum_{j=1}^{r}\|W_j^{n_j}u\|_{L^2_s} + \|u\|_{L^2_{-\infty}}\right)\\ &\quad + (\text{s.c.})\left(\sum_{j=1}^{r}\|W_j u\|_{L^2_{s+(n_j-1)\epsilon_0}} + \sum_{j=1}^{r}\|u\|_{L^2_{s+n_j\epsilon_0}}\right).\end{aligned} \tag{4.52}$$

Subtracting a small constant times $\sum_{j=1}^{r}\|W_j u\|_{L^2_{s+(n_j-1)\epsilon_0}} + \sum_{j=1}^{r}\|u\|_{L^2_{s+n_j\epsilon_0}}$ from both sides of (4.52) completes the proof. □

Lemma 4.5.14. *For all $s \in \mathbb{R}$,*

$$\sum_{j=1}^{r}\sum_{l=0}^{n_j-1}\|W_j^l u\|_{L^2_{s+\epsilon_0}} \lesssim \sum_{j=1}^{r}\|W_j^{n_j}u\|_{L^2_s} + \|u\|_{L^2_{-\infty}}, \quad u \in C_0^\infty(\Omega_1).$$

Proof. This follows immediately from Lemma 4.5.13. □

Theorem 4.5.5 is a statement about all distributions $u \in C_0^\infty(\mathfrak{M})'$. All of our results, up to this point, have been about functions $u \in C_0^\infty(\Omega_1)$. To make the move from smooth functions to distributions, we approximate each distribution by a smooth function. To do this, let $\hat{\psi} \in C_0^\infty(\mathbb{R}^n)$ equal 1 on a neighborhood of 0. For $\delta > 0$, let $\psi_\delta(D)$ denote the pseudo-differential operator

$$f \mapsto \left(\hat{\psi}(\delta\xi)\hat{f}(\xi)\right)^\vee. \tag{4.53}$$

Note that $\hat{\psi}(\delta\xi) \in S^{-\infty} = \bigcap_{m\in\mathbb{R}} S^m$, where S^m denotes the space of standard symbols from Definition 2.2.3; to indicate this, we say $\psi_\delta(D)$ is a pseudo-differential operator of order $-\infty$. Because of this, $\hat{\psi}(\delta\xi) : \mathscr{S}(\mathbb{R}^n)' \to C^\infty_{\mathrm{loc}}(\mathbb{R}^n)$. However, $\hat{\psi}(\delta\xi) \in S^0$, *uniformly in* $\delta \in (0,1]$; more precisely, $\{\hat{\psi}(\delta\xi) : \delta \in (0,1]\} \subset S^0$ is a bounded set. Also, $\lim_{\delta\to 0}\psi_\delta(D) = I$. We write $S^{(s,\delta)}$, perhaps with a subscript, to denote some standard pseudo-differential operator of order $-\infty$, such that $\{S^{(s,\delta)} : \delta \in (0,1]\}$ is a bounded set of pseudo-differential operators of order s. When we write $\widetilde{S}^{(s,\delta)}$, perhaps with a subscript, we mean a finite number of terms of that form. For example, if we write $\|\widetilde{S}^{(s,\delta)}u\|_{L^2}$, it means $\|S_1^{(s,\delta)}u\|_{L^2} + \cdots + \|S_Q^{(s,\delta)}u\|_{L^2}$, for some finite collection $S_1^{(s,\delta)}, \dots, S_Q^{(s,\delta)}$. Here the finite collection does not depend on the function or distribution u under consideration.

Lemma 4.5.15. *Let $\mathscr{P}$ be any partial differential operator with smooth coefficients. Suppose that for all $s \in \mathbb{R}$ and all $\phi_1, \phi_2 \in C_0^\infty(\Omega_1)$, with $\phi_1 \prec \phi_2$, and $\forall S^{(s,\delta)}$ as described above,*

$$\begin{aligned}\sum_{j=1}^{r}\sum_{l=0}^{n_j} \|W_j^l \phi_1 S^{(s,\delta)}u\|_{L^2} \lesssim \|\phi_2 \mathscr{P}u\|_{L^2_s} &+ \sum_{j=1}^{r}\sum_{l=0}^{n_j-1} \|W_j^l \phi_2 \widetilde{S}^{(s,\delta)}u\|_{L^2} \\ &+ \|u\|_{L^2_{-\infty}}, \quad \forall u \in \mathscr{S}(\mathbb{R}^n)'.\end{aligned} \tag{4.54}$$

Then, for all $s \in \mathbb{R}$ and $\phi_1, \phi_2 \in C_0^\infty(\Omega_1)$ with $\phi_1 \prec \phi_2$, we have

$$\|\phi_1 u\|_{L^2_{s+\epsilon_0}} \lesssim \|\phi_2 \mathscr{P}u\|_{L^2_s} + \|\phi_2 u\|_{L^2_{-\infty}}, \quad \forall u \in C_0^\infty(\Omega_1)'. \tag{4.55}$$

Proof. Recall that the term $\|W_j^l \phi_2 \widetilde{S}^{(s,\delta)}u\|_{L^2}$ really denotes a finite sum of such terms.

Because of the localizations ϕ_1 and ϕ_2 in (4.55), it suffices to prove (4.55) for $u \in \mathscr{S}(\mathbb{R}^n)'$. For the remainder of the proof, u will denote an arbitrary element of $\mathscr{S}(\mathbb{R}^n)'$.

Fix $\phi_1, \phi_2, \phi_3 \in C_0^\infty(\Omega_1)$ with $\phi_1 \prec \phi_3 \prec \phi_2$ and let $\eta_j \in C_0^\infty(\Omega_1)$, $j \in \mathbb{N}$, be such that $\phi_1 \prec \eta_0 \prec \eta_1 \prec \eta_2 \prec \cdots \prec \phi_3$.

Let $S_0^{(s,\delta)}$ be given. We claim that $\forall k \in \mathbb{N}_+$, there exists $\widetilde{S}_k^{(s-(k-1)\epsilon_0,\delta)}$ such that

$$\sum_{j=1}^{r}\sum_{l=0}^{n_j} \|W_j^l \eta_0 S_0^{(s,\delta)}u\|_{L^2} \lesssim \|\eta_k \mathscr{P}u\|_{L^2_s} + \sum_{j=1}^{r}\sum_{l=0}^{n_j-1} \|W_j^l \eta_k \widetilde{S}_k^{(s-(k-1)\epsilon_0,\delta)}u\|_{L^2} + \|u\|_{L^2_{-\infty}}. \tag{4.56}$$

We prove (4.56) by induction on k. The base case, $k = 1$, follows from the hypothesis (4.54). We assume (4.56) for $k \in \mathbb{N}_+$ and prove it for $k+1$. Fix $\tilde{\eta}_{k,1}, \tilde{\eta}_{k,2} \in C_0^\infty(\Omega_1)$ with $\eta_k \prec \tilde{\eta}_{k,1} \prec \tilde{\eta}_{k,2} \prec \eta_{k+1}$. By the inductive hypothesis and Lemma 4.5.14,

$$\begin{aligned}\sum_{j=1}^{r}\sum_{l=0}^{n_j} \|W_j^l \eta_0 S_0^{(s,\delta)}u\|_{L^2} &\lesssim \|\eta_k \mathscr{P}u\|_{L^2_s} + \sum_{j=1}^{r}\sum_{l=0}^{n_j-1} \|W_j^l \eta_k \widetilde{S}_k^{(s-(k-1)\epsilon_0,\delta)}u\|_{L^2} + \|u\|_{L^2_{-\infty}} \\ &\lesssim \|\eta_k \mathscr{P}u\|_{L^2_s} + \sum_{j=1}^{r}\sum_{l=0}^{n_j} \|W_j^l \eta_k \widetilde{S}_k^{(s-(k-1)\epsilon_0,\delta)}u\|_{L^2_{-\epsilon_0}} + \|u\|_{L^2_{-\infty}}.\end{aligned} \tag{4.57}$$

Consider, using the calculus of pseudo-differential operators,

$$\begin{aligned}\sum_{j=1}^{r}\sum_{l=0}^{n_j}\|W_j^l\eta_k\widetilde{S}_k^{(s-(k-1)\epsilon_0,\delta)}u\|_{L^2_{-\epsilon_0}} &\lesssim \sum_{j=1}^{r}\sum_{l=0}^{n_j}\|\tilde{\eta}_{k,1}\Lambda^{-\epsilon_0}W_j^l\widetilde{S}_k^{(s-(k-1)\epsilon_0,\delta)}u\|_{L^2}+\|u\|_{L^2_{-\infty}}\\ &\lesssim \sum_{j=1}^{r}\sum_{l=0}^{n_j}\|\tilde{\eta}_{k,1}W_j^l\widetilde{S}_{k+1}^{(s-k\epsilon_0,\delta)}u\|_{L^2}+\|u\|_{L^2_{-\infty}}\\ &= \sum_{j=1}^{r}\sum_{l=0}^{n_j}\|\tilde{\eta}_{k,1}W_j^l\tilde{\eta}_{k,2}\widetilde{S}_{k+1}^{(s-k\epsilon_0,\delta)}u\|_{L^2}+\|u\|_{L^2_{-\infty}}\\ &\lesssim \sum_{j=1}^{r}\sum_{l=0}^{n_j}\|W_j^l\tilde{\eta}_{k,2}\widetilde{S}_{k+1}^{(s-k\epsilon_0,\delta)}u\|_{L^2}+\|u\|_{L^2_{-\infty}}.\end{aligned}\tag{4.58}$$

The hypothesis (4.54) gives

$$\begin{aligned}&\sum_{j=1}^{r}\sum_{l=0}^{n_j}\|W_j^l\tilde{\eta}_{k,2}\widetilde{S}_{k+1}^{(s-k\epsilon_0,\delta)}u\|_{L^2}\\ &\quad\lesssim \|\eta_{k+1}\mathscr{P}u\|_{L^2_{s-k\epsilon_0}}+\sum_{j=1}^{r}\sum_{l=0}^{n_j-1}\|W_j^l\eta_{k+1}\widetilde{S}_{k+1}^{(s-k\epsilon_0,\delta)}u\|_{L^2}+\|u\|_{L^2_{-\infty}}.\end{aligned}\tag{4.59}$$

Plugging (4.58) and (4.59) into (4.57), we get

$$\sum_{j=1}^{r}\sum_{l=0}^{n_j}\|W_j^l\eta_0S_0^{(s,\delta)}u\|_{L^2}\lesssim\|\eta_{k+1}\mathscr{P}u\|_{L^2_s}+\sum_{j=1}^{r}\sum_{l=0}^{n_j-1}\|W_j^l\eta_{k+1}\widetilde{S}_{k+1}^{(s-k\epsilon_0,\delta)}u\|_{L^2}+\|u\|_{L^2_{-\infty}},$$

which completes the inductive step and therefore the proof of (4.56).

Taking k large in (4.56) implies

$$\sum_{j=1}^{r}\sum_{l=0}^{n_j}\|W_j^l\eta_0S_0^{(s,\delta)}u\|_{L^2}\lesssim\|\phi_3\mathscr{P}u\|_{L^2_s}+\|u\|_{L^2_{-\infty}}.$$

By Lemma 4.5.14, this implies

$$\|\eta_0S_0^{(s,\delta)}u\|_{L^2_{\epsilon_0}}\lesssim\|\phi_3\mathscr{P}u\|_{L^2_s}+\|u\|_{L^2_{-\infty}}.\tag{4.60}$$

We may now choose a particular $S_0^{(s,\delta)}$. Namely, we take $S_0^{(s,\delta)} := \psi_\delta(D)\Lambda^s$, where $\psi_\delta(D)$ is as defined in (4.53). Consider we have

$$\begin{aligned}\|\phi_1u\|_{L^2_{s+\epsilon_0}} &\lesssim \sup_{\delta>0}\|\eta_0\Lambda^s\psi_\delta(D)\phi_1u\|_{L^2_{\epsilon_0}}+\|u\|_{L^2_{-\infty}}\\ &= \sup_{\delta>0}\|\eta_0S_0^{(s,\delta)}u\|_{L^2_{\epsilon_0}}+\|u\|_{L^2_{-\infty}}\\ &\lesssim \|\phi_3\mathscr{P}u\|_{L^2_s}+\|u\|_{L^2_{-\infty}},\end{aligned}\tag{4.61}$$

where the last inequality follows from (4.60). Replacing u with ϕ_2u in (4.61), we obtain

$$\|\phi_1 u\|_{L^2_{s+\epsilon_0}} \lesssim \|\phi_3 \mathscr{P} u\|_{L^2_s} + \|\phi_2 u\|_{L^2_{-\infty}} \lesssim \|\phi_2 \mathscr{P} u\|_{L^2_s} + \|\phi_2 u\|_{L^2_{-\infty}},$$

completing the proof. □

Lemma 4.5.16. *Let $\phi_1, \phi_2, \phi_3 \in C_0^\infty(\Omega_1)$ with $\phi_1 \prec \phi_2 \prec \phi_3$. Then*

$$[\mathscr{L}, \operatorname{Mult}[\phi_1] S^{(s,\delta)}] = \sum_{j=1}^{r} \sum_{l=1}^{2n_j-1} \operatorname{Mult}[\phi_2] W_j^l \operatorname{Mult}[\phi_3] S_{j,l}^{(s,\delta)},$$

where $S^{(s,\delta)}$ and $S_{j,l}^{(s,\delta)}$ are pseudo-differential operators of order $-\infty$ which are pseudo-differential operators of order s, uniformly in $\delta \in (0,1]$.

Proof. This follows easily from the calculus of pseudo-differential operators (Theorem 2.2.8). □

Lemma 4.5.17. *Let $\phi_1, \phi_2 \in C_0^\infty(\Omega_1)$ with $\phi_1 \prec \phi_2$. Then, for $u \in \mathscr{S}(\mathbb{R}^n)'$,*

$$\begin{aligned} &|\langle [\operatorname{Mult}[\phi_1] S_0^{(s,\delta)}, \mathscr{L}] u, \operatorname{Mult}[\phi_1] S_0^{(s,\delta)} u \rangle| \\ &\quad \le (\text{l.c.}) \sum_{j=1}^{r} \sum_{l=0}^{n_j-1} \|W_j^l \phi_2 \widetilde{S}^{(s,\delta)} u\|_{L^2}^2 + (\text{s.c.}) \sum_{j=1}^{r} \sum_{l=0}^{n_j} \|W_j^l \phi_1 S_0^{(s,\delta)} u\|_{L^2}^2. \end{aligned}$$

Proof. Fix $\phi_1, \phi_2, \phi_3 \in C_0^\infty(\Omega_1)$ with $\phi_1 \prec \phi_2 \prec \phi_3$. We prove the result with ϕ_2 replaced by ϕ_3. Applying Lemma 4.5.16, we have

$$\begin{aligned} &|\langle [\operatorname{Mult}[\phi_1] S_0^{(s,\delta)}, \mathscr{L}] u, \operatorname{Mult}[\phi_1] S_0^{(s,\delta)} u \rangle| \\ &\quad = \left| \sum_{j=1}^{r} \sum_{l=0}^{2n_j-1} \langle \operatorname{Mult}[\phi_2] W_j^l \operatorname{Mult}[\phi_3] S_{j,l}^{(s,\delta)} u, \operatorname{Mult}[\phi_1] S_0^{(s,\delta)} u \rangle \right| \\ &\quad = \left| \sum_{j=1}^{r} \sum_{l=0}^{2n_j-1} \langle W_j^l \operatorname{Mult}[\phi_3] S_{j,l}^{(s,\delta)} u, \operatorname{Mult}[\phi_1] S_0^{(s,\delta)} u \rangle \right|. \end{aligned} \tag{4.62}$$

Since $W_j^* = -W_j + \operatorname{Mult}[f_j]$, where $f_j \in C_{\mathrm{loc}}^\infty(\mathbb{R}^n)$, the right-hand side of (4.62) is bounded by a sum of terms of the form

$$\begin{aligned} &\sum_{j=1}^{r} \sum_{l_1=0}^{n_j-1} \sum_{l_2=0}^{n_j} |\langle W_j^{l_1} \operatorname{Mult}[\phi_3] \widetilde{S}^{(s,\delta)} u, W_j^{l_2} \operatorname{Mult}[\phi_1] S_0^{(s,\delta)} u \rangle| \\ &\quad \le (\text{l.c.}) \sum_{j=1}^{r} \sum_{l=0}^{n_j-1} \|W_j^l \operatorname{Mult}[\phi_3] \widetilde{S}^{(s,\delta)} u\|_{L^2}^2 + (\text{s.c.}) \sum_{j=1}^{r} \sum_{l=0}^{n_j} \|W_j^l \operatorname{Mult}[\phi_1] S_0^{(s,\delta)} u\|_{L^2}^2, \end{aligned}$$

completing the proof. □

Proof of Theorem 4.5.5. We prove the result by verifying the conditions of Lemma 4.5.15. Let $u \in \mathscr{S}(\mathbb{R}^n)'$ and $\phi_1, \phi_2 \in C_0^\infty(\Omega_1)$ with $\phi_1 \prec \phi_2$. Then, using Lemma 4.5.17 and the Cauchy–Schwartz inequality,

$$
\begin{aligned}
\sum_{j=1}^{r}\sum_{l=0}^{n_j}\|W_j^l\phi_1 S_0^{(s,\delta)}u\|_{L^2}^2 &= |\langle \mathscr{L}\phi_1 S_0^{(s,\delta)}u, \phi_1 S_0^{(s,\delta)}u\rangle| \\
&\leq |\langle \phi_1 S_0^{(s,\delta)}\mathscr{L}u, \phi_1 S_0^{(s,\delta)}u\rangle| + |\langle [\mathscr{L}, \phi_1 S_0^{(s,\delta)}]u, \phi_1 S_0^{(s,\delta)}u\rangle| \\
&\leq (\text{l.c.})\left(\|\phi_1 S_0^{(s,\delta)}\mathscr{L}u\|_{L^2}^2 + \sum_{j=1}^{r}\sum_{l=0}^{n_j-1}\|W_j^l\phi_2\widetilde{S}^{(s,\delta)}u\|_{L^2}^2\right) \\
&\quad + (\text{s.c.})\sum_{j=1}^{r}\sum_{l=0}^{n_j}\|W_j^l\phi_1 S_0^{(s,\delta)}u\|_{L^2}^2.
\end{aligned} \tag{4.63}
$$

Subtracting (s.c.) $\sum_{j=1}^{r}\sum_{l=0}^{n_j}\|W_j^l\phi_1 S_0^{(s,\delta)}u\|_{L^2}^2$ from both sides of (4.63) and using

$$\|\phi_1 S_0^{(s,\delta)}\mathscr{L}u\|_{L^2}^2 \lesssim \|\phi_2\mathscr{L}u\|_{L_s^2}^2 + \|u\|_{L_{-\infty}^2},$$

we have

$$\sum_{j=1}^{r}\sum_{l=0}^{n_j}\|W_j^l\phi_1 S_0^{(s,\delta)}u\|_{L^2}^2 \lesssim \|\phi_2\mathscr{L}u\|_{L_s^2}^2 + \sum_{j=1}^{r}\sum_{l=0}^{n_j-1}\|W_j^l\phi_2\widetilde{S}^{(s,\delta)}u\|_{L^2}^2 + \|u\|_{L_{-\infty}^2}^2.$$

The result now follows from Lemma 4.5.15. □

4.5.2 Homogeneity

Understanding a fundamental solution for a hypoelliptic operator is particularly accessible when the operator is homogeneous under a dilation structure. In this section, we present these ideas.

Suppose we are given dilations on $\mathbb{R}^N$: $d = (d_1, \ldots, d_N)$, where $d_j \in \mathbb{N}_+$, that is, for $\delta > 0$ and $t \in \mathbb{R}^N$, $\delta^d t = (\delta^{d_1}t_1, \ldots, \delta^{d_N}t_N)$. We set $Q := d_1 + \cdots + d_N \in \mathbb{N}_+$, the so-called homogeneous dimension of $\mathbb{R}^N$ with respect to these dilations.

Definition 4.5.18. We say a partial differential operator with smooth coefficients, $\mathscr{P} : C_{\mathrm{loc}}^\infty(\mathbb{R}^N) \to C_{\mathrm{loc}}^\infty(\mathbb{R}^N)$, is homogeneous of degree $s \in \mathbb{R}$ if

$$\mathscr{P}(f(\delta^d t)) = \delta^s(\mathscr{P}f)(\delta^d t), \quad f \in C_{\mathrm{loc}}^\infty(\mathbb{R}^N).$$

Definition 4.5.19. For $K \in C_0^\infty(\mathbb{R}^N)'$ and $\delta > 0$, we define $K(\delta^d t) \in C_0^\infty(\mathbb{R}^N)'$ by

$$\int K(\delta^d t)f(t)\,dt := \delta^{-Q}\int K(t)f(\delta^{-d}t)\,dt, \quad f \in C_0^\infty(\mathbb{R}^N).$$

Note that when K agrees with integration against a locally integrable function, this definition agrees with the usual, pointwise, definition of the function $K(\delta^d t)$.

Definition 4.5.20. We say a distribution is homogeneous of degree $s \in \mathbb{R}$ if $K(\delta^d t) = \delta^s K(t)$, $\forall \delta > 0$.

The next result, due to Folland [90], is the key result of this section, which we state without proof. For a proof, the reader is referred to [90, Theorem 2.1] and also [99, Chapter 3, Section 3].

Theorem 4.5.21. *Let $\mathscr{P} : C^\infty_{\mathrm{loc}}(\mathbb{R}^N) \to C^\infty_{\mathrm{loc}}(\mathbb{R}^N)$ be a partial differential operator with smooth coefficients, which is homogeneous of degree $s \in (0, Q)$. Suppose $\mathscr{P}$ and $\mathscr{P}^*$ are both hypoelliptic. Then there exists a unique distribution $K \in C^\infty_0(\mathbb{R}^N)'$ such that K is homogeneous of degree $s - Q$ and $\mathscr{P}K = \delta_0$ (where δ_0 denotes the Dirac δ function at 0). This $K(t)$ agrees with a $C^\infty_{\mathrm{loc}}(\mathbb{R}^n)$ function on $\{t \neq 0\}$.*

Following (4.5) we define a homogeneous norm on $\mathbb{R}^N$ by

$$\|t\| = \left(\sum_{l=1}^{N} |t_l|^{(2\prod_{j=1}^{N} d_j)/d_l} \right)^{1/2\prod_{j=1}^{N} d_j} \approx \sum_{l=1}^{N} |t_l|^{1/d_l}.$$

Note that $\|\delta^d t\| = \delta \|t\|$, for $\delta > 0$. For a multi-index $\alpha \in \mathbb{N}^N$, define $\deg_d(\alpha) := \sum_{l=1}^{N} \alpha_l d_l \in \mathbb{N}$.

Proposition 4.5.22. *Let $K \in C^\infty_0(\mathbb{R}^N)'$ be homogeneous of degree $s - Q$, for $s \in (-\infty, Q]$, and let $\eta \in C^\infty_0(\mathbb{R}^n)$. Suppose $K(t)$ agrees with a $C^\infty_{\mathrm{loc}}(\mathbb{R}^n)$ function on $\{t \neq 0\}$. Then, $\forall \alpha \in \mathbb{N}^N$,*

$$\left|\partial_\xi^\alpha \widehat{\eta K}(\xi)\right| \le C_\alpha (1 + \|\xi\|)^{-s-\deg_d(\alpha)}. \tag{4.64}$$

Proof. We have

$$\partial_{\xi^\alpha} \widehat{\eta K}(\xi) = \int K(t)(-2\pi i t)^\alpha \eta(t) e^{-2\pi i t\cdot\xi}\, dt.$$

Note that $\{(-2\pi i t)^\alpha \eta(t) e^{-2\pi i t\cdot\xi} : \|\xi\| \le 1\} \subset C^\infty_0(\mathbb{R}^N)$ is a bounded set, and therefore

$$\sup_{\|\xi\|\le 1} \left|\partial_\xi^\alpha \widehat{\eta K}(\xi)\right| < \infty.$$

This proves (4.64) when $\|\xi\| \le 1$.

To prove (4.64) when $\|\xi\| \ge 1$, we will show, with $\eta_\delta(t) := \eta(\delta^d t)$,

$$\sup_{\|\xi\|=1} \sup_{\delta\in(0,1]} \left|\partial_\xi^\alpha \widehat{\eta_\delta K}(\xi)\right| < \infty. \tag{4.65}$$

First, we see why (4.65) implies (4.64) for $\|\xi\| \ge 1$. Note that by the homogeneity of K, we have

$$\widehat{\eta K}(\delta^{-d}\xi) = \int \eta(t)K(t)e^{-2\pi i(\delta^{-d}t)\cdot\xi}\,dt = \delta^s \int \eta(\delta^d t)K(t)e^{-2\pi i t\cdot\xi}\,dt = \delta^s \widehat{\eta_\delta K}(\xi).$$

Thus, using (4.65), we have, for $\delta \in (0,1]$,

$$\begin{aligned}\sup_{\|\xi\|=\delta^{-1}} \left|\partial_\xi^\alpha \widehat{\eta K}(\xi)\right| &= \delta^{\deg_d(\alpha)} \sup_{\|\xi\|=1} \left|\partial_\xi^\alpha (\widehat{\eta K}(\delta^{-d}\xi))\right| \\ &= \delta^{\deg_d(\alpha)+s} \sup_{\|\xi\|=1} \left|\partial_\xi^\alpha \widehat{\eta_\delta K}(\xi)\right| \lesssim \delta^{\deg_d(\alpha)+s}.\end{aligned}$$

This immediately implies (4.64) for $\|\xi\| \geq 1$.

We turn to proving (4.65). Fix $\phi \in C_0^\infty(\{t : \|t\| < 2\})$ with $\phi \equiv 1$ on $\{t : \|t\| \leq 3/2\}$. We have

$$\begin{aligned}\partial_\xi^\alpha \widehat{\eta_\delta K}(\xi) &= \int K(t)\eta_\delta(t)(-2\pi i t)^\alpha e^{-2\pi i t\cdot\xi}\,dt \\ &= \int K(t)\eta_\delta(t)(-2\pi i t)^\alpha \phi(t)e^{-2\pi i t\cdot\xi}\,dt \\ &\quad + \int K(t)\eta_\delta(t)(-2\pi i t)^\alpha (1-\phi(t))e^{-2\pi i t\cdot\xi}\,dt.\end{aligned} \tag{4.66}$$

Note that $\{\eta_\delta(t)\phi(t)(-2\pi i t)^\alpha e^{-2\pi i t\cdot\xi} : \|\xi\| = 1, \delta \in (0,1]\} \subset C_0^\infty(\mathbb{R}^N)$ is a bounded set, and therefore,

$$\sup_{\|\xi\|=1}\sup_{\delta\in(0,1]} \left|\int K(t)\eta_\delta(t)(-2\pi i t)^\alpha \phi(t)e^{-2\pi i t\cdot\xi}\,dt\right| < \infty. \tag{4.67}$$

Thus, it suffices to estimate the second term on the right-hand side of (4.66).

We claim that, $\forall \beta \in \mathbb{N}^N$, $t \neq 0$,

$$\left|\partial_t^\beta((-2\pi i t)^\alpha \eta_\delta(t)K(t))\right| \leq C_{\alpha,\beta}\|t\|^{-s-\deg_d(\beta)+\deg_d(\alpha)}. \tag{4.68}$$

Indeed, fix $D > 0$ such that $\operatorname{supp}(\eta) \subseteq \{t : \|t\| \leq D\}$. Then

$$\begin{aligned}\left|\partial_t^\beta((-2\pi i t)^\alpha \eta_\delta(t)K(t))\right| &\lesssim \sum_{\beta_1+\beta_2+\beta_3=\beta} \left|(\partial_t^{\beta_1}((-2\pi i t)^\alpha))(\partial_t^{\beta_2}\eta_\delta(t))(\partial_t^{\beta_3}K(t))\right| \\ &\lesssim \sum_{\beta_1+\beta_2+\beta_3} \|t\|^{\deg_d(\alpha)-\deg_d(\beta_1)}\delta^{\deg_d(\beta_2)}\chi_{\{\|t\|\leq D\delta^{-1}\}}\|t\|^{-\deg_d(\beta_3)} \\ &\lesssim \sum_{\beta_1+\beta_2+\beta_3} \|t\|^{\deg_d(\alpha)-\deg_d(\beta_1)-\deg_d(\beta_2)-\deg_d(\beta_3)},\end{aligned}$$

which proves (4.68).

We claim that, $\forall \beta \in \mathbb{N}^N$,

$$\left|\partial_t^\beta((-2\pi i t)^\alpha(1-\phi(t))\eta_\delta(t)K(t))\right| \leq C'_{\alpha,\beta}\|t\|^{-s-|\beta|+\deg_d(\alpha)}\chi_{\{\|t\|\geq 1\}}. \tag{4.69}$$

Indeed,

$$\begin{aligned}&\left|\partial_t^{\beta}((-2\pi it)^{\alpha}(1-\phi(t))\eta_\delta(t)K(t))\right|\\&\quad\lesssim\sum_{\beta_1+\beta_2=\beta}\left|(\partial_t^{\beta_1}(1-\phi(t)))(\partial_t^{\beta_2}((-2\pi it)^{\alpha}\eta_\delta(t)K(t)))\right|.\end{aligned}$$

When $\beta_1\neq 0$, $\partial_t^{\beta_1}(1-\phi(t))$ is supported on $\{3/2\leq\|t\|\leq 2\}$, and therefore, using (4.68), we have

$$\begin{aligned}&\left|(\partial_t^{\beta_1}(1-\phi(t)))(\partial_t^{\beta_2}((-2\pi it)^{\alpha}\eta_\delta(t)K(t)))\right|\\&\quad\lesssim\chi_{\{3/2\leq\|t\|\leq 2\}}\|t\|^{-s-\deg_d(\beta_2)+\deg_d(\alpha)}\lesssim\chi_{\{\|t\|\geq 1\}}.\end{aligned}$$

When $\beta_1=0$, then $\beta_2=\beta$. Using that $(1-\phi(t))$ is supported on $\{\|t\|\geq 3/2\}$ and (4.68), we have

$$\begin{aligned}&\left|(\partial_t^{\beta_1}(1-\phi(t)))(\partial_t^{\beta_2}((-2\pi it)^{\alpha}\eta_\delta(t)K(t)))\right|\\&\quad\left|((1-\phi(t)))(\partial_t^{\beta}((-2\pi it)^{\alpha}\eta_\delta(t)K(t)))\right|\\&\quad\lesssim\chi_{\{\|t\|\geq 3/2\}}\|t\|^{-s-\deg_d(\beta)+\deg_d(\alpha)}\lesssim\chi_{\{\|t\|\geq 3/2\}}\|t\|^{-s-|\beta|+\deg_d(\alpha)},\end{aligned}$$

where in the last inequality we have used $|\beta|\leq\deg_d(\beta)$. Combining the previous three equations proves (4.69).

Consider, for the second term on the right-hand side of (4.66), we have for $L\in\mathbb{N}$ sufficiently large

$$\begin{aligned}&\sup_{\|\xi\|=1}\sup_{\delta\in(0,1]}\left|\int K(t)\eta_\delta(t)(-2\pi it)^{\alpha}(1-\phi(x))e^{-2\pi it\cdot\xi}\,dt\right|\\&\quad=\sup_{\|\xi\|=1}\sup_{\delta\in(0,1]}(-|2\pi\xi|^2)^{-L}\left|\int(\triangle_t^{L}K(t)\eta_\delta(t)(-2\pi it)^{\alpha}(1-\phi(x)))e^{-2\pi it\cdot\xi}\,dt\right|\\&\quad\lesssim\sup_{\xi=1}\sup_{\delta\in(0,1]}(-|2\pi\xi|^2)^{-L}\int_{\|t\|\geq 1}\|t\|^{-s-2L+\deg_d(\alpha)}\,dt<\infty,\end{aligned}\tag{4.70}$$

where the second to last inequality used (4.69).

Combining (4.67) and (4.70) proves (4.65) and completes the proof. □

Remark 4.5.23. We use Proposition 4.5.22 in the following way. Suppose $K\in C_0^{\infty}(\mathbb{R}^N)'$ satisfies the hypotheses of Proposition 4.5.22 and fix $\eta\in C_0^{\infty}(B^N(a))$ and $\psi\in C_0^{\infty}(\Omega)$, where $\Omega\Subset\mathfrak{M}$ and $\mathfrak{M}$ is some manifold. We wish to see the distribution $\psi(x)\eta(t)K(t)$ as of the form $\check{b}(x,t,t)=\psi(x)\eta(t)K(t)$ for some $b(x,u,\xi)\in S^{-s}(a,\Omega,d)$. Indeed, take any $\eta'(t)\in C_0^{\infty}(\mathbb{R}^N)$ with $\eta\prec\eta'$ and set $\check{b}(x,u,t):=\psi(x)\eta(u)\eta'(t)K(t)$. Then Proposition 4.5.22 shows $b(x,u,\xi)\in S^{-s}(a,\Omega,d)$, and it is immediate to see that $\check{b}(x,t,t)=\psi(x)\eta(t)K(t)$.

4.5.3 Nilpotent Lie groups

An important model case of vector fields with formal degrees comes from graded nilpotent Lie groups. Following an idea of Rothschild and Stein [200], we will construct a parametrix for a general sub-Laplacian, by using a fundamental solution for a sub-Laplacian on a graded nilpotent Lie group. In this section, we introduce the basic background information regarding such groups. We refer the reader to the books [53] and [99] for more information.

Let $\mathfrak{g}$ be a finite-dimensional Lie algebra. Define $\mathfrak{g}^{(0)} = \mathfrak{g}$ and recursively set $\mathfrak{g}^{(k+1)} := [\mathfrak{g}, \mathfrak{g}^{(k)}]$.

Definition 4.5.24. We say $\mathfrak{g}$ is nilpotent of step $k \in \mathbb{N}$ if $\mathfrak{g}^{(k)} = \{0\}$. We say $\mathfrak{g}$ is nilpotent if $\mathfrak{g}$ is nilpotent of step k for some k.

Definition 4.5.25. For a Lie group G, we say G is nilpotent if its Lie algebra is nilpotent.

It is well known that if $\mathfrak{g}$ is a nilpotent Lie algebra, then there is a unique connected, simply connected, nilpotent Lie group G whose Lie algebra is $\mathfrak{g}$. In this case, the exponential map $\exp : \mathfrak{g} \to G$ is a diffeomorphism (see [53, Theorem 1.2.1]). In particular, as a manifold $G \cong \mathbb{R}^{\dim G}$. Given two elements of the Lie algebra $X, Y \in \mathfrak{g}$, we can compute the group multiplication $e^X e^Y$ via the Baker–Campbell–Hausdorff formula, namely, $e^X e^Y = e^{\mathrm{BCH}(X,Y)}$, where $\mathrm{BCH}(X, Y)$ is a finite sum due to the nilpotent hypothesis; see [53, Theorem 1.2.1]. Given a basis $V_1, \ldots, V_{\dim G}$ of $\mathfrak{g}$, we obtain global coordinates on G by $(v_1, \ldots, v_{\dim G}) \mapsto e^{v_1 V_1 + \cdots + v_{\dim G} V_{\dim G}}$, $\mathbb{R}^{\dim G} \xrightarrow{\sim} G$. In this coordinate system, the group multiplication can be computed explicitly via the Baker–Campbell–Hausdorff formula: the group product of $(v_1, \ldots, v_{\dim G})$ and $(u_1, \ldots, u_{\dim G})$ is given by the coefficients of $V_1, \ldots, V_{\dim G}$ in the formula $\mathrm{BCH}(v_1 V_1 + \cdots + v_{\dim G} V_{\dim G}, u_1 V_1 + \cdots u_{\dim G} V_{\dim G})$; this is a polynomial in v and u. Henceforth, we identify G with $\mathfrak{g}$ via the exponential map; therefore, $\mathfrak{g}$ has a group structure defined by the Baker–Campbell–Hausdorff formula. With this identification, 0 is the identity element, and the group inverse corresponds with multiplication by -1. If, further, we pick a basis for $\mathfrak{g}$, this identifies G with $\mathbb{R}^{\dim G}$ and $\mathbb{R}^{\dim G}$ has a group structure defined by the Baker–Campbell–Hausdorff formula. The group G is unimodular and with the above identification $G \cong \mathbb{R}^{\dim G}$, the Lebesgue measure is a Haar measure on G; see [53, Theorem 1.2.10].

By the Schwartz kernel theorem, right invariant operators $C_0^\infty(G) \to C_0^\infty(G)'$ are of the form $f(x) \mapsto K * f(x) = \int K(xy^{-1}) f(y)\, dy$ for some $K \in C_0^\infty(G)'$. Left invariant operators are of the form $f \mapsto f * K$.

Definition 4.5.26. We say a finite-dimensional, nilpotent Lie algebra $\mathfrak{g}$ is graded if

$$\mathfrak{g} = \oplus_{\mu=1}^{\nu} \mathfrak{v}_\mu,$$

where $[\mathfrak{v}_{\mu_1}, \mathfrak{v}_{\mu_2}] \subseteq \mathfrak{v}_{\mu_1+\mu_2}$ and we have taken $\mathfrak{v}_\mu = \{0\}$ for $\mu > \nu$.

Definition 4.5.27. We say a nilpotent Lie group is graded if its Lie algebra is graded.

Henceforth, fix a graded nilpotent Lie algebra $\mathfrak{g} = \oplus_{\mu=1}^{\nu} \mathfrak{v}_\mu$ and let G be the associated graded nilpotent Lie group. Define dilations on $\mathfrak{g}$ by $\delta \cdot X := \delta^\mu X$, for $\delta > 0$ and $X \in \mathfrak{v}_\mu$, and extend this by linearity. These dilations are Lie algebra automorphisms of $\mathfrak{g}$. We henceforth give $\mathfrak{g}$ this dilation structure. By the Baker–Campbell–Hausdorff formula, these dilations are also group automorphisms.

For each $\mu \in \{1, \ldots, \nu\}$, fix a basis $X_1^\mu, \ldots, X_{l_\mu}^\mu$ of $\mathfrak{v}_\mu$, so that $\{X_j^\mu : \mu \in \{1, \ldots, \nu\}, 1 \le j \le l_\mu\}$ is a basis for $\mathfrak{g}$, $l_\mu = \dim \mathfrak{v}_\mu$, and $\sum_\mu l_\mu = \dim G$. We write $t \in \mathbb{R}^{\dim G}$ as $t = (t_j^\mu)_{\mu \in \{1,\ldots,\nu\}, 1 \le j \le l_\mu}$ and identify $\mathbb{R}^{\dim G}$ with $\mathfrak{g} \cong G$ via $t \mapsto \sum t_j^\mu X_j^\mu$. Under this identification, the dilation structure on $\mathfrak{g}$ gives single-parameter dilations on $\mathbb{R}^{\dim G}$, via $\delta^d t := (\delta^\mu t_j^\mu)$; here $d \in \mathbb{N}^{\dim G}$ is the vector which equals μ in the coordinate corresponding to t_j^μ. We henceforth make this identification and treat both $\mathfrak{g}$ and G as $\mathbb{R}^{\dim G}$, with this dilation structure. Thus, $\mathbb{R}^{\dim G}$ has a group structure inherited from G. Set $Q = \sum_{\mu \in \{1,\ldots,\nu\}} \mu \dim \mathfrak{v}_\mu$, the homogeneous dimension of G with this dilation structure.

We identify $\mathfrak{g}$ with $T_0 G$, the tangent space to the identity of G. For $V \in \mathfrak{g}$, we identify V with the distribution which takes $f \in C_0^\infty(\mathbb{R}^n)'$ and computes the directional derivative at 0 given by Vf. There is a unique left invariant vector field V^L on G such that $V^L(0) = V$ and a unique right invariant vector field V^R with $V^R(0) = V$. In fact, $V^L f = f * V$ and $V^R f = V * f$. In particular, the elements $X_j^\mu \in \mathfrak{v}_\mu$ induce left invariant vector fields $X_j^{\mu,L}$ and right invariant vector fields $X_j^{\mu,R}$, which (in the above chosen coordinates) satisfy $X_j^{\mu,L}(0) = X_j^{\mu,R}(0) = \partial_{t_j^\mu}$. We have

$$X_j^{\mu,L} f = f * (\partial_{t_j^\mu} \delta_0), \quad X_j^{\mu,R} f = (\partial_{t_j^\mu} \delta_0) * f,$$

where δ_0 denotes the Dirac δ function at 0. We have $(V^L)^* = -V^L$ and $(V^R)^* = -V^R$.

Lemma 4.5.28. *Let $V \in \mathfrak{v}_\mu$. Then V^L and V^R are homogeneous differential operators of degree μ, in the sense of Definition* 4.5.18.

Proof. It suffices to prove the result for $V = X_j^\mu$. We have

$$\begin{aligned} X_j^{\mu,L}(f(\delta^d x)) &= \int f(\delta^d (xy^{-1}))(\partial_{t_j^\mu} \delta_0)(y)\, dy = \int f((\delta^d x)(\delta^d y)^{-1})(\partial_{t_j^\mu} \delta_0)(y)\, dy \\ &= \delta^{-Q} \int f((\delta^d x) y^{-1})(\partial_{t_j^\mu} \delta_0)(\delta^{-d} y)\, dy = \delta^{\mu - Q} \int f((\delta^d x) y^{-1}) \partial_{t_j^\mu}(\delta_0(\delta^{-d} y))\, dy \\ &= \delta^\mu \int f(\delta^d (x) y^{-1})(\partial_{t_j^\mu} \delta_0)(y)\, dy = \delta^\mu (X_j^{\mu,L} f)(\delta^d x). \end{aligned}$$

A similar computation proves the result for $X_j^{\mu,R}$. □

Let $(\widehat{W}, đ) = \{(\widehat{W}_1, đ_1), \ldots, (\widehat{W}_r, đ_r)\} \subset \mathfrak{g} \times \mathbb{N}_+$ satisfy $\widehat{W}_j \in \mathfrak{v}_{đ_j}$ such that $\widehat{W}_1, \ldots, \widehat{W}_r$ generate $\mathfrak{g}$ as a Lie algebra. Let $\widehat{W}_j^R$ be the right invariant vector fields associated with $\widehat{W}_j$. Fix $\kappa \in \mathbb{N}_+$ such that $đ_j$ divides κ for every j. We define a right invariant sub-Laplacian by

$$\mathscr{L}^R := \sum_{j=1}^{r} (-1)^{\kappa/đ_j} (\widehat{W}_j^R)^{2\kappa/đ_j} = \sum_{j=1} ((\widehat{W}_j^R)^*)^{\kappa/đ_j} (\widehat{W}_j^R)^{\kappa/đ_j}. \tag{4.71}$$

Proposition 4.5.29. *Suppose $2\kappa < Q$. Then there is a unique distribution $K \in C_0^\infty(G)'$ which is homogeneous of degree $2\kappa - Q$ such that $\mathscr{L}^R K = \delta_0$. This $K(t)$ agrees with a $C^\infty_{\mathrm{loc}}(\mathbb{R}^{\dim G})$ function on $\{t \neq 0\}$.*

Proof. $\mathscr{L}^R$ is subelliptic by Theorem 4.5.5 and is therefore hypoelliptic. By (4.71), $(\mathscr{L}^R)^* = \mathscr{L}^R$, and therefore $(\mathscr{L}^R)^*$ is also hypoelliptic. Lemma 4.5.28 implies that $\mathscr{L}^R$ is homogeneous of degree 2κ. The result follows from Theorem 4.5.21. □

A similar result holds for the left invariant sub-Laplacian, with the same proof.

4.5.4 Free nilpotent Lie algebras

We use a special nilpotent Lie algebra: the free nilpotent Lie algebra of step $\mathscr{K} \in \mathbb{N}$ on r generators, $\mathfrak{n}_{\mathscr{K},r}$. Informally, we take $\mathfrak{n}_{\mathscr{K},r}$ to be the nilpotent Lie algebra of step k, which is generated as a Lie algebra by r formal elements $\widehat{W}_1, \ldots, \widehat{W}_r$, but otherwise has as few relations as possible. Formally, $\mathfrak{n}_{\mathscr{K},r}$, with generators $\widehat{W}_1, \ldots, \widehat{W}_r$, is the unique[1] nilpotent Lie algebra of step $\mathscr{K}$ satisfying the following universal property: if $\mathfrak{g}$ is another nilpotent Lie algebra of step at most $\mathscr{K}$ and $\psi : \{\widehat{W}_1, \ldots, \widehat{W}_r\} \to \mathfrak{g}$ is any function, then there is a unique Lie algebra homomorphism which makes the following diagram commute:

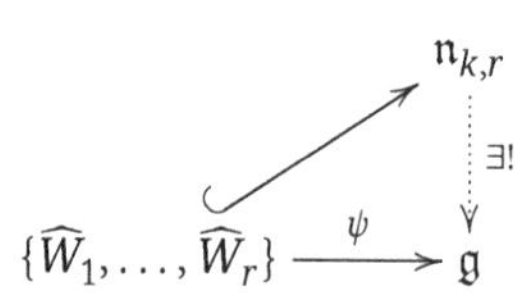

It is easy to show that such a Lie algebra exists. Furthermore, this Lie algebra can be given a gradation as follows. Assign to each $\widehat{W}_j$ a formal degree $đ_j \in \mathbb{N}_+$. Then there is a unique grading $\mathfrak{n}_{\mathscr{K},r} = \oplus_{\mu=1}^{\nu} \mathfrak{v}_\mu$, as in Definition 4.5.26, such that $\widehat{W}_j \in \mathfrak{v}_{đ_j}$.

1 The universal property determines $\mathfrak{n}_{\mathscr{K},r}$ uniquely up to isomorphisms of Lie algebras.

4.5.5 The calculus

In this section, we present a weak analog of the calculus of pseudo-differential operators.[2] Let $(W, \vec{d}) = \{(W_1, \vec{d}_1), \dots, (W_r, \vec{d}_r)\} \subset C^\infty_{\mathrm{loc}}(\mathfrak{M}; T\mathfrak{M}) \times \mathbb{N}_+$ be such that $\mathrm{Gen}((W, \vec{d}))$ is locally finitely generated. Fix relatively compact, open sets $\Omega_1 \Subset \Omega_2 \Subset \Omega_3 \Subset \mathfrak{M}$.

Pick a finite set $\mathcal{F} \subset \mathrm{Gen}((W, \vec{d}))$ so that $\mathrm{Gen}((W, \vec{d}))$ is finitely generated by $\mathcal{F}$ on Ω_3 and $(W, \vec{d}) \subseteq \mathcal{F}$. Set $M_0 := \max\{e : (Z, e) \in \mathcal{F}\}$, let $\mathcal{K} \geq 2M_0 - 1$, and let $\mathfrak{n}_{\mathcal{K},r}$ be the free nilpotent Lie algebra of step $\mathcal{K}$ with r generators $\widehat{W}_1, \dots, \widehat{W}_r$. We assign to each $\widehat{W}_j$ the formal degree $\vec{d}_j$ and give $\mathfrak{n}_{\mathcal{K},r}$ the associated grading as described in Section 4.5.4: $\mathfrak{n}_{\mathcal{K},r} = \oplus_{\mu=1}^{\nu} \mathfrak{v}_\mu$. We let $N_{\mathcal{K},r}$ be the connected, simply connected Lie group whose Lie algebra is $\mathfrak{n}_{\mathcal{K},r}$.

Let $q = \dim \mathfrak{n}_{\mathcal{K},r}$ and pick a basis of $\mathfrak{n}_{\mathcal{K},r}$, $\widehat{X}_1, \dots, \widehat{X}_q$, of the form

$$\widehat{X}_j = \mathrm{ad}(\widehat{W}_{l_1^j})\, \mathrm{ad}(\widehat{W}_{l_2^j}) \cdots \mathrm{ad}(\widehat{W}_{l_{L_j-1}^j}) \widehat{W}_{l_{L_j}^j},$$

for some choice of $l_1^j, \dots, l_{L_j}^j \in \{1, \dots, r\}$. Here, $\mathrm{ad}(V_1)V_2 = [V_1, V_2]$. Choose this basis in such a way that $\widehat{W}_1, \dots, \widehat{W}_r$ are elements of the basis. Note that $\widehat{X}_j \in \mathfrak{v}_{d_j}$ where $d_j := \vec{d}_{l_1^j} + \dots + \vec{d}_{l_{L_j}^j}$. We use this basis to identify $N_{\mathcal{K},r}$ and $\mathfrak{n}_{\mathcal{K},r}$ with $\mathbb{R}^q$ as in Section 4.5.3. For $t, s \in \mathbb{R}^q$ we write $t \times s$ for the group multiplication induced by $N_{\mathcal{K},r}$, and we write $\delta^d t$ for the dilations described in Section 4.5.3.

On $\mathfrak{M}$, we define vector fields $X_1, \dots, X_q$ as follows:

$$X_j := \mathrm{ad}(W_{l_1^j})\, \mathrm{ad}(W_{l_2^j}) \cdots \mathrm{ad}(W_{l_{L_j-1}^j}) W_{l_{L_j}^j}.$$

We set $(X, d) := \{(X_1, d_1), \dots, (X_q, d_q)\}$, and we set

$$\mathcal{F}_1 := \{(X_j, d_j) : d_j \leq M_0\}. \tag{4.72}$$

Lemma 4.5.30. *$\mathrm{Gen}((W, \vec{d}))$ is finitely generated by $\mathcal{F}_1$ and by (X, d) on Ω_3. Also, $(W, \vec{d}) \subseteq \mathcal{F}_1 \subseteq (X, d)$.*

Proof. By the above construction, $(W, \vec{d}) \subseteq \mathcal{F}_1 \subseteq (X, d) \subset \mathrm{Gen}((W, \vec{d}))$. Furthermore, since $\mathfrak{n}_{\mathcal{K},r}$ is the *free* nilpotent Lie algebra of step $\mathcal{K}$ with r generators and $M_0 \leq \mathcal{K}$, for every $(Z, e) \in \mathrm{Gen}((W, \vec{d}))$ with $e \leq M_0$, Z can be written as a linear combination, with constant coefficients, of elements of the form X_j where $d_j = e$. In particular, this is true for every element of $\mathcal{F}$. Since $\mathrm{Gen}((W, \vec{d}))$ is finitely generated by $\mathcal{F}$ on Ω_3, it follows that $\mathrm{Gen}((W, \vec{d}))$ is finitely generated by $\mathcal{F}_1$ on Ω_3. Since $\mathcal{F}_1 \subseteq (X, d)$, the same is true of (X, d). □

2 A much more powerful calculus of pseudo-differential operators is obtained in [2], though such results are not necessary for our purposes.

We let $\widehat{W}_j^L, \widehat{W}_j^R$ and $\widehat{X}_j^L, \widehat{X}_j^R$ be the left and right invariant vector fields corresponding to $\widehat{W}_j$ and $\widehat{X}_j$, respectively. By Lemma 4.5.28, $\widehat{W}_j^R$ is homogeneous of degree $\vec{d}_j$ and $\widehat{X}_j^R$ is homogeneous of degree d_j. The main result of this section is the following proposition.

Proposition 4.5.31. *If $a = a(\Omega_1, \Omega_2, (X, d)) > 0$ is sufficiently small, then the following holds. Suppose $b \in S^s(a, \Omega_1, d)$ and S is the (X, d), a pseudo-differential operator of order s supported in $\Omega_1 \times \Omega_2$ given by*

$$Sf(x) = \int f(e^{-t\cdot X}x)\check{b}(x, t, t)\, dt.$$

Let $b_{j,L}(x, u, \xi)$ and $b_{j,R}(x, u, \xi)$ be defined by $\check{b}_{j,L}(x, u, t) = \widehat{W}_j^L \check{b}(x, u, t)$ and $\check{b}_{j,R}(x, u, t) = \widehat{W}_j^R \check{b}(x, u, t)$, where $\widehat{W}_j^L$ and $\widehat{W}_j^R$ are acting in the t variable. Then

$$W_j Sf(x) = S_{j,L}f(x) + R_{j,L}f(x), \quad SW_jf(x) = S_{j,R}f(x) + R_{j,R}f(x),$$

where

$$S_{j,L}f(x) = \int f(e^{-t\cdot X}x)\check{b}_{j,L}(x, t, t)\, dt, \quad S_{j,R}f(x) = \int f(e^{-t\cdot X}x)\check{b}_{j,R}(x, t, t)\, dt,$$

$R_{j,L}$ and $R_{j,R}$ are (X, d), a pseudo-differential operators of order s supported in $\Omega_1 \times \Omega_2$, and $S_{j,L}$ and $S_{j,R}$ are (X, d), a pseudo-differential operators of order $s + d_j$ supported in $\Omega_1 \times \Omega_2$.

The rest of this section is devoted to the proof of Proposition 4.5.31. We prove only the result for SW_j as that is the only part we use in our later results and the proof for W_jS is similar.

Lemma 4.5.32. *Let $b(x, u, \xi) \in S^s(a, \Omega_2, d)$, for some $s \in \mathbb{R}$ and $a > 0$. Then:*

(i) *Suppose $g(t, x) \in C^\infty_{\mathrm{loc}}(B^q(a) \times \Omega_2)$ satisfies $\deg_d(g) \geq e$ for some $e \in \mathbb{N}$ (see Definition 4.2.1). Then $g(t, x)\check{b}(x, t, t) = \check{b}_g(x, t, t)$, where $b_g(x, u, \xi) \in S^{s-e}(a, \Omega_1, d)$.*

(ii) *For each $j \in \{1, \ldots, q\}$, treating each $\widehat{X}_j^R$ as a differential operator in the $t \in \mathbb{R}^q \cong N_{\mathcal{K},r}$ variable,*

$$\widehat{X}_j^R \check{b}(x, t, t) = \check{b}_{j,1}(x, t, t), \quad \partial_{t_j}\check{b}(x, t, t) = \check{b}_{j,2}(x, t, t),$$

where $b_{j,l}(x, u, \xi) \in S^{m+d_j}(a, \Omega_1, d)$, $\check{b}_{j,1}(x, u, t) = \widehat{X}_j^R \check{b}(x, u, t) + \check{r}_{j,1}(x, u, t)$ and $\check{b}_{j,2}(x, u, t) = \partial_{t_j}\check{b}(x, u, t) + \check{r}_{j,2}(x, u, t)$, where $r_{j,l}(x, u, \xi) \in S^m(a, \Omega_1, d)$.

Proof. By Proposition 4.1.14, we may write

$$\check{b}(x, u, t) = \sum_{j\in\mathbb{N}} 2^{js}\, \mathrm{Dil}^d_{2^j}(\varsigma_j)(x, u, t), \tag{4.73}$$

where $\{\varsigma_j : j \in \mathbb{N}\} \subset C_0^\infty(\Omega_1)\widehat{\otimes}C_0^\infty(B^q(a))\widehat{\otimes}\mathscr{S}(\mathbb{R}^q)$ is a bounded set and $\varsigma_j \in C_0^\infty(\Omega_1)\widehat{\otimes} C_0^\infty(B^q(a))\widehat{\otimes}\mathscr{S}_0(\mathbb{R}^q)$ for $j > 0$.

For (i), write $g(t,x) = \sum_{l=1}^{L} t^{\alpha_l} g_l(t,x)$, where $g_l(t,x) \in C^\infty_{\mathrm{loc}}(B^q(a) \times \Omega_2)$ and $\deg_d(\alpha_l) \geq e$. Then we take

$$\check{b}_g(x,u,t) := \sum_{l=1}^{L} g_l(u,x) t^{\alpha_l} \check{b}(x,u,t).$$

Thus, to prove (i), it suffices to show $\check{c}_l(x,u,t) := g_l(u,x)t^{\alpha_l}\check{b}(x,u,t)$ satisfies $c_l(x,u,\xi) \in S^{s-e}(a,\Omega_1,d)$. We have, by (4.73),

$$\check{c}_l(x,u,t) = \sum_{j\in\mathbb{N}} 2^{js} g(u,x)t^{\alpha_l}\,\mathrm{Dil}^d_{2^j}(\varsigma_j)(x,u,t) = \sum_{j\in\mathbb{N}} 2^{js-\deg_d(\alpha_l)}\,\mathrm{Dil}^d_{2^j}(\tilde{\varsigma}_j)(x,u,t),$$

where $\tilde{\varsigma}_j(x,u,t) := g(u,x)t^{\alpha_l}\varsigma_j(x,u,t)$. It follows immediately from the hypotheses on ς_j that $\{\tilde{\varsigma}_j : j \in \mathbb{N}\} \subset C_0^\infty(\Omega_1)\widehat{\otimes} C_0^\infty(B^q(a))\widehat{\otimes}\mathscr{S}(\mathbb{R}^q)$ is a bounded set and $\tilde{\varsigma}_j \in C_0^\infty(\Omega_1)\widehat{\otimes} C_0^\infty(B^q(a))\widehat{\otimes}\mathscr{S}_0(\mathbb{R}^q)$ for $j > 0$. Proposition 4.1.14 implies $c_l(x,u,\xi) \in S^{s-\deg_d(\alpha_l)}(a,\Omega_1,d) \subseteq S^{s-e}(a,\Omega_1,d)$, completing the proof of (i).

Turning to (ii), we prove the results for $\widehat{X}_j^R$ and ∂_{t_j} simultaneously. Indeed, if we let $V_j \in \{\widehat{X}_j^R, \partial_{t_j}\}$, then V_j is a vector field with polynomial coefficients which is homogeneous of degree d_j in the sense of Definition 4.5.18 (see Lemma 4.5.28). Note that $V\check{b}(x,t,t) = \check{b}_V(x,t,t)+r_V(x,t,t)$, where $\check{b}_V(x,u,t) = V_t(\check{b}(x,u,t))$, $r_V(x,u,\xi) = V_u b(x,u,\xi)$, and V_t and V_u denote V acting in the t and u variable, respectively. It is clear that $r_V \in S^m(a,\Omega_1,d)$. Thus, (ii) will follow if we show that $b_V \in S^{m+d_j}(a,\Omega_1,d)$. Using (4.73), we have

$$\check{b}_V(x,u,t) = \sum_{j\in\mathbb{N}} 2^{js} V_t\,\mathrm{Dil}^d_{2^j}(\varsigma_j)(x,u,t) = \sum_{j\in\mathbb{N}} 2^{js+d_j}\,\mathrm{Dil}^d_{2^j}(\tilde{\varsigma}_j)(x,u,t),$$

where $\tilde{\varsigma}_j(x,u,t) := V_t\varsigma_j(x,u,t)$ and we have used the homogeneity of V. Since V is a vector field with polynomial coefficients, it follows that $\{\tilde{\varsigma}_j : j \in \mathbb{N}\} \subset C_0^\infty(\Omega_1)\widehat{\otimes} C_0^\infty(B^q(a))\widehat{\otimes}\mathscr{S}(\mathbb{R}^q)$ is a bounded set and $\tilde{\varsigma}_j \in C_0^\infty(\Omega_1)\widehat{\otimes} C_0^\infty(B^q(a))\widehat{\otimes}\mathscr{S}_0(\mathbb{R}^q)$ for $j > 0$. Proposition 4.1.14 implies $b_V(x,u,\xi) \in S^{m+d_j}(a,\Omega_1,d)$, completing the proof. □

Lemma 4.5.33. *If $a = a(\Omega_1,\Omega_2,(X,d)) > 0$ is sufficiently small, the following holds. Let S be an (X,d), a pseudo-differential operator of order $s \in \mathbb{R}$ supported in $\Omega_1 \times \Omega_2$. Then, for $j \in \{1,\dots,q\}$, SX_j is an (X,d), a pseudo-differential operator of order $s + d_j$.*

Proof. This is a straightforward consequence of Theorem 4.3.3 and Lemma 4.3.4 (ii). Indeed, by Theorem 4.3.3, $S = \sum_{k\in\mathbb{N}} 2^{ks}E_k$, where $\{(E_k, 2^{-k}) : k \in \mathbb{N}\}$ is a bounded set of (X,d), a pseudo-differential operator scales supported in $\Omega_1 \times \Omega_2$. Thus, $SX_k = \sum_{k\in\mathbb{N}} 2^{k(s+d_j)}E_k 2^{-kd_j X_j} =: \sum_{k\in\mathbb{N}} 2^{k(s+d_j)}\widetilde{E}_k$. By Lemma 4.3.4 (ii), $\{(\widetilde{E}_k, 2^{-k}) : k \in \mathbb{N}\}$ is a bounded set of (X,d), a pseudo-differential operator scales supported in $\Omega_1 \times \Omega_2$, and therefore, by another application of Theorem 4.3.3, SX_k is an (X,d), a pseudo-differential operator of order $s + d_j$. □

Lemma 4.5.34. *If $a = a(\Omega_1, \Omega_2, (X, d)) > 0$ is sufficiently small, the following holds. Let $c \in C^\infty_{\text{loc}}(\Omega_3)$ and let S be an (X, d), a pseudo-differential operator of order $s \in \mathbb{R}$ supported in $\Omega_1 \times \Omega_2$. Then $S\,\text{Mult}[c]$ is an (X, d), a pseudo-differential operator of order s supported in $\Omega_1 \times \Omega_2$.*

Proof. If $Sf(x) = \int f(e^{-t\cdot X}x)\check{b}(x,t,t)$, then $S\,\text{Mult}[c]f(x) = \int f(e^{-t\cdot X})\check{b}_1(x,t,t)$, where $b_1(x,u,\xi) = c(e^{-u\cdot X}x)b(x,u,\xi)$. It follows easily from the definitions that if $b \in S^s(a,\Omega_1,d)$, then $b_1 \in S^s(a,\Omega_1,d)$. This completes the proof. □

Lemma 4.5.35. *If $a = a(\Omega_1, \Omega_2, (X, d)) > 0$ is sufficiently small, the following holds. Suppose $(Z, e) \in \text{Gen}((W, đ))$ and S is an (X, d), a pseudo-differential operator of order $s \in \mathbb{R}$ supported in $\Omega_1 \times \Omega_2$. Then SZ is an (X, d), a pseudo-differential operator of order $s + M_0$ supported in $\Omega_1 \times \Omega_2$.*

Proof. By Lemma 4.5.30, $\text{Gen}((W, đ))$ is finitely generated by $\mathcal{F}_1$ on Ω_3. Thus, since $\mathcal{F}_1$ is given by (4.72), we may write $Z = \sum_{d_j \le M_0} c_j X_j$, so $SZ = \sum_{d_j \le M_0} S\,\text{Mult}[c_j]X_j$. By Lemma 4.5.34, $S\,\text{Mult}[c_j]$ is an (X, d), a pseudo-differential operator of order s supported on $\Omega_1 \times \Omega_2$, and therefore by Lemma 4.5.33, $S\,\text{Mult}[c_j]X_j$ is an (X, d), a pseudo-differential operator of order $s + d_j$ supported on $\Omega_1 \times \Omega_2$. It follows that $SZ = \sum_{d_j \le M_0} S\,\text{Mult}[c_j]X_j$ is an (X, d), a pseudo-differential operator of order $s + M_0$ supported on $\Omega_1 \times \Omega_2$, completing the proof. □

Lemma 4.5.36. *If $a = a(\Omega_1, \Omega_2, (X, d)) > 0$ is sufficiently small, then for $1 \le j \le q$, $f \in C^\infty_{\text{loc}}(\Omega_3)$,*

$$(X_j f)(e^{-t\cdot X}x) = -\widehat{X}_j^R(f(e^{-t\cdot X}x)) + \sum_{\deg_d(\alpha) \ge \mathscr{K}+1-d_j} \sum_{k=1}^{q} h_{j,\alpha}^k(t,x)t^\alpha (X_k f)(e^{-t\cdot X}x),$$

where the sum $\sum_{\deg_d(\alpha) \ge \mathscr{K}+1-d_j}$ denotes a finite sum of such terms and $h_{j,\alpha}^k(t,x) \in C^\infty_{\text{loc}}(B^q(a) \times \Omega_2)$.

Proof. By Proposition 4.2.2 (iii) we know that

$$\widehat{X}_j^R f(e^{-t\cdot X}x) = \sum_{k=1}^{q} p_j^k(t,x)(X_k f)(e^{-t\cdot X}x),$$

for some $p_j^k(t,x) \in C^\infty_{\text{loc}}(B^q(a) \times \Omega_2)$. Thus, the result will follow if we can show that

$$\widehat{X}_j^R f(e^{-t\cdot X}x) - (X_j f)(e^{-t\cdot X}x) = O(|t|^{\mathscr{K}}). \tag{4.74}$$

We have, by the Baker–Campbell–Hausdorff formula and the definition of multiplication on $\mathbb{R}^q \cong N_{\mathscr{K},r}$,

$$\begin{aligned}(X_jf)(e^{-t\cdot X}x) &= \partial_{s_j}\big|_{s=0} f(e^{s\cdot X}e^{-t\cdot X}x) = \partial_{s_j}\big|_{s=0}(e^{-t\cdot X}e^{s\cdot X}f)(x)\\ &= \partial_{s_j}\big|_{s=0}(e^{((-s)\times t)^{-1}\cdot X}f)(x) + \partial_{s_j}\big|_{s=0}((e^{-t\cdot X}e^{s\cdot X}f)(x) - (e^{((-s)\times t)^{-1}\cdot X}f)(x))\\ &= \partial_{s_j}\big|_{s=0}(e^{-((-s)\times t)\cdot X}f)(x) + \partial_{s_j}\big|_{s=0}O((\|t\|+\|s\|)^{\mathscr{K}+1})\\ &= -\widehat{X}_j^R e^{-t\cdot X}f(x) + O(\|t\|^{\mathscr{K}+1-d_j}) = -\widehat{X}_j^R f(e^{-t\cdot X}x) + O(\|t\|^{\mathscr{K}+1-d_j}),\end{aligned}$$

where we have used the fact that $s^{-1} = -s$ in the group $N_{\mathscr{K},r} \cong \mathbb{R}^q$. This establishes (4.74) and completes the proof. □

Lemma 4.5.37. *If $a = a(\Omega_1, \Omega_2, (X, d)) > 0$ is sufficiently small, the following holds. Suppose $b(x, u, \xi) \in S^s(a, \Omega_1, d)$ and S is the (X, d), a pseudo-differential operator of order s supported in $\Omega_1 \times \Omega_2$ given by*

$$Sf(x) = \int f(e^{-t\cdot X}x)\check{b}(x, t, t)\, dt.$$

Then, for $(X_j, d_j) \in \mathcal{F}_1$,

$$SX_jf(x) = \int f(e^{-t\cdot X}x)\check{b}_j(x, t, t)\, dt + R_jf(x),$$

where $\check{b}_j(x, u, t) = \widehat{X}_j^R\check{b}(x, u, t)$ (where $\widehat{X}_j^R$ acts in the t variable) and R is an (X, d), a pseudo-differential operator of order s supported in $\Omega_1 \times \Omega_2$. Finally, $b_j \in S^{s+d_j}(a, \Omega, d)$.

Proof. Lemma 4.5.36 shows that

$$\begin{aligned}SX_jf(x) &= \int (X_jf)(e^{-t\cdot X}x)\check{b}(x, t, t)\, dt\\ &= -\int (\widehat{X}_j^R f(e^{-t\cdot X}x))\check{b}(x, t, t)\, dt + \sum_{k=1}^{q}\int (X_kf)(e^{-t\cdot X}x)g_j^k(t, x)\check{b}(x, t, t)\, dt\\ &= (\mathrm{I}) + (\mathrm{II}),\end{aligned}$$

where $g_j^k(t, x)$ is a finite sum of the form $\sum_{\deg_d(\alpha)\geq \mathscr{K}+1-d_j} h_{j,\alpha}^k(t, x)t^\alpha$ and $h_{j,\alpha}^k \in C^\infty_{\mathrm{loc}}(B^q(a)\times \Omega_2)$. Note that $\deg_d(g_j^k) \geq \mathscr{K} + 1 - d_j$.

By Lemma 4.5.32 (i), $g_j^k(t, x)\check{b}(x, t, t) = \check{b}_{j,k}(x, t, t)$, where

$$b_{j,k}(x, u, \xi) \in S^{s-\mathscr{K}-1+d_j}(a, \Omega_1, d).$$

Thus,

$$(\mathrm{II}) = \sum_{k=1}^{q} S_kX_kf(x),$$

where each S_k is an (X, d), a pseudo-differential operator of order $s-\mathscr{K}-1+d_j$ supported in $\Omega_1\times\Omega_2$. Lemma 4.5.35 shows that each S_kX_k is an (X, d), a pseudo-differential operator of order $s - \mathscr{K} - 1 + d_j + M_0 \leq s$ supported in $\Omega_1 \times \Omega_2$; here we have used $d_j \leq M_0$ since $(X_j, d_j) \in \mathcal{F}_1$ and we have used the fact that $\mathscr{K}$ was chosen to be $\geq 2M_0 - 1$. This shows that $(\mathrm{II}) = R_{j,1}f(x)$, where $R_{j,1}$ is an (X, d), a pseudo-differential operator of order s supported in $\Omega_1 \times \Omega_2$.

For (I), integrating by parts shows that

$$(\mathrm{I}) = \int f(e^{-t\cdot X}x)\widehat{X}_j^R \check{b}(x,t,t)\,dt.$$

Lemma 4.5.32 (ii) shows that $\widehat{X}_j^R \check{b}(x,t,t) = \check{b}_j(x,t,t) + \check{r}_{j,2}(x,t,t)$, where $r_{j,2} \in S^s(a, \Omega_1, d)$. This proves that (I) is of the desired form.

Finally, it follows from Lemma 4.5.32 (ii) that $b_j \in S^{s+d_j}(a, \Omega, d)$. □

Proof of Proposition 4.5.31. The result for SW_j follows immediately from Lemma 4.5.37, as $(W_j, \mathfrak{d}_j) \in \mathcal{F}_1$ (see Lemma 4.5.30). The proof for $W_j S$ is similar, and we leave it to the reader. □

4.5.6 The parametrix

In this section, we complete the proof of Theorem 4.5.1; we take all the same assumptions and notation as in that theorem. In particular, we have vector fields $(W, \mathfrak{d}) = \{(W_1, \mathfrak{d}_1), \ldots, (W_r, \mathfrak{d}_r)\} \subset C^\infty_{\mathrm{loc}}(\mathfrak{M}; T\mathfrak{M}) \times \mathbb{N}_+$ such that $\mathrm{Gen}((W, \mathfrak{d}))$ is locally finitely generated, and we are given $\kappa \in \mathbb{N}_+$ such that $\mathfrak{d}_j$ divides κ for every j.

We take $\mathfrak{n}_{\mathscr{K},r}$ as in Section 4.5.5, where we pick $\mathscr{K}$ so that $\mathscr{K} \geq 2\kappa + 1$. We let $\widehat{W}_j^R$ be as in Section 4.5.5, and we define $\mathscr{L}^R$ by (4.71). We define $(X, d) = \{(X_1, d_1), \ldots, (X_q, d_q)\}$ as in Section 4.5.5. In particular, $N_{\mathscr{K},r} \cong \mathbb{R}^q$.

With Q the homogeneous dimension of $N_{\mathscr{K},r} \cong \mathbb{R}^q$, we have $Q \geq \mathscr{K} > 2\kappa$. Thus, Proposition 4.5.29 shows that there is a unique distribution $K \in C_0^\infty(\mathbb{R}^q)'$ which is homogeneous of degree $2\kappa - Q$ such that $\mathscr{L}^R K = \delta_0$. This $K(t)$ agrees with a $C^\infty_{\mathrm{loc}}(\mathbb{R}^q)$ function on $\{t \neq 0\}$.

Take $a = a(\Omega_1, \Omega_2, (X, d)) > 0$ small and $\eta \in C_0^\infty(B^q(a))$ with $\eta = 1$ on a neighborhood of $0 \in \mathbb{R}^q$. For $\psi \in C_0^\infty(\Omega_1)$ we set

$$T_\psi f(x) := \int f(e^{-t\cdot X}x)\psi(x)\eta(t)K(t)\,dt.$$

Remark 4.5.23 shows that T_ψ is an (X, d), a pseudo-differential operator of order -2κ supported in $\Omega_1 \times \Omega_2$. We will show that this T_ψ satisfies the conclusions of Theorem 4.5.1. Note that we may take $a > 0$ as small as we like in the definition of T_ψ.

In what follows, we take all the same notation as in Section 4.5.5.

Lemma 4.5.38. *If $a = a(\Omega_1, \Omega_2, (X, d)) > 0$ is sufficiently small, the following holds. Suppose $b(x, u, \xi) \in S^s(a, \Omega_1, d)$ and let S be the (X, d), a pseudo-differential operator of order s supported in $\Omega_1 \times \Omega_2$ given by*

$$Sf(x) = \int f(e^{-t\cdot X}x)\check{b}(x,t,t)\,dt.$$

Then, for every $l \in \mathbb{N}$, $j \in \{1,\dots,r\}$,

$$SW_j^l f(x) = S_{j,l}f(x) + R_{j,l}f(x),$$

where $R_{j,l}$ is an (X,d), a pseudo-differential operator of order $s + d_j(l-1)$ supported in $\Omega_1 \times \Omega_2$ and $S_{j,l}$ is the (X,d), a pseudo-differential operator of order $s + d_j$ supported in $\Omega_1 \times \Omega_2$ given by

$$S_{j,l}f(x) = \int f(e^{t\cdot X}x)\check{b}_{j,l}(x,t,t)\,dt,$$

where $b_{j,l}(x,u,\xi) \in S^{s+d_j}(a,\Omega_1,d)$ is defined by $\check{b}_{j,l}(x,u,t) = (\widehat{W}_j^R)^l\check{b}(x,u,t)$, where $\widehat{W}_j^R$ is acting in the $t \in \mathbb{R}^q \cong N_{\mathscr{K},r}$ variable.

Proof. We proceed by induction on l. The base case, $l = 0$, is trivial (one can take $R_{j,0} = 0$ and $S_{j,0} = S$). We suppose we have the result for l and prove it for l replaced by $l+1$. By the inductive hypothesis, we have

$$SW_j^l f(x) = S_{j,l}W_j f(x) + R_{j,l}W_j f(x),$$

where $S_{j,l}$ and $R_{j,l}$ are as in the statement of the lemma. Proposition 4.5.31 applied to $S_{j,l}W_j$ shows that $S_{j,l}W_j$ is of the desired form. Lemma 4.5.33 (and the fact that $(W_j, d_j) \in (X,d)$) applied to $R_{j,l}W_j$ shows that $R_{j,l}W_j$ is an (X,d), a pseudo-differential operator of order $s + d_j(l-1) + d_j = s + d_j l$ supported in $\Omega_1 \times \Omega_2$ and therefore of the desired form. This completes the proof. □

Proof of Theorem 4.5.1. A simple computation shows that

$$\mathscr{L} = \sum_{j=1}^{r}(-1)^{\kappa/d_j}W_j^{2\kappa/d_j} + \sum_{j=1}^{r}\sum_{l=\kappa}^{2\kappa/d_j-1} c_{j,l}W_j^l,$$

where $c_{j,k} \in C^\infty_{\mathrm{loc}}(\mathfrak{M})$. Applying Lemma 4.5.34 once and Lemma 4.5.33 l times shows that

$$T_\psi c_{j,l}W_j^l$$

is an (X,d), a pseudo-differential operator of order $-2\kappa + d_j l$ supported in $\Omega_1 \times \Omega_2$. Thus,

$$T_\psi \sum_{j=1}^{r}\sum_{l=\kappa}^{2\kappa/d_j-1} c_{j,l}W_j^l$$

is an (X,d), a pseudo-differential operator of order -1 supported in $\Omega_1 \times \Omega_2$.

Lemma 4.5.38 shows that

$$T_\psi \sum_{j=1}^{r} (-1)^{\kappa/d_j} W_j^{2\kappa/d_j} = S + R,$$

where R is an (X, d), a pseudo-differential operator of order -1 supported on $\Omega_1 \times \Omega_2$, and

$$\begin{aligned} Sf(x) &= \int f(e^{-t\cdot X}x)\psi(x)\eta(t)(\mathscr{L}^R K(t))\,dt \\ &= \int f(e^{-t\cdot X}x)\psi(x)\eta(t)\delta_0(t)\,dt = \psi(x)f(x). \end{aligned}$$

Combining the above, we conclude $T_\psi \mathscr{L} \equiv \mathrm{Mult}[\psi]$, modulo (X, d), a pseudo-differential operators of order -1 supported on $\Omega_1 \times \Omega_2$, completing the proof. □

4.5.7 Limitations of the lifting procedure

The proof of Theorem 4.5.1, sometimes called the Rothschild–Stein lifting[3] procedure, is a powerful tool in the study of parametrices for partial differential operators based on vector fields. When it works, it provides detailed information which can be difficult or impossible to obtain by other methods. For example, in Theorem 4.5.1 we did not even require the underlying vector fields to satisfy Hörmander's condition.

Unfortunately, it is not possible to obtain a parametrix for every maximally subelliptic operator in this way: there are maximally subelliptic operators which have no corresponding lifted maximally subelliptic operator on a nilpotent Lie group. Since the goal of this text is to study maximally subelliptic operators in full generality, we are forced to move beyond the Rothschild–Stein lifting procedure; this is taken up in Chapter 8. However, Theorem 4.5.1 is a key tool in some of our methods for proving more general results. Thus, there are several situations we consider where the Rothschild–Stein lifting procedure is not directly applicable, but still plays an important role in the proof.

To see an example of an operator which is maximally subelliptic but does not have a corresponding lifted maximally subelliptic operator, we consider operators of the following form, where W_1, W_2 are Hörmander vector fields and $\lambda \in \mathbb{C}$:

$$\mathscr{L}_\lambda = W_1^2 + W_2^2 + i\lambda[W_1, W_2].$$

---

3 Unlike the proof due to Rothschild and Stein [200], we did not lift the operator to a nilpotent Lie group. Instead, following Goodman [99], we took a fundamental solution on a nilpotent Lie group and pushed it forward to a lower-dimensional space to obtain a parametrix. Thus, with this perspective, "lifting" is not the most appropriate word to use.

It is a result of Rothschild and Stein [200] that when $|\lambda| < 1$, $\mathscr{L}_\lambda$ is maximally subelliptic of degree 2 with respect to $(W,1) = \{(W_1,1),(W_2,1)\}$. However, the situation for $|\lambda| \geq 1$ is more subtle.

When W_1 and W_2 are the left invariant vector fields on the Heisenberg group described in Example 1.1.2 (iii), then it is a result of Folland and Stein [91] that $\mathscr{L}_\lambda$ is maximally subelliptic for all but a discrete set of $\lambda \in \mathbb{C}$; see also the presentation in [216, Chapter XIII]. On the other hand, Helffer [119] (building on work of Rothschild and Stein [200]) showed that $\mathbb{R}^2$ and the Heisenberg group are the only stratified Lie groups with this property: if W_1 and W_2 are generating left invariant vector fields on a stratified nilpotent Lie group of step higher than 2, then $\mathscr{L}_\lambda$ is only hypoelliptic for $|\lambda| < 1$.

Unlike the special case of stratified nilpotent Lie groups, there are choices of Hörmander vector fields which satisfy Hörmander's condition of arbitrarily high order, where $\mathscr{L}_\lambda$ is maximally subelliptic for all but a discrete set of λ. Indeed, on $\mathbb{R}^2$ we take $W_1 = \partial_x$ and $W_2 = x^k\partial_y$, where $k \in \mathbb{N}$ is fixed, so that W_1 and W_2 satisfy Hörmander's condition of order $k+1$ on $\mathbb{R}^2$. Then, except for a discrete set of $\lambda \in \mathbb{C}$, $\mathscr{L}_\lambda$ is maximally subelliptic of degree 2 with respect to $(W,1)$ on $\mathbb{R}^2$. When k is odd this follows from methods of Gilioli and Trèves [97] and when k is even this follows from methods of Menikoff [170]. Thus, when $\lambda \in \mathbb{C}$ is outside a discrete set, $\mathscr{L}_\lambda$ is maximally subelliptic of degree 2 with respect to $(W,1)$, but when $|\lambda| \geq 1$ and $k \geq 2$ it cannot be lifted to a maximally subelliptic operator on a nilpotent Lie group. Hence, the Rothschild–Stein lifting procedure destroys the maximally subellipticity of $\mathscr{L}_\lambda$ in this case.

A recent, elegant pseudo-differential calculus of Androulidakis, Moshen, and Yuncken [2] salvages many aspects of the Rothschild–Stein lifting procedure to the general setting. See Sections 4.6 and 8.8.

4.6 Further reading and references

In the single-parameter ($\nu = 1$) case, (X, d) pseudo-differential operators were first studied by Folland and Stein [91] and Rothschild and Stein [200]. In particular, Rothschild and Stein used them to obtain results similar to Theorem 4.5.1. Since these original works, similar operators have been used in many different contexts. For example, they were clarified by Goodman [99] and used by Rothschild [198], and a related theory with a calculus was developed by Christ, Geller, Głowacki, and Polin [49]. In the multi-parameter setting ($\nu > 1$), similar operators were used by the author in [220] and [221].

When $\kappa = 1$ and $d_j = 1$ for every j, the sub-Laplacian (4.39) was originally studied by Hörmander [125]; it is sometimes called Hörmander's sub-Laplacian. In this context, Hörmander proved Theorem 4.5.5. Kohn [142] gave a simpler proof of Hörmander's result, and it is this simpler proof that we use to prove Theorem 4.5.5. As mentioned earlier, the choice of ϵ_0 in Theorem 4.5.5 is not optimal. When $\kappa = 2$ and $d_j = 1$ for every j, it is a result of Rothschild and Stein [200] that one can take $\epsilon_0 = 2/m$, where $W_1, \ldots, W_r$ satisfy Hörmander's condition of order m on Ω_1, and this is optimal. See Corollary 8.2.5

for a further discussion of the optimal ϵ_0. For general κ and d_j, we presented the theory for $\mathscr{L}$ given by (4.39) using the same methods that were originally developed for $\kappa = 1$, $d_j = 1$. One of the early settings where this more general operator was explicitly used was in the work of Helffer and Nourrigat [121].

In the single-parameter setting, Androulidakis, Moshen, and Yuncken [2] recently introduced a pseudo-differential calculus which contains the parametrices for maximally subelliptic operators, based on the ideas of [3]. This is a powerful and general approach, which yields parametrices for all maximally subelliptic operators; however, it does not seem to directly apply to the multi-parameter theory we use in this text or to directly imply regularity theory for maximally subelliptic operators outside the setting of adapted L^2 Sobolev spaces. It is likely that these pseudo-differential operators could be combined with some methods of this text to give a pseudo-differential approach to some of the regularity theory of linear maximally subelliptic operators, though we do not pursue that here.

5 Singular integrals

Parametrices for elliptic partial differential operators are often constructed using the Fourier transform; see Proposition 2.2.27. Unfortunately, the Fourier transform is not a decisive tool when studying maximally subelliptic partial differential operators. Instead, we construct parametrices for maximally subelliptic operators by using a filtered algebra of singular integral operators. We present singular integral operators in three different settings, each of which is useful in a different way when studying maximally subelliptic operators.

Throughout this chapter, $\mathfrak{M}$ is a connected C^∞ manifold, endowed with a smooth, strictly positive density Vol. Fix $\Omega \Subset \mathfrak{M}$ a relatively compact, open set.

5.1 The three settings

In this section, we describe the three settings under which we introduce singular integral operators. We call these the **single-parameter** setting, the **multi-parameter Hörmander** setting, and the **general multi-parameter** setting.

5.1.1 The single-parameter setting

In this setting, we are given Hörmander vector fields with formal degrees $(W, \vec{d}) = \{(W_1, \vec{d}_1), \dots, (W_r, \vec{d}_r)\} \subset C^\infty_{\mathrm{loc}}(\mathfrak{M}; T\mathfrak{M}) \times \mathbb{N}_+$. For $t \in \mathbb{R}$, we will define a filtered algebra $\mathscr{A}^t(\Omega, (W, \vec{d}))$ consisting of operators in $\mathrm{Hom}(C_0^\infty(\mathfrak{M}), C_0^\infty(\mathfrak{M})')$, whose Schwartz kernels are supported in $\Omega \times \Omega$. These operators have the following properties:

- $\mathscr{A}^t(\Omega, (W, \vec{d}))$ is a filtered algebra: for $T \in \mathscr{A}^t(\Omega, (W, \vec{d}))$ and $S \in \mathscr{A}^s(\Omega, (W, \vec{d}))$, we have $TS \in \mathscr{A}^{t+s}(\Omega, (W, \vec{d}))$. See Proposition 5.8.6.
- If $t \leq s$, then $\mathscr{A}^t(\Omega, (W, \vec{d})) \subseteq \mathscr{A}^s(\Omega, (W, \vec{d}))$. See Proposition 5.8.2.
- Operators in $\mathscr{A}^0(\Omega, (W, \vec{d}))$ extend to bounded operators on $L^p(\mathfrak{M}, \mathrm{Vol})$. See Proposition 5.10.1.
- More generally, operators in $\mathscr{A}^t(\Omega, (W, \vec{d}))$ have good boundedness properties on function spaces adapted to $(W, \vec{d})$. See Theorem 6.2.10.
- Left parametrices for maximally subelliptic operators of degree κ are "locally" in $\mathscr{A}^{-\kappa}(\cdot, (W, \vec{d}))$. See Theorem 8.1.1 (vii).
- If $\psi \in C_0^\infty(\Omega)$, then $\mathrm{Mult}[\psi] \in \mathscr{A}^0(\Omega, (W, \vec{d}))$ and $\mathrm{Mult}[\psi] W_j \in \mathscr{A}^{\vec{d}_j}(\Omega, (W, \vec{d}))$.
- More generally, if $\psi \in C_0^\infty(\Omega)$ and α is an ordered multi-index, then $\mathrm{Mult}[\psi] W^\alpha \in \mathscr{A}^{\deg_{\vec{d}}(\alpha)}(\Omega, (W, \vec{d}))$. See Proposition 5.8.3.
- $\mathscr{A}^t(\Omega, (W, \vec{d}))$ depends on $(W, \vec{d})$ only though the weak equivalence class of $(W, \vec{d})$. See Proposition 5.8.12.

https://doi.org/10.1515/9783111085647-005

In fact, we will give four definitions of the filtered algebra $\mathscr{A}^t(\Omega,(W,d))$. We will define sets $\mathscr{A}_1^t(\Omega,(W,d))$, $\mathscr{A}_2^t(\Omega,(W,d))$, $\mathscr{A}_4^t(\Omega,(W,d))$, and for $K\in\mathbb{N}$, $\mathscr{A}_3^t(\Omega,(W,d),K)$. A main theorem (Theorem 5.2.12) is that these four definitions all give the same set: for all $K\in\mathbb{N}$,

$$\mathscr{A}_1^t(\Omega,(W,d))=\mathscr{A}_2^t(\Omega,(W,d))=\mathscr{A}_3^t(\Omega,(W,d),K)=\mathscr{A}_4^t(\Omega,(W,d)).$$

We let $\mathscr{A}^t(\Omega,(W,d))$ be the common value.

These single-parameter singular integrals are one of our main tools when proving properties of both linear and nonlinear maximally subelliptic PDEs in terms of function spaces adapted to (W,d).

5.1.2 The multi-parameter Hörmander setting

In this setting, we are given $\nu\in\mathbb{N}_+$ and for each $\mu\in\{1,\dots,\nu\}$ Hörmander vector fields with formal degrees $(W^\mu,d^\mu)=\{(W_1^\mu,d_1^\mu),\dots,(W_{r_\mu}^\mu,d_{r_\mu}^\mu)\}\subset C_{\mathrm{loc}}^\infty(\mathfrak{M};T\mathfrak{M})\times\mathbb{N}_+$. We make the following key assumption.

Key assumption: We assume $(W^1,d^1),\dots,(W^\nu,d^\nu)$ pairwise locally weakly approximately commute.

Set

$$\begin{aligned}(W,\vec{d})&=\{(W_1,\vec{d}_1),\dots,(W_r,\vec{d}_r)\}:=(W^1,d^1)\boxtimes(W^2,d^2)\boxtimes\cdots\boxtimes(W^\nu,d^\nu)\\&=\{(W_j^\mu,d_j^\mu\mathfrak{e}_\mu):\mu\in\{1,\dots,\nu\},1\le j\le r_\mu\}\subset C_{\mathrm{loc}}^\infty(\mathfrak{M};T\mathfrak{M})\times(\mathbb{N}^\nu\setminus\{0\}).\end{aligned}\tag{5.1}$$

For $t\in\mathbb{R}^\nu$, we will define a filtered algebra $\mathscr{A}^t(\Omega,(W,\vec{d}))$ consisting of operators in $\mathrm{Hom}(C_0^\infty(\mathfrak{M}),C_0^\infty(\mathfrak{M})')$, whose Schwartz kernels are supported in $\Omega\times\Omega$. These operators have the following properties:

- When $\nu=1$, $\mathscr{A}^t(\Omega,(W,\vec{d}))$ coincides with the single-parameter algebra of the same name in Section 5.1.1.
- $\mathscr{A}^t(\Omega,(W,\vec{d}))$ is a filtered algebra: for $T\in\mathscr{A}^t(\Omega,(W,\vec{d}))$ and $S\in\mathscr{A}^s(\Omega,(W,\vec{d}))$, we have $TS\in\mathscr{A}^{t+s}(\Omega,(W,\vec{d}))$. See Proposition 5.8.6.
- If $t\le s$ (coordinatewise), then $\mathscr{A}^t(\Omega,(W,\vec{d}))\subseteq\mathscr{A}^s(\Omega,(W,\vec{d}))$. See Proposition 5.8.2.
- Operators in $\mathscr{A}^0(\Omega,(W,\vec{d}))$ extend to bounded operators on $L^p(\mathfrak{M},\mathrm{Vol})$. See Theorem 6.2.10 and Proposition 6.2.13.
- More generally, operators in $\mathscr{A}^t(\Omega,(W,\vec{d}))$ have good boundedness properties on function spaces adapted to $(W,\vec{d})$. See Theorem 6.2.10.
- Operators in $\mathscr{A}^t(\Omega,(W,\vec{d}))$ are pseudo-local, i. e., their Schwartz kernels are smooth off the diagonal. See Theorem 5.8.18.
- Left parametrices for maximally subelliptic operators are locally in $\mathscr{A}(\cdot,(W,\vec{d}))$.
- If $\psi\in C_0^\infty(\Omega)$, then $\mathrm{Mult}[\psi]\in\mathscr{A}^0(\Omega,(W,\vec{d}))$ and $\mathrm{Mult}[\psi]W_j\in\mathscr{A}^{\vec{d}_j}(\Omega,(W,\vec{d}))$.
- More generally, if $\psi\in C_0^\infty(\Omega)$ and α is an ordered multi-index, then $\mathrm{Mult}[\psi]W^\alpha\in\mathscr{A}^{\deg_{\vec{d}}(\alpha)}(\Omega,(W,\vec{d}))$. See Proposition 5.8.3.

– $\mathscr{A}^t(\Omega, (W, \vec{d}))$ depends on $(W^\mu, \vec{d}^\mu)$ only though the weak equivalence class of $(W^\mu, \vec{d}^\mu)$. See Proposition 5.8.12.

Example 5.1.1. A main example to consider is the following. Suppose $(W^1, \vec{d}^1)$ are Hörmander vector fields with formal degrees. Suppose $(W^2, \vec{d}^2) = \{(W_1^2, 1), \dots, (W_{r_2}^2, 1)\}$ are such that $W_1^2(x), \dots, W_{r_2}^2(x)$ span $T_x\mathfrak{M}$, $\forall x \in \mathfrak{M}$, i. e., they satisfy Hörmander's condition of order 1. In this case, $(W^1, \vec{d}^1)$ and $(W^2, \vec{d}^2)$ locally weakly approximately commute – in fact, they locally strongly approximately commute; see Lemma 3.8.3. We will see that parametrices for partial differential operators which are maximally subelliptic of degree κ with respect to $(W^1, \vec{d}^1)$ are locally in $\mathscr{A}^{(-\kappa,0)}(\cdot, (W, \vec{d}))$ (see Section 5.11.2) and standard pseudo-differential operators of order m are locally in $\mathscr{A}^{(0,m)}(\cdot, (W, \vec{d}))$ (see Section 5.8.2). Thus, we have a pseudo-local algebra of singular integral operators containing both the parametrices for maximally subelliptic operators and the standard pseudo-differential operators.

To define $\mathscr{A}^t(\Omega, (W, \vec{d}))$, we generalize two of the definitions from the single-parameter setting in Section 5.1.1. Thus, we define two sets $\mathscr{A}_1^t(\Omega, (W, \vec{d}))$ and $\mathscr{A}_2^t(\Omega, (W, \vec{d}))$. A main theorem (see Theorem 5.2.18) is that these two sets are equal, and we denote their common value by $\mathscr{A}^t(\Omega, (W, \vec{d}))$.

These pseudo-local, multi-parameter singular integrals generalize the singular integrals from Section 5.1.1 to a multi-parameter setting. This will allow us to derive certain regularity properties of maximally subelliptic operators with respect to function spaces which are not adapted to the same vector fields, for example in the setting described in Example 3.5.5.

Unfortunately, the setting described in this section is not quantitatively invariant under rescalings (see Example 3.5.5). Because of this, the singular integrals operators described here are not general enough to prove the results we require for nonlinear maximally subelliptic PDEs. For this, we introduce a more general setting for our singular integrals, described in Section 5.1.3.

5.1.3 The general multi-parameter setting

In this setting, we are given $\nu \in \mathbb{N}_+$ and for each $\mu \in \{1, \dots, \nu\}$ vector fields with single-parameter formal degrees $(W^\mu, \vec{d}^\mu) = \{(W_1^\mu, \vec{d}_1^\mu), \dots, (W_{r_\mu}^\mu, \vec{d}_{r_\mu}^\mu)\} \subset C_{\mathrm{loc}}^\infty(\mathfrak{M}; T\mathfrak{M}) \times \mathbb{N}_+$. We make the following key assumption.

Key assumption: We assume $(W^1, \vec{d}^1), \dots, (W^\nu, \vec{d}^\nu)$ pairwise locally weakly approximately commute and for each $\mu \in \{1, \dots, \nu\}$, $\mathrm{Gen}((W^\mu, \vec{d}^\mu))$ is locally finitely generated on $\mathfrak{M}$.

Remark 5.1.2. By Proposition 3.4.14 (in the case $\nu = 1$ applied to each $(W^\mu, \vec{d}^\mu)$), the setting in Section 5.1.2 is a special case of the setting described here.

As in (5.1), set

$$(W, \vec{d}) = \{(W_1, \vec{d}_1), \ldots, (W_r, \vec{d}_r)\} := (W^1, d^1) \boxtimes (W^2, d^2) \boxtimes \cdots \boxtimes (W^\nu, d^\nu)$$
$$= \{(W_j^\mu, d_j^\mu \mathfrak{e}_\mu) : \mu \in \{1, \ldots, \nu\}, 1 \le j \le r_\mu\} \subset C^\infty_{\mathrm{loc}}(\mathfrak{M}; T\mathfrak{M}) \times (\mathbb{N}^\nu \setminus \{0\}).$$

For $t \in \mathbb{R}^\nu$, we will define a filtered algebra $\widetilde{\mathscr{A}}^t(\Omega, (W, \vec{d}))$ consisting of operators whose Schwartz kernels are supported in $\Omega \times \Omega$. These operators have the following properties:

- $\widetilde{\mathscr{A}}^t(\Omega, (W, \vec{d}))$ is a filtered algebra: for $T \in \widetilde{\mathscr{A}}^t(\Omega, (W, \vec{d}))$ and $S \in \widetilde{\mathscr{A}}^s(\Omega, (W, \vec{d}))$, we have $TS \in \widetilde{\mathscr{A}}^{t+s}(\Omega, (W, \vec{d}))$. See Proposition 5.8.6.
- If $t \le s$ (coordinatewise), then $\widetilde{\mathscr{A}}^t(\Omega, (W, \vec{d})) \subseteq \widetilde{\mathscr{A}}^s(\Omega, (W, \vec{d}))$. See Proposition 5.8.2.
- Operators in $\widetilde{\mathscr{A}}^0(\Omega, (W, \vec{d}))$ extend to bounded operators on $L^p(\mathfrak{M}, \mathrm{Vol})$. See Theorem 6.2.10 and Proposition 6.2.13.
- More generally, operators in $\widetilde{\mathscr{A}}^t(\Omega, (W, \vec{d}))$ have good boundedness properties on function spaces adapted to $(W, \vec{d})$. See Theorem 6.2.10.
- Left parametrices for maximally subelliptic operators are locally in $\widetilde{\mathscr{A}}(\cdot, (W, \vec{d}))$.
- If $\psi \in C_0^\infty(\Omega)$, then $\mathrm{Mult}[\psi] \in \widetilde{\mathscr{A}}^0(\Omega, (W, \vec{d}))$ and $\mathrm{Mult}[\psi] W_j \in \widetilde{\mathscr{A}}^{\vec{d}_j}(\Omega, (W, \vec{d}))$.
- More generally, if $\psi \in C_0^\infty(\Omega)$ and α is an ordered multi-index, then $\mathrm{Mult}[\psi] W^\alpha \in \widetilde{\mathscr{A}}^{\deg_{\vec{d}}(\alpha)}(\Omega, (W, \vec{d}))$. See Proposition 5.8.3.
- If $(W, \vec{d})$ satisfies the stronger hypotheses of Section 5.1.2, then $\mathscr{A}^t(\Omega, (W, \vec{d})) \subseteq \widetilde{\mathscr{A}}^t(\Omega, (W, \vec{d}))$.
- $\widetilde{\mathscr{A}}^t(\Omega, (W, \vec{d}))$ depends on (W^μ, d^μ) only though the weak equivalence class of (W^μ, d^μ). See Proposition 5.8.4.

To define $\widetilde{\mathscr{A}}^t(\Omega, (W, \vec{d}))$ we modify the definition from the multi-parameter Hörmander setting described in Section 5.1.2. Thus, we define two sets $\widetilde{\mathscr{A}}_1^t(\Omega, (W, \vec{d}))$ and $\widetilde{\mathscr{A}}_2^t(\Omega, (W, \vec{d}))$. A main theorem (see Theorem 5.2.30) is that these two sets are equal, and we denote their common value by $\widetilde{\mathscr{A}}^t(\Omega, (W, \vec{d}))$.

These generalized multi-parameter singular integrals are one of our main tools when proving regularity properties of nonlinear maximally subelliptic PDEs with respect to function spaces which are not adapted to the same vector fields (for example, when proving regularity properties of nonlinear maximally subelliptic PDEs with respect to the classical Zygmund–Hölder spaces).

5.2 The algebras of singular integrals

In this section, we define the filtered algebras of singular integrals in the three settings described in Section 5.1. The definitions in this section are somewhat complicated, and the reader may wish to understand them first in the elliptic setting; this is contained in Section 2.3.

In what follows, using the Schwartz kernel theorem, we identify $\mathrm{Hom}(C_0^\infty(\mathfrak{M}), C_0^\infty(\mathfrak{M})') \cong C_0^\infty(\mathfrak{M}\times\mathfrak{M})'$. Furthermore, we abuse notation and for a distribution $T(x,y) \in C_0^\infty(\mathfrak{M}\times\mathfrak{M})$, we write the pairing between $T(x,y)$ and $f(x,y) \in C_0^\infty(\mathfrak{M}\times\mathfrak{M})$ as integration:

$$\int T(x,y)f(x,y)\, d\,\mathrm{Vol}(x)\, d\,\mathrm{Vol}(y).$$

If, for some open set $U \subseteq \mathfrak{M}\times\mathfrak{M}$, $T\big|_U$ is given by integration against an $L^1_{\mathrm{loc}}(U, \mathrm{Vol}\otimes\mathrm{Vol})$ function, then we identify T with this function on that open set.

Thus, given an operator $T \in \mathrm{Hom}(C_0^\infty(\mathfrak{M}), C_0^\infty(\mathfrak{M})')$, we treat T as a distribution $T(x,y)$, and conversely. In particular, given $F(x,y) \in L^1_{\mathrm{loc}}(\mathfrak{M}\times\mathfrak{M}, \mathrm{Vol}\otimes\mathrm{Vol})$, we identify F with a distribution and therefore with an operator. In this way, we may treat functions of two variables as operators. For an operator $T \in \mathrm{Hom}(C_0^\infty(\mathfrak{M}), C_0^\infty(\mathfrak{M})')$, we write $T^* \in \mathrm{Hom}(C_0^\infty(\mathfrak{M}), C_0^\infty(\mathfrak{M})')$ for the formal $L^2(\mathfrak{M}, \mathrm{Vol})$ adjoint of T, which satisfies

$$\int f(x)\overline{(Tg)(x)}\, d\,\mathrm{Vol}(x) = \int (T^*f)(x)\overline{g(x)}\, d\,\mathrm{Vol}(x), \quad \forall f,g \in C_0^\infty(\mathfrak{M}).$$

5.2.1 The single-parameter setting

We take the setting described in Section 5.1.1; thus, we are given Hörmander vector fields with formal degrees $(W, \vec{d}) = \{(W_1, \vec{d}_1), \dots, (W_r, \vec{d}_r)\}$. As described in Section 5.1.1, we will introduce four sets of operators, which all turn out to be the same set: $\mathscr{A}_1^t(\Omega, (W, \vec{d}))$, $\mathscr{A}_2^t(\Omega, (W, \vec{d}))$, $\mathscr{A}_4^t(\Omega, (W, \vec{d}))$, and for $K \in \mathbb{N}$, $\mathscr{A}_3^t(\Omega, (W, \vec{d}), K)$. The definitions of $\mathscr{A}_3^t$ and $\mathscr{A}_4^t$ are the most similar to definitions which are usually seen in the literature, and we begin with those. For $j \in \mathbb{R}$, we write $2^{-j\vec{d}}W$ for $\{2^{-j\vec{d}_1}W_1, \dots, 2^{-j\vec{d}_r}W_r\}$.

Definition 5.2.1. We say a set $\mathcal{B} \subset C_0^\infty(\mathfrak{M}) \times \mathfrak{M} \times (0,1]$ is a bounded set of $(W, \vec{d})$ bump functions in Ω if:
- $\forall(\phi, x, 2^{-j}) \in \mathcal{B}$, we have $x \in \Omega$.
- $\forall(\phi, x, 2^{-j}) \in \mathcal{B}$, we have $\mathrm{supp}(\phi) \subseteq B_{(W,\vec{d})}(x, 2^{-j}) \cap \Omega$.
- For all ordered multi-indices α, there exists $C_\alpha \geq 0$ such that $\forall(\phi, x, 2^{-j}) \in \mathcal{B}$,

$$\sup_{z\in\mathfrak{M}} \left|(2^{-j\vec{d}}W)^\alpha \phi(z)\right| \leq C_\alpha \left(\mathrm{Vol}(B_{(W,\vec{d})}(x, 2^{-j})) \wedge 1\right)^{-1}.$$

Definition 5.2.2. For $t \in \mathbb{R}$, $K \geq 0$, we let $\mathscr{A}_3^t(\Omega, (W, \vec{d}), K) \subseteq \mathrm{Hom}(C_0^\infty(\mathfrak{M}), C_0^\infty(\mathfrak{M})')$ be the set of all $T \in \mathrm{Hom}(C_0^\infty(\mathfrak{M}), C_0^\infty(\mathfrak{M})')$ such that the following hold:
- $\mathrm{supp}(T) \subseteq \Omega \times \Omega$.
- (Growth condition) The Schwartz kernel of T agrees with a C^∞_{loc} function on $\{(x,z) \in \mathfrak{M}\times\mathfrak{M} : x \neq z\}$ and satisfies, for all ordered multi-indices α and β,

$$|W_x^\alpha W_z^\beta T(x,z)| \leq C_{\alpha,\beta} \frac{\rho_{(W,d)}(x,z)^{-((t+\deg_{\vec{d}}(\alpha)+\deg_{\vec{d}}(\beta))\vee K)}}{\mathrm{Vol}(B_{(W,d)}(x,\rho_{(W,d)}(x,z)))\wedge 1},$$

where W_x denotes the list of vector fields $W_1,\dots,W_r$ acting as partial differential operators in the x variable, and similarly for W_z.
- (Cancelation condition) For every bounded set of $(W,\vec{d})$ bump functions in Ω, $\mathcal{B} \subset C_0^\infty(\mathfrak{M}) \times \mathfrak{M} \times (0,1]$, and each ordered multi-index α, there exists $C_{\mathcal{B},\alpha} \geq 0$ with

$$\sup_{\substack{(\phi,z,2^{-j})\in\mathcal{B}\\ x\in\mathfrak{M}}} 2^{-j((t+\deg_{\vec{d}}(\alpha))\vee K)}(\mathrm{Vol}(B_{(W,d)}(x,2^{-j}))\wedge 1)|W^\alpha T\phi(x)| \leq C_{\mathcal{B},\alpha}$$

with the same estimate for T^* in place of T.

Remark 5.2.3. A priori, $\mathscr{A}_3^t(\Omega,(W,\vec{d}),K) \subseteq \mathscr{A}_3^t(\Omega,(W,\vec{d}),K+1)$. As a consequence of Theorem 5.2.12, we will see that $\mathscr{A}_3^t(\Omega,(W,\vec{d}),K)$ does not depend on K.

The estimates given in the growth condition of $\mathscr{A}_3^t(\Omega,(W,\vec{d}),K)$ are only sharp when $t+\deg_{\vec{d}}(\alpha)+\deg_{\vec{d}}(\beta) \geq K$: the definition actually implies better estimates than are assumed when $t+\deg_{\vec{d}}(\alpha)+\deg_{\vec{d}}(\beta) < K$. A similar remark holds for the cancelation condition. The next definition makes precise the estimates for lower-order derivatives. For the remainder of this chapter, fix $Q_1 = Q_1(\Omega) \geq n$ as in Corollary 3.3.9.

Definition 5.2.4. For $t \in \mathbb{R}$, we let $\mathscr{A}_4^t(\Omega,(W,\vec{d})) \subseteq \mathrm{Hom}(C_0^\infty(\mathfrak{M}), C_0^\infty(\mathfrak{M})')$ be the set of all $T \in \mathrm{Hom}(C_0^\infty(\mathfrak{M}), C_0^\infty(\mathfrak{M})')$ such that the following hold:
- $\mathrm{supp}(T) \subseteq \Omega\times\Omega$.
- (Growth condition) The Schwartz kernel of T agrees with a C_{loc}^∞ function on $\{(x,z) \in \mathfrak{M}\times\mathfrak{M} : x \neq z\}$ and satisfies, for all ordered multi-indices α and β and all $M \geq 0$ such that $t+\deg_{\vec{d}}(\alpha)+\deg_{\vec{d}}(\beta)+M > -Q_1$,

$$|W_x^\alpha W_z^\beta T(x,z)| \leq C_{\alpha,\beta} \frac{\rho_{(W,d)}(x,z)^{-(t+\deg_{\vec{d}}(\alpha)+\deg_{\vec{d}}(\beta)+M)}}{\mathrm{Vol}(B_{(W,d)}(x,\rho_{(W,d)}(x,z)))\wedge 1}.$$

- (Cancelation condition) For every bounded set of $(W,\vec{d})$ bump functions in Ω, $\mathcal{B} \subset C_0^\infty(\mathfrak{M}) \times \mathfrak{M} \times (0,1]$, each ordered multi-index α, and each $M \geq 0$ such that $t+\deg_{\vec{d}}(\alpha)+M > -Q_1$, there exists $C_{\mathcal{B},\alpha,M} \geq 0$ with

$$\sup_{\substack{(\phi,z,2^{-j})\in\mathcal{B}\\ x\in\mathfrak{M}}} 2^{-j(t+\deg_{\vec{d}}(\alpha)+M)}(\mathrm{Vol}(B_{(W,d)}(x,2^{-j}))\wedge 1)|W^\alpha T\phi(x)| \leq C_{\mathcal{B},\alpha,M}$$

with the same estimate for T^* in place of T.

Remark 5.2.5. In the definition of $\mathscr{A}_4^t(\Omega,(W,\vec{d}))$, because Ω is relatively compact, the conditions become stronger the smaller M is. For example, in the growth condition, if $t+\deg_{\vec{d}}(\alpha)+\deg_{\vec{d}}(\beta) > -Q_1$, then one may simply take $M=0$. In particular, if $t > -Q_1$, one does not need the parameter M at all. A similar remark holds for the cancelation condition.

For the definitions of $\mathscr{A}_1^t$ and $\mathscr{A}_2^t$, we require some additional definitions.

Definition 5.2.6. We say $\mathcal{E} \subset C_0^\infty(\Omega\times\Omega)\times(0,1]$ is a bounded set of $(W, \vec{d})$ pre-elementary operators supported in Ω if the following hold:
- $\bigcup_{(E,2^{-j})\in\mathcal{E}} \operatorname{supp}(E) \Subset \Omega \times \Omega$.
- For all ordered multi-indices α, β, for all $m \in \mathbb{N}$, there exists $C = C(\mathcal{E}, m, \alpha, \beta) \geq 0$ such that for all $(E, 2^{-j}) \in \mathcal{E}$,

$$\left|(2^{-j\vec{d}} W_x)^\alpha (2^{-j\vec{d}} W_z)^\beta E(x,z)\right| \leq C \frac{(1 + 2^j \rho_{(W,\vec{d})}(x,z))^{-m}}{\operatorname{Vol}(B_{(W,\vec{d})}(x, 2^{-j} + \rho_{(W,\vec{d})}(x,z))) \wedge 1}.$$

Remark 5.2.7. Definition 5.2.6 uses the quantity

$$\frac{(1 + 2^j \rho_{(W,\vec{d})}(x,z))^{-m}}{\operatorname{Vol}(B_{(W,\vec{d})}(x, 2^{-j} + \rho_{(W,\vec{d})}(x,z))) \wedge 1}.$$

It is important for what follows that this is essentially symmetric in x and z. Namely, for $x, z \in \Omega$,

$$\frac{(1 + 2^j \rho_{(W,\vec{d})}(x,z))^{-m}}{\operatorname{Vol}(B_{(W,\vec{d})}(x, 2^{-j} + \rho_{(W,\vec{d})}(x,z))) \wedge 1} \approx \frac{(1 + 2^j \rho_{(W,\vec{d})}(z,x))^{-m}}{\operatorname{Vol}(B_{(W,\vec{d})}(z, 2^{-j} + \rho_{(W,\vec{d})}(z,x))) \wedge 1}.$$

See Proposition 5.4.3.

Definition 5.2.8. We define the set of bounded sets of $(W, \vec{d})$ elementary operators supported in Ω, $\mathscr{G}_\Omega$, to be the largest set of subsets of $C_0^\infty(\Omega \times \Omega) \times (0,1]$ such that for all $\mathcal{E} \in \mathscr{G}_\Omega$:
- $\mathcal{E}$ is a bounded set of $(W, \vec{d})$ pre-elementary operators supported in Ω.
- $\forall (E, 2^{-j}) \in \mathcal{E}$,

$$E = \sum_{|\alpha|,|\beta|\leq 1} 2^{-(2-|\alpha|-|\beta|)j} (2^{-j\vec{d}} W)^\alpha E_{\alpha,\beta} (2^{-j\vec{d}} W)^\beta, \tag{5.2}$$

where $\{(E_{\alpha,\beta}, 2^{-j}) : (E, 2^{-j}) \in \mathcal{E}, |\alpha|, |\beta| \leq 1\} \in \mathscr{G}_\Omega$.

For $\mathcal{E} \in \mathscr{G}_\Omega$, we say $\mathcal{E}$ is a bounded set of $(W, \vec{d})$ elementary operators supported in Ω. We say $\mathcal{E}$ is a bounded set of $(W, \vec{d})$ elementary operators if there exists an $\Omega \Subset \mathfrak{M}$ such that $\mathcal{E}$ is a bounded set of $(W, \vec{d})$ elementary operators supported in Ω.

Remark 5.2.9. In Definition 5.2.8 (and in all similar definitions), we have defined $\mathscr{G}_\Omega$ to be the *largest* set satisfying certain axioms. That there is a largest such set follows from the fact that if $\{\mathscr{G}_\Omega^a\}$ is the collection of all sets satisfying the axioms, then $\bigcup_a \mathscr{G}_\Omega^a$ also satisfies the axioms. See Remark 2.3.6 for further comments on this definition.

Definition 5.2.10. For $t \in \mathbb{R}$, we let $\mathscr{A}_1^t(\Omega, (W, \vec{d})) \subseteq \mathrm{Hom}(C_0^\infty(\mathfrak{M}), C_0^\infty(\mathfrak{M})')$ be the set of all $T \in \mathrm{Hom}(C_0^\infty(\mathfrak{M}), C_0^\infty(\mathfrak{M})')$ such that $\mathrm{supp}(T) \subseteq \Omega \times \Omega$ and for every bounded set of $(W, \vec{d})$ elementary operators supported in Ω, $\mathcal{E}$,

$$\{(2^{-jt}TE, 2^{-j}) : (E, 2^{-j}) \in \mathcal{E}\}$$

is a bounded set of $(W, \vec{d})$ elementary operators supported in Ω.

Definition 5.2.11. For $t \in \mathbb{R}$, we let $\mathscr{A}_2^t(\Omega, (W, \vec{d})) \subseteq \mathrm{Hom}(C_0^\infty(\mathfrak{M}), C_0^\infty(\mathfrak{M})')$ be the set of all $T \in \mathrm{Hom}(C_0^\infty(\mathfrak{M}), C_0^\infty(\mathfrak{M})')$ such that there exists a bounded set of $(W, \vec{d})$ elementary operators supported in Ω, $\{(E_j, 2^{-j}) : j \in \mathbb{N}\}$, with $T = \sum_{j \in \mathbb{N}} 2^{jt} E_j$. See Proposition 5.5.10 for a description of the convergence of this sum.

The next theorem is our main theorem concerning the above definitions. It is proved in Section 5.7.

Theorem 5.2.12. *For every $K \in \mathbb{N}$, $t \in \mathbb{R}$,*

$$\mathscr{A}_1^t(\Omega, (W, \vec{d})) = \mathscr{A}_2^t(\Omega, (W, \vec{d})) = \mathscr{A}_3^t(\Omega, (W, \vec{d}), K) = \mathscr{A}_4^t(\Omega, (W, \vec{d})).$$

We denote their common value by $\mathscr{A}^t(\Omega, (W, \vec{d}))$.

Remark 5.2.13. The definition of $\mathscr{A}_4^t(\Omega, (W, \vec{d}))$ depends on Q_1 from Corollary 3.3.9. If Q_1 satisfies the conclusions of Corollary 3.3.9, then any $Q_1' \in [n, Q_1]$ also satisfies the conclusions in place of Q_1. However, the algebra defined by $\mathscr{A}_4^t(\Omega, (W, \vec{d}))$ does not depend on which such Q_1 is used: this follows from Theorem 5.2.12, as none of the other characterizations of $\mathscr{A}^t(\Omega, (W, \vec{d}))$ depend on the choice of Q_1.

5.2.2 The multi-parameter Hörmander setting

We take the setting described in Section 5.1.2. Thus, we are given ν families of Hörmander vector fields with formal degrees $(W^1, \vec{d}^1), \ldots, (W^\nu, \vec{d}^\nu)$ which pairwise locally weakly approximately commute, and we define $(W, \vec{d}) := (W^1, \vec{d}^1) \boxtimes (W^2, \vec{d}^2) \boxtimes \cdots \boxtimes (W^\nu, \vec{d}^\nu)$. Our goal is to define the algebra $\mathscr{A}^t(\Omega, (W, \vec{d}))$, for $t \in \mathbb{R}^\nu$. To do this, we generalize Definitions 5.2.10 and 5.2.11 to this setting. For $j \in \mathbb{R}^\nu$, we write $2^{-j\vec{d}}W$ for $\{2^{-j\cdot \vec{d}_1}W_1, \ldots, 2^{-j\cdot \vec{d}_r}W_r\}$.

For $x \in \mathfrak{M}$, $\delta > 0$, $j \in \mathbb{R}^\nu$, set

$$B_{2^{-j}}(x, \delta) := B_{(W,\vec{d})}(x, \delta 2^{-j}).$$

$B_{2^{-j}}(x, \delta)$ are metric balls, and we let $\rho_{2^{-j}}(x, z)$ be the corresponding metric. Note that $\rho_{2^{-j}}(x, z) = \rho_{(2^{-j\vec{d}}W, |\vec{d}|_1)}(x, z)$, where $|\vec{d}|_1$ denotes the list of single-parameter formal degrees $|\vec{d}_1|_1, \ldots, |\vec{d}_q|_1$.

Definition 5.2.14. We say $\mathcal{E} \subset C_0^\infty(\Omega \times \Omega) \times (0,1]^\nu$ is a bounded set of $(W, \vec{d})$ pre-elementary operators supported in Ω if the following hold:

- $\bigcup_{(E,2^{-j})\in\mathcal{E}} \operatorname{supp}(E) \Subset \Omega \times \Omega$.
- For all ordered multi-indices α, β, for all $m \in \mathbb{N}$, there exists $C = C(\mathcal{E}, \alpha, \beta, m) \geq 0$ such that $\forall (E, 2^{-j}) \in \mathcal{E}$,

$$\left|(2^{-j\vec{d}} W_x)^\alpha (2^{-j\vec{d}} W_z)^\beta E(x,z)\right| \leq C \frac{(1 + \rho_{2^{-j}}(x,z))^{-m}}{\operatorname{Vol}(B_{2^{-j}}(x, 1 + \rho_{2^{-j}}(x,z))) \wedge 1}.$$

Definition 5.2.15. We define the set of bounded sets of $(W, \vec{d})$ elementary operators supported in Ω, $\mathcal{G}_\Omega$, to be the largest set of subsets of $C_0^\infty(\Omega \times \Omega) \times (0,1]^\nu$ such that $\forall \mathcal{E} \in \mathcal{G}_\Omega$:

- $\mathcal{E}$ is a bounded set of $(W, \vec{d})$ pre-elementary operators supported in Ω.
- For all $(E, 2^{-j}) \in \mathcal{E}$, $\mu \in \{1, \ldots, \nu\}$,

$$E = \sum_{|\alpha_\mu|,|\beta_\mu|\leq 1} 2^{-(2-|\alpha_\mu|-|\beta_\mu|)j_\mu} (2^{-j_\mu d^\mu} W^\mu)^{\alpha_\mu} E_{\mu,\alpha_\mu,\beta_\mu} (2^{-j_\mu d^\mu} W^\mu)^{\beta_\mu}, \tag{5.3}$$

  where

$$\{(E_{\mu,\alpha_\mu,\beta_\mu}, 2^{-j}) : (E, 2^{-j}) \in \mathcal{E}, \mu \in \{1, \ldots, \nu\}, |\alpha_\mu|, |\beta_\mu| \leq 1\} \in \mathcal{G}_\Omega.$$

For $\mathcal{E} \in \mathcal{G}_\Omega$, we say $\mathcal{E}$ is a bounded set of $(W, \vec{d})$ elementary operators supported in Ω. We say $\mathcal{E}$ is a bounded set of $(W, \vec{d})$ elementary operators if there exists $\Omega \Subset \mathfrak{M}$ such that $\mathcal{E}$ is a bounded set of $(W, \vec{d})$ elementary operators supported in Ω.

Definition 5.2.16. For $t \in \mathbb{R}^\nu$, we let $\mathscr{A}_1^t(\Omega, (W, \vec{d})) \subseteq \operatorname{Hom}(C_0^\infty(\mathfrak{M}), C_0^\infty(\mathfrak{M})')$ be the set of all $T \in \operatorname{Hom}(C_0^\infty(\mathfrak{M}), C_0^\infty(\mathfrak{M})')$ such that $\operatorname{supp}(T) \subseteq \Omega \times \Omega$ and for every bounded set of $(W, \vec{d})$ elementary operators supported in Ω, $\mathcal{E}$,

$$\{(2^{-j\cdot t} TE, 2^{-j}) : (E, 2^{-j}) \in \mathcal{E}\}$$

is a bounded set of $(W, \vec{d})$ elementary operators supported in Ω.

Definition 5.2.17. For $t \in \mathbb{R}^\nu$, we let $\mathscr{A}_2^t(\Omega, (W, \vec{d})) \subseteq \operatorname{Hom}(C_0^\infty(\mathfrak{M}), C_0^\infty(\mathfrak{M})')$ be the set of all $T \in \operatorname{Hom}(C_0^\infty(\mathfrak{M}), C_0^\infty(\mathfrak{M})')$ such that there exists a bounded set of $(W, \vec{d})$ elementary operators supported in Ω, $\{(E_j, 2^{-j}) : j \in \mathbb{N}^\nu\}$, with $T = \sum_{j\in\mathbb{N}^\nu} 2^{j\cdot t} E_j$. See Proposition 5.5.10 for a description of the convergence of this sum.

The next theorem is our main theorem concerning the above definitions. It is proved in Section 5.7.

Theorem 5.2.18. *For every $t \in \mathbb{R}^\nu$,*

$$\mathscr{A}_1^t(\Omega, (W, \vec{d})) = \mathscr{A}_2^t(\Omega, (W, \vec{d})).$$

We denote their common value by $\mathscr{A}^t(\Omega, (W, \vec{d}))$.

Remark 5.2.19. It is straightforward to see that when $\nu = 1$, all of the above definitions coincide with the similar definitions in Section 5.2.1. This uses the fact that in the case $\nu = 1$, $2^j \rho_{(W,\vec{d})}(x,z) = \rho_{2^{-j}}(x,z)$ and $B_{(W,\vec{d})}(x, 2^{-j}\delta) = B_{2^{-j}}(x,\delta)$; these follow directly from the definitions. In particular, $\frac{(1+2^j\rho_{(W,\vec{d})}(x,z))^{-m}}{\mathrm{Vol}(B_{(W,\vec{d})}(x,2^{-j}+\rho_{(W,\vec{d})}(x,z)))\wedge 1} = \frac{(1+\rho_{2^{-j}}(x,z))^{-m}}{\mathrm{Vol}(B_{2^{-j}}(x,1+\rho_{2^{-j}}(x,z)))\wedge 1}$. Thus, the definitions here generalize the classical Calderón–Zygmund paradigm to this multi-parameter setting.

Remark 5.2.20. Definitions 5.2.16 and 5.2.17 generalize the definitions of $\mathscr{A}_1^t$ and $\mathscr{A}_2^t$ from Section 5.2.1 to the multi-parameter setting. However, we did not present conditions analogous to the more familiar $\mathscr{A}_3^t$ and $\mathscr{A}_4^t$ from Definitions 5.2.2 and 5.2.4. It is possible to do this to a certain extent; however, the definitions are a bit complicated and not useful in our applications. For a presentation of this in a more general context, see [220, Chapter 5].

5.2.3 The general multi-parameter setting

We take the setting described in Section 5.1.3. Thus, we are given ν sets of vector fields with single-parameter formal degrees, $(W^1, \vec{d}^1), \dots, (W^\nu, \vec{d}^\nu)$, which pairwise locally weakly approximately commute and such that $\mathrm{Gen}((W^\mu, \vec{d}^\mu))$ is locally finitely generated for each $\mu \in \{1, \dots, \nu\}$. We define $(W, \vec{d}) := (W^1, \vec{d}^1) \boxtimes (W^2, \vec{d}^2) \boxtimes \cdots \boxtimes (W^\nu, \vec{d}^\nu)$. Our goal is to define the algebra $\widetilde{\mathscr{A}}^t(\Omega, (W, \vec{d}))$. To do this, we present analogs of Definitions 5.2.16 and 5.2.17 in this more general setting. For $j \in \mathbb{R}^\nu$, we write $2^{-j\vec{d}}W$ for $\{2^{-j\cdot\vec{d}_1}W_1, \dots, 2^{-j\cdot\vec{d}_r}W_r\}$.

Definition 5.2.21. For $L \in \mathbb{N}$ and $U \subseteq \mathfrak{M}$ open, define the vector space

$$C_W^L(U) := \{f \in C(U) : W^\alpha f \in C(U), \forall |\alpha| \le L\},$$

with norm

$$\|f\|_{C_W^L(U)} := \sup_{x\in U} \sum_{|\alpha|\le L} |W^\alpha f(x)|.$$

With this norm, $C_W^L(U)$ is a Banach space.

Definition 5.2.22. We set

$$C_{W,\mathrm{loc}}^\infty(\mathfrak{M}) := \{f : \mathfrak{M} \to \mathbb{C} \mid f\big|_U \in C_W^L(U), \forall U \Subset \mathfrak{M} \text{ open}, \forall L \in \mathbb{N}\}.$$

We give $C_{W,\mathrm{loc}}^\infty(\mathfrak{M})$ the usual projective limit topology (similar to [238, Chapter 10, Example I]), making $C_{W,\mathrm{loc}}^\infty(\mathfrak{M})$ a Fréchet space.

Definition 5.2.23. We set

$$C^\infty_{W,0}(\mathfrak{M}) := \{f \in C^\infty_{W,\mathrm{loc}}(\mathfrak{M}) : \operatorname{supp}(f) \text{ is compact}\}.$$

We give $C^\infty_{W,0}(\mathfrak{M})$ the usual inductive limit topology (similar to [238, Chapter 13, Example II]), making $C^\infty_{W,0}(\mathfrak{M})$ an LF space.

Remark 5.2.24. We have $C^\infty_{\mathrm{loc}}(\mathfrak{M}) \hookrightarrow C^\infty_{W,\mathrm{loc}}(\mathfrak{M})$ and $C^\infty_0(\mathfrak{M}) \hookrightarrow C^\infty_{W,0}(\mathfrak{M})$, where the inclusions are continuous. If $W_1, \dots, W_r$ satisfy Hörmander's condition, then $C^\infty_{W,\mathrm{loc}}(\mathfrak{M}) = C^\infty_{\mathrm{loc}}(\mathfrak{M})$ and $C^\infty_{W,0}(\mathfrak{M}) = C^\infty_0(\mathfrak{M})$ with equality of topologies.

Remark 5.2.25. By Remark 5.2.24, $\operatorname{Hom}(C^\infty_{W,0}(\mathfrak{M}), C^\infty_{W,0}(\mathfrak{M})') \subseteq \operatorname{Hom}(C^\infty_0(\mathfrak{M}), C^\infty_0(\mathfrak{M})')$, and therefore we identify elements of $\operatorname{Hom}(C^\infty_{W,0}(\mathfrak{M}), C^\infty_{W,0}(\mathfrak{M})')$ with their Schwartz kernels.

Definition 5.2.26. We say $\mathcal{E} \subset \operatorname{Hom}(C^\infty_{W,0}(\mathfrak{M}), C^\infty_{W,\mathrm{loc}}(\mathfrak{M})) \times (0,1]^\nu$ is a bounded set of generalized $(W, \vec{d})$ pre-elementary operators supported in Ω if the following hold:

- $\bigcup_{(E,2^{-j})\in\mathcal{E}} \operatorname{supp}(E) \Subset \Omega \times \Omega$.
- For all ordered multi-indices α, β the operator

$$(2^{-j\vec{d}}W)^\alpha E(2^{-j\vec{d}}W)^\beta \tag{5.4}$$

  extends to a bounded operator $L^1(\mathfrak{M}, \mathrm{Vol}) \to L^1(\mathfrak{M}, \mathrm{Vol})$, and

$$\sup_{(E,2^{-j})\in\mathcal{E}} \|(2^{-j\vec{d}}W)^\alpha E(2^{-j\vec{d}}W)^\beta\|_{L^1(\mathfrak{M},\mathrm{Vol})\to L^1(\mathfrak{M},\mathrm{Vol})} < \infty. \tag{5.5}$$

  We henceforth identify (5.4) with this extended operator. Note that since $\operatorname{supp}(E) \Subset \mathfrak{M} \times \mathfrak{M}$, it follows that $(2^{-j\vec{d}}W)^\alpha E(2^{-j\vec{d}}W)^\beta f$ is defined for all $f \in L^1_{\mathrm{loc}}(\mathfrak{M}, \mathrm{Vol})$.
- For all ordered multi-indices α, β,

$$\sup_{(E,2^{-j})\in\mathcal{E}} \|(2^{-j\vec{d}}W)^\alpha E(2^{-j\vec{d}}W)^\beta\|_{L^\infty(\mathfrak{M},\mathrm{Vol})\to L^\infty(\mathfrak{M},\mathrm{Vol})} < \infty. \tag{5.6}$$

- For all ordered multi-indices α, β and for every countable set $\mathcal{E}' := \{(E_k, \delta_k) : k \in \mathbb{N}\} \subseteq \mathcal{E}$, set

$$\mathcal{T}_{\mathcal{E}',\alpha,\beta}\{f_k\}_{k\in\mathbb{N}} := \{(\delta_k^{\vec{d}}W)^\alpha E_k(\delta_k^{\vec{d}}W)^\beta\}_{k\in\mathbb{N}}. \tag{5.7}$$

  Then, for all $p \in (1,\infty)$, $q \in (1,\infty]$,

$$\sup_{\mathcal{E}'} \|\mathcal{T}_{\mathcal{E}',\alpha,\beta}\|_{L^p(\mathfrak{M},\mathrm{Vol};\ell^q(\mathbb{N}))\to L^p(\mathfrak{M},\mathrm{Vol};\ell^q(\mathbb{N}))} < \infty,$$

  where the supremum is taken over all such countable subsets $\mathcal{E}' \subseteq \mathcal{E}$.

- For $(E,2^{-j}) \in \mathcal{E}$, we assume E^* initially defined on $C^\infty_{W,\mathrm{loc}}(\mathfrak{M})'$ restricts to an operator in $\mathrm{Hom}(C^\infty_{W,0}(\mathfrak{M}), C^\infty_{W,\mathrm{loc}}(\mathfrak{M}))$. Furthermore, set

$$\begin{aligned}\mathcal{E}^* &:= \{(E^*,2^{-j}) : (E,2^{-j}) \in \mathcal{E}\}\\ &\subseteq \mathrm{Hom}(C^\infty_{W,0}(\mathfrak{M}), C^\infty_{W,\mathrm{loc}}(\mathfrak{M})) \times (0,1]^\nu.\end{aligned}$$

We assume that all of the above also holds with $\mathcal{E}$ replaced by $\mathcal{E}^*$.

Definition 5.2.27. We define the set of bounded sets of generalized $(W,\vec{d})$ elementary operators supported in Ω, $\widetilde{\mathcal{G}}_\Omega$, to be the largest set of subsets of $\mathrm{Hom}(C^\infty_{W,0}(\mathfrak{M}), C^\infty_{W,\mathrm{loc}}(\mathfrak{M}))$ such that $\forall \mathcal{E} \in \widetilde{\mathcal{G}}_\Omega$:

- $\mathcal{E}$ is a bounded set of generalized $(W,\vec{d})$ pre-elementary operators supported in Ω.
- For all $(E,2^{-j}) \in \mathcal{E}$, $\mu \in \{1,\dots,\nu\}$,

$$E = \sum_{|\alpha_\mu|,|\beta_\mu|\le 1} 2^{-(2-|\alpha_\mu|-|\beta_\mu|)j_\mu} (2^{-j_\mu d^\mu} W^\mu)^{\alpha_\mu} E_{\mu,\alpha_\mu,\beta_\mu} (2^{-j_\mu d^\mu} W^\mu)^{\beta_\mu},$$

where

$$\{(E_{\mu,\alpha_\mu,\beta_\mu}, 2^{-j}) : (E,2^{-j}) \in \mathcal{E}, \mu \in \{1,\dots,\nu\}, |\alpha_\mu|, |\beta_\mu| \le 1\} \in \widetilde{\mathcal{G}}_\Omega.$$

For $\mathcal{E} \in \widetilde{\mathcal{G}}_\Omega$, we say $\mathcal{E}$ is a bounded set of generalized $(W,\vec{d})$ elementary operators supported in Ω. We say $\mathcal{E}$ is a bounded set of generalized $(W,\vec{d})$ elementary operators if there exists $\Omega \Subset \mathfrak{M}$ such that $\mathcal{E}$ is a bounded set of generalized $(W,\vec{d})$ elementary operators supported in Ω.

Definition 5.2.28. For $t \in \mathbb{R}^\nu$, we let $\widetilde{\mathcal{A}}^t_1(\Omega,(W,\vec{d})) \subseteq \mathrm{Hom}(C^\infty_{W,0}(\mathfrak{M}), C^\infty_{W,0}(\mathfrak{M})')$ be the set of all $T \in \mathrm{Hom}(C^\infty_{W,0}(\mathfrak{M}), C^\infty_{W,0}(\mathfrak{M})')$ such that $\mathrm{supp}(T) \subseteq \Omega\times\Omega$ and for every bounded set of generalized $(W,\vec{d})$ elementary operators supported in Ω, $\mathcal{E}$,

$$\{(2^{-j\cdot t}TE, 2^{-j}) : (E,2^{-j}) \in \mathcal{E}\}$$

is a bounded set of generalized $(W,\vec{d})$ elementary operators supported in Ω.

Definition 5.2.29. For $t \in \mathbb{R}^\nu$, we let $\widetilde{\mathcal{A}}^t_2(\Omega,(W,\vec{d})) \subseteq \mathrm{Hom}(C^\infty_{W,0}(\mathfrak{M}), C^\infty_{W,0}(\mathfrak{M})')$ be the set of all $T \in \mathrm{Hom}(C^\infty_{W,0}(\mathfrak{M}), C^\infty_{W,0}(\mathfrak{M})')$ such that there exists a bounded set of generalized $(W,\vec{d})$ elementary operators supported in Ω, $\{(E_j,2^{-j}) : j \in \mathbb{N}^\nu\}$, with $T = \sum_{j\in\mathbb{N}^\nu} 2^{j\cdot t}E_j$. See Proposition 5.5.10 for a description of the convergence of this sum.

The next theorem is our main theorem concerning the above definitions. It is proved in Section 5.7.

Theorem 5.2.30. *For every* $t \in \mathbb{R}^\nu$,

$$\widetilde{\mathcal{A}}^t_1(\Omega,(W,\vec{d})) = \widetilde{\mathcal{A}}^t_2(\Omega,(W,\vec{d})).$$

We denote their common value by $\widetilde{\mathcal{A}}^t(\Omega,(W,\vec{d}))$.

Remark 5.2.31. The main difference between the definitions in this section and those in Section 5.2.2 is the definition of bounded sets of $(W, \vec{d})$ pre-elementary operators (Definition 5.2.14) versus the definition of bounded sets of generalized $(W, \vec{d})$ pre-elementary operators (Definition 5.2.26). Other than this one difference, the rest of the definitions are formally the same. Because of this, many of our proofs in the multi-parameter Hörmander setting will be formally the same as proofs of corresponding results in the general multi-parameter setting.

5.3 Notation

We will prove similar results in all three settings described in this chapter: the single-parameter Hörmander setting described in Section 5.1.1, the multi-parameter Hörmander setting described in Section 5.1.2, and the general multi-parameter setting described in Section 5.1.3. Most of our results in the single-parameter Hörmander setting are a special case of the results in the multi-parameter Hörmander setting. The results we prove in the multi-parameter Hörmander setting and the general multi-parameter setting are similar and have similar proofs, so we study these two cases simultaneously.

Thus, in this chapter, we will refer to two settings: the **Hörmander setting** (short for the multi-parameter Hörmander setting of Section 5.1.2) and the **general setting** (short for the multi-parameter general setting of Section 5.1.3).

In either setting, we are given $\nu \in \mathbb{N}_+$ and ν lists of vector fields with single-parameter formal degrees:

$$(W^\mu, d^\mu) = \{(W_1^\mu, d_1^\mu), \dots, (W_{r_\mu}^\mu, d_{r_\mu}^\mu)\} \subset C^\infty_{\mathrm{loc}}(\mathfrak{M}; T\mathfrak{M}) \times \mathbb{N}_+.$$

The assumptions in the two settings are the following.

Hörmander setting: We assume that each (W^μ, d^μ) are Hörmander vector fields with formal degrees and that $(W^1, d^1), \dots, (W^\nu, d^\nu)$ pairwise locally weakly approximately commute.

General setting: We assume that for each $\mu \in \{1, \dots, \nu\}$, $\mathrm{Gen}((W^\mu, d^\mu))$ is locally finitely generated on $\mathfrak{M}$ and that $(W^1, d^1), \dots, (W^\nu, d^\nu)$ pairwise locally weakly approximately commute.

Remark 5.1.2 shows that the Hörmander setting is a special case of the general setting.

In either setting, we define

$$\begin{aligned}(W, \vec{d}) = \{(W_1, \vec{d}_1), \dots, (W_r, \vec{d}_r)\} &:= (W^1, d^1) \boxtimes (W^2, d^2) \boxtimes \cdots \boxtimes (W^\nu, d^\nu)\\ &= \{(W_j^\mu, d_j^\mu \mathfrak{e}_\mu) : \mu \in \{1, \dots, \nu\}, 1 \le j \le r_\mu\} \subset C^\infty_{\mathrm{loc}}(\mathfrak{M}; T\mathfrak{M}) \times (\mathbb{N}^\nu \setminus \{0\}).\end{aligned}$$

When we consider the Hörmander setting, we will consider objects like bounded sets of $(W, \vec{d})$ pre-elementary operators supported in Ω (Definition 5.2.14), bounded sets of $(W, \vec{d})$ elementary operators supported in Ω (Definition 5.2.15), and $\mathscr{A}^t(\Omega, (W, \vec{d}))$. In the general setting, we will consider objects like bounded sets of generalized $(W, \vec{d})$ pre-elementary operators supported in Ω (Definition 5.2.26), bounded sets of generalized $(W, \vec{d})$ elementary operators supported in Ω (Definition 5.2.27), and $\widetilde{\mathscr{A}}^t(\Omega, (W, \vec{d}))$.

5.4 Pre-elementary operators

In this section we work in the Hörmander setting (see Section 5.3) and prove some results we need concerning bounded sets of $(W, \vec{d})$ pre-elementary operators (Definition 5.2.14). Bounded sets of $(W, \vec{d})$ pre-elementary operators are based on the quantity

$$\frac{(1 + \rho_{2^{-j}}(x, z))^{-m}}{\operatorname{Vol}(B_{2^{-j}}(x, 1 + \rho_{2^{-j}}(x, z))) \wedge 1}, \tag{5.8}$$

and many of the results of this section focus on a better understanding of this quantity.

We will sometimes allow some of the components of j in (5.8) to be equal to ∞, i. e., we will allow some of the components of 2^{-j} to be equal to 0. More precisely, we will consider $j \in (-\infty, \infty]^\nu \setminus \{(\infty, \dots, \infty)\}$. The definition of (5.8) in this more general setting is the same when some of the components of 2^{-j} equal 0. This corresponds to considering a setting with a lower number of parameters, where ν denotes the number of parameters, as the next lemma shows.

Lemma 5.4.1. *Let $j \in (-\infty, \infty]^\nu \setminus \{(\infty, \dots, \infty)\}$ and let $E = \{\mu \in \{1, \dots, \nu\} : j_\mu \neq \infty\}$. Let $j_E \in \mathbb{R}^{|E|}$ be the vector consisting of just the finite components of j. Then*

$$B_{2^{-j}}(x, \delta) = \widetilde{B}_{2^{-j_E}}(x, \delta), \tag{5.9}$$

$$\rho_{2^{-j}}(x, y) = \tilde{\rho}_{2^{-j_E}}(x, y), \tag{5.10}$$

$$\frac{(1 + \rho_{2^{-j}}(x, z))^{-m}}{\operatorname{Vol}(B_{2^{-j}}(x, 1 + \rho_{2^{-j}}(x, z))) \wedge 1} = \frac{(1 + \tilde{\rho}_{2^{-j_E}}(x, z))^{-m}}{\operatorname{Vol}(\widetilde{B}_{2^{-j_E}}(x, 1 + \tilde{\rho}_{2^{-j_E}}(x, z))) \wedge 1}, \tag{5.11}$$

where $\tilde{\rho}_{2^{-j_E}}$ and $\widetilde{B}_{2^{-j_E}}$ are obtained by replacing $(W, \vec{d})$ with $(W^E, \vec{d}^E)$ in the definition of $\rho_{2^{-j}}$ and $B_{2^{-j}}$, where $(W^E, \vec{d}^E)$ is as described in Section 3.5.2.

Proof. Indeed, directly from the definitions, since $(\delta 2^{-j})_\mu = 0$ for all $\mu \notin E$, we have

$$B_{2^{-j}}(x, \delta) = B_{(W, \vec{d})}(x, \delta 2^{-j}) = B_{(W^E, \vec{d}^E)}(x, \delta 2^{-j_E}) = \widetilde{B}_{2^{-j_E}}(x, \delta),$$

proving (5.9). Since $\rho_{2^{-j}}$ is the metric associated with the balls $B_{2^{-j}}$ and $\tilde{\rho}_{2^{-j_E}}$ is the metric associated with the balls $B_{2^{-j_E}}$, (5.10) follows from (5.9). Finally, (5.11) follows by combining (5.9) and (5.10). □

Remark 5.4.2. For $\emptyset \neq E \subseteq \{1,\dots,\nu\}$, $(W^E, \vec{d}^E)$ satisfies all the same hypotheses as $(W, \vec{d})$. Thus, in light of Lemma 5.4.1, many of the results in this section in the general case $j \in (-\infty,\infty]^\nu \setminus \{(\infty,\dots,\infty)\}$ follow from the special case $j \in \mathbb{R}^\nu$.

Proposition 5.4.3. *For $x, z \in \Omega$, $j \in (-\infty,\infty]^\nu \setminus \{(\infty,\dots,\infty)\}$, $m \in \mathbb{N}$,*

$$\frac{(1+\rho_{2^{-j}}(x,z))^{-m}}{\operatorname{Vol}(B_{2^{-j}}(x, 1+\rho_{2^{-j}}(x,z))) \wedge 1} \approx \frac{(1+\rho_{2^{-j}}(z,x))^{-m}}{\operatorname{Vol}(B_{2^{-j}}(z, 1+\rho_{2^{-j}}(z,x))) \wedge 1}, \tag{5.12}$$

where the implicit constants depend only on Ω and $(W, \vec{d})$.

Proof. By Remark 5.4.2, it suffices to prove the result when $j \in \mathbb{R}^\nu$. Since $\rho_{2^{-j}}(x,z) = \rho_{2^{-j}}(z,x)$, we only need to consider the denominator of both sides of (5.12). Using Theorem 3.5.4 (n), we have

$$\begin{aligned}
\operatorname{Vol}(B_{2^{-j}}(x, 1+\rho_{2^{-j}}(x,z))) \wedge 1 &\leq \operatorname{Vol}(B_{2^{-j}}(z, 2(1+\rho_{2^{-j}}(x,z)))) \wedge 1 \\
&= \operatorname{Vol}(B_{(W,\vec{d})}(x, 2(1+\rho_{2^{-j}}(x,z))2^{-j})) \wedge 1 \\
&\lesssim \operatorname{Vol}(B_{(W,\vec{d})}(x, (1+\rho_{2^{-j}}(x,z))2^{-j})) \wedge 1 \\
&= \operatorname{Vol}(B_{2^{-j}}(z, 1+\rho_{2^{-j}}(z,x))) \wedge 1.
\end{aligned}$$

By reversing the roles of x and z, we also have

$$\operatorname{Vol}(B_{2^{-j}}(z, 1+\rho_{2^{-j}}(z,x))) \wedge 1 \lesssim \operatorname{Vol}(B_{2^{-j}}(x, 1+\rho_{2^{-j}}(x,z))) \wedge 1,$$

completing the proof. □

Proposition 5.4.4. *For all $m \in \mathbb{N}$, there exist $m' = m'(m, \Omega, (W, \vec{d})) \in \mathbb{N}$ and $C = C(m, \Omega, (W, \vec{d})) \geq 0$ such that $\forall x, z \in \Omega$, $j, k \in (-\infty,\infty]^\nu \setminus \{(\infty,\dots,\infty)\}$,*

$$\begin{aligned}
&\int_\Omega \left(\frac{(1+\rho_{2^{-j}}(x,y))^{-m'}}{\operatorname{Vol}(B_{2^{-j}}(x, 1+\rho_{2^{-j}}(x,y))) \wedge 1} \right) \left(\frac{(1+\rho_{2^{-k}}(y,z))^{-m'}}{\operatorname{Vol}(B_{2^{-k}}(y, 1+\rho_{2^{-k}}(y,z))) \wedge 1} \right) d\operatorname{Vol}(y) \\
&\quad \leq C \frac{(1+\rho_{2^{-j\wedge k}}(x,z))^{-m}}{\operatorname{Vol}(B_{2^{-j\wedge k}}(x, 1+\rho_{2^{-j\wedge k}}(x,z))) \wedge 1}.
\end{aligned}$$

The proof of Proposition 5.4.4 requires several preliminary results, which will also be useful for some other purposes.

Lemma 5.4.5. *For all $x, z \in \mathfrak{M}$ and $j, k \in \mathbb{R}^\nu$, we have $2^{-|k-j|}\rho_{2^{-j}}(x,z) \leq \rho_{2^{-k}}(x,z)$.*

Proof. Directly from the definition, we have

$$2^{-|k-j|}\rho_{2^{-j}}(x,z) = \rho_{2^{|k-j|}2^{-j}}(x,z) \leq \rho_{2^{-k}}(x,z),$$

where the inequality follows from the fact that $2^{|k-j|}2^{-j} \geq 2^{-k}$, coordinatewise. □

Lemma 5.4.6. *There exists $C = C(\Omega, (W, \vec{d})) \geq 0$ such that for all $x, z \in \Omega$ and $j, k \in \mathbb{R}^\nu$, we have*

$$2^{-C|k-j|}(\mathrm{Vol}(B_{2^{-j}}(x, 1 + \rho_{2^{-j}}(x, z))) \wedge 1) \leq \mathrm{Vol}(B_{2^{-k}}(x, 1 + \rho_{2^{-k}}(x, z))) \wedge 1.$$

Proof. By Theorem 3.5.4 (n), there is a constant $D = D(\Omega, (W, \vec{d})) \geq 0$ such that for all $x \in \Omega$ and $r \in (0, \infty)^\nu$,

$$\mathrm{Vol}(B_{(W,\vec{d})}(x, 2r)) \wedge 1 \leq 2^D(\mathrm{Vol}(B_{(W,\vec{d})}(x, r)) \wedge 1). \tag{5.13}$$

We have, using Lemma 5.4.5 and (5.13),

$$\begin{aligned}\mathrm{Vol}(B_{2^{-j}}(x, 1 + \rho_{2^{-j}}(x, z))) \wedge 1 &= \mathrm{Vol}(B_{(W,\vec{d})}(x, (1 + \rho_{2^{-j}}(x, z))2^{-j})) \wedge 1\\ &\leq \mathrm{Vol}(B_{(W,\vec{d})}(x, (1 + 2^{|k-j|}\rho_{2^{-k}}(x, z))2^{|k-j|}2^{-k})) \wedge 1\\ &\leq \mathrm{Vol}(B_{(W,\vec{d})}(x, (1 + \rho_{2^{-k}}(x, z))2^{2|k-j|}2^{-k})) \wedge 1\\ &\leq 2^{2D|k-j|}(\mathrm{Vol}(B_{(W,\vec{d})}(x, (1 + \rho_{2^{-k}}(x, z))2^{-k})) \wedge 1)\\ &= 2^{2D|k-j|}(\mathrm{Vol}(B_{2^{-k}}(x, 1 + \rho_{2^{-k}}(x, z))) \wedge 1).\end{aligned}$$

Taking $C = 2D$ completes the proof. □

Lemma 5.4.7. *For all $m \in \mathbb{N}$, there exists $M = M(m, (W, \vec{d}), \Omega) \geq 0$ such that for all $x, z \in \Omega$ and $j, k \in \mathbb{R}^\nu$, we have*

$$2^{-M|j-k|}\frac{(1 + \rho_{2^{-k}}(x, z))^{-m}}{\mathrm{Vol}(B_{2^{-k}}(x, 1 + \rho_{2^{-k}}(x, z))) \wedge 1} \leq \frac{(1 + \rho_{2^{-j}}(x, z))^{-m}}{\mathrm{Vol}(B_{2^{-j}}(x, 1 + \rho_{2^{-j}}(x, z))) \wedge 1}.$$

Proof. We use Lemmas 5.4.5 and 5.4.6 to see that for $M = M(m, (W, \vec{d}), \Omega) > 0$ sufficiently large,

$$\begin{aligned}&2^{-M|j-k|}\frac{(1 + \rho_{2^{-k}}(x, z))^{-m}}{\mathrm{Vol}(B_{2^{-k}}(x, 1 + \rho_{2^{-k}}(x, z))) \wedge 1}\\ &\quad\leq \frac{(1 + 2^{|k-j|}\rho_{2^{-k}}(x, z))^{-m}}{2^{(M-m)|j-k|}(\mathrm{Vol}(B_{2^{-k}}(x, 1 + \rho_{2^{-k}}(x, z))) \wedge 1)}\\ &\quad\leq \frac{(1 + \rho_{2^{-j}}(x, z))^{-m}}{\mathrm{Vol}(B_{2^{-j}}(x, 1 + \rho_{2^{-j}}(x, z))) \wedge 1}.\end{aligned}$$

□

Lemma 5.4.8. *For all $m \in \mathbb{N}$, there exists $M = M(m, (W, \vec{d}), \Omega) \geq 0$ such that $\forall j \in \mathbb{R}^\nu$,*

$$\sum_{k \in \mathbb{Z}^\nu} 2^{-M|j-k|}\frac{(1 + \rho_{2^{-k}}(x, z))^{-m}}{\mathrm{Vol}(B_{2^{-k}}(x, 1 + \rho_{2^{-k}}(x, z))) \wedge 1} \leq \nu 2^\nu \frac{(1 + \rho_{2^{-j}}(x, z))^{-m}}{\mathrm{Vol}(B_{2^{-j}}(x, 1 + \rho_{2^{-j}}(x, z))) \wedge 1}.$$

Proof. We prove the result with M replaced by $M + 1$, where M is as in Lemma 5.4.7. Using Lemma 5.4.7, we have

$$\sum_{k\in\mathbb{Z}^\nu} 2^{-(M+1)|j-k|} \frac{(1+\rho_{2^{-k}}(x,z))^{-m}}{\operatorname{Vol}(B_{2^{-k}}(x,1+\rho_{2^{-k}}(x,z)))\wedge 1} \le \frac{(1+\rho_{2^{-j}}(x,z))^{-m}}{\operatorname{Vol}(B_{2^{-j}}(x,1+\rho_{2^{-j}}(x,z)))\wedge 1} \sum_{k\in\mathbb{Z}^\nu} 2^{-|j-k|}.$$

The result follows. □

In the rest of this section, for $j,k \in (-\infty,\infty]^\nu$, when we compute $|j-k|$ we use the convention $\infty-\infty = 0$. Whenever we compute $|j-k|$, j and k will be infinite in the same components, and because of our convention, $|j-k|$ will always be a finite number.

Lemma 5.4.9. *Fix $\emptyset \neq E \subseteq \{1,\dots,\nu\}$. Set $\mathcal{I}_E := \{k \in (\mathbb{Z}\cup\{\infty\})^\nu : k_\mu \neq \infty \text{ iff } \mu \in E\}$. For all $m \in \mathbb{N}$, there exists $M = M(m,(W,\vec{d}),\Omega) \ge 0$, such that for all $D > 0$, there exists $C_1 = C_1(D,m,(W,\vec{d}),\Omega) \ge 0$, such that $\forall x,z \in \Omega$, $\forall j \in (-\infty,\infty]^\nu$ with $j_\mu \neq \infty$ iff $\mu \in E$,*

$$\sum_{k\in\mathcal{I}_E} 2^{-M|k-j|} \frac{\chi_{\{\rho_{2^{-k}}(x,z)<D\}}}{\operatorname{Vol}(B_{2^{-k}}(x,1))\wedge 1} \le C_1 \frac{(1+\rho_{2^{-j}}(x,z))^{-m}}{\operatorname{Vol}(B_{2^{-j}}(x,1+\rho_{2^{-j}}(x,z)))\wedge 1}. \tag{5.14}$$

Conversely, for all $m \in \mathbb{N}$ and $D > 0$, there exists $C_2 = C_2(m,D,(W,\vec{d}),\Omega) \ge 0$ such that $\forall x,z \in \Omega$, $\forall j \in \mathbb{R}^\nu$,

$$\frac{(1+\rho_{2^{-j}}(x,z))^{-m}}{\operatorname{Vol}(B_{2^{-j}}(x,1+\rho_{2^{-j}}(x,z)))\wedge 1} \le C_2 \sum_{\substack{k\le j\\ k\in\mathcal{I}_E}} 2^{-(m/\sqrt{\nu})|k-j|} \frac{\chi_{\{\rho_{2^{-k}}(x,z)<D\}}}{\operatorname{Vol}(B_{2^{-k}}(x,1))\wedge 1}. \tag{5.15}$$

Proof. In light of Lemma 5.4.1 (see Remark 5.4.2), it suffices to consider only the case $E = \{1,\dots,\nu\}$. Henceforth, we assume $E = \{1,\dots,\nu\}$ and therefore $\mathcal{I}_E = \mathbb{Z}^\nu$.

We begin with (5.14). Note that by Theorem 3.5.4 (n) we have, $\forall x,z\in\Omega$,

$$\begin{aligned}\frac{\chi_{\{\rho_{2^{-k}}(x,z)<D\}}}{\operatorname{Vol}(B_{2^{-k}}(x,1))\wedge 1} &\lesssim \frac{\chi_{\{\rho_{2^{-k}}(x,z)<D\}}}{\operatorname{Vol}(B_{2^{-k}}(x,1+\rho_{2^{-k}}(x,z)))\wedge 1}\\ &\lesssim \frac{(1+\rho_{2^{-k}}(x,z))^{-m}}{\operatorname{Vol}(B_{2^{-k}}(x,1+\rho_{2^{-k}}(x,z)))\wedge 1},\end{aligned} \tag{5.16}$$

where the implicit constants depend on m and D. Using (5.16) and taking $M = M(m,(W,\vec{d}),\Omega) \ge 0$ as in Lemma 5.4.8, we have

$$\begin{aligned}\sum_{k\in\mathbb{Z}^\nu} 2^{-M|k-j|} \frac{\chi_{\{\rho_{2^{-k}}(x,z)<D\}}}{\operatorname{Vol}(B_{2^{-k}}(x,1))\wedge 1} &\lesssim \sum_{k\in\mathbb{Z}^\nu} 2^{-M|k-j|} \frac{(1+\rho_{2^{-k}}(x,z))^{-m}}{\operatorname{Vol}(B_{2^{-k}}(x,1+\rho_{2^{-k}}(x,z)))\wedge 1}\\ &\lesssim \frac{(1+\rho_{2^{-j}}(x,z))^{-m}}{\operatorname{Vol}(B_{2^{-j}}(x,1+\rho_{2^{-j}}(x,z)))\wedge 1},\end{aligned}$$

proving (5.14).

We turn to (5.15). Set $l := (\lfloor j_1 \rfloor, \dots, \lfloor j_\nu \rfloor) \in \mathbb{Z}^\nu$. By Lemma 5.4.7,

$$\frac{(1+\rho_{2^{-j}}(x,z))^{-m}}{\mathrm{Vol}(B_{2^{-j}}(x,1+\rho_{2^{-j}}(x,z)))\wedge 1} \approx \frac{(1+\rho_{2^{-l}}(x,z))^{-m}}{\mathrm{Vol}(B_{2^{-l}}(x,1+\rho_{2^{-l}}(x,z)))\wedge 1},$$

so it suffices to prove (5.15) with j replaced by l.

If $\rho_{2^{-l}}(x,z) < D$, then by Theorem 3.5.4 (n) we have

$$\frac{(1+\rho_{2^{-l}}(x,z))^{-m}}{\mathrm{Vol}(B_{2^{-l}}(x,1+\rho_{2^{-l}}(x,z)))\wedge 1} \approx \frac{\chi_{\{\rho_{2^{-l}}(x,z)<D\}}}{\mathrm{Vol}(B_{2^{-l}}(x,1))\wedge 1},$$

and (5.15) follows.

Otherwise, $\rho_{2^{-l}}(x,z) > D$. Pick $p \in \mathbb{N}_+$ such that $2^p(D/2) \le \rho_{2^{-l}}(x,z) < 2^{p+1}(D/2)$. Set $l' = (l_1-p, l_2-p, \dots, l_\nu-p) \in \mathbb{Z}^\nu$. Note that $l' \le l$, $|j-l'| \le \sqrt{\nu}p$, and $\rho_{2^{-l'}}(x,z) = 2^{-p}\rho_{2^{-l}}(x,z)$ and therefore

$$D/2 \le \rho_{2^{-l'}}(x,z) < D.$$

Thus, using the above, the fact that $\rho_{2^{-l}}(x,z) \approx 2^p D$, and Theorem 3.5.4 (n),

$$\begin{aligned}
\frac{(1+\rho_{2^{-l}}(x,z))^{-m}}{\mathrm{Vol}(B_{2^{-l}}(x,1+\rho_{2^{-l}}(x,z)))\wedge 1} &\approx \frac{2^{-mp}}{\mathrm{Vol}(B_{2^{-l}}(x,2^p\rho_{2^{-l'}}(x,z)))\wedge 1} \\
= \frac{2^{-mp}}{\mathrm{Vol}(B_{2^{-l'}}(x,\rho_{2^{-l'}}(x,z)))\wedge 1} &\approx \frac{2^{-mp}}{\mathrm{Vol}(B_{2^{-l'}}(x,1+\rho_{2^{-l'}}(x,z)))\wedge 1} \\
&\lesssim 2^{-(m/\sqrt{\nu})|j-l'|}\frac{\chi_{\{\rho_{2^{-l'}}(x,z)<D\}}}{\mathrm{Vol}(B_{2^{-l'}}(x,1+\rho_{2^{-l'}}(x,z)))\wedge 1},
\end{aligned}$$

completing the proof. □

Proof of Proposition 5.4.4. In the notation of this section, Proposition 3.10.1 can be rewritten as $\forall j', k' \in (-\infty,\infty]^\nu \setminus \{(\infty,\dots,\infty)\}$, $\forall x, z \in \Omega$,

$$\begin{aligned}
&\int_\Omega \left(\frac{\chi_{\{\rho_{2^{-j'}}(x,y)<1\}}}{\mathrm{Vol}(B_{2^{-j'}}(x,1))\wedge 1}\right)\left(\frac{\chi_{\{\rho_{2^{-k'}}(y,z)<1\}}}{\mathrm{Vol}(B_{2^{-k'}}(y,1))\wedge 1}\right) d\,\mathrm{Vol}(y) \\
&\quad\lesssim \frac{\chi_{\{\rho_{2^{-(j'\wedge k')}}(x,z)<2\}}}{\mathrm{Vol}(B_{2^{-(j'\wedge k')}}(x,1))\wedge 1}.
\end{aligned} \tag{5.17}$$

For $j, k \in (-\infty,\infty]^\nu \setminus \{(\infty,\dots,\infty)\}$ let $E := \{\mu \in \{1,\dots,\nu\} : j_\mu \ne \infty\}$ and $F := \{\mu \in \{1,\dots,\nu\} : k_\mu \ne \infty\}$. Let $\mathcal{I}_E$ and $\mathcal{I}_F$ be as in Lemma 5.4.9.

Using Lemma 5.4.9 and (5.17), we have by taking $m' = m'(m,\Omega,(W,\vec{d})) \in \mathbb{N}$ large, $\forall x, z \in \Omega$, $j, k \in (-\infty,\infty]^\nu \setminus \{(\infty,\dots,\infty)\}$, and using the convention $\infty - \infty = 0$,

$$\int_\Omega \frac{(1+\rho_{2^{-j}}(x,y))^{-m'}}{\mathrm{Vol}(B_{2^{-j}}(x,1+\rho_{2^{-j}}(x,y)))\wedge 1}\frac{(1+\rho_{2^{-k}}(y,z))^{-m'}}{\mathrm{Vol}(B_{2^{-k}}(y,1+\rho_{2^{-k}}(y,z)))\wedge 1}\,d\,\mathrm{Vol}(y)$$
$$\lesssim \sum_{\substack{j'\le j\\ j'\in\mathcal{I}_E}}\sum_{\substack{k'\le k\\ k'\in\mathcal{I}_F}} 2^{-(m'/\sqrt{\nu})|j'-j|-(m'/\sqrt{\nu})|k'-k|}$$
$$\times\int_\Omega \frac{\chi_{\{\rho_{2^{-j'}}(x,y)<1\}}}{\mathrm{Vol}(B_{2^{-j'}}(x,1))\wedge 1}\frac{\chi_{\{\rho_{2^{-k'}}(y,z)<1\}}}{\mathrm{Vol}(B_{2^{-k'}}(y,1))\wedge 1}\,d\,\mathrm{Vol}(y)$$
$$\lesssim \sum_{\substack{j'\le j\\ j'\in\mathcal{I}_E}}\sum_{\substack{k'\le k\\ k'\in\mathcal{I}_F}} 2^{-(m'/\sqrt{\nu})|j'-j|-(m'/\sqrt{\nu})|k'-k|}\frac{\chi_{\{\rho_{2^{-j'\wedge k'}}(x,z)<2\}}}{\mathrm{Vol}(B_{2^{-j'\wedge k'}}(x,1))\wedge 1}$$
$$\le 2^{\nu}\sum_{\substack{l\le j\wedge k\\ l\in\mathcal{I}_{E\cup F}}}\sum_{\substack{l\le l'\le j\vee k\\ l'\in\mathcal{I}_{E\cap F}}} 2^{-(m'\nu^{-2}/2)|j\wedge k-l|-(m'\nu^{-2}/2)|j\vee k-l'|}\frac{\chi_{\{\rho_{2^{-l}}(x,z)<2\}}}{\mathrm{Vol}(B_{2^{-l}}(x,1))\wedge 1}$$
$$\lesssim \sum_{\substack{l\le j\wedge k\\ l\in\mathcal{I}_{E\cup F}}} 2^{-(m'\nu^{-2}/2)|j\wedge k-l|}\frac{\chi_{\{\rho_{2^{-l}}(x,z)<2\}}}{\mathrm{Vol}(B_{2^{-l}}(x,1))\wedge 1}$$
$$\lesssim \frac{(1+\rho_{2^{-j\wedge k}}(x,z))^{-m}}{\mathrm{Vol}(B_{2^{-j\wedge k}}(x,1+\rho_{2^{-j\wedge k}}(x,z)))\wedge 1},$$

where in the first inequality, we used (5.15), in the second we used (5.17), in the third we took $l=j'\wedge k'$ and $l'=j'\vee k'$, and in the last we took m' large and applied (5.14). □

Lemma 5.4.9 can also be used to prove L^p boundedness properties of bounded sets of $(W,\vec{d})$ pre-elementary operators, and the next results show.

Lemma 5.4.10. *We have*

$$\sup_{x\in\Omega}\sup_{j\in\mathbb{R}^\nu}\int_\Omega \frac{(1+\rho_{2^{-j}}(x,z))^{-1}}{\mathrm{Vol}(B_{2^{-j}}(x,1+\rho_{2^{-j}}(x,z)))\wedge 1}\,d\,\mathrm{Vol}(z)<\infty, \tag{5.18}$$

$$\sup_{z\in\Omega}\sup_{j\in\mathbb{R}^\nu}\int_\Omega \frac{(1+\rho_{2^{-j}}(x,z))^{-1}}{\mathrm{Vol}(B_{2^{-j}}(x,1+\rho_{2^{-j}}(x,z)))\wedge 1}\,d\,\mathrm{Vol}(x)<\infty. \tag{5.19}$$

Proof. Proposition 5.4.3 shows that (5.18) and (5.19) are equivalent, so we prove only (5.18). Since $\mathrm{Vol}(\Omega)<\infty$, we have $\mathrm{Vol}(A\cap\Omega)\lesssim\mathrm{Vol}(A)\wedge 1$, where the implicit constant depends on Ω. Applying (5.15) with $m=1$, $D=1$, and $E=\{1,\dots,\nu\}$, we have

$$\int_\Omega \frac{(1+2^j\rho_{(W,d)}(x,z))^{-1}}{\mathrm{Vol}(B_{(W,d)}(x,2^{-j}+\rho_{(W,d)}(x,z)))\wedge 1}\,d\,\mathrm{Vol}(z)$$
$$\lesssim \sum_{\substack{k\le j\\ k\in\mathbb{Z}^\nu}} 2^{-|j-k|/\sqrt{\nu}}\int_\Omega \frac{\chi_{\{\rho_{2^{-l}}(x,z)<1\}}}{\mathrm{Vol}(B_{2^{-k}}(x,1))\wedge 1}\,d\,\mathrm{Vol}(z)$$

$$\lesssim \sum_{\substack{k\leq j\\ k\in\mathbb{Z}^\nu}} 2^{-|j-k|/\sqrt{\nu}} \frac{\operatorname{Vol}(B_{2^{-l}}(x,1)\cap\Omega)}{\operatorname{Vol}(B_{2^{-k}}(x,1))\wedge 1}$$
$$\lesssim \sum_{\substack{k\leq j\\ k\in\mathbb{Z}^\nu}} 2^{-|j-k|/\sqrt{\nu}} \lesssim 1,$$

completing the proof. □

Corollary 5.4.11. *Let $\mathcal{E}$ be a bounded set of $(W,\vec{d})$ pre-elementary operators supported in Ω. Then*

$$\sup_{p\in[1,\infty]} \sup_{(E,2^{-j})\in\mathcal{E}} \|E\|_{L^p(\mathfrak{M},\operatorname{Vol})\to L^p(\mathfrak{M},\operatorname{Vol})} < \infty. \tag{5.20}$$

Proof. Since, for $(E,2^{-j})\in\mathcal{E}$,

$$|E(x,z)| \lesssim \chi_\Omega(x)\chi_\Omega(z)\frac{(1+\rho_{2^{-j}}(x,z))^{-1}}{\operatorname{Vol}(B_{2^{-j}}(x,1+\rho_{2^{-j}}(x,z)))\wedge 1},$$

when $p=1$, (5.20) follows from (5.19), and when $p=\infty$, it follows from (5.18). The result for general p follows from the Riesz–Thorin interpolation theorem. □

Lemma 5.4.12. *Let $\mathcal{E}$ be a bounded set of $(W,\vec{d})$ pre-elementary operators supported in Ω. For every countable set $\mathcal{E}' := \{(E_k,\delta_k) : k\in\mathbb{N}\}\subseteq\mathcal{E}$, set*

$$\mathcal{T}_{\mathcal{E}'}\{f_k\}_{k\in\mathbb{N}} := \{E_k f_k\}_{k\in\mathbb{N}}.$$

Then, for all $p\in(1,\infty)$, $q\in(1,\infty]$,

$$\sup_{\mathcal{E}'} \|\mathcal{T}_{\mathcal{E}'}\|_{L^p(\mathfrak{M},\operatorname{Vol};\ell^q(\mathbb{N}))\to L^p(\mathfrak{M},\operatorname{Vol};\ell^q(\mathbb{N}))},$$

where the supremum is taken over all such countable subsets $\mathcal{E}'\subseteq\mathcal{E}$.

Proof. We will show that there exists $C_\mathcal{E}\geq 0$ such that $\forall (E,2^{-j})\in\mathcal{E}, f\in L^1_{\mathrm{loc}}(\mathfrak{M};\operatorname{Vol})$, we have

$$|Ef(x)| \leq C_\mathcal{E}\chi_\Omega(x)\mathcal{M}_{(W,\vec{d})}(\chi_\Omega f)(x), \tag{5.21}$$

where $\mathcal{M}_{(W,\vec{d})}$ is defined and studied in Section 3.11.2. Using (5.21), the result then immediately follows from Theorem 3.11.9.

To see (5.21), we apply (5.15) with $D=1$ and $m=1$ (and $E=\{1,\dots,\nu\}$) to see that for $(E,2^{-j})\in\mathcal{E}$,

$$
\begin{aligned}
|Ef(x)| &\lesssim \chi_\Omega(x) \int_\Omega \frac{(1+\rho_{2^{-j}}(x,z))^{-1}}{\mathrm{Vol}(B_{2^{-j}}(x, 1+\rho_{2^{-j}}(x,z))) \wedge 1} |f(z)|\, d\,\mathrm{Vol}(z) \\
&\lesssim \chi_\Omega(x) \sum_{\substack{k\leq j\\ k\in\mathbb{Z}^\nu}} 2^{-(1/\sqrt{\nu})|k-j|} \int_\Omega \frac{\chi_{\{\rho_{2^{-k}}(x,z)<1\}}}{\mathrm{Vol}(B_{2^{-k}}(x,1)) \wedge 1} |f(z)|\, d\,\mathrm{Vol}(z) \\
&\lesssim \chi_\Omega(x) \sum_{\substack{k\leq j\\ k\in\mathbb{Z}^\nu}} 2^{-(1/\sqrt{\nu})|k-j|} \mathcal{M}_{(W,\vec{d})}(\chi_\Omega f)(x) \lesssim \chi_\Omega(x) \mathcal{M}_{(W,\vec{d})}(\chi_\Omega f)(x),
\end{aligned}
$$

completing the proof. □

The expression

$$
\frac{(1+\rho_{2^{-j}}(x,z))^{-m}}{\mathrm{Vol}(B_{2^{-j}}(x, 1+\rho_{2^{-j}}(x,z))) \wedge 1}
$$

depends on the choices of $(W^1, \vec{d}^1), \ldots, (W^\nu, \vec{d}^\nu)$. Our next result shows that it does not depend on the weak equivalence class of these vector fields with formal degrees. Indeed, for each $\mu \in \{1, \ldots, \nu\}$, let

$$
(Z^\mu, \vec{d}^\mu) = \{(Z^\mu_1, \vec{d}^\mu_1), \ldots, (Z^\mu_{s_\mu}, \vec{d}^\mu_{s_\mu})\} \subset C^\infty_{\mathrm{loc}}(\mathfrak{M}; T\mathfrak{M}) \times \mathbb{N}_+
$$

be such that $(W^\mu, \vec{d}^\mu)$ and $(Z^\mu, \vec{d}^\mu)$ are locally weakly equivalent on $\mathfrak{M}$. By Lemma 3.14.11, $(Z^\mu, \vec{d}^\mu)$ pairwise locally weakly approximately commute, and by Proposition 3.14.14, each $(Z^\mu, \vec{d}^\mu)$ are Hörmander vector fields with formal degrees, that is, $(Z^1, \vec{d}^1), \ldots, (Z^\nu, \vec{d}^\nu)$ satisfy all the same hypotheses as $(W^1, \vec{d}^1), \ldots, (W^\nu, \vec{d}^\nu)$. We define $(Z, \vec{d}) := (Z^1, \vec{d}^1) \boxtimes (Z^2, \vec{d}^2) \boxtimes \cdots \boxtimes (Z^\nu, \vec{d}^\nu)$. We can define the various ingredients in

$$
\frac{(1+\rho_{2^{-j}}(x,z))^{-m}}{\mathrm{Vol}(B_{2^{-j}}(x, 1+\rho_{2^{-j}}(x,z))) \wedge 1}
$$

with either $(W, \vec{d})$ or $(Z, \vec{d})$. We write

$$
\frac{(1+\rho_{2^{-j},(W,\vec{d})}(x,z))^{-m}}{\mathrm{Vol}(B_{2^{-j},(W,\vec{d})}(x, 1+\rho_{2^{-j},(W,\vec{d})}(x,z))) \wedge 1}, \quad
\frac{(1+\rho_{2^{-j},(Z,\vec{d})}(x,z))^{-m}}{\mathrm{Vol}(B_{2\ j,(Z,\vec{d})}(x, 1+\rho_{2^{-j},(Z,\vec{d})}(x,z))) \wedge 1}
$$

for these two choices.

Proposition 5.4.13. *For each $m \in \mathbb{N}$,*

$$
\begin{aligned}
&\frac{(1+\rho_{2^{-j},(W,\vec{d})}(x,z))^{-m}}{\mathrm{Vol}(B_{2^{-j},(W,\vec{d})}(x, 1+\rho_{2^{-j},(W,\vec{d})}(x,z))) \wedge 1} \\
&\qquad \approx \frac{(1+\rho_{2^{-j},(Z,\vec{d})}(x,z))^{-m}}{\mathrm{Vol}(B_{2^{-j},(Z,\vec{d})}(x, 1+\rho_{2^{-j},(Z,\vec{d})}(x,z))) \wedge 1},
\end{aligned}
$$

for all $x, z \in \Omega$ and $j \in \mathbb{R}$, where the implicit constants do not depend on $x, z \in \Omega$ or $j \in \mathbb{R}$, but may depend on any of the other ingredients.

To prove Proposition 5.4.13, we use the next lemma.

Lemma 5.4.14. *For all $x, z \in \Omega$ and $j \in \mathbb{R}$,*

$$\rho_{2^{-j},(W,\vec{d})}(x,z) \approx \rho_{2^{-j},(Z,\vec{d})}(x,z). \tag{5.22}$$

Proof. For $r > 0$, we have $\rho_{r2^{-j},(W,\vec{d})}(x,z) = r^{-1}\rho_{2^{-j},(W,\vec{d})}(x,z)$, and similarly for $(W,\vec{d})$ replaced by $(Z,\vec{d})$. Thus, it suffices to prove the lemma when $j \in [0,\infty)^\nu$ and $j_\mu = 0$ for some μ. Without loss of generality, we assume $j_1 = 0$ and $j_2, \ldots, j_\nu \geq 0$.

Since Ω is relatively compact, it is relatively compact with respect to the metric $\rho_{(W^1,d^1)}$ (see Lemma 3.1.7). We have $\sup_{x,z\in\Omega}\rho_{(W^1,d^1)}(x,z) < \infty$. Therefore, for $x, z \in \Omega$,

$$\rho_{2^{-j},(W,\vec{d})}(x,z) \leq \rho_{(W^1,d^1)}(x,z) \lesssim 1.$$

We similarly have $\rho_{2^{-j},(Z,\vec{d})}(x,z) \lesssim 1$.

Fix $\xi_4 > 0$ as in Proposition 3.14.13. Suppose $\rho_{2^{-j},(W,\vec{d})}(x,z) < \xi_4 r$, where $r \in (0,1]$. Then by Proposition 3.14.13, $z \in B_{(W,\vec{d})}(x, \xi_4 r2^{-j}) \subseteq B_{(Z,\vec{d})}(x, r2^{-j})$, that is, $\rho_{2^{-j},(Z,\vec{d})}(x,z) < r$.

Since we have $\rho_{(Z,\vec{d}),2^{-j}}(x,z) \lesssim 1$, we conclude that $\rho_{2^{-j},(Z,\vec{d})}(x,z) \lesssim \rho_{2^{-j},(W,\vec{d})}(x,z)$. The reverse estimate follows by symmetry. □

Proof of Proposition 5.4.13. In light of Lemma 5.4.14, it suffices to show that for all $x, z \in \Omega$,

$$\begin{aligned}&\mathrm{Vol}(B_{2^{-j},(W,\vec{d})}(x, 1 + \rho_{2^{-j},(W,\vec{d})}(x,z))) \wedge 1\\ &\quad\approx \mathrm{Vol}(B_{2^{-j},(Z,\vec{d})}(x, 1 + \rho_{2^{-j},(Z,\vec{d})}(x,z))) \wedge 1.\end{aligned}$$

That is, we wish to show

$$\begin{aligned}&\mathrm{Vol}(B_{(W,\vec{d})}(x, (1 + \rho_{2^{-j},(W,\vec{d})}(x,z))2^{-j})) \wedge 1\\ &\quad\approx \mathrm{Vol}(B_{(Z,\vec{d})}(x, (1 + \rho_{2^{-j},(Z,\vec{d})}(x,z))2^{-j})) \wedge 1.\end{aligned} \tag{5.23}$$

By Lemma 5.4.14, $\rho_{2^{-j},(W,\vec{d})}(x,z) \approx \rho_{2^{-j},(Z,\vec{d})}(x,z)$, and therefore Theorem 3.5.4 (n) implies

$$\begin{aligned}&\mathrm{Vol}(B_{(W,\vec{d})}(x, (1 + \rho_{2^{-j},(W,\vec{d})}(x,z))2^{-j})) \wedge 1\\ &\quad\approx \mathrm{Vol}(B_{(W,\vec{d})}(x, (1 + \rho_{2^{-j},(Z,\vec{d})}(x,z))2^{-j})) \wedge 1.\end{aligned} \tag{5.24}$$

Proposition 3.14.14 implies

$$\begin{aligned}&\mathrm{Vol}(B_{(W,\vec{d})}(x, (1 + \rho_{2^{-j},(Z,\vec{d})}(x,z))2^{-j})) \wedge 1\\ &\quad\approx \mathrm{Vol}(B_{(Z,\vec{d})}(x, (1 + \rho_{2^{-j},(Z,\vec{d})}(x,z))2^{-j})) \wedge 1.\end{aligned} \tag{5.25}$$

Combining (5.24) and (5.25) yields (5.23) and completes the proof. □

5.5 Elementary operators

In this section we prove the basic results concerning $(W, \vec{d})$ elementary operators and generalized $(W, \vec{d})$ elementary operators. Most of our results hold for both $(W, \vec{d})$ elementary operators and generalized $(W, \vec{d})$ elementary operators with nearly identical proofs, so we state many of the results in this section in both cases. Because of this, many of the results in this section will be stated in both the Hörmander setting and the general setting (see Section 5.3). In the Hörmander setting, we will study bounded sets of $(W, \vec{d})$ elementary operators supported in Ω as in Definition 5.2.15, and in the general setting, we will study bounded sets of generalized $(W, \vec{d})$ elementary operators supported in Ω as in Definition 5.2.27.

Throughout this section Ω will denote a relatively compact, open set $\Omega \Subset \mathfrak{M}$.

Our first main result gives an equivalent characterization of bounded sets of (generalized) $(W, \vec{d})$ elementary operators. For this, recall the definition of a $(W^\mu, \vec{d}^\mu)$ partial differential operator in Ω of degree $\leq \kappa_\mu \in \mathbb{Z}$, as defined in Definition 3.12.1.

Proposition 5.5.1. Hörmander setting: *Let $\mathcal{G}'_\Omega$ be the largest set of subsets of $C_0^\infty(\Omega \times \Omega) \times (0,1]^\nu$ such that for all $\mathcal{E} \in \mathcal{G}'_\Omega$:*

- *$\mathcal{E}$ is a bounded set of $(W, \vec{d})$ pre-elementary operators supported in Ω.*
- *For each $\mu \in \{1, \dots, \nu\}$, there are finite sets depending on $\mathcal{E}$,*

$$\mathcal{F}_1^\mu = \{(\mathcal{P}_{1,1}^\mu, \kappa_{1,1}^\mu), \dots, (\mathcal{P}_{1,L_1}^\mu, \kappa_{1,L_1}^\mu)\},$$
$$\mathcal{F}_2^\mu = \{(\mathcal{P}_{2,1}^\mu, \kappa_{2,1}^\mu), \dots, (\mathcal{P}_{2,L_2}^\mu, \kappa_{2,L_2}^\mu)\},$$

where each $\mathcal{P}_{a,l}^\mu$ is a $(W^\mu, \vec{d}^\mu)$ partial differential operator in Ω of degree $\leq \kappa_{a,l}^\mu$, $\kappa_{a,l}^\mu \in \mathbb{N}$, such that $\forall (E, 2^{-j}) \in \mathcal{E}$,

$$E = \sum_{l_1=1}^{L_1} \sum_{l_2=1}^{L_2} 2^{-\epsilon_{1,l_1}^\mu j_\mu - \epsilon_{2,l_2}^\mu j_\mu} 2^{-j_\mu \kappa_{1,l_1}^\mu} \mathcal{P}_{1,l_1}^\mu E_{\mu,l_1,l_2} 2^{-j_\mu \kappa_{2,l_2}^\mu} \mathcal{P}_{2,l_2}^\mu, \tag{5.26}$$

where $\epsilon_{a,l}^\mu = \begin{cases} 1, & \kappa_{a,l}^\mu = 0, \\ 0, & \text{otherwise,} \end{cases}$ *and*

$$\{(E_{\mu,l_1,l_2}, 2^{-j}) : (E, 2^{-j}) \in \mathcal{E}, l \leq l_1 \leq L_1, 1 \leq l_2 \leq L_2\} \in \mathcal{G}'_\Omega.$$

Then $\mathcal{G}'_\Omega = \mathcal{G}_\Omega$, where $\mathcal{G}_\Omega$ is the set of bounded sets of elementary operators supported in Ω.

General setting: *We define $\widetilde{\mathcal{G}}'_\Omega$ as above, but with bounded sets of $(W, \vec{d})$ pre-elementary operators replaced by bounded sets of generalized $(W, \vec{d})$ pre-elementary operators and $C_0^\infty(\Omega \times \Omega)$ replaced by $\mathrm{Hom}(C_{W,0}^\infty(\mathfrak{M}), C_{W,\mathrm{loc}}^\infty(\mathfrak{M}))$. Then $\widetilde{\mathcal{G}}'_\Omega = \widetilde{\mathcal{G}}_\Omega$, where $\widetilde{\mathcal{G}}_\Omega$ is the set of bounded sets of generalized $(W, \vec{d})$ elementary operators supported in Ω.*

To prove Proposition 5.5.1, we require a few lemmas.

Lemma 5.5.2. Hörmander setting: *Suppose $\mathcal{E}$ is a bounded set of $(W, \vec{d})$ pre-elementary operators supported in Ω. Then the following are bounded sets of $(W, \vec{d})$ pre-elementary operators:*

(i) *for $\psi \in C^\infty_{\mathrm{loc}}(\Omega)$, the set*

$$\{(\mathrm{Mult}[\psi]E, 2^{-j}), (E\,\mathrm{Mult}[\psi], 2^{-j}) : (E, 2^{-j}) \in \mathcal{E}\},$$

(ii) *the set*

$$\{(aE_1 + bE_2, 2^{-j}) : (E_1, 2^{-j}), (E_2, 2^{-j}) \in \mathcal{E}, a, b \in \mathbb{C}, |a|, |b| \leq 1\},$$

(iii) *for α an ordered multi-index, the set*

$$\{((2^{-j\vec{d}}W)^\alpha E, 2^{-j}), (E(2^{-j\vec{d}}W)^\alpha, 2^{-j}) : (E, 2^{-j}) \in \mathcal{E}\},$$

(iv) *letting $\mathscr{P}^\mu$ be a $(W^\mu, \vec{d}^\mu)$ partial differential operator in Ω of degree $\leq \kappa_\mu \in \mathbb{Z}$, the set*

$$\{(2^{-j_\mu \kappa_\mu}\mathscr{P}^\mu E, 2^{-j}), (2^{-j_\mu \kappa_\mu} E\mathscr{P}^\mu, 2^{-j}) : (E, 2^{-j}) \in \mathcal{E}\},$$

(v) *letting $\mathscr{P}$ be a $(W, \vec{d})$ partial differential operator in Ω of degree $\leq \kappa \in \mathbb{Z}^\nu$, the set*

$$\{(2^{-j\cdot\kappa}\mathscr{P}E, 2^{-j}), (2^{-j\cdot\kappa}E\mathscr{P}, 2^{-j}) : (E, 2^{-j}) \in \mathcal{E}\}.$$

General setting: *The same result holds with "$(W, \vec{d})$ pre-elementary" replaced by "generalized $(W, \vec{d})$ pre-elementary" throughout.*

Proof. We begin with (i). In either setting, we have

$$\bigcup_{(E,2^{-j})\in\mathcal{E}} \mathrm{supp}(E) \Subset \Omega \times \Omega.$$

Thus, we may take $\phi \in C_0^\infty(\Omega)$ with $\phi \equiv 1$ on a large enough set that $\mathrm{Mult}[\phi]E = E = E\,\mathrm{Mult}[\phi]$ for $(E, 2^{-j}) \in \mathcal{E}$. Because of this, it suffices to replace ψ with $\phi\psi \in C_0^\infty(\Omega)$ in the statement of (i). In particular, it suffices to prove (i) in the special case where $\psi \in C_0^\infty(\Omega)$. In this case, the result follows easily from the definitions.

Item (ii) follows immediately from the definitions.

We now turn to (iii) and begin with the Hörmander setting. That

$$\{((2^{-j\vec{d}}W_x)^\alpha E(x,z), 2^{-j}), ((2^{-j\vec{d}}W_z)^\alpha E(x,z)) : (E, 2^{-j}) \in \mathcal{E}\} \tag{5.27}$$

is a bounded set of $(W, \vec{d})$ pre-elementary operators supported in Ω follows from the definition. Thus,

$$\{((2^{-j\vec{d}}W)^{\alpha}E, 2^{-j}) : (E, 2^{-j}) \in \mathcal{E}\}$$

is a bounded set of $(W, \vec{d})$ pre-elementary operators supported in Ω. Finally, note that

$$[E(2^{-j\vec{d}}W)^{\alpha}](x, z) = ((2^{-j\vec{d}}W_z)^{\alpha})^{\top} E(x, z).$$

Since $((2^{-j\vec{d}}W_z)^{\alpha})^{\top}$ can be written as finite linear combination, with coefficients bounded uniformly by 1, of terms of the form $\mathrm{Mult}[\psi](2^{-j\vec{d}}W_z)^{\beta}$, where $\psi \in C^{\infty}_{\mathrm{loc}}(\mathfrak{M})$ and $\deg_{\vec{d}}(\beta) \le \deg_{\vec{d}}(\alpha)$, it follows from (i), (ii), and (5.27) that

$$\{(E(2^{-j\vec{d}}W)^{\alpha}, 2^{-j}) : (E, 2^{-j}) \in \mathcal{E}\}$$

is a bounded set of $(W, \vec{d})$ pre-elementary operators supported in Ω, completing the proof of (iii) in the Hörmander setting.

Next we prove (iii) in the general setting. Fix an ordered multi-index γ. The result will follow once we show that for $(E, 2^{-j}) \in \mathcal{E}$ and

$$T_E \in \{(2^{-j\vec{d}}W)^{\gamma}E, E(2^{-j\vec{d}}W)^{\gamma}, [(2^{-j\vec{d}}W)^{\gamma}E]^{*}, [E(2^{-j\vec{d}}W)^{\gamma}]^{*}\}$$

we have

$$\sup_{(E,2^{-j})\in\mathcal{E}} \|(2^{-j\vec{d}}W)^{\alpha} T_E (2^{-j\vec{d}}W)^{\beta}\|_{L^p\to L^p} < \infty. \tag{5.28}$$

Also, for $\mathcal{E} := \{(E_k, 2^{-j_k}) : k \in \mathbb{N}\} \subseteq \mathcal{E}$, we set

$$\mathcal{T}_{\mathcal{E}',\alpha,\beta}\{f_k\}_{k\in\mathbb{N}} := \{(2^{-j_k\vec{d}}W)^{\alpha} T_E (2^{-j_k\vec{d}}W)^{\beta} f_k\}_{k\in\mathbb{N}}.$$

We require

$$\sup_{\mathcal{E}'\subseteq\mathcal{E}} \|\mathcal{T}_{\mathcal{E}',\alpha,\beta}\|_{L^p(\ell^q)\to L^p(\ell^q)} < \infty. \tag{5.29}$$

We turn to proving (5.28) and (5.29). Let δ be the ordered multi-index given by concatenating the lists α and γ so that $(2^{-j\vec{d}}W)^{\alpha}(2^{-j\vec{d}}W)^{\gamma} = (2^{-j\vec{d}}W)^{\delta}$. If $T_E = (2^{-j\vec{d}}W)^{\gamma}E$, then $(2^{-j\vec{d}}W)^{\alpha}T_E(2^{-j\vec{d}}W)^{\beta} = (2^{-j\vec{d}}W)^{\delta}E(2^{-j\vec{d}}W)^{\beta}$. In this case, (5.28) and (5.29) follow immediately from the definition of bounded sets of generalized $(W, \vec{d})$ pre-elementary operators (Definition 5.2.26). Similarly, if $T_E = E(2^{-j\vec{d}}W)^{\beta}$, then there is a δ with $(2^{-j\vec{d}}W)^{\alpha}T_E(2^{-j\vec{d}}W)^{\beta} = (2^{-j\vec{d}}W)^{\alpha}E(2^{-j\vec{d}}W)^{\delta}$ and again (5.28) and (5.29) follow immediately from the definition of bounded sets of generalized $(W, \vec{d})$ pre-elementary operators. Next we consider the case where $T_E = [(2^{-j\vec{d}}W)^{\gamma}E]^{*}$. In this case, we have, for some functions $f_{\delta} \in C^{\infty}_{\mathrm{loc}}(\mathfrak{M})$,

$$T_E = \sum_{\deg_{\vec{d}}(\delta)\leq \deg_{\vec{d}}(\gamma)} (E^* \operatorname{Mult}[f_\delta] 2^{-j\cdot(\deg_{\vec{d}}(\gamma)-\deg_{\vec{d}}(\delta))})(2^{-j\vec{d}}W)^\delta$$
$$=: \sum_{\deg_{\vec{d}}(\delta)\leq \deg_{\vec{d}}(\gamma)} E_\delta (2^{-j\vec{d}}W)^\delta. \tag{5.30}$$

The definition of bounded sets of generalized $(W, \vec{d})$ pre-elementary operators (Definition 5.2.26) shows that $\{(E^*, 2^{-j}) : (E, 2^{-j}) \in \mathcal{E}\}$ is a bounded set of generalized $(W, \vec{d})$ pre-elementary operators supported in Ω, and (i) and (ii) then imply

$$\{(E_\delta, 2^{-j}) : (E, 2^{-j}) \in \mathcal{E}, \deg_{\vec{d}}(\delta) \leq \deg_{\vec{d}}(\gamma)\}.$$

Thus, (5.30) shows that T_E is a finite sum of terms of the type for which we have already established (5.28) and (5.29), which establishes (5.28) and (5.29). A similar proof gives (5.28) and (5.29) when $T_E = [E(2^{-j\vec{d}}W)^\gamma]^*$, which completes the proof of (iii).

Item (iv) is the special case of (v) where $\kappa = \kappa_\mu \mathfrak{e}_\mu$.

Finally, we turn to (v); the proof is the same in both settings. When $\kappa \in \mathbb{Z}^\nu \setminus \mathbb{N}^\nu$, $\mathscr{P} = 0$ and the result is trivial; thus, we consider only the case $\kappa \in \mathbb{N}^\nu$. By the definition, $2^{-j\cdot\kappa}\mathscr{P}$ can be written as a finite linear combination, with coefficients bounded uniformly by 1, of terms of the form $\operatorname{Mult}[\psi](2^{-j\vec{d}}W)^\alpha$, where $\deg_{\vec{d}}(\alpha) \leq \kappa$ and $\psi \in C^\infty_{\mathrm{loc}}(\Omega)$. From this, the result follows from (i), (ii), and (iii). □

Lemma 5.5.3. Hörmander setting: *Let $\mathcal{E} \in \mathscr{G}'_\Omega$. Then*

$$\{(aE_1 + bE_2, 2^{-j}) : (E_1, 2^{-j}), (E_2, 2^{-j}) \in \mathcal{E}, a, b \in \mathbb{C}, |a|, |b| \leq 1\} \in \mathscr{G}_{\Omega'}.$$

General setting: *The same result holds with $\mathscr{G}'_\Omega$ replaced by $\widetilde{\mathscr{G}}'_\Omega$ throughout.*

Proof. This follows immediately from the definitions. □

Lemma 5.5.4. Hörmander setting: *Suppose $\mathcal{E} \in \mathscr{G}'_\Omega$ and $\mathscr{P}^{\mu_0}$ is a $(W^{\mu_0}, \vec{d}^{\mu_0})$ partial differential operator in Ω of degree $\leq \kappa_{\mu_0} \in \mathbb{Z}$. Then*

$$\{(2^{-j_{\mu_0}\kappa_{\mu_0}}\mathscr{P}^{\mu_0}E, 2^{-j}), (2^{-j_{\mu_0}\kappa_{\mu_0}}E\mathscr{P}^{\mu_0}, 2^{-j}) : (E, 2^{-j}) \in \mathcal{E}\} \in \mathscr{G}'_\Omega.$$

General setting: *The same result holds with $\mathscr{G}'_\Omega$ replaced by $\widetilde{\mathscr{G}}'_\Omega$ throughout.*

Proof. The proofs in the two settings are nearly identical, so we prove the result in the Hörmander setting. When $\kappa_{\mu_0} \in \mathbb{Z} \setminus \mathbb{N}$, $\mathscr{P}^{\mu_0} = 0$ and the result is trivial. We henceforth assume $\kappa_{\mu_0} \in \mathbb{N}$.

Without loss of generality, we take $\mu_0 = 1$, so that $\mathscr{P}^1$ is a $(W^1, \vec{d}^1)$ partial differential operator in Ω of degree $\leq \kappa_1$. We will show that

$$\mathcal{E}_1 := \{(2^{-j_1\kappa_1}\mathscr{P}^1E, 2^{-j}) : (E, 2^{-j}) \in \mathcal{E}\} \in \mathscr{G}'_\Omega.$$

The result with $2^{-j_1\kappa_1}\mathscr{P}^1 E$ replaced by $2^{-j_1\kappa_1}E\mathscr{P}^1$ has a similar proof and we leave it to the interested reader.

Lemma 5.5.2 (iv) shows that $\mathcal{E}_1$ is a bounded set of $(W,\vec{d})$ pre-elementary operators supported in Ω. Thus, the result will follow if we can show for each $\mu \in \{1,\dots,\nu\}$ that $2^{-j_1\kappa_1}\mathscr{P}^1 E$ can be written as a sum as in (5.26), where E_{μ,l_1,l_2} is "of the same form" as $2^{-j_1\kappa_1}\mathscr{P}^1 E$.

There are two cases to consider: $\mu = 1$ and $\mu \neq 1$. We begin with the case $\mu = 1$. By the definition of $\mathscr{G}'_\Omega$, there are finite sets

$$\mathcal{F}_1^1 = \{(\mathscr{P}_{1,1}^1,\kappa_{1,1}^1),\dots,(\mathscr{P}_{1,L_1}^1,\kappa_{1,L_1}^1)\},$$
$$\mathcal{F}_2^1 = \{(\mathscr{P}_{2,1}^1,\kappa_{2,1}^1),\dots,(\mathscr{P}_{2,L_2}^1,\kappa_{2,L_2}^1)\},$$

where each $\mathscr{P}_{a,l}^1$ is a (W^1,d^1) partial differential operator in Ω of degree $\le \kappa_{a,l}^1$, and

$$2^{-j_1\kappa_1}\mathscr{P}^1 E = \sum_{l_1=1}^{L_1}\sum_{l_2=1}^{L_2} 2^{-\epsilon_{1,l_1}^1 j_1-\epsilon_{2,l_2}^1 j_1}2^{-j_1(\kappa_{1,l_1}^1+\kappa_1)}\mathscr{P}_1\mathscr{P}_{1,l_1}^1 E_{1,l_1,l_2}2^{-j_1\kappa_{2,l_2}^1}\mathscr{P}_{2,l_2}^1,$$

where $\epsilon_{a,l}^1 = \begin{cases}1, & \kappa_{a,l}^1 = 0,\\ 0, & \text{otherwise,}\end{cases}$ and

$$\{(E_{1,l_1,l_2},2^{-j}) : (E,2^{-j}) \in \mathcal{E}, l \le l_1 \le L_1, 1 \le l_2 \le L_2\} \in \mathscr{G}'_\Omega.$$

Set $\tilde{\epsilon}_{a,l}^1 = \begin{cases}1, & \kappa_1+\kappa_{a,l}^1 = 0,\\ 0, & \text{otherwise.}\end{cases}$ Note that $\tilde{\epsilon}_{a,l}^1 \le \epsilon_{a,l}^1$, and we have

$$2^{-j_1\kappa_1}\mathscr{P}^1 E = 2^{-\tilde{\epsilon}_{1,l_1}^1 j_1-\epsilon_{2,l_2}^1 j_1}2^{-j_1(\kappa_{1,l_1}^1+\kappa_1)}\mathscr{P}_1\mathscr{P}_{1,l_1}^1(2^{-j_1(\epsilon_{a,l}^1-\tilde{\epsilon}_{a,l}^1)}E_{1,l_1,l_2})2^{-j_1\kappa_{2,l_2}^1}\mathscr{P}_{2,l_2}^1.$$

Lemma 5.5.3 shows that

$$\{(2^{-j_1(\epsilon_{a,l}^1-\tilde{\epsilon}_{a,l}^1)}E_{1,l_1,l_2},2^{-j}) : (E,2^{-j}) \in \mathcal{E}, 1 \le l_1 \le L_1, 1 \le l_2 \le L_2\} \in \mathscr{G}'_\Omega,$$

and Lemma 3.12.4 shows that $\mathscr{P}_1\mathscr{P}_{1,l_1}^1$ is a (W^1,d^1) partial differential operator in Ω of degree $\le \kappa_1+\kappa_{1,l_1}^1$. This sees $2^{-j_1\kappa_1}\mathscr{P}^1 E$ as being of the form (5.26), and completes the proof in the case $\mu = 1$.

Now suppose $\mu \neq 1$. Without loss of generality, we take $\mu = 2$. By the definition of $\mathscr{G}'_\Omega$, there are finite sets

$$\mathcal{F}_1^2 = \{(\mathscr{P}_{1,1}^2,\kappa_{1,1}^2),\dots,(\mathscr{P}_{1,L_1}^2,\kappa_{1,L_1}^2)\},$$
$$\mathcal{F}_2^2 = \{(\mathscr{P}_{2,1}^2,\kappa_{2,1}^2),\dots,(\mathscr{P}_{2,L_2}^2,\kappa_{2,L_2}^2)\},$$

where each $\mathscr{P}_{a,l}^2$ is a (W^2,d^2) partial differential operator in Ω of degree $\le \kappa_{a,l}^2$, and

$$2^{-j_1\kappa_1}\mathscr{P}^1 E = \sum_{l_1=1}^{L_1}\sum_{l_2=1}^{L_2} 2^{-\epsilon^2_{1,l_1}j_2-\epsilon^2_{2,l_2}j_2}2^{-j_2\kappa^2_{1,l_1}}2^{-j_1\kappa_1}\mathscr{P}_1\mathscr{P}^2_{1,l_1}E_{2,l_1,l_2}2^{-j_2\kappa^2_{2,l_2}}\mathscr{P}^1_{2,l_2},$$

where $\epsilon^2_{a,l} = \begin{cases} 1, & \kappa^2_{a,l} = 0, \\ 0, & \text{otherwise,} \end{cases}$ and

$$\{(E_{2,l_1,l_2}, 2^{-j}) : (E, 2^{-j}) \in \mathcal{E}, l \le l_1 \le L_1, 1 \le l_2 \le L_2\} \in \mathscr{G}'_\Omega.$$

Corollary 3.12.8 shows that $\mathscr{P}^1\mathscr{P}^2_{1,l_1}$ can be written as a finite sum of terms of the form $\widetilde{\mathscr{P}}^2_{1,l_1}\widetilde{\mathscr{P}}^1$, where $\widetilde{\mathscr{P}}^2_{1,l_1}$ is a $(W^2, \vec{d}^2)$ partial differential operator in Ω of degree $\le \kappa^2_{1,l_1}$ and $\widetilde{\mathscr{P}}^1$ is a $(W^1, \vec{d}^1)$ partial differential operator in Ω of degree $\le \kappa_1$. Thus, $2^{-j_1\kappa_1}\mathscr{P}^1 E$ can be written as a finite sum of operators terms of the form

$$2^{-\epsilon^2_{1,l_1}j_2-\epsilon^2_{2,l_2}j_2}2^{-j_2\kappa^2_{1,l_1}}\widetilde{\mathscr{P}}^2_{1,l_1}(2^{-j_1\kappa_1}\widetilde{\mathscr{P}}^1 E_{2,l_1,l_2})2^{-j_2\kappa^2_{2,l_2}}\mathscr{P}^1_{2,l_2}.$$

This sees $2^{-j_1\kappa_1}\mathscr{P}^1 E$ as being of the form (5.26), where each E_{2,l_1,l_2} is of the same form as $2^{-j_1\kappa_1}\mathscr{P}^1 E$. This completes the proof. □

Proof of Proposition 5.5.1. The proof in the Hörmander setting and the general setting are nearly identical, so we present only the proof in the Hörmander setting.

We begin with the containment $\mathscr{G}_\Omega \subseteq \mathscr{G}'_\Omega$. Note that $(2^{-j_\mu \vec{d}^\mu} W^\mu)^{\alpha_\mu} = 2^{-j_\mu \deg_{\vec{d}^\mu}(\alpha_\mu)}(W^\mu)^{\alpha_\mu}$ and $(W^\mu)^{\alpha_\mu}$ is a $(W^\mu, \vec{d}^\mu)$ partial differential operator in Ω of degree $\le \deg_{\vec{d}^\mu}(\alpha_\mu)$. Because of this, elements of $\mathscr{G}_\Omega$ satisfy the axioms of elements of $\mathscr{G}'_\Omega$. Since $\mathscr{G}'_\Omega$ is the largest set satisfying these properties, we have $\mathscr{G}_\Omega \subseteq \mathscr{G}'_\Omega$.

We turn to the reverse containment $\mathscr{G}'_\Omega \subseteq \mathscr{G}_\Omega$. Suppose $\mathcal{E} \in \mathscr{G}'_\Omega$. By our assumption $\mathcal{E}$ is a bounded set of $(W, \vec{d})$ pre-elementary operators supported in Ω. The proof will be complete if we can show that for all $(E, 2^{-j}) \in \mathcal{E}$ and $\mu \in \{1, \dots, \nu\}$, we have

$$E = \sum_{|\alpha_\mu|,|\beta_\mu|\le 1} 2^{-(2-|\alpha_\mu|-|\beta_\mu|)j_\mu}(2^{-j_\mu \vec{d}^\mu} W^\mu)^{\alpha_\mu}\widetilde{E}_{\mu,\alpha_\mu,\beta_\mu}(2^{-j_\mu \vec{d}^\mu} W^\mu)^{\beta_\mu}, \tag{5.31}$$

where

$$\{(\widetilde{E}_{\mu,\alpha_\mu,\beta_\mu}, 2^{-j}) : (E, 2^{-j}) \in \mathcal{E}, \mu \in \{1, \dots, \nu\}, |\alpha_\mu|, |\beta_\mu| \le 1\} \in \mathscr{G}'_\Omega,$$

as then elements of $\mathscr{G}'_\Omega$ will satisfy the axioms of $\mathscr{G}_\Omega$; since $\mathscr{G}_\Omega$ is the largest set satisfying this, we will have $\mathscr{G}'_\Omega \subseteq \mathscr{G}_\Omega$.

Fix $\mu \in \{1, \dots, \nu\}$. We write E as in (5.26); we will show that each term on the right-hand side of (5.26) is of the form (5.31), which will complete the proof. Fix $1 \le l_1 \le L_1$ and $1 \le l_2 \le L_2$.

Suppose $\kappa^{\mu}_{1,l_1} = \kappa^{\mu}_{2,l_2} = 0$. Then $\epsilon^{\mu}_{1,l_1} = \epsilon^{\mu}_{2,l_2} = 1$ and we have

$$\begin{aligned} &2^{-\epsilon^{\mu}_{1,l_1} j_\mu - \epsilon^{\mu}_{2,l_2} j_\mu} 2^{-j_\mu \kappa^{\mu}_{1,l_1}} \mathscr{P}^{\mu}_{1,l_1} E_{\mu,l_1,l_2} 2^{-j_\mu \kappa^{\mu}_{2,l_2}} \mathscr{P}^{\mu}_{2,l_2} \\ &\quad = 2^{-2j_\mu} (2^{-j_\mu \kappa^{\mu}_{1,l_1}} \mathscr{P}^{\mu}_{1,l_1} E_{\mu,l_1,l_2} 2^{-j_\mu \kappa^{\mu}_{2,l_2}} \mathscr{P}^{\mu}_{2,l_2}). \end{aligned} \tag{5.32}$$

By two applications of Lemma 5.5.4,

$$\{(2^{-j_\mu \kappa^{\mu}_{1,l_1}} \mathscr{P}^{\mu}_{1,l_1} E_{\mu,l_1,l_2} 2^{-j_\mu \kappa^{\mu}_{2,l_2}} \mathscr{P}^{\mu}_{2,l_2}, 2^{-j}) : (E, 2^{-j}) \in \mathcal{E}\} \in \mathscr{G}'_\Omega.$$

This shows that (5.32) is of the form (5.26) (with $|\alpha_\mu| = |\beta_\mu| = 0$). This completes the proof when $\kappa^{\mu}_{1,l_1} = \kappa^{\mu}_{2,l_2} = 0$.

Now suppose $\kappa^{\mu}_{1,l_1} \neq 0$. Then $\mathscr{P}^{\mu}_{1,l_1}$ can be written as a sum of terms of the form

$$W^{\mu}_k \widetilde{\mathscr{P}}^{\mu}_{k,1,l_1} \quad \text{and} \quad \widetilde{\mathscr{P}}^{0}_{1},$$

where $\widetilde{\mathscr{P}}^0_1$ is a (W^μ, d^μ) partial differential operator in Ω of degree ≤ 0 (i. e., multiplication by a $C^\infty_{\mathrm{loc}}(\Omega)$ function) and $\widetilde{\mathscr{P}}^{\mu}_{k,1,l_1}$ is a (W^μ, d^μ) partial differential operator in Ω of degree $\leq \kappa^{\mu}_{1,l_1} - d^{\mu}_k$.

If, in addition, $\kappa^{\mu}_{2,l_2} = 0$, then $\epsilon^{\mu}_{2,l_2} = 1$, and therefore

$$2^{-\epsilon^{\mu}_{1,l_1} j_\mu - \epsilon^{\mu}_{2,l_2} j_\mu} 2^{-j_\mu \kappa^{\mu}_{1,l_1}} \mathscr{P}^{\mu}_{1,l_1} E_{\mu,l_1,l_2} 2^{-j_\mu \kappa^{\mu}_{2,l_2}} \mathscr{P}^{\mu}_{2,l_2}$$

can be written as a sum of terms of the following forms:

$$2^{-j_\mu} (2^{-j_\mu d^{\mu}_k} W^{\mu}_k)(2^{-j_\mu(\kappa^{\mu}_{1,l_1} - d^{\mu}_k)} \widetilde{\mathscr{P}}^{\mu}_{k,1,l_1} E_{\mu,l_1,l_2} 2^{-j_\mu \kappa^{\mu}_{2,l_2}} \mathscr{P}^{\mu}_{2,l_2}), \tag{5.33}$$

$$2^{-2j_\mu} (2^{-j_\mu(\kappa^{\mu}_{1,l_1} - 1)} \widetilde{\mathscr{P}}^{0}_{1} E_{\mu,l_1,l_2} 2^{-j_\mu \kappa^{\mu}_{2,l_2}} \mathscr{P}^{\mu}_{2,l_2}). \tag{5.34}$$

Lemma 5.5.4 shows that

$$\begin{aligned} &\{(2^{-j_\mu(\kappa^{\mu}_{1,l_1} - d^{\mu}_k)} \widetilde{\mathscr{P}}^{\mu}_{k,1,l_1} E_{\mu,l_1,l_2} 2^{-j_\mu \kappa^{\mu}_{2,l_2}} \mathscr{P}^{\mu}_{2,l_2}, 2^{-j}) : (E, 2^{-j}) \in \mathcal{E}\} \in \mathscr{G}'_\Omega, \\ &\{(2^{-j_\mu(\kappa^{\mu}_{1,l_1} - 1)} \widetilde{\mathscr{P}}^{0}_{1} E_{\mu,l_1,l_2} 2^{-j_\mu \kappa^{\mu}_{2,l_2}} \mathscr{P}^{\mu}_{2,l_2}, 2^{-j}) : (E, 2^{-j}) \in \mathcal{E}\} \in \mathscr{G}'_\Omega. \end{aligned}$$

Thus, (5.33) is of the form (5.26) with $|\alpha_\mu| = 1$ and $|\beta_\mu| = 0$ and (5.34) is of the form (5.26) with $|\alpha_\mu| = |\beta_\mu| = 0$. This completes the proof when $\kappa^{\mu}_{1,l_1} \neq 0$ and $\kappa^{\mu}_{2,l_2} = 0$.

Finally, we turn to the case where $k^{\mu}_{2,l_2} \neq 0$. In this case $\mathscr{P}^{\mu}_{2,l_2}$ can be written as a sum of terms of the form

$$\widetilde{\mathscr{P}}^{\mu}_{m,1,l_2} W^{\mu}_m \quad \text{and} \quad \widetilde{\mathscr{P}}^{0}_{2},$$

where $\widetilde{\mathscr{P}}_2^0$ is a $(W^\mu, \vec{d}^\mu)$ partial differential operator in Ω of degree ≤ 0 (i. e., multiplication by a $C_{\mathrm{loc}}^\infty(\Omega)$ function) and $\widetilde{\mathscr{P}}^\mu_{m,1,l_2}$ is a $(W^\mu, \vec{d}^\mu)$ partial differential operator in Ω of degree $\leq \kappa^\mu_{2,l_2} - d^\mu_m$. Therefore,

$$2^{-\epsilon^\mu_{1,l_1} j_\mu - \epsilon^\mu_{2,l_2} j_\mu} 2^{-j_\mu \kappa^\mu_{1,l_1}} \mathscr{P}^\mu_{1,l_1} E_{\mu,l_1,l_2} 2^{-j_\mu \kappa^\mu_{2,l_2}} \mathscr{P}^\mu_{2,l_2}$$

can be written as a sum of terms of the following forms:

$$\begin{aligned} &(2^{-j_\mu d^\mu_k} W^\mu_k)(2^{-j_\mu(\kappa^\mu_{1,l_1} - d^\mu_k)} \widetilde{\mathscr{P}}^\mu_{k,1,l_1} E_{\mu,l_1,l_2} 2^{-j_\mu(\kappa^\mu_{2,l_2} - d^\mu_m)} \widetilde{\mathscr{P}}^\mu_{m,2,l_2})(2^{-j_\mu d^\mu_m} W^\mu_m) \\ &\quad =: (2^{-j_\mu d^\mu_k} W^\mu_k) \widetilde{E}_1 (2^{-j_\mu d^\mu_m} W^\mu_m), \end{aligned} \tag{5.35}$$

$$\begin{aligned} &2^{-j_\mu}(2^{-j_\mu d^\mu_k} W^\mu_k)(2^{-j_\mu(\kappa^\mu_{1,l_1} - d^\mu_k)} \widetilde{\mathscr{P}}^\mu_{k,1,l_1} E_{\mu,l_1,l_2} 2^{-j_\mu(\kappa^\mu_{2,l_2} - 1)} \widetilde{\mathscr{P}}^0_2) \\ &\quad =: 2^{-j_\mu}(2^{-j_\mu d^\mu_k} W^\mu_k) \widetilde{E}_2, \end{aligned} \tag{5.36}$$

$$\begin{aligned} &2^{-j_\mu}(2^{-j_\mu(\kappa^\mu_{1,l_1} - 1)} \widetilde{\mathscr{P}}^0_1 E_{\mu,l_1,l_2} 2^{-j_\mu(\kappa^\mu_{2,l_2} - d^\mu_m)} \widetilde{\mathscr{P}}^\mu_{m,2,l_2})(2^{-j_\mu d^\mu_m} W^\mu_m) \\ &\quad =: 2^{-j_\mu} \widetilde{E}_3 (2^{-j_\mu d^\mu_m} W^\mu_m), \end{aligned} \tag{5.37}$$

$$\begin{aligned} &2^{-2j_\mu}(2^{-j_\mu(\kappa^\mu_{1,l_1} - 1)} \widetilde{\mathscr{P}}^0_1 E_{\mu,l_1,l_2} 2^{-j_\mu(\kappa^\mu_{2,l_2} - 1)} \widetilde{\mathscr{P}}^0_2) \\ &\quad =: 2^{-j_\mu}(2^{-j_\mu d^\mu_k} W^\mu_k) \widetilde{E}_4. \end{aligned} \tag{5.38}$$

Lemma 5.5.4 shows that

$$\{(\widetilde{E}_1, 2^{-j}), (\widetilde{E}_2, 2^{-j}), (\widetilde{E}_3, 2^{-j}), (\widetilde{E}_4, 2^{-j}) : (E, 2^{-j}) \in \mathcal{E}\} \in \mathscr{G}'_\Omega.$$

Thus, (5.35) is of the form (5.26) with $|\alpha_\mu| = |\beta_\mu| = 1$, (5.36) is of the form (5.26) with $|\alpha_\mu| = 1$ and $|\beta_\mu| = 0$, (5.37) is of the form (5.26) with $|\alpha_\mu| = 0$ and $|\beta_\mu| = 1$, and (5.38) is of the form (5.26) with $|\alpha_\mu| = |\beta_\mu| = 0$. This completes the proof. □

With Proposition 5.5.1 in hand, we can prove the basic properties of elementary operators.

Proposition 5.5.5. Hörmander setting: *Let $\mathcal{E}$ be a bounded set of $(W, \vec{d})$ elementary operators supported in Ω.*

(a) $\{(aE_1 + bE_2, 2^{-j}) : (E_1, 2^{-j}), (E_2, 2^{-j}) \in \mathcal{E}, a, b \in \mathbb{C}, |a|, |b| \leq 1\}$ *is a bounded set of $(W, \vec{d})$ elementary operators supported in Ω.*

(b) *Suppose $\{a_l\}_{l \in \mathbb{N}}$ is a sequence of complex numbers with $\sum |a_l| < \infty$. Then*

$$\left\{\left(\sum_{l \in \mathbb{N}} a_l E_l, 2^{-j}\right) : \{(E_l, 2^{-j}) : l \in \mathbb{N}\} \subseteq \mathcal{E}\right\}$$

is a bounded set of $(W, \vec{d})$ elementary operators supported in Ω.

(c) *If $\psi \in C^\infty_{\mathrm{loc}}(\Omega)$, then*

$$\{(\mathrm{Mult}[\psi]E, 2^{-j}), (E\,\mathrm{Mult}[\psi], 2^{-j}) : (E, 2^{-j}) \in \mathcal{E}\}$$

is a bounded set of $(W, \vec{d})$ elementary operators supported in Ω.

(d) *$\mathcal{E}^* := \{(E^*, 2^{-j}) : (E, 2^{-j}) \in \mathcal{E}\}$ is a bounded set of $(W, \vec{d})$ elementary operators supported in Ω.*

(e) *Fix an ordered multi-index α. Then*

$$\{((2^{-j\vec{d}}W)^\alpha E, 2^{-j}), (E(2^{-j\vec{d}}W)^\alpha, 2^{-j}) : (E, 2^{-j}) \in \mathcal{E}\}$$

is a bounded set of $(W, \vec{d})$ elementary operators supported in Ω.

(f) *Suppose $\mu_0 \in \{1, \dots, \nu\}$ and $\mathscr{P}^{\mu_0}$ is a (W^{μ_0}, d^{μ_0}) partial differential operator in Ω of degree $\le \kappa_{\mu_0} \in \mathbb{Z}$. Then*

$$\{(2^{-j_{\mu_0}\kappa_{\mu_0}}\mathscr{P}^{\mu_0}E, 2^{-j}), (2^{-j_{\mu_0}\kappa_{\mu_0}}E\mathscr{P}^{\mu_0}, 2^{-j}) : (E, 2^{-j}) \in \mathcal{E}\}$$

is a bounded set of $(W, \vec{d})$ elementary operators supported in Ω.

(g) *Suppose $\mathscr{P}$ is a $(W, \vec{d})$ partial differential operator in Ω of degree $\le \kappa \in \mathbb{Z}^\nu$. Then*

$$\{(2^{-j\cdot\kappa}\mathscr{P}E, 2^{-j}), (2^{-j\cdot\kappa}E\mathscr{P}, 2^{-j}) : (E, 2^{-j}) \in \mathcal{E}\}$$

is a bounded set of $(W, \vec{d})$ elementary operators supported in Ω.

(h) *For each $N \in \mathbb{N}$ and $\mu \in \{1, \dots, \nu\}$, every $(E, 2^{-j}) \in \mathcal{E}$ can be written as*

$$E = \sum_{|\alpha_\mu| \le N} 2^{(|\alpha_\mu| - N)j_\mu}(2^{-j_\mu d^\mu}W^\mu)^{\alpha_\mu}E_{\mu,\alpha_\mu},$$

where $\{(E_{\mu,\alpha_\mu}, 2^{-j}) : (E, 2^{-j}) \in \mathcal{E}, \mu \in \{1, \dots, \nu\}, |\alpha_\mu| \le N\}$ is a bounded set of $(W, \vec{d})$ elementary operators supported in Ω. Similarly, we have

$$E = \sum_{|\alpha_\mu| \le N} 2^{(|\alpha_\mu| - N)j_\mu}\widetilde{E}_{\mu,\alpha_\mu}(2^{-j_\mu d^\mu}W^\mu)^{\alpha_\mu},$$

where $\{(\widetilde{E}_{\mu,\alpha_\mu}, 2^{-j}) : (E, 2^{-j}) \in \mathcal{E}, \mu \in \{1, \dots, \nu\}, |\alpha_\mu| \le N\}$ is a bounded set of $(W, \vec{d})$ elementary operators supported in Ω.

General setting: *All of the above holds with "$(W, \vec{d})$ elementary operators" replaced with "generalized $(W, \vec{d})$ elementary operators" throughout.*

To prove Proposition 5.5.5, we use the next lemma.

Lemma 5.5.6. Hörmander setting: *Let $\mathcal{E}$ be a bounded set of $(W, \vec{d})$ pre-elementary operators supported in Ω. Then*

$$\{(E^*, 2^{-j}) : (E, 2^{-j}) \in \mathcal{E}\}$$

is a bounded set of $(W, \vec{d})$ pre-elementary operators supported in Ω.

General setting: *The same result holds with $(W, \vec{d})$ pre-elementary operators replaced by generalized $(W, \vec{d})$ pre-elementary operators.*

Proof. In the general setting, this is part of the definition. In the Hörmander setting, since $E^*(x,z) = \overline{E(z,x)}$, this follows immediately from the definitions and Proposition 5.4.3. □

Proof of Proposition 5.5.5. The proofs in the general setting are nearly identical to the proofs in the Hörmander setting, so we prove only the results in the Hörmander setting. Items (a) and (b) follow easily from the definitions.

In light of Proposition 5.5.1, (f) follows from Lemma 5.5.4. Item (c) is the special case $\kappa_{\mu_0} = 0$ of (f). It suffices to prove (e) in the case $|\alpha| = 1$, in which case it is a special case of (f).

We now consider (g). When $\kappa \in \mathbb{Z}^\nu \setminus \{0\}$, $\mathscr{P} = 0$ and (g) is trivial. When $\kappa \in \mathbb{N}^\nu$, note that $2^{-j\cdot\kappa}\mathscr{P}$ can be written as a finite linear combination, with coefficients bounded by 1, of operators of the form $\mathrm{Mult}[\psi](2^{-j\vec{d}}W)^\alpha$, where $\psi \in C^\infty_{\mathrm{loc}}(\Omega)$ and $\deg_{\vec{d}}(\alpha) \le \kappa$. From here, (g) follows from (a), (c), and (e).

We turn to (h). By applying Definition 5.2.15 N times, we have

$$E = \sum_{|\alpha_\mu|,|\beta_\mu|\le N} 2^{(|\alpha_\mu|-N)j_\mu}2^{(|\beta_\mu|-N)j_\mu}(2^{-j_\mu d^\mu}W^\mu)^{\alpha_\mu}\widehat{E}_{\mu,\alpha_\mu,\beta_\mu}(2^{-j_\mu d^\mu}W^\mu)^{\beta_\mu}$$

$$=: \begin{cases} \sum_{|\alpha_\mu|\le N} 2^{(|\alpha_\mu|-N)j_\mu}(2^{-j_\mu d^\mu}W^\mu)^{\alpha_\mu}E_{\mu,\alpha_\mu}, \\ \sum_{|\beta_\mu|\le N} 2^{(|\beta_\mu|-N)j_\mu}\widetilde{E}_{\mu,\beta_\mu}(2^{-j_\mu d^\mu}W^\mu)^{\beta_\mu}, \end{cases}$$

where $\{(\widehat{E}_{\mu,\alpha_\mu,\beta_\mu}, 2^{-j}) : (E, 2^{-j}) \in \mathcal{E}, |\alpha_\mu|, |\beta_\mu| \le N\}$ is a bounded set of $(W, \vec{d})$ elementary operators supported in Ω. By (a) and (e),

$$\{(E_{\mu,\alpha_\mu}, 2^{-j}), (\widetilde{E}_{\mu,\beta_\mu}, 2^{-j}) : (E, 2^{-j}) \in \mathcal{E}, |\alpha_\mu|, |\beta_\mu| \le N\}$$

is a bounded set of elementary operators supported in Ω. This proves (h).

We turn to (d). By Lemma 5.5.6, $\mathcal{E}^*$ is a bounded set of $(W, \vec{d})$ pre-elementary operators supported in Ω. Thus, the result will follow if, for each $\mu \in \{1, \ldots, \nu\}$, we can write E^* in the form (5.26), where each E_{μ,l_1,l_2} is of the same form as E^*. Indeed, applying Proposition 5.5.1 to E, there are finite sets

$$\mathcal{F}_1^\mu = \{(\mathscr{P}^\mu_{1,1}, \kappa^\mu_{1,1}), \ldots, (\mathscr{P}^\mu_{1,L_1}, \kappa^\mu_{1,L_1})\},$$
$$\mathcal{F}_2^\mu = \{(\mathscr{P}^\mu_{2,1}, \kappa^\mu_{2,1}), \ldots, (\mathscr{P}^\mu_{2,L_2}, \kappa^\mu_{2,L_2})\},$$

where each $\mathscr{P}^\mu_{a,l}$ is a (W^μ, d^μ) partial differential operator in Ω of degree $\le \kappa^\mu_{a,l}$, $\kappa^\mu_{a,l} \in \mathbb{N}$, and such that $\forall (E, 2^{-j}) \in \mathcal{E}$,

$$E = \sum_{l_1=1}^{L_1}\sum_{l_2=1}^{L_2} 2^{-\epsilon^{\mu}_{1,l_1} j_\mu - \epsilon^{\mu}_{2,l_2} j_\mu} 2^{-j_\mu \kappa^{\mu}_{1,l_1}} \mathscr{P}^{\mu}_{1,l_1} \widehat{E}_{\mu,l_1,l_2} 2^{-j_\mu \kappa^{\mu}_{2,l_2}} \mathscr{P}^{\mu}_{2,l_2}, \tag{5.39}$$

where $\epsilon^{\mu}_{a,l} = \begin{cases} 1, & \kappa^{\mu}_{a,l} = 0, \\ 0, & \text{otherwise}, \end{cases}$ and

$$\{(\widehat{E}_{\mu,l_1,l_2}, 2^{-j}) : (E, 2^{-j}) \in \mathcal{E}, l \le l_1 \le L_1, 1 \le l_2 \le L_2\}$$

is a bounded set of $(W, \vec{d})$ elementary operators supported in Ω. Taking the adjoint of (5.39), we have

$$E^* = \sum_{l_1=1}^{L_1}\sum_{l_2=1}^{L_2} 2^{-\epsilon^{\mu}_{1,l_1} j_\mu - \epsilon^{\mu}_{2,l_2} j_\mu} 2^{-j_\mu \kappa^{\mu}_{2,l_2}} (\mathscr{P}^{\mu}_{2,l_2})^* \widehat{E}^*_{\mu,l_1,l_2} 2^{-j_\mu \kappa^{\mu}_{1,l_1}} (\mathscr{P}^{\mu}_{1,l_1})^*.$$

Since $(\mathscr{P}^{\mu}_{a,l})^*$ is a $(W, \vec{d})$ partial differential operator in Ω of degree $\le \kappa^{\mu}_{a,l}$, this sees E^* as being of the desired form, completing the proof. □

It will be convenient to have another characterization of bounded sets of (generalized) $(W, \vec{d})$ elementary operators, which is very similar to Proposition 5.5.1. It serves as a converse to Proposition 5.5.5 (h).

Corollary 5.5.7. Hörmander setting: *Let $\mathcal{G}''_\Omega$ be the largest set of subsets of $C_0^\infty(\Omega \times \Omega) \times (0,1]^\nu$ such that for all $\mathcal{E} \in \mathcal{G}''_\Omega$:*

- *$\mathcal{E}$ is a bounded set of $(W, \vec{d})$ pre-elementary operators supported in Ω.*
- *For each $\mu \in \{1, \dots, \nu\}$, there are finite sets depending on $\mathcal{E}$,*

$$\mathcal{F}^{\mu}_1 = \{(\mathscr{P}^{\mu}_{1,1}, \kappa^{\mu}_{1,1}), \dots, (\mathscr{P}^{\mu}_{1,L_1}, \kappa^{\mu}_{1,L_1})\},$$
$$\mathcal{F}^{\mu}_2 = \{(\mathscr{P}^{\mu}_{2,1}, \kappa^{\mu}_{2,1}), \dots, (\mathscr{P}^{\mu}_{2,L_2}, \kappa^{\mu}_{2,L_2})\},$$

where each $\mathscr{P}^{\mu}_{a,l}$ is a $(W^\mu, \vec{d}^\mu)$ partial differential operator in Ω of degree $\le \kappa^{\mu}_{a,l}$, $\kappa^{\mu}_{a,l} \in \mathbb{N}$, and such that $\forall (E, 2^{-j}) \in \mathcal{E}$,

$$E = \begin{cases} \sum_{l_1=1}^{L_1} 2^{-\epsilon^{\mu}_{1,l_1} j_\mu} 2^{-j_\mu \kappa^{\mu}_{1,l_1}} \mathscr{P}^{\mu}_{1,l_1} E_{\mu,l_1}, \\ \sum_{l_2=1}^{L_2} 2^{-\epsilon^{\mu}_{2,l_2} j_\mu} \widetilde{E}_{\mu,l_2} 2^{-j_\mu \kappa^{\mu}_{2,l_2}} \mathscr{P}^{\mu}_{2,l_2}, \end{cases} \tag{5.40}$$

where $\epsilon^{\mu}_{a,l} = \begin{cases} 1, & \kappa^{\mu}_{a,l} = 0, \\ 0, & \text{otherwise}, \end{cases}$ and

$$\{(E_{\mu,l_1}, 2^{-j}), (\widetilde{E}_{\mu,l_2}, 2^{-j}) : (E, 2^{-j}) \in \mathcal{E}, l \le l_1 \le L_1, 1 \le l_2 \le L_2\} \in \mathcal{G}''_\Omega.$$

Then $\mathcal{G}''_\Omega = \mathcal{G}_\Omega$, where $\mathcal{G}_\Omega$ is the set of bounded sets of elementary operators supported in Ω.

General setting: *We define* $\widetilde{\mathscr{G}}''_\Omega$ *as above, but with bounded sets of* $(W,\vec{d})$ *pre-elementary operators replaced by bounded sets of generalized* $(W,\vec{d})$ *pre-elementary operators and* $C_0^\infty(\Omega\times\Omega)$ *replaced by* $\mathrm{Hom}(C^\infty_{W,0}(\mathfrak{M}), C^\infty_{W,0}(\mathfrak{M})')$. *Then* $\widetilde{\mathscr{G}}''_\Omega = \widetilde{\mathscr{G}}_\Omega$, *where* $\widetilde{\mathscr{G}}_\Omega$ *is the set of bounded sets of generalized* $(W,\vec{d})$ *elementary operators supported in* Ω.

Proof. We prove only the result in the Hörmander setting. The proof in the general setting is nearly identical. It follows from Proposition 5.5.5 (h) with $N = 1$ that $\mathscr{G}_\Omega \subseteq \mathscr{G}''_\Omega$. Applying the definition of $\mathscr{G}''_\Omega$ twice, it follows that $\mathscr{G}''_\Omega \subseteq \mathscr{G}'_\Omega = \mathscr{G}_\Omega$. □

Proposition 5.5.8. *In the Hörmander setting (which is a special case of the general setting), if* $\mathcal{E}$ *is a bounded set of* $(W,\vec{d})$ *elementary operators supported in* Ω, *then* $\mathcal{E}$ *is a bounded set of generalized* $(W,\vec{d})$ *elementary operators supported in* Ω.

To prove Proposition 5.5.8, we require the next lemma.

Lemma 5.5.9. *In the Hörmander setting, if* $\mathcal{E}$ *is a bounded set of* $(W,\vec{d})$ *pre-elementary operators supported in* Ω, *then* $\mathcal{E}$ *is a bounded set of generalized* $(W,\vec{d})$ *pre-elementary operators supported in* Ω.

Proof. Fix ordered multi-indices α,β. We wish to show

$$\sup_{(E,2^{-j})\in\mathcal{E}} \|(2^{-j\vec{d}}W)^\alpha E(2^{-j\vec{d}}W)^\beta\|_{L^1(\mathfrak{M},\mathrm{Vol})\to L^1(\mathfrak{M},\mathrm{Vol})} < \infty,$$

$$\sup_{(E,2^{-j})\in\mathcal{E}} \|(2^{-j\vec{d}}W)^\alpha E(2^{-j\vec{d}}W)^\beta\|_{L^\infty(\mathfrak{M},\mathrm{Vol})\to L^\infty(\mathfrak{M},\mathrm{Vol})} < \infty,$$

$$\sup_{\mathcal{E}'} \|\mathcal{T}_{\mathcal{E}',\alpha,\beta}\|_{L^p(\mathfrak{M},\mathrm{Vol};\ell^q(\mathbb{N}))\to L^p(\mathfrak{M},\mathrm{Vol};\ell^q(\mathbb{N}))} < \infty,$$

where $\mathcal{T}_{\mathcal{E}',\alpha,\beta}$ is defined by (5.7) and the supremum is taken over all countable subsets $\mathcal{E}' \subseteq \mathcal{E}$. This will complete the proof since Lemma 5.5.6 shows that $\mathcal{E}^* := \{(E^*,2^{-j}) : (E,2^{-j}) \in \mathcal{E}\}$ is a bounded set of $(W,\vec{d})$ elementary operators supported in Ω, and therefore the above estimates also hold with $\mathcal{E}$ replaced by $\mathcal{E}^*$.

Lemma 5.5.2 (iii) shows that it suffices to prove the above estimates in the case where $|\alpha| = |\beta| = 0$. From here, the result follows from Corollary 5.4.11 and Lemma 5.4.12. □

Proof of Proposition 5.5.8. The definitions of bounded sets of $(W,\vec{d})$ elementary operators and bounded sets of generalized $(W,\vec{d})$ elementary operators are formally the same and only differ due to the difference between bounded sets of $(W,\vec{d})$ pre-elementary operators versus bounded sets of generalized $(W,\vec{d})$ pre-elementary operators. Thus, the result follows from Lemma 5.5.9. □

Proposition 5.5.10. Hörmander setting: *Let* $t \in \mathbb{R}^\nu$ *and let* $\{(E_j,2^{-j}) : j \in \mathbb{N}^\nu\}$ *be a bounded set of* $(W,\vec{d})$ *elementary operators supported in* Ω. *Then*

$$\sum_{j\in\mathbb{N}^\nu} 2^{j\cdot t}E_j$$

converges in the topology of bounded convergence on $\mathrm{Hom}(C^\infty_{\mathrm{loc}}(\mathfrak{M}), C^\infty_0(\mathfrak{M}))$*, in the topology of bounded convergence on* $\mathrm{Hom}(C^\infty_0(\mathfrak{M})', C^\infty_{\mathrm{loc}}(\mathfrak{M})')$*, and in* $C^\infty_0(\mathfrak{M}\times\mathfrak{M})'$*.*

General setting: *Let* $t \in \mathbb{R}^\nu$ *and let* $\{(E_j, 2^{-j}) : j \in \mathbb{N}^\nu\}$ *be a bounded set of generalized* $(W, \vec{d})$ *elementary operators supported in* Ω*. Then*

$$\sum_{j\in\mathbb{N}^\nu} 2^{j\cdot t} E_j$$

converges in the topology of bounded convergence on $\mathrm{Hom}(C^\infty_{W,\mathrm{loc}}(\mathfrak{M}), C^\infty_{W,0}(\mathfrak{M}))$*, in the topology of bounded convergence on* $\mathrm{Hom}(C^\infty_{W,0}(\mathfrak{M})', C^\infty_{W,\mathrm{loc}}(\mathfrak{M})')$*, and in* $C^\infty_0(\mathfrak{M}\times\mathfrak{M})'$*.*

Proof. In light of Lemma 5.5.9 and Remark 5.2.24, the result in the Hörmander setting follows from the result in the general setting. Thus, we consider only the general setting.

We begin by showing the convergence in the topology of bounded convergence on $\mathrm{Hom}(C^\infty_{W,\mathrm{loc}}(\mathfrak{M}), C^\infty_{W,0}(\mathfrak{M}))$. Let $\mathcal{B} \subset C^\infty_{W,\mathrm{loc}}(\mathfrak{M})$ be a bounded set. We will show that for every ordered multi-index α, the sum

$$\sum_{j\in\mathbb{N}^\nu} 2^{j\cdot t} W^\alpha E_j f(x) \tag{5.41}$$

converges absolutely and uniformly for $x \in \mathfrak{M}$, $f \in \mathcal{B}$. This will complete the proof since $W^\alpha E_j f(x)$ is continuous, by hypothesis, supported in Ω, and $\overline{\Omega}$ is compact.

Note that

$$\sum_{j\in\mathbb{N}^\nu} 2^{j\cdot t} W^\alpha E_j f = \sum_{j\in\mathbb{N}^\nu} 2^{j\cdot(t+\deg_{\vec{d}}(\alpha))}((2^{-j\vec{d}}W)^\alpha E_j) f.$$

Proposition 5.5.5 (e) shows that

$$\{((2^{-j\vec{d}}W)^\alpha E_j, 2^{-j}) : j \in \mathbb{N}^\nu\}$$

is a bounded set of generalized $(W, \vec{d})$ elementary operators supported in Ω. Thus, by replacing t with $t + \deg_{\vec{d}}(\alpha)$, it suffices to prove the convergence of (5.41) in the special case $|\alpha| = 0$.

Fix $\mu \in \{1, \dots, \nu\}$. We will show that

$$\sum_{\substack{j\in\mathbb{N}^\nu\\ j_\mu = |j|_\infty}} 2^{j\cdot t} W^\alpha E_j f(x) \tag{5.42}$$

converges absolutely and uniformly for $x \in \mathfrak{M}$, $f \in \mathcal{B}$; the result will then follow.

Take $N \ge |t|_1 + 1$. By Proposition 5.5.5 (h) we have

$$\sum_{\substack{j\in\mathbb{N}^\nu\\ j_\mu=|j|_\infty}} 2^{j\cdot t}W^\alpha E_j f(x) = \sum_{\substack{j\in\mathbb{N}^\nu\\ j_\mu=|j|_\infty}}\sum_{|\alpha_\mu|\le N} 2^{j\cdot t}2^{-(N-|\alpha_\mu|)j_\mu}E_{j,\mu,\alpha_\mu}(2^{-j_\mu d^\mu}W^\mu)^{\alpha_\mu}f(x)$$
$$= \sum_{\substack{j\in\mathbb{N}^\nu\\ j_\mu=|j|_\infty}}\sum_{|\alpha_\mu|\le N} 2^{-j_\mu}(2^{j\cdot t-(N-1)j_\mu-(\deg_{d^\mu}(\alpha_\mu)-|\alpha_\mu|)j_\mu}E_{j,\mu,\alpha_\mu})((W^\mu)^{\alpha_\mu}f)(x)$$
$$=: \sum_{\substack{j\in\mathbb{N}^\nu\\ j_\mu=|j|_\infty}}\sum_{|\alpha_\mu|\le N} 2^{-j_\mu}\widehat{E}_{j,\mu,\alpha_\mu}((W^\mu)^{\alpha_\mu}f)(x),$$

where $\{(E_{j,\mu,\alpha_\mu},2^{-j}) : j\in\mathbb{N}^\nu, |j|_\infty = j_\mu, |\alpha_\mu|\le N\}$ is a bounded set of generalized $(W,\vec{d})$ elementary operators supported in Ω. The choice of N and the fact that $|j|_\infty = j_\mu$ shows that $2^{j\cdot t-(N-1)j_\mu-(\deg_{d^\mu}(\alpha_\mu)-|\alpha_\mu|)j_\mu}\in(0,1]$, and therefore Proposition 5.5.5 (a) implies that $\{(\widehat{E}_{j,\mu,\alpha_\mu},2^{-j}) : j\in\mathbb{N}^\nu, |j|_\infty = j_\mu, |\alpha_\mu|\le N\}$ is a bounded set of generalized $(W,\vec{d})$ elementary operators supported in Ω. Since $\{(W^\mu)^{\alpha_\mu}f : f\in\mathcal{B}\}$ is a bounded subset of $C^\infty_{W,\mathrm{loc}}(\mathfrak{M})$, this shows that it suffices to prove the convergence of (5.42) in the special case where $t=-\mathfrak{e}_\mu$.

Let $\phi\in C_0^\infty(\mathfrak{M})$ equal 1 on Ω. Since $\operatorname{supp}(E_j)\subseteq\Omega$, we have $E_jf = E_j\phi f$. The definition of generalized $(W,\vec{d})$ pre-elementary operators shows that $\|E_jf\|_{L^\infty} = \|E_j\phi f\|_{L^\infty}\lesssim \|\phi f\|_{L^\infty}$. Thus, considering (5.42) in the special case $t=-\mathfrak{e}_\mu$, we have

$$\sum_{\substack{j\in\mathbb{N}^\nu\\ j_\mu=|j|_\infty}} |2^{-j_\mu}W^\alpha E_jf(x)| \lesssim \sum_{\substack{j\in\mathbb{N}^\nu\\ j_\mu=|j|_\infty}} 2^{-j_\mu}\|\phi f\|_{L^\infty} \lesssim \sum_{j\in\mathbb{N}^\nu} 2^{-|j|_\infty} < \infty,$$

completing the proof of the convergence in the topology of bounded convergence on $\operatorname{Hom}(C^\infty_{W,\mathrm{loc}}(\mathfrak{M}), C^\infty_{W,0}(\mathfrak{M}))$. The convergence in $C_0^\infty(\mathfrak{M}\times\mathfrak{M})'$ is an immediate consequence of this.

Finally, we turn to the convergence in the topology of bounded convergence on $\operatorname{Hom}(C^\infty_{W,0}(\mathfrak{M})', C^\infty_{W,\mathrm{loc}}(\mathfrak{M})')$. By Proposition 5.5.5 (d), $\{(E_j^*,2^{-j}) : j\in\mathbb{N}^\nu\}$ is a bounded set of generalized $(W,\vec{d})$ elementary operators supported in Ω, and therefore we have shown that $\sum_{j\in\mathbb{N}^\nu}2^{j\cdot t}E_j^*$ converges in the topology of bounded convergence on $\operatorname{Hom}(C^\infty_{W,\mathrm{loc}}(\mathfrak{M}), C^\infty_{W,0}(\mathfrak{M}))$. Thus, the claimed convergence in $\operatorname{Hom}(C^\infty_{W,0}(\mathfrak{M})', C^\infty_{W,\mathrm{loc}}(\mathfrak{M})')$ follows by duality. □

The next proposition shows that elementary operators form an algebra, and elementary operators at different scales are "almost orthogonal."

Proposition 5.5.11. Hörmander setting: *Suppose $\mathcal{E}$ is a bounded set of $(W,\vec{d})$ elementary operators supported in Ω. Then, for every $N\in\mathbb{N}$,*

$$\{(2^{N|j-k|}E_1E_2,2^{-j}),(2^{N|j-k|}E_1E_2,2^{-k}) : (E_1,2^{-j}),(E_2,2^{-k})\in\mathcal{E}\}$$

is a bounded set of $(W,\vec{d})$ elementary operators supported in Ω.

General setting: *The same result holds with "$(W, \vec{d})$ elementary operators" replaced with "generalized $(W, \vec{d})$ elementary operators" throughout.*

The proof of Proposition 5.5.11 requires several lemmas.

Lemma 5.5.12. Hörmander setting: *Let $\mathcal{E}$ be a bounded set of $(W, \vec{d})$ pre-elementary operators supported in Ω. Then, for all $m \in \mathbb{N}$ and all ordered multi-indices α, β, there exist $N = N(m, \alpha, \beta, \vec{d}) \in \mathbb{N}$, $C = C(m, \alpha, \beta, \mathcal{E}, (W, \vec{d})) \geq 0$ such that for all $(F_1, 2^{-j}), (F_2, 2^{-k}) \in \mathcal{E}$ and all $l \in \{j, k, j \wedge k\}$,*

$$\begin{aligned}2^{-N|j-k|}&|(2^{-l\vec{d}}W_x)^{\alpha}(2^{-l\vec{d}}W)^{\beta}[F_1F_2](x,z)|\\ &\leq C\frac{(1+\rho_{2^{-l}}(x,z))^{-m}}{\operatorname{Vol}(B_{2^{-l}}(x,1+\rho_{2^{-l}}(x,z)))\wedge 1}.\end{aligned} \tag{5.43}$$

General setting: *Let $\mathcal{E}$ be a bounded set of generalized $(W, \vec{d})$ pre-elementary operators supported in Ω. Then, for all ordered multi-indices α, β, there exists $N = N(\alpha, \beta, \vec{d})$ such that the following hold:*

- *For $p = 1, \infty$, $\sup_{\substack{(F_1,2^{-j})\in\mathcal{E}\\(F_2,2^{-k})\in\mathcal{E}}} \max_{l\in\{j,k,j\wedge k\}} 2^{-N|j-k|}\|(2^{-l\vec{d}}W)^{\alpha}F_1F_2(2^{-l\vec{d}}W)^{\beta}\|_{L^p\to L^p} < \infty$.*
- *For every countable set $\mathcal{E}' = \{(F_{m,1}, 2^{-j_m}), (F_{m,2}, 2^{-k_m}) : m \in \mathbb{N}\} \subseteq \mathcal{E}$ and every sequence $\{l_m\}_{m\in\mathbb{N}}$ with $l_m \in \{j_m, k_m, j_m \wedge k_m\}$, set*

$$\mathcal{T}_{\mathcal{E}',\alpha,\beta,\{l_m\}}\{f_m\}_{m\in\mathbb{N}} := \{2^{-N|j_m-k_m|}(2^{-l_m\vec{d}}W)^{\alpha}F_{m,1}F_{m,2}(2^{-l_m\vec{d}}W)^{\beta}f_m\}_{m\in\mathbb{N}}.$$

  Then, for $p \in (1, \infty)$, $q \in (1, \infty]$,

$$\sup_{\mathcal{E}',\{l_m\}}\|\mathcal{T}_{\mathcal{E}',\alpha,\beta,\{l_m\}}\|_{L^p(\ell^q)\to L^p(\ell^q)} < \infty,$$

  where the supremum is taken over all such choices of $\mathcal{E}'$ and $\{l_m\}$.
- *The above also hold with $\mathcal{E}$ replaced by $\{(E^*, 2^{-j}) : (E, 2^{-j}) \in \mathcal{E}\}$.*

Proof. We begin with the proof in the Hörmander setting. Let $N_1 := N + |\deg_{\vec{d}}(\alpha)| + |\deg_{\vec{d}}(\beta)|$. Then to prove (5.43), it suffices to show that

$$\begin{aligned}2^{-N_1|j-k|}&|(2^{-j\vec{d}}W_x)^{\alpha}(2^{-k\vec{d}}W)^{\beta}[F_1F_2](x,z)|\\ &\leq C\frac{(1+\rho_{2^{-l}}(x,z))^{-m}}{\operatorname{Vol}(B_{2^{-l}}(x,1+\rho_{2^{-l}}(x,z)))\wedge 1}.\end{aligned} \tag{5.44}$$

From the definition of $(W, \vec{d})$ pre-elementary operators supported in Ω (Definition 5.2.14), the left-hand side of (5.44) is zero unless $x, z \in \Omega$; so we henceforth assume $x, z \in \Omega$.

From the definitions, we have, using Proposition 5.4.4 and taking $m' = m'(m, \Omega, (W, \vec{d}))$ sufficiently large,

$$\begin{aligned}
&|(2^{-j\vec{d}} W_x)^{\alpha} (2^{-k\vec{d}} W)^{\beta} [F_1 F_2](x,z)| \\
&\lesssim \int_{\Omega} \left(\frac{(1+\rho_{2^{-j}}(x,y))^{-m'}}{\mathrm{Vol}(B_{2^{-j}}(x, 1+\rho_{2^{-j}}(x,y))) \wedge 1} \right) \left(\frac{(1+\rho_{2^{-k}}(y,z))^{-m'}}{\mathrm{Vol}(B_{2^{-k}}(y, 1+\rho_{2^{-k}}(y,z))) \wedge 1} \right) d\,\mathrm{Vol}(y) \\
&\lesssim \frac{(1+\rho_{2^{-j\wedge k}}(x,z))^{-m}}{\mathrm{Vol}(B_{2^{-j\wedge k}}(x, 1+\rho_{2^{-j\wedge k}}(x,z))) \wedge 1}.
\end{aligned}$$

Thus, we have

$$\begin{aligned}
&2^{-N_1|j-k|} |(2^{-j\vec{d}} W_x)^{\alpha} (2^{-k\vec{d}} W)^{\beta} [F_1 F_2](x,z)| \\
&\lesssim 2^{-N_1|j-k|} \frac{(1+\rho_{2^{-j\wedge k}}(x,z))^{-m}}{\mathrm{Vol}(B_{2^{-j\wedge k}}(x, 1+\rho_{2^{-j\wedge k}}(x,z))) \wedge 1}.
\end{aligned}$$

Taking $N_1 = N_1(m, (W, \vec{d}), \Omega)$ sufficiently large, (5.44) now follows from Lemma 5.4.7.

In the general setting, by taking $N = |\deg_{\vec{d}}(\alpha)| + |\deg_{\vec{d}}(\beta)|$, we see that it suffices to prove the result with N replaced by 0, $(2^{-l\vec{d}} W)^{\alpha}$ replaced by $(2^{-j\vec{d}} W)^{\alpha}$, $(2^{-l\vec{d}} W)^{\beta}$ replaced by $(2^{-k\vec{d}} W)^{\beta}$, $(2^{-l_m\vec{d}} W)^{\alpha}$ replaced by $(2^{-j_m\vec{d}} W)^{\alpha}$, and $(2^{-l_m\vec{d}} W)^{\beta}$ replaced by $(2^{-k_m\vec{d}} W)^{\beta}$. With these replacements, the result follows immediately from the definition of bounded sets of generalized $(W, \vec{d})$ pre-elementary operators (Definition 5.2.26). Here we are using the fact that the definition of bounded sets of generalized $(W, \vec{d})$ pre-elementary operators implies that $\{(E^*, 2^{-j}) : (E, 2^{-j}) \in \mathcal{E}\}$ is also a bounded set of generalized $(W, \vec{d})$ pre-elementary operators supported in Ω. □

Lemma 5.5.13. *Fix $\mu \in \{1, \dots, \nu\}$.*

Hörmander setting: *Let $\mathcal{E}_1$ be a bounded set of $(W, \vec{d})$ elementary operators supported in Ω and let $\mathcal{E}_2$ be a bounded set of $(W, \vec{d})$ pre-elementary operators supported in Ω. Then, for every $N \in \mathbb{N}$,*

$$\begin{aligned}
\Big\{&(2^{N|j-k|} E_1 E_2, 2^{-l}), (2^{N|j-k|} E_2 E_1, 2^{-l}) : (E_1, 2^{-j}) \in \mathcal{E}_1, (E_2, 2^{-k}) \in \mathcal{E}_2, \\
&l \in \{j, k, j \wedge k\}, j_\mu - k_\mu = |j-k|_\infty \Big\}
\end{aligned}$$

is a bounded set of $(W, \vec{d})$ pre-elementary operators supported in Ω.

General setting: *The same result holds with "$(W, \vec{d})$ elementary operators" and "$(W, \vec{d})$ pre-elementary operators" replaced with "generalized $(W, \vec{d})$ elementary operators" and "generalized $(W, \vec{d})$ pre-elementary operators," respectively, throughout.*

Proof. We begin with the Hörmander setting. We only prove that

$$\begin{aligned}
\Big\{&(2^{N|j-k|} E_1 E_2, 2^{-l}) : (E_1, 2^{-j}) \in \mathcal{E}_1, (E_2, 2^{-k}) \in \mathcal{E}_2, \\
&l \in \{j, k, j \wedge k\}, j_\mu - k_\mu = |j-k|_\infty \Big\}
\end{aligned}$$

is a bounded set of $(W, \vec{d})$ elementary operators supported in Ω. The result with E_1E_2 replaced by E_2E_1 has a similar proof, and we leave it to the reader.

That

$$\bigcup_{\substack{(E_1,2^{-j})\in\mathcal{E}_1\\(E_2,2^{-k})\in\mathcal{E}_2}} \operatorname{supp}(E_1E_2) \Subset \Omega\times\Omega$$

follows immediately from the definitions. Thus, the result will follow once we show that for every $N_0 \in \mathbb{N}$, $m \in \mathbb{N}$, and ordered multi-indices α, β, there exists $C = C(\alpha,\beta,N_0,m,\mathcal{E}_1,\mathcal{E}_2) \geq 0$ such that for all $(E_1,2^{-j}) \in \mathcal{E}_1$, $(E_2,2^{-k}) \in \mathcal{E}_2$ with $j_\mu - k_\mu = |j-k|_\infty$, for all $x, z \in \mathfrak{M}$ and $l \in \{j, k, j\wedge k\}$,

$$\begin{aligned} &2^{N_0|j-k|_\infty}\left|(2^{-l\vec{d}}W_x)^\alpha(2^{-l\vec{d}}W_z)^\beta[E_1E_2](x,z)\right| \\ &\quad\leq C\frac{(1+\rho_{2^{-l}}(x,z))^{-m}}{\operatorname{Vol}(B_{2^{-l}}(x,1+\rho_{2^{-l}}(x,z)))\wedge 1}. \end{aligned} \tag{5.45}$$

Fix $N = N(m,\alpha,\beta,N_0,\vec{d})$ large to be chosen later. We claim that it suffices to prove (5.45) with N_0 replaced by $N_0 - N$. Indeed, by Proposition 5.5.5 (h), we can write

$$E_1 = \sum_{|\alpha_\mu|\leq N} 2^{-j_\mu(N-|\alpha_\mu|)}E_{1,\alpha_\mu}(2^{-j_\mu d^\mu}W^\mu)^{\alpha_\mu},$$

where $\{(E_{1,\alpha_\mu},2^{-j}) : (E,2^{-j}) \in \mathcal{E}_1, |\alpha_\mu| \leq N\}$ is a bounded set of $(W,\vec{d})$ elementary operators supported in Ω. Thus,

$$\begin{aligned} E_1E_2 &= \sum_{|\alpha_\mu|\leq N} 2^{-j_\mu(N-|\alpha_\mu|)-(j_\mu-k_\mu)\deg_{d^\mu}(\alpha_\mu)}E_{1,\alpha_\mu}[(2^{-k_\mu d^\mu}W^\mu)^{\alpha_\mu}E_2] \\ &=: \sum_{|\alpha_\mu|\leq N} 2^{-N(j_\mu-k_\mu)}E_{1,\alpha_\mu}E_{2,\alpha_\mu,j,k}, \end{aligned} \tag{5.46}$$

where

$$E_{2,\alpha_\mu,j,k} := 2^{N(j_\mu-k_\mu)-j_\mu(N-|\alpha_\mu|)-(j_\mu-k_\mu)\deg_{d^\mu}(\alpha_\mu)}(2^{-k_\mu d^\mu}W^\mu)^{\alpha_\mu}E_2.$$

Since $|\alpha_\mu| \leq \deg_{d^\mu}(\alpha_\mu)$, it follows from Lemma 5.5.2 (ii) and (iii) that

$$\{(E_{2,\alpha_\mu,j,k},2^{-k}) : (E_2,2^{-k}) \in \mathcal{E}_2, |\alpha_\mu| \leq N, k \in [0,\infty)^\nu \text{ with } j_\mu - k_\mu = |j-k|_\infty\}$$

is a bounded set of $(W,\vec{d})$ pre-elementary operators supported in Ω.

Plugging (5.46) into (5.45) and using the fact that $j_\mu - k_\mu = |j-k|_\infty$, we see that it suffices to prove (5.45) with N_0 replaced by $N_0 - N$.

With N_0 replaced by $N_0 - N$, if N is sufficiently large, (5.45) follows from Lemma 5.5.12, completing the proof in the Hörmander case.

The proof in the general setting is very similar. As before, we only show that

$$\Big\{(2^{N|j-k|}E_1E_2, 2^{-l}) : (E_1, 2^{-j}) \in \mathcal{E}_1, (E_2, 2^{-k}) \in \mathcal{E}_2,$$
$$l \in \{j, k, j \wedge k\}, j_\mu - k_\mu = |j-k|_\infty\Big\}$$

is a bounded set of generalized $(W, \vec{d})$ elementary operators supported in Ω.

That

$$\bigcup_{\substack{(E_1,2^{-j})\in\mathcal{E}_1\\ (E_2,2^{-k})\in\mathcal{E}_2}} \operatorname{supp}(E_1E_2) \Subset \Omega \times \Omega$$

follows immediately from the definitions. Thus, the result will follow once we show that for every $N_0 \in \mathbb{N}$ we have for all ordered multi-indices α and β

$$\sup_{\substack{(E_1,2^{-j})\in\mathcal{E}_1\\ (E_2,2^{-k})\in\mathcal{E}_2\\ j_\mu-k_\mu=|j-k|_\infty}} \max_{l\in\{j,k,j\wedge k\}} \max_{p\in\{1,\infty\}} 2^{N_0|j-k|_\infty} \|(2^{-l\vec{d}}W)^\alpha E_1E_2(2^{-l\vec{d}}W)^\beta\|_{L^p\to L^p} < \infty$$

and for all countable sets we have

$$\mathcal{E}_1' = \{(E_{m,1}, 2^{-j_m}) : m \in \mathbb{N}\} \subseteq \mathcal{E}_1, \quad \mathcal{E}_2' = \{(E_{m,2}, 2^{-k_m}) : m \in \mathbb{N}\} \subseteq \mathcal{E}_2,$$

satisfying $(j_m)_\mu - (k_m)_\mu = |j_m - k_m|_\infty$. For every sequence $l_m \in \{j_m, k_m, j_m \wedge k_m\}$, set

$$\mathcal{T}_{N_0,\mathcal{E}_1',\mathcal{E}_2',\alpha,\beta,\{l_m\}}\{f_m\}_{m\in\mathbb{N}} := \{2^{N_0|j_m-k_m|_\infty}(2^{-l_m\vec{d}}W)^\alpha E_{m,1}E_{m,2}(2^{-l_m\vec{d}}W)^\beta f_m\}_{m\in\mathbb{N}}.$$

Then for $p \in (1,\infty)$, $q \in (1,\infty]$,

$$\sup_{\mathcal{E}_1',\mathcal{E}_2',\{l_m\}} \|\mathcal{T}_{N_0,\mathcal{E}_1',\mathcal{E}_2',\alpha,\beta,\{l_m\}}\|_{L^p(\mathfrak{M},\mathrm{Vol};\ell^q(\mathbb{N}))\to L^p(\mathfrak{M},\mathrm{Vol};\ell^q(\mathbb{N}))} < \infty,$$

where the supremum is taken over all such countable sets $\mathcal{E}_1'$ and $\mathcal{E}_2'$ and sequences $\{l_m\}$. We also require the above estimates with E_1E_2 replaced by $E_2^*E_1^*$ and $E_{m,1}E_{m,2}$ replaced by $E_{m,2}^*E_{m,1}^*$ throughout.

Just as in the proof of the Hörmander setting, by applying Proposition 5.5.5 (h) to E_1 and $E_{m,1}$, it suffices to prove the above estimates with N_0 replaced by N_0-N, where N is as large as we wish. By taking N sufficiently large, the estimates follow from Lemma 5.5.12. For the estimates with E_1E_2 replaced by $E_2^*E_1^*$ and $E_{m,1}E_{m,2}$ replaced by $E_{m,2}^*E_{m,1}^*$ throughout, note that Lemma 5.5.6 and Proposition 5.5.5 (d) show that E_1^*, E_2^*, $E_{1,m}^*$, and $E_{2,m}^*$ are of the same form as E_1, E_2, $E_{1,m}$, and $E_{2,m}$. Using this, a similar argument (reversing the roles of E_1 and E_2 and of $E_{1,m}$ and $E_{2,m}$) gives the desired estimates. We leave the details to the reader. □

Proof of Proposition 5.5.11. The proof in the general setting is nearly identical to the proof in the Hörmander setting, so we only prove the result in the Hörmander setting.

We prove that

$$\mathcal{E}_N := \{(2^{N|j-k|}E_1E_2, 2^{-j}) : (E_1, 2^{-j}), (E_2, 2^{-k}) \in \mathcal{E}\}$$

is a bounded set of $(W, \vec{d})$ elementary operators supported in Ω. The proof with $(2^{N|j-k|}E_1E_2, 2^{-j})$ replaced by $(2^{N|j-k|}E_1E_2, 2^{-k})$ is similar and we leave it to the interested reader.

Lemma 5.5.13 shows that $\mathcal{E}_N$ is a bounded set of $(W, \vec{d})$ pre-elementary operators supported in Ω. Thus, it suffices to show that $2^{N|j-k|}E_1E_2$ can be written in the form (5.3), where $E_{\mu,\alpha_\mu,\beta_\mu}$ is of the same form as $2^{N|j-k|}E_1E_2$, with a possibly different choice of N.

Fix $\mu \in \{1, \dots, \nu\}$. Letting $(E_1, 2^{-j}), (E_2, 2^{-k}) \in \mathcal{E}$ and applying Proposition 5.5.5 (h) with $N = 1$ to E_1 and E_2, we see that

$$\begin{aligned}
2^{N|j-k|}E_1E_2 &= \sum_{|\alpha_\mu|,|\beta_\mu|\le 1} 2^{N|j-k|-(1-|\alpha_\mu|)j_\mu-(1-|\beta_\mu|)k_\mu}(2^{-j_\mu d^\mu}W^\mu)^{\alpha_\mu}E_{1,\alpha_\mu}E_{2,\beta_\mu}(2^{-k_\mu d^\mu}W^\mu)^{\beta_\mu} \\
&= \sum_{|\alpha_\mu|,|\beta_\mu|\le 1} 2^{N|j-k|-(1-|\alpha_\mu|)j_\mu-(1-|\beta_\mu|)k_\mu-(k_\mu-j_\mu)\deg_{d^\mu}(\beta_\mu)} \\
&\qquad \times (2^{-j_\mu d^\mu}W^\mu)^{\alpha_\mu}E_{1,\alpha_\mu}E_{2,\beta_\mu}(2^{-j_\mu d^\mu}W^\mu)^{\beta_\mu} \\
&=: \sum_{|\alpha_\mu|,|\beta_\mu|\le 1} 2^{-(2-|\alpha_\mu|-|\beta_\mu|)}(2^{-j_\mu d^\mu}W^\mu)^{\alpha_\mu} \\
&\qquad \times 2^{(N+1-|\beta_\mu|+\deg_{d^\mu}(\beta_\mu))|j-k|}\widetilde{E}_{1,\alpha_\mu,\beta_\mu,k}E_{2,\beta_\mu}(2^{-j_\mu d^\mu}W^\mu)^{\beta_\mu},
\end{aligned}$$

where $\{(E_{1,\alpha_\mu}, 2^{-j}), (E_{2,\beta_\mu}, 2^{-k}) : (E_1, 2^{-j}), (E_2, 2^{-k}) \in \mathcal{E}, |\alpha_\mu|, |\beta_\mu| \le 1\}$ is a bounded set of $(W, \vec{d})$ elementary operators supported in Ω and

$$\widetilde{E}_{1,\alpha_\mu,\beta_\mu,k} := 2^{(1-|\beta_\mu|)(j_\mu-k_\mu)-(k_\mu-j_\mu)\deg_{d^\mu}(\beta_\mu)-|j-k|(1-|\beta_\mu|+\deg_{d^\mu}(\beta_\mu))}E_{1,\alpha_\mu}.$$

Since the above coefficient of $E_{1,\alpha}$ in the above equation is in $(0, 1]$, it follows from Proposition 5.5.5 (a) that $\{(\widetilde{E}_{1,\alpha_\mu,\beta_\mu,k}, 2^{-j}) : (E, 2^{-j}) \in \mathcal{E}, |\alpha_\mu|, |\beta_\mu| \le 1, k \in [0, \infty)^\nu\}$ is a bounded set of $(W, \vec{d})$ elementary operators supported in Ω.

Thus, we have written $2^{N|j-k|}E_1E_2$ in the form (5.3), where $E_{\mu,\alpha_\mu,\beta_\mu}$ is of the same form as $2^{N|j-k|}E_1E_2$, with N replaced by $N + 1 - |\beta_\mu| + \deg_{d^\mu}(\beta_\mu)$. This completes the proof. □

The notion of (generalized) $(W, \vec{d})$ elementary operators supported in Ω depends only on the weak equivalence class of $(W^1, d^1), \dots, (W^\nu, d^\nu)$, as we now discuss. For each $\mu \in \{1, \dots, \nu\}$, let

$$(Z^\mu, d^\mu) := \{(Z^\mu_1, d^\mu_1), \dots, (Z^\mu_{s_\mu}, d^\mu_{s_\mu})\} \subset C^\infty_{\mathrm{loc}}(\mathfrak{M}; T\mathfrak{M}) \times \mathbb{N}_+$$

be such that (Z^μ, d^μ) is locally weakly equivalent to (W^μ, d^μ) on $\mathfrak{M}$. Lemma 3.14.11 and Proposition 3.14.14 imply that in both the general setting and in the Hörmander setting,

$(Z^1, \vec{d}^1), \dots, (Z^\nu, \vec{d}^\nu)$ satisfies the same hypotheses as $(W^1, \vec{d}^1), \dots, (W^\nu, \vec{d}^\nu)$. We define $(Z, \vec{d}) := (Z^1, \vec{d}^1) \boxtimes (Z^2, \vec{d}^2) \boxtimes \cdots \boxtimes (Z^\nu, \vec{d}^\nu)$, i. e., $(Z, \vec{d})$ is to $(Z^1, \vec{d}^1), \dots, (Z^\nu, \vec{d}^\nu)$ as $(W, \vec{d})$ is to $(W^1, \vec{d}^1), \dots, (W^\nu, \vec{d}^\nu)$.

Proposition 5.5.14. Hörmander setting: *$\mathcal{E}$ is a bounded set of $(W, \vec{d})$ elementary operators supported in Ω if and only if $\mathcal{E}$ is a bounded set of $(Z, \vec{d})$ elementary operators supported in Ω.*

General setting: *$\mathcal{E}$ is a bounded set of generalized $(W, \vec{d})$ elementary operators supported in Ω if and only if $\mathcal{E}$ is a bounded set of generalized $(Z, \vec{d})$ elementary operators supported in Ω.*

To prove Proposition 5.5.14 we require a few lemmas.

Lemma 5.5.15. *In both the Hörmander setting and the general setting, the following holds. Let $\mathscr{P}$ be a partial differential operator on Ω with smooth coefficients. Let $D \in \mathbb{Z}^\nu$. Then the following are equivalent:*

(i) *$\mathscr{P}$ is a $(W, \vec{d})$ partial differential operator on Ω of degree $\le D$.*
(ii) *$\mathscr{P}$ is a $(Z, \vec{d})$ partial differential operator on Ω of degree $\le D$.*

In particular, $\mathscr{P}$ is a $(W^\mu, \vec{d}^\mu)$ partial differential operator on Ω of degree $\le \kappa_\mu$ if and only if $\mathscr{P}$ is a $(Z^\mu, \vec{d}^\mu)$ partial differential operator on Ω of degree $\le \kappa_\mu$.

Proof. If $D \in \mathbb{Z}^\nu \setminus \mathbb{N}^\nu$, then in either case $\mathscr{P} = 0$ and the result is trivial; thus, we henceforth assume $D \in \mathbb{N}^\nu$.

Let $(X^\mu, d^\mu) = \{(X_1^\mu, d_1^\mu), \dots, (X_{q_\mu}^\mu, d_{q_\mu}^\mu)\} \subset \operatorname{Gen}((W^\mu, \vec{d}^\mu))$ be such that $(W^\mu, \vec{d}^\mu) \subseteq (X^\mu, d^\mu)$ and $\operatorname{Gen}((W^\mu, \vec{d}^\mu))$ is finitely generated by (X^μ, d^μ) on Ω. Set

$$(X, \vec{d}) = \{(X_1, \vec{d}_1), \dots, (X_q, \vec{d}_q)\} := (X^1, d^1) \boxtimes (X^2, d^2) \boxtimes \cdots \boxtimes (X^\nu, d^\nu) \subset C^\infty_{\mathrm{loc}}(\mathfrak{M}; T\mathfrak{M}) \times (\mathbb{N}^\nu \setminus \{0\}).$$

Define $(Y, \vec{d})$ in the same way with $(W^\mu, \vec{d}^\mu)$ replaced with $(Z^\mu, \vec{d}^\mu)$ throughout. Lemma 3.14.12 shows that $(Y, \vec{d})$ and $(X, \vec{d})$ are strongly equivalent on Ω.

Since $(W, \vec{d}) \subseteq (X, \vec{d}) \subset \operatorname{Gen}((W, \vec{d}))$, (i) is equivalent to

$$\mathscr{P} = \sum_{\deg_{\vec{d}}(\alpha) \le D} \tilde{a}_\alpha(x) X^\alpha, \quad \tilde{a}_\alpha \in C^\infty_{\mathrm{loc}}(\Omega); \tag{5.47}$$

indeed, for each l, X_l may be written as a constant coefficient linear combination of W^β, where $\deg_{\vec{d}}(\beta) = \vec{d}_l$. Similarly, (ii) is equivalent to

$$\mathscr{P} = \sum_{\deg_{\vec{d}}(\alpha) \le D} \tilde{b}_\alpha(x) Y^\alpha, \quad \tilde{b}_\alpha \in C^\infty_{\mathrm{loc}}(\Omega). \tag{5.48}$$

Since $(X, \vec{d})$ and $(Y, \vec{d})$ are strongly equivalent on Ω, (5.47) is equivalent to (5.48), which completes the proof of (i) $\Leftrightarrow$ (ii).

That $\mathscr{P}$ is a $(W^\mu, \vec{d}^\mu)$ partial differential operator on Ω of degree $\leq \kappa_\mu$ if and only if $\mathscr{P}$ is a $(Z^\mu, \vec{d}^\mu)$ partial differential operator on Ω of degree $\leq \kappa_\mu$ follows by applying the above result in the case $\nu = 1$ to the lists $(W^\mu, \vec{d}^\mu)$ and $(Z^\mu, \vec{d}^\mu)$. □

Lemma 5.5.16. Hörmander setting: *The definition of bounded sets of $(W, \vec{d})$ pre-elementary operators supported in Ω (Definition 5.2.14) can be equivalently stated by replacing each instance of $(2^{-j\vec{d}}W)^\gamma$ by $2^{-j\cdot D}\mathscr{P}$, where $D \in \mathbb{N}^\nu$ and $\mathscr{P}$ is an arbitrary $(W, \vec{d})$ partial differential operator on Ω of degree $\leq D$. More precisely, $\mathcal{E} \subset C_0^\infty(\Omega \times \Omega) \times (0,1]^\nu$ is a bounded set of $(W, \vec{d})$ pre-elementary operators supported in Ω if and only if the following hold:*

- $\bigcup_{(E,2^{-j})\in\mathcal{E}} \operatorname{supp}(E) \Subset \Omega$.
- *For every $D_1, D_2 \in \mathbb{N}^\nu$, all $\mathscr{P}_1$ and $\mathscr{P}_2$ $(W, \vec{d})$ partial differential operators on Ω of degree $\leq D_1$ and $\leq D_2$, respectively, and all $m \in \mathbb{N}$, there exists $C \geq 0$ such that $\forall (E, 2^{-j}) \in \mathcal{E}$,*

$$\left|(2^{-j\cdot D_1}\mathscr{P}_{1,x})(2^{-j\cdot D_2}\mathscr{P}_{2,z})E(x,z)\right| \leq C\frac{(1+\rho_{2^{-j}}(x,z))^{-m}}{\operatorname{Vol}(B_{2^{-j}}(x, 1+\rho_{2^{-j}}(x,z))) \wedge 1}.$$

General setting: *The definition of bounded sets of generalized $(W, \vec{d})$ pre-elementary operators supported in Ω (Definition 5.2.26) can be equivalently stated by replacing each instance of $(2^{-j\vec{d}}W)^\gamma$ by $2^{-j\cdot D}\mathscr{P}$, where $D \in \mathbb{N}^\nu$ $\mathscr{P}$ is an arbitrary $(W, \vec{d})$ partial differential operator on Ω of degree $\leq D$.*

Proof. We proof the result in the Hörmander setting; the proof in the general setting is similar. Since $(2^{-j\vec{d}}W)^\gamma = 2^{-j\deg_{\vec{d}}(\gamma)}W^\gamma$, we see that for any ordered multi-index γ, $(2^{-j\vec{d}}W)^\gamma$ is of the form $2^{-j\cdot D}\mathscr{P}$, where $\mathscr{P}$ is a $(W, \vec{d})$ partial differential operator on Ω of degree $\leq D$. This shows that if $\mathcal{E}$ satisfies the above hypotheses, then $\mathcal{E}$ is a bounded set of $(W, \vec{d})$ pre-elementary operators supported in Ω.

For the converse, let $\mathcal{E}$ be a bounded set of $(W, \vec{d})$ pre-elementary operators supported in Ω. Then if $\mathscr{P}_1$ and $\mathscr{P}_2$ are as in the statement of the lemma, we may write $\mathscr{P}_1 = \sum_{\deg_{\vec{d}}(\alpha)\leq D_1} a_\alpha W^\alpha$, $\mathscr{P}_2 = \sum_{\deg_{\vec{d}}(\alpha)\leq D_2} b_\alpha W^\alpha$, with $a_\alpha, b_\alpha \in C^\infty_{\mathrm{loc}}(\Omega)$. Thus, for every $m \in \mathbb{N}$,

$$\begin{aligned}
\left|(2^{-j\cdot D_1}\mathscr{P}_{1,x})(2^{-j\cdot D_2}\mathscr{P}_{2,z})E(x,z)\right| &\leq \sum_{\substack{\deg_{\vec{d}}(\alpha)\leq D_1\\ \deg_{\vec{d}}(\beta)\leq D_2}} 2^{-j\cdot(D_1-\deg_{\vec{d}}(\alpha))-j\cdot(D_2\deg_{\vec{d}}(\beta))} \\
&\qquad \times \left|a_\alpha(x)b_\beta(z)(2^{-j\vec{d}}W_x)^\alpha(2^{-j\vec{d}}W_z)^\beta E(x,z)\right| \\
&\lesssim \sum_{\substack{\deg_{\vec{d}}(\alpha)\leq D_1\\ \deg_{\vec{d}}(\beta)\leq D_2}} \left|(2^{-j\vec{d}}W_x)^\alpha(2^{-j\vec{d}}W_z)^\beta E(x,z)\right| \\
&\lesssim \frac{(1+\rho_{2^{-j}}(x,z))^{-m}}{\operatorname{Vol}(B_{2^{-j}}(x, 1+\rho_{2^{-j}}(x,z))) \wedge 1},
\end{aligned}$$

where in the second to last inequality we have used $\bigcup_{(E,2^{-j})} \mathrm{supp}(E) \Subset \Omega$ and $a_\alpha, b_\beta \in C^\infty_{\mathrm{loc}}(\Omega)$. It follows that $\mathcal{E}$ satisfies the conditions of the lemma. □

Lemma 5.5.17. Hörmander setting: *$\mathcal{E}$ is a bounded set of $(W, \vec{d})$ pre-elementary operators supported in Ω if and only if $\mathcal{E}$ is a bounded set of $(Z, \vec{d})$ pre-elementary operators supported in Ω.*

General setting: *$\mathcal{E}$ is a bounded set of generalized $(W, \vec{d})$ pre-elementary operators supported in Ω if and only if $\mathcal{E}$ is a bounded set of generalized $(Z, \vec{d})$ pre-elementary operators supported in Ω.*

Proof. We begin with the Hörmander setting. In light of Lemma 5.5.16, the definition of bounded sets of $(W, \vec{d})$ pre-elementary operators depends on $(W, \vec{d})$ in two ways: in the definition of the bound $\frac{(1+\rho_{2^{-j}}(x,z))^{-m}}{\mathrm{Vol}(B_{2^{-j}}(x,1+\rho_{2^{-j}}(x,z)))\wedge 1}$ and in the way $(W, \vec{d})$ partial differential operators on Ω were used in Lemma 5.5.16. Lemma 5.5.15 shows that $(W, \vec{d})$ partial differential operators on Ω of degree $\leq D \in \mathbb{Z}^\nu$ are the same thing as $(Z, \vec{d})$ partial differential operators on Ω of degree $\leq D$. Proposition 5.4.13 shows that if one replaces $(W, \vec{d})$ with $(Z, \vec{d})$ one obtains an equivalent estimate in $\frac{(1+\rho_{2^{-j}}(x,z))^{-m}}{\mathrm{Vol}(B_{2^{-j}}(x,1+\rho_{2^{-j}}(x,z)))\wedge 1}$. This completes the proof in the Hörmander setting.

In the general setting, Lemma 5.5.16 shows that the definition of bounded sets of $(W, \vec{d})$ pre-elementary operators depends on $(W, \vec{d})$ only through the use of $(W, \vec{d})$ partial differential operators on Ω as described in that lemma. Lemma 5.5.15 shows that this class of partial differential operators is equivalently defined with $(Z, \vec{d})$, completing the proof in the general setting. □

Proof of Proposition 5.5.14. We only prove the result in the Hörmander setting; the proof in the general setting is nearly identical. Lemma 5.5.17 shows that bounded sets of $(W, \vec{d})$ pre-elementary operators are the same as bounded sets of $(Z, \vec{d})$ pre-elementary operators. Lemma 5.5.15 shows that $\mathscr{P}_\mu$ is a $(W^\mu, \vec{d}^\mu)$ partial differential operator in Ω of degree $\leq \kappa_\mu$ if and only if $\mathscr{P}_\mu$ is a $(Z^\mu, \vec{d}^\mu)$ partial differential operator in Ω of degree $\leq \kappa_\mu$. Using these remarks, the result follows immediately from Proposition 5.5.1. □

5.6 Pseudo-differential operators

In this section, we work in the same two settings described in Section 5.3: the Hörmander setting and the general setting.

In either setting, each $\mathrm{Gen}((W^\mu, \vec{d}^\mu))$ is finitely generated by some $(X^\mu, d^\mu) = \{(X_1^\mu, d_1^\mu), \dots, (X_{q_\mu}, d_{q_\mu}^\mu)\} \subset \mathrm{Gen}((W^\mu, \vec{d}^\mu))$ on Ω. In the general setting, this is part of the hypotheses; for the Hörmander setting, see Remark 5.1.2. We take (X^μ, d^μ) so that $(W^\mu, \vec{d}^\mu) \subseteq (X^\mu, d^\mu)$.

Set

$$(X, \vec{d}) = \{(X_1, \vec{d}_1), \dots, (X_q, \vec{d}_q)\} := (X^1, d^1) \boxtimes (X^2, d^2) \boxtimes \cdots \boxtimes (X^\nu, d^\nu)$$
$$= \{(X_j^\mu, d_j^\mu \mathfrak{e}_\mu) : \mu \in \{1, \dots, \nu\}, 1 \le j \le q_\mu\} \subset C^\infty_{\mathrm{loc}}(\mathfrak{M}; T\mathfrak{M}) \times (\mathbb{N}^\nu \setminus \{0\}).$$

By Remark 4.0.1, $(X, \vec{d})$ satisfies the assumptions of Chapter 4, with $\mathfrak{M}$ replaced by Ω. Fix open, relatively compact sets $\Omega_1 \Subset \Omega_2 \Subset \Omega$. It makes sense to talk about $(X, \vec{d})$, a pseudo-differential operators of order $t \in \mathbb{R}^\nu$, as defined in Definition 4.1.2. The main result of this section is that these pseudo-differential operators are elements of our algebras.

Proposition 5.6.1. *If $a = a(\Omega_1, \Omega_2, (X, \vec{d})) > 0$ is sufficiently small and T is an $(X, \vec{d})$, a pseudo-differential operator of order $t \in \mathbb{R}^\nu$ supported in $\Omega_1 \times \Omega_2$, then we have the following.*
Hörmander setting: *We have $T \in \mathscr{A}_2^t(\Omega, (W, \vec{d}))$.*
General setting: *We have $T \in \widetilde{\mathscr{A}}_2^t(\Omega, (W, \vec{d}))$.*

Corollary 5.6.2. *Let $\psi \in C_0^\infty(\Omega)$. Then we have the following.*
Hörmander setting: *We have* $\mathrm{Mult}[\psi] \in \mathscr{A}_2^t(\Omega, (W, \vec{d}))$.
General setting: *We have* $\mathrm{Mult}[\psi] \in \widetilde{\mathscr{A}}_2^t(\Omega, (W, \vec{d}))$.

Proof. Pick Ω_1 and Ω_2 with $\mathrm{supp}(\psi) \subseteq \Omega_1 \Subset \Omega_2 \Subset \Omega$. Example 4.1.6 shows that for any $a > 0$, $\mathrm{Mult}[\psi]$ is an $(X, \vec{d})$, a pseudo-differential operator of order 0 supported in $\Omega_1 \times \Omega_2$. From here, the result follows from Proposition 5.6.1. □

Proposition 5.6.1 is a simple consequence of the next result.

Proposition 5.6.3. *If $a = a(\Omega_1, \Omega_2, (X, \vec{d})) > 0$ is sufficiently small and $\mathcal{E}$ is a bounded set of $(X, \vec{d})$, a pseudo-differential operator scales supported in $\Omega_1 \times \Omega_2$, then we have the following.*
Hörmander setting: *$\mathcal{E}$ is a bounded set of $(W, \vec{d})$ elementary operators supported in Ω.*
General setting: *$\mathcal{E}$ is a bounded set of generalized $(W, \vec{d})$ elementary operators supported in Ω.*

Proof of Proposition 5.6.1. We only prove the result in the Hörmander setting; the proof in the general setting is nearly identical and we leave it to the reader. If $a = a(\Omega_1, \Omega_2, (X, \vec{d})) > 0$ is sufficiently small, then Theorem 4.3.3 shows that T may be written as $T = \sum_{j \in \mathbb{N}^\nu} 2^{j \cdot t} E_j$, where $\{(E_j, 2^{-j}) : j \in \mathbb{N}^\nu\}$ is a bounded set of $(W, \vec{d})$, a pseudo-differential operator scales supported in $\Omega_1 \times \Omega_2$. Proposition 5.6.3 implies $\{(E_j, 2^{-j}) : j \in \mathbb{N}^\nu\}$ is a bounded set of $(W, \vec{d})$ elementary operators supported in Ω, establishing that $T \in \mathscr{A}_2^t(\Omega, (W, \vec{d})) = \mathscr{A}^t(\Omega, (W, \vec{d}))$, completing the proof. □

The rest of this section is devoted to the proof of Proposition 5.6.3, which requires several lemmas. For these, we use the same notation as from Chapter 4. In particular, we decompose $t \in \mathbb{R}^q$ as $t = (t^1, \dots, t^\nu) \in \mathbb{R}^{q_1} \times \cdots \times \mathbb{R}^{q_\nu}$.

Lemma 5.6.4. *In the Hörmander setting, if $a = a(\Omega_1, \Omega_2, (X, \vec{d})) > 0$ is sufficiently small and $b(x,t) : \mathfrak{M} \times \mathbb{R}^q$ is a bounded measurable function with $\operatorname{supp}(b) \subseteq \Omega_1 \times B^q(a)$, then the following holds. For $j \in [0,\infty)^\nu$, set*

$$F_j f(x) := \int f(e^{-t^1 \cdot 2^{-j_1 d^1} X^1} \cdots e^{-t^\nu \cdot 2^{-j_\nu d^\nu} X^\nu}) b(x,t)\, dt.$$

Then the Schwartz kernel, $F_j(x,y)$, of F_j is a bounded measurable function and satisfies

$$|F_j(x,y)| \leq C \|b\|_{L^\infty} \frac{\chi_{B_{2^{-j}}(x,1)}(y) \chi_{\Omega_1}(x) \chi_{\Omega_2}(y)}{\operatorname{Vol}(B_{2^{-j}}(x,1)) \wedge 1},$$

where $C = C(\Omega_1, \Omega_2, (X, \vec{d})) \geq 0$.

Proof. The result for general $b(x,t)$ follows from the same result with $b(x,t) := \chi_{\Omega_1}(x)\chi_{B^q(a)}(t)$. In this special case, the result follows from Lemma 3.10.2 (i) and (iii). □

Lemma 5.6.5. *In the Hörmander setting, if $a = a(\Omega_1, \Omega_2, (X, \vec{d})) > 0$ is sufficiently small, the following holds. Let $\mathcal{B} \subset C_0^\infty(\Omega_1) \widehat{\otimes} C_0^\infty(B^q(a)) \widehat{\otimes} \mathcal{S}(\mathbb{R}^q)$ be a bounded set, and for $\varsigma \in \mathcal{B}$ and $j \in [0,\infty)^\nu$, set*

$$F_j f(x) := \int f(e^{-t^1 \cdot X^1} \cdots e^{-t^\nu \cdot X^\nu}) \operatorname{Dil}^{\vec{d}}_{2^j}(\varsigma)(x,t,t)\, dt.$$

Then the Schwartz kernel, $F_j(x,y)$, of F_j is a bounded measurable function satisfying, $\forall m \in \mathbb{N}$,

$$|F_j(x,y)| \leq C \chi_{\Omega_1}(x) \chi_{\Omega_2}(y) \frac{(1 + \rho_{2^{-j}}(x,y))^{-m}}{\operatorname{Vol}(B_{2^{-j}}(x, 1 + \rho_{2^{-j}}(x,y))) \wedge 1}, \tag{5.49}$$

where $C = C(\mathcal{B}, \Omega_1, \Omega_2, (X, \vec{d})) \geq 0$.

Proof. By Lemma 4.3.9, for $\varsigma \in \mathcal{B}$ there is a set

$$\{\gamma_k : k \in \mathbb{N}^\nu, k \leq j\} \subset C_0^\infty(\Omega_1 \times B^q(a) \times B^q(a))$$

such that

$$\operatorname{Dil}^{\vec{d}}_{2^j}(\varsigma)(x,t,t) = \sum_{\substack{k \leq j \\ k \in \mathbb{N}^\nu}} \operatorname{Dil}^{\vec{d}}_{2^k}(\gamma_k)(x,t,t), \tag{5.50}$$

and for every $M \in \mathbb{N}$,

$$\{2^{M|j-k|}\gamma_k : j \in [0,\infty)^\nu, k \le j, k \in \mathbb{N}^\nu, \varsigma \in \mathcal{B}\} \subset C_0^\infty(\Omega_1 \times B^q(a) \times B^q(a))$$

is a bounded set.

By Lemma 5.6.4, if for $k \le j$,

$$\begin{aligned} F_{j,k}f(x) &:= \int f(e^{-t^1\cdot X^1}\cdots e^{-t^\nu\cdot X^\nu})\,\mathrm{Dil}_{2^k}^{\vec{d}}(\gamma_k)(x,t,t)\,dt \\ &= \int f(e^{-t^1\cdot 2^{-k_1 d^1}X^1}\cdots e^{-t^\nu\cdot 2^{-k_\nu d^\nu}X^\nu})\gamma_k(x,2^{-k\vec{d}}t,t)\,dt, \end{aligned}$$

then $F_{j,k}$ is a measurable function and for every $M \in \mathbb{N}$,

$$|F_{j,k}(x,y)| \lesssim 2^{-M|k-j|}\frac{\chi_{B_{2^{-k}}(x,1)}(y)\chi_{\Omega_1}(x)\chi_{\Omega_2}(y)}{\mathrm{Vol}(B_{2^{-j}}(x,1)) \wedge 1}. \tag{5.51}$$

Using (5.50) and (5.51), we have, for every $m, M \in \mathbb{N}$, using Theorem 3.5.4 (n),

$$\begin{aligned} |F_j(x,y)| &\le \sum_{\substack{k\in\mathbb{N}^\nu \\ k\le j}} |F_{j,k}(x,y)| \lesssim \chi_{\Omega_1}(x)\chi_{\Omega_2}(y)\sum_{\substack{k\in\mathbb{N}^\nu \\ k\le j}} 2^{-M|j-k|}\frac{\chi_{B_{2^{-k}}(x,1)}(y)}{\mathrm{Vol}(B_{2^{-j}}(x,1)) \wedge 1} \\ &\lesssim \chi_{\Omega_1}(x)\chi_{\Omega_2}(y)\sum_{\substack{k\in\mathbb{N}^\nu \\ k\le j}} 2^{-M|j-k|}\frac{\chi_{B_{2^{-k}}(x,1)}(y)}{\mathrm{Vol}(B_{2^{-j}}(x,1+\rho_{2^{-k}}(x,y))) \wedge 1} \\ &\lesssim \chi_{\Omega_1}(x)\chi_{\Omega_2}(y)\sum_{\substack{k\in\mathbb{N}^\nu \\ k\le j}} 2^{-M|j-k|}\frac{(1+\rho_{2^{-k}}(x,y))^{-m}}{\mathrm{Vol}(B_{2^{-k}}(x,1+\rho_{2^{-k}}(x,y))) \wedge 1}. \end{aligned}$$

From here, (5.49) follows by taking M large (depending on m) and applying Lemma 5.4.8. □

Lemma 5.6.6. *If $a = a(\Omega_1, \Omega_2, (X, \vec{d})) > 0$ is sufficiently small and $\mathcal{E}$ is a bounded set of $(X, \vec{d})$, a pre-pseudo-differential operator scales supported in $\Omega_1 \times \Omega_2$, then we have the following.*

Hörmander setting: *$\mathcal{E}$ is a bounded set of $(W, \vec{d})$ pre-elementary operators supported in Ω.*

General setting: *$\mathcal{E}$ is a bounded set of generalized $(W, \vec{d})$ pre-elementary operators supported in Ω.*

Proof. In either case, since $\mathcal{E}$ is a bounded set of $(X, \vec{d})$, *a* pre-pseudo-differential operator scales supported in $\Omega_1 \times \Omega_2$, we have

$$\bigcup_{(E,2^{-j})\in\mathcal{E}} \mathrm{supp}(E) \subseteq \overline{\Omega_1 \times \Omega_2} \Subset \Omega \times \Omega. \tag{5.52}$$

We first consider the Hörmander setting. In light of (5.52), the result will follow once we show that for all ordered multi-indices α, β and all $m \in \mathbb{N}$,

$$\left|(2^{-j\vec{d}}W_x)^\alpha (2^{-j\vec{d}}W_z)^\beta E(x,z)\right| \lesssim \frac{(1+\rho_{2^{-j}}(x,z))^{-m}}{\operatorname{Vol}(B_{2^{-j}}(x,1+\rho_{2^{-j}}(x,z)))\wedge 1}. \tag{5.53}$$

Indeed, since $W_1,\ldots,W_r$ satisfy Hörmander's condition, not only does (5.53) give the desired bound, but when combined with (5.52) it also implies that $E \in C_0^\infty(\Omega\times\Omega)$.

Since

$$\begin{aligned}[(2^{-j\vec{d}}W)^\beta]^* &= (-1)^{|\beta|}(2^{-j\vec{d}}W)^\beta \\ &\quad + \sum_{\substack{\deg_{\vec{d}}(\gamma)\le\deg_{\vec{d}}(\beta)\\ |\gamma|<|\beta|}} 2^{-j\cdot(\deg_{\vec{d}}(\beta)-\deg_{\vec{d}}(\gamma))}(2^{-j\vec{d}}W)^\beta \operatorname{Mult}[g_{\beta,\gamma}],\end{aligned}$$

where $g_{\beta,\gamma} \in C^\infty_{\mathrm{loc}}(\mathfrak{M})$, to prove (5.53) it suffices to show that for all α, β, and m,

$$\left|[(2^{-j\vec{d}}W)^\alpha E(2^{-j\vec{d}}W)^\beta](x,z)\right| \lesssim \frac{(1+\rho_{2^{-j}}(x,z))^{-m}}{\operatorname{Vol}(B_{2^{-j}}(x,1+\rho_{2^{-j}}(x,z)))\wedge 1}. \tag{5.54}$$

When $|\alpha| = |\beta| = 0$, (5.54) follows directly from Lemma 5.6.5. For general α and β, by applying Lemma 4.3.4 (i) twice, we show that

$$\{((2^{-j\vec{d}}W)^\alpha E(2^{-j\vec{d}}W)^\beta, 2^{-j}) : (E,2^{-j}) \in \mathcal{E}\}$$

is a bounded set of $(X,\vec{d})$, a pre-pseudo-differential operator scales supported in $\Omega_1\times\Omega_2$. Thus, (5.54) for general α and β follows from (5.54) with $|\alpha| = |\beta| = 0$, which we have already shown. This completes the proof in the Hörmander case.

We turn to the general setting. In light of (5.52), the result will follow once we show that:

(a) $\mathcal{E} \subseteq \operatorname{Hom}(C^\infty_{W,\mathrm{loc}}(\mathfrak{M}), C^\infty_{W,0}(\mathfrak{M}))\times(0,1]^\nu$.

(b) For all α, β,

$$\sup_{(E,2^{-j})\in\mathcal{E}} \max_{p\in\{1,\infty\}} \|(2^{-j\vec{d}}W)^\alpha E(2^{-j\vec{d}}W)^\beta\|_{L^p\to L^p} < \infty. \tag{5.55}$$

(c) If $\mathcal{E}' := \{(E_k, 2^{-j_k}) : k\in\mathbb{N}\} \subseteq \mathcal{E}$ and for each α,β we set

$$\mathcal{T}_{\mathcal{E}',\alpha,\beta}\{f_k\}_{k\in\mathbb{N}} := \{(2^{-j\vec{d}}W)^\alpha E(2^{-j\vec{d}}W)^\beta f_k\}_{k\in\mathbb{N}},$$

then for $p\in(1,\infty)$, $q\in(1,\infty]$,

$$\sup_{\mathcal{E}'} \|\mathcal{T}_{\mathcal{E}',\alpha,\beta}\|_{L^p(\mathfrak{M},\mathrm{Vol};\ell^q(\mathbb{N}))\to L^p(\mathfrak{M},\mathrm{Vol};\ell^q(\mathbb{N}))} < \infty, \tag{5.56}$$

where the supremum is taken over all countable subsets $\mathcal{E}'\subseteq\mathcal{E}$.

(d) The above holds with $\mathcal{E}$ replaced by $\mathcal{E}^* := \{(E^*,2^{-j}) : (E,2^{-j})\in\mathcal{E}\}$.

We begin with (a). Let $(E, 2^{-j}) \in \mathcal{E}$. We wish to show that $E \in \mathrm{Hom}(C^\infty_{W,\mathrm{loc}}(\mathfrak{M}), C^\infty_{W,0}(\mathfrak{M}))$. In fact, we will show $E \in \mathrm{Hom}(C_{\mathrm{loc}}(\mathfrak{M}), C^\infty_{W,0}(\mathfrak{M}))$, where $C_{\mathrm{loc}}(\mathfrak{M})$ denotes the Fréchet space of continuous functions on $\mathfrak{M}$. First, we claim $E \in \mathrm{Hom}(C(\mathfrak{M}), C_{\mathrm{loc}}(\mathfrak{M}))$. Indeed, E is of the form

$$Ef(x) = \int f(\Gamma(t,x))\, \mathrm{Dil}^{\vec{d}}_{2^j}(\varsigma)(x,t,t)\, dt,$$

where $\Gamma(t,x) \in C^\infty_{\mathrm{loc}}(B^q(a) \times \Omega_1)$ is defined in (4.3) and $\varsigma \in C_0^\infty(\Omega)\widehat{\otimes} C_0^\infty(B^q(a))\widehat{\otimes}\mathscr{S}(\mathbb{R}^q)$. From this formula, it is immediate to verify that $E \in \mathrm{Hom}(C(\mathfrak{M}), C_{\mathrm{loc}}(\mathfrak{M}))$. Lemma 4.3.4 (i) shows that $W^\alpha E$ is of the same form as E for any α (here we are using the fact that the list W is a sublist of the list X). Thus, it follows that for every α, $W^\alpha E \in \mathrm{Hom}(C(\mathfrak{M}), C_{\mathrm{loc}}(\mathfrak{M}))$. This is equivalent to $E \in \mathrm{Hom}(C(\mathfrak{M}), C^\infty_{W,\mathrm{loc}}(\mathfrak{M}))$. Since (5.52) implies $\mathrm{supp}(E) \Subset \mathfrak{M} \times \mathfrak{M}$, we in fact have $E \in \mathrm{Hom}(C_{\mathrm{loc}}(\mathfrak{M}), C^\infty_{W,0}(\mathfrak{M}))$, completing the proof of (a).

Next we turn to (b) and (c). By applying Lemma 4.3.4 (i) twice, we show that

$$\{((2^{-j\vec{d}}W)^\alpha E (2^{-j\vec{d}}W)^\beta, 2^{-j}) : (E, 2^{-j}) \in \mathcal{E}\}$$

is a bounded set of $(X, \vec{d})$, a pre-pseudo-differential operator scales supported in $\Omega_1 \times \Omega_2$. Thus, (5.55) and (5.56) for general α and β follow from the special case where $|\alpha| = |\beta| = 0$. The case $|\alpha| = |\beta| = 0$ is Proposition 4.3.8.

Finally, to see (d), note that Lemma 4.3.11 (i) shows that $\mathcal{E}^*$ is of the same form as $\mathcal{E}$. Thus, the same proof gives the desired results for $\mathcal{E}^*$. □

Proof of Proposition 5.6.3. We prove only the result in the Hörmander setting. The proof in the general setting is nearly identical. We use the characterization of elementary operators given in Proposition 5.5.1. Lemma 5.6.6 shows that $\mathcal{E}$ is a bounded set of $(W, \vec{\mathfrak{d}})$ pre-elementary operators supported in Ω. Thus, it suffices to show, for $(E, 2^{-j}) \in \mathcal{E}$ and $\mu \in \{1, \ldots, \nu\}$, that E can be written in the form (5.26), where each E_{μ,l_1,l_2} is of the same form as E. Proposition 4.3.5 shows that E can be written as

$$E = \sum_{\substack{|\alpha_\mu| \le 1 \\ |\beta_\mu| \le 1}} 2^{-j\mu(2-|\alpha_\mu|-|\beta_\mu|)} (2^{-j_\mu d^\mu} X^\mu)^{\alpha_\mu} E_{\mu,\alpha_\mu,\beta_\mu} (2^{-j_\mu d^\mu} X^\mu)^{\beta_\mu},$$

where

$$\{(E_{\mu,\alpha_\mu,\beta_\mu}, 2^{-j}) : (E, 2^{-j}) \in \mathcal{E}, \mu \in \{1, \ldots, \nu\}, |\alpha_\mu|, |\beta_\mu| \le 1\}$$

is a bounded set of $(X, \vec{d})$, a pseudo-differential scales supported in $\Omega_1 \times \Omega_2$. Since $(2^{-j_\mu d^\mu} X^\mu)^{\alpha_\mu} = 2^{-j_\mu \deg_{d^\mu}(\alpha_\mu)} (X^\mu)^{\alpha_\mu}$ and $(X^\mu)^{\alpha_\mu}$ is a $(W^\mu, \vec{\mathfrak{d}}^\mu)$ partial differential operator in Ω of degree $\le \deg_{d^\mu}(\alpha_\mu)$ and similarly for $(2^{-j_\mu d^\mu} X^\mu)^{\beta_\mu}$, this sees E as being of the form (5.26) and completes the proof. □

5.7 Equivalence of the definitions

In this section, we prove Theorems 5.2.12, 5.2.18, and 5.2.30, i. e., that the various definitions we gave for the algebras of singular integrals are equivalent.

Proofs of Theorems 5.2.18 *and* 5.2.30. The proofs of Theorems 5.2.18 and 5.2.30 are nearly identical, so we prove only Theorem 5.2.30.

Suppose $T \in \widetilde{\mathscr{A}}_1^t(\Omega, (W, \vec{d}))$. Then, in particular, $T \in \operatorname{Hom}(C^\infty_{W,0}(\mathfrak{M}), C^\infty_{W,0}(\mathfrak{M})')$ and $\operatorname{supp}(T) \Subset \Omega \times \Omega$. Take $\psi \in C_0^\infty(\Omega)$ with $\psi \equiv 1$ on a large enough set so that $T \operatorname{Mult}[\psi] = T$.

By Corollary 5.6.2, we have $\operatorname{Mult}[\psi] \in \widetilde{\mathscr{A}}_2^0(\Omega, (W, \vec{d}))$, and therefore

$$\operatorname{Mult}[\psi] = \sum_{j \in \mathbb{N}^\nu} E_j, \tag{5.57}$$

where $\{(E_j, 2^{-j}) : j \in \mathbb{N}^\nu\}$ is a bounded set of $(W, \vec{d})$ elementary operators supported in Ω. By Proposition 5.5.10, the sum in (5.57) converges in the topology of bounded convergence on $\operatorname{Hom}(C^\infty_{W,\mathrm{loc}}(\mathfrak{M}), C^\infty_{W,0}(\mathfrak{M}))$. Thus, by the assumption that $T \in \operatorname{Hom}(C^\infty_{W,0}(\mathfrak{M}), C^\infty_{W,0}(\mathfrak{M})')$, we have

$$T = T \operatorname{Mult}[\psi] = T \sum_{j \in \mathbb{N}^\nu} E_j = \sum_{j \in \mathbb{N}^\nu} 2^{j\cdot t}(2^{-j\cdot t} T E_j).$$

Since $T \in \widetilde{\mathscr{A}}_1^t(\Omega, (W, \vec{d}))$, $\{(2^{-j\cdot t} T E_j, 2^{-j}) : j \in \mathbb{N}^\nu\}$ is a bounded set of $(W, \vec{d})$ elementary operators supported in Ω. This completes the proof that $T \in \widetilde{\mathscr{A}}_2^t(\Omega, (W, \vec{d}))$ and therefore $\widetilde{\mathscr{A}}_1^t(\Omega, (W, \vec{d})) \subseteq \widetilde{\mathscr{A}}_2^t(\Omega, (W, \vec{d}))$.

Suppose $T \in \widetilde{\mathscr{A}}_2^t(\Omega, (W, \vec{d}))$. Then

$$T = \sum_{j \in \mathbb{N}^\nu} 2^{j\cdot t} E_j, \tag{5.58}$$

where $\{(E_j, 2^{-j}) : j \in \mathbb{N}^\nu\}$ is a bounded set of $(W, \vec{d})$ elementary operators supported in Ω. We take the convention

$$E_j := 0, \quad j \in \mathbb{Z}^\nu \setminus \mathbb{N}^\nu. \tag{5.59}$$

By Proposition 5.5.10, the sum in (5.58) converges in the topology of bounded convergence on $\operatorname{Hom}(C^\infty_{W,\mathrm{loc}}(\mathfrak{M}), C^\infty_{W,0}(\mathfrak{M}))$.

Let $\mathcal{E}$ be a bounded set of $(W, \vec{d})$ elementary operators supported in Ω and let $(E, 2^{-k}) \in \mathcal{E}$. In particular, we have $E \in \operatorname{Hom}(C^\infty_{W,0}(\mathfrak{M}), C^\infty_{W,\mathrm{loc}}(\mathfrak{M}))$ and $\operatorname{supp}(E) \Subset \Omega$, and therefore $E \in \operatorname{Hom}(C^\infty_{W,0}(\mathfrak{M}), C^\infty_{W,0}(\mathfrak{M}))$. Thus, using the above described convergence of the sum (5.58) and the convention (5.59), we have

$$2^{-k\cdot t}TE = \sum_{j\in\mathbb{N}^\nu} 2^{(j-k)\cdot t}E_jE = \sum_{j\in\mathbb{N}^\nu} 2^{-|j-k|}(2^{(j-k)\cdot t+|j-k|}E_jE)$$
$$= \sum_{l\in\mathbb{Z}^\nu} 2^{-|l|}(2^{l\cdot t+|l|}E_{k+l}E). \tag{5.60}$$

Proposition 5.5.11 shows that

$$\{(2^{l\cdot t+|l|}E_{k+l}E, 2^{-k}) : l \in \mathbb{Z}^\nu, (E, 2^{-k}) \in \mathcal{E}\}$$

is a bounded set of generalized $(W, \vec{d})$ elementary operators supported in Ω, and therefore Proposition 5.5.5 (b) applied to the right-hand side of (5.60) shows that $\{(2^{-k\cdot t}TE, 2^{-k}) : (E, 2^{-k}) \in \mathcal{E}\}$ is a bounded set of generalized $(W, \vec{d})$ elementary operators supported in Ω. This shows that $T \in \widetilde{\mathscr{A}}_1^t(\Omega, (W, \vec{d}))$, and therefore $\widetilde{\mathscr{A}}_2^t(\Omega, (W, \vec{d})) \subseteq \widetilde{\mathscr{A}}_1^t(\Omega, (W, \vec{d}))$. □

Before we turn to the single-parameter case and prove Theorem 5.2.12, we introduce another way to check to see if an operator is in $\mathscr{A}^t(\Omega, (W, \vec{d}))$ or $\widetilde{\mathscr{A}}^t(\Omega, (W, \vec{d}))$. For this, we use the Hörmander setting and the general setting as described in Section 5.3.

Proposition 5.7.1. *Fix* $t \in \mathbb{R}^\nu$.

Hörmander setting: *Let* $T \in \mathrm{Hom}(C_0^\infty(\mathfrak{M}), C_0^\infty(\mathfrak{M}))$ *have* $\mathrm{supp}(T) \subseteq \Omega \times \Omega$. *Suppose that for every bounded set of* $(W, \vec{d})$ *elementary operators supported in* Ω, $\mathcal{E}$,

$$\{(2^{-j\cdot t}TE, 2^{-j}), (2^{-j\cdot t}ET, 2^{-j}) : (E, 2^{-j}) \in \mathcal{E}\}$$

is a bounded set of $(W, \vec{d})$ *pre-elementary operators supported in* Ω. *Then* $T \in \mathscr{A}^t(\Omega, (W, \vec{d}))$.

General setting: *Let* $T \in \mathrm{Hom}(C_{W,0}^\infty(\mathfrak{M}), C_{W,0}^\infty(\mathfrak{M}))$ *have* $\mathrm{supp}(T) \subseteq \Omega \times \Omega$. *Suppose that for every bounded set of generalized* $(W, \vec{d})$ *elementary operators supported in* Ω, $\mathcal{E}$,

$$\{(2^{-j\cdot t}TE, 2^{-j}), (2^{-j\cdot t}ET, 2^{-j}) : (E, 2^{-j}) \in \mathcal{E}\}$$

is a bounded set of generalized $(W, \vec{d})$ *pre-elementary operators supported in* Ω. *Then* $T \in \widetilde{\mathscr{A}}^t(\Omega, (W, \vec{d}))$.

Proof. We prove only the result in the Hörmander setting; the proof in the general setting is nearly identical. Let $\mathcal{E}$ be a bounded set of $(W, \vec{d})$ elementary operators supported in Ω. We will show that, for every $N \in \mathbb{N}$,

$$\mathcal{E}_N := \{(2^{N|j-k|_\infty-(j\wedge k)\cdot t}E_1TE_2, 2^{-j\wedge k}) : (E_1, 2^{-j}), (E_2, 2^{-k}) \in \mathcal{E}\} \tag{5.61}$$

is a bounded set of $(W, \vec{d})$ elementary operators supported in Ω.

First we see why this will complete the proof. Because Ω is pre-compact and $\mathrm{supp}(T) \subseteq \Omega \times \Omega$ is closed, we have $\mathrm{supp}(T) \Subset \Omega \times \Omega$. Take $\psi \in C_0^\infty(\Omega)$ with $\psi \equiv 1$ on a large enough set that $\mathrm{Mult}[\psi]T\,\mathrm{Mult}[\psi] = T$. By Corollary 5.6.2, there is a bounded set of $(W, \vec{d})$ elementary operators supported in Ω, $\{(E_j, 2^{-j}) : j \in \mathbb{N}^\nu\}$ with $T = \sum_{j\in\mathbb{N}^\nu} E_j$.

Proposition 5.5.10 shows that this sum converges in the topology of bounded convergence on $\mathrm{Hom}(C^\infty_{\mathrm{loc}}(\mathfrak{M}), C^\infty_0(\mathfrak{M}))$. Since $T \in \mathrm{Hom}(C^\infty_0(\mathfrak{M}), C^\infty_0(\mathfrak{M}))$, we have

$$T = \mathrm{Mult}[\psi] T \,\mathrm{Mult}[\psi] = \sum_{j,k\in\mathbb{N}^\nu} E_j T E_k. \tag{5.62}$$

Set, for $l \in \mathbb{N}^\nu$,

$$\begin{aligned}\widetilde{E}_l &:= \sum_{\substack{j,k\in\mathbb{N}^\nu\\ j\wedge k=l}} 2^{-l\cdot t} E_j T E_k = \sum_{\substack{j,k\in\mathbb{N}^\nu\\ j\wedge k=l}} 2^{-|j-k|_\infty}(2^{-l\cdot t+|j-k|_\infty} E_j T E_k)\\ &= \sum_{\substack{j',k'\in\mathbb{N}^\nu\\ j'\wedge k'=0}} 2^{-|j'-k'|_\infty}(2^{-l\cdot t+|j'-k'|_\infty} E_{l+j'} T E_{l+k'})\\ &=: \sum_{\substack{j',k'\in\mathbb{N}^\nu\\ j'\wedge k'=0}} 2^{-|j'-k'|_\infty}\widehat{E}_{l,j',k'}.\end{aligned} \tag{5.63}$$

It follows from our claim about (5.61) that $\{(\widehat{E}_{l,j',k'}, 2^{-l}) : j', k' \in \mathbb{N}^\nu\}$ is a bounded set of $(W, \vec{d})$ elementary operators supported in Ω. Thus, since $\sum_{\substack{j',k'\in\mathbb{N}^\nu\\ j'\wedge k'=0}} 2^{-|j'-k'|_\infty} < \infty$, Proposition 5.5.5 (b) implies that $\{(\widetilde{E}_l, 2^{-l}) : l \in \mathbb{N}^\nu\}$ is a bounded set of $(W, \vec{d})$ elementary operators supported in Ω. Plugging (5.63) into (5.62), we see that $T = \sum_{l\in\mathbb{N}^\nu} 2^{l\cdot t}\widetilde{E}_l$, and therefore $T \in \mathscr{A}^t_2(\Omega, (W, \vec{d})) = \mathscr{A}^t(\Omega, (W, \vec{d}))$, as desired.

We turn to proving that $\mathcal{E}_N$ from (5.61) is a bounded set of $(W, \vec{d})$ elementary operators supported in Ω.

First, we claim that $\mathcal{E}_N$ is a bounded set of $(W, \vec{d})$ pre-elementary operators supported in Ω. To do this, we separate $\mathcal{E}_N$ into two sets: when $|j-k|_\infty = j_\mu - k_\mu$ for some μ and when $|j-k|_\infty = k_\mu - j_\mu$ for some μ. Suppose $|j-k|_\infty = j_\mu - k_\mu$ for some μ. Then, by hypothesis,

$$\{(2^{-k\cdot t} T E_2, 2^{-k}) : k \in \mathbb{N}^\nu\}$$

is a bounded set of $(W, \vec{d})$ pre-elementary operators supported in Ω, and Lemma 5.5.13 then implies that for every $N \in \mathbb{N}$,

$$\{(2^{N|j-k|_\infty} 2^{-(j\wedge k)\cdot t} E_1 T E_2, 2^{-j\wedge k}) : j,k \in \mathbb{N}^\nu, j_\mu - k_\mu = |j-k|_\infty, \text{ for some } \mu\}$$

is a bounded set of $(W, \vec{d})$ pre-elementary operators supported in Ω. When $|j-k|_\infty = k_\mu - j_\mu$ for some μ, the same proof works with the roles of E_1 and E_2 reversed. This completes the proof that $\mathcal{E}_N$ is a bounded set of $(W, \vec{d})$ pre-elementary operators supported in Ω.

The result will follow once we show that for each $\mu \in \{1, \ldots, \nu\}$, $2^{N|j-k|_\infty} 2^{-t\cdot(j\wedge k)} E_1 T E_2$ can be written in the form (5.3) (with j replaced by $j \wedge k$), where each $E_{\mu,\alpha_\mu,\beta_\mu}$ is of the same form as $2^{N|j-k|_\infty}2^{-t\cdot(j\wedge k)}E_1TE_2$. Applying Proposition 5.5.5 (h) (with $N=1$) to E_1 and E_2, we have

$$
\begin{aligned}
2^{N|j-k|_\infty - t\cdot(j\wedge k)}E_1TE_2 &= \sum_{|\alpha_\mu|,|\beta_\mu|\leq 1} 2^{-(1-|\alpha_\mu|)j_\mu}2^{-(1-|\beta_\mu|)k_\mu}(2^{-j_\mu \vec{d}^\mu}W^\mu)^{\alpha_\mu} \\
&\qquad [2^{N|j-k|_\infty - t\cdot(j\wedge k)}\widetilde{E}_{1,\alpha_\mu}T\widehat{E}_{2,\beta_\mu}](2^{-k_\mu \vec{d}^\mu}W^\mu)^{\beta_\mu} \\
&=: \sum_{|\alpha_\mu|,|\beta_\mu|\leq 1} 2^{-(2-|\alpha_\mu|-|\beta_\mu|)(j_\mu\wedge k_\mu)}(2^{-(j_\mu\wedge k_\mu)\vec{d}^\mu}W^\mu)^{\alpha_\mu} \\
&\qquad [2^{N|j-k|_\infty - t\cdot(j\wedge k)}\widetilde{\widetilde{E}}_{1,\alpha_\mu,\beta_\mu,k}T\widehat{E}_{2,\beta_\mu}](2^{-(j_\mu\wedge k_\mu)\vec{d}^\mu}W^\mu)^{\beta_\mu},
\end{aligned}
\tag{5.64}
$$

where

$$\{(\widetilde{E}_{1,\alpha_\mu},2^{-j}),(\widehat{E}_{2,\beta_\mu},2^{-k}) : (E_1,2^{-j}),(E_2,2^{-k})\in\mathcal{E}, |\alpha_\mu|,|\beta_\mu|\leq 1\}$$

is a bounded set of $(W,\vec{d})$ elementary operators supported in Ω and

$$\widetilde{\widetilde{E}}_{1,\alpha_\mu,\beta_\mu,k} := 2^{-(j_\mu-(j_\mu\wedge k_\mu))(\deg_{\vec{d}^\mu}(\alpha_\mu)-(1-|\alpha_\mu|))-(k_\mu-(j_\mu\wedge k_\mu))(\deg_{\vec{d}^\mu}(\beta_\mu)-(1-|\beta_\mu|))}\widetilde{E}_{1,\alpha_\mu}.$$

Since the coefficient of $\widetilde{E}_{1,\alpha_\mu}$ in the above equation is in $(0,1]$, Proposition 5.5.5 (a) shows that

$$\{(\widetilde{\widetilde{E}}_{1,\alpha_\mu,\beta_\mu,k},2^{-j}) : (E_1,2^{-j}),(E_2,2^{-k})\in\mathcal{E}, |\alpha_\mu|,|\beta_\mu|\leq 1\}$$

is a bounded set of $(W,\vec{d})$ elementary operators supported in Ω. Thus, (5.64) shows that $2^{N|j-k|_\infty - t\cdot(j\wedge k)}E_1TE_2$ is an appropriate sum of derivatives of operators of the same form, as desired, completing the proof. □

The remainder of this section is devoted to the proof of Theorem 5.2.12; thus, we work in the single-parameter case and are given Hörmander vector fields with formal degrees $(W,d) = \{(W_1,d_1),\dots,(W_r,d_r)\}$. To prove the result, we show that for $t\in\mathbb{R}$, $K\geq 0$,

$$
\begin{aligned}
\mathscr{A}_1^t(\Omega,(W,d)) = \mathscr{A}_2^t(\Omega,(W,d)) &\subseteq \mathscr{A}_4^t(\Omega,(W,\vec{d})) \\
&\subseteq \mathscr{A}_3^t(\Omega,(W,d),K) \subseteq \mathscr{A}_2^t(\Omega,(W,d)).
\end{aligned}
\tag{5.65}
$$

That $\mathscr{A}_1^t(\Omega,(W,d)) = \mathscr{A}_2^t(\Omega,(W,d))$ is the special case $\nu=1$ of Theorem 5.2.18 (see Remark 5.2.19). That $\mathscr{A}_4^t(\Omega,(W,\vec{d}))\subseteq\mathscr{A}_3^t(\Omega,(W,d),K)$ follows immediately from the definitions. Thus, there are only two containments of (5.65) left to prove. In what follows, we will use Remark 5.2.19 freely. In particular, we have

$$\frac{(1+2^j\rho_{(W,d)}(x,z))^{-m}}{\mathrm{Vol}(B_{(W,d)}(x,2^{-j}+\rho_{(W,d)}(x,z)))\wedge 1} = \frac{(1+\rho_{2^{-j}}(x,z))^{-m}}{\mathrm{Vol}(B_{2^{-j}}(x,1+\rho_{2^{-j}}(x,z)))\wedge 1}. \tag{5.66}$$

This will allow us to apply results proved in the multi-parameter setting to this single-parameter setting.

Lemma 5.7.2. *Let $\mathcal{B}$ be a bounded set of $(W, \vec{d})$ bump functions in Ω and let $\mathcal{E}$ be a bounded set of $(W, \vec{d})$ pre-elementary operators supported in Ω. Then, for every ordered multi-index α, $\forall m \in \mathbb{N}$, there exists $C = C(\mathcal{B}, \mathcal{E}, \alpha, m, (W, \vec{d}), \Omega) \geq 0$ such that $\forall (E, 2^{-j}) \in \mathcal{E}$, $\forall (\phi, z, 2^{-k}) \in \mathcal{B}$,*

$$\left|(2^{-j\vec{d}} W)^{\alpha} E\phi(x)\right| \leq C \frac{(1 + \rho_{2^{-j\wedge k}}(x, z))^{-m}}{\operatorname{Vol}(B_{2^{-j\wedge k}}(x, 1 + \rho_{2^{-j\wedge k}}(x, z))) \wedge 1}. \tag{5.67}$$

Proof. By Lemma 5.5.2 (iii), $\{((2^{-j\vec{d}} W)^{\alpha} E, 2^{-j}) : (E, 2^{-j}) \in \mathcal{E}\}$ is a bounded set of $(W, \vec{d})$ pre-elementary operators supported in Ω. Thus, it suffices to prove (5.67) in the special case $|\alpha| = 0$.

For $(\phi, z, 2^{-k}) \in \mathcal{E}$, we have for every $m' \in \mathbb{N}$, using Theorem 3.3.1 (b),

$$\begin{aligned}
|\phi(y)| &\lesssim \frac{\chi_{B_{(W,\vec{d})}(z, 2^{-k})}(y)\chi_{\Omega}(y)}{\operatorname{Vol}(B_{(W,\vec{d})}(z, 2^{-k})) \wedge 1} \approx \frac{\chi_{B_{(W,\vec{d})}(z, 2^{-k})}(y)\chi_{\Omega}(y)}{\operatorname{Vol}(B_{(W,\vec{d})}(y, 2^{-k} + \rho_{(W,\vec{d})}(y, z))) \wedge 1} \\
&\lesssim \chi_{\Omega}(y) \frac{(1 + 2^{k}\rho_{(W,\vec{d})}(x, z))^{-m'}}{\operatorname{Vol}(B_{(W,\vec{d})}(x, 2^{-k} + \rho_{(W,\vec{d})}(x, z))) \wedge 1} \\
&= \chi_{\Omega}(y) \frac{(1 + \rho_{2^{-k}}(x, z))^{-m'}}{\operatorname{Vol}(B_{2^{-k}}(x, 1 + \rho_{2^{-k}}(x, z))) \wedge 1}.
\end{aligned} \tag{5.68}$$

Let $(E, 2^{-j}) \in \mathcal{E}$. For $x \notin \Omega$, the left-hand side of (5.67) equals zero. Thus, we may assume $x \in \Omega$. Also, we have $z \in \Omega$, by the definition of bounded sets of bump functions in Ω (Definition 5.2.1). Using (5.68), we have, by taking m' large, depending on m,

$$\begin{aligned}
|E\phi(x)| &\lesssim \int_{\Omega} \left|E(x, y)\phi(y)\right| \, d\operatorname{Vol}(y) \\
&\lesssim \int_{\Omega} \left(\frac{(1 + \rho_{2^{-j}}(x, y))^{-m'}}{\operatorname{Vol}(B_{2^{-j}}(x, 1 + \rho_{2^{-j}}(x, y))) \wedge 1}\right)\left(\frac{(1 + \rho_{2^{-k}}(y, z))^{-m'}}{\operatorname{Vol}(B_{2^{-k}}(y, 1 + \rho_{2^{-k}}(y, z))) \wedge 1}\right) d\operatorname{Vol}(y) \\
&\lesssim \frac{(1 + \rho_{2^{-j\wedge k}}(x, z))^{-m}}{\operatorname{Vol}(B_{2^{-j\wedge k}}(x, 1 + \rho_{2^{-j\wedge k}}(x, z))) \wedge 1},
\end{aligned}$$

where the last estimate follows from Proposition 5.4.4, completing the proof. □

Lemma 5.7.3. *Let $\mathcal{B}$ be a bounded set of $(W, \vec{d})$ bump functions in Ω and let $\mathcal{E}$ be a bounded set of $(W, \vec{d})$ elementary operators supported in Ω. Then, for every ordered multi-index α, $\forall m, N \in \mathbb{N}$, there exists $C = C(\mathcal{B}, \mathcal{E}, \alpha, m, N, (W, \vec{d}), \Omega) \geq 0$ such that $\forall (E, 2^{-j}) \in \mathcal{E}$, $(\phi, z, 2^{-k}) \in \mathcal{B}$,*

$$\left|(2^{-j\vec{d}} W)^{\alpha} E\phi(x)\right| \leq C 2^{-N(j - j\wedge k)} \frac{(1 + \rho_{2^{-j}}(x, z))^{-m}}{\operatorname{Vol}(B_{2^{-j}}(x, 1 + \rho_{2^{-j}}(x, z))) \wedge 1}. \tag{5.69}$$

Proof. When $j \le k$, this follows immediately from Lemma 5.7.2, so we consider only the case $k < j$. Using Lemma 5.4.7, take $M = M(m, (W, d), \Omega) \in \mathbb{N}$ so large that

$$2^{-M|j-k|} \frac{(1+\rho_{2^{-k}}(x,z))^{-m}}{\operatorname{Vol}(B_{2^{-k}}(x, 1+\rho_{2^{-k}}(x,z))) \wedge 1} \le \frac{(1+\rho_{2^{-j}}(x,z))^{-m}}{\operatorname{Vol}(B_{2^{-j}}(x, 1+\rho_{2^{-j}}(x,z))) \wedge 1}. \tag{5.70}$$

We apply Proposition 5.5.5 (h) with N replaced by $N + M$ to see that

$$E = \sum_{|\beta| \le (N+M)} 2^{-j(N+M-|\beta|)} E_\beta (2^{-jd} W)^\beta,$$

where $\{(E_\beta, 2^{-j}) : (E, 2^{-j}) \in \mathcal{E}, |\beta| \le N\}$ is a bounded set of (W, d) elementary operators supported in Ω. We have

$$2^{N(j-k)} E\phi = 2^{-(j-k)M} \sum_{|\beta| \le (N+M)} 2^{N(j-k)} 2^{-j(N-|\beta|)-kM} 2^{\deg_d(\beta)(k-j)} E_\beta (2^{-kd} W)^\beta \phi. \tag{5.71}$$

Since $k < j$ and $\deg_d(\beta) \ge |\beta|$, we have $2^{N(j-k)} 2^{-j(N-|\beta|)-kM} 2^{\deg_d(\beta)(k-j)-kM} \le 1$. Since $\{((2^{-kd} W)^\beta \phi, 2^{-k}) : (\phi, z, 2^{-k}) \in \mathcal{B}\}$ is a bounded set of (W, d) bump functions in Ω, plugging (5.71) into (5.69), we see that it instead suffices to prove

$$2^{-(j-k)M} |(2^{-jd} W)^\alpha E\phi(x)| \lesssim \frac{(1+\rho_{2^{-j}}(x,z))^{-m}}{\operatorname{Vol}(B_{2^{-j}}(x, 1+\rho_{2^{-j}}(x,z))) \wedge 1}, \tag{5.72}$$

where we apply (5.72) with E replaced by E_β and ϕ replaced by $(2^{-kd} W)^\beta \phi$.

Lemma 5.7.2 combined with (5.70) establishes (5.72), completing the proof. □

Proof of $\mathscr{A}_2^t(\Omega, (W, d)) \subseteq \mathscr{A}_4^t(\Omega, (W, d))$. Suppose $T \in \mathscr{A}_2^t(\Omega, (W, d))$. Then there is a bounded set of (W, d) elementary operators supported in Ω, $\{(E_j, 2^{-j}) : j \in \mathbb{N}\}$ with $T = \sum_{j \in \mathbb{N}} 2^{jt} E_j$, and this sum converges in the topology of bounded convergence on $\operatorname{Hom}(C^\infty_{\mathrm{loc}}(\mathfrak{M}), C_0^\infty(\mathfrak{M}))$ by Proposition 5.5.10. By assumption $\operatorname{supp}(T) \subseteq \Omega \times \Omega$, so we only need to show that T satisfies the growth and cancelation conditions from Definition 5.2.4.

We begin with the growth condition. Fix ordered multi-indices α and β and let $M \ge 0$ be such that $t + \deg_d(\alpha) + \deg_d(\beta) + M > -Q_1$. We will show that

$$\sum_{j \in \mathbb{N}} 2^{jt} W_x^\alpha W_y^\beta E_j(x,y)$$

converges absolutely and uniformly on compact subsets on $\{(x,y) \in \mathfrak{M} \times \mathfrak{M} : x \ne y\}$ and that we have

$$\sum_{j \in \mathbb{N}} 2^{jt} |W_x^\alpha W_y^\beta E_j(x,y)| \le C_{\alpha,\beta} \frac{\rho_{(W,d)}(x,y)^{-(t+\deg_d(\alpha)+\deg_d(\beta)+M)}}{\operatorname{Vol}(B_{(W,d)}(x, \rho_{(W,d)}(x,y))) \wedge 1}. \tag{5.73}$$

Since $T = \sum_{j\in\mathbb{N}} 2^{jt} E_j$, with convergence in distribution, and since $W_1, \ldots, W_r$ satisfy Hörmander's condition, it follows from (5.73) that $T(x,y)$ is C^∞_{loc} on $\{(x,y) : x \neq y\}$ and satisfies the growth condition.

We turn to proving the above convergence and the estimate (5.73). If $(x,y) \notin \Omega \times \Omega$, then $E_j(x,y) = 0$, $\forall j$. Thus, we may henceforth assume $x,y \in \Omega$. Set $t_0 := t + \deg_{\vec{d}}(\alpha) + \deg_{\vec{d}}(\beta)$. For any $m \in \mathbb{N}$ we have, using Theorem 3.3.1 (b),

$$\begin{aligned}
&\sum_{j\in\mathbb{N}} 2^{jt} |W_x^\alpha W_z^\beta E_j(x,y)| \\
&\quad = \sum_{j\in\mathbb{N}} 2^{jt_0} |(2^{-j\vec{d}} W_x)^\alpha (2^{-j\vec{d}} W_y)^\beta E_j(x,y)| \\
&\quad \lesssim \sum_{j\in\mathbb{N}} 2^{jt_0} \frac{(1 + 2^j \rho_{(W,\vec{d})}(x,y))^{-m}}{\mathrm{Vol}(B_{(W,\vec{d})}(x, 2^{-j} + \rho_{(W,\vec{d})}(x,y))) \wedge 1} \\
&\quad = \sum_{\substack{2^{-j} \le \rho_{(W,\vec{d})}(x,y) \\ j\in\mathbb{N}}} \cdot + \sum_{\substack{\rho_{(W,\vec{d})}(x,y) < 2^{-j} \\ j\in\mathbb{N}}} \cdot \\
&\quad \lesssim \sum_{\substack{2^{-j} \le \rho_{(W,\vec{d})}(x,y) \\ j\in\mathbb{N}}} 2^{j(t_0 - m)} \rho_{(W,\vec{d})}(x,y)^{-m} \left(\mathrm{Vol}(B_{(W,\vec{d})}(x, \rho_{(W,\vec{d})}(x,y))) \wedge 1\right)^{-1} \\
&\qquad + \sum_{\substack{\rho_{(W,\vec{d})}(x,y) < 2^{-j} \\ j\in\mathbb{N}}} 2^{jt_0} \left(\mathrm{Vol}(B_{(W,\vec{d})}(x, 2^{-j})) \wedge 1\right)^{-1} \\
&\quad =: (\mathrm{I}) + (\mathrm{II}).
\end{aligned}$$

By taking $m > t_0$, (I) is a geometric sum and we have

$$\begin{aligned}
(\mathrm{I}) &\lesssim \frac{\rho_{(W,\vec{d})}(x,y)^{-t_0}}{\mathrm{Vol}(B_{(W,\vec{d})}(x, \rho_{(W,\vec{d})}(x,y))) \wedge 1} \\
&= \frac{\rho_{(W,\vec{d})}(x,y)^{-(t + \deg_{\vec{d}}(\alpha) + \deg_{\vec{d}}(\beta))}}{\mathrm{Vol}(B_{(W,\vec{d})}(x, \rho_{(W,\vec{d})}(x,y))) \wedge 1}.
\end{aligned} \tag{5.74}$$

Since Ω is relatively compact, we have $\rho_{(W,\vec{d})}(x,y) \lesssim 1$, $\forall x,y \in \Omega$. Thus, (5.74) is even better than the desired bound (5.73).

We turn to (II) and let $\delta_1 = \delta_1((W,\vec{d}),\Omega) > 0$ be as in Corollary 3.3.9. By Corollary 3.3.9 (i),

$$\mathrm{Vol}(B_{(W,\vec{d})}(x,\delta)) \wedge 1 \approx \begin{cases} \mathrm{Vol}(B_{(W,\vec{d})}(x,\delta)) & \delta \in (0, \delta_1], \\ 1 & \delta \ge \delta_1. \end{cases} \tag{5.75}$$

Using (5.75), we have

$$(\mathrm{II}) \lesssim \sum_{\substack{\rho_{(W,d)}(x,y)<2^{-j}\le\delta_1 \\ j\in\mathbb{N}}} 2^{jt_0} \operatorname{Vol}(B_{(W,d)}(x,2^{-j}))^{-1} + \sum_{\rho_{(W,d)}(x,y)\vee\delta_1<2^{-j}} 2^{jt}$$
$$=: (\mathrm{III}) + (\mathrm{IV}).$$

For (IV), we have

$$(\mathrm{IV}) = \sum_{\rho_{(W,d)}(x,y)\vee\delta_1<2^{-j}} 2^{jt_0} \lesssim 1. \tag{5.76}$$

We claim that

$$1 \lesssim \frac{\rho_{(W,d)}(x,y)^{-t_0-M}}{\operatorname{Vol}(B_{(W,d)}(x,\rho_{(W,d)}(x,y)))\wedge 1}, \tag{5.77}$$

where we will use the fact that M is chosen so that $t_0 + M > -Q_1$. Indeed, if $\rho_{(W,d)}(x,y) \ge \delta_1 \approx 1$, then $\rho_{(W,d)}(x,y) \approx 1$ (since Ω is relatively compact), and (5.75) therefore shows that the right-hand side of (5.77) is ≈ 1. If $\rho_{(W,d)}(x,y) < \delta_1$, then (5.75) and Corollary 3.3.9 (i) and (ii) show that

$$\begin{aligned}
&\operatorname{Vol}(B_{(W,d)}(x,\rho_{(W,d)}(x,y)))\wedge 1 \approx \operatorname{Vol}(B_{(W,d)}(x,\rho_{(W,d)}(x,y))) \\
&\quad\lesssim (\rho_{(W,d)}(x,y)/\delta_1)^{Q_1}\operatorname{Vol}(B_{(W,d)}(x,\delta_1)) \approx \rho_{(W,d)}(x,y)^{Q_1} \lesssim \rho_{(W,d)}(x,y)^{-t_0-M},
\end{aligned}$$

where in the last estimate, we have used $\rho_{(W,d)}(x,y) \lesssim 1$ (since Ω is relatively compact). This establishes (5.77). Combining this with (5.76) gives

$$(\mathrm{IV}) \lesssim \frac{\rho_{(W,d)}(x,y)^{-t_0-M}}{\operatorname{Vol}(B_{(W,d)}(x,\rho_{(W,d)}(x,y)))\wedge 1},$$

as desired.

Finally, we turn to (III). If $\rho_{(W,d)}(x,y) \ge \delta_1$, then $(\mathrm{III}) = 0$, so we may assume $\rho_{(W,d)}(x,y) < \delta_1$. Using $t_0 + M > -Q_1$, $\rho_{(W,d)}(x,y) \lesssim 1$, and Corollary 3.3.9 (ii) we have

$$\begin{aligned}
(\mathrm{III}) &\lesssim \sum_{\substack{\rho_{(W,d)}(x,y)<2^{-j}\le\delta_1 \\ j\in\mathbb{N}}} 2^{jt_0}(2^j\rho_{(W,d)}(x,y))^{Q_1}\operatorname{Vol}(B_{(W,d)}(x,\rho_{(W,d)}(x,y)))^{-1} \\
&\lesssim \begin{cases} \frac{\rho_{(W,d)}(x,y)^{-t_0}}{\operatorname{Vol}(B_{(W,d)}(x,\rho_{(W,d)}(x,y)))}, & t_0 + Q_1 > 0, \\ \frac{\rho_{(W,d)}(x,y)^{Q_1}}{\operatorname{Vol}(B_{(W,d)}(x,\rho_{(W,d)}(x,y)))}, & t_0 + Q_1 < 0, \\ (-\log(\rho_{(W,d)}(x,y)))\frac{\rho_{(W,d)}(x,y)^{Q_1}}{\operatorname{Vol}(B_{(W,d)}(x,\rho_{(W,d)}(x,y)))}, & t_0 + Q_1 = 0 \end{cases} \\
&\lesssim \frac{\rho_{(W,d)}(x,y)^{-(t_0+M)}}{\operatorname{Vol}(B_{(W,d)}(x,\rho_{(W,d)}(x,y)))}\frac{\rho_{(W,d)}(x,y)^{-(t_0+M)}}{\operatorname{Vol}(B_{(W,d)}(x,\rho_{(W,d)}(x,y)))\wedge 1}.
\end{aligned}$$

This completes the proof of the growth condition.

We turn to the cancelation condition. Let $\mathcal{B}$ be a bounded set of bump functions in Ω. Fix an ordered multi-index α and $M \geq 0$ with $t+\deg_{\vec{d}}(\alpha)+M > -Q_1$. For $(\phi, z, 2^{-k}) \in \mathcal{B}$, using Lemma 5.7.3 and (5.66), we have, for every $m, N \in \mathbb{N}$, taking $\delta_1 > 0$ as above,

$$
\begin{aligned}
&|2^{-k(t+M)}(2^{-k\vec{d}}W)^{\alpha}T\phi(x)| \\
&\quad \leq \sum_{j\in\mathbb{N}} 2^{(j-k)t+(j-k)\deg_{\vec{d}}(\alpha)-kM}|(2^{-j\vec{d}}W)^{\alpha}E_j\phi(x)| \\
&\quad \lesssim \sum_{j\in\mathbb{N}} 2^{(j-k)t+(j-k)\deg_{\vec{d}}(\alpha)-kM-N(j-j\wedge k)} \frac{(1+2^{j\wedge k}\rho_{(W,\vec{d})}(x,z))^{-m}}{\mathrm{Vol}(B_{(W,\vec{d})}(x,2^{-j\wedge k}+\rho_{(W,\vec{d})}(x,z)))\wedge 1} \\
&\quad \lesssim \sum_{\substack{j\in\mathbb{N}\\ 2^{-j\wedge k}<\delta_1}} 2^{(j-k)(t+\deg_{\vec{d}}(\alpha))-kM-N(j-j\wedge k)}\,\mathrm{Vol}(B_{(W,\vec{d})}(x,2^{-j\wedge k}))^{-1} \\
&\qquad + \sum_{\substack{j\in\mathbb{N}\\ 2^{-j\wedge k}\geq\delta_1}} 2^{(j-k)(t+\deg_{\vec{d}}(\alpha))-kM-N(j-j\wedge k)} \\
&\quad =: (\mathrm{V}) + (\mathrm{VI}).
\end{aligned}
$$

We begin by estimating (VI). If $2^{-k} \geq \delta_1$, then (VI) $= 0$, so we assume $1 \geq 2^{-k} \geq \delta_1$, and therefore $2^{-k} \approx 1$. We have, by taking $N \geq -(t+\deg_{\vec{d}}(\alpha))$,

$$
\begin{aligned}
(\mathrm{VI}) &= \sum_{\substack{j\in\mathbb{N}\\ 2^{-j}\geq\delta_1}} 2^{j(t+\deg_{\vec{d}}(\alpha)-N)}2^{-k(t+\deg_{\vec{d}}(\alpha))-N(j\wedge k)-kM} \\
&\approx \sum_{\substack{j\in\mathbb{N}\\ 2^{-j}\geq\delta_1}} 2^{j(t+\deg_{\vec{d}}(\alpha)-N)} \lesssim 1 \leq (\mathrm{Vol}(B_{(W,\vec{d})}(x,2^{-k}))\wedge 1)^{-1},
\end{aligned}
$$

which is the desired bound.

We separate (V) into two terms:

$$
\begin{aligned}
(\mathrm{V}) &= \sum_{\substack{0\leq j\leq k\\ 2^{-j}<\delta_1}} 2^{(j-k)(t+\deg_{\vec{d}}(\alpha))-kM}\,\mathrm{Vol}(B_{(W,\vec{d})}(x,2^{-j}))^{-1} \\
&\quad + \sum_{\substack{j>k\\ 2^{-k}<\delta_1}} 2^{(j-k)(t+\deg_{\vec{d}}(\alpha)-N)-kM}\,\mathrm{Vol}(B_{(W,\vec{d})}(x,2^{-k}))^{-1} \\
&=: (\mathrm{VII}) + (\mathrm{VIII}).
\end{aligned}
$$

We turn to estimating (VIII). We have, by taking $N \geq -(t+\deg_{\vec{d}}(\alpha))$,

$$
\begin{aligned}
(\mathrm{VIII}) &= \sum_{j>k} 2^{(j-k)(t+\deg_{\vec{d}}(\alpha)-N)-kM}\,\mathrm{Vol}(B_{(W,\vec{d})}(x,2^{-k}))^{-1} \\
&\approx 2^{-kM}\,\mathrm{Vol}(B_{(W,\vec{d})}(x,2^{-k}))^{-1} \lesssim (\mathrm{Vol}(B_{(W,\vec{d})}(x,2^{-k}))\wedge 1)^{-1},
\end{aligned}
$$

which is the desired bound.

Finally, we estimate (VII). We have, using Corollary 3.3.9 (ii),

$$
\begin{aligned}
\text{(VII)} &= \sum_{\substack{0\leq j\leq k\\ 2^{-j}<\delta_1}} 2^{(j-k)(t+\deg_{d}(\alpha))-kM}\operatorname{Vol}(B_{(W,d)}(x,2^{-j}))^{-1}\\
&\lesssim \sum_{\substack{0\leq j\leq k\\ 2^{-j}\leq\delta_1}} 2^{(j-k)(t+\deg_{d}(\alpha))-kM}\left(\frac{2^{-k}\wedge\delta_1}{2^{-j}}\right)^{Q_1}\operatorname{Vol}(B_{(W,d)}(x,2^{-k}\wedge\delta_1))^{-1}\\
&\approx \sum_{\substack{0\leq j\leq k\\ 2^{-j}\leq\delta_1}} 2^{j(t+\deg_{d}(\alpha)+Q_1)}2^{k(-t-\deg_{d}(\alpha)-M-Q_1)}\operatorname{Vol}(B_{(W,d)}(x,2^{-k}\wedge\delta_1))^{-1}\\
&\lesssim \begin{cases} 2^{k(-t-\deg_{d}(\alpha)-M-Q_1)}\operatorname{Vol}(B_{(W,d)}(x,2^{-k}\wedge\delta_1))^{-1}, & t+\deg_{d}(\alpha)+Q_1<0,\\ 2^{-Mk}\operatorname{Vol}(B_{(W,d)}(x,2^{-k}\wedge\delta_1))^{-1}, & t+\deg_{d}(\alpha)+Q_1>0,\\ k2^{-Mk}\operatorname{Vol}(B_{(W,d)}(x,2^{-k}\wedge\delta_1))^{-1}, & t+\deg_{d}(\alpha)+Q_1=0\end{cases}\\
&\lesssim \operatorname{Vol}(B_{(W,d)}(x,2^{-k}\wedge\delta_1))^{-1}\lesssim(\operatorname{Vol}(B_{(W,d)}(x,2^{-k}))\wedge 1)^{-1},
\end{aligned}
$$

where the second to last estimate follows from $M\geq 0$ and $t+\deg_{d}(\alpha)+M+Q_1>0$ and the last estimate follows from (5.75). This completes the proof. □

We turn to final containment: $\mathscr{A}_3^t(\Omega,(W,d),K)\subseteq\mathscr{A}_2^t(\Omega,(W,d))$, $\forall K\geq 0$. For this, we will use Proposition 5.7.1 (in the case $\nu=1$). We begin with two lemmas.

Lemma 5.7.4. *Suppose $T\in\mathscr{A}_3^t(\Omega,(W,d),K)$ for some $K\in\mathbb{N}$ and suppose $\mathcal{E}$ is a bounded set of (W,d) elementary operators supported in Ω. Then*

$$\{(2^{-jt}TE,2^{-j}):(E,2^{-j})\in\mathcal{E}\}$$

is a bounded set of (W,d) pre-elementary operators supported in Ω.

Proof. Since $\operatorname{supp}(T)\subseteq\Omega\times\Omega$, Ω is relatively compact and $\operatorname{supp}(T)$ is closed in $\mathfrak{M}\times\mathfrak{M}$, we have $\operatorname{supp}(T)\Subset\Omega\times\Omega$. Combining this with $\bigcup_{(E,2^{-j})\in\mathcal{E}}\operatorname{supp}(E)\Subset\Omega\times\Omega$, it follows that $\bigcup_{(E,2^{-j})\in\mathcal{E}}\operatorname{supp}(TE)\Subset\Omega\times\Omega$.

Fix ordered multi-indices α,β and fix $m\in\mathbb{N}$. We wish to show that there exists $C\geq 0$ such that $\forall x,z\in\Omega$, $\forall(E,2^{-j})\in\mathcal{E}$,

$$2^{-jt}\left|(2^{-jd}W_x)^{\alpha}(2^{-jd}W_z)^{\beta}[TE](x,z)\right|\leq C\frac{(1+2^j\rho_{(W,d)}(x,z))^{-m}}{\operatorname{Vol}(B_{(W,d)}(x,2^{-j}+\rho_{(W,d)}(x,z)))\wedge 1}.$$

Equivalently, we will show that for all α,β,

$$
\begin{aligned}
&2^{-jt}\left|(2^{-jd}W_x)^{\alpha}[(2^{-jd}W_z)^{\beta}]^*[TE](x,z)\right| \\
&\quad\lesssim \frac{(1+2^j\rho_{(W,d)}(x,z))^{-m}}{\operatorname{Vol}(B_{(W,d)}(x,2^{-j}+\rho_{(W,d)}(x,z)))\wedge 1}.
\end{aligned} \tag{5.78}
$$

Using the fact that $[(2^{-dj}W_z)^{\beta}]^*E(y,z) = [E(2^{-dj}W)^{\beta}](y,z)$ and

$$
\{(E(2^{-dj}W)^{\beta},2^{-j}) : (E,2^{-j}) \in \mathcal{E}\}
$$

is a bounded set of (W,d) elementary operators supported in Ω (by Proposition 5.5.5 (e)), it suffices to prove (5.78) in the case $|\beta| = 0$. It follows easily from the definitions that $W_x^{\alpha}T(x,y) = W^{\alpha}T \in \mathscr{A}_3^t(\Omega,(W,d),K)$, and therefore it suffices to prove (5.78) in the case $|\alpha| = 0$, by replacing t with $t+\deg_d(\alpha)$.

We turn to proving (5.78) in the case $|\alpha| = |\beta| = 0$. Fix $M = M(m,\Omega,(W,d)) \in \mathbb{N}$ so large that Proposition 5.4.4 gives, $\forall x,z \in \Omega$,

$$
\begin{aligned}
&\int_{\Omega}\left(\frac{(1+2^j\rho_{(W,d)}(x,y))^{-M}}{\operatorname{Vol}(B_{(W,d)}(x,2^{-j}+\rho_{(W,d)}(x,y)))\wedge 1}\right)\left(\frac{(1+2^j\rho_{(W,d)}(y,z))^{-M}}{\operatorname{Vol}(B_{(W,d)}(y,2^{-j}+\rho_{(W,d)}(y,z)))\wedge 1}\right)d\operatorname{Vol}(y) \\
&\quad\lesssim \frac{(1+2^j\rho_{(W,d)}(x,z))^{-m}}{\operatorname{Vol}(B_{(W,d)}(x,2^{-j}+\rho_{(W,d)}(x,z)))\wedge 1}.
\end{aligned} \tag{5.79}
$$

Pick $N \in \mathbb{N}$ large enough that $t+N \geq K \vee M$. Applying Proposition 5.5.5 (h) with this choice of N, we may write

$$
E = \sum_{|\alpha|\leq N} 2^{-(N-|\alpha|)j}(2^{-jd}W)^{\alpha}E_{\alpha}, \tag{5.80}
$$

where $\{(E_{\alpha},2^{-j}) : (E,2^{-j}) \in \mathcal{E}, |\alpha| \leq N\}$ is a bounded set of (W,d) elementary operators supported in Ω.

Fix $x,z \in \Omega$ (all estimates which follow will be independent of x,z). Set $\mathcal{K} := \overline{\Omega}$ and let $\Omega_1 \Subset \mathfrak{M}$ be open with $\mathcal{K} \Subset \Omega_1$. Let $\phi_{x,2^{-j}} \in C_0^{\infty}(\Omega_1)$ be the function from Corollary 3.3.8. In particular, $\operatorname{supp}(\phi_{x,2^{-j}}) \subseteq B_{(W,d)}(x,2^{-j})$, $\phi_{x,2^{-j}} \equiv 1$ on $B_{(W,d)}(x,\xi_3^2 2^{-j})$, where $\xi_3 \approx 1$, and for every $N \in \mathbb{N}$,

$$
\sup_{\substack{x\in\mathcal{K}\\ j\in[0,\infty)}} \sum_{|\alpha|\leq N} \|(2^{-jd}W)^{\alpha}\phi_{x,2^{-j}}\|_{C(\mathfrak{M})} < \infty.
$$

Set $\psi_{\alpha}(y) := \phi_{x,2^{-j}}(y)E_{\alpha}(y,z)$, where we have suppressed the dependence of ψ_{α} on x and z. Let γ be an ordered multi-index. We have, $\forall y \in \operatorname{supp}(\psi) \subseteq B_{(W,d)}(x,2^{-j}) \cap \Omega$,

$$\begin{aligned}\left|(2^{-j\vec{d}}W)^{\gamma}\psi(y)\right| &\lesssim \frac{(1+2^{j}\rho_{(W,\vec{d})}(y,z))^{-m}}{\mathrm{Vol}(B_{(W,\vec{d})}(y,2^{-j}+\rho_{(W,\vec{d})}(y,z)))\wedge 1}\\ &\approx \frac{(1+2^{j}\rho_{(W,\vec{d})}(x,z))^{-m}}{\mathrm{Vol}(B_{(W,\vec{d})}(x,2^{-j}+\rho_{(W,\vec{d})}(x,z)))\wedge 1},\end{aligned} \tag{5.81}$$

where we have used $2^{-j}+\rho_{(W,\vec{d})}(x,z) \approx 2^{-j} \approx 2^{-j}\rho_{(W,\vec{d})}(y,z)$, $\forall y \in \mathrm{supp}(\psi) \subseteq B_{(W,\vec{d})}(x,2^{-j})$ and Theorem 3.3.1 (b) to see that for such y,

$$\begin{aligned}\mathrm{Vol}(B_{(W,\vec{d})}(y,2^{-j}+\rho_{(W,\vec{d})}(y,z)))\wedge 1 &\approx \mathrm{Vol}(B_{(W,\vec{d})}(y,2^{-j}))\wedge 1\\ &\approx \mathrm{Vol}(B_{(W,\vec{d})}(y,2^{-j}+\rho_{(W,\vec{d})}(x,z)))\wedge 1.\end{aligned}$$

Consider, using (5.80),

$$2^{-jt}\left|[TE](x,z)\right| \leq \sum_{|\alpha|\leq N} 2^{-j(t+N-|\alpha|)}\left|[T(2^{-j\vec{d}}W)^{\alpha}E_{\alpha}](x,z)\right|. \tag{5.82}$$

We bound each term on the right-hand side of (5.82) separately. Set $S_{\alpha} := TW^{\alpha}$. It follows easily from the definitions that $S_{\alpha} \in \mathscr{A}^{t}_{t+\deg_{\vec{d}}(\alpha)}(\Omega,(W,\vec{d}),K)$. We have, for $|\alpha|\leq N$,

$$\begin{aligned}&2^{-j(t+N-|\alpha|)}\left|[T(2^{-j\vec{d}}W)^{\alpha}E_{\alpha}](x,z)\right|\\ &\quad\leq 2^{-j(t+N)-j(\deg_{\vec{d}}(\alpha)-|\alpha|)}\left|[S_{\alpha}\psi_{\alpha}](x)\right|\\ &\qquad+2^{-j(t+N)-j(\deg_{\vec{d}}(\alpha)-|\alpha|)}\int_{\Omega}\left|S_{\alpha}(x,y)(1-\phi_{x,2^{-j}}(y))E_{\alpha}(y,z)\right|\,d\,\mathrm{Vol}(y)\\ &\quad=: \mathrm{(I)}+\mathrm{(II)}.\end{aligned}$$

We turn to bounding (I). Using the fact that $N+t\geq K$ and $\deg_{\vec{d}}(\alpha)-|\alpha|\geq 0$, the cancelation condition for S_{α} along with the bound (5.81) shows that

$$\mathrm{(I)} \lesssim \frac{(1+2^{j}\rho_{(W,\vec{d})}(x,z))^{-m}}{\mathrm{Vol}(B_{(W,\vec{d})}(x,2^{-j}+\rho_{(W,\vec{d})}(x,z)))\wedge 1},$$

as desired.

For (II) we use the growth condition for S_{α}. Note that $N+t+\deg_{\vec{d}}(\alpha)-|\alpha|\geq (t+\deg_{\vec{d}}(\alpha))\vee K$. Using that $\rho_{(W,\vec{d})}(x,y)\lesssim 1$, $\forall (x,y)\in \mathrm{supp}(S_{\alpha})\subseteq \Omega\times\Omega$, we have, for $\rho_{(W,\vec{d})}(x,y)\geq 2^{-j}\xi_3^2$ with $(x,y)\in\mathrm{supp}(S_{\alpha})$,

$$\begin{aligned}&2^{-j(t+N+\deg_{\vec{d}}(\alpha)-|\alpha|)}|S_{\alpha}(x,y)|\\ &\quad\lesssim 2^{-j(t+N+\deg_{\vec{d}}(\alpha)-|\alpha|)}\frac{\rho_{(W,\vec{d})}(x,y)^{-((t+\deg_{\vec{d}}(\alpha))\vee K)}}{\mathrm{Vol}(B_{(W,\vec{d})}(x,\rho_{(W,\vec{d})}(x,y)))\wedge 1}\\ &\quad\lesssim 2^{-j(t+N+\deg_{\vec{d}}(\alpha)-|\alpha|)}\frac{\rho_{(W,\vec{d})}(x,y)^{-(t+N+\deg_{\vec{d}}(\alpha)-|\alpha|)}}{\mathrm{Vol}(B_{(W,\vec{d})}(x,\rho_{(W,\vec{d})}(x,y)))\wedge 1}\end{aligned}$$

$$\begin{aligned}
&\lesssim \frac{(1+2^j\rho_{(W,d)}(x,y))^{-(t+N+\deg_d(\alpha)-|\alpha|)}}{\mathrm{Vol}(B_{(W,d)}(x,\rho_{(W,d)}(x,y)))\wedge 1} \\
&\lesssim \frac{(1+2^j\rho_{(W,d)}(x,y))^{-(t+N+\deg_d(\alpha)-|\alpha|)}}{\mathrm{Vol}(B_{(W,d)}(x,2^{-j}+\rho_{(W,d)}(x,y)))\wedge 1} \\
&\lesssim \frac{(1+2^j\rho_{(W,d)}(x,y))^{-M}}{\mathrm{Vol}(B_{(W,d)}(x,2^{-j}+\rho_{(W,d)}(x,y)))\wedge 1},
\end{aligned} \tag{5.83}$$

where the second to last estimate uses Theorem 3.3.1 (b) and the last estimate uses $t + N + \deg_d(\alpha) - |\alpha| \geq t + N \geq M$.

Because $(1 - \phi_{x,2^{-j}}(y))$ is supported on $\rho_{(W,d)}(x,y) \geq 2^{-j}\xi_3^2$, we use (5.83) to conclude that

$$\begin{aligned}
(\mathrm{II}) \lesssim \int_\Omega &\frac{(1+2^j\rho_{(W,d)}(x,y))^{-M}}{\mathrm{Vol}(B_{(W,d)}(x,2^{-j}+\rho_{(W,d)}(x,y)))\wedge 1} \\
&\times \frac{(1+2^j\rho_{(W,d)}(y,z))^{-M}}{\mathrm{Vol}(B_{(W,d)}(y,2^{-j}+\rho_{(W,d)}(y,z)))\wedge 1}\, d\,\mathrm{Vol}(y).
\end{aligned}$$

Combining this with (5.79) we obtain

$$(\mathrm{II}) \lesssim \frac{(1+2^j\rho_{(W,d)}(x,z))^{-m}}{\mathrm{Vol}(B_{(W,d)}(x,2^{-j}+\rho_{(W,d)}(x,z)))\wedge 1},$$

which is the desired bound, completing the proof. □

Lemma 5.7.5. *Let $T \in \mathscr{A}_3^t(\Omega,(W,d),K)$, for some $K \geq 0$. Then $T \in \mathrm{Hom}(C_0^\infty(\mathfrak{M}), C_0^\infty(\mathfrak{M}))$ and $T^* \in \mathscr{A}_1^t(\Omega,(W,d),K)$.*

Proof. A priori, $T \in \mathscr{A}_1^t(\Omega,(W,d),K) \subseteq \mathrm{Hom}(C_0^\infty(\mathfrak{M}), C_0^\infty(\mathfrak{M})')$. It follows from the cancelation condition that, in fact, $T \in \mathrm{Hom}(C_0^\infty(\mathfrak{M}), C_{\mathrm{loc}}^\infty(\mathfrak{M}))$. Since $\mathrm{supp}(T) \subseteq \Omega \times \Omega$ and Ω is relatively compact, we have $T \in \mathrm{Hom}(C_0^\infty(\mathfrak{M}), C_0^\infty(\mathfrak{M}))$. A similar argument, using the cancelation condition for T^*, shows that $T^* \in \mathrm{Hom}(C_0^\infty(\mathfrak{M}), C_0^\infty(\mathfrak{M}))$.

By Theorem 3.3.1 (b), $\mathrm{Vol}(B_{(W,d)}(x,\rho_{(W,d)}(x,z)) \wedge 1 \approx \mathrm{Vol}(B_{(W,d)}(z,\rho_{(W,d)}(x,z)) \wedge 1$ (see Proposition 5.4.3 for a similar argument). Using this, the growth condition for T^* follows immediately from the growth condition for T. Since the cancelation condition is symmetric in T and T^*, we have $T \in \mathscr{A}_3^t(\Omega,(W,d),K)$, as desired. □

Proof of $\mathscr{A}_3^t(\Omega,(W,d),K) \subseteq \mathscr{A}_2^t(\Omega,(W,d))$. Let $T \in \mathscr{A}_3^t(\Omega,(W,d),K)$. By Lemma 5.7.5, $T \in \mathrm{Hom}(C_0^\infty(\mathfrak{M}), C_0^\infty(\mathfrak{M}))$, so to show that $T \in \mathscr{A}_2^t(\Omega,(W,d))$, in the case $\nu = 1$ of Proposition 5.7.1, it suffices to show that for every bounded set of (W,d) elementary operators supported in Ω, $\mathcal{E}$,

$$\{(2^{-jt}ET, 2^{-j}), (2^{-jt}TE, 2^{-j}) : (E, 2^{-j}) \in \mathcal{E}\}$$

is a bounded set of $(W, \vec{d})$ pre-elementary operators supported in Ω.

Lemma 5.7.4 shows that

$$\{(2^{-jt}TE, 2^{-j}) : (E, 2^{-j}) \in \mathcal{E}\}$$

is a bounded set of $(W, \vec{d})$ pre-elementary operators supported in Ω. By Lemma 5.7.5, $T^* \in \mathscr{A}_3^t(\Omega, (W, \vec{d}), K)$, and therefore by Proposition 5.5.5 (d), Lemma 5.7.4 shows that

$$\{(2^{-jt}T^*E^*, 2^{-j}) : (E, 2^{-j}) \in \mathcal{E}\}$$

is a bounded set of $(W, \vec{d})$ pre-elementary operators supported in Ω. Lemma 5.5.6 shows that this implies that

$$\{(2^{-jt}ET, 2^{-j}) : (E, 2^{-j}) \in \mathcal{E}\}$$

is a bounded set of $(W, \vec{d})$ pre-elementary operators supported in Ω, completing the proof. □

5.8 Basic properties

Several basic properties of the singular integrals we are considering follow directly from the equivalences in Theorems 5.2.12, 5.2.18, and 5.2.30. As before, we consider the two settings described in Section 5.3. In the Hörmander setting, we will be studying the singular integrals in $\mathscr{A}^t(\Omega, (W, \vec{d}))$, and in the general setting, we will be studying the singular integrals in $\widetilde{\mathscr{A}}^t(\Omega, (W, \vec{d}))$. The single-parameter setting described in Section 5.1.1 is a special case of the Hörmander setting. See also the beginning of Section 5.5 for a further discussion of these two settings.

Proposition 5.8.1. Hörmander setting: *If $T \in \mathscr{A}^t(\Omega, (W, \vec{d}))$, then $T^* \in \mathscr{A}^t(\Omega, (W, \vec{d}))$.*
General setting: *If $T \in \widetilde{\mathscr{A}}^t(\Omega, (W, \vec{d}))$, then $T^* \in \widetilde{\mathscr{A}}^t(\Omega, (W, \vec{d}))$.*

Proof. We prove only the result in the Hörmander setting; a similar proof applies in the general setting. We use $T \in \mathscr{A}^t(\Omega, (W, \vec{d})) = \mathscr{A}_2^t(\Omega, (W, \vec{d}))$, so that there is a bounded set of $(W, \vec{d})$ elementary operators supported in Ω, $\{(E_j, 2^{-j}) : j \in \mathbb{N}^\nu\}$, with $T = \sum_{j\in\mathbb{N}^\nu} 2^{j\cdot t}E_j$. Then $T^* = \sum_{j\in\mathbb{N}^\nu} 2^{j\cdot t}E_j^*$. Proposition 5.5.5 (d) shows that $\{(E_j^*, 2^{-j}) : j \in \mathbb{N}^\nu\}$ is a bounded set of $(W, \vec{d})$ elementary operators supported in Ω. Thus, $T^* \in \mathscr{A}_2^t(\Omega, (W, \vec{d})) = \mathscr{A}^t(\Omega, (W, \vec{d}))$. □

Proposition 5.8.2. *Let $t, s \in \mathbb{R}^\nu$, with $t \leq s$ (i. e., $t_\mu \leq s_\mu$, $\forall \mu \in \{1, \ldots, \nu\}$). Then we have the following.*
Hörmander setting: *We have $\mathscr{A}^t(\Omega, (W, \vec{d})) \subseteq \mathscr{A}^s(\Omega, (W, \vec{d}))$.*
General setting: *We have $\widetilde{\mathscr{A}}^t(\Omega, (W, \vec{d})) \subseteq \widetilde{\mathscr{A}}^s(\Omega, (W, \vec{d}))$.*

Proof. In the Hörmander setting, this follows easily from $\mathscr{A}^{\cdot}(\Omega,(W,\vec{d})) = \mathscr{A}_2^{\cdot}(\Omega,(W,\vec{d}))$. Similarly, in the general setting this follows easily from $\widetilde{\mathscr{A}}^{\cdot}(\Omega,(W,\vec{d})) = \widetilde{\mathscr{A}}_2^{\cdot}(\Omega,(W,\vec{d}))$. □

Proposition 5.8.3. **Hörmander setting:** *Let $\psi \in C_0^\infty(\Omega)$ and let a be an ordered-multi index. Then* $\mathrm{Mult}[\psi]W^\alpha \in \mathscr{A}^{\deg_{\vec{d}}(\alpha)}(\Omega,(W,\vec{d}))$. *In particular,* $\mathrm{Mult}[\psi] \in \mathscr{A}^0(\Omega,(W,\vec{d}))$.
General setting: *The same result holds with $\mathscr{A}$ replaced by $\widetilde{\mathscr{A}}$.*

Proof. We prove only the result in the Hörmander setting; the proof in the general setting is nearly identical. Let $\mathcal{E}$ be a bounded set of $(W,\vec{d})$ elementary operators supported in Ω. For $(E,2^{-j}) \in \mathcal{E}$, we have

$$\mathrm{Mult}[\psi]W^\alpha E = 2^{j\cdot\deg_{\vec{d}}(\alpha)}\,\mathrm{Mult}[\psi](2^{-j\vec{d}}W)^\alpha E.$$

Thus, Proposition 5.5.5 (c) and (e) shows that

$$\{(2^{-j\cdot\deg_{\vec{d}}(\alpha)}\,\mathrm{Mult}[\psi]W^\alpha E, 2^{-j}) : (E,2^{-j}) \in \mathcal{E}\}$$

is a bounded set of $(W,\vec{d})$ elementary operators supported in Ω. It follows that $\mathrm{Mult}[\psi]W^\alpha \in \mathscr{A}_1^{\deg_{\vec{d}}(\alpha)}(\Omega,(W,\vec{d})) = \mathscr{A}^{\deg_{\vec{d}}(\alpha)}(\Omega,(W,\vec{d}))$. Taking $|\alpha| = 0$ shows $\mathrm{Mult}[\psi] \in \mathscr{A}^0(\Omega,(W,\vec{d}))$. □

Proposition 5.8.4. *In the Hörmander setting, $\mathscr{A}^t(\Omega,(W,\vec{d})) \subseteq \widetilde{\mathscr{A}}^t(\Omega,(W,\vec{d}))$.*

Proof. Let $T \in \mathscr{A}^t(\Omega,(W,\vec{d})) = \mathscr{A}_2^t(\Omega,(W,\vec{d}))$. Then there is a bounded set of $(W,\vec{d})$ elementary operators supported in Ω, $\{(E_j,2^{-j}) : j \in \mathbb{N}^\nu\}$, with $T = \sum_{j\in\mathbb{N}^\nu} 2^{j\cdot t}E_j$. Proposition 5.5.8 shows that $\{(E_j,2^{-j}) : j \in \mathbb{N}^\nu\}$ is a bounded set of $(W,\vec{d})$-generalized elementary operators supported on Ω, and therefore $T \in \widetilde{\mathscr{A}}_2^t(\Omega,(W,\vec{d})) = \widetilde{\mathscr{A}}^t(\Omega,(W,\vec{d}))$. □

Proposition 5.8.5. **Hörmander setting:** *Let $T \in \mathscr{A}^t(\Omega,(W,\vec{d}))$. Then $T \in \mathrm{Hom}(C_{\mathrm{loc}}^\infty(\mathfrak{M}), C_0^\infty(\mathfrak{M}))$ and $T \in \mathrm{Hom}(C_0^\infty(\mathfrak{M})', C_{\mathrm{loc}}^\infty(\mathfrak{M})')$.*
General setting: *Let $T \in \widetilde{\mathscr{A}}^t(\Omega,(W,\vec{d}))$. Then $T \in \mathrm{Hom}(C_{W,\mathrm{loc}}^\infty(\mathfrak{M}), C_{W,0}^\infty(\mathfrak{M}))$ and $T \in \mathrm{Hom}(C_{W,0}^\infty(\mathfrak{M})', C_{W,\mathrm{loc}}^\infty(\mathfrak{M})')$.*

Proof. In light of Proposition 5.8.4 and Remark 5.2.24, the result in the Hörmander setting is a special case of the result in the general setting. In the general setting, for $T \in \widetilde{\mathscr{A}}^t(\Omega,(W,\vec{d})) = \widetilde{\mathscr{A}}_2^t(\Omega,(W,\vec{d}))$, there is a bounded set of generalized $(W,\vec{d})$ elementary operators supported in Ω, $\{(E_j,2^{-j}) : j \in \mathbb{N}^\nu\}$ with $T = \sum_{j\in\mathbb{N}^\nu} 2^{j\cdot t}E_j$. Proposition 5.5.10 shows that this sum converges in both $\mathrm{Hom}(C_{W,\mathrm{loc}}^\infty(\mathfrak{M}), C_{W,0}^\infty(\mathfrak{M}))$ and $\mathrm{Hom}(C_{W,0}^\infty(\mathfrak{M})', C_{W,\mathrm{loc}}^\infty(\mathfrak{M})')$, completing the proof. □

Proposition 5.8.6. **Hörmander setting:** *$\mathscr{A}(\Omega,(W,\vec{d}))$ is a filtered algebra. More precisely, if $T \in \mathscr{A}^t(\Omega,(W,\vec{d}))$ and $S \in \mathscr{A}^s(\Omega,(W,\vec{d}))$, then $TS \in \mathscr{A}^{t+s}(\Omega,(W,\vec{d}))$.*
General setting: *$\widetilde{\mathscr{A}}(\Omega,(W,\vec{d}))$ is a filtered algebra. More precisely, if $T \in \widetilde{\mathscr{A}}^t(\Omega,(W,\vec{d}))$ and $S \in \widetilde{\mathscr{A}}^s(\Omega,(W,\vec{d}))$, then $TS \in \widetilde{\mathscr{A}}^{t+s}(\Omega,(W,\vec{d}))$.*

Proof. We prove only the result in the general setting; the proof in the Hörmander setting is nearly identical. Proposition 5.8.5 shows that TS makes sense as an operator in $\mathrm{Hom}(C^\infty_{W,\mathrm{loc}}(\mathfrak{M}), C^\infty_{W,0}(\mathfrak{M}))$. It now follows immediately from $\widetilde{\mathscr{A}}(\Omega,(W,\vec{d})) = \widetilde{\mathscr{A}}_1(\Omega,(W,\vec{d}))$ that $TS \in \widetilde{\mathscr{A}}^{t+s}(\Omega,(W,\vec{d}))$. □

Proposition 5.8.7. *Suppose $\Omega_1 \subseteq \Omega_2$ are open, relatively compact subsets of $\mathfrak{M}$.*
Hörmander setting: *We have $\mathscr{A}^t(\Omega_1,(W,\vec{d})) \subseteq \mathscr{A}^t(\Omega_2,(W,\vec{d}))$. Conversely, if $T \in \mathscr{A}^t(\Omega_2,(W,\vec{d}))$ satisfies $\mathrm{supp}(T) \subseteq \Omega_1\times\Omega_1$, then $T \in \mathscr{A}^t(\Omega_1,(W,\vec{d}))$.*
General setting: *We have $\widetilde{\mathscr{A}}^t(\Omega_1,(W,\vec{d})) \subseteq \widetilde{\mathscr{A}}^t(\Omega_2,(W,\vec{d}))$. Conversely, if $T \in \widetilde{\mathscr{A}}^t(\Omega_2,(W,\vec{d}))$ satisfies $\mathrm{supp}(T) \subseteq \Omega_1\times\Omega_1$, then $T \in \widetilde{\mathscr{A}}^t(\Omega_1,(W,\vec{d}))$.*

Proof. We only prove the result in the Hörmander setting; the proof in the general setting is similar. If $T \in \mathscr{A}^t(\Omega_1,(W,\vec{d})) = \mathscr{A}^t_2(\Omega_1,(W,\vec{d}))$, then there exists a bounded set of $(W,\vec{d})$ elementary operators supported in Ω_1, $\{(E_j,2^{-j}) : j\in\mathbb{N}^\nu\}$ with $T=\sum_{j\in\mathbb{N}^\nu} 2^{j\cdot t}E_j$. Since $\Omega_1\subseteq\Omega_2$, $\{(E_j,2^{-j}) : j\in\mathbb{N}^\nu\}$ is a bounded set of $(W,\vec{d})$ elementary operators supported in Ω_2, and therefore $T\in\mathscr{A}^t_2(\Omega_2,(W,\vec{d})) = \mathscr{A}^t(\Omega_2,(W,\vec{d}))$.

Conversely, suppose $T \in \mathscr{A}^t(\Omega_2,(W,\vec{d}))$ with $\mathrm{supp}(T)\subseteq\Omega_1\times\Omega_1$. Since $\mathrm{supp}(T)$ is closed and $\Omega_1 \Subset \mathfrak{M}$, we have $\mathrm{supp}(T)\Subset\Omega_1\times\Omega_1$. Take $\psi\in C^\infty_0(\Omega_1)$ with $T = \mathrm{Mult}[\psi]T\,\mathrm{Mult}[\psi]$. Since $T\in\mathscr{A}^t(\Omega_2,(W,\vec{d})) = \mathscr{A}^t_2(\Omega_2,(W,\vec{d}))$, there exists a bounded set of $(W,\vec{d})$ elementary operators supported in Ω_2, $\{(E_j,2^{-j}) : j\in\mathbb{N}^\nu\}$ with $T=\sum_{j\in\mathbb{N}^\nu}2^{j\cdot t}E_j$; by Proposition 5.5.10 this sum converges in the topology of bounded convergence on $\mathrm{Hom}(C^\infty_{\mathrm{loc}}(\mathfrak{M}), C^\infty_0(\mathfrak{M}))$. Thus, we have

$$T = \mathrm{Mult}[\psi]T\,\mathrm{Mult}[\psi] = \sum_{j\in\mathbb{N}^\nu} 2^{j\cdot t}\,\mathrm{Mult}[\psi]E_j\,\mathrm{Mult}[\psi].$$

It follows from Proposition 5.5.5 (c) that $\{(\mathrm{Mult}[\psi]E_j\,\mathrm{Mult}[\psi],2^{-j}) : j\in\mathbb{N}^\nu\}$ is a bounded set of $(W,\vec{d})$ elementary operators supported in Ω_2; however, since $\psi\in C^\infty_0(\Omega_1)$, the same proof shows that $\{(\mathrm{Mult}[\psi]E_j\,\mathrm{Mult}[\psi],2^{-j}) : j\in\mathbb{N}^\nu\}$ is a bounded set of $(W,\vec{d})$ elementary operators supported in Ω_1. Thus, $T\in\mathscr{A}^t_2(\Omega_1,(W,\vec{d})) = \mathscr{A}^t(\Omega_1,(W,\vec{d}))$. □

Remark 5.8.8. In light of Proposition 5.8.7, it makes sense to define

$$\mathscr{A}^t(W,\vec{d}) := \bigcup_{\Omega\Subset\mathfrak{M}} \mathscr{A}^t(\Omega,(W,\vec{d})),\qquad \widetilde{\mathscr{A}}^t(W,\vec{d}) := \bigcup_{\Omega\Subset\mathfrak{M}} \widetilde{\mathscr{A}}^t(\Omega,(W,\vec{d})),$$

where the unions are taken over all relatively compact open sets $\Omega\Subset\mathfrak{M}$.

Proposition 5.8.9. *Let α be an ordered multi-index and let $\phi\in C^\infty_{\mathrm{loc}}(\Omega)$.*
Hörmander setting: *If $T\in\mathscr{A}^t(\Omega,(W,\vec{d}))$, then*

$$\mathrm{Mult}[\phi]W^\alpha T,\, T\,\mathrm{Mult}[\phi]W^\alpha \in \mathscr{A}^{t+\deg_{\vec{d}}(\alpha)}(\Omega,(W,\vec{d})).$$

General setting: *If* $T \in \widetilde{\mathscr{A}}^t(\Omega, (W, \vec{d}))$, *then*

$$\mathrm{Mult}[\phi]W^\alpha T, T\,\mathrm{Mult}[\phi]W^\alpha \in \widetilde{\mathscr{A}}^{t+\deg_{\vec{d}}(\alpha)}(\Omega, (W, \vec{d})).$$

Proof. In either case, we have $\mathrm{supp}(T) \subseteq \Omega \times \Omega$, and since $\mathrm{supp}(T)$ is closed and $\Omega \Subset \mathfrak{M}$, we have $\mathrm{supp}(T) \Subset \Omega \times \Omega$. Take $\psi \in C_0^\infty(\Omega)$ so that $\psi \otimes \psi \equiv 1$ on a neighborhood of $\mathrm{supp}(T)$. Then $\mathrm{Mult}[\phi]W^\alpha T = \mathrm{Mult}[\phi\psi]W^\alpha T$ and $T\,\mathrm{Mult}[\phi]W^\alpha = T\,\mathrm{Mult}[\phi\psi]W^\alpha$. From here, the result follows from Propositions 5.8.3 and 5.8.6. □

Corollary 5.8.10. *Let* $\mathscr{P}$ *be a* $(W, \vec{d})$ *partial differential operator on* Ω *of degree* $\leq \kappa \in \mathbb{Z}^\nu$.
Hörmander setting: *If* $T \in \mathscr{A}^t(\Omega, (W, \vec{d}))$, *then* $\mathscr{P}T, T\mathscr{P} \in \mathscr{A}^{t+\kappa}(\Omega, (W, \vec{d}))$.
General setting: *If* $T \in \widetilde{\mathscr{A}}^t(\Omega, (W, \vec{d}))$, *then* $\mathscr{P}T, T\mathscr{P} \in \widetilde{\mathscr{A}}^{t+\kappa}(\Omega, (W, \vec{d}))$.

Proof. We prove only the result in the Hörmander setting; the proof in the general setting is nearly identical. If $\kappa \in \mathbb{Z}^\nu \setminus \mathbb{N}^\nu$, then $\mathscr{P} = 0$ and the result is trivial, so we henceforth assume $\kappa \in \mathbb{N}^\nu$.

From Definition 3.12.1, we see that $\mathscr{P}$ can be written as

$$\mathscr{P} = \sum_{\deg_{\vec{d}}(\alpha)\leq\kappa} \mathrm{Mult}[a_\alpha]W^\alpha, \quad a_\alpha \in C^\infty_{\mathrm{loc}}(\Omega).$$

Propositions 5.8.9 and 5.8.2 show that for $\deg_{\vec{d}}(\alpha) \leq \kappa$,

$$\mathrm{Mult}[a_\alpha]W^\alpha T, T\,\mathrm{Mult}[a_\alpha]W^\alpha \in \mathscr{A}^{t+\deg_{\vec{d}}(\alpha)}(\Omega, (W, \vec{d})) \subseteq \mathscr{A}^{t+\kappa}(\Omega, (W, \vec{d})).$$

The result follows. □

Proposition 5.8.11. *Let* $T \in C_0^\infty(\Omega \times \Omega)$.
Hörmander setting: *We have* $T \in \bigcap_{t\in\mathbb{R}^\nu} \mathscr{A}^t(\Omega, (W, \vec{d}))$.
General setting: *We have* $T \in \bigcap_{t\in\mathbb{R}^\nu} \widetilde{\mathscr{A}}^t(\Omega, (W, \vec{d}))$.

Proof. We prove only the result in the Hörmander setting. The proof in the general setting is similar. It is straightforward to verify that $\{(T, 1)\}$ is a bounded set of $(W, \vec{d})$ elementary operators supported in Ω. It follows that $T \in \mathscr{A}^t(\Omega, (W, \vec{d})) = \mathscr{A}_2^t(\Omega, (W, \vec{d}))$. □

Proposition 5.8.12. *For each* $\mu \in \{1, \ldots, \nu\}$ *let*

$$(Z^\mu, \vec{d}^\mu) = \{(Z_1^\mu, d_1^\mu), \ldots, (Z_{s_\mu}^\mu, d_{s_\mu}^\mu)\} \subset C^\infty_{\mathrm{loc}}(\mathfrak{M}; T\mathfrak{M}) \times \mathbb{N}_+$$

be such that $(Z^\mu, \vec{d}^\mu)$ *and* $(W^\mu, \vec{d}^\mu)$ *are locally weakly equivalent on* $\mathfrak{M}$. *Define* $(Z, \vec{d}) := (Z^1, \vec{d}^1)\boxtimes(Z^2, \vec{d}^2)\boxtimes\cdots\boxtimes(Z^\nu, \vec{d}^\nu)$, *i. e.,* $(Z, \vec{d})$ *is to* $(Z^\mu, \vec{d}^\mu)$, $1 \leq \mu \leq \nu$, *as* $(W, \vec{d})$ *is to* $(W^\mu, \vec{d}^\mu)$, $1 \leq \mu \leq \nu$.

Hörmander setting: *For every* $t \in \mathbb{R}^\nu$, $\mathscr{A}^t(\Omega, (W, \vec{d})) = \mathscr{A}^t(\Omega, (Z, \vec{d}))$.
General setting: *For every* $t \in \mathbb{R}^\nu$, $\widetilde{\mathscr{A}}^t(\Omega, (W, \vec{d})) = \widetilde{\mathscr{A}}^t(\Omega, (W, \vec{d}))$.

Proof. We prove only the result in the Hörmander setting; the proof in the general setting is nearly identical. Proposition 5.5.14 shows that $\mathcal{E}$ is a bounded set of $(W, \vec{d})$ elementary operators supported in Ω if and only if $\mathcal{E}$ is a bounded set of $(Z, \vec{d})$ elementary operators supported in Ω. From here, the result follows immediately from $\mathscr{A}^t(\Omega, (W, \vec{d})) = \mathscr{A}_2^t(\Omega, (W, \vec{d}))$ and $\mathscr{A}^t(\Omega, (Z, \vec{d})) = \mathscr{A}_2^t(\Omega, (Z, \vec{d}))$. □

The results of Section 5.6 show that our pseudo-differential operators give rise to operators in the algebras $\mathscr{A}$ and $\widetilde{\mathscr{A}}$. This is the source of some important examples of singular integrals in these algebras. We turn to presenting these ideas.

We take (X^μ, d^μ) and $(X, \vec{d})$ as in Section 5.6 and fix relatively compact, open sets $\Omega_1 \Subset \Omega_2 \Subset \Omega$.

Proposition 5.8.13. *If* $a = a(\Omega_1, \Omega_2, (X, \vec{d})) > 0$ *is sufficiently small and* T *is an* $(X, \vec{d})$, *a pseudo-differential operator of order* $t \in \mathbb{R}^\nu$ *supported in* $\Omega_1 \times \Omega_2$, *then we have the following.*
Hörmander setting: *We have* $T \in \mathscr{A}^t(\Omega, (W, \vec{d}))$.
General setting: *We have* $T \in \widetilde{\mathscr{A}}^t(\Omega, (W, \vec{d}))$.

Proof. In light of Theorems 5.2.18 and 5.2.30, this is a restatement of Proposition 5.6.1. □

Corollary 5.8.14. *If* $a = a(\Omega_1, \Omega_2, (X, \vec{d})) > 0$ *is sufficiently small and* T *is an* (X^μ, d^μ), *a pseudo-differential operator of order* $t_\mu \in \mathbb{R}$ *supported in* $\Omega_1 \times \Omega_2$, *then we have the following.*
Hörmander setting: *We have* $T \in \mathscr{A}^{t_\mu \mathfrak{e}_\mu}(\Omega, (W, \vec{d}))$.
General setting: *We have* $T \in \widetilde{\mathscr{A}}^{t_\mu \mathfrak{e}_\mu}(\Omega, (W, \vec{d}))$.

Proof. Proposition 4.4.1 shows that T is an $(X, \vec{d})$, a pseudo-differential operator of order $t_\mu \mathfrak{e}_\mu$. Thus, the result follows from Proposition 5.8.13. □

Corollary 5.8.15. *Fix* $\mu \in \{1, \ldots, \nu\}$ *and* $\kappa_\mu \in \mathbb{N}_+$ *such that* d_j^μ *divides* κ_μ *for all* $1 \le j \le r_\mu$. *Consider the sub-Laplacian*

$$\mathscr{L}_\mu := \sum_{j=1}^{r_\mu} ((W_j^\mu)^{\kappa_\mu / d_j^\mu})^* (W_j^\mu)^{\kappa_\mu / d_j^\mu}. \tag{5.84}$$

Let $\psi \in C_0^\infty(\Omega)$. *Then there exists* $T_{\mu,\psi}$, *with* $T_{\mu,\psi}\mathscr{L}_\mu = \mathrm{Mult}[\psi] + E$, *where we have the following.*
Hörmander setting: *We have* $T_{\mu,\psi} \in \mathscr{A}^{-2\kappa_\mu \mathfrak{e}_\mu}(\Omega, (W, \vec{d}))$ *and* $E \in \mathscr{A}^{-\mathfrak{e}_\mu}(\Omega, (W, \vec{d}))$.
General setting: *We have* $T_{\mu,\psi} \in \widetilde{\mathscr{A}}^{-2\kappa_\mu \mathfrak{e}_\mu}(\Omega, (W, \vec{d}))$ *and* $E \in \widetilde{\mathscr{A}}^{-\mathfrak{e}_\mu}(\Omega, (W, \vec{d}))$.

Proof. Fix $\Omega_1 \Subset \Omega_2 \Subset \mathfrak{M}$ open and relatively compact with $\psi \in C_0^\infty(\Omega_1)$. For $a > 0$ as small as we like, Theorem 4.5.1 shows that there is an (X^μ, d^μ), *a* pseudo-differential operator $T_{\mu,\psi}$ of order $-2\kappa_| mu$ supported in $\Omega_1 \times \Omega_2$ such that $T_{\mu,\psi}\mathscr{L}_\mu = \mathrm{Mult}[\psi] + E$, where E is an (X^μ, d^μ), *a* pseudo-differential operator of order -1 supported in $\Omega_1 \times \Omega_2$. By taking $a > 0$ sufficiently small, the result follows from Corollary 5.8.14. □

Remark 5.8.16. When $(W^\mu, \vec{d}^\mu)$ are Hörmander vector fields with formal degrees, then $\mathscr{L}_\mu$ as defined in (5.84) is maximally subelliptic with respect to $(W^\mu, \vec{d}^\mu)$. Corollary 5.8.15 is our first example of a more general result: parametrices for maximally subelliptic operators are in the algebras $\mathscr{A}$ and $\widetilde{\mathscr{A}}$. See Theorem 8.1.1 (i) ⇒ (vii).

Proposition 5.8.17. *The definitions of $\mathscr{A}$ and $\widetilde{\mathscr{A}}$ are invariant under diffeomorphisms. More precisely, suppose $\mathfrak{N}$ is another C^∞ manifold and $\Psi : \mathfrak{M} \xrightarrow{\sim} \mathfrak{N}$ is a C^∞_{loc} diffeomorphism. Set*

$$(\Psi_* W, \vec{d}) := \{(\Psi_* W_1, \vec{d}_1), \ldots, (\Psi_* W_r, \vec{d}_r)\} \subset C^\infty_{\mathrm{loc}}(\mathfrak{N}; T\mathfrak{N}) \times (\mathbb{N}^\nu \setminus \{0\}).$$

Then we have the following.

Hörmander setting: *We have $T \in \mathscr{A}^t(\Omega, (W, \vec{d}))$ if and only if $\Psi_* T \Psi^* \in \mathscr{A}^t(\Psi(\Omega), (\Psi_* W, \vec{d}))$.*

General setting: *We have $T \in \widetilde{\mathscr{A}}^t(\Omega, (W, \vec{d}))$ if and only if $\Psi_* T \Psi^* \in \widetilde{\mathscr{A}}^t(\Psi(\Omega), (\Psi_* W, \vec{d}))$.*

Proof. It is easy to see that all of our assumptions and definitions are invariant under diffeomorphisms, so this result follows just by tracing through the definitions. □

5.8.1 Pseudo-locality

In this section, we consider only the Hörmander setting and prove that operators in $\mathscr{A}^t(\Omega, (W, \vec{d}))$ are pseudo-local.[1] More precisely, we show the following theorem.

Theorem 5.8.18. *For all $t \in \mathbb{R}^\nu$, the operators in $\mathscr{A}^t(\Omega, (W, \vec{d}))$ are pseudo-local. More precisely, if $T \in \mathscr{A}^t(\Omega, (W, \vec{d}))$, then*

$$T(x, y)\big|_{\{(x,y): x \neq y\}} \in C^\infty_{\mathrm{loc}}(\{(x, y) \in \mathfrak{M} \times \mathfrak{M} : x \neq y\}).$$

The rest of this section is devoted to the proof of Theorem 5.8.18. We begin with a reduction. Since $W = \{W_1, \ldots, W_r\}$ satisfies Hörmander's condition, it suffices to show that $\forall \alpha, \beta$,

$$W_x^\alpha [W_z^\beta]^* T(x, z)$$

1 When $\nu = 1$, that the operators are pseudo-local follows from $\mathscr{A}^t(\Omega, (W, \vec{d})) = \mathscr{A}_4^t(\Omega, (W, \vec{d}))$, so this section is only interesting in the case $\nu \geq 2$.

is a bounded function on a neighborhood of each fixed $(x,z) \in \mathfrak{M} \times \mathfrak{M}$ with $x \neq z$. Furthermore, using Proposition 5.8.9,

$$W_x^\alpha [W_z^\beta]^* T = [W^\alpha T W^\beta](x,z) \in \mathscr{A}^{t+\deg_{\vec{d}}(\alpha)+\deg_{\vec{d}}(\beta)}(\Omega, (W, \vec{d})).$$

Thus, by replacing t with $t + \deg_{\vec{d}}(\alpha) + \deg_{\vec{d}}(\beta)$, we see that it suffices to show that $\forall T \in \mathscr{A}^t(\Omega, (W, \vec{d}))$,

$$T(x,z)\big|_{\{(x,z):x\neq z\}}$$

is a locally bounded function.

Fix $T \in \mathscr{A}^t(\Omega, (W, \vec{d}))$. Using the fact that $T \in \mathscr{A}_2^t(\Omega, (W, \vec{d}))$, we have $T = \sum_{j\in\mathbb{N}^\nu} 2^{j\cdot t} E_j$, where $\{(E_j, 2^{-j}) : j \in \mathbb{N}^\nu\}$ is a bounded set of $(W, \vec{d})$ elementary operators supported in Ω. We wish to bound

$$|T(x,z)| \leq \sum_{j\in\mathbb{N}^\nu} 2^{j\cdot t} |E_j(x,z)|.$$

Our assumptions are symmetric in the parameters $1, \ldots, \nu$, so it suffices to show that

$$\sum_{\substack{j\in\mathbb{N}^\nu \\ j_1\geq j_2\geq\cdots\geq j_\nu}} 2^{j\cdot t} |E_j(x,z)|$$

is locally bounded on $x \neq z$.

Set $2^{-0} = (1, 1, \ldots, 1) \in (0,1]^\nu$. In light of the above remarks, Theorem 5.8.18 follows immediately from the next lemma.

Lemma 5.8.19. *There exists $M = M(t) \geq 0$ such that $\forall x, z \in \Omega$, $x \neq z$,*

$$\sum_{\substack{j\in\mathbb{N}^\nu \\ j_1\geq j_2\geq\cdots\geq j_\nu}} 2^{j\cdot t} |E_j(x,z)| \lesssim \frac{\rho_{2^{-0}}(x,z)^{-M}}{\mathrm{Vol}(B_{2^{-0}}(x, 1 + \rho_{2^{-0}}(x,z))) \wedge 1}.$$

To help prove Lemma 5.8.19, we introduce the next lemma.

Lemma 5.8.20. *For every $N \in \mathbb{N}$, there exists $M = M(t,N) \geq 1$, such that for all $m \in \mathbb{N}$, there exists $C \geq 0$ such that for all $j \in \mathbb{N}^\nu$ with $j_1 \geq j_2 \geq \cdots \geq j_\nu$, $\forall x, z \in \Omega$,*

$$2^{j\cdot t} |E_j(x,z)| \leq C 2^{M j_\nu} 2^{-N j_1} \frac{(1 + 2^{j_\nu} \rho_{2^{-0}}(x,z))^{-m}}{\mathrm{Vol}(B_{2^{-0}}(x, 1 + \rho_{2^{-0}}(x,z))) \wedge 1}.$$

Proof. Since $W^\nu = \{W_1^\nu, \ldots, W_{r_\nu}^\nu\}$ satisfies Hörmander's condition on $\mathfrak{M}$, for all α_1, there exists $K(\alpha_1) \in \mathbb{N}$ such that

$$(W^1)^{\alpha_1} = \sum_{|\beta_\nu|\leq K(\alpha_1)} f_{\alpha_1,\beta_\nu} (W^\nu)^{\beta_\nu}, \quad f_{\alpha_1,\beta_\nu} \in C^\infty_{\mathrm{loc}}(\Omega). \tag{5.85}$$

Fix $\widetilde{N} = \widetilde{N}(N,t) \in \mathbb{N}$ large to be chosen later. By Proposition 5.5.5 (h), we have

$$E_j = \sum_{|\alpha_1|\leq \widetilde{N}} 2^{-(\widetilde{N}-|\alpha_1|)j_1}(2^{-j_1 d^1}W^1)^{\alpha_1} E_{j,\alpha_1},$$

where $\{(E_{j,\alpha_1}, 2^{-j}) : j_1 \geq j_2 \geq \cdots \geq j_\nu, |\alpha_1| \leq \widetilde{N}\}$ is a bounded set of $(W,\vec{d})$ elementary operators supported in Ω. Taking $M = M(\widetilde{N})$ large and using (5.85), we have

$$\begin{aligned}
E_j &= \sum_{|\alpha_1|\leq \widetilde{N}} \sum_{|\beta_\nu|\leq K(\alpha_1)} 2^{-(\widetilde{N}-|\alpha_1|)-j_1 \deg_{d^1}(\alpha_1)+j_\nu \deg_{d^\nu}(\beta_\nu)} \operatorname{Mult}[f_{\alpha_1,\beta_\nu}](2^{-j_\nu d^\nu}W^\nu)^{\beta_\nu} E_{j,\alpha_1} \\
&= \sum_{|\alpha_1|\leq \widetilde{N}} \sum_{|\beta_\nu|\leq K(\alpha_1)} 2^{-\widetilde{N}j_1+j_\nu M}\big[2^{-j_1(\deg_{d^1}(\alpha_1)-|\alpha_1|)-j_\nu(M-\deg_{d^\nu}(\beta_\nu))} \operatorname{Mult}[f_{\alpha_1,\beta_\nu}](2^{-j_\nu d^\nu}W^\nu)^{\beta_\nu} E_{j,\alpha_1}\big] \\
&=: \sum_{|\alpha_1|\leq \widetilde{N}} \sum_{|\beta_\nu|\leq K(\alpha_1)} 2^{-\widetilde{N}j_1+j_\nu M} E_{j,\alpha_1,\beta_\nu},
\end{aligned}$$

where if M is taken so large that $M - \deg_{d^\nu}(\beta_\nu) \geq 0$, $\forall |\beta_\nu| \leq K(\alpha_1)$ and all $|\alpha_1| \leq \widetilde{N}$, then $\{(E_{j,\alpha_1,\beta_\nu}, 2^{-j}) : j_1 \geq j_2 \geq \cdots \geq j_\nu, |\alpha_1| \leq \widetilde{N}, |\beta_\nu| \leq K(\alpha_1)\}$ is a bounded set of $(W,\vec{d})$ elementary operators supported in Ω (here we have used Proposition 5.5.5 (a), (c), and (e) and the fact that $|\alpha_1| \leq \deg_{d^1}(\alpha_1)$).

Thus, for every $m \in \mathbb{N}$,

$$2^{j\cdot t}|E_j(x,z)| \lesssim 2^{-\widetilde{N}j_1+Mj_\nu+|j|_\infty|t|_1} \frac{(1+\rho_{2^{-j}}(x,z))^{-m}}{\operatorname{Vol}(B_{2^{-j}}(x,1+\rho_{2^{-j}}(x,z)))\wedge 1}. \tag{5.86}$$

Note that since $j_1 \geq j_2 \geq \cdots \geq j_\nu$,

$$\rho_{2^{-j}}(x,z) \geq \rho_{2^{-(j_\nu,j_\nu,\ldots,j_\nu)}}(x,z) = 2^{j_\nu}\rho_{2^{-0}}(x,z). \tag{5.87}$$

Theorem 3.5.4 (n) implies that there exists $D = D(\Omega) \geq 0$ such that $\operatorname{Vol}(B_{(W,\vec{d})}(x,2\delta))\wedge 1 \leq 2^D(\operatorname{Vol}(B_{(W,\vec{d})}(x,\delta))\wedge 1)$, $\forall \delta \in (0,\infty)^\nu$, $x \in \Omega$. Combining this with (5.87), we have

$$\begin{aligned}
\operatorname{Vol}(B_{2^{-0}}(x,1+\rho_{2^{-0}}(x,z)))\wedge 1 &= \operatorname{Vol}(B_{(W,\vec{d})}(x,(1+\rho_{2^{-0}}(x,z))2^{-0}))\wedge 1 \\
&\leq \operatorname{Vol}(B_{(W,\vec{d})}(x,2^{2|j|_\infty}(1+\rho_{2^{-j}}(x,z))2^{-j}))\wedge 1 \\
&\leq 2^{2D|j|_\infty}\operatorname{Vol}(B_{(W,\vec{d})}(x,(1+\rho_{2^{-j}}(x,z))2^{-j}))\wedge 1 \\
&= 2^{2D|j|_\infty}\operatorname{Vol}(B_{2^{-j}}(x,1+\rho_{2^{-j}}(x,z)))\wedge 1.
\end{aligned} \tag{5.88}$$

Plugging (5.87) and (5.88) into (5.86) and using the fact that $|j|_\infty = j_1$, we obtain

$$2^{j\cdot t}|E_j(x,z)| \lesssim 2^{-(\widetilde{N}-2D-|t|_1)j_1+Mj_\nu} \frac{(1+2^{j_\nu}\rho_{2^{-0}}(x,z))^{-m}}{\operatorname{Vol}(B_{2^{-0}}(x,1+\rho_{2^{-0}}(x,z)))\wedge 1}.$$

Taking $\widetilde{N} \geq N + 2D + |t|_1$ completes the proof. □

Proof of Lemma 5.8.19. We take $N = \nu - 1$ in Lemma 5.8.20 and let $M \geq 1$ be as in that lemma. Then taking $m = M + 1$ in the statement of Lemma 5.8.20, we have, for $x, z \in \Omega$,

$$\begin{aligned}
\sum_{\substack{j \in \mathbb{N}^\nu \\ j_1 \geq j_2 \geq \cdots \geq j_\nu}} 2^{j \cdot t} |E_j(x,z)| &\lesssim \sum_{j_1 \geq j_2 \geq \cdots \geq j_\nu} 2^{-(\nu-1)j_1} 2^{Mj_\nu} \frac{(1 + 2^{j_\nu} \rho_{2^{-0}}(x,z))^{-(M+1)}}{\operatorname{Vol}(B_{2^{-0}}(x, 1 + \rho_{2^{-0}}(x,z))) \wedge 1} \\
&\lesssim \sum_{j_1 \geq j_2 \geq \cdots \geq j_\nu} 2^{-j_1 - j_2 - \cdots - j_{\nu-1}} 2^{Mj_\nu} \frac{(1 + 2^{j_\nu} \rho_{2^{-0}}(x,z))^{-(M+1)}}{\operatorname{Vol}(B_{2^{-0}}(x, 1 + \rho_{2^{-0}}(x,z))) \wedge 1} \\
&\lesssim \sum_{j_\nu \in \mathbb{N}} 2^{Mj_\nu} \frac{(1 + 2^{j_\nu} \rho_{2^{-0}}(x,z))^{-(M+1)}}{\operatorname{Vol}(B_{2^{-0}}(x, 1 + \rho_{2^{-0}}(x,z))) \wedge 1} \\
&= \sum_{\substack{j_\nu \in \mathbb{N} \\ 2^{j_\nu} \rho_{2^{-0}}(x,z) \leq 1}} \cdot + \sum_{\substack{j_\nu \in \mathbb{N} \\ 2^{j_\nu} \rho_{2^{-0}}(x,z) \geq 1}} \cdot =: (\mathrm{I}) + (\mathrm{II}).
\end{aligned}$$

We bound the terms (I) and (II) separately. For (I), we have

$$\begin{aligned}
(\mathrm{I}) &\lesssim \sum_{\substack{j_\nu \in \mathbb{N} \\ 2^{j_\nu} \rho_{2^{-0}}(x,z) \leq 1}} 2^{Mj_\nu} \frac{1}{\operatorname{Vol}(B_{2^{-0}}(x, 1 + \rho_{2^{-0}}(x,z))) \wedge 1} \\
&\approx \frac{\rho_{2^{-0}}(x,z)^{-M}}{\operatorname{Vol}(B_{2^{-0}}(x, 1 + \rho_{2^{-0}}(x,z))) \wedge 1},
\end{aligned}$$

as desired. For (II), we have

$$\begin{aligned}
(\mathrm{II}) &\lesssim \sum_{\substack{j_\nu \in \mathbb{N} \\ 2^{j_\nu} \rho_{2^{-0}}(x,z) \geq 1}} 2^{Mj_\nu} 2^{-(M+1)j_\nu} \frac{\rho_{2^{-0}}(x,z)^{-(M+1)}}{\operatorname{Vol}(B_{2^{-0}}(x, 1 + \rho_{2^{-0}}(x,z))) \wedge 1} \\
&\lesssim \frac{\rho_{2^{-0}}(x,z)^{-(M+1)}}{\operatorname{Vol}(B_{2^{-0}}(x, 1 + \rho_{2^{-0}}(x,z))) \wedge 1}.
\end{aligned}$$

This proves the result with M replaced by $M + 1$ (where we have used $\rho_{2^{-0}}(x,z) \lesssim 1$, $\forall x, z \in \Omega$, by the relative compactness of Ω). □

Remark 5.8.21. Not only did we prove that operators in $\mathscr{A}^t(\Omega, (W, \vec{d}))$ are pseudo-local, but Lemma 5.8.19 implies a bound on the kernel off of the diagonal. However, this estimate is not optimal. In particular, we did not specify M in the lemma. More seriously, even if the optimal M were found, this would not yield the optimal estimate. Even in the most simple non-trivial special cases with $\nu > 1$, it is somewhat involved to find the optimal estimate (see, for example, [229, Section 9.1]). Moreover, even if the optimal estimates were obtained, we would not know of any use for these estimates; in particular, they would not simplify any of our definitions or proofs.

5.8.2 Standard pseudo-differential operators

To understand the regularity properties of maximally subelliptic operators with respect to the standard Besov and Triebel–Lizorkin spaces, an important property of $\mathscr{A}$ and $\widetilde{\mathscr{A}}$ is that we may set them up in such a way to contain the standard pseudo-differential operators on $\mathfrak{M}$.[2]

For $\nu \geq 1$, we let $(W^1, \vec{d}^1), \ldots, (W^{\nu-1}, \vec{d}^{\nu-1}) \subset C^\infty_{\mathrm{loc}}(\mathfrak{M}; T\mathfrak{M}) \times \mathbb{N}_+$ satisfy the assumptions of this chapter. That is, we consider two settings.

Hörmander setting: We assume $(W^1, \vec{d}^1), \ldots, (W^{\nu-1}, \vec{d}^{\nu-1})$ pairwise locally weakly approximately commute and each $(W^\mu, \vec{d}^\mu)$ is a set of Hörmander vector fields with formal degrees on $\mathfrak{M}$.

General setting: We assume $(W^1, \vec{d}^1), \ldots, (W^{\nu-1}, \vec{d}^{\nu-1})$ pairwise locally weakly approximately commute and each $\operatorname{Gen}((W^\mu, \vec{d}^\mu))$ is locally finitely generated on $\mathfrak{M}$.

Let $(W^\nu, \vec{d}^\nu) = \{(W^\nu_1, 1), \ldots, (W^\nu_{r_\nu}, 1)\} \subset C^\infty_{\mathrm{loc}}(\mathfrak{M}; T\mathfrak{M}) \times \mathbb{N}_+$ be such that

$$\operatorname{span}\{W^\nu_1(x), \ldots, W^\nu_{r_\nu}(x)\} = T_x\mathfrak{M}, \quad \forall x \in \mathfrak{M}.$$

Lemma 5.8.22. *$(W^1, \vec{d}^1), \ldots, (W^\nu, \vec{d}^\nu)$ satisfy the hypotheses of this chapter. More precisely, we have the following.*

Hörmander setting: *$(W^1, \vec{d}^1), \ldots, (W^\nu, \vec{d}^\nu)$ pairwise locally weakly approximately commute and each $(W^\mu, \vec{d}^\mu)$ is a set of Hörmander vector fields with formal degrees on $\mathfrak{M}$.*

General setting: *$(W^1, \vec{d}^1), \ldots, (W^\nu, \vec{d}^\nu)$ pairwise locally weakly approximately commute and each $\operatorname{Gen}((W^\mu, \vec{d}^\mu))$ is locally finitely generated on $\mathfrak{M}$.*

Proof. Lemma 3.8.3 shows us that $(W^\nu, \vec{d}^\nu)$ and $(W^\mu, \vec{d}^\mu)$ locally strongly approximately commute, $\forall \mu \in \{1, \ldots, \nu\}$. Therefore, by Proposition 3.8.6 (b), $(W^\nu, \vec{d}^\nu)$ and $(W^\mu, \vec{d}^\mu)$ locally weakly approximately commute. Since $(W^\nu, \vec{d}^\nu)$ are Hörmander vector fields with formal degrees ($W^\nu_1, \ldots, W^\nu_{r_\nu}$ satisfy Hörmander's condition of order 1 on $\mathfrak{M}$), the result follows (see also the case $\nu = 1$ of Proposition 3.4.14). □

We define $(W, \vec{d}) := (W^1, \vec{d}^1) \boxtimes (W^2, \vec{d}^2) \boxtimes \cdots (W^\nu, \vec{d}^\nu)$.

Theorem 5.8.23. *Let $b(x, D)$ be a standard pseudo-differential operator of order $m \in \mathbb{R}$ on $\mathfrak{M}$ with $\operatorname{supp}(b(x, D)) \subseteq \Omega \times \Omega$. Then we have the following.*

Hörmander setting: *We have $b(x, D) \in \mathscr{A}^{m e_\nu}(\Omega, (W, \vec{d}))$.*

General setting: *We have $b(x, D) \in \widetilde{\mathscr{A}}^{m e_\nu}(\Omega, (W, \vec{d}))$.*

The rest of this section is devoted to the proof of Theorem 5.8.23.

---

2 A pseudo-differential operator of order $m \in \mathbb{R}$ on a manifold is a pseudo-local operator which when restricted to any local coordinate system equals a standard pseudo-differential operator of order m on $\mathbb{R}^n$. See [239, Chapter 1, Section 5] for a detailed definition.

Lemma 5.8.24. *Suppose $b(x,D)$ is a standard pseudo-differential operator of order $m \in \mathbb{R}$. Let $U \Subset \mathbb{R}^n$ and $V \Subset \mathfrak{M}$ be relatively compact open sets such that there is a C^∞_{loc} diffeomorphism $\Phi : U \xrightarrow{\sim} V$. Suppose* $\operatorname{supp}(b(x,D)) \Subset V \times V$. *Then we have the following.*
Hörmander setting: *We have $b(x,D) \in \mathscr{A}^{me_\nu}(V,(W,\vec{d}))$.*
General setting: *We have $b(x,D) \in \widetilde{\mathscr{A}}^{me_\nu}(V,(W,\vec{d}))$.*

Proof. We prove the result in the Hörmander setting; the proof in the general setting is nearly identical.

For each $\mu \in \{1,\dots,\nu\}$, set

$$\begin{aligned}(Z^\mu, d^\mu) &= \{(Z^\mu_1, d^\mu_1),\dots,(Z^\mu_{r_\mu}, d^\mu_{r_\mu})\}\\ &:= \{(\Phi^* W^\mu_1, d^\mu_1),\dots,(\Phi^* W^\mu_{r_\mu}, d^\mu_{r_\mu})\}\\ &\subset C^\infty_{\mathrm{loc}}(U;TU)\times\mathbb{N}_+.\end{aligned}$$

Set $(\partial,1) := \{(\partial_{x_1},1),\dots,(\partial_{x_n},1)\}$. Note that (Z^ν, d^ν) and $(\partial,1)$ are locally strongly equivalent on U (and are therefore locally weakly equivalent on U). We define two lists of vector fields with ν-parameter formal degrees:

$$\begin{aligned}(Z',\vec{d}') &:= (Z^1,d^1)\boxtimes(Z^2,d^2)\boxtimes\cdots\boxtimes(Z^{\nu-1},d^{\nu-1})\boxtimes(\partial,1),\\ (Z,\vec{d}) &:= (Z^1,d^1)\boxtimes(Z^2,d^2)\boxtimes\cdots\boxtimes(Z^\nu,d^\nu).\end{aligned}$$

By our assumption, $\Phi^* b(x,D)\Phi_*$ is a standard pseudo-differential operator of order m supported in $U \Subset \mathbb{R}^n$. Take $U_1 \Subset U_2 \Subset U_3 \Subset U$ open with $\operatorname{supp}(\Phi^* b(x,D)\Phi_*) \Subset U_1 \times U_1$. Proposition 4.1.7 implies that for any $a > 0$ sufficiently small, we may write $\Phi^* b(x,D)\Phi_* = T_a + R_a$, where T_a is a $(\partial,1)$, a pseudo-differential operator of order m supported in $U_1 \times U_2$ and $R_a \in C^\infty_0(U_2 \times U_2)$. By Corollary 5.8.14, if $a > 0$ is chosen sufficiently small, we have $T_a \in \mathscr{A}^{me_\nu}(U_3,(Z',\vec{d}'))$. Since $R_a \in C^\infty_0(U_2\times U_2)$, Proposition 5.8.11 implies that $R_a \in \mathscr{A}^{me_\nu}(U_3,(Z',\vec{d}'))$. Thus, $\Phi^* b(x,D)\Phi_* = T_a + R_a \in \mathscr{A}^{me_\nu}(U_3,(Z',\vec{d}'))$.

Since (Z^ν,d^ν) and $(\partial,1)$ are locally, weakly equivalent on U, Proposition 5.8.12 implies that $\mathscr{A}^{me_\nu}(U_3,(Z',\vec{d}')) = \mathscr{A}^{me_\nu}(U_3,(Z,\vec{d}))$. We conclude that $\Phi^* b(x,D)\Phi_* \in \mathscr{A}^{me_\nu}(U_3,(Z,\vec{d}))$. By Proposition 5.8.17, this is equivalent to $b(x,D) \in \mathscr{A}^{me_\nu}(\Phi(U_3),(W,\vec{d}))$. Since $\Phi(U_3) \subseteq V$, Proposition 5.8.7 shows that $b(x,D) \in \mathscr{A}^{me_\nu}(V,(W,\vec{d}))$, completing the proof. □

Proof of Theorem 5.8.23. We prove the result in the Hörmander setting; the proof in the general setting is nearly identical.

Let $\phi_1,\dots,\phi_L \in C^\infty_0(\mathfrak{M})$ be such that

$$\sum_{l=1}^{L}\phi_l \equiv 1 \quad \text{on } \Omega$$

and such that if $\operatorname{supp}(\phi_j) \cap \operatorname{supp}(\phi_k) \neq \emptyset$, then there exists $V_{j,k} \Subset \mathfrak{M}$ open and relatively compact with $\operatorname{supp}(\phi_j) \cup \operatorname{supp}(\phi_k) \Subset V_{j,k}$ and a C^∞_{loc} diffeomorphism $\Phi : U_{j,k} \xrightarrow{\sim} V_{j,k}$, where $U_{j,k} \Subset \mathbb{R}^n$ is open.

Since $\operatorname{supp}(b(x,D)) \subseteq \Omega \times \Omega$, we have $b(x,D) = \sum_{j,k} \operatorname{Mult}[\phi_j] b(x,D) \operatorname{Mult}[\phi_k]$. Thus, it suffices to show that for $j,k \in \{1,\dots L\}$, $\operatorname{Mult}[\phi_j] b(x,D) \operatorname{Mult}[\phi_k] \in \mathscr{A}^{me_\nu}(\Omega,(W,\vec{d}))$.

If $\operatorname{supp}(\phi_j) \cap \operatorname{supp}(\phi_k) \neq \emptyset$, then by Lemma 5.8.24, $\operatorname{Mult}[\phi_j] b(x,D) \operatorname{Mult}[\phi_k] \in \mathscr{A}^{me_\nu}(V_{j,k},(W,\vec{d}))$. Since $\operatorname{supp}(\operatorname{Mult}[\phi_j] b(x,D) \operatorname{Mult}[\phi_k]) \subseteq \Omega \cap V_{j,k}$, Proposition 5.8.7 shows that $\operatorname{Mult}[\phi_j] b(x,D) \operatorname{Mult}[\phi_k] \in \mathscr{A}^{me_\nu}(V_{j,k} \cap \Omega,(W,\vec{d})) \subseteq \mathscr{A}^{me_\nu}(\Omega,(W,\vec{d}))$, as desired.

If $\operatorname{supp}(\phi_j) \cap \operatorname{supp}(\phi_k) = \emptyset$, then since standard pseudo-differential operators are pseudo-local, we have $\operatorname{Mult}[\phi_j] b(x,D) \operatorname{Mult}[\phi_k] \in C^\infty_0(\Omega \times \Omega)$. Proposition 5.8.11 then implies that $\operatorname{Mult}[\phi_j] b(x,D) \operatorname{Mult}[\phi_k] \in \mathscr{A}^{me_\nu}(\Omega,(W,\vec{d}))$, completing the proof. □

5.9 An important subalgebra

We again consider the two settings described in Section 5.3. Fix $\nu_1 \in \{1,\dots,\nu-1\}$. Set

$$\begin{aligned}(Y,\vec{d}) &= \{(Y_1,\vec{d}_1),\dots,(Y_{s_1},\vec{d}_{s_1})\}\\ &:= (W^1,\vec{d}^1) \boxtimes (W^2,\vec{d}^2) \boxtimes \cdots (W^{\nu_1},\vec{d}^{\nu_1}) \subset C^\infty_{\mathrm{loc}}(\mathfrak{M};T\mathfrak{M}) \times (\mathbb{N}^{\nu_1} \setminus \{0\}).\end{aligned}$$

In either the Hörmander setting or the general setting, $(Y,\vec{d})$ satisfies the same assumptions as $(W,\vec{d})$, with ν replaced by ν_1. Set $\nu_2 = \nu - \nu_1$ and similarly define

$$\begin{aligned}(Z,\vec{d}) &= \{(Z_1,\vec{d}_1),\dots,(Z_{s_2},\vec{d}_{s_2})\}\\ &:= (W^{\nu_1+1},\vec{d}^{\nu_1+1}) \boxtimes (W^{\nu_1+2},\vec{d}^{\nu_1+2}) \boxtimes \cdots \boxtimes (W^{\nu},\vec{d}^{\nu}) \subset C^\infty_{\mathrm{loc}}(\mathfrak{M};T\mathfrak{M}) \times (\mathbb{N}^{\nu_2} \setminus \{0\}).\end{aligned}$$

We describe the thrust of this section in the Hörmander setting; though as we will see similar results hold in the general setting as well. For $t \in \mathbb{R}^{\nu_1}$, there are two very similar filtered algebras of singular integrals: $\mathscr{A}^t(\Omega,(Y,\vec{d}))$ and $\mathscr{A}^{(t,0_{\nu_2})}(\Omega,(W,\vec{d}))$. These algebras are not the same. In fact, in general we have

$$\mathscr{A}^t(\Omega,(Y,\vec{d})) \subsetneq \mathscr{A}^{(t,0_{\nu_2})}(\Omega,(W,\vec{d})), \quad \mathscr{A}^{(t,0_{\nu_2})}(\Omega,(W,\vec{d})) \subsetneq \mathscr{A}^t(\Omega,(Y,\vec{d})).$$

However, many important operators lie in the intersection:

$$\mathscr{A}^t(\Omega,(Y,\vec{d})) \bigcap \mathscr{A}^{(t,0_{\nu_2})}(\Omega,(W,\vec{d})).$$

An example of such operators can be obtained by using the theory described in Section 4.4 (see, for example, Corollary 5.8.14). We used this to show that parametrices for the sub-Laplacian can be viewed as elements of higher parameter algebras (Corollary 5.8.15). In fact, we will see in Section 5.11.2 that a similar result holds for all maximally subelliptic operators.

In this section, we introduce a filtered subalgebra of $\mathscr{A}^t(\Omega, (Y, \vec{d})) \bigcap \mathscr{A}^{(t,0_{\nu_2})}(\Omega, (W, \vec{d}))$ which contains many important examples and has convenient properties. Roughly speaking, this is the algebra of operators in $\mathscr{A}^t(\Omega, (Y, \vec{d}))$ which "approximately commute" with $(Z, \vec{d})$.

Definition 5.9.1. Hörmander setting: For $t \in \mathbb{R}^{\nu_1}$, let $\mathscr{D}^t(\Omega, (Y, \vec{d}), (Z, \vec{d})) \subseteq \mathscr{A}^t(\Omega, (Y, \vec{d}))$ be the set of those $T^{(t)} \in \mathscr{A}^t(\Omega, (Y, \vec{d}))$ such that for every $\kappa \in \mathbb{N}^{\nu_2}$ and every $(Z, \vec{d})$ partial differential operator on Ω of degree $\leq \kappa$, $\mathscr{P}^{(\kappa)}$, we have:

(1) $T^{(t)}\mathscr{P}^{(\kappa)}$ can be written as a finite sum of operators of the form

$$\widetilde{\mathscr{P}}^{(\kappa)}\widetilde{T}^{(t)},$$

where $\widetilde{\mathscr{P}}^{(\kappa)}$ is a $(Z, \vec{d})$ partial differential operator on Ω of degree $\leq \kappa$ and $\widetilde{T}^{(t)} \in \mathscr{A}^t(\Omega, (Y, \vec{d}))$.

(2) $\mathscr{P}^{(\kappa)}T^{(t)}$ can be written as a finite sum of operators of the form

$$\widetilde{T}^{(t)}\widetilde{\mathscr{P}}^{(\kappa)},$$

where $\widetilde{\mathscr{P}}^{(\kappa)}$ is a $(Z, \vec{d})$ partial differential operator on Ω of degree $\leq \kappa$ and $\widetilde{T}^{(t)} \in \mathscr{A}^t(\Omega, (Y, \vec{d}))$.

General setting: We define $\widetilde{\mathscr{D}}^t(\Omega, (Y, \vec{d}), (Z, \vec{d}))$ as above by replacing $\mathscr{A}$ with $\widetilde{\mathscr{A}}$ throughout.

It follows immediately from Proposition 5.8.7 that if $\Omega_1 \subseteq \Omega_2$, then $\mathscr{D}^t(\Omega_1, (Y, \vec{d}), (Z, \vec{d})) \subseteq \mathscr{D}^t(\Omega_2, (Y, \vec{d}), (Z, \vec{d}))$ and $\widetilde{\mathscr{D}}^t(\Omega_1, (Y, \vec{d}), (Z, \vec{d})) \subseteq \widetilde{\mathscr{D}}^t(\Omega_2, (Y, \vec{d}), (Z, \vec{d}))$. This justifies the next definition.

Definition 5.9.2. Hörmander setting: We set

$$\mathscr{D}^t((Y, \vec{d}), (Z, \vec{d})) := \bigcup_{\Omega \Subset \mathfrak{M}} \mathscr{D}^t(\Omega, (Y, \vec{d}), (Z, \vec{d})).$$

General setting: We set

$$\widetilde{\mathscr{D}}^t((Y, \vec{d}), (Z, \vec{d})) := \bigcup_{\Omega \Subset \mathfrak{M}} \widetilde{\mathscr{D}}^t(\Omega, (Y, \vec{d}), (Z, \vec{d})).$$

We begin by proving some very simple properties of these operators.

Proposition 5.9.3. Hörmander setting: *$\mathscr{D}^t(\Omega, (Y, \vec{d}), (Z, \vec{d}))$ is a filtered subalgebra of $\mathscr{A}^t(\Omega, (Y, \vec{d}))$. In other words, if $T \in \mathscr{D}^t(\Omega, (Y, \vec{d}), (Z, \vec{d}))$ and $S \in \mathscr{D}^s(\Omega, (Y, \vec{d}), (Z, \vec{d}))$, then $TS \in \mathscr{D}^{t+s}(\Omega, (Y, \vec{d}), (Z, \vec{d}))$.*

General setting: *The same result holds with $\mathscr{D}$ replaced by $\widetilde{\mathscr{D}}$ and $\mathscr{A}$ replaced by $\widetilde{\mathscr{A}}$ throughout.*

Proof. We prove only the result in the Hörmander setting; the proof in the general setting is similar. Suppose $T^{(t)} \in \mathscr{D}^t(\Omega, (Y, \vec{d}), (Z, \vec{d}))$ and $S^{(s)} \in \mathscr{D}^s(\Omega, (Y, \vec{d}), (Z, \vec{d}))$. First, we note by Proposition 5.8.6 that $T^{(t)}S^{(s)} \in \mathscr{A}^{(t+s)}(\Omega, (Y, \vec{d}))$. Let $\mathscr{P}^{(\kappa)}$ be a $(Z, \vec{d})$ partial differential operator on Ω of degree $\leq \kappa$. Applying the definition of $\mathscr{D}$ twice, we may write $\mathscr{P}^{(\kappa)}T^{(t)}S^{(s)}$ as a finite sum of operators of the form $\widetilde{T}^{(t)}\widetilde{S}^{(s)}\widetilde{\mathscr{P}}^{(\kappa)}$, where $\widetilde{\mathscr{P}}^{(\kappa)}$ is a $(Z, \vec{d})$ partial differential operator on Ω of degree $\leq \kappa$, $\widetilde{T}^{(t)} \in \mathscr{A}^{t}(\Omega, (Y, \vec{d}))$, and $\widetilde{S}^{(s)} \in \mathscr{A}^{s}(\Omega, (Y, \vec{d}))$. By Proposition 5.8.6, $\widetilde{T}^{(t)}\widetilde{S}^{(s)} \in \mathscr{A}^{t+s}(\Omega, (Y, \vec{d}))$. This shows that $\mathscr{P}^{(\kappa)}T^{(t)}S^{(s)}$ is a finite sum of operators of the correct form. A similar proof shows that $T^{(t)}S^{(s)}\mathscr{P}^{(\kappa)}$ is also a finite sum of operators of the correct form, which completes the proof that $TS \in \mathscr{D}^{t+s}(\Omega, (Y, \vec{d}), (Z, \vec{d}))$. □

Proposition 5.9.4. *Let $\beta \in \mathbb{Z}^{\nu_1}$ and let $\mathcal{Q}$ be a $(Y, \vec{d})$ partial differential operator on Ω of degree $\leq \beta$.*
Hörmander setting: *Let $T \in \mathscr{D}^{t}(\Omega, (Y, \vec{d}), (Z, \vec{d}))$. Then $\mathcal{Q}T, T\mathcal{Q} \in \mathscr{D}^{t+\beta}$.*
General setting: *Let $T \in \widetilde{\mathscr{D}}^{t}(\Omega, (Y, \vec{d}), (Z, \vec{d}))$. Then $\mathcal{Q}T, T\mathcal{Q} \in \widetilde{\mathscr{D}}^{t+\beta}$.*

Proof. We prove only the result in the Hörmander setting and only the result for $\mathcal{Q}T$; the proofs for the remaining parts are similar. First, we note that by Corollary 5.8.10, $\mathcal{Q}T \in \mathscr{A}^{t+\beta}(\Omega, (Y, \vec{d}))$. Let $\mathscr{P}^{(\kappa)}$ be a $(Z, \vec{d})$ partial differential operator on Ω of degree $\leq \kappa$. By Corollary 3.12.8, $\mathscr{P}^{(\kappa)}\mathcal{Q}$ can be written as a finite sum of operators of the form $\widetilde{\mathcal{Q}}^{(\beta)}\widetilde{\mathscr{P}}^{(\kappa)}$, where $\widetilde{\mathcal{Q}}^{(\beta)}$ is a $(Y, \vec{d})$ partial differential operator on Ω of degree $\leq \beta$ and $\widetilde{\mathscr{P}}^{(\kappa)}$ is a $(Z, \vec{d})$ partial differential operator on Ω of degree $\leq \kappa$. By the definition of $\mathscr{D}^{t}(\Omega, (Y, \vec{d}), (Z, \vec{d}))$, $\widetilde{\mathscr{P}}^{(\kappa)}T$ can be written as a finite sum of operators of the form $\widetilde{T}^{(t)}\widehat{\mathscr{P}}^{(\kappa)}$, where $\widetilde{T}^{(t)} \in \mathscr{A}^{t}(\Omega, (Y, \vec{d}))$ and $\widehat{\mathscr{P}}^{(\kappa)}$ is a $(Z, \vec{d})$ partial differential operator on Ω of degree $\leq \kappa$. Putting these together, we see that $\mathscr{P}^{(\kappa)}\mathcal{Q}T$ can be written as a finite sum of operators of the form $\widetilde{\mathcal{Q}}^{(\beta)}\widetilde{T}^{(t)}\widehat{\mathscr{P}}^{(\kappa)}$. Corollary 5.8.10 shows that $\widetilde{\mathcal{Q}}^{(\beta)}\widetilde{T}^{(t)} \in \mathscr{A}^{t+\beta}(\Omega, (Y, \vec{d}))$, which completes the proof that $\mathscr{P}^{(\kappa)}\mathcal{Q}T$ can be written as a sum of terms of the desired form.

Similarly, $\mathcal{Q}T\mathscr{P}^{(\kappa)}$ can be written as a finite sum of terms of the form $\widehat{\mathscr{P}}^{(\kappa)}\widetilde{\mathcal{Q}}^{(\beta)}\widetilde{T}^{(t)}$, and by Corollary 5.8.10, $\widetilde{\mathcal{Q}}^{(\beta)}\widetilde{T}^{(t)} \in \mathscr{A}^{t+\beta}(\Omega, (Y, \vec{d}))$, which shows that $\mathcal{Q}T\mathscr{P}^{(\kappa)}$ can be written as a sum of terms of the desired form.

It follows that $\mathcal{Q}T \in \mathscr{D}^{t+\beta}(\Omega, (Y, \vec{d}), (Z, \vec{d}))$, completing the proof. □

Proposition 5.9.5. Hörmander setting: *Let $T \in \mathscr{D}^{t}$. Then $T^{*} \in \mathscr{D}^{t}$.*
General setting: *Let $T \in \widetilde{\mathscr{D}}^{t}$. Then $T^{*} \in \widetilde{\mathscr{D}}^{t}$.*

Proof. We prove only the result in the Hörmander setting; the proof in the general setting is nearly identical. Suppose $T \in \mathscr{D}^{t}(\Omega, (Y, \vec{d}), (Z, \vec{d})) \subseteq \mathscr{A}^{t}(\Omega, (Y, \vec{d}))$. By Proposition 5.8.1, $T^{*} \in \mathscr{A}^{t}(\Omega, (Y, \vec{d}))$.

Let $\mathscr{P}$ be a $(Z, \vec{d})$ partial differential operator on Ω of degree $\leq \kappa$. By Lemma 3.12.5, $\mathscr{P}^{*}$ is also a $(Z, \vec{d})$ partial differential operator on Ω of degree $\leq \kappa$. Thus, $\mathscr{P}T^{*} = [T\mathscr{P}^{*}]$ can be written as a finite sum of terms of the form $[\widetilde{\mathscr{P}}\widetilde{T}]^{*} = \widetilde{T}^{*}\widetilde{\mathscr{P}}^{*}$, where $\widetilde{\mathscr{P}}$ is a $(Z, \vec{d})$ partial differential operator on Ω of degree $\leq \kappa$ and $\widetilde{T} \in \mathscr{A}^{t}(\Omega, (Y, \vec{d}))$. By Lemma 3.12.5, $\widetilde{\mathscr{P}}^{*}$ is a $(Z, \vec{d})$ partial differential operator on Ω of degree $\leq \kappa$, and by Proposition 5.8.1, $\widetilde{T}^{*} \in \mathscr{A}^{t}(\Omega, (Y, \vec{d}))$. This shows that $\mathscr{P}T^{*}$ is of the desired form.

A similar argument proves that $T^*\mathscr{P}$ is of the desired form, which completes the proof. □

We now present a main theorem regarding the algebras $\mathscr{D}^t$ and $\widetilde{\mathscr{D}}^t$.

Theorem 5.9.6. Hörmander setting: *For all* $t \in \mathbb{R}^{\nu_1}$, $\mathscr{D}^t(\Omega, (Y, \vec{d}), (Z, \vec{d})) \subseteq \mathscr{A}^{(t, 0_{\nu_2})}(\Omega, (W, \vec{d}))$.
General setting: *For all* $t \in \mathbb{R}^{\nu_1}$, $\widetilde{\mathscr{D}}^t(\Omega, (Y, \vec{d}), (Z, \vec{d})) \subseteq \widetilde{\mathscr{A}}^{(t, 0_{\nu_2})}(\Omega, (W, \vec{d}))$.

The rest of this section is devoted to the proof of Theorem 5.9.6.

To prove Theorem 5.9.6 in the Hörmander setting, we will need to study the interaction between $(Y, \vec{d})$ pre-elementary operators supported in Ω and $(W, \vec{d})$ pre-elementary operators supported in Ω.

$(W, \vec{d})$ pre-elementary operators are defined in terms of the quantity

$$\frac{(1 + \rho_{2^{-j}}(x, z))^{-m}}{\mathrm{Vol}(B_{2^{-j}}(x, 1 + \rho_{2^{-j}}(x, z))) \wedge 1},$$

where $j \in [0, \infty)^\nu$ and $\rho_{2^{-j}}$ and $B_{2^{-j}}$ are defined in terms of $(W, \vec{d})$ as in Section 5.1.2. Analogously, if for $k \in \mathbb{R}^{\nu_1}$ we define $\tilde{\rho}_{2^{-k}}$ and $\widetilde{B}_{2^{-k}}$ in the same way, but with $(W, \vec{d})$ replaced by $(Y, \vec{d})$, then $(Y, \vec{d})$ pre-elementary operators are defined in terms of the quantity

$$\frac{(1 + \tilde{\rho}_{2^{-k}}(x, z))^{-m}}{\mathrm{Vol}(\widetilde{B}_{2^{-k}}(x, 1 + \tilde{\rho}_{2^{-k}}(x, z))) \wedge 1}.$$

Lemma 5.4.1 shows that

$$\frac{(1 + \tilde{\rho}_{2^{-k}}(x, z))^{-m}}{\mathrm{Vol}(\widetilde{B}_{2^{-k}}(x, 1 + \tilde{\rho}_{2^{-k}}(x, z))) \wedge 1} = \frac{(1 + \rho_{2^{-(k,\infty)}}(x, z))^{-m}}{\mathrm{Vol}(B_{2^{-(k,\infty)}}(x, 1 + \rho_{2^{-(k,\infty)}}(x, z))) \wedge 1}. \tag{5.89}$$

For $j \in \mathbb{R}^\nu \cong \mathbb{R}^{\nu_1} \times \mathbb{R}^{\nu_2}$, we write $j = (j^1, j^2)$, where $j^1 \in \mathbb{R}^{\nu_1}$ and $j^2 \in \mathbb{R}^{\nu_2}$.

Lemma 5.9.7. *In the Hörmander setting, for every $m \in \mathbb{N}$, there exist $m' = m'(m, \Omega, (W, \vec{d})) \in \mathbb{N}$ and $C = C(m, \Omega, (W, \vec{d})) \geq 0$ such that $\forall x, z \in \Omega$, $j \in \mathbb{R}^\nu$, $k \in \mathbb{R}^{\nu_1}$, we have*

$$\begin{aligned}
&\int_\Omega \left(\frac{(1 + \rho_{2^{-j}}(x, y))^{-m'}}{\mathrm{Vol}(B_{2^{-j}}(x, 1 + \rho_{2^{-j}}(x, y))) \wedge 1} \right) \left(\frac{(1 + \tilde{\rho}_{2^{-k}}(y, z))^{-m'}}{\mathrm{Vol}(\widetilde{B}_{2^{-k}}(y, 1 + \tilde{\rho}_{2^{-k}}(y, z))) \wedge 1} \right) d\,\mathrm{Vol}(y) \\
&\quad \leq C \frac{(1 + \rho_{2^{-(j^1 \wedge k, j^2)}}(x, z))^{-m}}{\mathrm{Vol}(B_{2^{-(j^1 \wedge k, j^2)}}(x, 1 + \rho_{2^{-(j^1 \wedge k, j^2)}}(x, z))) \wedge 1},
\end{aligned} \tag{5.90}$$

$$\begin{aligned}
&\int_\Omega \left(\frac{(1 + \tilde{\rho}_{2^{-k}}(y, z))^{-m'}}{\mathrm{Vol}(\widetilde{B}_{2^{-k}}(y, 1 + \tilde{\rho}_{2^{-k}}(y, z))) \wedge 1} \right) \left(\frac{(1 + \rho_{2^{-j}}(x, y))^{-m'}}{\mathrm{Vol}(B_{2^{-j}}(x, 1 + \rho_{2^{-j}}(x, y))) \wedge 1} \right) d\,\mathrm{Vol}(y) \\
&\quad \leq C \frac{(1 + \rho_{2^{-(j^1 \wedge k, j^2)}}(x, z))^{-m}}{\mathrm{Vol}(B_{2^{-(j^1 \wedge k, j^2)}}(x, 1 + \rho_{2^{-(j^1 \wedge k, j^2)}}(x, z))) \wedge 1}.
\end{aligned} \tag{5.91}$$

Proof. The inequality (5.90) follows from Proposition 5.4.4 by replacing k in that proposition with (k, ∞) and applying (5.89). Similarly, (5.91) follows from Proposition 5.4.4 by taking j in that proposition to be (k, ∞), taking k in that proposition to be j, and applying (5.89). □

Lemma 5.9.8. Hörmander setting: *Let $\mathcal{E}_1$ be a bounded set of $(Y, \vec{d})$ elementary operators supported in Ω and let $\mathcal{E}_2$ be a bounded set of $(W, \vec{d})$ elementary operators supported in Ω. Then, $\forall N \in \mathbb{N}$, $m \in \mathbb{N}$, there exists $C = C(N, m, \mathcal{E}_1, \mathcal{E}_2) \geq 0$ such that for all $(E_1, 2^{-k}) \in \mathcal{E}_1$ and $(E_2, 2^{-j}) \in \mathcal{E}_2$ and all $x, y \in \Omega$,*

$$2^{N|j^1-k|}\big|[E_1E_2](x,y)\big| \leq C\frac{(1+\rho_{2^{-j}}(x,y))^{-m}}{\mathrm{Vol}(B_{2^{-j}}(x, 1+\rho_{2^{-j}}(x,y)))\wedge 1}, \tag{5.92}$$

$$2^{N|j^1-k|}\big|[E_2E_1](x,y)\big| \leq C\frac{(1+\rho_{2^{-j}}(x,y))^{-m}}{\mathrm{Vol}(B_{2^{-j}}(x, 1+\rho_{2^{-j}}(x,y)))\wedge 1}. \tag{5.93}$$

General setting: *Let $\mathcal{E}_1$ be a bounded set of generalized $(Y, \vec{d})$ elementary operators supported in Ω and let $\mathcal{E}_2$ be a bounded set of generalized $(W, \vec{d})$ elementary operators supported in Ω. Then, for all $N \in \mathbb{N}$, the following hold:*

- *For $p \in \{1, \infty\}$,*

$$\sup_{\substack{(E_1,2^{-k})\in\mathcal{E}_1\\(E_2,2^{-j})\in\mathcal{E}_2}} 2^{N|j^1-k|}\big(\|E_1E_2\|_{L^p\to L^p} + \|E_2E_1\|_{L^p\to L^p}\big) < \infty. \tag{5.94}$$

- *For all countable sets $\widehat{\mathcal{E}}_1 = \{(E_{1,l}, 2^{-k_l}) : l \in \mathbb{N}\} \subseteq \mathcal{E}_1$ and $\widehat{\mathcal{E}}_2 = \{(E_{2,l}, 2^{-j_l}) : l \in \mathbb{N}\} \subseteq \mathcal{E}_2$, if we set*

$$\mathcal{T}^1_{\widehat{\mathcal{E}}_1,\widehat{\mathcal{E}}_2,N}\{f_l\}_{l\in\mathbb{N}} := \{2^{N|j^1_l-k_l|}E_{1,l}E_{2,l}f_l\}_{l\in\mathbb{N}},$$

$$\mathcal{T}^2_{\widehat{\mathcal{E}}_1,\widehat{\mathcal{E}}_2,N}\{f_l\}_{l\in\mathbb{N}} := \{2^{N|j^1_l-k_l|}E_{2,l}E_{1,l}f_l\}_{l\in\mathbb{N}},$$

then, for all $p \in (1, \infty)$, $q \in (1, \infty]$,

$$\sup_{\widehat{\mathcal{E}}_1,\widehat{\mathcal{E}}_2}\big(\|\mathcal{T}^1_{\widehat{\mathcal{E}}_1,\widehat{\mathcal{E}}_2,N}\|_{L^p(\ell^q)\to L^p(\ell^q)} + \|\mathcal{T}^2_{\widehat{\mathcal{E}}_1,\widehat{\mathcal{E}}_2,N}\|_{L^p(\ell^q)\to L^p(\ell^q)}\big) < \infty.$$

Here, the supremum is taking over all such choices of $\widehat{\mathcal{E}}_1$ and $\widehat{\mathcal{E}}_2$.

Proof. We begin with the Hörmander setting and prove only (5.92); the proof for (5.93) is similar. Fix $M = M(m, (W, \vec{d}), \Omega)$ large, to be chosen later. We first claim that it suffices to prove (5.92) with N replaced by $-M$. For $(E_1, 2^{-k}) \in \mathcal{E}_1$ and $(E_2, 2^{-j}) \in \mathcal{E}_2$, we separate the proof of this claim into two cases.

The first case is when there is a $\mu \in \{1, \dots, \nu_1\}$ such that $k_\mu - j_\mu = |j^1 - k|_\infty$. We apply Proposition 5.5.5 (h) with N replaced by $N_1 := \lceil \sqrt{\nu_1}N + \sqrt{\nu_1}M \rceil$ to E_1 to see

$$\begin{aligned}
2^{N|j^1-k|}E_1E_2 &= 2^{N|j^1-k|} \sum_{|\alpha_\mu| \le N_1} 2^{(|\alpha_\mu|-N_1)k_\mu} E_{1,\alpha_\mu}(2^{-k_\mu \vec{d}^\mu} W^\mu)^{\alpha_\mu} E_2 \\
&= 2^{N|j^1-k|} \sum_{|\alpha_\mu| \le N_1} 2^{(|\alpha_\mu|-N_1)k_\mu - \deg_{\vec{d}^\mu}(\alpha_\mu)(k_\mu - j_\mu)} E_{1,\alpha-\mu}[(2^{-j_\mu \vec{d}^\mu} W^\mu)^{\alpha_\mu} E_2] \\
&= \sum_{|\alpha_\mu| \le N_1} 2^{-N_1|j^1-k|_\infty + N|j^1-k|} E_{1,\alpha_\mu}[2^{N_1|j^1-k|_\infty - (|\alpha_\mu|-N_1)k_\mu - \deg_{\vec{d}^\mu}(k_\mu - j_\mu)}(2^{-j_\mu \vec{d}^\mu} W^\mu)^{\alpha_\mu} E_2] \\
&=: 2^{-N_1|j^1-k|_\infty + N|j^1-k|} E_{1,\alpha_\mu} E_{2,\alpha_\mu},
\end{aligned}$$

where $\{(E_{1,\alpha_\mu}, 2^{-k}) : (E_1, 2^{-k}) \in \mathcal{E}_1, |\alpha_\mu| \le N_1\}$ is a bounded set of $(Y, \vec{d})$ elementary operators supported in Ω, and by Proposition 5.5.5 (a) and (e), $\{(E_{2,\alpha_\mu}, 2^{-j}) : (E_2, 2^{-j}) \in \mathcal{E}_2, k_\mu - j_\mu = |j^1 - k|_\infty, |\alpha_\mu| \le N_1\}$ is a bounded set of $(W, \vec{d})$ elementary operators supported in Ω (where we have used the fact that $2^{N_1|j^1-k|_\infty - (|\alpha_\mu|-N_1)k_\mu - \deg_{\vec{d}^\mu}(k_\mu - j_\mu)} \le 1$ in the sum). Since $2^{-N_1|j^1-k|_\infty + N|j^1-k|} \le 2^{-M|j^1-k|}$, we see that it suffices to prove (5.92) with N replaced by $-M$ in this case.

The second case is when there is a $\mu \in \{1, \dots, \nu_1\}$ such that $j_\mu - k_\mu = |j'^1 - k|_\infty$; the proof is similar to the first case. Indeed, we apply Proposition 5.5.5 (h) with N replaced by $N_1 := \lceil \sqrt{\nu_1}N + \sqrt{\nu_1}M \rceil$ to E_2 to see that

$$\begin{aligned}
2^{N|j^1-k|}E_1E_2 &= 2^{N|j^1-k|} \sum_{|\alpha_\mu| \le N_1} 2^{(|\alpha_\mu|-N_1)j_\mu} E_1(2^{-j_\mu \vec{d}^\mu} W^\mu)^{\alpha_\mu} E_{2,\alpha_\mu} \\
&= 2^{N|j^1-k|} \sum_{|\alpha_\mu| \le N_1} 2^{(|\alpha_\mu|-N_1)j_\mu - \deg_{\vec{d}^\mu}(\alpha_\mu)(j_\mu - k_\mu)} E_1(2^{-k_\mu \vec{d}^\mu} W^\mu)^{\alpha_\mu} E_{2,\alpha_\mu} \\
&= \sum_{|\alpha_\mu| \le N_1} 2^{-N_1|j^1-k| + N|j^1-k|} [2^{N_1|j-k|_\infty - (|\alpha_\mu|-N_1)j_\mu - \deg_{\vec{d}^\mu}(\alpha_\mu)(j_\mu - k_\mu)} E_1(2^{-k_\mu \vec{d}^\mu} W^\mu)^{\alpha_\mu}] E_{2,\alpha_\mu} \\
&=: \sum_{|\alpha_\mu| \le N_1} 2^{-N_1|j^1-k| + N|j^1-k|} E_{1,\alpha_\mu} E_{2,\alpha_\mu},
\end{aligned}$$

where $\{(E_{2,\alpha_\mu}, 2^{-j}) : (E_2, 2^{-j}) \in \mathcal{E}_2, |\alpha_\mu| \le N_1\}$ is a bounded set of $(W, \vec{d})$ elementary operators supported in Ω, and by Proposition 5.5.5 (a) and (e), $\{(E_{1,\alpha_\mu}, 2^{-k}) : (E_1, 2^{-k}) \in \mathcal{E}_1, j_\mu - k_\mu = |j^1 - k|_\infty, |\alpha_\mu| \le N_1\}$ is a bounded set of $(Y, \vec{d})$ elementary operators supported in Ω (where we have used the fact that $2^{N_1|j-k|_\infty - (|\alpha_\mu|-N_1)j_\mu - \deg_{\vec{d}^\mu}(\alpha_\mu)(j_\mu - k_\mu)} \le 1$ in the sum). Since $2^{-N_1|j^1-k|_\infty + N|j^1-k|} \le 2^{-M|j^1-k|}$, we see that it suffices to prove (5.92) with N replaced by $-M$.

We turn to proving (5.92) with N replaced by $-M$, where $M = M(m, (W, \vec{d}), \Omega)$ can be as large as we like. By Lemmas 5.4.7 and 5.9.7, we have, by taking $M = M(m, (W, \vec{d}), \Omega)$ sufficiently large, for all $x, y \in \Omega$,

$$
\begin{aligned}
2^{-M|j^1-k|}\big|[E_1E_2](x,y)\big| &\lesssim 2^{-M|j^1-k|}\frac{(1+\rho_{2^{-(j^1\wedge k,j^2)}}(x,y))^{-m}}{\mathrm{Vol}(B_{2^{-(j^1\wedge k,j^2)}}(x,1+\rho_{2^{-(j^1\wedge k,j^2)}}(x,y)))\wedge 1}\\
&\le 2^{-M|j-(j^1\wedge k,j^2)|}\frac{(1+\rho_{2^{-(j^1\wedge k,j^2)}}(x,y))^{-m}}{\mathrm{Vol}(B_{2^{-(j^1\wedge k,j^2)}}(x,1+\rho_{2^{-(j^1\wedge k,j^2)}}(x,y)))\wedge 1}\\
&\lesssim \frac{(1+\rho_{2^{-j}}(x,y))^{-m}}{\mathrm{Vol}(B_{2^{-j}}(x,1+\rho_{2^{-j}}(x,y)))\wedge 1},
\end{aligned}
$$

completing the proof in the Hörmander setting.

The proof in the general setting is similar and easier. Indeed, by the same argument as above, it suffices to prove the result with N replaced with 0. Once N is replaced with 0, the result follows immediately from the definition of generalized pre-elementary operators supported in Ω (Definition 5.2.26). □

Lemma 5.9.9. *Let $\kappa_1 \in \mathbb{N}^{\nu_1}$ and let $\mathscr{P}_1$ be a $(Y, \vec{d})$ partial differential operator on Ω of degree $\le \kappa_1$. Let $t \in \mathbb{R}^{\nu_1}$.*

Hörmander setting: *Suppose $T \in \mathscr{A}^t(\Omega, (Y, \vec{d}))$ and let $\mathcal{E}$ be a bounded set of $(W, \vec{d})$ elementary operators supported in Ω. Then, for every $m \in \mathbb{N}$, there exists $C \ge 0$ such that for all $(E, 2^{-j}) \in \mathcal{E}$ and all $(x, z) \in \Omega$,*

$$2^{-(j^1\cdot t+j^1\cdot\kappa_1)}\big|(\mathscr{P}_1TE)(x,z)\big| \le C\frac{(1+\rho_{2^{-j}}(x,z))^{-m}}{\mathrm{Vol}(B_{2^{-j}}(x,1+\rho_{2^{-j}}(x,z)))\wedge 1}, \tag{5.95}$$

$$2^{-(j^1\cdot t+j^1\cdot\kappa_1)}\big|(ET\mathscr{P}_1)(x,z)\big| \le C\frac{(1+\rho_{2^{-j}}(x,z))^{-m}}{\mathrm{Vol}(B_{2^{-j}}(x,1+\rho_{2^{-j}}(x,z)))\wedge 1}. \tag{5.96}$$

General setting: *Suppose $T \in \widetilde{\mathscr{A}}^t(\Omega, (Y, \vec{d}))$ and let $\mathcal{E}$ be a bounded set of $(W, \vec{d})$ elementary operators supported in Ω. Then:*

(i) *For $p \in \{1, \infty\}$,*

$$\sup_{(E,2^{-j})\in\mathcal{E}} 2^{-(j^1\cdot t+j^1\cdot\kappa_1)}\big(\|\mathscr{P}_1TE\|_{L^p\to L^p} + \|ET\mathscr{P}_1\|_{L^p\to L^p}\big) < \infty.$$

(ii) *For each countable set $\widetilde{\mathcal{E}} = \{(E_l, 2^{-j_l}) : l \in \mathbb{N}\} \subseteq \mathcal{E}$, we set*

$$
\begin{aligned}
\mathcal{T}^1_{\widetilde{\mathcal{E}}}\{f_l\}_{l\in\mathbb{N}} &:= \{2^{-(t\cdot j_l^1+\kappa_1\cdot j_l^1)}\mathscr{P}_1TE_lf_l\}_{l\in\mathbb{N}},\\
\mathcal{T}^2_{\widetilde{\mathcal{E}}}\{f_l\}_{l\in\mathbb{N}} &:= \{2^{-(t\cdot j_l^1+\kappa_1\cdot j_l^1)}E_lT\mathscr{P}_1f_l\}_{l\in\mathbb{N}}.
\end{aligned}
$$

Then, for $p \in (1, \infty)$, $q \in (1, \infty]$, we have

$$\sup_{\widetilde{\mathcal{E}}}\big(\|\mathcal{T}^1_{\widetilde{\mathcal{E}}}\|_{L^p(\ell^q)\to L^p(\ell^q)} + \|\mathcal{T}^2_{\widetilde{\mathcal{E}}}\|_{L^p(\ell^q)\to L^p(\ell^q)}\big) < \infty, \tag{5.97}$$

where the supremum is taken over all countable subsets of $\mathcal{E}$.

Proof. We begin with the Hörmander setting. First, we note that (5.96) follows from (5.95). Indeed,

$$\overline{[ET\mathscr{P}_1](x,z)} = [(ET\mathscr{P}_1)^*](z,x) = [\mathscr{P}_1^* T^* E^*](z,x).$$

By Lemma 3.12.5, $\mathscr{P}_1^*$ is a $(Y,\vec{d})$ partial differential operator on Ω of degree $\le \kappa_1$, by Proposition 5.8.1, $T^* \in \mathscr{A}^t(\Omega,(Y,\vec{d}))$, and by Proposition 5.5.5 (d), $\{(E^*,2^{-j}) : (E,2^{-j}) \in \mathcal{E}\}$ is a bounded set of $(W,\vec{d})$ elementary operators supported in Ω. Thus, once we prove (5.95), we will have

$$\begin{aligned}2^{-(j^1\cdot t+j^1\cdot\kappa_1)}|(ET\mathscr{P}_1)(x,z)| &\lesssim \frac{(1+\rho_{2^{-j}}(z,x))^{-m}}{\mathrm{Vol}(B_{2^{-j}}(z,1+\rho_{2^{-j}}(z,x)))\wedge 1}\\ &\approx \frac{(1+\rho_{2^{-j}}(x,z))^{-m}}{\mathrm{Vol}(B_{2^{-j}}(x,1+\rho_{2^{-j}}(x,z)))\wedge 1},\end{aligned}$$

where the last estimate follows from Proposition 5.4.3.

We turn to proving (5.95). By Corollary 5.8.10, $\mathscr{P}_1 T \in \mathscr{A}^{t+\kappa_1}(\Omega,(Y,\vec{d}))$. Thus, by replacing T with $\mathscr{P}_1 T$ and t with $t+\kappa_1$, it suffices to prove that for $p\in\{1,\infty\}$,

$$2^{-(j^1\cdot t)}|(TE)(x,z)| \le C\frac{(1+\rho_{2^{-j}}(x,z))^{-m}}{\mathrm{Vol}(B_{2^{-j}}(x,1+\rho_{2^{-j}}(x,z)))\wedge 1}. \tag{5.98}$$

Since $T\in\mathscr{A}^t(\Omega,(Y,\vec{d})) = \mathscr{A}_2^t(\Omega,(Y,\vec{d}))$, there is a bounded set of $(Y,\vec{d})$ elementary operators supported in Ω, $\{(E_k,2^{-k}) : k\in\mathbb{N}^{\nu_1}\}$, such that $T=\sum_{k\in\mathbb{N}^{\nu_1}} 2^{k\cdot t}E_k$. Let $x,z\in\Omega$. Using Lemma 5.9.8, we have, for every $m\in\mathbb{N}$,

$$\begin{aligned}|2^{-j^1\cdot t}TE(x,z)| &\le \sum_{k\in\mathbb{N}^{\nu_1}} |2^{(k-j^1)\cdot t}E_kE(x,z)|\\ &\lesssim \sum_{k\in\mathbb{N}^{\nu_1}} 2^{-|k-j^1|}\frac{(1+\rho_{2^{-j}}(x,z))^{-m}}{\mathrm{Vol}(B_{2^{-j}}(x,1+\rho_{2^{-j}}(x,z)))\wedge 1}\\ &\lesssim \frac{(1+\rho_{2^{-j}}(x,z))^{-m}}{\mathrm{Vol}(B_{2^{-j}}(x,1+\rho_{2^{-j}}(x,z)))\wedge 1},\end{aligned}$$

proving (5.98) and completing the proof in the Hörmander setting.

We now turn to the general setting. Using Corollary 5.8.10, we have $\mathscr{P}_1T, T\mathscr{P}_1 \in \widetilde{\mathscr{A}}^{t+\kappa_1}(\Omega,(Y,\vec{d}))$. Thus, by replacing T with either $\mathscr{P}_1T$ or $T\mathscr{P}_1$, it suffices to prove the result when $\kappa_1=0$ and $\mathscr{P}_1=\mathrm{Mult}[1]$. Since $T\in\widetilde{\mathscr{A}}^t(\Omega,(Y,\vec{d}))=\widetilde{\mathscr{A}}_2^t(\Omega,(Y,\vec{d}))$, there is a bounded set of generalized $(Y,\vec{d})$ elementary operators supported in Ω, $\{(E_k,2^{-k}) : k\in\mathbb{N}^{\nu_1}\}$, with $T=\sum_{k\in\mathbb{N}^{\nu_1}}2^{k\cdot t}E_k$. We take the convention that $E_k := 0$ for $k\in\mathbb{Z}^{\nu_1}\setminus\mathbb{N}^{\nu_1}$.

We begin with (i). We have, for $(E,2^{-j})\in\mathcal{E}$ and $p\in\{1,\infty\}$, using (5.94),

$$2^{-j^1\cdot t}\|TE\|_{L^p\to L^p} \le \sum_{k\in\mathbb{N}^{\nu_1}} 2^{(k-j^1)\cdot t}\|E_kE\|_{L^p\to L^p} \lesssim \sum_{k\in\mathbb{N}^{\nu_1}} 2^{-|k-j^1|} \lesssim 1.$$

A similar proof shows that $2^{-j^1\cdot t}\|ET\|_{L^p\to L^p}\lesssim 1$, completing the proof of (i).

We next turn to (ii) and prove only the result for $\mathcal{T}^1_{\widetilde{\mathcal{E}}}$; the proof for $\mathcal{T}^2_{\widetilde{\mathcal{E}}}$ is similar and we leave it to the reader. Let $\widetilde{\mathcal{E}}=\{(\widetilde{E}_l,2^{-j_l}):l\in\mathbb{N}\}\subseteq\mathcal{E}$ and consider for $p\in(1,\infty)$, $q\in(1,\infty]$,

$$\begin{aligned}\sup_{\widetilde{\mathcal{E}}\subseteq\mathcal{E}}\|\mathcal{T}^1_{\widetilde{\mathcal{E}}}\{f_l\}_{l\in\mathbb{N}}\|_{L^p(\ell^q)}&=\sup_{\widetilde{\mathcal{E}}\subseteq\mathcal{E}}\left\|\left\{\sum_{k\in\mathbb{N}^{\nu_1}}2^{t\cdot(k-j_l^1)}E_k\widetilde{E}_{j_l}f_l\right\}_{l\in\mathbb{N}}\right\|_{L^p(\ell^q)}\\&=\sup_{\widetilde{\mathcal{E}}\subseteq\mathcal{E}}\left\|\left\{\sum_{k\in\mathbb{N}^{\nu_1}}2^{-|k-j_l^1|}(2^{t\cdot(k-j_l^1)+|k-j_l^1|}2^{t\cdot(k-j_l^1)}E_k\widetilde{E}_{j_l}f_l)\right\}_{l\in\mathbb{N}}\right\|_{L^p(\ell^q)}\\&=\sup_{\widetilde{\mathcal{E}}\subseteq\mathcal{E}}\left\|\left\{\sum_{r\in\mathbb{Z}^{\nu_1}}2^{-|r|}(2^{t\cdot r+|r|}2^{t\cdot r}E_{j_l^1+r}\widetilde{E}_{j_l}f_l)\right\}_{l\in\mathbb{N}}\right\|_{L^p(\ell^q)}\\&\leq\sum_{r\in\mathbb{Z}^{\nu_1}}2^{-|r|}\sup_{\widetilde{\mathcal{E}}\subseteq\mathcal{E}}\|\{2^{t\cdot r+|r|}2^{t\cdot r}E_{j_l^1+r}\widetilde{E}_{j_l}f_l\}_{l\in\mathbb{N}}\|_{L^p(\ell^q)}\\&\lesssim\sum_{r\in\mathbb{Z}^{\nu_1}}2^{-|r|}\sup_{\widetilde{\mathcal{E}}\subseteq\mathcal{E}}\|\{f_l\}_{l\in\mathbb{N}}\|_{L^p(\ell^q)}\\&\lesssim\|\{f_l\}_{l\in\mathbb{N}}\|_{L^p(\ell^q)},\end{aligned}$$

where the second to last inequality follows from (5.97). This completes the proof of (ii). □

Lemma 5.9.10. *In both the Hörmander setting and the general setting, $(Y,\vec{d})$ and $(Z,\vec{d})$ locally weakly approximately commute.*

Proof. This follows from Corollary 3.8.11. Indeed, for $\mu_1\in\{1,\dots,\nu_1\}$, we take $\mathcal{S}_{\mu_1}:=(W^{\mu_1},\vec{d}^{\mu_1})$, and for $\mu_2\in\{1,\dots,\nu_2\}$, we take $\mathcal{T}_{\mu_2}:=(W^{\nu_1+\mu_2},\vec{d}^{\nu_1+\mu_2})$. Then Corollary 3.8.11 applies to $\mathcal{S}_1,\dots,\mathcal{S}_{\nu_1},\mathcal{T}_1,\dots,\mathcal{T}_{\nu_2}$. From that corollary, $\mathcal{S}_0$ equals $(Y,\vec{d})$ and $\mathcal{T}_0$ equals $(Z,\vec{d})$. Thus, $(Y,\vec{d})$ and $(Z,\vec{d})$ locally weakly approximately commute, completing he proof. □

Lemma 5.9.11. *In the general setting, $\widetilde{\mathscr{D}}^t(\Omega,(Y,\vec{d}),(Z,\vec{d}))\subseteq\mathrm{Hom}(C^\infty_{W,\mathrm{loc}}(\mathfrak{M}),C^\infty_{W,0}(\mathfrak{M}))$.*

Proof. First we claim

$$\widetilde{\mathscr{A}}^t(\Omega,(Y,\vec{d}))\subseteq\mathrm{Hom}(C^\infty_{W,\mathrm{loc}}(\mathfrak{M}),C^\infty_{Y,0}(\mathfrak{M})).\tag{5.99}$$

Indeed, Proposition 5.8.5 shows that for $\widetilde{T}\in\widetilde{\mathscr{A}}^t(\Omega,(Y,\vec{d}))$, we have $\widetilde{T}\in\mathrm{Hom}(C^\infty_{Y,\mathrm{loc}}(\mathfrak{M}),C^\infty_{Y,0}(\mathfrak{M}))$. Since $Y\subseteq W$, $C^\infty_{W,\mathrm{loc}}(\mathfrak{M})$ embeds continuously into $C^\infty_{Y,\mathrm{loc}}(\mathfrak{M})$ and therefore $\widetilde{T}\in\mathrm{Hom}(C^\infty_{W,\mathrm{loc}}(\mathfrak{M}),C^\infty_{Y,0}(\mathfrak{M}))$, establishing (5.99).

As a consequence of (5.99), for any $(Y,\vec{d})$ partial differential operator on Ω of degree $\leq\kappa_1\in\mathbb{Z}$, $\mathscr{P}_1^{(\kappa_1)}$, and any $\widetilde{T}\in\widetilde{\mathscr{A}}^t(\Omega,(Y,\vec{d}))$, we have

$$\mathscr{P}_1^{(\kappa_1)}\widetilde{T}\in\mathrm{Hom}(C^\infty_{W,\mathrm{loc}}(\Omega),C(\Omega)).\tag{5.100}$$

Let $T \in \widetilde{\mathscr{D}}^t(\Omega, (Y, \vec{d}), (Z, \vec{d}))$. We wish to show

$$T \in \operatorname{Hom}(C^\infty_{W,\mathrm{loc}}(\mathfrak{M}), C^\infty_{W,0}(\mathfrak{M})). \tag{5.101}$$

Since $T \in \widetilde{\mathscr{A}}^t(\Omega, (Y, \vec{d}))$, we have $\operatorname{supp}(T) \Subset \Omega \times \Omega$, and therefore (5.101) is equivalent to

$$T \in \operatorname{Hom}(C^\infty_{W,0}(\Omega), C^\infty_{W,\mathrm{loc}}(\Omega)). \tag{5.102}$$

To prove (5.102), we will show that for every ordered multi-index α,

$$W^\alpha T \in \operatorname{Hom}(C^\infty_{W,0}(\Omega), C(\Omega)). \tag{5.103}$$

We turn to proving (5.103). Let $\kappa := \deg_{\vec{d}}(\alpha) \in \mathbb{N}^\nu$ and write $\kappa = (\kappa_1, \kappa_2) \in \mathbb{N}^{\nu_1} \times \mathbb{N}^{\nu_2} \cong \mathbb{N}^\nu$. Note that W^α is a $(W, \vec{d})$ partial differential operator on Ω of degree $\leq \kappa$. Lemma 5.9.10 implies that $(Y, \vec{d})$ and $(Z, \vec{d})$ weakly approximately commute on Ω. Proposition 3.12.13 then implies that W^α can be written as a finite sum of terms of the form $\mathscr{P}_1^{(\kappa_1)} \mathscr{P}_2^{(\kappa_2)}$, where $\mathscr{P}_1^{(\kappa_1)}$ is a $(Y, \vec{d})$ partial differential operator on Ω of degree $\leq \kappa_1$ and $\mathscr{P}_2^{(\kappa_2)}$ is a $(Z, \vec{d})$ partial differential operator on Ω of degree $\leq \kappa_2$. Thus, to prove (5.103) it suffices to show that for such $\mathscr{P}_1^{(\kappa_1)}$ and $\mathscr{P}_2^{(\kappa_2)}$, we have

$$\mathscr{P}_1^{(\kappa_1)} \mathscr{P}_2^{(\kappa_2)} T \in \operatorname{Hom}(C^\infty_{W,0}(\Omega), C(\Omega)). \tag{5.104}$$

By the definition of $\widetilde{\mathscr{D}}^t(\Omega, (Y, \vec{d}), (Z, \vec{d}))$, $\mathscr{P}_1^{(\kappa_1)} \mathscr{P}_2^{(\kappa_2)} T$ can be written as a finite sum of terms of the form $\mathscr{P}_1^{(\kappa_1)} \widetilde{T} \widetilde{\mathscr{P}}_2^{(\kappa_2)}$, where $\widetilde{T} \in \widetilde{\mathscr{A}}^t(\Omega, (Y, \vec{d}))$ and $\widetilde{\mathscr{P}}_2^{(\kappa_2)}$ is a $(Z, \vec{d})$ partial differential operator on Ω of degree $\leq \kappa_2$. Thus, to prove (5.104) it suffices to show that for such $\widetilde{T}$ and $\widetilde{\mathscr{P}}_2^{(\kappa_2)}$,

$$\mathscr{P}_1^{(\kappa_1)} \widetilde{T} \widetilde{\mathscr{P}}_2^{(\kappa_2)} \in \operatorname{Hom}(C^\infty_{W,0}(\Omega), C(\Omega)). \tag{5.105}$$

Since $\widetilde{\mathscr{P}}_2^{(\kappa_2)}$ is a $(Z, \vec{d})$ partial differential operator on Ω of degree $\leq \kappa_2$, it is also a $(W, \vec{d})$ partial differential operator on Ω of degree $\leq (0_{\nu_1}, \kappa_2)$. Thus,

$$\widetilde{\mathscr{P}}_2^{(\kappa_2)} \in \operatorname{Hom}(C^\infty_{W,0}(\Omega), C^\infty_{W,0}(\Omega)). \tag{5.106}$$

Combining (5.100) and (5.106) gives (5.105) and completes the proof. □

Lemma 5.9.12. *Fix $t \in \mathbb{R}^{\nu_1}$.*
Hörmander setting: *Let $T^{(t)} \in \mathscr{D}^t(\Omega, (Y, \vec{d}), (Z, \vec{d}))$ and suppose $\mathcal{E}$ is a bounded set of $(W, \vec{d})$ elementary operators supported in Ω. Then*

$$\{(2^{-j^1 \cdot t} T^{(t)} E, 2^{-j}), (2^{-j^1 \cdot t} E T^{(t)}, 2^{-j}) : (E, 2^{-j}) \in \mathcal{E}\}$$

is a bounded set of $(W, \vec{d})$ pre-elementary operators supported in Ω.

General setting: *Let $T^{(t)} \in \widetilde{\mathscr{D}}^t(\Omega, (Y, \vec{d}), (Z, \vec{d}))$ and suppose $\mathcal{E}$ is a bounded set of generalized $(W, \vec{d})$ elementary operators supported in Ω. Then*

$$\{(2^{-j^1 \cdot t} T^{(t)} E, 2^{-j}), (2^{-j^1 \cdot t} E T^{(t)}, 2^{-j}) : (E, 2^{-j}) \in \mathcal{E}\} \tag{5.107}$$

is a bounded set of generalized $(W, \vec{d})$ pre-elementary operators supported in Ω.

Proof. In either setting, we have $\operatorname{supp}(T^{(t)}) \Subset \Omega \times \Omega$ and $\bigcup_{(E,2^{-j})\in\mathcal{E}} \operatorname{supp}(E) \Subset \Omega \times \Omega$. Therefore,

$$\bigcup_{(E,2^{-j})\in\mathcal{E}} \operatorname{supp}(2^{-j^1 \cdot t} T^{(t)} E) \Subset \Omega \times \Omega. \tag{5.108}$$

We turn to the Hörmander setting. We prove only that

$$\{(2^{-j^1 \cdot t} T^{(t)} E, 2^{-j}) : (E, 2^{-j}) \in \mathcal{E}\}$$

is a bounded set of $(W, \vec{d})$ pre-elementary operators supported in Ω. The proof with $T^{(t)}E$ replaced by $ET^{(t)}$ is similar, and we leave it to the reader.

In light of (5.108), it suffices to show that for all ordered multi-indices α and β,

$$\begin{aligned} &|2^{-j^1 \cdot t} (2^{-j\vec{d}} W_x)^\alpha (2^{-j\vec{d}} W_z)^\beta [T^{(t)} E](x, z)| \\ &\quad \leq C_{\alpha,\beta} \frac{(1 + \rho_{2^{-j}}(x, z))^{-m}}{\operatorname{Vol}(B_{2^{-j}}(x, 1 + \rho_{2^{-j}}(x, z))) \wedge 1}, \end{aligned} \tag{5.109}$$

for all $(E, 2^{-j}) \in \mathcal{E}$ and $x, z \in \Omega$.

Note that

$$\begin{aligned} &2^{-j^1 \cdot t} (2^{-j\vec{d}} W_x)^\alpha (2^{-j\vec{d}} W_z)^\beta [T^{(t)} E](x, z) \\ &\quad = 2^{-j^1 \cdot t} [(2^{-j\vec{d}} W)^\alpha T^{(t)} E((2^{-j\vec{d}} W)^\beta)^*](x, z). \end{aligned} \tag{5.110}$$

Since $((2^{-j\vec{d}} W)^\beta)^* = 2^{-j \cdot \deg_{\vec{d}}(\beta)} (W^\beta)^*$ and W^β is a $(W, \vec{d})$ partial differential operator on Ω of degree $\leq \deg_{\vec{d}}(\beta)$, Proposition 5.5.5 (g) and Lemma 3.12.5 imply that

$$\{(E((2^{-j\vec{d}} W)^\beta)^*, 2^{-j}) : (E, 2^{-j}) \in \mathcal{E}\}$$

is a bounded set of $(W, \vec{d})$ elementary operators supported in Ω. Combining this with (5.110) shows that it suffices to prove (5.109) in the special case where $|\beta| = 0$, i. e., to show that for all $x, z \in \Omega$ and $(E, 2^{-j}) \in \mathcal{E}$,

$$|2^{-j^1 \cdot t} [(2^{-j\vec{d}} W)^\alpha T^{(t)} E](x, z)| \lesssim \frac{(1 + \rho_{2^{-j}}(x, z))^{-m}}{\operatorname{Vol}(B_{2^{-j}}(x, 1 + \rho_{2^{-j}}(x, z))) \wedge 1}. \tag{5.111}$$

Since $(2^{-j\vec{d}}W)^{\alpha} = 2^{-j\cdot\deg_{\vec{d}}(\alpha)}W^{\alpha}$ and W^{α} is a $(W,\vec{d})$ partial differential operator on Ω of degree $\leq \deg_{\vec{d}}(\alpha)$, to prove (5.111), it suffices to show that for every $\mathscr{P}^{(\kappa)}$ a $(W,\vec{d})$ partial differential operator on Ω of degree $\leq \kappa \in \mathbb{N}^{\nu}$, we have for all $x, z \in \Omega$, for all $(E, 2^{-j}) \in \mathcal{E}$,

$$|2^{-j^1\cdot t - j\cdot\kappa}[\mathscr{P}^{(\kappa)}T^{(t)}E](x,z)| \lesssim \frac{(1+\rho_{2^{-j}}(x,z))^{-m}}{\operatorname{Vol}(B_{2^{-j}}(x,1+\rho_{2^{-j}}(x,z)))\wedge 1}. \tag{5.112}$$

Let $\kappa = (\kappa_1,\kappa_2) \in \mathbb{N}^{\nu_1}\times\mathbb{N}^{\nu_2} \cong \mathbb{N}^{\nu}$. Lemma 5.9.10 implies that $(Y,\vec{d})$ and $(Z,\vec{d})$ weakly approximately commute on Ω. Proposition 3.12.13 then implies that $\mathscr{P}^{(\kappa)}$ can be written as a finite sum of terms of the form $\mathscr{P}_1^{(\kappa_1)}\mathscr{P}_2^{(\kappa_2)}$, where $\mathscr{P}_1^{(\kappa_1)}$ is a $(Y,\vec{d})$ partial differential operator on Ω of degree $\leq \kappa_1$ and $\mathscr{P}_2^{(\kappa_2)}$ is a $(Z,\vec{d})$ partial differential operator on Ω of degree $\leq \kappa_2$. Thus, to prove (5.112), it suffices to show that for such $\mathscr{P}_1^{(\kappa_1)}$ and $\mathscr{P}_2^{(\kappa_2)}$, we have for all $x, z \in \Omega$, for all $(E, 2^{-j}) \in \mathcal{E}$,

$$\begin{aligned} &|2^{-j^1\cdot t}2^{-j^1\cdot\kappa_1}2^{-j^2\cdot\kappa_2}[\mathscr{P}_1^{(\kappa_1)}\mathscr{P}_2^{(\kappa_2)}T^{(t)}E](x,z)| \\ &\quad\lesssim \frac{(1+\rho_{2^{-j}}(x,z))^{-m}}{\operatorname{Vol}(B_{2^{-j}}(x,1+\rho_{2^{-j}}(x,z)))\wedge 1}. \end{aligned} \tag{5.113}$$

By the assumption that $T^{(t)} \in \widetilde{\mathscr{D}}^{t}(\Omega,(Y,\vec{d}),(Z,\vec{d}))$, $\mathscr{P}_2^{(\kappa_2)}T$ can be written as a finite sum of terms of the form $\widetilde{T}^{(t)}\widetilde{\mathscr{P}}_2^{(\kappa_2)}$, where $\widetilde{T}^{(t)} \in \mathscr{A}^{t}(\Omega,(Y,\vec{d}))$ and $\widetilde{\mathscr{P}}_2^{(\kappa_2)}$ is a $(Z,\vec{d})$ partial differential operator on Ω of degree $\leq \kappa_2$. Thus, to prove (5.113), it suffices to show that for such $\widetilde{T}^{(t)}$ and $\widetilde{\mathscr{P}}_2^{(\kappa_2)}$,

$$\begin{aligned} &|2^{-j^1\cdot t}2^{-j^1\cdot\kappa_1}[\mathscr{P}_1^{(\kappa_1)}\widetilde{T}^{(t)}2^{-j^2\cdot\kappa_2}\widetilde{\mathscr{P}}_2^{(\kappa_2)}E](x,z)| \\ &\quad\lesssim \frac{(1+\rho_{2^{-j}}(x,z))^{-m}}{\operatorname{Vol}(B_{2^{-j}}(x,1+\rho_{2^{-j}}(x,z)))\wedge 1}. \end{aligned} \tag{5.114}$$

Since $\widetilde{\mathscr{P}}_2^{(\kappa_2)}$ is a $(Z,\vec{d})$ partial differential operator on Ω of degree $\leq \kappa_2$, it is a $(W,\vec{d})$ partial differential operator on Ω of degree $\leq (0_{\nu_1},\kappa_2)$. Combining this with the fact that $j^2\cdot\kappa_2 = j\cdot(0_{\nu_1},\kappa_2)$, Proposition 5.5.5 (g) implies that

$$\{(2^{-j^2\cdot\kappa_2}\widetilde{\mathscr{P}}_2^{(\kappa_2)}E, 2^{-j}) : (E,2^{-j}) \in \mathcal{E}\}$$

is a bounded set of $(W,\vec{d})$ elementary operators supported in Ω. Thus, by replacing E with $2^{-j^2\cdot\kappa_2}\widetilde{\mathscr{P}}_2^{(\kappa_2)}E$, we see that to prove (5.114) it suffices to show that for all $x, z \in \Omega$, for all $(E,2^{-j}) \in \mathcal{E}$,

$$|2^{-j^1\cdot t}2^{-j^1\cdot\kappa_1}[\mathscr{P}_1^{(\kappa_1)}\widetilde{T}^{(t)}E](x,z)| \lesssim \frac{(1+\rho_{2^{-j}}(x,z))^{-m}}{\operatorname{Vol}(B_{2^{-j}}(x,1+\rho_{2^{-j}}(x,z)))\wedge 1}. \tag{5.115}$$

The estimate (5.115) follows from (5.95), completing the proof in the Hörmander setting.

We now turn to the general setting. In light of (5.108), it suffices to prove the following.

(A) For all $(E, 2^{-j}) \in \mathcal{E}$,

$$T^{(t)}E, ET^{(t)}, (T^{(t)}E)^*, (ET^{(t)})^* \in \operatorname{Hom}(C^{\infty}_{W,0}(\mathfrak{M}), C^{\infty}_{W,\mathrm{loc}}(\mathfrak{M})).$$

(B) For $p \in \{1, \infty\}$ and for $S_E \in \{T^{(t)}E, ET^{(t)}, (T^{(t)}E)^*, (ET^{(t)})^*\}$, for all ordered multi-indices α and β,

$$\sup_{(E,2^{-j})\in\mathcal{E}} 2^{-j^1\cdot t}\|(2^{-j\vec{d}}W)^{\alpha}S_E(2^{-j\vec{d}}W)^{\beta}\|_{L^p\to L^p} < \infty.$$

(C) For $p \in (1, \infty)$ and $q \in (1, \infty]$, the following holds. Let $\widetilde{\mathcal{E}} = \{(E_l, 2^{-j_l}) : l \in \mathbb{N}\} \subseteq \mathcal{E}$ be a countable set. For $r = 1, \dots, 4$, let $S_{r,l}$ be $2^{-j_l^1\cdot t}TE_l$ if $r = 1$, $2^{-j_l^1\cdot t}E_lT$ if $r = 2$, $(2^{-j_l^1\cdot t}TE_l)^*$ if $r = 3$, and $(2^{-j_l^1\cdot t}E_lT)^*$ if $r = 4$. Set, for ordered multi-indices α and β,

$$\mathcal{T}^r_{\widetilde{\mathcal{E}},\alpha,\beta}\{f_l\}_{l\in\mathbb{N}} := \{(2^{-j_l\vec{d}}W)^{\alpha}S_{r,l}(2^{-j_l\vec{d}}W)^{\beta}f_l\}_{l\in\mathbb{N}}.$$

Then

$$\sup_{\widetilde{\mathcal{E}}\subseteq\mathcal{E}} \|\mathcal{T}_{\widetilde{\mathcal{E}},\alpha,\beta}\|_{L^p(\ell^q)\to L^p(\ell^q)} < \infty,$$

where the supremum is taken over all such countable subsets of $\mathcal{E}$.

We turn to proving the above results. By Proposition 5.5.5 (d), $\{(E^*, 2^{-j}) : (E, 2^{-j}) \in \mathcal{E}\}$ is a bounded set of generalized $(W, \vec{d})$ elementary operators supported in Ω. By Proposition 5.9.5, $(T^{(t)})^* \in \widetilde{\mathscr{D}}^t(\Omega, (Y, \vec{d}), (Z, \vec{d}))$. Thus, $(T^{(t)}E)^*$ is of the same form as $ET^{(t)}$ and $(ET^{(t)})^*$ is of the same form as $T^{(t)}E$. Because of this, it suffices to prove (A) for $T^{(t)}E$ and $ET^{(t)}$, (B) when $S_E \in \{T^{(t)}E, ET^{(t)}\}$, and (C) for $r = 1, 2$.

We turn to (A). Since $\mathcal{E} \subseteq \operatorname{Hom}(C^{\infty}_{W,0}(\mathfrak{M}), C^{\infty}_{W,\mathrm{loc}}(\mathfrak{M})) \times (0, 1]^{\nu}$ (see Definition 5.2.26) and $\bigcup_{(E,2^{-j})\in\mathcal{E}} \operatorname{supp}(E) \Subset \Omega \times \Omega$, we have $\mathcal{E} \subseteq \operatorname{Hom}(C^{\infty}_{W,\mathrm{loc}}(\mathfrak{M}), C^{\infty}_{W,0}(\mathfrak{M})) \times (0, 1]^{\nu}$. Combining this with Lemma 5.9.11 implies that $T^{(t)}E, ET^{(t)} \in \operatorname{Hom}(C^{\infty}_{W,0}(\mathfrak{M}), C^{\infty}_{W,\mathrm{loc}}(\mathfrak{M}))$, for $(E, 2^{-j}) \in \mathcal{E}$, as desired.

For (B), the proof when $S_E = ET^{(t)}$ is similar to the proof when $S_E = T^{(t)}E$, so we only prove (B) when $S_E = T^{(t)}E$. Similarly, we only prove (C) for $r = 1$ (the proof for $r = 2$ being similar).

By Proposition 5.5.5 (e), $\{(E(2^{-j\vec{d}}W)^{\beta}, 2^{-j}) : (E, 2^{-j}) \in \mathcal{E}\}$ is a bounded set of generalized $(W, \vec{d})$ elementary operators supported in Ω. Thus, by replacing E with $E(2^{-j\vec{d}}W)^{\beta}$ it suffices to prove (B) and (C) in the special case $|\beta| = 0$. As in the proof in the Hörmander setting, we may write $(2^{-j\vec{d}}W)^{\alpha}T^{(t)}$ as a finite sum of terms of the form

$$2^{-j^1\cdot\kappa_1}\mathscr{P}_1^{(\kappa_1)}\widetilde{T}^{(t)}2^{-j^2\cdot\kappa_2}\widetilde{\mathscr{P}}_2^{(\kappa_2)},$$

where $\kappa = (\kappa_1, \kappa_2) := \deg_{\vec{d}}(\alpha) \in \mathbb{N}^\nu$, $\mathscr{P}_1^{(\kappa_1)}$ is a $(Y, \vec{d})$ partial differential operator on Ω of degree $\leq \kappa_1$, $\widetilde{\mathscr{P}}_2^{(\kappa_2)}$ is a $(Z, \vec{d})$ partial differential operator on Ω of degree $\leq \kappa_2$, and $\widetilde{T}^{(t)} \in \widetilde{\mathscr{A}}^t(\Omega, (Y, \vec{d}))$. Thus, it suffices to prove (B) and (C) with $(2^{-j\vec{d}} W)^\alpha T^{(t)} E$ replaced by

$$2^{-j^1 \cdot \kappa_1} \mathscr{P}_1^{(\kappa_1)} \widetilde{T}^{(t)} 2^{-j^2 \cdot \kappa_2} \widetilde{\mathscr{P}}_2^{(\kappa_2)} E,$$

where $\mathscr{P}_1^{(\kappa_1)}$, $\widetilde{T}^{(t)}$, and $\widetilde{\mathscr{P}}_2^{(\kappa_2)}$ are as described above.

Since $\widetilde{\mathscr{P}}_2^{(\kappa_2)}$ is a $(Z, \vec{d})$ partial differential operator on Ω of degree $\leq \kappa_2$, it is a $(W, \vec{d})$ partial differential operator on Ω of degree $\leq (0_{\nu_1}, \kappa_2)$. Combining this with the fact that $j^2 \cdot \kappa_2 = j \cdot (0_{\nu_1}, \kappa_2)$, Proposition 5.5.5 (g) implies that

$$\{(2^{-j^2 \cdot \kappa_2} \widetilde{\mathscr{P}}_2^{(\kappa_2)} E, 2^{-j}) : (E, 2^{-j}) \in \mathcal{E}\}$$

is a bounded set of generalized $(W, \vec{d})$ elementary operators supported in Ω. Thus, by replacing E with $2^{-j^2 \cdot \kappa_2} \widetilde{\mathscr{P}}_2^{(\kappa_2)} E$, we see that it suffices to prove (B) and (C) with $(2^{-j\vec{d}} W)^\alpha T^{(t)} E$ replaced by

$$2^{-j^1 \cdot \kappa_1} \mathscr{P}_1^{(\kappa_1)} \widetilde{T}^{(t)} E.$$

From here, the result follows from Lemma 5.9.9, completing the proof. □

Proof of Theorem 5.9.6. We begin with the Hörmander setting. Let $T \in \mathscr{D}^t(\Omega, (Y, \vec{d}), (Z, \vec{d})) \subseteq \mathscr{A}^t(\Omega, (Y, \vec{d}))$. By Proposition 5.8.5, $T \in \operatorname{Hom}(C_0^\infty(\Omega), C_0^\infty(\Omega))$, and by the definition of $\mathscr{A}^t(\Omega, (Y, \vec{d}))$ we have $\operatorname{supp}(T) \subseteq \Omega \times \Omega$. From here, that $T \in \mathscr{A}^{(t, 0_{\nu_2})}(\Omega, (W, \vec{d}))$ follows by combining Proposition 5.7.1 and Lemma 5.9.12.

The proof in the general setting is similar. Let $T \in \widetilde{\mathscr{D}}^t(\Omega, (Y, \vec{d}), (Z, \vec{d})) \subseteq \widetilde{\mathscr{A}}^t(\Omega, (Y, \vec{d}))$. By Lemma 5.9.11, $T \in \operatorname{Hom}(C_{W,0}^\infty(\Omega), C_{W,0}^\infty(\Omega))$. By the definition of $\widetilde{\mathscr{A}}^t(\Omega, (Y, \vec{d}))$, we have $\operatorname{supp}(T) \subseteq \Omega \times \Omega$. From here, that $T \in \widetilde{\mathscr{A}}^{(t, 0_{\nu_2})}(\Omega, (W, \vec{d}))$ follows by combining Proposition 5.7.1 and Lemma 5.9.12. □

5.9.1 Standard pseudo-differential operators

Let $(Y, 1) = \{(Y_1, 1), \ldots, (Y_\nu, 1)\} \subset C_{\mathrm{loc}}^\infty(\mathfrak{M}; T\mathfrak{M}) \times \mathbb{N}_+$ be such that

$$\operatorname{span}\{Y_1(x), \ldots, Y_\nu(x)\} = T_x\mathfrak{M}, \quad \forall x \in \mathfrak{M}.$$

Fix a relatively compact open set $\Omega \Subset \mathfrak{M}$ and $m \in \mathbb{R}$. Suppose $b(x, D)$ is a standard pseudo-differential operator of order m on $\mathfrak{M}$. Theorem 5.8.23 implies $b(x, D) \in \mathscr{A}^m(\Omega, (Y, 1))$. In this section we show that more is true: $b(x, D)$ lies in the algebras $\mathscr{D}^m$ and $\widetilde{\mathscr{D}}^m$. We turn to making this precise.

We consider the two settings of Section 5.3, so we are given a list $(W^1, \vec{d}^1), \dots, (W^\nu, \vec{d}^\nu)$ as in that section and set

$$(W, \vec{d}) = \{(W_1, \vec{d}_1), \dots, (W_r, \vec{d}_r)\} \\ := (W^1, \vec{d}^1) \boxtimes (W^2, \vec{d}^2) \boxtimes \cdots \boxtimes (W^\nu, \vec{d}^\nu) \subset C^\infty_{\mathrm{loc}}(\mathfrak{M}; T\mathfrak{M}) \times (\mathbb{N}^\nu \setminus \{0\}).$$

By Lemma 3.8.3, for each $\mu \in \{1, \dots, \nu\}$, $(W^\mu, \vec{d}^\mu)$ and $(Y, 1)$ strongly locally approximately commute, and therefore by Proposition 3.8.6 (b) they weakly locally approximately commute. It is evident that $(Y, 1)$ are Hörmander vector fields with formal degrees (indeed, $Y_1, \dots, Y_\nu$ satisfy Hörmander's condition of order 1 at every point). Thus, in either the Hörmander setting or the general setting, $(W^1, \vec{d}^1), \dots, (W^\nu, \vec{d}^\nu), (Y, 1)$ satisfy the same hypotheses as $(W^1, \vec{d}^1), \dots, (W^\nu, \vec{d}^\nu)$ with ν replaced by $\nu + 1$. In particular, in the Hörmander setting, it makes sense to talk about $\mathscr{D}^t(\Omega, (Y, 1), (W, \vec{d}))$, and in the general setting, it makes sense to talk about $\widetilde{\mathscr{D}}^t(\Omega, (Y, 1), (W, \vec{d}))$.

The main result of this setting is the next proposition.

Proposition 5.9.13. *Suppose $b(x, D)$ is a standard pseudo-differential operator of order $m \in \mathbb{R}$ with $\mathrm{supp}(b(x, D)) \subseteq \Omega \times \Omega$. Then we have the following.*
Hörmander setting: *We have $b(x, D) \in \mathscr{D}^m(\Omega, (Y, 1), (W, \vec{d}))$.*
General setting: *We have $b(x, D) \in \widetilde{\mathscr{D}}^m(\Omega, (Y, 1), (W, \vec{d}))$.*

Proof. We prove the result in the Hörmander setting; the proof in the general setting follows by replacing $\mathscr{A}$ and $\mathscr{D}$ with $\widetilde{\mathscr{A}}$ and $\widetilde{\mathscr{D}}$ throughout.

Let $\mathscr{P}^{(\kappa)}$ be a $(W, \vec{d})$ partial differential operator on Ω of degree $\le \kappa \in \mathbb{Z}^\nu$. We will show that:

(i) $b(x, D)\mathscr{P}^{(\kappa)}$ can be written as a finite sum of operators of the form

$$\widetilde{\mathscr{P}}^{(\kappa)} \tilde{b}(x, D),$$

where $\widetilde{\mathscr{P}}^{(\kappa)}$ is a $(W, \vec{d})$ partial differential operator on Ω of degree $\le \kappa$ and $\tilde{b}(x, D)$ is a standard pseudo-differential operator of order $m \in \mathbb{R}$ with $\mathrm{supp}(\tilde{b}(x, D)) \subseteq \Omega \times \Omega$.

(ii) $\mathscr{P}^{(\kappa)} b(x, D)$ can be written as a finite sum of operators of the form

$$\tilde{b}(x, D)\widetilde{\mathscr{P}}^{(\kappa)}, \tag{5.116}$$

where $\widetilde{\mathscr{P}}^{(\kappa)}$ is a $(W, \vec{d})$ partial differential operator on Ω of degree $\le \kappa$ and $\tilde{b}(x, D)$ is a standard pseudo-differential operator of order $m \in \mathbb{R}$ with $\mathrm{supp}(\tilde{b}(x, D)) \subseteq \Omega \times \Omega$.

To see why this will complete the proof, note that Theorem 5.8.23 implies that $b(x, D) \in \mathscr{A}^m(\Omega, (Y, 1))$ and $\tilde{b}(x, D) \in \mathscr{A}^m(\Omega, (Y, 1))$ for each of the above $\tilde{b}$. Thus, if (i) and (ii) hold, the conditions for $b(x, D) \in \mathscr{D}^m(\Omega, (Y, 1), (W, \vec{d}))$ are verified.

The proofs for (i) and (ii) are similar, so we prove only (ii). If $\kappa \in \mathbb{Z}^\nu \setminus \mathbb{N}^\nu$, then $\mathscr{P}^{(\kappa)} = 0$ and (ii) is trivial. We henceforth assume $\kappa \in \mathbb{N}^\nu$.

By definition $\mathscr{P}^{(\kappa)}$ (see Definition 3.12.1) can be written as a finite sum of terms of the form

$$\text{Mult}[g] W_{l_1} W_{l_2} \cdots W_{l_K}, \tag{5.117}$$

where $g \in C^\infty_{\text{loc}}(\Omega)$ and $\vec{d}_{l_1} + \cdots + \vec{d}_{l_K} \le \kappa$. It suffices to consider the case where $\mathscr{P}^{(\kappa)}$ is given by just one term of the form (5.117); we henceforth assume this.

Since $\text{supp}(b(x,D)) \subseteq \Omega \times \Omega$, $\text{supp}(b(x,D))$ is closed, and $\Omega \Subset \mathfrak{M}$, we have $b(x,D) \Subset \Omega \times \Omega$. Take $\phi \in C^\infty_0(\Omega)$ such that $\phi \otimes \phi \equiv 1$ on a neighborhood of $supp(b(x,D))$. It suffices to prove the result with $\mathscr{P}^{(\kappa)}$ replaced by $\text{Mult}[\phi]\mathscr{P}^{(\kappa)}$. That is, we may replace $\mathscr{P}^{(\kappa)}$ with

$$\text{Mult}[g\phi] W_{l_1} W_{l_2} \cdots W_{l_K}.$$

From here, the calculus of pseudo-differential operators shows that $\text{Mult}[\phi]\mathscr{P}^{(\kappa)} b(x,D)$ can be written as a finite sum of terms of the form

$$\tilde{b}(x,D)\,\text{Mult}[\tilde{g}] Z_{l_{k_1}} Z_{l_{k_2}} \cdots Z_{l_{k_R}},$$

where $\tilde{g} \in C^\infty_0(\Omega)$, $1 \le k_1 < k_2 < \cdots < k_R \le K$, and $\tilde{b}(x,D)$ is a standard pseudo-differential operator of order m. This is of the desired form (5.116), which completes the proof. □

5.10 L^p bounds in the single-parameter case

In the multi-parameter case, operators in $\mathscr{A}^0(\Omega,(W,\vec{d}))$ and, more generally, operators in $\widetilde{\mathscr{A}}^0(\Omega,(W,\vec{d}))$ are bounded on $L^p(\mathfrak{M},\text{Vol})$, $1 < p < \infty$. We will see this as part of our more general theory of Triebel–Lizorkin spaces in Chapter 6.[3] However, in the single-parameter case ($\nu = 1$), the operators in $\mathscr{A}^0(\Omega,(W,\vec{d}))$ fall under the general Calderón–Zygmund paradigm, and we can prove the L^p boundedness using standard Calderón–Zygmund theory. We present this here.

The setting is the single-parameter setting of Section 5.2.1. Thus, we are given Hörmander vector fields with formal degrees on $\mathfrak{M}$: $(W,\vec{d}) = \{(W_1,\vec{d}_1),\ldots,(W_r,\vec{d})\} \subset C^\infty_{\text{loc}}(\mathfrak{M};T\mathfrak{M}) \times \mathbb{N}_+$. As before, $\Omega \Subset \mathfrak{M}$ denotes a relatively compact, open set.

Proposition 5.10.1. *If $T \in \mathscr{A}^0(\Omega,(W,\vec{d}))$, then T extends to a bounded operator $T : L^p(\mathfrak{M},\text{Vol}) \to L^p(\mathfrak{M},\text{Vol})$, $1 < p < \infty$.*

3 The L^2 boundedness of operators in $\mathscr{A}^0(\Omega,(W,\vec{d}))$ and $\widetilde{\mathscr{A}}^0(\Omega,(W,\vec{d}))$ is easy to see from elementary considerations. See Remark 5.10.3.

By Theorem 3.3.1 (b), the balls $B_{(W,d)}(x,\delta)$ paired with the measure Vol locally give $\mathfrak{M}$ the structure of a space of homogeneous type. Since operators in $\mathscr{A}^0(\Omega,(W,\vec{d}))$ have Schwartz kernels with compact support, the classical Calderón–Zygmund theory can be applied to them. In particular, Proposition 5.10.1 follows from the $T(1)$ theorem on spaces of homogeneous type due to David, Journé, and Semmes [59]; here we use $\mathscr{A}^0(\Omega,(W,\vec{d})) = \mathscr{A}_4^0(\Omega,(W,\vec{d}))$. However, in later sections, we will apply Proposition 5.10.1 an infinite number of times, and we will require that the L^p bounds of the operator are uniform over this infinite number of applications. While it is possible to use the proof of the $T(1)$ theorem to deduce the required uniformity, it is easier to see from a more elementary proof, which we present. A uniform result is presented in Section 5.10.1.

Lemma 5.10.2. *If $T \in \mathscr{A}^0(\Omega,(W,\vec{d}))$, then T extends to a bounded operator*

$$T : L^2(\mathfrak{M},\mathrm{Vol}) \to L^2(\mathfrak{M},\mathrm{Vol}).$$

Proof. Since $T \in \mathscr{A}^0(\Omega,(W,\vec{d})) = \mathscr{A}_2^t(\Omega,(W,\vec{d}))$, there is a bounded set of $(W,\vec{d})$ elementary operators supported in Ω, $\{(E_j,2^{-j}) : j \in \mathbb{N}\}$, such that $T = \sum_{j\in\mathbb{N}} E_j$. The result will follow from the Cotlar–Stein lemma (Lemma 2.3.25) once we show that, $\forall j,k \in \mathbb{N}$,

$$\|E_jE_k^*\|_{L^2\to L^2}, \|E_k^*E_j\|_{L^2\to L^2} \lesssim 2^{-|j-k|}. \tag{5.118}$$

By Proposition 5.5.5 (d), $\{(E_k^*,2^{-k}) : k \in \mathbb{N}\}$ is a bounded set of $(W,\vec{d})$ elementary operators supported in Ω. Therefore, by Proposition 5.5.11,

$$\{(2^{|j-k|}E_jE_k^*,2^{-j}),(2^{|j-k|}E_k^*E_j,2^{-j}) : j,k \in \mathbb{N}\}$$

is a bounded set of $(W,\vec{d})$ elementary operators supported in Ω. Corollary 5.4.11 implies that

$$\sup_{j,k\in\mathbb{N}} \|2^{|j-k|}E_jE_k^*\|_{L^2\to L^2}, \sup_{j,k\in\mathbb{N}} \|2^{|j-k|}E_k^*E_j\|_{L^2\to L^2} < \infty.$$

This is equivalent to (5.118), completing the proof. □

Remark 5.10.3. In the multi-parameter setting, $\nu > 1$, it is straightforward to modify the proof of Lemma 5.10.2 to prove that operators in $\widetilde{\mathscr{A}}^0(\Omega,(W,\vec{d}))$ are bounded on L^2, and as a special case that operators in $\mathscr{A}^0(\Omega,(W,\vec{d}))$ are bounded in L^2. However, we will see this as part of more general results in Chapter 6.

Proof of Proposition 5.10.1. We will show that T extends to a bounded operator T : $L^p(\mathfrak{M},\mathrm{Vol}) \to L^p(\mathfrak{M},\mathrm{Vol})$, $1 < p \le 2$. Since $T^* \in \mathscr{A}^0(\Omega,(W,\vec{d}))$, by Proposition 5.8.1, the result for $2 \le p < \infty$ will follow by duality.

We use the proof method of [216, Chapter 1, Section 5]. For the remainder of the proof, we use $T \in \mathscr{A}^0(\Omega,(W,\vec{d})) = \mathscr{A}_4^0(\Omega,(W,\vec{d}))$. Since we already know that T extends to a bounded operator on L^2, by Lemma 5.10.2, by the methods of [216, Chapter 1, Section 5] it suffices to show that there exists $C \geq 0$ such that $\forall \delta > 0, y \in \Omega, \overline{y} \in B_{(W,d)}(y, \delta)$,

$$\int\limits_{B_{(W,d)}(y,2\delta)^c \cap \Omega} |T(x,y) - T(x,\overline{y})| \, d\,\mathrm{Vol}(x) \leq C. \tag{5.119}$$

For such $y, \overline{y}$, let $\gamma : [0,1] \to B_{(W,d)}(y, \delta)$ be an absolutely continuous path with $\gamma(0) = y$, $\gamma(1) = \overline{y}$, and $\gamma'(t) = \sum \delta^{d_l} a_l(t) W_l(\gamma(t))$, with $\| \sum |a_l|^2 \|_{L^\infty} < 1$. Note that for $x \in B_{(W,d)}(y, 2\delta)^c$, $\forall t \in [0,1]$, $\rho_{(W,d)}(x, \gamma(t)) \approx \rho_{(W,d)}(x,y)$, since $\gamma(t) \in B_{(W,d)}(y, \delta)$. Thus, for $x \in B_{(W,d)}(y, 2\delta)^c$ we have, using Theorem 3.3.1 (b),

$$\begin{aligned}
|T(x,y) - T(x,\overline{y})| &\leq \int_0^1 |\partial_t T(x, \gamma(t))| \, dt \leq \sum_{l=1}^r \int_0^1 |\delta^{d_l} (W_{l,y} T)(x, \gamma(t))| \, dt \\
&\lesssim \sum_{l=1}^r \delta^{d_l} \int_0^1 \frac{\rho_{(W,d)}(x, \gamma(t))^{-d_l}}{\mathrm{Vol}(B_{(W,d)}(x, \rho_{(W,d)}(x, \gamma(t)))) \wedge 1} \\
&\approx \sum_{l=1}^r \delta^{d_l} \frac{\rho_{(W,d)}(x,y)^{-d_l}}{\mathrm{Vol}(B_{(W,d)}(x, \rho_{(W,d)}(x,y))) \wedge 1}.
\end{aligned} \tag{5.120}$$

Using (5.120) and Theorem 3.3.1 (b), we have

$$\begin{aligned}
&\int\limits_{B_{(W,d)}(y,2\delta)^c \cap \Omega} |T(x,y) - T(x,\overline{y})| \, d\,\mathrm{Vol}(x) \\
&= \sum_{j=1}^\infty \int\limits_{(B_{(W,d)}(y,2^{j+1}\delta) \setminus B_{(W,d)}(y,2^j\delta)) \cap \Omega} |T(x,y) - T(x,\overline{y})| \, d\,\mathrm{Vol}(x) \\
&\lesssim \sum_{j=1}^\infty \sum_{l=1}^r \int\limits_{(B_{(W,d)}(y,2^{j+1}\delta) \setminus B_{(W,d)}(y,2^j\delta)) \cap \Omega} \frac{\delta^{d_l} \rho_{(W,d)}(x,y)^{-d_l}}{\mathrm{Vol}(B_{(W,d)}(y, \rho_{(W,d)}(x,y))) \wedge 1} \, d\,\mathrm{Vol}(x) \\
&\approx \sum_{j=1}^\infty \sum_{l=1}^r \int\limits_{(B_{(W,d)}(y,2^{j+1}\delta) \setminus B_{(W,d)}(y,2^j\delta)) \cap \Omega} \frac{2^{-jd_l}}{\mathrm{Vol}(B_{(W,d)}(y, 2^j\delta)) \wedge 1} \, d\,\mathrm{Vol}(x) \\
&\lesssim \sum_{j=1}^\infty \sum_{l=1}^r 2^{-jd_l} \frac{\mathrm{Vol}(B_{(W,d)}(y, 2^{j+1}\delta) \cap \Omega)}{\mathrm{Vol}(B_{(W,d)}(y, 2^j\delta)) \wedge 1} \\
&\approx \sum_{j=1}^\infty \sum_{l=1}^r 2^{-jd_l} \approx 1,
\end{aligned}$$

where we have used $\mathrm{Vol}(\Omega) < \infty$, since Ω is relatively compact. This completes the proof of (5.119). □

5.10.1 A uniform result

Fix $n \in \mathbb{N}$ and let $(Y, d\!\!\!/) = \{(Y_1, d\!\!\!/_1), \ldots, (Y_q, d\!\!\!/_q)\} \subset C^\infty(B^n(1); \mathbb{R}^n) \times \mathbb{N}_+$ be C^∞ vector fields with formal degrees on $B^n(1)$. We suppose:

– $\operatorname{span}\{Y_1(u), \ldots, Y_q(u)\} = T_u B^n(1)$, $\forall u \in B^n(1)$, uniformly in u in the sense that

$$\inf_{u \in B^n(1)} \max_{j_1, \ldots, j_n \in \{1, \ldots, q\}} \left|\det(Y_{j_1}(u)| \cdots |Y_{j_n}(u))\right| > 0. \tag{5.121}$$

– $[Y_j, Y_k] = \sum_{d\!\!\!/_l \leq d\!\!\!/_j + d\!\!\!/_k} c_{j,k}^l Y_l$, where $c_{j,k}^l \in C^\infty(B^n(1))$.

Fix $\eta \in (0, 1)$. If $a = a(\eta, (Y, d\!\!\!/)) > 0$ is sufficiently small, the following holds. Let T be a $(Y, d\!\!\!/), a$ pseudo-differential operator of order 0 supported in $B^n(\eta/2) \times B^n(\eta)$. Then, by Proposition 5.8.13, $T \in \mathscr{A}^0(B^n(1), (Y, d\!\!\!/))$, and therefore by Proposition 5.10.1, T extends to a bounded operator $T : L^p(B^n(1), \sigma_{\mathrm{Leb}}) \to L^p(B^n(1), \sigma_{\mathrm{Leb}})$, $1 < p < \infty$, where σ_{Leb} denotes the usual Lebesgue density on $\mathbb{R}^n$. In other words, for $1 < p < \infty$,

$$\|T\|_{L^p(B^n(1), \sigma_{\mathrm{Leb}}) \to L^p(B^n(1), \sigma_{\mathrm{Leb}})} \leq C_p, \tag{5.122}$$

for some constant $C_p \geq 0$. By keeping track of constants in the abovementioned proofs, we can be more explicit about what $C_p \geq 0$ and $a > 0$ can be chosen to depend on. Here we describe this dependence.

The constant *a*: The constant $a > 0$ can be chosen to depend only on upper bounds for η^{-1}, $(1 - \eta)^{-1}$, n, q, and $\max_{1 \leq j \leq q} \|Y_j\|_{C^1(B^n(1); \mathbb{R}^n)}$.

Since T is a $(Y, d\!\!\!/), a$ pseudo-differential operator of order 0 supported in $B^n(\eta/2) \times B^n(\eta)$, T can be written as

$$T = \sum_{j \in \mathbb{N}} E_j,$$

where

$$E_j f(x) = \int f(e^{-t \cdot Y} x) \operatorname{Dil}_{2^j}^{d}(\varsigma_j)(x, t, t)\, dt,$$

where $\{\varsigma_j : j \in \mathbb{N}\} \subset C_0^\infty(B^n(\eta/2)) \widehat{\otimes} C_0^\infty(B^q(a)) \widehat{\otimes} \mathscr{S}(\mathbb{R}^q)$ is a bounded set and $\varsigma_j \in C_0^\infty(B^n(\eta/2)) \widehat{\otimes} C_0^\infty(B^q(a)) \widehat{\otimes} \mathscr{S}_0(\mathbb{R}^q)$ for $j > 0$.

The constant C_p: There is a constant $L \in \mathbb{N}$, depending only on upper bounds for n and q, such that the constant $C_p \geq 0$ from (5.122) can be chosen to depend only on:

– upper bounds for n, q, η^{-1}, $(1 - \eta)^{-1}$, p, $(1 - p)^{-1}$, and $\max\{d\!\!\!/_j : 1 \leq j \leq q\}$,

– an upper bound for

$$\max_{1 \leq j \leq q} \|Y_j\|_{C^L(B^n(1); \mathbb{R}^n)},$$

- an upper bound for

$$\sum_{|\alpha|,|\beta|,|\gamma|\leq L} \sup_{x\in B^n(\eta/2)} \sup_{s\in B^q(a)} \sup_{t\in\mathbb{R}^q} \sup_{j\in\mathbb{N}} (1+|t|)^L \left|\partial_x^\alpha \partial_s^\beta \partial_t^\gamma \varsigma_j(x,s,t)\right|,$$

- a lower bound, > 0, for the left-hand side of (5.121).

5.10.2 Beyond Hörmander's condition

Proposition 5.10.1 concerned the case where $(W, \vec{d})$ were Hörmander vector fields with formal degrees, which is the setting to which the classical Calderón–Zygmund theory most directly applies. However, the uniformity described in Section 5.10.1 allows us to deduce the L^p boundedness of some operators outside of the setting of Hörmander's condition. At the end of this section, we use this to develop a Littlewood–Paley square function, which will be a crucial tool in our study of Sobolev spaces in Section 6.9.

Let $(X, d) = \{(X_1, d_1), \dots, (X_q, d_q)\} \subset C^\infty_{\mathrm{loc}}(\mathfrak{M}; T\mathfrak{M}) \times \mathbb{N}_+$ be C^∞_{loc} vector fields with formal degrees satisfying

$$[X_j, X_k] = \sum_{d_l \leq d_j + d_k} c^l_{j,k} X_l, \quad c^l_{j,k} \in C^\infty_{\mathrm{loc}}(\mathfrak{M}). \tag{5.123}$$

Note that we are not assuming that $X_1, \dots, X_q$ span the tangent space at any point, and we can therefore not directly use Proposition 5.10.1 to deduce the L^p boundedness of (X, d), a pseudo-differential operators of order 0.

Fix relatively compact, open sets $\Omega_3 \Subset \Omega_4 \Subset \mathfrak{M}$. The main result of this section is the following.

Proposition 5.10.4. *If $a = a(\Omega_3, \Omega_4, (X, d)) > 0$ is sufficiently small, the following holds. Let $\mathcal{E}$ be a bounded set of (X, d), a pseudo-differential operator scales supported in $\Omega_3 \times \Omega_4$. For $\mathcal{E}' = \{(E_j, 2^{-j}) : j \in \mathbb{N}\} \subseteq \mathcal{E}$, set $T = T[\mathcal{E}'] := \sum_{j\in\mathbb{N}} E_j$. Then, for $p \in (1, \infty)$, $T : L^p(\mathfrak{M}, \mathrm{Vol}) \to L^p(\mathfrak{M}, \mathrm{Vol})$ and*

$$\sup_{\substack{\{(E_j, 2^{-j}) : j\in\mathbb{N}\} \\ \subseteq \mathcal{E}}} \|T\|_{L^p(\mathfrak{M},\mathrm{Vol}) \to L^p(\mathfrak{M},\mathrm{Vol})} < \infty.$$

We turn to the proof of Proposition 5.10.4. The idea is the following: because of (5.123) the classical Frobenius theorem (Theorem 3.1.15) foliates $\mathfrak{M}$ into leaves, and $X_1, \dots, X_q$ span the tangent space at every point on the leaves. The idea is to apply Proposition 5.10.1 on each leaf (using the uniform version described in Section 5.10.1) to deduce the L^p boundedness of T. There is a difficulty with this proof outline, though: the foliation may have singular points (see Definition 3.1.21). In this case, the usual proofs of the Frobenius theorem do not give uniform control over the coordinate charts defining the foliation near a singular point, and we do not immediately obtain enough uniformity to deduce the L^p boundedness of T. Fortunately, Theorem 3.5.1 (in the case $\nu = 1$) does give the required uniformity for these coordinate charts.

Pick Ω_1, Ω_5 open and relatively compact and $\mathcal{K}$ compact with $\Omega_3 \Subset \Omega_4 \Subset \Omega_5 \Subset \mathcal{K} \Subset \Omega_1 \Subset \mathfrak{M}$. We apply Theorem 3.5.1 (in the case $\nu = 1$) with this choice of $\mathcal{K}$ and Ω_1 and with $(X, d) = (W, \vec{d})$. For $x \in \mathcal{K}$ and $\delta \in (0,1]$, let $\Phi_{x,\delta}$ and $N(x) = \dim \operatorname{span}\{X_1(x), \dots, X_q(x)\}$ be as in that theorem. We will use Proposition 3.9.1, with Ω_3 in that proposition replaced by Ω_4 and Ω_4 replaced by Ω_5 and with $\mathcal{K}$ playing the role of the set with the same name. We let $\eta_1 > 0$ be as in that proposition.

Lemma 5.10.5. *If $a = a(\Omega_3, \Omega_4, (X, d)) > 0$ is sufficiently small, the following holds. Let $\mathcal{E}$ be a bounded set of (X, d), a pseudo-differential operator scales supported in $\Omega_3 \times \Omega_4$. For each $p \in (1, \infty)$, there exists $D_p \geq 0$ such that for all $x \in \mathcal{K}$ and $f \in C_0^\infty(\Omega_1)$,*

$$\sup_{\substack{\{(E_j, 2^{-j}) : j \in \mathbb{N}\} \\ \subseteq \mathcal{E}}} \int_{B^{N(x)}(\eta_1/4)} |Tf(\Phi_{x,1}(u))|^p \, du \leq D_p \int_{B^{N(x)}(\eta_1)} |f(\Phi_{x,1}(u))|^p \, du, \tag{5.124}$$

where $T = T[\{(E_j, 2^{-j}) : j \in \mathbb{N}\}] = \sum_{j \in \mathbb{N}} E_j$, as in Proposition 5.10.4.

Proof. Since $\mathcal{E}$ is a bounded set of (X, d), a pseudo-differential operator scales, by Definition 4.3.1, for $(E_j, 2^{-j}) \in \mathcal{E}$, we can write

$$E_j f(x) = \int f(e^{-t\cdot X} x) \operatorname{Dil}^d_{2^j}(\varsigma_j)(x, t, t) \, dt, \tag{5.125}$$

where $\{\varsigma_j : (E_j, 2^{-j}) \in \mathcal{E}\} \subset C_0^\infty(\Omega_3) \widehat{\otimes} C_0^\infty(B^q(a)) \widehat{\otimes} \mathscr{S}(\mathbb{R}^q)$ is a bounded set and $\varsigma_j \in C_0^\infty(\Omega_3) \widehat{\otimes} C_0^\infty(B^q(a)) \widehat{\otimes} \mathscr{S}_0(\mathbb{R}^q)$ for $j > 0$.

We begin with the case where $N(x) = 0$, i. e., $X_1(x) = \cdots = X_q(x) = 0$, $\Phi_{x,1}(u) \equiv x$, $B^{N(x)}(\eta) = \{0\}$, and du denotes the point mass at $0 \in \mathbb{R}^0$. In this (degenerate) case, we will show that

$$|E_j f(x)| \lesssim 2^{-j} |f(x)|. \tag{5.126}$$

First we see why (5.126) yields (5.124) in this case. Indeed, by (5.126), we have $|Tf(x)| \lesssim \sum_{j \in \mathbb{N}} 2^{-j} |f(x)| \lesssim |f(x)|$. Therefore, we have

$$\begin{aligned} \sup_{\substack{\{(E_j, 2^{-j}) : j \in \mathbb{N}\} \\ \subseteq \mathcal{E}}} \int_{B^{N(x)}(\eta_1/4)} |Tf(\Phi_{x,1}(u))|^p \, du &= \sup_{\substack{\{(E_j, 2^{-j}) : j \in \mathbb{N}\} \\ \subseteq \mathcal{E}}} |Tf(x)|^p \, du \\ &\lesssim |f(x)|^p = \int_{B^{N(x)}(\eta_1)} |f(\Phi_{x,1}(u))|^p \, du, \end{aligned}$$

which yields (5.124) in this case. We turn to proving (5.126). Since $X_1(x) = \cdots = X_q(x) = 0$, we have $e^{-t\cdot X} x = x$. Thus, (5.125) can be rewritten as

$$E_j f(x) = f(x) \int \mathrm{Dil}^d_{2^j}(\varsigma_j)(x,t,t)\,dt,$$

so to prove (5.126), it suffices to show that

$$\left|\int \mathrm{Dil}^d_{2^j}(\varsigma_j)(x,t,t)\,dt\right| \lesssim 2^{-j}. \tag{5.127}$$

When $j = 0$, we have, by Definition 4.1.8 (ii),

$$\left|\mathrm{Dil}^d_{2^0}(\varsigma_j)(x,t,t)\,dt\right| = \left|\int \varsigma_0(x,t,t)\,dt\right| \lesssim \int (1+|t|)^{-q-1}\,dt \lesssim 1,$$

proving (5.127) when $j = 0$. We now assume $j > 0$, and therefore $\varsigma_j \in C_0^\infty(\Omega_3)\widehat{\otimes} C_0^\infty(B^q(a))\widehat{\otimes} \mathscr{S}_0(\mathbb{R}^q)$. By Corollary 4.1.19, we can write

$$\varsigma_j(x,s,t) = \triangle_t \triangle_t^{-1} \varsigma_j(x,s,t) = \sum_{k=1}^q \partial_{t_k}(-\partial_{t_k} \triangle_t^{-1} \varsigma_j(x,s,t)) =: \sum_{k=1}^q \partial_{t_k} \tilde{\varsigma}_{j,k}(x,s,t),$$

where $\{\tilde{\varsigma}_{j,k} : (E_j, 2^{-j}) \in \mathcal{E}, j > 0, k \in \{1,\dots,q\}\} \subset C_0^\infty(\Omega_3)\widehat{\otimes} C_0^\infty(B^q(a))\widehat{\otimes}\mathscr{S}(\mathbb{R}^q)$ is a bounded set. Thus, we have, integrating by parts and using Definition 4.1.8 (ii),

$$\begin{aligned}
\left|\int \mathrm{Dil}^d_{2^j}(\varsigma_j)(x,t,t)\,dt\right| &\le \sum_{k=1}^q \left|\int \mathrm{Dil}^d_{2^j}(\partial_{t_k}\tilde{\varsigma}_{j,k})(x,s,t)\Big|_{s=t}\,dt\right| \\
&= \sum_{k=1}^q 2^{-j d_k} \left|\int (\partial_{t_k} \mathrm{Dil}^d_{2^j}(\tilde{\varsigma}_{j,k})(x,s,t))\Big|_{s=t}\,dt\right| \\
&= \sum_{k=1}^q 2^{-j d_k} \left|\int (\partial_{s_k} \mathrm{Dil}^d_{2^j}(\tilde{\varsigma}_{j,k})(x,s,t))\Big|_{s=t}\,dt\right| \\
&= \sum_{k=1}^q 2^{-j d_k} \left|\int (\mathrm{Dil}^d_{2^j}(\partial_{s_k}\tilde{\varsigma}_{j,k})(x,s,t))\Big|_{s=t}\,dt\right| \\
&= \sum_{k=1}^q 2^{-j d_k} \left|\int (\partial_{s_k}\tilde{\varsigma}_{j,k})(x, 2^{-jd}t, t)\,dt\right| \\
&\lesssim \sum_{k=1}^q 2^{-j d_k} \int (1+|t|)^{-q-1}\,dt \lesssim \sum_{k=1}^q 2^{-j d_k} \lesssim 2^{-j}.
\end{aligned}$$

This establishes (5.127) and completes the proof in the case $N(x) = 0$.

It remains to prove the result when $N(x) \in \{1,\dots,n = \dim \mathfrak{M}\}$. For each $m \in \{1,\dots,n\}$, let $\phi_m \in C_0^\infty(B^m(\eta_1/2))$ satisfy $\phi_m \equiv 1$ on $B^m(\eta_1/4)$; note that we have only made $n = \dim \mathfrak{M} \approx 1$ such choices of ϕ_m. We will show that

$$\sup_{\substack{\{(E_j,2^{-j}):j\in\mathbb{N}\}\\ \subseteq\mathcal{E}}} \int_{B^{N(x)}(\eta_1/4)} |\phi_{N(x)}(u) Tf(\Phi_{x,1}(u))|^p\,du \le D_p \int_{B^{N(x)}(\eta_1)} |f(\Phi_{x,1}(u))|^p\,du, \tag{5.128}$$

and the result will follow.

Let $X_j^{x,1} := \Phi_{x,1}^* X_j$, as in Theorem 3.5.1. We will apply the results in Section 5.10.1 with $(Y, \vec{d}) = (X^{x,1}, d) = \{(X_1^{x,1}, d_1), \dots, (X_1^{x_1}, d_q)\}$ to obtain (5.128). Thus, we need to show that the assumptions in Section 5.10.1 hold uniformly for $x \in \mathcal{K}$. Pulling back (5.123) via $\Phi_{x,1}$, we have

$$[X_j^{x,1}, X_k^{x,1}] = \sum_{d_l \leq d_j + d_k} c_{j,k}^l \circ \Phi_{x,1} X_l.$$

By Theorem 3.5.1 (j), using the fact that $\Phi_{x,1}(B^{N(x)}(1)) \subseteq \Omega_1$ (see Theorem 3.5.1), we have, for every $L \in \mathbb{N}$,

$$\begin{aligned}\sup_{x \in \mathcal{K}} \|c_{j,k}^l \circ \Phi_{x,1}\|_{C^L(B^{N(x)}(1))} &\approx \sup_{x \in \mathcal{K}} \sum_{|\alpha| \leq L} \|(X^{x,1})^\alpha c_{j,k}^l \circ \Phi_{x,1}\|_{C(B^{N(x)}(1))} \\ &= \sup_{x \in \mathcal{K}} \sum_{|\alpha| \leq L} \|X^\alpha c_{j,k}^l\|_{C(\Phi_{x,1}(B^{N(x)}(1)))} \leq \sup_{x \in \mathcal{K}} \sum_{|\alpha| \leq L} \|X^\alpha c_{j,k}^l\|_{C(\Omega_1)} < \infty,\end{aligned}$$

where the $< \infty$ follows from the fact that $c_{j,k}^l \in C_{\mathrm{loc}}^\infty(\mathfrak{M})$ and $\Omega_1 \Subset \mathfrak{M}$. Combining this with Theorem 3.5.1 (g) and (h), we see that the vector fields $(Y, \vec{d}) := (X^{x,1}, d)$ satisfy the hypotheses of Section 5.10.1, uniformly for $x \in \mathcal{K}$.

With ς_j as in (5.125), for $x \in \mathcal{K}$, set

$$\varsigma_{j,x}(u, s, t) := \phi_{N(x)} \varsigma_j(\Phi_{x,1}(u), s, t). \tag{5.129}$$

Note that $\varsigma_{x,j} \in C_0^\infty(\Omega_3) \widehat{\otimes} C_0^\infty(B^{N(x)}(\eta_1/2)) \widehat{\otimes} \mathscr{S}(\mathbb{R}^q)$ and $\varsigma_{x,j} \in C_0^\infty(\Omega_3) \widehat{\otimes} C_0^\infty(B^{N(x)}(\eta_1/2)) \widehat{\otimes} \mathscr{S}_0(\mathbb{R}^q)$ for $j > 0$.

We claim, for all $L \in \mathbb{N}$,

$$\sum_{|\alpha|, |\beta|, |\gamma| \leq L} \sup_{\substack{x \in \mathcal{K} \\ N(x) \neq 0}} \sup_{\substack{u \in B^{N(x)}(1) \\ (E_j, 2^{-j}) \in \mathcal{E}}} \sup_{\substack{s \in B^q(a) \\ t \in \mathbb{R}^q}} (1 + |t|)^L |\partial_u^\alpha \partial_s^\beta \partial_t^\gamma \varsigma_{j,x}(u, s, t)| < \infty, \tag{5.130}$$

where $\varsigma_j = \varsigma_j(E_j, 2^{-j})$ is as in (5.125) and $\varsigma_{j,x} = \varsigma_{j,x}(\varsigma_j)$ is defined in (5.129). Indeed, by Theorem 3.5.1 (g) and (h), using the fact that $\Phi_{x,1}(B^{N(x)}(1)) \subseteq \Omega_1$ (see Theorem 3.5.1), we have

$$\begin{aligned}&\sum_{|\alpha|, |\beta|, |\gamma| \leq L} \sup_{\substack{x \in \mathcal{K} \\ N(x) \neq 0}} \sup_{\substack{u \in B^{N(x)}(1) \\ (E_j, 2^{-j}) \in \mathcal{E}}} \sup_{\substack{s \in B^q(a) \\ t \in \mathbb{R}^q}} (1 + |t|)^L |\partial_u^\alpha \partial_s^\beta \partial_t^\gamma \varsigma_{j,x}(u, s, t)| \\ &\lesssim \sum_{|\alpha|, |\beta|, |\gamma| \leq L} \sup_{\substack{x \in \mathcal{K} \\ N(x) \neq 0}} \sup_{\substack{u \in B^{N(x)}(1) \\ (E_j, 2^{-j}) \in \mathcal{E}}} \sup_{\substack{s \in B^q(a) \\ t \in \mathbb{R}^q}} (1 + |t|)^L |(X_u^{x,1})^\alpha \partial_s^\beta \partial_t^\gamma \varsigma_{j,x}(u, s, t)|\end{aligned}$$

$$= \sum_{|\alpha|,|\beta|,|\gamma|\leq L} \sup_{\substack{x\in\mathcal{K}\\ N(x)\neq 0}} \sup_{\substack{y\in\Phi_{x,1}(B^{N(x)}(1))\\ (E_j,2^{-j})\in\mathcal{E}}} \sup_{\substack{s\in B^q(a)\\ t\in\mathbb{R}^q}} (1+|t|)^L |X_y^\alpha \partial_s^\beta \partial_t^\gamma \varsigma_j(y,s,t)|$$

$$\leq \sum_{|\alpha|,|\beta|,|\gamma|\leq L} \sup_{\substack{y\in\Omega_1\\ (E_j,2^{-j})\in\mathcal{E}}} \sup_{\substack{s\in B^q(a)\\ t\in\mathbb{R}^q}} (1+|t|)^L |X_y^\alpha \partial_s^\beta \partial_t^\gamma \varsigma_j(y,s,t)| < \infty,$$

where the $< \infty$ follows from the fact that $\{\varsigma_j : (E_j, 2^{-j}) \in \mathcal{E}\} \subset C_0^\infty(\Omega_3)\widehat{\otimes}C_0^\infty(B^q(a))\widehat{\otimes}\mathscr{S}(\mathbb{R}^q)$ is a bounded set. This establishes (5.130).

For $(E_j, 2^{-j}) \in \mathcal{E}$, we have

$$\phi_{N(x)}(u)Ef(\Phi_{x,1}(u)) = \int f\circ\Phi_{x,1}(e^{-t\cdot X^{x,1}}x)\phi_{N(x)}(u)\,\mathrm{Dil}_{2^j}^d(\varsigma_{j,x})(u,t,t) =: E_{j,x}(f\circ\Phi_{x,1})(u),$$

where by (5.130), $\mathcal{E}_x := \{(E_x, 2^{-j}) : (E, 2^{-j}) \in \mathcal{E}\}$ is a bounded set of $(X^{x,1}, d)$, *a* pseudo-differential operator scales supported in $B^{N(x)}(\eta_1/2) \times B^{N(x)}(\eta_1)$, uniformly for $x \in \mathcal{K}$, provided $a > 0$ is sufficiently small as described in Section 5.10.1.

If $\{(E_{j,x}, 2^{-j}) : j \in \mathbb{N}\} \subseteq \mathcal{E}_x$, then by the above remarks, the results in Section 5.10.1 show that if $a > 0$ is sufficiently small (as described in that section, and therefore it can be chosen independent of $x \in \mathcal{K}$) and $T_x := \sum_{j\in\mathbb{N}} E_{j,x}$, then

$$\sup_{x\in\mathcal{K}} \sup_{\substack{\{(E_{j,x},2^{-j}):j\in\mathbb{N}\}\\ \subseteq\mathcal{E}}} \|T_x\|_{L^p(B^{N(x)},\sigma_{\mathrm{Leb}})\to L^p(B^{N(x)},\sigma_{\mathrm{Leb}})} < \infty. \tag{5.131}$$

Furthermore, since $\mathcal{E}_x$ is a bounded set of $(X^{x,1}, d)$, *a* pseudo-differential operator scales supported in $B^{N(x)}(\eta_1/2) \times B^{N(x)}(\eta_1)$, we have $\mathrm{supp}(T_x) \subseteq B^{N(x)}(\eta_1/2) \times B^{N(x)}(\eta_1)$. Combining this with (5.131), we have, for $f \in C_0^\infty(\Omega_1)$,

$$\int |\phi_{N(x)}(u)Tf(\phi_{x,1}(u))|^p\,du = \int |T_x(f\circ\Phi_{x,1})(u)|^p\,du \lesssim \int_{B^{N(x)}(1)} |f\circ\Phi_{x,1}(u)|^p\,du.$$

This establishes (5.128) and completes the proof. □

Proof of Proposition 5.10.4. In what follows, all implicit constants will be independent of the choice of $\{(E_j, 2^{-j}) : j \in \mathbb{N}\} \subseteq \mathcal{E}$ with $T = \sum_{j\in\mathbb{N}} E_j$.

Since T is an (X, d), *a* pseudo-differential operator supported in $\Omega_3 \times \Omega_4$, we have $\mathrm{supp}(T) \Subset \Omega_3 \times \Omega_4$. We will prove that, for $1 < p < \infty$,

$$\|Tf\|_{L^p(\Omega_3,\mathrm{Vol})} \lesssim \|f\|_{L^p(\mathfrak{M},\mathrm{Vol})}, \quad \forall f \in C_0^\infty(\Omega_4), \tag{5.132}$$

and the result will follow by the density of $C_0^\infty(\Omega_4)$ in $L^p(\Omega_4, \mathrm{Vol})$. We have, for $1 < p < \infty$,

$$\begin{aligned}
&\int_{\Omega_3} |Tf(x)|^p \, d\,\mathrm{Vol}(x) \\
&\quad \approx \int_{\Omega_5} \int_{B^{N(x)}(\eta_1/4)} |Tf(\Phi_{x,1}(u))|^p \, du \, d\,\mathrm{Vol}(x) && \text{Proposition 3.9.1} \\
&\quad \lesssim \int_{\Omega_5} \int_{B^{N(x)}(\eta_1)} |f \circ \Phi_{x,1}(u)| \, du \, d\,\mathrm{Vol}(x) && \text{Lemma 5.10.5} \\
&\quad \approx \int_{\Omega_4} |f(x)|^p \, d\,\mathrm{Vol}(x) && \text{Proposition 3.9.1.}
\end{aligned}$$

Formula (5.132) follows, completing the proof. □

A square function: We now use Proposition 5.10.4 to develop a Littlewood–Paley square function. Fix a compact set $\mathcal{K}_0 \Subset \Omega_3$. Fix $\psi \in C_0^\infty(\Omega_3)$ with $\psi \equiv 1$ on a neighborhood of $\mathcal{K}_0$. Let $a > 0$ be a small number; in particular small enough that Proposition 5.10.4 applies. Example 4.1.6 shows that $\mathrm{Mult}[\psi]$ is an (X, d), a pseudo-differential operator of order 0 supported in $\Omega_3 \times \Omega_4$, and therefore Theorem 4.3.3 shows that there is a bounded set of (X, d), a pseudo-differential operators supported in $\Omega_3 \times \Omega_4$, $\{(D_j, 2^{-j}) : j \in \mathbb{N}\}$, with $\mathrm{Mult}[\psi] = \sum_{j\in\mathbb{N}} D_j$.

Proposition 5.10.6. *Suppose $p \in (1, \infty)$. Then*

$$\left\| \left(\sum_{j\in\mathbb{N}} |D_j f|^2 \right)^{\frac{1}{2}} \right\|_{L^p(\mathfrak{M},\mathrm{Vol})} \leq C_p \|f\|_{L^p(\mathfrak{M},\mathrm{Vol})}, \quad \forall f \in L^p(\mathfrak{M}, \mathrm{Vol}). \tag{5.133}$$

Moreover, if $\mathrm{supp}(f) \subseteq \mathcal{K}_0$, we also have the reverse inequality,

$$\|f\|_{L^p(\mathfrak{M},\mathrm{Vol})} \leq C_p \left\| \left(\sum_{j\in\mathbb{N}} |D_j f|^2 \right)^{\frac{1}{2}} \right\|_{L^p(\mathfrak{M},\mathrm{Vol})}, \quad \forall f \in L^p(\mathcal{K}_0, \mathrm{Vol}). \tag{5.134}$$

Here, $C_p \geq 0$ does not depend on f, but may depend on any of the other ingredients in the proposition.

Before we prove Proposition 5.10.6, we present a corollary which is how we will use the proposition in our applications.

Corollary 5.10.7. *Let $\{\epsilon_j\}_{j\in\mathbb{N}}$ be a sequence of i. i. d. random variables of mean 0 taking values in ± 1 and fix $p \in (1, \infty)$. We have, for all $f \in L^p(\mathcal{K}_0, \mathrm{Vol})$,*

$$\left(\mathbb{E} \left\| \sum_{j\in\mathbb{N}} \epsilon_j D_j f \right\|^p_{L^p(\mathfrak{M},\mathrm{Vol})} \right)^{\frac{1}{p}} \approx \left\| \left(\sum_{j\in\mathbb{N}} |D_j f|^2 \right)^{\frac{1}{2}} \right\|_{L^p(\mathfrak{M},\mathrm{Vol})} \approx \|f\|_{L^p(\mathfrak{M},\mathrm{Vol})}.$$

Proof. It follows from the Khintchine inequality (Theorem 2.4.20) that

$$\left(\mathbb{E}\left\|\sum_{j\in\mathbb{N}}\epsilon_j D_j f\right\|^p_{L^p(\mathfrak{M},\mathrm{Vol})}\right)^{\frac{1}{p}} = \left\|\left(\mathbb{E}\left|\sum_{j\in\mathbb{N}}\epsilon_j D_j f\right|^p\right)^{\frac{1}{p}}\right\|_{L^p(\mathfrak{M},\mathrm{Vol})} \approx \left\|\left(\sum_{j\in\mathbb{N}}|D_j f|^2\right)^{\frac{1}{2}}\right\|_{L^p(\mathfrak{M},\mathrm{Vol})}.$$

It follows from Proposition 5.10.6 that

$$\left\|\left(\sum_{j\in\mathbb{N}}|D_j f|^2\right)^{\frac{1}{2}}\right\|_{L^p(\mathfrak{M},\mathrm{Vol})} \approx \|f\|_{L^p(\mathfrak{M},\mathrm{Vol})}. \qquad \square$$

We now turn to the proof of Proposition 5.10.6. A difficulty we must overcome is that D_j do not satisfy a nice version of the Calderón reproducing formula. However, we are able get around this by using the fact that D_j are "almost orthogonal."

For $j \in \mathbb{Z}\setminus\mathbb{N}$, set $D_j := 0$. For $M > 0$, set

$$U_M := \sum_{\substack{j\in\mathbb{N}\\ |l|\le M}} D_j D_{j+l}, \quad R_M := \sum_{\substack{j\in\mathbb{N}\\ |l|>M}} D_j D_{j+l}.$$

Note that $U_M + R_M = \sum_{j,k\in\mathbb{N}} D_j D_k = \mathrm{Mult}[\psi^2]$.

Proposition 5.10.8 (Calderón reproducing type formula). *Fix $p \in (1,\infty)$ and $\psi_1 \in C_0^\infty(\Omega_3)$ with $\psi \equiv 1$ on a neighborhood of* $\mathrm{supp}(\psi_1)$. *There exist $M = M(p)$ and $V_M : L^p(\mathfrak{M},\mathrm{Vol}) \to L^p(\mathfrak{M},\mathrm{Vol})$ such that* $\mathrm{Mult}[\psi_1]U_M V_M = V_M U_M \mathrm{Mult}[\psi_1] = \mathrm{Mult}[\psi_1]$.

To prove Proposition 5.10.8, we use the next lemma.

Lemma 5.10.9. *Fix $p \in (1,\infty)$. Then*

$$\lim_{M\to\infty} \|R_M\|_{L^p(\mathfrak{M},\mathrm{Vol})\to L^p(\mathfrak{M},\mathrm{Vol})} = 0.$$

Before we prove Lemma 5.10.9, we see how it implies Proposition 5.10.8.

Proof of Proposition 5.10.8. Recall that $U_M = \mathrm{Mult}[\psi^2] - R_M$; Lemma 5.10.9 shows that R_M is bounded on L^p, and therefore the same is true of U_M. Take $\psi_2 \in C_0^\infty(\Omega_3)$ with $\psi_1 \prec \psi_2 \prec \psi$. By Lemma 5.10.9, we may take $M = M(p)$ so large that $\|R_M \mathrm{Mult}[\psi_2]\|_{L^p\to L^p} < 1$. Define $V_M := \sum_{l=0}^\infty \mathrm{Mult}[\psi_2](R_M \mathrm{Mult}[\psi_2])^l = \sum_{l=0}^\infty (\mathrm{Mult}[\psi_2]R_M)^l \mathrm{Mult}[\psi_2]$, with convergence in the uniform operator topology on $\mathrm{Hom}(L^p, L^p)$. Using the fact that $\mathrm{Mult}[\psi_1]\mathrm{Mult}[\psi_2] = \mathrm{Mult}[\psi_1]$ and $\mathrm{Mult}[\psi_2]\mathrm{Mult}[\psi^2] = \mathrm{Mult}[\psi_2]$, we have

$$\begin{aligned}\mathrm{Mult}[\psi_1]U_M V_M &= \mathrm{Mult}[\psi_1](\mathrm{Mult}[\psi^2] - R_M)\left(\sum_{l=0}^\infty \mathrm{Mult}[\psi_2](R_M \mathrm{Mult}[\psi_2])^l\right)\\ &= \mathrm{Mult}[\psi_1]\left(\sum_{l=0}^\infty \mathrm{Mult}[\psi_2](R_M \mathrm{Mult}[\psi_2])^l - \sum_{l=0}^\infty R_M \mathrm{Mult}[\psi_2](R_M \mathrm{Mult}[\psi_2])^l\right)\end{aligned}$$

$$= \mathrm{Mult}[\psi_1]\left(\sum_{l=0}^{\infty}\mathrm{Mult}[\psi_2](R_M\,\mathrm{Mult}[\psi_2])^l - \sum_{l=1}^{\infty}\mathrm{Mult}[\psi_2](R_M\,\mathrm{Mult}[\psi_2])^l\right)$$
$$= \mathrm{Mult}[\psi_1]\,\mathrm{Mult}[\psi_2] = \mathrm{Mult}[\psi_1].$$

A similar proof gives $V_M U_M\,\mathrm{Mult}[\psi_1] = \mathrm{Mult}[\psi_1]$, which completes the proof. □

Lemma 5.10.9 follows by interpolating the next two lemmas.

Lemma 5.10.10. *For* $p \in (1,\infty)$, $M \geq 1$,

$$\|R_M\|_{L^p(\mathfrak{M},\mathrm{Vol})\to L^p(\mathfrak{M},\mathrm{Vol})} \leq C_p M.$$

Lemma 5.10.11. *For every* $N \geq 0$,

$$\|R_M\|_{L^2(\mathfrak{M},\mathrm{Vol})\to L^2(\mathfrak{M},\mathrm{Vol})} \leq C_M 2^{-NM}.$$

To prove Lemma 5.10.10, we require two more lemmas, the first of which implies (5.133).

Lemma 5.10.12. *Fix* $p \in (1,\infty)$. *Then*

$$\left\|\left(\sum_{j\in\mathbb{N}}|D_j f|^2\right)^{\frac{1}{2}}\right\|_{L^p(\mathfrak{M},\mathrm{Vol})} \leq C_p\|f\|_{L^p(\mathfrak{M},\mathrm{Vol})}, \quad \forall f \in L^p(\mathfrak{M},\mathrm{Vol}).$$

The same result holds with D_j *replaced by* D_j^* *throughout.*

Proof. In the proof that follows, we only use the fact that $\{(D_j, 2^{-j}) : j \in \mathbb{N}\}$ is a bounded set of (X, d), a pseudo-differential operator scales supported in $\Omega_3 \times \Omega_4$. By taking $a > 0$ small enough, Lemma 4.3.11 (ii) implies that $\{(D_j^*, 2^{-j}) : j \in \mathbb{N}\}$ is a bounded set of (X, d), a pseudo-differential operator scales supported in $\Omega_4 \times \Omega_5$. Thus, the same proof applies with D_j^* in place of D_j, so it suffices to consider only the result for D_j.

For any sequence $\{\epsilon_j\}_{j\in\mathbb{N}} \subseteq \{\pm1\}$, let $T = T[\{\epsilon_j\}_{j\in\mathbb{N}}] = \sum_{j\in\mathbb{N}}\epsilon_j D_j$. Since $\{(\epsilon_j D_j, 2^{-j}) : \epsilon_j \in \{\pm1\}, j \in \mathbb{N}\}$ is a bounded set of (X, d), a pseudo-differential operator scales supported in $\Omega_3 \times \Omega_4$, Proposition 5.10.4 implies that

$$\sup_{\{\epsilon_j\}_{j\in\mathbb{N}}\subseteq\{\pm1\}} \|T[\{\epsilon_j\}]\|_{L^p\to L^p} < \infty. \tag{5.135}$$

Now let $\{\epsilon_j\}_{j\in\mathbb{N}}$ be a sequence of i. i. d. random variables of mean 0 taking values ±1. Using the Khintchine inequality (Theorem 2.4.20) and (5.135), we have

$$\left\|\left(\sum_{j\in\mathbb{N}}|D_j f|^2\right)^{\frac{1}{2}}\right\|_{L^p} \approx \left\|\left(\mathbb{E}\left|\sum_{j\in\mathbb{N}}\epsilon_j D_j\right|^p\right)^{\frac{1}{p}}\right\|_{L^p} = \left(\mathbb{E}\left\|\sum_{j\in\mathbb{N}}\epsilon_j D_j f\right\|_{L^p}^p\right)^{\frac{1}{p}}$$
$$= \left(\mathbb{E}\|T[\{\epsilon_j\}]f\|_{L^p}^p\right)^{\frac{1}{p}} \lesssim \left(\mathbb{E}\|f\|_{L^p}^p\right)^{\frac{1}{p}} = \|f\|_{L^p},$$

completing the proof. □

Lemma 5.10.13. *For $p \in (1,\infty)$ and $M \geq 1$, the sum $U_M = \sum_{|l|\leq M}\sum_{j\in\mathbb{N}} D_j D_{j+l}$ converges in weak operator topology on* $\mathrm{Hom}(L^p, L^p)$ *and $\|U_M\|_{L^p\to L^p} \leq C_p M$.*

Proof. Let $q \in (1,\infty)$ be dual to p (i. e., $\frac{1}{p}+\frac{1}{q}=1$). We will show that for $f \in L^p$ and $g \in L^q$ with $\|g\|_{L^q}=1$, we have

$$\sum_{|l|\leq M}\sum_{j\in\mathbb{N}} |\langle g, D_j D_{j+l} f\rangle| \lesssim M\|f\|_{L^p},$$

which will complete the proof. Indeed, we have, using Lemma 5.10.12,

$$\begin{aligned}
\sum_{|l|\leq M}\sum_{j\in\mathbb{N}} |\langle g, D_j D_{j+l} f\rangle| &= \sum_{|l|\leq M}\sum_{j\in\mathbb{N}} |\langle D_j^* g, D_{j+l} f\rangle| \\
&\leq \sum_{|l|\leq M}\int \Big(\sum_{j\in\mathbb{N}} |D_j^* g|^2\Big)^{\frac{1}{2}} \Big(\sum_{j\in\mathbb{N}} |D_{j+l} f|^2\Big)^{\frac{1}{2}} \, d\,\mathrm{Vol} \\
&\leq \sum_{|l|\leq M}\int \Big(\sum_{j\in\mathbb{N}} |D_j^* g|^2\Big)^{\frac{1}{2}} \Big(\sum_{j\in\mathbb{N}} |D_j f|^2\Big)^{\frac{1}{2}} \, d\,\mathrm{Vol} \\
&\leq 2M \Big\|\Big(\sum_{j\in\mathbb{N}} |D_j^* g|^2\Big)^{\frac{1}{2}}\Big\|_{L^q} \Big\|\Big(\sum_{j\in\mathbb{N}} |D_j f|^2\Big)^{\frac{1}{2}}\Big\|_{L^p} \\
&\lesssim M\|g\|_{L^q}\|f\|_{L^p} = M\|f\|_{L^p},
\end{aligned}$$

as desired, completing the proof. □

Proof of Lemma 5.10.10. Since $R_M = \mathrm{Mult}[\psi^2] - U_M$, it suffices to prove the result with U_M in place of R_M, and this follows from Lemma 5.10.13. □

Proof of Lemma 5.10.11. Since $\{(D_j, 2^{-j}) : j \in \mathbb{N}\}$ is a bounded set of (X,d), *a* pseudo-differential operator scales supported in $\Omega_3\times\Omega_4$, if $a>0$ is small enough, Lemma 4.3.11 (ii) implies that $\{(D_j^*, 2^{-j}) : j \in \mathbb{N}\}$ is a bounded set of (X,d), *a* pseudo-differential operator scales supported in $\Omega_4\times\Omega_5$. Thus, if $a>0$ is small enough, Proposition 5.6.3 (with $(W,\vec{d}) = (X,d)$) implies that $\{(D_j, 2^{-j}), (D_j^*, 2^{-j}) : j \in \mathbb{N}\}$ is a bounded set of generalized (X,d) elementary operators supported in Ω_1. Thus, Proposition 5.5.11 implies that for every $N \in \mathbb{N}$,

$$\begin{gathered}
\{(2^{N|l|}D_j D_{j+l}, 2^{-j}), (2^{N|j_1-j_2|}D_{j_1}D_{j_2}^*, 2^{-j_1}), (2^{-N|l|}D_{j+l}^* D_j^*, 2^{-j}), \\
(2^{N|j_1-j_2|}D_{j_1}^* D_{j_2}, 2^{-j_1}) : j, j_1 \in \mathbb{N}, j_2, l \in \mathbb{Z}\}
\end{gathered}$$

is a bounded set of generalized (X,d) elementary operators supported in Ω_1. Thus, we have (see Definition 5.2.26), for every $N\in\mathbb{N}$, $j, j_1 \in \mathbb{N}$, and $j_2, l \in \mathbb{Z}$,

$$\begin{aligned}
\|D_j D_{j_1+l}\|_{L^2\to L^2} &\lesssim 2^{-N|l|}, \\
\|D_{j_1} D_{j_2}^*\|_{L^2\to L^2} &\lesssim 2^{-N|j_1-j_2|},
\end{aligned}$$

$$\|D_{j+l}^* D_j^*\|_{L^2\to L^2} \lesssim 2^{-N|l|},$$
$$\|D_{j_1}^* D_{j_2}\|_{L^2\to L^2} \lesssim 2^{-N|j_1-j_2|}.$$

It follows that for every $N \in \mathbb{N}$, there exists $C_N \geq 0$ such that for all $j_1, j_2 \in \mathbb{N}$ and $l_1, l_2 \in \mathbb{Z}$,

$$\|D_j D_{j_1+l_1} D_{j_2+l_2}^* D_{j_2}^*\|_{L^2\to L^2}, \|D_{j_1+l_1}^* D_{j_1}^* D_{j_2} D_{j_2+l_2}\|_{L^2\to L^2} \leq C_N 2^{-N \operatorname{diam}\{j_1, j_1+l_1, j_2+l_2, j_2\}},$$

where diam denotes the diameter of the set.

The Cotlar–Stein lemma (Lemma 2.3.25) applies to show that $\forall N \in \mathbb{N}$,

$$\|R_M\|_{L^2\to L^2} \lesssim \sup_{\substack{j_1\in\mathbb{N} \\ |l_1|>M}} \sum_{\substack{j_2\in\mathbb{N} \\ |l_2|>M}} 2^{-N \operatorname{diam}\{j_1, j_1+l_1, j_2+l_2, j_2\}/2} \lesssim 2^{-NM/2}.$$

Replacing N with $2N$ completes the proof. □

Proof of Proposition 5.10.6. Formula (5.133) follows from Lemma 5.10.12, so we prove only (5.134). Fix $\psi_1 \in C_0^\infty(\Omega_3)$, with $\psi_1 \equiv 1$ on a neighborhood of $\mathcal{K}_0$, and $\psi_1 \prec \psi$. Fix $p \in (1,\infty)$ and take $M = M(p)$ and V_M as in Proposition 5.10.8, so that $\operatorname{Mult}[\psi_1]U_M V_M = V_M U_M \operatorname{Mult}[\psi_1] = \operatorname{Mult}[\psi_1]$. Let $q \in (1,\infty)$ be dual to p (i. e., $\frac{1}{p} + \frac{1}{q} = 1$), so that $V_M^* : L^q \to L^q$ is bounded.

Let $g \in L^q(\mathfrak{M}, \operatorname{Vol})$ with $\|g\|_{L^q} = 1$. Let $f \in L^p(\mathcal{K}_0, \operatorname{Vol})$. We have

$$\begin{aligned}
&|\langle g, f\rangle| \\
&\quad = |\langle g, \operatorname{Mult}[\psi_1]f\rangle| && \text{using } \operatorname{Mult}[\psi_1]f = f \\
&\quad = |\langle g, V_M U_M \operatorname{Mult}[\psi_1]f\rangle| && \text{using } V_M U_M \operatorname{Mult}[\psi_1] = \operatorname{Mult}[\psi_1] \\
&\quad = |\langle V_M^* g, U_M f\rangle| && \text{using } \operatorname{Mult}[\psi_1]f = f \\
&\quad = \left| \sum_{|l|\leq M} \sum_{j\in\mathbb{N}} \langle V_M^* g, D_j D_{j+l} f\rangle \right| && \text{using Lemma 5.10.13} \\
&\quad \leq \sum_{|l|\leq M} \sum_{j\in\mathbb{N}} |\langle D_j^* V_M^* g, D_{j+l} f\rangle| \\
&\quad \leq \sum_{|l|\leq M} \int \sum_{j\in\mathbb{N}} |D_j^* V_M^* g||D_{j+l} f| \, d\operatorname{Vol} \\
&\quad \leq \sum_{|l|\leq M} \int \left(\sum_{j\in\mathbb{N}} |D_j^* V_M^* g|^2 \right)^{\frac{1}{2}} \left(\sum_{j\in\mathbb{N}} |D_{j+l} f|^2 \right)^{\frac{1}{2}} d\operatorname{Vol} && \text{using Cauchy–Schwartz} \\
&\quad \leq \sum_{|l|\leq M} \left\| \left(\sum_{j\in\mathbb{N}} |D_j^* V_M^* g|^2 \right)^{\frac{1}{2}} \right\|_{L^q} \left\| \left(\sum_{j\in\mathbb{N}} |D_j f|^2 \right)^{\frac{1}{2}} \right\|_{L^p} && \text{using Hölder's inequality}
\end{aligned}$$

$$\lesssim M\|V_M^* g\|_{L^q}\left\|\left(\sum_{j\in\mathbb{N}}|D_j f|^2\right)^{\frac{1}{2}}\right\|_{L^p} \qquad \text{using Lemma 5.10.12}$$

$$\lesssim \left\|\left(\sum_{j\in\mathbb{N}}|D_j f|^2\right)^{\frac{1}{2}}\right\|_{L^p} \qquad \text{using } V_M : L^q \to L^q, M \approx 1, \|g\|_{L^q} = 1.$$

Taking the supremum over all such g establishes (5.134) and completes the proof. □

5.11 Parametrices

The main reason the algebras $\widetilde{\mathscr{A}}^t(\Omega, (W, \vec{d}))$ and $\widetilde{\mathscr{A}}^t(\Omega, (W, \vec{d}))$ are useful to us is that parametrices for maximally subelliptic PDEs are locally in these algebras. In this section we present the main results we will use to deduce that the parametrices lie in these algebras.

As in the previous sections, we consider the Hörmander setting of Section 5.1.2 and the general setting of Section 5.1.3. The Schwartz kernels of operators in $\mathscr{A}^t(\Omega, (W, \vec{d}))$ and $\widetilde{\mathscr{A}}^t(\Omega, (W, \vec{d}))$ have compact support, which will usually not be the case for the parametrices we consider. To get around this issue we introduce the next definition; for this, recall that $\mathscr{A}^t(W, \vec{d}) = \bigcup_{\Omega\Subset\mathfrak{M}} \mathscr{A}^t(\Omega, (W, \vec{d}))$ and $\widetilde{\mathscr{A}}^t(W, \vec{d}) = \bigcup_{\Omega\Subset\mathfrak{M}} \widetilde{\mathscr{A}}^t(\Omega, (W, \vec{d}))$ (see Remark 5.8.8).

Definition 5.11.1. Let $t \in \mathbb{R}^\nu$.

Hörmander setting: We let $\mathscr{A}^t_{\mathrm{loc}}(W, \vec{d})$ be the set of all $T \in \mathrm{Hom}(C_0^\infty(\mathfrak{M}), C_0^\infty(\mathfrak{M})')$ such that $\forall \psi_1, \psi_2 \in C_0^\infty(\mathfrak{M})$, $\mathrm{Mult}[\psi_1] T\, \mathrm{Mult}[\psi_2] \in \mathscr{A}^t(W, \vec{d})$.

General setting: We let $\widetilde{\mathscr{A}}^t_{\mathrm{loc}}(W, \vec{d})$ be the set of all $T \in \mathrm{Hom}(C^\infty_{W,0}(\mathfrak{M}), C^\infty_{W,0}(\mathfrak{M})')$ such that $\forall \psi_1, \psi_2 \in C_0^\infty(\mathfrak{M})$, $\mathrm{Mult}[\psi_1] T\, \mathrm{Mult}[\psi_2] \in \widetilde{\mathscr{A}}^t(W, \vec{d})$.

Definition 5.11.2. Let $(Y, \vec{d})$, $(Z, \vec{d})$, and ν_1 be as in Section 5.9 and let $t \in \mathbb{R}^{\nu_1}$.

Hörmander setting: We let $\mathscr{D}^t_{\mathrm{loc}}((Y, \vec{d}), (Z, \vec{d}))$ be the set of all $T \in \mathrm{Hom}(C_0^\infty(\mathfrak{M}), C_0^\infty(\mathfrak{M})')$ such that $\forall \psi_1, \psi_2 \in C_0^\infty(\mathfrak{M})$, $\mathrm{Mult}[\psi_1] T\, \mathrm{Mult}[\psi_2] \in \mathscr{D}^t((Y, \vec{d}), (Z, \vec{d}))$ (see Definition 5.9.2).

General setting: We let $\widetilde{\mathscr{D}}^t_{\mathrm{loc}}((Y, \vec{d}), (Z, \vec{d}))$ be the set of all $T \in \mathrm{Hom}(C^\infty_{Y,0}(\mathfrak{M}), C^\infty_{Y,0}(\mathfrak{M})')$ such that $\forall \psi_1, \psi_2 \in C^\infty_{Y,0}(\mathfrak{M})$, $\mathrm{Mult}[\psi_1] T\, \mathrm{Mult}[\psi_2] \in \widetilde{\mathscr{D}}^t((Y, \vec{d}), (Z, \vec{d}))$ (see Definition 5.9.2).

Definition 5.11.3. Fix $t \in \mathbb{R}^\nu$ and $\mathbf{D}_1, \mathbf{D}_2 \in \mathbb{N}_+$.

Hörmander setting: We let $\mathscr{A}^t((W, \vec{d}); \mathbb{C}^{\mathbf{D}_1}, \mathbb{C}^{\mathbf{D}_2})$ be the set of all $T \in \mathrm{Hom}(C_0^\infty(\mathfrak{M}; \mathbb{C}^{\mathbf{D}_1}), C_0^\infty(\mathfrak{M}; \mathbb{C}^{\mathbf{D}_2})')$ which are $\mathbf{D}_2 \times \mathbf{D}_1$ matrices of operators in $\mathscr{A}^t(W, \vec{d})$. Similarly, we let $\mathscr{A}^t_{\mathrm{loc}}((W, \vec{d}); \mathbb{C}^{\mathbf{D}_1}, \mathbb{C}^{\mathbf{D}_2})$ be the set of all $T \in \mathrm{Hom}(C_0^\infty(\mathfrak{M}; \mathbb{C}^{\mathbf{D}_1}), C_0^\infty(\mathfrak{M}; \mathbb{C}^{\mathbf{D}_2})')$ which are $\mathbf{D}_2 \times \mathbf{D}_1$ matrices of operators in $\mathscr{A}^t_{\mathrm{loc}}(W, \vec{d})$. Also, for $\Omega \Subset \mathfrak{M}$ open and relatively compact, we let $\mathscr{A}^t(\Omega, (W, \vec{d}); \mathbb{C}^{\mathbf{D}_1}, \mathbb{C}^{\mathbf{D}_2})$ be the set of all $T \in \mathscr{A}^t((W, \vec{d}); \mathbb{C}^{\mathbf{D}_1}, \mathbb{C}^{\mathbf{D}_2})$ with $\operatorname{supp}(T) \subseteq \Omega \times \Omega$.

General setting: We let $\widetilde{\mathscr{A}}^t((W, \vec{d}); \mathbb{C}^{\mathbf{D}_1}, \mathbb{C}^{\mathbf{D}_2})$ be the set of all $T \in \operatorname{Hom}(C^\infty_{W,0}(\mathfrak{M}; \mathbb{C}^{\mathbf{D}_1}), C^\infty_{W,0}(\mathfrak{M}; \mathbb{C}^{\mathbf{D}_2})')$ which are $\mathbf{D}_2 \times \mathbf{D}_1$ matrices of operators in $\widetilde{\mathscr{A}}^t(W, \vec{d})$. Similarly, we let $\widetilde{\mathscr{A}}^t_{\mathrm{loc}}((W, \vec{d}); \mathbb{C}^{\mathbf{D}_1}, \mathbb{C}^{\mathbf{D}_2})$ be the set of all $T \in \operatorname{Hom}(C^\infty_{W,0}(\mathfrak{M}; \mathbb{C}^{\mathbf{D}_1}), C^\infty_{W,0}(\mathfrak{M}; \mathbb{C}^{\mathbf{D}_2})')$ which are $\mathbf{D}_2 \times \mathbf{D}_1$ matrices of operators in $\widetilde{\mathscr{A}}^t_{\mathrm{loc}}(W, \vec{d})$. Also, for $\Omega \Subset \mathfrak{M}$ open and relatively compact, we let $\widetilde{\mathscr{A}}^t(\Omega, (W, \vec{d}); \mathbb{C}^{\mathbf{D}_1}, \mathbb{C}^{\mathbf{D}_2})$ be the set of all $T \in \widetilde{\mathscr{A}}^t((W, \vec{d}); \mathbb{C}^{\mathbf{D}_1}, \mathbb{C}^{\mathbf{D}_2})$ with $\operatorname{supp}(T) \subseteq \Omega \times \Omega$.

We have a definition similar to Definition 5.11.3 for the algebras $\mathscr{D}$ and $\widetilde{\mathscr{D}}$.

Definition 5.11.4. Let $(Y, \vec{d})$, $(Z, \vec{d})$, and ν_1 be as in Section 5.9, let $t \in \mathbb{R}^{\nu_1}$, and let $\mathbf{D}_1, \mathbf{D}_2 \in \mathbb{N}_+$.

Hörmander setting: We let $\mathscr{D}^t((Y, \vec{d}), (Z, \vec{d}); \mathbb{C}^{\mathbf{D}_1}, \mathbb{C}^{\mathbf{D}_2})$ be the set of all

$$T \in \operatorname{Hom}(C^\infty_0(\mathfrak{M}; \mathbb{C}^{\mathbf{D}_1}), C^\infty_0(\mathfrak{M}; \mathbb{C}^{\mathbf{D}_2})')$$

which are $\mathbf{D}_2 \times \mathbf{D}_1$ matrices of operators in $\mathscr{D}^t((Y, \vec{d}), (Z, \vec{d}))$. Similarly, we let $\mathscr{D}^t_{\mathrm{loc}}((Y, \vec{d}), (Z, \vec{d}); \mathbb{C}^{\mathbf{D}_1}, \mathbb{C}^{\mathbf{D}_2})$ be the set of all $T \in \operatorname{Hom}(C^\infty_0(\mathfrak{M}; \mathbb{C}^{\mathbf{D}_1}), C^\infty_0(\mathfrak{M}; \mathbb{C}^{\mathbf{D}_2})')$ which are $\mathbf{D}_2 \times \mathbf{D}_1$ matrices of operators in $\mathscr{D}^t_{\mathrm{loc}}((Y, \vec{d}), (Z, \vec{d}))$. Also, for $\Omega \Subset \mathfrak{M}$ open and relatively compact, we let $\mathscr{D}^t(\Omega, (Y, \vec{d}), (Z, \vec{d}); \mathbb{C}^{\mathbf{D}_1}, \mathbb{C}^{\mathbf{D}_2})$ be the set of all $T \in \mathscr{D}^t((Y, \vec{d}), (Z, \vec{d}); \mathbb{C}^{\mathbf{D}_1}, \mathbb{C}^{\mathbf{D}_2})$ with $\operatorname{supp}(T) \subseteq \Omega \times \Omega$.

General setting: We let $\widetilde{\mathscr{D}}^t((Y, \vec{d}), (Z, \vec{d}); \mathbb{C}^{\mathbf{D}_1}, \mathbb{C}^{\mathbf{D}_2})$ be the set of all

$$T \in \operatorname{Hom}(C^\infty_{Y,0}(\mathfrak{M}; \mathbb{C}^{\mathbf{D}_1}), C^\infty_{Y,0}(\mathfrak{M}; \mathbb{C}^{\mathbf{D}_2})')$$

which are $\mathbf{D}_2 \times \mathbf{D}_1$ matrices of operators in $\widetilde{\mathscr{D}}^t((Y, \vec{d}), (Z, \vec{d}))$. Similarly, we let $\widetilde{\mathscr{D}}^t_{\mathrm{loc}}((Y, \vec{d}), (Z, \vec{d}); \mathbb{C}^{\mathbf{D}_1}, \mathbb{C}^{\mathbf{D}_2})$ be the set of all $T \in \operatorname{Hom}(C^\infty_{Y,0}(\mathfrak{M}; \mathbb{C}^{\mathbf{D}_1}), C^\infty_{Y,0}(\mathfrak{M}; \mathbb{C}^{\mathbf{D}_2})')$ which are $\mathbf{D}_2 \times \mathbf{D}_1$ matrices of operators in $\widetilde{\mathscr{D}}^t_{\mathrm{loc}}((Y, \vec{d}), (Z, \vec{d}))$. Also, for $\Omega \Subset \mathfrak{M}$ open and relatively compact, we let $\widetilde{\mathscr{D}}^t(\Omega, (Y, \vec{d}), (Z, \vec{d}); \mathbb{C}^{\mathbf{D}_1}, \mathbb{C}^{\mathbf{D}_2})$ be the set of all $T \in \widetilde{\mathscr{D}}^t((Y, \vec{d}), (Z, \vec{d}); \mathbb{C}^{\mathbf{D}_1}, \mathbb{C}^{\mathbf{D}_2})$ with $\operatorname{supp}(T) \subseteq \Omega \times \Omega$.

5.11.1 Parametrices via heat equations

The main way we will construct parametrices for general maximally subelliptic PDEs is via heat equations, and in this section, we describe how this works. The setting is the single-parameter setting of Section 5.1.1; thus, we are given Hörmander vector fields with formal degrees on $\mathfrak{M}$: $(W, \vec{d}) = \{(W_1, d_1), \ldots, (W_r, d_r)\} \subset C^\infty_{\mathrm{loc}}(\mathfrak{M}; T\mathfrak{M}) \times \mathbb{N}_+$.

Fix $\kappa, D \in \mathbb{N}_+$ and let $\mathscr{L} := \sum_{\deg_{\vec{d}}(\alpha) \leq 2\kappa} a_\alpha(x) W^\alpha$, where $a_\alpha \in C^\infty_{\mathrm{loc}}(\mathfrak{M}; \mathbb{M}^{D\times D}(\mathbb{C}))$. We think of $\mathscr{L}$ as a densely defined operator on $L^2(\mathfrak{M}, \mathrm{Vol}; \mathbb{C}^D)$ with dense domain $C^\infty_0(\mathfrak{M}; \mathbb{C}^D)$. We assume $\mathscr{L}$ is non-negative and symmetric, and we let $\overline{\mathscr{L}}$ be a non-negative, self-adjoint extension of $\mathscr{L}$. The goal of this section is to give conditions on the heat operator $e^{-t\overline{\mathscr{L}}}$ which ensure $\mathscr{L}$ has a two-sided parametrix in $\mathscr{A}_{\mathrm{loc}}^{-2\kappa}((W,\vec{d});\mathbb{C}^D,\mathbb{C}^D)$; because $\mathscr{L}\in\mathscr{A}_{\mathrm{loc}}^{2\kappa}((W,\vec{d});\mathbb{C}^D,\mathbb{C}^D)$ (see Propositions 5.8.3 and 5.8.2), this is the best possible situation.

Definition 5.11.5. We say $\mathcal{E}\subset C_{\mathrm{loc}}^\infty(\mathfrak{M}\times\mathfrak{M})\times(0,1]$ is a bounded set of locally $(W,\vec{d})$ pre-elementary operators if for every relatively compact, open set $\Omega\Subset\mathfrak{M}$, for all ordered multi-indices α and β, and for all $m\in\mathbb{N}$, there exists $C=C(\mathcal{E},\Omega,\alpha,\beta,m)\geq 0$ such that, $\forall(F,2^{-j})\in\mathcal{E}$, $\forall x,z\in\Omega$,

$$|(2^{-j\vec{d}}W_x)^\alpha(2^{-j\vec{d}}W_z)^\beta F(x,z)|\leq C\frac{(1+2^j\rho_{(W,\vec{d})}(x,z))^{-m}}{\mathrm{Vol}(B_{(W,\vec{d})}(x,2^{-j}+\rho_{(W,\vec{d})}(x,z)))\wedge 1}.$$

Remark 5.11.6. We define bounded sets of $(W,\vec{d})$ pre-elementary operators supported in Ω, bounded sets of $(W,\vec{d})$ elementary operators supported in Ω, and bounded sets of locally $(W,\vec{d})$ pre-elementary operators taking values in $\mathbb{M}^{D\times D}$ (instead of $\mathbb{C}$) in the obvious way. In this section, all such operators take values in $\mathbb{M}^{D\times D}$, and for $A\in\mathbb{M}^{D\times D}$ we write $|A|$ for the usual operator norm as an operator on $\mathbb{C}^D$.

The main result of this section is the next theorem.

Theorem 5.11.7. *Suppose*

$$\{(e^{-t\overline{\mathscr{L}}},t^{1/2\kappa}):t\in(0,1]\}$$

is a bounded set of locally $(W,\vec{d})$ pre-elementary operators. Then there exists $S\in\mathscr{A}_{\mathrm{loc}}^{-2\kappa}((W,\vec{d});\mathbb{C}^D,\mathbb{C}^D)$ such that

$$\mathscr{L}S,S\mathscr{L}\equiv I\mod C_{\mathrm{loc}}^\infty(\mathfrak{M}\times\mathfrak{M};\mathbb{M}^{D\times D}(\mathbb{C})).$$

There are many constructions of parametrices in the literature where the conclusions are similar to Theorem 5.11.7; see for example [200, 182, 140]. However, these references all assume that $2\kappa<Q_1$, where Q_1 is as in Corollary 3.3.9. This allows those references to use a definition like $\mathscr{A}_4^{-2\kappa}$ (Definition 5.2.4) with $M=0$. Here, we use our more general definitions to construct a parametrix no matter how large κ is. To prove Theorem 5.11.7 we require several preliminary results. Throughout this section, we let $\Omega\Subset\mathfrak{M}$ be a relatively compact, open set.

Lemma 5.11.8. *Let $\mathcal{E}_1$ be a bounded set of $(W,\vec{d})$ pre-elementary operators supported in Ω and let $\mathcal{E}_2$ be a bounded set of locally $(W,\vec{d})$ pre-elementary operators. Let $\psi\in C_{\mathrm{loc}}^\infty(\mathfrak{M})$ and let α be an ordered multi-index. Then:*

(i) $\{(\mathrm{Mult}[\psi](2^{-j\vec{d}}W)^\alpha F,2^{-j}),(F\,\mathrm{Mult}[\psi](2^{-j\vec{d}}W)^\alpha,2^{-j}):(F,2^{-j})\in\mathcal{E}_1\}$ *is a bounded set of $(W,\vec{d})$ pre-elementary operators supported in Ω.*

(ii) $\{(\mathrm{Mult}[\psi](2^{-j\vec{d}}W)^\alpha F,2^{-j}),(F\,\mathrm{Mult}[\psi](2^{-j\vec{d}}W)^\alpha,2^{-j}):(F,2^{-j})\in\mathcal{E}_2\}$ *is a bounded set of locally $(W,\vec{d})$ pre-elementary operators.*

Proof. This follows immediately from the definitions. □

Lemma 5.11.9. *For all ordered multi-indices α and β and all $m \in \mathbb{N}$, there exists $N \in \mathbb{N}$ such that the following holds. Let $\mathcal{E}_1$ be a bounded set of locally $(W, \vec{d})$ pre-elementary operators and let $\mathcal{E}_2$ be a bounded set of $(W, \vec{d})$ pre-elementary operators supported in Ω. Then there exists $C = C(\mathcal{E}_1, \mathcal{E}_2, m, \alpha, \beta, \Omega, (W, \vec{d})) \geq 0$ such that for all $x \in \Omega$, $z \in \mathfrak{M}$, $(F_1, 2^{-j_1}) \in \mathcal{E}_1$, $(F_2, 2^{-j_2}) \in \mathcal{E}_2$, $k \in \{j_1, j_2\}$, we have*

$$2^{-N|j_1-j_2|}\left|(2^{-k\vec{d}}W_x)^{\alpha}(2^{-k\vec{d}}W_z)^{\beta}[F_1F_2](x,z)\right| \leq C\frac{(1+2^k\rho_{(W,\vec{d})}(x,z))^{-m}}{\mathrm{Vol}(B_{(W,\vec{d})}(x,2^{-k}+\rho_{(W,\vec{d})}(x,z)))\wedge 1}. \tag{5.136}$$

Proof. Fix $\Omega_1 \Subset \mathfrak{M}$ with $\Omega \Subset \Omega_1$ and let $\psi \in C_0^\infty(\Omega_1)$ satisfy $\psi \equiv 1$ on Ω. Note that for $(F_1, 2^{-j_1}) \in \mathcal{E}_1$, $(F_2, 2^{-j_2}) \in \mathcal{E}_2$,

$$[F_1F_2](x,z) = [\widetilde{F}_1F_2](x,z), \quad x \in \Omega, z \in \mathfrak{M}, \tag{5.137}$$

where $\widetilde{F}_1 = \mathrm{Mult}[\psi]F_1\,\mathrm{Mult}[\psi]$. Also, directly from the definitions,

$$\{(\widetilde{F}_1, 2^{-j_1}) : (F_1, 2^{-j_1}) \in \mathcal{E}_1\}\bigcup \mathcal{E}_2$$

is a bounded set of $(W, \vec{d})$ pre-elementary operators supported in Ω_1. Lemma 5.5.12 implies (5.136) with F_1 replaced by $\widetilde{F}_1$ (here, we are using Remark 5.2.19), and the result follows by (5.137). □

Lemma 5.11.10. *Let $\mathcal{E}_1$ be a bounded set of locally $(W, \vec{d})$ pre-elementary operators, let $\mathcal{E}_2$ be a bounded set of $(W, \vec{d})$ elementary operators supported in Ω, and let $\psi \in C_0^\infty(\Omega)$. Then, for all $N \in \mathbb{N}$, the set*

$$\{(2^{Nj}\,\mathrm{Mult}[\psi]FE, 2^{-j}) : (F, 1) \in \mathcal{E}_1, (E, 2^{-j}) \in \mathcal{E}_2\}$$

is a bounded set of $(W, \vec{d})$ elementary operators supported in Ω.

Proof. Since $\bigcup_{(E,2^{-j})\in\mathcal{E}_2} \mathrm{supp}(E) \Subset \Omega$, we may pick $\psi_1 \in C_0^\infty(\Omega)$ which is equal to 1 on a large enough set that $\mathrm{Mult}[\psi_1]E = E$, $\forall (E, 2^{-j}) \in \mathcal{E}_2$. It follows immediately from the definitions that

$$\{(\mathrm{Mult}[\psi]F\,\mathrm{Mult}[\psi_1], 1) : (F, 1) \in \mathcal{E}_1\} \tag{5.138}$$

is a bounded set of $(W, \vec{d})$ pre-elementary operators supported in Ω. Since we are only considering those $(F, 2^{-j})$ with $j = 0$, it is in fact a bounded set of $(W, \vec{d})$ elementary operators supported in Ω: in (5.2), when $j = 0$, one may take $E_{0,0} = E$ and $E_{\alpha,\beta} = 0$ for $|\alpha|, |\beta| > 0$.

Because (5.138) is a bounded set of (W, d) elementary operators supported in Ω, Proposition 5.5.11 shows that for every $N \in \mathbb{N}$,

$$\begin{aligned}&\{(2^{Nj}\operatorname{Mult}[\psi]FE, 2^{-j}) : (F,1) \in \mathcal{E}_1, (E, 2^{-j}) \in \mathcal{E}_2\}\\&\quad = \{(2^{Nj}\operatorname{Mult}[\psi]F\operatorname{Mult}[\psi_1]E, 2^{-j}) : (F,1) \in \mathcal{E}_1, (E, 2^{-j}) \in \mathcal{E}_2\}\end{aligned}$$

is a bounded set of (W, d) elementary operators supported in Ω, completing the proof. □

Lemma 5.11.11. *For all ordered multi-indices α and β and all $m \in \mathbb{N}$, there exists $M = M(\alpha, \beta, m, (W, d), \Omega) \in \mathbb{N}$ such that the following holds. Let $\mathcal{E}$ be a bounded set of (W, d) pre-elementary operators supported in Ω. Let $F_t(x,z) = F(t,x,z) : (0,1] \times \mathfrak{M} \times \mathfrak{M} \to \mathbb{M}^{D\times D}$ be a measurable function such that $\{(F_t, t^{1/2\kappa}) : t \in (0,1]\}$ is a bounded set of locally (W, d) pre-elementary operators. Then there exists $C \geq 0$ such that, $\forall x \in \Omega$, $\forall z \in \mathfrak{M}$, $\forall (G, 2^{-j}) \in \mathcal{E}$, $\forall M_1 \geq M$,*

$$\begin{aligned}&\left|(2^{-jd}W_x)^\alpha (2^{-jd}W_z)^\beta \int_0^1 \left(\frac{t}{2^{-2\kappa j}} \wedge \frac{2^{-2\kappa j}}{t}\right)^{M_1} [F_t G](x,z)\right|\\&\quad \leq C 2^{-2\kappa j} \frac{(1 + 2^j \rho_{(W,d)}(x,z))^{-m}}{\operatorname{Vol}(B_{(W,d)}(x, 2^{-j} + \rho_{(W,d)}(x,z))) \wedge 1}.\end{aligned} \tag{5.139}$$

Proof. We may replace M_1 with M as the left-hand side of (5.139) is decreasing in M_1. By taking M sufficiently large and applying Lemma 5.11.9 (with $M - 2$ playing the role of N in that lemma), we have, for $x \in \Omega$ and $z \in \mathfrak{M}$,

$$\begin{aligned}&\left|(2^{-jd}W_x)^\alpha (2^{-jd}W_z)^\beta \int_0^1 \left(\frac{t}{2^{-2\kappa j}} \wedge \frac{2^{-2\kappa j}}{t}\right)^{M} [F_t G](x,z)\right|\\&\quad \lesssim \frac{(1 + 2^j \rho_{(W,d)}(x,z))^{-m}}{\operatorname{Vol}(B_{(W,d)}(x, 2^{-j} + \rho_{(W,d)}(x,z))) \wedge 1} \int_0^1 \left(\frac{t}{2^{-2\kappa j}} \wedge \frac{2^{-2\kappa j}}{t}\right)^2 dt.\end{aligned}$$

But,

$$\int_0^1 \left(\frac{t}{2^{-2\kappa j}} \wedge \frac{2^{-2\kappa j}}{t}\right)^2 dt = \int_0^{2^{-2\kappa j}} \frac{t^2}{2^{-4\kappa j}}\, dt + \int_0^1 \frac{2^{-4\kappa j}}{t^2}\, dt \lesssim 2^{-2\kappa j}.$$

This completes the proof. □

For the remainder of this section, we assume the hypotheses of Theorem 5.11.7.

Lemma 5.11.12. *For all ordered multi-indices α and β and all $m \in \mathbb{N}$, there exists $M = M(\alpha, \beta, m, (W, d), \Omega) \in \mathbb{N}$ such that the following holds. Let $\mathcal{E}$ be a bounded set of (W, d)*

pre-elementary operators supported in Ω. Then, $\forall M_1 \geq M$, $\forall M_2 \geq 4\kappa M_1$, $\forall |\gamma| \leq M_2$, γ *an ordered multi-index, there exists* $C \geq 0$ *such that* $\forall (F, 2^{-j}) \in \mathcal{E}$, $\forall x \in \Omega$, $\forall z \in \mathfrak{M}$,

$$\left| (2^{-j\vec{d}} W_x)^\alpha (2^{-j\vec{d}} W_z)^\beta \int_0^1 \left(\frac{t}{2^{-2\kappa j}}\right)^{M_1} [e^{-t\overline{\mathcal{L}}} (2^{-j\vec{d}} W)^\gamma 2^{-j(M_2-|\gamma|)} F](x,z)\, dt \right|$$
$$\leq C 2^{-2\kappa j} \frac{(1 + 2^j \rho_{(W,\vec{d})}(x,z))^{-m}}{\mathrm{Vol}(B_{(W,\vec{d})}(x, 2^{-j} + \rho_{(W,\vec{d})}(x,z))) \wedge 1}.$$

Proof. We separate the integral into $\int_0^1 = \int_0^{2^{-2\kappa j}} + \int_{2^{-2\kappa j}}^1$. We begin with the first integral. It follows from Lemma 5.11.8 (i) that (for any M_2 and $|\gamma| \leq M_2$),

$$\{((2^{-j\vec{d}} W)^\gamma 2^{-j(M_2-|\gamma|)} F, 2^{-j}) : (F, 2^{-j}) \in \mathcal{E}\}$$

is a bounded set of $(W, \vec{d})$ pre-elementary operators supported in Ω. Applying Lemma 5.11.11 with G replaced by $(2^{-j\vec{d}} W)^\gamma 2^{-j(M_2-|\gamma|)} F$ and F_t replaced by $e^{-t\overline{\mathcal{L}}}$, it follows that for every $m \in \mathbb{N}$, if M is sufficiently large, we have, with γ, M_1, M_2, and F as in the statement of the lemma,

$$\left| (2^{-j\vec{d}} W_x)^\alpha (2^{-j\vec{d}} W_z)^\beta \int_0^{2^{-2\kappa j}} \left(\frac{t}{2^{-2\kappa j}}\right)^{M_1} [e^{-t\overline{\mathcal{L}}} (2^{-j\vec{d}} W)^\gamma 2^{-j(M_2-|\gamma|)} F](x,z)\, dt \right|$$
$$= \left| (2^{-j\vec{d}} W_x)^\alpha (2^{-j\vec{d}} W_z)^\beta \int_0^{2^{-2\kappa j}} \left(\frac{t}{2^{-2\kappa j}} \wedge \frac{2^{-2\kappa j}}{t}\right)^{M_1} [e^{-t\overline{\mathcal{L}}} (2^{-j\vec{d}} W)^\gamma 2^{-j(M_2-|\gamma|)} F](x,z)\, dt \right|$$
$$\lesssim 2^{-2\kappa j} \frac{(1 + 2^j \rho_{(W,\vec{d})}(x,z))^{-m}}{\mathrm{Vol}(B_{(W,\vec{d})}(x, 2^{-j} + \rho_{(W,\vec{d})}(x,z))) \wedge 1},$$

as desired.

We now turn to the term $\int_{2^{-2\kappa j}}^1$. With $M_1 \geq M$ and $M_2 \geq 4\kappa M_1$, we have, for some measurable functions $B_{1,j}(t)$, $B_{2,j}(t)$ with $|B_{1,j}(t)|, |B_{2,j}(t)| \leq 1$, $\forall t$,

$$\left| (2^{-j\vec{d}} W_x)^\alpha (2^{-j\vec{d}} W_z)^\beta \int_{2^{-2\kappa j}}^1 \left(\frac{t}{2^{-2\kappa j}}\right)^{M_1} [e^{-t\overline{\mathcal{L}}} (2^{-j\vec{d}} W)^\gamma 2^{-j(M_2-|\gamma|)} F](x,z)\, dt \right|$$
$$= \left| (2^{-j\vec{d}} W_x)^\alpha (2^{-j\vec{d}} W_z)^\beta \int_{2^{-2\kappa j}}^1 \left(\frac{t}{2^{-2\kappa j}}\right)^{M_1} 2^{-j(\deg_{\vec{d}}(\gamma)+M_2-|\gamma|)} t^{-\deg_{\vec{d}}(\gamma)/2\kappa} \right.$$
$$\left. \times [2^{-t\overline{\mathcal{L}}} (t^{\vec{d}/2\kappa} W)^\gamma F](x,z)\, dt \right|$$

$$= \left| (2^{-j\vec{d}} W_x)^{\alpha} (2^{-j\vec{d}} W_z)^{\beta} \int_{2^{-2\kappa j}}^{1} \left(\frac{t}{2^{-2\kappa j}}\right)^{M_1} 2^{-jM_2} t^{-M_2/2\kappa} B_{1,j}(t) [2^{-t\overline{\mathscr{L}}} (t^{\vec{d}/2\kappa} W)^{\gamma} F](x,z)\, dt \right|$$

$$= \left| (2^{-j\vec{d}} W_x)^{\alpha} (2^{-j\vec{d}} W_z)^{\beta} \int_{2^{-2\kappa j}}^{1} \left(\frac{t}{2^{-2\kappa j}} \wedge \frac{2^{-2\kappa j}}{t}\right)^{M_1} B_{2,j}(t) [2^{-t\overline{\mathscr{L}}} (t^{\vec{d}/2\kappa} W)^{\gamma} F](x,z)\, dt \right|.$$

It follows from the hypotheses and Lemma 5.11.8 (ii) that

$$\{(B_{2,j}(t) e^{-t\overline{\mathscr{L}}} (t^{\vec{d}/2\kappa} W)^{\gamma}, t^{1/2\kappa}) : t \in (0,1], j \in [0,\infty)\}$$

is a bounded set of locally $(W, \vec{d})$ pre-elementary operators. From here, the desired estimate follows from Lemma 5.11.11, completing the proof. □

Lemma 5.11.13. *For $f \in C_0^\infty(\mathfrak{M}; \mathbb{C}^D)$, $\frac{d}{dt} e^{-t\overline{\mathscr{L}}} f = -\mathscr{L} e^{-t\overline{\mathscr{L}}} f = -e^{-t\overline{\mathscr{L}}} \mathscr{L} f$, where $\mathscr{L}$ is taken in the sense of distributions.*

Proof. Because $\overline{\mathscr{L}}$ agrees with $\mathscr{L}$ on $C_0^\infty(\mathfrak{M}; \mathbb{C}^D)$, we have $\frac{d}{dt} e^{-t\overline{\mathscr{L}}} f = -e^{-t\overline{\mathscr{L}}} \overline{\mathscr{L}} f = -e^{-t\overline{\mathscr{L}}} \mathscr{L} f$. Also, on the domain of $\overline{\mathscr{L}}$, $\overline{\mathscr{L}}$ agrees with $\mathscr{L}$ in the sense of distributions, so $\frac{d}{dt} e^{-t\overline{\mathscr{L}}} f = -\overline{\mathscr{L}} e^{-t\overline{\mathscr{L}}} f = -\mathscr{L} e^{-t\overline{\mathscr{L}}} f$. □

Lemma 5.11.14. *For all $\psi \in C_0^\infty(\Omega)$, for every $\mathcal{E}$ a bounded set of $(W, \vec{d})$ elementary operators supported in Ω, $\forall a \in \mathbb{N}$,*

$$\left\{ \left(2^{2\kappa j} \operatorname{Mult}[\psi] \int_0^1 \left(\frac{t}{2^{-2\kappa j}}\right)^{a} e^{-t\overline{\mathscr{L}}} E\, dt, 2^{-j} \right) : (E, 2^{-j}) \in \mathcal{E} \right\}$$

is a bounded set of $(W, \vec{d})$ pre-elementary operators supported in Ω.

Proof. It is clear that

$$\bigcup_{(E,2^{-j})\in\mathcal{E}} \operatorname{supp}\left(\operatorname{Mult}[\psi] \int_0^1 \left(\frac{t}{2^{-2\kappa j}}\right)^{a} e^{-t\overline{\mathscr{L}}} E\, dt \right) \Subset \Omega \times \Omega.$$

Fix ordered multi-indices α and β and $m \in \mathbb{N}$. The result will follow once we show that there exists $C \geq 0$ such that $\forall (E, 2^{-j}) \in \mathcal{E}$, $\forall x, z \in \Omega$,

$$\begin{aligned} &\left| (2^{-j\vec{d}} W_x)^{\alpha} (2^{-j\vec{d}} W_z)^{\beta} \int_0^1 \left(\frac{t}{2^{-2\kappa j}}\right)^{a} [e^{-t\overline{\mathscr{L}}} E](x,z)\, dt \right| \\ &\quad \leq C 2^{-2\kappa j} \frac{(1 + 2^j \rho_{(W,\vec{d})}(x,z))^{-m}}{\operatorname{Vol}(B_{(W,\vec{d})}(x, 2^{-j} + \rho_{(W,\vec{d})}(x,z))) \wedge 1}. \end{aligned} \tag{5.140}$$

First we claim that if (5.140) holds with a replaced by $a+1$ (for every $\mathcal{E}$ with C depending on $\mathcal{E}$), then it holds for a as well. Indeed, for $(E, 2^{-j}) \in \mathcal{E}$, Lemma 5.11.13 shows that $\frac{d}{dt} e^{-t\overline{\mathscr{L}}} E = -e^{-t\overline{\mathscr{L}}} \mathscr{L} E$. Thus, we have

$$\begin{aligned}\int_0^1 \left(\frac{t}{2^{-2\kappa j}}\right)^a e^{-t\overline{\mathscr{L}}} E\,dt &= \frac{2^{-2\kappa j}}{a+1} \int_0^1 \left(\frac{d}{dt}\left(\frac{t}{2^{-2\kappa j}}\right)^{a+1}\right) e^{-t\overline{\mathscr{L}}} E\,dt \\ &= \frac{2^{2\kappa j a}}{a+1} e^{-\overline{\mathscr{L}}} E + \frac{1}{a+1} \int_0^1 \left(\frac{t}{2^{-2\kappa j}}\right)^{a+1} e^{-t\overline{\mathscr{L}}} 2^{-2\kappa j} \mathscr{L} E\,dt.\end{aligned} \tag{5.141}$$

It follows from Lemma 5.11.10 that $\frac{2^{2\kappa j a}}{a+1} e^{-\overline{\mathscr{L}}} E$ satisfies the desired estimates in (5.140). For the second term on the right-hand side of (5.141), note that Proposition 5.5.5 (a), (c), and (e) implies that $\{(2^{-2\kappa j} \mathscr{L} E, 2^{-j}) : (E, 2^{-j}) \in \mathcal{E}\}$ is a bounded set of $(W, \vec{d})$ elementary operators supported in Ω. Thus, the second term on the right-hand side of (5.141) is of the same form as (5.140), but with a replaced by $a+1$.

The above discussion shows that it suffices to prove (5.140) for a sufficiently large (depending on α, β, m, $(W, \vec{d})$, and Ω). Let $M = M(\alpha, \beta, m, (W, \vec{d}), \Omega) \in \mathbb{N}$ be as in Lemma 5.11.12 and take $a \geq M$. We take $M_2 \geq 4\kappa a$. Applying Proposition 5.5.5 (h) with $N = M_2$, we see that

$$\int_0^1 \left(\frac{t}{2^{-2\kappa j}}\right)^a e^{-t\overline{\mathscr{L}}} E\,dt = \sum_{|\gamma| \leq M_2} \int_0^1 \left(\frac{t}{2^{-2\kappa j}}\right)^a e^{-t\overline{\mathscr{L}}} 2^{-j(M_2 - |\gamma|)} (2^{-j\vec{d}} W)^\gamma E_\gamma\,dt,$$

where $\{(E_\gamma, 2^{-j}) : (E, 2^{-j}) \in \mathcal{E}, |\gamma| \leq M_2\}$ is a bounded set of $(W, \vec{d})$ elementary operators supported in Ω. From here, (5.140) follows from Lemma 5.11.12. □

Lemma 5.11.15. *For all $\psi \in C_0^\infty(\Omega)$, for all bounded sets of $(W, \vec{d})$ elementary operators supported in Ω, $\mathcal{E}$, $\forall a, N \in \mathbb{N}$, $\forall c_1, c_2 \in \mathbb{C}$,*

$$\begin{aligned}\Bigg\{\Bigg(& c_1 2^{Nj} \operatorname{Mult}[\psi] e^{-\overline{\mathscr{L}}} E_1 + c_2 2^{2\kappa j} \operatorname{Mult}[\psi] \int_0^1 \left(\frac{t}{2^{-2\kappa j}}\right)^a e^{-t\overline{\mathscr{L}}} E_2\,dt, 2^{-j}\Bigg) \\ &: (E_1, 2^{-j}), (E_2, 2^{-j}) \in \mathcal{E}\Bigg\}\end{aligned} \tag{5.142}$$

is a bounded set of $(W, \vec{d})$ elementary operators supported in Ω.

Proof. Lemmas 5.11.10 and 5.11.14 show that (5.142) is a bounded set of $(W, \vec{d})$ pre-elementary operators supported in Ω. To show that it is a bounded set of $(W, \vec{d})$ elementary operators supported in Ω, we use the characterization given in Corollary 5.5.7. We do this by proving (5.40) (in the case $\nu = 1$) for the operators in (5.142), where $E_{\mu,l_1} = E_{l_1}$ and $\widetilde{E}_{\mu,l_2} = \widetilde{E}_{l_2}$ are a linear combination of operators of the same form as the ones in (5.142) (perhaps with different choices of c_1, c_2, N, a, and ψ). It then follows that the set in (5.142) is an element of $\mathcal{G}''_\Omega$ as defined in Corollary 5.5.7, which will complete the proof.

We begin with the E_{l_1} operators from (5.40). Let $\psi_2 \in C_0^\infty(\Omega)$ satisfy $\psi \prec \psi_2$. Note that for $(E, 2^{-j}) \in \mathcal{E}, E \in C_0^\infty(\mathfrak{M}\times\mathfrak{M}; \mathbb{C}^D)$, and Lemma 5.11.13 shows that $\frac{d}{dt} e^{-t\overline{\mathcal{L}}} E = -\mathcal{L} e^{-t\overline{\mathcal{L}}} E$. Using this, we have, for $(E_1, 2^{-j}), (E_2, 2^{-j}) \in \mathcal{E}$,

$$\begin{aligned}
&c_1 2^{Nj} \operatorname{Mult}[\psi] e^{-\overline{\mathcal{L}}} E_1 + c_2 2^{2\kappa j} \operatorname{Mult}[\psi] \int_0^1 \left(\frac{t}{2^{-2\kappa j}}\right)^a e^{-t\overline{\mathcal{L}}} E_2 \, dt \\
&= c_1 2^{Nj} \operatorname{Mult}[\psi] e^{-t\overline{\mathcal{L}}} E_1 + \frac{c_2}{a+1} \operatorname{Mult}[\psi] \int_0^1 \left(\frac{d}{dt}\left(\frac{t}{2^{-2\kappa j}}\right)^{a+1}\right) e^{-t\overline{\mathcal{L}}} E \, dt \\
&= c_1 2^{Nj} \operatorname{Mult}[\psi] e^{-\overline{\mathcal{L}}} E_1 + \frac{c_2}{a+1} 2^{2\kappa j(a+1)} \operatorname{Mult}[\psi] e^{-\overline{\mathcal{L}}} E_2 \\
&\quad + 2^{-2\kappa j} \operatorname{Mult}[\psi] \mathcal{L} \frac{c_2}{a+1} \int_0^1 \left(\frac{t}{2^{-2\kappa j}}\right)^{a+1} e^{-t\overline{\mathcal{L}}} E_2 \, dt \\
&= 2^{-j}\left(c_1 2^{(N+1)j} \operatorname{Mult}[\psi] e^{-\overline{\mathcal{L}}} E_1 + \frac{c_2}{a+1} 2^{j(2\kappa(a+1)+1)} \operatorname{Mult}[\psi] e^{-\overline{\mathcal{L}}} E_2\right) \\
&\quad + 2^{-2\kappa j} \operatorname{Mult}[\psi] \mathcal{L} \frac{c_2}{a+1} \int_0^1 \left(\frac{t}{2^{-2\kappa j}}\right)^{a+1} e^{-t\overline{\mathcal{L}}} E_2 \, dt.
\end{aligned} \tag{5.143}$$

The first term on the right-hand side of (5.143) is 2^{-j} times a linear combination of terms of the same form as the ones in (5.142) (with different choices of N) and is therefore of the desired form. For the second term, we use the fact that $\operatorname{Mult}[\psi]\mathcal{L}$ is a $D \times D$ matrix of $(W, \vec{d})$ partial differential operator in Ω of degree $\leq 2\kappa$, and therefore the second term on the right-hand side of (5.143) is of the desired form as well.

We now turn to the $\widetilde{E}_{l_2}$ terms from (5.40). For this, we apply Proposition 5.5.5 (h) with $N = 1$ to E_2 to see that

$$\begin{aligned}
&c_1 2^{Nj} \operatorname{Mult}[\psi] e^{-\overline{\mathcal{L}}} E_1 + c_2 2^{2\kappa j} \operatorname{Mult}[\psi] \int_0^1 \left(\frac{t}{2^{-2\kappa j}}\right)^a e^{-t\overline{\mathcal{L}}} E_2 \, dt \\
&= 2^{-j}(c_1 2^{(N+1)j} \operatorname{Mult}[\psi] e^{-\overline{\mathcal{L}}} E_1) \\
&\quad + \sum_{|\beta|\leq 1} 2^{j(|\beta|-1)} \left(c_2 2^{2\kappa j} \operatorname{Mult}[\psi] \int_0^1 \left(\frac{t}{2^{-2\kappa j}}\right)^a e^{-t\overline{\mathcal{L}}} E_\beta \, dt \right) (2^{-j\vec{d}} W)^\beta,
\end{aligned} \tag{5.144}$$

where $\{(E_\beta, 2^{-j}) : (E, 2^{-j}) \in \mathcal{E}, |\beta| \leq 1\}$ is a bounded set of $(W, \vec{d})$ elementary operators supported in Ω. The first term on the right-hand side of (5.144) is 2^{-j} times a term of the same form as the one in (5.142) (with different choices of N) and is therefore of the desired form. Using the fact that $(2^{-j\vec{d}} W)^\beta = 2^{-j \deg_{\vec{d}}(\beta)} W^\beta$ and W^β is a $(W, \vec{d})$ partial differential operator of degree $\leq \deg_{d}(\beta)$, we see that the second term on the right-hand side of (5.144) is also of the desired form, completing the proof. □

Corollary 5.11.16. *We have* $\int_0^1 e^{-t\overline{\mathscr{L}}}\,dt \in \mathscr{A}_{\mathrm{loc}}^{-2\kappa}((W,\vec{d});\mathbb{C}^D,\mathbb{C}^D)$.

Proof. Let $\psi_1,\psi_2 \in C_0^\infty(\mathfrak{M})$ and pick $\Omega \Subset \mathfrak{M}$ open and relatively compact such that $\psi_1,\psi_2 \in C_0^\infty(\Omega)$. We wish to show $\mathrm{Mult}[\psi_1](\int_0^1 e^{-t\overline{\mathscr{L}}}\,dt)\,\mathrm{Mult}[\psi_2] \in \mathscr{A}^{-2\kappa}(\Omega,(W,\vec{d});\mathbb{C}^D,\mathbb{C}^D)$. Let $\mathcal{E}$ be a bounded set of $(W,\vec{d})$ elementary operators supported in Ω. By Proposition 5.5.5 (c),

$$\{(\mathrm{Mult}[\psi_2]E,2^{-j}) : (E,2^{-j}) \in \mathcal{E}\}$$

is a bounded set of $(W,\vec{d})$ elementary operators supported in Ω. Thus, by Lemma 5.11.15 with $a=0$, $c_1=0$, and $c_2=1$,

$$\left\{\left(2^{2\kappa j}\,\mathrm{Mult}[\psi_1]\left(\int_0^1 e^{-t\overline{\mathscr{L}}}\,dt\right)\mathrm{Mult}[\psi_2]E,2^{-j}\right) : (E,2^{-j}) \in \mathcal{E}\right\}$$

is a bounded set of $(W,\vec{d})$ elementary operators supported in Ω. It follows that $\mathrm{Mult}[\psi_1](\int_0^1 e^{-t\overline{\mathscr{L}}}\,dt)\,\mathrm{Mult}[\psi_2] \in \mathscr{A}_2^{-2\kappa}(\Omega,(W,\vec{d});\mathbb{C}^D,\mathbb{C}^D) = \mathscr{A}^{-2\kappa}(\Omega,(W,\vec{d});\mathbb{C}^D,\mathbb{C}^D)$, completing the proof. □

Proof of Theorem 5.11.7. Set $S=\int_0^1 e^{-t\overline{\mathscr{L}}}\,dt$, so that $S \in \mathscr{A}_{\mathrm{loc}}^{-2\kappa}((W,\vec{d});\mathbb{C}^D,\mathbb{C}^D)$ by Corollary 5.11.16. Using Lemma 5.11.13, we have, for $f \in C_0^\infty(\mathfrak{M};\mathbb{C}^D)$,

$$\left(\int_0^1 e^{-t\overline{\mathscr{L}}}\,dt\right)\mathscr{L}f = -\int_0^1 \frac{d}{dt}e^{-t\overline{\mathscr{L}}}f\,dt = f - e^{-\overline{\mathscr{L}}}f,$$

that is, $S\mathscr{L} = I - e^{-\overline{\mathscr{L}}} \equiv I \mod C_{\mathrm{loc}}^\infty(\mathfrak{M}\times\mathfrak{M};\mathbb{M}^{D\times D})$. Also, for $f,g \in C_0^\infty(\mathfrak{M};\mathbb{C}^D)$, we have (thinking of $\mathscr{L}$ as being taken in the sense of distributions),

$$\langle f, S\mathscr{L}g\rangle_{L^2(\mathfrak{M},\mathrm{Vol};\mathbb{C}^D)} = \langle \mathscr{L}Sf, g\rangle_{L^2(\mathfrak{M},\mathrm{Vol};\mathbb{C}^D)},$$

and it follows that $\mathscr{L}S \equiv I \mod C_{\mathrm{loc}}^\infty(\mathfrak{M}\times\mathfrak{M};\mathbb{M}^{D\times D})$ as well. □

5.11.2 Parametrices and other geometries

In this section, we work in either the Hörmander setting or the general setting described in Section 5.3. However, in the general setting, we impose the **additional hypothesis** that $(W^1,\vec{d}^1)$ are Hörmander vector fields with formal degrees. Thus, in either setting, $(W^1,\vec{d}^1)$ are Hörmander vector fields with formal degrees. We assume $\nu \geq 2$.

Fix $\kappa_1, D \in \mathbb{N}_+$ and suppose $\mathscr{L}$ is an operator of the form

$$\mathscr{L} := \sum_{\deg_{d^1}(\alpha)\leq\kappa_1} a_\alpha(x)(W^1)^\alpha, \quad a_\alpha \in C^\infty_{\mathrm{loc}}(\mathfrak{M}; \mathbb{M}^{D\times D}(\mathbb{C})).$$

Suppose that there exists a parametrix for $\mathscr{L}$: $\mathscr{L}^{-1} \in \mathscr{A}^{-\kappa_1}_{\mathrm{loc}}((W^1, \vec{d}^1); \mathbb{C}^D, \mathbb{C}^D)$, that is, $\mathscr{L}^{-1}$ satisfies

$$\mathscr{L}\mathscr{L}^{-1}, \mathscr{L}^{-1}\mathscr{L} \equiv I \mod C^\infty_{\mathrm{loc}}(\mathfrak{M}\times\mathfrak{M}; \mathbb{M}^{D\times D}).$$

Note that despite our notation, $\mathscr{L}^{-1}$ is not assumed to be an exact inverse of $\mathscr{L}$, but just an inverse modulo infinitely smoothing operators.

Set

$$\begin{aligned}(Z, \vec{d}) = \{(Z_1, \vec{d}_1), \dots, (Z_s, \vec{d}_s)\} &:= (W^2, \vec{d}^2) \boxtimes (W^3, \vec{d}^3) \boxtimes \cdots \boxtimes (W^\nu, \vec{d}^\nu) \\ &\subset C^\infty_{\mathrm{loc}}(\mathfrak{M}; T\mathfrak{M}) \times (\mathbb{N}^{\nu-1} \setminus \{0\}).\end{aligned}$$

The main theorem of this section is the following.

Theorem 5.11.17. *Under the above hypotheses, we have the following.*
Hörmander setting: *We have* $\mathscr{L}^{-1} \in \mathscr{D}^{-\kappa_1}_{\mathrm{loc}}((W^1, \vec{d}^1), (Z, \vec{d}); \mathbb{C}^D, \mathbb{C}^D)$.
General setting: *We have* $\mathscr{L}^{-1} \in \widetilde{\mathscr{D}}^{-\kappa_1}_{\mathrm{loc}}((W^1, \vec{d}^1), (Z, \vec{d}); \mathbb{C}^D, \mathbb{C}^D)$.

Corollary 5.11.18. *Under the above hypotheses, we have the following.*
Hörmander setting: *We have* $\mathscr{L}^{-1} \in \mathscr{A}^{(-\kappa_1, 0_{\nu-1})}_{\mathrm{loc}}((W, \vec{d}); \mathbb{C}^D, \mathbb{C}^D)$.
General setting: *We have* $\mathscr{L}^{-1} \in \widetilde{\mathscr{A}}^{(-\kappa_1, 0_{\nu-1})}_{\mathrm{loc}}((W, \vec{d}); \mathbb{C}^D, \mathbb{C}^D)$.

Proof. We prove only the result in the Hörmander setting; the proof in the general setting is similar. Let $\psi_1, \psi_2 \in C^\infty_0(\mathfrak{M})$. By Theorem 5.11.17, $\mathrm{Mult}[\psi_1]\mathscr{L}^{-1}\mathrm{Mult}[\psi_2] \in \mathscr{D}^{-\kappa}((W^1, \vec{d}^1), (Z, \vec{d}); \mathbb{C}^D, \mathbb{C}^D)$. Therefore, by Theorem 5.9.6, $\mathrm{Mult}[\psi_1]\mathscr{L}^{-1}\mathrm{Mult}[\psi_2] \in \mathscr{A}^{(-\kappa_1, 0_{\nu_2})}((W, \vec{d}); \mathbb{C}^D, \mathbb{C}^D)$. Since $\psi_1, \psi_2 \in C^\infty_0(\mathfrak{M})$ were arbitrary, we have $\mathscr{L}^{-1} \in \mathscr{A}^{(-\kappa_1, 0_{\nu-1})}_{\mathrm{loc}}((W, \vec{d}); \mathbb{C}^D, \mathbb{C}^D)$, as desired, completing the proof. □

The rest of this section is devoted to the proof of Theorem 5.11.17.

Most of the proof will be the same in the Hörmander setting and the general setting, so we often do not specify which setting we are in (when unspecified, the reader may assume we are in the general setting, since the Hörmander setting is a special case of the general setting). We begin by introducing some notation to help with the proof.

Fix a relatively compact, open set $\Omega \Subset \mathfrak{M}$. For $\kappa \in \mathbb{N}$, $\mathscr{P}^{(\kappa)}_1$ and $\widetilde{\mathscr{P}}^{(\kappa)}_1$ denote an arbitrary operator of the form

$$\sum_{\deg_{d^1}(\alpha)\leq\kappa} b_\alpha(x)(W^1)^\alpha, \quad b_\alpha \in C^\infty_{\mathrm{loc}}(\Omega; \mathbb{M}^{D\times D}(\mathbb{C})).$$

When $\kappa \in \mathbb{Z} \setminus \mathbb{N}$, $\mathscr{P}_1^{(\kappa)} = 0 = \widetilde{\mathscr{P}}_1^{(\kappa)}$. In other words, $\mathscr{P}_1^{(\kappa)}$ or $\widetilde{\mathscr{P}}_1^{(\kappa)}$ are $D \times D$ matrices of $(W^1, \vec{d}^1)$ partial differential operators on Ω of degree $\leq \kappa$.

For $\kappa_2 \in \mathbb{Z}^{\nu-1}$ and $k \in \mathbb{Z}$, we let $\mathscr{P}_2^{(\kappa_2,k)}$ and $\widetilde{\mathscr{P}}_2^{(\kappa_2,k)}$ denote arbitrary (scalar-valued) $(Z, \vec{d})$ partial differential operators on Ω of order $\leq \kappa_2$ and length $\leq k$ (see Definition 3.12.1). Also, $\mathscr{P}_2^{(\kappa_2)}$ and $\widetilde{\mathscr{P}}_2^{(\kappa_2)}$ will denote arbitrary $(Z, \vec{d})$ partial differential operators on Ω of order $\leq \kappa_2$.

For $t \in \mathbb{R}$, $T^{(t)}$ and $\widetilde{T}^{(t)}$ denote arbitrary operators in $\mathscr{A}^t(\Omega, (W^1, \vec{d}^1); \mathbb{C}^D, \mathbb{C}^D)$. Also, $T^{(-\infty)}$ and $\widetilde{T}^{(-\infty)}$ denote an arbitrary element of $\bigcap_{t \in \mathbb{R}} \mathscr{A}^t(\Omega, (W^1, \vec{d}^1); \mathbb{C}^D, \mathbb{C}^D)$.

Remark 5.11.19. If $R \in C_0^\infty(\Omega \times \Omega; \mathbb{M}^{D \times D})$, then Proposition 5.8.11 implies that R is of the form $T^{(-\infty)}$.

The particular operator which $\mathscr{P}_1^{(\kappa)}$, $\widetilde{\mathscr{P}}_1^{(\kappa)}$, $\mathscr{P}_2^{(\kappa_2,k)}$, $\widetilde{\mathscr{P}}_2^{(\kappa_2,k)}$, $T^{(t)}$, or $\widetilde{T}^{(t)}$ denotes may change from line to line. All that is important is that it is of the correct form. For the remainder of the proof, fix $\psi_1, \psi_2 \in C_0^\infty(\Omega)$.

The basic idea of the proof of Theorem 5.11.17 is the following. If two operators A and B commute and A is invertible, then A^{-1} and B commute. Corollary 3.12.7 implies that $\mathscr{P}_2^{(\kappa_2)}$ and $\mathscr{L}$ "approximately commute" and we will use this to show that $\mathscr{P}_2^{(\kappa_2)}$ and $\mathrm{Mult}[\psi_1]\mathscr{L}^{-1}\mathrm{Mult}[\psi_2]$ "approximately commute," which will complete the proof. We turn to making this precise.

Lemma 5.11.20. *Suppose $\phi_1, \phi_2 \in C_0^\infty(\Omega)$ with $\phi_1 \prec \phi_2$. Then for $\kappa_2 \in \mathbb{Z}^{\nu-1}$ and $k \in \mathbb{Z}$, $\mathrm{Mult}[\phi_1]\mathscr{P}_2^{(\kappa_2,k)}$ is of the form $\widetilde{\mathscr{P}}_2^{(\kappa_2,k)}\mathrm{Mult}[\phi_2]$ and $\mathscr{P}_2^{(\kappa_2,k)}\mathrm{Mult}[\phi_1]$ is of the form $\mathrm{Mult}[\phi_2]\widetilde{\mathscr{P}}_2^{(\kappa_2,k)}$.*

Proof. Since $\mathscr{P}_2^{(\kappa_2,k)}$ is a partial differential operator, we have

$$\mathrm{Mult}[\phi_1]\mathscr{P}_2^{(\kappa_2,k)} = \mathrm{Mult}[\phi_1]\mathscr{P}_2^{(\kappa_2,k)}\mathrm{Mult}[\phi_2].$$

Since $\mathrm{Mult}[\phi_1]$ is a $(Z, \vec{d})$ partial differential operator of degree $\leq 0_{\nu-1}$ and length ≤ 0, Lemma 3.12.4 shows that $\mathrm{Mult}[\phi_1]\mathscr{P}_2^{(\kappa_2,k)} = \widetilde{\mathscr{P}}_2^{(\kappa_2,k)}$. This completes the proof for $\mathrm{Mult}[\phi_1]\mathscr{P}_2^{(\kappa_2,k)}$. A similar proof works for $\mathscr{P}_2^{(\kappa_2,k)}\mathrm{Mult}[\phi_1]$; we leave the details to the reader. □

Lemma 5.11.21. *There is a $K = K(\Omega) \in \mathbb{N}$ such that every operator of the form $\mathscr{P}_2^{(\kappa_2,k)}$ is a $(W^1, \vec{d}^1)$ partial differential operator on Ω of degree $\leq Kk$.*

Proof. When $k \leq -1$, $\mathscr{P}_2^{(\kappa_2,k)} = 0$ is trivial, so we assume $k \in \mathbb{N}$.

The hypothesis that $W_1^1, \dots, W_{r_1}^1$ are Hörmander vector fields and the compactness of $\overline{\Omega}$ imply that $W_1^1, \dots, W_{r_1}^1$ satisfy Hörmander's condition of order m on Ω for some $m \in \mathbb{N}$. Thus, every vector field $V \in C_{\mathrm{loc}}^\infty(\Omega; T\Omega)$ is a $(W^1, \vec{d}^1)$ partial differential operator on Ω of degree $\leq K := m \max_{1 \leq j \leq r_1}\{\vec{d}_j^1\}$.

By definition, $\mathscr{P}_2^{(\kappa_2,k)}$ can be written as a finite sum of operators of the form

$$\mathrm{Mult}[g]V_1V_2\cdots V_R,$$

where $g \in C^\infty_{\mathrm{loc}}(\Omega)$ and each $(V_j, \vec{d}_j) \in (Z, \vec{d})$ with $|\vec{d}_1|_1 + \cdots + |\vec{d}_R|_1 \leq k$. In particular, $R \leq k$. Thus,

$$\mathrm{Mult}[g]V_1V_2\cdots V_R$$

is a $(W^1, \vec{d}^1)$ partial differential operator on Ω of degree $\leq RK \leq kK$, completing the proof. □

Lemma 5.11.22. *Let $\kappa_2 \in \mathbb{N}^{\nu-1}$. Any operator of the form $T^{(-\kappa)}$ where $\kappa \geq \kappa_1$ is of the form*

$$\widetilde{T}^{(-\kappa_1)}\widetilde{\mathscr{P}}_2^{(\kappa_2)} \tag{5.145}$$

and of the form

$$\widetilde{\mathscr{P}}_2^{(\kappa_2)}\widetilde{T}^{(-\kappa_1)}. \tag{5.146}$$

In particular, any operator of the form $T^{(-\infty)}$ is of the forms (5.145) *and* (5.146).

Proof. Since $\mathrm{supp}(T^{(-\kappa)}) \Subset \Omega \times \Omega$, we may take $\psi \in C_0^\infty(\Omega)$ with $\psi \equiv 1$ on a large enough set that $T^{(-\kappa)}\,\mathrm{Mult}[\psi] = T^{(-\kappa)} = \mathrm{Mult}[\psi]T^{(-\kappa)}$. Since $\kappa_2 \in \mathbb{N}^{\nu-1}$, $\mathrm{Mult}[\psi]$ is of the form $\widetilde{\mathscr{P}}_2^{(\kappa_2)}$, and therefore

$$\widetilde{T}^{(-\kappa)} = \widetilde{T}^{(-\kappa)}\,\mathrm{Mult}[\psi] = \widetilde{T}^{(-\kappa)}\widetilde{\mathscr{P}}_2^{(\kappa_2)}$$

and

$$\widetilde{T}^{(-\kappa)} = \mathrm{Mult}[\psi]\widetilde{T}^{(-\kappa)} = \widetilde{\mathscr{P}}_2^{(\kappa_2)}\widetilde{T}^{(-\kappa)}.$$

Since $\kappa \geq \kappa_1$, Proposition 5.8.2 shows that $\widetilde{T}^{(-\kappa)}$ is of the form $\widetilde{T}^{(-\kappa_1)}$. □

Lemma 5.11.23. *Fix $\phi_0, \phi_1, \phi_2, \ldots \in C_0^\infty(\Omega)$ with*

$$\psi_1, \psi_2 \prec \phi_0 \prec \phi_1 \prec \phi_2 \prec \cdots.$$

(A) *For every $N \in \mathbb{N}$, any operator of the form $\mathscr{P}_2^{(\kappa_2)}\,\mathrm{Mult}[\psi_1]\mathscr{L}^{-1}\,\mathrm{Mult}[\psi_2]$ can be written as a finite sum of terms of the form*

$$\widetilde{T}^{(-\kappa_1)}\widetilde{\mathscr{P}}_2^{(\kappa_2)} \tag{5.147}$$

and

$$\widetilde{T}^{(-j)}\widetilde{\mathscr{P}}_2^{(\kappa_2,|\kappa_2|_1-N+j)}\,\mathrm{Mult}[\phi_N]\mathscr{L}^{-1}\,\mathrm{Mult}[\psi_2], \quad j = 0,1,\ldots,N. \tag{5.148}$$

(B) *For every* $N \in \mathbb{N}$, *any operator of the form* $\operatorname{Mult}[\psi_1]\mathcal{L}^{-1}\operatorname{Mult}[\psi_2]\mathcal{P}_2^{(\kappa_2)}$ *can be written as a finite sum of terms of the form*

$$\widetilde{\mathcal{P}}_2^{(\kappa_2)}\widetilde{T}^{(-\kappa_1)}$$

and

$$\operatorname{Mult}[\psi_1]\mathcal{L}^{-1}\operatorname{Mult}[\phi_N]\widetilde{\mathcal{P}}_2^{(\kappa_2,|\kappa_2|_1-N+j)}\widetilde{T}^{(-j)},\quad j=0,1,\dots,N.$$

Proof. The proofs of (A) and (B) are similar (they just involve reversing the roles of the left and right sides of the operators in question), so we prove only (A). When $\kappa_2 \in \mathbb{Z}^{\nu-1}\setminus\mathbb{N}^{\nu-1}$, then $\mathcal{P}_2^{(\kappa_2)}=0$ and the result is trivial. Thus, we henceforth assume $\kappa_2\in\mathbb{N}^{\nu-1}$.

We prove (A) by induction on N. We begin with the base case, $N=0$. Proposition 5.8.3 shows that $\operatorname{Mult}[\phi_0]$ is of the form $T^{(0)}$. Thus, since $\psi_1 \prec \phi_0$, using Lemma 5.11.20, we have

$$\begin{aligned}\mathcal{P}_2^{(\kappa_2)}\operatorname{Mult}[\psi_1]\mathcal{L}^{-1}\operatorname{Mult}[\psi_2] &= \mathcal{P}_2^{(\kappa_2)}\operatorname{Mult}[\psi_1]\operatorname{Mult}[\phi_0]\mathcal{L}^{-1}\operatorname{Mult}[\psi_2]\\ = \operatorname{Mult}[\phi_0]\widetilde{\mathcal{P}}_2^{(\kappa_2)}\operatorname{Mult}[\phi_0]\mathcal{L}^{-1}\operatorname{Mult}[\psi_2] &= T^{(0)}\widetilde{\mathcal{P}}_2^{(\kappa_2)}\operatorname{Mult}[\phi_0]\mathcal{L}^{-1}\operatorname{Mult}[\psi_2],\end{aligned}$$

thereby proving that $\mathcal{P}_2^{(\kappa_2)}\operatorname{Mult}[\psi_1]\mathcal{L}^{-1}\operatorname{Mult}[\psi_2]$ is of the form (5.148) with $j=0, N=0$.

We assume (A) for some $N\in\mathbb{N}$ and prove it for $N+1$. Thus, we wish to show that any term of the form (5.148) can be written as a finite sum of terms of the form (5.147) and (5.148) with N replaced by $N+1$.

We consider an arbitrary term of the form (5.147). For $j=0,1,\dots,N$, we have, by Lemma 5.11.20,

$$\begin{aligned}&T^{(-j)}\mathcal{P}_2^{(\kappa_2,|\kappa_2|_1-N+j)}\operatorname{Mult}[\phi_N]\mathcal{L}^{-1}\operatorname{Mult}[\psi_2]\\ &= T^{(-j)}\operatorname{Mult}[\phi_{N+1}]\widetilde{\mathcal{P}}^{(\kappa_2,|\kappa_2|_1-N+j)}\mathcal{L}^{-1}\operatorname{Mult}[\psi_2]\\ &= T^{(-j)}\operatorname{Mult}[\phi_{N+1}]\mathcal{L}^{-1}\widetilde{\mathcal{P}}^{(\kappa_2,|\kappa_2|_1-N+j)}\operatorname{Mult}[\psi_2]\\ &\quad + T^{(-j)}\operatorname{Mult}[\phi_{N+1}][\widetilde{\mathcal{P}}_2^{(\kappa_2,|\kappa_2|_1-N+j)},\mathcal{L}^{-1}]\operatorname{Mult}[\psi_2]\\ &=: \text{(I)} + \text{(II)}.\end{aligned}$$

We show that the terms (I) and (II) above are of the desired form separately. We begin with (I).

By the hypothesis that $\mathcal{L}^{-1}\mathcal{A}_{\mathrm{loc}}^{-\kappa_1}(W^1,d^1)$, we have $\operatorname{Mult}[\phi_{N+1}]\mathcal{L}^{-1}\operatorname{Mult}[\phi_0]=\widetilde{T}^{(-\kappa_1)}$. Thus, using Lemma 5.11.20, we have (with the operator $\widetilde{\mathcal{P}}^{(\kappa_2,|\kappa_2|_1-N+j)}$ possibly changing from line to line)

$$\begin{aligned}\text{(I)} &= T^{(-j)}\operatorname{Mult}[\phi_{N+1}]\mathcal{L}^{-1}\widetilde{\mathcal{P}}^{(\kappa_2,|\kappa_2|_1-N+j)}\operatorname{Mult}[\psi_2]\\ &= T^{(-j)}\operatorname{Mult}[\phi_{N+1}]\mathcal{L}^{-1}\operatorname{Mult}[\phi_0]\widetilde{\mathcal{P}}^{(\kappa_2,|\kappa_2|_1-N+j)}\\ &= T^{(-j)}\widetilde{T}^{(-\kappa_1)}\widetilde{\mathcal{P}}^{(\kappa_2,|\kappa_2|_1-N+j)}\\ &= \widetilde{T}^{(-\kappa_1-j)}\widetilde{\mathcal{P}}^{(\kappa_2)},\end{aligned}$$

where the last equality uses Proposition 5.8.6. Since $j \geq 0$, Proposition 5.8.2 shows $\widetilde{T}^{(-\kappa_1-j)}$ is of the form $\widetilde{T}^{(-\kappa_1)}$. Thus, (I) is of the form (5.147), as desired.

Turning to (II), we use the fact that $\mathscr{L}\mathscr{L}^{-1}, \mathscr{L}^{-1}\mathscr{L} \in C^{\infty}_{\mathrm{loc}}(\Omega \times \Omega; \mathbb{M}^{D\times D}(\mathbb{C}))$ and that $\mathscr{L}^{-1}$ is pseudo-local (see the growth condition in Definition 5.2.4). We see that for some $R_1, R_2 \in C^{\infty}_{\mathrm{loc}}(\mathfrak{M} \times \mathfrak{M}; \mathbb{M}^{D\times D})$,

$$\mathrm{Mult}[\phi_{N+1}]\mathscr{L}^{-1}\,\mathrm{Mult}[\phi_{N+2}]\mathscr{L} = \mathrm{Mult}[\phi_{N+1}] + \mathrm{Mult}[\phi_{N+3}]R_1\,\mathrm{Mult}[\phi_{N+3}], \tag{5.149}$$

$$\mathscr{L}\,\mathrm{Mult}[\phi_0]\mathscr{L}^{-1}\,\mathrm{Mult}[\psi_2] = \mathrm{Mult}[\psi_2] + \mathrm{Mult}[\phi_{N+3}]R_2\,\mathrm{Mult}[\phi_{N+3}]. \tag{5.150}$$

Using (5.149) and (5.150), we have

$$\begin{aligned}
(\mathrm{II}) &= T^{(-j)}\,\mathrm{Mult}[\phi_{N+1}]\big(\widetilde{\mathscr{P}}_2^{(\kappa_2,|\kappa_2|_1-N+j)}\mathscr{L}^{-1} - \mathscr{L}^{-1}\widetilde{\mathscr{P}}_2^{(\kappa_2,|\kappa_2|_1-N+j)}\big)\,\mathrm{Mult}[\psi_2] \\
&= T^{(-j)}\,\mathrm{Mult}[\phi_{N+1}]\mathscr{L}^{-1}\,\mathrm{Mult}[\phi_{N+2}]\mathscr{L}\big(\widetilde{\mathscr{P}}_2^{(\kappa_2,|\kappa_2|_1-N+j)}\mathscr{L}^{-1} - \mathscr{L}^{-1}\widetilde{\mathscr{P}}_2^{(\kappa_2,|\kappa_2|_1-N+j)}\big) \\
&\qquad \times \mathscr{L}\,\mathrm{Mult}[\phi_0]\mathscr{L}^{-1}\,\mathrm{Mult}[\psi_2] \\
&\quad - T^{(-j)}\,\mathrm{Mult}[\phi_{N+1}]\mathscr{L}^{-1}\,\mathrm{Mult}[\phi_{N+2}]\mathscr{L}\big(\widetilde{\mathscr{P}}_2^{(\kappa_2,|\kappa_2|_1-N+j)}\mathscr{L}^{-1} - \mathscr{L}^{-1}\widetilde{\mathscr{P}}_2^{(\kappa_2,|\kappa_2|_1-N+j)}\big) \\
&\qquad \times \mathrm{Mult}[\phi_{N+3}]R_2\,\mathrm{Mult}[\phi_{N+3}] \\
&\quad - T^{(-j)}\,\mathrm{Mult}[\phi_{N+3}]R_1\,\mathrm{Mult}[\phi_{N+3}]\big(\widetilde{\mathscr{P}}_2^{(\kappa_2,|\kappa_2|_1-N+j)}\mathscr{L}^{-1} - \mathscr{L}^{-1}\widetilde{\mathscr{P}}_2^{(\kappa_2,|\kappa_2|_1-N+j)}\big) \\
&\qquad \times \mathrm{Mult}[\psi_2] \\
&=: (\mathrm{III}) - (\mathrm{IV}) - (\mathrm{V}).
\end{aligned}$$

We study the terms (III) and (IV) + (V) separately.

We begin with (IV) + (V). Note that if $\xi_1, \xi_2 \in C_0^{\infty}(\Omega)$, we have

$$\mathrm{Mult}[\xi_1]\mathscr{L}^{-1}\,\mathrm{Mult}[\xi_2] = T^{(-\kappa_1)}, \tag{5.151}$$

and by Proposition 5.8.3,

$$\mathrm{Mult}[\xi_1]\mathscr{L} = T^{(\kappa_1)}. \tag{5.152}$$

Using Remark 5.11.19, if $R \in C^{\infty}_{\mathrm{loc}}(\Omega \times \Omega; \mathbb{M}^{D\times D}(\mathbb{C}))$, then

$$\mathrm{Mult}[\xi_1]R\,\mathrm{Mult}[\xi_2] = T^{(-\infty)}. \tag{5.153}$$

Also, if $K = K(\Omega) \in \mathbb{N}$ is as in Lemma 5.11.21, then, using Remark 3.12.2, $\widetilde{\mathscr{P}}_2^{(\kappa_2)} = \widetilde{\mathscr{P}}_2^{(\kappa_2,|\kappa_2|_1)}$ is a $(W^1, \vec{d}^1)$ partial differential operator on Ω of degree $\leq |\kappa_2|_1 K$, and therefore by Proposition 5.8.3,

$$\mathrm{Mult}[\xi_1]\widetilde{\mathscr{P}}_2^{(\kappa_2)} = T^{(|\kappa_2|_1 K)}. \tag{5.154}$$

Using (5.151), (5.152), (5.153), and (5.154), we have

$$\begin{aligned}
(\mathrm{IV}) + (\mathrm{V}) &= T^{(-j)} \operatorname{Mult}[\phi_{N+1}]\mathscr{L}^{-1} \operatorname{Mult}[\phi_{N+2}]\mathscr{L} \operatorname{Mult}[\phi_{N+3}]\widetilde{\mathscr{P}}_2^{(\kappa_2)} \operatorname{Mult}[\phi_{N+4}] \\
&\quad \times \mathscr{L}^{-1} \operatorname{Mult}[\phi_{N+3}]R_2 \operatorname{Mult}[\phi_{N+3}] \\
&\quad - T^{(-j)} \operatorname{Mult}[\phi_{N+1}]\mathscr{L}^{-1} \operatorname{Mult}[\phi_{N+2}]\mathscr{L} \operatorname{Mult}[\phi_{N+3}]\mathscr{L}^{-1} \operatorname{Mult}[\phi_{N+4}] \\
&\quad \times \widetilde{\mathscr{P}}_2^{(\kappa_2)} \operatorname{Mult}[\phi_{N+3}]R_2 \operatorname{Mult}[\phi_{N+3}] \\
&\quad + T^{(-j)} \operatorname{Mult}[\phi_{N+3}]R_1 \operatorname{Mult}[\phi_{N+3}]\widetilde{\mathscr{P}}_2^{(\kappa_2)} \operatorname{Mult}[\phi_{N+4}]\mathscr{L}^{-1} \operatorname{Mult}[\psi_2] \\
&\quad - T^{(-j)} \operatorname{Mult}[\phi_{N+3}]R_1 \operatorname{Mult}[\phi_{N+3}]\mathscr{L}^{-1} \operatorname{Mult}[\phi_0]\widetilde{\mathscr{P}}_2^{(\kappa_2)} \operatorname{Mult}[\psi_2] \\
&= T^{(-j)}T^{(-\kappa_1)}T^{(\kappa_1)}T^{(|\kappa_2|_1 K)}T^{(-\kappa_1)}T^{(-\infty)} + T^{(-j)}T^{(-\kappa_1)}T^{(\kappa_1)}T^{(-\kappa_1)}T^{(|\kappa_2|_1 K)}T^{(-\infty)} \\
&\quad + T^{(-j)}T^{(-\infty)}T^{(|\kappa_2|_1 K)}T^{(-\kappa_1)} + T^{(-j)}T^{(-\infty)}T^{(-\kappa_1)}T^{(|\kappa_2|_1 K)} \\
&= \widetilde{T}^{(-\infty)},
\end{aligned}$$

where the last equality follows from Proposition 5.8.6. Lemma 5.11.22 now shows that (IV) + (V) is of the form (5.147).

We turn to (III). Using the fact that $\mathscr{L}\mathscr{L}^{-1} = R_3$ and $\mathscr{L}^{-1}\mathscr{L} = R_4$, where $R_3, R_4 \in C^{\infty}_{\mathrm{loc}}(\mathfrak{M} \times \mathfrak{M}; \mathbb{M}^{D\times D}(\mathbb{C}))$,

$$\begin{aligned}
(\mathrm{III}) &= T^{(-j)} \operatorname{Mult}[\phi_{N+1}]\mathscr{L}^{-1} \operatorname{Mult}[\phi_{N+2}](\mathscr{L}\widetilde{\mathscr{P}}_2^{(\kappa_2,|\kappa_2|_1-N+j)} - \widetilde{\mathscr{P}}_2^{(\kappa_2,|\kappa_2|_1-N+j)}\mathscr{L}) \\
&\quad \times \operatorname{Mult}[\phi_0]\mathscr{L}^{-1} \operatorname{Mult}[\psi_2] \\
&\quad + T^{(-j)} \operatorname{Mult}[\phi_{N+1}]\mathscr{L}^{-1} \operatorname{Mult}[\phi_{N+2}]\mathscr{L}\widetilde{\mathscr{P}}_2^{(\kappa_2,|\kappa_2|_1-N+j)}R_4 \operatorname{Mult}[\phi_0]\mathscr{L}^{-1} \operatorname{Mult}[\psi_2] \\
&\quad - T^{(-j)} \operatorname{Mult}[\phi_{N+1}]\mathscr{L}^{-1} \operatorname{Mult}[\phi_{N+2}]R_3\widetilde{\mathscr{P}}_2^{(\kappa_2,|\kappa_2|_1-N+j)}\mathscr{L} \operatorname{Mult}[\phi_0]\mathscr{L}^{-1} \operatorname{Mult}[\psi_2] \\
&= (\mathrm{VI}) + (\mathrm{VII}) - (\mathrm{VIII}).
\end{aligned}$$

Using (5.151), (5.152), (5.153), and (5.154), we have

$$\begin{aligned}
(\mathrm{VII}) &= T^{(-j)} \operatorname{Mult}[\phi_{N+1}]\mathscr{L}^{-1} \operatorname{Mult}[\phi_{N+2}]\mathscr{L} \operatorname{Mult}[\phi_{N+3}]\widetilde{\mathscr{P}}_2^{(\kappa_2)} \operatorname{Mult}[\Phi_{N+4}]R_4 \operatorname{Mult}[\phi_0] \\
&\quad \times \mathscr{L}^{-1} \operatorname{Mult}[\psi_2] \\
&= T^{(-j)}T^{(-\kappa_1)}T^{(\kappa_1)}T^{(|\kappa_2|_1 K)}T^{(-\infty)}T^{(-\kappa_1)} = \widetilde{T}^{(-\infty)},
\end{aligned}$$

where the last equality follows from Proposition 5.8.6. Lemma 5.11.22 now shows that (VII) is of the form (5.147). A similar proof shows that (VIII) $= \widetilde{T}^{(-\infty)}$ and is therefore of the form (5.147) (by Lemma 5.11.22).

We turn to (VI). Corollary 3.8.11 with $\nu_1 = 1$, $\nu_2 = \nu - 1$, $\mathcal{S}_1 = (W^1, \vec{d}^1)$, and $\mathcal{T}_\mu = (W^{\mu-1}, \vec{d}^{\mu-1})$, $2 \le \mu \le \nu$, shows that $(W^1, \vec{d}^1)$ and $(Z, \vec{d})$ locally weakly approximately commute. Since $\widetilde{\mathscr{P}}_2^{(\kappa_2,|\kappa_2|_1-N+j)}$ is a *scalar-valued* $(Z, \vec{d})$ partial differential operator on Ω of degree $\le \kappa_2$ and length $\le |\kappa_2|_1 - N + j$ and $\mathscr{L}$ is a matrix of $(W^1, \vec{d}^1)$ partial differential operators on Ω of degree $\le \kappa_1$ and length $\le \kappa_1$ (see Remark 3.12.2), Theorem 3.12.6 implies that $\mathscr{L}\widetilde{\mathscr{P}}_2^{(\kappa_2,|\kappa_2|_1-N+j)} - \widetilde{\mathscr{P}}_2^{(\kappa_2,|\kappa_2|_1-N+j)}\mathscr{L}$ can be written as a finite sum of terms of the form

$$\mathscr{P}_1^{(\kappa_1,\kappa_1)}\mathscr{P}_2^{(\kappa_2,|\kappa_2|_1-N+j-1)} \quad \text{and} \quad \mathscr{P}_1^{(\kappa_1,\kappa_1-1)}\mathscr{P}_2^{(\kappa_2,|\kappa_2|_1-N+j-1)},$$

where each $\mathscr{P}_1^{(\kappa,k)}$ is a $D \times D$ matrix of (W^1, d^1) partial differential operators on Ω of degree $\leq \kappa$ and length $\leq k$. Since (W^1, d^1) is a single-parameter list, each $\mathscr{P}_1^{(\kappa,k)}$ is in fact a $D\times D$ matrix of (W^1, d^1) partial differential operators on Ω of degree $\leq \min\{\kappa, k\}$. Thus, $\mathscr{L}\widetilde{\mathscr{P}}_2^{(\kappa_2,|\kappa_2|_1-N+j)} - \widetilde{\mathscr{P}}_2^{(\kappa_2,|\kappa_2|_1-N+j)}\mathscr{L}$ can be written as a finite sum of terms of the form

$$\mathscr{P}_1^{(\kappa_1)}\mathscr{P}_2^{(\kappa_2,|\kappa_2|_1-N+j-1)} \quad \text{and} \quad \mathscr{P}_1^{(\kappa_1-1)}\mathscr{P}_2^{(\kappa_2,|\kappa_2|_1-N+j-1)}.$$

Thus (VI) can be written as a finite sum of terms of the form

$$\begin{aligned}
(\mathrm{IX}) &= T^{(-j)}\,\mathrm{Mult}[\phi_{N+1}]\mathscr{L}^{-1}\,\mathrm{Mult}[\phi_{N+2}]\mathscr{P}_1^{(\kappa_1)}\mathscr{P}_2^{(\kappa_2,|\kappa_2|_1-N+j-1)}\,\mathrm{Mult}[\phi_0]\mathscr{L}^{-1}\,\mathrm{Mult}[\psi_2],\\
(\mathrm{X}) &= T^{(-j)}\,\mathrm{Mult}[\phi_{N+1}]\mathscr{L}^{-1}\,\mathrm{Mult}[\phi_{N+2}]\mathscr{P}_1^{(\kappa_1-1)}\mathscr{P}_2^{(\kappa_2,|\kappa_2|_1-N+j)}\,\mathrm{Mult}[\phi_0]\mathscr{L}^{-1}\,\mathrm{Mult}[\psi_2].
\end{aligned}$$

By Proposition 5.8.3 we have, for $\xi \in C_0^\infty(\Omega)$,

$$\mathrm{Mult}[\xi]\mathscr{P}_1^{(k)} = T^{(k)}. \tag{5.155}$$

Using (5.151), (5.155), and Lemma 5.11.20, we have

$$\begin{aligned}
(\mathrm{IX}) &= T^{(-j)}\,\mathrm{Mult}[\phi_{N+1}]\mathscr{L}^{-1}\,\mathrm{Mult}[\phi_{N+2}]\mathscr{P}_1^{(\kappa_1)}\,\mathrm{Mult}[\phi_1]\widetilde{\mathscr{P}}_2^{(\kappa_2,|\kappa_2|_1-N+j-1)}\mathscr{L}^{-1}\,\mathrm{Mult}[\psi_2]\\
&= T^{(-j)}T^{(-\kappa_1)}T^{(\kappa_1)}\widetilde{\mathscr{P}}_2^{(\kappa_2,|\kappa_2|_1-N+j-1)}\,\mathrm{Mult}[\phi_{N+1}]\mathscr{L}^{-1}\,\mathrm{Mult}[\psi_2]\\
&= \widetilde{T}^{(-j)}\widetilde{\mathscr{P}}_2^{(\kappa_2,|\kappa_2|_1-N+j-1)}\,\mathrm{Mult}[\phi_{N+1}]\mathscr{L}^{-1}\,\mathrm{Mult}[\psi_2],
\end{aligned}$$

where the last equality uses Proposition 5.8.6. This is of the form (5.148) with N replaced by $N+1$, as desired.

Similarly, using (5.151), (5.155), and Lemma 5.11.20, we have

$$\begin{aligned}
(\mathrm{X}) &= T^{(-j)}\,\mathrm{Mult}[\phi_{N+1}]\mathscr{L}^{-1}\,\mathrm{Mult}[\phi_{N+2}]\mathscr{P}_1^{(\kappa_1-1)}\,\mathrm{Mult}[\phi_1]\\
&\quad\times \widetilde{\mathscr{P}}_2^{(\kappa_2,|\kappa_2|_1-N+j)}\,\mathrm{Mult}[\phi_{N+1}]\mathscr{L}^{-1}\,\mathrm{Mult}[\psi_2]\\
&= T^{(-j)}T^{(-\kappa_1)}T^{(\kappa_1-1)}\widetilde{\mathscr{P}}_2^{(\kappa_2,|\kappa_2|_1-N+j)}\,\mathrm{Mult}[\phi_{N+1}]\mathscr{L}^{-1}\,\mathrm{Mult}[\psi_2]\\
&= \widetilde{T}^{(-(j+1))}\widetilde{\mathscr{P}}_2^{(\kappa_2,|\kappa_2|_1-N+j)}\,\mathrm{Mult}[\phi_{N+1}]\mathscr{L}^{-1}\,\mathrm{Mult}[\psi_2],
\end{aligned}$$

where the last equality uses Proposition 5.8.6. This is of the form (5.148) with N replaced by $N+1$ and j replaced by $j+1$, as desired, completing the proof. □

Lemma 5.11.24. (A) $\mathscr{P}_2^{(\kappa_2)}\,\mathrm{Mult}[\psi_1]\mathscr{L}^{-1}\,\mathrm{Mult}[\psi_2]$ *can be written as a finite sum of terms of the form*

$$\widetilde{T}^{(-\kappa_1)}\widetilde{\mathscr{P}}_2^{(\kappa_2)}. \tag{5.156}$$

(B) $\mathrm{Mult}[\psi_1]\mathscr{L}^{-1}\mathrm{Mult}[\psi_2]\mathscr{P}_2^{(\kappa_2)}$ *can be written as a finite sum of terms of the form*

$$\widetilde{\mathscr{P}}_2^{(\kappa_2)}\widetilde{T}^{(-\kappa_1)}. \tag{5.157}$$

Proof. The proofs for (A) and (B) are similar (only the roles of the left- and right-hand sides of the operators are switched), so we prove only (A).

We use Lemma 5.11.23. Since (5.147) is already of the form (5.156), it suffices to show that if N is sufficiently large, any term of the form (5.148) is of the form (5.156).

Take $N := (K+1)|\kappa_2|_1$, where $K = K(\Omega)$ is as in Lemma 5.11.21. We claim

$$-j + K(|\kappa_2|_1 - N + j) - \kappa_1 \le -\kappa_1, \quad j = 0, 1, 2, \ldots, N. \tag{5.158}$$

Indeed, if $N - j \ge |\kappa_2|_1$, then $K(|\kappa_2|_1 - N + j) \le 0$, and (5.158) holds in this case. On the other hand, if $N - j \le |\kappa_2|_1$, then $j \ge N - |\kappa_2|_1 = K|\kappa_2|_1 \ge K(|\kappa_2|_1 - N + j)$, proving (5.158).

We consider a term of the form (5.148). By Lemma 5.11.21, $\widetilde{\mathscr{P}}_2^{(\kappa_2, |\kappa_2|_1 - N + j)}$ is a $(W^1, \vec{d}^1)$ partial differential operator on Ω of degree $\le K(|\kappa_2|_1 - N + j)$. Combining this with Corollary 5.8.10 shows that $\widetilde{T}^{(-j)}\widetilde{\mathscr{P}}_2^{(\kappa_2, |\kappa_2|_1 - N + j)} = \widetilde{T}^{(-j + K(|\kappa_2|_1 - N + j))}$. By hypothesis, $\mathrm{Mult}[\psi_1]\mathscr{L}^{-1}\mathrm{Mult}[\psi_2] \in \mathscr{A}^{-\kappa_1}(\Omega, (W^1, \vec{d}^1); \mathbb{C}^D, \mathbb{C}^D)$, and therefore by Proposition 5.8.6, $\widetilde{T}^{(-j + K(|\kappa_2|_1 - N + j))}\mathrm{Mult}[\psi_1]\mathscr{L}^{-1}\mathrm{Mult}[\psi_2] = \widetilde{T}^{(-j + (|\kappa_2|_1 - N + j) - \kappa_1)}$. Combining this with (5.158) and Proposition 5.8.2 shows that (5.148) is of the form $\widetilde{T}^{(-\kappa_1)}$. Lemma 5.11.22 shows that this is of the form (5.156), completing the proof. □

Proof of Theorem 5.11.17. Since $\Omega \Subset \mathfrak{M}$ was an arbitrary relatively compact, open set and $\psi_1, \psi_2 \in C_0^\infty(\Omega)$ were arbitrary, it suffices to show that

$$\begin{aligned} T &:= \mathrm{Mult}[\psi_1]\mathscr{L}^{-1}\mathrm{Mult}[\psi_2] \\ &\in \begin{cases} \mathscr{D}^{-\kappa_1}(\Omega, (W^1, \vec{d}^1), (Z, \vec{d}); \mathbb{C}^D, \mathbb{C}^D), & \text{in the Hörmander setting,} \\ \widetilde{\mathscr{D}}^{-\kappa_1}(\Omega, (W^1, \vec{d}^1), (Z, \vec{d}); \mathbb{C}^D, \mathbb{C}^D), & \text{in the general setting.} \end{cases} \end{aligned} \tag{5.159}$$

We think of T as a $D \times D$ matrix of operators.

Hörmander setting: By hypothesis, we have $T \in \mathscr{A}^{-\kappa_1}(\Omega, (W^1, \vec{d}^1); \mathbb{C}^D, \mathbb{C}^D)$. This along with Lemma 5.11.24 is exactly the statement that $T \in \mathscr{D}^{-\kappa_1}(\Omega, (W^1, \vec{d}^1), (Z, \vec{d}); \mathbb{C}^D, \mathbb{C}^D)$, which proves (5.159) and completes the proof in this case.

General setting: By hypothesis, we have $T \in \mathscr{A}^{-\kappa_1}(\Omega, (W^1, \vec{d}^1); \mathbb{C}^D, \mathbb{C}^D)$. Thus, Proposition 5.8.4 implies that $T \in \widetilde{\mathscr{A}}^{-\kappa_1}(\Omega, (W^1, \vec{d}^1); \mathbb{C}^D, \mathbb{C}^D)$. Lemma 5.11.24 shows that $\mathscr{P}_2^{(\kappa_2)}T$ can be written as a finite sum of terms of the form (5.156), while $T\mathscr{P}_2^{(\kappa_2)}$ can be written as a finite sum of terms of the form (5.157), where each $\widetilde{T}^{(-\kappa_1)} \in \mathscr{A}^{-\kappa_1}(\Omega, (W^1, \vec{d}^1); \mathbb{C}^D, \mathbb{C}^D)$. By Proposition 5.8.4, each $\widetilde{T}^{(-\kappa_1)} \in \widetilde{\mathscr{A}}^{-\kappa_1}(\Omega, (W^1, \vec{d}^1); \mathbb{C}^D, \mathbb{C}^D)$. This combining these remarks verifies that $T \in \widetilde{\mathscr{D}}^{-\kappa_1}(\Omega, (W^1, \vec{d}^1), (Z, \vec{d}); \mathbb{C}^D, \mathbb{C}^D)$, which proves (5.159) and completes the proof. □

5.12 Spectral multipliers

In this section, we briefly describe that spectral multipliers of Hörmander's sub-Laplacian are singular integrals in the sense of Section 5.2.1. This is taken from [220, Section 2.6], and we refer the reader there for further details.

In this section, we consider $\mathfrak{M}$ a smooth, **compact** manifold with smooth, strictly positive density Vol. Let $W_1, \ldots, W_r$ be Hörmander vector fields on $\mathfrak{M}$ and assign to each W_j the formal degree 1:

$$(W, 1) = \{(W_1, 1), \ldots, (W_r, 1)\}.$$

Consider Hörmander's sub-Laplacian:

$$\mathscr{L} = W_1^* W_1 + \cdots + W_r^* W_r,$$

where W_j^* denotes the formal $L^2(\mathfrak{M}, \mathrm{Vol})$ adjoint of W_j.

$\mathscr{L}$, with dense domain $C^\infty(\mathfrak{M})$, is clearly a non-negative symmetric operator on $L^2(\mathfrak{M}, \mathrm{Vol})$. By [220, Lemma 2.6.3] it is essentially self-adjoint: there is a unique self-adjoint extension. Henceforth, we identify $\mathscr{L}$ with this self-adjoint extension.

Let E be the spectral decomposition of $\mathscr{L}$, so that for a Borel measurable function $m : [0, \infty) \to \mathbb{C}$, we have

$$m(\mathscr{L}) = \int_{[0,\infty)} m(\lambda) dE(\lambda).$$

The eigenspace corresponding to eigenvalue 0 of $\mathscr{L}$ consists of only the constant functions (see [220, Lemma 2.6.2]). In other words, $E(\{0\})$ is equal to the orthogonal projection onto the constant functions.

The main theorem of this section is the following.

Theorem 5.12.1 (Theorem 2.6.6 of [220]). *Fix $t \in \mathbb{R}$ and let $m : [0, \infty) \to \mathbb{C}$ be a Borel measurable function with $m|_{(0,\infty)} \in C^\infty_{\mathrm{loc}}((0, \infty))$ which satisfies*

$$\sup_{\lambda>0} \lambda^{-t} \|(\lambda \partial_\lambda)^a m(\lambda)\| < \infty, \quad \forall a \in \mathbb{N}.$$

Then $m(\mathscr{L}) \in \mathscr{A}^{2t}(\mathfrak{M}, (W, 1))$.

Corollary 5.12.2. *For $s \in \mathbb{R}$ define*

$$\mathscr{L}^s = \int_{(0,\infty)} \lambda^s \, dE(\lambda).$$

Then $\mathscr{L}^s \in \mathscr{A}^{2s}(\mathfrak{M}, (W, 1))$.

Proof. The function

$$m(\lambda) = \begin{cases} \lambda^s & \text{if } \lambda > 0, \\ 0 & \text{if } \lambda = 0 \end{cases}$$

satisfies the hypotheses of Theorem 5.12.1 with $t = s$. □

Corollary 5.12.3. *For $s \in \mathbb{R}$, $(I + \mathscr{L})^s \in \mathscr{A}^{2s}(\mathfrak{M}, (W, 1))$.*

Proof. The function $m(\lambda) = (1 + \lambda)^s$ satisfies the hypotheses of Theorem 5.12.1 with $t = s$. □

We devote the rest of this section to a brief discussion of the proof of Theorem 5.12.1. We refer the reader to [220, Section 2.6] for a full proof. Let $\widehat{\mathscr{S}_0}(\mathbb{R}) = \{f \in \mathscr{S} : \partial_\xi^j f(0) = 0, \forall j \in \mathbb{N}\}$: the space of Schwartz functions which vanish to infinite order at 0. $\widehat{\mathscr{S}_0}(\mathbb{R})$ is a closed subspace of $\mathscr{S}(\mathbb{R})$, and we give it the subspace topology. Note that the Fourier transform is an isomorphism $\mathscr{S}_0(\mathbb{R}) \to \widehat{\mathscr{S}_0}(\mathbb{R})$, justifying the notation.

The key to Theorem 5.12.1 is the next proposition, which we state without proof. The proof can be found in [220, Section 2.6].

Proposition 5.12.4 (Proposition 2.6.11 of [220]). *Let $\mathcal{B} \subset \widehat{\mathscr{S}_0}(\mathbb{R})$ be a bounded set. Then*

$$\{(m(s^2\mathscr{L}), s) : s \in (0, 1], m \in \mathcal{B}\}$$

is a bounded set of $(W, 1)$ elementary operators supported in $\mathfrak{M}$.

Proof of Theorem 5.12.1. By [220, Lemma 2.6.4], the spectrum of $\mathscr{L}$ is discrete. Let $\lambda_0 > 0$ be the least non-zero eigenvalue of $\mathscr{L}$. Let $\psi \in C_0^\infty(\mathbb{R})$ be a non-negative function which equals 1 on $(-3\lambda_0/4, 3\lambda_0/4)$ and which equals 0 outside $(-7\lambda_0/8, 7\lambda_0/8)$. Define $\phi(\lambda) = \psi(\lambda) - \psi(4\lambda)$. Note that $1 = \psi(\lambda) + \sum_{j=1}^\infty \phi(2^{-2j}\lambda)$ and therefore

$$m(\mathscr{L}) = m(0)\psi(\mathscr{L}) + \sum_{j=1}^{\infty} \phi(2^{-2j}\mathscr{L})m(\mathscr{L}). \tag{5.160}$$

Note that $m(0)\psi(\mathscr{L}) = m(0)E(\{0\})$, where $E(\{0\})$ is projection onto the constant functions. It is straightforward to verify that the singleton set $\{(m(0)E(\{0\}), 1)\}$ is a bounded set of $(W, 1)$ elementary operators supported in $\mathfrak{M}$. From the assumptions on m and ϕ,

$$\{2^{-2jt}m(2^{2j}\lambda)\phi(\lambda) : j \in \mathbb{N}\} \subset \widehat{\mathscr{S}_0}(\mathbb{R})$$

is a bounded set. Proposition 5.12.4 implies that

$$\{(2^{-2jt}\phi(2^{-2j}\mathscr{L})m(\mathscr{L}), 2^{-j})\}$$

is a bounded set of $(W, 1)$ elementary operators supported in $\mathfrak{M}$.

Combining the above, (5.160) shows that

$$m(\mathscr{L}) = \sum_{j\in\mathbb{N}} 2^{2jt} E_j,$$

where $\{(E_j, 2^{-j}) : j \in \mathbb{N}\}$ is a bounded set of $(W, 1)$ elementary operators supported in $\mathfrak{M}$. We conclude that $m(\mathscr{L}) \in \mathscr{A}_2^{2t}(\mathfrak{M}, (W, 1)) = \mathscr{A}^{2t}(\mathfrak{M}, (W, 1))$, completing the proof. □

Remark 5.12.5. The above proof in the special case $m(\lambda) \equiv 1$ gives a decomposition of the identity operator:

$$I = \psi(\mathscr{L}) + \sum_{j=1}^{\infty} \phi(2^{-2j}\mathscr{L}) =: \sum_{j\in\mathbb{N}} D_j,$$

where $\{(D_j, 2^{-j}) : j \in \mathbb{N}\}$ is a bounded set of $(W, 1)$ elementary operators supported in $\mathfrak{M}$.

Remark 5.12.6. In this section, we focused on the case where the formal degrees are all equal to 1: $\vec{d}_1 = \vec{d}_2 = \cdots = \vec{d}_r = 1$. Recently, using the theory from this text, Lingxiao Zhang [254] obtained similar results for the more general sub-Laplacian (4.39) and other maximally subelliptic operators. To achieve this, Zhang utilized the short-time Gaussian type bounds for the heat equation established in Theorem 8.1.1 (ix). The proof of Proposition 5.12.4 given in [220, Section 2.6] does not extend to this more general setting, because that proof relies on the finite propagation speed (in terms of $\rho_{(W,1)}$) of the wave equation for $\mathscr{L}$ (which is a result of Melrose [169]) and hence cannot work for higher-order operators whose wave equations do not have finite speed of propagation.

5.13 Further reading and references

The *single-parameter* singular integrals in this chapter have a long history. They take their roots in the work on nilpotent Lie groups initiated by Folland and Stein [91] and Folland [90]. The general, quantitative study of Carnot-Carathéodory geometry due to Nagel, Stein, and Wainger [189] (see Section 3.3) shows that single-parameter Carnot–Carathéodory balls give rise to a space of homogeneous type in the sense of Coifman and Weiss [52, 59], and therefore it makes sense to pursue a theory of singular integrals on these spaces.

Singular integrals, in the single-parameter setting, of the type $\mathscr{A}^t(\Omega, (W, \vec{d}))$ (see Section 5.2.1) were first introduced under the name NIS operators by Nagel, Rosay, Stein, and Wainger [182], though an important precursor was introduced by Nagel and Stein [186]. These operators were later used by Chang, Nagel, and Stein [34] and Koenig [140]. These references all used a definition similar to $\mathscr{A}_4^t(\Omega, (W, \vec{d}))$ (see Definition 5.2.4) and restricted attention to $t > -Q_1$ so that M in Definition 5.2.4 could be taken equal to 0 (see Remark 5.2.5); thus, our definitions extend these ideas to all $t \in \mathbb{R}$ in a way that maintains that the operators form a filtered algebra (Proposition 5.8.6). In these references, an extra assumption was placed on the singular integrals: in [182, 34] it was assumed that there is a sequence $T_j \in C_0^\infty(\mathfrak{M} \times \mathfrak{M})$ satisfying the assumptions of Definition 5.2.4 uniformly in j and such that $T_j \to T$ in distribution. In [140], this was replaced with an a priori estimate. What we have used, instead of these extra assumptions, is the seemingly weaker, qualitative hypothesis that $T \in \mathrm{Hom}(C_0^\infty(\mathfrak{M}), C_0^\infty(\mathfrak{M})')$, without which we would not even be able to talk about the Schwartz kernel of T. By Theorem 5.2.12, the a priori stronger assumptions in [182, 34, 140] actually give rise to the same class of singular integrals. For example, the equality $\mathscr{A}_2^t(\Omega, (W, d)) = \mathscr{A}_4^t(\Omega, (W, d))$ gives such a sequence of smooth approximations (namely the partial sums of $T = \sum_{j \in \mathbb{N}} 2^{jt} E_j$). Elementary operators and the definitions of $\mathscr{A}_1^t(\Omega, (W, d))$ and $\mathscr{A}_2^t(\Omega, (W, d))$, on a compact manifold, were introduced by the author in [220, Chapter 2]. The single-parameter results in this text improve on the theory in [220, Chapter 2] in three ways: we do not assume the manifold is compact, we extend the definition of $\mathscr{A}_4^t$ to $t \le -Q_1$ (see Definition 5.2.4), and we introduced the fourth equivalent definition $\mathscr{A}_3^t$ (see Definition 5.2.2).

The general relationship between single-parameter singular integrals and the heat equation, described in Section 5.11.1, is classical. In the specific setting we are considering, regarding the singular integrals on Carnot–Carathéodory spaces, $\mathscr{A}_{\mathrm{loc}}^t$, it was used by Nagel and Stein in [188]. In this particular setting, though, parametrices for maximally subelliptic operators are not usually constructed via the heat equation. This is the case, for example, in [182, 34, 140], and more generally in [220, Theorem 2.4.8]. In those references, one starts with an a priori parametrix (guaranteed by the hypoellipticity of the operator and [238, Theorem 52.3]), and then one proves estimates on this parametrix via an a priori subelliptic estimate. See [220, Theorem 2.4.8] for an exposition of this method. The methods from Section 5.11.1 give an easier way to obtain good quantitative estimates on the operator, which we require for the study of nonlinear equations in Chapter 9.

We turn to describing some of the history of the singular integrals in the *multi-parameter Hörmander setting*, described in Section 5.2.2. This provides a theory of pseudo-local (see Section 5.8.1) singular integrals which form a filtered algebra (Proposition 5.8.6) and can be set up in such a way to contain both the parametrices for maximally subelliptic operators (see Section 5.11.2 and Theorem 8.1.1 (i) ⇒ (vii)) and the standard pseudo-differential operators (see Section 5.8.2).

Traditionally, the phrase "multi-parameter singular integrals" has often referred to the product theory of singular integrals, which are not pseudo-local in general. The product theory of singular integrals began with the work of R. Fefferman and Stein [86]. Another important early work is the work of Journé [136]. After these early works, many works followed. See, for example, [85, 37, 201, 180] for some works which have some relationship to the ideas in this chapter – this is just a small sampling of the many papers on the product theory of singular integrals.

The main idea of the product theory of singular integrals is that the ambient space is a product of spaces $\mathfrak{M} = \mathfrak{M}_1 \times \mathfrak{M}_2 \times \cdots \times \mathfrak{M}_\nu$, and the corresponding balls are of a product type $B((x_1, \ldots, x_\nu), (\delta_1, \ldots, \delta_\nu)) = B_1(x_1, \delta_1) \times \cdots \times B_\nu(x_\nu, \delta_\nu)$. Informally, we might say that the ν different geometries do not "overlap." This is the opposite situation to the one studied in Section 5.2.2, where the geometries all are on the same space, and completely "overlap."

The first papers which dealt with multi-parameter geometries which at least partially overlapped were on flag type singular integral operators. This began with the work of Müller, Ricci, and Stein [180, 181] on certain spectral multipliers on the Heisenberg group. The study of these flag kernels was later taken up by Nagel, Ricci, and Stein [184] and Nagel, Ricci, Stein, and Wainger [183]. Since these original works, several authors have studied these flag kernels; see, for example, [111], where end point Hardy space estimates were shown for the operators introduced in [180, 181].

In [220], the author put forth a theory which unified the above settings, at least when the underlying balls were assumed to be multi-parameter Carnot–Carathéodory balls. Special cases of this theory reproduced the product theory and the theory of flag kernels, which are inherently not pseudo-local. However, it was also shown (see [219, Section 5.1.4]) that in the multi-parameter Hörmander setting of Section 5.2.2, where the geometries completely overlap, the singular integrals described there are pseudo-local. In defining these algebras, [220] introduced the notion of elementary operators described in Definition 5.2.15. The algebra defined in Section 5.2.2 is essentially a special case of the general theory in [220]; however, by working in this special case we are able to make the theory considerably less technical and the main results have a cleaner statement.

Beyond [220] there have been a few other theories which introduce an algebra of pseudo-local, multi-parameter singular integrals which contain both the standard pseudo-differential operators and parametrices for certain subelliptic operators; see, for example, [185] and [227]. Neither of these theories seems general enough to attack general maximally subelliptic operators. We do not know the relationship between these theories and the general theory put forth in [220] or the theory from Section 5.2.2. A precursor to the general theory in [220], due to the author, developed a theory of multi-parameter convolution operators on a nilpotent Lie group which contained both the left and right invariant operators on the group [229]. This is essentially a special case of the setting in Section 5.2.2.

Even though the operators in Section 5.2.2 are pseudo-local, we did not define algebra in terms of pointwise estimates for the Schwartz kernel. This is in contrast to many of the previous theories for multi-parameter singular integrals. In fact, even when we proved the operators were pseudo-local (in Section 5.8.1), the estimates we obtained were far from optimal. In the special case studied in [229], a similar proof was used to obtain much sharper pointwise estimates for the Schwartz kernel. Even there, though, these estimates do not seem strong enough to define the algebra. In [220, Chapter 5] another definition of the algebra is given in terms of certain kinds of "test functions." This is perhaps closer to the idea of pointwise estimates for the Schwartz kernel, though it is still somewhat different.

The multi-parameter singular integrals in the general setting from Section 5.2.3 (which are not necessarily pseudo-local) seem to be new. The idea here is to axiomatize what we need from the theory in the Hörmander setting to be able to derive the function space estimates we require. The reason we must introduce these singular integrals is that the hypotheses are quantitatively invariant under the rescalings described in Section 3.5. We need this quantitative scale invariance when we move to studying nonlinear PDEs in Chapter 9. Alternatively, we could have introduced a theory similar to the one in [220, Chapter 5] which generalized the singular integrals from Section 5.2.2 and is appropriately quantitatively scale invariant, though this is somewhat technical, because one can no longer use the underlying compact support of the operator to deduce any quantitative estimates. Proceeding as in Section 5.2.3, which is sufficient for our purposes, is more straightforward.

Special cases of the theory in Section 5.2.3 include the flag kernels and product theory described above. See [220, Sections 4.1 and 4.2] for more details.

Spectral multipliers, like the ones described in Section 5.12, have been studied by many authors. In the group setting, see [127, 179, 1, 48]. A very general result of Stein [214] deals with the case where m is of Laplace transform type. The proof of Theorem 5.12.1, which can be found in [220, Section 2.6], uses the finite propagation speed of the wave equation for Hörmander's sub-Laplacian, which is a result of Melrose [169]. The proof also uses ideas of Sikora [211] and Coulhon and Duong [55].

6 Besov and Triebel–Lizorkin spaces

As described in Section 1.2 (see also Section 1.7.1), one of the main useful properties of elliptic operators is that they are locally left invertible modulo smooth functions on the classical Besov and Triebel–Lizorkin spaces; among these important spaces are the L^p Sobolev spaces ($1 < p < \infty$) and the Zygmund–Hölder spaces.

In this chapter, we generalize these spaces to achieve the same sort of left invertibility for maximally subelliptic operators. We present the results in two settings. Throughout this chapter, $\mathfrak{M}$ will be a connected C^∞ manifold of dimension $n \in \mathbb{N}_+$, endowed with a smooth, strictly positive density Vol.

The first setting is the single-parameter setting, where we are given $(W, \vec{d}) = \{(W_1, \vec{d}_1), \ldots, (W_r, \vec{d}_r)\} \subset C^\infty_{\mathrm{loc}}(\mathfrak{M}; T\mathfrak{M}) \times \mathbb{N}_+$, which are Hörmander vector fields with formal degrees. For $\mathcal{K} \Subset \mathfrak{M}$ a compact set, we define two scales of spaces of distributions supported in $\mathcal{K}$:

- The Besov spaces: for $s \in \mathbb{R}$, $p, q \in [1, \infty]$, $\mathscr{B}^s_{p,q}(\mathcal{K}, (W, \vec{d}))$.
- The Triebel–Lizorkin spaces: for $s \in \mathbb{R}$, $p \in (1, \infty)$, $q \in (1, \infty]$, $\mathscr{F}^s_{p,q}(\mathcal{K}, (W, \vec{d}))$.

The spaces are defined in such a way that each W_j acts as a differential operator of "degree" $\vec{d}_j$. Of particular interest to us are the spaces $\mathscr{F}^s_{p,2}(\mathcal{K}, (W, \vec{d}))$, which we see as a non-isotropic L^p Sobolev space, and $\mathscr{B}^s_{\infty,\infty}(\mathcal{K}, (W, \vec{d}))$, which we see as a non-isotropic Zygmund–Hölder space. These Zygmund–Hölder spaces play a key role in the study of nonlinear maximally subelliptic equations. The single-parameter Besov and Triebel–Lizorkin spaces are the right spaces in which to state the optimal regularity properties of maximally subelliptic operators.

Unfortunately the single-parameter theory described above is not powerful enough to deduce regularity properties of nonlinear maximally subelliptic equations with respect to other geometries. For example, when considering the classical (Euclidean) Besov and Triebel–Lizorkin spaces from Section 2.4, the single-parameter theory is not strong enough to deduce the regularity theory of fully nonlinear maximally subelliptic operators with respect to these classical spaces (as we will see, it is enough to deduce much of the regularity theory for linear operators). Instead, we introduce multi-parameter Besov and Triebel–Lizorkin spaces in the general setting of Section 5.1.3. These spaces generalize the single-parameter spaces described above. They give scales of spaces which can involve various geometries (for example the sub-Riemannian geometry associated with $(W, \vec{d})$ combined with the classical Euclidean geometry) and on which maximally subelliptic operators are essentially isomorphisms.

A main difficulty we face is that we do not have a simple Calderón reproducing formula in the generality in which we work.[1] We make do by repeatedly leveraging various

1 However, in some more specialized settings, we can recreate a Calderón reproducing formula; see the proof of Proposition 6.12.1.

https://doi.org/10.1515/9783111085647-006

types of almost orthogonality, though this makes the proofs somewhat more technical than in the classical setting. Also, for this same reason, we define the spaces by generalizing Proposition 2.4.18 (ii).

First, we describe some main ideas in a simple special case in Section 6.1. Then we state some of the main definitions and results of this chapter in Sections 6.2 and 6.3, before turning to the proofs in later sections.

6.1 Informal description of the norms

There are several subtleties involved in defining the Besov and Triebel–Lizorkin spaces. We defer those subtleties to later sections, and in this section informally describe the norms we use in a special case while ignoring some difficulties; the reader should consult Section 6.2 for rigorous and more general versions of the concepts described here.

We begin by reminding the reader of the definition of the classical Besov and Triebel–Lizorkin norms on $\mathbb{R}^n$, described in Section 2.4. Let $\hat{\phi} \in C_0^\infty(\mathbb{R}^n)$ be such that $\hat{\phi}(\xi) \equiv 1$ on a neighborhood of $0 \in \mathbb{R}^n$. For $j \geq 0$ set

$$\hat{\varsigma}_j(\xi) := \begin{cases} \hat{\phi}(\xi), & j = 0, \\ \hat{\phi}(\xi) - \hat{\phi}(2\xi), & j \geq 1. \end{cases}$$

For $j \geq 0$, define $D_j^1 : \mathscr{S}(\mathbb{R}^n)' \to \mathscr{S}(\mathbb{R}^n)'$ by

$$(D_j^1 f)^\wedge(\xi) := \hat{\varsigma}_j(2^{-j}\xi)\hat{f}(\xi). \tag{6.1}$$

We have

$$\|f\|_{\mathscr{B}^s_{p,q}(\mathbb{R}^n)} = \|\{2^{js} D_j^1 f\}_{j\in\mathbb{N}}\|_{\ell^q(\mathbb{N};L^p(\mathbb{R}^n))}, \quad f \in \mathscr{S}(\mathbb{R}^n)',$$
$$\|f\|_{\mathscr{F}^s_{p,q}(\mathbb{R}^n)} = \|\{2^{js} D_j^1 f\}_{j\in\mathbb{N}}\|_{L^p(\mathbb{R}^n;\ell^q(\mathbb{N}))}, \quad f \in \mathscr{S}(\mathbb{R}^n)'.$$

Formula (6.1) can be equivalently written as

$$D_j^1 f(x) = \int f(x-t)\,\mathrm{Dil}_{2^j}(\varsigma_j)(t)\,dt,$$

where $\mathrm{Dil}_{2^j}(\varsigma_j)(t) = 2^{nj}\varsigma_j(2^j t)$. Let $\eta \in C_0^\infty(\mathbb{R}^n)$ satisfy $\eta \equiv 1$ on a neighborhood of 0. Set

$$D_j^2 f(x) := \int f(x-t)\eta(t)\,\mathrm{Dil}_{2^j}(\varsigma_j)(t)\,dt.$$

It is not hard to see that we have

$$\|f\|_{\mathscr{B}^s_{p,q}(\mathbb{R}^n)} \approx \|\{2^{js} D_j^2 f\}_{j\in\mathbb{N}}\|_{\ell^q(\mathbb{N};L^p(\mathbb{R}^n))}, \quad f \in \mathscr{S}(\mathbb{R}^n)',$$
$$\|f\|_{\mathscr{F}^s_{p,q}(\mathbb{R}^n)} \approx \|\{2^{js} D_j^2 f\}_{j\in\mathbb{N}}\|_{L^p(\mathbb{R}^n;\ell^q(\mathbb{N}))}, \quad f \in \mathscr{S}(\mathbb{R}^n)'.$$

Finally, fix a compact set $\mathcal{K} \Subset \mathbb{R}^n$ and let $\psi \in C_0^\infty(\mathbb{R}^n)$ have $\psi \equiv 1$ on a neighborhood of $\mathcal{K}$. Set

$$\begin{aligned} D_j^3 f(x) &:= \psi(x) \int f(x-t)\eta(t) \operatorname{Dil}_{2^j}(\varsigma_j)(t)\, dt \\ &= \psi(x) \int f(e^{-t\cdot\partial}x)\eta(t) \operatorname{Dil}_{2^j}(\varsigma_j)(t)\, dt. \end{aligned} \tag{6.2}$$

We have

$$\|f\|_{\mathscr{B}^s_{p,q}(\mathbb{R}^n)} \approx \|\{2^{js} D_j^3 f\}_{j\in\mathbb{N}}\|_{\ell^q(\mathbb{N};L^p(\mathbb{R}^n))}, \quad f \in \mathscr{S}(\mathbb{R}^n)' \text{ with } \operatorname{supp}(f) \subseteq \mathcal{K},$$
$$\|f\|_{\mathscr{F}^s_{p,q}(\mathbb{R}^n)} \approx \|\{2^{js} D_j^3 f\}_{j\in\mathbb{N}}\|_{L^p(\mathbb{R}^n;\ell^q(\mathbb{N}))}, \quad f \in \mathscr{S}(\mathbb{R}^n)' \text{ with } \operatorname{supp}(f) \subseteq \mathcal{K}.$$

Let us now turn to the simple case where $\mathfrak{M} = \mathbb{R}^2$ and we are given the list of Hörmander vector fields with formal degrees $(W, \vec{d}) = \{(\partial_x, 1), (x\partial_y, 1)\}$, and fix a compact set $\mathcal{K} \Subset \mathbb{R}^2$. We wish to define the norms $\|f\|_{\mathscr{B}^s_{p,q}(W,\vec{d})}$ and $\|f\|_{\mathscr{F}^s_{p,q}(W,\vec{d})}$ when $\operatorname{supp}(f) \subseteq \mathcal{K}$. We will define these spaces so that $\partial_x, x\partial_y : \mathscr{B}^s_{p,q}(\mathcal{K}, (W, \vec{d})) \to \mathscr{B}^{s-1}_{p,q}(\mathcal{K}, (W, \vec{d}))$, and similarly for $\mathscr{F}^s_{p,q}(\mathcal{K}, (W, \vec{d}))$. In particular, we have $\partial_y = [\partial_x, x\partial_y] : \mathscr{B}^s_{p,q}(\mathcal{K}, (W, \vec{d})) \to \mathscr{B}^{s-2}_{p,q}(\mathcal{K}, (W, \vec{d}))$.

Let

$$(X, d) = \{(X_1, d_1), (X_2, d_2), (X_3, d_3)\} := \{(\partial_x, 1), (x\partial_y, 1), (\partial_y, 2)\};$$

see Example 3.3.6. Take $\hat{\phi} \in C_0^\infty(\mathbb{R}^3)$ as above, so that $\hat{\phi} \equiv 1$ on a neighborhood of 0. Define

$$\hat{\varsigma}_j(\xi_1, \xi_2, \xi_3) := \begin{cases} \hat{\phi}(\xi_1, \xi_2, \xi_3), & j = 0, \\ \hat{\phi}(\xi_1, \xi_2, \xi_3) - \hat{\phi}(2\xi_1, 2\xi_2, 2^2\xi_3) = \hat{\phi}(\xi) - \hat{\phi}(2^d\xi), & j \geq 1. \end{cases}$$

Fix $\eta \in C_0^\infty(\mathbb{R}^3)$ with $\eta \equiv 1$ on a neighborhood of 0 and η with small support. Fix $\psi \in C_0^\infty \mathbb{R}^2$ with $\psi \equiv 1$ on a neighborhood of $\mathcal{K}$. As an analog to (6.2), we set

$$D_j f(x) := \psi(x) \int f(e^{-t\cdot X}x)\eta(t) \operatorname{Dil}^d_{2^j}(\varsigma_j)(t)\, dt,$$

where

$$\operatorname{Dil}^d_{2^j}(\varsigma_j)(t_1, t_2, t_3) = 2^{4j}\varsigma_j(2^j t_1, 2^j t_2, 2^{2j} t_3);$$

see Definition 4.1.12. Note that because ψ has compact support and η has small support, the expression $e^{-t\cdot X}x$, $x \in \operatorname{supp}(\psi)$ and $t \in \operatorname{supp}(\eta)$ will always make sense, as long as the support of η is chosen sufficiently small – how small will, in general, depend on the vector fields X.

For f with $\mathrm{supp}(f) \Subset \mathcal{K}$, we set

$$\|f\|_{\mathscr{B}^s_{p,q}(W,d)} = \left\|\{2^{js} D_j f\}_{j\in\mathbb{N}}\right\|_{\ell^q(\mathbb{N};L^p(\mathbb{R}^2))}, \quad \mathrm{supp}(f) \subseteq \mathcal{K},$$
$$\|f\|_{\mathscr{F}^s_{p,q}(W,d)} = \left\|\{2^{js} D_j f\}_{j\in\mathbb{N}}\right\|_{L^p(\mathbb{R}^2;\ell^q(\mathbb{N}))}, \quad \mathrm{supp}(f) \subseteq \mathcal{K}.$$

We have made many choices, but we will show that the equivalence class of the above norms does not depend on these choices.

Away from $x = 0$, ∂_x and $x\partial_y$ span the tangent space to $\mathbb{R}^2$. Because of this (and the fact that ∂_x and $x\partial_y$ have formal degree 1), Theorem 6.6.7 shows that the norms $\|\cdot\|_{\mathscr{B}^s_{p,q}(W,d)}$ and $\|\cdot\|_{\mathscr{F}^s_{p,q}(W,d)}$ are locally equivalent to the standard norms $\|\cdot\|_{\mathscr{B}^s_{p,q}(\mathbb{R}^2)}$ and $\|\cdot\|_{\mathscr{F}^s_{p,q}(\mathbb{R}^2)}$, *away from* $x = 0$. However, as $x \to 0$, these norms are not comparable: near $x = 0$ and when $s > 0$, $\|\cdot\|_{\mathscr{B}^s_{p,q}(\mathbb{R}^2)}$ and $\|\cdot\|_{\mathscr{F}^s_{p,q}(\mathbb{R}^2)}$ are larger (stronger) norms than $\|\cdot\|_{\mathscr{B}^s_{p,q}(W,d)}$ and $\|\cdot\|_{\mathscr{F}^s_{p,q}(W,d)}$, respectively. Theorem 6.6.13 and Remark 6.6.14 describe how to compare these norms.

For example, in Corollary 6.2.14, we see that for $\kappa \in \mathbb{N}$ and $1 < p < \infty$,

$$\|f\|_{\mathscr{F}^\kappa_{p,2}(W,d)} \approx \sum_{|\alpha|\le\kappa} \|(\partial_x, x\partial_y)^\alpha f\|_{L^p(\mathbb{R}^2)}, \quad \mathrm{supp}(f) \subseteq \mathcal{K}. \tag{6.3}$$

From (6.3) it can be seen that, near $x = 0$, the best one can say is

$$\|f\|_{\mathscr{F}^\kappa_{p,2}(\mathbb{R}^2)} \lesssim \|f\|_{\mathscr{F}^{2\kappa}_{p,2}(W,d)} \lesssim \|f\|_{\mathscr{F}^{2\kappa}_{p,2}(\mathbb{R}^2)}. \tag{6.4}$$

One can think of (6.4) as reflecting the fact that $\partial_y = [\partial_x, x\partial_y]$ is treated as a differential operator of degree 2 with respect to the norms $\|\cdot\|_{\mathscr{F}^s_{p,2}(W,d)}$.

6.2 The single-parameter spaces

Let $(W, d) = \{(W_1, d_1), \dots, (W_r, d_r)\} \subset C^\infty_{\mathrm{loc}}(\mathfrak{M}; T\mathfrak{M}) \times \mathbb{N}_+$ be Hörmander vector fields with formal degrees on $\mathfrak{M}$. Fix a compact set $\mathcal{K} \Subset \mathfrak{M}$. Our goal is to define the following spaces consisting of distributions in $C^\infty_0(\mathfrak{M})'$ with support in $\mathcal{K}$:

- The Besov spaces: for $s \in \mathbb{R}$, $p, q \in [1,\infty]$, $\mathscr{B}^s_{p,q}(\mathcal{K}, (W, d))$.
- The Triebel–Lizorkin spaces: for $s \in \mathbb{R}$, $p \in (1,\infty)$, $q \in (1,\infty]$, $\mathscr{F}^s_{p,q}(\mathcal{K}, (W, d))$.

Notation 6.2.1. $\mathscr{V}$ will denote one of the spaces:

$$\{\ell^q(\mathbb{N}; L^p(\mathfrak{M}, \mathrm{Vol})) : p, q \in [1,\infty]\} \bigcup \{L^p(\mathfrak{M}, \mathrm{Vol}; \ell^q(\mathbb{N})) : p \in (1,\infty), q \in (1,\infty]\}.$$

See (2.28) and (2.29) for further details.

Notation 6.2.2. For $s \in \mathbb{R}$, $\mathscr{X}^s(\mathcal{K},(W,\vec{d}))$ will denote any one of the spaces (to be defined)

$$\{\mathscr{B}^s_{p,q}(\mathcal{K},(W,\vec{d})) : p,q \in [1,\infty]\} \bigcup \{\mathscr{F}^s_{p,q}(\mathcal{K},(W,\vec{d})) : p \in (1,\infty), q \in (1,\infty]\}.$$

We take

$$\mathscr{V} := \begin{cases} \ell^q(\mathbb{N}; L^p(\mathfrak{M},\mathrm{Vol})), & \text{when } \mathscr{X}^s(\mathcal{K},(W,\vec{d})) = \mathscr{B}^s_{p,q}(\mathcal{K},(W,\vec{d})), \\ L^p(\mathfrak{M},\mathrm{Vol};\ell^q(\mathbb{N})), & \text{when } \mathscr{X}^s(\mathcal{K},(W,\vec{d})) = \mathscr{F}^s_{p,q}(\mathcal{K},(W,\vec{d})). \end{cases}$$

Definition 6.2.3. Let $\mathcal{E}$ be a bounded set of $(W,\vec{d})$ elementary operators (see Definition 5.2.8). We set, for $f \in C_0^\infty(\mathfrak{M})'$ and $s \in \mathbb{R}$,

$$\|f\|_{\mathscr{V},s,\mathcal{E}} := \sup_{\{(E_j,2^{-j}):j\in\mathbb{N}\}\subseteq\mathcal{E}} \|\{2^{js}E_jf\}_{j\in\mathbb{N}}\|_{\mathscr{V}}.$$

This defines an extended[2] semi-norm on $C_0^\infty(\mathfrak{M})'$.

Motivated by Proposition 2.4.18 we make the next definition.

Definition 6.2.4. For $s \in \mathbb{R}$, we define $\mathscr{X}^s(\mathcal{K},(W,\vec{d}))$ to be the space of all those $f \in C_0^\infty(\mathfrak{M})'$ such that $\mathrm{supp}(f) \subseteq \mathcal{K}$ and for every bounded set of $(W,\vec{d})$ elementary operators, $\mathcal{E}$, we have $\|f\|_{\mathscr{V},s,\mathcal{E}} < \infty$.

It is clear from the definition that $\mathscr{X}^s(\mathcal{K},(W,\vec{d}))$ can be given the structure of a locally convex topological vector space. In fact, $\mathscr{X}^s(\mathcal{K},(W,\vec{d}))$ is a Banach space. We turn to introducing a norm on $\mathscr{X}^s(\mathcal{K},(W,\vec{d}))$ which achieves this.

Fix $\psi \in C_0^\infty(\mathfrak{M})$ with $\psi \equiv 1$ on a neighborhood of $\mathcal{K}$. Let $\Omega \Subset \mathfrak{M}$ be an open set with $\psi \in C_0^\infty(\Omega)$. By Proposition 5.8.3, $\mathrm{Mult}[\psi] \in \mathscr{A}^0(\Omega,(W,\vec{d})) = \mathscr{A}_2^0(\Omega,(W,\vec{d}))$. Thus, there exists a bounded set of $(W,\vec{d})$ elementary operators, $\mathcal{D}_0 = \{(D_j,2^{-j}) : j \in \mathbb{N}\}$ with $\mathrm{Mult}[\psi] = \sum_{j\in\mathbb{N}} D_j$ (see Proposition 5.5.10 for the convergence of this sum).

Proposition 6.2.5. $\|\cdot\|_{\mathscr{V},s,\mathcal{D}_0}$ *defines a norm on* $\mathscr{X}^s(\mathcal{K},(W,\vec{d}))$. *With respect to this norm,* $\mathscr{X}^s(\mathcal{K},(W,\vec{d}))$ *is a Banach space.*

Proposition 6.2.6. *The equivalence class of the norm* $\|\cdot\|_{\mathscr{V},s,\mathcal{D}_0}$ *does not depend on any of the choices we made above. More precisely, if* $\tilde{\psi} \in C_0^\infty(\mathfrak{M})$ *is such that* $\tilde{\psi} \equiv 1$ *on a neighborhood of* $\mathcal{K}$ *and* $\widetilde{\mathcal{D}} = \{(\widetilde{D}_j,2^{-j}) : j \in \mathbb{N}\}$ *is a bounded set of* $(W,\vec{d})$ *elementary operators with* $\mathrm{Mult}[\tilde{\psi}] = \sum_{j\in\mathbb{N}} \widetilde{D}_j$, *then*

$$\|f\|_{\mathscr{V},s,\mathcal{D}_0} \approx \|f\|_{\mathscr{V},s,\widetilde{\mathcal{D}}}, \quad \forall f \in \mathscr{X}^s(\mathcal{K},(W,\vec{d})).$$

2 It is an extended semi-norm in that it satisfies all the axioms of a semi-norm, except that it may take the value ∞.

Here, the implicit constants do not depend on f, but may depend on any of the other choices.

In light of Proposition 6.2.6, the next definition makes sense.

Definition 6.2.7. For $f \in \mathscr{X}^s(\mathcal{K},(W,\vec{d}))$, we write

$$\|f\|_{\mathscr{X}^s(W,\vec{d})} := \|f\|_{\mathscr{V},s,\mathcal{D}_0} = \|\{2^{js}D_jf\}_{j\in\mathbb{N}}\|_{\mathscr{V}}.$$

Remark 6.2.8. Only the equivalence class of the norm $\|\cdot\|_{\mathscr{X}^s(W,\vec{d})}$ is well-defined on $\mathscr{X}^s(\mathcal{K},(W,\vec{d}))$, though that is all we require. The notation $\|\cdot\|_{\mathscr{X}^s(W,\vec{d})}$ does not involve the compact set $\mathcal{K}$, since given two compact sets $\mathcal{K}$ and $\mathcal{K}'$, we may pick ψ so that $\psi\equiv 1$ on a neighborhood of $\mathcal{K}\cup\mathcal{K}'$, and we may therefore use the same norm on $\mathscr{X}^s(\mathcal{K},(W,\vec{d}))$ and $\mathscr{X}^s(\mathcal{K}',(W,\vec{d}))$.

Remark 6.2.9. Proposition 6.2.6 only shows that the equivalence class of the norm $\|f\|_{\mathscr{X}^s(W,\vec{d})}$ is well-defined for $f \in \mathscr{X}^s(\mathcal{K},(W,\vec{d}))$, not for all $f \in C_0^\infty(\mathfrak{M})'$ with $\operatorname{supp} f \subseteq \mathcal{K}$. In particular, we **do not claim** here that if $\|f\|_{\mathscr{X}^s(W,\vec{d})}<\infty$, then $f \in \mathscr{X}^s(\mathcal{K},(W,\vec{d}))$. This problem can be fixed by modifying the norm; see Remark 6.4.6 and also Section 6.10 for further comments.

Theorem 6.2.10. *Let Ω be an open set with $\Omega \Subset \mathcal{K}$. Then operators in $\mathscr{A}^t(\Omega,(W,\vec{d}))$ are bounded $\mathscr{X}^s(\mathcal{K},(W,\vec{d})) \to \mathscr{X}^{s-t}(\mathcal{K},(W,\vec{d}))$.*

The next two results show that the vector field W_j acts like a differential operator of degree $\vec{d}_j$ on these spaces.

Proposition 6.2.11. *For any ordered multi-index α, W^α defines a bounded operator $\mathscr{X}^s(\mathcal{K},(W,\vec{d})) \to \mathscr{X}^{s-\deg_{\vec{d}}(\alpha)}(\mathcal{K},(W,\vec{d}))$.*

Theorem 6.2.12. *Fix $\kappa \in \mathbb{N}_+$ such that $\vec{d}_j$ divides κ for $1\le j\le r$. Set $n_j := \kappa/\vec{d}_j \in \mathbb{N}_+$ and $\mathscr{L} := \sum_{j=1}^r (W_j^{n_j})^* W_j^{n_j}$, the sub-Laplacian (see Section 4.5). Then, for $f \in C_0^\infty(\mathfrak{M})'$, the following are equivalent:*

(i) $f \in \mathscr{X}^s(\mathcal{K},(W,\vec{d}))$.
(ii) *For every ordered multi-index α with $\deg_{\vec{d}}(\alpha)\le\kappa$, we have $W^\alpha f \in \mathscr{X}^{s-\kappa}(\mathcal{K},(W,\vec{d}))$.*
(iii) *$f \in \mathscr{X}^{s-\kappa}(\mathcal{K},(W,\vec{d}))$ and $W_j^{n_j} f \in \mathscr{X}^{s-\kappa}(\mathcal{K},(W,\vec{d}))$ for $1\le j\le r$.*
(iv) *$f, \mathscr{L}f \in \mathscr{X}^{s-2\kappa}(\mathcal{K},(W,\vec{d}))$.*

Furthermore, in this case we have

$$\|f\|_{\mathscr{X}^s(W,\vec{d})} \approx \sum_{\deg_{\vec{d}}(\alpha)\le\kappa} \|W^\alpha f\|_{\mathscr{X}^{s-\kappa}(W,\vec{d})} \approx \|f\|_{\mathscr{X}^{s-\kappa}(W,\vec{d})} + \sum_{j=1}^r \|W_j^{n_j} f\|_{\mathscr{X}^{s-\kappa}(W,\vec{d})}$$
$$\approx \|f\|_{\mathscr{X}^{s-2\kappa}(W,\vec{d})} + \|\mathscr{L}f\|_{\mathscr{X}^{s-2\kappa}(W,\vec{d})}.$$

One space of particular importance is $\mathscr{F}^s_{p,2}$ (for $1 < p < \infty$), which can be thought of as a non-isotropic L^p Sobolev space. This interpretation is justified by the next two results.

Proposition 6.2.13. *For $1 < p < \infty$, we have $\mathscr{F}^0_{p,2}(\mathcal{K}, (W, d)) = L^p(\mathcal{K}, \mathrm{Vol})$ and*

$$\|f\|_{\mathscr{F}^0_{p,2}} \approx \|f\|_{L^p(\mathcal{K},\mathrm{Vol})}, \quad \forall f \in L^p(\mathcal{K}, \mathrm{Vol}).$$

Corollary 6.2.14. *Fix $\kappa \in \mathbb{N}_+$ such that d_j divides κ for $1 \le j \le r$; set $n_j := \kappa / d_j \in \mathbb{N}_+$. Then, for $1 < p < \infty$ and $f \in C_0^\infty(\mathfrak{M})'$, the following are equivalent:*
(i) $f \in \mathscr{F}^\kappa_{p,2}(\mathcal{K}, (W, d))$.
(ii) *For every ordered multi-index α with $\deg_d(\alpha) \le \kappa$, we have $W^\alpha f \in L^p(\mathcal{K}, \mathrm{Vol})$.*
(iii) $f \in L^p(\mathcal{K}, \mathrm{Vol})$ *and* $W_j^{n_j} f \in L^p(\mathcal{K}, \mathrm{Vol})$ *for* $1 \le j \le r$.

Furthermore, in this case we have

$$\|f\|_{\mathscr{F}^\kappa_{p,2}(W,d)} \approx \sum_{\deg_d(\alpha) \le \kappa} \|W^\alpha f\|_{L^p} \approx \|f\|_{L^p} + \sum_{j=1}^r \|W_j^{n_j} f\|_{L^p}. \tag{6.5}$$

Also, if $\mathscr{L} := \sum_{j=1}^r (W_j^{n_j})^* W_j^{n_j}$ is the sub-Laplacian, then the following are equivalent:
(a) $f \in \mathscr{F}^{2\kappa}_{p,2}(\mathcal{K}, (W, d))$.
(b) $f, \mathscr{L} f \in L^p(\mathcal{K}, \mathrm{Vol})$.

Furthermore, in this case we have

$$\|f\|_{\mathscr{F}^{2\kappa}_{p,2}(W,d)} \approx \|f\|_{L^p} + \|\mathscr{L} f\|_{L^p}.$$

Proof. This follows by combining Proposition 6.2.13 and Theorem 6.2.12. □

All of the above results follow from more general results in the multi-parameter setting. See Section 6.3.

Finally, there is one last space which will be important for our applications, namely, the Zygmund–Hölder space, for $s > 0$,

$$\mathscr{C}^s(\mathcal{K}, (W, d)) := \mathscr{B}^s_{\infty,\infty}(\mathcal{K}, (W, d)).$$

This space plays the same role in the study of nonlinear maximally subelliptic equations that the classical Zygmund–Hölder spaces play for nonlinear elliptic equations. We defer discussion of these spaces to Chapter 7.

The classical Besov and Triebel–Lizorkin spaces are a special case of the above spaces: with the right choice of (W, d), these spaces exactly correspond to the classical spaces with the restriction that the distributions are supported in $\mathcal{K}$. See Section 6.6.1 for a further discussion.

See Section 6.12 for another approach to these function spaces using spectral multipliers.

6.3 The multi-parameter spaces

We work in the general setting described in Section 5.1.3. Thus, for some $\nu \in \mathbb{N}_+$ we are given ν lists of vector fields with formal degrees:

$$(W^\mu, d^\mu) = \{(W_1^\mu, d_1^\mu), \ldots, (W_{r_\mu}^\mu, d_{r_\mu}^\mu)\} \subset C_{\mathrm{loc}}^\infty(\mathfrak{M}; T\mathfrak{M}) \times \mathbb{N}_+.$$

We assume that $(W^1, d^1), \ldots, (W^\nu, d^\nu)$ pairwise locally weakly approximately commute and that for each $\mu \in \{1, \ldots, \nu\}$, $\mathrm{Gen}((W^\mu, d^\mu))$ is locally finitely generated on $\mathfrak{M}$. We define

$$\begin{aligned}(W, \vec{d}) &= \{(W_1, \vec{d}_1), \ldots, (W_r, \vec{d}_r)\} \\ &:= (W^1, d^1) \boxtimes (W^2, d^2) \boxtimes \cdots \boxtimes (W^\nu, d^\nu) \subset C_{\mathrm{loc}}^\infty(\mathfrak{M}; T\mathfrak{M}) \times (\mathbb{N}^\nu \setminus \{0\}).\end{aligned}$$

Fix $\mathcal{K} \Subset \mathfrak{M}$. The goal of this section is to define the following spaces consisting of distributions in $C_{W,0}^\infty(\mathfrak{M})'$ with support in $\mathcal{K}$:
- The Besov spaces: for $s \in \mathbb{R}^\nu$, $p, q, \in [1, \infty]$, $\mathscr{B}_{p,q}^s(\mathcal{K}, (W, \vec{d}))$.
- The Triebel–Lizorkin spaces: for $s \in \mathbb{R}^\nu$, $p \in (1, \infty)$, $q \in (1, \infty]$, $\mathscr{F}_{p,q}^s(\mathcal{K}, (W, \vec{d}))$.

Similar to Notations 6.2.1 and 6.2.2, we introduce the following.

Notation 6.3.1. $\mathscr{V}$ will denote one of the spaces:

$$\{\ell^q(\mathbb{N}^\nu; L^p(\mathfrak{M}, \mathrm{Vol})) : p, q \in [1, \infty]\} \bigcup \{L^p(\mathfrak{M}, \mathrm{Vol}; \ell^q(\mathbb{N}^\nu)) : p \in (1, \infty), q \in (1, \infty]\}.$$

For a sequence of distributions $\{f_j\}_{j\in\mathbb{N}} \subset C_0^\infty(\mathfrak{M})'$, when we consider $\|\{f_j\}_{j\in\mathbb{N}}\|_{\mathscr{V}}$ we set this equal to ∞ if any of the f_j does not agree with integration against an $L^1_{\mathrm{loc}}(\mathfrak{M}, \mathrm{Vol})$ function. If all f_j agree with integration against an $L^1_{\mathrm{loc}}(\mathfrak{M}, \mathrm{Vol})$ function, $\|\{f_j\}_{j\in\mathbb{N}}\|_{\mathscr{V}}$ has the usual definition (see (2.28) and (2.29)).

Notation 6.3.2. For $s \in \mathbb{R}$, $\mathscr{X}^s(\mathcal{K}, (W, \vec{d}))$ will denote any one of the spaces (to be defined)

$$\{\mathscr{B}_{p,q}^s(\mathcal{K}, (W, \vec{d})) : p, q \in [1, \infty]\} \bigcup \{\mathscr{F}_{p,q}^s(\mathcal{K}, (W, \vec{d})) : p \in (1, \infty), q \in (1, \infty]\}.$$

When $\mathscr{X}^s(\mathcal{K}, (W, \vec{d})) = \mathscr{B}_{p,q}^s(\mathcal{K}, (W, \vec{d}))$ we take $\mathscr{V}$ to be $\ell^q(\mathbb{N}^\nu; L^p(\mathfrak{M}, \mathrm{Vol}))$, and when $\mathscr{X}^s(\mathcal{K}, (W, \vec{d})) = \mathscr{F}_{p,q}^s(\mathcal{K}, (W, \vec{d}))$ we take $\mathscr{V}$ to be $L^p(\mathfrak{M}, \mathrm{Vol}; \ell^q(\mathbb{N}^\nu))$.

Definition 6.3.3. Let $\mathcal{E}$ be a bounded set of generalized $(W, \vec{d})$ elementary operators (see Definition 5.2.27). We set, for $f \in C_{W,0}^\infty(\mathfrak{M})'$ and $s \in \mathbb{R}^\nu$,

$$\|f\|_{\mathcal{V},s,\mathcal{E}} := \sup_{\{(E_j,2^{-j}):j\in\mathbb{N}^\nu\}\subseteq\mathcal{E}} \left\|\{2^{j\cdot s}E_jf\}\right\|_{\mathcal{V}}.$$

This defines an extended semi-norm on $C^\infty_{W,0}(\mathfrak{M})'$.

Remark 6.3.4. In Definition 6.3.3, we applied E_j to f, where $(E_j, 2^{-j}) \in \mathcal{E}$ and $f \in C^\infty_{W,0}(\mathfrak{M})'$. If one looks at the definition of generalized $(W, \vec{d})$ pre-elementary operators (Definition 5.2.26), E_j is only given as an element of $\mathrm{Hom}(C^\infty_{W,0}(\mathfrak{M}), C^\infty_{W,\mathrm{loc}}(\mathfrak{M}))$. However, as part of Definition 5.2.26 it is also assumed that $E_j^* \in \mathrm{Hom}(C^\infty_{W,0}(\mathfrak{M}), C^\infty_{W,\mathrm{loc}}(\mathfrak{M}))$. Thus, by duality, E_j extends to an operator in $\mathrm{Hom}(C^\infty_{W,\mathrm{loc}}(\mathfrak{M})', C^\infty_{W,0}(\mathfrak{M})')$. Combining this with the assumed compact support of E_j (also in Definition 5.2.26), we see that $E_j \in \mathrm{Hom}(C^\infty_{W,0}(\mathfrak{M})', C^\infty_{W,\mathrm{loc}}(\mathfrak{M})')$. In particular, E_jf makes sense for $f \in C^\infty_{W,0}(\mathfrak{M})'$.

Definition 6.3.5. For $s \in \mathbb{R}^\nu$, we define $\mathscr{X}^s(\mathcal{K}, (W, \vec{d}))$ to be the space of all those $f \in C^\infty_{W,0}(\mathfrak{M})'$ such that $\mathrm{supp}(f) \subseteq \mathcal{K}$ and for every bounded set of generalized $(W, \vec{d})$ elementary operators, $\mathcal{E}$, we have $\|f\|_{\mathcal{V},s,\mathcal{E}} < \infty$.

We next present a norm which turns $\mathscr{X}^s(\mathcal{K}, (W, \vec{d}))$ into a Banach space. Fix $\psi \in C_0^\infty(\mathfrak{M})$ with $\psi \equiv 1$ on a neighborhood of $\mathcal{K}$. Let $\Omega \Subset \mathfrak{M}$ be an open set with $\psi \in C_0^\infty(\Omega)$. By Proposition 5.8.3, $\mathrm{Mult}[\psi] \in \widetilde{\mathscr{A}}^0(\Omega, (W, \vec{d})) = \widetilde{\mathscr{A}}^0_2(\Omega, (W, \vec{d}))$. Thus, there exists a bounded set of generalized $(W, \vec{d})$ elementary operators, $\mathcal{D}_0 = \{(D_j, 2^{-j}) : j \in \mathbb{N}\}$ with $\mathrm{Mult}[\psi] = \sum_{j\in\mathbb{N}} D_j$ (see Proposition 5.5.10 for the convergence of this sum).

Proposition 6.3.6. *$\|\cdot\|_{\mathcal{V},s,\mathcal{D}_0}$ defines a norm on $\mathscr{X}^s(\mathcal{K}, (W, \vec{d}))$. With respect to this norm, $\mathscr{X}^s(\mathcal{K}, (W, \vec{d}))$ is a Banach space.*

Proposition 6.3.7. *The equivalence class of the norm $\|\cdot\|_{\mathcal{V},s,\mathcal{D}_0}$ does not depend on any of the choices we made above. More precisely, if $\tilde{\psi} \in C_0^\infty(\mathfrak{M})$ is such that $\tilde{\psi} \equiv 1$ on a neighborhood of $\mathcal{K}$ and $\widetilde{\mathcal{D}} = \{(\widetilde{D}_j, 2^{-j}) : j \in \mathbb{N}\}$ is a bounded set of generalized $(W, \vec{d})$ elementary operators with $\mathrm{Mult}[\tilde{\psi}] = \sum_{j\in\mathbb{N}} \widetilde{D}_j$, then*

$$\|f\|_{\mathcal{V},s,\mathcal{D}_0} \approx \|f\|_{\mathcal{V},s,\widetilde{\mathcal{D}}}, \quad \forall f \in \mathscr{X}^s(\mathcal{K}, (W, \vec{d})).$$

Here, the implicit constants do not depend on f, but may depend on any of the other choices.

In light of Proposition 6.3.7, the next definition makes sense.

Definition 6.3.8. For $f \in \mathscr{X}^s(\mathcal{K}, (W, \vec{d}))$, we write

$$\|f\|_{\mathscr{X}^s(W,\vec{d})} := \|f\|_{\mathcal{V},s,\mathcal{D}_0} = \left\|\{2^{j\cdot s}D_j\}_{j\in\mathbb{N}}\right\|_{\mathcal{V}}.$$

Remarks 6.2.8 and 6.2.9 also apply in this more general setting. In particular, we have the following.

Remark 6.3.9. The notation $\|\cdot\|_{\mathscr{X}^s(W,\vec{d})}$ does not involve the compact set $\mathcal{K}$, since given two compact sets $\mathcal{K}$ and $\mathcal{K}'$, we may pick ψ so that $\psi \equiv 1$ on a neighborhood of $\mathcal{K} \cup \mathcal{K}'$, and we may therefore use the same norm on $\mathscr{X}^s(\mathcal{K}, (W, \vec{d}))$ and $\mathscr{X}^s(\mathcal{K}', (W, \vec{d}))$.

Theorem 6.3.10. *Let Ω be an open set with $\Omega \subseteq \mathcal{K}$. Then operators in $\widetilde{\mathscr{A}}^t(\Omega, (W, \vec{d}))$ are bounded $\mathscr{X}^s(\mathcal{K}, (W, \vec{d})) \to \mathscr{X}^{s-t}(\mathcal{K}, (W, \vec{d}))$.*

The above results are proved in Section 6.5. Analogs of all the results in Section 6.2 hold in this more general setting as well, and will be presented throughout the chapter. In particular, similar to Proposition 6.2.11 and Theorem 6.2.12, W_j acts like a differential operator of degree $\vec{d}_j \in \mathbb{N}^\nu$ on the spaces $\mathscr{X}^s(\mathcal{K}, (W, \vec{d}))$ (see Corollary 6.5.11, Proposition 6.5.12, and Corollary 6.5.13). Similar to Proposition 6.2.13 and Corollary 6.2.14, $\mathscr{F}^s_{p,2}(\mathcal{K}, (W, \vec{d}))$ can be viewed as non-isotropic L^p Sobolev spaces for $1 < p < \infty$ (see Section 6.9).

Definition 6.3.11. For $N \in \mathbb{N}_+$, we define the vector-valued Besov ($\mathscr{B}^s_{p,q}(\mathcal{K}, (W, \vec{d}); \mathbb{C}^N)$) and Triebel–Lizorkin ($\mathscr{F}^s_{p,q}(\mathcal{K}, (W, \vec{d}); \mathbb{C}^N)$), by letting $\mathscr{X}^s(\mathcal{K}, (W, \vec{d}); \mathbb{C}^N)$ consist of those distributions $u = (u_1, \ldots, u_N)$ with each $u_j \in \mathscr{X}^s(\mathcal{K}, (W, \vec{d}))$. We set

$$\|u\|_{\mathscr{X}^s((W,\vec{d});\mathbb{C}^N)} := \sum_{j=1}^N \|u_j\|_{\mathscr{X}^s(W,\vec{d})}.$$

6.4 The main estimate

This section is devoted to the main technical estimate which is at the heart of many of the proofs in this chapter. Recall that in Section 6.3, we fixed $\psi \in C_0^\infty(\mathfrak{M})$ with $\psi \equiv 1$ on $\mathcal{K}$ and a bounded set of generalized $(W, \vec{d})$ elementary operators, $\mathcal{D}_0 = \{(D_j, 2^{-j}) : j \in \mathbb{N}^\nu\}$, with $\mathrm{Mult}[\psi] = \sum_{j\in\mathbb{N}^\nu} D_j$.

Notation 6.4.1. For any sequence of operators indexed by $j \in \mathbb{N}^\nu$, E_j, we define $E_j := 0$ for $j \in \mathbb{Z}^\nu \setminus \mathbb{N}^\nu$. For example, using this notation, we have $\mathrm{Mult}[\psi] = \sum_{j\in\mathbb{Z}^\nu} D_j$.

For $j, k, l \in \mathbb{Z}^\nu$ and $N \geq 1$, set

$$F_{N,j,k,l} := 2^{N|k|+|l|} D_{j+k} D_{j+k+l}.$$

Note that using Notation 6.4.1, $F_{N,j,k,l} = 0$ if $j + k \in \mathbb{Z}^\nu \setminus \mathbb{N}^\nu$ or $j + k + l \in \mathbb{Z}^\nu \setminus \mathbb{N}^\nu$. Set

$$\mathcal{D}_N := \{(F_{N,j,k,l}, 2^{-(j+k)}) : j, k, l \in \mathbb{Z}^\nu, j + k \in \mathbb{N}^\nu, |l| > |k|\}. \tag{6.6}$$

Lemma 6.4.2. *For $N \geq 1$, $\mathcal{D}_N$ is a bounded set of generalized $(W, \vec{d})$ elementary operators. Moreover, if $\mathcal{D}_0$ is a bounded set of generalized $(W, \vec{d})$ elementary operators supported in Ω, then so is $\mathcal{D}_N$.*

Proof. Let $\Omega \Subset \mathfrak{M}$ be open and relatively compact such that $\mathcal{D}_0$ is a bounded set of generalized $(W, \vec{d})$ elementary operators supported in Ω.

Since $|l| > |k|$ in the definition of $\mathcal{D}_N$, by Proposition 5.5.5 (a), it suffices to show that

$$\{(2^{(N+2)|l|}D_{j+k}D_{j+k+l}, 2^{-(j+k)}) : j, k, l \in \mathbb{Z}^\nu, j+k \in \mathbb{N}^\nu\}$$

is a bounded set of generalized $(W, \vec{d})$ elementary operators supported in Ω. This follows directly from Proposition 5.5.11, completing the proof. □

The next result is the main technical estimate which lies at the heart of many of the proofs in this chapter.

Theorem 6.4.3. *Fix $K_0 \Subset \mathbb{R}^\nu$ compact and set $N := \sup\{|s| : s \in K_0\} + 1$. Suppose $s \in K_0$, $f \in C^\infty_{W,0}(\mathfrak{M})'$ with $\operatorname{supp} f \subseteq \mathcal{K}$, and $\|f\|_{\mathcal{V},s,\mathcal{D}_N} < \infty$. Then, for every bounded set of generalized $(W, \vec{d})$ elementary operators $\mathcal{E}$, there exists $C = C(\mathcal{E}, K_0, \mathcal{V}) \geq 0$ such that*

$$\|f\|_{\mathcal{V},s,\mathcal{E}} \leq C\|f\|_{\mathcal{V},s,\mathcal{D}_0},$$

where if the right-hand side is finite, so is the left-hand side. Here, C does not depend on f, $\|f\|_{\mathcal{V},s,\mathcal{D}_N}$, or the particular $s \in K_0$.

Before we prove Theorem 6.4.3 we present three simple corollaries.

Corollary 6.4.4. *Let $s \in \mathbb{R}^\nu$ and let $\mathcal{E}$ be a bounded set of generalized $(W, \vec{d})$ elementary operators. Then $\|\cdot\|_{\mathcal{V},s,\mathcal{E}}$ defines a continuous semi-norm on $\mathscr{X}^s(\mathcal{K}, (W, \vec{d}))$. In particular, there is a constant $C \geq 0$ with*

$$\|f\|_{\mathcal{V},s,\mathcal{E}} \leq C\|f\|_{\mathcal{V},s,\mathcal{D}_0} = \|f\|_{\mathscr{X}^s(W,\vec{d})}, \quad \forall f \in \mathscr{X}^s(\mathcal{K}, (W, \vec{d})).$$

Furthermore, if $K_0 \Subset \mathbb{R}^\nu$ is a compact set, then the constant $C \geq 0$ can be chosen independent of $s \in K_0$.

Proof. Suppose $f \in \mathscr{X}^s(\mathcal{K}, (W, \vec{d}))$. Set $N := \sup\{|s| : s \in K_0\} + 1$. By Lemma 6.4.2, $\mathcal{D}_N$ is a bounded set of generalized $(W, \vec{d})$ elementary operators. Thus, by the definition of $\mathscr{X}^s(\mathcal{K}, (W, \vec{d}))$ (Definition 6.3.5), we have $\|f\|_{\mathcal{V},s,\mathcal{D}_N} < \infty$. Applying Theorem 6.4.3 we see that

$$\|f\|_{\mathcal{V},s,\mathcal{E}} \lesssim \|f\|_{\mathcal{V},s,\mathcal{D}_0} = \|f\|_{\mathscr{X}^s(W,\vec{d})},$$

as desired, completing the proof. □

Corollary 6.4.5. *For $f \in C^\infty_{W,0}(\mathfrak{M})'$, we have $f \in \mathscr{X}^s(\mathcal{K}, (W, \vec{d}))$ if and only if $\|f\|_{\mathcal{V},s,\mathcal{D}_0} + \|f\|_{\mathcal{V},s,\mathcal{D}_{|s|+1}} < \infty$ and $\operatorname{supp} f \subseteq \mathcal{K}$. Moreover, the norm $\|\cdot\|_{\mathcal{V},s,\mathcal{D}_0} + \|\cdot\|_{\mathcal{V},s,\mathcal{D}_{|s|+1}}$ is equivalent to the norm $\|\cdot\|_{\mathscr{X}^s(W,\vec{d})}$ on $\mathscr{X}^s(\mathcal{K}, (W, \vec{d}))$.*

Proof. Suppose $f \in \mathscr{X}^s(\mathcal{K}, (W, \vec{d}))$. Then by the definition of $\mathscr{X}^s(\mathcal{K}, (W, \vec{d}))$ (Definition 6.3.5) we have $\mathrm{supp}(f) \subseteq \mathcal{K}$ and for every bounded set of generalized $(W, \vec{d})$ elementary operators, $\mathcal{E}$, we have $\|f\|_{\mathscr{V},s,\mathcal{E}} < \infty$. Since $\mathcal{D}_0$ is a bounded set of generalized $(W, \vec{d})$ elementary operators by construction and $\mathcal{D}_{|s|+1}$ is a bounded set of generalized $(W, \vec{d})$ elementary operators by Lemma 6.4.2, we have $\|f\|_{\mathscr{V},s,\mathcal{D}_0} + \|f\|_{\mathscr{V},s,\mathcal{D}_{|s|+1}} < \infty$.

Conversely, suppose $\mathrm{supp} f \subseteq \mathcal{K}$ and $\|f\|_{\mathscr{V},s,\mathcal{D}_0} + \|f\|_{\mathscr{V},s,\mathcal{D}_{|s|+1}} < \infty$. Let $\mathcal{E}$ be a bounded set of generalized $(W, \vec{d})$ elementary operators. Applying Theorem 6.4.3 with $K_0 = \{s\}$, we have

$$\|f\|_{\mathscr{V},s,\mathcal{E}} \lesssim \|f\|_{\mathscr{V},s,\mathcal{D}_0} < \infty.$$

Since $\mathcal{E}$ was arbitrary, we see that $f \in \mathscr{X}^s(\mathcal{K}, (W, \vec{d}))$.

Finally, we show that the two norms are equivalent. Clearly, for $f \in \mathscr{X}^s(\mathcal{K}, (W, \vec{d}))$, we have

$$\|f\|_{\mathscr{X}^s(W,\vec{d})} = \|f\|_{\mathscr{V},s,\mathcal{D}_0} \leq \|f\|_{\mathscr{V},s,\mathcal{D}_0} + \|f\|_{\mathscr{V},s,\mathcal{D}_{|s|+1}}.$$

Conversely, Corollary 6.4.4 shows that for $f \in \mathscr{X}^s(\mathcal{K}, (W, \vec{d}))$,

$$\|f\|_{\mathscr{V},s,\mathcal{D}_0} + \|f\|_{\mathscr{V},s,\mathcal{D}_{|s|+1}} \lesssim \|f\|_{\mathscr{X}^s(W,\vec{d})},$$

completing the proof. □

Remark 6.4.6. Proposition 6.3.7 states that the equivalence class of the norm $\|f\|_{\mathscr{X}^s(W,\vec{d})}$ is well-defined for $f \in \mathscr{X}^s(\mathcal{K}, (W, \vec{d}))$. However, it does not make any claim for $f \notin \mathscr{X}^s(\mathcal{K}, (W, \vec{d}))$. In particular, **it is not claimed** that if $\|f\|_{\mathscr{X}^s(W,\vec{d})} < \infty$ and $\mathrm{supp}(f) \subseteq \mathcal{K}$, then $f \in \mathscr{X}^s(\mathcal{K}, (W, \vec{d}))$. However, if one replaces $\|f\|_{\mathscr{X}^s(W,\vec{d})}$ with the equivalent norm $\|f\|_{\mathscr{V},s,\mathcal{D}_0} + \|f\|_{\mathscr{V},s,\mathcal{D}_{|s|+1}}$, then Corollary 6.4.5 shows that if $\mathrm{supp}(f) \subseteq \mathcal{K}$ and the norm is finite, then $f \in \mathscr{X}^s(\mathcal{K}, (W, \vec{d}))$. See also Section 6.10 for some more comments on this issue.

Corollary 6.4.7. *Let $\Omega \Subset \mathfrak{M}$ be any relatively compact, open set with $\mathcal{K} \Subset \Omega$. Then for $f \in C^\infty_{W,0}(\mathfrak{M})'$ with* $\mathrm{supp}(f) \subseteq \mathcal{K}$, *the following are equivalent:*

(i) $f \in \mathscr{X}^s(\mathcal{K}, (W, \vec{d}))$.

(ii) *For every bounded set of generalized $(W, \vec{d})$ elementary operators supported in Ω, $\mathcal{E}$, we have*

$$\|f\|_{\mathscr{V},s,\mathcal{E}} < \infty.$$

Proof. (i) ⇒ (ii): This follows immediately from the definitions.

(ii) ⇒ (i): By construction, we may take $\mathcal{D}_0$ to be a bounded set of generalized $(W, \vec{d})$ elementary operators supported in Ω. By Lemma 6.4.2, $\mathcal{D}_N$ is then also a bounded set of generalized $(W, \vec{d})$ elementary operators supported in Ω, for every $N \geq 1$. Thus, if (ii) holds, then $\|f\|_{\mathscr{V},s,\mathcal{D}_0} + \|f\|_{\mathscr{V},s,\mathcal{D}_{|s|+1}} < \infty$ and Corollary 6.4.5 implies that (i) holds. □

Remark 6.4.8. Corollary 6.4.7 shows that the spaces $\mathscr{X}^s(\mathcal{K}, (W, \vec{d}))$ only depend on the choice of $(W, \vec{d})$ on an arbitrarily small neighborhood of $\mathcal{K}$. This justifies the fact that the notation $\mathscr{X}^s(\mathcal{K}, (W, \vec{d}))$ does not mention the ambient manifold $\mathfrak{M}$, as $\mathfrak{M}$ can be replaced by any open set containing $\mathcal{K}$, and we would obtain the same space $\mathscr{X}^s(\mathcal{K}, (W, \vec{d}))$.

We turn to the proof of Theorem 6.4.3. We begin with two lemmas.

Lemma 6.4.9. *Let $\mathcal{E}$ be a bounded set of generalized $(W, \vec{d})$ elementary operators and let*

$$\mathcal{E}_1 := \{(E_j^1, 2^{-j}) : j \in \mathbb{N}^\nu\}, \quad \mathcal{E}_2 := \{(E_j^2, 2^{-j}) : j \in \mathbb{N}^\nu\} \subseteq \mathcal{E}.$$

For $k \in \mathbb{Z}^\nu$, define $\mathcal{T}_k = \mathcal{T}_k(\mathcal{E}_1, \mathcal{E}_2)$ by

$$\mathcal{T}_k \{f_j\}_{j\in\mathbb{N}^\nu} := \{E_j^1 E_{j+k}^2 f_j\}_{j\in\mathbb{N}^\nu}.$$

Then $\mathcal{T}_k \in \mathrm{Hom}(\mathscr{V}, \mathscr{V})$ and there exists $C = C(\mathcal{E}, N, \mathcal{V}) \geq 0$ such that for all $k \in \mathbb{Z}^\nu$,

$$\sup_{\mathcal{E}_1,\mathcal{E}_2\subseteq\mathcal{E}} \|\mathcal{T}_k\|_{\mathscr{V}\to\mathscr{V}} \leq C 2^{-N|k|},$$

where the supremum is taken over all such $\mathcal{E}_1, \mathcal{E}_2 \subseteq \mathcal{E}$.

Proof. Proposition 5.5.11 shows that for every $N \in \mathbb{N}$,

$$\mathcal{E}_N := \{(2^{N|k|} E_1 E_2, 2^{-j}) : (E_1, 2^{-j}), (E_2, 2^{-(j+k)}) \in \mathcal{E}\}$$

is a bounded set of generalized $(W, \vec{d})$ elementary operators. For $\mathcal{F}_N := \{(F_j, 2^{-j}) : j \in \mathbb{N}^\nu\} \subseteq \mathcal{E}_N$, define $\mathcal{S} = \mathcal{S}(\mathcal{F}_N)$ by $\mathcal{S}\{f_j\}_{j\in\mathbb{N}^\nu} := \{F_j f_j\}_{j\in\mathbb{N}^\nu}$. Thus, we have

$$2^{N|k|} \sup_{\mathcal{E}_1,\mathcal{E}_2\subseteq\mathcal{E}} \|\mathcal{T}_k\|_{\mathscr{V}\to\mathscr{V}} \lesssim \sup_{\mathcal{F}_N\subseteq\mathcal{E}_N} \|\mathcal{S}\|_{\mathscr{V}\to\mathscr{V}} \lesssim 1,$$

where the $\lesssim 1$ follows from the fact that $\mathcal{E}_N$ is a bounded set of generalized $(W, \vec{d})$ elementary operators supported and we have used the definition of bounded sets of generalized $(W, \vec{d})$ pre-elementary operators (Definition 5.2.26). This completes the proof. □

Lemma 6.4.10 (A triangle inequality). *Suppose that for $j \in \mathbb{N}^\nu$, $k \in \mathbb{N}$, $f_{j,k} \in C_0^\infty(\mathfrak{M})'$ is such that*

$$\sum_{k\in\mathbb{N}} \|\{f_{j,k}\}_{j\in\mathbb{N}^\nu}\|_{\mathscr{V}} < \infty.$$

Then $\sum_{k\in\mathbb{N}} f_{j,k}$ *converges in* $C_0^\infty(\mathfrak{M})'$ *for every* j *and*

$$\left\| \left\{ \sum_{k\in\mathbb{N}} f_{j,k} \right\}_{j\in\mathbb{N}^\nu} \right\|_{\mathscr{V}} \le \sum_{k\in\mathbb{N}} \|\{f_{j,k}\}_{j\in\mathbb{N}^\nu}\|_{\mathscr{V}}. \tag{6.7}$$

Proof. Recall Notation 6.3.1, so that $\mathscr{V}$ is either $L^p(\mathfrak{M}, \mathrm{Vol}; \ell^q(\mathbb{N}^\nu))$ or $\ell^q(\mathbb{N}^\nu; L^p(\mathfrak{M}, \mathrm{Vol}))$. We present the proof when $p, q \neq \infty$; the case where either p or q is infinite is similar and easier.

Since $\sum_{k\in\mathbb{N}} \|\{f_{j,k}\}_{j\in\mathbb{N}^\nu}\|_{\mathscr{V}} < \infty$, we have, for every $j \in \mathbb{N}$, $\sum_{k\in\mathbb{N}} \|f_{j,k}\|_{L^p} < \infty$, which implies $\sum_{k\in\mathbb{N}} f_{j,k}$ converges in $L^p(\mathfrak{M}, \mathrm{Vol})$, and therefore in $C_0^\infty(\mathfrak{M})'$.

It remains to prove (6.7). Since $\sum_{k\in\mathbb{N}} f_{j,k}$ converges in $L^p(\mathfrak{M}, \mathrm{Vol})$ we may choose a sequence $N_l \to \infty$ such that $\sum_{k=0}^{N_l} f_{j,k} \xrightarrow{l\to\infty} \sum_{k\in\mathbb{N}} f_{j,k}$ pointwise almost everywhere. Applying Fatou's lemma and the triangle inequality, we see that

$$\begin{aligned}\left\| \left\{ \sum_{k\in\mathbb{N}} f_{j,k} \right\}_{j\in\mathbb{N}^\nu} \right\|_{\mathscr{V}} &\le \liminf_{l\to\infty} \left\| \left\{ \sum_{k=0}^{N_l} f_{j,k} \right\}_{j\in\mathbb{N}^\nu} \right\|_{\mathscr{V}} \\ &\le \liminf_{l\to\infty} \sum_{k=0}^{N_l} \|\{f_{j,k}\}_{j\in\mathbb{N}^\nu}\|_{\mathscr{V}} = \sum_{k\in\mathbb{N}} \|\{f_{j,k}\}_{j\in\mathbb{N}^\nu}\|_{\mathscr{V}},\end{aligned}$$

completing the proof. □

Proof of Theorem 6.4.3. Set $\mathcal{E}' := \mathcal{E} \cup \mathcal{D}_N$. It follows from Lemma 6.4.2 that $\mathcal{E}'$ is a bounded set of generalized $(W, \vec{d})$ elementary operators. Note that

$$\max\{\|f\|_{\mathscr{V},s,\mathcal{E}}, \|f\|_{\mathscr{V},s,\mathcal{D}_N}\} \le \|f\|_{\mathscr{V},s,\mathcal{E}'}. \tag{6.8}$$

Thus, it suffices to prove the result with $\mathcal{E}$ replaced by $\mathcal{E}'$.

Let $\{(E_j, 2^{-j}) : j \in \mathbb{N}^\nu\} \subseteq \mathcal{E}'$. Since $\mathrm{supp}(f) \subseteq \mathcal{K}$, we have

$$\begin{aligned}E_j f &= E_j \,\mathrm{Mult}[\psi]\, \mathrm{Mult}[\psi] f = E_j \left(\sum_{k\in\mathbb{N}^\nu} D_{j+k} \right) \left(\sum_{l\in\mathbb{Z}^\nu} D_{j+k+l} \right) f \\ &= \sum_{k,l\in\mathbb{Z}^\nu} E_j D_{j+k} D_{j+k+l} f,\end{aligned}$$

where the above sums all converge in the sense of $C_{W,0}^\infty(\mathfrak{M})'$ (see Proposition 5.5.10) and we are using the fact that each of $E_j, D_j \in \mathrm{Hom}(C_{W,0}^\infty(\mathfrak{M})', C_{W,\mathrm{loc}}^\infty(\mathfrak{M})')$ (see Remark 6.3.4). Thus, using Lemma 6.4.10, we have

$$\begin{aligned}\|\{2^{j\cdot s} E_j\}_{j\in\mathbb{N}^\nu}\|_{\mathscr{V}} &= \left\| \left\{ \sum_{k,l\in\mathbb{Z}^\nu} 2^{j\cdot s} E_j D_{j+k} D_{j+k+l} f \right\}_{j\in\mathbb{N}^\nu} \right\|_{\mathscr{V}} \\ &\le \sum_{k,l\in\mathbb{Z}^\nu} \|\{2^{j\cdot s} E_j D_{j+k} D_{j+k+l}\}_{j\in\mathbb{N}^\nu}\|_{\mathscr{V}},\end{aligned} \tag{6.9}$$

where if the right-hand side of (6.9) is finite, then the left-hand side is as well.

Fix $M \in \mathbb{N}$ large to be chosen later. We wish to bound the right-hand side of (6.9). To do so, we separate it into three terms:

$$(\mathrm{I}) := \sum_{\substack{k,l\in\mathbb{Z}^\nu \\ |l|\le M}} \cdot, \quad (\mathrm{II}) := \sum_{\substack{k,l\in\mathbb{Z}^\nu \\ |l|>M, |k|\le |l|}} \cdot, \quad (\mathrm{III}) := \sum_{\substack{k,l\in\mathbb{Z}^\nu \\ |l|>M, |l|>|k|}} \cdot.$$

Note that the right-hand side of (6.9) equals (I) + (II) + (III).

In what follows $A \lesssim B$ means $A \le CB$, where C may depend on $\mathscr{V}, K_0, \mathcal{E}, \mathcal{D}_0$, and $\mathcal{D}_N$, but not on k, l, M, f, or the particular choice of $\{(E_j, 2^{-j}) : j \in \mathbb{N}^\nu\} \subseteq \mathcal{E}'$.

We begin by estimating (I). We define a vector-valued operator, for $k \in \mathbb{Z}^\nu$, by

$$\mathcal{T}_k^1\{f_j\}_{j\in\mathbb{N}^\nu} := \{2^{|k|-k\cdot s}E_jD_{j+k}f_j\}_{j\in\mathbb{N}^\nu}.$$

Lemma 6.4.9 shows that $\|\mathcal{T}_k^1\|_{\mathscr{V}\to\mathscr{V}} \lesssim 1$. Thus, we have

$$\begin{aligned}(\mathrm{I}) &= \sum_{\substack{k,l\in\mathbb{Z}^\nu \\ |l|\le M}} 2^{-l\cdot s-|k|}\|\{(2^{|k|-k\cdot s}E_jD_{j+k})(2^{(j+k+l)\cdot s}D_{j+k+l})f\}_{j\in\mathbb{N}^\nu}\|_{\mathscr{V}} \\ &= \sum_{\substack{k,l\in\mathbb{Z}^\nu \\ |l|\le M}} 2^{-l\cdot s-|k|}\|\mathcal{T}_k^1\{2^{(j+k+l)\cdot s}D_{j+k+l}f\}_{j\in\mathbb{N}^\nu}\|_{\mathscr{V}} \\ &\lesssim \sum_{\substack{k,l\in\mathbb{Z}^\nu \\ |l|\le M}} 2^{-l\cdot s-|k|}\|\{2^{(j+k+l)\cdot s}D_{j+k+l}f\}_{j\in\mathbb{N}^\nu}\|_{\mathscr{V}} \\ &\le \sum_{\substack{k,l\in\mathbb{Z}^\nu \\ |l|\le M}} 2^{-l\cdot s-|k|}\|f\|_{\mathscr{V},s,\mathcal{D}_0} \lesssim 2^{M|s|}\|f\|_{\mathscr{V},s,\mathcal{D}_0}.\end{aligned} \tag{6.10}$$

We turn to (II). For $k, l \in \mathbb{Z}^\nu$ with $|k| \ge |l|$, we define the vector-valued operator

$$\mathcal{T}_{k,l}^2\{f_j\}_{j\in\mathbb{N}^\nu} := \{2^{|k|+|l|-(k+l)\cdot s}E_jD_{j+k}\}_{j\in\mathbb{N}^\nu}.$$

Using the fact that $|k| \ge |l|$, Lemma 6.4.9 shows that $\|\mathcal{T}_{k,l}^2\|_{\mathscr{V}\to\mathscr{V}} \lesssim 1$. Thus, we have

$$\begin{aligned}(\mathrm{II}) &= \sum_{\substack{k,l\in\mathbb{Z}^\nu \\ |l|>M,|k|\ge|l|}} 2^{-|k|-|l|}\|\{(2^{|k|+|l|-(k+l)\cdot s}E_jD_{j+k})(2^{(j+k+l)\cdot s}D_{j+k+l}f)\}_{j\in\mathbb{N}^\nu}\|_{\mathscr{V}} \\ &= \sum_{\substack{k,l\in\mathbb{Z}^\nu \\ |l|>M,|k|\ge|l|}} 2^{-|k|-|l|}\|\mathcal{T}_{k,l}^2\{2^{(j+k+l)\cdot s}D_{j+k+l}\}_{j\in\mathbb{N}^\nu}\|_{\mathscr{V}} \\ &\lesssim \sum_{\substack{k,l\in\mathbb{Z}^\nu \\ |l|>M,|k|\ge|l|}} 2^{-|k|-|l|}\|\{2^{(j+k+l)\cdot s}D_{j+k+l}\}_{j\in\mathbb{N}^\nu}\|_{\mathscr{V}} \\ &\le \sum_{\substack{k,l\in\mathbb{Z}^\nu \\ |l|>M,|k|\ge|l|}} 2^{-|k|-|l|}\|f\|_{\mathscr{V},s,\mathcal{D}_0} \lesssim 2^{-M}\|f\|_{\mathscr{V},s,\mathcal{D}_0}.\end{aligned} \tag{6.11}$$

Finally, we turn to (III). Define a vector-valued operator

$$\mathcal{T}^3\{f_j\}_{j\in\mathbb{N}^\nu} := \{E_j f_j\}_{j\in\mathbb{N}^\nu}.$$

Since $\mathcal{E}'$ is a bounded set of generalized $(W,\vec{d})$ elementary operators, it follows from the definition (see Definition 5.2.26) that $\|\mathcal{T}^3\|_{\mathcal{V}\to\mathcal{V}} \lesssim 1$. Thus, we have, using $2^{-N|k|-k\cdot s+|k|} \lesssim 1$, by the definition of N,

$$\begin{aligned}
\text{(III)} &= \sum_{\substack{k,l\in\mathbb{Z}^\nu\\ |l|>M,|l|>|k|}} 2^{-|k|-|l|}\big\|\{E_j(2^{(j+k)\cdot s}2^{-k\cdot s+|k|+|l|}D_{j+k}D_{j+k+l}f)\}_{j\in\mathbb{N}^\nu}\big\|_{\mathcal{V}}\\
&= \sum_{\substack{k,l\in\mathbb{Z}^\nu\\ |l|>M,|l|>|k|}} 2^{-|k|-|l|}\big\|\mathcal{T}^3\{2^{-N|k|-k\cdot s+|k|}2^{(j+k)\cdot s}F_{N,j,k,l}f\}_{j\in\mathbb{N}^\nu}\big\|_{\mathcal{V}}\\
&\lesssim \sum_{\substack{k,l\in\mathbb{Z}^\nu\\ |l|>M,|l|>|k|}} 2^{-|k|-|l|}\big\|\{2^{-N|k|-k\cdot s+|k|}2^{(j+k)\cdot s}F_{N,j,k,l}f\}_{j\in\mathbb{N}^\nu}\big\|_{\mathcal{V}}\\
&\lesssim \sum_{\substack{k,l\in\mathbb{Z}^\nu\\ |l|>M,|l|>|k|}} 2^{-|k|-|l|}\big\|\{2^{(j+k)\cdot s}F_{N,j,k,l}f\}_{j\in\mathbb{N}^\nu}\big\|_{\mathcal{V}}\\
&\lesssim \sum_{\substack{k,l\in\mathbb{Z}^\nu\\ |l|>M,|l|>|k|}} 2^{-|k|-|l|}\|f\|_{\mathcal{V},s,\mathcal{D}_N} \lesssim 2^{-M}\|f\|_{\mathcal{V},s,\mathcal{D}_N}.
\end{aligned}\tag{6.12}$$

Plugging (6.10), (6.11), and (6.12) into (6.9) we see that there exists $C \lesssim 1$ with

$$\|f\|_{\mathcal{V},s,\mathcal{E}'} \le C\big(2^{M|s|}\|f\|_{\mathcal{V},s,\mathcal{D}_0} + 2^{-M}\|f\|_{\mathcal{V},s,\mathcal{D}_N}\big).$$

Taking M so large that $C2^{-M} \le \frac{1}{2}$, using the hypothesis that $\|f\|_{\mathcal{V},s,\mathcal{D}_N} < \infty$, and using (6.8), we see that

$$\|f\|_{\mathcal{V},s,\mathcal{E}'} \le 2C2^{M|s|}\|f\|_{\mathcal{V},s,\mathcal{D}_0},$$

completing the proof. □

6.5 Some basic properties

In this section we prove some basic properties of the Besov and Triebel–Lizorkin spaces. Among other things, we prove the results stated in Section 6.3.

We begin with Proposition 6.3.7, that the equivalence class of the norm $\|\cdot\|_{\mathscr{X}^s(W,\vec{d})}$ is well-defined.

Proof of Proposition 6.3.7. The statement of the proposition is symmetric in $\mathcal{D}_0$ and $\widetilde{\mathcal{D}}$, so it suffices to show that

$$\|f\|_{\mathcal{V},s,\widetilde{\mathcal{D}}} \lesssim \|f\|_{\mathcal{V},s,\mathcal{D}_0}, \quad \forall f \in \mathscr{X}^s(\mathcal{K},(W,\vec{d})). \tag{6.13}$$

Since $\widetilde{\mathcal{D}}$ is a bounded set of generalized $(W,\vec{d})$ elementary operators, (6.13) follows directly from Corollary 6.4.4, completing the proof. □

Next, we turn to Theorem 6.3.10: the boundedness of operators in $\widetilde{\mathscr{A}}^t(\Omega,(W,\vec{d}))$.

Proof of Theorem 6.3.10. By assumption, $\Omega \Subset \mathcal{K}$ is open. Fix $T \in \widetilde{\mathscr{A}}^t(\Omega,(W,\vec{d}))$ and $f \in \mathscr{X}^s(\mathcal{K},(W,\vec{d}))$.

We wish to show $Tf \in \mathscr{X}^{s-t}(\mathcal{K},(W,\vec{d}))$. Clearly, $\operatorname{supp}(Tf) \subseteq \Omega \subseteq \mathcal{K}$. Let $\mathcal{E}$ be a bounded set of generalized $(W,\vec{d})$ elementary operators. Thus, there exists some $\Omega' \Subset \mathfrak{M}$ open such that $\mathcal{E}$ is a bounded set of generalized $(W,\vec{d})$ elementary operators supported in Ω'. By replacing Ω' with $\Omega' \cup \Omega$, we may assume $\Omega \subseteq \Omega'$.

Set $\mathcal{E}' := \{(2^{-j\cdot t}ET, 2^{-j}) : (E,2^{-j}) \in \mathcal{E}\}$.

We claim $\mathcal{E}'$ is a bounded set of generalized $(W,\vec{d})$ elementary operators supported in Ω'. Indeed, by Proposition 5.5.5 (d), $\{(E^*,2^{-j}) : (E,2^{-j}) \in \mathcal{E}\}$ is a bounded set of generalized $(W,\vec{d})$ elementary operators supported in Ω'. By Proposition 5.8.1 and Proposition 5.8.7, $T^* \in \widetilde{\mathscr{A}}^t(\Omega,(W,\vec{d})) \subseteq \widetilde{\mathscr{A}}^t(\Omega',(W,\vec{d})) = \widetilde{\mathscr{A}}^t_1(\Omega',(W,\vec{d}))$. Thus, $\{(2^{-j\cdot t}T^*E^*,2^{-j}) : (E,2^{-j}) \in \mathcal{E}\}$ is a bounded set of generalized $(W,\vec{d})$ elementary operators supported in Ω'. Therefore, by Proposition 5.5.5 (d), $\mathcal{E}'$ is a bounded set of generalized $(W,\vec{d})$ elementary operators supported in Ω' as desired.

Thus, we have

$$\begin{aligned}\|Tf\|_{\mathcal{V},s-t,\mathcal{E}} &= \sup_{\{(E_j,2^{-j}):j\in\mathbb{N}^\nu\}\subseteq\mathcal{E}} \|\{2^{(s-t)\cdot j}E_jTf\}_{j\in\mathbb{N}^\nu}\|_{\mathcal{V}} \\ &\le \sup_{\{(E_j',2^{-j}):j\in\mathbb{N}^\nu\}\subseteq\mathcal{E}'} \|\{2^{s\cdot j}E_j'f\}_{j\in\mathbb{N}^\nu}\|_{\mathcal{V}} = \|f\|_{\mathcal{V},s,\mathcal{E}'} \lesssim \|f\|_{\mathscr{X}^s(W,\vec{d})} < \infty,\end{aligned} \tag{6.14}$$

where the last $\lesssim$ uses Corollary 6.4.4 and the fact that $f \in \mathscr{X}^s(\mathcal{K},(W,\vec{d}))$.

Since $\mathcal{E}$ was an arbitrary bounded set of generalized $(W,\vec{d})$ elementary operators, we conclude $Tf \in \mathscr{X}^{s-t}(\mathcal{K},(W,\vec{d}))$. Moreover, applying (6.14) in the special case $\mathcal{E} = \mathcal{D}_0$, we see that

$$\|Tf\|_{\mathscr{X}^{s-t}(W,\vec{d})} = \|Tf\|_{\mathcal{V},s-t,\mathcal{D}_0} \lesssim \|f\|_{\mathscr{X}^s(W,\vec{d})},$$

proving that $T : \mathscr{X}^s(\mathcal{K},(W,\vec{d})) \to \mathscr{X}^{s-t}(\mathcal{K},(W,\vec{d}))$ is bounded. □

Next we turn to proving Proportion 6.3.6: that $\mathscr{X}^s(\mathcal{K},(W,\vec{d}))$ is a Banach space with norm $\|\cdot\|_{\mathcal{V},s,\mathcal{D}_0} = \|\cdot\|_{\mathscr{X}^s(W,\vec{d})}$. To do this, we introduce some new notation just for this section and several preliminary lemmas. For $\mu \in \{1,\dots,\nu\}$, set $\mathcal{I}_\mu := \{j \in \mathbb{N}^\nu : j_\mu = |j|_\infty\}$.

For $g \in C^\infty_{W,0}(\mathfrak{M})$, a bounded set of generalized $(W,\vec{d})$ elementary operators $\mathcal{E}$, $N \in \mathbb{R}$, and $\mu \in \{1,\dots,\nu\}$, set, for $f \in C^\infty_{W,0}(\mathfrak{M})'$,

$$\|f\|_{g,N,\mathcal{E}} := \sup_{\{(E_j,2^{-j}):j\in\mathbb{N}^\nu\}\subseteq\mathcal{E}} \sum_{j\in\mathbb{N}^\nu} 2^{N|j|}\left|\int g(x)E_jf(x)\,d\operatorname{Vol}(x)\right|,$$

$$\|f\|_{g,N,\mathcal{E},\mu} := \sup_{\{(E_j,2^{-j}):j\in\mathcal{I}_\mu\}\subseteq\mathcal{E}} \sum_{j\in\mathcal{I}_\mu} 2^{N|j|_\infty}\left|\int g(x)E_jf(x)\,d\operatorname{Vol}(x)\right|.$$

Lemma 6.5.1. *Let $\mathcal{E}$ be a bounded set of generalized $(W,\vec{d})$ elementary operators, $g \in C^\infty_{W,0}(\mathfrak{M})$, and $N \in \mathbb{R}$. Then, $\|\cdot\|_{g,N,\mathcal{E}}$ defines a continuous semi-norm on $\mathscr{X}^s(\mathcal{K},(W,\vec{d}))$. In other words,*

$$\|f\|_{g,N,\mathcal{E}} \le C_{g,N,\mathcal{E}}\|f\|_{\mathscr{V},s,\mathcal{D}_0} = C_{g,N,\mathcal{E}}\|f\|_{\mathscr{X}^s(W,\vec{d})}.$$

To prove Lemma 6.5.1 we begin with two easier lemmas.

Lemma 6.5.2. *Let $\mathcal{E}$ be a bounded set of generalized $(W,\vec{d})$ elementary operators, $g \in C^\infty_{W,0}(\mathfrak{M})$, and $N \le -|s|\sqrt{\nu}-1$. Then for $\mu \in \{1,\dots,\nu\}$, $\|\cdot\|_{g,N,\mathcal{E},\mu}$ defines a continuous semi-norm on $\mathscr{X}^s(\mathcal{K},(W,\vec{d}))$.*

Proof. Under the hypotheses on N we claim

$$(N+1)|j|_\infty \le s\cdot j. \tag{6.15}$$

Indeed, if $s\cdot j \ge 0$, (6.15) is clear. If $s\cdot j < 0$, then (6.15) is equivalent to $|s\cdot j| \le |N+1||j|_\infty = (-N-1)|j|_\infty$. By Cauchy–Schwartz we have $|s\cdot j| \le |s||j| \le \sqrt{\nu}|s||j|_\infty \le (-N-1)|j|_\infty$ as desired, completing the proof of (6.15).

Without loss of generality, we may replace $\mathcal{E}$ with

$$\mathcal{E}\cup\{(0,2^{-j}) : j\in\mathbb{N}^\nu\}. \tag{6.16}$$

We first prove the result when $\mathscr{X}^s(\mathcal{K},(W,\vec{d})) = \mathscr{B}^s_{p,q}(\mathcal{K},(W,\vec{d}))$ for some $p,q \in [1,\infty]$. We prove the result only in the case $q \in (1,\infty)$; the case where $q \in \{1,\infty\}$ is similar and easier, so we leave it to the reader. Let p' be dual to p ($\frac{1}{p'}+\frac{1}{p}=1$) and let q' be dual to q. We have, for $\{(E_j,2^{-j}) : j\in\mathcal{I}_\mu\}\subseteq\mathcal{E}$ and $f\in\mathscr{B}^s_{p,q}(\mathcal{K},(W,\vec{d}))$, using (6.15),

$$\begin{aligned}
&\sum_{j\in\mathcal{I}_\mu} 2^{N|j|_\infty}\left|\int g(x)E_jf(x)\,d\operatorname{Vol}(x)\right| \\
&\le \left(\sum_{j\in\mathcal{I}_\mu} 2^{-|j|_\infty q'}\right)^{\frac{1}{q'}}\left(\sum_{j\in\mathcal{I}_\mu}\left(2^{(N+1)|j|_\infty}\|g\|_{L^{p'}}\|E_jf\|_{L^p}\right)^q\right)^{\frac{1}{q}} \\
&\lesssim \left\|\{2^{(N+1)|j|_\infty}E_jf\}_{j\in\mathcal{I}_\mu}\right\|_{\ell^q(\mathcal{I}_\mu;L^p(\mathfrak{M},\operatorname{Vol}))} \lesssim \left\|\{2^{s\cdot j}E_jf\}_{j\in\mathbb{N}^\nu}\right\|_{\ell^q(\mathcal{I}_\mu;L^p(\mathfrak{M},\operatorname{Vol}))} \\
&\le \|f\|_{\mathscr{V},s,\mathcal{E}} \lesssim \|f\|_{\mathscr{B}^s_{p,q}(W,\vec{d})},
\end{aligned}$$

where the last estimate uses $f \in \mathscr{B}^s_{p,q}(\mathcal{K}, (W, \vec{d}))$ and Corollary 6.4.4 and the second to last estimate used the fact that we replaced $\mathcal{E}$ with (6.16). Taking the supremum over all $\{(E_j, 2^{-j}) : j \in \mathcal{I}_\mu\} \subseteq \mathcal{E}$, we see that

$$\|f\|_{g,N,\mathcal{E},\mu} \lesssim \|f\|_{\mathscr{B}^s_{p,q}(W,\vec{d})}, \quad \forall f \in \mathscr{B}^s_{p,q}(\mathcal{K}, (W, \vec{d})),$$

completing the proof in the case $\mathscr{X}^s(\mathcal{K}, (W, \vec{d})) = \mathscr{B}^s_{p,q}(\mathcal{K}, (W, \vec{d}))$.

We now turn to the case where $\mathscr{X}^s(\mathcal{K}, (W, \vec{d})) = \mathscr{F}^s_{p,q}(\mathcal{K}, (W, \vec{d}))$ for some $p \in (1, \infty)$, $q \in (1, \infty]$. We prove only the case where $q \in (1, \infty)$; the case where $q = \infty$ follows by an obvious modification, which we leave to the reader. Let p' be dual to p and let q' be dual to q. We have, for $\{(E_j, 2^{-j}) : j \in \mathcal{I}_\mu\} \subseteq \mathcal{E}$ and $f \in \mathscr{F}^s_{p,q}(\mathcal{K}, (W, \vec{d}))$, using (6.15),

$$\begin{aligned}
&\sum_{j \in \mathcal{I}_\mu} 2^{N|j|_\infty} \left| \int g(x) E_j f(x)\, d\,\mathrm{Vol}(x) \right| \\
&\quad \le \int |g(x)| \sum_{j \in \mathcal{I}_\mu} |E_j f(x)|\, d\,\mathrm{Vol}(x) \\
&\quad \le \int |g(x)| \Bigg(\sum_{j \in \mathcal{I}_\mu} 2^{-|j|_\infty q'} \Bigg)^{\frac{1}{q'}} \Bigg(\sum_{j \in \mathcal{I}_\mu} (2^{(N+1)|j|_\infty} |E_j f(x)|)^q \Bigg)^{\frac{1}{q}} d\,\mathrm{Vol}(x) \\
&\quad \lesssim \|g\|_{L^{p'}} \left\| \{2^{(N+1)|j|_\infty} E_j f\}_{j \in \mathbb{N}^\nu} \right\|_{L^p(\mathfrak{M}, \mathrm{Vol}; \ell^q(\mathcal{I}_\mu))} \\
&\quad \lesssim \left\| \{2^{s \cdot j} E_j f\}_{j \in \mathbb{N}^\nu} \right\|_{L^p(\mathfrak{M}, \mathrm{Vol}; \ell^q(\mathcal{I}_\mu))} \le \|f\|_{\mathscr{V}, s, \mathcal{E}} \lesssim \|f\|_{\mathscr{F}^s_{p,q}(W,\vec{d})},
\end{aligned}$$

where the last estimate uses $f \in \mathscr{F}^s_{p,q}(\mathcal{K}, (W, \vec{d}))$ and Corollary 6.4.4 and the second to last estimate used the fact that we replaced $\mathcal{E}$ with (6.16). Taking the supremum over all $\{(E_j, 2^{-j}) : j \in \mathcal{I}_\mu\} \subseteq \mathcal{E}$, we see that

$$\|f\|_{g,N,\mathcal{E},\mu} \lesssim \|f\|_{\mathscr{F}^s_{p,q}(W,\vec{d})}, \quad \forall f \in \mathscr{F}^s_{p,q}(\mathcal{K}, (W, \vec{d})),$$

completing the proof. □

Lemma 6.5.3. *Let $\mathcal{E}$ be a bounded set of generalized $(W, \vec{d})$ elementary operators, $g \in C^\infty_{W,0}(\mathfrak{M})$, and $N \in \mathbb{R}$. Then for $\mu \in \{1, \dots, \nu\}$, $\|\cdot\|_{g,N,\mathcal{E},\mu}$ defines a continuous semi-norm on $\mathscr{X}^s(\mathcal{K}, (W, \vec{d}))$.*

Proof. Take $M \in \mathbb{N}$ with $M \ge |s|\sqrt{\nu} + 1 + N$. Applying Proposition 5.5.5 (h), we have, for $(E_j, 2^{-j}) \in \mathcal{E}$,

$$E_j = \sum_{|\alpha_\mu| \le M} 2^{-(M - |\alpha_\mu|) j_\mu} (2^{-j_\mu d^\mu} W^\mu)^{\alpha_\mu} E_{j,\mu,\alpha_\mu},$$

where $\widetilde{\mathcal{E}} := \{(E_{j,\mu,\alpha_\mu}, 2^{-j}) : (E, 2^{-j}) \in \mathcal{E}, |\alpha_\mu| \le M\}$ is a bounded set of generalized $(W, \vec{d})$ elementary operators supported in Ω.

We have, for $f \in \mathscr{X}^s(\mathcal{K}, (W, \vec{d}))$, using the fact that $j_\mu = |j|_\infty$ for $j \in \mathcal{I}_\mu$,

$$\begin{aligned}
\|f\|_{g,N,\mathcal{E},\mu} &= \sup_{\{(E_j,2^{-j}):j\in\mathcal{I}_\mu\}\subseteq\mathcal{E}} \sum_{j\in\mathcal{I}_\mu} 2^{N|j|_\infty} \left|\int g(x)E_jf(x)\,d\operatorname{Vol}(x)\right| \\
&\leq \sum_{|\alpha_\mu|\leq M} \sup_{\{(E_j,2^{-j}):j\in\mathcal{I}_\mu\}\subseteq\mathcal{E}} \sum_{j\in\mathcal{I}_\mu} 2^{(N-M)|j|_\infty+(|\alpha_\mu|-\deg_{d^\mu}(\alpha_\mu))|j|_\infty} \\
&\quad\times\left|\int g(x)(W^\mu)^{\alpha_\mu} E_{j,\mu,\alpha_\mu} f(x)\,d\operatorname{Vol}(X)\right| \\
&\leq \sum_{|\alpha_\mu|\leq M} \sup_{\{(E_j,2^{-j}):j\in\mathcal{I}_\mu\}\subseteq\mathcal{E}} \sum_{j\in\mathcal{I}_\mu} 2^{-(|s|\sqrt{\nu}+1)|j|_\infty} \\
&\quad\times\left|\int (((W^\mu)^{\alpha_\mu})^* g(x))E_{j,\mu,\alpha_\mu} f(x)\,d\operatorname{Vol}(X)\right| \\
&\leq \sum_{|\alpha_\mu|\leq M} \|f\|_{((W^\mu)^{\alpha_\mu})^*g,-(|s|\sqrt{\nu}+1),\widetilde{\mathcal{E}},\mu} \lesssim \|f\|_{\mathscr{X}^s(W,\vec{d})},
\end{aligned}$$

where the last estimate used Lemma 6.5.2. This completes the proof. □

Proof of Lemma 6.5.1. It follows directly from the definitions that

$$\|f\|_{g,N,\mathcal{E}} \leq \sum_{\mu\in\{1,\dots,\nu\}} \|f\|_{g,(\sqrt{\nu}N)\vee 0,\mathcal{E},\mu}.$$

Thus, the result follows immediately from Lemma 6.5.3. □

Lemma 6.5.4. *For $g \in C^\infty_{W,0}(\mathfrak{M})$, the map $\lambda_g : C^\infty_{W,0}(\mathfrak{M})' \to \mathbb{C}$ given by $\lambda_g(f) = \int fg$ restricts to a bounded linear functional on $\mathscr{X}^s(\mathcal{K}, (W, \vec{d}))$.*

Proof. Let $f \in \mathscr{X}^s(\mathcal{K}, (W, \vec{d}))$. Since $f = \operatorname{Mult}[\psi]f = \sum_{j\in\mathbb{N}^\nu} D_jf$ (with convergence in $C^\infty_{W,\mathrm{loc}}(\mathfrak{M})'$ by Proposition 5.5.10), we see that

$$\begin{aligned}
|\lambda_g(f)| &= \left|\sum_{j\in\mathbb{N}^\nu} \int g(x)D_jf(x)\,d\operatorname{Vol}(x)\right| \\
&\leq \sum_{j\in\mathbb{N}^\nu} \left|\int g(x)D_jf(x)\right| = \|f\|_{g,0,\mathcal{D}_0} \lesssim \|f\|_{\mathscr{X}^s(W,\vec{d})},
\end{aligned}$$

where the last estimate follows from Lemma 6.5.1, completing the proof. □

Proof of Proportion 6.3.6. First we show that $\|\cdot\|_{\mathscr{V},s,\mathcal{D}_0}$ is a norm on $\mathscr{X}^s(\mathcal{K}, (W, \vec{d}))$ (as opposed to merely a semi-norm). Indeed, for $f \in \mathscr{X}^s(\mathcal{K}, (W, \vec{d}))$, we have $f = \operatorname{Mult}[\psi]f = \sum_{j\in\mathbb{N}^\nu} D_jf$, with convergence in $C^\infty_{W,\mathrm{loc}}(\mathfrak{M})'$ by Proposition 5.5.10. If $\|f\|_{\mathscr{V},s,\mathcal{D}_0} = 0$, then $D_jf = 0$ for every $j \in \mathbb{N}^\nu$, and therefore $f = 0$. We conclude that $\|\cdot\|_{\mathscr{V},s,\mathcal{D}_0}$ is a norm on $\mathscr{X}^s(\mathcal{K}, (W, \vec{d}))$.

It remains to show that $\mathscr{X}^s(\mathcal{K},(W,\vec{d}))$ is complete with respect to this norm. Let $\{f_l\}_{l\in\mathbb{N}} \subset \mathscr{X}^s(\mathcal{K},(W,\vec{d}))$ be a Cauchy sequence. Our goal is to show that f_l converges in $\mathscr{X}^s(\mathcal{K},(W,\vec{d}))$.

Let $\Omega \Subset \mathfrak{M}$ be open with $\mathcal{K} \Subset \Omega$. Set $\mathcal{K}_1 := \overline{\Omega}$, so that $\mathcal{K}_1$ is compact. This makes $C^\infty_{W,0}(\mathcal{K}_1)$ (the space of $C^\infty_{W,0}(\mathfrak{M})$ functions supported in $\mathcal{K}_1$) a Fréchet space. For $f \in C^\infty_{W,0}(\mathfrak{M})'$ and $g \in C^\infty_{W,0}(\mathcal{K}_1)$, set $\mu_f(g) = \int gf$. It follows from Lemma 6.5.4 that $\{\mu_{f_l}(g)\}_{l\in\mathbb{N}} \subset \mathbb{C}$, for every $g \in C^\infty_{W,0}(\mathcal{K}_1)$. Set $\mu(g) := \lim_{l\to\infty}\mu_{f_l}(g)$. By the Banach–Steinhaus theorem (see [206, Theorem 2.8]), $\mu : C^\infty_{W,0}(\mathcal{K}_1) \to \mathbb{C}$ is continuous. In particular, $\mu : C^\infty_{W,0}(\Omega) \to \mathbb{C}$ is continuous, and therefore $\mu(g) = \int fg$ for some $f \in C^\infty_{W,0}(\Omega)'$. Since $\mu_{f_l}(g) \to \mu(g)$, $\forall g \in C^\infty_{W,0}(\Omega)$, we have $f_l \to f$ in $C^\infty_{W,0}(\Omega)'$.

Since $\operatorname{supp} f_l \subseteq \mathcal{K}$ for every l, we have $\operatorname{supp} f \subseteq \mathcal{K} \Subset \Omega$. Therefore, f extends to an element of $C^\infty_{W,\mathrm{loc}}(\mathfrak{M})' \subseteq C^\infty_{W,0}(\mathfrak{M})'$ and $f_l \to f$ in $C^\infty_{W,0}(\mathfrak{M})'$.

Take $N := |s|+1$ and define $\mathcal{D}_N$ as in (6.6). Let $\mathcal{E} := \mathcal{D}_N \cup \mathcal{D}_0$. For each $(E,2^{-j}) \in \mathcal{E}$, by Remark 6.3.4, $E \in \operatorname{Hom}(C^\infty_{W,0}(\mathfrak{M})', C^\infty_{W,0}(\mathfrak{M})')$, and therefore $Ef_l \to f$ in $C^\infty_{W,0}(\mathfrak{M})'$.

The proof will be complete once we show that $f \in \mathscr{X}^s(\mathcal{K},(W,\vec{d}))$ and $f_l \xrightarrow{l\to\infty} f$ in $\mathscr{X}^s(\mathcal{K},(W,\vec{d}))$.

Depending on whether $\mathscr{X}^s(\mathcal{K},(W,\vec{d}))$ equals $\mathscr{B}^s_{p,q}(\mathcal{K},(W,\vec{d}))$ or $\mathscr{F}^s_{p,q}(\mathcal{K},(W,\vec{d}))$, $\mathscr{V}$ is equal to either $\ell^q(\mathbb{N}^\nu; L^p(\mathfrak{M},\mathrm{Vol}))$ or $L^p(\mathfrak{M},\mathrm{Vol}; \ell^q(\mathbb{N}^\nu))$ (see Notation 6.3.2). In either case, let $p \in [1,\infty]$ be the corresponding p. We claim, for every $(E,2^{-j}) \in \mathcal{E}$, that Ef_l is Cauchy in $L^p(\mathfrak{M},\mathrm{Vol})$. Indeed, using Corollary 6.4.4, we have

$$2^{-j\cdot s}\|E(f_a - f_b)\|_{L^p} \le \|f_a - f_b\|_{\mathscr{V},s,\mathcal{D}_0} + \|f_a - f_b\|_{\mathscr{V},s,\mathcal{D}_N} \lesssim \|f_a - f_b\|_{\mathscr{V},a,\mathcal{D}_0}.$$

Since $\{f_l\}_{l\in\mathbb{N}}$ is Cauchy in $\|\cdot\|_{\mathscr{V},s,\mathcal{D}_0}$, it follows that $\{Ef_l\}_{l\in\mathbb{N}}$ is Cauchy in $L^p(\mathfrak{M},\mathrm{Vol})$. Since we already have $Ef_l \to Ef$ in $C^\infty_{W,0}(\mathfrak{M})'$, it follows that $Ef \in L^p(\mathfrak{M},\mathrm{Vol})$ and $Ef_l \to Ef$ in $L^p(\mathfrak{M},\mathrm{Vol})$.

Since $Ef_l \to Ef$ in L^p for every $(E,2^{-j}) \in \mathcal{E}$ and $\mathcal{E}$ is countable, there exists a subsequence f_{l_k} such that $Ef_{l_k} \to Ef$ pointwise almost everywhere for every $(E,2^{-j}) \in \mathcal{E}$. We have, for any $\mathcal{E}' := \{(E_j,2^{-j}) : j \in \mathbb{N}^\nu\} \subseteq \mathcal{E}$ (with constants independent of the choice of $\mathcal{E}'$), using Fatou's lemma,

$$\begin{aligned}\|\{2^{j\cdot s}E_jf\}_{j\in\mathbb{N}^\nu}\|_{\mathscr{V}} &\le \liminf_{k\to\infty}\|\{2^{j\cdot s}E_jf_{l_k}\}_{j\in\mathbb{N}^\nu}\|_{\mathscr{V}} \\ &\le \liminf_{k\to\infty}\|f_{l_k}\|_{\mathscr{V},s,\mathcal{E}} \lesssim \liminf_{k\to\infty}\|f_{l_k}\|_{\mathscr{V},s,\mathcal{D}_0} \lesssim 1,\end{aligned}$$

where the second to last estimate follows from Corollary 6.4.4 and the final estimate follows from the fact that f_{l_k} is Cauchy in $\|\cdot\|_{\mathscr{V},s,\mathcal{D}_0}$. Taking the supremum over all such $\mathcal{E}' \subseteq \mathcal{E}$, we see that

$$\|f\|_{\mathscr{V},s,\mathcal{D}_0} + \|f\|_{\mathscr{V},s,\mathcal{D}_N} \le 2\|f\|_{\mathscr{V},s,\mathcal{E}} < \infty.$$

By Corollary 6.4.5, we have $f \in \mathscr{X}^s(\mathcal{K},(W,\vec{d}))$.

Similarly, using the fact that $\mathcal{D}_0 \subseteq \mathcal{E}$ and Fatou's lemma, we have

$$\begin{aligned}\limsup_{m\to\infty} \|f - f_m\|_{\mathscr{V},s,\mathcal{D}_0} &= \limsup_{m\to\infty} \big\|\{2^{j\cdot s} D_j(f - f_m)\}_{j\in\mathbb{N}^\nu}\big\|_{\mathscr{V}} \\ &\leq \limsup_{m\to\infty} \liminf_{k\to\infty} \big\|\{2^{j\cdot s} D_j(f_{l_k} - f_m)\}_{j\in\mathbb{N}^\nu}\big\|_{\mathscr{V}} \\ &= \limsup_{m\to\infty} \liminf_{k\to\infty} \|f_{l_k} - f_m\|_{\mathscr{V},s,\mathcal{D}_0} = 0.\end{aligned}$$

Thus, $f_l \to f$ in $\mathscr{X}^s(\mathcal{K}, (W, \vec{d}))$, completing the proof. □

Proposition 6.5.5. *Let $C^\infty_{W,0}(\mathcal{K})$ be the space of those $C^\infty_{W,0}(\mathfrak{M})$ functions supported in $\mathcal{K}$. Then $C^\infty_{W,0}(\mathcal{K}) \subseteq \mathscr{X}^s(\mathcal{K}, (W, \vec{d}))$ (for every $s \in \mathbb{R}^\nu$).*

Proof. Let $f \in C^\infty_{W,0}(\mathcal{K})$ and let $\mathcal{E}$ be a bounded set of generalized $(W, \vec{d})$ elementary operators supported in Ω for some $\Omega \Subset \mathfrak{M}$. We will show that

$$\|f\|_{\mathscr{V},s,\mathcal{E}} < \infty, \tag{6.17}$$

which will complete the proof.

Set $\mathcal{I}_\mu := \{j \in \mathbb{N}^\nu : j_\mu = |j|_\infty\}$ and set

$$\mathscr{V}_\mu := \begin{cases} \ell^q(\mathcal{I}_\mu; L^p(\mathfrak{M}, \mathrm{Vol})) & \text{if } \mathscr{V} = \ell^q(\mathbb{N}^\nu; L^p(\mathfrak{M}, \mathrm{Vol})), \\ L^p(\mathfrak{M}, \mathrm{Vol}; \ell^q(\mathcal{I}_\mu)) & \text{if } \mathscr{V} = L^p(\mathfrak{M}, \mathrm{Vol}; \ell^q(\mathbb{N}^\nu)). \end{cases} \tag{6.18}$$

We have

$$\begin{aligned}\|f\|_{\mathscr{V},s,\mathcal{E}} &= \sup_{\{(E_j,2^{-j}):j\in\mathbb{N}^\nu\}\subseteq\mathcal{E}} \big\|\{2^{j\cdot s} E_j f\}_{j\in\mathbb{N}^\nu}\big\|_{\mathscr{V}} \\ &\leq \sum_{\mu\in\{1,\dots,\nu\}} \sup_{\{(E_j,2^{-j}):j\in\mathcal{I}_\mu\}\subseteq\mathcal{E}} \big\|\{2^{j\cdot s} E_j f\}_{j\in\mathcal{I}_\mu}\big\|_{\mathscr{V}_\mu}.\end{aligned} \tag{6.19}$$

Let $N := \lceil |s|\sqrt{\nu}\rceil + 1$. In light of (6.19), to show (6.17) it suffices to show that for $\mu \in \{1, \dots, \nu\}$,

$$\sup_{\{(E_j,2^{-j}):j\in\mathcal{I}_\mu\}\subseteq\mathcal{E}} \big\|\{2^{(N-1)j_\mu} E_j f\}_{j\in\mathcal{I}_\mu}\big\|_{\mathscr{V}_\mu} < \infty, \tag{6.20}$$

where we have used the fact that $j_\mu = |j|_\infty$ for $j \in \mathcal{I}_\mu$.

Fix $\mu \in \{1, \dots, \nu\}$. For $(E_j, 2^{-j}) \in \mathcal{E}$, we apply Proposition 5.5.5 (h) to see that

$$\begin{aligned}E_j &= \sum_{|\alpha_\mu|\leq N} 2^{-(N-|\alpha_\mu|)j_\mu} E_{j,\mu,\alpha_\mu} (2^{-j_\mu d^\mu} W^\mu)^{\alpha_\mu} \\ &= \sum_{|\alpha_\mu|\leq N} 2^{-Nj_\mu} 2^{-(\deg_{d^\mu}(\alpha_\mu)-|\alpha_\mu|)j_\mu} E_{j,\mu,\alpha_\mu} (W^\mu)^{\alpha_\mu},\end{aligned} \tag{6.21}$$

where $\mathcal{E}' := \{(E_{j,\mu,\alpha_\mu}, 2^{-j}) : (E_j, 2^{-j}) \in \mathcal{E}, |\alpha_\mu| \leq N\}$ is a bounded set of generalized $(W, \vec{d})$ elementary operators supported in Ω.

Since $\deg_{\vec{d}^\mu}(\alpha_\mu) - |\alpha_\mu| \geq 0$ for all α_μ, using (6.21), we see that

$$\begin{aligned}
&\sup_{\{(E_j,2^{-j}):j\in\mathcal{I}_\mu\}\subseteq\mathcal{E}} \|\{2^{(N-1)j_\mu}E_jf\}_{j\in\mathcal{I}_\mu}\|_{\mathscr{V}_\mu} \\
&\quad\leq \sum_{|\alpha_\mu|\leq N} \sup_{\{(E_j,2^{-j}):j\in\mathcal{I}_\mu\}\subseteq\mathcal{E}} \|\{2^{-j_\mu}E_{j,\mu,\alpha_\mu}(W^\mu)^{\alpha_\mu}f\}_{j\in\mathcal{I}_\mu}\|_{\mathscr{V}_\mu} \\
&\quad\lesssim \sup_{\{(E_j',2^{-j}):j\in\mathcal{I}_\mu\}\subseteq\mathcal{E}'} \|\{2^{-j_\mu}E_j'(W^\mu)^{\alpha_\mu}f\}_{j\in\mathcal{I}_\mu}\|_{\mathscr{V}_\mu}.
\end{aligned} \tag{6.22}$$

Let $p \in [1,\infty]$ be as in (6.18). Using the definition of bounded sets of generalized $(W,\vec{d})$ pre-elementary operators (Definition 5.2.26), we have

$$\begin{aligned}
&\sup_{\{(E_j',2^{-j}):j\in\mathcal{I}_\mu\}\subseteq\mathcal{E}'} \|\{2^{-j_\mu}E_j'(W^\mu)^{\alpha_\mu}f\}_{j\in\mathcal{I}_\mu}\|_{\mathscr{V}_\mu} \\
&\quad\lesssim \|\{2^{-j_\mu}(W^\mu)^{\alpha_\mu}f\}_{j\in\mathcal{I}_\mu}\|_{\mathscr{V}_\mu} \\
&\quad= \|\{2^{-|j|_\infty}(W^\mu)^{\alpha_\mu}f\}_{j\in\mathcal{I}_\mu}\|_{\mathscr{V}_\mu} \lesssim \|(W^\mu)^{\alpha_\mu}f\|_{L^p(\mathfrak{M},\mathrm{Vol})} < \infty.
\end{aligned} \tag{6.23}$$

Combining (6.22) and (6.23) gives (6.20) and completes the proof. □

Proposition 6.5.6. *Suppose $\mathcal{K}_1, \mathcal{K}_2 \Subset \mathfrak{M}$ are two compact sets with $\mathcal{K}_1 \subseteq \mathcal{K}_2$. Then $\mathscr{X}^s(\mathcal{K}_1,(W,\vec{d})) \subseteq \mathscr{X}^s(\mathcal{K}_2,(W,\vec{d}))$ is a closed subspace (and the subspace topology on $\mathscr{X}^s(\mathcal{K}_1,(W,\vec{d}))$ agrees with its original topology). Furthermore, $\mathscr{X}^s(\mathcal{K}_1,(W,\vec{d}))$ consists of precisely those $f \in \mathscr{X}^s(\mathcal{K}_2,(W,\vec{d}))$ such that* $\mathrm{supp}(f) \subseteq \mathcal{K}_1$.

Proof. It follows directly from Definition 6.3.5 that for $f \in C^\infty_{W,0}(\mathfrak{M})'$, the following are equivalent:
- $f \in \mathscr{X}^s(\mathcal{K}_2,(W,\vec{d}))$ and $\mathrm{supp}(f) \subseteq \mathcal{K}_1$.
- $f \in \mathscr{X}^s(\mathcal{K}_1,(W,\vec{d}))$.

This proves that $\mathscr{X}^s(\mathcal{K}_1,(W,\vec{d})) \subseteq \mathscr{X}^s(\mathcal{K}_2,(W,\vec{d}))$ and $\mathscr{X}^s(\mathcal{K}_1,(W,\vec{d}))$ consists of precisely those $f \in \mathscr{X}^s(\mathcal{K}_2,(W,\vec{d}))$ such that $\mathrm{supp}(f) \subseteq \mathcal{K}_1$.

By Remark 6.3.9 we may choose the same norm on $\mathscr{X}^s(\mathcal{K}_2,(W,\vec{d}))$ and $\mathscr{X}^s(\mathcal{K}_1,(W,\vec{d}))$ so that the subspace topology on $\mathscr{X}^s(\mathcal{K}_1,(W,\vec{d}))$ agrees with its original topology. In particular, since $\mathscr{X}^s(\mathcal{K}_1,(W,\vec{d}))$ is complete (Proposition 6.3.6), we see that $\mathscr{X}^s(\mathcal{K}_1,(W,\vec{d}))$ is a closed subspace of $\mathscr{X}^s(\mathcal{K}_2,(W,\vec{d}))$. □

In light of Proposition 6.5.6, the next definition makes sense.

Definition 6.5.7. For $s \in \mathbb{R}^\nu$, we set

$$\mathscr{X}^s_{\mathrm{cpt}}(W,\vec{d}) := \bigcup_{\substack{\mathcal{K}\Subset\mathfrak{M}\\ \mathcal{K}\text{ compact}}} \mathscr{X}^s(\mathcal{K},(W,\vec{d})).$$

Remark 6.5.8. Proposition 6.5.6 can be restated as saying that the following are equivalent:
- $u \in \mathscr{X}^s(\mathcal{K}, (W, \vec{d}))$.
- $u \in \mathscr{X}^s_{\mathrm{cpt}}(W, \vec{d})$ and $\mathrm{supp}(u) \subseteq \mathcal{K}$.

The next results show that each vector field W_j acts as a differential operator of degree $\vec{d}_j \in \mathbb{N}^\nu$.

Proposition 6.5.9. *Let $\Omega \Subset \mathfrak{M}$ be an open set with $\mathcal{K} \Subset \Omega$. Let $\kappa \in \mathbb{Z}^\nu$ and let $\mathscr{P}$ be a $(W, \vec{d})$ partial differential operator on Ω of degree $\le \kappa$. Then, for every $s \in \mathbb{R}^\nu$, $\mathscr{P} : \mathscr{X}^s(\mathcal{K}, (W, \vec{d})) \to \mathscr{X}^{s-\kappa}(\mathcal{K}, (W, \vec{d}))$ and is a bounded operator.*

Proof. If $\kappa \in \mathbb{Z}^\nu \setminus \mathbb{N}^\nu$, then $\mathscr{P} = 0$ and the result is trivial. We henceforth assume $\kappa \in \mathbb{N}^\nu$.

Take $\phi \in C_0^\infty(\Omega)$ with $\phi \equiv 1$ on a neighborhood of $\mathcal{K}$. For $f \in \mathscr{X}^s(\mathcal{K}, (W, \vec{d}))$ we have $\mathrm{supp}(f) \subseteq \mathcal{K}$ and therefore $\mathscr{P}f = \mathrm{Mult}[\phi]\mathscr{P}f$. Thus, it suffices to prove the result with $\mathscr{P}$ replaced by $\mathrm{Mult}[\phi]\mathscr{P}$.

By definition (see Definition 3.12.1), $\mathrm{Mult}[\phi]\mathscr{P}$ can be written as a finite sum of operators of the form $\mathrm{Mult}[\phi g_\alpha]W^\alpha$, where $g_\alpha \in C^\infty_{\mathrm{loc}}(\Omega)$ and $\deg_{\vec{d}}(\alpha) \le \kappa$. By Propositions 5.8.3 and 5.8.2, $\mathrm{Mult}[\phi g_\alpha]W^\alpha \in \widetilde{\mathscr{A}}^{\deg_{\vec{d}}(\alpha)}(\Omega, (W, \vec{d})) \subseteq \widetilde{\mathscr{A}}^\kappa(\Omega, (W, \vec{d}))$. We conclude that $\mathrm{Mult}[\phi]\mathscr{P} \in \widetilde{\mathscr{A}}^\kappa(\Omega, (W, \vec{d}))$.

Set $\mathcal{K}_1 := \overline{\Omega} \Subset \mathfrak{M}$. Since $\mathrm{Mult}[\phi]\mathscr{P} \in \widetilde{\mathscr{A}}^\kappa(\Omega, (W, \vec{d}))$, by Theorem 6.3.10, $\mathrm{Mult}[\phi]\mathscr{P} : \mathscr{X}^s(\mathcal{K}_1, (W, \vec{d})) \to \mathscr{X}^{s-\kappa}(\mathcal{K}_1, (W, \vec{d}))$ is a bounded operator. By Proposition 6.5.6, $\mathscr{X}^s(\mathcal{K}, (W, \vec{d}))$ is a closed subspace of $\mathscr{X}^s(\mathcal{K}_1, (W, \vec{d}))$, and therefore $\mathrm{Mult}[\phi]\mathscr{P} : \mathscr{X}^s(\mathcal{K}, (W, \vec{d})) \to \mathscr{X}^{s-\kappa}(\mathcal{K}_1, (W, \vec{d}))$ is a bounded operator.

By Proposition 6.5.6, the topology on $\mathscr{X}^{s-\kappa}(\mathcal{K}, (W, \vec{d}))$ agrees with its topology as a subspace of $\mathscr{X}^{s-\kappa}(\mathcal{K}_1, (W, \vec{d}))$. Thus, to complete the proof, it suffices to show that $\mathrm{Mult}[\phi]\mathscr{P} : \mathscr{X}^s(\mathcal{K}, (W, \vec{d})) \to \mathscr{X}^{s-\kappa}(\mathcal{K}, (W, \vec{d}))$. Take $f \in \mathscr{X}^s(\mathcal{K}, (W, \vec{d}))$. We have already shown that $\mathrm{Mult}[\phi]\mathscr{P}f \in \mathscr{X}^{s-\kappa}(\mathcal{K}_1, (W, \vec{d}))$. Since $\mathrm{supp}(f) \subseteq \mathcal{K}$ and $\mathscr{P}$ is a partial differential operator, we have $\mathrm{supp}(\mathrm{Mult}[\phi]\mathscr{P}f) \subseteq \mathcal{K}$. Proposition 6.5.6 implies that $\mathrm{Mult}[\phi]\mathscr{P}f \in \mathscr{X}^{s-\kappa}(\mathcal{K}, (W, \vec{d}))$, completing the proof. □

Corollary 6.5.10. *Let $\phi \in C^\infty_{\mathrm{loc}}(\mathfrak{M})$. Then $\mathrm{Mult}[\phi]$ is a bounded operator $\mathscr{X}^s(\mathcal{K}, (W, \vec{d})) \to \mathscr{X}^s(\mathcal{K}, (W, \vec{d}))$.*

Proof. For any $\Omega \Subset \mathfrak{M}$, $\mathrm{Mult}[\phi]$ is a $(W, \vec{d})$ partial differential operator on Ω of degree ≤ 0. Thus, the result follows immediately from Proposition 6.5.9. □

Corollary 6.5.11. *For any ordered multi-index α,*

$$W^\alpha : \mathscr{X}^s(\mathcal{K}, (W, \vec{d})) \to \mathscr{X}^{s-\deg_{\vec{d}}(\alpha)}(\mathcal{K}, (W, \vec{d}))$$

is a bounded operator. In particular, $W_j : \mathscr{X}^s(\mathcal{K}, (W, \vec{d})) \to \mathscr{X}^{s-\vec{d}_j}(\mathcal{K}, (W, \vec{d}))$ is a bounded operator.

Proof. Let $\Omega \Subset \mathfrak{M}$ be open with $\mathcal{K} \Subset \Omega$. Then W^α is a $(W, \vec{d})$ partial differential operator on Ω of degree $\leq \deg_{\vec{d}}(\alpha)$. Thus, the claimed boundedness of W^α follows directly from Proposition 6.5.9. The claim about W_j is the special case $|\alpha| = 1$ of the claim about W^α. □

Proposition 6.5.12. *Fix $\mu \in \{1, \dots, \nu\}$ and $\kappa_\mu \in \mathbb{N}_+$ such that d_j^μ divides κ_μ for $1 \leq j \leq r_\mu$. Set $n_j^\mu := \kappa_\mu / d_j^\mu \in \mathbb{N}_+$ and let*

$$\mathcal{L}_\mu := \sum_{j=1}^{r_\mu} ((W_j^\mu)^{n_j^\mu})^* (W_j^\mu)^{n_j^\mu}. \tag{6.24}$$

For $s \in \mathbb{R}^\nu$ and $f \in C_{W,0}^\infty(\mathfrak{M})'$, the following are equivalent:
(i) $f \in \mathscr{X}^s(\mathcal{K}, (W, \vec{d}))$.
(ii) *For all ordered multi-indices α_μ with $\deg_{\vec{d}^\mu}(\alpha_\mu) \leq \kappa_\mu$, we have*

$$(W^\mu)^{\alpha_\mu} f \in \mathscr{X}^{s - \kappa_\mu \mathfrak{e}_\mu}(\mathcal{K}, (W, \vec{d})).$$

(iii) $f \in \mathscr{X}^{s-\kappa_\mu \mathfrak{e}_\mu}(\mathcal{K}, (W, \vec{d}))$ *and* $(W_j^\mu)^{n_j^\mu} f \in \mathscr{X}^{s-\kappa_\mu \mathfrak{e}_\mu}(\mathcal{K}, (W, \vec{d}))$ *for* $1 \leq j \leq r_\mu$.
(iv) $f, \mathcal{L}_\mu f \in \mathscr{X}^{s-2\kappa_\mu \mathfrak{e}_\mu}(\mathcal{K}, (W, \vec{d}))$.

Furthermore, we have, for all $f \in \mathscr{X}^s(\mathcal{K}, (W, \vec{d}))$,

$$\begin{aligned} \|f\|_{\mathscr{X}^s(W,\vec{d})} &\approx \sum_{\deg_{\vec{d}^\mu}(\alpha_\mu) \leq \kappa_\mu} \|(W^\mu)^{\alpha_\mu} f\|_{\mathscr{X}^{s-\kappa_\mu}(W,\vec{d})} \\ &\approx \|f\|_{\mathscr{X}^{s-\kappa_\mu \mathfrak{e}_\mu}(W,\vec{d})} + \sum_{j=1}^{r_\mu} \|(W_j^\mu)^{n_j^\mu} f\|_{\mathscr{X}^{s-\kappa_\mu \mathfrak{e}_\mu}(W,\vec{d})} \\ &\approx \|f\|_{\mathscr{X}^{s-2\kappa_\mu \mathfrak{e}_\mu}(W,\vec{d})} + \|\mathcal{L}_\mu f\|_{\mathscr{X}^{s-2\kappa_\mu \mathfrak{e}_\mu}(W,\vec{d})}, \end{aligned} \tag{6.25}$$

where the implicit constants do not depend on f (but may depend on any of the other ingredients).

Proof. (i) ⇒ (ii): Suppose $f \in \mathscr{X}^s(\mathcal{K}, (W, \vec{d}))$ and $\deg_{\vec{d}^\mu}(\alpha_\mu) \leq \kappa_\mu$. Since $(W^\mu)^{\alpha_\mu}$ is a $(W, \vec{d})$ partial differential operator of degree $\leq \deg_{\vec{d}^\mu}(\alpha_\mu)\mathfrak{e}_\mu \leq \kappa_\mu \mathfrak{e}_\mu$ on Ω (for any Ω open with $\mathcal{K} \Subset \Omega$), Proposition 6.5.9 shows that $(W^\mu)^{\alpha_\mu} f \in \mathscr{X}^{s-\kappa_\mu \mathfrak{e}_\mu}(\mathcal{K}, (W, \vec{d}))$ and

$$\sum_{\deg_{\vec{d}^\mu}(\alpha_\mu) \leq \kappa_\mu} \|(W^\mu)^{\alpha_\mu} f\|_{\mathscr{X}^{s-\kappa_\mu}(W,\vec{d})} \lesssim \|f\|_{\mathscr{X}^s(W,\vec{d})}. \tag{6.26}$$

(ii) ⇒ (iii): Suppose (ii) holds. Since $(W_j^\mu)^{n_j^\mu} = (W^\mu)^{\alpha_\mu}$, for some $\deg_{\vec{d}^\mu}(\alpha_\mu) = \kappa_\mu$, (iii) holds, and moreover,

$$\begin{aligned}\|f\|_{\mathscr{X}^{s-\kappa_\mu e_\mu}(W,\vec{d})} + \sum_{j=1}^{r_\mu}\|(W_j^\mu)^{n_j^\mu} f\|_{\mathscr{X}^{s-\kappa_\mu e_\mu}(W,\vec{d})} \\ \lesssim \sum_{\deg_{d^\mu}(\alpha_\mu)\le\kappa_\mu} \|(W^\mu)^{\alpha_\mu} f\|_{\mathscr{X}^{s-\kappa_\mu}(W,\vec{d})}.\end{aligned} \tag{6.27}$$

(iii) $\Rightarrow$ (iv): Suppose (iii) holds. Fix $\Omega \Subset \mathfrak{M}$ open with $\mathcal{K} \Subset \Omega$. $(W_j^\mu)^{n_j^\mu}$ is a $(W,\vec{d})$ partial differential operator on Ω of degree $\le \kappa_\mu e_\mu$, and therefore by Lemma 3.12.5, $((W_j^\mu)^{n_j^\mu})^*$ is a $(W,\vec{d})$ partial differential operator on Ω of degree $\le \kappa_\mu e_\mu$. Thus, since (by assumption) $(W_j^\mu)^{n_j^\mu} f \in \mathscr{X}^{s-\kappa_\mu e_\mu}(\mathcal{K},(W,\vec{d}))$, Proposition 6.5.9 shows that $((W_j^\mu)^{n_j^\mu})^*(W_j^\mu)^{n_j^\mu} f \in \mathscr{X}^{s-2\kappa_\mu e_\mu}(\mathcal{K},(W,\vec{d}))$ and

$$\|((W_j^\mu)^{n_j^\mu})^*(W_j^\mu)^{n_j^\mu} f\|_{\mathscr{X}^{s-2\kappa_\mu e_\mu}(W,\vec{d})} \lesssim \|(W_j^\mu)^{n_j^\mu} f\|_{\mathscr{X}^{s-\kappa_\mu e_\mu}(W,\vec{d})}.$$

Using the formula for $\mathscr{L}_\mu$, we conclude that $\mathscr{L}_\mu f \in \mathscr{X}^{s-2\kappa_\mu e_\mu}$ and

$$\|\mathscr{L}_\mu f\|_{\mathscr{X}^{s-2\kappa_\mu e_\mu}(W,\vec{d})} \lesssim \|(W_j^\mu)^{n_j^\mu} f\|_{\mathscr{X}^{s-\kappa_\mu e_\mu}(W,\vec{d})}. \tag{6.28}$$

By assumption, we have $f \in \mathscr{X}^{s-\kappa_\mu e_\mu}(\mathcal{K},(W,\vec{d})) \subseteq \mathscr{X}^{s-2\kappa_\mu e_\mu}(\mathcal{K},(W,\vec{d}))$, which completes the proof of (iv). Moreover,

$$\|f\|_{\mathscr{X}^{s-2\kappa_\mu e_\mu}(W,\vec{d})} \le \|f\|_{\mathscr{X}^{s-\kappa_\mu e_\mu}(W,\vec{d})}. \tag{6.29}$$

By combining (6.28) and (6.29), we obtain

$$\begin{aligned}\|f\|_{\mathscr{X}^{s-2\kappa_\mu e_\mu}(W,\vec{d})} + \|\mathscr{L}_\mu f\|_{\mathscr{X}^{s-2\kappa_\mu e_\mu}(W,\vec{d})} \\ \lesssim \|f\|_{\mathscr{X}^{s-\kappa_\mu e_\mu}(W,\vec{d})} + \sum_{j=1}^{r_\mu}\|(W_j^\mu)^{n_j^\mu} f\|_{\mathscr{X}^{s-\kappa_\mu e_\mu}(W,\vec{d})}.\end{aligned} \tag{6.30}$$

(iv) $\Rightarrow$ (i): We claim the following.

Claim: Under the assumptions of (iv), for $L \in \{0,1,\dots,2\kappa_\mu\}$, we have

$$f \in \mathscr{X}^{s-2\kappa_\mu e_\mu + L e_\mu}(\mathcal{K},(W,\vec{d}))$$

and

$$\|f\|_{\mathscr{X}^{s-2\kappa_\mu e_\mu + L e_\mu}(W,\vec{d})} \lesssim \|f\|_{\mathscr{X}^{s-2\kappa_\mu e_\mu}(W,\vec{d})} + \|\mathscr{L}_\mu f\|_{\mathscr{X}^{s-2\kappa_\mu e_\mu}(W,\vec{d})}. \tag{6.31}$$

We prove the claim by induction on L. The base case, $L = 0$, follows immediately from the assumption (iv). Thus, we assume the claim for $L \in \{0,1,\dots,2\kappa_\mu - 1\}$ and prove it for $L+1$.

Fix $\Omega \Subset \mathfrak{M}$ open with $\mathcal{K} \Subset \Omega$. Let $\mathcal{K}_1 := \overline{\Omega}$, so that $\mathcal{K}_1$ is compact with $\mathcal{K} \Subset \mathcal{K}_1$. Fix $\phi \in C_0^\infty(\Omega)$ with $\phi \equiv 1$ on $\mathcal{K}$. By Corollary 5.8.15 there exists $T \in \mathscr{A}^{-2\kappa_\mu \mathfrak{e}_\mu}(\Omega, (W, \vec{d}))$ with

$$T\mathscr{L}_\mu = \mathrm{Mult}[\phi] + R,$$

where $R \in \widetilde{\mathscr{A}}^{-\mathfrak{e}_\mu}(\Omega, (W, \vec{d}))$.

By Proposition 6.5.6 and assumption (iv),

$$\mathscr{L}_\mu f \in \mathscr{X}^{s-2\kappa_\mu \mathfrak{e}_\mu}(\mathcal{K}, (W, \vec{d})) \subseteq \mathscr{X}^{s-2\kappa_\mu \mathfrak{e}_\mu}(\mathcal{K}_1, (W, \vec{d})).$$

Thus, by Theorem 6.3.10, using the fact that $T \in \mathscr{A}^{-2\kappa_\mu \mathfrak{e}_\mu}(\Omega, (W, \vec{d}))$ and $\mathrm{supp}(f) \subseteq \mathcal{K}$, we have

$$f + Rf = \mathrm{Mult}[\phi]f + Rf = T\mathscr{L}_\mu f \in \mathscr{X}^{s}(\mathcal{K}_1, (W, \vec{d})),$$

with

$$\|f + Rf\|_{\mathscr{X}^{s}(W,\vec{d})} \lesssim \|\mathscr{L}_\mu f\|_{\mathscr{X}^{s-2\kappa_\mu \mathfrak{e}_\mu}(W,\vec{d})}. \tag{6.32}$$

Using the inductive hypothesis and Proposition 6.5.6, we have

$$f \in \mathscr{X}^{s-2\kappa_\mu \mathfrak{e}_\mu + L\mathfrak{e}_\mu}(\mathcal{K}, (W, \vec{d})) \subseteq \mathscr{X}^{s-2\kappa_\mu \mathfrak{e}_\mu + L\mathfrak{e}_\mu}(\mathcal{K}_1, (W, \vec{d})).$$

Thus, since $R \in \widetilde{\mathscr{A}}^{-\mathfrak{e}_\mu}(\Omega, (W, \vec{d}))$, applying Theorem 6.3.10, we have

$$Rf \in \mathscr{X}^{s-2\kappa_\mu \mathfrak{e}_\mu + (L+1)\mathfrak{e}_\mu}(\mathcal{K}_1, (W, \vec{d}))$$

with

$$\begin{aligned} \|Rf\|_{\mathscr{X}^{s-2\kappa_\mu \mathfrak{e}_\mu+(L+1)\mathfrak{e}_\mu}(W,\vec{d})} &\lesssim \|f\|_{\mathscr{X}^{s-2\kappa_\mu \mathfrak{e}_\mu+L\mathfrak{e}_\mu}(W,\vec{d})} \\ &\lesssim \|f\|_{\mathscr{X}^{s-2\kappa_\mu \mathfrak{e}_\mu}(W,\vec{d})} + \|\mathscr{L}_\mu f\|_{\mathscr{X}^{s-2\kappa_\mu \mathfrak{e}_\mu}(W,\vec{d})}, \end{aligned} \tag{6.33}$$

where the last estimate uses the inductive hypothesis.

Combining the above, we see that $f = (f + Rf) - Rf \in \mathscr{X}^{s-2\kappa_\mu \mathfrak{e}_\mu+(L+1)\mathfrak{e}_\mu}(\mathcal{K}_1, (W, \vec{d}))$. Since $\mathrm{supp}(f) \subseteq \mathcal{K}$, Proposition 6.5.6 implies that $f \in \mathscr{X}^{s-2\kappa_\mu \mathfrak{e}_\mu+(L+1)\mathfrak{e}_\mu}(\mathcal{K}, (W, \vec{d}))$. Finally, combining (6.32) and (6.33), we obtain (6.31) with L replaced by $L + 1$. This completes the proof of the claim.

Applying the claim with $L = 2\kappa_\mu$ shows that (i) holds, and moreover,

$$\|f\|_{\mathscr{X}^{s}(W,\vec{d})} \lesssim \|f\|_{\mathscr{X}^{s-2\kappa_\mu \mathfrak{e}_\mu}(W,\vec{d})} + \|\mathscr{L}_\mu f\|_{\mathscr{X}^{s-2\kappa_\mu \mathfrak{e}_\mu}(W,\vec{d})}. \tag{6.34}$$

Thus, we have completed the proof of the equivalence of the conditions. Moreover, (6.25) follows by combining (6.26), (6.27), (6.30), and (6.34). □

Corollary 6.5.13. *For each $\mu \in \{1, \dots, \nu\}$, let $\kappa_\mu \in \mathbb{N}$ be such that $\vec{d}^\mu_j$ divides κ_μ for $1 \leq j \leq r_\mu$ (we allow the possibility $\kappa_\mu = 0$). Set $\kappa := (\kappa_1, \dots, \kappa_\nu) \in \mathbb{N}^\nu$. Then, for $s \in \mathbb{R}^\nu$ and $f \in C^\infty_{W,0}(\mathfrak{M})'$, the following are equivalent:*

(i) $f \in \mathscr{X}^s(\mathcal{K}, (W, \vec{d}))$.

(ii) *For every ordered multi-index α with $\deg_{\vec{d}}(\alpha) \leq \kappa$, we have $W^\alpha f \in \mathscr{X}^{s-\kappa}(\mathcal{K}, (W, \vec{d}))$.*

Furthermore, for $f \in \mathscr{X}^s(\mathcal{K}, (W, \vec{d}))$, we have

$$\|f\|_{\mathscr{X}^s(W,\vec{d})} \approx \sum_{\deg_{\vec{d}}(\alpha) \leq \kappa} \|W^\alpha f\|_{\mathscr{X}^{s-\kappa}(W,\vec{d})}, \tag{6.35}$$

where the implicit constants do not depend on f (but may depend on any of the other ingredients).

Proof. (i) ⇒ (ii): By Corollary 6.5.11, for $\deg_{\vec{d}}(\alpha) \leq \kappa$,

$$W^\alpha : \mathscr{X}^s(\mathcal{K}, (W, \vec{d})) \to \mathscr{X}^{s-\deg_{\vec{d}}(\alpha)}(\mathcal{K}, (W, \vec{d})) \subseteq \mathscr{X}^{s-\kappa}(\mathcal{K}, (W, \vec{d}))$$

is bounded, and therefore (i) ⇒ (ii) and moreover

$$\sum_{\deg_{\vec{d}}(\alpha) \leq \kappa} \|W^\alpha f\|_{\mathscr{X}^{s-\kappa}(W,\vec{d})} \lesssim \|f\|_{\mathscr{X}^s(W,\vec{d})}. \tag{6.36}$$

(ii) ⇒ (i): Suppose (ii) holds. For each $\mu \in \{1, \dots, \nu+1\}$ set $\kappa^\mu = (0, 0, \dots, 0, \kappa_\mu, \kappa_{\mu+1}, \dots, \kappa_\nu) \in \mathbb{N}^\nu$, with $\kappa^{\nu+1} = 0 \in \mathbb{N}^\nu$.

Claim: For $\mu \in \{1, \dots, \nu + 1\}$ and for every ordered multi-index α with $\deg_{\vec{d}}(\alpha) \leq \kappa^\mu$, we have $W^\alpha f \in \mathscr{X}^{s-\kappa^\mu}(\mathcal{K}, (W, \vec{d}))$ and

$$\sum_{\deg_{\vec{d}}(\alpha) \leq \kappa^\mu} \|W^\alpha f\|_{\mathscr{X}^{s-\kappa^\mu}(W,\vec{d})} \lesssim \sum_{\deg_{\vec{d}}(\alpha) \leq \kappa} \|W^\alpha f\|_{\mathscr{X}^{s-\kappa}(W,\vec{d})}. \tag{6.37}$$

We prove the claim by induction on μ. The base case, $\mu = 1$, follows immediately from hypothesis (ii) ((6.37) being trivial in the case $\mu = 1$). We assume the result for some $\mu \in \{1, \dots, \nu\}$ and prove it for $\mu + 1$.

If $\kappa_\mu = 0$, then $\kappa^\mu = \kappa^{\mu+1}$ and there is nothing to prove. Thus, we assume $\kappa_\mu \neq 0$. Fix β with $\deg_{\vec{d}}(\beta) \leq \kappa^{\mu+1}$. For any α_μ with $\deg_{d^\mu}(\alpha_\mu) \leq \kappa_\mu$, $(W^\mu)^{\alpha_\mu} W^\beta$ is of the form W^γ for some γ with $\deg_{\vec{d}}(\gamma) \leq \kappa^\mu$. Thus, by the inductive hypothesis, we have $(W^\mu)^{\alpha_\mu} W^\beta f \in \mathscr{X}^{s-\kappa^\mu}(\mathcal{K}, (W, \vec{d}))$ and

$$\sum_{\deg_{d^\mu}(\alpha_\mu) \leq \kappa_\mu} \|(W^\mu)^{\alpha_\mu} W^\beta f\|_{\mathscr{X}^{s-\kappa^\mu}(W,\vec{d})} \lesssim \sum_{\deg_{\vec{d}}(\alpha) \leq \kappa} \|W^\alpha f\|_{\mathscr{X}^{s-\kappa}(W,\vec{d})}.$$

Proposition 6.5.12 (ii) ⇒ (i) implies that $W^\beta f \in \mathscr{X}^{s-\kappa^{\mu+1}}(\mathcal{K}, (W, \vec{d}))$. Moreover, (6.25) implies that

$$\begin{aligned}\|W^{\beta}f\|_{\mathscr{X}^{s-\kappa^{\mu+1}}(W,\vec{d})} &\lesssim \sum_{\deg_{d^{\mu}}(\alpha_{\mu})\le\kappa_{\mu}} \|(W^{\mu})^{\alpha_{\mu}}W^{\beta}f\|_{\mathscr{X}^{s-\kappa^{\mu}}(W,\vec{d})}\\ &\lesssim \sum_{\deg_{\vec{d}}(\alpha)\le\kappa} \|W^{\alpha}f\|_{\mathscr{X}^{s-\kappa}(W,\vec{d})}.\end{aligned} \tag{6.38}$$

Since β was arbitrary with $\deg_{\vec{d}}(\beta) \le \kappa^{\mu+1}$, summing (6.38) over such β completes the proof of the claim.

Applying the claim with $\mu = \nu + 1$ shows that (i) holds, and moreover,

$$\|f\|_{\mathscr{X}^{s}(W,\vec{d})} \lesssim \sum_{\deg_{\vec{d}}(\alpha)\le\kappa} \|W^{\alpha}f\|_{\mathscr{X}^{s-\kappa}(W,\vec{d})}. \tag{6.39}$$

Finally, (6.35) follows by combining (6.36) and (6.39). □

Proposition 6.5.14. *For each $\mu \in \{1,\dots,\nu\}$, let*

$$(Z^{\mu}, d^{\mu}) = \{(Z_1^{\mu}, d_1^{\mu}),\dots,(Z_{s_{\mu}}^{\mu}, d_{s_{\mu}}^{\mu})\} \subset C^{\infty}_{\mathrm{loc}}(\mathfrak{M}; T\mathfrak{M}) \times \mathbb{N}_+$$

be such that (Z^{μ}, d^{μ}) and (W^{μ}, d^{μ}) are locally weakly equivalent on $\mathfrak{M}$. Define $(Z, \vec{d}) = (Z^1, d^1)\boxtimes(Z^2, d^2)\boxtimes\cdots(Z^{\nu}, d^{\nu})$, i. e., $(Z, \vec{d})$ is to (Z^{μ}, d^{μ}), $1 \le \mu \le \nu$, as $(W, \vec{d})$ is to (W^{μ}, d^{μ}), $1 \le \mu \le \nu$. Then $\mathscr{X}^{s}(\mathcal{K}, (W, \vec{d})) = \mathscr{X}^{s}(\mathcal{K}, (Z, \vec{d}))$, with equivalent norms.

Proof. By Proposition 5.5.14, $\mathcal{E}$ is a bounded set of generalized $(W, \vec{d})$ elementary operators if and only if $\mathcal{E}$ is a bounded set of generalized $(Z, \vec{d})$ elementary operators. Using this, the result follows directly from the definitions. □

Proposition 6.5.15. *The spaces $\mathscr{X}^{s}(\mathcal{K}, (W, \vec{d}))$ are invariant under diffeomorphisms. More precisely, suppose $\mathfrak{N}$ is another C^{∞} manifold and $\Psi : \mathfrak{M} \xrightarrow{\sim} \mathfrak{N}$ is a $C^{\infty}_{\mathrm{loc}}$ diffeomorphism. Set*

$$(\Psi_* W, \vec{d}) := \{(\Psi_* W_1, \vec{d}_1),\dots,(\Psi_* W_r, \vec{d}_r)\} \subset C^{\infty}_{\mathrm{loc}}(\mathfrak{N}; T\mathfrak{N}) \times (\mathbb{N}^{\nu} \setminus \{0\}).$$

Then

$$\mathscr{X}^{s}(\mathcal{K}, (W, \vec{d})) = \mathscr{X}^{s}(\Psi(\mathcal{K}), (\Psi_* W, \vec{d}))$$

and

$$\|f\|_{\mathscr{X}^{s}(W,\vec{d})} \approx \|\Psi_* f\|_{\mathscr{X}^{s}(\Psi_* W,\vec{d})}, \quad \forall f \in \mathscr{X}^{s}(\mathcal{K}, (W, \vec{d})). \tag{6.40}$$

Proof. It is easy to see that all of our assumptions and definitions are invariant under diffeomorphisms, so this result follows just by tracing through the definitions. □

Proposition 6.5.16. *Fix $p \in [1,\infty]$ such that $\mathscr{X}^{s} = \mathscr{B}^{s}_{p,q}$ or $\mathscr{X}^{s} = \mathscr{F}^{s}_{p,q}$ (see Notation 6.3.2). Then, $\forall s \in (0,\infty)^{\nu}$, we have*

$$\mathscr{X}^s(\mathcal{K},(W,\vec{d})) \subseteq L^p(\mathcal{K},\mathrm{Vol}) \subseteq \mathscr{X}^{-s}(\mathcal{K},(W,\vec{d})), \tag{6.41}$$

and the inclusions are continuous.

Proof. We begin with the second inclusion in (6.41). Let $f \in L^p(\mathcal{K},\mathrm{Vol})$ and let $\mathcal{E}$ be a bounded set of generalized $(W,\vec{d})$ elementary operators. Interpolation of (5.5) and (5.6) (with $|\alpha| = |\beta| = 0$) shows that

$$\sup_{(E,2^{-j})\in\mathcal{E}} \|E\|_{L^p(\mathfrak{M},\mathrm{Vol})\to L^p(\mathfrak{M},\mathrm{Vol})} < \infty.$$

Let $\{(E_j,2^{-j}) : j \in \mathbb{N}^\nu\} \subseteq \mathcal{E}$. We have

$$\begin{aligned}\|\{2^{-j\cdot s}E_jf\}_{j\in\mathbb{N}^\nu}\|_{\mathscr{V}} &\le \|\{2^{-j\cdot s}E_jf\}_{j\in\mathbb{N}^\nu}\|_{\ell^1(\mathbb{N}^\nu;L^p(\mathfrak{M},\mathrm{Vol}))} = \sum_{j\in\mathbb{N}^\nu} 2^{-j\cdot s}\|E_jf\|_{L^p(\mathfrak{M},\mathrm{Vol})}\\ &\lesssim \sum_{j\in\mathbb{N}^\nu} 2^{-j\cdot s}\|f\|_{L^p(\mathfrak{M},\mathrm{Vol})} \lesssim \|f\|_{L^p(\mathfrak{M},\mathrm{Vol})}.\end{aligned} \tag{6.42}$$

Taking the supremum of the left-hand side of (6.42) over $\{(E_j,2^{-j}) : j \in \mathbb{N}^\nu\} \subseteq \mathcal{E}$ shows that

$$\|f\|_{\mathscr{V},s,\mathcal{E}} \lesssim \|f\|_{L^p(\mathfrak{M},\mathrm{Vol})} < \infty. \tag{6.43}$$

This shows that $f \in \mathscr{X}^{-s}(\mathcal{K},(W,\vec{d}))$, proving the second containment in (6.41). Moreover, setting $\mathcal{E} = \mathcal{D}_0$ in (6.43) shows that $\|f\|_{\mathscr{X}^{-s}(W,\vec{d})} \lesssim \|f\|_{L^p}$, proving that the inclusion is continuous.

We turn to the first inclusion in (6.41). Let $f \in \mathscr{X}^s(\mathcal{K},(W,\vec{d}))$. Since $\mathrm{supp}(f) \subseteq \mathcal{K}$, it suffices to show that $f \in L^p(\mathfrak{M},\mathrm{Vol})$ and $\|f\|_{L^p(\mathfrak{M},\mathrm{Vol})} \lesssim \|f\|_{\mathscr{X}^s(W,\vec{d})}$. Since $\mathrm{supp}(f) \subseteq \mathcal{K}$, we have $f = \mathrm{Mult}[\psi]f = \sum_{j\in\mathbb{N}^\nu} D_jf$ with convergence in $C^\infty_{W,0}(\mathfrak{M})'$ (see Proposition 5.5.10). Since $f \in \mathscr{X}^s(\mathcal{K},(W,\vec{d}))$, we have $D_jf \in L^p(\mathfrak{M},\mathrm{Vol})$ and $\|D_jf\|_{L^p} \lesssim 2^{-j\cdot s}\|f\|_{\mathscr{X}^s(W,\vec{d})}$. This implies that $f = \sum_{j\in\mathbb{N}^\nu} D_jf$ converges in $L^p(\mathfrak{M},\mathrm{Vol})$, and moreover,

$$\|f\|_{L^p} \le \sum_{j\in\mathbb{N}^\nu} \|D_jf\|_{L^p} \lesssim \sum_{j\in\mathbb{N}^\nu} 2^{-j\cdot s}\|f\|_{\mathscr{X}^s(W,\vec{d})} \lesssim \|f\|_{\mathscr{X}^s(W,\vec{d})},$$

completing the proof. □

6.5.1 Smooth functions are dense

Recall Notation 6.2.2, which says that $\mathscr{X}^s$ is either of the form $\mathscr{B}^s_{p,q}$, for some $p,q \in [1,\infty]$, or of the form $\mathscr{F}^s_{p,q}$, for some $p \in (1,\infty)$, $q \in (1,\infty]$; for the remainder of this section, fix such p,q. The main result of this section shows that if $q \neq \infty$, then "smooth" functions are dense in $\mathscr{X}^s_{\mathrm{cpt}}(W,\vec{d})$.

Most of the proofs of this section are stated in the general setting of this chapter, but we do obtain a better result in the special case of the Hörmander setting (see Section 5.3 for the distinction between the general setting and the Hörmander setting). The main result of this section is the next proposition.

Proposition 6.5.17. *Suppose $q \neq \infty$. Fix $s \in \mathbb{R}^\nu$. Let $\mathcal{K} \Subset \mathfrak{M}$ be compact and let $\Omega \Subset \mathfrak{M}$ be an open, relatively compact set with $\mathcal{K} \Subset \Omega$.*
Hörmander setting: *$\forall f \in \mathscr{X}^s(\mathcal{K},(W,\vec{d}))$, there exists a sequence $\{f_L\}_{L\in\mathbb{N}} \subset C_0^\infty(\Omega)$ such that $f_L \xrightarrow{L\to\infty} f$ in $\mathscr{X}^s(\overline{\Omega},(W,\vec{d}))$.*
General setting: *$\forall f \in \mathscr{X}^s(\mathcal{K},(W,\vec{d}))$, there exists a sequence*

$$\{f_L\}_{L\in\mathbb{N}} \subset \bigcap_{t\in\mathbb{R}^\nu} \mathscr{X}^t(\overline{\Omega},(W,\vec{d}))$$

such that $f_L \xrightarrow{L\to\infty} f$ in $\mathscr{X}^s(\overline{\Omega},(W,\vec{d}))$.

The rest of this section is devoted to the proof of Proposition 6.5.17. We take $\mathcal{K} \Subset \Omega \Subset \mathfrak{M}$ as in that proposition and fix $s \in \mathbb{R}^\nu$. Throughout this section, $p \in [1,\infty]$ and $q \in [1,\infty)$ are as above and we insist $q \neq \infty$. We define $\mathscr{V}$ as in Notation 6.3.2.

Lemma 6.5.18. *Suppose $\mathcal{E}$ is a bounded set of generalized $(W,\vec{d})$ elementary operators and let $\{(E_{j,k},2^{-j}) : j,l \in \mathbb{N}^\nu\} \subseteq \mathcal{E}$. Then, $\forall f \in \mathscr{X}^s(\mathcal{K},(W,\vec{d}))$,*

$$\lim_{L\to\infty} \sum_{l\in\mathbb{Z}^\nu} 2^{-|l|} \|\{2^{j\cdot s}\chi_{\{|j|_\infty>L\}} E_{j,l} f\}_{j\in\mathbb{N}^\nu}\|_{\mathscr{V}} = 0, \tag{6.44}$$

where

$$\chi_{\{|j|_\infty>L\}} = \begin{cases} 1, & |j|_\infty > L, \\ 0, & \textit{otherwise.} \end{cases}$$

Proof. We prove the result in the case where $\mathscr{V} = L^p(\mathfrak{M},\mathrm{Vol};\ell^q(\mathbb{N}^\nu))$, $p \in (1,\infty)$; the proof in the case where $\mathscr{V} = \ell^q(\mathbb{N}^\nu;L^p(\mathfrak{M},\mathrm{Vol}))$, $p \in [1,\infty]$, is similar and easier.

Note that

$$|2^{j\cdot s}\chi_{\{|j|_\infty>L\}} E_{j,l} f| \le 2^{j\cdot s}|E_{j,l} f|,$$

and by Corollary 6.4.4,

$$\begin{aligned}\sum_{j\in\mathbb{Z}^\nu} 2^{-|l|} \|\{2^{j\cdot s}|E_{j,l} f|\}_{j\in\mathbb{N}^\nu}\|_{\mathscr{V}} &\le \sum_{j\in\mathbb{Z}^\nu} 2^{-|l|}\|f\|_{\mathscr{V},s,\mathcal{E}} \\ &\lesssim \sum_{l\in\mathbb{Z}^\nu} 2^{-|l|}\|f\|_{\mathscr{X}^s(W,\vec{d})} < \infty.\end{aligned}$$

Thus, Lebesgue's dominated convergence theorem can be applied to the left-hand side of (6.44) to show that (using $p < \infty$)

$$\begin{aligned}\lim_{L\to\infty} \sum_{l\in\mathbb{Z}^\nu} 2^{-|l|} \| \{2^{j\cdot s} \chi_{\{|j|_\infty > L\}} E_{j,l} f\}_{j\in\mathbb{N}^\nu} \|_{\mathscr{V}} &= \lim_{L\to\infty} \sum_{l\in\mathbb{Z}^\nu} 2^{-|l|} \left\| \left(\sum_{|j|_\infty > L} |2^{j\cdot s} E_{j,l} f|^q \right)^{\frac{1}{q}} \right\|_{L^p(\mathfrak{M},\mathrm{Vol})} \\ &= \sum_{l\in\mathbb{Z}^\nu} 2^{-|l|} \left\| \left(\lim_{L\to\infty} \sum_{|j|_\infty > L} |2^{j\cdot s} E_{j,l} f|^q \right)^{\frac{1}{q}} \right\|_{L^p(\mathfrak{M},\mathrm{Vol})} = 0,\end{aligned}$$

where we have used the fact that if $\sum_{j\in\mathbb{N}^\nu} |a_j| < \infty$, for a sequence $a_j \in \mathbb{C}$, then $\lim_{L\to\infty} \sum_{|j|_\infty > L} |a_j| = 0$. □

As before, we take $\psi \in C_0^\infty(\Omega)$ with $\psi \equiv 1$ on $\mathcal{K}$. We let $\mathcal{D}_0 = \{(D_j, 2^{-j}) : j \in \mathbb{N}^\nu\}$ be a bounded set of generalized $(W, \vec{d})$ elementary operators supported in Ω with $\mathrm{Mult}[\psi] = \sum_{j\in\mathbb{N}^\nu} D_j$. For $L \in \mathbb{N}$, set $P_L := \sum_{|j|_\infty \le L} D_j$. Using Definition 5.2.29, we have $P_L \in \bigcap_{t\in\mathbb{R}^\nu} \mathscr{A}_2^t(\Omega, (W, \vec{d})) = \bigcap_{t\in\mathbb{R}^\nu} \widetilde{\mathscr{A}}^t(\Omega, (W, \vec{d}))$.

Lemma 6.5.19. *We have,* $\forall f \in \mathscr{X}^s(\mathcal{K}, (W, \vec{d}))$, $P_L f \xrightarrow{L\to\infty} f$ *in* $\mathscr{X}^s(\overline{\Omega}, (W, \vec{d}))$.

Proof. Let $f \in \mathscr{X}^s(\mathcal{K}, (W, \vec{d}))$. In particular, since $\mathrm{supp}(f) \subseteq \mathcal{K}$, we have $f = \mathrm{Mult}[\psi] f$ and therefore

$$(I - P_L) f = (\mathrm{Mult}[\psi] - P_L) f = \sum_{|j|_\infty > L} f.$$

By Proposition 6.5.6, $f \in \mathscr{X}^s(\mathcal{K}, (W, \vec{d})) \subseteq \mathscr{X}^s(\overline{\Omega}, (W, \vec{d}))$. Since $P_L \in \widetilde{\mathscr{A}}^0(\Omega, (W, \vec{d}))$, Theorem 6.3.10 shows that $P_L f \in \mathscr{X}^s(\overline{\Omega}, (W, \vec{d}))$, $\forall L \in \mathbb{N}$.

Thus, to prove the result it suffices to show that

$$\left\| \sum_{|k|_\infty > L} D_k f \right\|_{\mathscr{X}^s(W,\vec{d})} \xrightarrow{L\to\infty} 0. \tag{6.45}$$

Fix $\tilde{\psi} \in C_0^\infty(\mathfrak{M})$ with $\tilde{\psi} \equiv 1$ on a neighborhood of $\overline{\Omega}$. Let $\{(\widetilde{D}_j, 2^{-j}) : j \in \mathbb{N}^\nu\}$ be a bounded set of generalized $(W, \vec{d})$ elementary operators with $\sum_{j\in\mathbb{N}^\nu} \widetilde{D}_j = \mathrm{Mult}[\tilde{\psi}]$. We take the convention $D_j = 0 = \widetilde{D}_j$ for $j \in \mathbb{Z}^\nu \setminus \mathbb{N}^\nu$. Set, for $j, l \in \mathbb{Z}^\nu$, $E_{j+l,l} := 2^{|l|} 2^{-l\cdot s} \widetilde{D}_j D_{j+l}$. Proposition 5.5.11 implies that

$$\{(E_{j,l}, 2^{-j}) : j \in \mathbb{N}^\nu, l \in \mathbb{Z}^\nu\}$$

is a bounded set of generalized $(W, \vec{d})$ elementary operators.

Using the definition of $\|\cdot\|_{\mathscr{X}^s(W,\vec{d})}$ (see Definition 6.3.8), we have

$$\begin{aligned}\Big\|\sum_{|k|_\infty>L} D_k f\Big\|_{\mathscr{X}^s(W,\vec{d})} &\approx \Big\|\Big\{\sum_{|k|_\infty>L} 2^{j\cdot s}\widetilde{D}_j D_k f\Big\}_{j\in\mathbb{N}^\nu}\Big\|_{\mathscr{V}} \\ &= \Big\|\Big\{\sum_{l\in\mathbb{Z}^\nu} 2^{-|l|}2^{(j+l)\cdot s}\chi_{\{|j+l|_\infty>L\}}E_{j+l,l}f\Big\}_{j\in\mathbb{N}^\nu}\Big\|_{\mathscr{V}} \\ &\leq \sum_{l\in\mathbb{Z}^\nu} 2^{-|l|}\|\{2^{j\cdot s}\chi_{\{|j|_\infty>L\}}E_{j,l}f\}_{j\in\mathbb{N}^\nu}\|_{\mathscr{V}} \xrightarrow{L\to\infty} 0,\end{aligned}$$

where the final limit uses Lemma 6.5.18. This establishes (6.45) and completes the proof. □

Proof of Proposition 6.5.17. We take $f_L := P_L f$. Lemma 6.5.19 shows that $f_L \to f$ in $\mathscr{X}^s(\overline{\Omega},(W,\vec{d}))$.

In the general setting, we use Proposition 6.5.6 to see that $f \in \mathscr{X}^s(\mathcal{K},(W,\vec{d})) \subseteq \mathscr{X}^s(\overline{\Omega},(W,\vec{d}))$. Since $P_L \in \bigcap_{t\in\mathbb{R}^\nu}\mathscr{A}^t(\Omega,(W,\vec{d}))$, Theorem 6.3.10 shows that $f_L = P_L f \in \bigcap_{t\in\mathbb{R}^\nu}\mathscr{X}^t(\overline{\Omega},(W,\vec{d}))$, $\forall L\in\mathbb{N}$, completing the proof in the general setting.

In the Hörmander setting, we note that we may take $\{(D_j,2^{-j}) : j\in\mathbb{N}^\nu\}$ to be a bounded set of $(W,\vec{d})$ elementary operators supported in Ω (see Corollary 5.6.2). In particular, $D_j\in C_0^\infty(\Omega\times\Omega)$ and therefore $P_L\in C_0^\infty(\Omega\times\Omega)$. It follows that $f_L = P_L f\in C_0^\infty(\Omega)$, completing the proof. □

6.6 The single-parameter spaces, revisited

We return to the single-parameter setting of Section 6.2; thus, we are given Hörmander vector fields with single-parameter formal degrees: $(W,\vec{d}) = \{(W_1,\vec{d}_1),\dots,(W_r,\vec{d}_r)\}\subset C^\infty_{\mathrm{loc}}(\mathfrak{M};T\mathfrak{M})\times\mathbb{N}_+$. This is a special case of the general multi-parameter setting covered in Section 6.3. The next result shows that the spaces defined in this single-parameter setting are a special case of the spaces defined in the general multi-parameter setting covered in Section 6.3.

Proposition 6.6.1. *For $s\in\mathbb{R}$, $\mathscr{X}^s(\mathcal{K},(W,\vec{d}))$ as defined in Definition 6.2.4 is the same as $\mathscr{X}^s(\mathcal{K},(W,\vec{d}))$ as defined in Definition 6.3.5. More precisely, in the setting of this section, $C_0^\infty(\mathfrak{M})' = C^\infty_{W,0}(\mathfrak{M})'$, and for $f\in C_0^\infty(\mathfrak{M})'$ the following are equivalent:*

(i) *Fix $\Omega\Subset\mathfrak{M}$ relatively compact and open with $\mathcal{K}\Subset\Omega$. Then, for every bounded set of $(W,\vec{d})$ elementary operators supported in Ω, $\mathcal{E}$, we have $\|f\|_{\mathscr{V},s,\mathcal{E}}<\infty$.*
(ii) *For every bounded set of $(W,\vec{d})$ elementary operators, $\mathcal{E}$, $\|f\|_{\mathscr{V},s,\mathcal{E}}<\infty$.*
(iii) *For every bounded set of generalized $(W,\vec{d})$ elementary operators, $\mathcal{E}$, $\|f\|_{\mathscr{V},s,\mathcal{E}}<\infty$.*

Furthermore, in the definition of the norm $\|\cdot\|_{\mathscr{X}^s(W,d)} = \|\cdot\|_{\mathscr{V},s,\mathcal{D}_0}$, we may take the set $\mathcal{D}_0$ to be a bounded set of $(W,\vec{d})$ elementary operators.

Proof. It follows from Remark 5.2.24 that $C_0^\infty(\mathfrak{M})' = C_{W,0}^\infty(\mathfrak{M})'$. Since every bounded set of $(W, \vec{d})$ elementary operators is a bounded set of generalized $(W, \vec{d})$ elementary operators, by Proposition 5.5.8, we have (ii) ⇒ (iii). Also, (ii) ⇒ (i) is trivial.

Fix $\Omega \Subset \mathfrak{M}$ as in (i). Let $\psi \in C_0^\infty(\Omega)$ satisfy $\psi \equiv 1$ on a neighborhood of $\mathcal{K}$. Proposition 5.8.3 shows that $\operatorname{Mult}[\psi] \in \mathscr{A}^0(\Omega, (W, \vec{d})) = \mathscr{A}_2^0(\Omega, (W, \vec{d}))$. In particular, there exists a bounded set of $(W, \vec{d})$ elementary operators supported in Ω, $\mathcal{D}_0 = \{(D_j, 2^{-j}) : j \in \mathbb{N}\}$ such that $\operatorname{Mult}[\psi] = \sum_{j\in\mathbb{N}} D_j$. This shows that we may take $\mathcal{D}_0$ in the definition of the norm $\|\cdot\|_{\mathscr{X}^s(W,\vec{d})} = \|\cdot\|_{\mathscr{V},s,\mathcal{D}_0}$ to be a bounded set of $(W, \vec{d})$ elementary operators.

Finally, we turn to showing (i) ⇒ (iii). Suppose $f \in C_0^\infty(\mathfrak{M})'$ is such that (i) holds. To show that (iii) holds, Corollary 6.4.5 shows that it suffices to show $\|f\|_{\mathscr{V},s,\mathcal{D}_0} < \infty$ and $\|f\|_{\mathscr{V},s,\mathcal{D}_N} < \infty$ for all $N \geq 1$, where $\mathcal{D}_N$ is defined in (6.6). Since $\mathcal{D}_0$ is a bounded set of $(W, \vec{d})$ elementary operators supported in Ω, (i) implies $\|f\|_{\mathscr{V},s,\mathcal{D}_0} < \infty$. Similarly, $\|f\|_{\mathscr{V},s,\mathcal{D}_N} < \infty$ will follow once we show that $\mathcal{D}_N$ is a bounded set of $(W, \vec{d})$ elementary operators supported in Ω, for every $N \geq 1$. Since $|l| > |k|$ in the definition of $\mathcal{D}_N$ (see (6.6)), by Proposition 5.5.5 (a), it suffices to show that

$$\{(2^{(N+2)|l|} D_{j+k} D_{j+k+l}, 2^{-(j+k)}) : j, k, l \in \mathbb{Z}, j + k \in \mathbb{N}\} \tag{6.46}$$

is a bounded set of $(W, \vec{d})$ elementary operators supported in Ω (recall that $D_{j+k+l} = 0$ if $j + k + l < 0$; see Notation 6.4.1). It follows directly from Proposition 5.5.11 that the set in (6.46) is a bounded set of $(W, \vec{d})$ elementary operators supported in Ω, completing the proof. □

Remark 6.6.2. By a proof very similar to the one of Proposition 6.6.1, one can see that in the multi-parameter Hörmander setting of Section 5.1.2 (which is a special case of the general multi-parameter setting described in this chapter), the spaces $\mathscr{X}^s(\mathcal{K}, (W, \vec{d}))$ can be defined using bounded sets of $(W, \vec{d})$ elementary operators, instead of bounded sets of generalized $(W, \vec{d})$ elementary operators.

In light of Proposition 6.6.1, many of our main results concerning the single-parameter spaces follow from more general results concerning multi-parameter spaces. For example, we can now prove many of the results from Section 6.2.

Proof of Proposition 6.2.5. In light of Proposition 6.6.1, this is a special case of Proposition 6.3.6. □

Proof of Proposition 6.2.6. Since every bounded set of $(W, \vec{d})$ elementary operators is a bounded set of generalized $(W, \vec{d})$ elementary operators, by Proposition 5.5.8, this is a special case of Proposition 6.3.7. □

Proof of Theorem 6.2.10. By Proposition 5.8.4, $\mathscr{A}^t(\Omega, (W, \vec{d})) \subseteq \widetilde{\mathscr{A}}^t(\Omega, (W, \vec{d}))$. Thus, in light of Proposition 6.6.1, this result is a special case of Theorem 6.3.10. □

Proof of Proposition 6.2.11. In light of Proposition 6.6.1, this is a special case of Corollary 6.5.11. □

Proof of Theorem 6.2.12. In light of Proposition 6.6.1, this is a special case of Proposition 6.5.12. □

We defer the proof of Proposition 6.2.13 to Section 6.9.

6.6.1 The classical spaces

The Besov and Triebel–Lizorkin spaces defined in this chapter generalize the classical Besov and Triebel–Lizorkin spaces on $\mathbb{R}^n$ as described in Section 2.4. For $s \in \mathbb{R}$, we write $\mathscr{X}^s$ to denote any one of $\mathscr{B}^s_{p,q}$, $p, q \in [1, \infty]$, or $\mathscr{F}^s_{p,q}$, $p \in (1, \infty)$, $q \in (1, \infty]$. Thus, $\mathscr{X}^s(\mathbb{R}^n)$ denotes one of the spaces $\mathscr{B}^s_{p,q}(\mathbb{R}^n)$ or $\mathscr{F}^s_{p,q}(\mathbb{R}^n)$ from Section 2.4.

We begin by stating and proving a result on $\mathbb{R}^n$, after which we will turn to a more general result on manifolds.

Proposition 6.6.3. *Let* $\mathfrak{M} = \mathbb{R}^n$ *(endowed with* Vol = *Lebesgue measure) and let* $\mathcal{K} \Subset \mathbb{R}^n$ *be a compact subset. Set* $(\partial, 1) := \{(\partial_{x_1}, 1), (\partial_{x_n}, 1)\} \subset C^\infty_{\mathrm{loc}}(\mathbb{R}^n; T\mathbb{R}^n) \times \mathbb{N}_+$. *Then, for* $s \in \mathbb{R}$,

$$\mathscr{X}^s(\mathcal{K}, (\partial, 1)) = \{f \in \mathscr{X}^s(\mathbb{R}^n) : \mathrm{supp}(f) \subseteq \mathcal{K}\}. \tag{6.47}$$

Moreover,

$$\|f\|_{\mathscr{X}^s(\partial,1)} \approx \|f\|_{\mathscr{X}^s(\mathbb{R}^n)}, \quad \forall f \in \mathscr{X}^s(\mathcal{K}, (\partial, 1)), \tag{6.48}$$

where the implicit constants do not depend on f, *but may depend on any of the other ingredients.*

Proof. First we claim

$$\rho_{(\partial,1)}(x, y) \approx |x - y|, \quad \forall x, y \in \mathbb{R}^n. \tag{6.49}$$

Indeed, $\rho_{(\partial,1)}(x, y)$ and $|x - y|$ are both translation invariant, so it suffices to prove (6.49) in the case $x = 0$. It also follows easily from the definitions that for $\delta > 0$, $\rho_{(\partial,1)}(\delta x, \delta y) = \delta \rho_{(\partial,1)}(x, y)$ and of course $|\delta x - \delta y| = \delta |x - y|$. Thus, to show (6.49) it suffices to show that

$$\rho_{(\partial,1)}(0, y) \approx 1, \quad \forall y \in \mathbb{R}^n \text{ with } |y| = 1. \tag{6.50}$$

Since $y \mapsto \rho_{(\partial,1)}(0, y)$ is continuous (see Lemma 3.1.7) and $\{y \in \mathbb{R}^n : |y| = 1\}$ is compact, (6.50) follows, completing the proof of (6.49).

Let $\Omega \Subset \mathbb{R}^n$ be open and relatively compact. In light of (6.49), it follows easily from the definitions that if $\mathcal{E}$ is a bounded set of $(\partial, 1)$ elementary operators supported on Ω, then $\mathcal{E}$ is a bounded set of elementary operators in the sense of Definition 2.3.5. Thus, if $f \in \mathscr{X}^s(\mathbb{R}^n)$, it follows from Proposition 2.4.18 that for every bounded set of $(\partial, 1)$ elementary operators supported in Ω, $\mathcal{E}$, we have

$$\|f\|_{\mathcal{V},s,\mathcal{E}} < \infty.$$

Since $\Omega \Subset \mathbb{R}^n$ was arbitrary, we see that if we also have $\mathrm{supp}(f) \subseteq \mathcal{K}$, then $f \in \mathscr{X}^s(\mathcal{K}, (\partial, 1))$. This proves the $\supseteq$ part of (6.47).

Moreover, if $\psi \in C_0^\infty(\mathbb{R}^n)$ satisfies $\psi \equiv 1$ on a neighborhood of $\mathcal{K}$ and $\mathcal{D}_0 = \{(D_j, 2^{-j}) : j \in \mathbb{N}\}$ is a bounded set of $(\partial, 1)$ elementary operators with $\mathrm{Mult}[\psi] = \sum_{j\in\mathbb{N}} D_j$, then Proposition 2.4.11 shows that

$$\|f\|_{\mathscr{X}^s(\partial,1)} = \|f\|_{\mathcal{V},s,\mathcal{D}_0} \lesssim \|f\|_{\mathscr{X}^s(\mathbb{R}^n)}. \tag{6.51}$$

Suppose $f \in \mathscr{X}^s(\mathcal{K}, (\partial, 1))$. The same proof as in Proposition 6.4.3 and Corollary 6.4.4 shows that $\mathcal{E}$ in those results can be taken to instead be a bounded set of elementary operators as in Definition 2.3.5 (see also the proof of Proposition 2.4.11). Thus, we have, for any such bounded set of elementary operators, $\mathcal{E}$,

$$\|f\|_{\mathcal{V},s,\mathcal{E}} \lesssim \|f\|_{\mathscr{X}^s(\partial,1)} < \infty. \tag{6.52}$$

We conclude that $f \in \mathscr{X}^s(\mathbb{R}^n)$ (see Proposition 2.4.18). This proves the $\subseteq$ part of (6.47).

In particular, if we let $\widetilde{\mathcal{D}}_0 = \{(\widetilde{D}_j, 2^{-j}) : j \in \mathbb{N}\}$, where $\widetilde{D}_j$ is the operator D_j from Section 2.4, then $\widetilde{\mathcal{D}}_0$ is a bounded set of elementary operators and (6.52) with $\mathcal{E} = \widetilde{\mathcal{D}}_0$ implies

$$\|f\|_{\mathscr{X}^s(\mathbb{R}^n)} = \|f\|_{\mathcal{V},s,\widetilde{\mathcal{D}}_0} \lesssim \|f\|_{\mathscr{X}^s(\partial,1)}. \tag{6.53}$$

Combining (6.51) and (6.53) yields (6.48) and completes the proof. □

There is a more general version of Proposition 6.6.3 on an arbitrary connected, smooth manifold, which we now present. To begin, we need to define the classical Besov and Triebel–Lizorkin spaces on an arbitrary manifold, and we restrict our attention to distributions with compact support. We denote these spaces by, for $\mathcal{K} \Subset \mathfrak{M}$,

$$\mathscr{B}^s_{p,q,\mathrm{std}}(\mathcal{K}), \quad \mathscr{F}^s_{p,q,\mathrm{std}}(\mathcal{K}).$$

As before, we let $\mathscr{X}^s_{\mathrm{std}}(\mathcal{K})$ denote any one of $\mathscr{B}^s_{p,q,\mathrm{std}}(\mathcal{K})$, $p, q \in [1, \infty]$, or $\mathscr{F}^s_{p,q,\mathrm{std}}(\mathcal{K})$, $p \in (1, \infty)$, $q \in (1, \infty]$.

Definition 6.6.4. $\mathscr{X}^s_{\mathrm{std}}(\mathcal{K})$ consists of those $f \in C_0^\infty(\mathfrak{M})'$ with $\mathrm{supp}(f) \subseteq \mathcal{K}$ such that for each $x \in \mathcal{K}$, there is a C^∞_{loc} diffeomorphism $\Phi_x : B^n(1) \xrightarrow{\sim} \Phi_x(B^n(1))$, where $\Phi_x(B^n(1))$ is an open neighborhood of x, and $\psi_x \in C_0^\infty(\Phi_x(B^n(1)))$ with $\psi_x \equiv 1$ on a neighborhood of x such that $\Phi_x^* \psi_x f \in \mathscr{X}^s(\mathbb{R}^n)$. We set

$$\mathscr{X}^s_{\mathrm{std,cpt}}(\mathfrak{M}) := \bigcup_{\substack{\mathcal{K}\Subset\mathfrak{M}\\ \mathcal{K}\text{ compact}}} \mathscr{X}^s_{\mathrm{std}}(\mathcal{K}). \tag{6.54}$$

We wish to endow $\mathscr{X}^s_{\mathrm{std}}(\mathcal{K})$ with a norm. For each $x \in \mathfrak{M}$, let $\Phi_x : B^n(1) \xrightarrow{\sim} \Phi_x(B^n(1))$ be a C^∞_{loc} diffeomorphism such that $\Phi_x(0) = x$ and $\Phi_x(B^n(1)) \Subset \mathfrak{M}$ is open and relatively compact. Cover $\mathcal{K}$ by a finite collection of open sets of the form $\Phi_{x_1}(B^n(1)), \dots, \Phi_{x_L}(B^n(1))$. Take $\psi_j \in C_0^\infty(\Phi_{x_j}(B^n(1)))$ such that $\sum_{j=1}^L \psi_j \equiv 1$ on a neighborhood of $\mathcal{K}$.

Definition 6.6.5. For $f \in \mathscr{X}^s_{\mathrm{std}}(\mathcal{K})$ we define the norm

$$\|f\|_{\mathscr{X}^s_{\mathrm{std}}(\mathcal{K})} := \sum_{j=1}^{L} \|\Phi_{x_j}^* \psi_j f\|_{\mathscr{X}^s(\mathbb{R}^n)}.$$

With this norm, $\mathscr{X}^s_{\mathrm{std}}(\mathcal{K})$ is a Banach space. We state the next simple lemma without proof.

Lemma 6.6.6. *For $f \in C_0^\infty(\mathfrak{M})'$, the following are equivalent:*
- $f \in \mathscr{X}^s_{\mathrm{std}}(\mathcal{K})$.
- $\mathrm{supp}(f) \subseteq \mathcal{K}$ *and for each* $1 \le j \le L$, $\Phi_{x_j}^* \psi_j f \in \mathscr{X}^s(\mathbb{R}^n)$.

Moreover, on $\mathscr{X}^s_{\mathrm{std}}(\mathcal{K})$, the equivalence class of the norm $\|\cdot\|_{\mathscr{X}^s_{\mathrm{std}}(\mathcal{K})}$ does not depend on any of the choices we made.

We can now present the main result of this section, which is a generalization of Proposition 6.6.3.

Theorem 6.6.7. *Let $(Y,1) = \{(Y_1,1), \dots, (Y_s,1)\} \subset C^\infty_{\mathrm{loc}}(\mathfrak{M}; T\mathfrak{M}) \times \mathbb{N}_+$ be such that*

$$\mathrm{span}\{Y_1(x), \dots, Y_s(x)\} = T_x\mathfrak{M}, \quad \forall x \in \mathfrak{M}. \tag{6.55}$$

Then

$$\mathscr{X}^s_{\mathrm{std}}(\mathcal{K}) = \mathscr{X}^s(\mathcal{K}, (Y,1)),$$

with equality of topologies.

To prove Theorem 6.6.7, we require a lemma.

Lemma 6.6.8. *For $f \in C_0^\infty(\mathfrak{M})'$ with $\mathrm{supp}(f) \subseteq \mathcal{K}$, the following are equivalent:*
(i) $f \in \mathscr{X}^s(\mathcal{K}, (Y,1))$.
(ii) $\psi_j f \in \mathscr{X}^s(\mathcal{K}, (Y,1))$, *for* $1 \le j \le L$.

Moreover, for $f \in \mathscr{X}^s(\mathcal{K}, (Y,1))$,

$$\|f\|_{\mathscr{X}^s(Y,1)} \approx \sum_{j=1}^{L} \|\psi_j f\|_{\mathscr{X}^s(Y,1)}. \tag{6.56}$$

Proof. (i) ⇒ (ii): Suppose $f \in \mathscr{X}^s(\mathcal{K}, (Y,1))$. By Corollary 6.5.10, $\mathrm{Mult}[\psi_j] : \mathscr{X}^s(\mathcal{K}, (Y,1)) \to \mathscr{X}^s(\mathcal{K}, (Y,1))$ is a bounded operator. Thus, $\psi_j f \in \mathscr{X}^s(\mathcal{K}, (Y,1))$ and

$$\sum_{j=1}^{L} \|\psi_j f\|_{\mathscr{X}^s(Y,1)} \lesssim \|f\|_{\mathscr{X}^s(Y,1)}. \tag{6.57}$$

(ii) ⇒ (i): Suppose (ii) holds. Then, since $\sum_{j=1}^{L} \psi_j \equiv 1$ on $\mathcal{K}$, we have $f = \sum_{j=1}^{L} \psi_j f \in \mathscr{X}^s(\mathcal{K}, (Y, 1))$. Moreover, by the triangle inequality,

$$\|f\|_{\mathscr{X}^s(Y,1)} \le \sum_{j=1}^{L} \|\psi_j f\|_{\mathscr{X}^s(Y,1)}. \tag{6.58}$$

Combining (6.57) and (6.58) yields (6.56) and completes the proof. □

Proof of Theorem 6.6.7. In light of Lemmas 6.6.6 and 6.6.8, to prove the result it suffices to show that for $1 \le j \le L$,

$$\psi_j f \in \mathscr{X}^s(\mathcal{K}, (Y, 1)) \iff \Phi_{X_j}^* \psi_j f \in \mathscr{X}^s(\mathbb{R}^n), \tag{6.59}$$

$$\|\psi_j f\|_{\mathscr{X}^s(Y,1)} \approx \|\Phi_{X_j} \psi_j f\|_{\mathscr{X}^s(\mathbb{R}^n)}. \tag{6.60}$$

Set $\mathcal{K}_j := \operatorname{supp}(\psi_j f) \Subset \Phi_{X_j}(B^n(1))$. By Proposition 6.5.6,

$$\psi_j f \in \mathscr{X}^s(\mathcal{K}, (Y, 1)) \iff \psi_j f \in \mathscr{X}^s(\mathcal{K}_j, (Y, 1)). \tag{6.61}$$

By Proposition 6.5.15 (see also Remark 6.4.8), we have

$$\psi_j f \in \mathscr{X}^s(\mathcal{K}_j, (Y, 1)) \iff \Phi_{X_j}^* \psi_j f \in \mathscr{X}^s(\Phi_{X_j}^{-1}(\mathcal{K}_j), (\Phi_{X_j}^* Y, 1)), \tag{6.62}$$

$$\|\psi_j f\|_{\mathscr{X}^s(Y,1)} \approx \|\Phi_{X_j}^* \psi_j f\|_{\mathscr{X}^s(\Phi_{X_j}^* Y,1)}. \tag{6.63}$$

The hypothesis (6.55) implies

$$\operatorname{span}\{\Phi_{X_j}^* Y_1(u), \ldots \Phi_{X_j^*} Y_s(u)\} = T_u B^n(1), \quad \forall u \in B^n(1).$$

It follows that $(\Phi_{X_j}^* Y, 1)$ and $(\partial, 1)$ are locally strongly equivalent on $B^n(1)$, and therefore they are locally weakly equivalent on $B^n(1)$. Thus, by Proposition 6.5.14,

$$\Phi_{X_j}^* \psi_j f \in \mathscr{X}^s(\Phi_{X_j}^{-1}(\mathcal{K}_j), (\Phi_{X_j}^* Y, 1)) \iff \Phi_{X_j}^* \psi_j f \in \mathscr{X}^s(\Phi_{X_j}^{-1}(\mathcal{K}_j), (\partial, 1)), \tag{6.64}$$

$$\|\Phi_{X_j}^* \psi_j f\|_{\mathscr{X}^s(\Phi_{X_j}^* Y,1)} \approx \|\Phi_{X_j}^* \psi_j f\|_{\mathscr{X}^s(\partial,1)}. \tag{6.65}$$

Since $\operatorname{supp}(\Phi_{X_j}^* \psi_j f) \subseteq \Phi_{X_j}^{-1}(\mathcal{K}_j) \Subset B^n(1)$, Proposition 6.6.3 (see also Remark 6.4.8) shows

$$\Phi_{X_j}^* \psi_j f \in \mathscr{X}^s(\Phi_{X_j}^{-1}(\mathcal{K}_j), (\partial, 1)) \iff \Phi_{X_j}^* \psi_j f \in \mathscr{X}^s(\mathbb{R}^n), \tag{6.66}$$

$$\|\Phi_{X_j}^* \psi_j f\|_{\mathscr{X}^s(\partial,1)} \approx \|\Phi_{X_j}^* \psi_j f\|_{\mathscr{X}^s(\mathbb{R}^n)}. \tag{6.67}$$

Combining (6.61), (6.62), (6.63), (6.64), (6.65), (6.66), and (6.67) proves (6.59) and (6.60) and completes the proof. □

6.6.2 Comparing single-parameter spaces

Let $(W, \vec{d}) = \{(W_1, \vec{d}_1), \dots, (W_r, \vec{d}_r)\}$ and $(Z, \vec{d}) = \{(Z_1, \vec{d}_1), \dots, (Z_v, \vec{d}_v)\}$ be two lists of Hörmander vector fields with formal degrees on $\mathfrak{M}$. In this section, we compare the spaces $\mathscr{X}^{s_1}(\mathcal{K}, (W, \vec{d}))$ and $\mathscr{X}^{s_2}(\mathcal{K}, (Z, \vec{d}))$.

Definition 6.6.9. For $x \in \mathfrak{M}$ and $1 \le j \le v$, define

$$\begin{aligned}\lambda(x, (W, \vec{d}), (Z_j, \vec{d}_j)) := \inf\{\lambda' > 0 :& \exists \text{ a neighborhood } U \text{ of } x \text{ such that } Z_j|_U \\ & \text{is in the } C^\infty_{\mathrm{loc}}(U) \text{ module generated by} \\ & \{X_1 : (X_1, d_1) \in \mathrm{Gen}((W, \vec{d})), d_1 \le \lambda' \vec{d}_j\}\}.\end{aligned} \tag{6.68}$$

Set

$$\lambda(x, (W, \vec{d}), (Z, \vec{d})) := \max_{1 \le j \le v} \lambda(x, (W, \vec{d}), (Z, \vec{d})). \tag{6.69}$$

Lemma 6.6.10. *We have $\lambda(x, (W, \vec{d}), (Z, \vec{d})) < \infty$ and the infimum in (6.68) is achieved (i. e., $\inf$ can be replaced with $\min$).*

Proof. Because $W_1, \dots, W_r$ satisfy Hörmander's condition on $\mathfrak{M}$, given a relatively compact neighborhood U of x, there exists $M \in \mathbb{N}$ such that

$$\operatorname{span}\{X_1(y) : (X_1, d_1) \in \mathrm{Gen}((W, \vec{d})), d_1 \le M\} = T_y\mathfrak{M}, \quad \forall y \in U.$$

Thus, $\lambda(x, (W, \vec{d}), (Z, \vec{d})) \le M < \infty$.

To see that the infimum is achieved, it suffices to show that the set over which we are taking the infimum is closed, that is, we wish to show that the following set is open:

$$\begin{aligned}\Lambda := \{\lambda' > 0 :& \exists \text{ a neighborhood } U \text{ of } x \text{ such that } Z_j|_U \text{ is not in the } C^\infty_{\mathrm{loc}}(U) \text{ module} \\ & \text{generated by } \{X_1 : (X_1, d_1) \in \mathrm{Gen}((W, \vec{d})), d_1 \le \lambda' \vec{d}_j\}\}.\end{aligned}$$

Indeed, since each $d_1 \in \mathbb{N}_+$, for $(X_1, d_1) \in \mathrm{Gen}((W, \vec{d}))$, if $\lambda' \in \Lambda$, then $\lambda' + \epsilon \in \Lambda$ for all ϵ sufficiently small. This shows Λ is open and completes the proof. □

Corollary 6.6.11. *There is an open neighborhood, $\Omega \Subset \mathfrak{M}$, of x such that $(W, \vec{d})$ weakly $\lambda(x, (W, \vec{d}), (Z, \vec{d}))$-controls $(Z, \vec{d})$ on Ω.*

Proof. Since the infimum in (6.68) is achieved, by Lemma 6.6.10, and since $(Z, \vec{d})$ is a finite set, there is a neighborhood Ω of x such that for $1 \le j \le v$, $Z_j|_\Omega$ is in the $C^\infty_{\mathrm{loc}}(\Omega)$ module generated by $\{X_1 : (X_1, d_1) \in \mathrm{Gen}((W, \vec{d})), d_1 \le \lambda \vec{d}_j\}$, where $\lambda := \lambda(x, (W, \vec{d}), (Z, \vec{d}))$. In other words, $(W, \vec{d})$ weakly λ-controls $(Z, \vec{d})$ on Ω (see Definition 3.14.3). This completes the proof. □

Remark 6.6.12. In fact, $\lambda(x, (W, d), (Z, d))$ is chosen to be the least number such that Corollary 6.6.11 holds. In many important cases, this implies our results are sharp. See Section 6.7.1.

The main result of this section is the next theorem. Informally, it allows us to trade derivatives with respect to (Z, d) for derivatives with respect to (W, d), with the cost of a factor of $\lambda(x, (W, d), (Z, d))$.

Theorem 6.6.13. *For every $x \in \mathfrak{M}$, there exists a relatively compact neighborhood $\Omega_0 \Subset \mathfrak{M}$ of x such that if $\mathcal{K} := \overline{\Omega_0}$ and $\lambda := \lambda(x, (W, d), (Z, d))$, then:*

(A) *For all $s_0, \epsilon > 0$,*

$$\mathscr{X}^{\lambda s_0+\epsilon}(\mathcal{K}, (W, d)) \subseteq \mathscr{X}^{s_0}((Z, d), \mathcal{K}), \tag{6.70}$$

and the inclusion is continuous.

(B) *For all $s_0, \epsilon > 0$,*

$$\mathscr{X}^{-(s_0/\lambda)+\epsilon}(\mathcal{K}, (Z, d)) \subseteq \mathscr{X}^{-s_0}(\mathcal{K}, (W, d)), \tag{6.71}$$

and the inclusion is continuous.

Remark 6.6.14. Theorem 6.6.13 involves a "loss" of ϵ derivatives, for an arbitrarily small $\epsilon > 0$. It is not clear if this loss is necessary or merely an artifact of the proof, and in some situations we have similar results without such a loss (see, for example, Proposition 6.6.21 and Section 6.7). It seems especially likely that the loss is not needed in the special case $\mathscr{X}^s = \mathscr{F}^s_{p,2}$, and it seems likely that methods similar to Proposition 6.8.8 could be used to prove such a result; however, we do not pursue that here. When (W, d) and (Z, d) weakly locally approximately commute, then we do have for $s_0 > 0$

$$\|f\|_{\mathscr{F}^{s_0}_{p,2}(Z,d)} \lesssim \|f\|_{\mathscr{F}^{\lambda s_0}_{p,2}(W,d)},$$

$$\|f\|_{\mathscr{F}^{-s_0}_{p,2}(Z,d)} \lesssim \|f\|_{\mathscr{F}^{-(s_0/\lambda)}_{p,2}(W,d)},$$

for all $f \in C_0^\infty(\mathfrak{M})$ with $\mathrm{supp}(f) \subseteq \mathcal{K}$. This can be seen by combining Proposition 6.8.8 and Theorem 6.7.2. We do not use this result, though, so we do not present the full proof.

Before we prove Theorem 6.6.13 we introduce an important corollary: the comparison between the Besov and Triebel–Lizorkin spaces associated with (W, d) and the standard Besov and Triebel–Lizorkin spaces (see Definition 6.6.4).

Definition 6.6.15. For $x \in \mathfrak{M}$, set

$$\begin{aligned}\lambda_{\mathrm{std}}(x, (W, d)) := \min\{\, &\max\{d_1, \dots, d_n\} : (X_1, d_1), \dots, (X_n, d_n) \in \mathrm{Gen}((W, d)),\\ &\text{satisfy } \mathrm{span}\{X_1(x), \dots, X_n(x)\} = T_x\mathfrak{M}\}.\end{aligned}$$

Also, set

$$\Lambda_{\mathrm{std}}(x,(W,\vec{d}))^{-1} := \max\{\vec{d}_j : W_j \text{ is not identically zero on any neighborhood of } x\}.$$

Example 6.6.16. If $\vec{d}_1 = \cdots = \vec{d}_r = 1$, then $\lambda_{\mathrm{std}}(x,(W,\vec{d})) = m$, where $W_1,\ldots,W_r$ satisfy Hörmander's condition of order m at x (see Definition 1.1.1). In this special case, $\Lambda_{\mathrm{std}}(x,(W,\vec{d})) = 1$.

Lemma 6.6.17. *If $(Y,1) = \{(Y_1,1),\ldots,(Y_\nu,1)\} \subset C^\infty_{\mathrm{loc}}(\mathfrak{M};T\mathfrak{M})\times\mathbb{N}_+$ are such that $Y_1,\ldots,Y_\nu$ span the tangent space at every point of $\mathfrak{M}$, then $\lambda_{\mathrm{std}}(x,(W,\vec{d})) = \lambda(x,(W,\vec{d}),(Y,1))$ and $\Lambda_{\mathrm{std}}(x,(W,\vec{d})) = \lambda(x,(Y,1),(W,\vec{d}))$.*

Proof. This follows easily from the definitions. □

Corollary 6.6.18. *For every $x \in \mathfrak{M}$, there exists a relatively compact neighborhood $\Omega_0 \Subset \mathfrak{M}$ of x such that if $\mathcal{K} = \overline{\Omega_0}$, $\lambda_{\mathrm{std}} := \lambda_{\mathrm{std}}(x,(W,\vec{d}))$, and $\Lambda_{\mathrm{std}} := \Lambda_{\mathrm{std}}(x,(W,\vec{d}))$, then we have, for all $s_0,\epsilon > 0$,*

$$\begin{aligned}
\mathscr{X}^{\lambda_{\mathrm{std}}s_0+\epsilon}(\mathcal{K},(W,\vec{d})) &\subseteq \mathscr{X}^{s_0}_{\mathrm{std}}(\mathcal{K}),\\
\mathscr{X}^{\Lambda_{\mathrm{std}}s_0+\epsilon}_{\mathrm{std}}(\mathcal{K}) &\subseteq \mathscr{X}^{s_0}(\mathcal{K},(W,\vec{d})),\\
\mathscr{X}^{-(s_0/\lambda_{\mathrm{std}})+\epsilon}_{\mathrm{std}}(\mathcal{K}) &\subseteq \mathscr{X}^{-s_0}(\mathcal{K},(W,\vec{d})),\\
\mathscr{X}^{-(s_0/\Lambda_{\mathrm{std}})+\epsilon}(\mathcal{K},(W,\vec{d})) &\subseteq \mathscr{X}^{-s_0}_{\mathrm{std}}(\mathcal{K}),
\end{aligned}$$

and the inclusions are all continuous.

Proof. Pick $(Y,1) = \{(Y_1,1),\ldots,(Y_n,1)\} \subset C^\infty_{\mathrm{loc}}(\mathfrak{M};T\mathfrak{M})\times\mathbb{N}_+$ such that $Y_1(x),\ldots,Y_n(x)$ form a basis for $T_x\mathfrak{M}$. Thus, there is an open neighborhood $\mathfrak{N} \subseteq \mathfrak{M}$ of x such that $Y_1(y),\ldots,Y_n(y)$ form a basis for $T_y\mathfrak{N}$, $\forall y \in \mathfrak{N}$. By Remark 6.4.8, it suffices to prove the result with $\mathfrak{M}$ replaced by $\mathfrak{N}$; in particular, we may assume $Y_1,\ldots,Y_n$ span the tangent space at every point.

Combining Lemma 6.6.17 and Theorem 6.6.13 shows that the result holds with $\mathscr{X}^s_{\mathrm{std}}(\mathcal{K})$ replaced by $\mathscr{X}^s(\mathcal{K},(Y,1))$ throughout. Theorem 6.6.7 shows that $\mathscr{X}^s(\mathcal{K},(Y,1)) = \mathscr{X}^s_{\mathrm{std}}(\mathcal{K})$, with equality of topologies, completing the proof. □

The rest of this section is devoted to the proof of Theorem 6.6.13. By Corollary 6.6.11, we may take $\Omega \Subset \mathfrak{M}$ an open, relatively compact neighborhood of x, such that $(W,\vec{d})$ weakly λ-controls $(Z,\vec{d})$ on Ω. Let $\Omega_0 \Subset \Omega$ be an open neighborhood of x and set $\mathcal{K} := \overline{\Omega_0} \Subset \Omega$. For the remainder of the section, we write $A \lesssim B$ to mean $A \le C_\epsilon B$, where $C_\epsilon \ge 0$ may depend on $\epsilon > 0$.

Lemma 6.6.19. *Let $\mathcal{E}_1$ be a bounded set of $(W,\vec{d})$ elementary operators supported in Ω and let $\mathcal{E}_2$ be a bounded set of $(Z,\vec{d})$ elementary operators supported in Ω. Then, for every $N \in \mathbb{N}$, there exists $L = L(N) \in \mathbb{N}$, $\mathcal{E}_1' = \mathcal{E}_1'(N)$ a bounded set of $(W,\vec{d})$ elementary* *operators supported in* Ω, *and* $\mathcal{E}_2' = \mathcal{E}_2'(N)$ *a bounded set of* $(Z, \vec{d})$ *elementary operators supported in* Ω *such that the following holds* $\forall (E_1, 2^{-j_1}) \in \mathcal{E}_1, (E_2, 2^{-j_2}) \in \mathcal{E}_2$:

$$2^{N|j_2 \vee (\lambda j_1) - \lambda j_1|} E_1 E_2 = \sum_{K=1}^{L} E_1^K E_2^K, \tag{6.72}$$

where $(E_1^K, 2^{-j_1}) \in \mathcal{E}_1'$ *and* $(E_2^K, 2^{-j_2}) \in \mathcal{E}_2'$. *Similarly,*

$$2^{N|j_2 \vee (\lambda j_1) - \lambda j_1|} E_2 E_1 = \sum_{K=1}^{L} F_2^K F_1^K, \tag{6.73}$$

where $(F_1^K, 2^{-j_1}) \in \mathcal{E}_1'$ *and* $(F_2^K, 2^{-j_2}) \in \mathcal{E}_2'$.

Proof. The proofs of (6.72) and (6.73) are similar, so we prove only (6.72). It suffices to prove (6.72) in the case $N = 1$, as the result for general N follows from repeated application of the result for $N = 1$.

By picking $\mathcal{E}_1'$ and $\mathcal{E}_2'$ such that

$$\mathcal{E}_1 \subseteq \mathcal{E}_1', \quad \mathcal{E}_2 \subseteq \mathcal{E}_2', \quad \{(0, 2^{-j}) : j \geq 0\} \subseteq \mathcal{E}_1' \cap \mathcal{E}_2',$$

(6.72) is trivial when $\lambda j_1 \geq j_2$. Thus, we turn to proving (6.72) in the case where $N = 1$ and $j_2 > \lambda j_1$.

By Proposition 5.5.5 (h) (with $N = 1$), we can write

$$\begin{aligned} 2^{j_2 - \lambda j_1} E_1 E_2 &= 2^{j_2 - \lambda j_1} \left(2^{-j_2} E_1 E_{2,0} + \sum_{l=1}^{\nu} E_1 2^{-j_2 \vec{d}_l} Z_l E_{2,l} \right) \\ &= 2^{-\lambda j_1} E_1 E_{2,0} + \sum_{l=1}^{\nu} 2^{(j_2 - \lambda j_1)(1 - \vec{d}_l)} E_1 2^{-\lambda j_1 \vec{d}_l} Z_l E_{2,l}, \end{aligned} \tag{6.74}$$

where $\{(E_{2,l}, 2^{-j_2}) : l \in \{0, \dots, \nu\}, (E_2, 2^{-j_2}) \in \mathcal{E}_2\}$ is a bounded set of $(Z, \vec{d})$ elementary operators supported in Ω. We set $\widetilde{E}_{2,0} := 2^{-\lambda j_1} E_{2,0}$ and $\widetilde{E}_{2,l} := 2^{(j_2 - \lambda j_1)(1 - \vec{d}_l)} E_{2,l}$ for $l = 1, \dots, \nu$. Then, since $2^{-\lambda j_1}, 2^{(j_2 - \lambda j_1)(1 - \vec{d}_l)} \leq 1$, Proposition 5.5.5 (a) shows that

$$\{(\widetilde{E}_{2,l}, 2^{-j_2}) : (E_2, 2^{-j_2}) \in \mathcal{E}_{,2} > \lambda j_1\}$$

is a bounded set of $(Z, \vec{d})$ elementary operators supported in Ω.

Since $(W, \vec{d})$ weakly λ-controls $(Z, \vec{d})$ on Ω (by the choice of Ω), it follows from the definitions that Z_l is a $(W, \vec{d})$ partial differential operator on Ω of degree $\leq \lambda \vec{d}_l$. Thus, if we set $\widetilde{E}_{1,l} := E_1 2^{-\lambda \vec{d}_l} Z_l$, it follows from Proposition 5.5.5 (f) that $\{(\widetilde{E}_{1,l}, 2^{-j_1}) : (E_2, 2^{-j}) \in \mathcal{E}_1, l \in \{1, \dots, \nu\}\}$ is a bounded set of $(W, \vec{d})$ elementary operators supported in Ω; see also Remark 3.12.3.

Combining the above discussion with (6.74) shows that

$$2^{j_2-\lambda j_1}E_1E_2 = E_1\widetilde{E}_{2,0} + \sum_{l=1}^{\nu} \widetilde{E}_{1,l}\widetilde{E}_{2,l},$$

completing the proof. □

Recall that by Notation 6.2.2, $\mathscr{V}$ is either $L^p(\mathfrak{M}, \mathrm{Vol}; \ell^q(\mathbb{N}))$ or $\ell^q(\mathbb{N}; L^p(\mathfrak{M}, \mathrm{Vol}))$. Set

$$\widetilde{\mathscr{V}} := \begin{cases} L^p(\mathfrak{M}, \mathrm{Vol}; \ell^q(\mathbb{N}^2)), & \text{if } \mathscr{V} = L^p(\mathfrak{M}, \mathrm{Vol}; \ell^q(\mathbb{N})), \\ \ell^q(\mathbb{N}^2; L^p(\mathfrak{M}, \mathrm{Vol})), & \text{if } \mathscr{V} = \ell^q(\mathbb{N}; L^p(\mathfrak{M}, \mathrm{Vol})). \end{cases} \tag{6.75}$$

We fix these $p, q \in [1, \infty]$ for the remainder of the section.

Lemma 6.6.20. *For every $\epsilon > 0$ and every sequence of measurable functions $\{f_{j_1,j_2}\}_{j_1,j_2\in\mathbb{N}}$,*

$$\left\| \left\{ \sum_{j_1\in\mathbb{N}} f_{j_1,j_2} \right\}_{j_2\in\mathbb{N}} \right\|_{\mathscr{V}} \lesssim \|\{2^{\epsilon j_1} f_{j_1,j_2}\}_{(j_1,j_2)\in\mathbb{N}^2}\|_{\widetilde{\mathscr{V}}}, \tag{6.76}$$

$$\|\{f_{j_1,j_2}\}_{(j_1,j_2)\in\mathbb{N}^2}\|_{\widetilde{\mathscr{V}}} \leq \sum_{j_1\in\mathbb{N}} \|\{f_{j_1,j_2}\}_{j_2\in\mathbb{N}}\|_{\mathscr{V}}. \tag{6.77}$$

Proof. For (6.76), we use

$$\sum_{j_1\in\mathbb{N}} 2^{-\epsilon j_1}|b_{j_1}| \lesssim \sup_{j_1\in\mathbb{N}} |b_{j_1}| \leq \left(\sum_{j_1\in\mathbb{N}} |b_{j_1}|^q \right)^{\frac{1}{q}}. \tag{6.78}$$

Thus, if $\mathscr{V} = \ell^q(\mathbb{N}; L^p(\mathfrak{M}, \mathrm{Vol}))$, we have

$$\left\| \left\{ \sum_{j_1\in\mathbb{N}} f_{j_1,j_2} \right\}_{j_2\in\mathbb{N}} \right\|_{\mathscr{V}} \lesssim \left\| \left\{ \left(\sum_{j_1\in\mathbb{N}} |2^{\epsilon j_1} f_{j_1,j_2}|^q \right)^{\frac{1}{q}} \right\}_{j_2\in\mathbb{N}} \right\|_{\mathscr{V}} = \|\{2^{\epsilon j_1} f_{j_1,j_2}\}_{(j_1,j_2)\in\mathbb{N}^2}\|_{\widetilde{\mathscr{V}}},$$

as desired. Similarly, if $\mathscr{V} = L^p(\mathfrak{M}, \mathrm{Vol}; \ell^q(\mathbb{N}))$, we have, using the triangle inequality and (6.78),

$$\begin{aligned} \left\| \left\{ \sum_{j_1\in\mathbb{N}} f_{j_1,j_2} \right\}_{j_2\in\mathbb{N}} \right\|_{\mathscr{V}} &\leq \sum_{j_1\in\mathbb{N}} \|\{f_{j_1,j_2}\}_{j_2\in\mathbb{N}}\|_{\mathscr{V}} \\ &\lesssim \left(\sum_{j_1\in\mathbb{N}} \|\{2^{\epsilon j_1} f_{j_1,j_2}\}_{j_2\in\mathbb{N}}\|_{\mathscr{V}}^q \right)^{\frac{1}{q}} = \|\{2^{\epsilon j_1} f_{j_1,j_2}\}_{(j_1,j_2)\in\mathbb{N}^2}\|_{\widetilde{\mathscr{V}}}, \end{aligned}$$

completing the proof of (6.76).

When $\mathscr{V} = \ell^q(\mathbb{N}; L^p(\mathfrak{M}, \mathrm{Vol}))$, (6.77) follows immediately from the inequality $\|\cdot\|_{\ell^q(\mathbb{N})} \leq \|\cdot\|_{\ell^1(\mathbb{N})}$ applied in the j_1 variable. When $\mathscr{V} = L^p(\mathfrak{M}, \mathrm{Vol}; \ell^q(\mathbb{N}))$, again using $\|\cdot\|_{\ell^q(\mathbb{N})} \leq \|\cdot\|_{\ell^1(\mathbb{N})}$ and the triangle inequality, we have

$$\|\{f_{j_1,j_2}\}_{(j_1,j_2)\in\mathbb{N}^2}\|_{\widetilde{\mathscr{V}}} \leq \left\| \left\{ \sum_{j_1\in\mathbb{N}} |f_{j_1,j_2}| \right\}_{j_2\in\mathbb{N}} \right\|_{\mathscr{V}} \leq \sum_{j_1\in\mathbb{N}} \|\{f_{j_1,j_2}\}_{j_2\in\mathbb{N}}\|_{\mathscr{V}},$$

completing the proof. □

Proof of Theorem 6.6.13 (A). Set $\epsilon_1 := \epsilon/(1+\lambda) > 0$. Let $\tilde{s} := s_0 + \epsilon_1 > 0$. Set $s_1 := \epsilon_1 + \lambda\tilde{s} = \epsilon + \lambda s_0$ and $s_2 := s_0 - \tilde{s} = -\epsilon_1$. Let $\vec{s} := (s_1, s_2)$. Note that

$$\vec{s} + (-\lambda\tilde{s}, \tilde{s}) = (\epsilon_1, s_0). \tag{6.79}$$

Let $\mathcal{E}_2$ be a bounded set of $(Z, \vec{d})$ elementary operators supported in Ω. Let $\{(E_{j_2}, 2^{-j_2}) : j_2 \in \mathbb{N}\} \subseteq \mathcal{E}_2$. In the estimates which follow, implicit constants will not depend on the particular choice of $\{(E_{j_2}, 2^{-j_2}) : j_2 \in \mathbb{N}\} \subseteq \mathcal{E}_2$.

Let $\{(D_{j_1}, 2^{-j_1}) : j_1 \in \mathbb{N}\}$ be a bounded set of $(W, \vec{d})$ elementary operators supported in Ω such that $\sum_{j_1\in\mathbb{N}} D_{j_1} = \mathrm{Mult}[\psi]$, where $\psi \in C_0^\infty(\Omega)$ satisfies $\psi \equiv 1$ on $\mathcal{K}$.

We have, for $f \in \mathscr{X}^{-\lambda s_0+\epsilon}((W, \vec{d}), \mathcal{K})$, using (6.76) and (6.79),

$$\begin{aligned}
\|\{2^{j_2 s_0} E_{j_2} f\}_{j_2\in\mathbb{N}}\|_{\mathscr{V}} &= \left\| \left\{ \sum_{j_1\in\mathbb{N}} 2^{j_2 s_0} E_{j_2} D_{j_1} f \right\}_{j_2\in\mathbb{N}} \right\|_{\mathscr{V}} \\
&\lesssim \|\{2^{j\cdot(\epsilon_1,s_0)} E_{j_2} D_{j_1} f\}_{j\in\mathbb{N}^2}\|_{\widetilde{\mathscr{V}}} = \|\{2^{j\cdot\vec{s}} 2^{j\cdot(-\lambda\tilde{s},\tilde{s})} E_{j_2} D_{j_1} f\}_{j\in\mathbb{N}^2}\|_{\widetilde{\mathscr{V}}} \\
&= \|\{2^{j\cdot\vec{s}} 2^{\tilde{s}(j_2-\lambda j_1)} E_{j_2} D_{j_1} f\}_{j\in\mathbb{N}^2}\|_{\widetilde{\mathscr{V}}} \\
&\leq \|\{2^{j\cdot\vec{s}} 2^{\lceil\tilde{s}\rceil|(j_2\vee(\lambda j_1))-\lambda j_1|} E_{j_2} D_{j_1} f\}_{j\in\mathbb{N}^2}\|_{\widetilde{\mathscr{V}}}.
\end{aligned} \tag{6.80}$$

By Lemma 6.6.19,

$$2^{\lceil\tilde{s}\rceil|(j_2\vee(\lambda j_1))-\lambda j_1|} E_{j_2} D_{j_1} = \sum_{K=1}^{L} \widetilde{E}_{j_2}^K \widetilde{F}_{j_1}^K, \tag{6.81}$$

where $\{(\widetilde{E}_{j_2}^K, 2^{-j_2}) : (E_{j_2}, 2^{-j_2}) \in \mathcal{E}_2, j_1 \in \mathbb{N}, 1 \leq L \leq K\}$ is a bounded set of $(Z, \vec{d})$ elementary operators supported in Ω and $\{(\widetilde{F}_{j_1}^K, 2^{-j_1}) : (E_{j_2}, 2^{-j_2}) \in \mathcal{E}_2, j_1 \in \mathbb{N}, 1 \leq L \leq K\}$ is a bounded set of $(W, \vec{d})$ elementary operators supported in Ω, where $L \approx 1$. Here, $\widetilde{E}_{j_2}^K$ depends on j_1 and $\widetilde{F}_{j_1}^K$ depends on $\widetilde{E}_{j_2}^K$, though we suppress this dependence.

Plugging (6.81) into (6.80), we see that

$$\begin{aligned}\left\|\{2^{j_2 s_0} E_{j_2} f\}_{j_2\in\mathbb{N}}\right\|_{\mathcal{V}} &\lesssim \sum_{K=1}^{L} \left\|\{2^{j\cdot\vec{s}}\widetilde{E}_{j_2}^K \widetilde{F}_{j_1}^K f\}_{j\in\mathbb{N}^2}\right\|_{\widetilde{\mathcal{V}}} \\ &= \sum_{K=1}^{L} \left\|\{2^{-\epsilon_1 j_2} 2^{j_1 s_1}\widetilde{E}_{j_2}^K \widetilde{F}_{j_1}^K f\}_{j\in\mathbb{N}^2}\right\|_{\widetilde{\mathcal{V}}} \\ &\le \sum_{K=1}^{L}\sum_{j_2\in\mathbb{N}} 2^{-\epsilon_1 j_2} \left\|\{2^{j_1 s_1}\widetilde{E}_{j_2}^K \widetilde{F}_{j_1}^K f\}_{j_1\in\mathbb{N}}\right\|_{\mathcal{V}},\end{aligned} \tag{6.82}$$

where the last inequality uses (6.77).

Since $\{(\widetilde{E}_{j_2}^K, 2^{-j_2}) : (E_{j_2}, 2^{-j_2}) \in \mathcal{E}_2, j_1 \in \mathbb{N}, 1 \le L \le K\}$ is a bounded set of $(Z, \vec{d})$ elementary operators supported in Ω, it follows from Corollary 5.4.11 and Lemma 5.4.12 that

$$\left\|\{2^{j_1 s_1}\widetilde{E}_{j_2}^K \widetilde{F}_{j_1}^K f\}_{j_1\in\mathbb{N}}\right\|_{\mathcal{V}} \lesssim \left\|\{2^{j_1 s_1}\widetilde{F}_{j_1}^K f\}_{j_1\in\mathbb{N}}\right\|_{\mathcal{V}}. \tag{6.83}$$

Since $\{(\widetilde{F}_{j_1}^K, 2^{-j_1}) : (E_{j_2}, 2^{-j_2}) \in \mathcal{E}_2, j_1 \in \mathbb{N}, 1 \le L \le K\}$ is a bounded set of $(W, \vec{d})$ elementary operators supported in Ω, Theorem 6.4.3 shows that

$$\left\|\{2^{j_1 s_1}\widetilde{F}_{j_1}^K f\}_{j_1\in\mathbb{N}}\right\|_{\mathcal{V}} \lesssim \|f\|_{\mathcal{X}^{s_1}(W,\vec{d})} = \|f\|_{\mathcal{X}^{\epsilon+\lambda s_0}(W,\vec{d})}. \tag{6.84}$$

Plugging (6.83) and (6.84) into (6.82) and using the fact that $L \approx 1$, we obtain

$$\left\|\{2^{j_2 s_0} E_{j_2} f\}_{j_2\in\mathbb{N}}\right\|_{\mathcal{V}} \lesssim \sum_{K=1}^{L}\sum_{j_2\in\mathbb{N}} 2^{-\epsilon_1 j_2}\|f\|_{\mathcal{X}^{\epsilon+\lambda s_0}(W,\vec{d})} \approx \|f\|_{\mathcal{X}^{\epsilon+\lambda s_0}(W,\vec{d})}. \tag{6.85}$$

Taking the supremum of (6.85) over $\{(E_{j_2}, 2^{-j_2}) : j_2 \in \mathbb{N}\} \subseteq \mathcal{E}_2$ shows that

$$\|f\|_{\mathcal{V},s_0,\mathcal{E}_2} \lesssim \|f\|_{\mathcal{X}^{\epsilon+\lambda s_0}(W,\vec{d})} < \infty. \tag{6.86}$$

Since $\mathcal{E}_2$ was an arbitrary bounded set of $(W, \vec{d})$ elementary operators supported in Ω, it follows from Proposition 6.6.1 (i) ⇒ (ii) that $f \in \mathcal{X}^{s_0}(\mathcal{K}, (Z, \vec{d}))$. This completes the proof of the containment (6.70).

To see that the inclusion is continuous, let $\widetilde{\mathcal{D}} = \{(\widetilde{D}_{j_2}, 2^{-j_2}) : j_2 \in \mathbb{N}\}$ be a bounded set of $(Z, \vec{d})$ elementary operators supported in Ω with $\sum_{j_2\in\mathbb{N}} \widetilde{D}_{j_2} = \mathrm{Mult}[\psi]$. Applying (6.86) with $\mathcal{E}_2 = \widetilde{\mathcal{D}}$ shows that

$$\|f\|_{\mathcal{X}^{s_0}(Z,\vec{d})} \approx \|f\|_{\mathcal{V},s_0,\widetilde{\mathcal{D}}} \lesssim \|f\|_{\mathcal{X}^{\epsilon+\lambda s_0}(W,\vec{d})}.$$

Thus, the inclusion map is bounded, completing the proof. □

Proof of Theorem 6.6.13 (B). Take $\epsilon_1 > 0$ so small that $\epsilon_1(1+\frac{1}{\lambda}) \le \epsilon$ and $\tilde{s} := (s_0-\epsilon_1)/\lambda > 0$. Set $s_2 := \epsilon_1 - \tilde{s} = (1+1/\lambda)\epsilon_1 - s_0/\lambda \le \epsilon - s_0/\lambda$. Set $s_1 := -\epsilon_1 = \lambda\tilde{s} - s_0$, and let $\vec{s} := (s_1, s_2)$. Note that

$$\vec{s} + (-\lambda\tilde{s}, \tilde{s}) = (-s_0, \epsilon_1). \tag{6.87}$$

Let $\mathcal{E}_1$ be a bounded set of $(W, \vec{d})$ elementary operators supported in Ω. Let $\{(E_{j_1}, 2^{-j_1}) : j_1 \in \mathbb{N}\} \subseteq \mathcal{E}_1$. In the estimates which follow, implicit constants will not depend on the particular choice of $\{(E_{j_1}, 2^{-j_1}) : j_1 \in \mathbb{N}\} \subseteq \mathcal{E}_1$.

Let $\{(D_{j_2}, 2^{-j_2}) : j_2 \in \mathbb{N}\}$ be a bounded set of $(Z, \vec{d})$ elementary operators supported in Ω such that $\sum_{j_2 \in \mathbb{N}} D_{j_2} = \mathrm{Mult}[\psi]$, where $\psi \in C_0^\infty(\Omega)$ satisfies $\psi \equiv 1$ on $\mathcal{K}$.

We have, for $f \in \mathscr{X}^{-(s_0/\lambda)+\epsilon}(\mathcal{K}, (Z, \vec{d}))$, using (6.76) and (6.87),

$$\begin{aligned}
\|\{2^{-j_1 s_0} E_{j_1} f\}_{j_1 \in \mathbb{N}}\|_{\mathcal{V}} &= \left\| \left\{ \sum_{j_2 \in \mathbb{N}} 2^{-j_1 s_0} E_{j_1} D_{j_2} f \right\}_{j_1 \in \mathbb{N}} \right\|_{\mathcal{V}} \\
&\lesssim \|\{2^{j \cdot (-s_0, \epsilon_1)} E_{j_1} D_{j_2} f\}_{j \in \mathbb{N}^2}\|_{\widetilde{\mathcal{V}}} = \|\{2^{j \cdot \vec{s} + j \cdot (-\lambda \tilde{s}, \tilde{s})} E_{j_1} D_{j_2} f\}_{j \in \mathbb{N}^2}\|_{\widetilde{\mathcal{V}}} \\
&= \|\{2^{j \cdot \vec{s} + \tilde{s}(j_2 - \lambda j_1)} E_{j_1} D_{j_2} f\}_{j \in \mathbb{N}^2}\|_{\widetilde{\mathcal{V}}} \\
&\leq \|\{2^{j \cdot \vec{s} + \lceil \tilde{s} \rceil |(j_2 \vee (\lambda j_1)) - \lambda j_1|} E_{j_1} D_{j_2} f\}_{j \in \mathbb{N}^2}\|_{\widetilde{\mathcal{V}}}.
\end{aligned} \tag{6.88}$$

By Lemma 6.6.19,

$$2^{\lceil \tilde{s} \rceil |(j_2 \vee (\lambda j_1)) - \lambda j_1|} E_{j_1} D_{j_2} = \sum_{K=1}^{L} \widetilde{E}_{j_1}^K \widetilde{F}_{j_2}^K, \tag{6.89}$$

where $\{(\widetilde{E}_{j_1}^K, 2^{-j_1}) : (E_{j_1}, 2^{-j_1}) \in \mathcal{E}_1, j_2 \in \mathbb{N}, 1 \leq L \leq K\}$ is a bounded set of $(W, \vec{d})$ elementary operators supported in Ω and $\{(\widetilde{F}_{j_2}^K, 2^{-j_2}) : (E_{j_1}, 2^{-j_1}) \in \mathcal{E}_1, j_2 \in \mathbb{N}, 1 \leq L \leq K\}$ is a bounded set of $(Z, \vec{d})$ elementary operators supported in Ω, where $L \approx 1$. Here, $\widetilde{E}_{j_1}^K$ depends on j_2 and $\widetilde{F}_{j_2}^K$ depends on $\widetilde{E}_{j_1}^K$, though we suppress this dependence.

Plugging (6.89) into (6.88), we obtain

$$\begin{aligned}
\|\{2^{-j_1 s_0} E_{j_1} f\}_{j_1 \in \mathbb{N}}\|_{\mathcal{V}} &\lesssim \sum_{K=1}^{L} \|\{2^{j \cdot \vec{s}} \widetilde{E}_{j_1}^K \widetilde{F}_{j_2}^K f\}_{j \in \mathbb{N}^2}\|_{\widetilde{\mathcal{V}}} \\
&= \sum_{K=1}^{L} \|\{2^{-\epsilon_1 j_1} 2^{j_2 s_2} \widetilde{E}_{j_1}^K \widetilde{F}_{j_2}^K f\}_{j \in \mathbb{N}^2}\|_{\widetilde{\mathcal{V}}} \\
&\leq \sum_{K=1}^{L} \sum_{j_1 \in \mathbb{N}} 2^{-\epsilon_1 j_1} \|\{2^{j_2 s_2} \widetilde{E}_{j_1}^K \widetilde{F}_{j_2}^K f\}_{j_2 \in \mathbb{N}}\|_{\mathcal{V}},
\end{aligned} \tag{6.90}$$

where the last inequality uses (6.77).

Since $\{(\widetilde{E}_{j_1}^K, 2^{-j_1}) : (E_{j_1}, 2^{-j_1}) \in \mathcal{E}_1, j_2 \in \mathbb{N}, 1 \leq L \leq K\}$ is a bounded set of $(W, \vec{d})$ elementary operators supported in Ω, it follows from Corollary 5.4.11 and Lemma 5.4.12 that

$$\|\{2^{j_2 s_2} \widetilde{E}_{j_1}^K \widetilde{F}_{j_2}^K f\}_{j_2 \in \mathbb{N}}\|_{\mathcal{V}} \lesssim \|\{2^{j_2 s_2} \widetilde{F}_{j_2}^K f\}_{j_2 \in \mathbb{N}}\|_{\mathcal{V}}. \tag{6.91}$$

Since $\{(\widetilde{F}_{j_2}^K, 2^{-j_2}) : (E_{j_1}, 2^{-j_1}) \in \mathcal{E}_1, j_1 \in \mathbb{N}, 1 \leq L \leq K\}$ is a bounded set of $(W, \vec{d})$ elementary operators supported in Ω, Theorem 6.4.3 shows that

$$\left\|\{2^{j_2 s_2}\widetilde{F}_{j_2}^K f\}_{j_2\in\mathbb{N}}\right\|_{\mathscr{V}} \lesssim \|f\|_{\mathscr{X}^{s_2}(Z,\vec{d})} = \|f\|_{\mathscr{X}^{-(s_0/\lambda)+\epsilon}(Z,\vec{d})}. \tag{6.92}$$

Plugging (6.91) and (6.92) into (6.90) and using the fact that $L \approx 1$, we obtain

$$\left\|\{2^{-j_1 s_0}E_{j_1}f\}_{j_1\in\mathbb{N}}\right\|_{\mathscr{V}} \lesssim \sum_{K=1}^{L}\sum_{j_1\in\mathbb{N}} 2^{-\epsilon_1 j_1}\|f\|_{\mathscr{X}^{-(s_0/\lambda)+\epsilon}(Z,\vec{d})} \approx \|f\|_{\mathscr{X}^{-(s_0/\lambda)+\epsilon}(Z,\vec{d})}. \tag{6.93}$$

Taking the supremum of (6.93) over $\{(E_{j_1}, 2^{-j_1}) : j_1 \in \mathbb{N}\} \subseteq \mathcal{E}_1$ shows that

$$\|f\|_{\mathscr{V},-s_0,\mathcal{E}_1} \lesssim \|f\|_{\mathscr{X}^{-(s_0/\lambda)+\epsilon}(Z,\vec{d})} < \infty. \tag{6.94}$$

Since $\mathcal{E}_1$ was an arbitrary bounded set of $(Z,\vec{d})$ elementary operators supported in Ω, it follows from Proposition 6.6.1 (i) $\Rightarrow$ (ii) that $f \in \mathscr{X}^{-s_0}(\mathcal{K},(W,\vec{d}))$. This completes the proof of the containment (6.71).

To see that the inclusion is continuous, let $\widetilde{\mathcal{D}} = \{(\widetilde{D}_{j_1}, 2^{-j_1}) : j_1 \in \mathbb{N}\}$ be a bounded set of $(W,\vec{d})$ elementary operators supported in Ω with $\sum_{j_1\in\mathbb{N}} \widetilde{D}_{j_1} = \mathrm{Mult}[\psi]$. Applying (6.94) with $\mathcal{E}_1 = \widetilde{\mathcal{D}}$ shows that

$$\|f\|_{\mathscr{X}^{-s_0}(W,\vec{d})} \approx \|f\|_{\mathscr{V},-s_0,\widetilde{\mathcal{D}}} \lesssim \|f\|_{\mathscr{X}^{-(s_0/\lambda)+\epsilon}(Z,\vec{d})}.$$

Thus, the inclusion map is bounded, completing the proof. □

6.6.3 Boundedness of operators in $\mathscr{D}$

Let $(W,\vec{d}) = \{(W_1,\vec{d}_1),\ldots,(W_r,\vec{d}_r)\}$ and $(Z,\vec{d}) = \{(Z_1,\vec{d}_1),\ldots,(Z_v,\vec{d}_v)\}$ be two lists of Hörmander vector fields with formal degrees on $\mathfrak{M}$. We assume $(W,\vec{d})$ and $(Z,\vec{d})$ locally weakly approximately commute. Because $(W,\vec{d})$ and $(Z,\vec{d})$ locally weakly approximately commute, it makes sense to talk of the filtered algebra $\mathscr{D}^t(\Omega,(W,\vec{d}),(Z,\vec{d})) \subseteq \mathscr{A}^t(\Omega,(W,\vec{d}))$ as defined in Section 5.9. Informally speaking, $\mathscr{D}^t(\Omega,(W,\vec{d}),(Z,\vec{d}))$ consists of those operators in $\mathscr{A}^t(\Omega,(W,\vec{d}))$ which "approximately commute" with the vector fields in $(Z,\vec{d})$. This is enough to show that the operators in $\mathscr{D}^\kappa(\Omega,(W,\vec{d}),(Z,\vec{d}))$, for $\kappa > 0$, have good mapping properties on the spaces $\mathscr{X}^s(\mathcal{K},(Z,\vec{d}))$, as the next result shows.

Proposition 6.6.21. *For every $x \in \mathfrak{M}$, there exists a relatively compact, open neighborhood $\Omega_0 \Subset \mathfrak{M}$ of x such that if $\lambda := \lambda(x,(W,\vec{d}),(Z,\vec{d}))$ (see Definition 6.6.9) and $\mathcal{K} := \overline{\Omega_0}$, then for every $\kappa > 0$ and $T \in \mathscr{D}^\kappa(\Omega_0,(W,\vec{d}),(Z,\vec{d}))$, $T : \mathscr{X}^{s-\kappa/\lambda}(\mathcal{K},(Z,\vec{d})) \to \mathscr{X}^s(\mathcal{K},(Z,\vec{d}))$ is bounded.*

Proposition 6.6.21 in particular shows that standard pseudo-differential operators have the expected mapping properties on the spaces $\mathscr{X}^s(\mathcal{K},(Z,\vec{d}))$.

Corollary 6.6.22. *For every $x \in \mathfrak{M}$, there exists a relatively compact, open neighborhood $\Omega_0 \Subset \mathfrak{M}$ of x such that if $\Lambda_{\mathrm{std}} := \Lambda_{\mathrm{std}}(x, (Z, \vec{d}))$ (see Definition 6.6.15) and $\mathcal{K} := \overline{\Omega_0}$, then for every $\kappa > 0$ and $b(x, D)$ a standard pseudo-differential operator of order κ with* $\operatorname{supp}(b(x, D)) \subseteq \Omega \times \Omega$, $b(x, D) : \mathscr{X}^{s-\kappa/\Lambda_{\mathrm{std}}}(\mathcal{K}, (Z, \vec{d})) \to \mathscr{X}^{s}(\mathcal{K}, (Z, \vec{d}))$ *is bounded.*

Proof. Pick $Y_1, \ldots, Y_n \in C^{\infty}_{\mathrm{loc}}(\mathfrak{M}; T\mathfrak{M})$ such that $Y_1(x), \ldots, Y_n(x)$ form a basis for $T_x\mathfrak{M}$. By continuity, there is a neighborhood $\mathfrak{N} \subseteq \mathfrak{M}$ of x such that $Y_1(y), \ldots, Y_n(y)$ form a basis for $T_y\mathfrak{N}$, for all $y \in \mathfrak{N}$. By Remark 6.4.8 it suffices to prove the result with $\mathfrak{M}$ replaced by $\mathfrak{N}$. We henceforth make this replacement.

Let $(Y, 1) = \{(Y_1, 1), \ldots, (Y_n, 1)\}$. By Lemma 6.6.17, $\Lambda_{\mathrm{std}}(x, (Z, \vec{d})) = \lambda(x, (Y, 1), (Z, \vec{d}))$. Let Ω_0 and $\mathcal{K}$ be as in Proposition 6.6.21 with $(W, \vec{d})$ replaced by $(Y, 1)$. By Proposition 5.9.13 we have $b(x, D) \in \mathscr{D}^{\kappa}(\Omega_0, (Y, 1), (Z, \vec{d}))$. From here, Proposition 6.6.21 completes the proof. □

An important application of Proposition 6.6.21 is to give regularity properties of maximally subelliptic operators with respect to Besov and Triebel–Lizorkin spaces corresponding to other geometries. In particular, it gives sharp regularity of maximally subelliptic operators with respect the standard Besov and Triebel–Lizorkin spaces. See Section 8.2.2.

The rest of this section is devoted to the proof of Proposition 6.6.21. Fix $x \in \mathfrak{M}$, for which we wish to prove Proposition 6.6.21. By Corollary 6.6.11 there is a neighborhood $\Omega \Subset \mathfrak{M}$ such that $(W, \vec{d})$ weakly $\lambda = \lambda(x, (W, \vec{d}), (Z, \vec{d}))$-controls $(Z, \vec{d})$ on Ω. Let $\Omega_0 \Subset \Omega$ be an open neighborhood of x and let $\mathcal{K} := \overline{\Omega_0} \Subset \Omega$. These are the Ω_0 and $\mathcal{K}$ for which we prove Proposition 6.6.21.

We recall Notation 6.4.1, where for any operator, E_j, indexed by $j \in \mathbb{N}$, we set $E_j := 0$ for $j < 0$.

Lemma 6.6.23. *Let $\mathcal{F}$ be a bounded set of $(W, \vec{d})$ elementary operators supported in Ω and let $\mathcal{E}$ be a bounded set of $(Z, \vec{d})$ elementary operators supported in Ω. Let*

$$\mathcal{E}_1' = \{(E_j^1, 2^{-j}) : j \in \mathbb{N}\}, \quad \mathcal{E}_2' = \{(E_j^2, 2^{-j}) : j \in \mathbb{N}\} \subseteq \mathcal{E}, \tag{6.95}$$

$$\mathcal{F}' = \{(F_k, 2^{-k}) : k \in \mathbb{N}\} \subseteq \mathcal{F}. \tag{6.96}$$

Then, for every $M \in \mathbb{N}$ and $s \in \mathbb{R}$,

$$\sup_{\substack{\mathcal{E}_1', \mathcal{E}_2' \\ \mathcal{F}'}} \sup_{k<0, l \in \mathbb{Z}} 2^{M|k|} \left\| \{E_j^1 F_{\lceil \frac{j+l}{\lambda} \rceil + k} 2^{s(j+l)} E_{j+l}^2 f\}_{j \in \mathbb{N}} \right\|_{\mathscr{V}} \lesssim \|f\|_{\mathscr{X}^s(Z, \vec{d})}, \quad \forall f \in \mathscr{X}^s(\mathcal{K}, (Z, \vec{d})), \tag{6.97}$$

where the first supremum is taken over all such subsets $\mathcal{E}_1', \mathcal{E}_2' \subseteq \mathcal{E}$ and $\mathcal{F}' \subseteq \mathcal{F}$.

Proof. Let $\mathcal{E}_1', \mathcal{E}_2'$ be as in (6.95) and let $\mathcal{F}'$ be as in (6.96). In what follows, the estimates will be independent of the particular choice of $\mathcal{E}_1', \mathcal{E}_2', \mathcal{F}'$.

Fix $M_0 \geq 2M/\lambda + 2M\lambda$. By Lemma 6.6.19, we may write

$$F_{\lceil \frac{j+l}{\lambda} \rceil + k} E^2_{j+l} = 2^{-M_0 |(j+l) \vee ((\lceil \frac{j+l}{\lambda} \rceil + k)\lambda) - \lambda(\lceil \frac{j+l}{\lambda} \rceil + k)|} \sum_{K=1}^{L} \widetilde{F}^K_{\lceil \frac{j+l}{\lambda} \rceil + k} \widetilde{E}^K_{j+l},$$

where $L \approx 1$, $\{(\widetilde{F}^K_k, 2^{-k}) : (F_k, 2^{-k}) \in \mathcal{F}, (E^2_j, 2^{-j}) \in \mathcal{E}, K = 1, \ldots, L\}$ is a bounded set of $(W, \vec{d})$ elementary operators supported in Ω and $\{(\widetilde{E}^K_j, 2^{-j}) : (F_k, 2^{-k}) \in \mathcal{F}, (E^2_j, 2^{-j}) \in \mathcal{E}, K = 1, \ldots, L\}$ is a bounded set of $(Z, \vec{d})$ elementary operators supported in Ω.

For $k < 0$, $M_0 |(j+l) \vee ((\lceil \frac{j+l}{\lambda} \rceil + k)\lambda) - \lambda(\lceil \frac{j+l}{\lambda} \rceil + k)| \geq M|k| - \lambda$, and therefore, for $k < 0$, since $\lambda \approx 1$, we have (with bounds independent of $k < 0$ and $l \in \mathbb{Z}$)

$$2^{M|k|} \left\| \{E^1_j F_{\lceil \frac{j+l}{\lambda} \rceil + k} 2^{s(j+l)} E^2_{j+l} f\}_{j \in \mathbb{N}} \right\|_{\mathcal{V}} \lesssim \sum_{K=1}^{L} \left\| \{E^1_j \widetilde{F}^K_{\lceil \frac{j+l}{\lambda} \rceil + k} 2^{s(j+l)} \widetilde{E}^K_{j+l} f\}_{j \in \mathbb{N}} \right\|_{\mathcal{V}}. \tag{6.98}$$

Using the fact that $\mathcal{E}$ is a bounded set of $(Z, \vec{d})$ elementary operators supported in Ω and that $\{(\widetilde{F}^K_k, 2^{-k}) : (F_k, 2^{-k}) \in \mathcal{F}, (E^2_j, 2^{-j}) \in \mathcal{E}, K = 1, \ldots, L\}$ is a bounded set of $(W, \vec{d})$ elementary operators supported in Ω, it follows from Corollary 5.4.11 and Lemma 5.4.12 that

$$\begin{aligned} \sum_{K=1}^{L} \left\| \{E^1_j \widetilde{F}^K_{\lceil \frac{j+l}{\lambda} \rceil + k} 2^{s(j+l)} \widetilde{E}^K_{j+l} f\}_{j \in \mathbb{N}} \right\|_{\mathcal{V}} &\lesssim \sum_{K=1}^{L} \left\| \{\widetilde{F}^K_{\lceil \frac{j+l}{\lambda} \rceil + k} 2^{s(j+l)} \widetilde{E}^K_{j+l} f\}_{j \in \mathbb{N}} \right\|_{\mathcal{V}} \\ &\lesssim \sum_{K=1}^{L} \left\| \{2^{s(j+l)} \widetilde{E}^K_{j+l} f\}_{j \in \mathbb{N}} \right\|_{\mathcal{V}}. \end{aligned} \tag{6.99}$$

Using the fact that $\{(\widetilde{E}^K_j, 2^{-j}) : (F_k, 2^{-k}) \in \mathcal{F}, (E^2_j, 2^{-j}) \in \mathcal{E}, K = 1, \ldots, L\}$ is a bounded set of $(Z, \vec{d})$ elementary operators supported in Ω, it follows from Corollary 6.4.4 and the fact that $L \approx 1$ that for all $f \in \mathscr{X}^s(\mathcal{K}, (Z, \vec{d}))$,

$$\sum_{K=1}^{L} \left\| \{2^{s(j+l)} \widetilde{E}^K_{j+l} f\}_{j \in \mathbb{N}} \right\|_{\mathcal{V}} \lesssim \sum_{K=1}^{L} \|f\|_{\mathscr{X}^s(Z, \vec{d})} \approx \|f\|_{\mathscr{X}^s(Z, \vec{d})}. \tag{6.100}$$

Combining (6.98), (6.99), and (6.100) yields (6.97) and completes the proof. □

Lemma 6.6.24. *Let $\kappa > 0$ and $T \in \mathscr{A}^{-\kappa}(\Omega, (W, \vec{d}))$. Let $\mathcal{E}$ be a bounded set of $(Z, \vec{d})$ elementary operators supported in Ω. Let*

$$\mathcal{E}'_1 = \{(E^1_j, 2^{-j}) : j \in \mathbb{N}\}, \quad \mathcal{E}'_2 = \{(E^2_j, 2^{-j}) : j \in \mathbb{N}\} \subseteq \mathcal{E}. \tag{6.101}$$

Then, for every $s \in \mathbb{R}$,

$$\sup_{l \in \mathbb{Z}} \sup_{\mathcal{E}'_1, \mathcal{E}'_2} \left\| \{E^1_j T 2^{(j+l)s} E^2_{j+l} f\}_{j \in \mathbb{N}} \right\|_{\mathcal{V}} \lesssim \|f\|_{\mathscr{X}^{s-\kappa/\lambda}}. \tag{6.102}$$

Proof. Using $T \in \mathscr{A}^{-\kappa}(\Omega, (W, \vec{d})) = \mathscr{A}_2^{-\kappa}(\Omega, (W, \vec{d}))$, we may write $T = \sum_{k\in\mathbb{N}} 2^{-k\kappa} F_k$, where $\{(F_k, 2^{-k}) : k \in \mathbb{N}\}$ is a bounded set of $(W, \vec{d})$ elementary operators supported in Ω. By Proposition 5.5.10, this sum converges in the topology of bounded convergence on $\mathrm{Hom}(C_{\mathrm{loc}}^{\infty}(\mathfrak{M}), C_0^{\infty}(\mathfrak{M}))$. In particular, with $\mathcal{E}_1'$ and $\mathcal{E}_2'$ as in (6.101), since $E_j^1, E_{j+l}^2 \in C_0^{\infty}(\Omega \times \Omega)$, we have, for any $f \in C_0^{\infty}(\mathfrak{M})'$,

$$E_j^1 T 2^{(j+l)s} E_{j+l}^2 f = \sum_{k\in\mathbb{Z}} E_j^1 2^{-(k+\lceil \frac{j+l}{\lambda} \rceil)\kappa} F_{\lceil \frac{j+l}{\lambda} \rceil + k} 2^{(j+l)s} E_{j+l}^2 f. \tag{6.103}$$

In what follows, the implicit constants will be independent of the particular choice of $\mathcal{E}_1'$ and $\mathcal{E}_2'$ in (6.101) and independent of $l \in \mathbb{Z}$. Note that $2^{\lceil \frac{j+l}{\lambda} \rceil \kappa} \approx 2^{\kappa \frac{j+l}{\lambda}}$. Thus, we have, for $f \in \mathscr{X}^{s-\kappa/\lambda}(\mathcal{K}, (Z, \vec{d}))$, using (6.103),

$$\begin{aligned} &\big\| \{E_j^1 T 2^{(j+l)s} E_{j+l}^2 f\}_{j\in\mathbb{N}} \big\|_{\mathscr{V}} \\ &\quad \le \sum_{k\in\mathbb{Z}} \big\| \{E_j^1 2^{-(k+\lceil \frac{j+l}{\lambda} \rceil)\kappa} F_{\lceil \frac{j+l}{\lambda} \rceil + k} 2^{(j+l)s} E_{j+l}^2 f\}_{j\in\mathbb{N}} \big\|_{\mathscr{V}} \\ &\quad \approx \sum_{k\in\mathbb{Z}} 2^{-k\kappa} \big\| \{E_j^1 F_{\lceil \frac{j+l}{\lambda} \rceil + k} 2^{(j+l)(s-\kappa/\lambda)} E_{j+l}^2 f\}_{j\in\mathbb{N}} \big\|_{\mathscr{V}}. \end{aligned} \tag{6.104}$$

We separate the sum on the right-hand side of (6.104) into two parts: $k \ge 0$ and $k < 0$. For $k \ge 0$, using Corollary 5.4.11 and Lemma 5.4.12 we have

$$\begin{aligned} \big\| \{E_j^1 F_{\lceil \frac{j+l}{\lambda} \rceil + k} 2^{(j+l)(s-\kappa/\lambda)} E_{j+l}^2 f\}_{j\in\mathbb{N}} \big\|_{\mathscr{V}} &\lesssim \big\| \{F_{\lceil \frac{j+l}{\lambda} \rceil + k} 2^{(j+l)(s-\kappa/\lambda)} E_{j+l}^2 f\}_{j\in\mathbb{N}} \big\|_{\mathscr{V}} \\ &\lesssim \big\| \{2^{(j+l)(s-\kappa/\lambda)} E_{j+l}^2 f\}_{j\in\mathbb{N}} \big\|_{\mathscr{V}} \lesssim \|f\|_{\mathscr{X}^{s-\kappa/\lambda}(Z,\vec{d})}, \end{aligned}$$

where the last estimate follows from Corollary 6.4.4. Thus,

$$\begin{aligned} &\sum_{k\ge 0} 2^{-k\kappa} \big\| \{E_j^1 F_{\lceil \frac{j+l}{\lambda} \rceil + k} 2^{(j+l)(s-\kappa/\lambda)} E_{j+l}^2 f\}_{j\in\mathbb{N}} \big\|_{\mathscr{V}} \\ &\quad \lesssim \sum_{k\ge 0} 2^{-k\kappa} \|f\|_{\mathscr{X}^{s-\kappa/\lambda}(Z,\vec{d})} \lesssim \|f\|_{\mathscr{X}^{s-\kappa/\lambda}(Z,\vec{d})}, \end{aligned} \tag{6.105}$$

where we have used $\kappa > 0$.

For $k < 0$, we apply Lemma 6.6.23 with $M = \kappa + 1$ and s replaced by $s - \kappa/\lambda$ to see that

$$\begin{aligned} &\sum_{k< 0} 2^{-k\kappa} \big\| \{E_j^1 F_{\lceil \frac{j+l}{\lambda} \rceil + k} 2^{(j+l)(s-\kappa/\lambda)} E_{j+l}^2 f\}_{j\in\mathbb{N}} \big\|_{\mathscr{V}} \\ &\quad \lesssim \sum_{k< 0} 2^{-|k|} \|f\|_{\mathscr{X}^{s-\kappa/\lambda}(Z,\vec{d})} \lesssim \|f\|_{\mathscr{X}^{s-\kappa/\lambda}(Z,\vec{d})}. \end{aligned} \tag{6.106}$$

Combining (6.104), (6.105), and (6.106) yields (6.102) and completes the proof. □

Lemma 6.6.25. *Let $t \in \mathbb{R}$ and $T \in \mathscr{D}^t(\Omega, (W, \vec{d}), (Z, \vec{d}))$. Let $\mathcal{E}$ be a bounded set of $(Z, \vec{d})$ elementary operators supported in Ω. Then, for every $M \in \mathbb{N}$ and $(E_j, 2^{-j}), (F_k, 2^{-k}) \in \mathcal{E}$, we have*

$$2^{M|j-k|} E_j T F_k = \sum_{K=1}^{L} \widetilde{E}_j^K T_K \widetilde{F}_k^K,$$

where $L = L(M, (Z, \vec{d})) \in \mathbb{N}$ does not depend on sE, $T_L \in \mathscr{A}^t(\Omega, (W, \vec{d}))$ for $K = 1, \dots L$ and

$$\{(\widetilde{E}_j^K, 2^{-j}), (\widetilde{F}_k^K, 2^{-k}) : (E_j, 2^{-j}) \in \mathcal{E}, (F_k, 2^{-k}) \in \mathcal{E}, K = 1, \dots, L\}$$

is a bounded set of $(Z, \vec{d})$ elementary operators supported in Ω.

Proof. There are two cases to consider: $j \ge k$ and $k \ge j$. The proof in the two cases are similar, so we prove only the case $j \ge k$.

Applying Proposition 5.5.5 (h) to E_j, we see that

$$\begin{aligned} E_j &= \sum_{|\alpha| \le M} 2^{-(M-|\alpha|)j} \widetilde{E}_{j,\alpha} (2^{-j\vec{d}} Z)^\alpha \\ &= \sum_{|\alpha| \le M} 2^{-(M-|\alpha|)j} \widetilde{E}_{j,\alpha} 2^{-j \deg_{\vec{d}}(\alpha)} Z^\alpha, \end{aligned}$$

where $\{(\widetilde{E}_{j,\alpha}, 2^{-j}) : (E_j, 2^{-j}) \in \mathcal{E}\}$ is a bounded set of $(Z, \vec{d})$ elementary operators supported in Ω.

Fix $|\alpha| \le M$. By the definition of $\mathscr{D}^t(\Omega, (W, \vec{d}), (Z, \vec{d}))$ (see Definition 5.9.1), we can write $Z^\alpha T$ as a finite sum of terms of the form $\widetilde{T}^{(t)} \mathscr{P}^{(\deg_{\vec{d}}(\alpha))}$, where $\widetilde{T}^{(t)} \in \mathscr{A}^t(\Omega, (W, \vec{d}))$ and $\mathscr{P}^{(\deg_{\vec{d}}(\alpha))}$ is a $(Z, \vec{d})$ partial differential operator on Ω of degree $\le \deg_{\vec{d}}(\alpha)$. Thus, it suffices to show, for $|\alpha| \le M$ and such $\widetilde{T}^{(t)}$ and $\mathscr{P}^{(\deg_{\vec{d}}(\alpha))}$, that

$$2^{M|j-k|} \widetilde{E}_{j,\alpha} 2^{-(M-|\alpha|)j} \widetilde{T}(t) 2^{-j \deg_{\vec{d}}(\alpha)} \mathscr{P}^{(\deg_{\vec{d}}(\alpha))} F_k = \widehat{E}_j \widetilde{T}^{(t)} \widetilde{F}_k, \tag{6.107}$$

where

$$\{(\widehat{E}_j, 2^{-j}), (\widetilde{F}_k, 2^{-k}) : (E_j, 2^{-j}) \in \mathcal{E}, (F_k, 2^{-k}) \in \mathcal{E}\} \tag{6.108}$$

is a bounded set of $(Z, \vec{d})$ elementary operators supported in Ω.

Indeed, we take

$$\begin{aligned} \widehat{E}_j &:= 2^{M|j-k|-(M-|\alpha|)j - \deg_{\vec{d}}(\alpha)(j-k)} \widetilde{E}_{j,\alpha}, \\ \widetilde{F}_k &:= 2^{-k \deg_{\vec{d}}(\alpha)} \mathscr{P}^{(\deg_{\vec{d}}(\alpha))} F_k. \end{aligned}$$

Clearly, (6.107) holds with these choices. Since $2^{M|j-k|-(M-|\alpha|)j-\deg_{\vec{d}}(\alpha)(j-k)} \le 1$, it follows from Proposition 5.5.5 (a) and (f) that (6.108) is a bounded set of $(Z, \vec{d})$ elementary operators supported in Ω, completing the proof. □

Proof of Proposition 6.6.21. Let $\mathcal{E}$ be a bounded set of $(Z, \vec{d})$ elementary operators supported in Ω, let $\psi \in C_0^\infty(\Omega)$ be such that $\psi \equiv 1$ on a neighborhood of $\mathcal{K}$, and let $\{(D_j, 2^{-j}) : j \in \mathbb{N}\}$ be a bounded set of $(Z, \vec{d})$ elementary operators supported in Ω such that $\sum_{j\in\mathbb{N}} D_j = \mathrm{Mult}[\psi]$.

Let $\mathcal{E}' = \{(E_j, 2^{-j}) : j \in \mathbb{N}\} \subseteq \mathcal{E}$. In what follows the implicit constants will not depend on the particular choice of $\mathcal{E}'$. We first claim, for $M \in \mathbb{N}$,

$$\begin{aligned}&\sup_{\substack{\mathcal{E}'\\ l\in\mathbb{Z}}} 2^{M|l|}\big\|\{E_j T 2^{(j+l)s} D_{j+l} f\}_{j\in\mathbb{N}}\big\|_{\mathcal{V}} \\ &\quad\lesssim \|f\|_{\mathscr{X}^{s-\kappa/\lambda}(Z,\vec{d})}, \quad \forall f \in \mathscr{X}^{s}(\mathcal{K},(Z,\vec{d})),\end{aligned} \tag{6.109}$$

where the supremum is taken over all such $\mathcal{E}' \subseteq \mathcal{E}$. Indeed, to see (6.109), note that by Lemma 6.6.25, we have

$$2^{M|l|} E_j T D_{j+l} = \sum_{K=1}^{L} \widetilde{E}_j^K T_K \widetilde{F}_{j+l}^K,$$

where $T_K \in \mathscr{A}^{-\kappa}(\Omega,(W,\vec{d}))$, $L \approx 1$, and $\{(\widetilde{E}_j^K, 2^{-j}), (\widetilde{F}_k^K, 2^{-k}) : (E_j, 2^{-j}) \in \mathcal{E}, l \in \mathbb{Z}, K = 1, \dots L\}$ is a bounded set of $(Z, \vec{d})$ elementary operators supported in Ω. Thus, to prove (6.109), it suffices to show that for $K = 1, \dots, L$,

$$\big\|\{\widetilde{E}_j^K T_K 2^{(j+l)s} \widetilde{F}_{j+l}^K f\}_{j\in\mathbb{N}}\big\|_{\mathcal{V}} \lesssim \|f\|_{\mathscr{X}^{s-\kappa/\lambda}(Z,\vec{d})}, \quad \forall f \in \mathscr{X}^{s}(\mathcal{K},(Z,\vec{d})). \tag{6.110}$$

But (6.110) follows from Lemma 6.6.24, which completes the proof of (6.109).

With (6.109) in hand, we turn to proving the proposition. Recall that $\mathrm{Mult}[\psi] = \sum_{j\in\mathbb{N}} D_j$, where by Proposition 5.5.10, this sum converges in the topology of bounded convergence on $\mathrm{Hom}(C_0^\infty(\mathfrak{M})', C_{\mathrm{loc}}^\infty(\mathfrak{M})')$. By Proposition 5.8.5, $T \in \mathscr{D}^{-\kappa}(\Omega,(W,\vec{d}),(Z,\vec{d})) \subseteq \mathscr{A}^{-\kappa}(\Omega,(W,\vec{d})) \subseteq \mathrm{Hom}(C_0^\infty(\mathfrak{M})', C_{\mathrm{loc}}^\infty(\mathfrak{M})')$. Since we also have $E_j \in C_0^\infty(\Omega\times\Omega)$, we have, for all $f \in \mathscr{X}^{s-\kappa/\lambda}(\mathcal{K},(Z,\vec{d})) \subseteq C_0^\infty(\mathfrak{M})'$,

$$E_j T f = E_j T\,\mathrm{Mult}[\psi] f = \sum_{l\in\mathbb{Z}} E_j T D_{j+l} f,$$

with convergence in $C_0^\infty(\mathfrak{M})$. Thus, we have, using (6.109) with $M \geq |s| + 1$, for all $f \in \mathscr{X}^{s-\kappa/\lambda}(\mathcal{K},(Z,\vec{d}))$,

$$\begin{aligned}\big\|\{2^{js} E_j T f\}_{j\in\mathbb{N}}\big\|_{\mathcal{V}} &= \left\|\left\{\sum_{l\in\mathbb{Z}} 2^{js} E_j T D_{j+l} f\right\}_{j\in\mathbb{N}}\right\|_{\mathcal{V}} \\ &\leq \sum_{l\in\mathbb{Z}} \big\|\{2^{js} E_j T D_{j+l} f\}_{j\in\mathbb{N}}\big\|_{\mathcal{V}} \\ &\leq \sum_{l\in\mathbb{Z}} 2^{-|l|} 2^{(|s|+1)|l|} \big\|\{E_j T 2^{(j+l)s} D_{j+l} f\}_{j\in\mathbb{N}}\big\|_{\mathcal{V}} \\ &\lesssim \sum_{l\in\mathbb{Z}} 2^{-|l|} \|f\|_{\mathscr{X}^{s-\kappa/\lambda}(Z,\vec{d})} \lesssim \|f\|_{\mathscr{X}^{s-\kappa/\lambda}(Z,\vec{d})}.\end{aligned} \tag{6.111}$$

Taking the supremum of (6.111) over all such $\mathcal{E}' \subseteq \mathcal{E}$, we see that

$$\|Tf\|_{\mathcal{V},s,\mathcal{E}} \lesssim \|f\|_{\mathscr{X}^{s-\kappa/\lambda}(Z,\vec{d})} < \infty. \tag{6.112}$$

In particular, since $\mathcal{E}$ was an arbitrary bounded set of $(Z,\vec{d})$ elementary operators supported in Ω and $\mathrm{supp}(Tf) \subseteq \Omega_0 \subseteq \mathcal{K}$, Proposition 6.6.1 (i) $\Rightarrow$ (iii) implies $Tf \in \mathscr{X}^s(\mathcal{K},(Z,\vec{d}))$. This shows that $T : \mathscr{X}^{s-\kappa/\lambda}(\mathcal{K},(Z,\vec{d})) \to \mathscr{X}^s(\mathcal{K},(Z,\vec{d}))$. To show the map is bounded, we apply (6.112) with $\mathcal{E} = \mathcal{D}_0$ to see that

$$\|Tf\|_{\mathscr{X}^s(Z,\vec{d})} \approx \|Tf\|_{\mathcal{V},s,\mathcal{D}_0} \lesssim \|f\|_{\mathscr{X}^{s-\kappa/\lambda}(Z,\vec{d})},$$

completing the proof. □

6.7 Trading derivatives

We return to the general multi-parameter setting described in Section 6.3. Fix $\nu_1 \in \{1,\dots,\nu-1\}$ and $\nu_2 = \nu - \nu_1$. Set

$$\begin{aligned}(Y^\mu, \vec{d}^\mu) &:= (W^\mu, \vec{d}^\mu), \quad 1 \le \mu \le \nu_1,\\ (Y,\vec{d}) = \{(Y_1,\vec{d}_1),\dots,(Y_{\nu_Y},\vec{d}_{\nu_Y})\} &:= (Y^1,\vec{d}^1) \boxtimes (Y^2,\vec{d}^2) \boxtimes \cdots \boxtimes (Y^{\nu_1},\vec{d}^{\nu_1})\\ &\subset C^\infty_{\mathrm{loc}}(\mathfrak{M}; T\mathfrak{M}) \times (\mathbb{N}^{\nu_1} \setminus \{0\}).\end{aligned}$$

Similarly define

$$\begin{aligned}(Z^\mu, \vec{d}^\mu) &:= (W^{\nu_1+\mu}, \vec{d}^{\nu_1+\mu}), \quad 1 \le \mu \le \nu_2,\\ (Z,\vec{d}) = \{(Z_1,\vec{d}_1),\dots,(Z_{\nu_Z},\vec{d}_{\nu_Z})\} &:= (Z^1,\vec{d}^1) \boxtimes (Z^2,\vec{d}^2) \boxtimes \cdots \boxtimes (Z^{\nu_2},\vec{d}^{\nu_2})\\ &\subset C^\infty_{\mathrm{loc}}(\mathfrak{M}; T\mathfrak{M}) \times (\mathbb{N}^{\nu_2} \setminus \{0\}).\end{aligned}$$

In short, $(Y,\vec{d})$ is to $(Y^1,\vec{d}^1),\dots,(Y^{\nu_1},\vec{d}^{\nu_1})$ as $(Z,\vec{d})$ is to $(Z^1,\vec{d}^1),\dots,(Z^{\nu_2},\vec{d}^{\nu_2})$ and as $(W,\vec{d})$ is to $(W^1,\vec{d}^1),\dots,(W^\nu,\vec{d}^\nu)$.

Fix $\mathcal{K} \Subset \Omega \Subset \mathfrak{M}$, where $\mathcal{K}$ is compact and Ω is open and relatively compact. Fix a $\nu_2 \times \nu_1$ matrix λ, with components in $[0,\infty]$. If $(Y,\vec{d})$ weakly λ-controls $(Z,\vec{d})$ on Ω, then (informally) each vector field Z_j counts as a $(Y,\vec{d})$ derivative of "degree" $\lambda^\top(\vec{d}_j)$. We turn to making this idea precise using the multi-parameter Besov and Triebel–Lizorkin spaces. For $j \in [0,\infty)^\nu$, we write $j = (j^Y, j^Z) \in [0,\infty)^{\nu_1} \times [0,\infty)^{\nu_2}$.

Notation 6.7.1. For $j^Y \in [0,\infty)^{\nu_1}$, to compute $\lambda(j^Y) \in [0,\infty]^{\nu_2}$, we use the convention $\infty \cdot 0 = 0$. However, for $s \in [0,\infty)^{\nu_2}$, when we compute $\lambda^\top(s)$, we use the convention $\infty \cdot 0 = 0$. When we take the dot product of two elements of $[0,\infty]^{\nu_1}$, we use the convention $\infty \cdot 0 = 0$, but when we take the dot product of two elements of $[0,\infty]^{\nu_2}$, we use the convention $\infty \cdot 0 = \infty$. Thus, $\lambda(j^Y) \cdot s = j^Y \cdot \lambda^\top(s)$.

The main result of this section is the next theorem.

Theorem 6.7.2. *Suppose $(Y,\vec{d})$ weakly λ-controls $(Z,\vec{d})$ on Ω. Then, for $s \in \mathbb{R}^\nu$ and $\tilde{s} \in [0,\infty)^{\nu_2}$ such that $\lambda^\top(\tilde{s})$ is not equal to ∞ in any component, we have*

$$\mathscr{X}^s(\mathcal{K},(W,\vec{d})) \subseteq \mathscr{X}^{s+(-\lambda^\top(\tilde{s}),\tilde{s})}(\mathcal{K},(W,\vec{d})), \tag{6.113}$$

and the inclusion is continuous.

To prove Theorem 6.7.2, we introduce a lemma.

Lemma 6.7.3. *Suppose $(Y,\vec{d})$ weakly λ-controls $(Z,\vec{d})$ on Ω. Let $\mathcal{E}$ be a bounded set of generalized $(W,\vec{d})$ elementary operators supported in Ω. Then, $\forall N \geq 0$,*

$$\{(2^{N|(j^Z\vee\lambda(j^Y))-\lambda(j^Y)|}E_j, 2^{-j}) : (E_j,2^{-j}) \in \mathcal{E}\}$$

is a bounded set of generalized $(W,\vec{d})$ elementary operators supported in Ω.

Proof. We will show that

$$\mathcal{E}' := \{(2^{|(j^Z\vee\lambda(j^Y))-\lambda(j^Y)|_\infty}E_j, 2^{-j}) : (E_j,2^{-j}) \in \mathcal{E}\}$$

is a bounded set of generalized $(W,\vec{d})$ elementary operators supported in Ω. Repeated application of this proves the proposition.

We consider ν_2 sets: for $\mu \in \{1,\dots,\nu_2\}$, set

$$\mathcal{E}'_\mu := \{(2^{j^Z_\mu-\lambda(j^Y)_\mu}E_j, 2^{-j}) : (E_j,2^{-j}) \in \mathcal{E}, j^Z_\mu - \lambda(j^Y)_\mu = |(j^Z \vee \lambda(j^Y)) - \lambda(j^Y)|_\infty\}.$$

Note that $\mathcal{E}' \subseteq \mathcal{E} \cup \mathcal{E}'_1 \cup \dots \cup \mathcal{E}'_{\nu_2}$. Thus, it suffices to show that for each $\mu \in \{1,\dots,\nu_2\}$, $\mathcal{E}'_\mu$ is a bounded set of generalized $(W,\vec{d})$ elementary operators supported in Ω.

Let $(E_j,2^{-j}) \in \mathcal{E}'_\mu$, so that $j^Z_\mu - \lambda(j^Y)_\mu = |(j^Z\vee\lambda(j^Y))-\lambda(j^Y)|_\infty$. In particular, $\lambda(j^Y)_\mu \neq \infty$. By Proposition 5.5.5 (h) (in the case $N=1$), using the fact that $(Z^\mu,\vec{d}^\mu) = (W^{\mu+\nu_1}, \vec{d}^{\mu+\nu_1})$, we can write

$$\begin{aligned}2^{j^Z_\mu-\lambda(j^Y)_\mu}E_j &= 2^{j^Z_\mu-\lambda(j^Y)_\mu}\sum_{|\alpha_\mu|\leq 1} 2^{-(1-|\alpha_\mu|)j^Z_\mu}(2^{-j^Z_\mu}Z^\mu)^{\alpha_\mu}E_{j,\alpha_\mu} \\ &= 2^{-\lambda(j^Y)_\mu}E_{j,0} + \sum_{|\alpha_\mu|=1} 2^{(j^Z_\mu-\lambda(j^Y)_\mu)(1-\deg_{\vec{d}^\mu}(\alpha_\mu))}2^{-\lambda(j^Y)_\mu\deg_{\vec{d}^\mu}(\alpha_\mu)}(Z^\mu)^{\alpha_\mu}E_{j,\alpha_\mu},\end{aligned} \tag{6.114}$$

where $\{(E_{j,\alpha_\mu},2^{-j}) : (E_j,2^{-j}) \in \mathcal{E}, |\alpha_\mu| \leq 1\}$ is a bounded set of generalized $(W,\vec{d})$ elementary operators supported in Ω.

Since $2^{-\lambda(j^Y)_\mu} \in [0,1]$, Proposition 5.5.5 (a) shows that $\{(2^{-\lambda(j^Y)_\mu}E_{j,0},2^{-j}) : (E_j,2^{-j}) \in \mathcal{E}\}$ is a bounded set of generalized $(W,\vec{d})$ elementary operators supported in Ω. Combining this with the fact that for $|\alpha_\mu| = 1$, $2^{(j^Z_\mu - \lambda(j^Y)_\mu)(1-\deg_{\vec{d}^\mu}(\alpha_\mu))} \in [0,1]$, by (6.114) and Proposition 5.5.5 (a), to show that $\mathcal{E}'_\mu$ is a bounded set of generalized $(W, \vec{d})$ elementary operators supported in Ω, it suffices to show for $|\alpha_\mu| = 1$ that

$$\{(2^{-\lambda(j^Y)_\mu \deg_{\vec{d}^\mu}(\alpha_\mu)}(Z^\mu)^{\alpha_\mu} E_{j,\alpha_\mu}, 2^{-j}) : (E, 2^{-j}) \in \mathcal{E}\} \tag{6.115}$$

is a bounded set of $(W, \vec{d})$ elementary operators supported in Ω.

Consider, for some $1 \le k \le \nu_Z$,

$$\begin{aligned} 2^{-\lambda(j^Y)_\mu \deg_{\vec{d}^\mu}(\alpha_\mu)}(Z^\mu)^{\alpha_\mu} E_{j,\alpha_\mu} &= 2^{-\lambda(j^Y)\cdot \vec{d}_k} Z_k E_{j,\alpha_\mu} \\ &= 2^{-j^Y\cdot\lambda^\intercal(\vec{d}_k)} Z_k E_{j,\alpha_\mu} = 2^{-j\cdot(\lambda^\intercal(\vec{d}_k), 0_{\nu_2})} Z_k E_{j,\alpha_\mu}. \end{aligned} \tag{6.116}$$

Since $(Y, \vec{d})$ weakly λ-controls $(Z, \vec{d})$ on Ω, it follows from the definitions that $\mathbb{Z}_k$ is a $(Y, \vec{d})$ partial differential operator on Ω of degree $\le \lambda^\intercal(\vec{d}_k)$, and therefore it is a $(W, \vec{d})$ partial differential operator on Ω of degree $\le (\lambda^\intercal(\vec{d}_k), 0_{\nu_2})$. Thus, by Proposition 5.5.5 (g),

$$\{(2^{-j\cdot(\lambda^\intercal(\vec{d}_k), 0_{\nu_2})} Z_k E_{j,\alpha_\mu}, 2^{-j}) : (E_j, 2^{-j}) \in \mathcal{E}\}$$

is a bounded set of generalized $(W, \vec{d})$ elementary operators supported in Ω. Combining this with (6.116) shows that (6.115) is a bounded set of generalized $(W, \vec{d})$ elementary operators supported in Ω, which completes the proof. □

Proof of Theorem 6.7.2. Let $\mathcal{E}$ be a bounded set of generalized $(W, \vec{d})$ elementary operators supported in Ω. Set

$$\widetilde{\mathcal{E}} := \{(2^{|\tilde{s}||(j^Z \vee \lambda(j^Y)) - \lambda(j^Y)|} E_j, 2^{-j}) : (E_j, 2^{-j}) \in \mathcal{E}\}.$$

By Lemma 6.7.3, $\widetilde{\mathcal{E}}$ is a bounded set of generalized $(W, \vec{d})$ elementary operators supported in Ω.

Let $f \in \mathscr{X}^s(\mathcal{K}, (W, \vec{d}))$. We have, using the fact that $\tilde{s}$ is non-negative in every component,

$$\begin{aligned} \|f\|_{\mathscr{V}, s+(-\lambda^\intercal(\tilde{s}), \tilde{s}), \mathcal{E}} &= \sup_{\{(E_j, 2^{-j}) : j \in \mathbb{N}^\nu\} \subseteq \mathcal{E}} \|\{2^{j\cdot s} 2^{(-\lambda^\intercal(\tilde{s}), \tilde{s})\cdot j} E_j f\}_{j\in\mathbb{N}^\nu}\|_{\mathscr{V}} \\ &= \sup_{\{(E_j, 2^{-j}) : j \in \mathbb{N}^\nu\} \subseteq \mathcal{E}} \|\{2^{j\cdot s} 2^{\tilde{s}\cdot(j^Z - \lambda(j^Y))} E_j f\}_{j\in\mathbb{N}^\nu}\|_{\mathscr{V}} \\ &\le \sup_{\{(E_j, 2^{-j}) : j \in \mathbb{N}^\nu\} \subseteq \mathcal{E}} \|\{2^{j\cdot s} 2^{|\tilde{s}||(j^Z \vee \lambda^\intercal(j^Y)) - \lambda(j^Y)|} E_j f\}_{j\in\mathbb{N}^\nu}\|_{\mathscr{V}} \\ &\le \sup_{\{(\widetilde{E}_j, 2^{-j}) : j \in \mathbb{N}^\nu\} \subseteq \widetilde{\mathcal{E}}} \|\{2^{j\cdot s} \widetilde{E}_j f\}_{j\in\mathbb{N}^\nu}\|_{\mathscr{V}} = \|f\|_{\mathscr{V}, s, \widetilde{\mathcal{E}}} \lesssim \|f\|_{\mathscr{X}^s(W, \vec{d})} < \infty, \end{aligned} \tag{6.117}$$

where the final $\lesssim$ follows from Corollary 6.4.4.

Since $\mathcal{E}$ was an arbitrary bounded set of generalized $(W, \vec{d})$ elementary operators supported in Ω and $\mathcal{K} \Subset \Omega$, it follows from Corollary 6.4.7 that $f \in \mathscr{X}^{s+(-\lambda^\top(\vec{s}),\vec{s})}(\mathcal{K}, (W, \vec{d}))$. This completes the proof of the containment (6.113). To see that the inclusion is continuous, we apply (6.117) with $\mathcal{E} = \mathcal{D}_0$ (where $\mathcal{D}_0$ is a bounded set of generalized $(W, \vec{d})$ elementary operators supported in Ω as in Definition 6.3.8) to see that for all $f \in \mathscr{X}^s(\mathcal{K}, (W, \vec{d}))$,

$$\|f\|_{\mathscr{X}^{s+\left(-\lambda^\top(\vec{s}),\vec{s}\right)}(W,\vec{d})} \approx \|f\|_{\mathscr{V},s+\left(-\lambda^\top(\vec{s}),\vec{s}\right),\mathcal{D}_0} \lesssim \|f\|_{\mathscr{X}^s(W,\vec{d})}.$$

This shows that the inclusion map is bounded and completes the proof. □

6.7.1 Sharpness

In some of our most important applications, Theorem 6.7.2 is sharp; we describe that in this section.

Suppose $(Y, \vec{d}) = \{(Y_1, \vec{d}_1), \dots, (Y_{r_1}, \vec{d}_{r_1})\} \subset C^\infty_{\mathrm{loc}}(\mathfrak{M}; T\mathfrak{M}) \times \mathbb{N}_+$ and $(Z, \vec{d}) = \{(Z_1, \vec{d}_1), \dots, (Z_{r_2}, \vec{d}_{r_2})\} \subset C^\infty_{\mathrm{loc}}(\mathfrak{M}; T\mathfrak{M}) \times \mathbb{N}_+$ are two sets of Hörmander vector fields with formal degrees such that $(Y, \vec{d})$ and $(Z, \vec{d})$ locally weakly approximately commute on $\mathfrak{M}$. We set $(W, \vec{d}) := (Y, \vec{d}) \boxtimes (Z, \vec{d})$.

By Corollary 6.6.11, for each $x \in \mathfrak{M}$, there is an open neighborhood $\Omega \Subset \mathfrak{M}$ of x such that $(Y, \vec{d})$ weakly $\lambda(x, (Y, \vec{d}), (Z, \vec{d}))$-control $(Z, \vec{d})$ on Ω. Theorem 6.7.2 therefore applies in this setting to any $\mathcal{K} \Subset \mathfrak{M}$. The next definition gives a setting in which Theorem 6.7.2 is sharp.

Definition 6.7.4. For $x \in \mathfrak{M}$, we say $\lambda(x, (Y, \vec{d}), (Z, \vec{d}))$ is sharp if for every open neighborhood $U \subseteq \mathfrak{M}$ of x, there exist $(V, \vec{d}) \in \mathrm{Gen}((Z, \vec{d}))$ and $y \in U$ such that

$$V(y) \notin \mathrm{span}\{X_1(y) : (X_1, d_1) \in \mathrm{Gen}((Y, \vec{d})), d_1 < \lambda \vec{d}\}. \tag{6.118}$$

Remark 6.7.5. Since $(Y, \vec{d})$ weakly $\lambda(x, (Y, \vec{d}), (Z, \vec{d}))$-controls $(Z, \vec{d})$ on Ω (Corollary 6.6.11), it follows from Lemma 3.14.8 that $(Y, \vec{d})$ weakly $\lambda(x, (Y, \vec{d}), (Z, \vec{d}))$-controls $\mathrm{Gen}((Z, \vec{d}))$ on Ω. In other words, $\mathrm{Gen}((Y, \vec{d}))$ strongly $\lambda(x, (Y, \vec{d}), (Z, \vec{d}))$-controls $\mathrm{Gen}((Z, \vec{d}))$ on Ω. This implies that for every $y \in \Omega$, (6.118) holds if $<$ is replaced by $\leq$. This helps to explain the terminology "is sharp."

Example 6.7.6. Suppose $(Y, 1) = \{(Y_1, 1), \dots, (Y_{r_1}, 1)\}$ is such that $Y_1, \dots, Y_{r_1}$ span the tangent space at every point. Then $\lambda(x, (Y, 1), (Z, \vec{d}))$ and $\lambda(x, (Z, \vec{d}), (Y, 1))$ are both sharp. This shows that $\lambda_{\mathrm{std}}(x, (Z, \vec{d}))$ and $\Lambda_{\mathrm{std}}(x, (Z, \vec{d}))$ as defined in Definition 6.6.15 are both "sharp." See Lemma 6.6.17.

Proposition 6.7.7. *For every $x \in \mathfrak{M}$, there is an open neighborhood $\Omega_0 \Subset \mathfrak{M}$ of x such that for every $s \in \mathbb{R}^2$ and $\tilde{s} > 0$, we have, with $\lambda := \lambda(x, (Y, \vec{d}), (Z, \vec{d}))$,*

$$\|f\|_{\mathscr{X}^{s+(-\lambda(\tilde{s}),\tilde{s})}(W,\vec{d})} \le C_{x,s,\tilde{s},\mathscr{X}^s} \|f\|_{\mathscr{X}^s(W,\vec{d})}, \quad \forall f \in \mathscr{X}^s(\overline{\Omega_0}, (W, \vec{d})). \tag{6.119}$$

Furthermore, if $\lambda(x, (Y, \vec{d}), (Z, \vec{d}))$ is sharp, then (6.119) is sharp in the following sense. Suppose there exist $s \in \mathbb{R}^2$, $\tilde{s} > 0$, $\epsilon_1, \epsilon_2 \ge 0$, a space $\mathscr{X}_0^s$ of the form

$$\mathscr{X}_0^s \in \{\mathscr{B}^s_{p,q} : p, q \in [1, \infty]\} \bigcup \{\mathscr{F}^s_{p,q} : p \in (1, \infty), q \in (1, \infty]\}, \tag{6.120}$$

a neighborhood U of x, and $C \ge 0$ such that

$$\|f\|_{\mathscr{X}_0^{s+(-\lambda(\tilde{s}),\tilde{s})+(\epsilon_1,\epsilon_2)}(W,\vec{d})} \le C\|f\|_{\mathscr{X}_0^s(W,\vec{d})}, \quad \forall f \in C_0^\infty(U). \tag{6.121}$$

Then $\epsilon_1 = \epsilon_2 = 0$.

The bound (6.119) is a special case of inclusion map in Theorem 6.7.2 being continuous; thus, we need only to prove the sharpness part of Proposition 6.7.7. The rest of this section is devoted to the proof of this sharpness. Fix $x \in \mathfrak{M}$, $s \in \mathbb{R}^2$, $\tilde{s} > 0$, $\epsilon_1, \epsilon_2 \ge 0$, $\mathscr{X}_0^s$, and $U \Subset \mathfrak{M}$ as in Proposition 6.7.7, such that (6.121) holds. We assume $\lambda(x, (Y, \vec{d}), (Z, \vec{d}))$ is sharp. Our goal is to show $\epsilon_1 = \epsilon_2 = 0$. Recall that $\mathscr{X}_0^s$ is of the form (6.120). Fix $p \in [1, \infty]$ so that $\mathscr{X}_0^s$ is equal to either $\mathscr{B}^s_{p,q}$ or $\mathscr{F}^s_{p,q}$, for some $q \in [1, \infty]$.

By Corollary 6.6.11, there is a neighborhood $\Omega \Subset U$ of x such that $(Y, \vec{d})$ weakly $\lambda := \lambda(x, (Y, \vec{d}), (Z, \vec{d}))$-controls $(Z, \vec{d})$ on Ω. Fix $\Omega_0 \Subset \Omega_1 \Subset \Omega_2 \Subset \Omega$ open with $x \in \Omega_0$. Using Proposition 3.4.14, in the case $\nu = 1$, $\mathrm{Gen}((Y, \vec{d}))$ is finitely generated on Ω by some $(X^1, d^1) = \{(X_1^1, d_1^1), \dots, (X_{q_1}^1, d_{q_1}^1)\} \subseteq \mathrm{Gen}((Y, \vec{d}))$ and $\mathrm{Gen}((Z, \vec{d}))$ is finitely generated on Ω by some $(X^2, d^2) = \{(X_1^2, d_1^2), \dots, (X_{q_2}^2, d_{q_2}^2)\} \subseteq \mathrm{Gen}((Z, \vec{d}))$. Since $\lambda(x, (Y, \vec{d}), (Z, \vec{d}))$ is sharp, by possibly adding an element to (X^2, d^2), we may assume $(X_1^2, d_1^2) = (V, \vec{d})$, where $(V, \vec{d}) \in \mathrm{Gen}((Z, \vec{d}))$ is as in Definition 6.7.4 with $U = \Omega_0$. We will be considering pseudo-differential operators corresponding to $(X^1, d^1), (X^2, d^2)$ as in Chapter 4. To this end, define $(X, \vec{d})$ as in (4.1), that is,

$$\begin{aligned}(X, \vec{d}) = \{(X_1, \vec{d}_1), \dots, (X_q, \vec{d}_q)\} &:= (X^1, d^1) \boxtimes (X^2, d^2) \\ &\subset C^\infty_{\mathrm{loc}}(\mathfrak{M}; T\mathfrak{M}) \times (\mathbb{N}^2 \setminus \{0\}).\end{aligned}$$

Note that $q = q_1 + q_2$. For $t \in \mathbb{R}^q$, we write $t = (t^1, t^2) \in \mathbb{R}^{q_1} \times \mathbb{R}^{q_2} = \mathbb{R}^q$.

Fix $a > 0$ small and $N \in \mathbb{N}$ large, to be chosen later. Let $\psi \in C_0^\infty(\Omega_1)$ and $\varsigma \in C_0^\infty(B^q(a))$ be such that

$$\int (t^1)^{\alpha_1} \varsigma(t^1, t^2)\, dt^1 \equiv 0, \quad \int (t^2)^{\alpha_2} \varsigma(t^1, t^2)\, dt^2 \equiv 0, \quad \forall |\alpha_1|, |\alpha_2| \le N. \tag{6.122}$$

Define, for $j \geq 0$, with $\lambda := \lambda(x, (Y, \vec{d}), (Z, \vec{d}))$,

$$T_j f(x) := \int f(e^{-t^1 \cdot X^1} e^{-t^2 \cdot X^2} x) \psi(x) \operatorname{Dil}^{\vec{d}}_{2^{(j,\lambda j)}}(\varsigma)(t)\, dt, \tag{6.123}$$

where $\operatorname{Dil}^{\vec{d}}_{2^{(j,\lambda j)}}(\varsigma)(t)$ is defined as in Definition 4.1.12 (here we are taking $f(x, s, t) = \varsigma(t)$ in Definition 4.1.12).

The proof of Proposition 6.7.7 follows by combining the next two lemmas.

Lemma 6.7.8. *Suppose $\epsilon_1 + \epsilon_2 > 0$. Then if $a > 0$ is sufficiently small and $N \in \mathbb{N}$ is sufficiently large, we have*

$$\lim_{j\to\infty} \|T_j T_j\|_{L^p(\mathfrak{M},\mathrm{Vol}) \to L^p(\mathfrak{M},\mathrm{Vol})} = 0.$$

Lemma 6.7.9. *For every $a > 0$ and $N \in \mathbb{N}$, there exist a choice of ψ and ς as above with*

$$\limsup_{j\to\infty} \|T_j T_j\|_{L^p(\mathfrak{M},\mathrm{Vol}) \to L^p(\mathfrak{M},\mathrm{Vol})} > 0.$$

Proof of Proposition 6.7.7. In light of Lemmas 6.7.8 and 6.7.9, we must have $\epsilon_1 + \epsilon_2 \leq 0$. Since $\epsilon_1, \epsilon_2 \geq 0$, this implies $\epsilon_1 = \epsilon_2 = 0$, completing the proof. □

We begin by proving Lemma 6.7.8; this lemma does not require that $\lambda(x, (Y, \vec{d}), (Z, \vec{d}))$ be sharp. We begin with two auxiliary lemmas.

Lemma 6.7.10. *Fix $\nu \in \mathbb{R}^2$. If $N = N(\nu, (X, \vec{d})) \in \mathbb{N}$ is sufficiently large and $a = a(\Omega_1, \Omega_2, (X, \vec{d})) > 0$ is sufficiently small, then $2^{(j,\lambda j)\cdot \nu} T_j$ is an $(X, \vec{d})$, a pseudo-differential operator of order ν supported in $\Omega_1 \times \Omega_2$, uniformly for $j \geq 0$. More precisely, if T_j is associated with the kernel $\check{b}_j(x, s, t)$ as in Definition* 4.1.2, *then $\{b_j : j \geq 0\} \subset S^m(a, \Omega_1, \vec{d})$ is a bounded set, where $S^m(a, \Omega_1, \vec{d})$ is defined in Definition* 4.1.1 *and is given the obvious locally convex topology. Here, $b_j(x, s, \xi)$ is the Fourier transform of $\check{b}_j(x, s, t)$ in the t variable.*

Proof. Take $\varsigma_0 \in C_0^\infty(B^q(a))$ such that $\varsigma \prec \varsigma_0$. Thus, if we set

$$\check{b}_j(x, s, t) := \psi(x)\varsigma_0(s)(2^{(j,\lambda j)\cdot \nu} \operatorname{Dil}^{\vec{d}}_{2^{(j,\lambda j)}}(\varsigma)(t)),$$

then

$$T_j f(x) = \int f(e^{-t^1 \cdot X^1} e^{-t^2 \cdot X^2} x) \check{b}_j(x, t, t)\, dt.$$

Thus, if $b_j(x, s, \xi)$ is the Fourier transform of $\check{b}_j(x, s, t)$ in the t variable, we wish to show that $\{b_j : j \geq 0\} \subset S^m(a, \Omega_1, \vec{d})$ is a bounded set. In other words, we wish to show that the conditions of Definition 4.1.1 hold uniformly in $j \geq 0$.

Note that

$$\bigcup_{j\geq 0}\{(x,s)\in \mathfrak{M}\times B^q(a) : \exists \xi\in \mathbb{R}^q \text{ with } b_j(x,s,\xi)\neq 0\} \subseteq \operatorname{supp}(\psi(x)\varsigma_0(s)) \Subset \Omega_1\times B^q(a).$$

Thus, Definition 4.1.1 (i) holds uniformly in $j\geq 0$.

All that remains to show is that Definition 4.1.1 (ii) holds uniformly in $j\geq 0$. Since $b_j(x,s,\xi) = \psi(x)\varsigma_0(s)2^{(j,\lambda j)\cdot \nu}\hat{\varsigma}(\xi)$, where $\hat{\varsigma}$ is the Fourier transform of ς, to show Definition 4.1.1 (ii) holds uniformly in $j\geq 0$, it suffices to show that, $\forall \alpha,\beta$,

$$\begin{aligned} &|\partial_{\xi_1}^{\alpha}\partial_{\xi_2}^{\beta} 2^{(j,\lambda j)\cdot \nu}\hat{\varsigma}(2^{-jd^1}\xi^1, 2^{-j\lambda d^2}\xi^2)| \\ &\quad \leq C_{\alpha,\beta}(1+\|\xi^1\|_1)^{\nu_1-\deg_{d^1}(\alpha)}(1+\|\xi^2\|_2)^{\nu_2-\deg_{d^2}(\beta)}, \end{aligned} \tag{6.124}$$

where $C_{\alpha,\beta}$ does not depend on $j\geq 0$.

To prove (6.124), we use the following elementary inequality. For $w_1\in\mathbb{R}$, $L\geq 0$ with $L\geq |w_1|$, we have

$$2^{j_1 w_1}(2^{-j_1}\|\xi^1\|_1)^{w_1\vee 0}(1+2^{-j_1}\|\xi^1\|_1)^{-L} \lesssim (1+\|\xi^1\|_1)^{w_1}, \tag{6.125}$$

where the implicit constant does not depend on $j_1\geq 0$ or ξ^1, but may depend on any of the other ingredients. Indeed, to see (6.125), we consider three cases. When $w_1\geq 0$, it is immediate that the left-hand side of (6.125) is $\lesssim \|\xi^1\|_1^{w_1}\lesssim (1+\|\xi^1\|_1)^{w_1}$, establishing (6.125) in this case. When $w_1<0$ and $2^{-j_1}\|\xi^1\|_1\leq 1$, we have, using the fact that $j\geq 0$,

$$\begin{aligned} 2^{j_1 w_1}(2^{-j_1}\|\xi^1\|_1)^{w_1\vee 0}(1+2^{-j_1}\|\xi^1\|_1)^{-L} &\lesssim 2^{j_1 w_1} = 2^{-j_1|w_1|} \\ &\lesssim \|\xi^1\|_1^{-|w_1|}\wedge 1 = \|\xi^1\|_1^{w_1}\wedge 1 \approx (1+\|\xi^1\|_1)^{w_1}, \end{aligned}$$

establishing (6.125) in this case. Finally, if $w_1<0$ and $2^{-j_1}\|\xi^1\|_1>1$, we have, using the fact that $L\geq |w_1|$,

$$\begin{aligned} 2^{jw_1}(2^{-j_1}\|\xi^1\|_1)^{w_1\vee 0}(1+2^{-j_1}\|\xi^1\|_1)^{-L} &\lesssim 2^{j_1 w_1}2^{Lj_1}\|\xi^1\|_1^{-L} \\ &= \|\xi^1\|_1^{w_1}(\|\xi^1\|_1^{-w_1}2^{-j_1 w_1})(2^{Lj_1}\|\xi^1\|_1^{-L}) \\ &\leq \|\xi^1\|_1^{w_1} \lesssim (1+\|\xi^1\|_1^{w_1}), \end{aligned}$$

where the final estimate uses the fact that $\|\xi_1\|_1\geq 2^{-j_1}\geq 1$ in this case. This completes the proof of (6.125).

By (6.122), by taking $N\in\mathbb{N}$ large, $\hat{\varsigma}(\xi^1,\xi^2)$ vanishes to as high an order as we like at $\xi^1=0$ and at $\xi^2=0$. Since $\hat{\varsigma}\in\mathscr{S}(\mathbb{R}^q)$ as well, we see that by taking $N=N(\nu,\vec{d})\in\mathbb{N}$ large, we have, $\forall\alpha,\beta,L$,

$$|\partial_{\xi^1}^{\alpha}\partial_{\xi^2}^{\beta}\hat{\varsigma}(\xi^1,\xi_2)| \lesssim \|\xi^1\|_1^{(\nu_1-\deg_{d^1}(\alpha))\vee 0}\|\xi^2\|_2^{(\nu_2-\deg_{d^2}(\beta))\vee 0}(1+\|\xi^1\|_1)^{-L}(1+\|\xi^2\|_2)^{-L}. \tag{6.126}$$

Thus, using (4.6), we have, for every α, β, L,

$$\begin{aligned}
&|\partial_{\xi_1}^{\alpha}\partial_{\xi_2}^{\beta} 2^{(j,\lambda j)\cdot v}\hat{\varsigma}(2^{-jd^1}\xi^1, 2^{-j\lambda d^2}\xi^2)| \\
&\lesssim 2^{j(v_1-\deg_{d^1}(\alpha))}2^{\lambda j(v_2-\deg_{d^2}(\beta))}\|2^{-jd^1}\xi^1\|_1^{(v_1-\deg_{d^1}(\alpha))\vee 0} \\
&\quad\times \|2^{-\lambda j d^2}\xi^2\|_2^{(v_2-\deg_{d^2}(\beta))\vee 0}(1+\|2^{-jd^1}\xi^1\|_1)^{-L}(1+\|2^{-\lambda j d^2}\xi^2\|_2)^{-L} \\
&= 2^{j(v_1-\deg_{d^1}(\alpha))}(2^{-j}\|\xi^1\|_1)^{(v_1-\deg_{d^1}(\alpha))\vee 0}(1+2^{-j}\|\xi^1\|_1)^{-L} \\
&\quad\times 2^{\lambda j(v_2-\deg_{d^2}(\beta))}(2^{-\lambda j}\|\xi^2\|_2)^{(v_2-\deg_{d^2}(\beta))\vee 0}(1+2^{-\lambda j}\|\xi^2\|_2)^{-L} \\
&\lesssim (1+\|\xi^1\|_1)^{v_1-\deg_{d^1}(\alpha)}(1+\|\xi^2\|_2)^{v_2-\deg_{d^2}(\beta)},
\end{aligned}$$

where the final estimate follows by taking $L = L(v_1, v_2, \alpha, \beta)$ large and applying (6.125) twice (once with $j_1 = j$ and $\|\xi^1\|_1 = \|\xi^1\|_1$ and once with $j_1 = \lambda j$ and $\|\xi^1\|_1$ replaced by $\|\xi^2\|_2$). This establishes (6.124) and completes the proof. □

Lemma 6.7.11. *If $a > 0$ is sufficiently small, the following holds. For $v \in \mathbb{R}^2$, if $N = N(v) \in \mathbb{N}$ is sufficiently large, then we have, $\forall w \in \mathbb{R}^2$,*

$$\|2^{v\cdot(j,\lambda j)}T_j f\|_{\mathscr{X}_0^w(W,\vec{d})} \lesssim \|f\|_{\mathscr{X}_0^{w+v}(W,\vec{d})}, \quad \forall f \in C_0^\infty(\Omega), \tag{6.127}$$

where the implicit constant depends on neither $j \geq 0$ nor f.

Proof. By Lemma 6.7.10, $2^{v\cdot(j,\lambda j)}T_j$ is an $(X, \vec{d}), a$ pseudo-differential operator of order v supported in $\Omega_1 \times \Omega_2$, uniformly for $j \geq 0$. Thus, by Proposition 5.8.13, $2^{v\cdot(j,\lambda j)}T_j \in \widetilde{\mathscr{A}}^v(\Omega, (W, \vec{d}))$, uniformly in $j \geq 0$, and therefore by Proposition 6.3.10,

$$\|2^{v\cdot(j,\lambda j)}T_j\|_{\mathscr{X}_0^{w+v}(\overline{\Omega},(W,\vec{d}))\to\mathscr{X}_0^{w}(\overline{\Omega},(W,\vec{d}))} \lesssim 1. \tag{6.128}$$

By Proposition 6.5.5, $C_0^\infty(\Omega) \subseteq C_{W,0}^\infty(\Omega) \subseteq \mathscr{X}_0^{w+v}(\overline{\Omega}, (W, \vec{d}))$. Thus, (6.127) follows from (6.128), completing the proof. □

Proof of Lemma 6.7.8. The assumption of the lemma is that (6.121) holds for some $\epsilon_1, \epsilon_2 \geq 0$ with $\epsilon_1 + \epsilon_2 > 0$. We first claim that we may assume $\epsilon_1, \epsilon_2 > 0$. Indeed, if $\epsilon_1 = 0$ and $\epsilon_2 > 0$, then by (6.119) (which we have already seen as a consequence of Theorem 6.7.2), (6.121) also holds with $\tilde{s}$ replaced by $\tilde{s} + \epsilon'$, ϵ_1 replaced by $\epsilon_1 + \lambda\epsilon'$, and ϵ_2 replaced by $\epsilon_2 - \epsilon'$. Taking $\epsilon' \in (0, \epsilon_2)$ shows that we may assume $\epsilon_1, \epsilon_2 > 0$. Similarly, if $\epsilon_2 = 0$ and $\epsilon_1 > 0$, then by (6.119), (6.121) also holds with $\tilde{s}$ replaced by $\tilde{s} - \epsilon'$, ϵ_1 replaced by $\epsilon_1 - \lambda\epsilon'$, and ϵ_2 replaced by $\epsilon_2 + \epsilon'$. Thus, taking $\epsilon' > 0$ so small that $\epsilon_1 - \lambda\epsilon' > 0$ and $\tilde{s} - \epsilon' > 0$ shows that we may assume $\epsilon_1, \epsilon_2 > 0$. Set $\epsilon_1' := \epsilon_1/2 > 0$ and $\epsilon_2' := \epsilon_2/2 > 0$.

We henceforth assume (6.121) holds for some $\tilde{s} > 0$ and $\epsilon_1, \epsilon_2 > 0$. Fix $\psi_0 \in C_0^\infty(\mathfrak{M})$ with $\psi_0 \equiv 1$ on a neighborhood of $\overline{\Omega}$. Let $\{(D_j, 2^{-j}) : j \in \mathbb{N}^2\}$ be a bounded set of $(W, \vec{d})$ elementary operators satisfying $\sum_{j\in\mathbb{N}^2} D_j = \mathrm{Mult}[\psi_0]$.

We let $\mathscr{V}_0$ be to $\mathscr{X}_0^s$ as $\mathscr{V}$ is to $\mathscr{X}^s$ (see Notation 6.3.2), that is, if $\mathscr{X}_0^s = \mathscr{B}_{p,q}^s$, then $\mathscr{V}_0$ is $\ell^q(\mathbb{N}^2; L^p(\mathfrak{M}, \mathrm{Vol}))$, and if $\mathscr{X}_0^s = \mathscr{F}_{p,q}^s$, then $\mathscr{V}_0$ is $L^q(\mathfrak{M}, \mathrm{Vol}; \ell^p(\mathbb{N}^2))$. By a proof very similar to Lemma 6.6.20, we have the following elementary inequalities:

$$\|\{f_j\}_{j\in\mathbb{N}^2}\|_{L^p(\mathfrak{M},\mathrm{Vol};\ell^1(\mathbb{N}^2))} \lesssim \|\{2^{(\epsilon_1',\epsilon_2')\cdot j} f_j\}\|_{\mathscr{V}_0}, \tag{6.129}$$

$$\|\{f_j\}_{j\in\mathbb{N}^2}\|_{\mathscr{V}_0} \leq \|\{f_j\}_{j\in\mathbb{N}^2}\|_{\ell^1(\mathbb{N}^2;L^p(\mathfrak{M},\mathrm{Vol}))}. \tag{6.130}$$

By the formula for T_j, if $a > 0$ is sufficiently small, we have $T_j : C_0^\infty(\mathfrak{M}) \to C_0^\infty(\Omega_1)$. In particular, for $f \in C_0^\infty(\mathfrak{M})$, by Proposition 5.5.10, $T_j T_j f = \mathrm{Mult}[\psi_0] T_j T_j f = \sum_{k\in\mathbb{N}} D_k T_j T_j f$, with convergence in $C_0^\infty(\mathfrak{M})$.

By taking $N = N(s, \tilde{s}, \lambda, \epsilon_1, \epsilon_2) \in \mathbb{N}$ sufficiently large and applying Lemma 6.7.11 and (6.121), we have, for $f \in C_0^\infty(\Omega_2)$,

$$\begin{aligned}
\|T_j T_j f\|_{L^p} &= \Big\|\sum_{k\in\mathbb{N}} D_k T_j T_j f\Big\|_{L^p} \\
&\lesssim \|\{2^{k\cdot(\epsilon_1',\epsilon_2')} D_k T_j T_j f\}\|_{\mathscr{V}_0} && \text{by (6.129)} \\
&\approx \|T_j T_j f\|_{\mathscr{X}_0^{(\epsilon_1',\epsilon_2')}(W,\vec{d})} \\
&= 2^{-(s+(\epsilon_1',\epsilon_2'))\cdot(j,\lambda j)} \|2^{(s+(\epsilon_1',\epsilon_2')+(-\lambda(\tilde{s}),\tilde{s}))\cdot(j,\lambda j)} T_j T_j f\|_{\mathscr{X}_0^{(\epsilon_1',\epsilon_2')}(W,\vec{d})} \\
&\lesssim 2^{-(s+(\epsilon_1',\epsilon_2'))\cdot(j,\lambda j)} \|T_j f\|_{\mathscr{X}_0^{s+(\epsilon_1,\epsilon_2)+(-\lambda(\tilde{s}),\tilde{s})}(W,\vec{d})} && \text{by Lemma 6.7.11} \\
&\lesssim 2^{-(s+(\epsilon_1',\epsilon_2'))\cdot(j,\lambda j)} \|T_j f\|_{\mathscr{X}_0^{s}(W,\vec{d})} && \text{by (6.121)} \\
&\lesssim 2^{-(\epsilon_1'/2,\epsilon_2'/2)\cdot(j,\lambda j)} \|f\|_{\mathscr{X}^{(-\epsilon_1'/2,-\epsilon_2'/2)}(W,\vec{d})} && \text{by Lemma 6.7.11} \\
&\approx 2^{-(\epsilon_1'/2,\epsilon_2'/2)\cdot(j,\lambda j)} \|\{2^{-(\epsilon_1'/2,\epsilon_2'/2)\cdot k} D_k f\}_{k\in\mathbb{N}^2}\|_{\mathscr{V}_0} \\
&\leq 2^{-(\epsilon_1'/2,\epsilon_2'/2)\cdot(j,\lambda j)} \sum_{k\in\mathbb{N}^2} 2^{-(\epsilon_1'/2,\epsilon_2'/2)\cdot k} \|D_k f\|_{L^p} && \text{by (6.130)} \\
&\lesssim 2^{-(\epsilon_1'/2,\epsilon_2'/2)\cdot(j,\lambda j)} \sum_{k\in\mathbb{N}^2} 2^{-(\epsilon_1'/2,\epsilon_2'/2)\cdot k} \|f\|_{L^p} && \text{by Corollary 5.4.11} \\
&\lesssim 2^{-(\epsilon_1'/2,\epsilon_2'/2)\cdot(j,\lambda j)} \|f\|_{L^p}.
\end{aligned}$$

Since T_j is an $(X, \vec{d})$, a elementary operator supported in $\Omega_1 \times \Omega_2$ and the above equation holds for all $f \in C_0^\infty(\Omega_2)$, we conclude that $\|T_j T_j\|_{L^p\to L^p} \lesssim 2^{-(\epsilon_1'/2,\epsilon_2'/2)\cdot(j,\lambda j)}$. The result follows. □

We now turn to the proof of Lemma 6.7.9, where the sharpness of $\lambda = \lambda(x, (Y, \vec{d}), (Z, \vec{d}))$ is essential. Recall that we have chosen $(X_1^2, d_1^2) = (V, \vec{d})$, where $(V, \vec{d}) \in \mathrm{Gen}((Z, \vec{d}))$ is as in Definition 6.7.4 with $U = \Omega_0$. Thus, there is a point $x_0 \in \Omega_0$ such that

$$X_1^2(x_0) \notin \mathrm{span}\{X_k^1(x_0) : d_k^1 < \lambda d_1^2\}. \tag{6.131}$$

For $\delta \in (0,1]$, let $\Phi_{x,\delta} : B^n(1) \to \Omega$ be the map from Theorem 3.3.7 when applied with $\mathcal{K} = \{x_0\}$, $\Omega_1 = \Omega_1$, $(W, \vec{d}) = (Y, \vec{d})$, and $(X, d) = (X^1, d^1)$.

Lemma 6.7.12. *For every $L \in \mathbb{N}$, we have*

$$\sup_{\delta\in(0,1]} \|\Phi^*_{x_0,\delta}\delta^{d^1_j} X^1_j\|_{C^L(B^n(1);\mathbb{R}^n)} < \infty, \quad 1 \le j \le q_1, \tag{6.132}$$

$$\sup_{\delta\in(0,1]} \|\Phi^*_{x_0,\delta}\delta^{\lambda d^2_j} X^2_j\|_{C^L(B^n(1);\mathbb{R}^n)} < \infty, \quad 1 \le j \le q_2. \tag{6.133}$$

Furthermore, there exists a sequence $\delta_l \downarrow 0$ such that

$$\Phi^*_{x_0,\delta_l}\delta_l^{d^1_j} X^1_j \xrightarrow{l\to\infty} Y^1_j \in C^\infty(B^n(1);\mathbb{R}^n), \quad 1 \le j \le q_1, \tag{6.134}$$

$$\Phi^*_{x_0,\delta_l}\delta_l^{\lambda d^2_j} X^2_j \xrightarrow{l\to\infty} Y^2_j \in C^\infty(B^n(1);\mathbb{R}^n), \quad 1 \le j \le q_2, \tag{6.135}$$

with convergence in $C^\infty(B^n(1);\mathbb{R}^n)$. Finally, we have $Y^2_1(0) \neq 0$, and by possibly reordering (X^1, d^1), we also have $Y^1_1(0) \neq 0$.

Proof. The estimate (6.132) is Theorem 3.3.7 (j).

Next, we establish (6.133). By the choice of Ω, $(Y, \vec{d})$ weakly λ-controls $(Z, \vec{d})$ on Ω. Thus, Lemma 3.14.9 (v) and (vi) implies (X^1, d^1) strongly λ-controls (X^2, d^2) on Ω. In other words, for $1 \le j \le q_2$, we may write

$$X^2_j = \sum_{d^1_k \le \lambda d^2_j} c^k_j X^1_k, \quad c^k_j \in C^\infty_{\mathrm{loc}}(\Omega).$$

Thus, for $\delta \in (0,1]$ and $1 \le j \le q_2$, we have

$$\Phi^*_{x_0,\delta}\delta^{\lambda d^2_j} X^2_j = \sum_{d^1_k \le \lambda d^2_j} \delta^{\lambda d^2_j - d^1_j}(c^k_j \circ \Phi_{x_0,\delta})\Phi^*_{x_0,\delta}\delta^{d^1_k} X^1_k. \tag{6.136}$$

In light of (6.132) and (6.136), to prove (6.133) it suffices to show that for every $L \in \mathbb{N}$,

$$\sup_{\delta\in(0,1]} \|c^k_j \circ \Phi_{x_0,\delta}\|_{C^L(B^n(1);\mathbb{R}^n)} < \infty, \quad 1 \le j \le q_2, 1 \le k \le q_1. \tag{6.137}$$

By Theorem 3.3.7 (m), (6.137) is equivalent to

$$\sup_{\delta\in(0,1]} \sum_{|\alpha|\le L} \|(\Phi^*_{x_0,\delta}\delta^{d_1} X^1)^\alpha c^k_j \circ \Phi_{x_0,\delta}\|_{C(B^n(1);\mathbb{R}^n)} < \infty. \tag{6.138}$$

But $(\Phi^*_{x_0,\delta}\delta^{d_1} X^1)^\alpha c^k_j \circ \Phi_{x_0,\delta} = ((\delta^{d_1} X^1)^\alpha c^k_j) \circ \Phi_{x_0,\delta}$. Thus, (6.138) is equivalent to

$$\sup_{\delta\in(0,1]} \sum_{|\alpha|\le L} \|(\delta^{d^1} X^1)^\alpha c^k_j\|_{C(\Phi_{x_0,\delta}(B^n(1));\mathbb{R}^n)} < \infty. \tag{6.139}$$

Theorem 3.3.7 shows that $\Phi_{x_0,\delta}(B^n(1)) \subset \Omega_1$, and therefore, using $\delta \in (0,1]$,

$$\sup_{\delta\in(0,1]} \sum_{|\alpha|\leq L} \|(\delta^{d^1}X^1)^\alpha c_j^k\|_{C(\Phi_{x_0,\delta}(B^n(1));\mathbb{R}^n)} \leq \sum_{|\alpha|\leq L} \|(X^1)^\alpha c_j^k\|_{C(\Omega_1;\mathbb{R}^n)} < \infty,$$

where the $< \infty$ follows from the fact that $c_j^k \in C^\infty_{\mathrm{loc}}(\Omega)$ and $\Omega_1 \Subset \Omega$. This establishes (6.139) and completes the proof of (6.133).

By (6.132) and (6.133),

$$\{\Phi^*_{x_0,\delta}\delta^{d_k^1}X_k^1, \Phi^*_{x_0,\delta}\delta^{\lambda d_j^2}X_j^2 : \delta \in (0,1], 1 \leq k \leq q_1, 1 \leq j \leq q_2\} \subset C^\infty(B^n(1);\mathbb{R}^n)$$

is a bounded set and therefore relatively compact. Thus, there exists a sequence $\delta_l \downarrow 0$ such that (6.134) and (6.135) hold.

We claim $Y_1^2(0) \neq 0$. For contradiction, suppose $Y_1^2(0) = 0$. Thus, $\Phi^*_{x_0,\delta}\delta_l^{\lambda d_1^2}X_1^2(0) \xrightarrow{l\to\infty} 0$. By Theorem 3.3.7 (k) and (j), for every $\epsilon > 0$, there exists $L \in \mathbb{N}$ such that $l \geq L$ implies

$$(\Phi^*_{x_0,\delta_l}\delta_l^{\lambda d_1^2}X_1^2)(0) = \sum_{k=1}^{q_1} a_l^{\epsilon,k}(\Phi^*_{x_0,\delta}\delta_l^{d_k^1}X_k^1)(0), \quad |a_l^{k,\epsilon}| < \epsilon. \tag{6.140}$$

Since $\Phi_{x_0,\delta_l}(0) = x_0$, (6.140) is equivalent to

$$\delta_l^{\lambda d_1^2}X_1^2(x_0) = \sum_{k=1}^{q_1} a_l^{\epsilon,k}\delta_l^{d_k^1}X_k^1(x_0).$$

We therefore have

$$X_1^2(x_0) - \sum_{\lambda d_1^2\leq d_k^1} a_l^{\epsilon,k}\delta_l^{d_k^1-\lambda d_1^2}X_k^1(x_0) \in \operatorname{span}\{X_k(x_0) : d_k^1 < \lambda d_1^2\}.$$

Taking $\epsilon \downarrow 0$ and $l \to \infty$ and using the fact that $|a_l^{\epsilon,k}| < \epsilon$ and that finite-dimensional vector spaces are closed, we see that

$$X_1^2(x_0) \in \operatorname{span}\{X_k^1(x_0) : d_k^1 < \lambda d_1^2\}.$$

This contradicts (6.131) and completes the proof that $Y_1^2(0) \neq 0$.

Finally, we claim that, by possibly reordering (X^1, d^1), we have $Y_1^1(0) \neq 0$. Indeed, by taking $\lim_{l\to\infty}$ of Theorem 3.3.7 (k), we see

$$\max_{k_1,\ldots,k_n\in\{1,\ldots,q_1\}} \left|\det(Y_{k_1}^1(0)|\cdots|Y_{k_n}^1(0))\right| > 0.$$

In particular, $Y_k^1(0) \neq 0$ for some $k \in \{1,\ldots,q_1\}$. Thus, by possibly reordering (X^1, d^1) to switch (X_k^1, d_k^1) and (X_1^1, d_1^1), we have $Y_1^1(0) \neq 0$. □

For the rest of this section, we take $\delta_l \downarrow 0$ and $Y^1 = (Y_1^1, \dots, Y_{q_1}^1)$ and $Y^2 = (Y_1^2, \dots, Y_{q_2}^2)$ as in Lemma 6.7.12. In particular, $Y_1^1(0) \neq 0$ and $Y_1^2(0) \neq 0$.

Given $a > 0$ small and $N \in \mathbb{N}$, pick $f_0 \in C_0^\infty(B^n(1/2))$ and $\varsigma \in C_0^\infty(B^q(a))$ satisfying (6.122) such that

$$\int f_0(e^{-t^1\cdot Y^1} e^{-t^2\cdot Y^2} e^{-t^3\cdot Y^1} e^{-t^4\cdot Y^2} 0)\varsigma(t^1,t^2)\varsigma(t^3,t^4)\, dt^2\, dt^2\, dt^3\, dt^4 \neq 0.$$

Since $Y_1^1(0) \neq 0$ and $Y_1^2(0) \neq 0$, this is always possible. Note that by continuity, we have, for u in some open neighborhood of $0 \in \mathbb{R}^n$,

$$\int f_0(e^{-t^1\cdot Y^1} e^{-t^2\cdot Y^2} e^{-t^3\cdot Y^1} e^{-t^4\cdot Y^2} u)\varsigma(t^1,t^2)\varsigma(t^3,t^4)\, dt^2\, dt^2\, dt^3\, dt^4 \neq 0. \tag{6.141}$$

Take $\psi \in C_0^\infty(\Omega_1)$ with $\psi \equiv 1$ on a neighborhood of x_0. Define T_j as in (6.123) with this choice of ψ and ς. We will show that $\limsup_{j\to\infty} \|T_jT_j\|_{L^p\to L^p} > 0$, which will prove Lemma 6.7.9, since $a > 0$ and $N \in \mathbb{N}$ were arbitrary.

Define $j_l \uparrow \infty$ by $2^{-j_l} = \delta_l$. Set

$$F_l := \Phi_{x_0,\delta_l}^* T_{j_l} T_{j_l} (\Phi_{x_0,\delta_l})_*.$$

Lemma 6.7.13. *For every $l \in \mathbb{N}$,*

$$\|F_l\|_{L^p(B^n(1),\sigma_{\mathrm{Leb}})\to L^p(B^n(1),\sigma_{\mathrm{Leb}})} \lesssim \|T_{j_l}T_{j_l}\|_{L^p(\mathfrak{M},\mathrm{Vol})\to L^p(\mathfrak{M},\mathrm{Vol})},$$

where σ_{Leb} denotes the usual Lebesgue density on $\mathbb{R}^n$.

Proof. Write $S_l := T_{j_l}T_{j_l}$. Let $h_{x_0,\delta_l} \in C^\infty(B^n(1))$ be as in Theorem 3.3.7 (n), so that $\Phi_{x_0,\delta_l}^* \mathrm{Vol} = \Lambda(x_0,\delta_l) h_{x_0,\delta_l}\sigma_{\mathrm{Leb}}$, where $\Lambda(x_0,\delta_l)$ is as in Theorem 3.3.7 (n). In particular, by Theorem 3.3.7 (n), we have $h_{x_0,\delta} \approx 1$. Note that for $g \in C_0^\infty(B^n(1))$, we have $\mathrm{supp}((\Phi_{x_0,\delta})_* g) \subseteq \Phi_{x_0,\delta_l}(B^n(1))$. Thus, we have, for $g \in C_0^\infty(B^n(1))$,

$$\begin{aligned}
\|F_l g\|_{L^p(B^n(1),\sigma_{\mathrm{Leb}})} &\approx \|F_l g\|_{L^p(B^n(1),h_{x_0,\delta}\sigma_{\mathrm{Leb}})} \\
&= \Lambda(x_0,\delta_l)^{-1/p} \|F_l g\|_{L^p(B^n(1),\Phi_{x_0,\delta_l}^* \mathrm{Vol})} \\
&= \Lambda(x_0,\delta_l)^{-1/p} \|\Phi_{x_0,\delta_l}^* S_l (\Phi_{x_0,\delta_l})_* g\|_{L^p(B^n(1),\Phi_{x_0,\delta_l}^* \mathrm{Vol})} \\
&= \Lambda(x_0,\delta_l)^{-1/p} \|S_l (\Phi_{x_0,\delta_l})_* g\|_{L^p(\Phi_{x_0,\delta}(B^n(1)),\mathrm{Vol})} \\
&\leq \Lambda(x_0,\delta_l)^{-1/p} \|S_l\|_{L^p(\mathfrak{M},\mathrm{Vol})\to L^p(\mathfrak{M},\mathrm{Vol})} \|(\Phi_{x_0,\delta_l})_* g\|_{L^p(\mathfrak{M},\mathrm{Vol})} \\
&= \Lambda(x_0,\delta_l)^{-1/p} \|S_l\|_{L^p(\mathfrak{M},\mathrm{Vol})\to L^p(\mathfrak{M},\mathrm{Vol})} \|(\Phi_{x_0,\delta_l})_* g\|_{L^p(\Phi_{x_0,\delta}(B^n(1)),\mathrm{Vol})} \\
&= \Lambda(x_0,\delta_l)^{-1/p} \|S_l\|_{L^p(\mathfrak{M},\mathrm{Vol})\to L^p(\mathfrak{M},\mathrm{Vol})} \|g\|_{L^p(B^n(1),\Lambda(x_0,\delta_l)h_{x_0,\delta_l}\sigma_{\mathrm{Leb}})} \\
&= \|S_l\|_{L^p(\mathfrak{M},\mathrm{Vol})\to L^p(\mathfrak{M},\mathrm{Vol})} \|g\|_{L^p(B^n(1),h_{x_0,\delta_l}\sigma_{\mathrm{Leb}})} \\
&\approx \|S_l\|_{L^p(\mathfrak{M},\mathrm{Vol})\to L^p(\mathfrak{M},\mathrm{Vol})} \|g\|_{L^p(B^n(1),\sigma_{\mathrm{Leb}})}.
\end{aligned}$$

The result follows. □

Lemma 6.7.14. *The limit* $\lim_{l\to\infty} F_l f_0(u)$ *exists (pointwise) for all* $u \in B^n(1)$ *and is non-zero for* u *in a neighborhood of* 0. *Here,* $f_0 \in C_0^\infty(B^n(1/2))$ *is the particular function chosen so that* (6.141) *holds.*

Proof. Consider, for $u \in B^n(1)$,

$$\begin{aligned} F_l f_0(u) = \int & f_0 \circ \Phi_{x_0,\delta_l^*}(e^{-t^1\cdot X^1} e^{-t^2\cdot X^2} e^{-t^3\cdot X^1} e^{-t^4\cdot X^2} \Phi_{x_0,\delta_l}(u)) \\ & \psi(\Phi_{x_0,\delta_l}(u))\psi(e^{-t^3\cdot X^1} e^{-t^4\cdot X^2} \Phi_{x_0,\delta_l}(u)) \operatorname{Dil}^{\vec{d}}_{2^{(j_l,\lambda j_l)}}(\varsigma)(t^1,t^2) \\ & \operatorname{Dil}^{\vec{d}}_{2^{(j_l,\lambda j_l)}}(\varsigma)(t^3,t^4)\, dt^1\, dt^2\, dt^3\, dt^4. \end{aligned} \tag{6.142}$$

By Theorem 3.3.7, $\Phi_{x_0,\delta_l}(B^n(1)) \subseteq B_{(X,d)}(x_0,\delta_l)$. Therefore, for any neighborhood V of x_0, since $\delta_l \downarrow 0$, if l is sufficiently large, $\Phi_{x_0,\delta_l}(B^n(1)) \subseteq V$; this uses Lemma 3.1.7. Since $\psi \equiv 1$ on a neighborhood of x_0, this shows that for l sufficiently large, $\psi(\Phi_{x_0,\delta_l}(u)) \equiv 1$. Similarly, since $j_l \uparrow \infty$, for l large the integral in (6.142) is only over t^3 and t^4 small. Thus, for l sufficiently large, we also have $\psi(e^{-t^3\cdot X^1} e^{-t^4\cdot X^2} \Phi_{x_0,\delta_l}(u)) \equiv 1$ on the domain of integration of (6.142).

Thus, for l large, we have (using $\delta_l = 2^{-j_l}$)

$$\begin{aligned} F_l f_0(u) &= \int f_0 \circ \Phi^{-1}_{x_0,\delta}(e^{-t^1\cdot X^1} e^{-t^2\cdot X^2} e^{-t^3\cdot X^1} e^{-t^4\cdot X^2} \Phi_{x_0,\delta_l}(u)) \\ &\qquad \operatorname{Dil}^{\vec{d}}_{2^{(j_l,\lambda j_l)}}(\varsigma)(t^1,t^2) \operatorname{Dil}^{\vec{d}}_{2^{(j_l,\lambda j_l)}}(\varsigma)(t^3,t^4)\, dt^1\, dt^2\, dt^3\, dt^4 \\ &= \int f_0 \circ \Phi^{-1}_{x_0,\delta}(e^{-t^1\cdot 2^{-d^1 j_l} X^1} e^{-t^2\cdot 2^{-\lambda d^2 j_l} X^2} e^{-t^3\cdot 2^{-d^1 j_l} X^1} e^{-t^4\cdot 2^{-\lambda d^2 j_l} X^2} \Phi_{x_0,\delta_l}(u)) \\ &\qquad \varsigma(t^1,t^2)\varsigma(t^3,t^4)\, dt^1\, dt^2\, dt^3\, dt^4 \\ &= \int f_0(e^{-t^1\cdot \Phi^*_{x,\delta_l}\delta_l^{d^1} X^1} e^{-t^2\cdot \Phi^*_{x,\delta_l}\delta_l^{\lambda d^2} X^2} e^{-t^3\cdot \Phi^*_{x,\delta_l}\delta_l^{d^1} X^1} e^{-t^4\cdot \Phi^*_{x,\delta_l}\delta_l^{\lambda d^2} X^2} u) \\ &\qquad \varsigma(t^1,t^2)\varsigma(t^3,t^4)\, dt^1\, dt^2\, dt^3\, dt^4 \\ &\xrightarrow{l\to\infty} \int f_0(e^{-t^1\cdot Y^1} e^{-t^2\cdot Y^2} e^{-t^3\cdot Y^1} e^{-t^4\cdot Y^2} u)\varsigma(t^1,t^2)\varsigma(t^3,t^4)\, dt^1\, dt^2\, dt^3\, dt^4, \end{aligned}$$

where the last line follows from Lemma 6.7.12. Thus the limit exists pointwise, and by (6.141) is non-zero for u in a neighborhood of 0. □

Proof of Lemma 6.7.9. Suppose, for contradiction, that $\lim_{j\to\infty} \|T_j T_j\|_{L^p(\mathfrak{M},\mathrm{Vol})\to L^p(\mathfrak{M},\mathrm{Vol})} = 0$. Then by Lemma 6.7.13, $\lim_{l\to\infty} \|F_l\|_{L^p(B^n(1),\sigma_{\mathrm{Leb}})\to L^p(B^n(1),\sigma_{\mathrm{Leb}})} = 0$. In particular, $\lim_{l\to\infty} \|F_l f_0\|_{L^p(B^n(1),\sigma_{\mathrm{Leb}})} = 0$. Thus, there is a subsequence $l_k \to \infty$ such that $F_{l_k} f_0 \xrightarrow{k\to\infty} 0$, pointwise almost everywhere. This contradicts Lemma 6.7.14 and completes the proof. □

6.8 Adding parameters

Suppose $\nu \geq 2$. Set

$$\begin{aligned}(Y^\mu, \vec{d}^\mu) &:= (W^\mu, \vec{d}^\mu), \quad 1 \leq \mu \leq \nu - 1,\\ (Y, \vec{d}) = \{(Y_1, \vec{d}_1), \ldots, (Y_{\nu_Y}, \vec{d}_{\nu_Y})\} &:= (Y^1, \vec{d}^1) \boxtimes (Y^2, \vec{d}^2) \boxtimes \cdots (Y^{\nu-1}, \vec{d}^{\nu-1})\\ &\subset C^\infty_{\mathrm{loc}}(\mathfrak{M}; T\mathfrak{M}) \times (\mathbb{N}^{\nu-1} \setminus \{0\}).\end{aligned}$$

$(Y, \vec{d})$ satisfies all the same assumptions as $(W, \vec{d})$, but with ν replaced by $\nu-1$. For $s \in \mathbb{R}^\nu$, we have two different scales of spaces: $\mathscr{X}^s(\mathcal{K}, (Y, \vec{d}))$ and $\mathscr{X}^{(s,0)}(\mathcal{K}, (W, \vec{d}))$. These spaces are similar, but unless $\mathscr{X}^\bullet = \mathscr{F}^\bullet_{p,2}$, they are not the same (for a discussion of the case $\mathscr{X}^\bullet = \mathscr{F}^\bullet_{p,2}$, see Proposition 6.8.2).

Fix a compact set $\mathcal{K} \Subset \mathfrak{M}$ and $s \in \mathbb{R}^{\nu-1}$. The main results of this section are the following.

Proposition 6.8.1. *For every $\epsilon > 0$, we have*

$$\mathscr{X}^{(s,\epsilon)}(\mathcal{K}, (W, \vec{d})) \subseteq \mathscr{X}^s(\mathcal{K}, (Y, \vec{d})) \subseteq \mathscr{X}^{(s,-\epsilon)}(\mathcal{K}, (W, \vec{d})),$$

and the inclusions are continuous.

Proposition 6.8.2. *For $1 < p < \infty$,*

$$\mathscr{F}^{(s,0)}_{p,2}(\mathcal{K}, (W, \vec{d})) = \mathscr{F}^s_{p,2}(\mathcal{K}, (Y, \vec{d})),$$

with equality of topologies.

We begin with the proof of Proposition 6.8.1, for which we need to introduce some notation and preliminary lemmas.

Fix $\psi', \psi \in C^\infty_0(\mathfrak{M})$, with $\psi', \psi \equiv 1$ on a neighborhood of $\mathcal{K}$. Let $\mathcal{D}'_0 = \{(D'_{j'}, 2^{-j'}) : j' \in \mathbb{N}^{\nu-1}\}$ be a bounded set of generalized $(Y, \vec{d})$ elementary operators with $\sum_{j' \in \mathbb{N}^{\nu-1}} D'_{j'} = \mathrm{Mult}[\psi']$, and let $\mathcal{D}_0 = \{(D_j, 2^{-j}) : j \in \mathbb{N}^\nu\}$ be a bounded set of generalized $(W, \vec{d})$ elementary operators with $\sum_{j \in \mathbb{N}^\nu} D_j = \mathrm{Mult}[\psi]$. Given $j \in \mathbb{N}^\nu$, we write $j = (j', j_\nu) \in \mathbb{N}^{\nu-1} \times \mathbb{N}$. As in Notation 6.3.2, if $\mathscr{X}^\bullet = \mathscr{B}^\bullet_{p,q}$, we set $\mathscr{V} = \ell^q(\mathbb{N}^\nu; L^p(\mathfrak{M}, \mathrm{Vol}))$ and $\mathscr{V}' = \ell^q(\mathbb{N}^{\nu-1}; L^p(\mathfrak{M}, \mathrm{Vol}))$. If $\mathscr{X}^\bullet = \mathscr{F}^\bullet_{p,q}$, we set $\mathscr{V} = L^p(\mathfrak{M}, \mathrm{Vol}; \ell^q(\mathbb{N}^\nu))$ and $\mathscr{V}' = L^p(\mathfrak{M}, \mathrm{Vol}; \ell^q(\mathbb{N}^{\nu-1}))$. Throughout this section, we fix $p, q \in [1, \infty]$ as above.

Lemma 6.8.3. *Let $\epsilon > 0$.*

(I) *For $\{f_j\}_{j \in \mathbb{N}^\nu} \in \mathscr{V}$,*

$$\sum_{j_\nu \in \mathbb{N}} 2^{-\epsilon j_\nu} \big\| \{f_j\}_{j' \in \mathbb{N}^{\nu-1}} \big\|_{\mathscr{V}'} \lesssim_\epsilon \big\| \{f_j\}_{j \in \mathbb{N}^\nu} \big\|_{\mathscr{V}}.$$

(II) *For* $\{f_{j'}\}_{j'\in\mathbb{N}^{\nu-1}} \in \mathscr{V}'$ *and* $\epsilon > 0$,

$$\|\{2^{-\epsilon j_\nu} f_{j'}\}_{j\in\mathbb{N}^\nu}\|_{\mathscr{V}} \lesssim_\epsilon \|\{f_{j'}\}_{j'\in\mathbb{N}^{\nu-1}}\|_{\mathscr{V}'}.$$

Proof. (I): When $\mathscr{V} = \ell^q(L^p)$, (I) follows immediately. When $\mathscr{V} = L^p(\ell^q)$, we have

$$\begin{aligned}\sum_{j\in\mathbb{N}^\nu} 2^{-\epsilon j_\nu} \|\{f_j\}_{j'\in\mathbb{N}^{\nu-1}}\|_{\mathscr{V}'} &\lesssim \sup_{j_\nu\in\mathbb{N}} \|\{f_j\}_{j'\in\mathbb{N}^{\nu-1}}\|_{\mathscr{V}'} \\ &\le \left\|\left\{\sup_{j_\nu\in\mathbb{N}} |f_j|\right\}_{j'\in\mathbb{N}^{\nu-1}}\right\|_{\mathscr{V}'} \le \|\{f_j\}_{j\in\mathbb{N}^\nu}\|_{\mathscr{V}},\end{aligned}$$

as desired.

(II): When $\mathscr{V} = L^p(\ell^q)$, (II) follows immediately. When $\mathscr{V} = \ell^q(L^p)$, we have

$$\begin{aligned}\|\{2^{-\epsilon j_\nu} f_{j'}\}_{j\in\mathbb{N}^\nu}\|_{\mathscr{V}} &\le \left\|\left\{\sum_{j_\nu\in\mathbb{N}} 2^{-\epsilon j_\nu} |f_{j'}|\right\}_{j'\in\mathbb{N}^{\nu-1}}\right\|_{\mathscr{V}'} \\ &\le \sum_{j_\nu\in\mathbb{N}} 2^{-\epsilon j_\nu} \|\{f_{j'}\}_{j'\in\mathbb{N}^{\nu-1}}\|_{\mathscr{V}'} \lesssim \|\{f_{j'}\}_{j'\in\mathbb{N}^{\nu-1}}\|_{\mathscr{V}'},\end{aligned}$$

as desired. □

Lemma 6.8.4. *Let* $\epsilon > 0$.

(I) *Let* $\mathcal{E}'$ *be a bounded set of generalized* $(W, \vec{d})$ *pre-elementary operators. Then*

$$\sup_{\{(E'_{j'}, 2^{-j'}):j'\in\mathbb{N}^{\nu-1}\}\subseteq\mathcal{E}'} \sum_{j_\nu\in\mathbb{N}} \|\{2^{-j_\nu\epsilon} E'_{j'} f_j\}_{j'\in\mathbb{N}^{\nu-1}}\|_{\mathscr{V}'} \lesssim_\epsilon \|\{f_j\}_{j\in\mathbb{N}^\nu}\|_{\mathscr{V}},$$

for all $\{f_j\}_{j\in\mathbb{N}^\nu} \in \mathscr{V}$.

(II) *Let* $\mathcal{E}$ *be a bounded set of generalized* $(W, \vec{d})$ *pre-elementary operators. Then*

$$\sup_{\{(E_j, 2^{-j}):j\in\mathbb{N}^\nu\}\subseteq\mathcal{E}} \|\{2^{-j_\nu\epsilon} E_j f_{j'}\}_{j\in\mathbb{N}^\nu}\|_{\mathscr{V}} \lesssim_\epsilon \|\{f_{j'}\}_{j'\in\mathbb{N}^{\nu-1}}\|_{\mathscr{V}'},$$

for all $\{f_{j'}\}_{j'\in\mathbb{N}^{\nu-1}} \in \mathscr{V}'$.

Proof. (I): Using Definition 5.2.26 and Lemma 6.8.3 (I), we have

$$\begin{aligned}&\sup_{\{(E'_{j'}, 2^{-j'}):j'\in\mathbb{N}^{\nu-1}\}\subseteq\mathcal{E}'} \sum_{j_\nu\in\mathbb{N}} \|\{2^{-j_\nu\epsilon} E'_{j'} f_j\}_{j'\in\mathbb{N}^{\nu-1}}\|_{\mathscr{V}'} \\ &\quad\le \sum_{j_\nu\in\mathbb{N}} \sup_{\{(E'_{j'}, 2^{-j'}):j'\in\mathbb{N}^{\nu-1}\}\subseteq\mathcal{E}'} \|\{2^{-j_\nu\epsilon} E'_{j'} f_j\}_{j'\in\mathbb{N}^{\nu-1}}\|_{\mathscr{V}'} \\ &\quad\lesssim \sum_{j_\nu\in\mathbb{N}} \|\{2^{-j_\nu\epsilon} f_j\}_{j'\in\mathbb{N}^{\nu-1}}\|_{\mathscr{V}'} \lesssim \|\{f_j\}_{j\in\mathbb{N}^\nu}\|_{\mathscr{V}}.\end{aligned}$$

(II): Set $g_j := 2^{-j_\nu \epsilon} f_{j'}$. Then by Lemma 6.8.3 (II), $\{g_j\}_{j\in\mathbb{N}^\nu} \in \mathscr{V}$ and $\|\{g_j\}_{j\in\mathbb{N}^\nu}\|_{\mathscr{V}} \lesssim \|\{f_{j'}\}_{j'\in\mathbb{N}^{\nu-1}}\|_{\mathscr{V}'}$. Using this and Definition 5.2.26, we have

$$\sup_{\{(E_j,2^{-j}):j\in\mathbb{N}^\nu\}\subseteq\mathcal{E}} \|\{2^{-j_\nu\epsilon} E_j f_{j'}\}_{j\in\mathbb{N}^\nu}\|_{\mathscr{V}} \quad \sup_{\{(E_j,2^{-j}):j\in\mathbb{N}^\nu\}\subseteq\mathcal{E}} \|\{E_j g_j\}_{j\in\mathbb{N}^\nu}\|_{\mathscr{V}}$$
$$\lesssim \|\{g_j\}_{j\in\mathbb{N}^\nu}\|_{\mathscr{V}} \lesssim \|\{f_{j'}\}_{j'\in\mathbb{N}^{\nu-1}}\|_{\mathscr{V}'}. \qquad \square$$

Lemma 6.8.5. *For all $\epsilon > 0$, $\mathscr{X}^{(s,\epsilon)}(\mathcal{K},(W,\vec{d})) \subseteq C^\infty_{Y,0}(\mathfrak{M})'$ and $\forall f \in \mathscr{X}^{(s,\epsilon)}(\mathcal{K},(W,\vec{d}))$, $\sum_{j\in\mathbb{N}^\nu} D_j f = f$ with convergence in $C^\infty_{Y,0}(\mathfrak{M})'$.*

Proof. A priori, Definition 6.3.5 gives $\mathscr{X}^{(s,\epsilon)}(\mathcal{K},(W,\vec{d})) \subseteq C^\infty_{W,0}(\mathfrak{M})'$ and Proposition 5.5.10 shows

$$f = \mathrm{Mult}[\psi]f = \sum_{j\in\mathbb{N}^\nu} D_j f, \quad \forall f \in \mathscr{X}^{(s,\epsilon)}(\mathcal{K},(W,\vec{d})), \tag{6.143}$$

with convergence in $C^\infty_{W,0}(\mathfrak{M})'$. We will use the fact that $\epsilon > 0$ to show that the sum in (6.143) converges in $C^\infty_{Y,0}(\mathfrak{M})'$, which will complete the proof.

Fix $g \in C^\infty_{Y,0}(\mathfrak{M})$ and let $N = N(s) \in \mathbb{N}$ be so large that $|j'|_\infty N \geq |j'\cdot s|$ for every $j' \in \mathbb{N}^{\nu-1}$. We will show that, for all $j \in \mathbb{N}^\nu$,

$$|\langle g, D_j f\rangle| \lesssim \sum_{|\alpha|\leq N} \|Y^\alpha g\|_{L^q} 2^{-j'\cdot s - |j'|_\infty N - j_\nu\epsilon} \|f\|_{\mathscr{X}^{(s,\epsilon)}(W,\vec{d})}. \tag{6.144}$$

Once we show (6.144), it will immediately follow that the sum in (6.143) converges in $C^\infty_{Y,0}(\mathfrak{M})'$, completing the proof.

For $j \in \mathbb{N}^\nu$, let $\mu \in \{1,\dots,\nu-1\}$ be such that $|j'|_\infty = j_\mu$. Using Proposition 5.5.5 (h), we can write

$$\begin{aligned} D_j &= \sum_{|\alpha_\mu|\leq N} 2^{(|\alpha_\mu|-N)j_\mu} (2^{-j_\mu d^\mu} W^\mu)^{\alpha_\mu} E_{j,\mu,\alpha_\mu} \\ &= \sum_{|\alpha_\mu|\leq N} 2^{(|\alpha_\mu|-N)j_\mu} (2^{-j'_\mu d^\mu} Y^\mu)^{\alpha_\mu} E_{j,\mu,\alpha_\mu}, \end{aligned}$$

where $\{(E_{j,\mu,\alpha_\mu}, 2^{-j}) : j \in \mathbb{N}^\nu, \mu \in \{1,\dots,\nu-1\}, |\alpha_\mu| \leq N\}$ is a bounded set of generalized $(W,\vec{d})$ elementary operators. Since $f \in \mathscr{X}^{(s,\epsilon)}(\mathcal{K},(W,\vec{d}))$, it follows from Corollary 6.4.4 that $\|E_{j_\mu,\mu,\alpha_\mu} f\|_{L^p(\mathfrak{M},\mathrm{Vol})} \lesssim 2^{-j\cdot(s,\epsilon)} \|f\|_{\mathscr{X}^{(s,\epsilon)}(W,\vec{d})}$.

Thus,

$$\begin{aligned} |\langle g, D_j f\rangle| &\leq \sum_{|\alpha_\mu|\leq N} 2^{(|\alpha_\mu|-N)j_\mu} |\langle g, (2^{-j'_\mu d^\mu} Y^\mu)^{\alpha_\mu} E_{j,\mu,\alpha_\mu} f\rangle| \\ &= \sum_{|\alpha_\mu|\leq N} 2^{(|\alpha_\mu|-N-\deg_{d^\mu}(\alpha_\mu))j_\mu} |\langle ((2^{-j'_\mu d^\mu} Y^\mu)^{\alpha_\mu})^* g, E_{j,\mu,\alpha_\mu} f\rangle| \end{aligned}$$

$$\lesssim 2^{-Nj_\mu}\Big(\sum_{|\beta_\mu|\leq N}\|(Y^\mu)^{\beta_\mu}g\|_{L^q}\Big)\Big(\max_{|\alpha_\mu|\leq N}\|E_{j,\mu,\alpha_\mu}f\|_{L^p}\Big)$$
$$\lesssim 2^{-N|j'|_\infty}2^{-j'\cdot s-j_\nu\epsilon}\sum_{|\beta_\mu|\leq N}\|(Y^\mu)^{\beta_\mu}g\|_{L^q}\|f\|_{\mathscr{X}^{(s,\epsilon)}(W,\vec{d})},$$

establishing (6.144) and completing the proof. □

Lemma 6.8.6. *Let $\mathcal{E}$ be a bounded set of generalized $(W,\vec{d})$ elementary operators and let $\mathcal{E}'$ be a bounded set of generalized $(Y,\vec{d})$ elementary operators. Then, for every $N\in\mathbb{N}$, there exists $L=L(N)\in\mathbb{N}$ such that $\forall(E_j,2^{-j})\in\mathcal{E}$, $(E'_{k'},2^{-k'})\in\mathcal{E}'$,*

$$E_jE'_{k'}=\sum_{K=1}^{L}2^{-N|k'-j'|}F_{j,K}F'_{k',K}\tag{6.145}$$

and

$$E'_{k'}E_j=\sum_{K=1}^{L}2^{-N|k'-j'|}\widetilde{F}'_{k',K}\widetilde{F}_{j,K},\tag{6.146}$$

where

$$\{(F_{j,K},2^{-j}),(\widetilde{F}_{j,K},2^{-j}):(E_j,2^{-j})\in\mathcal{E},(E'_{k'},2^{-k'})\in\mathcal{E}',1\leq K\leq L\}$$

is a bounded set of generalized $(W,\vec{d})$ elementary operators and

$$\{(F'_{k',K},2^{-k'}),(\widetilde{F}'_{k',K},2^{-k'}):(E_j,2^{-j})\in\mathcal{E},(E'_{k'},2^{-k'})\in\mathcal{E}',1\leq K\leq L\}$$

is a bounded set of generalized $(Y,\vec{d})$ elementary operators.

Proof. We prove only (6.145); the proof of (6.146) is similar and we leave it to the reader.

We will prove (6.145) with $2^{-N|k'-j'|}$ replaced by $2^{-N|k'-j'|_\infty}$; the result will then follow by increasing N and using Proposition 5.5.5 (a).

Pick $\mu\in\{1,\dots,\nu-1\}$ such that $|j'_\mu-k'_\mu|=|j'-k'|_\infty$. When $k'_\mu\geq j'_\mu$, we apply Proposition 5.5.5 (h) to $E'_{k'}$ to see that

$$E'_{k'}=\sum_{|\alpha_\mu|\leq N}2^{(|\alpha_\mu|-N)k'_\mu}(2^{-k'_\mu d^\mu}Y^\mu)^{\alpha_\mu}F'_{k',\alpha_\mu}$$
$$=\sum_{|\alpha_\mu|\leq N}2^{(|\alpha_\mu|-N)k'_\mu}(2^{-k'_\mu d^\mu}W^\mu)^{\alpha_\mu}F'_{k',\alpha_\mu}$$

where $\{(F'_{k',\alpha_\mu},2^{-k'}):(E'_{k'},2^{-k'})\in\mathcal{E}'\}$ is a bounded set of $(Y,\vec{d})$ elementary operators and we have used $(W^\mu,d^\mu)=(Y^\mu,d^\mu)$ for $1\leq\mu\leq\nu-1$. Thus, we have

$$E_j E'_{k'} = 2^{-N(k'_\mu - j'_\mu)} \sum_{|\alpha_\mu| \le N} [2^{-Nj'_\mu + (|\alpha_\mu| - \deg_{\vec{d}^\mu}(\alpha_\mu))(k'_\mu - j'_\mu)} E_j (2^{-j'_\mu d^\mu} W^\mu)^{\alpha_\mu}] F'_{k',\alpha_\mu}$$
$$=: 2^{-N|k'-j'|_\infty} \sum_{|\alpha_\mu| \le N} F_{j,\alpha_\mu} F'_{k',\alpha_\mu},$$

where by Proposition 5.5.5 (e) and (a), $\{(F_{j,\alpha_\mu}, 2^{-j}) : (E_j, 2^{-j}) \in \mathcal{E}, (E'_{k'}, 2^{-k'}) \in \mathcal{E}', |\alpha_\mu| \le N\}$ is a bounded set of generalized $(W, \vec{d})$ elementary operators. This shows that $E_j E'_{k'}$ is of the form (6.145).

If $k'_\mu < j'_\mu$, then a similar proof works where we apply Proposition 5.5.5 (h) to E_j and Proposition 5.5.5 (e) and (a) to $E'_{k'}$; we leave the details to the interested reader. □

Proof of Proposition 6.8.1. In the proof which follows, for an operator indexed by $j \in \mathbb{N}^\nu$, E_j (for example, $E_j = D_j$), we take the convention $E_j = 0$ for $j \in \mathbb{Z}^\nu \setminus \{0\}$. We use a similar convention for operators indexed by $j' \in \mathbb{N}^{\nu-1}$.

We begin with the inclusion $\mathscr{X}^{(s,\epsilon)}(\mathcal{K}, (W, \vec{d})) \hookrightarrow \mathscr{X}^s(\mathcal{K}, (Y, \vec{d}))$. Let $\mathcal{E}'$ be a bounded set of generalized $(Y, \vec{d})$ elementary operators and fix $f \in \mathscr{X}^{(s,\epsilon)}$. We will show that

$$\|f\|_{\mathscr{V}',s,\mathcal{E}'} \lesssim \|f\|_{\mathscr{X}^{(s,\epsilon)}(W,\vec{d})}. \tag{6.147}$$

Lemma 6.8.5 shows that $f \in C^\infty_{Y,0}(\mathfrak{M})'$, and therefore the left-hand side of (6.147) is defined. Since $\mathcal{E}'$ was arbitrary and $\operatorname{supp}(f) \subseteq \mathcal{K}$, (6.147) implies $f \in \mathscr{X}^s(\mathcal{K}, (Y, \vec{d}))$. Moreover, by taking $\mathcal{E}' = \mathcal{D}'_0$, (6.147) implies $\|f\|_{\mathscr{X}^s(Y,\vec{d})} \lesssim \|f\|_{\mathscr{X}^{(s,\epsilon)}(W,\vec{d})}$, proving that the inclusion $\mathscr{X}^{(s,\epsilon)}(\mathcal{K}, (W, \vec{d})) \hookrightarrow \mathscr{X}^s(\mathcal{K}, (Y, \vec{d}))$ is continuous.

Thus, to prove the inclusion $\mathscr{X}^{(s,\epsilon)}(\mathcal{K}, (W, \vec{d})) \hookrightarrow \mathscr{X}^s(\mathcal{K}, (Y, \vec{d}))$, it suffices to prove (6.147). Fix $\{(E_{j'}, 2^{-j'}) : j' \in \mathbb{N}^{\nu-1}\} \subseteq \mathcal{E}$. We will show that

$$\|\{2^{-j'\cdot s} E'_{j'} f\}_{j' \in \mathbb{N}^{\nu-1}}\|_{\mathscr{V}'} \lesssim \|f\|_{\mathscr{X}^{(s,\epsilon)}(W,\vec{d})}. \tag{6.148}$$

Taking the supremum of (6.148) over $\{(E_{j'}, 2^{-j'}) : j' \in \mathbb{N}^{\nu-1}\} \subseteq \mathcal{E}$ establishes (6.147). Thus, we turn to establishing (6.148).

Lemma 6.8.5 shows that $f = \sum_{j \in \mathbb{N}^\nu} D_j f$ with convergence in $C^\infty_{Y,0}(\mathfrak{M})'$, so

$$\begin{aligned} \|\{2^{-j'\cdot s} E'_{j'} f\}_{j' \in \mathbb{N}^{\nu-1}}\|_{\mathscr{V}'} &= \left\| \left\{ 2^{-j'\cdot s} E'_{j'} \sum_{j_\nu \in \mathbb{N}} \sum_{l \in \mathbb{Z}^{\nu-1}} D_{(j'+l, j_\nu)} f \right\}_{j' \in \mathbb{N}^{\nu-1}} \right\|_{\mathscr{V}'} \\ &\le \sum_{l \in \mathbb{Z}^{\nu-1}} \sum_{j_\nu \in \mathbb{N}} \|\{2^{-j'\cdot s} E'_{j'} D_{(j'+l, j_\nu)} f\}_{j' \in \mathbb{N}^{\nu-1}}\|_{\mathscr{V}'}. \end{aligned} \tag{6.149}$$

Take $N > |s|$. By Lemma 6.8.6, we may write (with $K \approx 1$)

$$E'_{j'} D_{(j'+l, j_\nu)} = \sum_{L=1}^{K} 2^{-N|l|} F'_{j',L} F_{(j'+l, j_\nu), L}, \tag{6.150}$$

where $\{(F'_{j',L}, 2^{-j'}) : (E'_{j'}, 2^{-j'}) \in \mathcal{E}', j_\nu \in \mathbb{N}, l \in \mathbb{Z}^{\nu-1}, 1 \le L \le K\}$ is a bounded set of generalized $(Y, \vec{d})$ elementary operators and $\{(F_{k,L}, 2^{-k}) : (E'_{j'}, 2^{-j'}) \in \mathcal{E}', k \in \mathbb{N}^\nu, 1 \le L \le K\}$ is a bounded set of generalized $(W, \vec{d})$ elementary operators.

Plugging (6.150) into (6.149), we see that

$$\begin{aligned}
&\|\{2^{-j'\cdot s} E'_{j'} f\}_{j'\in\mathbb{N}^{\nu-1}}\|_{\mathcal{V}'} \\
&\quad\le \sum_{l\in\mathbb{Z}^{\nu-1}} \sum_{j_\nu\in\mathbb{N}} \sum_{L=1}^{K} 2^{-N|l|-s\cdot l} \|\{2^{-j_\nu\epsilon} F'_{j',L} 2^{(s,\epsilon)\cdot(j'+l',j_\nu)} F_{(j'+l',j_\nu),L} f\}_{j'\in\mathbb{N}^{\nu-1}}\|_{\mathcal{V}'} \\
&\quad\lesssim \sum_{l\in\mathbb{Z}^{\nu-1}} \sum_{L=1}^{K} 2^{-N|l|-s\cdot l} \|\{2^{(s,\epsilon)\cdot(j'+l',j_\nu)} F_{(j'+l',j_\nu),L} f\}_{j\in\mathbb{N}^\nu}\|_{\mathcal{V}} && \text{Lemma 6.8.4 (I)} \\
&\quad\lesssim \sum_{l\in\mathbb{Z}^{\nu-1}} \sum_{L=1}^{K} 2^{-N|l|-s\cdot l} \|f\|_{\mathscr{X}^{(s,\epsilon)}(W,\vec{d})} && \text{Corollary 6.4.4} \\
&\quad\lesssim \|f\|_{\mathscr{X}^{(s,\epsilon)}(W,\vec{d})},
\end{aligned}$$

establishing (6.148) and completing the proof of the inclusion $\mathscr{X}^{(s,\epsilon)}(\mathcal{K}, (W, \vec{d})) \hookrightarrow \mathscr{X}^{s}(\mathcal{K}, (Y, \vec{d}))$.

We turn to the inclusion $\mathscr{X}^{s}(\mathcal{K}, (Y, \vec{d})) \hookrightarrow \mathscr{X}^{(s,-\epsilon)}(\mathcal{K}, (W, \vec{d}))$. Let $\mathcal{E}$ be a bounded set of generalized $(W, \vec{d})$ elementary operators and let $f \in \mathscr{X}^{s}(\mathcal{K}, (Y, \vec{d}))$. We will show that

$$\|f\|_{\mathcal{V},(s,-\epsilon),\mathcal{E}} \lesssim \|f\|_{\mathscr{X}^{s}(Y,\vec{d})}. \tag{6.151}$$

Since $f \in \mathscr{X}^{s}(\mathcal{K}, (Y, \vec{d})) \subseteq C^\infty_{Y,0}(\mathfrak{M})' \subseteq C^\infty_{W,0}(\mathfrak{M})'$, the left-hand side of (6.151) makes sense. Since $\mathcal{E}$ was arbitrary and $\operatorname{supp}(f) \subseteq \mathcal{K}$, (6.151) implies $f \in \mathscr{X}^{(s,-\epsilon)}(\mathcal{K}, (W, \vec{d}))$. Moreover, by taking $\mathcal{E} = \mathcal{D}_0$, (6.151) gives $\|f\|_{\mathscr{X}^{(s,-\epsilon)}(W,\vec{d})} \lesssim \|f\|_{\mathscr{X}^{s}(Y,\vec{d})}$, establishing the continuous inclusion $\mathscr{X}^{s}(\mathcal{K}, (Y, \vec{d})) \hookrightarrow \mathscr{X}^{(s,-\epsilon)}(\mathcal{K}, (W, \vec{d}))$. We turn to proving (6.151).

Let $\{(E_j, 2^{-j}) : j \in \mathbb{N}^\nu\} \subseteq \mathcal{E}$. We will show that

$$\|\{2^{-j\cdot(s,-\epsilon)} E_j f\}_{j\in\mathbb{N}^\nu}\|_{\mathcal{V}} \lesssim \|f\|_{\mathscr{X}^{s}(Y,\vec{d})}. \tag{6.152}$$

Taking the supremum of (6.152) over $\{(E_j, 2^{-j}) : j \in \mathbb{N}^\nu\} \subseteq \mathcal{E}$ establishes (6.151). Thus, we turn to establishing (6.152).

Proposition 5.5.10 shows $f = \operatorname{Mult}[\psi']f = \sum_{j'\in\mathbb{N}^{\nu-1}} D'_{j'} f$ with convergence in $C^\infty_{Y,0}(\mathfrak{M})'$; therefore, this sum also converges in the weaker sense of $C^\infty_{W,0}(\mathfrak{M})'$. Thus,

$$\begin{aligned}
\|\{2^{\,j\cdot(s,-\epsilon)} E_j f\}_{j\in\mathbb{N}^\nu}\|_{\mathcal{V}} &= \left\| \left\{ 2^{-j\cdot(s,-\epsilon)} E_j \sum_{l\in\mathbb{Z}^{\nu-1}} D'_{j'+l} f \right\}_{j\in\mathbb{N}^\nu} \right\|_{\mathcal{V}} \\
&\le \sum_{l\in\mathbb{Z}^{\nu-1}} \|\{2^{-j\cdot(s,-\epsilon)} E_j D'_{j'+l} f\}_{j\in\mathbb{N}^\nu}\|_{\mathcal{V}}.
\end{aligned} \tag{6.153}$$

Take $N > |s|$. By Lemma 6.8.6, we may write (with $K \approx 1$)

$$E_j D'_{j'+l} = \sum_{L=1}^{K} 2^{-N|l|} F_{j,L} F'_{j'+l,L}, \tag{6.154}$$

where $\{(F_{j,L}, 2^{-j}) : (E_j, 2^{-j}) \in \mathcal{E}, l \in \mathbb{Z}^{\nu-1}, 1 \le L \le K\}$ is a bounded set of generalized $(W, \vec{d})$ elementary operators and $\{(F'_{k',L}, 2^{-k'}) : (E_j, 2^{-j}) \in \mathcal{E}, k' \in \mathbb{N}^{\nu-1}, 1 \le L \le K\}$ is a bounded set of generalized $(Y, \vec{d})$ elementary operators.

Plugging (6.154) into (6.153), we see that

$$\begin{aligned}
&\|\{2^{-j\cdot(s,-\epsilon)} E_j f\}_{j\in\mathbb{N}^\nu}\|_{\mathcal{V}} \\
&\le \sum_{l\in\mathbb{Z}^{\nu-1}} \sum_{L=1}^{K} 2^{-N|l|+l\cdot s} \|\{2^{-j_\nu \epsilon} F_{j,L} 2^{-(j'+l')\cdot s} F'_{j'+l,L} f\}_{j\in\mathbb{N}^\nu}\|_{\mathcal{V}} \\
&\lesssim \sum_{l\in\mathbb{Z}^{\nu-1}} \sum_{L=1}^{K} 2^{-|l|(N-|s|)} \|\{2^{s\cdot(j'+l)} F'_{j'+l',L} f\}_{j'\in\mathbb{N}^{\nu-1}}\|_{\mathcal{V}'} && \text{Lemma 6.8.4(II)} \\
&\lesssim \sum_{l\in\mathbb{Z}^{\nu-1}} \sum_{L=1}^{K} 2^{-|l|(N-|s|)} \|f\|_{\mathscr{X}^s(Y,\vec{d})} && \text{Corollary 6.4.4} \\
&\lesssim \|f\|_{\mathscr{X}^s(Y,\vec{d})},
\end{aligned}$$

establishing (6.152) and completing the proof. □

The remainder of this section is devoted to the proof of Proposition 6.8.2. This is achieved by combining the next two results.

Lemma 6.8.7. *Fix $\epsilon_0 > 0$ and $s = (s_1, \ldots, s_\nu) \in \mathbb{R}^\nu$. Suppose*

$$f \in \bigcap_{\epsilon\in(0,\epsilon_0]} \mathscr{X}^{(s_1-\epsilon, s_2-\epsilon,\ldots,s_\nu-\epsilon)}(\mathcal{K}, (W, \vec{d}))$$

and $\|f\|_{\mathcal{V},s,\mathcal{D}_0} < \infty$. Then $f \in \mathscr{X}^s(\mathcal{K}, (W, \vec{d}))$. This result holds even in the case $\nu = 1$.

Proposition 6.8.8. *One may choose $\mathcal{D}_0$ and $\mathcal{D}'_0$ such that the following holds. Fix $s \in \mathbb{R}^{\nu-1}$ and $p \in (1, \infty)$. For $f \in C_0^\infty(\mathfrak{M})'$ such $\|f\|_{L^p(\mathfrak{M},\mathrm{Vol};\ell^2(\mathbb{N}^{\nu-1})),s,\mathcal{D}'_0} < \infty$. Then*

$$\|f\|_{L^p(\mathfrak{M},\mathrm{Vol};\ell^2(\mathbb{N}^{\nu-1})),s,\mathcal{D}'_0} \approx \|f\|_{L^p(\mathfrak{M},\mathrm{Vol};\ell^2(\mathbb{N}^{\nu})),(s,0),\mathcal{D}_0}, \tag{6.155}$$

where the implicit constants depend on $p \in (1, \infty)$ and the choices of $\mathcal{D}_0$ and $\mathcal{D}'_0$. Here, the left-hand side of (6.155) *is finite if and only if the right-hand side is.*

First we see why the above two results give Proposition 6.8.2.

Proof of Proposition 6.8.2. Since

$$\|f\|_{\mathscr{F}^{(s,0)}_{p,2}(W,\vec{d})} \approx \|f\|_{L^p(\mathfrak{M},\mathrm{Vol};\ell^2(\mathbb{N}^\nu)),(s,0),\mathcal{D}_0}, \quad \forall f \in \mathscr{F}^{(s,0)}_{p,2}(W,\vec{d}),$$
$$\|f\|_{\mathscr{F}^{s}_{p,2}(Y,\vec{d})} \approx \|f\|_{L^p(\mathfrak{M},\mathrm{Vol};\ell^2(\mathbb{N}^{\nu-1})),s,\mathcal{D}_0'}, \quad \forall f \in \mathscr{F}^{s}_{p,2}(Y,\vec{d}),$$

once we establish $\mathscr{F}^{(s,0)}_{p,2}(\mathcal{K},(W,\vec{d})) = \mathscr{F}^{s}_{p,2}(\mathcal{K},(Y,\vec{d}))$, it follows immediately from (6.155) that the topologies are equal.

Let $f \in \mathscr{F}^{(s,0)}_{p,2}(\mathcal{K},(W,\vec{d}))$ and fix $\Omega \Subset \mathfrak{M}$ open and relatively compact such that $\mathcal{K} \Subset \Omega$ and ψ and ψ' are equal to 1 on a neighborhood of $\overline{\Omega}$. Fix $\epsilon > 0$. By Proposition 6.5.17 there is a sequence $\{f_L\}_{L\in\mathbb{N}} \subset \mathscr{F}^{(s,\epsilon)}_{p,2}((W,\vec{d}),\overline{\Omega})$ such that $f_L \xrightarrow{L\to\infty} f$ in $\mathscr{F}^{(s,0)}_{p,2}(\overline{\Omega},(W,\vec{d}))$. By Proposition 6.8.1, we have $f_L \in \mathscr{F}^{s}_{p,2}(\overline{\Omega},(Y,\vec{d}))$, $\forall L \in \mathbb{N}$. Therefore, using (6.155), we have

$$\begin{aligned}\|f_L - f_{L'}\|_{\mathscr{F}^{s}_{p,2}(Y,\vec{d})} &\approx \|f_L - f_{L'}\|_{L^p(\mathfrak{M},\mathrm{Vol};\ell^2(\mathbb{N}^{\nu-1})),s,\mathcal{D}_0'}\\ &\approx \|f_L - f_{L'}\|_{L^p(\mathfrak{M},\mathrm{Vol};\ell^2(\mathbb{N}^{\nu})),(s,0),\mathcal{D}_0} \approx \|f_L - f_{L'}\|_{\mathscr{F}^{(s,0)}_{p,2}(W,\vec{d})}.\end{aligned}$$

We conclude that $\{f_L\}_{L\in\mathbb{N}}$ is Cauchy in $\mathscr{F}^{s}_{p,2}(\overline{\Omega},(Y,\vec{d}))$. Since $\mathscr{F}^{s}_{p,2}(\overline{\Omega},(Y,\vec{d}))$ is complete (Proposition 6.3.6), we conclude that $f \in \mathscr{F}^{s}_{p,2}(\overline{\Omega},(Y,\vec{d}))$. Since $\mathrm{supp}(f) \subseteq \mathcal{K}$, Proposition 6.5.6 shows that $f \in \mathscr{F}^{s}_{p,2}(\mathcal{K},(Y,\vec{d}))$. This establishes the containment $\mathscr{F}^{(s,0)}_{p,2}(\mathcal{K},(W,\vec{d})) \subseteq \mathscr{F}^{s}_{p,2}(\mathcal{K},(Y,\vec{d}))$.

Suppose $f \in \mathscr{F}^{s}_{p,2}(\mathcal{K},(Y,\vec{d}))$. Then by (6.155) we have

$$\|f\|_{L^p(\mathfrak{M},\mathrm{Vol};\ell^2(\mathbb{N}^{\nu})),(s,0),\mathcal{D}_0} \approx \|f\|_{L^p(\mathfrak{M},\mathrm{Vol};\ell^2(\mathbb{N}^{\nu-1})),s,\mathcal{D}_0'} \approx \|f\|_{\mathscr{F}^{s}_{p,2}(Y,\vec{d})} < \infty.$$

Proposition 6.8.1 shows that

$$f \in \bigcap_{\epsilon>0} \mathscr{F}^{(s,-\epsilon)}_{p,2}(\mathcal{K},(W,\vec{d})) \subseteq \bigcap_{\epsilon>0} \mathscr{F}^{(s,0)-(\epsilon,\epsilon,\ldots,\epsilon)}_{p,2}(\mathcal{K},(W,\vec{d})).$$

Using the above two equations combined with Lemma 6.8.7 shows that

$$f \in \mathscr{F}^{(s,0)}_{p,2}(\mathcal{K},(W,\vec{d})).$$

This establishes the containment $\mathscr{F}^{s}_{p,2}(\mathcal{K},(Y,\vec{d})) \subseteq \mathscr{F}^{(s,0)}_{p,2}(\mathcal{K},(W,\vec{d}))$ and completes the proof. □

Proof of Lemma 6.8.7. Let $\mathcal{E}$ be a bounded set of generalized $(W,\vec{d})$ elementary operators. We wish to show $\|f\|_{\mathcal{V},s,\mathcal{E}} < \infty$. Since $f \in \bigcap_{\epsilon\in(0,\epsilon_0]} \mathscr{X}^{(s_1-\epsilon,s_2-\epsilon,\ldots,s_\nu-\epsilon)}(\mathcal{K},(W,\vec{d}))$, Corollary 6.4.4 with $K_0 = [s_1-\epsilon_0,s_1]\times\cdots\times[s_\nu-\epsilon_0,s_\nu]$ shows that there exists $C \geq 0$ such that for every $\epsilon \in (0,\epsilon_0]$,

$$\|f\|_{\mathcal{V},s-(\epsilon,\epsilon,\ldots,\epsilon),\mathcal{E}} \leq C\|f\|_{\mathcal{V},s-(\epsilon,\epsilon,\ldots,\epsilon),\mathcal{D}_0} \leq C\|f\|_{\mathcal{V},s,\mathcal{D}_0}.$$

Thus, by the monotone convergence theorem,[3] we have, for $\{(E_j, 2^{-j}) \in \mathcal{E} : j \in \mathbb{N}^\nu\} \subseteq \mathcal{E}$,

$$\begin{aligned}\|\{2^{j\cdot s}E_j f\}_{j\in\mathbb{N}^\nu}\|_{\mathscr{V}} &= \lim_{\epsilon\downarrow 0}\|\{2^{j\cdot(s-(\epsilon,\ldots,\epsilon))}E_j f\}_{j\in\mathbb{N}^\nu}\|_{\mathscr{V}} \\ &\le \lim_{\epsilon\downarrow 0}\|f\|_{\mathscr{V},s-(\epsilon,\epsilon,\ldots,\epsilon),\mathcal{E}} \le C\|f\|_{\mathscr{V},s,\mathcal{D}_0}.\end{aligned}$$

Therefore, taking the supremum over all $\{(E_j, 2^{-j}) \in \mathcal{E} : j \in \mathbb{N}^\nu\} \subseteq \mathcal{E}$, we see that

$$\|f\|_{\mathscr{V},s,\mathcal{E}} \le C\|f\|_{\mathscr{V},s,\mathcal{D}_0} < \infty. \tag{6.156}$$

Since $f \in \mathscr{X}^{(s_1-\epsilon_0,s_2-\epsilon_0,\ldots,s_\nu-\epsilon_0)}(\mathcal{K},(W,\vec{d}))$, we have $\operatorname{supp}(f) \subseteq \mathcal{K}$. Combining this with (6.156) and using the fact that $\mathcal{E}$ was an arbitrary bounded set of generalized $(W,\vec{d})$ elementary operators, we obtain $f \in \mathscr{X}^s(\mathcal{K},(W,\vec{d}))$, completing the proof. □

We close this section by proving Proposition 6.8.8. For this, we must introduce the choices of $\mathcal{D}_0$ and $\mathcal{D}_0'$ we use.

Fix open sets

$$\mathcal{K} \Subset \Omega_1^1 \Subset \Omega_2^1 \Subset \Omega_1^2 \Subset \Omega_2^2 \Subset \cdots \Subset \Omega_1^\nu \Subset \Omega_2^\nu \Subset \Omega_1 \Subset \Omega_2 \Subset \Omega \Subset \mathfrak{M}.$$

For each $\mu \in \{1,\ldots,\nu\}$, let $\psi_\mu \in C_0^\infty(\Omega_1^\mu)$ satisfy $\psi_\mu \equiv 1$ on a neighborhood of $\overline{\Omega_2^{\mu-1}}$ if $\mu \ge 2$ and $\psi_1 \equiv 1$ on a neighborhood of $\mathcal{K}$.

By hypothesis, for each $\mu \in \{1,\ldots,\nu\}$, $\operatorname{Gen}((W^\mu, \vec{d}^\mu))$ is locally finitely generated. Pick

$$(X^\mu, d^\mu) := \{(X_1^\mu, d_1^\mu), \ldots, (X_{q_\mu}^\mu, d_{q_\mu}^\mu)\} \subset \operatorname{Gen}((W^\mu, \vec{d}^\mu))$$

such that $\operatorname{Gen}((W^\mu, \vec{d}^\mu))$ is finitely generated by (X^μ, d^μ) on Ω. Set

$$(X, \vec{d}) := (X^1, d^1) \boxtimes (X^2, d^2) \boxtimes \cdots \boxtimes (X^\nu, d^\nu) \subset C_{\mathrm{loc}}^\infty(\mathfrak{M}; T\mathfrak{M}) \times (\mathbb{N}^\nu \setminus \{0\}),$$

and similarly,

$$(X', \vec{d}') := (X^1, d^1) \boxtimes (X^2, d^2) \boxtimes \cdots \boxtimes (X^{\nu-1}, d^{\nu-1}) \subset C_{\mathrm{loc}}^\infty(\mathfrak{M}; T\mathfrak{M}) \times (\mathbb{N}^{\nu-1} \setminus \{0\}).$$

Note that $(X, \vec{d})$ is to $(W, \vec{d})$ as $(X', \vec{d}')$ is to $(Y, \vec{d})$.

Let $a_\mu > 0$ be a small number to be chosen later. Example 4.1.6 shows that $\operatorname{Mult}[\psi_\mu]$ is an (X^μ, d^μ), a_μ partial differential operator of order 0 supported in $\Omega_1^\mu \times \Omega_2^\mu$. Therefore, by Theorem 4.3.3, we can write $\operatorname{Mult}[\psi_\mu] = \sum_{j_\mu \in \mathbb{N}} D_{j_\mu}^\mu$, where $\{(D_{j_\mu}^\mu, 2^{-j_\mu}) : j_\mu \in \mathbb{N}\}$ is a bounded set of (X^μ, d^μ), a pseudo-differential operator scales supported in $\Omega_1^\mu \times \Omega_2^\mu$.

3 We use the monotone convergence theorem when $\mathscr{V} = L^p(\ell^q)$ or $\mathscr{V} = \ell^q(L^p)$, where $p, q \in [1,\infty)$. If either p or q is infinite, one can combine an elementary argument with the monotone convergence theorem to obtain the same result.

Define

$$D'_{j'} := D^{\nu-1}_{j_{\nu-1}} D^{\nu-2}_{j_{\nu-2}} \cdots D^1_{j_1}, \quad D_j = D^{\nu}_{j_\nu} D^{\nu-1}_{j_{\nu-1}} \cdots D^1_{j_1} = D^{\nu}_{j_\nu} D'_{j'}.$$

Note that

$$\begin{aligned} \sum_{j\in\mathbb{N}^\nu} D_j &= \Big(\sum_{j_\nu\in\mathbb{N}} D^{\nu}_{j_\nu}\Big)\Big(\sum_{j_{\nu-1}\in\mathbb{N}} D^{\nu-1}_{j_{\nu-1}}\Big)\cdots\Big(\sum_{j_1\in\mathbb{N}} D^1_{j_1}\Big) \\ &= \mathrm{Mult}[\psi_\nu]\,\mathrm{Mult}[\psi_{\nu-1}]\cdots\mathrm{Mult}[\psi_1] = \mathrm{Mult}[\psi_1]. \end{aligned}$$

Similarly,

$$\sum_{j'\in\mathbb{N}^{\nu-1}} D'_{j'} = \mathrm{Mult}[\psi_{\nu-1}]\,\mathrm{Mult}[\psi_{\nu-2}]\cdots\mathrm{Mult}[\psi_1] = \mathrm{Mult}[\psi_1].$$

Set

$$\mathcal{D}_0 := \{(D_j, 2^{-j}) : j \in \mathbb{N}^\nu\}, \quad \mathcal{D}'_0 := \{(D'_{j'}, 2^{-j'}) : j' \in \mathbb{N}^{\nu-1}\}.$$

Remark 6.8.9. Note that $\mathcal{D}'_0$ is of the same form as $\mathcal{D}_0$, but with ν replaced by $\nu - 1$.

Lemma 6.8.10. *Let $a > 0$ be a given small number. If $a_\mu = a_\mu(a, (X, \vec{d}), \Omega^1_1, \Omega^1_2, \ldots, \Omega^1_\nu, \Omega^2_\nu, \Omega_1, \Omega_2) > 0$ is sufficiently small, for $\mu \in \{1, \ldots, \nu\}$, then $\mathcal{D}_0$ is a bounded set of $(X, \vec{d})$, a pseudo-differential operator scales supported in $\Omega_1 \times \Omega_2$ and $\mathcal{D}'_0$ is a bounded set of $(X', \vec{d}')$, a pseudo-differential operator scales supported in $\Omega_1 \times \Omega_2$.*

Proof. The result for $\mathcal{D}'_0$ is the same as the result for $\mathcal{D}_0$, with ν replaced by $\nu - 1$, so it suffices to prove only the result for $\mathcal{D}_0$.

By construction, for each $\mu \in \{1, \ldots, \nu\}$, $\{(D^\mu_{j_\mu}, 2^{-j_\mu}) : j_\mu \in \mathbb{N}\}$ is a bounded set of (X^μ, d^μ), a_μ pseudo-differential operator scales supported in $\Omega^1_\mu \times \Omega^2_\mu$. Thus, by Definition 4.3.1, we may write

$$D^\mu_{j_\mu} f(x) = \int f(e^{-t^\mu \cdot X^\mu} x)\, \mathrm{Dil}^{d^\mu}_{2^{j_\mu}}(\varsigma^\mu_{j_\mu})(x, t^\mu, t^\mu),$$

where $\varsigma^\mu_{j_\mu} \in C^\infty_0(\Omega^\mu_1)\widehat{\otimes} C^\infty_0(B^{q_\mu}(a_\mu))\widehat{\otimes}\mathscr{S}_0(\mathbb{R}^{q_\mu})$ if $j_\mu \neq 0$ and

$$\{\varsigma^\mu_{j_\mu} : j_\mu \in \mathbb{N}\} \subset C^\infty_0(\Omega^\mu_1)\widehat{\otimes} C^\infty_0(B^{q_\mu}(a_\mu))\widehat{\otimes}\mathscr{S}(\mathbb{R}^{q_\mu})$$

is a bounded set.

For $j \in \mathbb{N}^\nu$, set

$$\begin{aligned} \varsigma_j(x, (s^1, \ldots, s^\nu), (t^1, \ldots, t^\nu)) &= \varsigma^1_{j_1}(e^{-s^2\cdot X^2} e^{-s^3\cdot X^3} \cdots e^{-s^\nu\cdot X^\nu} x, s^1, t^1) \\ &\quad\times \varsigma^2_{j_2}(e^{-s^3\cdot X^3} e^{-s^4\cdot X^4} \cdots e^{-s^\nu\cdot X^\nu} x, s^2, t^2) \\ &\quad\times \cdots \varsigma^\nu_{j_\nu}(x, s^\nu, t^\nu). \end{aligned}$$

We only consider $s^{\mu} \in B^{q_{\mu}}(a_{\mu})$, where we may take $a_{\mu} > 0$ small. Thus, the above expression makes sense. Moreover, since $\Omega_1 \mu \Subset \Omega_1$, we see that if $a_{\mu} > 0$ is sufficiently small for each $\mu \in \{1, \dots, \nu\}$, then

$$\{\varsigma_j : j \in \mathbb{N}^{\nu}\} \subset C_0^{\infty}(\Omega_1)\widehat{\otimes} C_0^{\infty}(B^q(a))\widehat{\otimes}\mathscr{S}(\mathbb{R}^q)$$

is a bounded set. Furthermore, since $\varsigma_{j_\mu}^{\mu} \in C_0^{\infty}(\Omega_1^{\mu})\widehat{\otimes} C_0^{\infty}(B^{q_{\mu}}(a_{\mu}))\widehat{\otimes}\mathscr{S}_0(\mathbb{R}^{q_{\mu}})$ if $j_{\mu} \neq 0$, it follows from the definitions that $\varsigma_j \in C_0^{\infty}(\Omega_1)\widehat{\otimes} C_0^{\infty}(B^q(a))\widehat{\otimes}\mathscr{S}_{\{\mu : j_{\mu} \neq 0\}}(\mathbb{R}^q)$.

Finally, we have

$$\begin{aligned} D_j f(x) &= D_{j_\nu}^{\nu} D_{j_{\nu-1}}^{\nu-1} \cdots D_{j_1}^{1} f(x) \\ &= \int f(e^{-t^1 \cdot X^1} e^{-t^2 \cdot X^2} \cdots e^{-t^{\nu} \cdot X^{\nu}} x) \operatorname{Dil}_{2^j}^{\vec{d}}(\varsigma)(x,t,t)\, dt. \end{aligned}$$

Combining the above, we see that $\mathcal{D}_0$ is a bounded set of $(X, \vec{d})$, a pseudo-differential operator scales supported in $\Omega_1 \times \Omega_2$, completing the proof. □

Corollary 6.8.11. *If $a_{\mu} = a_{\mu}(a, (X, \vec{d}), \Omega_1^1, \Omega_2^1, \dots, \Omega_{\nu}^1, \Omega_{\nu}^2, \Omega_1, \Omega_2, \Omega) > 0$ is sufficiently small, for $\mu \in \{1, \dots, \nu\}$, then $\mathcal{D}_0$ is a bounded set of generalized $(W, \vec{d})$ elementary operators supported in Ω and $\mathcal{D}_0'$ is a bounded set of generalized $(Y, \vec{d})$ elementary operators supported in Ω.*

Proof. Given any $a > 0$, by choosing $a_{\mu} > 0$ small enough, $\mu \in \{1, \dots, \nu\}$, Lemma 6.8.10 shows that $\mathcal{D}_0$ is a $(X, \vec{d})$, a pseudo-differential operator scales supported in $\Omega_1 \times \Omega_2$ and $\mathcal{D}_0'$ is a bounded set of $(X', \vec{d}')$, a pseudo-differential operator scales supported in $\Omega_1 \times \Omega_2$. From here, by taking $a = a((X, \vec{d}), \Omega_1, \Omega_2) > 0$ small enough, the result follows from Proposition 5.6.3. □

To prove Proposition 6.8.8, we need a well-known multi-parameter version of the Khintchine inequality.

Theorem 6.8.12 (Multi-parameter Khintchine inequality). *Let $\{\epsilon_j^{\mu}\}_{j \in \mathbb{N}, \mu \in \{1,\dots,\nu\}}$ be i. i. d. random variables of mean zero taking values in ± 1. For $j = (j_1, \dots, j_{\nu}) \in \mathbb{N}^{\nu}$, set $\epsilon_j := \epsilon_{j_1}^1 \epsilon_{j_2}^2 \cdots \epsilon_{j_\nu}^{\nu}$. Fix $p \in (1, \infty)$. Then for any sequence of complex numbers $\{a_j\}_{j \in \mathbb{N}^{\nu}}$ with $\sum |a_j|^2 < \infty$, we have*

$$\Big(\sum_{j \in \mathbb{N}^{\mu}} |a_j|^2\Big)^{\frac{1}{2}} \approx \Big(\mathbb{E}\Big|\sum_{j \in \mathbb{N}^{\nu}} \epsilon_j a_j\Big|^p\Big)^{\frac{1}{p}},$$

where the implicit constants depend on $p \in (0, \infty)$.

See [213, Appendix D] for a proof of Theorem 6.8.12.

Proof of Proposition 6.8.8. We use Corollary 5.10.7 with $\mathcal{K}_0 \Subset \Omega_3 \Subset \Omega_4$ replaced by $\overline{\Omega_2^{\nu-1}} \Subset \Omega_1^\nu \Subset \Omega_2^\nu$. The results of that corollary hold with $\{(D_j, 2^{-j}) : j \in \mathbb{N}\}$ replaced by $\{(D_{j_\nu}^\nu, 2^{-j_\nu}) : j_\nu \in \mathbb{N}\}$.

Let $\{\epsilon_j^\mu\}_{j\in\mathbb{N},\mu\in\{1,\ldots,\nu\}}$ be i. i. d. random variables of mean zero taking values in ± 1. For $j = (j_1, \ldots, j_\nu) \in \mathbb{N}^\nu$, set $\epsilon_j := \epsilon_{j_1}^1 \epsilon_{j_2}^2 \cdots \epsilon_{j_\nu}^\nu$ and $\epsilon_{j'} := \epsilon_{j_1}^1 \epsilon_{j_2}^2 \cdots \epsilon_{j_{\nu-1}}^{\nu-1}$. By Corollary 5.10.7, we have, for every $g \in L^p(\Omega_2^{\nu-1}, \mathrm{Vol})$,

$$\mathbb{E}\left\| \sum_{j_\nu \in \mathbb{N}} \epsilon_{j_\nu}^\nu D_{j_\nu}^\nu g \right\|_{L^p}^p \approx \|g\|_{L^p}^p. \tag{6.157}$$

Let $f \in C_0^\infty(\mathfrak{M})'$ be such that $\|f\|_{L^p(\mathfrak{M},\mathrm{Vol};\ell^2(\mathbb{N}^{\nu-1})),s,\mathcal{D}_0'} < \infty$. Using Theorem 6.8.12, we have

$$\begin{aligned}\left(\mathbb{E}\left\| \sum_{j'\in\mathbb{N}^{\nu-1}} \epsilon_{j'} 2^{j'\cdot s} D_{j'}' f \right\|_{L^p}\right)^{\frac{1}{p}} &= \left\| \left(\mathbb{E}\left| \sum_{j'\in\mathbb{N}^{\nu-1}} \epsilon_{j'} 2^{j'\cdot s} D_{j'}' f \right|^p\right)^{\frac{1}{p}} \right\|_{L^p} \\ &\approx \|f\|_{L^p(\mathfrak{M},\mathrm{Vol};\ell^2(\mathbb{N}^{\nu-1})),s,\mathcal{D}_0'} < \infty.\end{aligned}$$

In particular, $\sum_{j'\in\mathbb{N}^{\nu-1}} \epsilon_{j'} 2^{j'\cdot s} D_{j'}' f \in L^p(\mathfrak{M}, \mathrm{Vol})$, $\mathbb{E}$-almost surely. We also know that $\mathrm{supp}(D_{j_{\nu-1}}^{\nu-1}) \subseteq \Omega_1^{\nu-1} \times \Omega_2^{\nu-1}$, and since $D_{j'}' = D_{j_{\nu-1}}^{\nu-1} D_{j_{\nu-2}}^{\nu-2} \cdots D_{j_1}^1$, we have

$$\mathrm{supp}\left(\sum_{j'\in\mathbb{N}^{\nu-1}} \epsilon_{j'} 2^{j'\cdot s} D_{j'}' f\right) \subseteq \Omega_2^{\nu-1}.$$

Thus, we may apply (6.157) with $g = \sum_{j'\in\mathbb{N}^{\nu-1}} \epsilon_{j'} 2^{j'\cdot s} D_{j'}' f$. We have

$$\begin{aligned}&\|f\|_{L^p(\mathfrak{M},\mathrm{Vol};\ell^2(\mathbb{N}^{\nu-1})),s,\mathcal{D}_0'} \\ &\quad = \left\| \left(\sum_{j'\in\mathbb{N}^{\nu-1}} |2^{j'\cdot s} D_{j'}' f|^2 \right) \right\|_{L^p} \\ &\quad \approx \left\| \left(\mathbb{E}\left| \sum_{j'\in\mathbb{N}^{\nu-1}} \epsilon_{j'} 2^{j'\cdot s} D_{j'}' f \right|^p\right)^{\frac{1}{p}} \right\|_{L^p} && \text{by Theorem 6.8.12} \\ &\quad = \left(\mathbb{E}\left\| \sum_{j'\in\mathbb{N}^{\nu-1}} \epsilon_{j'} 2^{j'\cdot s} D_{j'}' f \right\|_{L^p}^p\right)^{\frac{1}{p}} \\ &\quad \approx \left(\mathbb{E}\left\| \sum_{j_\nu\in\mathbb{N}} \sum_{j'\in\mathbb{N}^{\nu-1}} \epsilon_{j_\nu}^\nu \epsilon_{j'} 2^{j'\cdot s} D_{j_\nu}^\nu D_{j'}' f \right\|_{L^p}^p\right)^{\frac{1}{p}} && \text{by (6.157)}\end{aligned}$$

$$= \left(\mathbb{E} \left\| \sum_{j \in \mathbb{N}^\nu} \epsilon_j 2^{j' \cdot s} D_j f \right\|_{L^p}^p \right)^{\frac{1}{p}}$$

$$= \left\| \left(\mathbb{E} \left| \sum_{j \in \mathbb{N}^\nu} \epsilon_j 2^{j' \cdot s} D_j f \right|^p \right)^{\frac{1}{p}} \right\|_{L^p}$$

$$\approx \left\| \left(\sum_{j \in \mathbb{N}^\nu} |2^{j' \cdot s} D_j f|^2 \right) \right\|_{L^p} \qquad \text{by Theorem 6.8.12}$$

$$= \|f\|_{L^p(\mathfrak{M}, \mathrm{Vol}; \ell^2(\mathbb{N}^{\nu-1})), (s,0), \mathcal{D}_0},$$

completing the proof. □

Along with Proposition 6.8.8, the same proof ideas give the following useful corollary.

Corollary 6.8.13. *For* $a_1, \ldots, a_\nu > 0$ *sufficiently small, we have the following. For* $f \in L^p(\mathfrak{M}, \mathrm{Vol})$ *with* $\mathrm{supp}(f) \subseteq \mathcal{K}$, *we have*

$$\|f\|_{L^p(\mathfrak{M}, \mathrm{Vol}; \ell^2(\mathbb{N}^\nu)), 0, \mathcal{D}_0} \approx \|f\|_{L^p(\mathfrak{M}, \mathrm{Vol})}.$$

This corollary holds even when $\nu = 1$.

Proof. We prove the result by induction on ν. For the base case, $\nu = 1$, we use Corollary 5.10.7 with $\mathcal{K}_0 \Subset \Omega_3 \Subset \Omega_4$ replaced by $\mathcal{K} \Subset \Omega_1^1 \Subset \Omega_2^1$, with $D_{j_1}^1$ playing the role of D_j in that result, to see that, for $f \in L^p(\mathcal{K}, \mathrm{Vol})$,

$$\left\| \left(\sum_{j_1 \in \mathbb{N}} |D_{j_1}^1 f|^2 \right)^{\frac{1}{2}} \right\|_{L^p} \approx \|f\|_{L^p},$$

which is exactly the claimed result when $\nu = 1$.

The inductive step is the case $s = 0$ of Proposition 6.8.8 (see also Remark 6.8.9). □

6.9 Sobolev spaces

An important special case of the Triebel–Lizorkin spaces are the spaces $\mathscr{F}_{p,2}^{s}(\mathcal{K}, (W, \vec{d}))$, $1 < p < \infty$, which we think of as L^p Sobolev spaces adapted to $(W, \vec{d})$, consisting of distributions supported in $\mathcal{K}$. The key to understanding this is the next proposition and its corollary.

Proposition 6.9.1. *For* $1 < p < \infty$, $\mathscr{F}_{p,2}^{0}(\mathcal{K}, (W, \vec{d})) = L^p(\mathcal{K}, \mathrm{Vol})$, *and the corresponding norms are equivalent.*

Proof. Let $\mathcal{D}_0$ be as in Corollary 6.8.13. Since (see Proposition 6.3.7)

$$\|\cdot\|_{\mathscr{F}^0_{p,2}(W,\vec{d})} \approx \|\cdot\|_{L^p(\mathfrak{M},\mathrm{Vol};\ell^2(\mathbb{N}^\nu)),0,\mathcal{D}_0},$$

Corollary 6.8.13 shows that the norms $\|\cdot\|_{L^p}$ and $\|\cdot\|_{\mathscr{F}^0_{p,2}(W,\vec{d})}$ are equivalent on $L^p(\mathcal{K},\mathrm{Vol}) \bigcap \mathscr{F}^0_{p,2}(\mathcal{K},(W,\vec{d}))$. Thus, once we have the equality of the spaces, the equivalence of the norms follows from Corollary 6.8.13.

First we prove the result in the special case where $\mathcal{K} = \overline{\Omega_1}$, for some open set $\Omega_1 \Subset \mathfrak{M}$. For $f \in L^p(\mathcal{K},\mathrm{Vol})$, by Corollary 6.8.13 we have

$$\|f\|_{L^p(\mathcal{K},\mathrm{Vol})} \approx \|f\|_{L^p(\mathfrak{M},\mathrm{Vol};\ell^2(\mathbb{N}^\nu)),s,\mathcal{D}_0}.$$

By Proposition 6.5.5, $C_0^\infty(\Omega_1) \subseteq \mathscr{F}^0_{p,2}(\mathcal{K},(W,\vec{d}))$, and therefore for $f \in C_0^\infty(\Omega_1)$, we have

$$\|f\|_{L^p} \approx \|f\|_{L^p(\mathfrak{M},\mathrm{Vol};\ell^2(\mathbb{N}^\nu)),0,\mathcal{D}_0} \approx \|f\|_{\mathscr{F}^0_{p,2}(W,\vec{d})}. \tag{6.158}$$

Using the fact that $C_0^\infty(\Omega_1)$ is dense in $L^p(\mathcal{K},\mathrm{Vol})$ and that $\mathscr{F}^0_{p,2}(\mathcal{K},(W,\vec{d}))$ is complete (Proposition 6.3.6), we have

$$L^p(\mathcal{K},\mathrm{Vol}) = \overline{C_0^\infty(\Omega_1)} \subseteq \mathscr{F}^0_{p,2}(\mathcal{K},(W,\vec{d})),$$

where $\overline{C_0^\infty(\Omega_1)}$ denotes the completion of $C_0^\infty(\Omega_1)$ with respect to any of the equivalent norms (6.158).

Conversely, we wish to show $\mathscr{F}^0_{p,2}(\mathcal{K},(W,\vec{d})) \subseteq L^p(\mathcal{K},\mathrm{Vol})$. Fix $f \in \mathscr{F}^0_{2,p}(\mathcal{K},(W,\vec{d}))$. Let $\Omega \Subset \mathfrak{M}$ be an open, relatively compact set and $\epsilon > 0$. By Proposition 6.5.17, there exists a sequence $\{f_L\}_{L\in\mathbb{N}} \subset \mathscr{F}^{(\epsilon,\epsilon,\ldots,\epsilon)}_{s,2}(\overline{\Omega},(W,\vec{d}))$ such that $f_L \xrightarrow{L\to\infty} f$ in $\mathscr{F}^0_{p,2}(\overline{\Omega},(W,\vec{d}))$. Proposition 6.5.16 shows that $f_L \in L^p(\overline{\Omega},\mathrm{Vol})$ for every L. Thus, since the sequence f_L is Cauchy in the norm $\|\cdot\|_{\mathscr{F}^0_{p,2}}$, it is also Cauchy in the equivalent norm $\|\cdot\|_{L^p}$. We conclude that f_L converges in $L^p(\overline{\Omega},\mathrm{Vol})$, and therefore $f \in L^p(\mathfrak{M},\mathrm{Vol})$. Since $\mathrm{supp}(f) \subseteq \mathcal{K}$, we conclude that $f \in L^p(\mathcal{K},\mathrm{Vol})$, as desired. This completes the proof in the case where $\mathcal{K} = \overline{\Omega_1}$.

For general $\mathcal{K}$, pick $\Omega_1 \Subset \mathfrak{M}$ open with $\mathcal{K} \Subset \Omega_1$ and set $\mathcal{K}_1 := \overline{\Omega_1}$. Proposition 6.5.6 shows that

$$f \in \mathscr{F}^0_{p,2}(\mathcal{K},(W,\vec{d})) \iff f \in \mathscr{F}^0_{p,2}(\mathcal{K}_1,(W,\vec{d})) \text{ and } \mathrm{supp}(f) \subseteq \mathcal{K},$$

and it is a standard fact that

$$f \in L^p(\mathcal{K},\mathrm{Vol}) \iff f \in L^p(\mathcal{K}_1,\mathrm{Vol}) \text{ and } \mathrm{supp}(f) \subseteq \mathcal{K}.$$

Thus, the equality $L^p(\mathcal{K},\mathrm{Vol}) = \mathscr{F}^0_{p,2}(\mathcal{K},(W,\vec{d}))$ follows from the already proved equality $L^p(\mathcal{K}_1,\mathrm{Vol}) = \mathscr{F}^0_{p,2}(\mathcal{K}_1,(W,\vec{d}))$, completing the proof. □

Proof of Proposition 6.2.13. In light of Proposition 6.6.1, this is a special case of Proposition 6.9.1. □

Corollary 6.9.2. *Fix* $p \in (1,\infty)$. *For each* $\mu \in \{1,\ldots,\nu\}$, *let* $\kappa_\mu \in \mathbb{N}$ *be such that* d_j^μ *divides* κ_μ *for* $1 \le j \le r_\mu$ *(we allow the possibility* $\kappa_\mu = 0$*). Set* $\kappa := (\kappa_1,\ldots,\kappa_\nu) \in \mathbb{N}^\nu$. *Then, for* $s \in \mathbb{R}^\nu$ *and* $f \in C^\infty_{W,0}(\mathfrak{M})'$, *the following are equivalent:*
(i) $f \in \mathscr{F}^\kappa_{p,2}(\mathcal{K},(W,\vec{d}))$.
(ii) *For every ordered multi-index* α *with* $\deg_{\vec{d}}(\alpha) \le \kappa$, *we have* $W^\alpha f \in L^p(\mathcal{K},\mathrm{Vol})$.

Furthermore, for $f \in \mathscr{F}^\kappa_{p,2}(\mathcal{K},(W,\vec{d}))$, *we have*

$$\|f\|_{\mathscr{F}^\kappa_{p,2}(W,\vec{d})} \approx \sum_{\deg_{\vec{d}}(\alpha)\le\kappa} \|W^\alpha f\|_{L^p},$$

where the implicit constants do not depend on f *(but may depend on any of the other ingredients).*

Proof. In light of Proposition 6.9.1, this follows immediately from Corollary 6.5.13 with $s = \kappa$. □

6.10 Distributions of finite norm

In Definition 6.3.8, we defined the norm $\|f\|_{\mathscr{X}^s(W,\vec{d})} := \|f\|_{\mathscr{V},s,\mathcal{D}_0}$. However, if $f \in C^\infty_{W,0}(\mathfrak{M})'$ satisfies $\mathrm{supp}(f) \subseteq \mathcal{K}$ and $\|f\|_{\mathscr{V},s,\mathcal{D}_0} < \infty$, we have not shown that $f \in \mathscr{X}^s(\mathcal{K},(W,\vec{d}))$. Remark 6.4.6 shows that we can replace $\|\cdot\|_{\mathscr{V},s,\mathcal{D}_0}$ with a different norm which is equivalent on $\mathscr{X}^s(\mathcal{K},(W,\vec{d}))$ and whose finiteness does imply belonging to the space $\mathscr{X}^s(\mathcal{K},(W,\vec{d}))$.

For some of our applications, the norm $\|\cdot\|_{\mathscr{V},s,\mathcal{D}_0}$ is the most convenient choice. The main result of this section shows that if $s \in (0,\infty)^\nu$, then finiteness of this norm does imply belonging to the space $\mathscr{X}^s(\mathcal{K},(W,\vec{d}))$.

Proposition 6.10.1. *Fix* $s \in (0,\infty)^\nu$. *Let* $f \in C^\infty_{W,0}(\mathfrak{M})'$ *with* $\mathrm{supp}(f) \subseteq \mathcal{K}$. *Then the following are equivalent:*
(i) $f \in \mathscr{X}^s(\mathcal{K},(W,\vec{d}))$.
(ii) $\|f\|_{\mathscr{V},s,\mathcal{D}_0} < \infty$.

Proof. (i)⇒(ii) is trivial, so we prove only the reverse implication. Thus, suppose $f \in C^\infty_{W,0}(\mathfrak{M})'$ with $\mathrm{supp}(f) \subseteq \mathcal{K}$ and (ii) holds. Recall that $\mathcal{D}_0 = \{(D_j, 2^{-j}) : j \in \mathbb{N}^\nu\}$ is a bounded set of generalized $(W,\vec{d})$ elementary operators with $\sum_{j\in\mathbb{N}^\nu} D_j = \mathrm{Mult}[\psi]$, where $\psi \in C^\infty_0(\mathfrak{M})$ satisfies $\psi \equiv 1$ on a neighborhood of $\mathcal{K}$.

$\mathscr{V}$ is equal to either $L^p(\mathfrak{M},\mathrm{Vol};\ell^q(\mathbb{N}^\nu))$ or $\ell^q(\mathbb{N}^\nu;L^p(\mathfrak{M},\mathrm{Vol}))$ (see Notation 6.3.1). In either case, fix this $p,q \in [1,\infty]$.

We claim $f \in L^p(\mathfrak{M}, \mathrm{Vol})$. Indeed, since $\mathrm{supp}(f) \subseteq \mathcal{K}$, we have $\sum_{j\in\mathbb{N}^\nu} D_j f = \mathrm{Mult}[\psi]f = f$, with convergence in $C^\infty_{W,\mathrm{loc}}(\mathfrak{M})'$, by Proposition 5.5.10. We also have

$$\|\{2^{j\cdot s} D_j f\}_{j\in\mathbb{N}^\nu}\|_{\mathcal{V}} = \|f\|_{\mathcal{V},s,\mathcal{D}_0} < \infty.$$

In particular, this implies

$$\|D_j f\|_{L^p(\mathfrak{M},\mathrm{Vol})} \lesssim 2^{-j\cdot s}. \tag{6.159}$$

Thus, $\sum_{j\in\mathbb{N}^\nu} \|D_j f\|_{L^p} < \infty$, and therefore $f = \sum_{j\in\mathbb{N}^\nu} D_j f$ converges in the norm topology on $L^p(\mathfrak{M}, \mathrm{Vol})$, showing that $f \in L^p(\mathfrak{M}, \mathrm{Vol})$.

Set

$$\epsilon_0 := \frac{\min_{\mu\in\{1,\ldots,\nu\}} s_\mu}{2} > 0.$$

We will show $\forall \epsilon \in (0, \epsilon_0]$ that $f \in \mathscr{X}^{s-\epsilon(1,1,\ldots,1)}(\mathcal{K}, (W, \vec{d}))$, and (i) will then follow from Lemma 6.8.7. For the remainder of the proof, ϵ will denote an arbitrary element of $(0, \epsilon_0]$.

Fix a bounded set of $(W, \vec{d})$ of generalized $(W, \vec{d})$ elementary operators, $\mathcal{E}$. In what follows, $(E_j, 2^{-j})$ will denote an arbitrary element of $\mathcal{E}$. We write $A \lesssim B$ for $A \leq CB$, where $C \geq 0$ does not depend on $(E_j, 2^{-j}) \in \mathcal{E}$ or on the variable $k \in \mathbb{N}^\nu$. However, C may depend on f.

We claim that, $\forall N \in \mathbb{N}$,

$$\|E_j D_k\|_{L^p(\mathfrak{M},\mathrm{Vol})\to L^p(\mathfrak{M},\mathrm{Vol})} \lesssim 2^{-N|j_1-k_1|}2^{-N|j_2-k_2|}\cdots 2^{-N|j_\nu-k_\nu|}. \tag{6.160}$$

Indeed, by Proposition 5.5.11, for every $N_0 \in \mathbb{N}$, $\{(2^{N_0|j-k|}E_j D_k, 2^{-j}) : (E_j, 2^{-j}) \in \mathcal{E}, k \in \mathbb{N}^\nu\}$ is a bounded set of generalized $(W, \vec{d})$ elementary operators and therefore by Definition 5.2.26,

$$\|E_j D_k\|_{L^p(\mathfrak{M},\mathrm{Vol})\to L^p(\mathfrak{M},\mathrm{Vol})} \lesssim 2^{-N_0|j-k|}.$$

The estimate (6.160) follows by taking $N_0 \geq N\nu$.

Fix $\delta = \delta(\epsilon) \in (0,1)$ large enough that if we set $c_\mu := \frac{s_\mu-\epsilon}{\delta s_\mu}$, we have $c_\mu \in (0,1)$; this is always possible since $\frac{s_\mu-\epsilon}{s_\mu} \in (0,1)$ by the choice of ϵ. By (6.160) with N replaced by $N/(1-\delta)$ and the already proved fact that $f \in L^p$, we have

$$\|E_j D_k f\|_{L^p} \lesssim 2^{-\frac{N}{1-\delta}|j_1-k_1|}2^{-\frac{N}{1-\delta}|j_2-k_2|}\cdots 2^{-\frac{N}{1-\delta}|j_\nu-k_\nu|}. \tag{6.161}$$

By (6.159) and the fact that $\sup_{(E_j,2^{-j})\in\mathcal{E}} \|E_j\|_{L^p\to L^p} \lesssim 1$ (see Definition 5.2.26), we have

$$\|E_j D_k f\|_{L^p} \lesssim 2^{-k\cdot s}. \tag{6.162}$$

Combining (6.161) and (6.162), we have

$$\|E_j D_k f\|_{L^p} \lesssim 2^{-k_1\delta s_1} 2^{-k_2\delta s_2} \cdots 2^{-k_\nu \delta s_\nu} 2^{-N|j_1-k_1|} 2^{-N|j_2-k_2|} \cdots 2^{-N|j_\nu-k_\nu|}. \tag{6.163}$$

We claim that, $\forall \epsilon \in (0, \epsilon_0]$,

$$\|E_j f\|_{L^p(\mathfrak{M},\mathrm{Vol})} \leq C_\epsilon 2^{-j\cdot(s-\epsilon(1,1,\ldots,1))}. \tag{6.164}$$

Indeed, using (6.163) with $N \geq \max_{\mu\in\{1,\ldots,\nu\}}(s_\mu - \epsilon)/(1-c_\mu)$, we have

$$\begin{aligned}
\|E_j f\|_{L^p} &= \Big\| E_j \sum_{k\in\mathbb{N}^\nu} D_k f \Big\|_{L^p} \leq \sum_{k\in\mathbb{N}^\nu} \|E_j D_k f\|_{L^p} \\
&= \sum_{F\subseteq\{1,\ldots,\nu\}} \sum_{\substack{k_\mu \geq c_\mu j_\mu, \mu\in F \\ k_\mu < c_\mu j_\mu, \mu\notin F}} \|E_j D_k f\|_{L^p} \\
&\lesssim \sum_{F\subseteq\{1,\ldots,\nu\}} \sum_{\substack{k_\mu \geq c_\mu j_\mu, \mu\in F \\ k_\mu < c_\mu j_\mu, \mu\notin F}} \Big(\prod_{\mu\in F} 2^{-k_\mu\delta_\mu s_\mu}\Big)\Big(\prod_{\mu\notin F} 2^{-N(j_\mu-k_\mu)}\Big) \\
&= \sum_{F\subseteq\{1,\ldots,\nu\}} \Big(\prod_{\mu\in F} \sum_{k_\mu\geq c_\mu j_\mu} 2^{-k_\mu\delta_\mu s_\mu}\Big)\Big(\prod_{\mu\notin F} \sum_{k_\mu < c_\mu j_\mu} 2^{-N(j_\mu-k_\mu)}\Big) \\
&\lesssim \sum_{F\subseteq\{1,\ldots,\nu\}} \Big(\prod_{\mu\in F} 2^{-c_\mu\delta_\mu j_\mu s_\mu}\Big)\Big(\prod_{\mu\notin F} 2^{-N(1-c_\mu)j_\mu}\Big) \\
&\leq \Big(\prod_{\mu\in F} 2^{-(s_\mu-\epsilon)j_\mu}\Big)\Big(\prod_{\mu\notin F} 2^{-(s_\mu-\epsilon)j_\mu}\Big) \\
&\lesssim 2^{-j\cdot(s-\epsilon(1,1,\ldots,1))},
\end{aligned}$$

proving (6.164).

Now let $\{(E_j, 2^{-j}) : j \in \mathbb{N}^\nu\} \subseteq \mathcal{E}$. In what follows, implicit constants will not depend on the particular subset of $\mathcal{E}$. Since $p, q \geq 1$, we have $\|\cdot\|_{L^p(\ell^q)} \leq \|\cdot\|_{L^p(\ell^1)} \leq \|\cdot\|_{\ell^1(L^p)}$ and $\|\cdot\|_{\ell^q(L^p)} \leq \|\cdot\|_{\ell^1(L^p)}$. Thus, we have $\|\cdot\|_{\mathscr{V}} \leq \|\cdot\|_{\ell^q(\mathbb{N}^\nu; L^p(\mathfrak{M},\mathrm{Vol}))}$. Using this and (6.164) with ϵ replaced by $\epsilon/2$, we have

$$\begin{aligned}
\big\|\{2^{j\cdot(s-\epsilon(1,1,\ldots,1))} E_j f\}_{j\in\mathbb{N}^\nu}\big\|_{\mathscr{V}} &\leq \sum_{j\in\mathbb{N}^\nu} 2^{j\cdot(s-\epsilon(1,1,\ldots,1))} \|E_j f\|_{L^p(\mathfrak{M},\mathrm{Vol})} \\
&\lesssim \sum_{j\in\mathbb{N}^\nu} 2^{-j\cdot(\epsilon/2)(1,1,\ldots,1)} \lesssim 1.
\end{aligned}$$

Taking the supremum over all $\{(E_j, 2^{-j}) : j \in \mathbb{N}^\nu\} \subseteq \mathcal{E}$ shows that $\|f\|_{\mathscr{V},s-\epsilon(1,1,\ldots,1),\mathcal{E}} < \infty$. Since $\mathrm{supp}(f) \subseteq \mathcal{K}$, this implies $f \in \mathscr{X}^{s-\epsilon(1,1,\ldots,1)}(\mathcal{K}, (W, \vec{d}))$.

Since $\epsilon \in (0, \epsilon_0]$ was arbitrary, we have $f \in \bigcap_{\epsilon\in(0,\epsilon_0]} \mathscr{X}^{s-\epsilon(1,1,\ldots,1)}(\mathcal{K}, (W, \vec{d}))$. Since $\|f\|_{\mathscr{V},s,\mathcal{D}_0} < \infty$, Lemma 6.8.7 implies $f \in \mathscr{X}^s(\mathcal{K}, (W, \vec{d}))$, completing the proof. □

6.11 An explicit choice of norm

Fix $\mathcal{K} \Subset \mathfrak{M}$, compact. For $f \in \mathscr{X}^s(\mathcal{K}, (W, \vec{d}))$, in Definition 6.2.7, we defined the norm

$$\|f\|_{\mathscr{X}^s(W,\vec{d})} := \|f\|_{\mathscr{V},s,\mathcal{D}_0}, \quad f \in \mathscr{X}^s(\mathcal{K}, (W, \vec{d})), \tag{6.165}$$

where $\mathcal{D}_0 = \{(D_j, 2^{-j}) : j \in \mathbb{N}^\nu\}$ is a bounded set of $(W, \vec{d})$ generalized elementary operators with $\sum_{j\in\mathbb{N}^\nu} D_j = \mathrm{Mult}[\psi]$ and $\psi \in C_0^\infty(\mathfrak{M})$ is such that $\psi \equiv 1$ on a neighborhood of $\mathcal{K}$.

In the above definition, (6.165), several choices were made. Proposition 6.2.6 shows that the equivalence class of the norm does not depend on any of these choices; however, the norm itself does depend on the choices. Thus, if one is only interested in qualitative results, the particular choice of norm is not relevant. However, when we study nonlinear maximally subelliptic equations, we will require quantitative estimates, and then the particular choice of norm does matter.

In this section, we describe a particular choice of $\mathcal{D}_0$, along with all the choices which must be made in the definition of the norm. This will allow us to be explicit about the exact norm (not just the equivalence class of the norm) when we turn to quantitative estimates.

Fix relatively compact open sets $\Omega_0, \Omega_1, \Omega_2$, and Ω with $\mathcal{K} \Subset \Omega_0 \Subset \Omega_1 \Subset \Omega_2 \Subset \Omega \Subset \mathfrak{M}$. Fix $\psi \in C_0^\infty(\Omega_1)$ with $\psi \equiv 1$ on $\overline{\Omega_0}$. The final choice of norm will depend on $\mathcal{K}, \Omega_0, \Omega_1$, and Ω_2 (along with other choices we will make precise) but not on ψ or Ω.

Recall that our goal is to write

$$\mathrm{Mult}[\psi] = \sum_{j\in\mathbb{N}^\nu} D_j,$$

where $\mathcal{D}_0 = \{(D_j, 2^{-j}) : j \in \mathbb{N}^\nu\}$ is a bounded set of $(W, \vec{d})$ generalized elementary operators with $\sum_{j\in\mathbb{N}^\nu} D_j = \mathrm{Mult}[\psi]$, for a specific choice of $\mathcal{D}_0$.

By assumption, each $\mathrm{Gen}((W^\mu, \vec{d}^\mu))$ is finitely generated by some $(X^\mu, d^\mu) = \{(X_1^\mu, d_1^\mu), \ldots, (X_{q_\mu}^\mu, d_{q_\mu}^\mu)\} \subset \mathrm{Gen}((W^\mu, \vec{d}^\mu))$ on Ω with $(W^\mu, \vec{d}^\mu) \subseteq (X^\mu, d^\mu)$. Set

$$(X, \vec{d}) = \{(X_1, \vec{d}_1), \ldots, (X_q, \vec{d}_q)\} := (X^1, d^1) \boxtimes (X^2, d^2) \boxtimes \cdots \boxtimes (X^\nu, d^\nu).$$

In particular, $(W, \vec{d}) \subseteq (X, \vec{d}) \subseteq \mathrm{Gen}((W, \vec{d}))$ and by Proposition 3.8.8 (a), $\mathrm{Gen}((W, \vec{d}))$ is finitely generated by $(X, \vec{d})$ on Ω; therefore, by Remark 3.4.12, we have, for $1 \le j, k \le q$,

$$[X_j, X_k] = \sum_{\vec{d}_l \le \vec{d}_j + \vec{d}_l} c_{j,k}^l X_l, \quad c_{j,k}^l \in C_{\mathrm{loc}}^\infty(\Omega).$$

Fix $a > 0$ a small number. Throughout the construction, we may shrink $a > 0$ a finite number of times; we will see that how small a must be depends only on $\Omega_0, \Omega_1, \Omega_2$, and $(X, \vec{d})$.

Fix $\eta \in C_0^\infty(B^q(a))$ with $\eta \equiv 1$ on a neighborhood of 0. For each $\mu \in \{1,\ldots,\nu\}$, let $\rho_\mu(t^\mu) \in \mathscr{S}(\mathbb{R}^{q_\mu})$ be such that $\hat{\rho}_\mu(\xi^\mu) \equiv 1$ on a neighborhood of $0 \in \mathbb{R}^{q_\mu}$, where $\hat{\rho}_\mu$ denotes the Fourier transform of ρ_μ. Note that ρ_μ can be chosen to depend only on q_μ and η can be chosen to depend only a and q.

For $F \subseteq \{1,\ldots,\nu\}$, define $\tau_F(t) \in \mathscr{S}(\mathbb{R}^q)$ by

$$\hat{\tau}_F(\xi) = \left[\prod_{\mu\notin F} \hat{\rho}_\mu(\xi^\mu)\right]\left[\prod_{\mu\in F}(\hat{\rho}_\mu(\xi^\mu) - \hat{\rho}_\mu(2^{d^\mu}\xi_\mu))\right],$$

where $2^{d^\mu}\xi^\mu$ is defined in (4.4) and $\hat{\tau}_F$ denotes the Fourier transform of τ_F. Since $\hat{\rho}_\mu(\xi^\mu) - \hat{\rho}_\mu(2^{d^\mu}\xi^\mu) \equiv 0$ for $|\xi^\mu|$ small, we have $\tau_F \in \mathscr{S}_F(\mathbb{R}^q)$, i. e.,

$$\int_{\mathbb{R}^{q_\mu}} (t^\mu)^\alpha \tau_F(t)\, dt^\mu \equiv 0, \quad \forall \mu \in F, \alpha \in \mathbb{N}^{q_\mu}. \tag{6.166}$$

For $j \in \mathbb{N}^\nu$, set

$$\varsigma_j(x,s,t) := \psi(x)\eta(s)\tau_{\{\mu:j_\mu\neq 0\}}(t).$$

As in (4.3), set

$$\Gamma(t,x) = \Gamma((t^1,\ldots,t^\nu),x) := e^{-t^1\cdot X^1}e^{-t^2\cdot X^2}\cdots e^{-t^\nu\cdot X^\nu}x. \tag{6.167}$$

If $a > 0$ is sufficiently small, then $\Gamma(t,x)$ is defined for $x \in \Omega_1$ and $t \in B^q(a)$ and satisfies

$$x \in \Omega_1, t \in B^q(a) \Rightarrow \Gamma(t,x) \in \Omega_2, \tag{6.168}$$

$$\Gamma(t,x) \in \mathcal{K}, t \in B^q(a) \Rightarrow x \in \Omega_0. \tag{6.169}$$

For $j \in \mathbb{N}^\nu$, set

$$\begin{aligned} D_j f(x) &:= \psi(x)\int f(\Gamma(t,x))\,\mathrm{Dil}_{2^j}^{\vec{d}}(\varsigma_j)(x,t,t)\,dt \\ &= \psi(x)\int f(\Gamma(t,x))\eta(t)\,\mathrm{Dil}_{2^j}^{\vec{d}}(\tau_{\{\mu:j_\mu\neq0\}})(t)\,dt, \end{aligned} \tag{6.170}$$

$$P_j := \sum_{\substack{k\in\mathbb{N}^\nu\\ k\leq j}} D_j.$$

Proposition 6.11.1. *With D_j and P_j as above, if $a = a((X,\vec{d}),\mathcal{K},\Omega_0,\Omega_1,\Omega_2) > 0$ is sufficiently small, we have:*

(i) *$\{(D_j, 2^{-j}) : j \in \mathbb{N}^\nu\}$ is a bounded set of generalized $(W,\vec{d})$ elementary operators supported in Ω.*

(ii) $\{(P_j, 2^{-j}) : j \in \mathbb{N}^\nu\}$ *is a bounded set of generalized* $(W, \vec{d})$ *pre-elementary operators supported in* Ω.
(iii) $\sum_{j\in\mathbb{N}^\nu} D_j = \mathrm{Mult}[\psi]$; *see Proposition* 5.5.10 *for the convergence of this sum.*
(iv) *If* $\mathrm{supp}(f) \subseteq \mathcal{K}$, *then*

$$D_j f(x) = \int f(\Gamma(t,x))\eta(t)\,\mathrm{Dil}_{2^j}^{\vec{d}}(\tau_{\{\mu:j_\mu\neq 0\}})(t)\,dt;$$

in particular, $D_j f$ *does not depend on the choice of* ψ.

In light of Proposition 6.11.1, the next definition makes sense and yields an explicit choice of norm in the equivalence class of $\|\cdot\|_{\mathscr{X}^s(W,\vec{d})}$.

Definition 6.11.2. For $f \in C_0^\infty(\mathfrak{M})'$ with $\mathrm{supp}(f) \subseteq \mathcal{K}$ and $s \in \mathbb{R}^\nu$, we set

$$\|f\|_{\mathscr{X}^s(W,\vec{d}),(X,\vec{d}),a,\Omega_0,\Omega_1,\Omega_2,\mathrm{Vol}} := \|f\|_{\mathscr{V},s,\{(D_j,2^{-j}):j\in\mathbb{N}^\nu\}}.$$

The rest of this section is devoted to the proof of Proposition 6.11.1. To prove it, we introduce the following more precise version.

Proposition 6.11.3. *If* $a = a(\Omega_1, \Omega_2, \mathcal{K}) > 0$ *is sufficiently small, then:*
(i) $\sum_{j\in\mathbb{N}^\nu} D_j = \mathrm{Mult}[\psi]$ *with convergence in the topology of bounded convergence on* $\mathrm{Hom}(C^\infty_{\mathrm{loc}}(\mathfrak{M}), C_0^\infty(\Omega_1))$ *and in the weak operator topology on* $\mathrm{Hom}(C_0^\infty(\mathfrak{M})', C^\infty_{\mathrm{loc}}(\Omega_1)')$, *where* $C^\infty_{\mathrm{loc}}(\Omega_1)'$ *is given the weak topology.*
(ii) $\{(D_j, 2^{-j}) : j \in \mathbb{N}^\nu\}$ *is a bounded set of* $(X, \vec{d}), a$ *pseudo-differential operator scales supported in* $\Omega_1 \times \Omega_2$.
(iii) $\{(P_j, 2^{-j}) : j \in \mathbb{N}^\nu\}$ *is a bounded set of* $(X, \vec{d}), a$ *pre-pseudo-differential operator scales supported in* $\Omega_1 \times \Omega_2$.

Proof. (ii): Since $\{\varsigma_j : j \in \mathbb{N}^\nu\} = \{\psi \otimes \eta \otimes \tau_F : F \subseteq \{1,\dots,\nu\}\} \subset C_0^\infty(\Omega_1)\widehat{\otimes}C_0^\infty(B^q(a))\widehat{\otimes}\mathscr{S}(\mathbb{R}^q)$ is a finite set, it is a bounded set. It follows from (6.166) that $\varsigma_j = \psi \otimes \eta \otimes \tau_{\{\mu:j_\mu\neq 0\}} \in C_0^\infty(\Omega_1)\widehat{\otimes}C_0^\infty(B^q(a))\widehat{\otimes}\mathscr{S}_{\{\mu:j_\mu\neq 0\}}(\mathbb{R}^q)$. From here, (ii) follows immediately from the definitions (see Definition 4.3.1 (ii)).

(i): By (ii), $\{(D_j, 2^{-j}) : j \in \mathbb{N}^\nu\}$ is a bounded set of $(X, \vec{d}), a$ pseudo-differential operator scales supported in $\Omega_1 \times \Omega_2$. Thus, by Theorem 4.3.3, if $a > 0$ is sufficiently small, $\sum_{j\in\mathbb{N}^\nu} D_j$ converges in the topology of bounded convergence on $\mathrm{Hom}(C^\infty_{\mathrm{loc}}(\mathfrak{M}), C_0^\infty(\Omega_1))$ and in the weak operator topology on $\mathrm{Hom}(C_0^\infty(\mathfrak{M})', C^\infty_{\mathrm{loc}}(\Omega_1)')$, where $C^\infty_{\mathrm{loc}}(\Omega_1)'$ is given the weak topology. Thus, to prove (i) it suffices to show that

$$\sum_{j\in\mathbb{N}^\nu} D_j f(x) = \psi(x) f(x), \quad \forall f \in C_0^\infty(\mathfrak{M}). \tag{6.171}$$

By Remark 4.1.4,

$$\begin{aligned}\sum_{j\in\mathbb{N}^\nu} D_jf(x) &= \int f(\Gamma(t,x))\Big(\sum_{j\in\mathbb{N}^\nu} \mathrm{Dil}^{\vec{d}}_{2^j}(\varsigma_j)(x,t,t)\Big)\,dt \\ &= \iint f(\Gamma(t,x))\Big(\sum_{j\in\mathbb{N}^\nu} \hat{\varsigma}_j(x,t,2^{-j\vec{d}}\xi)\Big)e^{2\pi i t\cdot\xi}\,dt\,d\xi.\end{aligned} \tag{6.172}$$

Because of this, to prove (6.171), it suffices to show that

$$\sum_{j\in\mathbb{N}^\nu} \hat{\varsigma}_j(x,s,\xi) = \psi(x)\eta(s). \tag{6.173}$$

Indeed, once we have (6.173), then (6.172) together with the fact that $\eta(0)=1$ yields

$$\begin{aligned}\sum_{j\in\mathbb{N}^\nu} D_jf(x) &= \iint f(\Gamma(t,x))\psi(x)\eta(t)e^{2\pi i t\cdot\xi}\,dt\,d\xi \\ &= \iint f(\Gamma(t,x))\psi(x)\eta(t)\delta_0(t)\,dt = f(\Gamma(0,x))\psi(x)\eta(0) = f(x)\psi(x).\end{aligned}$$

We complete the proof of (i) by proving (6.173). In fact, we have, for $j\in\mathbb{N}^\nu$,

$$\begin{aligned}\sum_{\substack{k\le j\\ k\in\mathbb{N}^\nu}} \hat{\varsigma}_k(x,s,2^{-k\vec{d}}\xi) &= \psi(x)\eta(s)\prod_{\mu\in\{1,\dots,\nu\}}\Big(\hat{\rho}_\mu(\xi^\mu) + \sum_{1\le k_\mu\le j_\mu}(\hat{\rho}_\mu(2^{-k_\mu d^\mu}\xi^\mu) - \hat{\rho}_\mu(2^{-(k_\mu-1)d^\mu}\xi^\mu))\Big) \\ &= \psi(x)\eta(s)\hat{\rho}_1(2^{-j_1 d^1}\xi^1)\cdots\hat{\rho}_\nu(2^{-j_\nu d^\nu}\xi^\nu).\end{aligned} \tag{6.174}$$

Since each $\hat{\rho}_\mu(\xi^\mu)=1$ for ξ^μ small, the right-hand side of (6.174) clearly tends to $\psi(x)\eta(s)$ as $j_1,\dots,j_\nu\to\infty$, proving (6.173) and completing the proof of (i).

(iii): By (6.174), we have

$$\sum_{\substack{k\le j\\ k\in\mathbb{N}^\nu}} \hat{\varsigma}_k(x,s,2^{-k\vec{d}}\xi) = \psi(x)\eta(s)\hat{\rho}_1(2^{-j_1 d^1}\xi^1)\cdots\hat{\rho}_\nu(2^{-j_\nu d^\nu}\xi^\nu) = \hat{\tau}_\emptyset(x,s,2^{-j\vec{d}}\xi).$$

Thus,

$$P_jf(x) = \int f(\Gamma(t,x))\,\mathrm{Dil}^{\vec{d}}_{2^j}(\tau_\emptyset)(x,t,t)\,dt.$$

Since $\tau_\emptyset\in C_0^\infty(\Omega_1)\widehat{\otimes}C_0^\infty(B^q(a))\widehat{\otimes}\mathscr{S}(\mathbb{R}^q)$, if $a>0$ is sufficiently small, (iii) follows immediately from the definition of bounded sets of $(X,\vec{d})$, a pre-pseudo-differential operator scales supported in $\Omega_1\times\Omega_2$ (see Definition 4.3.1 (i)). □

Proof of Proposition 6.11.1. (i): In light of Proposition 6.11.3 (ii), this follows from Proposition 5.6.3.

(ii): In light of Proposition 6.11.3 (iii), this follows from Lemma 5.6.6.

(iii): This follows from Proposition 6.11.3 (i).

(iv): By (6.169), if $\mathrm{supp}(f) \subseteq \mathcal{K}$, then the integral

$$\int f(\Gamma(t,x))\eta(t)\,\mathrm{Dil}_{2^j}^{\vec{d}}(\tau_{\{\mu:j_\mu\neq 0\}})(t)\,dt$$

is supported on $x \in \Omega_0$. For such x, $\psi(x) = 1$, and therefore,

$$\begin{aligned} D_j f(x) &= \psi(x)\int f(\Gamma(t,x))\eta(t)\,\mathrm{Dil}_{2^j}^{\vec{d}}(\tau_{\{\mu:j_\mu\neq 0\}})(t)\,dt \\ &= \int f(\Gamma(t,x))\eta(t)\,\mathrm{Dil}_{2^j}^{\vec{d}}(\tau_{\{\mu:j_\mu\neq 0\}})(t)\,dt, \end{aligned}$$

completing the proof. □

6.11.1 The unit scale

In many of our main results, we require quantitative estimates at the multi-parameter unit scale described in Section 3.15.1. Here, $\mathfrak{M} = B^n(1)$ and $\mathrm{Vol} = h\sigma_{\mathrm{Leb}}$, where $h \in C^\infty(B^n(1);(0,\infty))$ with $\inf_{u\in B^n(1)} h(u) > 0$.

Letting $(W^1,\vec{d}^1),\ldots,(W^\nu,\vec{d}^\nu)$ and $(X^1,d^1),\ldots,(X^\nu,d^\nu)$ be as in Section 3.15.1, we set $(W,\vec{d}) := (W^1,\vec{d}^1)\boxtimes\cdots\boxtimes(W^\nu,\vec{d}^\nu)$ and $(X,\vec{d}) := (X^1,d^1)\boxtimes\cdots\boxtimes(X^\nu,d^\nu)$. For $\sigma \in (0,1)$ and $f \in C_0^\infty(B^n(1))'$ with $\mathrm{supp}(f) \subseteq \overline{B^n(\sigma)}$, we define

$$\|f\|_{\mathscr{X}^s(W,\vec{d}),(X,\vec{d}),a,\sigma,h} := \|f\|_{\mathscr{X}^s(W,\vec{d}),(X,\vec{d}),a,B^n(\sigma+(1-\sigma)/5),B^n(\sigma+2(1-\sigma)/5),B^n(\sigma+3(1-\sigma)/5),\mathrm{Vol}}.$$

In our applications, we will usually take $\sigma = 7/8$.

The main idea is that when we prove results in this setting, the estimates will only depend on $(W,\vec{d})$, $(X,\vec{d})$, and h through multi-parameter unit-admissible constants as described in Definition 3.15.1.

6.11.2 Scaling

The norm at the unit scale in Section 6.11.1 plays a particular role in some scaling estimates.

Fix compact sets $\mathcal{K}_0, \mathcal{K} \Subset \mathfrak{M}$ and relatively compact, open sets $U, \Omega_1, \Omega_2 \Subset \mathfrak{M}$, with $\mathcal{K}_0 \Subset U \Subset \mathcal{K} \Subset \Omega_1 \Subset \Omega_2 \Subset \mathfrak{M}$. Let $(W^1,\vec{d}^1),\ldots,(W^\nu,\vec{d}^\nu),(X^1,d^1),\ldots,(X^\nu,d^\nu) \subset C^\infty_{\mathrm{loc}}(\mathfrak{M};T\mathfrak{M})$, and $\lambda \in (0,\infty)^\nu$ be as in Section 3.15.2, with these choices of $\mathcal{K}$, Ω_1, and Ω_2. As usual, set $(W,\vec{d}) := (W^1,\vec{d}^1)\boxtimes\cdots\boxtimes(W^\nu,\vec{d}^\nu)$ and $(X,\vec{d}) := (X^1,d^1)\boxtimes\cdots\boxtimes(X^\nu,d^\nu)$.

Consider the following norm from Section 6.11:

$$\|f\|_{\mathscr{X}^s(W,\vec{d}),(X,\vec{d}),a,\Omega_0,\Omega_1,\Omega_2,\mathrm{Vol}}, \quad f \in \mathscr{X}^s(\mathcal{K},(W,\vec{d})),$$

defined in Definition 6.11.2. For $\delta \in (0,1]$, we also have the norm

$$\|f\|_{\mathscr{X}^s(\delta^{\lambda\vec{d}}W,\vec{d}),(\delta^{\lambda\vec{d}}X,\vec{d}),a,\Omega_0,\Omega_1,\Omega_2,\mathrm{Vol}}, \quad f \in \mathscr{X}^s(\mathcal{K},(W,\vec{d})).$$

These norms are both in the equivalence class of $\|f\|_{\mathscr{X}^s(W,\vec{d})}$ and therefore, for $f \in \mathscr{X}^s(\mathcal{K},(W,\vec{d}))$,

$$\|f\|_{\mathscr{X}^s(W,\vec{d}),(X,\vec{d}),a,\Omega_0,\Omega_1,\Omega_2,\mathrm{Vol}} \approx_\delta \|f\|_{\mathscr{X}^s(\delta^{\lambda\vec{d}}W,\vec{d}),(\delta^{\lambda\vec{d}}X,\vec{d}),a,\Omega_0,\Omega_1,\Omega_2,\mathrm{Vol}},$$

where the implicit constants depend on δ. When we turn to use scaling, as in Theorems 3.15.3 and 3.15.5 to prove quantitative results, the estimates are often in terms of $\|f\|_{\mathscr{X}^s(\delta^{\lambda\vec{d}}W,\vec{d}),(\delta^{\lambda\vec{d}}X,\vec{d}),a,\Omega_0,\Omega_1,\Omega_2,\mathrm{Vol}}$ instead of $\|f\|_{\mathscr{X}^s(W,\vec{d}),(X,\vec{d}),a,\Omega_0,\Omega_1,\Omega_2,\mathrm{Vol}}$. This is due to the next proposition.

Proposition 6.11.4. *Fix $\sigma \in (0,1)$. If $a = a(\mathcal{K},(X,\vec{d}),\Omega_0,\Omega_1,\Omega_2,\sigma) > 0$ is sufficiently small, the following holds. Let $\Phi_{x,\delta}$ be as in Theorem 3.15.3, with $\mathcal{K}$ replaced by $\mathcal{K}_0$ and Ω_1 replaced by U, and let $(W^{\mu,x,\delta},\vec{d}^{\mu}), (X^{\mu,x,\delta},d^{\mu}) \subset C^\infty(B^n(1);TB^n(1)) \times \mathbb{N}_+$ and $h_{x,\delta} \in C^\infty(B^n(1))$ be as in that theorem. Set $(W^{x,\delta},\vec{d}) := (W^{1,x,\delta},\vec{d}^1) \boxtimes \cdots \boxtimes (W^{\nu,x,\delta},\vec{d}^\nu)$ and $(X^{x,\delta},\vec{d}) := (X^{1,x,\delta},\vec{d}^1) \boxtimes \cdots \boxtimes (X^{\nu,x,\delta},\vec{d}^\nu)$. Fix $\phi \in C_0^\infty(B^n(\sigma))$. Then, for any $u \in C_0^\infty(\mathfrak{M})'$, $x \in \mathcal{K}_0$, $\delta \in (0,1]$, and $s \in \mathbb{R}^\nu$, we have*[4]

$$\|\phi\Phi_{x,\delta}^* u\|_{\mathscr{X}^s(W^{x,\delta},\vec{d}),(X^{x,\delta},\vec{d}),a,\sigma,h_{x,\delta}} = \Lambda(x,\delta)^{-1/p}\|(\phi\circ\Phi_{x,\delta}^{-1})u\|_{\mathscr{X}^s(\delta^{\lambda\vec{d}}W,\vec{d}),(\delta^{\lambda\vec{d}}X,\vec{d}),a,\Omega_0,\Omega_1,\Omega_2,\mathrm{Vol}},$$

where $\Lambda(x,\delta)$ is given by

$$\Lambda(x,\delta) := \max_{j_1,\ldots,j_n\in\{1,\ldots,q_1\}} \mathrm{Vol}(x)(\delta^{d_{j_1}^1}X_{j_1}^1(x),\ldots,\delta^{d_{j_n}^1}X_{j_n}^1(x)) \tag{6.175}$$

and we use the convention $\Lambda(x,\delta)^{1/\infty} = 1$.

Proof. Let $\widehat{D}_j^{x,\delta}$ be the operator D_j from (6.170), with $(X,\vec{d})$ replaced by $(X^{x,\delta},\vec{d})$. Similarly, let D_j^δ be the operator from (6.170) with $(X,\vec{d})$ replaced by $(\delta^{\lambda d}X,\vec{d})$. Recall that τ_F is chosen to depend only on $q_1,\ldots,q_\nu$, and η is chosen to depend only on a and q. Thus, we choose the *same* τ_F and η in the definitions of $\widehat{D}_j^{x,\delta}$ and D_j^δ.

By Proposition 6.11.1 (iv), we have

$$\begin{aligned}
&\widehat{D}_j^{x,\delta}(\phi\Phi_{x,\delta}^* u)(v)\\
&= \int (\phi\Phi_{x,\delta}^* u)(e^{-t^1\cdot X^{1,x,\delta}}e^{-t^2\cdot X^{2,x,\delta}}\cdots e^{-t^\nu\cdot X^{\nu,x,\delta}}v)\eta(t)\,\mathrm{Dil}_{2^j}^{\vec{d}}(\tau_{\{\mu:j_\mu\neq 0\}})(t)\,dt \qquad (6.176)\\
&= \int ((\phi\circ\Phi_{x,\delta}^{-1})u)(e^{-t^1\cdot\delta^{d^1}X^1}e^{-t^2\cdot(\lambda_2\delta)^{d^2}X^2}\cdots e^{-t^\nu\cdot(\lambda_\nu\delta)^{d^\nu}X^\nu}\Phi_{x,\delta}(v))\eta(t)\,\mathrm{Dil}_{2^j}^{\vec{d}}(\tau_{\{\mu:j_\mu\neq 0\}})(t)\,dt\\
&= [D_j^\delta((\phi\circ\Phi_{x,\delta})u)](\Phi_{x,\delta}(v))\\
&= \Phi_{x,\delta}^* D_j^\delta((\phi\circ\Phi_{x,\delta}^{-1})u)(v),
\end{aligned}$$

4 Here, $p \in [1,\infty]$ is the p for which $\mathscr{X}^s = \mathscr{F}_{p,q}^s$ or $\mathscr{X}^s = \mathscr{B}_{p,q}^s$.

where the final equality uses Proposition 6.11.1 (iv) again.

We complete the proof in the case where $\mathscr{X}^s = \mathscr{F}^s_{p,q}$. A similar proof works in the case where $\mathscr{X}^s = \mathscr{B}^s_{p,q}$; we leave the details to the reader. Using (6.176) and $\Phi^*_{x,\delta}\operatorname{Vol} = \Lambda(x,\delta)h_{x,\delta}\sigma_{\mathrm{Leb}}$, we have

$$
\begin{aligned}
&\|\phi\Phi^*_{x,\delta}u\|_{\mathscr{X}^s(W^{x,\delta},\vec{d}),(X^{x,\delta},\vec{d}),a,\sigma,h_{x,\delta}} \\
&\quad = \big\|\{\widehat{D}_j^{x,\delta}(\phi\Phi^*_{x,\delta}u)\}_{j\in\mathbb{N}^\nu}\big\|_{L^p(B^n(1),h_{x,\delta}\sigma_{\mathrm{Leb}};\ell^q(\mathbb{N}^\nu))} \\
&\quad = \Lambda(x,\delta)^{-1/p}\big\|\{\widehat{D}_j^{x,\delta}(\phi\Phi^*_{x,\delta}u)\}_{j\in\mathbb{N}^\nu}\big\|_{L^p(B^n(1),\Phi^*_{x,\delta}\operatorname{Vol};\ell^q(\mathbb{N}^\nu))} \\
&\quad = \Lambda(x,\delta)^{-1/p}\big\|\{\Phi^*_{x,\delta}D_j^{\delta}((\phi\circ\Phi^{-1}_{x,\delta})u)\}_{j\in\mathbb{N}^\nu}\big\|_{L^p(B^n(1),\Phi^*_{x,\delta}\operatorname{Vol};\ell^q(\mathbb{N}^\nu))} \\
&\quad = \Lambda(x,\delta)^{-1/p}\big\|\{D_j^{\delta}((\phi\circ\Phi^{-1}_{x,\delta})u)\}_{j\in\mathbb{N}^\nu}\big\|_{L^p(\mathfrak{M},\operatorname{Vol};\ell^q(\mathbb{N}^\nu))} \\
&\quad = \Lambda(x,\delta)^{-1/p}\|(\phi\circ\Phi^{-1}_{x,\delta})u\|_{\mathscr{X}^s(\delta^{\lambda\vec{d}}W,\vec{d}),(\delta^{\lambda\vec{d}}X,\vec{d}),a,\Omega_0,\Omega_1,\Omega_2,\operatorname{Vol}}.
\end{aligned}
$$

□

6.12 A spectral definition

In the single-parameter setting, the Besov and Triebel–Lizorkin spaces can sometimes be equivalently defined in terms of spectral multipliers for Hörmander's sub-Laplacian. We outline this in the simplest setting.

In this section, we assume $\mathfrak{M}$ is a smooth, **compact** manifold with smooth strictly positive density Vol. On $\mathfrak{M}$ we are given Hörmander vector fields $W_1,\ldots,W_r$. We assign to each W_j the formal degree 1:

$$(W,1) = \{(W_1,1),\ldots,(W_r,1)\}.$$

Let $\mathscr{L} = W_1^*W_1 + \cdots + W_r^*W_r$: Hörmander's sub-Laplacian. As discussed in Section 5.12, $\mathscr{L}$ is essentially self-adjoint, and we identify $\mathscr{L}$ with its self-adjoint extension.

In Remark 5.12.5 we present a decomposition of the identity operator

$$I = \sum_{j\in\mathbb{N}} D_j,$$

where $\mathcal{D}_0 := \{(D_j,2^{-j}) : j\in\mathbb{N}\}$ is a bounded set of $(W,1)$ elementary operators supported in $\mathfrak{M}$. Moreover, there is $\psi\in C_0^\infty(\mathbb{R})$ with $\psi\equiv 1$ on a neighborhood of 0 such that if $\phi(\lambda) = \psi(\lambda) - \psi(4\lambda)$, then

$$D_j = \begin{cases} \psi(\mathscr{L}) & \text{if } j = 0, \\ \phi(2^{-2j}\mathscr{L}) & \text{if } j \geq 1. \end{cases}$$

The Besov, $\mathscr{B}^s_{p,q}(\mathfrak{M},(W,1))$, and Triebel–Lizorkin, $\mathscr{F}^s_{p,q}(\mathfrak{M},(W,1))$, spaces can be equivalently defined using these spectral multipliers, as the next result shows. We use Notations 6.2.1 and 6.2.2 and Definition 6.2.3.

Proposition 6.12.1. *Fix* $s \in \mathbb{R}$. *For* $f \in C_0^\infty(\mathfrak{M})'$ *the following are equivalent:*

(i) $f \in \mathscr{X}^s(\mathfrak{M}, (W, 1))$,

(ii) $\|f\|_{\mathscr{V},s,\mathcal{D}_0} < \infty$.

Moreover, in this case we have $\|f\|_{\mathscr{X}^s(W,1)} \approx \|f\|_{\mathscr{V},s,\mathcal{D}_0}$.

Proof. (i) ⇔ (ii): This follows from a simple reprise of the proof of Proposition 2.4.18. The key is that due to the spectral theorem we have an analog of the Calderón reproducing formula (Lemma 2.4.10). Indeed, for $C_0 \in \mathbb{N}$ define

$$\widetilde{D}_j^{C_0} := \sum_{\substack{|k-j| \le C_0 \\ k \in \mathbb{N}}} D_k.$$

Then if $C_0 = C_0(\psi)$ is sufficiently large, we have $D_j = D_j \widetilde{D}_j^{C_0}$, $\forall j$. Using this and replacing Lemma 2.4.16 with Lemma 6.4.9, the proof of Proposition 2.4.18 goes through nearly unchanged to establish (i) ⇔ (ii).

That $\|f\|_{\mathscr{X}^s(W,1)} \approx \|f\|_{\mathscr{V},s,\mathcal{D}_0}$ is a special case of Proposition 6.2.6. □

Remark 6.12.2. We presented the above in the special case where $\mathfrak{d}_1 = \cdots = \mathfrak{d}_r = 1$. Similar results for more general formal degrees and more general operators $\mathscr{L}$ have recently been obtained by Lingxiao Zhang [254] using the theory from this text. See also Remark 5.12.6.

6.13 Further questions

In this chapter, we have restricted attention to those results which are most directly useful for our study of maximally subelliptic PDEs. There are a number of very natural questions which we did not pursue. We mention a few of those here.

- We restricted our attention to $p, q \in [1, \infty]$ for the Besov spaces $\mathscr{B}_{p,q}^s$ and to $p \in (1, \infty)$, $q \in (1, \infty]$ for the Triebel–Lizorkin spaces $\mathscr{F}_{p,q}^s$. In particular, we did not study the local Hardy spaces $\mathscr{F}_{p,2}^0$ where $p \in (0, 1]$ or bounded mean oscillation (bmo) $\mathscr{F}_{\infty,2}^0$,[5] though the local Hardy and bmo spaces have been studied by other authors on more general spaces of homogeneous type (see the references in Section 6.14). It would be interesting to extend the values of p and q for which our spaces are defined.
- In the definitions of $\mathscr{B}_{p,q}^s$ and $\mathscr{F}_{p,q}^s$, we used the space $\ell^q(\mathbb{N}^\nu)$. When, for example, $\nu = 2$, one could replace $\ell^q(\mathbb{N}^2)$ in the definitions with $\ell^{q_1}(\mathbb{N}; \ell^{q_1}(\mathbb{N}))$ and obtain

5 In the classical setting, when studying $\mathscr{F}_{\infty,q}^s$ one needs a modification of the definition; see [244, Section 2.3.4]. To study such spaces in the generality of this text would likely require a similar modification.

a more general definition. One could even mix the definitions of $\mathscr{B}^s_{p,q}$ and $\mathscr{F}^s_{p,q}$ by using the space $\ell^{q_1}(\mathbb{N}; L^p(\mathfrak{M}, \mathrm{Vol}; \ell^{q_2}(\mathbb{N})))$. In this way one could create many more scales of multi-parameter function spaces which are naturally associated with our multi-parameter geometries. It would be interesting to have a general theory of such function spaces.

- There is an extensive interpolation and embedding theory for the classical Besov and Triebel–Lizorkin spaces. There are many interesting questions relating to what extent these results generalize to the spaces studied here.
- The classical Besov and Triebel–Lizorkin spaces can be defined using the heat equation; see [244, Section 2.12.2]. In light of Theorem 8.1.1 (i) ⇒ (ix), it seems likely that in the single-parameter case $\nu = 1$, the same methods of this chapter could be used to prove that the norms $\|\cdot\|_{\mathscr{B}^s_{p,q}(W,d)}$ and $\|\cdot\|_{\mathscr{F}^s_{p,s}(W,d)}$ can be equivalently defined using the heat equation for any non-negative self-adjoint extension of $\mathscr{P}^*\mathscr{P}$, where $\mathscr{P}$ is any maximally subelliptic operator.
- To what extent is the loss of ϵ derivatives in Theorem 6.6.13 necessary? See Remark 6.6.14.
- In Section 6.10 it was shown that if $s \in (0,\infty)^\nu$, the Besov and Triebel–Lizorkin spaces consisted of precisely those distributions of finite norm (see Proposition 6.10.1). If $s \in \mathbb{R}^\nu$, then by Remark 6.4.6 one has a similar result with a more complicated norm, but it is unclear if one can use the simpler norm used in Proposition 6.10.1. Can Proposition 6.10.1 be extended to all $s \in \mathbb{R}^\nu$? By being more precise about the norm, this can be done in some cases; see Proposition 6.12.1.

6.14 Further reading and references

There is a vast body of literature[6] on Besov ($\mathscr{B}^s_{p,q}$) and Triebel–Lizorkin ($\mathscr{F}^s_{p,q}$) spaces, mostly in the single-parameter case $\nu = 1$, where $s \in \mathbb{R}$. We are interested in settings where we can develop concrete characterizations of these spaces. For the most part, this literature can be separated into two nearly distinct categories:

(i) Theories which define $\mathscr{B}^s_{p,q}$ and/or $\mathscr{F}^s_{p,q}$ for all $s \in \mathbb{R}$. These are usually based off of $\mathbb{R}^n$ as in the classical theory (see [244] for a detailed history) or more generally based on some sort of Lie groups (see [92, 90, 207] for some early examples of this).[7]

(ii) Theories which define $\mathscr{B}^s_{p,q}$ and/or $\mathscr{F}^s_{p,q}$ for $|s|$ small (e. g., $|s| < 1$). With the restriction that $|s|$ be small, one can define these function spaces for much more general underlying spaces. There has been much work on this topic when the underlying

6 Because the literature on function spaces is so extensive, we make no attempt to cite every important work. Rather, we hope to give the reader a starting point for further reading in any topic of interest.

7 Much work has been done on the Hardy spaces $\mathscr{F}^0_{p,2}$, $p \in (0,1]$. Even though we did not address Hardy spaces in this chapter, many of the methods developed for Hardy spaces are important for studying the Besov and Triebel–Lizorkin spaces in generality.

space is a space of homogeneous type. This began with the work of Han and Sawyer [106], under some additional hypotheses on the space of homogeneous type. Some other early important works are [114, 115, 112, 113]. A breakthrough by Auscher and Hytönen [5, 6] allowed the creation of an appropriate Littlewood–Paley theory on general spaces of homogeneous type using wavelets, leading to a theory of Besov and Triebel–Lizorkin spaces (for $|s|$ small) on general spaces of homogeneous type. See the recent papers [110, 107, 108, 116, 138, 23, 24, 22, 21, 95, 96, 117] for various approaches to this theory.

The distinction between (i) and (ii) comes from the following observation. On $\mathbb{R}^n$ (or more generally a Lie group) one can make sense of what it means to have high-order moments vanish (or, similarly, to make sense of high-order differences). On a general space of homogeneous type, it is only clear how to define first moments vanishing (similarly, first-order differences).

The setting of this chapter lies between the two categories described above. We do not have any sort of group structure with which to work, though (at least when $\nu = 1$) we are working on a (local) space of homogeneous type (see Remark 3.3.2). At this intermediate level of generality, we are able to develop the Besov and Triebel–Lizorkin spaces for all s. The idea is that, though there is no underlying group structure (even locally), there is an intrinsic and quantitative notion of high levels of smoothness adapted to our geometry (see Definition 5.2.21). This is in contrast to a general space of homogeneous type, where one can only define smoothness up to Lipschitz regularity. Elementary operators are appropriately scaled high-order derivatives of elementary operators up to controllable error terms (see Proposition 5.5.5 (h)): this is the analog we use of having "high-order moments vanish." See Example 2.3.8 for a description of why this can be viewed as an analog of having high-order moments vanish. See Section 7.7 for the analog we use of high-order differences.

As described in Section 6.9, the spaces $\mathscr{F}^s_{p,2}(\mathcal{K}, (W, \vec{d}))$ can be viewed as L^p Sobolev spaces. From this perspective, these generalize the non-isotropic Sobolev spaces which were first used outside of the group setting by Rothschild and Stein [200, Section 16], where the definition was restricted to the special case $s \in \mathbb{N}$. These Sobolev spaces were later used in [182, 140]. A special case of the spaces $\mathscr{F}^s_{p,2}(\mathcal{K}, (W, \vec{d}))$ described in this chapter was defined in [220, Section 2.10], where the connection with the sub-Laplacian was made more explicit.

In the multi-parameter setting, $\nu \geq 2$, there has been considerable work studying Besov and Triebel–Lizorkin spaces when the underlying space is a product of single-parameter spaces. Early important work on this in $\mathbb{R}^{n_1} \times \mathbb{R}^{n_2}$ was due to Chang and R. Fefferman [35–37], Journé [136], and Pipher [197]; see also [160, 68, 70, 71, 41]. For some results beyond the Euclidean setting see, for example, [109, 43, 42, 148, 159]. Beyond the product setting, less is known. For the flag setting, see for example [111, 72]. For spaces on $\mathbb{R}^n$ with two different underlying homogeneities, see for example [69, 67, 66]. Lu, Shen, and Zhang [158] studied the boundedness of the multi-parameter singular integrals from

[220] on Hardy spaces and local Hardy spaces, results which have many similarities to Theorem 6.3.10.

The Sobolev spaces $\mathscr{F}^{s}_{p,2}(\mathcal{K},(W,\vec{d}))$, in the multi-parameter setting, were defined and studied in [220, Section 5.1.1], and were defined in an even more general setting in [221].

7 Zygmund–Hölder spaces

As we saw in Section 2.8, the classical Zygmund–Hölder spaces $\mathscr{C}^s(\mathbb{R}^n) = \mathscr{B}^s_{\infty,\infty}(\mathbb{R}^n)$, where $s > 0$, play an important role in the study of nonlinear elliptic PDEs. For maximally subelliptic operators, an analogous role is played by $\mathscr{B}^s_{\infty,\infty}(\mathcal{K}, (W, \vec{d}))$, where $s \in (0, \infty)^\nu$. In this chapter, we develop the theory of this space, which will be useful when we turn to studying nonlinear equations in Chapter 9.

Throughout this chapter, $\mathfrak{M}$ is a connected C^∞ manifold of dimension $n \in \mathbb{N}_+$.[1] We work in the general setting described in Section 5.1.3. Thus, for some $\nu \in \mathbb{N}_+$ we are given ν lists of vector fields with formal degrees:

$$(W^\mu, \vec{d}^\mu) = \{(W^\mu_1, \vec{d}^\mu_1), \dots, (W^\mu_{r_\mu}, \vec{d}^\mu_{r_\mu})\} \subset C^\infty_{\mathrm{loc}}(\mathfrak{M}; T\mathfrak{M}) \times \mathbb{N}_+.$$

We assume that $(W^1, \vec{d}^1), \dots, (W^\nu, \vec{d}^\nu)$ pairwise locally weakly approximately commute and that for each $\mu \in \{1, \dots, \nu\}$, $\mathrm{Gen}((W^\mu, \vec{d}^\mu))$ is locally finitely generated on $\mathfrak{M}$. We define

$$\begin{aligned}(W, \vec{d}) &= \{(W_1, \vec{d}_1), \dots, (W_r, \vec{d}_r)\} \\ &:= (W^1, \vec{d}^1) \boxtimes (W^2, \vec{d}^2) \boxtimes \cdots \boxtimes (W^\nu, \vec{d}^\nu) \subset C^\infty_{\mathrm{loc}}(\mathfrak{M}; T\mathfrak{M}) \times (\mathbb{N}^\nu \setminus \{0\}).\end{aligned}$$

In addition, we make the **additional assumption** that $W_1, \dots, W_r$ satisfy Hörmander's condition on $\mathfrak{M}$.

Throughout this chapter, $\mathcal{K}$ will denote a compact set and Ω will denote an open, relatively compact set, with $\mathcal{K} \Subset \Omega \Subset \mathfrak{M}$.

Definition 7.0.1. For $s \in (0, \infty)^\nu$ we define the Zygmund–Hölder spaces with respect to $(W, \vec{d})$ by $\mathscr{C}^s(\mathcal{K}, (W, \vec{d})) := \mathscr{B}^s_{\infty,\infty}(\mathcal{K}, (W, \vec{d}))$ and $\mathscr{C}^s(\mathcal{K}, (W, \vec{d}); \mathbb{C}^N) := \mathscr{B}^s_{\infty,\infty}(\mathcal{K}, (W, \vec{d}); \mathbb{C}^N)$ with equality of norms:

$$\|\cdot\|_{\mathscr{C}^s(W,\vec{d})} := \|\cdot\|_{\mathscr{B}^s_{\infty,\infty}(W,\vec{d})}, \quad \|\cdot\|_{\mathscr{C}^s((W,\vec{d});\mathbb{C}^N)} := \|\cdot\|_{\mathscr{B}^s_{\infty,\infty}((W,\vec{d});\mathbb{C}^N)}.$$

As in Definition 6.5.7, we set

$$\mathscr{C}^s_{\mathrm{cpt}}(W, \vec{d}) := \bigcup_{\substack{\mathcal{K} \Subset \mathfrak{M} \\ \mathcal{K} \text{ compact}}} \mathscr{C}^s(\mathcal{K}, (W, \vec{d})).$$

We define the classical Zygmund–Hölder spaces consisting of functions with compact support by $\mathscr{C}^s_{\mathrm{std}}(\mathcal{K}) := \mathscr{B}^s_{\infty,\infty,\mathrm{std}}(\mathcal{K})$ and

---

1 In this chapter, the only L^p space we consider is L^∞. Because of this, it does not matter what strictly positive, smooth density Vol we use on $\mathfrak{M}$: they all give rise to the same norm $\|\cdot\|_{L^\infty(\mathfrak{M})}$.

https://doi.org/10.1515/9783111085647-007

$$\mathscr{C}^s_{\mathrm{std,cpt}}(\mathfrak{M}) := \bigcup_{\substack{\mathcal{K}\Subset\mathfrak{M}\\ \mathcal{K}\text{ compact}}} \mathscr{C}^s_{\mathrm{std}}(\mathcal{K}).$$

The main purpose of this section is to develop the parts of the theory of $\mathscr{C}^s(\mathcal{K},(W,\vec{d}))$ which are useful when studying nonlinear PDEs. The key results are:

- $\mathscr{C}^s(\mathcal{K},(W,\vec{d}))$ forms an algebra (Proposition 7.4.1).
- For appropriately smooth functions $F(x,\zeta)$, where $x\in\mathfrak{M}$ and $\zeta\in\mathbb{R}^N$, we have

$$F(x,u(x))\in\mathscr{C}^s(\mathcal{K},(W,\vec{d})),\quad \forall u\in\mathscr{C}^s(\mathcal{K},(W,\vec{d});\mathbb{R}^N).$$

  See Section 7.5.
- In the single-parameter case, $\nu = 1$, we present a difference characterization of $\mathscr{C}^s(\mathcal{K},(W,d))$ and describe the connection with the classical Hölder spaces with respect to the metric $\rho_{(W,d)}(x,y)$. See Sections 7.7 and 7.7.1.

7.1 The norm

For $s\in(0,\infty)^\nu$ and $f\in\mathscr{C}^s(\mathcal{K},(W,\vec{d}))$, the norm $\|f\|_{\mathscr{C}^s(W,\vec{d})}$ is defined as follows. Let $\psi\in C_0^\infty(\Omega)$ satisfy $\psi\equiv 1$ on a neighborhood of $\mathcal{K}$. By Proposition 5.8.3 we may decompose $\mathrm{Mult}[\psi]=\sum_{j\in\mathbb{N}^\nu}D_j$, where $\{(D_j,2^{-j}):j\in\mathbb{N}^\nu\}$ is a bounded set of generalized $(W,\vec{d})$ elementary operators supported in Ω. Then

$$\|f\|_{\mathscr{C}^s(W,\vec{d})}=\|f\|_{\mathscr{B}^s_{\infty,\infty}(W,\vec{d})}=\sup_{j\in\mathbb{N}^\nu}2^{j\cdot s}\|D_jf\|_{L^\infty}.$$

Proposition 6.3.7 shows that the equivalence class of this norm does not depend on any of the choices we have made, and Proposition 6.10.1 shows that

$$\mathscr{C}^s(\mathcal{K},(W,\vec{d}))=\{f\in C^\infty_{W,0}(\mathfrak{M})':\mathrm{supp}(f)\subseteq\mathcal{K},\|f\|_{\mathscr{C}^s(W,\vec{d})}<\infty\}.\tag{7.1}$$

We use the flexibility that we may choose any such $\{(D_j,2^{-j}):j\in\mathbb{N}^\nu\}$, and we will use the particular choice from Section 6.11, that is, we choose Ω_0, Ω_1, and Ω_2 as in that section, we let $a>0$ be small, and we define D_j as in (6.170). The main result of this section is that with this choice, we have the following proposition.

Proposition 7.1.1. *Fix $\psi\in C_0^\infty(\Omega)$ with $\psi\equiv 1$ on a neighborhood of $\mathcal{K}$. There exists a bounded set of generalized $(W,\vec{d})$ elementary operators supported in Ω, $\{(D_j,2^{-j}):j\in\mathbb{N}^\nu\}$ such that the following holds. Set $P_j:=\sum_{k\le j}D_j$. Then:*

(i) $\sum_{j\in\mathbb{N}^\nu}D_j=\mathrm{Mult}[\psi]$; *see Proposition* 5.5.10 *for the convergence of this sum.*

(ii) $\forall j\in\mathbb{N}^\nu$, *we have* $P_j,D_j\in C_0^\infty(\Omega\times\Omega)$.

(iii) $\forall s\in(0,\infty)^\nu$, $\forall f\in\mathscr{C}^s(\mathcal{K},(W,\vec{d}))$,

$$\sum_{j\in\mathbb{N}^\nu}D_jf=\psi f,$$

with convergence in $C(\mathfrak{M})$, *that is,* $P_j f \to f$ *in* $C(\mathfrak{M})$ *as* $j_1, \ldots, j_\nu \to \infty$.

(iv) $\{(P_j, 2^{-j}) : j \in \mathbb{N}^\nu\}$ *is a bounded set of generalized* $(W, \vec{d})$ *pre-elementary operators supported in* Ω.

Remark 7.1.2. Proposition 7.1.1 (ii) is the reason we assumed the additional assumption in this chapter that $W_1, \ldots, W_r$ satisfy Hörmander's condition on $\mathfrak{M}$.

Remark 7.1.3. We take, for $f \in \mathscr{C}^s(\mathcal{K}, (W, \vec{d}))$,

$$\|f\|_{\mathscr{C}^s(W,\vec{d})} := \|f\|_{\ell^\infty(\mathbb{N}^\nu; L^\infty(\mathfrak{M})),s,\mathcal{D}_0} = \sup_{j\in\mathbb{N}^\nu} 2^{j\cdot s}\|D_j f\|_{L^\infty(\mathfrak{M})}$$
$$= \|f\|_{\mathscr{B}^s_{\infty,\infty}(W,\vec{d}),(X,\vec{d}),a,\Omega_0,\Omega_1,\Omega_2,\mathrm{Vol}}.$$

See Definition 6.11.2. More generally, for $f \in \mathscr{C}^s(\mathcal{K}, (W, \vec{d}); \mathbb{C}^N)$, we take

$$\|f\|_{\mathscr{C}^s((W,\vec{d});\mathbb{C}^N)} := \|f\|_{\ell^\infty(\mathbb{N}^\nu; L^\infty(\mathfrak{M};\mathbb{C}^N)),s,\mathcal{D}_0} = \sup_{j\in\mathbb{N}^\nu} 2^{j\cdot s}\|D_j f\|_{L^\infty(\mathfrak{M};\mathbb{C}^N)}.$$

Before we prove Proposition 7.1.1, we state a corollary which is the main way Proposition 7.1.1 will be used.

Corollary 7.1.4. *Let* D_j *and* P_j *be as in Proposition* 7.1.1. *Then:*

(i) *For every ordered multi-index* α,

$$\sup_{j\in\mathbb{N}^\nu}\|(2^{-j\vec{d}}W)^\alpha P_j\|_{L^\infty(\mathfrak{M})\to L^\infty(\mathfrak{M})} < \infty, \quad \sup_{j\in\mathbb{N}^\nu}\|(2^{-j\vec{d}}W)^\alpha D_j\|_{L^\infty(\mathfrak{M})\to L^\infty(\mathfrak{M})} < \infty.$$

(ii) *For each* $\mu \in \{1, \ldots, \nu\}$ *and* $M \in \mathbb{N}$, *we can write*

$$D_j = \sum_{|\alpha_\mu|\le M} 2^{-(M-|\alpha_\mu|)j_\mu} E_{j,\alpha_\mu,\mu,M}(2^{-j_\mu d^\mu} W^\mu)^{\alpha_\mu}, \tag{7.2}$$

where

$$\sup_{j\in\mathbb{N}^\nu}\sup_{|\alpha_\mu|\le M}\|E_{j,\alpha_\mu,\mu,M}\|_{L^\infty(\mathfrak{M})\to L^\infty(\mathfrak{M})} < \infty. \tag{7.3}$$

Proof. (i): Since $\{(D_j, 2^{-j}) : j \in \mathbb{N}^\nu\}$ is a bounded set of generalized $(W, \vec{d})$ elementary operators supported in Ω, it is a bounded set of generalized $(W, \vec{d})$ pre-elementary operators supported in Ω (see Definition 5.2.27). Also, by Proposition 7.1.1 (iv), $\{(P_j, 2^{-j}) : j \in \mathbb{N}^\nu\}$ is a bounded set of generalized $(W, \vec{d})$ pre-elementary operators supported in Ω. From here, (i) follows from the definition of bounded sets of generalized $(W, \vec{d})$ pre-elementary operators (see Definition 5.2.26).

(ii): Because $\{(D_j, 2^{-j}) : j \in \mathbb{N}^\nu\}$ is a bounded set of generalized $(W, \vec{d})$ elementary operators supported in Ω, Proposition 5.5.5 (h) shows that we may write D_j as in (7.2), where $\mathcal{E}_{\mu,M} := \{(E_{j,\alpha_\mu,\mu,M}, 2^{-j}) : j \in \mathbb{N}^\nu, |\alpha_\mu| \leq M\}$ is a bounded set of generalized $(W, \vec{d})$ elementary operators supported in Ω. In particular, $\mathcal{E}_{\mu,M}$ is a bounded set of generalized $(W, \vec{d})$ pre-elementary operators supported in Ω (see Definition 5.2.27). From here, (7.3) follows from the definition of bounded sets of generalized $(W, \vec{d})$ pre-elementary operators (see Definition 5.2.26). □

Proof of Proposition 7.1.1. By Proposition 6.11.1 (i), $\{(D_j, 2^{-j}) : j \in \mathbb{N}^\nu\}$ is a bounded set of generalized $(W, \vec{d})$ elementary operators supported in Ω. Item (iv) follows from Proposition 6.11.1 (ii). Item (i) follows from Proposition 6.11.1 (iii).

(ii): Since each P_j is given by the finite sum $P_j = \sum_{k \leq j} D_j$, it suffices to show that $D_j \in C_0^\infty(\Omega_1 \times \Omega_2)$. We use the formula for D_j given in (6.170). By Proposition 6.11.3 (ii) we already know that $\{(D_j, 2^{-j}) : j \in \mathbb{N}^\nu\}$ is a bounded set of $(X, \vec{d})$, a pseudo-differential operator scales supported in $\Omega_1 \times \Omega_2$. In particular, $\operatorname{supp}(D_j) \Subset \Omega_1 \times \Omega_2$, for each $j \in \mathbb{N}^\nu$. Thus, it suffices to show that $D_j \in C_{\mathrm{loc}}^\infty(\mathfrak{M} \times \mathfrak{M})$. By Proposition 3.8.8 (a), $\operatorname{Gen}((W, \vec{d}))$ is finitely generated by $(X, \vec{d})$ on Ω. In particular, since $W_1, \dots, W_r$ satisfy Hörmander's condition, $X_1(x), \dots, X_q(x)$ must span $T_x\mathfrak{M}$, $\forall x \in \Omega$. By the formula for $\Gamma(t, x)$ (see (6.167)), we have $\partial_{t_j}\big|_{t=0}\Gamma(t, x) = -X_j(x)$, for $1 \leq j \leq q$. In particular, for $x \in \Omega$, the $q \times n$ matrix $d_t\Gamma(t, x)\big|_{t=0}$ has full rank. The function $\operatorname{Dil}_{2^j}^{\vec{d}}(\varsigma_j)(x, t, t)$ is supported in $x \in \Omega_1$ and $t \in B^q(a)$. Since $\Omega_1 \Subset \Omega$, if $a > 0$ is sufficiently small, we may apply the change of variables $u = \Gamma(t, x)$ in the t variable to see

$$D_j f(x) = \int f(\Gamma(t, x)) \operatorname{Dil}_{2^j}^{\vec{d}}(\varsigma_j)(x, t, t)\, dt = \int f(u) K_j(x, u)\, du,$$

where $K_j(x, u) \in C_{\mathrm{loc}}^\infty(\mathfrak{M} \times \mathfrak{M})$, completing the proof of (ii).

(iii): Let $f \in \mathscr{C}^s(\mathcal{K}, (W, \vec{d}))$. It follows from (i) that $\sum_{j \in \mathbb{N}^\nu} D_j f = \psi f$ with convergence in $C_{\mathrm{loc}}^\infty(\mathfrak{M})'$. Thus, to complete the proof of (iii) it suffices to show that $\sum_{j \in \mathbb{N}^\nu} D_j f$ converges in $C(\mathfrak{M})$. It follows from (ii) that $D_j f \in C_0^\infty(\mathfrak{M}) \subseteq C(\mathfrak{M})$ for every $j \in \mathbb{N}^\nu$, and therefore it suffices to show that $\sum_{j \in \mathbb{N}^\nu} D_j f$ converges in $L^\infty(\mathfrak{M})$. Since $\{(D_j, 2^{-j}) : j \in \mathbb{N}^\nu\}$ is a bounded set of generalized $(W, \vec{d})$ elementary operators, we have, by Corollary 6.4.4 (with $\mathscr{V} = \ell^\infty(\mathbb{N}^\nu; L^\infty(\mathfrak{M}))$, $\mathscr{X}^s = \mathscr{B}^s_{\infty,\infty}$, and $\mathcal{E} = \{(D_j, 2^{-j}) : j \in \mathbb{N}^\nu\}$),

$$\sup_{j \in \mathbb{N}^\nu} 2^{j \cdot s} \|D_j f\|_{L^\infty} \lesssim \|f\|_{\mathscr{C}^s(W, \vec{d})}.$$

Since $s \in (0, \infty)^\nu$, it follows that $\sum_{j \in \mathbb{N}^\nu} D_j f$ converges in $L^\infty(\mathfrak{M})$, completing the proof. □

Remark 7.1.5. In Section 7.7, we will use the above choice of D_j in the single-parameter setting ($\nu = 1$). Thus, we are given Hörmander vector fields with formal degrees $(W, \vec{d})$, and we pick $(X, d) = \{(X_1, d_1), \dots, (X_q, d_q)\} \subset \operatorname{Gen}((W, \vec{d}))$ such that $\operatorname{Gen}((W, \vec{d}))$ is finitely generated by (X, d) on Ω. The construction is:

- Pick any $\psi \in C_0^\infty(\Omega)$ with $\psi \equiv 1$ on a neighborhood of $\mathcal{K}$.

- Pick any $\rho \in \mathscr{S}(\mathbb{R}^q)$ with $\hat{\rho}(\xi) \equiv 1$ on a neighborhood of 0.
- For $a > 0$ sufficiently small (how small depends on $\operatorname{supp}(\psi)$), pick any $\eta \in C_0^\infty(B^q(a))$ with $\eta \equiv 1$ on a neighborhood of 0.
- For $j \in \mathbb{N}$, we define $\varsigma_j \in \mathscr{S}(\mathbb{R}^q)$ by

$$\hat{\varsigma}_j(\xi) := \begin{cases} \hat{\rho}(\xi), & j = 0 \\ \hat{\rho}(\xi) - \hat{\rho}(2^d \xi), & j \geq 1. \end{cases}$$

  Note that $\varsigma_j \in \mathscr{S}_0(\mathbb{R}^q)$ for $j \geq 1$ and $\varsigma_j = \varsigma_1$ for all $j \geq 1$.
- Set

$$D_j f(x) = \psi(x) \int f(e^{-t \cdot X} x) \eta(t) \operatorname{Dil}_{2^j}^d(\varsigma_j)(t)\, dt.$$

7.2 Decomposition into smooth functions

Similar to the classical setting on $\mathbb{R}^n$ (see Section 2.5.1), one of the most convenient ways to study functions in $\mathscr{C}^s(\mathcal{K}, (W, \vec{d}))$ is to decompose them into a sum of smooth functions. The next proposition gives such a decomposition.

Proposition 7.2.1. *Let $s \in (0, \infty)^\nu$. For $f \in C_0^\infty(\mathfrak{M})'$, the following are equivalent:*
(i) $f \in \mathscr{C}^s(\mathcal{K}, (W, \vec{d}))$.
(ii) $\operatorname{supp} f \subseteq \mathcal{K}$ *and there exists a sequence $\{f_j\}_{j \in \mathbb{N}^\nu} \subset C_0^\infty(\mathfrak{M})$ such that for every ordered multi-index α*

$$\sup_{j \in \mathbb{N}^\nu} 2^{j \cdot s} \|(2^{-j\vec{d}} W)^\alpha f_j\|_{L^\infty(\mathfrak{M})} < \infty$$

and $f = \sum_{j \in \mathbb{N}^\nu} f_j$ with convergence in $C(\mathfrak{M})$.

In this case, there exists $M = M(s, \nu) \in \mathbb{N}$ such that

$$\|f\|_{\mathscr{C}^s(W, \vec{d})} \leq C \sum_{|\alpha| \leq M} \sup_{j \in \mathbb{N}^\nu} 2^{j \cdot s} \|(2^{-j\vec{d}} W)^\alpha f_j\|_{L^\infty(\mathfrak{M})}. \tag{7.4}$$

Furthermore, f_j may be chosen such that $f_j \in C_0^\infty(\Omega)$, for every $j \in \mathbb{N}^\nu$, and for every ordered multi-index α,

$$\sup_{j \in \mathbb{N}^\nu} 2^{j \cdot s} \|(2^{-j\vec{d}} W)^\alpha f_j\|_{L^\infty(\mathfrak{M})} \leq C_\alpha \|f\|_{\mathscr{C}^s(W, \vec{d})}. \tag{7.5}$$

Here, C and C_α do not depend on f, but may depend on any other ingredient in the proposition.

Proof. Suppose (i) holds, i. e., $f \in \mathscr{C}^s(\mathcal{K}, (W, \vec{d}))$. Then by definition $\mathrm{supp}(f) \subseteq \mathcal{K}$. Let D_j be as in Proposition 7.1.1 and set $f_j := D_j f$. By Proposition 7.1.1 (ii), $D_j \in C_0^\infty(\Omega \times \Omega)$ and therefore $f_j \in C_0^\infty(\Omega)$. By Proposition 7.1.1 (iii), $\sum_{j\in\mathbb{N}^\nu} f_j = f$, with convergence in $C(\mathfrak{M})$. Finally, let α be an ordered multi-index and set $E_{j,\alpha} := (2^{-j\vec{d}} W)^\alpha D_j$. Since $\{(D_j, 2^{-j}) : j \in \mathbb{N}^\nu\}$ is a bounded set of generalized $(W, \vec{d})$ elementary operators supported in Ω, Proposition 5.5.5 (e) implies that $\mathcal{E}_\alpha := \{(E_{j,\alpha}, 2^{-j}) : j \in \mathbb{N}^\nu\}$ is also a bounded set of generalized $(W, \vec{d})$ elementary operators supported in Ω. Thus, by Corollary 6.4.4, we have

$$\begin{aligned}\sup_{j\in\mathbb{N}^\nu} 2^{j\cdot s}\|(2^{-j\vec{d}} W)^\alpha f_j\|_{L^\infty(\mathfrak{M})} &= \sup_{j\in\mathbb{N}^\nu} 2^{j\cdot s}\|E_{j,\alpha} f\|_{L^\infty(\mathfrak{M})} \\ &= \|f\|_{\ell^\infty(\mathbb{N}^\nu;L^\infty(\mathfrak{M})),s,\mathcal{E}_\alpha} \lesssim \|f\|_{\mathscr{B}^s_{\infty,\infty}(W,\vec{d})} = \|f\|_{\mathscr{C}^s(W,\vec{d})}.\end{aligned}$$

Thus, (ii) holds, and moreover we have $f_j \in C_0^\infty(\Omega)$, $\forall j \in \mathbb{N}^\nu$, and (7.5) holds.

Suppose (ii) holds. Fix $M \geq \nu(|s|_\infty + 1)$. We will show that for all $j, k \in \mathbb{N}^\nu$,

$$\|D_j f_k\|_{L^\infty} \lesssim 2^{-M|(j\vee k)-k|_\infty} \sum_{|\alpha|\leq M} \|(2^{-k\vec{d}} W)^\alpha f_k\|_{L^\infty}. \tag{7.6}$$

First, we see why (7.6) will complete the proof. Indeed, we then have

$$\begin{aligned}\|D_j f\|_{L^\infty} &\leq \sum_{k\in\mathbb{N}^\nu} \|D_j f_k\|_{L^\infty} \\ &\lesssim \left(\sup_{k'\in\mathbb{N}^\nu} 2^{k'\cdot s} \sum_{|\alpha|\leq M} \|(2^{-k'\vec{d}} W)^\alpha f_{k'}\|_{L^\infty}\right) \sum_{k\in\mathbb{N}^\nu} 2^{-M|(j\vee k)-k|_\infty} 2^{-k\cdot s}.\end{aligned} \tag{7.7}$$

Also,

$$\begin{aligned}\sum_{k\in\mathbb{N}^\nu} 2^{-M|(j\vee k)-k|_\infty} 2^{-k\cdot s} &\leq \sum_{k\in\mathbb{N}^\nu} 2^{-|(j\vee k)-k|_1} 2^{-(j\vee k)\cdot s} \\ &= \prod_{\mu\in\{1,\ldots,\nu\}} \sum_{k_\mu\in\mathbb{N}} 2^{-((j_\mu\vee k_\mu)-k_\mu)} 2^{-(j_\mu\vee k_\mu)s_\mu} \\ &\leq \prod_{\mu\in\{1,\ldots,\nu\}} \left(\sum_{k_\mu\leq j_\mu} 2^{-(j_\mu-k_\mu)} 2^{-j_\mu s_\mu} + \sum_{k_\mu\geq j_\mu} 2^{-k_\mu s_\mu}\right) \\ &\lesssim \prod_{\mu\in\{1,\ldots,\nu\}} 2^{-j_\mu s_\mu} = 2^{-j\cdot s}.\end{aligned} \tag{7.8}$$

Combining (7.7) and (7.8) establishes (7.4). In particular, since $\|f\|_{\mathscr{C}^s(W,\vec{d})} < \infty$ and $\mathrm{supp} f \subseteq \mathcal{K}$, (7.1) shows that $f \in \mathscr{C}^s(\mathcal{K}, (W, \vec{d}))$ and therefore (i) holds.

Thus, we will complete the proof by proving (7.6). If $|(j \vee k) - k|_\infty = 0$, then since $\sup_{j\in\mathbb{N}^\nu} \|D_j\|_{L^\infty\to L^\infty} < \infty$ (Corollary 7.1.4 (i)), we have $\|D_j f_k\|_{L^\infty} \lesssim \|f_k\|_{L^\infty}$, proving (7.6) in this case.

We turn to proving (7.6) when $|(j\vee k)-k|_\infty > 0$. Fix $\mu \in \{1,\dots,\nu\}$ such that $j_\mu - k_\mu = |(j\vee k)-j|_\infty$. Letting $E_{j,\alpha_\mu,\mu,M}$ be as in Corollary 7.1.4 (ii), so that $\|E_{j,\alpha_\mu,\mu,M}\|_{L^\infty\to L^\infty} \lesssim 1$, we have

$$\begin{aligned}\|D_j f_k\|_{L^\infty} &\le \sum_{|\alpha_\mu|\le M} 2^{-j_\mu(M-|\alpha_\mu|)}\|E_{j,\alpha_\mu,\mu,M}(2^{-j_\mu d^\mu}W^\mu)^{\alpha_\mu} f_k\|_{L^\infty}\\ &\lesssim \sum_{|\alpha_\mu|\le M} 2^{-j_\mu(M-|\alpha_\mu|)}2^{-(j_\mu-k_\mu)\deg_{d^\mu}(\alpha_\mu)}\|(2^{-k_\mu d^\mu}W^\mu)^{\alpha_\mu} f_k\|_{L^\infty}\\ &\le \sum_{|\alpha_\mu|\le M} 2^{-M(j_\mu-k_\mu)}\|(2^{-k_\mu d^\mu}W^\mu)^{\alpha_\mu} f_k\|_{L^\infty}\\ &\le 2^{-M|(j\vee k)-k|_\infty}\sum_{|\alpha|\le M}\|(2^{-k\vec{d}}W)^{\alpha} f_k\|_{L^\infty},\end{aligned}$$

establishing (7.6) and completing the proof. □

Corollary 7.2.2. *For $s \in (0,\infty)^\nu$, $\mathscr{C}^s(\mathcal{K},(W,\vec{d})) \subseteq C(\mathfrak{M})$ and the inclusion is continuous.*

Proof. For $f \in \mathscr{C}^s(\mathcal{K},(W,\vec{d}))$, Proposition 7.2.1 shows that $f = \sum_{j\in\mathbb{N}^\nu} f_j$, where $f_j \in C_0^\infty(\mathfrak{M}) \subseteq C(\mathfrak{M})$, with convergence in $C(\mathfrak{M})$. This proves that $f \in C(\mathfrak{M})$. Moreover, by choosing f_j so that (7.5) holds, we have

$$\|f\|_{C(\mathfrak{M})} \le \sum_{j\in\mathbb{N}^\nu}\|f_j\|_{L^\infty} \lesssim \sum_{j\in\mathbb{N}^\nu} 2^{-j\cdot s}\|f\|_{\mathscr{C}^s(W,\vec{d})} \lesssim \|f\|_{\mathscr{C}^s(W,\vec{d})},$$

proving that the inclusion map is continuous. □

7.3 Bounds of some sums

In this chapter, we require some elementary estimates for sums, which we present in this section. The two main estimates we require are the next two lemmas.

Lemma 7.3.1. *Let $s \in (0,\infty)^\nu$ and $M > |s|_1$. Then, $\forall k \in \mathbb{N}^\nu$,*

$$\sum_{j,k'\in\mathbb{N}^\nu} 2^{-M|(k\vee k'\vee j)-(k'\vee j)|_\infty}2^{-k'\cdot s}2^{-j\cdot s} \le C_{s,M,\nu}2^{-k\cdot s}.$$

Lemma 7.3.2. *Let $s \in (0,\infty)^\nu$, $t > 0$, $M_1 > |s|_1$, and $M_2 > t$. Then, for all $k \in \mathbb{N}^\nu$ and $l \in \mathbb{N}$,*

$$\sum_{\substack{j,k'\in\mathbb{N}^\nu\\ l'\in\mathbb{N}}} 2^{-M_1|(k\vee k'\vee j)-(k'\vee j)|_\infty}2^{-M_2|(l\vee l')-l'|}2^{-k'\cdot s}2^{-j\cdot s}2^{-l't} \le C_{s,t,M_1,M_2,\nu}2^{-k\cdot s}2^{-lt}.$$

We begin by proving two easier lemmas in the single-parameter case.

Lemma 7.3.3. *Let $s > 0$ and $M > s$. Then, $\forall k \in \mathbb{N}$,*

$$\sum_{k' \in \mathbb{N}} 2^{-M|(k \vee k') - k'|} 2^{-k's} \leq C_{s,M} 2^{-ks}.$$

Proof. We have

$$\sum_{k' \geq k} 2^{-M|(k \vee k') - k'|} 2^{-k's} = \sum_{k' \geq k} 2^{-k's} \lesssim 2^{-ks} \tag{7.9}$$

and

$$\begin{aligned}\sum_{k' \leq k} 2^{-M|(k \vee k') - k'|} 2^{-k's} &= \sum_{k' \leq k} 2^{-M(k-k')} 2^{-k's} \\ &\leq 2^{-ks} \sum_{k' \in \mathbb{N}} 2^{-(M-s)|k-k'|} \lesssim 2^{-ks}.\end{aligned} \tag{7.10}$$

Combining (7.9) and (7.10) completes the proof. □

Lemma 7.3.4. *Let $s > 0$ and $M > s$. Then, $\forall k \in \mathbb{N}$,*

$$\sum_{j,k' \in \mathbb{N}} 2^{-M|(k \vee k' \vee j) - (k' \vee j)|} 2^{-k's} 2^{-js} \leq C_{s,M} 2^{-ks}.$$

Proof. The sum is symmetric in j and k', so it suffices to estimate just the sum $\sum_{j \leq k'}$. We have

$$\begin{aligned}&\sum_{j \leq k'} 2^{-M|(k \vee k' \vee j) - (k' \vee j)|} 2^{-k's} 2^{-js} \\ &\quad \leq \left(\sum_{j \in \mathbb{N}} 2^{-js} \right) \left(\sum_{k' \in \mathbb{N}} 2^{-M|(k \vee k') - k'|} 2^{-ks} \right) \lesssim 2^{-ks},\end{aligned}$$

where the final estimate follows from Lemma 7.3.3, completing the proof. □

Proof of Lemma 7.3.1. Pick $M_1, \ldots, M_\nu$ with $M = \sum_{\mu=1}^{\nu} M_\mu$ and $M_\mu > s_\mu$. We have, using Lemma 7.3.4,

$$\begin{aligned}&\sum_{j,k' \in \mathbb{N}^\nu} 2^{-M|(k \vee k' \vee j) - (k' \vee j)|_\infty} 2^{-k' \cdot s} 2^{-j \cdot s} \\ &\quad \leq \prod_{\mu \in \{1,\ldots,\nu\}} \left(\sum_{j_\mu, k'_\mu \in \mathbb{N}} 2^{-M_\mu |(k_\mu \vee k'_\mu \vee j_\mu) - (k'_\mu \vee j_\mu)|} 2^{-k'_\mu s_\mu} 2^{-j_\mu s_\mu} \right) \\ &\quad \lesssim \prod_{\mu \in \{1,\ldots,\nu\}} 2^{-k_\mu s_\mu} = 2^{-k \cdot s},\end{aligned}$$

completing the proof. □

Proof of Lemma 7.3.2. Applying Lemmas 7.3.3 and 7.3.1, we have

$$\begin{aligned}
&\sum_{\substack{j,k'\in\mathbb{N}^\nu\\ l'\in\mathbb{N}}} 2^{-M_1|(k\vee k'\vee j)-(k'\vee j)|_\infty}2^{-M_2|(l\vee l')-l'|}2^{-k'\cdot s}2^{-j\cdot s}2^{-l't}\\
&\quad=\Bigg(\sum_{j,k'\in\mathbb{N}^\nu} 2^{-M_1|(k\vee k'\vee j)-(k'\vee j)|_\infty}2^{-k'\cdot s}2^{-j\cdot s}\Bigg)\Bigg(\sum_{l'\in\mathbb{N}} 2^{-M_2|(l\vee l')-l'|}2^{-l't}\Bigg)\\
&\quad\lesssim 2^{-k\cdot s}2^{-lt},
\end{aligned}$$

completing the proof. □

7.4 Algebra

When considering nonlinear PDEs, an important property of the space $\mathscr{C}^s(\mathcal{K},(W,\vec{d}))$ is that it forms an algebra. Indeed, we have the following.

Proposition 7.4.1. *Fix $s\in(0,\infty)^\nu$. For $f,g\in\mathscr{C}^s(\mathcal{K},(W,\vec{d}))$, we have $fg\in\mathscr{C}^s(\mathcal{K},(W,\vec{d}))$, and moreover,*

$$\|fg\|_{\mathscr{C}^s(W,\vec{d})}\le C\|f\|_{\mathscr{C}^s(W,\vec{d})}\|g\|_{L^\infty}+\|f\|_{L^\infty}\|g\|_{\mathscr{C}^s(W,\vec{d})},\tag{7.11}$$

where $C=C(s,\mathcal{K},(W,\vec{d}))\ge 0$.

The estimate (7.11) is called a *tame* estimate and such estimates are very important in the study of nonlinear PDEs (see Remark 7.5.4). It is somewhat easier to prove the weaker non-tame estimate:

$$\|fg\|_{\mathscr{C}^s(W,\vec{d})}\lesssim\|f\|_{\mathscr{C}^s(W,\vec{d})}\|g\|_{\mathscr{C}^s(W,\vec{d})}.\tag{7.12}$$

We defer the proof of the tame estimate (7.11) to Theorem 7.5.2 (iii), and in this section prove only the easier non-tame estimate (7.12), as a warm-up for the more difficult proofs which follow. The proof of (7.11) uses a more complicated para-paraproduct decomposition which we introduce to study more general compositions.

Before we prove (7.12), we need some new notation.

Notation 7.4.2 (The product rule). We often consider derivatives of the form $(2^{-j\vec{d}}W)^\alpha$, where α is an ordered multi-index. Recall that this means that $\alpha=(\alpha_1,\dots,\alpha_L)$ is a list with $\alpha_j\in\{1,\dots,r\}$. We then have

$$(2^{-j\vec{d}}W)^\alpha=(2^{-j\cdot\vec{d}_{\alpha_1}}W_{\alpha_1})\cdots(2^{-j\cdot\vec{d}_{\alpha_L}}W_{\alpha_L}).$$

We write $\sum_{\beta_1\dot{+}\beta_2=\alpha}$ to denote the sum over non-overlapping sublists β_1,β_2 of α which combine to make the full list α. For example, $\sum_{\beta_1\dot{+}\beta_2=(1,2)}$ sums over the choices

$$\{\beta_1 = (), \beta_2 = (1,2)\}, \{\beta_1 = (1), \beta_2 = (2)\}, \{\beta_1 = (2), \beta_2 = (1)\}, \{\beta_1 = (1,2), \beta_2 = ()\}.$$

We take this sum with repetitions, so for example $\sum_{\beta_1 \dot{+} \beta_2 = (1,1)}$ sums over

$$\{\beta_1 = (), \beta_2 = (1,1)\}, \{\beta_1 = (1), \beta_2 = (1)\}, \{\beta_1 = (1), \beta_2 = (1)\}, \{\beta_1 = (1,1), \beta_2 = ()\}.$$

With this notation, we have the product rule:

$$(2^{-j\vec{d}} W)^{\alpha}(fg) = \sum_{\beta_1 \dot{+} \beta_2 = \alpha} ((2^{-j\vec{d}} W)^{\beta_1} f)((2^{-j\vec{d}} W)^{\beta_2} g).$$

We similarly define sums with more terms, for example $\sum_{\beta_1 \dot{+} \beta_2 \dot{+} \cdots \dot{+} \beta_R = \alpha}$, and obtain a similar product rule.

Proof of the algebra property and (7.12). By Corollary 7.2.2, $\mathscr{C}^s(\mathcal{K}, (W, \vec{d})) \subseteq C(\mathfrak{M})$, so the product fg is defined pointwise. Without loss of generality, we assume $\|f\|_{\mathscr{C}^s(W,\vec{d})} = \|g\|_{\mathscr{C}^s(W,\vec{d})} = 1$.

By Proposition 7.2.1, we may decompose $f = \sum_{k \in \mathbb{N}^\nu} f_k$ and $g = \sum_{l \in \mathbb{N}^\nu} g_l$, where for every ordered multi-index α,

$$2^{k\cdot s} \|(2^{-k\vec{d}} W)^{\alpha} f_k\|_{L^\infty}, 2^{l\cdot s} \|(2^{-l\vec{d}} W)^{\alpha} g_l\|_{L^\infty} \lesssim 1.$$

Fix $M > |s|_1$. We will show that

$$\|D_j(f_k g_l)\|_{L^\infty} \lesssim 2^{-M|(j\vee k\vee l)-(k\vee l)|} 2^{-k\cdot s} 2^{-l\cdot s}. \tag{7.13}$$

First, we see why (7.13) completes the proof. Indeed, we have, using (7.13) and Lemma 7.3.1,

$$\|D_j fg\|_{L^\infty} \le \sum_{k,l \in \mathbb{N}^\nu} \|D_j f_k g_l\|_{L^\infty} \lesssim \sum_{k,l \in \mathbb{N}^\nu} 2^{-M|(j\vee k\vee l)-(k\vee l)|} 2^{-k\cdot s} 2^{-l\cdot s} \lesssim 2^{-j\cdot s}.$$

This establishes (7.12). Moreover, since $\operatorname{supp}(f), \operatorname{supp}(g) \subseteq \mathcal{K}$, we have $\operatorname{supp}(fg) \subseteq \mathcal{K}$. Since (7.12) implies $\|fg\|_{\mathscr{C}^s(W,\vec{d})} < \infty$, it follows from (7.1) that $fg \in \mathscr{C}^s(\mathcal{K}, (W, \vec{d}))$.

We turn to proving (7.13). If $k \vee l = j \vee k \vee l$, then using $\|D_j\|_{L^\infty \to L^\infty} \lesssim 1$ (see Corollary 7.1.4 (i)), we have

$$\|D_j(f_k g_l)\|_{L^\infty} \lesssim \|f_k g_l\|_{L^\infty} \le \|f_k\|_{L^\infty} \|g_l\|_{L^\infty} \lesssim 2^{-k\cdot s} 2^{-l\cdot s},$$

proving (7.13) in this case.

Suppose $|(j \vee k \vee l) - (k \vee l)|_\infty > 0$. Pick $\mu \in \{1, \dots, \nu\}$ such that $j_\mu - (k_\mu \vee l_\mu) = |(j \vee k \vee l) - (k \vee l)|_\infty$. Letting $E_{j,\alpha_\mu,\mu,M}$ be as in Corollary 7.1.4 (ii), we have $\|E_{j,\alpha_\mu,\mu,M}\|_{L^\infty \to L^\infty} \lesssim 1$ and (using Notation 7.4.2 and Corollary 7.1.4 (ii))

$$\begin{aligned}
\|D_j(f_k g_l)\|_{L^\infty} &\leq \sum_{|\alpha_\mu|\leq M} 2^{-j_\mu(M-|\alpha_\mu|)} \|E_{j,\alpha_\mu,\mu,M}(2^{-j_\mu \vec{d}^\mu} W^\mu)^{\alpha_\mu}(f_k g_l)\|_{L^\infty} \\
&\lesssim \sum_{|\alpha_\mu|\leq M} 2^{-j_\mu(M-|\alpha_\mu|)} \|(2^{-j_\mu \vec{d}^\mu} W^\mu)^{\alpha_\mu}(f_k g_l)\|_{L^\infty} \\
&\lesssim \sum_{|\alpha_\mu|\leq M} \sum_{\beta_1^\mu+\beta_2^\mu=\alpha_\mu} 2^{-j_\mu(M-|\alpha_\mu|)} \|(2^{-j_\mu \vec{d}^\mu} W^\mu)^{\beta_1^\mu} f_k\|_{L^\infty} \|(2^{-j_\mu \vec{d}^\mu} W^\mu)^{\beta_2^\mu} g_l\|_{L^\infty} \\
&= \sum_{|\alpha_\mu|\leq M} \sum_{\beta_1^\mu+\beta_2^\mu=\alpha_\mu} 2^{-j_\mu(M-|\alpha_\mu|)} 2^{-(j_\mu-k_\mu)\deg_{\vec{d}^\mu}(\beta_\mu^1)} 2^{-(j_\mu-l_\mu)\deg_{\vec{d}^\mu}(\beta_\mu^2)} \\
&\qquad\qquad \times \|(2^{-k_\mu \vec{d}^\mu} W^\mu)^{\beta_1^\mu} f_k\|_{L^\infty} \|(2^{-l_\mu \vec{d}^\mu} W^\mu)^{\beta_2^\mu} g_l\|_{L^\infty} \\
&\lesssim \sum_{|\alpha_\mu|\leq M} \sum_{\beta_1^\mu+\beta_2^\mu=\alpha_\mu} 2^{-j_\mu(M-|\alpha_\mu|)} 2^{-(j_\mu-k_\mu)\deg_{\vec{d}^\mu}(\beta_\mu^1)} 2^{-(j_\mu-l_\mu)\deg_{\vec{d}^\mu}(\beta_\mu^2)} \\
&\lesssim 2^{-(j_\mu-(k_\mu\vee l_\mu))M} = 2^{-|(j\vee k\vee l)-(k\vee l)|_\infty M},
\end{aligned}$$

where in the final $\lesssim$ we have used $\deg_{\vec{d}^\mu}(\beta_\mu^1) + \deg_{\vec{d}^\mu}(\beta_\mu^2) = \deg_{\vec{d}^\mu}(\alpha_\mu) \geq |\alpha_\mu|$. This proves (7.13) and completes the proof. □

7.5 Compositions

Fix $N \in \mathbb{N}_+$. For functions $u \in \mathscr{C}^s(\mathcal{K}, (W, \vec{d}); \mathbb{R}^N)$, we wish to consider compositions of the form $F(x, u(x))$, where $F(x, \zeta) : \mathfrak{M} \times \mathbb{R}^N \to \mathbb{C}$ is a sufficiently smooth function.[2]

Informally, the class of functions $F(x, \zeta)$ we consider are $\mathscr{C}^s(\mathcal{K}, (W, \vec{d}))$ in the x variable and $\mathscr{C}^t(\mathbb{R}^N)$ in the ζ variable (see Definition 2.5.1). To rigorously define this product Zygmund–Hölder space, let $\psi \in C_0^\infty(\Omega)$ satisfy $\psi \equiv 1$ on a neighborhood of $\mathcal{K}$ and let $\{(D_j, 2^{-j}) : j \in \mathbb{N}^\nu\}$ be a bounded set of generalized $(W, \vec{d})$ elementary operators supported in Ω with $\sum_{j\in\mathbb{N}^\nu} D_j = \mathrm{Mult}[\psi]$ (we will usually take D_j from Proposition 7.1.1). Let $\{(\widetilde{D}_l, 2^{-l}) : l \in \mathbb{N}^\nu\}$ be given by (2.30) with n replaced by N; in particular, it is a bounded set of elementary operators on $\mathbb{R}^N$ in the sense of Definition 2.3.5 and $\sum_{l\in\mathbb{N}} \widetilde{D}_l = I$. Set $\widehat{D}_{j,l} := D_j \otimes \widetilde{D}_l$.

Definition 7.5.1. For $s \in (0, \infty)^\nu$ and $t > 0$, we let $\mathscr{C}^{s,t}(\mathcal{K} \times \mathbb{R}^N, (W, \vec{d}) \boxtimes \nabla_{\mathbb{R}^N})$ denote the space of all $F \in C_0^\infty(\mathfrak{M} \times \mathbb{R}^N)'$ such that the following norm is finite:

$$\|F\|_{\mathscr{C}^{s,t}((W,\vec{d})\boxtimes\nabla_{\mathbb{R}^N})} := \sup_{\substack{j\in\mathbb{N}^\nu \\ l\in\mathbb{N}}} 2^{j\cdot s} 2^{lt} \|\widehat{D}_{j,l} F\|_{L^\infty(\mathfrak{M}\times\mathbb{R}^N)} < \infty.$$

2 Here, we are considering functions $u(x)$ taking values in $\mathbb{R}^N$ instead of $\mathbb{C}^N$ (as was used in most of the rest of the text). There is no real difference for our purposes: functions taking values in $\mathbb{R}^N$ can be viewed as functions taking values in $\mathbb{C}^N$, while functions taking values in $\mathbb{C}^N$ can be viewed as functions taking values in $\mathbb{R}^{2N}$. The use of $\mathbb{R}^N$ in this section makes the notation a little simpler.

For $M \in \mathbb{N}_+$, we define the vector-valued space $\mathscr{C}^{s,t}(\mathcal{K}\times\mathbb{R}^N, (W,\vec{d})\boxtimes\nabla_{\mathbb{R}^N};\mathbb{C}^M)$ to consist of those distributions $F=(F_1,\dots,F_M)$ with each $F_k \in \mathscr{C}^{s,t}(\mathcal{K}\times\mathbb{R}^N, (W,\vec{d})\boxtimes\nabla_{\mathbb{R}^N})$. We set

$$\|F\|_{\mathscr{C}^{s,t}((W,\vec{d})\boxtimes\nabla_{\mathbb{R}^N};\mathbb{C}^M)} := \sum_{k=1}^{M} \|F_k\|_{\mathscr{C}^{s,t}((W,\vec{d})\boxtimes\nabla_{\mathbb{R}^N})}$$

and

$$\mathscr{C}^{s,t}_{\mathrm{cpt}}((W,\vec{d})\boxtimes\nabla_{\mathbb{R}^N}) := \bigcup_{\substack{\mathcal{K}\Subset\mathfrak{M}\\ \mathcal{K}\text{ compact}}} \mathscr{C}^{s,t}(\mathcal{K}\times\mathbb{R}^N, (W,\vec{d})\boxtimes\nabla_{\mathbb{R}^N}). \tag{7.14}$$

The main result of this section is the next theorem.

Theorem 7.5.2. *Fix $s\in(0,\infty)^\nu$ and $t>0$. Let $F\in\mathscr{C}^{s,t}(\mathcal{K}\times\mathbb{R}^N,(W,\vec{d})\boxtimes\nabla_{\mathbb{R}^N})$, $u,v\in\mathscr{C}^s(\mathcal{K},(W,\vec{d});\mathbb{R}^N)$, and $w\in\mathscr{C}^s(\mathcal{K},(W,\vec{d}))$. Then:*

(i) *If $t>\lfloor|s|_1\rfloor+1+\nu$, then*

$$F(x,u(x)+\zeta)\in\mathscr{C}^{s,t-\lfloor|s|_1\rfloor-1-\nu}(\mathcal{K}\times\mathbb{R}^N,(W,\vec{d})\boxtimes\nabla_{\mathbb{R}^N})$$

and

$$\begin{aligned}&\|F(x,u(x)+\zeta)\|_{\mathscr{C}^{s,t-\lfloor|s|_1\rfloor-1-\nu}((W,\vec{d})\boxtimes\nabla_{\mathbb{R}^N})}\\ &\quad\lesssim \|F\|_{\mathscr{C}^{s,t}((W,\vec{d})\boxtimes\nabla_{\mathbb{R}^N})}\big(1+\|u\|_{\mathscr{C}^s(W,\vec{d})}\big)\big(1+\|u\|_{L^\infty}\big)^{\lfloor|s|_1\rfloor+\nu}.\end{aligned}$$

(ii) *If $t>\lfloor|s|_1\rfloor+1+\nu$, then*

$$F(x,u(x))\in\mathscr{C}^s(\mathcal{K},(W,\vec{d}))$$

and

$$\begin{aligned}&\|F(x,u(x))\|_{\mathscr{C}^s(W,\vec{d})}\\ &\quad\lesssim \|F\|_{\mathscr{C}^{s,t}((W,\vec{d})\boxtimes\nabla_{\mathbb{R}^N})}\big(1+\|u\|_{\mathscr{C}^s(W,\vec{d})}\big)\big(1+\|u\|_{L^\infty}\big)^{\lfloor|s|_1\rfloor+\nu}.\end{aligned}$$

(iii) *We have $u\cdot v\in\mathscr{C}^s(\mathcal{K},(W,\vec{d}))$ and*

$$\|u\cdot v\|_{\mathscr{C}^s(W,\vec{d})}\lesssim \|u\|_{\mathscr{C}^s(W,\vec{d})}\|v\|_{L^\infty}+\|u\|_{L^\infty}\|v\|_{\mathscr{C}^s(W,\vec{d})}.$$

(iv) *If $t>\lfloor|s|_1\rfloor+1+\nu$ we have $F(x,u(x)+\zeta)w(x)\in\mathscr{C}^{s,t-\lfloor|s|_1\rfloor-1-\nu}(\mathcal{K}\times\mathbb{R}^N,(W,\vec{d})\boxtimes\nabla_{\mathbb{R}^N})$ and*

$$\begin{aligned}&\|F(x,u(x)+\zeta)w(z)\|_{\mathscr{C}^{s,t-\lfloor|s|_1\rfloor-1-\nu}((W,\vec{d})\boxtimes\nabla_{\mathbb{R}^N})}\\ &\quad\lesssim \|F\|_{\mathscr{C}^{s,t}((W,\vec{d})\boxtimes\nabla_{\mathbb{R}^N})}\big(\|w\|_{\mathscr{C}^s(W,\vec{d})}+\|u\|_{\mathscr{C}^s(W,\vec{d})}\|w\|_{L^\infty}\big)\big(1+\|u\|_{L^\infty}\big)^{\lfloor|s|_1\rfloor+\nu}.\end{aligned}$$

(v) *If* $t > \lfloor |s|_1 \rfloor + 2 + \nu$,

$$\|F(x,u(x)+\zeta) - F(x,v(x)+\zeta)\|_{\mathscr{C}^{s,t-\lfloor|s|_1\rfloor-2-\nu}((W,\vec{d})\boxtimes\nabla_{\mathbb{R}^N})}$$
$$\lesssim \|F\|_{\mathscr{C}^{s,t}((W,\vec{d})\boxtimes\nabla_{\mathbb{R}^N})}\big(\|u-v\|_{\mathscr{C}^s(W,\vec{d})} + (\|u\|_{\mathscr{C}^s(W,\vec{d})} + \|v\|_{\mathscr{C}^s(W,\vec{d})})\|u-v\|_{L^\infty}\big)$$
$$\times \big(1 + \|u\|_{L^\infty} + \|v\|_{L^\infty}\big)^{\lfloor|s|_1\rfloor+\nu}.$$

(vi) *If* $t > \lfloor |s|_1 \rfloor + 2 + \nu$, *then*

$$\|F(x,u(x)) - F(x,v(x))\|_{\mathscr{C}^s(W,\vec{d})}$$
$$\lesssim \|F\|_{\mathscr{C}^{s,t}((W,\vec{d})\boxtimes\nabla_{\mathbb{R}^N})}\big(\|u-v\|_{\mathscr{C}^s(W,\vec{d})} + (\|u\|_{\mathscr{C}^s(W,\vec{d})} + \|v\|_{\mathscr{C}^s(W,\vec{d})})\|u-v\|_{L^\infty}\big)$$
$$\times \big(1 + \|u\|_{L^\infty} + \|v\|_{L^\infty}\big)^{\lfloor|s|_1\rfloor+\nu}.$$

(vii) *Fix* $L \in \mathbb{N}_+$ *and suppose* $\partial_\zeta^\beta F(x,0) \equiv 0$, $\forall |\beta| < L$ *and* $t > \lfloor |s|_1 \rfloor + 1 + \nu + L$. *Then*

$$\|F(x,u(x))\|_{\mathscr{C}^s(W,\vec{d})} \lesssim \|F\|_{\mathscr{C}^{s,t}((W,\vec{d})\boxtimes\nabla_{\mathbb{R}^N})}\big(1+\|u\|_{L^\infty}\big)^{\lfloor|s|_1\rfloor+\nu+1}\|u\|_{L^\infty}^{L-1}\|u\|_{\mathscr{C}^s(W,\vec{d})}.$$

(viii) *Fix* $L \in \mathbb{N}$, $L \geq 2$ *and suppose* $\partial_\zeta^\beta F(x,0) \equiv 0$, $\forall 0 < |\beta| < L$ *and* $t > \lfloor |s|_1 \rfloor + 1 + \nu + L$. *Then*

$$\|F(x,u(x)) - F(x,v(x))\|_{\mathscr{C}^s(W,\vec{d})}$$
$$\lesssim \|F\|_{\mathscr{C}^{s,t}((W,\vec{d})\boxtimes\nabla_{\mathbb{R}^N})}\big(1+\|u\|_{L^\infty}+\|v\|_{L^\infty}\big)^{\lfloor|s|_1\rfloor+\nu+1}\big(\|u\|_{L^\infty}+\|v\|_{L^\infty}\big)^{L-2}$$
$$\times \big(\|u\|_{\mathscr{C}^s(W,\vec{d})} + \|v\|_{\mathscr{C}^s(W,\vec{d})}\big)\|u-v\|_{L^\infty}$$
$$+ \|F\|_{\mathscr{C}^{s,t}((W,\vec{d})\boxtimes\nabla_{\mathbb{R}^N})}\big(\|u\|_{L^\infty}+\|v\|_{L^\infty}\big)^{L-1}\|u-v\|_{\mathscr{C}^s(W,\vec{d})}.$$

The implicit constants do not depend on F, u, v, *or* w, *but may depend on any of the other ingredients in the result.*

Remark 7.5.3. Theorem 7.5.2 (vi) says, roughly, that the map $u \mapsto F(x,u(x))$ is Lipschitz in u. This will allow us to use arguments similar to the contraction mapping principle when studying fully nonlinear maximally subelliptic PDEs.

Remark 7.5.4. All the estimates in Theorem 7.5.2 are *tame* in the sense that each term on the right-hand side of the inequalities only has one factor of $\|\cdot\|_{\mathscr{C}^s(W,\vec{d})}$; all the other factors are $\|\cdot\|_{L^\infty}$. See also Remark 2.6.5. Tame estimates play an essential role in the proof in Section 9.2.4.

Remark 7.5.5. It is likely that Theorem 7.5.2 assumes too much regularity in the ζ variable (i. e., the result is probably not sharp in t). However, finding the sharp result seems to be difficult. See also Remark 2.6.4.

Remark 7.5.6. In Definition 7.5.1, we have used the fact that $W_1, \ldots, W_r$ satisfy Hörmander's condition, and therefore $C^\infty_{W,0}(\mathfrak{M}) = C^\infty_0(\mathfrak{M})$ (see Remark 5.2.24). Thus, in the proof which follows, it makes sense to consider $\widehat{D}_{j,l}F = (D_j \otimes \widetilde{D}_l)F$, for any $F \in C_0^\infty(\mathfrak{M} \times \mathbb{R}^N)' \cong C_0^\infty(\mathfrak{M})' \widehat{\otimes} C_0^\infty(\mathbb{R}^N)'$.

7.5.1 Properties of the product Zygmund–Hölder spaces

The space $\mathscr{C}^{s,t}(\mathcal{K} \times \mathbb{R}^N, (W, \vec{d}) \boxtimes \nabla_{\mathbb{R}^N})$ is closely related to the $(\nu+1)$-parameter Zygmund–Hölder spaces $\mathscr{C}^{(s,t)}$ associated with the vector fields with $(\nu+1)$-parameter formal degrees:

$$(W, \vec{d}) \boxtimes \{(\partial_{x_1}, 1), \ldots, (\partial_{x_n}, 1)\}$$

The difference is that in $\mathscr{C}^{s,t}(\mathcal{K} \times \mathbb{R}^N, (W, \vec{d}) \boxtimes \nabla_{\mathbb{R}^N})$ we do not require that the functions $F(x, \zeta)$ have compact support in the ζ variable. This is possible since the vector fields $\partial_{x_1}, \ldots, \partial_{x_n}$ are easy to study at both large and small scales (since they are homogeneous under the usual global family of dilations on $\mathbb{R}^n$), while general smooth vector fields can only be easily studied as small scales.

The proofs that we have exhibited for $\mathscr{C}^s(\mathcal{K}, (W, \vec{d}))$ can be easily modified to prove similar results for $\mathscr{C}^{s,t}(\mathcal{K} \times \mathbb{R}^N, (W, \vec{d}) \boxtimes \nabla_{\mathbb{R}^N})$. We state some of these results without proof here.

Proposition 7.5.7. *For $s \in (0, \infty)^\nu$ and $t > 0$, the following are equivalent:*

(i) $F \in \mathscr{C}^{s,t}(\mathcal{K} \times \mathbb{R}^N, (W, \vec{d}) \boxtimes \nabla_{\mathbb{R}^N})$.

(ii) $\operatorname{supp}(F) \subseteq \mathcal{K} \times \mathbb{R}^N$ *and for every bounded set of generalized $(W, \vec{d})$ elementary operators, $\mathcal{E}_1$, and every bounded set of elementary operators on $\mathbb{R}^N$, $\mathcal{E}_2$ (see Definition 2.3.5), we have*

$$\sup_{\{(E_j, 2^{-j}) : j \in \mathbb{N}^\nu\} \subseteq \mathcal{E}_1} \; \sup_{\{(\widetilde{E}_l, 2^{-l}) : l \in \mathbb{N}\} \subseteq \mathcal{E}_2} \; \sup_{\substack{j \in \mathbb{N}^\nu \\ l \in \mathbb{N}}} 2^{j \cdot s} 2^{lt} \|(E_j \otimes \widetilde{E}_l) F\|_{L^\infty(\mathfrak{M} \times \mathbb{R}^N)} < \infty.$$

Proposition 7.5.8. *For $s \in (0, \infty)^\nu$ and $t > 0$, $\mathscr{C}^{s,t}(\mathcal{K} \times \mathbb{R}^N, (W, \vec{d}) \boxtimes \nabla_{\mathbb{R}^N})$ is a Banach space when endowed with the norm $\|\cdot\|_{\mathscr{C}^{s,t}((W,\vec{d}) \boxtimes \nabla_{\mathbb{R}^N})}$.*

Proposition 7.5.9. *Let $s \in (0, \infty)^\nu$ and $t > 0$. For $F \in C_0^\infty(\mathfrak{M} \times \mathbb{R}^N)'$, the following are equivalent:*

(i) $F \in \mathscr{C}^{s,t}(\mathcal{K} \times \mathbb{R}^N, (W, \vec{d}) \boxtimes \nabla_{\mathbb{R}^N})$.

(ii) $\operatorname{supp}(F) \subseteq \mathcal{K} \times \mathbb{R}^N$ *and there exists a sequence $\{F_{j,l}\}_{\substack{j \in \mathbb{N}^\nu \\ l \in \mathbb{N}}} \subset C^\infty_{\mathrm{loc}}(\mathfrak{M} \times \mathbb{R}^N)$ such that for every ordered multi-index α and every multi-index β,*

$$\sup_{\substack{j \in \mathbb{N}^\nu \\ l \in \mathbb{N}}} 2^{j \cdot s} 2^{lt} \|(2^{-j\vec{d}} W_x)^\alpha (2^{-l} \partial_\zeta)^\beta F_{j,l}(x, \zeta)\|_{L^\infty(\mathfrak{M} \times \mathbb{R}^N)} < \infty,$$

with $F = \sum_{j,l} F_{j,l}$, where the sum converges in $C(\mathfrak{M} \times \mathbb{R}^N)$.

In this case, there exists $M = M(s,t,\nu) \in \mathbb{N}$ such that

$$\|F\|_{\mathscr{C}^{s,t}((W,\vec{d})\boxtimes\nabla_{\mathbb{R}^N})} \leq C \sum_{|\alpha|,|\beta|\leq M} \sup_{\substack{j\in\mathbb{N}^\nu\\ l\in\mathbb{N}}} 2^{j\cdot s}2^{lt}\|(2^{-j\vec{d}}W_x)^\alpha(2^{-l}\partial_\zeta)^\beta F_{j,l}(x,\zeta)\|_{L^\infty(\mathfrak{M}\times\mathbb{R}^N)}.$$

Furthermore, $F_{j,l}$ can be chosen such that $\operatorname{supp}(F_{j,l}) \subseteq \Omega\times\mathbb{R}^N$, $\forall j,l$, *and for each ordered multi-index α and multi-index β,*

$$\sup_{\substack{j\in\mathbb{N}^\nu\\ l\in\mathbb{N}}} 2^{j\cdot s}2^{lt}\|(2^{-j\vec{d}}W_x)^\alpha(2^{-l}\partial_\zeta)^\beta F_{j,l}(x,\zeta)\|_{L^\infty(\mathfrak{M}\times\mathbb{R}^N)} \leq C_{\alpha,\beta}\|F\|_{\mathscr{C}^{s,t}((W,\vec{d})\boxtimes\nabla_{\mathbb{R}^N})}. \tag{7.15}$$

Here, C and $C_{\alpha,\beta}$ do not depend on F or $F_{j,l}$, but may depend on any of the other ingredients in the result.

Proposition 7.5.10. *For $s \in (0,\infty)^\nu$ and $t > 0$, $\mathscr{C}^{s,t}(\mathcal{K}\times\mathbb{R}^N,(W,\vec{d})\boxtimes\nabla_{\mathbb{R}^N}) \subseteq C(\mathfrak{M}\times\mathbb{R}^N)$ and the inclusion is continuous.*

Proposition 7.5.11. *Let $s \in (0,\infty)^\nu$, $\beta \in \mathbb{N}^N$ a multi-index, and $t > |\beta|$. Then $F(x,\zeta) \mapsto \partial_\zeta^\beta F(x,\zeta)$ is a continuous map $\mathscr{C}^{s,t}(\mathcal{K}\times\mathbb{R}^N,(W,\vec{d})\boxtimes\nabla_{\mathbb{R}^N}) \mapsto \mathscr{C}^{s,t-|\beta|}(\mathcal{K}\times\mathbb{R}^N,(W,\vec{d})\boxtimes\nabla_{\mathbb{R}^N})$.*

We include one final result in this section, which is not an analog of a previous result, but instead a simple consequence of Proposition 7.5.9.

Corollary 7.5.12. *For $s \in (0,\infty)^\nu$ and $t > 0$, we have:*
(i) *The map $F(x,\zeta) \mapsto F(x,0)$ is continuous $\mathscr{C}^{s,t}(\mathcal{K}\times\mathbb{R}^N,(W,\vec{d})\boxtimes\nabla_{\mathbb{R}^N}) \to \mathscr{C}^s(\mathcal{K},(W,\vec{d}))$.*
(ii) *The map $F(x,\zeta_1,\dots,\zeta_N) \mapsto F(x,\zeta_1,\dots,\zeta_{N-1},0)$ is continuous $\mathscr{C}^{s,t}(\mathcal{K}\times\mathbb{R}^N,(W,\vec{d})\boxtimes\nabla_{\mathbb{R}^N}) \to \mathscr{C}^{s,t}(\mathcal{K}\times\mathbb{R}^{N-1},(W,\vec{d})\boxtimes\nabla_{\mathbb{R}^{N-1}})$.*
(iii) *For $x_0 \in \mathfrak{M}$, the map $F(x,\zeta) \mapsto F(x_0,\zeta)$ is continuous $\mathscr{C}^{s,t}(\mathcal{K}\times\mathbb{R}^N,(W,\vec{d})\boxtimes\nabla_{\mathbb{R}^N}) \to \mathscr{C}^t(\mathbb{R}^N)$.*

Proof. Let $F \in \mathscr{C}^{s,t}(\mathcal{K}\times\mathbb{R}^N,(W,\vec{d})\boxtimes\nabla_{\mathbb{R}^N})$. By Proposition 7.5.10, $F \in C(\mathfrak{M}\times\mathbb{R}^N)$, so the functions $F(x,0) \in C(\mathfrak{M})$ and $F(x,\zeta_1,\dots,\zeta_{N-1},0)$ are defined pointwise. Proposition 7.5.9 shows that we may write $F = \sum_{\substack{j\in\mathbb{N}^\nu\\ l\in\mathbb{N}}} F_{j,l}$, where $F_{j,l} \in C^\infty_{\mathrm{loc}}(\mathfrak{M}\times\mathbb{R}^N)$ and for every ordered multi-index α and every multi-index $\beta \in \mathbb{N}^N$ we have

$$\|(2^{-j\vec{d}}W_x)^\alpha(2^{-l}\partial_\zeta)^\beta F_{j,l}(x,\zeta)\|_{L^\infty(\mathfrak{M}\times\mathbb{R}^N)} \lesssim 2^{-j\cdot s}2^{-lt}\|F\|_{\mathscr{C}^{s,t}((W,\vec{d})\boxtimes\nabla_{\mathbb{R}^N})}. \tag{7.16}$$

(i): Set $G_j(x) := \sum_{l\in\mathbb{N}} F_{j,l}(x,0)$ so that $F(x,0) = \sum_{j\in\mathbb{N}^\nu} G_j(x)$. Using (7.16) with $\beta = 0$, we have, for every ordered multi-index α,

$$\begin{aligned}\|(2^{-j\vec{d}}W)^\alpha G_j\|_{L^\infty(\mathfrak{M})} &\leq \sum_{l\in\mathbb{N}}\|(2^{-j\vec{d}}W_x)^\alpha F_{j,l}(x,\zeta)\|_{L^\infty(\mathfrak{M}\times\mathbb{R}^N)}\\ &\lesssim \sum_{l\in\mathbb{N}} 2^{-j\cdot s}2^{-lt}\|F\|_{\mathscr{C}^{s,t}((W,\vec{d})\boxtimes\nabla_{\mathbb{R}^N})} \lesssim 2^{-j\cdot s}\|F\|_{\mathscr{C}^{s,t}((W,\vec{d})\boxtimes\nabla_{\mathbb{R}^N})}.\end{aligned} \tag{7.17}$$

Combining (7.17) with the fact that $\mathrm{supp}(F(\cdot,0)) \subseteq \mathcal{K}$ (since $\mathrm{supp}(F) \subseteq \mathcal{K}\times\mathbb{R}^N$), Proposition 7.2.1 shows that $F(\cdot,0) \in \mathscr{C}^s(\mathcal{K},(W,\vec{d}))$ with $\|F(\cdot,0)\|_{\mathscr{C}^s(W,\vec{d})} \lesssim \|F\|_{\mathscr{C}^{s,t}((W,\vec{d})\boxtimes\nabla_{\mathbb{R}^N})}$, completing the proof.

(ii): Let $\zeta' = (\zeta_1,\dots,\zeta_{N-1})$. It follows directly from (7.16) that for every ordered multi-index α and every multi-index $\beta' \in \mathbb{N}^{N-1}$, we have

$$\|(2^{-j\vec{d}}W_x)^\alpha(2^{-l}\partial_{\zeta'})^{\beta'}F_{j,l}(x,\zeta',0)\|_{L^\infty(\mathfrak{M}\times\mathbb{R}^N)} \lesssim 2^{-j\cdot s}2^{-lt}\|F\|_{\mathscr{C}^{s,t}((W,\vec{d})\boxtimes\nabla_{\mathbb{R}^N})}.$$

Since $F(x,\zeta',0) = \sum_{\substack{j\in\mathbb{N}^\nu\\ l\in\mathbb{N}}} F_{j,l}(x,\zeta',0)$ and $\mathrm{supp}(F(x,\zeta',0)) \subseteq \mathcal{K}\times\mathbb{R}^{N-1}$, since $\mathrm{supp}(F) \subseteq \mathcal{K}\times\mathbb{R}^N$, Proposition 7.5.9 implies $F(x,\zeta',0) \in \mathscr{C}^{s,t}(\mathcal{K}\times\mathbb{R}^{N-1},(W,\vec{d})\boxtimes\nabla_{\mathbb{R}^{N-1}})$ and $\|F(x,\zeta',0)\|_{\mathscr{C}^{s,t}((W,\vec{d})\boxtimes\nabla_{\mathbb{R}^{N-1}})} \lesssim \|F\|_{\mathscr{C}^{s,t}((W,\vec{d})\boxtimes\nabla_{\mathbb{R}^N})}$, completing the proof.

(iii): Set $G_l(\zeta) := \sum_{j\in\mathbb{N}^\nu} F_{j,l}(x_0,\zeta)$ so that $F(x_0,\zeta) = \sum_{l\in\mathbb{N}} G_l(\zeta)$. Using (7.16) with $|\alpha| = 0$, we have, for every multi-index β,

$$\begin{aligned}\|(2^{-lt}\partial_\zeta)^\beta G_l\|_{L^\infty(\mathbb{R}^N)} &\le \sum_{j\in\mathbb{N}^\nu}\|(2^{-j\vec{d}}W_x)^\alpha F_{j,l}(x,\zeta)\|_{L^\infty(\mathfrak{M}\times\mathbb{R}^N)}\\ &\lesssim \sum_{j\in\mathbb{N}^\nu} 2^{-j\cdot s}2^{-lt}\|F\|_{\mathscr{C}^{s,t}((W,\vec{d})\boxtimes\nabla_{\mathbb{R}^N})} \lesssim 2^{-lt}\|F\|_{\mathscr{C}^{s,t}((W,\vec{d})\boxtimes\nabla_{\mathbb{R}^N})}.\end{aligned} \tag{7.18}$$

Combining (7.18) with Proposition 2.5.16 shows that $\|F(x_0,\zeta)\|_{\mathscr{C}^t(\mathbb{R}^N)} \lesssim \|F\|_{\mathscr{C}^{s,t}((W,\vec{d})\boxtimes\nabla_{\mathbb{R}^N})}$, completing the proof. □

7.5.2 Proof of the composition theorem

In this section, we prove Theorem 7.5.2. As we will see, the main difficulties are contained in Theorem 7.5.2 (i); the rest of Theorem 7.5.2 follows as an easy consequence. To prove Theorem 7.5.2 (i), we use the next lemma.

Lemma 7.5.13. *Fix $s \in (0,\infty)^\nu$ and $t > 0$. Set $M_1 := \lfloor |s|_1\rfloor + 1$ and $M_2 := \lfloor t\rfloor + 1$. Suppose that for each $j,k \in \mathbb{N}^\nu$ and $l \in \mathbb{N}$, $H_{j,k,l}(x,\zeta) \in C^\infty_{\mathrm{loc}}(\mathfrak{M}\times\mathbb{R}^N)$ are such that $\forall\mu \in \{1,\dots,\nu\}$, $|\alpha_\mu| \le M_1$, and $|\beta| \le M_2$,*

$$\|(2^{-(k_\mu\vee j_\mu)d^\mu}W^\mu)^{\alpha_\mu}(2^{-l}\partial_\zeta)^\beta H_{j,k,l}\|_{L^\infty(\mathfrak{M}\times\mathbb{R}^N)} \le A2^{-k\cdot s}2^{-j\cdot s}2^{-lt},$$

for some constant $A \ge 0$. Set $H(x,\zeta) := \sum_{j,k\in\mathbb{N}^\nu}\sum_{l\in\mathbb{N}} H_{j,k,l}(x,\zeta)$ and suppose $\mathrm{supp}(H) \subseteq \mathcal{K}\times\mathbb{R}^N$. *Then $H \in \mathscr{C}^{s,t}(\mathcal{K}\times\mathbb{R}^N,(W,\vec{d})\boxtimes\nabla_{\mathbb{R}^N})$ and*

$$\|H\|_{\mathscr{C}^{s,t}((W,\vec{d})\boxtimes\nabla_{\mathbb{R}^N})} \le C_{s,t,(W,\vec{d}),\mathcal{K},N}A.$$

Proof. We will show that for $j, k, k' \in \mathbb{N}^\nu$ and $l, l' \in \mathbb{N}$,

$$\|\widehat{D}_{k,l}H_{j,k',l'}\|_{L^\infty(\mathfrak{M}\times\mathbb{R}^N)} \lesssim A2^{-M_1|(j\vee k\vee k')-(j\vee k')|_\infty}2^{-M_2|(l\vee l')-l'|}2^{-j\cdot s}2^{-k'\cdot s}2^{-l't}. \tag{7.19}$$

First, we see why (7.19) will complete the proof. Indeed, using (7.19) we have

$$\begin{aligned}\|\widehat{D}_{k,l}H\|_{L^\infty(\mathfrak{M}\times\mathbb{R}^N)} &\le \sum_{\substack{j,k'\in\mathbb{N}^\nu\\ l'\in\mathbb{N}}}\|\widehat{D}_{k,l}H_{j,k',l'}\|_{L^\infty(\mathfrak{M}\times\mathbb{R}^N)}\\ &\lesssim A\sum_{\substack{j,k'\in\mathbb{N}^\nu\\ l'\in\mathbb{N}}}2^{-M_1|(j\vee k\vee k')-(j\vee k')|_\infty}2^{-M_2|(l\vee l')-l'|}2^{-j\cdot s}2^{-k'\cdot s}2^{-l't}\\ &\lesssim A2^{-k\cdot s}2^{-lt},\end{aligned}\tag{7.20}$$

where in the final estimate we have used $M_1 > |s|_1$, $M_2 > t$ and applied Lemma 7.3.2. By hypothesis, we have $\operatorname{supp}(H) \subseteq \mathcal{K}\times\mathbb{R}^N$, and (7.20) implies $\|H\|_{\mathscr{C}^{s,t}((W,\vec{d})\boxtimes\nabla_{\mathbb{R}^N})} \lesssim A$, which completes the proof.

We turn to proving (7.19), which we separate into several cases. Suppose $|(k\vee k'\vee j)-(k'\vee j)|_\infty = 0$ and $|(l\vee l')-l'| = 0$. By Corollary 7.1.4 (i) and Lemma 2.3.26, we have

$$\|\widehat{D}_{k,l}\|_{L^\infty(\mathfrak{M}\times\mathbb{R}^N)\to L^\infty(\mathfrak{M}\times\mathbb{R}^N)} = \|D_k\|_{L^\infty(\mathfrak{M})\to L^\infty(\mathfrak{M})}\|\widetilde{D}_l\|_{L^\infty(\mathbb{R}^N)\to L^\infty(\mathbb{R}^N)} \lesssim 1.$$

Thus,

$$\|\widehat{D}_{k,l}H_{j,k',l'}\|_{L^\infty} \lesssim \|H_{j,k',l'}\|_{L^\infty} \le A2^{-j\cdot s}2^{-k'\cdot s}2^{-l't},$$

completing the proof of (7.19) in this case.

Next we consider the case where $|(k\vee k'\vee j)-(k'\vee j)|_\infty > 0$. Pick $\mu \in \{1,\dots,\nu\}$ such that $k_\mu - (k'_\mu\vee j_\mu) = |(k\vee k'\vee j)-(k'\vee j)|_\infty$. Let E_{k,α_μ,μ,M_1} be as in Corollary 7.1.4 (ii). We consider two subcases: $|(l\vee l')-l'| = 0$ and $|(l\vee l')-l'| > 0$. If $|(l\vee l')-l'| = 0$, we use $\|\widetilde{D}_l\|_{L^\infty(\mathbb{R}^N)\to L^\infty(\mathbb{R}^N)} \lesssim 1$ (Lemma 2.3.26) to see that

$$\begin{aligned}\|\widehat{D}_{k,l}H_{j,k',l'}\|_{L^\infty} &\le \sum_{|\alpha_\mu|\le M_1}2^{-(M_1-|\alpha_\mu|)k_\mu}\|(E_{k,\alpha_\mu,\mu,M_1}\otimes\widetilde{D}_l)(2^{-k_\mu d^\mu}W^\mu)^{\alpha_\mu}H_{j,k',l'}\|_{L^\infty}\\ &\lesssim \sum_{|\alpha_\mu|\le M_1}2^{-(M_1-|\alpha_\mu|)k_\mu}\|(2^{-k_\mu d^\mu}W^\mu)^{\alpha_\mu}H_{j,k',l'}\|_{L^\infty}\\ &\lesssim \sum_{|\alpha_\mu|\le M_1}2^{-(M_1-|\alpha_\mu|)k_\mu}2^{-\deg_{d^\mu}(\alpha_\mu)(k_\mu-(k'_\mu\vee j_\mu))}\|(2^{-(k'_\mu\vee j_\mu)d^\mu}W^\mu)^{\alpha_\mu}H_{j,k',l'}\|_{L^\infty}\\ &\lesssim 2^{-M_1(k_\mu-(k'_\mu\vee j_\mu))}A2^{-k'\cdot s}2^{-j\cdot s}2^{-l't}\\ &= A2^{-M_1|(j\vee k\vee k')-(j\vee k')|_\infty}2^{-k'\cdot s}2^{-j\cdot s}2^{-l't},\end{aligned}$$

where we have used $\deg_{d^\mu}(\alpha_\mu) \ge |\alpha_\mu|$, establishing (7.19). If $|(l\vee l')-l'| > 0$, we apply Proposition 2.3.18 (d) to write

$$\widetilde{D}_l = \sum_{|\beta|\leq M_2} 2^{-l(M_2-|\beta|)} \widetilde{E}_{l,\beta} (2^{-l}\partial_\zeta)^\beta, \tag{7.21}$$

where $\{(\widetilde{E}_{l,\beta}, 2^{-l}) : l \in \mathbb{N}, |\beta| \leq M_2\}$ is a bounded set of elementary operators on $\mathbb{R}^N$, and therefore, by Lemma 2.3.26, $\|\widetilde{E}_{l,\beta}\|_{L^\infty(\mathbb{R}^N)\to L^\infty(\mathbb{R}^N)} \lesssim 1$. Thus, we have

$$\begin{aligned}
&\|\widehat{D}_{k,l} H_{j,k',l'}\|_{L^\infty} \\
&\quad\leq \sum_{|\alpha_\mu|\leq M_1} \sum_{|\beta|\leq M_2} 2^{-(M_1-|\alpha_\mu|)k_\mu} 2^{-l(M_2-|\beta|)} \|(E_{k,\alpha_\mu,\mu,M_1} \otimes \widetilde{E}_{l,\beta})(2^{-k_\mu \vec{d}^\mu} W^\mu)^{\alpha_\mu} (2^{-l}\partial_\zeta)^\beta H_{j,k',l'}\|_{L^\infty} \\
&\quad\lesssim \sum_{|\alpha_\mu|\leq M_1} \sum_{|\beta|\leq M_2} 2^{-(M_1-|\alpha_\mu|)k_\mu} 2^{-l(M_2-|\beta|)} \|(2^{-k_\mu \vec{d}^\mu} W^\mu)^{\alpha_\mu} (2^{-l}\partial_\zeta)^\beta H_{j,k',l'}\|_{L^\infty} \\
&\quad\lesssim \sum_{|\alpha_\mu|\leq M_1} \sum_{|\beta|\leq M_2} 2^{-(M_1-|\alpha_\mu|)k_\mu} 2^{-\deg_{\vec{d}^\mu}(\alpha_\mu)(k_\mu-(k'_\mu\vee j_\mu))} 2^{-l(M_2-|\beta|)} 2^{-|\beta|(l-l')} \\
&\qquad\qquad \times \|(2^{-(k'_\mu\vee j_\mu)\vec{d}^\mu} W^\mu)^{\alpha_\mu} (2^{-l'}\partial_\zeta)^\beta H_{j,k',l'}\|_{L^\infty} \\
&\quad\lesssim A 2^{-M_1(k_\mu-(k'_\mu\vee j_\mu))} 2^{-M_2(l-l')} A 2^{-k'\cdot s} 2^{-j\cdot s} 2^{-l't} \\
&\quad= A 2^{-M_1|(j\vee k\vee k')-(j\vee k')|_\infty} 2^{-M_2|(l\vee l')-l'|} 2^{-j\cdot s} 2^{-k'\cdot s} 2^{-l't},
\end{aligned}$$

where we have used $\deg_{\vec{d}^\mu}(\alpha_\mu) \geq |\alpha_\mu|$, establishing (7.19).

Finally, we consider the case where $|(k\vee k'\vee j) - (k'\vee j)|_\infty = 0$ and $|(l\vee l') - l'| > 0$. As in the previous case, we use (7.21), where $\|\widetilde{E}_{l,\beta}\|_{L^\infty\to L^\infty} \lesssim 1$. Using $\|D_k\|_{L^\infty\to L^\infty} \lesssim 1$ (Corollary 7.1.4 (i)), we have

$$\begin{aligned}
\|\widehat{D}_{k,l} H_{j,k',l'}\|_{L^\infty} &\leq \sum_{|\beta|\leq M_2} 2^{-l(M_2-|\beta|)} \|(D_k \otimes \widetilde{E}_{l,\beta})(2^{-l}\partial_\zeta)^\beta H_{j,k',l'}\|_{L^\infty} \\
&\lesssim \sum_{|\beta|\leq M_2} 2^{-l(M_2-|\beta|)} 2^{-|\beta|(l-l')} \|(2^{-l'}\partial_\zeta)^\beta H_{j,k',l'}\|_{L^\infty} \\
&\lesssim 2^{-M_2(l-l')} A 2^{-j\cdot s} 2^{-k'\cdot s} 2^{-l't} = A 2^{-M_2|(l\vee l')-l'|} 2^{-j\cdot s} 2^{-k'\cdot s} 2^{-l't},
\end{aligned}$$

establishing (7.19) and completing the proof. □

To prove Theorem 7.5.2 (i), we wish to decompose $F(x, u(x) + \zeta)$ as in Lemma 7.5.13. Let $F \in \mathscr{C}^{s,t}(\mathcal{K} \times \mathbb{R}^N, (W, \vec{d}) \boxtimes \nabla_{\mathbb{R}^N})$ and $u \in \mathscr{C}^s(\mathcal{K}, (W, \vec{d}))$. For $E \subseteq \{1, \ldots, \nu\}$, let $\mathfrak{e}_E = \sum_{\mu\in E} \mathfrak{e}_\mu$, where $\mathfrak{e}_1, \ldots, \mathfrak{e}_\nu$ is the standard basis for $\mathbb{R}^\nu$. Decompose

$$F = \sum_{\substack{k\in\mathbb{N}^\nu \\ l\in\mathbb{N}}} F_{k,l} \tag{7.22}$$

as in Proposition 7.5.9, with $F_{k,l} \in C^\infty_{\mathrm{loc}}(\mathfrak{M} \times \mathbb{R}^N)$ satisfying (7.15). The decomposition of $F(x, u(x) + \zeta)$ we use is the following:

$$
\begin{aligned}
F(x,u(x)+\zeta) &= \lim_{J\to\infty} F(x,P_{(J,J,\ldots,J)}u(x)+\zeta) \\
&= \sum_{j\in\mathbb{N}^\nu}\sum_{E\subseteq\{\mu:j_\mu\neq 0\}} (-1)^{|E|}F(x,P_{j-\mathfrak{e}_E}u(x)+\zeta) \qquad (7.23)\\
&= \sum_{\substack{j,k\in\mathbb{N}^\nu\\ l\in\mathbb{N}}}\sum_{E\subseteq\{\mu:j_\mu\neq 0\}} (-1)^{|E|}F_{k,l}(x,P_{j-\mathfrak{e}_E}u(x)+\zeta)\\
&=: \sum_{\substack{j,k\in\mathbb{N}^\nu\\ l\in\mathbb{N}}} H_{j,k,l}(x,\zeta).
\end{aligned}
$$

Note that $H_{j,k,l}\in C^\infty_{\mathrm{loc}}(\mathfrak{M}\times\mathbb{R}^N)$.

The decomposition (7.23) is the decomposition to which we wish to apply Lemma 7.5.13. Thus, we wish to estimate the functions $H_{j,k,l}(x,\zeta)$. Before we enter into details, we describe how we will study $H_{j,k,l}$ in the special case $\nu = 2$ to help give the reader an overview of how we will proceed. Suppose $g(\sigma_1,\sigma_2)$ is a function of two variables. Then

$$
g(1,1)-g(1,0)-g(0,1)+g(0,0) = \iint_{[0,1]^2} \partial_{\sigma_1}\partial_{\sigma_2} g(\sigma_1,\sigma_2)\,d\sigma.
$$

In the case $\nu=2$, when $j\in\mathbb{N}^2_+$, we write $H_{j,k,l}(x,\zeta)$ as

$$
\begin{aligned}
H_{j,k,l}(x,\zeta) &= F_{k,l}(x,P_{(j_1,j_2)}u(x)+\zeta) - F_{k,l}(x,P_{(j_1-1,j_2)}u(x)+\zeta)\\
&\quad - F_{k,l}(x,P_{(j_1,j_2-1)}u(x)+\zeta) + F_{k,l}(x,P_{(j_1-1,j_2-1)}u(x)+\zeta)\\
&= \iint_{[0,1]^2} \partial_{\sigma_1}\partial_{\sigma_2} F_{k,l}\Big(x, P_{j_1-1,j_2-1}u(x) \qquad (7.24)\\
&\qquad + \sigma_1 D_{j_1,j_2-1}u(x) + \sigma_2 D_{j_1-1,j_2}u(x) + \sigma_1\sigma_2 D_{j_1,j_2}u(x)+\zeta\Big)\,d\sigma.
\end{aligned}
$$

Thus, to estimate $H_{j,k,l}$, it suffices to instead estimate the above integrand. By the chain rule, this integrand involves the factors $D_{j_1-1,j_2}u$, $D_{j_1,j_2-1}u$, or $D_{j_1,j_2}u$. If $u\in\mathscr{C}^s(\mathcal{K},(W,\vec{d}))$, then these factors give a gain of $2^{-j\cdot s}$, which is the key to our estimates. If $j_1\geq 1$, but $j_2=0$, we instead use

$$
\begin{aligned}
H_{j,k,l}(x,\zeta) &= F_{k,l}(x,P_{(j_1,0)}u(x)+\zeta) - F_{k,l}(x,P_{(j_1-1,0)}u(x)+\zeta)\\
&= \int_{[0,1]} \partial_{\sigma_1} F_{k,l}(x,P_{j_1-1,0}u(x)+\sigma_1 D_{j_1,0}u(x)+\zeta)\,d\sigma_1.
\end{aligned}
$$

Again the chain rule gives a gain of $2^{-j\cdot s}$. A similar formula works when $j_2\geq 1$ and $j_1=0$. Finally, when $j=(j_1,j_2)=(0,0)$, $2^{-j\cdot s}=1$, so we can directly use the formula $H_{j,k,l}(x,\zeta)=F_{k,l}(x,P_0u(x)+\zeta)$.

Notation 7.5.14. For a function $G(x,\zeta) \in C^\infty_{\mathrm{loc}}(\mathfrak{M} \times \mathbb{R}^N)$ and $L \in \mathbb{N}_+$, we write $\mathbf{D}^L_\zeta G(x,\zeta)$ for the L-th derivative of $G(x,\zeta)$ in the ζ variable. Thus, $\mathbf{D}^L_\zeta G(x,\zeta)$ is an L-linear form on $\mathbb{R}^N$. We write $\|\mathbf{D}^L_\zeta G(x,\zeta)\|_{L^\infty}$ for the $L^\infty(\mathfrak{M} \times \mathbb{R}^N)$ norm of $\mathbf{D}^L_\zeta G(x,\zeta)$ thought of as taking values in the space of L-linear forms on $\mathbb{R}^N$ with the usual norm. In other words, $\|\mathbf{D}^L_\zeta G(x,\zeta)\|_{L^\infty}$ is the least constant such that for all $v_1,\dots,v_L \in \mathbb{R}^N$

$$\|(\mathbf{D}^L_\zeta G(x,\zeta))[v_1,\dots,v_L]\|_{L^\infty(\mathfrak{M}\times\mathbb{R}^N)} \leq \|\mathbf{D}^L_\zeta G(x,\zeta)\|_{L^\infty}|v_1|\cdots|v_L|.$$

When $L = 1$, we may identify $\mathbf{D}_\zeta G(x,\zeta)$ with a vector in $\mathbb{R}^N$. Thus, it makes sense to consider

$$\|\mathbf{D}_\zeta G(x,\zeta)\|_{\mathscr{C}^{s,t}((W,\vec{d})\boxtimes\nabla_{\mathbb{R}^N})} := \|\mathbf{D}_\zeta G(x,\zeta)\|_{\mathscr{C}^{s,t}((W,\vec{d})\boxtimes\nabla_{\mathbb{R}^N};\mathbb{R}^N)}.$$

Lemma 7.5.15. *Fix $s \in (0,\infty)^\nu$, $t \in \mathbb{R}$, and $A \geq 0$. Fix $k \in \mathbb{N}^\nu$ and $l \in \mathbb{N}$ and suppose we are given $G_{k,l}(x,\zeta) \in C^\infty_{\mathrm{loc}}(\mathfrak{M} \times \mathbb{R}^N)$ satisfying, for all α and β,*

$$\|(2^{-k\vec{d}}W)^\alpha(2^{-l}\partial_\zeta)G_{k,l}\|_{L^\infty(\mathfrak{M}\times\mathbb{R}^N)} \leq C_{\alpha,\beta}A_0 2^{-k\cdot s}2^{-lt}. \tag{7.25}$$

Let $u \in L^\infty(\mathfrak{M};\mathbb{R}^N)$. For $E \subseteq \{1,\dots,\nu\}$, let $a_E \in [0,1]$. For $j \in \mathbb{N}^\nu$, set

$$u_j := \begin{cases} P_{j-\mathfrak{e}_{\{\mu:j_\mu\neq 0\}}}u + \sum_{E\subsetneq\{\mu:j_\mu\neq 0\}} a_E D_{j-\mathfrak{e}_E}u, & j \neq 0,\\ P_0 u, & j = 0.\end{cases}$$

Then, for all $\mu \in \{1,\dots,\nu\}$, α_μ, and β,

$$\begin{aligned}&\|(2^{-(j_\mu\vee k_\mu)d^\mu}W^\mu)^{\alpha_\mu}(2^{-l}\partial_\zeta)^\beta G_{k,l}(x,u_j(x)+\zeta)\|_{L^\infty(\mathfrak{M}\times\mathbb{R}^N)}\\ &\quad\leq \widetilde{C}_{\mu,\alpha_\mu,\beta}A_0 2^{-k\cdot s}2^{-l(t-|\alpha_\mu|)}(1+\|u\|_{L^\infty})^{|\alpha_\mu|},\end{aligned} \tag{7.26}$$

where $\widetilde{C}_{\mu,\alpha_\mu,\beta}$ does not depend on A_0, $j,k \in \mathbb{N}^\nu$, $l \in \mathbb{N}$, $a_E \in [0,1]$, or $u \in L^\infty(\mathfrak{M};\mathbb{R}^N)$.

Proof. By Corollary 7.1.4 (i), we have, for all α,

$$\|(2^{-j\vec{d}}W)^\alpha u_j\|_{L^\infty} \lesssim 1. \tag{7.27}$$

Consider, using Notations 7.4.2 and 7.5.14,

$$\begin{aligned}&(2^{-(j_\mu\vee k_\mu)d^\mu}W^\mu)^{\alpha_\mu}(2^{-l}\partial_\zeta)^\beta G_{k,l}(x,u_j(x)+\zeta)\\ &= ((2^{-(j_\mu\vee k_\mu)d^\mu}W^\mu_x)^{\alpha_\mu}(2^{-l}\partial_\zeta)^\beta G_{k,l})(x,u_j(x)+\zeta)\\ &\quad+\sum_{\substack{\alpha_\mu^1+\alpha_\mu^2\\ 1\leq L\leq|\alpha_\mu^2|}}\sum_{\substack{\gamma_1+\gamma_2+\dots+\gamma_L\\=\alpha_\mu^2}}(((2^{-(j_\mu\vee k_\mu)d^\mu}W^\mu_x)^{\alpha_\mu^1}(2^{-l}\partial_\zeta)^\beta\mathbf{D}^L_\zeta G_{k,l})(x,u_j(x)+\zeta))\\ &\qquad[(2^{-(j_\mu\vee k_\mu)d^\mu}W^\mu)^{\gamma_1}u_j(x),\dots,(2^{-(j_\mu\vee k_\mu)d^\mu}W^\mu)^{\gamma_L}u_j(x)].\end{aligned} \tag{7.28}$$

For the first term on the right-hand side of (7.28), we have, using (7.25),

$$\begin{aligned}
&\|((2^{-(j_\mu\vee k_\mu)\vec{d}^\mu}W_x^\mu)^{\alpha_\mu}(2^{-l}\partial_\zeta)^\beta G_{k,l})(x,u_j(x)+\zeta)\|_{L^\infty}\\
&\leq \|(2^{-k_\mu\vec{d}^\mu}W^\mu)^{\alpha_\mu}(2^{-l}\partial_\zeta)^\beta G_{k,l}\|_{L^\infty}\\
&\lesssim A_0 2^{-k\cdot s}2^{-lt},
\end{aligned}$$

which is even better than the desired bound (7.26).

We turn to bounding the sum on the right-hand side of (7.28); we bound each term of the sum separately. We have, using $L\leq|\alpha_\mu^2|\leq|\alpha_\mu|$, (7.25), and (7.27),

$$\begin{aligned}
&\Big\|(((2^{-(j_\mu\vee k_\mu)\vec{d}^\mu}W_x^\mu)^{\alpha_\mu^1}(2^{-l}\partial_\zeta)^\beta \mathbf{D}_\zeta^L G_{k,l})(x,u_j(x)+\zeta))\\
&\qquad[(2^{-(j_\mu\vee k_\mu)\vec{d}^\mu}W^\mu)^{\gamma_1}u_j(x),\ldots,(2^{-(j_\mu\vee k_\mu)\vec{d}^\mu}W^\mu)^{\gamma_L}u_j(x)]\Big\|_{L^\infty}\\
&\leq 2^{lL}\|(2^{-k_\mu\vec{d}^\mu}W^\mu)^{\alpha_\mu^1}(2^{-l}\partial_\zeta)^\beta(2^{-l}\mathbf{D}_\zeta)^L G_{k,l}\|_{L^\infty}\prod_{s=1}^{L}\|(2^{-j_\mu\vec{d}^\mu}W^\mu)^{\gamma_s}u_j\|_{L^\infty}\\
&\lesssim 2^{lL}A_0 2^{-k\cdot s}2^{-lt}\|u\|_{L^\infty}^L\leq A_0 2^{-k\cdot s}2^{-l(t-|\alpha_\mu|)}(1+\|u\|_{L^\infty})^{|\alpha_\mu|},
\end{aligned}$$

completing the proof. □

Lemma 7.5.16. *Let $H_{j,k,l}$ be as in* (7.23)*, where $u\in\mathscr{C}^s(\mathcal{K},(W,\vec{d});\mathbb{R}^N)$. Then, for all $\mu\in\{1,\ldots,\nu\}$, α_μ, and β,*

$$\begin{aligned}
&\|(2^{-(k_\mu\vee j_\mu)\vec{d}^\mu}W^\mu)^{\alpha_\mu}(2^{-l}\partial_\zeta)^\beta H_{j,k,l}(x,\zeta)\|_{L^\infty}\\
&\leq C_{\mu,\alpha_\mu,\beta}\|F\|_{\mathscr{C}^{s,t}((W,\vec{d})\boxtimes\nabla_{\mathbb{R}^N})}2^{-k\cdot s}2^{-j\cdot s}2^{-l(t-|\alpha_\mu|-\nu)}\\
&\quad\times(1+\|u\|_{\mathscr{C}^s((W,\vec{d});\mathbb{R}^N)})(1+\|u\|_{L^\infty})^{|\alpha_\mu|+\nu-1},
\end{aligned}\tag{7.29}$$

where $C_{\mu,\alpha_\mu,\beta}\geq 0$ does not depend on $j,k\in\mathbb{N}^\nu$, $l\in\mathbb{N}$, F, or u.

Proof. Before delving into details, we encourage the reader to review (7.24), which contains the main formula we use in a simple special case.

By (7.23), we have

$$H_{j,k,l}(x,\zeta)=\sum_{E\subseteq\{\mu:j_\mu\neq 0\}}F_{k,l}(x,P_{j-\mathfrak{e}_E}u(x)+\zeta),$$

where $F_{k,l}$ satisfy (7.15). Note that this implies that $F_{k,l}$ satisfies the assumptions of $G_{k,l}$ in Lemma 7.5.15 with $A_0=\|F\|_{\mathscr{C}^{s,t}((W,\vec{d})\boxtimes\nabla_{\mathbb{R}^N})}$.

First we consider the case $j=0$, so that $H_{0,k,l}(x,\zeta)=F_{k,l}(x,P_0u(x)+\zeta)$. Applying Lemma 7.5.15 with $j=0$ gives

$$\begin{aligned}&\|(2^{-k_\mu d^\mu} W^\mu)^{\alpha_\mu}(2^{-l}\partial_\zeta)F_{k,l}(x,P_0u(x)+\zeta)\|_{L^\infty}\\&\quad\lesssim \|F\|_{\mathscr{C}^{s,t}((W,\vec{d})\boxtimes\nabla_{\mathbb{R}^N})}2^{-k\cdot s}2^{-l(t+|\alpha_\mu|)}(1+\|u\|_{L^\infty})^{|\alpha_\mu|},\end{aligned}$$

which is even better than the desired estimate (7.29).

Let $j\in\mathbb{N}^\nu\setminus\{0\}$. Without loss of generality, we assume $\{\mu : j_\mu\neq 0\}=\{1,\dots,\nu_0\}$ for some $1\leq\nu_0\leq\nu$. We write $\sigma=(\sigma_1,\dots,\sigma_{\nu_0})\in\mathbb{R}^{\nu_0}$. For $E\subseteq\{1,\dots,\nu_0\}$, we write $\sigma_E:=\prod_{\mu\in E}\sigma_\mu$ and $E^c:=\{1,\dots,\nu_0\}\setminus E$. We have

$$\begin{aligned}H_{j,k,l}(x,\zeta)&=\sum_{E\subseteq\{1,\dots,\nu_0\}}(-1)^{|E|}F_{k,l}(x,P_{j-\mathfrak{e}_E}u(x)+\zeta)\\&=\int_{[0,1]^{\nu_0}}\partial_{\sigma_1}\partial_{\sigma_2}\cdots\partial_{\sigma_{\nu_0}}\left(F_{k,l}\left(x,P_{j-\mathfrak{e}_{\{1,\dots,\nu_0\}}}u(x)+\sum_{E\subsetneq\{1,\dots,\nu_0\}}\sigma_{E^c}D_{j-\mathfrak{e}_E}u(x)+\zeta\right)\right)d\sigma.\end{aligned}$$

Thus, to prove (7.29) it suffices to show that for all $\sigma\in[0,1]^{\nu_0}$,

$$\begin{aligned}&\left\|(2^{-(j_\mu\vee k_\mu)d^\mu}W^\mu)^{\alpha_\mu}(2^{-l}\partial_\zeta)^\beta\partial_{\sigma_1}\cdots\partial_{\sigma_{\nu_0}}F_{k,l}\left(x,P_{j-\mathfrak{e}_{\{1,\dots,\nu_0\}}}u(x)+\sum_{E\subsetneq\{1,\dots,\nu_0\}}\sigma_{E^c}D_{j-\mathfrak{e}_E}u(x)+\zeta\right)\right\|_{L^\infty}\\&\quad\lesssim\|F\|_{\mathscr{C}^{s,t}((W,\vec{d})\boxtimes\nabla_{\mathbb{R}^N})}2^{-k\cdot s}2^{-j\cdot s}2^{-l(t-|\alpha_\mu|-\nu_0)}(1+\|u\|_{\mathscr{C}^s((W,\vec{d});\mathbb{R}^N)})(1+\|u\|_{L^\infty})^{|\alpha_\mu|+\nu-1}.\end{aligned}\tag{7.30}$$

Let $1\leq L\leq\nu_0$ and $E_1,\dots,E_L,\widetilde{E}_1,\dots,\widetilde{E}_L\subseteq\{1,\dots,\mu_0\}$. Then, since $\nu_0\geq 1$,

$$\partial_{\sigma_1}\cdots\partial_{\sigma_{\nu_0}}F_{k,l}\left(x,P_{j-\mathfrak{e}_{\{1,\dots,\nu_0\}}}u(x)+\sum_{E\subsetneq\{1,\dots,\nu_0\}}\sigma_{E^c}D_{j-\mathfrak{e}_E}u(x)+\zeta\right)$$

can be written as a finite sum of terms of the form

$$(\mathbf{D}_\zeta^L F_{k,l})\left(x,P_{j-\mathfrak{e}_{\{1,\dots,\nu_0\}}}u(x)+\sum_{E\subsetneq\{1,\dots,\nu_0\}}\sigma_{E^c}D_{j-\mathfrak{e}_E}u(x)+\zeta\right)[\sigma_{E_1}D_{j-\mathfrak{e}_{\widetilde{E}_1}}u,\dots,\sigma_{E_L}D_{j-\mathfrak{e}_{\widetilde{E}_L}}u].$$

Since each $\sigma_{E_s}\in[0,1]$, to prove (7.30) it suffices to show that

$$\begin{aligned}&\left\|(2^{-(j_\mu\vee k_\mu)d^\mu}W^\mu)^{\alpha_\mu}(2^{-l}\partial_\zeta)^\beta\left[(\mathbf{D}_\zeta^L F_{k,l})\left(x,P_{j-\mathfrak{e}_{\{1,\dots,\nu_0\}}}u(x)+\sum_{E\subsetneq\{1,\dots,\nu_0\}}\sigma_{E^c}D_{j-\mathfrak{e}_E}u(x)+\zeta\right)\right.\right.\\&\qquad\left.\left.[D_{j-\mathfrak{e}_{\widetilde{E}_1}}u,\dots,D_{j-\mathfrak{e}_{\widetilde{E}_L}}u]\right]\right\|_{L^\infty}\\&\quad\lesssim\|F\|_{\mathscr{C}^{s,t}((W,\vec{d})\boxtimes\nabla_{\mathbb{R}^N})}2^{-k\cdot s}2^{-j\cdot s}2^{-l(t-|\alpha_\mu|-\nu_0)}\\&\qquad\times(1+\|u\|_{\mathscr{C}^s((W,\vec{d});\mathbb{R}^N)})(1+\|u\|_{L^\infty})^{|\alpha_\mu|+\nu-1}.\end{aligned}\tag{7.31}$$

Using Notations 7.4.2 and 7.5.14, we have

$$
\begin{aligned}
&\Big\| (2^{-(j_\mu \vee k_\mu) d^\mu} W^\mu)^{\alpha_\mu} (2^{-l}\partial_\zeta)^\beta \Big[(\mathbf{D}_\zeta^L F_{k,l})\Big(x, P_{j-\mathfrak{e}_{\{1,\dots,\nu_0\}}} u(x) + \sum_{E \subsetneq \{1,\dots,\nu_0\}} \sigma_{E^c} D_{j-\mathfrak{e}_E} u(x) + \zeta \Big) \\
&\qquad [D_{j-\mathfrak{e}_{\tilde{E}_1}} u, \dots, D_{j-\mathfrak{e}_{\tilde{E}_L}} u] \Big] \Big\|_{L^\infty} \\
&\leq \sum_{\substack{\gamma_0+\gamma_1+\dots+\gamma_L \\ = \alpha_\mu}} \Big\| (2^{-(j_\mu \vee k_\mu) d^\mu} W^\mu)^{\gamma_0} (2^{-l}\partial_\zeta)^\beta \\
&\qquad (\mathbf{D}_\zeta^L F_{k,l})\Big(x, P_{j-\mathfrak{e}_{\{1,\dots,\nu_0\}}} u(x) + \sum_{E \subsetneq \{1,\dots,\nu_0\}} \sigma_{E^c} D_{j-\mathfrak{e}_E} u(x) + \zeta \Big) \Big\|_{L^\infty} \\
&\qquad \times \prod_{s=1}^{L} \| (2^{-(j_\mu \vee k_\mu) d^\mu} W^\mu)^{\gamma_s} D_{j-\mathfrak{e}_{\tilde{E}_s}} u \|_{L^\infty}.
\end{aligned} \tag{7.32}
$$

We bound each term in the sum in (7.32) separately.

By (7.15), the components of the tensor $(2^{-l}\mathbf{D}_\zeta)^L F_{k,l}$ satisfy the assumptions of $G_{k,l}$ in Lemma 7.5.15 with $A_0 = \|F\|_{\mathscr{C}^{s,t}((W,\vec{d}) \boxtimes \nabla_{\mathbb{R}^N})}$. Thus, Lemma 7.5.15 implies

$$
\begin{aligned}
&\Big\| (2^{-(j_\mu \vee k_\mu) d^\mu} W^\mu)^{\alpha_\mu} (2^{-l}\partial_\zeta)^\beta (\mathbf{D}_\zeta^L F_{k,l})\Big(x, P_{j-\mathfrak{e}_{\{1,\dots,\nu_0\}}} u(x) + \sum_{E \subsetneq \{1,\dots,\nu_0\}} \sigma_{E^c} D_{j-\mathfrak{e}_E} u(x) + \zeta \Big) \Big\|_{L^\infty} \\
&\lesssim 2^{Ll} \|F\|_{\mathscr{C}^{s,t}((W,\vec{d}) \boxtimes \nabla_{\mathbb{R}^N})} 2^{-k\cdot s} 2^{-l(t-|\gamma_0|)} (1 + \|u\|_{L^\infty})^{|\gamma_0|} \\
&\leq \|F\|_{\mathscr{C}^{s,t}((W,\vec{d}) \boxtimes \nabla_{\mathbb{R}^N})} 2^{-k\cdot s} 2^{-l(t-|\alpha_\mu|-\nu)} (1 + \|u\|_{L^\infty})^{|\alpha_\mu|},
\end{aligned} \tag{7.33}
$$

where in the last estimate we have used $L \leq \nu_0 \leq \nu$ and $|\gamma_0| \leq |\alpha_\mu|$.

Since $|j - (j - \mathfrak{e}_{\tilde{E}_1})|_\infty \leq 1$, Proposition 5.5.5 (a) and (e) implies that

$$
\{((2^{-j_\mu d^\mu} W^\mu)^{\gamma_1} D_{j-\mathfrak{e}_{\tilde{E}_1}}, 2^{-j}) : j \in \mathbb{N}^\nu\}
$$

is a bounded set of generalized $(W, \vec{d})$ elementary operators, and therefore by Corollary 6.4.4,

$$
\begin{aligned}
\| (2^{-(j_\mu \vee k_\mu) d^\mu} W^\mu)^{\gamma_1} D_{j-\mathfrak{e}_{\tilde{E}_1}} u \|_{L^\infty} &\leq \| (2^{-j_\mu d^\mu} W^\mu)^{\gamma_1} D_{j-\mathfrak{e}_{\tilde{E}_1}} u \|_{L^\infty} \\
&\lesssim 2^{-j\cdot s} \|u\|_{\mathscr{C}^s((W,\vec{d});\mathbb{R}^N)}.
\end{aligned} \tag{7.34}
$$

We use Corollary 7.1.4 (i) to see that

$$\begin{aligned}
&\prod_{s=2}^{L} \|(2^{-(j_\mu \vee k_\mu) d^\mu} W^\mu)^{\gamma_s} D_{j-\mathfrak{e}_{\tilde{E}_s}} u\|_{L^\infty} \\
&\quad \le \prod_{s=2}^{L} \|(2^{-j_\mu d^\mu} W^\mu)^{\gamma_s} D_{j-\mathfrak{e}_{\tilde{E}_s}} u\|_{L^\infty} \\
&\quad \lesssim \prod_{s=2}^{L} \|u\|_{L^\infty} \le (1+\|u\|_{L^\infty})^{\nu-1},
\end{aligned} \tag{7.35}$$

where in the last estimate we have used $L \le \nu_0 \le \nu$.

Plugging (7.33), (7.34), and (7.35) into (7.32) yields (7.31) and completes the proof. □

Proof of Theorem 7.5.2. (i): Take $H_{j,k,l}$ as in (7.23), so that $F(x,u(x)+\zeta) = \sum_{\substack{j,k\in\mathbb{N}^\nu \\ l\in\mathbb{N}}} H_{j,k,l}(x,\zeta)$. Set $M_1 := \lfloor |s|_1 \rfloor + 1$ and $M_2 := \lfloor t \rfloor - \lfloor |s|_1 \rfloor - \nu$. For $\mu \in \{1,\dots,\nu\}$, $|\alpha_\mu| \le M_1$, and $|\beta| \le M_2$, Lemma 7.5.16 implies

$$\begin{aligned}
&\|(2^{-(k_\mu \vee j_\mu) d^\mu} W^\mu)^{\alpha_\mu} (2^{-l}\partial_\zeta)^\beta H_{j,k,l}\|_{L^\infty} \\
&\quad \lesssim \|F\|_{\mathscr{C}^{s,t}((W,\vec{d})\boxtimes\nabla_{\mathbb{R}^N})} 2^{-k\cdot s} 2^{-j\cdot s} 2^{-l(t-|\alpha_\mu|-\nu)} \\
&\qquad \times (1+\|u\|_{\mathscr{C}^s((W,\vec{d});\mathbb{R}^N)})(1+\|u\|_{L^\infty})^{|\alpha_\mu|+\nu-1} \\
&\quad \lesssim \|F\|_{\mathscr{C}^{s,t}((W,\vec{d})\boxtimes\nabla_{\mathbb{R}^N})} 2^{-k\cdot s} 2^{-j\cdot s} 2^{-l(t-\lfloor |s|_1 \rfloor-1-\nu)} \\
&\qquad \times (1+\|u\|_{\mathscr{C}^s((W,\vec{d});\mathbb{R}^N)})(1+\|u\|_{L^\infty})^{\lfloor |s|_1 \rfloor+\nu}.
\end{aligned}$$

This shows that $H_{j,k,l}$ satisfies the assumptions of Lemma 7.5.13 with t replaced by $t - \lfloor |s|_1 \rfloor - 1 - \nu$ and $A = \|F\|_{\mathscr{C}^{s,t}((W,\vec{d})\boxtimes\nabla_{\mathbb{R}^N})}(1+\|u\|_{\mathscr{C}^s((W,\vec{d});\mathbb{R}^N)})(1+\|u\|_{L^\infty})^{\lfloor |s|_1 \rfloor+\nu}$. The conclusion of Lemma 7.5.13, using $F(x,u(x)+\zeta) = \sum_{\substack{j,k\in\mathbb{N}^\nu \\ l\in\mathbb{N}}} H_{j,k,l}(x,\zeta)$, is exactly the conclusion (i).

(ii): Since $F(x,u(x)) = F(x,u(x)+\zeta)\big|_{\zeta=0}$, (ii) follows by combining (i) and Corollary 7.5.12 (i).

(iii): We first consider the case where $\|u\|_{L^\infty} = \|v\|_{L^\infty} = 1$. Fix $\psi_0 \in C_0^\infty(\mathfrak{M})$ with $\psi_0 \equiv 1$ on $\mathcal{K}$ and $\phi(\zeta_1,\zeta_2) \in C_0^\infty(\mathbb{R}^N \times \mathbb{R}^N)$ with $\phi(\zeta_1,\zeta_2) \equiv 1$ if $|\zeta_1|, |\zeta_2| \le 1$. Set $F(x,\zeta) := \psi_0(x)\phi(\zeta_1,\zeta_2)\zeta_1\cdot\zeta_2$. It follows immediately from Proposition 7.5.9 that $F \in \mathscr{C}^{s,t}(\mathrm{supp}(\phi)\times\mathbb{R}^{2N}, (W,\vec{d})\boxtimes\nabla_{\mathbb{R}^{2N}})$, $\forall t>0$.

If $\|u\|_{L^\infty} = \|v\|_{L^\infty} = 1$, then $u(x)\cdot v(x) = F(x,u(x),v(x))$ and (ii) implies

$$\begin{aligned}
\|u\cdot v\|_{\mathscr{C}^s(W,\vec{d})} &= \|F(x,u(x),v(x))\|_{\mathscr{C}^s(W,\vec{d})} \\
&\lesssim \|F\|_{\mathscr{C}^{s,\lfloor |s|_1 \rfloor+2+\nu}((W,\vec{d})\boxtimes\nabla_{\mathbb{R}^N})}(1+\|u\|_{\mathscr{C}^s(W,\vec{d})}+\|v\|_{\mathscr{C}^s(W,\vec{d})}) \\
&\quad \times (1+\|u\|_{L^\infty}+\|v\|_{L^\infty}) \\
&\lesssim 1+\|u\|_{\mathscr{C}^s(W,\vec{d})}+\|v\|_{\mathscr{C}^s(W,\vec{d})}.
\end{aligned} \tag{7.36}$$

For general (not identically zero) $u, v \in \mathscr{C}^s(\mathcal{K}, (W, \vec{d}); \mathbb{R}^N)$, we apply (7.36) to $u/\|u\|_{L^\infty}$ and $v/\|v\|_{L^\infty}$ to see that

$$\begin{aligned}\|u \cdot v\|_{\mathscr{C}^s(W,\vec{d})} &\lesssim \|u\|_{L^\infty}\|v\|_{L^\infty} + \|u\|_{\mathscr{C}^s(W,\vec{d})}\|v\|_{L^\infty} + \|u\|_{L^\infty}\|v\|_{\mathscr{C}^s(W,\vec{d})} \\ &\lesssim \|u\|_{\mathscr{C}^s(W,\vec{d})}\|v\|_{L^\infty} + \|u\|_{L^\infty}\|v\|_{\mathscr{C}^s(W,\vec{d})},\end{aligned}$$

where the final inequality uses Corollary 7.2.2.

(iv): First we prove the result with $\|w\|_{L^\infty} = 1$. Let $\phi(\zeta') \in C_0^\infty(\mathbb{R})$ satisfy $\phi \equiv 1$ for $|\zeta'| \leq 1$. Set

$$\widetilde{F}(x, \zeta, \zeta') := F(x, \zeta)\phi(\zeta')\zeta'.$$

It follows easily from Proposition 7.5.9 that $\widetilde{F} \in \mathscr{C}^{s,t}(\mathcal{K} \times \mathbb{R}^{N+1}, (W, \vec{d}) \boxtimes \nabla_{\mathbb{R}^{N+1}})$ and $\|\widetilde{F}\|_{\mathscr{C}^{s,t}((W,\vec{d})\boxtimes\nabla_{\mathbb{R}^{N+1}})} \lesssim \|F\|_{\mathscr{C}^{s,t}((W,\vec{d})\boxtimes\nabla_{\mathbb{R}^N})}$.

When $\|w\|_{L^\infty} = 1$, we have $F(x, u(x) + \zeta)w(x) = \widetilde{F}(x, u(x) + \zeta, w(x) + \zeta')\big|_{\zeta'=0}$. Thus, applying (i) and Corollary 7.5.12 (ii), we have

$$\begin{aligned}&\|F(x, u(x) + \zeta)w(x)\|_{\mathscr{C}^{s,t-\lfloor|s|_1\rfloor-1-\nu}((W,\vec{d})\boxtimes\nabla_{\mathbb{R}^N})} \\ &\quad = \big\|\widetilde{F}(x, u(x) + \zeta, w(x) + \zeta')\big|_{\zeta'=0}\big\|_{\mathscr{C}^{s,t-\lfloor|s|_1\rfloor-1-\nu}((W,\vec{d})\boxtimes\nabla_{\mathbb{R}^N})} \\ &\quad \lesssim \big\|\widetilde{F}(x, u(x) + \zeta, w(x) + \zeta')\big\|_{\mathscr{C}^{s,t-\lfloor|s|_1\rfloor-1-\nu}((W,\vec{d})\boxtimes\nabla_{\mathbb{R}^{N+1}})} \\ &\quad \lesssim \|\widetilde{F}\|_{\mathscr{C}^{s,t}((W,\vec{d})\boxtimes\nabla_{\mathbb{R}^{N+1}})}\big(1 + \|u\|_{\mathscr{C}^s(W,\vec{d})} + \|w\|_{\mathscr{C}^s(W,\vec{d})}\big)\big(1 + \|u\|_{L^\infty} + \|w\|_{L^\infty}\big)^{\lfloor|s|_1\rfloor+\nu} \\ &\quad \lesssim \|F\|_{\mathscr{C}^{s,t}((W,\vec{d})\boxtimes\nabla_{\mathbb{R}^N})}\big(1 + \|u\|_{\mathscr{C}^s(W,\vec{d})} + \|w\|_{\mathscr{C}^s(W,\vec{d})}\big)\big(1 + \|u\|_{L^\infty}\big)^{\lfloor|s|_1\rfloor+\nu}.\end{aligned} \tag{7.37}$$

For general (not identically zero) $w \in \mathscr{C}^s(\mathcal{K}, (W, \vec{d}))$, we apply (7.37) to $w/\|w\|_{L^\infty}$ to see that

$$\begin{aligned}&\|F(x, u(x) + \zeta)w(x)\|_{\mathscr{C}^{s,t-\lfloor|s|_1\rfloor-1-\nu}((W,\vec{d})\boxtimes\nabla_{\mathbb{R}^N})} \\ &\quad \lesssim \|F\|_{\mathscr{C}^{s,t}((W,\vec{d})\boxtimes\nabla_{\mathbb{R}^N})}\big(\|w\|_{L^\infty} + \|w\|_{L^\infty}\|u\|_{\mathscr{C}^s(W,\vec{d})} + \|w\|_{\mathscr{C}^s(W,\vec{d})}\big)\big(1 + \|u\|_{L^\infty}\big)^{\lfloor|s|_1\rfloor+\nu} \\ &\quad \lesssim \|F\|_{\mathscr{C}^{s,t}((W,\vec{d})\boxtimes\nabla_{\mathbb{R}^N})}\big(\|w\|_{L^\infty}\|u\|_{\mathscr{C}^s(W,\vec{d})} + \|w\|_{\mathscr{C}^s(W,\vec{d})}\big)\big(1 + \|u\|_{L^\infty}\big)^{\lfloor|s|_1\rfloor+\nu},\end{aligned}$$

where the final inequality uses Corollary 7.2.2.

(v): Using (iv) and Proposition 7.5.11, we have

$$\begin{aligned}&\big\|F(x, u(x) + \zeta) - F(x, v(x) + \zeta)\big\|_{\mathscr{C}^{s,t-\lfloor|s|_1\rfloor-2-\nu}((W,\vec{d})\boxtimes\nabla_{\mathbb{R}^N})} \\ &\quad = \left\|\int_0^1 \mathbf{D}_\zeta F(x, \sigma u(x) + (1-\sigma)v(x) + \zeta)(u(x) - v(x))\, d\sigma\right\|_{\mathscr{C}^{s,t-\lfloor|s|_1\rfloor-2-\nu}((W,\vec{d})\boxtimes\nabla_{\mathbb{R}^N})} \\ &\quad \leq \int_0^1 \big\|\mathbf{D}_\zeta F(x, \sigma u(x) + (1-\sigma)v(x) + \zeta)(u(x) - v(x))\big\|_{\mathscr{C}^{s,t-\lfloor|s|_1\rfloor-2-\nu}((W,\vec{d})\boxtimes\nabla_{\mathbb{R}^N})}\, d\sigma\end{aligned}$$

$$
\begin{aligned}
&\le \sup_{\sigma\in[0,1]} \left\|\mathbf{D}_\zeta F(x,\sigma u(x)+(1-\sigma)v(x)+\zeta)(u(x)-v(x))\right\|_{\mathscr{C}^{s,t-\lfloor|s|_1\rfloor-2-\nu}((W,\vec{d})\boxtimes\nabla_{\mathbb{R}^N})}\\
&\lesssim \|\mathbf{D}_\zeta F\|_{\mathscr{C}^{s,t-1}((W,\vec{d})\boxtimes\nabla_{\mathbb{R}^N})}\big(\|u-v\|_{\mathscr{C}^s(W,\vec{d})}+\|\sigma u+(1-\sigma)v\|_{\mathscr{C}^s(W,\vec{d})}\|u-v\|_{L^\infty}\big)\\
&\quad\times\big(1+\|\sigma u+(1-\sigma)v\|_{L^\infty}\big)^{\lfloor|s|_1\rfloor+\nu}\\
&\lesssim \|F\|_{\mathscr{C}^{s,t}((W,\vec{d})\boxtimes\nabla_{\mathbb{R}^N})}\big(\|u-v\|_{\mathscr{C}^s(W,\vec{d})}+\big(\|u\|_{\mathscr{C}^s(W,\vec{d})}+\|v\|_{\mathscr{C}^s(W,\vec{d})}\big)\|u-v\|_{L^\infty}\big)\\
&\quad\times\big(1+\|u\|_{L^\infty}+\|v\|_{L^\infty}\big)^{\lfloor|s|_1\rfloor+\nu}.
\end{aligned}
$$

(vi): Since $F(x,u(x))-F(x,v(x)) = F(x,u(x)+\zeta)-F(x,v(x)+\zeta)\big|_{\zeta=0}$, this follows from (v) and Corollary 7.5.12 (i).

(vii): We prove (vii) by induction on L. We begin with the base case, $L=1$. When $L=1$, using the fact that $F(x,0)\equiv 0$, we have $F(x,u(x))=F(x,u(x))-F(x,0)$. From here, the result when $L=1$ follows from (vi) with $v=0$.

We assume (vi) for some $L\in\mathbb{N}_+$ and prove it for $L+1$; thus, we assume the hypotheses with L replaced by $L+1$. Using the fact that $F(x,0)\equiv 0$ and using (iii), we have

$$
\begin{aligned}
\|F(x,u(x))\|_{\mathscr{C}^s(W,\vec{d})} &= \|F(x,u(x))-F(x,v(x))\|_{\mathscr{C}^s(W,\vec{d})}\\
&= \left\|\int_0^1 \mathbf{D}_\zeta F(x,\sigma u(x))u(x)\,d\sigma\right\|_{\mathscr{C}^s(W,\vec{d})}\\
&\le \int_0^1 \left\|\mathbf{D}_\zeta F(x,\sigma u(x))u(x)\right\|_{\mathscr{C}^s(W,\vec{d})}\,d\sigma\\
&\le \sup_{\sigma\in[0,1]}\left\|\mathbf{D}_\zeta F(x,\sigma u(x))u(x)\right\|_{\mathscr{C}^s(W,\vec{d})}\\
&\lesssim \sup_{\sigma\in[0,1]}\left\|\mathbf{D}_\zeta F(x,\sigma u(x))\right\|_{\mathscr{C}^s(W,\vec{d})}\|u\|_{L^\infty}\\
&\quad+\sup_{\sigma\in[0,1]}\|\mathbf{D}_\zeta F(x,\sigma u(x))\|_{L^\infty}\|u\|_{\mathscr{C}^s(W,\vec{d})}\\
&=: \text{(I)}+\text{(II)},
\end{aligned}
\tag{7.38}
$$

where when we applied (iii) to the various summands of the matrix multiplication $\mathbf{D}_\zeta F(x,\sigma u(x))u(x)$.

To bound (I) we use the inductive hypothesis applied to $\mathbf{D}_\zeta F$ with t replaced by $t-1>\lfloor|s|_1\rfloor+1+\nu+L$ and Proposition 7.5.11 to see

$$
\begin{aligned}
\text{(I)} &= \sup_{\sigma\in[0,1]}\left\|\mathbf{D}_\zeta F(x,\sigma u(x))\right\|_{\mathscr{C}^s(W,\vec{d})}\|u\|_{L^\infty}\\
&\lesssim \sup_{\sigma\in[0,1]}\|\mathbf{D}_\zeta F\|_{\mathscr{C}^{s,t-1}((W,\vec{d})\boxtimes\nabla_{\mathbb{R}^N})}\big(1+\|\sigma u\|_{L^\infty}\big)^{\lfloor|s|_1\rfloor+\nu+1}\|\sigma u\|_{L^\infty}^{L-1}\|\sigma u\|_{\mathscr{C}^s(W,\vec{d})}\|u\|_{L^\infty}\\
&\lesssim \|F\|_{\mathscr{C}^{s,t}((W,\vec{d})\boxtimes\nabla_{\mathbb{R}^N})}\big(1+\|u\|_{L^\infty}\big)^{\lfloor|s|_1\rfloor+\nu+1}\|u\|_{L^\infty}^{L}\|u\|_{\mathscr{C}^s(W,\vec{d})}.
\end{aligned}
\tag{7.39}
$$

For (II) we use the fact that $\partial_\zeta^\beta \mathbf{D}_\zeta F(x,0) \equiv 0$, $\forall |\beta| < L$, the fact that $t > L+1$, and Propositions 7.5.11 and 7.5.10 to see that

$$\begin{aligned}\sup_{\sigma\in[0,1]} \left\|\mathbf{D}_\zeta F(x,\sigma u(x))\right\|_{L^\infty} &\lesssim \sup_{\sigma\in[0,1]} \sum_{|\beta|=L} \left\|\partial_\zeta^L \mathbf{D}_\zeta F(x,\zeta)\right\|_{L^\infty} \|\sigma u(x)\|_{L^\infty}^L \\ &\lesssim \|F\|_{\mathscr{C}^{s,t}((W,\vec{d})\boxtimes\nabla_{\mathbb{R}^N})} \|u\|_{L^\infty}^L.\end{aligned}$$

We conclude

$$(\mathrm{II}) \lesssim \|F\|_{\mathscr{C}^{s,t}((W,\vec{d})\boxtimes\nabla_{\mathbb{R}^N})} \|u\|_{L^\infty}^L \|u\|_{\mathscr{C}^s(W,\vec{d})}. \tag{7.40}$$

Using (7.39) and (7.40) to bound the right-hand side of (7.38) completes the inductive step.

(viii): We have, using (iii),

$$\begin{aligned}&\|F(x,u(x)) - F(x,v(x))\|_{\mathscr{C}^s(W,\vec{d})} \\ &= \left\| \int_0^1 \mathbf{D}_\zeta F(x,\sigma u(x) + (1-\sigma)v(x))(u(x)-v(x))\, d\sigma \right\|_{\mathscr{C}^s(W,\vec{d})} \\ &\le \int_0^1 \left\|\mathbf{D}_\zeta F(x,\sigma u(x) + (1-\sigma)v(x))(u(x)-v(x))\right\|_{\mathscr{C}^s(W,\vec{d})} d\sigma \\ &\le \sup_{\sigma\in[0,1]} \left\|\mathbf{D}_\zeta F(x,\sigma u(x) + (1-\sigma)v(x))(u(x)-v(x))\right\|_{\mathscr{C}^s(W,\vec{d})} \\ &\lesssim \sup_{\sigma\in[0,1]} \|\mathbf{D}_\zeta F(x,\sigma u(x) + (1-\sigma)v(x))\|_{\mathscr{C}^s(W,\vec{d})} \|u-v\|_{L^\infty} \\ &\quad + \sup_{\sigma\in[0,1]} \|\mathbf{D}_\zeta F(x,\sigma u(x) + (1-\sigma)v(x))\|_{L^\infty} \|u-v\|_{\mathscr{C}^s(W,\vec{d})} \\ &=: (\mathrm{III}) + (\mathrm{IV}),\end{aligned} \tag{7.41}$$

where we applied (iii) to the various summands of the matrix multiplication $\mathbf{D}_\zeta F(x,\sigma u(x) + (1-\sigma)v(x))(u(x)-v(x))$.

For (III), we have $\partial_\zeta^\beta \mathbf{D}_\zeta F(x,0) \equiv 0$, $\forall |\beta| \le L-1$, and we may therefore apply (vii) with L replaced by $L-1$ and t replaced by $t-1$ to see that

$$\begin{aligned}&\sup_{\sigma\in[0,1]} \left\|\mathbf{D}_\zeta F(x,\sigma u(x) + (1-\sigma)v(x))\right\|_{\mathscr{C}^s(W,\vec{d})} \\ &\lesssim \sup_{\sigma\in[0,1]} \|\mathbf{D}_\zeta F\|_{\mathscr{C}^{s,t-1}((W,\vec{d})\boxtimes\nabla_{\mathbb{R}^N})} \big(1 + \|\sigma u + (1-\sigma)v\|_{L^\infty}\big)^{\lfloor |s|_1 \rfloor + \nu + 1} \\ &\quad \times \|\sigma u + (1-\sigma)v\|_{L^\infty}^{L-2} \|\sigma u + (1-\sigma)v\|_{\mathscr{C}^s(W,\vec{d})} \\ &\lesssim \|F\|_{\mathscr{C}^{s,t}((W,\vec{d})\boxtimes\nabla_{\mathbb{R}^N})} \big(1 + \|u\|_{L^\infty} + \|v\|_{L^\infty}\big)^{\lfloor |s|_1 \rfloor + \nu + 1} \\ &\quad \times \big(\|u\|_{L^\infty} + \|v\|_{L^\infty}\big)^{L-2} \big(\|u\|_{\mathscr{C}^s(W,\vec{d})} + \|v\|_{\mathscr{C}^s(W,\vec{d})}\big),\end{aligned}$$

where the final estimate used Proposition 7.5.11. It follows that

$$\begin{aligned}(\mathrm{III}) &\lesssim \|F\|_{\mathscr{C}^{s,t}((W,\vec{d})\boxtimes\nabla_{\mathbb{R}^N})}\big(1+\|u\|_{L^\infty}+\|v\|_{L^\infty}\big)^{\lfloor|s|_1\rfloor+\nu+1} \\ &\quad\times\big(\|u\|_{L^\infty}+\|v\|_{L^\infty}\big)^{L-2}\big(\|u\|_{\mathscr{C}^s(W,\vec{d})}+\|v\|_{\mathscr{C}^s(W,\vec{d})}\big)\|u-v\|_{L^\infty}.\end{aligned} \tag{7.42}$$

For (IV), we use the fact that $\partial_\zeta^\beta \mathbf{D}_\zeta f(x,0)\equiv 0$, $\forall|\beta|<L-1$, to see that

$$\begin{aligned}&\sup_{\sigma\in[0,1]}\big\|\mathbf{D}_\zeta F(x,\sigma u(x)+(1-\sigma)v(x))\big\|_{L^\infty} \\ &\quad\lesssim \sup_{\sigma\in[0,1]}\sum_{|\beta|=L-1}\big\|\partial_\zeta^\beta\mathbf{D}_\zeta F(x,\sigma u(x)+(1-\sigma)v(x))\big\|_{L^\infty}\big\|\sigma u+(1-\sigma)v\big\|_{L^\infty}^{L-1} \\ &\quad\lesssim \|F\|_{\mathscr{C}^{s,t}((W,\vec{d})\boxtimes\nabla_{\mathbb{R}^N})}\big(\|u\|_{L^\infty}+\|v\|_{L^\infty}\big)^{L-1},\end{aligned}$$

where in the last inequality we used $t>L$ and Propositions 7.5.11 and 7.5.10. It follows that

$$(\mathrm{IV})\lesssim \|F\|_{\mathscr{C}^{s,t}((W,\vec{d})\boxtimes\nabla_{\mathbb{R}^N})}\big(\|u\|_{L^\infty}+\|v\|_{L^\infty}\big)^{L-1}\|u-v\|_{\mathscr{C}^s(W,\vec{d})}. \tag{7.43}$$

Using (7.42) and (7.43) to bound the right-hand side of (7.41) completes the proof. □

7.5.3 The classical product Zygmund–Hölder spaces

Just as in Section 6.6.1, we can understand classical product Zygmund–Hölder spaces in terms of the framework of this section.

Definition 7.5.17. Fix $\mathcal{K}\Subset\mathfrak{M}$ compact. For $s,t>0$, we let $\mathscr{C}^{s,t}_{\mathrm{std}}(\mathcal{K}\times\mathbb{R}^N)$ be the space consisting of those $F\in C_0^\infty(\mathfrak{M}\times\mathbb{R}^N)'$ with $\mathrm{supp}(F)\subseteq\mathcal{K}\times\mathbb{R}^N$ such that for each $x\in\mathcal{K}$, there exists a C^∞_{loc} diffeomorphism $\Phi_x :\xrightarrow{\sim}\Phi_x(B^n(1))$, where $\Phi_x(B^n(1))$ is an open neighborhood of x and $\psi_x\in C_0^\infty(\Phi_x(B^n(1)))$ with $\psi_x\equiv 1$ on a neighborhood of x such that $\Phi_x^*\psi_x F\in\mathscr{C}^{s,t}(\mathbb{R}^n\times\mathbb{R}^N)$.

Remark 7.5.18. In Definition 7.5.17, and in the rest of this section, we used Φ_x^* to denote the pullback in the first variable. So, for example, $\Phi_x^*\psi_x F$ could be more explicitly written as $(\Phi_x\otimes\mathrm{Id}_{\mathbb{R}^N})^*\psi_x F$, where $\mathrm{Id}_{\mathbb{R}^N}:\mathbb{R}^N\to\mathbb{R}^N$ denotes the identity map.

Similar to Section 6.6.1, $\mathscr{C}^{s,t}_{\mathrm{std}}(\mathcal{K}\times\mathbb{R}^N)$ can be endowed with a norm which turns it into a Banach space. Indeed, for each $x\in\mathfrak{M}$, let $\Phi_x : B^n(1)\xrightarrow{\sim}\Phi_x(B^n(1))$ be a C^∞_{loc} diffeomorphism such that $\Phi_x(B^n(1))\subseteq\mathfrak{M}$ is open and $x=\Phi_x(0)$. Cover $\mathcal{K}$ by a finite collection of open sets of the form $\Phi_{x_1}(B^n(1)),\dots,\Phi_{x_L}(B^n(1))$. Take $\psi_j\in C_0^\infty(\Phi_{x_j}(B^n(1)))$ such that $\sum_{j=1}^L\psi_j\equiv 1$ on a neighborhood of $\mathcal{K}$. For $F\in\mathscr{C}^{s,t}_{\mathrm{std}}(\mathcal{K}\times\mathbb{R}^N)$, we define the norm

$$\|F\|_{\mathscr{C}^{s,t}_{\mathrm{std}}(\mathcal{K}\times\mathbb{R}^N)} := \sum_{j=1}^{L} \|\Phi^*_{x_j}\psi_j F\|_{\mathscr{C}^{s,t}(\mathbb{R}^n\times\mathbb{R}^N)}.$$

With this norm, $\mathscr{C}^{s,t}_{\mathrm{std}}(\mathcal{K}\times\mathbb{R}^N)$ is a Banach space, and the equivalence class of the norm does not depend on any of the choices we made. We set

$$\mathscr{C}^{s,t}_{\mathrm{std,cpt}}(\mathfrak{M}\times\mathbb{R}^N) := \bigcup_{\mathcal{K}\Subset\mathfrak{M}\,\mathcal{K}\text{ compact}} \mathscr{C}^{s,t}_{\mathrm{std}}(\mathcal{K}\times\mathbb{R}^N).$$

We have the following analog of Theorem 6.6.7, which has a similar proof, which we omit.

Theorem 7.5.19. *Let* $(Y,1) = \{(Y_1,1),\dots,(Y_s,1)\} \subset C^\infty_{\mathrm{loc}}(\mathfrak{M};T\mathfrak{M})\times\mathbb{N}_+$ *be such that*

$$\operatorname{span}\{Y_1(x),\dots,Y_s(x)\} = T_x\mathfrak{M}, \quad \forall x\in\mathfrak{M}.$$

Then

$$\mathscr{C}^{s,t}_{\mathrm{std}}(\mathcal{K}\times\mathbb{R}^N) = \mathscr{C}^{s,t}(\mathcal{K}\times\mathbb{R}^N,(Y,1)\boxtimes\nabla_{\mathbb{R}^N}),$$

with equality of topologies.

7.6 Adding parameters

Using the results from Section 6.8, we can relate single-parameter Zygmund–Hölder spaces with higher-parameter Zygmund–Hölder spaces, as the next result shows.

Proposition 7.6.1. *Let* $(Y,\mathsf{d}),(Z,\mathsf{d}) \subset C^\infty_{\mathrm{loc}}(\mathfrak{M};T\mathfrak{M})\times\mathbb{N}_+$ *be two lists of Hörmander vector fields with formal degrees on* $\mathfrak{M}$. *Suppose* (Y,d) *and* (Z,d) *locally weakly approximately commute on* $\mathfrak{M}$ *and set* $(W,\vec{\mathsf{d}}) := (Y,\mathsf{d})\boxtimes(Z,\mathsf{d}) \subset C^\infty_{\mathrm{loc}}(\mathfrak{M};T\mathfrak{M})\times(\mathbb{N}^2\setminus\{0\})$. *Then,* $\forall\epsilon>0$, $\exists\epsilon'>0$, $\forall s>\epsilon$, $\forall t>0$,

$$\mathscr{C}^{s}(\mathcal{K},(Z,\mathsf{d})) \subseteq \mathscr{C}^{(\epsilon',s-\epsilon)}(\mathcal{K},(W,\vec{\mathsf{d}})), \tag{7.44}$$

$$\mathscr{C}^{s,t}(\mathcal{K}\times\mathbb{R}^N,(Z,\mathsf{d})\boxtimes\nabla_{\mathbb{R}^N}) \subseteq \mathscr{C}^{(\epsilon',s-\epsilon),t}(\mathcal{K}\times\mathbb{R}^N,(W,\vec{\mathsf{d}})\boxtimes\nabla_{\mathbb{R}^N}), \tag{7.45}$$

and $\forall\epsilon>0$, $\forall s,t>0$,

$$\mathscr{C}^{(\epsilon,s)}(\mathcal{K},(W,\vec{\mathsf{d}})) \subseteq \mathscr{C}^{s}(\mathcal{K},(Z,\mathsf{d})), \tag{7.46}$$

$$\mathscr{C}^{(\epsilon,s),t}(\mathcal{K}\times\mathbb{R}^N,(W,\vec{\mathsf{d}})\boxtimes\nabla_{\mathbb{R}^N}) \subseteq \mathscr{C}^{s,t}(\mathcal{K}\times\mathbb{R}^N,(Z,\mathsf{d})\boxtimes\nabla_{\mathbb{R}^N}), \tag{7.47}$$

where the inclusions are continuous. Moreover, if $\Omega\Subset\mathfrak{M}$ *is open and relatively compact with* $\mathcal{K}\Subset\Omega$ *and* (Z,d) *weakly* λ*-controls* (Y,d) *on* Ω, *then we may take* $\epsilon'=\epsilon/2\lambda$.

Proof. We prove (7.44) and (7.46) by applying results from Chapter 6. The proofs for (7.45) and (7.47) are similar and follow by a straightforward adaptation of these previous results to the similar spaces $\mathscr{C}^{\bullet,\bullet}(\mathcal{K}\times\mathbb{R}^N,(W,\vec{d})\boxtimes\nabla_{\mathbb{R}^N})$; we leave the details of this adaptation to the reader. In this proof, all inclusions of spaces are continuous.

For (7.46), we have, using Proposition 6.8.1,

$$\mathscr{C}^{(\epsilon,s)}(\mathcal{K},(W,\vec{d}))=\mathscr{B}^{(\epsilon,s)}_{\infty,\infty}(\mathcal{K},(W,\vec{d}))\subseteq\mathscr{B}^{s}_{\infty,\infty}(\mathcal{K},(Z,\vec{d}))=\mathscr{C}^{s}(\mathcal{K},(Z,\vec{d})).$$

For (7.44) we take $\lambda>0$ as in the statement of the proposition (such a λ always exists by Hörmander's condition and the relative compactness of Ω). Set $\epsilon':=\epsilon/2\lambda$. Proposition 6.8.1 gives

$$\mathscr{C}^{s}(\mathcal{K},(Z,\vec{d}))=\mathscr{B}^{s}_{\infty,\infty}(\mathcal{K},(Z,\vec{d}))\subseteq\mathscr{B}^{(-\epsilon',s)}_{\infty,\infty}(\mathcal{K},(W,\vec{d})).\tag{7.48}$$

Applying Theorem 6.7.2 with $\tilde{s}=\epsilon/\lambda$, we have

$$\begin{aligned}\mathscr{B}^{(-\epsilon',s)}_{\infty,\infty}(\mathcal{K},(W,\vec{d}))&\subseteq\mathscr{B}^{(-\epsilon',s)+(\epsilon/\lambda,-\epsilon)}_{\infty,\infty}(\mathcal{K},(W,\vec{d}))\\&=\mathscr{B}^{(\epsilon',s-\epsilon)}_{\infty,\infty}(\mathcal{K},(W,\vec{d}))=\mathscr{C}^{(\epsilon',s-\epsilon)}(\mathcal{K},(W,\vec{d})).\end{aligned}\tag{7.49}$$

Combining (7.48) and (7.49) yields (7.44) and completes the proof. □

7.7 Difference characterization

Proposition 2.5.3 gives an elementary difference characterization of the classical spaces $\mathscr{C}^s(\mathbb{R}^n)$. In this section, we present a similar difference characterization of the spaces $\mathscr{C}^s(\mathcal{K},(W,\vec{d}))$ in the single-parameter[3] setting $\nu=1$, where $(W,\vec{d})$ are Hörmander vector fields with (single-parameter) formal degrees. In Section 7.7.1 we use this characterization to describe the connection between the spaces $\mathscr{C}^s(\mathcal{K},(W,\vec{d}))$ and more elementary Hölder type spaces.

The Hölder spaces, $C^{0,r}$ for $r\in[0,1]$, can be defined on any metric space, because they are defined in terms of first-order differences. However, similar to Proposition 2.5.3, to characterize $\mathscr{C}^s(\mathcal{K},(W,\vec{d}))$ we need to use higher-order differences. This is a difficulty, since higher-order differences generally require some kind of group structure, and we have no such structure in our setting. Instead, we proceed by considering higher-order differences of functions restricted to sufficiently regular paths, as we make precise below.

3 The restriction to the single-parameter case is probably not necessary, and is only for simplicity.

Definition 7.7.1. For $k \in \mathbb{N}$ and $l > 0$ we let $L_k^\infty([0,l])$ be defined as

$$L_k^\infty([0,l]) = \begin{cases} L^\infty([0,l]), & k = 0, \\ C^{k-1,1}([0,l]), & k \geq 1. \end{cases}$$

Fix a finite collection

$$(X,d) = \{(X_1,d_1),\dots,(X_q,d_q)\} \subset C^\infty_{\mathrm{loc}}(\mathfrak{M};T\mathfrak{M}) \times \mathbb{N}_+.$$

Definition 7.7.2. For $l \in \mathbb{N}_+$ and $\delta > 0$, we let $\mathcal{P}^l_\delta(\mathfrak{M},(X,d))$ denote the set of those absolutely continuous paths $\gamma : [0,l] \to \mathfrak{M}$ such that

$$\frac{d}{dt}\gamma(t) = \sum_{j=1}^q a_j(t) X_j(\gamma(t)), \quad \text{a. e.,}$$

where $a_j \in L^\infty_{l-1}([0,l])$ and

$$\|\partial_t^{l'} a_j\|_{L^\infty([0,l])} < \delta^{d_j \vee (l'+1)}, \quad 0 \leq l' \leq l-1. \tag{7.50}$$

For $l = 0$, we let $\mathcal{P}^0_\delta(\mathfrak{M},(X,d))$ consist of all paths $\gamma : \{0\} \to \mathfrak{M}$, i. e., $\mathcal{P}^0_\delta(\mathfrak{M},(X,d)) \cong \mathfrak{M}$.

Definition 7.7.3. For $K \in \mathbb{N}_+$ and $g : [0,K] \to \mathbb{C}$, we set $\mathrm{Diff}(g)(t) := g(t+1) - g(t) : [0,K-1] \to \mathbb{C}$.

Remark 7.7.4. If $g : [0,K] \to \mathbb{C}$, then $\mathrm{Diff}^K(g) : \{0\} \to \mathbb{C}$. In particular, it makes sense to consider $\mathrm{Diff}^K(g)(0)$.

Definition 7.7.5. For $l \in \mathbb{N}$, $\gamma : [0,l] \to \mathfrak{M}$, and $f : \mathfrak{M} \to \mathbb{C}$, we set

$$\triangle^l_\gamma f := \mathrm{Diff}^l(f \circ \gamma)(0) \in \mathbb{C}.$$

Definition 7.7.6. For $l \in \mathbb{N}_+$, $s \in (0,l)$, and $f \in C(\mathfrak{M})$, we set

$$\|f\|_{\mathscr{C}^s_l(X,d)} := \sum_{l'=0}^{l} \sup_{\delta \in (0,\infty)} \sup_{\gamma \in \mathcal{P}^{l'}_\delta(\mathfrak{M},(X,d))} \delta^{-sl'/l} |\triangle^{l'}_\gamma f|$$

and

$$\mathscr{C}^s_l(\mathfrak{M},(X,d)) := \{f \in C(\mathfrak{M}) : \|f\|_{\mathscr{C}^s_l(X,d)} < \infty\}.$$

Proposition 7.7.7. *For $l \in \mathbb{N}_+$ and $s \in (0,l)$, $\mathscr{C}^s_l(\mathfrak{M},(X,d))$ is an algebra and $\forall f,g \in \mathscr{C}^s_l(\mathfrak{M},(X,d))$*

$$\|fg\|_{\mathscr{C}^s_l(X,d)} \leq C_l \|f\|_{\mathscr{C}^s_l(X,d)} \|g\|_{\mathscr{C}^s_l(X,d)},$$

where C_l is a constant depending only on l.

Proof. Fix $0 \le l' \le l$ and $\gamma \in \mathcal{P}_\delta^{l'}(\mathfrak{M}, (X, d))$. We wish to show

$$|\triangle_\gamma^{l'}(fg)| \lesssim \delta^{sl'/l} \|f\|_{\mathscr{C}_l^s(X,d)} \|g\|_{\mathscr{C}_l^s(X,d)}, \tag{7.51}$$

and this will complete the proof.

We have

$$\begin{aligned}\triangle_\gamma^{l'}(fg) &= \mathrm{Diff}^{l'}((f \circ \gamma)(g \circ \gamma))(0) \\ &= \sum_{k=0}^{l'} \binom{l'}{k} (\mathrm{Diff}^{l'-k}(f \circ \gamma))(k)(\mathrm{Diff}^k(g \circ \gamma))(0).\end{aligned} \tag{7.52}$$

For $0 \le k \le l'$, we set $\gamma_k(t) := \gamma(t + k)$. Then $\gamma_k \in \mathcal{P}_\delta^{l'-k}(\mathfrak{M}, (X, d))$, and therefore

$$|\mathrm{Diff}^{l'-k}(f \circ \gamma)(k)| = |\triangle_{\gamma_k}^{l'-k} f| \le \delta^{s(l'-k)/l} \|f\|_{\mathscr{C}_l^s(X,d)}. \tag{7.53}$$

Also, $\gamma \in \mathcal{P}_\delta^{l'}(\mathfrak{M}, (X, d)) \subseteq \mathcal{P}_\delta^k(\mathfrak{M}, (X, d))$ and $\mathrm{Diff}^k(g \circ \gamma)(0) = \triangle_\gamma^k g$. Therefore,

$$|\mathrm{Diff}^k(g \circ \gamma)(0)| = |\triangle_\gamma^k g| \le \delta^{sk/l} \|g\|_{\mathscr{C}_l^s(X,d)}. \tag{7.54}$$

Plugging (7.53) and (7.54) into (7.52) establishes (7.51) and completes the proof. □

Now let

$$(W, \vec{d}) = \{(W_1, \vec{d}_1), \ldots, (W_r, \vec{d}_r)\} \subset C_{\mathrm{loc}}^\infty(\mathfrak{M}; T\mathfrak{M}) \times \mathbb{N}_+$$

be Hörmander vector fields with (single-parameter) formal degrees. Recall that $\mathcal{K} \Subset \mathfrak{M}$ denotes a compact set and $\Omega \Subset \mathfrak{M}$ denotes a relatively compact, open set with $\mathcal{K} \Subset \Omega$. Fix a new relatively compact, open set $\widetilde{\Omega} \Subset \mathfrak{M}$ with $\Omega \Subset \widetilde{\Omega}$. By Proposition 3.4.14 (in the case $\nu = 1$), we may pick a finite collection

$$(X, d) = \{(X_1, d_1), \ldots, (X_q, d_q)\} \subset \mathrm{Gen}((W, \vec{d}))$$

such that $\mathrm{Gen}((W, \vec{d}))$ is finitely generated by (X, d) on $\widetilde{\Omega}$.

Theorem 7.7.8. *Suppose* $\mathrm{Gen}((W, \vec{d}))$ *is finitely generated by* $(X, d) \subset \mathrm{Gen}((W, \vec{d}))$ *on* $\widetilde{\Omega}$. *Then, for all* $l \in \mathbb{N}_+$ *and* $s \in (0, l)$,

$$\mathscr{C}^s(\mathcal{K}, (W, \vec{d})) = \{f \in \mathscr{C}_l^s(\mathfrak{M}, (X, d)) : \mathrm{supp}(f) \subseteq \mathcal{K}\}.$$

Moreover, these two spaces have equivalent norms: for $f \in \mathscr{C}^s(\mathcal{K}, (W, \vec{d}))$,

$$\|f\|_{\mathscr{C}^s(W,\vec{d})} \approx \|f\|_{\mathscr{C}_l^s(X,d)},$$

where the implicit constants do not depend on f*, but may depend on any of the other ingredients.*

The rest of this section is devoted to the proof of Theorem 7.7.8. We henceforth assume $\mathrm{Gen}((W, \vec{d}))$ is finitely generated by $(X, d) \subset \mathrm{Gen}((W, \vec{d}))$ on $\widetilde{\Omega}$. We begin with the easier containment $\mathscr{C}^s(\mathcal{K}, (W, \vec{d})) \subseteq \mathscr{C}^s_l(\mathfrak{M}, (X, d))$. For this, the next two lemmas are useful.

Lemma 7.7.9. *Let $l \in \mathbb{N}_+$. Then, for all $\delta \in (0, 1]$, $h \in C_0^\infty(\mathfrak{M})$, $\gamma \in \mathcal{P}^l_\delta(\mathfrak{M}, (X, d))$, and $0 \le l' \le l$,*

$$\partial_t^{l'}(h \circ \gamma)(t) = \sum_{|\beta| \le l'} a_\beta^{l'}(t)(X^\beta h)(\gamma(t)), \tag{7.55}$$

where for all $0 \le k \le l - l'$,

$$\|\partial_t^k a_\beta^{l'}\|_{L^\infty([0,l])} \le C_{k,q,l}\delta^{\deg_d(\beta)\vee(k+l')}.$$

Proof. We prove the result by induction on l'. The base case, $l' = 0$, follows by taking $a_0^0 \equiv 1$. Let $l \ge l' \ge 1$. We assume the result for $l' - 1$ and prove it for l'. We have, letting $a_j(t)$ be as in Definition 7.7.2,

$$\begin{aligned}\partial_t^{l'} h \circ \gamma(t) = \partial_t \partial_t^{l'-1} h \circ \gamma(t) &= \partial_t \sum_{|\beta| \le l'-1} a_\beta^{l'-1}(t)(X^\beta h)(\gamma(t)) \\ &= \sum_{|\beta| \le l'-1} (\partial_t a_\beta^{l'-1})(t)(X^\beta h)(\gamma(t)) + \sum_{|\beta| \le l'-1} \sum_{j=1}^{q} a_\beta^{l'-1} a_j(t)(X^\beta X_j h)(\gamma(t)) \\ &:= (\mathrm{I}) + (\mathrm{II}).\end{aligned}$$

We wish to show that (I) and (II) are of the desired form (7.55).

For (I), for $0 \le k \le l - l'$ we have $k + 1 \le l - l' + 1 = l - (l' - 1)$, and therefore by the inductive hypothesis,

$$\|\partial_t^k \partial_t a_\beta^{l'-1}\|_{L^\infty} = \|\partial_t^{k+1} a_\beta^{l'-1}\|_{L^\infty} \lesssim \delta^{\deg_d(\beta)\vee(k+1+(l'-1))} = \delta^{\deg_d(\beta)\vee(k+l')},$$

establishing that (I) is of the form (7.55).

For (II), we have $X^\beta X_j = X^{\tilde{\beta}}$, with $\deg_d(\tilde{\beta}) = \deg_d(\beta) + d_j$, and we therefore wish to show that for all $0 \le k \le l - l'$,

$$\|\partial_t^k (a_\beta^{l'-1}(t) a_j(t))\|_{L^\infty} \lesssim \delta^{(\deg_d(\beta)+d_j)\vee(k+l')}. \tag{7.56}$$

Using (7.50) and the inductive hypothesis, we have, for $0 \le k \le l - l'$,

$$\begin{aligned}\|\partial_t^k (a_\beta^{l'-1} a_j)\|_{L^\infty} &\lesssim \sum_{k_1=0}^{k} \|\partial_t^{k_1} a_\beta^{l'-1}\|_{L^\infty} \|\partial_t^{k-k_1} a_j\|_{L^\infty} \\ &\lesssim \sum_{k_1=0}^{k} \delta^{\deg_d(\beta)\vee(k_1+l'-1)} \delta^{d_j \vee(k-k_1+1)} \lesssim \delta^{(\deg_d(\beta)+d_j)\vee(k+l')}.\end{aligned}$$

This establishes (7.56) and proves (II) is of the desired form, completing the proof. □

Lemma 7.7.10. *Fix $l \in \mathbb{N}$. Then for all $h \in C_0^\infty(\mathfrak{M})$, $\delta \in (0,\infty)$, and $\gamma \in \mathcal{P}_\delta^l(\mathfrak{M},(X,d))$,*

$$|\triangle_\gamma^l h| \lesssim \left(\sum_{|\beta|\leq l} \delta^{\deg_d(\beta)\vee l}\|X^\beta h\|_{L^\infty}\right) \wedge \|h\|_{L^\infty}.$$

Proof. Clearly, $|\triangle_\gamma^l h| \lesssim \|h\|_{L^\infty}$. Also, for $\delta > 1$,

$$\sum_{|\beta|\leq l} \delta^{\deg_d(\beta)\vee l}\|X^\beta h\|_{L^\infty} \geq \|h\|_{L^\infty}.$$

Thus, it suffices to assume that $\delta \in (0,1]$ and to prove that

$$|\triangle_\gamma^l h| \lesssim \sum_{|\beta|\leq l} \delta^{\deg_d(\beta)\vee l}\|X^\beta h\|_{L^\infty}. \tag{7.57}$$

Consider

$$\operatorname{Diff}(h\circ\gamma)(t) = \int_0^1 \partial_t(h\circ\gamma)(t+u)\,du.$$

Therefore,

$$\triangle_\gamma^l h = \operatorname{Diff}^l(h\circ\gamma)(0) = \int_{[0,1]^l} (\partial_t^l(h\circ\gamma))(u_1+u_2+\cdots+u_l)\,du,$$

so with a_β^l as in Lemma 7.7.9, we have

$$|\triangle_\gamma^l h| \leq \|\partial_t^l(h\circ\gamma)\|_{L^\infty([0,l])} \leq \sum_{|\beta|\leq l}\|a_\beta^l\|_{L^\infty}\|X^\beta h\|_{L^\infty} \lesssim \sum_{|\beta|\leq l} \delta^{\deg_d(\beta)\vee l}\|X^\beta h\|_{L^\infty}.$$

This establishes (7.57) and completes the proof. □

We are prepared to prove the easy half of Theorem 7.7.8.

Proposition 7.7.11. *Let $l \in \mathbb{N}_+$ and $s \in (0,l)$. Then*

$$\|f\|_{\mathscr{C}_l^s(X,d)} \lesssim \|f\|_{\mathscr{C}^s(W,\vec{d})}, \quad \forall f \in \mathscr{C}^s(\mathcal{K},(W,\vec{d})), \tag{7.58}$$

and $\mathscr{C}^s(\mathcal{K},(W,\vec{d})) \subseteq \mathscr{C}_l^s(\mathfrak{M},(X,d))$.

Proof. We proceed by induction on $l \in \mathbb{N}_+$. Suppose we know the result holds $\forall l' \in \mathbb{N}_+$ with $l' < l$; we will prove the result for l. Because of how we have set up this induction, we do not require a base case (the base case $l = 1$ is included in the inductive step).

Suppose $f \in \mathscr{C}^s(\mathcal{K},(W,\vec{d}))$. By Corollary 7.2.2, $f \in C(\mathfrak{M})$ and $\|f\|_{C(\mathfrak{M})} \lesssim \|f\|_{\mathscr{C}^s(W,\vec{d})}$. For $0 < l' < l$, we have, by the inductive hypothesis,

$$\sup_{\substack{\delta\in(0,\infty)\\ \gamma\in\mathcal{P}_\delta^{l'}(\mathfrak{M},(X,d))}} \delta^{-sl'/l}|\triangle_\gamma^{l'}f| \le \|f\|_{\mathscr{C}_{l'}^{sl'/l}(X,d)} \lesssim \|f\|_{\mathscr{C}^{sl'/l}(W,\vec{d})} \le \|f\|_{\mathscr{C}^s(W,\vec{d})}.$$

Thus, to prove (7.58) it suffices to show that for $\delta > 0$, $\gamma \in \mathcal{P}_\delta^l(\mathfrak{M}, (X, d))$,

$$|\triangle_\gamma^l f| \lesssim \delta^s \|f\|_{\mathscr{C}^s(W,\vec{d})}. \tag{7.59}$$

By Proposition 7.2.1 we may write $f = \sum_{j\in\mathbb{N}} f_j$, where $f_j \in C_0^\infty(\Omega)$ satisfy, for all α,

$$\left\|(2^{-j\vec{d}}W)^\alpha f_j\right\|_{L^\infty} \lesssim 2^{-js}\|f\|_{\mathscr{C}^s(W,\vec{d})}.$$

Since $(X, d) \subset \mathrm{Gen}((W, \vec{d}))$, we therefore have, for all β,

$$\left\|(2^{-jd}X)^\beta f_j\right\|_{L^\infty} \lesssim 2^{-js}\|f\|_{\mathscr{C}^s(W,\vec{d})}. \tag{7.60}$$

By Lemma 7.7.10 and (7.60) we have

$$\begin{aligned}
|\triangle_\gamma^l f_j| &\lesssim \left(\sum_{|\beta|\le l} \delta^{\deg_d(\beta)\vee l}\left\|X^\beta f_j\right\|_{L^\infty}\right) \wedge \|f_j\|_{L^\infty} \\
&\lesssim \left(\left(2^{-js}\sum_{|\beta|\le l} \delta^{\deg_d(\beta)\vee l} 2^{j\deg_d(\beta)}\right) \wedge 2^{-js}\right)\|f\|_{\mathscr{C}^s(W,\vec{d})} \\
&\lesssim 2^{-js}((2^j\delta)^l \wedge 1)\|f\|_{\mathscr{C}^s(W,\vec{d})}.
\end{aligned}$$

Hence, using $l > s$, we have

$$\begin{aligned}
|\triangle_\gamma^l f| &\le \sum_{j\in\mathbb{N}}|\triangle_\gamma^l f_j| \lesssim \|f\|_{\mathscr{C}^s(W,\vec{d})}\sum_{j\in\mathbb{N}} 2^{-js}((\delta 2^j)^l \wedge 1) \\
&\le \|f\|_{\mathscr{C}^s(W,\vec{d})}\left(\sum_{2^{-j}\ge\delta} 2^{j(l-s)}\delta^l + \sum_{2^{-j}\le\delta} 2^{-js}\right) \\
&\lesssim \|f\|_{\mathscr{C}^s(W,\vec{d})}(\delta^s + \delta^s) \lesssim \|f\|_{\mathscr{C}^s(W,\vec{d})}\delta^s.
\end{aligned}$$

This establishes (7.59) and completes the proof of (7.58). The containment $\mathscr{C}^s(\mathcal{K}, (W, \vec{d})) \subseteq \mathscr{C}_l^s(\mathfrak{M}, (X, d))$ is an immediate consequence of (7.58). □

We turn to the other half of Theorem 7.7.8: the containment

$$\{f \in \mathscr{C}_l^s(\mathfrak{M}, (X, d)) : \mathrm{supp}(f) \subseteq \mathcal{K}\} \subseteq \mathscr{C}^s(\mathcal{K}, (W, \vec{d})).$$

The basic proof idea is the same as in Lemma 2.5.12, but here the operators D_j are more complicated: see Section 7.1. To make the same proof outline work requires a number of technical estimates. To do this, we introduce several lemmas.

Suppose $U \subseteq \mathbb{R}^q \times \widetilde{\Omega}$ is an open neighborhood of $\{0\} \times \widetilde{\Omega}$ and $g(t,x) \in C^\infty_{\mathrm{loc}}(U)$. Then, by Definition 4.2.1 (with $\mathfrak{M}$ replaced by $\widetilde{\Omega}$), it makes sense to ask if $\deg_d(g) \geq d_0$ for $d_0 \in \mathbb{Z}$. Since this is the single-parameter case ($\nu = 1$), we can define $\deg_d(g) \in \mathbb{N}$ to be the least number $e \in \mathbb{N}$ such that $\deg_d(g) \geq d_0 \Leftrightarrow e \geq d_0$, $\forall d_0 \in \mathbb{Z}$.

Lemma 7.7.12. *Let $U \subseteq \mathbb{R}^q \times \widetilde{\Omega}$ be an open neighborhood of $\{0\} \times \widetilde{\Omega}$ and let $\mathcal{K}_0 \Subset \widetilde{\Omega}$ be compact. Fix $l \in \mathbb{N}_+$. Then if $a_0 = a_0(\mathcal{K}_0, U, (X,d), l) > 0$ is sufficiently small, the following holds. Suppose $\gamma : [0,l] \to \mathcal{K}_0$ satisfies, for some $t \in B^q(a_0)$, $g_1, \ldots, g_q \in C^\infty_{\mathrm{loc}}(U)$, $k \in \{1, \ldots, q\}$, $j \in (0,\infty)$ with $2^{-j} \leq a_0$, $j' \in [0,j]$,*

$$\frac{d}{ds}\gamma(s) = \sum_{p=1}^{q} 2^{-jd_k} g_k^p(2^{-j'd}t + 2^{-jd_k}s e_k, \gamma(s)) X_p(\gamma(s)), \tag{7.61}$$

where $\deg_d(g_k^p) \geq d_p - d_k$. Then there exists $A = A(\mathcal{K}_0, (X,d), l, g_1, \ldots, g_q) \geq 0$, not depending on t, j, or j', such that $\gamma \in \mathcal{P}^l_{A2^{-j'}}(\mathfrak{M}, (X,d))$.

Proof. Note that $a_0 > 0$ must be sufficiently small for the right-hand side of (7.61) to even make sense because g_k^p is only defined on U. In particular, $g(t,x)$ is defined for all $x \in \mathcal{K}_0$ for t sufficiently small (independent of $x \in \mathcal{K}_0$).

The conclusion that $\gamma \in \mathcal{P}^l_{A2^{-j}}(\mathfrak{M}, (X,d))$, for some $A \lesssim 1$, is equivalent to stating that for $0 \leq l' \leq l-1$,

$$\sup_{t \in B^q(a_0)} \sup_{0 \leq s \leq l} \left|\partial_s^{l'} 2^{-jd_k} g_k^p(2^{-j'd}t + 2^{-jd_k}s e_k, \gamma(s))\right| \lesssim 2^{-j'\left(d_p \vee (l'+1)\right)}. \tag{7.62}$$

To prove (7.62), we prove a more general estimate. Let $g \in C^\infty_{\mathrm{loc}}(U)$ satisfy $\deg_d(g) \geq d_0 - d_k$ for some $d_0 \in \mathbb{N}$. We will show for all $l_1, l_2 \in \mathbb{N}$ with $0 \leq l_1 + l_2 \leq l$,

$$\sup_{t \in B^q(a_0)} \sup_{0 \leq s_1, s_2 \leq l} \left|\partial_{s_1}^{l_1}\partial_{s_2}^{l_2} 2^{-jd_k} g(2^{-j'd}t + 2^{-jd_k}s_1 e_k, \gamma(s_2))\right| \lesssim 2^{-j'\left(d_0 \vee (l_1+l_2+1)\right)}. \tag{7.63}$$

Indeed, setting $s_1 = s_2 = s$ in (7.63) and summing over all $l_1 + l_2 = l'$ proves (7.62).

We prove (7.63) by induction on l_2. We begin with the base case, $l_2 = 0$. By the definition of $\deg_d(g) \geq d_0 - d_k$, we may write $g(t,x) = \sum_{p=1}^{P} t^{\alpha_p} h_p(t,x)$, where $P \geq 1$, $h_p \in C^\infty_{\mathrm{loc}}(U)$, and $\deg_d(\alpha_p) \geq d_0 - d_k$. It suffices to prove (7.63) when g is given by one such term, so we assume

$$g(t,x) = t^\alpha h(t,x),$$

where $h \in C^\infty_{\mathrm{loc}}(U)$ and $\deg_d(\alpha) \geq d_0 - d_k$.

We have, for some constants $c_{l_1'}^{l_1}$,

$$\begin{aligned}
&\partial_{s_1}^{l_1} 2^{-jd_k} g(2^{-j'd}t + 2^{-jd_k}s_1\mathfrak{e}_k, \gamma(s_2)) \\
&\quad = \partial_{s_1}^{l_1}[2^{-jd_k}(2^{-j'd}t + 2^{-jd_k}s\mathfrak{e}_k)^{\alpha} h(2^{-j'd}t + 2^{-jd_k}s_1\mathfrak{e}_k, \gamma(s_2))] \\
&\quad = \sum_{l_1'=0}^{l_1\wedge\alpha_k} c_{l_1'}^{l_1} 2^{-j(1+l_1)d_k}(2^{-jd}t + 2^{-jd_k}s_1\mathfrak{e}_k)^{\alpha - l_1'\mathfrak{e}_k}(\partial_{t_k}^{l_1-l_1'}h)(2^{-jd}t + 2^{-jd_k}s_1\mathfrak{e}_k, \gamma(s_2)).
\end{aligned} \tag{7.64}$$

In the above expression, we are only considering $|t| < a_0 \lesssim 1$, $|s| \le l \lesssim 1$, and $2^{-j'} \lesssim 2^{-j}$. Therefore, for any multi-index β,

$$|(2^{-j'd}t + 2^{-jd_k}s\mathfrak{e}_k)^{\beta}| \lesssim 2^{-j'\deg_d(\beta)}. \tag{7.65}$$

Since $\gamma(s_2) \in \mathcal{K}_0$, $|2^{-j'd}t + 2^{-jd_k}s\mathfrak{e}_k| \le (l+1)a_0$, and $h \in C_{\mathrm{loc}}^{\infty}(U)$, if $a_0 > 0$ is sufficiently small, we have by compactness

$$|(\partial_t^{l_1-l_1'}h)(2^{-j'd}t + 2^{-jd_k}s_1\mathfrak{e}_k, \gamma(s_2))| \lesssim 1. \tag{7.66}$$

Plugging (7.65) and (7.66) into (7.64) and using $\deg_d(\alpha) \ge d_0 - d_k$ and $2^{-j} \lesssim 2^{-j'}$, we have

$$\begin{aligned}
|\partial_{s_1}^{l_1} 2^{-jd_k} g(2^{-j'd}t + 2^{-jd_k}s_1\mathfrak{e}_k, \gamma(s_2))| &\lesssim \sum_{l_1'=0}^{l_1\wedge\alpha_k} 2^{-j(1+l_1)d_k} 2^{-j'\deg_d(\alpha - l_1'\mathfrak{e}_k)} \\
&\lesssim 2^{-j(1+l_1)d_k} \wedge \left(\sum_{l_1'=0}^{l_1} 2^{-j(1+l_1)d_k} 2^{-j'(d_0-d_k-l_1'd_k)}\right) \\
&\lesssim 2^{-j'(1+l_1)d_k} \wedge 2^{-j'd_0} \lesssim 2^{-j'(d_0\vee(1+l_1))}.
\end{aligned}$$

This proves (7.63) in the base case $l_2 = 0$.

Let $l_2 \ge 1$. We assume that (7.63) holds for all lesser values of l_2. Applying the inductive hypothesis, we have, for all $0 \le l_2' < l_2$,

$$\begin{aligned}
&\sup_{0\le s_2\le l} |\partial_{s_2} 2^{-jd_k} g(2^{-j'd}t + 2^{-jd_k}s_2\mathfrak{e}_k, \gamma(s_2))| \\
&\quad \lesssim \sup_{0\le s_1,s_2\le l} \sum_{l_{1,1}+l_{2,1}=l_2'} |\partial_{s_1}^{l_{1,1}}\partial_{s_2}^{l_{2,1}} 2^{-jd_k} g(2^{-j'd}t + 2^{-jd_k}s_1\mathfrak{e}_k, \gamma(s_2))| \\
&\quad \lesssim 2^{-j'(d_0\vee(l_2'+1))}.
\end{aligned} \tag{7.67}$$

Consider, by (7.61),

$$\begin{aligned}
&\partial_{s_2} 2^{-jd_k} g(2^{-j'd}t + 2^{-jd_k}s_1\mathfrak{e}_k, \gamma(s_2)) \\
&\quad = \sum_{p=1}^{q} 2^{-2jd_k} g_k^p(2^{-j'd}t + 2^{-jd_k}s_2\mathfrak{e}_k, \gamma(s_2))(X_p g)(2^{-j'd}t + 2^{-jd_k}s_1\mathfrak{e}_k, \gamma(s_2)).
\end{aligned} \tag{7.68}$$

We have $\deg_d(g_k^p) \geq d_p - d_k$ by hypothesis and $\deg_d(X_p g) \geq \deg_d(g) \geq d_0 - d_k$. Thus, we can apply (7.67) to $2^{-jd_k} g_k^p(2^{-j'd}t + 2^{-jd_k}s_2\mathfrak{e}_k, \gamma(s_2))$ (with d_0 replaced by d_p) and the inductive hypothesis to $2^{-jd_k}(X_p g)(2^{-j'd}t + 2^{-jd_k}s_1\mathfrak{e}_k, \gamma(s_2))$. Applying these to (7.68) shows that

$$\begin{aligned}
&|\partial_{s_1}^{l_1}\partial_{s_2}^{l_2} 2^{-jd_k} g(2^{-j'd}t + 2^{-jd_k}s_1\mathfrak{e}_k, \gamma(s_2))| \\
&\quad \lesssim \sum_{l_{2,1}+l_{2,2}=l_2-1} |2^{-jd_k}\partial_{s_2}^{l_{2,1}} g_k^p(2^{-j'd}t + 2^{-jd_k}s_2\mathfrak{e}_k, \gamma(s_2))| \\
&\qquad \times |2^{-jd_k}\partial_{s_1}^{l_1}\partial_{s_2}^{l_{2,2}}(X_p g)(2^{-j'd}t + 2^{-jd_k}s_1\mathfrak{e}_k, \gamma(s_2))| \\
&\quad \lesssim \sum_{l_{2,1}+l_{2,2}=l_2-1} 2^{-j'(d_p\vee(l_{2,1}+1))}2^{-j'(d_0\vee(l_1+l_{2,2}+1))} \\
&\quad \lesssim 2^{-j'(d_0\vee(l_1+l_2+1))}.
\end{aligned}$$

This establishes (7.63) and completes the proof. □

Lemma 7.7.13. *Fix $l \in \mathbb{N}_+$. If $a_0 = a_0(\widetilde{\Omega}, \Omega, (X, d), l) > 0$ is sufficiently small, the following holds. There exists $A = A(\widetilde{\Omega}, \Omega, (X, d), l) \geq 0$ such that for $x \in \Omega$, $t \in B^q(a_0)$, $j \in (0, \infty)$ with $2^{-j} \leq a_0$, $j' \in [0, j]$, and $k \in \{1, \dots, q\}$, if we set*

$$\gamma_{x,t,j,j'}(s) := \exp(-(2^{-j'd}t + 2^{-jd_k}s\mathfrak{e}_k)\cdot X)x,$$

then $\gamma_{x,t,j,j'} \in \mathcal{P}^l_{A2^{-j'}}(\mathfrak{M}, (X, d))$.

Proof. By Remark 3.4.12, we have

$$[X_j, X_k] = \sum_{d_l \leq d_j + d_k} c_{j,k}^l X_l, \quad c_{j,k}^l \in C^\infty_{\mathrm{loc}}(\Omega).$$

In other words, (X, d) satisfies the hypotheses of Chapter 4 with $\mathfrak{M}$ replaced by $\widetilde{\Omega}$ and $\nu = 1$. Thus, we can apply Proposition 4.2.2 to understand the exponential map $\Gamma(t, x) = e^{-t\cdot X}x$. Fix $\mathcal{K}_0$ compact with $\Omega \Subset \mathcal{K}_0 \Subset \widetilde{\Omega}$. If $a_0 > 0$ is sufficiently small, then $\gamma_{x,t,j,j'}(s) \in \mathcal{K}_0$, $\forall s \in [0, l]$.

Proposition 4.2.2 (iv) implies

$$\frac{d}{ds}\gamma_{x,t,j,j'}(s) = -2^{-jd_k}X_k(\gamma_{x,t,j,j'}) + \sum_{p=1}^{q} 2^{-jd_k} g_{4,k}^p(2^{-j'd}t + 2^{-jd_k}s\mathfrak{e}_k, \gamma_{x,t,j,j'}(s))X_p(\gamma_{x,t,j,j'}),$$

where $g_{4,k}^p(t, x)$ is as in that result. In particular, $\deg_d(g_{4,k}^p) \geq d_p - d_k$. From here, the result follows from Lemma 7.7.12. □

For a function $F : \mathbb{R}^q \to \mathbb{C}$ and $v \in \mathbb{R}^q$, we define $\mathrm{Diff}_v F : \mathbb{R}^q \to \mathbb{C}$ by

$$\mathrm{Diff}_v F(t) := F(t + v) - F(t). \tag{7.69}$$

If F is only defined on a subset of $\mathbb{R}^q$, then $\operatorname{Diff}_v F(t)$ is only defined for t where (7.69) makes sense.

Lemma 7.7.14. *Fix $l \in \mathbb{N}_+$ and $s \in (0,l)$. There exists $a_0 = a_0(\widetilde{\Omega}, \Omega, (X,d), l) > 0$ such that the following holds. There exists $C \geq 0$ such that for all $j \in (0,\infty)$ with $2^{-j} \leq a_0$, $j' \in [0,j]$, $k \in \{1,\dots,q\}$, $0 \leq l' \leq l$, $f \in \mathscr{C}_l^s(\mathfrak{M},(X,d))$,*

$$\sup_{\substack{t\in B^q(a_0)\\ x\in\Omega}} \left|\operatorname{Diff}^{l'}_{2^{(j'-j)d_k}\mathfrak{e}_k} f(e^{-2^{-j'}d t\cdot X}x)\right| \leq C 2^{-j's l'/l} \|f\|_{\mathscr{C}_l^s(X,d)}.$$

Here, $\operatorname{Diff}_{2^{(j'-j)d_k}\mathfrak{e}_k}$ is acting on $f(e^{-2^{-j'}d t\cdot X}x)$ in the $t \in \mathbb{R}^q$ variable.

Proof. For $t \in B^q(a_0)$ and $x \in \Omega$, we set $\gamma(s) := \exp(-(2^{-j'}d t + 2^{-jd_k} s\mathfrak{e}_k)X)$. Lemma 7.7.13 shows that there exists $A \lesssim 1$ with $\gamma \in \mathcal{P}^l_{A2^{-j'}}(\mathfrak{M},(X,d)) \subseteq \mathcal{P}^{l'}_{A2^{-j'}}(\mathfrak{M},(X,d))$. Thus,

$$\begin{aligned}\left|\operatorname{Diff}^{l'}_{2^{(j'-j)d_k}\mathfrak{e}_k} f(e^{-2^{-j'}d t\cdot X}x)\right| &= \left|(\operatorname{Diff}^{l'}(f\circ\gamma))(0)\right| = \left|\triangle^{l'}_\gamma f\right| \\ &\leq (A2^{-j'})^{sl'/l}\|f\|_{\mathscr{C}_l^s(X,d)} \lesssim 2^{-j'sl'/l}\|f\|_{\mathscr{C}_l^s(X,d)}.\end{aligned}$$

□

Lemma 7.7.15. *Fix $l \in \mathbb{N}_+$ and $s \in (0,l)$. If $a_0 = a_0(\widetilde{\Omega}, \Omega, (X,d), l) > 0$ is sufficiently small, the following holds. Let $\mathcal{B} \subset C_0^\infty(B^q(a_0))$ be a bounded set. Then there exists $C \geq 0$ such that for all $j \in (0,\infty)$ with $2^{-j} \leq a_0$, $j' \in [0,j]$, $0 \leq l' \leq l$, $k \in \{1,\dots,q\}$, and $f \in \mathscr{C}_l^s(\mathfrak{M},(W,\vec{d}))$,*

$$\sup_{x\in\Omega}\sup_{\eta,\varsigma\in\mathcal{B}} \left|\int f(e^{-t\cdot X}x)\operatorname{Diff}^{l'}_{-2^{-jd_k}\mathfrak{e}_k}(\eta(t)\operatorname{Dil}^d_{2^{-j'}}(\varsigma)(t))\,dt\right| \leq C2^{-j'sl'/l}\|f\|_{\mathscr{C}_l^s(X,d)}.$$

Proof. Let $a_0 > 0$ be as in Lemma 7.7.14. Applying that lemma, we have, for $x \in \Omega$ and $\eta, \varsigma \in \mathcal{B}$,

$$\begin{aligned}&\left|\int f(e^{-t\cdot X}x)\operatorname{Diff}^{l'}_{-2^{-jd_k}\mathfrak{e}_k}(\eta(t)\operatorname{Dil}^d_{2^{-j'}}(\varsigma)(t))\,dt\right| \\ &\quad= \left|\int f(e^{-2^{-j'}d t\cdot X}x)\operatorname{Diff}^{l'}_{-2^{(j'-j)d_k}\mathfrak{e}_k}(\eta(2^{-j'}d t)\varsigma(t))\,dt\right| \\ &\quad= \left|\int (\operatorname{Diff}^{l'}_{2^{(j'-j)d_k}\mathfrak{e}_k} f(e^{-2^{-j'}d t\cdot X}x))\eta(2^{-j'}d t)\varsigma(t)\,dt\right| \\ &\quad\lesssim 2^{-j'sl'/l}\|f\|_{\mathscr{C}_l^s(X,d)}\int\left|\eta(2^{-j'}d t)\varsigma(t)\right|dt \\ &\quad\lesssim 2^{-j'sl'/l}\|f\|_{\mathscr{C}_l^s(X,d)}.\end{aligned}$$

□

Lemma 7.7.15 required $\varsigma \in C_0^\infty(B^q(a_0))$; however, we require a similar result for $\varsigma \in \mathscr{S}(\mathbb{R}^q)$. This can be achieved by decomposing the case of $\varsigma \in \mathscr{S}(\mathbb{R}^q)$ into a sum of the cases where $\varsigma \in C_0^\infty(B^q(a_0))$. This decomposition is described in the next lemma.

Lemma 7.7.16. *Fix $a_0 > 0$. Let $\mathcal{B}_1 \subset C_0^\infty(B^q(a_0))$ and $\mathcal{B}_2 \subset \mathscr{S}(\mathbb{R}^q)$ be bounded sets. Then, for all $\eta \in \mathcal{B}_1$, $\varsigma \in \mathcal{B}_2$, $j \in [0,\infty)$, there exists*

$$\{\gamma_j : j' \in \mathbb{N}, j' \le j\} \subset C_0^\infty(B^q(a_0))$$

such that

$$\eta(t)\,\mathrm{Dil}^d_{2^j}(\varsigma) = \eta \sum_{\substack{j' \in \mathbb{N}\\ j' \le j}} \mathrm{Dil}^d_{2^{j'}}(\gamma_{j'})(t). \tag{7.70}$$

Furthermore, for every $M \in \mathbb{N}$,

$$\{2^{M|j-j'|}\gamma_{j'} : j \in [0,\infty), j' \le j, \varsigma \in \mathcal{B}_2, \eta \in \mathcal{B}_1\} \subset C_0^\infty(B^q(a_0)) \tag{7.71}$$

is a bounded set.

Proof. Fix $\tilde{\eta} \in C_0^\infty(B^q(a_0))$ such that $\tilde{\eta} \equiv 1$ on $\bigcup_{\eta \in \mathcal{B}_1} \mathrm{supp}(\eta)$ and such that $\tilde{\eta} \equiv 1$ on a neighborhood of 0. Fix $j \in \mathbb{N}$. For $j' \in \mathbb{N}$ with $j' \le j$, let

$$\sigma_{j'}(t) := \begin{cases} \tilde{\eta}(t) - \tilde{\eta}(2^d t), & j' + 1 \le j, \\ \tilde{\eta}(t), & j' + 1 > j. \end{cases}$$

Note that

$$\tilde{\eta}(t) = \sum_{j'=0}^{j} \sigma_{j'}(2^{j'd} t) \tag{7.72}$$

and $\sigma_{j'}(t) = 0$ if $j' \le j - 1$ and $|t|$ is sufficiently small (independent of j, j'). For $\varsigma \in \mathcal{B}_2$, set

$$\gamma_{j'}(t) := \sigma_{j'}(t)\,\mathrm{Dil}^d_{2^{j-j'}}(\varsigma)(t).$$

It follows immediately from the definitions that

$$\{\gamma_{j'} : j \in [0,\infty), j' \le j, (j-j') \le 1, \varsigma \in \mathcal{B}_2\}$$

is a bounded set. So, to prove (7.71) is a bounded set, it suffices to consider the case where $j - j' > 1$.

For $j - j' > 1$, since $\sigma_{j'}(t) = 0$ for $|t|$ small, using the Schwartz bounds for ς we have, for all $m \in \mathbb{N}$ and multi-indices $\alpha \in \mathbb{N}^q$,

$$|\partial_t^\alpha \gamma_{j'}(t)| \lesssim \chi_{\{|t|\approx 1\}}(1 + |2^{(j-j')d} t|)^{-m-|\alpha| \max d_k} 2^{(j-j')|\alpha| \max d_k} \lesssim 2^{-m|j-j'|}. \tag{7.73}$$

By the definition of $\gamma_{j'}$, there is a compact set $\mathcal{K}_0 \Subset B^q(a_0)$ independent of j, j', and ς such that $\mathrm{supp}(\gamma_{j'}) \subseteq \mathcal{K}_0$. Combining this with (7.73) shows that $\forall M \in \mathbb{N}$,

$$\{2^{M|j-j'|}\gamma_{j'} : j \in [0,\infty), j' \leq j, j' \in \mathbb{N}, j-j' \geq 1, \varsigma \in \mathcal{B}\} \subset C_0^\infty(B^q(a_0))$$

is a bounded set. This completes the proof that (7.71) is a bounded set.

Using (7.72), we have

$$\eta(t)\operatorname{Dil}_{2^j}^d(\varsigma)(t) = \eta(t)\tilde{\eta}(t)\operatorname{Dil}_{2^j}^d(\varsigma)(t) = \eta(t)\sum_{\substack{j'\in\mathbb{N}\\ j'\leq j}}\operatorname{Dil}_{2^{j'}}^d(\gamma_{j'})(t),$$

establishing (7.70) and completing the proof. □

Lemma 7.7.17. *Fix $l \in \mathbb{N}_+$ and $s \in (0,l)$. If $a_0 = a_0(\widetilde{\Omega}, \Omega, (X,d), l) > 0$ is sufficiently small, the following holds. Let $\mathcal{B}_1 \subset C_0^\infty(B^q(a_0))$ and $\mathcal{B}_2 \subset \mathscr{S}(\mathbb{R}^q)$ be bounded sets. Then there exists $C \geq 0$ such that for all $j \in [0,\infty)$ with $2^{-j} \leq a_0$, $\eta \in \mathcal{B}_1$, $\varsigma \in \mathcal{B}_2$, $0 \leq l' \leq l$, $k \in \{1,\dots,q\}$, and $f \in \mathscr{C}_l^s(\mathfrak{M},(X,d))$,*

$$\sup_{x\in\Omega}\sup_{\substack{\eta\in\mathcal{B}_1\\ \varsigma\in\mathcal{B}_2}}\left|\int f(2^{-t\cdot X}x)\operatorname{Diff}_{-2^{-jd_k}\mathfrak{e}_k}^{l'}(\eta(t)\operatorname{Dil}_{2^j}^d(\varsigma)(t))\,dt\right| \leq C2^{-jsl'/l}\|f\|_{\mathscr{C}_l^s(X,d)}.$$

Proof. By Lemma 7.7.16, for $\eta \in \mathcal{B}_1$ and $\varsigma \in \mathcal{B}_2$, we have

$$\eta(t)\operatorname{Dil}_{2^j}^d(\varsigma)(t) = \eta(t)\sum_{\substack{j'\in\mathbb{N}\\ j'\leq j}}\operatorname{Dil}_{2^{j'}}^d(\gamma_{j'})(t),$$

where $\forall M \in \mathbb{N}$,

$$\{2^{M|j-j'|}\gamma_{j'} : j \in [0,\infty), j' \leq j, j' \in \mathbb{N}, \varsigma \in \mathcal{B}_2, \eta \in \mathcal{B}_1\} \subset C_0^\infty(B^q(a_0))$$

is a bounded set.

By taking $M > s$ and applying Lemma 7.7.15, we have, for $x \in \Omega$,

$$\begin{aligned}
&\left|\int f(e^{-t\cdot X}x)\operatorname{Diff}_{-2^{-jd_k}\mathfrak{e}_k}^{l'}(\eta(t)\operatorname{Dil}_{2^j}^d(\varsigma)(t))\,dt\right| \\
&\quad\leq \sum_{\substack{j'\in\mathbb{N}\\ j'\leq j}}\left|\int f(e^{-t\cdot X}x)\operatorname{Diff}_{-2^{-jd_k}\mathfrak{e}_k}^{l'}(\eta(t)\operatorname{Dil}_{2^{j'}}^d(\gamma_{j'})(t))\,dt\right| \\
&\quad\lesssim \sum_{\substack{j'\in\mathbb{N}\\ j'\leq j}} 2^{-M|j-j'|}2^{-j'sl'/l}\|f\|_{\mathscr{C}_l^s(X,d)} \lesssim 2^{-jsl'/l}\|f\|_{\mathscr{C}_l^s(X,d)}.
\end{aligned}$$

□

Lemma 7.7.18. *Fix $l \in \mathbb{N}_+$ and $s \in (0,l)$. If $a_1 = a_1(\widetilde{\Omega}, \Omega, (X,d), l) > 0$ is sufficiently small, the following holds. Let $\mathcal{B}_1 \subset C_0^\infty(B^q(a_1))$ and $\mathcal{B}_2 \subset \mathscr{S}(\mathbb{R}^q)$ be bounded sets. Then there* *exists $C \geq 0$ such that for all $j \in [0,\infty)$, $\eta \in \mathcal{B}_1$, $\varsigma \in \mathcal{B}_2$, $k \in \{1,\ldots,q\}$,*

$$\sup_{x\in\Omega}\sup_{\substack{\eta\in\mathcal{B}_1\\ \varsigma\in\mathcal{B}_2}}\left|\int f(2^{-t\cdot X}x)\eta(t)\,\mathrm{Dil}^d_{2^j}[\mathrm{Diff}^l_{-\mathfrak{e}_k}\varsigma](t)\,dt\right| \leq C2^{-js}\|f\|_{\mathscr{C}^s_l(X,d)}.$$

Proof. We take $a_1 = a_0/2$, where a_0 is as in Lemma 7.7.17.

If $2^{-j} \geq a_1/l$, then $2^{-j} \approx 1$ and we have

$$\begin{aligned}
&\sup_{x\in\Omega}\sup_{\substack{\eta\in\mathcal{B}_1\\ \varsigma\in\mathcal{B}_2}}\left|\int f(2^{-t\cdot X}x)\eta(t)\,\mathrm{Dil}^d_{2^j}[\mathrm{Diff}^l_{-\mathfrak{e}_k}\varsigma](t)\,dt\right|\\
&\quad\lesssim \|f\|_{C(\mathfrak{M})}\sup_{\varsigma\in\mathcal{B}_2}\int\left|\mathrm{Dil}^d_{2^j}[\mathrm{Diff}^l_{-\mathfrak{e}_k}\varsigma](t)\right|dt\\
&\quad= \|f\|_{C(\mathfrak{M})}\sup_{\varsigma\in\mathcal{B}_2}\int\left|\mathrm{Diff}^l_{-\mathfrak{e}_k}\varsigma(t)\right|dt\\
&\quad\lesssim \|f\|_{C(\mathfrak{M})} \leq \|f\|_{\mathscr{C}^s_l(X,d)} \approx 2^{-js}\|f\|_{\mathscr{C}^s_l(X,d)}.
\end{aligned}$$

Thus we may henceforth assume $2^{-j} < a_1/l$.

For $v \in \mathbb{R}^q$ and $F : \mathbb{R}^q \to \mathbb{C}$, we set $\tau_v F(t) := F(t+v)$. We have

$$\mathrm{Diff}_v(FG) = (\tau_v F)(\mathrm{Diff}_v G) + (\mathrm{Diff}_v F)G.$$

In other words,

$$F\,\mathrm{Diff}_v G = \mathrm{Diff}_v((\tau_{-v}F)G) - (\mathrm{Diff}_v(\tau_{-v}F))G.$$

Therefore, for $l \in \mathbb{N}_+$,

$$F\,\mathrm{Diff}^l_v G = \sum_{l'=0}^{l}\binom{l}{l'}(-1)^{l-l'}\,\mathrm{Diff}^{l'}_v((\mathrm{Diff}^{l-l'}_v\,\tau^l_{-v}F)G). \tag{7.74}$$

Using (7.74) with $F = \eta$ and $G = \mathrm{Dil}^d_\varsigma(2^j)$ we have

$$\begin{aligned}
&\int f(e^{-t\cdot X}x)\eta(t)\,\mathrm{Dil}^d_{2^j}[\mathrm{Diff}^l_{-e_k}\varsigma](t)\,dt\\
&\quad= \int f(e^{-t\cdot X}x)\eta(t)\,\mathrm{Diff}^l_{-2^{-jd_k}\mathfrak{e}_k}(\mathrm{Dil}^d_{2^j}(\varsigma)(t))\,dt\\
&\quad= \sum_{l'=0}^{l}\binom{l}{l'}(-1)^{l-l'}\int f(2^{-t\cdot X}x)\,\mathrm{Diff}^{l'}_{-2^{-jd_k}\mathfrak{e}_k}[\eta_{j,l,l'}(t)\,\mathrm{Dil}^d_{2^j}(\varsigma)(t)]\,dt,
\end{aligned} \tag{7.75}$$

where

$$\eta_{j,l,l'} := \mathrm{Diff}^{l-l'}_{-2^{-jd_k}\mathfrak{e}_k}\,\tau^l_{2^{-jd_k}\mathfrak{e}_k}\eta. \tag{7.76}$$

In light of (7.75), it suffices to show that for $0 \le l' \le l$, $x \in \Omega$,

$$\left|\int f(e^{-t\cdot X}x)\,\mathrm{Diff}^{l'}_{-2^{-jd_k}\mathfrak{e}_k}[\eta_{j,l,l'}(t)\,\mathrm{Dil}^{d}_{2^j}(\varsigma)(t)]\,dt\right| \lesssim 2^{-js}\|f\|_{\mathscr{C}^s_l(X,d)}. \tag{7.77}$$

We claim that

$$\{2^{(l-l')jd_k}\eta_{j,l,l'} : 2^{-j} \le a_1/l, \eta \in \mathcal{B}_1, 0 \le l' \le l\} \subset C_0^\infty(B^q(a_0)) \tag{7.78}$$

is a bounded set. It is clear from (7.76) (using $2^{-j} < a_1/l$) that $\mathrm{supp}(\eta_{j,l,l'}) \subseteq \mathrm{supp}(\eta) + B^q(a_1)$. Since $a_1 = a_0/2$ and $\eta \in \mathcal{B}_1 \subset C_0^\infty(B^q(a_1))$ is a bounded set, this shows

$$\bigcup_{\eta\in\mathcal{B}_1}\bigcup_{\substack{2^{-j}<a_1/l\\0\le l'\le l}} \mathrm{supp}(\eta_{j,l,l'}) \Subset B^q(a_0). \tag{7.79}$$

From here, the case $l' = l$ of (7.78) follows immediately from the fact that $\mathcal{B}_1 \subset C_0^\infty(B^q(a_1))$ is a bounded set. For $0 \le l' < l$, we have

$$\begin{aligned}\eta_{j,l,l'}(t) &= \mathrm{Diff}^{l-l'}_{-2^{-jd_k}\mathfrak{e}_k}\tau^{l}_{2^{-jd_k}\mathfrak{e}_k}\eta(t)\\ &= 2^{-j(l-l')d_k}\int\limits_{[0,1]^{l-l'}}(\partial_{t_k}^{l-l'}\eta)\left(t+\sum_{p=1}^{l-l'}2^{-jd_k}u_p\mathfrak{e}_k\right)du.\end{aligned}$$

It follows that for every multi-index α, we have

$$|\partial_t^\alpha \eta_{j,l,l'}(t)| \lesssim 2^{-j(l-l')d_k}.$$

Combining this with (7.79) proves that (7.78) is a bounded set. Thus, applying Lemma 7.7.17 with η replaced by $2^{(l-l')jd_k}\eta_{j,l,l'}$, we have

$$\begin{aligned}&\left|\int f(e^{-t\cdot X}x)\,\mathrm{Diff}^{l'}_{-2^{-jd_k}\mathfrak{e}_k}[\eta_{j,l,l'}(t)\,\mathrm{Dil}^{d}_{2^j}(\varsigma)(t)]\,dt\right|\\ &\quad\lesssim 2^{-(l-l')jd_k}2^{-jsl'/l}\|f\|_{\mathscr{C}^s_l(X,d)}\\ &\quad\lesssim 2^{-sj(l-l')/l}2^{-jsl'/l}\|f\|_{\mathscr{C}^s_l(X,d)} = 2^{-js}\|f\|_{\mathscr{C}^s_l(X,d)},\end{aligned}$$

where in the final inequality we have used $s/l < 1 \le d_k$. This establishes (7.77) and completes the proof. □

We are prepared to choose the operators D_j with which to define the norm

$$\|f\|_{\mathscr{C}^s(W,\vec{d})} = \sup_{j\in\mathbb{N}} 2^{js}\|D_jf\|_{L^\infty}. \tag{7.80}$$

We follow the construction in Section 7.1; see Remark 7.1.5. Let $a_1 = a_1(\widetilde{\Omega}, \Omega, (X,d), l) > 0$ be a small number, in particular so small that Lemma 7.7.18 holds. Let $\eta \in C_0^\infty(B^q(a_1))$ equal 1 on a neighborhood of 0. Take $\rho \in \mathscr{S}(\mathbb{R}^q)$ with $\hat{\rho}(\xi) \equiv 1$ on a neighborhood of $\xi = 0$ and such that $\operatorname{supp}(\hat{\rho}) \subset B^q(1)$. Define $\varsigma_0, \varsigma_1 \in \mathscr{S}(\mathbb{R}^q)$ by

$$\hat{\varsigma}_0(\xi) := \hat{\rho}(\xi), \quad \hat{\varsigma}_1(\xi) = \hat{\rho}(\xi) - \hat{\rho}(2^d \xi).$$

Note that $\operatorname{supp}(\hat{\varsigma}_1) \subset B^q(1)$ and $\hat{\varsigma}_1(\xi) \equiv 1$ for ξ on a neighborhood of 0. Define

$$D_j f(x) := \begin{cases} \psi(x) \int f(e^{-t\cdot X}x)\eta(t)\varsigma_0(t)\,dt, & j = 0, \\ \psi(x) \int f(e^{-t\cdot X}x)\eta(t)\operatorname{Dil}_{2^j}^d(\varsigma_1)(t)\,dt, & j \geq 1. \end{cases}$$

Proposition 6.11.3 (see also Remark 7.1.5) implies that if $a_1 > 0$ is sufficiently small in the above construction, then the norm $\|f\|_{\mathscr{C}^s(W,\vec{d})}$ can be defined by (7.80).

Lemma 7.7.19. *Fix $l \in \mathbb{N}_+$ and $s \in (0, l)$. With the above choice of D_j,*

$$\|D_j f\|_{L^\infty} \lesssim 2^{-js} \|f\|_{\mathscr{C}_l^s(X,d)}, \quad \forall f \in \mathscr{C}_l^s(X,d). \tag{7.81}$$

Proof. For $j = 0$, we use $\|D_0\|_{L^\infty \to L^\infty} \lesssim 1$ (see Corollary 7.1.4 (i)) to see that

$$\|D_0 f\|_{L^\infty} \lesssim \|f\|_{L^\infty} \leq \|f\|_{\mathscr{C}_l^s(X,d)},$$

establishing (7.81) in the case $j = 0$.

For $j > 0$, we apply Lemma 2.5.11 to write[4]

$$\varsigma_1 = \sum_{k=1}^q \operatorname{Diff}_{-e_k}^l \varsigma_{1,k},$$

where $\varsigma_{1,k} \in \mathscr{S}(\mathbb{R}^q)$. Applying Lemma 7.7.18 and using $\operatorname{supp}(D_j f) \subseteq \operatorname{supp}(\psi) \subseteq \Omega$, we have

$$\begin{aligned} \|D_j f\|_{L^\infty} &= \sup_{x \in \Omega} |D_j f(x)| \\ &\leq \sum_{k=1}^q \sup_{x \in \Omega} \left| \int f(e^{-t\cdot X}x)\eta(t)\operatorname{Dil}_{2^j}^d[\operatorname{Diff}_{-e_k}^l \varsigma_{1,k}](t)\,dt \right| \\ &\lesssim 2^{-js} \|f\|_{\mathscr{C}_l^s(X,d)}, \end{aligned}$$

completing the proof. □

4 Lemma 2.5.11 was stated using $\operatorname{Diff}_{e_k}^l$ but the same proof yields the result with $\operatorname{Diff}_{-e_k}^l$.

Proposition 7.7.20. *Fix $l \in \mathbb{N}_+$ and $s \in (0, l)$. Then*

$$\{f \in \mathscr{C}_l^s(X, d) : \operatorname{supp}(f) \subseteq \mathcal{K}\} \subseteq \mathscr{C}^s(\mathcal{K}, (W, \vec{d})), \tag{7.82}$$

and for $f \in \mathscr{C}_l^s(X, d)$ with $\operatorname{supp}(f) \subseteq \mathcal{K}$, we have

$$\|f\|_{\mathscr{C}^s(W,\vec{d})} \lesssim \|f\|_{\mathscr{C}_l^s(X,d)}. \tag{7.83}$$

Proof. Inequality (7.83) follows directly from Lemma 7.7.19 (see (7.80)). The containment (7.82) follows from inequality (7.83) (see (7.1)). □

Proof of Theorem 7.7.8. This follows by combining Propositions 7.7.11 and 7.7.20. □

7.7.1 Hölder spaces

Because the spaces $\mathscr{C}_1^s(\mathfrak{M}, (X, d))$ involve only first-order differences, they are equal to more standard Hölder spaces, as we will make precise in this section. We can use this observation and Theorem 7.7.8 to connect the Zygmund–Hölder spaces $\mathscr{C}^s(\mathcal{K}, (W, \vec{d}))$ with standard Hölder spaces in some cases.

We begin by defining the standard Hölder spaces. For this, let $(X, d) = \{(X_1, d_1), \ldots, (X_q, d_q)\} \subset C^\infty_{\mathrm{loc}}(\mathfrak{M}; T\mathfrak{M}) \times \mathbb{N}_+$ be any finite collection of smooth vector fields with single-parameter formal degrees.

Definition 7.7.21. For $s \in [0, 1]$ we set

$$\|f\|_{C^{0,s}(X,d)} := \|f\|_{C(\mathfrak{M})} + \sup_{\substack{x,y\in\mathfrak{M}\\ x\neq y}} \rho_{(X,d)}(x,y)^{-s}|f(x) - f(y)|$$

and

$$C^{0,s}(\mathfrak{M}, (X, d)) := \{f \in C(\mathfrak{M}) : \|f\|_{C^{0,r}(X,d)} < \infty\}.$$

Here, $\rho_{(X,d)}$ is the extended metric defined in (1.14).

We will sometimes be interested in the special case where $d_1 = d_2 = \cdots d_q = 1$. We denote this case by

$$(X, 1) = \{(X_1, 1), \ldots, (X_q, 1)\}.$$

In this special case, it is possible to extend Definition 7.7.21 to the spaces $C^{m,s}(\mathfrak{M}, (X, 1))$ for $m \in \mathbb{N}$, as the next definition shows.

Definition 7.7.22. For $m \in \mathbb{N}$ and $s \in [0, 1]$, we set

$$C^{m,s}(\mathfrak{M}, (X, 1)) := \{f \in C(\mathfrak{M}) : X^\alpha f \in C^{0,s}(\mathfrak{M}, (X, 1)), \forall|\alpha| \leq m\}$$

and

$$\|f\|_{C^{m,s}(X,1)} := \sum_{|\alpha|\le m} \|X^\alpha\|_{C^{0,s}(X,1)}.$$

For the remainder of this section, let

$$(W, \vec{d}) = \{(W_1, \vec{d}_1), \ldots, (W_r, \vec{d}_r)\} \subset C^\infty_{\mathrm{loc}}(\mathfrak{M}; T\mathfrak{M}) \times \mathbb{N}_+$$

be Hörmander vector fields with (single-parameter) formal degrees. We let $(W, 1)$ denote the special case where $\vec{d}_1 = \cdots = \vec{d}_r = 1$. The main result of this section is the next theorem.

Theorem 7.7.23. *Fix $s \in (0,1)$. Then:*
(i) *$\{f \in C^{0,s}(\mathfrak{M}, (W, \vec{d})) : \operatorname{supp}(f) \subseteq \mathcal{K}\} = \mathscr{C}^s(\mathcal{K}, (W, \vec{d}))$, with equivalent norms.*
(ii) *For $m \in \mathbb{N}$, $\{f \in C^{m,s}(\mathfrak{M}, (W,1)) : \operatorname{supp}(f) \subseteq \mathcal{K}\} = \mathscr{C}^{m+s}(\mathcal{K}, (W,1))$, with equivalent norms.*

The rest of this section is devoted to the proof of Theorem 7.7.23.

Lemma 7.7.24. *For $s \in (0,1)$, $C^{0,s}(\mathfrak{M}, (X, d)) = \mathscr{C}^s_1(\mathfrak{M}, (X, d))$, with equivalent norms.*

Proof. We will show that

$$\|f\|_{C^{0,s}(X,d)} \approx \|f\|_{\mathscr{C}^s_1(X,d)}, \quad \forall f \in C(\mathfrak{M}),$$

where if one side is infinite, the other side is as well. The result will then follow.

By unraveling the definitions we have

$$\begin{aligned}\sup_{\delta\in(0,\infty)} \sup_{\gamma\in\mathcal{P}^0_\delta(\mathfrak{M},(X,d))} |\triangle^0_\gamma f| &= \sup_{\delta\in(0,\infty)} \sup_{\gamma\in\mathcal{P}^0_\delta(\mathfrak{M},(X,d))} |f(\gamma(0))| \\ &= \sup_{x\in\mathfrak{M}} |f(x)| = \|f\|_{C(\mathfrak{M})}.\end{aligned}$$

Thus, it suffices to show that

$$\sup_{\delta\in(0,\infty)} \sup_{\gamma\in\mathcal{P}^1_\delta(\mathfrak{M},(X,d))} \delta^{-s}|\triangle^1_\gamma f| \approx \sup_{\substack{x,y\in\mathfrak{M}\\ x\ne y}} \rho_{(X,d)}(x,y)^{-s}|f(x) - f(y)|. \tag{7.84}$$

We begin with the $\lesssim$ part of (7.84). Let $\delta \in (0,\infty)$ and $\gamma \in \mathcal{P}^1_\delta(\mathfrak{M}, (X, d))$. If $\gamma(1) = \gamma(0)$, then the left-hand side of (7.84) equals 0 and there is nothing to show. Otherwise, by the definition of $\rho_{(X,d)}$ (see (1.14)), we have $\rho_{(X,d)}(\gamma(0), \gamma(1)) < \sqrt{q}\delta$. Therefore,

$$\delta^{-s}|\triangle^1_\gamma f| = \delta^{-s}|f(\gamma(1)) - f(\gamma(0))| \lesssim \sup_{\substack{x,y\in\mathfrak{M}\\ x\ne y}} \rho_{(X,d)}(x,y)^{-s}|f(x) - f(y)|,$$

taking the supremum over such δ and γ proves the $\lesssim$ part of (7.84).

We turn to the $\gtrsim$ part of (7.84). Let $x, y \in \mathfrak{M}$ with $x \neq y$. If $\rho_{(X,d)}(x,y) = \infty$, then the right-hand side of (7.84) equals 0 and there is nothing to show. Otherwise, take $\delta > 0$ such that $\rho_{(X,d)}(x,y) < \delta$. By the definition of $\rho_{(X,d)}$ (see (1.14)), there is an absolutely continuous path $\gamma : [0,1] \to \mathfrak{M}$ such that $\gamma(0) = x$, $\gamma(1) = y$, and $\gamma'(t) = \sum_{j=1}^{q} a_j(t) X_j(\gamma(t))$, with

$$\left\| \sum_{j=1}^{q} |a_j|^2 \right\|_{L^\infty([0,1])} < 1.$$

This implies $\gamma \in \mathcal{P}_\delta^1(\mathfrak{M}, (X, d))$, so

$$\delta^{-s}|f(x) - f(y)| \le \sup_{\delta \in (0,\infty)} \sup_{\gamma \in \mathcal{P}_\delta^1(\mathfrak{M},(X,d))} \delta^{-s}|\triangle_\gamma^1 f|.$$

Taking the supremum over all δ with $\rho_{(X,d)}(x,y) < \delta$ proves the $\gtrsim$ part of (7.84) (in fact proves $\ge$). □

Remark 7.7.25. In the definition of $\mathcal{P}_\delta^1(\mathfrak{M}, (X, d))$ (Definition 7.7.2), if we replace (7.50) with the slight modification

$$\left\| \sum_{j=1}^{q} |\delta^{-(d_j \vee (l'+1))} \partial_t^{l'} a_j|^2 \right\|_{L^\infty} < 1,$$

then one obtains the same spaces $\mathscr{C}_l^s(\mathfrak{M}, (X, d))$, with an equivalent (though not equal) norm. With this change, one has equality of norms in Lemma 7.7.24, instead of merely equivalence of norms.

Fix a relatively compact open set $\widetilde{\Omega} \Subset \mathfrak{M}$ with $\mathcal{K} \Subset \Omega \Subset \widetilde{\Omega} \Subset \mathfrak{M}$.

Lemma 7.7.26. *Fix $s \in (0,1)$. Suppose* $\mathrm{Gen}((W, \vec{d}))$ *is finitely generated by* (X, d) *on* $\widetilde{\Omega}$ *and* $(W, \vec{d}) \subseteq (X, d)$. *Then*

$$\{f \in C^{0,s}(\mathfrak{M}, (W, \vec{d})) : \mathrm{supp}(f) \subseteq \mathcal{K}\} = \{f \in C^{0,s}(\mathfrak{M}, (X, d)) : \mathrm{supp}(f) \subseteq \mathcal{K}\},$$

with equivalence of norms.

Proof. Let $f \in C(\mathfrak{M})$ with $\mathrm{supp}(f) \subseteq \mathcal{K}$ and let $x, y \in \mathfrak{M}$ with $x \neq y$. We will show that

$$\begin{aligned} &\|f\|_{C(\mathfrak{M})} + \rho_{(W,\vec{d})}(x,y)^{-s}|f(x) - f(y)| \\ &\approx \|f\|_{C(\mathfrak{M})} + \rho_{(X,d)}(x,y)^{-s}|f(x) - f(y)|, \end{aligned} \tag{7.85}$$

with constants independent of f, x, and y. Taking the supremum over such x and y completes the proof.

If $x \notin \mathcal{K}$ and $y \notin \mathcal{K}$, then both sides of (7.85) equal $\|f\|_{C(\mathfrak{M})}$. Thus, it suffices to consider the case where at least one of x or y is in $\mathcal{K}$. Without loss of generality, we assume $x \in \mathcal{K}$.

Set $\mathcal{K}_0 := \overline{\Omega}$ and pick Ω_1 open with $\mathcal{K}_0 \Subset \Omega_1 \Subset \widetilde{\Omega}$. We apply Theorem 3.3.7 with $\mathcal{K}$ replaced by $\mathcal{K}_0$ and this choice of Ω_0 to $(W, \vec{d})$ and (X, d). Let $\xi_3 \in (0,1]$ be as in that theorem.

Suppose $\rho_{(X,d)}(x,y) < \xi_3$. Then, by Theorem 3.3.7 (b) for $\delta \in (0,1]$,

$$\rho_{(X,d)}(x,y) < \xi_3\delta \Rightarrow \rho_{(W,\vec{d})}(x,y) < \delta \Rightarrow \rho_{(X,d)}(x,y) < \delta.$$

We conclude that $\rho_{(X,d)}(x,y) \approx \rho_{(W,\vec{d})}(x,y)$, and (7.85) follows in this case.

Suppose $\rho_{(X,d)}(x,y) \geq \xi_3$; in particular, $\rho_{(X,d)}(x,y) \gtrsim 1$. Since $(W, \vec{d}) \subseteq (X, d)$, we also have $\rho_{(W,\vec{d})}(x,y) \geq \rho_{(X,d)}(x,y) \gtrsim 1$. Thus,

$$\rho_{(W,\vec{d})}(x,y)^{-s}|f(x) - f(y)| \lesssim \|f\|_{C(\mathfrak{M})}, \quad \rho_{(X,d)}(x,y)^{-s}|f(x) - f(y)| \lesssim \|f\|_{C(\mathfrak{M})}.$$

Formula (7.85) follows, completing the proof. □

Proof of Theorem 7.7.23. Item (i) follows by combining Theorem 7.7.8 and Lemmas 7.7.24 and 7.7.26.

We turn to (ii), where $\vec{d}_1 = \cdots = \vec{d}_r = 1$. In this case, for an ordered multi-index α, we have $\deg_{\vec{d}}(\alpha) = |\alpha|$. Let $f \in C_0^\infty(\mathfrak{M})'$ with $\mathrm{supp}(f) \subseteq \mathcal{K}$. Applying (i) and Theorem 6.2.12 (with s replaced by $s+m$, $\mathscr{X}^s = \mathscr{B}^s_{\infty,\infty} = \mathscr{C}^s$, and $\kappa = m$), we see that $f \in \mathscr{C}^{m+s}(\mathcal{K}, (W,1))$ if and only if for all $|\alpha| \leq m$, $W^\alpha f \in \mathscr{C}^s(\mathcal{K}, (W,1)) = C^{0,s}(\mathfrak{M}, (W,1))$ if and only if $f \in C^{m,s}(\mathfrak{M}, (W,1))$. Moreover, in this case,

$$\|f\|_{\mathscr{C}^{m+s}(W,1)} \approx \sum_{|\alpha|\leq m} \|W^\alpha f\|_{\mathscr{C}^s(W,1)} \approx \sum_{|\alpha|\leq m} \|W^\alpha f\|_{C^{0,s}(W,1)} = \|f\|_{C^{m,s}(W,1)},$$

completing the proof. □

8 Linear maximally subelliptic operators

In this chapter, we study linear maximally subelliptic operators with smooth coefficients. Similar to the case of elliptic operators, linear maximally subelliptic operators with constant coefficients are a decisive tool for studying fully nonlinear maximally subelliptic operators and for studying linear maximally subelliptic operators with non-smooth coefficients. See Chapter 9 for more details.

Throughout this chapter, $\mathfrak{M}$ will be a connected C^∞ manifold endowed with a smooth, strictly positive density Vol. Let

$$(W, \vec{d}) = \{(W_1, \vec{d}_1), \ldots, (W_r, \vec{d}_r)\} \subset C^\infty_{\mathrm{loc}}(\mathfrak{M}; T\mathfrak{M}) \times \mathbb{N}_+$$

be Hörmander vector fields with (single-parameter) formal degrees on $\mathfrak{M}$.

Fix $\kappa \in \mathbb{N}_+$ such that $\vec{d}_j$ divides κ for $1 \leq j \leq r$ and set $n_j := \kappa/\vec{d}_j \in \mathbb{N}_+$. Fix $\mathbf{D}_1, \mathbf{D}_2 \in \mathbb{N}_+$. We consider partial differential operators of the form

$$\mathscr{P} := \sum_{\deg_{\vec{d}}(\alpha) \leq \kappa} a_\alpha(x) W^\alpha, \quad a_\alpha \in C^\infty_{\mathrm{loc}}(\mathfrak{M}; \mathbb{M}^{\mathbf{D}_1 \times \mathbf{D}_2}(\mathbb{C})). \tag{8.1}$$

We repeat here the definition of maximal subellipticity (Definition 1.1.7), as it is the central concept of this chapter (and even of this text).

Definition 8.0.1. We say that $\mathscr{P}$ given by (8.1) is maximally subelliptic of degree κ with respect to $(W, \vec{d})$ on $\mathfrak{M}$ if for every relatively compact, open set $\Omega \Subset \mathfrak{M}$, there exists $C_\Omega \geq 0$ such that for all $f \in C^\infty_0(\Omega; \mathbb{C}^{\mathbf{D}_2})$,

$$\sum_{j=1}^{r} \|W_j^{n_j} f\|_{L^2(\mathfrak{M},\mathrm{Vol};\mathbb{C}^{\mathbf{D}_2})} \leq C_\Omega (\|\mathscr{P} f\|_{L^2(\mathfrak{M},\mathrm{Vol};\mathbb{C}^{\mathbf{D}_1})} + \|f\|_{L^2(\mathfrak{M},\mathrm{Vol};\mathbb{C}^{\mathbf{D}_2})}). \tag{8.2}$$

When $\mathfrak{M}$ is clear from the context, we just say $\mathscr{P}$ is maximally subelliptic of degree κ with respect to $(W, \vec{d})$.

Remark 8.0.2. The density Vol does not play a role in Definition 8.0.1: one obtains an equivalent definition with any choice of smooth, strictly positive density.

8.1 The main result

The main result of this chapter is a theorem which includes several equivalent characterizations of maximally subelliptic operators (in terms of regularity properties, heat equations, and parametrices).

Using the L^2 structure on $(\mathfrak{M}, \mathrm{Vol})$, we may take the formal $L^2(\mathfrak{M}, \mathrm{Vol})$ adjoint of W_j:

$$W_j^* = -W_j + \mathrm{Mult}[f_j], \quad f_j \in C^\infty_{\mathrm{loc}}(\mathfrak{M}). \tag{8.3}$$

https://doi.org/10.1515/9783111085647-008

Letting $\mathscr{P}$ be as in (8.1), we can also take its formal adjoint. Using (8.3), we have

$$\mathscr{P}^*\mathscr{P} = \sum_{\deg_{d}(\alpha),\deg_{d}(\beta)\leq\kappa} b_{\alpha,\beta}(x)W^{\alpha}W^{\beta}, \quad b_{\alpha,\beta} \in C^{\infty}_{\mathrm{loc}}(\mathfrak{M}; \mathbb{M}^{\mathbf{D}_2\times\mathbf{D}_2}(\mathbb{C})). \tag{8.4}$$

In light of (8.4), it makes sense to ask whether $\mathscr{P}^*\mathscr{P}$ is maximally subelliptic of degree 2κ with respect to (W, d). We also treat $\mathscr{P}^*\mathscr{P}$ as a densely defined operator on $L^2(\mathfrak{M}, \mathrm{Vol}; \mathbb{C}^{\mathbf{D}_2})$, with dense domain $C_0^{\infty}(\mathfrak{M}; \mathbb{C}^{\mathbf{D}_2})$. Thought of in this way, $\mathscr{P}^*\mathscr{P}$ is a nonnegative, symmetric, densely defined operator. We are now prepared to state the main result of this chapter.

Theorem 8.1.1. *Let $(W, d) = \{(W_1, d_1), \ldots, (W_r, d_r)\} \subset C^{\infty}_{\mathrm{loc}}(\mathfrak{M}; T\mathfrak{M})\times\mathbb{N}_+$ be Hörmander vector fields with formal degrees on $\mathfrak{M}$ and let $\mathscr{P}$ be of the form* (8.1)*. The following are equivalent:*

(i) *$\mathscr{P}$ is maximally subelliptic of degree κ with respect to (W, d) on $\mathfrak{M}$.*

(ii) *$\mathscr{P}_0 := \sum_{\deg_{d}(\alpha)=\kappa} a_{\alpha}(x)W^{\alpha}$ is maximally subelliptic of degree κ with respect to (W, d) on $\mathfrak{M}$. Recall that $\mathscr{P} = \sum_{\deg_{d}(\alpha)\leq\kappa} a_{\alpha}(x)W^{\alpha}$, so that $\mathscr{P}_0$ consists of just those terms of degree precisely κ.*

(iii) *$\forall x_0 \in \mathfrak{M}$, there exists an open neighborhood $U \subseteq \mathfrak{M}$ of x_0 such that $\mathscr{P}$ is maximally subelliptic of degree κ with respect to (W, d) on U.*

(iv) *$\forall x_0 \in \mathfrak{M}$, there exists an open neighborhood $U \subseteq \mathfrak{M}$ of x_0 such that the frozen coefficient operator $\sum_{\deg_{d}(\alpha)\leq\kappa} a_{\alpha}(x_0)W^{\alpha}$ is maximally subelliptic of degree κ with respect to (W, d) on U.*

(v) *For any scale of spaces $\mathscr{X}^s$ of the form*

$$\mathscr{X}^s \in \{\mathscr{B}^s_{p,q} : p, q \in [1,\infty]\}\bigcup\{\mathscr{F}^s_{p,q} : p \in (1,\infty), q \in (1,\infty]\},$$

we have the following. Let $\phi_1, \phi_2, \phi_3 \in C_0^{\infty}(\mathfrak{M})$ with $\phi_1 \prec \phi_2 \prec \phi_3$. Then, $\forall s \in \mathbb{R}$,

$$\begin{aligned}&\phi_3\mathscr{P}u \in \mathscr{X}^s_{\mathrm{cpt}}((W, d); \mathbb{C}^{\mathbf{D}_1}) \Rightarrow \phi_1 u \in \mathscr{X}^{s+\kappa}_{\mathrm{cpt}}((W, d); \mathbb{C}^{\mathbf{D}_2}),\\ &\forall u \in C_0^{\infty}(\mathfrak{M})'.\end{aligned} \tag{8.5}$$

Moreover, for every $N \geq 0$ there exists $C = C(s, N, \mathscr{X}^s, (W, d), \phi_1, \phi_2, \phi_3) \geq 0$ such that

$$\begin{aligned}&\|\phi_1 u\|_{\mathscr{X}^{s+\kappa}((W,d);\mathbb{C}^{\mathbf{D}_2})} \leq C(\|\phi_3\mathscr{P}u\|_{\mathscr{X}^{s}((W,d);\mathbb{C}^{\mathbf{D}_1})} + \|\phi_2 u\|_{\mathscr{X}^{s-N}((W,d);\mathbb{C}^{\mathbf{D}_2})}),\\ &\forall u \in C_0^{\infty}(\mathfrak{M}; \mathbb{C}^{\mathbf{D}_2})' \text{ with } \phi_3\mathscr{P}u \in \mathscr{X}^s_{\mathrm{cpt}}(W, d).\end{aligned} \tag{8.6}$$

(vi) *$\mathscr{P}^*\mathscr{P}$ is maximally subelliptic of degree 2κ with respect to (W, d) on $\mathfrak{M}$ (see* (8.4)*).*

(vii) *There exists $T \in \mathscr{A}^{-\kappa}_{\mathrm{loc}}((W, d); \mathbb{C}^{\mathbf{D}_2}, \mathbb{C}^{\mathbf{D}_1})$ (see Definition* 5.11.3*) such that*

$$T\mathscr{P} \equiv I \mod C^{\infty}_{\mathrm{loc}}(\mathfrak{M}\times\mathfrak{M}; \mathbb{M}^{\mathbf{D}_2\times\mathbf{D}_2}(\mathbb{C})).$$

(viii) *There exists* $S \in \mathscr{A}_{\mathrm{loc}}^{-2\kappa}((W, \vec{d}); \mathbb{C}^{\mathbf{D}_2}, \mathbb{C}^{\mathbf{D}_2})$ *such that*

$$S\mathscr{P}^*\mathscr{P}, \mathscr{P}^*\mathscr{P}S \equiv I \mod C_{\mathrm{loc}}^{\infty}(\mathfrak{M} \times \mathfrak{M}; \mathbb{M}^{\mathbf{D}_2 \times \mathbf{D}_2}(\mathbb{C})).$$

(ix) *For every non-negative self-adjoint extension* $\overline{\mathscr{L}}$ *of* $\mathscr{P}^*\mathscr{P}$ *and all* $t > 0$, *the Schwartz kernel of* $e^{-t\overline{\mathscr{L}}}$ *is a* $C_{\mathrm{loc}}^{\infty}(\mathfrak{M} \times \mathfrak{M}; \mathbb{M}^{\mathbf{D}_2 \times \mathbf{D}_2}(\mathbb{C}))$ *function (denoted by* $e^{-t\overline{\mathscr{L}}}(x,y)$*), which satisfies the following. For every* $\mathcal{K} \Subset \mathfrak{M}$ *compact, there exists* $c > 0$ *such that for all ordered multi-indices* α, β, *for all* $s \in \mathbb{N}$, *there exists* $C \geq 0$, $\forall t \in (0,1]$, $\forall x, y \in \mathcal{K}$,

$$\left|\partial_t^s W_x^{\alpha} W_y^{\beta} e^{-t\overline{\mathscr{L}}}(x,y)\right| \leq C\left(\rho_{(W,\vec{d})}(x,y) + t^{\frac{1}{2\kappa}}\right)^{-\deg_{\vec{d}}(\alpha) - \deg_{\vec{d}}(\beta) - 2\kappa s}$$
$$\times \exp\left(-c\left(\frac{\rho_{(W,\vec{d})}(x,y)^{2\kappa}}{t}\right)^{\frac{1}{2\kappa-1}}\right)\left(\mathrm{Vol}(B_{(W,\vec{d})}(x, \rho_{(W,\vec{d})}(x,y) + t^{\frac{1}{2\kappa}})) \wedge 1\right)^{-1}.$$

(x) *There exists a non-negative self-adjoint extension* $\overline{\mathscr{L}}$ *of* $\mathscr{P}^*\mathscr{P}$ *satisfying the conclusions of (ix).*

(xi) *For every non-negative self-adjoint extension* $\overline{\mathscr{L}}$ *of* $\mathscr{P}^*\mathscr{P}$,

$$\{(e^{-t\overline{\mathscr{L}}}, t^{1/2\kappa}) : t \in (0,1]\}$$

is a bounded set of locally $(W, \vec{d})$ *pre-elementary operators.*

(xii) *There exists a non-negative self-adjoint extension* $\overline{\mathscr{L}}$ *of* $\mathscr{P}^*\mathscr{P}$ *satisfying the conclusions of (xi).*

This chapter is devoted to proving Theorem 8.1.1 and understanding some of its consequences. The proof of Theorem 8.1.1 is completed in Section 8.5.

Remark 8.1.2. Item (v) is a priori stronger than (i) in several ways. Indeed, fix a relatively compact, open set $\Omega \Subset \mathfrak{M}$ and choose ϕ_1, ϕ_2, ϕ_3 as in (v) with $\phi_1 \equiv 1$ on Ω. Then, applying (v) with $\mathscr{X}^s = \mathscr{F}_{p,2}^s$ for $p \in (1, \infty)$ and $N = s = 0$ and using Corollary 6.2.14 shows that $\forall f \in C_0^{\infty}(\Omega; \mathbb{C}^{\mathbf{D}_2})$,

$$\sum_{\deg_{\vec{d}}(\alpha) \leq \kappa} \|W^{\alpha} f\|_{L^p(\mathfrak{M}, \mathrm{Vol}; \mathbb{C}^{\mathbf{D}_2})} \lesssim \|\mathscr{P} f\|_{L^p(\mathfrak{M}, \mathrm{Vol}; \mathbb{C}^{\mathbf{D}_1})} + \|f\|_{L^p(\mathfrak{M}, \mathrm{Vol}; \mathbb{C}^{\mathbf{D}_2})}. \tag{8.7}$$

Specializing (8.7) to $p = 2$, we obtain an a priori stronger inequality than (8.2). Thus, (v) improves the left-hand side of (8.2), generalizes (8.2) from $p = 2$ to all $p \in (1, \infty)$, extends (8.2) beyond just the L^2 Sobolev spaces $\mathscr{F}_{2,2}^s$ to all Besov and Triebel–Lizorkin spaces, and localizes (8.2) using the cut-off functions ϕ_1, ϕ_2, and ϕ_3.

Remark 8.1.3. The estimate defining maximal subellipticity, (8.2), is in some sense the strongest possible local estimate one can assume. Indeed, due to the form (8.1) of $\mathscr{P}$, it follows immediately that the reverse inequality in (8.7) holds. Thus, if $\mathscr{P}$ is maximally subelliptic of degree κ with respect to $(W, \vec{d})$ on $\mathfrak{M}$, $\Omega \Subset \mathfrak{M}$ is open and relatively compact, and $p \in (1, \infty)$, we have $\forall f \in C_0^\infty(\Omega; \mathbb{C}^{\mathbf{D}_2})$,

$$\sum_{\deg_{\vec{d}}(\alpha) \leq \kappa} \|W^\alpha f\|_{L^p(\mathfrak{M},\mathrm{Vol};\mathbb{C}^{\mathbf{D}_2})} \approx \|\mathscr{P} f\|_{L^p(\mathfrak{M},\mathrm{Vol};\mathbb{C}^{\mathbf{D}_1})} + \|f\|_{L^p(\mathfrak{M},\mathrm{Vol};\mathbb{C}^{\mathbf{D}_2})}.$$

Remark 8.1.4. Recently, another condition equivalent to maximal subellipticity, in terms of the representation theory of certain nilpotent Lie groups, was established by Androulidakis, Moshen, and Yuncken [2]. See Section 8.8.

Remark 8.1.5. Though Theorem 8.1.1 is not a local theorem, the main arguments used to prove it are. Because of this, there is a more general result, with the same proof, for vector bundles over $\mathfrak{M}$, whereas Theorem 8.1.1 is just stated for trivial vector bundles. See Section 8.6 for more details.

8.2 Further regularity properties

Combining Theorem 8.1.1 with the results from Chapters 5 and 6, we can deduce sharp regularity properties of maximally subelliptic operators with respect to the Besov and Triebel–Lizorkin spaces corresponding to other geometries. We separate this into three parts:

- Regularity properties of maximally subelliptic operators with respect to the function spaces corresponding to different lists of Hörmander vector fields with formal degrees. See Section 8.2.1.
- Sharp regularity properties of maximally subelliptic operators with respect to the standard Besov and Triebel–Lizorkin spaces on $\mathfrak{M}$. In particular, we see that maximally subelliptic operators are subelliptic and prove the sharp subelliptic gain. This is a special case of the previous point. See Section 8.2.2.
- Sharp regularity properties of maximally subelliptic operators with respect to multi-parameter function spaces. See Section 8.2.3. Even if one is only interested in single-parameter function spaces, the results on multi-parameter function spaces are essential for understanding the regularity properties of fully nonlinear maximally subelliptic equations in Section 9.1.1 (see the discussion in Section 1.8). They are also useful for obtaining some single-parameter results about linear operators (for example, Corollary 8.2.9).

8.2.1 Single-parameter function spaces

Proposition 8.2.1. *Let $\mathscr{P}$ be given by* (8.1) *and suppose $\mathscr{P}$ is maximally subelliptic of degree κ with respect to $(W, \vec{d})$ on $\mathfrak{M}$. Let $(Z, \vec{d}) = \{(Z_1, \vec{d}_1), \ldots, (Z_\nu, \vec{d}_\nu)\} \subset C^\infty_{\mathrm{loc}}(\mathfrak{M}; T\mathfrak{M}) \times \mathbb{N}_+$ be another set of Hörmander vector fields with single-parameter formal degrees on $\mathfrak{M}$.*

Suppose $(W, \vec{d})$ *and* $(Z, \vec{d})$ *locally weakly approximately commute on* $\mathfrak{M}$. *Fix* $x_0 \in \mathfrak{M}$ *and let* $\lambda := \lambda(x_0, (W, \vec{d}), (Z, \vec{d})) > 0$ *be as in Definition* 6.6.9. *Then there exists a relatively compact, open neighborhood* $\Omega \Subset \mathfrak{M}$ *of* x_0 *such that the following holds. For every family of spaces* $\mathscr{X}^s$ *of the form*

$$\mathscr{X}^s \in \{\mathscr{B}^s_{p,q} : p, q \in [1, \infty]\} \bigcup \{\mathscr{F}^s_{p,q} : p \in (1, \infty), q \in (1, \infty]\} \tag{8.8}$$

and for every $\phi_1, \phi_2, \phi_3 \in C_0^\infty(\Omega)$ *with* $\phi_1 \prec \phi_2 \prec \phi_3$, *we have*

$$\begin{aligned} &\phi_3 \mathscr{P} u \in \mathscr{X}^s_{\mathrm{cpt}}(Z, \vec{d}) \Rightarrow \phi_1 u \in \mathscr{X}^{s+\kappa/\lambda}_{\mathrm{cpt}}(Z, \vec{d}), \\ &\forall u \in C_0^\infty(\mathfrak{M}; \mathbb{C}^{\mathbf{D}_2})'. \end{aligned} \tag{8.9}$$

Moreover, for every $N \geq 0$ *there exists* $C = C(s, N, \mathscr{P}, \mathscr{X}^s, (W, \vec{d}), (Z, \vec{d}), \phi_1, \phi_2, \phi_3) \geq 0$ *such that*

$$\begin{aligned} &\|\phi_1 u\|_{\mathscr{X}^{s+\kappa/\lambda}(Z,\vec{d})} \leq C(\|\phi_3 \mathscr{P} u\|_{\mathscr{X}^s(Z,\vec{d})} + \|\phi_2 u\|_{\mathscr{X}^{s-N}(Z,\vec{d})}), \\ &\forall u \in C_0^\infty(\mathfrak{M}; \mathbb{C}^{\mathbf{D}_2})' \text{ with } \phi_3 \mathscr{P} u \in \mathscr{X}^s_{\mathrm{cpt}}(Z, \vec{d}). \end{aligned} \tag{8.10}$$

Furthermore, if $\lambda(x_0, (W, \vec{d}), (Z, \vec{d}))$ *is sharp (see Definition* 6.7.4*), this result is optimal in the following sense. Suppose* $\mathscr{P}$ *is any partial differential operator of the form* (8.1) *(not necessarily maximally subelliptic) and suppose that there exists a neighborhood* Ω' *of* x_0, *a scale of spaces* $\mathscr{X}^s_0$ *of the form* (8.8), $s \in \mathbb{R}$, $C \geq 0$, *and* $\epsilon \geq 0$ *such that* $\forall u \in C_0^\infty(\Omega'; \mathbb{C}^{\mathbf{D}_2})$ *we have*

$$\|u\|_{\mathscr{X}_0^{s+\kappa/\lambda+\epsilon}(Z,\vec{d})} \leq C(\|\mathscr{P} u\|_{\mathscr{X}_0^s(Z,\vec{d})} + \|u\|_{\mathscr{X}_0^s(Z,\vec{d})}). \tag{8.11}$$

Then $\epsilon = 0$.

Proof. By Theorem 8.1.1 (i) ⇒ (viii), there exists $S \in \mathscr{A}^{-2\kappa}_{\mathrm{loc}}((W, \vec{d}); \mathbb{C}^{\mathbf{D}_2}, \mathbb{C}^{\mathbf{D}_2})$ such that

$$S\mathscr{P}^*\mathscr{P}, \mathscr{P}^*\mathscr{P}S \equiv I \mod C^\infty_{\mathrm{loc}}(\mathfrak{M} \times \mathfrak{M}; \mathbb{M}^{\mathbf{D}_2 \times \mathbf{D}_2}(\mathbb{C})).$$

By Theorem 5.11.17, $S \in \mathscr{D}^{-2\kappa}_{\mathrm{loc}}((W, \vec{d}), (Z, \vec{d}); \mathbb{C}^{\mathbf{D}_2}, \mathbb{C}^{\mathbf{D}_2})$. Set $T := S\mathscr{P}^*$ so that T satisfies

$$T\mathscr{P} \equiv I \mod C^\infty_{\mathrm{loc}}(\mathfrak{M} \times \mathfrak{M}; \mathbb{M}^{\mathbf{D}_2 \times \mathbf{D}_2}(\mathbb{C})). \tag{8.12}$$

By Proposition 5.9.4, $T \in \mathscr{D}^{-\kappa}_{\mathrm{loc}}((W, \vec{d}), (Z, \vec{d}); \mathbb{C}^{\mathbf{D}_2}, \mathbb{C}^{\mathbf{D}_1})$. In particular, since

$$T \in \mathscr{D}^{-\kappa}_{\mathrm{loc}}((W, \vec{d}), (Z, \vec{d}); \mathbb{C}^{\mathbf{D}_2}, \mathbb{C}^{\mathbf{D}_1}) \subseteq \mathscr{A}^{-\kappa}_{\mathrm{loc}}((W, \vec{d}); \mathbb{C}^{\mathbf{D}_2}, \mathbb{C}^{\mathbf{D}_1}),$$

T is pseudo-local, i. e., its Schwartz kernel, $T(x, y)$, is C^∞_{loc} on $x \neq y$ (see, for example, the growth condition in Definition 5.2.2).

Let $\Omega_0 \Subset \mathfrak{M}$ be an open neighborhood of x_0 as guaranteed by Proposition 6.6.21 (with x replaced by x_0 in that proposition). Set $\mathcal{K}_1 := \overline{\Omega_0}$. By Proposition 6.6.21, we have, $\forall \gamma_1, \gamma_2 \in C_0^\infty(\Omega_0)$,

$$\operatorname{Mult}[\gamma_1] T \operatorname{Mult}[\gamma_2] : \mathscr{X}^s(\mathcal{K}_1, (Z, \vec{d})) \to \mathscr{X}^{s+\kappa/\lambda}(\mathcal{K}_1, (Z, \vec{d})) \text{ is bounded.} \tag{8.13}$$

Fix $\Omega \Subset \Omega_0$ an open neighborhood of x_0 and let $\mathcal{K} := \overline{\Omega} \Subset \Omega_0$. This is the Ω for which we prove the result. Let $\Psi_1, \Psi_2, \Psi_3 \in C_0^\infty(\Omega_0)$, with $\Psi_1 \prec \Psi_2 \prec \Psi_3$, and $\Psi_1 \equiv 1$ on $\mathcal{K}$.

Note that $\phi_1 \prec \phi_2 \prec \phi_3 \prec \Psi_1$, and therefore (8.9) and (8.10) are equivalent with u replaced by $\Psi_1 u$. We henceforth make this replacement, so that $\Psi_2 u = u$.

Let R_1, R_2 denote elements of $C_{\mathrm{loc}}^\infty(\mathfrak{M} \times \mathfrak{M}; \mathbb{M}^{\mathbf{D}_2 \times \mathbf{D}_2})$ (which depend on ϕ_3). We have, using the pseudo-locality of T and (8.12),

$$\begin{aligned} \phi_1 u = \phi_1 \Psi_2 u &= \phi_1 T \mathscr{P} \Psi_2 u + \phi_1 R_1 \Psi_2 u = \phi_1 T \phi_3 \mathscr{P} \Psi_2 u + \phi_1 R_2 \Psi_2 u \\ &= \phi_1 T \phi_3 \mathscr{P} u + \phi_1 R_2 \Psi_2 u = (\operatorname{Mult}[\phi_1] T \operatorname{Mult}[\Psi_2]) \phi_3 \mathscr{P} u + \phi_1 R_2 \Psi_2 u. \end{aligned} \tag{8.14}$$

Using (8.14), we can prove (8.9). Indeed, suppose $\phi_3 \mathscr{P} u \in \mathscr{X}_{\mathrm{cpt}}^s(Z, \vec{d})$; since $\operatorname{supp}(\phi_3) \subseteq \mathcal{K}_1$, Remark 6.5.8 shows that $\phi_3 \mathscr{P} u \in \mathscr{X}^s(\mathcal{K}_1, (Z, \vec{d}))$. Therefore, (8.13) implies $(\operatorname{Mult}[\phi_1] T \operatorname{Mult}[\Psi_2]) \phi_3 \mathscr{P} u \in \mathscr{X}^{s+\kappa/\lambda}(\mathcal{K}_1, (Z, \vec{d}))$. Also, since $R_2 \in C_{\mathrm{loc}}^\infty(\mathfrak{M} \times \mathfrak{M}; \mathbb{M}^{\mathbf{D}_2 \times \mathbf{D}_2})$, we have $\phi_1 R_2 \Psi_2 u \in C_{\mathrm{loc}}^\infty(\Omega) \subseteq \mathscr{X}^{s+\kappa/\lambda}(\mathcal{K}_1, (Z, \vec{d}))$, where the $\subseteq$ follows from Proposition 6.5.5. Combining the above with (8.14), we conclude $\phi_1 u \in \mathscr{X}^{s+\kappa/\lambda}(\mathcal{K}_1, (Z, \vec{d})) \subseteq \mathscr{X}_{\mathrm{cpt}}^{s+\kappa/\lambda}(Z, \vec{d})$, as desired.

We turn to (8.10). Let $\psi \in C_0^\infty(\Omega)$ with $\phi_2 \prec \psi \prec \phi_3$. Applying (8.14) with ϕ_3 replaced by ψ we obtain (with a possibly different choice of $R_2 \in C_{\mathrm{loc}}^\infty(\mathfrak{M} \times \mathfrak{M}; \mathbb{M}^{\mathbf{D}_2 \times \mathbf{D}_2})$), for $u \in \mathscr{X}^s(\mathcal{K}, (Z, \vec{d}))$,

$$\begin{aligned} \|\phi_1 u\|_{\mathscr{X}^{s+\kappa/\lambda}(Z,\vec{d})} &\leq \|\operatorname{Mult}[\phi_1] T \operatorname{Mult}[\Psi_2] \psi \mathscr{P} u\|_{\mathscr{X}^{s+\kappa/\lambda}(Z,\vec{d})} + \|\phi_1 R_2 \Psi_2 u\|_{\mathscr{X}^{s+\kappa/\lambda}(Z,\vec{d})} \\ &\lesssim \|\psi \mathscr{P} u\|_{\mathscr{X}^s(Z,\vec{d})} + \|\Psi_2 u\|_{\mathscr{X}^{s-N}(Z,\vec{d})}, \end{aligned} \tag{8.15}$$

where the final estimate used (8.13) for the first term, and for the second term we used the fact that $\operatorname{Mult}[\phi_1] R_2 \operatorname{Mult}[\Psi_2] \in C_0^\infty(\Omega_0 \times \Omega_0; \mathbb{M}^{\mathbf{D}_2 \times \mathbf{D}_2}(\mathbb{C})) \subseteq \mathscr{A}^{-\kappa/\lambda - N}(\Omega_0, (Z, \vec{d}); \mathbb{C}^{\mathbf{D}_2}, \mathbb{C}^{\mathbf{D}_2})$ (Proposition 5.8.11) and Theorem 6.2.10. Replacing u with $\phi_2 u$ in (8.15)[1] shows that

$$\begin{aligned} \|\phi_1 u\|_{\mathscr{X}^{s+\kappa/\lambda}(Z,\vec{d})} &\lesssim \|\psi \mathscr{P} u\|_{\mathscr{X}^s(Z,\vec{d})} + \|\phi_2 u\|_{\mathscr{X}^{s-N}(Z,\vec{d})} \\ &= \|\psi \phi_3 \mathscr{P} u\|_{\mathscr{X}^s(Z,\vec{d})} + \|\phi_2 u\|_{\mathscr{X}^{s-N}(Z,\vec{d})} \lesssim \|\phi_3 \mathscr{P} u\|_{\mathscr{X}^s(Z,\vec{d})} + \|\phi_2 u\|_{\mathscr{X}^{s-N}(Z,\vec{d})}, \end{aligned}$$

where the last estimate used Corollary 6.5.10. This completes the proof of (8.10).

Finally we turn to the sharpness. Suppose $\lambda(x_0, (W, \vec{d}), (Z, \vec{d}))$ is sharp. For contradiction, suppose (8.11) holds for some $\epsilon > 0$ and some choice of $\mathscr{X}_0^s$, $C \geq 0$, and Ω'. By

1 Since $\phi_3 \mathscr{P} u \in \mathscr{X}_{\mathrm{cpt}}^s(Z, \vec{d})$, the first part of the proof with ϕ_1 replaced with ϕ_2 implies $\phi_2 u \in \mathscr{X}_{\mathrm{cpt}}^{s+\kappa/\lambda}(Z, \vec{d}) \subseteq \mathscr{X}_{\mathrm{cpt}}^s(Z, \vec{d})$.

Corollary 6.6.11, we may shrink Ω' so that (W, d) weakly $\lambda = \lambda(x_0, (W, d), (Z, d))$-controls (Z, d) on Ω'. Let $(Y, \vec{d})$ be the set of vector fields with two parameter degrees given by

$$\begin{aligned}(Y, \vec{d}) &:= \{(W_1, (d_1, 0)), \ldots, (W_r, (d_r, 0))\} \bigcup \{(Z_1, (0, d_1)), \ldots, (Z_\nu, (0, d_\nu))\} \\ &\subset C^\infty_{\mathrm{loc}}(\mathfrak{M}; T\mathfrak{M}) \times (\mathbb{N}^2 \setminus \{0\}).\end{aligned}$$

Using Proposition 6.8.1, using our assumption (8.11), and using $\epsilon > 0$, we have, for $u \in C_0^\infty(\Omega'; \mathbb{C}^{\mathbf{D}_2})$,

$$\begin{aligned}&\|u\|_{\mathscr{X}_0^{(\kappa+\lambda\epsilon/4,s)+(-\lambda(\kappa/\lambda+\epsilon/2),\kappa/\lambda+\epsilon/2)+(\lambda\epsilon/8,\epsilon/2)}(Y,\vec{d})} \\ &= \|u\|_{\mathscr{X}_0^{(-\lambda\epsilon/8,s+\kappa/\lambda+\epsilon)}(Y,\vec{d})} \\ &\lesssim \|u\|_{\mathscr{X}_0^{s+\kappa/\lambda+\epsilon}(Z,d)} \lesssim \|\mathscr{P}u\|_{\mathscr{X}_0^{s}(Z,d)} + \|u\|_{\mathscr{X}_0^{s}(Z,d)} \\ &\lesssim \|\mathscr{P}u\|_{\mathscr{X}_0^{(\lambda\epsilon/4,s)}(Y,\vec{d})} + \|u\|_{\mathscr{X}_0^{(\lambda\epsilon/4,s)}} \lesssim \|u\|_{\mathscr{X}_0^{(\lambda\epsilon/4+\kappa,s)}(Y,\vec{d})},\end{aligned} \tag{8.16}$$

where in the last estimate we used Proposition 6.5.9.

Combining (8.16) with Proposition 6.7.7 shows $\epsilon = 0$, a contradiction, completing the proof. □

Proposition 8.2.1 assumed that (W, d) and (Z, d) locally weakly approximately commute, which covers the most important examples. Without this assumption, we can still obtain a weaker result.

Proposition 8.2.2. *Let $\mathscr{P}$ be given by* (8.1) *and suppose $\mathscr{P}$ is maximally subelliptic of degree κ with respect to (W, d) on $\mathfrak{M}$. Let $(Z, d) = \{(Z_1, d_1), \ldots, (Z_\nu, d_\nu)\} \subset C^\infty_{\mathrm{loc}}(\mathfrak{M}; T\mathfrak{M}) \times \mathbb{N}_+$ be another set of Hörmander vector fields with single-parameter formal degrees on $\mathfrak{M}$. Fix $x_0 \in \mathfrak{M}$ and let $\lambda := \lambda(x_0, (W, d), (Z, d)) > 0$ be as in Definition* 6.6.9. *Then there exists a relatively compact, open neighborhood $\Omega \Subset \mathfrak{M}$ of x_0 such that the following holds. For every family of spaces $\mathscr{X}^s$ of the form*

$$\mathscr{X}^s \in \{\mathscr{B}^s_{p,q} : p, q \in [1, \infty]\} \bigcup \{\mathscr{F}^s_{p,q} : p \in (1, \infty), q \in (1, \infty]\}$$

and for every $\phi_1, \phi_2, \phi_3 \in C_0^\infty(\Omega)$ with $\phi_1 \prec \phi_2 \prec \phi_3$, we have, $\forall \epsilon > 0$,

$$\begin{aligned}&\phi_3 \mathscr{P} u \in \mathscr{X}^0_{\mathrm{cpt}}(Z, d) \Rightarrow \phi_1 u \in \mathscr{X}^{-\epsilon+\kappa/\lambda}_{\mathrm{cpt}}(Z, d), \\ &\forall u \in C_0^\infty(\mathfrak{M}; \mathbb{C}^{\mathbf{D}_2})'.\end{aligned} \tag{8.17}$$

Moreover, for every $N \geq 0$ there exists $C = C(\epsilon, N, \mathscr{P}, \mathscr{X}^s, (W, d), (Z, d), \phi_1, \phi_2, \phi_3) \geq 0$ such that

$$\begin{aligned}&\|\phi_1 u\|_{\mathscr{X}^{-\epsilon+\kappa/\lambda}(Z,d)} \leq C(\|\phi_3 \mathscr{P} u\|_{\mathscr{X}^0(Z,d)} + \|\phi_2 u\|_{\mathscr{X}^{-N}(Z,d)}), \\ &\forall u \in C_0^\infty(\mathfrak{M}; \mathbb{C}^{\mathbf{D}_2})' \text{ with } \phi_3 \mathscr{P} u \in \mathscr{X}^0_{\mathrm{cpt}}(Z, d).\end{aligned} \tag{8.18}$$

Proof. Let $\Omega \Subset \mathfrak{M}$ be the relatively compact, open neighborhood of x_0 called Ω_0 in Theorem 6.6.13 (with x_0 playing the role of x in that theorem). Set $\mathcal{K} := \overline{\Omega}$. Fix $\epsilon > 0$ and suppose $u \in C_0^\infty(\mathfrak{M}; \mathbb{C}^{D_2})'$ with $\phi_3 u \in \mathscr{X}^0_{\mathrm{cpt}}(Z, \vec{d})$. Then, since $\operatorname{supp}(\phi_3) \subseteq \mathcal{K}$, Remark 6.5.8 shows that $\phi_3 u \in \mathscr{X}^0(Z, \vec{d})\mathcal{K}$.

The result is stronger the smaller ϵ is, so we may assume $\epsilon \in (0, \kappa/\lambda)$. By Theorem 6.6.13 (B) with $s_0 = \lambda\epsilon/2$ and ϵ replaced by $\epsilon/2$, we have

$$\begin{aligned} &\phi_3 \mathscr{P} u \in \mathscr{X}^{-\lambda\epsilon/2}(\mathcal{K}, (W, \vec{d})), \\ &\|\phi_3 \mathscr{P} u\|_{\mathscr{X}^{-\lambda\epsilon/2}(W,\vec{d})} \lesssim \|\phi_3 \mathscr{P} u\|_{\mathscr{X}^0(Z,\vec{d})}. \end{aligned} \tag{8.19}$$

Applying Theorem 8.1.1 (i) $\Rightarrow$ (v) we see that

$$\begin{aligned} &\phi_1 u \in \mathscr{X}^{\kappa-\lambda\epsilon/2}(\mathcal{K}, (W, \vec{d})), \\ &\|\phi_1 u\|_{\mathscr{X}^{\kappa-\lambda\epsilon/2}(W,\vec{d})} \lesssim \|\phi_3 \mathscr{P} u\|_{\mathscr{X}^{-\lambda\epsilon/2}(W,\vec{d})} + \|\phi_2 u\|_{\mathscr{X}^{-(N+\epsilon)\lambda}(W,\vec{d})}. \end{aligned} \tag{8.20}$$

Using $\kappa - \lambda\epsilon/2 = \lambda(\kappa/\lambda - \epsilon) + \lambda\epsilon/2$, Theorem 6.6.13 (A) with $s_0 = -\epsilon + \kappa/\lambda > 0$ and ϵ replaced by $\lambda\epsilon/2$ shows that

$$\begin{aligned} &\phi_1 u \in \mathscr{X}^{\kappa/\lambda-\epsilon}(\mathcal{K}, (Z, \vec{d})) \subseteq \mathscr{X}^{\kappa/\lambda-\epsilon}_{\mathrm{cpt}}(Z, \vec{d}), \\ &\|\phi_1 u\|_{\mathscr{X}^{\kappa/\lambda-\epsilon}(Z,\vec{d})} \lesssim \|\phi_1 u\|_{\mathscr{X}^{\kappa-\lambda\epsilon/2}(W,\vec{d})}. \end{aligned} \tag{8.21}$$

This establishes (8.17). For (8.18), combining (8.19), (8.20), and (8.21), we have

$$\|\phi_1 u\|_{\mathscr{X}^{\kappa/\lambda-\epsilon}(Z,\vec{d})} \lesssim \|\phi_3 \mathscr{P} u\|_{\mathscr{X}^0(Z,\vec{d})} + \|\phi_2 u\|_{\mathscr{X}^{-(N+\epsilon)\lambda}(W,\vec{d})}. \tag{8.22}$$

Applying Theorem 6.6.13 (B) with $s_0 = (N + \epsilon)\lambda$, we have

$$\|\phi_2 u\|_{\mathscr{X}^{-(N+\epsilon)\lambda}(W,\vec{d})} \lesssim \|\phi_2 u\|_{\mathscr{X}^{-N}(Z,\vec{d})}. \tag{8.23}$$

Combining (8.22) and (8.23) yields (8.18) and completes the proof. □

8.2.2 Standard function spaces

As a simple consequence of Proposition 8.2.1, in this section we will obtain the sharp regularity properties of maximally subelliptic operators with respect to the standard Besov and Triebel–Lizorkin spaces on $\mathfrak{M}$.

For this we recall the spaces, for $\mathcal{K} \Subset \mathfrak{M}$ compact,

$$\mathscr{B}^s_{p,q,\mathrm{std}}(\mathcal{K}), \quad \mathscr{F}^s_{p,q,\mathrm{std}}(\mathcal{K}),$$

defined in Section 6.6.1 (see Definition 6.6.4). These are nothing more than the usual Besov and Triebel–Lizorkin spaces consisting of distributions supported in $\mathcal{K}$. We also recall the positive numbers $\lambda_{\mathrm{std}}(x,(W,\vec{d}))$ and $\Lambda_{\mathrm{std}}(x,(W,\vec{d}))$ from Definition 6.6.15. In the important special case where $\vec{d}_1 = \vec{d}_2 = \cdots = \vec{d}_r = 1$, then $\lambda_{\mathrm{std}}(x,(W,\vec{d})) = m$, where $W_1,\dots,W_r$ satisfy Hörmander's condition of order m at x, and $\Lambda_{\mathrm{std}}(x,(W,\vec{d})) = 1$. See Example 6.6.16. Finally, if $\mathscr{X}^s_{\mathrm{std}}$ is one of $\mathscr{B}^s_{p,q,\mathrm{std}}$ or $\mathscr{F}^s_{p,q,\mathrm{std}}$, recall the definition of $\mathscr{X}^s_{\mathrm{std,cpt}}(\mathfrak{M})$ given in (6.54) – the space of those distributions in $\mathscr{X}^s_{\mathrm{std}}(\mathcal{K})$ for some $\mathcal{K} \Subset \mathfrak{M}$.

Corollary 8.2.3. *Let $\mathscr{P}$ be given by (8.1) and suppose $\mathscr{P}$ is maximally subelliptic of degree κ with respect to $(W,\vec{d})$ on $\mathfrak{M}$. Fix $x_0 \in \mathfrak{M}$. There exists a relatively compact, open neighborhood $\Omega \Subset \mathfrak{M}$ of x_0 such that the following holds with $\lambda := \lambda_{\mathrm{std}}(x_0,(W,\vec{d}))$. For every family of spaces $\mathscr{X}^s_{\mathrm{std}}$ of the form*

$$\mathscr{X}^s_{\mathrm{std}} \in \{\mathscr{B}^s_{p,q,\mathrm{std}} : p,q \in [1,\infty]\} \bigcup \{\mathscr{F}^s_{p,q,\mathrm{std}} : p \in (1,\infty), q \in (1,\infty]\} \tag{8.24}$$

and for every $\phi_1,\phi_2,\phi_3 \in C_0^\infty(\Omega)$ with $\phi_1 \prec \phi_2 \prec \phi_3$, we have

$$\phi_3 \mathscr{P} u \in \mathscr{X}^s_{\mathrm{std,cpt}}(\mathfrak{M}) \Rightarrow \phi_1 u \in \mathscr{X}^{s+\kappa/\lambda}_{\mathrm{std,cpt}}(\mathfrak{M}),$$
$$\forall u \in C_0^\infty(\mathfrak{M};\mathbb{C}^{\mathrm{D}_2})'.$$

Moreover, for every $N \geq 0$ there exists $C = C(s,N,\mathscr{P},\mathscr{X}^s_{\mathrm{std}},(W,\vec{d}),\phi_1,\phi_2,\phi_3) \geq 0$ such that

$$\|\phi_1 u\|_{\mathscr{X}^{s+\kappa/\lambda}_{\mathrm{std}}} \leq C\big(\|\phi_3 \mathscr{P} u\|_{\mathscr{X}^s_{\mathrm{std}}} + \|\phi_2 u\|_{\mathscr{X}^{s-N}_{\mathrm{std}}}\big),$$
$$\forall u \in C_0^\infty(\mathfrak{M};\mathbb{C}^{\mathrm{D}_2})' \text{ with } \phi_3 \mathscr{P} u \in \mathscr{X}^s_{\mathrm{std,cpt}}(\mathfrak{M}).$$

Furthermore, this result is optimal in the following sense. Suppose $\mathscr{P}$ is any partial differential operator of the form (8.1) (not necessarily maximally subelliptic) and suppose that there exist a neighborhood Ω' of x_0, a scale of spaces $\mathscr{X}^s_{0,\mathrm{std}}$ of the form (8.24), $s \in \mathbb{R}$, $C \geq 0$, and $\epsilon \geq 0$ such that $\forall u \in C_0^\infty(\Omega';\mathbb{C}^{\mathrm{D}_2})$ we have

$$\|u\|_{\mathscr{X}^{s+\kappa/\lambda+\epsilon}_{0,\mathrm{std}}} \leq C\big(\|\mathscr{P} u\|_{\mathscr{X}^s_{0,\mathrm{std}}} + \|u\|_{\mathscr{X}^s_{0,\mathrm{std}}}\big).$$

Then $\epsilon = 0$.

To prove Corollary 8.2.3 (and Corollary 8.2.8 below), we use the next lemma.

Lemma 8.2.4. *Fix $x_0 \in \mathfrak{M}$. There exist an open neighborhood $\mathfrak{N} \subseteq \mathfrak{M}$ of x_0 and vector fields $Z_1,\dots,Z_n \in C^\infty_{\mathrm{loc}}(\mathfrak{N};T\mathfrak{N})$ such that if $(Z,1) = \{(Z_1,1),\dots,(Z_n,1)\}$, we have:*

(i) $\mathrm{span}\{Z_1(y),\dots,Z_n(y)\} = T_y\mathfrak{N}$, $\forall y \in \mathfrak{N}$.
(ii) $\lambda_{\mathrm{std}}(x_0,(W,\vec{d})) = \lambda(x_0,(W,\vec{d}),(Z,1))$ *and* $\Lambda_{\mathrm{std}}(x_0,(W,\vec{d})) = \lambda(x_0,(Z,1),(W,\vec{d}))$.
(iii) $\lambda(x_0,(W,\vec{d}),(Z,1))$ *and* $\lambda(x_0,(Z,1),(W,\vec{d}))$ *are sharp.*
(iv) $(W,\vec{d})$ *and* $(Z,1)$ *locally weakly approximately commute on* $\mathfrak{N}$.
(v) *For every compact set $\mathcal{K} \Subset \mathfrak{N}$ and every space $\mathscr{X}^s_{\mathrm{std}}$ of the form (8.24), we have*

$$\mathscr{X}^s_{\mathrm{std}}(\mathcal{K}) = \mathscr{X}^s(\mathcal{K},(Z,1)),$$

with equivalent norms. In particular, $\mathscr{X}^s_{\mathrm{std,cpt}}(\mathfrak{N}) = \mathscr{X}^s_{\mathrm{cpt}}(\mathfrak{N},(Z,1))$.

Proof. Pick $Z_1, \dots, Z_n \in C^\infty_{\mathrm{loc}}(\mathfrak{M}; T\mathfrak{M})$ such that $Z_1(x_0), \dots, Z_n(x_0)$ form a basis for $T_{x_0}\mathfrak{M}$. By continuity, there is a neighborhood $\mathfrak{N} \subseteq \mathfrak{M}$ of x_0 such that (i) holds. Lemma 6.6.17 with $\mathfrak{M}$ replaced by $\mathfrak{N}$ implies (ii). Item (iii) follows from the definitions (see also Example 6.7.6). Lemma 3.8.3 implies that $(W, \vec{d})$ and $(Z, 1)$ locally strongly approximately commute on $\mathfrak{N}$, and therefore Proposition 3.8.6 (b) implies they locally weakly approximately commute on $\mathfrak{N}$; item (iv) follows. Item (v) is an immediate consequence of Theorem 6.6.7. □

Proof of Corollary 8.2.3. Let $\mathfrak{N}$ and $(Z, 1)$ be as in Lemma 8.2.4. In light of Lemma 8.2.4, the result is an immediate consequence of Proposition 8.2.1 with $(Z, \vec{d})$ replaced by $(Z, 1)$. □

Corollary 8.2.3 shows that maximally subelliptic operators are subelliptic and gives the maximal possible subelliptic gain. Indeed, letting L^p_s denote the standard L^p Sobolev space of order $s \in \mathbb{R}$, we have the following.

Corollary 8.2.5. *Let $\mathscr{P}$ be given by* (8.1) *and suppose $\mathscr{P}$ is maximally subelliptic of degree κ with respect to $(W, \vec{d})$ on $\mathfrak{M}$. Fix $x_0 \in \mathfrak{M}$. There exists a relatively compact, open neighborhood $\Omega \Subset \mathfrak{M}$ of x_0 such that the following holds with $\lambda := \lambda_{\mathrm{std}}(x_0, (W, \vec{d}))$. For every $p \in (1, \infty)$, $s \in \mathbb{R}$, and $\phi_1, \phi_2, \phi_3 \in C^\infty_0(\Omega)$ with $\phi_1 \prec \phi_2 \prec \phi_3$, we have*

$$\phi_3 \mathscr{P} u \in L^p_s \Rightarrow \phi_1 u \in L^p_{s+\kappa/\lambda}, \quad \forall u \in C^\infty_0(\mathfrak{M}; \mathbb{C}^{\mathbf{D}_2})'.$$

Moreover, for every $N \geq 0$, there exists $C = C(p, s, N, \mathscr{P}, (W, \vec{d}), \phi_1, \phi_2, \phi_3) \geq 0$ such that

$$\|\phi_1 u\|_{L^p_{s+\kappa/\lambda}} \leq C(\|\phi_3 \mathscr{P} u\|_{L^p_s} + \|\phi_2 u\|_{L^p_{s-N}}), \quad \forall u \in C^\infty_0(\mathfrak{M}; \mathbb{C}^{\mathbf{D}_2})',$$

where if the right-hand side is finite, so is the left-hand side. Furthermore, this result is optimal in the following sense. Suppose $\mathscr{P}$ is any partial differential operator of the form (8.1) *(not necessarily maximally subelliptic) and suppose that there exist a neighborhood Ω' of x_0, $p \in (1, \infty)$, $s \in \mathbb{R}$, $C \geq 0$, and $\epsilon \geq 0$ such that $\forall u \in C^\infty_0(\Omega'; \mathbb{C}^{\mathbf{D}_2})$ we have*

$$\|u\|_{L^p_{s+\kappa/\lambda+\epsilon}} \leq C(\|\mathscr{P} u\|_{L^p_s} + \|u\|_{L^p_s}).$$

Then $\epsilon = 0$.

Proof. This is the special case of Corollary 8.2.3 with $\mathscr{X}^s_{\mathrm{std}} = \mathscr{F}^s_{p,2,\mathrm{std}}$, since when restricted to distributions with fixed compact support, the spaces $\mathscr{F}^s_{p,2,\mathrm{std}}$ and L^p_s are equal and have equivalent norms. □

Remark 8.2.6. Corollary 8.2.5 shows that Corollary 8.2.3 implies maximally subelliptic operators are subelliptic and gives the sharp subelliptic gain. Corollary 8.2.3 does more than this: it gives the sharp gain on Besov and Triebel–Lizorkin spaces other than just the standard Sobolev spaces. For example, Corollary 8.2.3 gives the sharp gain on the classical Zygmund–Hölder spaces $\mathscr{B}^s_{\infty,\infty,\mathrm{std}}$.

Remark 8.2.7. If $\mathscr{P}$ is given by (8.1), then it follows from Corollary 8.2.5 that the following are equivalent:

- $\mathscr{P}$ is elliptic of degree κ near x_0.
- $\mathscr{P}$ is maximally subelliptic of degree κ with respect to $(W, \vec{d})$ and $\operatorname{span}\{W_j(x_0) : \vec{d}_j = 1\} = T_{x_0}\mathfrak{M}$.

In particular, even in the special case where $\vec{d}_1 = \vec{d}_2 = \cdots = \vec{d}_r = 1$, the only way an operator of the form (8.1) can be elliptic of degree κ is if $W_1, \dots, W_r$ satisfy Hörmander's condition of order 1. Moreover, Corollary 8.2.5 shows that the higher the order of Hörmander's condition, the more degenerate the operator will be.

Corollary 8.2.8. *Let $\widehat{\mathscr{P}}$ be a partial differential operator of order κ on $\mathfrak{M}$, with smooth coefficients. Fix $x_0 \in \mathfrak{M}$ and suppose $\widehat{\mathscr{P}}$ is of order κ elliptic at x_0. Let $\Lambda := \Lambda_{\mathrm{std}}(x_0, (W, \vec{d}))$. Then there exists a relatively compact, open neighborhood $\Omega \Subset \mathfrak{M}$ of x_0 such that the following holds. For every family of spaces $\mathscr{X}^s$ of the form*

$$\mathscr{X}^s \in \{\mathscr{B}^s_{p,q} : p, q \in [1, \infty]\} \bigcup \{\mathscr{F}^s_{p,q} : p \in (1, \infty), q \in (1, \infty]\} \tag{8.25}$$

and for every $\phi_1, \phi_2, \phi_3 \in C_0^\infty(\Omega)$ with $\phi_1 \prec \phi_2 \prec \phi_3$, we have

$$\phi_3 \widehat{\mathscr{P}} u \in \mathscr{X}^s_{\mathrm{cpt}}(W, \vec{d}) \Rightarrow \phi_1 u \in \mathscr{X}^{s+\kappa/\Lambda}_{\mathrm{cpt}}(W, \vec{d}),$$
$$\forall u \in C_0^\infty(\mathfrak{M}; \mathbb{C}^{\mathbf{D}_2})'.$$

Moreover, for every $N \geq 0$ there exists $C = C(s, N, \widehat{\mathscr{P}}, \mathscr{X}^s, (W, \vec{d}), \phi_1, \phi_2, \phi_3) \geq 0$ such that

$$\|\phi_1 u\|_{\mathscr{X}^{s+\kappa/\Lambda}(W,\vec{d})} \leq C(\|\phi_3 \widehat{\mathscr{P}} u\|_{\mathscr{X}^s(W,\vec{d})} + \|\phi_2 u\|_{\mathscr{X}^{s-N}(W,\vec{d})}),$$
$$\forall u \in C_0^\infty(\mathfrak{M}; \mathbb{C}^{\mathbf{D}_2})' \text{ with } \phi_3 \widehat{\mathscr{P}} u \in \mathscr{X}^s_{\mathrm{cpt}}(W, \vec{d}).$$

Furthermore, this result is optimal in the following sense. Let $\widehat{\mathscr{P}}$ be any partial differential operator of order κ on $\mathfrak{M}$, with smooth coefficients (not necessarily elliptic). Suppose that there exist a neighborhood Ω' of x_0, a scale of spaces $\mathscr{X}^s_0$ of the form (8.25), $s \in \mathbb{R}$, $C \geq 0$, and $\epsilon \geq 0$ such that $\forall u \in C_0^\infty(\Omega'; \mathbb{C}^{\mathbf{D}_2})$ we have

$$\|u\|_{\mathscr{X}_0^{s+\kappa/\Lambda+\epsilon}(W,\vec{d})} \leq C(\|\widehat{\mathscr{P}} u\|_{\mathscr{X}_0^s(W,\vec{d})} + \|u\|_{\mathscr{X}_0^s(W,\vec{d})}).$$

Then $\epsilon = 0$.

Proof. Let $\mathfrak{N}$ and $(Z, 1)$ be as in Lemma 8.2.4. By possibly shrinking $\mathfrak{N}$, we may assume $\widehat{\mathscr{P}}$ is elliptic on $\mathfrak{N}$. In particular, this implies that $\widehat{\mathscr{P}}$ is maximally subelliptic of degree κ with respect to $(Z, 1)$ on $\mathfrak{N}$. By Remark 6.4.8, it suffices to prove the result with $\mathfrak{M}$ replaced by $\mathfrak{N}$. Lemma 8.2.4 shows that Proposition 8.2.1 applies with $\mathscr{P}$ replaced by $\widehat{\mathscr{P}}$, $(W, \vec{d})$ replaced by $(Z, 1)$, $(Z, \vec{d})$ replaced by $(W, \vec{d})$, $\mathfrak{M}$ replaced by $\mathfrak{N}$, and λ replaced by Λ. The conclusions of Proposition 8.2.1 with these replacements are exactly the conclusions of the corollary. □

As described in those results, Corollaries 8.2.3 and 8.2.5 are sharp. However, we can give a more precise version of Corollary 8.2.5, and if we allow a loss of ϵ derivatives, a more precise version of Corollary 8.2.3 can also be given. This is contained in the next result.

Corollary 8.2.9. *Let $\mathscr{P}$ be given by (8.1) and suppose $\mathscr{P}$ is maximally subelliptic of degree κ with respect to $(W, \vec{d})$ on $\mathfrak{M}$. Fix $\phi_1, \phi_2, \phi_3 \in C_0^\infty(\mathfrak{M})$ with $\phi_1 \prec \phi_2 \prec \phi_3$. Then, $\forall s \in \mathbb{R}$, we have the following:*

(i) *For every $p \in (1,\infty)$, we have*

$$\phi_3 \mathscr{P} u \in L_s^p \Rightarrow \phi_1 W^\alpha u \in L_s^p,$$
$$\forall u \in C_0^\infty(\mathfrak{M})', \quad \forall \deg_{\vec{d}}(\alpha) \leq \kappa.$$

Moreover, for every $N \geq 0$ there exists $C = C(s, N, p, \mathscr{P}, (W, \vec{d}), \phi_1, \phi_2, \phi_3) \geq 0$ with

$$\sum_{\deg_{\vec{d}}(\alpha) \leq \kappa} \|\phi_1 W^\alpha u\|_{L_s^p} \leq C(\|\phi_3 \mathscr{P} u\|_{L_s^p} + \|\phi_2 u\|_{L_{s-N}^p}), \quad \forall u \in C_0^\infty(\mathfrak{M})',$$

where if the right-hand side is finite, so is the left-hand side.

(ii) *Let $\mathscr{X}_{\mathrm{std}}^s$ be any space of the form (8.24). Then, $\forall \epsilon > 0$,*

$$\phi_3 \mathscr{P} u \in \mathscr{X}_{\mathrm{std,cpt}}^s \Rightarrow \phi_1 W^\alpha u \in \mathscr{X}_{\mathrm{std,cpt}}^{s-\epsilon},$$
$$\forall u \in C_0^\infty(\mathfrak{M})', \quad \forall \deg_{\vec{d}}(\alpha) \leq \kappa.$$

Moreover, for every $N \geq 0$ there exists $C = C(s, N, \epsilon, \mathscr{X}_{\mathrm{std}}^s, \mathscr{P}, (W, \vec{d}), \phi_1, \phi_2, \phi_3) \geq 0$ with

$$\sum_{\deg_{\vec{d}}(\alpha) \leq \kappa} \|\phi_1 W^\alpha u\|_{\mathscr{X}_{\mathrm{std}}^s} \leq C(\|\phi_3 \mathscr{P} u\|_{\mathscr{X}_{\mathrm{std}}^{s+\epsilon}} + \|\phi_2 u\|_{\mathscr{X}_{\mathrm{std}}^{s-N}}), \quad \forall u \in C_0^\infty(\mathfrak{M})',$$

where if the right-hand side is finite, so is the left-hand side.

We defer the proof of Corollary 8.2.9 to Section 8.2.3, where we see it as a consequence of more general multi-parameter results.

Remark 8.2.10. It would be interesting to establish Corollary 8.2.9 (ii) with $\epsilon = 0$. It seems likely that this is possible: in the special case where $\mathscr{X}_{\mathrm{std}}^s = \mathscr{F}_{p,2,\mathrm{std}}^s$, this is exactly Corollary 8.2.9 (i). Also, Corollary 8.2.3 makes it seem likely that Corollary 8.2.9 (ii) holds with $\epsilon = 0$.

8.2.3 Multi-parameter function spaces

Essential to our study of the regularity properties of fully nonlinear maximally subelliptic equations (with respect to, for example, the standard Zygmund–Hölder spaces) are the regularity properties of linear maximally subelliptic operators with respect to multi-parameter Besov and Triebel–Lizorkin spaces. We present the relevant multi-parameter results in this section.

For each $\mu \in \{1,\dots,\nu\}$, let $(W^\mu, \vec{d}^\mu) = \{(W_1^\mu, \vec{d}_1^\mu),\dots,(W_{r_\mu}^\mu, \vec{d}_{r_\mu}^\mu)\} \subset C_{\mathrm{loc}}^\infty(\mathfrak{M}; T\mathfrak{M}) \times \mathbb{N}_+$ be C_{loc}^∞ vector fields with formal degrees. We assume that:

- $(W^1, \vec{d}^1)$ are Hörmander vector fields with formal degrees on $\mathfrak{M}$.
- For $2 \le \mu \le \nu$, $\mathrm{Gen}((W^\mu, \vec{d}^\mu))$ is locally finitely generated on $\mathfrak{M}$.
- $(W^1, \vec{d}^1),\dots,(W^\nu, \vec{d}^\nu)$ pairwise locally weakly approximately commute on $\mathfrak{M}$.

Remark 8.2.11. The above assumptions are exactly the assumptions of the main multi-parameter setting described in Section 3.15.

Define $(W, \vec{d}) := (W^1, \vec{d}^1) \boxtimes (W^2, \vec{d}^2) \boxtimes \cdots \boxtimes (W^\nu, \vec{d}^\nu)$. Since $(W^1, \vec{d}^1)$ are Hörmander vector fields with formal degrees, Proposition 3.4.14 (in the case $\nu = 1$) implies $\mathrm{Gen}((W^1, \vec{d}^1))$ is locally finitely generated on $\mathfrak{M}$. In particular, $(W, \vec{d})$ satisfies all the hypotheses of Section 5.1.3, and it therefore makes sense to talk about singular integrals like those in $\widetilde{\mathscr{A}}^s_{\mathrm{loc}}(W, \vec{d})$ (for $s \in \mathbb{R}^\nu$) and the ν-parameter Besov and Triebel–Lizorkin spaces with respect to $(W, \vec{d})$ as defined in Section 6.3.

Fix $\kappa \in \mathbb{N}_+$ such that $\vec{d}_j^1$ divides κ for $1 \le j \le r_1$. We consider partial differential operators of the form

$$\mathscr{P} := \sum_{\deg_{\vec{d}^1}(\alpha) \le \kappa} a_\alpha(x)(W^1)^\alpha, \quad a_\alpha \in C_{\mathrm{loc}}^\infty(\mathfrak{M}; \mathbb{M}^{\mathbf{D}_1 \times \mathbf{D}_2}(\mathbb{C})). \tag{8.26}$$

Proposition 8.2.12. *Let $\mathscr{P}$ be given by* (8.26) *and suppose $\mathscr{P}$ is maximally subelliptic of degree κ with respect to $(W^1, \vec{d}^1)$ on $\mathfrak{M}$. Then for any scale of spaces $\mathscr{X}^s$, $s \in \mathbb{R}^\nu$, of the form*

$$\mathscr{X}^s \in \{\mathscr{B}^s_{p,q} : p, q \in [1,\infty]\} \bigcup \{\mathscr{F}^s_{p,q} : p \in (1,\infty), q \in (1,\infty]\},$$

the following holds. Let $\phi_1, \phi_2, \phi_3 \in C_0^\infty(\mathfrak{M})$ with $\phi_1 \prec \phi_2 \prec \phi_3$. Then, $\forall s \in \mathbb{R}^\nu$,

$$\begin{aligned} &\phi_3 \mathscr{P} u \in \mathscr{X}^s_{\mathrm{cpt}}(W, \vec{d}) \Rightarrow \phi_1 u \in \mathscr{X}^{s+\kappa e_1}_{\mathrm{cpt}}(W, \vec{d}), \\ &\forall u \in C_0^\infty(\mathfrak{M}; \mathbb{C}^{\mathbf{D}_2})'. \end{aligned} \tag{8.27}$$

Moreover, $\forall \vec{N} \in [0,\infty)^\nu$, there exists $C = C(s, \vec{N}, \mathscr{P}, \mathscr{X}^s, (W, \vec{d}), \phi_1, \phi_2, \phi_3) \ge 0$ such that

$$\begin{aligned} &\|\phi_1 u\|_{\mathscr{X}^{s+\kappa e_1}(W,\vec{d})} \le C(\|\phi_3 \mathscr{P} u\|_{\mathscr{X}^s(W,\vec{d})} + \|\phi_2 u\|_{\mathscr{X}^s(W,\vec{d})}), \\ &\forall u \in C_0^\infty(\mathfrak{M}; \mathbb{C}^{\mathbf{D}_2})' \text{ with } \phi_3 \mathscr{P} u \in \mathscr{X}^s_{\mathrm{cpt}}(W, \vec{d}). \end{aligned} \tag{8.28}$$

Proof. By Theorem 8.1.1 (i) ⇒ (viii), there exists $S \in \mathscr{A}_{\mathrm{loc}}^{-2\kappa}((W^1, \vec{d}^1); \mathbb{C}^{\mathbf{D}_2}, \mathbb{C}^{\mathbf{D}_2})$ such that

$$S\mathscr{P}^*\mathscr{P}, \mathscr{P}^*\mathscr{P}S \equiv I \mod C_{\mathrm{loc}}^{\infty}(\mathfrak{M} \times \mathfrak{M}; \mathbb{M}^{\mathbf{D}_2 \times \mathbf{D}_2}(\mathbb{C})).$$

By Corollary 5.11.18, we have $S \in \widetilde{\mathscr{A}}_{\mathrm{loc}}^{-2\kappa e_1}((W, \vec{d}); \mathbb{C}^{\mathbf{D}_2}, \mathbb{C}^{\mathbf{D}_2})$. Set $T := S\mathscr{P}^*$ so that T satisfies

$$T\mathscr{P} \equiv I \mod C_{\mathrm{loc}}^{\infty}(\mathfrak{M} \times \mathfrak{M}; \mathbb{M}^{\mathbf{D}_2 \times \mathbf{D}_2}(\mathbb{C})). \tag{8.29}$$

It follows from Corollary 5.8.10 that $T \in \mathscr{A}_{\mathrm{loc}}^{-\kappa}((W^1, \vec{d}^1); \mathbb{C}^{\mathbf{D}_2}, \mathbb{C}^{\mathbf{D}_1}) \bigcap \widetilde{\mathscr{A}}_{\mathrm{loc}}^{-\kappa e_1}((W, \vec{d}); \mathbb{C}^{\mathbf{D}_2}, \mathbb{C}^{\mathbf{D}_1})$. In particular, since $T \in \mathscr{A}_{\mathrm{loc}}^{-\kappa}((W^1, \vec{d}^1); \mathbb{C}^{\mathbf{D}_2}, \mathbb{C}^{\mathbf{D}_1})$, T is pseudo-local, i. e., its Schwartz kernel, $T(x, y)$, is $C_{\mathrm{loc}}^{\infty}$ on $x \neq y$ (see, for example, the growth condition in Definition 5.2.2).

Set $\mathcal{K} := \mathrm{supp}(\phi_3)$. Let $\Psi_1, \Psi_2, \Psi_3 \in C_0^{\infty}(\mathfrak{M})$, with $\Psi_1 \prec \Psi_2 \prec \Psi_3$, and $\Psi_1 \equiv 1$ on $\mathcal{K}$. Take $\Omega_0 \Subset \mathfrak{M}$ open and relatively compact with $\Psi_3 \in C_0^{\infty}(\Omega_0)$ and set $\mathcal{K}_1 := \overline{\Omega_0}$. Note that $\phi_1 \prec \phi_2 \prec \phi_3 \prec \Psi_1$, and therefore (8.27) and (8.28) are equivalent with u replaced by $\Psi_1 u$. We henceforth make this replacement, so that $\Psi_2 u = u$.

Let R_1, R_2 denote elements of $C_{\mathrm{loc}}^{\infty}(\mathfrak{M} \times \mathfrak{M}; \mathbb{M}^{\mathbf{D}_2 \times \mathbf{D}_2})$ (which depend on ϕ_3). We have, using the pseudo-locality of T and (8.29),

$$\begin{aligned} \phi_1 u &= \phi_1 \Psi_2 u = \phi_1 T\mathscr{P}\Psi_2 u + \phi_1 R_1 \Psi_2 u = \phi_1 T \phi_3 \mathscr{P}\Psi_2 u + \phi_1 R_2 \Psi_2 u \\ &= \phi_1 T \phi_3 \mathscr{P} u + \phi_1 R_2 \Psi_2 u = (\mathrm{Mult}[\phi_1] T \,\mathrm{Mult}[\Psi_2]) \phi_3 \mathscr{P} u + \phi_1 R_2 \Psi_2 u. \end{aligned} \tag{8.30}$$

Using (8.30), we can prove (8.27). Indeed, suppose $\phi_3 \mathscr{P} u \in \mathscr{X}_{\mathrm{cpt}}^{s}(W, \vec{d})$. Then, since $\mathrm{supp}(\phi_3) \subseteq \mathcal{K}_1$, Remark 6.5.8 implies $\phi_3 \mathscr{P} u \in \mathscr{X}^{s}(\mathcal{K}_1, (W, \vec{d}))$. Since $\mathrm{Mult}[\phi_1] T \,\mathrm{Mult}[\Psi_2] \in \widetilde{\mathscr{A}}^{-\kappa e_1}(\Omega_0, (W, \vec{d}); \mathbb{C}^{\mathbf{D}_2}, \mathbb{C}^{\mathbf{D}_1})$, Theorem 6.3.10 implies that $(\mathrm{Mult}[\phi_1] T \,\mathrm{Mult}[\Psi_2]) \phi_3 \mathscr{P} u \in \mathscr{X}^{s+\kappa e_1}(\mathcal{K}_1, (W, \vec{d}))$. Also, since $R_2 \in C_{\mathrm{loc}}^{\infty}(\mathfrak{M} \times \mathfrak{M}; \mathbb{M}^{\mathbf{D}_2 \times \mathbf{D}_2})$, we have $\phi_1 R_2 \Psi_2 u \in C_{\mathrm{loc}}^{\infty}(\Omega) \subseteq \mathscr{X}^{s+\kappa e_1}(\mathcal{K}_1, (W, \vec{d}))$, where the $\subseteq$ follows from Proposition 6.5.5. Combining the above with (8.14), we conclude that $\phi_1 u \in \mathscr{X}^{s+\kappa e_1}(\mathcal{K}_1, (W, \vec{d})) \subseteq \mathscr{X}_{\mathrm{cpt}}^{s+\kappa e_1}(W, \vec{d})$, as desired.

We turn to (8.28). Let $\psi \in C_0^{\infty}(\Omega)$ with $\phi_2 \prec \psi \prec \phi_3$. Applying (8.30) with ϕ_3 replaced by ψ we obtain (with a possibly different choice of $R_2 \in C_{\mathrm{loc}}^{\infty}(\mathfrak{M} \times \mathfrak{M}; \mathbb{M}^{\mathbf{D}_2 \times \mathbf{D}_2})$), for $u \in \mathscr{X}^{s}(\mathcal{K}, (W, \vec{d}))$,

$$\begin{aligned} \|\phi_1 u\|_{\mathscr{X}^{s+\kappa e_1}(W, \vec{d})} &\leq \|\mathrm{Mult}[\phi_1] T \,\mathrm{Mult}[\Psi_2] \psi \mathscr{P} u\|_{\mathscr{X}^{s+\kappa e_1}(W, \vec{d})} \\ &\quad + \|\phi_1 R_2 \Psi_2 u\|_{\mathscr{X}^{s+\kappa e_1}(W, \vec{d})} \\ &\lesssim \|\psi \mathscr{P} u\|_{\mathscr{X}^{s}(W, \vec{d})} + \|\Psi_2 u\|_{\mathscr{X}^{s-\vec{N}}(W, \vec{d})}, \end{aligned} \tag{8.31}$$

where the final estimate used $\mathrm{Mult}[\phi_1] T \,\mathrm{Mult}[\Psi_2] \in \widetilde{\mathscr{A}}^{-\kappa e_1}(\Omega_0, (W, \vec{d}); \mathbb{C}^{\mathbf{D}_2}, \mathbb{C}^{\mathbf{D}_1})$ and Theorem 6.3.10 for the first term, and for the second term we used the fact that $\mathrm{Mult}[\phi_1] R_2 \,\mathrm{Mult}[\Psi_2] \in C_0^{\infty}(\Omega_0 \times \Omega_0; \mathbb{M}^{\mathbf{D}_2 \times \mathbf{D}_2}(\mathbb{C})) \subseteq \widetilde{\mathscr{A}}^{-\kappa e_1 - \vec{N}}(\Omega_0, (W, \vec{d}); \mathbb{C}^{\mathbf{D}_2}, \mathbb{C}^{\mathbf{D}_2})$ (Proposition 5.8.11) and Theorem 6.3.10. Replacing u with $\phi_2 u$ in (8.31)[2] shows that

---

2 Since $\phi_3 \mathscr{P} u \in \mathscr{X}_{\mathrm{cpt}}^{s}(W, \vec{d})$, the first part of the proof with ϕ_1 replaced by ϕ_2 shows that $\phi_2 u \in \mathscr{X}^{s+\kappa e_1}(\mathcal{K}, (W, \vec{d})) \subseteq \mathscr{X}^{s}(\mathcal{K}, (W, \vec{d}))$, and therefore (8.31) applies.

$$\begin{aligned}\|\phi_1 u\|_{\mathscr{X}^{s+\kappa e_1}(W,\vec{d})} &\lesssim \|\psi \mathscr{P} u\|_{\mathscr{X}^s(W,\vec{d})} + \|\phi_2 u\|_{\mathscr{X}^{s-\vec{N}}(W,\vec{d})} \\ &= \|\psi\phi_3 \mathscr{P} u\|_{\mathscr{X}^s(W,\vec{d})} + \|\phi_2 u\|_{\mathscr{X}^{s-\vec{N}}(W,\vec{d})} \\ &\lesssim \|\phi_3 \mathscr{P} u\|_{\mathscr{X}^s(W,\vec{d})} + \|\phi_2 u\|_{\mathscr{X}^{s-\vec{N}}(W,\vec{d})},\end{aligned}$$

where the last estimate used Corollary 6.5.10. This completes the proof. □

Using results from Chapter 6, we can deduce new regularity estimates for maximally subelliptic operators directly from Proposition 8.2.12. We present two such estimates in the next two corollaries.

Corollary 8.2.13. *Let $\mathscr{P}$ be given by* (8.26) *and suppose $\mathscr{P}$ is maximally subelliptic of degree κ with respect to $(W^1, \vec{d}^1)$ on $\mathfrak{M}$. Let $(Y, \vec{d}) := (W^2, \vec{d}^2) \boxtimes (W^3, \vec{d}^3) \boxtimes \cdots (W^\nu, \vec{d}^\nu)$. Fix $\phi_1, \phi_2, \phi_3 \in C_0^\infty(\mathfrak{M})$ with $\phi_1 \prec \phi_2 \prec \phi_3$. Then, for all $s \in \mathbb{R}^{\nu-1}$, $p \in (1,\infty)$,*

$$\begin{aligned}&\phi_3 \mathscr{P} u \in \mathscr{F}^s_{p,2,\mathrm{cpt}}(Y,\vec{d}) \Rightarrow \phi_1 (W^1)^\alpha u \in \mathscr{F}^s_{p,2,\mathrm{cpt}}(Y,\vec{d}), \\ &\forall u \in C_0^\infty(\mathfrak{M})', \quad \forall \deg_{\vec{d}^1}(\alpha) \le \kappa.\end{aligned} \tag{8.32}$$

Moreover, for every $\vec{N} \in [0,\infty)^{\nu-1}$, there exists $C = C(s, \mathscr{P}, p, (W,\vec{d}), \phi_1, \phi_2) \ge 0$ such that

$$\sum_{\deg_{\vec{d}^1}(\alpha)\le\kappa} \|\phi_1 (W^1)^\alpha u\|_{\mathscr{F}^s_{p,2}(Y,\vec{d})} \le C\left(\|\phi_3 \mathscr{P} u\|_{\mathscr{F}^s_{p,2}(Y,\vec{d})} + \|\phi_2 u\|_{\mathscr{F}^{s-\vec{N}}_{p,2}(Y,\vec{d})}\right), \tag{8.33}$$

for all $u \in C_0^\infty(\mathfrak{M})'$ with $\phi_3 \mathscr{P} u \in \mathscr{F}^s_{p,2,\mathrm{cpt}}(Y,\vec{d})$.

Proof. For $\mathcal{K} \Subset \mathfrak{M}$ compact, Proposition 6.8.2 shows that

$$\mathscr{F}^s_{p,2}(\mathcal{K}, (Y,\vec{d})) = \mathscr{F}^{(0,s)}_{p,2}(\mathcal{K}, (W,\vec{d})),$$

with equivalent norms. In particular, this shows that $\mathscr{F}^s_{p,2,\mathrm{cpt}}(Y,\vec{d}) = \mathscr{F}^{(0,s)}_{p,2,\mathrm{cpt}}(W,\vec{d})$.

Fix $\phi_{1.5} \in C_0^\infty(\mathfrak{M})$ with $\phi_1 \prec \phi_{1.5} \prec \phi_2$. Suppose $\phi_3 \mathscr{P} u \in \mathscr{F}^s_{p,2,\mathrm{cpt}}(Y,\vec{d}) = \mathscr{F}^{(0,s)}_{p,2,\mathrm{cpt}}(W,\vec{d})$. Proposition 8.2.12 implies $\phi_{1.5} u \in \mathscr{F}^{(\kappa,s)}_{p,2,\mathrm{cpt}}(W,\vec{d})$ and moreover we also have $\phi_3 u \in \mathscr{F}^s_{p,2,\mathrm{cpt}}(Y,\vec{d}) = \mathscr{F}^{(0,s)}_{p,2,\mathrm{cpt}}(W,\vec{d})$, so Proposition 8.2.12 also implies that

$$\begin{aligned}\|\phi_{1.5} u\|_{\mathscr{F}^{(\kappa,s)}_{p,2}(W,\vec{d})} &\lesssim \|\phi_3 \mathscr{P} u\|_{\mathscr{F}^{(0,s)}_{p,2}(W,\vec{d})} + \|\phi_2 u\|_{\mathscr{F}^{(0,s)-(0,\vec{N})}_{p,2}(W,\vec{d})} \\ &\approx \|\phi_3 \mathscr{P} u\|_{\mathscr{F}^s_{p,2}(Y,\vec{d})} + \|\phi_2 u\|_{\mathscr{F}^{s-\vec{N}}_{p,2}(Y,\vec{d})}.\end{aligned} \tag{8.34}$$

Since $\phi_{1.5} u \in \mathscr{F}^{(\kappa,s)}_{p,2,\mathrm{cpt}}(W,\vec{d})$, Corollaries 6.5.10 and 6.5.11 imply, for $\deg_{\vec{d}^1}(\alpha) \le \kappa$,

$$\begin{aligned}\phi_1 (W^1)^\alpha u = \phi_1 (W^1)^\alpha \phi_{1.5} u &\in \mathscr{F}^{(\kappa-\deg_{\vec{d}^1}(\alpha),s)}_{p,2,\mathrm{cpt}}(W,\vec{d}) \\ &\subseteq \mathscr{F}^{(0,s)}_{p,2,\mathrm{cpt}}(W,\vec{d}) = \mathscr{F}^s_{p,2,\mathrm{cpt}}(Y,\vec{d}),\end{aligned}$$

establishing (8.32), and moreover

$$\sum_{\deg_{d^1}(\alpha)\leq\kappa} \|\phi_1(W^1)^\alpha u\|_{\mathscr{F}^s_{p,2}(Y,\vec{d})} \approx \sum_{\deg_{d^1}(\alpha)\leq\kappa} \|\phi_1(W^1)^\alpha u\|_{\mathscr{F}^{(0,s)}_{p,2}(W,\vec{d})} \lesssim \|\phi_{1.5}u\|_{\mathscr{F}^{(\kappa,s)}_{p,2}(W,\vec{d})}. \tag{8.35}$$

Combining (8.34) and (8.35) yields (8.33) and completes the proof. □

Corollary 8.2.13 only concerned the Sobolev spaces $\mathscr{F}^s_{p,2}$. If one is willing to "lose ϵ derivatives," then it is sometimes possible to prove a similar result for more general spaces, as the next result shows.

Corollary 8.2.14. *Consider the special case where $\nu = 2$ and suppose (W^2, d^2) are Hörmander vector fields with formal degrees. Let $\mathscr{P}$ be given by (8.26) and suppose $\mathscr{P}$ is maximally subelliptic of degree κ with respect to (W^1, d^1) on $\mathfrak{M}$. Then, for any scale of spaces $\mathscr{X}^s$, $s \in \mathbb{R}$, of the form*

$$\mathscr{X}^s \in \{\mathscr{B}^s_{p,q} : p, q \in [1,\infty]\} \bigcup \{\mathscr{F}^s_{p,q} : p \in (1,\infty), q \in (1,\infty]\},$$

the following holds. Let $\phi_1, \phi_2, \phi_3 \in C_0^\infty(\mathfrak{M})$ with $\phi_1 \prec \phi_2 \prec \phi_3$ and $\epsilon > 0$. We have

$$\begin{aligned}&\phi_3\mathscr{P}u \in \mathscr{X}^s_{\mathrm{cpt}}(W^2, d^2) \Rightarrow \phi_1(W^1)^\alpha u \in \mathscr{X}^{s-\epsilon}_{\mathrm{cpt}}(W^2, d^2),\\ &\forall u \in C_0^\infty(\mathfrak{M})', \quad \forall \deg_{d^1}(\alpha) \leq \kappa.\end{aligned} \tag{8.36}$$

Moreover, $\forall N \geq 0$, there exists $C = C(s, N, \epsilon, \mathscr{P}, \mathscr{X}^s, (W^1, d^1), (W^2, d^2), \phi_1, \phi_2) \geq 0$ such that

$$\sum_{\deg_{d^1}(\alpha)\leq\kappa} \|\phi_1(W^1)^\alpha u\|_{\mathscr{X}^{s-\epsilon}(W^2,d^2)} \leq C(\|\phi_3\mathscr{P}u\|_{\mathscr{X}^s(W^2,d^2)} + \|\phi_2 u\|_{\mathscr{X}^{s-N}(W^2,d^2)}), \tag{8.37}$$

$\forall u \in C_0^\infty(\mathfrak{M})'$ with $\phi_3\mathscr{P}u \in \mathscr{X}^s_{\mathrm{cpt}}(W^2, d^2)$.

Proof. Set $\mathcal{K} := \mathrm{supp}(\phi_3)$ and let $\Omega \Subset \mathfrak{M}$ be an open, relatively compact set with $\mathcal{K} \subseteq \Omega$. Since $W^2_1, \ldots, W^2_{r_2}$ satisfy Hörmander's condition on $\mathfrak{M}$, it follows directly from the definitions that (W^2, d^2) weakly λ-controls (W^1, d^1) on Ω for some $\lambda > 0$. Set $\epsilon_1 := \epsilon/2\lambda > 0$.

Fix $\phi_{1.5} \in C_0^\infty(\mathfrak{M})$ with $\phi_1 \prec \phi_{1.5} \prec \phi_2$. Suppose $\phi_3\mathscr{P}u \in \mathscr{X}^s(\mathcal{K}, (W^2, d^2))$. Then, by Proposition 6.8.1, we have $\phi_3\mathscr{P}u \in \mathscr{X}^s(\mathcal{K}, (W^2, d^2)) \subseteq \mathscr{X}^{(-\epsilon_1,s)}(\mathcal{K}, (W, \vec{d}))$. Proposition 8.2.12 implies that $\phi_{1.5}u \in \mathscr{X}^{(-\epsilon_1+\kappa,s)}(\mathcal{K}, (W, \vec{d}))$ and moreover, Propositions 8.2.12 and 6.8.1 imply that

$$\begin{aligned}\|\phi_{1.5}u\|_{\mathscr{X}^{(-\epsilon_1+\kappa,s)}(W,\vec{d})} &\lesssim \|\phi_3\mathscr{P}u\|_{\mathscr{X}^{(-\epsilon_1,s)}(W,\vec{d})} + \|\phi_2 u\|_{\mathscr{X}^{[(-\epsilon_1, s) - (0,N)][(W,\vec{d})]}}\\ &\lesssim \|\phi_3\mathscr{P}u\|_{\mathscr{X}^s(W^2,d^2)} + \|\phi_2 u\|_{\mathscr{X}^{s-N}(W^2,d^2)}.\end{aligned} \tag{8.38}$$

Since $\phi_{1.5}u \in \mathscr{X}^{(-\epsilon_1+\kappa,s)}(\mathcal{K},(W,\vec{d}))$, Corollaries 6.5.10 and 6.5.11 imply, for $\deg_{\vec{d}^1}(\alpha) \leq \kappa$,

$$\begin{aligned}\phi_1(W^1)^\alpha u = \phi_1(W^1)^\alpha \phi_{1.5}u &\in \mathscr{X}^{(-\epsilon_1+\kappa-\deg_{\vec{d}^1}(\alpha),s)}(\mathcal{K},(W,\vec{d}))\\ &\subseteq \mathscr{X}^{(-\epsilon_1,s)}(\mathcal{K},(W,\vec{d}))\end{aligned}$$

and

$$\sum_{\deg_{\vec{d}^1}(\alpha)\leq\kappa} \|\phi_1(W^1)^\alpha u\|_{\mathscr{X}^{(-\epsilon_1,s)}(W,\vec{d})} \lesssim \|\phi_{1.5}u\|_{\mathscr{X}^{(-\epsilon_1+\kappa,s)}(W,\vec{d})}. \tag{8.39}$$

By Theorem 6.7.2, we have, for $\deg_{\vec{d}^1}(\alpha) \leq \kappa$,

$$\phi_1(W^1)^\alpha u \in \mathscr{X}^{(-\epsilon_1,s)}(\mathcal{K},(W,\vec{d})) \subseteq \mathscr{X}^{(-\epsilon_1,s)+(\epsilon/\lambda,-\epsilon)}(\mathcal{K},(W,\vec{d})) = \mathscr{X}^{(\epsilon_1,s-\epsilon)}(\mathcal{K},(W,\vec{d}))$$

and

$$\sum_{\deg_{\vec{d}^1}(\alpha)\leq\kappa} \|\phi_1(W^1)^\alpha u\|_{\mathscr{X}^{(\epsilon_1,s-\epsilon)}(W,\vec{d})} \lesssim \sum_{\deg_{\vec{d}^1}(\alpha)\leq\kappa} \|\phi_1(W^1)^\alpha u\|_{\mathscr{X}^{(-\epsilon_1,s)}(W,\vec{d})}. \tag{8.40}$$

Finally, by Proposition 6.8.1 we have, for $\deg_{\vec{d}^1}(\alpha) \leq \kappa$,

$$\phi_1(W^1)^\alpha u \in \mathscr{X}^{(\epsilon_1,s-\epsilon)}(\mathcal{K},(W,\vec{d})) \subseteq \mathscr{X}^{s-\epsilon}(\mathcal{K},(W^2,\vec{d}^2)),$$

establishing (8.36), and moreover

$$\sum_{\deg_{\vec{d}^1}(\alpha)\leq\kappa} \|\phi_1(W^1)^\alpha u\|_{\mathscr{X}^{s-\epsilon}(W^2,\vec{d}^2)} \lesssim \sum_{\deg_{\vec{d}^1}(\alpha)\leq\kappa} \|\phi_1(W^1)^\alpha u\|_{\mathscr{X}^{(\epsilon_1,s-\epsilon)}(W,\vec{d})}. \tag{8.41}$$

Combining (8.38), (8.39), (8.40), and (8.41) gives (8.37) and completes the proof. □

Proof of Corollary 8.2.9. Set $\mathcal{K} := \mathrm{supp}(\phi_2) \Subset \mathfrak{M}$ and let $\mathfrak{N} \Subset \mathfrak{M}$ be an open and relatively compact subset of $\mathfrak{M}$ with $\mathcal{K} \Subset \mathfrak{N}$. Let $(Z,1) := \{(Z_1,1),\dots,(Z_\nu,1)\} \subset C^\infty_{\mathrm{loc}}(\mathfrak{N};T\mathfrak{N}) \times \mathbb{N}_+$ be such that $\mathrm{span}\{Z_1(x),\dots,Z_\nu(x)\} = T_x\mathfrak{N}$, $\forall x \in \mathfrak{N}$. Lemma 3.8.3 shows $(Z,1)$ and $(W,\vec{d})$ locally strongly approximately commute on $\mathfrak{N}$, and therefore Proposition 3.8.6 (b) implies they locally weakly approximately commute on $\mathfrak{N}$.

Theorem 6.6.7 shows that $\mathscr{X}^s_{\mathrm{std}}(\mathcal{K}) = \mathscr{X}^s(\mathcal{K},(Z,1))$, with equivalent norms. We take $\nu = 2$ and replace $(W^1,\vec{d}^1)$ with $(W,\vec{d})$ and $(W^2,\vec{d}^2)$ with $(Z,1)$. Using the above remarks, (i) follows from Corollary 8.2.13 and (ii) follows from Corollary 8.2.14. □

Remark 8.2.15. When $(W^1,\vec{d}^1) = (W^2,\vec{d}^2)$, the conclusion of Corollary 8.2.14 with $\epsilon = 0$ follows immediately from Theorem 8.1.1 (i) ⇒ (v).

8.3 A priori estimates

The key to proving Theorem 8.1.1 is combining scaling techniques with a priori estimates which take place on the unit ball. Because we combine these results with scaling techniques, the results in this section will be applied an infinite number of times in the proof of Theorem 8.1.1. Thus, it is important to be explicit about what our estimates depend on, to later show that they remain uniform over this infinite number of applications.

We suppose we are given Hörmander vector fields with formal degrees on $B^n(1)$,

$$(W, \vec{d}) = \{(W_1, \vec{d}_1), \dots, (W_r, \vec{d}_r)\} \subset C^\infty(B^n(1); TB^n(1)),$$

and we assume $W_1, \dots, W_r$ satisfy Hörmander's condition of order $m \in \mathbb{N}_+$ on $B^n(1)$. We make all the same assumptions as in Section 3.2. Fix a strictly positive smooth function $h(u) \in C^\infty(B^n(1))$ with $\inf_{u\in B^n(1)} h(u) > 0$; we will sometimes endow $B^n(1)$ with the measure $h(u)\sigma_{\mathrm{Leb}}$, where σ_{Leb} denotes the usual Lebesgue density on $B^n(1)$. We will also sometimes endow $B^n(1)$ with the usual Lebesgue density σ_{Leb}.

Fix $\kappa \in \mathbb{N}_+$ such that $\vec{d}_j$ divides κ, for $1 \le j \le r$. Set $n_j := \kappa/\vec{d}_j \in \mathbb{N}_+$. Fix $D \in \mathbb{N}_+$. The basic object of study in this section is a partial differential operator of the form

$$\mathscr{L} := \sum_{\deg_{\vec{d}}(\alpha), \deg_{\vec{d}}(\beta) \le \kappa} b_{\alpha,\beta}(x) W^\alpha W^\beta, \quad b_{\alpha,\beta} \in C^\infty(B^n(1); \mathbb{M}^{D\times D}(\mathbb{C})). \tag{8.42}$$

Definition 8.3.1. We say an operator $\mathscr{L}$ of the form (8.42) is maximally subelliptic of type 2 of degree 2κ at the unit scale with respect to $(W, \vec{d})$ and $h\sigma_{\mathrm{Leb}}$ if there exists $A \ge 0$ such that $\forall f \in C_0^\infty(B^n(1); \mathbb{C}^D)$,

$$\begin{aligned}\sum_{j=1}^r \|W_j^{n_j} f\|^2_{L^2(B^n(1), h\sigma_{\mathrm{Leb}}; \mathbb{C}^D)} \\ \le A\big(|\langle \mathscr{L} f, f\rangle_{L^2(B^n(1), h\sigma_{\mathrm{Leb}}; \mathbb{C}^D)}| + \|f\|^2_{L^2(B^n(1), h\sigma_{\mathrm{Leb}}; \mathbb{C}^D)}\big).\end{aligned} \tag{8.43}$$

See Remark 8.5.3 and Corollary 8.5.4 for some comments on the relationship between maximal subellipticity of type 2 and maximal subellipticity.

The main result of this section (Proposition 8.3.6) shows that maximally subelliptic operators of type 2 at the unit scale are subelliptic; we use this result to conclude some important corollaries. As mentioned before, it is important to be explicit about what our estimates depend on. For this, we introduce the next definition.

Definition 8.3.2. For a parameter ι, we say C is an ι-$\mathscr{L}$-unit-admissible constant if C can be chosen to depend only on:

- Anything an (ι, κ)-unit-admissible constant can depend on in Definition 3.2.1.
- An upper bound for D and the constant $A \ge 0$ from (8.43).

– An upper bound for

$$\|h\|_{C^L(B^n(1))} + \max_{\deg_{\vec{d}}(\alpha),\deg_{\vec{d}}(\beta)\leq\kappa} \|b_{\alpha,\beta}\|_{C^L(B^n(1);\mathbb{M}^{D\times D})},$$

where $L \in \mathbb{N}_+$ can be chosen to depend only on ι and upper bounds for m, κ, n, r, and $\max\{\vec{d}_1,\dots,\vec{d}_r\}$.
– A lower bound > 0 for $\inf_{u\in B^n(1)} h(u)$.

We say C is an $\mathscr{L}$-unit-admissible constant if it is a 0-$\mathscr{L}$-unit-admissible constant. For another parameter ι_0, we say $C = C(\iota_0)$ is an ι-$\mathscr{L}$-unit-admissible constant if C is an ι-$\mathscr{L}$-unit-admissible constant which may also depend on ι_0.

In this section, we write $A \lesssim B$ to mean $A \leq CB$, where $C = C(\iota_0) \geq 0$ is an ι-$\mathscr{L}$-unit-admissible constant, for appropriate choices of ι and ι_0, which will usually be clear from the context. When the choices of ι and ι_0 are not clear, we will be explicit. All that matters for our applications, though, is that the estimates do not depend on the particular choice of $\mathscr{L}$, $(W,\vec{d})$, or h, except through the quantities discussed in Definition 8.3.2.

8.3.1 Some preliminary estimates

Set

$$\epsilon_0 := \left(m\max\{\vec{d}_1,\dots,\vec{d}_r\}\right)^{-1} > 0; \tag{8.44}$$

recall that $W_1,\dots,W_r$ satisfy Hörmander's condition of order $m \in \mathbb{N}_+$ on $B^n(1)$. For $s \in \mathbb{R}$, we let $\|\cdot\|_{L^2_s}$ denote the standard L^2 Sobolev norm of order s on $\mathbb{R}^n$.

Lemma 8.3.3. *Fix $\sigma \in (0,1)$ and $\kappa \in \mathbb{N}_+$ such that $\vec{d}_j$ divides κ for $1 \leq j \leq r$. Then:*
(i) $\forall f \in C_0^\infty(B^n(1-\sigma);\mathbb{C}^D)$,

$$\sum_{\deg_{\vec{d}}(\alpha)\leq\kappa} \|W^\alpha f\|_{L^2(B^n(1),\sigma_{\mathrm{Leb}};\mathbb{C}^D)} \approx \sum_{j=1}^r \|W_j^{n_j} f\|_{L^2(B^n(1),\sigma_{\mathrm{Leb}};\mathbb{C}^D)} + \|f\|_{L^2(B^n(1),\sigma_{\mathrm{Leb}};\mathbb{C}^D)}.$$

(ii) $\forall f \in C_0^\infty(B^n(1-\sigma);\mathbb{C}^D)$,

$$\sum_{\deg_{\vec{d}}(\alpha)\leq\kappa-1} \|W^\alpha f\|_{L^2(B^n(1),\sigma_{\mathrm{Leb}};\mathbb{C}^D)} \lesssim \sum_{\deg_{\vec{d}}(\alpha)\leq\kappa} \|W^\alpha f\|_{L^2_{-\epsilon_0}},$$

$$\sum_{\deg_{\vec{d}}(\alpha)\leq\kappa-1} \|W^\alpha f\|_{L^2_{\epsilon_0}} \lesssim \sum_{\deg_{\vec{d}}(\alpha)\leq\kappa} \|W^\alpha f\|_{L^2(B^n(1),\sigma_{\mathrm{Leb}};\mathbb{C}^D)}.$$

(iii) $\forall f \in C_0^\infty(B^n(1-\sigma);\mathbb{C}^D)$,

$$\|f\|_{L^2_{\kappa\epsilon_0}} \lesssim \sum_{\deg_{\vec{d}}(\alpha)\leq\kappa} \|W^\alpha f\|_{L^2(B^n(1),\sigma_{\mathrm{Leb}};\mathbb{C}^D)}.$$

Here, the implicit constants are κ-unit-admissible constants as in Definition 3.2.1 *that can also depend on a lower bound* > 0 *for* σ.

Remark 8.3.4. The choice of ϵ_0 in Lemma 8.3.3 (ii) and (iii) is not optimal. However, finding the optimal choice is not important for our purposes: all we use at this point is the fact that $\epsilon_0 > 0$ depends only on m and $d_1, \dots, d_r$. When we combine this with scaling and techniques from harmonic analysis, we will obtain optimal results like those discussed in Section 8.2.2. Thus, the non-optimal Lemma 8.3.3 is an important tool in later proving optimal results.

Remark 8.3.5. In the important special case $d_1 = \cdots = d_r = 1$, Lemma 8.3.3 (ii) (with a different choice of $\epsilon_0 > 0$) follows from Proposition 4.5.8, which has a much simpler and much more elementary proof. This is one of the main ways in which working with general $d_1, \dots, d_r$ is harder than the case where $d_1 = \cdots = d_r = 1$. Even in the general case, one can easily prove Lemma 8.3.3 (iii) (with a different choice of ϵ_0) using Proposition 4.5.8, though we do not know of a direct, elementary proof of Lemma 8.3.3 (ii).

Proof of Lemma 8.3.3. We prove the result without keeping track of the dependence of the constants. The dependence follows by keeping track of constants in the results we use, which we leave to the reader.

The estimate (i) follows from Corollary 6.2.14 (in particular, (6.5)), with $\mathfrak{M} = B^n(1)$, $\mathcal{K} := \overline{B^n(1-\sigma)}$, and $\mathrm{Vol} = \sigma_{\mathrm{Leb}}$.

The estimates (ii) and (iii) follow from a more general claim. Namely, for $s \in \mathbb{R}$ and $\mu \in \{0, \dots, \kappa\}$,

$$\sum_{\deg_d(\alpha) \le \kappa - \mu} \|W^\alpha f\|_{L^2_s} \lesssim \sum_{\deg_d(\alpha) \le \kappa} \|W^\alpha f\|_{L^2_{s-\mu\epsilon_0}}, \quad \forall f \in C_0^\infty(B^n(1-\sigma)). \tag{8.45}$$

Indeed, (ii) follows from the cases $s = 0$, $\mu = 1$ and $s = \epsilon_0$, $\mu = 1$ of (8.45), while (iii) is the special case $s = \kappa\epsilon_0$, $\mu = \kappa$. We turn to proving (8.45). Throughout this proof, f will denote an arbitrary element of $C_0^\infty(B^n(1-\sigma))$.

We consider the Hörmander vector fields with formal degrees on $B^n(1)$ given by $(\partial, 1) = \{(\partial_{x_1}, 1), \dots, (\partial_{x_n}, 1)\}$. Lemma 3.8.3 implies (W, d) and $(\partial, 1)$ locally strongly approximately commute on $B^n(1)$, and therefore Proposition 3.8.6 (b) implies they locally weakly approximately commute on $B^n(1)$. We define the list of vector fields with two-parameter formal degrees, $(Y, \vec{d})$, by

$$(Y, \vec{d}) = \{(W_1, (d_1, 0)), \dots, (W_r, (d_r, 0))\} \bigcup \{(\partial_{x_1}, (0,1)), \dots, (\partial_{x_n}, (0,1))\}.$$

$(Y, \vec{d})$ satisfies all the assumptions of Section 5.1.2 and therefore satisfies the assumptions of Section 5.1.3 (see Remark 5.1.2). In particular, all of the results concerning the function spaces from Chapter 6 apply with $(W, \vec{d})$ replaced by $(Y, \vec{d})$.

By Theorem 6.6.7 (see also Definition 6.6.4), we have, $\forall s \in \mathbb{R}$,

$$\|f\|_{L^2_s} \approx \|f\|_{\mathscr{F}^s_{2,2,\mathrm{std}}} \approx \|f\|_{\mathscr{F}^s_{2,2}(\partial,1)}. \tag{8.46}$$

Applying Proposition 6.8.8, we have, $\forall s \in \mathbb{R}$,

$$\|f\|_{\mathscr{F}^s_{2,2}(\partial,1)} \approx \|f\|_{\mathscr{F}^{(0,s)}_{2,2}(Y,\vec{d})}. \tag{8.47}$$

Thus, using (8.46), (8.47), and Corollary 6.5.11, we have

$$\begin{aligned} \sum_{\deg_{\vec{d}}(\alpha)\le\kappa-\mu} \|W^\alpha f\|_{L^2_s} &\approx \sum_{\deg_{\vec{d}}(\alpha)\le\kappa-\mu} \|W^\alpha f\|_{\mathscr{F}^{(0,s)}_{2,2}(Y,\vec{d})} \\ &\lesssim \|f\|_{\mathscr{F}^{(\kappa-\mu,s)}_{2,2}(Y,\vec{d})}. \end{aligned} \tag{8.48}$$

Because $W_1,\dots,W_r$ satisfy Hörmander's condition of order m, for each $j = 1,\dots,n$ we have

$$\partial_{x_j} = \sum_k c_j^k X_k, \quad (X_k, d_k) \in \mathrm{Gen}((W,\vec{d})),\ d_k \le \epsilon_0^{-1},\ c_j^k \in C^\infty_{\mathrm{loc}}(B^n(1)).$$

This shows that $(W,\vec{d})$ weakly ϵ_0^{-1}-controls $(\partial, 1)$ on $B^n(1)$ (see Definition 3.14.3). By Theorem 6.7.2, we have

$$\|f\|_{\mathscr{F}^{(\kappa-\mu,s)}_{2,2}(Y,\vec{d})} = \|f\|_{\mathscr{F}^{(\kappa,s-\mu\epsilon_0)+(-\mu,\mu\epsilon_0)}_{2,2}(Y,\vec{d})} \lesssim \|f\|_{\mathscr{F}^{(\kappa,s-\mu\epsilon_0)}_{2,2}(Y,\vec{d})}. \tag{8.49}$$

By Proposition 6.5.12 (in particular, (6.25)), we have, $\forall s \in \mathbb{R}$,

$$\|f\|_{\mathscr{F}^{(\kappa,s-\mu\epsilon_0)}_{2,2}(Y,\vec{d})} \approx \sum_{\deg_{\vec{d}}(\alpha)\le\kappa} \|W^\alpha f\|_{\mathscr{F}^{(0,s-\mu\epsilon_0)}_{2,2}(Y,\vec{d})}. \tag{8.50}$$

Applying (8.46) and (8.47), we have, $\forall s \in \mathbb{R}$,

$$\begin{aligned} \sum_{\deg_{\vec{d}}(\alpha)\le\kappa} \|W^\alpha f\|_{\mathscr{F}^{(0,s-\mu\epsilon_0)}_{2,2}(Y,\vec{d})} &\approx \sum_{\deg_{\vec{d}}(\alpha)\le\kappa} \|W^\alpha f\|_{\mathscr{F}^{s-\mu\epsilon_0}_{2,2}(\partial,1)} \\ &\approx \sum_{\deg_{\vec{d}}(\alpha)\le\kappa} \|W^\alpha f\|_{L^2_{s-\mu\epsilon_0}}. \end{aligned} \tag{8.51}$$

Combining (8.48), (8.49), (8.50), and (8.51) establishes (8.45) and completes the proof. □

8.3.2 Subelliptic estimates

In this section we present the subelliptic estimate which lies at the heart of the proof of Theorem 8.1.1. Throughout this section, we assume that $\mathscr{L}$ is a partial differential operator of the form (8.42) and we assume that $\mathscr{L}$ is maximally subelliptic of type 2 of degree 2κ at the unit scale with respect to $(W, \vec{d})$ and $h\sigma_{\mathrm{Leb}}$ (see Definition 8.3.1). Let $\epsilon_0 > 0$ be as in (8.44). The main result of this section is the next proposition.

Proposition 8.3.6. *$\mathscr{L}$ is subelliptic on $B^n(1)$ in the following sense. For every $s \in \mathbb{R}$ and $\phi_1, \phi_2 \in C_0^\infty(B^n(1))$ with $\phi_1 \prec \phi_2$, we have*

$$\|\phi_1 f\|_{L^2_{s+\kappa\epsilon_0}} \lesssim \|\phi_2 \mathscr{L} f\|_{L^2_s} + \|\phi_2 f\|_{L^2_{-\infty}}, \quad \forall f \in C_0^\infty(B^n(1); \mathbb{C}^D)', \tag{8.52}$$

where if the right-hand side is finite, so is the left-hand side. More precisely, for every $s \in \mathbb{R}$, $N \in \mathbb{N}$, there exists an (s, N)-$\mathscr{L}$-unit-admissible constant $C = C(\phi_1, \phi_2) \geq 0$ (see Definition 8.3.2) such that $\forall f \in C_0^\infty(B^n(1); \mathbb{C}^D)'$,

$$\|\phi_1 f\|_{L^2_{s+\kappa\epsilon_0}} \leq C(\|\phi_2 \mathscr{L} f\|_{L^2_s} + \|\phi_2 f\|_{L^2_{-N}}).$$

Remark 8.3.7. The gain of $\kappa\epsilon_0$ on the left-hand side of (8.52) is not optimal. Similar to Remark 8.3.4, finding the optimal gain is not important for our applications: any subelliptic gain will allow us to deduce optimal estimates for maximally subelliptic operators as in Section 8.2.2. In particular, if $\mathscr{L}$ is, in addition, known to be symmetric and non-negative, then Corollary 8.5.4 shows that $\mathscr{L}$ is maximally subelliptic. The optimal subelliptic gain then follows from Corollary 8.2.5.

Before we prove Proposition 8.3.6, we state two important results which follow from it.

Corollary 8.3.8. *For all $s > 0$ and $\phi_1, \phi_2 \in C_0^\infty(B^n(1))$ with $\phi_1 \prec \phi_2$, there exists an s-$\mathscr{L}$-unit-admissible constant $C = C(\phi_1, \phi_2) \geq 0$ with*

$$\|\phi_1 f\|_{L^2_s} \leq C \sum_{j=0}^{\lceil \frac{s}{\kappa\epsilon_0} \rceil} \|\phi_2 \mathscr{L}^j f\|_{L^2(B^n(1), \sigma_{\mathrm{Leb}}; \mathbb{C}^D)}, \quad \forall f \in C_0^\infty(B^n(1); \mathbb{C}^D)',$$

where if the right-hand side is finite, so is the left-hand side.

Proposition 8.3.9. *Think of $\mathscr{L}$ as a densely defined operator on $L^2(B^n(1), h\sigma_{\mathrm{Leb}}; \mathbb{C}^D)$ with dense domain $C_0^\infty(B^n(1); \mathbb{C}^D)$. Suppose $\mathscr{L}$ is non-negative and symmetric. Then there is an $\mathscr{L}$-unit-admissible constant $C \geq 0$ such that the following holds. For every $\epsilon \in (0, 1/2]$, $\forall u(t, x) \in C^\infty_{\mathrm{loc}}((-1, 1) \times B^n(1); \mathbb{C}^D))$ such that $\partial_t u(t, x) = -\epsilon^{-2\kappa} \mathscr{L} u(t, x)$, $\forall k \in [1, 1/2\epsilon]$,*

$$\begin{aligned} &\|\partial_t u\|_{L^2((-1+k\epsilon, 1-k\epsilon) \times B^n(1/2), dt \times d(h\sigma_{\mathrm{Leb}}); \mathbb{C}^D)} \\ &\quad \leq C\epsilon^{-2\kappa} \|u\|_{L^2((-1+(k-1)\epsilon, 1-(k-1)\epsilon) \times B^n(1), dt \times d(h\sigma_{\mathrm{Leb}}); \mathbb{C}^D)}. \end{aligned}$$

Remark 8.3.10. The measure $h\sigma_{\mathrm{Leb}}$ plays a secondary role in the above results. Indeed, because h is smooth, it is bounded below, and all its derivatives are bounded, in most estimates it is not important whether one uses σ_{Leb} or $h\sigma_{\mathrm{Leb}}$. For example, since h is bounded above and below by $\mathscr{L}$-unit-admissible constants, we have

$$\|f\|_{L^2(B^n(1),h\sigma_{\mathrm{Leb}})} \approx \|f\|_{L^2(B^n(1),\sigma_{\mathrm{Leb}})}, \quad \forall f \in L^2(B^n(1),\sigma_{\mathrm{Leb}}). \tag{8.53}$$

One important place h plays a role is in Proposition 8.3.9: $\mathscr{L}$ being symmetric and non-negative on $L^2(B^n(1),h\sigma_{\mathrm{Leb}};\mathbb{C}^D)$ is not the same thing as $\mathscr{L}$ being symmetric and non-negative on $L^2(B^n(1),\sigma_{\mathrm{Leb}};\mathbb{C}^D)$. As described at the start of Section 8.3, the above results will be applied an infinite number of times, and h will vary over these applications, though we will be able to estimate h uniformly across these applications. In this section, every L^2 inner product will be taken with respect to the measure $h\sigma_{\mathrm{Leb}}$.

We turn to proving Propositions 8.3.6 and 8.3.9 and Corollary 8.3.8.

We begin with the proof of Proposition 8.3.6. Note that the conclusions of Proposition 8.3.6 remain unchanged if we replace $W_1,\dots,W_r$ with $\phi_3 W_1,\dots,\phi_3 W_r$, where $\phi_3 \in C_0^\infty(B^n(1))$ with $\phi_2 \prec \phi_3$. With this replacement, $W_1,\dots,W_r$ are standard pseudo-differential operators of order 1. Because of this, we may treat $W_1,\dots,W_r$ as standard pseudo-differential operators of order 1 in the proofs which follow. Similarly, we can view the coefficients $b_{\alpha,\beta}$ in (8.42) as elements of $C_0^\infty(B^n(1);\mathbb{M}^{D\times D}(\mathbb{C}))$ rather than as elements of the larger space $C^\infty(B^n(1);\mathbb{M}^{D\times D}(\mathbb{C}))$.

As in Section 4.5.1, we write $S^{(s,\delta)}$, perhaps with a subscript to denote a standard pseudo-differential operator of order $-\infty$ such that $\{S^{(s,\delta)} : \delta \in (0,1]\}$ is a bounded set of standard pseudo-differential operators of order s. When we write $\widetilde{S}^{(s,\delta)}$ we mean a finite sum of such terms. For example, if we write $\|\widetilde{S}^{(s,\delta)}u\|_{L^2}$, it means $\|S_1^{(s,\delta)}u\|_{L^2} + \cdots + \|S_Q^{(s,\delta)}u\|_{L^2}$, for some finite collection $S_1^{(s,\delta)},\dots,S_Q^{(s,\delta)}$. Here the finite collection does not depend on the function or distribution u under consideration.

Lemma 8.3.11. *Suppose $\mathscr{Q}$ is a partial differential operator on $B^n(1)$ with smooth coefficients taking values in $\mathbb{M}^{D\times D}(\mathbb{C})$. Suppose that for all $\phi_1,\phi_2 \in C_0^\infty(B^n(1))$ with $\phi_1 \prec \phi_2$, $\forall s \in \mathbb{R}$,*

$$\begin{aligned}&\sum_{\deg_{\vec{d}}(\alpha)\le\kappa} \|W^\alpha \phi_1 S^{(s,\delta)} f\|_{L^2(B^n(1),\sigma_{\mathrm{Leb}};\mathbb{C}^D)}\\ &\quad\lesssim \|\phi_2 \mathscr{Q} f\|_{L^2_s} + \sum_{\deg_{\vec{d}}(\alpha)\le\kappa} \|W^\alpha \phi_2 \widetilde{S}^{(s,\delta)} f\|_{L^2_{-\epsilon_0}} + \|f\|_{L^2_{-\infty}},\end{aligned} \tag{8.54}$$

$\forall f \in \mathscr{S}(\mathbb{R}^n;\mathbb{C}^D)'$. Then, $\forall \phi_1,\phi_2 \in C_0^\infty(B^n(1))$ with $\phi_1 \prec \phi_2$, $\forall s \in \mathbb{R}$,

$$\|\phi_1 f\|_{L^2_{s+\kappa\epsilon_0}} \lesssim \|\phi_2 \mathscr{Q} f\|_{L^2_s} + \|\phi_2 f\|_{L^2_{-\infty}}, \tag{8.55}$$

$\forall f \in C_0^\infty(B^n(1);\mathbb{C}^D)'$.

Proof. Recall that $\|W^\alpha \phi_2 \widetilde{S}^{(s,\delta)} f\|_{L^2_{-\epsilon_0}}$ denotes a finite sum of such terms. In this proof we write L^2 for the space $L^2(B^n(1),\sigma_{\mathrm{Leb}};\mathbb{C}^D)$. Because of the localizations ϕ_1 and ϕ_2 in (8.55), it suffices to prove (8.55) for $f \in \mathscr{S}(\mathbb{R}^n;\mathbb{C}^D)'$; henceforth, f denotes an arbitrary element of $\mathscr{S}(\mathbb{R}^n;\mathbb{C}^D)'$.

Fix $\phi_1, \phi_2, \phi_{1.5} \in C_0^\infty(B^n(1))$ with $\phi_1 \prec \phi_{1.5} \prec \phi_2$. For $j \in \mathbb{N}$, let $\eta_j \in C_0^\infty(B^n(1))$ satisfy $\phi_1 \prec \eta_0 \prec \eta_1 \prec \eta_2 \prec \cdots \prec \phi_{1.5}$.

Let $S_0^{(s,\delta)}$ be given. We claim that, $\forall k \in \mathbb{N}$, there exist $\widetilde{S}_k^{(s-k\epsilon_0,\delta)}$ such that

$$\begin{aligned}&\sum_{\deg_d(\alpha)\leq\kappa} \|W^\alpha \eta_0 S_0^{(s,\delta)} f\|_{L^2}\\ &\quad\lesssim \|\eta_k \mathscr{Q} f\|_{L^2_s} + \sum_{\deg_d(\alpha)\leq\kappa} \|W^\alpha \eta_k \widetilde{S}_k^{(s-k\epsilon_0)} f\|_{L^2} + \|f\|_{L^2_{-\infty}}.\end{aligned} \tag{8.56}$$

We prove (8.56) by induction on k. The base case, $k = 0$, is trivial. We assume (8.56) for some $k \in \mathbb{N}$ and prove it for $k+1$. Fix $\tilde{\eta}_{k,1}, \tilde{\eta}_{k,2} \in C_0^\infty(B^n(1))$ with $\eta_k \prec \tilde{\eta}_{k,1} \prec \tilde{\eta}_{k,2} \prec \eta_{k+1}$.

By the inductive hypothesis and (8.54) (with ϕ_1 replaced by η_k and ϕ_2 replaced by $\tilde{\eta}_{k,1}$), we have

$$\begin{aligned}&\sum_{\deg_d(\alpha)\leq\kappa} \|W^\alpha \eta_0 S_0^{(s,\delta)} f\|_{L^2}\\ &\quad\lesssim \|\eta_k \mathscr{Q} f\|_{L^2_s} + \sum_{\deg_d(\alpha)\leq\kappa} \|W^\alpha \eta_k \widetilde{S}^{(s-k\epsilon_0,\delta)} f\|_{L^2} + \|f\|_{L^2_{-\infty}}\\ &\quad\lesssim \|\tilde{\eta}_{k,1} \mathscr{Q} f\|_{L^2_s} + \sum_{\deg_d(\alpha)\leq\kappa} \|W^\alpha \tilde{\eta}_{k,1} \widetilde{S}^{(s-k\epsilon_0,\delta)} f\|_{L^2_{-\epsilon_0}} + \|f\|_{L^2_{-\infty}}\\ &\quad\lesssim \|\eta_{k+1} \mathscr{Q} f\|_{L^2_s} + \sum_{\deg_d(\alpha)\leq\kappa} \|W^\alpha \tilde{\eta}_{k,1} \widetilde{S}^{(s-k\epsilon_0,\delta)} f\|_{L^2_{-\epsilon_0}} + \|f\|_{L^2_{-\infty}}.\end{aligned}$$

Thus, to prove the inductive step, it suffices to show that

$$\sum_{\deg_d(\alpha)\leq\kappa} \|W^\alpha \tilde{\eta}_{k,1} \widetilde{S}^{(s-k\epsilon_0,\delta)} f\|_{L^2_{-\epsilon_0}} \lesssim \sum_{\deg_d(\alpha)\leq\kappa} \|W^\alpha \eta_{k+1} \widetilde{S}^{(s-(k+1)\epsilon_0,\delta)} f\|_{L^2}. \tag{8.57}$$

Using the calculus of pseudo-differential operators (Theorem 2.2.8), we have, with Λ^s as in Definition 2.2.10,

$$\begin{aligned}&\sum_{\deg_d(\alpha)\leq\kappa} \|W^\alpha \tilde{\eta}_{k,1} \widetilde{S}^{(s-k\epsilon_0,\delta)} f\|_{L^2_{-\epsilon_0}} \lesssim \sum_{\deg_d(\alpha)\leq\kappa} \|\tilde{\eta}_{k,2} \Lambda^{-\epsilon_0} W^\alpha \tilde{\eta}_{k,1} \widetilde{S}^{(s-k\epsilon_0,\delta)} f\|_{L^2}\\ &\quad\lesssim \sum_{\deg_d(\alpha)\leq\kappa} \|\tilde{\eta}_{k,2} W^\alpha \widetilde{S}^{(s-(k+1)\epsilon_0,\delta)} f\|_{L^2} = \sum_{\deg_d(\alpha)\leq\kappa} \|\tilde{\eta}_{k,2} W^\alpha \eta_{k+1} \widetilde{S}^{(s-(k+1)\epsilon_0,\delta)} f\|_{L^2}\\ &\quad\lesssim \sum_{\deg_d(\alpha)\leq\kappa} \|W^\alpha \eta_{k+1} \widetilde{S}^{(s-(k+1)\epsilon_0,\delta)} f\|_{L^2},\end{aligned}$$

establishing (8.57) and completing the proof of (8.56).

Taking k large in (8.56) and using $\eta_k \prec \phi_{1.5}$ shows that

$$\sum_{\deg_d(\alpha)\leq\kappa} \|W^\alpha \eta_0 S_0^{(s,\delta)} f\|_{L^2} \lesssim \|\phi_{1.5} \mathscr{Q} f\|_{L^2_s} + \|f\|_{L^2_{-\infty}}.$$

By Lemma 8.3.3 (iii), this implies

$$\|\eta_0 S_0^{(s,\delta)} f\|_{L^2_{\epsilon_0\kappa}} \lesssim \|\phi_{1.5}\mathcal{Q}f\|_{L^2_s} + \|f\|_{L^2_{-\infty}}. \tag{8.58}$$

We now choose a particular $S_0^{(s,\delta)}$. Namely, we take $S_0^{(s,\delta)} = \Lambda^s\psi_\delta(D)\,\mathrm{Mult}[\phi_1]$, where $\hat{\psi} \in C_0^\infty(\mathbb{R}^n)$ equals 1 on a neighborhood of 0 and $\psi_\delta(D)$ is given by (4.53). We have, using (8.58),

$$\begin{aligned}\|\phi_1 f\|_{L^2_{s+\epsilon_0\kappa}} &\lesssim \sup_{\delta>0} \|\eta_0\Lambda^s\psi_\delta(D)\phi_1 f\|_{L^2_{\kappa\epsilon_0}} + \|f\|_{L^2_{-\infty}} \\ &= \sup_{\delta>0} \|\eta_0 S_0^{(s,\delta)} f\|_{L^2_{\kappa\epsilon_0}} + \|f\|_{L^2_{-\infty}} \lesssim \|\phi_{1.5}\mathcal{Q}f\|_{L^2_s} + \|f\|_{L^2_{-\infty}}.\end{aligned} \tag{8.59}$$

Replacing f with $\phi_2 f$ in (8.59) and using $\phi_1 \prec \phi_{1.5} \prec \phi_2$, we have

$$\|\phi_1 f\|_{L^2_{s+\epsilon_0\kappa}} \lesssim \|\phi_{1.5}\mathcal{Q}f\|_{L^2_s} + \|\phi_2 f\|_{L^2_{-\infty}} \lesssim \|\phi_2\mathcal{Q}f\|_{L^2_s} + \|\phi_2 f\|_{L^2_{-\infty}},$$

completing the proof. □

Lemma 8.3.12. *Let $\phi_1, \phi_2, \phi_3 \in C_0^\infty(B^n(1))$ satisfy $\phi_1 \prec \phi_2 \prec \phi_3$. Then*

$$[\mathcal{L}, \mathrm{Mult}[\phi_1]S^{(s,\delta)}] = \sum_{\substack{\deg_d(\alpha),\deg_d(\beta)\le\kappa \\ \deg_d(\alpha)+\deg_d(\beta)\le 2\kappa-1}} \mathrm{Mult}[\phi_2]W^\alpha W^\beta\,\mathrm{Mult}[\phi_3]S_{\alpha,\beta}^{(s,\delta)},$$

where $S^{(s,\delta)}$ and $S_{j,l}^{(s,\delta)}$ are pseudo-differential operators of order $-\infty$ which are pseudo-differential operators of order s, uniformly in $\delta \in (0,1]$.

Proof. This follows easily from the calculus of pseudo-differential operators (Theorem 2.2.8). □

We recall the large constant (l.c.) and small constant (s.c.) notation defined in Section 4.5.1; see the discussion surrounding (4.40).

Lemma 8.3.13. *For $\phi_1, \phi_2 \in C_0^\infty(B^n(1))$ with $\phi_1 \prec \phi_2$, we have*

$$\begin{aligned}&|\langle[\mathrm{Mult}[\phi_1]S_0^{(s,\delta)}, \mathcal{L}]u, \phi_1 S_0^{(s,\delta)}u\rangle_{L^2(B^n(1),h\sigma_{\mathrm{Leb}};\mathbb{C}^D)}| \\ &\quad\le (\mathrm{l.c.})\sum_{\deg_d(\alpha)\le\kappa}\|W^\alpha\phi_2\widetilde{S}^{(s,\delta)}u\|^2_{L^2_{-\epsilon_0}} + (\mathrm{s.c.})\sum_{\deg_d(\alpha)\le\kappa}\|W^\alpha\phi_1 S_0^{(s,\delta)}u\|^2_{L^2(B^n(1),\sigma_{\mathrm{Leb}};\mathbb{C}^D)},\end{aligned}$$

for all $u \in \mathscr{S}(\mathbb{R}^n;\mathbb{C}^D)'$.

Furthermore, if $u \in \mathscr{S}(\mathbb{R}^n;\mathbb{C}^D)$, we may replace $S_0^{(s,\delta)}$ with $S^{(s)}$ and $\widetilde{S}^{(s,\delta)}$ with $\widetilde{S}^{(s)}$.

Proof. We prove only the result for $u \in \mathscr{S}(\mathbb{R}^n;\mathbb{C}^D)'$. The same proof applies for $u \in \mathscr{S}(\mathbb{R}^n;\mathbb{C}^D)$ (by replacing $S_0^{(s,\delta)}$ with $S^{(s)}$ and $\widetilde{S}^{(s,\delta)}$ with $\widetilde{S}^{(s)}$ throughout the proof). When

$u \in \mathscr{S}(\mathbb{R}^n;\mathbb{C}^D)'$ we require that the pseudo-differential operators be (qualitatively) infinitely smoothing to guarantee that we are working with smooth functions; this is not necessary when $u \in \mathscr{S}(\mathbb{R}^n;\mathbb{C}^D)$.

Fix $\phi_1, \phi_2, \phi_3 \in C_0^\infty(B^n(1))$ with $\phi_1 \prec \phi_2 \prec \phi_3$. We prove the result with ϕ_2 replaced by ϕ_3. Applying Lemma 8.3.12, we have

$$\begin{aligned}
&\left|\langle [\mathrm{Mult}[\phi_1] S_0^{(s,\delta)}, \mathscr{L}] u, \phi_1 S_0^{(s,\delta)} u\rangle_{L^2(B^n(1),h\sigma_{\mathrm{Leb}};\mathbb{C}^D)}\right| \\
&= \left| \sum_{\substack{\deg_d(\alpha),\deg_d(\beta)\le\kappa \\ \deg_d(\alpha)+\deg_d(\beta)\le 2\kappa-1}} \langle \phi_2 W^\alpha W^\beta \phi_3 S_{\alpha,\beta}^{(s,\delta)} u, \phi_1 S_0^{(s,\delta)} u\rangle_{L^2(B^n(1),h\sigma_{\mathrm{Leb}};\mathbb{C}^D)}\right|.
\end{aligned} \tag{8.60}$$

By the calculus of pseudo-differential operators, we see that the right-hand side of (8.60) is bounded by a finite sum of terms of the form

$$\begin{aligned}
&\sum_{\substack{\deg_d(\alpha),\deg_d(\beta)\le\kappa \\ \deg_d(\alpha)+\deg_d(\beta)\le 2\kappa-1}} \left|\langle W^\beta \phi_3 \widetilde{S}_{\alpha,\beta}^{(s,\delta)} u, W^\alpha \phi_1 S_0^{(s,\delta)} u\rangle_{L^2(B^n(1),h\sigma_{\mathrm{Leb}};\mathbb{C}^D)}\right| \\
&\quad \le \left| \sum_{\substack{\deg_d(\alpha)\le\kappa \\ \deg_d(\beta)\le\kappa-1}} \cdot \right| + \left| \sum_{\substack{\deg_d(\alpha)\le\kappa-1 \\ \deg_d(\beta)\le\kappa}} \cdot \right| =: (\mathrm{I}) + (\mathrm{II}).
\end{aligned}$$

We begin with (I). Using the Cauchy–Schwartz inequality, (8.53), and Lemma 8.3.3 (ii), we have

$$\begin{aligned}
(\mathrm{I}) &\le (\mathrm{l.c.}) \sum_{\deg_d(\beta)\le\kappa-1} \|W^\beta \phi_3 \widetilde{S}_{\alpha,\beta}^{(s,\delta)} u\|_{L^2(B^n(1),\sigma_{\mathrm{Leb}})}^2 + (\mathrm{s.c.}) \sum_{\deg_d(\alpha)\le\kappa} \|W^\alpha \phi_1 S_0^{(s,\delta)} u\|_{L^2(B^n(1),\sigma_{\mathrm{Leb}})}^2 \\
&\le (\mathrm{l.c.}) \sum_{\deg_d(\beta)\le\kappa} \|W^\beta \phi_3 \widetilde{S}_{\alpha,\beta}^{(s,\delta)} u\|_{L^2_{-\epsilon_0}}^2 + (\mathrm{s.c.}) \sum_{\deg_d(\alpha)\le\kappa} \|W^\alpha \phi_1 S_0^{(s,\delta)} u\|_{L^2(B^n(1),\sigma_{\mathrm{Leb}})}^2,
\end{aligned}$$

which is the desired bound.

We turn to (II). Let Λ^s be as in Definition 2.2.10 and let $S^{(-\epsilon_0)}$ denote a standard pseudo-differential operator of order $-\epsilon_0$. Using the Cauchy–Schwartz inequality, the calculus of pseudo-differential operators, (8.53), and Lemma 8.3.3 (ii), we have

$$\begin{aligned}
(\mathrm{II}) &\le \sum_{\substack{\deg_d(\alpha)\le\kappa-1 \\ \deg_d(\beta)\le\kappa}} \left|\langle S^{(-\epsilon_0)} W^\beta \phi_3 \widetilde{S}_{\alpha,\beta}^{(s,\delta)} u, \Lambda^{\epsilon_0} W^\alpha \phi_1 S_0^{(s,\delta)} u\rangle_{L^2(B^n(1),h\sigma_{\mathrm{Leb}};\mathbb{C}^D)}\right| \\
&\le (\mathrm{l.c.}) \sum_{\deg_d(\beta)\le\kappa} \|W^\beta \phi_3 \widetilde{S}_{\alpha,\beta}^{(s,\delta)} u\|_{L^2_{-\epsilon_0}}^2 + (\mathrm{s.c.}) \sum_{\deg_d(\alpha)\le\kappa-1} \|W^\alpha \phi_1 S_0^{(s,\delta)} u\|_{L^2_{\epsilon_0}}^2 \\
&\le (\mathrm{l.c.}) \sum_{\deg_d(\beta)\le\kappa} \|W^\beta \phi_3 \widetilde{S}_{\alpha,\beta}^{(s,\delta)} u\|_{L^2_{-\epsilon_0}}^2 + (\mathrm{s.c.}) \sum_{\deg_d(\alpha)\le\kappa} \|W^\alpha \phi_1 S_0^{(s,\delta)} u\|_{L^2(B^n(1),\sigma_{\mathrm{Leb}})}^2,
\end{aligned}$$

which is the desired bound, completing the proof. □

Proof of Proposition 8.3.6. We prove the proposition by verifying the conditions of Lemma 8.3.11. Let $\phi_1, \phi_2 \in C_0^\infty(B^n(1))$ with $\phi_1 \prec \phi_2$. Using (8.53), Lemma 8.3.3 (i), the Cauchy–Schwartz inequality, Lemma 8.3.13, and Lemma 8.3.3 (ii), we have, for $u \in \mathscr{S}(\mathbb{R}^n;\mathbb{C}^D)'$,

$$\begin{aligned}
&\sum_{\deg_{\vec{d}}(\alpha)\le\kappa} \|W^\alpha \phi_1 S_0^{(s,\delta)} u\|^2_{L^2(B^n(1),\sigma_{\mathrm{Leb}})} \\
&\approx \sum_{j=1}^r \|W_j^{n_j}\phi_1 S_0^{(s,\delta)}u\|^2_{L^2(B^n(1),\sigma_{\mathrm{Leb}})} + \|\phi_1 S_0^{(s,\delta)}u\|^2_{L^2(B^n(1),\sigma_{\mathrm{Leb}})} \\
&\lesssim \sum_{j=1}^r \|W_j^{n_j}\phi_1 S_0^{(s,\delta)}u\|^2_{L^2(B^n(1),h\sigma_{\mathrm{Leb}})} + \|\phi_2 S_0^{(s,\delta)}u\|^2_{L^2(B^n(1),\sigma_{\mathrm{Leb}})} \\
&\lesssim |\langle \mathscr{L}\phi_1 S_0^{(s,\delta)}u, \phi_1 S_0^{(s,\delta)}u\rangle_{L^2(B^n(1),h\sigma_{\mathrm{Leb}};\mathbb{C}^D)}| + \|\phi_2 S_0^{(s,\delta)}u\|^2_{L^2(B^n(1),\sigma_{\mathrm{Leb}})} \\
&\lesssim |\langle \phi_1 S_0^{(s,\delta)}\mathscr{L}u, \phi_1 S_0^{(s,\delta)}u\rangle_{L^2(B^n(1),h\sigma_{\mathrm{Leb}};\mathbb{C}^D)}| \\
&\quad + |\langle [\mathscr{L}, \mathrm{Mult}[\phi_1]S_0^{(s,\delta)}]u, \phi_1 S_0^{(s,\delta)}u\rangle_{L^2(B^n(1),h\sigma_{\mathrm{Leb}};\mathbb{C}^D)}| + \|\phi_2 S_0^{(s,\delta)}u\|^2_{L^2(B^n(1),\sigma_{\mathrm{Leb}})} \\
&\le (\mathrm{l.c.})\Big(\|\phi_1 S_0^{(s,\delta)}\mathscr{L}u\|^2_{L^2(B^n(1),\sigma_{\mathrm{Leb}})} + \sum_{\deg_{\vec{d}}(\alpha)\le\kappa}\|W^\alpha\phi_2\widetilde{S}^{(s,\delta)}u\|^2_{L^2_{-\epsilon_0}}\Big) \\
&\quad + (\mathrm{s.c.})\left(\sum_{\deg_{\vec{d}}(\alpha)\le\kappa}\|W^\alpha\phi_1 S_0^{(s,\delta)}u\|^2_{L^2(B^n(1),\sigma_{\mathrm{Leb}})}\right).
\end{aligned}$$

Subtracting $(\mathrm{s.c.})(\sum_{\deg_{\vec{d}}(\alpha)\le\kappa}\|W^\alpha\phi_1 S_0^{(s,\delta)}u\|^2_{L^2(B^n(1),\sigma_{\mathrm{Leb}})})$ from both sides of the above equation yields

$$\begin{aligned}
&\sum_{\deg_{\vec{d}}(\alpha)\le\kappa} \|W^\alpha \phi_1 S_0^{(s,\delta)} u\|^2_{L^2(B^n(1),\sigma_{\mathrm{Leb}})} \\
&\quad\lesssim \|\phi_1 S_0^{(s,\delta)}\mathscr{L}u\|^2_{L^2(B^n(1),\sigma_{\mathrm{Leb}})} + \sum_{\deg_{\vec{d}}(\alpha)\le\kappa}\|W^\alpha\phi_2\widetilde{S}^{(s,\delta)}u\|^2_{L^2_{-\epsilon_0}}.
\end{aligned}\tag{8.61}$$

Combining

$$\|\phi_1 S_0^{(s,\delta)}\mathscr{L}u\|^2_{L^2(B^n(1),\sigma_{\mathrm{Leb}})} \lesssim \|\phi_2\mathscr{L}u\|^2_{L^2_s} + \|u\|_{L^2_{-\infty}}$$

with (8.61) shows that

$$\begin{aligned}
&\sum_{\deg_{\vec{d}}(\alpha)\le\kappa} \|W^\alpha \phi_1 S_0^{(s,\delta)} u\|^2_{L^2(B^n(1),\sigma_{\mathrm{Leb}})} \\
&\quad\lesssim \|\phi_2\mathscr{L}u\|^2_{L^2_s} + \sum_{\deg_{\vec{d}}(\alpha)\le\kappa}\|W^\alpha\phi_2\widetilde{S}^{(s,\delta)}u\|^2_{L^2_{-\epsilon_0}} + \|u\|^2_{L^2_{-\infty}}.
\end{aligned}$$

Lemma 8.3.11 applies to complete the proof. □

Proof of Corollary 8.3.8. Let L^2 denote $L^2(B^n(1), \sigma_{\mathrm{Leb}}; \mathbb{C}^D)$. Let $\eta_j \in C_0^\infty(B^n(1))$, $j \in \mathbb{N}$, be a sequence with $\phi_1 = \eta_0 \prec \eta_1 \prec \eta_2 \prec \cdots \prec \phi_2$. Then, using Proposition 8.3.6, we have, for $N \in \mathbb{N}$,

$$\begin{aligned}
\|\phi_1 f\|_{L^2_s} = \|\eta_0 f\|_{L^2_s} &\lesssim \|\eta_1 \mathscr{L} f\|_{L^2_{s-\kappa\epsilon_0}} + \|\eta_1 f\|_{L^2} \\
&\lesssim \|\eta_2 \mathscr{L}^2 f\|_{L^2_{s-2\kappa\epsilon_0}} + \|\eta_2 \mathscr{L} f\|_{L^2} + \|\eta_2 f\|_{L^2} \\
&\lesssim \cdots \\
&\lesssim \|\eta_N \mathscr{L}^N f\|_{L^2_{s-N\kappa\epsilon_0}} + \|\eta_N \mathscr{L}^{N-1} f\|_{L^2} + \|\eta_N \mathscr{L}^{N-2} f\|_{L^2} + \cdots + \|\eta_N f\|_{L^2}.
\end{aligned}$$

Taking $N = \lceil \frac{s}{\kappa\epsilon_0} \rceil$ and using $\eta_N \prec \phi_2$ completes the proof. □

All that remains in this section is to prove Proposition 8.3.9. We henceforth assume that $\mathscr{L}$ is symmetric and non-negative on $L^2(B^n(1), h\sigma_{\mathrm{Leb}}; \mathbb{C}^D)$ with dense domain $C_0^\infty(B^n(1); \mathbb{C}^D)$. We continue to assume that $\mathscr{L}$ is maximally subelliptic of type 2 of degree 2κ at the unit scale with respect to $(W, \vec{d})$ and $h\sigma_{\mathrm{Leb}}$ (see Definition 8.3.1). In particular, since $\mathscr{L}$ is non-negative, this implies there exists $A \geq 0$ with

$$\sum_{j=1}^{r} \|W_j^{n_j} f\|^2_{L^2(B^n(1), h\sigma_{\mathrm{Leb}}; \mathbb{C}^D)} \leq A(\langle \mathscr{L} f, f\rangle_{L^2(B^n(1), h\sigma_{\mathrm{Leb}}; \mathbb{C}^D)} + \|f\|^2_{L^2(B^n(1), h\sigma_{\mathrm{Leb}}; \mathbb{C}^D)}), \tag{8.62}$$

$\forall f \in C_0^\infty(B^n(1); \mathbb{C}^D)$.

Lemma 8.3.14. *For all $s \in \mathbb{R}$ and $\phi_1, \phi_2 \in C_0^\infty(B^n(1))$ with $\phi_1 \prec \phi_2$, there exist s, ϕ_1, ϕ_2-$\mathscr{L}$-unit-admissible constants $C_1, D_1 \geq 0$ such that $\forall f \in C_0^\infty(B^n(1); \mathbb{C}^D)$,*

$$\begin{aligned}
\sum_{\deg_{\vec{d}}(\alpha) \leq \kappa} \|W^\alpha \phi_1 f\|^2_{L^2_s} \leq C_1 \operatorname{Re}\langle \phi_2 \Lambda^s \phi_1 \mathscr{L} f, \phi_2 \Lambda^s \phi_1 f\rangle_{L^2(B^n(1), h\sigma_{\mathrm{Leb}}; \mathbb{C}^D)} \\
+ D_1 \sum_{\deg_{\vec{d}}(\alpha) \leq \kappa} \|W^\alpha \phi_2 f\|^2_{L^2_{s-\epsilon_0}}.
\end{aligned}$$

Proof. In this proof, we use the (l.c.) and (s.c.) notation, but with an added twist. For real numbers $A_1, A_2 \in \mathbb{R}$, $B_1, B_2 \geq 0$, we write $A_1 \leq A_2 + (\text{l.c.})B_1 + (\text{s.c.})B_2$ to mean that for all $\epsilon > 0$, there exists $C_\epsilon > 0$ with $A_1 \leq A_2 + C_\epsilon B_1 + \epsilon B_2$. Throughout this proof, f will denote an arbitrary element of $C_0^\infty(B^n(1); \mathbb{C}^D)$.

Using the calculus of pseudo-differential operators (Theorem 2.2.8), we have, with $\phi_3 \in C_0^\infty(B^n(1))$, with $\phi_2 \prec \phi_3$, and using (8.53) and Lemma 8.3.3 (i) and (ii), where $S^{(s)}$ and $\widetilde{S}^{(s)}$, perhaps with a subscript, are standard pseudo-differential operators of order s which may change from line to line,

$$\begin{aligned}
\sum_{\deg_{\vec{d}}(\alpha) \leq \kappa} \|W^\alpha \phi_1 f\|_{L^2_s} &\approx \sum_{\deg_{\vec{d}}(\alpha) \leq \kappa} \|\phi_2 \Lambda^s W^\alpha \phi_1 f\|_{L^2(\sigma_{\mathrm{Leb}})} \\
&\lesssim \sum_{\deg_{\vec{d}}(\alpha) \leq \kappa} \|W^\alpha \phi_2 \Lambda^s \phi_1 f\|_{L^2(\sigma_{\mathrm{Leb}})} + \sum_{\deg_{\vec{d}}(\alpha) \leq \kappa - 1} \|W^\alpha \phi_3 S^{(s)}_\alpha \phi_1 f\|_{L^2(\sigma_{\mathrm{Leb}})}
\end{aligned}$$

$$
\begin{aligned}
&\lesssim \sum_{j=1}^{r} \|W_j^{n_j}\phi_2\Lambda^s\phi_1 f\|_{L^2(\sigma_{\mathrm{Leb}})} + \sum_{\deg_{\vec{d}}(\alpha)\le\kappa-1} \|W^\alpha \phi_2 \widetilde{S}_\alpha^{(s)}\phi_1 f\|_{L^2(\sigma_{\mathrm{Leb}})} \qquad (8.63)\\
&\lesssim \sum_{j=1}^{r} \|W_j^{n_j}\phi_2\Lambda^s\phi_1 f\|_{L^2(\sigma_{\mathrm{Leb}})} + \sum_{\deg_{\vec{d}}(\alpha)\le\kappa} \|W^\alpha \phi_2 \widetilde{S}_\alpha^{(s)}\phi_1 f\|_{L^2_{-\epsilon_0}}\\
&\lesssim \sum_{j=1}^{r} \|W_j^{n_j}\phi_2\Lambda^s\phi_1 f\|_{L^2(\sigma_{\mathrm{Leb}})} + \sum_{\deg_{\vec{d}}(\alpha)\le\kappa} \|\widetilde{S}_\alpha^{(s)} W^\alpha \phi_2 f\|_{L^2_{-\epsilon_0}}\\
&\lesssim \sum_{j=1}^{r} \|W_j^{n_j}\phi_2\Lambda^s\phi_1 f\|_{L^2(h\sigma_{\mathrm{Leb}})} + \sum_{\deg_{\vec{d}}(\alpha)\le\kappa} \|W^\alpha \phi_2 f\|_{L^2_{s-\epsilon_0}}.
\end{aligned}
$$

Using Lemma 8.3.13 (with ϕ_1 in that lemma replaced by ϕ_2) and the calculus of pseudo-differential operators, we have, for $\phi_3 \in C_0^\infty(B^n(1))$ with $\phi_3 \prec \phi_2$,

$$
\begin{aligned}
&|\langle [\mathscr{L}, \phi_2\Lambda^s\phi_1] f, \phi_2\Lambda^s\phi_1 f\rangle_{L^2(h\sigma_{\mathrm{Leb}})}|\\
&\quad = |\langle [\mathscr{L}, \phi_2(\Lambda^s\phi_1)]\phi_2 f, \phi_2(\Lambda^s\phi_1)\phi_2 f\rangle_{L^2(h\sigma_{\mathrm{Leb}})}|\\
&\quad \le (\mathrm{l.c.}) \sum_{\deg_{\vec{d}}(\alpha)\le\kappa} \|W^\alpha\phi_3 \widetilde{S}^{(s,\delta)}\phi_2 f\|_{L^2_{-\epsilon_0}}^2 + (\mathrm{s.c.}) \sum_{\deg_{\vec{d}}(\alpha)\le\kappa} \|W^\alpha \phi_2(\Lambda^s\phi_1)\phi_2 f\|_{L^2}^2 \qquad (8.64)\\
&\quad \le (\mathrm{l.c.}) \sum_{\deg_{\vec{d}}(\alpha)\le\kappa} \|W^\alpha\phi_2 f\|_{L^2_{s-\epsilon_0}}^2 + (\mathrm{s.c.}) \sum_{\deg_{\vec{d}}(\alpha)\le\kappa} \|W^\alpha \phi_2\Lambda^s\phi_1 f\|_{L^2}^2.
\end{aligned}
$$

Turning to the first term on the right-hand side of (8.63), using (8.62), (8.53), Lemma 8.3.3 (iii), and (8.64), we have

$$
\begin{aligned}
&\sum_{j=1}^{r} \|W_j^{n_j}\phi_2\Lambda^s\phi_1 f\|_{L^2(h\sigma_{\mathrm{Leb}})}^2\\
&\quad \lesssim \mathrm{Re}\langle \mathscr{L}\phi_2\Lambda^s\phi_1 f, \phi_2\Lambda^s\phi_1 f\rangle_{L^2(h\sigma_{\mathrm{Leb}})} + \|\phi_2\Lambda^s\phi_1 f\|_{L^2(h\sigma_{\mathrm{Leb}})}^2\\
&\quad \lesssim \mathrm{Re}\langle \mathscr{L}\phi_2\Lambda^s\phi_1 f, \phi_2\Lambda^s\phi_1 f\rangle_{L^2(h\sigma_{\mathrm{Leb}})} + \sum_{\deg_{\vec{d}}(\alpha)\le\kappa} \|W^\alpha\phi_2\Lambda^s\phi_1 f\|_{L^2_{-\kappa\epsilon_0}}^2\\
&\quad \lesssim \mathrm{Re}\langle \mathscr{L}\phi_2\Lambda^s\phi_1 f, \phi_2\Lambda^s\phi_1 f\rangle_{L^2(h\sigma_{\mathrm{Leb}})} + \sum_{\deg_{\vec{d}}(\alpha)\le\kappa} \|W^\alpha\phi_2 f\|_{L^2_{s-\kappa\epsilon_0}}^2 \qquad (8.65)\\
&\quad = \mathrm{Re}\langle \phi_2\Lambda^s\phi_1\mathscr{L} f, \phi_2\Lambda^s\phi_1 f\rangle_{L^2(h\sigma_{\mathrm{Leb}})} + \mathrm{Re}\langle [\mathscr{L}, \phi_2\Lambda^s\phi_1] f, \phi_2\Lambda^s\phi_1 f\rangle_{L^2(h\sigma_{\mathrm{Leb}})}\\
&\qquad + \sum_{\deg_{\vec{d}}(\alpha)\le\kappa} \|W^\alpha\phi_2 f\|_{L^2_{s-\kappa\epsilon_0}}^2\\
&\quad \le \mathrm{Re}\langle \phi_2\Lambda^s\phi_1\mathscr{L} f, \phi_2\Lambda^s\phi_1 f\rangle_{L^2(h\sigma_{\mathrm{Leb}})} + (\mathrm{l.c.}) \sum_{\deg_{\vec{d}}(\alpha)\le\kappa} \|W^\alpha\phi_2 f\|_{L^2_{s-\epsilon_0}}^2\\
&\qquad + (\mathrm{s.c.}) \sum_{\deg_{\vec{d}}(\alpha)\le\kappa} \|W^\alpha\phi_1 f\|_{L^2_s}^2.
\end{aligned}
$$

Combining (8.63) and (8.65) shows that there exists a constant $C \geq 0$ such that

$$\begin{aligned}\sum_{\deg_d(\alpha)\leq\kappa} \|W^\alpha \phi_1 f\|_{L^2_s} \leq C\,\mathrm{Re}\langle \phi_2\Lambda^s\phi_1\mathscr{L}f, \phi_2\Lambda^s\phi_1 f\rangle_{L^2(h\sigma_{\mathrm{Leb}})} \\ + (\text{l.c.})\sum_{\deg_d(\alpha)\leq\kappa} \|W^\alpha\phi_2 f\|^2_{L^2_{s-\epsilon_0}} + (\text{s.c.})\sum_{\deg_d(\alpha)\leq\kappa} \|W^\alpha\phi_1 f\|^2_{L^2_s}.\end{aligned} \tag{8.66}$$

Subtracting (s.c.) $\sum_{\deg_d(\alpha)\leq\kappa} \|W^\alpha\phi_1 f\|^2_{L^2_s}$ from both sides of (8.66) completes the proof. □

For the next lemmas, we consider functions $u(t,x)$, where $t \in \mathbb{R}$ and $x \in \mathbb{R}^n$. We let Λ_x^s denote the pseudo-differential operator in the x variable with symbol $(1+|\xi|^2)^{s/2}$, where ξ is dual to x. We use the space $L^2(\mathbb{R}, dt; L^2_s)$ to denote the Hilbert space with norm

$$\|u\|^2_{L^2(\mathbb{R},dt;L^2_s)} := \int_{-\infty}^{\infty} \|u(t,\cdot)\|^2_{L^2_s(\mathbb{R}^n)}\,dt.$$

Lemma 8.3.15. *Let $u(t,x) \in C^\infty_{\mathrm{loc}}((-1,1)\times B^n(1);\mathbb{C}^D)$ and let $\phi_1,\phi_2 \in C_0^\infty(B^n(1))$ and $\psi_1,\psi_2 \in C_0^\infty((-1,1))$ satisfy $\phi_1 \prec \phi_2$ and $\psi_1 \prec \psi_2$. Then, for $s \in \mathbb{R}$,*

$$\left|\int_{-\infty}^{\infty} |\psi_1(t)|^2\,\mathrm{Re}\langle \phi_2\Lambda_x^s\phi_1\partial_t u(t,\cdot), \phi_2\Lambda_x^s\phi_1 u(t,\cdot)\rangle_{L^2(B^n(1),h\sigma_{\mathrm{Leb}};\mathbb{C}^D)}\,dt\right| \lesssim \|\psi_2\phi_1 u\|^2_{L^2(\mathbb{R},dt;L^2_s)}.$$

Proof. We have

$$\begin{aligned}&\left|\int_{-\infty}^{\infty} |\psi_1(t)|^2\,\mathrm{Re}\langle \phi_2\Lambda_x^s\phi_1\partial_t u(t,\cdot), \phi_2\Lambda_x^s\phi_1 u(t,\cdot)\rangle_{L^2(h\sigma_{\mathrm{Leb}})}\,dt\right| \\ &= \frac{1}{2}\left|\int_{-\infty}^{\infty} |\psi_1(t)|^2\partial_t\|\phi_2\Lambda_x^s\phi_1 u(t,\cdot)\|^2_{L^2(h\sigma_{\mathrm{Leb}})}\,dt\right| \\ &= \frac{1}{2}\left|\int_{-\infty}^{\infty} (\partial_t|\psi_1(t)|^2)\|\phi_2\Lambda_x^s\phi_1 u(t,\cdot)\|^2_{L^2(h\sigma_{\mathrm{Leb}})}\,dt\right| \\ &\lesssim \int_{-\infty}^{\infty} |\psi_2(t)|^2\|\phi_1(\cdot)u(t,\cdot)\|^2_{L^2_s}\,dt = \|\psi_2\phi_1 u\|^2_{L^2(\mathbb{R},dt;L^2_s)}.\end{aligned}$$

□

Lemma 8.3.16. *Let $u(t,x) \in C^\infty_{\mathrm{loc}}((-1,1)\times B^n(1);\mathbb{C}^D)$ satisfy $\partial_t u = -\mathscr{L}u$. Let $\phi_1,\phi_2 \in C_0^\infty(B^n(1))$ and $\psi_1,\psi_2 \in C_0^\infty((-1,1))$ satisfy $\phi_1 \prec \phi_2$ and $\psi_1 \prec \psi_2$. Then, for $s \in \mathbb{R}$,*

$$\sum_{\deg_d(\alpha)\leq\kappa} \|\psi_1 W^\alpha\phi_1 u\|_{L^2(\mathbb{R},dt;L^2_s)} \lesssim \sum_{\deg_d(\alpha)\leq\kappa} \|\psi_2 W^\alpha\phi_2 u\|_{L^2(\mathbb{R},dt;L^2_{s-\epsilon_0})}.$$

Proof. We have, using Lemma 8.3.14,

$$\begin{aligned}
\sum_{\deg_d(\alpha)\le\kappa} \|\psi_1 W^\alpha \phi_1 u\|^2_{L^2(\mathbb{R},dt;L^2_s)} &= \int_{-\infty}^{\infty} |\psi_1(t)|^2 \sum_{\deg_d(\alpha)\le\kappa} \|W^\alpha \phi_1 u(t,\cdot)\|^2_{L^2_s}\,dt \\
&\le C_1 \int_{-\infty}^{\infty} |\psi_1(t)|^2 \operatorname{Re}\langle \phi_2\Lambda^s_X\phi_1\mathscr{L}u(t,\cdot), \phi_2\Lambda^s_X\phi_1 u(t,\cdot)\rangle_{L^2(h\sigma_{\mathrm{Leb}})}\,dt \\
&\quad + D_1 \int_{-\infty}^{\infty} |\psi_1(t)|^2 \sum_{\deg_d(\alpha)\le\kappa} \|W^\alpha \phi_2 u(t,\cdot)\|^2_{L^2_{s-\epsilon_0}}\,dt \qquad (8.67)\\
&= -C_1 \int_{-\infty}^{\infty} |\psi_1(t)|^2 \operatorname{Re}\langle \phi_2\Lambda^s_X\phi_1\partial_t u(t,\cdot), \phi_2\Lambda^s_X\phi_1 u(t,\cdot)\rangle_{L^2(h\sigma_{\mathrm{Leb}})}\,dt \\
&\quad + D_1 \sum_{\deg_d(\alpha)\le\kappa} \|\psi_1 W^\alpha \phi_2 u\|^2_{L^2(\mathbb{R},dt;L^2_{s-\epsilon_0})}. \\
&\lesssim \|\psi_2\phi_1 u\|^2_{L^2(\mathbb{R},dt;L^2_s)} + \sum_{\deg_d(\alpha)\le\kappa} \|\psi_1 W^\alpha \phi_2 u\|^2_{L^2(\mathbb{R},dt;L^2_{s-\epsilon_0})},
\end{aligned}$$

where the last inequality follows from Lemma 8.3.15. Since $\psi_1 \prec \psi_2$, we have

$$\sum_{\deg_d(\alpha)\le\kappa} \|\psi_1 W^\alpha \phi_2 u\|_{L^2(\mathbb{R},dt;L^2_{s-\epsilon_0})} \lesssim \sum_{\deg_d(\alpha)\le\kappa} \|\psi_2 W^\alpha \phi_2 u\|^2_{L^2(\mathbb{R},dt;L^2_{s-\epsilon_0})},$$

and therefore the second term on the right-hand side of (8.67) has the desired bound.

We finish the proof by estimating the first term on the right-hand side of (8.67). Using Lemma 8.3.3 (iii) and the calculus of pseudo-differential operators, we have, for $f \in C^\infty_{\mathrm{loc}}(B^n(1))$,

$$\begin{aligned}
\|\phi_1 f\|_{L^2_s} \approx \|\phi_2\Lambda^s\phi_1 f\|_{L^2(\sigma_{\mathrm{Leb}})} &\lesssim \sum_{\deg_d(\alpha)\le\kappa} \|W^\alpha\phi_2\Lambda^s\phi_1 f\|_{L^2_{-\kappa\epsilon_0}} \\
&\lesssim \sum_{\deg_d(\alpha)\le\kappa} \|\widetilde{S}^{(s)} W^\alpha \phi_2 f\|_{L^2_{-\kappa\epsilon_0}} \lesssim \sum_{\deg_d(\alpha)\le\kappa} \|W^\alpha\phi_2 f\|_{L^2_{s-\kappa\epsilon_0}} \\
&\lesssim \sum_{\deg_d(\alpha)\le\kappa} \|W^\alpha\phi_2 f\|_{L^2_{s-\epsilon_0}}.
\end{aligned}$$

Thus, we have

$$\|\psi_2\phi_1 u\|_{L^2(\mathbb{R},dt;L^2_s)} \lesssim \sum_{\deg_d(\alpha)\le\kappa} \|\psi_2 W^\alpha \phi_2 u\|^2_{L^2(\mathbb{R},dt;L^2_{s-\epsilon_0})},$$

which shows that the first term on the right-hand side of (8.67) also has the desired bound, completing the proof. □

Lemma 8.3.17. *Let $u(t,x) \in C^\infty_{\mathrm{loc}}((-1,1)\times B^n(1);\mathbb{C}^D)$ satisfy $\partial_t u = -\mathscr{L}u$. Let $\phi_1,\phi_2 \in C^\infty_0(B^n(1))$ and $\psi_1,\psi_2 \in C^\infty_0((-1,1))$ satisfy $\phi_1 \prec \phi_2$ and $\psi_1 \prec \psi_2$. Then, for $s \in \mathbb{R}$,*

$$\|\psi_1\phi_1 u\|_{L^2(\mathbb{R},dt;L^2_s)} \lesssim \|\psi_2\phi_2 u\|_{L^2(\mathbb{R}\times B^n(1),dt\times(h\sigma_{\mathrm{Leb}}))}.$$

Proof. Pick sequences $\{\eta_j\}_{j\in\mathbb{N}} \subset C_0^\infty(B^n(1))$ and $\{\gamma_j\}_{j\in\mathbb{N}} \subset C_0^\infty((-1,1))$ with $\phi_1 = \eta_0 \prec \eta_1 \prec \eta_2 \prec \cdots \prec \phi_2$ and $\psi_1 = \gamma_0 \prec \gamma_1 \prec \gamma_2 \prec \cdots \prec \psi_2$. By repeated application of Lemma 8.3.16, we have, for every $N \in \mathbb{N}$,

$$\begin{aligned}
\|\psi_1\phi_1 u\|_{L^2(\mathbb{R},dt;L^2_s)} &\le \sum_{\deg_d(\alpha)\le\kappa} \|\gamma_0 W^\alpha \eta_0 u\|_{L^2(\mathbb{R},dt;L^2_s)} \\
&\lesssim \sum_{\deg_d(\alpha)\le\kappa} \|\gamma_1 W^\alpha \eta_1 u\|_{L^2(\mathbb{R},dt;L^2_{s-\epsilon_0})} \lesssim \sum_{\deg_d(\alpha)\le\kappa} \|\gamma_2 W^\alpha \eta_2 u\|_{L^2(\mathbb{R},dt;L^2_{s-2\epsilon_0})} \\
&\lesssim \cdots \lesssim \sum_{\deg_d(\alpha)\le\kappa} \|\gamma_N W^\alpha \eta_N u\|_{L^2(\mathbb{R},dt;L^2_{s-N\epsilon_0})} \\
&\lesssim \|\gamma_N \eta_N u\|_{L^2(\mathbb{R},dt;L^2_{s+\kappa-N\epsilon_0})} \lesssim \|\psi_2\phi_2 u\|_{L^2(\mathbb{R},dt;L^2_{s+\kappa-N\epsilon_0})}.
\end{aligned}$$

Taking N so large that $N\epsilon_0 \ge s+\kappa$ and using (8.53) completes the proof. □

Lemma 8.3.18. *Let $u(t,x) \in C^\infty_{\mathrm{loc}}((-1,1)\times B^n(1);\mathbb{C}^D)$ satisfy $\partial_t u = -\mathscr{L}u$. Then*

$$\|\partial_t u\|_{L^2((-1/2,1/2)\times B^n(1/2),dt\times d(h\sigma_{\mathrm{Leb}});\mathbb{C}^D)} \lesssim \|u\|_{L^2((-1,1)\times B^n(1),dt\times d(h\sigma_{\mathrm{Leb}});\mathbb{C}^D)}.$$

Proof. Let $\phi_1,\phi_2 \in C_0^\infty(B^n(1))$ and $\psi_1,\psi_2 \in C_0^\infty((-1,1))$ satisfy $\phi_1 \prec \phi_2$, $\psi_1 \prec \psi_2$, $\phi_1 \equiv 1$ on $B^n(1/2)$, and $\psi_1 \equiv 1$ on $B^n(-1/2,1/2)$. Using (8.53) and Lemma 8.3.17, we have

$$\begin{aligned}
\|\partial_t u\|_{L^2((-1/2,1/2)\times B^n(1/2),dt\times(h\sigma_{\mathrm{Leb}}))} &= \|\mathscr{L}u\|_{L^2((-1/2,1/2)\times B^n(1/2),dt\times(h\sigma_{\mathrm{Leb}}))} \\
&\approx \|\mathscr{L}u\|_{L^2((-1/2,1/2)\times B^n(1/2),dt\times\sigma_{\mathrm{Leb}})} \\
&\lesssim \|\psi_1\phi_1 u\|_{L^2(\mathbb{R},dt;L^2_{2\kappa})} \lesssim \|\psi_2\phi_2 u\|_{L^2(\mathbb{R}\times B^n(1),dt\times d(h\sigma_{\mathrm{Leb}}))} \\
&\lesssim \|u\|_{L^2((-1,1)\times B^n(1),dt\times d(\sigma_{\mathrm{Leb}}))}.
\end{aligned}$$
□

Lemma 8.3.19. *There exists an $\mathscr{L}$-unit-admissible constant $C \ge 0$ such that $\forall \epsilon \in (0,1/2]$, $\forall u \in C^\infty_{\mathrm{loc}}((-\epsilon^{-2\kappa},\epsilon^{-2\kappa})\times B^n(1);\mathbb{C}^D)$ satisfying $\partial_t u = -\mathscr{L}u$, $\forall k \in [1,(2\epsilon)^{-1}]$, we have*

$$\begin{aligned}
&\|\partial_t u\|_{L^2((-\epsilon^{-2\kappa}+(k\epsilon)\epsilon^{-2\kappa},\epsilon^{-2\kappa}-(k\epsilon)\epsilon^{-2\kappa})\times B^n(1/2),dt\times d(h\sigma_{\mathrm{Leb}});\mathbb{C}^D)} \\
&\quad \le C\|u\|_{L^2((-\epsilon^{-2\kappa}+((k-1)\epsilon)\epsilon^{-2\kappa},\epsilon^{-2\kappa}-((k-1)\epsilon)\epsilon^{-2\kappa})\times B^n(1),dt\times d(h\sigma_{\mathrm{Leb}});\mathbb{C}^D)}.
\end{aligned} \tag{8.68}$$

Proof. Note that

$$\begin{aligned}
&(-\epsilon^{-2\kappa}+((k-1)\epsilon)\epsilon^{-2\kappa},\epsilon^{-2\kappa}-((k-1)\epsilon)\epsilon^{-2\kappa}) \setminus (-\epsilon^{-2\kappa}+(k\epsilon)\epsilon^{-2\kappa},\epsilon^{-2\kappa}-(k\epsilon)\epsilon^{-2\kappa}) \\
&\quad = (-\epsilon^{-2\kappa}+((k-1)\epsilon)\epsilon^{-2\kappa},-\epsilon^{-2\kappa}+(k\epsilon)\epsilon^{-2\kappa}] \bigcup [\epsilon^{-2\kappa}-(k\epsilon)\epsilon^{-2\kappa},\epsilon^{-2\kappa}-((k-1)\epsilon)\epsilon^{-2\kappa}),
\end{aligned}$$

which are two intervals of length $\epsilon^{-2\kappa+1} \ge \epsilon^{-1} \ge 2$. Thus, we may cover $(\epsilon^{-2\kappa}+(k\epsilon)\epsilon^{-2\kappa}$, $\epsilon^{-2\kappa}-(k\epsilon)\epsilon^{-2\kappa})$ by a finite collection of intervals of the form $(t_i-1/2,t_i+1/2)$, $i=1,\dots,L$, where $t_i \in (\epsilon^{-2\kappa}+(k\epsilon)\epsilon^{-2\kappa},\epsilon^{-2\kappa}-(k\epsilon)\epsilon^{-2\kappa})$, $(t_i-1,t_i+1) \subseteq (-\epsilon^{-2\kappa}+((k-1)\epsilon)\epsilon^{-2\kappa},\epsilon^{-2\kappa}-((k-

$1)\epsilon)\epsilon^{-2\kappa})$, and no point lies in more than four of the intervals $(t_i - 1, t_i + 1)$. Lemma 8.3.18 shows that

$$\|\partial_t u\|_{L^2((t_i-1/2,t_i+1/2)\times B^n(1/2),dt\times d(h\sigma_{\mathrm{Leb}}))} \lesssim \|u\|_{L^2((t_i-1,t_i+1)\times B^n(1),dt\times d(h\sigma_{\mathrm{Leb}}))}.$$

We conclude that

$$\begin{aligned}
&\|\partial_t u\|^2_{L^2((-\epsilon^{-2\kappa}+(k\epsilon)\epsilon^{-2\kappa},\epsilon^{-2\kappa}-(k\epsilon)\epsilon^{-2\kappa})\times B^n(1/2),dt\times d(h\sigma_{\mathrm{Leb}}))} \\
&\quad \le \sum_{i=1}^{L} \|\partial_t u\|^2_{L^2((t_i-1/2,t_i+1/2)\times B^n(1/2),dt\times d(h\sigma_{\mathrm{Leb}}))} \\
&\quad \lesssim \sum_{i=1}^{L} \|u\|^2_{L^2((t_i-1,t_i+1)\times B^n(1),dt\times d(h\sigma_{\mathrm{Leb}}))} \\
&\quad \lesssim \|u\|^2_{L^2((-\epsilon^{-2\kappa}+((k-1)\epsilon)\epsilon^{-2\kappa},\epsilon^{-2\kappa}-((k-1)\epsilon)\epsilon^{-2\kappa})\times B^n(1),dt\times d(h\sigma_{\mathrm{Leb}}))}.
\end{aligned}$$ □

Proof of Proposition 8.3.9. Set $u_\epsilon(t,x) := u(\epsilon^{2\kappa}t, x)$, so that $\partial_t u_\epsilon = -\mathscr{L}u$. Lemma 8.3.19 implies that (8.68) holds with u replaced by u_ϵ. The result follows. □

8.3.3 Exponential estimates

To prove Theorem 8.1.1 (ix) we need to conclude exponential estimates. Remarkably, Proposition 8.3.9 combined with some technical scaling techniques implies the required exponential estimates. In fact, in the proof of Proposition 8.3.9, we used simple scaling techniques in the t variable applied to Lemma 8.3.18. Thus, we will obtain the required exponential estimates by combining scaling techniques (in both variables (t,x)) with the result at the unit scale Lemma 8.3.18. The main result of this section is the following.

Proposition 8.3.20. *Let $\mathscr{L}$ be given by* (8.42). *Suppose that $\mathscr{L}$, thought of as a densely defined operator on $L^2(B^n(1), h\sigma_{\mathrm{Leb}}; \mathbb{C}^D)$ with dense domain $C_0^\infty(B^n(1);\mathbb{C}^D)$, is symmetric and non-negative and suppose that $\mathscr{L}$ is maximally subelliptic of type 2 of degree 2κ at the unit scale with respect to $(W, \vec{d})$ and $h\sigma_{\mathrm{Leb}}$. Then there exists an $\mathscr{L}$-unit-admissible constant $c > 0$ such that for every ordered multi-index α, there exists an α-$\mathscr{L}$-unit-admissible constant $C_\alpha \ge 0$ such that the following holds. If $u(t,x) \in C^\infty_{\mathrm{loc}}((-1,1)\times B^n(1);\mathbb{C}^D)$ satisfies $\partial_t u(t,x) = -\mathscr{L}u(t,x)$ and $u(t,x) \equiv 0$ for $t \in (-1,0]$, then $\forall t \in (-1/4, 1/4)$,*

$$\sup_{x\in B^n(1/4)} |W^\alpha u(t,x)| \le C_\alpha \exp(-ct^{-\frac{1}{2\kappa-1}})\|u\|_{L^2((-1,1)\times B^n(1);\mathbb{C}^D)}.$$

We prove Proposition 8.3.20 by proving $u(t,x)$ has a certain Gevrey regularity in the t variable. This idea originates in the work of Jerison and Sánchez-Calle [134], who use it to study the heat kernel associated with Hörmander's sub-Laplacian – see Section 8.10 for more details on the history of this proof.

The rest of this section is devoted to the proof of Proposition 8.3.20; we henceforth assume $\mathscr{L}$ is as given in that proposition. The heart of the proof is a technical, quantitative scaling result.

Fix $0 < \epsilon_1 < \epsilon_2 \leq 1$ and set $\epsilon := \epsilon_2 - \epsilon_1$; assume $\epsilon \in (0, 1/2]$. It is important for the proof that all estimates in this section are independent of the particular choice of ϵ_1, ϵ_2. Informally, we wish to apply Theorem 3.3.7 with $\mathcal{K} = \overline{B^n(\epsilon_1)}$ and $\Omega_1 = B^n(\epsilon_2)$; however, if one proceeds in this way without care, then the estimates so obtained depend on ϵ_1 and ϵ_2. Instead, we prove an analog of Theorem 3.3.7 where all of the estimates are $\mathscr{L}$-unit-admissible and, in particular, do not depend on ϵ_1 or ϵ_2. Like Theorem 3.3.7, this result follows easily from Theorem 3.6.5.

First set

$$(X, d) = \{(X_1, d_1), \dots, (X_q, d_q)\} := \Big\{(Z, e) \in \mathrm{Gen}((W, \vec{d})) : e \leq m \max_{1 \leq j \leq r} \vec{d}_j\Big\}.$$

In particular, since $W_1, \dots, W_r$ satisfy Hörmander's condition of order m on $B^n(1)$, we have

$$\mathrm{span}\{X_1(x), \dots, X_q(x)\} = T_x B^n(1), \quad \forall x \in B^n(1).$$

Moreover, by the definition of unit-admissible constants (in particular, see (3.4)), we have

$$\inf_{x \in B^n(1)} \max_{j_1, \dots, j_n \in \{1, \dots, q\}} \big|\det(X_{j_1}(x)| \cdots |X_{j_n}(x))\big| \gtrsim 1. \tag{8.69}$$

Set $\mathrm{Vol} := h\sigma_{\mathrm{Leb}}$ and

$$\begin{aligned} \Lambda(x, \delta) &:= \max_{j_1, \dots, j_n \in \{1, \dots, q\}} \mathrm{Vol}(x)(\delta^{d_{j_1}} X_{j_1}(x), \dots, \delta^{d_{j_n}} X_{j_n}(x)) \\ &= \max_{j_1, \dots, j_n \in \{1, \dots, q\}} h(x) \big|\det(X_{j_1}(x)| \cdots |X_{j_n}(x))\big|. \end{aligned} \tag{8.70}$$

The main scaling result we use is the next proposition.

Proposition 8.3.21. *There exist $\mathscr{L}$-unit-admissible constants $c_1, c_2 \in (0, 1]$, not depending on ϵ_1 or ϵ_2, such that the following hold:*
(a) $B_{(X,d)}(x, 2c_1\epsilon) \subseteq B^n(\epsilon_2)$, $\forall x \in B^n(\epsilon_1)$.
(b) $\forall \delta \in (0, 6c_1\epsilon]$, $\forall x \in B^n(\epsilon_1)$,

$$\mathrm{Vol}(B_{(X,d)}(x, 2\delta)) \lesssim \mathrm{Vol}(B_{(X,d)}(x, \delta)).$$

For every $x \in B^n(\epsilon_1)$, $\delta \in (0, c_1\epsilon]$, there exists $\Phi_{x,\delta} : B^n(1) \to B_{(X,d)}(x, \delta)$ such that:
(c) $\Phi_{x,\delta}(0) = x$.
(d) *$\Phi_{x,\delta}$ is a smooth coordinate system, that is, $\Phi_{x,\delta}(B^n(1)) \subseteq B^n(1)$ is open and $\Phi_{x,\delta} : B^n(1) \to \Phi_{x,\delta}(B^n(1))$ is a C^∞_{loc} diffeomorphism.*
(e) $B_{(X,d)}(x, c_2\delta) \subseteq \Phi_{x,\delta}(B^n(1/2)) \subseteq \Phi_{x,\delta}(B^n(1)) \subseteq B_{(X,d)}(x, \delta)$.

Set $W_j^{x,\delta} := \Phi_{x,\delta}^* \delta^{d_j} W_j$ *and* $X_k^{x,\delta} := \delta^{d_k} X_k$. *Let* $(W^{x,\delta}, \vec{d}) = \{(W_1^{x,\delta}, \vec{d}_1), \dots, (W_r^{x,\delta}, \vec{d}_r)\}$ *and* $(X^{x,\delta}, d) = \{(X_1^{x,\delta}, d_1), \dots, (X_q^{x,\delta}, d_q)\}$.

(f) $W_j^{x,\delta}$ *and* $X_k^{x,\delta}$ *are in* $C^\infty(B^n(1); \mathbb{R}^n)$ *in the quantitative sense that* $\forall L \in \mathbb{N}$, *there exists an* L-$\mathscr{L}$-*unit-admissible constant* C_L *such that for* $1 \le j \le r$, $1 \le k \le q$,

$$\sup_{\substack{x \in B^n(\epsilon_1)\\ \delta \in (0, c_1\epsilon]}} \|W_j^{x,\delta}\|_{C^L(B^n(1);\mathbb{R}^n)} \le C_L, \quad \sup_{\substack{x \in B^n(\epsilon_1)\\ \delta \in (0, c_1\epsilon]}} \|X_k^{x,\delta}\|_{C^L(B^n(1);\mathbb{R}^n)} \le C_L.$$

(g) $X_1^{x,\delta}(u), \dots, X_q^{x,\delta}(u)$ *span* $T_u B^n(1)$ *uniformly in the sense that*

$$\inf_{\substack{x \in B^n(\epsilon_1)\\ \delta \in (0, c_1\epsilon]}} \max_{j_1, \dots, j_n \in \{1, \dots, q\}} \inf_{u \in B^n(1)} \left|\det(X_{j_1}^{x,\delta}(u)| \cdots |X_{j_n}^{x,\delta}(u))\right| \gtrsim 1.$$

(h) $(W^{x,\delta}, \vec{d})$ *are Hörmander vector fields with formal degrees at the unit scale and all of the estimates defining* L-*unit-admissible constants with respect to* $(W^{x,\delta}, \vec{d})$ *as in Definition* 3.2.1 *can be chosen to be* L-$\mathscr{L}$-*unit-admissible constants.*

(i) $\forall x \in B^n(\epsilon_1)$, $\delta \in (0, c_1\epsilon]$, $L \in \mathbb{N}$,

$$\|f\|_{C^L(B^n(1))} \approx \sum_{|\alpha| \le L} \|(X^{x,\delta})^\alpha f\|_{C(B^n(1))}, \quad \forall f \in C(B^n(1)).$$

For $x \in B^n(\epsilon_1)$ *and* $\delta \in (0, c_1\epsilon]$ *define* $h_{x,\delta} \in C^\infty(B^n(1))$ *by* $\Phi_{x,\delta}^* \operatorname{Vol} = \Lambda(x,\delta) h_{x,\delta} \sigma_{\mathrm{Leb}}$. *Then:*

(j) $\forall x \in B^n(\epsilon_1)$, $\delta \in (0, c_1\epsilon]$, $u \in B^n(1)$, $h_{x,\delta}(u) \approx 1$. *Furthermore, for every* $L \in \mathbb{N}$,

$$\sup_{\substack{x \in B^n(\epsilon_1)\\ \delta \in (0, c_1\epsilon]}} \|h_{x,\delta}\|_{C^L(B^n(1))} \lesssim 1,$$

where the implicit constant is an L-$\mathscr{L}$-*unit-admissible constant.*

Proof. We prove this result by applying Theorem 3.6.5 with x_0 replaced by $x \in B^n(\epsilon_1)$ and $X_1, \dots, X_q$ replaced by $\delta^{d_1} X_1, \dots, \delta^{d_q} X_q$ for $\delta \in (0, c_3\epsilon]$, where $c_3 \in (0,1]$ is a small $\mathscr{L}$-unit-admissible constant to be chosen later.

The central aspect of this proof is to show that L-admissible constants and L, Vol-admissible constants (as in Definition 3.6.4) in this application of Theorem 3.6.5 are L-$\mathscr{L}$-unit-admissible constants. Thus, we must verify that each of the quantities in Definition 3.6.4 can be bounded by L-$\mathscr{L}$-unit-admissible constants.

By the Picard–Lindelöf theorem, if $c_3 \in (0,1]$ is sufficiently small (depending only on m, q, and $\max_{1 \le j \le q} \|X_j\|_{C(B^n(1);\mathbb{R}^n)}$), then for $\delta \in (0, c_3\epsilon]$, $\delta^{d_1} X_1, \dots, \delta^{d_q} X_q$ satisfy $C(x, 1, B^n(1))$, $\forall x \in B^n(\epsilon_1)$ (see Definition 3.6.3). Indeed, $\delta^{d_j} \le (c_3\epsilon)^{d_j} \le c_3\epsilon$. Therefore, by taking $c_3 > 0$ small, each $\delta^{d_k} X_k$ is a small linear combination of $\epsilon\partial_{x_1}, \dots, \epsilon\partial_{x_n}$. Since $\epsilon_1 + \epsilon = \epsilon_2 \le 1$, it follows that $\delta^{d_1} X_1, \dots, \delta^{d_q} X_q$ satisfy $C(x, 1, B^n(1))$. In particular, we make take $\eta = 1$ in our application of Theorem 3.6.5. Moreover, by the same argument, by taking $c_3 > 0$ small, we have, for $x \in B^n(\epsilon_1)$ and $\delta \in (0, c_3\epsilon]$,

$$B_{\delta^d X}(x) = B_{(X,d)}(x, \delta) \subseteq B^n(\epsilon_2).$$

We will choose $c_1 \le c_3/2$, and therefore (a) holds.

Lemma A.2.11 shows that $\tau_0 > 0$ can be chosen to be an $\mathcal{L}$-unit-admissible constant.

In the application of Theorem 3.6.5, we take $\xi = 1$. We claim that

$$[X_j, X_k] = \sum_{d_l \le d_j + d_k} c_{j,k}^l X_l, \quad c_{j,k}^l \in C^\infty(B^n(1)), \tag{8.71}$$

and $\forall L \in \mathbb{N}$, $\|c_{j,k}^l\|_{C^L(B^n(1))} \lesssim 1$. Indeed, if $d_j + d_k \le m \max_{1\le p\le r} \check{d}_p$, then $([X_j, X_k], d_j + d_k) \in (X, d)$, and therefore (8.71) holds with one of the $c_{j,k}^l$ equal to 1 and the rest equal to 0. If $d_j + d_k > m \max_{1\le p\le r} \check{d}_p$, then we use $\|X_p\|_{C^L(B^n(1))} \lesssim 1$ and (8.69) to write

$$[X_j, X_k] = \sum_{l=1}^{q} c_{j,k}^l X_l, \quad c_{j,k}^l \in C^\infty(B^n(1)),$$

where $\forall L \in \mathbb{N}$, $\|c_{j,k}^l\|_{C^L(B^n(1))} \lesssim 1$. Since $d_j + d_k > m \max_{1\le p\le r} \check{d}_p \ge \max\{d_l : 1 \le l \le q\}$, (8.71) follows.

Multiplying (8.71) by $\delta^{d_j + d_k}$, we see that

$$[\delta^{d_j} X_j, \delta^{d_k} X_k] = \sum_{l=1}^{q} c_{j,k}^{l,\delta} \delta^{d_l} X_l,$$

where

$$c_{j,k}^{l,\delta} = \begin{cases} \delta^{d_l + d_k - d_l} c_{j,k}^l, & d_l \le d_j + d_k, \\ 0, & d_l > d_j + d_k. \end{cases}$$

Consider, $\forall \delta \in (0, c_3\epsilon] \subseteq (0, 1]$ and $x \in B^n(1)$, we have, for any ordered multi-index α,

$$\begin{aligned} \max_{j,k,l} \|(\delta^d X)^\alpha c_{j,k}^{l,\delta}\|_{C(B_{\delta^d X}(x))} &= \max_{j,k,l} \|(\delta^d X)^\alpha c_{j,k}^{l,\delta}\|_{C(B_{(X,d)}(x,\delta))} \\ &\le \max_{d_l \le d_j + d_k} \|X^\alpha c_{j,k}^l\|_{C(B^n(1))} \lesssim 1, \end{aligned}$$

where the implicit constant is an α-$\mathcal{L}$-unit-admissible constant.

Consider

$$\mathrm{Lie}_{\delta^{d_k} X_k} \mathrm{Vol} = (\delta^{d_k} X_k h)\sigma_{\mathrm{Leb}} + \delta^{d_k} h \,\mathrm{div}(X_k)\sigma_{\mathrm{Leb}} =: f_{k,\delta}\sigma_{\mathrm{Leb}}.$$

We have, for $L \in \mathbb{N}$,

$$\sum_{|\alpha|\le L} \|(\delta^d X)^\alpha f_{j,\delta}\|_{C(B_{\delta^d X}(x))} \lesssim \sum_{|\alpha|\le L} \|X^\alpha X_j h\|_{C(B^n(1))} + \sum_{|\alpha|\le L} \|X^\alpha (h\,\mathrm{div}(X_k))\|_{C(B^n(1))} \lesssim 1.$$

We use $\zeta = 1$ in Theorem 3.6.5, so that (after reordering the vector fields) the left-hand side of (3.16) equals 1. This is always possible: see the remarks following (3.16). In Theorem 3.6.5, the vector fields are reordered so that the first n vector fields satisfy (3.16). This reordering depends on x and δ. We proceed without reordering the vector fields, so that $X_1, \ldots, X_n$ in that theorem are really of the form $\delta^{d_{j_1(x,\delta)}} X_{j_1(x,\delta)}, \ldots, \delta^{d_{j_n(x,\delta)}} X_{j_n(x,\delta)}$ for some $j_1(x,\delta), \ldots, j_n(x,\delta) \in \{1, \ldots, q\}$.

It follows from the above that L-admissible and L, Vol-admissible constants (as in Definition 3.6.4) are L-$\mathscr{L}$-unit-admissible constants in our application of Theorem 3.6.5. We let $\Psi_{x,\delta}$ be the map Φ from Theorem 3.6.5 when applied with $x_0 = x \in B^n(\epsilon_1)$ and $X_1, \ldots, X_q$ replaced by $\delta^{d_1} X_1, \ldots, \delta^{d_q} X_q$, where $\delta \in (0, c_3\epsilon]$, and we let $\eta_1 \in (0,1]$ be the 1-admissible constant from that theorem. We set $\Phi_{x,\delta}(t) := \Psi_{x,\delta}(\eta_1 t) : B^n(1) \xrightarrow{\sim} \Phi_{x,\delta}(B^n(1))$.

Item (c) follows from the definition of $\Phi_{x,\delta}(t)$ (see (3.17)). Item (d) follows from Theorem 3.6.5, (a), and (b).

For (e), we let $c_2 = \xi_2$, where ξ_2 is as in Theorem 3.6.5. Then, by Theorem 3.6.5 (c), we have

$$\begin{aligned} B_{(X,d)}(x, c_2\delta) &= B_{(c_2\delta)^d X}(x) \subseteq B_{\xi_2 \delta^d X}(x) \subseteq \Phi_{x,\delta}(B^n(1/2)) \\ &= \Psi_{x,\delta}(B^n(1/2)) \subseteq \Phi_{x,\delta}(B^n(1)) = \Psi_{x,\delta}(B^n(1)) \subseteq B_{\delta^d X}(x) \\ &= B_{(X,d)}(x, \delta). \end{aligned} \tag{8.72}$$

Thus, (e) holds (by picking $c_1 \le c_3$).

Since $\Phi_{x,\delta}^* \delta^{d_k} X_k(u) = \eta_1^{-1} \Psi_{x,\delta}^* \delta^{d_k} X_k(\eta_1 u) = \eta_1^{-1} Y_k(\eta_1 u)$, where Y_k is as in Theorem 3.6.5 and $\eta_1 \approx 1$, it follows from Theorem 3.6.5 (e) that $\|X_k^{x,\delta}\|_{C^L(B^n(1);\mathbb{R}^n)} \lesssim 1$. Since each $W_j^{x,\delta}$ is of the form $X_k^{x,\delta}$ for some k, (f) follows.

By Theorem 3.6.5 (d), we have

$$\inf_{u \in B^n(\eta)} \max_{j_1, \ldots, j_n \in \{1, \ldots, q\}} \left|\det(Y_{j_1}(u) | \cdots | Y_{j_n}(u))\right| \ge \sigma_n,$$

where $\sigma_n > 0$ depends only on n. Since $X_k^{x,\delta}(u) = \eta_1^{-1} Y_k(\eta_1 u)$ and $\eta_1 \approx 1$, (g) follows.

(h): Since $(X, d) = \{(Z, e) \in \mathrm{Gen}((W, \vec{d})) : e \le m \max \vec{d}_j\}$, it follows that

$$(X^{x,\delta}, d) = \{(Z^{x,\delta}, e) \in \mathrm{Gen}((W^{x,\delta}, \vec{d})) : e \le m \max \vec{d}_j\}.$$

Thus, (h) follows from (f) and (g).

Again using the fact that $X_k^{x,\delta}(u) = \eta_1^{-1} Y_k(\eta_1 u)$ and $\eta_1 \approx 1$, (i) follows from Theorem 3.6.5 (f).

(j): Let $\tilde{h}$ be the function h from Theorem 3.6.5. Since $\Phi_{x,\delta}(t) = \Psi_{x,\delta}(\eta_1 t)$, we see that $h_{x,\delta}(t) = \eta_1^{-n} \tilde{h}(\eta_1 t)$. Since $\eta_1 \approx 1$, (j) follows from Theorem 3.6.5 (g) and (h).

All that remains to show is (b). This will follow once we show that if $c_1 \in (0,1]$ is a sufficiently small $\mathscr{L}$-unit-admissible constant, then for $\delta \in (0, 12c_1\epsilon]$, we have

$$\mathrm{Vol}(B_{(X,d)}(x,\delta)) \approx \Lambda(x,\delta). \tag{8.73}$$

It is immediately clear that $\Lambda(x, 2\delta) \lesssim \Lambda(x,\delta)$ (see (8.70)), and therefore (b) follows from (8.73).

We turn to proving (8.73). By (e) we have, for $\delta \in (0, c_3\epsilon]$ and $x \in B^n(\epsilon_1)$,

$$\mathrm{Vol}(B_{(X,d)}(x, c_2\delta)) \le \mathrm{Vol}(\Phi_{x,\delta}(B^n(1))) \le \mathrm{Vol}(B_{(X,d)}(x,\delta)). \tag{8.74}$$

By (j),

$$\mathrm{Vol}(\Phi_{x,\delta}(B^n(1))) = \Lambda(x,\delta) \int_{B^n(1)} h_{x,\delta}(t)\, dt \approx \Lambda(x,\delta). \tag{8.75}$$

Combining (8.74) and (8.75) shows that

$$\begin{aligned} \Lambda(x,\delta) &\lesssim \mathrm{Vol}(B_{(X,d)}(x,\delta)), \quad \delta \in (0, c_3\epsilon], \\ \mathrm{Vol}(B_{(X,d)}(x,\delta)) &\lesssim \Lambda(x,\delta/c_2) \approx \Lambda(x,\delta), \quad \delta \in (0, c_2c_3\epsilon]. \end{aligned}$$

Thus, so long as $c_1 \le \frac{c_2c_3}{12}$, (8.73) follows, completing the proof. □

Let $c_1, c_2 \in (0,1]$ be the unit-$\mathscr{L}$-admissible constants in Proposition 8.3.21. Henceforth, we fix $\delta := c_1\epsilon \approx \epsilon$ and we apply Proposition 8.3.21 with this choice of δ. Let $\Phi_{x,\delta}$, $(W^{x,\delta}, \vec{d})$, $(X^{x,\delta}, d)$, and $\mathrm{Vol}_{x,\delta} := \Phi_{x,\delta}^*(h\sigma_{\mathrm{Leb}}) = \Lambda(x,\delta)h_{x,\delta}\sigma_{\mathrm{Leb}}$ be as in that proposition.

Set, for $x \in B^n(\epsilon_1)$,

$$\mathscr{L}^{x,\epsilon} := \Phi_{x,\delta}^* \epsilon^{2\kappa} \mathscr{L}(\Phi_{x,\delta})_*.$$

Note that $\mathscr{L}^{x,\epsilon}$ is a partial differential operator with smooth coefficients on $B^n(1)$.

Lemma 8.3.22. *With $\mathscr{L} = \mathscr{L}^{x,\epsilon}$, $h = h_{x,\delta}$, and $(W, \vec{d}) = (W^{x,\delta}, \vec{d})$, the hypotheses of Proposition 8.3.9 hold, and $\mathscr{L}^{x,\epsilon}$-unit-admissible constants can be taken to be $\mathscr{L}$-unit-admissible constants. In particular, the conclusion of Proposition 8.3.9 holds for $\mathscr{L}^{x,\epsilon}$ and $h_{x,\delta}$, where $C \ge 0$ is an $\mathscr{L}$-unit-admissible constant.*

Proof. That the vector fields with formal degree $(W^{x,\delta}, \vec{d})$ are of the correct form is Proposition 8.3.21 (h). That $h_{x,\delta}$ satisfies the correct estimates is Proposition 8.3.21 (j).

We next show that $\mathscr{L}^{x,\epsilon}$ is of the correct form. Multiplying (8.42) by $\epsilon^{2\kappa}$ and using $\delta = c_1\epsilon$, we have

$$\epsilon^{2\kappa}\mathscr{L} = \sum_{\deg_{\vec{d}}(\alpha), \deg_{\vec{d}}(\beta) \le \kappa} \epsilon^{2\kappa - \deg_{\vec{d}}(\alpha) - \deg_{\vec{d}}(\beta)} c_1^{\deg_{\vec{d}}(\alpha) + \deg_{\vec{d}}(\beta)} b_{\alpha,\beta}(x) (\delta^{\vec{d}} W)^\alpha (\delta^{\vec{d}} W)^\beta.$$

Therefore,

$$\mathscr{L}^{x,\epsilon} = \sum_{\deg_{d}(\alpha),\deg_{d}(\beta)\leq\kappa} b^{x,\epsilon}_{\alpha,\beta}(W^{x,\delta})^{\alpha}(W^{x,\delta})^{\beta},$$

where

$$b^{x,\epsilon}_{\alpha,\beta} = \epsilon^{2\kappa-\deg_{d}(\alpha)-\deg_{d}(\beta)} c_1^{\deg_{d}(\alpha)+\deg_{d}(\beta)} b_{\alpha,\beta}\circ\Phi_{x,\delta}.$$

We wish to show that, for every $L\in\mathbb{N}$,

$$\|b^{x,\epsilon}_{\alpha,\beta}\|_{C^L(B^n(1))} \lesssim 1, \tag{8.76}$$

where the implicit constant is an L-$\mathscr{L}$-unit-admissible constant. By Proposition 8.3.21 (i), we have, using $0<\epsilon, c_1, \delta\leq 1$,

$$\begin{aligned}\|b^{x,\epsilon}_{\alpha,\beta}\|_{C^L(B^n(1))} &\approx \sum_{|\alpha|\leq L} \|(X^{x,\delta})^{\alpha} b^{x,\epsilon}_{\alpha,\beta}\|_{C(B^n(1))}\\ &= \sum_{|\alpha|\leq L} \epsilon^{2\kappa-\deg_{d}(\alpha)-\deg_{d}(\beta)} c_1^{\deg_{d}(\alpha)+\deg_{d}(\beta)} \|(\delta^{d}X)^{\alpha} b_{\alpha,\beta}\|_{C(\Phi_{x,\delta}(B^n(1)))}\\ &\leq \sum_{|\alpha|\leq L} \|X^{\alpha} b_{\alpha,\beta}\|_{C(B^n(1))} \lesssim 1,\end{aligned}$$

establishing (8.76).

We require that $\mathscr{L}^{x,\epsilon}$ is symmetric and non-negative on $L^2(B^n(1), h_{x,\delta}\sigma_{\mathrm{Leb}}; \mathbb{C}^D)$ when thought of as an operator with dense domain $C_0^\infty(B^n(1);\mathbb{C}^D)$. By assumption, $\mathscr{L}$ is symmetric and non-negative, and therefore, $\forall f_1, f_2\in C_0^\infty(B^n(1);\mathbb{C}^D)$, we have

$$\langle \epsilon^{2\kappa}\mathscr{L}f_1, f_2\rangle_{L^2(h\sigma_{\mathrm{Leb}})} = \langle f_1, \epsilon^{2\kappa}\mathscr{L}f_2\rangle_{L^2(h\sigma_{\mathrm{Leb}})}, \tag{8.77}$$

$$\langle \epsilon^{2\kappa}\mathscr{L}f_1, f_1\rangle_{L^2(h\sigma_{\mathrm{Leb}})} \geq 0. \tag{8.78}$$

Let $g_1, g_2\in C_0^\infty(B^n(1);\mathbb{C}^D)$ and set $f_1 := g_1\circ\Phi_{x,\delta}$ and $f_2 := g_2\circ\Phi_{x,\delta}$. By Proposition 8.3.21 (d), $f_1, f_2\in C_0^\infty(B^n(1);\mathbb{C}^D)$. Plugging this choice of f_1, f_2 into (8.77) and (8.78), we obtain

$$\langle \mathscr{L}^{x,\epsilon}g_1, g_2\rangle_{L^2(\mathrm{Vol}_{x,\delta})} = \langle g_1, \mathscr{L}^{x,\epsilon}g_2\rangle_{L^2(\mathrm{Vol}_{x,\delta})}, \tag{8.79}$$

$$\langle \mathscr{L}^{x,\epsilon}g_1, g_1\rangle_{L^2(\mathrm{Vol}_{x,\delta})} \geq 0. \tag{8.80}$$

Dividing (8.79) and (8.80) by $\Lambda(x,\delta)$ shows that

$$\langle \mathscr{L}^{x,\epsilon}g_1, g_2\rangle_{L^2(h_{x,\delta}\sigma_{\mathrm{Leb}})} = \langle g_1, \mathscr{L}^{x,\epsilon}g_2\rangle_{L^2(h_{x,\delta}\sigma_{\mathrm{Leb}})},$$

$$\langle \mathscr{L}^{x,\epsilon}g_1, g_1\rangle_{L^2(h_{x,\delta}\sigma_{\mathrm{Leb}})} \geq 0.$$

Therefore, $\mathscr{L}^{x,\epsilon}$ is symmetric and non-negative, as desired.

Finally, we wish to show that $\mathscr{L}^{x,\epsilon}$ is maximally subelliptic of type 2 of degree 2κ at the unit scale with respect to $(W^{x,\delta}, \vec{d})$ and $h_{x,\delta}\sigma_{\mathrm{Leb}}$ and where the constant in the definition of maximally subelliptic of type 2 (see Definition 8.3.1) can be taken to be an $\mathscr{L}$-unit-admissible constant.

By assumption, $\mathscr{L}$ is symmetric, non-negative, and maximally subelliptic of type 2 of degree 2κ at the unit scale with respect to $(W, \vec{d})$ and $h\sigma_{\mathrm{Leb}}$. In other words, for $f \in C_0^\infty(B^n(1); \mathbb{C}^D)$, we have

$$\sum_{j=1}^r \|W_j^{n_j} f\|_{L^2(h\sigma_{\mathrm{Leb}})}^2 \leq A(\langle \mathscr{L}f, f\rangle_{L^2(h\sigma_{\mathrm{Leb}})} + \|f\|_{L^2(h\sigma_{\mathrm{Leb}})}^2), \tag{8.81}$$

where, by definition, $A \geq 0$ is an $\mathscr{L}$-unit-admissible constant. Multiplying both sides of (8.81) by $\delta^{2\kappa}$ and using $\delta \approx \epsilon$ and $0 < \epsilon \leq 1$, we see that

$$\begin{aligned}\sum_{j=1}^r \|(\delta^{d_j} W_j)^{n_j} f\|_{L^2(h\sigma_{\mathrm{Leb}})}^2 &\lesssim (\langle \epsilon^{2\kappa} \mathscr{L}f, f\rangle_{L^2(h\sigma_{\mathrm{Leb}})} + \epsilon^{2\kappa}\|f\|_{L^2(h\sigma_{\mathrm{Leb}})}^2) \\ &\leq (\langle \epsilon^{2\kappa} \mathscr{L}f, f\rangle_{L^2(h\sigma_{\mathrm{Leb}})} + \|f\|_{L^2(h\sigma_{\mathrm{Leb}})}^2).\end{aligned} \tag{8.82}$$

Let $g \in C_0^\infty(B^n(1); \mathbb{C}^D)$ and let $f := g \circ \Phi_{x,\delta}$. By Proposition 8.3.21 (d), $f \in C_0^\infty(B^n(1); \mathbb{C}^D)$. Plugging this choice of f into (8.82), we obtain

$$\sum_{j=1}^r \|(W_j^{x,\delta})^{n_j} g\|_{L^2(\mathrm{Vol}_{x,\delta})}^2 \lesssim (\langle \mathscr{L}^{x,\epsilon} g, g\rangle_{L^2(\mathrm{Vol}_{x,\delta})} + \|g\|_{L^2(\mathrm{Vol}_{x,\delta})}^2). \tag{8.83}$$

Dividing (8.83) by $\Lambda(x, \delta)$ gives

$$\sum_{j=1}^r \|(W_j^{x,\delta})^{n_j} g\|_{L^2(h_{x,\delta}\sigma_{\mathrm{Leb}})}^2 \lesssim (\langle \mathscr{L}^{x,\epsilon} g, g\rangle_{L^2(h_{x,\delta}\sigma_{\mathrm{Leb}})} + \|g\|_{L^2(h_{x,\delta}\sigma_{\mathrm{Leb}})}^2),$$

completing the proof. □

For the rest of this section, let $u(t,y) \in C_{\mathrm{loc}}^\infty(B^n(1) \times (-1,1); \mathbb{C}^D)$ satisfy $\partial_t u = -\mathscr{L}u$ and $u(t,y) \equiv 0$ for $t \in (-1, 0]$.

Lemma 8.3.23. *For $k \in [1, 1/2\epsilon]$ and $x \in B^n(\epsilon_1)$,*

$$\begin{aligned}&\|\partial_t u\|_{L^2((-1+k\epsilon, 1-k\epsilon)\times B_{(X,d)}(x,\epsilon c_2 c_1), dt\times d(h\sigma_{\mathrm{Leb}}); \mathbb{C}^D)} \\ &\quad\lesssim \epsilon^{-2\kappa} \|u\|_{L^2((-1+(k-1)\epsilon, 1-(k-1)\epsilon)\times B_{(X,d)}(x,\epsilon c_1), dt\times d(h\sigma_{\mathrm{Leb}}); \mathbb{C}^D)}.\end{aligned} \tag{8.84}$$

Proof. Fix $x \in B^n(\epsilon_1)$ and set $u_{x,\epsilon}(t,y) := u(t, \Phi_{x,\delta}(y))$. Note that $u_{x,\epsilon}$ satisfies $\partial_t u_{x,\epsilon}(t,y) = -\epsilon^{-2\kappa}\mathscr{L}^{x,\epsilon} u_{x,\epsilon}(t,y)$, where $\mathscr{L}^{x,\epsilon}$ is acting in the $y \in B^n(1)$ variable. Also, by Proposition 8.3.21 (d), $u_{x,\epsilon} \in C_{\mathrm{loc}}^\infty((-1,1) \times B^n(1); \mathbb{C}^D)$.

Lemma 8.3.22 shows that Proposition 8.3.9 applies to $u_{x,\epsilon}$ and the constant $C \geq 0$ from Proposition 8.3.9 is an $\mathscr{L}$-unit-admissible constant. In other words, for $k \in [1, 1/2\epsilon]$,

$$\begin{aligned} &\|\partial_t u_{x,\epsilon}\|_{L^2((-1+k\epsilon,1-k\epsilon)\times B^n(1/2),dt\times d(h_{x,\delta}\sigma_{\mathrm{Leb}}))} \\ &\quad \lesssim \epsilon^{-2\kappa}\|u_{x,\epsilon}\|_{L^2((-1+(k-1)\epsilon,1-(k-1)\epsilon)\times B^n(1),dt\times d(h_{x,\delta}\sigma_{\mathrm{Leb}}))}. \end{aligned} \tag{8.85}$$

Multiplying both sides of (8.85) by $\Lambda(x,\delta)^{\frac{1}{2}}$, we obtain

$$\begin{aligned} &\|\partial_t u_{x,\epsilon}\|_{L^2((-1+k\epsilon,1-k\epsilon)\times B^n(1/2),dt\times d\,\mathrm{Vol}_{x,\delta})} \\ &\quad \lesssim \epsilon^{-2\kappa}\|u_{x,\epsilon}\|_{L^2((-1+(k-1)\epsilon,1-(k-1)\epsilon)\times B^n(1),dt\times d\,\mathrm{Vol}_{x,\delta})}. \end{aligned} \tag{8.86}$$

Changing variables, (8.86) is equivalent to

$$\begin{aligned} &\|\partial_t u\|_{L^2((-1+k\epsilon,1-k\epsilon)\times\Phi_{x,\delta}(B^n(1/2)),dt\times d(h\sigma_{\mathrm{Leb}}))} \\ &\quad \lesssim \epsilon^{-2\kappa}\|u\|_{L^2((-1+(k-1)\epsilon,1-(k-1)\epsilon)\times\Phi_{x,\delta}(B^n(1)),dt\times d(h\sigma_{\mathrm{Leb}}))}. \end{aligned} \tag{8.87}$$

By Proposition 8.3.21 (e) and the definition $\delta = c_1\epsilon$, we have

$$B_{(X,d)}(x, c_1c_2\epsilon) \subseteq \Phi_{x,\delta}(B^n(1/2)) \subseteq \Phi_{x,\delta}(B^n(1)) \subseteq B_{(X,d)}(x, c_1\epsilon).$$

Thus, (8.87) implies (8.84), completing the proof. □

The next lemma has a standard proof (see, for example, [216, page 32]), though we include the proof to make clear how all our estimates involve only $\mathscr{L}$-unit-admissible constants.

Lemma 8.3.24. *There is a finite collection $x_1, \ldots, x_L \in B^n(\epsilon_1)$ and an $\mathscr{L}$-unit-admissible constant $M \in \mathbb{N}_+$ such that $\{B_{(X,d)}(x_i, \epsilon c_1c_2) : i = 1, \ldots, L\}$ is a cover for $B^n(\epsilon_1)$, no point lies in more than M of the balls $B_{(X,d)}(x_i, \epsilon c_1)$, $i = 1, \ldots L$, and for each i, $B_{(X,d)}(x_i, \epsilon c_1) \subseteq B^n(\epsilon_2)$.*

Proof. The keys to this proof are the doubling estimate Proposition 8.3.21 (b) and the fact that the balls $B_{(X,d)}(x,\delta)$ are metric balls. Let $B_{(X,d)}(x_i, (\epsilon c_1c_2)/2)$, $i = 1, \ldots, L$, be a maximal disjoint subcollection of $\{B_{(X,d)}(x, (\epsilon c_1c_2)/2) : x \in B^n(\epsilon_1)\}$. Note that $\bigcup_{i=1}^L B_{(X,d)}(x_i, \epsilon c_1c_2) \supseteq B^n(\epsilon_1)$, as if there were $y \in B^n(\epsilon_1) \setminus \bigcup_{i=1}^L B_{(X,d)}(x_0, \epsilon c_1c_2)$, then $B_{(X,d)}(y, (\epsilon c_1c_2)/2)$ would be disjoint from $B_{(X,d)}(x_i, (\epsilon c_1c_2)/2)$, $i = 1, \ldots L$, contradicting maximality.

Now suppose $y \in B_{(X,d)}(x_{i_1}, \epsilon c_1) \cap \cdots \cap B_{(X,d)}(x_{i_M}, \epsilon c_1)$; we wish to show that $M \lesssim 1$. Since $y \in B_{(X,d)}(x_{i_l}, \epsilon c_1)$, we have

$$B_{(X,d)}(x_{i_l}, (\epsilon c_1c_2)/2) \subseteq B_{(X,d)}(x_{i_l}, \epsilon c_1) \subseteq B_{(X,d)}(y, 2\epsilon c_1). \tag{8.88}$$

Because the balls $B_{(X,d)}(x_{i_l}, (\epsilon c_1c_2/2))$ are disjoint, (8.88) implies that

$$\mathrm{Vol}(B_{(X,d)}(y, 2\epsilon c_1)) \geq \sum_{l=1}^{M} \mathrm{Vol}(B_{(X,d)}(x_{i_l}, (\epsilon c_1 c_2)/2)). \tag{8.89}$$

Since $B_{(X,d)}(y, \epsilon c_1) \subseteq B_{(X,d)}(x_{i_l}, 2\epsilon c_1)$, repeated application of Proposition 8.3.21 (b) shows that

$$\begin{aligned}\mathrm{Vol}(B_{(X,d)}(x_{i_l}, (\epsilon c_1 c_2)/2)) &\approx \mathrm{Vol}(B_{(X,d)}(x_{i_l}, 2\epsilon c_1)) \\ &\geq \mathrm{Vol}(B_{(X,d)}(y, \epsilon c_1)) \approx \mathrm{Vol}(B_{(X,d)}(y, 2\epsilon c_1)).\end{aligned} \tag{8.90}$$

Plugging (8.90) into (8.89) shows that

$$\mathrm{Vol}(B_{(X,d)}(y, 2\epsilon c_1)) \gtrsim M\, \mathrm{Vol}(B_{(X,d)}(y, 2\epsilon c_1)),$$

which implies $M \lesssim 1$, as desired.

Finally, it follows from the fact that $x_i \in B^n(\epsilon_1)$ and Proposition 8.3.21 (a) that $B_{(X,d)}(x_i, \epsilon c_1) \subseteq B^n(\epsilon_2)$. □

Lemma 8.3.25. *For $k \in [1, 1/2\epsilon]$,*

$$\begin{aligned}&\|\partial_t u\|_{L^2((-1+k\epsilon,1-k\epsilon)\times B^n(\epsilon_1),dt\times d(h\sigma_{\mathrm{Leb}});\mathbb{C}^D)} \\ &\quad \lesssim \epsilon^{-2\kappa} \|u\|_{L^2((-1+(k-1)\epsilon,1-(k-1)\epsilon)\times B^n(\epsilon_2),dt\times d(h\sigma_{\mathrm{Leb}});\mathbb{C}^D)}.\end{aligned}$$

Proof. Take $x_1, \ldots, x_L \in B^n(\epsilon_1)$ as in Lemma 8.3.24. Using the fact that $B_{(X,d)}(x_i, \epsilon c_1 c_2)$ is a cover for $B^n(\epsilon_1)$, we have

$$\begin{aligned}&\|\partial_t u\|^2_{L^2((-1+k\epsilon,1-k\epsilon)\times B^n(\epsilon_1),dt\times d(h\sigma_{\mathrm{Leb}}))} \\ &\quad \leq \sum_{i=1}^{L} \|\partial_t u\|^2_{L^2((-1+k\epsilon,1-k\epsilon)\times B_{(X,d)}(x_i,\epsilon c_1 c_2),dt\times d(h\sigma_{\mathrm{Leb}}))} \\ &\quad \lesssim \sum_{i=1}^{L} \epsilon^{-4\kappa} \|u\|^2_{L^2((-1+(k-1)\epsilon,1-(k-1)\epsilon)\times B_{(X,d)}(x_i,\epsilon c_1),dt\times d(h\sigma_{\mathrm{Leb}}))},\end{aligned} \tag{8.91}$$

where the last estimate uses Lemma 8.3.23. Using the fact that no point lies in more than $M \lesssim 1$ of the balls $B_{(X,d)}(x_i, c_1\epsilon) \subseteq B^n(\epsilon_2)$, we have

$$\begin{aligned}&\sum_{i=1}^{L} \epsilon^{-4\kappa} \|u\|^2_{L^2((-1+(k-1)\epsilon,1-(k-1)\epsilon)\times B_{(X,d)}(x_i,\epsilon c_1),dt\times d(h\sigma_{\mathrm{Leb}}))} \\ &\quad \lesssim \epsilon^{-4\kappa} \|u\|^2_{L^2((-1+(k-1)\epsilon,1-(k-1)\epsilon)\times B^n(\epsilon_2),dt\times d(h\sigma_{\mathrm{Leb}}))}.\end{aligned} \tag{8.92}$$

Combining (8.91) and (8.92) and taking square roots completes the proof. □

Lemma 8.3.26. *For $l \in \mathbb{N}_+$ and $k \in \{1, \ldots, l\}$, we have*

$$\begin{aligned}&\|\partial_t u\|_{L^2((-1+\frac{k}{2l},1-\frac{k}{2l})\times B^n(1-\frac{k}{2l}),dt\times(h\sigma_{\mathrm{Leb}});\mathbb{C}^D)} \\ &\quad \lesssim l^{2\kappa} \|u\|_{L^2((-1+\frac{k-1}{2l},1-\frac{k-1}{2l})\times B^n(1-\frac{k-1}{2l}),dt\times d(h\sigma_{\mathrm{Leb}});\mathbb{C}^D)}.\end{aligned}$$

Proof. Apply Lemma 8.3.25 with $\epsilon_1 = 1 - \frac{k}{2l}$ and $\epsilon_2 = 1 - \frac{k-1}{2l}$, so that $\epsilon = \epsilon_2 - \epsilon_1 = \frac{1}{2l}$. The result follows. □

Lemma 8.3.27. *There exists an $\mathscr{L}$-unit-admissible constant $R_1 \geq 0$ such that $\forall l \in \mathbb{N}$,*

$$\begin{aligned}&\|\partial_t^l u\|_{L^2((-1/2,1/2)\times B^n(1/2),dt\times d(h\sigma_{\mathrm{Leb}});\mathbb{C}^D)}\\&\leq R_1^l (l!)^{2\kappa}\|u\|_{L^2((-1,1)\times B^n(1),dt\times d(h\sigma_{\mathrm{Leb}});\mathbb{C}^D)}.\end{aligned}$$

Proof. The case $l = 0$ is trivial, so we assume $l \geq 1$. We claim that there exists an $\mathscr{L}$-unit-admissible constant $C_1 \geq 0$ such that $\forall k \in \{1, \ldots, l\}$,

$$\begin{aligned}&\|\partial_t^k u\|_{L^2((-1+\frac{k}{2l},1-\frac{k}{2l})\times B^n(1-\frac{k}{2l}),dt\times d(h\sigma_{\mathrm{Leb}}))}\\&\leq C_1 l^{2\kappa}\|\partial_t^{k-1}u\|_{L^2((-1+\frac{k-1}{2l},1-\frac{k-1}{2l})\times B^n(1-\frac{k-1}{2l}),dt\times d(h\sigma_{\mathrm{Leb}}))}.\end{aligned} \tag{8.93}$$

Indeed, (8.93) follows by applying Lemma 8.3.26 to $\partial_t^{k-1}u$, which satisfies the same hypotheses as u (in particular, $\partial_t\partial_t^{k-1}u = -\mathscr{L}\partial_t^{k-1}u$).

Applying (8.93) l times gives

$$\|\partial_t^l u\|_{L^2((-1/2,1/2)\times B^n(1/2),dt\times d(h\sigma_{\mathrm{Leb}}))} \leq C_1^l l^{2\kappa l}\|u\|_{L^2((-1,1)\times B^n(1),dt\times d(h\sigma_{\mathrm{Leb}}))}.$$

The result now follows by Stirling's approximation. □

Lemma 8.3.28. *Let $v(t,x) \in C^\infty_{\mathrm{loc}}((0,1)\times B^n(1);\mathbb{C}^D)$. Then, for every ordered multi-index α, there exist an $\mathscr{L}$-unit-admissible constant $N = N(\alpha) \in \mathbb{N}$ and an α-$\mathscr{L}$-unit-admissible constant $C_\alpha \geq 0$ such that*

$$\sup_{\substack{t\in(-1/4,1/4)\\x\in B^n(1/4)}} |W_x^\alpha v(t,x)| \leq C_\alpha \sum_{r_1=0}^{N(\alpha)}\sum_{r_2=0}^{N(\alpha)} \|\partial_t^{r_1}\mathscr{L}^{r_2}v\|_{L^2((-1/2,1/2)\times B^n(1/2),dt\times d(h\sigma_{\mathrm{Leb}});\mathbb{C}^D)}.$$

Proof. Pick $\phi_1, \phi_2 \in C_0^\infty(B^n(3/8))$ with $\phi_1 \prec \phi_2$ and $\phi_1 \equiv 1$ on $B^n(1/4)$ and pick $\psi_1, \psi_2 \in C_0^\infty((-3/8,3/8))$ with $\psi_1 \prec \psi_2$ and $\psi_1 \equiv 1$ on $(-1/4,1/4)$. By the Sobolev embedding theorem there exists $s = s(|\alpha|, n) > 0$ (s depending only on $|\alpha|$ and n) so that the following holds:

$$\begin{aligned}\sup_{\substack{t\in(-1/4,1/4)\\x\in B^n(1/4)}} |W_x^\alpha v(t,x)| &\leq \sup_{t,x}|W_x^\alpha \phi_1(x)\psi_1(t)v(t,x)|\\&\lesssim \|\phi_1\psi_1 v\|_{L^2_s(\mathbb{R}\times\mathbb{R}^n)} \lesssim \sum_{r_1=0}^{\lceil s\rceil}\|\psi_2\phi_1\partial_t^{r_1}v(t,\cdot)\|_{L^2(\mathbb{R},dt;L^2_s(\mathbb{R}^n))}.\end{aligned} \tag{8.94}$$

Corollary 8.3.8 implies that

$$\|\psi_2\phi_1\partial_t^{r_1}v(t,\cdot)\|_{L^2(\mathbb{R},dt;L_s^2(\mathbb{R}^n))} \lesssim \sum_{r_2=0}^{\lceil \frac{s}{\kappa\epsilon_0}\rceil} \|\psi_2\phi_1\partial_t^{r_1}\mathscr{L}^{r_2}v\|_{L^2(\mathbb{R}\times\mathbb{R}^n,dt\times d\sigma_{\mathrm{Leb}})}. \tag{8.95}$$

Combining (8.94) and (8.95), using the support of ϕ_2 and ψ_2, and using (8.53), we have

$$\begin{aligned}\sup_{\substack{t\in(-1/4,1/4)\\ x\in B^n(1/4)}} |W_x^\alpha v(t,x)| &\lesssim \sum_{r_1=0}^{\lceil s\rceil}\sum_{r_2=0}^{\lceil \frac{s}{\kappa\epsilon_0}\rceil} \|\psi_2\phi_1\partial_t^{r_1}\mathscr{L}^{r_2}v\|_{L^2(\mathbb{R}\times\mathbb{R}^n,dt\times d\sigma_{\mathrm{Leb}})}\\ &\lesssim \sum_{r_1=0}^{\lceil s\rceil}\sum_{r_2=0}^{\lceil \frac{s}{\kappa\epsilon_0}\rceil} \|\partial_t^{r_1}\mathscr{L}^{r_2}v\|_{L^2((-1/2,1/2)\times B^n(1/2),dt\times d(h\sigma_{\mathrm{Leb}}))}.\end{aligned}$$

Taking $N(\alpha) := \max\{\lceil s\rceil, \lceil \frac{s}{\kappa\epsilon_0}\rceil\}$ completes the proof. □

Lemma 8.3.29. *There exists an $\mathscr{L}$-unit-admissible constant $R_2 > 0$ such that for all ordered multi-indices α, there exists an α-$\mathscr{L}$-unit-admissible constant $C_\alpha \geq 0$ such that for all $l \in \mathbb{N}$,*

$$\sup_{\substack{t\in(-1/4,1/4)\\ x\in B^n(1/4)}} |\partial_t^l W^\alpha u(t,x)| \leq C_\alpha R_2^l (l!)^{2\kappa} \|u\|_{L^2((-1,1)\times B^n(1),dt\times d(h\sigma_{\mathrm{Leb}});\mathbb{C}^D)}. \tag{8.96}$$

Proof. By Lemma 8.3.28, there exists an $\mathscr{L}$-unit-admissible constant $N(\alpha) \in \mathbb{N}$ such that

$$\begin{aligned}\sup_{\substack{t\in(-1/4,1/4)\\ x\in B^n(1/4)}} |\partial_t^l W^\alpha u(t,x)| &\lesssim \sum_{r_1=0}^{N(\alpha)}\sum_{r_2=0}^{N(\alpha)} \|\partial_t^{l+r_1}\mathscr{L}^{r_2}u\|_{L^2((-1/2,1/2)\times B^n(1/2),dt\times d(h\sigma_{\mathrm{Leb}}))}\\ &= \sum_{r_1=0}^{N(\alpha)}\sum_{r_2=0}^{N(\alpha)} \|\partial_t^{l+r_1+r_2}u\|_{L^2((-1/2,1/2)\times B^n(1/2),dt\times d(h\sigma_{\mathrm{Leb}}))},\end{aligned} \tag{8.97}$$

where in the equality we have used $\partial_t u = -\mathscr{L}u$. Applying Lemma 8.3.27, we have

$$\begin{aligned}&\sum_{r_1=0}^{N(\alpha)}\sum_{r_2=0}^{N(\alpha)} \|\partial_t^{l+r_1+r_2}u\|_{L^2((-1/2,1/2)\times B^n(1/2),dt\times d(h\sigma_{\mathrm{Leb}}))}\\ &\quad\lesssim \|u\|_{L^2((-1,1)\times B^n(1),dt\times d(h\sigma_{\mathrm{Leb}}))} \sum_{r_1=0}^{N(\alpha)}\sum_{r_2=0}^{N(\alpha)} ((l+r_1+r_2)!)^{2\kappa} R_1^{l+r_1+r_2}\\ &\quad\lesssim \|u\|_{L^2((-1,1)\times B^n(1),dt\times d(h\sigma_{\mathrm{Leb}}))} R_1^l \sum_{r_1=0}^{N(\alpha)}\sum_{r_2=0}^{N(\alpha)} ((l+r_1+r_2)!)^{2\kappa}.\end{aligned} \tag{8.98}$$

We use the fact that for $l \in \mathbb{N}_+$ and $p \in \mathbb{N}$,

$$(l+p)! \leq (l+p)^p(l!) \leq C_{p,1}l^p(l!) \leq C_{p,2}2^l(l!) \leq C_{p,3}^l(l!),$$

where each $C_{p,j} \geq 0$ depends only on p. Thus, there exists an $\mathscr{L}$-unit-admissible constant $R_2 > 0$ such that

$$R_1^l \sum_{r_1=0}^{N(\alpha)} \sum_{r_2=0}^{N(\alpha)} ((l + r_1 + r_2)!)^{2\kappa} \lesssim R_2^l (l!)^{2\kappa}. \tag{8.99}$$

Combining (8.97), (8.98), and (8.99) completes the proof. □

The next lemma is [134, Lemma 2] (when the parameter $\gamma = 2$) and [118, Lemma 6] (for general $\gamma > 1$). We include the proof for completeness.

Lemma 8.3.30. *Let $f \in C^{\infty}_{\mathrm{loc}}((-1/4, 1/4))$ satisfy $f(t) = 0$ for $t \in (-1/4, 0]$. If there exist $\gamma > 1$, $C \geq 0$, and $R > 0$ such that*

$$|\partial_t^l f(t)| \leq CR^l (l!)^{\gamma}, \quad \forall t \in (0, 1/4), l \in \mathbb{N}, \tag{8.100}$$

then

$$|f(t)| \leq Ce^{\gamma-1} \exp\{-(\gamma - 1)e^{-1}(Rt)^{-\frac{1}{\gamma-1}}\}, \quad t \in (-1/4, 1/4).$$

Proof. The result for $C = 0$ is trivial. For $C > 0$, by replacing f with f/C, we see that it suffices to prove the result for $C = 1$. We henceforth assume $C = 1$. Thus, we wish to show that

$$|f(t)| \leq e^{\gamma-1} \exp\{-(\gamma - 1)e^{-1}(Rt)^{-\frac{1}{\gamma-1}}\}, \quad t \in (-1/4, 1/4). \tag{8.101}$$

The inequality (8.101) is trivial for $t \in (-1/4, 0]$ (because $f(t) = 0$ for $t \leq 0$), so we assume $t > 0$.

If $0 \leq e^{-1}(Rt)^{-1/(\gamma-1)} < 1$, then the right-hand side of (8.101) ≥ 1, so (8.101) follows from the case $l = 0$ of (8.100) (recall that we are assuming $C = 1$).

Otherwise, pick $l \in \mathbb{N}$ such that $l+1 \leq e^{-1}(Rt)^{-1/(\gamma-1)} < l+2$, so that $(Rt)^{1/(\gamma-1)} \leq \frac{1}{e(l+1)}$. We have

$$\begin{aligned}
|f(t)| &\leq \int_0^t \frac{(t-s)^l}{l!} |f^{(l+1)}(s)| \, ds \leq R^{l+1}((l+1)!)^{\gamma} \frac{t^{l+1}}{(l+1)!} \\
&= R^{l+1}((l+1)!)^{\gamma-1} t^{l+1} \leq [(Rt)^{\frac{1}{\gamma-1}}(l+1)]^{(\gamma-1)(l+1)} \leq e^{-(\gamma-1)((l+2)-1)} \\
&\leq \exp\{-(\gamma - 1)(e^{-1}(Rt)^{-1/(\gamma-1)} - 1)\},
\end{aligned}$$

which gives (8.101), completing the proof. □

Proof of Proposition 8.3.20. Lemma 8.3.29 shows that (8.96) holds. Thus, Lemma 8.3.30 applies to the function $t \mapsto W^{\alpha}u(t, x)$ to show that

$$\sup_{x\in B^n(1/4)} |W^\alpha u(t,x)| \le C_\alpha e^{2\kappa-1}\exp\{-(2\kappa-1)e^{-1}R_2^{-\frac{1}{2\kappa-1}}t^{-\frac{1}{2\kappa-1}}\}\|u\|_{L^2((-1,1)\times B^n(1),dt\times d(h\sigma_{\mathrm{Leb}}))},$$

for $t \in (-1/4, 1/4)$, where C_α and R_2 are the constants from Lemma 8.3.29. The result now follows with $c := (2\kappa-1)e^{-1}R_2^{-\frac{1}{2\kappa-1}}$ and C_α replaced by $C_\alpha e^{2\kappa-1}$. □

Remark 8.3.31. There are many proofs in harmonic analysis that work for more general quasi-metric balls instead of metric balls. However, the proof of Proposition 8.3.20 used the fact that the balls $B_{(X,d)}(x,\delta)$ are metric balls in an essential way. Quasi-metric balls arise if we allow d_j to be elements of $(0,1)$. Since we take $d_j \in \mathbb{N}_+$, we always have $d_j \ge 1$ and $d_k \ge 1$, and the balls $B_{(W,d)}(x,\delta)$ and $B_{(X,d)}(x,\delta)$ are metric balls. The part of the proof of Proposition 8.3.20 that used the fact that $d_k \ge 1$, $k = 1,\dots,q$, is when we chose $c_1 > 0$ so that $B_{(X,d)}(x, 2c_1\epsilon) \subseteq B^n(\epsilon_2)$, for all $x \in B^n(\epsilon_1)$; recall that $\epsilon = \epsilon_2 - \epsilon_1$. If some of the d_k were less than 1, then choosing such a $c_1 > 0$ might not be possible.

8.4 Heat equations

In this section, we present the desired heat kernel bounds. We work on a connected C^∞ manifold $\mathfrak{M}$ with smooth, strictly positive density Vol. Let

$$(W, d) = \{(W_1, d_1), \dots, (W_r, d_r)\} \subset C^\infty_{\mathrm{loc}}(\mathfrak{M}; T\mathfrak{M}) \times \mathbb{N}_+$$

be Hörmander vector fields with (single-parameter) formal degrees on $\mathfrak{M}$.

Fix $D \in \mathbb{N}_+$ and, as in the beginning of this section, let $\kappa \in \mathbb{N}_+$ be such that d_j divides κ for $1 \le j \le r$. Set $n_j := \frac{\kappa}{d_j} \in \mathbb{N}_+$. In this section, we consider partial differential operators of the form

$$\mathcal{L} = \sum_{\deg_d(\alpha),\deg_d(\beta)\le\kappa} b_{\alpha,\beta}(x) W^\alpha W^\beta, \quad b_{\alpha,\beta} \in C^\infty_{\mathrm{loc}}(\mathfrak{M}; \mathbb{M}^{D\times D}(\mathbb{C})). \tag{8.102}$$

Definition 8.4.1. We say that $\mathcal{L}$ given by (8.102) is maximally subelliptic of type 2 of degree 2κ with respect to (W, d) on $(\mathfrak{M}, \mathrm{Vol})$ if for every relatively compact, open set $\Omega \Subset \mathfrak{M}$, there exists $C_\Omega \ge 0$ such that for all $f \in C_0^\infty(\Omega; \mathbb{C}^{\mathbf{D}_2})$,

$$\sum_{j=1}^r \|W_j^{n_j} f\|^2_{L^2(\mathfrak{M},\mathrm{Vol};\mathbb{C}^{\mathbf{D}_2})} \le C_\Omega(|\langle \mathcal{L}f, f\rangle_{L^2(\mathfrak{M},\mathrm{Vol};\mathbb{C}^D)}| + \|f\|^2_{L^2(\mathfrak{M},\mathrm{Vol};\mathbb{C}^{\mathbf{D}_2})}).$$

When $(\mathfrak{M}, \mathrm{Vol})$ is also clear from the context, we say $\mathcal{P}$ is maximally subelliptic of type 2 of degree κ with respect to (W, d).

Assumption. Throughout this section, we assume that $\mathcal{L}$, when thought of as a densely defined operator on $L^2(\mathfrak{M}, \mathrm{Vol}; \mathbb{C}^D)$ with dense domain $C_0^\infty(\mathfrak{M}; \mathbb{C}^D)$, is symmetric and non-negative. Let $\overline{\mathcal{L}}$ be any non-negative, self-adjoint extension of $\mathcal{L}$.

Theorem 8.4.2. *Suppose $\mathscr{L}$ given by* (8.102) *is maximally subelliptic of type 2 of degree 2κ with respect to* $(W, \vec{d})$ *on* $(\mathfrak{M}, \mathrm{Vol})$ *(and is symmetric and non-negative as in the above assumption). Then, for $t > 0$, the Schwartz kernel of $e^{-t\overline{\mathscr{L}}}$ is a smooth function, which we denote by $e^{-t\overline{\mathscr{L}}}(x,y) \in C^\infty_{\mathrm{loc}}(\mathfrak{M}\times\mathfrak{M}; \mathbb{M}^{D\times D}(\mathbb{C}))$. Furthermore, for every $\mathcal{K} \Subset \mathfrak{M}$, there exists $c = c(\mathcal{K}) > 0$ such that for all ordered multi-indices α and β and all $s \in \mathbb{N}$, there exists $C = C(\mathcal{K}, s, \alpha, \beta)$ such that $\forall t > 0$, $\forall x, y \in \mathcal{K}$,*

$$\begin{aligned}\left|\partial_t^s W_x^\alpha W_y^\beta e^{-t\overline{\mathscr{L}}}(x,y)\right| \leq C\big((\rho_{(W,\vec{d})}(x,y) + t^{\frac{1}{2\kappa}}) \wedge 1\big)^{-\deg_{\vec{d}}(\alpha)-\deg_{\vec{d}}(\beta)-2\kappa s}\\ \times \exp\left(-c\left(\frac{\rho_{(W,\vec{d})}(x,y)^{2\kappa}}{t}\right)^{\frac{1}{2\kappa-1}}\right)\\ \times \big(\mathrm{Vol}(B_{(W,\vec{d})}(x, \rho_{(W,\vec{d})}(x,y) + t^{\frac{1}{2\kappa}})) \wedge 1\big)^{-1}.\end{aligned} \tag{8.103}$$

The rest of this section is devoted to the proof of Theorem 8.4.2. Fix $\mathcal{K} \Subset \Omega_1 \Subset \Omega_2 \Subset \mathfrak{M}$, with $\mathcal{K}$ compact and Ω_1 and Ω_2 open and relatively compact. Theorem 3.3.7 applies to give for each $x \in \mathcal{K}$ and $\delta \in (0,1]$ a map $\Phi_{x,\delta}$ as in that theorem. Set, for $x \in \mathcal{K}$ and $\delta \in (0,1]$,

$$\mathscr{L}^{x,\delta} := \Phi_{x,\delta}^* \delta^{2\kappa} \mathscr{L} (\Phi_{x,\delta})_*. \tag{8.104}$$

$\mathscr{L}^{x,\delta}$ is a partial differential operator on $B^n(1)$. Let $(X, d) = \{(X_1, d_1), \ldots, (X_q, d_q)\} \subset \mathrm{Gen}((W, \vec{d}))$ be as in Theorem 3.3.7, i. e., $(W, \vec{d}) \subseteq (X, d)$ and $\mathrm{Gen}((W, \vec{d}))$ is finitely generated by (X, d) on Ω_2.

We prove Theorem 8.4.2 in three steps. In the first step, in Section 8.4.1, we prove results "at the unit scale" in the case where $\mathscr{L}$ is a partial differential operator on $B^n(1)$; this step uses the a priori estimates from Section 8.3. In the second step, we prove abstract results taking place on $B_{(X,d)}(x, \delta)$ for some fixed $x \in \mathcal{K}$ and $\delta \in (0,1]$. We do this by changing variables via Theorem 3.3.7 to translate the result to $B^n(1)$ and apply the result from the first step to $\mathscr{L}^{x,\delta}$; importantly, the estimates in this step will not depend on $x \in \mathcal{K}$ or $\delta \in (0,1]$. Finally, in the third step we use the results from the second step to conclude Theorem 8.4.2.

Remark 8.4.3. Note that for f in the domain of $\overline{\mathscr{L}}$, $\overline{\mathscr{L}}f = \mathscr{L}f$, where $\mathscr{L}f$ is taken in the sense of distributions. Indeed, for such f and for $g \in C_0^\infty(\mathfrak{M}; \mathbb{C}^D)$, we have

$$\langle \overline{\mathscr{L}}f, g\rangle_{L^2(\mathfrak{M},\mathrm{Vol})} = \langle f, \overline{\mathscr{L}}g\rangle_{L^2(\mathfrak{M},\mathrm{Vol})} = \langle f, \mathscr{L}g\rangle_{L^2(\mathfrak{M},\mathrm{Vol})}.$$

Since $\mathscr{L}$ is symmetric, this shows that $\overline{\mathscr{L}}f = \mathscr{L}f$.

Remark 8.4.4. For some comments on the relationship between maximal subellipticity and maximal subellipticity of type 2, see Remark 8.5.3 and Corollary 8.5.4.

8.4.1 Step I: the unit scale

In this section, we work on $B^n(1)$ and use the same setting as in Section 8.3. Thus, we are given

$$(W, \vec{d}) = \{(W_1, \vec{d}_1), \dots, (W_r, \vec{d}_r)\} \subset C^\infty(B^n(1); TB^n(1)),$$

a strictly positive smooth function $h \in C^\infty(B^n(1))$ with $\inf_{u \in B^n(1)} h(u) > 0$, and an operator $\mathscr{L}$ of the form (8.42) which is maximally subelliptic of type 2 of degree 2κ at the unit scale with respect to $(W, \vec{d})$ and $h\sigma_{\mathrm{Leb}}$, in the sense that Definition 8.3.1 holds with these choices. We assume all the same assumptions as in Section 8.3 and define ι-$\mathscr{L}$-unit-admissible constants as in Definition 8.3.2.

Fix a constant $\Lambda > 0$; none of the constants in our estimates will depend on the choice of Λ. We will often endow $B^n(1)$ with the measure $\mathrm{Vol}_0 := \Lambda h \sigma_{\mathrm{Leb}}$. For $U \subseteq B^n(1)$ open, let $\mathscr{R}_U : L^2(B^n(1), \mathrm{Vol}_0) \to L^2(U, \mathrm{Vol}_0)$ denote the restriction operator $\mathscr{R}_U : f \mapsto f\big|_U$.

8.4.1.1 On diagonal bounds

Fix $D \in \mathbb{N}_+$ and suppose $(\mathfrak{M}_1, \mu_1)$ is a σ-finite measure space. Let

$$T : L^2(\mathfrak{M}, \mu_1; \mathbb{C}^{D_0}) \to L^2(B^n(1), \mathrm{Vol}_0; \mathbb{C}^D)$$

be a bounded linear map such that $\forall L \in \mathbb{N}_+$, $\exists C_L \geq 0$ with

$$\sum_{j=0}^{L} \|\mathscr{L}^j T\|_{L^2(\mathfrak{M}_1, \mu_1) \to L^2(B^n(1), \mathrm{Vol}_0)} \leq C_L.$$

Remark 8.4.5. When $\Lambda = 1$, we have in mind $T = e^{-t\overline{\mathscr{L}}}$, where $t \approx 1$ and $\overline{\mathscr{L}}$ is any non-negative self-adjoint extension of $\mathscr{L}$.

Proposition 8.4.6. *Under the above conditions, $\mathscr{R}_{B^n(1/2)} T : L^2(\mathfrak{M}_1, \mu_1; \mathbb{C}^D) \to L^2(B^n(1/2), \mathrm{Vol}_0; \mathbb{C}^D)$ is given by integration against a measurable function $T(x,y) : B^n(1/2) \times \mathfrak{M}_1 \to \mathbb{M}^{D \times D_0}(\mathbb{C})$. More precisely, for all $f \in L^2(B^n(1/2), \mathrm{Vol}_0; \mathbb{C}^D)$ and $g \in L^2(\mathfrak{M}_1, \mu_1; \mathbb{C}^{D_0})$, we have*

$$\int f(x) \cdot (Tg)(x)\, d\,\mathrm{Vol}_0(x) = \iint f(x) \cdot (T(x,y) g(y))\, d\,\mathrm{Vol}_0(x)\, d\mu_1(y).$$

Furthermore, for every $M \in \mathbb{N}$, there exists $L \in \mathbb{N}$ depending only on upper bounds for M, n, m, and $\max\{\vec{d}_j : 1 \leq j \leq r\}$ and an M-$\mathscr{L}$-unit-admissible constant $E_M = E_M(C_L) \geq 0$ such that

$$\sup_{x \in B^n(1/2)} \sum_{\deg_{\vec{d}}(\alpha) \leq M} \|W_x^\alpha T(x, \cdot)\|_{L^2(\mathfrak{M}_1, \mu_1; \mathbb{M}^{D \times D_0})} \leq E_M \Lambda^{-\frac{1}{2}}.$$

Proof. Fix $f \in L^2(\mathfrak{M}_1, \mu_1; \mathbb{C}^D)$. Since $\mathrm{Vol}_0 = \Lambda h\sigma_{\mathrm{Leb}}$ and $\Lambda > 0$ is a constant, we have, $\forall j \in \mathbb{N}$, $\mathscr{L}^j Tf \in L^2(B^n(1), \mathrm{Vol}_1; \mathbb{C}^D) = L^2(B^n(1), h\sigma_{\mathrm{Leb}}; \mathbb{C}^D)$. Corollary 8.3.8 applies to show that $Tf \in C^\infty_{\mathrm{loc}}(B^n(1); \mathbb{C}^D)$. Furthermore, fix $\phi_1, \phi_2 \in C_0^\infty(B^n(1))$ with $\phi_1 \equiv 1$ on $B^n(1/2)$ and $\phi_1 \prec \phi_2$. Then we have, for $s = s(n) \in (0, \infty)$ sufficiently large, by the Sobolev embedding theorem and Corollary 8.3.8, with $L(s, M) := \lceil \frac{s+M}{\kappa\epsilon_0} \rceil$,

$$\begin{aligned}
&\sup_{x \in B^n(1/2)} \sum_{\deg_{\vec{d}}(\alpha) \le M} |W^\alpha Tf(x)| \lesssim \|\phi_1 Tf\|_{L^2_{s+M}} \\
&\lesssim \sum_{j=0}^{L(s,M)} \|\phi_2 \mathscr{L}^j Tf\|_{L^2(B^n(1), h\sigma_{\mathrm{Leb}})} = \Lambda^{-\frac{1}{2}} \sum_{j=0}^{L(s,M)} \|\phi_2 \mathscr{L}^j Tf\|_{L^2(B^n(1), \mathrm{Vol}_0)} \\
&\le C_{L(s,M)} \Lambda^{-\frac{1}{2}} \|f\|_{L^2(\mathfrak{M}_1, \mu_1)}.
\end{aligned}$$

Taking the supremum over all such f with $\|f\|_{L^2(\mathfrak{M}_1,\mu_1)} = 1$ yields the result. □

8.4.1.2 Off-diagonal bounds

Let $(\mathfrak{M}_1, \mu_1)$ be a σ-finite measure space and fix $D_0 \in \mathbb{N}$. Let $\sigma > 0$ and for $t \in [0, \sigma)$ suppose that

$$T_t : L^2(\mathfrak{M}_1, \mu_1; \mathbb{C}^{D_0}) \to L^2(B^n(1), \mathrm{Vol}_0; \mathbb{C}^D)$$

is a bounded linear map and there is a dense subspace $\mathscr{D} \subseteq L^2(\mathfrak{M}_1, \mu_1; \mathbb{C}^{D_0})$ such that the following hold:

- For all $f \in \mathscr{D}$, $T_t f \in C^\infty_{\mathrm{loc}}([0,\sigma) \times B^n(1); \mathbb{C}^D)$, where we are treating $T_t f(x)$ as a function of $(t, x) \in [0, \sigma) \times B^n(1)$.
- For all $f \in \mathscr{D}$, $T_t f(x)$ vanishes to infinite order in t as $t \downarrow 0$.
- For all $f \in \mathscr{D}$, $(\partial_t + \sigma^{-1}\mathscr{L})T_t f = 0$.
- $\|T_t\|_{L^2(\mathfrak{M}_1,\mu_1;\mathbb{C}^{D_0}) \to L^2(B^n(1),\mathrm{Vol}_0;\mathbb{C}^D)} \le 1$, $\forall t \in [0, \sigma)$.

Remark 8.4.7. When $\Lambda = \sigma = 1$, informally an example of T_t can be given as follows. Let U and V be disjoint open sets and let $\overline{\mathscr{L}}$ be any non-negative self-adjoint extension of $\mathscr{L}$. Then T_t can be taken to be the operator $f \mapsto e^{-t\overline{\mathscr{L}}} f\big|_U$, where we restrict our attention to $\mathrm{supp}(f) \subseteq V$. Since $U \cap V = \emptyset$, if $f \in C_0^\infty(V)$, $e^{-t\overline{\mathscr{L}}} f\big|_U$ vanishes to infinite order as $t \downarrow 0$.

Proposition 8.4.8. *Suppose $\mathscr{L}$, when thought of as a densely defined operator on $L^2(B^n(1), \mathrm{Vol}_0; \mathbb{C}^D)$ with dense domain $C_0^\infty(B^n(1); \mathbb{C}^D)$, is symmetric and non-negative (and maximally subelliptic of type 2 of degree 2κ at the unit scale with respect to $(W, \vec{d})$ and $h\sigma_{\mathrm{Leb}}$, as assumed above). Then, for every $t \in [0, \sigma/4)$, $\mathscr{R}_{B^n(1/4)} T_t$ is given by integration against a measurable function $T_t(x, y) : B^n(1/4) \times \mathfrak{M}_1 \to \mathbb{M}^{D \times D_0}(\mathbb{C})$. More precisely, for all $f \in L^2(B^n(1/4), \mathrm{Vol}_0; \mathbb{C}^D)$ and $g \in L^2(\mathfrak{M}_1, \mu_1; \mathbb{C}^{D_0})$, for $t \in [0, \sigma/4)$ we have*

$$\int f(x) \cdot (T_t g)(x)\, d\,\mathrm{Vol}_0(x) = \iint f(x) \cdot (T_t(x, y) g(y))\, d\,\mathrm{Vol}(x) d\mu_1(y).$$

Furthermore, there exists an $\mathscr{L}$-unit-admissible constant $c > 0$ such that for all $s \in \mathbb{N}$ and all ordered multi-indices α and β, there exists an (s, α)-$\mathscr{L}$-unit-admissible constant $D_{s,\alpha} \geq 0$ such that for $t \in [0, \sigma/4)$,

$$\begin{aligned} &\sup_{x \in B^n(1/4)} \|\partial_t^s W_x^\alpha T_t(x, \cdot)\|_{L^2(\mathfrak{M}_1, \mu_1; \mathbb{M}^{D \times D_0})} \\ &\leq D_{s,\alpha} \sigma^{-s} \Lambda^{-\frac{1}{2}} \exp(-c\sigma^{\frac{1}{2\kappa-1}} t^{-\frac{1}{2\kappa-1}}). \end{aligned} \tag{8.105}$$

Proof. Since $\mathrm{Vol}_0 = \Lambda h \sigma_{\mathrm{Leb}}$ and $\Lambda > 0$ is constant, $\mathscr{L}$, when thought of as a densely defined operator on $L^2(B^n(1), h\sigma_{\mathrm{Leb}}; \mathbb{C}^D)$ with dense domain $C_0^\infty(B^n(1); \mathbb{C}^D)$, is symmetric and non-negative.

Because $\partial_t^s T_t = (-1)^s \sigma^{-s} \mathscr{L}^s T_t$ and due to the form of $\mathscr{L}$, it suffices to prove (8.105) for $s = 0$. Fix $f \in \mathscr{D}$ and set

$$u(t, x) := \begin{cases} T_{\sigma t} f(x), & t \in [0, 1), \\ 0, & t \in (-1, 0). \end{cases}$$

Note that $u \in C_{\mathrm{loc}}^\infty((-1, 1) \times B^n(1); \mathbb{C}^D)$ and $\partial_t u(t, x) = -\mathscr{L} u(t, x)$. Proposition 8.3.20 implies that there exists an $\mathscr{L}$-unit-admissible constant $c > 0$ such that for $t \in [0, 1/4)$,

$$\begin{aligned} \sup_{x \in B^n(1/4)} |W_x^\alpha u(t, x)| &\lesssim \exp(-ct^{-\frac{1}{2\kappa-1}}) \|u\|_{L^2((-1,1) \times B^n(1), dt \times h\sigma_{\mathrm{Leb}})} \\ &= \Lambda^{-\frac{1}{2}} \exp(-ct^{-\frac{1}{2\kappa-1}}) \|T_{\sigma t} f\|_{L^2((-1,1) \times B^n(1), dt \times d\,\mathrm{Vol}_0)} \\ &\leq \Lambda^{-\frac{1}{2}} \exp(-ct^{-\frac{1}{2\kappa-1}}) \|f\|_{L^2(\mathfrak{M}_1, \mu_1)}. \end{aligned}$$

Taking the supremum over all $f \in \mathscr{D}$ with $\|f\|_{L^2(\mathfrak{M}_1, \mu_1)}$ yields, for $t \in [0, 1/4)$,

$$\sup_{x \in B^n(1/4)} \|W_x^\alpha T_{\sigma t}(x, \cdot)\|_{L^2(\mathfrak{M}_1, \mu_1; \mathbb{M}^{D \times D_0})} \lesssim \Lambda^{-\frac{1}{2}} \exp(-ct^{-\frac{1}{2\kappa-1}}).$$

From here the result follows. □

8.4.2 Step II: a single point and scale

We return to the main setting of this section on an abstract manifold $\mathfrak{M}$ with smooth, strictly positive density Vol and Hörmander vector fields with formal degrees $(W, \vec{d}) = \{(W_1, \vec{d}_1), \ldots, (W_r, \vec{d}_r)\}$. We let $\mathscr{L}$ be given by (8.42) and assume that $\mathscr{L}$ is symmetric, non-negative, and maximally subelliptic of type 2 of degree 2κ with respect to $(W, \vec{d})$ on $(\mathfrak{M}, \mathrm{Vol})$.

The main thrust of this section is that Propositions 8.4.6 and 8.4.8 apply to $\mathscr{L}^{x,\delta}$, *uniformly* for $x \in \mathcal{K}$ and $\delta \in (0, 1]$, where $\mathscr{L}^{x,\delta}$ is defined in (8.104). Indeed, we have the following proposition.

Proposition 8.4.9. *$\mathscr{L}^{x,\delta}$ satisfies the hypotheses of Section 8.4.1 uniformly for $x \in \mathcal{K}$ and $\delta \in (0,1]$, and ι-$\mathscr{L}^{x,\delta}$-unit-admissible constants (as in Definition 8.3.2) can be chosen independent of $x \in \mathcal{K}$ and $\delta \in (0,1]$. More precisely, we have the following:*

(a) *Let $W_j^{x,\delta} : \Phi_{x,\delta}^* \delta^{d_j} W_j$ and set $(W^{x,\delta}, \vec{d}) := \{(W_1^{x,\delta}, \vec{d}_1), \ldots, (W_r^{x,\delta}, \vec{d}_r)\}$. Then $(W^{x,\delta}, \vec{d})$ are Hörmander vector fields at the unit scale, uniformly for $x \in \mathcal{K}$, $\delta \in (0,1]$.*

(b) *We give $B^n(1)$ the density $\Phi_{x,\delta}^* \mathrm{Vol} = \Lambda(x,\delta) h_{x,\delta} \sigma_{\mathrm{Leb}}$, where $\Lambda(x,\delta) > 0$ and $h_{x,\delta} \in C^\infty(B^n(1))$ are as in Theorem 3.3.7. Then*

$$\inf_{\substack{x \in B^n(1) \\ \delta \in (0,1]}} \inf_{u \in B^n(1)} h_{x,\delta}(u) > 0$$

and for every $L \in \mathbb{N}$

$$\sup_{\substack{x \in B^n(1) \\ \delta \in (0,1]}} \|h_{x,\delta}\|_{C^L(B^n(1))} < \infty.$$

(c) *$\mathscr{L}^{x,\delta}$ can be written in the form*

$$\mathscr{L}^{x,\delta} = \sum_{\deg_{\vec{d}}(\alpha), \deg_{\vec{d}}(\beta) \le \kappa} b_{\alpha,\beta}^{x,\delta} (W^{x,\delta})^\alpha (W^{x,\delta})^\beta,$$

where $b_{\alpha,\beta}^{x,\delta} \in C^\infty(B^n(1); \mathbb{M}^{D\times D}(\mathbb{C}))$ satisfy, for every $L \in \mathbb{N}$,

$$\sup_{\substack{x \in B^n(1) \\ \delta \in (0,1]}} \|b_{\alpha,\beta}^{x,\delta}\|_{C^L(B^n(1))} < \infty.$$

(d) *$\mathscr{L}^{x,\delta}$ is symmetric and non-negative when thought of as a densely defined operator on $L^2(B^n(1), \Phi_{x,\delta}^* \mathrm{Vol}; \mathbb{C}^D)$ with dense domain $C_0^\infty(B^n(1); \mathbb{C}^D)$ is symmetric and non-negative.*

(e) *$\mathscr{L}^{x,\delta}$ is maximally subelliptic of type 2 of order 2κ at the unit scale with respect to $(W^{x,\delta}, \vec{d})$ and $h_{x,\delta}\sigma_{\mathrm{Leb}}$, uniformly in $x \in \mathcal{K}$ and $\delta \in (0,1]$ in the sense that there exists $A \ge 0$ such that $\forall x \in \mathcal{K}$, $\forall \delta \in (0,1]$, $\forall f \in C_0^\infty(B^n(1); \mathbb{C}^D)$,*

$$\begin{aligned}&\sum_{j=1}^r \|(W_j^{x,\delta})^{n_j} f\|_{L^2(B^n(1), h_{x,\delta}\sigma_{\mathrm{Leb}}; \mathbb{C}^D)}^2 \\ &\quad \le A\big(|\langle \mathscr{L}^{x,\delta} f, f\rangle_{L^2(B^n(1), h_{x,\delta}\sigma_{\mathrm{Leb}}; \mathbb{C}^D)}| + \|f\|_{L^2(B^n(1), h_{x,\delta}\sigma_{\mathrm{Leb}}; \mathbb{C}^D)}^2\big).\end{aligned}$$

Proof. (a): This follows from Theorem 3.3.7 (l).

(b): This follows from Theorem 3.3.7 (n).

(c): Let $f, g, \in C_0^\infty(B^n(1); \mathbb{C}^D)$. By Theorem 3.3.7 (h), $f \circ \Phi_{x,\delta}, g \circ \Phi_{x,\delta} \in C_0^\infty(\mathfrak{M}; \mathbb{C}^D)$. Thus, by changing variables and using the fact that $\mathscr{L}$ is symmetric, we have

$$\begin{aligned}\langle \mathscr{L}^{x,\delta} f, g\rangle_{L^2(B^n(1),\Phi^*_{x,\delta}\mathrm{Vol})} &= \delta^{2\kappa}\langle \Phi^*_{x,\delta}\mathscr{L}(\Phi_{x,\delta})_* f, g\rangle_{L^2(B^n(1),\Phi^*_{x,\delta}\mathrm{Vol})}\\ &= \delta^{2\kappa}\langle \mathscr{L}(f\circ\Phi_{x,\delta}), g\circ\Phi_{x,\delta}\rangle_{L^2(B^n(1),\mathrm{Vol})}\\ &= \delta^{2\kappa}\langle f\circ\Phi_{x,\delta}, \mathscr{L}(g\circ\Phi_{x,\delta})\rangle_{L^2(B^n(1),\mathrm{Vol})}\\ &= \langle f, \mathscr{L}^{x,\delta} g\rangle_{L^2(B^n(1),\Phi^*_{x,\delta}\mathrm{Vol})}.\end{aligned} \tag{8.106}$$

We conclude that $\mathscr{L}^{x,\delta}$ is symmetric. Plugging $f = g$ into (8.106) and using the fact that $\mathscr{L}$ is non-negative, we see that

$$\langle \mathscr{L}^{x,\delta} f, f\rangle_{L^2(B^n(1),\Phi^*_{x,\delta}\mathrm{Vol})} = \delta^{2\kappa}\langle \mathscr{L}(f\circ\Phi_{x,\delta}), f\circ\Phi_{x,\delta}\rangle_{L^2(B^n(1),\mathrm{Vol})} \geq 0,$$

and we conclude that $\mathscr{L}^{x,\delta}$ is non-negative.

(e): Let $C_{\Omega_1} \geq 0$ be the constant from Definition 8.4.1, when applied to $\mathscr{L}$. Let $f \in C_0^\infty(B^n(1);\mathbb{C}^D)$, $x \in \mathcal{K}$, and $\delta \in (0,1]$. Set $u := f\circ\Phi_{x,\delta}$. By Theorem 3.3.7 (h) and the fact that $\Phi_{x,\delta}(B^n(1)) \subseteq \Omega_1$ (as stated in Theorem 3.3.7), we see that $u \in C_0^\infty(\Omega_1;\mathbb{C}^D)$. Thus, we have

$$\begin{aligned}\sum_{j=1}^r \|(W_j^{x,\delta})^{n_j} f\|^2_{L^2(B^n(1),h\sigma_{\mathrm{Leb}})} &= \Lambda^{-1}\sum_{j=1}^r \|(\Phi^*_{x,\delta}\delta^{d_j}W_j)^{n_j}\Phi^*_{x,\delta}u\|^2_{L^2(B^n(1),\Phi^*_{x,\delta}\mathrm{Vol})}\\ &= \Lambda^{-1}\delta^{2\kappa}\sum_{j=1}^r \|W_j^{n_j}u\|^2_{L^2(\mathfrak{M},\mathrm{Vol})}\\ &\leq C_{\Omega_1}\Lambda^{-1}\delta^{2\kappa}\big(|\langle \mathscr{L}u,u\rangle_{L^2(\mathfrak{M},\mathrm{Vol})}| + \|u\|^2_{L^2(\mathfrak{M},\mathrm{Vol})}\big)\\ &\leq C_{\Omega_1}\Lambda^{-1}\big(|\langle \delta^{2\kappa}\mathscr{L}u,u\rangle_{L^2(\mathfrak{M},\mathrm{Vol})}| + \|u\|^2_{L^2(\mathfrak{M},\mathrm{Vol})}\big)\\ &= C_{\Omega_1}\Lambda^{-1}\big(|\langle \mathscr{L}^{x,\delta}f,f\rangle_{L^2(B^n(1),\Phi^*_{x,\delta}\mathrm{Vol})}| + \|f\|^2_{L^2(B^n(1),\Phi^*_{x,\delta}\mathrm{Vol})}\big)\\ &= C_{\Omega_1}\big(|\langle \mathscr{L}^{x,\delta}f,f\rangle_{L^2(B^n(1),h_{x,\delta}\sigma_{\mathrm{Leb}})}| + \|f\|^2_{L^2(B^n(1),h_{x,\delta}\sigma_{\mathrm{Leb}})}\big),\end{aligned}$$

completing the proof. □

For the remainder of this section, fix $x_0 \in \mathcal{K}$ and $\delta_0 \in (0,1]$. Importantly, the estimates which follow are all uniform in $x_0 \in \mathcal{K}$ and $\delta_0 \in (0,1]$. For $U \subseteq B_{(X,d)}(x_0,\delta_0)$ open, let $\mathscr{R}_U : L^2(B_{(X,d)}(x_0,\delta_0),\mathrm{Vol};\mathbb{C}^D) \to L^2(U,\mathrm{Vol};\mathbb{C}^D)$ denote the restriction map $\mathscr{R}_U : f \mapsto f|_U$.

8.4.2.1 On-diagonal bounds

Fix $D_0 \in \mathbb{N}_+$ and suppose that $(\mathfrak{M}_1,\mu_1)$ is a σ-finite measure space and

$$T_{x_0,\delta_0} : L^2(\mathfrak{M}_1,\mu_1;\mathbb{C}^{D_0}) \to L^2(B_{(X,d)}(x_0,\delta_0),\mathrm{Vol};\mathbb{C}^D)$$

is a bounded linear map such that $\forall L \in \mathbb{N}$, $\exists C_L \geq 0$ with

$$\sum_{j=0}^{L} \|(\delta_0^{2\kappa}\mathscr{L})^j T_{x_0,\delta_0}\|_{L^2(\mathfrak{M}_1,\mu_1;\mathbb{C}^{D_0})\to L^2(B_{(X,d)}(x_0,\delta_0),\mathrm{Vol};\mathbb{C}^D)} \leq C_L.$$

Remark 8.4.10. One should think of T_{x_0,δ_0} as $e^{-t\overline{\mathscr{L}}}$, where $t \approx \delta^{2\kappa}$ and $\overline{\mathscr{L}}$ is any non-negative self-adjoint extension of $\mathscr{L}$.

Let $\xi_3 > 0$ be the constant from Theorem 3.3.7; note that ξ_3 does not depend on the choice of $x_0 \in \mathcal{K}$ or $\delta_0 \in (0,1]$.

Proposition 8.4.11. *Under the above conditions,* $\mathscr{R}_{B_{(X,d)}(x_0,\xi_3\delta_0)} T_{x_0,\delta_0} : L^2(\mathfrak{M}_1,\mu_1;\mathbb{C}^{D_0}) \to L^2(B_{(X,d)}(x_0,\delta_0\xi_3),\mathrm{Vol};\mathbb{C}^D)$ *is given by integration against a measurable function*

$$T_{x_0,\delta_0}(x,y) : B_{(X,d)}(x_0,\xi_3\delta_0) \times \mathfrak{M}_1 \to \mathbb{M}^{D\times D_0(\mathbb{C})}$$

in the sense that for all $f \in L^2(B_{(X,d)}(x_0,\xi_3\delta_0),\mathrm{Vol};\mathbb{C}^D)$ *and* $g \in L^2(\mathfrak{M}_1,\mu_1;\mathbb{C}^{D_0})$, *we have*

$$\begin{aligned} &\int f(x)\cdot (T_{x_0,\delta_0} g)(x)\, d\,\mathrm{Vol}(x) \\ &\quad = \iint f(x)\cdot (T_{x_0,\delta_0}(x,y) g(y))\, d\,\mathrm{Vol}(x) d\mu_1(y). \end{aligned} \tag{8.107}$$

Furthermore, for every $M \in \mathbb{N}$, *there exists* $L = L(M) \in \mathbb{N}$ *and* $E_M = E_M(C_L) \geq 0$, *not depending on* $x_0 \in \mathcal{K}$ *or* $\delta_0 \in (0,1]$, *such that*

$$\begin{aligned} &\sup_{x\in B_{(W,\vec{d})}(x_0,\xi_3\delta_0)} \sum_{\deg_{\vec{d}}(\alpha)\leq M} \|(\delta_0^{\vec{d}} W_x)^\alpha T_{x_0,\delta_0}(x,\cdot)\|_{L^2(\mathfrak{M}_1,\mu_1;\mathbb{M}^{D\times D_0})} \\ &\quad \leq E_M(\mathrm{Vol}(B_{(W,\vec{d})}(x_0,\delta_0))\wedge 1)^{-\frac{1}{2}}. \end{aligned} \tag{8.108}$$

Proof. Proposition 8.4.9 shows that Proposition 8.4.6 applies with $(W,\vec{d})$ replaced by $(W^{x_0,\delta_0},\vec{d})$, $\mathscr{L}$ replaced by $\mathscr{L}^{x_0,\delta_0}$, T replaced by $\Phi^*_{x_0,\delta}T_{x_0,\delta_0}$, and Vol_0 replaced by $\mathrm{Vol}_{x_0,\delta_0} := \Phi^*_{x_0,\delta_0}\mathrm{Vol} = \Lambda(x_0,\delta_0)h_{x_0,\delta_0}\sigma_{\mathrm{Leb}}$. Moreover, all of the estimates obtained from Proposition 8.4.6 are uniform in $x_0 \in \mathcal{K}$ and $\delta \in (0,1]$.

Thus, there is a measurable function $S(u,y) : B^n(1/2) \times \mathfrak{M}_1 \to \mathbb{M}^{D\times D_0}(\mathbb{C})$ such that for all $f \in L^2(B^n(1/2),\mathrm{Vol}_{x_0,\delta_0})$ and $g \in L^2(\mathfrak{M}_1,\mu_1;\mathbb{C}^{D_0})$, we have

$$\begin{aligned} &\int f(u)\cdot (\Phi^*_{x_0,\delta_0}T_{x_0,\delta_0} g)(u)\, d\,\mathrm{Vol}_{x_0,\delta_0}(u) \\ &\quad = \iint f(u)\cdot (S(u,y) g(y))\, d\,\mathrm{Vol}_{x_0,\delta_0}(u) d\mu_1(y), \end{aligned} \tag{8.109}$$

which satisfies

$$\begin{aligned} &\sup_{u\in B^n(1/2)} \sum_{\deg_{\vec{d}}(\alpha)\leq M} \|(W_u^{x_0,\delta_0})^\alpha S(u,\cdot)\|_{L^2(\mathfrak{M}_1,\mu_1)} \\ &\quad \lesssim \Lambda(x_0,\delta_0)^{-\frac{1}{2}} \leq (\Lambda(x_0,\delta_0)\wedge 1)^{-\frac{1}{2}} \approx (\mathrm{Vol}(B_{(W,\vec{d})}(x_0,\delta_0))\wedge 1)^{-\frac{1}{2}}, \end{aligned} \tag{8.110}$$

where the final estimate follows from Theorem 3.3.7 (d). Setting $T_{x_0,\delta_0}(x,y) := S(\Phi_{x_0,\delta_0}^{-1}(x),y)$, changing variables $x = \Phi_{x_0,\delta_0}(u)$ in (8.109), and using $B_{(X,d)}(x_0,\xi_3\delta_0) \subseteq \Phi_{x_0,\delta_0}(B^n(1/2))$ (by Theorem 3.3.7 (i)) shows that (8.107) holds. Similarly, again using $B_{(X,d)}(x_0,\xi_3\delta_0) \subseteq \Phi_{x_0,\delta_0}(B^n(1/2))$, (8.108) follows from (8.110). □

8.4.2.2 Off-diagonal bounds

Suppose $(\mathfrak{M}_1,\mu_1)$ is a σ-finite measure space. Fix $D_0 \in \mathbb{N}_+$ and $\sigma > 0$. For $t \in [0,\sigma)$, suppose that

$$T_{t,x_0,\delta_0} : L^2(\mathfrak{M}_1,\mu_1;\mathbb{C}^{D_0}) \to L^2(B_{(X,d)}(x_0,\delta_0),\mathrm{Vol};\mathbb{C}^D)$$

is a bounded linear map and there is a dense subspace $\mathscr{D} \subseteq L^2(\mathfrak{M},\mu_1)$ such that the following hold:

- For all $f \in \mathscr{D}$, $T_{t,x_0,\delta_0}f \in C^\infty_{\mathrm{loc}}([0,\sigma) \times B_{(X,d)}(x_0,\delta_0);\mathbb{C}^D)$, where we are treating $T_{t,x_0,\delta_0}f(x)$ as a function of $(t,x) \in [0,\sigma) \times B_{(X,d)}(x_0,\delta_0)$.
- For all $f \in \mathscr{D}$, $T_{t,x_0,\delta_0}f(x)$ vanishes to infinite order in t as $t \downarrow 0$.
- For all $f \in \mathscr{D}$, $(\partial_t + \sigma^{-1}\delta_0^{2\kappa}\mathscr{L})T_{t,x_0,\delta_0}f = 0$.
- We have $\|T_{t,x_0,\delta_0}\|_{L^2(\mathfrak{M}_1,\mu_1;\mathbb{C}^{D_0})\to L^2(B_{(X,d)}(x_0,\delta_0),\mathrm{Vol};\mathbb{C}^D)} \leq 1$, $\forall t \in [0,\sigma)$.

Proposition 8.4.12. *For every $t \in [0,\sigma/4)$, $\mathscr{R}_{\Phi_{x_0,\delta_0}(B^n(1/4))}T_{t,x_0,\delta_0}$ is given by integration against a measurable function $T_{t,x_0,\delta_0}(x,y) : \Phi_{x_0,\delta}(B^n(1/4)) \times \mathfrak{M}_1 \to \mathbb{M}^{D\times D_0}(\mathbb{C})$. More precisely, for all $f \in L^2(\Phi_{x_0,\delta_0}(B^n(1/4)),\mathrm{Vol};\mathbb{C}^D)$ and $g \in L^2(\mathfrak{M}_1,\mu_1)$, for $t \in [0,\sigma/4)$, we have*

$$\begin{aligned}&\int f(x)\cdot(T_{t,x_0,\delta_0}g)(x)\,d\mathrm{Vol}(x)\\ &\quad= \iint f(x)\cdot(T_{t,x_0,\delta_0}(x,y)g(y))\,d\mathrm{Vol}(x)d\mu_1(y).\end{aligned} \tag{8.111}$$

Furthermore, there exists a constant $c = c(\mathcal{K},(W,d)) > 0$ such that for all $s \in \mathbb{N}$ and all ordered multi-indices α and β, there exists $D_{s,\alpha} \geq 0$ such that for $t \in [0,\sigma/4)$,

$$\begin{aligned}&\sup_{x\in\Phi_{x_0,\delta_0}(B^n(1/4))} \|\partial_t^s(\delta_0^d W_x)^\alpha T_{t,x_0,\delta_0}(x,\cdot)\|_{L^2(\mathfrak{M}_1,\mu_1;\mathbb{M}^{D\times D_0})}\\ &\quad\leq D_{s,\alpha}\sigma^{-s}(\mathrm{Vol}(B_{(W,d)}(x_0,\delta))\wedge 1)^{-\frac{1}{2}}\exp(-c\sigma^{\frac{1}{2\kappa-1}}t^{-\frac{1}{2\kappa-1}}).\end{aligned} \tag{8.112}$$

Here $c > 0$ and $D_{s,\alpha} \geq 0$ do not depend on $x_0 \in \mathcal{K}$, $\delta_0 \in (0,1]$, or T_{t,x_0,δ_0}.

Proof. Let $S_t := \Phi_{x_0,\delta}^* T_{t,x_0,\delta_0} : L^2(\mathfrak{M}_1,\mu_1;\mathbb{C}^{D_0}) \to L^2(B^n(1),\Phi_{x_0,\delta_0}^*\mathrm{Vol})$. Proposition 8.4.9 shows that Proposition 8.4.8 applies with (W,d) replaced by (W^{x_0,δ_0},d), $\mathscr{L}$ replaced by $\mathscr{L}^{x_0,\delta_0}$, T_t replaced by S_t, and Vol_0 replaced by $\mathrm{Vol}_{x_0,\delta_0} := \Phi_{x_0,\delta_0}^*\mathrm{Vol} = \Lambda(x_0,\delta_0)h_{x_0,\delta_0}\sigma_{\mathrm{Leb}}$. Moreover, all of the estimates obtained from Proposition 8.4.8 are uniform in $x_0 \in \mathcal{K}$ and $\delta \in (0,1]$.

Proposition 8.4.8 implies that for $t \in [0, \sigma/4)$ there is a measurable function $S_t : B^n(1/4) \times \mathfrak{M}_1 \to \mathbb{M}^{D\times D_0}(\mathbb{C})$ such that for all $f \in L^2(B^n(1/4), \mathrm{Vol}_0; \mathbb{C}^D)$ and $g \in L^2(\mathfrak{M}_1, \mu_1; \mathbb{C}^{D_0})$, we have

$$\begin{aligned}&\int f(u) \cdot (S_t g)(u)\, d\,\mathrm{Vol}_{x_0,\delta_0}(u)\\ &\quad = \iint f(u) \cdot (S_t(u,y) g(y))\, d\,\mathrm{Vol}_{x_0,\delta_0}(u) d\mu_1(y).\end{aligned} \tag{8.113}$$

Furthermore, Proposition 8.4.8 shows that there exists $c \gtrsim 1$ such that for all $s \in \mathbb{N}$, for all ordered multi-indices α and β, and for $t \in [0, \sigma/4)$,

$$\begin{aligned}&\sup_{u\in B^n(1/4)} \|\partial_t^s (W_u^{x_0,\delta_0})^\alpha S_t(u,\cdot)\|_{L^2(\mathfrak{M}_1,\mu_1;\mathbb{M}^{D\times D_0})}\\ &\lesssim \sigma^{-s} \Lambda(x_0,\delta_0)^{-\frac{1}{2}} \exp(-c\sigma^{\frac{1}{2\kappa-1}} t^{-\frac{1}{2\kappa-1}})\\ &\lesssim \sigma^{-s} (\Lambda(x_0,\delta_0) \wedge 1)^{-\frac{1}{2}} \exp(-c\sigma^{\frac{1}{2\kappa-1}} t^{-\frac{1}{2\kappa-1}})\\ &\lesssim \sigma^{-s} (\mathrm{Vol}(B_{(W,d)}(x_0,\delta_0)) \wedge 1)^{-\frac{1}{2}} \exp(-c\sigma^{\frac{1}{2\kappa-1}} t^{-\frac{1}{2\kappa-1}}),\end{aligned} \tag{8.114}$$

where the final estimate used Theorem 3.3.7 (d).

Setting $T_{t,x_0,\delta_0}(x,y) := S_t(\Phi_{x_0,\delta_0}^{-1}(x), y)$ and changing variables $x = \Phi_{x_0,\delta_0}(u)$ in (8.113) establishes (8.111). Similarly, (8.112) follows from (8.114). □

8.4.3 Step III: all scales

In this section, we use the results from Section 8.4.2 to complete the proof of Theorem 8.4.2.

We separate the proof into three parts:

- Large time: $\forall t > 0$, the Schwartz kernel of $e^{-t\overline{\mathscr{L}}}$ is a $C_{\mathrm{loc}}^\infty(\mathfrak{M}\times\mathfrak{M}; \mathbb{M}^{D\times D}(\mathbb{C}))$ function, which we denote by $e^{-t\overline{\mathscr{L}}}(x,y)$. Furthermore, $\forall \delta_2 > 0$, $\forall s \in \mathbb{N}$, $\forall \alpha, \beta$, $\exists C_{\mathcal{K},\delta_2,s,\alpha,\beta} \geq 0$ such that $\forall t \geq \delta_2^{2\kappa}$, $x, y \in \mathcal{K}$, we have

$$|\partial_t^s W_x^\alpha W_y^\beta e^{-t\overline{\mathscr{L}}}(x,y)| \leq C_{\mathcal{K},\delta_2,s,\alpha,\beta}. \tag{8.115}$$

- On-diagonal bounds: $\forall s \in \mathbb{N}$, $\forall \alpha, \beta$, $\exists C_{\mathcal{K},s,\alpha,\beta} \geq 0$ such that $\forall x, y \in \mathcal{K}$, $t > 0$ with $\rho_{(X,d)}(x,y) \leq 16 t^{\frac{1}{2\kappa}} \leq 1$, we have

$$|\partial_t^s W_x^\alpha W_y^\beta e^{-t\overline{\mathscr{L}}}(x,y)| \leq C_{\mathcal{K},s,\alpha,\beta} \frac{t^{-(\deg_d(\alpha)+\deg_d(\beta)+2\kappa s)/2\kappa}}{\mathrm{Vol}(B_{(W,d)}(x, t^{\frac{1}{2\kappa}})) \wedge 1}. \tag{8.116}$$

- Off-diagonal bounds: There exists $c_1 = c_1(\mathcal{K}) > 0$ such that $\forall s \in \mathbb{N}$, $\forall \alpha, \beta$, $\exists C_{\mathcal{K},s,\alpha,\beta} \geq 0$ such that $\forall x, y \in \mathcal{K}$, $t > 0$ with $16 t^{\frac{1}{2\kappa}} \leq \rho_{(X,d)}(x,y) \leq 1$, we have

$$\begin{aligned}|\partial_t^s W_x^\alpha W_y^\beta e^{-t\mathscr{L}}(x,y)| \le C_{\mathcal{K},s,\alpha,\beta}\rho_{(X,d)}(x,y)^{-\deg_{\vec{d}}(\alpha)-\deg_{\vec{d}}(\beta)-2\kappa s} \\ \times \exp\left(-c_1\left(\frac{\rho_{(X,d)}(x,y)^{2\kappa}}{t}\right)^{\frac{1}{2\kappa-1}}\right) \\ \times (\mathrm{Vol}(B_{(W,\vec{d})}(x,\rho_{(X,d)}(x,y)))\wedge 1)^{-1}.\end{aligned} \tag{8.117}$$

Before we prove the above estimates we see how they imply Theorem 8.4.2. For this, we need the next lemma.

Lemma 8.4.13. *For all* $x,y\in\mathcal{K}$, $\rho_{(W,\vec{d})}(x,y)\approx\rho_{(X,d)}(x,y)$.

Proof. This lemma follows easily from the case $j=0$, $\nu=1$ of Lemma 5.4.14,[3] but we include the proof in this simple special case. Since $(W,\vec{d})\subseteq(X,d)$, we have $\rho_{(X,d)}(x,y)\le\rho_{(W,\vec{d})}(x,y)$, $\forall x,y\in\mathfrak{M}$.

Let $x,y\in\mathcal{K}$. Theorem 3.3.7 (b) shows that if $\delta\in(0,1]$ and $\rho_{(X,d)}(x,y)<\xi_3\delta$, then $\rho_{(W,\vec{d})}(x,y)<\delta$. Thus, if $\rho_{(X,d)}(x,y)<\xi_3$, we have $\rho_{(W,\vec{d})}(x,y)<\xi_3^{-1}\rho_{(X,d)}(x,y)$. Finally, if $\rho_{(X,d)}(x,y)\ge\xi_3$, since $\mathcal{K}$ is compact with respect to $\rho_{(W,\vec{d})}$ (see Lemma 3.1.7), we have $\rho_{(W,\vec{d})}(x,y)\lesssim 1\lesssim\rho_{(X,d)}(x,y)$, completing the proof. □

Proof of Theorem 8.4.2. By the large time estimates, we have $e^{-t\mathscr{L}}(x,y)\in C^\infty_{\mathrm{loc}}(\mathfrak{M}\times\mathfrak{M};\mathbb{M}^{D\times D}(\mathbb{C}))$, for $t>0$. Thus, it suffices to prove (8.103). Fix $\mathcal{K}\Subset\mathfrak{M}$, $x,y\in\mathcal{K}$, and $t>0$. We separate the proof of (8.103) into the following three cases: $16t^{\frac{1}{2\kappa}}\le 1\wedge\rho_{(X,d)}(x,y)$, $\rho_{(X,d)}(x,y)\le 16t^{\frac{1}{2\kappa}}$, and $16t^{\frac{1}{2\kappa}}\ge 1$. In all cases, we use Lemma 8.4.13, so that $\rho_{(X,d)}(x,y)\approx\rho_{(W,\vec{d})}(x,y)$.

$16t^{\frac{1}{2\kappa}}\le\rho_{(X,d)}(x,y)\wedge 1$: In this case, we have

$$\rho_{(X,d)}(x,y)\approx\rho_{(X,d)}(x,y)+t^{\frac{1}{2\kappa}}\approx\rho_{(W,\vec{d})}(x,y)+t^{\frac{1}{2\kappa}}. \tag{8.118}$$

Since $\mathcal{K}$ is compact with respect to $\rho_{(W,\vec{d})}$ (Lemma 3.1.7), we have $\rho_{(W,\vec{d})}(x,y)\lesssim 1$, and therefore since we also have $t\le 1$,

$$\rho_{(W,\vec{d})}(x,y)+t^{\frac{1}{2\kappa}}\approx(\rho_{(W,\vec{d})}(x,y)+t^{\frac{1}{2\kappa}})\wedge 1. \tag{8.119}$$

Combining (8.118) with Theorem 3.3.7 (f), we see that

$$\begin{aligned}&\mathrm{Vol}(B_{(W,\vec{d})}(x,\rho_{(X,d)}(x,y)))\wedge 1\\ &\quad\approx\mathrm{Vol}(B_{(W,\vec{d})}(x,\rho_{(W,\vec{d})}(x,y)+t^{\frac{1}{2\kappa}}))\wedge 1.\end{aligned} \tag{8.120}$$

Let $c_1=c_1(\mathcal{K})>0$ be as in (8.117) and take $c=c(\mathcal{K})>0$ such that

$$c\rho_{(W,\vec{d})}(x,y)^{\frac{2\kappa}{2\kappa-1}}\le c_1\rho_{(X,d)}(x,y)^{\frac{2\kappa}{2\kappa-1}}. \tag{8.121}$$

3 In fact, Lemma 5.4.14 implies that $\rho_{(W,\vec{d})}$ and $\rho_{(X,d)}$ are equivalent on any compact subset of $\mathfrak{M}$.

Using (8.118), (8.119), (8.120), and (8.121) to bound (8.117) yields (8.103), completing the proof in this case. We use this choice of $c = c(\mathcal{K}) > 0$ in the rest of the proof.

$\rho_{(X,d)}(x,y) \leq 16t^{\frac{1}{2\kappa}} \leq 1$: In this case, we have

$$t^{\frac{1}{2\kappa}} \approx \rho_{(X,d)}(x,y) + t^{\frac{1}{2\kappa}} \approx \rho_{(W,d)}(x,y) + t^{\frac{1}{2\kappa}}, \tag{8.122}$$

$$t^{\frac{1}{2\kappa}} \approx (\rho_{(X,d)}(x,y) + t^{\frac{1}{2\kappa}}) \wedge 1 \approx (\rho_{(W,d)}(x,y) + t^{\frac{1}{2\kappa}}) \wedge 1. \tag{8.123}$$

Combining (8.122) with Theorem 3.3.7 (f), we see that

$$\mathrm{Vol}(B_{(W,d)}(x, t^{\frac{1}{2\kappa}})) \wedge 1 \approx \mathrm{Vol}(B_{(W,d)}(x, \rho_{(W,d)}(x,y) + t^{\frac{1}{2\kappa}})) \wedge 1. \tag{8.124}$$

Finally, in this case we have $\rho_{(W,d)}(x,y) \approx \rho_{(X,d)}(x,y) \leq 16t^{\frac{1}{2\kappa}}$, and therefore

$$\exp\left(-c\left(\frac{\rho_{(W,d)}(x,y)^{2\kappa}}{t}\right)^{\frac{1}{2\kappa-1}}\right) \gtrsim 1. \tag{8.125}$$

Using (8.123), (8.124), and (8.125) to bound (8.116) yields (8.103), completing the proof in this case.

$16t^{\frac{1}{2\kappa}} \geq 1$: In this case, we have

$$(\rho_{(W,d)}(x,y) + t^{\frac{1}{2\kappa}}) \wedge 1 \approx 1. \tag{8.126}$$

Clearly,

$$(\mathrm{Vol}(B_{(W,d)}(x, \rho_{(W,d)}(x,y) + t^{\frac{1}{2\kappa}})) \wedge 1)^{-1} \geq 1. \tag{8.127}$$

Since $\mathcal{K}$ is compact with respect to $\rho_{(W,d)}$ (see Lemma 3.1.7), we have $\rho_{(W,d)}(x,y) \lesssim 1$. Since $t \gtrsim 1$, we see that

$$\exp\left(-c\left(\frac{\rho_{(W,d)}(x,y)^{2\kappa}}{t}\right)^{\frac{1}{2\kappa-1}}\right) \approx 1. \tag{8.128}$$

Combining (8.126), (8.127), and (8.128) shows that the right-hand side of (8.103) is $\gtrsim 1$. Therefore, (8.103) follows from (8.115) with $\delta_2 = \frac{1}{16}$. □

We turn to proving the estimates (8.115), (8.116), and (8.117).

Lemma 8.4.14. *For $t > 0$, the Schwartz kernel of $e^{-t\overline{\mathcal{L}}}$ is an $L^1_{\mathrm{loc}}(\mathfrak{M} \times \mathfrak{M}, \mathrm{Vol} \otimes \mathrm{Vol}; \mathbb{M}^{D\times D}(\mathbb{C}))$ function $e^{-t\overline{\mathcal{L}}}(x,y)$ which satisfies $\forall \alpha$, $\exists C_{\mathcal{K},\alpha} \geq 0$ such that $\forall x \in \mathcal{K}$, $\forall t > 0$, we have*

$$\begin{aligned}&\|W_x^\alpha e^{-(t/2)\overline{\mathcal{L}}}(x,\cdot)\|_{L^2(\mathfrak{M},\mathrm{Vol};\mathbb{M}^{D\times D})}, \|W_x^\alpha e^{-(t/2)\overline{\mathcal{L}}}(\cdot,x)\|_{L^2(\mathfrak{M},\mathrm{Vol};\mathbb{M}^{D\times D})}\\ &\leq C_{\mathcal{K},\alpha}(t^{\frac{1}{2\kappa}} \wedge 1)^{-\deg_d(\alpha)}(\mathrm{Vol}(B_{(W,d)}(x, t^{\frac{1}{2\kappa}} \wedge 1)) \wedge 1)^{-\frac{1}{2}}.\end{aligned}$$

Proof. Set $t_1 := t \wedge 1$. Note that, $\forall L \in \mathbb{N}$, $\exists C_L \geq 0$ (C_L depending only on L), such that

$$\begin{aligned}\sum_{j=0}^{L} &\|(t_1\mathscr{L})^j e^{-(t/2)\overline{\mathscr{L}}}\|_{L^2(\mathfrak{M},\mathrm{Vol})\to L^2(\mathfrak{M},\mathrm{Vol})} \\ &= \sum_{j=0}^{L} \|(t_1\overline{\mathscr{L}})^j e^{-(t/2)\overline{\mathscr{L}}}\|_{L^2(\mathfrak{M},\mathrm{Vol})\to L^2(\mathfrak{M},\mathrm{Vol})} \leq C_L,\end{aligned} \tag{8.129}$$

where we have used the fact that $\overline{\mathscr{L}}$ agrees with $\mathscr{L}$ on the domain of $\overline{\mathscr{L}}$, where $\mathscr{L}$ is taken in the sense of distributions (see Remark 8.4.3).

Fix $x_0 \in \mathcal{K}$. By restricting (8.129) to $B_{(X,d)}(x_0, t_1^{\frac{1}{2\kappa}})$, we see that

$$\sum_{j=0}^{L} \|(t_1\mathscr{L})^j e^{-(t/2)\overline{\mathscr{L}}}\|_{L^2(\mathfrak{M},\mathrm{Vol})\to L^2(B_{(X,d)}(x_0,t_1^{\frac{1}{2\kappa}}),\mathrm{Vol})} \leq C_L. \tag{8.130}$$

Letting $\mathscr{R}_U : f \mapsto f|_U$ denote the restriction map, Proposition 8.4.11 (with $\delta_0 = t_1^{\frac{1}{2\kappa}}$ and $T_{x_0,\delta_0} = e^{-(t/2)\overline{\mathscr{L}}}$) shows that $\mathscr{R}_{B_{(X,d)}(x_0,\xi_3 t_1^{\frac{1}{2\kappa}})} e^{-(t/2)\overline{\mathscr{L}}}$ is given by integration against a measurable function $e^{-(t/2)\overline{\mathscr{L}}}(x,y)$ satisfying, for every α,

$$\begin{aligned}\sup_{x\in B_{(X,d)}(x_0,\xi_3 t_1^{\frac{1}{2\kappa}})} &\|(t_1^{\frac{d}{2\kappa}} W_x)^\alpha e^{-(t/2)\overline{\mathscr{L}}}(x,\cdot)\|_{L^2(\mathfrak{M},\mathrm{Vol})} \\ &\lesssim (\mathrm{Vol}(B_{(W,d)}(x_0, t_1^{\frac{1}{2\kappa}})) \wedge 1)^{-\frac{1}{2}},\end{aligned} \tag{8.131}$$

where $\xi_3 = \xi_3(\mathcal{K}) > 0$. In particular, $e^{-(t/2)\overline{\mathscr{L}}}(x,y) \in L^1_{\mathrm{loc}}(B_{(X,d)}(x_0, \xi_3 t_1^{\frac{1}{2\kappa}}) \times \mathfrak{M}, \mathrm{Vol} \otimes \mathrm{Vol}; \mathbb{M}^{D\times D})$ and $\forall \phi_1 \in C_0^\infty(B_{(X,d)}(x_0, \xi_3 t_1^{\frac{1}{2\kappa}}); \mathbb{C}^D)$, $\forall \phi_2 \in C_0^\infty(\mathfrak{M}; \mathbb{C}^D)$

$$\begin{aligned}\int &\phi_1(x) \cdot (e^{-(t/2)\overline{\mathscr{L}}}\phi_2)(x)\, d\,\mathrm{Vol}(x) \\ &= \iint \phi_1(x) \cdot (e^{-(t/2)\overline{\mathscr{L}}}(y)\phi_2(y))\, d\,\mathrm{Vol}(x) d\,\mathrm{Vol}(y).\end{aligned}$$

Since $B_{(X,d)}(x_0, \xi_3 t^{\frac{1}{2\kappa}})$ is an open neighborhood of x_0, $x_0 \in \mathcal{K}$ is arbitrary, and $\mathcal{K} \Subset \mathfrak{M}$ is arbitrary, it follows that the Schwartz kernel of $e^{-(t/2)\overline{\mathscr{L}}}$ is an $L^1_{\mathrm{loc}}(\mathfrak{M}\times\mathfrak{M}, \mathrm{Vol}\otimes\mathrm{Vol}; \mathbb{M}^{D\times D}(\mathbb{C}))$ function. Moreover, by setting $x = x_0$ in (8.131), we have

$$\begin{aligned}&\|W_x^\alpha e^{-(t/2)\overline{\mathscr{L}}}(x,\cdot)\|_{L^2(\mathfrak{M},\mathrm{Vol};\mathbb{M}^{D\times D})} \\ &\lesssim (t^{\frac{1}{2\kappa}} \wedge 1)^{-\deg_d(\alpha)} (\mathrm{Vol}(B_{(W,d)}(x, t^{\frac{1}{2\kappa}} \wedge 1)) \wedge 1)^{-\frac{1}{2}}.\end{aligned}$$

Since $e^{-(t/2)\overline{\mathscr{L}}}$ is self-adjoint, the same estimate holds for $\|W_x^\alpha e^{-(t/2)\overline{\mathscr{L}}}(\cdot,x)\|_{L^2(\mathfrak{M},\mathrm{Vol};\mathbb{M}^{D\times D})}$, completing the proof. □

Using the fact that $e^{-t\overline{\mathscr{L}}} = e^{-(t/2)\overline{\mathscr{L}}} \circ e^{-(t/2)\overline{\mathscr{L}}}$, we have, for $t > 0$ and all α, β, using Lemma 8.4.14 and the Cauchy–Schwartz inequality,

$$\begin{aligned} &|W_x^\alpha W_y^\beta e^{-t\overline{\mathscr{L}}}(x,y)| \\ &= \left|\int \left(W_x^\alpha e^{-(t/2)\overline{\mathscr{L}}}(x,z)\right)\left(W_y^\beta e^{-(t/2)\overline{\mathscr{L}}}(z,y)\right) d\,\mathrm{Vol}(z)\right| \\ &\le \|W_x^\alpha e^{-(t/2)\overline{\mathscr{L}}}(x,\cdot)\|_{L^2(\mathfrak{M},\mathrm{Vol})} \|W_y^\beta e^{-(t/2)\overline{\mathscr{L}}}(\cdot,y)\|_{L^2(\mathfrak{M},\mathrm{Vol})}. \end{aligned} \tag{8.132}$$

Proof of (8.115). Because $\partial_t^s e^{-t\overline{\mathscr{L}}} = (-1)^s \overline{\mathscr{L}}^s e^{-t\overline{\mathscr{L}}} = (-1)^s \mathscr{L} e^{-t\overline{\mathscr{L}}}$, where $\mathscr{L}$ is taken in the sense of distributions (see Remark 8.4.3), and due to the form of $\mathscr{L}$, it suffices to prove (8.115) in the case $s = 0$.

Let $\delta_2 > 0$; the estimates which follow depend on δ_2. For $t \ge \delta_2^{2\kappa}$ and $x, y \in \mathcal{K}$, we have by Lemma 8.4.14 and (8.132)

$$\begin{aligned} &|W_x^\alpha W_y^\beta e^{-t\overline{\mathscr{L}}}(x,y)| \\ &\le \|W_x^\alpha e^{-(t/2)\overline{\mathscr{L}}}(x,\cdot)\|_{L^2(\mathfrak{M},\mathrm{Vol})} \|W_y^\beta e^{-(t/2)\overline{\mathscr{L}}}(\cdot,y)\|_{L^2(\mathfrak{M},\mathrm{Vol})} \\ &\lesssim (t^{\frac{1}{2\kappa}} \wedge 1)^{-\deg_d(\alpha)-\deg_d(\beta)} (\mathrm{Vol}(B_{(W,d)}(x, t^{\frac{1}{2\kappa}} \wedge 1)) \wedge 1)^{-\frac{1}{2}} \\ &\quad \times (\mathrm{Vol}(B_{(W,d)}(y, t^{\frac{1}{2\kappa}} \wedge 1)) \wedge 1)^{-\frac{1}{2}}. \end{aligned} \tag{8.133}$$

Since $t \ge \delta_2^{2\kappa}$, we have $t^{\frac{1}{2\kappa}} \wedge 1 \approx 1$. Also, letting $\delta_1 = \delta_1(\Omega_1) > 0$ be as in Corollary 3.3.9 (recall that $\mathcal{K} \Subset \Omega_1 \Subset \mathfrak{M}$ and Ω_1 is open), using repeated application of Theorem 3.3.7 (f), we have, for $z \in \mathcal{K}$,

$$\begin{aligned} \mathrm{Vol}(B_{(W,d)}(z, t^{\frac{1}{2\kappa}} \wedge 1)) \wedge 1 &\ge \mathrm{Vol}(B_{(W,d)}(z, \delta_2 \wedge 1)) \wedge 1 \\ &\approx \mathrm{Vol}(B_{(W,d)}(z, \delta_1)) \wedge 1 \approx 1, \end{aligned}$$

where the final $\approx$ used Corollary 3.3.9. Thus, the right-hand side of (8.133) is $\lesssim 1$, completing the proof. □

Corollary 8.4.15. *We have* $e^{-t\overline{\mathscr{L}}}(x,y) \in C^\infty_{\mathrm{loc}}(\mathfrak{M} \times \mathfrak{M}; \mathbb{M}^{D\times D})$.

Proof. The estimate (8.115) shows that for every α and β, $W_x^\alpha W_y^\beta e^{-t\overline{\mathscr{L}}}(x,y)$ is a locally bounded function on $\mathfrak{M} \times \mathfrak{M}$. The result follows, since $W_1, \ldots, W_r$ satisfy Hörmander's condition. □

Proof of (8.116). Because $\partial_t^s e^{-t\overline{\mathscr{L}}} = (-1)^s \overline{\mathscr{L}}^s e^{-t\overline{\mathscr{L}}} = (-1)^s \mathscr{L} e^{-t\overline{\mathscr{L}}}$, where $\mathscr{L}$ is taken in the sense of distributions (see Remark 8.4.3), and due to the form of $\mathscr{L}$, it suffices to prove (8.116) in the case $s = 0$.

Suppose $x, y \in \mathcal{K}$ with $\rho_{(X,d)}(x,y) \le 16 t^{\frac{1}{2\kappa}} \le 1$. By Lemma 8.4.14 and (8.132), using $t^{\frac{1}{2\kappa}} < 1$, we have

$$\begin{aligned}
&\left|W_x^\alpha W_y^\beta e^{-t\overline{\mathscr{L}}}(x,y)\right| \\
&\leq \left\|W_x^\alpha e^{-(t/2)\overline{\mathscr{L}}}(x,\cdot)\right\|_{L^2(\mathfrak{M},\mathrm{Vol})}\left\|W_y^\beta e^{-(t/2)\overline{\mathscr{L}}}(\cdot,y)\right\|_{L^2(\mathfrak{M},\mathrm{Vol})} \\
&\lesssim t^{-(\deg_{\vec{d}}(\alpha)+\deg_{\vec{d}}(\beta))/2\kappa}\left(\mathrm{Vol}(B_{(W,\vec{d})}(x,t^{\frac{1}{2\kappa}}))\wedge 1\right)^{-\frac{1}{2}} \\
&\quad\times\left(\mathrm{Vol}(B_{(W,\vec{d})}(y,t^{\frac{1}{2\kappa}}))\wedge 1\right)^{-\frac{1}{2}}.
\end{aligned} \tag{8.134}$$

Since $t^{\frac{1}{2\kappa}} \geq \rho_{(X,d)}(x,y)/16$, we have $B_{(X,d)}(x,t^{\frac{1}{2\kappa}}) \subseteq B_{(X,d)}(y,48t^{\frac{1}{2\kappa}})$. Thus, by Theorem 3.3.7 (d) and repeated application of Theorem 3.3.7 (f), we have

$$\begin{aligned}
\mathrm{Vol}(B_{(W,\vec{d})}(x,t^{\frac{1}{2\kappa}}))\wedge 1 &\approx \mathrm{Vol}(B_{(X,d)}(x,t^{\frac{1}{2\kappa}}))\wedge 1 \\
&\leq \mathrm{Vol}(B_{(X,d)}(y,48t^{\frac{1}{2\kappa}}))\wedge 1 \approx \mathrm{Vol}(B_{(X,d)}(y,t^{\frac{1}{2\kappa}}))\wedge 1 \\
&\approx \mathrm{Vol}(B_{(W,\vec{d})}(y,t^{\frac{1}{2\kappa}}))\wedge 1.
\end{aligned} \tag{8.135}$$

Using (8.135) to estimate the right-hand side of (8.134) gives the case $s = 0$ of (8.116) and completes the proof. □

Lemma 8.4.16. *Fix $\delta > 0$ and $x_0 \in \mathfrak{M}$. Fix $f \in C_0^\infty(\mathfrak{M} \setminus \overline{B_{(X,d)}(x_0,\delta)};\mathbb{C}^D)$. For $t \in \mathbb{R}$ and $x \in B_{(X,d)}(x_0,\delta)$, set*

$$u(t,x) = \begin{cases} e^{-t\overline{\mathscr{L}}}f(x), & t \geq 0, \\ 0, & t < 0. \end{cases}$$

Then $u \in C^\infty_{\mathrm{loc}}(\mathbb{R}\times B_{(X,d)}(x_0,\delta);\mathbb{C}^D)$ and $u(t,x)$ vanishes to infinite order in t as $t \to 0$.

Proof. $B_{(X,d)}(x_0,\delta) \subseteq \mathfrak{M}$ is an open set (see Lemma 3.1.7) and therefore it is a submanifold. Fix $y_0 \in B_{(X,d)}(x_0,\delta)$ and let $U \subseteq B_{(X,d)}(x_0,\delta)$ be an open neighborhood of y_0 such that there exists a C^∞_{loc} diffeomorphism $\Phi : B^n(2) \xrightarrow{\sim} U$ with $\Phi(0) = y_0$. We give $B^n(2)$ the density $h\sigma_{\mathrm{Leb}} := \Phi^*\,\mathrm{Vol}$. Note that $h \in C^\infty_{\mathrm{loc}}(B^n(2))$ and $h(v) > 0$, $\forall v \in B^n(2)$.

We will show that $u(t,\Phi(v)) \in C^\infty_{\mathrm{loc}}(\mathbb{R}\times B^n(1/2);\mathbb{C}^D)$ and vanishes to infinite order as $t \to 0$. This will complete the proof since $y_0 \in B_{(X,d)}(x_0,\delta)$ was arbitrary and $\Phi(B^n(1/2))$ is an open neighborhood of y_0.

For $t > 0$, we have $\partial_t u(t,x) = -\overline{\mathscr{L}}u(t,x) = -\mathscr{L}u(t,x)$ (see Remark 8.4.3). Thus, $\partial_t u(t,\Phi(v)) = -\Phi^*\mathscr{L}\Phi_* u(t,\Phi(v))$, where $\Phi^*\mathscr{L}\Phi_*$ is a partial differential operator on $B^n(2)$, acting in the v variable.

Since $\mathscr{L}$ is maximally subelliptic of type 2 of degree 2κ with respect to $(W,\vec{d})$ on $(\mathfrak{M},\mathrm{Vol})$, it follows immediately from the definitions that $\Phi^*\mathscr{L}\Phi_*$ is maximally subelliptic of type 2 of degree 2κ with respect to $(\Phi^*W,\vec{d}) := \{(\Phi^*W_1,\vec{d}_1),\ldots,(\Phi^*W_r,\vec{d}_r)\}$ on $(B^n(2),\Phi^*\,\mathrm{Vol})$. In particular, by restricting our attention to $B^n(1)$, we see that $\Phi^*\mathscr{L}\Phi_*$ is maximally subelliptic of type 2 of degree 2κ at the unit scale with respect to $(\Phi^*W,\vec{d})$ and $h\sigma_{\mathrm{Leb}}$.

Fix $M \in \mathbb{N}$. We begin by showing $u(t,\Phi(v)) \in C^\infty((0,M)\times B^n(1/2);\mathbb{C}^D)$. Fix $\phi_1,\phi_2 \in C_0^\infty(B^n(1))$ with $\phi_1 \prec \phi_2$ and $\phi_1 \equiv 1$ on $B^n(3/4)$. Corollary 8.3.8 applied in the v variable implies that $\forall s>0$, $\forall k \in \mathbb{N}$, $\exists N(s) \in \mathbb{N}$, $C_s \geq 0$ such that

$$\begin{aligned}
&\|\phi_1(\cdot)\partial_t^k u(t,\Phi(\cdot))\|_{L^2((0,M);L^2_s(\mathbb{R}^n))}\\
&\quad = \|\phi_1(\cdot)(\Phi^*\mathscr{L}\Phi_*)^k u(t,\Phi(\cdot))\|_{L^2((0,M);L^2_s(\mathbb{R}^n))}\\
&\quad \leq C_s \sum_{j=k}^{N(s)+k} \|\phi(\cdot)(\Phi^*\mathscr{L}\Phi_*)^j u(t,\Phi(\cdot))\|_{L^2((0,M)\times\mathbb{R}^n, dt\times d h\sigma_{\mathrm{Leb}})}.
\end{aligned} \tag{8.136}$$

But, using Remark 8.4.3,

$$\begin{aligned}
(\Phi^*\mathscr{L}\Phi_*)^j u(t,\Phi(v)) &= (\mathscr{L}^j u)(t,\Phi(v)) = (\mathscr{L}^j e^{-t\overline{\mathscr{L}}} f)(\Phi(v))\\
&= (\overline{\mathscr{L}}^j e^{-t\overline{\mathscr{L}}} f)(\Phi(v)) = (e^{-t\overline{\mathscr{L}}}\,\overline{\mathscr{L}}^j f)(\Phi(v)) = (e^{-t\overline{\mathscr{L}}}\mathscr{L}^j f)(\Phi(v)).
\end{aligned}$$

Thus, changing variables, we have

$$\begin{aligned}
&\|\phi(\cdot)(\Phi^*\mathscr{L}\Phi_*)^j u(t,\Phi(\cdot))\|_{L^2((0,M)\times\mathbb{R}^n, dt\times dh\sigma_{\mathrm{Leb}})}\\
&\quad \leq \|e^{-t\overline{\mathscr{L}}}\mathscr{L}^j f\|_{L^2((0,M)\times\mathfrak{M}, dt\times d\,\mathrm{Vol})} \leq \|\mathscr{L}^j f\|_{L^2((0,M)\times\mathfrak{M}, dt\times d\,\mathrm{Vol})}\\
&\quad = \left(\int_0^M \|\mathscr{L}^j f\|^2_{L^2(\mathfrak{M},\mathrm{Vol})}\right)^{\frac{1}{2}} < \infty,
\end{aligned} \tag{8.137}$$

where we have used $\mathscr{L}^j f \in C_0^\infty(\mathfrak{M};\mathbb{C}^D) \subseteq L^2(\mathfrak{M},\mathrm{Vol};\mathbb{C}^D)$. Combining (8.136) and (8.137), we see that $\forall M,k \in \mathbb{N}$, $\forall s>0$,

$$\|\phi_1(\cdot)\partial_t^k u(t,\Phi(\cdot))\|_{L^2((0,M);L^2_s(\mathbb{R}^n))} < \infty.$$

Using the fact that $\phi_1 \equiv 1$ on $B^n(3/4) \supseteq B^n(1/2)$, taking k and s large, and applying the Sobolev embedding theorem, we see that $u(t,\Phi(v)) \in C^\infty((0,M)\times B^n(1/2);\mathbb{C}^D)$. In particular, since M was arbitrary, we obtain $u(t,\Phi(v)) \in C^\infty_{\mathrm{loc}}((0,\infty)\times B^n(1/2);\mathbb{C}^D)$.

Note that for $x \in B_{(X,d)}(x_0,\delta)$, $u(0,x) = e^{-0\overline{\mathscr{L}}} f(x) = f(x) = 0$. Therefore, $u(t,\Phi(v)) \equiv 0$ for $t \leq 0$. Thus, to complete the proof, it suffices to show that for every $k \in \mathbb{N}$,

$$\lim_{t\downarrow 0} \partial_t^k u(t,\Phi(\cdot)) = 0,$$

where the limit is taken in the Fréchet space $C^\infty(B^n(1/2))$. To prove this, fix $k \in \mathbb{N}$ and let $\{t_j\}_{j\in\mathbb{N}} \subset (0,1)$ be any sequence with $t_j \to 0$. We will show that there is a subsequence t_{j_l} such that

$$\lim_{l\to\infty} (\partial_t^k u)(t_{j_l},\Phi(\cdot)) = 0,$$

where the limit is taken in $C^\infty(B^n(1/2))$; this will complete the proof. Since $u(t,\Phi(v)) \in C^\infty((0,1)\times B^n(1/2))$, $\{u(t_j,\Phi(\cdot)) : j \in \mathbb{N}\} \subset C^\infty(B^n(1/2))$ is a bounded set, so there is a subsequence t_{j_l} such that

$$\lim_{l\to\infty} (\partial_t^k u)(t_{j_l}, \Phi(\cdot)) = u_0(\cdot),$$

where $u_0 \in C^\infty(B^n(1/2))$, with convergence in $C^\infty(B^n(1/2))$. Thus, to complete the proof we wish to show that $u_0 = 0$. This will follow if we show that

$$\lim_{t\downarrow 0} \partial_t^k u(t,\Phi(\cdot)) = 0, \tag{8.138}$$

with convergence in the sense of $C_0^\infty(B^n(1/2);\mathbb{C}^D)'$.

Let $\phi \in C_0^\infty(B^n(1/2);\mathbb{C}^D)$. We have

$$\begin{aligned}
\lim_{t\downarrow 0} \int \overline{\phi(v)}\partial_t^k u(t,\Phi(v))h(v)\,dv &= \lim_{t\downarrow 0} \int \overline{\phi(\Phi^{-1}(x))}\partial_t^k u(t,x)\,d\operatorname{Vol}(x) \\
&= \lim_{t\downarrow 0} \int \overline{\phi(\Phi^{-1}(x))}\partial_t^k (e^{-t\mathscr{L}} f)(x)\,d\operatorname{Vol}(x) \\
&= \lim_{t\downarrow 0} \int \overline{\phi(\Phi^{-1}(x))}(-1)^k (e^{-t\mathscr{L}}\mathscr{L}^k f)(x)\,d\operatorname{Vol}(x) \\
&= \int \overline{\phi(\Phi^{-1}(x))}(-1)^k (\mathscr{L}^k f)(x)\,d\operatorname{Vol}(x) = 0,
\end{aligned}$$

where in the final equality we have used the fact that $\phi(\Phi^{-1}(\cdot))$ is supported in $B_{(X,d)}(x_0,\delta)$ and f is supported in $\mathfrak{M} \setminus \overline{B_{(X,d)}(x_0,\delta)}$. This establishes (8.138) and completes the proof. □

Lemma 8.4.17. *There exists $c_2 = c_2(\mathcal{K}) > 0$ such that $\forall \alpha$, $\exists C_{\mathcal{K},\alpha} \geq 0$ such that for all $\delta \in (0,1]$, $x \in \mathcal{K}$, and $t \in [0,\delta^{2\kappa}/4)$,*

$$\begin{aligned}
&\|W_x^\alpha e^{-t\mathscr{L}}(x,\cdot)\|_{L^2(\mathfrak{M}\setminus\overline{B_{(X,d)}(x,\delta)},\operatorname{Vol};\mathbb{M}^{D\times D})}, \|W_x^\alpha e^{-t\mathscr{L}}(\cdot,x)\|_{L^2(\mathfrak{M}\setminus\overline{B_{(X,d)}(x,\delta)},\operatorname{Vol};\mathbb{M}^{D\times D})} \\
&\quad \leq C_{\mathcal{K},\alpha}\delta^{-\deg_d(\alpha)} \exp(-c_2\delta^{\frac{2\kappa}{2\kappa-1}} t^{-\frac{1}{2\kappa-1}})(\operatorname{Vol}(B_{(W,d)}(x,\delta))\wedge 1)^{-\frac{1}{2}}.
\end{aligned}$$

Proof. Fix $\delta_0 \in (0,1]$ and $x_0 \in \mathcal{K}$. Let $\mathscr{R}_{B_{(X,d)}(x_0,\delta_0)} : f \mapsto f\big|_{B_{(X,d)}(x_0,\delta_0)}$ denote the restriction map. Define, for $t \geq 0$,

$$T_{t,x_0,\delta_0} := \mathscr{R}_{B_{(X,d)}(x_0,\delta_0)} e^{-t\mathscr{L}} : L^2(\mathfrak{M}\setminus\overline{B_{(X,d)}(x_0,\delta_0)},\operatorname{Vol};\mathbb{C}^D) \to L^2(B_{(X,d)}(x_0,\delta_0);\mathbb{C}^D).$$

We will apply Proposition 8.4.12 with this choice of T_{t,x_0,δ_0}, $(\mathfrak{M}_1,\mu_1) = (\mathfrak{M}\setminus\overline{B_{(X,d)}(x_0,\delta_0)}, \operatorname{Vol})$, $\sigma = \delta_0^{2\kappa}$, $\mathscr{D} = C_0^\infty(\mathfrak{M}\setminus\overline{B_{(X,d)}(x_0,\delta_0)})$, and $D_0 = D$.

We begin by verifying the hypotheses of Proposition 8.4.12. Clearly,

$$\|T_{t,x_0,\delta_0}\|_{L^2(\mathfrak{M}\setminus\overline{B_{(X,d)}(x_0,\delta_0)},\operatorname{Vol})\to L^2(B_{(X,d)}(x_0,\delta_0),\operatorname{Vol})} \leq 1.$$

For $f \in \mathscr{D}$, it follows immediately from Lemma 8.4.16 that $T_{t,x_0,\delta_0}f \in C^\infty_{\mathrm{loc}}([0,\sigma) \times B_{(X,d)}(x_0,\delta_0);\mathbb{C}^D)$ and vanishes to infinite order as $t \downarrow 0$. Finally, for $f \in \mathscr{D}$, using Remark 8.4.3 and the fact that $B_{(X,d)}(x_0,\delta_0) \subseteq \mathfrak{M}$ is open (Lemma 3.1.7), we have

$$\begin{aligned}\partial_t T_{t,x_0,\delta_0}f &= -\mathscr{R}_{B_{(X,d)}(x_0,\delta_0)}\overline{\mathscr{L}}e^{-t\overline{\mathscr{L}}}f = -\mathscr{R}_{B_{(X,d)}(x_0,\delta_0)}\mathscr{L}e^{-t\overline{\mathscr{L}}}f \\ &= -\mathscr{L}\mathscr{R}_{B_{(X,d)}(x_0,\delta_0)}e^{-t\overline{\mathscr{L}}}f = -\mathscr{L}T_{t,x_0,\delta_0}f = -\sigma^{-1}\delta_0^{2\kappa}\mathscr{L}T_{t,x_0,\delta_0}.\end{aligned}$$

Thus, all of the hypotheses of Proposition 8.4.12 are satisfied, so the proposition implies that there exists $c_2 = c_2(\mathcal{K}) > 0$ such that the following holds:

$$\begin{aligned}&\sup_{x\in\Phi_{x_0,\delta_0}(B^n(1/4))} \|(\delta_0^d W_x)^\alpha e^{-t\overline{\mathscr{L}}}(x,\cdot)\|_{L^2(\mathfrak{M}\setminus\overline{B_{(X,d)}(x_0,\delta_0)},\mathrm{Vol})} \\ &\quad= \sup_{x\in\Phi_{x_0,\delta_0}(B^n(1/4))} \|(\delta_0^d W_x)^\alpha T_{t,x_0,\delta_0}(x,\cdot)\|_{L^2(\mathfrak{M}\setminus\overline{B_{(X,d)}(x_0,\delta_0)},\mathrm{Vol})} \\ &\quad\lesssim (\mathrm{Vol}(B_{(W,d)}(x_0,\delta_0))\wedge 1)^{-\frac{1}{2}}\exp(-c_2\delta_0^{\frac{2\kappa}{2\kappa-1}}t^{-\frac{1}{2\kappa-1}}).\end{aligned}$$

Setting $x = \Phi_{x_0,\delta_0}(0) = x_0$ in the above equation and using the fact that $(\delta_0^d W)^\alpha = \delta_0^{\deg_d(\alpha)}W^\alpha$ proves the desired bound for $\|W_x^\alpha e^{-t\overline{\mathscr{L}}}(x,\cdot)\|_{L^2(\mathfrak{M}\setminus\overline{B_{(X,d)}(x,\delta)},\mathrm{Vol};\mathbb{M}^{D\times D})}$ (with x replaced by x_0 and δ replaced by δ_0). Since $e^{-t\overline{\mathscr{L}}}$ is self-adjoint, the desired bound for $\|W_x^\alpha e^{-t\overline{\mathscr{L}}}(\cdot,x)\|_{L^2(\mathfrak{M}\setminus\overline{B_{(X,d)}(x,\delta)},\mathrm{Vol};\mathbb{M}^{D\times D})}$ also follows. □

Lemma 8.4.18. *Suppose $a, b, c > 0$ are given. Then*

$$\sup_{x\in\mathcal{K}}\sup_{\substack{\delta',\delta\in(0,\infty)\\ \delta'\leq\delta}} \left(\frac{\delta}{\delta'}\right)^a \frac{(\mathrm{Vol}(B_{(W,d)}(x,\delta))\wedge 1)^{\frac{1}{2}}}{(\mathrm{Vol}(B_{(W,d)}(x,\delta'))\wedge 1)^{\frac{1}{2}}}\exp\left(-c\left(\frac{\delta}{\delta'}\right)^b\right) < \infty.$$

Proof. It follows from repeated application of Theorem 3.3.7 (f) that there exists $p = p(\mathcal{K}) > 0$ and $C_{\mathcal{K}} \geq 0$ such that $\forall x \in \mathcal{K}$, $\delta, \delta' \in (0,\infty)$ with $\delta' \leq \delta$,

$$\begin{aligned}\mathrm{Vol}(B_{(W,d)}(x,\delta))\wedge 1 &= \mathrm{Vol}(B_{(W,d)}(x,(\delta/\delta')\delta'))\wedge 1 \\ &\leq C_{\mathcal{K}}(\delta/\delta')^p(\mathrm{Vol}(B_{(W,d)}(x,\delta))\wedge 1).\end{aligned}$$

Thus, we have

$$\begin{aligned}&\left(\frac{\delta}{\delta'}\right)^a \frac{(\mathrm{Vol}(B_{(W,d)}(x,\delta))\wedge 1)^{\frac{1}{2}}}{(\mathrm{Vol}(B_{(W,d)}(x,\delta'))\wedge 1)^{\frac{1}{2}}}\exp\left(-c\left(\frac{\delta}{\delta'}\right)^b\right) \\ &\quad\leq C_{\mathcal{K}}^{\frac{1}{2}}\left(\frac{\delta}{\delta'}\right)^{a+p/2}\exp\left(-c\left(\frac{\delta}{\delta'}\right)^b\right).\end{aligned}$$

From here, the result follows. □

Proof of (8.117). Because $\partial_t^s e^{-t\overline{\mathscr{L}}} = (-1)^s \overline{\mathscr{L}}^s e^{-t\overline{\mathscr{L}}} = (-1)^s \mathscr{L} e^{-t\overline{\mathscr{L}}}$, where $\mathscr{L}$ is taken in the sense of distributions (see Remark 8.4.3), and due to the form of $\mathscr{L}$, it suffices to prove (8.117) in the case $s = 0$.

Fix $x, y \in \mathcal{K}$ and $t > 0$ with $16t^{\frac{1}{2\kappa}} \le \rho_{(X,d)}(x,y) \le 1$. Set $\delta := \rho_{(X,d)}(x,y)/3$. Note that $t < \delta^{2\kappa}/8$ and $\overline{B_{(X,d)}(x,\delta)} \cap \overline{B_{(X,d)}(y,\delta)} = \emptyset$. Using (8.132), we have

$$
\begin{aligned}
|W_x^\alpha W_y^\beta e^{-t\overline{\mathscr{L}}}(x,y)| &= \Big|\int (W_x^\alpha e^{-(t/2)\overline{\mathscr{L}}}(x,z))\,(W_y^\beta e^{-(t/2)\overline{\mathscr{L}}}(z,y))\,d\,\mathrm{Vol}(z)\Big| \\
&\le \int\limits_{z\in\mathfrak{M}\setminus B_{(X,d)}(x,\delta)} |(W_x^\alpha e^{-(t/2)\overline{\mathscr{L}}}(x,z))(W_y^\beta e^{-(t/2)\overline{\mathscr{L}}}(z,y))|\,d\,\mathrm{Vol}(z) \\
&\quad + \int\limits_{z\in\mathfrak{M}\setminus B_{(X,d)}(y,\delta)} |(W_x^\alpha e^{-(t/2)\overline{\mathscr{L}}}(x,z))(W_y^\beta e^{-(t/2)\overline{\mathscr{L}}}(z,y))|\,d\,\mathrm{Vol}(z) \\
&=: (\mathrm{I}) + (\mathrm{II}).
\end{aligned}
$$

We estimate only (I); the estimate for (II) follows by a similar proof by reversing the roles of x and y.

In the next equation we use $t < \delta^{2\kappa}/4 < 1$. We also freely use Theorem 3.3.7 (f). Applying the Cauchy–Schwartz inequality and Lemmas 8.4.17 and 8.4.14, we have, with $c_2 > 0$ as in Lemmas 8.4.17 and $c_3 := c_2/9 > 0$,

$$
\begin{aligned}
(\mathrm{I}) &\le \|W_x^\alpha e^{-(t/2)\overline{\mathscr{L}}}(x,\cdot)\|_{L^2(\mathfrak{M}\setminus\overline{B_{(X,d)}(x,\delta)})} \|W_y^\beta e^{-(t/2)\overline{\mathscr{L}}}(\cdot,y)\|_{L^2(\mathfrak{M},\mathrm{Vol})} \\
&\lesssim \delta^{-\deg_d(\alpha)} \exp(-c_2\delta^{\frac{2\kappa}{2\kappa-1}}(t/2)^{-\frac{1}{2\kappa-1}})(\mathrm{Vol}(B_{(W,d)}(x,\delta))\wedge 1)^{-\frac12} \\
&\quad \times (t^{1/2\kappa})^{-\deg_d(\beta)}(\mathrm{Vol}(B_{(W,d)}(y,t^{\frac{1}{2\kappa}}))\wedge 1)^{-\frac12} \\
&\approx \rho_{(X,d)}(x,y)^{-\deg_d(\alpha)} t^{-\deg_d(\beta)/2\kappa} \exp\left(-c_3\left(\frac{\rho_{(X,d)}(x,y)^{2\kappa}}{t}\right)^{\frac{1}{2\kappa-1}}\right) \\
&\quad \times (\mathrm{Vol}(B_{(W,d)}(x,\rho_{(X,d)}(x,y)))\wedge 1)^{-\frac12}(\mathrm{Vol}(B_{(X,d)}(y,t^{\frac{1}{2\kappa}}))\wedge 1)^{-\frac12} \\
&= \rho_{(X,d)}(x,y)^{-\deg_d(\alpha)-\deg_d(\beta)}(\mathrm{Vol}(B_{(W,d)}(x,\rho_{(X,d)}(x,y)))\wedge 1)^{-\frac12} \\
&\quad \times (\mathrm{Vol}(B_{(W,d)}(y,\rho_{(X,d)}(x,y)))\wedge 1)^{-\frac12} \exp\left(-(c_3/2)\left(\frac{\rho_{(X,d)}(x,y)^{2\kappa}}{t}\right)^{\frac{1}{2\kappa-1}}\right) \\
&\quad \times \left\{\frac{(\mathrm{Vol}(B_{(W,d)}(y,\rho_{(X,d)}(x,y)))\wedge 1)^{\frac12}}{(\mathrm{Vol}(B_{(W,d)}(y,t^{\frac{1}{2\kappa}}))\wedge 1)^{\frac12}}\left(\frac{\rho_{(X,d)}(x,y)}{t^{\frac{1}{2\kappa}}}\right)^{\deg_d(\beta)}\right. \\
&\quad \left.\exp\left(-(c_3/2)\left(\frac{\rho_{(X,d)}(x,y)^{2\kappa}}{t}\right)^{\frac{1}{2\kappa-1}}\right)\right\}.
\end{aligned}
$$

By Lemma 8.4.18, the term inside $\{\cdot\}$ is $\lesssim 1$. Thus, we have, with $c_1 = c_3/2$,

$$
\begin{aligned}
(\mathrm{I}) &\lesssim \rho_{(X,d)}(x,y)^{-\deg_d(\alpha)-\deg_d(\beta)}\left(\operatorname{Vol}(B_{(W,d)}(x,\rho_{(X,d)}(x,y)))\wedge 1\right)^{-\frac{1}{2}} \\
&\quad\times\left(\operatorname{Vol}(B_{(W,d)}(y,\rho_{(X,d)}(x,y)))\wedge 1\right)^{-\frac{1}{2}}\exp\left(-c_1\left(\frac{\rho_{(X,d)}(x,y)^{2\kappa}}{t}\right)^{\frac{1}{2\kappa-1}}\right).
\end{aligned}
\tag{8.139}
$$

Finally, using Theorem 3.3.7 (d) and (f), we have

$$
\begin{aligned}
&\operatorname{Vol}(B_{(W,d)}(x,\rho_{(X,d)}(x,y)))\wedge 1 \approx \operatorname{Vol}(B_{(X,d)}(x,\rho_{(X,d)}(x,y)))\wedge 1 \\
&\leq \operatorname{Vol}(B_{(X,d)}(y,3\rho_{(X,d)}(x,y)))\wedge 1 \lesssim \operatorname{Vol}(B_{(X,d)}(y,\rho_{(X,d)}(x,y)))\wedge 1 \\
&\approx \operatorname{Vol}(B_{(W,d)}(y,\rho_{(X,d)}(x,y)))\wedge 1.
\end{aligned}
\tag{8.140}
$$

Using (8.140) to bound the right-hand side of (8.139) gives the desired bound in the case $s = 0$ of (8.117) and completes the proof. □

8.5 Proof of the main result

In this section, we use the theory developed up to this point to complete the proof of Theorem 8.1.1.

Proof of Theorem 8.1.1 *except for (ii) and (iv).* Here we prove the equivalence of all parts of Theorem 8.1.1 except for (ii) and (iv). We do this by proving the following implications:

$$
\begin{array}{ccccccccccc}
(i) & \Rightarrow & (ix) & \Rightarrow & (x) & & & & & & \\
 & & \Downarrow & & \Downarrow & & & & & & \\
 & & (xi) & \Rightarrow & (xii) & \Rightarrow & (viii) & \Rightarrow & (vii) & \Rightarrow & (v) \Rightarrow (i)
\end{array}
$$

$$
(viii) \Rightarrow (vi) \Rightarrow (vii) \qquad\qquad (i) \Rightarrow (iii) \Rightarrow (i).
$$

(i) ⇒ (ix): It follows immediately from the definitions that if $\mathscr{P}$ is maximally subelliptic of degree κ with respect to $(W, \vec{d})$ on $\mathfrak{M}$ (see Definition 8.0.1), then $\mathscr{L} := \mathscr{P}^*\mathscr{P}$ is maximally subelliptic of type 2 of degree 2κ with respect to $(W, \vec{d})$ on $(\mathfrak{M}, \mathrm{Vol})$ (see Definition 8.4.1), where the formal adjoint $\mathscr{P}^*$ is taken in the sense of $L^2(\mathfrak{M}, \mathrm{Vol})$. Clearly, $\mathscr{L}$ is symmetric and non-negative when treated as a densely defined operator on $L^2(\mathfrak{M}, \mathrm{Vol}; \mathbb{C}^{\mathbf{D}_2})$ with dense domain $C_0^\infty(\mathfrak{M}; \mathbb{C}^{\mathbf{D}_2})$. Item (ix) now follows from Theorem 8.4.2. We make one comment. In the estimate in Theorem 8.4.2, it is written that $(\rho_{(W,d)}(x,y) + t^{\frac{1}{2\kappa}}) \wedge 1$, while in (ix) there is no $\wedge 1$. Because (ix) is restricted to $t \in (0,1]$ and $x, y \in \mathcal{K}$ and $\mathcal{K}$ is compact with respect to $\rho_{(W,d)}$ (Lemma 3.1.7), we have $\rho_{(W,d)}(x,y) + t^{\frac{1}{2\kappa}} \lesssim 1$. Thus, one may remove this $\wedge 1$ in the statement of Theorem 8.4.2 in the case we are considering. Item (ix) follows.

(ix) ⇒ (x) and (xi) ⇒ (xii): Since $\mathscr{P}^*\mathscr{P}$ is a symmetric, non-negative operator, it has at least one non-negative, self-adjoint extension (for example, the Friedrichs extension). Thus, both (ix) ⇒ (x) and (xi) ⇒ (xii) are immediate.

(ix) ⇒ (xi) and (x) ⇒ (xii): This follows immediately from the definitions.

(xii) ⇒ (viii): This follows from Theorem 5.11.7.

(viii) ⇒ (vii): Let $S \in \mathscr{A}_{\mathrm{loc}}^{-2\kappa}((W, \vec{d}); \mathbb{C}^{\mathbf{D}_2}, \mathbb{C}^{\mathbf{D}_2})$ be the operator from (viii) and set $T := S\mathscr{L}^*$. Clearly, $T\mathscr{P} \equiv I \mod C_{\mathrm{loc}}^{\infty}(\mathfrak{M} \times \mathfrak{M}; \mathbb{M}^{\mathbf{D}_2\times\mathbf{D}_2}(\mathbb{C}))$. It follows from Corollary 5.8.10 that $T \in \mathscr{A}_{\mathrm{loc}}^{-\kappa}((W, \vec{d}); \mathbb{C}^{\mathbf{D}_2}, \mathbb{C}^{\mathbf{D}_1})$, which establishes (vii).

(vii) ⇒ (v): Let $T \in \mathscr{A}_{\mathrm{loc}}^{-\kappa}((W, \vec{d}); \mathbb{C}^{\mathbf{D}_2}, \mathbb{C}^{\mathbf{D}_1})$ be the operator from (vii). Let $\mathcal{K} := \mathrm{supp}(\phi_2)$ and fix $\Psi_1, \Psi_2, \Psi_3 \in C_0^{\infty}(\mathfrak{M})$ with $\Psi_1 \prec \Psi_2 \prec \Psi_3$ and $\Psi_1 \equiv 1$ on $\mathcal{K}$. Fix $\Omega_1 \Subset \mathfrak{M}$ open and relatively compact with $\Psi_3 \in C_0^{\infty}(\Omega_1)$ and set $\mathcal{K}_1 := \overline{\Omega_1} \Subset \mathfrak{M}$.

Formulas (8.5) and (8.6) remain unchanged if u is replaced by $\Psi_1 u$; we henceforth make this replacement, so that we may assume $\Psi_2 u = u$.

Let R_1, R_2 denote elements of $C_{\mathrm{loc}}^{\infty}(\mathfrak{M} \times \mathfrak{M}; \mathbb{M}^{\mathbf{D}_2\times\mathbf{D}_2})$. Using the fact that $T\mathscr{P} \equiv I \mod C_{\mathrm{loc}}^{\infty}(\mathfrak{M} \times \mathfrak{M}; \mathbb{M}^{\mathbf{D}_2\times\mathbf{D}_2})$ and the pseudo-locality of T (see, for example, the growth condition in Definition 5.2.2), we have

$$\begin{aligned} \phi_1 u &= \phi_1 \Psi_2 u = \phi_1 T\mathscr{P}\Psi_2 u + \phi_1 R_1 \Psi_2 u = \phi_1 T\phi_3 \mathscr{P}\Psi_2 u + \phi_1 R_2 \Psi_2 u \\ &= \phi_1 T \phi_3 \mathscr{P} u + \phi_1 R_2 \Psi_2 u = \phi_1 T \Psi_2 \phi_3 \mathscr{P} u + \phi_1 R_2 \Psi_2 u. \end{aligned} \tag{8.141}$$

We now prove (8.5). If $\phi_3 \mathscr{P} u \in \mathscr{X}_{\mathrm{cpt}}^{s}(W, \vec{d})$, then by Remark 6.5.8 (since $\mathrm{supp}(\phi_2) \subseteq \mathcal{K}_1$), we have $\phi_3 \mathscr{P} u \in \mathscr{X}^{s}(\mathcal{K}_1, (W, \vec{d}))$. By assumption, we have $\mathrm{Mult}[\phi_1] T \,\mathrm{Mult}[\Psi_2] \in \mathscr{A}^{-\kappa}(\Omega_1, (W, \vec{d}))$, and therefore by Theorem 6.2.10, we have

$$\phi_1 T \Psi_2 \phi_3 \mathscr{P} u = (\mathrm{Mult}[\phi_1] T \,\mathrm{Mult}[\Psi_2]) \phi_3 \mathscr{P} u \in \mathscr{X}^{s+\kappa_1}(\mathcal{K}_1, (W, \vec{d})). \tag{8.142}$$

For any distribution $u \in C_0^{\infty}(\mathfrak{M}; \mathbb{C}^{\mathbf{D}_2})'$, we have $\phi_1 R_2 \Psi_2 u \in C_0^{\infty}(\mathfrak{M})$ and $\mathrm{supp}(\phi_1 R_2 \Psi_2 u) \subseteq \mathcal{K}_1$. Thus, by Proposition 6.5.5 (see also Remark 5.2.24), we have

$$\phi_1 R_2 \Psi_2 u \in \mathscr{X}^{s+\kappa}(\mathcal{K}_1, (W, \vec{d})). \tag{8.143}$$

Combining (8.141), (8.142), and (8.143), we conclude that $\phi_1 u \in \mathscr{X}^{s+\kappa}(\mathcal{K}_1, (W, \vec{d})) \subseteq \mathscr{X}_{\mathrm{cpt}}^{s+\kappa}(W, \vec{d})$, completing the proof of (8.5).

We turn to (8.6). Let $\psi \in C_0^{\infty}(\mathfrak{M})$ satisfy $\phi_2 \prec \psi \prec \phi_3$. We begin by assuming $u \in \mathscr{X}^{s}(\mathcal{K}_1, (W, \vec{d}))$ – we will later replace u with $\phi_2 u$. Applying (8.141) with ϕ_3 replaced by ψ we have (with a different choice of R_2)

$$\begin{aligned} \|\phi_1 u\|_{\mathscr{X}^{s+\kappa}(W,d)} &\le \|\phi_1 T \Psi_2 \psi \mathscr{P} u\|_{\mathscr{X}^{s+\kappa}(W,d)} + \|\phi_1 R_2 \Psi_2 u\|_{\mathscr{X}^{s+\kappa}(W,d)} \\ &= \|(\mathrm{Mult}[\phi_1] T \,\mathrm{Mult}[\Psi_2])(\psi \mathscr{P} u)\|_{\mathscr{X}^{s+\kappa}(W,d)} \\ &\quad + \|(\mathrm{Mult}[\phi_1] R_2 \,\mathrm{Mult}[\Psi_3]) \Psi_2 u\|_{\mathscr{X}^{s+\kappa}(W,d)}. \end{aligned} \tag{8.144}$$

By assumption, $\mathrm{Mult}[\phi_1] T \,\mathrm{Mult}[\Psi_2] \in \mathscr{A}^{-\kappa}(\Omega_1, (W, \vec{d}))$. Since $\mathrm{Mult}[\phi_1] R_2 \,\mathrm{Mult}[\Psi_3] \in C_0^{\infty}(\Omega_1 \times \Omega_1)$, Proposition 5.8.11 shows that $\mathrm{Mult}[\phi_1] R_2 \,\mathrm{Mult}[\Psi_3] \in \mathscr{A}^{-\kappa-N}(\Omega_1, (W, \vec{d}))$. Applying Theorem 6.2.10 to (8.144), we conclude that

$$\|\phi_1 u\|_{\mathscr{X}^{s+\kappa}(W,\vec{d})} \lesssim \|\psi \mathscr{P} u\|_{\mathscr{X}^{s}(W,\vec{d})} + \|\Psi_2 u\|_{\mathscr{X}^{s-N}(W,\vec{d})}. \tag{8.145}$$

Now suppose $u \in C_0^\infty(\mathfrak{M};\mathbb{C}^{\mathbf{D}_2})'$ with $\phi_3\mathscr{P}u \in \mathscr{X}^s_{\mathrm{cpt}}(W,\vec{d})$; by (8.5) with ϕ_1 replaced by ϕ_2, we have $\phi_2 u \in \mathscr{X}^{s+\kappa}_{\mathrm{cpt}}(W,\vec{d}) \subseteq \mathscr{X}^{s}_{\mathrm{cpt}}(W,\vec{d})$. Applying (8.145) to $\phi_2 u$ and using $\phi_1 \prec \phi_2 \prec \psi \prec \phi_3 \prec \Psi_2$, we have

$$\begin{aligned}\|\phi_1 u\|_{\mathscr{X}^{s+\kappa}(W,\vec{d})} &\lesssim \|\psi\mathscr{P}u\|_{\mathscr{X}^{s}(W,\vec{d})} + \|\phi_2 u\|_{\mathscr{X}^{s-N}(W,\vec{d})}\\ &= \|\psi\phi_3\mathscr{P}u\|_{\mathscr{X}^{s}(W,\vec{d})} + \|\phi_2 u\|_{\mathscr{X}^{s-N}(W,\vec{d})}\\ &\lesssim \|\phi_3\mathscr{P}u\|_{\mathscr{X}^{s}(W,\vec{d})} + \|\phi_2 u\|_{\mathscr{X}^{s-N}(W,\vec{d})},\end{aligned}$$

where the final estimate used Corollary 6.5.10. This completes the proof of (v).

(v) ⇒ (i): Fix a relatively compact, open set $\Omega \Subset \mathfrak{M}$ and $\phi_1,\phi_2,\phi_3 \in C_0^\infty(\mathfrak{M})$ with $\phi_1 \prec \phi_2 \prec \phi_3$ and $\phi_1 \equiv 1$ on $\overline{\Omega}$. Let $\mathcal{K} := \mathrm{supp}(\phi_2)$. By Proposition 6.5.5 (see also Remark 5.2.24), we have $C_0^\infty(\Omega) \subseteq \mathscr{F}^s_{2,2}(\mathcal{K},(W,\vec{d}))$, $\forall s \in \mathbb{R}$. Applying (8.6) with $s = N = 0$ we see that $\forall u \in C_0^\infty(\Omega)$,

$$\begin{aligned}\|u\|_{\mathscr{F}^{\kappa}_{2,2}(W,\vec{d})} &= \|\phi_1 u\|_{\mathscr{F}^{\kappa}_{2,2}(W,\vec{d})} \lesssim \|\phi_3\mathscr{P}u\|_{\mathscr{F}^{0}_{2,2}(W,\vec{d})} + \|\phi_2 u\|_{\mathscr{F}^{0}_{2,2}(W,\vec{d})}\\ &= \|\mathscr{P}u\|_{\mathscr{F}^{0}_{2,2}(W,\vec{d})} + \|u\|_{\mathscr{F}^{0}_{2,2}(W,\vec{d})} \approx \|\mathscr{P}u\|_{L^2(\mathfrak{M},\mathrm{Vol})} + \|u\|_{L^2(\mathfrak{M},\mathrm{Vol})},\end{aligned} \tag{8.146}$$

where the last ≈ step uses Proposition 6.2.13. By Corollary 6.2.14 (in particular, (6.5)), we have

$$\sum_{j=1}^{r}\|W_j^{n_j}u\|_{L^2(\mathfrak{M},\mathrm{Vol})} \lesssim \|u\|_{\mathscr{F}^{\kappa}_{2,2}(W,\vec{d})}. \tag{8.147}$$

Combining (8.146) and (8.147) establishes (i).

We have so far established the equivalence (i) ⇔ (ix) ⇔ (x) ⇔ (xi) ⇔ (xii) ⇔ (viii) ⇔ (vii) ⇔ (v).

(viii) ⇒ (vi): This follows from the already proved implication (vii) ⇒ (i) applied to the operator $\mathscr{P}^*\mathscr{P}$.

(vi) ⇒ (vii): Suppose (vi) holds. Using the already proved implication (i) ⇒ (vii) applied to the operator $\mathscr{P}^*\mathscr{P}$ shows that there is $S \in \mathscr{A}^{-2\kappa}_{\mathrm{loc}}((W,\vec{d});\mathbb{C}^{\mathbf{D}_2},\mathbb{C}^{\mathbf{D}_2})$ such that $S\mathscr{P}^*\mathscr{P} \equiv I \mod C^\infty_{\mathrm{loc}}(\mathfrak{M}\times\mathfrak{M};\mathbb{M}^{\mathbf{D}_2\times\mathbf{D}_2}(\mathbb{C}))$. Set $T := S\mathscr{L}^*$. Clearly, $T\mathscr{P} \equiv I \mod C^\infty_{\mathrm{loc}}(\mathfrak{M}\times\mathfrak{M};\mathbb{M}^{\mathbf{D}_2\times\mathbf{D}_2}(\mathbb{C}))$. It follows from Corollary 5.8.10 that $T \in \mathscr{A}^{-\kappa}_{\mathrm{loc}}((W,\vec{d});\mathbb{C}^{\mathbf{D}_2},\mathbb{C}^{\mathbf{D}_1})$, which establishes (vii).

(i) ⇒ (iii): This is obvious.

(iii) ⇒ (i): Suppose (iii) holds and fix a relatively compact, open set $\Omega \Subset \mathfrak{M}$. For each $x \in \mathfrak{M}$, by hypothesis, there is an open neighborhood U_x of x such that $\mathscr{P}$ is maximally subelliptic of degree κ with respect to $(W,\vec{d})$ on U_x. The collection $\{U_x : x \in \mathfrak{M}\}$ is an open cover for the compact set $\overline{\Omega}$; let $U_{x_1},\dots,U_{x_L}$ be a finite subcover.

For $j = 1,\dots,L$, pick $\phi_{1,j} \in C_0^\infty(U_{x_j})$ such that $\sum_{j=1}^{L}\phi_{1,j} \equiv 1$ on Ω. Let $\phi_{2,j} \in C_0^\infty(U_{x_j})$ be such that $\phi_{1,j} \prec \phi_{2,j}$.

Let $f \in C_0^\infty(\Omega;\mathbb{C}^{\mathbf{D}_2})$. We use the already proved implication (i) ⇒ (v) with $\mathfrak{M}$ replaced by U_{x_j} and Proposition 6.5.5 (see also Remark 5.2.24) to see that

$$\|\phi_{1,j}f\|_{\mathscr{F}^{\kappa}_{2,2}(W,d)} \lesssim \|\phi_{2,j}\mathscr{P}f\|_{\mathscr{F}^{0}_{2,2}(W,d)} + \|\phi_{2,j}f\|_{\mathscr{F}^{0}_{2,2}(W,d)}$$
$$\approx \|\phi_{2,j}\mathscr{P}f\|_{L^2(\mathfrak{M},\mathrm{Vol})} + \|\phi_{2,j}f\|_{L^2(\mathfrak{M},\mathrm{Vol})} \lesssim \|\mathscr{P}f\|_{L^2(\mathfrak{M},\mathrm{Vol})} + \|f\|_{L^2(\mathfrak{M},\mathrm{Vol})},$$

where use have also used Proposition 6.2.13. Thus, using Corollary 6.2.14 (in particular, (6.5)), we have

$$\sum_{k=1}^{r}\|W_k^{n_k}f\|_{L^2} \lesssim \|f\|_{\mathscr{F}^{\kappa}_{2,2}(W,d)} \leq \sum_{j=1}^{L}\|\phi_{1,j}f\|_{\mathscr{F}^{\kappa}_{2,2}(W,d)} \lesssim \|\mathscr{P}f\|_{L^2} + \|f\|_{L^2}.$$

This establishes (i). □

To complete the proof of Theorem 8.1.1 it remains to show that (ii) and (iv) are equivalent to the other conditions.

Lemma 8.5.1. *Suppose we have two partial differential operators of the form* (8.1)*:*

$$\mathscr{P}_1 := \sum_{\deg_{d}(\alpha)\leq\kappa} b_\alpha(x)W^\alpha, \quad \mathscr{P}_2 := \sum_{\deg_{d}(\alpha)\leq\kappa} c_\alpha(x)W^\alpha,$$

where $b_\alpha, c_\alpha \in C^\infty_{\mathrm{loc}}(\mathfrak{M}; \mathbb{M}^{\mathbf{D}_1\times\mathbf{D}_2}(\mathbb{C}))$. Suppose $b_\alpha = c_\alpha$ when $\deg_{d}(\alpha) = \kappa$. Then $\mathscr{P}_1$ is maximally subelliptic of degree κ with respect to (W, d) on $\mathfrak{M}$ if and only if $\mathscr{P}_2$ is maximally subelliptic of degree κ with respect to (W, d) on $\mathfrak{M}$.

Proof. By symmetry, we only need to show that if $\mathscr{P}_1$ is maximally subelliptic of degree κ with respect to (W, d) on $\mathfrak{M}$, then $\mathscr{P}_2$ is maximally subelliptic of degree κ with respect to (W, d) on $\mathfrak{M}$. Thus, we suppose $\mathscr{P}_1$ is maximally subelliptic of degree κ with respect to (W, d) on $\mathfrak{M}$.

Fix a relatively compact open set $\Omega \Subset \mathfrak{M}$ and let $\phi_1, \phi_2 \in C_0^\infty(\mathfrak{M})$ satisfy $\phi_1 \prec \phi_2$ and $\phi_1 \equiv 1$ on Ω. In what follows, f will denote an element of $C_0^\infty(\Omega)$, and we will freely use the fact that $C_0^\infty(\Omega) \subseteq \mathscr{F}^{s}_{2,2}(\overline{\Omega}, (W, d))$, $\forall s \in \mathbb{R}$ (see Proposition 6.5.5 and Remark 5.2.24).

For any $s \in \mathbb{R}$, using the already proved implication Theorem 8.1.1 (i) ⇒ (v), we have, $\forall f \in C_0^\infty(\Omega)$,

$$\begin{aligned}
\|f\|_{\mathscr{F}^{s+\kappa}_{2,2}(W,d)} &= \|\phi_1 f\|_{\mathscr{F}^{s+\kappa}_{2,2}(W,d)} \\
&\lesssim \|\phi_2\mathscr{P}_1 f\|_{\mathscr{F}^{s}_{2,2}(W,d)} + \|\phi_2 f\|_{\mathscr{F}^{s}_{2,2}(W,d)} \\
&\leq \|\phi_2\mathscr{P}_2 f\|_{\mathscr{F}^{s}_{2,2}(W,d)} + \|\phi_2(\mathscr{P}_1 - \mathscr{P}_2)f\|_{\mathscr{F}^{s}_{2,2}(W,d)} + \|\phi_2 f\|_{\mathscr{F}^{s}_{2,2}(W,d)} \\
&\lesssim \|\mathscr{P}_2 f\|_{\mathscr{F}^{s}_{2,2}(W,d)} + \|f\|_{\mathscr{F}^{s+\kappa-1}_{2,2}(W,d)},
\end{aligned} \tag{8.148}$$

where the last estimate used Proposition 6.5.9 and Corollary 6.5.10.

We claim, for $j \in \{0, \dots, \kappa\}$, that

$$\|f\|_{\mathscr{F}^{\kappa}_{2,2}(W,d)} \lesssim \|\mathscr{P}_2 f\|_{\mathscr{F}^{0}_{2,2}(W,d)} + \|f\|_{\mathscr{F}^{\kappa-j}_{2,2}(W,d)}. \tag{8.149}$$

We prove (8.149) by induction on j. The base case, $j = 0$, is trivial.

Suppose we have (8.149) for some $j \in \{0, \dots, \kappa - 1\}$. We wish to prove it for $j + 1$. Applying the inductive hypothesis and (8.148) with $s = -j$, we have

$$\begin{aligned}\|f\|_{\mathscr{F}^{\kappa}_{2,2}(W,\vec{d})} &\lesssim \|\mathscr{P}_2 f\|_{\mathscr{F}^{0}_{2,2}(W,\vec{d})} + \|f\|_{\mathscr{F}^{\kappa-j}_{2,2}(W,\vec{d})}\\ &\lesssim \|\mathscr{P}_2 f\|_{\mathscr{F}^{0}_{2,2}(W,\vec{d})} + \|\mathscr{P}_2 f\|_{\mathscr{F}^{-j}_{2,2}(W,\vec{d})} + \|f\|_{\mathscr{F}^{\kappa-j-1}_{2,2}(W,\vec{d})}\\ &\lesssim \|\mathscr{P}_2 f\|_{\mathscr{F}^{0}_{2,2}(W,\vec{d})} + \|f\|_{\mathscr{F}^{\kappa-j-1}_{2,2}(W,\vec{d})},\end{aligned}$$

completing the proof of (8.149).

Applying (8.149) and using Corollary 6.2.14 (in particular, (6.5)) and Proposition 6.2.13, we have

$$\sum_{k=1}^{r} \|W_k^{n_k} f\|_{L^2} \lesssim \|f\|_{\mathscr{F}^{\kappa}_{2,2}(W,\vec{d})} \lesssim \|\mathscr{P}_2 f\|_{\mathscr{F}^{0}_{2,2}} + \|f\|_{\mathscr{F}^{0}_{2,2}} \approx \|\mathscr{P}_2 f\|_{L^2} + \|f\|_{L^2}.$$

We conclude that $\mathscr{P}_2$ is maximally subelliptic of degree κ with respect to $(W, \vec{d})$ on $\mathfrak{M}$, completing the proof. □

Proof the equivalence of Theorem 8.1.1 *(ii).* (i) ⇔ (ii): This is a special case of Lemma 8.5.1. □

Lemma 8.5.2. *Suppose we have two partial differential operators of the form* (8.1)*:*

$$\mathscr{P}_1 := \sum_{\deg_{\vec{d}}(\alpha)\leq\kappa} b_\alpha(x) W^\alpha, \quad \mathscr{P}_2 := \sum_{\deg_{\vec{d}}(\alpha)\leq\kappa} c_\alpha(x) W^\alpha,$$

where $b_\alpha, c_\alpha \in C^\infty_{\mathrm{loc}}(\mathfrak{M}; \mathbb{M}^{\mathbf{D}_1\times\mathbf{D}_2}(\mathbb{C}))$. *Fix* $x_0 \in \mathfrak{M}$ *and suppose* $b_\alpha(x_0) = c_\alpha(x_0)$, $\forall \alpha$. *Then the following are equivalent:*

(a) *There exists a neighborhood* U *of* x_0 *such that* $\mathscr{P}_1$ *is maximally subelliptic of degree* κ *with respect to* $(W, \vec{d})$ *on* U.

(b) *There exists a neighborhood* U *of* x_0 *such that* $\mathscr{P}_2$ *is maximally subelliptic of degree* κ *with respect to* $(W, \vec{d})$ *on* U.

Proof. By symmetry, it suffices to prove (a) ⇒ (b). Suppose $\mathscr{P}_1$ is maximally subelliptic of degree κ with respect to $(W, \vec{d})$ on U, where U is a neighborhood of x_0. Let $U_1 \Subset U$ be a neighborhood of x_0 and let $\phi_1, \phi_2 \in C_0^\infty(U)$, with $\phi_1 \prec \phi_2$ and $\phi_1 \equiv 1$ on U_1.

We will freely use the fact that $C_0^\infty(U) \subseteq \mathscr{F}^{s}_{2,2}(\overline{U}, (W, \vec{d}))$, $\forall s \in \mathbb{R}$ (see Proposition 6.5.5 and Remark 5.2.24).

Using Corollary 6.2.14 (in particular, (6.5)) and the already proved implication Theorem 8.1.1 (i) ⇒ (v), we have, for $f \in C_0^\infty(U_1; \mathbb{C}^{\mathbf{D}_2})$,

$$\sum_{\deg_{d}(\alpha)\leq\kappa}\|W^{\alpha}f\|_{L^2}\approx\|f\|_{\mathcal{F}^{\kappa}_{2,2}(W,d)}=\|\phi_1 f\|_{\mathcal{F}^{\kappa}_{2,2}(W,d)}$$
$$\lesssim\|\phi_2\mathcal{P}_1 f\|_{\mathcal{F}^{0}_{2,2}(W,d)}+\|\phi_2 f\|_{\mathcal{F}^{0}_{2,2}(W,d)}=\|\mathcal{P}_1 f\|_{\mathcal{F}^{0}_{2,2}(W,d)}+\|f\|_{\mathcal{F}^{0}_{2,2}(W,d)}$$
$$\approx\|\mathcal{P}_1 f\|_{L^2}+\|f\|_{L^2}.$$

Thus, there exists $C>0$ with

$$\sum_{\deg_{d}(\alpha)\leq\kappa}\|W^{\alpha}f\|_{L^2}\leq C(\|\mathcal{P}_1 f\|_{L^2}+\|f\|_{L^2}).$$

Take $V\subseteq U_1$ a neighborhood of x_0 so small that

$$\sup_{x\in V}\max_{\deg_{d}(\alpha)\leq\kappa}\|b_{\alpha(x)}-c_{\alpha}(x)\|_{\mathbb{M}^{D_1\times D_2}}\leq\frac{1}{2C}.$$

It follows that for $f\in C_0^{\infty}(V;\mathbb{C}^{D_2})$,

$$\sum_{\deg_{d}(\alpha)\leq\kappa}\|W^{\alpha}f\|_{L^2}\leq\frac{1}{2}\sum_{\deg_{d}(\alpha)\leq\kappa}\|W^{\alpha}f\|_{L^2}+C(\|\mathcal{P}_2 f\|_{L^2}+\|f\|_{L^2}).$$

We conclude that $\forall f\in C_0^{\infty}(V;\mathbb{C}^{D_2})$,

$$\sum_{j=1}^{r}\|W_j^{n_j}f\|_{L^2}\leq\sum_{\deg_{d}(\alpha)\leq\kappa}\|W^{\alpha}f\|_{L^2}\leq 2C(\|\mathcal{P}_2 f\|_{L^2}+\|f\|_{L^2}).$$

We have shown that $\mathcal{P}_2$ is maximally subelliptic of degree κ with respect to (W,d) on V, completing the proof. □

Proof the equivalence of Theorem 8.1.1 *(iv).* (iii) ⇔ (iv) This is a special case of Lemma 8.5.2. □

Remark 8.5.3. The exact relationship between maximal subellipticity and maximal subellipticity of type 2 is not immediately clear. It is obvious that $\mathcal{P}$ is maximally subelliptic of degree κ with respect to (W,d) on $\mathfrak{M}$ if and only if $\mathcal{P}^*\mathcal{P}$ is maximally subelliptic of type 2 of degree 2κ with respect to (W,d) on $(\mathfrak{M},\mathrm{Vol})$, where the formal adjoint $\mathcal{P}^*$ is taken with respect to $L^2(\mathfrak{M},\mathrm{Vol})$. It seems likely that the methods of [2] can be used to show that every operator which is maximally subelliptic of type 2 is maximally subelliptic; however, we do not pursue this here. We content ourselves with the next result, which follows easily from the results we have already shown.

Corollary 8.5.4. *Suppose $\mathcal{L}$ is a partial differential operator of the form* (8.102) *and satisfies:*

- *$\mathcal{L}$ is maximally subelliptic of type* 2 *of degree 2κ with respect to (W,d) on $(\mathfrak{M},\mathrm{Vol})$.*
- *$\mathcal{L}$ is symmetric and non-negative, when thought of as a densely defined operator on $L^2(\mathfrak{M},\mathrm{Vol};\mathbb{C}^D)$ with dense domain $C_0^{\infty}(\mathfrak{M};\mathbb{C}^D)$.*

Then $\mathcal{L}$ is maximally subelliptic of degree 2κ with respect to (W,d) on $\mathfrak{M}$.

Proof. Let $\overline{\mathscr{L}}$ be any non-negative, self-adjoint extension of $\mathscr{L}$; for example, we may take $\overline{\mathscr{L}}$ to be the Friedrichs extension.

Fix a compact set $\mathcal{K} \Subset \mathfrak{M}$. Since $\mathcal{K}$ is compact with respect to $\rho_{(W,\vec{d})}$ (see Lemma 3.1.7), we have $\rho_{(W,\vec{d})}(x,y) \lesssim 1$, $\forall x,y \in \mathcal{K}$. In particular, for $x,y \in \mathcal{K}$ and $t \in (0,1]$, we have

$$\rho_{(W,\vec{d})}(x,y) + t^{\frac{1}{2\kappa}} \approx (\rho_{(W,\vec{d})}(x,y) + t^{\frac{1}{2\kappa}}) \wedge 1. \tag{8.150}$$

By Theorem 8.4.2 and (8.150), we have $e^{-t\overline{\mathscr{L}}}(x,y) \in C^\infty_{\mathrm{loc}}(\mathfrak{M}\times\mathfrak{M}; \mathbb{M}^{D\times D}(\mathbb{C}))$ and there exists $c > 0$ such that for all ordered multi-indices α, β, there exists $C_{\alpha,\beta}$ such that $\forall x,y \in \mathcal{K}$ and $t \in (0,1]$,

$$\begin{aligned}
\left|W_x^\alpha W_y^\beta e^{-t\overline{\mathscr{L}}}(x,y)\right| \leq{}& C_{\alpha,\beta}(\rho_{(W,\vec{d})}(x,y) + t^{\frac{1}{2\kappa}})^{-\deg_{\vec{d}}(\alpha)-\deg_{\vec{d}}(\beta)} \\
&\times \exp\left(-c\left(\frac{\rho_{(W,\vec{d})}(x,y)^{2\kappa}}{t}\right)^{\frac{1}{2\kappa-1}}\right) \\
&\times \left(\mathrm{Vol}(B_{(W,\vec{d})}(x, \rho_{(W,\vec{d})}(x,y) + t^{\frac{1}{2\kappa}})) \wedge 1\right)^{-1}.
\end{aligned}$$

It follows that $\{(e^{-t\overline{\mathscr{L}}}, t^{1/2\kappa}) : t \in (0,1]\}$ is a bounded set of locally $(W, \vec{d})$ pre-elementary operators. By Theorem 5.11.7, there exists $S \in \mathscr{A}^{-2\kappa}_{\mathrm{loc}}((W,\vec{d}); \mathbb{C}^D, \mathbb{C}^D)$ such that

$$\mathscr{L}S, S\mathscr{L} \equiv I \mod C^\infty_{\mathrm{loc}}(\mathfrak{M}\times\mathfrak{M}; \mathbb{M}^{D\times D}(\mathbb{C})).$$

Theorem 8.1.1 (vii) $\Rightarrow$ (i) now implies that $\mathscr{L}$ is maximally subelliptic of degree 2κ with respect to $(W, \vec{d})$ on $\mathfrak{M}$. □

8.6 Vector bundles

Theorem 8.1.1 (along most other results in this text) is stated for trivial vector bundles over $\mathfrak{M}$. However, there is an analogous result for arbitrary vector bundles over $\mathfrak{M}$. The idea is that while Theorem 8.1.1 is not (strictly speaking) a local theorem, the main methods to prove it are local. Thus, a similar result holds for arbitrary vector bundles with the same proof. In this section, we formulate this version on vector bundles. Elliptic operators on vector bundles have many applications to geometry; we state the below results for maximally subelliptic operators on vector bundles in hopes that they may find similar applications.

Let $\mathfrak{M}$ be a connected C^∞ manifold of dimension $n \geq 1$ and let $(W,\vec{d}) = \{(W_1, \vec{d}_1), \dots, (W_r, \vec{d}_r)\} \subset C^\infty_{\mathrm{loc}}(\mathfrak{M}; T\mathfrak{M}) \times \mathbb{N}_+$ be Hörmander vector fields with formal degrees on $\mathfrak{M}$. Fix $\kappa \in \mathbb{N}_+$ such that $\vec{d}_j$ divides κ for $j = 1, \dots, r$ and set $n_j := \kappa/\vec{d}_j \in \mathbb{N}_+$.

Let $\pi_1 : \mathcal{V}_1 \to \mathfrak{M}$ and $\pi_2 : \mathcal{V}_2 \to \mathfrak{M}$ be smooth, complex vector bundles over $\mathfrak{M}$ of finite ranks $\mathbf{D}_1$ and $\mathbf{D}_2$, respectively.

Fix a smooth connection ∇ on $\pi_2 : \mathcal{V}_2 \to \mathfrak{M}$ and smooth inner products $\langle\cdot,\cdot\rangle_{1,x}$ and $\langle\cdot,\cdot\rangle_{2,x}$ on $\mathcal{V}_1$ and $\mathcal{V}_2$, respectively. Most of the results and definitions which follow do not depend on the choice of connection or the choices of inner products (see Remarks 8.6.1, 8.6.3, 8.6.5, and 8.6.6). For an ordered multi-index $\alpha = (\alpha_1, \alpha_2, \dots, \alpha_L)$, we write

$$\nabla_W^\alpha = \nabla_{W_{\alpha_1}} \nabla_{W_{\alpha_2}} \cdots \nabla_{W_{\alpha_L}}.$$

Since $\mathcal{V}_1$ and $\mathcal{V}_2$ have inner products, and therefore norms, it makes sense to consider $L^p(\mathfrak{M}, \mathrm{Vol}; \mathcal{V}_1)$ and $L^p(\mathfrak{M}, \mathrm{Vol}; \mathcal{V}_2)$: the Banach spaces of L^p sections of $\mathcal{V}_1$ and $\mathcal{V}_2$. We write $\underline{\mathrm{Hom}}(\mathcal{V}_2, \mathcal{V}_1)$ for the smooth vector bundle over $\mathfrak{M}$ whose fiber at $x \in \mathfrak{M}$ is given by $\underline{\mathrm{Hom}}(\mathcal{V}_2, \mathcal{V}_1)_x = \mathrm{Hom}(\mathcal{V}_{2,x}, \mathcal{V}_{1,x})$. We write $\mathrm{Hom}(\mathcal{V}_2, \mathcal{V}_1)$ for the smooth vector bundle over $\mathfrak{M} \times \mathfrak{M}$ whose fiber at $(x,y) \in \mathfrak{M} \times \mathfrak{M}$ is given by $\mathrm{Hom}(\mathcal{V}_2, \mathcal{V}_1)_{(x,y)} = \mathrm{Hom}(\mathcal{V}_{2,y}, \mathcal{V}_{1,x})$.

We consider partial differential operators of the form

$$\mathscr{P} = \sum_{\deg_{\vec{d}}(\alpha) \le \kappa} a_\alpha \nabla_W^\alpha, \quad a_\alpha \in C_{\mathrm{loc}}^\infty(\mathfrak{M}; \underline{\mathrm{Hom}}(\mathcal{V}_2, \mathcal{V}_1)). \tag{8.151}$$

Remark 8.6.1. While (8.151) uses the connection, ∇, the class of operators covered by (8.151) does not depend on ∇. Indeed, it is straightforward to verify that $\mathscr{P}$ is of the form (8.151) if and only if the following holds. For every $x \in \mathfrak{M}$, let $U \subseteq \mathfrak{M}$ be an open neighborhood of x with local trivializations $\phi_1 : \pi_1^{-1}(U) \xrightarrow{\sim} U \times \mathbb{C}^{\mathbf{D}_1}$, $\phi_2 : \pi_2^{-1}(U) \xrightarrow{\sim} U \times \mathbb{C}^{\mathbf{D}_2}$. We assume $(\phi_1)_* \mathscr{P} \phi_2^* = \sum_{\deg_{\vec{d}}(\alpha)\le\kappa} a_\alpha^{\phi_1,\phi_2} W^\alpha$, where $a_\alpha^{\phi_1,\phi_2} \in C_{\mathrm{loc}}^\infty(U; \mathbb{M}^{\mathbf{D}_1 \times \mathbf{D}_2}(\mathbb{C}))$.

Definition 8.6.2. We say $\mathscr{P}$ given by (8.151) is maximally subelliptic of degree κ with respect to $(W, \vec{d})$ on $\mathfrak{M}$ if the following holds. For every relatively compact, open set $\Omega \Subset \mathfrak{M}$, there exists $C_\Omega \ge 0$ such that $\forall f \in C_0^\infty(\Omega; \mathcal{V}_2)$,

$$\sum_{j=1}^r \|\nabla_{W_j}^{n_j} f\|_{L^2(\mathfrak{M},\mathrm{Vol};\mathcal{V}_2)} \le C_\Omega \big(\|\mathscr{P} f\|_{L^2(\mathfrak{M},\mathrm{Vol};\mathcal{V}_1)} + \|f\|_{L^2(\mathfrak{M},\mathrm{Vol};\mathcal{V}_2)} \big). \tag{8.152}$$

Remark 8.6.3. For a fixed operator $\mathscr{P}$, whether $\mathscr{P}$ is maximally subelliptic does not depend on the choice of connection, ∇, even though the expression (8.152) involves the connection. To see this, fix a relatively compact, open set $\Omega \Subset \mathfrak{M}$; we consider $f \in C_0^\infty(\mathfrak{M}, \mathrm{Vol}; \mathcal{V}_2)$. A simple integration by parts shows that for $l \in \mathbb{N}_+$,

$$\|\nabla_{W_j}^l f\|_{L^2(\mathfrak{M},\mathrm{Vol};\mathcal{V}_2)}^2 \le (\mathrm{l.c.}) \|\nabla_{W_j}^{l+1} f\|_{L^2(\mathfrak{M},\mathrm{Vol};\mathcal{V}_2)}^2 + (\mathrm{s.c.}) \|\nabla_{W_j}^{l-1} f\|_{L^2(\mathfrak{M},\mathrm{Vol};\mathcal{V}_2)}^2.$$

Using this, a straightforward induction argument shows that

$$\|\nabla_{W_j}^{n_j} f\|_{L^2(\mathfrak{M},\mathrm{Vol};\mathcal{V}_2)} + \|f\|_{L^2(\mathfrak{M},\mathrm{Vol};\mathcal{V}_2)} \approx \sum_{l=0}^{n_j} \|\nabla_{W_j}^l f\|_{L^2(\mathfrak{M},\mathrm{Vol};\mathcal{V}_2)},$$

and therefore (8.152) is equivalent to

$$\sum_{j=1}^{r}\sum_{l=0}^{n_j}\|\nabla_{W_j}^{l}f\|_{L^2(\mathfrak{M},\mathrm{Vol};\mathcal{V}_2)} \lesssim \|\mathscr{P}f\|_{L^2(\mathfrak{M},\mathrm{Vol};\mathcal{V}_1)} + \|f\|_{L^2(\mathfrak{M},\mathrm{Vol};\mathcal{V}_2)}. \tag{8.153}$$

A simple computation shows that if $\widetilde{\nabla}$ is another smooth connection on $\pi_2 : \mathcal{V}_2 \to \mathfrak{M}$, then

$$\sum_{j=1}^{r}\sum_{l=0}^{n_j}\|\nabla_{W_j}^{l}f\|_{L^2(\mathfrak{M},\mathrm{Vol};\mathcal{V}_2)} \approx \sum_{j=1}^{r}\sum_{l=0}^{n_j}\|\widetilde{\nabla}_{W_j}^{l}f\|_{L^2(\mathfrak{M},\mathrm{Vol};\mathcal{V}_2)}, \tag{8.154}$$

where the implicit constants do not depend on f, but do depend on $\Omega \Subset \mathfrak{M}$ and the various other choices we have made. Combining (8.154) and (8.153) shows that (8.152) does not depend on the choice of connection.

Using their inner products, we may identify $\mathcal{V}_1$ and $\mathcal{V}_2$ with their duals, and we do so henceforth. We will be considering operators $T : C_0^\infty(\mathfrak{M};\mathcal{V}_1) \to C_{\mathrm{loc}}^\infty(\mathfrak{M};\mathcal{V}_2)$. We say such an operator is given by integration against a function $T(x,y) \in L^1_{\mathrm{loc}}(\mathfrak{M}\times\mathfrak{M}, \mathrm{Vol}\times\mathrm{Vol}; \mathrm{Hom}(\mathcal{V}_1,\mathcal{V}_2))$ if $\forall f \in C_0^\infty(\mathfrak{M};\mathcal{V}_1)$ and $g \in C_0^\infty(\mathfrak{M};\mathcal{V}_2)$ we have

$$\int \langle g(x), Tf(x)\rangle_{2,x}\, d\,\mathrm{Vol}(x) = \iint \langle g(x), T(x,y)f(y)\rangle_{2,x}\, d\,\mathrm{Vol}(x)\, d\,\mathrm{Vol}(y).$$

We write $T \equiv S \mod C_{\mathrm{loc}}^\infty(\mathfrak{M}\times\mathfrak{M};\mathrm{Hom}(\mathcal{V}_1,\mathcal{V}_2))$ if $T - S$ is given by integration against an element of $C_{\mathrm{loc}}^\infty(\mathfrak{M}\times\mathfrak{M};\mathrm{Hom}(\mathcal{V}_1,\mathcal{V}_2))$.

For an element $E \in C_{\mathrm{loc}}^\infty(\mathfrak{M}\times\mathfrak{M};\mathrm{Hom}(\mathcal{V}_1,\mathcal{V}_2))$ it makes sense to consider $|E(x,y)|$, the operator norm of $E(x,y) : \mathcal{V}_{1,y} \to \mathcal{V}_{2,x}$, where the norms on $\mathcal{V}_{1,y}$ and $\mathcal{V}_{2,x}$ are given by their inner products. Using this, one can generalize all of the definitions concerning $\mathscr{A}^t(\Omega,(W,\vec{d}))$ to define objects like $\mathscr{A}^t(\Omega,(W,\vec{d});\mathrm{Hom}(\mathcal{V}_1,\mathcal{V}_2))$; wherever we have used W^α, one instead uses ∇_W^α. It is perhaps more simple to proceed locally to obtain an equivalent definition:

Definition 8.6.4. For $t \in \mathbb{R}$, we let $\mathscr{A}_{\mathrm{loc}}^t((W,\vec{d});\mathrm{Hom}(\mathcal{V}_1,\mathcal{V}_2))$ be the space of those $T : C_0^\infty(\mathfrak{M};\mathcal{V}_1) \to C_{\mathrm{loc}}^\infty(\mathfrak{M};\mathcal{V}_2)$ such that the following holds. For every $x \in \mathfrak{M}$, let $U \subseteq \mathfrak{M}$ be a neighborhood of x with local trivializations $\phi_1 : \pi_1^{-1}(U) \xrightarrow{\sim} U \times \mathbb{C}^{\mathbf{D}_1}$, $\phi_2 : \pi_2^{-1}(U) \xrightarrow{\sim} U\times\mathbb{C}^{\mathbf{D}_2}$. We assume $(\phi_2)_* T\phi_1^* \in \mathscr{A}_{\mathrm{loc}}^t((W,\vec{d});\mathbb{C}^{\mathbf{D}_1},\mathbb{C}^{\mathbf{D}_2})$, where we have replaced $\mathfrak{M}$ with U in the definition of $\mathscr{A}_{\mathrm{loc}}^t((W,\vec{d});\mathbb{C}^{\mathbf{D}_1},\mathbb{C}^{\mathbf{D}_2})$.

Remark 8.6.5. Definition 8.6.4 does not depend on the choices of connection or inner products.

Similarly, one can define the Besov and Triebel–Lizorkin spaces of sections in $\mathcal{V}$ for $\mathcal{K} \Subset \mathfrak{M}$ compact $\mathscr{B}_{p,q}^s(\mathcal{K},(W,\vec{d});\mathcal{V})$ and $\mathscr{F}_{p,q}^s(\mathcal{K},(W,\vec{d});\mathcal{V})$, either by a simple reprise of our definitions or by using a partition of unity and reducing to local trivializations. One obtains the same spaces and equivalent norms either way.

Remark 8.6.6. Using the inner products on $\mathcal{V}_1$ and $\mathcal{V}_2$, given $\mathscr{P}$ of the form (8.151), it makes sense to consider the formal adjoint $\mathscr{P}^*$, in the sense of operators $L^2(\mathfrak{M}, \mathrm{Vol}; \mathcal{V}_2) \to L^2(\mathfrak{M}, \mathrm{Vol}; \mathcal{V}_1)$. This adjoint depends on the choices of inner products on $\mathcal{V}_1$ and $\mathcal{V}_2$; however, Theorem 8.6.7 holds for any choice of smooth inner products.

We are prepared to state the vector bundle analog of Theorem 8.1.1. We present analogs of all parts of Theorem 8.1.1 except for Theorem 8.1.1 (iv) as the process of freezing coefficients cannot be done intrinsically on a vector bundle.

Theorem 8.6.7. *In the setting above, let $\mathscr{P}$ be of the form* (8.151). *The following are equivalent:*

(a) *$\mathscr{P}$ is maximally subelliptic of degree κ with respect to $(W, \vec{d})$ on $\mathfrak{M}$.*

(b) *$\mathscr{P}_0 := \sum_{\deg_{\vec{d}}(\alpha)=\kappa} a_\alpha \nabla_W^\alpha$ is maximally subelliptic of degree κ with respect to $(W, \vec{d})$ on $\mathfrak{M}$.*

(c) *$\forall x_0 \in \mathfrak{M}$, there exists an open neighborhood $U \subseteq \mathfrak{M}$ of x_0 such that $\mathscr{P}$ is maximally subelliptic of degree κ with respect to $(W, \vec{d})$ on U.*

(d) *For any scale of spaces $\mathscr{X}^s$ of the form*

$$\mathscr{X}^s \in \{\mathscr{B}^s_{p,q} : p, q \in [1, \infty]\} \bigcup \{\mathscr{F}^s_{p,q} : p \in (1, \infty), q \in (1, \infty]\},$$

we have the following. Let $\phi_1, \phi_2, \phi_3 \in C_0^\infty(\mathfrak{M})$ with $\phi_1 \prec \phi_2 \prec \phi_3$. Then $\forall s \in \mathbb{R}$,

$$\phi_3 \mathscr{P} u \in \mathscr{X}^s_{\mathrm{cpt}}((W, \vec{d}); \mathcal{V}_1) \Rightarrow \phi_1 u \in \mathscr{X}^{s+\kappa}_{\mathrm{cpt}}((W, \vec{d}); \mathcal{V}_2),$$
$$\forall u \in C_0^\infty(\mathfrak{M})'.$$

Moreover, for every $N \geq 0$, there exists $C = C(s, N, \mathscr{X}^s, (W, \vec{d}), \phi_1, \phi_2, \phi_3) \geq 0$ such that

$$\|\phi_1 u\|_{\mathscr{X}^{s+\kappa}((W,\vec{d});\mathcal{V}_2)} \leq C(\|\phi_3 \mathscr{P} u\|_{\mathscr{X}^s((W,\vec{d});\mathcal{V}_1)} + \|\phi_2 u\|_{\mathscr{X}^{s-N}((W,\vec{d});\mathcal{V}_2)}),$$
$$\forall u \in C_0^\infty(\mathfrak{M}; \mathcal{V}_2)' \text{ with } \phi_3 \mathscr{P} u \in \mathscr{X}^s_{\mathrm{cpt}}(W, \vec{d}).$$

(e) *$\mathscr{P}^* \mathscr{P}$ is maximally subelliptic of degree 2κ with respect to $(W, \vec{d})$ on $\mathfrak{M}$.*

(f) *There exists $T \in \mathscr{A}^{-\kappa}_{\mathrm{loc}}((W, \vec{d}); \mathrm{Hom}(\mathcal{V}_1, \mathcal{V}_2))$ such that*

$$T\mathscr{P} \equiv I \mod C^\infty_{\mathrm{loc}}(\mathfrak{M} \times \mathfrak{M}; \mathrm{Hom}(\mathcal{V}_2, \mathcal{V}_2)).$$

(g) *There exists $S \in \mathscr{A}^{-2\kappa}_{\mathrm{loc}}((W, \vec{d}); \mathrm{Hom}(\mathcal{V}_2, \mathcal{V}_2))$ such that*

$$S\mathscr{P}^* \mathscr{P}, \mathscr{P}^* \mathscr{P} S \equiv I \mod C^\infty_{\mathrm{loc}}(\mathfrak{M} \times \mathfrak{M}; \mathrm{Hom}(\mathcal{V}_2, \mathcal{V}_2)).$$

(h) *For every non-negative self-adjoint extension $\overline{\mathscr{L}}$ of $\mathscr{P}^* \mathscr{P}$ and all $t > 0$, $e^{-t\overline{\mathscr{L}}}$ is given by integration against a $C^\infty_{\mathrm{loc}}(\mathfrak{M} \times \mathfrak{M}; \mathrm{Hom}(\mathcal{V}_2, \mathcal{V}_2))$ function (denoted by $e^{-t\overline{\mathscr{L}}}(x, y)$) which satisfies the following. For every $\mathcal{K} \Subset \mathfrak{M}$ compact, there exists $c > 0$ such that* *for all ordered multi-indices α, β, for all $s \in \mathbb{N}$, there exists $C \geq 0$ such that $\forall t \in (0,1]$, $\forall x, y \in \mathcal{K}$,*

$$|\partial_t^s W_x^\alpha W_y^\beta e^{-t\overline{\mathcal{L}}}(x,y)| \leq C(\rho_{(W,d)}(x,y) + t^{\frac{1}{2\kappa}})^{-\deg_d(\alpha)-\deg_d(\beta)-2\kappa s}$$
$$\times \exp\left(-c\left(\frac{\rho_{(W,d)}(x,y)^{2\kappa}}{t}\right)^{\frac{1}{2\kappa-1}}\right)(\mathrm{Vol}(B_{(W,d)}(x, \rho_{(W,d)}(x,y) + t^{\frac{1}{2\kappa}})) \wedge 1)^{-1}.$$

(i) *There exists a non-negative self-adjoint extension $\overline{\mathcal{L}}$ of $\mathcal{P}^* \mathcal{P}$ satisfying the conclusions of (h).*

(j) *For every non-negative self-adjoint extension $\overline{\mathcal{L}}$ of $\mathcal{P}^* \mathcal{P}$, $e^{-t\overline{\mathcal{L}}}$ is given by integration against a $C^\infty_{\mathrm{loc}}(\mathfrak{M} \times \mathfrak{M}; \mathrm{Hom}(\mathcal{V}_2, \mathcal{V}_2))$ function (denoted by $e^{-t\overline{\mathcal{L}}}(x,y)$), which satisfies the following. For every $\mathcal{K} \Subset \mathfrak{M}$, $\forall \alpha, \beta$ ordered multi-indices, $\forall l \in \mathbb{N}$, $\exists C \geq 0$ such that $\forall t \in (0,1]$, $\forall x, y \in \mathcal{K}$,*

$$|\nabla_{W_x}^\alpha \nabla_{W_y}^\beta e^{-t\overline{\mathcal{L}}}(x,y)| \leq Ct^{-(\deg_d(\alpha)-\deg_d(\beta))/2\kappa}(1 + t^{-\frac{1}{2\kappa}}\rho_{(W,d)}(x,y))^{-l}$$
$$\times (\mathrm{Vol}(B_{(W,d)}(x, \rho_{(W,d)}(x,y) + t^{\frac{1}{2\kappa}})) \wedge 1)^{-1}.$$

(k) *There exists a non-negative self-adjoint extension $\overline{\mathcal{L}}$ of $\mathcal{P}^* \mathcal{P}$ satisfying the conclusions of (j).*

We also have analogs of the regularity properties from Section 8.2, though those results are local in nature, so the results on arbitrary vector bundles easily follow from the corresponding results on trivial vector bundles.

8.7 Quantitative regularity estimates

The regularity results in Theorem 8.1.1 (v) and more generally in Section 8.2 are qualitative in the sense that we have not been precise about what quantities the various constants depend on. For example, in (8.6), if ϕ_1 and ϕ_2 have small support, we wish to understand what the right estimate is to incorporate this dependence in a quantitative way. This is important when we turn to applications to nonlinear equations in Chapter 9.

In this section, we describe how to use the methods of this chapter combined with the scaling ideas in Section 3.15.2 to deduce the correct quantitative results.

8.7.1 The unit scale

We begin by describing a quantitative version of Proposition 8.2.12 at the unit scale. We work in the setting of Section 3.15.1. In particular, we assume $\mathfrak{M} = B^n(1)$ and $\mathrm{Vol} = h\sigma_{\mathrm{Leb}}$, where $h \in C^\infty(B^n(1); (0,\infty))$ with $\inf_u h(u) > 0$. We assume we are given $(W^1, \vec{d}^1), \dots, (W^\nu, \vec{d}^\nu)$ and $(X^1, d^1), \dots, (X^\nu, d^\nu)$ satisfying the hypotheses of Section 3.15.1. These choices satisfy the hypotheses of Section 8.2.3 in a quantitative way. As described in Section 3.15.1, $(W^1, \vec{d}^1)$ are Hörmander vector fields with formal degrees, and therefore it makes sense to ask whether a partial differential operator is maximally subelliptic with respect to $(W^1, \vec{d}^1)$.

The quantitative estimates are terms of the norm $\|f\|_{\mathscr{X}^s(W,\vec{d}),(X,\vec{d}),a,7/8,h}$ described in Section 6.11.1; here $a \gtrsim 1$ can be chosen to be a 0-multi-parameter unit-admissible constant as in Definition 3.15.1.

Fix $\kappa \in \mathbb{N}_+$ such that $\vec{d}_j^1$ divides κ, $1 \le j \le r_1$. Set $n_j := \kappa / \vec{d}_j^1$. Fix $\mathbf{D}_1, \mathbf{D}_2 \in \mathbb{N}_+$. We consider a partial differential operator of the form

$$\mathscr{P} = \sum_{\deg_{\vec{d}^1}(\alpha) \le \kappa} (W^1)^\alpha, \quad a_\alpha \in C^\infty(B^n(1); \mathbb{M}^{\mathbf{D}_1 \times \mathbf{D}_2}(\mathbb{C})). \tag{8.155}$$

We assume that $\mathscr{P}$ is maximally subelliptic of degree κ with respect to $(W^1, \vec{d}^1)$ in the sense that $\exists A \ge 0$ such that $\forall f \in C_0^\infty(B^n(1))$,

$$\sum_{j=1}^{r_1} \|(W_j^1)^{n_j} f\|_{L^2(B^n(1), h\sigma_{\mathrm{Leb}})} \le A(\|\mathscr{P} f\|_{L^2(B^n(1), h\sigma_{\mathrm{Leb}})} + \|f\|_{L^2(B^n(1), h\sigma_{\mathrm{Leb}})}).$$

Proposition 8.7.1. *For any scale of spaces $\mathscr{X}^s$, $s \in \mathbb{R}^\nu$, of the form*

$$\mathscr{X}^s \in \{\mathscr{B}_{p,q}^s : p, q \in [1, \infty]\} \bigcup \{\mathscr{F}_{p,q}^s : p \in (1, \infty), q \in (1, \infty]\},$$

the following holds. Let $\mathcal{B} \subset C_0^\infty(B^n(7/8))$ be a bounded set. Then, $\forall s \in \mathbb{R}^\nu$, $\forall \vec{N} \in [0, \infty)^\nu$, there exists $C \ge 0$ such that for all $\phi_1, \phi_2, \phi_3 \in \mathcal{B}$ with $\phi_1 \prec \phi_2 \prec \phi_3$,

$$\begin{aligned} \|\phi_1 u\|_{\mathscr{X}^{s+\kappa e_1}(W,\vec{d}),(X,\vec{d}),a,7/8,h} \le C\Big(&\|\phi_3 \mathscr{P} u\|_{\mathscr{X}^s(W,\vec{d}),(X,\vec{d}),a,7/8,h} \\ &+ \|\phi_2 u\|_{\mathscr{X}^{s-\vec{N}}(W,\vec{d}),(X,\vec{d}),a,7/8,h}\Big), \end{aligned} \tag{8.156}$$

$\forall u \in C_0^\infty(B^n(1); \mathbb{C}^{\mathbf{D}_2})'$ with $\phi_3 \mathscr{P} u \in \mathscr{X}_{\mathrm{cpt}}^s(W, \vec{d})$. Moreover, there exists $L \in \mathbb{N}$, depending only on $\max_{1 \le j \le q} |\vec{d}_j|_1$, κ, s, $\vec{N}$, and $\mathscr{X}^s$, such that $C = C(A, \max_\alpha \|a_\alpha\|_{C^L(B^n(1))}, \mathcal{B}, \mathbf{D}_1, \mathbf{D}_2) \ge 0$ is a $(\kappa, s, \vec{N}, \mathscr{X}^s)$-multi-parameter unit-admissible constant as in Definition 3.15.1.

Except for the dependence of C, Proposition 8.7.1 is a special case of Proposition 8.2.12. One can prove Proposition 8.7.1 by keeping careful track of the estimates in the proof of Proposition 8.2.12; we leave the details to the reader (see, for example, Theorem 3.15.5).

We state a simple corollary of Proposition 8.7.1 which is important for our applications.

Corollary 8.7.2. *For any scale of spaces $\mathscr{X}^s$, $s \in \mathbb{R}^\nu$, of the form*

$$\mathscr{X}^s \in \{\mathscr{B}_{p,q}^s : p, q \in [1, \infty]\} \bigcup \{\mathscr{F}_{p,q}^s : p \in (1, \infty), q \in (1, \infty]\},$$

the following holds. Let $p \in [1,\infty]$ be such that $\mathscr{X}^s = \mathscr{B}^s_{p,q}$ or $\mathscr{X}^s = \mathscr{F}^s_{p,q}$ and let $\mathcal{B} \subset C_0^\infty(B^n(7/8))$ be a bounded set. Then, $\forall s \in \mathbb{R}^\nu$, there exists $C \geq 0$ such that for all $\phi_1, \phi_2 \in \mathcal{B}$ with $\phi_1 \prec \phi_2$,

$$\|\phi_1 u\|_{\mathscr{X}^{s+\kappa e_1}(W,\vec{d}),(X,\vec{d}),a,7/8,h} \leq C\Big(\|\phi_2 \mathscr{P} u\|_{\mathscr{X}^{s}(W,\vec{d}),(X,\vec{d}),a,7/8,h} + \|\phi_2 u\|_{L^p(B^n(1),h\sigma_{\mathrm{Leb}})}\Big), \tag{8.157}$$

$\forall u \in C_0^\infty(B^n(1);\mathbb{C}^{\mathbf{D}_2})'$ with $\phi_2 \mathscr{P} u \in \mathscr{X}^s_{\mathrm{cpt}}(W,\vec{d})$ and $\phi_2 u \in L^p(B^n(1), h\sigma_{\mathrm{Leb}})$. Moreover, there exists $L \in \mathbb{N}$, depending only on $\max_{1\leq j\leq q}|\vec{d}_j|_1$, κ, s, and $\mathscr{X}^s$, such that $C = C(A, \max_\alpha \|a_\alpha\|_{C^L(B^n(1))}, \mathcal{B}, \mathbf{D}_1, \mathbf{D}_2) \geq 0$ is a $(\kappa, s, \mathscr{X}^s)$-multi-parameter unit-admissible constant as in Definition 3.15.1.

Proof. For notational simplicity, in this proof we write $\|\cdot\|_{\mathscr{X}^s(W,\vec{d})}$ to mean $\|\cdot\|_{\mathscr{X}^s(W,\vec{d}),(X,\vec{d}),a,7/8,h}$.

Define $\vec{N} \in [0,\infty)^\nu$ given by $\vec{N}_\mu := (s_\mu + 1) \vee 0$. For $\phi_1, \phi_2 \in \mathcal{B}$ with $\phi_1 \prec \phi_2$, let $\phi_{1.5} \in C_0^\infty(B^n(7/8))$ be such that $\phi_1 \prec \phi_{1.5} \prec \phi_2$ and $\{\phi_{1.5} : \phi_1, \phi_2 \in \mathcal{B}\} \subset C_0^\infty(B^n(7/8))$ is a bounded set. Proposition 8.7.1 shows that

$$\|\phi_1 u\|_{\mathscr{X}^{s+\kappa e_1}(W,\vec{d})} \lesssim \|\phi_2 \mathscr{P} u\|_{\mathscr{X}^{s}(W,\vec{d})} + \|\phi_{1.5} u\|_{\mathscr{X}^{s-\vec{N}}(W,\vec{d})}. \tag{8.158}$$

Since $s_\mu - \vec{N}_\mu \leq -1$, $\forall \mu$, Proposition 6.5.16 shows that

$$\|\phi_{1.5} u\|_{\mathscr{X}^{s-\vec{N}}(W,\vec{d})} \leq \|\phi_{1.5} u\|_{\mathscr{X}^{(-1,-1,\dots,-1)}(W,\vec{d})} \lesssim \|\phi_{1.5} u\|_{L^p} \lesssim \|\phi_2 u\|_{L^p}, \tag{8.159}$$

where keeping track of estimates in the proof of Proposition 6.5.16 shows that the implicit constant is of the desired form. Combining (8.158) and (8.159) completes the proof. □

8.7.2 A small scale

We now turn to a quantitative version of Proposition 8.2.12 which takes place on a small scale. We work in the same setting as Proposition 8.2.12 (i. e., the main multi-parameter setting of Section 3.15). Thus, we are given vector fields with single-parameter formal degrees $(W^1, \vec{d}^1), \dots, (W^\nu, \vec{d}^\nu) \subset C^\infty_{\mathrm{loc}}(\mathfrak{M}; T\mathfrak{M})$ satisfying the hypotheses given in Section 3.15. In particular, $(W^1, \vec{d}^1) = \{(W^1_1, \vec{d}^1_1), \dots, (W^1_{r_1}, \vec{d}^1_{r_1})\}$ are Hörmander vector fields with formal degrees on $\mathfrak{M}$.

Fix $\mathcal{K} \Subset \Omega_0 \Subset \Omega_1 \Subset \Omega_2 \Subset \mathfrak{M}$, with $\mathcal{K}$ compact and Ω_0, Ω_1, and Ω_2 open and relatively compact. Pick $(X^1, d^1), \dots, (X^\nu, d^\nu) \subset C^\infty_{\mathrm{loc}}(\mathfrak{M}; T\mathfrak{M})$ and $\lambda = (1, \lambda_2, \dots, \lambda_\nu) \in (0,\infty)^\nu$ as in Section 3.15.2 (with the above choices of $\mathcal{K}$, Ω_1, and Ω_2). Let $\Phi_{x,\delta}$ be as in Theorem 3.15.3, with these choices. As usual, we set $(W, \vec{d}) := (W^1, \vec{d}^1) \boxtimes \cdots \boxtimes (W^\nu, \vec{d}^\nu)$ and $(X, \vec{d}) := (X^1, d^1) \boxtimes \cdots \boxtimes (X^\nu, d^\nu)$.

Proposition 8.7.3. *There exists $a > 0$ such that the following holds. Fix $\kappa \in \mathbb{N}_+$ such that $\vec{d}_j^1$ divides κ for $1 \le j \le r_1$. Let $\mathscr{P}$ be given by (8.26) and suppose $\mathscr{P}$ is maximally subelliptic of degree κ with respect to $(W^1, \vec{d}^1)$ on $\mathfrak{M}$. Then, for any scale of spaces $\mathscr{X}^s$, $s \in \mathbb{R}^\nu$, of the form*

$$\mathscr{X}^s \in \{\mathscr{B}^s_{p,q} : p, q \in [1,\infty]\} \bigcup \{\mathscr{F}^s_{p,q} : p \in (1,\infty), q \in (1,\infty]\},$$

we have the following. Let $\Phi_{x,\delta}$ be as in Theorem 3.15.3. Fix $\phi_1, \phi_2, \phi_3 \in C_0^\infty(B^n(7/8))$ with $\phi_1 \prec \phi_2 \prec \phi_3$. Then, for every $s \in \mathbb{R}^\nu$ and $\vec{N} \in [0,\infty)^\nu$, there exists $C \ge 0$ such that the following holds $\forall x \in \mathcal{K}$, $\delta \in (0,1]$:

$$\begin{aligned} &\|(\phi_1 \circ \Phi_{x,\delta}^{-1})u\|_{\mathscr{X}^{s+\kappa e_1}(\delta^{\lambda\vec{d}} W, \vec{d}),(\delta^{\lambda \vec{d}} X, \vec{d}),a,\Omega_0,\Omega_1,\Omega_2,\mathrm{Vol}} \\ &\quad \le C\Big(\delta^\kappa \|(\phi_3 \circ \Phi_{x,\delta}^{-1})\mathscr{P}u\|_{\mathscr{X}^{s}(\delta^{\lambda\vec{d}} W, \vec{d}),(\delta^{\lambda \vec{d}} X, \vec{d}),a,\Omega_0,\Omega_1,\Omega_2,\mathrm{Vol}} \\ &\qquad + \|(\phi_2 \circ \Phi_{x,\delta}^{-1})u\|_{\mathscr{X}^{s-\vec{N}}(\delta^{\lambda\vec{d}} W, \vec{d}),(\delta^{\lambda \vec{d}} X, \vec{d}),a,\Omega_0,\Omega_1,\Omega_2,\mathrm{Vol}}\Big), \end{aligned} \tag{8.160}$$

$\forall u \in C_0^\infty(\mathfrak{M}; \mathbb{C}^{\mathbf{D}_2})'$ with $(\phi_3 \circ \Phi_{x,\delta}^{-1})\mathscr{P}u \in \mathscr{X}^s(\mathcal{K}, (W, \vec{d}))$. Here, $C \ge 0$ does not depend on $x \in \mathcal{K}$ or $\delta \in (0,1]$, but may depend on any of the other ingredients.

Remark 8.7.4. In light of Theorem 3.3.7 (i) one should think of $\Phi_{x,\delta}(B^n(7/8))$ as being comparable to $B_{(W^1,\vec{d}^1)}(x,\delta)$ and $B_{(X^1,\vec{d}^1)}(x,\delta)$. Thus, the estimate (8.160) is taking place at scale $\approx \delta$ with respect to $\rho_{(W^1,\vec{d}^1)}$ (alternatively, $\rho_{(X^1,\vec{d}^1)}$).

Proof. Set $\mathscr{P}^{x,\delta} := \Phi_{x,\delta}^* \delta^\kappa \mathscr{P}(\Phi_{x,\delta})_*$. Using Proposition 6.11.4 and the fact that $\Phi_{x,\delta}^* \delta^\kappa \mathscr{P}u = \mathscr{P}^{x,\delta}\Phi_{x,\delta}^* u$, (8.160) is equivalent to

$$\begin{aligned} &\|\phi_1 \Phi_{x,\delta}^* u\|_{\mathscr{X}^{s+\kappa e_1}(W^{x,\delta},\vec{d}),(X^{x,\delta},\vec{d}),a,7/8,h_{x,\delta}} \\ &\quad \le C\Big(\|\phi_3 \mathscr{P}^{x,\delta}\Phi_{x,\delta}^* u\|_{\mathscr{X}^{s}(W^{x,\delta},\vec{d}),(X^{x,\delta},\vec{d}),a,7/8,h_{x,\delta}} \\ &\qquad + \|\phi_2 \Phi_{x,\delta}^* u\|_{\mathscr{X}^{s-\vec{N}}(W^{x,\delta},\vec{d}),(X^{x,\delta},\vec{d}),a,7/8,h_{x,\delta}}\Big), \end{aligned} \tag{8.161}$$

where $(W^{x,\delta}, \vec{d})$ and $(X^{x,\delta}, \vec{d})$ are as in Proposition 6.11.4. We will prove (8.161) by applying Proposition 8.7.1 with $(W, \vec{d})$, $(X, \vec{d})$, and h replaced by $(W^{x,\delta}, \vec{d})$, $(X^{x,\delta}, \vec{d})$, and $h_{x,\delta}$, respectively. Here, $x \in \mathcal{K}$ and $\delta \in (0,1]$. We use the same scaling ideas as in Section 3.3.1.

Theorem 3.15.3 shows that $(W^{x,\delta}, \vec{d})$, $(X^{x,\delta}, \vec{d})$, and $h_{x,\delta}$ satisfy the hypotheses of Proposition 8.7.1 uniformly for $x \in \mathcal{K}$ and $\delta \in (0,1]$. Thus, it remains to show that $\mathscr{P}^{x,\delta}$ also satisfies the hypotheses uniformly in $x \in \mathcal{K}$ and $\delta \in (0,1]$.

First we must show that $\mathscr{P}^{x,\delta}$ is of the form (8.155). We have

$$\begin{aligned} \mathscr{P}^{x,\delta} = \Phi_{x,\delta}^* \delta^\kappa \mathscr{P}(\Phi_{x,\delta})_* &= \sum_{\deg_{\vec{d}^1}(\alpha) \le \kappa} \delta^{\kappa - \deg_{\vec{d}^1}(\alpha)} (a_\alpha \circ \Phi_{x,\delta})(W^{1,x,\delta})^\alpha \\ &=: \sum_{\deg_{\vec{d}^1}(\alpha) \le \kappa} a_\alpha^{x,\delta} (W^{1,x,\delta})^\alpha. \end{aligned}$$

We claim, for every $L \in \mathbb{N}$ and $\deg_{d^1}(\alpha) \leq \kappa$,

$$\sup_{\substack{x \in \mathcal{K} \\ \delta \in (0,1]}} \|a_\alpha^{x,\delta}\|_{C^L(B^n(1);\mathbb{M}^{\mathbf{D}_1\times\mathbf{D}_2})} < \infty. \tag{8.162}$$

Indeed, Theorem 3.3.7 (m) gives

$$\begin{aligned}
\|a_\alpha^{x,\delta}\|_{C^L(B^n(1);\mathbb{M}^{\mathbf{D}_1\times\mathbf{D}_2})} &\approx \sum_{\deg_{d^1}(\beta)\leq L} \|(X^{1,x,\delta})^\beta a_\alpha^{x,\delta}\|_{C(B^n(1))} \\
&= \sum_{\deg_{d^1}(\beta)\leq L} \delta^{\kappa-\deg_{d^1}(\alpha)} \|(X^{1,x,\delta})^\beta a_\alpha \circ \Phi_{x,\delta}\|_{C(B^n(1))} \\
&= \sum_{\deg_{d^1}(\beta)\leq L} \delta^{\kappa-\deg_{d^1}(\alpha)+\deg_{d^1}(\beta)} \|(X^1)^\beta a_\alpha\|_{C(\Phi_{x,\delta}(B^n(1)))} \\
&\leq \sum_{\deg_{d^1}(\beta)\leq L} \|(X^1)^\beta a_\alpha\|_{C(\Omega_1)} \lesssim 1,
\end{aligned}$$

where the $\lesssim 1$ uses the relative compactness of Ω_1. This proves (8.162) and shows $\mathscr{P}^{x,\delta}$ is of the form (8.155) uniformly for $x \in \mathcal{K}$ and $\delta \in (0,1]$.

Finally, we must show that $\mathscr{P}^{x,\delta}$ is maximally subelliptic of degree κ with respect to $(W^{1,x,\delta}, \vec{d}^1)$, uniformly in $x \in \mathcal{K}$ and $\delta \in (0,1]$. By the assumption that $\mathscr{P}$ is maximally subelliptic of degree κ with respect to $(W^1, \vec{d}^1)$ on $\mathfrak{M}$, there exists $A \geq 0$ such that $\forall f \in C_0^\infty(\Omega; \mathbb{C}^{\mathbf{D}_2})$, we have, with $n_j := \kappa/\vec{d}_j^1 \in \mathbb{N}_+$,

$$\sum_{j=1}^r \|(W_j^1)^{n_j} f\|_{L^2(\mathfrak{M},\mathrm{Vol};\mathbb{C}^{\mathbf{D}_2})} \leq A(\|\mathscr{P}f\|_{L^2(\mathfrak{M},\mathrm{Vol};\mathbb{C}^{\mathbf{D}_1})} + \|f\|_{L^2(\mathfrak{M},\mathrm{Vol};\mathbb{C}^{\mathbf{D}_2})}). \tag{8.163}$$

Multiplying (8.163) by δ^κ, we have

$$\begin{aligned}
\sum_{j=1}^r \|(\delta^{\vec{d}_j^1} W_j^1)^{n_j} f\|_{L^2(\mathfrak{M},\mathrm{Vol};\mathbb{C}^{\mathbf{D}_2})} &\leq A(\|\delta^\kappa \mathscr{P}f\|_{L^2(\mathfrak{M},\mathrm{Vol};\mathbb{C}^{\mathbf{D}_1})} + \delta^\kappa \|f\|_{L^2(\mathfrak{M},\mathrm{Vol};\mathbb{C}^{\mathbf{D}_2})}) \\
&\leq A(\|\delta^\kappa \mathscr{P}f\|_{L^2(\mathfrak{M},\mathrm{Vol};\mathbb{C}^{\mathbf{D}_1})} + \|f\|_{L^2(\mathfrak{M},\mathrm{Vol};\mathbb{C}^{\mathbf{D}_2})}).
\end{aligned} \tag{8.164}$$

For $g \in C_0^\infty(B^n(1); \mathbb{C}^{\mathbf{D}_2})$, we have $f := g \circ \Phi_{x,\delta}^{-1} \in C_0^\infty(\Omega_1; \mathbb{C}^{\mathbf{D}_2})$. Plugging this choice of f into (8.164) and using $\Phi_{x,\delta}^* \mathrm{Vol} = \Lambda(x,\delta) h_{x,\delta} \sigma_{\mathrm{Leb}}$ (where $\Lambda(x,\delta)$ is given by (6.175)), we have

$$\begin{aligned}
&\sum_{j=1}^r \|(W_j^{1,x,\delta})^{n_j} g\|_{L^2(B^n(1),\Lambda(x,\delta)h_{x,\delta}\sigma_{\mathrm{Leb}};\mathbb{C}^{\mathbf{D}_2})} \\
&\quad \leq A(\|\mathscr{P}^{x,\delta} g\|_{L^2(B^n(1),\Lambda(x,\delta)h_{x,\delta}\sigma_{\mathrm{Leb}};\mathbb{C}^{\mathbf{D}_1})} + \|g\|_{L^2(B^n(1),\Lambda(x,\delta)h_{x,\delta}\sigma_{\mathrm{Leb}};\mathbb{C}^{\mathbf{D}_2})}).
\end{aligned} \tag{8.165}$$

Multiplying both sides of (8.165) by $\Lambda(x,\delta)^{-\frac{1}{2}}$ gives

$$\sum_{j=1}^{r} \|(W_j^{1,x,\delta})^{n_j} g\|_{L^2(B^n(1),h_{x,\delta}\sigma_{\mathrm{Leb}};\mathbb{C}^{D_2})} \tag{8.166}$$
$$\leq A(\|\mathscr{P}^{x,\delta} g\|_{L^2(B^n(1),h_{x,\delta}\sigma_{\mathrm{Leb}};\mathbb{C}^{D_1})} + \|g\|_{L^2(B^n(1),h_{x,\delta}\sigma_{\mathrm{Leb}};\mathbb{C}^{D_2})}).$$

Since A does not depend on $x \in \mathcal{K}$ or $\delta \in (0,1]$, (8.166) shows that $\mathscr{P}^{x,\delta}$ is maximally subelliptic of degree κ with respect to $(W^{1,x,\delta}, \vec{d}^1)$, uniformly in $x \in \mathcal{K}$ and $\delta \in (0,1]$. This completes the proof that the assumptions of Proposition 8.7.1 hold uniformly for $x \in \mathcal{K}$ and $\delta \in (0,1]$. Thus, (8.161) follows from Proposition 8.7.1, which completes the proof. □

8.8 Representation theory and Rockland's condition

Probably the most familiar characterization of elliptic partial differential operators comes via the Fourier transform. Let $\mathscr{E}$ be a constant coefficient partial differential operator on $\mathbb{R}^n$ which is homogeneous of degree κ:

$$\mathscr{E} = \sum_{|\alpha|=\kappa} a_\alpha \partial_x^\alpha, \quad a_\alpha \in \mathbb{C}.$$

Since $\mathscr{E}$ is translation invariant on $\mathbb{R}^n$, conjugating it with the Fourier transform gives a one-dimensional operator for each $\xi \in \mathbb{R}^n$:

$$\widehat{\mathscr{E}}(\xi) = (-2\pi i)^\kappa \sum_{|\alpha|=\kappa} a_\alpha \xi^\alpha.$$

Then $\mathscr{E}$ is elliptic if and only if $\widehat{\mathscr{E}}(\xi) \neq 0$, $\forall \xi \neq 0$.

Thinking of the Fourier transform as the unitary representation theory of $\mathbb{R}^n$, this characterization of ellipticity has a particularly elegant generalization to nilpotent Lie groups. Let G be a connected, simply connected, graded, nilpotent Lie group (see Section 4.5.3), with Lie algebra $\mathfrak{g}$ with grading $\mathfrak{g} = \oplus_{\mu=1}^{\nu} \mathfrak{v}_\mu$ (see Definition 4.5.26). As in Section 4.5.3, the grading induces dilations on the Lie group. Fix $(W, \vec{d}) = \{(W_1, \vec{d}_1), \dots, (W_r, \vec{d}_r)\} \subset \mathfrak{g} \times \mathbb{N}_+$ such that $W_1, \dots, W_r$ generate $\mathfrak{g}$ and $W_j \in \mathfrak{v}_{\vec{d}_j}$. Fix $\kappa \in \mathbb{N}_+$ such that $\vec{d}_j$ divides κ for each $1 \leq j \leq r$. We identify $W_1, \dots, W_r$ with the left invariant vector fields on G. Let $\mathscr{P}$ be a left invariant partial differential operator which is homogeneous of degree κ with respect to the dilations. Note that $\mathscr{P}$ is of the form

$$\mathscr{P} = \sum_{\deg_{\vec{d}}(\alpha)=\kappa} a_\alpha W^\alpha, \quad a_\alpha \in \mathbb{C}.$$

Theorem 8.8.1. *Let $\mathscr{P}$ be as above. Then the following are equivalent:*
- *$\mathscr{P}$ is hypoelliptic.*
- *$\mathscr{P}$ is maximally subelliptic of degree κ with respect to $(W, \vec{d})$ on G.*
- *For every non-trivial irreducible unitary representation π of G, the operator $\pi(\mathscr{P})$ is injective in the space of C^∞ vectors of the representation.*

Theorem 8.8.1 was first proved in the case of the Heisenberg group by Rockland [205], who conjectured the general result; see also [171, 78]. Theorem 8.8.1 was proved in general by Helffer and Nourrigat [121]; see also [120, 9]. Helffer and Nourrigat, in turn, conjectured an extension of Theorem 8.8.1 to general partial differential operators on manifolds like the ones studied in this chapter [122]. This more general conjecture was recently proved by Androulidakis, Moshen, and Yuncken [2], who construct parametrices for maximally subelliptic operators using a pseudo-differential calculus based on the ideas of [3].

8.9 Positive definite forms

Let $(W, \vec{d}) = \{(W_1, \vec{d}_1), \ldots, (W_r, \vec{d}_r)\}$ be Hörmander vector fields with formal degrees on $\mathfrak{M}$. Fix $\kappa \in \mathbb{N}_+$ such that $\vec{d}_j$ divides κ for every j; set $n_j := \kappa/\vec{d}_j \in \mathbb{N}_+$. In this section, we describe the following result.

Proposition 8.9.1. *Let $a(x) = (a_{i,j}(x)) \in C^\infty_{\mathrm{loc}}(\mathfrak{M}; \mathbb{M}^{r\times r}(\mathbb{C}))$ be a smooth, self-adjoint matrix which is strictly positive definite at every point. For $\deg_{\vec{d}}(\alpha) < \kappa$, let $b_\alpha(x) \in C^\infty_{\mathrm{loc}}(\mathfrak{M}; \mathbb{C})$. Then the operator*

$$\mathscr{L}_a u := \sum_{1\leq i,j\leq r} a_{i,j}(x) W_i^{n_i} W_j^{n_j} u(x) + \sum_{\deg_{\vec{d}}(\alpha)<\kappa} b_\alpha(x) W^\alpha u(x)$$

is maximally subelliptic of degree 2κ with respect to $(W, \vec{d})$ on $\mathfrak{M}$.

Example 8.9.2. Proposition 8.9.1 is easiest to understand when $\vec{d}_j = 1$, $\forall j$. As usual, we denote this special case by $(W, 1)$. In this case, we may take $\kappa \in \mathbb{N}_+$ to be any positive integer. As special cases of Proposition 8.9.1 we have, for a smooth, self-adjoint, strictly positive definite matrix $(a_{i,j}(x))$ and for any smooth functions $b_\alpha(x)$:

- The operator

$$\sum_{1\leq i,j\leq r} a_{i,j}(x) W_i W_j u(x) + \sum_{j=1}^{r} b_j(x) W_j u(x) + b_0(x) u(x)$$

  is maximally subelliptic of degree 2 with respect to $(W, 1)$ on $\mathfrak{M}$.
- More generally, for every $l \in \mathbb{N}_+$, the operator

$$\sum_{1\leq i,j\leq r} a_{i,j}(x) W_i^l W_j^l u(x) + \sum_{|\alpha|<2l} W^\alpha u(x)$$

  is maximally subelliptic of degree $2l$ with respect to $(W, 1)$ on $\mathfrak{M}$.

It is possible to prove Proposition 8.9.1 by adapting the methods we used for the sub-Laplacian in Section 4.5. However, it is perhaps instructive to instead see it as a simple consequence of Theorem 8.1.1.

Lemma 8.9.3. *Let $a = (a_{i,j}) \in \mathbb{M}^{r\times r}(\mathbb{C})$ be a constant, self-adjoint matrix which is strictly positive definite and for $\deg_{d}(\alpha) < 2\kappa$ let $b_\alpha \in \mathbb{C}$. Then the operator*

$$\mathscr{L}_a u := \sum_{1\leq i,j\leq r} a_{i,j} W_i^{n_i} W_j^{n_j} u(x) + \sum_{\deg_{d}(\alpha)<\kappa} b_\alpha W^\alpha u(x)$$

is maximally subelliptic of degree 2κ with respect to (W, d) on $\mathfrak{M}$.

Proof of Proposition 8.9.1. This follows immediately from Lemma 8.9.3 and Theorem 8.1.1 (iv) ⇒ (i). □

Proof of Lemma 8.9.3. Let $A = (a_{i,j})$ and $C = A^{\frac{1}{2}}$. Define

$$C\nabla_W := C \begin{bmatrix} W_1^{n_1} \\ W_2^{n_2} \\ \vdots \\ W_r^{n_r} \end{bmatrix}.$$

Since C is invertible (because it is strictly positive definite), it follows immediately from the definitions that $C\nabla_W$ is maximally subelliptic of degree κ with respect to (W, d) on $\mathfrak{M}$. Theorem 8.1.1 (i) ⇒ (vi) shows that

$$\sum_{i,j} (W_i^{n_i})^* a_{i,j} W_j^{n_j} = (C\nabla_W)^*(C\nabla_W)$$

is maximally subelliptic of degree 2κ with respect to (W, d) on $\mathfrak{M}$. From here, Theorem 8.1.1 (i) ⇔ (ii) completes the proof. □

The proof given above for Proposition 8.9.1 can be adapted to show that other operators are maximally subelliptic. For example, the proof goes through nearly unchanged to show the following.

Proposition 8.9.4. *Set $\mathbf{D} := \#\{\alpha : \deg_{d}(\alpha) = \kappa\}$. Let $a(x) = (a_{\alpha,\beta}(x)) \in C^\infty_{\mathrm{loc}}(\mathfrak{M}; \mathbb{M}^{\mathbf{D}\times\mathbf{D}}(\mathbb{C}))$ be a smooth, self-adjoint matrix which is strictly positive definite at every point. For $\deg_{d}(\alpha) < \kappa$, let $b_\alpha(x) \in C^\infty_{\mathrm{loc}}(\mathfrak{M}; \mathbb{C})$. Then the operator*

$$\widehat{\mathscr{L}_a} u := \sum_{\deg_{d}(\alpha),\deg_{d}(\beta)=\kappa} a_{\alpha,\beta}(x) W^\alpha W^\beta u + \sum_{\deg_{d}(\alpha)<2\kappa} b_\alpha(x) W^\alpha u$$

is maximally subelliptic of degree 2κ with respect to (W, d) on $\mathfrak{M}$.

8.10 Further reading and references

One of the important early works introducing maximally subelliptic operators was the influential work of Hörmander [125]. Let $W_1, \ldots, W_r$ be Hörmander vector fields;[4] in [125] the operator $W_1 + W_2^2 + W_3^2 + \cdots + W_r^2$ was shown to be subelliptic; it was later shown by Rothschild and Stein [200] that this operator is maximally subelliptic with respect to $\{(W_1, 2), (W_2, 1), (W_3, 1), \ldots, (W_r, 1)\}$. The methods of [125] were simplified by Kohn [142], and it is Kohn's more simple approach that we use to prove the a priori estimates in Sections 4.5.1 and 8.3.2. The first time that the defining property of maximal subellipticity, (8.2), was explicitly stated in a non-trivial special case seems to be in the paper of Folland and Stein [91, Theorem 9.4]. The general case was first explicitly described by Helffer and Nourrigat; see, for example, [122].

The results of Hörmander and Kohn took place in L^2, and the subelliptic gain shown was not sharp. To move to L^p, $1 < p < \infty$, and to prove the sharp subelliptic gain, singular integral techniques were introduced by Folland and Stein [91], Folland [90], and Rothschild and Stein [200]. These references also address Zygmund–Hölder spaces, though they are referred to as Lipschitz spaces. These references were tied to the analysis on nilpotent Lie groups as was used in Section 4.5; because of this, they cannot be used to study general maximally subelliptic operators (see Section 4.5.7). See the text [16] for an excellent history of many parts of this subject, along with many of the original motivations for studying these operators. See [88] for a detailed treatment of left invariant maximally subelliptic operators on homogeneous Lie groups (called Rockland operators in that setting), where much more is known.

To move beyond methods involving nilpotent Lie groups, motivated by questions in several complex variables, Nagel, Rosay, Stein, and Wainger [182] introduced (single-parameter) singular integrals like the ones in Section 5.2.1; they showed that certain maximally subelliptic operators from several complex variables have parametrices which were singular integral operators. These singular integrals were later used for similar purposes by Chang, Nagel, and Stein [34] and Koenig [140]; see also the work of Christ [46, 47] and Lu [154] for similar ideas. The methods of these papers were generalized to study general maximally subelliptic operators $\mathscr{P}$ such that $\mathscr{P}^*$ was also maximally subelliptic by the author [220, Theorem 2.4.8]. Unfortunately, these methods have some drawbacks: they require that $\mathscr{P}$ and $\mathscr{P}^*$ both be a priori maximally subelliptic, and it is much harder to keep track of the quantitative bounds in these proofs. This is why we used the heat kernel in this paper to prove Theorem 8.1.1: it is more general (we only needed $\mathscr{P}$ to be maximally subelliptic) and it provides better quantitative control. See [192, 61, 62, 199, 141, 218] for some other papers relating maximal subellipticity to several complex variables.

4 Hörmander vector fields and Hörmander's sub-Laplacian are so named because of Hörmander's original work [125].

The above described singular integrals are based on the underlying Carnot–Carathéodory geometry. The connection between this Carnot–Carathéodory geometry and maximal subellipticity was made clear in the setting of nilpotent Lie groups by Folland and Stein [91] and Folland [90]. This connection on general manifolds began with the work of Nagel, Stein, and Wainger [189]. Two other important early papers connecting Carnot–Carathéodory geometry with similar partial differential operators were those of C. Fefferman and Sánchez-Calle [84] and C. Fefferman and Phong [82].

Let $(W, d\!\!\!\!/\,) = \{(W_1, d\!\!\!\!/\,_1), \ldots, (W_r, d\!\!\!\!/\,_r)\}$ be Hörmander vector fields with formal degrees and fix $\kappa \in \mathbb{N}_+$ such that $d\!\!\!\!/\,_j$ divides κ for $1 \leq j \leq r$. Set $n_j := \kappa / d\!\!\!\!/\,_j \in \mathbb{N}_+$. The gradient operator $\nabla_{\kappa,W} : f \mapsto (W_1^{n_1} f, \ldots, W_r^{n_r} f)$ is clearly maximally subelliptic of degree κ with respect to $(W, d\!\!\!\!/\,)$. Set

$$\mathscr{L} := \nabla_{\kappa,W}^* \nabla_{\kappa,W} = \sum_{j=1}^{r} (W_j^{n_j})^* W_j^{n_j},$$

the sub-Laplacian. Let $\overline{\mathscr{L}}$ be any non-negative, self-adjoint extension of $\mathscr{L}$. Theorem 8.1.1 (i) ⇒ (ix) gives estimates on the heat kernel $e^{-t\overline{\mathscr{L}}}$. Such estimates have a long history; most work has been done in the setting of left invariant vector fields on a Lie group, often a homogeneous Lie group. When $d\!\!\!\!/\,_j = 1$, for all j, and $\kappa = 1$ (so $n_j = 1$ for all j), heat kernel estimates date back to works of Gårding [93] and Nelson [190]; see also [90]. The method of using Gevrey regularity (like we use in Theorem 8.1.1) to obtain exponential heat kernel bounds (on more general manifolds, not just Lie groups) was introduced by Jerison and Sánchez-Calle [134], and they attribute this idea to E. M. Stein. For a further history of related ideas, see [131].

For general $d\!\!\!\!/\,_j$, but in the setting of nilpotent Lie groups, heat kernels were studied by Folland and Stein [92], Dziubański and Hulanicki [80], Dziubański, Hebisch, and Zienkiewicz [79] (who studied more general left invariant maximally subelliptic operators), and Hebisch [118], who used Gevrey regularity like we do in Theorem 8.1.1. See [234, 235, 236] for a different approach working with more general maximally subelliptic operators on a Lie group. The works above are just a small sampling of the many papers devoted to heat kernel estimates for operators like $\overline{\mathscr{L}}$, an active area of research.

When $d\!\!\!\!/\,_j = 1, 1 \leq j \leq r$, and $\kappa = 1$, there is another approach to understanding the heat kernel $e^{-t\overline{\mathscr{L}}}$ and parametrices for $\mathscr{L}$ which can be quite useful. Melrose [169] showed that the wave equation for $\mathscr{L}$ has finite speed of propagation (with respect to the metric $\rho_{(W,1)}$). Using this, one can deduce sharp exponential estimates for the heat equation and also study more general multipliers for $\overline{\mathscr{L}}$; see, for example, [211]. See [220, Section 2.6] for a further discussion. Unfortunately, when studying general maximally subelliptic operators, we must deal with operators whose wave equations do not have finite speed of propagation, so this method is not available to us.

As described above, many parts of Theorem 8.1.1 in various special cases are known. But it seems that Theorem 8.1.1, in the generality in which it is stated, is new. Similarly, special cases of the regularity results in Section 8.2 are known, but the results there are new in the generality in which they are stated.

There are connections between maximal subellipticity and probability theory. This was first described by Malliavin [161], though we do not use those techniques in this text. See [193] for an exposition.

Maximally subelliptic operators have been used to great effect in the study of the stochastically forced Navier–Stokes equation. This began with the work of E and Mattingly [81], and many other works followed; see, for example, [104].

We do not address boundary value problems at all in this text, because such questions are not yet well understood at the level of generality addressed here. Some interesting results have been obtained in special cases. See, for example, [132, 133, 94].

We also do not address Poincaré type inequalities, but there have been many papers on Poincaré type inequalities for Hörmander vector fields, beginning with the influential work of Jerison [129]. After Jerison's original work, hundreds of papers followed.

The index theory for maximally subelliptic operators has recently been studied by Moshen [172], using the pseudo-differential calculus of [2].

The above is only a partial overview of the vast theory which has arisen from Hörmander's work [125], which has been cited thousands of times.

9 Nonlinear maximally subelliptic equations

In Chapter 8 we obtained the sharp regularity properties for linear maximally subelliptic operators. These results are strong enough that we can use them to deduce sharp regularity properties of fully nonlinear maximally subelliptic equations.

We study nonlinear equations as described in Section 1.1.2. Throughout this chapter, $\mathfrak{M}$ will be a connected, smooth manifold. Given Hörmander vector fields with formal degrees on $\mathfrak{M}$, $(W, \vec{d}) = \{(W_1, \vec{d}_1), \ldots, (W_r, \vec{d}_r)\} \subset C^\infty_{\mathrm{loc}}(\mathfrak{M}; T\mathfrak{M}) \times \mathbb{N}_+$, we take $\kappa \in \mathbb{N}_+$ such that $\vec{d}_j$ divides κ, for $1 \leq j \leq r$. Fix $\mathbf{D}_1, \mathbf{D}_2 \in \mathbb{N}_+$; as in (1.7), we set

$$N := \mathbf{D}_2 \times \#\{\alpha \text{ an ordered multi-index} : \deg_{\vec{d}}(\alpha) \leq \kappa\}. \tag{9.1}$$

Given a sufficiently smooth function $u = (u_1, \ldots, u_{\mathbf{D}_2}) : \mathfrak{M} \to \mathbb{R}^{\mathbf{D}_2}$, we define

$$\{W^\alpha u\}_{\deg_{\vec{d}}(\alpha) \leq \kappa} : \mathfrak{M} \to \mathbb{R}^N$$

to be the vector whose components are given by $W^\alpha u_j$, $\deg_{\vec{d}}(\alpha) \leq \kappa$ and $1 \leq j \leq \mathbf{D}_2$. Let $F(x, \zeta) : \mathfrak{M} \times \mathbb{R}^N \to \mathbb{R}^{\mathbf{D}_1}$ be a function. The goal of this chapter is to study the interior regularity theory of nonlinear equations written in the form[1]

$$F(x, \{W^\alpha u(x)\}_{\deg_{\vec{d}}(\alpha) \leq \kappa}) = g(x) \tag{9.2}$$

under the assumption that (9.2) is maximally subelliptic in the sense of Definition 1.1.11.

9.1 Main qualitative results

The main result of this chapter (Theorem 9.2.1) is a quantitative result. However, it is easiest to initially understand this result through its qualitative consequences. In both the single- and multi-parameter settings we will use the Zygmund spaces $\mathscr{C}^{\vec{\mathbf{s}}}(W, \vec{d})$ from Definition 7.0.1 and the product Zygmund spaces $\mathscr{C}^{\vec{\mathbf{s}},\mathbf{t}}((W, \vec{d}) \boxtimes \nabla_{\mathbb{R}^N})$ from Definition 7.5.1.

Definition 9.1.1. Fix $x_0 \in \mathfrak{M}$.

- For a distribution $u \in C_0^\infty(\mathfrak{M})'$, we say $u \in \mathscr{C}^{\vec{\mathbf{s}}}(W, \vec{d})$ near x_0 if there exists $\phi \in C_0^\infty(\mathfrak{M})$ with $\phi \equiv 1$ on a neighborhood of x_0 such that $\phi u \in \mathscr{C}^{\vec{\mathbf{s}}}_{\mathrm{cpt}}(W, \vec{d})$.
- For a distribution $F \in C_0^\infty(\mathfrak{M} \times \mathbb{R}^N)'$, we say $F(x, \zeta) \in \mathscr{C}^{\vec{\mathbf{s}},\mathbf{t}}((W, \vec{d}) \boxtimes \nabla_{\mathbb{R}^N})$ near x_0 if there exists $\phi \in C_0^\infty(\mathfrak{M})$ with $\phi \equiv 1$ on a neighborhood of x_0 such that $\phi F \in \mathscr{C}^{\vec{\mathbf{s}},\mathbf{t}}_{\mathrm{cpt}}((W, \vec{d}) \boxtimes \nabla_{\mathbb{R}^N})$; recall from (7.14) that this only requires that $\phi(x)F(x, \zeta)$ has compact support in the x variable, not the ζ variable.

1 By Hörmander's condition, *every* fully nonlinear equation can be written in the form (9.2).

https://doi.org/10.1515/9783111085647-009

9.1.1 Single-parameter results

As in the introduction to this chapter, we let

$$(W, \vec{d}) = \{(W_1, \vec{d}_1), \ldots, (W_r, \vec{d}_r)\} \subset C^\infty_{\mathrm{loc}}(\mathfrak{M}; T\mathfrak{M}) \times \mathbb{N}_+$$

be Hörmander vector fields with single-parameter formal degrees on $\mathfrak{M}$. We fix $\kappa \in \mathbb{N}_+$ such that $\vec{d}_j$ divides κ, $\forall j$; we define N as in (9.1).

We consider the following PDE, where $F(x, \zeta) : \mathfrak{M} \times \mathbb{R}^N \to \mathbb{R}^{\mathbf{D}_1}$, $g(x) : \mathfrak{M} \to \mathbb{R}^{\mathbf{D}_1}$, and $u(x) : \mathfrak{M} \to \mathbb{R}^{\mathbf{D}_2}$ are given:

$$F(x, \{W^\alpha u(x)\}_{\deg_{\vec{d}}(\alpha) \le \kappa}) = g(x). \tag{9.3}$$

Fix $x_0 \in \mathfrak{M}$.

Main assumptions: Throughout this section, we assume that

$$F(x, \{W^\alpha u(x)\}_{\deg_{\vec{d}}(\alpha) \le \kappa})$$

is maximally subelliptic at (x_0, u) of degree κ with respect to $(W, \vec{d})$ (see Definition 1.1.11), and we assume (9.3) holds for x in a neighborhood of x_0.

Theorem 9.1.2. *Let* $\mathbf{s} > \mathbf{r} > 0$, $\mathbf{t} > 2\lfloor \mathbf{s} \rfloor + 6$. *Suppose that:*
- $F \in \mathscr{C}^{\mathbf{s},\mathbf{t}}((W, \vec{d}) \boxtimes \nabla_{\mathbb{R}^N})$ *near* x_0,
- $g \in \mathscr{C}^{\mathbf{s}}(W, \vec{d})$ *near* x_0,
- $u \in \mathscr{C}^{\mathbf{r}+\kappa}(W, \vec{d})$ *near* x_0.

Then $u \in \mathscr{C}^{\mathbf{s}+\kappa}(W, \vec{d})$ *near* x_0.

Proof. This is the case $\nu = 1$ of Theorem 9.1.14, below. □

Theorem 9.1.2 gives the sharp regularity of fully nonlinear maximally subelliptic PDEs in terms of the adapted Zygmund–Hölder spaces $\mathscr{C}^{\mathbf{s}}(W, \vec{d})$.

Remark 9.1.3. Theorem 9.1.2 is likely not sharp with respect to regularity of $F(x, \zeta)$ in the ζ variable. That is, it likely holds for some smaller values of $\mathbf{t}$. See Remark 2.6.4. The parameter $\mathbf{r} > 0$ (i. e., the a priori regularity of u) does not play a role in the conclusion of Theorem 9.1.2; however, the assumption $u \in \mathscr{C}^{\mathbf{r}+\kappa}(W, \vec{d})$ near x_0 does ensure that all aspects of (9.3) are defined in the classical sense. Theorem 9.1.2 is clearly sharp with respect to all the other ingredients.

Remark 9.1.4. If (9.3) holds on a relatively compact open set Ω, then the estimates for u blow up as one approaches the boundary of Ω. This blowup is made precise in Section 9.3.

Corollary 9.1.5. *Let* $\mathbf{r} > 0$. *Suppose that:*
- $F(x, \zeta) \in C^\infty_{\mathrm{loc}}(\mathfrak{M} \times \mathbb{R}^N)$ *for* x *near* x_0,

– $g \in C^\infty_{\mathrm{loc}}$ *near* x_0,
– $u \in \mathscr{C}^{\mathbf{r}+\kappa}$ *near* x_0.

Then $u \in C^\infty_{\mathrm{loc}}$ *near* x_0.

Proof. By the maximal subellipticity hypothesis, the linear partial differential operator

$$v \mapsto d_\zeta F(x_0, \{W^\alpha u(x_0)\}_{\deg_{\vec{d}}(\alpha)\le\kappa})\{W^\alpha v\}_{\deg_{\vec{d}}(\alpha)\le\kappa}$$

is maximally subelliptic of degree κ with respect to $(W, \vec{d})$ on a neighborhood of x_0. By continuity, it follows that there is a neighborhood V of x_0 such that for all $x_1 \in V$,

$$v \mapsto d_\zeta F(x_1, \{W^\alpha u(x_1)\}_{\deg_{\vec{d}}(\alpha)\le\kappa})\{W^\alpha v\}_{\deg_{\vec{d}}(\alpha)\le\kappa}$$

is maximally subelliptic at of degree κ with respect to $(W, \vec{d})$ on a neighborhood of x_1; see Lemma 8.5.2 for a similar continuity argument. That is,

$$F(x, \{W^\alpha u(x)\}_{\deg_{\vec{d}}(\alpha)\le\kappa})$$

is maximally subelliptic at (x_1, u) of degree κ with respect to $(W, \vec{d})$, $\forall x_1 \in V$. Fix $U \Subset V$ with $x_0 \in U$ and $\phi \in C_0^\infty(V)$ with $\phi \equiv 1$ on a neighborhood of x_0. By possibly shrinking U, we may apply Theorem 9.1.2 to each $x_1 \in \overline{U}$, and using the compactness of $\overline{U}$ and a partition of unity we see that $\phi u \in \mathscr{C}^{\mathbf{s}+\kappa}(\overline{U}, (W, \vec{d}))$, $\forall \mathbf{s} > 0$. The result follows. □

It is also of interest to understand the regularity theory with respect to the Zygmund–Hölder spaces adapted to different vector fields with formal degrees. The next theorem gives such a result, which is sharp up to a "loss of ϵ derivatives."

Proposition 9.1.6. *Fix* $\mathbf{s} > \mathbf{r} > 0$, $\mathbf{t} > 2\lfloor \mathbf{s} \rfloor + 8$. *Let* $(Z, \vec{d}) = \{(Z_1, \vec{d}_1), \ldots, (Z_\nu, \vec{d}_\nu)\} \subset C^\infty_{\mathrm{loc}}(\mathfrak{M}; T\mathfrak{M}) \times \mathbb{N}_+$ *be another set of Hörmander vector fields with single-parameter formal degrees on* $\mathfrak{M}$. *Let* $\lambda := \lambda(x_0, (W, \vec{d}), (Z, \vec{d})) > 0$ *be as in Definition* 6.6.9. *Suppose that:*
– $(W, \vec{d})$ *and* $(Z, \vec{d})$ *locally weakly approximately commute on* $\mathfrak{M}$,
– $F \in \mathscr{C}^{\mathbf{s},\mathbf{t}}((Z, \vec{d}) \boxtimes \nabla_{\mathbb{R}^N})$ *near* x_0,
– $g \in \mathscr{C}^{\mathbf{s}}(Z, \vec{d})$ *near* x_0,
– $W^\alpha u \in \mathscr{C}^{\mathbf{r}}(Z, \vec{d})$ *near* x_0, $\forall \deg_{\vec{d}}(\alpha) \le \kappa$.

Then, $\forall \epsilon > 0$, *we have:*
(I) $u \in \mathscr{C}^{\mathbf{s}+\kappa/\lambda-\epsilon}(Z, \vec{d})$ *near* x_0, *and*
(II) $W^\alpha u \in \mathscr{C}^{\mathbf{s}-\epsilon}(Z, \vec{d})$ *near* x_0, $\forall \deg_{\vec{d}}(\alpha) \le \kappa$.

We see Proposition 9.1.6 as a corollary of a multi-parameter result in Section 9.1.2.

Proposition 8.2.1 gives the same gain of κ/λ derivatives for linear maximally subelliptic PDEs (without the loss of ϵ), and it is shown there that this gain is sharp. Thus,

Proposition 9.1.6 is sharp up to the endpoint $\epsilon = 0$. It would be interesting to determine whether Proposition 9.1.6 holds with $\epsilon = 0$.

Remark 9.1.7. In Proposition 9.1.6, (I) is a consequence of (II) by Proposition 8.2.1; however, we will prove (I) directly.

A special case of Proposition 9.1.6 gives the sharp (up to a loss of $\epsilon > 0$ derivatives) for fully nonlinear maximally subelliptic PDEs with respect to the classical Zygmund–Hölder spaces.

Corollary 9.1.8. *Fix* $\mathbf{s} > \mathbf{r} > 0$ *and* $\mathbf{t} > 2\lfloor \mathbf{s} \rfloor + 8$. *Let* $\lambda := \lambda_{\mathrm{std}}(x_0, (W, \vec{d}))$ *(see Definition* 6.6.15*). Suppose that:*
- $F \in \mathscr{C}^{\mathbf{s},\mathbf{t}}_{\mathrm{std}}$ *near* x_0,
- $g \in \mathscr{C}^{\mathbf{s}}_{\mathrm{std}}$ *near* x_0,
- $W^{\alpha} u \in \mathscr{C}^{\mathbf{r}}_{\mathrm{std}}$ *near* x_0, $\forall \deg_{\vec{d}}(\alpha) \leq \kappa$.

Then, $\forall \epsilon > 0$, *we have:*
(I) $u \in \mathscr{C}^{\mathbf{s}+\kappa/\lambda-\epsilon}_{\mathrm{std}}$ *near* x_0, *and*
(II) $W^{\alpha} u \in \mathscr{C}^{\mathbf{s}-\epsilon}_{\mathrm{std}}$ *near* x_0, $\forall \deg_{\vec{d}}(\alpha) \leq \kappa$.

Proof. Let $\mathfrak{N}$ and $(Z, 1)$ be as in Lemma 8.2.4. In light of Lemma 8.2.4, the result is an immediate consequence of Proposition 9.1.6 with $(Z, \vec{d})$ replaced by $(Z, 1)$. □

9.1.1.1 Qualitative Schauder estimates

The above results immediately imply corresponding qualitative Schauder estimate type results. We describe these qualitative results here and discuss more quantitative versions (which describe the blowup near the boundary) in Section 9.3.1.

Let $\mathscr{P}u(x) = \sum_{\deg_{\vec{d}}(\alpha) \leq \kappa} a_{\alpha}(x) W^{\alpha} u(x)$, where $a_{\alpha} : \mathfrak{M} \to \mathbb{M}^{\mathbf{D}_1 \times \mathbf{D}_2}(\mathbb{R})$. Fix $x_0 \in \mathfrak{M}$ and define the frozen coefficient operator

$$\mathscr{P}_{x_0} := \sum_{\deg_{\vec{d}}(\alpha) \leq \kappa} a_{\alpha}(x_0) W^{\alpha}.$$

We make the following assumption.

Main assumption: There exists a neighborhood Ω of x_0 such that $\mathscr{P}_{x_0}$ is maximally subelliptic of degree κ with respect to $(W, \vec{d})$ on Ω.

Furthermore, we suppose we are given functions $u : \mathfrak{M} \to \mathbb{R}^{\mathbf{D}_2}$ and $g : \mathfrak{M} \to \mathbb{R}^{\mathbf{D}_1}$ satisfying

$$\mathscr{P}u = g.$$

Example 9.1.9. Suppose $(W, \vec{d}) = (W, 1) = \{(W_1, 1), \ldots, (W_r, 1)\}$, i. e., each $\vec{d}_j = 1$. Consider the operator

$$u \mapsto \sum_{i,j} a_{i,j}(x) W_i W_j u(x).$$

If the $r \times r$ matrix $(a_{i,j}(x_0))$ is strictly positive definite, then the main assumption above is satisfied for this operator. See Section 8.9 for this and generalizations.

Corollary 9.1.10. *Let* $\mathbf{s} > \mathbf{r} > 0$. *Suppose that:*
- $a_\alpha \in \mathscr{C}^{\mathbf{s}}(W, \vec{d})$ *near* x_0, $\forall \deg_{\vec{d}}(\alpha) \leq \kappa$,
- $g \in \mathscr{C}^{\mathbf{s}}(W, \vec{d})$ *near* x_0,
- $u \in \mathscr{C}^{\mathbf{r}+\kappa}(W, \vec{d})$ *near* x_0.

Then $u \in \mathscr{C}^{\mathbf{s}+\kappa}(W, \vec{d})$ *near* x_0.

Corollary 9.1.11. *Fix* $\mathbf{s} > \mathbf{r} > 0$. *Let* $(Z, \vec{d}) = \{(Z_1, \vec{d}_1), \ldots, (Z_\nu, \vec{d}_\nu)\} \subset C^\infty_{\mathrm{loc}}(\mathfrak{M}; T\mathfrak{M}) \times \mathbb{N}_+$ *be another set of Hörmander vector fields with single-parameter formal degrees on* $\mathfrak{M}$. *Let* $\lambda := \lambda(x_0, (W, \vec{d}), (Z, \vec{d})) > 0$ *be as in Definition* 6.6.9. *Suppose that:*
- $(W, \vec{d})$ *and* $(Z, \vec{d})$ *locally weakly approximately commute on* $\mathfrak{M}$,
- $a_\alpha \in \mathscr{C}^{\mathbf{s}}(Z, \vec{d})$ *near* x_0, $\forall \deg_{\vec{d}}(\alpha) \leq \kappa$,
- $g \in \mathscr{C}^{\mathbf{s}}(Z, \vec{d})$ *near* x_0,
- $W^\alpha u \in \mathscr{C}^{\mathbf{r}}(Z, \vec{d})$ *near* x_0, $\forall \deg_{\vec{d}}(\alpha) \leq \kappa$.

Then, $\forall \epsilon > 0$, *we have:*
- $u \in \mathscr{C}^{\mathbf{s}+\kappa/\lambda-\epsilon}(Z, \vec{d})$ *near* x_0, *and*
- $W^\alpha u \in \mathscr{C}^{\mathbf{s}-\epsilon}(Z, \vec{d})$ *near* x_0, $\forall \deg_{\vec{d}}(\alpha) \leq \kappa$.

Corollary 9.1.12. *Fix* $\mathbf{s} > \mathbf{r} > 0$. *Let* $\lambda := \lambda_{\mathrm{std}}(x_0, (W, \vec{d}))$ *(see Definition* 6.6.15*). Suppose that:*
- $a_\alpha \in \mathscr{C}^{\mathbf{s}}_{\mathrm{std}}$ *near* x_0, $\forall \deg_{\vec{d}}(\alpha) \leq \kappa$,
- $g \in \mathscr{C}^{\mathbf{s}}_{\mathrm{std}}$ *near* x_0,
- $W^\alpha u \in \mathscr{C}^{\mathbf{r}}_{\mathrm{std}}$ *near* x_0, $\forall \deg_{\vec{d}}(\alpha) \leq \kappa$.

Then, $\forall \epsilon > 0$, *we have:*
- $u \in \mathscr{C}^{\mathbf{s}+\kappa/\lambda-\epsilon}_{\mathrm{std}}$ *near* x_0, *and*
- $W^\alpha u \in \mathscr{C}^{\mathbf{s}-\epsilon}_{\mathrm{std}}$ *near* x_0, $\forall \deg_{\vec{d}}(\alpha) \leq \kappa$.

Each of the above corollaries is a nearly immediate consequence of the corresponding fully nonlinear result above. We include the proof reducing Corollary 9.1.10 to Theorem 9.1.2; the other proofs are similar, and we leave them to the interested reader.

Proof of Corollary 9.1.10. Fix a small, compact neighborhood $\mathcal{K} \Subset \mathfrak{M}$ of x_0 and let $\psi \in C^\infty_0(\mathfrak{M})$ be such that $\psi \equiv 1$ on $\mathcal{K}$ with small enough support that $\psi u \in \mathscr{C}^{\mathbf{r}+\kappa}_{\mathrm{cpt}}(W, \vec{d})$. Using Corollaries 7.2.2 and 6.5.11, we have, for $\deg_{\vec{d}}(\alpha) \leq \kappa$,

$$\|W^\alpha u\|_{L^\infty(\mathcal{K})} \lesssim \|W^\alpha \psi u\|_{\mathscr{C}^{\mathbf{r}}(W,\vec{d})} \lesssim \|\psi u\|_{\mathscr{C}^{\mathbf{r}+\kappa}(W,\vec{d})} < \infty.$$

Dividing u and g by $\sum_{\deg_{d}(\alpha)\leq\kappa}\|W^{\alpha}u\|_{L^{\infty}(\mathcal{K})}$ we see that we may assume $\|W^{\alpha}u\|_{L^{\infty}(\mathcal{K})}\leq 1$, $\forall\deg_{d}(\alpha)\leq\kappa$.

Fix $\phi(\zeta)\in C_0^{\infty}(\mathbb{R}^N)$ with $\phi(\zeta)\equiv 1$ on a neighborhood of $\{\zeta\in\mathbb{R}^N : |\zeta_\alpha|\leq 2, \forall\deg_{d}(\alpha)\leq\kappa\}$. Define

$$F(x,\zeta):=\sum_{\deg_{d}(\alpha)\leq\kappa} a_\alpha(x)\zeta_\alpha\phi(\zeta).$$

Clearly, $F\in\mathscr{C}^{\mathbf{s},2\lfloor\mathbf{s}\rfloor+7}((W,d)\boxtimes\nabla_{\mathbb{R}^N})$ near x_0.

Note that

$$F(x,\{W^{\alpha}u(x)\}_{\deg_{d}(\alpha)\leq\kappa})=\mathscr{P}u(x),\quad x\text{ near }x_0. \tag{9.4}$$

Also,

$$d_\zeta F(x_0,\{W^{\alpha}u(x_0)\}_{\deg_{d}(\alpha)\leq\kappa})\{W^{\alpha}\}_{\deg_{d}(\alpha)\leq\kappa}=\mathscr{P}_{x_0},$$

so the assumption of the maximal subellipticity of $\mathscr{P}_{x_0}$ gives that $F(x,\{W^{\alpha}u(x)\}_{\deg_{d}(\alpha)\leq\kappa})$ is maximally subelliptic at (x_0,u) of degree κ with respect to (W,d). We have shown that all the hypotheses of Theorem 9.1.2 hold for (9.4). The conclusion of Theorem 9.1.2 in this case is $u\in\mathscr{C}^{\mathbf{s}+\kappa}(W,d)$ near x_0, completing the proof. □

Remark 9.1.13. The proof above of Corollary 9.1.10 does not fully exploit the fact that $\mathscr{P}$ is a linear partial differential operator. By exploiting this, following a proof very similar to Theorem 9.1.14, one can weaken the a priori assumption that $u\in\mathscr{C}^{\mathbf{r}+\kappa}(W,d)$ near x_0. To push this idea to its logical conclusion would involve introducing considerable notation, and we do not do so here since the main focus of this chapter is to study fully nonlinear equations. See Remark 2.8.6.

9.1.2 Multi-parameter results

Fix $\nu\in\mathbb{N}_+$ and let $(W^1,d^1),\dots,(W^\nu,d^\nu)\subset C^{\infty}_{\mathrm{loc}}(\mathfrak{M};T\mathfrak{M})\times\mathbb{N}_+$ be finite lists of vector fields with formal degrees satisfying the hypotheses of the main multi-parameter setting described in Section 3.15. Namely, we assume that:
- (W^1,d^1) are Hörmander vector fields with formal degrees on $\mathfrak{M}$.
- $\mathrm{Gen}((W^\mu,d^\mu))$ is locally finitely generated for $\mu\in\{2,\dots,\nu\}$.
- $(W^1,d^1),\dots,(W^\nu,d^\nu)$ pairwise weakly locally approximately commute on $\mathfrak{M}$.

Set $(W,\vec{d}):=(W^1,d^1)\boxtimes(W^2,d^2)\boxtimes\cdots\boxtimes(W^\nu,d^\nu)\subset C^{\infty}_{\mathrm{loc}}(\mathfrak{M};T\mathfrak{M})\times(\mathbb{N}^\nu\setminus\{0\})$. Fix $\kappa\in\mathbb{N}_+$ such that d^1_j divides κ, $\forall j$. Let $\mathbf{D}_1,\mathbf{D}_2\in\mathbb{N}_+$ and set

$$N:=\mathbf{D}_2\times\#\{\alpha\text{ an ordered multi-index}:\deg_{d^1}(\alpha)\leq\kappa\}. \tag{9.5}$$

We consider the following PDE, where $F(x,\zeta) : \mathfrak{M} \times \mathbb{R}^N \to \mathbb{R}^{\mathbf{D}_1}$, $g(x) : \mathfrak{M} \to \mathbb{R}^{\mathbf{D}_1}$, and $u(x) : \mathfrak{M} \to \mathbb{R}^{\mathbf{D}_2}$ are given:

$$F(x, \{(W^1)^\alpha u(x)\}_{\deg_{\vec{d}^1}(\alpha)\leq\kappa}) = g(x). \tag{9.6}$$

Fix $x_0 \in \mathfrak{M}$.

Main assumptions: We assume that

$$F(x, \{(W^1)^\alpha u(x)\}_{\deg_{\vec{d}^1}(\alpha)\leq\kappa})$$

is maximally subelliptic at (x_0, u) of degree κ with respect to $(W^1, \vec{d}^1)$, and we assume (9.3) holds for x in a neighborhood of x_0.

Theorem 9.1.14. *Fix $\vec{\mathbf{r}}, \vec{\mathbf{s}} \in (0,\infty)^\nu$ and $\mathbf{t} > 2\lfloor|\vec{\mathbf{s}}|_1\rfloor + 4 + 2\nu$. Suppose that:*
- *$F \in \mathscr{C}^{\vec{\mathbf{s}},\mathbf{t}}((W,\vec{d}) \boxtimes \nabla_{\mathbb{R}^N})$ near x_0,*
- *$g \in \mathscr{C}^{\vec{\mathbf{s}}}(W,\vec{d})$ near x_0,*
- *$u \in \mathscr{C}^{\vec{\mathbf{r}}+\kappa e_1}(W,\vec{d})$ near x_0.*

Then $u \in \mathscr{C}^{\vec{\mathbf{s}}+\kappa e_1}(W,\vec{d})$ near x_0.

Proof. This is a qualitative version of the (stronger) quantitative result Theorem 9.2.1. Indeed, the hypotheses of this theorem are invariant under diffeomorphisms. By picking a coordinate chart $\Phi : B^n(2) \xrightarrow{\sim} \Phi(B^n(2)) \subseteq \mathfrak{M}$, where $\Phi(0) = x_0$ and $\Phi(B^n(2))$ is a small neighborhood of $x_0 \in \mathfrak{M}$, writing the above assumptions in this coordinate system, and restricting attention to $B^n(1)$, we see that all the hypotheses of Theorem 9.2.1 hold. The conclusion of Theorem 9.2.1 implies the conclusion of this theorem. □

We next present the reduction of Proposition 9.1.6 to Theorem 9.1.14. This combines several ideas from previous chapters.

Proof of Proposition 9.1.6. We rename $(W, \vec{d})$ to be $(W^1, \vec{d}^1)$, set $(W^2, \vec{d}^2) := (Z, \vec{d})$, and let $(W, \vec{d}) := (W^1, \vec{d}^1) \boxtimes (W^2, \vec{d}^2)$. By the assumptions of Proposition 9.1.6, we may pick $\psi_1, \psi_2 \in C_0^\infty(\mathfrak{M})$, with $\psi_1 \prec \psi_2$, with $\psi_1 \equiv 1$ on a neighborhood of x_0, and with small enough support that if $\mathcal{K} := \operatorname{supp}(\psi_2)$, we have

$$\psi_1(x)F(x,\zeta) \in \mathscr{C}^{\mathbf{s},\mathbf{t}}(\mathcal{K}\times\mathbb{R}^N, (W^2,\vec{d}^2)\boxtimes\nabla_{\mathbb{R}^N}), \tag{9.7}$$

$$\psi_1 g \in \mathscr{C}^{\mathbf{s}}(\mathcal{K}, (W^2,\vec{d}^2)), \tag{9.8}$$

$$\psi_2(W^1)^\alpha u \in \mathscr{C}^{\mathbf{r}}(\mathcal{K}, (W^2,\vec{d}^2)), \quad \forall \deg_{\vec{d}^1}(\alpha) \leq \kappa. \tag{9.9}$$

Note that for $\deg_{\vec{d}^1}(\alpha) \leq \kappa$, we have

$$(W^1)^\alpha \psi_1 u = \psi_2 (W^1)^\alpha \psi_1 u = \sum_{\deg_{\vec{d}^1}(\beta)\leq\kappa} g_\alpha^\beta \psi_2 (W^1)^\beta u,$$

where $g_\alpha^\beta \in C_0^\infty(\mathfrak{M})$. Therefore, Corollary 6.5.10 combined with (9.9) implies that

$$(W^1)^\alpha \psi_1 u \in \mathscr{C}^{\mathbf{r}}(\mathcal{K}, (W^2, \vec{d}^2)), \quad \forall \deg_{\vec{d}^1}(\alpha) \le \kappa. \tag{9.10}$$

The result is stronger the smaller $\epsilon > 0$ is, so we may assume $\epsilon \in (0, \mathbf{r})$. Proposition 7.6.1 applied to (9.7), (9.8), and (9.10) shows that there exists $\epsilon' > 0$ with

$$\begin{aligned} \psi_1(x)F(x,\zeta) &\in \mathscr{C}^{(\epsilon',\mathbf{s}-\epsilon),\mathbf{t}}(\mathcal{K} \times \mathbb{R}^N, (W, \vec{d}) \boxtimes \nabla_{\mathbb{R}^N}), \\ \psi_1 g &\in \mathscr{C}^{(\epsilon',\mathbf{s}-\epsilon)}(\mathcal{K}, (W, \vec{d})), \\ (W^1)^\alpha \psi_1 u &\in \mathscr{C}^{(\epsilon',\mathbf{r}-\epsilon)}(\mathcal{K}, (W, \vec{d})), \quad \forall \deg_{\vec{d}^1}(\alpha) \le \kappa. \end{aligned}$$

Proposition 6.5.12 shows that $\psi_1 u \in \mathscr{C}^{(\epsilon'+\kappa,\mathbf{r}-\epsilon)}(\mathcal{K}, (W, \vec{d}))$.

Since $\psi_1 \equiv 1$ on a neighborhood of x_0, the above shows that Theorem 9.1.14 applies to give that there exists $\tilde{\phi} \in C_0^\infty(\mathfrak{M})$, with $\tilde{\phi} \equiv 1$ on a neighborhood of x_0, such that

$$\tilde{\phi} u \in \mathscr{C}_{\mathrm{cpt}}^{(\epsilon'+\kappa,\mathbf{s}-\epsilon)}(W, \vec{d}). \tag{9.11}$$

(II): Take $\phi \in C_0^\infty(\mathfrak{M})$ with $\phi \prec \tilde{\phi}$ and $\phi \equiv 1$ on a neighborhood of x_0. Using (9.11), Corollaries 6.5.10 and 6.5.11 show that for all $\deg_{\vec{d}^1}(\alpha) \le \kappa$,

$$\phi (W^1)^\alpha u = \phi (W^1)^\alpha \tilde{\phi} u \in \mathscr{C}_{\mathrm{cpt}}^{(\epsilon'+\kappa-\deg_{\vec{d}^1}(\alpha),\mathbf{s}-\epsilon)}(W, \vec{d}) \subseteq \mathscr{C}_{\mathrm{cpt}}^{(\epsilon',\mathbf{s}-\epsilon)}(W, \vec{d}).$$

From here, Proposition 7.6.1 implies that for all $\deg_{\vec{d}^1}(\alpha) \le \kappa$,

$$\phi (W^1)^\alpha u \in \mathscr{C}_{\mathrm{cpt}}^{(\epsilon',\mathbf{s}-\epsilon)}(W, \vec{d}) \subseteq \mathscr{C}_{\mathrm{cpt}}^{\mathbf{s}-\epsilon}(W^2, \vec{d}^2),$$

completing the proof of (II).

(I): Letting $\lambda := \lambda(x_0, (W^1, \vec{d}^1), (W^2, \vec{d}^2)) > 0$ be as in Definition 6.6.9, Corollary 6.6.11 shows that $(W^1, \vec{d}^1)$ weakly λ-controls $(W^2, \vec{d}^2)$ on Ω' for some relatively compact open neighborhood $\Omega' \Subset \mathfrak{M}$ of x_0. Take $\phi \in C_0^\infty(\Omega')$ with $\phi \prec \tilde{\phi}$ and $\phi \equiv 1$ on a neighborhood of x_0 and set $\mathcal{K}_1 := \mathrm{supp}(\phi) \Subset \Omega'$. Corollary 6.5.10 and (9.11) combine to show that

$$\phi u = \phi \tilde{\phi} u \in \mathscr{C}_{\mathrm{cpt}}^{(\epsilon'+\kappa,\mathbf{s}-\epsilon)}(W, \vec{d}),$$

and therefore, since $\mathrm{supp}(\phi) \subseteq \mathcal{K}_1$, $\phi u \in \mathscr{C}^{(\epsilon'+\kappa,\mathbf{s}-\epsilon)}(\mathcal{K}_1, (W, \vec{d}))$. By Theorem 6.7.2 with $\tilde{\mathbf{s}} = \kappa/\lambda$, we have

$$\begin{aligned} \phi u \in \mathscr{C}^{(\epsilon'+\kappa,\mathbf{s}-\epsilon)}(\mathcal{K}_1, (W, \vec{d})) &\subseteq \mathscr{C}^{(\epsilon'+\kappa,\mathbf{s}-\epsilon)+(-\kappa,\kappa/\lambda)}(\mathcal{K}_1, (W, \vec{d})) \\ &= \mathscr{C}^{(\epsilon',\mathbf{s}+\kappa/\lambda-\epsilon)}(\mathcal{K}_1, (W, \vec{d})). \end{aligned}$$

From here, Proposition 7.6.1 implies

$$\phi u \in \mathscr{C}^{(\epsilon',\mathbf{s}+\kappa/\lambda-\epsilon)}(\mathcal{K}_1,(W,\vec{d})) \subseteq \mathscr{C}^{\mathbf{s}+\kappa/\lambda-\epsilon}(\mathcal{K}_1,(W^2,\vec{d}^2)),$$

completing the proof. □

9.1.2.1 Qualitative Schauder estimates

Similar to Corollary 9.1.10, Theorem 9.1.14 immediately gives a qualitative Schauder type estimate. We state this result here; the proof is similar to the proof of Corollary 9.1.10, and we leave it to the interested reader.

Let $\mathscr{P}u(x) = \sum_{\deg_{\vec{d}^1}(\alpha)\leq\kappa} a_\alpha(x)(W^1)^\alpha u(x)$, where $a_\alpha : \mathfrak{M} \to \mathbb{M}^{\mathbf{D}_1\times\mathbf{D}_2}(\mathbb{R})$. Fix $x_0 \in \mathfrak{M}$ and define the frozen coefficient operator

$$\mathscr{P}_{x_0} := \sum_{\deg_{\vec{d}^1}(\alpha)\leq\kappa} a_\alpha(x_0)(W^1)^\alpha.$$

We make the following assumption.

Main assumption: There exists a neighborhood Ω of x_0 such that $\mathscr{P}_{x_0}$ is maximally subelliptic of degree κ with respect to $(W,\vec{d})$ on Ω.

Furthermore, we suppose that we are given functions $u : \mathfrak{M} \to \mathbb{R}^{\mathbf{D}_2}$ and $g : \mathfrak{M} \to \mathbb{R}^{\mathbf{D}_1}$ satisfying

$$\mathscr{P}u = g.$$

Corollary 9.1.15. *Let $\vec{\mathbf{s}},\vec{\mathbf{r}} \in (0,\infty)^\nu$. Suppose that:*
- $a_\alpha \in \mathscr{C}^{\vec{\mathbf{s}}}(W,\vec{d})$ *near* x_0, $\forall \deg_{\vec{d}^1}(\alpha) \leq \kappa$,
- $g \in \mathscr{C}^{\vec{\mathbf{s}}}(W,\vec{d})$ *near* x_0,
- $u \in \mathscr{C}^{\vec{\mathbf{r}}+\kappa}(W,\vec{d})$ *near* x_0.

Then $u \in \mathscr{C}^{\vec{\mathbf{s}}+\kappa}(W,\vec{d})$ near x_0.

Proof. This is an immediate consequence of Theorem 9.1.14; see the proof of Corollary 9.1.10 for a similar reduction. □

9.2 Main quantitative result

The main results of this chapter follow from a quantitative version of Theorem 9.1.14, which we state and prove in this section. We work in the multi-parameter unit scale setting described in Section 3.15.1. Thus, we work on $B^n(1)$ endowed with the density $\mathrm{Vol} = h\sigma_{\mathrm{Leb}}$, where $h \in C^\infty(B^n(1);\mathbb{R})$, with $\inf_{x\in B^n(1)} h(x) > 0$. Fix $\nu \geq 1$; we assume that for each $\mu \in \{1,\ldots,\nu\}$ we are given

$$(W^\mu,\vec{d}^\mu) := \{(W_1^\mu,\vec{d}_1^\mu),\ldots,(W_{r_\mu}^\mu,\vec{d}_{r_\mu}^\mu)\} \subset C^\infty(B^n(1);TB^n(1)) \times \mathbb{N}_+,$$
$$(X^\mu,d^\mu) = \{(X_1^\mu,d_1^\mu),\ldots,(X_{q_\mu}^\mu,d_{q_\mu}^\mu)\} \subset \mathrm{Gen}((W^\mu,\vec{d}^\mu))$$

with $(W^\mu, \vec{d}^\mu) \subseteq (X^\mu, d^\mu)$ and such that the assumptions (i) and (ii) from Section 3.15.1 hold. We will speak of ι-multi-parameter unit-admissible constants as in Definition 3.15.1.

Fix $\mathbf{D}_1, \mathbf{D}_2, \kappa \in \mathbb{N}_+$ such that $\vec{d}_j^1$ divides κ, $\forall j$; define N as in (9.5). Fix $\vec{\mathbf{r}}, \vec{\mathbf{s}} \in (0,\infty)^\nu$ and $\sigma > 0$. Suppose that:

$$\begin{aligned} F(x,\zeta) &\in \mathscr{C}^{\vec{\mathbf{s}},2\lfloor|\vec{\mathbf{s}}|_1\rfloor+4+2\nu+\sigma}(\overline{B^n(7/8)} \times \mathbb{R}^N, (W,\vec{d}) \boxtimes \nabla_{\mathbb{R}^N}; \mathbb{R}^{\mathbf{D}_1}), \\ u &\in \mathscr{C}^{\vec{\mathbf{r}}+\kappa e_1}(\overline{B^n(7/8)}, (W,\vec{d}); \mathbb{R}^{\mathbf{D}_2}), \\ g &\in \mathscr{C}^{\vec{\mathbf{s}}}(\overline{B^n(7/8)}, (W,\vec{d}); \mathbb{R}^{\mathbf{D}_1}). \end{aligned}$$

Suppose also that we have the following:

$$F(x, \{(W^1)^\alpha u(x)\}_{\deg_{\vec{d}^1}(\alpha)\le\kappa}) = g(x), \quad \forall x \in B^n(3/4). \tag{9.12}$$

Because the results in this section are quantitative, it is important to use an explicit choice of norm for $\|\cdot\|_{\mathscr{C}^{\vec{\mathbf{s}}}(W,\vec{d})}$ and $\|\cdot\|_{\mathscr{C}^{\vec{\mathbf{s}},t}((W,\vec{d})\boxtimes\nabla_{\mathbb{R}^N})}$, rather than just work with an equivalence class of norms as we usually do. The choice we use is described in Section 9.2.1; for the rest of this section, we use this choice.

Fix $C_0 \ge 0$ such that

$$\|F\|_{\mathscr{C}^{\vec{\mathbf{s}},2\lfloor|\vec{\mathbf{s}}|_1\rfloor+4+2\nu+\sigma}((W,\vec{d})\boxtimes\nabla_{\mathbb{R}^N})}, \|u\|_{\mathscr{C}^{\vec{\mathbf{r}}+\kappa e_1}(W,\vec{d})}, \|g\|_{\mathscr{C}^{\vec{\mathbf{s}}}(W,\vec{d})} \le C_0.$$

Set $n_j := \kappa/\vec{d}_j^1 \in \mathbb{N}_+$ and define $\mathscr{P}$ by

$$\mathscr{P}v := d_\zeta F(0, \{(W^1)^\alpha u(0)\}_{\deg_{\vec{d}^1}(\alpha)\le\kappa})\{(W^1)^\alpha v\}_{\deg_{\vec{d}^1}(\alpha)\le\kappa}. \tag{9.13}$$

We assume $\mathscr{P}$ is maximally subelliptic in the quantitative sense that there exists $A \ge 0$ with

$$\begin{aligned} &\sum_{j=1}^{r_1} \|(W_j^1)^{n_j} f\|_{L^2(B^n(1),h\sigma_{\mathrm{Leb}};\mathbb{C}^{\mathbf{D}_2})} \\ &\quad \le A(\|\mathscr{P}f\|_{L^2(B^n(1),h\sigma_{\mathrm{Leb}};\mathbb{C}^{\mathbf{D}_1})} + \|f\|_{L^2(B^n(1),h\sigma_{\mathrm{Leb}};\mathbb{C}^{\mathbf{D}_2})}), \end{aligned} \tag{9.14}$$

for all $f \in C_0^\infty(B^n(1);\mathbb{C}^{\mathbf{D}_2})$.

We apply Theorem 3.15.5 with $\sigma = 7/8$; let $\Phi_{x,\delta} : B^n(1) \xrightarrow{\sim} \Phi_{x,\delta}(B^n(1)) \subseteq B_{(X^1,d^1)}(x,\delta) \subseteq B^n(1)$ be as in that theorem.[2]

Our main nonlinear regularity theorem is the following

2 Since the vector fields $X_1^1, \dots, X_{q_1}^1$ are only defined on $B^n(1)$, we have $B_{(X^1,d^1)}(x,\delta) \subseteq B^n(1)$, $\forall \delta > 0$, by the definition of $B_{(X^1,d^1)}(x,\delta)$ – see (1.12) and (1.13).

Theorem 9.2.1. *Suppose all of the above assumptions hold; in particular, u satisfies* (9.12). *There exists an* $(\vec{\mathbf{s}}, \vec{\mathbf{r}}, \kappa, \sigma)$*-multi-parameter unit-admissible constant* $\delta = \delta(A, C_0, \mathbf{D}_1, \mathbf{D}_2) \in (0,1)$ *such that the following holds. Let* $\psi_0 \in C_0^\infty(B^n(2/3))$. *Then* $(\psi_0 \circ \Phi_{0,\delta})u \in \mathscr{C}^{\vec{\mathbf{s}}+\kappa e_1}(\overline{B^n(7/8)}, (W, \vec{d}); \mathbb{C}^{\mathbf{D}_2})$ *and*

$$\|(\psi_0 \circ \Phi_{0,\delta}^{-1})u\|_{\mathscr{C}^{\vec{\mathbf{s}}+\kappa e_1}(W,\vec{d})} \leq C_5,$$

where $C_5 = C_5(A, C_0, \mathbf{D}_1, \mathbf{D}_2, \psi_0) \geq 0$ *is an* $(\vec{\mathbf{s}}, \vec{\mathbf{r}}, \kappa, \sigma)$*-multi-parameter unit-admissible constant.*

Remark 9.2.2. Theorem 9.2.1 is a generalization of the corresponding result for elliptic equations: Theorem 2.8.7. The reader interested in learning the proof of Theorem 9.2.1 might find it useful to first understand the proof of Theorem 2.8.7, which has the same outline, though many of the technicalities become much simpler.

To ease notation, in the proof of Theorem 9.2.1, we write

$$\mathfrak{W}_1^\kappa u(x) := \{(W^1)^\alpha u(x)\}_{\deg_{\vec{d}^1}(\alpha)\leq\kappa}.$$

We prove Theorem 9.2.1 in three steps of increasing generality:

- **Step I:** We first prove Theorem 9.2.1 in the case of a perturbation of a linear operator, which is the heart of the proof. More precisely, when (9.12) is of the form

$$\mathscr{P}u(x) = R_1(x, \mathfrak{W}_1^\kappa u(x)) + R_2(x, \mathfrak{W}_1^\kappa u(x)), \tag{9.15}$$

  where $\mathscr{P}$ is maximally subelliptic as above, $\|R_1\|_{\mathscr{C}^{\vec{\mathbf{s}},\lfloor|\vec{\mathbf{s}}|_1\rfloor+2+\nu+\sigma}((W,\vec{d})\boxtimes\nabla_{\mathbb{R}^N})}$ is small, $R_2 \in \mathscr{C}^{\vec{\mathbf{s}},\lfloor|\vec{\mathbf{s}}|_1\rfloor+3+\nu+\sigma}((W,\vec{d})\boxtimes\nabla_{\mathbb{R}^N})$ with $R_2(x,0) \equiv 0$ and $d_\zeta R_2(x,0) \equiv 0$, and $\|u\|_{\mathscr{C}^{\vec{\mathbf{r}}+\kappa e_1}(W,\vec{d})}$ is small. Because R_2 vanishes to second order at $\zeta = 0$ and $\|u\|_{\mathscr{C}^{\vec{\mathbf{r}}+\kappa e_1}(W,\vec{d})}$ is small, the right-hand side of (9.15) is small. We may therefore study the equation as a perturbation of a linear maximally subelliptic equation, and contraction mapping type arguments can be used to prove the desired regularity.
- **Step II:** We next prove Theorem 9.2.1 in the case where (9.12) is of the form

$$G(\mathfrak{W}_1^\kappa u(x)) + R_3(x, \mathfrak{W}_1^\kappa u(x)) = 0, \tag{9.16}$$

  where $\|R_3\|_{\mathscr{C}^{\vec{\mathbf{s}},\lfloor|\vec{\mathbf{s}}|_1\rfloor+2+\nu+\sigma}((W,\vec{d})\boxtimes\nabla_{\mathbb{R}^N})}$ is small, $G \in \mathscr{C}^{\lfloor|s|_1\rfloor+3+\nu+\sigma}(\mathbb{R}^N)$, and $\|u\|_{\mathscr{C}^{\vec{\mathbf{r}}+\kappa e_1}(W,\vec{d})}$ is small. We assume that

$$v \mapsto d_\zeta G(\mathfrak{W}_1^\kappa u(0))\mathfrak{W}_1^\kappa v$$

  is maximally subelliptic, as described above. We will see that it is a straightforward calculus exercise to reduce this case to the previous one.
- **Step III:** Finally, we prove Theorem 9.2.1 by reducing it to Step II. We do this by showing that the hypotheses of Step II hold if we “zoom in” around $x = 0$. This zooming

is achieved by rescaling via Theorem 3.15.5, which retains all of the hypotheses in a quantitative way.

Remark 9.2.3. In Step III of the above outline, we require more regularity in the ζ variable than in the previous steps. This is because the reduction from Step III to Step II loses some regularity in the ζ variable. This loss is probably not necessary, though we do not pursue that here.

Remark 9.2.4. In this section, we will apply several results from previous chapters. Not all of these results are stated in a quantitative way. However, by keeping track of the proofs of these results, one can prove that all estimates can be taken to be in terms of multi-parameter unit-admissible constants (when applied in the setting of this section). We will freely use this fact without mentioning it in the proofs which follow. We will write $A \lesssim B$, where $A \leq CB$ and C is an appropriate multi-parameter unit-admissible constant and $A \approx B$ for $A \lesssim B$ and $B \lesssim A$. It will be clear from the context exactly what C depends on; in particular, we will make the dependence explicit in the statement of results and in the proofs the dependence will usually be the same as in the statement of the result.

Throughout this section, $\vec{\mathbf{r}}, \vec{\mathbf{s}} \in (0,\infty)^\nu$, $\sigma > 0$, $\mathbf{D}_1, \mathbf{D}_2 \in \mathbb{N}_+$, $\kappa \in \mathbb{N}_+$ such that $\vec{d}_j^1$ divides κ, $\forall 1 \leq j_1 \leq r_1$ and $n_j := \kappa / \vec{d}_j^1 \in \mathbb{N}_+$, $N \in \mathbb{N}_+$, and $(W, \vec{d})$ are as described above.

9.2.1 Vector fields and norms

Let $\lambda = (1, \lambda_1, \dots, \lambda_\nu) \in (0,\infty)^\nu$ be as in Section 3.15.3. For $\delta \in (0,1]$ and $x \in \overline{B^n(7/8)}$, we define

$$
\begin{aligned}
(W^{\mu,x,\delta}, \vec{d}^\mu) &:= \{(W_1^{\mu,x,\delta}, \vec{d}_1^\mu), \dots, (W_{r_\mu}^{\mu,x,\delta}, \vec{d}_{r_\mu}^\mu)\} \\
&\subset C^\infty(B^n(1); TB^n(1)) \times \mathbb{N}_+, \\
(X^{\mu,x,\delta}, d^\mu) &:= \{(X_1^{\mu,x,\delta}, d_1^\mu), \dots, (X_{q_\mu}^{\mu,x,\delta}, d_{q_\mu}^\mu)\} \\
&\subset C^\infty(B^n(1); TB^n(1)) \times \mathbb{N}_+,
\end{aligned}
$$

and $h_{x,\delta} \in C^\infty(B^n(1))$ as in Theorem 3.15.5 with $\sigma = 7/8$. We set

$$
\begin{aligned}
(W^{x,\delta}, \vec{d}) &:= (W^{1,x,\delta}, \vec{d}^1) \boxtimes (W^{2,x,\delta}, \vec{d}^2) \boxtimes \cdots \boxtimes (W^{\nu,x,\delta}, \vec{d}^\nu) \\
&\subset C^\infty(B^n(1); TB^n(1)) \times (\mathbb{N}^\nu \setminus \{0\}), \\
(X^{x,\delta}, \vec{d}) &:= (X^{1,x,\delta}, d^1) \boxtimes (X^{2,x,\delta}, d^2) \boxtimes \cdots \boxtimes (X^{\nu,x,\delta}, d^\nu) \\
&\subset C^\infty(B^n(1); TB^n(1)) \times (\mathbb{N}^\nu \setminus \{0\}).
\end{aligned}
$$

Remark 9.2.5. For each $x \in \overline{B^n(7/8)}$ and $\delta \in (0,1]$, $(W^{x,\delta}, \vec{d})$, $(X^{x,\delta}, \vec{d})$, and $h_{x,\delta}$ can be viewed as data for the multi-parameter unit scale setting of Section 3.15.1. Importantly, Theorem 3.15.5 shows that multi-parameter unit-admissible constants (as in Definition 3.15.1) with respect to $(W^{x,\delta}, \vec{d})$, $(X^{x,\delta}, \vec{d})$, and $h_{x,\delta}$ can be chosen to be multi-parameter unit-admissible constants with respect to $(W, \vec{d})$, $(X, \vec{d})$, and h. In short, if we prove a result in terms of $(W, \vec{d})$, $(X, \vec{d})$, and h, we may apply that result to $(W^{x,\delta}, \vec{d})$, $(X^{x,\delta}, \vec{d})$, and $h_{x,\delta}$, and multi-parameter unit-admissible constants will be preserved. In this section, all results will be in terms of multi-parameter unit-admissible constants, and this allows us to freely scale those results.

For each $\delta \in (0,1]^\nu$, we also consider the vector fields with formal degrees:

$$\begin{aligned}
(\delta^{\lambda\vec{d}} W, \vec{d}) &:= (\delta^{\lambda_1 d^1} W^1, d^1) \boxtimes (\delta^{\lambda_2 d_2} W^2, d^2) \boxtimes \cdots \boxtimes (\delta^{\lambda_\nu \nu} W^\nu, d^\nu) \\
&\subset C^\infty(B^n(1); TB^n(1)) \times (\mathbb{N}^\nu \setminus \{0\}), \\
(\delta^{\lambda\vec{d}} X, \vec{d}) &:= (\delta^{\lambda_1 d^1} W^1, d^1) \boxtimes (\delta^{\lambda_2 d_2} X^2, d^2) \boxtimes \cdots \boxtimes (\delta^{\lambda_\nu \nu} X^\nu, d^\nu) \\
&\subset C^\infty(B^n(1); TB^n(1)) \times (\mathbb{N}^\nu \setminus \{0\}),
\end{aligned}$$

in particular, $(W^{x,\delta}, \vec{d}) = (\Phi^*_{x,\delta}\delta^{\lambda\vec{d}} W, \vec{d})$ and $(X^{x,\delta}, \vec{d}) = (\Phi^*_{x,\delta}\delta^{\lambda\vec{d}} X, \vec{d})$.

We define the operator D_j, $j \in \mathbb{N}^\nu$, as in Section 6.11, with $\mathcal{K} = \overline{B^n(7/8)}$, $\Omega_0 = B^n(15/16)$, and $a > 0$ a small, 0-multi-parameter unit-admissible constant. Note that D_j depends on $(X, \vec{d})$. For $x \in \overline{B^n(7/8)}$ and $\delta \in (0,1]$ we define $D_j^{x,\delta}$ in the same way with $(X, \vec{d})$ replaced by $(X^{x,\delta}, \vec{d})$, and we define D_j^δ with $(X, \vec{d})$ replaced by $(\delta^{\lambda\vec{d}} X, \vec{d})$.

We define the following norms: for a function $f \in L^\infty(B^n(1))$,

$$\begin{aligned}
\|f\|_{\mathscr{C}^{\vec{s}}(W,\vec{d})} &:= \sup_{j\in\mathbb{N}^\nu} 2^{j\cdot\vec{s}} \|D_j f\|_{L^\infty} = \|f\|_{\mathscr{B}^{\vec{s}}_{\infty,\infty}(W,\vec{d}),(X,\vec{d}),a,\sigma,h}, \\
\|f\|_{\mathscr{C}^{\vec{s}}(\delta^{\lambda\vec{d}} W,\vec{d})} &:= \sup_{j\in\mathbb{N}^\nu} 2^{j\cdot\vec{s}} \|D_j^\delta f\|_{L^\infty} = \|f\|_{\mathscr{B}^{\vec{s}}_{\infty,\infty}(\delta^{\lambda\vec{d}} W,\vec{d}),(\delta^{\lambda\vec{d}} X,\vec{d}),a,\sigma,h}, \\
\|f\|_{\mathscr{C}^{\vec{s}}(W^{x,\delta},\vec{d})} &:= \sup_{j\in\mathbb{N}^\nu} 2^{j\cdot\vec{s}} \|D_j^{x,\delta} f\|_{L^\infty} = \|f\|_{\mathscr{B}^{\vec{s}}_{\infty,\infty}(W^{x,\delta},\vec{d}),(X^{x,\delta},\vec{d}),a,\sigma,h}.
\end{aligned}$$

For a function $F(x, \zeta) \in L^\infty(B^n(1) \times \mathbb{R}^N)$, with $\widetilde{D}_l$ as in Section 7.5,

$$\begin{aligned}
\|F\|_{\mathscr{C}^{\vec{s},\mathtt{t}}((W,\vec{d})\boxtimes\nabla_{\mathbb{R}^N})} &:= \sup_{j\in\mathbb{N}^\nu} \sup_{l\in\mathbb{N}} 2^{j\cdot\vec{s}+l\mathtt{t}} \|(D_j \otimes \widetilde{D}_l) F\|_{L^\infty}, \\
\|F\|_{\mathscr{C}^{\vec{s},\mathtt{t}}((\delta^{\lambda\vec{d}} W,\vec{d})\boxtimes\nabla_{\mathbb{R}^N})} &:= \sup_{j\in\mathbb{N}^\nu} \sup_{l\in\mathbb{N}} 2^{j\cdot\vec{s}+l\mathtt{t}} \|(D_j^\delta \otimes \widetilde{D}_l) F\|_{L^\infty}, \\
\|F\|_{\mathscr{C}^{\vec{s},\mathtt{t}}((W^{x,\delta},\vec{d})\boxtimes\nabla_{\mathbb{R}^N})} &:= \sup_{j\in\mathbb{N}^\nu} \sup_{l\in\mathbb{N}} 2^{j\cdot\vec{s}+l\mathtt{t}} \|(D_j^{x,\delta} \otimes \widetilde{D}_l) F\|_{L^\infty}.
\end{aligned}$$

The above definitions all extend to all $\vec{s} \in \mathbb{R}^\nu$, though we use the notation $\mathscr{B}^{\vec{s}}_{\infty,\infty}$ instead of $\mathscr{C}^{\vec{s}}$ to make it clear that we are not working with a Zygmund–Hölder space. That is, we set, for $\vec{\mathbf{s}} \in \mathbb{R}^\nu$,

$$\|f\|_{\mathscr{B}^{\vec{\mathbf{s}}}_{\infty,\infty}(W,\vec{d})} := \sup_{j\in\mathbb{N}^\nu} 2^{j\cdot\vec{\mathbf{s}}}\|D_j f\|_{L^\infty} = \|f\|_{\mathscr{B}^{\vec{\mathbf{s}}}_{\infty,\infty}(W,\vec{d}),(X,\vec{d}),a,\sigma,h},$$

$$\|f\|_{\mathscr{B}^{\vec{\mathbf{s}}}_{\infty,\infty}(\delta^{\lambda\vec{d}}W,\vec{d})} := \sup_{j\in\mathbb{N}^\nu} 2^{j\cdot\vec{\mathbf{s}}}\|D^\delta_j f\|_{L^\infty} = \|f\|_{\mathscr{B}^{\vec{\mathbf{s}}}_{\infty,\infty}(\delta^{\lambda\vec{d}}W,\vec{d}),(\delta^{\lambda\vec{d}}X,\vec{d}),a,\sigma,h},$$

$$\|f\|_{\mathscr{B}^{\vec{\mathbf{s}}}_{\infty,\infty}(W^{x,\delta},\vec{d})} := \sup_{j\in\mathbb{N}^\nu} 2^{j\cdot\vec{\mathbf{s}}}\|D^{x,\delta}_j f\|_{L^\infty} = \|f\|_{\mathscr{B}^{\vec{\mathbf{s}}}_{\infty,\infty}(W^{x,\delta},\vec{d}),(X^{x,\delta},\vec{d}),a,\sigma,h}.$$

The norms $\|\cdot\|_{\mathscr{C}^{\vec{\mathbf{s}}}(W,\vec{d})}$ and $\|\cdot\|_{\mathscr{C}^{\vec{\mathbf{s}}}(W^{x,\delta},\vec{d})}$ take place at the "unit scale," and because of this we will be able to directly prove quantitative estimates in terms of these norms. When $\delta > 0$ is small, we cannot directly prove quantitative results about $\|f\|_{\mathscr{C}^{\vec{\mathbf{s}}}(\delta^\lambda W,\vec{d})}$. However, the next two results allow us to indirectly conclude quantitative estimates about these norms.

Lemma 9.2.6. *Fix $c_0 \in (0,1]$ and $\vec{\mathbf{s}} \in (0,\infty)^\nu$. There exists an $\vec{\mathbf{s}}$-multi-parameter unit-admissible constant $C_1 = C_1(c_0) \geq 1$ such that $\forall \delta \in [c_0, 1]$, $\forall f \in \mathscr{C}^{\vec{\mathbf{s}}}(\overline{B^n(7/8)}, (W,\vec{d}))$,*

$$C_1^{-1}\|f\|_{\mathscr{C}^{\vec{\mathbf{s}}}(W,\vec{d})} \leq \|f\|_{\mathscr{C}^{\vec{\mathbf{s}}}(\delta^\lambda W,\vec{d})} \leq C_1\|f\|_{\mathscr{C}^{\vec{\mathbf{s}}}(W,\vec{d})}.$$

Proof. This follows from a simple reprise of the proof of Proposition 6.3.7, keeping track of constants in that proof. □

Lemma 9.2.7. *For $x \in \overline{B^n(7/8)}$, $\delta \in (0,1]$, $\vec{\mathbf{s}} \in (0,\infty)^\nu$, and $\phi \in C_0^\infty(B^n(7/8))$, we have*

$$\|\phi\Phi^*_{x,\delta}f\|_{\mathscr{C}^{\vec{\mathbf{s}}}(W^{x,\delta},\vec{d})} = \|(\phi\circ\Phi^{-1}_{x,\delta})f\|_{\mathscr{C}^{\vec{\mathbf{s}}}(\delta^\lambda W,\vec{d})},$$

$\forall f \in C_0^\infty(B^n(1))'$.

Proof. This is the case $p = \infty$ of Proposition 6.11.4. □

9.2.2 Bump functions

A key element of Step I is the existence of bump functions at every sub-Riemannian scale, which is due to Nagel and Stein [187].

We apply Theorem 3.15.5 (with $\sigma = 7/8$); in this section we use all the same notation as in that theorem. In particular, we let $\xi_3 \in (0,1]$ be the 0-multi-parameter unit-admissible constant from that theorem.

Theorem 9.2.8. *There exist 0-multi-parameter unit-admissible constants $\xi_4, \xi_5 \in (0,1]$ with $\xi_5 < \xi_4 < \xi_3$ such that $\forall x \in B^n(7/8)$ and $\delta \in (0,1]$, there exists $\phi_{x,\delta} \in C_0^\infty(B^n(1))$ and, setting $\xi_6 := \xi_5^4/\xi_4^3 \in (0,\xi_5)$, we have $\forall x \in B^n(7/8)$, $\delta \in (0,1]$:*
(i) $\operatorname{supp}(\phi_{x,\delta}) \subseteq B_{(W^1,d^1)}(x,\xi_4\delta)$.
(ii) $0 \leq \phi_{x,\delta} \leq 1$.

(iii) $\phi_{x,\delta} \equiv 1$ *on a neighborhood of* $\overline{B_{(W^1,\vec{d}^1)}(x, \xi_5\delta)}$.
(iv) $\phi_{x,\frac{\xi_5}{\xi_4}\delta} \prec \phi_{x,\delta}$.
(v) *For* $\xi_0 \in (0,1]$, $\operatorname{supp}(\phi_{x,\xi_0\delta}) \subseteq \Phi_{x,\delta}(B^n(1/2))$. *In particular,*

$$\operatorname{supp}(\phi_{x,\frac{\xi_4}{\xi_5}\xi_6\delta}), \operatorname{supp}(\phi_{x,(\frac{\xi_4}{\xi_5})^2\xi_6\delta}), \operatorname{supp}(\phi_{x,(\frac{\xi_4}{\xi_5})^3\xi_6\delta}) \subseteq \Phi_{x,\delta}(B^n(1/2)).$$

(vi) *For every* α,

$$\sup_{\substack{x\in B^n(7/8)\\ \delta\in(0,1]}} \|(\delta^{\lambda\vec{d}}X)^\alpha \phi_{x,\delta}\|_{C(B^n(1))} \leq C_\alpha,$$

where C_α *is an* α*-multi-parameter unit-admissible constant.*
(vii) *For* $\xi_0 \in (0,1]$ *and* $L \in \mathbb{N}$,

$$\|\phi_{x,\xi_0\delta} \circ \Phi_{x,\delta}\|_{C^L(B^n(1))} \leq C_L \xi_0^{-L\max_{1\leq j\leq q_1} d_j^1},$$

where $C_L \geq 0$ *is an* L*-multi-parameter unit-admissible constant.*
(viii) *There exists a* 0*-multi-parameter unit-admissible constant* $N_1 \in \mathbb{N}_+$ *such that* $\forall \vec{\mathbf{s}} \in (0,\infty)^\nu$, *there exists* $C_{\vec{\mathbf{s}}} \geq 0$ *an* $\vec{\mathbf{s}}$*-multi-parameter unit-admissible constant such that* $\forall x \in B^n(3/4)$ *and* $\delta \in (0,1]$ *with* $B_{(W^1,\vec{d}^1)}(x,\delta) \subseteq B^n(3/4)$, *there exist* $x_1, \dots, x_{N_1} \in B^n(3/4)$ *with*

$$B_{(W^1,\vec{d}^1)}\left(x_j, \frac{\xi_4^3\xi_6}{\xi_5^3}\delta\right) \subseteq B^n(3/4), \tag{9.17}$$

$$\|\phi_{x,\delta}u\|_{\mathscr{C}^{\vec{\mathbf{s}}}(\delta^{\lambda\vec{d}}W,\vec{d})} \leq C_{\vec{\mathbf{s}}} \sum_{j=1}^{N_1} \|\phi_{x_j,\xi_6\delta}u\|_{\mathscr{C}^{\vec{\mathbf{s}}}(\delta^{\lambda\vec{d}}W,\vec{d})}, \tag{9.18}$$

for all $u \in C_0^\infty(B^n(1))'$.
(ix) *There exist* 0*-multi-parameter unit-admissible constants* $N_2 \in \mathbb{N}_+$ *and* $\delta_1 \in (0,1]$ *and* $x_1, \dots, x_{N_2} \in B^n(3/4)$ *with* $B_{(W^1,\vec{d}^1)}(x_j,\delta_1) \subseteq B^n(3/4)$, $1 \leq j \leq N_2$, *such that* $\forall \vec{\mathbf{s}} \in (0,\infty)^\nu$, $\forall \psi \in C_0^\infty(B^n(2/3))$, *there exists an* $\vec{\mathbf{s}}$*-multi-parameter unit-admissible constant* $C_{\vec{\mathbf{s}}} = C_{\vec{\mathbf{s}}}(\psi) \geq 0$ *such that*

$$\|\psi u\|_{\mathscr{C}^{\vec{\mathbf{s}}}(W,\vec{d})} \leq C_{\vec{\mathbf{s}}} \sum_{j=1}^{N_2} \|\phi_{x_j,\delta_1}u\|_{\mathscr{C}^{\vec{\mathbf{s}}}(\delta_1^{\lambda\vec{d}}W,\vec{d})}. \tag{9.19}$$

(x) *Fix* $\xi_0 \in (0,1]$ *and* $\vec{\mathbf{s}} \in (0,\infty)^\nu$. *Then*

$$\|\phi_{x,\xi_0\delta}u\|_{\mathscr{C}^{\vec{\mathbf{s}}}(\delta^{\lambda\vec{d}}W,\vec{d})} \approx \|\phi_{x,\xi_0\delta}u\|_{\mathscr{C}^{\vec{\mathbf{s}}}((\xi_0\delta)^{\lambda\vec{d}}W,\vec{d})},$$

$\forall u \in C_0^\infty(B^n(1))'$, *where the implicit constants are* $\vec{\mathbf{s}}$*-multi-parameter unit-admissible constants which depend on* ξ_0.

Proof. With $\xi_3, \delta_0 \in (0,1]$ as in Theorem 3.15.5, we set $\xi_4 := (\xi_3/2) \wedge (\delta_0/2)$. By Lemma 3.2.4 and Theorem 3.15.5, there exists a 0-multi-parameter unit-admissible constant $\eta_2 \in (0,1/2]$ so small that $\forall x \in B^n(7/8)$ and $\delta \in (0,1]$,

$$B^n(\eta_2) \subseteq B_{(\xi_4^{d^1} W^{1,x,\delta}, d^1)}(0,1) = B_{(W^{1,x,\delta}, d^1)}(0, \xi_4). \tag{9.20}$$

Using the fact that $\Phi_{x,\delta}(B_{(W^{1,x,\delta},d^1)}(0,\xi_3)) = B_{(W^1,d^1)}(x, \xi_3\delta)$ (since $\Phi_{x,\delta}^* \delta^{d^1} W^1 = W^{1,x,\delta}$, by definition), (9.20) and Theorem 3.15.5 (f) show that

$$B^n(\eta_2) \subseteq B_{(W^{1,x,\delta},d^1)}(0,\xi_4) \subseteq B_{(W^{1,x,\delta},d^1)}(0,\xi_3) \subseteq B^n(1/2). \tag{9.21}$$

Take $\phi_0 \in C_0^\infty(B^n(\eta_2))$ such that $\phi_0 \equiv 1$ on $\overline{B^n(\eta_2/2)}$ and $0 \le \phi_0 \le 1$; note that ϕ_0 depends only on η_2 and n – it does not depend on x or δ.

By the Picard–Lindelöf theorem and Theorem 3.15.5 (h) (with $L = 1$), we make take $\xi_5 \in (0, \xi_4)$ a multi-parameter unit-admissible constant so small that $B_{(W^{1,x,\delta},d^1)}(0,\xi_5) \subseteq B^n(\eta_2/4)$, $\forall x \in B^n(7/8)$, $\delta \in (0,1]$. In particular, we have

$$\phi_0 \equiv 1 \text{ on a neighborhood of } \overline{B_{(W^{1,x,\delta},d^1)}(0,\xi_5)} \subseteq B^n(\eta_4/2), \tag{9.22}$$

$\forall x \in B^n(7/8)$, $\delta \in (0,1]$.

Set $\phi_{x,\delta} := \phi_0 \circ \Phi_{x,\delta}^{-1}$. Since ϕ_0 is supported on $B^n(\eta_2) \subseteq B_{(W^{1,x,\delta},d^1)}(x,\xi_4)$ and $\Phi_{x,\delta}(B_{(W^{1,x,\delta},d^1)}(x,\xi_4)) = B_{(\delta^{d^1} W^1, d^1)}(x,\xi_4) = B_{(W^1,d^1)}(x,\xi_4\delta)$, (i) follows. Since $0 \le \phi_0 \le 1$, (ii) follows immediately from the definition. Applying $\Phi_{x,\delta}$ to (9.22) shows that

$$\phi_{x,\delta} \equiv 1 \text{ on a neighborhood of } \overline{B_{(W^1,d^1)}(x,\xi_5\delta)},$$

establishing (iii).

(iv): Using (i) we have

$$\operatorname{supp}(\phi_{x,\frac{\xi_5}{\xi_4}\delta}) \subseteq B_{(W^1,d^1)}(x, \xi_5\delta).$$

Applying (iii) shows that $\phi_{x,\delta} \equiv 1$ on a neighborhood of $\operatorname{supp}(\phi_{x,\frac{\xi_5}{\xi_4}\delta})$, establishing (iv).

(v): For $\xi_0 \in (0,1]$, we have, using (i) and Theorem 3.15.5 (f),

$$\operatorname{supp}(\phi_{x,\xi_0\delta}) \subseteq B_{(W^1,d^1)}(x, \xi_4\xi_0\delta) \subseteq B_{(W^1,d^1)}(x,\xi_3\delta) \subseteq \Phi_{x,\delta}(B^n(1/2)),$$

establishing (v).

(vi): We have, for $x \in B^n(7/8)$ and $\delta \in (0,1]$, using the fact that $\Phi_{x,\delta}^* \delta^{d^1} X^1 = X^{1,x,\delta}$,

$$\begin{aligned} \|(\delta^{\lambda\vec{d}} X)^\alpha \phi_{x,\delta}\|_{C(B^n(1))} &= \|((X^{x,\delta})^\alpha \phi_0) \circ \Phi_{x,\delta}^{-1}\|_{C(B^n(1))} \\ &= \|(X^{x,\delta})^\alpha \phi_0\|_{C(\Phi_{x,\delta}(B^n(1)))} \lesssim 1, \end{aligned}$$

where in the last estimate, we have used the fact that $\phi_0 \in C_0^\infty(B^n(\eta_2))$ is a fixed function and we have used Theorem 3.15.5 (h).

(vii): Using Theorem 3.15.5 (j), we have

$$\begin{aligned}\|\phi_{x,\xi_0\delta} \circ \Phi_{x,\delta}\|_{C^L(B^n(1))} &\approx \sum_{|\alpha| \leq L} \|(X^{1,x,\delta})^\alpha \phi_{x,\xi_0\delta} \circ \Phi_{x,\delta}\|_{C(B^n(1))} \\ &= \sum_{|\alpha| \leq L} \xi_0^{-\deg_{d^1}(\alpha)} \|((\xi_0\delta)^{d^1} X^1)^\alpha \phi_{x,\xi_0\delta}\|_{C(B^n(1))} \lesssim \sum_{|\alpha| \leq L} \xi_0^{-\deg_{d^1}(\alpha)},\end{aligned}$$

where the final estimate used (vi). Item (vii) follows.

(viii): Item (viii) is trivial when the right-hand side of (9.18) is infinite, so we assume the right-hand side is finite. The statement of (viii) has N_1 independent of x and δ, but by allowing repetitions in $x_1, \dots, x_{N_1}$, it suffices to prove (viii) with $N_1 \leq C_1$, where C_1 is a 0-multi-parameter unit-admissible constant (that is, N_1 may depend on x and δ, so long as it is bounded by a 0-multi-parameter unit-admissible constant).

Let $\delta \in (0, 1]$ and $x \in B^n(3/4)$ be such that $B_{(W^1,d^1)}(x, \delta) \subseteq B^n(3/4)$. We will pick $x_1, \dots, x_{N_1} \in B_{(W^1,d^1)}(x, \xi_4\delta)$. Note that we will then have, by the choices of ξ_6 and ξ_4, for $1 \leq j \leq N_1$,

$$\begin{aligned}B_{(W^1,d^1)}\left(x_j, \frac{\xi_4^3\xi_6}{\xi_5^3}\delta\right) &= B_{(W^1,d^1)}(x_j, \xi_5\delta) \subseteq B_{(W^1,d^1)}(x, 2\xi_4\delta) \\ &\subseteq B_{(W^1,d^1)}(x, \delta) \subseteq B^n(3/4),\end{aligned} \tag{9.23}$$

and therefore (9.17) follows. In particular, $x_j \in B^n(3/4)$.

We turn to choosing $x_1, \dots, x_{N_1}$. Pick from the collection

$$\left\{B_{(W^1,d^1)}\left(y, \frac{\xi_5\xi_6}{8}\delta\right) : y \in B_{(W^1,d^1)}(x, \xi_4\delta)\right\}$$

a maximal disjoint subcollection:

$$B_{(W^1,d^1)}\left(x_1, \frac{\xi_5\xi_6}{8}\delta\right), \dots, B_{(W^1,d^1)}\left(x_{N_1}, \frac{\xi_5\xi_6}{8}\right). \tag{9.24}$$

By maximality,

$$B_{(W^1,d^1)}\left(x_1, \frac{\xi_5\xi_6}{2}\delta\right), \dots, B_{(W^1,d^1)}\left(x_{N_1}, \frac{\xi_5\xi_6}{2}\right)$$

is a cover for $B_{(W^1,d^1)}(x, \xi_4\delta)$. Indeed, suppose that there were a $y \in B_{(W^1,d^1)}(x, \xi_4\delta) \setminus \bigcup B_{(W^1,d^1)}(x_j, \frac{\xi_5\xi_6}{2}\delta)$. Then $B_{(W^1,d^1)}(y, \frac{\xi_5\xi_6}{8}\delta)$ would be disjoint from the collection (9.24), contradicting maximality.

We claim $N_1 \leq C_0$, where $C_0 \geq 0$ is a 0-multi-parameter unit-admissible constant. Indeed, (9.23) implies each $x_j \in B^n(3/4)$, and therefore since $2\xi_4 \leq \delta_0$, repeated application of Theorem 3.15.5 (c) gives

$$\begin{aligned}\mathrm{Vol}\left(B_{(W^1,d^1)}\left(x_j,\frac{\xi_5\xi_6}{8}\delta\right)\right) &\gtrsim \mathrm{Vol}(B_{(W^1,d^1)}(x_j,2\xi_4\delta)) \\ &\geq \mathrm{Vol}(B_{(W^1,d^1)}(x,\xi_4\delta)),\end{aligned} \tag{9.25}$$

where in the final estimate we used $B_{(W^1,d^1)}(x,\xi_4\delta) \subseteq B_{(W^1,d^1)}(x_j,2\xi_4\delta)$.

Since $\xi_5\xi_6/8 \leq \xi_4$, we have $B_{(W^1,d^1)}(x_j,\frac{\xi_5\xi_6}{8}\delta) \subseteq B_{(W^1,d^1)}(x,2\xi_4)$. Combining this with the disjointness of the balls in (9.24) and Theorem 3.15.5 (c), we have

$$\begin{aligned}\mathrm{Vol}(B_{(W^1,d^1)}(x,\xi_4\delta)) &\gtrsim \mathrm{Vol}(B_{(W^1,d^1)}(x,2\xi_4\delta)) \\ &\geq \sum_{j=1}^{N_1}\mathrm{Vol}\left(B_{(W^1,d^1)}\left(x_j,\frac{\xi_5\xi_6}{8}\delta\right)\right) \gtrsim N_1\,\mathrm{Vol}(B_{(W^1,d^1)}(x,\xi_4\delta)),\end{aligned}$$

where the last estimate used (9.25). We conclude that $N_1 \lesssim 1$, as desired.

Using (ii), we have

$$0 \leq \sum_{j=1}^{N_1}\phi_{x_j,\xi_6\delta} \leq N_1.$$

Moreover, by (iii) we have $\phi_{x_j,\xi_6\delta} \equiv 1$ on $B_{(W^1,d^1)}(x_j,\xi_5\xi_6\delta)$, and since $B_{(W^1,d^1)}(x_j,\xi_5\xi_6\delta/2)$, $1 \leq j \leq N_1$, is a cover for $B_{(W^1,d^1)}(x,\xi_4\delta)$, we have

$$1 \leq \sum_{j=1}^{N_1}\phi_{x_j,\xi_6\delta} \text{ on } B_{(W^1,d^1)}(x,\xi_4\delta). \tag{9.26}$$

Set

$$g_{x,\delta} := \frac{\phi_{x,\delta}}{\sum_{j=1}^{N_1}\phi_{x_j,\xi_6\delta}}. \tag{9.27}$$

By (9.26), (i), and (ii), we have $0 \leq g_{x,\delta} \leq 1$. Moreover, using (vi) we also have, $\forall L \in \mathbb{N}$,

$$\sum_{|\alpha|\leq L}\|(\delta^{d^1}X^1)^\alpha g_{x,\delta}\|_{C(B^n(1))} \leq C_L,$$

where $C_L \geq 0$ is an L-multi-parameter unit-admissible constant. In particular, we have

$$\sum_{|\alpha|\leq L}\|(X^{1,x,\delta})^\alpha(g_{x,\delta}\circ\Phi_{x,\delta})\|_{C(B^n(1))} \leq \sum_{|\alpha|\leq L}\|(\delta^{d^1}X^1)^\alpha g_{x,\delta}\|_{C(B^n(1))} \leq C_L.$$

By Theorem 3.15.5 (j), this gives

$$\|g_{x,\delta}\circ\Phi_{x,\delta}^{-1}\|_{C^L} \lesssim_L 1, \tag{9.28}$$

where the implicit constant is an L-multi-parameter unit-admissible constant.

We claim $\operatorname{supp}(\phi_{x_j,\xi_6\delta} \circ \Phi_{x,\delta}) \subseteq B^n(1/2)$. Indeed, we have, by (i) and the fact that $x_j \in B_{(W^1,d^1)}(x, \xi_4\delta)$,

$$\begin{aligned}\operatorname{supp}(\phi_{x_j,\xi_6\delta}) &\subseteq B_{(W^1,d^1)}(x_j, \xi_6\xi_4\delta)\\ &\subseteq B_{(W^1,d^1)}(x_j, \xi_4\delta) \subseteq B_{(W^1,d^1)}(x, 2\xi_4\delta) \subseteq B_{(W^1,d^1)}(x, \xi_3\delta).\end{aligned}$$

Therefore,

$$\operatorname{supp}(\phi_{x_j,\xi_6\delta} \circ \Phi_{x,\delta}) \subseteq B_{(W^{1,x,\delta},d^1)}(x, \xi_3) \subseteq B^n(1/2), \tag{9.29}$$

where the final containment uses Theorem 3.15.5 (f).

We have

$$\begin{aligned}\|\phi_{x,\delta}u\|_{\mathscr{C}^{\vec{s}}(\delta^{\lambda\vec{d}}W,\vec{d})} &= \left\| \left(\sum_{j=1}^{N_1} \phi_{x_j,\xi_6\delta} \right) g_{x,\delta}u \right\|_{\mathscr{C}^{\vec{s}}(\delta^{\lambda\vec{d}}W,\vec{d})}\\ \sum_{j=1}^{N_1} &\|g_{x,\delta}\phi_{x_j,\xi_6\delta}u\|_{\mathscr{C}^{\vec{s}}(\delta^{\lambda\vec{d}}W,\vec{d})}\\ &= \sum_{j=1}^{N_1} \|(g_{x,\delta} \circ \Phi_{x,\delta})(\phi_{x_j,\xi_6\delta} \circ \Phi_{x,\delta})u \circ \Phi_{x,\delta}\|_{\mathscr{C}^{\vec{s}}(W^{x,\delta},\vec{d})},\end{aligned} \tag{9.30}$$

where the last line uses Lemma 9.2.7 and the fact that $\phi_{x,\xi_6\delta} \circ \Phi_{x,\delta} \in C_0^\infty(B^n(1/2))$ (which uses (9.29)).

By (9.28), the C^L norm of $g_{x,\delta} \circ \Phi_{x,\delta}$ is bounded by an L-multi-parameter unit-admissible constant. Thus, it follows from Corollary 6.5.10 that

$$\|(g_{x,\delta} \circ \Phi_{x,\delta})w\|_{\mathscr{C}^{\vec{s}}(W^{x,\delta},\vec{d})} \leq C_{\vec{s}}\|w\|_{\mathscr{C}^{\vec{s}}(W^{x,\delta},\vec{d})}, \tag{9.31}$$

$\forall w \in \mathscr{C}^{\vec{s}}(\overline{B^n(7/8)}, (W^{x,\delta}, \vec{d}))$, where $C_{\vec{s}} \geq 0$ is an $\vec{s}$-multi-parameter unit-admissible constant.

At the start of the proof to this part, we reduced to the case where the right-hand side of (9.18) is finite. Therefore, by Lemma 9.2.7 (and using (9.29)), we have

$$\|(\phi_{x_j,\xi_6\delta} \circ \Phi_{x,\delta})u \circ \Phi_{x,\delta}\|_{\mathscr{C}^{\vec{s}}(W^{x,\delta},\vec{d})} < \infty.$$

Combining this with (9.29), Proposition 6.10.1 shows that

$$(\phi_{x_j,\xi_6\delta} \circ \Phi_{x,\delta})u \circ \Phi_{x,\delta} \in \mathscr{C}^{\vec{s}}(\overline{B^n(7/8)}, (W^{x,\delta}, \vec{d})).$$

Thus, (9.31) applies to show that

$$\begin{aligned}&\|(g_{x,\delta} \circ \Phi_{x,\delta})(\phi_{x_j,\xi_6\delta} \circ \Phi_{x,\delta})u \circ \Phi_{x,\delta}\|_{\mathscr{C}^{\vec{s}}(W^{x,\delta},\vec{d})}\\ &\quad \leq C_{\vec{s}}\|(\phi_{x_j,\xi_6\delta} \circ \Phi_{x,\delta})u \circ \Phi_{x,\delta}\|_{\mathscr{C}^{\vec{s}}(W^{x,\delta},\vec{d})}.\end{aligned} \tag{9.32}$$

Plugging (9.32) into (9.30) gives

$$\|\phi_{x,\delta}u\|_{\mathscr{C}^{\vec{s}}(\delta^{\lambda\vec{d}}W,\vec{d})} \leq C_{\vec{s}} \sum_{j=1}^{N_1} \|(\phi_{x_j,\xi_6\delta} \circ \Phi_{x,\delta})u \circ \Phi_{x,\delta}\|_{\mathscr{C}^{\vec{s}}(W^{x,\delta},\vec{d})}. \tag{9.33}$$

Lemma 9.2.7 (also using (9.29)) shows that (9.33) is equivalent to (9.18), which completes the proof of (viii).

(ix): This is a simple reprise of some of the ideas in the proof of (viii). As in the proof of (viii), by allowing repetitions in $x_1, \ldots, x_{N_2}$, it suffices to prove the result with $N_2 \leq C_0$, where $C_0 \geq 0$ is a 0-multi-parameter unit-admissible constant. Also, the result is trivial when the right-hand side of (9.19) is infinite, so we assume that the right-hand side is finite.

By the Picard–Lindelöf theorem and the case $L = 1$ of Theorem 3.15.5 (h), we may take $\delta_0 \in (0, 1]$ to be a 0-multi-parameter unit-admissible constant so small that

$$B_{(W^1,d^1)}(x, \delta_1) \subseteq B^n(3/4), \quad \forall x \in B^n(2/3). \tag{9.34}$$

Pick from the collection

$$\{B_{(W^1,d^1)}(y, \xi_5\delta_1/8) : y \in B^n(2/3)\}$$

a maximal disjoint subcollection

$$B_{(W^1,d^1)}(x_1, \xi_5\delta_1/8), \ldots, B_{(W^1,d^1)}(x_{N_2}, \xi_5\delta_1/8). \tag{9.35}$$

As in the proof of (viii), maximality implies that

$$B_{(W^1,d^1)}(x_1, \xi_5\delta_1/2), \ldots, B_{(W^1,d^1)}(x_{N_2}, \xi_5\delta_1/2)$$

is a cover for $B^n(2/3)$.

Since $\xi_5 \leq \xi_4 \leq \delta_0/2$, we have $\xi_5\delta_1/2 \in (0, \delta_0]$. In particular, it follows from Theorem 3.15.5 (b) that

$$\mathrm{Vol}(B_{(W^1,d^1)}(x_j, \xi_5\delta_1/8)) \approx \Lambda(x_j, \xi_5\delta_1/8) \approx 1,$$

where the final $\approx$ used the formula for Λ and the fact that $\xi_5\delta_2/2 \approx 1$. Thus, using the disjointness of the balls in (9.35) and using (9.34), we have

$$1 \approx \mathrm{Vol}(B^n(3/4)) \geq \sum_{j=1}^{N_2} \mathrm{Vol}(B_{(W^1,d^1)}(x_j, \xi_5\delta_1/8)) \gtrsim N_2.$$

We conclude $N_2 \lesssim 1$, i. e., $N_2 \leq C_0$, where $C_0 \geq 0$ is a 0-multi-parameter unit-admissible constant.

By (ii), $0 \leq \sum_{j=1}^{N_2} \phi_{x_j,\delta_1} \leq N_2$. Since $\phi_{x_j,\delta_1} \equiv 1$ on $B_{(W^1,d^1)}(x_j, \xi_5\delta_1)$ (by (iii)) and $B_{(W^1,d^1)}(x_j, \xi_5\delta_1/2)$ form a cover for $B^n(2/3)$, we have $1 \leq \sum_{j=1}^{N_2} \phi_{x_j,\delta_1}$ on $B^n(2/3)$. Since $\delta_1 \approx 1$, (vii) shows that $\|\phi_{x_j,\delta_1}\|_{C^L(B^n(1))} \lesssim_L 1$, for every $L \in \mathbb{N}$, where the implicit constant is an L-multi-parameter unit-admissible constant.

For $\psi \in C_0^\infty(B^n(2/3))$, we set

$$g_\psi := \frac{\psi}{\sum_{j=1}^{N_2} \phi_{x_j,\delta}}.$$

By the above remarks, we have $g_\psi \in C_0^\infty(B^n(2/3))$ and $\|g_\psi\|_{C^L(B^n(1))} \leq C_L$, where $C_L = C_L(\psi) \geq 0$ is an L-multi-parameter unit-admissible constant.

Thus, we have

$$\|\psi u\|_{\mathscr{C}^{\vec{s}}(W,\vec{d})} = \left\| \sum_{j=1}^{N_2} g_\psi \phi_{x_j,\delta_1} u \right\|_{\mathscr{C}^{\vec{s}}(W,\vec{d})} \leq \sum_{j=1}^{N_2} \|g_\psi \phi_{x_j,\delta_1} u\|_{\mathscr{C}^{\vec{s}}(W,\vec{d})}. \tag{9.36}$$

Since $\operatorname{supp}(\phi_{x_j,\delta_1}) \subseteq B_{(W^1,d^1)}(x_j, \xi_4\delta_1) \subseteq B^n(3/4)$, by (9.34), as we are assuming $\|\phi_{x_j,\delta_1} u\|_{\mathscr{C}^{\vec{s}}(W,\vec{d})} < \infty$, Proposition 6.10.1 implies that $\phi_{x_j,\delta_1} u \in \mathscr{C}^{\vec{s}}(\overline{B^n(3/4)}, (W, \vec{d}))$. Thus, using Corollary 6.5.10, we have

$$\|g_\psi \phi_{x_j,\delta_1} u\|_{\mathscr{C}^{\vec{s}}(W,\vec{d})} \leq C_{\vec{s}} \|\phi_{x_j,\delta_1} u\|_{\mathscr{C}^{\vec{s}}(W,\vec{d})}, \tag{9.37}$$

where $C_{\vec{s}} = C_{\vec{s}}(\psi) \geq 0$ is an L-multi-parameter unit-admissible constant. Using Lemma 9.2.6 and the fact that $\delta_1 \approx 1$, we have

$$\|\phi_{x_j,\delta_1} u\|_{\mathscr{C}^{\vec{s}}(W,\vec{d})} \approx_{\vec{s}} \|\phi_{x_j,\delta_1} u\|_{\mathscr{C}^{\vec{s}}(\delta_1^{\lambda\vec{d}} W,\vec{d})}. \tag{9.38}$$

Combining (9.36), (9.37), and (9.38) yields (9.19) and completes the proof.

(x): Fix $\gamma_1, \gamma_2 \in [\xi_0, 1]$. We will show that

$$\|\phi_{x,\xi_0\delta} u\|_{\mathscr{C}^{\vec{s}}((\gamma_1\delta)^{\lambda\vec{d}} W,\vec{d})} \lesssim \|\phi_{x,\xi_0\delta} u\|_{\mathscr{C}^{\vec{s}}((\gamma_2\delta)^{\lambda\vec{d}} W,\vec{d})}. \tag{9.39}$$

The result will then follow by applying (9.39) with $\gamma_1 = \xi_0$ and $\gamma_2 = 1$ and with $\gamma_1 = 1$ and $\gamma_2 = \xi_0$.

Suppose that

$$\|\phi_{x,\xi_0\delta} u\|_{\mathscr{C}^{\vec{s}}((\gamma_2\delta)^{\lambda\vec{d}} W,\vec{d})} \leq B < \infty. \tag{9.40}$$

We will show that $\|\phi_{x,\xi_0\delta} u\|_{\mathscr{C}^{\vec{s}}((\gamma_1\delta)^{\lambda\vec{d}} W,\vec{d})} \lesssim B$, which will establish (9.39) and complete the proof.

By (i), we have $\operatorname{supp}(\phi_{x,\xi_0\delta}) \subseteq B_{(W^1,d^1)}(x, \xi_0\xi_4\delta) = B_{(W^1,d^1)}(x, \gamma_2\delta\xi_4\xi_0/\gamma_2)$. Therefore,

$$\operatorname{supp}(\phi_{x,\xi_0\delta} \circ \Phi_{x,\gamma_2\delta}) \subseteq B_{(W^{1,x,\gamma_2\delta},d^1)}(0, \xi_0\xi_4/\gamma_2) \subseteq B_{(W^{1,x,\gamma_2\delta},d^1)}(0, \xi_4) \subseteq B^n(7/8).$$

Using this and Lemma 9.2.7, we see that (9.40) is equivalent to

$$\|(\phi_{x,\xi_0\delta} \circ \Phi_{x,\gamma_2\delta})u \circ \Phi_{x,\gamma_2\delta}\|_{\mathscr{C}^{\vec{s}}(W^{1,x,\gamma_2\delta},\vec{d})} \lesssim B. \tag{9.41}$$

By Proposition 6.10.1, this implies that $(\phi_{x,\xi_0\delta}\circ\Phi_{x,\gamma_2\delta})u\circ\Phi_{x,\gamma_2\delta} \in \mathscr{C}^{\vec{s}}(\overline{B^n(7/8)}, (W^{1,x,\gamma_2\delta}, \vec{d}))$.

Proposition 7.2.1 shows that (9.41) implies that there exists a sequence $\hat{v}_j \in C^\infty(B^n(1))$, $j \in \mathbb{N}^\nu$, such that

$$\sum_{j\in\mathbb{N}^\nu} \hat{v}_j = (\phi_{x,\xi_0\delta} \circ \Phi_{x,\gamma_2\delta})u \circ \Phi_{x,\gamma_2\delta}$$

and

$$\|(2^{-j\vec{d}} W^{x,\gamma_2\delta})^\alpha \hat{v}_j\|_{L^\infty(B^n(1))} \lesssim_\alpha 2^{-j\cdot s}B, \quad \forall\alpha.$$

Letting $v_j := \hat{v}_j \circ \Phi^{-1}_{x,\gamma_2\delta}$, we have $v_j \in C_0^\infty(\mathfrak{M})$, $\sum_{j\in\mathbb{N}^\nu} v_j = \phi_{x,\xi_0\delta}u$, and

$$\|((\gamma_2\delta)^{\lambda\vec{d}} 2^{-j\vec{d}} W)^\alpha v_j\|_{L^\infty} \lesssim_\alpha 2^{-j\cdot s}B.$$

Since $\gamma_2 \approx \gamma_1$, this implies that

$$\|((\gamma_1\delta)^{\lambda\vec{d}} 2^{-j\vec{d}} W)^\alpha v_j\|_{L^\infty} \lesssim_\alpha 2^{-j\cdot s}B.$$

Letting $\tilde{v}_j := v_j \circ \Phi_{x,\gamma_1,\delta}$, we see that $\tilde{v}_j \in C^\infty(B^n(1))$ with

$$\|(2^{-j\vec{d}} W^{x,\gamma_1\delta})^\alpha \tilde{v}_j\|_{L^\infty(B^n(1))} \lesssim_\alpha 2^{-j\cdot s}B, \quad \forall\alpha,$$

and $\sum_{j\in\mathbb{N}^\nu} \tilde{v}_j = (\phi_{x,\xi_0\delta} \circ \Phi_{x,\gamma_1\delta})u \circ \Phi_{x,\gamma_1\delta}$. By Proposition 7.2.1, this implies that

$$\|(\phi_{x,\xi_0\delta} \circ \Phi_{x,\gamma_1\delta})u \circ \Phi_{x,\gamma_1\delta}\|_{\mathscr{C}^{\vec{s}}(W^{1,x,\gamma_1\delta},\vec{d})} \lesssim B.$$

As before, Lemma 9.2.7 implies that this is equivalent to

$$\|\phi_{x,\xi_0\delta}u\|_{\mathscr{C}^{\vec{s}}((\gamma_1\delta)^{\lambda\vec{d}} W,\vec{d})} \lesssim B,$$

completing the proof. □

9.2.3 Scaled estimates

As described in Section 9.2.1, we may apply the previous results in this text to the norms $\|\cdot\|_{\mathscr{C}^{\vec{s}}(W^{x,\delta},\vec{d})}$ and all estimates will be multi-parameter unit-admissible constants. This is not true of the norms $\|\cdot\|_{\mathscr{C}^{\vec{s}}(\delta^{\lambda\vec{d}} W,\vec{d})}$; when previous proofs are applied to these norms, the estimates blow up as $\delta \downarrow 0$. However, by using Lemma 9.2.7 we can conclude results about the norms $\|\cdot\|_{\mathscr{C}^{\vec{s}}(\delta^{\lambda\vec{d}}W,\vec{d})}$. In this section, we use this idea to rephrase several earlier results in the text in terms of the norms $\|\cdot\|_{\mathscr{C}^{\vec{s}}(\delta^{\lambda\vec{d}}W,\vec{d})}$. We let $\phi_{x,\delta}$ and ξ_4, ξ_5, and ξ_6 be as in Theorem 9.2.8.

Recall that $\kappa \in \mathbb{N}_+$ is such that d_j^1 divides κ, $\forall 1 \le j \le r_1$, and $n_j = \kappa/d_j^1 \in \mathbb{N}_+$. Let $\mathscr{P}$ be a partial differential operator of the form

$$\mathscr{P} = \sum_{\deg_{d^1}(\alpha)\le\kappa} a_\alpha (W^1)^\alpha,$$

where $a_\alpha \in \mathbb{M}^{\mathbf{D}_1\times\mathbf{D}_2}(\mathbb{R})$ are constant matrices. Fix $B \ge 0$ such that $\max_\alpha |a_\alpha| \le B$.

Proposition 9.2.9. *Suppose $\mathscr{P}$ is maximally subelliptic of degree κ with respect to (W^1, d^1), in the sense that $\exists A \ge 0$ with*

$$\sum_{j=1}^{r_1} \|(W_j^1)^{n_j} f\|_{L^2(B^n(1),h\sigma_{\mathrm{Leb}})} \le A(\|\mathscr{P} f\|_{L^2(B^n(1),h\sigma_{\mathrm{Leb}})} + \|f\|_{L^2(B^n(1),h\sigma_{\mathrm{Leb}})}),$$

$\forall f \in C_0^\infty(B^n(31/32); \mathbb{C}^{\mathbf{D}_2})$. Fix $\vec{s} \in \mathbb{R}^\nu$. For every $x \in B^n(7/8)$ and $\delta \in (0,1]$, we have

$$\begin{aligned} &\|\phi_{x,\xi_6\delta} u\|_{\mathscr{B}^{\vec{s}+\kappa e_1}_{\infty,\infty}(\delta^{\lambda\vec{d}}W,\vec{d})} \\ &\quad \le C(\|\phi_{x,\frac{\xi_4\xi_6}{\xi_5}\delta}\delta^{2\kappa}\mathscr{P}^*\mathscr{P}u\|_{\mathscr{B}^{\vec{s}-\kappa e_1}_{\infty,\infty}(\delta^{\lambda\vec{d}}W,\vec{d})} + \|\phi_{x,\frac{\xi_4\xi_6}{\xi_5}\delta} u\|_{L^\infty}), \end{aligned} \tag{9.42}$$

$\forall u \in C_0^\infty(B^n(1); \mathbb{C}^{\mathbf{D}_2})'$ with $\mathscr{P}^*\mathscr{P}u \in \mathscr{B}^{\vec{s}-\kappa e_1}_{\infty,\infty}(\overline{B^n(7/8)}, (W,\vec{d}))$, where $C = C(A, B, \mathbf{D}_1, \mathbf{D}_2) \ge 0$ is an $(\vec{s},\kappa)$-multi-parameter unit-admissible constant.

Proof. By Theorem 8.1.1 (i) $\Rightarrow$ (iv), $\mathscr{P}^*\mathscr{P}$ is maximally subelliptic of degree 2κ with respect to $(W, \vec{d})$ in the sense that

$$\sum_{j=1}^{r_1} \|(W_j^1)^{2n_j} f\|_{L^2(B^n(1),h\sigma_{\mathrm{Leb}})} \lesssim \|\mathscr{P}^*\mathscr{P} f\|_{L^2(B^n(1),h\sigma_{\mathrm{Leb}})} + \|f\|_{L^2(B^n(1),h\sigma_{\mathrm{Leb}})}, \tag{9.43}$$

$\forall f \in C_0^\infty(B^n(15/16))$.

For $\xi_0 \in \{\xi_6, \xi_4\xi_6/\xi_5\}$, set $\psi_{x,\xi_0\delta} := \phi_{x,\xi_0\delta} \circ \Phi_{x,\delta}$. Theorem 9.2.8 (v) and (vii) shows that $\psi_{x,\xi_0\delta} \in C_0^\infty(B^n(1))$ and for every $L \in \mathbb{N}$, $\|\psi_{x,\xi_0\delta}\|_{C^L} \lesssim_L 1$, where the implicit constant is an L-multi-parameter unit-admissible constant. Theorem 9.2.8 (iv) shows that $\psi_{x,\xi_6\delta} \prec \psi_{x,\frac{\xi_4\xi_6}{\xi_5}\delta}$.

From here, the proof is very similar to the proof of Proposition 8.7.3, using Corollary 8.7.2; we describe the main idcas. Sct $\mathscr{L}^{x,\delta} := \Phi_{x,\delta}^* \delta^{2\kappa}\mathscr{P}^*\mathscr{P}(\Phi_{x,\delta})_*$. By Lemma 9.2.7, (9.42) follows from

$$\|\psi_{x,\xi_6\delta} v\|_{\mathscr{B}^{\vec{s}+\kappa e_1}_{\infty,\infty}(W^{x,\delta},\vec{d})} \lesssim \|\psi_{x,\frac{\xi_4\xi_6}{\xi_5}\delta}\mathscr{L}^{x,\delta} v\|_{\mathscr{B}^{\vec{s}-\kappa e_1}_{\infty,\infty}(W^{x,\delta},\vec{d})} + \|\psi_{x,\frac{\xi_4\xi_6}{\xi_5}\delta} v\|_{L^\infty}, \tag{9.44}$$

for all $v \in C_0^\infty(B^n(1))'$ with $\psi_{x,\frac{\xi_4\xi_6}{\xi_5}\delta} v \in L^\infty(B^n(1))$.

To prove (9.44), we will apply Corollary 8.7.2. Just as in Section 3.3.1 and the proof of Proposition 8.7.3, we have

$$\mathscr{L}^{x,\delta} = \sum_{\deg_{\vec{d}^1}(\alpha)\leq 2\kappa} b_\alpha^{x,\delta} (W^{1,x,\delta})^\alpha,$$

where $b_\alpha^{x,\delta} \in C^\infty(B^n(1); \mathbb{M}^{\mathbf{D}_2\times\mathbf{D}_2})$ and for every $L \in \mathbb{N}$, $\|b_\alpha^{x,\delta}\|_{C^L} \lesssim_L 1$. Also, following the proof in Section 3.3.1 (or, equivalently, the proof of Proposition 8.7.3), (9.43) implies that

$$\sum_{j=1}^{r_1} \|(W_j^{1,x,\delta})^{2n_j} f\|_{L^2(B^n(1),h_{x,\delta}\sigma_{\mathrm{Leb}})} \lesssim \|\mathscr{L}^{x,\delta} f\|_{L^2(B^n(1),h_{x,\delta}\sigma_{\mathrm{Leb}})} + \|f\|_{L^2(B^n(1),h_{x,\delta}\sigma_{\mathrm{Leb}})},$$

$\forall f \in C_0^\infty(B^n(1))$, where $h_{x,\delta}$ is as in Theorem 3.15.5.

Thus, we have verified that all the hypotheses of Corollary 8.7.2 hold (when applied to $\mathscr{L}^{x,\delta}$) and that result (with $p = \infty$) implies (9.44) and completes the proof. □

Proposition 9.2.10. *Let $E_\infty \in C^\infty(B^n(1) \times B^n(1))$ and fix $\xi_0 \in (0,1]$. Then, $\forall \vec{s} \in \mathbb{R}^\nu$, there exists $L \in \mathbb{N}$, $C_{\vec{s}} \geq 0$, such that $\forall x \in B^n(7/8)$, $\delta \in (0,1]$,*

$$\|\phi_{x,\xi_0\delta} E_\infty u\|_{\mathscr{B}^{\vec{s}}_{\infty,\infty}(\delta^{\lambda\vec{d}} W,\vec{d})} \leq C_{\vec{s}} \|E_\infty\|_{C^L(B^n(1)\times B^n(1))} \|u\|_{L^\infty}, \tag{9.45}$$

$\forall u \in L^\infty(B^n(1))$, where $L = L(\vec{s}) \in \mathbb{N}$ is a 0-multi-parameter unit-admissible constant and $C_{\vec{s}} = C_{\vec{s}}(\xi_0) \geq 0$ is an $\vec{s}$-multi-parameter unit-admissible constant.

Proof. Since the left-hand side of (9.45) is increasing in $\vec{s}$, it suffices to prove the result for $\vec{s} \in (0,\infty)^\nu$, when $\|\cdot\|_{\mathscr{B}^{\vec{s}}_{\infty,\infty}(\delta^{\lambda\vec{d}} W,\vec{d})} = \|\cdot\|_{\mathscr{C}^{\vec{s}}(\delta^{\lambda\vec{d}} W,\vec{d})}$.

It follows from Theorem 9.2.8 (vi) (and using $(W,\vec{d}) \subseteq (X,\vec{d})$) that, $\forall L \in \mathbb{N}$,

$$\sum_{\deg_{\vec{d}}(\alpha)\leq L} \|(\delta^{\lambda\vec{d}} W)^\alpha \phi_{x,\xi_0\delta} E_\infty u\|_{L^\infty(B^n(1))} \lesssim_L \|E_\infty\|_{C^L} \|u\|_{L^\infty}.$$

Thus, $\forall L \in \mathbb{N}$,

$$\sum_{\deg_{\vec{d}}(\alpha)\leq L} \|(W^{x,\delta})^\alpha \Phi_{x,\delta}^*(\phi_{x,\xi_0\delta} E_\infty u)\|_{L^\infty(B^n(1))} \lesssim_L \|E_\infty\|_{C^L} \|u\|_{L^\infty}.$$

By Proposition 7.2.1 (see (7.4)), it follows that for some multi-parameter unit-admissible constant $L = L(\vec{s}) \in \mathbb{N}$, we have

$$\|\Phi_{x,\delta}^*(\phi_{x,\xi_0\delta} E_\infty u)\|_{\mathscr{C}^{\vec{s}}(W^{x,\delta},\vec{d})} \lesssim \|E_\infty\|_{C^L} \|u\|_{L^\infty}. \tag{9.46}$$

Lemma 9.2.7 shows that (9.46) is equivalent to (9.45), completing the proof. □

Proposition 9.2.11. *Fix* $\xi_0 \in (0,1]$. *Then,* $\forall \vec{\mathbf{s}} \in \mathbb{R}^\nu$, $\exists C_{\vec{\mathbf{s}}} \geq 0$ *such that* $\forall x \in B^n(7/8)$, $\delta \in (0,1]$,

$$\|\phi_{x,\xi_0\frac{\xi_5}{\xi_4}\delta}\delta^\kappa \mathscr{P} u\|_{\mathscr{B}^{\vec{\mathbf{s}}}_{\infty,\infty}(\delta^{\lambda\vec{d}}W,\vec{d})} \leq C_{\vec{\mathbf{s}}} B \|\phi_{x,\xi_0\delta} u\|_{\mathscr{B}^{\vec{\mathbf{s}}+\kappa e_1}_{\infty,\infty}(\delta^{\lambda\vec{d}}W,\vec{d})},$$

$\forall u \in C_0^\infty(B^n(1);\mathbb{C}^{\mathbf{D}_2})'$ *with* $\phi_{x,\xi_0\delta} u \in \mathscr{B}^{\vec{\mathbf{s}}+\kappa e_1}_{\infty,\infty}(W,\vec{d})$. *Similarly,*

$$\|\phi_{x,\xi_0\frac{\xi_5}{\xi_4}\delta}\delta^\kappa \mathscr{P}^* v\|_{\mathscr{B}^{\vec{\mathbf{s}}}_{\infty,\infty}(\delta^{\lambda\vec{d}}W,\vec{d})} \leq C_{\vec{\mathbf{s}}} B \|\phi_{x,\xi_0\delta} v\|_{\mathscr{B}^{\vec{\mathbf{s}}+\kappa e_1}_{\infty,\infty}(\delta^{\lambda\vec{d}}W,\vec{d})},$$

$\forall u \in C_0^\infty(B^n(1);\mathbb{C}^{\mathbf{D}_1})'$ *with* $\phi_{x,\xi_0\delta} u \in \mathscr{B}^{\vec{\mathbf{s}}+\kappa e_1}_{\infty,\infty}(W,\vec{d})$. *Here,* $C_{\vec{\mathbf{s}}} = C_{\vec{\mathbf{s}}}(\xi_0,\mathbf{D}_1,\mathbf{D}_2) \geq 0$ *is an* $(\vec{\mathbf{s}},\kappa)$*-multi-parameter unit-admissible constant.*

Proof. Without loss of generality, we take $B = 1$ and it suffices to consider only the case $\mathbf{D}_1 = \mathbf{D}_2 = 1$.

Both $\mathscr{P}$ and $\mathscr{P}^*$ are of the form

$$\widetilde{\mathscr{P}} = \sum_{\deg_{\vec{d}^1}(\alpha)\leq\kappa} b_\alpha (W^1)^\alpha,$$

where $b_\alpha \in C^\infty(B^n(1))$ and for every L, $\|b_\alpha\|_{C^L(B^n(1))} \lesssim_L 1$, where the implicit constant is an L-multi-parameter unit-admissible constant.

For $x \in B^n(7/8)$, $\delta \in (0,1]$, set

$$\widetilde{\mathscr{P}}^{x,\delta} := \Phi_{x,\delta}^* \delta^\kappa \widetilde{\mathscr{P}} (\Phi_{x,\delta})_* = \sum_{\deg_{\vec{d}^1}(\alpha)\leq\kappa} b_\alpha^{x,\delta}(W^{1,x,\delta}),$$

where $b_\alpha^{x,\delta} = \delta^{\kappa-\deg_{\vec{d}^1}(\alpha)} b_\alpha \circ \Phi_{x,\delta}$. Using Theorem 3.15.5 (j), we have, $\forall L \in \mathbb{N}$,

$$\begin{aligned}\|b_\alpha^{x,\delta}\|_{C^L(B^n(1))} &\approx_L \sum_{|\beta|\leq L} \|(X^{1,x,\delta})^\alpha b_\alpha^{x,\delta}\|_{C(B^n(1))} \\ &= \sum_{|\beta|\leq L} \delta^{\beta+\kappa-\deg_{\vec{d}^1}(\alpha)} \|(X^1)^\alpha b_\alpha\|_{C(\Phi_{x,\delta}(B^n(1)))} \lesssim_L 1.\end{aligned}$$

We also have, by Theorem 9.2.8 (v) and (vii), $\psi_{x,\xi_0\frac{\xi_5}{\xi_4}\delta} := \phi_{x,\xi_0\frac{\xi_5}{\xi_4}\delta} \circ \Phi_{x,\delta} \in C_0^\infty(B^n(1))$, and for every L, $\|\psi_{x,\xi_0\frac{\xi_5}{\xi_4}\delta}\|_{C^L} \lesssim_L 1$.

Combining the above and using the fact that $\psi_{x,\xi_0\frac{\xi_5}{\xi_4}\delta} \prec \psi_{x,\xi_0}$ (by Theorem 9.2.8 (iv)), it follows from Proposition 6.5.9 that

$$\begin{aligned}\|\psi_{x,\xi_0\frac{\xi_5}{\xi_4}\delta}\widetilde{\mathscr{P}}^{x,\delta} v\|_{\mathscr{B}^{\vec{\mathbf{s}}}_{\infty,\infty}(W^{x,\delta},\vec{d})} &= \|\psi_{x,\xi_0\frac{\xi_5}{\xi_4}\delta}\widetilde{\mathscr{P}}^{x,\delta}\psi_{x,\xi_0\delta} v\|_{\mathscr{B}^{\vec{\mathbf{s}}}_{\infty,\infty}(W^{x,\delta},\vec{d})} \\ &\lesssim_{\vec{\mathbf{s}},\kappa} \|\psi_{x,\xi_0\delta} v\|_{\mathscr{B}^{\vec{\mathbf{s}}+\kappa e_1}_{\infty,\infty}(W^{x,\delta},\vec{d})},\end{aligned} \tag{9.47}$$

$\forall v \in C_0^\infty(B^n(1))'$ with $\psi_{x,\xi_0\delta} v \in \mathscr{B}^{\vec{\mathbf{s}}+\kappa e_1}_{\infty,\infty}(W^{x,\delta})$.

We have, using Lemma 9.2.7 and (9.47),

$$\|\phi_{x,\xi_0\frac{\xi_5}{\xi_4}\delta}\delta^{\kappa}\widetilde{\mathscr{P}}u\|_{\mathscr{B}^{\vec{s}}_{\infty,\infty}(\delta^{\lambda\vec{d}}W,\vec{d})} = \|\phi_{x,\xi_0\frac{\xi_5}{\xi_4}\delta}\widetilde{\mathscr{P}}^{x,\delta}\Phi^*_{x,\delta}u\|_{\mathscr{B}^{\vec{s}}_{\infty,\infty}(W^{x,\delta},\vec{d})}$$
$$\lesssim_{\vec{s},\kappa} \|\psi_{x,\xi_0\delta}\Phi^*_{x,\delta}u\|_{\mathscr{B}^{\vec{s}+\kappa e_1}_{\infty,\infty}(W^{x,\delta},\vec{d})} = \|\phi_{x,\xi_0\delta}u\|_{\mathscr{B}^{\vec{s}+\kappa e_1}_{\infty,\infty}(\delta^{\lambda\vec{d}}W,\vec{d})},$$

completing the proof. □

Lemma 9.2.12. *Let $\vec{\mathbf{s}} \in (0,\infty)^\nu$ and $\mathbf{t} > 0$. Then, for $u \in \mathscr{C}^{\vec{\mathbf{s}}}(\overline{B^n(7/8)},(W,\vec{d}))$ and $\psi \in C_0^\infty(B^n(7/8))$,*

$$\|\psi\Phi^*_{x,\delta}u\|_{\mathscr{C}^{\vec{s}}(W^{x,\delta},\vec{d})} \leq C_{\vec{s}}\|\psi\|_{C^L}\|u\|_{\mathscr{C}^{\vec{s}}(W,\vec{d})}, \tag{9.48}$$

where $L \in \mathbb{N}$ and $C_{\vec{s}} \geq 0$ are $\vec{\mathbf{s}}$-multi-parameter unit-admissible constants. Similarly, for $F \in \mathscr{C}^{\vec{\mathbf{s}},\mathbf{t}}((W,\vec{d}) \boxtimes \nabla_{\mathbb{R}^N})$ and $\psi \in C_0^\infty(B^n(7/8))$,

$$\|\psi(\cdot)F(\Phi_{x,\delta}(\cdot),\zeta)\|_{\mathscr{C}^{\vec{s},\mathbf{t}}((W^{x,\delta},\vec{d})\boxtimes\nabla_{\mathbb{R}^N})} \leq C_{\vec{s},\mathbf{t}}\|\psi\|_{C^L}\|F\|_{\mathscr{C}^{\vec{s},\mathbf{t}}((W,\vec{d})\boxtimes\nabla_{\mathbb{R}^N})}, \tag{9.49}$$

where $L \in \mathbb{N}$ and $C_{\vec{\mathbf{s}},\mathbf{t}} = C_{\vec{\mathbf{s}},\mathbf{t}}(N) \geq 0$ are $(\vec{\mathbf{s}},\mathbf{t})$-multi-parameter unit-admissible constants.

Proof. Let $u \in \mathscr{C}^{\vec{s}}(\overline{B^n(7/8)},(W,\vec{d}))$. By Proposition 7.2.1 we may write $u = \sum_{j\in\mathbb{N}^\nu} u_j$, where $u_j \in C_0^\infty(B^n(1))$ and for every α,

$$\sup_{j\in\mathbb{N}^\nu} 2^{j\cdot s}\|(2^{-j\vec{d}}W)^\alpha u_j\|_{L^\infty(B^n(1))} \lesssim_{\vec{s},\alpha} \|u\|_{\mathscr{C}^s(W,\vec{d})}.$$

Thus, we have

$$\begin{aligned}&\sup_{j\in\mathbb{N}^\nu} 2^{j\cdot s}\|(2^{-j\vec{d}}W^{x,\delta})^\alpha\Phi^*_{x,\delta}u_j\|_{L^\infty(B^n(1))}\\&\quad\leq \sup_{j\in\mathbb{N}^\nu} 2^{j\cdot s}\delta^{\deg_{d^1}(\alpha)}\|(2^{-j\vec{d}}W)^\alpha u_j\|_{L^\infty(B^n(1))} \lesssim_{\vec{s},\alpha} \|u\|_{\mathscr{C}^s(W,\vec{d})}.\end{aligned} \tag{9.50}$$

It follows that for every $L \in \mathbb{N}$,

$$\sum_{|\alpha|\leq L}\sup_{j\in\mathbb{N}^\nu} 2^{j\cdot s}\|(2^{-j\vec{d}}W^{x,\delta})^\alpha\psi\Phi^*_{x,\delta}u_j\|_{L^\infty(B^n(1))} \lesssim_{\vec{s},L} \|u\|_{\mathscr{C}^s(W,\vec{d})}.$$

Since $\psi\Phi^*_{x,\delta}u = \sum_{j\in\mathbb{N}^\nu}\psi\Phi^*_{x,\delta}u_j$, (9.48) follows from Proposition 7.2.1 (in particular, (7.4)).

The same proof gives (9.49), by replacing Proposition 7.2.1 with Proposition 7.5.9. □

Corollary 9.2.13. *Fix $\xi_0 \in (0,1]$ and $s \in (0,\infty)^\nu$. Then, $\forall x \in B^n(7/8)$, $\delta \in (0,1]$, $u \in \mathscr{C}^{\vec{\mathbf{s}}}(\overline{B^n(7/8)},(W,\vec{d}))$,*

$$\|\phi_{x,\xi_0\delta}u\|_{\mathscr{C}^{\vec{s}}(\delta^{\lambda\vec{d}}W,\vec{d})} \leq C_{\vec{s}}\|u\|_{\mathscr{C}^{\vec{s}}(W,\vec{d})}, \tag{9.51}$$

where $C_{\vec{\mathbf{s}}} = C_{\vec{\mathbf{s}}}(\xi_0) \geq 0$ is an $\vec{\mathbf{s}}$-multi-parameter unit-admissible constant.

Proof. By Lemma 9.2.7, (9.51) is equivalent to

$$\|(\phi_{x,\xi_0\delta}\circ\Phi_{x,\delta})\Phi_{x,\delta}^{*}u\|_{\mathscr{C}^{\vec{s}}(W^{x,\delta},\vec{d})}\leq C_{\vec{s}}\|u\|_{\mathscr{C}^{\vec{s}}(W,\vec{d})}. \tag{9.52}$$

Formula (9.52) follows from (9.48) with $\psi=\phi_{x,\xi_0\delta}\circ\Phi_{x,\delta}$, where we use Theorem 9.2.8 (v) and (vii) to see $\phi_{x,\xi_0\delta}\circ\Phi_{x,\delta}\in C_0^{\infty}(B^n(1/2))$ and $\|\phi_{x,\xi_0\delta}\circ\Phi_{x,\delta}\|_{C^L}\lesssim_L 1$, $\forall L\in\mathbb{N}$. □

Proposition 9.2.14. *Fix $\vec{s}\in(0,\infty)^{\nu}$, $\mathbf{t}>0$, $\xi_0\in(0,1]$, $N\in\mathbb{N}_+$. Let $F(x,\zeta)\in\mathscr{C}^{\vec{s},\mathbf{t}}(\overline{B^N(7/8)\times\mathbb{R}^N},(W,\vec{d})\boxtimes\nabla_{\mathbb{R}^N})$ and $u\in\mathscr{C}^{\vec{s}}(\overline{B^n(7/8)},(W,\vec{d});\mathbb{R}^N)$. Then, $\forall x\in B^n(7/8)$ and $\delta\in(0,1]$, the following hold:*

(i) *If $\mathbf{t}>\lfloor|\vec{s}|_1\rfloor+1+\nu$, then*

$$\begin{aligned}&\|\phi_{x,\xi_0\delta}(\cdot)F(\cdot,\phi_{x,\xi_0\delta}(\cdot)u(\cdot))\|_{\mathscr{C}^{\vec{s}}(\delta^{\lambda\vec{d}}W,\vec{d})}\\&\quad\leq C_{\vec{s},\mathbf{t}}\|F\|_{\mathscr{C}^{\vec{s},\mathbf{t}}((W,\vec{d})\boxtimes\nabla_{\mathbb{R}^N})}(1+\|\phi_{x,\xi_0\delta}u\|_{\mathscr{C}^{\vec{s}}(\delta^{\lambda\vec{d}}W,\vec{d})})(1+\|\phi_{x,\xi_0\delta}u\|_{L^\infty})^{\lfloor|s|_1\rfloor+\nu},\end{aligned}$$

where $C_{\vec{s},\mathbf{t}}=C_{\vec{s},\mathbf{t}}(\xi_0,N)\geq 0$ is an $(\vec{s},\mathbf{t})$-multi-parameter unit-admissible constant.

(ii) *Fix $L\in\mathbb{N}_+$ and suppose that $\partial_\zeta^{\beta}F(x,0)\equiv 0$, $\forall|\beta|<L$, and $\mathbf{t}>\lfloor|\vec{s}|_1\rfloor+1+\nu+L$. Then*

$$\begin{aligned}&\|\phi_{x,\xi_0\delta}(\cdot)F(\cdot,\phi_{x,\xi_0\delta}(\cdot)u(\cdot))\|_{\mathscr{C}^{\vec{s}}(\delta^{\lambda\vec{d}}W,\vec{d})}\\&\quad\leq C_{\vec{s},\mathbf{t},L}\|F\|_{\mathscr{C}^{\vec{s},\mathbf{t}}((W,\vec{d})\boxtimes\nabla_{\mathbb{R}^N})}(1+\|\phi_{x,\xi_0\delta}u\|_{L^\infty})^{\lfloor|s|_1\rfloor+\nu+1}\|\phi_{x,\xi_0\delta}u\|_{L^\infty}^{L-1}\|\phi_{x,\xi_0\delta}u\|_{\mathscr{C}^{\vec{s}}(\delta^{\lambda\vec{d}}W,\vec{d})},\end{aligned}$$

where $C_{\vec{s},\mathbf{t},L}=C_{\vec{s},\mathbf{t},L}(\xi_0,N)\geq 0$ is an $(\vec{s},\mathbf{t},L)$-multi-parameter unit-admissible constant.

Proof. The proofs of (i) and (ii) are nearly identical: we pull back the setting via $\Phi_{x,\delta}$ and then apply the corresponding part of Theorem 7.5.2.

Indeed, for (i) we apply Lemma 9.2.7 and Theorem 7.5.2 (ii) to see that, with $\psi_{x,\xi_0\delta}:=\phi_{x,\xi_0\delta}\circ\Phi_{x,\delta}$,

$$\begin{aligned}&\|\phi_{x,\xi_0\delta}(\cdot)F(\cdot,\phi_{x,\xi_0\delta}(\cdot)u(\cdot))\|_{\mathscr{C}^{\vec{s}}(\delta^{\lambda\vec{d}}W,\vec{d})}\\&\quad=\|\psi_{x,\xi_0\delta}(\cdot)F(\Phi_{x,\delta}(\cdot),\psi_{x,\xi_0\delta}(\cdot)\Phi_{x,\delta}^{*}u(\cdot))\|_{\mathscr{C}^{\vec{s}}(W^{x,\delta},\vec{d})}\\&\quad\lesssim_{\vec{s},\mathbf{t}}\|\psi_{x,\xi_0\delta}(\cdot)F(\Phi_{x,\delta}(\cdot),\zeta)\|_{\mathscr{C}^{\vec{s},\mathbf{t}}((W^{x,\delta},\vec{d})\boxtimes\nabla_{\mathbb{R}^N})}(1+\|\psi_{x,\xi_0\delta}\Phi_{x,\delta}^{*}u\|_{\mathscr{C}^{\vec{s}}(W^{x,\delta},\vec{d})})\\&\qquad\times(1+\|\psi_{x,\xi_0\delta}\Phi_{x,\delta}^{*}u\|_{L^\infty})^{\lfloor|s|_1\rfloor+\nu}.\end{aligned}$$

Note that

$$\|\psi_{x,\xi_0\delta}\Phi_{x,\delta}^{*}u\|_{L^\infty}=\|\phi_{x,\xi_0\delta}u\|_{L^\infty},$$

and by Lemma 9.2.7 we have

$$\|\psi_{x,\xi_0\delta}\Phi_{x,\delta}^{*}u\|_{\mathscr{C}^{\vec{s}}(W^{x,\delta},\vec{d})}=\|\phi_{x,\xi_0\delta}u\|_{\mathscr{C}^{\vec{s}}(\delta^{\lambda\vec{d}}W,\vec{d})}.$$

Thus, to complete the proof of (i) it suffices to show that

$$\|\psi_{x,\xi_0\delta}(\cdot)F(\Phi_{x,\delta}(\cdot),\zeta)\|_{\mathscr{C}^{\vec{\mathbf{s}},\mathbf{t}}((W^{x,\delta},\vec{d})\boxtimes\nabla_{\mathbb{R}^N})} \lesssim_{\vec{\mathbf{s}},\mathbf{t}} \|F\|_{\mathscr{C}^{\vec{\mathbf{s}},\mathbf{t}}((W,\vec{d})\boxtimes\nabla_{\mathbb{R}^N})}. \tag{9.53}$$

The estimate (9.53) follows from (9.49) with $\psi = \psi_{x,\xi_0\delta} = \phi_{x,\xi_0\delta}\circ\Phi_{x,\delta}$, where we use Theorem 9.2.8 (v) and (vii) to see that $\phi_{x,\xi_0\delta}\circ\Phi_{x,\delta}\in C_0^\infty(B^n(1/2))$ and $\|\phi_{x,\xi_0\delta}\circ\Phi_{x,\delta}\|_{C^L}\lesssim_L 1$, $\forall L\in\mathbb{N}$.

The proof for (ii) is the same, but with Theorem 7.5.2 (ii) replaced by Theorem 7.5.2 (vii). □

9.2.4 Step I: Perturbation of a linear operator

Let

$$\mathscr{P} = \sum_{\deg_{\vec{d}^1}(\alpha)\le\kappa} a_\alpha (W^1)^\alpha, \quad a_\alpha\in\mathbb{M}^{\mathbf{D}_1\times\mathbf{D}_2}(\mathbb{R})$$

be maximally subelliptic of degree κ with respect to $(W^1,\vec{d}^1)$ in the sense that

$$\sum_{j=1}^{r_1}\|(W_j^1)^{n_j}u\|_{L^2(B^n(1),h\sigma_{\mathrm{Leb}})} \le A(\|\mathscr{P}u\|_{L^2(B^n(1),h\sigma_{\mathrm{Leb}})} + \|u\|_{L^2(B^n(1),h\sigma_{\mathrm{Leb}})}),$$

$\forall u\in C_0^\infty(B^n(31/32);\mathbb{C}^{\mathbf{D}_2})$. Here $a_\alpha\in\mathbb{M}^{\mathbf{D}_1\times\mathbf{D}_2}(\mathbb{R})$ are constant matrices. Fix $B\ge 0$ with $|a_\alpha|\le B$, $\forall\alpha$.

Let $\epsilon_1\in(0,1]$ be a small number to be chosen later and fix $C_1\ge 0$. Suppose that $R_1\in\mathscr{C}^{\vec{\mathbf{s}},\lfloor|\vec{\mathbf{s}}|_1\rfloor+2+\nu+\sigma}(\overline{B^n(7/8)}\times\mathbb{R}^N,(W,\vec{d})\boxtimes\nabla_{\mathbb{R}^N};\mathbb{R}^{\mathbf{D}_1})$ and $R_2\in\mathscr{C}^{\vec{\mathbf{s}},\lfloor|\vec{\mathbf{s}}|_1\rfloor+3+\nu+\sigma}(\overline{B^n(7/8)}\times\mathbb{R}^N,(W,\vec{d})\boxtimes\nabla_{\mathbb{R}^N};\mathbb{R}^{\mathbf{D}_1})$ with

$$\|R_1\|_{\mathscr{C}^{\vec{\mathbf{s}},\lfloor|\vec{\mathbf{s}}|_1\rfloor+2+\nu+\sigma}((W,\vec{d})\boxtimes\nabla_{\mathbb{R}^N})}\le\epsilon_1,\quad \|R_2\|_{\mathscr{C}^{\vec{\mathbf{s}},\lfloor|\vec{\mathbf{s}}|_1\rfloor+3+\nu+\sigma}((W,\vec{d})\boxtimes\nabla_{\mathbb{R}^N})}\le C_1.$$

We suppose that

$$\partial_\zeta^\beta R_2(x,0)\equiv 0,\quad \forall|\beta|\le 1.$$

Let $u\in\mathscr{C}^{\vec{\mathbf{r}}+\kappa e_1}(\overline{B^n(7/8)},(W,\vec{d});\mathbb{R}^{\mathbf{D}_2})$ with

$$\|u\|_{\mathscr{C}^{\vec{\mathbf{r}}+\kappa e_1}(W,\vec{d})}\le\epsilon_1$$

such that u satisfies the following nonlinear PDE:

$$\mathscr{P}u(x) = R_1(x,\mathfrak{W}_1^\kappa u(x)) + R_2(x,\mathfrak{W}_1^\kappa u(x)),\quad \forall x\in B^n(13/16).$$

Proposition 9.2.15. *There exists $\epsilon_1 \in (0,1]$ sufficiently small such that if the above holds, then $\forall \psi_0 \in C_0^\infty(B^n(2/3))$, we have*

$$\psi_0 u \in \mathscr{C}^{\vec{s}+\kappa\epsilon_1}(\overline{B^n(7/8)}, (W, \vec{d}); \mathbb{R}^{\mathbf{D}_2})$$

and

$$\|\psi_0 u\|_{\mathscr{C}^{\vec{s}+\kappa\epsilon_1}(W,\vec{d})} \lesssim C_2.$$

Here, $\epsilon_1 = \epsilon_1(C_1, A, B, \mathbf{D}_1, \mathbf{D}_2) \in (0,1]$ and $C_2 = C_2(C_1, A, B, \mathbf{D}_1, \mathbf{D}_2, \psi_0) \geq 0$ are $(\kappa, \vec{\mathbf{s}}, \vec{\mathbf{r}}, \sigma)$-multi-parameter unit-admissible constants.

We use Remark 9.2.4 freely throughout this section. That is, we will apply results from previous chapters to the setting of this section; all the estimates from those results are in terms of appropriate multi-parameter unit-admissible constants, as can be seen by keeping track of the estimates in the proofs. Throughout the proof, we shrink $\epsilon_1 > 0$ a finite number of times (while still keeping it an appropriate admissible constant).

Without loss of generality, we assume that $\vec{\mathbf{s}} \geq \vec{\mathbf{r}}$, as we may replace $\vec{\mathbf{r}}$ with $\vec{\mathbf{r}} \wedge \vec{\mathbf{s}}$. Fix $\psi_1, \psi_2, \psi_3 \in C_0^\infty(B^n(13/16))$ with $\psi_3 \prec \psi_2 \prec \psi_1$ and $\psi_3 \equiv 1$ on a neighborhood of $\overline{B^n(3/4)}$.

Since $\mathscr{P}$ is maximally subelliptic, Theorem 8.1.1 (i) $\Rightarrow$ (viii) implies that there exists $T \in \mathscr{A}_{\mathrm{loc}}^{-2\kappa}((W^1, \vec{d}^1); \mathbb{C}^{\mathbf{D}_2}, \mathbb{C}^{\mathbf{D}_2})$ (where we are using the ambient manifold $B^n(31/32)$) such that

$$T\mathscr{P}^*\mathscr{P}, \mathscr{P}^*\mathscr{P}T \equiv I \mod C_{\mathrm{loc}}^\infty(B^n(31/32) \times B^n(31/32); \mathbb{M}^{\mathbf{D}_2\times\mathbf{D}_2}(\mathbb{C})). \tag{9.54}$$

Proposition 5.8.4 shows that $T \in \widetilde{\mathscr{A}}_{\mathrm{loc}}^{-2\kappa}((W^1, \vec{d}^1); \mathbb{C}^{\mathbf{D}_2}, \mathbb{C}^{\mathbf{D}_2})$. Corollary 5.11.18 shows that $T \in \widetilde{\mathscr{A}}_{\mathrm{loc}}^{-2\kappa\epsilon_1}((W, \vec{d}); \mathbb{C}^{\mathbf{D}_2}, \mathbb{C}^{\mathbf{D}_2})$.

We define

$$\begin{aligned} V_\infty &:= \psi_2 T\mathscr{P}^*\psi_1(R_1(x, \mathfrak{W}_1^\kappa u(x)) + R_2(x, \mathfrak{W}_1^\kappa u(x))) \\ &= \psi_2 T\mathscr{P}^*\psi_1\mathscr{P}u \\ &= \psi_2 T\mathscr{P}^*\mathscr{P}u + \psi_2 T\mathscr{P}^*(1-\psi_1)\mathscr{P}u = \psi_2 u + \psi_2 e_\infty, \end{aligned} \tag{9.55}$$

where by the pseudo-locality of T (see the growth condition in Definition 5.2.4) and (9.54), we have $e_\infty \in C_0^\infty(B^n(7/8))$. Moreover (using $\mathrm{supp}(u) \subseteq \overline{B^n(7/8)}$, Proposition 5.8.11, and Theorem 6.3.10),

$$\|e_\infty\|_{\mathscr{C}^{\vec{r}+\kappa\epsilon_1}(W,\vec{d})}, \|e_\infty\|_{\mathscr{C}^{\vec{s}+\kappa\epsilon_1}(W,\vec{d})} \lesssim \|u\|_{\mathscr{C}^{\vec{r}+\kappa\epsilon_1}(W,\vec{d})} \lesssim \epsilon_1.$$

Since $T\mathscr{P}^*\psi_1\mathscr{P} \in \widetilde{\mathscr{A}}_{\mathrm{loc}}^{0_\nu}((W, \vec{d}); \mathbb{C}^{\mathbf{D}_2}, \mathbb{C}^{\mathbf{D}_2})$ (see Corollary 5.8.10) and using the fact that $\mathrm{supp}(u) \subseteq \overline{B^n(7/8)}$, Theorem 6.3.10 shows that

$$\|V_\infty\|_{\mathscr{C}^{\vec{r}+\kappa\epsilon_1}(W,\vec{d})} = \|\psi_2 T\mathscr{P}^*\mathscr{P}u\|_{\mathscr{C}^{\vec{r}+\kappa\epsilon_1}(W,\vec{d})} \lesssim \|u\|_{\mathscr{C}^{\vec{r}+\kappa\epsilon_1}(W,\vec{d})} \lesssim \epsilon_1,$$

so there exists $C_3 \approx 1$ with

$$\|V_\infty\|_{\mathscr{C}^{\vec{\mathrm{r}}+\kappa e_1}(W,\vec{d})} \leq C_3\epsilon_1. \tag{9.56}$$

Set $H := u - V_\infty$. We have $H \in \mathscr{C}^{\vec{\mathrm{r}}+\kappa e_1}(\overline{B^n(7/8)}, (W, \vec{d}))$ and

$$\|H\|_{\mathscr{C}^{\vec{\mathrm{r}}+\kappa e_1}(W,\vec{d})} \leq \|u\|_{\mathscr{C}^{\vec{\mathrm{r}}+\kappa e_1}(W,\vec{d})} + \|V_\infty\|_{\mathscr{C}^{\vec{\mathrm{r}}+\kappa e_1}(W,\vec{d})} \lesssim \epsilon_1. \tag{9.57}$$

Also, since $\psi_3 H = \psi_3 e_\infty$, using Corollary 6.5.10, we have

$$\|\psi_3 H\|_{\mathscr{C}^{\vec{\mathrm{s}}+\kappa e_1}(W,\vec{d})} \lesssim \|e_\infty\|_{\mathscr{C}^{\vec{\mathrm{s}}+\kappa e_1}(W,\vec{d})} \lesssim \epsilon_1. \tag{9.58}$$

For $V \in \mathscr{C}^{\vec{\mathrm{r}}+\kappa e_1}(\overline{B^n(7/8)}, (W, \vec{d}))$, set

$$\mathscr{T}(V) := \mathrm{Mult}[\psi_2] T \mathscr{P}^* \psi_1(\cdot)\big(R_1(\cdot, \mathfrak{W}_1^\kappa(V+H)(\cdot)) + R_2(\cdot, \mathfrak{W}_1^\kappa(V+H)(\cdot))\big).$$

It follows from the fact that $H \in \mathscr{C}^{\vec{\mathrm{r}}+\kappa e_1}(\overline{B^n(7/8)}, (W, \vec{d}))$, Theorem 6.3.10, and Theorem 7.5.2 (ii) that $\mathscr{T} : \mathscr{C}^{\vec{\mathrm{r}}+\kappa e_1}(\overline{B^n(7/8)}, (W, \vec{d})) \to \mathscr{C}^{\vec{\mathrm{r}}+\kappa e_1}(\overline{B^n(7/8)}, (W, \vec{d}))$. We also have $\mathscr{T}(V_\infty) = V_\infty$, by (9.55).

Let $D \in [C_3 \vee 1, 1/\epsilon_1]$, to be chosen later; we will choose $D \approx 1$. Define

$$\mathscr{M}_{D,\epsilon_1,\vec{\mathrm{r}}} := \big\{V \in \mathscr{C}^{\vec{\mathrm{r}}+\kappa e_1}(\overline{B^n(7/8)}, (W, \vec{d})) : \|V\|_{\mathscr{C}^{\vec{\mathrm{r}}+\kappa e_1}(W,\vec{d})} \leq D\epsilon_1\big\}.$$

Note that $\mathscr{M}_{D,\epsilon_1,\vec{\mathrm{r}}}$ is a closed subspace of the Banach space $\mathscr{C}^{\vec{\mathrm{r}}+\kappa e_1}(\overline{B^n(7/8)}, (W, \vec{d}))$ (see Proposition 6.3.6), and therefore it is a complete metric space with distance induced by $\|\cdot\|_{\mathscr{C}^{\vec{\mathrm{r}}+\kappa e_1}(W,\vec{d})}$. Furthermore, since $D \geq C_3$, (9.56) shows that $V_\infty \in \mathscr{M}_{D,\epsilon_1,\vec{\mathrm{r}}}$. We will show that $\mathscr{T} : \mathscr{M}_{D,\epsilon_1,\vec{\mathrm{r}}} \to \mathscr{M}_{D,\epsilon_1,\vec{\mathrm{r}}}$ and that it is a strict contraction (with appropriate choices of D and ϵ_1); it will then follow that V_∞ is the unique fixed point of $\mathscr{T}$ in $\mathscr{M}_{D,\epsilon_1,\vec{\mathrm{r}}}$.

For $V \in \mathscr{M}_{D,\epsilon_1,\vec{\mathrm{r}}}$, we have, using Corollary 6.5.11, (9.57), and $D \geq 1$,

$$\begin{aligned}\|\mathfrak{W}_1^\kappa(V+H)\|_{\mathscr{C}^{\vec{\mathrm{r}}}(W,\vec{d})} &\lesssim \|V+H\|_{\mathscr{C}^{\vec{\mathrm{r}}+\kappa e_1}(W,\vec{d})} \\ &\leq \|V\|_{\mathscr{C}^{\vec{\mathrm{r}}+\kappa e}(W,\vec{d})} + \|H\|_{\mathscr{C}^{\vec{\mathrm{r}}+\kappa e}(W,\vec{d})} \lesssim D\epsilon_1 + \epsilon_1 \lesssim D\epsilon_1.\end{aligned} \tag{9.59}$$

In particular, using Corollary 7.2.2, we have

$$\|\mathfrak{W}_1^\kappa(V+H)\|_{L^\infty} \lesssim \|\mathfrak{W}_1^\kappa(V+H)\|_{\mathscr{C}^{\vec{\mathrm{r}}}(W,\vec{d})} \lesssim D\epsilon_1. \tag{9.60}$$

Using Theorem 6.3.10, Theorem 7.5.2 (ii) applied to R_1, and Theorem 7.5.2 (vii) with $L = 2$ applied to R_2, for $V \in \mathscr{M}_{D,\epsilon_1,\vec{\mathrm{r}}}$, we have

$$\begin{aligned}&\|\mathscr{T}(V)\|_{\mathscr{C}^{\vec{\mathrm{r}}+\kappa e_1}(W,\vec{d})} \\ &= \big\|\psi_2 T \mathscr{P}^* \psi_1\big(R_1(x, \mathfrak{W}_1^\kappa(V+H)(x)) + R_2(x, \mathfrak{W}_1^\kappa(V+H)(x))\big)\big\|_{\mathscr{C}^{\vec{\mathrm{r}}+\kappa e_1}(W,\vec{d})} \\ &\lesssim \big\|R_1(x, \mathfrak{W}_1^\kappa(V+H)(x))\big\|_{\mathscr{C}^{\vec{\mathrm{r}}}(W,\vec{d})} + \big\|R_2(x, \mathfrak{W}_1^\kappa(V+H)(x))\big\|_{\mathscr{C}^{\vec{\mathrm{r}}}(W,\vec{d})}\end{aligned}$$

$$\begin{aligned}
&\lesssim \|R_1\|_{\mathscr{C}^{\vec{\mathbf{r}},\lfloor|\vec{\mathbf{r}}|_1\rfloor+1+\nu+\sigma}((W,\vec{d})\boxtimes\nabla_{\mathbb{R}^N})}\big(1+\|\mathfrak{W}_1^K(V+H)\|_{\mathscr{C}^{\vec{\mathbf{r}}}(W,\vec{d})}\big)\big(1+\|\mathfrak{W}_1^K(V+H)\|_{L^\infty}\big)^{\lfloor|\vec{\mathbf{r}}|_1\rfloor+\nu} \\
&\quad + \|R_2\|_{\mathscr{C}^{\vec{\mathbf{r}},\lfloor|\vec{\mathbf{r}}|_1\rfloor+3+\nu+\sigma}((W,\vec{d})\boxtimes\nabla_{\mathbb{R}^N})}\big(1+\|\mathfrak{W}_1^K(V+H)\|_{L^\infty}\big)^{\lfloor|\vec{\mathbf{r}}|_1\rfloor+\nu+1} \\
&\qquad \times \|\mathfrak{W}_1^K(V+H)\|_{L^\infty}\|\mathfrak{W}_1^K(V+H)\|_{\mathscr{C}^{\vec{\mathbf{r}}}(W,\vec{d})} \\
&\lesssim \epsilon_1(1+D\epsilon_1)(1+D\epsilon_1)^{\lfloor|\vec{\mathbf{r}}|_1\rfloor+1} + (1+D\epsilon_1)^{\lfloor|\vec{\mathbf{r}}|_1\rfloor+\nu+1}(D\epsilon_1)^2 \lesssim \epsilon_1 + D^2\epsilon_1^2,
\end{aligned}$$

where in the second to last inequality we used (9.59) and (9.60), and in the last we used $D \leq 1/\epsilon_1$. That is, there exists $C_4 \approx 1$ with

$$\|\mathscr{T}(V)\|_{\mathscr{C}^{\vec{\mathbf{r}}+\kappa\epsilon_1}(W,\vec{d})} \leq C_4(\epsilon_1 + D^2\epsilon_1^2), \quad \forall V \in \mathscr{M}_{D,\epsilon_1,\vec{\mathbf{r}}}. \tag{9.61}$$

We take $D := \max\{C_3, 2C_4, 1\}$, so that $D \approx 1$, and we take $\epsilon_1 \in (0, 1/D^2]$ to be chosen later. Then $C_4(\epsilon_1 + \epsilon_1^2 D^2) \leq D\epsilon_1$ and (9.61) implies

$$\|\mathscr{T}(V)\|_{\mathscr{C}^{\vec{\mathbf{r}}+\kappa\epsilon_1}(W,\vec{d})} \leq D\epsilon_1, \quad \forall V \in \mathscr{M}_{D,\epsilon_1,\vec{\mathbf{r}}}.$$

That is, $\mathscr{T} : \mathscr{M}_{D,\epsilon_1,\vec{\mathbf{r}}} \to \mathscr{M}_{D,\epsilon_1,\vec{\mathbf{r}}}$.

Now that we have chosen $D \approx 1$, (9.59) and (9.60) become

$$\|\mathfrak{W}_1^K(V+H)\|_{\mathscr{C}^{\vec{\mathbf{r}}}(W,\vec{d})}, \|\mathfrak{W}_1^K(V+H)\|_{L^\infty} \lesssim \epsilon_1, \quad \forall V \in \mathscr{M}_{D,\epsilon_1,\vec{\mathbf{r}}}. \tag{9.62}$$

Using (9.62), Theorem 6.3.10, Theorem 7.5.2 (vi) applied to R_1, and Theorem 7.5.2 (viii) with $L = 2$ applied to R_2, we have

$$\begin{aligned}
&\|\mathscr{T}(V_1) - \mathscr{T}(V_2)\|_{\mathscr{C}^{\vec{\mathbf{r}}+\kappa\epsilon_1}(W,\vec{d})} \\
&\quad = \|\psi_2 T\mathscr{P}^*\psi_1(R_1(x,\mathfrak{W}_1^K(V_1+H)(x)) - R_1(x,\mathfrak{W}_1^K(V_2+H)(x)) \\
&\qquad + R_2(x,\mathfrak{W}_1^K(V_1+H)(x)) - R_2(x,\mathfrak{W}_1^K(V_2+H)(x)))\|_{\mathscr{C}^{\vec{\mathbf{r}}+\kappa\epsilon_1}(W,\vec{d})} \\
&\quad \lesssim \|R_1(x,\mathfrak{W}_1^K(V_1+H)(x)) - R_1(x,\mathfrak{W}_1^K(V_2+H)(x))\|_{\mathscr{C}^{\vec{\mathbf{r}}}(W,\vec{d})} \\
&\qquad + \|R_2(x,\mathfrak{W}_1^K(V_1+H)(x)) - R_2(x,\mathfrak{W}_1^K(V_2+H)(x))\|_{\mathscr{C}^{\vec{\mathbf{r}}}(W,\vec{d})} \\
&\quad \lesssim \|R_1\|_{\mathscr{C}^{\vec{\mathbf{r}},\lfloor|\vec{\mathbf{r}}|_1\rfloor+2+\nu+\sigma}((W,\vec{d})\boxtimes\nabla_{\mathbb{R}^N})}\Big(\|\mathfrak{W}_1^K(V_1-V_2)\|_{\mathscr{C}^{\vec{\mathbf{r}}}(W,\vec{d})} \\
&\qquad + \big(\|\mathfrak{W}_1^K(V_1+H)\|_{\mathscr{C}^{\vec{\mathbf{r}}}(W,\vec{d})} + \|\mathfrak{W}_1^K(V_2+H)\|_{\mathscr{C}^{\vec{\mathbf{r}}}(W,\vec{d})}\big)\|\mathfrak{W}_1^K(V_1-V_2)\|_{L^\infty}\Big) \\
&\qquad\quad \times \big(1+\|\mathfrak{W}_1^K(V_1+H)\|_{L^\infty} + \|\mathfrak{W}_1^K(V_2+H)\|_{L^\infty}\big)^{\lfloor|\vec{\mathbf{r}}|_1\rfloor+\nu} \\
&\qquad + \|R_2\|_{\mathscr{C}^{\vec{\mathbf{r}},\lfloor|\vec{\mathbf{r}}|_1\rfloor+3+\nu+\sigma}((W,\vec{d})\boxtimes\nabla_{\mathbb{R}^N})}\big(1+\|\mathfrak{W}_1^K(V_1+H)\|_{L^\infty} + \|\mathfrak{W}_1^K(V_2+H)\|_{L^\infty}\big)^{\lfloor|\vec{\mathbf{r}}|_1\rfloor+\nu+1} \\
&\qquad\quad \times \big(\|\mathfrak{W}_1^K(V_1+H)\|_{\mathscr{C}^{\vec{\mathbf{r}}}(W,\vec{d})} + \|\mathfrak{W}_1^K(V_2+H)\|_{\mathscr{C}^{\vec{\mathbf{r}}}(W,\vec{d})}\big)\|\mathfrak{W}_1^K(V_1-V_2)\|_{L^\infty} \\
&\qquad + \|R_2\|_{\mathscr{C}^{\vec{\mathbf{r}},\lfloor|\vec{\mathbf{r}}|_1\rfloor+3+\nu+\sigma}((W,\vec{d})\boxtimes\nabla_{\mathbb{R}^N})}\Big(\|\mathfrak{W}_1^K(V_1+H)\|_{L^\infty} \\
&\qquad\qquad\qquad + \|\mathfrak{W}_1^K(V_2+H)\|_{L^\infty}\Big)\|\mathfrak{W}_1^K(V_1-V_2)\|_{\mathscr{C}^{\vec{\mathbf{r}}}(W,\vec{d})}
\end{aligned}$$

$$\lesssim \epsilon_1\big(\|V_1 - V_2\|_{\mathscr{C}^{\vec{\mathbf{r}}+\kappa e_1}(W,\vec{d})} + \epsilon_1\|\mathfrak{W}_1^{\kappa}(V_1 - V_2)\|_{L^\infty}\big)(1+\epsilon_1)^{\lfloor|\vec{\mathbf{r}}|_1\rfloor+\nu}$$
$$+ (1+\epsilon_1)^{\lfloor|\vec{\mathbf{r}}|_1\rfloor+\nu+1}\epsilon_1\|\mathfrak{W}_1^{\kappa}(V_1 - V_2)\|_{L^\infty}$$
$$+ \epsilon_1\|\mathfrak{W}_1^{\kappa}(V_1 - V_2)\|_{L^\infty}$$
$$\lesssim \epsilon_1\|V_1 - V_2\|_{\mathscr{C}^{\vec{\mathbf{r}}+\kappa e_1}(W,\vec{d})} + \epsilon_1\|\mathfrak{W}_1^{\kappa}(V_1 - V_2)\|_{L^\infty}$$
$$\lesssim \epsilon_1\|V_1 - V_2\|_{\mathscr{C}^{\vec{\mathbf{r}}+\kappa e_1}(W,\vec{d})},$$

where we have used Corollaries 7.2.2 and 6.5.11 to see that

$$\|\mathfrak{W}_1^{\kappa}(V_1 - V_2)\|_{L^\infty} \lesssim \|\mathfrak{W}_1^{\kappa}(V_1 - V_2)\|_{\mathscr{C}^{\vec{\mathbf{r}}}(W,\vec{d})} \lesssim \|V_1 - V_2\|_{\mathscr{C}^{\vec{\mathbf{r}}+\kappa e_1}(W,\vec{d})}.$$

We conclude that $\exists C_5 \approx 1$ with

$$\|\mathscr{T}(V_1) - \mathscr{T}(V_2)\|_{\mathscr{C}^{\vec{\mathbf{r}}+\kappa e_1}(W,\vec{d})} \le C_5\epsilon_1\|V_1 - V_2\|_{\mathscr{C}^{\vec{\mathbf{r}}+\kappa e_1}(W,\vec{d})}.$$

In particular, if $\epsilon_1 \in (0, \min\{1/2C_5, 1/D^2\}]$, we have

$$\|\mathscr{T}(V_1) - \mathscr{T}(V_2)\|_{\mathscr{C}^{\vec{\mathbf{r}}+\kappa e_1}(W,\vec{d})} \le \frac{1}{2}\|V_1 - V_2\|_{\mathscr{C}^{\vec{\mathbf{r}}+\kappa e_1}(W,\vec{d})},$$

and therefore $\mathscr{T} : \mathscr{M}_{D,\epsilon_1,\vec{\mathbf{r}}} \to \mathscr{M}_{D,\epsilon_1,\vec{\mathbf{r}}}$ is a strict contraction with unique fixed point V_∞.

Set $V_0 := 0$ and for $j \ge 0$ set $V_{j+1} := \mathscr{T}(V_j)$. We have established $V_j \in \mathscr{M}_{D,\epsilon_1,\vec{\mathbf{r}}}$, $\forall j$, and the contraction mapping principle shows that $V_j \xrightarrow{j\to\infty} V_\infty$ in $\mathscr{C}^{\vec{\mathbf{r}}+\kappa e_1}(\overline{B^n(7/8)}, (W,\vec{d}))$. Since $V_j \in \mathscr{M}_{D,\epsilon_1,\vec{\mathbf{r}}}$, we have

$$\|V_j\|_{\mathscr{C}^{\vec{\mathbf{r}}+\kappa e_1}(W,\vec{d})} \lesssim \epsilon_1,$$

and therefore just as in (9.62), we have

$$\|\mathfrak{W}_1^{\kappa}(V_j + H)\|_{\mathscr{C}^{\vec{\mathbf{r}}}(W,\vec{d})}, \|\mathfrak{W}_1^{\kappa}(V_j + H)\|_{L^\infty} \lesssim \|V_j\|_{\mathscr{C}^{\vec{\mathbf{r}}+\kappa e_1}(W,\vec{d})} + \epsilon_1 \lesssim \epsilon_1, \tag{9.63}$$

$\forall j \in \mathbb{N}$. Similarly,

$$\|\mathfrak{W}_1^{\kappa} V_j\|_{\mathscr{C}^{\vec{\mathbf{r}}}(W,\vec{d})}, \|\mathfrak{W}_1^{\kappa} V_j\|_{L^\infty} \lesssim \|V_j\|_{\mathscr{C}^{\vec{\mathbf{r}}+\kappa e_1}(W,\vec{d})} \lesssim \epsilon_1, \tag{9.64}$$

$\forall j \in \mathbb{N}$.

Lemma 9.2.16. *For $\psi' \in C_0^\infty(B^n(13/16))$ with $\psi' \prec \psi_3$, we have*

$$\psi' V_j \in \mathscr{C}^{\vec{\mathbf{s}}+\kappa e_1}(\overline{B^n(7/8)}, (W,\vec{d})), \quad \forall j \in \mathbb{N}.$$

Proof. We prove the lemma by induction on j. Since $V_0 = 0$, the base case, $j = 0$, is trivial. We assume the result for j and prove it for $j+1$.

Take $\psi'', \psi''' \in C_0^\infty(B^n(13/16))$ with $\psi' \prec \psi'' \prec \psi''' \prec \psi_3$. By the inductive hypothesis, we have $\psi''' V_j \in \mathscr{C}^{\vec{s}+\kappa e_1}(\overline{B^n(7/8)}, (W, \vec{d}))$ and we wish to show $\psi' V_j \in \mathscr{C}^{\vec{s}+\kappa e_1}(\overline{B^n(7/8)}, (W, \vec{d}))$.

We have

$$\begin{aligned}
\psi' V_{j+1} &= \psi' \mathscr{T}(V_j) \\
&= \psi' T \mathscr{P}^* \psi_1 (R_1(x, \mathfrak{W}_1^\kappa (V_j + H)(x)) + R_2(x, \mathfrak{W}_1^\kappa (V_j + H)(x))) \\
&= \psi' T \mathscr{P}^* (1 - \psi'') \psi_1 (R_1(x, \mathfrak{W}_1^\kappa (V_j + H)(x)) + R_2(x, \mathfrak{W}_1^\kappa (V_j + H)(x))) \\
&\quad + \psi' T \mathscr{P}^* \psi'' (R_1(x, \mathfrak{W}_1^\kappa (V_j + H)(x)) + R_2(x, \mathfrak{W}_1^\kappa (V_j + H)(x))) \\
&=: (\mathrm{I}) + (\mathrm{II}).
\end{aligned}$$

Since $\psi' \prec \psi''$ and $T \in \mathscr{A}_{\mathrm{loc}}^{-2\kappa}((W^1, \vec{d}^1); \mathbb{C}^{D_2}, \mathbb{C}^{D_2})$ is pseudo-local (see the growth condition in Definition 5.2.4), we have $\mathrm{Mult}[\psi'] T \mathscr{P}^* \mathrm{Mult}[(1 - \psi'')\psi_1] \in C_0^\infty(B^n(13/16) \times B^n(13/16))$. It follows that

$$\begin{aligned}
(\mathrm{I}) &= \psi' T \mathscr{P}^* (1 - \psi'') \psi_1 (R_1(x, \mathfrak{W}_1^\kappa (V_j + H)(x)) + R_2(x, \mathfrak{W}_1^\kappa (V_j + H)(x))) \\
&\in C_0^\infty(B^n(13/16)) \subseteq \mathscr{C}^{\vec{s}+\kappa e_1}(\overline{B^n(7/8)}, (W, \vec{d})),
\end{aligned}$$

where the $\subseteq$ uses Proposition 6.5.5. So to complete the proof, it suffices to show that $(\mathrm{II}) \in \mathscr{C}^{\vec{s}+\kappa e_1}(\overline{B^n(7/8)}, (W, \vec{d}))$.

We have

$$(\mathrm{II}) = \psi' T \mathscr{P}^* \psi'' (R_1(x, \mathfrak{W}_1^\kappa \psi''' (V_j + H)(x)) + R_2(x, \mathfrak{W}_1^\kappa \psi''' (V_j + H)(x))).$$

By the inductive hypothesis, $\psi''' V_j \in \mathscr{C}^{\vec{s}+\kappa e_1}(\overline{B^n(7/8)}, (W, \vec{d}))$, and since $\psi''' H = \psi''' \psi_3 H = \psi''' \psi_3 e_\infty \in C_0^\infty(B^n(13/16))$, we also have $\psi''' H \in \mathscr{C}^{\vec{s}+\kappa e_1}(\overline{B^n(7/8)}, (W, \vec{d}))$. Corollary 6.5.11 shows that $\mathfrak{W}_1^\kappa \psi''' (V_j + H) \in \mathscr{C}^{\vec{s}}(\overline{B^n(7/8)}, (W, \vec{d}))$ and Theorem 7.5.2 (ii) implies that $R_1(x, \mathfrak{W}_1^\kappa \psi''' (V_j + H)(x)) + R_2(x, \mathfrak{W}_1^\kappa \psi''' (V_j + H)(x)) \in \mathscr{C}^{\vec{s}}(\overline{B^n(7/8)}, (W, \vec{d}))$. Since $\mathrm{Mult}[\psi'] T \mathscr{P}^* \mathrm{Mult}[\psi''] \in \widetilde{\mathscr{A}}^{-\kappa e_1}(B^n(7/8), (W, \vec{d}))$, Theorem 6.3.10 implies that $(\mathrm{II}) \in \mathscr{C}^{\vec{s}+\kappa e_1}(\overline{B^n(7/8)}, (W, \vec{d}))$, as desired, completing the proof. □

We use the functions $\phi_{x,\delta}$ and the constants ξ_4, ξ_5, and ξ_6 from Theorem 9.2.8. Fix $x_0 \in B^n(3/4)$ and $\delta \in (0, 1]$ with $B_{(W^1, \vec{d}^1)}(x_0, (\xi_4^3 \xi_6 / \xi_5^3)\delta) \subseteq B^n(3/4)$. We have, by Proposition 9.2.9 (and using Lemma 9.2.16),

$$\|\phi_{x_0, \xi_6 \delta} V_{j+1}\|_{\mathscr{C}^{\vec{s}+\kappa e_1}(\delta^{\lambda \vec{d}} W, \vec{d})} \lesssim \|\phi_{x_0, \frac{\xi_4 \xi_6}{\xi_5} \delta} \delta^{2\kappa} \mathscr{P}^* \mathscr{P} V_{j+1}\|_{\mathscr{B}_{\infty,\infty}^{\vec{s}-\kappa e_1}(\delta^{\lambda \vec{d}} W, \vec{d})} + \|\phi_{x_0, \frac{\xi_4 \xi_6}{\xi_5} \delta} V_{j+1}\|_{L^\infty}. \tag{9.65}$$

We estimate the two terms on the right-hand side of (9.65).

By (9.64) we have

$$\|\phi_{x_0, \frac{\xi_4 \xi_6}{\xi_5} \delta} V_{j+1}\|_{L^\infty} \le \|V_{j+1}\|_{L^\infty} \lesssim \epsilon_1. \tag{9.66}$$

To bound the first term on the right-hand side of (9.65), we use the fact that $\mathrm{supp}(\phi_{x_0,\frac{\xi_4\xi_6}{\xi_5}\delta}) \subseteq B_{(W^1,d^1)}(x_0,(\xi_4^3\xi_6/\xi_5^3)\delta) \subseteq B^n(3/4)$ and therefore $\phi_{x_0,\frac{\xi_4\xi_6}{\xi_5}\delta} \prec \psi_2$ to see that

$$\begin{aligned}
&\phi_{x_0,\frac{\xi_4\xi_6}{\xi_5}\delta}\mathscr{P}^*\mathscr{P}\,\mathrm{Mult}[\psi_2]T\mathscr{P}^*\,\mathrm{Mult}[\psi_1] \\
&\quad= \phi_{x_0,\frac{\xi_4\xi_6}{\xi_5}\delta}\mathscr{P}^*\mathscr{P}T\mathscr{P}^*\,\mathrm{Mult}[\psi_1] \\
&\quad= \phi_{x_0,\frac{\xi_4\xi_6}{\xi_5}\delta}\mathscr{P}^*\,\mathrm{Mult}[\psi_1] + \phi_{x_0,\frac{\xi_4\xi_6}{\xi_5}\delta}E_\infty\,\mathrm{Mult}[\psi_1],
\end{aligned}$$

where $E_\infty \in C_0^\infty(B^n(7/8)\times B^n(7/8))$ with $\|E_\infty\|_{C^L} \lesssim_L 1$, $\forall L \in \mathbb{N}$. Thus, we have

$$\begin{aligned}
&\|\phi_{x_0,\frac{\xi_4\xi_6}{\xi_5}\delta}\delta^{2\kappa}\mathscr{P}^*\mathscr{P}V_{j+1}\|_{\mathscr{B}^{\vec{\mathfrak{s}}-\kappa e_1}_{\infty,\infty}(\delta^{\lambda\vec{d}}W,\vec{d})} \\
&= \|\phi_{x_0,\frac{\xi_4\xi_6}{\xi_5}\delta}\delta^{2\kappa}\mathscr{P}^*\mathscr{P}\mathscr{T}(V_j)\|_{\mathscr{B}^{\vec{\mathfrak{s}}-\kappa e_1}_{\infty,\infty}(\delta^{\lambda\vec{d}}W,\vec{d})} \\
&= \|\phi_{x_0,\frac{\xi_4\xi_6}{\xi_5}\delta}\delta^{2\kappa}\mathscr{P}^*\mathscr{P}\psi_2T\mathscr{P}^*\psi_1(R_1(x,\mathfrak{W}_1^\kappa(V_j+H)(x))+R_2(x,\mathfrak{W}_1^\kappa(V_j+H)(x)))\|_{\mathscr{B}^{\vec{\mathfrak{s}}-\kappa e_1}_{\infty,\infty}(\delta^{\lambda\vec{d}}W,\vec{d})} \\
&\le \|\phi_{x_0,\frac{\xi_4\xi_6}{\xi_5}\delta}\delta^{2\kappa}\mathscr{P}^*\psi_1(R_1(x,\mathfrak{W}_1^\kappa(V_j+H)(x))+R_2(x,\mathfrak{W}_1^\kappa(V_j+H)(x)))\|_{\mathscr{B}^{\vec{\mathfrak{s}}-\kappa e_1}_{\infty,\infty}(\delta^{\lambda\vec{d}}W,\vec{d})} \\
&\quad+ \|\phi_{x_0,\frac{\xi_4\xi_6}{\xi_5}\delta}\delta^{2\kappa}E_\infty\psi_1(R_1(x,\mathfrak{W}_1^\kappa(V_j+H)(x))+R_2(x,\mathfrak{W}_1^\kappa(V_j+H)(x)))\|_{\mathscr{B}^{\vec{\mathfrak{s}}-\kappa e_1}_{\infty,\infty}(\delta^{\lambda\vec{d}}W,\vec{d})} \\
&=: (\mathrm{III}) + (\mathrm{IV}).
\end{aligned} \tag{9.67}$$

To bound (IV), we use Proposition 9.2.10 to see that, for some $L \approx 1$,

$$\begin{aligned}
(\mathrm{IV}) &\lesssim \delta^{2\kappa}\|E_\infty\|_{C^L}\|R_1(x,\mathfrak{W}_1^\kappa(V_j+H)(x))+R_2(x,\mathfrak{W}_1^\kappa(V_j+H)(x))\|_{L^\infty} \\
&\lesssim \delta^{2\kappa}(\|R_1(x,\mathfrak{W}_1^\kappa(V_j+H)(x))\|_{L^\infty}+\|R_2(x,\mathfrak{W}_1^\kappa(V_j+H)(x))\|_{L^\infty}).
\end{aligned} \tag{9.68}$$

We have, using Proposition 7.5.10,

$$\begin{aligned}
\|R_1(x,\mathfrak{W}_1^\kappa(V_j+H)(x))\|_{L^\infty} &\le \|R_1\|_{L^\infty} \\
&\lesssim \|R_1\|_{\mathscr{C}^{\vec{\mathfrak{s}},\lfloor|\vec{\mathfrak{s}}|_1\rfloor+2+\nu+\sigma}((W,\vec{d})\boxtimes\nabla_{\mathbb{R}^N})} \le \epsilon_1.
\end{aligned} \tag{9.69}$$

Since $R_2(x,0)\equiv 0$, we have, using Propositions 7.5.10 and 7.5.11,

$$\begin{aligned}
\|R_2(x,\mathfrak{W}_1^\kappa(V_j+H)(x))\|_{L^\infty} &\lesssim \sum_{|\beta|\le 1}\|\partial_\zeta^\beta R_2(x,\zeta)\|_{L^\infty}\|\mathfrak{W}_1^\kappa(V_j+H)\|_{L^\infty} \\
&\lesssim \sum_{|\beta|\le 1}\|\partial_\zeta^\beta R_2\|_{\mathscr{C}^{\vec{\mathfrak{s}},\lfloor|\vec{\mathfrak{s}}|_1\rfloor+2+\nu+\sigma}((W,\vec{d})\boxtimes\nabla_{\mathbb{R}^N})}\|\mathfrak{W}_1^\kappa(V_j+H)\|_{L^\infty} \\
&\lesssim \|R_2\|_{\mathscr{C}^{\vec{\mathfrak{s}},\lfloor|\vec{\mathfrak{s}}|_1\rfloor+3+\nu+\sigma}((W,\vec{d})\boxtimes\nabla_{\mathbb{R}^N})}\|\mathfrak{W}_1^\kappa(V_j+H)\|_{L^\infty} \\
&\lesssim \|\mathfrak{W}_1^\kappa(V_j+H)\|_{L^\infty} \lesssim \epsilon_1,
\end{aligned} \tag{9.70}$$

where the final estimate used (9.63). Using (9.69) and (9.70) to estimate the right-hand side of (9.68) shows that

$$(\mathrm{IV}) \lesssim \delta^{2\kappa}\epsilon_1 \lesssim \epsilon_1. \tag{9.71}$$

For (III), using the fact that $\phi_{x_0,\frac{\xi_4\xi_6}{\xi_5}\delta} \prec \phi_{x_0,\frac{\xi_4^2\xi_6}{\xi_5^2}\delta}$ (see Theorem 9.2.8 (iv)), we have

$$\begin{aligned}
(\mathrm{III}) &= \Big\|\phi_{x_0,\frac{\xi_4\xi_6}{\xi_5}\delta}\delta^{2\kappa}\mathscr{P}^*\psi_1\Big(R_1(x,\phi_{x_0,\frac{\xi_4^2\xi_6}{\xi_5^2}\delta}(x)\mathfrak{W}_1^\kappa(V_j+H)(x)) \\
&\quad + R_2(x,\phi_{x_0,\frac{\xi_4^2\xi_6}{\xi_5^2}\delta}(x)\mathfrak{W}_1^\kappa(V_j+H)(x))\Big)\Big\|_{\mathscr{B}^{\mathfrak{s}-\kappa e_1}_{\infty,\infty}(\delta^{\lambda\vec{d}}W,\vec{d})} \\
&\lesssim \delta^\kappa\Big\|\phi_{x_0,\frac{\xi_4^2\xi_6}{\xi_5^2}\delta}\Big(R_1(x,\phi_{x_0,\frac{\xi_4^2\xi_6}{\xi_5^2}\delta}(x)\mathfrak{W}_1^\kappa(V_j+H)(x)) \\
&\quad + R_2(x,\phi_{x_0,\frac{\xi_4^2\xi_6}{\xi_5^2}\delta}(x)\mathfrak{W}_1^\kappa(V_j+H)(x))\Big)\Big\|_{\mathscr{C}^{\mathfrak{s}}(\delta^{\lambda\vec{d}}W,\vec{d})} \\
&\le \delta^\kappa\Big\|\phi_{x_0,\frac{\xi_4^2\xi_6}{\xi_5^2}\delta}R_1(x,\phi_{x_0,\frac{\xi_4^2\xi_6}{\xi_5^2}\delta}(x)\mathfrak{W}_1^\kappa(V_j+H)(x))\Big\|_{\mathscr{C}^{\mathfrak{s}}(\delta^{\lambda\vec{d}}W,\vec{d})} \\
&\quad + \delta^\kappa\Big\|\phi_{x_0,\frac{\xi_4^2\xi_6}{\xi_5^2}\delta}R_2(x,\phi_{x_0,\frac{\xi_4^2\xi_6}{\xi_5^2}\delta}(x)\mathfrak{W}_1^\kappa(V_j+H)(x))\Big\|_{\mathscr{C}^{\mathfrak{s}}(\delta^{\lambda\vec{d}}W,\vec{d})} \\
&=: (\mathrm{V}) + (\mathrm{VI}),
\end{aligned} \tag{9.72}$$

where we used Proposition 9.2.11 and $\phi_{x_0,\frac{\xi_4^2\xi_6}{\xi_5^2}\delta} \prec \psi_1$ (since $\mathrm{supp}(\phi_{x_0,\frac{\xi_4^2\xi_6}{\xi_5^2}\delta}) \subseteq B_{(W^1,d^1)}(x_0, (\xi_4^3\xi_6/\xi_5^3)\delta) \subseteq B^n(3/4)$) (see Theorem 9.2.8 (i)).

Before we estimate (V) and (VI), we need some preliminary estimates. First, using Corollary 9.2.13 and the fact that $\phi_{x_0,\frac{\xi_4^3\xi_6}{\xi_5^3}\delta} \prec \psi_3$ (since $\mathrm{supp}(\phi_{x_0,\frac{\xi_4^3\xi_6}{\xi_5^3}\delta}) \subseteq B_{(W^1,d^1)}(x_0, (\xi_4^3\xi_6/\xi_5^3)\delta) \subseteq B^n(3/4)$) (see Theorem 9.2.8 (i)), we have

$$\begin{aligned}
\Big\|\phi_{x_0,\frac{\xi_4^3\xi_6}{\xi_5^3}\delta}H\Big\|_{\mathscr{C}^{\mathfrak{s}+\kappa e_1}(\delta^{\lambda\vec{d}}W,\vec{d})} &= \Big\|\phi_{x_0,\frac{\xi_4^3\xi_6}{\xi_5^3}\delta}\psi_3 H\Big\|_{\mathscr{C}^{\mathfrak{s}+\kappa e_1}(\delta^{\lambda\vec{d}}W,\vec{d})} \\
&\lesssim \|\psi_3 H\|_{\mathscr{C}^{\mathfrak{s}+\kappa e_1}(W,\vec{d})} \lesssim \epsilon_1,
\end{aligned} \tag{9.73}$$

where the final estimate used (9.58). Next, using Proposition 9.2.11 and (9.73), we have

$$\begin{aligned}
&\delta^\kappa\Big\|\phi_{x_0,\frac{\xi_4^2\xi_6}{\xi_5^2}\delta}\mathfrak{W}_1^\kappa(V_j+H)\Big\|_{\mathscr{C}^{\mathfrak{s}}(\delta^{\lambda\vec{d}}W,\vec{d})} \\
&\quad\lesssim \Big\|\phi_{x_0,\frac{\xi_4^3\xi_6}{\xi_5^3}\delta}(V_j+H)\Big\|_{\mathscr{C}^{\mathfrak{s}+\kappa e_1}(\delta^{\lambda\vec{d}}W,\vec{d})} \\
&\quad\le \Big\|\phi_{x_0,\frac{\xi_4^3\xi_6}{\xi_5^3}\delta}V_j\Big\|_{\mathscr{C}^{\mathfrak{s}+\kappa e_1}(\delta^{\lambda\vec{d}}W,\vec{d})} + \Big\|\phi_{x_0,\frac{\xi_4^3\xi_6}{\xi_5^3}\delta}H\Big\|_{\mathscr{C}^{\mathfrak{s}+\kappa e_1}(\delta^{\lambda\vec{d}}W,\vec{d})} \\
&\quad\lesssim \Big\|\phi_{x_0,\frac{\xi_4^3\xi_6}{\xi_5^3}\delta}V_j\Big\|_{\mathscr{C}^{\mathfrak{s}+\kappa e_1}(\delta^{\lambda\vec{d}}W,\vec{d})} + \epsilon_1.
\end{aligned} \tag{9.74}$$

Finally, since $0 \leq \phi_{x_0,\frac{\xi_4^2\xi_6}{\xi_5^2}\delta} \leq 1$ (Theorem 9.2.8 (ii)), we have, using (9.63),

$$\Big\|\phi_{x_0,\frac{\xi_4^2\xi_6}{\xi_5^2}\delta}\mathfrak{W}_1^\kappa(V_j+H)\Big\|_{L^\infty} \leq \|\mathfrak{W}_1^\kappa(V_j+H)\|_{L^\infty} \leq \epsilon_1. \tag{9.75}$$

For the following estimates of (V) and (VI), we freely use (9.73), (9.74), and (9.75).

For (V) we have, by Proposition 9.2.14 (i),

$$\begin{aligned}
(\mathrm{V}) &= \delta^\kappa\Big\|\phi_{x_0,\frac{\xi_4^2\xi_6}{\xi_5^2}\delta}R_1(x,\phi_{x_0,\frac{\xi_4^2\xi_6}{\xi_5^2}\delta}(x)\mathfrak{W}_1^\kappa(V_j+H)(x))\Big\|_{\mathscr{C}^{\vec{s}}(\delta^{\lambda\vec{d}}W,\vec{d})}\\
&\lesssim \|R_1\|_{\mathscr{C}^{s,\lfloor|s|_1\rfloor+1+\nu+\sigma}((W,\vec{d})\boxtimes\nabla_{\mathbb{R}^N})}\Big(\delta^\kappa+\delta^\kappa\Big\|\phi_{x_0,\frac{\xi_4^2\xi_6}{\xi_5^2}\delta}\mathfrak{W}_1^\kappa(V_j+H)\Big\|_{\mathscr{C}^{\vec{s}}(\delta^{\lambda\vec{d}}W,\vec{d})}\Big)\\
&\quad\times\Big(1+\Big\|\phi_{x_0,\frac{\xi_4^2\xi_6}{\xi_5^2}\delta}\mathfrak{W}_1^\kappa(V_j+H)\Big\|_{L^\infty}\Big)^{\lfloor|s|_1\rfloor+\nu}\\
&\lesssim \epsilon_1\Big(1+\Big\|\phi_{x_0,\frac{\xi_4^3\xi_6}{\xi_5^3}\delta}V_j\Big\|_{\mathscr{C}^{\vec{s}+\kappa e_1}(\delta^{\lambda\vec{d}}W,\vec{d})}+\epsilon_1\Big)(1+\epsilon_1)^{\lfloor|s|_1\rfloor+\nu}\\
&\lesssim \epsilon_1+\epsilon_1\Big\|\phi_{x_0,\frac{\xi_4^3\xi_6}{\xi_5^3}\delta}V_j\Big\|_{\mathscr{C}^{\vec{s}+\kappa e_1}(\delta^{\lambda\vec{d}}W,\vec{d})}.
\end{aligned}\tag{9.76}$$

For (VI) we have, by Proposition 9.2.14 (ii) with $L=2$,

$$\begin{aligned}
&\delta^\kappa\Big\|\phi_{x_0,\frac{\xi_4^2\xi_6}{\xi_5^2}\delta}R_2(x,\phi_{x_0,\frac{\xi_4^2\xi_6}{\xi_5^2}\delta}(x)\mathfrak{W}_1^\kappa(V_j+H)(x))\Big\|_{\mathscr{C}^{\vec{s}}(\delta^{\lambda\vec{d}}W,\vec{d})}\\
&\lesssim \|R_2\|_{\mathscr{C}^{\vec{s},\lfloor|\vec{s}|_1\rfloor+3+\nu+\sigma}((W,\vec{d})\boxtimes\nabla_{\mathbb{R}^N})}\Big(1+\Big\|\phi_{x_0,\frac{\xi_4^2\xi_6}{\xi_5^2}\delta}(x)\mathfrak{W}_1^\kappa(V_j+H)\Big\|_{L^\infty}\Big)^{\lfloor|s|_1\rfloor+\nu+1}\\
&\quad\times\Big\|\phi_{x_0,\frac{\xi_4^2\xi_6}{\xi_5^2}\delta}\mathfrak{W}_1^\kappa(V_j+H)\Big\|_{L^\infty}\delta^\kappa\Big\|\phi_{x_0,\frac{\xi_4^2\xi_6}{\xi_5^2}\delta}\mathfrak{W}_1^\kappa(V_j+H)\Big\|_{\mathscr{C}^{\vec{s}}(\delta^{\lambda\vec{d}}W,\vec{d})}\\
&\lesssim (1+\epsilon_1)^{\lfloor|s|_1\rfloor+\nu+1}\epsilon_1\Big(\Big\|\phi_{x_0,\frac{\xi_4^3\xi_6}{\xi_5^3}\delta}V_j\Big\|_{\mathscr{C}^{\vec{s}+\kappa e_1}(\delta^{\lambda\vec{d}}W,\vec{d})}+\epsilon_1\Big)\\
&\lesssim \epsilon_1^2+\epsilon_1\Big\|\phi_{x_0,\frac{\xi_4^3\xi_6}{\xi_5^3}\delta}V_j\Big\|_{\mathscr{C}^{\vec{s}+\kappa e_1}(\delta^{\lambda\vec{d}}W,\vec{d})}.
\end{aligned}\tag{9.77}$$

Using (9.76) and (9.77) to bound the right-hand side of (9.72) shows that

$$(\mathrm{III}) \lesssim \epsilon_1+\epsilon_1\Big\|\phi_{x_0,\frac{\xi_4^3\xi_6}{\xi_5^3}\delta}V_j\Big\|_{\mathscr{C}^{\vec{s}+\kappa e_1}(\delta^{\lambda\vec{d}}W,\vec{d})}. \tag{9.78}$$

Using (9.71) and (9.78) to bound the right-hand side of (9.67) shows that

$$\|\phi_{x_0,\frac{\xi_4\xi_6}{\xi_5}\delta}\delta^{2\kappa}\mathscr{P}^*\mathscr{P}V_{j+1}\|_{\mathscr{B}^{\vec{s}-\kappa e_1}_{\infty,\infty}(\delta^{\lambda\vec{d}}W,\vec{d})} \lesssim \epsilon_1+\epsilon_1\Big\|\phi_{x_0,\frac{\xi_4^3\xi_6}{\xi_5^3}\delta}V_j\Big\|_{\mathscr{C}^{\vec{s}+\kappa e_1}(\delta^{\lambda\vec{d}}W,\vec{d})}. \tag{9.79}$$

Using (9.79) and (9.66) to bound the right-hand side of (9.65) and using Theorem 9.2.8 (x) shows that

$$\begin{aligned}\|\phi_{x_0,\xi_6\delta}V_{j+1}\|_{\mathscr{C}^{\mathfrak{s}+\kappa\epsilon_1}(\delta^{\lambda\vec{d}}W,\vec{d})} &\lesssim \epsilon_1+\epsilon_1\Big\|\phi_{x_0,\frac{\xi_4^3\xi_6}{\xi_5^3}\delta}V_j\Big\|_{\mathscr{C}^{\mathfrak{s}+\kappa\epsilon_1}(\delta^{\lambda\vec{d}}W,\vec{d})}\\ &\approx \epsilon_1+\epsilon_1\Big\|\phi_{x_0,\frac{\xi_4^3\xi_6}{\xi_5^3}\delta}V_j\Big\|_{\mathscr{C}^{\mathfrak{s}+\kappa\epsilon_1}(((\xi_4^3\xi_6/\xi_5^3)\delta)^{\lambda\vec{d}}W,\vec{d})}.\end{aligned} \tag{9.80}$$

Set

$$Q_j := \sup_{\substack{\delta\in(0,1]\\ B_{(W^1,d^1)}(x,\delta)\subseteq B^n(3/4)}} \|\phi_{x,\delta}V_j\|_{\mathscr{C}^{\mathfrak{s}+\kappa\epsilon_1}(\delta^{\lambda\vec{d}}W,\vec{d})} \in [0,\infty],$$

where the supremum is taken over all such x and δ. Since $V_0=0$, we have $Q_0=0$. For $x_0\in B^n(3/4)$ and $\delta\in(0,1]$ such that $B_{(W^1,d^1)}(x_0,(\xi_4^3\xi_6/\xi_5^3)\delta)\subseteq B^n(3/4)$, (9.80) implies that

$$\|\phi_{x_0,\xi_6\delta}V_{j+1}\|_{\mathscr{C}^{\mathfrak{s}+\kappa\epsilon_1}(\delta^{\lambda\vec{d}}W,\vec{d})} \lesssim \epsilon_1+\epsilon_1 Q_j. \tag{9.81}$$

Take $x\in B^n(3/4)$, $\delta\in(0,1]$ such that $B_{(W^1,d^1)}(x,\delta)\subseteq B^n(3/4)$. Take $N_1\approx 1$ and $x_1,\dots,x_{N_1}$ as in Theorem 9.2.8 (viii). Because (9.17) holds, (9.81) implies

$$\|\phi_{x_j,\xi_6\delta}V_{j+1}\|_{\mathscr{C}^{\mathfrak{s}+\kappa\epsilon_1}(\delta^{\lambda\vec{d}}W,\vec{d})} \lesssim \epsilon_1+\epsilon_1 Q_j,\quad 1\le j\le N_1. \tag{9.82}$$

Plugging (9.82) into (9.18) shows that

$$\|\phi_{x,\delta}V_{j+1}\|_{\mathscr{C}^{\mathfrak{s}+\kappa\epsilon_1}(\delta^{\lambda\vec{d}}W,\vec{d})} \lesssim \sum_{j=1}^{N_1}\|\phi_{x_j,\xi_6\delta}V_{j+1}\|_{\mathscr{C}^{\mathfrak{s}+\kappa\epsilon_1}(\delta^{\lambda\vec{d}}W,\vec{d})} \lesssim \epsilon_1+\epsilon_1 Q_j.$$

Taking the supremum over such x and δ shows that

$$Q_{j+1}\lesssim \epsilon_1+\epsilon_1 Q_j.$$

That is, $\exists C_6\approx 1$ with

$$Q_{j+1}\le C_6(\epsilon_1+\epsilon_1 Q_j).$$

We take $\epsilon_1 := \min\{1/2C_6, 1/2C_5, 1/D^2\}$. Then we have

$$Q_{j+1}\le \frac{1}{2}+\frac{Q_j}{2}. \tag{9.83}$$

Since $Q_0=0$, a simple induction using (9.83) shows that $Q_j\le 1$, $\forall j\in\mathbb{N}$.

Finally, we take $\psi_0 \in C_0^\infty(B^n(2/3))$. Take $x_1, \dots, x_{N_2}$ and $\delta_1 \in (0,1]$ as in Theorem 9.2.8 (ix). In particular, we have $B_{(W^1,d^1)}(x_j, \delta_1) \subseteq B^n(3/4)$, $\delta_1 \approx 1$, and $N_2 \approx 1$. By (9.19), we have

$$\|\psi_0 V_j\|_{\mathscr{C}^{\vec{s}+\kappa e_1}(W,\vec{d})} \lesssim \sum_{j=1}^{N_2} \|\phi_{x_j,\delta_1} V_j\|_{\mathscr{C}^{\vec{s}+\kappa e_1}(\delta_1^{\lambda \vec{d}} W,\vec{d})} \lesssim Q_j \lesssim 1.$$

That is, for $l \in N^\nu$,

$$\|D_l \psi_0 V_j\|_{L^\infty} \lesssim 2^{-l\cdot(\vec{s}+\kappa e_1)}. \tag{9.84}$$

Since $V_j \to V_\infty$ in $\mathscr{C}^{\vec{r}+\kappa e_1}(\overline{B^n(7/8)}, (W, \vec{d}))$, Corollary 7.2.2 shows that $V_j \to V_\infty$ in L^∞. Since each D_l is bounded on L^∞ (see (5.6)), taking the limit as $j \to \infty$ in (9.84) shows that

$$\|D_l \psi_0 V_\infty\|_{L^\infty} \lesssim 2^{-l\cdot(\vec{s}+\kappa e_1)}, \quad \forall l \in \mathbb{N}^\nu. \tag{9.85}$$

Since $\operatorname{supp}(\psi_0 V_j) \subseteq \overline{B^n(7/8)}$, (9.85) implies that $\psi_0 V_\infty \in \mathscr{C}^{\vec{s}+\kappa e_1}(\overline{B^n(7/8)}, (W, \vec{d}))$ with

$$\|\psi_0 V_\infty\|_{\mathscr{C}^{\vec{s}+\kappa e_1}(W,\vec{d})} \lesssim 1. \tag{9.86}$$

Since $\psi_0 \prec \psi_3$, we have $\psi_0 H = \psi_0 \psi_3 H = \psi_0 \psi_3 e_\infty \in C_0^\infty(B^n(7/8))$. Using this, Proposition 6.5.5 shows that $\psi_0 H \in \mathscr{C}^{\vec{s}+\kappa e_1}(\overline{B^n(7/8)}, (W, \vec{d}))$. Moreover, using Corollary 6.5.10 and (9.58), we have

$$\|\psi_0 H\|_{\mathscr{C}^{\vec{s}+\kappa e_1}(W,\vec{d})} = \|\psi_0 \psi_3 H\|_{\mathscr{C}^{\vec{s}+\kappa e_1}(W,\vec{d})} \lesssim \|\psi_3 H\|_{\mathscr{C}^{\vec{s}+\kappa e_1}(W,\vec{d})} \lesssim \epsilon_1. \tag{9.87}$$

Since $u = H + V_\infty$, we have $\psi_0 u \in \mathscr{C}^{\vec{s}+\kappa e_1}(\overline{B^n(7/8)}, (W, \vec{d}))$, and using (9.86) and (9.87) we see that

$$\|\psi_0 u\|_{\mathscr{C}^{\vec{s}+\kappa e_1}(W,\vec{d})} \leq \|\psi_0 H\|_{\mathscr{C}^{\vec{s}+\kappa e_1}(W,\vec{d})} + \|\psi_0 V_\infty\|_{\mathscr{C}^{\vec{s}+\kappa e_1}(W,\vec{d})} \lesssim 1,$$

completing the proof.

9.2.5 Step II: A simpler form

In this section, we present a special case of Proposition 9.2.15 whose hypotheses will prove easier to verify in Step III (Section 9.2.7).

Let $\epsilon_2 \in (0,1]$ be a small number, to be chosen later. Suppose

$$R_3(x,\zeta) \in \mathscr{C}^{\vec{s},\lfloor|\vec{s}|_1\rfloor+2+\nu+\sigma}(\overline{B^n(7/8)} \times \mathbb{R}^N, (W, \vec{d}) \boxtimes \nabla_{\mathbb{R}^N}; \mathbb{R}^{\mathbf{D}_1})$$

with

$$\|R_3\|_{\mathscr{C}^{s,\lfloor|\vec{s}|_1\rfloor+2+\nu+\sigma}((W,\vec{d})\boxtimes\nabla_{\mathbb{R}^N})} \leq \epsilon_2. \tag{9.88}$$

Let $G(\zeta) \in \mathscr{C}^{\lfloor|s|_1\rfloor+3+\nu+\sigma}(\mathbb{R}^N;\mathbb{R}^{\mathbf{D}_1})$ and take $C_3 \geq 0$ with

$$\|G\|_{\mathscr{C}^{\lfloor|s|_1\rfloor+3+\nu+\sigma}(\mathbb{R}^N)} \leq C_3$$

and $u \in \mathscr{C}^{\vec{\mathbf{r}}+\kappa e_1}(\overline{B^n(7/8)},(W,\vec{d});\mathbb{R}^{\mathbf{D}_2})$ with

$$\|u\|_{\mathscr{C}^{\vec{\mathbf{r}}+\kappa e_1}(W,\vec{d})} \leq \epsilon_2.$$

We assume $dG(\mathfrak{W}_1^\kappa u(0))\mathfrak{W}_1^\kappa$ is maximally subelliptic of degree κ with respect to $(W,\vec{d})$ in the sense that there exists $A_2 \geq 0$ such that

$$\begin{aligned}&\sum_{j=1}^{r_1}\|(W_j^1)^{n_j}v\|_{L^2(B^n(1),h\sigma_{\mathrm{Leb}})}\\&\quad\leq A_2\big(\|dG(\mathfrak{W}_1^\kappa u(0))\mathfrak{W}_1^\kappa v\|_{L^2(B^n(1),h\sigma_{\mathrm{Leb}})}+\|v\|_{L^2(B^n(1),h\sigma_{\mathrm{Leb}})}\big),\end{aligned} \tag{9.89}$$

for all $v \in C_0^\infty(B^n(1);\mathbb{C}^{\mathbf{D}_2})$. Finally, we suppose u satisfies the following nonlinear PDE:

$$G(\mathfrak{W}_1^\kappa u(x)) + R_3(x,\mathfrak{W}_1^\kappa u(x)) = 0, \quad \forall x \in B^n(13/16). \tag{9.90}$$

Proposition 9.2.17. *There exists $\epsilon_2 \in (0,1]$ sufficiently small such that if the above holds, then $\forall\psi_0 \in C_0^\infty(B^n(2/3))$ we have*

$$\psi_0 u \in \mathscr{C}^{\vec{\mathbf{s}}+\kappa e_1}((W,\vec{d});\mathbb{R}^{\mathbf{D}_2})$$

and $\exists C_4 \geq 0$ with

$$\|\psi_0 u\|_{\mathscr{C}^{\vec{\mathbf{s}}+\kappa e_1}(W,\vec{d})} \leq C_4.$$

Here, $\epsilon_2 = \epsilon_2(C_3,A_2,\mathbf{D}_1,\mathbf{D}_2) \in (0,1]$ and $C_4 = C_4(C_3,A_2,\mathbf{D}_1,\mathbf{D}_2,\psi_0) \geq 0$ are $(\kappa,\vec{\mathbf{s}},\vec{\mathbf{r}},\sigma)$-multi-parameter unit-admissible constants.

Proof. Lemma 8.3.3 (i) combined with the fact that $h \approx 1$ shows that

$$\begin{aligned}&\sum_{\deg_{\vec{d}^1}(\alpha)\leq\kappa}\|(W^1)^\alpha v\|_{L^2(B^n(1),h\sigma_{\mathrm{Leb}})}\\&\quad\approx\sum_{j=1}^{r_1}\|(W_j^1)^{n_j}v\|_{L^2(B^n(1),h\sigma_{\mathrm{Leb}})}+\|v\|_{L^2(B^n(1),h\sigma_{\mathrm{Leb}})},\end{aligned}$$

$\forall v \in C_0^\infty(B^n(31/32);\mathbb{C}^{\mathbf{D}_2})$. Combining this with (9.89) shows that there exists $A_3 \approx 1$ with

$$\begin{aligned}&\sum_{\deg_{\vec{d}^1}(\alpha)\leq\kappa}\|(W^1)^\alpha v\|_{L^2(B^n(1),h\sigma_{\mathrm{Leb}})}\\&\leq A_3(\|dG(\mathfrak{W}_1^\kappa u(0))\mathfrak{W}_1^\kappa v\|_{L^2(B^n(1),h\sigma_{\mathrm{Leb}})}+\|v\|_{L^2(B^n(1),h\sigma_{\mathrm{Leb}})}),\end{aligned}\tag{9.91}$$

$\forall v\in C_0^\infty(B^n(31/32);\mathbb{C}^{\mathbf{D}_2})$.

We claim that if $\epsilon_2>0$ is sufficiently small, then

$$\begin{aligned}&\sum_{\deg_{\vec{d}^1}(\alpha)\leq\kappa}\|(W^1)^\alpha v\|_{L^2(B^n(1),h\sigma_{\mathrm{Leb}})}\\&\leq 2A_3(\|dG(0)\mathfrak{W}_1^\kappa v\|_{L^2(B^n(1),h\sigma_{\mathrm{Leb}})}+\|v\|_{L^2(B^n(1),h\sigma_{\mathrm{Leb}})}),\end{aligned}\tag{9.92}$$

$\forall v\in C_0^\infty(B^n(31/32);\mathbb{C}^{\mathbf{D}_2})$. To see (9.92), we first use Corollaries 7.2.2 and 6.5.11 to see that

$$\|\mathfrak{W}_1^\kappa u\|_{L^\infty}\lesssim\|\mathfrak{W}_1^\kappa u\|_{\mathscr{C}^{\vec{\mathbf{r}}}(W,\vec{d})}\lesssim\|u\|_{\mathscr{C}^{\vec{\mathbf{r}}+\kappa e_1}(W,\vec{d})}\lesssim\epsilon_2.\tag{9.93}$$

Thus, since $\lfloor|s|_1\rfloor+3+\nu+\sigma>2$, we have

$$\begin{aligned}|dG(\mathfrak{W}_1^\kappa u(0))-dG(0)|&\lesssim\sum_{|\beta|\leq 2}\|\partial_\zeta^\beta G\|_{L^\infty(\mathbb{R}^N)}|\mathfrak{W}_1^\kappa u(0)|\\&\lesssim\|G\|_{\mathscr{C}^{\lfloor|s|_1\rfloor+3+\nu+\sigma}(\mathbb{R}^N)}\|\mathfrak{W}_1^\kappa u\|_{L^\infty}\lesssim\epsilon_2.\end{aligned}\tag{9.94}$$

Plugging (9.94) into (9.91) we conclude that for some $C_5\approx 1$, we have

$$\begin{aligned}\|\mathfrak{W}_1^\kappa v\|_{L^2(B^n(1),h\sigma_{\mathrm{Leb}})}&\leq A_3(\|dG(0)\mathfrak{W}_1^\kappa v\|_{L^2(B^n(1),h\sigma_{\mathrm{Leb}})}+\|v\|_{L^2(B^n(1),h\sigma_{\mathrm{Leb}})})\\&\quad+A_3C_5\epsilon_2\|\mathfrak{W}_1^\kappa v\|_{L^2(B^n(1),h\sigma_{\mathrm{Leb}})},\end{aligned}$$

$\forall v\in C_0^\infty(B^n(31/32);\mathbb{C}^{\mathbf{D}_2})$. Taking $\epsilon_2\in(0,1\wedge 1/2A_3C_5]$ gives (9.92).

Next, using $\lfloor|s|_1\rfloor+3+\nu+\sigma>1$ and (9.93), we have

$$\begin{aligned}|G(\mathfrak{W}_1^\kappa u(0))-G(0)|&\lesssim\sum_{|\beta|\leq 1}\|\partial_\zeta^\beta G\|_{L^\infty}|\mathfrak{W}_1^\kappa u(0)|\\&\lesssim\|G\|_{\mathscr{C}^{\lfloor|s|_1\rfloor+3+\nu+\sigma}(\mathbb{R}^N)}\|\mathfrak{W}_1^\kappa u\|_{L^\infty}\lesssim\epsilon_2.\end{aligned}\tag{9.95}$$

Also, using (9.90) and Proposition 7.5.10 gives

$$\begin{aligned}|G(\mathfrak{W}_1^\kappa u(0))|&\leq|G(\mathfrak{W}_1^\kappa u(0))+R_3(0,\mathfrak{W}_1^\kappa u(0))|+\|R_3\|_{L^\infty}\\&=0+\|R_3\|_{L^\infty}\lesssim\|R_3\|_{\mathscr{C}^{s,\lfloor|s|_1\rfloor+2+\nu+\sigma}((W,\vec{d})\boxtimes\nabla_{\mathbb{R}^N})}\lesssim\epsilon_2.\end{aligned}\tag{9.96}$$

Combining (9.95) and (9.96) gives

$$|G(0)|\lesssim\epsilon_2.\tag{9.97}$$

Define $E(\zeta)$ by $G(\zeta)=G(0)+d_\zeta G(0)\zeta+E(\zeta)$, so that $E(0)=0$, $d_\zeta E(0)=0$, $E\in\mathscr{C}^{\lfloor|s|_1\rfloor+3+\nu+\sigma}(\mathbb{R}^N)$, and

$$\|E\|_{\mathscr{C}^{\lfloor|s|_1\rfloor+3+\nu+\sigma}(\mathbb{R}^N)} \lesssim 1. \tag{9.98}$$

Set $\mathscr{P} := d_\zeta G(0)\mathfrak{W}_1^\kappa$. Rewriting (9.90), we have

$$\mathscr{P}u(x) = -G(0) - R_3(x, \mathfrak{W}_1^\kappa u(x)) - E(\mathfrak{W}_1^\kappa u(x)), \quad \forall x \in B^n(13/16). \tag{9.99}$$

Fix $\psi_1 \in C_0^\infty(B^n(7/8))$ with $\psi_1 \equiv 1$ on $B^n(13/16)$. Set

$$R_1(x,\zeta) := -\psi_1(x)(G(0) + R_3(x,\zeta)), \quad R_2(x,\zeta) := -\psi_1(x)E(\zeta).$$

Then (9.99) implies

$$\mathscr{P}u(x) = R_1(x, \mathfrak{W}_1^\kappa u(x)) + R_2(x, \mathfrak{W}_1^\kappa u(x)), \quad \forall x \in B^n(13/16). \tag{9.100}$$

Furthermore, using (9.97) and (9.88), we have

$$\|R_1\|_{\mathscr{C}^{s,\lfloor|s|_1\rfloor+2+\nu+\sigma}((W,\vec{d})\boxtimes\nabla_{\mathbb{R}^N})} \lesssim |G(0)| + \|R_3\|_{\mathscr{C}^{s,\lfloor|s|_1\rfloor+2+\nu+\sigma}((W,\vec{d})\boxtimes\nabla_{\mathbb{R}^N})} \lesssim \epsilon_2,$$

and using (9.98),

$$\|R_2\|_{\mathscr{C}^{s,\lfloor|s|_1\rfloor+3+\nu+\sigma}((W,\vec{d})\boxtimes\nabla_{\mathbb{R}^N})} \lesssim \|E\|_{\mathscr{C}^{\lfloor|s|_1\rfloor+3+\nu+\sigma}(\mathbb{R}^N)} \lesssim 1.$$

That is, there is $C_6 \approx 1$ with

$$\|R_1\|_{\mathscr{C}^{s,\lfloor|s|_1\rfloor+2+\nu+\sigma}((W,\vec{d})\boxtimes\nabla_{\mathbb{R}^N})} \le C_6\epsilon_2, \quad \|R_2\|_{\mathscr{C}^{s,\lfloor|s|_1\rfloor+3+\nu+\sigma}((W,\vec{d})\boxtimes\nabla_{\mathbb{R}^N})} \le C_6.$$

Also note that

$$|d_\zeta G(0)| \le \|G\|_{\mathscr{C}^{\lfloor|s|_1\rfloor+3+\nu+\sigma}(\mathbb{R}^N)} \lesssim 1,$$

i. e., there exists $C_7 \approx 1$ with

$$|d_\zeta G(0)| \le C_7.$$

Since $E(0) = 0$ and $d_\zeta E(0) = 0$, we have $\partial_\zeta^\beta R_2(x,0) \equiv 0$, $\forall |\beta| \le 1$.

Let $\epsilon_1 = \epsilon_1(C_6, 2A_3, C_7, \mathbf{D}_1, \mathbf{D}_2) \in (0,1]$ be the $(\kappa, \vec{\mathbf{s}}, \vec{\mathbf{r}}, \sigma)$-multi-parameter unit-admissible constant of the same name from Proposition 9.2.15. Taking $\epsilon_2 := \min\{\epsilon_1/C_6, 1\}$, $\mathscr{P}$, R_1, and R_2 satisfy all the hypotheses of Proposition 9.2.15 (the maximal subellipticity of $\mathscr{P}$ is (9.92)). The conclusions of Proposition 9.2.15, in this case, are exactly the conclusions of this proposition. □

9.2.6 Scaled decompositions of Zygmund–Hölder functions

To complete the proof of Theorem 9.2.1, we require scaled analogs of Propositions 7.2.1 and 7.5.9. In this section, we use only the scaling maps $\Phi_{0,\delta}$; even though the results can be generalized to work with all $\Phi_{x,\delta}$, we only require them with $x = 0$ and we restrict our attention to that case for simplicity. Let $\lambda = (1, \lambda_2, \dots, \lambda_\nu) \in (0,\infty)^\nu$ be as in Section 3.15.3.

Proposition 9.2.18. *Fix $J \geq 0$, $\vec{\mathbf{s}} \in (0,\infty)^\nu$, $\mathbf{t} > 0$, and $\psi \in C_0^\infty(B^n(7/8))$.*

(i) *Suppose that for each $j \in \mathbb{N}^\nu$, $u_j \in C^\infty_{\mathrm{loc}}(B^n(1))$ satisfies, $\forall L \in \mathbb{N}$,*

$$\sum_{|\alpha|\leq L} \|(2^{-(j+\lambda J)\vec{d}} W)^\alpha u_j\|_{L^\infty(B_{(X^1,d^1)}(0,2^{-J}))} \leq C_L 2^{-j\cdot\vec{\mathbf{s}}} E,$$

for some constants $C_L, E \geq 0$. Set

$$\hat{u}(y) := \psi(y) \sum_{j\in\mathbb{N}^\nu} u_j(\Phi_{0,2^{-J}}(y)).$$

Then $\hat{u} \in \mathscr{C}^{s}(\overline{B^n(7/8)}, (W^{0,2^{-J}}, \vec{d}))$ and

$$\|\hat{u}\|_{\mathscr{C}^{\vec{\mathbf{s}}}(W^{0,2^{-J}},\vec{d})} \leq C_{\vec{\mathbf{s}}} E,$$

where there exists a multi-parameter unit-admissible constant $L = L(\vec{\mathbf{s}}) \in \mathbb{N}$ such that $C_{\vec{\mathbf{s}}} = C_{\vec{\mathbf{s}}}(C_L, \psi)$ is an $\vec{\mathbf{s}}$-multi-parameter unit-admissible constant.

(ii) *Suppose that for $j \in \mathbb{N}^\nu$ and $k \in \mathbb{N}$, $H_{j,k} \in C^\infty_{\mathrm{loc}}(B^n(1) \times \mathbb{R}^N)$ satisfies, $\forall L \in \mathbb{N}$,*

$$\sum_{|\alpha|,|\beta|\leq L} \|(2^{-(j+\lambda J)\vec{d}} W)^\alpha (\partial_\zeta)^\beta H_{j,k}(x,\zeta)\|_{L^\infty(B_{(X^1,d^1)}(0,2^{-J})\times\mathbb{R}^N)} \leq C_L E 2^{-j\cdot\vec{\mathbf{s}}-k\mathbf{t}}, \tag{9.101}$$

for some constants $C_L, E \geq 0$. Set

$$\hat{H}(y,\zeta) := \psi(y) \sum_{\substack{j\in\mathbb{N}^\nu\\ k\in\mathbb{N}}} H_{j,k}(\Phi_{0,2^{-J}}(y), \zeta).$$

Then $\hat{H} \in \mathscr{C}^{\vec{\mathbf{s}},\mathbf{t}}(\overline{B^n(7/8)} \times \mathbb{R}^N, (W^{0,2^{-J}}, \vec{d}) \boxtimes \nabla_{\mathbb{R}^N})$ and

$$\|\hat{H}\|_{\mathscr{C}^{\vec{\mathbf{s}},\mathbf{t}}((W^{0,2^{-J}},\vec{d})\boxtimes\nabla_{\mathbb{R}^N})} \leq C_{\vec{\mathbf{s}},\mathbf{t}} E,$$

where there exists a multi-parameter unit-admissible constant $L = L(\vec{\mathbf{s}}, \mathbf{t}) \in \mathbb{N}$ such that $C_{\vec{\mathbf{s}},\mathbf{t}} = C_{\vec{\mathbf{s}},\mathbf{t}}(C_L, N, \psi)$ is an $\vec{\mathbf{s}}, \mathbf{t}$-multi-parameter unit-admissible constant.

Proof. The proofs for (i) and (ii) are similar; we begin with the proof of (ii).

Set $\hat{H}_{j,k}(y,\zeta) := \psi(y) H_{j,k}(\Phi_{0,2^{-J}}(y,\zeta))$. Since $\psi \in C_0^\infty(B^n(7/8))$ is a fixed function, $\Phi^*_{0,2^{-J}} 2^{-J\lambda\cdot\vec{d}} W = W^{0,2^{-J}}$ (Theorem 3.15.5), and $\|W_l^{\mu,0,2^{-J}}\|_{C^L} \lesssim 1$, $\forall l, \mu \in \{1,\dots,\nu\}$ (see Theorem 3.15.5 (h)), we have

$$
\begin{aligned}
&\sum_{|\alpha|,|\beta|\le L} \|(2^{-j\vec{d}}W^{0,2^{-J}})^{\alpha}(2^{-k}\partial_\zeta)^{\beta}\hat{H}_{j,k}(y,\zeta)\|_{L^\infty(B^n(1)\times\mathbb{R}^N)}\\
&\quad\lesssim_L \sum_{|\alpha|,|\beta|\le L} \|(2^{-j\vec{d}}W^{0,2^{-J}})^{\alpha}(2^{-k}\partial_\zeta)^{\beta}H_{j,k}(\Phi_{0,2^{-J}}(y),\zeta)\|_{L^\infty(B^n(1)\times\mathbb{R}^N)}\\
&\quad= \sum_{|\alpha|,|\beta|\le L} \|(2^{-(j+\lambda J)\vec{d}}W)^{\alpha}(2^{-k}\partial_\zeta)^{\beta}H_{j,k}(x,\zeta)\|_{L^\infty(\Phi_{0,2^{-J}}(B^n(1))\times\mathbb{R}^N)}\\
&\quad\le \sum_{|\alpha|,|\beta|\le L} \|(2^{-(j+\lambda J)\vec{d}}W)^{\alpha}(2^{-k}\partial_\zeta)^{\beta}H_{j,k}(x,\zeta)\|_{L^\infty(B_{(X^1,d^1)}(0,2^{-J})\times\mathbb{R}^N)}\\
&\quad\le C_L 2^{-j\cdot\vec{\mathbf{s}}-k\cdot\mathbf{t}}E,
\end{aligned}
$$

where in the second to last inequality we used $\Phi_{0,2^{-J}}(B^n(1)) \subseteq B_{(X^1,d^1)}(0,2^{-J})$ (see Theorem 3.15.5) and in the last inequality we used the assumption (9.101). Since $\hat{H} = \sum_{j,k}\hat{H}_{j,k}$, the result follows from Proposition 7.5.9.

The proof for (i) is similar, where there is no ζ variable and we use Proposition 7.2.1 in place of Proposition 7.5.9. □

Lemma 9.2.19. *Fix $\vec{\mathbf{s}} \in (0,\infty)^\nu$. Let $u \in C^\infty_{\mathrm{loc}}(B^n(1))$ and $J \in \mathbb{N}$. Suppose that, $\forall L \in \mathbb{N}$,*

$$\sum_{|\beta|\le L} \|(2^{-Jd^1}W^1)^{\beta}u\|_{L^\infty(B_{(X^1,d^1)}(0,2^{-J}))} \le C_L E, \tag{9.102}$$

for some constants $C_L, E \ge 0$. Fix $\psi \in C_0^\infty(B^n(7/8))$ and set

$$\hat{u}(y) := \psi(y)u(\Phi_{0,2^{-J}}(y)).$$

Then $u \in \mathscr{C}^{\vec{\mathbf{s}}}(\overline{B^n(7/8)}, (W^{0,2^{-J}}, \vec{d}))$ and

$$\|\hat{u}\|_{\mathscr{C}^{\vec{\mathbf{s}}}(W^{0,2^{-J}},\vec{d})} \le C_{\vec{\mathbf{s}}}E, \tag{9.103}$$

where there exists a multi-parameter unit-admissible constant $L = L(\vec{\mathbf{s}}) \in \mathbb{N}$ such that $C_{\vec{\mathbf{s}}} = C_{\vec{\mathbf{s}}}(C_L,\psi)$ is an $\vec{\mathbf{s}}$-multi-parameter unit-admissible constant.

Proof. Using the fact that $\psi \in C_0^\infty(B^n(7/8))$ is a fixed function, $\Phi_{0,2^{-J}}^{*}2^{-J\cdot d^1}W^1 = W^{1,0,2^{-J}}$ (Theorem 3.15.5), and $\|W_l^{1,0,2^{-J}}\|_{C^L} \lesssim 1$, $\forall l$ (see Theorem 3.15.5 (h)), we have

$$
\begin{aligned}
\sum_{|\beta|\le L} \|(W^{1,0,2^{-J}})^{\beta}\hat{u}\|_{L^\infty(B^n(1))} &\lesssim_L \sum_{|\beta|\le L} \|(W^{1,0,2^{-J}})^{\beta}u\circ\Phi_{0,2^{-J}}\|_{L^\infty(B^n(1))}\\
&= \sum_{|\beta|\le L} \|(2^{-Jd^1}W^1)^{\beta}u\|_{L^\infty(\Phi_{0,2^{-J}}(B^n(1)))}\\
&\le \sum_{|\beta|\le L} \|(2^{-Jd^1}W^1)^{\beta}u\|_{L^\infty(B_{(X^1,d^1)}(0,2^{-J}))}\\
&\le C_L E,
\end{aligned}
$$

where the second to last inequality uses $\Phi_{0,2^{-J}}(B^n(1)) \subseteq B_{(X^1,d^1)}(0,2^{-J})$ (see Theorem 3.15.5) and in the last inequality we used the assumption (9.102).

Using the fact that each $X_l^{1,0,2^{-J}}$ is a commutator of the $W_l^{1,0,2^{-J}}$ vector fields (see Theorem 3.15.5), it follows that

$$\sum_{|\beta|\leq L} \|(X^{1,0,2^{-J}})^{\beta}\hat{u}\|_{L^\infty(B^n(1))} \lesssim_L E. \tag{9.104}$$

By Theorem 3.15.5 (j), (9.104) implies that

$$\|\hat{u}\|_{C^L(B^n(1))} \lesssim_L E. \tag{9.105}$$

Using Theorem 3.15.5 (h), (9.105) implies that

$$\sum_{|\alpha|\leq L} \|(W^{0,2^{-J}})^{\alpha}\hat{u}\|_{L^\infty(B^n(1))} \lesssim_L E.$$

Using Proposition 7.2.1 (with only one term) and the fact that $\operatorname{supp}(\hat{u}) \subseteq B^n(7/8)$, it follows that $\hat{u} \in \mathscr{C}^{\vec{s}}(\overline{B^n(7/8)}, (W^{0,2^{-J}}, \vec{d}))$ (for every $\vec{s} \in (0,\infty)^\nu$) and (9.103) holds. □

9.2.7 Step III: Completion of the proof

To prove Theorem 9.2.1, we "zoom in" near $0 \in \mathbb{R}^n$ by using the scaling maps $\Phi_{0,\delta}$. We will see that if we zoom in sufficiently close to zero, then (9.12) is of the form covered by Proposition 9.2.17.

First, note that it suffices to prove Theorem 9.2.1 with $g = 0$ by replacing $F(x,\zeta)$ with $F(x,\zeta) - g(x)$; we henceforth assume that $g = 0$.

Take $J \in \mathbb{N}$ large, to be chosen later. Set

$$u_J := \sum_{\substack{0\leq j_1<J\\ j_2,\ldots,j_\nu\in\mathbb{N}}} D_j u, \quad v_J := 2^{\kappa J} \sum_{\substack{j_1\geq J\\ j_2,\ldots,j_\nu\in\mathbb{N}}} D_j u.$$

Note that

$$u_J + 2^{-\kappa J} v_J = \sum_{j\in\mathbb{N}^\nu} D_j u = u. \tag{9.106}$$

Set

$$\begin{aligned} G_J(\zeta) &:= F(0, \{(W^1)^\alpha u_J(0)\}_{\deg_{d^1}(\alpha)\leq\kappa} + \{2^{-(\kappa-\deg_{d^1}(\alpha))J}\zeta_\alpha\}_{\deg_{d^1}(\alpha)\leq\kappa}),\\ H_J(x,\zeta) &:= F(x, \{(W^1)^\alpha u_J(x)\}_{\deg_{d^1}(\alpha)\leq\kappa} + \{2^{-(\kappa-\deg_{d^1}(\alpha))J}\zeta_\alpha\}_{\deg_{d^1}(\alpha)\leq\kappa})\\ &\quad - F(0, \{(W^1)^\alpha u_J(0)\}_{\deg_{d^1}(\alpha)\leq\kappa} + \{2^{-(\kappa-\deg_{d^1}(\alpha))J}\zeta_\alpha\}_{\deg_{d^1}(\alpha)\leq\kappa}). \end{aligned}$$

Consider

$$G_J(\{(2^{-J\vec{d}^1}W^1)^{\alpha}v_J(x)\}_{\deg_{\vec{d}^1}(\alpha)\le\kappa}) + H_J(x,\{(2^{-J\vec{d}^1}W^1)^{\alpha}v_J(x)\}_{\deg_{\vec{d}^1}(\alpha)\le\kappa}) \\ = F(x,\{(W^1)^{\alpha}u(x)\}_{\deg_{\vec{d}^1}(\alpha)\le\kappa}).$$

Thus, (9.12) shows that (using $g = 0$)

$$G_J(\{(2^{-J\vec{d}^1}W^1)^{\alpha}v_J(x)\}_{\deg_{\vec{d}^1}(\alpha)\le\kappa}) \\ + H_J(x,\{(2^{-J\vec{d}^1}W^1)^{\alpha}v_J(x)\}_{\deg_{\vec{d}^1}(\alpha)\le\kappa}) = 0, \quad \forall x\in B^n(3/4). \tag{9.107}$$

We will pull back (9.107) by $\Phi_{0,2^{-J}}$. Then by taking $J \in \mathbb{N}$ sufficiently large, we will see that Proposition 9.2.17 applies to the pulled back equation, which will complete the proof. To see that the pulled back equation satisfies the conditions of Proposition 9.2.17, we require several preliminary results. In these results, $\vec{\mathbf{r}},\vec{\mathbf{s}}\in(0,\infty)^{\nu}$, $\mathbf{t}>0$, and $\kappa\in\mathbb{N}_+$.

Lemma 9.2.20. *Fix $\psi\in C_0^{\infty}(B^n(7/8))$ and $u\in\mathscr{C}^{\vec{\mathbf{r}}+\kappa e_1}(\overline{B^n(7/8)},(W,\vec{d}))$. For $J\in[0,\infty)$ and $\deg_{\vec{d}^1}(\alpha)\le\kappa$, set*

$$\begin{aligned}
\overline{u}_{J,\alpha}(y) &:= \sum_{\substack{0\le j_1<J\\ j_2,\dots,j_\nu\in\mathbb{N}}} ((W^1)^{\alpha}D_ju)(\Phi_{0,2^{-J}}(y)) = ((W^1)^{\alpha}u_J)(\Phi_{0,2^{-J}}(y)),\\
\tilde{u}_{J,\alpha}(y) &:= \sum_{\substack{0\le j_1<J\\ j_2,\dots,j_\nu\in\mathbb{N}}} [((W^1)^{\alpha}D_ju)(\Phi_{0,2^{-J}}(y)) - ((W^1)^{\alpha}D_ju)(0)]\\
&= \overline{u}_{J,\alpha}(y) - \overline{u}_{J,\alpha}(0),\\
\overline{v}_J(y) &:= \sum_{\substack{j_1\ge J\\ j_2,\dots,j_\nu\in\mathbb{N}}} 2^{J\kappa}D_ju(\Phi_{0,2^{-J}}(y)) = v_J(\Phi_{0,2^{-J}}(y)).
\end{aligned}$$

Then

$$\|\psi\overline{u}_{J,\alpha}\|_{\mathscr{C}^{\vec{\mathbf{s}}}(W^{0,2^{-J}},\vec{d})} \le C_1\|u\|_{\mathscr{C}^{\vec{\mathbf{r}}+\kappa e_1}(W,\vec{d})}, \tag{9.108}$$

$$\|\psi\tilde{u}_{J,\alpha}\|_{\mathscr{C}^{\vec{\mathbf{s}}}(W^{0,2^{-J}},\vec{d})} \le C_2 2^{-J(\vec{\mathbf{r}}_1\wedge 1)}\|u\|_{\mathscr{C}^{\vec{\mathbf{r}}+\kappa e_1}(W,\vec{d})}, \tag{9.109}$$

$$\|\psi\overline{v}_J\|_{\mathscr{C}^{\vec{\mathbf{r}}+\kappa e_1}(W^{0,2^{-J}},\vec{d})} \le C_3 2^{-J\vec{\mathbf{r}}_1}\|u\|_{\mathscr{C}^{\vec{\mathbf{r}}+\kappa e_1}(W,\vec{d})}. \tag{9.110}$$

Here, $C_1 = C_1(\psi)$, $C_2 = C_2(\psi)\ge 0$ are $(\vec{\mathbf{r}},\vec{\mathbf{s}},\kappa)$-multi-parameter unit-admissible constants and $C_3 = C_3(\psi)\ge 0$ is an $(\vec{\mathbf{r}},\kappa)$-multi-parameter unit-admissible constant.

Proof. We begin by proving (9.108) and (9.109). For any ordered multi-indices α and β, it follows from Proposition 5.5.5 (e) that

$$\{((2^{-j\vec{d}^1}X^1)^{\beta}(2^{-j\vec{d}^1}W^1)^{\alpha}D_j, 2^{-j}) : j\in\mathbb{N}^{\nu}\}$$

is a bounded set of generalized $(W, \vec{d})$ elementary operators, and therefore Corollary 6.4.4 shows that

$$\|(2^{-jd^1}X^1)^\beta(2^{-jd^1}W)^\alpha D_j u\|_{L^\infty} \lesssim_{\alpha,\beta} 2^{-j\cdot(\vec{\mathbf{r}}+\kappa e_1)}\|u\|_{\mathscr{C}^{\vec{\mathbf{r}}+\kappa e_1}(W,\vec{d})}. \tag{9.111}$$

Letting $\deg_{d^1}(\alpha) \le \kappa$, letting β be any ordered multi-index, and letting $j \in \mathbb{N}^\nu$ such that $j_1 < J$, we have, using (9.111),

$$\begin{aligned}
&\|(2^{-Jd^1}X^1)^\beta(W^1)^\alpha D_j u\|_{L^\infty}\\
&= 2^{(j_1-J)\deg_{d^1}(\beta)}2^{j_1\deg_{d^1}(\alpha)}\|(2^{-j_1d^1}X^1)^\beta(2^{-j_1d^1}W^1)^\alpha D_j u\|_{L^\infty}\\
&\lesssim 2^{(j_1-J)\deg_{d^1}(\beta)}2^{j_1\deg_{d^1}(\alpha)}2^{-j\cdot(\vec{\mathbf{r}}+\kappa e_1)}\|u\|_{\mathscr{C}^{\vec{\mathbf{r}}+\kappa e_1}(W,\vec{d})}\\
&\lesssim 2^{(j_1-J)\deg_{d^1}(\beta)}2^{-j\cdot\vec{\mathbf{r}}}\|u\|_{\mathscr{C}^{\vec{\mathbf{r}}+\kappa e_1}(W,\vec{d})}.
\end{aligned} \tag{9.112}$$

Thus,

$$\begin{aligned}
&\|(2^{-Jd^1}X^1)^\beta(W^1)^\alpha u_J\|_{L^\infty}\\
&= \left\|(2^{-Jd^1}X^1)^\beta(W^1)^\alpha \sum_{\substack{0\le j_1<J\\ j_2,\ldots,j_\nu\in\mathbb{N}}} D_j u\right\|_{L^\infty}\\
&\le \sum_{\substack{0\le j_1<J\\ j_2,\ldots,j_\nu\in\mathbb{N}}} \|(2^{-Jd^1}X^1)^\beta(W^1)^\alpha D_j u\|_{L^\infty}\\
&\lesssim \sum_{\substack{0\le j_1<J\\ j_2,\ldots,j_\nu\in\mathbb{N}}} 2^{(j_1-J)\deg_{d^1}(\beta)}2^{-j\cdot\vec{\mathbf{r}}}\|u\|_{\mathscr{C}^{\vec{\mathbf{r}}+\kappa e_1}(W,\vec{d})}\\
&\lesssim \begin{cases} \|u\|_{\mathscr{C}^{\vec{\mathbf{r}}+\kappa e_1}(W,\vec{d})}, & |\beta| = 0,\\ 2^{-J(\vec{\mathbf{r}}_1\wedge 1)}\|u\|_{\mathscr{C}^{\vec{\mathbf{r}}+\kappa e_1}(W,\vec{d})}, & |\beta| > 0.\end{cases}
\end{aligned} \tag{9.113}$$

In particular, (9.113) combined with Lemma 9.2.19 (using the fact that every $(2^{-Jd^1}W^1)^{\beta'}$ is of the form $(2^{-Jd^1}X)^\beta$ for some β) shows that (9.108) holds.

Let $x \in B_{(W^1,d^1)}(0, 2^{-J})$. Thus, there exists an absolutely continuous curve $\gamma : [0,1] \to B_{(X^1,d^1)}(0, 2^{-J})$ such that $\gamma(0) = 0$, $\gamma(1) = x$, and $\gamma'(t) = \sum_{l=1}^{q_1} a_l(t)2^{-Jd_l^1}X_l^1(\gamma(t))$, with $\|\sum|a_l|^2\|_{L^\infty([0,1])} \le 1$. Hence, we have

$$\begin{aligned}
&|(W^1)^\alpha D_j u(x) - (W^1)^\alpha D_j u(0)|\\
&= |(W^1)^\alpha D_j u(\gamma(1)) - (W^1)^\alpha D_j u(\gamma(0))|\\
&\le \int_0^1 \left|\frac{d}{dt}(W^1)^\alpha D_j u(\gamma(t))\right| dt
\end{aligned}$$

$$
\begin{aligned}
&\lesssim \sum_{l=1}^{q_1} \int_0^1 \left|2^{-J d^1} X_l^1 (W^1)^{\alpha} D_j u(\gamma(t))\right| dt \\
&\lesssim \sum_{l=1}^{q_1} \left\|2^{-J d^1} X_l^1 (W^1)^{\alpha} D_j u\right\|_{L^\infty} \\
&\lesssim 2^{(j_1-J)} 2^{-j\cdot\vec{\mathrm{r}}} \|u\|_{\mathscr{C}^{\vec{\mathrm{r}}+\kappa e_1}(W,\vec{d})},
\end{aligned}
$$

where the final inequality used (9.112). Thus,

$$
\begin{aligned}
&\Bigg\| \sum_{\substack{j_1<J\\ j_2,\dots,j_\nu\in\mathbb{N}}} [(W^1)^{\alpha} D_j u(\cdot) - (W^1)^{\alpha} D_j u(0)] \Bigg\|_{L^\infty(B_{(X^1,d^1)}(0,2^{-J}))} \\
&\quad \lesssim \sum_{\substack{j_1<J\\ j_2,\dots,j_\nu\in\mathbb{N}}} 2^{(j_1-J)} 2^{-j\cdot\vec{\mathrm{r}}} \|u\|_{\mathscr{C}^{\vec{\mathrm{r}}+\kappa e_1}(W,\vec{d})} \lesssim 2^{-J(\vec{\mathrm{r}}_1\wedge 1)} \|u\|_{\mathscr{C}^{\vec{\mathrm{r}}+\kappa e_1}(W,\vec{d})}.
\end{aligned}
\tag{9.114}
$$

By (9.113) (using the fact that every $(2^{-J d^1} W^1)^{\beta'}$ is of the form $(2^{-J d^1} X)^{\beta}$ for some β) we also have, for any $|\beta'| > 0$,

$$
\begin{aligned}
&\Bigg\| (2^{-J d^1} W^1)^{\beta'} \sum_{\substack{j_1<J\\ j_2,\dots,j_\nu\in\mathbb{N}}} [(W^1)^{\alpha} D_j u(\cdot) - (W^1)^{\alpha} D_j u(0)] \Bigg\|_{L^\infty(B_{(X^1,d^1)}(0,2^{-J}))} \\
&= \Bigg\| (2^{-J d^1} W^1)^{\beta'} \sum_{\substack{j_1<J\\ j_2,\dots,j_\nu\in\mathbb{N}}} (W^1)^{\alpha} D_j u(\cdot) \Bigg\|_{L^\infty(B_{(X^1,d^1)}(0,2^{-J}))} \\
&\lesssim 2^{-J(\vec{\mathrm{r}}_1\wedge 1)} \|u\|_{\mathscr{C}^{\vec{\mathrm{r}}+\kappa e_1}(W,\vec{d})}.
\end{aligned}
\tag{9.115}
$$

Combining (9.114) and (9.115) shows that the hypotheses of Lemma 9.2.19 hold for $\tilde{u}_{J,\alpha}$ with $E = 2^{-J(\vec{\mathrm{r}}_1\wedge 1)}$, and the conclusion of Lemma 9.2.19 is (9.109).

We turn to (9.110). We have

$$
\overline{v}_J = \sum_{j\in\mathbb{N}^\nu} 2^{J\kappa} D_{j+J e_1} u(\Phi_{0,2^{-J}}(y)).
$$

Thus, by Proposition 9.2.18 (i) with $E = 2^{-J\vec{\mathrm{r}}_1}$, to prove (9.110) it suffices to show that $\forall L \in \mathbb{N}$,

$$
\sum_{|\beta|\le L} \left\|(2^{-(j+J\lambda)\vec{d}} W)^{\beta} 2^{J\kappa} D_{j+J e_1} u\right\|_{L^\infty} \lesssim_L 2^{-j\cdot(\vec{\mathrm{r}}+\kappa e_1)} 2^{-J\vec{\mathrm{r}}_1} \|u\|_{\mathscr{C}^{\vec{\mathrm{r}}+\kappa e_1}(W,\vec{d})}.
\tag{9.116}
$$

Using (9.111) with $|\beta| = 0$ and α replaced by β, we have

$$\begin{aligned}
&\sum_{|\beta|\le L} \|(2^{-(j+J\lambda)\vec{d}} W)^{\beta} 2^{J\kappa} D_{j+J\mathfrak{e}_1} u\|_{L^\infty} \\
&\quad = \sum_{|\beta|\le L} 2^{-J(0,\lambda_2,\ldots,\lambda_\nu)\cdot \deg_{\vec{d}}(\beta)} 2^{J\kappa} \|(2^{-(j+J\mathfrak{e}_1)\vec{d}} W)^{\beta} D_{j+J\mathfrak{e}_1} u\|_{L^\infty} \\
&\quad \lesssim_L 2^{-(j+J\mathfrak{e}_1)\cdot(\vec{\mathbf{r}}+\kappa\mathfrak{e}_1)} 2^{J\kappa} \|u\|_{\mathscr{C}^{\vec{\mathbf{r}}+\kappa\mathfrak{e}_1}(W,\vec{d})} \\
&\quad \le 2^{-j\cdot(\vec{\mathbf{r}}+\kappa\mathfrak{e}_1)} 2^{-J\vec{\mathbf{r}}_1} \|u\|_{\mathscr{C}^{\vec{\mathbf{r}}+\kappa\mathfrak{e}_1}(W,\vec{d})},
\end{aligned}$$

establishing (9.116) and completing the proof. □

Our next proposition will be used to show that the pullback of the H_J function in (9.107) is small if J is large.

Proposition 9.2.21. *Let* $\mathbf{t} > \lfloor |\vec{\mathbf{s}}|_1 \rfloor + 2 + \nu$, $u \in \mathscr{C}^{\vec{\mathbf{r}}+\kappa\mathfrak{e}_1}(\overline{B^n(7/8)}, (W,\vec{d}))[\mathbb{R}^{\mathbf{D}_2}]$, *and* $F \in \mathscr{C}^{\vec{\mathbf{s}},\mathbf{t}}(\overline{B^n(7/8)} \boxtimes \nabla_{\mathbb{R}^N})$. *For* $J \in \mathbb{N}$, $\psi \in C_0^\infty(B^n(7/8))$, *and* $c_\alpha \in [0,1]$ *for* $\deg_{\vec{d}^1}(\alpha) \le \kappa$, *set*

$$\begin{aligned}
\hat{H}_J^c(y,\zeta) := \psi(y)\Bigg[F\Bigg(\Phi_{0,2^{-J}}(y), \sum_{\substack{j_1<J \\ j_2,\ldots,j_\nu \in \mathbb{N}}} \mathfrak{W}_1^\kappa D_j u(\Phi_{0,2^{-J}}(y)) + \{c_\alpha \zeta_\alpha\}_{\deg_{\vec{d}^1}(\alpha)\le\kappa} \Bigg) \\
- F\Bigg(0, \sum_{\substack{j_1<J \\ j_2,\ldots,j_\nu \in \mathbb{N}}} \mathfrak{W}_1^\kappa D_j u(0) + \{c_\alpha \zeta_\alpha\}_{\deg_{\vec{d}^1}(\alpha)\le\kappa} \Bigg)\Bigg].
\end{aligned}$$

Then $\hat{H}_J^c \in \mathscr{C}^{\vec{\mathbf{s}},\mathbf{t}-\lfloor|\vec{\mathbf{s}}|_1\rfloor-2-\nu}(\overline{B^n(7/8)} \times \mathbb{R}^N, (W^{0,2^{-J}},\vec{d}) \boxtimes \nabla_{\mathbb{R}^N})$ *and*

$$\begin{aligned}
&\|\hat{H}_J^c\|_{\mathscr{C}^{\vec{\mathbf{s}},\mathbf{t}-\lfloor|\vec{\mathbf{s}}|_1\rfloor-2-\nu}((W^{0,2^{-J}},\vec{d})\boxtimes\nabla_{\mathbb{R}^N})} \\
&\quad \le C\|F\|_{\mathscr{C}^{\vec{\mathbf{s}},\mathbf{t}}((W,\vec{d})\boxtimes\nabla_{\mathbb{R}^N})} \left(1 + \|u\|_{\mathscr{C}^{\vec{\mathbf{r}}+\kappa\mathfrak{e}_1}(W,\vec{d})}\right)^{\lfloor|\vec{\mathbf{s}}|_1\rfloor+\nu+2} \\
&\quad \times \left(2^{-J(\vec{\mathbf{r}}_1\wedge 1)} + 2^{-J(\min_\mu \lambda_\mu(\vec{\mathbf{s}}_\mu\wedge 1))}\right).
\end{aligned}$$

Here, $C = C(\psi, \mathbf{D}_2) \ge 0$ *is an* $(\vec{\mathbf{r}},\vec{\mathbf{s}},\mathbf{t})$*-multi-parameter unit-admissible constant.*

We separate the main part of the proof of Proposition 9.2.21 into the next two lemmas.

Lemma 9.2.22. *With the same setup as Proposition* 9.2.21, *set*

$$\begin{aligned}
\hat{H}_{1,J}(y,\zeta) := \psi(y)\Bigg[F\Bigg(\Phi_{0,2^{-J}}(y), \sum_{\substack{j_1<J \\ j_2,\ldots,j_\nu \in \mathbb{N}}} \mathfrak{W}_1^\kappa D_j u(\Phi_{0,2^{-J}}(y)) + \zeta \Bigg) \\
- F\Bigg(\Phi_{0,2^{-J}}(y), \sum_{\substack{j_1<J \\ j_2,\ldots,j_\nu \in \mathbb{N}}} \mathfrak{W}_1^\kappa D_j u(0) + \zeta \Bigg)\Bigg].
\end{aligned}$$

Then $\widehat{H}_{1,J} \in \mathscr{C}^{\vec{\mathbf{s}},\mathbf{t}-\lfloor|\vec{\mathbf{s}}|_1\rfloor-2-\nu}(\overline{B^n(7/8)} \times \mathbb{R}^N, (W^{0,2^{-J}}, \vec{d}) \boxtimes \nabla_{\mathbb{R}^N})$ *and*

$$\begin{aligned}
&\|\widehat{H}_{1,J}\|_{\mathscr{C}^{\vec{\mathbf{s}},\mathbf{t}-\lfloor|\vec{\mathbf{s}}|_1\rfloor-2-\nu}((W^{0,2^{-J}},\vec{d})\boxtimes\nabla_{\mathbb{R}^N})} \\
&\quad \leq C2^{-J(\vec{\mathbf{r}}_1\wedge 1)}\|F\|_{\mathscr{C}^{\vec{\mathbf{s}},\mathbf{t}}((W,\vec{d})\boxtimes\nabla_{\mathbb{R}^N})}\big(1+\|u\|_{\mathscr{C}^{\vec{\mathbf{r}}+\kappa e_1}(W,\vec{d})}\big)^{\lfloor|\vec{\mathbf{s}}|_1\rfloor+\nu+1}\|u\|_{\mathscr{C}^{\vec{\mathbf{r}}+\kappa e_1}(W,\vec{d})}.
\end{aligned}$$

Here, $C = C(\psi, \mathbf{D}_2) \geq 0$ *is an* $(\vec{\mathbf{r}}, \vec{\mathbf{s}}, \mathbf{t})$*-multi-parameter unit-admissible constant.*

Proof. Take $\tilde{\psi} \in C_0^\infty(B^n(7/8))$ with $\psi \prec \psi$. Then

$$\begin{aligned}
\widehat{H}_{1,J}(y,\zeta) = {}& \psi(y)F(\Phi_{0,2^{-J}}(y), \{\tilde{\psi}(y)\overline{u}_{J,a}(y)\}_{\deg_{\vec{d}^1}(a)\leq\kappa} + \zeta) \\
&- \psi(y)F(\Phi_{0,2^{-J}}(y), \{\tilde{\psi}(y)\overline{u}_{J,a}(0)\}_{\deg_{\vec{d}^1}(a)\leq\kappa} + \zeta),
\end{aligned}$$

where $\overline{u}_{J,a}$ is as in Lemma 9.2.20. Lemma 9.2.12 (in particular, (9.49)) shows that

$$\psi(y)F(\Phi_{0,2^{-J}}(y), \zeta) \in \mathscr{C}^{\vec{\mathbf{s}},\mathbf{t}}(\overline{B^n(7/8)} \times \mathbb{R}^N, (W^{0,2^{-J}}, \vec{d}) \boxtimes \nabla_{\mathbb{R}^N})$$

with

$$\|\psi(y)F(\Phi_{0,2^{-J}}(y), \zeta)\|_{\mathscr{C}^{\vec{\mathbf{s}},\mathbf{t}}((W^{0,2^{-J}},\vec{d})\boxtimes\nabla_{\mathbb{R}^N})} \lesssim \|F\|_{\mathscr{C}^{\vec{\mathbf{s}},\mathbf{t}}((W,\vec{d})\boxtimes\nabla_{\mathbb{R}^N})}.$$

Applying Theorem 7.5.2 (v) with F replaced by $\psi(y)F(\Phi_{0,2^{-J}}(y), \zeta)$, u replaced by $\{\tilde{\psi}(y)\overline{u}_{J,a}(y)\}_{\deg_{\vec{d}^1}(a)\leq\kappa}$, v replaced by $\{\tilde{\psi}(y)\overline{u}_{J,a}(0)\}_{\deg_{\vec{d}^1}(a)\leq\kappa}$, and $(W, \vec{d})$ replaced by $(W^{0,2^{-J}}, \vec{d})$ shows that $H_{1,J} \in \mathscr{C}^{\vec{\mathbf{s}},\mathbf{t}-\lfloor|\vec{\mathbf{s}}|_1\rfloor-2-\nu}(\overline{B^n(7/8)} \times \mathbb{R}^N, (W, \vec{d}) \boxtimes \nabla_{\mathbb{R}^N})$, and using the above remarks shows that

$$\begin{aligned}
&\|\widehat{H}_{1,J}\|_{\mathscr{C}^{\vec{\mathbf{s}},\mathbf{t}-\lfloor|\vec{\mathbf{s}}|_1\rfloor-2-\nu}((W,\vec{d})\boxtimes\nabla_{\mathbb{R}^N})} \\
&\quad \lesssim \|F\|_{\mathscr{C}^{\vec{\mathbf{s}},\mathbf{t}}((W,\vec{d})\boxtimes\nabla_{\mathbb{R}^N})}\Big(\big\|\{\tilde{\psi}(y)\overline{u}_{J,a}(y) - \tilde{\psi}(y)\overline{u}_{J,a}(0)\}_{\deg_{\vec{d}^1}(a)\leq\kappa}\big\|_{\mathscr{C}^{\vec{\mathbf{s}}}(W^{0,2^{-J}},\vec{d})} \\
&\qquad + \big(\|\{\tilde{\psi}(y)\overline{u}_{J,a}(y)\}_{\deg_{\vec{d}^1}(a)\leq\kappa}\|_{\mathscr{C}^{\vec{\mathbf{s}}}(W,\vec{d})} + \|\{\tilde{\psi}(y)\overline{u}_{J,a}(0)\}_{\deg_{\vec{d}^1}(a)\leq\kappa}\|_{\mathscr{C}^{\vec{\mathbf{s}}}(W,\vec{d})}\big) \\
&\qquad \times \big\|\{\tilde{\psi}(y)\overline{u}_{J,a}(y) - \tilde{\psi}(y)\overline{u}_{J,a}(0)\}_{\deg_{\vec{d}^1}(a)\leq\kappa}\big\|_{L^\infty}\Big) \\
&\qquad \times \Big(1 + \big\|\{\tilde{\psi}(y)\overline{u}_{J,a}(y)\}_{\deg_{\vec{d}^1}(a)\leq\kappa}\big\|_{L^\infty} + \big\|\{\tilde{\psi}(y)\overline{u}_{J,a}(0)\}_{\deg_{\vec{d}^1}(a)\leq\kappa}\big\|_{L^\infty}\Big)^{\lfloor|s|_1\rfloor+\nu}.
\end{aligned} \tag{9.117}$$

By (9.108), we have

$$\|\{\tilde{\psi}(y)\overline{u}_{J,a}(y)\}_{\deg_{\vec{d}^1}(a)\leq\kappa}\|_{\mathscr{C}^{\vec{\mathbf{s}}}(W,\vec{d})} \lesssim \|u\|_{\mathscr{C}^{\vec{\mathbf{r}}+\kappa e_1}}. \tag{9.118}$$

Therefore, using Corollary 7.2.2,

$$\begin{aligned}
&\big\|\{\tilde{\psi}(y)\overline{u}_{J,a}(y)\}_{\deg_{\vec{d}^1}(a)\leq\kappa}\big\|_{L^\infty} \\
&\quad \lesssim \|\{\tilde{\psi}(y)\overline{u}_{J,a}(y)\}_{\deg_{\vec{d}^1}(a)\leq\kappa}\|_{\mathscr{C}^{\vec{\mathbf{s}}}(W,\vec{d})} \lesssim \|u\|_{\mathscr{C}^{\vec{\mathbf{r}}+\kappa e_1}}.
\end{aligned} \tag{9.119}$$

By taking $\hat{\psi} \in C_0^\infty(B^n(7/8))$ with $\tilde{\psi}(0) = 1$, we have

$$\begin{aligned}(9.120)\quad &\|\{\tilde{\psi}(y)\overline{u}_{J,a}(0)\}_{\deg_{\vec{d}1}(a)\leq\kappa}\|_{\mathscr{C}^{\vec{s}}(W,\vec{d})}\\ &\lesssim \|\tilde{\psi}\|_{\mathscr{C}^{\vec{s}}(W,\vec{d})}\left\|\{\hat{\psi}(y)\overline{u}_{J,a}(y)\}_{\deg_{\vec{d}1}(a)\leq\kappa}\right\|_{L^\infty}\\ &\lesssim \|u\|_{\mathscr{C}^{\vec{r}+\kappa e_1}},\end{aligned}$$

where in the last inequality we have used (9.119) (with $\tilde{\psi}$ replaced by $\hat{\psi}$) and the fact that $\tilde{\psi} \in C_0^\infty(B^n(7/8))$ is fixed. Using Corollary 7.2.2 and (9.120), we have

$$\begin{aligned}(9.121)\quad \left\|\{\tilde{\psi}(y)\overline{u}_{J,a}(0)\}_{\deg_{\vec{d}1}(a)\leq\kappa}\right\|_{L^\infty} &\lesssim \|\{\tilde{\psi}(y)\overline{u}_{J,a}(0)\}_{\deg_{\vec{d}1}(a)\leq\kappa}\|_{\mathscr{C}^{\vec{s}}(W,\vec{d})}\\ &\lesssim \|u\|_{\mathscr{C}^{\vec{r}+\kappa e_1}}.\end{aligned}$$

Using (9.109), we have

$$\begin{aligned}(9.122)\quad &\left\|\{\tilde{\psi}(y)\overline{u}_{J,a}(y) - \tilde{\psi}(y)\overline{u}_{J,a}(0)\}_{\deg_{\vec{d}1}(a)\leq\kappa}\right\|_{\mathscr{C}^{\vec{s}}(W^{0,2^{-J}},\vec{d})}\\ &= \left\|\{\tilde{\psi}(y)\tilde{u}_{J,a}(y)\}_{\deg_{\vec{d}1}(a)\leq\kappa}\right\|_{\mathscr{C}^{\vec{s}}(W^{0,2^{-J}},\vec{d})} \lesssim 2^{-J(\vec{r}_1\wedge 1)}\|u\|_{\mathscr{C}^{\vec{r}+\kappa e_1}}.\end{aligned}$$

Therefore, using Corollary 7.2.2,

$$\begin{aligned}(9.123)\quad &\left\|\{\tilde{\psi}(y)\overline{u}_{J,a}(y) - \tilde{\psi}(y)\overline{u}_{J,a}(0)\}_{\deg_{\vec{d}1}(a)\leq\kappa}\right\|_{L^\infty}\\ &\lesssim \left\|\{\tilde{\psi}(y)\overline{u}_{J,a}(y) - \tilde{\psi}(y)\overline{u}_{J,a}(0)\}_{\deg_{\vec{d}1}(a)\leq\kappa}\right\|_{\mathscr{C}^{\vec{s}}(W^{0,2^{-J}},\vec{d})}\\ &\lesssim 2^{-J(\vec{r}_1\wedge 1)}\|u\|_{\mathscr{C}^{\vec{r}+\kappa e_1}}.\end{aligned}$$

Using (9.118), (9.119), (9.120), (9.121), (9.122), and (9.123) to bound the right-hand side of (9.117) shows that

$$\begin{aligned}&\|\hat{H}_{1,J}\|_{\mathscr{C}^{\vec{s},\mathbf{t}-\lfloor|\vec{s}|_1\rfloor-2-\nu}((W,\vec{d})\boxtimes\nabla_{\mathbb{R}^N})}\\ &\lesssim \|F\|_{\mathscr{C}^{\vec{s},\mathbf{t}}((W,\vec{d})\boxtimes\nabla_{\mathbb{R}^N})}\left(2^{-J(\vec{r}_1\wedge 1)}\|u\|_{\mathscr{C}^{\vec{r}+\kappa e_1}(W,\vec{d})} + 2^{-J(\vec{r}_1\wedge 1)}\|u\|^2_{\mathscr{C}^{\vec{r}+\kappa e_1}(W,\vec{d})}\right)\\ &\quad\times\left(1 + \|u\|_{\mathscr{C}^{\vec{r}+\kappa e_1}(W,\vec{d})}\right)^{\lfloor|s|_1\rfloor+\nu}\\ &= \|F\|_{\mathscr{C}^{\vec{s},\mathbf{t}}((W,\vec{d})\boxtimes\nabla_{\mathbb{R}^N})}2^{-J(\vec{r}_1\wedge 1)}\|u\|_{\mathscr{C}^{\vec{r}+\kappa e_1}(W,\vec{d})}\left(1 + \|u\|_{\mathscr{C}^{\vec{r}+\kappa e_1}(W,\vec{d})}\right)^{\lfloor|s|_1\rfloor+\nu+1}.\end{aligned}$$ □

Lemma 9.2.23. *With the same setup as Proposition* 9.2.21, *for $\zeta_0 \in \mathbb{R}^N$ fixed, set*

$$\hat{H}_{2,J}(y,\zeta) := \psi(y)(F(\Phi_{0,2^{-J}}(y), \zeta_0 + \zeta) - F(\Phi_{0,2^{-J}}(0), \zeta_0 + \zeta)).$$

Then $\hat{H}_{2,J} \in \mathscr{C}^{\vec{s},\mathbf{t}}(\overline{B^n(7/8)} \times \mathbb{R}^N, (W^{0,2^{-J}}, \vec{d}) \boxtimes \nabla_{\mathbb{R}^N})$ and

$$\|\hat{H}_{2,J}\|_{\mathscr{C}^{\vec{s},\mathbf{t}}((W^{0,2^{-J}},\vec{d})\boxtimes\nabla_{\mathbb{R}^N})} \leq C2^{-J\min_\mu(\lambda_\mu(\vec{s}_\mu\wedge 1))}\|F\|_{\mathscr{C}^{\vec{s},\mathbf{t}}((W,\vec{d})\boxtimes\nabla_{\mathbb{R}^N})}.$$

Here, $C = C(\psi, N) \geq 0$ is an $(\vec{s},\mathbf{t})$-multi-parameter unit-admissible constant.

Proof. Since $\|\hat{H}_{2,J}(y,\zeta)\|_{\mathscr{C}^{\vec{\mathbf{s}},\mathbf{t}}((W^{0,2^{-J}},\vec{d})\boxtimes\nabla_{\mathbb{R}^N})} = \|\hat{H}_{2,J}(y,\zeta-\zeta_0)\|_{\mathscr{C}^{\vec{\mathbf{s}},\mathbf{t}}((W^{0,2^{-J}},\vec{d})\boxtimes\nabla_{\mathbb{R}^N})}$, it suffices to prove the result with $\zeta_0 = 0$; we henceforth assume $\zeta_0 = 0$. By Proposition 7.5.9, we may decompose $F = \sum_{j\in\mathbb{N}^\nu, k\in\mathbb{N}} F_{j,k}$, where $F_{j,k} \in C^\infty_{\mathrm{loc}}(B^n(1)\times\mathbb{R}^N)$ satisfy, $\forall \alpha,\beta$,

$$\|(2^{-j\vec{d}}W)^\alpha(2^{-k}\partial_\zeta)^\beta F_{j,k}\|_{L^\infty} \lesssim_{\alpha,\beta} 2^{-j\cdot\vec{\mathbf{s}}-k\mathbf{t}}\|F\|_{\mathscr{C}^{\vec{\mathbf{s}},\mathbf{t}}((W,\vec{d})\boxtimes\nabla_{\mathbb{R}^N})}. \tag{9.124}$$

Recall that $\lambda = (\lambda_1,\dots,\lambda_\nu) \in (0,\infty)^\nu$, with $\lambda_1 = 1$. For $J\in\mathbb{N}$, $j\in\mathbb{N}^\nu$, set

$$\begin{aligned}\mathscr{J}(J,j) := \{j' \in \mathbb{N}^\nu : {}& j_\mu + \lambda_\mu J \le j'_\mu < j_\mu + \lambda_\mu J + 1 \text{ if } j_\mu > 0,\\ & 0 \le j'_\mu < \lambda_\mu J + 1 \text{ if } j_\mu = 0\},\end{aligned}$$

and set

$$H_{J,j,k}(x,\zeta) := \sum_{j'\in\mathscr{J}(J,j)} F_{j',k}(x,\zeta) - F_{j',k}(0,\zeta).$$

Note that $\sum_{\substack{j\in\mathbb{N}^\nu\\ k\in\mathbb{N}}} H_{J,j,k}(x,\zeta) = F(x,\zeta) - F(0,\zeta)$ and therefore

$$\hat{H}_{2,J}(y,\zeta) = \psi(y)\sum_{\substack{j\in\mathbb{N}^\nu\\ k\in\mathbb{N}}} H_{J,j,k}(\Phi_{0,2^{-J}}(y),\zeta).$$

Thus, by Proposition 9.2.18 (ii) with $E = 2^{-J(\min_\mu \lambda_\mu(\vec{\mathbf{s}}_\mu\wedge 1))}$, to prove the result it suffices to show that for every α,β,

$$\begin{aligned}&\|(2^{-(j+J\lambda)\vec{d}}W)^\alpha(2^{-k}\partial_\zeta)^\beta H_{J,j,k}(x,\zeta)\|_{L^\infty(B_{(X^1,d^1)}(0,2^{-J})\times\mathbb{R}^N)}\\ &\quad\lesssim_{\alpha,\beta} 2^{-j\cdot\vec{\mathbf{s}}-k\mathbf{t}}2^{-J(\min_\mu \lambda_\mu(\vec{\mathbf{s}}_\mu\wedge 1))}\|F\|_{\mathscr{C}^{\vec{\mathbf{s}},\mathbf{t}}((W,\vec{d})\boxtimes\nabla_{\mathbb{R}^N})}.\end{aligned}\tag{9.125}$$

We first establish (9.125) in the case where either $|\alpha|>0$ or $|j|>0$. Pick μ_0 such that either $j_{\mu_0}>0$ or $\deg_{\vec{d}}(\alpha)_{\mu_0}>0$. We have, using (9.124) and the definition of $\mathscr{J}(J,j)$,

$$\begin{aligned}&\|(2^{-(j+J\lambda)\vec{d}}W)^\alpha(2^{-k}\partial_\zeta)^\beta H_{J,j,k}(x,\zeta)\|_{L^\infty(B^n(1)\times\mathbb{R}^N)}\\ &\le 2\sum_{j'\in\mathscr{J}(J,j)}\|(2^{-(j+J\lambda)\vec{d}}W)^\alpha(2^{-k}\partial_\zeta)^\beta F_{j',k}(x,\zeta)\|_{L^\infty(B^n(1)\times\mathbb{R}^N)}\\ &= 2\sum_{j'\in\mathscr{J}(J,j)} 2^{-(j+\lambda J-j')\cdot\deg_{\vec{d}}(\alpha)}\|(2^{-j'\vec{d}}W)^\alpha(2^{-j}\partial_\zeta)^\beta F_{j',k}\|_{L^\infty(B^n(1)\times\mathbb{R}^N)}\\ &\lesssim_{\alpha,\beta}\sum_{j'\in\mathscr{J}(J,j)} 2^{-(j+\lambda J-j')\cdot\deg_{\vec{d}}(\alpha)}2^{-j'\cdot\vec{\mathbf{s}}-k\mathbf{t}}\|F\|_{\mathscr{C}^{\vec{\mathbf{s}},\mathbf{t}}((W,\vec{d})\boxtimes\nabla_{\mathbb{R}^N})}\end{aligned}$$

$$\approx 2^{-k\mathbf{t}}\|F\|_{\mathscr{C}^{\vec{\mathbf{s}},\mathbf{t}}((W,\vec{d})\boxtimes\nabla_{\mathbb{R}^N})}\left[\prod_{\mu:j_\mu>0} 2^{-(j_\mu+\lambda_\mu J)\vec{\mathbf{s}}_\mu}\right]$$
$$\times\left[\prod_{\mu:j_\mu=0}\sum_{0\leq j'_\mu<\lambda_\mu J+1} 2^{-(J\lambda_\mu-j'_\mu)\mathfrak{e}_\mu \deg_{\vec{d}}(\alpha)_\mu}2^{-j'_\mu\vec{\mathbf{s}}_\mu}\right]$$
$$\lesssim 2^{-k\mathbf{t}}\|F\|_{\mathscr{C}^{\vec{\mathbf{s}},\mathbf{t}}((W,\vec{d})\boxtimes\nabla_{\mathbb{R}^N})}\left[\prod_{\mu:j_\mu>0} 2^{-(j_\mu+\lambda_\mu J)\vec{\mathbf{s}}_\mu}\right]\left[\prod_{\substack{\mu:j_\mu=0\\ \deg_{\vec{d}}(\alpha)_\mu\geq 1}} 2^{-J\lambda_\mu(\vec{\mathbf{s}}_\mu\wedge 1)}\right]$$
$$\lesssim 2^{-k\mathbf{t}-j\cdot\vec{\mathbf{s}}}2^{-\lambda_\mu J(\vec{\mathbf{s}}_\mu\wedge 1)}\|F\|_{\mathscr{C}^{\vec{\mathbf{s}},\mathbf{t}}((W,\vec{d})\boxtimes\nabla_{\mathbb{R}^N})},$$

establishing (9.125) in this case.

Finally, we prove (9.125) in the case where $|\alpha|=0$ and $|j|=0$. Let $x\in B_{(X^1,d^1)}(0,2^{-J})$; then there exists an absolutely continuous $\gamma:[0,1]\to B^n(1)$ with $\gamma(0)=0$, $\gamma(1)=x$, $\gamma'(t)=\sum_{l=1}^{q_1}2^{-Jd_l^1}a_l(t)X_l^1(\gamma(t))$, where $\|\sum_l|a_l|^2\|_{L^\infty([0,1])}<1$. Using the fact that $\lambda_1=1$ and therefore $j_1'<J+1$, $\forall j'\in\mathscr{J}(J,0)$, and (9.124), we have

$$\begin{aligned}
&|(2^{-k}\partial_\zeta)^\beta H_{J,0,l}(x,\zeta)|\\
&\quad\leq\sum_{j'\in\mathscr{J}(J,0)}|(2^{-k}\partial_\zeta)^\beta F_{j',k}(x,\zeta)-(2^{-k}\partial_\zeta)^\beta F_{j',k}(0,\zeta)|\\
&\quad\leq\sum_{j'\in\mathscr{J}(J,0)}\int_0^1\left|(2^{-k}\partial_\zeta)^\beta\frac{d}{dt}F_{j',k}(\gamma(t),\zeta)\right|dt\\
&\quad\lesssim\sum_{j'\in\mathscr{J}(J,0)}\sum_{l=1}^{q_1}2^{-(J-j_1')d_l^1}\int_0^1\left|2^{-j_1'd_l^1}X_l^1(2^{-k}\partial_\zeta)^\beta F_{j',k}(\gamma(t),\zeta)\right|dt\\
&\quad\lesssim\sum_{j'\in\mathscr{J}(J,0)}\sum_{l=1}^{q_1}2^{-(J-j_1')d_l^1}\|2^{-j_1'd_l^1}X_l^1(2^{-k}\partial_\zeta)^\beta F_{j',k}(x,\zeta)\|_{L^\infty}\\
&\quad\lesssim_\beta\sum_{j'\in\mathscr{J}(J,0)}2^{-(J-j_1')}2^{-j'\cdot\vec{\mathbf{s}}}2^{-k\mathbf{t}}\|F\|_{\mathscr{C}^{\vec{\mathbf{s}},\mathbf{t}}((W,\vec{d})\boxtimes\nabla_{\mathbb{R}^N})}\\
&\quad\lesssim 2^{-k\mathbf{t}}2^{-J(\vec{\mathbf{s}}_1\wedge 1)}\|F\|_{\mathscr{C}^{\vec{\mathbf{s}},\mathbf{t}}((W,\vec{d})\boxtimes\nabla_{\mathbb{R}^N})}=2^{-k\mathbf{t}}2^{-J\lambda_1(\vec{\mathbf{s}}_1\wedge 1)}\|F\|_{\mathscr{C}^{\vec{\mathbf{s}},\mathbf{t}}((W,\vec{d})\boxtimes\nabla_{\mathbb{R}^N})}.
\end{aligned}\tag{9.126}$$

Taking the supremum of (9.126) over $x\in B_{(X^1,d^1)}(0,2^{-J})$ establishes (9.125) in this case and completes the proof. □

Proof of Proposition 9.2.21. We denote the case where all $c_\alpha=1$ by $\hat{H}_J^1$. Since

$$\|\hat{H}_J^c\|_{\mathscr{C}^{\vec{\mathbf{s}},\mathbf{t}-\lfloor|\vec{\mathbf{s}}|_1\rfloor-2-\nu}((W^{0,2^{-J}},\vec{d})\boxtimes\nabla_{\mathbb{R}^N})}\lesssim\|\hat{H}_J^1\|_{\mathscr{C}^{\vec{\mathbf{s}},\mathbf{t}-\lfloor|\vec{\mathbf{s}}|_1\rfloor-2-\nu}((W^{0,2^{-J}},\vec{d})\boxtimes\nabla_{\mathbb{R}^N})},$$

it suffices to prove the result for $\hat{H}_J^1$. We have

$$\hat{H}_J^1(y,\zeta)=\hat{H}_{1,J}(y,\zeta)+\hat{H}_{2,J}(y,\zeta),$$

where

$$\hat{H}_{1,J}(y,\zeta) := \psi(y)\Bigg[F\Bigg(\Phi_{0,2^{-J}}(y), \sum_{\substack{j_1<J\\ j_2,\ldots,j_\nu\in\mathbb{N}}} \mathfrak{W}_1^\kappa D_j u(\Phi_{0,2^{-J}}(y)) + \zeta\Bigg) - F\Bigg(\Phi_{0,2^{-J}}(y), \sum_{\substack{j_1<J\\ j_2,\ldots,j_\nu\in\mathbb{N}}} \mathfrak{W}_1^\kappa D_j u(0) + \zeta\Bigg)\Bigg],$$

$$\hat{H}_{2,J}(y,\zeta) := \psi(y)\Bigg[F\Bigg(\Phi_{0,2^{-J}}(y), \sum_{\substack{j_1<J\\ j_2,\ldots,j_\nu\in\mathbb{N}}} \mathfrak{W}_1^\kappa D_j u(0) + \zeta\Bigg) - F\Bigg(0, \sum_{\substack{j_1<J\\ j_2,\ldots,j_\nu\in\mathbb{N}}} \mathfrak{W}_1^\kappa D_j u(0) + \zeta\Bigg)\Bigg].$$

Lemma 9.2.22 shows that

$$\begin{aligned}&\|\hat{H}_{1,J}\|_{\mathscr{C}^{\vec{\mathbf{s}},\mathbf{t}-\lfloor|\vec{\mathbf{s}}|_1\rfloor-2-\nu}((W^{0,2^{-J}},\vec{d})\boxtimes\nabla_{\mathbb{R}^N})}\\ &\quad\lesssim 2^{-J(\vec{\mathbf{r}}_1\wedge 1)}\|F\|_{\mathscr{C}^{\vec{\mathbf{s}},\mathbf{t}}((W,\vec{d})\boxtimes\nabla_{\mathbb{R}^N})}\big(1+\|u\|_{\mathscr{C}^{\vec{\mathbf{r}}+\kappa e_1}(W,\vec{d})}\big)^{\lfloor|\vec{\mathbf{s}}|_1\rfloor+\nu+1}\|u\|_{\mathscr{C}^{\vec{\mathbf{r}}+\kappa e_1}(W,\vec{d})}.\end{aligned}$$

Lemma 9.2.23 shows that

$$\|\hat{H}_{2,J}\|_{\mathscr{C}^{\vec{\mathbf{s}},\mathbf{t}}((W^{0,2^{-J}},\vec{d})\boxtimes\nabla_{\mathbb{R}^N})} \lesssim 2^{-J\min_\mu(\lambda_\mu(\vec{\mathbf{s}}_\mu\wedge 1))}\|F\|_{\mathscr{C}^{\vec{\mathbf{s}},\mathbf{t}}((W,\vec{d})\boxtimes\nabla_{\mathbb{R}^N})}.$$

The result follows. □

Proof of Theorem 9.2.1. As mentioned before, by replacing $F(x,\zeta)$ with $F(x,\zeta)-g(x)$, it suffices to consider the case $g=0$. Since δ and C_5 in the statement of the theorem are allowed to depend on C_0, in this proof we have

$$\|F\|_{\mathscr{C}^{\vec{\mathbf{s}},2\lfloor|\vec{\mathbf{s}}|_1\rfloor+4+2\nu+\sigma}((W,\vec{d})\boxtimes\nabla_{\mathbb{R}^N})}, \|u\|_{\mathscr{C}^{\vec{\mathbf{r}}+\kappa e_1}(W,\vec{d})} \lesssim 1.$$

Fix $\psi\in C_0^\infty(B^n(7/8))$ with $\psi\equiv 1$ on a neighborhood of $B^n(13/16)$. We consider only $J\in\mathbb{N}_+$ large enough that $B_{(X^1,d^1)}(0,2^{-J})\subseteq B^n(13/16)$; by the Picard–Lindelöf theorem how large J must be for $B_{(X^1,d^1)}(0,2^{-J})\subseteq B^n(13/16)$ can be chosen to depend only on an upper bound for $\sum_{l=1}^{q_1}\|X_l^1\|_{C^1}$ and therefore may be bounded by a 0-multi-parameter unit-admissible constant.

Set

$$\begin{aligned}&\hat{u}_J(y) := \psi(y)u_J(\Phi_{0,2^{-J}}(y)), \quad \hat{v}_J(y) := \psi(y)u_J(\Phi_{0,2^{-J}}(y)),\\ &\hat{H}_J(y,\zeta) := \psi(y)H_J(\Phi_{0,2^{-J}}(y),\zeta).\end{aligned}$$

Since $\psi(y) \equiv 1$ on $B^n(13/16)$ and $\Phi_{0,2^{-J}}(B^n(1)) \subseteq B_{(X^1,\vec{d}^1)}(0,2^{-J}) \subseteq B^n(13/16)$ (see Theorem 3.15.5), it follows from (9.107) that

$$G_J(\{(W^{1,0,2^{-J}})^\alpha \hat{v}_J(y)\}) + \hat{H}_J(y, \{(W^{1,0,2^{-J}})^\alpha \hat{v}_J(y)\}) = 0, \quad \forall y \in B^n(13/16). \tag{9.127}$$

We will show that for J sufficiently large, (9.127) satisfies the hypotheses of Proposition 9.2.17.

Using Corollary 7.5.12 (iii), we have

$$\begin{aligned}\|G_J\|_{\mathscr{C}^{\lfloor|\vec{s}|_1\rfloor+3+\nu+\sigma}(\mathbb{R}^N)} &= \|F(0,\cdot)\|_{\mathscr{C}^{\lfloor|\vec{s}|_1\rfloor+3+\nu+\sigma}(\mathbb{R}^N)} \\ &\lesssim \|F\|_{\mathscr{C}^{\vec{s},\lfloor|\vec{s}|_1\rfloor+3+\nu+\sigma}((W,\vec{d})\boxtimes\nabla_{\mathbb{R}^N})} \le \|F\|_{\mathscr{C}^{\vec{s},2\lfloor|\vec{s}|_1\rfloor+4+2\nu+\sigma}((W,\vec{d})\boxtimes\nabla_{\mathbb{R}^N})} \lesssim 1.\end{aligned}$$

That is, $\exists C_3 \approx 1$ with

$$\|G_J\|_{\mathscr{C}^{\lfloor|\vec{s}|_1\rfloor+3+\nu+\sigma}(\mathbb{R}^N)} \le C_3. \tag{9.128}$$

Set

$$\widehat{\mathscr{P}}_J w := dG_J(\{(W^{1,0,2^{-J}})^\alpha \hat{v}_J(0)\}_{\deg_{\vec{d}^1}(\alpha)})\{(W^{1,0,2^{-J}})^\alpha w\}_{\deg_{\vec{d}^1}(\alpha)\le\kappa}. \tag{9.129}$$

We wish to show that $\widehat{\mathscr{P}}_J$ is maximally subelliptic of degree κ with respect to $(W^{1,0,2^{-J}}, \vec{d}^1)$, *uniformly* in J in a quantitative way.

Let $\mathscr{P}$ be given by (9.13). Then $\mathscr{P}$ is of the form

$$\mathscr{P} = \sum_{\deg_{\vec{d}^1}(\alpha)\le\kappa} a_\alpha (W^1)^\alpha,$$

where $a_\alpha \in \mathbb{M}^{\mathbf{D}_1\times\mathbf{D}_2}(\mathbb{R})$ are constant matrices; a_α are defined by

$$dG(\{(W^1)^\alpha u(0)\})\{\zeta_\alpha\}_{\deg_{\vec{d}^1}(\alpha)\le\kappa} = \sum_{\deg_{\vec{d}^1}(\alpha)\le\kappa} a_\alpha \zeta_\alpha.$$

Using the fact that $(W^1)^\alpha u_J(0) + 2^{-\kappa J}(W^1)^\alpha v_J(0) = (W^1)^\alpha u(0)$ (see (9.106)), we have

$$\begin{aligned}&dG_J(\{(2^{-J\vec{d}^1} W^1)^\alpha v_J(0)\})\{\zeta_\alpha\}_{\deg_{\vec{d}^1}(\alpha)\le\kappa} \\ &\quad = dG(\{(W^1)^\alpha u(0)\})\{2^{-(\kappa-\deg_{\vec{d}^1}(\alpha))J}\zeta_\alpha\}_{\deg_{\vec{d}^1}(\alpha)\le\kappa} \\ &\quad = \sum_{\deg_{\vec{d}^1}(\alpha)\le\kappa} 2^{-(\kappa-\deg_{\vec{d}^1}(\alpha))J} a_\alpha\zeta_\alpha.\end{aligned} \tag{9.130}$$

Since $\Phi_{0,2^{-J}}(0) = 0$ and $\Phi_{0,2^{-J}}^* 2^{-J d_l^1} W_l^1 = W_l^{1,0,2^{-J}}$ (see Theorem 3.15.5), we have

$$(W^{1,0,2^{-J}})^\alpha \hat{v}_J(0) = (2^{-J\vec{d}^1} W^1)^\alpha v_J(0). \tag{9.131}$$

Plugging (9.131) into (9.130) and using (9.129), we have

$$\begin{aligned}\widehat{\mathscr{P}}_J w &:= dG_J(\{(W^{1,0,2^{-J}})^\alpha \hat{v}_J(0)\}_{\deg_{\vec{d}^1}(\alpha)})\{(W^{1,0,2^{-J}})^\alpha w\}_{\deg_{\vec{d}^1}(\alpha)\le\kappa} \\ &= dG_J(\{(2^{-J\vec{d}^1}W^1)^\alpha v_J(0)\})\{(W^{1,0,2^{-J}})^\alpha w\}_{\deg_{\vec{d}^1}(\alpha)\le\kappa} \\ &= \sum_{\deg_{\vec{d}^1}(\alpha)\le\kappa} 2^{-(\kappa-\deg_{\vec{d}^1}(\alpha))J} a_\alpha (W^{1,0,2^{-J}})^\alpha w.\end{aligned}$$

Thus, we have

$$\begin{aligned}\Phi_{0,2^{-J}}^* 2^{-\kappa J}\mathscr{P}(\Phi_{0,2^{-J}})_* &= \Phi_{0,2^{-J}}^* \sum_{\deg_{\vec{d}^1}(\alpha)\le\kappa} 2^{-(\kappa-\deg_{\vec{d}^1}(\alpha))J} a_\alpha (2^{-J\vec{d}^1}W^1)^\alpha (\Phi_{0,2^{-J}})_* \\ &= \sum_{\deg_{\vec{d}^1}(\alpha)\le\kappa} 2^{-(\kappa-\deg_{\vec{d}^1}(\alpha))J} a_\alpha (W^{1,0,2^{-J}})^\alpha = \widehat{\mathscr{P}}_J.\end{aligned} \tag{9.132}$$

Multiplying (9.14) by $2^{-\kappa J}$ gives, $\forall f \in C_0^\infty(B^n(1);\mathbb{C}^{\mathbf{D}_2})$,

$$\begin{aligned}&\sum_{j=1}^{r_1} \|(2^{-J d_j^1} W_j^1)^{n_j} f\|_{L^2(B^n(1),h\sigma_{\mathrm{Leb}};\mathbb{C}^{\mathbf{D}_2})} \\ &\quad\le A(\|2^{-\kappa J}\mathscr{P} f\|_{L^2(B^n(1),h\sigma_{\mathrm{Leb}};\mathbb{C}^{\mathbf{D}_1})} + 2^{-\kappa J}\|f\|_{L^2(B^n(1),h\sigma_{\mathrm{Leb}};\mathbb{C}^{\mathbf{D}_2})}).\end{aligned} \tag{9.133}$$

For $g \in C_0^\infty(B^n(1);\mathbb{C}^{\mathbf{D}_2})$ we apply (9.133) to $g \circ \Phi_{0,2^{-J}}^{-1} \in C_0^\infty(B^n(1);\mathbb{C}^{\mathbf{D}_2})$ and use (9.132) to see that

$$\begin{aligned}&\sum_{j=1}^{r_1} \|(W^{1,0,2^{-J}})^{n_j} g\|_{L^2(B^n(1),\Lambda(0,2^{-J})h_{0,2^{-J}}\sigma_{\mathrm{Leb}};\mathbb{C}^{\mathbf{D}_2})} \\ &\quad\le A(\|\widehat{\mathscr{P}}_J g\|_{L^2(B^n(1),\Lambda(0,2^{-J})h_{0,2^{-J}}\sigma_{\mathrm{Leb}};\mathbb{C}^{\mathbf{D}_1})} + 2^{-\kappa J}\|g\|_{L^2(B^n(1),\Lambda(0,2^{-J})h_{0,2^{-J}}\sigma_{\mathrm{Leb}};\mathbb{C}^{\mathbf{D}_2})}) \\ &\quad\le A(\|\widehat{\mathscr{P}}_J g\|_{L^2(B^n(1),\Lambda(0,2^{-J})h_{0,2^{-J}}\sigma_{\mathrm{Leb}};\mathbb{C}^{\mathbf{D}_1})} + \|g\|_{L^2(B^n(1),\Lambda(0,2^{-J})h_{0,2^{-J}}\sigma_{\mathrm{Leb}};\mathbb{C}^{\mathbf{D}_2})}),\end{aligned} \tag{9.134}$$

where $\Lambda(0,2^{-J})$ and $h_{0,2^{-J}}$ are as in Theorem 3.15.5.

Multiplying both sides of (9.134) by $\Lambda(0,2^{-J})^{-1/2}$, we obtain

$$\begin{aligned}&\sum_{j=1}^{r_1} \|(W^{1,0,2^{-J}})^{n_j} g\|_{L^2(B^n(1),h_{0,2^{-J}}\sigma_{\mathrm{Leb}};\mathbb{C}^{\mathbf{D}_2})} \\ &\quad\le A(\|\widehat{\mathscr{P}}_J g\|_{L^2(B^n(1),h_{0,2^{-J}}\sigma_{\mathrm{Leb}};\mathbb{C}^{\mathbf{D}_1})} + \|g\|_{L^2(B^n(1),h_{0,2^{-J}}\sigma_{\mathrm{Leb}};\mathbb{C}^{\mathbf{D}_2})}).\end{aligned} \tag{9.135}$$

That is, $\widehat{\mathscr{P}}_J$ is maximally subelliptic of degree κ with respect to $(W^{1,0,2^{-J}}, \vec{d}^1)$ (and using the measure $h_{0,2^{-J}}\sigma_{\mathrm{Leb}}$) with the *same constant* A as in our assumption (9.14).

Theorem 3.15.5 shows that multi-parameter unit-admissible constants with respect to $(W^{0,2^{-J}}, \vec{d})$ and $h_{0,2^{-J}}$ can be taken to be multi-parameter unit-admissible constants with respect to $(W, \vec{d})$ and h. Thus, we may apply Proposition 9.2.17 with $(W, \vec{d})$ and h replaced by $(W^{0,2^{-J}}, \vec{d})$ and $h_{0,2^{-J}}$ to find a $(\kappa, \vec{\mathbf{s}}, \vec{\mathbf{r}}, \sigma)$-multi-parameter unit-admissible constant $\epsilon_2 = \epsilon_2(C_3, A, \mathbf{D}_1, \mathbf{D}_2) \in (0,1]$ as in that result.

We have already shown that G_J satisfies the assumptions of G in Proposition 9.2.17 (see (9.128) and (9.135)). Thus, we will be able to apply Proposition 9.2.17 to (9.127) once we show that if J is sufficiently large, then

$$\|\hat{H}_J\|_{\mathscr{C}^{\vec{\mathbf{s}},\lfloor|\vec{\mathbf{s}}|_1\rfloor+2+\nu+\sigma}((W^{0,2^{-J}},\vec{d})\boxtimes\nabla_{\mathbb{R}^N})} \leq \epsilon_2, \tag{9.136}$$

$$\|\hat{v}_J\|_{\mathscr{C}^{\vec{\mathbf{r}}+\kappa e_1}(W^{0,2^{-J}},\vec{d})} \leq \epsilon_2. \tag{9.137}$$

Proposition 9.2.21 with $c_a = 2^{-(\kappa-\deg_{\vec{d}^1}(a))}$ shows that

$$\begin{aligned}
&\|\hat{H}_J\|_{\mathscr{C}^{\vec{\mathbf{s}},\lfloor|\vec{\mathbf{s}}|_1\rfloor+2+\nu+\sigma}((W^{0,2^{-J}},\vec{d})\boxtimes\nabla_{\mathbb{R}^N})} \\
&\quad = \|\hat{H}_J\|_{\mathscr{C}^{\vec{\mathbf{s}},2\lfloor|\vec{\mathbf{s}}|_1\rfloor+4+2\nu+\sigma-\lfloor|\vec{\mathbf{s}}|_1\rfloor-2-\nu}((W^{0,2^{-J}},\vec{d})\boxtimes\nabla_{\mathbb{R}^N})} \\
&\quad \lesssim \|F\|_{\mathscr{C}^{\vec{\mathbf{s}},2\lfloor|\vec{\mathbf{s}}|_1\rfloor+4+2\nu+\sigma}((W,\vec{d})\boxtimes\nabla_{\mathbb{R}^N})}\big(1+\|u\|_{\mathscr{C}^{\vec{\mathbf{r}}+\kappa e_1}(W,\vec{d})}\big)^{\lfloor|\vec{\mathbf{s}}|_1\rfloor+\nu+2} \\
&\qquad \times \big(2^{-J(\vec{\mathbf{r}}_1\wedge 1)}+2^{-J(\min_\mu \lambda_\mu(s_\mu\wedge 1))}\big) \\
&\quad \lesssim \big(2^{-J(\vec{\mathbf{r}}_1\wedge 1)}+2^{-J(\min_\mu \lambda_\mu(s_\mu\wedge 1))}\big).
\end{aligned} \tag{9.138}$$

The estimate (9.110) shows that

$$\|\hat{v}_J\|_{\mathscr{C}^{\vec{\mathbf{r}}+\kappa e_1}(W^{0,2^{-J}},\vec{d})} \lesssim 2^{-J\vec{\mathbf{r}}_1}\|u\|_{\mathscr{C}^{\vec{\mathbf{r}}+\kappa e_1}(W,\vec{d})} \lesssim 2^{-J\vec{\mathbf{r}}_1}. \tag{9.139}$$

The estimates (9.138) and (9.139) show that if we take $J = J(A, C_0, \mathbf{D}_1, \mathbf{D}_2) \in \mathbb{N}$ to be a sufficiently large $(\vec{\mathbf{s}}, \vec{\mathbf{r}}, \kappa, \sigma)$-multi-parameter unit-admissible constant, then (9.136) and (9.137) hold.

For such a J, Proposition 9.2.17 applies to (9.127) to show that for $\psi_0 \in C_0^\infty(B^n(2/3))$, $\psi_0\hat{v}_J \in \mathscr{C}^{\vec{\mathbf{s}}+\kappa e_1}(\overline{B^n(7/8)}, (W^{0,2^{-J}}, \vec{d}))$ and

$$\big\|\psi_0\hat{v}_J\big\|_{\mathscr{C}^{\vec{\mathbf{s}}+\kappa e_1}(W^{0,2^{-J}},\vec{d})} \lesssim 1. \tag{9.140}$$

Since $\psi \equiv 1$ on a neighborhood of $\overline{B^n(13/16)}$, we have $\psi_0\psi = \psi_0$ and therefore (9.140) is equivalent to

$$\big\|\psi_0\Phi^*_{0,2^{-J}}v_J\big\|_{\mathscr{C}^{\vec{\mathbf{s}}+\kappa e_1}(W^{0,2^{-J}},\vec{d})} \lesssim 1. \tag{9.141}$$

The estimate (9.108) with $|\alpha| = 0$, $\vec{\mathbf{s}}$ replaced with $\vec{\mathbf{s}} + \kappa e_1$ and ψ replaced with ψ_0 shows that

$$\big\|\psi_0\Phi^*_{0,2^{-J}}u_J\big\|_{\mathscr{C}^{\vec{\mathbf{s}}+\kappa e_1}(W^{0,2^{-J}},\vec{d})} \lesssim 1. \tag{9.142}$$

Using (9.106), (9.141) and (9.142) imply

$$\|\psi_0 \Phi_{0,2^{-J}}^{*} u\|_{\mathscr{C}^{\tilde{s}+\kappa e_1}(W^{0,2^{-J}},\vec{d})} \lesssim 1. \tag{9.143}$$

By Lemma 9.2.7, (9.143) is equivalent to

$$\|(\psi_0 \circ \Phi_{0,2^{-J}}^{-1})u\|_{\mathscr{C}^{\tilde{s}+\kappa e_1}(2^{-J\lambda}W,\vec{d})} \lesssim 1. \tag{9.144}$$

Since $J \approx 1$, Lemma 9.2.6 shows that (9.144) implies

$$\|(\psi_0 \circ \Phi_{0,2^{-J}}^{-1})u\|_{\mathscr{C}^{\tilde{s}+\kappa e_1}(W,\vec{d})} \lesssim 1.$$

Taking $\delta = 2^{-J}$ completes the proof. □

9.3 Weighted estimates near the boundary

Let $\Omega \Subset \mathfrak{M}$ be a relatively compact, open set. If a nonlinear maximally subelliptic equation of the form (9.3) holds on Ω, then the estimates obtained for u in Theorem 9.1.2 (on $\|u\|_{\mathscr{C}^{s+\kappa}(W,\vec{d})}$) blow up near the boundary of Ω. To make this blowup precise, we introduce weighted versions of the norms introduced in Section 7.7. See Section 2.8.4 for a description of these ideas in the simpler special case of elliptic PDEs.

We take the same basic setting as in Section 9.1.1, so that $(W,\vec{d}) = \{(W_1,\vec{d}_1),\dots,(W_r,\vec{d}_r)\} \subset C^{\infty}_{\mathrm{loc}}(\mathfrak{M};T\mathfrak{M}) \times \mathbb{N}_+$ are Hörmander vector fields with single-parameter formal degrees on $\mathfrak{M}$. We set, for $x \in \mathfrak{M}$,

$$\mathrm{dist}_{(W,\vec{d})}(x,\partial\Omega) := \inf\{\rho_{(W,\vec{d})}(x,y) : y \in \partial\Omega\}.$$

Fix $\mathcal{K} \Subset \Omega_1 \Subset \mathfrak{M}$, with $\mathcal{K}$ compact and Ω_1 open and relatively compact with $\Omega \Subset \mathcal{K}$. By Proposition 3.4.14 (in the case $\nu = 1$), we may pick a finite collection

$$(X,d) = \{(X_1,d_1),\dots,(X_q,d_q)\} \subset \mathrm{Gen}((W,\vec{d}))$$

such that $\mathrm{Gen}((W,\vec{d}))$ is finitely generated by (X,d) on Ω_1 and $(W,\vec{d}) \subseteq (X,d)$.

Definition 9.3.1. For $l \in \mathbb{N}_+$, $\mathbf{s} \in (0,l)$, and $f \in C_{\mathrm{loc}}(\Omega)$, we define the weighted interior norm:

$$\|f\|_{\widehat{\mathscr{C}}_l^{\mathbf{s}}(\Omega,(W,\vec{d}),(X,d))} := \sum_{l'=0}^{l} \sup_{\delta\in(0,\infty)} \sup_{\gamma\in\mathcal{P}_\delta^{l'}(\Omega,(X,d))} \mathrm{dist}_{(W,\vec{d})}(\gamma(0),\partial\Omega)^{\mathbf{s}l'/l}\delta^{-\mathbf{s}l'/l}|\triangle_\gamma^{l'} f|.$$

Fix $\mathbf{s} > \mathbf{r} > 0$, $\sigma > 0$, and $\mathcal{K} \Subset \mathfrak{M}$ compact with $\Omega \Subset \mathcal{K}$. With $F(x,\zeta) \in \mathscr{C}^{\mathbf{s},3\lfloor \mathbf{s}\rfloor+8+\sigma}(\mathcal{K}\times\mathbb{R}^N,(W,\vec{d})\boxtimes\nabla_{\mathbb{R}^N};\mathbb{R}^{\mathbf{D}_1})$, $u \in \mathscr{C}^{\mathbf{r}+\kappa}(\mathcal{K},(W,\vec{d});\mathbb{R}^{\mathbf{D}_2})$, and $g \in \mathscr{C}^{\mathbf{s}}(\mathcal{K},(W,\vec{d});\mathbb{R}^{\mathbf{D}_1})$ given, we assume

$$F(x, \{W^\alpha u(x)\}_{\deg_{d}(\alpha)\le\kappa}) = g(x), \quad x \in \Omega. \tag{9.145}$$

Since the functions involved are continuous, this equality in fact holds on $\overline{\Omega}$.

Finally, we assume that $F(x, \{W^\alpha u(x)\}_{\deg_{d}(\alpha)\le\kappa})$ is maximally subelliptic in the following sense.

Maximally subelliptic assumption: For each $x_0 \in \Omega$, set

$$\mathscr{P}_{x_0} v := d_\zeta F(x_0, \{W^\alpha u(x_0)\}_{\deg_{d}(\alpha)\le\kappa})\{W^\alpha v\}_{\deg_{d}(\alpha)\le\kappa}. \tag{9.146}$$

We assume that there exists $A \ge 0$ such that $\forall f \in C_0^\infty(\Omega; \mathbb{C}^{\mathbf{D}_2})$, $\forall x_0 \in \Omega$,

$$\sum_{j=1}^{r} \|W_j^{n_j} f\|_{L^2(\mathfrak{M},\mathrm{Vol};\mathbb{C}^{\mathbf{D}_2})} \le A\big(\|\mathscr{P}_{x_0} f\|_{L^2(\mathfrak{M},\mathrm{Vol};\mathbb{C}^{\mathbf{D}_1})} + \|f\|_{L^2(\mathfrak{M},\mathrm{Vol};\mathbb{C}^{\mathbf{D}_2})}\big). \tag{9.147}$$

Theorem 9.3.2. *Under the above hypotheses, $\forall l \in \mathbb{N}_+$ with $l > \mathrm{s} + \kappa$, there exists $C \ge 0$ such that*

$$\|u\|_{\mathscr{C}_l^{\mathrm{s}+\kappa}(\Omega,(W,d),(X,d))} \le C.$$

Here, $C \ge 0$ can be chosen such that the dependence of C on F, u, and g is only in terms of $A \ge 0$ and upper bounds for $\|F\|_{\mathscr{C}^{\mathrm{s},3\lfloor \mathrm{s}\rfloor+8+\sigma}((W,d)\boxtimes\nabla_{\mathbb{R}^N})}$, $\|u\|_{\mathscr{C}^{\mathrm{r}+\kappa}(W,d)}$, and $\|g\|_{\mathscr{C}^{\mathrm{s}}(W,d)}$. C may depend on any of the other ingredients in the result.

The rest of this section is devoted to the proof of Theorem 9.3.2, which uses ideas very similar to the scaling ideas used in Section 9.2.7; see also Section 2.8.4, where the same proof is carried out in the much simpler elliptic setting. In the proof, we use the scaling maps $\Phi_{x,\delta}$, $x \in \mathcal{K}$, $\delta \in (0,1]$, from Theorem 3.3.7, with the current choices of $\mathcal{K}$ and Ω_1. Let

$$\begin{aligned}
(W^{x,\delta}, d) &= \{(W_1^{x,\delta}, d_1), \ldots, (W_r^{x,\delta}, d_r)\} \\
&:= \{(\Phi_{x,\delta}^* \delta^{d_1} W_1, d_1), \ldots, (\Phi_{x,\delta}^* \delta^{d_r} W_r, d_r)\} \subset C^\infty(B^n(1); TB^n(1)) \times \mathbb{N}_+, \\
(X^{x,\delta}, d) &= \{(X_1^{x,\delta}, d_1), \ldots, (X_q^{x,\delta}, d_q)\} \\
&:= \{(\Phi_{x,\delta}^* \delta^{d_1} X_1, d_1), \ldots, (\Phi_{x,\delta}^* \delta^{d_q} X_q, d_q)\} \subset C^\infty(B^n(1); TB^n(1)) \times \mathbb{N}_+
\end{aligned}$$

be as in Theorem 3.3.7. Since $(W^{x,\delta}, d)$ and $(X^{x,\delta}, d)$ are Hörmander vector fields at the unit scale, we may define the quantitative norms $\|f\|_{\mathscr{C}^{\mathrm{s}}(W^{x,\delta},d)}$ and $\|F\|_{\mathscr{C}^{\mathrm{s},\mathrm{t}}((W^{x,\delta},d)\boxtimes\nabla_{\mathbb{R}^N})}$ as in Section 9.2.1; henceforth when we refer to these norms we use the particular choices from Section 9.2.1. The norms $\|f\|_{\mathscr{C}^{\mathrm{s}}(W,d)}$ and $\|F\|_{\mathscr{C}^{\mathrm{s},\mathrm{t}}((W,d)\boxtimes\nabla_{\mathbb{R}^N})}$ can be any choice from the equivalence class of norms – the choice does not affect the conclusion of Theorem 9.3.2.

In Theorem 9.3.2, by replacing $F(x,\zeta)$ with $F(x,\zeta)-g(x)$, it suffices to prove the result in the case $g=0$; we henceforth assume $g=0$. Take $\hat{\psi}\in C_0^\infty(\Omega_1)$ with $\hat{\psi}\equiv 1$ on a neighborhood of $\mathcal{K}$. Let $\{(D_j,2^{-j}) : j\in\mathbb{N}\}$ be a bounded set of $(W,\vec{d})$ elementary operators supported in Ω_1 with $\sum_{j\in\mathbb{N}} D_j=\mathrm{Mult}[\hat{\psi}]$ (this is always possible since by Proposition 5.8.3, $\mathrm{Mult}[\tilde{\psi}]\in\mathscr{A}^0(\Omega_1,(W,\vec{d}))=\mathscr{A}_2^0(\Omega_1,(W,\vec{d}))$).

For $J\in[0,\infty)$, set

$$u_J:=\sum_{0\le j<J} D_j u,\quad v_J:=2^{\kappa J}\sum_{j\ge J} D_j u$$

so that

$$u_J+2^{-\kappa J}v_J=\mathrm{Mult}[\hat{\psi}]u=u. \tag{9.148}$$

Also, set

$$F_J(x,\zeta):=F(x,\{W^\alpha u_J(x)\}_{\deg_{\vec{d}}(\alpha)\le\kappa}+\{2^{-(\kappa-\deg_{\vec{d}}(\alpha))}\zeta_\alpha\}_{\deg_{\vec{d}}(\alpha)\le\kappa}).$$

Using (9.148) and the fact that we are taking $g=0$, (9.145) implies

$$\begin{aligned} &F_J(x,\{(2^{-J\vec{d}}W)^\alpha v_J(x)\}_{\deg_{\vec{d}}(\alpha)\le\kappa}) \\ &\quad=F(x,\{W^\alpha u(x)\}_{\deg_{\vec{d}}(\alpha)\le\kappa})=0,\quad \forall x\in\Omega. \end{aligned} \tag{9.149}$$

We require the following lemmas, which are simple analogs of the corresponding results in Section 9.2.7.

Lemma 9.3.3. *Fix $\psi\in C_0^\infty(B^n(7/8))$, $\mathbf{r},\mathbf{s}>0$, and $u\in\mathscr{C}^{\mathbf{r}+\kappa}(\mathcal{K},(W,\vec{d}))$. For $J\in[0,\infty)$ and $\deg_{\vec{d}}(\alpha)\le\kappa$, set*

$$\begin{aligned} \bar{u}_{x,J,\alpha}(y)&:=\sum_{0\le j<J}(W^\alpha D_j u)(\Phi_{x_0,2^{-J}}(y))=(W^\alpha u_J)(\Phi_{x_0,2^{-J}}(y)),\\ \bar{v}_{x,J}(y)&:=\sum_{j\ge J}2^{\kappa J}D_j u(\Phi_{x_0,2^{-J}}(y))=v_J(\Phi_{x_0,2^{-J}}(y)). \end{aligned}$$

Then, for $x\in\Omega$,

$$\begin{aligned} \|\psi\bar{u}_{x,J,\alpha}\|_{\mathscr{C}^{\mathbf{s}}(W^{x_0,2^{-J}},\vec{d})}&\lesssim\|u\|_{\mathscr{C}^{\mathbf{r}+\kappa}(W,\vec{d})},\quad\forall\deg_{\vec{d}}(\alpha)\le\kappa,\\ \|\psi\bar{v}_{x,J}\|_{\mathscr{C}^{\mathbf{r}+\kappa}(W^{x_0,2^{-J}},\vec{d})}&\lesssim\|u\|_{\mathscr{C}^{\mathbf{r}+\kappa}(W,\vec{d})}. \end{aligned}$$

Proof. This is a simple reprise of Lemma 9.2.20. □

Lemma 9.3.4. *Fix $\psi\in C_0^\infty(B^n(7/8))$, $\mathbf{s}>0$, $\mathbf{t}>\lfloor\mathbf{s}\rfloor+2$. If $F\in\mathscr{C}^{\mathbf{s},\mathbf{t}}(\mathcal{K}\times\mathbb{R}^N,(W,\vec{d})\boxtimes\nabla_{\mathbb{R}^N})$, we have, $\forall J\ge 0$,*

$$\begin{aligned} &\|\psi(y)F_J(\Phi_{x,2^{-J}}(y),\zeta)\|_{\mathscr{C}^{\mathbf{s},\mathbf{t}-\lfloor\mathbf{s}\rfloor-2}((W^{x_0,2^{-J}},\vec{d})\boxtimes\nabla_{\mathbb{R}^N})}\\ &\quad\lesssim\|F\|_{\mathscr{C}^{\mathbf{s},\mathbf{t}}((W,\vec{d})\boxtimes\nabla_{\mathbb{R}^N})}\left(1+\|u\|_{\mathscr{C}^{\mathbf{r}+\kappa}(W,\vec{d})}\right)^{\lfloor\mathbf{s}\rfloor+2}. \end{aligned}$$

Proof. By a proof just like the one in Lemma 9.2.12 for (9.49), we have

$$\|\psi(y)F(\Phi_{x,2^{-J}}(y),\zeta)\|_{\mathscr{C}^{s,t}((W^{x_0,2^{-J}},\vec{d})\boxtimes\nabla_{\mathbb{R}^N})} \lesssim \|F\|_{\mathscr{C}^{s,t}((W,\vec{d})\boxtimes\nabla_{\mathbb{R}^N})}. \tag{9.150}$$

Pick $\tilde{\psi} \in C_0^\infty(B^n(7/8))$ with $\psi \prec \tilde{\psi}$. Set

$$\overline{G}_{x,J}(y,\zeta) := F(\Phi_{x,2^{-J}}(y), \{\tilde{\psi}(y)W^\alpha u_J(\Phi_{x,2^{-J}}(y))\}_{\deg_{\vec{d}}(\alpha)\leq\kappa}).$$

By Theorem 7.5.2 (i), we have

$$\begin{aligned}
&\|\psi(y)\overline{G}_J(\Phi_{x,2^{-J}}(y),\zeta)\|_{\mathscr{C}^{s,t-\lfloor s\rfloor-2}((W^{x_0,2^{-J}},\vec{d})\boxtimes\nabla_{\mathbb{R}^N})}\\
&\quad\lesssim \|\psi(y)F(\Phi_{x,2^{-J}}(y),\zeta)\|_{\mathscr{C}^{s,t}((W^{x_0,2^{-J}},\vec{d})\boxtimes\nabla_{\mathbb{R}^N})}\big(1+\big\|\{\tilde{\psi}W^\alpha u_J\circ\Phi_{x,2^{-J}}\}_{\deg_{\vec{d}}(\alpha)\leq\kappa}\big\|_{\mathscr{C}^s(W^{x_0,2^{-J}},\vec{d})}\big)\\
&\quad\quad\times\big(1+\big\|\{\tilde{\psi}W^\alpha u_J\circ\Phi_{x,2^{-J}}\}_{\deg_{\vec{d}}(\alpha)\leq\kappa}\big\|_{L^\infty}\big)^{\lfloor s\rfloor+1}.
\end{aligned} \tag{9.151}$$

By Lemma 9.3.3, we have

$$\big\|\{\tilde{\psi}W^\alpha u_J\circ\Phi_{x,2^{-J}}\}_{\deg_{\vec{d}}(\alpha)\leq\kappa}\big\|_{\mathscr{C}^s(W^{x_0,2^{-J}},\vec{d})} \lesssim \|u\|_{\mathscr{C}^{r+\kappa}(W,\vec{d})}. \tag{9.152}$$

Using Corollary 7.2.2 and (9.152), we have

$$\begin{aligned}
&\big\|\{\tilde{\psi}W^\alpha u_J\circ\Phi_{x,2^{-J}}\}_{\deg_{\vec{d}}(\alpha)\leq\kappa}\big\|_{L^\infty}\\
&\quad\lesssim \big\|\{\tilde{\psi}W^\alpha u_J\circ\Phi_{x,2^{-J}}\}_{\deg_{\vec{d}}(\alpha)\leq\kappa}\big\|_{\mathscr{C}^s(W^{x_0,2^{-J}},\vec{d})} \lesssim \|u\|_{\mathscr{C}^{r+\kappa}(W,\vec{d})}.
\end{aligned} \tag{9.153}$$

Using (9.150), (9.152), and (9.153) to bound the right-hand side of (9.151) shows that

$$\begin{aligned}
&\|\psi(y)\overline{G}_J(\Phi_{x,2^{-J}}(y),\zeta)\|_{\mathscr{C}^{s,t-\lfloor s\rfloor-2}((W^{x_0,2^{-J}},\vec{d})\boxtimes\nabla_{\mathbb{R}^N})}\\
&\quad\lesssim \|F\|_{\mathscr{C}^{s,t}((W,\vec{d})\boxtimes\nabla_{\mathbb{R}^N})}\big(1+\|u\|_{\mathscr{C}^{r+\kappa}(W,\vec{d})}\big)^{\lfloor s\rfloor+2}.
\end{aligned} \tag{9.154}$$

Finally, since $\psi \prec \tilde{\psi}$, we have

$$\psi(y)F_J(\phi_{x,2^{-J}}(y),\zeta) = \psi(y)\overline{G}_J(y,\{2^{-J(\kappa-\deg_{\vec{d}}(\alpha))}\zeta_\alpha\}_{\deg_{\vec{d}}(\alpha)\leq\kappa}).$$

Thus, using (9.154), we have

$$\begin{aligned}
&\|\psi(y)F_J(\Phi_{x,2^{-J}}(y),\zeta)\|_{\mathscr{C}^{s,t-\lfloor s\rfloor-2}((W^{x_0,2^{-J}},\vec{d})\boxtimes\nabla_{\mathbb{R}^N})}\\
&\quad\lesssim \|\psi(y)\overline{G}_J(\Phi_{x,2^{-J}}(y),\zeta)\|_{\mathscr{C}^{s,t-\lfloor s\rfloor-2}((W^{x_0,2^{-J}},\vec{d})\boxtimes\nabla_{\mathbb{R}^N})}\\
&\quad\lesssim \|F\|_{\mathscr{C}^{s,t}((W,\vec{d})\boxtimes\nabla_{\mathbb{R}^N})}\big(1+\|u\|_{\mathscr{C}^{r+\kappa}(W,\vec{d})}\big)^{\lfloor s\rfloor+2}.
\end{aligned}$$

□

Proof of Theorem 9.3.2. Fix $\xi_0 \in (0,1]$ a small number to be chosen later; we will eventually choose $\xi_0 \approx 1$. For $\delta > 0$ and $0 \le l' \le l$, let $\gamma \in \mathcal{P}^{l'}_{\xi_0\delta}(\Omega, (X, d))$ and set $x_0 := \gamma(0) \in \Omega$. Let $\xi_3 \in (0,1]$, $\xi_3 \approx 1$, be as in Theorem 3.3.7 and set $2^{-J_0} := (\mathrm{dist}_{(W,d)}(x_0, \partial\Omega) \wedge 1)\xi_3$. Since $\overline{\Omega}$ is compact, we have (using Lemma 3.1.7), $\mathrm{dist}_{(W,d)}(x_0, \partial\Omega) \lesssim 1$ and therefore $2^{-J_0} \approx \mathrm{dist}_{(W,d)}(x_0, \partial\Omega)$. Note that $J_0 \in [0, \infty)$.

To prove the statement of the theorem (using the fact that we will choose $\xi_0 \approx 1$), it suffices to show that

$$(2^{-J_0(\mathrm{s}+\kappa)/l}\delta^{-(\mathrm{s}+\kappa)/l})^{l'}|\triangle^{l'}_\gamma u| \lesssim 1. \tag{9.155}$$

If $\delta \ge 2^{-J_0}$, then using Corollary 7.2.2, we have

$$(2^{-J_0(\mathrm{s}+\kappa)/l}\delta^{-(\mathrm{s}+\kappa)/l})^{l'}|\triangle^{l'}_\gamma u| \le |\triangle^{l'}_\gamma u| \lesssim \|u\|_{L^\infty} \lesssim \|u\|_{\mathscr{C}^{\mathrm{r}+\kappa}(W,d)} \lesssim 1,$$

establishing (9.155). If $l' = 0$, then δ and 2^{-J_0} do not play a role in (9.155) and the same proof shows (9.155) holds.

We turn to establishing (9.155) in the remaining case where $2^{-J_0} > \delta$ and $1 \le l' \le l$. Using Theorem 3.3.7 (i), we have

$$\begin{aligned}\Phi_{x_0,2^{-J_0}}(B^n(1)) &\subseteq B_{(X,d)}(x, 2^{-J_0}) = B_{(X,d)}(x, (\mathrm{dist}_{(W,d)}(x_0, \partial\Omega) \wedge 1)\xi_3)\\ &\subseteq B_{(W,d)}(x, \mathrm{dist}_{(W,d)}(x_0, \partial\Omega)) \subseteq \Omega.\end{aligned} \tag{9.156}$$

Pulling back (9.149) via $\Phi_{x_0,2^{-J_0}}$ and using (9.156), we have

$$F_{J_0}(\Phi_{x_0,2^{-J_0}}(y), \{(W^{x_0,2^{-J_0}})^\alpha v_{J_0}(\Phi_{x_0,2^{-J_0}}(y))\}_{\deg_d(\alpha)\le\kappa}) = 0, \quad \forall y \in B^n(1). \tag{9.157}$$

Take $\psi \in C_0^\infty(B^n(7/8))$ such that $\psi \equiv 1$ on a neighborhood of $\overline{B^n(3/4)}$. Set

$$\begin{aligned}\widehat{F}_{x_0,J_0}(y, \zeta) &:= \psi(y)F_{J_0}(\Phi_{x_0,J_0}(y), \zeta),\\ \hat{v}_{x_0,J_0}(y) &:= \psi(y)v_{J_0}(\Phi_{x_0,2^{-J_0}}(y)).\end{aligned} \tag{9.158}$$

Since $\psi \equiv 1$ on $B^n(3/4)$, (9.157) shows that

$$\widehat{F}_{x_0,J_0}(y, \{(W^{x_0,2^{-J_0}})^\alpha \hat{v}_{x_0,J_0}(y)\}_{\deg_d(\alpha)\le\kappa}) = 0, \quad \forall y \in B^n(3/4). \tag{9.159}$$

By Lemma 9.3.3, we have

$$\|\hat{v}_{x_0,J_0}\|_{\mathscr{C}^{\mathrm{r}+\kappa}(W^{x_0,2^{-J_0}},d)} \lesssim \|u\|_{\mathscr{C}^{\mathrm{r}+\kappa}(W,d)} \lesssim 1, \tag{9.160}$$

and by Lemma 9.3.4,

$$\|\widehat{F}_{x_0,J_0}\|_{\mathscr{C}^{\mathrm{s},2\lfloor\mathrm{s}\rfloor+6+\sigma}((W^{x_0,2^{-J_0}},d)\boxtimes\nabla_{\mathbb{R}^N})} \lesssim \|F\|_{\mathscr{C}^{\mathrm{s},3\lfloor\mathrm{s}\rfloor+8+\sigma}((W,d)\boxtimes\nabla_{\mathbb{R}^N})} \lesssim 1. \tag{9.161}$$

Our goal is to apply Theorem 9.2.1 to (9.159). To do so, we need to show that (9.159) is maximally subelliptic uniformly in x_0. That is, define

$$\widehat{\mathscr{P}}_{x_0,J_0} := d_\zeta \widehat{F}_{x_0,J_0}(0, \{(W^{x_0,2^{-J_0}})^\alpha \hat{v}_{J_0}(0)\}_{\deg_{\vec{d}}(\alpha)\leq\kappa})\{(W^{x_0,2^{-J_0}})^\alpha\}_{\deg_{\vec{d}}(\alpha)\leq\kappa}.$$

We wish to show $\widehat{\mathscr{P}}_{x_0,J_0}$ is maximally subelliptic of degree κ with respect to $(W^{x_0,2^{-J_0}}, \vec{d})$, uniformly in $x_0 \in \Omega$.

Recall $\mathscr{P}_{x_0}$ from (9.146). From the definition, we know that $\mathscr{P}_{x_0}$ is of the form

$$\mathscr{P}_{x_0} = \sum_{\deg_{\vec{d}}(\alpha)\leq\kappa} a_{\alpha,x_0} W^\alpha,$$

where $a_{\alpha,x_0} \in \mathbb{M}^{\mathbf{D}_1\times\mathbf{D}_2}(\mathbb{R})$ are constant matrices, and we have

$$d_\zeta F(x_0, \{W^\alpha u(x_0)\}_{\deg_{\vec{d}}(\alpha)\leq\kappa})\{\zeta_\alpha\}_{\deg_{\vec{d}}(\alpha)\leq\kappa} = \sum_{\deg_{\vec{d}}(\alpha)\leq\kappa} a_{\alpha,x_0}\zeta_\alpha.$$

By (9.148) we have $W^\alpha u_{J_0}(x_0) + 2^{-\kappa J_0} W^\alpha v_{J_0}(x_0) = W^\alpha u(x_0)$, and therefore,

$$\begin{aligned} &dF_{J_0}(x_0, \{(2^{-J\vec{d}} W)^\alpha v_{J_0}(x_0)\}_{\deg_{\vec{d}}(\alpha)\leq\kappa})\{\zeta_\alpha\}_{\deg_{\vec{d}}(\alpha)\leq\kappa} \\ &\quad = d_\zeta F(x_0, \{W^\alpha u(x_0)\}_{\deg_{\vec{d}}(\alpha)\leq\kappa})\{2^{-(\kappa-\deg_{\vec{d}}(\alpha))J_0}\zeta_\alpha\}_{\deg_{\vec{d}}(\alpha)\leq\kappa} \\ &\quad = \sum_{\deg_{\vec{d}}(\alpha)\leq\kappa} 2^{-(\kappa-\deg_{\vec{d}}(\alpha))J_0} a_{\alpha,x_0}\zeta_\alpha. \end{aligned} \tag{9.162}$$

Since $\Phi_{x_0,2^{-J_0}}(0) = x_0$ (see Theorem 3.3.7 (g)), $\Phi^*_{x_0,2^{-J_0}} 2^{-J_0 \vec{d}_l} W_l = W_l^{x_0,2^{-J_0}}$, and $\psi \equiv 1$ on a neighborhood of 0, we have

$$(W^{x_0,2^{-J_0}})^\alpha \hat{v}_{x_0,J_0}(0) = (2^{-J\vec{d}} W)^\alpha v_{J_0}(x_0). \tag{9.163}$$

Using (9.163), (9.162), and again using $\Phi_{x_0,2^{-J_0}}(0) = x_0$, we have

$$\begin{aligned} \widehat{\mathscr{P}}_{x_0,J_0} w &= d_\zeta \widehat{F}_{x_0,J_0}(0, \{(W^{x_0,2^{-J_0}})^\alpha \hat{v}_{x_0,J_0}(0)\}_{\deg_{\vec{d}}(\alpha)\leq\kappa})\{(W^{x_0,2^{-J_0}})^\alpha w\}_{\deg_{\vec{d}}(\alpha)\leq\kappa} \\ &= dF_{J_0}(x_0, \{(2^{-J\vec{d}} W)^\alpha v_{J_0}(x_0)\}_{\deg_{\vec{d}}(\alpha)\leq\kappa})\{(W^{x_0,2^{-J_0}})^\alpha w\}_{\deg_{\vec{d}}(\alpha)\leq\kappa} \\ &= \sum_{\deg_{\vec{d}}(\alpha)\leq\kappa} 2^{-(\kappa-\deg_{\vec{d}}(\alpha))J_0} a_{\alpha,x_0}(W^{x_0,2^{-J_0}})^\alpha w. \end{aligned} \tag{9.164}$$

Thus, we have

$$\begin{aligned} &\Phi^*_{x_0,2^{-J_0}} 2^{-\kappa J_0} \mathscr{P}_{x_0}(\Phi_{x_0,2^{-J_0}})_* \\ &\quad = \sum_{\deg_{\vec{d}}(\alpha)\leq\kappa} 2^{-(\kappa-\deg_{\vec{d}}(\alpha))J_0} a_{\alpha,x_0}\Phi^*_{x_0,2^{-J_0}}(2^{-J_0\vec{d}} W)^\alpha(\Phi_{x_0,2^{-J_0}})_* \\ &\quad = \sum_{\deg_{\vec{d}}(\alpha)\leq\kappa} 2^{-(\kappa-\deg_{\vec{d}}(\alpha))J_0} a_{\alpha,x_0}(W^{x_0,2^{-J_0}})^\alpha \\ &\quad = \widehat{\mathscr{P}}_{x_0,J_0}, \end{aligned} \tag{9.165}$$

where the final equality used (9.164).

Multiplying (9.147) by $2^{-\kappa J_0}$ gives, $\forall f \in C_0^\infty(\Omega;\mathbb{C}^{\mathbf{D}_2})$,

$$\begin{aligned}\sum_{j=1}^{r}&\|(2^{-d_j J_0} W_j)^{n_j} f\|_{L^2(\mathfrak{M},\mathrm{Vol};\mathbb{C}^{\mathbf{D}_2})}\\ &\leq A(\|2^{-\kappa J_0}\mathscr{P}_{x_0} f\|_{L^2(\mathfrak{M},\mathrm{Vol};\mathbb{C}^{\mathbf{D}_1})} + 2^{-\kappa J_0}\|f\|_{L^2(\mathfrak{M},\mathrm{Vol};\mathbb{C}^{\mathbf{D}_2})})\\ &\leq A(\|2^{-\kappa J_0}\mathscr{P}_{x_0} f\|_{L^2(\mathfrak{M},\mathrm{Vol};\mathbb{C}^{\mathbf{D}_1})} + \|f\|_{L^2(\mathfrak{M},\mathrm{Vol};\mathbb{C}^{\mathbf{D}_2})}).\end{aligned} \tag{9.166}$$

For $g \in C_0^\infty(B^n(1);\mathbb{C}^{\mathbf{D}_2})$ we have $g \circ \Phi_{x_0,2^{-J_0}}^{-1} \in C_0^\infty(\Omega;\mathbb{C}^{\mathbf{D}_2})$ (see (9.156)). Applying (9.166) with f replaced by $g \circ \Phi_{x_0,2^{-J_0}}^{-1}$, using (9.165), and letting $h_{x_0,2^{-J_0}}$ and $\Lambda(x_0,2^{-J_0})$ be as in Theorem 3.3.7, we have

$$\begin{aligned}\sum_{j=1}^{r}&\|(W_j^{x_0,2^{-J_0}})^{n_j} g\|_{L^2(B^n(1),\Lambda(x_0,2^{-J_0})h_{x_0,2^{-J_0}}\sigma_{\mathrm{Leb}};\mathbb{C}^{\mathbf{D}_2})}\\ &\leq A\Big(\|\widehat{\mathscr{P}}_{x_0,2^{-J_0}} g\|_{L^2(B^n(1),\Lambda(x_0,2^{-J_0})h_{x_0,2^{-J_0}}\sigma_{\mathrm{Leb}};\mathbb{C}^{\mathbf{D}_1})}\\ &\qquad + \|g\|_{L^2(B^n(1),\Lambda(x_0,2^{-J_0})h_{x_0,2^{-J_0}}\sigma_{\mathrm{Leb}};\mathbb{C}^{\mathbf{D}_2})}\Big).\end{aligned} \tag{9.167}$$

Multiplying both sides of (9.167) by $\Lambda(x_0,2^{-J_0})^{-\frac{1}{2}}$, we see that

$$\begin{aligned}\sum_{j=1}^{r}&\|(W_j^{x_0,2^{-J_0}})^{n_j} g\|_{L^2(B^n(1),h_{x_0,2^{-J_0}}\sigma_{\mathrm{Leb}};\mathbb{C}^{\mathbf{D}_2})}\\ &\leq A(\|\widehat{\mathscr{P}}_{x_0,2^{-J_0}} g\|_{L^2(B^n(1),h_{x_0,2^{-J_0}}\sigma_{\mathrm{Leb}};\mathbb{C}^{\mathbf{D}_1})} + \|g\|_{L^2(B^n(1),h_{x_0,2^{-J_0}}\sigma_{\mathrm{Leb}};\mathbb{C}^{\mathbf{D}_2})}).\end{aligned} \tag{9.168}$$

That is, $\widehat{\mathscr{P}}_{x_0,2^{-J_0}}$ is maximally subelliptic of degree κ with respect to $(W^{x_0,2^{-J_0}}, \vec{d})$, with constant A from (9.147).

In light of (9.159), (9.160), (9.161), (9.168), and Theorem 3.3.7 (j) and (n), Theorem 9.2.1 applies to (9.159) (with $(W,\vec{d})$ and h replaced by $(W^{x_0,2^{-J_0}},\vec{d})$ and $h_{x_0,2^{-J_0}}$, respectively). Let $\psi_0 \in C_0^\infty(B^n(2/3))$ satisfy $\psi_0 \equiv 1$ on $B^n(1/2)$. The conclusion of Theorem 9.2.1 in this case then shows that there exists $\delta_1 \in (0,1]$, with $\delta_1 \approx 1$, such that

$$\|(\psi_0 \circ \widehat{\Phi}_{0,\delta_1}^{-1})\hat{v}_{x_0,J_0}\|_{\mathscr{C}^{s+\kappa}(W^{x_0,2^{-J_0}},\vec{d})} \lesssim 1, \tag{9.169}$$

where $\widehat{\Phi}_{0,\delta_1}$ is the map Φ_{0,δ_1} from Theorem 3.15.5 when applied to $(W^{x_0,2^{-J_0}},\vec{d})$ and $(X^{x_0,2^{-J_0}}, d)$ with $\sigma = 7/8$ (so $\nu = 1$ in this application).

Recall that our goal is to prove (9.155). Since $\gamma \in \mathcal{P}_{\xi_0\delta}^{l'}(\Omega,(X,d))$ and $\gamma(0) = x_0$, it follows immediately from Definition 7.7.2 that $\gamma([0,l']) \subseteq B_{(X,d)}(x_0, ql'\xi_0\delta)$. We have the flexibility to choose ξ_0 as small as we like, as long as we maintain $\xi_0 \approx 1$. We take $\xi_0 \leq \xi_3/ql \approx 1$. Then, since $l' \leq l$ and $\delta < 2^{-J_0}$, we have (using Theorem 3.3.7 (i))

$$\gamma([0,l']) \subseteq B_{(X,d)}(x_0,\xi_3\delta) \subseteq B_{(X,d)}(x_0,\xi_3 2^{-J_0}) \subseteq \Phi_{x_0,2^{-J_0}}(B^n(1/2)).$$

Because of this, it makes sense to define

$$\hat{\gamma}(t) := \Phi^{-1}_{x_0,2^{-J_0}} \circ \gamma(t) : [0,l'] \to B^n(1/2).$$

By the definition of $\gamma \in \mathcal{P}^{l'}_{\xi_0\delta}(\Omega,(X,d))$ (see Definition 7.7.2), we have

$$\gamma'(t) = \sum_{j=1}^{q} a_j(t) X_j(\gamma(t)) = \sum_{j=1}^{q} 2^{J_0 d_j} a_j(t) 2^{-J_0 d_j} X_j(\gamma(t)), \quad a.e., \tag{9.170}$$

where

$$\|\partial_t^{l''} a_j\|_{L^\infty([0,l'])} < (\xi_0\delta)^{d_j \vee (l''+1)}, \quad 1 \le j \le q,\ 0 \le l'' \le l'-1. \tag{9.171}$$

Since $\Phi^*_{x_0,2^{-J_0}} 2^{-J_0 d_j} X_j = X_j^{x_0,2^{-J_0}}$, (9.170) shows that

$$\hat{\gamma}'(t) = \sum_{j=1}^{q} 2^{J_0 d_j} a_j(t) X_j^{x_0,2^{-J_0}}(\gamma(t)),$$

where (9.171) implies

$$\|\partial_t^{l''} 2^{-J_0 d_j} a_j\|_{L^\infty([0,l'])} < 2^{J_0 d_j} (\xi_0\delta)^{d_j \vee (l''+1)} \le (2^{J_0}\xi_0\delta)^{d_j \vee (l''+1)},$$
$$1 \le j \le q, 0 \le l'' \le l'-1.$$

We conclude that $\hat{\gamma} \in \mathcal{P}^{l'}_{2^{J_0}\xi_0\delta}(B^n(1/2),(X^{x_0,2^{-J_0}},d))$.

The estimate (9.155) is equivalent to

$$(2^{J_0}\delta)^{-l'(\mathrm{s}+\kappa)/l} \left|\triangle^{l'}_{\hat{\gamma}} u \circ \Phi_{x_0,2^{-J_0}}\right| \lesssim 1. \tag{9.172}$$

By (9.148), to prove (9.172) it suffices to establish the next two estimates:

$$(2^{J_0}\delta)^{-l'(\mathrm{s}+\kappa)/l} \left|\triangle^{l'}_{\hat{\gamma}} u_{J_0} \circ \Phi_{x_0,2^{-J_0}}\right| \lesssim 1, \tag{9.173}$$

$$(2^{J_0}\delta)^{-l'(\mathrm{s}+\kappa)/l} \left|\triangle^{l'}_{\hat{\gamma}} v_{J_0} \circ \Phi_{x_0,2^{-J_0}}\right| \lesssim 1. \tag{9.174}$$

We first establish the easier of these two estimates: (9.173). Fix $\tilde{\psi} \in C_0^\infty(B^n(7/8))$ with $\tilde{\psi} \equiv 1$ on $B^n(1/2)$. Using Theorem 7.7.8 and Lemma 9.3.3 (with s replaced with $\mathrm{s}+\kappa$), we have

$$\begin{aligned} &\|\tilde{\psi} u_{J_0} \circ \Phi_{x_0,2^{-J_0}}\|_{\mathscr{C}_l^{\mathrm{s}+\kappa}(X^{x_0,2^{-J_0}},d)} \\ &\quad \lesssim \|\tilde{\psi} u_{J_0} \circ \Phi_{x_0,2^{-J_0}}\|_{\mathscr{C}^{\mathrm{s}+\kappa}(W^{x_0,2^{-J_0}},d)} \lesssim \|u\|_{\mathscr{C}^{r+\kappa}(W,d)} \lesssim 1. \end{aligned} \tag{9.175}$$

Since $\hat{\gamma} \in \mathcal{P}^{l'}_{2^{J_0}\xi_0\delta}(B^n(1/2), (X^{x_0,2^{-J_0}}, d))$ and $\xi_0 \approx 1$, (9.175) implies

$$(2^{J_0}\delta)^{-l'(\mathrm{s}+\kappa)/l} |\triangle^{l'}_{\hat{\gamma}} \tilde{\psi} u_{J_0} \circ \Phi_{x_0,2^{-J_0}}| \leq \|\tilde{\psi} u_{J_0} \circ \Phi_{x_0,2^{-J_0}}\|_{\mathscr{C}^{\mathrm{s}+\kappa}_l(X^{x_0,2^{-J_0}},d)} \lesssim 1. \tag{9.176}$$

We have already established that $\hat{\gamma}([0,l']) \subseteq B^n(1/2)$. Since $\tilde{\psi} \equiv 1$ on $B^n(1/2)$, we have

$$\triangle^{l'}_{\hat{\gamma}} \tilde{\psi} u_{J_0} \circ \Phi_{x_0,2^{-J_0}} = \triangle^{l'}_{\hat{\gamma}} u_{J_0} \circ \Phi_{x_0,2^{-J_0}}. \tag{9.177}$$

Combining (9.176) and (9.177) establishes (9.173).

By taking $\xi_0 \approx 1$ sufficiently small, we will show that a similar approach works for (9.174). We have, by Theorem 7.7.8 and (9.169),

$$\begin{aligned} &\|(\psi_0 \circ \widehat{\Phi}^{-1}_{0,\delta_1})\hat{v}_{x_0,J_0}\|_{\mathscr{C}^{\mathrm{s}+\kappa}_l(X^{x_0,2^{-J_0}},d)} \\ &\quad \lesssim \|(\psi_0 \circ \widehat{\Phi}^{-1}_{0,\delta_1})\hat{v}_{x_0,J_0}\|_{\mathscr{C}^{\mathrm{s}+\kappa}(W^{x_0,2^{-J_0}},d)} \lesssim 1. \end{aligned} \tag{9.178}$$

Since $\hat{\gamma} \in \mathcal{P}^{l'}_{2^{J_0}\xi_0\delta}(B^n(1/2), (X^{x_0,2^{-J_0}}, d))$ and $\xi_0 \approx 1$, (9.178) implies

$$\begin{aligned} &(2^{J_0}\delta)^{-l'(\mathrm{s}+\kappa)/l} |\triangle^{l'}_{\hat{\gamma}} (\psi_0 \circ \widehat{\Phi}^{-1}_{0,\delta_1})\hat{v}_{x_0,J_0}| \\ &\quad \leq \|(\psi_0 \circ \widehat{\Phi}^{-1}_{0,\delta_1})\hat{v}_{x_0,J_0}\|_{\mathscr{C}^{\mathrm{s}+\kappa}_l(X^{x_0,2^{-J_0}},d)} \lesssim 1. \end{aligned} \tag{9.179}$$

Since $\psi_0 \equiv 1$ on $B^n(1/2)$, we have $\psi_0 \circ \widehat{\Phi}^{-1}_{0,\delta_1} \equiv 1$ on $\widehat{\Phi}_{0,\delta_1}(B^n(1/2))$. Let $\hat{\xi}_3 \in (0,1]$, $\hat{\xi}_3 \approx 1$, be the ξ_3 constant from Theorem 3.15.5. Using Theorem 3.15.5 (f), we have

$$\psi_0 \circ \Phi^{-1}_{0,\delta_1} \equiv 1 \text{ on } B_{(X^{x_0,2^{-J_0}},d)}(0, \hat{\xi}_3\delta_1) \subseteq \widehat{\Phi}_{0,\delta_1}(B^n(1/2)). \tag{9.180}$$

We take $\xi_0 \approx 1$ so small that $\xi_0 \leq \hat{\xi}_3\delta_1/ql \approx 1$. Then, since $\hat{\gamma} \in \mathcal{P}^{l'}_{2^{J_0}\xi_0\delta}(B^n(1/2), (X^{x_0,2^{-J_0}}, d))$, $l' \leq l$, and $2^{-J_0} > \delta$, we have

$$\hat{\gamma}([0,l']) \subseteq B_{(X^{x_0,2^{-J_0}},d)}(0, ql\xi_0 2^{J_0}\delta) \subseteq B_{(X^{x_0,2^{-J_0}},d)}(0, \hat{\xi}_3\delta_1). \tag{9.181}$$

Combining (9.180) and (9.181) shows that

$$\triangle^{l'}_{\hat{\gamma}} (\psi_0 \circ \widehat{\Phi}^{-1}_{0,\delta_1})\hat{v}_{x_0,J_0} = \triangle^{l'}_{\hat{\gamma}} \hat{v}_{x_0,J_0}. \tag{9.182}$$

Since $\psi \equiv 1$ on $B^n(3/4)$ and we have already established that $\hat{\gamma}([0,l']) \subseteq B^n(1/2)$, using (9.158), we have

$$\triangle^{l'}_{\hat{\gamma}} \hat{v}_{x_0,J_0} = \triangle^{l'}_{\hat{\gamma}} v_{J_0} \circ \Phi_{x_0,2^{-J_0}}. \tag{9.183}$$

Plugging (9.183) and (9.182) into (9.179) establishes (9.174) and completes the proof. □

9.3.1 Weighted Schauder estimates near the boundary

We take the same setting as above, but we now assume that our partial differential operator is linear and has the form

$$\mathscr{P}u(x) = \sum_{\deg_d(\alpha)\leq\kappa} a_\alpha(x)W^\alpha u(x),$$

where $a_\alpha \in \mathscr{C}^{\mathbf{s}}(\mathcal{K},(W,\vec{d});\mathbb{M}^{\mathbf{D}_1\times\mathbf{D}_2}(\mathbb{R}))$.

Maximally subelliptic assumption: For each $x_0 \in \Omega$, define the frozen coefficient operator

$$\mathscr{P}_{x_0}v := \sum_{\deg_{\vec{d}}(\alpha)\leq\kappa} a_\alpha(x_0)W^\alpha u.$$

We assume that there exists $A \geq 0$ such that $\forall f \in C_0^\infty(\Omega;\mathbb{C}^{\mathbf{D}_2})$, $\forall x_0 \in \Omega$,

$$\sum_{j=1}^r \|W_j^{n_j}f\|_{L^2(\mathfrak{M},\mathrm{Vol};\mathbb{C}^{\mathbf{D}_2})} \leq A\big(\|\mathscr{P}_{x_0}f\|_{L^2(\mathfrak{M},\mathrm{Vol};\mathbb{C}^{\mathbf{D}_1})} + \|f\|_{L^2(\mathfrak{M},\mathrm{Vol};\mathbb{C}^{\mathbf{D}_2})}\big).$$

Theorem 9.3.5. *Under the above hypotheses, suppose that* $u \in \mathscr{C}^{\mathbf{r}+\kappa}(\mathcal{K},(W,\vec{d});\mathbb{R}^{\mathbf{D}_2})$ *and* $g \in \mathscr{C}^{\mathbf{s}}(\mathcal{K},(W,\vec{d});\mathbb{R}^{\mathbf{D}_1})$ *such that*

$$\mathscr{P}u(x) = g(x), \quad \forall x \in \Omega.$$

Then, $\forall l \in \mathbb{N}_+$ *with* $l > \mathbf{s}$, *we have*

$$\|u\|_{\widehat{\mathscr{C}}_l^{\mathbf{s}+\kappa}(\Omega,(W,d),(X,d))} \leq C\big(\|u\|_{\mathscr{C}^{\mathbf{r}+\kappa}(W,d)} + \|g\|_{\mathscr{C}^{\mathbf{s}}(W,d)}\big).$$

Here, $C \geq 0$ *does not depend on* u *or* g, *but may depend on any of the other ingredients in the result.*

Remark 9.3.6. Theorem 9.3.5 can likely be improved in several ways by using the methods of this text, though we do not do so here. See Remark 9.1.13.

Proof of Theorem 9.3.5. By replacing u and g with $u/(\|u\|_{\mathscr{C}^{\mathbf{r}+\kappa}(W,d)} + \|g\|_{\mathscr{C}^{\mathbf{s}}(W,d)})$ and $g/(\|u\|_{\mathscr{C}^{\mathbf{r}+\kappa}(W,d)} + \|g\|_{\mathscr{C}^{\mathbf{s}}(W,d)})$, respectively, we see that we may assume $\|u\|_{\mathscr{C}^{\mathbf{r}+\kappa}(W,d)}$, $\|g\|_{\mathscr{C}^{\mathbf{s}}(W,d)} \leq 1$ and in this case it suffices to prove

$$\|u\|_{\widehat{\mathscr{C}}_l^{\mathbf{s}+\kappa}(\Omega,(W,d),(X,d))} \lesssim 1. \tag{9.184}$$

Using Corollaries 7.2.2 and 6.5.11, we have, $\forall \deg_{\vec{d}}(\alpha) \leq \kappa$,

$$\|W^\alpha u\|_{L^\infty} \leq \|W^\alpha u\|_{\mathscr{C}^{\mathbf{r}}(W,d)} \lesssim \|u\|_{\mathscr{C}^{\mathbf{r}+\kappa}(W,d)} \leq 1.$$

We conclude that there exists $C \approx 1$ with $\|W^\alpha u\|_{L^\infty} \le C, \forall \deg_{d}(\alpha) \le \kappa$. Fix $\phi(\zeta) \in C_0^\infty(\mathbb{R}^N; \mathbb{R})$ with $\phi(\zeta) \equiv 1$ on a neighborhood of $\{\zeta : |\zeta_\alpha| \le 2C, \forall \deg_{d}(\alpha) \le \kappa\}$. Define

$$F(x,\zeta) := \sum_{\deg_{d}(\alpha)\le\kappa} a_\alpha(x)\zeta_\alpha\phi(\zeta).$$

Note that $F(x,\zeta) \in \mathscr{C}^{\mathbf{s},3\lfloor\mathbf{s}\rfloor+9}(\mathcal{K}\times\mathbb{R}^N, (W, d) \boxtimes \nabla_{\mathbb{R}^N}; \mathbb{R}^{\mathbf{D}_1})$ with $\|F\|_{\mathscr{C}^{\mathbf{s},3\lfloor\mathbf{s}\rfloor+9}((W,d)\boxtimes\nabla_{\mathbb{R}^N})} \lesssim 1$.

Since $|W^\alpha u(x)| \le C, \forall \deg_{d}(\alpha) \le \kappa, \forall x$, we have

$$\mathscr{P}u(x) = F(x, \{W^\alpha u(x)\}_{\deg_{d}(\alpha)\le\kappa}),$$

and therefore,

$$F(x, \{W^\alpha u(x)\}_{\deg_{d}(\alpha)\le\kappa}) = g(x), \quad \forall x \in \Omega. \tag{9.185}$$

Again using the fact that $|W^\alpha u(x)| \le C, \forall \deg_{d}(\alpha) \le \kappa, \forall x$, we have

$$d_\zeta F(x, \{W^\alpha u(x)\}_{\deg_{d}(\alpha)\le\kappa})\{\zeta_\alpha\}_{\deg_{d}(\alpha)\le\kappa} = \sum_{\deg_{d}(\alpha)\le\kappa} a_\alpha(x)\zeta_\alpha, \quad \forall x \in \Omega.$$

Therefore,

$$\mathscr{P}_{x_0} = d_\zeta F(x_0, \{W^\alpha u(x_0)\}_{\deg_{d}(\alpha)\le\kappa})\{W^\alpha\}_{\deg_{d}(\alpha)\le\kappa}.$$

We have established that all the hypotheses of Theorem 9.3.2 hold for (9.185). The conclusion of Theorem 9.3.2 in this case is exactly (9.184), completing the proof. □

9.4 Examples

In this section, we present some examples of fully nonlinear maximally subelliptic equations, which generalize ideas from the classical elliptic theory.

9.4.1 Second-order equations

Many classical examples of fully nonlinear elliptic equations are second-order equations whose linearization is given by a positive definite form. These equations often have generalizations to the maximally subelliptic setting. On $\mathbb{R}^n$, consider a nonlinear partial differential operator given by

$$u \mapsto F(x, \{\partial_x^\alpha u(x)\}_{|\alpha|\le 2}), \tag{9.186}$$

where F is a real-valued function. If we consider F as a function of $x \in \mathbb{R}^n$ and ζ, then ζ is a vector in $\mathbb{R}^{n^2+n+1}$,

$$\zeta = (\zeta_0, \zeta_1, \dots, \zeta_n, \zeta_{1,1}, \dots, \zeta_{i,j}, \dots, \zeta_{n,n}),$$

where, for example, the coordinate $\zeta_{i,j}$ corresponds to $\partial_{x_i}\partial_{x_j} u$. For a fixed u and x_0, if the $n \times n$ matrix

$$\begin{pmatrix} \partial_{\zeta_{1,1}} F(x_0, \{\partial_x^\alpha u(x_0)\}_{|\alpha|\le 2}) & \cdots & \partial_{\zeta_{1,n}} F(x_0, \{\partial_x^\alpha u(x_0)\}_{|\alpha|\le 2}) \\ \vdots & \ddots & \vdots \\ \partial_{\zeta_{n,1}} F(x_0, \{\partial_x^\alpha u(x_0)\}_{|\alpha|\le 2}) & \cdots & \partial_{\zeta_{n,n}} F(x_0, \{\partial_x^\alpha u(x_0)\}_{|\alpha|\le 2}) \end{pmatrix}$$

is symmetric and strictly positive definite, then (9.186) is elliptic at (x_0, u).

A similar phenomenon occurs in the maximally subelliptic setting. Let $(W, 1) = \{(W_1, 1), \dots, (W_r, 1)\}$ be Hörmander vector fields on a smooth, connected manifold $\mathfrak{M}$, each paired with formal degree 1. We consider a nonlinear partial differential operator given by

$$u \mapsto F(x, \{W^\alpha u(x)\}_{|\alpha|\le 2}), \tag{9.187}$$

where F is a real-valued function. Similar to the above, we consider $F = F(x, \zeta)$ as a function of $x \in \mathfrak{M}$ and $\zeta \in \mathbb{R}^{r^2+r+1}$,

$$\zeta = (\zeta_0, \zeta_1, \dots, \zeta_r, \zeta_{1,1}, \dots, \zeta_{i,j}, \dots, \zeta_{r,r}),$$

where, for example, the coordinate $\zeta_{i,j}$ corresponds to $W_i W_j u$.

Proposition 9.4.1. *Fix $x_0 \in \mathfrak{M}$ and $u : \mathfrak{M} \to \mathbb{R}$. If the $r \times r$ matrix*

$$\begin{pmatrix} \partial_{\zeta_{1,1}} F(x_0, \{W^\alpha u(x_0)\}_{|\alpha|\le 2}) & \cdots & \partial_{\zeta_{1,r}} F(x_0, \{W^\alpha u(x_0)\}_{|\alpha|\le 2}) \\ \vdots & \ddots & \vdots \\ \partial_{\zeta_{r,1}} F(x_0, \{W^\alpha u(x_0)\}_{|\alpha|\le 2}) & \cdots & \partial_{\zeta_{r,r}} F(x_0, \{W^\alpha u(x_0)\}_{|\alpha|\le 2}) \end{pmatrix} \tag{9.188}$$

is symmetric and strictly positive definite, then (9.187) *is maximally subelliptic at* (x_0, u) *of degree* 2 *with respect to* $(W, 1)$ *in the sense of Definition* 1.1.11.

Proof. By the definition (Definition 1.1.11) the maximal subellipticity of (9.187) at (x_0, u) is equivalent to the maximal subellipticity of the linear operator:

$$\begin{aligned} v \mapsto\ & \nabla_\zeta F(x_0, \{W^\alpha u(x_0)\}_{|\alpha|\le 2}) \cdot (v, W_1 v, \dots, W_r v, W_1 W_1 v, \dots, W_i W_j v, \dots, W_r W_r v) \\ &= \partial_{\zeta_0} F(x_0, \{W^\alpha u(x_0)\}_{|\alpha|\le 2}) v + \sum_{j=1}^{r} \partial_{\zeta_j} F(x_0, \{W^\alpha u(x_0)\}_{|\alpha|\le 2}) W_j v \\ &\quad + \sum_{j,k=1}^{r} \partial_{\zeta_{j,k}} F(x_0, \{W^\alpha u(x_0)\}_{|\alpha|\le 2}) W_j W_k v. \end{aligned} \tag{9.189}$$

The maximal subellipticity of (9.189) follows from Lemma 8.9.3. □

Corollary 9.4.2. *Let F, u, and g satisfy $F(x, \{W^\alpha u(x)\}_{|\alpha|\leq 2}) = g(x)$ and fix $x_0 \in \mathfrak{M}$ and* $\mathbf{s} > \mathbf{r} > 0$, $\mathbf{t} > 2\lfloor \mathbf{s} \rfloor + 6$. *Suppose that:*
- $F \in \mathscr{C}^{\mathbf{s},\mathbf{t}}((W,1) \boxtimes \nabla_{\mathbb{R}^N})$ *near* x_0,
- $g \in \mathscr{C}^{\mathbf{s}}(W,1)$ *near* x_0,
- $u \in \mathscr{C}^{\mathbf{r}+2}(W,1)$ *near* x_0,
- *the matrix given in* (9.188) *is symmetric and strictly positive definite.*

Then $u \in \mathscr{C}^{\mathbf{s}+2}(W,1)$ *near* x_0.

Proof. This follows by combining Theorem 9.1.2 and Proposition 9.4.1. □

Corollary 9.4.3. *Let F, u, and g satisfy $F(x, \{W^\alpha u(x)\}_{|\alpha|\leq 2}) = g(x)$ and fix $x_0 \in \mathfrak{M}$ and* $\mathbf{r} > 0$. *Suppose that:*
- $F \in C^\infty_{\mathrm{loc}}(\mathfrak{M} \times \mathbb{R}^n)$,
- $g \in C^\infty_{\mathrm{loc}}(\mathfrak{M})$ *near* x_0,
- $u \in \mathscr{C}^{\mathbf{r}+2}(W,1)$ *near* x_0,
- *the matrix given in* (9.188) *is symmetric and strictly positive definite.*

Then $u \in C^\infty_{\mathrm{loc}}(\mathfrak{M})$ *near* x_0.

Proof. This follows by combining Corollary 9.1.5 and Proposition 9.4.1. □

Remark 9.4.4. Similarly, we can apply the many other regularity results in this chapter to immediately obtain other regularity results about solutions to these second-order, maximally subelliptic equations. For example, Corollary 9.1.8 gives regularity properties in terms of the classical Zygmund–Hölder spaces.

Remark 9.4.5. In this text we only address classical solutions to fully nonlinear equations; this is because our main theorems deal with general fully nonlinear equations (of arbitrary order), so we cannot talk about weak solutions. In various settings, the theory of second-order elliptic equations has been extended to many types of weak solutions; see, for example, [25]. There has been some work on weak solutions for maximally subelliptic equations; see, for example, [168]. See also the references given in Section 9.5. It seems likely that much more can be done in parallel with elliptic theory, though we do not pursue that here.

9.4.2 The Monge–Ampère equation

An important classical example of a second-order fully nonlinear elliptic equation is the Monge–Ampère equation; see, e. g., [87]. For a function $u : \mathbb{R}^n \to \mathbb{R}$, let $(\partial_{x_i}\partial_{x_j} u)$ denote the $n \times n$ matrix with the (i,j) component equal to $\partial_{x_i}\partial_{x_j} u$. In the Monge–Ampère equation, one considers the Hessian:

$$u \mapsto \det(\partial_{x_i}\partial_{x_j} u). \tag{9.190}$$

When u is convex, this equation is elliptic. When u is at least C^2, one can rephrase this as saying that if $(\partial_{x_i}\partial_{x_j}u(x_0)) > 0$, i. e., if the matrix is positive definite, then (9.190) is elliptic at (x_0, u).

In [157] and [57] a generalization of convexity and the Monge–Ampère equation were introduced (and their relationship discussed) in the setting of stratified nilpotent Lie groups. Several papers followed, extending this notion to other settings; see [56, 58, 101, 102, 137, 245, 248, 32, 7, 168]. In this section, we describe this generalized Monge–Ampère equation in the general setting of Hörmander vector fields on a manifold and show that when the function is "convex" (in the sense of positive "Hessian"), the Monge–Ampère equation is maximally subelliptic. This is a special case of the more general second-order results described above.

As in Section 9.4.1, let $(W, 1) = \{(W_1, 1), \dots, (W_r, 1)\}$ be Hörmander vector fields, each paired with formal degree 1, on a connected smooth manifold $\mathfrak{M}$. For a real-valued function $u : \mathfrak{M} \to \mathbb{R}$, let $\mathscr{W}_{i,j}u(x) := \frac{1}{2}(W_iW_j + W_jW_i)u(x)$ and let $\mathscr{W}u := (\mathscr{W}_{i,j}u(x))$ denote the $r\times r$ matrix with the (i, j) component equal to $\mathscr{W}_{i,j}u(x)$. The Monge–Ampére equation in [57] takes the form

$$u \mapsto \det \mathscr{W}u. \tag{9.191}$$

Proposition 9.4.6. *Fix $x_0 \in \mathfrak{M}$ and suppose that $\mathscr{W}u(x_0) > 0$, i. e., $\mathscr{W}u(x_0)$ is a strictly positive definite matrix. Then* (9.191) *is maximally subelliptic at (x_0, u) of degree* 2 *with respect to $(W, 1)$ in the sense of Definition* 1.1.11.

Proof. Let $\mathscr{W}^{i,j}u$ denote the (i, j) component of the inverse matrix $(\mathscr{W}u)^{-1}$. Using the matrix identity

$$\left.\frac{d}{d\epsilon}\right|_{\epsilon=0} \det(A + \epsilon B) = \det(A)\operatorname{tr}(A^{-1}B), \tag{9.192}$$

we see that the linearization of (9.191) with respect to u at the point x_0 is the operator

$$v \mapsto \sum_{1\le i,j\le r} (\det \mathscr{W}u(x_0))\mathscr{W}^{i,j}u(x_0)\mathscr{W}_{i,j}v. \tag{9.193}$$

The matrix with (i, j) component equal to $(\det \mathscr{W}u(x_0))\mathscr{W}^{i,j}u(x_0)$ is strictly positive definite, and Lemma 8.9.3 implies that (9.193) is maximally subelliptic of degree 2 with respect to $(W, 1)$ on $\mathfrak{M}$, which completes the proof. □

In light of Proposition 9.4.6, the conclusions of Corollaries 9.4.2 and 9.4.3 hold when $\det \mathscr{W}u(x_0) > 0$. See Corollaries 9.4.11 and 9.4.12 for a precise statement of this in a more general setting.

Remark 9.4.7. As discussed in Remark 9.4.5, we only address classical solutions to the generalized Monge–Ampère equation (9.191). This is because we deduce results about (9.191) by applying the general fully nonlinear theory, where we cannot talk about weak solutions in general. For the classical Monge–Ampère equation (9.190) one can use the special form to talk about weak solutions; see the discussion of Alexandrov solutions in Chapter 2 of [87]. Similar ideas have been developed in the setting of Carnot groups; see, for example, [157]. It would be interesting to generalize these ideas to the general maximally subelliptic setting.

9.4.3 Higher-order Monge–Ampère equations

The generalized Monge–Ampère equation in Section 9.4.2 can be further generalized to higher-order equations in at least two different ways, which we present in this section.

Let $(W, \vec{d}) = \{(W_1, \vec{d}_1), \ldots, (W_r, \vec{d}_r)\}$ be Hörmander vector fields with formal degrees on $\mathfrak{M}$. Fix $\kappa \in \mathbb{N}_+$ such that $\vec{d}_j$ divides κ for every j and set $n_j := \kappa/\vec{d}_j \in \mathbb{N}_+$.

The first generalization: For a real-valued function $u : \mathfrak{M} \to \mathbb{R}$, let $\mathscr{W}[\kappa]u(x)$ denote the symmetric $r \times r$ real matrix with the (i,j) component equal to

$$\mathscr{W}[\kappa]_{i,j}u(x) := \frac{1}{2}(W_i^{n_i} W_j^{n_j} u(x) + W_j^{n_j} W_i^{n_i} u(x)).$$

We consider a higher-order generalized Monge–Ampère equation given by

$$u \mapsto \det \mathscr{W}[\kappa]u(x). \tag{9.194}$$

Remark 9.4.8. When $\vec{d}_j = 1$, $\forall 1 \leq j \leq r$, and $\kappa = 1$, then (9.194) is the same as (9.191). However, by taking $\kappa > 1$, (9.194) gives new higher-order operators, even in the special case where $\vec{d}_j = 1$, $\forall 1 \leq j \leq r$.

Proposition 9.4.9. *Fix $x_0 \in \mathfrak{M}$ and suppose that $\mathscr{W}[\kappa]u(x_0) > 0$, i. e., $\mathscr{W}[\kappa]u(x_0)$ is a strictly positive definite matrix. Then* (9.191) *is maximally subelliptic at (x_0, u) of degree 2κ with respect to $(W, \vec{d})$ in the sense of Definition* 1.1.11.

Proof. Let $\mathscr{W}[\kappa]^{i,j}u$ denote the (i,j) component of the inverse matrix $(\mathscr{W}[\kappa]u)^{-1}$. Using (9.192) we see that the linearization of (9.194) with respect to u at the point x_0 is the operator

$$v \mapsto \sum_{1 \leq i,j \leq r} (\det \mathscr{W}[\kappa]u(x_0))\mathscr{W}[\kappa]^{i,j}u(x_0)v. \tag{9.195}$$

The matrix with the (i,j) component equal to $(\det \mathscr{W}[\kappa]u(x_0))\mathscr{W}[\kappa]^{i,j}(x_0)$ is strictly positive definite, and Lemma 8.9.3 implies that (9.195) is maximally subelliptic of degree 2κ with respect to $(W, \vec{d})$ on $\mathfrak{M}$, which completes the proof. □

The second generalization: Set

$$\mathbf{D} := \#\{\alpha : \deg_{\vec{d}}(\alpha) = \kappa\}. \tag{9.196}$$

For a real-valued function $u : \mathfrak{M} \to \mathbb{R}$, let $\widehat{\mathscr{W}}[\kappa]u(x)$ denote the symmetric $\mathbf{D} \times \mathbf{D}$ real matrix with the (α, β) component equal to

$$\widehat{\mathscr{W}}[\kappa]_{\alpha,\beta}u(x) := \frac{1}{2}(W^\alpha W^\beta u(x) + W^\beta W^\alpha u(x)), \quad \deg_{\vec{d}}(\alpha) = \deg_{\vec{d}}(\beta) = \kappa.$$

We consider a higher-order generalized Monge–Ampère equation given by

$$u \mapsto \det \widehat{\mathscr{W}}[\kappa]u(x). \tag{9.197}$$

Proposition 9.4.10. *Fix $x_0 \in \mathfrak{M}$ and suppose $\widehat{\mathscr{W}}[\kappa]u(x_0) > 0$, i. e., $\widehat{\mathscr{W}}[\kappa]u(x_0)$ is a strictly positive definite matrix. Then* (9.197) *is maximally subelliptic at (x_0, u) of degree 2κ with respect to $(W, \vec{d})$ in the sense of Definition* 1.1.11.

Proof. This follows just as in the proof of Proposition 9.4.9 by replacing Lemma 8.9.3 with Proposition 8.9.4. □

Corollary 9.4.11. *Fix $\mathbf{s} > \mathbf{r} > 0$ and $x_0 \in \mathfrak{M}$. Suppose that:*
- $\mathscr{W}[\kappa]u(x_0) > 0$,
- $u \in \mathscr{C}^{\mathbf{r}+2\kappa}(W, \vec{d})$ *near* x_0,
- $\det \mathscr{W}[\kappa]u \in \mathscr{C}^{\mathbf{s}}(W, \vec{d})$ *near* x_0.

Then $u \in \mathscr{C}^{\mathbf{s}+2\kappa}(W, \vec{d})$ near x_0. The same result holds with $\mathscr{W}$ replaced by $\widehat{\mathscr{W}}$ throughout.

Proof. This follows by combining Theorem 9.1.2 and Propositions 9.4.9 and 9.4.10. □

Corollary 9.4.12. *Fix $\mathbf{r} > 0$ and $x_0 \in \mathfrak{M}$. Suppose that:*
- $\mathscr{W}[\kappa]u(x_0) > 0$,
- $u \in \mathscr{C}^{\mathbf{r}+2\kappa}(W, \vec{d})$ *near* x_0,
- $\det \mathscr{W}[\kappa]u \in C^\infty$ *near* x_0.

Then $u \in C^\infty$ near x_0. The same result holds with $\mathscr{W}$ replaced by $\widehat{\mathscr{W}}$ throughout.

Proof. This follows by combining Corollary 9.1.5 and Propositions 9.4.9 and 9.4.10. □

Remark 9.4.13. Similarly, we can apply the other regularity theorems in this chapter to immediately deduce results about (9.194) when either $\mathscr{W}[\kappa]u(x_0) > 0$ or $\widehat{\mathscr{W}}[\kappa]u(x_0) > 0$. For example, Corollary 9.1.8 gives regularity properties in terms of the classical Zygmund–Hölder spaces.

9.4.4 Higher-order equations

Just as the generalized Monge–Ampère equation from Section 9.4.2 can be seen as a special case of the more general second-order equations from Section 9.4.1, the higher-order generalized Monge–Ampère equations from Section 9.4.3 can be seen as a special case of higher-order operators whose maximal subellipticity is nearly immediate. The main idea is that the definition of a nonlinear partial differential operator being maximally subelliptic (Definition 1.1.11) is just that a frozen coefficient version of its linearization is maximally subelliptic. When this linearization is given by a positive definite quadratic form, the maximal subellipticity can often be deduced from results like Propositions 8.9.1 and 8.9.4. In this section we present two such results, which generalize the two types of higher-order Monge–Ampère equations from Section 9.4.3.

Let $(W, \vec{d}) = \{(W_1, \vec{d}_1), \dots, (W_r, \vec{d}_r)\}$ be Hörmander vector fields with formal degrees on $\mathfrak{M}$. Fix $\kappa \in \mathbb{N}_+$ such that $\vec{d}_j$ divides κ, $\forall 1 \leq j \leq r$, and set $n_j := \kappa/\vec{d}_j \in \mathbb{N}_+$. Consider a nonlinear partial differential operator of the form

$$u \mapsto F(x, \{W^\beta u(x)\}_{\deg_{\vec{d}}(\beta)\leq 2\kappa}), \tag{9.198}$$

where F is a real-valued function. We consider $F = F(x, \zeta)$ as a function of $x \in \mathfrak{M}$ and $\zeta = \{\zeta_\alpha\}_{\deg_{\vec{d}}(\alpha)\leq\kappa}$, where $\zeta_\alpha \in \mathbb{R}$.

For each $1 \leq i, j \leq r$ define the ordered multi-index

$$\alpha_{i,j} := (\underbrace{i, i, \dots, i}_{n_i \text{ terms}}, \underbrace{j, j, \dots, j}_{n_j \text{ terms}}).$$

Note that $\deg_{\vec{d}}(\alpha_{i,j}) = 2\kappa$ and $W^{\alpha_{i,j}} = W_i^{n_i} W_j^{n_j}$.

Our first result generalizes the higher-order Monge–Ampère operators from Proposition 9.4.9.

Proposition 9.4.14. *Fix $x_0 \in \mathfrak{M}$ and $u : \mathfrak{M} \to \mathbb{R}$. Suppose that:*

(i) *For each $\deg_{\vec{d}}(\alpha) = 2\kappa$ such that $\alpha \neq \alpha_{i,j}$, $\forall i, j$, we have $\partial_{\zeta_\alpha} F(x_0, \{W^\beta u(x_0)\}_{\deg_{\vec{d}}(\beta)\leq 2\kappa}) = 0$.*

(ii) *The $r \times r$ matrix*

$$\begin{pmatrix} \partial_{\zeta_{\alpha_{1,1}}} F(x_0, \{W^\beta u(x_0)\}_{\deg_{\vec{d}}(\beta)\leq 2\kappa}) & \cdots & \partial_{\zeta_{\alpha_{1,r}}} F(x_0, \{W^\beta u(x_0)\}_{\deg_{\vec{d}}(\beta)\leq 2\kappa}) \\ \vdots & \ddots & \vdots \\ \partial_{\zeta_{\alpha_{r,1}}} F(x_0, \{W^\beta u(x_0)\}_{\deg_{\vec{d}}(\beta)\leq 2\kappa}) & \cdots & \partial_{\zeta_{\alpha_{r,r}}} F(x_0, \{W^\beta u(x_0)\}_{\deg_{\vec{d}}(\beta)\leq 2\kappa}) \end{pmatrix}$$

is symmetric and strictly positive definite.

Then (9.198) *is maximally subelliptic at (x_0, u) of degree 2κ with respect to $(W, \vec{d})$ in the sense of Definition* 1.1.11.

Remark 9.4.15. The assumption (i) in Proposition 9.4.14 is somewhat artificial, and only serves to make it easy to recognize the maximal subellipticity of (9.198). Note that (i) automatically holds if $F(x,\zeta)$ does not depend on any of the coordinates ζ_α, where $\deg_{d}(\alpha) = 2\kappa$ and $\alpha \neq \alpha_{i,j}$ for all $1 \leq i,j \leq r$. For example, this is the case in the higher-order generalized Monge–Ampère equation (9.194).

Proof of Proposition 9.4.14. By the definition (Definition 1.1.11) the maximal subellipticity of (9.198) at (x_0,u) is equivalent to the maximal subellipticity of the linear operator:

$$\begin{aligned} v \mapsto \nabla_\zeta F(x_0, \{W^\beta u(x_0)\}_{\deg_{d}(\beta)\leq 2\kappa}) \cdot \{W^\alpha v\}_{\deg_{d}(\alpha)\leq 2\kappa} \\ = \sum_{1\leq i,j\leq r} \partial_{\zeta_{\alpha_{i,j}}} F(x_0, \{W^\beta u(x_0)\}_{\deg_{d}(\beta)\leq 2\kappa}) W^{n_i} W^{n_j} v \\ + \sum_{\deg_{d}(\alpha)<2\kappa} \partial_{\zeta_\alpha} F(x_0, \{W^\beta u(x_0)\}_{\deg_{d}(\beta)\leq 2\kappa}) W^\alpha v. \end{aligned} \tag{9.199}$$

The maximal subellipticity of (9.199) follows from Lemma 8.9.3. □

Our next result generalizes the higher-order Monge–Ampère operators from Proposition 9.4.10. Define $\mathbf{D} \in \mathbb{N}_+$ by (9.196). For an ordered multi-index α (i. e., a list of elements of $\{1,\dots,r\}$), we write $\alpha = \gamma_1 \oplus \gamma_2$ if α is the concatenation of the lists γ_1 and γ_2. In particular, we have $W^{\gamma_1\oplus\gamma_2} = W^{\gamma_1}W^{\gamma_2}$.

Proposition 9.4.16. *Fix $x_0 \in \mathfrak{M}$ and $u : \mathfrak{M} \to \mathbb{R}$. Suppose that:*
(i) *For each $\deg_{d}(\alpha) = 2\kappa$ such that $\alpha \neq \gamma_1 \oplus \gamma_2$ for every $\deg_{d}(\gamma_1) = \deg_{d}(\gamma_2) = \kappa$, we have $\partial_{\zeta_\alpha} F(x_0, \{W^\beta u(x_0)\}_{\deg_{d}(\beta)\leq 2\kappa}) = 0$.*
(ii) *The $\mathbf{D} \times \mathbf{D}$ matrix*

$$\left(\partial_{\zeta_{\gamma_1\oplus\gamma_2}} F(x_0, \{W^\beta u(x_0)\}_{\deg_{d}(\beta)\leq 2\kappa})\right)_{\deg_{d}(\gamma_1),\deg_{d}(\gamma_2)=\kappa}$$

is symmetric and strictly positive definite.

Then (9.198) *is maximally subelliptic at* (x_0,u) *of degree* 2κ *with respect to* (W,d) *in the sense of Definition* 1.1.11.

Remark 9.4.17. The assumption (i) of Proposition 9.4.16 holds automatically if $d_j = 1$, $\forall 1 \leq j \leq r$, since there are no such ordered multi-indices α in that case. The only way that (i) might not hold is if d_j are not all equal.

Proof of Proposition 9.4.16. This follows just as in the proof of Proposition 9.4.14 by replacing Lemma 8.9.3 with Proposition 8.9.4. □

Corollary 9.4.18. *Let F, u, and g satisfy $F(x, \{W^\alpha u(x)\}_{\deg_{d}(\alpha)\leq 2\kappa}) = g(x)$ and fix $x_0 \in \mathfrak{M}$ and $\mathbf{s} > \mathbf{r} > 0$, $\mathbf{t} > 2\lfloor \mathbf{s} \rfloor + 6$. Suppose that:*
- *$F \in \mathscr{C}^{\mathbf{s},\mathbf{t}}((W,d) \boxtimes \nabla_{\mathbb{R}^N})$ near x_0,*
- *$g \in \mathscr{C}^{\mathbf{s}}(W,d)$ near x_0,*

- $u \in \mathscr{C}^{\mathbf{r}+2\kappa}(W, \vec{d})$ *near* x_0,
- *the assumptions of either Proposition* 9.4.14 *or Proposition* 9.4.16 *hold.*

Then $u \in \mathscr{C}^{\mathbf{s}+2\kappa}(W, \vec{d})$ *near* x_0.

Proof. This follows by combining Theorem 9.1.2 and Propositions 9.4.14 and 9.4.16 □

Corollary 9.4.19. *Let* F, u, *and* g *satisfy* $F(x, \{W^\alpha u(x)\}_{\deg_{\vec{d}}(\alpha)\leq 2\kappa}) = g(x)$ *and fix* $x_0 \in \mathfrak{M}$ *and* $\mathbf{r} > 0$. *Suppose that:*
- $F \in C^\infty_{\mathrm{loc}}(\mathfrak{M} \times \mathbb{R}^n)$,
- $g \in C^\infty_{\mathrm{loc}}(\mathfrak{M})$ *near* x_0,
- $u \in \mathscr{C}^{\mathbf{r}+2\kappa}(W, 1)$ *near* x_0,
- *the assumptions of either Proposition* 9.4.14 *or Proposition* 9.4.16 *hold.*

Then $u \in C^\infty_{\mathrm{loc}}(\mathfrak{M})$ *near* x_0.

Proof. This follows by combining Corollary 9.1.5 and Propositions 9.4.14 and 9.4.16 □

Remark 9.4.20. Similarly, we can apply the many other regularity results in this chapter to immediately obtain other regularity results about solutions to these higher-order, maximally subelliptic equations. For example, Corollary 9.1.8 gives regularity properties in terms of the classical Zygmund–Hölder spaces.

9.5 Further reading and references

There has been much less work on nonlinear maximally subelliptic equations than there has been for linear maximally subelliptic equations. Nearly all the work has been on operators of order 2, whose linearizations are similar to the sub-Laplacian (and where the Rothschild–Stein lifting theorem applies; cf. Section 4.5.7). In contrast, we have studied general fully nonlinear equations. Because previous studies work with equations of a more special form, they are often able to introduce notions of weak solutions and are therefore able to deal with less regularity than we do. However, it seems likely that the methods of this chapter can, when applied to operators of a special form, be used to study weak solutions (for example, see Remark 9.1.13). Our results on weighted estimates near the boundary (in Section 9.3) also seem to be new.

Much of the previous work has been on quasi-linear maximally subelliptic equations, usually of the special form

$$\sum_{j=1}^{r} W_j^* A_j(x, u, W_1 u, \dots, W_r u) = f(x, u, W_1 u, \dots, W_r u), \tag{9.200}$$

where $W_1, \dots, W_r$ are Hörmander vector fields; this is a second-order quasi-linear equation. Except for a few exceptions, this is usually done in the context of Lie groups. The earliest work on quasi-linear maximally subelliptic equations of the form (9.200) was due to Xu [251], Capogna, Danielli, and Garofalo [30], and Capogna [28]. More recent results in the same vein include [31, 73, 76, 77, 74, 75, 64, 65, 89, 253, 164, 163, 165, 176–178, 29, 162, 202, 203, 19, 249]. For some other quasi-linear maximally subelliptic equations see [156]. The methods of these papers are unrelated to the methods of this text.

Beyond the quasi-linear setting, there has been some work on second-order fully nonlinear equations where weak solutions can be considered [168, 8, 11] and on linear operators with non-smooth coefficients [155, 18, 15, 17]. There is also a large body of literature on degenerate elliptic equations, some of which is closely related to maximally subellipticity. A recent paper which shares some themes of this chapter is [139].

Nonlinear maximally subelliptic equations appear in the geometry of several complex variables. See, for example, [145, 126, 173], where the Levi curvature equations are studied, related to domains of holomorphy, and where a real hypersurface evolves through a motion involving the Levi curvature (here, the vector fields themselves are taken to be nonlinear), and [130], where an analog of the Yamabe problem for CR manifolds is studied. Nonlinear maximally subelliptic equations have also had applications to finance [146] and have been connected to the study of human vision and image processing [196, 51, 13, 50].

As mentioned earlier, the proof due to Simon [212] was an inspiration for the proof of Theorem 9.2.1.

A Canonical coordinates

In this appendix, we discuss and prove the results from Section 3.6, which form the basis of the proofs of every main theorem of this text.

The central result (Theorem A.2.4) is a slight generalization of Theorem 3.6.5 where we work with C^2_{loc} vector fields instead of C^∞_{loc} vector fields. The results in this section are a simplified special case of the results in [228]; see Section A.5 for details on the history of these results as well as references to some other, stronger results.

A.1 Basic notation

Throughout this appendix, $\mathfrak{M}$ will denote a C^2 manifold of dimension $n \in \mathbb{N}_+$. Given $X = \{X_1, \ldots, X_q\} \subset C^1_{\mathrm{loc}}(\mathfrak{M}; T\mathfrak{M})$, a finite set of C^1_{loc} vector fields on $\mathfrak{M}$, we set

$$B_X(x, \delta) := B_{\delta X_1, \ldots, \delta X_q}(x), \quad x \in \mathfrak{M}, \delta > 0,$$

where $B_{\delta X_1, \ldots, \delta X_q}(x)$ is defined by (1.12).

Given a C^1_{loc} vector field $Y \in C^1_{\mathrm{loc}}(\mathfrak{M}; T\mathfrak{M})$ and a function $f : \mathfrak{M} \to \mathbb{C}$, we define

$$Yf(x) := \left.\frac{d}{dt}\right|_{t=0} f(e^{tY}x). \tag{A.1}$$

If we say $Yf(x)$ exists, this means that this derivative exists in the classical sense, $\forall x$. Note that this allows us to define $Yf(x)$ for x such that $e^{tY}x$ is defined for t small; in particular, $f(x)$ need not be defined on an open set containing x. For $\alpha = (\alpha_1, \ldots, \alpha_L) \in \{1, \ldots, q\}^L$, we define $X^\alpha f = X_{\alpha_1}(X_{\alpha_2}(\cdots(X_{\alpha_L} f)))$, and we say $X^\alpha f$ exists if at each stage the derivatives in the sense of (A.1) exist.

Since $\mathfrak{M}$ is merely a C^2 manifold, it does not make sense to talk about functions of regularity any higher than C^2_{loc}. However, it does make sense to talk about functions which are of higher regularity *with respect to the vector fields X*. To describe this, we generalize the Banach spaces $C^L_X(\mathfrak{M})$ from Definition 5.2.21 to the setting of C^1_{loc} vector fields.

Definition A.1.1. For $L \in \mathbb{N}$ and $U \subseteq \mathfrak{M}$ open, define the vector space

$$C^L_X(U) := \{f \in C(U) : X^\alpha f \text{ exists and } X^\alpha f \in C(U), \forall |\alpha| \le L\},$$

with norm

$$\|f\|_{C^L_X(U)} := \sup_{x \in U} \sum_{|\alpha| \le L} |X^\alpha f(x)|.$$

https://doi.org/10.1515/9783111085647-010

With this norm, $C^L_X(U)$ is a Banach space. We make the same definition for any set $U \subseteq \mathfrak{M}$ such that $e^{tX_j}x \in U$ for all $x \in U, 1 \leq j \leq q$, and $t \in \mathbb{R}$ sufficiently small (how small may depend on x and j). We set $C^\infty_X(U) := \bigcap_L C^L_X(U)$.

Likewise, we define $C^\infty_{X,\mathrm{loc}}(\mathfrak{M})$ just as in Definition 5.2.22; in particular, even though the manifold is only C^2, it makes sense to talk about functions which are infinitely smooth with respect to the vector fields X.

Let

$$\mathcal{I}(n,q) := \{(i_1,\dots,i_n) : i_j \in \{1,\dots q\}\} = \{1,\dots,q\}^n,$$
$$\mathcal{I}_0(n,q) := \{(i_1,\dots,i_n) \in \mathcal{I}(n,q) : i_1 < i_2 < \dots < i_n\}.$$

Given $J = (j_1,\dots,j_n) \in \mathcal{I}(n,q)$, we write $X_J = (X_{j_1},\dots,X_{j_n})$ and $\bigwedge X_J := X_{j_1} \wedge X_{j_2} \wedge \dots \wedge X_{j_n}$. If we also have $K \in \mathcal{I}(n,q)$ and $\bigwedge X_J(x_0) \neq 0$, then Definition 3.6.2 defines the function

$$\frac{\bigwedge X_K(x)}{\bigwedge X_J(x)}$$

for x near x_0.

A.2 The main results

Let $\mathfrak{M}$ be a C^2 manifold of dimension $n \in \mathbb{N}_+$ and let $X = \{X_1,\dots,X_q\} \subset C^1_{\mathrm{loc}}(\mathfrak{M}; T\mathfrak{M})$ be a finite set of C^1_{loc} vector fields on $\mathfrak{M}$ such that $\forall x \in \mathfrak{M}$, $T_x\mathfrak{M} = \mathrm{span}\{X_1(x),\dots,X_q(x)\}$. An important aspect of all the assumptions and estimates which follow is that they are quantitatively invariant under arbitrary C^2_{loc} diffeomorphisms. See Remark A.2.3.

Fix $x_0 \in \mathfrak{M}$ and $\xi \in (0,1]$. We write

$$[X_j, X_k] = \sum_{l=1}^{q} c^l_{j,k} X_l, \tag{A.2}$$

and we assume $c^l_{j,k} \in C(B_X(x_0,\xi))$; since $X_1,\dots,X_q$ span the tangent space at every point, it is easy to see that $B_X(x_0,\xi)$ is open in $\mathfrak{M}$ – see Lemma A.2.14.

Fix $\zeta \in (0,1]$. Let $J_0 \in \mathcal{I}(n,q)$ be such that $\bigwedge X_{J_0}(x_0) \neq 0$ and moreover

$$\max_{J \in \mathcal{I}(n,q)} \left| \frac{\bigwedge X_J(x_0)}{\bigwedge X_{J_0}(x_0)} \right| \leq \zeta^{-1}; \tag{A.3}$$

see Definition 3.6.2 for the definition of this quotient. Note that this is always possible: we may always choose J_0 so that the left-hand side of (A.3) equals 1, though it is sometimes convenient to have the flexibility to allow it to be greater than 1. Without loss of generality, we reorder $X_1,\dots,X_q$ so that $J_0 = (1,\dots,n)$.

As in Section 3.6, we require two technical quantities associated with X:

- Let $\eta \in (0,1]$ be such that X_{J_0} satisfies $\mathcal{C}(x_0, \eta, \mathfrak{M})$. See Definition 3.6.3.
- Let $\tau_0 \in (0,1]$ be such that for $\tau \in (0, \tau_0]$ the following holds: if $z \in B_{X_{J_0}}(x_0, \xi)$ is such that X_{J_0} satisfies $\mathcal{C}(z, \tau, B_{X_{J_0}}(x_0, \xi))$, $t \in B^n(\tau)$ is such that $e^{t \cdot X_{J_0}} z = z$, and $X_1(z), \ldots, X_n(z)$ are linearly independent, then $t = 0$.

Such an η and τ_0 always exist (by possibly shrinking $\xi > 0$ – see Proposition A.2.10); however, our quantitative estimates will depend on η and τ_0. See Section A.2.1 for a further discussion of these quantities.

Definition A.2.1. We say C is a 0-admissible constant if C can be chosen to depend only on upper bounds for q, ζ^{-1}, ξ^{-1}, and $\|c_{j,k}^l\|_{C(B_{X_{J_0}}(x_0,\xi))}$, $1 \le j,k,l \le q$.

Definition A.2.2. For $m \in \mathbb{N}_+$, if we say C is an m-admissible constant, this means that we assume $c_{j,k}^l \in C^m_{X_{J_0}}(B_{X_{J_0}}(x_0, \xi))$ for $1 \le j,k,l \le q$. C is allowed to depend on anything a 0-admissible constant can depend on and upper bounds for η^{-1}, τ_0^{-1}, m, and $\|c_{j,k}^l\|_{C^m_X(B_{X_{J_0}}(x_0,\xi))}$.

For $m \in \mathbb{N}$, we write $A \lesssim_m B$ for $A \le CB$, where $C \ge 0$ is an m-admissible constant. We write $A \approx_m B$ for $A \lesssim_m B$ and $B \lesssim_m A$.

Remark A.2.3. An important aspect of the admissible constants in Definitions A.2.1 and A.2.2 is that they are quantitative invariant under arbitrary C^2 diffeomorphisms. For example, because each X_j is C^1_{loc}, in local coordinates we may write $X_j = \sum_k a_j^k \partial_{x_k}$, where $a_j^k \in C^1$. However, our admissible constants do not depend on any sort of upper bound of $\|a_j^k\|_{C^1}$, because this norm depends on the choice of coordinate system. Similarly, since $\operatorname{span}\{X_1(x_0), \ldots, X_q(x_0)\} = T_{x_0}\mathfrak{M}$, we may choose $j_1, \ldots, j_n$ such that $|\det(a_{j_l}^k)_{l,k=1,\ldots,n}| > 0$; however, admissible constants do not depend on a lower bound of this quantity since it depends on the choice of coordinate system.

Because X_{J_0} satisfies $\mathcal{C}(x_0, \eta, \mathfrak{M})$, by hypothesis, we may define the map, for $t \in B^n(\eta)$,

$$\Phi(t) := e^{t \cdot X_{J_0}} x_0 = e^{t_1 X_1 + \cdots + t_n X_n} x_0. \tag{A.4}$$

Let $\eta_0 := \min\{\eta, \xi\}$ so that $\Phi : B^n(\eta_0) \to B_{X_{J_0}}(x_0, \xi)$. Note that, a priori, Φ is C^1_{loc} since $X_1, \ldots, X_n$ are C^1_{loc}.

Theorem A.2.4. *There exists a 0-admissible constant $\chi \in (0, \xi]$ such that:*

(a) $\forall y \in B_{X_{J_0}}(x_0, \chi)$, $\bigwedge X_{J_0}(y) \neq 0$.

(b) $\forall y \in B_{X_{J_0}}(x_0, \chi)$,

$$\max_{J \in \mathcal{I}(n,q)} \left| \frac{\bigwedge X_J(y)}{\bigwedge X_{J_0}(y)} \right| \approx_0 1. \tag{A.5}$$

(c) $\forall \chi' \in (0, \chi]$, $B_{X_{J_0}}(x_0, \chi')$ *is an open subset of* $\mathfrak{M}$.

For the rest of the theorem, we assume $c_{j,k}^l \in C^1_{X_{J_0}}(B_{X_{J_0}}(x_0, \xi))$, *for all* $1 \le j, k, l \le q$. *There exist 1-admissible constants* $\eta_1, \xi_1, \xi_2 \in (0, 1]$ *such that:*

(d) $\Phi(B^n(\eta_1))$ *is an open subset of* $B_{X_{J_0}}(x_0, \chi)$ *and is therefore an open subset of* $\mathfrak{M}$.

(e) $\Phi : B^n(\eta_1) \to \Phi(B^n(\eta_1))$ *is a* C^2_{loc} *diffeomorphism.*

(f) $B_X(x_0, \xi_2) \subseteq B_{X_{J_0}}(x_0, \xi_1) \subseteq \Phi(B^n(\eta_1/2)) \subseteq \Phi(B^n(\eta_1)) \subseteq B_{X_{J_0}}(x_0, \chi) \subseteq B_X(x_0, \xi)$.

Let $Y_j := \Phi^* X_j$ *and write* $Y_{J_0} = (I + A)\nabla$, *where* Y_{J_0} *denotes the column vector of vector fields* $Y_{J_0} = [Y_1, Y_2, \ldots, Y_n]^\top$, ∇ *denotes the gradient in* $\mathbb{R}^n$ *thought of as a column vector, and* $A \in C_{\mathrm{loc}}(B^n(\eta_1); \mathbb{M}^{n\times n})$.

(g) $A(0) = 0$ *and* $\sup_{t \in B^n(\eta_1)} \|A(t)\|_{\mathbb{M}^{n\times n}} \le \frac{1}{2}$.

(h) *For* $1 \le j \le q$ *and* $m \in \mathbb{N}_+$,

$$\|Y_j\|_{C^m(B^n(\eta_1);\mathbb{R}^n)} \lesssim_m 1.$$

(i) *We have the following equivalence of norms for* $f \in C(B^n(\eta_1))$ *and* $m \in \mathbb{N}_+$*:*

$$\|f\|_{C^m(B^n(\eta_1))} \approx_{(m-1)\vee 0} \|f\|_{C^m_{Y_{J_0}}(B^n(\eta_1))} \approx_{(m-1)\vee 0} \|f\|_{C^m_Y(B^n(\eta_1))}.$$

Remark A.2.5. In the context of Lie groups, the coordinates given by Φ in (A.4) are sometimes called canonical coordinates of the first kind.

Remark A.2.6. The regularity given in Theorem A.2.4 (h) is not sharp – indeed it "loses one derivative." Sharp results are obtained in terms of Zygmund–Hölder spaces in [224] and [230]. In [230] it is shown that to obtain these sharp results one cannot use the map Φ given by (A.4). Instead, a more complicated argument involving elliptic PDEs is used.

Remark A.2.7. In Theorem A.2.4 (and in the rest of the appendix), if we write an inequality which implies $\|f\|_{C^m(U)} < \infty$, we take this to mean that $f \in C^m(U)$ and that the inequality holds, and similarly for any other norm. See, for example, Theorem A.2.4 (h) and (i).

A.2.1 More on the assumptions

We further consider the constants $\eta > 0$ and $\tau_0 > 0$ used in Theorem A.2.4. First, we present two examples which show why these constants cannot be dispensed with in our definition of m-admissible constants (Definition A.2.2), and then we state a result which shows that such constants always exist.

Example A.2.8. This example demonstrates the importance of η. Let $\mathfrak{M} = \mathbb{R}$, $q = 1$, $x_0 > 0$, and $X_1 = x^2\partial_x$. In this case, $\eta > 0$ can be taken no larger than $1/x_0$, i. e., X_1

satisfies $\mathcal{C}(x_0, x_0^{-1}, \mathbb{R})$ but does not satisfy $\mathcal{C}(x_0, \eta', \mathbb{R})$ for any $\eta' > x_0^{-1}$ (because the ODE $\dot{y}(t) = y(t)^2$, $y(0) = x_0$ exists only for $t < x_0^{-1}$). If Theorem A.2.4 held with constants independent of η (and therefore independent of x_0), then we could conclude X_1 satisfied $\mathcal{C}(x_0, \eta', \mathbb{R})$ for some η' independent of x_0. This is because the condition $\mathcal{C}$ is invariant under a change of coordinates, and we can therefore check it in the coordinate system given by Φ in Theorem A.2.4. This is a contradiction, showing that η must play a role in the quantitative estimates in Theorem A.2.4.

Example A.2.9. This example demonstrates the importance of τ_0 and also shows its topological nature. The point of τ_0 is to ensure the map Φ in Theorem A.2.4 is injective. Let $\mathfrak{M} = S^1$, $q = 1$, $x_0 \in S^1$, and $X_1 = K\partial_\theta$ for some large constant $K \geq 1$. For this example, we must take $\tau_0 \leq 2\pi/K$. If the constants in Theorem A.2.4 did not depend on τ_0, they would also not depend on K. We could then conclude that τ_0 could be taken independent of K – this is because τ_0 is invariant under a chance of coordinates, and we can check it in the coordinate system given by Φ in Theorem A.2.4. This shows that τ_0 must play a role in the quantitative estimates in Theorem A.2.4.

The next result is a generalization of Proposition 3.6.7 and shows that such an $\eta > 0$ and $\tau_0 > 0$ always exist.

Proposition A.2.10. *Let $X_1, \ldots, X_q \in C^1_{\mathrm{loc}}(\mathfrak{M}; T\mathfrak{M})$ be C^1_{loc} vector fields on the C^2 manifold $\mathfrak{M}$. Fix an open set $U \subseteq \mathfrak{M}$ and a compact set $\mathcal{K} \Subset U$. Then:*
(i) *$\exists \eta > 0$ such that $\forall x_0 \in \mathcal{K}$, X satisfies $\mathcal{C}(x_0, \eta, U)$.*
(ii) *$\exists \tau_0 > 0$ such that $\forall \theta \in S^{q-1}$, if $x \in \mathcal{K}$ is such that $\theta_1 X_1(x) + \cdots + \theta_q X_q(x) \neq 0$, then $\forall r \in (0, \tau_0]$,*

$$e^{r\theta_1 X_1 + \cdots + r\theta_q X_q} x \neq x.$$

Proposition A.2.10 is a local result, so it suffices to instead prove the following local version.

Lemma A.2.11. *Fix $U \subseteq \mathbb{R}^n$ open and fix $\mathcal{K} \Subset U$ compact. Let $X_1, \ldots, X_q \in C^1(U; TU)$ be C^1 vector fields on U. Then:*
(i) *$\exists \eta > 0$ such that $\forall x_0 \in \mathcal{K}$, X satisfies $\mathcal{C}(x_0, \eta, U)$.*
(ii) *$\exists \tau_0 > 0$ such that $\forall \theta \in S^{q-1}$, if $x \in \mathcal{K}$ is such that $\theta_1 X_1(x) + \cdots + \theta_q X_q(x) \neq 0$, then $\forall r \in (0, \tau_0]$,*

$$e^{r\theta_1 X_1 + \cdots + r\theta_q X_q} x \neq x.$$

Moreover, $\eta > 0$ and $\tau_0 > 0$ can be chosen to depend only on n, the distance from $\mathcal{K}$ to ∂U, and upper bounds for q and $\max_{1 \leq j \leq q} \|X_j\|_{C^1(U;\mathbb{R}^n)}$.

Before we prove Lemma A.2.11, we need another standard lemma.

Lemma A.2.12. *Suppose $Z \in C^1(V;TV)$ is a C^1 vector field on an open set $V \subseteq \mathbb{R}^n$. Then there exists $\delta > 0$, depending only on n, such that if $\|Z\|_{C^1(V;\mathbb{R}^n)} \le \delta$, then there does not exist $x \in V$ with:*
- $e^{tZ}x \in V$, $\forall t \in [0,1]$.
- $e^{Z}x = x$.
- $Z(x) \neq 0$.

Proof. Suppose the lemma does not hold and we have an $x \in V$ and Z satisfying the above conditions. In the proof of this lemma, we use big-O notation – the implicit constants will only depend on n. Differentiating the identity

$$\frac{d}{dt}e^{tZ}x = Z(e^{tZ}x),$$

we obtain

$$\frac{d^2}{dt^2}e^{tZ}x = O\left(\delta\left|\frac{d}{dt}e^{tZ}x\right|\right). \tag{A.6}$$

Thus, by Grönwall's inequality,

$$\frac{d}{dt}e^{tZ}x = O\left(\left|\frac{d}{dt}\right|_{t=0} e^{tZ}x\right|\right) = O(|Z(x)|), \quad t \le 1. \tag{A.7}$$

Plugging (A.7) into (A.6), we see that

$$\frac{d^2}{dt^2}e^{tZ}x = O(\delta|Z(x)|). \tag{A.8}$$

Integrating (A.8), we obtain

$$\frac{d}{dt}e^{tZ}x = Z(x) + O(\delta|t||Z(x)|). \tag{A.9}$$

Integrating (A.9), we have

$$x = e^{Z}x = x + Z(x) + O(\delta|Z(x)|),$$

which is impossible if δ is sufficiently small (since $Z(x) \neq 0$). This completes the proof. □

Proof of Lemma A.2.11. Item (i) is a standard consequence of the Picard–Lindelöf theorem, so we only prove (ii). Take $\delta = \delta(n) > 0$ as in Lemma A.2.12 and take $\delta_1 > 0$ so small that $\forall y \in \mathcal{K}$, $t \in B^q(\delta_1)$, we have $e^{t_1X_1+t_2X_2+\cdots+t_qX_q}y \in U$; by the Picard–Lindelöf theorem τ_0, can be chosen to depend only on n, the distance from $\mathcal{K}$ to ∂U, and upper bounds for q and $\max_{1\le j\le q}\|X_j\|_{C^1(U;\mathbb{R}^n)}$. Set $C := \max_{1\le j\le q}\|X_j\|_{C^1(U;\mathbb{R}^n)}$ and let $\tau_0 := \min\{\delta_1, \delta/qC\}$. Now, (ii) follows from Lemma A.2.12. □

Remark A.2.13. Proposition A.2.10 shows that $\eta > 0$ and $\tau_0 > 0$ as in Theorem A.2.4 always exist. However, the proof of Lemma A.2.11 relies on the C^1 norms of the vector fields $X_1, \ldots, X_q$ in some fixed coordinate system. Theorem A.2.4 provides a coordinate system in which we have good bounds on the C^1 norms of these vector fields, but does not a priori assume the existence of such a coordinate system. By working with η and τ_0 in Definition A.2.2 we are able to present a theorem which does not assume any sort of quantitative bounds on the C^1 norms of $X_1, \ldots, X_q$. This allows us to apply Theorem A.2.4 to some vector fields which may be a priori rough or large in the coordinate system in which they are given, and Theorem A.2.4 provides a coordinate system where they are not rough or large.

We close this section with a lemma which shows that $B_X(x_0, \xi)$ is open in $\mathfrak{M}$.

Lemma A.2.14. *Let $\mathfrak{M}$ be an n-dimensional C^2 manifold and fix $x \in \mathfrak{M}$ and $\delta > 0$. Suppose $V_1, \ldots, V_q \in C^1_{\mathrm{loc}}(\mathfrak{M}; T\mathfrak{M})$ are C^1_{loc} vector fields which span the tangent space at every point of $B_V(x, \delta)$. Then $B_V(x, \delta)$ is open.*

Proof. Fix $x_0 \in B_V(x, \delta)$ and pick $j_1, \ldots, j_n \in \{1, \ldots, q\}$ such that $V_{j_1}(x_0), \ldots, V_{j_n}(x_0)$ form a basis for $T_{x_0}\mathfrak{M}$. Define

$$\Psi(t) = e^{t_1 V_{j_1} + \cdots + t_n V_{j_n}} x_0$$

so that $d\Psi(0) = (V_{j_1}(x_0) | \cdots | V_{j_n}(x_0))$ and therefore $d\Psi(0)$ is non-singular. Standard theorems show that Ψ is C^1_{loc}. It is clear that for $\delta' > 0$ sufficiently small, $\Psi(B^n(\delta')) \subseteq B_V(x, \delta)$, and the inverse function theorem shows that for $\delta' > 0$ sufficiently small, $\Psi(B^n(\delta')) \subseteq B_V(x, \delta)$ is open. Since $x_0 = \Psi(0)$, $\Psi(B^n(\delta'))$ is an open neighborhood of x_0 in $B_V(x, \delta)$. Since $x_0 \in B_V(x, \delta)$ was arbitrary, we conclude $B_V(x, \delta)$ is open. □

A.2.2 Densities

Under the same setting as Theorem A.2.4, let $\chi \in (0, \xi]$ be as in that theorem. In most of the applications in this text, we are given a density,[1] Vol, on $B_{X_{J_0}}(x_0, \chi)$, and we need to measure certain sets with respect to this density.

Let Vol be a C^1_{loc} density on the manifold $B_{X_{J_0}}(x_0, \chi)$. Suppose

$$\mathrm{Lie}_{X_j} \mathrm{Vol} = f_j \mathrm{Vol}, \quad 1 \le j \le n,\ f_j \in C(B_{X_{J_0}}(x_0, \chi)). \tag{A.10}$$

Our goal is to understand $\Phi^* \mathrm{Vol}$ and $\mathrm{Vol}(B_X(x_0, \xi_2))$, where Φ and ξ_2 are as in Theorem A.2.4.

1 In the rest of this text, Vol is assumed to be strictly positive, though we make no such assumption in this appendix.

Definition A.2.15. We say C is a $\{0;\mathrm{Vol}\}$-admissible constant if C is a 0-admissible constant (as in Definition A.2.1), which is also allowed to depend on upper bounds for $\|f_j\|_{C(B_{X_{J_0}}(x_0,\chi))}$, $1 \le j \le n$.

Definition A.2.16. For $m \in \mathbb{N}_+$, if we say C is an $\{m;\mathrm{Vol}\}$-admissible constant, this means that we assume $f_j \in C^m_{X_{J_0}}(B_{X_{J_0}}(x_0,\chi))$ and C is an m-admissible constant (as in Definition A.2.2) which is also allowed to depend on upper bounds for $\|f_j\|_{C^m_{X_{J_0}}(B_{X_{J_0}}(x_0,\chi))}$, $1 \le j \le n$.

We write $A \lesssim_{\{m;\mathrm{Vol}\}} B$ for $A \le CB$, where $C \ge 0$ is an $\{m;\mathrm{Vol}\}$-admissible constant. We write $A \approx_{\{m;\mathrm{Vol}\}B}$ for $A \lesssim_{\{m;\mathrm{Vol}\}} B$ and $B \lesssim_{\{m;\mathrm{Vol}\}} A$.

To help understand Vol, we introduce a distinguished density, Vol_0, on $B_{X_{J_0}}(x_0,\chi)$:

$$\mathrm{Vol}_0(Z_1,\dots,Z_n) := \left|\frac{Z_1 \wedge Z_2 \wedge \cdots \wedge Z_n}{X_1 \wedge X_2 \wedge \cdots \wedge X_n}\right|. \tag{A.11}$$

Note that Vol_0 is defined since $X_1 \wedge X_2 \wedge \cdots \wedge X_n = \bigwedge X_{J_0}$ is never zero on $B_{X_{J_0}}(x_0,\chi)$ by Theorem A.2.4 (a); Vol_0 is clearly a density.

Theorem A.2.17. *There exists $g \in C(B_{X_{J_0}}(x_0,\chi))$ such that* $\mathrm{Vol} = g\,\mathrm{Vol}_0$ *and:*

(i) *We have $g(x) \approx_{\{0;\mathrm{Vol}\}} g(x_0) = \mathrm{Vol}(X_1,\dots,X_n)(x_0)$, $\forall x \in B_{X_{J_0}}(x_0,\chi)$. In particular, g always has the same sign, and is either never zero or always zero.*

(ii) *We have the following regularity on g:*

$$\|g\|_{C^m_{X_{J_0}}} \lesssim_{\{(m-1)\vee 0;\mathrm{Vol}\}} |\mathrm{Vol}(X_1,\dots,X_n)(x_0)|.$$

Define $h \in C^1(B^n(\eta_1))$ by $\Phi^*\,\mathrm{Vol} = h\sigma_{\mathrm{Leb}}$.

(iii) *We have $h(t) \approx_{\{0;\mathrm{Vol}\}} \mathrm{Vol}(X_1,\dots,X_n)(x_0)$, $\forall t \in B^n(\eta_1)$. In particular, h always has the same sign and is either never zero or always zero.*

(iv) *We have the following regularity on h:*

$$\|h\|_{C^m(B^n(\eta_1))} \lesssim_{\{m;\mathrm{Vol}\}} |\mathrm{Vol}(X_1,\dots,X_n)(x_0)|.$$

Corollary A.2.18. *Let $\xi_2 \in (0,1]$ be as in Theorem* A.2.4. *Then*

$$\mathrm{Vol}(B_{X_{J_0}}(x_0,\xi_2)) \approx_{\{1;\mathrm{Vol}\}} \mathrm{Vol}(B_X(x_0,\xi_2)) \approx_{\{1;\mathrm{Vol}\}} \mathrm{Vol}(X_1,\dots,X_n)(x_0), \tag{A.12}$$

and therefore

$$\begin{aligned} |\mathrm{Vol}(B_{X_{J_0}}(x_0,\xi_2)| &\approx_{\{1;\mathrm{Vol}\}} |\mathrm{Vol}(B_X(x_0,\xi_2))| \\ &\approx_{\{1;\mathrm{Vol}\}} |\mathrm{Vol}(X_1,\dots,X_n)(x_0)| \approx_0 \max_{j_1,\dots,j_n\in\{1,\dots,q\}} |\mathrm{Vol}(X_{j_1},\dots,X_{j_n})(x_0)|. \end{aligned} \tag{A.13}$$

A.3 Qualitative consequences

For the main applications of this text, we require the quantitative nature of Theorem A.2.4 (see, e. g., Remark 3.6.1). However, Theorem A.2.4 also has interesting qualitative consequences: it provides a coordinate system in which the given C^1_{loc} vector fields may have a higher level of regularity.

Let $X_1, \dots, X_q \in C^1_{\mathrm{loc}}(\mathfrak{M}; T\mathfrak{M})$ be C^1_{loc} vector fields on the C^2 manifold, $\mathfrak{M}$, of dimension n, which span the tangent space at every point.

Theorem A.3.1 (The local theorem). *Fix $x_0 \in \mathfrak{M}$. The following are equivalent:*

(i) *There are an open neighborhood $V \subseteq \mathfrak{M}$ of x_0 and a C^2 diffeomorphism $\Phi : U \to V$, where $U \subseteq \mathbb{R}^n$ is open, such that $\Phi^* X_1, \dots, \Phi^* X_q \in C^\infty(U; \mathbb{R}^n)$.*

(ii) *Reorder the vector fields so that $X_1(x_0), \dots, X_n(x_0)$ are linearly independent. There is an open neighborhood $V \subseteq \mathfrak{M}$ of x_0 such that:*
- *$[X_j, X_k] = \sum_{k=1}^n \hat{c}^l_{j,k} X_l$, $1 \le j,k \le n$, where $\hat{c}^l_{j,k} \in C^\infty_X(V)$.*
- *For $n+1 \le j \le q$, $X_j = \sum_{k=1}^n b^k_j X_k$, where $b^k_j \in C^\infty_X(V)$.*

(iii) *There exists an open neighborhood $V \subseteq \mathfrak{M}$ of x_0 such that $[X_j, X_k] = \sum_{l=1}^q c^l_{j,k} X_l$, $1 \le j,k \le q$, where $c^l_{j,k} \in C^\infty_X(V)$.*

Remark A.3.2. Theorem A.3.1 (ii) and (iii) are similar, though have different advantages. In (ii), because $X_1, \dots, X_n$ are a basis for the tangent space of $\mathfrak{M}$ near x_0, the functions $\hat{c}^l_{j,k}$ and b^k_j are uniquely determined (so long as V is chosen sufficiently small). Moreover, one can directly check to see if (ii) holds by computing these functions and this computation can be done in any coordinate system. If $q > n$, $X_1, \dots, X_q$ are linearly dependent and $c^l_{j,k}$ in (iii) are not uniquely determined; (iii) only asks that there exists a choice of $c^l_{j,k}$ satisfying the conditions in (iii). Despite this lack of uniqueness, it is often more convenient to use the setting of (iii) since it is symmetric in $X_1, \dots, X_q$.

Proof of Theorem A.3.1. (i) $\Rightarrow$ (ii): Let U, V, x_0, and Φ be as in (i). Without loss of generality, assume $0 \in U$ and $\Phi(0) = x_0$. Reorder $X_1, \dots, X_q$ so that $X_1(x_0), \dots, X_n(x_0)$ are linearly independent and let $Y_j := \Phi^* X_j$ so that $Y_j \in C^\infty(U; \mathbb{R}^n)$, $1 \le j \le q$. Note that $Y_1(0), \dots, Y_n(0)$ form a basis for the tangent space $T_0 U$. Let $\eta > 0$ be so small that $B^n(2\eta) \subseteq U$ and $Y_1, \dots, Y_n$ form a basis for the tangent space at every point of $B^n(2\eta)$. Since $Y_1, \dots, Y_q \in C^\infty(B^n(\eta))$, it follows immediately from the definitions that $C^\infty_Y(B^n(\eta)) \subseteq C^\infty(B^n(\eta))$. Furthermore, since $Y_1, \dots, Y_n$ form a basis for the tangent space at every point of $B^n(2\eta)$, it is easy to see that $C^\infty(B^n(\eta)) \subseteq C^\infty_Y(B^n(\eta))$. We conclude that

$$C^\infty_Y(B^n(\eta)) = C^\infty(B^n(\eta)). \tag{A.14}$$

Using the fact that $Y_1, \dots, Y_n \in C^\infty(B^n(2\eta))$ form a basis for the tangent space at every point of $B^n(2\eta)$, we have, for $1 \le j,k \le q$,

$$[Y_j, Y_k] = \sum_{l=1}^{n} \tilde{c}_{j,k}^{l} Y_l, \quad \tilde{c}_{j,k}^{l} \in C^{\infty}(B^n(\eta)). \tag{A.15}$$

By (A.14), we have $\tilde{c}_{j,k}^{l} \in C_Y^{\infty}(B^n(\eta))$. Pushing (A.15) forward via Φ shows that $[X_j, X_k] = \sum_{l=1}^{n} \hat{c}_{j,k}^{l} X_l$, with $\hat{c}_{j,k}^{l} = \tilde{c}_{j,k}^{l} \circ \Phi^{-1} \in C_{\Phi_* Y}^{\infty}(\Phi(B^n(\eta))) = C_X^{\infty}(\Phi(B^n(\eta)))$.

Again using the fact that $Y_1, \dots, Y_n \in C^{\infty}(B^n(2\eta))$ form a basis for the tangent space at every point of $B^n(2\eta)$, we have, for $n+1 \le j \le q$,

$$Y_j = \sum_{k=1}^{n} \tilde{b}_j^k Y_k, \quad \tilde{b}_j^k \in C^{\infty}(B^n(\eta)). \tag{A.16}$$

By (A.14), we have $\tilde{b}_j^k \in C_Y^{\infty}(B^n(\eta))$. Pushing (A.16) forward via Φ shows that $X_j = \sum_{k=1}^{n} b_j^k X_k$, where $b_j^k = \tilde{b}_j^k \circ \Phi^{-1} \in C_{\Phi_* Y}^{\infty}(\Phi(B^n(\eta))) = C_X^{\infty}(\Phi(B^n(\eta)))$. This completes the proof of (ii) with $V = \Phi(B^n(\eta))$.

(ii) $\Rightarrow$ (iii): Suppose (ii) holds. We wish to show that for $1 \le j, k \le q$,

$$[X_j, X_k] = \sum_{l=1}^{q} c_{j,k}^{l}, \quad c_{j,k}^{l} \in C_X^{\infty}(V), \tag{A.17}$$

where V is as in (ii). For $1 \le j, k \le n$, (A.17) is contained in (ii). We prove (A.17) for $n+1 \le j, k \le q$; the remaining cases when $1 \le j \le n$ and $n+1 \le k \le q$ or $n+1 \le j \le q$ and $1 \le k \le n$ are similar and easier. Consider

$$\begin{aligned}
[X_j, X_k] &= \left[\sum_{l_1=1}^{n} b_j^{l_1} X_{l_1}, \sum_{l_2=1}^{n} b_k^{l_2} X_{l_2} \right] \\
&= \sum_{l_1, l_2=1}^{n} \left(b_j^{l_1} (X_{l_1} b_k^{l_2}) X_{l_2} - b_k^{l_2} (X_{l_2} b_j^{l_1}) X_{l_1} + \sum_{i=1}^{n} b_j^{l_1} b_k^{l_2} \hat{c}_{l_1, l_2}^{i} X_i \right).
\end{aligned}$$

We are given $b_j^k \in C_X^{\infty}(V)$ and $\hat{c}_{k_1,k_2}^{l} \in C_X^{\infty}(V)$; it follows that $X_l b_j^k \in C_X^{\infty}(V)$. From here, (A.17) follows immediately from the easily shown fact that $C_X^{\infty}(V)$ is an algebra, completing the proof of (iii).

(iii) $\Rightarrow$ (i): This is a consequence of Theorem A.2.4. We make a few comments to this end. As discussed in Section A.2.1, $\eta > 0$ and $\tau_0 > 0$ as in the hypotheses of Theorem A.2.4 always exist (see also Remark A.2.13). Take $\xi > 0$ so small that $B_X(x_0, \xi) \subseteq V$ and take $J_0 \in \mathcal{I}(n, q)$ so that (A.3) holds with $\zeta = 1$. We have, directly from the definitions,

$$c_{j,k}^{l} \in C_X^{\infty}(V) \subseteq C_X^{\infty}(B_X(x_0, \xi)) \subseteq C_{X_{J_0}}^{\infty}(B_{X_{J_0}}(x_0, \xi)).$$

In particular, all the hypotheses of Theorem A.2.4 hold and m-admissible constants as in Definition A.2.2 are finite for every $m \in \mathbb{N}_+$. This yields a map Φ as in Theorem A.2.4. This map satisfies the conclusions of (i) with $U = B^n(\eta_1)$; in particular, it follows from Theorem A.2.4 (h) that $\Phi^* X_1, \dots, \Phi^* X_q \in C^{\infty}(B^n(\eta_1); \mathbb{R}^n)$. □

Theorem A.3.1 has the following global consequence.

Theorem A.3.3 (The global theorem). *The following are equivalent:*
(i) *There exists a C^∞ atlas on $\mathfrak{M}$, compatible with its C^2 structure, such that $X_1, \ldots, X_q$ are C^∞_{loc} with respect to this atlas.*
(ii) *For each $x_0 \in \mathfrak{M}$, any of the three equivalent conditions from Theorem* A.3.1 *hold for this choice of x_0.*
(iii) $[X_j, X_k] = \sum_{l=1}^{q} c_{j,k}^{l} X_l$, $1 \leq j, k \leq q$, *where $c_{j,k}^{l} \in C^\infty_{X,\mathrm{loc}}(\mathfrak{M})$.*

Furthermore, under these conditions, the C^∞ structure on $\mathfrak{M}$ induced by the atlas in (i) is unique, in the sense that if there is another C^∞ atlas on $\mathfrak{M}$, compatible with its C^2 structure and such that $X_1, \ldots, X_q$ are also C^∞ with respect to this second atlas, then the identity map $\mathfrak{M} \to \mathfrak{M}$ is a C^∞_{loc} diffeomorphism between these two C^∞ manifold structures on $\mathfrak{M}$.

Remark A.3.4. Recall that for $m \in \mathbb{N} \cup \{\infty\}$, a C^m *atlas* of dimension $n \in \mathbb{N}$ on a topological space M is a collection $\{(\psi_\alpha, V_\alpha) : \alpha \in \mathcal{I}\}$, where $\mathcal{I}$ is an index set, $\{V_\alpha : \alpha \in \mathcal{I}\}$ is an open cover of M, $\psi_\alpha : V_\alpha \to \psi(V_\alpha)$ is a homeomorphism where $\psi_\alpha(V_\alpha) \subseteq \mathbb{R}^n$ is open, and the transition functions

$$\psi_\beta \circ \psi_\alpha^{-1} : \psi_\alpha(V_\alpha \cap V_\beta) \to \psi_\beta(V_\alpha \cap V_\beta)$$

are C^m_{loc}.

The uniqueness of the C^∞ structure described in Theorem A.3.3 follows from the next lemma.

Lemma A.3.5. *Let M and N be two n-dimensional C^∞ manifolds, suppose $X_1, \ldots, X_q$ are C^∞_{loc} vector fields on M which span the tangent space at every point, and let $Z_1, \ldots, Z_q$ be C^∞_{loc} vector fields on N. Let $\Psi : M \to N$ be a C^2_{loc} diffeomorphism such that $\Psi_* X_j = Z_j$. Then Ψ is a C^∞_{loc} diffeomorphism.*

Proof. We begin with the case where M and N are open subsets of $\mathbb{R}^n$; we use x to denote points in M and y to denote points in N.

Fix a point $x_0 \in M$. We wish to show that Ψ is C^∞_{loc} on a neighborhood of x_0. Reorder $X_1, \ldots, X_q$ so that $X_1(x_0), \ldots, X_n(x_0)$ are linearly independent and reorder $Z_1, \ldots, Z_q$ in the same way so that we still have $\Psi_* X_j = Z_j$. Since $X_1(x_0), \ldots, X_n(x_0)$ form a basis of $T_{x_0}M$, we may pick an open neighborhood U of x_0 so that $X_1(x), \ldots, X_n(x)$ form a basis for $T_x M$ space at every $x \in U$.

Define two $n \times n$ matrices $\mathscr{X}(x) := (X_1(x)| \cdots |X_n(x))$ and $\mathscr{Z}(y) = (Z_1(y)| \cdots |Z_n(y))$. By hypothesis, we have $\mathscr{X} \in C^\infty_{\mathrm{loc}}(M; \mathbb{M}^{n\times n})$ and $\mathscr{Z} \in C^\infty_{\mathrm{loc}}(N; \mathbb{M}^{n\times n})$. Since $\Psi_* X_j = Z_j$, we have the matrix equation

$$d\Psi(x)\mathscr{X}(x) = \mathscr{Z}(\Psi(x)). \tag{A.18}$$

Since $X_1, \dots, X_n$ span the tangent space at every point of U, the matrix $\mathscr{X}(x)$ is invertible, $\forall x \in U$, and we have $\mathscr{X}(x)^{-1} \in C^\infty_{\mathrm{loc}}(U; \mathbb{M}^{n\times n})$. From (A.18), we obtain

$$d\Psi(x) = \mathscr{Z}(\Psi(x))\mathscr{X}(x)^{-1}. \tag{A.19}$$

We prove, for all $m \in \mathbb{N}$, $m \geq 2$, that $\Psi \in C^m_{\mathrm{loc}}(U)$ by induction on m. The base case, $m = 2$, follows from the hypothesis that Ψ is a C^2_{loc} diffeomorphism. Suppose $\Psi \in C^m_{\mathrm{loc}}(U)$. Then by (A.19) it follows that $d\Psi \in C^m_{\mathrm{loc}}(U)$, and therefore $\Psi \in C^{m+1}_{\mathrm{loc}}(U)$. This achieves the inductive step and completes the proof in the case where both M and N are open subsets of $\mathbb{R}^n$.

We now turn to the general case where M and N are C^∞_{loc} manifolds. Since M and N are C^∞ manifolds, they have associated C^∞ atlases, $\{(\phi_\alpha, V_\alpha)\}$ and $\{(\psi_\beta, W_\beta)\}$; see Remark A.3.4. Saying that Ψ is C^∞_{loc} is equivalent to the statement that $\forall \alpha, \beta$,

$$\Psi_{\alpha,\beta} := \psi_\beta \circ \Psi \circ \phi_\alpha^{-1} : \phi_\alpha(V_\alpha \cap \Psi^{-1}(W_\beta)) \to \psi_\beta(\Psi(V_\alpha) \cap W_\beta)$$

is C^∞_{loc}.

By hypothesis, we have

$$(\Psi_{\alpha,\beta})_*(\phi_\alpha)_* X_j = (\psi_\beta)_* Z_j.$$

Since $(\phi_\alpha)_* X_1, \dots, (\phi_\alpha)_* X_q$ and $(\psi_\beta)_* Z_1, \dots, (\psi_\beta)_* Z_q$ are C^∞_{loc} by hypothesis and $(\phi_\alpha)_* X_1, \dots, (\phi_\alpha)_* X_q$ span the tangent space at every point of $\phi_\alpha(V_\alpha)$ by hypothesis, it follows from the above case where M and N are open subsets of $\mathbb{R}^n$ that $\Psi_{\alpha,\beta}$ is a C^∞_{loc} diffeomorphism, completing the proof. □

Proof of Theorem A.3.3. (ii) ⇒ (i): Under the condition (ii), for each $x \in \mathfrak{M}$, there exist open sets $U_x \subseteq \mathbb{R}^n$, $V_x \subseteq \mathfrak{M}$ with $x \in V_x$, and a C^2_{loc} diffeomorphism $\Phi_x : U_x \to V_x$ such that if $Y_j^x := \Phi_x^* X_j$, then $Y_j^x \in C^\infty_{\mathrm{loc}}(U_x; \mathbb{R}^n)$; note that $Y_1^x, \dots, Y_q^x$ span the tangent space to U_x at every point.

We wish to show that the collection $\{(\Phi_x^{-1}, V_x) : x \in \mathfrak{M}\}$ forms a C^∞ atlas on $\mathfrak{M}$ in the sense of Remark A.3.4. Once that is shown, (i) will follow since X_j will be C^∞_{loc} with respect to this atlas by definition and the atlas is clearly compatible with the C^2 structure on $\mathfrak{M}$. All that needs to be shown is that the transition functions are C^∞_{loc}. Take $x_1, x_2 \in \mathfrak{M}$ and consider the transition function $\Psi_{x_1,x_2} := \Phi_{x_2}^{-1} \circ \Phi_{x_1} : U_{x_1} \cap \Phi_{x_1}^{-1}(V_{x_2}) \to U_{x_2} \cap \Phi_{x_2}^{-1}(V_{x_1})$; we wish to show that Ψ_{x_1,x_2} is C^∞_{loc}. We already know that Ψ_{x_1,x_2} is a C^2_{loc} diffeomorphism and $(\Psi_{x_1,x_2})_* Y_j^{x_1} = Y_j^{x_2}$. It now follows from Lemma A.3.5 that Ψ_{x_1,x_2} is a C^∞_{loc} diffeomorphism, completing the proof of (i).

(i) ⇒ (iii): Suppose (i) holds, so that we may regard $\mathfrak{M}$ as a smooth manifold and $X_1, \dots, X_q$ as smooth vector fields on $\mathfrak{M}$. Since $X_1, \dots, X_q$ span the tangent space at every point, a simple partition of unity argument shows that we may write $[X_j, X_k] = \sum_{l=1}^q c_{j,k}^l X_l$, where $c_{j,k}^l \in C^\infty_{\mathrm{loc}}(\mathfrak{M})$. Since $X_1, \dots, X_q \in C^\infty_{\mathrm{loc}}(\mathfrak{M}; T\mathfrak{M})$, it follows easily from the definitions that $c_{j,k}^l \in C^\infty_{X,\mathrm{loc}}(\mathfrak{M})$, completing the proof of (iii).

(iii) ⇒ (ii): This is obvious.

Finally, as mentioned before, the uniqueness of the $C^{\infty}_{\mathrm{loc}}$ manifold structure as described in the theorem is an immediate consequence of Lemma A.3.5. □

Remark A.3.6. Theorems A.3.1 and A.3.3 deal with $C^{\infty}_{\mathrm{loc}}$ functions. Similar results hold for a finite level of smoothness provided one uses appropriate Zygmund–Hölder spaces; see [224, 230]. There are also analogous results for real analyticity; see [222].

Remark A.3.7. Though we do not use results like Theorems A.3.1 and A.3.3 in the main part of this text, it is possible they could be useful when studying PDEs. For example, one might be given vector fields whose coefficients come from the solution to a PDE which are not a priori known to be smooth. In such a situation, it is possible that Theorems A.3.1 and A.3.3 could give a manifold structure in which the vector fields are smooth, giving access to the theory of this text.

A.4 Proofs

We turn to the proofs of the main results of this appendix: Theorems A.2.4 and A.2.17 and Corollary A.2.18. The heart of the proof is the study of a certain ODE which arises in canonical coordinates; this is presented in Section A.4.1. We also require a quantitative special case of the inverse function theorem, described in Section A.4.2. With those preparations in hand, we prove Theorem A.2.4 in Section A.4.3. Finally, we prove the result about densities (Theorem A.2.17 and Corollary A.2.18) in Section A.4.4.

A.4.1 An ODE

The proof of Theorem A.2.4 rests on the study of the following ODE, defined for an $n \times n$ matrix $A(u)$, depending on $u \in B^n(\eta)$ for some $\eta > 0$. Write $u = r\theta$, $r > 0$, $\theta \in S^{n-1}$. The ODE is

$$\frac{\partial}{\partial r} rA(r\theta) = -A(r\theta)^2 - C(r\theta)A(r\theta) - C(r\theta), \tag{A.20}$$

where $C(u) \in C(B^n(\eta); \mathbb{M}^{n\times n}(\mathbb{R}))$ is a given function.

A.4.1.1 Derivation of the ODE

Let $X_1, \dots, X_n \in C^1_{\mathrm{loc}}(\mathfrak{M}; T\mathfrak{M})$ be C^1_{loc} vector fields on a C^2 manifold $\mathfrak{M}$ of dimension $n \in \mathbb{N}_+$. Fix $x \in \mathfrak{M}$ and $\epsilon > 0$ and suppose that:

- $X_1, \dots, X_n$ span the tangent space at every point of $\mathfrak{M}$.
- $\Phi(u) := e^{u_1X_1+u_2X_2+\cdots+u_nX_n}x$ exists for $u \in B^n(\epsilon)$.

Write $[X_j, X_k] = \sum_{l=1}^n c_{j,k}^l X_l$. Since $X_1, \dots, X_n$ form a basis for the tangent space of $\mathfrak{M}$ at every point, the functions $c_{j,k}^l \in C_{\mathrm{loc}}(\mathfrak{M})$ are uniquely determined. Classical theorems from the field of ODEs show that Φ is C^1_{loc} since $X_1, \dots, X_n$ are.

Let $U \subseteq \mathfrak{M}$ and $V \subseteq B^n(\epsilon)$ be open sets such that $\Phi|_V : V \to U$ is a C^1_{loc} diffeomorphism. Let $Y_j := \Phi|_V^* X_j$ so that Y_j is a C^0_{loc} vector field on V. Write

$$Y_j = \partial_{u_j} + \sum_{k=1}^n a_j^k(u)\partial_{u_k}, \tag{A.21}$$

where $a_j^k \in C_{\mathrm{loc}}(V)$. Let $A(u)$ denote the $n \times n$ matrix with (j,k) component $a_j^k(u)$ and let $C(u)$ denote the $n \times n$ matrix with (j,k) component $\sum_l u_l c_{j,l}^k \circ \Phi(u)$. We write u in polar coordinates as $u = r\theta$, $r \geq 0$.

Proposition A.4.1. *In the above setting, $A(u)$ satisfies the differential equation*

$$\frac{\partial}{\partial r} rA(r\theta) = -A(r\theta)^2 - C(r\theta)A(r\theta) - C(r\theta). \tag{A.22}$$

In particular, the derivative $\frac{\partial}{\partial r} rA(r\theta)$ exists in the classical sense.

We begin by proving Proposition A.4.1 in the smooth case.

Lemma A.4.2. *Proposition* A.4.1 *holds in the special case where $\mathfrak{M}$ is a C^∞ manifold and $X_1, \dots, X_n$ are C^∞_{loc} vector fields on $\mathfrak{M}$.*

Proof. When $X_1, \dots, X_n$ are C^∞_{loc}, then Φ is C^∞_{loc} and $\Phi|_V : V \to U$ is a C^∞_{loc} diffeomorphism. It follows that $Y_1, \dots, Y_n$ are C^∞_{loc} vector fields on V. Furthermore,

$$[Y_j, Y_k] = \sum_l \tilde{c}_{j,k}^l Y_l, \quad \text{where } \tilde{c}_{j,k}^l := c_{j,k}^l \circ \Phi. \tag{A.23}$$

Note that $d\Phi(r\theta)r\partial_r = rd\Phi(r\theta)\partial_r = r\theta \cdot X(\Phi(r\theta))$, since $\Phi(r\theta) = e^{r(\theta \cdot X)}x$, and we are identifying the X with the vector of vector fields $X = (X_1, \dots, X_n)$. Writing this in Cartesian coordinates, we have

$$\sum_{j=1}^n u_j \partial_{u_j} = \sum_{j=1}^n u_j Y_j. \tag{A.24}$$

Taking the Lie bracket of Y_i with (A.24) and using (A.23), we obtain

$$\begin{aligned}\sum_{j=1}^n ((Y_i u_j)\partial_{u_j} + u_j[Y_i, \partial_{u_j}]) &= \sum_{j=1}^n ((Y_i u_j)Y_j + u_j[Y_i, Y_j]) \\ &= \sum_{j=1}^n \left((Y_i u_j)Y_j + u_j \sum_{l=1}^n \tilde{c}_{i,j}^l(u) Y_l \right).\end{aligned} \tag{A.25}$$

We rewrite (A.25) as

$$\begin{aligned}&\left(\sum_{j=1}^{n} u_j[\partial_{u_j}, Y_i - \partial_{u_i}]\right) + Y_i - \partial_{u_i}\\&\quad = -\left(\sum_{j=1}^{n}((Y_i - \partial_{u_i})(u_j))(Y_j - \partial_{u_j})\right) - \sum_{j=1}^{n}\sum_{l=1}^{n} u_j\tilde{c}_{i,j}^{l}(u)Y_l.\end{aligned} \tag{A.26}$$

Plugging (A.21) into (A.26), we have

$$\begin{aligned}&\sum_{j=1}^{n}\sum_{k=1}^{n} u_j(\partial_{u_j}a_i^k)\partial_{u_k} + \sum_{k=1}^{n} a_i^k\partial_{u_k}\\&\quad = -\sum_{k=1}^{n}\sum_{j=1}^{n} a_i^j a_j^k\partial_{u_k} - \sum_{k=1}^{n}\sum_{j=1}^{n} u_j\tilde{c}_{i,j}^{k}\partial_{u_k} - \sum_{l=1}^{n}\sum_{k=1}^{n}\sum_{j=1}^{n} u_j\tilde{c}_{i,j}^{l}a_l^k\partial_{u_k}.\end{aligned} \tag{A.27}$$

Taking the ∂_{u_k} component of (A.27) and writing $1 + \sum_{j=1}^{n} u_j\partial_{u_j} = \partial_r r$, we have

$$\partial_r r a_i^k = -\sum_{j=1}^{n} a_i^j a_j^k - \sum_{j=1}^{n} u_j\tilde{c}_{i,j}^{k} - \sum_{l=1}^{n}\left(\sum_{j=1}^{n} u_j\tilde{c}_{i,j}^{l}\right)a_l^k.$$

This is exactly (A.22) and completes the proof. □

Proof of Proposition A.4.1. By a classical theorem of Whitney, there is a C^∞ manifold structure on $\mathfrak{M}$, compatible with its C^2 manifold structure, so we may assume $\mathfrak{M}$ is a C^∞ manifold. Pick relatively compact open sets $\tilde{V} \Subset V$ and $\tilde{U} \Subset U$ with $\Phi|_{\tilde{V}} : \tilde{V} \to \tilde{U}$ a C^1_{loc} diffeomorphism. Fix $u_0 \in \tilde{V}$. We will prove the result with V replaced by $B^n(u_0, \delta_0)$ for some $\delta_0 > 0$, and the result will follow since the conclusion is local.

Fix $\epsilon' \in (0, \epsilon)$ so large that $\tilde{V} \subseteq B^n(\epsilon')$. Let X_j^σ be smooth vector fields on $\mathfrak{M}$ such that $X_j^\sigma \to X_j$ in C^1_{loc} as $\sigma \downarrow 0$. Define

$$\Phi_\sigma(u) := e^{u_1X_1^\sigma + \cdots + u_nX_n^\sigma}x.$$

Then, for σ sufficiently small, $\Phi_\sigma(u)$ is defined for $u \in B^n(\epsilon')$ and $X_1^\sigma, \ldots, X_n^\sigma$ form a basis for the tangent space at every point of a neighborhood of the closure of $\Phi_\sigma(B^n(\epsilon'))$. Thus, we may write $[X_i^\sigma, X_j^\sigma] = \sum_k c_{i,k}^{k,\sigma}X_k^\sigma$, with $c_{i,j}^{k,\sigma} \to c_{i,j}^k$ in C^0_{loc} as $\sigma \downarrow 0$. Also, $\Phi_\sigma \to \Phi$ in $C^1_{\mathrm{loc}}(B^n(\epsilon'))$ as $\sigma \downarrow 0$, by standard theorems.

For σ sufficiently small, $|\det d\Phi_\sigma(u_0)| \geq \frac{1}{2}|\det d\Phi(u_0)| > 0$. The inverse function theorem shows that there is a $\delta_0 > 0$, independent of σ, such that for σ small, $\Phi_\sigma|_{B^n(u_0,\delta_0)}$ is a diffeomorphism onto its image.

Define A_σ and C_σ in the obvious way on $B^n(u_0, \delta_0)$, by using the vector fields $X_1^\sigma, \ldots, X_n^\sigma$ in place of $X_1, \ldots, X_n$. We have $A_\sigma \to A$ and $C_\sigma \to C$ in $C^0_{\mathrm{loc}}(B^n(u_0, \delta_0))$. Furthermore, by Lemma A.4.2, $\partial_r r A_\sigma = -A_\sigma^2 - C_\sigma A_\sigma - C_\sigma$, i. e., for $r_0\theta \in B^n(u_0, \delta_0)$ and r near r_0, we have

$$rA_\sigma(r\theta) = r_0 A_\sigma(r_0\theta) + \int_{r_0}^{r} -A_\sigma(s\theta)^2 - C_\sigma(s\theta)A_\sigma(s\theta) - C_\sigma(s\theta)\, ds. \tag{A.28}$$

Taking the limit as $\sigma \downarrow 0$ in (A.28) shows that

$$rA(r\theta) = r_0 A(r_0\theta) + \int_{r_0}^{r} -A(s\theta)^2 - C(s\theta)A(s\theta) - C(s\theta)\, ds.$$

It follows that $\partial_r rA$ exists in the classical sense and $\partial_r rA = -A^2 - CA - C$, completing the proof. □

A.4.1.2 Existence, uniqueness, and regularity

In this section, we describe the existence, uniqueness, and regularity of solutions to (A.20) satisfying $A(0) = 0$. If C is a matrix, we write $|C|$ for the usual matrix norm.

Proposition A.4.3. *Let $C \in C(B^n(\eta); \mathbb{M}^{n\times n})$ be given with $C(0) = 0$. Suppose that $|C(x)| \leq D|x|$, for $x \in B^n(\eta)$. Then if $\eta \leq (10D)^{-1}$, there exists a unique $A \in C^0_{\mathrm{loc}}(B^n(\eta); \mathbb{M}^{n\times n})$ with $A(0) = 0$ satisfying* (A.20). *This unique solution satisfies*

$$|A(x)| \leq \frac{5}{8} D|x| \quad \text{and} \quad |A(x)| \leq \frac{1}{16}, \quad \forall x \in B^n(\eta). \tag{A.29}$$

Furthermore, for this solution A and $m \in \mathbb{N}$,

$$C \in C^m(B^n(\eta); \mathbb{M}^{n\times n}) \implies A \in C^m(B^n(\eta); \mathbb{M}^{n\times n}) \tag{A.30}$$

and

$$\|A\|_{C^m(B^n(\eta);\mathbb{M}^{n\times n})} \leq K_{n,m}, \tag{A.31}$$

where $K_{n,m}$ can be chosen to depend only on n, m, and an upper bound for $\|C\|_{C^m(B^n(\eta);\mathbb{M}^{n\times n})}$.

The rest of this section is devoted to the proof of Proposition A.4.3. Define $\mathscr{T} : C_{\mathrm{loc}}(B^n(\eta); M^{n\times n}) \to C_{\mathrm{loc}}(B^n(\eta); \mathbb{M}^{n\times n})$ by

$$\mathscr{T}(A)(x) := \int_0^1 -A(sx)^2 - C(sx)A(sx) - C(sx)\, ds.$$

The relevance of $\mathscr{T}$ is the following lemma.

Lemma A.4.4. *$A \in C_{\mathrm{loc}}(B^n(\eta); \mathbb{M}^{n\times n})$ is a solution to* (A.20) *if and only if $\mathscr{T}(A) = A$. Also, writing $x = r\theta$, we have the following formula for $\mathscr{T}$ when $r > 0$:*

$$\mathscr{T}(A)(r\theta) = \frac{1}{r}\int_0^r -A(s\theta)^2 - C(s\theta)A(s\theta) - C(s\theta)\,ds. \tag{A.32}$$

Proof. Formula (A.32) follows from a straightforward change of variables in the definition of $\mathscr{T}$. It follows from (A.32) that $A \in C_{\mathrm{loc}}(B^n(\eta); \mathbb{M}^{n\times n})$ is a solution to (A.20) if and only if $\mathscr{T}(A) = A$. □

We begin with the uniqueness claim in Proposition A.4.3. Suppose $A_1, A_2 \in C_{\mathrm{loc}}(B^n(\eta); \mathbb{M}^{n\times n})$ are two solutions to (A.20) with $A_1(0) = A_2(0) = 0$. By Lemma A.4.4 we have $\mathscr{T}(A_1) = A_1$ and $\mathscr{T}(A_2) = A_2$. We first claim that $|A_j(x)| = O(|x|)$ as $|x| \to 0$, for $j = 1, 2$; we prove this for A_1 and the same will hold for A_2 by symmetry. Set $F(r) := \sup_{|x|\le r} |A_1(x)|$, note that $F : [0, \eta) \to \mathbb{R}$ is continuous increasing, and consider $F(0) = 0$. Since $\mathscr{T}(A_1) = A_1$ and $|C(sx)| \le Ds|x|$ by assumption, we have

$$\begin{aligned} |A_1(x)| &\le \int_0^1 F(s|x|)^2 + Ds|x|F(s|x|) + Ds|x|\,ds \\ &\le F(|x|)^2 + \frac{1}{2}D|x|F(|x|) + \frac{1}{2}D|x|, \end{aligned}$$

so $F(r) \le F(r)^2 + \frac{1}{2}DrF(r) + \frac{1}{2}Dr$, and therefore $F(r)(1 - F(r)) \le \frac{1}{2}DrF(r) + \frac{1}{2}Dr$. Taking r so small that $F(r) \le \frac{1}{2}$, we have $F(r) \le \frac{3}{2}Dr$. Thus, $|A_1(x)| = O(|x|)$ as $x \to 0$, as desired.

Writing x in polar coordinates $x = r\theta$ and using (A.32), we have, for $r > 0$,

$$\big|r(A_1(r\theta) - A_2(r\theta))\big| \le \int_0^r \big|s(A_1(s\theta) - A_2(s\theta))\big|\big(s^{-1}|A_1(s\theta)| + s^{-1}|A_2(s\theta)| + s^{-1}|C(s\theta)|\big)\,ds.$$

Using the fact that $|A_1(s\theta)|, |A_2(s\theta)|, |C(s\theta)| = O(s)$, the integral form of Grönwall's inequality shows that $A_1(r\theta) = A_2(r\theta)$ for $r > 0$ and therefore $A_1 = A_2$. This completes the proof of uniqueness.

We now turn to the existence part of Proposition A.4.3, which we prove using the contraction mapping principle. Let

$$\begin{aligned} \mathscr{M} := \Big\{ & A \in C(B^n(\eta); \mathbb{M}^{n\times n}) : A(0) = 0, \ \sup_{0\ne x\in B^n(\eta)} \frac{1}{|x|}|A(x)| < \infty, \\ & \sup_{x\in B^n(\eta)} |A(x)| \le \frac{1}{10} \Big\}. \end{aligned}$$

We give $\mathscr{M}$ the metric

$$d(A,B) := \sup_{0 \neq x \in B^n(\eta)} \frac{1}{|x|} \big|A(x) - B(x)\big|.$$

With this metric, $\mathscr{M}$ is a complete metric space.

Lemma A.4.5. *$\mathscr{T} : \mathscr{M} \to \mathscr{M}$ and $\forall A, B \in \mathscr{M}$, $d(\mathscr{T}(A), \mathscr{T}(B)) \leq \frac{1}{5} d(A,B)$. Also, $d(\mathscr{T}(0), 0) \leq D/2$.*

Proof. Let $A \in \mathscr{M}$. It follows easily from the definition of $\mathscr{T}$ that $\mathscr{T}(A)$ is continuous. For $x \in B^n(\eta)$, using $\eta < (10D)^{-1}$, we have

$$\begin{aligned} \big|\mathscr{T}(A)(x)\big| &\leq \int_0^1 \|A\|_{C^0}^2 + Ds|x|\|A\|_{C^0} + Ds|x| \, ds \\ &\leq \frac{1}{100} + \frac{D}{2}\eta\frac{1}{10} + \frac{D}{2}\eta \\ &\leq \frac{1}{100} + \frac{1}{200} + \frac{1}{20} \leq \frac{1}{10}. \end{aligned} \tag{A.33}$$

Also, for $0 \neq x \in B^n(\eta)$,

$$\frac{1}{|x|}\big|\mathscr{T}(0)(x)\big| \leq \frac{1}{|x|}\int_0^1 Ds|x| \, ds \leq \frac{1}{2}D, \tag{A.34}$$

so $\mathscr{T}(0) \in \mathscr{M}$ with $d(\mathscr{T}(0), 0) \leq D/2$.

Finally, for $A, B \in \mathscr{M}$, $0 \neq x \in B^n(\eta)$, we have

$$\begin{aligned} &\frac{1}{|x|}\big|\mathscr{T}(A)(x) - \mathscr{T}(B)(x)\big| \\ &\quad \leq \frac{1}{|x|}\int_0^1 \big|A(sx) - B(sx)\big|\big(\big|A(sx)\big| + \big|B(sx)\big| + \big|C(sx)\big|\big) \, ds \\ &\quad \leq \frac{1}{|x|}\int_0^1 s|x|d(A,B)\left(\frac{1}{5} + Ds|x|\right) ds \leq \int_0^1 sd(A,B)\left(\frac{1}{5} + \frac{s}{10}\right) ds \\ &\quad \leq d(A,B)\left(\frac{1}{10} + \frac{1}{30}\right) \leq \frac{1}{5}d(A,B). \end{aligned} \tag{A.35}$$

Putting $0 = B$ in (A.35) and using (A.34) shows $\sup_{0 \neq x \in B^n(\eta)} \frac{1}{|x|}|\mathscr{T}(A)(x)| < \infty$. Combining this with (A.33) shows that $\mathscr{T} : \mathscr{M} \to \mathscr{M}$. Further, (A.35) with arbitrary $A, B \in \mathscr{M}$ shows that $d(\mathscr{T}(A), \mathscr{T}(B)) \leq \frac{1}{5}d(A,B)$, and this completes the proof. □

By Lemma A.4.5, $\mathscr{T} : \mathscr{M} \to \mathscr{M}$ is a strict contraction, and the contraction mapping principle applies to show that if $A_0 = 0$, $A_a = \mathscr{T}(A_{a-1})$, $a \geq 1$, then $A_a \to A_\infty$ in $\mathscr{M}$, where $\mathscr{T}(A_\infty) = A_\infty$. A_∞ is the desired solution to (A.20) (see Lemma A.4.4).

Also, for $a \in \mathbb{N} \cup \{\infty\}$, we have, using Lemma A.4.5, for $0 \neq x \in B^n(\eta)$,

$$\begin{aligned}\frac{1}{|x|}|A_a(x)| &\leq d(A_a, 0) \leq \sum_{b=0}^{a-1} d(\mathcal{T}^{b+1}(0), \mathcal{T}^b(0)) \\ &\leq \sum_{b=0}^{a-1} 5^{-b} d(\mathcal{T}(0), 0) \leq \frac{5}{8} D.\end{aligned} \tag{A.36}$$

In particular, for $x \in B^n(\eta)$, we have $|A_\infty(x)| \leq \frac{5}{8}D|x|$. Also, since $\eta \leq (10D)^{-1}$, it follows that $|A_\infty(x)| \leq \frac{1}{16}$; this establishes (A.29).

It remains to prove the regularity properties of A_∞ in terms of the regularity of C, i. e., to establish (A.30) and (A.31) for A_∞. To do this, we use a modification of the contraction mapping principle due to Izzo.

Lemma A.4.6 (Izzo's contraction mapping principle [128]). *Suppose (M, d) is a metric space and $\{\mathcal{Q}_a\}_{a \in \mathbb{N}}$ is a sequence of contractions on M for which there exists $c < 1$ with*

$$d(\mathcal{Q}_a(x), \mathcal{Q}_a(y)) \leq c d(x, y), \quad \forall x, y \in M, a \in \mathbb{N}.$$

Suppose $\exists x_\infty \in M$ with $\lim_{a\to\infty} \mathcal{Q}_a(x_\infty) = x_\infty$. Let $x_0 \in M$ be arbitrary and define x_a recursively by $x_{a+1} = \mathcal{Q}_a(x_a)$. Then $\lim_{a\to\infty} x_a = x_\infty$.

Proof. We include a slightly modified version of the proof in [128]. For each $a \in \mathbb{N}$,

$$\begin{aligned}d(x_{a+1}, x_\infty) &= d(\mathcal{Q}_a(x_a), x_\infty) \\ &\leq d(\mathcal{Q}_a(x_a), \mathcal{Q}_a(x_\infty)) + d(\mathcal{Q}_a(x_\infty), x_\infty) \\ &\leq c d(x_a, x_\infty) + d(\mathcal{Q}_a(x_\infty), x_\infty).\end{aligned} \tag{A.37}$$

First we claim that the sequence $d(x_a, x_\infty)$ is bounded. Since $\mathcal{Q}_a(x_\infty) \to x_\infty$, $\exists N$, $a \geq N \Rightarrow d(\mathcal{Q}_a(x_\infty), x_\infty) < 1 - c$. Suppose $d(x_a, x_\infty)$ is not bounded; then $\exists a \geq N$ with $\max\{d(x_a, x_\infty), 1\} \leq d(x_{a+1}, x_\infty)$. Applying this to (A.37), we have $d(x_{a+1}, x_\infty) \leq c d(x_a, x_\infty) + d(\mathcal{Q}_a(x_\infty), x_\infty) < c d(x_{a+1}, x_\infty) + 1 - c$, so $d(x_{a+1}, x_\infty) < 1 \leq d(x_{a+1}, x_\infty)$, a contradiction. Thus, the sequence $d(x_a, x_\infty)$ is bounded.

Since $\mathcal{Q}_a(x_\infty) \to x_\infty$, (A.37) implies

$$\limsup_{a\to\infty} d(x_a, x_\infty) \leq c \limsup_{a\to\infty} d(x_a, x_\infty).$$

Since $\limsup_{a\to\infty} d(x_a, x_\infty) < \infty$, this gives $\limsup_{a\to\infty} d(x_a, x_\infty) = 0$, completing the proof. □

To establish (A.30) and (A.31) we prove the following when $C \in C^m(B^n(\eta); \mathbb{M}^{n\times n})$:

(I) $A_a \to A_\infty$ in $C^m(B^n(\eta); \mathbb{M}^{n\times n})$,

(II) $\|A_\infty\|_{C^m(B^n(\eta);\mathbb{M}^{n\times n})} \lesssim 1$, where the implicit constant can be chosen to depend only on n, m, and an upper bound for $\|C\|_{C^m(B^n(\eta);\mathbb{M}^{n\times n})}$.

We prove (I) and (II) by induction on m. The base case, $m = 0$, was established above: (I) was proved by the contraction mapping principle and (II) follows from (A.29).

Let $m \geq 1$, assume (I) and (II) hold with m replaced by $m-1$, and let $C \in C^m(B^n(\eta);\mathbb{M}^{n\times n})$. Fix a multi-index β with $|\beta| = m$. We will show (under our inductive hypothesis) that:

(a) $A_a \in C^m(B^n(\eta);\mathbb{M}^{n\times n})$, $\forall a \in \mathbb{N}$.
(b) $\exists B_\infty \in C(B^n(\eta);\mathbb{M}^{n\times n})$ such that $\partial_x^\beta A_a \xrightarrow{a\to\infty} B_\infty$ in $C(B^n(\eta);\mathbb{M}^{n\times n})$.
(c) $\|B_\infty\|_{C(B^n(\eta);\mathbb{M}^{n\times n})} \lesssim 1$, where the implicit constant can be chosen to depend only on n, m, and an upper bound for $\|C\|_{C^m(B^n(\eta);\mathbb{M}^{n\times n})}$.

To see why the above will complete the proof, note that (b) along with the fact that $A_a \to A_\infty$ in $C(B^n(\eta);\mathbb{M}^{n\times n})$ gives $\partial_x^\beta A_\infty = B_\infty$. Since β was an arbitrary multi-index with $|\beta| = m$, (b) along with the inductive hypothesis gives (I) and (c) along with the inductive hypothesis gives (II).

Having shown them to be sufficient, we turn to proving (a), (b), and (c).

Since $A_a = \mathscr{T}^a(0)$, (a) follows from the easily established fact that when $C \in C^m(B^n(\eta);\mathbb{M}^{n\times n})$, $\mathscr{T} : C^m(B^n(\eta);\mathbb{M}^{n\times n}) \to C^m(B^n(\eta);\mathbb{M}^{n\times n})$. To establish (b) and (c), we use Lemma A.4.6.

For $M_1, M_2 \in C^m(B^n(\eta);\mathbb{M}^{n\times n})$, define $R_\beta(M_1, M_2)$ by

$$\partial_x^\beta(M_1M_2)(x) = (\partial_x^\beta M_1(x))M_2(x) + M_1(x)(\partial_x^\beta M_2(x)) + R_\beta(M_1, M_2)(x),$$

that is,

$$R_\beta(M_1, M_2) = \sum_{\substack{\beta_1+\beta_2=\beta\\ \beta_1,\beta_2\neq 0}} \binom{\beta_1}{\beta} (\partial_x^{\beta_1} M_1)(\partial_x^{\beta_2} M_2).$$

By the inductive hypothesis, we have the following limits in $C(B^n(\eta);\mathbb{M}^{n\times n})$:

$$R_\beta(A_a, A_a) \xrightarrow{a\to\infty} R_\beta(A_\infty, A_\infty), \quad R_\beta(C, A_a) \xrightarrow{a\to\infty} R_\beta(C, A_\infty), \tag{A.38}$$

and the following estimates for $a \in \mathbb{N} \cup \{\infty\}$:

$$\|R_\beta(A_a, A_a)\|_{C(B^n(\eta);\mathbb{M}^{n\times n})}, \|R_\beta(C, A_a)\|_{C(B^n(\eta);\mathbb{M}^{n\times n})} \lesssim 1. \tag{A.39}$$

For $a \in \mathbb{N} \cup \{\infty\}$, $B \in C(B^n(\eta); \mathbb{M}^{n\times n})$, set

$$\mathscr{Q}_a(B)(x) := -\int_0^1 s^{|\beta|}\Big[B(sx)A_a(sx) + A_a(sx)B(sx) + C(sx)B(sx) + (\partial_x^\beta C)(sx)A_a(sx) + (\partial_x^\beta C)(sx) + R_\beta(A_a, A_a)(sx) + R_\beta(C, A_a)(sx)\Big]\, ds.$$

Lemma A.4.7. *For $a \in \mathbb{N} \cup \{\infty\}$, $\mathscr{Q}_a : C(B^n(\eta); \mathbb{M}^{n\times n}) \to C(B^n(\eta); \mathbb{M}^{n\times n})$ and satisfies*

$$\|\mathscr{Q}_a(B_1) - \mathscr{Q}_a(B_2)\|_{C(B^n(\eta);\mathbb{M}^{n\times n})} \le \frac{1}{8}\|B_1 - B_2\|_{C(B^n(\eta);\mathbb{M}^{n\times n})}, \tag{A.40}$$

$\forall B_1, B_2 \in C(B^n(\eta); \mathbb{M}^{n\times n})$. Furthermore, $\forall B \in C(B^n(\eta); \mathbb{M}^{n\times n})$, $\mathscr{Q}_a(B) \xrightarrow{a\to\infty} \mathscr{Q}_\infty(B)$, with convergence in $C(B^n(\eta); \mathbb{M}^{n\times n})$. Finally, $\|\mathscr{Q}_a(0)\|_{C(B^n(\eta);\mathbb{M}^{n\times n})} \lesssim 1$.

Proof. It follows immediately from the definitions, the inductive hypothesis, and the assumption that $C \in C^m(B^n(\eta); \mathbb{M}^{n\times n})$ that $\mathscr{Q}_a : C(B^n(\eta); \mathbb{M}^{n\times n}) \to C(B^n(\eta); \mathbb{M}^{n\times n})$. It follows from (A.38) and the fact that $A_a \to A_\infty$ in $C(B^n(\eta); \mathbb{M}^{n\times n})$ that $\mathscr{Q}_a(B) \xrightarrow{a\to\infty} \mathscr{Q}_\infty(B)$. It follows from (A.39), $\|C\|_{C^{B^n(\eta)}(\mathbb{M}^{n\times n})} \lesssim 1$, and $\|A_a\|_{C(B^n(\eta);\mathbb{M}^{n\times n})} \lesssim 1$ that $\|\mathscr{Q}_a(0)\|_{C(B^n(\eta);\mathbb{M}^{n\times n})} \lesssim 1$.

We turn to (A.40). Using (A.36) and the assumption $|C(x)| \le D|x|$, we have, for $x \in B^n(\eta)$,

$$\begin{aligned}
|\mathscr{Q}_a(B_1)(x) - \mathscr{Q}_a(B_2)(x)| &\le \int_0^1 s^{|\beta|}(2|A_a(sx)| + |C(sx)|)|B_1(sx) - B_2(sx)|\, ds \\
&\le \|B_1 - B_2\|_{C(B^n(\eta);\mathbb{M}^{n\times n})} \int_0^1 \frac{9}{4} D|x| s\, ds \\
&\le \frac{9}{8} D\eta \|B_1 - B_2\|_{C(B^n(\eta);\mathbb{M}^{n\times n})} \le \frac{9}{80}\|B_1 - B_2\|_{C(B^n(\eta);\mathbb{M}^{n\times n})},
\end{aligned}$$

establishing (A.40) and completing the proof. □

For $a \in \mathbb{N}$, set $B_a := \partial_x^\beta A_a$; note that $B_a \in C(B^n(\eta); \mathbb{M}^{n\times n})$ since we have already shown that $A_a \in C^m(B^n(\eta); \mathbb{M}^{n\times n})$. Also, $B_{a+1} = \partial_x^\beta A_{a+1} = \partial_x^\beta \mathscr{T}(A_a) = \mathscr{Q}_a(B_a)$.

Lemma A.4.7 shows that $\mathscr{Q}_\infty$ is a strict contraction on $C(B^n(\eta); \mathbb{M}^{n\times n})$, and the contraction mapping principle guarantees that it has a unique fixed point $B_\infty \in C(B^n(\eta); \mathbb{M}^{n\times n})$. Lemma A.4.7 shows that $\mathscr{Q}_a(B_\infty) \xrightarrow{a\to\infty} \mathscr{Q}_\infty(B_\infty) = B_\infty$. This, along with Lemma A.4.7, shows that Lemma A.4.6 applies to give $B_a \xrightarrow{a\to\infty} B_\infty$ in $C(B^n(\eta); \mathbb{M}^{n\times n})$. Since $B_a = \partial_x^\beta A_a$, this establishes (b).

Finally, to prove (c), note that B_∞ is the fixed point of the strict contraction $\mathscr{Q}_\infty$, so $\mathscr{Q}_\infty^a(0) \to B_\infty$. Hence, using (A.40),

$$\begin{aligned}\|B_\infty\|_{C(B^n(\eta);\mathbb{M}^{n\times n})} &\le \sum_{a=0}^{\infty} \|\mathscr{Q}_\infty^{a+1}(0) - \mathscr{Q}_\infty^{a}(0)\|_{C(B^n(\eta);\mathbb{M}^{n\times n})} \\ &\le \sum_{a=0}^{\infty} 8^{-a}\|\mathscr{Q}_\infty(0) - 0\|_{C(B^n(\eta);\mathbb{M}^{n\times n})} = \frac{8}{7}\|\mathscr{Q}_\infty(0)\|_{C(B^n(\eta);\mathbb{M}^{n\times n})} \lesssim 1,\end{aligned}$$

where the last inequality uses Lemma A.4.7.

A.4.2 An inverse function theorem

We require a quantitative version of a special case of the inverse function theorem that does not follow from the standard statement of that theorem, though we achieve it by keeping track of constants in a standard proof. We present it here.

Fix $\eta > 0$, let $Y_1, \dots, Y_n \in C^1(B^n(\eta); T\mathbb{R}^n)$ be vector fields on $B^n(\eta)$, and suppose they satisfy

$$\inf_{u\in B^n(\eta)} |\det(Y_1(u)|\cdots|Y_n(u))| \ge c_0 > 0.$$

Take $C_0 > 0$ such that $\|Y_j\|_{C^1(B^n(\eta);\mathbb{R}^n)} \le C_0$, $\forall 1 \le j \le n$. Define

$$\Psi_u(v) := e^{v_1Y_1+\cdots+v_nY_n}u.$$

Proposition A.4.8. *There exist $\kappa = \kappa(C_0, c_0, n) > 0$ and $\Delta_0 = \Delta_0(C_0, c_0, n, \eta) > 0$ such that for all $\delta \in (0, \Delta_0]$ and $u \in B^n(\kappa\delta)$, $v \mapsto \Psi_u(v)$ is defined and injective on $v \in B^n(\delta)$. Furthermore, $B^n(\kappa\delta) \subseteq \Psi_u(B^n(\delta))$.*

The rest of this section is devoted to the proof of Proposition A.4.8.

Lemma A.4.9. *Let $\delta_0 > 0$, let $F \in C^1(B^n(\delta_0); \mathbb{R}^n)$, and suppose $dF(0)$ is non-singular and $\sup_{x\in B^n(\delta_0)} \|dF(0)^{-1}dF(x) - I\|_{\mathbb{M}^{n\times n}} \le \frac{1}{2}$. Then $F(B^n(\delta_0)) \subseteq \mathbb{R}^n$ is open and $F : B^n(\delta_0) \to F(B^n(\delta_0))$ is a C^1_{loc} diffeomorphism. Furthermore, $F(B^n(\delta_0)) \supseteq B^n(F(0), \kappa_0\delta_0)$, where*

$$\kappa_0 := \|d(F^{-1})\|^{-1}_{C(F(B^n(\delta_0));\mathbb{M}^{n\times n})} \ge c_n|\det dF(0)|\|F\|^{-(n-1)}_{C^1(B^n(\delta_0);\mathbb{R}^n)}, \tag{A.41}$$

where $c_n > 0$ can be chosen to depend only on n.

Proof. We first show that F is injective. Fix $y \in \mathbb{R}^n$ and set $\phi(x) = x + dF(0)^{-1}(y - F(x))$. Note that $F(x) = y \Leftrightarrow \phi(x) = x$. Also, $\forall x \in B^n(\delta_0)$, $\|d\phi(x)\|_{\mathbb{M}^{n\times n}} = \|I - dF(0)^{-1}dF(x)\|_{\mathbb{M}^{n\times n}} \le \frac{1}{2}$. Hence, $|\phi(x_1) - \phi(x_2)| \le \frac{1}{2}|x_1 - x_2|$, and there is at most one solution to $\phi(x) = x$. In other words, there is at most one solution to $F(x) = y$, proving that F is injective.

Since $\|dF(0)^{-1}dF(x) - I\|_{\mathbb{M}^{n\times n}} \le \frac{1}{2}$, $\forall x \in B^n(\delta_0)$, it follows that $dF(x)$ is invertible, $\forall x \in B^n(\delta_0)$. Combining this with the fact that F is injective, the inverse function theorem shows that $F(B^n(\delta_0))$ is open and $F : B^n(\delta_0) \to F(B^n(\delta_0))$ is a C^1_{loc} diffeomorphism.

Next we prove the lower bound for κ_0 given in (A.41). In what follows, we use $A \lesssim B$ to denote $A \le C_n B$, where C_n can be chosen to depend only on n. Since $\|dF(0)^{-1}dF(x) - I\|_{\mathbb{M}^{n\times n}} \le \frac{1}{2}$ by assumption, we have

$$\inf_{x\in B^n(\delta_0)} \left|\det dF(x)\right| \gtrsim \left|\det dF(0)\right|. \tag{A.42}$$

Also, $\forall x \in B^n(\delta_0)$,

$$\|(dF(x))^{-1}\|_{\mathbb{M}^{n\times n}} \lesssim \left|\det dF(x)\right|^{-1} \|dF\|^{n-1}_{C(B^n(\delta_0);\mathbb{M}^{n\times n})},$$

as can be seen by the cofactor representation of $dF(x)^{-1}$. Hence,

$$\sup_{x\in B^n(\delta_0)} \|(dF(x))^{-1}\|_{\mathbb{M}^{n\times n}} \lesssim \Big(\inf_{y\in B^n(\delta_0)} \left|\det dF(y)\right|\Big)^{-1} \|dF\|^{n-1}_{C(B^n(\delta_0);\mathbb{M}^{n\times n})},$$

and therefore

$$\begin{aligned}\|d(F^{-1})\|_{C(F(B^n(\delta_0));\mathbb{M}^{n\times n})} &\lesssim \Big(\inf_{x\in B^n(\delta_0)} \left|\det dF(x)\right|\Big)^{-1} \|dF\|^{n-1}_{C(B^n(\delta_0);\mathbb{M}^{n\times n})}\\ &\lesssim \Big(\inf_{x\in B^n(\delta_0)} \left|\det dF(x)\right|\Big)^{-1} \|F\|^{n-1}_{C^1(B^n(\delta_0);\mathbb{M}^{n\times n})}.\end{aligned} \tag{A.43}$$

Combining (A.42) and (A.43) yields the inequality in (A.41).

Finally, we prove that $F(B^n(\delta_0)) \supseteq B^n(F(0), \kappa_0\delta_0)$. Take $\epsilon > 0$ to be the largest ϵ so that $B^n(F(0), \epsilon) \subseteq F(B^n(\delta_0))$; note that $\epsilon > 0$ by the inverse function theorem. The proof will be complete once we show $\epsilon \ge \delta_0\kappa_0$. Suppose, for contradiction, $\epsilon < \delta_0\kappa_0$. We have, by the mean value theorem,

$$F^{-1}(B^n(F(0), \epsilon)) \subseteq B^n(0, \epsilon\|dF^{-1}\|_{C(F(B^n(\delta_0));\mathbb{M}^{n\times n})}).$$

Thus, if $\epsilon < \kappa_0\delta_0$, then $F^{-1}(B^n(F(0), \epsilon)) \Subset B^n(0, \delta_0)$, which contradicts the maximality of ϵ and completes the proof. □

Lemma A.4.10. *Let Y_j, C_0, n, η, and Ψ be as in Proposition* A.4.8. *There exists $\delta_1 = \delta_1(C_0, n, \eta) > 0$ such that $\forall u \in B^n(\eta/2)$, Ψ_u is defined on $B^n(\delta_1)$ and satisfies*

$$\|\Psi_u\|_{C^1(B^n(\delta_1);\mathbb{R}^n)} \le C(C_0, n) \tag{A.44}$$

and $\forall u \in B^n(\eta/2)$, $v \in B^n(\delta_1)$,

$$\|d_v\Psi_u(v) - d_v\Psi_u(0)\|_{\mathbb{M}^{n\times n}} \le C(C_0, n)|v|, \tag{A.45}$$

where $C(C_0, n) \ge 0$ can be chosen to depend only on C_0 and n.

Proof. The existence of $\delta_1 > 0$ so that $\forall u \in B^n(\eta/2)$, $\Psi_u(v)$ is defined for $v \in B^n(\delta_1)$ and (A.44) are classical theorems from the field of ODEs. Thus, we prove only (A.45). We write $A \lesssim B$ for $A \leq CB$, where C can be chosen to depend only on C_0 and n. Using the equation $\partial_r \Psi_u(rv) = v \cdot Y(\Psi_u(r,v))$, we have

$$\Psi_u(v) = \int_0^1 v \cdot Y(\Psi_u(sv))\, ds.$$

Since $d_v\Psi_u(0) = (Y_1(u)|\cdots|Y_n(u))$, we have, $\forall u \in B^n(\eta/2)$, $v \in B^n(\delta_1)$,

$$\Psi_u(v) - (d_v\Psi_u(0))v = \int_0^1 v \cdot \big(Y(\Psi_u(sv)) - Y(\Psi_u(0))\big)\, ds. \tag{A.46}$$

Applying d_v to (A.46) and using the chain rule, we have, $\forall u \in B^n(\eta/2)$, $v \in B^n(\delta_1)$,

$$\begin{aligned}\|d_v\Psi_u(v) - d_v\Psi_u(0)\|_{\mathbb{M}^{n\times n}} &= \left\| \int_0^1 Y(\Psi_u(sv)) - Y(\Psi_u(0)) + sv^\top dY(\Psi_u(sv))d\Psi_u(sv)\, ds \right\|_{\mathbb{M}^{n\times n}} \\ &\lesssim |v| \|Y \circ \Psi_u\|_{C^1(B^n(\delta_1);\mathbb{M}^{n\times n})} + |v| \|Y\|_{C^1(B^n(\eta);\mathbb{M}^{n\times n})} \|\Psi_u\|_{C^1(B^n(\delta_1);\mathbb{R}^n)} \\ &\lesssim |v| \|Y\|_{C^1(B^n(\eta);\mathbb{M}^{n\times n})} \|\Psi_u\|_{C^1(B^n(\delta_1);\mathbb{R}^n)} \lesssim |v|,\end{aligned}$$

where we have written $Y(u)$ for the $\mathbb{M}^{n\times n}$-valued function $(Y_1(u)|\cdots|Y_n(u))$ and used (A.44). □

Proof of Proposition A.4.8. In what follows, we write $A \lesssim B$ for $A \leq CB$, where C can be chosen to depend only on n, C_0, and c_0, and we write $A \lesssim_\eta B$ if C can also depend on η. By taking $\delta_1 \gtrsim_\eta 1$ as in Lemma A.4.10, $\Psi_u(v)$ is defined for all $u \in B^n(\eta/2)$ and $v \in B^n(\delta_1)$. For such u, since $d\Psi_u(0) = (Y_1(u)|\cdots|Y_n(u))$ and $|\det(Y_1(u)|\cdots|Y_n(0))| \gtrsim 1$, we have $\|d_v\Psi_u(0)^{-1}\|_{\mathbb{M}^{n\times n}} \lesssim 1$. Hence, using (A.45), for $u \in B^n(\eta/2)$ $v \in B^n(\delta_1)$, we have

$$\left\|d_v\Psi_u(0)^{-1}d_v\Psi_u(v) - I\right\|_{\mathbb{M}^{n\times n}} \lesssim \left\|d_v\Psi_u(v) - d_v\Psi_u(0)\right\|_{\mathbb{M}^{n\times n}} \lesssim |v|.$$

Thus, if $\delta_2 \gtrsim_\eta 1$ is sufficiently small, we have, for all $u \in B^n(\eta/2)$ and $v \in B^n(\delta_2)$,

$$\left\|d_v\Psi_u(0)^{-1}d_v\Psi_u(v) - I\right\|_{\mathbb{M}^{n\times n}} \leq \frac{1}{2}. \tag{A.47}$$

Using (A.47), Lemma A.4.9 implies that for $u \in B^n(\eta/2)$, $\Psi_u : B^n(\delta_2) \to \Psi_u(B^n(\delta_2))$ is a C^1_{loc} diffeomorphism. Set $\kappa := \frac{1}{2}\|d(\Psi_u^{-1})\|^{-1}_{C(\Psi_u(B^n(\delta_2));\mathbb{M}^{n\times n})}$; note the extra factor of $\frac{1}{2}$ in κ as related to the definition of κ_0 in (A.41). The estimate (A.41) implies $\kappa \gtrsim 1$.

Take $\Delta_0 < \delta_2$, $\Delta_0 \gtrsim_\eta 1$ sufficiently small so that $\kappa\Delta_0 < \eta/2$. Then, for $\delta \in (0, \Delta_0]$ and $u \in B^n(\kappa\delta)$, Lemma A.4.9 shows that

$$\Psi_u(B^n(\delta)) \supseteq B^n(\Psi_u(0), 2\kappa\delta) = B^n(u, 2\kappa\delta) \supseteq B^n(0, \kappa\delta),$$

which completes the proof. □

A.4.3 Proof of the main result

We turn to the proof of Theorem A.2.4, which we separate into two parts: when $X_1, \ldots, X_q$ are linearly independent (i. e., when $n = q$) and more generally when $X_1, \ldots, X_q$ may be linearly dependent (i. e., when $q \geq n$).

A.4.3.1 Linearly independent

We start by proving Theorem A.2.4 in the special case where $n = q$, i. e., when $X_1, \ldots, X_q$ form a basis for the tangent space at every point of $\mathfrak{M}$. Note that in this case, $X_{J_0} = X$, so we may replace X_{J_0} with X throughout the statement of Theorem A.2.4.

We take $\chi = \xi$. By assumption, $X_1, \ldots, X_n$ form a basis for the tangent space at every point of $\mathfrak{M}$, in particular at every point of $B_X(x_0, \xi)$, and (a) follows. Item (b) is trivial when $n = q$ since the left-hand side of (A.5) equals 1 in this case. Since $X_1, \ldots, X_n$ span the tangent space at every point, Lemma A.2.14 shows that all balls of the form $B_X(x_0, \chi')$ are open, which establishes (c). With (a), (b), and (c) proved, we henceforth assume that $c_{j,k}^l \in C^1_{X_{J_0}}(B_{X_{J_0}}(x_0, \xi)) = C^1_X(B_X(x_0, \xi))$, $1 \leq j, k, l \leq n$.

Consider the map $\Phi : B^n(\eta_0) \to B_X(x_0, \xi)$ defined in (A.4), which we a priori know to be C^1_{loc}. Clearly, $d\Phi(0)\partial_{t_j} = X_j(x_0)$. Since $X_1(x_0), \ldots, X_n(x_0)$ form a basis of the tangent space $T_{x_0}B_X(x_0, \xi)$, the inverse function theorem shows that there exists a (non-admissible) $\delta > 0$ such that $\Phi : B^n(\delta) \to \Phi(B^n(\delta))$ is a C^1_{loc} diffeomorphism. Let $\widehat{Y}_j := \Phi|_{B^n(\delta)}^* X_j$, so that $\widehat{Y}_j$ is a C^0_{loc} vector field on $B^n(\delta)$. Write $\widehat{Y}_j = \partial_{t_j} + \sum_{k=1}^n \hat{a}_j^k(t)\partial_{t_k}$. Let $\widehat{A}(t) \in C_{\text{loc}}(B^n(\delta); \mathbb{M}^{n\times n})$ denote the $n \times n$ matrix with (j, k) component $\hat{a}_j^k(t)$ and let $C(t) \in C(B^n(\eta_0); \mathbb{M}^{n\times n})$ denote the $n \times n$ matrix with (j, k) component equal to $\sum_{l=1}^n t_l c_{j,l}^k \circ \Phi(t)$.

Proposition A.4.11. *Write t in polar coordinates, $t = r\theta$, and consider the differential equation*

$$\frac{\partial}{\partial r} rA(r\theta) = -A(r\theta)^2 - C(r\theta)A(r\theta) - C(r\theta), \tag{A.48}$$

defined for $A : B^n(\eta_0) \to \mathbb{M}^{n\times n}$. There exists a constant $\eta' \in (0, \eta_0]$, which is a 0-admissible constant which can also depend on a lower bound for $\eta > 0$, such that there exists a unique continuous solution $A \in C_{\text{loc}}(B^n(\eta'); \mathbb{M}^{n\times n})$ to (A.48) with $A(0) = 0$. Moreover, this solution lies in $C^1(B^n(\eta'); \mathbb{M}^{n\times n})$ and satisfies

$$\|A(t)\|_{\mathbb{M}^{n\times n}} \lesssim_0 |t| \quad \text{and} \quad \|A(t)\|_{\mathbb{M}^{n\times n}} \leq \frac{1}{2}, \quad \forall t \in B^n(\eta'). \tag{A.49}$$

For $m \in \mathbb{N}$, if $c_{j,k}^l \circ \Phi \in C^m(B^n(\eta'))$ with $\|c_{j,k}^l \circ \Phi\|_{C^m(B^n(\eta'))} \leq D_m$, $\forall j,k,l$, then $A \in C^m(B^n(\eta'); \mathbb{M}^{n\times n})$ and there exists a constant $C_m \geq 0$, which depends only on n, m, and D_m, such that $\|A\|_{C^m(B^n(\eta');\mathbb{M}^{n\times n})} \leq C_m$. Finally, $\widehat{A}\big|_{B^n(\min\{\eta',\delta\})} = A\big|_{B^n(\min\{\eta',\delta\})}$.

Proof. Note that by the definition of $C(t)$ and 0-admissible constants, we have $\|C(t)\|_{\mathbb{M}^{n\times n}} \lesssim |t|$. Also, $\widehat{A}$ satisfies (A.48) on $B^n(\delta)$ by Proposition A.4.1. Since $d\Phi(0)\partial_{t_j} = X_j(x_0)$, we have $\widehat{A}(0) = 0$. With these remarks in hand, the proposition (except for the claim $A \in C^1(B^n(\eta); \mathbb{M}^{n\times n})$) follows directly from Proposition A.4.3.

The claim that $A \in C^1(B^n(\eta'); \mathbb{M}^{n\times n})$ can be seen as follows. First note that we may assume $\eta' < \eta_0$, as if $\eta' = \eta_0$ we can replace η' with $\eta_0/2$. Since $c_{j,k}^l \in C^1_{X_{J_0}}(B_{X_{J_0}}(x_0,\xi)) = C^1_X(B_X(x_0,\xi))$ and $X_1,\dots,X_n$ span the tangent space at every point of $B_X(x_0,\xi)$, it follows that $c_{j,k}^l \in C^1_{\mathrm{loc}}(B_X(x_0,\xi))$. Since $\Phi : B^n(\eta_0) \to B_X(x_0,\xi)$ is a priori known to be C^1_{loc}, we have $c_{j,k}^l \circ \Phi \in C^1_{\mathrm{loc}}(B^n(\eta_0))$, and therefore $c_{j,k}^l \circ \Phi \in C^1(B^n(\eta'))$. We conclude that $C \in C^1(B^n(\eta'); \mathbb{M}^{n\times n})$, and it follows from Proposition A.4.3 that $A \in C^1(B^n(\eta'); \mathbb{M}^{n\times n})$. □

Fix $\eta' > 0$ and A as in Proposition A.4.11. Write $a_j^k(t)$ for the (j,k) component of $A(t)$ and set $Y_j := \partial_{t_j} + \sum_{k=1}^n a_j^k(t)\partial_{t_k}$. Note that $Y_1,\dots,Y_n$ are C^1 vector fields on $B^n(\eta')$. By Proposition A.4.11, $Y_j\big|_{B^n(\min\{\eta',\delta\})} = \widehat{Y}_j\big|_{B^n(\min\{\eta',\delta\})}$. Since δ is not admissible, we think of δ as being much smaller than η', so Y_j should be thought of as extending $\widehat{Y}_j$.

Proposition A.4.12. *We have, $\forall t \in B^n(\eta')$, $d\Phi(t)Y_j(t) = X_j(\Phi(t))$, $1 \leq j \leq n$.*

Proof. Fix $\theta \in S^{n-1}$ and set

$$r_1 := \sup\{r \in [0,\eta') : d\Phi(r'\theta)Y_j(r'\theta) = X_j(\Phi(r'\theta)), 0 \leq r' \leq r, 1 \leq j \leq n\}.$$

We wish to show that $r_1 = \eta'$, and this will complete the proof since $\theta \in S^{n-1}$ was arbitrary. Suppose, for contradiction, $r_1 < \eta'$. Since $Y_j\big|_{B^n(\min\{\eta',\delta\})} = \widehat{Y}_j\big|_{B^n(\min\{\eta',\delta\})}$ and $d\Phi(u)\widehat{Y}_j(u) = X_j(\Phi(u))$, we know $r_1 > 0$. By continuity, we have

$$d\Phi(r_1\theta)Y_j(r_1\theta) = X_j(\Phi(r_1\theta)). \tag{A.50}$$

By assumption, $X_1(\Phi(r_1\theta)),\dots,X_n(\Phi(r_1\theta))$ span $T_{\Phi(r_1\theta)}\mathfrak{M}$; combining this with (A.50) shows that $d\Phi(r_1\theta)$ is non-singular and therefore the inverse function theorem applies to Φ at the point $r_1\theta$. Thus, there exists a neighborhood V of $r_1\theta$ such that $\Phi : V \to \Phi(V)$ is a C^1_{loc} diffeomorphism. Pick $0 < r_2 < r_3 < r_1 < r_4 < \eta'$ such that $\{r'\theta : r_2 \leq r' \leq r_4\} \subset V$.

Let $\widetilde{Y}_j := \Phi\big|_V^* X_j$. By the choice of r_1, we have

$$\widetilde{Y}_j(r'\theta) = Y_j(r'\theta), \quad r_2 \leq r' \leq r_3,\ 1 \leq j \leq n. \tag{A.51}$$

Write $\widetilde{Y}_j(t) = \partial_{t_j} + \sum_{k=1}^n \tilde{a}_j^k(t)\partial_{t_k}$ and let $\widetilde{A}(t)$ denote the matrix with (j,k) component $\tilde{a}_j^k(t)$. By (A.51), we have $\widetilde{A}(r'\theta) = A(r'\theta)$ for $r_2 \leq r' \leq r_3$. Proposition A.4.1 shows that $\widetilde{A}$ satisfies (A.48). Away from $r = 0$, (A.48) is a standard ODE that both A and $\widetilde{A}$ satisfy.

Thus, by standard uniqueness theorems (using, for example, Grönwall's inequality) we have $\widetilde{A}(r'\theta) = A(r'\theta)$ for $r_2 \le r' \le r_4$, that is, $Y_j(r'\theta) = \widetilde{Y}_j(r'\theta)$, $r_2 \le r' \le r_4$, $1 \le j \le n$. We therefore have $d\Phi(r'\theta)Y_j(r'\theta) = d\Phi(r'\theta)\widetilde{Y}_j(r'\theta) = X_j(\Phi(r'\theta))$ for $r_2 \le r' \le r_4$. Since $r_2 < r_1 < r_4$, this contradicts our choice of r_1, completing the proof. □

Lemma A.4.13. *$\Phi : B^n(\eta') \to B_X(x_0, \xi)$ is C^2_{loc}.*

Proof. Since we already know $\Phi : B^n(\eta') \to \mathfrak{M}$ is C^1_{loc}, it suffices to show that the map $u \mapsto d\Phi(u)$, $u \in B^n(\eta')$, is C^1_{loc}. We have already remarked that $Y_1, \ldots, Y_n$ are C^1 (see Proposition A.4.11). Since $Y = (I + A)\nabla$, with $\|A(t)\|_{\mathbb{M}^{n\times n}} \le \frac{1}{2}$, $\forall t$ (see (A.49)), we conclude that $Y_1, \ldots, Y_n$ are a basis for the tangent space at every point of $B^n(\eta')$. Also, using Proposition A.4.12, $d\Phi(u)Y_j(u) = X_j(\Phi(u)) \in C^1_{\mathrm{loc}}$, since $X_j \in C^1_{\mathrm{loc}}$ and $\Phi \in C^1_{\mathrm{loc}}$. Since $Y_1, \ldots, Y_n$ are C^1 and a basis for the tangent space at every point, we conclude that $u \mapsto d\Phi(u)$ is C^1_{loc}, and therefore Φ is C^2_{loc}, completing the proof. □

Proposition A.4.14. *For $m \in \mathbb{N}$ and $\eta'' \in (0, \eta']$, we have, for any function f,*

$$\|f\|_{C^m(B^n(\eta''))} \approx_{(m-1)\vee 0} \|f\|_{C^m_Y(B^n(\eta''))} \tag{A.52}$$

and

$$\|Y_j\|_{C^m(B^n(\eta'');\mathbb{R}^n)} \lesssim_m 1. \tag{A.53}$$

Furthermore, for $m \in \mathbb{N}$, $1 \le j, k, l \le n$, we have

$$\|c^l_{j,k} \circ \Phi\|_{C^m(B^n(\eta'))} \lesssim_m 1. \tag{A.54}$$

Proof. Rewrite (A.53) as the equivalent statement

$$\|A\|_{C^m(B^n(\eta'');\mathbb{M}^{n\times n})} \lesssim_m 1. \tag{A.55}$$

We prove (A.52), (A.55), and (A.54) by induction on m.

Consider the base case, $m = 0$. Since $\|f\|_{C^0(B^n(\eta''))} = \|f\|_{C^0_Y(B^n(\eta''))}$, (A.52) is trivial when $m = 0$. The estimate (A.49) gives $\|A(t)\|_{\mathbb{M}^{n\times n}} \le \frac{1}{2}$, $\forall t \in B^n(\eta')$, and the case $m = 0$ of (A.55) follows. Finally, using the definition of 0-admissible constants (Definition A.2.1), we have

$$\begin{aligned}\|c^l_{j,k} \circ \Phi\|_{C^0(B^n(\eta'))} &= \|c^l_{j,k} \circ \Phi\|_{C(B^n(\eta'))} \\ &= \|c^l_{j,k}\|_{C(\Phi(B^n(\eta')))} \le \|c^l_{j,k}\|_{C(B_X(x_0,\xi))} \lesssim_0 1,\end{aligned}$$

establishing the case $m = 0$ of (A.54).

Let $m \ge 1$. We assume (A.52), (A.55), and (A.54) for $m - 1$ and prove them for m. In particular, the inductive hypothesis for (A.55) is equivalent to

$$\|Y_j\|_{C^{m-1}(B^n(\eta'');\mathbb{R}^n)} \lesssim_{m-1} 1, \quad 1 \le j \le n. \tag{A.56}$$

(A.52): Using (A.56) it follows immediately from the definitions that

$$\|f\|_{C_Y^m(B^n(\eta''))} \lesssim_{m-1} \|f\|_{C^m(B^n(\eta''))}.$$

For the reverse inequality, note that $Y = (I+A)\nabla$ and (A.49) gives $\sup_{t\in B^n(\eta')} \|A(t)\|_{\mathbb{M}^{n\times n}} \le \frac{1}{2}$, so $I + A$ is invertible and $\nabla = (I + A)^{-1}Y$. In particular, since (A.56) holds, we may write any ∂_X^β, with $|\beta| \le m$, as a linear combination of Y^α, $|\alpha| \le m$, with coefficients in $C(B^n(\eta''))$, which are $\lesssim_{m-1} 1$. It follows that

$$\|f\|_{C^m(B^n(\eta''))} \lesssim_{m-1} \|f\|_{C_Y^m(B^n(\eta''))},$$

establishing (A.52).

(A.54): Since $\Phi^* X_j = Y_j$, we have

$$Y^\alpha(g \circ \Phi) = (X^\alpha g) \circ \Phi. \tag{A.57}$$

Using (A.52) and (A.57), we have

$$\begin{aligned}\|c_{j,k}^l \circ \Phi\|_{C^m(B^n(\eta'))} &\approx_{m-1} \|c_{j,k}^l \circ \Phi\|_{C_Y^m(B^n(\eta'))}\\ &= \sum_{|\alpha|\le m} \|Y^\alpha(c_{j,k}^l \circ \Phi)\|_{C(B^n(\eta'))} = \sum_{|\alpha|\le m} \|X^\alpha c_{j,k}^l\|_{C(\Phi(B^n(\eta')))}\\ &\le \sum_{|\alpha|\le m} \|X^\alpha c_{j,k}^l\|_{C(B_X(x_0,\xi))} = \|c_{j,k}^l\|_{C_X^m(B_X(x_0,\xi))} \lesssim_m 1,\end{aligned}$$

where we have used the fact that $\Phi(B^n(\eta')) \subseteq B_X(x_0, \xi)$. This establishes (A.54).

(A.55): The estimate (A.54) gives $\|C\|_{C^m(B^n(\eta');\mathbb{M}^{n\times n})} \lesssim_m 1$. Proposition A.4.11 gives $\|A\|_{C^m(B^n(\eta');\mathbb{M}^{n\times n})} \lesssim_m 1$, which implies (A.55). □

Proposition A.4.15. *There exists a 1-admissible constant $\eta_1 \in (0, \eta']$ such that $\Phi|_{B^n(\eta_1)}$ is injective. Furthermore, $\Phi(B^n(\eta_1)) \subseteq B_X(x_0, \xi)$ is open and $\Phi : B^n(\eta_1) \to \Phi(B^n(\eta_1))$ is a C^2_{loc} diffeomorphism.*

Proof. Consider the maps, defined for $u, v \in \mathbb{R}^n$ sufficiently small, given by

$$\Psi_u(v) = e^{v_1 Y_1 + \cdots + v_n Y_n} u.$$

Since $Y = (I + A)\nabla$ and $\sup_{t\in B^n(\eta')} \|A(t)\|_{\mathbb{M}^{n\times n}} \le \frac{1}{2}$ (see (A.49)), we have

$$\left|\det(Y_1(t)| \cdots |Y_n(t))\right| \ge c_n > 0, \quad \forall t \in B^n(\eta'),$$

where $c_n > 0$ can be chosen to depend only on n. Furthermore, Proposition A.4.14 gives $\|Y_j\|_{C^1(B^n(\eta');\mathbb{R}^n)} \lesssim 1$.

Take $\Delta_0, \kappa > 0$ as in Proposition A.4.8 (with η' playing the role of η in that proposition). In light of the above comments, Δ_0 and κ can be taken to be 1-admissible constants. Set $\delta_1 := \min\{\Delta_0, \tau_0, 1\}$ so that $\delta_1 > 0$ is a 1-admissible constant; see Section A.2 for the definition of $\tau_0 > 0$. Let $\eta_1 := \min\{\delta_1\kappa, \eta'\} > 0$, so that η_1 is also a 1-admissible constant.

We claim $\Phi|_{B^n(\eta_1)}$ is injective. Let $u_1, u_2 \in B^n(\eta_1)$ be such that $\Phi(u_1) = \Phi(u_2)$; we wish to show $u_1 = u_2$. By Proposition A.4.8 there exists $v \in B^n(\delta_1)$ such that $u_2 = \Psi_{u_1}(v)$, i. e., $u_2 = e^{v\cdot Y}u_1$. Since $d\Phi(u)Y_j(u) = X_j(\Phi(u))$ (Proposition A.4.12), it follows that

$$\Phi(u_1) = \Phi(u_2) = \Phi(e^{v\cdot Y}u_1) = e^{v\cdot X}\Phi(u_1).$$

We have assumed that $X_1, \dots, X_n$ are linearly independent everywhere. Also, X satisfies $\mathcal{C}(\Phi(u_1), \delta_1, B_X(x_0, \xi))$ because Y satisfies $\mathcal{C}(u_1, \delta_1, B^n(\eta'))$ (by Proposition A.4.8). Hence, by the definition of τ_0, we have $v = 0$. We conclude that $u_2 = e^{v\cdot Y}u_1 = u_1$, and therefore Φ is injective.

Proposition A.4.12 shows that $d\Phi(u)Y_j(u) = X_j(\Phi(u))$, and we have assumed that $X_1, \dots, X_n$ form a basis for the tangent space at every point of $\mathfrak{M}$. It follows that $d\Phi(u)$ is non-singular, $\forall u \in B^n(\eta')$. The inverse function theorem applies to show that $\Phi : B^n(\eta') \to B_X(x_0, \xi)$ is an open map and is locally a C^1_{loc} diffeomorphism. In particular, $\Phi(B^n(\eta_1))$ is open. Hence, since $\Phi|_{B^n(\eta_1)}$ is injective, locally a C^1_{loc} diffeomorphism, and C^2_{loc} (Lemma A.4.13), we conclude that $\Phi : B^n(\eta_1) \to \Phi(B^n(\eta_1))$ is a C^2_{loc} diffeomorphism. □

Henceforth, we let η_1 be the 1-admissible constant from Proposition A.4.15.

Lemma A.4.16. *There exists a 1-admissible constant $\xi_1 \in (0, \xi]$ such that $B_X(x_0, \xi_1) \subseteq \Phi(B^n(\eta_1/2))$.*

Proof. Fix $\xi_1 \in (0, \xi]$ to be chosen later and suppose that $y \in B_X(x_0, \xi_1)$. Thus, there exists an absolutely continuous $\gamma : [0,1] \to B_X(x_0, \xi)$ with $\gamma(0) = x_0$, $\gamma(1) = y$, $\gamma'(t) = \sum_{j=1}^n b_j(t)\xi_1 X_j(\gamma(t))$, with $\| \sum |b_j(t)|^2 \|_{L^\infty([0,1])} < 1$. Define

$$t_0 := \sup\{t \in [0,1] : \gamma(t') \in \Phi(B^n(\eta_1/2)), \forall 0 \le t' \le t\}.$$

We want to show that by taking $\xi_1 > 0$ to be a sufficiently small 1-admissible constant, we have $t_0 = 1$ and $y = \gamma(1) \in \Phi(B^n(\eta_1/2))$. Note that $t_0 \ge 0$ since $\gamma(0) = x_0 = \Phi(0)$.

Suppose that $t_0 < 1$ or $t_0 = 1$ and $\gamma(1) \notin \Phi(B^n(\eta_1/2))$. In either case, we have $|\Phi^{-1}(\gamma(t_0))| = \eta_1/2$. Using the fact that $\|Y_j\|_{C^0(B^n(\eta_1);\mathbb{R}^n)} \lesssim_0 1$ and $\Phi(0) = x_0$, we have

$$\eta_1/2 = |\Phi^{-1}(\gamma(t_1))| = \left| \int_0^{t_0} \frac{d}{dt}\Phi^{-1} \circ \gamma(t)\, dt \right|$$
$$= \left| \int_0^{t_0} \sum_{j=1}^n b_j(t)\xi_1 Y_j(\Phi^{-1} \circ \gamma(t))\, dt \right| \lesssim_0 \xi_1.$$

This is a contradiction if ξ_1 is a sufficiently small 1-admissible constant, completing the proof. □

Taking $\chi = \xi$ and $\xi_1 = \xi_2$ and using the fact that $X = X_{J_0}$ and $Y = Y_{J_0}$, the above results combine to complete the proof of Theorem A.2.4 in the case $n = q$.

A.4.3.2 Linearly dependent

We prove Theorem A.2.4 in the general case $q \geq n$. We take the same setting and notation as in Theorem A.2.4.

We start with some general comments about Lie derivatives. Let $Z, W_1, \ldots, W_n \in C^1_{\mathrm{loc}}(\mathfrak{M}; T\mathfrak{M})$ be C^1_{loc} vector fields on $\mathfrak{M}$. The Lie derivative of $W_1 \wedge W_2 \wedge \cdots \wedge W_n$ with respect to Z is given by

$$\begin{aligned}\mathrm{Lie}_Z(W_1 \wedge W_2 \wedge \cdots \wedge W_n) = {} & [Z, W_1] \wedge W_2 \wedge W_3 \wedge \cdots \wedge W_n \\ & + W_1 \wedge [Z, W_2] \wedge W_3 \wedge \cdots \wedge W_n + \cdots \\ & + W_1 \wedge W_2 \wedge \cdots \wedge W_{n-1} \wedge [Z, W_n].\end{aligned} \tag{A.58}$$

Let $V_1, \ldots, V_n \in C^1_{\mathrm{loc}}(\mathfrak{M}; T\mathfrak{M})$ be C^1_{loc} vector fields which span the tangent space near a point $x_0 \in \mathfrak{M}$. Near x_0, using Definition 3.6.2, we may define a real-valued function by

$$x \mapsto \frac{W_1(x) \wedge W_2(x) \wedge \cdots \wedge W_n(x)}{V_1(x) \wedge V_2(x) \wedge \cdots \wedge V_n(x)}.$$

The derivative of this function with respect to Z is exactly what one would expect, as the next lemma shows.

Lemma A.4.17. *We have*

$$\begin{aligned}Z\frac{W_1 \wedge W_2 \wedge \cdots \wedge W_n}{V_1 \wedge V_2 \wedge \cdots \wedge V_n} = {} & \frac{\mathrm{Lie}_Z(W_1 \wedge W_2 \wedge \cdots \wedge W_n)}{V_1 \wedge V_2 \wedge \cdots \wedge V_n} \\ & - \frac{W_1 \wedge W_2 \wedge \cdots \wedge W_n}{V_1 \wedge V_2 \wedge \cdots \wedge V_n} \frac{\mathrm{Lie}_Z(V_1 \wedge V_2 \wedge \cdots \wedge V_n)}{V_1 \wedge V_2 \wedge \cdots \wedge V_n}.\end{aligned}$$

Proof. Let $\mathfrak{V} := V_1 \wedge V_2 \wedge \cdots \wedge V_n$ and $\mathfrak{W} := W_1 \wedge W_2 \wedge \cdots \wedge W_n$. Let ν be any C^1_{loc} n-form which is non-zero near x_0, so that by Definition 3.6.2,

$$\frac{\mathfrak{W}}{\mathfrak{V}} = \frac{\nu(\mathfrak{W})}{\nu(\mathfrak{V})}.$$

Because ν is non-zero near x_0 (and the space of n-forms is one-dimensional at each point), we may write $\mathrm{Lie}_Z \nu = f\nu$ for some continuous function f near x_0. Using [147, Proposition 18.9], we have

$$Z\nu(\mathfrak{W}) = (\mathrm{Lie}_Z \nu)(\mathfrak{W}) + \nu(\mathrm{Lie}_Z \mathfrak{W}) = f\nu(\mathfrak{W}) + \nu(\mathrm{Lie}_Z \mathfrak{W}),$$

and similarly with $\mathfrak{W}$ replaced by $\mathfrak{V}$. We conclude that near x_0,

$$
\begin{aligned}
Z\frac{\mathfrak{W}}{\mathfrak{V}} = Z\frac{\nu(\mathfrak{W})}{\nu(\mathfrak{V})} &= \frac{Z\nu(\mathfrak{W})}{\nu(\mathfrak{V})} - \frac{\nu(\mathfrak{W})}{\nu(\mathfrak{V})}\frac{Z\nu(\mathfrak{V})}{\nu(\mathfrak{V})} \\
&= \frac{f\nu(\mathfrak{W}) + \nu(\mathrm{Lie}_Z \mathfrak{W})}{\nu(\mathfrak{V})} - \frac{\nu(\mathfrak{W})}{\nu(\mathfrak{V})}\frac{f\nu(\mathfrak{V}) + \nu(\mathrm{Lie}_Z \mathfrak{V})}{\nu(\mathfrak{V})} \\
&= \frac{\nu(\mathrm{Lie}_Z \mathfrak{W})}{\nu(\mathfrak{V})} - \frac{\nu(\mathfrak{W})}{\nu(\mathfrak{V})}\frac{\nu(\mathrm{Lie}_Z \mathfrak{V})}{\nu(\mathfrak{V})} = \frac{\mathrm{Lie}_Z \mathfrak{W}}{\mathfrak{V}} - \frac{\mathfrak{W}}{\mathfrak{V}}\frac{\mathrm{Lie}_Z \mathfrak{V}}{\mathfrak{V}},
\end{aligned}
$$

completing the proof. □

Lemma A.4.18. *For $J \in \mathcal{I}(n,q)$, $1 \le j \le n$,*

$$\mathrm{Lie}_{X_j} \bigwedge X_J = \sum_{K \in \mathcal{I}_0(n,q)} g_{j,J}^K \bigwedge X_K, \quad \text{on } B_{X_{J_0}}(x_0, \xi),$$

where, $\forall m \in \mathbb{N}$,

$$\|g_{j,J}^K\|_{C^m_{X_{J_0}}(B_{X_{J_0}}(x_0,\xi))} \lesssim_m 1. \tag{A.59}$$

Proof. Let $J = (j_1, \dots, j_n)$. We have, using (A.58) and (A.2),

$$
\begin{aligned}
\mathrm{Lie}_{X_j} \bigwedge X_J &= \mathrm{Lie}_{X_j}(X_{j_1} \wedge X_{j_2} \wedge \cdots \wedge X_{j_n}) \\
&= \sum_{l=1}^{n} X_{j_1} \wedge X_{j_2} \wedge \cdots \wedge X_{j_{l-1}} \wedge [X_j, X_{j_l}] \wedge X_{j_{l+1}} \wedge \cdots \wedge X_{j_n} \\
&= \sum_{l=1}^{n}\sum_{k=1}^{q} c_{j,j_l}^k X_{j_1} \wedge X_{j_2} \wedge \cdots \wedge X_{j_{l-1}} \wedge X_k \wedge X_{j_{l+1}} \wedge \cdots \wedge X_{j_n}.
\end{aligned}
$$

The result follows from the anti-commutativity of $\wedge$ and the definition of m-admissible constants (see Definitions A.2.1 and A.2.2). □

Lemma A.4.19. *Let $\chi' \in (0, \xi]$. Suppose that for all $y \in B_{X_{J_0}}(x_0, \chi')$, $\bigwedge X_{J_0}(y) \neq 0$. Then, for $J \in \mathcal{I}(n,q)$, $1 \le j \le n$,*

$$X_j \frac{\bigwedge X_J}{\bigwedge X_{J_0}} = \sum_{K \in \mathcal{I}_0(n,q)} g_{j,J}^K \frac{\bigwedge X_K}{\bigwedge X_{J_0}} - \sum_{K \in \mathcal{I}_0(n,q)} g_{j,J_0}^K \frac{\bigwedge X_J}{\bigwedge X_{J_0}} \frac{\bigwedge X_K}{\bigwedge X_{J_0}} \quad \text{on } B_{X_{J_0}}(x_0, \chi'),$$

where $g_{j,J}^K$ are the functions from Lemma A.4.18.

Proof. This follows by combining Lemmas A.4.17 and A.4.18. □

Lemma A.4.20. *Fix $C, u_0 > 0$ and let $u_{u_0,C}(t)$ be the unique solution to*

$$\frac{d}{dt} u_{u_0,C}(t) = C(u_{u_0,C}(t) + u_{u_0,C}(t)^2), \quad u_{u_0,C}(0) = u_0,$$

defined on some maximum interval $[0, R_{u_0,C})$. *Let* $F(t)$ *be a non-negative, absolutely continuous function defined on* $[0, R')$ *with* $R' \le R_{u_0,C}$ *satisfying*

$$\frac{d}{dt}F(t) \le C(F(t) + F(t)^2), \quad F(0) \le u_0,$$

where the above inequality holds almost everywhere. Then, for $t \in [0, R')$, $F(t) \le u_{u_0,C}(t)$.

Proof. This is a special case of the Bihari–LaSalle inequality. □

Lemma A.4.21. *There exists a* 0*-admissible constant* $\chi \in (0, \xi]$ *such that the following holds. Suppose* $\gamma : [0, \chi] \to B_{X_{J_0}}(x_0, \xi)$ *is an absolutely continuous curve satisfying* $\gamma(0) = x_0$, $\gamma'(t) = \sum_{j=1}^n X_j(\gamma(t))$ *almost everywhere and* $\| \sum |a_j(t)|^2 \|_{L^\infty([0,\chi])} < 1$. *Suppose further that for some* $\chi' \in (0, \chi]$, $\bigwedge X_{J_0}(\gamma(t)) \neq 0$ *for* $t \in [0, \chi']$. *Then*

$$\sup_{\substack{J \in \mathcal{I}(n,q) \\ t \in [0,\chi']}} \left| \frac{\bigwedge X_J(\gamma(t))}{\bigwedge X_{J_0}(\gamma(t))} \right| \lesssim_0 1. \tag{A.60}$$

Here, the implicit constant depends on neither χ' *nor* γ.

Proof. Let $\chi \in (0, \xi]$ be a 0-admissible constant to be chosen later and let γ and $\chi' \in (0, \chi]$ be as in the statement of the lemma. We wish to show that if χ is chosen to be a sufficiently small 0-admissible constant (which forces χ' to be small), then (A.60) holds.

Set

$$F(t) := \sum_{J \in \mathcal{I}_0(n,q)} \left| \frac{\bigwedge X_J(\gamma(t))}{\bigwedge X_{J_0}(\gamma(t))} \right|^2.$$

We will show that if χ is a sufficiently small 0-admissible constant, then $F(t) \lesssim_0 1$, $\forall t \in [0, \chi']$, and this will complete the proof.[2]

Using Lemma A.4.19, we have, for almost every t,

$$\begin{aligned}
\frac{d}{dt}F(t) &= \sum_{J \in \mathcal{I}_0(n,q)} 2 \frac{\bigwedge X_J(\gamma(t))}{\bigwedge X_{J_0}(\gamma(t))} \sum_{j=1}^n a_j(t) \left(X_j \frac{\bigwedge X_J}{\bigwedge X_{J_0}} \right)(\gamma(t)) \\
&= \sum_{\substack{J \in \mathcal{I}_0(n,q) \\ K \in \mathcal{I}_0(n,q)}} \sum_{j=1}^n 2a_j(t) \frac{\bigwedge X_J(\gamma(t))}{\bigwedge X_{J_0}(\gamma(t))} \\
&\quad \times \left(g_{j,J}^K(\gamma(t)) \frac{\bigwedge X_K(\gamma(t))}{\bigwedge X_{J_0}(\gamma(t))} - g_{j,J_0}^K(\gamma(t)) \frac{\bigwedge X_J(\gamma(t))}{\bigwedge X_{J_0}(\gamma(t))} \frac{\bigwedge X_K(\gamma(t))}{\bigwedge X_{J_0}(\gamma(t))} \right) \\
&\lesssim_0 F(t) + F(t)^{3/2} \lesssim_0 F(t) + F(t)^2.
\end{aligned} \tag{A.61}$$

2 Here we are using the fact that $\forall K \in \mathcal{I}(n,q)$, either $\bigwedge X_K \equiv 0$ or $\exists J \in \mathcal{I}_0(n,q)$ with $\bigwedge X_K = \pm \bigwedge X_J$, by the basic properties of wedge products.

Also, we have, using (A.3),

$$F(0) = \sum_{J \in \mathcal{I}_0(n,q)} \left| \frac{\bigwedge X_J(x_0)}{\bigwedge X_{J_0}(x_0)} \right|^2 \lesssim_0 1. \tag{A.62}$$

Using (A.61) and (A.62) we see that there exist 0-admissible constants $C, u_0 > 0$ such that

$$\frac{d}{dt} F(t) \leq C(F(t) + F(t)^2), \quad F(0) \leq u_0,$$

where the inequality holds almost everywhere.

Standard theorems from the field of ODEs show that if $\chi_0 = \chi_0(C, u_0) > 0$ is chosen sufficiently small, then the unique solution $u(t)$ to

$$\frac{d}{dt} u(t) = C(u(t) + u(t)^2), \quad u(0) = u_0,$$

exists for $t \in [0, \chi_0]$ and satisfies $u(t) \leq 2u_0$, $\forall t \in [0, \chi_0]$. We take $\chi := \min\{\chi_0, \xi\} \in (0, \chi]$. For this choice of χ, which is 0-admissible since C and u_0 are 0-admissible, Lemma A.4.20 shows that $F(t) \leq 2u_0 \lesssim_0 1$, $\forall t \in [0, \chi']$, completing the proof. □

Proposition A.4.22. *There exists a 0-admissible constant $\chi \in (0, \xi]$ such that $\forall y \in B_{X_{J_0}}(x_0, \chi)$, $\bigwedge X_{J_0}(y) \neq 0$ and*

$$\sup_{\substack{J \in \mathcal{I}(n,q) \\ y \in B_{X_{J_0}}(x_0, \chi)}} \left| \frac{\bigwedge X_J(y)}{\bigwedge X_{J_0}(y)} \right| \lesssim_0 1. \tag{A.63}$$

Proof. Take χ as in Lemma A.4.21. First we claim that $\forall y \in B_{X_{J_0}}(x_0, \chi)$, $\bigwedge X_{J_0}(y) \neq 0$. Fix $y \in B_{X_{J_0}}(x_0, \chi)$, so that there exists an absolutely continuous $\gamma : [0, \chi] \to B_{X_{J_0}}(x_0, \chi)$ with $\gamma(0) = x_0$, $\gamma(\chi) = y$, $\gamma'(t) = \sum_{j=1}^n a_j(t) X_j(\gamma(t))$ almost everywhere, and $\| \sum |a_j(t)|^2 \|_{L^\infty([0,\chi])} < 1$. We will show that $\forall t \in [0, \chi]$ we have $\bigwedge X_{J_0}(\gamma(t)) \neq 0$, and it will then follow that $\bigwedge X_{J_0}(y) = \bigwedge X_{J_0}(\gamma(\chi)) \neq 0$.

Suppose not, so that $\bigwedge X_{J_0}(\gamma(t)) = 0$ for some $t \in [0, \chi]$, and let $t_0 := \inf\{t \in [0, \chi] : \bigwedge X_{J_0}(\gamma(t)) = 0\}$, so that $\bigwedge X_{J_0}(\gamma(t_0)) = 0$ but $\bigwedge X_{J_0}(\gamma(t)) \neq 0$, $\forall t \in [0, t_0)$. Note that $t_0 > 0$ since $\bigwedge X_{J_0}(\gamma(t)) = \bigwedge X_{J_0}(x_0) \neq 0$.

Let ν be a C^1_{loc} n-form, defined on a neighborhood of $\gamma(t_0)$ and which is non-zero at $\gamma(t_0)$. We have

$$\lim_{t \uparrow t_0} \nu(\bigwedge X_{J_0})(\gamma(t)) = 0, \quad \lim_{t \uparrow t_0} \nu(\bigwedge X_J(\gamma(t))) \neq 0,$$

by continuity, the fact that $X_1, \ldots, X_q$ span the tangent space at every point (and in particular at $\gamma(t_0)$), and the fact that ν is non-zero at $\gamma(t_0)$. We conclude, using Definition 3.6.2, that

$$\lim_{t\uparrow t_0}\max_{J\in\mathcal{I}(n,q)}\left|\frac{\bigwedge X_J(\gamma(t))}{\bigwedge X_{J_0}(\gamma(t))}\right|=\lim_{t\uparrow t_0}\max_{J\in\mathcal{I}(n,q)}\left|\frac{\nu(\bigwedge X_J)(\gamma(t))}{\nu(\bigwedge X_{J_0})(\gamma(t))}\right|=\infty. \tag{A.64}$$

Take any $\chi'\in(0,t_0)$. We know that $\forall t\in[0,\chi']$, $\bigwedge X_{J_0}(\gamma(t))\neq 0$. Lemma A.4.21 implies that

$$\sup_{\substack{J\in\mathcal{I}(n,q)\\ t\in[0,\chi']}}\left|\frac{\bigwedge X_J(\gamma(t))}{\bigwedge X_{J_0}(\gamma(t))}\right|\lesssim_0 1.$$

Since $\chi'\in(0,t_0)$ was arbitrary, we have

$$\sup_{\substack{J\in\mathcal{I}(n,q)\\ t\in[0,t_0)}}\left|\frac{\bigwedge X_J(\gamma(t))}{\bigwedge X_{J_0}(\gamma(t))}\right|\lesssim_0 1.$$

This contradicts (A.64) and completes the proof that $\bigwedge X_{J_0}(y)\neq 0$, $\forall y\in B_{X_{J_0}}(x_0,\chi)$.

To prove (A.63), take $y\in B_{X_{J_0}}(x_0,\chi)$. Then there exists an absolutely continuous $\gamma:[0,\chi]\to B_{X_{J_0}}(x_0,\chi)\subseteq B_{X_{J_0}}(x_0,\xi)$ with $\gamma(0)=x_0$, $\gamma(\chi)=y$, $\gamma'(t)=\sum_{j=1}^n a_j(t)X_j(\gamma(t))$ almost everywhere, and $\|\sum|a_j(t)|^2\|_{L^\infty([0,\chi])}<1$. We have already shown that $\bigwedge X_{J_0}(\gamma(t))\neq 0$, $\forall t\in[0,\chi]$. Lemma A.4.21 gives

$$\max_{J\in\mathcal{I}(n,q)}\left|\frac{\bigwedge X_J(y)}{\bigwedge X_{J_0}(y)}\right|=\max_{J\in\mathcal{I}(n,q)}\left|\frac{\bigwedge X_J(\gamma(\chi))}{\bigwedge X_{J_0}(\gamma(\chi))}\right|\lesssim_0 1.$$

Since $y\in B_{X_{J_0}}(x_0,\chi)$ was arbitrary, (A.63) follows. □

For the remainder of this section, we fix $\chi\in(0,\xi]$ as in Proposition A.4.22.

Lemma A.4.23. *$B_{X_{J_0}}(x_0,\chi)\subseteq\mathfrak{M}$ is open.*

Proof. By Proposition A.4.22, $X_1,\dots,X_n$ form a basis for the tangent space at every point of $B_{X_{J_0}}(x_0,\chi)$. The result now follows from Lemma A.2.14. □

Lemma A.4.24. *For $m\in\mathbb{N}$, $J\in\mathcal{I}(n,q)$,*

$$\left\|\frac{\bigwedge X_J}{\bigwedge X_{J_0}}\right\|_{C^m_{X_{J_0}}(B_{X_{J_0}}(x_0,\chi))}\lesssim_{(m-1)\vee 0} 1. \tag{A.65}$$

Proof. We prove (A.65) by induction on m. The base case, $m=0$, follows from Proposition A.4.22. Let $m\geq 1$; we assume (A.65) for $m-1$ and prove it for m.

We have, directly from Definition A.1.1,

$$\left\|\frac{\bigwedge X_J}{\bigwedge X_{J_0}}\right\|_{C^m_{X_{J_0}}(B_{X_{J_0}}(x_0,\chi))}=\left\|\frac{\bigwedge X_J}{\bigwedge X_{J_0}}\right\|_{C^{m-1}_{X_{J_0}}(B_{X_{J_0}}(x_0,\chi))}+\sum_{j=1}^n\left\|X_j\frac{\bigwedge X_J}{\bigwedge X_{J_0}}\right\|_{C^{m-1}_{X_{J_0}}(B_{X_{J_0}}(x_0,\chi))}. \tag{A.66}$$

The first term on the right-hand side of (A.66) is $\lesssim_{(m-2)\vee 0} 1$ by the inductive hypothesis, so we focus only on the second term. We have, using Lemma A.4.19 and letting $C_{m,n} \geq 0$ be a constant which depends only on m and n,

$$\begin{aligned}
&\left\| X_j \frac{\bigwedge X_J}{\bigwedge X_{J_0}} \right\|_{C^{m-1}_{X_{J_0}}(B_{X_{J_0}}(x_0,\chi))} \\
&\leq C_{m,n} \sum_{K\in\mathcal{I}_0(n,q)} \|g^K_{j,J}\|_{C^{m-1}_{X_{J_0}}(B_{X_{J_0}}(x_0,\xi))} \left\| \frac{\bigwedge X_K}{\bigwedge X_{J_0}} \right\|_{C^{m-1}_{X_{J_0}}(B_{X_{J_0}}(x_0,\chi))} \\
&\quad + C_{m,n} \sum_{K\in\mathcal{I}_0(n,q)} \|g^K_{j,J_0}\|_{C^{m-1}_{X_{J_0}}(B_{X_{J_0}}(x_0,\xi))} \left\| \frac{\bigwedge X_J}{\bigwedge X_{J_0}} \right\|_{C^{m-1}_{X_{J_0}}(B_{X_{J_0}}(x_0,\chi))} \left\| \frac{\bigwedge X_K}{\bigwedge X_{J_0}} \right\|_{C^{m-1}_{X_{J_0}}(B_{X_{J_0}}(x_0,\chi))} \\
&\lesssim_{m-1} 1,
\end{aligned}$$

where the last inequality follows from the bounds described in (A.59) and the inductive hypothesis. This completes the inductive step and the proof. □

Lemma A.4.25. *For $1 \leq k \leq q$, $1 \leq l \leq n$, there exist $\tilde{b}^l_k \in C(B_{X_{J_0}}(x_0,\chi))$ such that*

$$X_k = \sum_{l=1}^{n} \tilde{b}^l_k X_l, \tag{A.67}$$

where for $m \in \mathbb{N}$,

$$\|\tilde{b}^l_k\|_{C^m_{X_{J_0}}(B_{X_{J_0}}(x_0,\chi))} \lesssim_{(m-1)\vee 0} 1. \tag{A.68}$$

Proof. For $1 \leq k \leq n$ this is trivial (take $\tilde{b}^l_k = 1$ if $k = l$ and $\tilde{b}^l_k = 0$ if $k \neq l$); however, the proof which follows deals with all $1 \leq k \leq q$ simultaneously.

For $1 \leq k \leq q$, $1 \leq l \leq n$, let $J(l,k) = (1,2,\ldots,l-1,k,l+1,\ldots,n) \in \mathcal{I}(n,q)$. We have, by Cramer's rule,

$$X_k = \sum_{l=1}^{n} \frac{\bigwedge X_{J(l,k)}}{\bigwedge X_{J_0}} X_l.$$

Now, the result follows from Lemma A.4.24. □

Lemma A.4.26. *For $1 \leq i,j,l \leq n$, $\exists \hat{c}^l_{i,j} \in C(B_{X_{J_0}}(x_0,\chi))$ such that $[X_i,X_j] = \sum_{l=1}^{n} \hat{c}^l_{i,j} X_l$, where for $m \in \mathbb{N}$*

$$\|\hat{c}^l_{i,j}\|_{C^m_{X_{J_0}}(B_{X_{J_0}}(x_0,\chi))} \lesssim_m 1.$$

Proof. For $1 \le i,j \le n$, using (A.2) and Lemma A.4.25, we have

$$[X_i, X_j] = \sum_{k=1}^{q} c_{i,j}^k X_k = \sum_{l=1}^{n} \left(\sum_{k=1}^{q} c_{i,j}^k \tilde{b}_k^l \right) X_l.$$

Setting $\hat{c}_{i,j}^l := \sum_{k=1}^q c_{i,j}^k \tilde{b}_k^l$, the result follows from Lemma A.4.25 and the definition of m-admissible constants (Definitions A.2.1 and A.2.2). □

Proposition A.4.22 and Lemmas A.4.23 and A.4.26 show that the (already proved) case $n = q$ of Theorem A.2.4 applies to $X_1, \dots, X_n$ with ξ replaced by χ and $\mathfrak{M}$ replaced by $B_{X_{J_0}}(x_0, \chi)$.[3] In light of Lemma A.4.26, any constants which are m-admissible in the sense of this application of the case of $n = q$ of Theorem A.2.4 are m-admissible in the sense of this proof. Thus, from the case $n = q$ of Theorem A.2.4 we obtain 1-admissible constants $\xi_1, \eta_1 > 0$ and a map $\Phi : B^n(\eta_1) \to B_{X_{J_0}}(x_0, \chi)$ as in that theorem. Most of the case $q \ge n$ of Theorem A.2.4 has either already been shown above or follows immediately from this application of the case $q = n$. All that remains to show are (b), (c), there exists $\xi_2 > 0$ as in (f), (h) for $n + 1 \le j \le q$, and the second equivalence in (i).

Proof of (b). That

$$\max_{J \in \mathcal{I}(n,q)} \left| \frac{\bigwedge X_J(y)}{\bigwedge X_{J_0}(y)} \right| \ge 1, \quad \forall y \in B_{X_{J_0}}(x_0, \chi)$$

is clear (by taking $J = J_0$). That

$$\max_{J \in \mathcal{I}(n,q)} \left| \frac{\bigwedge X_J(y)}{\bigwedge X_{J_0}(y)} \right| \lesssim_0 1, \quad \forall y \in B_{X_{J_0}}(x_0, \chi)$$

is Proposition A.4.22. □

Proof of (c). It follows from Proposition A.4.22 that $X_1, \dots, X_n$ form a basis for the tangent space at every point of $B_{X_{J_0}}(x_0, \chi)$. From this, (c) follows immediately from Lemma A.2.14. □

That there exists a 1-admissible constant $\xi_2 > 0$ as in (f) follows by applying the next lemma with $\zeta_1 = \xi_1$.

Lemma A.4.27. *Fix $\zeta_1 \in (0, \chi]$. Then there is a 0-admissible constant $\zeta_2 > 0$ (which also depends on ζ_1) such that $B_X(x_0, \zeta_2) \subseteq B_{X_{J_0}}(x_0, \zeta_1)$.*

Proof. Let $\zeta_2 \in (0, \zeta_1]$. We will pick ζ_2 at the end of the proof. Suppose $y \in B_X(x_0, \zeta_2)$, so that there is an absolutely continuous $\gamma : [0,1] \to B_X(x_0, \zeta_2)$ with $\gamma(0) = x_0$, $\gamma(1) = y$, $\gamma'(t) = \sum_{j=1}^q a_j(t) \zeta_2 X_j(\gamma(t))$, and $\| \sum |a_j(t)|^2 \|_{L^\infty([0,1])} < 1$. Let

3 When we proved Theorem A.2.4 in the case $n = q$, we took $\xi = \chi$.

$$t_0 := \sup\{t \in [0,1] : \gamma(t') \in B_{X_{J_0}}(x_0, \zeta_1/2), \forall t' \in [0,t]\}.$$

We wish to show that if $\zeta_2 = \zeta_2(\zeta_1) > 0$ is taken to be a sufficiently small 0-admissible constant, then we have $t_0 = 1$ and $y = \gamma(1) \in B_{X_{J_0}}(x_0, \zeta_1)$.

In fact, we will prove $\gamma(t_0) \in B_{X_{J_0}}(x_0, \zeta_1/2)$. The result will then follow as if $t_0 < 1$, the fact that $B_{X_{J_0}}(x_0, \zeta_1/2)$ is open (see (c)) and the fact that γ is continuous show that $\gamma(t') \in B_{X_{J_0}}(x_0, \zeta_1/2)$ for $t' \in [0, t_0 + \epsilon)$ for some $\epsilon > 0$, which contradicts the choice of t_0.

We turn to proving $\gamma(t_0) \in B_{X_{J_0}}(x_0, \zeta_2/2)$. We have, using Lemma A.4.25,

$$\begin{aligned}\gamma'(t) &= \sum_{k=1}^{q} a_k(t)\zeta_2 X_l(\gamma(t)) = \sum_{l=1}^{n}\left(\sum_{k=1}^{q} a_k(t)\zeta_2 \tilde{b}_k^l(t)\right) X_l(\gamma(t)) \\ &=: \sum_{l=1}^{n} \tilde{a}_l(t)\frac{\zeta_1}{2} X_l(\gamma(t)),\end{aligned}$$

where by Lemma A.4.25, $\| \sum |\tilde{a}_l(t)|^2 \|_{L^\infty([0,1])} \lesssim_0 \frac{\zeta_2}{\zeta_1}$. Thus, by taking $\zeta_2 = \zeta_2(\zeta_1) > 0$ to be a sufficiently small 0-admissible constant, we have $\| \sum |\tilde{a}_l(t)|^2 \|_{L^\infty([0,1])} < 1$. It follows that for this choice of $\zeta_2 > 0$, we have $\gamma(t_0) \in B_{X_{J_0}}(x_0, \zeta_1/2)$, completing the proof. □

Proof of (h) for $n+1 \le j \le q$. The application of Theorem A.2.4 in the case $n = q$ already established (h) for $1 \le j \le n$. Set $b_k^l := \tilde{b}_k^l \circ \Phi$, where $\tilde{b}_k^l$ is as in Lemma A.4.25. Since $Y_j = \Phi^* X_j$, we have, for any ordered multi-index α,

$$Y^\alpha b_j^k = (X^\alpha \tilde{b}_j^k) \circ \Phi.$$

It follows that for every m,

$$\begin{aligned}\|b_j^k\|_{C^m_{Y_{J_0}}(B^n(\eta_1))} &= \sum_{|\alpha|\le m} \|Y_{J_0}^\alpha b_j^k\|_{C(B^n(\eta_1))} \\ &= \sum_{|\alpha|\le m} \|(X_{J_0}^\alpha \tilde{b}_j^k) \circ \Phi\|_{C(B^n(\eta_1))} = \sum_{|\alpha|\le m} \|X_{J_0}^\alpha \tilde{b}_j^k\|_{C(\Phi(B^n(\eta_1)))} \\ &\le \sum_{|\alpha|\le m} \|X_{J_0}^\alpha \tilde{b}_j^k\|_{C(B_{X_{J_0}}(x_0,\chi))} = \|\tilde{b}_j^k\|_{C^m_{X_{J_0}}(B_{X_{J_0}}(x_0,\chi))} \lesssim_{(m-1)\vee 0} 1,\end{aligned} \tag{A.69}$$

where the last inequality uses (A.68).

The first equivalence in (i) (which we have already established in the case $n = q$ of Theorem A.2.4) shows that

$$\|b_j^k\|_{C^m(B^n(\eta_1))} \approx_{(m-1)\vee 0} \|b_j^k\|_{C^m_{Y_{J_0}}(B^n(\eta_1))}. \tag{A.70}$$

Combining (A.69) and (A.70) shows that

$$\|b_j^k\|_{C^m(B^n(\eta_1))} \lesssim_{(m-1)\vee 0} 1. \tag{A.71}$$

Pulling back (A.67) via Φ shows that

$$Y_j = \sum_{l=1}^{n} b_k^l Y_l \quad \text{on } B^n(\eta_1). \tag{A.72}$$

The application of Theorem A.2.4 in the case $n = q$ already established (h) for $1 \leq j \leq n$. Combining this with (A.71) and (A.72) establishes (h) for $n + 1 \leq j \leq q$, completing the proof. □

Proof of (i). The case $n = q$ of Theorem A.2.4 already established the first equivalence, so it suffices to show that, $\forall m \in \mathbb{N}$,

$$\|f\|_{C^m_{Y_{J_0}}(B^n(\eta_1))} \leq \|f\|_{C^m_Y(B^n(\eta_1))}, \tag{A.73}$$

$$\|f\|_{C^m_Y(B^n(\eta_1))} \lesssim_{(m-1)\vee 0} \|f\|_{C^m(B^n(\eta_1))}. \tag{A.74}$$

Since Y_{J_0} is a sublist of Y, (A.73) follows immediately from the definitions. The estimate (A.74) follows from (h), completing the proof. □

A.4.4 Densities

In this section, we prove the results from Section A.2.2. Recall the density Vol_0 from (A.11):

$$\mathrm{Vol}_0(x)(Z_1(x), \ldots, Z_n(x)) := \left| \frac{Z_1(x) \wedge Z_2(x) \wedge \cdots \wedge Z_n(x)}{X_1(x) \wedge X_2(x) \wedge \cdots \wedge X_n(x)} \right|. \tag{A.75}$$

Throughout this section, we use the same notation as in Theorem A.2.4.

Lemma A.4.28. *We have* $\mathrm{Vol}_0(X_1, \ldots, X_n) \equiv 1$ *and for* $j_1, \ldots, j_n \in \{1, \ldots, q\}$, $\mathrm{Vol}_0(X_{j_1}, \ldots, X_{j_n}) \lesssim_0 1$.

Proof. It follows directly from the definition (A.75) that $\mathrm{Vol}_0(X_1, \ldots, X_n) \equiv 1$. It follows from Theorem A.2.4 (b) that $\mathrm{Vol}_0(X_{j_1}, \ldots, X_{j_n}) \lesssim_0 1$. □

Lemma A.4.29. *Let V and W be n-dimensional real vector spaces and let $A : W \to V$ be an invertible linear transformation. Let $v_1, \ldots, v_n$ be a basis for V and let $w_1, \ldots, w_n \in W$. Then*

$$\frac{Aw_1 \wedge Aw_2 \wedge \cdots \wedge Aw_n}{v_1 \wedge v_2 \wedge \cdots \wedge v_n} = \frac{w_1 \wedge w_2 \wedge \cdots \wedge w_n}{A^{-1}v_1 \wedge A^{-1}v_2 \wedge \cdots \wedge A^{-1}v_n}.$$

Proof. Let Z_1, Z_2 be one-dimensional real vector spaces and let $B : Z_1 \to Z_2$ be an invertible linear transformation. Let $z_1 \in Z_1$ and $0 \neq z_2 \in Z_2$. We claim that

$$\frac{Bz_1}{z_2} = \frac{z_1}{B^{-1}z_2}. \tag{A.76}$$

Indeed, let $\lambda_2 : Z_2 \to \mathbb{R}$ be any non-zero linear functional and set $\lambda_1 := \lambda_2 \circ B : Z_1 \to \mathbb{R}$ so that λ_1 is a non-zero linear functional. We have

$$\frac{Bz_1}{z_2} = \frac{\lambda_2(Bz_1)}{\lambda_2(z_2)} = \frac{\lambda_1(z_1)}{\lambda_1(B^{-1}z_2)} = \frac{z_1}{B^{-1}z_2}.$$

Applying (A.76) in the case $Z_1 = \bigwedge^n W$, $Z_2 = \bigwedge^n V$, and $B : Z_1 \to Z_2$ given by $B(w_1 \wedge w_2 \wedge \cdots \wedge w_n) = Aw_1 \wedge Aw_2 \wedge \cdots \wedge Aw_n$ completes the proof. □

Lemma A.4.30. *For* $1 \leq j \leq n$, $\mathrm{Lie}_{X_j} \mathrm{Vol}_0 = f_j^0 \mathrm{Vol}_0$, *where* $f_j^0 \in C(B_{X_{J_0}}(x_0, \chi))$. *Furthermore, for* $m \in \mathbb{N}$,

$$\|f_j^0\|_{C^m_{X_{J_0}}(B_{X_{J_0}}(x_0,\chi))} \lesssim_m 1. \tag{A.77}$$

Proof. Set $\phi_t(x) := e^{tX_j}x$ so that $\mathrm{Lie}_{X_j} \mathrm{Vol}_0 = \frac{\partial}{\partial_t}\big|_{t=0} \phi_t^* \mathrm{Vol}_0$. We write $d\phi_t(x)$ to denote the differential of $\phi_t(x)$ in the x variable. Using Lemma A.4.29, we have for t small

$$\begin{aligned}
(\phi_t^* \mathrm{Vol}_0)(x)(Z_1, \ldots, Z_n) &= \mathrm{Vol}_0(\Phi_t(x))(d\phi_t(x)Z_1(x), \ldots, d\phi_t(x)Z_n(x)) \\
&= \left|\frac{d\phi_t(x)Z_1(x) \wedge \cdots \wedge d\phi_t(x)Z_n(x)}{X_1(\phi_t(x)) \wedge \cdots \wedge X_n(\phi_t(x))}\right| \\
&= \left|\frac{Z_1(x) \wedge \cdots \wedge Z_n(x)}{d\phi_t(x)^{-1}X_1(\phi_t(x)) \wedge \cdots \wedge d\phi_t(x)^{-1}X_n(\phi_t(x))}\right| \\
&= \left|\frac{Z_1(x) \wedge \cdots \wedge Z_n(x)}{\phi_t^* X_1(x) \wedge \cdots \wedge \phi_t^* X_n(x)}\right|.
\end{aligned} \tag{A.78}$$

Fix $x \in B_{X_{J_0}}(x_0, \chi)$. We claim that the sign of

$$\frac{Z_1(x) \wedge \cdots \wedge Z_n(x)}{\phi_t^* X_1(x) \wedge \cdots \wedge \phi_t^* X_n(x)}$$

does not change for t small. To this end, let θ be a C^1_{loc} n-form which is non-zero near x. Since $X_1(x) \wedge X_(x) \wedge \cdots \wedge X_n(x) \neq 0$ (Theorem A.2.4 (a)), we have $\theta(x)(X_1(x) \wedge \cdots \wedge X_n(x)) \neq 0$, so by continuity, for t small, $\theta(x)(\phi_t^* X_1(x) \wedge \cdots \wedge \phi_t^* X_n(x)) \neq 0$. We conclude that for t sufficiently small,

$$\frac{Z_1(x) \wedge \cdots \wedge Z_n(x)}{\phi_t^* X_1(x) \wedge \cdots \wedge \phi_t^* X_n(x)} = \frac{\theta(x)(Z_1(x) \wedge \cdots \wedge Z_n(x))}{\theta(x)(\phi_t^* X_1(x) \wedge \cdots \wedge \phi_t^* X_n(x))}$$

does not change sign, and is either never zero or always zero.

Set, for t small,

$$\epsilon := \mathrm{sgn} \frac{Z_1(x) \wedge \cdots \wedge Z_n(x)}{\phi_t^* X_1(x) \wedge \cdots \wedge \phi_t^* X_n(x)},$$

and in the case that the quantity inside sgn equals zero, the choice of ϵ does not matter. By the above discussion, ϵ does not depend on t (for t small). By (A.78), using the functions $g_{j,J}^K$ from Lemmas A.4.18 and A.4.19, we have

$$\begin{aligned}
&\frac{\partial}{\partial t}\Big|_{t=0}(\phi_t^* \operatorname{Vol}_0)(x)(Z_1(x),\dots,Z_n(x))\\
&\quad= \frac{\partial}{\partial t}\Big|_{t=0}\left|\frac{Z_1(x)\wedge\cdots\wedge Z_n(x)}{\phi_t^*X_1(x)\wedge\cdots\wedge\phi_t^*X_n(x)}\right|\\
&\quad= \frac{\partial}{\partial t}\Big|_{t=0}\epsilon\frac{Z_1(x)\wedge\cdots\wedge Z_n(x)}{\phi_t^*X_1(x)\wedge\cdots\wedge\phi_t^*X_n(x)}\\
&\quad= \frac{\partial}{\partial t}\Big|_{t=0}\epsilon\frac{\theta(x)(Z_1(x)\wedge\cdots\wedge Z_n(x))}{\theta(x)(\phi_t^*X_1(x)\wedge\cdots\wedge\phi_t^*X_n(x))}\\
&\quad= -\epsilon\frac{\theta(x)(Z_1(x)\wedge\cdots\wedge Z_n(x))}{\theta(x)(X_1(x)\wedge\cdots\wedge X_n(x))^2}\frac{\partial}{\partial t}\Big|_{t=0}\theta(x)(\phi_t^*X_1(x)\wedge\cdots\wedge\phi_t^*(x)X_n(x))\\
&\quad= -\epsilon\frac{\theta(x)(Z_1(x)\wedge\cdots\wedge Z_n(x))}{\theta(x)(X_1(x)\wedge\cdots\wedge X_n(x))}\frac{\theta(x)(\operatorname{Lie}_{X_j}(X_1\wedge\cdots\wedge X_n))}{\theta(x)(X_1(x)\wedge\cdots\wedge X_n(x))}\\
&\quad= -\epsilon\frac{Z_1(x)\wedge\cdots\wedge Z_n(x)}{X_1(x)\wedge\cdots\wedge X_n(x)}\frac{\operatorname{Lie}_{X_j}(X_1\wedge\cdots\wedge X_n)}{X_1(x)\wedge\cdots\wedge X_n(x)}\\
&\quad= -\frac{\operatorname{Lie}_{X_j}(X_1\wedge\cdots\wedge X_n)}{X_1(x)\wedge\cdots\wedge X_n(x)}\operatorname{Vol}_0(x)(Z_1(x),\dots,Z_n(x))\\
&\quad= -\sum_{K\in\mathcal{I}_0(n,q)} g_{j,J_0}^K(x)\frac{\bigwedge X_K(x)}{\bigwedge X_{J_0}(x)}\operatorname{Vol}_0(x)(Z_1(X),\dots,Z_n(x)).
\end{aligned}$$

We conclude that

$$f_j^0 = -\sum_{K\in\mathcal{I}_0(n,q)} g_{j,J_0}^K\frac{\bigwedge X_K}{\bigwedge X_{J_0}}.$$

The estimate (A.77) now follows from Lemma A.4.24 and (A.59). □

Let $\sigma_0 := \Phi^*\operatorname{Vol}_0$, so that σ_0 is a density on $B^n(\eta_1)$. Define a function h_0 by $\sigma_0 = h_0\sigma_{\text{Leb}}$, so that $h_0 \in C_{\text{loc}}(B^n(\eta_1))$.

Lemma A.4.31. *We have* $h_0(t) = \det(I + A(t))^{-1}$, *where* $A(t)$ *is the matrix from Theorem* A.2.4. *In particular,* $h_0(t) \approx_0 1$, $\forall t \in B^n(\eta_1)$, *and for* $m \in \mathbb{N}$,

$$\|h_0\|_{C^m(B^n(\eta_1))} \lesssim_m 1. \tag{A.79}$$

Proof. Because $\|A(t)\|_{\mathbb{M}^{n\times n}} \le \frac{1}{2}$, $\forall t \in B^n(\eta_1)$ (Theorem A.2.4 (g)), $I + A(t)$ is invertible, and moreover,

$$\left|\det(I + A(t))^{-1}\right| = \det(I + A(t))^{-1} \approx_0 1, \quad \forall t \in B^n(\eta_1). \tag{A.80}$$

We have

$$\begin{aligned}h_0(t) &= \sigma_0(t)(\partial_{t_1},\ldots,\partial_{t_n})\\ &= \sigma_0(t)((I+A(t))^{-1}Y_1(t),\ldots,(I+A(t))^{-1}Y_n(t))\\ &= |\det(I+A(t))^{-1}|\sigma_0(t)(Y_1(t),\ldots,Y_n(t))\\ &= \det(I+A(t))^{-1}\operatorname{Vol}_0(\Phi(t))(X_1(\Phi(t)),\ldots,X_n(\Phi(t)))\\ &= \det(I+A(t))^{-1}.\end{aligned}$$

It now follows from (A.80) that $h_0(t) \approx_0 1$.

Recall that $Y_{J_0} = (I+A)\nabla$, so Theorem A.2.4 (h) implies that, $\forall m \in \mathbb{N}$,

$$\|A\|_{C^m(B^n(\eta_1);\mathbb{M}^{n\times n})} \lesssim_m 1. \tag{A.81}$$

Combining (A.81) with the fact that $\|A(t)\|_{\mathbb{M}^{n\times n}} \leq \frac{1}{2}, \forall t \in B^n(\eta_1)$ (Theorem A.2.4 (g)), shows that, $\forall m \in \mathbb{N}$,

$$\|(I+A)^{-1}\|_{C^m(B^n(\eta_1);\mathbb{M}^{n\times n})} \lesssim_m 1,$$

and therefore

$$\|h_0\|_{C^m(B^n(\eta_1))} = \|\det(I+A)^{-1}\|_{C^m(B^n(\eta_1))} \lesssim_m 1. \qquad \square$$

We turn to studying the density Vol from Section A.2.2; thus, we use the functions f_j from (A.10). Because Vol_0 is a non-zero density on $B_{X_{J_0}}(x_0,\chi)$, there is a unique $g \in C_{\mathrm{loc}}(B_{X_{J_0}}(x_0,\chi))$ such that $\operatorname{Vol} = g\operatorname{Vol}_0$.

Lemma A.4.32. *For* $1 \leq j \leq n$, $X_j g = (f_j - f_j^0)g$.

Proof. Using Lemma A.4.30, we have

$$\begin{aligned}f_j g \operatorname{Vol}_0 &= f_j \operatorname{Vol} = \operatorname{Lie}_{X_j}\operatorname{Vol} = \operatorname{Lie}_{X_j}(g\operatorname{Vol}_0)\\ &= (X_j g)\operatorname{Vol}_0 + g\operatorname{Lie}_{X_j}\operatorname{Vol}_0 = (X_j g)\operatorname{Vol}_0 + g f_j^0\operatorname{Vol}_0.\end{aligned}$$

The result follows. $\square$

Proof of Theorem A.2.17 *(i).* Note that $g(x_0) = g(x_0)\operatorname{Vol}_0(x_0)(X_1(x_0),\ldots,X_n(x_0)) = \operatorname{Vol}(x_0)(X_1(x_0),\ldots,X_n(x_0))$, by definition, so it suffices to show that $g(x) \approx_{\{0;\mathrm{Vol}\}} g(x_0)$, for $x \in B_{X_{J_0}}(x_0,\chi)$.

Fix $x \in B_{X_{J_0}}(x_0,\chi)$, so that there exists an absolutely continuous $\gamma : [0,1] \to B_{X_{J_0}}(x_0,\chi)$ with $\gamma(0) = x_0$, $\gamma(1) = x$, $\gamma'(t) = \sum_{j=1}^n a_j(t)\chi X_j(\gamma(t))$, and $\|\sum |a_j(t)|^2\|_{L^\infty([0,1])} < 1$. We have, using Lemma A.4.32,

$$\frac{d}{dt}g(\gamma(t)) = \sum_{j=1}^{n} \chi a_j(t) X_j g(\gamma(t)) = \sum_{j=1}^{n} \chi a_j(t)(f_j(\gamma(t)) - f_j^0(\gamma(t)))g(\gamma(t)).$$

That is, $g(\gamma(t))$ satisfies a linear ODE. Solving this ODE we have

$$g(x) = g(\gamma(t)) = e^{\int_0^1 \sum_{j=1}^{n} \chi a_j(t)(f_j(\gamma(s)) - f_j^0(\gamma(s)))\,ds} g(x_0). \tag{A.82}$$

The estimate (A.77) implies $\|f_j^0\|_{C(B_{X_{J_0}}(x_0,\chi))} \lesssim_0 1$. Using this and the definition of $\{0;\mathrm{Vol}\}$-admissible constants (Definition A.2.15), (A.82) gives $g(x) \approx_{\{0;\mathrm{Vol}\}} g(x_0)$, completing the proof. □

Proof of Theorem A.2.17 *(ii).* We prove the result by induction on $m \in \mathbb{N}$. The base case, $m = 0$, follows from Theorem A.2.17 (i). Let $m \in \mathbb{N}_+$. We assume the result for $m - 1$ and prove it for m. We have

$$\|g\|_{C^m_{X_{J_0}}(B_{X_{J_0}}(x_0,\chi))} = \|g\|_{C^{m-1}_{X_{J_0}}(B_{X_{J_0}}(x_0,\chi))} + \sum_{j=1}^{n} \|X_j g\|_{C^{m-1}_{X_{J_0}}(B_{X_{J_0}}(x_0,\chi))}. \tag{A.83}$$

The first term on the right-hand side of (A.83) is $\lesssim_{\{m-1;\mathrm{Vol}\}} |\mathrm{Vol}(X_1,\ldots,X_n)(x_0)|$ by the inductive hypothesis. For the second term, using Lemma A.4.32, we have, for a constant $C_{m,n} \geq 0$ depending only on m and n, for $1 \leq j \leq n$,

$$\begin{aligned}
\|X_j g\|_{C^{m-1}_{X_{J_0}}(B_{X_{J_0}}(x_0,\chi))} &= \|(f_j - f_j^0)g\|_{C^{m-1}_{X_{J_0}}(B_{X_{J_0}}(x_0,\chi))} \\
&\leq C_{m,n}\|f_j - f_j^0\|_{C^{m-1}_{X_{J_0}}(B_{X_{J_0}}(x_0,\chi))}\|g\|_{C^{m-1}_{X_{J_0}}(B_{X_{J_0}}(x_0,\chi))} \\
&\lesssim_{\{m-1;\mathrm{Vol}\}} |\mathrm{Vol}(X_1,\ldots,X_n)(x_0)|,
\end{aligned} \tag{A.84}$$

where the final inequality follows from the inductive hypothesis, (A.77), and the definition of $\{m-1;\mathrm{Vol}\}$-admissible constants (Definitions A.2.15 and A.2.16). Using (A.84) to bound the second term on the right-hand side of (A.83) completes the proof. □

Lemma A.4.33. *Let $h(t)$ be as in Theorem* A.2.17. *Then $h(t) = h_0(t)g \circ \Phi(t)$.*

Proof. We have

$$h\sigma_{\mathrm{Leb}} = \Phi^*\,\mathrm{Vol} = \Phi^* g\,\mathrm{Vol}_0 = (g \circ \Phi)\Phi^*\,\mathrm{Vol}_0 = (g \circ \Phi)h_0\sigma_{\mathrm{Leb}},$$

completing the proof. □

Proof of Theorem A.2.17 *(iii).* Since $h(t) = h_0(t)g \circ \Phi(t)$ (Lemma A.4.33), this follows from Theorem A.2.17 (i) and Lemma A.4.31. □

Proof of Theorem A.2.17 *(iv).* Using Theorem A.2.4 (i) and the fact that $Y_j = \Phi^* X_j$, we have, for $m \in \mathbb{N}$,

$$
\begin{aligned}
\|g \circ \Phi\|_{C^m(B^n(\eta_1))} &\approx_{(m-1)\vee 0} \|g \circ \Phi\|_{C^m_{Y_{J_0}}(B^n(\eta_1))} \\
&= \sum_{|\alpha|\le m} \|Y_{J_0}^{\alpha}(g \circ \Phi)\|_{C(B^n(\eta_1))} = \sum_{|\alpha|\le m} \|X_{J_0}^{\alpha} g\|_{C(\Phi(B^n(\eta_1)))} \\
&\le \sum_{|\alpha|\le m} \|X_{J_0}^{\alpha} g\|_{C(B_{X_{J_0}}(x_0,\chi))} = \|g\|_{C^m_{X_{J_0}}(B_{X_{J_0}}(x_0,\chi))} \\
&\lesssim_{\{(m-1)\vee 0;\mathrm{Vol}\}} |\mathrm{Vol}(X_1,\dots,X_n)(x_0)|,
\end{aligned}
\tag{A.85}
$$

where the final inequality uses Theorem A.2.17 (ii). Since $h(t) = h_0(t) g \circ \Phi(t)$ (Lemma A.4.33), combining (A.85) and (A.79) completes the proof. □

Having completed the proof of Theorem A.2.17, we turn to Corollary A.2.18. To facilitate this, we introduce a corollary of Theorem A.2.4.

Corollary A.4.34. *Let $\eta_1, \xi_1, \xi_2 > 0$ be as in Theorem* A.2.4. *There exist* 1*-admissible constants* $0 < \eta_2 \le \eta_1$ *and* $0 < \xi_4 \le \xi_3 \le \xi_2$ *such that*

$$
\begin{aligned}
B_X(x_0,\xi_4) &\subseteq B_{X_{J_0}}(x_0,\xi_3) \subseteq \Phi(B^n(\eta_2/2)) \subseteq \Phi(B^n(\eta_2)) \\
&\subseteq B_{X_{J_0}}(x_0,\xi_2) \subseteq B_X(x_0,\xi_2) \subseteq B_{X_{J_0}}(x_0,\xi_1) \subseteq \Phi(B^n(\eta_1/2)) \\
&\subseteq \Phi(B^n(\eta_1)) \subseteq B_{X_{J_0}}(x_0,\chi) \subseteq B_X(x_0,\xi).
\end{aligned}
$$

Proof. After obtaining η_1, ξ_1, and ξ_2 from Theorem A.2.4, we apply Theorem A.2.4 again with ξ replaced by ξ_2 to obtain η_2, ξ_3, and ξ_4 as in the statement of the corollary. □

Proof of Corollary A.2.18. We have

$$
\begin{aligned}
\mathrm{Vol}(B_{X_{J_0}}(x_0,\xi_2)) &= \int_{B_{X_{J_0}}(x_0,\xi_2)} d\,\mathrm{Vol} \\
&= \int_{\Phi^{-1}(B_{X_{J_0}}(x_0,\xi_2))} d\Phi^* \,\mathrm{Vol} = \int_{\Phi^{-1}(B_{X_{J_0}}(x_0,\xi_2))} h(t)\,dt \\
&\approx_{\{0;\mathrm{Vol}\}} \sigma_{\mathrm{Leb}}(\Phi^{-1}(B_{X_{J_0}}(x_0,\xi_2)))\,\mathrm{Vol}(X_1,\dots,X_n)(x_0),
\end{aligned}
\tag{A.86}
$$

where the last estimate uses Theorem A.2.17 (iii). By Corollary A.4.34 and the fact that $\eta_1, \eta_2 > 0$ are 1-admissible constants, we have

$$
1 \approx_1 \sigma_{\mathrm{Leb}}(B^n(\eta_2)) \le \sigma_{\mathrm{Leb}}(\Phi^{-1}(B_{X_{J_0}}(x_0,\xi_2))) \le \sigma_{\mathrm{Leb}}(B^n(\eta_1)) \approx_1 1. \tag{A.87}
$$

Combining (A.86) and (A.87) shows that $\mathrm{Vol}(B_{X_{J_0}}(x_0,\xi_2)) \approx_{\{1;\mathrm{Vol}\}} \mathrm{Vol}(X_1,\dots,X_n)(x_0)$. The same proof works with $B_{X_{J_0}}(x_0,\xi_2)$ replaced by $B_X(x_0,\xi_2)$, which completes the proof of (A.12).

The first two estimates in (A.13) follow from (A.12). All that remains to prove (A.13) is to show that

$$\left|\mathrm{Vol}(X_1,\ldots,X_n)(x_0)\right| \approx_0 \max_{j_1,\ldots,j_n\in\{1,\ldots,q\}} \left|\mathrm{Vol}(X_{j_1},\ldots,X_{j_n})(x_0)\right|.$$

Using Lemma A.4.28, we have

$$\begin{aligned}\left|\mathrm{Vol}(X_1,\ldots,X_n)(x_0)\right| &= \left|g(x_0)\,\mathrm{Vol}_0(X_1,\ldots,X_n)(x_0)\right| = |g(x_0)| \\ &\approx_0 |g(x_0)| \max_{j_1,\ldots,j_n\in\{1,\ldots q\}} \left|\mathrm{Vol}_0(X_{j_1},\ldots,X_{j_n})(x_0)\right| \\ &= \max_{j_1,\ldots,j_n\in\{1,\ldots q\}} \left|g(x_0)\,\mathrm{Vol}_0(X_{j_1},\ldots,X_{j_n})(x_0)\right| \\ &= \max_{j_1,\ldots,j_n\in\{1,\ldots q\}} \left|\mathrm{Vol}(X_{j_1},\ldots,X_{j_n})(x_0)\right|,\end{aligned}$$

completing the proof. □

A.4.5 Proof of Theorem 3.6.5

Theorem 3.6.5 follows easily from Theorems A.2.4 and A.2.17. We make a few comments on this.

In Theorem 3.6.5 (b), Φ is claimed to be a C^∞_{loc} diffeomorphism, while in Theorem A.2.4 (e), Φ is only shown to be a C^2_{loc} diffeomorphism. However, in the setting of Theorem 3.6.5, since $X_1,\ldots,X_n$ are C^∞_{loc} vector fields, classical theorems from the field of ODEs show that the map Φ given by (3.17) is C^∞_{loc}.

The function h in Theorem 3.6.5 is obtained by taking the function h from Theorem A.2.17 and dividing it by

$$\max_{j_1,\ldots,j_n\in\{1,\ldots,q\}} \mathrm{Vol}(X_{j_1},\ldots,X_{j_n})(x_0).$$

In the setting of Theorem 3.6.5, Vol is assumed to be strictly positive and therefore Corollary A.2.18 gives

$$\max_{j_1,\ldots,j_n\in\{1,\ldots,q\}} \mathrm{Vol}(X_{j_1},\ldots,X_{j_n})(x_0) \approx_0 \left|\mathrm{Vol}(X_1,\ldots,X_n)(x_0)\right|. \tag{A.88}$$

Using (A.88), Theorem 3.6.5 (g) follows from Theorem A.2.17 (iii) and Theorem 3.6.5 (h) follows from Theorem A.2.17 (iv).

The rest of Theorem 3.6.5 follows directly from Theorem A.2.4.

A.5 Further reading and references

The first result similar to Theorem A.2.4 was due to Nagel, Stein, and Wainger [189] and was essentially the same as Theorem 3.3.7; around the same time similar results were proved by C. Fefferman and Sánchez-Calle [84]. The methods of [189] are based on studying the Taylor series in t of the map (A.4). This leads to several issues: the methods are not sufficient for studying multi-parameter balls as in Section 3.5, they require that the vector fields be smooth in some given coordinate system, and they are not quantitatively invariant under a change of coordinates.

Motivated by questions involving the smoothing properties of Radon transforms, Tao and Wright [232] studied two-parameter Carnot–Carathéodory balls, $B(x, (\delta_1, \delta_2))$, where $\delta_1, \delta_2 \in (0, 1]$ were assumed to be "weakly comparable," i. e., $\delta_1^N \lesssim \delta_2$ and $\delta_2^N \lesssim \delta_1$ for some N. Even though, as was noted in that paper, the methods of [189] were sufficient to obtain the results they needed, Tao and Wright replaced a key step in the proof of [189] with the study of the ODE (A.20). That this ODE appears in canonical coordinates is classical (see [44, p. 155] for a similar ODE), though [232] seems to be the first reference which used it for the quantitative study of Carnot–Carathéodory balls, and we have followed their derivation.

In [219], building on this idea of Tao and Wright, the remaining Taylor series methods from [189] were replaced with techniques from the field of ODEs. This allowed for the study of multi-parameter balls (without any weak-comparability hypothesis) as in Section 3.5, though these results still required the vector fields to be given in some coordinate system with good a priori estimates on their coefficients.

Theorems A.2.4 and A.2.17 were proved in [228], and we have closely followed the proofs in that reference. [228] is the first paper on these coordinate systems where all the main results are quantitatively invariant under arbitrary C^2_{loc} diffeomorphisms: in the previously mentioned references, estimates were given in terms of the C^m norms of the coefficients of the vector fields in some fixed coordinate system. Thus, the vector fields had to be a priori "smooth" and "not large" in some fixed coordinate system; the concepts of vector fields being "smooth" and "not large" are not invariant under a C^2 change of coordinates. Theorems A.2.4 and A.2.17, on the other hand, remain completely unchanged if one pushes the entire setting forward under a C^2_{loc} diffeomorphism. This is because the notions of admissible constants (in Definitions A.2.1, A.2.2, A.2.15, and A.2.16) do not depend on any choice of coordinate system. The results in [228] are a bit more involved than Theorems A.2.4 and A.2.17; they work with some more general function spaces which are useful for proving sharp results (see the next paragraph), but are not useful for the applications in this text. We refer the reader to [228] for the more general results proved there.

Theorem A.2.4 "loses one derivative" and hence is not sharp. Indeed, one expects the vector fields $Y_1, \ldots, Y_q$ to be one derivative smoother than the functions $c_{i,j}^k$, i. e., one expects Theorem A.2.4 (h) with $\lesssim_m$ replaced by $\lesssim_{(m-1)\vee 0}$. Unfortunately, as shown in [230], this is not true. However, if one chooses a *different* map Φ and works with Zygmund–Hölder spaces instead of C^m spaces, then such an improved result is obtained in [224, 230], where methods from the area of elliptic PDEs are used to obtain sharp results. A result similar to Theorem A.2.4 in the real analytic category is obtained in [222].

Unlike [189, 232], Theorem A.2.4 works with C^1_{loc} vector fields, instead of C^∞_{loc} vector fields; see [230] for similar results for vector fields which are $\mathscr{C}^s$ for some $s > 0$. Results in the single-parameter case, similar to Theorem 3.3.7 but working with non-smooth vector fields, were obtained by Montanari and Morbidelli [174]. A result similar to Proposition A.4.1 in the case where ϵ was assumed to be small was obtained in [175, Appendix A]. A result very similar to Proposition A.4.8 can be found in [175, Theorem 4.5].

Theorems A.3.1 and A.3.3 are strengthened to a finite level of smoothness (in terms of Zygmund–Hölder spaces) in [224, 230] and an analogous result for real analyticity was obtained in [222]. These results are perhaps reminiscent of the celebrated results of DeTurck and Kazdan [63] regarding a coordinate system in which a Riemannian metric tensor has optimal regularity; for further comments on this, see [230].

Results like Theorem 3.3.7 have been used many times in the theory of several complex variables (see, e. g., [182, 140, 34, 188, 218, 40, 39, 38]). Because of this, it is often of interest to obtain a result like Theorem A.2.4, but where the map Φ is holomorphic, because if Φ is not holomorphic, rescalings will destroy the complex nature of the problem. Such general holomorphic coordinate systems were studied in [223].

Bibliography

[1] G. Alexopoulos, *Spectral multipliers on Lie groups of polynomial growth*, Proc. Am. Math. Soc. **120** (1994), no. 3, 973–979. MR1172944

[2] I. Androulidakis, O. Mohsen, and R. Yuncken, *A pseudodifferential calculus for maximally hypoelliptic operators and the Helffer-Nourrigat conjecture*, preprint, arXiv:2201.12060, 2022.

[3] I. Androulidakis and G. Skandalis, *Pseudodifferential calculus on a singular foliation*, J. Noncommut. Geom. **5** (2011), no. 1, 125–152. MR2746653

[4] N. Aronszajn and K. T. Smith, *Theory of Bessel potentials. I*, Ann. Inst. Fourier (Grenoble) **11** (1961), 385–475. MR143935

[5] P. Auscher and T. Hytönen, *Orthonormal bases of regular wavelets in spaces of homogeneous type*, Appl. Comput. Harmon. Anal. **34** (2013), no. 2, 266–296. MR3008566

[6] P. Auscher and T. Hytönen, *Addendum to Orthonormal bases of regular wavelets in spaces of homogeneous type [Appl. Comput. Harmon. Anal. 34(2) (2013) 266–296] [MR3008566]*, Appl. Comput. Harmon. Anal. **39** (2015), no. 3, 568–569. MR3398952

[7] M. Bardi and F. Dragoni, *Convexity and semiconvexity along vector fields*, Calc. Var. Partial Differ. Equ. **42** (2011), no. 3-4, 405–427. MR2846261

[8] M. Bardi and A. Goffi, *New strong maximum and comparison principles for fully nonlinear degenerate elliptic PDEs*, Calc. Var. Partial Differ. Equ. **58** (2019), no. 6, Paper No. 184, 20. MR4018316

[9] R. Beals, *Opérateurs invariants hypoelliptiques sur un groupe de Lie nilpotent*, Séminaire Goulaouic-Schwartz 1976/1977: Équations aux dérivées partielles et analyse fonctionnelle, Exp. No. 19, École Polytechnique, 1977, p. 8. MR0494313

[10] A. Besikovitch, *Sur la nature des fonctions à carré sommable et des ensembles mesurables*, Fundam. Math. **1** (1923), no. 4, 172–195.

[11] T. Bhattacharya and A. Mohammed, *On a strong maximum principle for fully nonlinear subelliptic equations with Hörmander condition*, Calc. Var. Partial Differ. Equ. **60** (2021), no. 1, Paper No. 9, 20. MR4197073

[12] A. Bonfiglioli and R. Fulci, *Topics in noncommutative algebra*, Lecture Notes in Mathematics, vol. 2034, Springer, Heidelberg, 2012, The theorem of Campbell, Baker, Hausdorff and Dynkin. MR2883818

[13] U. Boscain, R. A. Chertovskih, J. P. Gauthier, and A. O. Remizov, *Hypoelliptic diffusion and human vision: a semidiscrete new twist*, SIAM J. Imaging Sci. **7** (2014), no. 2, 669–695. MR3196939

[14] G. Bourdaud and M. L. de Cristoforis, *Functional calculus in Hölder-Zygmund spaces*, Trans. Am. Math. Soc. **354** (2002), no. 10, 4109–4129. MR1926867

[15] M. Bramanti, *Schauder estimates for parabolic and elliptic nondivergence operators of Hörmander type*, Subelliptic PDE's and applications to geometry and finance, Lect. Notes Semin. Interdiscip. Mat., vol. 6, Semin. Interdiscip. Mat. (S.I.M.), Potenza, 2007, pp. 69–82. MR2384636

[16] M. Bramanti, *An invitation to hypoelliptic operators and Hörmander's vector fields*, SpringerBriefs in Mathematics, Springer, Cham, 2014. MR3154431

[17] M. Bramanti and L. Brandolini, *Schauder estimates for parabolic nondivergence operators of Hörmander type*, J. Differ. Equ. **234** (2007), no. 1, 177–245. MR2298970

[18] M. Bramanti, L. Brandolini, E. Lanconelli, and F. Uguzzoni, *Non-divergence equations structured on Hörmander vector fields: heat kernels and Harnack inequalities*, Mem. Am. Math. Soc. **204** (2010), no. 961, vi+123. MR2604962

[19] M. Bramanti and M. Zhu, *L^p and Schauder estimates for nonvariational operators structured on Hörmander vector fields with drift*, Anal. PDE **6** (2013), no. 8, 1793–1855. MR3198585

[20] H. Brezis and F. Browder, *Partial differential equations in the 20th century*, Adv. Math. **135** (1998), no. 1, 76–144. MR1617413

[21] H.-Q. Bui, T. A. Bui, and X. T. Duong, *Weighted Besov and Triebel-Lizorkin spaces associated with operators and applications*, Forum Math. Sigma **8** (2020), Paper No. e11, 95. MR4070730

https://doi.org/10.1515/9783111085647-011

[22] T. A. Bui and X. T. Duong, *Sharp weighted estimates for square functions associated to operators on spaces of homogeneous type*, J. Geom. Anal. **30** (2020), no. 1, 874–900. MR4058541

[23] T. A. Bui, X. T. Duong, and F. K. Ly, *Maximal function characterizations for new local Hardy-type spaces on spaces of homogeneous type*, Trans. Am. Math. Soc. **370** (2018), no. 10, 7229–7292. MR3841848

[24] T. A. Bui, X. T. Duong, and F. K. Ly, *Maximal function characterizations for Hardy spaces on spaces of homogeneous type with finite measure and applications*, J. Funct. Anal. **278** (2020), no. 8, 108423, 55. MR4056995

[25] L. A. Caffarelli and X. Cabré, *Fully nonlinear elliptic equations*, American Mathematical Society Colloquium Publications, vol. 43, American Mathematical Society, Providence, RI, 1995. MR1351007

[26] A.-P. Calderón, *Lebesgue spaces of differentiable functions and distributions*, Proc. Sympos. Pure Math., Vol. IV, American Mathematical Society, Providence, R.I., 1961, pp. 33–49. MR0143037

[27] A. P. Calderon and A. Zygmund, *On the existence of certain singular integrals*, Acta Math. **88** (1952), 85–139. MR52553

[28] L. Capogna, *Regularity of quasi-linear equations in the Heisenberg group*, Commun. Pure Appl. Math. **50** (1997), no. 9, 867–889. MR1459590

[29] L. Capogna and G. Citti, *Regularity for subelliptic PDE through uniform estimates in multi-scale geometries*, Bull. Math. Sci. **6** (2016), no. 2, 173–230. MR3510691

[30] L. Capogna, D. Danielli, and N. Garofalo, *An embedding theorem and the Harnack inequality for nonlinear subelliptic equations*, Commun. Partial Differ. Equ. **18** (1993), no. 9-10, 1765–1794. MR1239930

[31] L. Capogna and N. Garofalo, *Regularity of minimizers of the calculus of variations in Carnot groups via hypoellipticity of systems of Hörmander type*, J. Eur. Math. Soc. **5** (2003), no. 1, 1–40. MR1961133

[32] L. Capogna, S. D. Pauls, and J. T. Tyson, *Convexity and horizontal second fundamental forms for hypersurfaces in Carnot groups*, Trans. Am. Math. Soc. **362** (2010), no. 8, 4045–4062. MR2608394

[33] C. Carathéodory, *Untersuchungen über die Grundlagen der Thermodynamik*, Math. Ann. **67** (1909), no. 3, 355–386. MR1511534

[34] D.-C. Chang, A. Nagel, and E. M. Stein, *Estimates for the $\bar{\partial}$-Neumann problem in pseudoconvex domains of finite type in* $\mathbf{C}^2$, Acta Math. **169** (1992), no. 3-4, 153–228. MR1194003

[35] S.-Y. A. Chang and R. Fefferman, *A continuous version of duality of* H^1 *with BMO on the bidisc*, Ann. of Math. (2) **112** (1980), no. 1, 179–201. MR584078

[36] S.-Y. A. Chang and R. Fefferman, *The Calderón-Zygmund decomposition on product domains*, Am. J. Math. **104** (1982), no. 3, 455–468. MR658542

[37] S.-Y. A. Chang and R. Fefferman, *Some recent developments in Fourier analysis and* H^p*-theory on product domains*, Bull. Am. Math. Soc. (N.S.) **12** (1985), no. 1, 1–43. MR766959

[38] P. Charpentier and Y. Dupain, *Extremal bases, geometrically separated domains and applications*, Algebra Anal. **26** (2014), no. 1, 196–269. MR3234809

[39] P. Charpentier and Y. Dupain, *Estimates for the Bergman and Szegö projections for pseudoconvex domains of finite type with locally diagonalizable Levi form*, Publ. Mat. **50** (2006), no. 2, 413–446. MR2273668

[40] P. Charpentier and Y. Dupain, *Geometry of pseudo-convex domains of finite type with locally diagonalizable Levi form and Bergman kernel*, J. Math. Pures Appl. (9) **85** (2006), no. 1, 71–118. MR2200592

[41] J. Chen, W. Ding, and G. Lu, *Boundedness of multi-parameter pseudo-differential operators on multi-parameter local Hardy spaces*, Forum Math. **32** (2020), no. 4, 919–936. MR4116647

[42] L. Chen, G. Lu, and X. Luo, *Boundedness of multi-parameter Fourier multiplier operators on Triebel-Lizorkin and Besov-Lipschitz spaces*, Nonlinear Anal. **134** (2016), 55–69. MR3462617

[43] P. Chen, X. T. Duong, J. Li, L. A. Ward, and L. Yan, *Product Hardy spaces associated to operators with heat kernel bounds on spaces of homogeneous type*, Math. Z. **282** (2016), no. 3-4, 1033–1065. MR3473656

[44] C. Chevalley, *Theory of Lie groups. I*, Princeton Mathematical Series, vol. 8, Princeton University Press, Princeton, N.J., 1946. MR0015396

[45] W.-L. Chow, *Über Systeme von linearen partiellen Differentialgleichungen erster Ordnung*, Math. Ann. **117** (1939), 98–105. MR1880

[46] M. Christ, *Regularity properties of the $\overline{\partial}_b$ equation on weakly pseudoconvex CR manifolds of dimension* 3, J. Am. Math. Soc. **1** (1988), no. 3, 587–646. MR928903

[47] M. Christ, *Estimates for fundamental solutions of second-order subelliptic differential operators*, Proc. Am. Math. Soc. **105** (1989), no. 1, 166–172. MR953002

[48] M. Christ, *L^p bounds for spectral multipliers on nilpotent groups*, Trans. Am. Math. Soc. **328** (1991), no. 1, 73–81. MR1104196

[49] M. Christ, D. Geller, P. Głowacki, and L. Polin, *Pseudodifferential operators on groups with dilations*, Duke Math. J. **68** (1992), no. 1, 31–65. MR1185817

[50] G. Citti, B. Franceschiello, G. Sanguinetti, and A. Sarti, *Sub-Riemannian mean curvature flow for image processing*, SIAM J. Imaging Sci. **9** (2016), no. 1, 212–237. MR3463049

[51] G. Citti and A. Sarti, *A cortical based model of perceptual completion in the roto-translation space*, J. Math. Imaging Vis. **24** (2006), no. 3, 307–326. MR2235475

[52] R. R. Coifman and G. Weiss, *Analyse harmonique non-commutative sur certains espaces homogènes*, Lecture Notes in Mathematics, Vol. 242, Springer-Verlag, Berlin-New York, 1971, Étude de certaines intégrales singulières. MR0499948

[53] L. J. Corwin and F. P. Greenleaf, *Representations of nilpotent Lie groups and their applications. Part I*, Cambridge Studies in Advanced Mathematics, vol. 18, Cambridge University Press, Cambridge, 1990, Basic theory and examples. MR1070979

[54] M. Cotlar, *A combinatorial inequality and its applications to L^2-spaces*, Rev. Mat. Cuyana **1** (1955), 41–55 (1956). MR80263

[55] T. Coulhon and X. T. Duong, *Riesz transforms for* $1 \le p \le 2$, Trans. Am. Math. Soc. **351** (1999), no. 3, 1151–1169. MR1458299

[56] D. Danielli, N. Garofalo, D. M. Nhieu, and F. Tournier, *The theorem of Busemann-Feller-Alexandrov in Carnot groups*, Commun. Anal. Geom. **12** (2004), no. 4, 853–886. MR2104079

[57] D. Danielli, N. Garofalo, and D.-M. Nhieu, *Notions of convexity in Carnot groups*, Commun. Anal. Geom. **11** (2003), no. 2, 263–341. MR2014879

[58] D. Danielli, N. Garofalo, and D.-M. Nhieu, *On the best possible character of the L^Q norm in some a priori estimates for non-divergence form equations in Carnot groups*, Proc. Am. Math. Soc. **131** (2003), no. 11, 3487–3498. MR1991760

[59] G. David, J.-L. Journé, and S. Semmes, *Opérateurs de Calderón-Zygmund, fonctions para-accrétives et interpolation*, Rev. Mat. Iberoam. **1** (1985), no. 4, 1–56. MR850408

[60] G. David and J.-L. Journé, *A boundedness criterion for generalized Calderón-Zygmund operators*, Ann. of Math. (2) **120** (1984), no. 2, 371–397. MR763911

[61] M. Derridj, *Regularité pour $\overline{\partial}$ dans quelques domaines faiblement pseudo-convexes*, J. Differ. Geom. **13** (1978), no. 4, 559–576 (1979). MR570218

[62] M. Derridj, *Microlocalisation et estimations pour $\overline{\partial}_b$ dans quelques hypersurfaces pseudoconvexes*, Invent. Math. **104** (1991), no. 3, 631–642. MR1106752

[63] D. M. DeTurck and J. L. Kazdan, *Some regularity theorems in Riemannian geometry*, Ann. Sci. Éc. Norm. Supér. (4) **14** (1981), no. 3, 249–260. MR644518

[64] G. Di Fazio, A. Domokos, M. S. Fanciullo, and J. J. Manfredi, *Subelliptic Cordes estimates in the Grušin plane*, Manuscr. Math. **120** (2006), no. 4, 419–433. MR2245893

[65] G. Di Fazio and P. Zamboni, *Hölder continuity for quasilinear subelliptic equations in Carnot Carathéodory spaces*, Math. Nachr. **272** (2004), 3–10. MR2079757

[66] W. Ding, *The boundedness of composition operators on Triebel-Lizorkin and Besov spaces with different homogeneities*, Acta Math. Sin. Engl. Ser. **30** (2014), no. 6, 933–948. MR3200443

[67] W. Ding and G. Lu, *Duality of multi-parameter Triebel-Lizorkin spaces associated with the composition of two singular integral operators*, Trans. Am. Math. Soc. **368** (2016), no. 10, 7119–7152. MR3471087

[68] W. Ding and G. Lu, *Boundedness of inhomogeneous Journé's type operators on multi-parameter local Hardy spaces*, Nonlinear Anal. **197** (2020), 111816, 31. MR4073512

[69] W. Ding, G. Lu, and Y. Zhu, *Multi-parameter Triebel-Lizorkin spaces associated with the composition of two singular integrals and their atomic decomposition*, Forum Math. **28** (2016), no. 1, 25–42. MR3441104

[70] W. Ding, G. Lu, and Y. Zhu, *Discrete Littlewood-Paley-Stein characterization of multi-parameter local Hardy spaces*, Forum Math. **31** (2019), no. 6, 1467–1488. MR4026464

[71] W. Ding, G. Lu, and Y. Zhu, *Multi-parameter local Hardy spaces*, Nonlinear Anal. **184** (2019), 352–380. MR3925053

[72] Y. Ding, G. Z. Lu, and B. L. Ma, *Multi-parameter Triebel-Lizorkin and Besov spaces associated with flag singular integrals*, Acta Math. Sin. Engl. Ser. **26** (2010), no. 4, 603–620. MR2591640

[73] A. Domokos, *Differentiability of solutions for the non-degenerate p-Laplacian in the Heisenberg group*, J. Differ. Equ. **204** (2004), no. 2, 439–470. MR2085543

[74] A. Domokos and J. J. Manfredi, *Nonlinear subelliptic equations*, Manuscr. Math. **130** (2009), no. 2, 251–271. MR2545517

[75] A. Domokos and J. J. Manfredi, *On the regularity of nonlinear subelliptic equations*, Around the research of Vladimir Maz'ya. II, Int. Math. Ser. (N. Y.), vol. 12, Springer, New York, 2010, pp. 145–157. MR2676172

[76] A. Domokos and J. J. Manfredi, *$C^{1,\alpha}$-regularity for p-harmonic functions in the Heisenberg group for p near 2*, The p-harmonic equation and recent advances in analysis, Contemp. Math., vol. 370, Amer. Math. Soc., Providence, RI, 2005, pp. 17–23. MR2126699

[77] A. Domokos and J. J. Manfredi, *Subelliptic Cordes estimates*, Proc. Am. Math. Soc. **133** (2005), no. 4, 1047–1056. MR2117205

[78] A. S. Dynin, *An algebra of pseudodifferential operators on the Heisenberg groups. Symbolic calculus*, Dokl. Akad. Nauk SSSR **227** (1976), no. 4, 792–795. MR0423432

[79] J. Dziubański, W. Hebisch, and J. Zienkiewicz, *Note on semigroups generated by positive Rockland operators on graded homogeneous groups*, Stud. Math. **110** (1994), no. 2, 115–126. MR1279987

[80] J. Dziubański and A. Hulanicki, *On semigroups generated by left-invariant positive differential operators on nilpotent Lie groups*, Stud. Math. **94** (1989), no. 1, 81–95. MR1008240

[81] W. E and J. C. Mattingly, *Ergodicity for the Navier-Stokes equation with degenerate random forcing: finite-dimensional approximation*, Commun. Pure Appl. Math. **54** (2001), no. 11, 1386–1402. MR1846802

[82] C. Fefferman and D. H. Phong, *Subelliptic eigenvalue problems*, Conference on harmonic analysis in honor of Antoni Zygmund, Vol. I, II (Chicago, Ill., 1981), Wadsworth Math. Ser., Wadsworth, Belmont, CA, 1983, pp. 590–606. MR730094

[83] C. Fefferman and E. M. Stein, *Some maximal inequalities*, Am. J. Math. **93** (1971), 107–115. MR284802

[84] C. L. Fefferman and A. Sánchez-Calle, *Fundamental solutions for second order subelliptic operators*, Ann. of Math. (2) **124** (1986), no. 2, 247–272. MR855295

[85] R. Fefferman, *Harmonic analysis on product spaces*, Ann. of Math. (2) **126** (1987), no. 1, 109–130. MR898053

[86] R. Fefferman and E. M. Stein, *Singular integrals on product spaces*, Adv. Math. **45** (1982), no. 2, 117–143. MR664621

[87] A. Figalli, *The Monge-Ampère equation and its applications*, Zurich Lectures in Advanced Mathematics, European Mathematical Society (EMS), Zürich, 2017. MR3617963

[88] V. Fischer and M. Ruzhansky, *Quantization on nilpotent Lie groups*, Progress in Mathematics, vol. 314, Birkhäuser/Springer, [Cham], 2016. MR3469687

[89] A. Föglein, *Partial regularity results for subelliptic systems in the Heisenberg group*, Calc. Var. Partial Differ. Equ. **32** (2008), no. 1, 25–51. MR2377405

[90] G. B. Folland, *Subelliptic estimates and function spaces on nilpotent Lie groups*, Ark. Mat. **13** (1975), no. 2, 161–207. MR494315

[91] G. B. Folland and E. M. Stein, *Estimates for the $\bar{\partial}_b$ complex and analysis on the Heisenberg group*, Commun. Pure Appl. Math. **27** (1974), 429–522. MR367477

[92] G. B. Folland and E. M. Stein, *Hardy spaces on homogeneous groups*, Mathematical Notes, vol. 28, Princeton University Press, Princeton, N.J.; University of Tokyo Press, Tokyo, 1982. MR657581

[93] L. Gårding, *Vecteurs analytiques dans les représentations des groups de Lie*, Bull. Soc. Math. Fr. **88** (1960), 73–93. MR119104

[94] N. Garofalo and E. Lanconelli, *Existence and nonexistence results for semilinear equations on the Heisenberg group*, Indiana Univ. Math. J. **41** (1992), no. 1, 71–98. MR1160903

[95] A. G. Georgiadis, G. Kerkyacharian, G. Kyriazis, and P. Petrushev, *Atomic and molecular decomposition of homogeneous spaces of distributions associated to non-negative self-adjoint operators*, J. Fourier Anal. Appl. **25** (2019), no. 6, 3259–3309. MR4029179

[96] A. G. Georgiadis and G. Kyriazis, *Embeddings between Triebel-Lizorkin spaces on metric spaces associated with operators*, Anal. Geom. Metric Spaces **8** (2020), no. 1, 418–429. MR4193446

[97] A. Gilioli and F. Trèves, *An example in the solvability theory of linear PDE's*, Am. J. Math. **96** (1974), 367–385. MR355285

[98] K. R. Goodearl and R. B. Warfield, Jr., *An introduction to noncommutative Noetherian rings, second ed.*, London Mathematical Society Student Texts, vol. 61, Cambridge University Press, Cambridge, 2004. MR2080008

[99] R. W. Goodman, *Nilpotent Lie groups: structure and applications to analysis*, Lecture Notes in Mathematics, Vol. 562, Springer-Verlag, Berlin-New York, 1976. MR0442149

[100] V. Guillemin, *Lecture notes*, http://math.mit.edu/~arita/18.101/, 2008, prepared by Ana Rita Pires, Accessed: 2017-05-02.

[101] C. E. Gutiérrez and A. Montanari, *Maximum and comparison principles for convex functions on the Heisenberg group*, Commun. Partial Differ. Equ. **29** (2004), no. 9-10, 1305–1334. MR2103838

[102] C. E. Gutiérrez and A. Montanari, *On the second order derivatives of convex functions on the Heisenberg group*, Ann. Sc. Norm. Super. Pisa, Cl. Sci. (5) **3** (2004), no. 2, 349–366. MR2075987

[103] U. Haagerup, *The best constants in the Khintchine inequality*, Stud. Math. **70** (1981), no. 3, 231–283 (1982). MR654838

[104] M. Hairer and J. C. Mattingly, *Ergodicity of the 2D Navier-Stokes equations with degenerate stochastic forcing*, Ann. of Math. (2) **164** (2006), no. 3, 993–1032. MR2259251

[105] R. S. Hamilton, *The inverse function theorem of Nash and Moser*, Bull. Am. Math. Soc. (N.S.) **7** (1982), no. 1, 65–222. MR656198

[106] Y. Han and E. T. Sawyer, *Littlewood-Paley theory on spaces of homogeneous type and the classical function spaces*, Mem. Am. Math. Soc. **110** (1994), no. 530, vi+126. MR1214968

[107] Y. Han, Y. Han, and J. Li, *Criterion of the boundedness of singular integrals on spaces of homogeneous type*, J. Funct. Anal. **271** (2016), no. 12, 3423–3464. MR3558247

[108] Y. Han, Y. Han, and J. Li, *Geometry and Hardy spaces on spaces of homogeneous type in the sense of Coifman and Weiss*, Sci. China Math. **60** (2017), no. 11, 2199–2218. MR3714572

[109] Y. Han, J. Li, and G. Lu, *Multiparameter Hardy space theory on Carnot-Carathéodory spaces and product spaces of homogeneous type*, Trans. Am. Math. Soc. **365** (2013), no. 1, 319–360. MR2984061

[110] Y. Han, J. Li, and L. A. Ward, *Hardy space theory on spaces of homogeneous type via orthonormal wavelet bases*, Appl. Comput. Harmon. Anal. **45** (2018), no. 1, 120–169. MR3790058

[111] Y. Han, G. Lu, and E. Sawyer, *Flag Hardy spaces and Marcinkiewicz multipliers on the Heisenberg group*, Anal. PDE **7** (2014), no. 7, 1465–1534. MR3293443

[112] Y. Han, D. Müller, and D. Yang, *Littlewood-Paley characterizations for Hardy spaces on spaces of homogeneous type*, Math. Nachr. **279** (2006), no. 13-14, 1505–1537. MR2269253

[113] Y. Han, D. Müller, and D. Yang, *A theory of Besov and Triebel-Lizorkin spaces on metric measure spaces modeled on Carnot-Carathéodory spaces*, Abstr. Appl. Anal. (2008), Art. ID 893409, 250. MR2485404

[114] Y. Han and D. Yang, *New characterizations and applications of inhomogeneous Besov and Triebel-Lizorkin spaces on homogeneous type spaces and fractals*, Diss. Math. (Rozprawy Mat.) **403** (2002), 102. MR1926534

[115] Y. Han and D. Yang, *Some new spaces of Besov and Triebel-Lizorkin type on homogeneous spaces*, Stud. Math. **156** (2003), no. 1, 67–97. MR1961062

[116] Z. He, Y. Han, J. Li, L. Liu, D. Yang, and W. Yuan, *A complete real-variable theory of Hardy spaces on spaces of homogeneous type*, J. Fourier Anal. Appl. **25** (2019), no. 5, 2197–2267. MR4014799

[117] Z. He, F. Wang, D. Yang, and W. Yuan, *Wavelet characterization of Besov and Triebel-Lizorkin spaces on spaces of homogeneous type and its applications*, Appl. Comput. Harmon. Anal. **54** (2021), 176–226. MR4237423

[118] W. Hebisch, *Sharp pointwise estimate for the kernels of the semigroup generated by sums of even powers of vector fields on homogeneous groups*, Stud. Math. **95** (1989), no. 1, 93–106. MR1024276

[119] B. Helffer, *Partial differential equations on nilpotent groups*, Lie group representations, III (College Park, Md., 1982/1983), Lecture Notes in Math., vol. 1077, Springer, Berlin, 1984, pp. 210–253. MR765555

[120] B. Helffer and F. Nourrigat, *Hypoellipticité pour des groupes nilpotents de rang de nilpotence* 3, Commun. Partial Differ. Equ. **3** (1978), no. 8, 643–743. MR499352

[121] B. Helffer and J. Nourrigat, *Caracterisation des opérateurs hypoelliptiques homogènes invariants à gauche sur un groupe de Lie nilpotent gradué*, Commun. Partial Differ. Equ. **4** (1979), no. 8, 899–958. MR537467

[122] B. Helffer and J. Nourrigat, *Hypoellipticité maximale pour des opérateurs polynômes de champs de vecteurs*, Progress in Mathematics, vol. 58, Birkhäuser Boston, Inc., Boston, MA, 1985. MR897103

[123] L. Hörmander, *Linear partial differential operators*, Die Grundlehren der mathematischen Wissenschaften, Bd. 116, Academic Press, Inc., Publishers, New York; Springer-Verlag, Berlin-Göttingen-Heidelberg, 1963. MR0161012

[124] L. Hörmander, *Pseudo-differential operators*, Commun. Pure Appl. Math. **18** (1965), 501–517. MR180740

[125] L. Hörmander, *Hypoelliptic second order differential equations*, Acta Math. **119** (1967), 147–171. MR222474

[126] G. Huisken and W. Klingenberg, *Flow of real hypersurfaces by the trace of the Levi form*, Math. Res. Lett. **6** (1999), no. 5-6, 645–661. MR1739222

[127] A. Hulanicki and J. W. Jenkins, *Almost everywhere summability on nilmanifolds*, Trans. Am. Math. Soc. **278** (1983), no. 2, 703–715. MR701519

[128] A. J. Izzo, *C^r convergence of Picard's successive approximations*, Proc. Am. Math. Soc. **127** (1999), no. 7, 2059–2063. MR1486736

[129] D. Jerison, *The Poincaré inequality for vector fields satisfying Hörmander's condition*, Duke Math. J. **53** (1986), no. 2, 503–523. MR850547

[130] D. Jerison and J. M. Lee, *A subelliptic, nonlinear eigenvalue problem and scalar curvature on CR manifolds*, Microlocal analysis (Boulder, Colo., 1983), Contemp. Math., vol. 27, Amer. Math. Soc., Providence, RI, 1984, pp. 57–63. MR741039

[131] D. Jerison and A. Sánchez-Calle, *Subelliptic, second order differential operators*, Complex analysis, III (College Park, Md., 1985–86), Lecture Notes in Math., vol. 1277, Springer, Berlin, 1987, pp. 46–77. MR922334

[132] D. S. Jerison, *The Dirichlet problem for the Kohn Laplacian on the Heisenberg group. I*, J. Funct. Anal. **43** (1981), no. 1, 97–142. MR639800

[133] D. S. Jerison, *The Dirichlet problem for the Kohn Laplacian on the Heisenberg group. II*, J. Funct. Anal. **43** (1981), no. 2, 224–257. MR633978

[134] D. S. Jerison and A. Sánchez-Calle, *Estimates for the heat kernel for a sum of squares of vector fields*, Indiana Univ. Math. J. **35** (1986), no. 4, 835–854. MR865430

[135] B. Jessen, J. Marcinkiewicz, and A. Zygmund, *Note on the differentiability of multiple integrals*, Fundam. Math. **25** (1935), no. 1, 217–234.

[136] J.-L. Journé, *Calderón-Zygmund operators on product spaces*, Rev. Mat. Iberoam. **1** (1985), no. 3, 55–91. MR836284

[137] P. Juutinen, G. Lu, J. J. Manfredi, and B. Stroffolini, *Convex functions on Carnot groups*, Rev. Mat. Iberoam. **23** (2007), no. 1, 191–200. MR2351130

[138] G. Kerkyacharian and P. Petrushev, *Heat kernel based decomposition of spaces of distributions in the framework of Dirichlet spaces*, Trans. Am. Math. Soc. **367** (2015), no. 1, 121–189. MR3271256

[139] H. Koch, A. Rüland, and W. Shi, *The variable coefficient thin obstacle problem: higher regularity*, Adv. Differ. Equ. **22** (2017), no. 11-12, 793–866. MR3692912

[140] K. D. Koenig, *On maximal Sobolev and Hölder estimates for the tangential Cauchy-Riemann operator and boundary Laplacian*, Am. J. Math. **124** (2002), no. 1, 129–197. MR1879002

[141] K. D. Koenig, *Maximal hypoellipticity for the $\bar{\partial}$-Neumann problem*, Adv. Math. **282** (2015), 128–219. MR3374525

[142] J. J. Kohn, *Lectures on degenerate elliptic problems*, Pseudodifferential operators with applications (Bressanone, 1977), Liguori, Naples, 1978, pp. 89–151. MR660652

[143] J. J. Kohn, *Hypoellipticity and loss of derivatives*, Ann. of Math. (2) **162** (2005), no. 2, 943–986, With an appendix by Makhlouf Derridj and David S. Tartakoff. MR2183286

[144] J. J. Kohn and L. Nirenberg, *An algebra of pseudo-differential operators*, Commun. Pure Appl. Math. **18** (1965), 269–305. MR176362

[145] E. Lanconelli and A. Montanari, *On a class of fully nonlinear PDEs from complex geometry*, Recent trends in nonlinear partial differential equations. I. Evolution problems, Contemp. Math., vol. 594, Amer. Math. Soc., Providence, RI, 2013, pp. 231–242. MR3155912

[146] E. Lanconelli, A. Pascucci, and S. Polidoro, *Linear and nonlinear ultraparabolic equations of Kolmogorov type arising in diffusion theory and in finance*, Nonlinear problems in mathematical physics and related topics, II, Int. Math. Ser. (N. Y.), vol. 2, Kluwer/Plenum, New York, 2002, pp. 243–265. MR1972000

[147] J. M. Lee, *Introduction to smooth manifolds*, Graduate Texts in Mathematics, vol. 218, Springer-Verlag, New York, 2003. MR1930091

[148] F. Liao, Z. Liu, and X. Zhang, *Smooth decompositions of Triebel-Lizorkin and Besov spaces on product spaces of homogeneous type*, J. Funct. Spaces (2014), Art. ID 396418, 15. MR3246448

[149] J. E. Littlewood and R. E. A. C. Paley, *Theorems on Fourier Series and Power Series*, J. Lond. Math. Soc. **6** (1931), no. 3, 230–233. MR1574750

[150] J. E. Littlewood and R. E. A. C. Paley, *Theorems on Fourier series and power series (II)*, Proc. Lond. Math. Soc. (2) **42** (1936), no. 1, 52–89. MR1577045

[151] J. E. Littlewood and R. E. A. C. Paley, *Theorems on Fourier Series and Power Series (III)*, Proc. Lond. Math. Soc. (2) **43** (1937), no. 2, 105–126. MR1575588

[152] P. I. Lizorkin, *Operators connected with fractional differentiation, and classes of differentiable functions*, Tr. Mat. Inst. Steklova **117** (1972), 212–243, 345, Studies in the theory of differentiable functions of several variables and its applications, IV. MR0370166

[153] P. I. Lizorkin, *Properties of functions in the spaces $\Lambda^r_{p,\theta}$*, Tr. Mat. Inst. Steklova **131** (1974), 158–181, 247, Studies in the theory of differentiable functions of several variables and its applications, V. MR0361756

[154] G. Lu, *Existence and size estimates for the Green's functions of differential operators constructed from degenerate vector fields*, Commun. Partial Differ. Equ. **17** (1992), no. 7-8, 1213–1251. MR1179284

[155] G. Lu, *Weighted Poincaré and Sobolev inequalities for vector fields satisfying Hörmander's condition and applications*, Rev. Mat. Iberoam. **8** (1992), no. 3, 367–439. MR1202416

[156] G. Lu, *Embedding theorems into Lipschitz and BMO spaces and applications to quasilinear subelliptic differential equations*, Publ. Mat. **40** (1996), no. 2, 301–329. MR1425620

[157] G. Lu, J. J. Manfredi, and B. Stroffolini, *Convex functions on the Heisenberg group*, Calc. Var. Partial Differ. Equ. **19** (2004), no. 1, 1–22. MR2027845

[158] G. Lu, J. Shen, and L. Zhang, *Multi-parameter hardy spaces theory and endpoint estimates for multi-parameter singular integrals*, Mem. Amer. Math. Soc., **281** (2023), no. 1388, vi+87 pp. ISBN: 978-1-4704-5537-8; 978-1-4704-7321-1

[159] G. Z. Lu and Y. Y. Xiao, *Atomic decomposition and boundedness criterion of operators on multi-parameter Hardy spaces of homogeneous type*, Acta Math. Sin. Engl. Ser. **28** (2012), no. 7, 1329–1346. MR2928481

[160] G. Z. Lu and Y. P. Zhu, *Singular integrals and weighted Triebel-Lizorkin and Besov spaces of arbitrary number of parameters*, Acta Math. Sin. Engl. Ser. **29** (2013), no. 1, 39–52. MR3001008

[161] P. Malliavin, *Stochastic calculus of variation and hypoelliptic operators*, Proceedings of the International Symposium on Stochastic Differential Equations (Res. Inst. Math. Sci., Kyoto Univ., Kyoto, 1976), Wiley, New York-Chichester-Brisbane, 1978, pp. 195–263. MR536013

[162] J. J. Manfredi and G. Mingione, *Regularity results for quasilinear elliptic equations in the Heisenberg group*, Math. Ann. **339** (2007), no. 3, 485–544. MR2336058

[163] S. Marchi, *$C^{1,\alpha}$ local regularity for the solutions of the p-Laplacian on the Heisenberg group for $2 \leq p < 1+\sqrt{5}$*, Z. Anal. Anwend. **20** (2001), no. 3, 617–636. MR1863937

[164] S. Marchi, *Erratum to: "$C^{1,\alpha}$ local regularity for the solutions of the p-Laplacian on the Heisenberg group for $2 \leq p < 1+\sqrt{5}$" [Z. Anal. Anwendungen **20** (2001), no. 3, 617–636; MR1863937 (2002i:35037)]*, Z. Anal. Anwend. **22** (2003), no. 2, 471–472. MR2000279

[165] S. Marchi, *$C^{1,\alpha}$ local regularity for the solutions of the p-Laplacian on the Heisenberg group. The case $1+\frac{1}{\sqrt{5}} < p \leq 2$*, Comment. Math. Univ. Carol. **44** (2003), no. 1, 33–56. MR2045844

[166] J. Marcinkiewicz, *Sur les séries de fourier*, Fundam. Math. **27** (1936), no. 1, 38–69.

[167] J. Marcinkiewicz, *Sur l'interpolation d'operations*, C. R. Acad. Sci. Paris. **208** (1939), 1272–1273.

[168] V. Martino and A. Montanari, *Nonsmooth solutions for a class of fully nonlinear PDE's on Lie groups*, Nonlinear Anal. **126** (2015), 115–130. MR3388873

[169] R. Melrose, *Propagation for the wave group of a positive subelliptic second-order differential operator*, Hyperbolic equations and related topics (Katata/Kyoto, 1984), Academic Press, Boston, MA, 1986, pp. 181–192. MR925249

[170] A. Menikoff, *Some examples of hypoelliptic partial differential equations*, Math. Ann. **221** (1976), no. 2, 167–181. MR481452

[171] K. G. Miller, *Hypoellipticity on the Heisenberg group*, J. Funct. Anal. **31** (1979), no. 3, 306–320. MR531133

[172] O. Mohsen, *On the index of maximally hypoelliptic differential operators*, preprint, arXiv:2201.13049, 2022.

[173] A. Montanari, *Real hypersurfaces evolving by Levi curvature: smooth regularity of solutions to the parabolic Levi equation*, Commun. Partial Differ. Equ. **26** (2001), no. 9-10, 1633–1664. MR1865940

[174] A. Montanari and D. Morbidelli, *Nonsmooth Hörmander vector fields and their control balls*, Trans. Am. Math. Soc. **364** (2012), no. 5, 2339–2375. MR2888209

[175] A. Montanari and D. Morbidelli, *A Frobenius-type theorem for singular Lipschitz distributions*, J. Math. Anal. Appl. **399** (2013), no. 2, 692–700. MR2996747

[176] S. Mukherjee, *On local Lipschitz regularity for quasilinear equations in the Heisenberg group*, Nonlinear Anal. **212** (2021), Paper No. 112453, 24. MR4273846

[177] S. Mukherjee and X. Zhong, *$C^{1,\alpha}$-regularity for variational problems in the Heisenberg group*, Anal. PDE **14** (2021), no. 2, 567–594. MR4241808

[178] S. Mukherjee and X. Zhong, *$C^{1,\alpha}$-regularity for variational problems in the Heisenberg group*, Anal. PDE **14** (2021), no. 2, 567–594. MR4241808

[179] D. Müller, *On Riesz means of eigenfunction expansions for the Kohn-Laplacian*, J. Reine Angew. Math. **401** (1989), 113–121. MR1018056

[180] D. Müller, F. Ricci, and E. M. Stein, *Marcinkiewicz multipliers and multi-parameter structure on Heisenberg (-type) groups. I*, Invent. Math. **119** (1995), no. 2, 199–233. MR1312498

[181] D. Müller, F. Ricci, and E. M. Stein, *Marcinkiewicz multipliers and multi-parameter structure on Heisenberg (-type) groups. II*, Math. Z. **221** (1996), no. 2, 267–291. MR1376298

[182] A. Nagel, J.-P. Rosay, E. M. Stein, and S. Wainger, *Estimates for the Bergman and Szegö kernels in $\mathbb{C}^2$*, Ann. of Math. (2) **129** (1989), no. 1, 113–149. MR979602

[183] A. Nagel, F. Ricci, E. Stein, and S. Wainger, *Singular integrals with flag kernels on homogeneous groups,I*, Rev. Mat. Iberoam. **28** (2012), no. 3, 631–722. MR2949616

[184] A. Nagel, F. Ricci, and E. M. Stein, *Singular integrals with flag kernels and analysis on quadratic CR manifolds*, J. Funct. Anal. **181** (2001), no. 1, 29–118. MR1818111
[185] A. Nagel, F. Ricci, E. M. Stein, and S. Wainger, *Algebras of singular integral operators with kernels controlled by multiple norms*, Mem. Am. Math. Soc. **256** (2018), no. 1230, vii+141. MR3862599
[186] A. Nagel and E. M. Stein, *Lectures on pseudodifferential operators: regularity theorems and applications to nonelliptic problems*, Mathematical Notes, vol. 24, Princeton University Press, Princeton, N.J.; University of Tokyo Press, Tokyo, 1979. MR549321
[187] A. Nagel and E. M. Stein, *Differentiable control metrics and scaled bump functions*, J. Differ. Geom. **57** (2001), no. 3, 465–492. MR1882665
[188] A. Nagel and E. M. Stein, *The $\Box_b$-heat equation on pseudoconvex manifolds of finite type in $\mathbb{C}^2$*, Math. Z. **238** (2001), no. 1, 37–88. MR1860735
[189] A. Nagel, E. M. Stein, and S. Wainger, *Balls and metrics defined by vector fields. I. Basic properties*, Acta Math. **155** (1985), no. 1-2, 103–147. MR793239
[190] E. Nelson, *Analytic vectors*, Ann. of Math. (2) **70** (1959), 572–615. MR107176
[191] L. I. Nicolaescu, *Lectures on the geometry of manifolds*, second ed., World Scientific Publishing Co. Pte. Ltd., Hackensack, NJ, 2007. MR2363924
[192] J. Nourrigat, *Hypoellipticité maximale pour le système de Cauchy-Riemann induit*, Goulaouic-Meyer-Schwartz Seminar, 1981/1982, École Polytech., Palaiseau, 1982, pp. Exp. No. X, 11. MR671607
[193] D. Nualart, *The Malliavin calculus and related topics*, second ed., Probability and its Applications (New York), Springer-Verlag, Berlin, 2006. MR2200233
[194] R. E. A. C. Paley, *A remarkable series of orthogonal functions (i)*, Proc. Lond. Math. Soc. **2** (1932), no. 1, 241–264.
[195] J. Peetre, *Sur les espaces de Besov*, C. R. Acad. Sci. Paris Sér. A-B **264** (1967), A281–A283. MR218887
[196] J. Petitot and Y. Tondut, *Vers une neurogéométrie. Fibrations corticales, structures de contact et contours subjectifs modaux*, Math. Inform. Sci. Humaines (1999), no. 145, 5–101. MR1697185
[197] J. Pipher, *Journé's covering lemma and its extension to higher dimensions*, Duke Math. J. **53** (1986), no. 3, 683–690. MR860666
[198] L. Rothschild, *A criterion for hypoellipticity of operators constructed from vector fields*, Commun. Partial Differ. Equ. **4** (1979), no. 6, 645–699. MR532580
[199] L. Rothschild, *Nonexistence of optimal L^2 estimates for the boundary Laplacian operator on certain weakly pseudoconvex domains*, Commun. Partial Differ. Equ. **5** (1980), no. 8, 897–912. MR583606
[200] L. Rothschild and E. M. Stein, *Hypoelliptic differential operators and nilpotent groups*, Acta Math. **137** (1976), no. 3-4, 247–320. MR436223
[201] F. Ricci and E. M. Stein, *Multiparameter singular integrals and maximal functions*, Ann. Inst. Fourier (Grenoble) **42** (1992), no. 3, 637–670. MR1182643
[202] D. Ricciotti, *p-Laplace equation in the Heisenberg group*, SpringerBriefs in Mathematics, Springer, [Cham]; BCAM Basque Center for Applied Mathematics, Bilbao, 2015, Regularity of solutions, BCAM SpringerBriefs. MR3444525
[203] D. Ricciotti, *On the $C^{1,\alpha}$ regularity of p-harmonic functions in the Heisenberg group*, Proc. Am. Math. Soc. **146** (2018), no. 7, 2937–2952. MR3787355
[204] M. Riesz, *Sur les fonctions conjuguées*, Math. Z. **27** (1928), no. 1, 218–244. MR1544909
[205] C. Rockland, *Hypoellipticity on the Heisenberg group-representation-theoretic criteria*, Trans. Am. Math. Soc. **240** (1978), 1–52. MR486314
[206] W. Rudin, *Functional analysis*, second ed., International Series in Pure and Applied Mathematics, McGraw-Hill, Inc., New York, 1991. MR1157815
[207] K. Saka, *Besov spaces and Sobolev spaces on a nilpotent Lie group*, Tohoku Math. J. (2) **31** (1979), no. 4, 383–437. MR558675
[208] J. Schur, *Bemerkungen zur Theorie der beschränkten Bilinearformen mit unendlich vielen Veränderlichen*, J. Reine Angew. Math. **140** (1911), 1–28. MR1580823

[209] R. T. Seeley, *Singular integrals on compact manifolds*, Am. J. Math. **81** (1959), 658–690. MR110022

[210] R. T. Seeley, *Refinement of the functional calculus of Calderón and Zygmund*, Nederl. Akad. Wetensch. Proc. Ser. A 68=Indag. Math. **27** (1965), 521–531. MR0226450

[211] A. Sikora, *Riesz transform, Gaussian bounds and the method of wave equation*, Math. Z. **247** (2004), no. 3, 643–662. MR2114433

[212] L. Simon, *Schauder estimates by scaling*, Calc. Var. Partial Differ. Equ. **5** (1997), no. 5, 391–407. MR1459795

[213] E. M. Stein, *Singular integrals and differentiability properties of functions*, Princeton Mathematical Series, No. 30, Princeton University Press, Princeton, N.J., 1970. MR0290095

[214] E. M. Stein, *Topics in harmonic analysis related to the Littlewood-Paley theory*, Annals of Mathematics Studies, No. 63, Princeton University Press, Princeton, N.J.; University of Tokyo Press, Tokyo, 1970. MR0252961

[215] E. M. Stein, *The development of square functions in the work of A. Zygmund*, Bull. Am. Math. Soc. (N.S.) **7** (1982), no. 2, 359–376. MR663787

[216] E. M. Stein, *Harmonic analysis: real-variable methods, orthogonality, and oscillatory integrals*, Princeton Mathematical Series, vol. 43, Princeton University Press, Princeton, NJ, 1993, With the assistance of Timothy S. Murphy, Monographs in Harmonic Analysis, III. MR1232192

[217] E. M. Stein, *Calderón and Zygmund's theory of singular integrals*, Harmonic analysis and partial differential equations (Chicago, IL, 1996), Chicago Lectures in Math., Univ. Chicago Press, Chicago, IL, 1999, pp. 1–26. MR1731194

[218] E. M. Stein, *The $\Box_b$ heat equation and multipliers via the wave equation*, Math. Z. **263** (2009), no. 4, 861–886. MR2551602

[219] E. M. Stein, *Multi-parameter Carnot-Carathéodory balls and the theorem of Frobenius*, Rev. Mat. Iberoam. **27** (2011), no. 2, 645–732. MR2848534

[220] E. M. Stein, *Multi-parameter singular integrals*, Annals of Mathematics Studies, vol. 189, Princeton University Press, Princeton, NJ, 2014. MR3241740

[221] E. M. Stein, *Sobolev spaces associated to singular and fractional Radon transforms*, Rev. Mat. Iberoam. **33** (2017), no. 2, 633–748. MR3651019

[222] E. M. Stein, *Coordinates adapted to vector fields III: real analyticity*, Asian J. Math. **24** (2020), no. 6, 1029–1078. MR4312758

[223] E. M. Stein, *Sub-Hermitian geometry and the quantitative Newlander-Nirenberg theorem*, Adv. Math. **368** (2020), 107137, 87. MR4085140

[224] E. M. Stein, *Coordinates adapted to vector fields II: sharp results*, Am. J. Math. **143** (2021), no. 6, 1791–1840. MR4349133

[225] E. M. Stein and B. Street, *Multi-parameter singular Radon transforms III: Real analytic surfaces*, Adv. Math. **229** (2012), no. 4, 2210–2238. MR2880220

[226] E. M. Stein and B. Street, *Multi-parameter singular Radon transforms II: The L^p theory*, Adv. Math. **248** (2013), 736–783. MR3107526

[227] E. M. Stein and P.-L. Yung, *Pseudodifferential operators of mixed type adapted to distributions of k-planes*, Math. Res. Lett. **20** (2013), no. 6, 1183–1208. MR3228630

[228] B. Stovall and B. Street, *Coordinates adapted to vector fields: canonical coordinates*, Geom. Funct. Anal. **28** (2018), no. 6, 1780–1862. MR3881835

[229] B. Street, *An algebra containing the two-sided convolution operators*, Adv. Math. **219** (2008), no. 1, 251–315. MR2435424

[230] B. Street and L. Yao, *Improving the regularity of vector fields*, J. Funct. Anal. **283** (2022), no. 5, Paper No. 109537, 75. MR4429573

[231] H. J. Sussmann, *Orbits of families of vector fields and integrability of distributions*, Trans. Am. Math. Soc. **180** (1973), 171–188. MR321133

[232] T. Tao and J. Wright, *L^p improving bounds for averages along curves*, J. Am. Math. Soc. **16** (2003), no. 3, 605–638. MR1969206

[233] M. E. Taylor, *Partial differential equations III. Nonlinear equations*, second ed., Applied Mathematical Sciences, vol. 117, Springer, New York, 2011. MR2744149

[234] A. F. M. ter Elst and D. W. Robinson, *Subcoercivity and subelliptic operators on Lie groups. I. Free nilpotent groups*, Potential Anal. **3** (1994), no. 3, 283–337. MR1290669

[235] A. F. M. ter Elst and D. W. Robinson, *Subcoercivity and subelliptic operators on Lie groups. II. The general case*, Potential Anal. **4** (1995), no. 3, 205–243. MR1331833

[236] A. F. M. ter Elst, D. W. Robinson, and A. Sikora, *Heat kernels and Riesz transforms on nilpotent Lie groups*, Colloq. Math. **74** (1997), no. 2, 191–218. MR1477562

[237] E. C. Titchmarsh, *On Conjugate Functions*, Proc. Lond. Math. Soc. (2) **29** (1928), no. 1, 49–80. MR1575323

[238] F. Tréves, *Topological vector spaces, distributions and kernels*, Academic Press, New York-London, 1967. MR0225131

[239] F. Trèves, *Introduction to pseudodifferential and Fourier integral operators. Vol. 1*, Plenum Press, New York-London, 1980, Pseudodifferential operators, The University Series in Mathematics. MR597144

[240] H. Triebel, *Spaces of distributions of Besov type on Euclidean n-space. Duality, interpolation*, Ark. Mat. **11** (1973), 13–64. MR348483

[241] H. Triebel, *Interpolation theory, function spaces, differential operators*, North-Holland Mathematical Library, vol. 18, North-Holland Publishing Co., Amsterdam-New York, 1978. MR503903

[242] H. Triebel, *Theory of function spaces. II*, Monographs in Mathematics, vol. 84, Birkhäuser Verlag, Basel, 1992. MR1163193

[243] H. Triebel, *Theory of function spaces. III*, Monographs in Mathematics, vol. 100, Birkhäuser Verlag, Basel, 2006. MR2250142

[244] H. Triebel, *Theory of function spaces*, Modern Birkhäuser Classics, Birkhäuser/Springer Basel AG, Basel, 2010, Reprint of 1983 edition [MR0730762], Also published in 1983 by Birkhäuser Verlag [MR0781540]. MR3024598

[245] N. S. Trudinger, *On Hessian measures for non-commuting vector fields*, Pure Appl. Math. Q. **2** (2006), no. 1, Special Issue: In honor of John H. Coates. Part 1, 147–161. MR2217569

[246] A. Unterberger and J. Bokobza, *Les opérateurs de Calderon-Zygmund précisés*, C. R. Acad. Sci. Paris **259** (1964), 1612–1614. MR176360

[247] A. Unterberger and J. Bokobza, *Les opérateurs de Calderon-Zygmund précisés*, C. R. Acad. Sci. Paris **260** (1965), 34–37. MR176361

[248] C. Wang, *Viscosity convex functions on Carnot groups*, Proc. Am. Math. Soc. **133** (2005), no. 4, 1247–1253. MR2117228

[249] J. Wang and P. Niu, *Optimal partial regularity for weak solutions of nonlinear sub-elliptic systems in Carnot groups*, Nonlinear Anal. **72** (2010), no. 11, 4162–4187. MR2606775

[250] H. Weyl, *Singuläre integralgleichungen mit besonderer berücksichtigung des fourierschen integraltheorems*, W. Fr. Kaestner, 1908.

[251] C. J. Xu, *Regularity for quasilinear second-order subelliptic equations*, Commun. Pure Appl. Math. **45** (1992), no. 1, 77–96. MR1135924

[252] O. Zariski and P. Samuel, *Commutative algebra. Vol. II*, The University Series in Higher Mathematics, D. Van Nostrand Co., Inc., Princeton, N.J.–Toronto-London-New York, 1960. MR0120249

[253] A. Zatorska-Goldstein, *Very weak solutions of nonlinear subelliptic equations*, Ann. Acad. Sci. Fenn., Math. **30** (2005), no. 2, 407–436. MR2173373

[254] L. Zhang, *Spectral multipliers for maximally subelliptic operators*, preprint, arXiv:2302.07086, 2023.

[255] A. Zygmund, *Smooth functions*, Duke Math. J. **12** (1945), 47–76. MR12691

Symbol Index

https://doi.org/10.1515/9783111085647-012

Index

https://doi.org/10.1515/9783111085647-013

De Gruyter Studies in Mathematics

Volume 92
Min Ru
Minimal Surfaces through Nevanlinna Theory, 2023
ISBN 978-3-11-099982-2, e-ISBN 978-3-11-098955-7, e-ISBN (ePUB) 978-3-11-098996-0

Volume 46/2
Steven Lord, Edward McDonald, Fedor Sukochev, Dmitriy Zanin
Singular Traces. Volume 2: Trace Formulas, 2nd Edition, 2023
ISBN 978-3-11-070000-8, e-ISBN 978-3-11-070017-6, e-ISBN (ePUB) 978-3-11-070024-4

Volume 91
Johnny Henderson, Rodica Luca
Boundary Value Problems for Second-Order Finite Difference Equations and Systems, 2023
ISBN 978-3-11-103931-2, e-ISBN 978-3-11-104037-0, e-ISBN (ePUB) 978-3-11-104045-5

Volume 90
William M. McGovern
Representation Theory and Geometry of the Flag Variety, 2022
ISBN 978-3-11-076690-5, e-ISBN 978-3-11-076694-3, e-ISBN (ePUB) 978-3-11-076696-7

Volume 89
Marco Abate
Holomorphic Dynamics on Hyperbolic Riemann Surfaces, 2022
ISBN 978-3-11-060105-3, e-ISBN 978-3-11-060197-8, e-ISBN (ePUB) 978-3-11-059874-2

Volume 88
Joel L. Schiff
Topics in Complex Analysis, 2022
ISBN 978-3-11-075769-9, e-ISBN 978-3-11-075782-8, e-ISBN (ePUB) 978-3-11-075787-3

Volume 66
Igor V. Nikolaev
Noncommutative Geometry. A Functorial Approach, 2nd Edition, 2022
ISBN 978-3-11-078860-0, e-ISBN 978-3-11-078870-9, e-ISBN (ePUB) 978-3-11-078881-5

Volume 87
René Erlin Castillo
Fourier Meets Hilbert and Riesz. An Introduction to the Corresponding Transforms, 2022
ISBN 978-3-11-078405-3, e-ISBN 978-3-11-078409-1, e-ISBN (ePUB) 978-3-11-078412-1

www.ingramcontent.com/pod-product-compliance
Ingram Content Group UK Ltd.
Pitfield, Milton Keynes, MK11 3LW, UK
UKHW050954290926
15470UKWH00006B/1822

9 783111 085173